D1366425

ATLAS OF IMMUNOLOGY

SECOND EDITION

ATLAS OF IMMUNOLOGY

SECOND EDITION

Julius M. Cruse, B.A., B.S., D.Med.Sc., M.D., Ph.D., Dr. *h.c.*, F.A.A.M., F.R.S.H., F.R.S.M.
Professor of Pathology
Director of Immunopathology and Transplantation Immunology
Director of Graduate Studies in Pathology
Department of Pathology
Associate Professor of Medicine and Associate Professor of Microbiology
Distinguished Professor of the History of Medicine
University of Mississippi Medical Center
Investigator of the Wilson Research Foundation
Mississippi Methodist Rehabilitation Center
Jackson, Mississippi

Robert E. Lewis, B.S., M.S., Ph.D., F.R.S.H., F.R.S.M.
Professor of Pathology
Director of Immunopathology and Transplantation Immunology
Department of Pathology
University of Mississippi Medical Center
Investigator of the Wilson Research Foundation
Mississippi Methodist Rehabilitation Center
Jackson, Mississippi

CRC PRESS

Boca Raton London New York Washington, D.C.

Library of Congress Cataloging-in-Publication Data

Cruse, Julius M., 1937-
Atlas of immunology / Julius M. Cruse, Robert E. Lewis.—2nd ed.
p. cm.
Includes index.
ISBN 0-8493-1567-0 (alk. paper)
1. Immunology—Atlases. I. Lewis, R.E. (Robert Edwin), 1947-II. Title.

QR182.C78 2003
616.07'9'0222—dc22 2003058464

International Standard Book Number 0-8493-1567-0
Library of Congress Card Number 2003058464
Printed in the United States of America 1 2 3 4 5 6 7 8 9 0
Printed on acid-free paper

Dedicated to

A. Wallace Conerly, M.D.
on the occasion of his retirement as
Vice Chancellor for Health Affairs and Dean of the School of Medicine
at The University of Mississippi Medical Center, Jackson, Mississippi.
Truly a person of vision, Dr. Conerly's selfless dedication to medical education and research has inspired all of us who are beneficiaries of his generosity in creating a better community, state, and nation through medical research and improved patient care.

Editorial Staff

CRC Press Editorial Staff

Authors

Julius M. Cruse, B.A., B.S., D.Med.Sc., M.D., Ph.D., Dr.h.c., is Professor of Pathology, Director of Immunopathology and Transplantation Immunology, Director of Graduate Studies in Pathology, Associate Professor of Medicine, Associate Professor of Microbiology, and Distinguished Professor of the History of Medicine at the University of Mississippi Medical Center in Jackson. Formerly, Dr. Cruse was Professor of Immunology and of Biology in the University of Mississippi Graduate School. Dr. Cruse graduated in 1958, earning B.A. and B.S. degrees in chemistry with honors from the University of Mississippi. He was a Fulbright Fellow in the University of Graz (Austria) medical faculty, where he wrote a thesis on Russian tickborne encephalitis virus and received a D.Med.Sc. degree *summa cum laude* in 1960. On his return to the United States, he entered the M.D./Ph.D. program at the University of Tennessee College of Medicine, Memphis, completing his M.D. degree in 1964 and Ph.D. in pathology (immunopathology) in 1966. Dr. Cruse also trained in pathology at the University of Tennessee Center for the Health Sciences, Memphis.

Dr. Cruse is a member of numerous professional societies, including the American Association of Immunologists (Historian), the American Society for Investigative Pathology, the American Society for Histocompatibility and Immunogenetics (Historian; Member of Council, 1997–1999; formerly Chairman, Publications Committee [1987–1995]), the Societé Francaise d'Immunologie, the Transplantation Society, and the Society for Experimental Biology and Medicine, among many others. He is a Fellow of the American Academy of Microbiology, a Fellow of the Royal Society of Health (U.K.) and a Fellow of the Royal Society of Medicine (London). He received the Doctor of Divinity, *honoris causa*, in 1999 from The General Theological Seminary of the Episcopal Church, New York City.

Dr. Cruse's research has centered on transplantation and tumor immunology, autoimmunity, MHC genetics in the pathogenesis of AIDS, and neuroendocrine immune interactions. He has received many research grants during his career and is presently funded by the Wilson Research Foundation for neuroendocrine–immune system interactions in patients with spinal cord injuries. He is the author of more than 250 publications in scholarly journals and 38 books, and has directed dissertation and thesis research for more than 40 graduate students during his career. He is editor-in-chief of the international journals *Immunologic Research, Experimental and Molecular Pathology*, and *Transgenics*. He was chief editor of the journal *Pathobiology* from 1982 to 1998 and was founder of *Immunologic Research*, *Transgenics*, and *Pathobiology*.

Robert E. Lewis, B.A., M.S., Ph.D., is Professor of Pathology and Director of Immunopathology and Transplantation Immunology in the Department of Pathology at the University of Mississippi Center in Jackson. Dr. Lewis received his B.A. and M.S. degrees in microbiology from the University of Mississippi and earned his Ph.D. in pathology (immunopathology) from the University of Mississippi Medical Center. Following specialty postdoctoral training at several medical institutions, Dr. Lewis has risen through the academic ranks from instructor to professor at the University of Mississippi Medical Center.

Dr. Lewis is a member of numerous professional societies, including the American Association of Immunologists, the American Society for Investigative Pathology, the Society for Experimental Biology and Medicine, the American Society for Microbiology, the Canadian Society for Immunology, and the American Society for Histocompatibility and Immunogenetics (Chairman, Publications Committee; member of Board of Directors), among numerous others. He is a Fellow of the Royal Society of Health of Great Britain and a Fellow of the Royal Society of Medicine (U.K.). Dr. Lewis has been the recipient of a number of research grants in his career and is currently funded by the Wilson Research Foundation for his research on neuroendocrine–immune system interaction in patients with spinal cord injuries.

Dr. Lewis has authored or coauthored more than 120 papers and 150 abstracts and has made numerous scientific presentations at both the national and international levels. In addition to neuroendocrine–immune interactions, his current research also includes immunogenetic aspects of AIDS progression. Dr. Lewis is a founder, senior editor, and deputy editor-in-chief of *Immunologic Research* and *Transgenics*, and is senior editor and deputy editor-in-chief of *Experimental and Molecular Pathology*. He was senior editor and deputy editor-in-chief of *Pathobiology* from 1982 to 1998.

Preface

The splendid reception of the first edition of this book in 1999 convinced both the authors and the publisher to prepare a second edition. The 4 years since this atlas first appeared have witnessed an exponential increase in immunological information emanating from more than 130 journals devoted to the subject. The *Journal of Immunology* is published twice monthly in an effort to accommodate an ever-increasing demand for immunological information among researchers spanning all fields of biomedicine. Besides the unprecedented advances in knowledge of cell receptors and signal transduction pathways, an avalanche of new information has been gleaned from contemporary research concerning cytokines and chemokines, with special reference to their structure and function. This edition has not only been thoroughly updated but also contains five new chapters on comparative immunology, autoimmunity, vaccines and immunization, therapeutic immunology, and diagnostic immunology.

The *Atlas of Immunology* is designed to provide a pictorial reference and serve as a primary resource as the most up-to-date and thorough, illustrated treatise available in the complex science of immunology. The book contains more than 1100 illustrations and depicts essentially every concept of importance in understanding the subject of immunology. It is addressed to immunologists and non-immunologists alike, including students, researchers, practitioners, and basic biomedical scientists. Use of the book does not require prior expertise. Some of the diagrams illustrate basic concepts, while others are designed for the specialist interested in a more detailed treatment of the subject matter of immunology. The group of illustrations is relatively complete and eliminates the need to refer to another source. The subject matter ranges from photographs of historical figures to molecular structures of recently characterized cell receptors, chemokines, and cytokines, the major histocompatibility complex molecules, immunoglobulins, and molecules of related interest to immunologists.

The subject matter is divided into chapters that follow an outline which correlates with a standard immunology textbook. This provides for a logical and sequential presentation and gives the reader ready access to each part of the subject matter as it relates to the other parts of the publication. These descriptive illustrations provide the reader with a concise and thorough understanding of basic immunological concepts that often intersect the purview of other basic and clinical scientific disciplines. A host of new illustrations, such as cellular adhesions molecules, is presented in a manner that facilitates better understanding of their role in intercellular and immune reactions. Figures that are pertinent to all of the immunological subspecialties, such as transplantation, autoimmunity, immunophysiology, immunopathology, antigen presentation, and the T cell receptor, to name a few, may be found in this publication. Those individuals with a need for ready access to a visual image of immunological information will want this book to be readily available on their bookshelf. No other publication provides the breadth or detail of illustrated immunological concepts as may be found in the *Atlas of Immunology*, Second Edition.

Acknowledgments

Although many individuals have offered help or suggestions in the preparation of this book, several deserve special mention. We are very grateful to Dr. Michael Hughson, chairman, Department of Pathology, University of Mississippi Medical Center, Jackson, for his support of our academic endeavors at this institution. Dr. Fredrick H. Shipkey, professor emeritus of pathology at the University of Mississippi Medical Center, provided valuable assistance in selecting and photographing appropriate surgical pathology specimens to illustrate immunological lesions. We express genuine appreciation to Dr. Edwin Eigenbrodt and Dr. Marsha Eigenbrodt for many photomicrographs. We thank Professor Albert Wahba for offering constructive criticism related to a number of the chemical structures, and express genuine appreciation to Dr. Virginia Lockard for providing the electron micrographs that appear in the book. We also thank Dr. Robert Peace for a case of Job's syndrome, Dr. Ray Shenefelt for the photomicrograph of cytomegalovirus, Dr. Jonathan Fratkin for photomicrographs of eye and muscle pathology, Dr. C.J. Chen for VKH photographs, Dr. Howard Shulman for GVH photographs, Dr. David DeBauche for providing an illustration of the Philadelphia chromosome, and Dorothy Whitcomb for the history photographs. We thank Dr. G. Reid Bishop of Mississippi College for his generous contribution of molecular models of cytokines and other configurations critical to immunology.

We express genuine appreciation to Julia Peteet, Dr. Huan Wang, Dr. Smaroula Dilioglou, Robert Morrison, Debra Small, Bill Buhner, and Sam Pierce for their dedicated efforts in helping us to complete this publication in a timely manner and making valuable editorial contributions. We also appreciate the constructive criticisms of Patsy Foley, B.S., M.T., C.H.T., C.H.S.; Jay Holliday, B.S., M.T., C.H.T.; Paula Hymel, B.S., M.T., C.H.T; Kevin Beason, B.S., M.T., C.H.T.; Maxine Crawford, B.S., C.H.T., Shawn Clinton, B.S., M.T., C.H.T.; and Susan Touchstone, B.S., M.T., S.B.B., C.H.T., C.H.S. We are most grateful to Joanna LaBresh of R&D Corp. for providing the artwork and for sharing with us a number of schematic diagrams of immunological molecules and concepts owned by R&D. It is a pleasure also to express our genuine gratitude to INOVA Diagnostics Inc., especially to Carol Peebles, for permitting us to use their photomicrographs of immunological concepts for the autoimmunity chapter and to Cell Marque Corp., especially to Michael Lacey, M.D., for providing figures for the chapter on diagnostic immunology. We would also like to commend the individuals at CRC Press — Judith Spiegel, Ph.D., Editor; Fequiere Vilsaint, former Editor; Pat Roberson, Production Manager; Amy Rodriguez, Project Editor; and all members of their staff — for their professionalism and unstinting efforts to bring this book to publication. To these individuals, we offer our grateful appreciation.

Special thanks are expressed to Dr. Daniel W. Jones, Dean of Medicine and Vice Chancellor for Health Affairs, University of Mississippi Medical Center, and his predecessor, Dr. A. Wallace Conerly, for their unstinting support of our research and academic endeavors. Their wise and enthusiastic leadership has facilitated our task and made it enjoyable. We are also most grateful to Mark Adams, CEO of Mississippi Methodist Rehabilitation Center, for his unconditional allegiance and encouragement of clinical research on spinal cord injury patients.

The authors' cellular immunology research has been funded through the generous support of the Wilson Research Foundation, Mississippi Methodist Rehabilitation Center, Jackson.

Illustration Credits

Structural image for the front cover, crystal structure of a cytokine/receptor complex: x-ray diffraction image of IL-6 (beta sign) chain provided through the courtesy of Research Collaboratory for Structural Bioinformatics, 2003. H.M. Berman, J. Westbrook, Z. Feng, G. Gilliland, T.N. Bhat, H. Weissing, I.N. Shindyalov, P.E. Bourne: *The Protein Data Bank, Nucleic Acids Research* 28 pp. 235–242 (2000).

Photograph of Dr. A. Wallace Conerly on dedication page furnished through the courtesy of Jay Ferchaud, Department of Public Relations, University of Mississippi Medical Center, Jackson.

Figures 1.1, 1.5, 1.8, 1.11, 1.12, 1.14, 1.15, 1.18, 1.19, 1.20, 1.21, 1.40, 1.41, 1.42, 1.46, and 1.49 reprinted from U.S. National Library of Medicine.

Figures 1.2, 1.3, 1.4, 1.6, 1.9, 1.10, 1.13, 1.16, 1.17, 1.22 through 1.30, 1.32 through 1.39, 1.43, 1.44, 1.45, 1.47, 1.48 and 1.50 through 1.54 reprinted from Whitcomb, D., *Immunology to 1980*. University of Wisconsin, Center of Health Sciences Library, Madison, 1985.

Figure 1.8 courtesy of the Cruse collection; adapted from Hemmelweit, F., *Collected Papers of Paul Ehrlich*, Pergamon Press, Tarrytown, NY, 1956–1960.

Figure 1.55 reprinted with permission of Jerry Berndt.

Figure 1.56 compliments of Professor Dr. Rolf Zinkernagel, Institute of Pathology, University of Zurich.

Figure 2.4 adapted from Lachman, P.J., Keith, P.S., Rosen, F.S., and Walport, M.J., *Clinical Aspects of Immunology,* 5th ed., Vol. 1993, p. 203, Fig. 112. With permission.

Figures 2.4, 7.85, and 11.18 redrawn from Lachmann, P.J., *Clinical Aspects of Immunology.* Blackwell Scientific Publications, Cambridge, MA, 1993. Reprinted by permission of Blackwell Science, Inc.

Figures 2.5, 2.6, 2.9, 2.19, 2.52, 4.4, 4.14, 4.17, 7.80, 7.83, 9.24, 9.25, 9.26, 9.29, 9.30, 9.37, 10.4, 10.5, 10.6, 10.10, 10.19, 10.24, 10.26, 10.30, 10.35, 10.37, 10.39, 10.40, 19.8, 19.9, 19.10, 20.16, 20.19, 20.22, 21.4, 21.5, 21.6, 21.7, 22.25, 23.25, 23.29, 23.40, and 23.41 reprinted from Protein Data Bank. Abola, E.E., Bernstein, F.C., Bryant, S.H., Koetzle, T.F., and Weng, J., In: *Crystallographic Database: Information Content, Software Systems, Scientific Applications.* Allen, F.H., Bergerhoff, G., and Wievers, R., Eds. Data Commission of the International Union of Crystallography, Bonn/Cambridge/Chester, 1987, pp. 107–132. Bernstein, F.C., Koetzle, T.F., Williams, G.J.B., Meyer, E.F. Jr., Bride, M.D., Rogers, J.R., Kennard, O., Simanouchi, T., and Tasumi, M., The protein data bank: a computer-based archival file for macromolecular structures, *Journal of Molecular Biology,* 112:535–542, 1977. These images are part of the Swiss-3D Image Collection. Manuel C. Peitsch, Geneva Biomedical Research Institute, Glaxo Wellcome R&D, Geneva, Switzerland.

Figures 2.16, 2.39, 2.78, 2.85, 4.18, 4.19, 9.22, 9.23, 9.28, 9.38, and 9.39 redrawn from Barclay, A.N., Birkeland, M.L., Brown, M.H., Beyers, A.D., Davis, S.J., Somoza, C., and Williams, A.F., *The Leucocyte Antigen Facts Book,* Academic Press, Orlando, FL, 1993.

Figures 2.30, 2.31, 2.41, 2.50, 2.68, 2.69, 2.71, 2.75, 2.76, 2.84, 2.85, 2.90, 2.94, 2.95, 2.98, 2.99, 2.103, 9.7, and 21.33 through 21.47 compliments of Marsha L. Eigenbrodt, MD, MPH, formerly assistant professor, Department of Medicine, and Edwin H. Eigenbrodt, MD, formerly professor of pathology, University of Mississippi Medical Center.

Figure 2.33 reprinted from Deutsch, M. and Weinreb, A., Apparatus for high-precision repetitive sequential optical measurement of lifting cells, *Cytometry* 16:214–226, 1994. Adapted by permission of Wiley-Liss, Inc., a subsidiary of John Wiley & Sons, Inc., and Marder, O., et al., Effect of interleukin-1α, interleukin-1β, and tumor necrosis factor-α on the intercellular fluorescein fluorescence polarization of human lung fibroblasts, *Pathobiology* 64(3):123–130.

Figure 2.44 redrawn from Ravetch, J.V. and Kinet, J.P., Fc receptors, *Annual Review of Immunology* 9:462, 1991.

Figures 2.79, 7.23, 7.25, 7.28, 12.2, 12.18, 12.22, 12.28 and 15.9 redrawn from Murray, P.R., *Medical Microbiology,* Mosby-Yearbook, St. Louis, MO, 1994.

Figure 2.86 redrawn from Tedder, T.F., Structure of the gene encoding the human B lymphocyte differentiation antigen CD20(B1), *Journal of Immunology* 142(7):2567, 1989.

Figures 3.2, 6.19, 21.20, and 22.5. Redrawn from Bellanti, J.A., *Immunology II.* W.B. Saunders Co., Philadelphia, PA, 1978.

Figure 24.04 ©1997 by Facts and Comparisons. Adapted with permission from *Immunofacts: Vaccines and Immunologic Drugs.* Facts and Comparisons, St. Louis, MO, a Wolters Kluwer Company, 1996.

Figure 4.12 reprinted from Janeway, C.A. Jr. and Travers, P., *Immunobiology: The Immune System in Health and Disease.* 3rd ed., pp. 4–5, 1997. Reprinted by permission of Routledge/Taylor & Francis Books, Inc.

Figure 4.22 reprinted with permission from *Nature.* Bjorkmam, P.J., Saper, M.A., Samraoui, B., Bennet, W.A.S., Strominger, J.L., and Wiley, D.C., Structure of the human class I histocompatibility antigen, HLA-A2, 329:506–512. ©1987 Macmillan Magazines, Ltd.

Figures 6.6, 6.9, and 27.50 redrawn from Paul, W.E., *Fundamental Immunology*, 3rd ed., Raven Press, New York, 1993.

Figure 7.1 redrawn from Hunkapiller, T. and Hood, L., Diversity of the immunoglobulin gene superfamily, *Advances in Immunology* 44:1–62, 1989.

Figure 7.11 courtesy of Mike Clark, PhD, Division of Immunology, Cambridge University.

Figure 7.12 reprinted with permission from *Nature.* Harris, L.F., Larson, S.E., Hasel, K.W., Day, J., Greenwood, A., and McPherson, A., The three-dimensional structure of an intact monoclonal antibody for canine lymphoma, 360(6402):369–372. ©1992 Macmillan Magazines Ltd.

Figures 7.20, 7.22 through 7.25, 7.30 through 7.32, and 6.52 for immunoglobulins redrawn from Oppenheim, J., Rosenstreich, D.L., and Peter, M., *Cellular Function Immunity and Inflammation*, Elsevier Science, New York, 1984.

Figure 7.35 redrawn from Capra, J.D. and Edmundson, A.B., The antibody combining site, *Scientific American* 236:50–54, 1977. ©George V. Kelvin/ *Scientific American.*

Figure 7.43 courtesy of Dr. Leon Carayannopoulos, Department of Microbiology, University of Texas, Southwestern Medical School, Dallas.

Figure 7.48 adapted from Kang, C. and Kohler, H., *Immunoregulation and Autoimmunity*, Vol. 3, Cruse, J.M. and Lewis, R.E. Eds., 226 S Karger, Basel, Switzerland, 1986.

Figure 7.66 reprinted with permission from Raghavan, M., et al., Analysis of the pH dependence of the neonatal Fc receptor/immunoglobulin G interaction using antibody and receptor variants, *Biochemistry* 34:14,469–14,657. ©1995 American Chemical Society; and from Junghans, R.P., Finally, the Brambell receptor (FcRB), *Immunologic Research* 16:29–57.

Figure 7.67 reprinted with permission from *Nature.* Brambel, F.W.R., Hemmings, W.A., and Morris, I.G., A theoretical model of gamma globulin catabolism, 203:1352–1355. © 1987 Macmillan Magazines, Ltd.

Figure 7.68 reprinted from Brambell, F.W.R., The transmission of immunity from mother to young and the catabolism of immunoglobulins, *The Lancet*, ii:1087–1093, 1966.

Figure 7.74 adapted from Haber, E., Quertermous, T., Matsueda, G.R., and Runge, M.S., Innovative approaches to plasminogen activator therapy, *Science* 243:52–56. ©1989 American Association for the Advancement of Science.

Figure 7.84 redrawn from Conrad, D.H., Keegan, A.D., Kalli, K.R., Van Dusen, R., Rao, M., and Levine, A.D., Superinduction of low affinity IgE receptors on murine B lymphocytes by LPS and interleukin-4, *Journal of Immunology* 141:1091–1097, 1988.

Figures 8.7, 8.13, 8.16, 8.18, and 8.21 redrawn from Eisen, H., *Immunology*, Lippincott-Raven Publishers, New York, pp. 371, 373, 385–386, 1974.

Figures 8.23 and 8.24 reprinted from Kabat, E.A., *Structural Concepts in Immunology and Immunochemistry,* Holt, Rinehart & Winston, New York, 1968.

Figure 9.2 reprinted from *Atlas of Tumor Pathology*, 2nd Series, Fascicle 13, Armed Forces Institute of Pathology.

Figures 9.3 and 9.8 reprinted from *Atlas of Tumor Pathology*, 3rd Series, Fascicle 21, Armed Forces Institute of Pathology.

Figures 9.4, 9.5, and 9.6 reprinted from Muller-Hermelink, H.K., Marina, M., and Palestra, G., Pathology of thymic epithelial tumors. In: *The Human Thymus. Current Topics in Pathology.* Muller-Hermelink, H.K., Ed. 1986; 75:207–268.

Figure 9.10 reprinted from van Wijingaert, F.P., Kendall, M.D., Schuurmann, H.J., Rademakers, L.H., Kater, L., Heterogeneity of epithelial cells in the human thymus. An ultrastructural study, *Cell and Tissue Research* 227–237, 1984.

Figure 9.13 reprinted from Lo, D., Reilly, C.R., DeKoning, J., Laufer, T.M., and Glimcher, L.H., Thymic stromal cell specialization of the T cell receptor repertoire, *Immunologic Research* 16(1):3–14, 1997.

Figure 9.27 adapted from Werner, K. and Ferrara, J., *Immunologic Research* 15(1), 1996, p.52.

Figures 9.33 and 17.90 redrawn from Davis, M.M., T cell receptor gene diversity and selection, *Annual Review of Biochemistry* 59:477, 1990.

Figure 10.28 redrawn from Ealick, S.E., Cook, W.J., and Vijay-Kumar, S., Three-dimensional structure of recombinant human interferon-γ, *Science* 252:698–702. ©1991 American Association for the Advancement of Science.

Figure 10.39 adapted from Rifkin, D.B. et al., *Thrombosis and Haemostasis* 1993:70, 177–179.

Figure 11.4 redrawn from Arlaud, G.J., Colomb, M.G., and Gagnon, J., A functional model of the human C1 complex. *Immunology Today* 8:107–109, 1987.

Figures 11.9 and 11.11 redrawn from Podack, E.R., Molecular mechanisms of cytolysis by complement and cytolytic lymphocytes, *Journal of Cellular Biochemistry* 30:133–70, 1986.

Figure 11.12 redrawn from Rooney, I.A., Oglesby, T.J., and Atkinson, J.P., Complement in human reproduction: activation and control, *Immunologic Research* 12(3): 276–294, 1993.

Figure 11.19 redrawn from Kinoshita, T., *Complement Today*. Cruse, J.M. and Lewis, R.E., Eds., 48 S. Karger, Basel Switzerland, 1993.

Figure 12.26 reprinted from Shwartzman, G. Phenomenon of Local Tissue Reactivity and Its Immunological, Pathological, and Clinical Significance, Paul B. Hoeber, Publisher (Lippincott-Raven Publishers), New York, 1937, p. 275.

Figures 14.2, 14.3, and 14.6 through 14.53 are furnished courtesy of INOVA Corp. and Ms. Carol Peebles, San Diego, CA.

Figures 16.5 and 16.22 adapted from Vengelen-Tyler, V., Ed., *Technical Manual,* 12th ed., American Association of Blood Banks, Bethesda, MD, 1996, pp. 231, 282.

Figures 16.12, 16.24 and 16.26 reprinted from Daniels, G., *Human Blood Groups*, Blackwell Science Ltd., Oxford, UK, pp.13, 271, 432, 1995.

Figures 16.33 and 16.11 adapted from Walker, R.H., Ed., *Technical Manual,* 11th ed., American Association of Blood Banks, Bethesda, MD, 1993, pp 242, 281.

Figures 17.8 and 17.69 redrawn from Cotran, R.S., Kumar, V., and Robbins, S.L., *Robbins Pathologic Basis of Disease,* B. Saunders Co., Philadelphia, PA, 1989.

Figures 17.23 and 17.87 reprinted from Valenzuela, R., Bergfeld, W.F., and Deodhar, S.D., *Immunofluorescent Patterns in Skin Diseases*, American Society of Clinical Pathologists Press, Chicago, IL, 1984. With permission of the ASCP Press.

Figures 17.27 and 17.28 adapted from Edmundson, A.B., Ely, K.R., Abola, E.E., Schiffer, M., and Panagotopoulos, N., Rotational allomerism and divergent evolution of domains in immunoglobulin light chains, *Biochemistry* 14:3953–3961. ©1975 American Chemical Society.

Figure 16.33 reprinted from Clemetson, K.J., Glycoproteins of the platelet plasma membrane, in *Platelet Membrane Glycoproteins,* George, J.N., Norden, A.T., and Philips, D.R., Eds., Plenum Press, New York, 1985, pp. 61-86, with permission.

Figure 17.46 redrawn from Stites, D.P., *Basic and Clinical Immunology*, Appleton & Lange, East Norwalk, CT, 1991.

Figure 17.112 redrawn from Dieppe, P.A., Bacon, P.A., Bamji, A.N., and Watt, I., *Atlas of Clinical Rheumatology,* Lea & Febiger, Philadelphia, PA, 1986.

Figure 21.17 adapted from *Splits, Associated Antigens and Inclusions,* PEL-FREEZE Clinical Systems, Brown Deer, WI, 1992.

Figures 21.48 through 21.62 compliments of Howard M. Shulman, MD, professor of pathology, University of Washington, Member Fred Hutchinson Cancer Research Center, Seattle.

Figures 22.1 through 22.3 reprinted from *Monoclonal Antiadhesion Molecules,* Nov. 8, 1994. Seikagaku Corp.

Figure 23.17 reproduced from *Journal of Cell Biology.* Tilney, L.G. and Portnoy, D.A., Actin filaments and the growth, movement, and spread of the intracellular bacterial parasite, *Listeria monocytogenes*, 1989, 109:1597–1608, with permission of the Rockefeller University Press.

Figure 22.23 adapted from Immunoscintigraphy (nude mouse) with a ^{131}I-labelled monoclonal antibody, photographs prepared by Hachmann, H. and Steinstraesser, A., Radiochemical Laboratory, Hrechst, Frankfort, Germany. With permission.

Figures 22.31 and 23.32 reprinted from Seifer, M. and Standring, D.N., Assembly and antigenicity of hepatitis B virus core particles, *Intervirology* 38:47–62, 1995.

Figure 23.36 courtesy of Farr-Jones, S., University of California at San Francisco.

Figures 27.3 and 27.4 redrawn from Hudson, L. and Hay, F.C., *Practical Immunology*, Blackwell Scientific Publications, Cambridge, MA, 1989.

Figure 27.10 redrawn from Miller, L.E., *Manual of Laboratory Immunology*, Lea & Febiger, Mal-vern, PA, 1991.

Figure 27.19 redrawn from Elek, S.D., *Staphyloccocus Pyogenes and its Relation to Disease*, E&S Livingstone Ltd., Edinburgh and London, 1959.

Figures 28.1 through 28.33 are furnished courtesy of Cell Marque Corp. and Dr. Michael Lacey, Hot Springs, AR.

Table of Contents

Chapter 1
History of Immunology 1

Chapter 2
Molecules, Cells, and Tissues of the Immune Response 25

Chapter 3
Antigens and Immunogens 105

Chapter 4
Major Histocompatibility Complex 127

Chapter 5
Antigen Processing and Presentation 145

Chapter 6
B Lymphocyte Development and Immunoglobulin Genes 157

Chapter 7
Immunoglobulin Synthesis, Properties, Structure, and Function 173

Chapter 8
Antigen–Antibody Interactions 227

Chapter 9
The Thymus and T Lymphocytes 251

Chapter 10
Cytokines and Chemokines 285

Chapter 11
The Complement System 321

Chapter 12
Types I, II, III, and IV Hypersensitivity 347

Chapter 13
Immunoregulation and Immunologic Tolerance 379

Chapter 14
Autoimmunity 387

Chapter 15
Mucosal Immunity 437

Chapter 16
Immunohematology 447

Chapter 17
Immunological Diseases and Immunopathology 469

Chapter 18
Immunodeficiencies: Congenital and Acquired 535

Chapter 19
Acquired Immune Deficiency Syndrome (AIDS) 555

Chapter 20
Immunosuppression 567

Chapter 21
Transplantation Immunology 579

Chapter 22
Tumor Immunology 613

Chapter 23
Immunity against Microorganisms 631

Chapter 24
Vaccines and Immunization 685

Chapter 25
Therapeutic Immunology 699

Chapter 26
Comparative Immunology 707

Chapter 27
Immunological Methods and Molecular Techniques 723

Chapter 28
Diagnostic Immunohistochemistry 777

Index 797

1 History of Immunology

The metamorphosis of immunology from a curiosity of medicine associated with vaccination to a modern science focused at the center of basic research in molecular medicine is chronicled here. The people and events that led to this development are no less fascinating than the subject itself. A great number of researchers in diverse areas of medicine and science contributed to building the body of knowledge we now possess. It is possible to name only a few here, but we owe a debt to them all. We are standing on the shoulders of giants, and in remembering their achievements we come to understand better the richness of our inheritance.

Resistance against infectious disease agents was the principal concern of bacteriologists and pathologists establishing the basis of classical immunology in the latter half of the 19th and early 20th centuries. Variolation was practiced for many years prior to Edward Jenner's famous studies proving that inoculation with cowpox could protect against subsequent exposure to smallpox. This established him as the founder of immunology. He contributed the first reliable method of conferring lasting immunity to a major contagious disease. Following the investigations by Louis Pasteur on immunization against anthrax, chicken cholera, and rabies, and Robert Koch's studies on hypersensitivity in tuberculosis, their disciples continued research on immunity against infectious disease agents. Emil von Behring and Paul Ehrlich developed antitoxin against diphtheria, while Elie Metchnikoff studied phagocytosis and cellular reactions in immunity. Hans Buchner described a principle in the blood later identified by Jules Bordet as alexine or complement. Bordet and Octave Gengou went on to develop the complement fixation test that was useful to assay antigen–antibody reactions. Karl Landsteiner described the ABO blood groups of man in 1900, followed by his elegant studies establishing the immunochemical basis of antigenic specificity.

Charles Robert Richet and Paul Jules Portier in the early 1900s attempted to immunize dogs against toxins in the tentacles of sea anemones but inadvertently induced a state of hypersusceptibility, which they termed anaphylaxis. Since that time, many other hypersensitivity and allergic phenomena that are closely related to immune reactions have been described. Four types of hypersensitivity reactions are now recognized as contributory mechanisms in the production of immunological diseases. From the early 1900s until the 1940s, immunochemistry was a predominant force maintaining that antibody was formed through a template mechanism. With the discovery of immunological tolerance by Peter Medawar in the 1940s, David Talmage's cell selection theory, and Frank M. Burnet's clonal selection theory of acquired immunity, it became apparent that a selective theory based on genetics was more commensurate with the facts than was the earlier template theory of the immunochemists. With the elucidation of immunoglobulin structure by Rodney Robert Porter and Gerald Edelman, among others, in the late 1950s and 1960s, modern immunology emerged at the frontier of medical research. Jean Baptiste Dausset described human histocompatibility antigens, and transplantation immunology developed into a major science, making possible the successful transplantation of organs. Bone marrow transplants became an effective treatment for severe combined immunodeficiency and related disorders. The year 1960 marked the beginning of a renaissance in cellular immunology, and the modern era dates from that time. Many subspecialties of immunology are now recognized and include such diverse topics as molecular immunology (immunochemistry), immunobiology, immunogenetics, immunopathology, tumor immunology, transplantation, comparative immunology, immunotoxicology, and immunopharmacology. Thus, it is apparent that immunology is only at the end of the beginning and has bright prospects for the future as evidenced by the exponential increase in immunologic literature in recent years.

In 1948, Astrid Elsa Fagraeus established the role of the plasma cell in antibody formation. The fluorescence antibody technique developed by Albert Coons was a major breakthrough for the identification of antigen in tissues and subsequently demonstrated antibody synthesis by individual cells. While attempting to immunize chickens in which the bursa of Fabricius had been removed, Bruce Glick et al. noted that antibody production did not take place. This was the first evidence of bursa-dependent antibody formation. Robert A. Good immediately realized the significance of this finding for immunodeficiencies of childhood. He and his associates in Minneapolis and J.F.A.P. Miller in England went on to show the role of the thymus in the immune response, and various investigators began to search for bursa equivalence in man and other animals. Thus, the immune system of many species was found to have distinct bursa-dependent, antibody-synthesizing, and thymus-dependent cell-mediated limbs. In 1959, James Gowans proved that lymphocytes actually recirculate. In 1966, Tzvee Nicholas Harris et al. demonstrated clearly that lymphocytes could form antibodies. In 1966 and 1967, Claman et al., David et al., and Mitchison et al. showed that T and B lymphocytes cooperate with one another in the production of an immune response. Various phenomena such as the switch from forming one class of immunoglobulin to another by B cells were demonstrated to be dependent upon a signal from T cells activating B cells to change from IgM to IgG or IgA production. B cells stimulated by antigen in which no T cell signal was given continued to produce IgM antibody. Such antigens were referred to as thymus-independent antigens, and others requiring T cell participation as thymus-dependent antigens. Mitchison et al. described a subset of T lymphocytes demonstrating helper activity, i.e., helper T cells. In 1971, Gershon and Condo described suppressor T cells. Suppressor T cells have been the subject of much investigation but have eluded confirmation by the techniques of molecular biology. Baruj Benacerraf et al. demonstrated the significant role played by gene products of the major histocompatibility complex in the specificity and regulation of T cell-dependent immune response. Jerne described the network theory of immunity in which antibodies formed against idiotypic specificities of antibody molecules followed by the formation of antiidiotypic antibodies constitutes a significant additional immunoregulatory process for immune system function. This postulate has been proved valid by numerous investigators. Tonegawa et al. and Leder et al. identified and cloned the genes that code for variable and constant diversity in antibody-combining sites. In 1975, George Kohler and Cesar Milstein successfully produced monoclonal antibodies by hybridizing mutant myeloma cells with antibody-producing B cells (hybridoma technique). The B cells conferred the antibody-producing capacity while the myeloma cells provided the capability for endless reproduction. Monoclonal antibodies are the valuable homogeneous products of hybridomas that have widespread application in diagnostic laboratory medicine.

Rhazes (Abu Bakr Muhammad ibn Zakariya) (865–932) Persian philosopher and alchemist who described measles and smallpox as different diseases. He also was a proponent of the theory that immunity is acquired. Rhazes is often cited as the premier physician of Islam.

Girolamo Fracastoro (1478–1553) A physician who was born in Verona and educated in Padua. His interests ranged from poetry to geography. He proposed the theory of acquired immunity and was a leader in the early theories of contagion. *Syphilis sive Morbus Gallicus*, 1530; *De Sympathia et Antipathia Rerum*, 1546; *De Contagione*, 1546.

Smallpox cartoon, artist unknown, from the Clement C. Fry collection. Yale Medical Library, contributed by Jason S. Zielonka, published in *J. Hist. Med.* 27:447, 1972. Legend translated: smallpox-disfigured father says, "How shameful that your pretty little children should call my children stupid and should run away, refusing to play with them as friends …." Meanwhile, the children lament: "Father dear, it appears to be your fault that they're avoiding us. To tell the truth, it looks as though you should have inoculated us against smallpox."

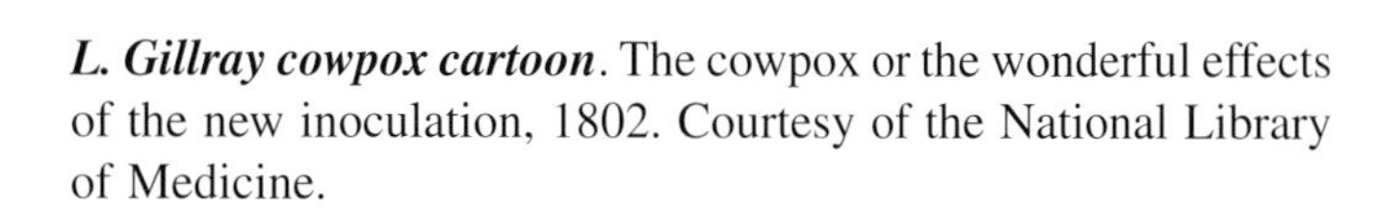

L. Gillray cowpox cartoon. The cowpox or the wonderful effects of the new inoculation, 1802. Courtesy of the National Library of Medicine.

Lady Mary Wortley Montagu (1689–1762) Often credited as the first to introduce inoculation as a means of preventing smallpox in England in 1722. After observing the practice in Turkey where her husband was posted as Ambassador to the Turkish Court, she had both her young son and daughter inoculated and interested the Prince and Princess of Wales in the practice. Accounts of inoculation against smallpox are found in her *Letters*, 1777. Robert Halsband authored a biography, *The Life of Lady Mary Wortley Montagu,* Clarendon, Oxford, 1956.

Edward Jenner (1749–1822) Often termed the founder of immunology for his contribution of the first reliable method of providing lasting immunity to a major contagious disease. He studied medicine under John Hunter and for most of his career was a country doctor in Berkeley in southern England. It was common knowledge in the country that an eruptive skin disease of cattle (cowpox) and a similar disease in horses called "grease" conferred immunity to smallpox on those who cared for the animals and caught the infection from them. Jenner carefully observed and recorded 23 cases. The results of his experiments were published, establishing his claim of credit for initiating the technique of vaccination. He vaccinated an 8-year-old boy, James Phipps, with matter taken from the arm of the milkmaid, Sara Nelmes, who was suffering from cowpox. After the infection subsided, he inoculated the child with smallpox and found that the inoculation had no effect. His results led to widespread adoption of vaccination in England and elsewhere in the world, ultimately leading to eradication of smallpox.

Louis Pasteur (1822–1895) French. Father of immunology. One of the most productive scientists of modern times, Pasteur's contributions included the crystallization of L- and D-tartaric acid disproving the theory of spontaneous generation, and studies of diseases in wine, beer, and silkworms. He successfully immunized sheep and cattle against anthrax, terming the technique "vaccination" in honor of Jenner. He used attenuated bacteria and viruses for vaccination. Pasteur produced a vaccine for rabies by drying the spinal cord of rabbits and using the material to prepare a series of 14 injections of increasing virulence. A child's life (Joseph Meister) was saved by this treatment. *Les Maladies des Vers a Soie,* 1865; *Etudes sur le Vin,* 1866; *Etudes sur la Biere,* 1876; *Oeuvres,* 1922–1939.

Julius Cohnheim (1839–1884) German experimental pathologist who was the first proponent of inflammation as a vascular phenomenon. *Lectures on General Pathology,* 1889.

Heinrich Hermann Robert Koch (1843–1910) German bacteriologist awarded the Nobel Prize in 1905 for his work on tuberculosis. Koch made many contributions to the field of bacteriology. Along with his postulates for proof of etiology, Koch instituted strict isolation and culture methods in bacteriology. He studied the life cycle of anthrax and discovered both the *Vibrio cholerae* and the tubercle bacillus. The Koch phenomenon and Koch-Weeks bacillus both bear his name.

Elie Metchnikoff (1845–1916) Born at Ivanovska, Ukraine, where he was a student of zoology with a very special interest in comparative embryology. He earned a Ph.D. degree at the University of Odessa where he also served as professor of zoology. He studied phagocytic cells of starfish larvae in 1884 in a marine laboratory in Italy. This served as the basis for his cellular phagocytic theory of immunity. On leaving Russia for political reasons, he accepted a position at the Institut Pasteur in Paris where he extended his work on the defensive role of phagocytes and championed his cellular theory of immunity. He also made numerous contributions to immunology and bacteriology. He shared the 1908 Nobel Prize in Physiology or Medicine with Paul Ehrlich "in recognition for their work on immunity." *Lecons sur le Pathologie de l'Inflammation*, 1892; *L'Immunite dans les Maladies Infectieuses*, 1901; *Etudes sur la Nature Humaine*, 1903.

Alexandre Besredka (1870–1940) Parisian immunologist who worked with Metchnikoff at the Pasteur Institute. He was born in Odessa. He contributed to studies of local immunity, anaphylaxis, and antianaphylaxis. *Anaphylaxie et Antianaphylaxie,* 1918; *Histoire d'une Idee: L'Oeuvre de Metchnikoff,* 1921; *Etudes sur l'Immunite dans les Maladies Infectieuses,* 1928.

Paul Ehrlich (1854–1915) Born in Silesia, Germany, and graduated as a doctor of medicine from the University of Leipzig. His scientific work included three areas of investigation. He first became interested in stains for tissues and cells and perfected some of the best ones to demonstrate the tubercle bacillus and leukocytes in blood. His first immunological studies were begun in 1890 when he was an assistant at the Institute for Infectious Disease under Robert Koch. After first studying the antibody response to the plant toxins abrin and ricin, Ehrlich published the first practical technique to standardize diphtheria toxin and antitoxin preparations in 1897. He proposed the first selective theory of antibody formation known as the "side chain theory" which stimulated much research by his colleagues in an attempt to disprove it. He served as director of his own institute in Frankfurt-am-Main where he published papers with a number of gifted colleagues, including Dr. Julius Morgenroth, on immune hemolysis and other immunological subjects. He also conducted a number of studies on cancer and devoted the final phase of his career to the development of chemotherapeutic agents for the treatment of disease. He shared the 1908 Nobel Prize with Metchnikoff for their studies on immunity. Fruits of these labors led to treatments for trypanosomiasis and syphilis (Salvarsan, "the magic bullet"). *Collected Studies on Immunity,* 1906; *Collected Papers of Paul Ehrlich,* 3 vols., 1957.

August von Wassermann (1866–1925) German physician who, with Neisser and Bruck, described the first serological test for syphilis, i.e., the Wassermann reaction. *Handbook der Pathogenen Mikroorganismen* (with Kolle), 1903.

Hans Buchner (1850–1902) German bacteriologist who was a professor of hygiene in Munich in 1894. He discovered complement. Through his studies of normal serum and its bactericidal effects, he became an advocate of the humoral theory of immunity.

Svante Arrhenius (1859–1927) Photographed with Paul Ehrlich, 1903. Coined the term "immunochemistry" and hypothesized that antigen–antibody complexes are reversible. He was awarded the Nobel Prize in Chemistry, 1903. *Immunochemistry,* 1907.

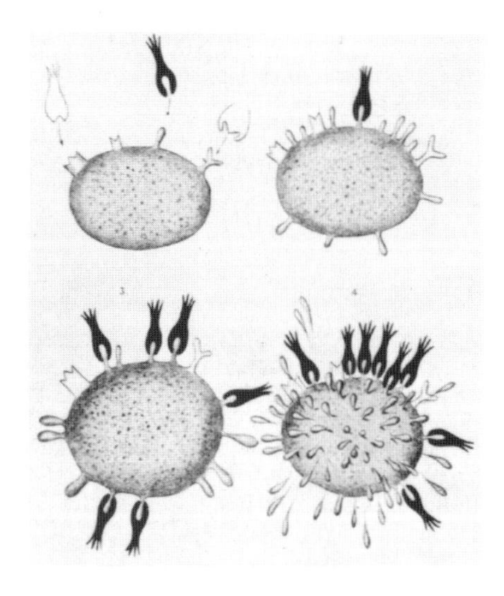

DIAGRAMMATIC REPRESENTATION OF THE SIDE-CHAIN THEORY
(PLATES I AND II)

Fig. 1 "The groups [the haptophore group of the side-chain of the cell and that of the food-stuff or the toxin] must be adapted to one another, *e.g.*, as male and female screw (PASTEUR), or as lock and key (E. FISCHER)."

Fig. 2 ". . . the first stage in the toxic action must be regarded as being the union of the toxin by means of its haptophore group to a special side-chain of the cell protoplasm."

Fig. 3 "The side-chain involved, so long as the union lasts, cannot exercise its normal, physiological, nutritive function . . ."

Fig. 4 "We are therefore now concerned with a defect which, according to the principles so ably worked out by . . . Weigert, is . . . [overcorrected] by regeneration."

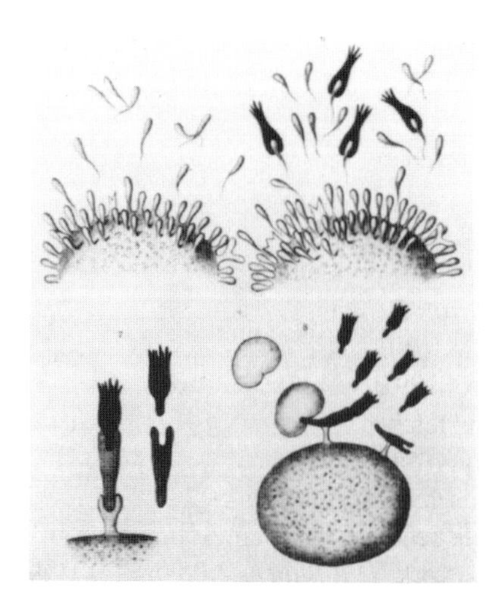

DIAGRAMMATIC REPRESENTATION OF THE SIDE-CHAIN THEORY
(*cont.*)

Fig. 5 ". . . the antitoxins represent nothing more than the side-chains reproduced in excess during regeneration and therefore pushed off from the protoplasm—thus coming to exist in a free state."

Fig. 6 [The free side-chains (circulating antitoxins) unite with the toxins and thus protect the cell.]

Fig. 7 ". . . two haptophore groups must be ascribed to the 'immune-body' [haemolytic amboceptor], one having a strong affinity for a corresponding haptophore group of the red blood corpuscles, . . . and another . . which . . . becomes united with the 'complement' . . ."

Fig. 8 "If a cell . . . has, with the assistance of an appropriate side-chain, fixed to itself a giant [protein] molecule . . . there is provided [only] one of the conditions essential for the cell nourishment. Such . . . molecules . . are not available until . . . they have been split into smaller fragments This will be . . . attained if . . . the 'tentacle' . . . possesses . . . a second haptophore group adapted to take to itself ferment-like material . . ."

Ehrlich side chain theory The first selective theory of antibody synthesis developed by Paul Ehrlich in 1900. Although elaborate in detail, the essential feature of the theory was that cells of the immune system possess the genetic capability to react to all known antigens and that each cell on the surface bears receptors with surface haptophore side chains. On combination with antigen, the side chains would be cast off into the circulation and new receptors would replace the old ones. These cast-off receptors represented antibody molecules in the circulation. Although far more complex than this explanation, the importance of the theory was in the amount of research stimulated to try to disprove it. Nevertheless, it was the first effort to account for the importance of genetics in immune responsiveness at a time when Mendel's basic studies had not even yet been "rediscovered" by De Vries.

Jules Jean Baptiste Vincent Bordet (1870–1961) Belgian physician who graduated with a doctor of medicine degree from the University of Brussels. He was preparateur in Metchnikoff's laboratory at the Institut Pasteur from 1894 to 1901, where he discovered immune hemolysis and elucidated the mechanisms of complement-mediated bacterial lysis. He and Gengou described complement fixation and pointed to its use in the diagnosis of infectious diseases. Their technique was subsequently used by von Wassermann to develop a complement-fixation test for syphilis which enjoyed worldwide popularity. His debates with Paul Ehrlich on the nature of antigen–antibody–complement interactions stimulated much useful research. He was awarded the Nobel Prize in Physiology or Medicine for his studies on immunity, 1919. *Traite de l'Immunite dans les Maladies Infectieuses,* 1920.

Emil Adolph von Behring (1854–1917) German bacteriologist who worked at the Institute for Infectious Disease in Berlin with Kitasato and Wernicke from 1890 to 1892 and demonstrated that circulating antitoxins against diphtheria and tetanus conferred immunity. He demonstrated that the passive administration of antitoxin, i.e., serum containing antitoxin could facilitate recovery. This represented the beginning of serum therapy, especially for diphtheria. He received the first Nobel Prize in Physiology or Medicine in 1901 for this work. *Die Blutserumtherapie,* 1902; *Gesammelte Abhandlungen,* 1915; *Behring, Gestalt und Werk,* 1940; *Emil von Behring zum Gedachtnis,* 1942.

Shibasaburo Kitasato (1892–1931) Codiscoverer with Emil von Behring of antitoxin antibodies.

Karl Landsteiner (1868–1943) Viennese pathologist and immunologist who later worked at the Rockefeller Institute for Medical Research in New York. Received the Nobel Prize in 1930 "for his discovery of the human blood groups." He was the first to infect monkeys with poliomyelitis and syphilis to allow controlled studies of these diseases. He established the immunochemical specificity of synthetic antigens and haptens. Landsteiner felt his most important contribution was in the area of antibody–hapten interactions. *Die Spezifizität der serologiochen Reaktionen,* 1933; *The Specificity of Serological Reactions,* 1945.

Charles Robert Richet (1850–1935) Parisian physician who became professor of physiology at the University of Paris. He was interested in the physiology of toxins and, with Portier, discovered anaphylaxis, for which he was awarded the Nobel Prize in Physiology or Medicine in 1913. He and Portier discovered anaphylaxis in dogs exposed to the toxins of murine invertebrates to which they had been previously sensitized. Thus, an immune-type reaction that was harmful rather than protective was demonstrated. Experimental anaphylaxis was later shown to be similar to certain types of hypersensitivity, which lent clinical as well as theoretical significance to the discovery. *L'Anaphylaxie,* 1911.

Paul Jules Portier (1866–1962) French physiologist who, with Richet, was the first to describe anaphylaxis.

Clemens Freiherr von Pirquet (1874–1929) Viennese physician who coined the term "allergy" and described serum sickness and its pathogenesis. He also developed a skin test for tuberculosis. He held academic appointments at Vienna, Johns Hopkins, and Breslau, and returned to Vienna in 1911 as director of the University Children's Clinic. *Die Serumkrenkheit* (with Schick), 1905; *Klinische Studien über Vakzination und Vakzinale Allergie,* 1907; *Allergy,* 1911.

Gaston Ramon (1886–1963) French immunologist who perfected the flocculation assay for diphtheria toxin.

Bela Schick (1877–1967) Austro-Hungarian pediatrician whose work with von Pirquet resulted in the discovery and description of serum sickness. He developed the test for diphtheria that bears his name. *Die Serumkrankheit* (with Pirquet), 1905.

Arthur Fernandez Coca (1875–1959) American allergist and immunologist. He was a major force in allergy and immunology. He named atopic antibodies and was a pioneer in the isolation of allergens. Together with Robert A. Cooke, Coca classified allergies in humans.

Robert Anderson Cooke (1880–1960) American immunologist and allergist who was instrumental in the founding of several allergy societies. With Coca he classified allergies in humans. Cooke also pioneered skin test methods and desensitization techniques.

Felix Haurowitz (1896–1988) A noted protein chemist from Prague who later came to the U.S. He investigated the chemistry of hemoglobins. In 1930 (with Breinl) he advanced the instruction theory of antibody formation. *Chemistry and Biology of Proteins,* 1950; *Immunochemistry and Biosynthesis of Antibodies,* 1968.

Jacques Oudin (1908–1986) French immunologist who was director of analytical immunology at the Pasteur Institute, Paris. His accomplishments include discovery of idiotypy and the agar single diffusion method antigen–antibody assay.

Almroth Edward Wright (1861–1947) British pathologist and immunologist who graduated with a doctor of medicine degree from Trinity College, Dublin, in 1889. He became professor of pathology at the Army Medical School in Netley in 1892. He became associated with the Institute of Pathology at St. Mary's Hospital Medical School, London, in 1902. Together with Douglas, he formulated a theory of opsonins and perfected an antitoxoid inoculation system. Wright studied immunology in Frankfurt-am-Main under Paul Ehrlich and made important contributions to the immunology of infectious diseases and immunization. He played a significant role in the founding of the American Association of Immunologists. His published works include *Pathology and Treatment of War Wounds,* 1942; *Researches in Clinical Physiology,* 1943; *Studies in Immunology,* 2 vols., 1944.

Carl Prausnitz-Giles (1876–1963) German physician from Breslau who conducted extensive research on allergies. He and Küstner successfully transferred food allergy with serum. This became the basis for the Prausnitz-Küstner test. He worked at the State Institute for Hygiene in Breslau and spent time at the Royal Institute for Public Health in London earlier in the century. In 1933, he left Germany to practice medicine on the Isle of Wight.

Nicolas Maurice Arthus (1862–1945) Paris physician. He studied venoms and their physiological effects; he was the first to describe local anaphylaxis (the Arthus reaction) in 1903. Arthus investigated the local necrotic lesion resulting from a local antigen–antibody reaction in an immunized animal. *De l'Anaphylaxie a l'Immunite*, 1921.

Albert Calmette (1863–1933) French physician who was subdirector of the Institut Pasteur in Paris. In a popular book published in 1920, *Bacillary Infection and Tuberculosis*, he emphasized the necessity of separating tuberculin reactivity from anaphylaxis. Together with Guerin, he perfected BCG vaccine and also investigated snake venom and plague serum.

Michael Heidelberger (1888–1991) American; a founder of immunochemistry. He began his career as an organic chemist. His contributions to immunology include the perfection of quantitative immunochemical methods and the immunochemical characterization of pneumococcal polysaccharides. Heidelberger's contributions to immunologic research are legion. During his career he received the Lasker Award, the National Medal of Science, the Behring Award, the Pasteur Medal, and the French Legion of Honor. *Lectures on Immunochemistry,* 1956.

Arne W. Tiselius (1902–1971) Swedish chemist who was educated at the University of Uppsala where he also worked in research. In 1934, he was at the Institute for Advanced Study in Princeton, worked for the Swedish National Research Council in 1946, and became president of the Nobel Foundation in 1960. Awarded the Nobel Prize in Chemistry in 1948, together with Elvin A. Kabat he perfected the electrophoresis technique and classified antibodies as γ globulins. He also developed synthetic blood plasmas.

Elvin Abraham Kabat (1914–2000) American immunochemist. With Tiselius he was the first to separate immunoglobulins electrophoretically. He also demonstrated that γ globulins can be distinguished as 7S or 19S. Other contributions include research on antibodies to carbohydrates, the antibody combining site, and the discovery of immunoglobulin chain variable regions. He received the National Medal of Science. *Experimental Immunochemistry* (with Mayer), 1948; *Blood Group Substances: Their Chemistry and Immunochemistry; Structural Concepts in Immunology and Immunochemistry,* 1956.

Henry Hallett Dale (1875–1968) British investigator who made a wide range of scientific contributions including work on the chemistry of nerve impulse transmissions, the discovery of histamine, and the development of the Schultz-Dale test for anaphylaxis. He received a Nobel Prize in 1935.

John Richardson Marrack (1899–1976) British physician who served as professor of chemical pathology at Cambridge and at London Hospital. He hypothesized that antibodies are bivalent, labeled antibodies with colored dyes, and proposed a lattice theory of antigen–antibody complex formation in fundamental physicochemical studies.

William Dameshek (1900–1969) Noted Russian–American hematologist who was among the first to understand autoimmune hemolytic anemias. He spent many years as editor-in-chief of the journal *Blood.*

Orjan Thomas Gunnersson Ouchterlony (1914–) Swedish bacteriologist who developed the antibody detection test that bears his name. Two-dimensional double diffusion with subsequent precipitation patterns is the basis of the assay. *Handbook of Immunodiffusion and Immunoelectrophoresis,* 1968.

Merrill Chase (1905–) American immunologist who worked with Karl Landsteiner at the Rockefeller Institute for Medical Research, New York. He investigated hypersensitivity, including delayed type hypersensitivity and contact dermatitis. He was the first to demonstrate the passive transfer of tuberculin and contact hypersensitivity and also made contributions in the fields of adjuvants and quantitative methods.

Philip Levine (1900–1987) Russian-American immunohematologist. With Landsteiner, he conducted pioneering research on blood group antigens, including discovery of the MNP system. His work contributed much to transfusion medicine and transplantation immunobiology.

Jules Freund (1890–1960) Hungarian physician who later worked in the U.S. He made many contributions to immunology, including work on antibody formation, studies on allergic encephalomyelitis, and the development of Freund's adjuvant. He received the Lasker Award in 1959.

Hans Zinsser (1878–1940) A leading American bacteriologist and a Columbia, Stanford, and Harvard educator whose work in immunology included hypersensitivity research and plague immunology, formulation of the unitarian theory of antibodies, and demonstration of differences between tuberculin and anaphylactic hypersensitivity. His famous text, *Microbiology* (with Hiss), 1911, has been through two dozen editions since its first appearance.

Max Theiler (1899–1972) South African virologist who received the Nobel Prize in 1951 "for his development of vaccines against yellow fever."

Gregory Shwartzman (1896–1965) Russian–American microbiologist who described local and systemic reactions that follow injection of bacterial endotoxins. The systemic Shwartzman reaction, a nonimmunologic phenomenon, is related to disseminated intravascular coagulation. The local Shwartzman reaction in skin resembles the immunologically based Arthus reaction in appearance. *Phenomenon of Local Tissue Reactivity and Its Immunological and Clinical Significance,* 1937.

Robin Coombs (1921–) British pathologist and immunologist who is best known for the Coombs' test as a means for detecting immunoglobulin on the surface of a patient's red blood cells. The test was developed in the 1940s to demonstrate autoantibodies on the surface of red blood cells that failed to cause agglutination of these erythrocytes. It is a test for autoimmune hemolytic anemia. He also contributed much to serology, immunohematology, and immunopathology. *The Serology of Conglutination and Its Relation to Disease*, 1961; *Clinical Aspects of Immunology* (with Gell), 1963.

Albert Hewett Coons (1912–1978) American immunologist and bacteriologist who was an early leader in immunohistochemistry with the development of fluorescent antibodies. Coons received the Lasker medal in 1959, the Ehrlich prize in 1961, and the Behring prize in 1966.

Pierre Grabar (1898–1986) French-educated immunologist, born in Kiev, who served as chef de service at the Institut Pasteur and as director of the National Center for Scientific Research, Paris. Best known for his work with Williams in the development of immunoelectrophoresis. He studied antigen–antibody interactions and developed the "carrier" theory of antibody function. He was instrumental in reviving European immunology in the era after World War II.

Herman Nathaniel Eisen (1918–) American physician whose research contributions range from equilibrium dialysis (with Karush) to the mechanism of contact dermatitis.

Milan Hasek (1925–1985) Czechoslovakian scientist whose contributions to immunology include investigations of immunologic tolerance and the development of chick embryo parabiosis. Hasek also made fundamental contributions to transplantation biology.

Gustav Joseph Victor Nossal (1931–) Australian immunologist whose seminal works have concentrated on antibodies and their formation. He served as director of the Walter and Eliza Hall Institute of Medical Research in Melbourne. *Antibodies and Immunity*, 1969; *Antigens, Lymphoid Cells and the Immune Response* (with Ada), 1971.

Ernest Witebsky (1901–1969) German–American immunologist and bacteriologist who made significant contributions to transfusion medicine and to concepts of autoimmune diseases. He was a direct descendent of the Ehrlich school of immunology, having worked at Heidelberg with Hans Sachs, Ehrlich's principal assistant, in 1929. He came to Mt. Sinai Hospital in New York in 1934 and became professor at the University of Buffalo in 1936, where he remained until his death. A major portion of his work on autoimmunity was the demonstration, with Noel R. Rose, of experimental autoimmune thyroiditis.

Noel Richard Rose (1927–) American immunologist and authority on autoimmune diseases who first discovered, with Witebsky, experimental autoimmune thyroiditis. His subsequent contributions to immunology are legion. He has authored numerous books and edited leading journals in the field.

Peter Alfred Gorer (1897–1961) British pathologist who was professor at Guy's Hospital Medical School, London, where he made major discoveries in transplantation genetics. With Snell, he discovered the H-2 murine histocompatibility complex. Most of his work was in transplantation genetics. He identified antigen II and described its association with tumor rejection. *The Gorer Symposium*, 1985.

Peter Brian Medawar (1915–1987) British transplantation biologist who earned his Ph.D. at Oxford in 1935, where he served as lecturer in zoology. He was subsequently a professor of zoology at Birmingham (1947) and at University College, London, in 1951. He became director of the Medical Research Council in 1962 and of the Clinical Research Center at Northwick Park in 1971. Together with Billingham and Brent, he made seminal discoveries in transplantation. He shared the 1960 Nobel Prize in Physiology or Medicine with Sir MacFarlane Burnet.

Ray David Owen (1915–) American geneticist who described erythrocyte mosaicism in dizygotic cattle twins. This discovery of reciprocal erythrocyte tolerance contributed to the concept of immunologic tolerance. The observation that cattle twins, which shared a common fetal circulation, were chimeras and could not reject transplants of each other's tissues later in life provided the groundwork for Burnet's ideas about tolerance and Medawar's work in transplantation.

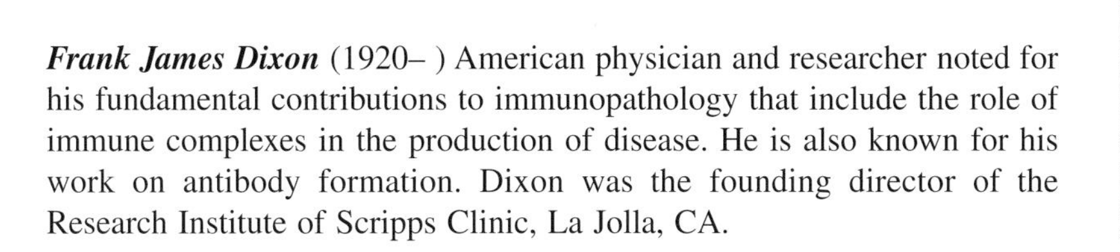

Frank James Dixon (1920–) American physician and researcher noted for his fundamental contributions to immunopathology that include the role of immune complexes in the production of disease. He is also known for his work on antibody formation. Dixon was the founding director of the Research Institute of Scripps Clinic, La Jolla, CA.

Niels Kaj Jerne (1911–1994) Immunologist, born in London and educated at Leiden and Copenhagen, who shared the Nobel Prize in 1984 with Kohler and Milstein for his contributions to immune system theory. These include his selective theory of antibody formation, the functional network of interacting antibodies and lymphocytes, and distinction of self from nonself by T lymphocytes. He studied antibody synthesis and avidity, perfected the hemolytic plaque assay, developed natural selection theory of antibody formation, and formulated the idiotypic network theory. He was named director of the Paul Ehrlich Institute in Frankfurt-am-Main in 1966 and director of the Basel Institute for Immunology in 1969.

David Wilson Talmage (1919–) American physician and investigator who in 1956 developed the cell selection theory of antibody formation. His work was a foundation for Burnet's subsequent clonal selection theory. After training in immunology with Taliaferro in Chicago where he became a professor in 1952, Talmage subsequently became chairman of microbiology in 1963, dean of medicine in 1968, and director of the Webb-Waring Institute in Denver in 1973. In addition to his investigations of antibody formation, he also studied heart transplantation tolerance. *The Chemistry of Immunity in Health and Disease* (with Cann), 1961.

Joshua Lederberg (1925–) American biochemist who made a significant contribution to immunology with his work on the clonal selection theory of antibody formation. He received a Nobel Prize in 1958 (with Beadle and Tatum) for genetic recombination and organization of genetic material in bacteria.

Henry Sherwood Lawrence (1916–) American immunologist. While studying type IV hypersensitivity and contact dermatitis, he discovered transfer factor. *Cellular and Humoral Aspects of Delayed Hypersensitivity,* 1959.

Jan Gosta Waldenström (1906–1996) Swedish physician who described macroglobulinemia, which now bears his name. He received the Gairdner Award in 1966.

Daniel Bovet (1907–1992) Primarily a pharmacologist and physiologist, Bovet received the Nobel Prize in 1957 for his contributions to the understanding of the role histamine plays in allergic reactions and the development of antihistamines. *Structure Chimique et Activite Pharmacodynamique des Medicaments du Systeme Nerveux Vegetatif*, 1948; *Curare and Curare-Like Agents*, 1959.

Frank MacFarlane Burnet (1899–1985) Australian virologist and immunologist who shared the Nobel Prize with Peter B. Medawar in 1960 for the discovery of acquired immunological tolerance. Burnet was a theoretician who made major contributions to the developing theories of self tolerance and clonal selection in antibody formation. Burnet and Fenner's suggested explanation of immunologic tolerance was tested by Medawar et al., who confirmed the hypothesis in 1953 using inbred strains of mice. *Production of Antibodies* (with Fenner), 1949; *Natural History of Infectious Diseases,* 1953; *Clonal Selection Theory of Antibody Formation,* 1959; *Autoimmune Diseases* (with Mackay), 1962; *Cellular Immunology,* 1969; *Changing Patterns* (autobiography), 1969.

George Davis Snell (1903–1996) American geneticist who shared the 1980 Nobel Prize in Physiology or Medicine with Jean Dausset and Baruj Benacerraf "for their work on genetically determined structures of the cell surface that regulate immunologic reactions." Snell's major contributions were in the field of mouse genetics including discovery of the H-2 locus and the development of congenic mice. He made many seminal contributions to transplantation genetics and received the Gairdner Award in 1976. *Histocompatibility* (with Dausset and Nathenson), 1976.

Jean Baptiste Gabriel Dausset (1916–) French physician and investigator. He pioneered research on the HLA system and the immunogenetics of histocompatibility. For this work, he shared a Nobel Prize with Benacerraf and Snell in 1980. He made numerous discoveries in immunogenetics and transplantation biology. *Immunohematologie, Biologique et Clinique*, 1956; *HLA and Disease* (with Svejaard), 1977.

Baruj Benacerraf (1920–) American immunologist born in Caracas, Venezuela. His multiple contributions include the carrier effect in delayed hypersensitivity, lymphocyte subsets, MHC, and Ir immunogenetics, for which he shared the Nobel Prize in 1980 with Jean Dausset and George Snell. Benacerraf et al. demonstrated that immune response Ir genes control an animal's response to a given antigen. These genes were localized in the I region of the MHC. *Textbook of Immunology* (with E. Unanue), 1979.

Henry George Kunkel (1916–1983) American physician and immunologist. The primary focus of his work was immunoglobulins. He characterized myeloma proteins as immunoglobulins and rheumatoid factor as an autoantibody. He also discovered IgA and idiotypy and contributed to immunoglobulin structure and genetics. Kunkel received the Lasker Award and the Gairdner Award. A graduate of Johns Hopkins Medical School, he served as professor of medicine at the Rockefeller Institute for Medical Research.

Astrid Elsa Fagraeus-Wallbom (1913–) Swedish investigator noted for her doctoral thesis which provided the first clear evidence that immunoglobulins are made in plasma cells. In 1962, she became chief of the Virus Department of the National Bacteriological Laboratory, and in 1965, professor of immunology at the Karolinska Institute in Stockholm. She also investigated cell membrane antigens and contributed to the field of clinical immunology. *Antibody Production in Relation to the Development of Plasma Cells*, 1948.

Rosalyn Sussman Yalow (1921–) American investigator who shared the 1977 Nobel Prize with Guillemin and Schally for her endocrinology research and perfection of the radioimmunoassay technique. With Berson, Yalow made an important discovery of the role antibodies play in insulin-resistant diabetes. Her technique provided a test to estimate nanogram or picogram quantities of various types of hormones and biologically active molecules, thereby advancing basic and clinical research.

J.F.A.P. Miller (1931–) Proved the role of the thymus in immunity while investigating gross leukemia in neonatal mice.

Robert Alan Good (1922–2003) American immunologist and pediatrician who has made major contributions to studies on the ontogeny and phylogeny of the immune response. Much of his work focused on the role of the thymus and the bursa of Fabricius in immunity. He and his colleagues demonstrated the role of the thymus in the education of lymphocytes. *The Thymus in Immunobiology*, 1964; *Phylogeny of Immunity*, 1966.

James Gowans (1924–) British physician and investigator whose principal contribution to immunology was the demonstration that lymphocytes recirculate via the thoracic duct, which radically changed understanding of the role lymphocytes play in immune reactions. He also investigated lymphocyte function. Gowans was made director of the MRC Cellular Immunobiology Unit, Oxford, in 1963.

Rodney Robert Porter (1917–1985) British biochemist who received the Nobel Prize in 1972, with Gerald Edelman, for their studies of antibodies and their chemical structure. Porter cleaved antibody molecules with the enzyme papain to yield Fab and Fc fragments. He suggested that antibodies have a four-chain structure. Fab fragments were shown to have the antigen-binding sites, whereas the Fc fragment conferred the antibody's biological properties. He also investigated the sequence of complement genes in the MHC. *Defense and Recognition*, 1973.

Gerald Maurice Edelman (1929–) American investigator who was professor at the Rockefeller University and shared the Nobel Prize in 1972 with Porter for their work on antibody structure. Edelman was the first to demonstrate that immunoglobulins are composed of light and heavy polypeptide chains. He also did pioneering work with the Bence-Jones protein, cell adhesion molecules, immunoglobulin amino acid sequence, and neurobiology.

Richard K. Gershon (1932–1983) One of the first to demonstrate the suppressor role of the T cell. The suppressor T cell was described as a subpopulation of lymphocytes that diminish or suppress antibody formation by B cells or down-regulate the ability of T lymphocytes to mount a cellular immune response. The inability to confirm the presence of receptor molecules on their surface has cast a cloud over the suppressor cell; however, functional suppressor cell effects are indisputable.

Kimishige Ishizaka (1925–) ***and Terako Ishizaka*** discovered IgE and have contributed to elucidation of its function.

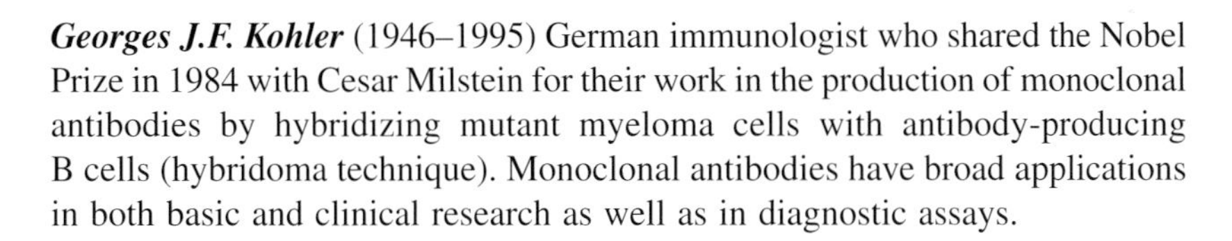

Georges J.F. Kohler (1946–1995) German immunologist who shared the Nobel Prize in 1984 with Cesar Milstein for their work in the production of monoclonal antibodies by hybridizing mutant myeloma cells with antibody-producing B cells (hybridoma technique). Monoclonal antibodies have broad applications in both basic and clinical research as well as in diagnostic assays.

Cesar Milstein (1927–2002) Immunologist born in Argentina who worked in the U.K. He shared the 1984 Nobel Prize with Kohler for their production of monoclonal antibodies by hybridizing mutant myeloma cells with antibody-producing B cells (hybridoma technique). The production of monoclonal antibodies by hybridoma technology revolutionized immunological research.

Susumu Tonegawa (1939–) Japanese-born immunologist working in the U.S. He received the Nobel Prize in 1987 for his research on immunoglobulin genes and antibody diversity. Tonegawa and many colleagues were responsible for the discovery of immunoglobulin gene C, V, J, and D regions and their rearrangement.

E. Donnall Thomas and Joseph E. Murray Recipients of the 1990 Nobel Prize for Physiology or Medicine for their work during the 1950s and 1960s on reducing the risk of organ rejection by the body's immune system. Murray performed the first successful organ transplant in the world, which was a kidney from one identical twin to another, at the Peter Bent Brigham Hospital in Boston in 1954. Two years later, Thomas was the first to perform a successful transplant of bone marrow, which he achieved by administering a drug that prevented rejection. The two doctors have made significant discoveries that "have enabled the development of organ and cell transplantation into a method for the treatment of human disease," said the Nobel Assembly in its citation for the prize.

Rolf Zinkernagel (right) (1944–) ***and Peter Doherty (left)*** (1940–) Recipients of the 1996 Nobel Prize for Physiology or Medicine for their demonstration of MHC restriction. In an investigation of how T lymphocytes protect mice against lymphocytic choriomeningitis virus (LCMV) infection, they found that T cells from mice infected by the virus killed only infected target cells expressing the same major histocompatibility complex (MHC) class I antigens but not those expressing a different MHC allele. In their study, murine cytotoxic T cells (CTL) would only lyse virus-infected target cells if the effector and target cells were H-2 compatible. This significant finding had broad implications, demonstrating that T cells did not recognize the virus directly but only in conjunction with MHC molecules.

2 Molecules, Cells, and Tissues of the Immune Response

The generation of an immune response of either the innate or acquired variety requires the interaction of specific molecules, cells, and tissues. This chapter provides an overview of these structures with brief descriptions, enhanced by schematic representations and light and electron micrographs of those elements whose interactions yield a highly tailored immune response that is critical to survival of the species. Many of the molecules of immunity are described in subsequent chapters. Adhesion molecules that are important in bringing cells together in the generation of immune responses, of directing cellular traffic through vessels or interaction of cells with matrix are presented here.

All lymphocytes in the body are derived from stem cells in the bone marrow. Those cells destined to become T cells migrate to the thymus where they undergo maturation and education prior to their residence in the peripheral lymphoid tissues. B cells undergo maturation in the bone marrow following their release. Both B and T cells occupy specific areas in the peripheral lymphoid tissues. Depictions of the thymus, lymph nodes, spleen, and other lymphoid organs are presented to give the reader a visual concept of immune system structure and development. The various cells involved in antigen presentation and development of an immune response are followed by a description of cells involved in effector immune functions. Understanding the molecules, cells, and tissues described in the pages that follow prepares the reader to appreciate the novel and fascinating interactions of these molecules and cells in the body tissues and organs which permit the generation of a highly specific immune response. Immunity may perform many vital functions; for example, the elimination of invading microbes, the activation of amplification mechanisms such as the complement pathway, or the development of protective antibodies or cytotoxic T cells that prevent the development of potentially fatal infectious diseases. By contrast, the immune system may generate responses that lead to hypersensitivity or tissue injury and disease. In either case, the process is fascinating and commands the attention and respect of the reader for Nature's incomparable versatility.

Immune: Natural or acquired resistance to a disease. Either a subclinical infection with the causative agent or deliberate immunization with antigenic substances prepared from it may render a host immune. Because of immunological memory, the immune state is heightened upon second exposure of individuals to an immunogen. A subject may become immune as a consequence of having experienced and recovered from an infectious disease.

CAM (cell adhesion molecules): Cell-selective proteins that promote adhesion of cells to one another and are calcium independent. They are believed to help direct migration of cells during embryogenesis. The majority of lymphocytes and monocytes express this antigen which is not found on other cells. The "humanized" antibodies specific for this epitope are termed Campath-1H. See CD52 in Chapter 11.

Cell adhesion molecules (CAMs) on the cell surface facilitate the binding of cells to each other in tissues as well as in cell-to-cell interaction. Most are grouped into protein families that include the integrins, selectins, mucin-like proteins, and the immunoglobulin superfamily.

Immune cell motility: Migration of immune cells is a principal host defense mechanism for the recruitment of leukocytes to inflammatory sites in the development of cell-mediated immunity. The induction of migratory responses follows the interaction of signal molecules with plasma membrane receptors, initiating cytoskeletal reorganization and changes in cell shape. Motile responses may be random, chemokinetic, chemotactic, or haptotactic. Random migration of unstimulated motility in chemokinetic migration, i.e., stimulated random movement of cells without a stimulus gradient, are motile responses that are not consistently directional. By contrast, responses that are directional include those that are chemotactic and haptotactic. They take place when cells are subjected to a signal gradient, and the cells migrate toward an increasing concentration of the stimulus. The various motile responses may participate in the mobilization of immune cells to sites of inflammation.

The immune system includes the molecules, cells, tissues, and organs that are associated with adaptive immunity such as the host defense mechanisms.

Immune system anatomy: The lymphocyte is the cell responsible for immune response specificity. The human mature lymphoid system is comprised of 2×10^{12}

lymphocytes together with various accessory cells that include epithelial cells, monocytes/macrophages, and other antigen-presenting cells. Accessory cells are a requisite for both maturation and effective functioning of lymphocytes. The thymus is the site of maturation of T cells and the bone marrow is the maturation site of B cells. These two tissues comprise the primary lymphoid organs. The secondary lymphoid organs consist of the cervical lymph nodes, ancillary lymph nodes, spleen, mesenteric lymph nodes, and inguinal lymph nodes. Mature lymphocytes migrate from the central lymphoid organs by way of the blood vessels to the secondary or peripheral tissues and organs, where they respond to antigen. Peripheral lymphoid tissues comprise the spleen, lymph nodes, and mucosa-associated lymphoid tissue (MALT) which is associated with the respiratory, genitourinary, and gastrointestinal tracts, making up 50% of the lymphoid cells of the body. The mucosa-associated lymphoid system consists of the adenoids, tonsils, and mucosa-associated lymphoid cells of the respiratory, genitourinary, gastrointestinal tracts, and Peyer's patches in the gut.

Immunity refers to a state of acquired or innate resistance or protection from a pathogenic microorganism or its products or from the effect of toxic substances such as snake or insect venom.

Cluster of differentiation (CD): The designation of antigens on the surface of leukocytes by their reactions with clusters of monoclonal antibodies. The antigens are designated as clusters of differentiation (CDs).

CD (cluster of differentiation): See cluster of differentiation. Molecular weights for the CD designations in this book are given for reduced conditions.

CD antigens are a cluster of differentiation antigens identified by monoclonal antibodies. The CD designation refers to a cluster of monoclonal antibodies, all of which have identical cellular reaction patterns and identify the same molecular species. Anti-CD refers to antiidiotype and should not be employed to name CD monoclonal antibodies. The CD designation was subsequently used to describe the recognized molecule but it had to be clarified by using the terms *antigen* or *molecule*. CD nomenclature is used by most investigators to designate leukocyte surface molecules. Provisional clusters are designated as CDw. CD antigen is a molecule of the cell membrane that is employed to differentiate human leukocyte subpopulations based upon their interaction with monoclonal antibody. The monoclonal antibodies that interact with the same membrane molecule are grouped into a common cluster of differentiation or CD.

CD molecules are cell surface molecules found on immune system cells that are designated "cluster of differentiation" or CD followed by a number such as CD33.

CMI is the abbreviation for cell-mediated immunity.

Immunoblast: Lymphoblast.

Immunochemistry is that branch of immunology concerned with the properties of antigens, antibodies, complement, T cell receptors, MHC molecules, and all the molecules and receptors that participate in immune interactions *in vivo* and *in vitro*. Immunochemistry aims to identify active sites in immune responses and define the forces that govern antigen–antibody interaction. It is also concerned with the design of new molecules such as catalytic antibodies and other biological catalysts. Also called molecular immunology.

Immunocompetent is an adjective that describes a mature functional lymphocyte that is able to recognize a specific antigen and mediate an immune response. It also may refer to the immune status of a human or other animal to indicate that the individual is capable of responding immunologically to an immunogenic challenge.

Selectins are a group of cell adhesion molecules (CAMs) that are glycoproteins and play an important role in the relationship of circulating cells to the endothelium. The members of this surface molecule family have three separate structural motifs. They have a single N-terminal (extracellular) lectin motif preceding a single epidermal growth factor repeat and various short consensus repeat homology units. They are involved in lymphocyte migration. These carbohydrate-binding proteins facilitate adhesion of leukocytes to endothelial cells. There is a single-chain transmembrane glycoprotein in each of the selectin molecules with a similar modular structure that includes an extracellular calcium-dependent lectin domain. The three separate groups of selectins include L-selectin (CD62L), expressed on leukocytes, P-selectin (CD62P), expressed on platelets and activated endothelium, and E-selectin (CD62E), expressed on activated endothelium. Under shear forces their characteristic structural motif is comprised of an N-terminal lectin domain, a domain with homology to epidermal growth factor (EGF), and various complement regulatory protein repeat sequences. See also E-selectin, L-selectin, P-selectin, and CD62.

Immunocyte, literally *immune cell*, is a term sometimes used by pathologists to describe plasma cells in stained tissue sections, e.g., in the papillary or reticular dermis in erythema multiforme.

Mucins are heavily glycosylated serine- and threonine-rich proteins that serve as ligands for selectins.

Immunocytochemistry refers to the visual recognition of target molecules, tissues, and cells through the specific reaction of antibody with antigen by using antibodies labeled with indicator molecules. By tagging an antibody with a fluorochrome, color-producing enzyme, or metallic particle, the target molecules can be identified.

MEL-14 is a selectin on the surface of lymphocytes significant in lymphocyte interaction with endothelial cells of peripheral lymph nodes. Selectins are important for adhesion despite shear forces associated with circulating blood. MEL-14 is lost from the surface of both granulocytes and T lymphocytes following their activation. MEL-14 combines with phosphorylated oligosaccharides.

MEL-14 antibody identifies a gp90 receptor that permits lymphocyte binding to peripheral lymph node high endothelial venules. Immature double-negative thymocytes comprise cells that vary from high to low in MEL-14 content. The gp90 MEL-14 epitope is a glycoprotein on murine lymph node lymphocyte surfaces. MEL-14 antibody prevents these lymphocytes from binding to postcapillary venules. The gp90 MEL-14 is apparently a lymphocyte homing receptor that directs these cells to lymph nodes in preference to lymphoid tissue associated with the gut.

Immunologic (or immunological) is an adjective referring to those aspects of a subject that fall under the purview of the scientific discipline of immunology.

An **immunological reaction** is an *in vivo* or *in vitro* response of lymphoid cells to an antigen they have never previously encountered or to an antigen for which they are already primed or sensitized. An immunological reaction may consist of antibody formation, cell-mediated immunity, or immunological tolerance. The humoral antibody and cell-mediated immune reactions may mediate either protective immunity or hypersensitivity, depending on various conditions.

A **microenvironment** is an organized, local interaction among cells that provides an interactive, dynamic, structural, or functional compartmentalization. The microenvironment may facilitate or regulate cell and molecular interactions through biologically active molecules. Microenvironments may exert their influence at the organ, tissue, cellular, or molecular levels. In the immune system, they include the thymic cortex and the thymic medulla, which are distinct; the microenvironment of lymphoid nodules; and a microenvironment of B cells in a lymphoid follicle, among others.

Microfilaments are cellular organelles that comprise a network of fibers of about 60 Å in diameter present beneath the membranes of round cells, occupying protrusions of the cells, or extending down microprojections such as microvilli. They are found as highly organized and prominent bundles of filaments, concentrated in regions of surface activity during motile processes or endocytosis. Microfilaments consist mainly of actin, a globular 42-kDa protein. In media of appropriate ionic strength, actin polymerizes in a double array to form microfilaments which are critical for cell movement, phagocytosis, fusion of phagosome and lysosome, and other important functions of cells belonging to the immune and other systems.

An **immunologically activated cell** is the term for an immunologically competent cell following its interaction with antigen. This response may be expressed either as lymphocyte transformation, immunological memory, cell-mediated immunity, immunologic tolerance, or antibody synthesis.

Bystander effects are indirect, nonantigen-specific phenomena that result in polyclonal responses. In contrast to antigen-specific interactions, bystander effects are the result of cellular interactions that take place without antigen recognition or under conditions where antigen and receptors for antigen are not involved. Bystander effects are phenomena linked to the specific immune response in that they do not happen on their own but only in connection with a specific response. Cells not directly involved in the antigen-specific response are transstimulated or "carried along" in the response.

Innocent bystander refers to a cell that is fatally injured during an immune response specific for a different cell type.

Bystander lysis refers to tissue cell lysis that is nonspecific. The tissue cells are not the specific targets during an immune response but are killed as innocent bystanders because of their close proximity to the site where nonspecific factors are released near the actual target of the immune response. Bystander lysis may occur by the Fas/FasL pathway depending on the polarity and kinetics of FasL surface expression and downregulation after TCR engagement. This cytotoxicity pathway may give rise to bystander lysis of Fas^+ target cells.

An **immunologically competent cell** is a lymphocyte, such as a B cell or T cell, that can recognize and respond to a specific antigen.

An **immunologist** is a person who makes a special study of immunology.

Immunology is that branch of biomedical science concerned with the response of the organism to immunogenic (antigenic) challenge, the recognition of self from nonself, and all the biological (*in vivo*), serological (*in vitro*), physical, and chemical aspects of immune phenomena.

Immunophysiology refers to the physiologic basis of immunologic processes.

LCAM is the abbreviation for leukocyte cell adhesion molecule.

A **neural cell adhesion molecule-L1 (NCAM-L1)** is a member of the Ig gene superfamily. Although originally identified in the nervous sytem, NCAM-L1 is also expressed in hematopoietic and epithelial cells. It may function in cell–cell and cell–matrix interactions. NCAM-L1 can support homophilic NCAM-L1–NCAM-L1 and integrin cell-binding. It can also bind with high affinity to the neural proteoglycan eurocan. NCAM-L1 promotes neurite outgrowth by functioning in neurite extension.

Activated leukocyte cell adhesion molecule (ALCAM/CD166) is a member of the immunoglobulin (Ig) gene superfamily. It is expressed by activated leukocytes and lymphocyte antigen CD6. The extracellular region of ALCAM contains five Ig-like domains. The N-terminal Ig domain binds specifically to CD6. ALCAM-CD6 interactions have been implicated in T cell development and regulation of T cell function. ALCAM may also play a role in progression of human melanoma.

Ligand refers to a molecule or part of a molecule that binds or forms a complex with another molecule such as a cell surface receptor. A ligand is any molecule that a receptor recognizes.

Cell surface receptors and ligands: Activation of caspases via ligand binding to cell surface receptors involves the TNF family of receptors and ligands. These receptors contain an 80-amino acid death domain (DD) that through homophilic interactions recruits adaptor proteins to form a signaling complex on the cytosolic surface of the receptor. The signaling induced by the ligand binding to the receptor appears to involve trimerization. Based on x-ray crystallography, the trimeric ligand has three equal faces; a receptor monomer interacts at each of the three junctions formed by the three faces. Thus, each receptor polypeptide contacts two ligands. The bringing together of three receptors, thereby orienting the intracellular DDs, appears to be the critical feature for signaling by these receptors. The adaptor proteins recruited to the aligned receptor DDs recruit either caspases or other signaling proteins. The exact mechanism by which recruitment of caspases-8 to the DD-induced complex causes activation of caspases-8 is not clear.

Homing receptors are molecules on a cell surface that direct traffic of that cell to a precise location in other tissues or organs. For example, lymphocytes bear surface receptors that facilitate their attachment to high endothelial cells of postcapillary venules in lymph nodes. Adhesion molecules present on lymphocyte surfaces enable lymphocytes to recirculate and home to specific tissues. Homing receptors bind to ligands termed *addressins* found on endothelial cells in affected vessels.

Adhesion molecules are extracellular matrix proteins that attract leukocytes from the circulation. For example, T and B lymphocytes possess lymph node homing receptors on their membranes that facilitate passage through high endothelial venules. Neutrophils migrate to areas of inflammation in response to endothelial leukocyte adhesion molecule-1 (ELAM-1) stimulated by TNF and IL-1 on the endothelium of vessels. B and T lymphocytes that pass through high endothelial venules have lymph node homing receptors. Adhesion molecules mediate cell adhesion to their surroundings and to neighboring cells. In the immune system, adhesion molecules are critical to most aspects of leukocyte function, including lymphocyte recirculation through lymphoid organs, leukocyte recruitment into inflammatory sites, antigen-specific recognition, and wound healing. The five principal structural families of adhesion molecules are (1) integrins, (2) immunoglobulin superfamily (IgSF) proteins, (3) selectins, (4) mucins, and (5) cadherins.

Neuropilin is a cell-surface protein that is a receptor for the collapsin/semaphorin family of neuronal guidance proteins.

Adhesion molecule assays: Cell adhesion molecules are cell-surface proteins involved in the binding of cells to each other, to endothelial cells, or to the extracellular matrix. Specific signals produced in response to wounding and infection control the expression and activation of the adhesion molecule. The interactions and responses initiated by the binding of these adhesion molecules to their receptors/ligands play important roles in the mediation of the inflammatory and immune reaction. The immediate response to a vessel wall injury is the adhesion of platelets to the injury site and the growth, by further aggregation of platelets, of a mass which tends to obstruct (often incompletely) the lumen of the damaged vessel. This platelet mass is called a **hemostatic plug**. The exposed basement membranes at the sites of injury are the substrate for platelet adhesion, but deeper tissue components may have a similar effect. Far from being static, the hemostatic plug has a continuous tendency to break up with new masses reformed immediately at the original site.

Integrins are a family of cell membrane glycoproteins that are heterodimers comprised of α and β chain subunits. They serve as extracellular matrix glycoprotein receptors. They identify the RGD sequence of the β subunit, which consists of the arginine-glycine-aspartic acid tripeptide that occasionally also includes serine. The RGD sequence serves as a receptor recognition signal. Extracellular matrix glycoproteins, for which integrins serve as receptors, include fibronectin, C3, and lymphocyte function-associated antigen 1 (LFA-1), among other proteins. Differences in the β chain serve as the basis for division of integrins into three categories. Each category has distinctive α chains. The β chain provides specificity. The same 95-kDa β chain is found in one category of integrins that includes LFA-1, p150,95, and complement receptor 3 (CR3). The same 130-kDa β chain is shared among VLA-1, VLA-2, VLA-3, VLA-4, VLA-5, VLA-6, and integrins found in chickens. A 110-kDa β chain is shared in common by another category that includes the vitronectin receptor and platelet glycoprotein IIb/IIIa. There are four repeats of 40 amino acid residues in the β chain extracellular domains. There are 45 amino acid residues in the β chain intracellular domains. The principal function of integrins is to link the cytoskeleton to extracellular ligands. They also participate in wound healing, cell migration, killing of target cells, and in phagocytosis. Leukocyte adhesion deficiency syndrome occurs when the β subunit of LFA-1 and Mac-1 is missing. VLA proteins facilitate binding of cells to collagen (VLA-1, VLA-2, and VLA-3), laminin (VLA-1, VLA-2, and VLA-6), and fibronectin (VLA-3, VLA-4, and VLA-5). The cell-to-cell contacts formed by integrins are critical for many aspects of the immune response such as antigen presentation, leukocyte-mediated cytotoxicity, and myeloid cell phagocytosis. Integrins comprise an essential part of an adhesion receptor cascade that guides leukocytes from the bloodstream across endothelium and into injured tissue in response to chemotactic signals.

Substrate adhesion molecules (SAM) are extracellular molecules that share a variety of sequence motifs with other adhesion molecules. Most prominent among these are segments similar to the type III repeats of fibronectin and immunoglobulin-like domains. In contrast to other morphoregulatory molecules, SAMs do not have to be made by the cells that bind them. SAMs can link and influence the behavior of one another. Examples include glycoproteins, collagens, and proteoglycans.

During chemokine-induced lymphocyte polarization, the cytoskeletal protein **moesin** is important for the redistribution of adhesion molecules to the cellular uropod.

Homing-cell adhesion molecule (H-CAM) is also known as CD44, $gp^{90\ hermes}$, GP85/Pgp-1, and ECMRIII. It is a lymphocyte transmembrane glycoprotein with a molecular weight of 85 to 95 kDa and is expressed in macrophages, granulocytes, fibroclasts, endothelial cells, and epithelial cells. H-CAM has been found to bind to extracellular matrix molecules such as collagen and hyaluronic acid. H-CAM is also an important signal transduction protein during lymphocyte adhesion as it has been demonstrated that phosphorylation by kinase C and acylation by acyl-transferases enhance H-CAM's interaction with cytoskeletal proteins.

Immunoreceptor tyrosine-based activation motif (ITAM): Amino acid sequences in the intracellular portion of signal-transducing cell surface molecules that are sites of tyrosine phosphorylation and of association with tyrosine kinases and phosphotyrosine-binding proteins that participate in signal transduction. Examples include Igα, Igβ, CD3 chains, and several Ig Fc receptors. Following receptor–ligand binding and phosphorylation, docking sites are formed for other molecules that participate in maintaining cell-activating signal transduction mechanisms.

ITAMs: Abbreviation for immunoreceptor tyrosine-based activation motifs.

Immunoreceptor tyrosine-based inhibition motif (ITIM): Motifs with effects that oppose those of immunoreceptor tyrosine-based activation motifs (ITAMs). These amino acids in the cytoplasmic tail of transmembrane molecules bind phosphate groups added by tyrosine kinases. This six-amino acid (isoleucine-X-tyrosine-X-X-leucine) motif is present in the cytoplasmic tails of immune system inhibitory receptors that include Fc RIIB on B lymphocytes and the killer inhibitory receptor (KIR) on the NK cells. Following receptor–ligand binding and phosphorylation on their tyrosine residue, a docking site is formed for protein tyrosine phosphatases that inhibit other signal transduction pathways, thereby negatively regulating cell activation.

Adhesion receptors (Figure 2.1) are proteins in cell membranes that facilitate the interaction of cells with

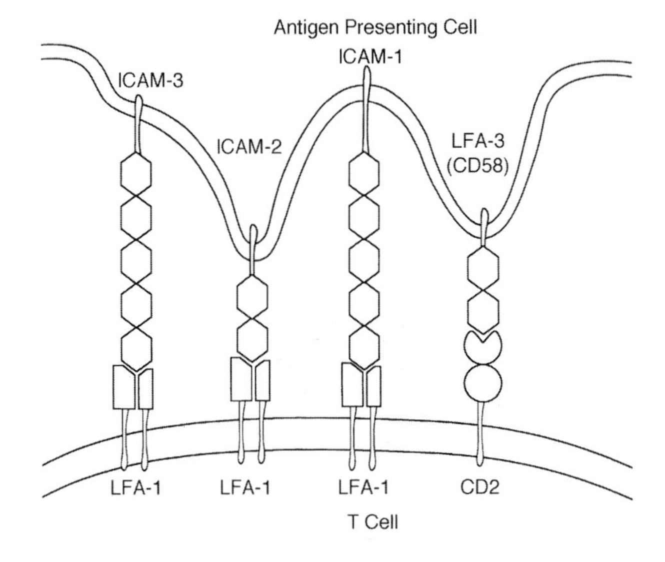

FIGURE 2.1 Adhesion receptors.

matrix. They play a significant role in adherence and chemoattraction in cell migration. They are divided into three groups that include the immunoglobulin superfamily which contains the T cell receptor/CD3, CD4, CD8, MHC class I, MHC class II, sCD2/LFA-2, LFA-3/CD58, ICAM-1, ICAM-2, and VCAM-2. The second group of adhesion receptors is made up of the integrin family which contains LFA-1, Mac-1, p150,95, VLA-5, VLA-4/LPAM-1, LPAM-2, and LPAM-3. The third family of adhesion receptors consists of selectin molecules that include Mel-14/LAM-1, ELAM-1, and CD62.

Heparin is a glycosaminoglycan comprised of two types of disaccharide repeating units. One is comprised of D-glucosamine and D-glucuronic acid, whereas the other is comprised of D-glucosamine and L-iduronic acid. Heparin is extensively sulfated and is an anticoagulant. It unites with an antithrombin III which can unite with and block numerous coagulation factors. It is produced by mast cells and endothelial cells. It is found in the lungs, liver, skin, and gastrointestinal mucosa. Because of its anticoagulant properties, heparin is useful for treatment of thrombosis and phlebitis.

Heparan sulfate is a glycosaminoglycan that resembles heparin and is comprised of the same disaccharide repeating unit. Yet, it is a smaller molecule and less sulfated than heparin. An extracellular substance, heparan sulfate is present in the lungs, arterial walls, and on numerous cell surfaces.

Reactive oxygen intermediates (ROIs) are highly reactive compounds that include superoxide anion (O_2), hydroxyl radicals (OH), and hydrogen peroxide (H_2O_2) that are produced in cells and tissues. Phagocytes use ROIs to form oxyhalides that injure ingested microorganisms. Release from cells may induce inflammatory responses leading to tissue injury.

4-1BB is a TNF receptor family molecule that binds specifically to 4-1BB ligand.

4-1BB ligand (4-1BBL) is a TNF family molecule that binds to 4-1BB.

Integrin family of leukocyte adhesive proteins: The CD11/CD18 family of molecules.

Integrins, HGF/SF activation of: Integrins and growth factor receptors can share common signaling pathways. Each type of receptor can impact the signal and ultimate response of the other. An example of a growth factor that has been shown to influence members of the integrin family of cell adhesion receptors is hepatocyte growth factor/scatter factor (HGF/SF). HGF/SF is a multifunctional cytokine that promotes mitogenesis, migration, invasion, and morphogenesis. HGF/SF-dependent signaling can modulate integrin function by promoting aggregation and cell adhesion.

Morphogenic responses to HGF/SF are dependent on adhesive events. HGF/SF-induced effects occur via signaling of the MET tyrosine kinase receptor, following ligand binding. HGF/SF binding to MET leads to enhanced integrin-mediated B cell and lymphoma cell adhesion. Blocking experiments with monoclonal antibodies directed against integrin subunits indicate that $\alpha_4\ \beta_1$ and $\alpha_5\ \beta_1$ integrins on hematopoietic progenitor cells are activated by HGF/SF to induce adhesion to fibronectin. The HGF/SF-dependent signal transduction pathway can also induce ligand-binding activity in functionally inactive $\alpha_v\ \beta_3$ integrins. These effects elicited by HGF/SF highlight the importance of growth factor regulation of integrin function in both normal and tumor cells.

LFA-1, LFA-2, LFA-3: See leukocyte functional antigens.

Lymphocyte function-associated antigen-1 (LFA-1) (Figure 2.2) is a glycoprotein comprised of a 180-kDa α chain and a 95-kDa β chain expressed on lymphocyte and phagocytic cell membranes. LFA-1's ligand is the intercellular adhesion molecule 1 (ICAM-1). It facilitates natural killer cell and cytotoxic T cell interaction with target cells. Complement receptor 3 and p150,95 share the same specificity of the 769-amino acid residue β chain found

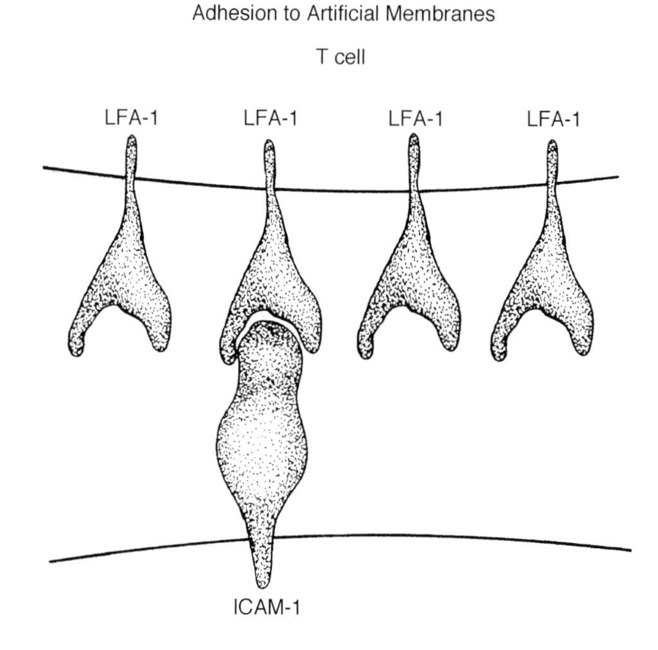

FIGURE 2.2 Lymphocyte function-associated antigen-1.

in LFA-1. A gene on chromosome 16 encodes the α chain whereas a gene on chromosome 21 encodes the β chain. This leukocyte integrin (β_2) adhesion molecule has a critical role in adhesion of leukocytes to each other and to other cells as well as microbial recognition by phagocytes. FLA-1 binds not only ICAM-1 but also ICAM-2 or ICAM-3. LFA-1-dependent cell adhesion is dependent on temperature, magnesium, and cytoskeleton. LFA-1 induces costimulatory signals that are believed to be significant in leukocyte function. LFA-1 function is critical to most aspects of the immune response. Also referred to as CD11a/CD18. See CD11a and CD18.

Lymphocyte function-associated antigen-2 (LFA-2): See CD2.

LFA-2 is a T cell antigen that is the receptor molecule for sheep red cells and is also referred to as the T11 antigen. The molecule has a 50-kDa mol wt. The antigen also seems to be involved in cell adherence, probably binding LFA-3 as its ligand.

LFA-3 is a 60-kDa polypeptide chain expressed on the surfaces of B cells, T cells, monocytes, granulocytes, platelets, fibroblasts, and endothelial cells of vessels. LFA-3 is the ligand for CD2 and is encoded by genes on chromosome 1 in man.

Lymphocyte function-associated antigen-3 (LFA-3) (Figure 2.3 and Figure 2.4) is a 60-kDa polypeptide chain expressed on the surfaces of B cells, T cells, monocytes, granulocytes, platelets, fibroblasts, and endothelial cells of vessels. LFA-3 or CD58 is expressed either as a transmembrane or a lipid-linked cell surface protein. The transmembrane form consists of a 188-amino acid extracellular region, a 23-amino acid transmembrane hydrophobic region, and a 12-amino acid intracellular hydrophilic region ending in the C-terminus. LFA-3 expression by antigen-presenting cells that include dendritic cells, macrophages, and B lymphocytes point to a possible role in regulating the immune response.

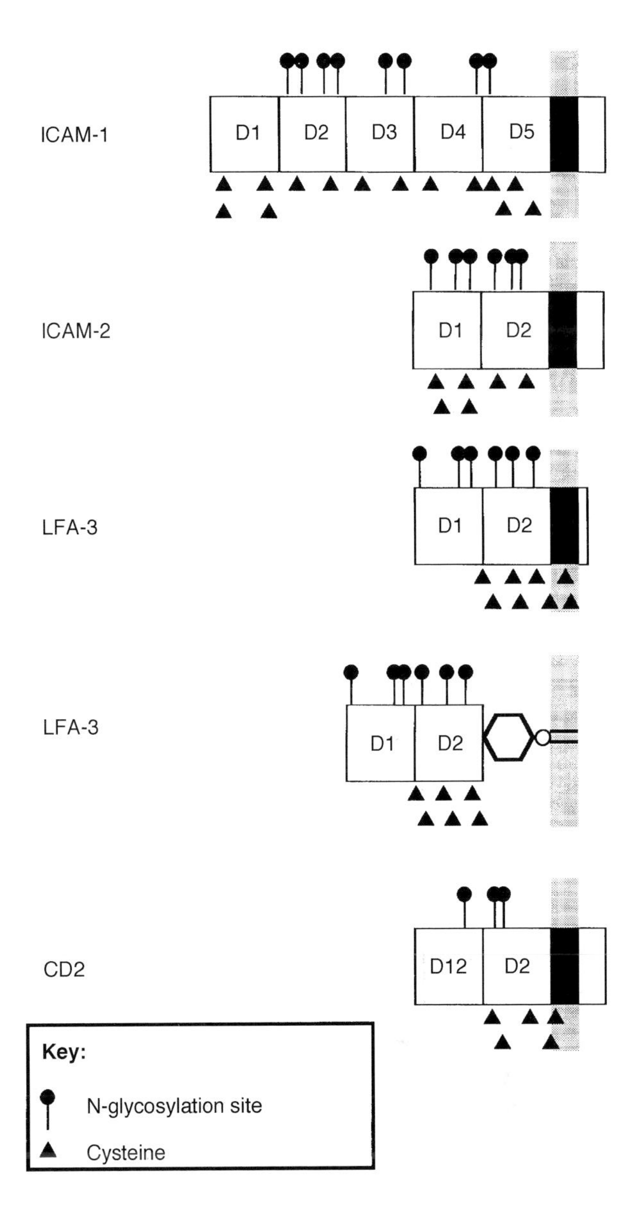

FIGURE 2.4 Immunoglobulin superfamily adhesion receptors.

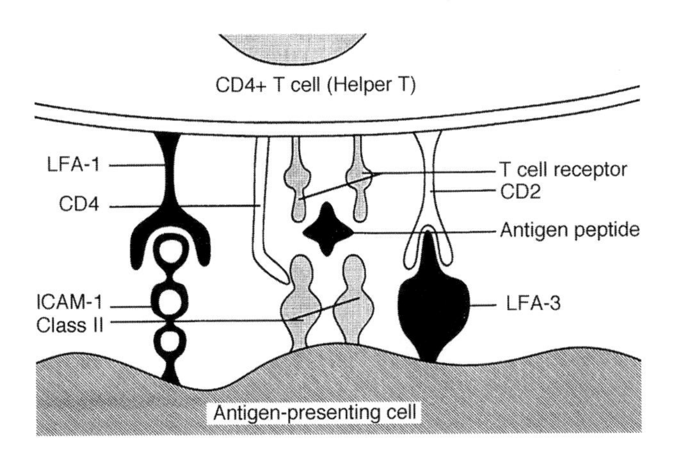

FIGURE 2.3 Lymphocyte function-associated antigen-3.

Intercellular adhesion molecule-1 (ICAM-1) (Figure 2.8) is a 90-kDa cellular membrane glycoprotein that occurs in multiple cell types including dendritic cells and endothelial cells. It is the lymphocyte function-associated antigen-1 (LFA-1) ligand. The LFA-1 molecules on cytotoxic T lymphocytes (CTL) interact with ICAM-1 molecules found on CTL target cells. Interferon γ, tumor necrosis factor, and IL-1 can elevate ICAM-1 expression. ICAM-1 is a member of the immunoglobulin gene superfamily of cell adhesion molecules. It plays a major role in the inflammatory response and in T cell-mediated host responses serving as

FIGURE 2.5 CD2 ribbon diagram. Resolution 2.0 Å.

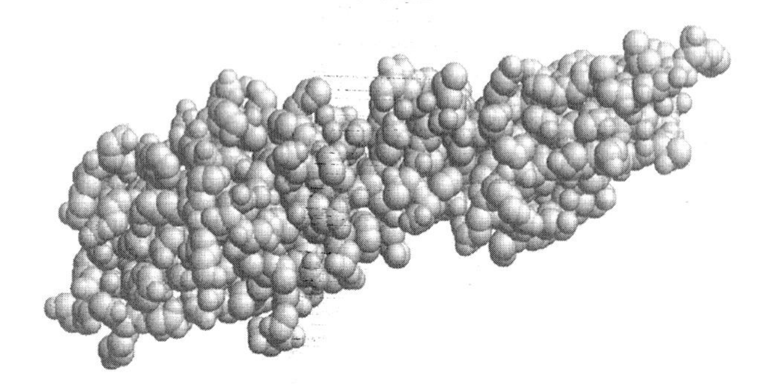

FIGURE 2.6 CD2 space fill. Resolution 2.0 Å.

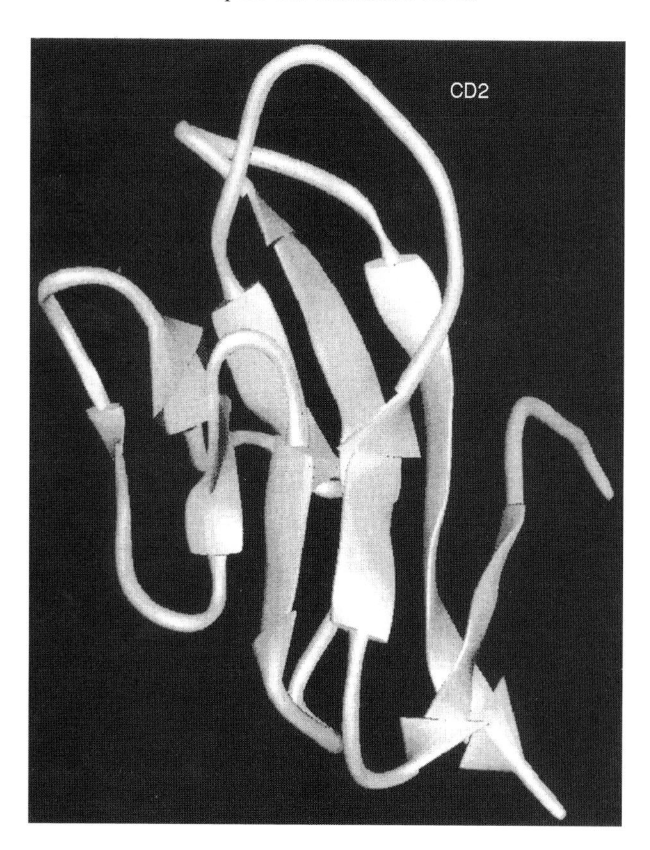

FIGURE 2.7 CD2 ribbon structure.

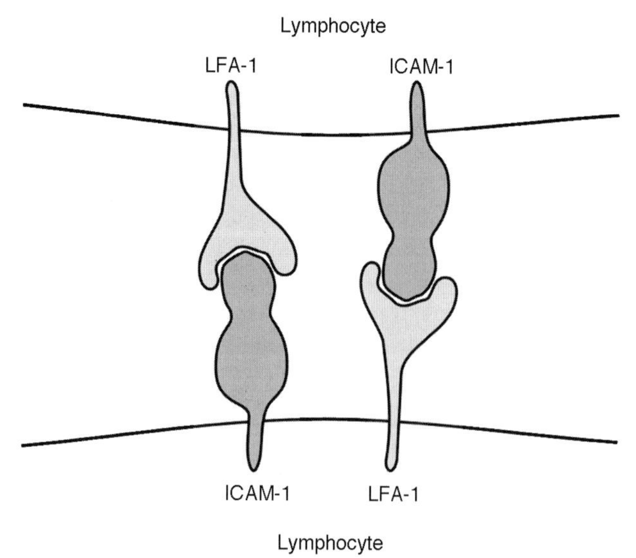

FIGURE 2.8 ICAM-1.

FIGURE 2.9 ICAM-2. Ribbon diagram. Resolution 2.2 Å.

a costimulatory molecule on antigen-presenting cells to activate MHC class II restricted T cells and on other types of cells in association with MHC class I to activate cytotoxic T cells. On endothelial cells, it facilitates migration of activated leukocytes to the site of injury. It is the cellular receptor for a subgroup of rhinoviruses.

ICAM-1 (intercellular adhesion molecule-1) is a γ interferon-induced protein which is needed for the migration of polymorphonuclear neutrophils into areas of inflammation.

Intercellular adhesion molecule-2 (ICAM-2) (Figure 2.9) is a protein that is a member of the immunoglobulin superfamily that is important in cellular interactions. It is a cell surface molecule that serves as a ligand for leukocyte integrins. ICAM-2 facilitates lymphocytes binding to antigen-presenting cells or to endothelial cells. It binds to LFA-1, a T lymphocyte integrin.

ICAM-2: See intercellular adhesion molecule.

Intercellular adhesion molecule-3 (ICAM-3) is a leukocyte cell surface molecule that plays a critical role in the interaction of T lymphocytes with antigen presenting cells. The interaction of the T lymphocyte with an antigen presenting cell through union of ICAM-1, ICAM-2, and ICAM-3 with LFA-1 molecules is also facilitated by the interaction of the T-cell surface molecule CD2 with LFA-3 present on antigen-presenting cells.

ICAM-3: See intercellular adhesion molecule-3.

Very late activation antigens (VLA molecules) are β-1 integrins that all have the CD19 β chain in common. They were originally described on T lymphocytes grown in long-term culture but were subsequently found on additional types of leukocytes and on cells other than blood cells. VLA proteins facilitate leukocyte adherence to vascular endothelium and extracellular matrix. Resting T lymphocytes express VLA-4, VLA-5, and VLA-6. VLA-4 is expressed on multiple cells that include thymocytes, lymphocytes in blood, B and T cell lines, monocytes, NK cells, and eosinophils. The extracellular matrix ligand for VLA-4 and VLA-5 is fibronectin, and for VLA-6 it is laminin. The binding of these molecules to their ligands gives T lymphocytes costimulator signals. VLA-5 is present on monocytes, memory T lymphocytes, platelets, and fibroblasts. It facilitates B and T cell binding to fibronectin. VLA-6, which is found on platelets, T cells, thymocytes, and monocytes, mediates platelet adhesion to laminin. VLA-3, a laminin receptor, binds collagen and identifies fibronectin. It is present on B cells, the thyroid, and the renal glomerulus. Platelet VLA-2 binds to collagen only, whereas endothelial cell VLA-2 combines with collagen and laminin. Lymphocytes bind through VLA-4 to high endothelial venules and to endothelial cell surface proteins (VCAM-1) in areas of inflammation. VLA-1, which is present on activated T cells, monocytes, melanoma cells, and smooth muscle cells, binds collagen and laminin.

VLA receptors refer to a family of integrin receptors found on cell surfaces. They consist of α and β transmembrane chain heterodimers. There is a VLA-binding site at the arginine-glycine-aspartamine sequences of vitronectin and fibronectin. VLA receptors occur principally on T lymphocytes. They also bind laminin and collagen. They participate in cell–extracellular matrix interactions.

Vascular cell adhesion molecule-1 (VCAM-1) (Figure 2.10 and Figure 2.11) is a molecule that binds lymphocytes and monocytes. It is found on activated endothelial cells, dendritic cells, tissue macrophages, bone marrow fibroblasts, and myoblasts. VCAM-1 belongs to the immunoglobulin gene superfamily and is a ligand for VLA-4 (integrin α4/β1) and integrin α4/β7. It plays an important role in leukocyte recruitment to inflammatory sites and facilitates lymphocyte, eosinophil, and monocyte adhesion to activated endothelium. It participates in lymphocyte–dendritic cell interaction in the immune response.

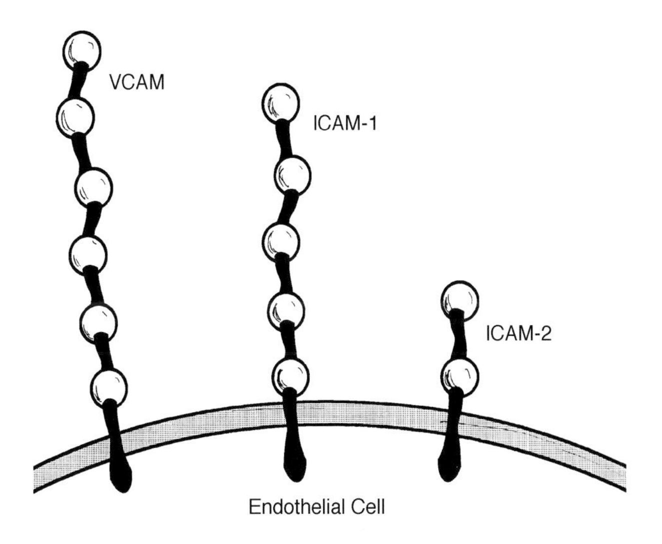

FIGURE 2.10 VCAM-1 bound to an endothelial cell.

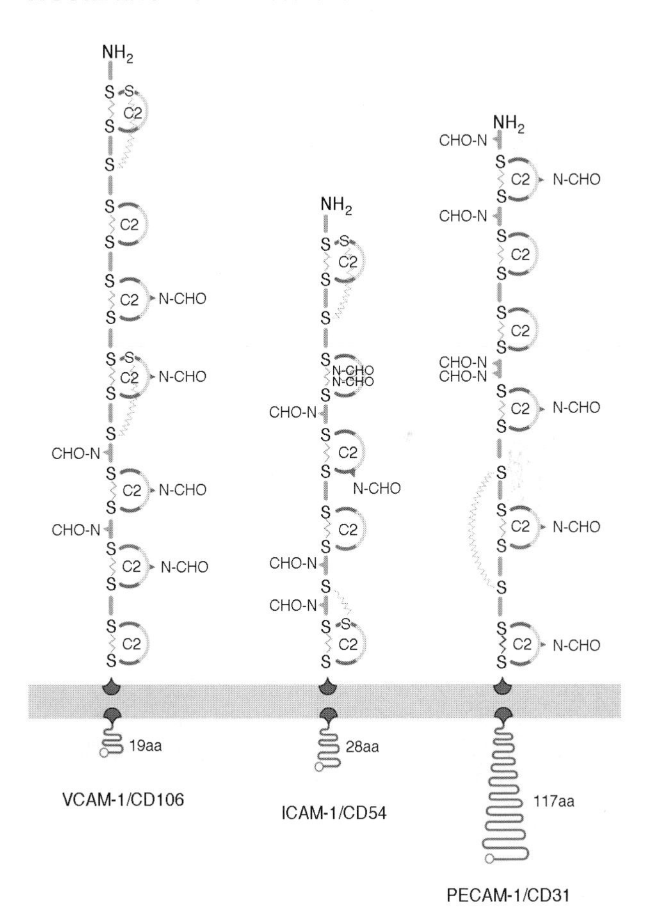

FIGURE 2.11 Schematic representation of VCAM-1.

Platelet endothelial cell adhesion molecule-1 (PECAM-1) (CD31) is an antigen that is a single-chain membrane glycoprotein with a 140-kDa mol wt. It is found on granulocytes, monocytes, macrophages, B cells, platelets, and endothelial cells. Although it is termed gpIIa′, it is different from the CD29 antigen. At present the function of CD31 is unknown. It may be an adhesion molecule.

PECAM (CD31): An immunoglobulin-like molecule present on leukocytes and at endothelial cell junctions. These molecules participate in leukocyte–endothelial cell interactions, as during an inflammatory response.

Endothelial leukocyte adhesion molecule-1 (ELAM-1) facilitates focal adhesion of leukocytes to blood vessel walls. It is induced by endotoxins and cytokines and belongs to the adhesion molecule family. ELAM-1 is considered to play a significant role in the pathogenesis of atherosclerosis and infectious and autoimmune diseases. Neutrophil and monocyte adherence to endothelial cells occurs during inflammation *in vivo* where there is leukocyte margination and migration to areas of inflammation. Endothelial cells activated by IL-1 and TNF synthesize ELAM-1, at least in culture. A 115-kDa chain and a 100-kDa chain comprise the ELAM-1 molecule.

ELAM-1 (endothelial leukocyte adhesion molecule-1) is a glycoprotein of the endothelium that facilitates adhesion of neutrophils. Structurally, it has an epidermal growth factor-like domain, a lectin-like domain, amino acid sequence homology with complement-regulating proteins, and six tandem-repeated motifs. Tumor necrosis factor, interleukin-1, and substance P induce its synthesis. Its immunoregulatory activities include attraction of neutrophils to inflammatory sites and mediating cell adhesion by sialyl-Lewis X, a carbohydrate ligand. It acts as an adhesion molecule or addressin for T lymphocytes that home to the skin.

Endothelin is a peptide comprised of 21 amino acid residues that is derived from aortic endothelial cells and is a powerful vasoconstrictor. A gene on chromosome 6 encodes the molecule. It produces an extended pressor response, stimulates release of aldosterone, inhibits release of renin, and impairs renal excretion. It is elevated in myocardial infarction and cardiogenic shock, major abdominal surgery, pulmonary hypertension, and uremia. It may have a role in the development of congestive heart failure.

Gatekeeper effect refers to contraction of endothelium mediated by IgE, permitting components of the blood to gain access to the extravascular space as a consequence of increased vascular permeability.

Fibronectin is an adhesion-promoting dimeric glycoprotein found abundantly in the connective tissue and basement membrane. The tetrapeptide Arg-Gly-Asp-Ser facilitates cell adhesion to fibrin, Clq, collagens, heparin, and types I-, II-, III-, V-, and VI-sulfated proteoglycans. Fibronectin is also present in plasma and on normal cell surfaces. Approximately 20 separate fibronectin chains are known. They are produced from the fibronectin gene by alternative splicing of the RNA transcript. Fibronectin is comprised of two 250-kDa subunits joined near their carboxy-terminal ends by disulfide bonds. The amino acid residues in the subunits vary in number from 2145 to 2445. Fibronectin is important in contact inhibition, cell movement in embryos, cell-substrate adhesion, inflammation, and wound healing. It may also serve as an opsonin.

Fibrinogen is one of the largest plasma proteins and has a mol wt of 330 to 340 kDa, comprising more than 3000 amino acid residues. The concentration in the plasma ranges between 200 and 500 mg/l00 ml. The molecule contains 3% carbohydrate, about 28 to 29 disulfide linkages, and one free sulfhydryl group. Fibrinogen exists as a dimer and can be split into two identical sets comprising three different polypeptide chains. Fibrinogen is susceptible to enzymatic cleavage by a variety of enzymes. The three polypeptide chains of fibrinogen are designated Aα, Bβ, and γ. By electron microscopy the dried fibrinogen molecule shows a linear arrangement of three nodules, 50 to 70 Å in diameter, connected by a strand about 15 Å thick.

Fibrinopeptides are released by the conversion of fibrinogen into fibrin. Thrombin splits fragments from the N-terminal region of Aα and Bβ chains of fibrinogen. The split fragments are called fibrinopeptide A and B, respectively, and are released in the fluid phase. They may be further degraded and may apparently have vasoactive functions. The release rate of fibrinopeptide A exceeds that of fibrinopeptide B and this differential release may play a role in the propensity of nascent fibrin to polymerize.

Fibrin is a protein responsible for the coagulation of blood. It is formed through the degradation of fibrinogen into fibrin monomers. Polymerization of the nascent fibrin molecules (comprising the α, β, and γ chains) occurs by end-to-end as well as lateral interactions. The fibrin polymer is envisaged as having two chains of the triad structure lying side by side in a staggered fashion in such a way that two terminal nodules are associated with the central nodule of a third molecule. The chains may also be twisted around each other. The fibrin polymer thus formed is stabilized under the action of a fibrin-stabilizing factor, another component of the coagulation system. Fibrinogen may also be degraded by plasmin. In this process, a number of intermediates, designated as fragments X, Y, D, and E, are formed. These fragments interfere with polymerization of fibrin by binding to nascent intact fibrin molecules, thus causing a defective and unstable polymerization. Fibrin itself is also cleaved by plasmin into similar

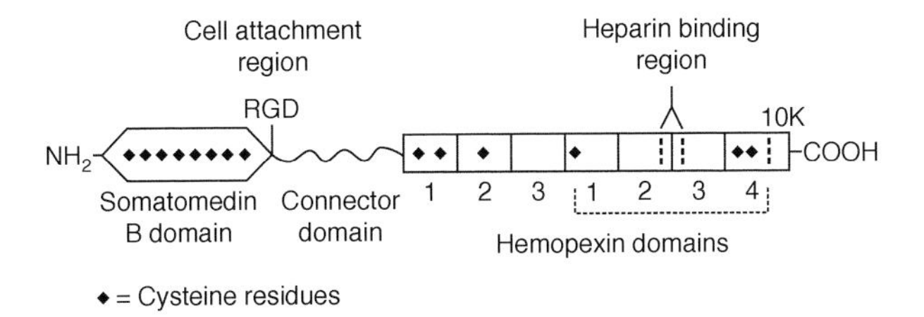

FIGURE 2.12 Vitronectin.

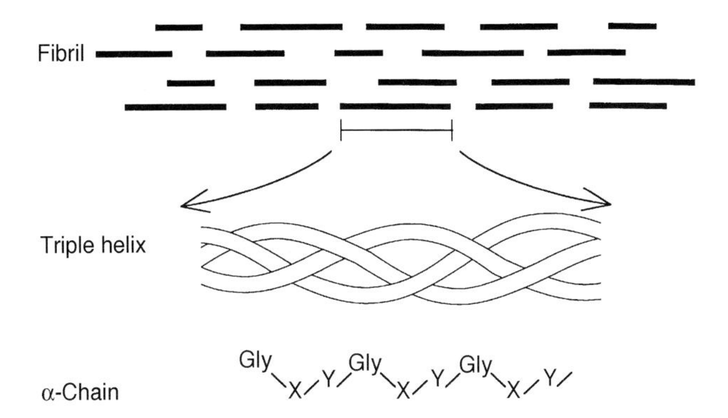

FIGURE 2.13 Collagen.

but shorter fragments collectively designated fibrin degradation products. Of course, any excess of such fragments will impair the normal coagulation process — an event with serious clinical significance. Abzymes, such as thromboplastin activator linked to an antibody specific for antigens in fibrin that are not present in fibrinogen, are used clinically to lyse fibrin clots obstructing coronary arteries in myocardial infarction patients.

Vitronectin (Figure 2.12), a cell adhesion molecule that is a 65-kDa glycoprotein, is found in the serum at a concentration of 20 mg/l. It combines with coagulation and fibrinolytic proteins and with C5b67 complex to block its insertion into lipid membranes. Vitronectin appears in the basement membrane, together with fibronectin in proliferative vitreoretinopathy. It decreases nonselective lysis of autologous cells by insertion of soluble C5b67 complexes from other cell surfaces. Vitronectin is also called *epibolin* and *protein S*.

In plasma, 65-kDa and 75-kDa glycoproteins that facilitate adherence of cells as well as the ability of cells to spread and to differentiate are known as **serum spreading factors**.

Collagen (Figure 2.13) is a 285-kDa extracellular matrix protein that contains proline, hydroxyproline, lysine, hydroxylysine, and glycine 30%. The structure consists of a triple helix of 95-kDa polypeptides forming a tropocollagen molecule that is resistant to proteases. Collagen types other than IV form fibrils with quarter stagger overlap between molecules that provide a fibrillar structure which resists tension. Several types of collagen have been described and most of them can be crosslinked through lysine side chain.

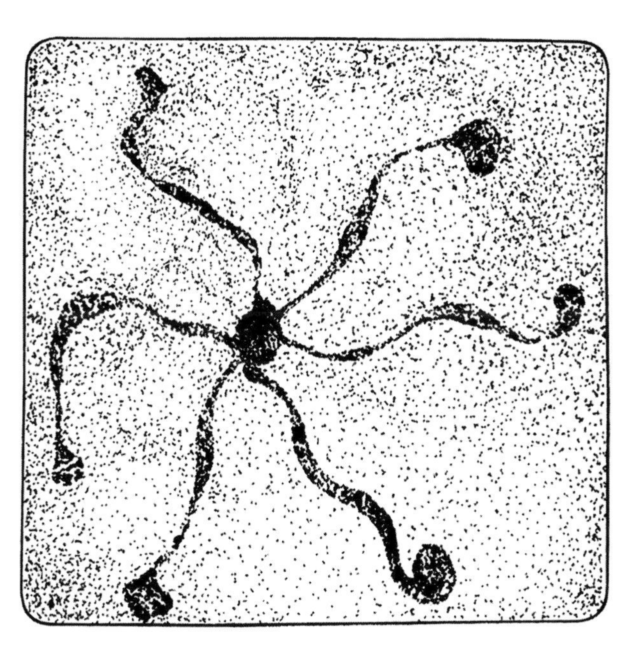

FIGURE 2.14 Tenascin.

Tenascin (Figure 2.14) is a matrix protein produced by embryonic mesenchymal cells. It facilitates epithelial tissue differentiation and consists of six 210-kDa proteins that are all alike.

Laminin (Figure 2.15) is a relatively large (820-kDa) basement membrane glycoprotein comprised of three polypeptide subunits. It belongs to the integrin receptor family which includes a 400-kDa α heavy chain and two 200-kDa light chains designated β-1 and β-2. By electron microscopy the molecule is arranged in the form of a cross. The domain structures of the α and β chains resemble one another. There are six primary domains. Domains I and II have repeat sequences forming α helices. Domains III and V are comprised of cysteine-rich repeating sequences. The globular regions are comprised of domains IV and VI. There is an additional short cysteine-rich α domain between domains I and II in the β-1 chain. There is a relatively large globular segment linked to the C-terminal of domain I, designated the "foot" in the α chain. Five "toes" on the foot contain repeat sequences. Laminins have biological functions and characteristics that include facilitation of cellular adhesion and linkage to other basement membrane constituents such as collagen type IV, heparan, and glycosaminoglycans. Laminins also facilitate neurite regeneration, an

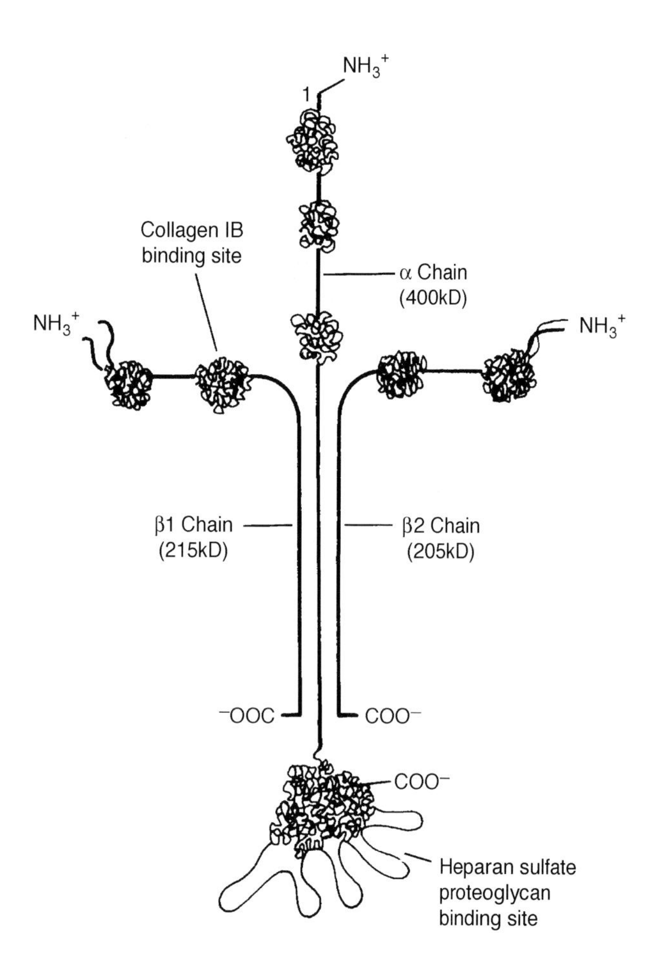

FIGURE 2.15 Laminin.

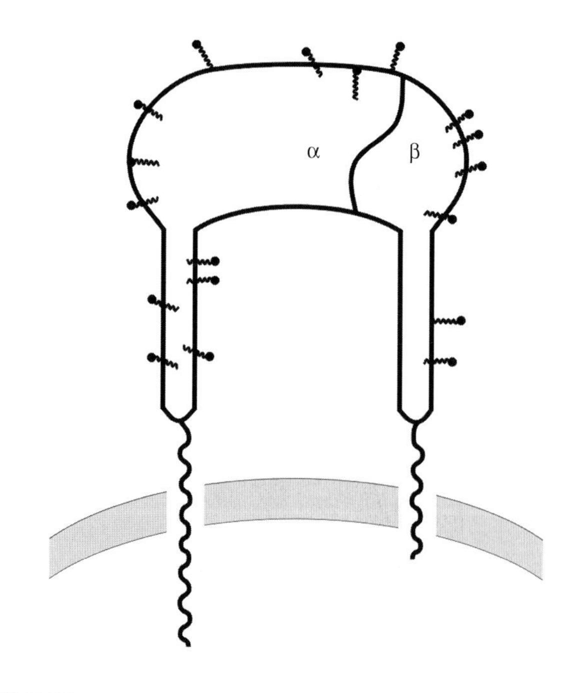

FIGURE 2.16 Mac-1.

activity associated with the foot of the molecule. There is more than one form of laminin, each representing different gene products, even though they possess a high degree of homology. S-laminin describes a form found only in synaptic and nonmuscle basal lamina. This is a single 190-kDa polypeptide (in the reduced form) and is greater than 1000 kDa in the nonreduced form. It is associated with the development or stabilization of synapses. S-laminin is homologous to the β-1 chain of laminin. Laminin facilitates cell attachment and migration. It plays a role in differentiation and metastasis and is produced by macrophages, endothelial cells, epithelial cells, and Schwann cells.

The **laminin receptor** is a membrane protein comprised of two disulfide bond-linked subunits, one relatively large and one relatively small. Its function appears to be for attachment of cells and for the outgrowth of neurites. It may share structural similarities with fibronectin and vitronectin, both of which are also integrins.

MAC-1 (Figure 2.16) is found on mononuclear phagocytes, neutrophils, NK cells, and mast cells. It is an integrin molecule comprised of an alpha chain (CD11b) linked noncovalently to a beta chain (CD18) that is the same as the beta chains of LFA-1 and of p150,95. MAC-1 facilitates phagocytosis of microbes that are coded with iC3b. It also facilitates neutrophil and monocyte adherence to the endothelium.

CD11: A "family" of three leukocyte-associated single chain molecules that has been identified in recent years (sometimes referred to as the LFA/Mac-1 family). They all consist of two polypeptide chains; the larger of these chains (α) is different for each member of the family; the smaller chain (β) is common to all three molecules (see CD11a, CD11b, CD11c).

CD11a: α chain of the LFA-1 molecule with a 180-kDa mol wt. It is present on leukocytes, monocytes, macrophages, and granulocytes but negative on platelets. LFA-1 binds the intercellular adhesion molecules ICAM-1 (CD54), ICAM-2, and ICAM-3.

A **human T lymphocyte** encircled by a ring of sheep red blood cells is referred to as an E rosette. This was used previously as a method to enumerate T lymphocytes (Figure 2.17).

GlyCAM-1 is a molecule resembling mucin that is present on high endothelial venules in lymphoid tissues. L-selectin molecules on lymphocytes in the peripheral blood bind GlyCAM-1 molecules, causing the lymphocytes to exit the blood circulation and circulate into the lymphoid tissues.

LAM-1 (leukocyte adhesion molecule-1) is a homing protein found on membranes, which combines with target

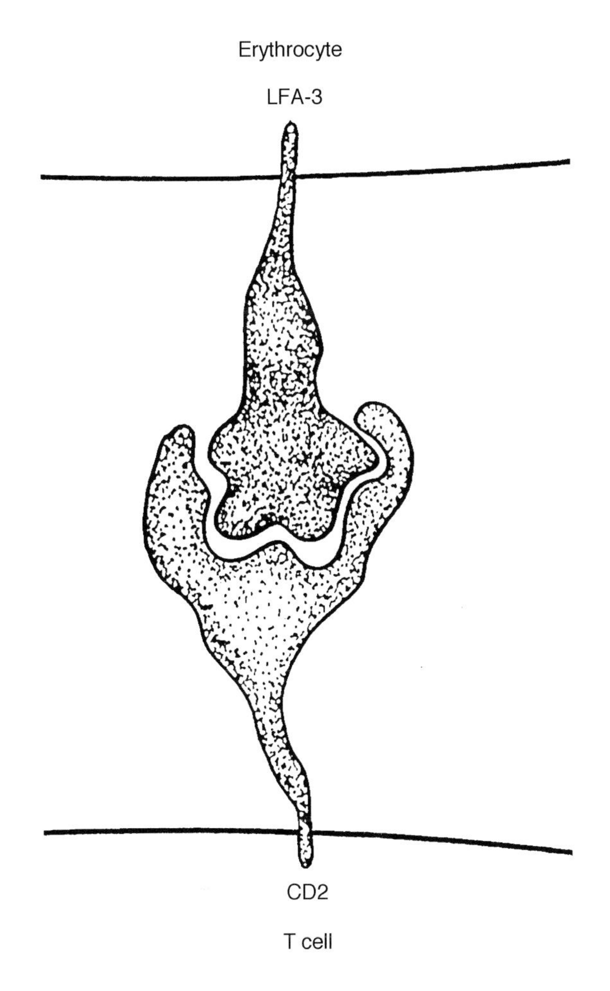

FIGURE 2.17 E rosette.

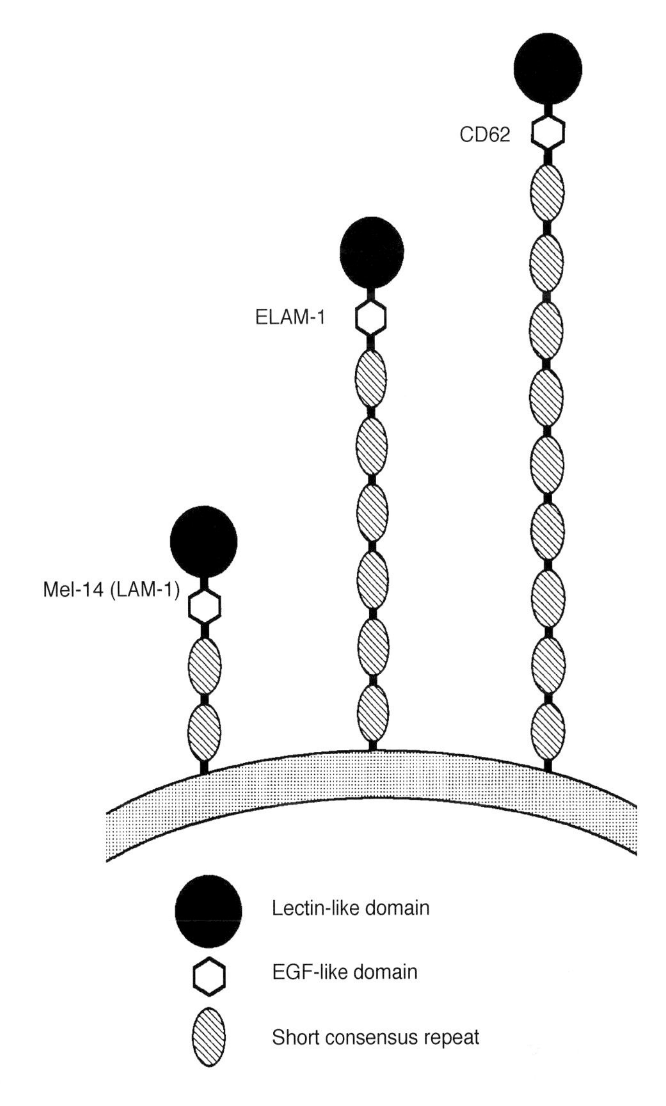

FIGURE 2.18 Selectins.

cell-specific glycoconjugates. It helps to regulate migration of leukocytes through lymphocyte binding to high endothelial venules and to regulate neutrophil adherence to endothelium at inflammatory sites.

CD44 is a transmembrane molecule with 80- to 90-kDa mol wt. It is found on some white and red cells and is weakly expressed on platelets. It functions probably as a homing receptor. CD44 is a receptor on cells for hyaluronic acid; it binds to hyaluronate. It mediates leukocyte adhesion. CD44 is a ubiquitous multistructural and multifunctional cell surface glycoprotein that participates in adhesive cell-to-cell and cell-to-matrix interactions. It also plays a role in cell migration and cell homing. Its main ligand is hyaluronic acid (HA), hyaluronate, hyaluronan. CD44 is expressed by numerous cell types of lymphohematopoietic origin including erythrocytes, T and B lymphocytes, natural killer cells, macrophages, Kupffer cells, dendritic cells, and granulocytes. It is also expressed in other types of cells such as fibroblasts and CNS cells. Besides hyaluronic acid, CD44 also interacts with other ECM ligands such as collagen, fibronectin, and laminin. In addition to function stated above, CD44 facilitates lymph node homing via binding to high endothelial venules, presentation of chemokines or growth factors to migrating cells, and growth signal transmission. CD44 concentration may be observed in areas of intensive cell migration and proliferation as in wound healing, inflammation, and carcinogenesis. Many cancer cells and their metastases express high levels of CD44. It may be used as a diagnostic or prognostic marker for selected human malignant diseases.

E-selectin (CD62E) (Figure 2.18) is a molecule found on activated endothelial cells, which recognizes sialylated Lewis X and related glycans. Its expression is associated with acute cytokine-mediated inflammation.

CD62E is a 140-kDa antigen present on endothelium. CD62E is endothelium leuckocyte adhesion molecule (ELAM). It mediates neutrophil rolling on the endothelium. It also binds sialyl-Lewis X.

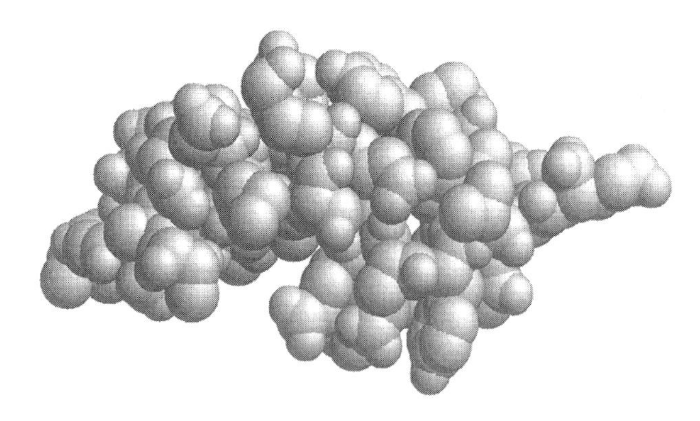

FIGURE 2.19 P-selectin NMR.

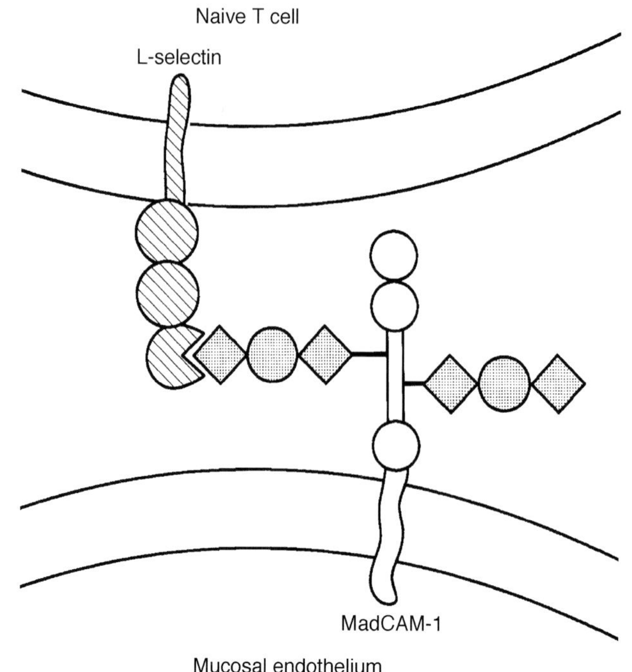

FIGURE 2.20 MadCAM-1.

P-selectin (CD62P) (Figure 2.19) is a molecule found in the storage granules of platelets and the Weibel-Palade bodies of endothelial cells. Ligands are sialylated Lewis X and related glycans. P-selectins are involved in the binding of leukocytes to endothelium and platelets to monocytes in areas of inflammation.

Weibel-Palade bodies are P-selectin granules found in endothelial cells. P-selectin is translocated rapidly to the cell surface following activation of an endothelial cell by such mediators as histamine and C5a.

CD62P is a 75- to 80-kDa antigen present on endothelial cells, platelets, and megakaryocytes. CD62P is an adhesion molecule that binds sialyl-Lewis X. It is a mediator of platelet interaction with monocytes and neutrophils. It also mediates neutrophil rolling on the endothelium. It is also referred to as P-selectin, GMP-140, PADGEM, or LECAM-3.

L-selectin (CD62L) is an adhesion molecule of the selectin family found on lymphocytes that is responsible for the homing of lymphocytes to lymph node high endothelial venules where it binds to CD34 and GlyCAM-1. This induces the migration of lymphocytes into tissues. L-selectin is also found on neutrophils where it acts to bind the cells to activated endothelium early in the inflammatory process.

CD62L, a 150-kDa antigen present on B and T cells, monocytes, and NK cells, is a leukocyte adhesion molecule (LAM). It mediates cell rolling on the endothelium. It also binds CD34 and GlyCAM. CD62L is also referred to as L-selectin, LECAM-1, or LAM-1.

Addressin is a molecule such as a peptide or protein that serves as a homing device to direct a molecule to a specific location (an example is ELAM-1). Lymphocytes from Peyer's patches home to mucosal endothelial cells bearing ligands for the lymphocyte homing receptor. Mucosal addressin cell adhesion molecule-1 (MadCAM) is the Peyer's patch addressin in the intestinal wall that links to the integrin $\alpha 4\beta 7$ on T lymphocytes that home to the intestine. Thus, endothelial cell addressins in separate anatomical locations bind to lymphocyte homing receptors leading to organ-specific lymphocyte homing.

Vascular addressins are mucin-like molecules on endothelial cells that bind selected leukocytes to particular anatomical cites.

LPAM-1 is a combination of $\alpha 4$ and $\beta 7$ integrin chains that mediate the binding of lymphocytes to the high endothelial venules of Peyer's patch in mice. The addressin for LPAM-1 is MadCAM-1.

MadCAM-1 (Figure 2.20) facilitates access of lymphocytes to the mucosal lymphoid tissue, as in the gastrointestinal tract.

Cadherins are one of four specific families of cell adhesion molecules that enable cells to interact with their environment. Cadherins help cells to communicate with other cells in immune surveillance, extravasation, trafficking, tumor metastasis, wound healing, and tissue localization. Cadherins are calcium dependent. The five different cadherins include N-cadherin, P-cadherin, T-cadherin, V-cadherin, and E-cadherin. Cytoplasmic domains of cadherins may interact with proteins of the cytoskeleton. They may bind to other receptors based on homophilic specificity, but they still depend on intracellular interactions linked to the cytoskeleton.

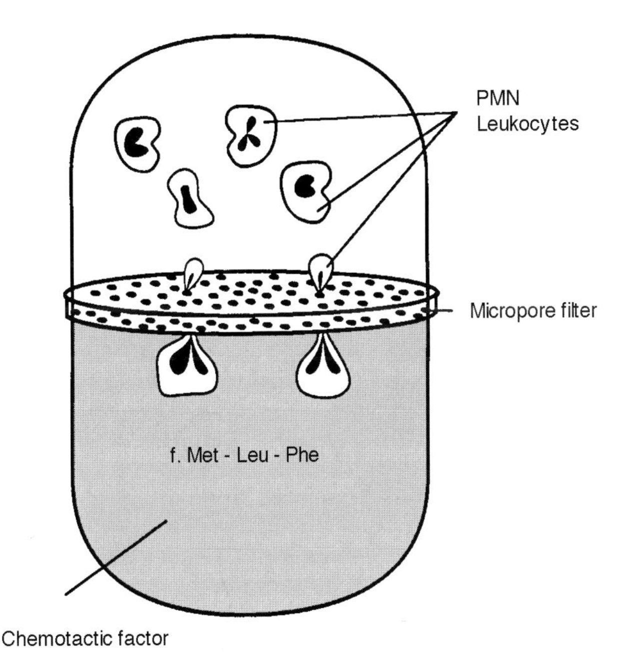

FIGURE 2.21 Chemotaxis.

E-cadherin and its associated cytoplasmic proteins α-, β-, and γ-catenin play an important role in epithelial cell–cell adhesion and in the maintenance of tissue architecture. Downregulation or mutation of the E-cadherin/catenin genes can disrupt intercellular adhesion, which may lead to cellular transformation and tumor progression.

Chemotaxis (Figure 2.21) is the process whereby chemical substances direct cell movement and orientation. The orientation and movement of cells in the direction of a chemical's concentration gradient are *positive chemotaxis* , whereas movement away from the concentration gradient is termed *negative chemotaxis* . Substances that induce chemotaxis are referred to as *chemotaxins* and are often small molecules, such as C5a, formyl peptides, lymphokines, bacterial products, leukotriene B4, etc., that induce positive chemotaxis of polymorphonuclear neutrophils, eosinophils, and monocytes. These cells move into inflammatory agents by chemotaxis. A dual chamber device called a *Boyden chamber* is used to measure chemotaxis in which phagocytic cells in culture are separated from a chemotactic substance by a membrane. The number of cells on the filter separating the cell chamber from the chemotaxis chamber reflects the chemotactic influence of the chemical substance for the cells.

Chemotactic factors include substances of both endogenous and exogenous origin. Among them are bacterial extracts, products of tissue injury, chemical substances, various proteins, and secretory products of cells. The most important among them are those generated from complement and described as anaphylatoxins. This name is related to their concurrent ability of stimulating the release of mediators from mast cells. Some chemotactic factors act specifically in directing migration of certain cell types. Others have a broader spectrum of activity. Many of them have additional activities besides acting as chemotactic factors. Such effects of aggregation and adhesion of cells, discharge of lysosomal enzymes, and phagocytosis by phagocytic cells may be concurrently stimulated. Participation in various immunologic phenomena such as cell triggering of cell–cell interactions is known for certain chemotactic factors. The structure of chemotactic factors and even the active region in their molecules have been determined in many instances. However, advances in the clarification of their mechanism of action have been facilitated by the use of synthetic oligopeptides with chemotactic activities. The specificity of such compounds depends both on the nature of the amino acid sequence and the position of amino acids in the peptide chain. Methionine at the NH2-terminal is essential for chemotactic activity. Formylation of Met leads to a 3,000- to 30,000-fold increase in activity. The second position from the NH2-terminal is also essential, and Leu, Phe, and Met in this position are essentially equivalent. Positively charged His and negatively charged Glu in this position are significantly less active, substantiating the role of a neutral amino acid in the second position at the N-terminal.

Directed migration of cells, known as chemotaxis, is mediated principally by the complement components C5a and C5a-des Arg. Neutrophil chemoattractants also include bacterial products such as *N*-formyl methionyl peptides, fibrinolysis products, oxidized lipids such as leukotriene B_4, and stimulated leukocyte products. Interleukin 8 is chemotactic for polymorphonuclear neutrophils. Chemokines that are chemotactic for polymorphonuclear neutrophils include epithelial cell-derived neutrophil activating peptide (ENA-78), neutrophil activating peptide 2 (NAP-2), growth-related oncogene (GRO-α, GRO-β, and GRO-γ), and macrophage inflammatory protein-2α and β (MIP-2α and MIP-2β). Polypeptides with chemotactic activity mainly for mononuclear cells (β chemokine) include monocyte chemoattractant protein-1, 2, and 3 (MCP-1, MCP-2, and MCP-3), macrophage inflammatory protein-1 (MIP-1) α and β, and RANTES. These chemotactic factors are derived from both inflammatory and noninflammatory cells including neutrophils, macrophages, smooth muscle cells, fibroblasts, epithelial cells, and endothelial cells. MCP-1 participates in the recruitment of monocytes in various pathologic or physiologic conditions. Neutrophil chemotaxis assays are performed using the microchamber technique. Chemotactic assays are also useful to reveal the presence of chemotaxis inhibitors in serum.

Chemotactic receptors are specific cellular receptors for chemotactic factors. In bacteria, such receptors are designated sensors and signalers and are associated with various transport mechanisms. The cellular receptors for chemotactic factors have not been isolated and characterized. In leukocytes, the chemotactic receptor appears to activate a serine proesterase enzyme, which sets in motion the sequence of events related to cell locomotion. The receptors appear specific for the chemotactic factors under consideration, and apparently the same receptors mediate all types of cellular responses inducible by a given chemotactic factor. However, these responses can be dissociated from each other, suggesting that binding to the putative receptor initiates a series of parallel, interdependent, and coordinated biochemical events leading to one or another type of response. Using a synthetic peptide *N*-formylmethionyl-leucyl-phenylalanine, about 2000 binding sites have been demonstrated per PMN leukocyte. The binding sites are specific, have a high affinity for the ligand, and are saturable. Competition for the binding sites is shown only by the parent or related compounds; the potency of the latter varies. Positional isomers may inhibit binding. Full occupancy of the receptors is not required for a maximal response, and occupancy of only 10 to 20% of them is sufficient. The presence of spare receptors may enhance the sensitivity in the presence of small concentrations of chemotactic factors and may contribute to the detection of a gradient. There also remains the possibility that some substances with chemotactic activity do not require specific binding sites on cell membranes.

Chemokinesis refers to the determination of the rate of movement or random motion of cells by chemical substances in the environment. The direction of cellular migration is determined by chemotaxis, not chemokinesis.

Leukotaxis is chemotaxis of leukocytes.

A **Boyden chamber** (Figure 2.22) is a two-compartment structure used in the laboratory to assay chemotaxis. The two chambers in the apparatus are separated by a micropore filter. The cells to be tested are placed in the upper chamber and a chemotactic agent such as F-met-leu-phe is placed in the lower chamber. As cells in the upper chamber settle to the filter surface, they migrate through the pores if the agent below chemoattracts them. On staining of the filter, cell migration can be evaluated.

EMF-1 (embryo fibroblast protein-1) is a chemokine of the α family (CXC family). It has been found in chicken fibroblasts and mononuclear cells, yet no human or murine homolog is known. Cultured chick embryo fibroblasts (CEFs) abundantly express the avian gene 9E3/CEF-4.

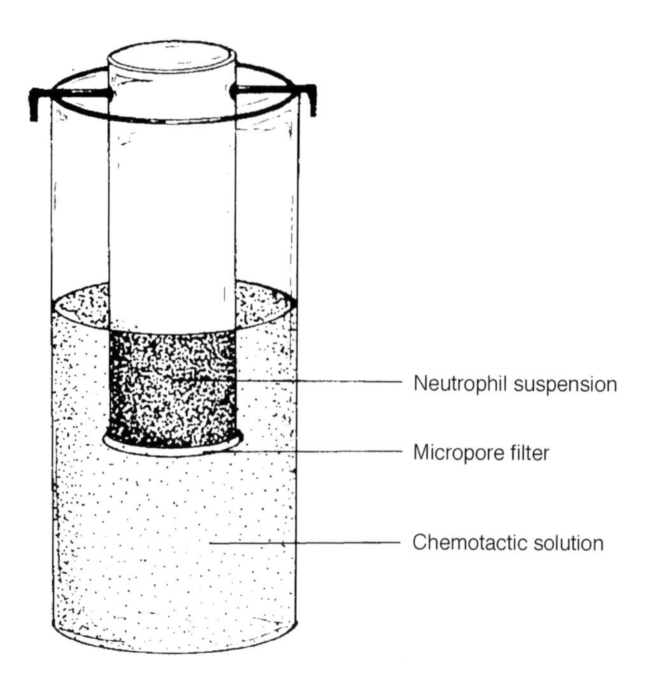

FIGURE 2.22 Boyden chamber.

The EMF-1 gene was isolated from RSV-transformed CEF identified by differential screening of a cDNA library. EMF-1 is characterized as a chemokine because its sequence resembles that of CTAP-III and PF4. RSV-infected cells represent the tissue source. Fibroblasts and mononuclear cells are the target cells. Expression of EMF-1 together with high collagen levels and in wounded tissues suggests that it has a role in the wound response and/or repair. EMF-1 is chemotactic for chicken peripheral blood mononuclear cells.

ENA-78 (epithelial derived neutrophil attractant-78) is a chemokine of the α family (CXC family) and is related to NAP-2, GRO-α, and IL-8. Tissue sources include epithelial cells and platelets. Neutrophils are the target cells. ENA-78 is increased in peripheral blood, synovial fluid, and synovial tissue from rheumatoid arthritis patients. ENA-78 mRNA levels are elevated in acutely rejecting human renal allografts compared with renal allografts that are not being rejected.

A **chemotactic peptide** is a peptide that attracts cell migration such as formyl-methionyl-leucyl-phenylalanine.

Chemotactic deactivation represents the reduced chemotactic responsiveness to a chemotactic agent caused by prior incubation of leukocytes with the same agent but in the absence of a concentration gradient. It can be tested by adding first the chemotactic factor to the upper chamber, washing, and then testing the response to the chemotactic factor placed in the lower chamber (no gradient being present). The mechanism of deactivation has been postulated as obstruction of the membrane channels involved in cation fluxes. Deactivation phenomena are used to discriminate between chemokinetic factors which enhance random migration and true chemotactic factors which cause directed migration. Only true chemotactic factors are able to induce deactivation.

A **chemoattractant** is a substance that attracts leukocytes and which may induce significant physiologic alterations in cells that express receptors for them.

Formyl-methionyl-leucyl-phenylalanine (F-Met-Leu-Phe) is a synthetic peptide that is a powerful chemotactic attractant for leukocytes, facilitating their migration. It may also induce neutrophil degranulation. This peptide resembles chemotactic factors released from bacteria. Following interaction with neutrophils, leukocyte migration is enhanced and complement receptor 3 molecules are increased in the cell membrane.

f-Met peptides are bacterial tripeptides such as formyl-Met-Leu-Phe that are chemotactic for inflammatory cells, inducing leukocyte migration.

ACT-2 is a human homolog of murine MIP-1b that chemoattracts monocytes but prefers activated $CD4^+$ cells to $CD8^+$ cells. T cells and monocytes are sources of ACT-2.

Adhesins are bacterial products that split proteins. They combine with human epithelial cell glycoprotein or glycolipid receptors, which could account for the increased incidence of pulmonary involvement attributable to *Pseudomonas aeruginosa* in patients who are intubated.

Annexins (lipocortins) are proteins with a highly conserved core region comprised of four or eight repeats of about 70 amino acid residues and a highly variable N-terminal region. The core region mediates Ca^{2+}-dependent binding to phospholipid membranes and forms a Ca^{2+} channel-like structure. Physical and structural features of annexin proteins suggest that they regulate many aspects of cell membrane function, including membrane trafficking, signal transduction, and cell–matrix interactions. Their actions resemble some of those of glucocorticoids, including antiinflammatory, antiedema, and immunosuppressive effects.

Apolipoprotein (APO-E) is a plasma protein involved in many functions including lipid transport, tissue repair, and the regulation of cellular growth and proliferation. There are three major isoforms of APO-E encoded by the epsilon 2, 3, or 4 alleles (APO-E2, APO-E3, APO-E4). APO-E3 is the most common variant. There is much interest in the APO-E4 variant as it may be implicated in Alzheimer's disease. Other APO-E polymorphisms have been implicated in disorders of lipid metabolism and heart disease.

β cells are insulin secreting cells in the islet of Langerhans of the pancreas.

β-pleated sheet is a protein configuration in which the β sheet polypeptide chains are extended and have a 35-nm axial distance. Hydrogen bonding between NH and CO groups of separate polypeptide chains stabilize the molecules. Adjacent molecules may be either parallel or antiparallel. The β-pleated sheet configuration is characteristic of amyloidosis and is revealed by Congo red staining followed by polarizing light microscopy, which yields an apple-green birefringence and ultrastructurally consists of nonbranching fibrils.

β barrel: See β sheet.

α-1 antitrypsin (A1AT): A glycoprotein in circulating blood that blocks trypsin, chymotrypsin, and elastase, among other enzymes. The gene on chromosome 14 encodes 25 separate allelic forms which differ according to electrophoretic mobility. The PiMM phenotype is physiologic. The PiZZ phenotype is the most frequent form of the deficiency which is associated with emphysema, cirrhosis, hepatic failure, and cholelithiasis, with an increased incidence of hepatocellular carcinoma. It is treated with prolastin. Adenoviruses may be employed to transfer the A1AT gene to lung epithelial cells, after which A1AT mRNA and functioning A1AT become demonstrable.

α helix: A spiral or coiled structure present in many proteins and polypeptides. It is defined by intrachain hydrogen bonds between -CO and -NH groups that hold the polypeptide chain together in a manner that results in 3.6 amino acid residues per helical turn. There is a 1.5-Å rise for each residue. The helix has a pitch of 5.4 Å. The helical backbone is formed by a peptide group and the α carbon. Hydrogen bonds link each -CO group to the -NH group of the fourth residue forward in the chain. The α helix may be left- or right-handed. Right-handed α helices are the ones found in proteins.

Chaperones are a group of proteins that includes BiP, a protein that binds the immunoglobulin heavy chain. Chaperones aid the proper folding of oligomeric protein complexes. They prevent incorrect conformations or enhance correct ones. Chaperones are believed to combine with the surfaces of proteins exposed during intermediate folding and to restrict further folding to the correct conformations. They take part in transmembrane targeting of selected proteins. Chaperones hold some proteins that are to be inserted into membranes in intermediate conformation in the cytoplasm until they interact with the target membrane. Besides BiP, they include heat shock proteins 70 and 90 and nucleoplasmins.

The **coagulation system** is a cascade of interaction among 12 proteins in blood serum that culminates in the generation of fibrin, which prevents bleeding from blood vessels whose integrity has been interrupted.

The **clotting system** is a mixture of cells, their fragments, zymogens, zymogen activation products and naturally occurring inhibitors, adhesive and structural proteins, phospholipids, lipids, cyclic and noncyclic nucleotides, hormones, and inorganic cations, all of which normally maintain blood flow. With disruption of the monocellular layer of endothelial cells lining a vessel wall, the subendothelial layer is exposed, bleeding occurs, and a cascade of events is initiated that leads to clot formation. These homeostatic reactions lead to formation of the primary platelet plug followed by a clot that mainly contains crosslinked fibrin (secondary hemostasis). After the blood vessel is repaired, the clot is dissolved by fibrinolysis. See also coagulation system.

Hageman factor (HF) is a zymogen in plasma that is activated by contact with a surface or by the kallikrein system at the beginning of the intrinsic pathway of blood coagulation. This is an 80-kDa plasma glycoprotein which, following activation, is split into an α and β chain. When activated, this substance is a serine protease that transforms prekallikrein into kallikrein. HF is coagulation factor XII.

Endocrine: An adjective describing regulatory molecules such as hormones that reach target cells from cells that are the site of their synthesis through the bloodstream.

F-actin: Actin molecules in a dual-stranded helical polymer. Together with the tropomyosin–tropinin regulatory complex, it constitutes the thin filaments of skeletal muscle.

Hemophilia is an inherited coagulation defect attributable to blood clotting factor VIII, factor IX, or factor XI deficiency. Hemophilia A patients are successfully maintained by the administration of exogenous factor VIII, which is now safe. Before mid-1985, factor products were a source of several cases of AIDS transmission when factor VIII was extracted from the blood of HIV-positive subjects by accident. Hemophilia B patients are treated with factor IX. Hemophilia A and B are cross linked, but hemophilia C is autosomal.

Lectins are glycoproteins that bind to specific sugars and oligosaccharides and link to glycoproteins or glycolipids on the cell surface. They can be extracted from plants or seeds, as well as from other sources. They are able to agglutinate cells such as erythrocytes through recognition of specific oligosaccharides and occasionally will react with a specific monosaccharide. Many lectins also function as mitogens and induce lymphocyte transformation, during which a small resting lymphocyte becomes a large blast cell that may undergo mitosis. Well-known mitogens used in experimental immunology include phytohemagglutinin, pokeweed mitogen, and concanavalin A.

Lectin-like receptors are macrophage and monocyte surface structures that bind sugar residues. The ability of these receptors to anchor polysaccharides and glycoproteins facilitates attachment during phagocytosis of microorganisms. Steroid hormones elevate the number of these cell-surface receptors.

Ischemia is deficient blood supply to a tissue as a consequence of vascular obstruction.

Isoforms are different versions of a protein encoded by alleles of a gene or by different but closely related genes.

Prekallikrein is a kallikrein precursor. The generation of kallikrein from prekallikrein can activate the intrinsic mechanism of blood coagulation.

Protein kinase C (PKC) is a serine/threonine kinase that Ca^{2+} activates in the cytoplasm of cells. It participates in T and B cell activation and is a receptor for phorbol ester that acts by signal transduction, leading to hormone secretion, enzyme secretion, neurotransmitter release, and mediation of inflammation. It is also involved in lipogenesis and gluconeogenesis. PKC participates also in differentiation of cells and in tumor promotion.

Protein S is a 69-kDa plasma protein that is vitamin-K dependent and serves as a cofactor for activated protein C. It occurs as an active single chain protein or as a dimeric protein that is disulfide-linked and inactive. Protein S, in the presence of phospholipid, facilitates protein C inactivation of factor Va and combines with C4b-binding proteins. Protein S deficiency, which is transmitted as an autosomal dominant, is characterized clinically by deep vein thrombosis, pulmonary thrombosis, and thrombophlebitis. Laurell rocket electrophoresis is used to assay protein S.

Phosphatase is an enzyme that deletes phosphate groups from protein amino acid residue side chains. Lymphocyte protein phosphatases control signal transduction and transcription factor activity. Protein phosphatases may show specificity for either phosphotyrosine residues or phosphoserine and phosphothreonine residues.

Small G proteins are monomeric G proteins, including Ras, that function as intracellular signaling molecules downstream of many transmembrane signaling events. In their active form they bind GTP and hydrolyze it to GDP to become inactive.

Stress proteins are characterized into major families generally according to molecular weight. Within a family, heat shock proteins show a high degree of sequence homology throughout the phylogenetic spectrum and are among the most highly conserved proteins in nature. Heat shock protein 70 from mycobacteria and humans reveals 50% sequence homology. In spite of this homology, there are subtle differences in the functions, inducibility, and cellular location among related heat shock proteins for a given species. Even though major stress proteins accumulate to very high levels in stressed cells, they are present at low to moderate levels in unstressed cells pointing to the fact that they play a role in normal cells. In addition to increased synthesis, many heat shock proteins change their intracellular distribution in response to stress. An important characteristic of heat shock proteins is their capacity to function as molecular chaperones, which describes their capacity to bind to denatured proteins, preventing their aggregation, and this helps explain the function of heat shock proteins under normal conditions and in stress situations.

Ubiquitin is a 7-kDa protein found free in the blood or bound to cytoplasmic, nuclear, or membrane proteins united through isopeptide bonds to numerous lysine residues. Ubiquitin combines with a target protein and marks it for degradation. It is a 76-amino acid residue polypeptide found in all eukaryotes, but not in prokaryotes. Ubiquitin is found in chromosomes covalently linked to histones, although the function is unknown. It is present on the lymphocyte homing receptor gp90Mel-14.

Ubiquitination is the covalent linkage of several copies of ubiquitin, a small polypeptide, to a protein. Protein that has been ubiquitinated is marked for proteolytic degradation by proteasomes, which is involved in class 1 MHC antigen processing and presentation.

Zymogen refers to the inactive state in which an enzyme may be synthesized. Proteolytic cleavage of the zymogen may lead to active enzyme formation.

Adaptor proteins are critical linkers between receptors and downstream signaling pathways that serve as bridges or scaffolds for recruitment of other signaling molecules. They are functionally heterogeneous, yet share an SH domain that permits interaction with phosphotyrosine residues formed by receptor-associated tyrosine kinases. During lymphocyte activation, they may be phosphorylated on tyrosine residues, which enables them to combine with other homology-2 (SH2) domain-containing proteins. LAT, SLP-76, and Grb-2 are examples of adaptor molecules that participate in T cell activation.

Neuropilin is a cell-surface protein that is a receptor for the collapsin/semaphorin family of neuronal guidance proteins.

Heat shock proteins (hsp): A restricted number of highly conserved cellular proteins that increase during metabolic stress such as exposure to heat. Heat shock proteins affect protein assembly into protein complexes, proper protein folding, protein uptake into cellular organelles, and protein sorting. The main group of hsps are 70-kDa proteins. Heat shock (stress) proteins are expressed by many pathogens and are classified into four families based on molecular size, i.e., hsp90, hsp70, hsp60, and small hsp (<40 kDa). Mycobacterial hsp65 antibodies are found in rheumatoid arthritis, atherosclerosis, multiple sclerosis, Alzheimer's disease, and Parkinson's disease. EIA is the method of choice for their detection, but there is no known clinical significance for hsp antibodies.

Antiheat shock protein antibodies have a broad phylogenetic distribution and share sequence similarities in molecules derived from bacteria, humans, or other animals. They play a significant role in inflammation. Heat shock proteins of mycobacteria are important in the induction of adjuvant arthritis by these microorganisms. It is found that 40% of SLE and 10 to 20% of RA patients have antibodies of IgM, IgG, and IgA classes to a 73-kDa protein of the hsp70 group. RA synovial fluid contains T lymphocytes that react with a 65-kDa mycobacterial heat shock protein. The significance of these observations of immune reactivity to heat shock proteins remains to be determined.

Heat shock protein antibodies: Antibodies of the IgM, IgG, and IgA classes specific for a 73-kDa chaperonin that belongs to the hsp70 family are present in the sera of approximately 40% of systemic lupus erythematosus patients and in 10 to 20% of individuals with rheumatoid arthritis. Antibodies specific for the 65-kDa heat shock protein derived from mycobacteria shows specificity for rheumatoid synovium. RA synovial fluid T cells specific for a 65-kDa mycobacterial heat shock protein have been reported to be inversely proportional to the disease duration.

Diacylglycerol (DAG), a substance formed by the action of phospholipase C-γ on inositol phospholipids that serves as an intracellular signaling molecule. DAG activates cytosolic protein kinase C, which further propagates the signal.

Insulin-like growth factors consist of IGF-I and IGF-II which are prohormones with M_r of 9K and 14K, respectively. IGF-I is a 7.6-kDa side-chain polypeptide hormone that resembles proinsulin structurally. It is formed by the liver and by fibroblasts. IGF-I is the sole effector of growth hormone activity. It is a primary growth regulator that is age dependent. It is expressed in juvenile life but declines after puberty. Circulating IGFs are not free in the plasma but are associated with binding proteins that may have the function of limiting the bioavailability of circulating IGFs, which may be a means of controlling growth factor activity. IGF-II is present mainly during the embryonic and fetal stages of mammalian development in various tissues. It is also present in the circulating plasma in association with binding proteins, reaching its highest level in the fetal circulation and declining following birth. IGF-II is important for growth of the whole organism.

Insulin-like growth factor-II (IGF-II) is a fetal growth factor that is expressed at high levels in many tissues during fetal and early postnatal development but only in the central nervous system thereafter.

N-linked oligosaccharide is covalently linked to asparagine residues in protein molecules. N-linked oligosaccharide manifests a core structure that is branched and comprised of two *N*-acetylglucosamine residues and three mannose residues. There are three types that differ on the basis of their exterior branches: (1) high-mannose oligosaccharide reveals two to six additional mannose residues linked to the polysaccharide core; (2) complex oligosaccharide comprised of two to five terminal branches that consist of *N*-acetylglucosamine, galactose, often *N*-acetylneuraminic acid, and occasionally fucose or another sugar; and (3) hybrid molecules that reveal characteristics of high-mannose and complex oligosaccharides.

hCG (human choriogonadotrophic hormone) is a glycoprotein comprised of lactose and hexosamine that is synthesized by syncytiotrophoblast, fetal kidney, and liver-selected tumors. It may be measured by radioimmunoassay or enzyme-linked immunosorbent assay (ELISA). It is elevated in patients with various types of tumors such as carcinoma of the liver, stomach, breast, pancreas, lungs, kidneys, and renal cortex, as well as conditions such as lymphoma, leukemia, melanoma, and seminoma.

The **fluid mosaic model** (Figure 2.23) is a fluid lipid molecular bilayer in the plasma membrane and organelle membranes of cells. This structure permits membrane proteins and glycoproteins to float. The lipid molecules are situated in a manner that arranges the polar heads toward outer surfaces and their hydrophobic side chains projecting into the interior. There can be lateral movement of molecules in the bilayer plain, or they may rotate on their long axis. This is the Singer–Nicholson "fluid mosaic." The bilayer consists of glycolipids and phospholipids. Amphipathic lipids and globular proteins are spaced throughout the membrane. The fluid consistency permits movement of the proteins, glycoprotein, and receptors laterally.

The **cytoskeleton** is a framework of cytoskeletal filaments present in the cell cytoplasm. They maintain the cell's internal arrangement, shape, and motility. This framework interacts with the membrane of the cell and with organelles in the cytoplasm. Microtubules, microfilaments, and intermediate filaments constitute the varieties of cytoskeletal filaments. Microtubules help to determine cell shape by polymerizing and depolymerizing. They are 24-nm diameter hollow tubes whose walls are comprised of protofilaments that contain α and β tubulin dimers. The 7.5-nm diameter microfilaments are actin polymers. In addition to their interaction with myosin filaments in muscle contraction, actin filaments may affect movement or cell shape through polymerization and depolymerization. Microfilaments participate in cytoplasmic streaming, ruffling of membranes, and phagocytosis. They may be responsible for limiting protein mobility in the cell membrane. The proteins of the 10-nm intermediate filaments differ according to the cells in which they occur. Vimentin intermediate filaments occur in macrophages, lymphocytes, and endothelial cells, whereas desmin occurs in muscle and epithelial cells containing keratin.

The **endoplasmic reticulum** (Figure 2.24) is a structure in the cytoplasm comprised of parallel membranes that are connected to the nuclear membranes. Lipids and selected proteins are synthesized in this organelle. The membrane is continuous and convoluted. Electron microscopy reveals

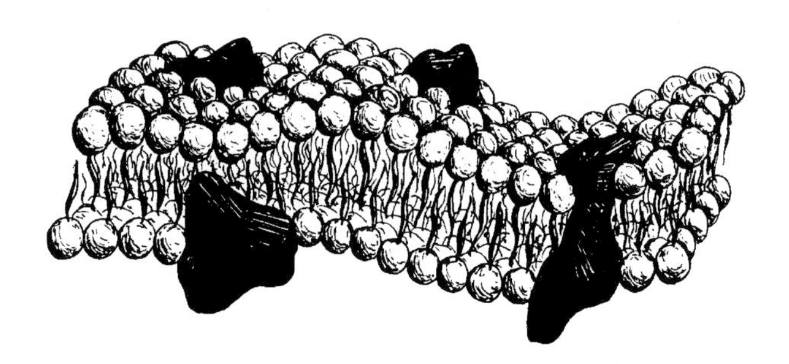

FIGURE 2.23 Fluid mosaic model.

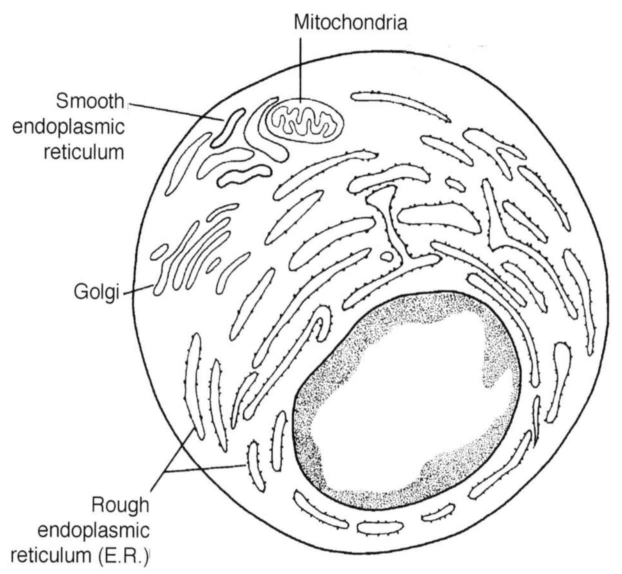

FIGURE 2.24 Eukaryotic cell.

rough endoplasmic reticulum, which contains ribosomes on the side exposed to the cytoplasm and smooth endoplasmic reticulum without ribosomes. Fatty acids and phospholipids are synthesized and metabolized in smooth endoplasmic reticulum. Selected membrane and organelle proteins, as well as secreted proteins, are synthesized in the rough endoplasmic reticulum. Cells such as plasma cells that produce antibodies or other specialized secretory proteins have abundant rough endoplasmic reticulum in the cytoplasm. Following formation, proteins move from the rough endoplasmic reticulum to the Golgi complex. They may be transported in vesicles that form from the endoplasmic reticulum and fuse with Golgi complex membranes. Once the secreted protein reaches the endoplasmic reticulum lumen, it does not have to cross any further barriers prior to exit from the cell.

The **Golgi apparatus** consists of a stack of vesicles enclosed by membranes found within a cell and serves as a site of glycosylation and packaging of secreted proteins. It is part of the GERL complex.

Golgi complex: Tubular cytoplasmic structures that participate in protein secretion. The complex consists of flattened membranous sacs on top of each other termed *cisternae*. These are also associated with spherical vesicles. Proteins arriving from the rough endoplasmic reticulum are processed in the Golgi complex and sent elsewhere in the cell. Proteins handled in this manner include those secreted constitutively, such as immunoglobulins; those of the membrane; those that are stored in secretory granules to be released on command; and lysosomal enzymes.

A **lysosome** (Figure 2.25) is a cytoplasmic organelle enclosed by a membrane that contains multiple hydrolytic enzymes including ribonuclease, deoxyribonuclease, phosphatase, glycosidase, collagenase, arylsulfatase, and cathespins. These hydrolytic enzymes may escape into a phagosome or to the outside. Lysosomes occur in numerous cells but are especially prominent in neutrophils and macrophages. The enzymes are critical for intracellular digestion. They may autolyze dead cells. Lysosomes participate in antigen processing by the class II MHC pathway. See also phagosome and phagocytosis.

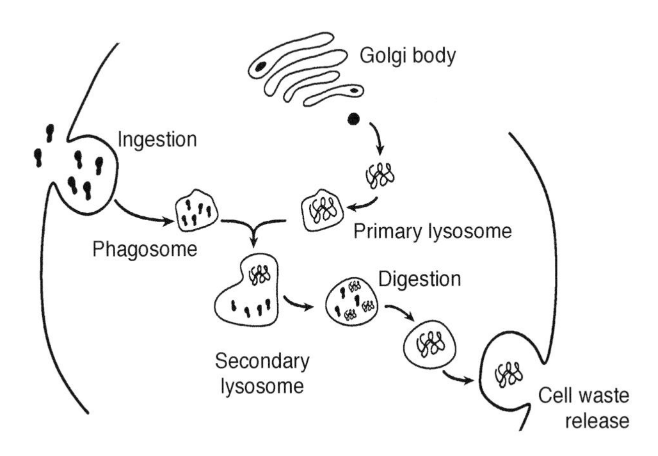

FIGURE 2.25 Lysosome.

Primary lysosome refers to a lysosome that has not yet fused with a phagosome.

LAMP 1 is a lysosomal membrane protein (CD107a).

LAMP 2 is a lysosomal membrane protein designated CD107b.

A **ribosome** is a subcellular organelle in the cytoplasm of a cell that is a site of amino acid incorporation in the process of protein synthesis.

An **endosome** is a 0.1 to 0.2 μm intracellular vesicle produced by endocytosis. Extracellular proteins are internalized in this structure during antigen processing. The endosome has an acidic pH and contains proteolytic enzymes that degrade proteins into peptides that bind to class II MHC molecules. MIIC, a subset of class II MHC-rich endosomes, has a critical role in antigen processing and presentation by the class II pathway.

The **mitochondria** are cytoplasmic organelles that are sites of metabolism in cells in aerobic eukaryotic cells where respiration, electron transport, oxidative phosphorylation, and citric acid cycle reactions occur. Mitochondria possess DNA and ribosomes.

Clathrin is the principal protein enclosing numerous coated vesicles. The molecular structure consists of three 180-kDa heavy chains and three 30- to 35-kDa light chains arranged into typical lattice structures comprised of pentagons or hexagons. These structures encircle the vesicles.

Microtubules: These organelles are hollow, cylindrical fibers of about 240 Å in diameter, radiating from the center of eukaryotic cells, including lymphocytes, phagocytes, and mast cells, in all directions toward the plasma membrane. The mitotic spindle is comprised of them. Microtubules form a sturdy cytoskeleton. They originate from the centriole, a structure occupying the concavity of the nucleus. Microtubules provide orientation of gross membrane activities, associate directly or indirectly with granules to enable their contact and fusion with endocytic vesicles, and direct reorganization of the cell membrane. Although not critical for the cell movement of chemotaxis, they are needed for "fine tuning" of cell locomotion. The major component of microtubules is tubulin, a dimeric protein.

Coated pit refers to a depression in the cell membrane coated with clathrin. Hormones such as insulin and epidermal growth factor may bind to their receptors in the coated pit or migrate toward the pit following binding of the ligand at another site. After the aggregation of

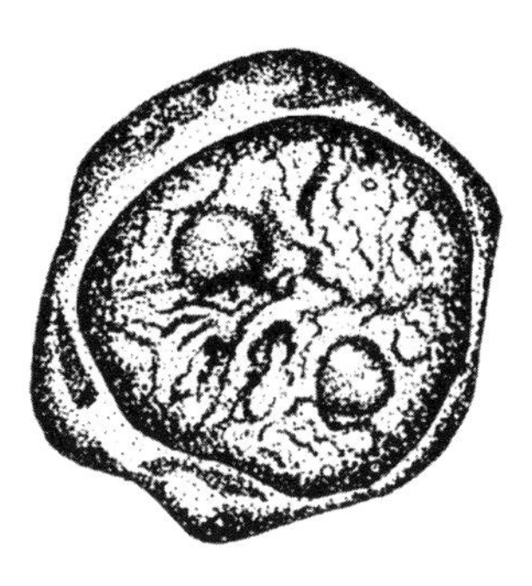

FIGURE 2.26 Stem cell.

complexes of receptor and ligand in the coated pits, they invaginate and bud off as coated vesicles containing the receptor–ligand complexes. These structures, called receptosomes, migrate into the cell by endocytosis. Following association with GERL structures, they fuse with lysosomes where receptors and ligands are degraded.

Coated vesicles are vesicles in the cytoplasm usually encircled by a coat of protein-containing clathrin molecules. They originate from coated pits and are important for protein secretion and receptor-mediated endocytosis. Coated vesicles convey receptor–macromolecule complexes from an extracellular to an intracellular location. Clathrin-coated vesicles convey proteins from one intracellular organelle to another. See also coated pit.

Stem cells (Figure 2.26) are relatively large cells with a cytoplasmic rim that stains with methyl green pyronin and a nucleus that has thin chromatin strands and contains nucleoli that are pyroninophilic. They are found in hematopoietic tissues such as the bone marrow. These stem cells are a part of the colony-forming unit (CFU) pool that indicates that individual cells are able to differentiate and proliferate under favorable conditions. The stem cells, CFU-S which are pluripotent, are capable of differentiating into committed precursor cells of the granulocyte and monocyte lineage (CFU-C), of erythropoietic lineage (CFU-E and BFU-E), and of megakaryocyte lineage (CFU-Mg). Lymphocytes, like other hematopoietic cells, are generated in the bone marrow. The stem cell compartment is composed of a continuum of cells that includes the most primitive with the greatest capacity for self-renewal and the least evidence of cell cycle activity to the most committed with a lesser capacity for self-renewal and the most evidence of cell cycle activity. Stem cells are precursor cells that are multipotential with the capacity to yield differentiated cell types with different functions and phenotypes. The proliferative capacity of stem cells is, however, limited.

CD34 is a vascular addressin present on lymph node high endothelial venules.

CD34 is a molecule (105- to 120-kDa mol wt) that is a single chain transmembrane glycoprotein present on immature hematopoietic cells and endothelial cells as well as bone marrow stromal cells. Three classes of CD34 epitopes have been defined by differential sensitivity to enzymatic cleavage with neuraminidase and with glycoprotease from *Pasteurella haemolytica*. Its gene is on chromosome 1. CD34 is the ligand for L-selectin (CD62L).

A **pluripotent stem cell** is a continuously dividing, undifferentiated bone marrow cell that has progeny consisting of additional stem cells together with cells of multiple separate lineages. Bone marrow hematopoietic stem cells may develop into cells of the myeloid, lymphoid, and erythroid lineages.

Colony-forming unit (CFU) comprises the hematopoietic stem cell and the progeny cells that derive from it. Mature (end-stage) hematopoietic cells in the blood are considered to develop from one CFU. Some progenitor cells are precursors of erythrocytes, others are precursors of polymorphonuclear leukocytes and monocytes, and still others are megakaryocyte and platelet precursors.

CFU-S (colony-forming units, spleen) refers to a mixed-cell population considered to contain the ideal stem cell that is pluripotent and capable of proliferating and renewing itself.

Colony-forming units, spleen (CFU-S): A hematopoietic precursor cell that can produce a tiny nodule in the spleen of mice that have been lethally irradiated. These small nodular areas are sites of cellular proliferation. Each arises from a single cell or colony-forming unit. The CFU-S form colonies of pluripotent stem cells.

Totipotent means having the potential for developing in various specialized ways in response to external/internal stimuli; of a cell or part.

Stem-cell factor (SCF) is a bone marrow stromal cell transmembrane protein that binds to c-Kit, a signaling receptor found on developing B cells and other developing leukocytes. SCF is a substance that promotes growth of hematopoietic precursor cells and is encoded by the murine SI gene. It serves as a ligand for the tyrosine kinase receptor family protooncogene termed c-*kit*. It apparently has a role in embryogenesis in cells linked to migratory patterns of hematopoietic stem cells, melanoblasts, and germ cells.

Embryonic stem (ES) cells are murine embryonic cells that are immortal in culture and retain the capacity to give rise to all cell lineages. They may be altered genetically *in vitro* and introduced into mouse blastocysts to give rise to mutant murine lines. Genes may be deleted in ES cells by homologous recombination to produce mutant ES cells that can give rise to gene knock-out mice.

An **erythroid progenitor** is an immature cell that leads to the production of erythrocytes and megakaryocytes but no other blood cell types.

Erythropoiesis refers to the formation of erythrocytes or red blood cells.

Erythropoietin is a 46-kDa glycoprotein produced by the kidney, more specifically by cells adjacent to the proximal renal tubules, based on the presence of substances such as heme in the kidneys which are oxygen sensitive. It stimulates red blood cell production and combines with erythroid precursor receptors to promote mature red cell development. Erythropoietin formation is increased by hypoxia. It is useful in the treatment of various types of anemia.

Hematopoiesis is the development of the cellular elements of the blood including erythrocytes, leukocytes, and platelets from pluripotent stem cells in the bone marrow in fetal liver. It is regulated by various cytokine growth factors synthesized by bone marrow stromal cells, T cells, or other types of cells.

Hematopoietic-inducing microenvironment (HIM) refers to an anatomical location in which the cells and cellular factors requisite for the generation development of hematopoietic cells may be found.

Hematopoietic lineage is a series of cells that develops from hematopoietic stem cells and which yields mature blood elements.

Hematopoietic system refers to those tissues and cells that generate the peripheral blood cells.

A **hemocytoblast** is a bone marrow stem cell.

Common lymphoid progenitors are stem cells from which all lymphocytes are derived. Pluripotent hematopoietic stem cells give rise to these progenitors.

Leukocytes are white blood cells. The principal types of leukocytes in the peripheral blood of man include polymorphonuclear neutrophils, eosinophils and basophils (granulocytes), and lymphocytes and monocytes.

Leukocytosis is an increase above normal of the peripheral blood leukocytes as reflected by a total white blood cell count of greater than 11,000/mm^3 of blood. This occurs frequently with acute infection.

Leukopenia is the reduction below normal of the number of white blood cells in the peripheral blood.

Leukocyte activation, the first step in activation, is adhesion through surface receptors on the cell. Stimulus recognition is also mediated through membrane-bound receptors. An inducible endothelial–leukocyte adhesion molecule that provides a mechanism for leukocyte-vessel wall adhesion has been described. Surface adherent leukocytes undergo a large prolonged respiratory burst. NADPH oxidase, which utilizes hexose monophosphate shunt-generated NADPH, catalyzes the respiratory burst. Both Ca^{2+} and protein kinase C play a key role in the activation pathway. CR3 facilitates the ability of phagocytes to bind and ingest opsonized particles. Molecules found to be powerful stimulators of PMN activity include recombinant 1FN-γ, granulocyte-macrophage colony-stimulating factor, TNF, and lymphotoxin.

Leukocyte adhesion molecules are facilitators of vascular endothelium aggregation, chemotaxis, cytotoxicity, biding of iC3b-coated particles, lymphocyte proliferation, and phagocytosis. The three main families of leukocyte adhesion molecules include the selectins, integrins, and immunoglobulin superfamily. Leukocyte adhesion deficiencies are partial or complete inherited deficiencies of cell surface expression of CD18 and CD11a–c. These deficiencies prevent granulocytes from migrating to extravascular sites of inflammation, leading to recurrent infections and possibly death.

LCA (leukocyte common antigen): See leukocyte common antigen, CD45.

CD45 is an antigen that is a single-chain glycoprotein referred to as the *leukocyte common antigen* (or "T200"). It consists of at least five high molecular weight glycoproteins present on the surface of the majority of human leukocytes (mol wts: 180, 190, 205, and 220 kDa). The different isoforms arise from a single gene via alternative mRNA splicing. The variation between the isoforms is all in the extracellular region. The larger (700-amino acid) intracellular portion is identical in all isoforms and has protein tyrosine phosphatase activity. It can potentially interact with intracellular protein kinases such as $p56^{lck}$, that may be involved in triggering cell activation. By dephosphorylating proteins, CD45 would act in an opposing fashion to a protein kinase. It facilitates signaling through B and T cell antigen receptors. Leukocyte common antigen (LCA) is present on all leukocyte surfaces. It is a transmembrane tyrosine phosphatase that is expressed in various isoforms on different types of cells, including the different subtypes of T cells. The isoforms are designated by CD45R followed by the exon whose presence gives rise to distinctive antibody-binding patterns.

CD45RB is a molecule which consists of four isoforms of CD45 (sequence encoded by exon B) with mol wts of 220, 205, and 190 kDa that is found on B cells, subsets of T cells, monocytes, macrophages, and granulocytes.

B220 is a form of CD45, a protein tyrosine phosphatase.

CD45R is a subfamily that is now divided into three isoforms: CD45RO, CD45RA, and CD45RB. The designation CD45R has been maintained for those antibodies that have not been tested on appropriate transfectants.

CD45RA is a 220,205-kDa isoform of CD45 (sequence encoded by exon A) that is found on B cells, monocytes, and a subtype of T cells. T cells expressing this isotype are naïve or virgin T cells and nonprimed $CD4^+$ and $CD8^+$ cells.

CD45RO is a 180-kDa isoform of CD45 (sequence not encoded by either exons A, B, or C), that is found on T cells, subset of B cells, monocytes, and macrophages. T cells expressing this antigen are T memory cells or primed T cells.

Leukocyte adhesion molecule-1 is a homing protein found on membranes, which combines with target cell specific glycoconjugates. It helps to regulate migration of leukocytes through lymphocytes binding to high endothelial venules and to regulate neutrophil adherence to endothelium at inflammatory sites.

Leukocyte adhesion proteins are membrane-associated dimeric glycoproteins comprised of a unique α subunit and a shared 95-kDa β subunit involved in cell-to-cell interactions. They include LFA-1, which is found on lymphocytes, neutrophils, and monocytes; Mac-1, which is found on neutrophils, eosinophils, NK cells, and monocytes; and p150,95, which is common to all leukocytes.

Leu-CAM: Leukocyte cell adhesion molecules.

L-plastin (LPL) is a 65-kDa actin-bundling protein, also called *fimbrin*, that is expressed in leukocytes, embryonic endoderm, and transformed cells. LPL localizes to phagocytic cups, phagosomes, and podosomes in phagocytes but its role is unclear. LPL is believed to be important in the formation and stabilization of F-actin filaments during phagocytosis.

Leukocyte chemotaxis inhibitors are humoral factors that inhibit the chemotaxis of leukocytes. They play a role in the regulation of inflammatory responses of both immune and nonimmune origin.

Leukocyte functional antigens (LFAs) are cell adhesion molecules that include LFA-1, a β_2 integrin; LFA-2, an immunoglobulin superfamily member; and LFA-3, an immunoglobulin superfamily member now designated CD58. LFA-1 facilitates T cell adhesion to endothelial cells and antigen-presenting cells.

Leukocyte integrins: See leukocyte functional antigens.

Mononuclear cells are leukocytes with single, round nuclei such as lymphocytes and macrophages, in contrast to polymorphonuclear leukocytes. Thus, the term refers to the mononuclear phagocytic system or to lymphocytes.

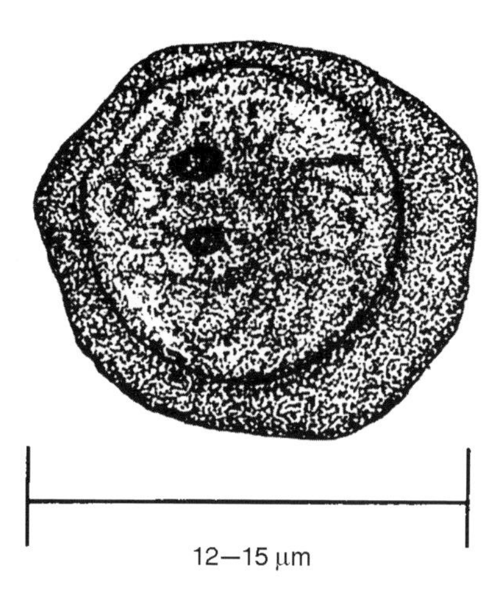

FIGURE 2.27 Lymphoblast.

A **progenitor cell** no longer contains the capacity for self-renewal and is committed to the generation of a specific cell lineage.

A **lymphoid progenitor cell** is a cell belonging to the lymphoid lineage that gives rise to lymphocytes.

A **lymphoblast** (Figure 2.27) is a relatively large cell of the lymphocyte lineage that bears a nucleus with fine chromatin and basophilic nucleoli. They form frequently following antigenic or mitogenic challenge of lymphoid cells, which leads to enlargement and division to produce effector lymphocytes that are active in immune reactions. The Epstein–Barr virus (EBV) is commonly used to transform B cells into B lymphoblasts in tissue culture to establish B lymphoblast cell lines. The lymphocyte that has enlarged to create a lymphoblast has an increased rate of synthesis of RNA and protein.

A **lymphocyte** (Figure 2.28 through Figure 2.33) is a round cell that measures 7 to 12 μm and contains a round to ovoid nucleus that may be indented. The chromatin is densely packed and stains dark blue with Romanowsky stains. Small lymphocytes contain a thin rim of robin's egg blue cytoplasm; a few azurophilic granules may be present. Large lymphocytes have more cytoplasm and a similar nucleus. Electron microscopy reveals villi that cover most of the cell surface. Lymphocytes are divided into two principal groups termed B and T lymphocytes. They are distinguished not on morphology but on the expression of distinctive surface molecules that have precise roles in immune reaction. In addition, natural killer cells, which are large granular lymphocytes, comprise a small percentage of the lymphocyte population. Lymphocytes express variable cell surface receptors for antigen.

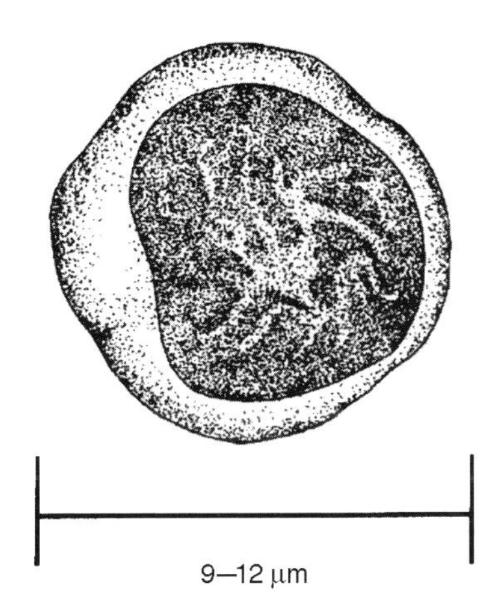

FIGURE 2.28 Lymphocyte.

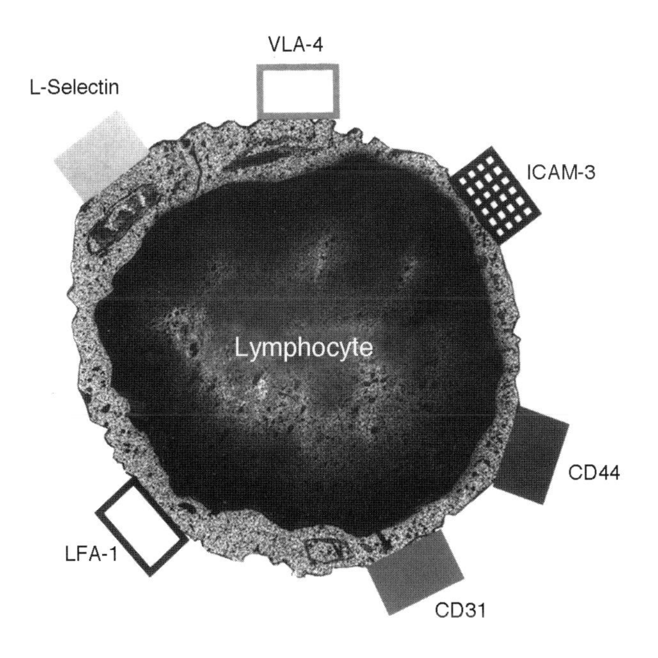

FIGURE 2.29 Lymphocyte.

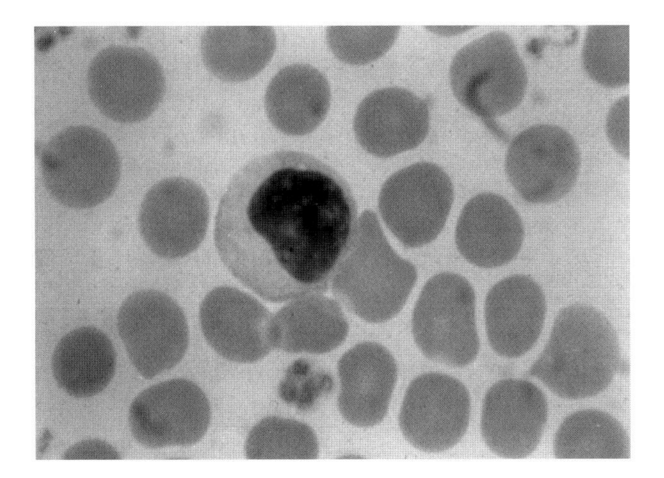

FIGURE 2.30 Lymphocyte in peripheral blood.

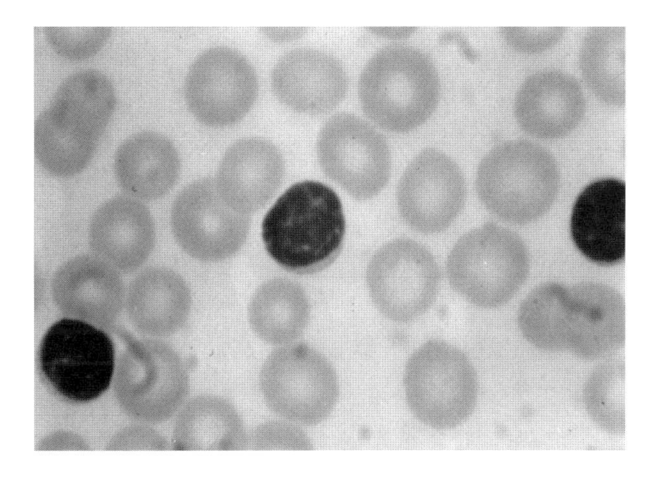

FIGURE 2.31 Small lymphocyte in peripheral blood.

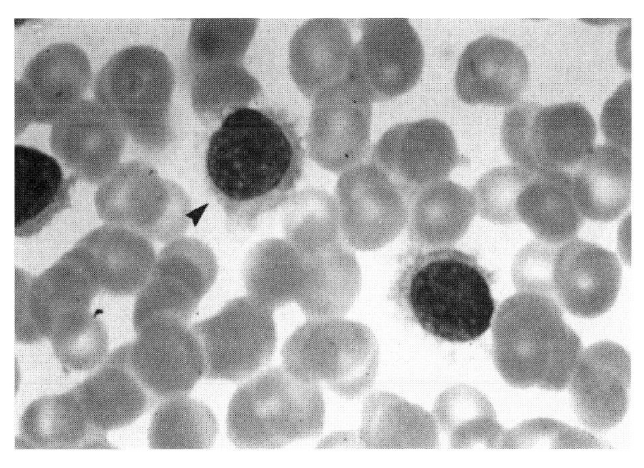

FIGURE 2.32 Lymphocytes in peripheral blood.

A **lymphoid cell** is a cell of the lymphoid system. The classic lymphoid cell is the lymphocyte.

Lymphoid cell series: (1) Cell lineages whose members morphologically resemble lymphocytes, their progenitors, and their progeny. (2) Organized tissues of the body in which the predominant cell type is the lymphocyte or cells of the lymphoid cell lineage. These include the lymph nodes, thymus, spleen, and gut-associated lymphoid tissue, among others.

Lymphopenia is a decrease below normal in the number of lymphocytes in the peripheral blood.

Lymphopoiesis is the differentiation of hematopoietic stem cells into lymphocytes.

Lymphoreticular is an adjective describing the system composed of lymphocytes and monocyte-macrophages, as well as the stromal elements that support them. The thymus, lymph nodes, spleen, tonsils, bone marrow, Peyer's patches,

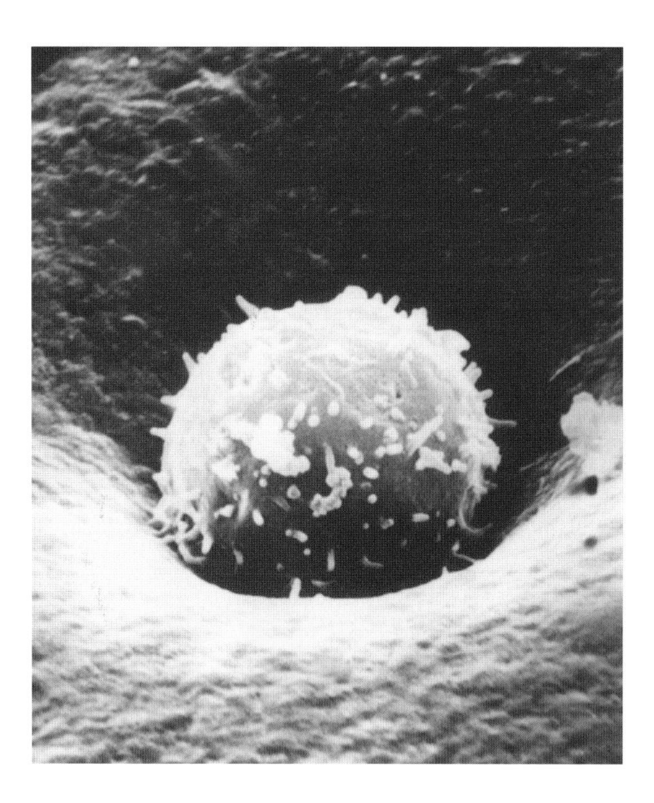

FIGURE 2.33 Lymphocyte scanning electron micrograph.

and avian bursa of Fabricius comprise the lymphoreticular tissues.

Lymphorrhages are accumulations of lymphocytes in inflamed muscle in selected muscle diseases such as myasthenia gravis.

Lymphoid lineage refers to lymphocytes of all varieties and the bone cells that are their precursors.

A **small lymphocyte** is one of the five types of leukocytes in the peripheral blood that measures 6 to 8 μm in diameter. In Wright's and Giemsa-stained blood smears, the nucleus stains dark blue and is encircled by a narrow rim of robin's egg blue cytoplasm. Even though most of the lymphocytes look alike, they differ greatly in origin and function. They differ in other features as well. By light microscopy, T and B lymphocytes and the E rosette subpopulations look the same. However, they have different phenotypic surface markers and differ greatly in function.

A **long-lived lymphocyte** is a small lymphocyte derived principally from the thymus that survives for months to years without dividing. These are in contrast to short-lived lymphocytes.

A **large lymphocyte** is 12 μm or greater in diameter.

An **effector lymphocyte** is a lymphocyte activated through either specific or nonspecific mechanisms to carry out a certain function in the immune response. Examples of effector lymphocytes include the NK cell, the tumor-infiltrating lymphocyte (TIF), the lymphokine activated killer (LAK) cell, cytotoxic T lymphocyte, helper T lymphocyte, and suppressor T cell. Most commonly, the term signifies a T lymphocyte capable of mediating cytotoxicity, suppression, or helper function.

A **nonadherent cell** is a cell that fails to stick to a surface such as a culture flask. A lymphocyte is an example of a nonadherent cell, whereas macrophages readily adhere to the glass surface of a tissue culture flask.

Paracortex is a T lymphocyte thymus-dependent area beneath and between lymph node cortex follicles.

An **activated lymphocyte** is a lymphocyte whose cell surface receptors have interacted with a specific antigen or with a mitogen such as phytohemagglutinin, concanavalin A, or staphylococcal protein A. The morphologic appearance of activated (or stimulated) lymphocytes is characteristic, and in this form the cells are called immunoblasts. The cells increase in size from 15 to 30 mm in diameter; show increased cytoplasmic basophilia; and develop vacuoles, lysosomes, and ribosomal aggregates. Pinocytotic vesicles are present on the cell membrane. The nucleus contains very little chromatin, which is limited to a thin marginal layer, and the nucleolus becomes conspicuous. The array of changes that follow stimulation is called transformation. Such cells are called transformed cells. An activated B lymphocyte may synthesize antibody molecules, whereas an activated T cell may mediate a cellular immune reaction.

LAMP 1 and LAMP 2 refer to lysosome-associated membrane proteins that are complex molecular complexes involved in maintaining lysosomal membrane integrity in cytotoxic lymphocytes and platelets.

By light microscopy, **resting lymphocytes** appear as a distinct and homogeneous population of round cells, each with a large, spherical or slightly kidney-shaped nucleus which occupies most of the cell and is surrounded by a narrow rim of basophilic cytoplasm with occasional vacuoles. The nucleus usually has a poorly visible single indentation and contains densely packed chromatin. Occasionally, nucleoli can be distinguished. The small lymphocyte variant, which is the predominant morphologic form, is slightly larger than an erythrocyte. Larger lymphocytes, ranging between 10 and 20 μm in diameter, are difficult to differentiate from monocytes. They have more cytoplasm and may show azurophilic granules. Intermediate-size forms between the two are described. By phase contrast microscopy, living lymphocytes show a feeble motility with ameboid movements that give the cells a hand-mirror shape. The mirror handle is called a uropod. In large lymphocytes, mitochondria and lysosomes are better visualized, and some cells show a spherical, birefringent, 0.5-μm diameter inclusion called a *Gall body*. Lymphocytes do not spread on surfaces. The different classes of lymphocytes cannot be distinguished by light microscopy. By scanning electron microscopy, B lymphocytes sometimes show a hairy (rough) surface, but this is apparently an artifact. Electron microscopy does not provide additional information except for visualization of the cellular organelles which are not abundant. This suggests that the small, resting lymphocytes are end-stage cells. However, under appropriate stimulation, they are capable of considerable morphologic changes.

Nonspecific T lymphocyte helper factor is a soluble factor released by $CD4^+$ helper T lymphocytes that nonspecifically activates other lymphocytes.

Diversity refers to the presence of numerous lymphocytes with different antigenic specificities in a subject to create a lymphocyte repertoire that is large and varied. Diversity, which is critical to adaptive immune responsiveness, is a consequence of structural variability in antigen-binding sites of lymphocyte receptors for antigen (antibodies and TCRs).

Emperipolesis is the intrusion or penetration of a lymphocyte into the cytoplasm of another cell followed by passage through the cell. Emperipolesis also describes the movement of one cell within another cell's cytoplasm.

Lymphocytosis is an elevated number of peripheral blood lymphocytes.

Theliolymphocytes are small lymphocytes associated with intestinal epithelial cells; also termed as intraepithelial lymphocyte.

Peripheral blood mononuclear cells are lymphocytes and monocytes in the peripheral blood that may be isolated by Ficoll Hypaque density centrifugation.

Lymphocyte receptor repertoire: All of the highly variable antigen receptors of B and T lymphocytes.

Productive rearrangement refers to lymphocyte receptor chain rearrangement in the appropriate reading frame for the receptor chain in question.

Lymphocyte trafficking is a process that is critical for interaction of the lymphocyte surface antigen receptor with epitopes. There is continuous migration of lymphocytes from the blood into lymphoid and nonlymphoid organs and back again to the blood by way of the lymphatics and venules. Lymphocytes remain in the blood circulation for approximately 30 min on each passage. Lymphocytes in the blood circulation are exchanged approximately 48 times per day, and about 5×10^{11} lymphocytes leave the blood circulation each day. Lymphocyte migration is regulated during entry, transit, and exit. Since there are only a few immunocompetent lymphocytes specific for each antigen, lymphocyte trafficking increases the probability of interaction between the lymphocyte and the epitope for which it is specific. Several adhesion molecules participate in receptor–ligand interactions involved in the entry of lymphocytes into lymphoid organs through endothelial venules. See also lymphocytes, circulating (or recirculating).

Lymphocytotrophic is the property of possessing a special attraction or affinity for lymphocytes. Examples include the attraction of the Epstein–Barr virus for B lymphocytes and the affinity of human immunodeficiency virus (HIV) for the helper/inducer (CD4) T lymphocyte.

In immunology, the term **naïve** refers to B and T lymphocytes that have not been exposed to antigen. Also called unprimed or virgin.

A **naïve lymphocyte** is a mature T or B lymphocyte that has never been exposed to antigen and is not derived from antigen-stimulated mature lymphocyte. Exposure of naïve lymphocytes to antigen leads to their differentiation into effector lymphocytes such as antibody-secreting B cells or helper T cells and cytolytic T lymphocytes (CTLs). Lymphocytes that migrate from the central lymphoid organs are naïve, i.e., naïve T cells from the thymus and naïve B cells from the bone marrow. The surface markers and recirculation patterns of naïve lymphocytes differ from those of lymphocytes activated previously.

Round cells is a term used by pathologists to describe mononuclear cells, especially lymphocytes, infiltrating tissues.

Short-lived lymphocytes are lymphocytes with a life span of 4 to 5 d, in contrast to long-lived lymphocytes which may last from months to years in the blood circulation.

Circulating lymphocytes, the lymphocytes present in the systemic circulation, represent a mixture of cells derived from different sources: (1) B and T cells exiting from bone marrow and thymus on their way to seed the peripheral lymphoid organs, (2) lymphocytes exiting the lymph nodes via lymphatics, collected by the thoracic duct and discharged into the superior vena cava, and (3) lymphocytes derived from direct discharge into the vascular sinuses of the spleen. About 70% of cells in the circulating pool are recirculating; that is, they undergo a cycle during which they exit the systemic circulation to return to lymphoid follicles, lymph nodes, and spleen, and start the cycle again. The cells in this recirculating pool are mostly long-lived mature T cells. About 30% of the lymphocytes of the intravascular pool do not recirculate. They comprise mostly short-lived immature T cells, which either live their life span intravascularly or are activated and exit the intravascular space. The exit of lymphocytes into the spleen occurs by direct discharge from the blood vessels. In the lymph nodes and lymphoid follicles, the exit of lymphocytes occurs through specialized structures, the postcapillary venules. These differ from other venules in that they have a tall endothelial covering. The exiting lymphocytes percolate through the endothelial cells, a mechanism whose real significance is not known. A number of agents such as cortisone or the bacterium *Bordetella pertussis* increase the extravascular exit of lymphocytes and prevent their return to circulation. The lymphocytes travel back and forth between the blood and lymph nodes or the spleen's marginal sinuses. Within 24 to 48 h they return via the lymphatics to the thoracic duct where they then reenter the blood.

The term **recirculating pool** refers to the continuous recirculation of T and B lymphocytes between the blood and lymph compartments.

Recirculation of lymphocytes is the continuous transport of lymphocytes from the blood to secondary lymphoid tissues, to lymph, and back into the blood. Traffic to the spleen represents an exception since lymphocytes only enter and exit the spleen via the blood.

Lymphocyte homing is the directed migration of circulating lymphocyte subsets to specific tissue locations. It is regulated by adhesion molecules, termed homing receptors, termed addressins, expressed on specific tissues, in different vascular beds. Selected T cells that home specifically to intestinal lymphoid tissues such as Peyer's patches are directed by binding of VLA-4 integrin on their surfaces MadCAM addressin on the endothelium of Peyer's patches.

Lymphocyte activation (Figure 2.34) follows stimulation of lymphocytes *in vitro* by antigen or mitogen which renders them metabolically active. Activated lymphocytes may undergo transformation or blastogenesis.

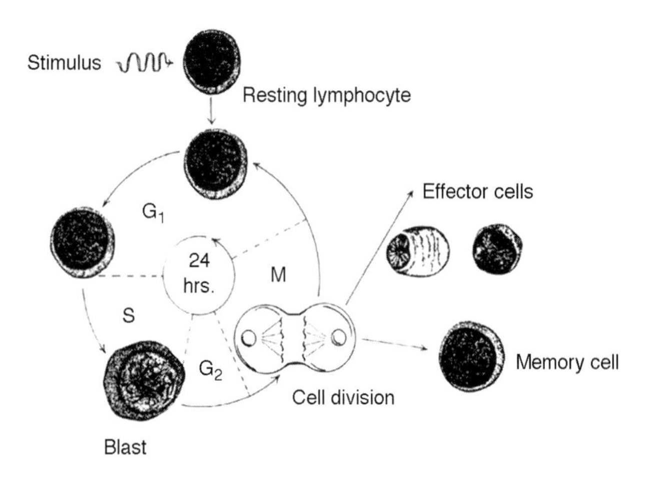

FIGURE 2.34 Lymphocyte activation.

Receptor-associated tyrosine kinases are molecules of the Src-family with which lymphocyte antigen receptors associate. They bind to the tails of receptors through their SH-2 domains.

Signal transduction is a process whereby signals received on the cell surface, such as by the binding of antigen to its receptor, are transmitted into the nucleus of the cell, resulting in altered gene expression.

SH-2 domain: See Src-family tyrosine kinases.

Src homology-2 (SH-2) domain: A 100-amino acid residue three-dimensional domain structure found in numerous signaling proteins that allows specific noncovalent interactions with other proteins by linking to phosphotyrosines. There is a unique binding specificity for each SH-2 domain that is determined by amino acid residues adjacent to the phosphotyrosine on the target protein. SH-2 domains serve as important sites of protein interaction during early signaling events in T and B lymphocytes.

Scr homology-3 (SH-3) domain: A 60-amino acid residue three-dimensional domain structure found in numerous signaling proteins that facilitates the binding of proteins to one another. SH-3 domains bind to proline residues and function in concert with SH-2 domains on the same protein molecules.

Lymphocyte chemotaxis: Lymphocytes are a heterogeneous motile cell population. Both T and B cells recirculate continuously between the blood and the lymphoid tissues. This recirculating cell population consists of naïve small lymphocytes which are not in the cell cycle. Once lymphocytes recognize antigen, their migration behavior changes. They enter the cell cycle and exit the recirculatory pool. An adhesion phenotype changes with loss of L-selectin and loss of affinity for the high endothelial venule (HEV) cells of lymphoid tissue. They increase expression and activity of various other adhesion molecules, which prevents their attaching to the endothelium at sites of inflammation, clustering around antigen-presenting cells, and interacting with target cells for cytotoxicity. Rather than continuing to monitor the environment for antigen, the lymphocyte changes to a cell that mediates effector functions. IL-2 and IL-15 are both excellent chemotactic factors for activated T lymphocytes. IL-16 is also a T cell attractant with selective activity for $CD4^+$. Several chemokines including both α and β chemokines have activity. B cells as well as T cells respond better to attractants following their activation. NK cells activated with IL-2 can respond to chemoattractants including several chemokines such as MIP-1 α, MCP-1 RANTES, and IL-8.

Activation protein-1 (AP-1): DNA-binding transcription factors composed of dimers of two proteins linked to each other through a shared structural motif termed a *leucine zipper*. An example of an AP-1 factor is one comprised of Fos and Jun proteins. Among the many different genes of the immune system in which AP-1 exerts transcriptional regulation of cytokine genes.

AP-1: Transcription factors, some of which have a role in lymphocyte activation. A transcription factor that binds the IL-2 promoter thereby regulating induction of the IL-2 gene. Immediately following T cell stimulation, c-*fos* mRNA is increased, and the c-*fos* gene product combines with the c-*jun* gene product to form AP-1. A similar series of events occurs following B cell stimulation; however, the genes regulated by B cell AP-1 are not known.

Activation refers to the stimulation of lymphocytes or macrophages to increase their functional activity, or the initiation of the multicomponent complement cascade in serum consisting of a series of enzyme-substrate reactions leading to the generation of functionally active effector molecules.

Activation phase refers to a stage in the adaptive immune response following recognition that is associated with lymphocyte proliferation and differentiation into effector cells.

Agonist peptides are peptide antigens that activate specific lymphocytes, enabling them to synthesize cytokines and to proliferate.

Antagonists are peptides whose sequence is closely related to that of an agonist peptide. They inhibit the response of a cloned T cell line specific for the agonist peptide. A molecule that interferes with the function of a receptor as a consequence of binding to it.

Blastogenesis is the activation of small lymphocytes to form blast cells. A blast cell is a relatively large cell that is greater than 8 μm in diameter with abundant RNA in the cytoplasm, a nucleus containing loosely arranged chromatin, and a prominent nucleolus. Blast cells are active in synthesizing DNA and contain numerous polyribosomes in the cytoplasm.

Blk: See tyrosine kinase.

CD40: An integral membrane glycoprotein that has a mol wt of 48/44 kDa and is also referred to as gp50. The antigen shares similarities with many nerve growth factor receptors. The CD40 antigen is expressed on peripheral blood and tonsillar B cells from the pre-B cell stage until the plasma cell stage where it is lost. It is also expressed on B cell leukemias and lymphomas, some carcinomas and interdigitating cells, and weakly on monocytes. It has been shown that the CD40 antigen is active in B cell proliferation. The CD40 ligand is gp39. CD40 binds CD40-L, the CD40 ligand. It is the receptor for the costimulatory signal for B cells. Its interaction with CD40 ligand on T cells induces B cell proliferation. CD40 belongs to the TNF-receptor family of molecules.

Second signals refer to the second of two signals required to activate lymphocytes. Lymphocyte activation requires the recognition of antigen by an antigen-specific leukocyte receptor, either in the soluble form by the B cell surface immunoglobulin receptor or complexed to an MHC molecule on an antigen-presenting cell by the αβ heterodimer of the T cell receptor complex. Following this first signal, lymphocytes do not become fully activated and are either turned off or become unresponsive to subsequent receptor stimulation, or they undergo apoptosis. A second signal is required to induce a productive immune response. The second signal enhances lymphocyte proliferation and promotes cell survival and/or prevents lymphocyte receptor unresponsiveness. Second signals may either potentiate signals transduced by TCR ligation and initiate enhanced proliferation or not only facilitate antigen-driven lymphocyte proliferation but also inhibit the induction of lymphocyte unresponsiveness and/or programmed cell death. These latter costimulatory signals activate intracellular pathways different from those induced by the antigen-receptor complex. Different surface molecules can provide second signals.

Tyrosine kinase is an enzyme that phosphorylates proteins on tyrosine residues. Enzymes of this family play a critical role in T and B cell activation. Lck, Fyn, and ZAP-70 are the principal tyrosine kinases critical for T cell activation, whereas Blk, Fyn, Lyn, and Syk are the main tyrosine kinases that are critical for B cell activation.

Blast transformation refers to the activation of small lymphocytes to form blast cells.

CD40-L: A 39-kDa antigen present on activated $CD4^+$ T cells. It is the ligand for CD40. It is also called T-BAM or gp39.

eph receptors and ephrins: The eph family of receptors is the largest known subfamily of receptor tyrosine kinases. The ligands are called ephrins. The ephrin/eph interactions are important in development, especially in cell–cell interactions involved in nervous system patterning (axon guidance) and possibly in cancer.

Ephrin/eph: Endothelial cells destined to become arteries express ephrin-B2, while the cognate receptor, eph B4, is expressed on endothelial cells destined to become veins. The ephrin/eph family of cell-surface proteins is important in the cell–cell recognition and signaling of nervous system patterning. Their specific location on venous vs. arterial endothelial cells suggests that the formation of a vascular system may be appreciably more complicated than predicted.

B7: A homodimeric immunoglobulin superfamily protein whose expression is restricted to the surface of cells that stimulate growth of T lymphocytes. The ligand for B7 is CD28. B7 is expressed by accessory cells and is important in costimulatory mechanisms. Some APCs may upregulate expression of B7 following activation by various stimuli including IFN-α, endotoxin, and MHC class II binding. B7 is also termed BB1, B7.1, or CD80.

A **B7.1 costimulatory molecule** is a 60-kDa protein that serves as a costimulatory ligand for CD28 but as an inhibitory ligand upon interacting with CTLA-4 molecules. Also called CD80.

A **B7.2 costimulatory molecule** is a costimulatory molecule whose sequence resembles that of B7. Dendritic cells, monocytes, activated T cells, and activated B lymphocytes may express B7.2, which is an 80-kDa protein that serves as a costimulatory ligand for CD28 but as an inhibitory ligand upon interacting with CTLA-4 molecules. Also called CD86.

A **CD40 ligand** is a molecule on T cells, which interacts with CD40 on B cell proliferation.

Phorbol ester(s): Esters of phorbol alcohol (4,9,12-β, 13,20-pentahydroxy-1,6-tigliadien-3-on) found in croton oil and myristic acid. Phorbol myristate acetate (PMA), which is of interest to immunologists, is a phorbol ester that is 12-*O*-tetradecanoylphorbol-13-acetate (TPA). This is a powerful tumor promoter that also exerts pleotrophic effects on cells in culture, such as stimulation of macromolecular synthesis and cell proliferation, induction of prostaglandin formation, alteration in the morphology and permeability of cells, and disappearance of surface fibronectin. PMA also acts on leukocytes. It links to and stimulates protein kinase C. This leads to threonine and serine residue phosphorylation in the transmembrane protein cytoplasmic domains such as in the CD2 and CD3 molecules. These events enhance interleukin-2 receptor expression on T cells and facilitate their proliferation in the presence of interleukin-1 as well as TPA. Mast cells, polymorphonuclear leukocytes, and platelets may all degranulate in the presence of TPA.

Activation-induced cell death (AICD): A phenomenon first observed in T hybridomas which die within 24 h of stimulation. It was also observed *in vivo* following systemic stimulation by bacterial sAgs or peptide antigens. It represents a heightened sensitivity of recently stimulated cells to apoptosis induced by TCR crosslinking, linked to the cell cycle. It can also eliminate T cells immediately at the time of initial stimulation, especially in virus infected individuals. In clonal exhaustion, AICD can lead to the complete elimination of all antigen-reactive cells and could represent the basis for high-dose tolerance.

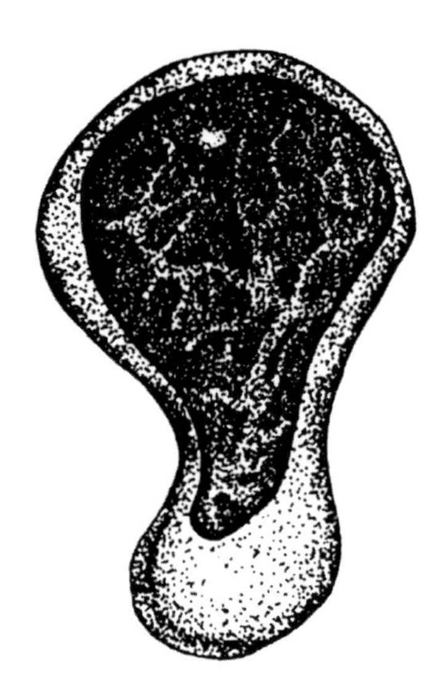

FIGURE 2.35 Uropod.

Phosphatidylinositol bisphosphate (PIP$_2$) is a membrane-associated phospholipid that is cleaved by a phospholipase C-γ to yield the signaling molecules discylglycerol and inositol trisphosphate.

Uropod (Figure 2.35) describes lymphocyte cytoplasm extending as an elongated tail or pseudopod in locomotion. The uropod may resemble the handle of a hand mirror. The plasma membrane covers the uropod cytoplasm.

T cells (Figure 2.36) are derived from hematopoietic precursors that migrate to the thymus where they undergo differentiation which continues thereafter to completion in the various lymphoid tissues throughout the body or during their circulation to and from these sites. T cells primarily are involved in the control of immune responses by providing specific cells capable of helping or suppressing these responses. They also have a number of other functions related to cell-mediated immune phenomena.

T cell: See T lymphocyte.

αβ T cells are T lymphocytes that express αβ chain heterodimers on their surface. The vast majority of T cells are of the αβ variety. T lymphocytes that express an antigen receptor compromised of α and β polypeptide chains. This population, to which most T cells belong, includes all those that recognize peptide antigen presented by MHC class I and class II molecules.

Regulatory T cells are T lymphocytes that can inhibit T cell responses. Suppressor T cells are an example of regulatory T cells.

Rosette refers to cells of one type surrounding a single cell of another type. In immunology, it was used as an early method to enumerate T cells, i.e., in the formation of E rosettes in which CD2 markers on human T lymphocytes

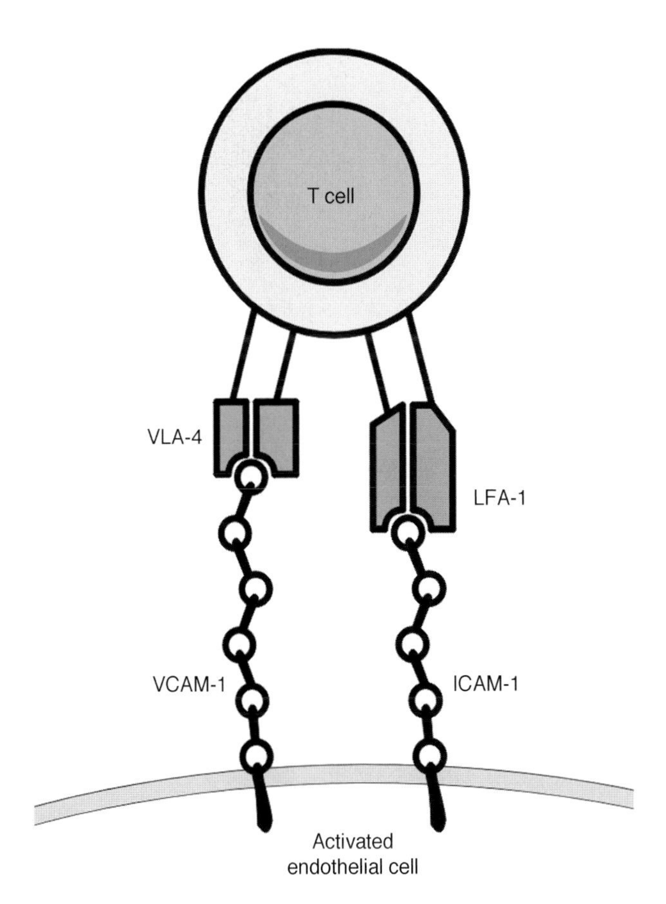

FIGURE 2.36 T cell.

adhere to LFA-3 molecules on sheep red cells surrounding them to give a rosette arrangement. Another example was the use of the EAC rosette, consisting of erythrocytes coated with antibody and complement which surrounded a B cell bearing Fc receptors or complement receptors on its surface.

Veto cells (Figure 2.37 and Figure 2.38) comprise a proposed population of cells suggested to facilitate maintenance of self-tolerance through veto of autoimmune responses by T cells. A "veto cell" would neutralize the function of an autoreactive T lymphocyte. A T cell identifies itself as an autoreactive lymphocyte by recognizing the surface antigen on the veto cell. No special receptors with specificity for the autoreactive T lymphocyte are required for the veto cell to render the T lymphocyte nonfunctional. Contemporary research suggests the existence of a veto cell that can eliminate cytotoxic T lymphocyte (CTL) precursors reactive against allogeneic class I major histocompatibility complex (MHC) molecules or against antigens presented in association with self class I MHC.

NF-κB is a transcription factor comprised of two chains of 50 kDa and 65 kDa. Under physiologic conditions it is found in the cytosol where it is bound to a third chain termed $I_{\kappa}B$, an inhibitor of $NF_{\kappa}B$ transcription.

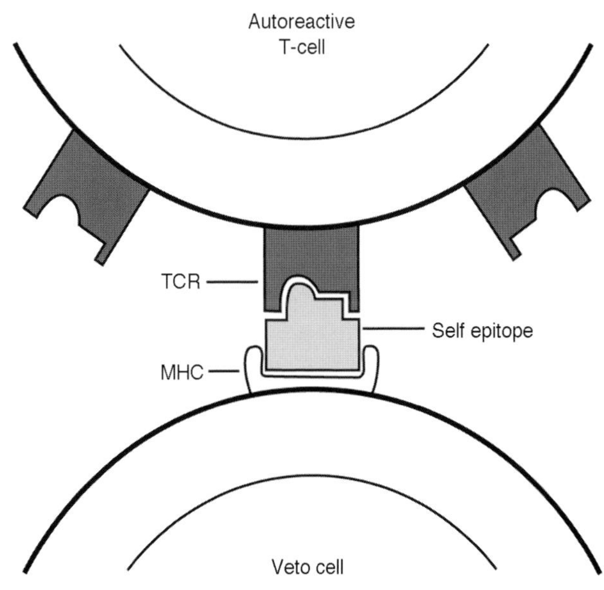

FIGURE 2.37 Veto effect.

B cells are the B lymphocytes that derive from the fetal liver in the early embryonal stages of development and from the bone marrow thereafter. In birds, maturation takes place in the bursa of Fabricius, a lymphoid structure derived from an outpouching of the hindgut near the cloaca. In mammals, maturation is in the bone marrow. Plasma cells that synthesize antibody develop from precursor B cells.

A **receptor** is a molecular configuration on a cell surface or macromolecule, which combines with molecules that are complementary to it. Examples include enzyme-substrate reactions, the T cell receptor, and membrane-bound immunoglobulin receptors of B cells. It is usually a transmembrane molecule that binds to a ligand on the cell surface, leading to biochemical changes within the cell.

Adrenergic receptors are structures on the surfaces of various types of cells that are designated α or β and interact with adrenergic drugs.

An **agonist** is a molecule that combines with a receptor and enables it to function.

In immunology, **humoral** refers to the antibody limb of the immune response, in contrast to the cell-mediated limb, together with the action of complement. Thus, immunity based on antibodies or antibodies and complement is produced and referred to as humoral immunity. Humoral immunity of the antibody type represents the products of the B cell system.

Antigen receptors: Cell surface immunoglobulin for B cells and T cell receptor for T cells. A single antigen specificity is expressed on the surface of each lymphocyte.

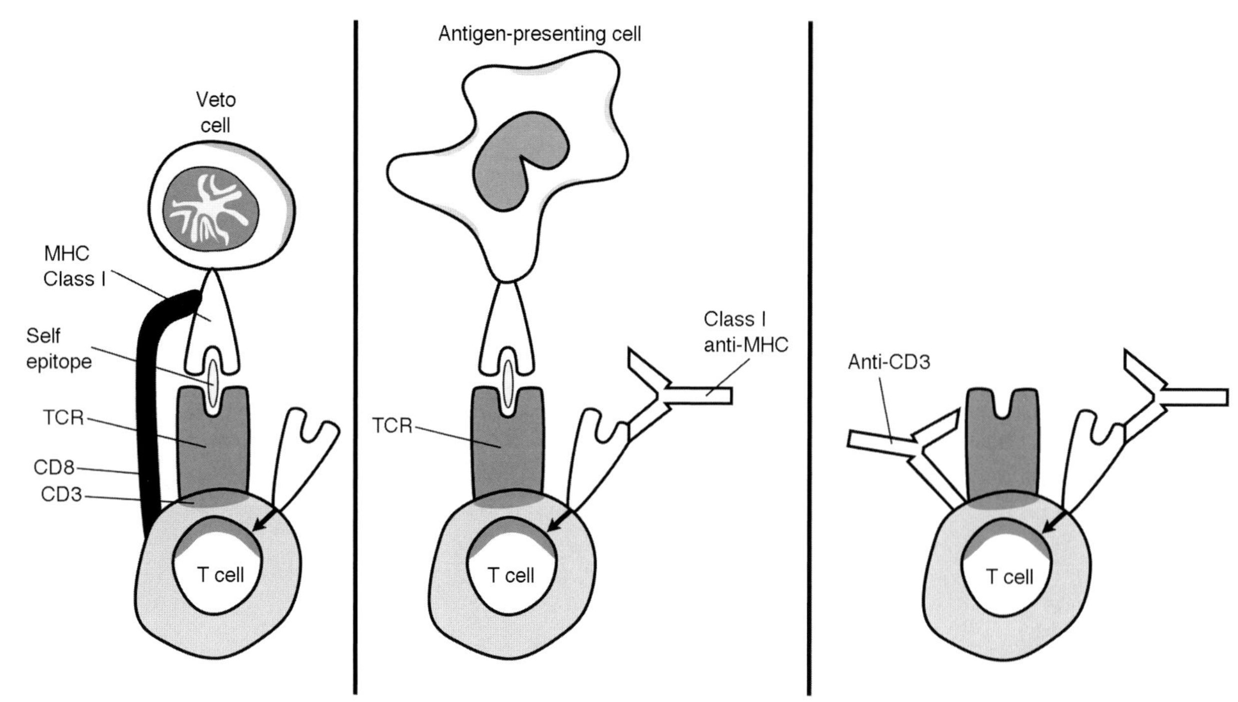

FIGURE 2.38 Veto effect.

Antigen-specific cells are antigen-binding cells such as B lymphocytes that recognize antigen with a unique antigen receptor comprising surface immunoglobulin. Monoclonal antibodies recognizing a single clonotype of T cell receptor can be used to identify antigen-specific cells and responses using this clone. Fluorescence-activated cell sorting can also be used to identify antigen-specific cells.

CD22 (Figure 2.39) is a molecule with a130- and b140-kDa mol wt that is expressed in the cytoplasm of B cells of the pro-B and pre-B cell stage and on the cell surface of mature B cells with surface Ig. The antigen is lost shortly before the terminal plasma cell phase. The molecule has five extracellular immunoglobulin domains and shows homology with myelin adhesion glycoprotein and with N-CAM (CD56). It participates in B cell adhesion to monocytes and T cells. It also is called BL-CAM.

Plasma cells (Figure 2.40 and Figure 2.41) are antibody-producing cells. Immunoglobulins are present in their cytoplasm, and secretion of immunoglobulin by plasma cells has been directly demonstrated *in vitro*. Increased levels of immunoglobulins in some pathologic conditions are associated with increased numbers of plasma cells and, conversely, their number at antibody-producing sites increases following immunization. Plasma cells develop from B cells and are large spherical or ellipsoidal cells 10 to 20 μm in size. Mature plasma cells have abundant cytoplasm, which stain deep blue with Wright's stain, and have an eccentrically located, round or oval nucleus, usually surrounded by

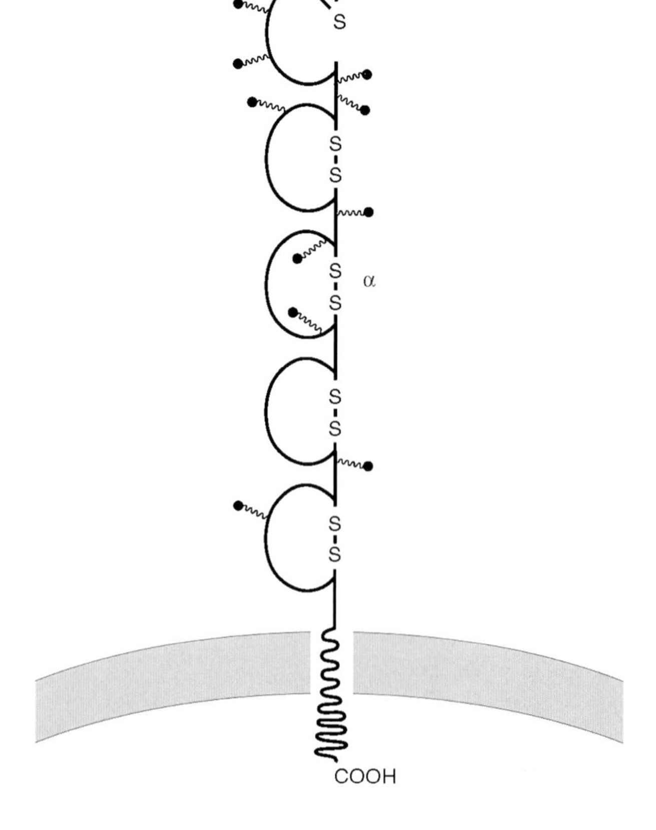

FIGURE 2.39 CD22.

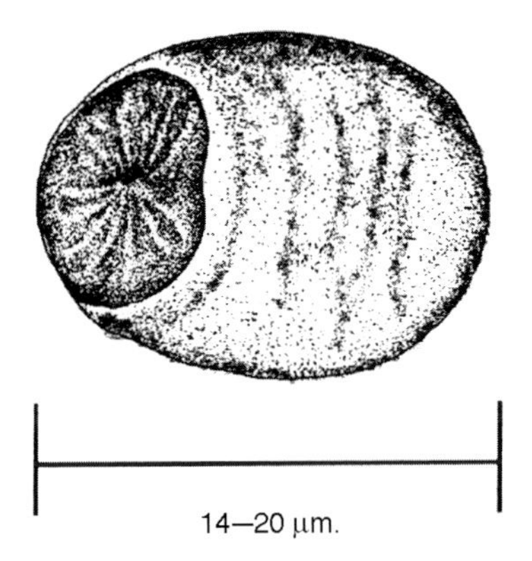

FIGURE 2.40 Plasma cell.

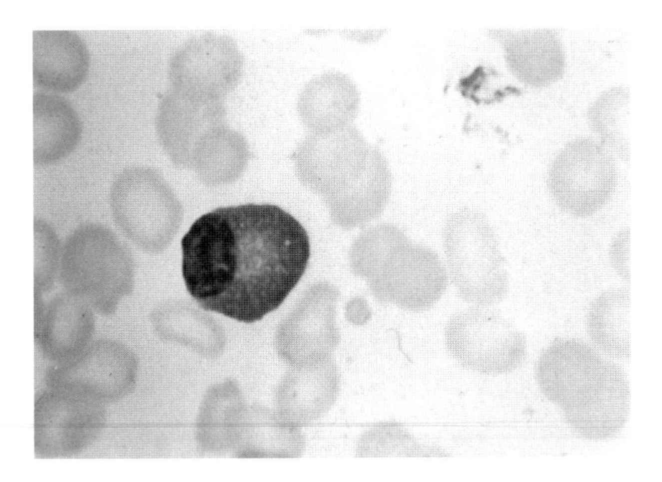

FIGURE 2.41 Plasma cell in peripheral blood.

a well-defined perinuclear clear zone. The nucleus contains coarse and clumped masses of chromatin, often arranged in a cartwheel fashion. The nuclei of normal, mature plasma cells have no nucleoli, but those of neoplastic plasma cells such as those seen in multiple myeloma have conspicuous nucleoli. The cytoplasm of normal plasma cells has conspicuous Golgi complex and rough endoplasmic reticulum and frequently contains vacuoles. The nuclear to cytoplasmic ratio is 1:2. By electron microscopy, plasma cells show very abundant endoplasmic reticulum, indicating extensive and active protein synthesis. Plasma cells do not express surface immunoglobulin or complement receptors, which distinguishes them from B lymphocytes.

Cartwheel nucleus is a descriptor for the arrangement of chromatin in the nucleus of a typical plasma cell based on the more and less electron-dense areas observed by electron microscopy. Euchromatin makes up the less electron-dense spokes of the wheel, whereas heterochromatin makes up the more electron-dense areas.

A **Russell body** is a sphere or globule in the endoplasmic reticulum of some plasma cells. These immunoglobulin-containing structures are stained pink by eosin.

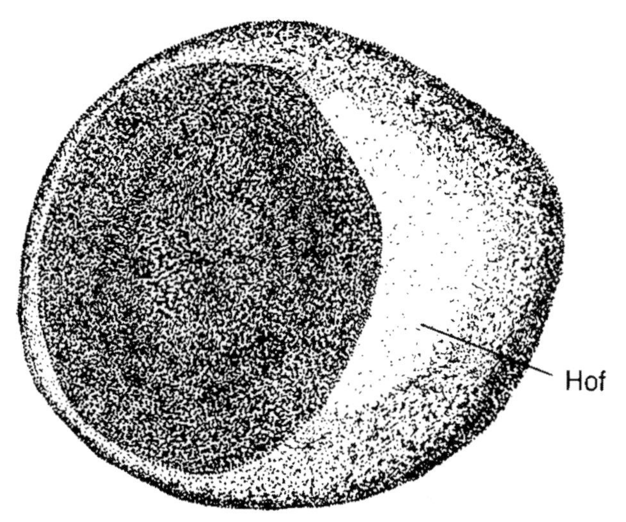

FIGURE 2.42 Hof.

A **clone** is a cell or organism that develops from a single progenitor cell and has exactly the same genotype and phenotype of the parent cell. Malignant proliferation of a clone of plasma cell in multiple myeloma represents a type of monoclonal gammopathy. The fusion of an antibody-producing B cell with a mutant myeloma cell *in vitro* by the action of polyethylene glycol to form a hybridoma that is immortal and produces monoclonal antibody is an example of the *in vitro* production of a clone.

Polyclonal means originating from multiple clones.

An **end cell** is a cell such as a mature plasma cell that is at the termination point in that cell line's maturation pattern. End cells do not further divide. They represent the final product of maturation.

Hof (Figure 2.42) is a German word for courtyard which refers to the perinuclear clear zone adjacent to the nucleus in plasma cells. Lymphoblasts and Reed–Sternberg cells may also exhibit a hof.

A **null cell** is a lymphocyte that does not manifest any markers of T or B cells, including cluster of differentiation (CD) antigens or surface immunoglobulins. Approximately 20% of peripheral lymphocytes are null cells. They play a role in antibody-dependent cell-mediated cytotoxicity (ADCC). They may be the principal cell in certain malignancies such as acute lymphocytic leukemia of children. The three types of null cells include (1) undifferentiated stem cells that may mature into T or B lymphocytes, (2) cells with labile IgG and high-affinity Fc receptors that are resistant to trypsin, and (3) large granular lymphocytes that constitute NK and K cells. The **null cell compartment** comprises 37% of the bone marrow lymphocytes, i.e., they do not have any of the markers characteristic of B or T cell lineage. They may differentiate into either B or T cells upon appropriate induction, the mechanism of which is unknown. Some null cells differentiate into killer (K) cells by developing Fc and complement receptors. The NK cells are also present in this cell population. Null, K, and NK cells, like committed lymphocytes, also migrate to the peripheral lymphoid organs such as spleen and lymph nodes or to the thymus, but they represent only a very small fraction of the total cells present there. At all locations, the null cells are part of the rapidly renewed pool of immature cells with a short life span (5 to 6 d). The null cells which have been committed to the T cell lineage migrate to the thymus to continue their differentiation.

Natural killer (NK) cells (Figure 2.43) attack and destroy certain virus-infected cells. They constitute an important part of the natural immune system, do not require prior contact with antigen, and are not MHC restricted by the major histocompatibility complex (MHC) antigens. NK cells are lymphoid cells of the natural immune system that express cytotoxicity against various nucleated cells, including tumor cells and virus-infected cells. NK cells, killer (K) cells, or ADCC cells induce lysis through the action of antibody. Immunologic memory is not involved, as previous contact with antigen is not necessary for NK cell activity. The NK cell is approximately 15 μm in diameter and has a kidney-shaped nucleus with several — often three — large cytoplasmic granules. The cells are also called large granular lymphocytes (LGL). In addition to the ability to kill selected tumor cells and some virus-infected cells, they also participate in ADCC by anchoring antibody to the cell surface through an Fcγ receptor. Thus, they are able to destroy antibody-coated nucleated cells. NK cells are believed to represent a significant part of the natural immune defense against spontaneously developing neoplastic cells and against infection by viruses. NK cell activity is measured by a ^{51}Cr release assay employing the K562 erythroleukemia cell line as a target.

NK cell: See natural killer cells.

Large granular lymphocytes (LGL): See natural killer cells.

LGL (large granular lymphocyte or null cell): These lymphocytes do not express B and T cell markers, but they have Fc receptors for IgG on their surface. They comprise approximately 3.5% of lymphocytes and originate in the bone marrow. The LGLs include natural killer cells which comprise 70% of LGLs, and killer cells which mediate ADCC.

Fragmentins are serine esterases present in cytotoxic T cell and natural killer cell cytoplasmic granules. The introduction of fragmentins into the cytosol of a cell causes apoptosis as the DNA in the nucleus is fragmented into 200 base pair multimers. Also called granzymes.

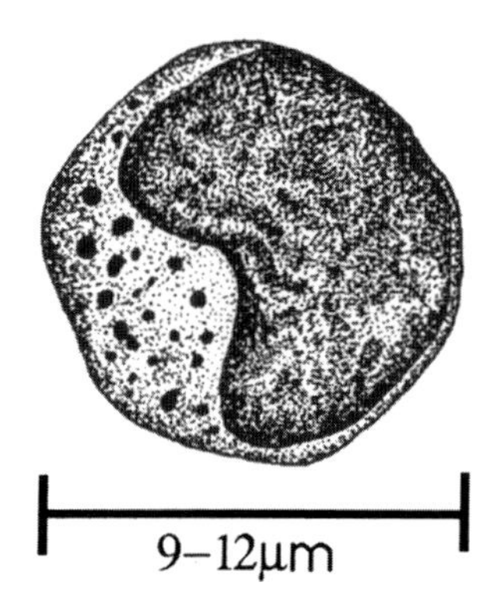

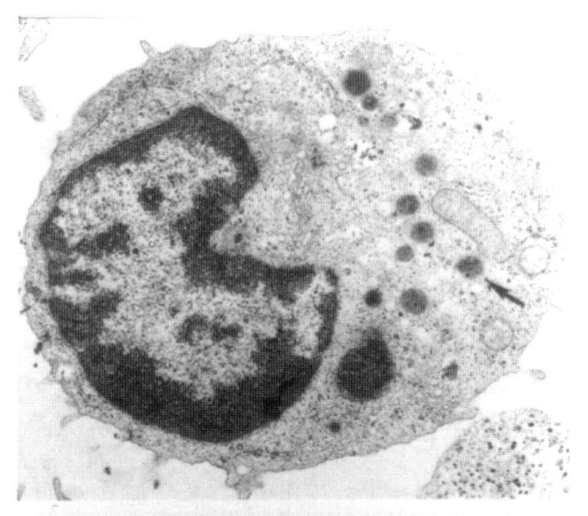

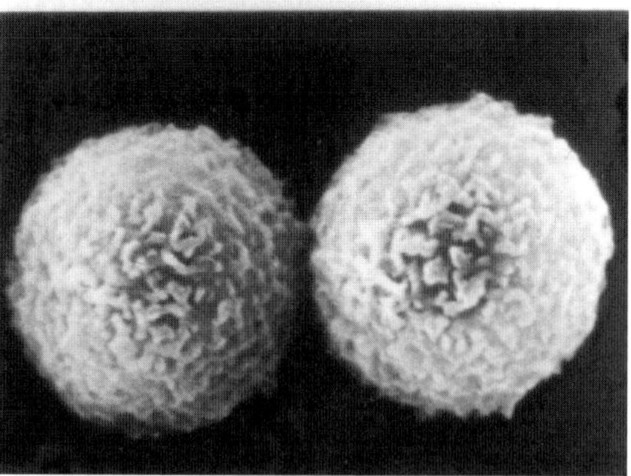

FIGURE 2.43 NK (Natural Killer) cell schematic representation. Transmission and scanning EM of human large granular lymphocyte.

Granzymes: Proteases released from LGL and cytotoxic T lymphocytes (CTL) granules that contribute to fatal injury of target cells subjected to the cytotoxic action of perforin. Antigranzyme antibodies inhibit target cell lysis. Serine esterase enzymes are present in the granules of cytotoxic T lymphocytes and NK cells. Granzymes induce target cell apoptosis after entering the cytosol. Also called fragmentins. Serine esterase enzymes are present in the granules of cytotoxic T lymphocytes and NK cells. Granzymes induce target cell apoptosis after entering the cytosol.

Killer activatory receptors (KARs) are NK cell or cytotoxic T cell surface receptors that can activate killing by these cell types.

NK 1.1 is a natural killer (NK) cell alloantigen identified in selected inbred mouse strains such as C57BL/6 mice.

NK1-T cell is a lymphocyte that shares certain characteristics with T cells, such as T cell receptors, in addition to those of natural killer cells.

NK1.1 CD4 T cells are a minor T cell subset that expresses the NK1.1 marker, a molecule usually present on NK cells. NK1.1 T lymphocytes also express α:β T cell receptors of limited diversity and either the coreceptor molecule CD4 or no coreceptor. They are present in abundance in the liver and synthesize cytokines soon after stimulation.

NK-T cell is a lymphoid cell that is intermediate between T lymphocytes and NK cells with respect to morphology and granule content. They may be $CD4^{-}8^{-}$ or $CD4^{+}8^{-}$, weakly expressed αβ TCR with an invariant α chain and highly restricted β chain specificity. They have a powerful capacity to synthesize IL-4. Many of these T cell receptors recognize antigens presented by the nonclassical MHC-like molecule CD1. Their surface NK1.1 receptor is lectin-like and is believed to recognize microbial carbohydrates.

LAG-3 is an NK cell activation molecule that is closely related to CD4 in its structure. It is a type I integral membrane protein and a member of the Ig superfamily with an Ig V-region-like domain and three Ig-C2 region-like domains. Its gene colocalizes with but is distinct from the CD4 gene on mouse chromosome 6. LAG-3 is expressed on activated T and NK cells, but not on resting lymphocytes. LAG-3 could be a coreceptor for a putative activation receptor.

CD16 (Figure 2.44) is an antigen that is also known as the low-affinity Fc receptor for complexed IgG-Fc_{γ}RIII. It is expressed on NK cells, granulocytes, (neutrophils), and macrophages. Structural differences in the CD16 antigen from granulocytes and NK cells have been reported. This apparent polymorphism suggests two different genes for the Fc_{γ}RIII molecule in polymorphonuclear leukocytes (PMN) and in NK cells. The CD16 molecule in NK cells has a transmembrane form, whereas it is phosphatidylinositol (PI)-linked in granulocytes. CD16 mediates phagocytosis. It is the functional receptor structure for performing ADCC. CD16 is also termed Fc_{γ}RIII.

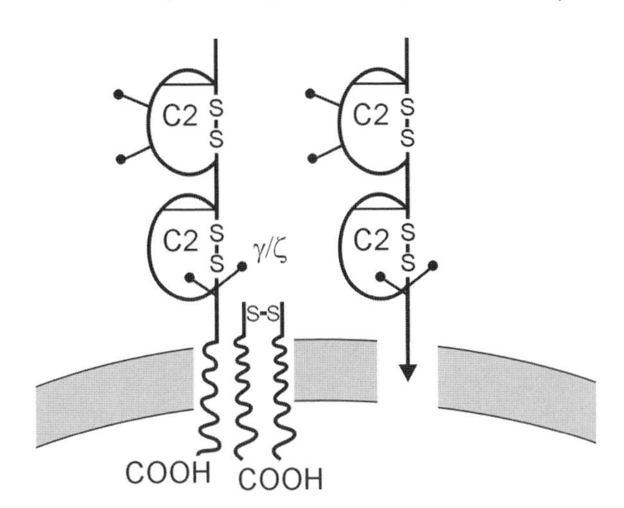

FIGURE 2.44 CD16.

CD56 (Figure 2.45) is a 220/135-kDa molecule that is an isoform of NCAM. It is used as a marker of NK cells, but it is also present on neuroectodermal cells.

CD57 is a 110-kDa myeloid-associated glycoprotein that is recognized by the antibody HNK1. It is a marker for

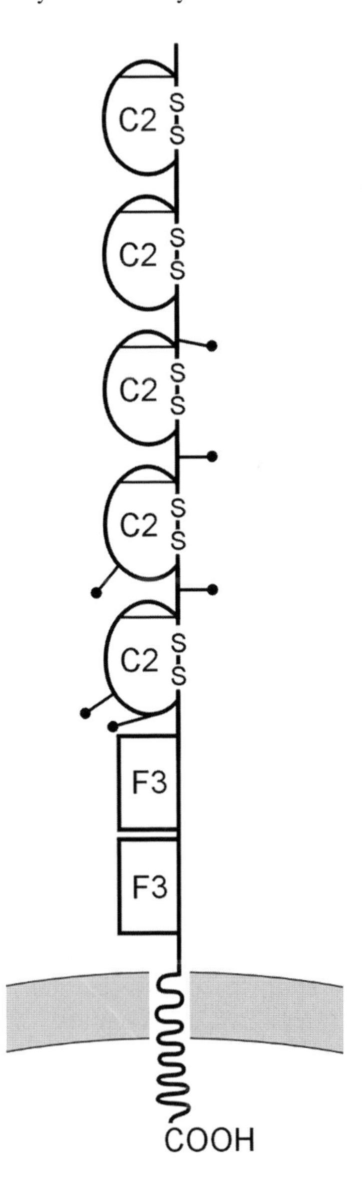

FIGURE 2.45 CD56.

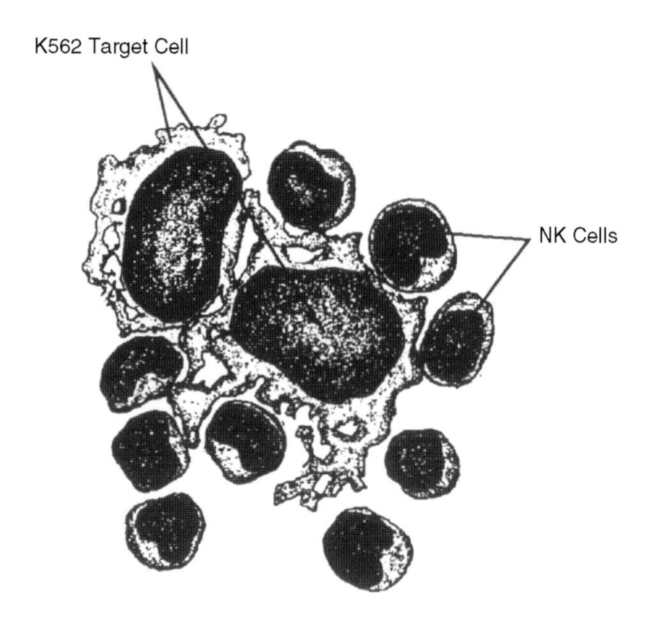

FIGURE 2.46 Schematic representation of a K562 target cell bound to NK cells.

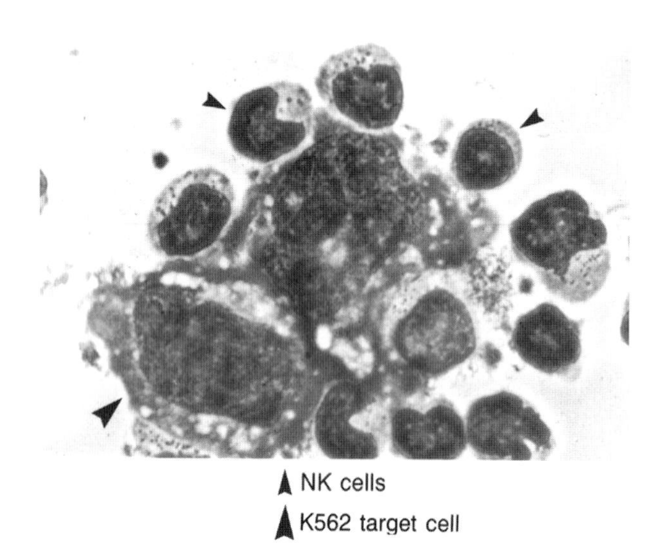

FIGURE 2.47 K562 target cells (large arrow) bound to NK cells (small arrows).

NK cells, but it is also found on some T and B cells. It is an oligosaccharide present on multiple cell surface glycoproteins.

K562 cells (Figure 2.46 and Figure 2.47) are a chronic myelogenous leukemia cell line that serves as a target cell in a ^{51}Cr release assay of natural killer (NK) cells. Following incubation of NK cells with ^{51}Cr-labeled target K562 cells, the amount of chromium released into the supernatant is measured, and the cytotoxicity is determined by use of a formula.

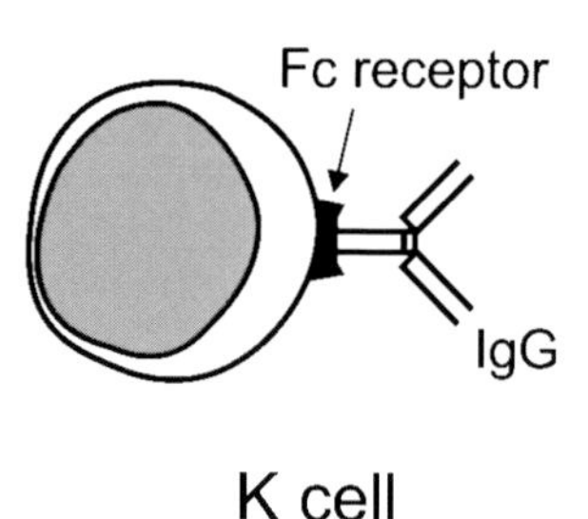

FIGURE 2.48 Schematic representation of a K cell.

K (killer) cells (Figure 2.48), also called null cells, have lymphocyte-like morphology but functional characteristics different from those of B and T cells. They are involved in a particular form of immune response: ADCC, killing target cells coated with IgG antibodies. A K cell is an Fc-bearing killer cell that has an effector function in mediating ADCC. An IgG antibody molecule binds through its Fc region to the K cell's Fc receptor. Following contact with a target cell bearing antigenic determinants on its surface for which the Fab regions of the antibody molecule attached to the K cell are specific, the lymphocyte-like K cell releases lymphokines that destroy the target. This represents a type of immune effector function in which cells and antibody participate. Besides K cells, other cells that mediate antibody-dependent cell-mediated cytotoxicity include natural killer (NK) cells, cytotoxic T cells, neutrophils, and macrophages. A **killer cell (K cell)** is a large granular lymphocyte bearing Fc receptors on its surface for IgG, which makes it capable of mediating ADCC. Complement is not involved in the reaction. Antibody may attach through its Fab regions to target cell epitopes and link to the killer cell through attachment of its Fc region to the K cell's Fc receptor, thereby facilitating cytolysis of the target by the killer cell, or an IgG antibody may first link via its Fc region to the Fc receptor on the killer cell surface and direct the K cell to its target. Cytolysis is induced by insertion of perforin polymer in the target cell membrane in a manner that resembles the insertion of C9 polymers in a cell membrane in complement-mediated lysis. Perforin is showered on the target cell membrane following release from the K cell.

Macrophages (Figure 2.49 and Figure 2.50) are mononuclear phagocytic cells derived from monocytes in the blood that were produced from stem cells in the bone marrow. These cells have a powerful, although nonspecific, role in immune defense. These intensely phagocytic cells contain lysosomes and exert microbicidal action against microbes they ingest. They also have effective tumoricidal activity. They may take up and degrade both protein and polysaccharide antigens and present them to T lymphocytes in the context of major histocompatibility complex class II molecules. They interact with both T and B lymphocytes in

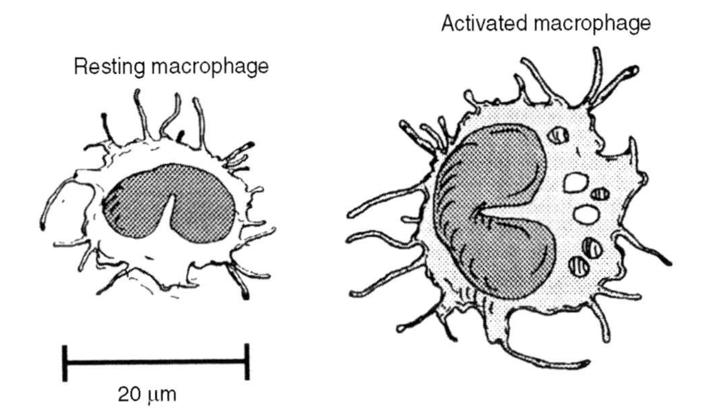

FIGURE 2.49 Schematic representation of a resting macrophage vs. an activated macrophage.

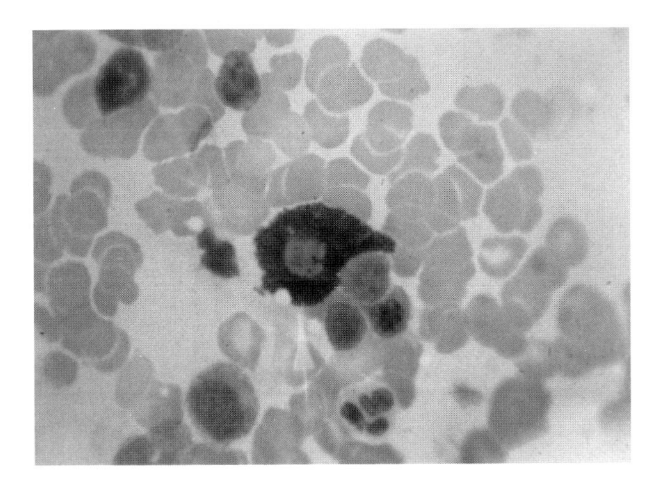

FIGURE 2.50 Macrophage-histiocyte in bone marrow.

immune reactions. They are frequently found in areas of epithelium, mesothelium, and blood vessels. Macrophages have been referred to as adherent cells since they readily adhere to glass and plastic and may spread on these surfaces and manifest chemotaxis. They have receptors for Fc and C3b on their surfaces, stain positively for nonspecific esterase and peroxidase, and are Ia antigen positive when acting as accessory cells that present antigen to $CD4^+$ lymphocytes in the generation of an immune response. Monocytes, which may differentiate into macrophages when they migrate into the tissues, make up 3 to 5% of leukocytes in the peripheral blood. Macrophages that are tissue-bound may be found in the lung alveoli, as microglial cells in the central nervous system, as Kupffer cells in the liver, as Langerhans cells in the skin, and as histiocytes in connective tissues, as well as macrophages in lymph nodes and peritoneum. Multiple substances are secreted by macrophages including complement components C1 through C5, factors B and D, properdin, C3b inactivators, and β-1H. They also produce monokines such as interleukin-1, acid hydrolase, proteases, lipases, and numerous other substances.

IFN-γ activates macrophages to increase their capacity to kill intracellular microorganisms. Macrophages are known by different names according to the tissue in which they are found, such as the microglia of the central nervous system, the Kupffer cells of the liver, alveolar macrophages of the lung, and osteoclasts in the bone.

ANAE (a-naphthyl acetate esterase): See nonspecific esterase.

An **activated macrophage** has been stimulated in some manner or by some substance to increase its functional efficiency with respect to phagocytosis, intracellular bactericidal activity, or lymphokine, i.e., IL-1 production. A lymphokine-activated mononuclear phagocyte is double the size of resting macrophages. MHC class II antigen surface expression is elevated, and lysosomes increase. The latter changes facilitate antimicrobial defense.

Multiple processes are involved in macrophage activation. These include an increase in size and number of cytoplasmic granules and a spread of membrane ruffling. Functional alterations include elevated metabolism and transport of amino acids and glucose; increased enzymatic activity; an elevation in prostaglandins, cGMP, plasminogen activator, intracellular calcium ions, phagocytosis, pinocytosis; and the ability to lyse bacteria and tumor cells.

Nitric oxide (NO) and **reactive oxygen species (ROS):** Reactive oxygen species (ROS) break down to form or generate free radicals. Cells possess elaborate systems to scavenge free radicals. When free radicals exceed the capacity of these systems, however, cells die. Cell death induced by free radicals has characteristics of both apoptosis and necrosis. The most compelling observation that cell death resulting from free radicals is related to the apoptotic process is found at the level of the mitochondria. The antiapoptotic protein, Bcl-2, inhibits cell death in response to free radicals. The mechanisms involved, however, are not fully understood. The radical-induced cell death may involve the mitochondrial permeability transition pore. Bcl-2 has been observed to be located near the permeability transition pore in the mitochondrial membrane. Nitirc oxide (NO) is produced by iNOS, eNOS, and nNOS. NO is a biological signaling molecule that elicits numerous biochemical responses. Reports have suggested that NO can affect key proteins or signaling pathways involved in apoptosis. Given the rapidly expanding roles for signaling through the generation of NO, it may be that NO has important influences on apoptosis.

A stimulated macrophage is one that has been activated *in vivo* or *in vitro*. The term *activated macrophage* is preferred.

Inducible NO synthase (iNOS) is a mechanism of macrophages or various other cells to activate NO synthesis in response to numerous different stimuli. This represents a principal mechanism of host resistance against murine intracellular infection and may exist in humans as well.

Nitric oxide (NO) is a biologic molecule with multiple effects, including an important role in intracellular signaling and functioning in macrophages as a powerful microbicidal agent against ingested microorganisms. NO is a neurotransmitter and an agent that maintains hemodynamic stability. Its role in human host defense has been controversial. Nitrite has been generated in human macrophage cultures in response to TNF-α and GM-CSF together with avirulent microbacterial strains. High levels of nitric oxide synthesis have been shown in response to a select group of stimuli.

Superoxide dismutase is an enzyme that defends an organism against oxygen-free radicals by catalyzing the interaction of superoxide anions with hydrogen ions to yield hydrogen peroxide and oxygen.

An **adherent cell** is a cell such as a macrophage (mononuclear phagocyte) that attaches to the wall of a culture flask, thereby facilitating the separation of such cells from B and T lymphocytes which are not adherent.

The **epithelioid cell** derives from the monocyte–macrophage system. Peripheral blood monocytes made adherent to cellophane strips and implanted into the subcutaneous tissue of an experimental animal develop into epithelioid cells. Conversion of the macrophage to an epithelioid cell is not preceded by a mitotic division of the macrophage. On the contrary, epithelioid cells are able to divide, resulting in round, small daughter cells which mature in 2 to 4 d, gaining structural and functional characteristics of young macrophages. Material that is taken up by macrophages but cannot be further processed prevents the conversion of epithelioid cells. The life span of the epithelioid cell is from 1 to 4 weeks.

The **epithelioid cell** is a particular type of cell characteristic of some types of granulomas such as in tuberculosis, sarcoidosis, leprosy, etc. The cell has poorly defined cellular outlines; cloudy, abundant eosinophilic cytoplasm; and an elongated and pale nucleus. By electron microscopy, the cell shows a few short and slender pseudopodia and well-developed cellular organelles. Mitochondria are generally elongated, the Golgi complex is prominent, and lysosomal dense bodies are scattered throughout the cytoplasm. Strands of endoplasmic reticulum, free ribosomes, and fibrils are present in the ground substance.

Apolipoprotein E is a 33-kDa protein produced by nonactivated macrophages but not monocytes. It binds low-density lipids as well as high-density cholesterol esters.

The term **resident macrophage** refers to a macrophage normally present at a tissue location without being induced to migrate there.

Tissue-fixed macrophage: Histiocyte.

Angry macrophage is a term sometimes used to refer to activated macrophages.

Macrophage immunity: Cellular immunity.

A **granuloma** is a tissue reaction characterized by altered macrophages (epithelioid cells), lymphocytes, and fibroblasts. These cells form microscopic masses of mononuclear cells. Giant cells form from some of these fused cells. Granulomas may be of the foreign body type, such as those surrounding silica or carbon particles, or of the immune type that encircle particulate antigens derived from microorganisms. Activated macrophages trap antigen, which may cause T cells to release lymphokines, causing more macrophages to accumulate. This process isolates the microorganism. Granulomas appear in cases of tuberculosis and develop under the influence of helper T cells that react against *Mycobacterium tuberculosis.* Some macrophages and epithelioid cells fuse to form multinucleated giant cells in immune granulomas. There may also be occasional neutrophils and eosinophils. Necrosis may develop. It is a delayed type of hypersensitivity reaction that persists as a consequence of the continuous presence of foreign body or infection.

Macrophage/monocyte chemotaxis: Macrophages and monocytes are strongly adherent cells and have a rate of locomotion slower than that of neutrophils. They mount a chemotactic response to microorganisms formyl-Met-Leu-Phe, C5a, C5a des Arg, leukotriene B_4, platelet-activating factor, thrombin, and elastin. Tumors in man and animals may produce an inhibitor that causes monocytes or macrophages to migrate poorly in chemotaxis assays.

Nitric oxide synthetase is an enzyme or family of enzymes that synthesizes vasoactive and microbicidal compound nitric oxide from L-arginine. The activation of macrophages by microorganisms of cytokines can induce a form of this enzyme.

A **histiocyte** is a tissue macrophage that is fixed in tissues such as connective tissues. Histiocytes are frequently around blood vessels and are actively phagocytic. They may be derived from monocytes in the circulating blood.

A **mannose receptor** is a lectin or carbohydrate-binding receptor on macrophages that binds mannose and fucose residues on the cell walls of microorganisms, thereby facilitating their phagocytosis.

Mannose-binding lectin (MBL) is a protein in the plasma that binds mannose residues on bacterial cell walls, thereby acting as opsonin to facilitate phagocytosis of the bacterium by macrophages. A surface receptor for Clq on macrophage surfaces also binds MBL and facilitates phagocytosis of the opsonized microorganisms.

Tingible body refers to nuclear debris present in macrophages of lymph node, spleen, and tonsil germinal centers, as well as in the dome of the appendix.

Tingible body macrophages are phagocytic cells that engulf apoptotic B cells which are formed in large numbers at the height of a germinal center response.

Phagocytosis is an important clearance mechanism for the removal and disposition of foreign agents and particles or damaged cells. Macrophages, monocytes, and polymorphonuclears cells are phagocytic cells. In special circumstances, other cells such as fibroblasts may show phagocytic properties; these are called facultative phagocytes.

Phagocytosis may involve nonimmunologic or immunologic mechanisms. Nonimmunologic phagocytosis refers to the ingestion of inert particles such as latex particles or of other particles that have been modified by chemical treatment or coated with protein. Damaged cells are also phagocytized by nonimmunologic mechanisms. Damaged cells may become coated with immunoglobulin or other proteins which facilitate their recognition.

Phagocytosis of microorganisms involves several steps: attachment, internalization, and digestion. After attachment, the particle is engulfed within a membrane fragment and a phagocytic vacuole is formed. The vacuole fuses with the primary lysosome to form the phagolysosome, in which the lysosomal enzymes are discharged and the enclosed material is digested. Remnants of indigestible material can be recognized subsequently as residual bodies. Polymorphonuclear neutrophils (PMNs), eosinophils, and macrophages play an important role in defending the host against microbial infection. PMNs and occasional eosinophils appear first in response to acute inflammation, followed later by macrophages. Chemotactic factors are released by actively multiplying microbes. These chemotactic factors are powerful attractants for phagocytic cells which have specific membrane receptors for the factors. Certain pyogenic bacteria may be destroyed soon after phagocytosis as a result of oxidative reactions. However, certain intracellular microorganisms such as *Mycobacteria* or *Listeria* are not killed merely by ingestion and may remain viable unless there is adequate cell-mediated immunity induced by γ interferon activation of macrophages.

Phagocytic dysfunction may be due to either extrinsic or intrinsic defects. The extrinsic variety encompasses opsonin deficiencies secondary to antibody or complement factor deficiencies, suppression of phagocytic cell numbers by immunosuppressive agents, corticosteroid-induced interference with phagocytic function, neutropenia, or abnormal neutrophil chemotaxis. Intrinsic phagocytic dysfunction is related to deficiencies in enzymatic killing of engulfed microorganisms. Examples of the intrinsic disorders include chronic granulomatous disease, myeloperoxidase deficiency, and glucose-6-phosphate dehydrogenase deficiency. Consequences of phagocytic dysfunction include increased susceptibility to bacterial infections but not to viral or protozoal infections. Selected phagocytic function disorders may be associated with severe fungal infections. Severe bacterial infections associated with phagocytic dysfunction range from mild skin infections to fatal systemic infections.

A **phagosome** is a phagocytic membrane-limited vesicle in a phagocyte that contains phagocytized material which is digested by lysosomal enzymes that enter the vesicle after fusion with lysosomes in the cytoplasm.

A **phagolysosome** is a cytoplasmic vesicle with a limiting membrane produced by the fusion of a phagosome with a lysosome. Substances within a phagolysosome are digested by hydrolysis.

A **microglial cell** is a phagocytic cell in the central nervous system. It is a bone marrow-derived perivascular cell of the mononuclear phagocyte system. In the central nervous system, it may act as an antigen-presenting cell, functioning in an MHC class II-restricted manner.

Tuftsin is a leukokinin globulin-derived substance that enhances phagocytosis. It is a tetrapeptide comprised of Thr-Lys-Pro-Arg. The leukokinin globulin from which it is derived represents immunoglobulin Fc receptor residues 289 through 292. Tuftsin is formed in the spleen. Its actions include neutrophil and macrophage chemotaxis, enhancing phagocyte motility, and promoting oxidative metabolism. It also facilitates antigen processing.

Facultative phagocytes are cells such as fibroblasts that may show phagocytic properties under special circumstances.

Armed macrophages are macrophages bearing surface IgG or IgM cytophilic antibodies or T cell lymphokines that render them capable of inducing antigen-specific cytotoxicity.

Pinocytosis refers to the uptake by a cell of small liquid droplets, minute particles, and solutes.

Endocytosis is a mechanism whereby substances are taken into a cell from the extracellular fluid through plasma membrane vesicles. This is accomplished by either pinocytosis or receptor-facilitated endocytosis. In pinocytosis, extracellular fluid is captured within a plasma membrane vesicle. In receptor-facilitated endocytosis, extracellular ligands bind to receptors, and coated pits and coated vesicles facilitate internalization. Clathrin-coated vesicles become uncoated and fuse to form endosomes. Ligand and receptor dissociate within the endosomes, and the receptor returns to the cell surface. Endosomes fused with lysosomes form secondary lysosomes where ligand degradation occurs. Low-density lipoproteins are handled in this manner.

An endocytic vesicle is a membrane structure derived from the plasma membrane that transports extracellular material into cells.

Endogenous means resulting from conditions within a cell or organism, rather than externally caused; derived internally.

A **Kupffer cell** is a liver macrophage that has become fixed as a mononuclear phagocytic cell in the liver sinusoids. It is an integral part of the mononuclear phagocyte (reticuloendothelial) system. Monocytes become attached to the interior surfaces of liver sinusoids where they develop into macrophages. They have CR1 and CR2 receptors, surface Fc receptors, and MHC class II molecules. They are actively phagocytic and remove foreign substances from the blood as they flow through the liver. Under certain disease conditions, they may phagocytize erythrocytes, leading to the deposition of hemosiderin particles derived from hemoglobin breakdown products.

CD9 is a single-chain protein with a mol wt of 24 kDa that is present on pre-B cells, monocytes, granulocytes, and platelets. Antibodies against the molecule can cause platelet aggregation. The CD9 antigen has protein kinase activity. It may be significant in aggregation and activation of platelets.

CD13 is an antigen that is a single-chain membrane glycoprotein with a mol wt of 130 kDa. It is present on monocytes, granulocytes, some macrophages, and connective tissue. CD13 has been shown to be aminopeptidase-*N*. It functions as a zinc metalloproteinase.

CD33 is an antigen that is a single-chain transmembrane glycoprotein, with a mol wt of 67 kDa. It is restricted to myeloid cells and is found on early progenitor cells, monocytes, myeloid leukemias, and weakly on some granulocytes.

An accessory cell (Figure 2.51) is a cell such as a dendritic cell or Langerhans cell, monocyte, or macrophage that facilitates T cell responses to protein antigens. B cells may also act as antigen-presenting cells, thereby serving an accessory cell function.

Accessory molecules are molecules other than the antigen receptor and major histocompatibility complex (MHC) that participate in cognitive, activation, and effector functions

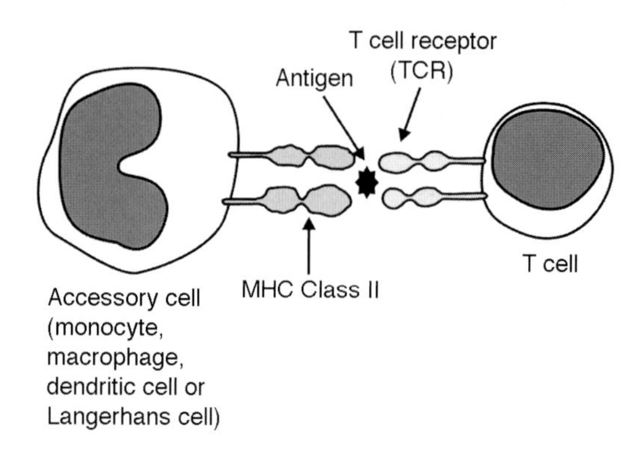

FIGURE 2.51 Accessory cell.

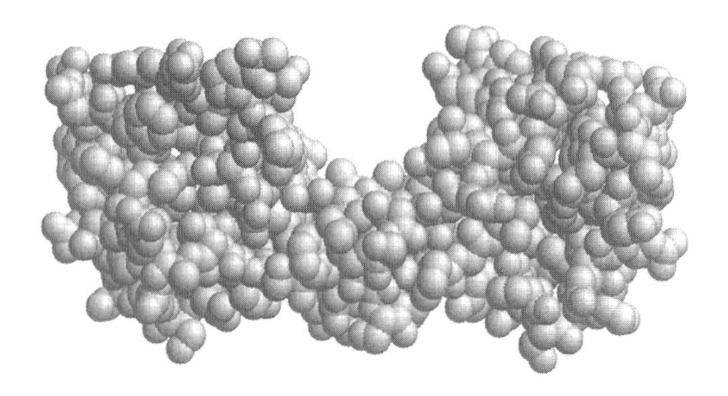

FIGURE 2.52 MIP-1α. NMR.

of T lymphocyte responsiveness. Adhesion molecules facilitating the interaction of T lymphocytes with other cells that signal transducing molecules which participate in T cell activation or migration are classified as accessory molecules.

Macrophage-activating factor (MAF) is a lymphokine such as γ interferon that accentuates the ability of macrophages to kill microbes and tumor cells. MAF is a lymphokine that enhances a macrophage's phagocytic activity and bactericidal and tumoricidal properties.

Macrophage chemotactic factor (MCF) refers to cytokines that act together with macrophages to induce facilitating migration. Among these substances are interleukins and interferons.

Macrophage chemotactic and activating factor (MCAF) is a chemoattractant and activator of macrophages produced by fibroblasts, monocytes, and endothelial cells as a result of exogenous stimuli and endogenous cytokines such as TNF, IL-1, and PDGF. It also has a role in activating monocytes to release an enzyme that is cytostatic for some tumor cells. MCAF also has a role in ELAM-1 and CD11 a and b surface expression in monocytes and is a potent degranulator of basophils.

Macrophage inflammatory protein-1-α (MIP-1) (Figure 2.52) is an endogeneous fever-inducing substance that binds heparin and is resistant to *cyclo*-oxygenase inhibition. Macrophages stimulated by endotoxin may secrete this protein, termed MIP-1, which differs from tumor necrosis factor (TNF) and IL-1 as well as other endogenous pyrogens because its action is not associated with prostaglandin synthesis. It appears indistinguishable from hematopoietic stem cell inhibitor and may function in growth regulation of hematopoietic cells.

Macrophage inflammatory peptide-2 (MIP-2) is an IL-8 type II receptor competitor and chemoattractant that is also involved in hemopoietic colony formation as a costimulator. It also degranulates murine neutrophils. The inflammatory activities of MIP-2 are very similar to those of IL-8.

Plasminogen is the inactive precursor of the proteolytic enzyme plasmin. Several serine proteases such as urokinase convert it to active plasmin. It is a β globulin widely distributed in tissue, body fluids, and plasma. Plasminogen is a single-chain monomeric molecule. Plasminogen activation occurs in two stages: The Glu-plasminogen activation begins with removal of two peptides at the N-terminus of the molecule and conversion to Lys-plasminogen. The second step involves the rapid conversion of Lys-plasminogen to Lys-plasmin.

TPA is the abbreviation for tissue-plasminogen activator.

A **plasminogen activator** is an enzyme produced by macrophages that converts plasminogen to plasmin which degrades fibrin.

Macrophage cytophilic antibody (Figure 2.53) is an antibody that becomes anchored to the Fc receptors on macrophage surfaces. This cytophilic antibody can be demonstrated by the immunocytoadherence test.

Macrophage functional assays are tests of macrophage function that include the following: (1) Chemotaxis — macrophages are placed in one end of a Boyden chamber and a chemoattractant is added to the other end. Macrophage migration toward the chemoattractant is assayed. (2) Lysis — macrophages acting against radiolabeled tumor cells or bacterial cells in suspension can be measured after suitable incubation by measuring the radioactivity of the supernatant. (3) Phagocytosis — radioactivity of macrophages that have ingested a radiolabeled target can be assayed.

Alveolar macrophage (Figure 2.54) is a macrophage in the lung alveoli that may remove inhaled particulate matter.

A **veiled cell** (Figure 2.55) is a mononuclear phagocytic cell that serves as an antigen-presenting cell. It is found in the afferent lymphatics and in the marginal sinus. It may manifest IL-2 receptors in the presence of GM-CSF.

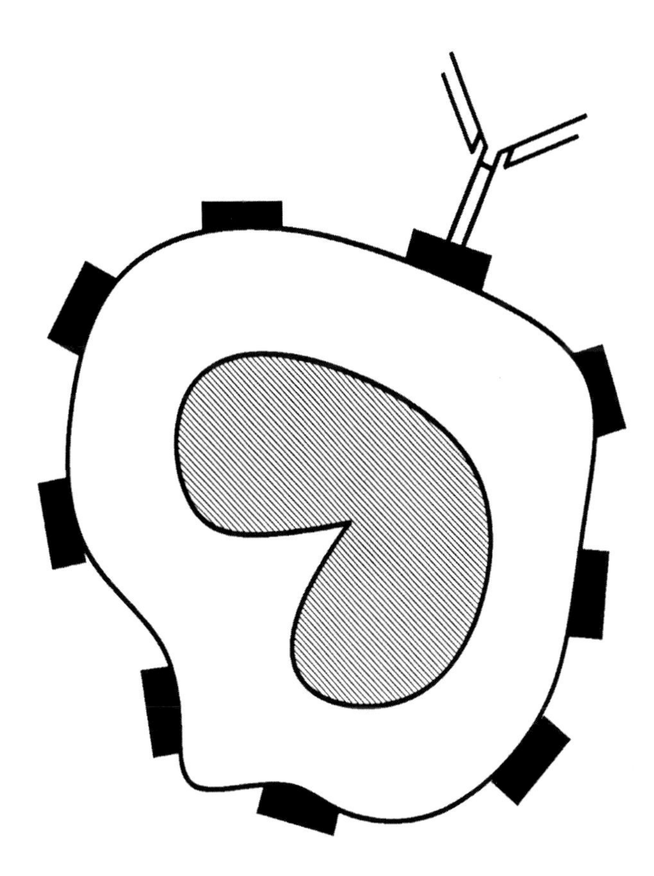

FIGURE 2.53 Macrophage cytophilic antibody.

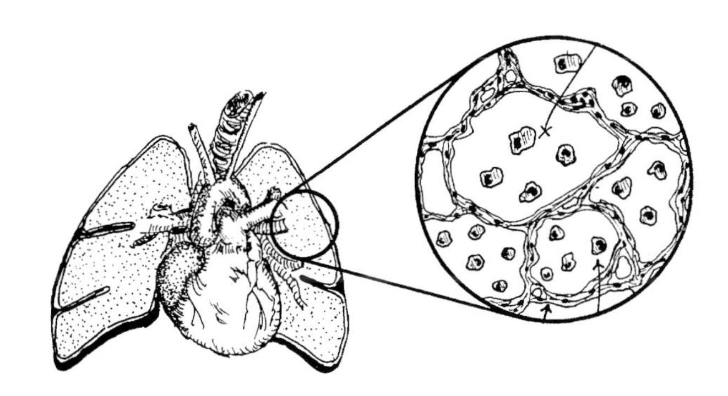

FIGURE 2.54 Alveolar macrophages.

Dendritic cells (DC) are mononuclear phagocytic cells found in the skin as Langerhans cells, in the lymph nodes as interdigitating cells, in the paracortex as veiled cells in the marginal sinus of afferent lymphatics, and as mononuclear phagocytes in the spleen where they present antigen to T lymphocytes. Dendritic reticular cells may have nonspecific esterase, Birbeck granules, endogenous peroxidase, possibly CD1, complement receptors CR1 and CR3, and Fc receptors. Dendritic cells (DC) are sentinels of the immune system. They originate from a bone marrow progenitor, travel through the blood, and are seeded into nonlymphoid tissues. DC capture and process exogenous antigens for presentation as peptide–MHC complexes at the cell surface and then migrate via the blood and afferent

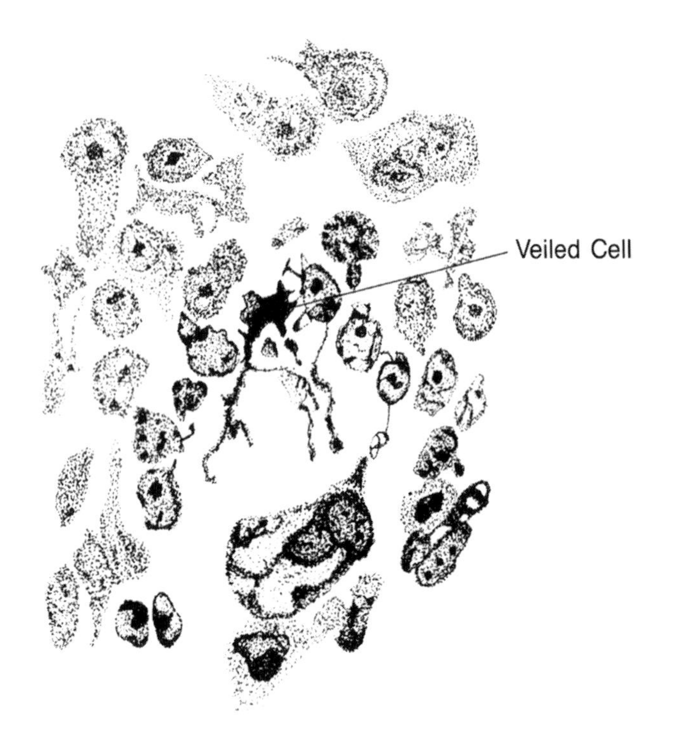

FIGURE 2.55 Veiled cell.

lymph to secondary lymph nodes. In the lymph nodes, they interact with T lymphocytes to facilitate activation of helper and killer T cells. DC have been named according to their appearance and distribution in the body (see Table 1). During the past decade, DC have been further characterized by lineage, by maturation stage, by functional and phenotype characteristics of these stages, and by mechanisms involved in migration and function. DC are being considered as adjuvants in immunization protocols for antiviral or antitumor immunity. Immature DC are defined by cell surface markers that represent functional capacity. They express the chemokine receptors CCR-1, CCR-2, CCR-5, CCR-6 (only $CD34^+$ HPC-derived DC), and CXCR-1, that are commonly thought to allow DC to migrate in response to inflammatory chemokines expressed by inflamed tissues. Immature DC are phagocytic and have a high level of macropinocytosis, allowing them to efficiently process and present antigen on class I molecules. Expression of Fc_γ (CD64) and the mannose receptors allow efficient capture of IgG immune complexes and antigens that expose mannose or fucose residues. The expression of E-cadherin allows DC to interact with tissue cells and remain in the tissues until activated. Following antigen processing, DC are remodeled. Fc and mannose receptors are downregulated, and there is disappearance of acidic intracellular compartments, resulting in a loss of endocytic activity. During this maturation process, the level of MHC class II molecules and costimulatory molecules is unregulated, and there is a change in chemokine receptor expression. Maturing DC home to T cell areas of secondary lymph nodes, where they present antigen to naïve T cells. *In vitro* culture of DC with CD40L, LPS, and TNF-α generates mature DC. These cells are very good stimulators of allogeneic T cell proliferation. The DC–T cell interaction is thought to be a two-way interaction. Evidence suggests that T cells interact with DC through CD40 ligation to enhance DC viability and their T cell stimulatory ability. Addition of CD40L induces DC to produce IL-12, which is known to support Th1 responses. LPS stimulation generates a weaker *in vitro* immune response than the CD40L-stimulated DC.

A **circulating dendritic cell** is a dendritic cell that has taken up antigen and is migrating to secondary lymphoid tissue such as a lymph node.

Dendritic epidermal cell: Mouse epidermal cells that are Thy-1^+, MHC class II molecule negative, and possess γδ T cell receptor associated with CD3. It is believed to be a variety of bone marrow-derived T lymphocyte that is separate from Langerhans cells in the skin.

Immature dendritic cells are cells that only exit the various body tissues in which they are present in response to an inflammatory mediator or an infection.

Interdigitating reticular cells: See dendritic cells.

Interstitial dendritic cells are found in most organs such as heart, lungs, liver, kidneys, and gastrointestinal tract.

Langerhans cells (Figure 2.56) are dendritic-appearing accessory cells interspersed between cells of the upper layer of the epidermis. They can be visualized by gold chloride impregnation of unfixed sections and show dendritic

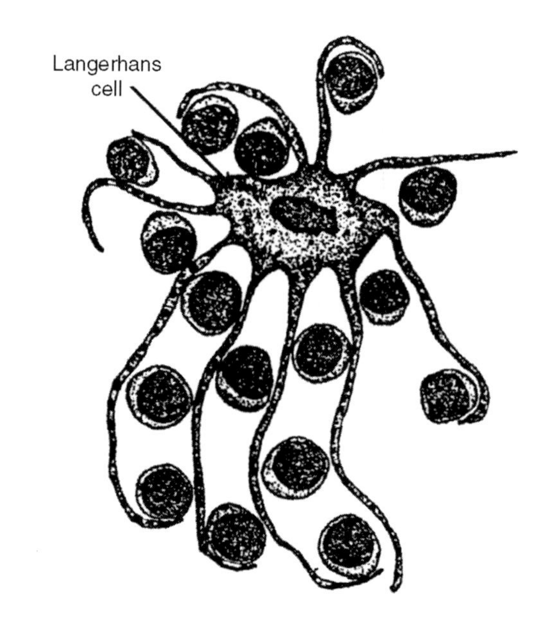

FIGURE 2.56 Langerhans cell.

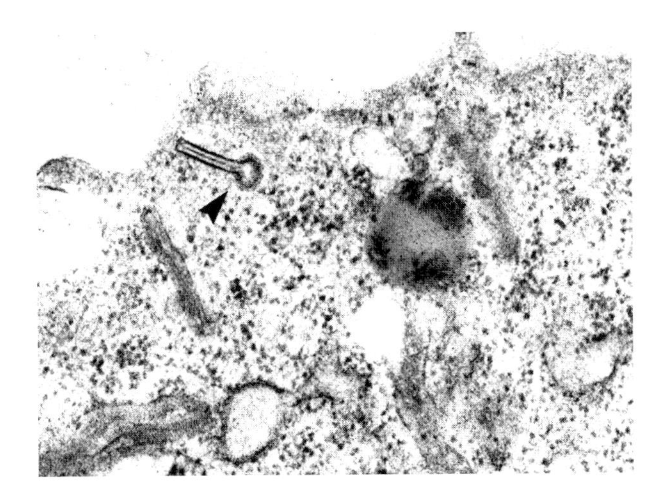

FIGURE 2.57 Birbeck granules.

processes but no intercellular bridges. By electron microscopy, they lack tonofibrils or desmosomes, have indented nuclei, and contain tennis racket-shaped Birbeck granules which are relatively small vacuoles, round to rectangular and measuring 10 nm. Following their formation from stem cells in the bone marrow, Langerhans cells migrate to the epidermis and then to the lymph nodes where they are described as dendritic cells, based upon their thin cytoplasmic processes that course between adjacent cells. Langerhans cells express both class I and class II histocompatibility antigens, as well as C3b receptors and IgG Fc receptors on their surfaces. They function as antigen-presenting cells. Epidermal Langerhans cells express complement receptors 1 and 3, Fc_γ receptors, and fluctuating quantities of CD1. Dendritic cells do not express Fc_γ receptors or CD1. Langerhans cells in the lymph nodes are found in the deep cortex. Epidermal Langerhans cells are important in the development of delayed-type hypersensitivity through the uptake of antigen in the skin and transport of it to the lymph nodes. Veiled cells in the lymph are indistinguishable from Langerhans cells.

Birbeck granules (Figure 2.57) are 10- to 30-nm diameter round cytoplasmic vesicle present in the cytoplasm of Langerhans cells in the epidermis.

The **reticuloendothelial system (RES)** is a former term for the mononuclear phagocyte system that includes Kupffer cells lining the sinusoids of the liver as well as macrophages of the spleen and lymph nodes (Figure 2.58). Aschoff introduced the term to describe cells that could take up and retain vital dyes and particles that had been injected into the body. In addition to macrophages, less active phagocytic cells such as fibroblasts and endothelial cells were also included in the original definition. The principal function of the mononuclear phagocyte system is to remove unwelcome particles from the blood. RES activity can be measured by the elimination rate of

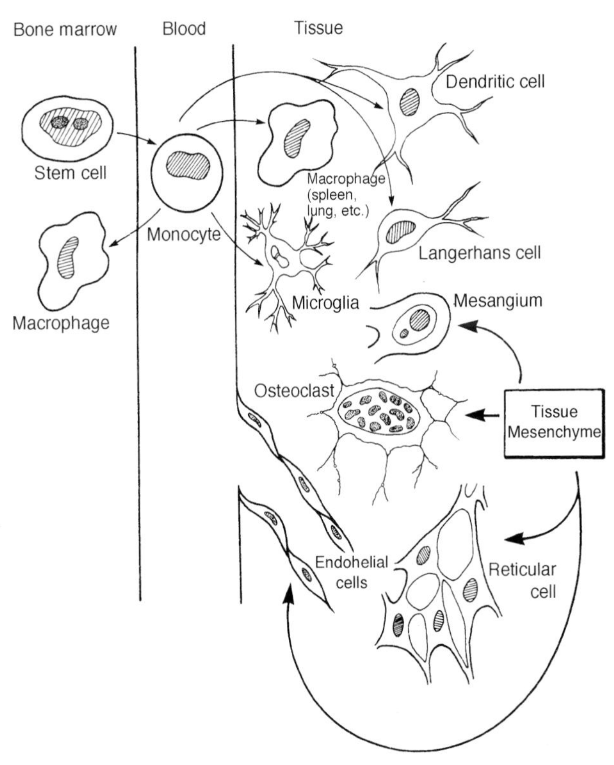

FIGURE 2.58 Reticuloendothelial system (RES).

radiolabeled molecules or cells such as albumin or erythrocytes coated with antibody.

The **mononuclear phagocyte system** (Figure 2.59) consists of mononuclear cells with pronounced phagocytic ability that are distributed extensively in lymphoid and other organs. *Mononuclear phagocyte system* should be used in place of the previously popular *reticuloendothelial system* to describe this group of cells. Mononuclear phagocytes originate from stem cells in the bone marrow that first differentiate into monocytes which appear in the blood for approximately 24 h or more with final differentiation into macrophages in the tissues. Macrophages usually occupy perivascular areas. Liver macrophages are termed Kupffer cells, whereas those in the lung are alveolar macrophages. The microglia represent macrophages of the central nervous system, whereas histiocytes represent macrophages of connective tissue. Tissue stem cells are monocytes that have wandered from the blood into the tissues and may differentiate into macrophages. Mononuclear phagocytes have a variety of surface receptors that enable them to bind carbohydrates or such protein molecules as C3 via complement receptor 1 and complement receptor 3, and IgG and IgE through Fc_γ and Fc_ε receptors. The surface expression of MHC class II molecules enables both monocytes and macrophages to serve as antigen-presenting cells to $CD4^+$ T lymphocytes. Mononuclear phagocytes secrete a rich array of molecular substances with various functions. A few of these are interleukin-1;

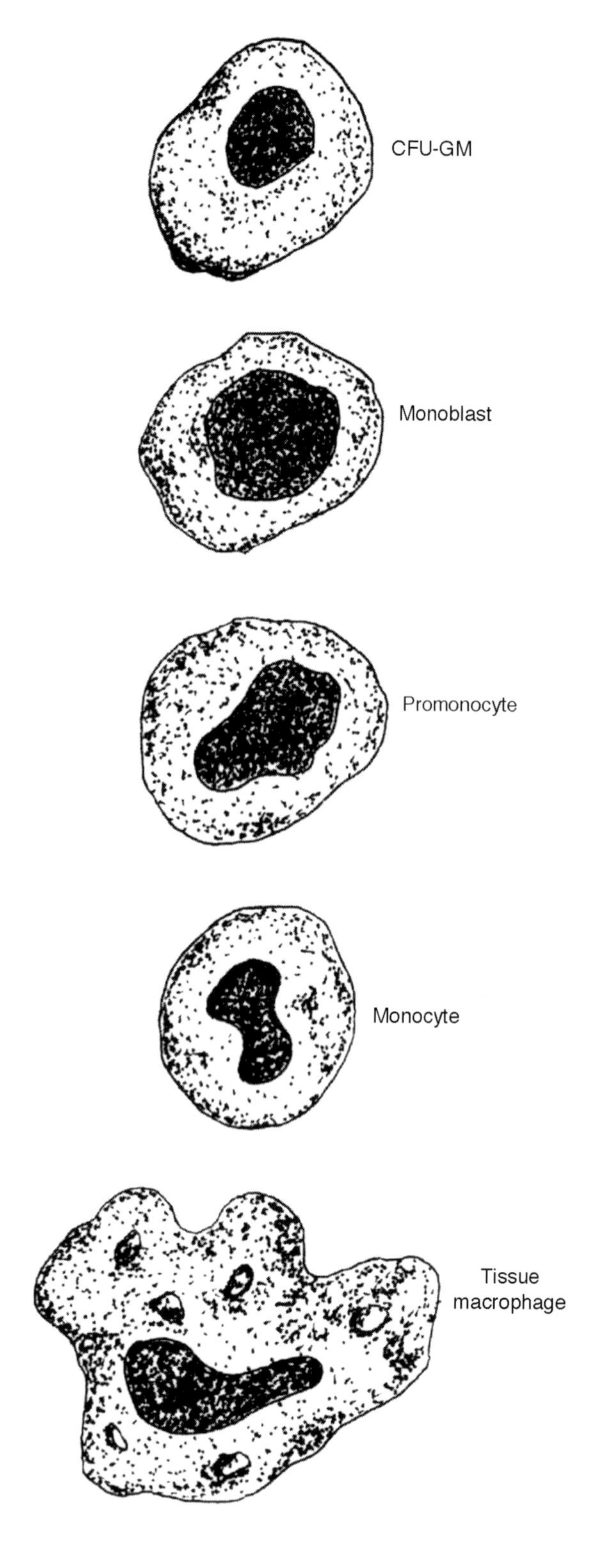

FIGURE 2.59 Mononuclear phagocyte system.

tumor necrosis factor α; interleukin-6; C2, C3, C4, and factor B complement proteins; prostaglandins; leukotrienes; and other substances.

Nonspecific esterase (α naphthyl acetate esterase) is an enzyme of mononuclear phagocytes and lymphocytes that is demonstrable by cytochemical staining. There is diffuse granular staining of the cytoplasm of mononuclear phagocytes which may help to identify them. Some human T cells are positive for nonspecific esterase but appear as one or several small localized dots within the T cell.

Mononuclear phagocyte: See mononuclear phagocyte system.

Scavenger receptors are structures on macrophages and other cell types that bind a variety of ligands and delete them from the blood. Scavenger receptors are especially abundant on Kupffer cells of the liver.

Saccharated iron oxide is a colloidal iron oxide employed to investigate the phagocytic capacity of mononuclear phagocytes.

Thorotrast (thorium dioxide ^{32}THOT) is a radiocontrast medium that yields α particles. It is no longer in use since neoplasia have been attributed to the substance. It is removed by the reticuloendothelial (mononuclear phagocyte) system. It induced hepatic angiosarcoma and also cholangiocarcinoma and hepatocellular carcinoma in some patients who received it. It has been known to produce other neoplasms. In immunology, it has been used in experimental animal studies involving the blockade of the reticuloendothelial system.

Myeloid cell series is an immature bone marrow cell (myeloblast) that is a precursor of the polymorphonuclear leukocyte series. This 18-μm diameter cell has a relatively large nucleus with finely distributed chromatin and two conspicuous nucleoli. The cytoplasm is basophilic when stained. During maturation, the cytoplasm becomes populated with large azurophilic primary granules, representing the promyelocyte stage. Later, the specific or secondary granules appear, representing the myelocyte stage. The nucleoli vanish as the nuclear chromatin forms dense aggregates. The chromatin in the nucleus condenses, and the cells no longer divide at this metamyelocyte stage of development. The nucleus assumes a sausage-like configuration known as a band. This subsequently develops into a three-lobed polymorphonuclear leukocyte, which develops into the neutrophils, eosinophils, and basophils that constitute myeloid cells. These latter three types are present in normal peripheral blood.

Granulocyte refers to the three types of polymorphonuclear leukocytes that differ mainly because of the staining properties of their cytoplasmic granules. The three types are classified as neutrophils, eosinophils, and basophils. They are all mature myeloid-series cells and have different functions. Granulocytes constitute 58 to 71% of the leukocytes in the blood circulation. See the individual cells for details.

Polymorphonuclear leukocytes (PMNs) (Figure 2.60 and Figure 2.61) are white blood cells with lobulated nuclei that are often trilobed. These cells are of the myeloid cell lineage and in the mature form can be differentiated into neutrophils, eosinophils, and basophils. This distinction is based on the staining characteristics of their cytoplasmic specific or secondary granules. These cells, which measure approximately 13 μm in diameter, are active in acute inflammatory responses.

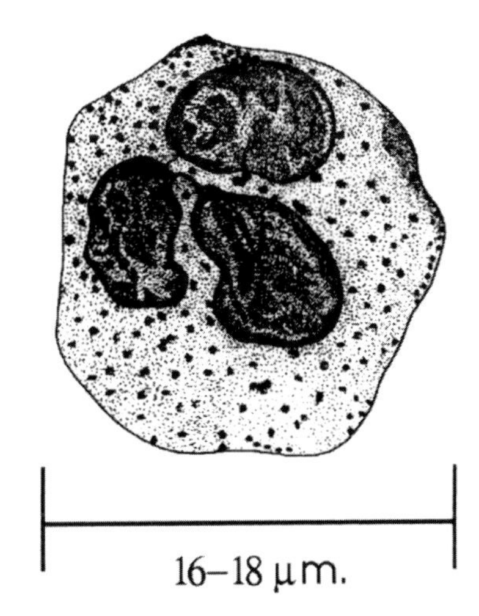

FIGURE 2.60 Schematic representation of a PMN cell.

PMN is the abbreviation for polymorphonuclear leukocyte.

A **neutrophil leukocyte** (Figure 2.62) is a peripheral blood polymorphonuclear leukocyte derived from the myeloid lineage. Neutrophils comprise 40 to 75% of the total white blood count numbering 2500 to 7500 cells/mm^3. They are phagocytic cells and have a multilobed nucleus and azurophilic and specific granules that appear lilac following staining with Wright's or Giemsa stains. They may be attracted to a local site by such chemotactic factors as C5a. They are the principal cells of acute inflammation and actively phagocytize invading microorganisms. Besides serving as the first line of cellular defense infection, they participate in such reactions as the uptake of antigen–antibody complexes in the Arthus reaction.

A **neutrophil** expresses Fc receptors and can participate in antibody-dependent cell-mediated cytotoxicity. It has the capacity to phagocytize microorganisms and digest them enzymatically.

Neutrophils chemotaxis: See chemotaxis and chemotactic factors.

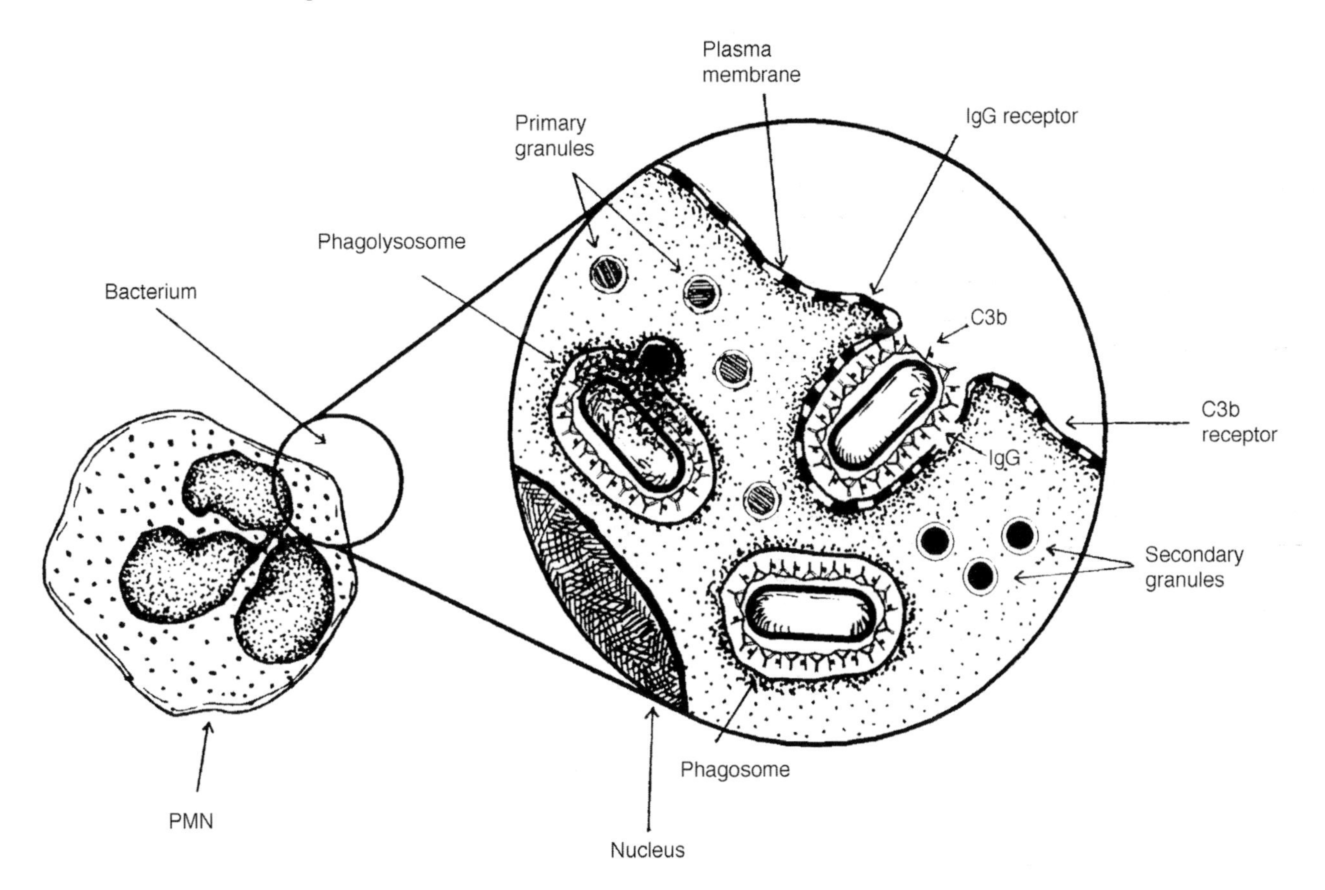

FIGURE 2.61 Polymorphonuclear leukocyte (PMN).

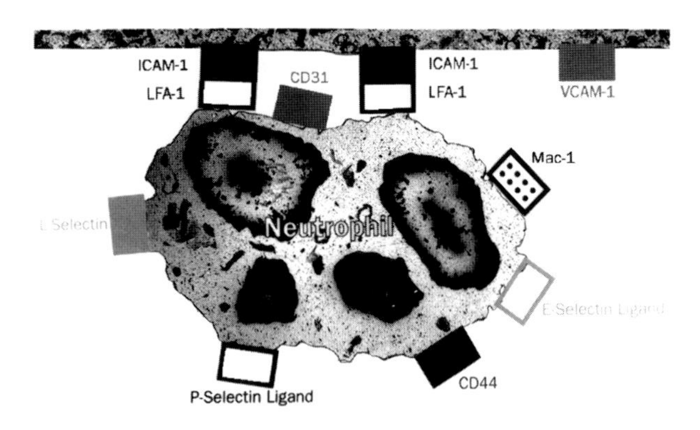

FIGURE 2.62 Neutrophil leukocyte.

Neutrophil microbicidal assay is a test that assesses the capacity of polymorphonuclear neutrophil leukocytes to kill intracellular bacteria.

Neutropenia refers to a diminished number of polymorphonuclear neutrophilic leukocytes in the peripheral blood circulation.

Myeloperoxidase is an enzyme present in the azurophil granules of neutrophilic leukocytes which catalyzes peroxidation of many microorganisms. Myeloperoxidase, in conjunction with hydrogen peroxidase and halide, has a bactericidal effect.

Primary granule: Azurophil granule.

Secondary granule is a structure in the cytoplasm of polymorphonuclear leukocytes which contains vitamin B_{12}-binding protein, lysozyme, and lactoferrin in neutrophils. Cationic peptides are present in eosinophil secondary granules. Histamine, platelet-activating factor, and heparin are present in the secondary granules of basophils.

A **tertiary granule** is a structure in the cytoplasm of polymorphonuclear neutrophils (PMNs) in which complement receptor 3 precursor, acid hydrolase, and gelatinase are located.

A **specific granule** is a secondary granule in the cytoplasm of polymorphonuclear leukocytes which contains lysozyme, vitamin B_{12}-binding protein, neutral proteases, and lactoferrin. It is smaller and fuses with phagosomes more quickly than does the azurophil granule.

Respiratory burst is a process used by neutrophils and monocytes to kill certain pathogenic microorganisms. It involves increased oxygen consumption with the generation of hydrogen peroxide and superoxide anions. This occurs also in macrophages that kill tumor cells. It is an abrupt elevation in oxygen consumption, which is followed by metabolic events in neutrophils and mononuclear cells preceding bacteriolysis. Partial reduction of oxygen by this process provides microbicidal oxidants. The initial event is a one-electron reduction of oxygen by membrane-bound oxidase to form a superoxide. The hexose monophosphate shunt reaction that accompanies this reduction liberates an H^+ which unites with the oxygen to produce H_2O_2.

NAP: Neutrophil alkaline phosphatase.

NAP-2 (neutrophil activating protein-2): A chemokine of the α family (CSC family). NAP-2 is a proteolytic fragment of platelet basic protein (PBP) corresponding to amino acids 25 to 94. CTAP-III and LA-PF4 or βTG released from activated platelets are inactive NAP-2 precursors. Leukocytes and leukocyte-derived proteases convert the inactive precursors into NAP-2 by proteolytic cleavage at the N-terminus. Platelets represent the tissue source, whereas neutrophils, basophils, eosinophils, fibroblasts, natural killer (NK) cells, megakaryocytes and endothelial cells.

A **superoxide anion** is a free radical formed by the addition of an electron to an oxygen molecule, causing it to become highly reactive. This takes place in inflammation or is induced by ionizing radiation. It is formed by reduction of molecular oxygen in polymorphonuclear neutrophils (PMNs) and mononuclear phagocytes. The hexose monophosphate shunt activation pathway enhances superoxide anion generation. Superoxide anion interacts with protons, additional superoxide anions, and hydrogen peroxide. Oxidation of one superoxide anion and reduction of another may lead to the formation of oxygen and hydrogen peroxide. Superoxide dismutase, found in phagocytes, catalyzes this reaction. Injury induced by superoxide anion is associated with age-related degeneration. It may also serve as a mutagen with implications for carcinogenesis. The superoxide anion plays a pivotal role in the ability of mononuclear phagocytes and neutrophils to kill microorganisms through their oxidative microbicidal function.

Oxygen-dependent killing: This is activated by a powerful oxidative burst that culminates in the formation of hydrogen peroxide and other antimicrobial substances. In addition to this oxygen-dependent killing mechanism, phagocytized intracellular microbes may be the targets of toxic substances released from granules into the phagosome leading to microbial cell death by an oxygen-independent mechanism. For oxygen-dependent killing of microbes, membranes of specific granules and phagosomes fuse. This permits interaction of NADPH oxidase with cytochrome b. With the aid of quinone, this combination reduces oxygen to superoxide anion O_2. In the presence of a catalyst superoxide dismutase, superoxidase ion is converted to hydrogen peroxide. The clinical relevance of this process is illustrated by chronic granulomatous

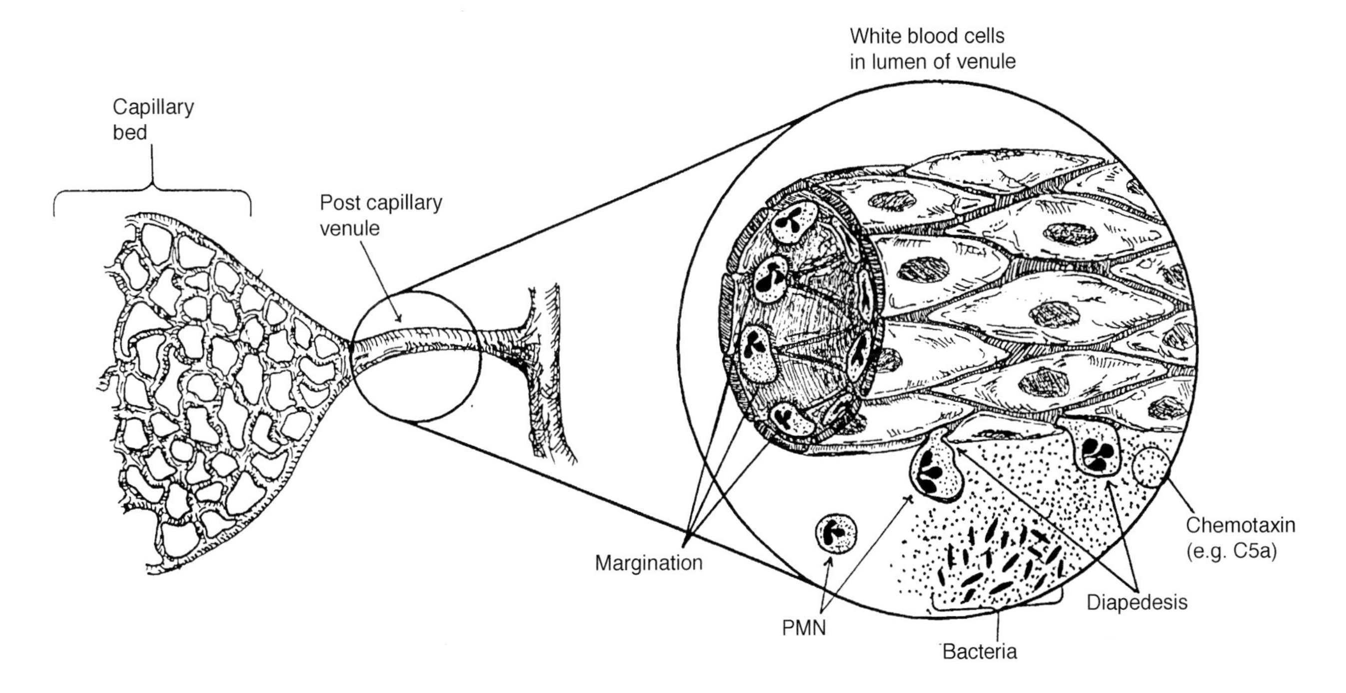

FIGURE 2.63 Diapedesis and margination.

disease (CGD) in children who fail to form superoxide anions. The patients have diminished cytochrome b, even though phagocytosis is normal. They have impaired ability to oxidize NADPH and destroy bacteria through the oxidative pathway. The oxidative mechanism kills microbes through a complex process. Hydrogen peroxide, together with myeloperoxidase, transforms chloride ions into hypochlorous ions that kill microorganisms. Azurophil granule fusion releases myeloperodase to the phagolysosome. Some microorganisms such as pneumococci may themselves form hydrogen peroxide.

Oxygen-independent killing: Following adherence of opsonized microbes to the neutrophil plasma membrane, lysozyme and lactoferrin are discharged from specific granules into phagosomes with which they have fused. Antimicrobial cationic proteins reach phagosomes from azurophil granules. These proteins kill Gram-negative microbes by interrupting their cell membrane integrity. They are far less effective against Gram-positive microorganisms.

SOD is an abbreviation for superoxide dismutase.

Diapedesis (Figure 2.63 and Figure 2.64) refers to cell migration from the interior of small vessels into tissue spaces as a consequence of constriction of endothelial cells in the wall.

Margination refers to the adherence of leukocytes in the peripheral blood to the endothelium of vessel walls. Approximately 50% of polymorphonuclear neutrophils marginate at one time. During inflammation, there is margination of leukocytes, followed by their migration out of the vessels.

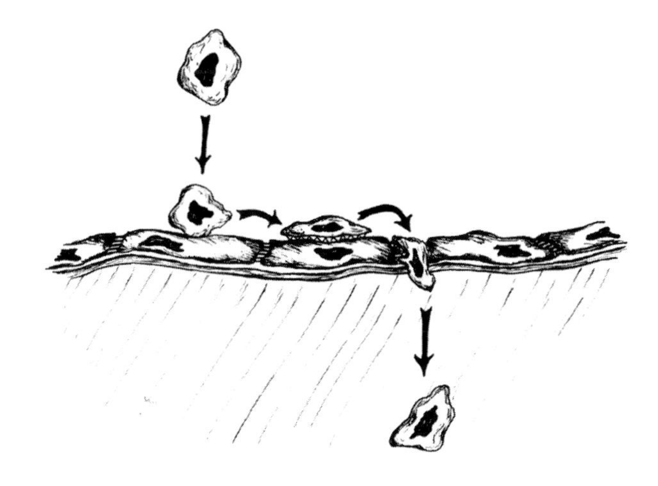

FIGURE 2.64 Diapedesis.

Phagocytes (Figure 2.65 and Figure 2.66) are cells such as mononuclear phagocytes and polymorphonuclear neutrophils that ingest and frequently digest particles such as bacteria, blood cells, and carbon particles, among many other particulate substances.

Surface phagocytosis refers to the facilitation of phagocytosis when microorganisms become attached to the surfaces of tissues, blood clots, or leukocytes.

Zippering is a mechanism in phagocytosis in which the phagocyte membrane cores the particle by a progressive

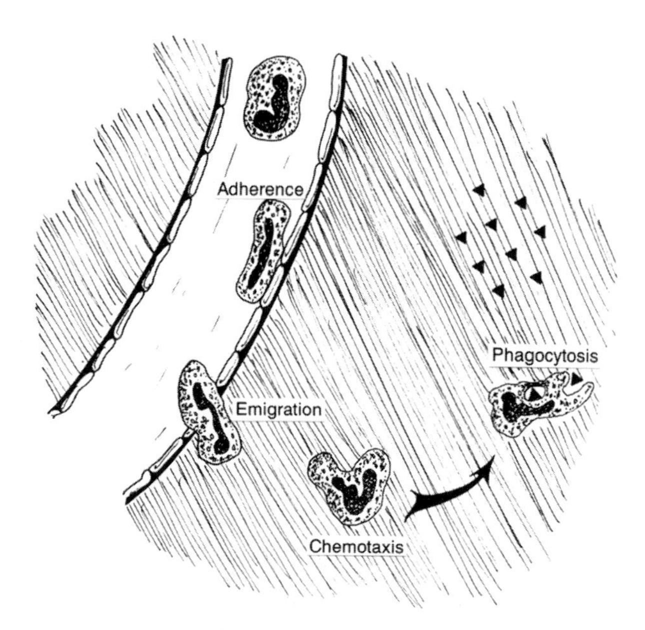

FIGURE 2.65 Phagocytosis.

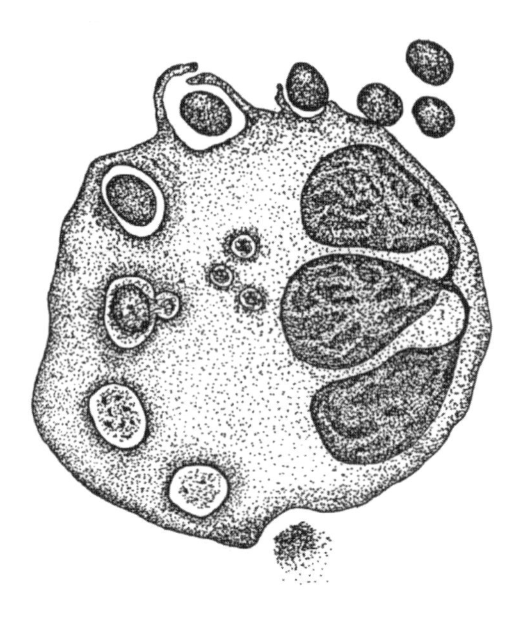

FIGURE 2.66 Schematic representation of phagocytosis.

adhesive interaction. Evidence in support of this process comes from experiments in which capped B cells are only partially internalized, whereas those coated uniformly with anti-IgG opsonizing antibody are engulfed fully.

Reactive nitrogen intermediates are very cytotoxic antimicrobial substances produced when oxygen and nitrogen combine within phagocytes such as neutrophils and macrophages.

Opsonin is a substance that binds to bacteria, erythrocytes, or other particles to increase their susceptibility to phagocytosis. Opsonins include antibodies such as IgG3, IgG1, and IgG2 that are specific for epitopes on the particle surface. Following interaction, the Fc region of the antibody becomes anchored to Fc receptors on phagocyte surfaces, thereby facilitating phagocytosis of the particles. In contrast to these so-called heat-stable antibody opsonins are the heat-labile products of complement activation such as C3b or C3bi, which are linked to particles by transacylation with the C3 thiolester. C3b combines with complement receptor 1 and C3bi combines with complement receptor 3 on phagocytic cells. Other substances that act as opsonins include the basement membrane constituent, fibronectin.

Opsonization is the facilitation of the phagocytosis of microorganisms or other particles such as erythrocytes through the coating of their surface with either immune or nonimmune opsonins. Antibody, such as IgG molecules, and complement fragments may opsonize extracellular bacteria or other microorganisms, rendering them susceptible to destruction by neutrophils and macrophages through phagocytosis.

In **opsonophagocytosis**, antibodies and/or complement, mainly C3, serve as opsonins by binding to epitopes on microorganisms and increasing their susceptibility to phagocytosis by polymorphonuclear leukocytes, especially neutrophils. Serum bactericidal activity and phagocytic killing are two principal mechanisms in host defense against bacteria. Opsonic antimicrobial antibodies are critical for optimal functioning of phagocytes in the uptake and containment of bacteria.

Toll-like receptors are receptors on the surfaces of phagocytes and other cells that signal the activation of macrophages responding to microbial products such as endotoxin in the natural or innate immune response. They are structurally homologous and share signal transduction pathways with the type I IL-1 receptor.

TLR1-10: See Toll-like receptors.

Pseudopodia are membrane extensions from motile and phagocytic cells.

Catalase is an enzyme present in activated phagocytes that causes degradation of hydrogen peroxide and superoxide dismutase.

Cationic proteins are phagocytic cell granule constituents that have antimicrobial properties.

A **phagolysosome** is a cytoplasmic vesicle with a limiting membrane produced by the fusion of a phagosome with a lysosome. Substances within a phagolysosome are digested by hydrolysis.

A **secondary lysosome** is a lysosome that has united with a phagosome.

A **suppressor macrophage** is a macrophage activated by its response to an infection or neoplasm in the host from which it was derived. It is able to block immunologic reactivity *in vitro* through production of prostaglandins, oxygen radicals, or other inhibitors produced through arachidonic acid metabolism.

Defensins are widely reactive antimicrobial cationic proteins present in polymorphonuclear neutrophilic leukocyte granules. They block cell transport activities and are lethal for Gram-positive and Gram-negative microorganisms. These peptides are rich in cystecine and are found in the skin and in neutrophil granules that function as broad-spectrum antibiotics that kill numerous bacteria and fungi. The inflammatory cytokines IL-1 and TNF facilitate synthesis of defensins. Defensins (human neutrophil proteins 1 to 4) are amphipathic, carbohydrate-free, cytotoxic membrane-active antimicrobial molecules. Three of the defensin peptides (HP-1, HP-2, and HP-3) are nearly identical in sequence. By contrast, the sequences of HP-4, H-HP-5, and HP-6 are very different. HP-1 and HP-2 are chemotactic for monocytes. High concentrations of HP-1 to HP-4 (25 to 200μg/ml) manifest antimicrobial and/or viricidal properties *in vitro*. HP-4 has the greatest defensin activity and HP-3 the least.

Eosinophils (Figure 2.67 and Figure 2.68) are polymorphonuclear leukocytes identified as brilliant reddish-orange refractile granules in Wright- or Giemsa-stained preparations by staining of secondary granules in the leukocyte cytoplasm. Cationic peptides are released from these secondary granules when an eosinophil interacts with a target cell and may lead to death of the target. Eosinophils make up 2 to 5% of the total white blood cells in man. After a brief residence in the circulation, eosinophils migrate into tissues by passing between the lining endothelial cells. It is believed that they do not return to the circulation. The distribution corresponds mainly to areas exposed to external environment such as skin,

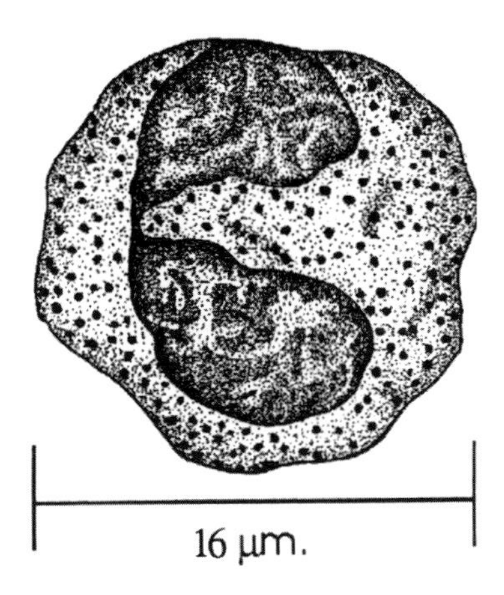

FIGURE 2.67 Eosinophil with segmented nucleus.

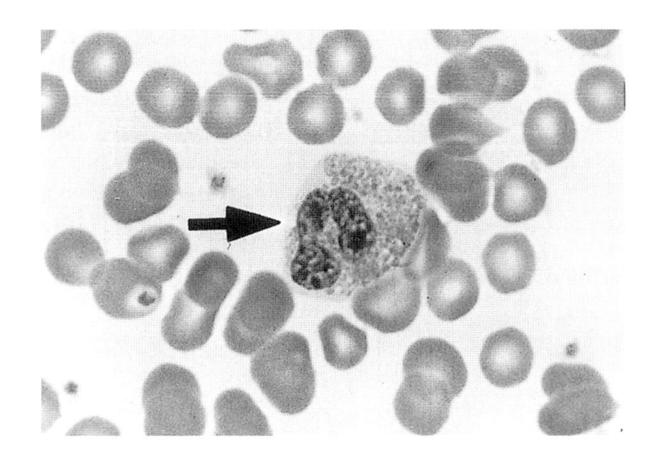

FIGURE 2.68 Eosinophil in peripheral blood.

mucosa of the bronchi, and gastrointestinal tract. Eosinophils are elevated during allergic reactions, especially type I immediate hypersensitivity responses, and are also elevated in individuals with parasitic infestations.

The **eosinophil differentiation factor** is a 20-kDa cytokine synthesized by some activated CD4$^+$ T lymphocytes and by activated mast cells. Formerly, it was called T cell replacing factor or B cell growth factor II. It facilitates B cell growth and differentiation into cells that secrete IgA. The eosinophil differentiation factor is costimulator with IL-2 and IL-4 of B cell growth and differentiation. IL-5 also stimulates eosinophil growth and differentiation. It activates mature eosinophils to render them capable of killing helminths. Through IL-5, T lymphocytes exert a regulatory effect on inflammation mediated by eosinophils. Because of its action in promoting eosinophil differentiation, it has been called eosinophil differentiation factor (EDF). IL-5 can facilitate B cell differentiation into plaque-forming cells of IgM and IgG classes. In parasitic diseases, IL-5 leads to eosinophilia.

Eosinophil chemotactic factors are mast cell granule peptides that induce eosinophil chemotaxis. These include two tetrapeptides: Val-Gly-Ser-Glu and Ala-Gly-Ser-Glu. Histamine also induces eosinophil chemotaxis.

Eosinophil cationic protein (ECP) is an eosinophil granule basic, single-chain, zinc-containing protein that manifests cytotoxic, helminthotoxic, ribonuclease, and bactericidal properties. ECP, major basic protein (MBP), eosinophil-derived neurotoxin (EDN), and eosinophil peroxidase (EPO) are the four main eosinophil granule proteins. ECP and MBP induce the release of preformed histamine and synthesis of vasoactive and proinflammatory mediators (PGD2) from activated human heart mast cells. Acute graft rejection and atopic dermatitis patients manifest provated ECP whereas systemic sclerosis patients develop increased serum levels of MBP. MBP is able to act as a cell stimulant and as a toxin. ECP, EPO, EDN, and MBP are versatile in their biological activities which include the capacity to activate other cells including basophils, neutrophils, and platelets. Both EIA and flow cytometry have been used to assay intracellular eosinophil proteins in eosinophils from bone marrow and peripheral blood.

Eosinophil granule major basic protein (EGMBP) is a polypeptide, rich in arginine, which is released from eosinophil granules. It is a powerful toxin for helminths and selected mammalian cells. EGMBP has a significant role in late-phase reactions in allergy and asthma and in late-phase skin reactions to allergens such as dust mites. EGMBP is believed to be significant in endomyocardial injury inducted by cardiac localization of eosinophil granule proteins. IL-3, GM-CSF, and IL-5 are eosinophilopoietic cytokines that activate eosinophils and facilitate their survival. These cytokines enhance eosinophil activation in the airways of patients with bronchial asthma, which leads to epithelial injury. IFN-α inhibits the release of EGMBP in hypereosinophilic syndrome patients. Immunohistochemistry is used to measure EGMBP from eosinophils infiltrating skin lesions of atopic dermatitis.

Eosinophil and neutrophil chemotactic activities: Chemotactic factors for eosinophils and neutrophils (ECA–NCA) are present in bronchoalveolar lavage fluid (BALF) of selected patients with asthma. ECA induces early atopic dermatitis lesions and is induced by transepidermal permeation of mite allergen. Eosinophils also participate in renal and liver allograft rejection as reflected by eosinophil cationic protein assays. Eosinophil major basic protein interacts with IL-1 and transforming growth factor-β to upregulate lung fibroblast to synthesize IL-6 cytokine. When stimulated by C5a, eosinophils produce increased levels of H_2O_2 as assayed by chemiluminescence (CL). Eosinophil activation may also be assayed by flow cytometry.

Defective adherence and the migration of neutrophils can be a cause of increased susceptibility to bacterial infection in neonates. The chemotactic cytokine, neutrophil-activating peptide ENA-78, is a proinflammatory polypeptide that shares sequence similarity with IL-8 and GRO-α. ENA-78 is a powerful upregulator of Mac-1 cell surface expression. It is found in cystic fibrosis lung and its mRNA levels are elevated in human pulmonary inflammation. Flow cytometry and nitroblue tetrazolium reduction can be used to access neutrophil activation.

Leukocidin is a bacterial toxin produced especially by staphylococci that is cytolytic. It is toxic principally for polymorphonuclear leukocytes and, to a lesser extent, for monocytes. It contains an F and an S component that combine with the cell membrane causing altered permeability. Less than toxic doses interfere with locomotion of polymorphonuclear neutrophils.

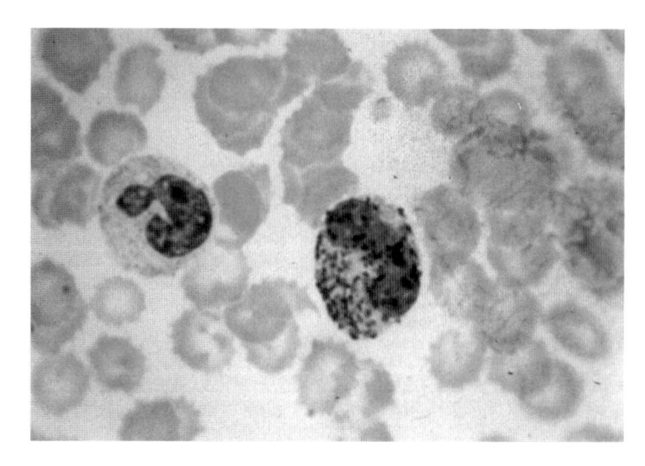

FIGURE 2.69 Basophil and neutrophil in peripheral blood.

Basophilic is an adjective that refers to an affinity of cells or tissues for basic stains leading to a bluish tint.

Basophils (Figure 2.69) are polymorphonuclear leukocytes of the myeloid lineage with distinctive basophilic secondary granules in the cytoplasm that frequently overlie the nucleus. These granules are storage depots for heparin, histamine, platelet-activating factor, and other pharmacological mediators of immediate hypersensitivity. Degranulation of the cells with release of these pharmacological mediators takes place following crosslinking by allergen or antigen of Fab regions of IgE receptor molecules bound through Fc receptors to the cell surface. They comprise less than 0.5% of peripheral blood leukocytes. Following crosslinking of surface-bound IgE molecules by specific allergen or antigen, granules are released by exocytosis. Substances liberated from the granules are pharmacological mediators of immediate (type I) anaphylactic hypersensitivity.

Basophil-derived kallikrein (BK-A) represents the only known instance where an activator of the kinin system is generated directly from a primary immune reaction. The molecule is a high molecular weight enzyme with arginine esterase activity. It is stored in the producing cells in a preformed state. Its release depends on basophil–IgE interactions with antigen and parallels the release of histamine.

Mast cells (Figures 2.70 and 2.71) are a normal component of the connective tissue that plays an important role in immediate (type I) hypersensitivity and inflammatory reactions by secreting a large variety of chemical mediators from storage sites in their granules upon stimulation. Their anatomical location at mucosal and cutaneous surfaces and about venules in deeper tissues is related to this role.

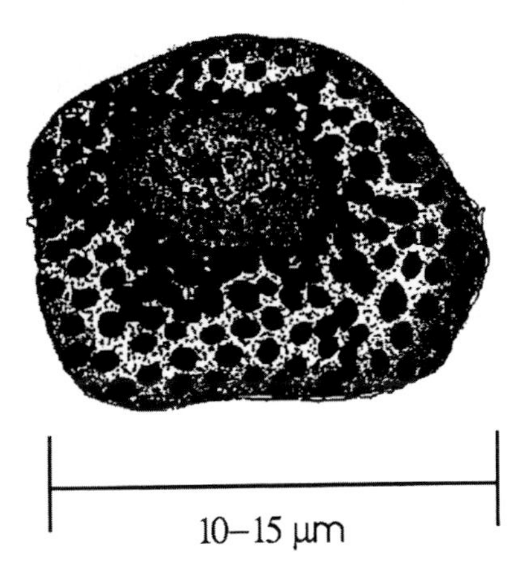

FIGURE 2.70 Mast cell.

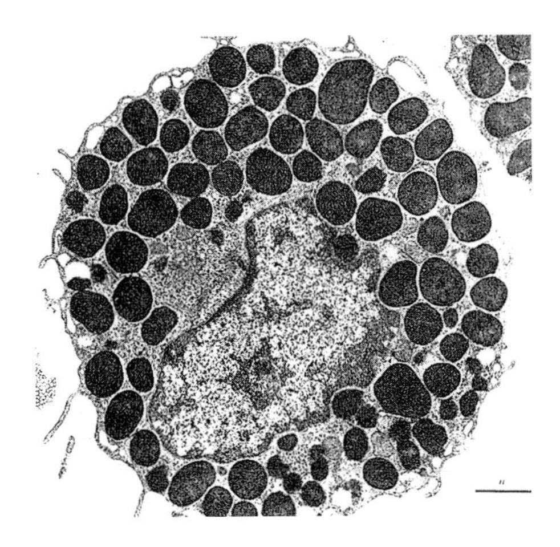

FIGURE 2.71 Mast cell in peripheral blood.

They can be identified easily by their characteristic granules, which stain metachromatically. The size and shape of mast cells vary, i.e., 10 to 30 μm in diameter. In adventitia of large vessels, they are elongated; in loose connective tissue, they are round or oval; and the shape in fibrous connective tissue may be angular. On their surfaces, they have Fc receptors for IgE. Crosslinking by either antigen for which the IgE Fab regions are specific or by anti-IgE or antireceptor antibody leads to degranulation with the release of pharmacological mediators of immediate hypersensitivity from their storage sites in the mast cell granules. Leukotrienes, prostaglandins, and platelet-activating factor are also produced and released following Fc_ε receptor crosslinking. Mast cell granules are approximately 0.5 μm in diameter and are electron dense. They contain many biologically active compounds, of which the most important are heparin, histamine, serotonin, and a variety of enzymes. Histamine is stored in the granule as a complex with heparin or serotonin. Mast cells also contain proteolytic enzymes such as plasmin and also hydroxylase, β glucuronidase, phosphatase, and a high uronidase inhibitor, to mention only the most important. Zinc, iron, and calcium are also found. Some substances released from mast cells are not stored in a preformed state but are synthesized following mast cell activation. These represent secondary mediators as opposed to the preformed primary mediators. Mast cell degranulation involves adenylate cyclase activation with rapid synthesis of cyclic AMP, protein kinase activation, phospholipid methylation, and serine esterase activation. Mast cells of the gastrointestinal and respiratory tracts that contain chondroitin sulfate produce leukotriene C_4, whereas connective tissue mast cells that contain heparin produce prostaglandin D_2.

Acyclic adenosine monophosphate (cAMP): Adenosine 3′,5′-(hydrogen phosphate). A critical regulator within cells. It is produced through the action of adenylate cyclase on adenosine triphosphate. It activates protein kinase C. It serves as a "second messenger" when hormones activate cells. Elevated cAMP concentrations in mast cells diminish their response to degranulation signals.

Adenosine is normally present in the plasma in a concentration of 0.03 μ*M* in man and 0.04 μ*M* in the dog. In various clinical states associated with hypoxia, the adenosine level increases five- to tenfold, suggesting that it may play a role in the release of mediators. Experimentally, adenosine is a powerful potentiator of mast cell function. The incubation of mast cells with adenosine does not induce the release of mediators. However, by preincubation with adenosine and subsequent challenge with a mediator-releasing agent, the response is markedly enhanced.

Exocytosis refers to the release of intracellular vesicle contents to the exterior of the cell. The vesicles make their way to the plasma membrane with which they fuse to permit the contents to be released to the external environment. Examples include immunoglobulin released from plasma cells and mast cell degranulation, which releases histamine and other pharmacological mediators of anaphylaxis to the exterior of the cell. Cytokines may also be released from cells by this process.

Exogenous means externally caused rather than resulting from conditions within the organism; derived externally.

Acyclic guanosine monophosphate (cGMP): Guanosine cyclic 3′,5′-(hydrogen phosphate). A cAMP antagonist produced by the action of guanylatecyclase on guanosine triphosphate. Elevated cGMP concentrations in mast cells accentuate their response to degranulation signals.

88 Monocytes (Figure 2.72 through Figure 2.76) are mononuclear phagocytic cells in the blood that are derived from promonocytes in the bone marrow. Following a relatively brief residence in the blood, they migrate into the tissues and are transformed into macrophages. They are less mature than macrophages, as suggested by fewer surface receptors, cytoplasmic organelles, and enzymes than the latter. Monocytes are larger than polymorphonuclear leukocytes, are actively phagocytic, and constitute 2 to 10% of the total white blood cell count in humans. The monocyte in the blood circulation is 15 to 25 μm in diameter. It has grayish-blue cytoplasm that contains lysosomes with enzymes such as acid phosphatase, arginase cachetepsin, collagenase, deoxyribonuclease, lipase, glucosidase, and plasminogen activator. The cell has a reniform nucleus with delicate lace-like chromatin. The monocyte has surface receptors such as the Fc receptor for IgG and a receptor for CR3. It is actively phagocytic and plays a significant role in antigen processing. Monocyte numbers are elevated in both benign and malignant conditions. Certain infections stimulate a reactive type of monocytosis, such as in tuberculosis, brucellosis, HIV-1 infection, and malaria.

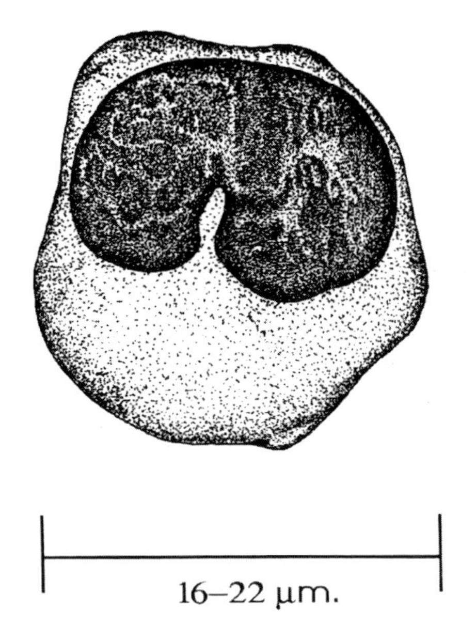

FIGURE 2.72 Monocyte.

The **monocyte–phagocyte system** is a system of cells that provides nonspecific immunity and is dependent on the activity of the monocyte/macrophage lineage cells, which are especially prominent in the spleen.

A **trophoblast** consists of a layer of cells in the placenta that synthesizes immunosuppressive agents. These cells are in contact with the lining of the uterus.

A **blast cell** is a relatively large cell that is greater than 8 μm in diameter with abundant RNA in the cytoplasm, a

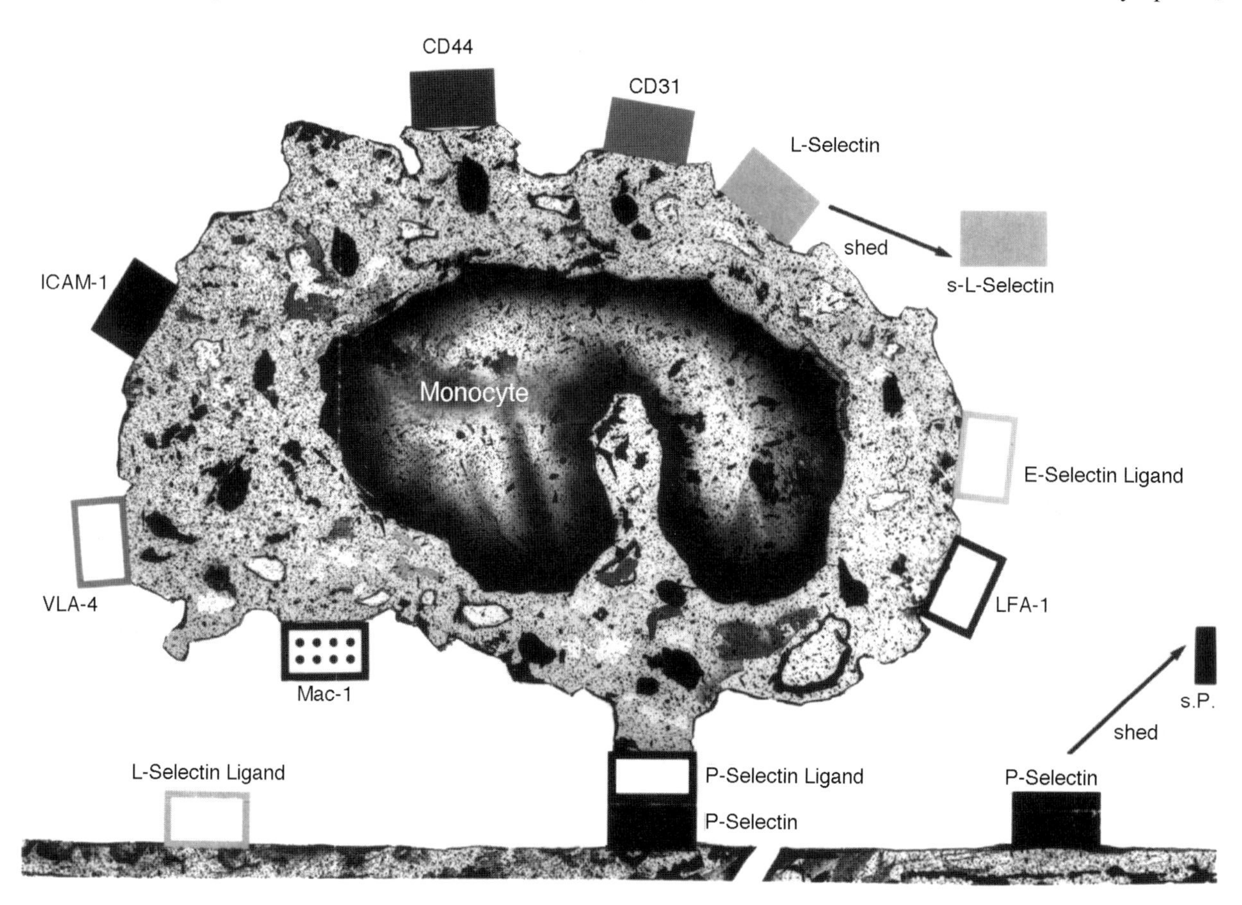

FIGURE 2.73 Monocyte.

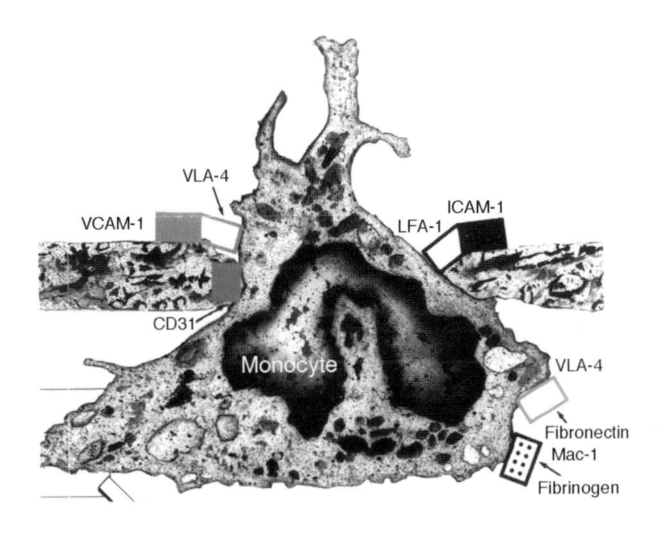

FIGURE 2.74 Monocyte.

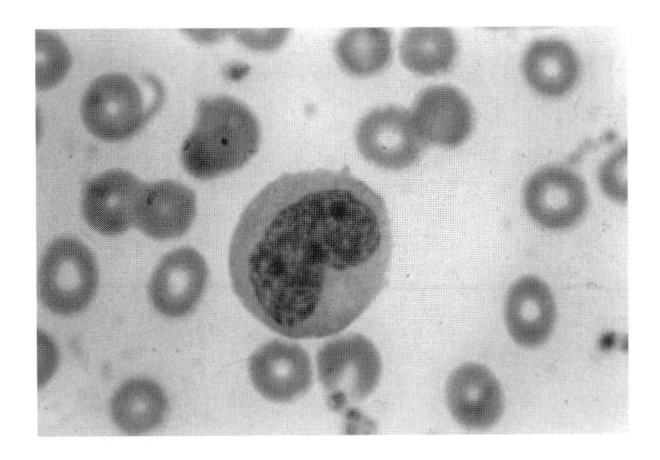

FIGURE 2.75 Monocyte in peripheral blood.

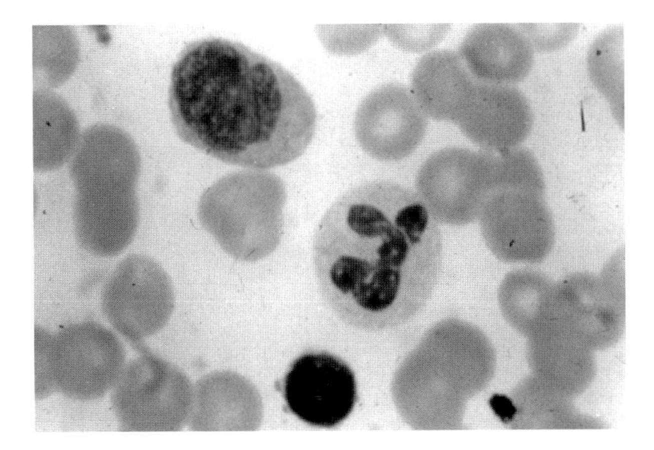

FIGURE 2.76 Monocyte, lymphocyte, and polymorphonuclear neutrophil.

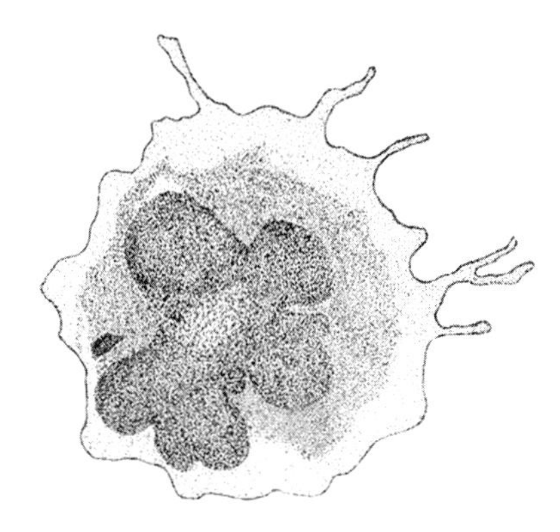

FIGURE 2.77 Megakaryocyte.

nucleus with loosely arranged chromatin, and a prominent nucleolus. Blast cells are active in synthesizing DNA and contain numerous polyribosomes in the cytoplasm.

Megakaryocytes (Figure 2.77) are relatively large bone-marrow giant cells that are multinuclear and from which blood platelets are derived by the breaking up of membrane-bound cytoplasm to produce the thrombocytes.

A **platelet** is a small (3 μm in diameter) round disk that is derived from bone marrow megakaryocytes but is present in the blood. Platelets function in blood clotting by releasing thromboplastin. They also harbor serotonin and histamine, which may be released during type I (anaphylactic hypersensitivity) reactions. Complement receptor 1 (CR1) is present on the platelets of mammals other than primates and is significant for immune adherence.

A **thrombocyte** is a blood platelet.

PAF is the abbreviation for platelet-activating factor.

Platelet-derived growth factor (PDGF) is a low molecular weight protein derived from human platelets that acts as a powerful connective tissue mitogen, causing fibroblast and intimal smooth muscle proliferation. It also induces vasoconstriction and chemotaxis and activates intracellular enzymes. PDGF plays an important role in atherosclerosis and fibroproliferative lesions such as glomerulonephritis, pulmonary fibrosis, myelofibrosis, and other processes. It is comprised of a two-chain, i.e., A or B, dimer. It can be an AA or BB homodimer or an AB heterodimer. Human PDGF-AA is a 26.5-kDa A-chain homodimeric protein comprised of 250 amino acid residues, whereas PDGF-BB is a 25-kDa B-chain homodimeric protein comprised of 218 amino acid residues. In addition to platelets, PDGF is released by activated mononuclear cells, endothelial cells, smooth muscle cells, and fibroblasts. It plays a physiologic role in wound repair and processes requiring accumulation of connective tissue. The three known polypeptide dimers of PDGF include AA, AB, and BB that bind to α or β dimeric tyrosine kinase receptors.

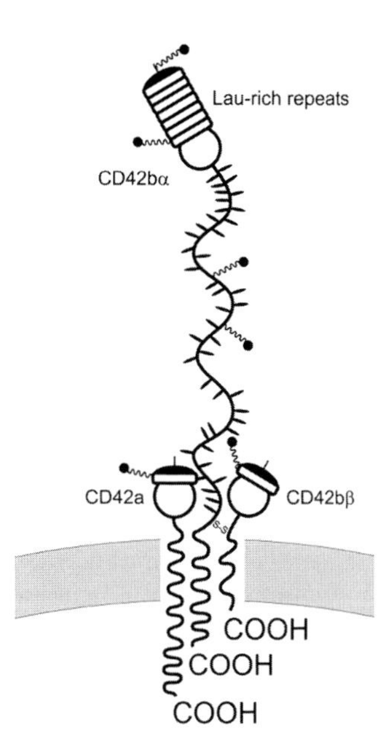

FIGURE 2.78 CD42.

Platelet-derived growth factor receptor (PDGF-R) is a glycoprotein in the membrane that has five extracellular domains that resemble those of immunoglobulins. It also has a kinase insert in the cytoplasm. The receptor protein must undergo a conformational change for signal transduction. A gene on chromosome 4q11 encodes PDGF-R.

CD42a is an antigen equivalent to glycoprotein IX that is a single-chain membrane glycoprotein with a 23-kDa mol wt. It is found on megakaryocytes and platelets. CD42a forms a noncovalent complex with CD4b (gpIb) which acts as a receptor for von Willebrand factor. It is absent or reduced in the Bernard–Soulier syndrome.

CD42b is an antigen equivalent to glycoprotein Ib, which is a two-chain membrane glycoprotein with a molecular weight of 170 kDa. CD42b has an α-chain of 135 kDa and a β-chain of 23 kDa. It is found on platelets and megakaryocytes. CD42b forms a noncovalent complex with CD42a (gpIX) which acts as a receptor for von Willebrand factor. The antigen is absent or reduced in the Bernard–Soulier syndrome.

CD42c (Figure 2.78) is a 22-kDa antigen found on platelets and megakaryocytes. It is also referred to as GPIB-β.

CD42d is an 85-kDa antigen present mainly on platelets and megakaryocytes. It is also referred to as GPV.

β lysin is a thrombocyte-derived antibacterial protein that is effective mainly against Gram-positive bacteria. It is released when blood platelets are disrupted, as occurs during clotting. β lysin acts as a nonantibody humoral substance that contributes to nonspecific immunity.

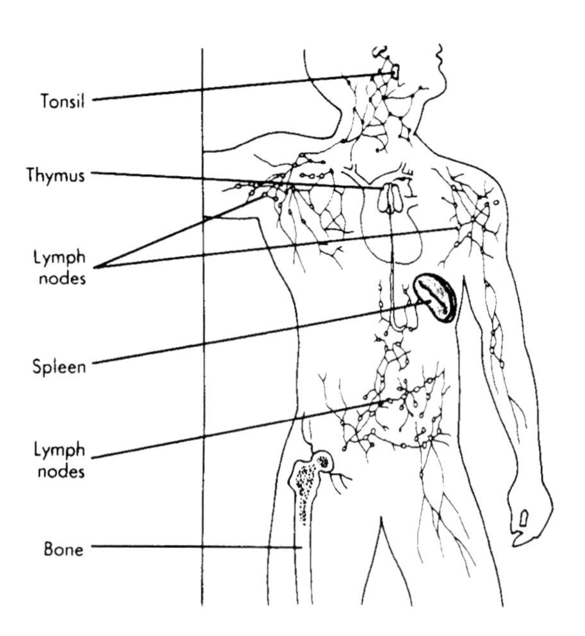

FIGURE 2.79 Lymphoid tissue.

Lymphoid system refers to the lymphoid organs and the lymphatic vessels.

Lymphoid tissues (Figure 2.79) are tissues that include the lymph nodes, spleen, thymus, Peyer's patches, tonsils, bursa of Fabricius in birds, and other lymphoid organs in which the predominant cell type is the lymphocyte.

Bone marrow is soft tissue within bone cavities that contains hematopoietic precursor cells and hematopoietic cells that are maturing into erythrocytes, the five types of leukocytes, and thrombocytes. Whereas red marrow is hemopoietic and is present in developing bone, ribs, vertebrae, and long bones, some of the red marrow may be replaced by fat and become yellow marrow.

Bone marrow cells are stem cells from which the formed elements of the blood, including erythrocytes, leukocytes, and platelets, are derived. B lymphocyte and T lymphocyte precursors are abundant. The B lymphocytes and pluripotent stem cells in bone marrow are important for reconstitution of an irradiated host. Bone marrow transplants are useful in the treatment of aplastic anemia, leukemias, and immunodeficiencies. Patients may donate their own marrow for subsequent bone marrow autotransplantation if they are to receive intensive doses of irradiation.

A **stromal cell** is a cell of nonhematopoietic origin that facilitates the growth and differentiation of hematopoietic cells.

Stromal cells are sessile cells that form an interconnected network that gives an organ structural integrity but also provides a specific inductive microenvironment that facilitates differentiation and maturation of incoming precursor

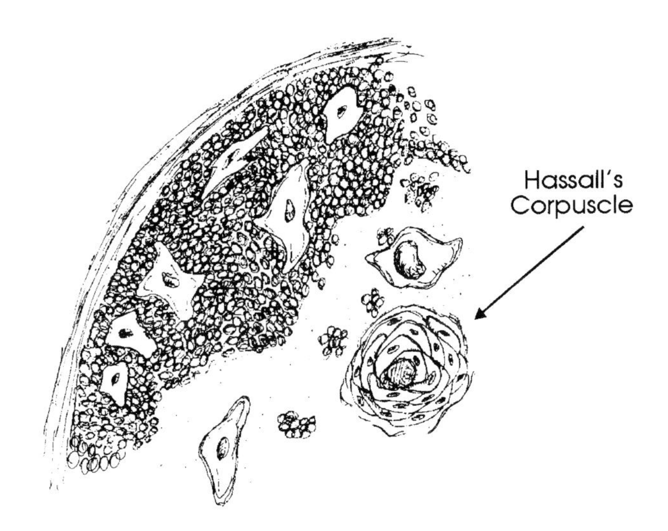

FIGURE 2.80 Histology of the thymus.

cells. Stromal cells and their organization are fully as complex as the cells whose development they regulate. For example, stromal cells of the thymus have been the best characterized with respect to their role in T lymphocyte maturation.

Adenoids are mucosa-associated lymphoid tissues located in the nasal cavity.

BALT: See bronchial-associated lymphoid tissue.

The **thymus** (Figure 2.80) is a triangular, bilobed structure enclosed in a thin fibrous capsule and located retrosternally. Each lobe is subdivided by prominent trabeculae into interconnecting lobules and each lobule comprises two histologically and functionally distinct areas — cortex and medulla. The cortex consists of a mesh of epithelial–reticular cells enclosing densely packed large lymphocytes. It has no germinal centers. The epithelial cell component is of endodermal origin; the lymphoid cells are of mesenchymal origin. The prothymocytes, which migrate from the bone marrow to the subcapsular regions of the cortex, are influenced by this microenvironment which directs their further development. The process of education is exerted by hormonal substances produced by the thymic epithelial cells. The cortical cells proliferate extensively. Parts of these cells are short-lived and die. The surviving cells acquire characteristics of thymocytes.

Annexin V binding: In normal, nonapoptotic cells, PS is segregated to the inner leaflet of the plasma membrane. During early stages of apoptosis, this asymmetry collaspes and PS becomes exposed on the outer surface cells. Annexin V is a protein that preferentially binds PS in a calcium-dependent manner. Binding of annexin V, in conjunction with dye exclusion (e.g., propidium iodide) to establish membrane integrity, can be used to identify apoptotic cells.

Programmed cell death is apoptosis in which death is activated from within the dying cell. It occurs in lymphocytes deprived of growth factors or costimulators. It is also known as "death by neglect" or "passive cell death" in which mitochondrial cytochrome c is released into the cytoplasm, caspase-9 is activated, and apoptosis is initiated.

Apoptosis is programmed cell death in which the chromatin becomes condensed and the DNA is degraded. The immune system employs apoptosis for clonal deletion of cortical thymocytes by antigen in immunologic tolerance. A healthy organism is an exquisitely integrated collection of differentiated cells, which maintain a balance between life and death. Some cells are irreplaceable; some cells complete their functions and are then sacrificed; and some cells live a finite lifetime to be replaced by yet another generation. A failure of cells to fulfill their destiny has catastrophic consequences for the organism. Apoptosis is the last phase of a cell's destiny. It is the controlled disassembly of a cell. When apoptosis occurs on schedule, neighboring cells and, more important, the organism itself, are not adversely affected. Apoptosis gone awry, however, has dire effects. When apoptosis occurs in irreplaceable cells, as in some neurodegenerative disorders, functions critical to the organism are lost. When cells fail to undergo apoptosis after serving their purpose, as in some autoimmune disorders, escaped cells adversely affect the organism. When cells become renegade and resist apoptosis, as in cancer, the outlaw cells create a dire situation for the organism. Mistiming of, or errors in, apoptosis can have devastating consequences on development. Apoptotic fidelity is, therefore, critical to the well-being of an organism. The process of apoptosis (programmed cell death) is regulated by signals generated when cytokines bind to their receptors. There are two types of cytokine-induced signals. The first initiates apoptosis. Cytokines producing an inductive signal include TNFα, FAS/APO-1 ligand, and TRAIL/APO-2 ligand. The second is an inhibitory signal that suppresses apoptosis. Cytokines producing inhibitory signals include those required for cell survival.

Apoptosis proceeds through cleavage of vital intracellular proteins. Caspases are inactive until a signal initiates activation of one, starting a cascade in which a series of other caspases are proteolytically activated. Although both signaling processes affect caspase activation, the mechanism differs.

Apoptosis is characterized by degradation of nuclear DNA, degeneration and condensation of nuclei, and phagocytosis of cell residue. Proliferating cells often undergo apoptosis as a natural process, and proliferating lymphocytes manifest rapid apoptosis during development and during immune responses. In contrast to the internal death program of apoptosis, necrosis describes death from without.

Caspases are closely related cysteine proteases that cleave protein substrates at the C-terminal sides of aspartic acid residues and represent components of enzymatic cascades that lead to apoptotic cell death. There are two pathways of activation of caspases in lymphocyte. One involves mitochondrial permeability changes in growth factor–deprived cells and the other involves signals from death receptors in the plasma membrane.

Cytosolic aspartate-specific proteases (CASPases): These are responsible for the deliberate disassembly of a cell into apoptotic bodies. Caspases are present as inactive proenzymes, most of which are activated by proteolytic cleavage. Caspase-8, caspase-9, and caspase-3 are situated at the pivotal junctions in apoptotic pathways. Caspase-8 initiates disassembly in response to extracellular apoptosis-inducing ligands and is activated in a complex associated with the receptor's cytoplasmic death domains. Caspase-9 activates disassembly in response to agents or insults that trigger release of cytochrome c from the mitochondria and is activated when complexed with dATP, APAF-1, and extramitochondrial cytochrome c. Caspase-3 appears to amplify caspase-8 and caspase-9 signals into a full-fledged commitment to disassembly. Both caspase-8 and caspase-9 can activate caspase-3 by proteolytic cleavage, and caspase-3 may then cleave vital cellular proteins or activate additional caspase by proteolytic cleavage. Many other caspases have been described. Caspases are a group of proteases that proteolytically disassemble the cell. Caspases are present in healthy cells as inactive proforms. During apoptosis, most caspases are activated by proteolytic cleavage. Caspase-9, however, may be active without being proteolytically cleaved. Activation is through autoproteolysis or cleavage by other caspases. Cleavage of caspases generates a pro-domain fragment and subunits of approximately 20 and 10 kDa. Active caspases appear to be tetramers consisting of two identical 20-kDa subunits and two identical 10-kDa subunits. Detection of either the 20- or 10-kDa subunit by immunoblotting may imply activation of the caspase. Colorimetric and fluorometric assays using fluorogenic peptide substrates can be used to measure caspase activity in apoptotic cells.

Caspases cleave substrate proteins at the carboxyl terminus of specific aspartates. Tetrameric peptides with fluorometric or colorimetric groups at the carboxyl terminal have been used to determine the Km of caspases. Although there is preference for peptides with a certain amino acids (aa) sequence, the aa sequence can have some variance. Caspases also have overlapping preferences for the tetrameric aa sequence (i.e., the same substrates can be cleaved by multiple caspases although one caspase may have a lower Km). Peptides containing groups that form covalent bonds with the cysteine residing at the active site of the caspase are often used to inhibit caspase activities.

Caspase substrates: The specificity of caspases translates into an order disassembly of cells by proteolytic cleavage of specific cellular protein. The paradigm substrate for caspase cleavage is PARP (polyADP-ribose polymerase). During apoptosis, intact 121-kDa PARP is cleaved by caspases into fragments of approximately 84 and 23 kDa. Generation of these fragments tends to be an inductor of apoptosis. Cleavage inactivates the enzymatic activity of PARP.

FLIP/FLAM is highly homologous to caspase-8. It does not, however, contain the active site required for proteolytic activity. FLIP appears to compete with caspase-8 binding to the cytosolic receptor complex, thereby preventing the activation of the caspase cascade in response to members of the TNF family of ligands. The exact *in vivo* influence of the IAP family of protein on apoptosis is not clear.

Apoptosis, caspase pathway: A group of intracellular proteases called caspases are responsible for the deliberate disassembly of the cell into apoptotic bodies during apoptosis. Caspases are present as inactive proenzymes that are activated by proteolytic cleavage. Caspases 8, 9, and 3 are situated at pivotal junctions in apoptosis pathways. Caspase-8 initiates disassembly in response to extracellular apoptosis-inducing ligands and is activated in a complex associated with the cytoplasmic death domain of many cell surface receptors for the ligands. Caspase-9 activates disassembly in response to agents or insults that trigger the release of cytochrome c from mitochondria and is activated when complexed with apoptotic protease activating factor 1 (APAF-1) and extramitochondrial cytochrome c. Caspase-3 appears to amplify caspase-8 and caspase-9 initiation signals into full-fledged commitment to disassembly. Caspase-8 and caspase-9 activate caspase-3 by proteolytic cleavage, and caspase-3 then cleaves vital cellular proteins or other caspases.

Cytochrome c: Suppression of the antiapoptotic members or activation of the proapoptotic members of the Bcl-2 family leads to altered mitochondrial membrane permeability resulting in release of cytochrome c into the cytosol. In the cytosol or on the surface of the mitochondria, cytochrome c is bound by the protein Apaf-1 (apoptotic protease activating factor) which also binds caspase-9 and dATP. Binding of cytochrome c triggers activation of caspase-9, which then accelerates apotosis by activating other caspases. Release of cytochrome c from mitochondria has been established by determining the distribution of cytochrome c in subcellular fractions of cells treated or untreated to induce apoptosis. Cytochrome c was primarily in the mitochondria-containing fractions obtained from healthy, nonapoptotic cells in the cytosolic nonmitochondria-containing fractions obtained from apoptotic cells. Using mitochondria-enriched fractions from mouse liver, rat liver, or cultured cells, it has been shown that release of cytochrome c from mitochondira is greatly accelerated by addition of Bax, fragments of Bid, and by cell extracts.

Immunotoxin-induced apoptosis: Immunotoxins are cytotoxic agents usually assembled as recombinant fusion proteins composed of a targeting domain and a toxin. The targeting domain controls the specificity of action and is usually derived from an antibody Fv fragment, a growth factor, or a soluble receptor. The protein toxins are obtained from bacteria, e.g., *Pseudomonas* endotoxin (PE) or diptheria toxin (DT), or from plants, e.g., ricin. Immunotoxins have been studied as treatments for cancer, graft-vs.-host disease, autoimmune disease, and AIDS.

The bacterial toxins PE and DT act via the ADP-ribosylation of elongation factor 2, thereby inactivating it. This results in the arrest of protein synthesis and subsequent cell death. These toxins can also induce apoptosis, although the mechanism is unknown. Two common features of apoptotic cell death are the activation of a group of cysteine proteases called caspases and the caspase-catalyzed cleavage of so-called "death substrates" such as the nuclear repair enzyme poly (ADP-ribose) polymerase (PARP).

Apoptosis and necrosis are two major processes by which cells die. Apoptosis is the ordered disassembly of the cell from within. Disassembly creates changes in the phospholipid content of the plasma membrane outer leaflet. PS is exposed on the outer leaflet and phagocytic cells that recognize this change may engulf the apoptotic cell or cell-derived, membrane-limited apoptotic bodies. Necrosis normally results from a severe cellular insult. Both internal organelle and plasma membrane integrity are lost, resulting in spilling of cytosolic and organellar contents into the surrounding environment. Immune cells are attached to the area and begin producing cytokines that generate an inflammatory response. Thus, cell death in the absence of an inflammatory response may be the best way to distinguish apoptosis from necrosis. Other techniques which have been used to distinguish apoptosis from necrosis in cultured cells and in tissue sections include detecting PS at the cell surface with annexin V binding, DNA laddering, and staining cleaved DNA fragments which contain characteristic ends. At the extremes, apoptosis and necrosis clearly involve different molecular mechanisms. It is not clear whether or not there is cellular death involving the molecular mechanisms of both apoptosis and necrosis. Cell death induced by free radicals, however, may have characteristics of apoptosis and necrosis.

Positive induction apoptosis: There are two central pathways that lead to apoptosis: (1) positive induction by ligand binding to plasma membrane receptor and (2) negative induction by loss of suppressor activity. Each leads to activation of cysteine proteases with homology to IL-1β converting enzyme (ICE) (i.e., caspases). Positive induction involves ligands related to TNF. Ligands are typically trimeric and bind to cell surface receptors causing aggregation (trimerization) of cell surface receptors. Receptor oligomerization orients their cytosolic-death domains into a configuration that recruits adaptor proteins. The adaptor complex recruits caspase-8. Caspase-8 is activated, and the cascade of caspase-mediated disassembly proceeds.

Negative induction apoptosis: Negative induction of apoptosis by loss of a suppressor activity involves the mitochondria. Release of cytochrome c from the mitochondria into the cytosol serves as a trigger to activate caspases. Permeability of the mitochondrial outer membrane is essential to initiation of apoptosis through this pathway. Proteins belonging to the Bcl-2 family appear to regulate the membrane permeability to ions and possibly to cytochrome c as well. Although these proteins can themselves form channels in membranes, the actual molecular mechanisms by which they regulate mitochondrial permeability and the solutes that are released are less clear. The Bcl-2 family is composed of a large group of antiapoptosis members that, when overexpressed, prevent apoptosis and a large group of proapoptosis members that when overexpressed induce apoptosis. The balance between the antiapoptotic and proapoptotic Bcl-2 family members may be critical in determining if a cell undergoes apoptosis. Thus, the suppressor activity of the antiapoptotic Bcl-2 family appears to be negated by the proapoptotic members. Many members of the proapoptotic Bcl-2 family are present in cells at levels sufficient to induce apoptosis. However, these members do not induce apoptosis because their activity is maintained in a latent form. Bax is present in the cytosol of live cells. After an appropriate signal, Bax undergoes a conformational change and moves to the mitochondrial membrane where it causes release of mitochondrial cytochrome c into the cytosol. BID is also present in the cytosol of live cells. After cleavage by caspase-8, it moves to the mitochondria where it causes release of cytochrome c possibly by altering the conformation of Bax. Similarly, BAK appears to undergo a conformational change that converts it from an inactive to an active state. Thus, understanding the molecular mechanisms responsible for regulating the Bcl-2 family activities creates the potential for pharmaceutical intervention to control apoptosis. The viability of many cells is dependent on a constant or intermittent supply of cytokines or growth factors. In the absence of an apoptosis-suppressing cytokine, cells may undergo apoptosis. Bad is a proapoptotic member of the Bcl-2 family and is sequestered in the cytosol when cytokines are present. Cytokine binding can activate PI3 kinase, which phosporylates Akt/PKB, which in turn phosphorylates Bad. Phosphorylated Bad is sequestered in the cytosol by the 14-3-3 protein. Removal of the cytokine turns the kinase pathway off, the phosphorylation state of Bad shifts to the dephosphorylated form, and dephosphorylated Bad causes release of cytochrome c from the mitochondria.

APO-1 is a synonym for *fas* gene. FAS membrane protein ligation has been shown to initiate apoptosis. This is the reverse action of bcl-2 protein, which blocks apoptosis.

Apoptosis, suppressors: The induction of apoptosis or progression through the process of apoptosis is inhibited by a group of proteins called *inhibitors of apoptosis*. These proteins contain a BIR (baculovirus IAP repeat) domain near the amino-terminus. The BIR domain can bind some caspases. Many members of the IAP family of proteins block proteolytic activation of caspase-3 and caspase-7. XIAP, cIAP-1, and cIAP-2 appear to block cytochrome c–induced activation of caspase-9, thereby preventing initiation of the caspase cascade. Since cIAP-1 and cIAP-2 were first identified as components in the cytosolic death domain-induced complex associated with the TNF family of receptors, they may inhibit apoptosis by additional mechanisms.

Death domains are protein molecular structures involved in protein to protein interactions. They were first recognized in proteins encoded by genes involved in programmed cell death or apoptosis.

Fas (AP0-1/CD95) is a member of the TNF receptor superfamily. Fas Ligand (FasL) is a member of the TNF family of type 2 membrane proteins. Soluble FasL can be produced by proteolysis of membrane-associated Fas. Ligation of Fas by FasL or anti-Fas antibody can induce apoptotic cell death in cells expressing Fas. Fas is expressed on selected cells, including T cells, and renders them susceptible to apoptotic death mediated by cells expressing Fas ligand.

FasL/Fas toxicity: Cytotoxic sequence that commences with crosslinking of target cell Fas by FasL on the effector cell. This does not require macromolecular synthesis or extracellular calcium. FasL crosslinks Fas triggering signals, which leads to a target cell apoptotic response. Fas crosslinking leads to activation of intracellular caspases.

Fas ligand: Binding to Fas initiates the death pathway of apoptosis in the Fas-bearing cell. Fas-mediated killing of T lymphocytes is critical for the maintenance of self-tolerance. *Fas* gene mutations can lead to systemic autoimmune disease. Fas Ligand is a member of the TNF family of proteins expressed on the cell surface of activated T lymphocytes. Binding Fas ligand to Fas initiates the signaling pathway that leads to apoptotic cell death of the cell expressing Fas.

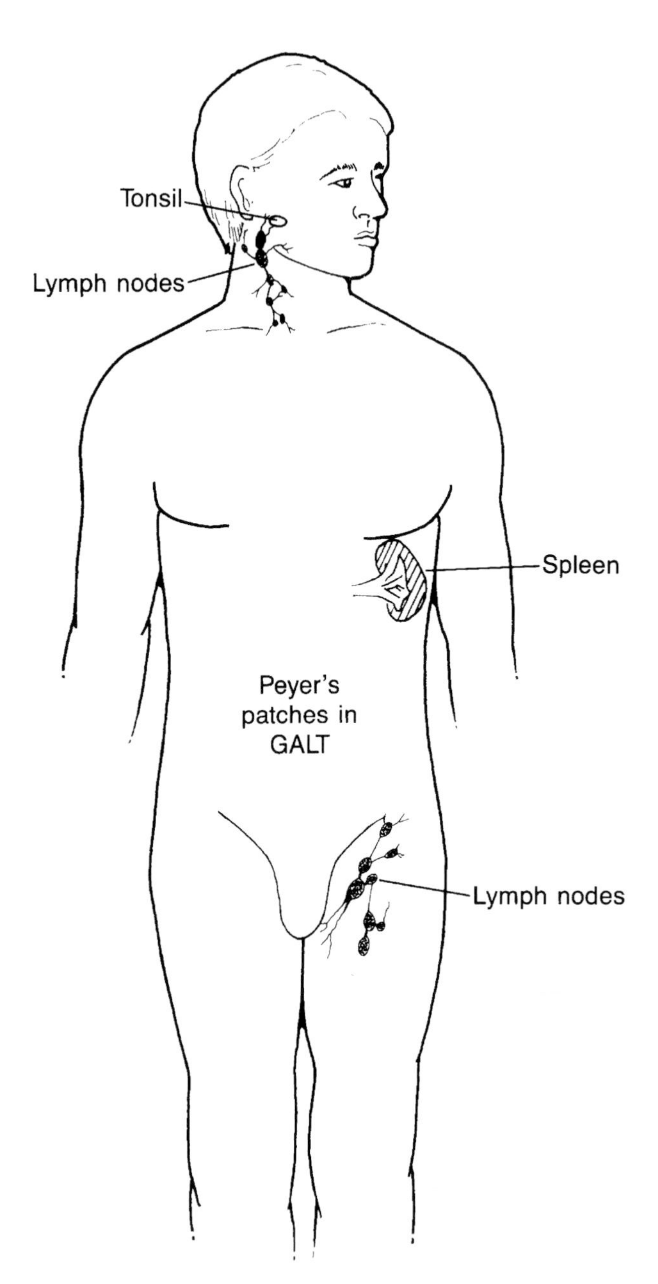

FIGURE 2.81 Peripheral lymphoid organs.

Peripheral lymphoid organs (Figure 2.81) are not required for ontogeny of the immune response. They include the lymph nodes, spleen, tonsils, Peyer's patches, and mucosal-associated lymphoid tissues in which immune responses are induced in contrast to the thymus, a central lymphoid organ in which lymphocytes develop.

Ikaros is a requisite transcription factor for all lineages of lymphoid cells to develop.

Afferent lymphatic vessels are the channels that transport lymph, which may contain antigens draining from

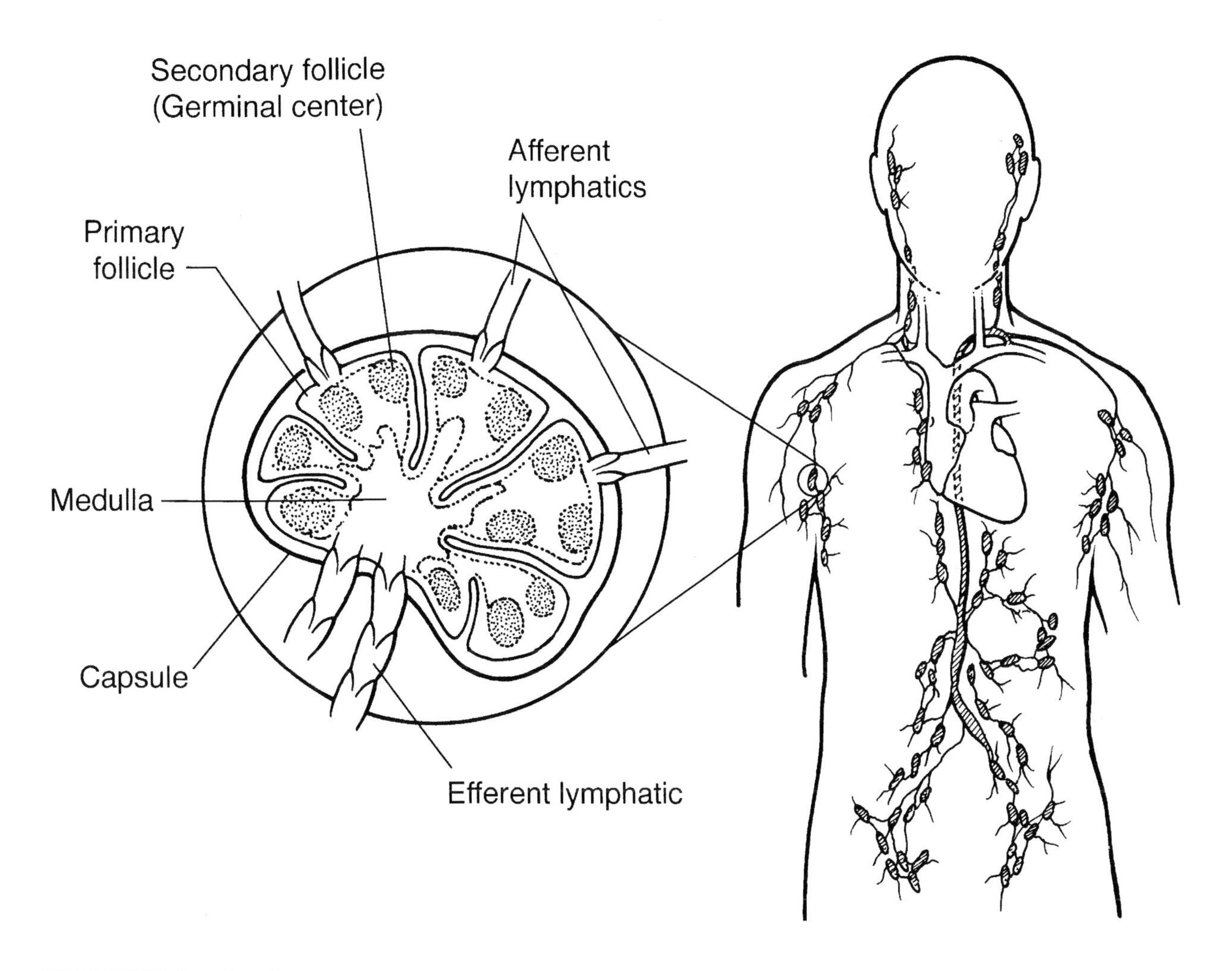

FIGURE 2.82 Lymph node.

connective tissue or from sites of infection in many anatomical locations, to the lymph nodes.

The **thoracic duct** is a canal that leads from the cisterna chyli, a dilated segment of the thoracic duct at its site of origin in the lumbar region, to the left subclavian vein.

Cisterna chyli: See thoracic duct.

Thoracic duct drainage is the deliberate removal of lymphocytes through drainage of lymph from the thoracic duct with a catheter.

A **lymph node** (Figure 2.82 to Figure 2.85) is a relatively small, i.e., 0.5-cm, secondary lymphoid organ that is a major site of immune reactivity. It is surrounded by a capsule and contains lymphocytes, macrophages, and dendritic cells in a loose reticulum environment. Lymph enters this organ from afferent lymphatics at the periphery, percolates through the node until it reaches the efferent lymphatics, exits at the hilus, and circulates to central lymph nodes and finally to the thoracic duct. The lymph node is divided into cortex and medulla. The superficial cortex contains B lymphocytes in follicles, and the deep cortex

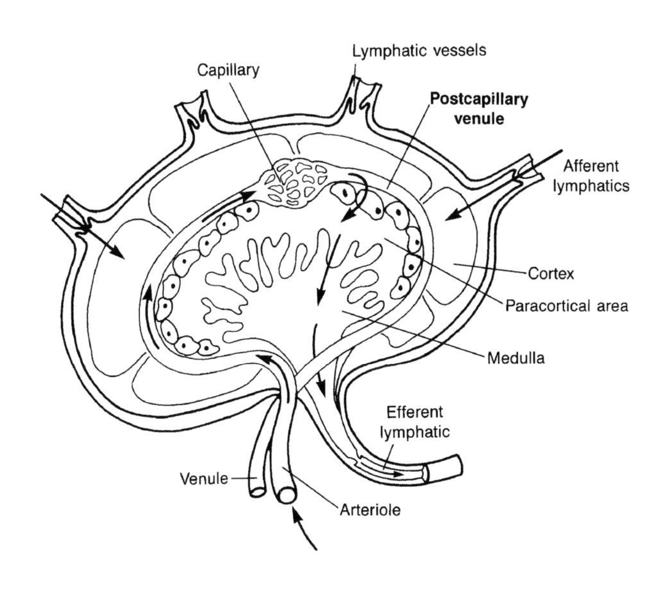

FIGURE 2.83 Structure of a lymph node.

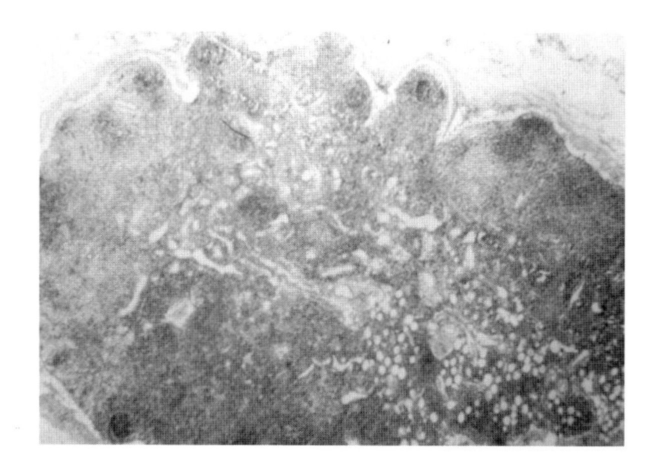

FIGURE 2.84 Lymph node (low power).

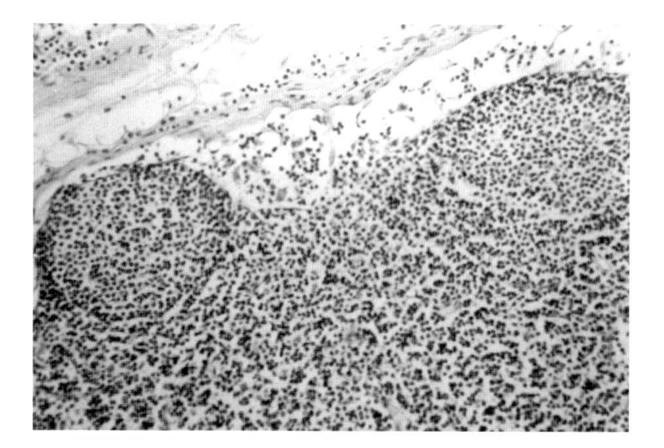

FIGURE 2.85 Subcapsular sinus lymph node.

is comprised of T lymphocytes. Differentiation of the specific cells continues in these areas and is driven by antigen and thymic hormones. Conversion of B cells into plasma cells occurs chiefly in the medullary region where enclosed lymphocytes are protected from undesirable influences by a macrophage sleeve. The postcapillary venules from which lymphocytes exit the lymph node are also located in the medullary region. Macrophages and follicular dendritic cells interact with antigen molecules that are transported to the lymph node in the lymph. Reticulum cells form medullary cords and sinuses in the central region. T lymphocytes percolate through the lymph nodes. They enter from the blood at the postcapillary venules of the deep cortex. They then enter the medullary sinuses and pass out of the node through the efferent lymphatics. T cells that interact with antigens are detained in the lymph node, which may be a site of major immunologic reactivity. The lymph node is divided into B and T lymphocyte regions. Individuals with B-cell or T-cell immunodeficiencies may reveal an absence of either lymphocyte type in the areas of the lymph node normally occupied by that cell population. The lymph node acts as a filter and may be an important site for phagocytosis and the initiation of immune responses.

Nasopharyngeal-associated lymphoreticular tissue (NALT) includes the palatine and nasopharyngeal tonsils (adenoids) which are mostly covered by a squamous epithelium. The palatine tonsils usually contain 10 to 20 crypts, which increase their surface area. The deeper regions of these crypts contain M cells that may take up encountered antigens. The tonsils contain all major classes of antigen presenting cells, including dendritic and Langerhans cells, macrophages, class I positive B cells, and antigen-retaining follicular dendritic cells in B cell germinal centers. Approximately one half of tonsillar cells are B lymphocytes situated mainly in follicles containing germinal centers. There is predominance of IgG blasts in germinal centers and of plasma cells in the parafollicular area. Approximately 40% of tonsillar cells are T cells, and more than 98% express the αβ TCR. Higher CD4:CD8 ratios are found in tonsils compared with peripheral blood. The tonsils reveal not only features of mucosal inductive sites, but also characteristics of effector sites with high numbers of plasma cells. The role of tonsils in host mucosal immunity following intranasal immunization remains to be determined.

Primary lymphoid organs are sites of maturation of B or T lymphocytes. The primary lymphoid organs for B lymphocytes in avian species is the bursa of Fabricius, whereas it is the bone marrow in adult mammals. By contrast, T-cell development occurs in the thymus of all vertebrates. Stem-cell maturation in primary lymphoid organs occurs without stimulation by antigen.

Lymph is the extracellular fluid that circulates in the lymphatic system vessels. Its composition resembles that of tissue fluids, although there is less protein in lymph than in plasma. Lymph in the mesentery contains fat, and lymph draining the intestine and liver often possesses more protein than does other lymph. The principal cell type in the lymph is the small lymphocyte, with rare large lymphocytes, monocytes, and macrophages. Occasional red cells and eosinophils are present. Coagulation factors are also present in lymph.

The **lymphatic system** is a network of lymphoid channels that transports lymph, a tissue fluid derived from the blood. It collects extracellular fluid from the periphery and channels it via the thoracic duct to the blood circulation. Lymph nodes at the intersection of the lymphatic vessels trap and retain antigens from the lymph. Situated at lymphatic vessel intersections are lymph nodes, Peyer's patches, and other organized lymphoid structures except the spleen, which communicates directly with the blood. The lymphatic system's main functions include the concentration of antigen from various body locations into a few lymphoid organs, the circulation of lymphocytes through lymphoid organs to permit antigen to interact with antigen-specific cells, and carrying antibody and immune effector cells to the blood circulation and tissue.

Lymphatics are vessels that transport the interstitial fluid called lymph to lymph nodes and away from them, directing it to the thoracic duct from where it reenters the blood stream.

Lymphatic vessels are thinly walled channels through which lymph and cells of the lymphatic system move through secondary lymphoid tissues such as lymph nodes, except for the spleen, to the thoracic duct which joins the blood circulation.

An **efferent lymphatic vessel** is the channel through which lymph and lymphocytes exit a lymph node in transit to the blood.

Lymphoid follicle: See lymphoid nodules.

Lymphoid nodules (or **follicles**) are aggregates of lymphoid cells present in the loose connective tissue supporting the respiratory and digestive membranes. They are also present in the spleen and may develop beneath any mucous membrane as a result of antigenic stimulation. They are poorly defined at birth. Characteristic lymphoid nodules are round and nonencapsulated. They may occur as isolated structures or may be confluent, such as in the tonsils, pharynx, and nasopharynx. In the tongue and pharynx, they form a characteristic structure referred to as Waldeyer's ring. In the terminal ileum, they form oblong patches termed Peyer's patches. The lymphoid nodules contain B and T cells and macrophages. Plasma cells in submucosal sites synthesize IgA, which is released in secretions.

Lymphoid organs are organized lymphoid tissues in which numerous lymphocytes interact with nonlymphoid stroma. The thymus and bone marrow are the primary lymphoid organs where lymphocytes are formed. The principal secondary lymphoid tissues where adaptive immune responses are initiated include the lymph nodes, spleen, and mucosa-associated lymphoid tissues, including the tonsils and Peyer's patches.

The term **lymphoid** is an adjective that describes tissues such as the lymph node, thymus, and spleen that contain a large population of lymphocytes.

Lymph gland: More correctly referred to as *lymph node*.

Central lymphoid organs are requisite organs for the development of the lymphoid and, therefore, of the immune system. These include the thymus, bone marrow, and bursa of Fabricus (also termed *primary lymphoid organs*). They are sites of lymphocyte development. Human T lymphocytes mature in the thymus, whereas B lymphocytes develop in the bone marrow.

A **generative lymphoid organ** is an organ in which lymphocytes arise from immature precursor cells. The principal generative lymphoid organ for T cells is the thymus, and for B cells the bone marrow.

Large pyroninophilic blast cells stain positively with methyl green pyronin stain. They are found in thymus-dependent areas of lymph nodes and other peripheral lymphoid tissues.

The **cortex** is the outer or peripheral layer of an organ.

Hyperplasia: An increase in the cell number of an organ that leads to an increase in the organ size. It is often linked to a physiological reaction to a stimulus and is reversible.

Follicles are circular or oval areas of lymphocytes in lymphoid tissues rich in B cells. They are present in the cortex of lymph nodes and in the splenic white pulp. Primary follicles contain B lymphocytes that are small and medium sized. They are demonstrable in lymph nodes prior to antigenic stimulation. Once a lymph node is stimulated by antigen, secondary follicles develop. They contain large B lymphocytes in the germinal centers where tingible body macrophages (those phagocytizing nuclear particles) and follicular dendritic cells are present.

Follicular center cells are B lymphocytes in germinal centers (secondary follicles).

Follicular hyperplasia is lymph node enlargement associated with an increase in follicle size and number. Germinal centers are usually present in the follicles. Follicular hyperplasia is often a postinfection reactive process in lymph nodes.

Germinal follicle: See germinal center.

Primary nodule: See primary follicle.

Appendix, vermiform is a gut-associated lymphoid tissue situated at the ileocecal junction of the gastrointestinal tract.

The **medulla** is the innermost or central region of an organ. The central area of a thymic lobe represents the thymic medulla, which is rich in antigen-presenting cells derived from the bone marrow and medullary epithelial cells. Macrophages and plasma cells are rich in the lymph node medulla through which lymph passes en route to the efferent lymphatics.

The **medullary cord** is a region of the lymph node medulla composed of macrophages as well as plasma cells and lies between the lymphatic sinusoids.

The **medullary sinuses** are potential cavities in the lymph node medulla that receive lymph prior to its entering efferent lymphatics.

Lymphadenitis is lymph node inflammation often caused by microbial (bacterial or viral) infection.

Benign lymphadenopathy refers to lymph node enlargement that is not associated with malignant neoplasms. Histologic types of benign lymphadenopathy include nodular, granulomatous, sinusoidal, paracortical, diffuse or obliterative, mixed, and depleted, whereas clinical states are associated with each histologic pattern.

Dermatopathic lymphadenitis is a benign lymph node hyperplasia that follows skin inflammation or infection.

A **draining lymph node** collects fluid draining through lymphatic channels from an anatomical site of infection before returning to the blood circulation.

Lymphadenopathy is lymph node enlargement due to any of several causes. Lymphadenopathies are reactive processes in lymph nodes due to various exogenous and endogenous stimulants. Possible etiologies include microorganisms, autoimmune diseases, immunodeficiencies, foreign bodies, tumors, and medical procedures. Lymphadenitis is reserved for lymph node enlargement caused by microorganisms, whereas lymphadenopathy applies to all other etiologies of lymph node enlargement. Lymphadenopathies are divided into reactive lymphadenopathies, lymphadenopathies associated with clinical syndromes, vascular lymphadenopathies, foreign body lymphadenopathies, and lymph node inclusions. In benign lymphadenopathy, there is variability of germinal center size, no invasion of the capsule or fat, mitotic activity confined to germinal centers, and localization in the cortex and nonhomogenous follicle distribution.

Thymus-dependent areas are regions of peripheral lymphoid tissues occupied by T lymphocytes. Specifically, these include the paracortical areas of lymph nodes, the zone between nodules and Peyer's patches, and the center of splenic Malpighian corpuscles. These regions contain small lymphocytes derived from the circulating cells that reach these areas by passing through high endothelial venules. Proof that these anatomical sites are thymus-dependent areas is provided by the demonstration that animals thymectomized as neonates do not have lymphocytes in these areas. Likewise, humans or animals with thymic hypoplasia or congenital aplasia of the thymus reveal no T cells in these areas.

A **primary follicle** is a densely packed accumulation of resting B lymphocytes and a network of follicular dendritic cells from a secondary lymphoid organ such as the lymph node cortex or the splenic white pulp where resting unstimulated B cells develop into germinal centers when stimulated by antigen.

Postcapillary venules are relatively small blood vessels lined with cuboidal epithelium through which blood circulates after it exits the capillaries and before it enters the veins. It is a frequent site of migration of lymphocytes and inflammatory cells into tissues during inflammation. Recirculating lymphocytes migrate from the blood to the lymph through high endothelial venules of lymph nodes.

Mantle zone refers to the rim of B lymphocytes that encircles lymphoid follicles. The function and role of mantle zone lymphocytes remain to be determined.

Reticular cells are stroma or framework cells which, together with reticular fibers, constitute the lymphoid tissue framework of lymph nodes, spleen, and bone marrow.

Reticulum cell: See reticular cell.

The first event in an adaptive immune response when antigen-specific lymphocytes bind to antigens is termed the **recognition phase.** This phase often occurs in secondary lymphoid tissues such as lymph nodes or spleen where antigens and naïve lymphocytes are present.

The **pharyngeal tonsils** are lymphoid follicles found in the roof and posterior wall of the nasopharynx. They are similar to Peyer's patches in the small intestine. Mucosal lymphoid follicles are rich in IgA-producing B cells that may be found in germinal centers.

Secondary lymphoid organs are the structures that include the lymph nodes, spleen, gut-associated lymphoid tissues, and tonsils where T and B lymphocytes interact with antigen-presenting accessory cells such as macrophages, resulting in the generation of an immune response.

Secondary lymphoid tissues are tissues in which immune responses are generated. They include lymph nodes, spleen, and mucosa-associated lymphoid tissues. Lymph nodes and spleen are also often referred to as *secondary lymphoid organs*.

Follicular dendritic cells (Figure 2.86) manifest narrow cytoplasmic processes that interdigitate between densely populated areas of B lymphocytes in lymph node follicles and in spleen. Antigen–antibody complexes adhere to the surfaces of follicular dendritic cells and are not generally endocytosed but are associated with the formation of germinal centers. These cells are bereft of class II histocompatibility molecules, although Fc receptors, complement receptor 1, and complement receptor 2 molecules are demonstrable on their surfaces. They display antigens on their surface for B-cell recognition and participate in the activation and selection of B cells expressing high-affinity membrane immunoglobulin during affinity maturation.

FIGURE 2.86 Follicular dendritic cell.

CD21 (Figure 2.87) is an antigen with a 145-kDa mol wt that is expressed on B cells and even more strongly on follicular dendritic cells. It appears when surface Ig is expressed after the pre-B cell stage and is lost during early stages of terminal B cell differentiation to the final plasma cell stage. CD21 is coded for by a gene found on chromosome 1 at band q32. The antigen functions as a receptor for the C3d complement component and also for Epstein–Barr virus. CD21, together with CD19 and CD81, constitutes the coreceptor for B cells. It is also termed CR2.

Germinal centers (Figure 2.88 to Figure 2.90) develop in lymph node and lymphoid aggregates within primary follicles of lymphoid tissues following antigenic stimulation. The mixed-cell population in the germinal center is comprised of B lymphoblasts (both cleaved and transformed lymphocytes), follicular dendritic cells, and numerous tingible body-containing macrophages. Germinal centers seen in various pathologic states include "burned out" germinal centers comprised of accumulations of pale histiocytes and scattered immunoblasts; "progressively transformed" centers that show a "starry sky" pattern containing epithelioid histiocytes, dendritic reticulum cells, increased T lymphocytes, and mantle zone lymphocytes; and "regressively transformed" germinal centers that are relatively small, with few lymphocytes, and which reveal an onion-skin layering of dendritic reticulum cells, vascular endothelial cells, and fibroblasts. The mantle zone (Figure 2.91) is a dense area of lymphocytes that encircles a germinal center.

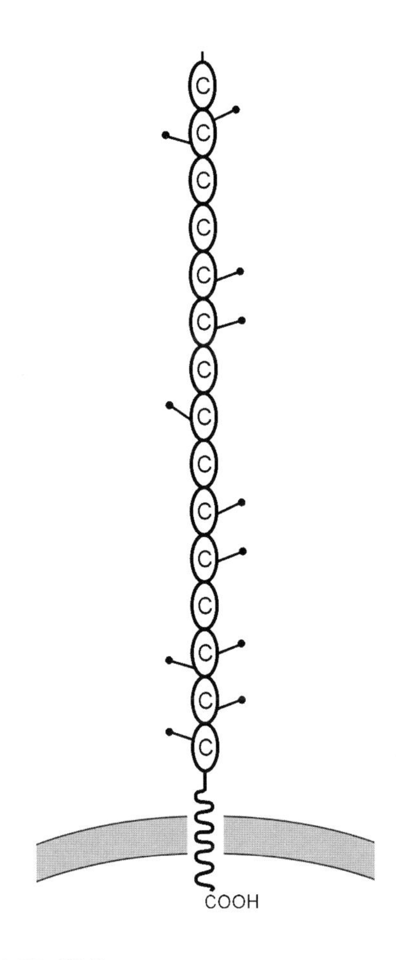

FIGURE 2.87 CD21.

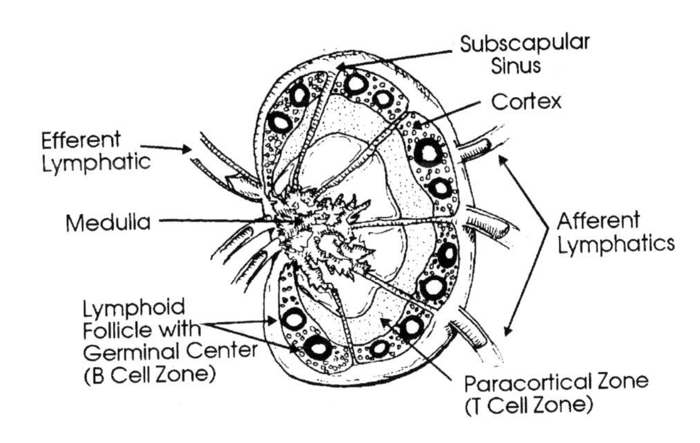

FIGURE 2.88 Germinal center.

Centroblasts are large, rapidly dividing cells in germinal centers. These cells, in which somatic hypermutation is thought to take place, give rise to memory and antibody-secreting B cells.

The **dark zone** is that part of a germinal center in secondary lymphoid tissue in which centroblasts undergo rapid division.

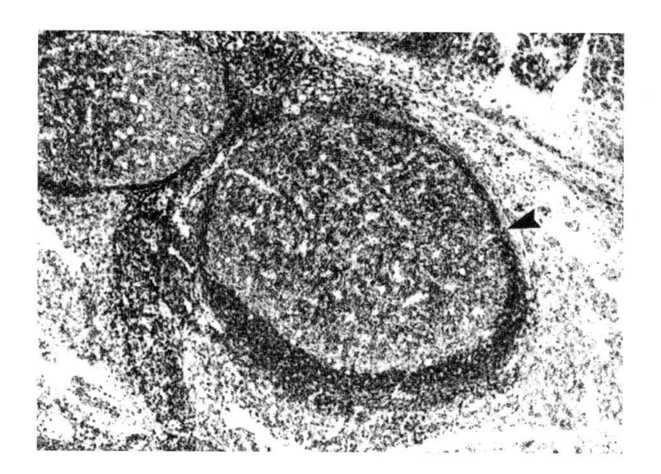

FIGURE 2.89 Germinal center.

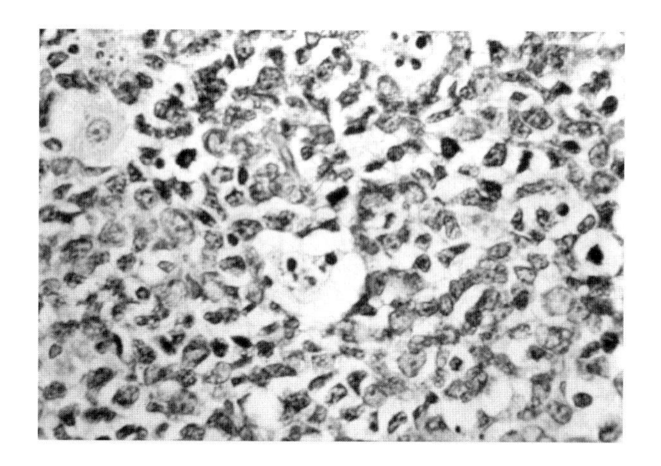

FIGURE 2.90 Tingible body macrophages in germinal center.

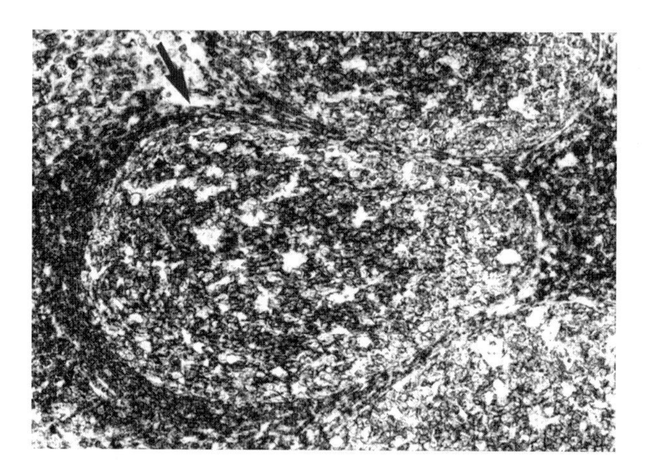

FIGURE 2.91 Mantle zone.

The **light zone** is a region of a germinal center in secondary lymphoid tissue containing centrocytes that are not dividing but are interacting with follicular dendritic cells.

Lymphocytopenic center: See germinal center.

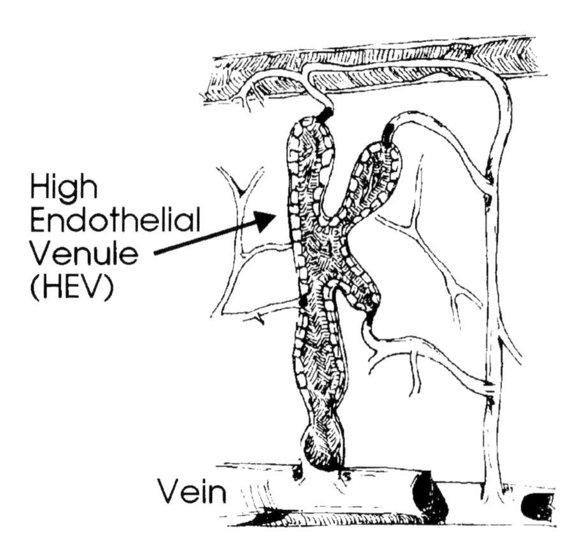

FIGURE 2.92 High endothelial venules (HEV).

A **secondary follicle** is an area in a peripheral lymphoid organ where a germinal center is located, usually associated with a secondary immune response more than with a primary one. It forms a ring of concentrically packed B lymphocytes surrounding a germinal center.

Centrocytes are small B lymphocytes in germinal centers that arise from centroblasts. They may give rise to antibody-secreting plasma cells, or to memory B lymphocytes, or may undergo apoptosis, depending on the interaction of their receptor with antigen.

Mantle refers to a dense zone of lymphocytes that encircles a germinal center.

High endothelial venules (HEV) are postcapillary venules of lymph node paracortical areas. They also occur in Peyer's patches, which are part of the gut-associated lymphoid tissue (GALT). Their specialized columnar cells bear receptors for antigen-primed lymphocytes. They signal lymphocytes to leave the peripheral blood circulation. A homing receptor for circulating lymphocytes is found in lymph nodes.

High endothelial postcapillary venules (Figure 2.92) are lymphoid organ vessels that are especially designed for circulating lymphocytes to gain access into the parenchyma of the organ. They contain cuboidal endothelium which permits lymphocytes to pass between the cells into the tissues. Lymphocyte recirculation from the blood to the lymph occurs through these vessel walls.

The **spleen** (Figure 2.93 to Figure 2.95) is an encapsulated organ in the abdominal cavity that has important immunologic and nonimmunologic functions. Vessels and nerves enter the spleen at the hilum, as in lymph nodes,

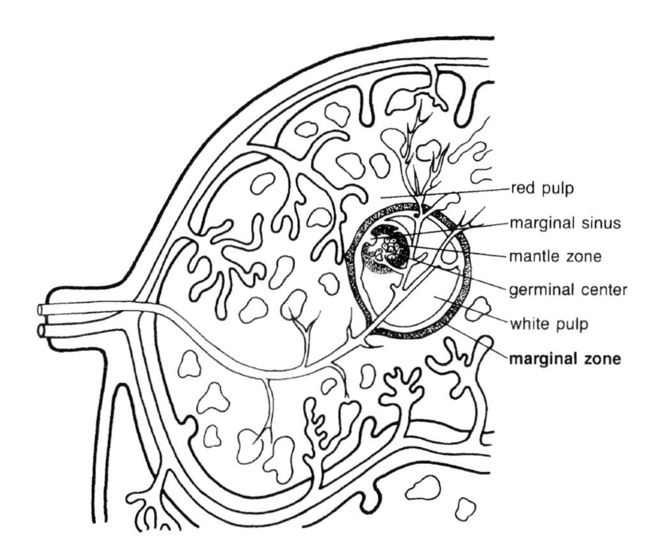

FIGURE 2.93 Spleen.

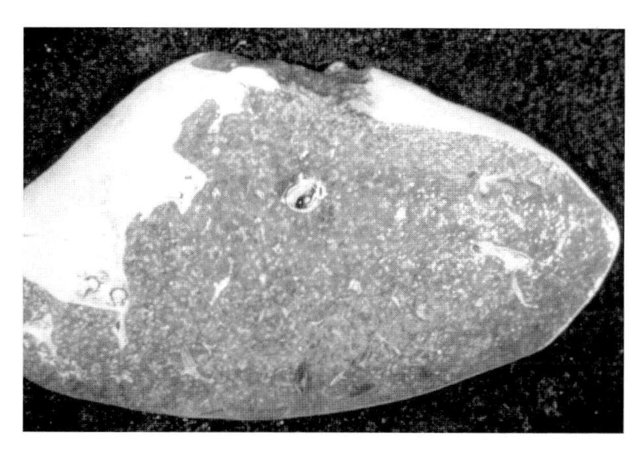

FIGURE 2.94 Spleen, gross specimen.

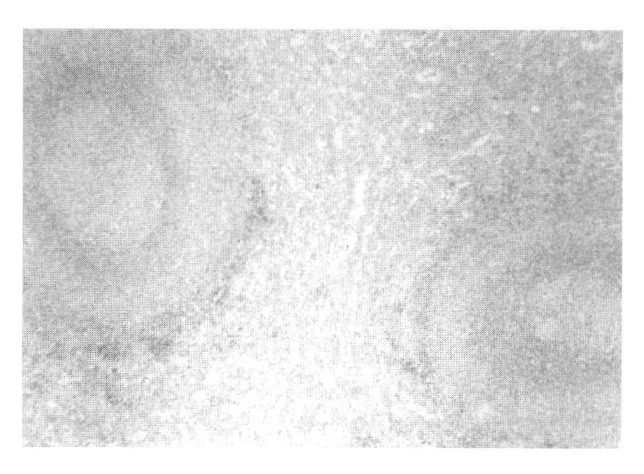

FIGURE 2.95 Spleen follicle with central follicular artery.

and travel part of their course within the fibrous trabeculae that emerge from the capsule. The splenic parenchyma has two regions that are functionally and histologically distinct. The white pulp consists of a thick layer of lymphocytes surrounding the arteries that have left the trabeculae. It forms a periarterial sheath that contains mainly T cells. The sheaths then expand along their course to form well-developed lymphoid nodules called *Malpighian corpuscles*. The red pulp consists of a mesh of reticular fibers continuous with the collagen fibers of the trabeculae. These fibers enclose an open system of sinusoids that drain into small veins and are lined by endothelial cells with reticular properties. The endothelium is discontinuous, leaving small slits through which cells have to pass during transit. Within the sinusoidal mesh are red blood cells, macrophages, lymphocytes, and plasma cells. The red pulp between adjacent sinusoids forms the pulp cords, sometimes called the *cords of Billroth*. The marginal zone consists of a poorly defined area between the white and the red pulp where the periarterial sheath and the lymphoid nodules merge. The blood vessels branch, and at the periphery of the marginal zone the blood empties into the pulp. Lymphocytes of the marginal zone are mainly T cells. They surround the periphery of the lymphoid nodules that comprise B cells. In this marginal region, the T and B cells contact each other. Some B cells may convert into immunoblasts. Further maturation to plasma cells occurs in the red pulp. Active follicles contain germinal centers in which lymphoblasts may be generated. They are discharged into sinusoids, and plasma cells may form. The spleen also contains dendritic cells that have long cytoplasmic extensions. Dendritic cells serve as antigen-presenting cells, interacting with lymphocytes. The spleen filters blood as the lymph nodes filter the lymph. The spleen is active in the formation of antibodies against intravenously administered particulate antigens. It has numerous additional functions, including the sequestration and destruction of senescent red blood cells, platelets, and lymphocytes.

Cords of Billroth are splenic medullary cords.

The **B cell corona** is the zone of splenic white pulp comprised mainly of B cells.

Red pulp (Figure 2.96) describes areas of the spleen comprised of the cords of Billroth and sinusoids. It is comprised of vascular sinusoids, with interspersed large numbers of macrophages, erythrocytes, dendritic cells, a few lymphocytes, and plasma cells. Macrophages in the red pulp ingest microorganisms, foreign particles, and injured red blood cells.

White pulp (Figure 2.97) refers to the periarteriolar lymphatic sheaths encircled by small lymphocytes, which are mainly T cells that surround germinal centers comprised of B lymphocytes and B lymphoblasts in normal splenic tissue. Following interaction of B cells in the germinal

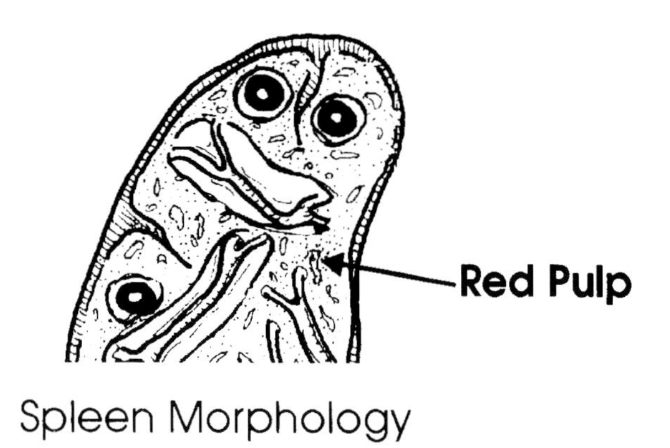

FIGURE 2.96 Red pulp.

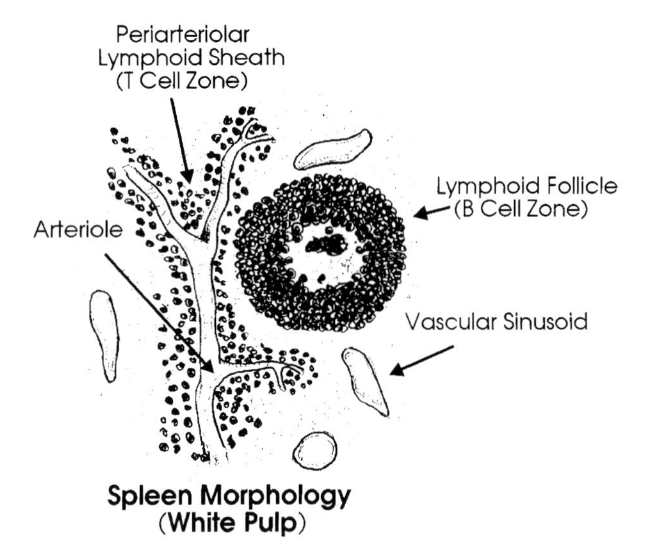

FIGURE 2.97 White pulp.

center with antigen in the blood, a primary immune response is generated within 24 h revealing immunoblastic proliferation and enlargement of germinal centers.

Periarteriolar lymphoid sheath is the thymus-dependent region in the splenic white pulp that is comprised mainly of T cells. There is lymphocyte cuffing of these small arterioles of the spleen adjacent to lymphoid follicles. Two thirds of the PALS T cells are $CD4^+$ and one third are $CD8^+$. During humoral immune responses against protein antigens, B cells are activated at the PALS and follicle interface and then made to migrate into the follicles to produce germinal centers.

The **marginal zone** is an exterior layer of lymphoid follicles of the spleen where T and B lymphocytes are loosely arranged encircling the periarterial lymphatic sheath. When antigens are injected intravenously, macrophages in this area actively phagocytize them. Marginal zone macrophages are especially adept at trapping polysaccharide antigens on their surfaces, where they may persist for long periods and be recognized by specific B cells or conveyed into follicles.

FIGURE 2.98 Mucosa of ileum with Peyer's patch.

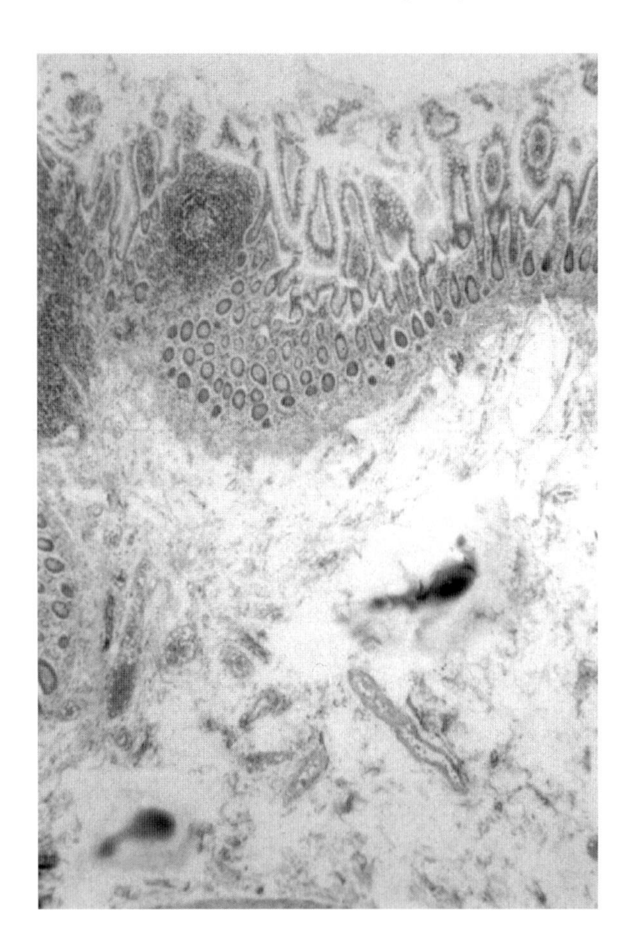

FIGURE 2.99 Peyer's patch (higher magnification).

Peyer's patches (Figure 2.98 and Figure 2.99) are lymphoid tissues in the submucosa of the small intestine. They are comprised of lymphocytes, plasma cells, germinal centers, and thymus-dependent areas. Peyer's patches are sites

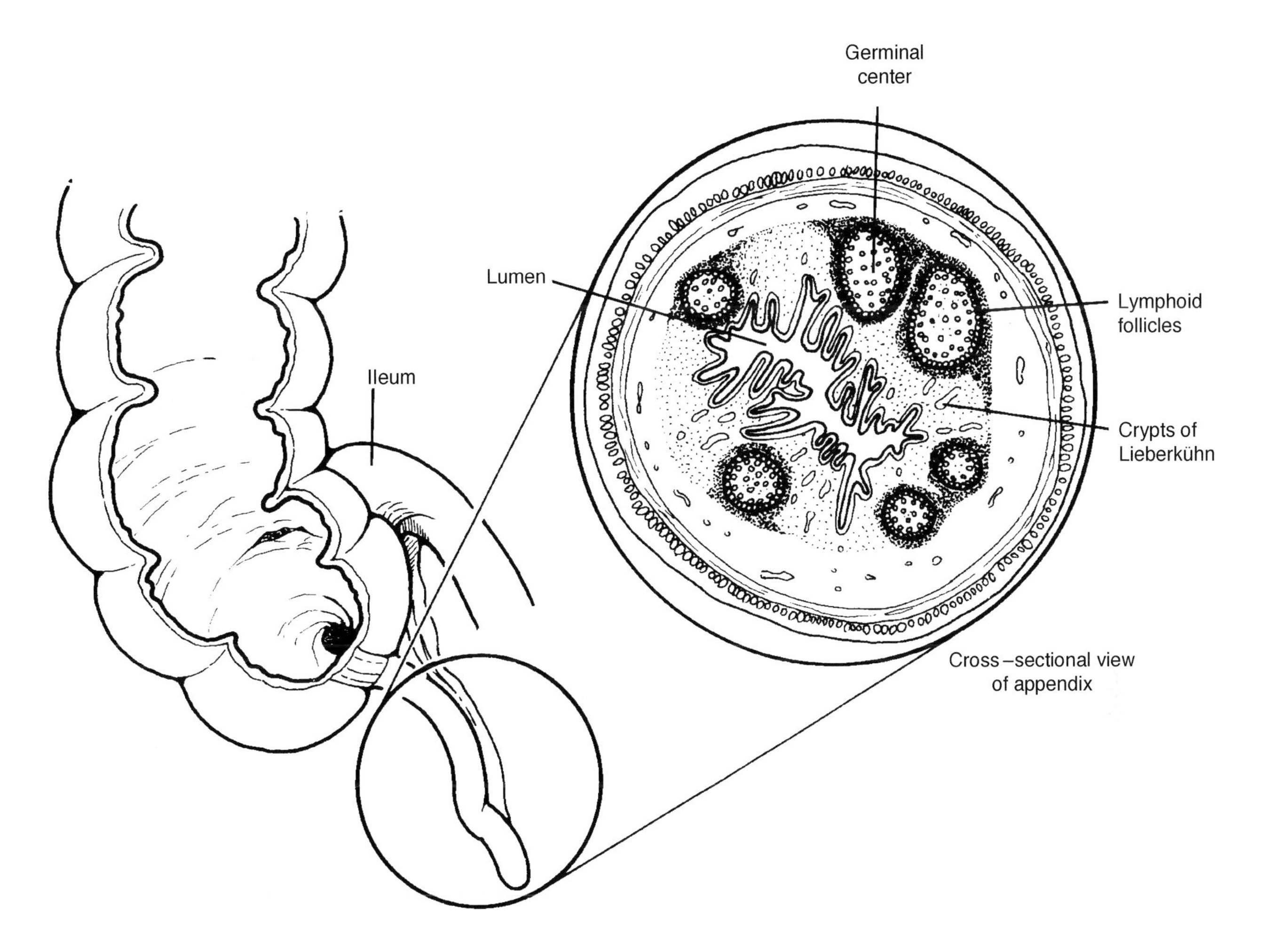

FIGURE 2.100 Appendix.

where immune responses to ingested antigens may be induced.

Vermiform appendix (Figure 2.100) is a lymphoid organ situated at the ileocecal junction of the gastrointestinal tract.

Waldeyer's ring (Figure 2.101) describes a circular arrangement of lymphoid tissue comprised of tonsils and adenoids encircling the pharynx–oral cavity junction.

Tonsils (Figure 2.102 and Figure 2.103) are lymphoid tissue masses at the intersection of the oral cavity and the pharynx, i.e., in the oropharynx. Tonsils contain mostly B lymphocytes and are classified as secondary lymphoid organs. There are several types of tonsils designated as palatine, flanked by the palatoglossal and palatopharyngeal arches; the pharyngeal, which are adenoids in the posterior pharynx; and the lingual, at the tongue's base.

Caecal tonsils are lymphoid aggregates containing germinal centers found in the gut wall in birds, specifically in the wall of the caecum.

Angiogenesis refers to the formation of new blood vessels under the influence of several protein factors produced by both natural and adaptive immune system cells. It may be associated with chronic inflammation. It is the formation of new vessels by sprouting of new capillaries from existing vessels — a fundamental phenomenon in diseases such as atherosclerosis, cancer, or diabetes and in physiological conditions such as the menstrual cycle and pregnancy. Angiogenesis is closely related to vasculogenesis, the formation of the vascular network from the stem cells in the embryo. In each case, the controlling mechanisms are the paracrine regulation of tyrosine kinase receptors, primarily on endothelial cells.

Angiogenesis factor: A macrophage-derived protein that facilitates neovascularization through stimulation of vascular endothelial cell growth. Among the five angiogenesis factors known, basic fibroblast growth factor may facilitate neovascularization in type IV delayed-hypersensitivity responses.

Angiogenic factors: Fibroblast growth factors (FGF) and vascular endothelial growth factors (VEGF) are endothelial cell mitogens. The key factors are the five VEGFs, the

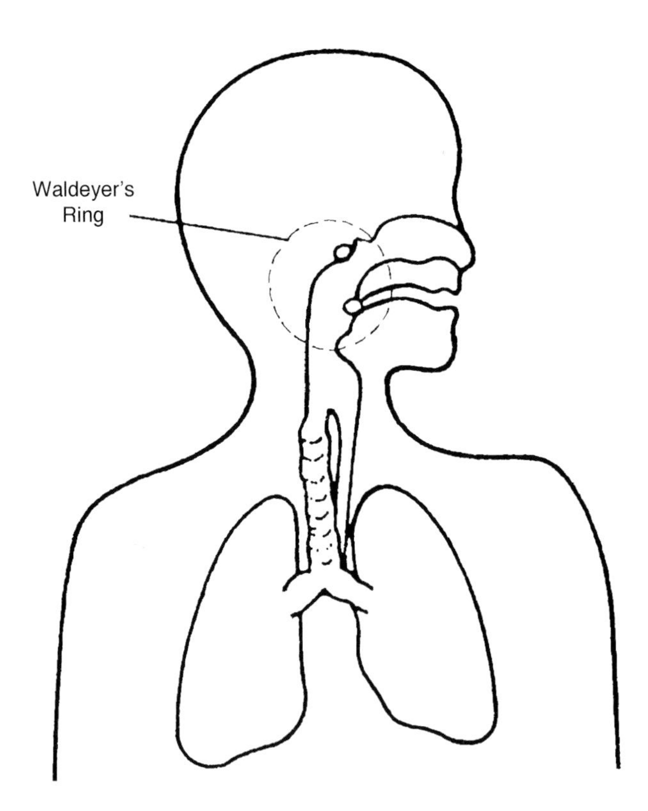

FIGURE 2.101 Waldeyer's ring.

VEGF receptors, VEGF-R1, -R2, and -R3, and placental growth factors (PIGF). In addition, several newer factors, such as the angiopoietins, ephrins, leptin, and chemokines, have been shown to be important in angiogenesis.

Angiogenin is a 14.4-kDa protein belonging to a family of proteins called the RNAse superfamily. Angiogenin acts as a potent inducer of blood vessel formation and is known to possess ribonuclease activity. If the ribonuclease activity is blocked, the angiogenic properties of angiogenin also appear to be inhibited. Angiogenin mRNA has been identified in a wide range of cell types.

Angiopoietins/Tie2: Two endothelial cell-specific tyrosine kinase receptors are Tie1 and Tie2. The ligands for Tie2 are angiopoietin-1 and angiopoietin-2 (Ang1 and Ang2). The ligand(s) for Tie1 has not been identified.

Interstitial fluid is the fluid present in the spaces between cells of an organ or tissue.

Transudation is the movement of electrolytes, fluid, and proteins of low molecular weight from the intravascular space to the extravascular space, as in inflammation.

Viscosity refers to the physical consistency of a fluid such as blood serum based on the size, shape, and conformation of its molecules. Molecular charge, sensitivity to temperature, and hydrostatic state affect viscosity.

Neuropeptides are substances that associate the nervous system with the inflammatory response. Neuropeptides serve as inflammatory mediators released from neurons in response to local tissue injury. They include substance P, vasoactive intestinal peptide, somatostatin, and calcitonin gene-related peptide. Multiple immunomodulatory activities have been attributed to these substances.

A **vasoactive intestinal peptide (VIP)** is a neuropeptide comprised of 28 residues that is a member of the secretin–glycogen group of molecules found in nerve fibers of blood vessels, in smooth muscle, and in upper respiratory tract glands. It activates adenylate cyclase and produces vasodilatation. It increases cardiac output, glycogenolysis, and bronchodilation, while preventing release of macromolecules from mucous-secreting glands. A deficiency of VIP aggravates bronchial asthma. Asthmatic patients usually do not have VIP. VIP may be increased in pancreatic islet G cell tumors.

Inflammation is a defense reaction of living tissue to injury. The literal meaning of the word is burning, and it originates from the cardinal symptoms of *rubor*, *calor*, *tumor*, and *dolor*, the Latin terms equivalent to redness, heat, swelling, and pain, respectively. It is beneficial for the host and essential for survival of the species, although in some cases the response is exaggerated and may be itself injurious.

Inflammation is the result of multiple interactions which have as a first objective localization of the process and removal of the irritant. This is followed by a period of repair. Inflammation is not necessarily of immunologic nature, although immunologic reactions are among the immediate causes inducing inflammation, and the immunologic status of the host determines the intensity of the inflammatory response. Inflammation tends to be less intense in infants whose immune system is not fully mature.

The causes of inflammation are numerous and include living microorganisms such as pathogenic bacteria and animal parasites which act mainly by the chemical poisons they produce and less by mechanical irritation; viruses which become offenders after they have multiplied in the host and cause cell damage; and fungi which grow at the surface of the skin but produce little or no inflammation in the dermis. Other causes of inflammation include physical agents such as trauma, thermal and radiant energy, and chemical agents which represent a large group of exogenous or endogenous causes which include immunologic offenders. See inflammatory response.

Late-phase reaction (LPR) is an inflammatory response that begins approximately 5 to 8 hours after exposure to antigen in IgE-mediated allergic diseases. In addition to inflammation, there is pruritus and minor cellular infiltration.

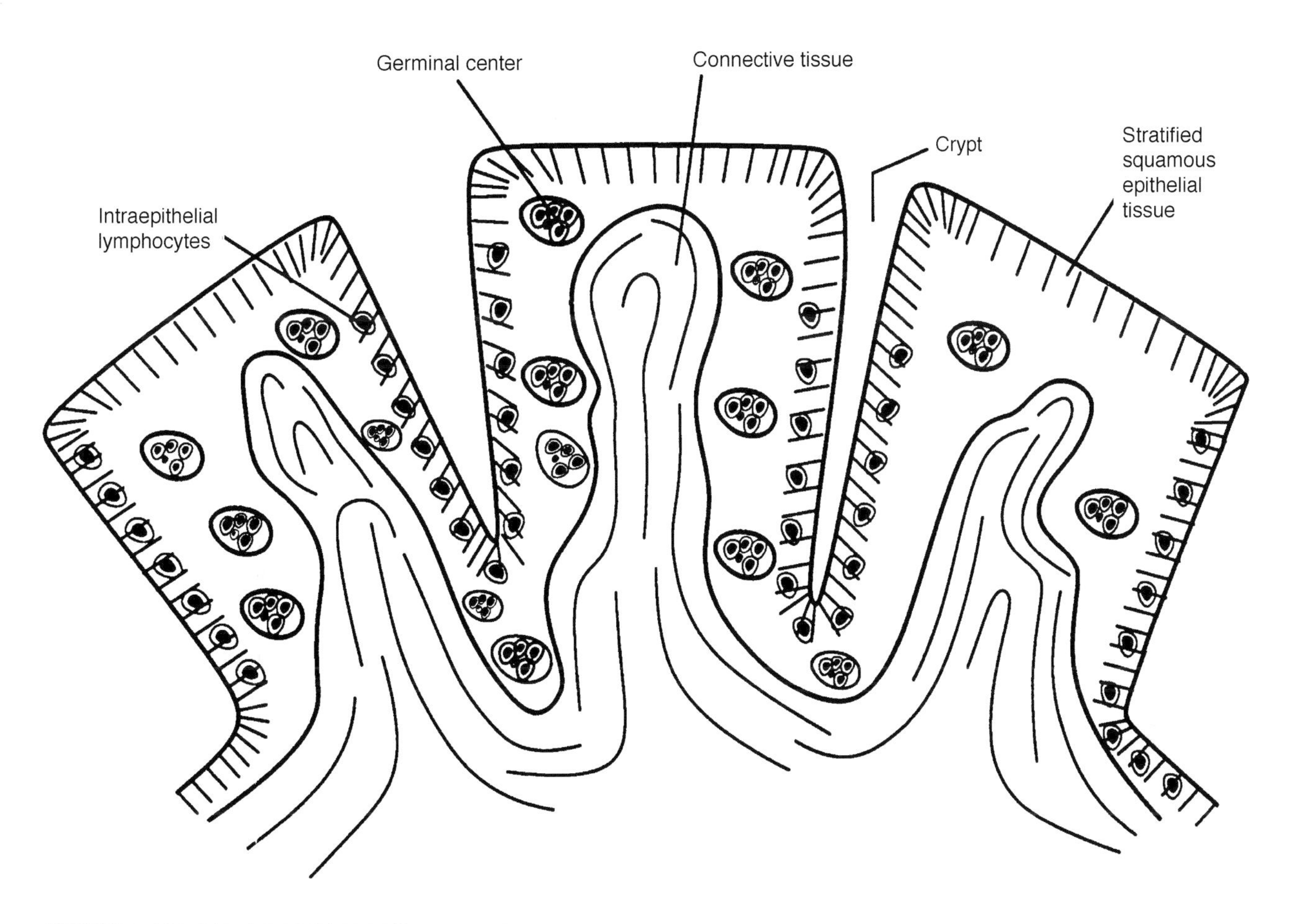

FIGURE 2.102 Histology of the tonsil.

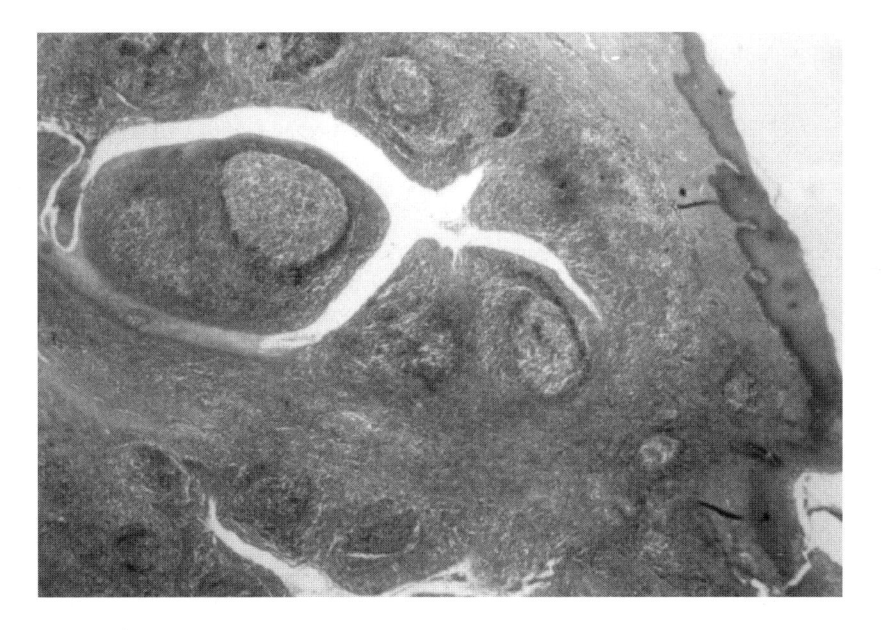

FIGURE 2.103 Tonsil.

Asthmatic patients may produce a delayed or secondary response following antigenic challenge, which involves the release of histamines from neutrophils. This induces secondary degranulation of mast cells and basophils, which stimulates bronchiole hyperreactivity. Whereas prostaglandin and then PGD_2 are produced in the primary response, they are not formed in the late-phase reaction. Cold air, ozone, viruses, or other irritants may induce the late-phase reaction. This condition may be treated with β adrenergic aerosols. It is resistant to treatment with antihistamine. Repeated episodes of late-phase reactivity can produce tissue injury.

LPR is the abbreviation for late-phase reaction.

The **wheal and flare reaction** is an immediate hypersensitivity, IgE-mediated (in man), reaction to an antigen. Application of antigen by a scratch test in a hypersensitive individual may be followed by erythema, which is the red flare, and edema, which is the wheal. Atopic subjects who have a hereditary component to their allergy experience the effects of histamine and other vasoactive amine released from mast cell granules following crosslinking of surface IgE molecules by antigen or allergen.

Acute inflammation is a reaction of sudden onset marked by the classic symptoms of pain, heat, redness, swelling, and loss of function. There is dilatation of arterioles, capillaries, and venules with increased permeability and blood flow. There is exudation of fluids, including plasma proteins, and migration of leukocytes into the inflammatory site. Inflammation is a localized protective response induced by injury or destruction of tissues. It is designed to destroy, dilute, or wall off both the offending agent and the injured tissue.

An **acute inflammatory response** represents an early defense mechanism to contain an infection and prevent its spread from the initial focus. When microbes multiply in host tissues, two principal defense mechanisms mounted against them are antibodies and leukocytes. The three major events in acute inflammation are: (1) dilation of capillaries to increase blood flow; (2) changes in the microvasculature structure, leading to escape of plasma proteins and leukocytes from the circulation; and (3) leukocyte emigration from the capillaries and accumulation at the site of injury. Widening of interendothelial cell junctions of venules or injury of endothelial cells facilitates the escape of plasma proteins from the vessels. Neutrophils attached to the endothelium through adhesion molecules escape the microvasculature and are attracted to sites of injury by chemotactic agents. This is followed by phagocytosis of microorganisms, which may lead to their intracellular destruction. Activated leukocytes may produce toxic metabolites and proteases that injure endothelium and tissues when they are released. Activation of the third complement component (C3) is also a critical step in inflammation.

Multiple chemical mediators of inflammation derived from either plasma or cells have been described. Mediators and plasma proteins such as complement are present as precursors that require activation to become biologically active. Mediators such as histamine and mast cells, derived from cells are present as precursors in intracellular granules. Following activation, these substances are secreted. Other mediators such as prostaglandins may be synthesized following stimulation. These mediators are quickly activated by enzymes or other substances such as antioxidants. A chemical mediator may also cause a target cell to release a secondary mediator with a similar or opposing action.

Besides histamine, other preformed chemical mediators in cells include serotonin and lysosomal enzymes. Those that are newly synthesized include prostaglandins, leukotrienes, platelet-activating factors, cytokines, and nitric oxide. Chemical mediators in plasma include complement fragments C3a and C5a and the C5b–g sequence. Three plasma-derived factors including kinins, complement, and clotting factors are involved in inflammation. Bradykinin is produced by activation of the kinin system. It induces arteriolar dilation and increased venule permeability through contraction of endothelial cells and extravascular smooth muscle. Activation of bradykinin precursors involves activated factor XII (Hageman factor) generated by its contact with injured tissues.

During clotting, fibrinopeptides produced during the conversion of fibrinogen to fibrin increase vascular permeability and are chemotactic for leukocytes. The fibrinolytic system participates in inflammation through the kinin system. Products produced during arachidonic acid metabolism also affect inflammation. These include prostaglandins and leukotrienes, which can mediate essentially every aspect of acute inflammation.

Inflammatory cells are cells of the blood and tissues that participate in acute and chronic inflammatory reactions. These include polymorphonuclear neutrophils, eosinophils, and macrophages.

An **inflammatory macrophage** is found in peritoneal exudate induced by thioglycolate broth or mineral oil injection into the peritoneal cavity of an experimental animal.

Contact system: A system of proteins in the plasma that engages in sequential interactions following contact with surfaces of particles that bear a negative charge, such as glass, or with substances such as lipopolysaccharides, collagen, etc. Bradykinin is produced through their sequential interaction. C1 inhibitor blocks the contact system. Anaphylactic shock, endotoxin shock, and inflammation are processes in which the contact system has a significant role.

Immune inflammation is the reaction to injury mediated by an adaptive immune response to antigen. Neutrophils and macrophages responding to T-cell cytokines may compromise inflammatory cellular infiltrate.

Exudate is composed of fluid-containing cells and cellular debris that have escaped from blood vessels and have been deposited in tissues or on tissue surfaces as a consequence of inflammation. In contrast to a transudate, an exudate is characterized by a high content of proteins, cells, or solid material derived from cells.

Exudation refers to the passage of blood cells and fluid containing serum proteins from the blood into the tissues during inflammation.

Extravasation is cell or fluid movement from the interior of blood vessels to the exterior.

An **inflammatory mediator** is a substance that participates in an inflammatory reaction.

***N*-formylmethionine** is an amino acid that initiates all bacterial proteins, but no mammalian proteins other than those produced within mitochondria. It alerts the innate immune system to infection of the host. Neutrophils express specific receptors for *N*-formylmethionine-containing peptides. These receptors mediate neutrophil activation.

Edema is tissue swelling as a result of fluid extravasation from the intravascular space.

Fibrosis is the formation of fibrous tissue, as in repair or replacement of parenchymatous elements. It is a process that leads to the development of a type of scar tissue at a site of chronic inflammation.

G proteins are proteins that bind guanosine triphosphate (GTP) and are converted to quanosine diphosphate (GDP) during cell signal tranduction. These heterotrimeric proteins are active when GTP occupies the guanine binding site and inactive when it anchors GDP. The two types of G proteins include the trimeric (α, β, γ) receptor-associated G protein and the small G proteins including Ras and Raf which function downstream of numerous transmembrane signaling events. Trimeric GTP-binding proteins are associated with parts of numerous cell surface receptors in the cytoplasm, including chemokine receptors.

The **G protein-coupled receptor family** refers to receptors for lipid inflammatory mediators, hormones, and chemokines which employ associated trimeric G proteins for intracellular signaling.

Guanine nucleotide exchange factors (GEFs) are proteins that can disengage bound GDP from small G proteins. This permits GTP to bind and activate the G protein.

GEF: See guanine nucleotide exchange factor

Fatty acids and immunity: Dietary lipids exert significant effects on antigen-specific and nonspecific immunity. These effects are related to total and fat-derived energy intake, synthesis of multiple eicosanoids, and alterations in cell membrane content. Eicosanoids are biologic mediators with multiple effects on immune cells. Their oversynthesis contributes to the development of chronic and acute inflammatory, autoimmune, atherosclerotic, and neoplastic diseases. Feeding a fish oil (*n*-3 PUFA) diet leads to recovery of splenic T cell blastogenesis, diminished secretion of PGEZ by splenic cells, and diminished splenic suppressor activity. PUFAs–immune system interactions are guided by the rate at which they are converted to eicosanoids. Dietary *n*-3 PUFAs diminish autoimmune, inflammatory, and atherosclerotic disease severity by diminishing the synthesis of *n*-6 PUFA-derived eicosanoids and cytokines. The increased eicosanoids found in shock and trauma can induce immunosuppression in humans. Supplementation of the diet with *n*-3 PUFA may protect from immunosuppression following trauma. Dietary fish oil supplements also improve joint tenderness in rheumatoid arthritis patients. Dietary supplements of *n*-3 PUFAs as linolenic acid or fish oil significantly inhibit the mixed lymphocyte reaction, which reflects graft survival. Linoleic acid is the only fatty acid needed to facilitate proliferation and maturation of immunoglobulin-secreting cells. B cell function may be suppressed by *n*-3 PUFAs by displacing linoleic acid and arachidonic acid. Cell membrane lipids play a critical role in both primary and secondary immune responses against an immunogenic challenge or to an infection.

An **acute-phase response (APR)** (Figure 2.104) is a nonspecific response by an individual stimulated by interleukin-1, interleukin-6, interferons, and tumor necrosis factor. C-reactive protein may show a striking rise within a few hours. Infection, inflammation, tissue injury, and, very infrequently, neoplasm may be associated with APR. The liver produces acute-phase proteins at an accelerated rate, and the endocrine system is affected with elevated gluconeogenesis, impaired thyroid function, and other changes. Immunologic and hematopoietic system changes include hypergammaglobulinemia and leukocytosis with a shift to the left. There is diminished formation of albumin, elevated ceruloplasmin, and diminished zinc and iron. Cellular elements may also be produced in addition to the acute phase proteins.

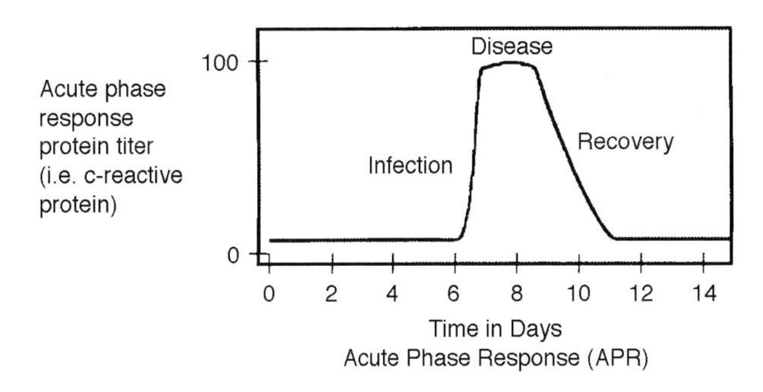

FIGURE 2.104 Acute phase response.

Acute phase proteins are plasma proteins that increase in concentration during inflammation. They are active during early phases of host defense against infection. They include mannose-binding protein (MBP), C-reactive protein (CRPJ), fibrinogen, selected complement components, and interferons.

Acute-phase reactants are serum proteins that increase during acute inflammation. These proteins, which migrate in the α-1 and α-2 electrophoretic regions, include α-1 antitrypsin, α-1 glycoprotein, amyloid A&P, antithrombin III, C-reactive protein, C1 esterase inhibitor, C3 complement, ceruloplasmin, fibrinogen, haptoglobin, orosomucoid, plasminogen, and transferrin. Most of the acute phase reactant proteins are synthesized in the liver and increase soon after infection as part of the systemic inflammatory response syndrome (SIRS). Inflammatory cytokines upregulate these molecules, including interleukin-6 (IL-6) and tumor necrosis factor (TNF). Acute phase reactants participate in the natural or innate response to microorganisms.

Pentraxins are a family of acute-phase plasma proteins comprised of five identical globular subunits. **Pentraxin family** refers to a category of glycoproteins in the blood that has a cyclic pentameric symmetric structure. It includes C-reactive protein, serum amyloid P, and complement C1.

Early induced responses are nonadaptive host responses induced by infectious agents early in infection. Their inductive phase differentiates them from innate immunity and their failure to involve clonal selection of antigen-specific lymphocytes distinguishes them from adaptive immunity.

Innate defense system is general or nonspecific and does not require previous exposure to the offending pathogen (or closely related organisms).

Innate immunity is a natural or native immunity that is present from birth and is designed to protect the host from injury or infection without previous contact with the infectious agent. It includes such factors as protection by the skin, mucous membranes, lysozyme in tears, stomach acid, and numerous other factors. Phagocytes, natural killer cells, complement, and cytokines represent key participants in natural innate immunity.

Native immunity is genetically determined host responsiveness that prevents healthy humans from becoming infected under normal circumstances by selected microorganisms that usually infect animals. This may be altered in the case of profound immunosuppression of humans, as in the case of acquired immune deficiency syndrome in which humans become infected with microorganisms such as *Mycobacterium avium intracellulare.*

Natural immunity: Innate immune mechanisms that do not depend upon previous exposure to an antigen. Among the numerous factors that contribute to natural resistance are the skin, mucous membranes, and other barriers to infection; lysozyme in tears and other antibacterial molecules; and natural killer (NK) cells.

Herd immunity: Nonspecific factors, as well as specific immunity, may have a significant role in resistance of a group (herd) of humans or other animals against an infectious disease agent. Elimination of reservoirs of the disease agent may be as important as specific immunity in diminishing disease incidence among individuals. "Herd immunity" also means that an epidemic will not follow infection of a single member of the herd or group if other members are immune to that particular infectious agent.

Lactoferrin is a protein that combines with iron and competes with microorganisms for it. This represents a nonantibody humoral substance that contributes to the body's natural defenses against infection. It is present in polymorphonuclear neutrophil granules as well as in milk. By combining with iron molecules, it deprives bacterial cells of this needed substance.

Lactoperoxidase is an enzyme present in milk and saliva that may be inhibitory to a number of microorganisms and serves as a nonantibody humoral substance that contributes to nonspecific immunity. Its mechanism of action resembles that of myeloperoxidase.

Nonsterile immunity: See premunition.

Acquired immunity is protective resistance against an infectious agent generated as a consequence of infection with a specific microorganism or as a result of deliberate immunization. See also primary immune response and secondary immune response.

Cellular immunity: See cell-mediated immunity.

Cellular immunology is the study of cells involved in immune phenomena.

Active immunity is protection attained as a consequence of clinical or subclinical infection or deliberate immunization with an infectious agent or its products. It is a type of adaptive immunity in which lymphocytes are activated in response to a foreign antigen to which they have been exposed. Compare with passive immunity.

Artificially acquired immunity is the use of deliberate active or passive immunization or vaccination to elicit protective immunity as opposed to immunity which results from unplanned and coincidental exposure to antigenic materials, including microorganisms in the environment.

Specific immunity refers to an immune state in which antibody or specifically sensitized or primed lymphocytes recognize an antigen and react with it. By contrast, immunologically competent cells may interact with antigen to produce specific immunosuppression termed *immunologic tolerance*.

Humoral immune response is a host defense mediated by antibody molecules found in the plasma, lymph, and tissue fluids. This type of immunity protects against extracellular bacteria in foreign micromolecules. Humoral immunity may be transferred passively with antibodies or serum containing antibodies.

Humoral antibody is found in the blood plasma, lymph, and other body fluids. Humoral antibody, together with complement, mediates humoral immunity which is based upon soluble effector molecules.

Protective immunity refers to both natural, nonspecific immune mechanisms and actively acquired specific immunity that result in the defense of a host against a particular pathogenic microorganism. Protective immunity may be induced either by active immunization with a vaccine prepared from antigens of a pathogenic microorganism or by experiencing either a subclinical or clinical infection with the pathogenic microorganism.

Protective antigens are the antigenic determinants of a pathogenic microorganism that stimulate an immune response that can protect a host against an infection by that microorganism. Thus, these particular antigenic specificities can be used for prophylactic immunization in vaccines to immunize susceptible hosts against possible future infections.

Functional immunity: See protective immunity.

Preemptive immunity refers to resistance shown by virus-infected cells to superinfection with a different virus.

Artificial passive immunity refers to the transfer of immunoglobulins from an immune individual to a nonimmune, susceptible recipient.

Passive immunity is a form of acquired immunity induced by the transfer of immune serum containing specific antibodies or of sensitized lymphoid cells from an immune to a nonimmune recipient host. Examples of passive immunity are the transfer of IgG antibodies across the placenta from mother to fetus or the ingestion of colostrum-containing antibodies by an infant. Antitoxins generated to protect against diphtheria or tetanus toxins represent a second example of passive humoral immunity, as used in the past. Specifically sensitized lymphoid cells transferred from an immune to a previously nonimmune recipient is termed *adoptive immunization*. The passive transfer of antibodies in immune serum can be used for the temporary protection of individuals exposed to certain infectious disease agents. They may be injected with hyperimmune globulin.

Gravity and immunity: Space flight has been associated with the development of neutrophilia, slight T cell lymphopenia, and diminished blastogenic responsiveness of T cells in postflight blood samples. Eosinophilia has been noted. Lymphopenia was marked by a decreased numbers of T cells and natural killer (NK) cells. Changes have also been observed in postflight concentrations of immunoglobulins, complement components, lysozyme, interferon, and α_2-macroglobulin. There is a modest depression in cell-mediated immunity after both short and long space flights. Human NK cells diminished and revealed decreased cytotoxic activity after both long and short space flights. Delayed-type dermal hypersensitivity (DTH) reactions decreased or even disappeared during prolonged residence in space. IgA and IgM rose but IgG remained constant during long space flights. No defects in humoral immunity have been noted. IL-2 and interferon synthesis by lymphocytes was significantly decreased in both humans and rodents after long flights.

Exercise and immunity: Exercise leads to altered distribution and function of immunocompetent cells. This is related in part to changes in hormone release, blood flow distribution, and other factors that affect immune system function. Vigorous exercise leads to an immediate leucocytosis. Exercise-induced immunosuppression or immunoenhancement may affect disease risk. Absolute numbers of $CD3^+$, $CD8^+$, and $CD16^+/C56^+$ (natural killer) cells increase after exercise. B lymphocytes also rise with acute exercise. But these increases return to preexercise levels within a few hours following its cessation. The $CD4^+/CD8^+$ lymphocyte helper/suppressor ratio diminishes soon after exercise, attributable to increased CD8 counts. Acute exercise is followed by increase in the concentration and *in vitro* cytolytic activity of $CD16^+/CD56^+$ cells, but exhaustive exercise leads to a decrease in the cytolytic activity of NK cells. Exercise also leads to cytokine-m lease, such as the elevation of IL-6 but not of IL-1. After exercise, TNF-a and GMCSF are essentially undetectable. IL-2 levels also decrease following exercise. Increased urinary concentrations of IFN-y, TNF-a, and IL-6 have been demonstrated following long distance running. Exercise has little functional impact on immune effector cells. The response of lymphocytes to T cell mitogen such as PHA and con A is diminished immediately after exercise but returns to normal within 24 hours. The proliferative response to B cell mitogens such as LPS is mixed. T and B cell mitogens such as PWM increase following exercise. Antibody synthesis is not much affected by limited exercise. IgG, IgM, and IgA levels as well as the ability to synthesize antibody to tetanus toxoid antigen are not compromised by exercise. Exercise prior to exposure to infection diminishes morbidity or mortality, yet exercise during an infection produces the reverse effect. Prolonged intense exercise is followed by some immunosuppression. The immune parameters altered by physical exercise are related to the neuroendocrine changes such as those that occur in response to physical or psychological stress.

Nonspecific immunity: Refers to mechanisms such as phagocytosis that nonspecifically remove invading microorganisms, as well as the action of chemical and physical barriers to infection such as acid in the stomach and the skin. Other nonspecific protective factors include lysozyme, β lysin, and interferon. Nonspecific or natural immunity does not depend on immunologic memory. Natural killer cells represent an important part of the natural immune cell system. Phagocytosis of invading microorganisms by polymorphonuclear neutrophils and monocytes represents another important aspect of nonspecific immunity.

Nutrition and immunity: The most frequent cause of immunodeficiency worldwide is malnutrition. Protein-energy malnutrition has an adverse effect on immunity, increases the frequency of opportunistic infections, and leads to lymphoid tissue atrophy with reduction in the size of the thymus and depletion of lymphoid cells in thymus dependent areas of lymph nodes, as well as a loss of lymphoid cells around small vessels in the spleen. It leads to delayed hypersensitivity responses in the skin to both new and recall antigens. The helper/suppressor ratio is significantly decreased, and lymphocyte proliferation and synthesis of DNA are diminished. Serum antibody responses are usually unaffected in protein-energy malnutrition. Phagocytosis is affected as a consequence of decreased complement, component C3, factor B, and total hemolytic activity; ingestion by phagocytes is intact and metabolic destruction of microorganisms is decreased as is synthesis of various cytokines including interleukin-2 and interferon-γ. Deficiency of pyridoxin, folic acid, vitamin A, vitamin C, and vitamin E lead to impaired cell-mediated immunity and diminished antibody responses. Vitamin B_6 deficiency leads to decreased lymphocyte stimulation responses to mitogens. A moderate increase in vitamin A enhances immune responses. Zinc deficiency leads to lymphoid atrophy, decreased cutaneous delayed hypersensitivity responses and allograft rejection, and diminished thymic hormone activity. Iron deficiency is the most common nutritional problem worldwide. It leads to impaired lymphocyte proliferation in response to mitogens and antigens and a low response to tetanus toxoid and herpes simplex antigens. Copper-deficient animals have fewer antibody forming cells compared to healthy controls. Dietary deficiencies of selected amino acids diminish antibody responses, but in other states an amino acid imbalance may enhance selected antibody responses, perhaps reflecting alterations in suppressor cells.

Vitamin A and immunity: A deficiency of vitamin A compromises acquired, adaptive, antigen-specific immunity. The deficiency has been linked to atrophy of thymus, spleen, lymph nodes, and Peyer's patches pointing to major alterations of immune effector cell mechanisms. Vitamin A deficiency is also associated with impaired ability to form an antibody response to T cell-dependent antigens such as tetanus toxoid, proteins, and viral infections. It is also linked to decreased antibody responsiveness to T cell-independent antigens such as pneumococcal polysaccharide and meningococcal polysaccharide. Vitamin A deficiency also compromises natural innate immunity since it is necessary for maintenance of mucosal surfaces, the first line of defense against infection. Immune effector cells that mediate nonspecific immunity include polymorphonuclear cells, macrophages, and natural killer (NK) cells. Neutrophil phagocytosis is diminished by Vitamin A deficiency, and viral infections are more severe because of diminished cytolytic activity by NK cells.

Immunologically, **vitamin A** may serve as an adjuvant to elevate antibody responses to soluble protein antigens in mice. The adjuvant effect is produced whether vitamin A is given orally or parenterally.

Vitamin B and immunity: B complex vitamins differ greatly in chemical structure and biological actions. Vitamin B_6 deficiency induces marked changes in immune function, especially the thymus. Thymic hormone activity is diminished and lymphopenia occurs. Vitamin B_6 deficiency suppresses delayed cutaneous hypersensitivity responses, primary and secondary T-cell mediator cytotoxicity, and skin graft rejection. It also impairs humoral immunity. The number of circulating lymphocytes is decreased. Folate and vitamin B_{12} deficiencies are linked to diminished host resistance and impaired lymphocyte function. Pantothenic acid deficiency suppresses humoral antibody responses to antigens. Thiamin, biotin, and riboflavin deficiencies have a moderate interfering action on immune function. Riboflavin deficiency diminishes humoral antibody formation in response to antigen. Intake of micronutrients, including the B complex vitamins two to three times higher than the U.S. RDA recommended dose, helps to maintain optimal immune function in healthy elderly adults.

Vitamin C and immunity: Ascorbic acid (vitamin C) is needed for the cells, tissues, and organs of the body to function properly. It is an antioxidant and a cofactor in many hydroxylating reactions. The immune system is sensitive to the level of vitamin C intake. Leukocytes have high concentrations of ascorbate that is used rapidly during infection and phagocytosis, which points to vitamin C's role in immunity. Vitamin C facilitates neutrophil chemotaxis and migration, induces interferon synthesis, maintains mucous membrane integrity, and has a role in the expression of delay type hypersensitivity. High dose vitamin C supplementation is believed to increase T and B lymphocyte proliferation. It diminishes nonspecific extracellular free radical injury and autotoxicity after the oxidative burst activity of stimulated neutrophils. It further enhances immune function indirectly by maintaining optimal levels of vitamin E.

Vitamin D and immunity: Calcitriol, the hormonal form of vitamin D, has a significant regulatory role in cell differentiation and proliferation of the immune system. It mediates its action through specific intracellular Vitamin D_3 receptors (VDRs). Among calcitriol's numerous effects on the immune system are the inhibition of cytokine release from monocytes; the prolongation of skin allograft survival in mice; the inhibition of autoimmune encephalomyelitis and thyroiditis in mice; the potentiation of murine primary immune responses; the restoration of defective macrophage and lymphocyte functions in vitamin D deficient rickets patients; and restoration of lymphocyte proliferation IL-2 synthesis in human dialysis patients, among many others.

Vitamin E and immunity: Vitamin E is required by the immune system. It is a major antioxidant that protects cell membranes from free radical attack. It is effective in preventing biological injury by immunoenhancement. Vitamin E in high doses diminishes $CD8^+$ T cells and increases $CD4^+$:$CD8^+$ T cell ratio; increases total lymphocyte count and stimulates cytotoxic cells, natural killer cells, phagocytosis by macrophages, and mitogen responsiveness. The immunostimulatory action of vitamin E renders it useful for therapeutic enhancement of the immune response in patients. Vitamin E's effect on the immune system depends on its interaction with other antioxidant and preoxidant nutrients, polyunsaturated fatty acids, and other factors that affect the immune response, including age and stress. Vitamin E stimulation of immunity is particularly important in the elderly in whom infectious disease and tumor incidence increase with age. Vitamin E facilitates host defense by inhibiting increases in tissue prostaglandin synthesis from arachidonic acid during infection. *In vitro*, vitamin E has been shown to stimulate IL-2 and interferon-γ by mitogen-stimulated lymphocytes. Vitamin E prevents lipid peroxidation of cell membranes, which may be a mechanism to enhance immune responses and phagocytosis.

Zinc is an element of great significance to the immune system as well as to other nonantigen-specific host defenses. The interleukins of the immune system play a role in zinc distribution and metabolism in the body. As a constituent of the active site in multiple metalloenzymes, zinc is critical in chemical prothesis within lymphocytes and leukocytes. Its role in the reproduction of cells is of critical significance for immunological reactions since nucleic acid synthesis depends, in part, on zinc metalloenzymes. Zinc facilitates cell membrane modification and stabilization. Zinc deficiency is associated with reversible dysfunction of T lymphocytes in man. It causes atrophy of the thymus and other lymphoid organs and is associated with diminished numbers of lymphocytes in the T cell areas of lymphoid tissues. There is also lymphopenia. Anergy develops in zinc-deficient patients; this signifies disordered cell-mediated immunity as a consequence of the deficiency. There is also a decrease in the synthesis of antibodies to T cell-dependent antigens. In zinc deficiency, there is a selective decline in the number of $CD4^+$ helper T cells and a strikingly decreased proliferative response to phytomitogens including PHA. Thymic hormonal function requires zinc. Deficiency of this element is also associated with decreased formation of monocytes and macrophages and with altered chemotaxis of granulocytes. Wound healing is impaired in these individuals who also show greatly increased susceptibility to infectious diseases, which are especially severe when they do develop.

Zinc and immunity: Zinc is found in all tissues and fluids of the body. It is mainly an intracellular ion with over 95% found within cells; 60 to 80% of the cellular zinc is located in the cytosol. About 85% of the total body zinc is present in the skeletal muscle and bone. Zinc is absorbed all along the small intestine but is taken up primarily in the jejunum. Its function falls into three categories: catalytic, structural, and regulatory. Catalytic roles are present in all six classes of enzymes. Over 50 different enzymes require zinc for normal activity, e.g., zinc metalloenzymes, in which its role is usually structural. Regulation of gene expression is also a significant biochemical function of zinc. Classic symptoms of zinc deficiency in experimental animals include retarded growth, depressed immune function, skin lesions, depressed appetite, skeletal abnormalities, and impaired reproduction. In humans, zinc deficiency causes severe growth retardation and sexual immaturity. Acute zinc toxicity may occur with intakes in the range of 1 to 2 g, which leads to gastric distress, dizziness, and nausea. High chronic intakes from supplements (150 to 300 mg/d) may impair immune function and reduce concentrations of high-density lipoprotein cholesterol. High intakes of zinc have been used to treat Wilson's disease, a copper accumulation disorder. Zinc is believed to induce synthesis of metallothionein in the intestinal mucosal cells. Though relatively nontoxic, chronic use of zinc supplements may induce nutrient imbalances and physiological effects.

Diagnosis of zinc deficiency is difficult because of the lack of a sensitive specific indicator of zinc status. Stress, infection, food intake, short-term fasting, and the individual's hormonal status all influence plasma zinc levels. Zinc deficiency is best assessed by using a combination of dietary, static, and functional signs of depletion. Red meat and shellfish constitute the best food sources of zinc.

Iron and immunity: Iron has two effects on immune function: (1) Micronutrients, such as iron, are redistributed in the body during infection and (2) dietary iron deficiency produces iron deficiency anemia and decreases immunocompetence. Both iron deficiency and iron excess compromise the immune system.

A **eukaryote** is a cell or organism with a real nucleus containing chromosomes encircled by a nuclear membrane.

Genome: All genetic information that is contained in a cell or in a gamete. It is the total genetic material found in the haploid set of chromosomes.

Germ line refers to unaltered genetic material that is transmitted from one generation to the next through gametes. An individual's germ line genes are those present in the zygote from which it arose. It refers to unrearranged genes rather than those rearranged for the production of immunoglobulin or T cell receptor (TCR) molecules.

Genomic DNA is found in the chromosomes. See DNA.

DNA library refers to a gene library or clone library comprised of multiple nucleotide sequences that are representative of all sections of the DNA in a particular genome. It is a random assemblage of DNA fragments from one organism, linked to vectors and cloned in an appropriate host. This prevents any individual sequence from being systematically excluded. Adjacent clones will overlap, and cloning large fragments helps to ensure that the library will contain all sequences. The DNA to be investigated is reduced to fragments by enzymatic or mechanical treatment, and the fragments are linked to appropriate vectors such as plasmids or viruses. The altered vectors are then introduced into host cells. This is followed by cloning. Transcribed DNA fragments termed *exons* and nontranscribed DNA fragments termed *introns* or spacers are part of the gene library. A probe may be used to screen a gene library to locate specific DNA sequences.

The term **gene bank** is a synonym for DNA library.

Hairpin loop describes the looped structure of hairpin DNA.

Kilobase (kb): 1000 DNA or RNA base pairs.

That segment of a strand of DNA responsible for coding is known as an **exon.** This continuous DNA sequence in a gene encodes the amino acid sequence of the gene product. Exons are buttressed on both ends by introns, which are noncoding regions of DNA. The coding sequence is transcribed in mature mRNA and subsequently translated into proteins. Exons produce folding regions, functional regions, domains, and subdomains. Introns, which are junk DNA, are spliced out. They constitute the turns or edges of secondary structure.

An **intron** is a structural gene segment that is not transcribed into RNA. Introns have no known function and are believed to be derived from "junk" DNA.

Intervening sequence: See intron.

An **inverted repeat** is a complementary sequence segments on a single strand of DNA. It is a palindrome when an inverted repeat's halves are placed side by side.

A **palindrome** is a DNA segment with dyad symmetrical structure. When read from 5′ to the 3′, it reveals an equivalent sequence whether read from forward or backward or from left or right. The base sequence in one strand is identical to the sequence in the second strand.

TATA box: An oligonucleotide sequence comprised of thymidine-adenine-thymidine-adenine found in numerous genes that are transcribed often or rapidly.

DNA polymerase is an enzyme that catalyzes DNA synthesis from deoxyribonucleotide triphosphate by employing a template of either single- or double-stranded DNA. This is termed DNA-dependent (direct) DNA polymerase in contrast to RNA-dependent (direct) DNA polymerase which employs an RNA template for DNA synthesis.

DNA polymerase I is DNA-dependent DNA polymerase whose principal function is in DNA repair and synthesis. It catalyzes DNA synthesis in the 5′ to 3′ sense. It also has a proofreading function (3′ → 5′ exonuclease) and a 5′ → 3′ exonuclease activity.

DNA polymerase II is DNA-dependent DNA polymerase in prokaryotes. It catalyzes DNA synthesis in the 5′ to 3′ sense, has a proofreading function (3′ → 5′ exonuclease), and is thought to play a role in DNA repair.

DNA polymerase III is DNA-dependent DNA polymerase in prokaryotes that catalyzes DNA synthesis in the 5′ to 3′ sense. It is the principal synthetic enzyme in DNA replication. It has a proofreading function (3′ → 5′ exonuclease) and 5′ → 3′ exonuclease activity.

DNA-dependent RNA polymerase is an enzyme that participates in DNA transcription. With DNA as a template, it catalyzes RNA synthesis from the ribonucleoside-5′-triphosphates.

DNA nucleotidylexotransferase (terminal deoxynucleotidyl-transferase [TdT]): DNA polymerase that randomly catalyzes deoxynucleotide addition to the 3′-OH end of a DNA strand in the absence of a template. It can also be employed to add homopolymer tails. Immature T and B lymphocytes contain TdT. The thymus is rich in TdT, which is also present in the bone marrow. TdT inserts a few nucleotides in T cell receptor gene and immunoglobulin gene segments at the V-D, D-J, and V-J junctions. This enhances sequence diversity.

DNA ligase is an enzyme that joins DNA strands during repair and replication. It serves as a catalyst in phosphodiester, binding between the 3′-OH and the 5′-PO_4 of the phosphate backbone of DNA.

Deoxyribonuclease is an endonuclease that catalyzes DNA hydrolysis.

Deoxyribonuclease I is an enzyme that catalyzes DNA hydrolysis to a mono- and oligonucleotides mixture comprised of fragments terminating in a 5′-phosphoryl nucleotide.

Deoxyribonuclease II is an enzyme that catalyzes DNA hydrolysis to a mono- and oligonucleotides mixture comprised of fragments terminating in a 3′-phosphoryl nucleotide.

Gene conversion is the recombination between two homologous genes in which a local segment of one gene is replaced by homologous segment of a second gene. In avian species and lagomorphs, gene conversion facilitates immunoglobulin receptor diversity principally through homologous inactive V gene segments exchanging short sequences with an active, rearranged variable-region gene.

Gene mapping refers to gene localization or gene order. Gene localization can be in relationship to other genes or to a chromosomal band. The term may also refer to the ordering of gene segments.

Pseudogene refers to a sequence of DNA that is similar to a sequence of a true gene but does not encode a protein due to defects that inhibit gene expression. Thus, pseudogenes represent nonusable or junk DNA. They may result from duplicated genes and have several defects as mutations accumulate.

Genetic code refers to the codons, i.e., nucleotide triplets, correlating with amino acid residues in protein synthesis. The nucleotide linear sequence in mRNA is translated into the amino acid residue sequence.

An **open reading frame (ORF)** is a length of RNA or DNA that encodes a protein and may signal the identification of a protein not described previously. An ORF begins with a start codon and does not contain a termination codon, but it ends at a stop codon.

An **unidentified reading frame (URF)** is an open reading frame (ORF) that does not correlate with a defined protein.

Nonproductive rearrangement refers to rearrangements in which gene segments are joined out of phase leading to failure to preserve the triplet-reading frame for translation. Nonproductive rearrangements of gene segments encoding T and B cell receptors lead to failure to encode a protein because the coding sequences are in wrong translational reading frame.

Complementarity is a genetic term that indicates the requirement for more than one gene to express a trait.

Homologous recombination describes the exchange of DNA fragments between two DNA molecules or chromatids of paired chromosomes (during crossover) at the site of identical nucleotide sequences.

Nuclear matrix proteins (NMPs) are substances that organize the nuclear chromatin. They are associated with DNA replication and RNA synthesis in hormone receptor binding. Antibodies against NMPs react with nuclear mitotic apparatus protein (NUMA). Thus, much of the nuclear matrix is devoted to formation of the mitotic apparatus (MA). NMPs participate in the cellular events that result in programmed cell death or apoptosis.

A **genotype** is an organism's genetic makeup. It constitutes the combined genetic constituents inherited from both parents and refers to the alleles present at one or more specific loci. When referring to microinjected transgenics, mice can be homozygous (transgene is present on only one chromosome in a pair), hemizygous (transgene is present on only one chromosome in a pair), or wild type (transgene is not present on either chromosome in a pair). When referring to knockout mice, the correct term is *homozygous* (sometimes called null mice).

Dominant phenotype: Trait manifested in an individual who is heterozygous at the gene locus of interest.

A **karyotype** is the number and shape of chromosomes within a cell. The karyotype may be characteristic for a particular species.

Haploid refers to a single copy of each autosome and one sex chromosome. This constitutes one set of unpaired chromosomes in a nucleus. The adjective may also refer to a cell containing this number of chromosomes.

Diploid is a descriptor to indicate dual copies of each autosome and two sex chromosomes in a cell nucleus.

The diploid cell has twice the number of chromosomes in a haploid cell.

An **allele** is one of the alternative forms of a gene at a single locus on a chromosome that encodes the phenotypic features of a certain inherited characteristic. The presence of multiple alleles such as at the MHC locus leads to polymorphism.

Pseudoalleles are genes that are closely linked but distinct. They have functional similarity and act as alleles in complementation investigations. However, crossover studies may separate them.

Highly polymorphic describes genes with many alleles and for which most subjects in a population are heterzygotes.

Allelic dropout: In the amplification of a DNA segment by the polymerase chain reaction, one of the alleles may not be amplified, leading to the false impression that the allele is absent. The phenomenon takes place at 82 to 90°C in the thermocycler.

Heterozygosity refers to the presence of different alleles at one or more loci on homologous chromosomes.

Heterozygous is the descriptor for individuals possessing two different alleles of a particular gene.

GATA-2 gene is a gene that encodes a transcription factor that is requisite for the development of lymphoid, erythroid, and myeloid hematopoietic cell lineages.

Genetic polymorphism refers to variation in a population attributable to the existence of two or more alleles of a gene.

A **codon** is a three-adjacent nucleotide sequence mRNA that acts as a coding unit for a specific amino acid during protein synthesis. The codon controls which amino acid is incorporated into the protein molecule at a certain position in the polypeptide chain. Out of 64 codons, 61 encode amino acids and 3 act as termination codons.

Consensus sequence refers to the typical nucleic acid or protein sequence where a nucleotide or amino acid residue present at each position is that found most often during comparison of numerous similar sequences in a specific molecular region.

Ras: One of a group of 21-kDa guanine nucleotide-binding proteins with intrinsic GTPase activity that participates in numerous different signal transduction pathways in a variety of cells. Ras gene mutations may be associated with tumor transformation. Ras is attracted to the plasma membrane by tyrosine phosphorylated adaptor proteins during T lymphocyte activation, where GDP-GTP exchange factors activated. GTP-Ras then activates the MAP kinase cascade that results in *fos* gene expression and assembly of AP-1 transcription factor. See small G proteins.

Polygenic is an adjective that refers to several nonallelic genes encoding the same or similar proteins. It may also signify any trait attributable to inheritance of more than one gene.

Recombinant DNA is the physical union of two or more strands of available DNA to form another DNA strand. The term describes the exchange of DNA during meiosis, mitosis, or gene conversion. It may also refer to DNA strands produced *in vitro*.

Recombinant DNA technology is the technique of isolating genes from one organism and purifying and reproducing them in another organism. This is often accomplished through ligation of genomic or cDNA into a plasmid or viral vector where replication of DNA takes place.

Gene cloning refers to the use of recombinant DNA technology to replicate genes or their fragments.

Restriction endonuclease are bacterial products that identify and combine with a short sequence of DNA. The enzyme acts as molecular scissors by cleaving the DNA at either the recognition site or at another location. Restriction endonucleases catalyze degradation of foreign DNA. They recognize precise base sequences of DNA and cut it into relatively few fragments termed *restriction fragments*. There are three major types of restriction endonucleases. Type I enzymes identify specific base sequences but cut the DNA elsewhere, i.e., approximately 1000 bp from the recognition site. Type II enzymes identify specific base sequences and cut the DNA either within or adjacent to these sequences. Type III endonucleases identify specific base sequences and cut the DNA approximately 25 bp from the recognition site.

An **isoschizomer** is one of several restriction endonucleases derived from different organisms that identify the same DNA base sequence for cleavage but do not always cleave DNA at the same location in the sequence. Target sequence methylation affects the action of isoschizomers which are valuable in investigations of DNA methylation.

A **restriction map** is a diagram of either a linear or circular molecule of DNA that indicates the points where one or more restriction enzymes would cleave the DNA. DNA is first digested with restriction endonucleases, which split the DNA into fragments that can be separated by gel electrophoresis. Size determination is accomplished by comparison with DNA fragments of known size.

RFLP (restriction fragment length polymorphism): Denotes local DNA sequence variations of man or other animals that may be revealed by the use of restriction endonucleases. These enzymes cut double-stranded DNA at points where they recognize a very specific oligonucleotide sequence, resulting in DNA fragments of different lengths that are unique to each individual animal or person. The fragments of different sizes are separated by electrophoresis. The technique is useful for a variety of purposes such as identifying genes associated with neurologic diseases (e.g., myotonic dystrophy) which are inherited as autosomal dominant genes or in documenting chimerism. The fragments may also be used as genetic markers to help identify the inheritance patterns of particular genes.

DNA fingerprinting is a method to demonstrate short, tandem-repeated, highly specific genomic sequences known as minisatellites. There is only a 1-in-30-billion probability that two persons would have the identical DNA fingerprint. It has greater specificity than RFLP analysis. Each individual has a different number of repeats. The insert-free wild-type M13 bacteriophage identifies the hypervariable minisatellites. The sequence of DNA that identifies the differences is confined to two clusters of 15-bp repeats in the protein III gene of the bacteriophage. The specificity of this probe, known as the Jeffries probe, renders it applicable to human genome mapping, parentage testing, and forensic science.

RNA may also be split into fragments by enzymatic digestion followed by electrophoresis. A characteristic pattern for that molecule is produced, which aids in identifying it.

DNA laddering: Endonucleases are activated during apoptosis. Activated endonucleases nick genomic DNA at internucleosomal sites to produce DNA fragments. Not all cell types generate uniformly nicked genomic DNA during apoptosis. Following gel electrophoresis, DNA fragments migrate in a pattern resembling a ladder with individual bands differing by approximately 200 bp.

TdT is an abbreviation for terminal deoxynucleotidyl transferase.

Terminal transferase: See DNA nucleotidyl exotransferase.

An **allophenic mouse** is a tetraparental, chimeric mouse whose genetic makeup is derived from four separate parents. It is produced by the association of two early eight cell embryos that differ genetically. A single blastocyst forms, it is placed in a pseudopregnant female uterus, and is permitted to develop to term. Tetraparental mice are widely used in immunological research.

3 Antigens and Immunogens

An infectious agent, whether a bacterium, fungus, virus, or parasite, contains an abundance of substances capable of inducing an immune response. These are called immunogens or antigens. Specifically, an immunogen is a substance capable of stimulating B cell, T cell, or both limbs of the immune response. An antigen is a substance that reacts with the products of an immune response stimulated by a specific immunogen, including both antibodies and/or T lymphocyte receptors. The "traditional" definition of antigen more correctly refers to an immunogen. A complete antigen is one that both induces an immune response and reacts with the products of it, whereas an incomplete antigen or hapten is unable to induce an immune response alone but is able to react with its products, e.g., antibodies. Haptens could be rendered immunogenic by covalently linking them to a carrier molecule.

Following the administration of an antigen (immunogen) to a host animal, antibody synthesis and/or cell-mediated immunity or immunologic tolerance may result. To be immunogenic, a substance usually needs to be foreign, although some autoantigens represent an exception. They should usually have a molecular weight of at least 10,000 kDa and be either proteins or polysaccharides. Nevertheless, immunogenicity depends upon the genetic capacity of the host to respond rather than merely depending upon the antigenic properties of an injected immunogen.

The specific parts of antigen molecules that elicit immune reactivity are known as antigenic determinants or epitopes. Even the earliest investigators in immunology recognized that small molecular weight substances such as simple chemicals could react with the products of an immune response but were not themselves immunogenic. These were termed haptens. Thus, a hapten is a relatively small molecule which by itself is unable to elicit an immune response when injected into an animal but is capable of reacting *in vitro* with an antibody specific for it. However, a hapten may be covalently linked to a carrier macromolecule such as a foreign protein that renders it immunogenic, and can form new antigenic determinants. Haptens often have highly reactive chemical groupings which permit them to autocouple with a substance such as a tissue protein. This type of reaction occurs in individuals who develop contact hypersensitivity to poison ivy or poison oak.

An **antigenic determinant** (Figure 3.1) interacts with the specific antigen-binding site in the variable region of an antibody molecule known as a paratope. The excellent fit between epitope and paratope is based on their three-dimensional interaction and noncovalent union. An antigenic determinant or epitope may also react with a T cell receptor for which it is specific. A single antigen molecule may have several different epitopes. Whereas an epitope interacts with the antigen binding region of an antibody molecule or with the T cell receptor, a separate region of the antigen that combines with class II MHC molecules is known as an agretope.

Antigenic determinants may be either conformational or linear. A conformational determinant is produced by spatial juxtaposition during folding of amino acid residues from different segments of the linear amino acid sequence. Conformational determinants are usually associated with natural rather than denatured proteins. A linear determinant is one produced by adjacent amino acid residues in the covalently sequenced proteins. They are usually available for interaction with an antibody only following denaturation of a protein and are not customarily in the native configuration. Antigenic determinants or epitopes are sometimes called immunodominant groups. In contrast to the natural antigens that constitute part of a microbe, one derived exclusively by laboratory synthesis and not obtained from living cells is termed a **synthetic antigen** (Figure 3.2). Synthetic polypeptide antigens have a backbone consisting of amino acids that usually include lysine. Side chains of different amino acids are attached directly to the backbone and then elongated with a homopolymer or, conversely, attached via the homopolymer. Side chains have contributed much to our knowledge of epitope structure and function. Side chains have well-defined specificities determined by the particular arrangement, number, and nature of the amino acid components of the molecule, and they may be made more complex by further coupling to haptens or derivatized with various compounds. The size of the molecule is less critical with synthetic antigens than with natural antigens. Thus, molecules as small as those of *p*-azobenzenearsonate coupled to three L-lysine residues (mol wt 750 kDa) or even of *p*-azobenzenearsonate-*N*-acetyl-L-tyrosine (mol wt 451 kDa) may be immunogenic. Specific antibodies are markedly stereospecific, and there is no crossreaction between them, e.g., poly-D-alanyl and poly-L-alanyl determinants. Studies employing synthetic antigens demonstrated the significance of aromatic, charged amino acid residues in proving the ability of synthetic polypeptides to induce an immune response.

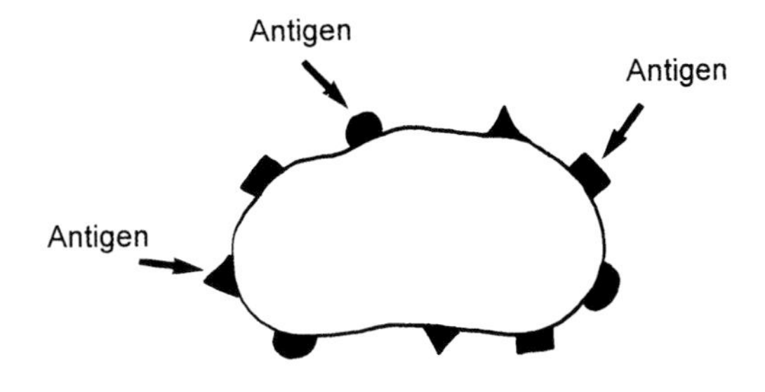

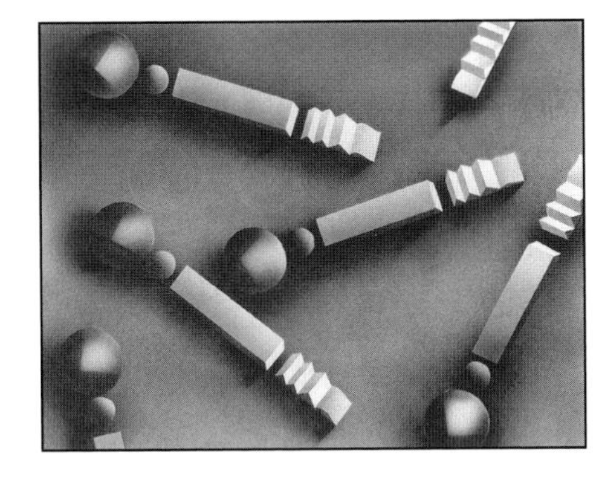

FIGURE 3.1 Schematic of antigenic determinants (epitopes).

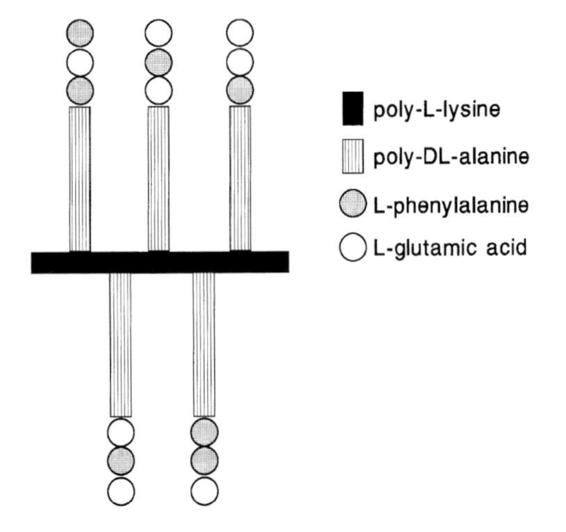

FIGURE 3.2 Synthetic polypeptide antigen with multichain copolymer (phe, G)-A-L.

An antigen molecule has two or more epitopes (or antigenic determinants) per molecule. Epitopes consists of approximately six amino acids or six monosaccharides. Epitopes that stimulate a greater antibody response than others are referred to as immunodominant epitopes.

The principal chemical features of antigens include their large size, complexity, and ability to be degraded by enzymes within phagocytes. Most antigens are of 10,000 kDa or greater molecular weight. Exceptions include such substances as insulin with 5700 kDa. Antigenicity is usually more easily demonstrated with molecules of greater molecular weight. However, size alone does not make the molecule antigenic. It must have a certain amount of internal structural complexity. Linear polymers of polylysine comprised of a repeating simple structure are not antigenic. The majority of protein antigens contain 20 different amino acids in an assorted arrangement. Oligosaccharides comprised of both monosaccharides and complex sugars are antigenic. The antigen must also be degradable by phagocytes to be antigenic. Antigen processing includes enzymatic digestion to prepare soluble macromolecular antigen. Substances such as D-amino acid polypeptides that cannot be degraded in phagocytes are not antigenic even though they might otherwise possess the characteristics of antigens. Foreignness is another characteristic that is critical for antigenicity. We do not respond to our own self-antigens because we are immunologically tolerant of them. During development, the body becomes tolerant of self-antigens as well as foreign antigens that may have been artificially introduced into the host prior to development of the immune system. The latter situation describes the induction of actively acquired immunologic tolerance, which was discovered by Medawar, Billingham, and Brent in the 1950s in studies on skin grafting. They inoculated fetal or newborn mice with cells of a different mouse strain prior to development of immunologic competence in the recipients. Once the recipient mice reached maturity, they were able to accept skin grafts from the donor strain without rejecting them. Since that time, many studies have been conducted defining the nature of T cell and B cell tolerance. In general, T cells are rendered tolerant with lower doses of antigen and for longer periods of time than are B cells. B cell tolerance is for relatively brief duration and requires much greater quantities of antigen than does T cell tolerance. Tolerance is a type of antigen-induced specific immunosuppression and antigen must remain in contact with immunocompetent cells for the tolerant state to be maintained. Tolerance induction is favored by the route of administration and the physical nature of the injected antigen. For example, the intravenous route of injection of solubilized antigen favors tolerance induction. By contrast, the injection of antigen in particulate form into the skin favors the development of immunity. An antigen that induces tolerance is often referred to as a tolerogen.

To mount an immune response to an antigen, a host must have appropriate immune response (Ir) genes. This has been proven in animal studies in which inbred strain-2 guinea pigs were shown to be responders whereas strain-13 were not. Lymphocyte proteins in man encoded by Ir genes include the class II MHC molecules designated DP, DQ, and DR that are found on human B cells and macrophages. This enables recognition among B cells, T cells, and macrophages. Antigens may be classified as either T cell dependent (TD) or T cell independent (TI). As shown in Figure 3.1, TD antigens are much more complex than TI antigens, are usually proteins, stimulate a full complement of immunoglobulins with all five classes represented, elicit an anamnestic or memory response, and are present in most pathogenic microorganisms. This ensures that an effective immune response can be generated in a host infected with these pathogens. By contrast, the simpler TI antigens are often polysaccharides or lipopolysaccharides that elicit only an IgM response and fail to stimulate an anamnestic response compared with T cell dependent antigens.

A **homopolymer** is a molecule comprised of repeating units of only one amino acid.

An **antigen** is a substance that reacts with the products of an immune response stimulated by a specific immunogen, including both antibodies and/or T lymphocyte receptors. It is presently considered to be one of many kinds of substances with which an antibody molecule or T cell receptor may bind. These include sugars, lipids, intermediary metabolites, autocoids, hormones, complex carbohydrates, phospholipids, nucleic acids, and proteins. By contrast, the traditional definition of antigen is a substance that may stimulate B and/or T cell limbs of the immune response and react with the products of that response, including immunoglobulin antibodies and/or specific receptors on T cells. (See definition of immunogen.) The traditional definition of antigen more correctly refers to an immunogen. A complete antigen is one that both induces an immune response and reacts with its products, whereas an incomplete antigen or hapten is unable to induce an immune response alone but is able to react with its products, e.g., antibodies. Haptens could be rendered immunogenic by covalently linking them to a carrier molecule. Following the administration of an antigen (immunogen) to a host animal, antibody synthesis and/or cell-mediated immunity or immunologic tolerance may result. To be immunogenic a substance usually needs to be foreign, although some autoantigens represent an exception. They should usually have a mol wt of at least 1000 kDa and be either proteins or polysaccharides. Nevertheless, immunogenicity depends also upon the genetic capacity of the host to respond to, rather than merely upon the antigenic properties of, an injected immunogen.

A **superantigen** is an antigen such as a bacterial toxin that is capable of stimulating many $CD4^+$ T cells, leading to the release of relatively large quantities of cytokines. Selected bacterial toxins may stimulate all T lymphocytes in the body that contain a certain family of V β T cell receptor genes. Superantigens may induce proliferation of 10% of $CD4^+$ T cells by combining with the T cell receptor V β and to the MHC HLA-DR α-1 domain. Superantigens are thymus-dependent (TD) antigens that do not require phagocytic processing. Instead of fitting into the T cell receptor (TCR) internal groove where a typical processed peptide antigen fits, superantigens bind to the external region of the $\alpha\beta$ TCR and simultaneously link to DP, DQ, or DR molecules on antigen-presenting cells. Superantigens react with multiple TCR molecules whose peripheral structure is similar. Thus, they stimulate multiple T cells that augment a protective T and B cell antibody response. This enhanced responsiveness to antigens such as toxins produced by staphylococci and streptococci is an important protective mechanism in the infected individual. Several staphylococcal enterotoxins are superantigens and may activate many T cells, resulting in the release of large quantities of cytokines that provoke pathophysiologic manifestations resembling endotoxin shock.

Abrin is a powerful toxin and lectin used in immunological research by Paul Ehrlich (*circa* 1900). It is extracted from the seeds of the jequirity plant and causes agglutination of erythrocytes.

An **alloantigen** is an antigen present in some members or strains of a species but not in others. Alloantigens include blood group substances on erythrocytes and histocompatibility antigens present in grafted tissues that stimulate an alloimmune response in the recipient not possessing them, as well as various proteins and enzymes. Two animals of a given species are said to be allogeneic with respect to each other. Alloantigens are commonly products of polymorphic genes.

Bovine serum albumin (BSA) is from the blood sera of cows and has been used extensively as an antigen in experimental immunologic research.

A **soluble antigen** is an antigen solubilized in an aqueous medium.

Toxins are poisons that are usually immunogenic and stimulate production of antibodies, termed antitoxins, with the ability to neutralize the harmful effects of the particular toxins eliciting their synthesis. The general groups of toxins include: (1) bacterial toxins — those produced by microorganisms such as those causing tetanus, diphtheria, botulism, and gas gangrene, as well as toxins of staphylococci, (2) phytotoxins — plant toxins such as ricin of the castor bean, crotein, and abrin derived from the Indian licorice seed, Gerukia, and (3) zootoxins — snake, spider, scorpion, bee, and wasp venoms.

Supratypic antigen is an inclusive term to describe an antigenic mosaic that can be separated into smaller but related parts called inclusions, splits, and subtypic antigens. Bw4 and Bw6 are classic examples of supertypic antigens. This implies that an antibody that detects Bw4 will also react with all antigens associated with Bw4, and an antibody that detects Bw6 will also react with all antigens associated with Bw6.

Supratypic antigen: See public antigen.

Venom is a poisonous or toxic substance produced by selected species such as snakes, arthropods, and bees. The poison is transmitted to the recipient through a bite or sting. At least 100 species of fish inject venoms that are hazardous to humans. These include sharks, rays, catfish, weever fish, scorpion fish, and stargazers.

T-dependent antigen: See thymus-dependent antigen.

TFA antigens are antigens in rabbits and rats that result from changes in liver cell components as a consequence of exposure to the anesthetic halothane. Rats that were administered halothane intraperitoneally expressed maximum amounts of the 100-, 76-, 59- and 57-kDa antigens after 12 h. The antigens were still detectable after 7 d. TFA antigen expression varies in humans as a consequence of variability of hepatic cytochrome P-450 isoenzyme profiles.

T-independent antigen: See thymus-independent antigen.

Biochemical sequestration refers to antigenic determinants hidden in a molecule which may be unable to act as immunogens or to react with antibody. Structural alterations in the molecule may render them identifiable and capable of serving as immunogens.

HSA is the abbreviation for human serum albumin, an antigen commonly used in experimental immunology.

Tetanus toxoid consists of formaldehyde-detoxified toxins of *Clostridium tetani*. It is an immunizing preparation to protect against tetanus. Individuals with increased likelihood of developing tetanus as a result of a deep, penetrating wound with a rusty nail or other contaminated instrument are immunized by subcutaneous inoculation. The preparation is available in both fluid and adsorbable forms. It is included in a mixture with diphtheria toxoid and pertussis vaccine and is known as DTP, DPT, or triple vaccine. It is employed routinely to immunize children less than 6 years old.

An **antigenic peptide** is a peptide that is able to induce an immune response and one that complexes with MHC, thereby permitting its recognition by a T cell receptor.

Ovalbumin (OA) is a protein derived from avian egg albumin (the egg white). It has been used extensively as an antigen in experimental immunology.

Antigenic profile is the total antigenic content, structure, or distribution of epitopes of a cell or tissue.

End-binders are selected anticarbohydrate specific antibodies that bind the ends of oligosaccharide antigens, in contrast to those that bind the sides of these molecules.

A **pneumococcal polysaccharide** is a polysaccharide found in the *Streptococcus pneumoniae* capsule that is a type-specific antigen. It is a virulence factor. Serotypes of this microorganism are based upon different specificities in the capsular polysaccharide which is comprised of oligosaccharide-repeating units. Glucose and glucuronic acid are the repeating units in type III polysaccharide.

A **capsular polysaccharide** is a constituent of the protective coating around a number of bacteria such as the pneumococcus (*Streptococcus pneumoniae*), which is a polysaccharide chemically, and stimulates the production of antibodies specific for its epitopes. In addition to the pneumococcus, other microorganisms such as Streptococci and certain Bacillus species have polysaccharide capsules.

SSS III: One of more than 70 types of specific soluble substances comprising the polysaccharide in capsules of *Streptococcus pneumoniae*, commonly known as the pneumococcus. It was used extensively by Michael Heidelberger and associates in perfecting the quantitative precipitation reaction.

A **crossreacting antigen** is an antigen that interacts with an antibody synthesized following immunogenic challenge with a different antigen. Epitopes shared between these two antigens or epitopes with a similar stereochemical configuration may account for this type of crossreactivity. The presence of the same or of a related epitope between bacterial cells, red blood cells, or other types of cells may crossreact with an antibody produced against either of them.

Carbohydrate antigens: The best known carbohydrate antigen is the specific soluble substance of the capsule of *Streptococcus pneumoniae* which is immunogenic in humans. Heidelberger developed the first effective vaccines against purified pneumococcal polysaccharide in the early 1940s which were used in the treatment of pneumonia caused by these microorganisms, yet interest in the vaccine waned as antibiotics were developed for treatment. With increased resistance of bacteria to antibiotics, however, there is a renewed interest in immunization with polysaccharide-based vaccines. Polysaccharides alone are relatively poor immunogens, especially in infants and in immunocompromised hosts. Pneumococci have 84 distinct serotypes, which further complicates the matter. Polysaccharides are classified as T cell-independent immunogens. They fail to induce immunologic memory, which is needed for booster responses in an immunization protocol. Only a few B cell clones are activated, leading to restricted yet polyclonal heterogeneity. The majority of polysaccharides can induce tolerance or unresponsiveness and fail to induce delayed-type hypersensitivity. Polysaccharide immunogenicity increases with molecular weight. Those polysaccharides that are less than 50 kDa are nonimmunogenic. Thus, immunization with purified polysaccharides has not been as effective as desired. The covalent linkage of a polysaccharide or of its epitopes to a protein carrier to form a conjugate vaccine has facilitated enhancement of immunogenecity in both humans and other animals, and induces immunologic memory. Antibodies formed against the conjugate vaccine are protective and bind the capsular polysaccharide from which they were derived. An example of this type of immunogen is HIV polysaccharide linked to tetanous toxoid, which has been successful in infant immunization.

Immunogenic carbohydrates: Carbohydrates are important in various immunological processes that include opsonization and phagocytosis of microorganisms, and cell activation and differentiation. They exert their effects through interactions with carbohydrate-binding proteins or lectins that have a widespread distribution in mammalian tissues, including the immune system. Immunogenic carbohydrates are usually large polymers of glucose (glucans and lentinans), mannose (mannans), xylose, (hemicelluloses), fructose (levans), or mixtures of these sugars. Complex carbohydrates may stimulate the immune system by activating macrophages with fungal glycans or stimulating T cells with lentinan. Acemannin activates macrophages and T cells, thereby influencing both cellular and humoral immunity. Glucans stimulate immunity against bacterial diseases by activating macrophages and stimulating their lysosomal and phagocytic activity. Complex carbohydrates may activate the immune systems of patients or experimental animals with neoplastic diseases. Some mannans and glucans are powerful anticancer agents. Lentinin derived from an edible mushroom has an antineoplastic effect against several different allogeneic and syngeneic tumors without mediating cytotoxicity of the tumor cells. Mannans derived from yeast also have a significant antitumor effect, as do levans that activate not only macrophages but B and T cells as well. Pectin is a gallactose-containing carbohydrate concentrated in citrus and has an antitumor effect.

In immunology, the term **repeating units** refers to antigens in which macromolecular configurations are repeated, such as the repeating units of β-1,4-glucose-β-1,3-glucuronic acid in type III pneumococcus polysaccharide. Polysaccharide antigens in the cell walls of Gram-negative bacteria also contain repeating structures. Antigens with this type of configuration are often thymus-independent antigens.

Biovin antigens are *Salmonella* O antigens. These carbohydrate–lipid–protein complexes withstand trichloroacetic acid treatment.

A **differentiation antigen** is an epitope that appears at various stages of development or in separate tissues. This cell surface antigenic determinant is present only on cells of a particular lineage and at a particular developmental stage; it may serve as an immunologic marker.

Immunopotency is the capability of a part of an antigen molecule to function as an epitope and induce the synthesis of specific antibodies.

Serum albumin is the principal protein in serum or plasma. It is soluble in water as well as in partially concentrated salt solution such as 50% saturated ammonium sulfate solution. It is coagulated by heat. It accounts for much of the plasma colloidal osmotic pressure. Serum albumin functions as a transport protein for fatty acids, bilirubin, or other large organic anions. It also carries selected hormones, including cortisol and thyroxine, and many drugs. It is formed in the liver, and levels in the serum decrease when there is protein malnutrition or significant liver and kidney disease. When the pH is neutral, albumin has a negative charge, causing its rapid movement toward the anode during electrophoresis. It is comprised of a single 585-amino acid residue chain and has a concentration of 35 to 55 mg/ml. Bovine serum albumin (BSA) and selected other serum albumins have been used as experimental immunogens in immunologic research.

BSA is the abbreviation for bovine serum albumin.

Lipopolysaccharide (LPS): See endotoxin.

An **eclipsed antigen** is an antigen such as one from a parasite, which so closely resembles host antigens that it fails to stimulate an immune response.

Embryonic antigens are protein or carbohydrate antigens synthesized during embryonic and fetal life that are either absent or formed in only minute quantities in normal adult subjects. Alpha fetoprotein (AFP) and carcinoembryonic antigen (CEA) are fetal antigens that may be synthesized once again in large amounts in individuals with certain tumors. Their detection and level during the course of the disease and following surgery to remove a tumor, reducing the antigen level, may serve as a diagnostic and prognostic indicator of the disease process. Blood group antigens, such as the iI, which are reversed in their levels of expression in the fetus and in the adult, may show a reemergence of i antigen in adults in patients with thalassemia and hypoplastic anemia. Cold autoagglutinins specific for i may be found in infectious mononucleosis patients. Common acute lymphoblastic leukemia antigen (CD10) is rarely found on peripheral blood cells of normal subjects, whereas as CALLA′ cells coexpressing IgM and CD19 molecules may be found in fetal bone marrow and peripheral blood samples. CD10 may be expressed in children with common acute lymphoblastic leukemia.

An **exoantigen** is a released antigen.

Glycosylphosphatidylinositol- (GPI)-linked membrane antigens comprise a class of cell surface antigens attached to the membrane by glycosylphosphatidylinositol. Monoclonal antibody studies indicate that in both human and murine subjects GTI-linked antigens are capable of stimulating T cells and sometimes B cells. Structurally they are diverse. See also Ly-5 and Qa-2.

Halothane antigens are antigenic determinants that result from the action of halothane on rabbit and rat liver cell components. Also referred to as TFA antigens.

D-amino acid polymers are synthetic peptides (and polypeptides) that are antigenic. They are found very infrequently in living organisms.

Heat-aggregated protein antigen refers to the partial denaturation of a protein antigen by mild heating. This diminishes the protein's solubility but causes it to express new epitopes. An example is the greater reactivity of rheumatoid factor (e.g., IgM anti-IgG autoantibody) with heat-aggregated γ globulin than with unheated γ globulin.

A **heteroantigen** is an antigen that induces an immune response in a species other than the one from which it was derived.

Heterogenetic antigen: See heterophile antigen.

A **heterologous antigen** is a crossreacting antigen.

Functional antigen: See protective antigen.

A **homologous antigen** is an antigen (immunogen) that stimulates the synthesis of an antibody and reacts specifically with it.

Hot antigen suicide: The labeling of an antigen with a powerful radioisotope such as ^{131}I proves lethal upon contact with antigen-binding cells that have receptors specific for it. This leads to a failure to synthesize antibodies specific for that antigen, provided the antigen-binding and antibody-synthesizing cells are one and the same. Hot antigen suicide supports the clonal selection theory of antibody formation.

Inaccessible antigens: See hidden determinant (epitopes).

Hemocyanin is a blood pigment that transports oxygen in invertebrates. In immunology, hemocyanin of the keyhole limpet has been widely used as an experimental antigen.

Keyhole limpet hemocyanin (KLH) is a traditional antigen widely used as a carrier in studies on immune responsiveness to haptens. This respiratory pigment containing copper is found in mollusks and crustaceans. It is usually immunogenic in vertebrate animals.

KLH is the abbreviation for keyhole limpet hemocyanin.

Nontissue-specific antigen is an antigen that is not confined to a single organ but is distributed in more than one normal tissue or organ, such as nuclear antigens.

Surface antigens are epitopes on a cell surface such as the bacterial antigens Vi and O.

Whereas an **artificial antigen** is prepared by chemical modification of a natural antigen, a synthetic antigen (Figure 3.2) is derived exclusively by laboratory synthesis and not obtained from living cells.

Polyclonal activators are agents that stimulate numerous lymphocytes without regard to their antigen specificity. The proliferating cells produce products such as immunoglobulins or cytokines. The proliferating cells lack cell-to-cell collaboration, which reduces mutations on polyclonal activators. Polyclonal activators use different activation pathways than antigen-stimulated lymphocytes, although pathways overlap and interact. There are numerous polyclonal activators that have different mechanisms of action on cells of immune systems. Among the polyclonal activators are mitogenic lectins that bind polysaccharides of surface structures such as PHA, Con-A, and PWM; bacterial cell wall products that bind lymphocyte receptors, such as lipopolysaccharide (LPS); calcium ionophores that alter calcium signals; phosphorylation modifiers that increase stimulatory phosphorylation events, such as phorbol myristate acetate; antigen-receptor ligands that bind to nonvariable parts of the receptor mechanism, such as staphylococcal protein A; ligands for costimulatory molecules that bind to lymphocytes and favor stimulation, such as anti-CD40 and CD40L; and transforming agents that infect cells and cause continued growth or activation, such as HIV. Monoclonal activators can be used to access the functional status of a particular lymphocyte population. Some polyclonal activators require more than one type of cell and may even induce inhibition and inactivation lymphocytes. Numerous interactions occur between antigen-specific and polyclonal forms of lymphocyte activation and inactivation. Costimulation and coinhibition regulates specific immune responses. Defects in signals in these various interactions, especially defects in signaling associated with antagonism and coinhibition, may predispose to autoimmune disease.

Phorbol ester(s) are esters of phorbol alcohol (4,9,12-b,13,20-pentahydroxy-1,6-tigliadien-3-on) found in croton oil and myristic acid. Phorbol myristate acetate (PMA), which is of interest to immunologists, is a phorbol ester that is 12-*O*-tetradecanoylphorbol-13-acetate (TPA). This is a powerful tumor promoter that also exerts pleotrophic effects on cells in culture, such as stimulation of macromolecular synthesis and cell proliferation, induction of prostaglandin formation, alteration in the morphology and permeability of cells, and disappearance of surface fibronectin. PMA also acts on leukocytes. It links to and stimulates protein kinase C. This leads to threonine and serine residue phosphorylation in the transmembrane protein cytoplasmic domains such as in the CD2 and CD3 molecules. These events enhance interleukin-2 receptor expression on T cells and facilitate their proliferation in the presence of interleukin-1 as well as TPA. Mast cells, polymorphonuclear leukocytes, and platelets may all degranulate in the presence of TPA.

Dextrans are polysaccharides of high molecular weight comprised of D-glucohomopolymers with α glycoside linkages, principally α-1,6 bonds. Dextrans serve as murine B lymphocyte mitogens. Some dextrans may also serve as thymus-independent antigens. Dextrans of relatively low molecular weight have been used as plasma expanders.

Synthetic polypeptide antigens have a backbone consisting of amino acids that usually include lysine. Side chains of different amino acids are attached directly to the backbone and then elongated with a homopolymer, or conversely, attached via the homopolymer. They have contributed much to our knowledge of epitope structure and function. They have well-defined specificities determined by the particular arrangement, number, and nature of the amino acid components of the molecule, and they may be made more complex by further coupling to haptens or derivatized with various compounds. The size of the molecule is less critical with synthetic antigens than with natural antigens. Thus, molecules as small as those of *p*-azobenzenearsonate coupled to three L-lysine residues (mol wt 750 kDa) or even of *p*-azobenzenearsonate-*N*-acetyl-L-tyrosine (mol wt 451) may be immunogenic. Specific antibodies are markedly stereospecific, and there is no crossreaction between, e.g., poly-D-alanyl and poly-L-alanyl determinants. Studies employing synthetic antigens demonstrated the significance of aromatic, charged amino acid residues in proving the ability of synthetic polypeptides to induce an immune response.

An **immunogen** is a substance that is able to induce a humoral antibody and/or cell-mediated immune response rather than immunological tolerance. The term *immunogen* is sometimes used interchangeably with *antigen*, yet the term signifies the ability to stimulate an immune response as well as react with the products of it, e.g., antibody. By contrast, antigen is reserved by some to mean a substance that reacts with an antibody. The principal immunogens are proteins and polysaccharides, whereas lipids may serve as haptens.

Immunogenicity is the ability of an antigen serving as an immunogen to induce an immune response in a particular species of recipient. Immunogenicity depends on a number of physical and chemical characteristics of the immunogen (antigen), as well as on the genetic capacity of the host response.

Immunogenic is an adjective that denotes the capacity to induce humoral antibody and/or cell-mediated immune responsiveness but not immunological tolerance. Immunogenicity depends on characteristics of the immunogen and on the injected animal's genetic capacity to respond to the immunogen. To be immunogenic, a substance must be foreign to the recipient. An immunogen that is of significant molecular size and complexity, as well as host factors such as previous exposure to the immunogen and immunocompetence, are all critical factors in immunogenicity.

Antigenicity is the property of a substance that renders it immunogenic or capable of stimulating an immune response. Antigenicity was more commonly used in the past to refer to what is now known as immunogenicity, although the two are still used interchangeably. An antigen is considered by many to be a substance that reacts with the products of immunogenic stimulation. It is a substance that combines specifically with antibodies formed or receptors of T cells stimulated during an immune response.

Antigenic refers to the ability of a substance to induce an immune response and to react with its products, which include antibodies and T lymphocyte receptors. The term *antigenic* has been largely replaced by *immunogenic*.

An **antigenic determinant** (Figure 3.3) is the site on an antigen molecule that is termed an epitope and interacts

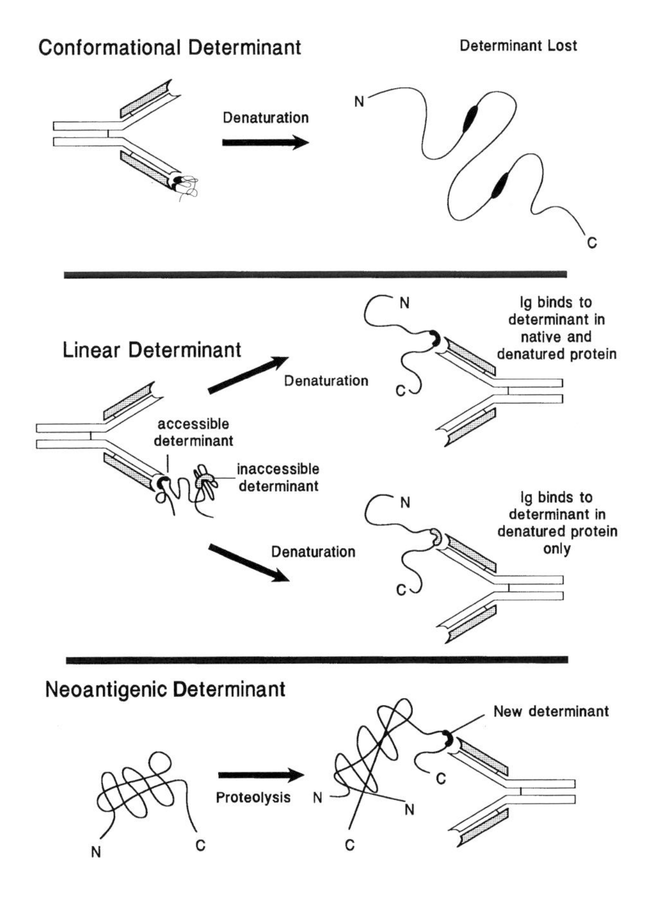

FIGURE 3.3 Antigenic determinants.

with the specific antigen-binding site in the variable region of an antibody molecule known as a paratope. The excellent fit between epitope and paratope is based on their three-dimensional interaction and noncovalent union. An antigenic determinant or epitope may also react with a T cell receptor for which it is specific. A lone antigen molecule may have several different epitopes available for reaction with antibody or T cell receptors.

A **conformational determinant** is an epitope composed of amino acid residues that are not contiguous and represent separated parts of the linear sequence of amino acids that are brought into proximity to one another by folding of the molecule. A conformational determinant is dependent on a three-dimensional structure. Conformational determinants are therefore usually associated with native rather than denatured proteins. Antibodies specific for linear determinants and others specific for conformational determinants may give clues as to whether a protein is denatured or native, respectively.

An **epitope** is an antigenic determinant. It is the simplest form or smallest structural area on a complex antigen molecule that can combine with an antibody or T lymphocyte receptor. It must be at least 1 kDa to elicit an antibody response. A smaller molecule such as a hapten may induce an immune response if combined with a carrier protein molecule. Multiple epitopes may be found on large nonpolymeric molecules. Based on x-ray crystallography, epitopes consist of prominently exposed "hill and ridge" regions that manifest surface rigidity. Antigenicity is diminished in more flexible sites.

Conformational epitopes are discontinuous determinants on a protein antigen formed from several separate regions in the primary sequence of a protein brought together by protein folding. Antibodies that bind conformational epitopes bind only native folded proteins.

Continuous epitopes are linear antigenic determinants on proteins that are contiguous in amino acid sequence and do not require folding of a protein into its native conformation for antibody to bind with it.

Cooperative determinant: Carrier determinant.

Cryptodeterminant: See hidden determinant.

Discontinuous epitopes: See conformational epitopes.

Epitope spreading refers to increased diversity of the response to autoantigens with time. Also termed determinant spreading or antigen spreading.

An **epitype** is a family or group of related epitopes.

A **hidden determinant** is an epitope on a cell or molecule that is unavailable for interaction with either lymphocyte receptors or the antigen-binding region of antibody molecules because of stereochemical factors. These hidden or cryptic determinants neither react with lymphocyte or antibody receptors nor induce an immune response unless an alteration in the molecule's steric configuration causes the epitope to be exposed.

Immunodominance is a descriptor for the immune response-generating capacity of that part of an epitope on an antigen molecule that serves as an immunodominant tip or area which provides the principal binding energy for reaction with a paratope on an antibody molecule or with a T cell receptor for antigen. The hapten portion of a hapten–carrier complex is often the immunodominant part of the molecule. Immunodominance refers to the region of an antigenic determinant that is the principal binding site for antibody.

An **immunodominant epitope** is the antigenic determinant on an antigen molecule that binds or fits best with the antibody or T cell receptor specific for it.

Immunodominant site: Refer to immunodominant epitope.

An **inducer determinant** is a hapten determinant.

Linear determinants are antigenic determinants produced by adjacent amino acid residues in the covalent sequence in proteins. Linear determinants of six amino acids interact with specific antibody. Occasionally, linear determinants may be on the surface of a native folded protein, but they are more commonly unavailable in the native configuration and only become available for interaction with antibody upon denaturation of the protein.

A **linear epitope** is an antigenic determinant of a protein molecule recognized by an antibody that consists of a linear sequence of amino acids within the protein's primary structure.

An **oligosaccharide determinant** is an epitope or antigenic determinant of a polysaccharide hapten that consists of relatively few, i.e., two to seven, pentoses, hexoses, or heptoses united by glycoside linkages.

A **sequential determinant** is an epitope whose specificity is determined by the sequence of several residues within the antigenic determinant rather than by the molecular configuration of the antigen molecule. A peptide segment of approximately six amino acid residues represents the sequential determinant structure.

Nonsequential epitopes are antigenic determinants that are widely separated in the primary sequence of the polypeptide chain but are near one another in the tertiary structure of the molecule.

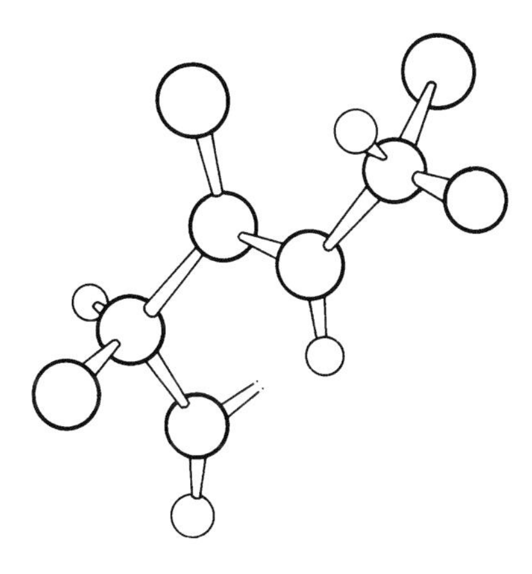

FIGURE 3.4 Primary structure.

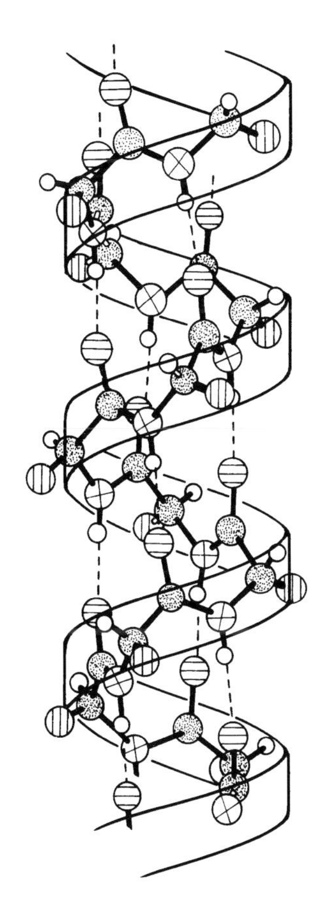

FIGURE 3.5 Secondary structure.

Determinant groups (or epitopes) are chemical mosaics found on macromolecular antigens that induce an immune response.

Primary structure (Figure 3.4) refers to a polypeptide or protein molecule's linear amino acid sequence.

Secondary structure (Figure 3.5) is based on polypeptide chain or polynucleotide strand folding along the axis or backbone of a molecule. This is a consequence of the formation of intramolecular hydrogen bonds joining carbonyl oxygen and amide nitrogen atoms. Secondary structure is based on the local spatial organization of polypeptide chain segments or polynucleotide strands irrespective of the structure of side chains or of the relationship of the segments to one another.

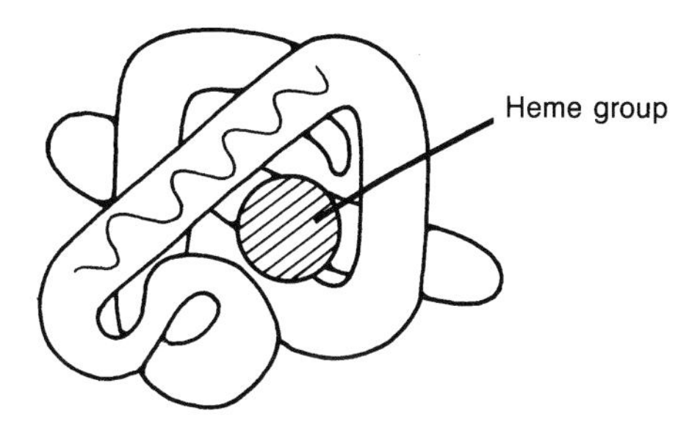

FIGURE 3.6 Tertiary structure.

FIGURE 3.7 Quaternary structure.

Tertiary structure (Figure 3.6) describes the folding of a polypeptide chain as a result of the interactions of its amino acid side chains which may be situated either near or distant along the chain. This three-dimensional folding occurs in globular proteins. Tertiary structure also refers to the spatial arrangement of protein atoms irrespective of their relationship to atoms in adjacent molecules.

Quaternary structure (Figure 3.7) refers to four components that are associated with one another. Two or more folded polypeptide chains packed into a configuration such as a tetramer. Quaternary antigenic determinants may be difficult to demonstrate in such structures as hemoglobulin molecules.

Denaturation is changing a protein's secondary and tertiary structure (coiling and folding) to produce a configuration that is either uncoiled or coiled more randomly. Whereas storage causes slow denaturation, heating or chemical treatment may induce more rapid denaturation of native protein molecules. Denaturation diminishes protein solubility and often abrogates the molecule's biologic activity. New or previously unexposed epitopes may be revealed as a consequence of denaturation.

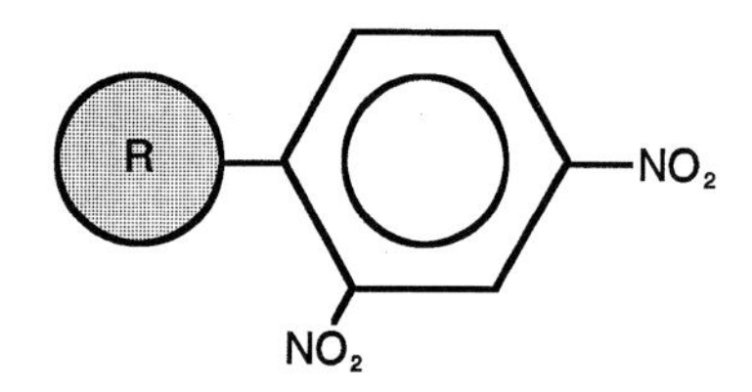

FIGURE 3.8 2,4-dinitrophenyl (DNP).

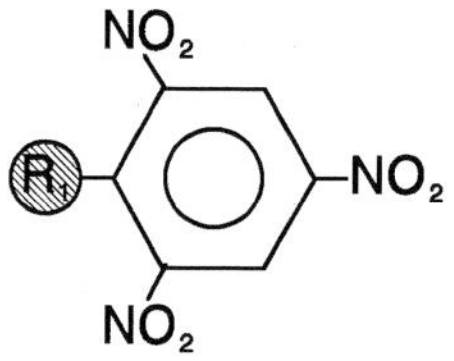

NO2
NO2
NO2

FIGURE 3.9 Trinitrophenyl group.

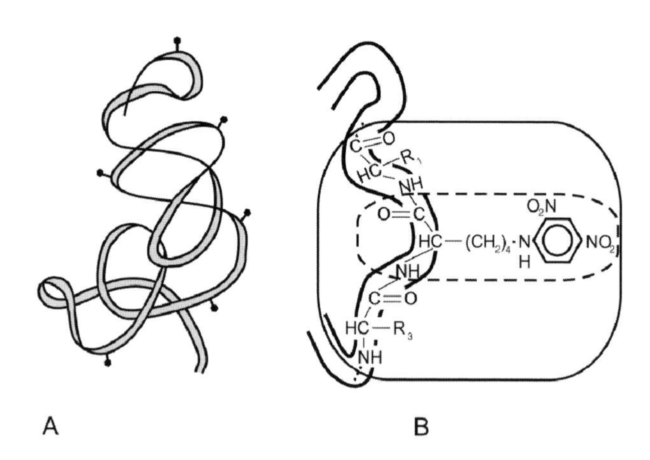

FIGURE 3.8A (a) Conjugated protein with substituents. (b) Broken line refers to the haptenic group and the solid line refers to the antigenic determinant.

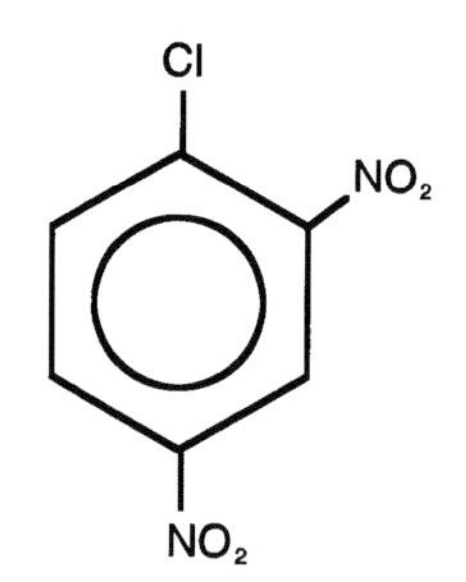

1-chloro-2,4-dinitrobenzene

FIGURE 3.10 Dinitrochlorobenzene.

2,4-dinitrophenyl (DNP) group (Figure 3.8 and Figure 3.8A): These are groups that may serve as haptens after they are chemically linked to –NH2 groups of proteins that interact with chlorodinitrobenzene, 2,4-dinitrobenzene sulphonic acid, or dinitrofluorobenzene. These protein carrier–DNP hapten antigens are useful as experimental immunogens. Antibodies specific for the DNP hapten, which are generated through immunization with the carrier–hapten complex, interact with low molecular weight substances that contain the DNP groups.

DNP: See dinitrophenyl group.

DNCB: See dinitrochlorobenzene.

DNBS (2,4-dinitrobenzene sulfonate) is a substance employed to generate dinitrophenylated proteins that are used as experimental antigens. Chemically, DNBS reacts principally with lysine residue-free ε amino groups if an alkaline pH is maintained.

A **trinitrophenyl (picryl) group** (Figure 3.9) is a chemical grouping that may serve as a hapten when it is linked to a protein through –NH2 groups by reaction with picryl chloride or trinitrobenzene sulfonic acid.

TNP: Abbreviation for trinitrophenyl group.

Dinitrochlorobenzene (DNCB) (Figure 3.10) is a substance employed to test an individual's capacity to develop a cell-mediated immune reaction. A solution of DNCB is applied to the skin of an individual not previously sensitized against this chemical, where it acts as a hapten, interacting with proteins of the skin. Reexposure of this same individual to a second application of DNCB 2 weeks after the first challenge results in a T cell-mediated delayed-type hypersensitivity (contact dermatitis) reaction. Persons with impaired delayed-type hypersensitivity or cell-mediated immunity might reveal an impaired response. The 2,4-dinitro-1-chlorobenzene interacts with free -amino terminal groups in polypeptide chains, as well as with side chains of lysine, tyrosine, histidine, cysteine, or other amino acid residues.

A **hapten** is a relatively small molecule which by itself is unable to elicit an immune response when injected into an animal but is capable of reacting *in vitro* with an antibody specific for it. Only if complexed to a carrier molecule prior to administration can a hapten induce an immune response. Haptens usually bear only one epitope. Pneumococcal polysaccharide is an example of a larger molecule that may act as a hapten in rabbits but as a complete antigen in humans.

A **carrier** is an immunogenic macromolecular protein, such as ovalbumin, to which an incomplete antigen termed a hapten may be conjugated either *in vitro* or *in vivo.* Whereas the hapten is not immunogenic by itself, conjugation to the carrier molecule renders it immunogenic.

When self proteins are appropriately modified by the hapten, they may serve as a carrier *in vivo*, a mechanism operative in allergy to drugs.

Carrier effect: To achieve a secondary immune response to a hapten, both the hapten and the carrier used in the initial immunization must be employed.

(TG)AL: Tyrosine and glutamic acid polymers fasten as side chains to a poly-L-lysine backbone through alanine residues. This substance is a synthetic antigen.

(Phe,G)AL is a poly-L-lysine backbone to which side chains of phenylalanine and glutamic acid short polymers are linked by alanine residues.

Conjugate usually refers to the covalent bonding of a protein carrier with a hapten, or it may refer to the labeling of a molecule such as an immunoglobulin with fluorescein isothiocyanate, ferritin, or an enzyme used in the enzyme-linked immunoabsorbent assay.

Conjugated antigen: See conjugate.

Carrier specificity refers to an immune response, either humoral antibody or cell-mediated immunity, that is specific for the carrier portion of a hapten–carrier complex that has been used as an immunogen. The carrier-specific part of the immune response does not react with the hapten either by itself or conjugated to a different carrier.

Schlepper is a name used by Landsteiner to refer to large macromolecules that serve as carriers for simple chemical molecules serving as haptens. The immunization of rabbits or other animals with a hapten–carrier complex leads to the formation of antibodies specific for the hapten as well as the carrier. T cells were later shown to be carrier specific and B cells hapten specific. Carriers are conjugated to haptens through covalent linkages such as the diazo linkage.

Hapten X: See CD15.

NIP (4-hydroxy,5-iodo,3-nitrophenylacetyl) is used as a hapten in experimental immunology.

A **hybrid hapten** is a hydrophobic type of hapten that lies within the folds of a protein carrier away from the aqueous solvent, creating a new spatial structure.

NP (4-hydroxy,3-nitrophenylacetyl) is a hapten used in experimental immunology that exhibits limited crossreactivity with NIP.

Hapten–carrier conjugate refers to the combination of a small molecule covalently linked to a large immunogenic carrier molecule.

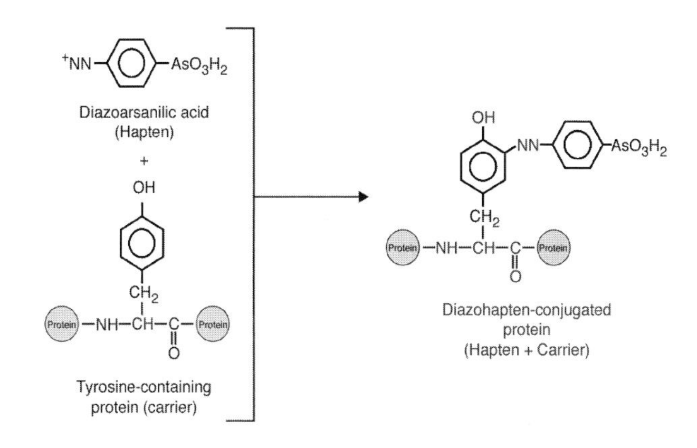

FIGURE 3.11 Hapten conjugates.

Hapten conjugate response: The response to **hapten conjugates** (Figure 3.11) requires two populations of lymphocytes, T and B cells. The cells producing the antibodies are derived from B cells. T cells act as helpers in this process. B cell preparations, depleted of T cells, cannot respond to hapten conjugates. The T cells are responsive to the carrier portion of the conjugate, although in some cases they also recognize the hapten. The influence of the carrier on the ensuing response is called carrier effect. The experimental design for demonstrating the carrier effect involves adoptive transfer of hapten-sensitive B cells and of T cells primed with one or another carrier. The primed cells are those which have already had a past opportunity to encounter the antigen.

Dinitrofluorobenzene (2,4-dinitro-1-fluorobenzene) (DNFB) is a chemical employed to prepare hapten–carrier conjugates. It inserts the 2,4-dinitrophenyl group into molecules containing free $-NH_2$ groups. When placed on the skin, it leads to contact hypersensitivity.

The **hapten inhibition test** is an assay for serological characterization or elucidation of the molecular structure of an epitope by blocking the antigen-binding site of an antibody specific for the epitope with a defined hapten.

Diazotization (Figure 3.12) is a method to introduce the diazo group ($-N^+{\equiv}N^-$) into a molecule. Landsteiner used this technique extensively in coupling low molecular weight chemicals acting as haptens to protein macromolecules serving as carriers. Aromatic amine derivatives can be coupled to side chains of selected amino acid residues to prepare protein–hapten conjugates, which, when used to immunize experimental animals such as rabbits, stimulate the synthesis of antibodies. Some of these antibodies are specific for the hapten, which by itself is unable to stimulate an immune response. First an aromatic amine reacts with nitrous acid generated through the combination of sodium nitrite with HCl. The diazonium salt is then

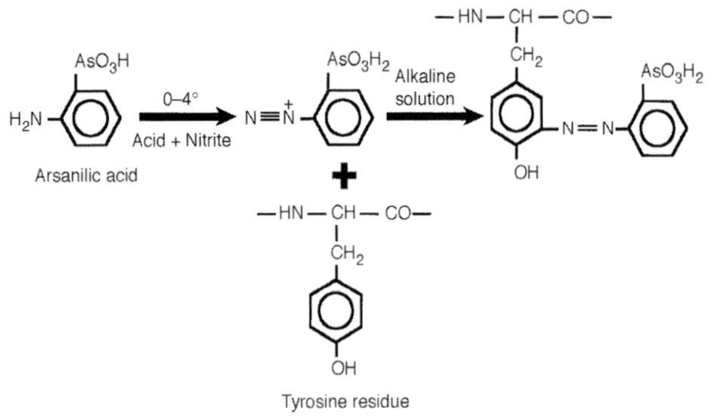

FIGURE 3.12 Diazotization.

combined with the protein at a pH that is slightly alkaline. The reaction products include monosubstituted tyrosine and histidine and also lysine residues that are disubstituted.

A **diazo salt** is a diazonium salt prepared by diazotization from an arylamine to yield a product with a diazo group. Diazotization has been widely used in the preparation of hapten–carrier conjugates for use in experimental immunology.

An **azoprotein** is produced by joining a substance to a protein through a diazo linkage –N = N–. Karl Landsteiner (in early 1900s) made extensive use of diazotization to prepare hapten–protein conjugates to define immunochemical specificity. See also diazo reaction.

An **antigen binding site** is the location on an antibody molecule where an antigenic determinant or epitope combines with it. The antigen-binding site is located in a cleft bordered by the N-terminal variable regions of heavy and light chain parts of the Fab region. Also called paratope. The liver has important immunologic functions by virtue of its mass of Kupffer cells, which represent the major part (90%) of the body's phagocytic capacity. Antigens escaping the intestinal barrier by passage through the liver are removed through a process called **antigen clearance**. The liver's anatomical position at the border between the splanchnic and systemic circulations substantiates its function as a filter for noxious substances, whether antigen or otherwise. The same removal mechanism is operative during liver passage in situations in which antigen circulates in the blood.

Antigen clearance: Exogenous clearance is a principal immune system function. Fixed mononuclear phagocytic system cells are the principal mechanism to eliminate antigen. The mechanism of removal depends on the biological and physical chemical properties of the antigen, its mode of presentation, and its capacity to induce a specific humoral or cellular immune response.

The simultaneous injection of two closely related antigens may lead to suppression or a decrease of the immune response to one of them compared to the antigen's ability to elicit an immune response if injected alone. This is known as **antigenic competition.** Proteins that are TD antigens are the ones with which **antigenic competition** occurs. The phenomenon has been claimed to be due in part to the competition by antigenic peptides for one binding site on class II MHC molecules. Antigenic competition was observed in the early days of vaccination when it was found that the immune response of a host to the individual components of a vaccine might be less than if they had been injected individually.

Antigen unmasking, the exposure of tissue antigens using an antigen unmasking solution based on a citric acid formula, is highly effective at revealing antigens in formalin-fixed, paraffin-embedded tissue sections when used in combination with a high-temperature treatment procedure.

Original antigenic sin: The immune response against a virus to which an individual was previously exposed, such as a parental strain, may be greater than it is against the immunizing agent, such as type A influenza virus variant. This concept is known as the doctrine of original antigenic sin.

When an individual is exposed to an antigen that is similar but not identical to an antigen to which he was previously exposed by either infection or immunization, the immune response to the second exposure is still directed against the first antigen. This was first noticed in the influenza virus infection. Due to antigenic drift and antigenic shift in influenza virus, reinfection with an antigenically altered strain generates a secondary immune response that is specific for the influenza virus strain that produced an earlier infection. Other highly immunogenic epitopes on a second and subsequent viruses are ignored.

Antigenic variation represents a mechanism whereby selected viruses, bacteria, and animal parasites may evade the host antibody or T cell immune response, thereby permitting antigenically altered etiologic agents of disease to produce a renewed infection. The variability among infectious disease agents is of critical significance in the development of effective vaccines. Antigenic variation affects the surface antigens of the viruses, bacteria, or animal parasite in which it occurs. By the time the host has developed a protective immune response against the antigens originally present, the latter will have been replaced in a few surviving microorganisms by new antigens to which the host is not immune, thereby permitting survival of the microorganism or animal parasite and its evasion of the host immune response. Thus from these few surviving viruses, bacteria, or animal parasites, a new

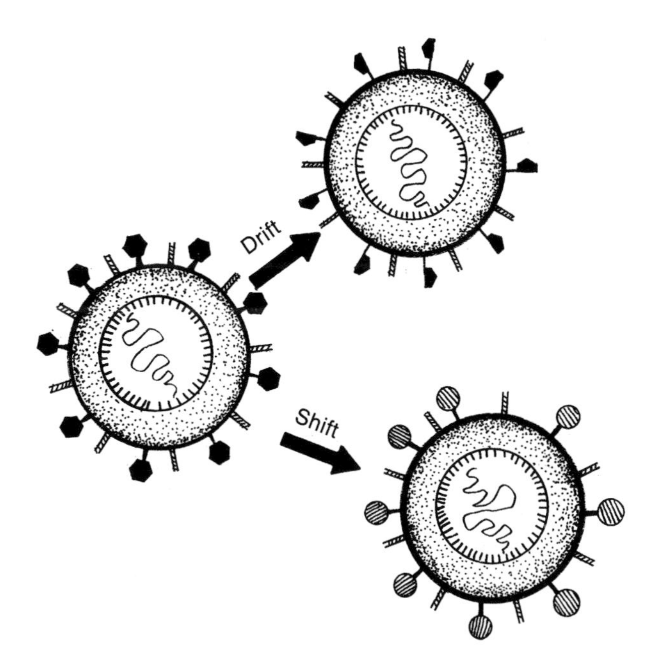

FIGURE 3.13 Antigenic drift and shift.

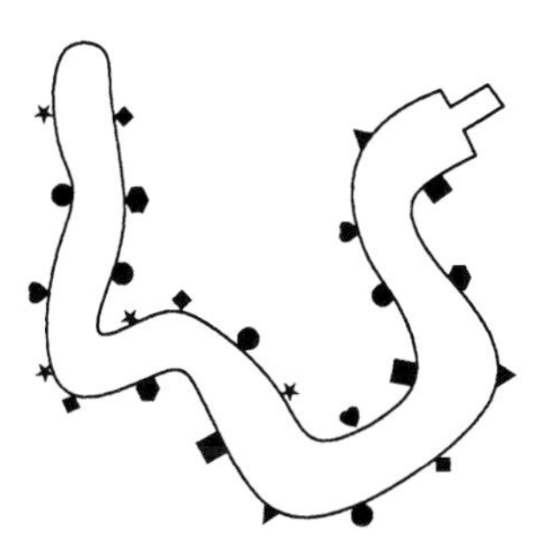

FIGURE 3.14 Thymus-dependent antigen.

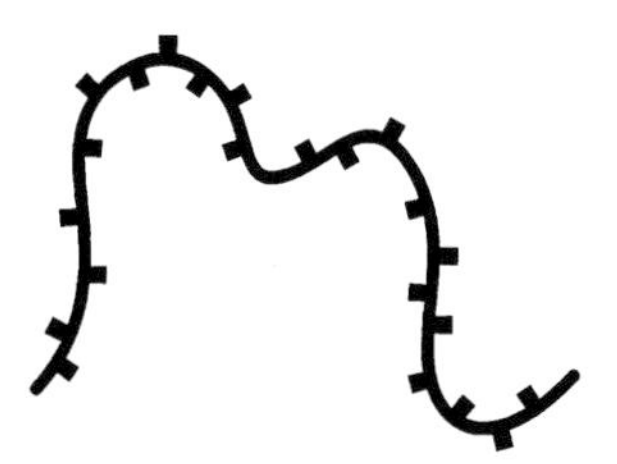

FIGURE 3.15 Thymus-independent antigen.

population of infectious agents is produced. This cycle may be repeated, thereby obfuscating the protective effects of the immune response.

Antigenic drift (Figure 3.13) refers to spontaneous variation, as in influenza virus, expressed as relatively minor differences exemplified by slow antigenic changes from one year to the next. Antigenic drift is believed to be due to mutation of the genes encoding the hemagglutinin or the neuraminidase components. Antigenic variants represent those viruses that have survived exposure to the host's neutralizing antibodies. Minor alterations in a viral genome might occur every few years, especially in influenza A subtypes that are made up of H1, H2, and H3 hemagglutinins and N1 and N2 neuraminidases. Antigenic shifts follow point mutations of DNA encoding these hemagglutinins and neuraminidases.

Antigenic shift (Figure 3.13) also describes a major antigenic change in which a strain with distinctive new antigens may appear, such as Asian or A2 influenza in 1957. Antigenic variants of type A influenza virus are known as subtypes. Influenza virus antigenic shift is attributable mainly to alterations in the hemagglutinin antigens with less frequent alterations in the neuraminidase antigens. The appearance of a new type A influenza virus signals the addition of a new epitope, even though several original antigenic determinants are still present. In contrast to antigenic drift, antigenic shift involves a principal alteration in a genome attributable to gene rearrangement between two related microorganisms. Since antigenic shift involves the acquisition of totally new antigens against which the host population is not immune, this alteration may lead to an epidemic of significant proportions.

Antigenic mosaicism refers to antigenic variation first discovered in pathogenic *Neisseria*. It is the result of genetic transformation between gonococcal strains. This is also observed in penicillin resistance of several bacterial species where the resistant organism contains DNA from a host commensal organism.

A **thymus-dependent (TD) antigen** (Figure 3.14) is an immunogen that requires T lymphocyte cooperation for B cells to synthesize specific antibodies. Presentation of TD antigen to T cells must be in the context of MHC class II molecules. TD antigens include proteins, polypeptides, hapten–carrier complexes, erythrocytes, and many other antigens that have diverse epitopes.

A **thymus-independent (TI) antigen** (Figure 3.15) is an immunogen that can stimulate B cells to synthesize antibodies without participation by T cells. These antigens are less complex than TD antigens. They are often polysaccharides that contain repeating epitopes or lipopolysaccharides (LPS) derived from Gram-negative microorganisms. TI antigens induce IgM synthesis by B lymphocytes without cooperation by T cells. They also do not stimulate immunological memory. Murine TI antigens are classified as either TI-1 or TI-2 antigens. LPS, which activate murine B cells without participation by T or other cells, are typical TI-1 antigens. Low concentrations of LPS stimulate synthesis of specific antigen, whereas high concentrations activate essentially all B cells to grow and differentiate. TI-2 antigens include polysaccharides, glycolipids, and

	T Cell-Dependent Antigen	T Cell-Independent Antigen
Structural Properties	Complex	Simple
Chemistry	Proteins; protein-nucleoprotein conjugates; glycoproteins; lipoproteins	Polysaccharide of pneumococcus; dextran polyvinyl pyrolidone; bacterial lipo-polysaccharide
Antibody class-induced	IgG, IgM, IgA (+IgD and IgE)	IgM
Immunological memory response	Yes	No
Present in most pathogenic microbes	Yes	No

FIGURE 3.16 Comparison of T cell-dependent with T cell-independent antigens.

nucleic acids. When T lymphocytes and macrophages are depleted, no antibody response develops against them.

An ***xid* gene** is an X-chromosome mutation designated *xid*. When homozygous it leads to diminished responsiveness to some thymus-independent antigens, limited decrease in responsiveness to thymus-dependent antigens, and defective terminal differentiation of B cells. Many autoimmune features are diminished when the *xid* gene is bred into autoimmune mouse strains. These include reduced anti-DNA antibody levels and diminished renal disease, together with increased survival.

A **private antigen** (Figure 3.16) is (1) an antigen confined to one MHC molecule; (2) an antigenic specificity restricted to a few individuals; (3) a tumor antigen restricted to a specific chemically induced tumor; (4) a low-frequency epitope present on red blood cells of fewer than 0.1% of the population, i.e., I, Pt^a, By, Bp^a, etc.; and (5) an HLA antigen encoded by one allele such as HLA-B27.

A **private specificity** is an epitope found on a protein encoded by a single allele. Thus, it is found only on one member of a group of proteins, such as alloantigens of the major histocompatibility complex, even though it may also apply to other alloantigenic systems.

A **public antigen (supratypic antigen)** (Figure 3.17) is an epitope which several distinct or private antigens have in common. A public antigen is one such blood group antigen that is present in greater than 99.9% of a population. It is detected by the indirect antiglobulin (Coombs' test). Examples include Ve, Ge, Jr, Gy^a, and OK^a. Antigens that occur frequently but are not public antigens include MNs, Lewis, Duffy, P, etc. In blood banking, there is a problem finding a suitable unit of blood for a tranfusion to recipients who have developed antibodies against public antigens.

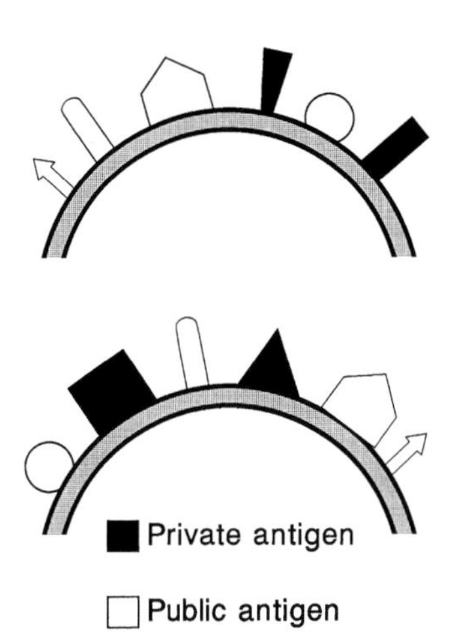

FIGURE 3.17 Public and private antigens.

A **heterophile antigen** is an antigen (epitope) present in divergent animal species, plants, and bacteria that manifest broad crossreactivity with antibodies of the heterophile group. Heterophile antigens induce the formation of heterophile antibodies when introduced into a species where they are absent. Heterophile antigens are often carbohydrates.

Forssman antigen is a heterophile or heterogenetic glycolipid antigen that stimulates the synthesis of antisheep hemolysin in rabbits. Its broad phylogenetic distribution spans both animal and plant kingdoms. The antigen is present in guinea pig and horse organs, but not in their red blood cells. In sheep, it is found exclusively in erythrocytes. Forssman antigen occurs in both red blood cells and organs in chickens. It is also present in goats, ostriches, mice, dogs, cats, spinach, *Bacillus anthracis*, and on the gastrointestinal mucosa of a limited number of people. It is absent in rabbits, rats, cows, pigs, cuckoos, beans, and *Salmonella typhi*. Forssman substance is ceramide tetrasaccharide. The Forssman antigen contains *N*-acylsphingo-sine (ceramide), galactose, and *N*-acetylgalactosamine. As originally defined, it is present in guinea pig kidney, is heat stable, and is alcohol soluble. Forssman antigen-containing tissue is effective in absorbing the homologous antibody from serum. Antibodies to the Forssman antigen occur in the sera of patients recovering from infectious mononucleosis.

Forssman antibody is an antibody specific for the Forssman (heterogenetic) antigen. Human serum may contain Forssman antibody as a natural antibody.

A **booster response** is the secondary antibody response produced during immunization of subjects primed by earlier exposure to the same antigen. Also called an anamnestic response and secondary response.

Immunization is the deliberate administration of an antigen or immunogen to induce active immunity that is often protective, as in the case of immunization against antigenic products of infectious disease agents.

The **inductive phase** is the time between antigen administration and detection of immune reactivity.

Inoculation is the introduction of an immunogen into an animal to induce immunity, usually to protect against an infectious disease agent.

Transformation is a heritable alteration in a cell as a consequence of investigative manipulation.

1. Lymphocyte transformation: The stimulation of a resting lymphocyte with a lectin, antigen, or lymphokine to undergo blast transformation associated with the cell's division, proliferation, and differentiation. This process can be assayed quantitatively by adding ^{3}H (thymidine) to the cell culture which becomes incorporated into the DNA.
2. Genetic transformation by DNA: Nonvirulent living pneumococci can become virulent after taking up DNA from dead pneumococci by transformation.
3. Cells can undergo neoplastic transformation in culture and acquire the capacity for unrestricted proliferation, thereby resembling neoplastic cells.

Immunize refers to the deliberate administration of an antigen or immunogen for the purpose of inducing active immunity that is often protective, as in the case of immunization against antigenic products of infectious disease agents. As a consequence of contact between the antigen or immunogen and immunologically competent cells of the host, specific antibodies and specifically reactive immune lymphoid cells are induced to confer a state of immunity.

Challenge refers to antigen deliberately administered to induce an immune reaction in an individual previously exposed to that antigen to determine the state of immunity.

ImD50 is the antigen (or vaccine) dose capable of successfully immunizing 50% of a particular animal test population.

Active immunization is the induction of an immune response either through exposure to an infectious agent or by deliberate immunization with products of the microorganism inducing the disease to develop protective immunity. A clinical disease or subclinical infection or vaccination may be used to induce the desired protective effect. Booster immunization injections given at intervals after primary exposure may lead to long-lasting immunity through the activation of immunological memory cells.

Parenteral refers to administration or injection of a substance into the animal body by any route except the alimentary tract.

Lymphocyte activation refers to the stimulation of lymphocytes *in vitro* by antigen or mitogen which renders them metabolically active. Activated lymphocytes may undergo transformation or blastogenesis.

The **two-signal hypothesis** is the concept that lymphocyte activation requires two separate signals, the first mediated by antigen and the second by either microbial products or consituents of the natural or innate immune response to microorganisms. The first signal mediated by antigen guarantees that the immune response will be specific. The second signal induced by microorganisms or innate immune responses ensures that immune responses are induced when required to defend against microorganisms or other offending agents but not against self antigens or harmless components. Signal 2 is now known as costimulation and is frequently mediated by professional antigen-presenting cell membrane molecules including B7 proteins.

An **immune response** is the reaction of the animal body to challenge by an immunogen. This is expressed as antibody production and/or cell-mediated immunity or immunologic tolerance. Immune response may follow stimulation by a wide variety of agents such as pathogenic microorganisms, tissue transplants, or other antigenic substances deliberately introduced for one purpose or another. Infectious agents may also induce inflammatory reactions characterized by the production of chemical mediators at the site of injury. In addition to this adaptive immune response described above is the innate or natural immunity that is present from birth and is designed to protect the host from injury or infection without previous contact with the infectious agent or antigen. It includes such protective factors as the skin, mucous membranes, lysozymen, tears, stomach acid, and numerous other factors. Phagocytes, natural killer cells, and complement represent key participants in natural innate immune responses.

Negative phase is the decrease in antibody titer immediately following injection of a second or booster dose of antigen to an animal previously given a primary injection of the same antigen. Following this initial drop of preformed antibody in the circulation, there is a rapid and pronounced rise in antibody titer, representing immunologic memory.

Antigen-specific suppressor cells (antigen-specific TS cells) can be demonstrated both in humoral and cell-mediated immunity. The TS cells active in the humoral response can be generated after priming with the carrier to be used in subsequent experiments with hapten–carrier conjugates. These TS cells can suppress the hapten-specific IgM and IgG antibody response if recipient animals are immunized with the hapten coupled to the homologous carrier. This type of suppression may have a differential effect on IgM and IgG antibody responses according to the time frame in which TS cells are administered to the recipient animal. The early IgG response is relatively independent of T cell function and, accordingly, less susceptible to TS cell effects. The late IgM and IgG responses are more T cell dependent and, accordingly, more susceptible to TS cell inhibition.

Primed refers to a lymphoid cell or an intact animal that has been exposed once to a specific antigen and which mounts a rapid and heightened response upon second exposure to this same antigen. Products of the reaction may be manifested as either increased antibody production or heightened cell-mediated immunity against that antigen.

A **primed lymphocyte** is one that has interacted with an antigen *in vivo* or *in vitro*.

Responder animals: Guinea pigs immunized by DNP-PLL or DNP-GL produce significant amounts of anti-DNP antibodies and develop a delayed type hypersensitivity reaction to the conjugate. Conversely, nonresponder animals produce little or no anti-DNP antibodies and do not develop delayed-type hypersensitivity. The capacity to respond to DNP-PLL antigen is a function of a dominant autosomal gene. All guinea pigs of strain-13 are nonresponders, whereas strain-2 guinea pigs and (2 × 13) F_1 hybrids are responders. The immune response gene present in strain-2 but absent in strain-13 is called the *PLL gene*.

Polygenic inheritance is phenotypic inheritance based on genetic variation at multiple loci. Numerous genetic loci may contribute to the inherited phenotype. Certain forms of immune responsiveness and susceptibility to selected diseases may be influenced by polygenic inheritance.

The **primary immune response** is the animal body's adaptive immune response to its first contact with antigen (Figure 3.18). It is characterized by a lag period of a few days following antigen administration before antibodies or specific T lymphocytes are detectable. The antibody produced consists mostly of IgM and is of relatively low titer and low affinity. This is in contrast to the secondary immune response in which the latent period is relatively brief and IgG is the predominant antibody. The most important event that occurs in the primary response is the activation of memory cells that recognize antigen immediately on their second encounter with it, leading to a secondary response. A similar pattern is followed in cell-mediated responses.

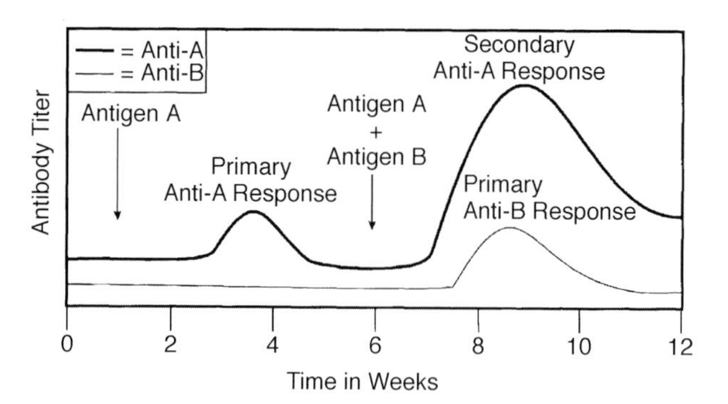

FIGURE 3.18 Primary immune response.

The **primary response** is the first response to an immunogen to which the recipient has not previously been exposed. It is principally an IgM response and generates immunologic memory.

A **nonresponder** is an animal that fails to generate an immune response following antigenic challenge. This may be genetically based, as in strain-13 guinea pigs which reveal unresponsiveness to selected antigens based on genetic factors.

The **secondary immune response** describes a heightened antibody response following second exposure to antigen in animals that have been primed by previous contact with the same antigen. The secondary immune response depends upon immunological memory learned from the first encounter with antigen. It is characterized by a steep and rapid rise in antibody titer, usually of the IgG class, accompanied by potent cell-mediated immunity. Protein and glycoprotein immunogens stimulate this type of response. The rapid rise in antibody synthesis is followed by a gradual exponential decline in titer. This is also known as the booster response observed following administration of antigens subsequent to the secondary exposure.

The **secondary response** is the anamnestic or memory immune response that occurs following second exposure to an antigen in an individual who has been sensitized by a primary immunizing dose of the same antigen. The secondary response occurs more rapidly and is of much greater magnitude than the primary response because of the memory cells that respond to the second challenge. Also called booster response.

Secondary antibody response: IgM appears before IgG in the primary antibody response. The inoculation of an immunogen into an experimental animal which has been primed by previous immunization with the same immunogen produces antibodies following the secondary immunogenic challenge that develop more rapidly, last longer, and reach a higher titer than in the primary response. Antibodies produced in the secondary response are predominantly IgG and reach levels that are tenfold or greater than those in the primary antibody response.

Iccosomes (immune complex coated antibodies) are small pieces of cell membrane derived from follicular dendritic cell processes and laden with immune complexes. These immune complex-coated particles fragment off follicular dendritic cell processes in lymphoid follicles early in a secondary or subsequent antibody response.

Immunoenhancement is the process of increasing or contributing to the level of immune response by various specific and nonspecific means such as immunization.

Immunopotentiation is facilitation of the immune response usually with the aid of adjuvants such as muramyl dipeptide, Freund's adjuvant, synthetic polynucleotides, or other agents. Biological response modifiers, cloned cytokines, and purified immunoglobulins have been used as immunopotentiating agents.

A **booster** is a second administration of immunogen to an individual primed months or years previously by a primary injection of the same immunogen. The purpose is to deliberately induce a secondary or anamnestic or memory immune response to facilitate protection against an infectious disease agent.

Booster injection refers to the administration of a second inoculation of an immunizing preparation, such as a vaccine, to which the individual has been previously exposed. The booster inoculation elicits a recall or anamnestic response through stimulation of memory cells that have encountered the same antigen previously. Booster injections are given after the passage of time sufficient for a primary immune response specific for the immunogen to have developed. Booster injections are frequently given to render the subject immune prior to the onset of a particular disease or to protect the individual when exposed to subjects infected with the infectious disease agent against which immunity is desired.

Booster phenomenon is an expansion in the diameter of a tuberculin reaction following the administration of a subsequent PPD skin test for tuberculosis. This is usually greater than 6 mm and shows an increase in size from below 10 mm to greater than 10 mm in diameter following the secondary challenge. A positive test suggests an increased immunologic recall as a consequence of either previous infection with *Mycobacterium tuberculosis* or other mycobacteria. It is seen in older subjects with previous *M. tuberculosis* infections who fail to convert to active disease.

An **anamnestic immune response** (Figure 3.19) is one that is accentuated and occurs following exposure of immunocompetent cells to an immunogen to which they have been exposed before. Commonly called the secondary or anamnestic response, it occurs rapidly, i.e., within hours following secondary immunogen inoculation and does not have the lag period observed with primary immunization. Immunologic memory is involved in the production of this response which generally consists of IgG antibodies of high titer and high affinity. There may also be heightened T cell (cell-mediated) immune reactivity. It is

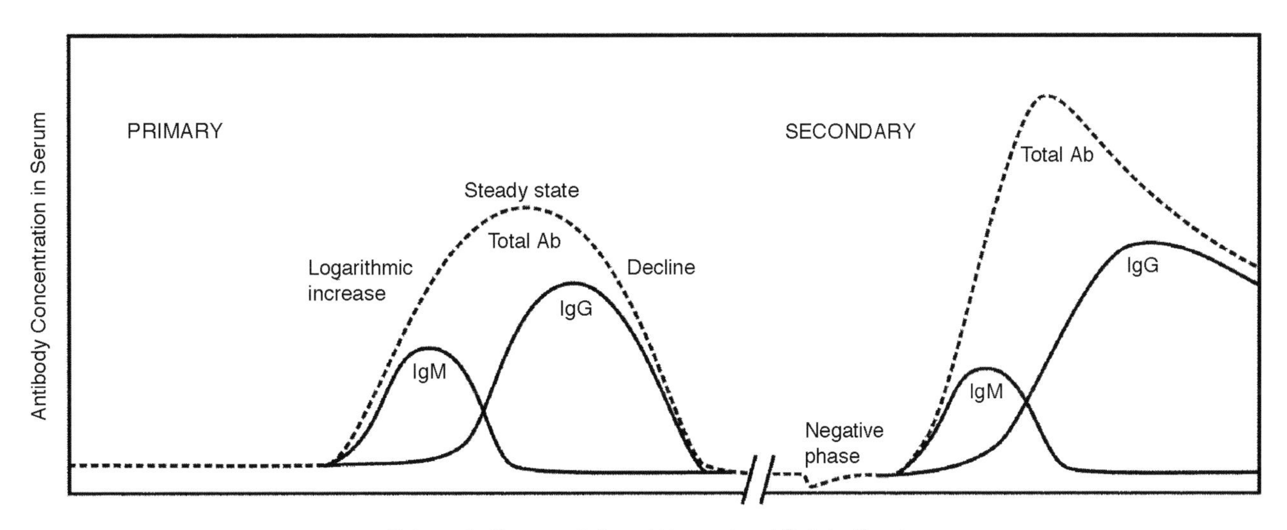

FIGURE 3.19 Anamnestic immune response.

also called memory or booster response or secondary immune response.

Anamnesis is immunologic memory. The term refers to the elevated immune response following secondary or tertiary administration of immunogen to a recipient previously primed or sensitized to the immunogen, i.e., the secondary response.

Anamnestic describes the recall response of immunologic memory that results in a rapid rise in antibody production following reexposure to the same antigen.

Memory is the capacity of the adaptive immune system to respond more rapidly, more effectively, and with greater magnitude to a second (or subsequent) exposure to an immunogen compared to the response to the primary exposure to the immunogen.

Memory cells are immunocompetent T and B lymphocytes that have the ability to mount an accentuated response to antigen compared to that of virgin immunocompetent cells because of their previous exposure to the antigen through immunization or infection.

Memory lymphocytes are lymphocytes of either the B or T type that respond rapidly with an enhanced memory or recall response to second or subsequent exposure to an antigen to which they were primed previously. Antigen-stimulation of naïve lymphocytes leads to the production of memory B and T cells that persist in a functionally dormant state years following antigen elimination.

Memory T cells are long-lived antigen-specific T lymphocytes that respond in an immediate and exaggerated manner to induce a heightened immune response to a specific antigen.

A **tertiary immune response** (Figure 3.20) is one that is induced by a third (second booster) administration of antigen. It closely resembles the secondary (or booster) immune response.

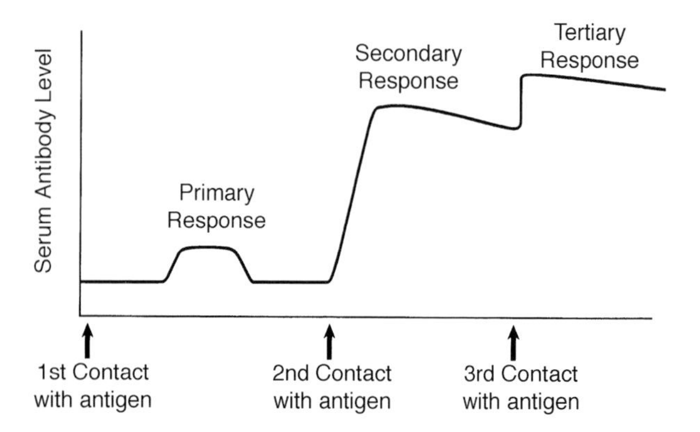

FIGURE 3.20 Tertiary immune response.

A **tertiary response** is the consequence of injecting an immunogen into an animal for the third time.

Tertiary immunization is the immune response following injection of the same immunogen for the third time.

An **adjuvant** (Figure 3.21) is a substance that facilitates or enhances the immune response to an antigen with which it is combined. Various types of adjuvants have been described, including Freund's complete and incomplete adjuvants, aluminum compounds, and muramyl dipeptide. Some of these act by forming a depot in tissues from which an antigen is slowly released. In addition, Freund's adjuvant attracts a large number of cells to the area of antigen deposition to provide increased immune responsiveness to it. Modern adjuvants include such agents as muramyl dipeptide. The ideal adjuvant is one that is biodegradable with elimination from the tissues once its immunoenhancing activity has been completed. An adjuvant nonspecifically facilitates an immune response to antigen. It usually combines with the immunogen but is sometimes given prior to or following antigen administration. Adjuvants represent a heterogenous class of compounds capable of augmenting the humoral or cell-mediated immune response to a given antigen. They are widely used in experimental work and for therapeutic purposes in vaccines. Adjuvants comprise compounds of mineral nature, products of microbial origin, and synthetic compounds. The primary effect of some adjuvants is postulated to be the retention of antigen at the inoculation site so that the immunogenic stimulus persists for a longer period of time. However, the mechanism by which adjuvants augment the immune response is poorly understood. The macrophage may be the target and mediator of action of some adjuvants, whereas others may require T cells for their response augmenting effect. Adjuvants such as LPS may act directly on B lymphocytes. They enhance activation of T lymphocytes by facilitating the accumulation and activation of accessory cells at a site of antigen exposure. Adjuvants facilitate expression by the accessory cell of T cell-activating costimulators and cytokines and are believed to prolong the expression of peptide-MHC complexes on the surface of antigen-presenting cells (APCs).

An **immunologic adjuvant** is a substance that enhances an immune response, either humoral or cellular or both, to an immunogen (antigen).

Alums are aluminum salts employed to adsorb and precipitate protein antigens from solution, followed by the use of the precipitated antigen as an immunogen which forms a depot in animal tissues. See also aluminum adjuvant.

An **adjuvant granuloma** is a tissue reaction that occurs at a local site following the injection of such adjuvant materials as Freund's complete adjuvant or alum, both of

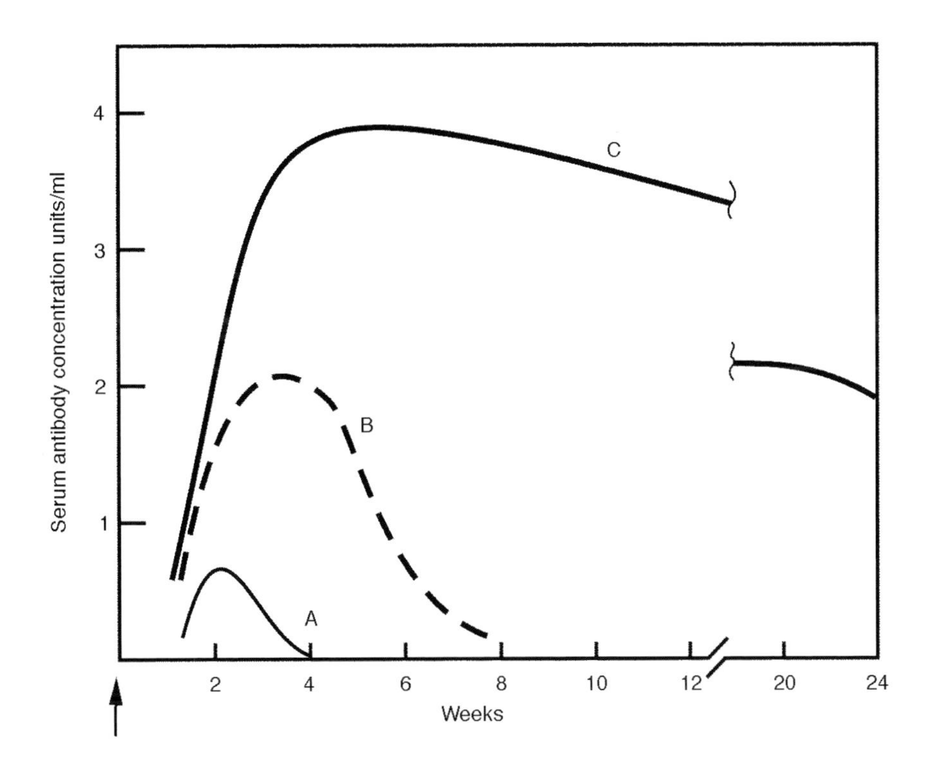

FIGURE 3.21 Effect of adjuvants. Schematic representation of the quantities of antibodies formed by rabbits following a single injection of a soluble protein antigen →, such as bovine gammaglobulin in dilute physiologic saline solution (A), adsorbed on precipitated alum (B), or incorporated into Freund's complete adjuvant (C).

which have been used extensively in immunologic research in past years.

Alum-precipitated antigen is a soluble protein antigen such as a toxoid adsorbed to aluminum salts during precipitation from solution. The aluminum salt acts as an adjuvant that facilitates an immune response to such antigens as diphtheria and tetanus toxoids. Soluble protein antigen is combined with 1% potassium aluminum sulfate, and sodium hydroxide is added until floccules are produced.

An **alum granuloma** is tissue reaction in the form of a granuloma produced at the local site of intramuscular or subcutaneous inoculation of a protein antigen precipitated from solution by an aluminum salt acting as adjuvant. Slow release of antigen from the granuloma has been considered to facilitate enhanced antibody synthesis to the antigen.

A **liposome** is a spherical lipid vesicle comprised of 5-nm phospholipid bilayers that enclose one or several aqueous units. These micropheres can be produced by dispersing phospholipid mixtures with or without sterols in aqueous solution. The liposomes produced consist of concentric phospholipid bilayers. Thus, they represent ideal models of cell membranes into which antigens may be embedded to induce an immune response. They also have been used to deliver drugs. They may or may not fuse with the cell membrane, and there are other problems associated with their use, such as whether or not they will leave the circulatory system or can be phagocytized by reticuloendothelial cells. Liposomes may act as immunologic adjuvants when antigens are incorporated into them. Liposomes serving as biological membrane models have also been used in studies on complement-mediated lysis.

Alum-precipitated toxoid: See alum-precipitated antigen.

Triton X-100 is a quaternary ammonium salt that is surface active. Chemically, it is isooctyl phenoxy polyethoxy ethanol. It has been used as a detergent, emulsifier, wetting agent, and surfactant.

An **aluminum adjuvant** is an aluminum-containing substance that has a powerful capacity to adsorb and precipitate protein antigens from solution. The use of these preparations as immunogen causes depot formation in the tissues at the site of inoculation from which the antigen is slowly released, thereby facilitating greater antibody production than if the antigen is dissipated and rapidly lost from the body. Substances used extensively in the past for this purpose include aluminum hydroxide gel, aluminum sulfate, and ammonium alum, as well as potassium alum.

Aluminum hydroxide gel [$Al(OH)_3$] was widely used in the past as an immunologic adjuvant by reacting antigen with 2% hydrated $AL(OH)_3$ to adsorb and precipitate the protein antigen from solution. See also aluminum adjuvant.

Tapioca adjuvant (historical) is an immunologic adjuvant consisting of starch granules to which molecular antigen was absorbed. This permitted the adjuvant–antigen complex to form a depot in the tissues from which the antigen was slowly released to stimulate a sustained antibody response.

APT (alum-precipitated toxoid): See alum-precipitated antigen.

Tween is a nonionic detergent.

Tween 80® is polyoxyethylene sorbitan monooleate. An emulsifying agent used in cultures of mycobacteria and in water-in-oil-in-water emulsion adjuvants as a stabilizing agent.

ISCOMs are immune stimulatory complexes of antigen bound in a matrix of lipid that serves as an adjuvant and facilitates uptake of antigen into the cytoplasm of a cell following plasma membrane–lipid fusion.

Aquaphor® is an emulsifying preparation of lanolin used extensively in the past to prepare the water-in-oil emulsion immunologic adjuvants of the Freund type.

Aqueous adjuvants: Freund's adjuvants are not used in man because of the ease with which hypersensitivity is induced and the unpredictability of the local reaction. A number of water-soluble synthetic components comprising active moieties of the mycobacterial cell wall have been synthesized in search of more adequate adjuvants. One of them is muramyl dipeptide (MDP), which is active when administered by the oral route. MDP in water is extremely active as an adjuvant and is not generally toxic when administered at high doses. It is neither mitogenic or immunogenic nor antigenic and is rapidly eliminated from the animal body. The simplicity of its chemical structure allows the study of the targets of its action in the immune system. Other synthetic adjuvant compounds are polynucleotides such as poly-inosine-poly-cytidine (poly-I:C), whose structure is similar to that of native nucleotides. Their mechanism of action involves signals which are rapidly received by the immune system since such compounds are destroyed in 5 to 10 min by the nucleases of the serum. The prevalent concept is that adjuvants have a number of other regulatory activities on the immune response, and the term *adjuvant* may be replaced by *immunoregulatory molecule*.

Arlacel® A is a mannide monooleate used as an emulsifier to stabilize water-in-oil emulsions employed as adjuvants, e.g., Freund's adjuvant, in experimental immunology.

Chemical adjuvants that are used for immunopotentiation include the polynucleotide poly-I:C and poly-A:U, vitamin D_3, dextran sulphate, inulin, dimethyl, dioctadecyl ammonium bromide (DDA), avridine; carbohydrate polymers similar to mannan, and trehalose dimycolate, among others. Two of the newer chemical adjuvants include polyphosphazines (initially introduced as slow release-promoting agents) and a *Leishmania* protein LeIF.

Depot-forming adjuvants are substances that facilitate an immune response by holding an antigen at the injection site following inoculation. They facilitate the slow release of antigen over an extended period and help attract macrophages to the site of antigen deposition. To be effective, they must be administered together with the antigen. Water-in-oil emulsion of the Freund type, as well as aluminum salt (aluminum hydroxide) adjuvants, are examples of depot-forming adjuvants. In the past, depot-forming and centrally acting adjuvants were distinguished. However, adjuvant action depends upon far more complicated cellular and molecular mechanisms than the simplistic views of depot formation advanced in the past.

Drakeol 6VR® is a purified light mineral oil employed to prepare water-in-oil emulsion adjuvants.

A **copolymer** is a polymer such as a polypeptide comprised of at least two separate chemical specificities, such as two different amino acids.

A **polynucleotide** is a linear polymer comprised of greater than ten nucleotides joined by 3′,5′-phosphodiester bonds. Double-stranded DNA chains that may serve as adjuvants when inoculated with antigens.

Incomplete Freund's adjuvant (IFA) is lightweight mineral oil without mycobacteria which, when combined with aqueous-phase antigen as a water-in-oil emulsion, enhances the humoral or antibody (B cell) limb of the immune response. It does not facilitate T cell-mediated immune responsiveness.

Mycobacterial adjuvants are substances that have long been used to enhance both humoral and cellular immune responses to antigen. Killed, dried mycobacteria, including *Mycobacterium tuberculosis* among other strains, are ground and suspended in lightweight mineral oil. With the aid of an emulsifying agent such as Arlacel A, added antigen in aqueous medium is incorporated to produce a water-in-oil emulsion. This mixture is used for immunization. The mycobacteria are especially effective in stimulating cell-mediated immunity to the antigen. The administration of this adjuvant without antigen may induce adjuvant arthritis in rats. Incorporation of normal tissues such as thyroid or adrenal into Freund's complete adjuvant may induce autoimmune disease if reinoculated into the animal of origin or other members of the strain with the same genetic background.

Mycobacterial peptidoglycolipid is a constituent of the wax D fraction of mycobacteria. The wax D fraction of *Mycobacterium tuberculosis* var. *hominis* contains the adjuvant principle associated with such mycobacterial adjuvants as Freund's adjuvant. By electron microscopy, the peptidoglycolipid has a homogeneous, intertwined filamentous structure.

Wax D is a high molecular weight (70 kDa) glycolipid and peptidoglycolipid extracted from wax fractions of *Mycobacterium tuberculosis* isolated from man. It is soluble in chloroform and insoluble in boiling acetone and has the adjuvant properties which *M. tuberculosis* organisms add to Freund's adjuvant preparations. Thus, wax D can be used to replace mycobacteria in adjuvant preparation to enhance cellular and humoral immune responsiveness to antigen.

Pertussis adjuvant contains killed *Bordetella pertussis* microorganisms that have been mixed with antigen to enhance antibody production. Pertussis adjuvant particularly facilitates IgE synthesis in rats or other animals. When used as a component in the triple vaccine, i.e., diphtheria–pertussis–tetanus (DPT) preparation used for childhood immunization, the killed *B. pertussis* microorganisms not only stimulate antibodies that protect against whooping cough but also facilitate antibody synthesis against both the diphtheria and tetanus toxoid preparations, thereby serving as an immunologic adjuvant.

Silica adjuvants: In the past, silica crystals and hydrated aluminum silicate (bentonite) were occasionally used to enhance the immune response to certain antigens. They were considered to have a central action on the immune system rather than serving as depot adjuvants, yet bentonite can delay the distribution of antigen from its site of inoculation by surface adsorption.

Squalene is a substance synthesized in the body when cholesterol is converted to fat. It is present in selected cosmetics and foods. Squalene has immunologic adjuvant properties and has been used in experimental vaccines as an adjuvant to facilitate the response to an immunogen. Adjuvants are incorporated into certain vaccines to improve the immune response to the vaccine constituents. Not all vaccines contain adjuvants. Some do not require them to induce protective immunity in recipients. Aluminum compounds are common adjuvants in licensed vaccines. Some vaccines administered to military personnel during the Gulf War, including the anthrax, botulinum toxoid, hepatitis B, and tetanus–diphtheria vaccines, contained aluminum compounds as adjuvants. Antibodies to artificial squalene found in the blood sera of Gulf War veterans have been interpreted by some investigators to imply that squalene or antibodies to squalene may have contributed to these illnesses. These investigators believe that squalene was used as an adjuvant, a claim that has been denied by the U.S. Department of Defense. Squalene is not approved by the FDA for use as an adjuvant in anthrax or any other vaccine.

Multiple-emulsion adjuvant is a water-in-oil-in-water emulsion adjuvant.

Solubilized water-in-oil adjuvant is a water-in-oil emulsion adjuvant comprised of a small volume aqueous phase compared to the volume of oil. A mixture of aqueous and oil phases results in an emulsion which is stabilized by the addition of emulsifying agents.

Double-emulsion adjuvant is a water-in-oil-in-water emulsion adjuvant.

Freund's adjuvant is a water-in-oil emulsion that facilitates or enhances an immune response to antigen that has been incorporated into the adjuvant. There are two forms. Freund's complete adjuvant (CFA) consists of lightweight mineral oil that contains killed, dried mycobacteria. Antigen in an aqueous phase is incorporated into the oil phase containing mycobacteria with the aid of an emulsifying agent such as Arlacel A. This emulsion is then used as the immunogen. Freund's incomplete adjuvant (IFA) differs from the complete form only in that it does not contain mycobacteria. In both cases, the augmenting effect depends on and parallels the magnitude of the local inflammatory lesion, essentially a nonnecrotic monocytic reaction with fibrous encapsulation. Whereas the complete form facilitates stimulation of both T and B limbs of the immune response, the incomplete variety enhances antibody formation but does not stimulate cell-mediated immunity except for transient Jones–Mote reactivity. The adjuvant principle in mycobacteria is the cell wall wax D fraction. CFA does not potentiate the immune response to the so-called thymus-independent antigens such as pneumococcal polysaccharide or polyvinylpyrrolidone. CFA may be combined with normal tissues and injected into animals of the type supplying the tissue to induce autoimmune diseases such as thyroiditis, allergic encephalomyelitis, or adjuvant arthritis.

Freund's complete adjuvant: See Freund's adjuvant.

Freund's incomplete adjuvant: See Freund's adjuvant.

CFA is an abbreviation for complete Freund's adjuvant.

Immunologic competency is the capacity of an animal's immune system to generate a response to an immunogen.

Complete Freund's adjuvant (CFA) consists of a suspension of killed and dried mycobacteria suspended in lightweight mineral oil, which when combined with an

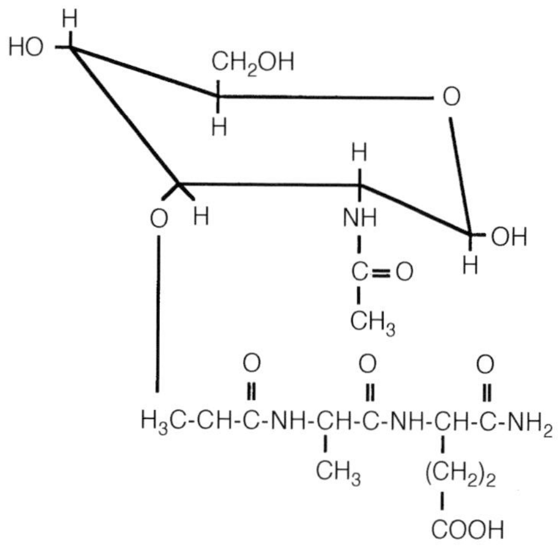

H_3C-CH-C-NH-CH-C-NH-CH-C-NH_2

N-acetylmuramyl-L-alanyl-D-isoglutamine

FIGURE 3.22 Muramyl dipeptide (MDP).

aqueous-phase antigen as a water-in-oil emulsion, enhances immunogenicity. CFA facilitates stimulation of both humoral (B cell) and cell-mediated (T cell) limbs of the immune response.

Incomplete Freund's adjuvant (IFA) is a lightweight mineral oil, which without mycobacteria, when combined with aqueous-phase antigen, such as a water-in-oil emulsion, enhances the humoral or antibody (B cell) limb of the immune response. It does not facilitate T-cell mediated immune responsiveness.

Muramyl dipeptide (MDP) (*N*-acetyl-muramyl-L-alanyl-D-isoglutamine) (Figure 3.22) is the active principle responsible for the immunologic adjuvant properties of complete Freund's adjuvant. It is an extract of the peptidoglycan of the cell walls of mycobacteria in complete Freund's adjuvant that has the immunopotentiating property of inducing delayed-type hypersensitivity and boosting antibody responses. It induces fever and lyses blood platelets and may produce a temporary leukopenia. However, purified derivatives without adverse side effects have been prepared for use as immunologic adjuvants and may prove useful for use in human vaccines.

MDP: Muramyl dipeptide.

4 Major Histocompatibility Complex

The first recognition of major histocompatibility complex (MHC) genes was based on their ability to encode proteins that serve as identity markers on tissues and cells that have been transplanted into an incompatible recipient. Their recognition by the recipient's lymphocytes leads to prompt rejection. The preoccupation of investigators with the role of histocompatibility antigens in transplantation obscured their real purpose of serving as identity markers on cells interacting with T cells carrying out specific immune functions through their own T cell receptors. In fact, T lymphocytes recognize antigens only in the context of MHC molecules.

Histocompatibility means tissue compatibility as in the transplantation of tissues or organs from one member to another of the same species, i.e., an allograft, or from one species to another, a xenograft. The genes that encode antigens which should match if a tissue or organ graft is to survive in the recipient are located in the MHC region. These are genes that encode the major, as opposed to minor, histocompatibility antigens that are expressed on cell membranes. The MHC gene is located on the short arm of chromosome 6 in man and of chromosome 17 in the mouse (Figure 4.1 and Figure 4.2). Class I (Figure 4.3 and Figure 4.4) and class II MHC antigens are important in tissue transplantation. The greater the match between donor and recipient, the more likely the transplant is to survive. For example, a six-antigen match implies sharing of two HLA-A antigens, two HLA-B antigens, and two HLA-DR antigens between donor and recipient. Even though antigenically dissimilar grafts may survive when a powerful immunosuppressive drug such as cyclosporine is used, the longevity of the graft is still improved by having as many antigens to match as possible.

Calnexin (Figure 4.5) is an 88-kDa membrane molecule that combines with newly formed α chains that also interact with nascent β_2-microglobulin. Calnexin maintains partial folding of the MHC class I molecule in the endoplasmic reticulum. It also interacts with MHC class II molecules, T cell receptors, and immunoglobulins that are partially folded. Class III MHC genes encode secreted proteins such as complement components, tumor necrosis factor, and soluble stem proteins. In humans, the A, B, and C regions encode class I MHC molecules, whereas the DP, DQ, and DR regions encode class II molecules. The C4, CD2, and Bf regions encode class III molecules in humans. Class I and class II MHC antigens are important in tissue transplantation. The greater the match between donor and recipient, the more likely the transplant is to survive. Even though antigenically dissimilar grafts may survive when a powerful immunosuppressive drug such as cyclosporine is used, the longevity of the graft is still improved by having as many antigens to match as possible.

Calreticulin is a soluble protein structurally related to calnexin and present in the endoplasmic reticulum, which facilitates proper folding of MHC molecules and other glycoproteins. It is a molecular chaperone that binds to MHC class I, MHC class II, as well as other immunoglobulin-like domain containing proteins, including antigen receptors for both T and B cells.

The **MHC** is a locus on a chromosome comprised of multiple genes that encode histocompatibility antigens that are cell surface glycoproteins, which play a significant role in interaction among immune cells. MHC genes encode both class I and class II MHC antigens. These antigens play critical roles in interactions among immune system cells, such as class II antigen participation in antigen presentation by macrophages to $CD4^+$ lymphocytes; the participation of class I MHC antigens in cytotoxicity mediated by $CD8^+$ T lymphocytes against target cells such as those infected by viruses, as well as various other immune reactions. MHC genes are very polymorphic and also encode a third category termed class III molecules that include complement proteins C2, C4, and factor B; P-450 cytochrome 21-hydroxylase; tumor necrosis factor; and lymphotoxin. The MHC locus in man is designated HLA, in the mouse as H2, in the chicken as B, in the dog as DLA, in the guinea pig as GPLA, and in the rat as RT1. The mouse and human MHC loci are the most widely studied. When organs are transplanted across major MHC locus differences between donor and recipient, graft rejection is prompt.

Major histocompatibility system: See major histocompatibility complex.

A **histocompatibility locus** is the specific site on a chromosome where the histocompatibility genes that encode histocompatibility antigens are located. There are major histocompatibility loci such as HLA in the human and H-2 in the mouse, across which incompatible grafts are rejected within 1 to 2 weeks. There are also several minor histocompatibility loci with more subtle antigenic differences, across which only slow, low-level graft rejection reactions occur.

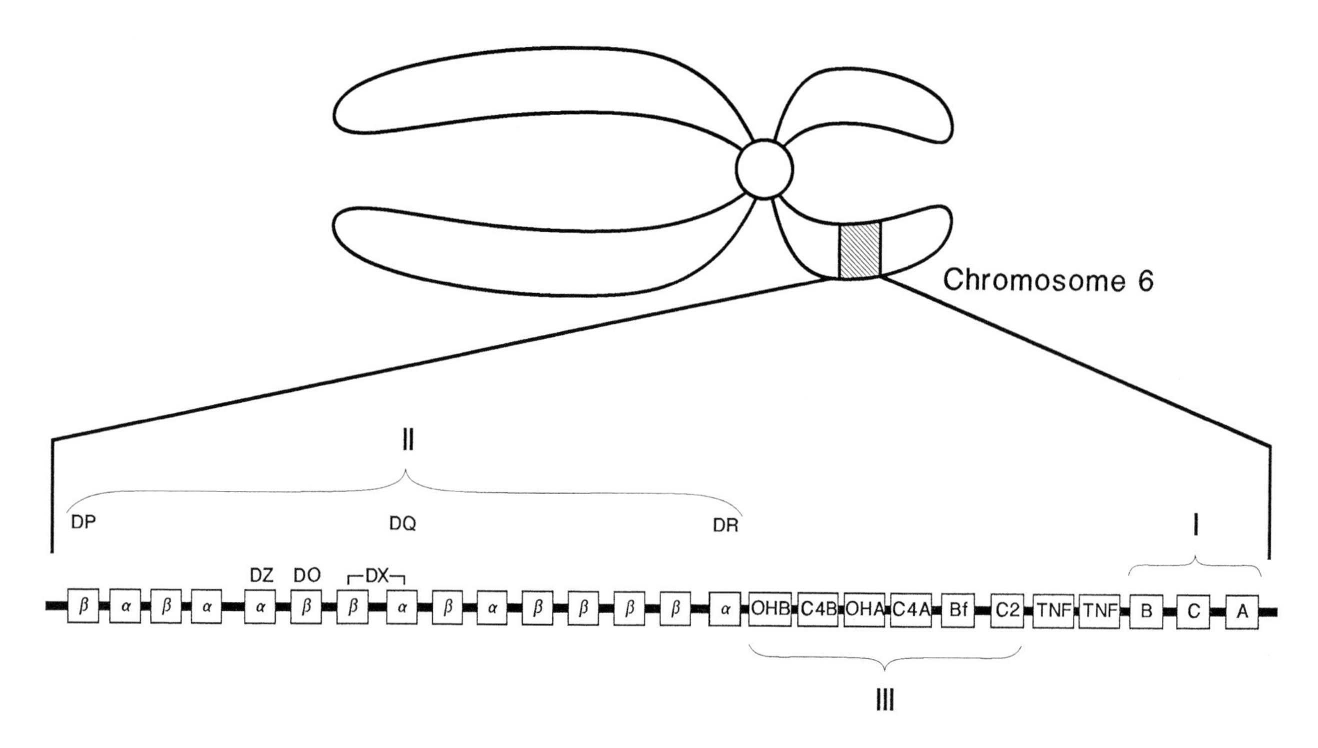

FIGURE 4.1 HLA is an abbreviation for human leukocyte antigen. The HLA locus in humans is found on the short arm of chromosome 6. The class I region consists of HLA-A, HLA-B, and HLA-C loci, and the class II region consists of the D region which is subdivided into HLA-DP, HLA-DQ, and HLA-DR subregions.

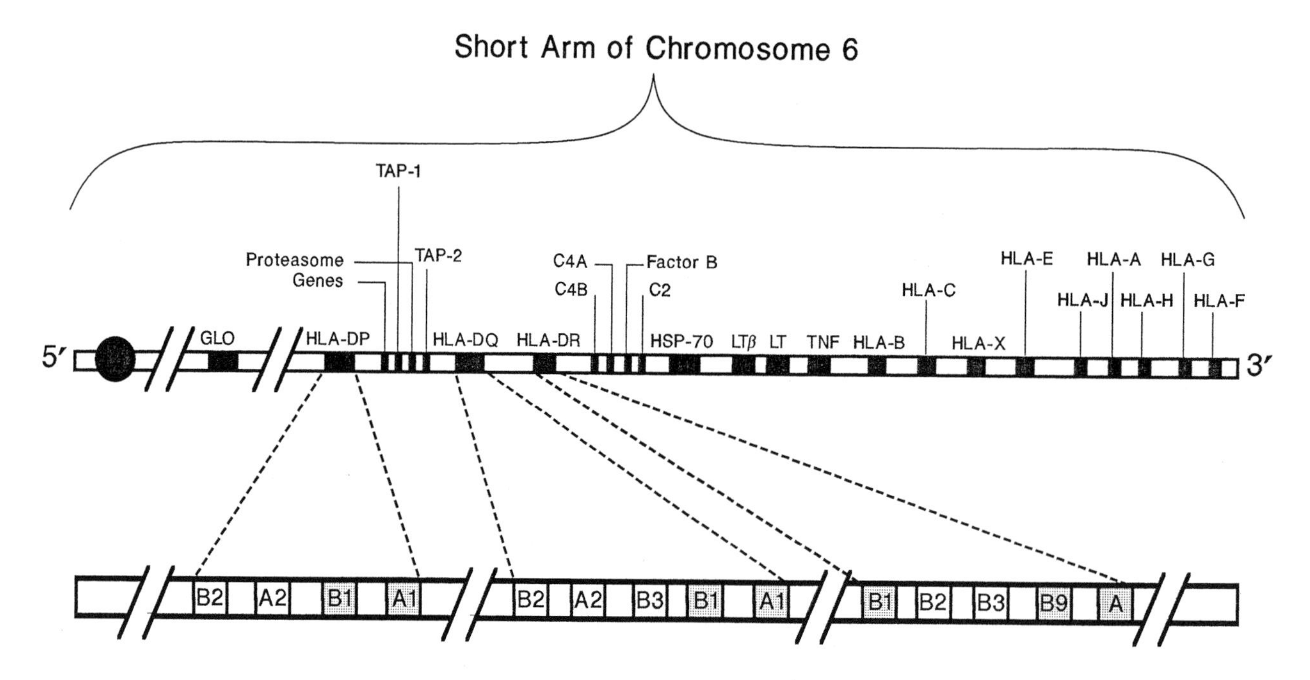

FIGURE 4.2 MHC genes encode the major histocompatibility antigens that are expressed on cell membranes. MHC genes in the mouse are located at the H-2 locus on chromosome 17, whereas the MHC genes in man are located at the HLA locus on the short arm of chromosome 6.

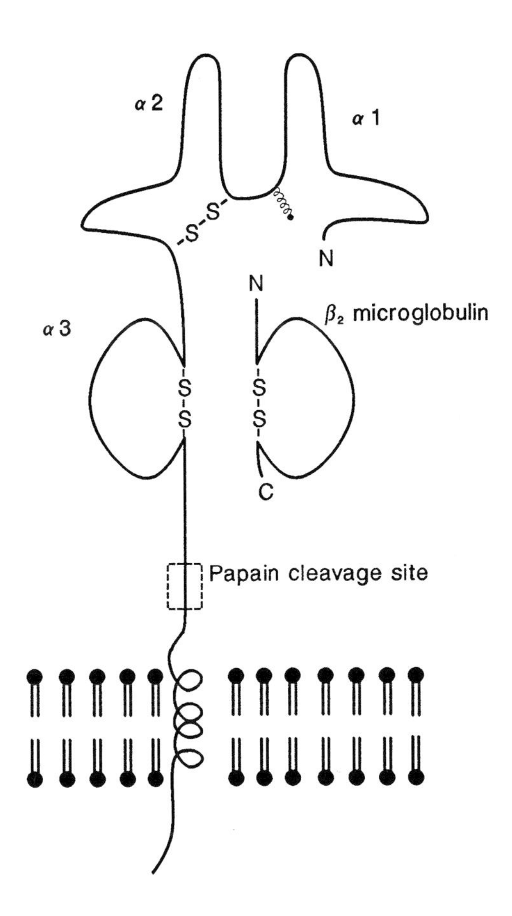

FIGURE 4.3 Class I MHC molecules are glycoproteins that play an important role in the interactions among cells of the immune system.

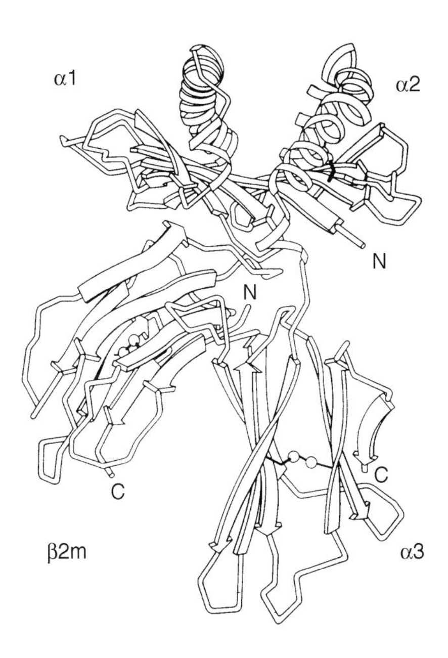

FIGURE 4.4 Schematic representation of the three-dimensional structure of the external domains of a human class I HLA molecule based on x-ray crystallographic analysis. The β strands are depicted as thick arrows and the α helices as spiral ribbons. Disulfide bonds are shown as two interconnected spheres. The α_1 and α_2 domains interact to form the peptide-binding cleft. Note the immunoglobulin-fold structure of the α_3 and the β_2 microglobulin.

A **histocompatibility antigen** is one of a group of genetically encoded antigens present on tissue cells of an animal that provoke a rejection response if the tissue containing them is transplanted to a genetically dissimilar recipient. These antigens are detected by typing lymphocytes on which they are expressed. They are encoded in the human by genes at the HLA locus on the short arm of chromosome 6. In the mouse, they are encoded by genes at the H-2 locus on chromosome 17.

MHC genes are major histocompatility complex genes. They are genes that encode the major, as opposed to minor, histocompatibility antigens that are expressed on cell membranes. MHC genes in the mouse are located at the H-2 locus on chromosome 17, whereas the MHC genes in man are located at the HLA locus on the short arm of chromosome 6.

A **centiMorgan (cM)** is a chromosomal unit of physical distance that corresponds to a 1% recombination frequency between two genes that are closely linked. Also termed a map unit.

A **locus** is the precise location of a gene on a chromosome of a gene or other chromosome marker; also the DNA at that position. The use of locus is sometimes restricted to mean regions of DNA that are expressed.

Loci is the plural of locus.

MHC is the abbreviation for major histocompatibility complex.

Codominantly expressed refers to gene expression when both alleles at one locus are expressed in approximately equal amounts in heterzygotes. The highly polymorphic MHC genes as well as most other genes manifest this property.

***bm* mutants:** Mouse H-2 *bm* mutants have served as elegant genetic tools to perform a detailed analysis of H-2 gene structure and its relationship to function of the gene products. The mutants are especially useful where intact animals are needed to investigate a particular process. They have aided the assignment of diverse biological and immunologic functions to single H-2 genes. They also influence mating preferences as evidenced by the ability of mice to distinguish parental animals from those carrying *bm* mutations by urine odor.

The **B complex** is the major histocompatibility complex (MHC) in chickens. Genes at these loci determine MHC class I and II antigens and erythrocyte antigens.

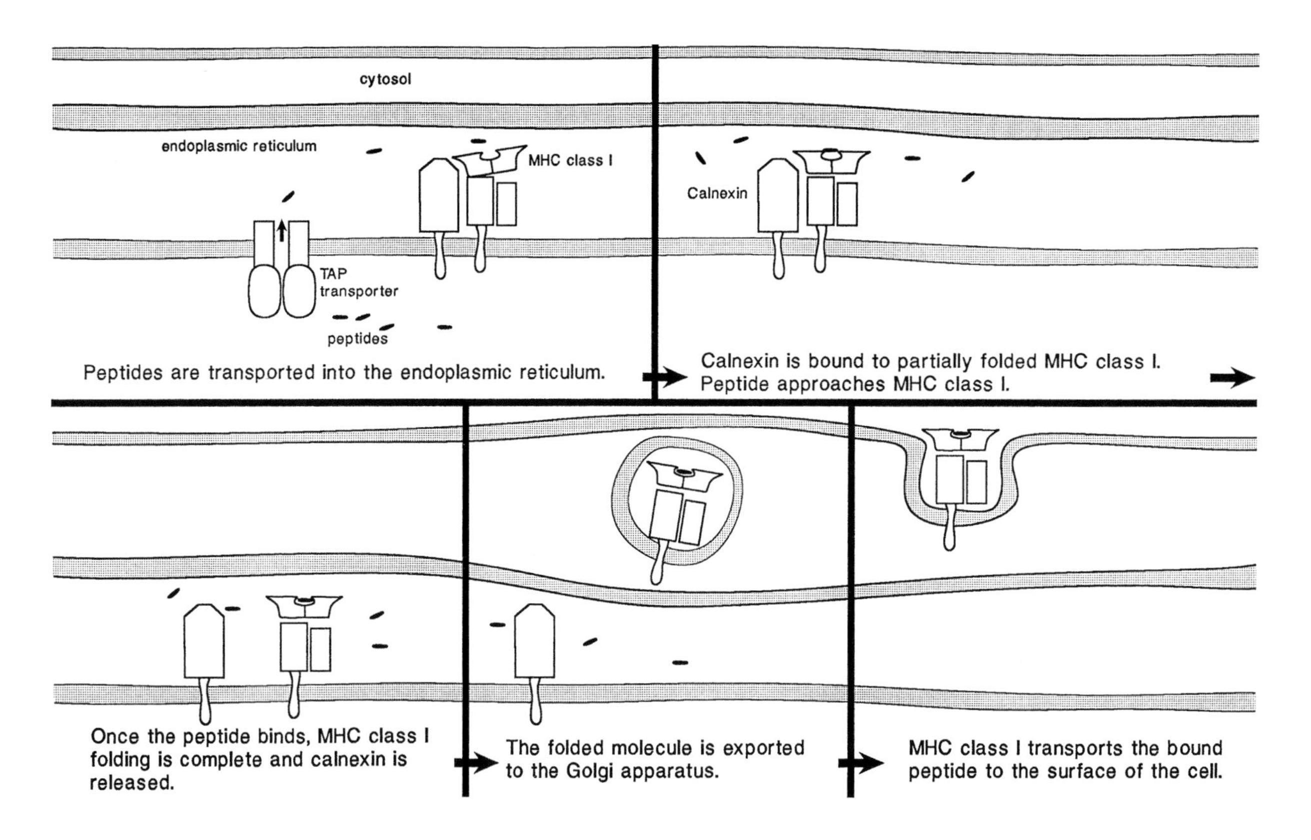

FIGURE 4.5 Calnexin is an 88 kD membrane molecule that combines with newly formed a chains that also interact with nascent b2-microglobulin. Calnexin maintains partial folding of the MHC-class I molecule in the endoplasmic reticulum. It also interacts with MHC class II molecules, T cell receptors, and immunoglobulins that are partially folded.

B genes are situated within the MHC locus on the short arm of chromosome 6. They are termed MHC class III genes. TNF-α and TNF-β genes are situated between the C2 and HLA-B genes. Another gene, designated FD, lies between the *Bf* and *C4a* genes. C2 and B complete primary structures have been deduced from cDNA and protein sequences. C2 is comprised of 732 residues and is an 81-kDa molecule, whereas B contains 739 residues and is an 83-kDa molecule. Both proteins have a three domain. Certain HLA alleles occur in a higher frequency in individuals with particular diseases than in the general population. This type of data permits estimation of the relative risk of developing a disease with every one known HLA allele. For example, there is a strong association between ankylosing spondylitis, which is an autoimmune disorder involving the vertebral joints and the class I MHC allele HLA-B27. There is a strong association between products of the polymorphic class II alleles HLA-DR and DQ and certain autoimmune diseases since class II MHC molecules are of great importance in the selection and activation of $CD4^+$T lymphocytes which regulate immune responses against protein antigens. For example, 95% of Caucasians with insulin-dependent (type I) diabetes mellitus have HLA-DR3 or HLA-DR4, or both. There is also a strong association of HLA-DR4 with rheumatoid arthritis. Numerous other examples exist and are the targets of current investigations, especially in extended studies employing DNA probes. Calculation of the relative risk (RR) and absolute risk (AR) can be found in this atlas under the definitions of those terms.

The **B locus** is the major histocompatibility locus in the chicken.

The **class IB genes** are linked to the MHC class I region that code for class I-like α chains. These genes that encode molecules on the cell surface that associate with β_2 microglobulin vary in their cell surface expression and tissue distribution from one species to another. An individual animal may have multiple class IB molecules. One such molecule has a role in presentation of peptides bearing N-formulated amino termini. Other class IB molecules may also be active in antigen presentation.

The **class I region** is that part of the major histocompatibility complex which contains the MHC class I heavy-chain genes.

Public specificity refers to the specificity of an epitope encoded by two or more alleles of an alloantigenic system. It is most frequently used to refer to major histocompatibility complex determinants such as MHC alleles where a single epitope is shared by multiple HLA molecules. In this system, antigenic products encoded by more than one allele at a single locus may carry public specificities. These specificities may also be encoded by alleles at loci that are separate but related. Thus, epitopes that represent public specificities are shared by two or more proteins in a particular group. For the specificity to be public it must be found on at least two proteins, but may be on multiple ones.

***Ir* genes** are immune response genes. Class II MHC genes that control immune responses are found in the Ir region. These genes govern the ability of an animal to respond immunologically to any particular antigen.

The **gene conversion hypothesis** describes the method by which alternative sequences can be introduced into the MHC genes without reciprocal crossover events. This mechanism could account for the incredible polymorphism of alleles. The alternative sequences may include those found in the class I-like genes and pseudogenes present on chromosome 6. Hypothetically, gene conversion was an evolutionary event as well as an ongoing one, giving rise to new mutations, and therefore new alleles within a population.

MHC molecules are major histocompatibility complex-encoded glycoproteins that are highly polymorphic. They complex with immunogenic peptides for antigen presentation to T lymphocytes. They fall into two classes designated MHC class I and MHC class II, each of which has a different role in the immune response. Also termed major histocompatibility antigens that serve as the principal alloantigens and play a principal role in tissue transplant rejection.

MHC functions: MHC class I molecules were originally identified as strong histocompatibility antigens and MHC class II glycoproteins were first described as *Ir* gene products. MHC class I and class II molecules have a central immunological function of focusing $CD8^+$ and $CD4^+$ T lymphocytes, respectively, to the surface of appropriate lymphocytes. MHC's principal role in immunity is to alert T cells to alterations in the surface integrity of other cells in order that they may be dealt with in a manner appropriate for maintenance of the interior milieu. MHC class I and class II molecules define self to the immune system. They represent a self-surveillance complex.

MHC peptide tetramers refer to a molecular complex fastened together by fluorescent streptavidin which possesses four binding sites for biotin that is attached to the tail of the MHC molecule. This makes it possible to stain specific T lymphocytes in any species.

MHC recombinant mice have a crossover within the MHC.

MHC peptide-binding specificity: The highly polymorphic membrane glycoprotein MHC molecules bind peptide fragments of proteins and display them for recognition by T lymphocytes. Their *raison d'etre* is to present self antigens in the thymus for tolerance induction and foreign antigens in the periphery for immune responsiveness. The organism distinguishes self from nonself through MHC peptide recognition by T cells.

Class I MHC molecules are glycoproteins that play an important role in the interactions among cells of the immune system. Class I molecules occur on essentially all nucleated cells of the animal body but are absent from trophoblast cells and sperm. The cell membrane of T lymphocytes is rich in class I molecules that are comprised of two distinct polypeptide chains, i.e., a 44-kDa α (heavy) chain and a 12-kDa β chain (β_2 microglobulin). There is a 40-kDa core polypeptide in the human α chain that has one N-linked oligosaccharide. Approximately 75% of the α chain is extracellular, including the amino terminus and the oligosaccharide group. The membrane portion is an abbreviated hydrophobic segment. The cytoplasm contains the 30-amino acid residue that comprises the carboxy terminus. The β_2 microglobulin component is neither linked to the cell surface nor to the α chain by covalent bonds. Its association with the α chain is noncovalent. Class I molecules consist of four parts that include an extracellular amino terminal peptide-binding site, an immunoglobulin (Ig)-like region, a transmembrane segment, and a cytoplasmic portion. The main function of MHC molecules is to bind foreign peptides to form a complex that T cells can recognize. The class I molecular site that binds protein antigens is a 180-amino acid residue segment at the class I α chain's amino terminus. The α-3 segment of the heavy chain contains approximately 90 amino acid residues between the α-2 segment's carboxy terminal end and the point of entrance into the plasma membrane. The α-3 segment joins the plasma membrane through a short connecting region and spans the membrane as a segment of 25 hydrophobic amino acid residues. This stabilizes the α chain of MHC class I in the membrane. The carboxy terminal region emerges as a 30-amino acid stretch that is present in the cytoplasm. Class I histocompatibility antigens are products of the MHC locus. HLA-A, -B, and -C genes located in the MHC region on the short arm of chromosome 6 in man encode these molecules. K, D, and L genes located on chromosome 17 in the H-2 complex in mice encode murine class I MHC antigens. The Tla complex situated near H-2 encodes additional class I molecules in mice. In T cell-mediated cytotoxicity, $CD8^+$ T lymphocytes kill antigen-bearing target cells. The cytotoxic T lymphocytes play a significant role in resistance to viral infection. Class I MHC molecules present viral antigens to CD8+ T lymphocytes as a viral peptide class I molecular complex, which is transported to the infected cell surface. Cytotoxic CD8+ T cells recognize this and lyse the target before the virus can replicate, thereby stopping the infection.

MHC class I molecules are those major histocompatibility complex polymorphic glycoproteins that present cytosol generated peptides to CD8+ T lymphocytes. They are heterodimers comprised of a class I heavy chain associated with β_2 microglobulin.

Leukocyte groups: Leukocytes may be grouped according to their surface antigens such as MHC class I and class II antigens. These surface antigens may be detected by several techniques that include the microlymphotoxicity assay and DNA typing.

A **major histocompatibility complex (MHC) molecule** is an MHC locus encoded heterodimeric membrane protein used by T lymphocytes to recognize antigen. They are divided into two structural types, designated class I and class II. Most nucleated cells of the body express class I MHC molecules that bind peptides derived from cytosolic proteins and are recognized by CD8+ T lymphocytes. The distribution of class II MHC molecules is more restricted, being confined mostly to professional antigen-presenting cells that bind peptides from enocytosed proteins that are recognized by CD4+ T cells.

Human leukocyte antigen (HLA) is the MHC in humans that contains the genes that encode the polymorphic MHC class I and class II molecules as well as other important genes.

Class I antigen is an MHC antigen found on nucleated cells on multiple tissues. In man, class I antigens are encoded by genes at A, B, and C loci, and in mice by genes at D and K loci.

***HLA* nonclassical class I genes** are located within the MHC class I region that encode products that can associate with β_2 microglobulin. However, their function and tissue distribution are different from those of HLA-A, -B, and -C molecules. Examples include HLA-E, -F, and -G. Of these, only HLA-G is expressed on the cell surface. It is uncertain whether or not these HLA molecules are involved in peptide-binding and presentation like classical class I molecules.

HLA-H is a pseudogene found in the MHC class I region that is structurally similar to HLA-A but is nonfunctional due to the absence of a cysteine residue at position 164 in its protein product and the deletion of the codon 227 nucleotide.

β_2 microglobulin (β_2M) is a thymic epithelium-derived polypeptide that is 11.8 kDa and makes up part of the MHC class I molecules that appear on the surfaces of nucleated cells. It is noncovalently linked to the MHC class I polypeptide chain. It promotes maturation of T lymphocytes and serves as a chemotactic factor. β_2 M makes up part of the peptide–antigen class I-(β_2M) complex involved in antigen presentation to cytotoxic T lymphocytes. Nascent β_2 M facilitates the formation of antigenic complexes that can stimulate T lymphocytes. This monomorphic polypeptide accumulates in the serum in renal dialysis patients and may lead to β_2 microglobulin (β_2 M)$_2$-induced amyloidosis. The extracellular protein β_2 M is encoded by a nonpolymorphic gene outside the MHC. It is homologous structurally to an immunoglobulin domain and is invariant among all class I molecules.

In the mouse, the **I region** is the DNA segment of the major histocompatibility complex where the gene that encodes MHC class II molecules are located. The 250-kb I region consists of I-A and I-E subregions. The genes designated pseudo $A_\beta 3$, $A_\beta 2$, $A_\beta 1$, and A_β are located in the 175-kb I-A subregion. The genes designated $E_\beta 2$ and E_α are located in the 75-kb I-E subregion. $E_\beta 1$ is located where the I-A and I-E subregions join. The S region contains the $E_\beta 3$ gene.

I-J is an area of the murine I region that has been predicted to encode for a suppressor cell antigen. Recombinant mouse strains were used to show that the I-J gene locus maps to a region between the I-A and I-E loci of the H-2 complex. Monoclonal antibodies against I-J react with antigen-specific suppressor of T cells which secrete suppressive factors. The anti-I-J specific antibodies react with an epitope among the suppressor cell factors. I-J determinant is associated with a 13-kDa protein known as the glycosylation-inhibiting factor, which has suppressive activity.

Ia antigens (immune-associated antigen) are products of MHC I region genes that encode murine cellular antigens. In man, the equivalent antigens are designated HLA-DR, which are encoded by MHC class II genes. B lymphocytes, monocytes, and activated T lymphocytes express Ia antigens.

The **Tla complex** consists of genes that map to the MHC region telomeric to H2 loci on chromosome 17 in mice. These genes encode MHC class I proteins such as Qa and Tla that have no known immune function. Qa and Tla proteins that closely resemble H2 MHC class I proteins in sequence, associate noncovalently with β-2 microglobulin. Expression of Qa and Tla proteins, unlike expression of MHC H-2 class I proteins, is limited to only selected mouse cells. For example, only hepatocytes express Q10 protein, only selected lymphocyte subpopulations such as activated T lymphocytes express Qa-2 proteins, and T lymphocytes express Tla proteins. Thus, Qa and Tla class I molecules differ in structure and expression from the remaining MHC class I genes and proteins in the mouse. This could

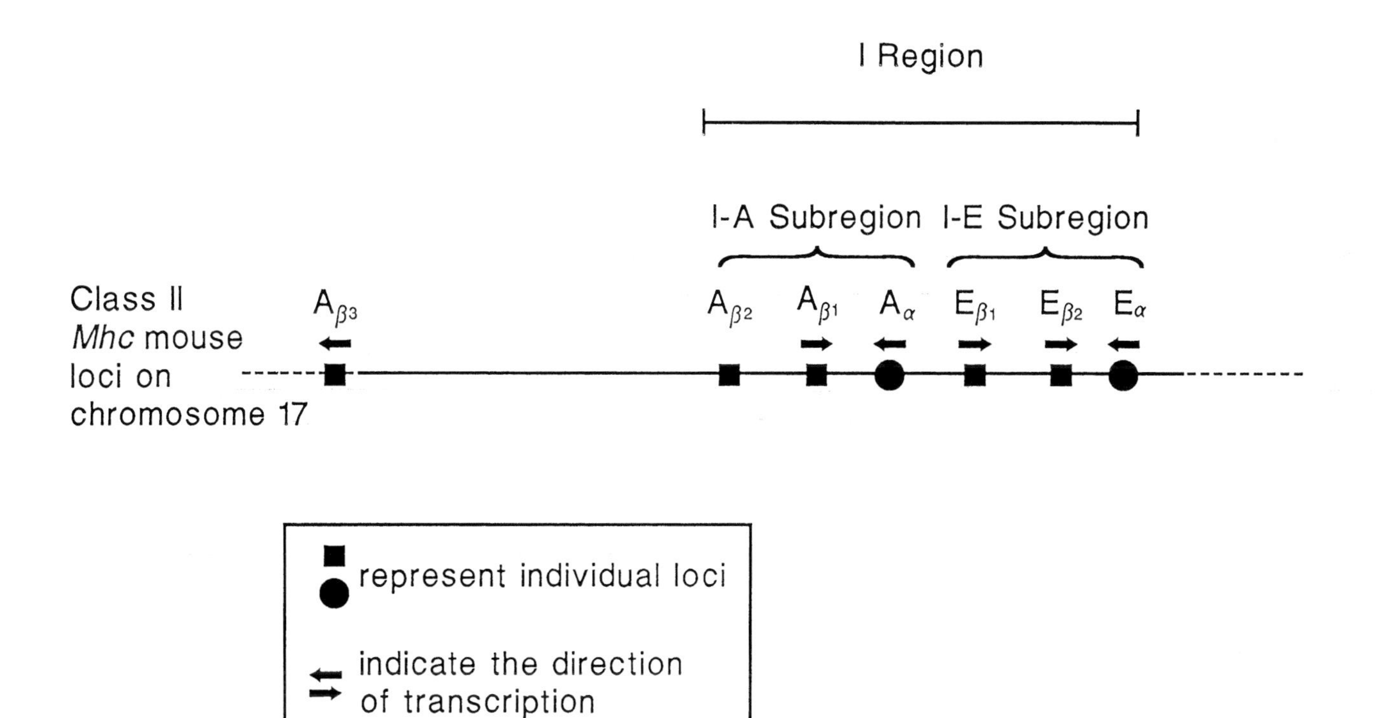

FIGURE 4.6 In the mouse, the I region is the DNA segment of the major histocompatibility complex where the genes that encode MHC class II molecules are located.

account for their failure to function in antigen presentation to T lymphocytes.

TL (thymic-leukemia antigen): An epitope on the thymocyte membrane of TL^+ mice. As the T lymphocytes mature, this antigen disappears, but resurfaces if leukemia develops. TL antigens are specific and are normally present on the cell surface of thymocytes of certain mouse strains. They are encoded by a group of structural genes located at the Tla locus, in the linkage group IX, very close to the D pole of the H-2 locus on chromosome 17. There are three structural TL genes, one of which has two alleles. The TL antigens are numbered from 1 to 4 specifying four antigens: TL.1, TL.2, TL.3, and TL.4. TL.3 and TL.4 are mutually exclusive. Their expression is under the control of regulatory genes, apparently located at the same Tla locus. Normal mouse thymocytes belong to three phenotypic groups: TL^-, TL.2, and TL.1,2,3. Development of leukemia in the mouse induces a restructuring of the TL surface antigens of thymocytes with expression of TL.1 and TL.2 in TL^- cells, expression of TL.1 in TL.2 cells, and expression of TL.4 in both TL^- and TL.2 cells. When normal thymic cells leave the thymus, the expression of TL antigen ceases. Thus, thymocytes are TL^+ (except the TL^- strains) and the peripheral T cells are TL^-. In transplantation experiments, TL^+ tumor cells undergo antigenic modulation. Tumor cells exposed to the homologous antibody stop expressing the antigen and thus escape lysis when subsequently exposed to the same antibody plus complement.

Tla antigen is a murine MHC class I histocompatibility antigen encoded by genes that are situated near the Qa region on chromosome 17. Thymocytes may express products of up to 6 alleles. Leukemia cells may aberrantly express Tla antigens.

In the mouse, the **I region** (Figure 4.6) is the DNA segment of the major histocompatibility complex where the genes that encode MHC class II molecules are located. The 250-kb I region consists of I-A and I-E subregions. The genes designated pseudo $A_{\beta}3$, $A_{\beta}2$, $A_{\beta}1$, and A_{α} are located in the 175-kb I-A subregion. The genes designated $E_{\beta}2$ and E_{α} are located where the I-A and I-E subregions join. The S region contains the $E_{\beta}3$ gene. Ia antigens are products of the MHC I region genes that encode murine cellular antigens. B lymphocytes, monocytes, and activated T lymphocytes express Ia antigens.

Qa region: See Tla complex.

The **Qa** locus is a subregion of the murine MHC located on the telomeric side of the H2 complex in a 1.5-cm stretch of DNA. The Qa region is a part of the Tla complex which also contains Tla. The Qa region is comprised of 220 kb which encode for class I MHC α chains that associate noncovalently with β_2 microglobulin. The Qa region is comprised of eight to ten genes designated as Q1, Q2, Q3, etc.

Qa antigens are class I histocompatibility antigens in mice designated Qa1, Qa2, and Qa10. They are encoded by genes in the QA region of the H-2 complex on the telomeric side. Lymphoid cells express Qa1 and Qa2 antigens, whereas hepatic cells express Q10 antigens. The Qa represents one of two regions of the Tla complex, with Tla representing the second region of this complex.

Qa-2 antigen is encoded by MHC class I genes and is produced in a GPI-linked form as well as in a soluble secreted form. It shows little polymorphism and is tissue specific. Anti-Qa-2 mAbs are potent T cell activators in the presence of PMA and a crosslinking anti-Ig.

K region: The K as well as the D region refer to class I MHC segments in the murine genome.

MHC class IB molecules are encoded by the MHC that are less polymorphic than the MHC class I and MHC class II molecules and which present a restricted set of antigens.

MHC class II transactivator (CIITA) is a protein that activates MHC class II gene transcription. The gene that encodes this molecule is one of several defective genes in bare lymphocyte syndrome characterized by a lack of MHC class II molecules on all cells.

The **MHC class II region** is comprised of three subregions designated DR, DQ, and DP. Multiple genetic loci are present in each of these. DN (previously DZ) and DO subregions are each comprised of one genetic locus. Each class II HLA molecule is comprised of one α and one β chain that constitute a heterodimer. Genes within each subregion encode a particular class II molecule's α and β chains. Class II genes that encode α chains are designated A, and class II genes that encode β gene are designated B. A number is used following A or B if a particular subregion contains two or more A or B genes.

The **MHC class II compartment (MIIC)** is a cellular site where MHC class II molecules concentrate, interact with HLA-DM, and bind antigenic peptides prior to migrating to the cell surface.

The **class II region** is that part of the major histocompatibility complex containing the MHC class II α- and β-chain genes.

DO and DM are MHC class II-like molecules. DO gene expression occurs exclusively in the thymus and on B lymphocytes. DNα and DOβ chains pair to produce the DO molecule. Information related to DM gene products awaits further investigation.

Class II MHC molecules are glycoprotein histocompatibility antigens that play a critical role in immune system cellular interactions. Each class II MHC molecule is comprised of a 32- to 34-kDa α chain and a 29- to 32-kDa β chain, each of which possess N-linked oligosaccharide groups, amino termini that are extracellular, and carboxy termini that are intracellular. Approximately 70% of both α and β chains are extracellular. Separate MHC genes encode the class II molecule α and β chains, which are polymorphic. Class II molecules resemble class I molecules structurally as revealed by class II molecule nucleotide and amino acid sequences. Class II MHC molecules consist of a peptide-binding region, a transmembrane segment, and an intracytoplasmic portion. The extracellular portion of α and β chains consist of α-1 and α-2 and β-1 and β-2 segments, respectively. The α-1 and α-2 segments constitute the peptide-binding region and consist of approximately 90 amino acid residues each. The immunoglobulin-like region is comprised of α-2 and β-2 segments that are folded into immunoglobulin domains in the class II molecule. The transmembrane region consists of approximately 25 hydrophobic amino acid residues. The transmembrane portion ends with a group of basic amino acid residues immediately followed by hydrophilic tails that extend into the cytoplasm and constitute the carboxy terminal ends of the chains. The α chain is more heavily glycosylated than is the β chain. Of the five exons in the α genes, one encodes the signal sequence and two code for the extracellular domains. The transmembrane domain and a portion of the 3′ untranslated segment are encoded by a fourth exon. The remaining part of the 3′ untranslated region is coded for by a fifth exon. Six exons are present in the β genes. They resemble α gene exons 1 through 3. The transmembrane domain and a portion of the cytoplasmic domain are encoded by a fourth exon, the cytoplasmic domain is coded for by the fifth exon, and the sixth exon encodes the 3′ region that is untranslated. B lymphocytes, macrophages, or other accessory cells express MHC class II antigens. γ-Interferon or other agents may induce an aberrant expression of class II antigen by other types of cells. Antigen-presenting cells (APC) such as macrophages present antigen at the cell surface to immunoreactive $CD4^+$ helper/inducer T cells in the context of MHC class II antigens. For appropriate presentation, the peptide must bind securely to the MHC class II molecules. Those that do not fail to elicit an immune response. Following interaction of the peptide and the $CD4^+$ helper T lymphocyte receptor, the CD4 cell is activated, interleukin-2 (IL-2) is released, and the immune response is initiated. In man, the class II antigens, DR, DP, and DQ are encoded by HLA-D region genes. In the mouse, class II antigens, designated as Ia antigens, are encoded by I region genes. The I invariant chain (Ii) represents an essentially nonpolymorphic polypeptide chain that is associated with MHC class II molecules of man and mouse.

Class II antigens are MHC histocompatibility antigens with limited distribution on such cells as B lymphocytes and macrophages. In man, these antigens are encoded by genes at the DR, DP, and DQ loci.

Class II transactivator (CIITA): See MHC class II transactivator.

The **S region** is the chromosomal segment of the murine MHC where genes are located that encode MHC class III molecules, including complement component C2, factor B, C4A (Slp), and C4B (Ss). The S locus within this region is the site of genes that encode a 200-kDa protein termed Ss (serum substance) that corresponds to C4 in the serum of man. Also within the S region can be found the gene for Slp (sex limited protein), a protein usually found only in male mice, the gene for 21-hydroxylase, an enzyme with no known immune function, and the gene for a serum β globulin. This term likewise refers to the chromosomal segment that lies between HLA-B and HLA-D where the genes encoding the corresponding human MHC class III molecules are situated.

Class III molecules are substances that include factors B, C2, and C4 that are encoded by genes in the MHC region. Although adjacent to class I and class II molecules that are important in histocompatibility, C3 genes are not important in this regard. The 100-kb region is located between HLA-B and HLA-D loci on the short arm of chromosome 6 in man and between the I and H-2D regions on chromosome 17 in mice. The genes encoding C4 and P-450 21-hydroxylase are closely linked.

Complotype is an MHC class III haplotype. It is a precise arrangement of linked alleles of MHC class III genes that encode C2, factor B, C4A, and C4B MHC class III molecules in man. Caucasians have 12 ordinary complotypes that may be in positive linkage disequilibrium.

Relative risk (RR) refers to the association of a particular disease with a certain HLA antigen. This represents the chance a person with the disease-associated HLA antigen has of developing the disease compared with that of a person who does not possess that antigen. Relative risk is calculated as follows:

where

p^+ = number of patients possessing a particular HLA antigen

c^- = number of controls not possessing the particular HLA antigen

p^- = number of patients not possessing the particular HLA antigen

c^+ = number of controls possessing the particular antigen

The higher the relative risk (above 1), the greater the antigen's frequency in the patient population.

MHC disease associations: See HLA disease association.

The **RhLA locus** is the major histocompatibility locus in the rhesus monkey.

MHC congenic mice are mice that differ only at the MHC complex.

Killer cell immunoglobulin-like receptors (KIRs) are NK cell receptors that bind to MHC class I molecules and transmit either activating or inhibitory signals to the T cell.

Killer inhibitory receptors (KIRs) are natural killer cell receptors that recognize self class I MHC molecules and transmit inhibitory signals that block activation of NK cell cytolytic processes. These receptors prevent NK cells from killing normal host cells expressing class I MHC molecules, yet allow lysis of virus-infected cells whose class I MHC expression has been suppressed. The inhibitory receptors fall into several classes that include immunoglobulin superfamily members, heterodimers of CD94 and selectin, and Ly49. These various receptors possess cytoplasmic tails bearing immunoreceptor tyrosine inhibition motifs (ITIMs) that participate in initiating inhibitory signal pathways.

GPLA is the guinea pig MHC. There are two loci that encode class I MHC molecules (*GPLA-B* and *GPLA-S*). The *GPLA-Ia* locus encodes class II MHC molecules. Complement protein factor B, C2, and C4 are encoded by other loci. The genes are *BF* (factor B), C2, C4, *Ia*, *B,* and *S*.

MHC mutant mice are mutant at one or more loci.

Coisogenic refers to inbred mouse strains that have an identical genotype except for a difference at one genetic locus. A point mutation in an inbred strain provides the opportunity to develop a coisogenic strain by inbreeding the mouse in which the mutation occurred. The line carrying the mutation is coisogenic with the line not expressing the mutation. Considering the problems associated with developing coisogenic lines, congenic mouse strains were developed as an alternative. Congenic strains of inbred mice (Figure 4.7) are believed to be genetically identical except for a difference at one genetic locus. Congenic strains are produced by crossing a donor strain and a background strain. Repeated backcrossing is made to the background strain and selecting in each generation for heterozygosity at a certain locus. Following 12 to 14 backcrosses, the progeny are inbred through brother–sister matings to yield a homozygous inbred strain. Mutation and genetic linkage may lead to random differences at a few other loci in the congenic strain. Designations for congenic strains consist of the symbol for the background strain followed by a period and then the symbol for the donor strain.

H-2 (Figure 4.8) designates the major histocompatibility complex in the mouse. H-2 genes are located on chromosome 17. They encode somatic cell surface antigens as well as the host *Ir* genes. Each of these has a length of 600 bp. There are four regions in the H-2 complex designated K, I, S, and D. K region genes encode class I histocompatibility molecules designated K. I region genes encode class II histocompatibility molecules designated I-A and I-E. S region genes encode class III molecules designated C2, C4, factor B, and P-450 cytochrome (21-hydroxylase). D region genes encode class I histocompatibility molecules designated D and L. Antigens that represent the H-2 type of a particular inbred strain of mice are encoded by H-2 alleles. Thus, differences in the antigenic structure between inbred mice of differing H-2 alleles is of critical importance in the acceptance or rejection of tissue grafts exchanged between them. K, D, and L subregions of H-2 correspond to A, B, and C subregions of HLA in humans. The I-A and I-E regions are equivalent to the human HLA-D region.

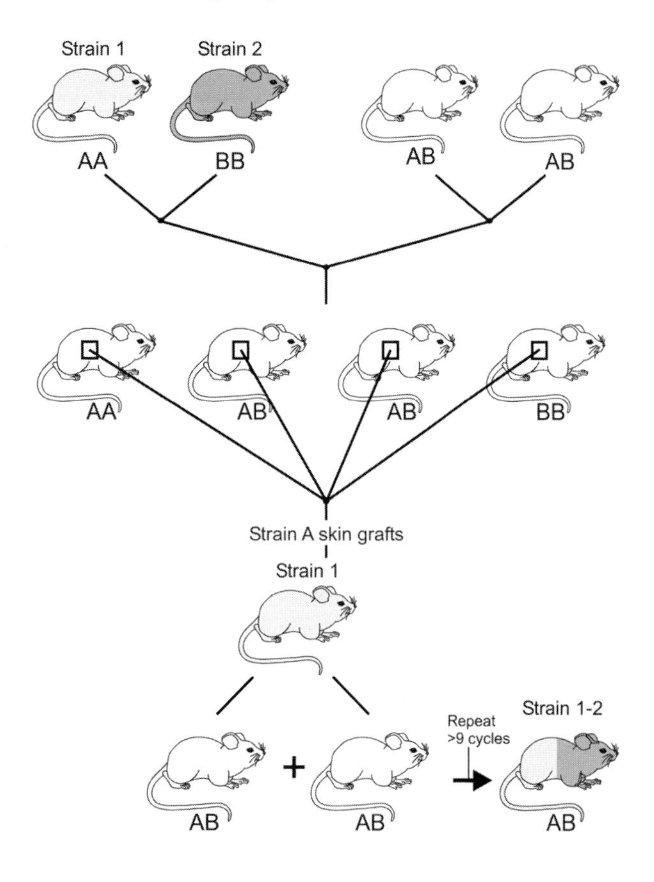

FIGURE 4.7 Congenic strains.

H-2 complex: See H2 histocompatibility system.

The **H-2 locus** is the mouse major histocompatibility region on chromosome 17.

H-2 restriction is MHC restriction involving the murine H-2 MHC.

The **H-2I region** is a murine H-2 MHC region where the genes encoding class II molecules are found.

H-2D and H-2K are murine H-2 MHC loci whose products are class I antigens. Both H-2D and H-2K loci have multiple alleles.

H-2L is the murine class I histocompatibility antigen found on spleen cells that serves as a target epitope in graft rejection.

HLA is the abbreviation for human leukocyte antigen. The HLA histocompatibility system in humans represents a complex of MHC class I molecules distributed on essentially all nucleated cells of the body and MHC

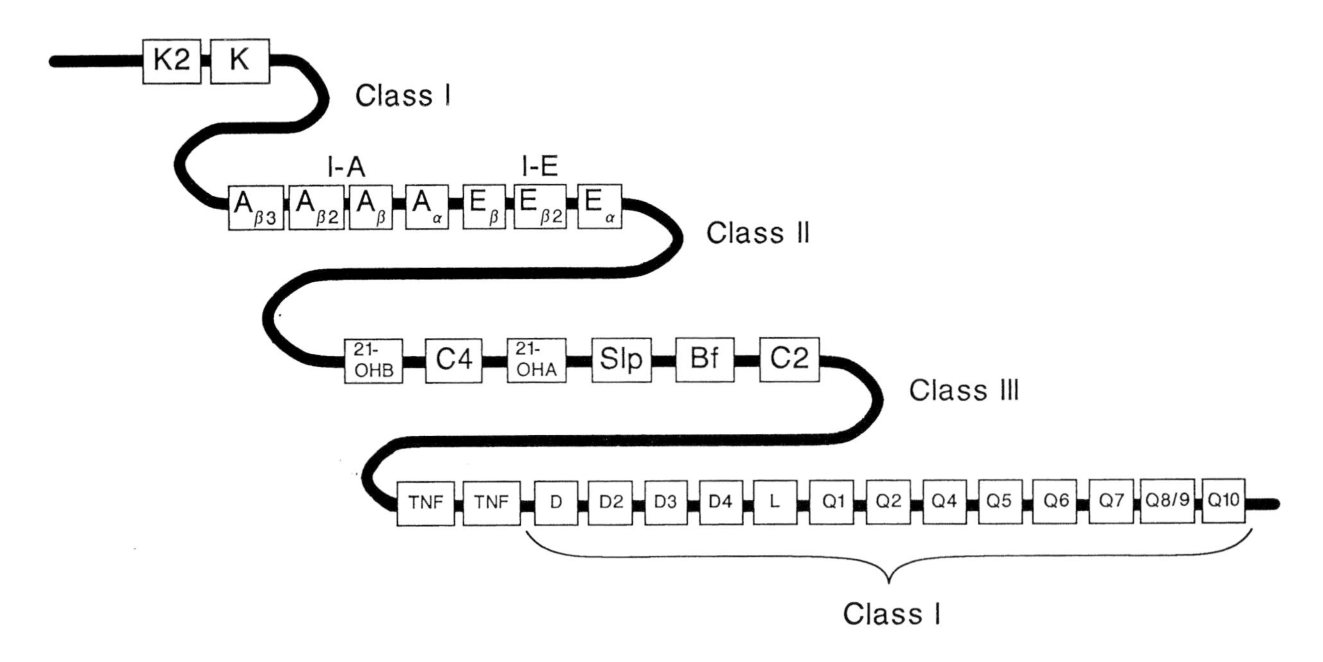

FIGURE 4.8 H-2 histocompatibility system is the major histocompatibility complex in the mouse.

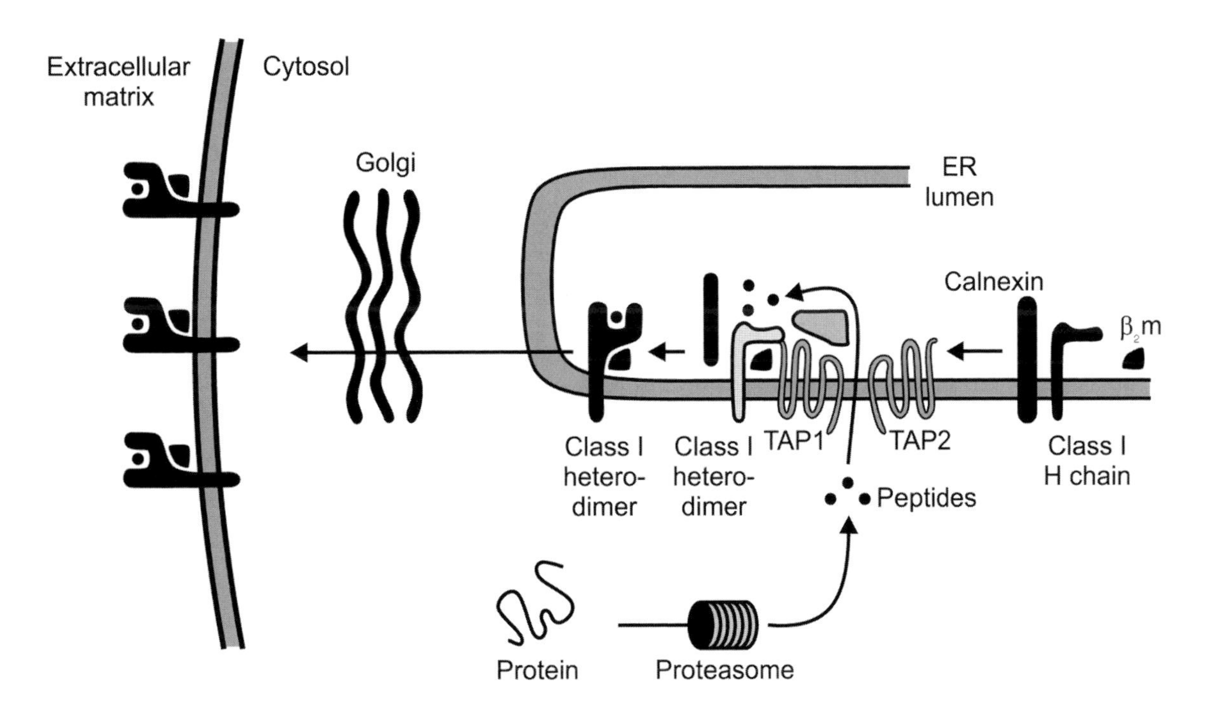

FIGURE 4.9 Class I MHC assembly.

class II molecules that are distributed on B lymphocytes, macrophages, and a few other cell types. These are encoded by genes at the major histocompatibility complex. The HLA locus in humans is found on the short arm of chromosome 6. This has now been well defined, and in addition to encoding surface isoantigens, genes at the HLA locus also encode *Ir* genes. The class I region consists of HLA-A, HLA-B, and HLA-C loci and the class II region consists of the D region which is subdivided into HLA-DP, HLA-DQ, and HLA-DR subregions. Class II molecules play an important role in the induction of an immune response since antigen-presenting cells must complex an exogenous antigen with class II molecules to present it to $CD4^+$ T lymphocytes in the presence of interleukin-1. Class I molecules are important in presentation of intracellular antigen to $CD8^+$ T lymphocytes as well as in effector functions of target cells (Figure 4.9). Class III molecules encoded by genes located between those that encode class I and class II molecules include C2, BF, C4a, and C4b. Class I and class II molecules play an important role in the transplantation of organs and tissues. HLA-C molecules act as inhibitors of the lytic capacity of natural killer (NK) cells and non-MHC-restricted T cells. The microlymphocytotoxicity assay is used for HLA-A, -B, -C, -DR, and -DQ typing but is gradually being replaced by molecular (DNA) typing. The primed lymphocyte test is used for DP typing. Uppercase letters designate individual HLA loci, as in HLA-B, and alleles are designated by numbers, as in HLA-B*0701.

HLA-A is a class I histocompatibility antigen in humans. It is expressed on nucleated cells of the body. Tissue typing to identify an individual's HLA-A antigens employs lymphocytes.

The **HLA locus** is the major histocompatibility locus in man.

HLA allelic variation: (Figure 4.20) Genomic analysis has identified specific individual allelic variants to explain HLA associations with rheumatoid arthritis, type I diabetes mellitus, multiple sclerosis, and celiac disease. There is a minimum of six α and eight β genes in distinct clusters, termed HLA-DR, -DQ, and -DP within the HLA class II genes. DO and DN class II genes are related but mapped outside the DR, DQ, and DP regions. There are two types of dimers along the HLA cell surface HLA-DR class II molecules. The dimers are made up of either DR α polypeptide associated with $DR_{\beta}1$ polypeptide or DR with $DR_{\beta}2$ polypeptide. Structural variation in class II gene products is linked to functional features of immune recognition leading to individual variations in histocompatibility, immune recognition, and susceptibility to disease. There are two types of structural variations which include variation among DP, DQ, and DR products in primary amino acid sequence by as much as 35% and individual variation attributable to different allelic forms of class II genes. The class II polypeptide chain possesses domains which are specific structural subunits containing variable sequences that distinguish among class II α genes

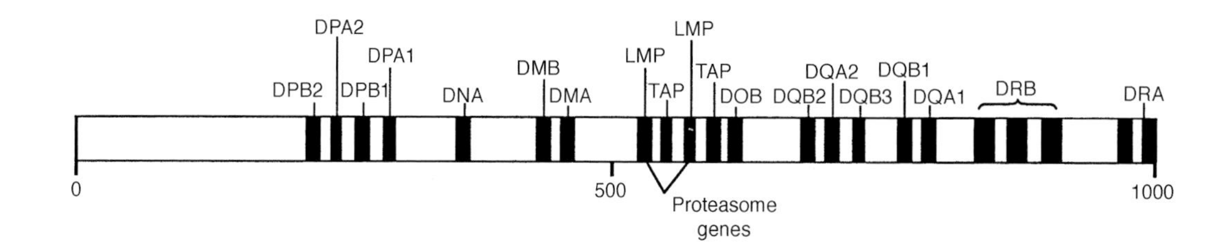

FIGURE 4.10 MHC class II.

or class II β genes. These allelic variation sites have been suggested to form epitopes which represent individual structural differences in immune recognition.

HLA class III: See MHC genes and class III MHC molecules.

HLA-D region (Figure 4.10) refers to the human MHC class II region comprised of three subregions designated DR, DQ, and DP. Multiple genetic loci are present in each of these. DN (previously DZ) and DO subregions are each comprised of one genetic locus. Each class II HLA molecule is comprised of one α and one β chain that constitute a heterodimer. Genes within each subregion encode a particular class II molecule's α and β chains. Class II genes that encode α chains are designated A, whereas class II genes that encode β chain are designated B. A number is used following A or B if a particular subregion contains two or more A or B genes.

Homozygous typing cells (HTCs) are cells obtained from a subject who is homozygous at the HLA-D locus. HTCs facilitate MLR typing of the human D locus.

HLA disease association: Certain HLA alleles occur in a higher frequency in individuals with particular diseases than in the general population. This type of data permits estimation of the relative risk of developing a disease with every known HLA allele. For example, there is a strong association between ankylosing spondylitis, which is an autoimmune disorder involving the vertebral joints, and the class I MHC allele, HLA-B27. There is a strong association between products of the polymorphic class II alleles HLA-DR and -DQ and certain autoimmune diseases since class II MHC molecules are of great importance in the selection and activation of $CD4^+$ T lymphocytes which regulate the immune responses against protein antigens. For example, 95% of Caucasians with insulin-dependent (type I) diabetes mellitus have HLA-DR3 or HLA-DR4, or both. There is also a strong association of HLA-DR4 with rheumatoid arthritis. Numerous other examples exist and are the targets of current investigations, especially in extended studies employing DNA probes. Calculation of the relative risk (RR) and absolute risk (AR) can be found in this atlas under the definitions of those terms.

The **HLA-DP subregion** is a site of two sets of genes designated HLA-DPA1 and HLA-DPB1 and the pseudogenes HLA-DPA2 and HLA-DPB2. DPα and DPβ chains encoded by the corresponding genes DPA1 and DPB1 unite to produce the DPαβ molecule. DP antigen or type is determined principally by the very polymorphic DPβ chain in contrast to the much less polymorphic DPα chain. DP molecules carry DPw1 to DPw6 antigens.

The **HLA-DQ subregion** contains two sets of genes, designated DQA1 and DQB1, and DQA2 and DQB2. DQA2 and DQB2 are pseudogenes. DQα and DQβ chains, encoded by DQA1 and DQB1 genes, unite to produce the DQαβ molecule. Although both DQα and DQβ chains are polymorphic, the DQβ chain is the principal factor in determining the DQ antigen or type. DQαβ molecules carry DQw1 to DQw9 specificities.

The **HLA-DR subregion** is the site of one HLA-DRA gene. Although DRB gene number varies with DR type, there are usually three DRB genes, termed DRB1, DRB2, and DRB3 (or DRB4). The DRB2 pseudogene is not expressed.

The DRα chain, encoded by the DRA gene, can unite with products of DRB1 and DRB3 (or DRB4) genes, which are the DRβ-1 and DRβ-3 (or DRβ-4) chains. This yields two separate DR molecules, DRαβ-1 and DRαβ-3 (or DRαβ-4). The DRβ chain determines the DR antigen (DR type) since it is very polymorphic, whereas the DRα chain is not. DRαβ-1 molecules carry DR specificities DR1 to DRw18. Yet, DRαβ-3 molecules carry the DRw52, and the DRαβ-4 molecules carry the DRw53 specificity.

HLA-DR antigenic specificities reflect the epitopes on DR gene products. Selected specificities have been mapped to define loci. HLA serologic typing requires the identification of a prescribed antigenic determinant on a particular HLA molecular product. One typing specificity can be present on many different molecules. Different alleles at the same locus may encode these various HLA molecules. Monoclonal antibodies are now used to recognize certain antigenic determinants shared by various molecules bearing the same HLA typing specificity. Monoclonal antibodies have been employed to recognize specific class II alleles with disease associations.

HLA-DM (Figure 4.11 to Figure 4.13) facilitates the loading of antigenic peptides onto MHC class II moleules. As a result of proteolysis of the invariant chain a small fragment called the class II-associated invariant chain peptide, or CLIP, remains bound to the MHC class II molecule. CLIP peptide is replaced by antigenic peptides, but in the absence of HLA-DM this does not occur. The HLA-DM molecule must therefore play some part in removal of the CLIP peptide and in the loading of antigenic peptides.

HLA nonclassical class I genes include genes located within the MHC class I region that encode products that can be associated with β_2 microglobulin. However, their function and tissue distribution are different from those of HLA-A, -B, and -C molecules. Examples include HLA-E, -F, and -G. Of these, only HLA-G is expressed on the cell surface. It is uncertain whether or not these HLA molecules are involved in peptide binding and presentation like classical class I molecules.

HLA-G is a nonclassical class I HLA antigen which exhibits a certain amount of polymorphism with the most extensive variability in the α-2 domain. It is found on trophoblasts, i.e., placenta cells and trophoblastic neoplasms. HLA-G is expressed only on cells such as placental extravillous cytotrophoblasts and choriocarcinoma that fail to express HLA-A, -B, and -C antigens. HLA-G expression is most pronounced during the first trimester of pregnancy. Trophoblast cells expressing HLA-G at the maternal–fetal junction may protect the semiallogeneic fetus from "rejection." Prominent HLA-G expression suggests maternal immune tolerance.

HLA-E is an HLA class I nonclassical molecule.

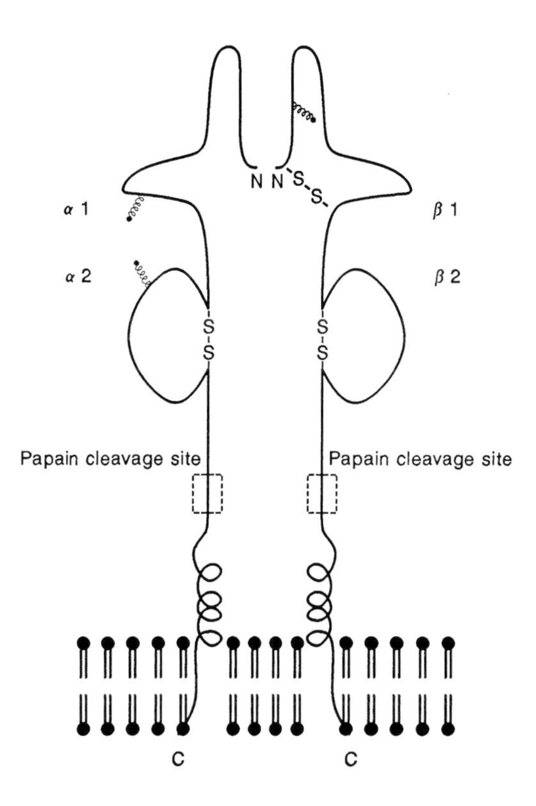

FIGURE 4.11 Class II MHC molecules are glycoprotein histocompatibility antigens that play a critical role in immune system cellular interactions. Each class II MHC molecule is comprised of a 32- to 34-kDa α chain and a 29- to 32-kDa β chain, each of which possesses N-linked oligosaccharide groups, amino termini that are extracellular, and carboxyl termini that are intracellular. Approximately 70% of both α and β chains are extracellular.

HLA-F is an HLA class I nonclassical molecule.

The ***C2* and *B* genes** are situated within the MHC locus on the short arm of chromosome 6. They are termed MHC class III genes. TNF-α and TNF-β genes are situated between the *C2* and *HLA-B* genes. Another gene, designated FD, lies between the *Bf* and *C4a* genes. C2 and B complete primary structures have been deduced from cDNA and protein sequences. C2 is comprised of 732 residues and is an 81-kDa molecule, whereas B contains 739 residues and is an 83-kDa molecule. Both proteins have a three-domain globular structure. During C3 convertase formation, the amino terminal domains C2b or Ba are split off. They contain consensus repeats that are present in CR1, CR2, H, DAF, and C4bp, which all combine with C3 and/or C4 fragments and regulate C3 convertases. The amino acid sequences of the C2 and B consensus repeats are known. C2b contains sites significant for C2 binding to C4b. Ba, resembling C2b, manifests binding sites significant in C3 convertase assembly. Available evidence indicates that C2b possesses a C4b-binding site and that Ba contains a corresponding C3b-binding site. In considering assembly and decay of C3 convertases,

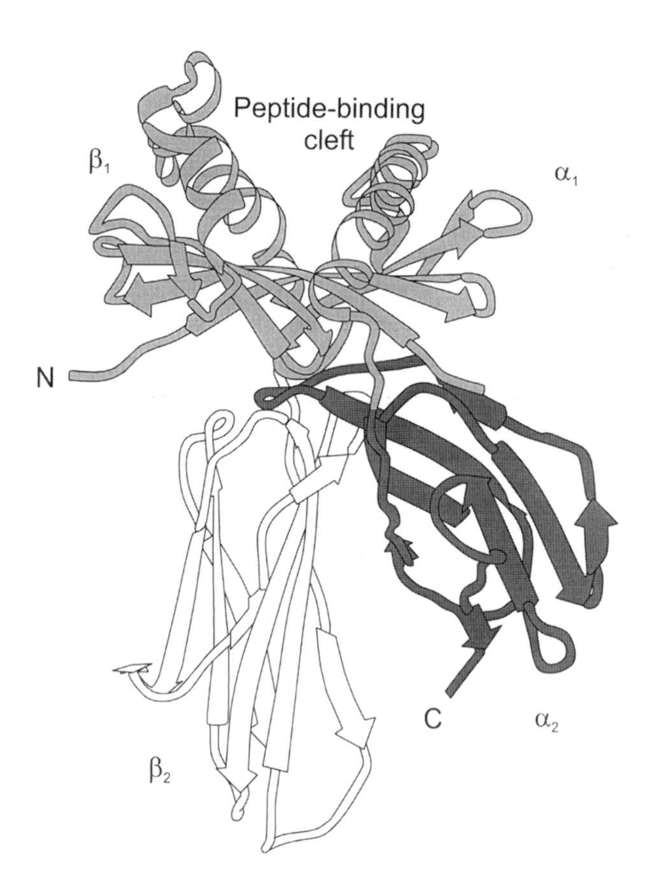

FIGURE 4.12 MHC class II molecular structure.

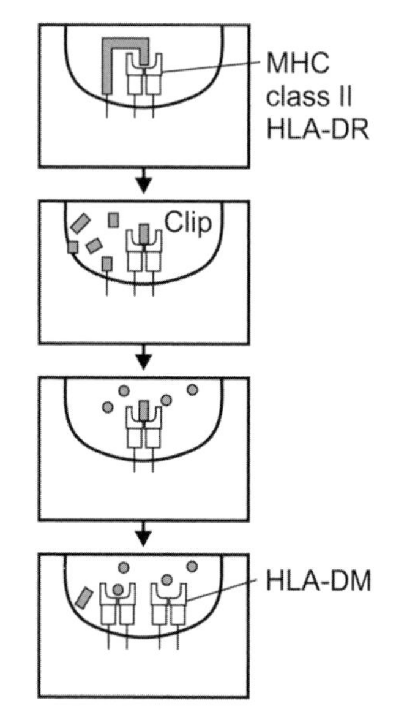

FIGURE 4.13 HLA-DM.

initial binding of the three-domain structures C2 or B to activator-bound C4b or C3b, respectively, requires one affinity site on the C2b/Ba domain and another on one of the remaining two domains. A transient change in C2a and Bb conformation results from C2 or B cleavage by C1s or D. This leads to greater binding affinity and Mg^{+2} sequestration, and acquisition of proteolytic activity for C3, C2a, or Bb dissociation leads to C3 convertase decay. Numerous serum-soluble and membrane-associated regulatory proteins control the rate of formation and association of C3 convertases.

Certain HLA alleles occur in a higher frequency in individuals with particular diseases than in the general population. This type of data permits estimation of the relative risk of developing a disease with every known HLA allele. For example, there is a strong association between ankylosing spondylitis, which is an autoimmune disorder involving the vertebral joints, and the class I MHC allele, HLA-B27. There is a strong association between products of the polymorphic class II alleles HLA-DR and -DQ and certain autoimmune diseases since class II MHC molecules are of great importance in the selection and activation of $CD4^+$ T lymphocytes which regulate immune responses against protein antigens. For example, 95% of Caucasians with insulin-dependent type I diabetes mellitus have HLA-DR3 or HLA-DR4, or both. There is also a strong association of HLA-DR4 with rheumatoid arthritis. Numerous other examples exist and are the targets of current investigations, especially in extended studies employing DNA probes. Calculation of the relative risk (RR) and absolute risk (AR) can be found in this atlas under definitions of those terms.

Genomic analysis has identified specific individual **allelic variants** to explain HLA associations with rheumatoid arthritis, type I diabetes, multiple sclerosis, and celiac disease. There are a minimum of six α and eight β genes in distinct clusters, termed HLA-DR, -DQ, and -DP within the HLA class II genes. DO and DN class II genes are related but mapped outside the DR, DQ, and DP regions. There are two types of dimers along the HLA cell-surface HLA-DR class II molecules. The dimers are made up of either DRα-polypeptide associated with DRβ_1-polypeptide or DR with DRβ_2-polypeptide. Structural variation in class II gene products is linked to functional features of immune recognition leading to individual variations in histocompatibility, immune recognition, and susceptibility to disease. There are two types of structural variations which include variations among DP, DQ, and DR products in primary amino acid sequence by as much as 35% and individual variation attributable to different allelic forms of class II genes. The class II polypeptide chain possesses domains which are specific structural subunits containing variable sequences that distinguish among class IIα genes or class IIβ genes. These allelic variation sites have been suggested to form epitopes that represent individual structural differences in immune recognition.

MHC restriction (Figure 4.14) is the recognition of antigen in the context of either class I or class II molecules

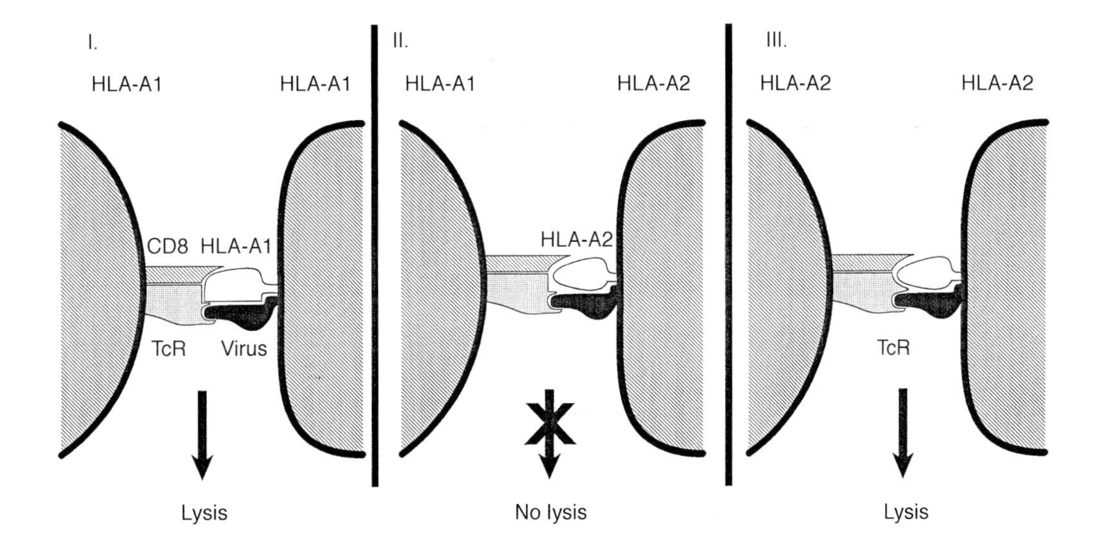

FIGURE 4.14 MHC restriction.

by the T cell receptor for antigen. In the afferent limb of the immune response, when antigen is being presented at the surface of a macrophage or other antigen-presenting cell to CD4+ T lymphocytes, this presentation must be in the context of MHC class II molecules for the CD4+ lymphocyte to recognize the antigen and proliferate in response to it. By contrast, cytotoxic (CD8+) T lymphocytes recognize foreign antigen such as viral antigens on infected target cells only in the context of class I MHC molecules. Once this recognition system is in place, the cytotoxic T cell can fatally injure the target cell through release of perforin molecules that penetrate the target cell surface.

Major histocompatibility complex restriction: See MHC restriction.

CD4 (Figure 4.15) is a single-chain glycoprotein, also referred to as the T4 antigen, that has a 56-kDa mol wt and is present on approximately two thirds of circulating human T cells, including most T cells of helper/inducer type. The antigen is also found on human monocytes and macrophages. The molecule is a receptor for gp120 of HIV-1 and HIV-2 (AIDS viruses). This antigen binds to class II MHC molecules on APC and may stabilize antigen-presenting cell and T cell interactions. It is physically associated with the intracellular tyrosine protein kinase, known as p561ck, which phosphorylates nearby proteins. This antigen is thereby relaying a signal to the cells. Crosslinking of CD4 may induce activation of this enzyme and phosphorylation of CD3.

The **CD4 molecule** exists as a monomer and contains four immunoglobulin-like domains. The first domains of CD4 form a rigid rod-like structure that is linked to the two carboxyl-terminal domains by a flexible link. The binding site for MHC class II molecules is thought to involve the D1 and D2 domains of CD4.

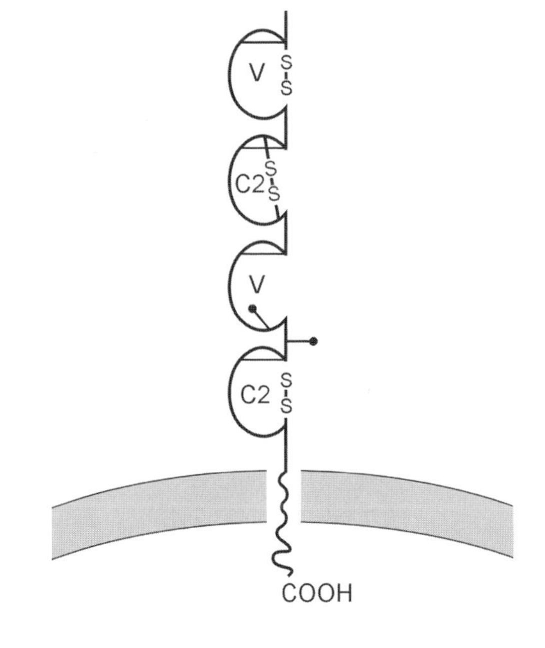

FIGURE 4.15 Structure of CD4.

CD8 (Figure 4.16 and Figure 4.17) is an antigen, also referred to as the T8 antigen, that has a 32- to 34-kDa mol wt. The CD8 antigen consists of two polypeptide chains α and β which may exist in the combination α/α homodimer or α/β heterodimers. Most antibodies are against the α-chain. This antigen binds to class I MHC molecules on APC and may stabilize APC/class I cell interactions.

The **CD8 molecule** is a heterodimer of an α- and β-chain that are covalently associated by a disulfide bond. The two chains of the dimer have similar structures, each having a single domain resembling an immunoglobulin variable domain and a stretch of peptide believed to be in a relatively extended conformation (Figure 4.18).

Each set of alleles is referred to as a **haplotype** (Figure 4.19). An individual inherits one haplotype from the mother and one haplotype from the father. In an outbred population, the offspring are generally heterozygous at many loci and will express both maternal and paternal MHC alleles. The alleles are therefore codominantly expressed, that is, both maternal and paternal gene products are expressed in the same cells. In inbred mice, however, each H-2 locus is homozygous because the maternal and paternal haplotypes are identical and all offspring express identical haplotypes.

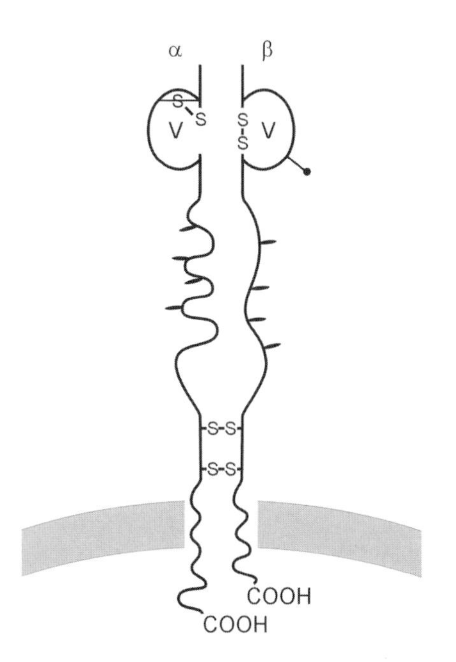

FIGURE 4.16 Structure of CD8.

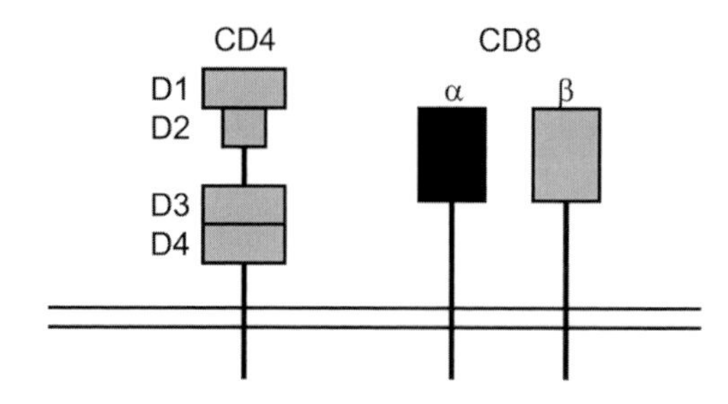

FIGURE 4.18 The outline structure of the CD4 and CD8 coreceptor molecules.

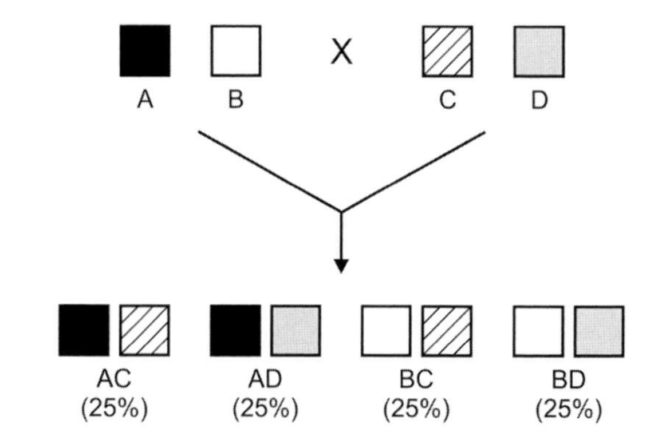

FIGURE 4.19 Each set of alleles is referred to as a haplotype.

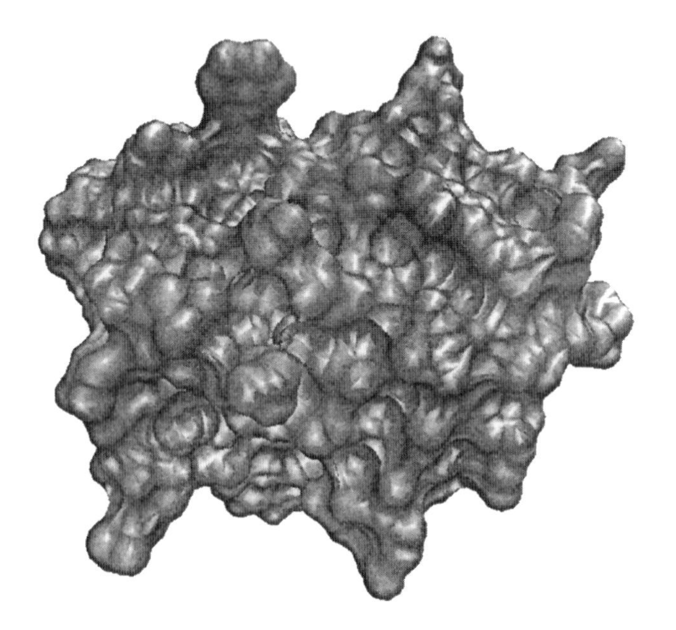

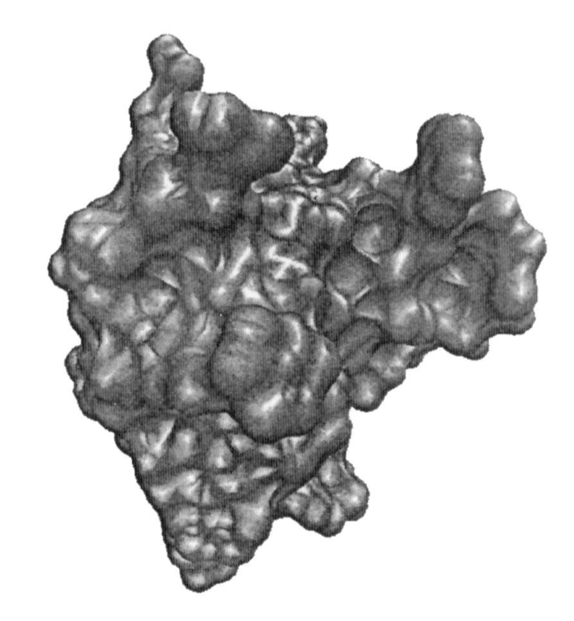

FIGURE 4.17 This structure consists of the N-terminal 114 residues of CD8. These residues make up a single immunoglobulin axis which coincides with a crystallographic twofold axis.

Name	Previous equivalents	Molecular characteristics
HLA-A	-	Class I α-chain
HLA-B	-	Class I α-chain
HLA-C	-	Class I α-chain
HLA-E	E,"6.2"	associated with class I 6.2-kB Hind III fragment
HLA-F	F, "5.4"	associated with class I 5.4-kB Hind III fragment
HLA-G	G,"6.0"	associated with class 6.0 Hind III fragment
HLA-H	H, AR, "12.4"	Class I pseudogene associated with 5.4-kB Hind III fragment
HLA-J	cda 12	Class I pseudogene associated with 5.9-kB Hind III fragment
HLA-K	HLA-70	Class I pseudogene associated with 7.0-kB Hind III fragment
HLA-L	HLA-92	Class I pseudogene associated with 9.2-kB Hind III fragment
HLA-DRA	DRα	DR α chain
HLA-DRB1	DRβI,DR1B	DR β1 chain determining specificities DR1, DR2, DR3, DR4, DR5, etc.
HLA-DRB2	DRβII	pseudogene with DR β-like sequences
HLA-DRB3	DRβIII, DR3B	DR β3 chain determining DR52 and Dw24, Dw25, Dw26 specificities
HLA-DRB4	DRβIV, DR4B	DR β4 chain determining DR53
HLA-DRB5	DRβIII	DR β5 chain determining DR51
HLA-DRB6	DRBX, DRBσ	DRB pseudogene found on DR1, DR2, and DR10 haplotypes
HLA-DRB7	DRBψ1	DRB pseudogene found on DR4, DR7 and DR9 haplotypes
HLA-DRB8	DRBψ2	DRB pseudogene found on DR4, DR7 and DR9 haplotypes
HLA-DRB9	M4.2 β exon	DRB pseudogene, isolated fragment
HLA-DQA1	DQα1, DQ1A	DQ α chain as expressed
HLA-DQB1	DQβ1, DQ1B	DQ β chain as expressed
HLA-DQA2	DXα, DQ2A	DQ α-chain-related sequence, not known to be expressed
HLA-DQB2	DXβ, DQ2B	DQ β-chain-related sequence, not known to be expressed
HLA-DQB3	DVβ,DQB3	DQ β-chain-related sequence, not known to be expressed
HLA-DOB	DOβ	DO β chain
HLA-DMA	RING6	DM α chain
HLA-DMB	RING7	DM β chain
HLA-DNA	DZα, DOα	DN α chain
HLA-DPA1	DPα1, DP1A	DP α chain as expressed
HLA-DPB1	DPβ2, DP2B	DP β chain as expressed
HLA-DPA2	DPα2, DP2A	DP α-chain-related pseudogene
HLA-DPB2	DPβ2, DP2B	DP β-chain-related pseudogene
TAP-1	RING4, Y3, PSF1	ABC (ATP binding cassette) transporter
TAP-2	RING11, Y1, PSF2	ABC (ATP binding cassette) transporter
LMP2	RING12	Proteasome-related sequence
LMP7	RING10	Proteasome-related sequence

FIGURE 4.20 Names for genes in the HLA region.

5 Antigen Processing and Presentation

T lymphocytes recognize antigens only in the context of self-MHC molecules on the surface of accessory cells. During processing, intact protein antigens are degraded into peptide fragments. Most epitopes that T cells recognize are peptide chain fragments. B cells and T cells often recognize different epitopes of an antigen leading to both antibody and cell-mediated immune responses. Before antigen can bind to MHC molecules, it must be processed into peptides in the intracellular organelles. $CD4^+$ helper T lymphocytes recognize antigens in the context of class II MHC molecules, a process known as class II MHC restriction. By contrast, $CD8^+$ cytotoxic T lymphocytes recognize antigens in the context of class I molecules, which is known as class I MHC restriction. Following the generation of peptides by proteolytic degradation in antigen-presenting cells, peptide–MHC complexes are presented on the surface of antigen-presenting cells where they may be recognized by T lymphocytes. Antigens derived from either intracellular or extracellular proteins may be processed to produce peptides from either self or foreign proteins that are presented by surface MHC molecules to T cells. In the class II MHC processing pathway, professional antigen-presenting cells such as macrophages, dendritic cells, or B lymphocytes incorporate extracellular proteins into endosomes where they are processed (Figure 5.1 and Figure 5.2). Enzymes within the vesicles of the endosomal pathway cleave proteins in the acidic environment.

Class II MHC heterodimeric molecules, united with invariant chain, are shifted to endosomal vesicles from the endoplasmic reticulum. Following cleavage of the invariant chain, DM molecules remove a tiny piece of invariant chain from the MHC molecules' peptide-binding groove. Following complexing of extracellular-derived peptide with the class II MHC molecule, the MHC–peptide complex is transported to the cell surface where presentation to $CD4^+$ T cells occurs. Proteins in the cytosol, such as those derived from viruses, may be processed through the class I MHC route of antigen presentation. The multiprotein complex in the cytoplasm, known as the proteasome effects, involve proteolytic degradation of proteins in the cytoplasm to yield many of the peptides that are presented by class I MHC molecules. TAP molecules transport peptides from the cytoplasm to the endoplasmic reticulum where they interact and bind to class I MHC dimeric molecules. Once the class I MHC molecules have become stabilized through peptide binding, the complex leaves the endoplasmic reticulum entering the Golgi apparatus en route to the surface of the cell. Thus, mechanisms are provided through MHC-restricted antigen presentation to guarantee that peptides derived from extracellular microbial proteins can be presented by class II MHC molecules to $CD4^+$ helper T cells and that peptides derived from intracellular microbes can be presented by class I MHC molecules to $CD8^+$ cytotoxic T lymphocytes. The generation of microbial peptides produced through antigen processing to combine with self MHC molecules is critical to the development of an appropriate immune response.

Antigen presentation (Figure 5.3 and Figure 5.4) is the expression of antigen molecules on the surface of a dendritic cell, macrophage, or other antigen-presenting cell in association with MHC class II molecules when the antigen is being presented to a $CD4^+$ T helper lymphocyte or in association with cell surface MHC class I molecules when presentation is to $CD8^+$ cytotoxic T lymphocytes (Figure 4.3). Antigen-presenting cells, known also as accessory cells, include dendritic cells, macrophages, and Langerhans cells of the skin as well as B lymphocytes. Target cells such as fibroblasts present antigen to $CD8^+$ cytotoxic T lymphocytes. Mononuclear phagocytes ingest proteins and split them into peptides in endosomes. These eight to ten amino acid residue peptides link to cell surface MHC class II molecules. For appropriate presentation, it is essential that peptides bind securely to the MHC class II molecules since those that do not bind or are bound only weakly are not presented and fail to elicit an immune response. Following interaction of the presented antigen and MHC class II molecules with the $CD4^+$ helper T cell receptor, the $CD4^+$ lymphocyte is activated, interleukin-2 is released, and IL-2 receptors are expressed on the $CT4^+$ lymphocyte surface. The IL-2 produced by the activated cell stimulates its own receptors as well as those of mononuclear phagocytes, increasing their microbicidal activity. IL-2 also stimulates B cells to synthesize antibodies. Whereas B cells may recognize a protein antigen in its native state, T lymphocytes recognize the peptides that result from antigen processing.

Antigen processing is the degradation of proteins into peptides capable of binding to MHC molecules for presentation to T lymphocytes. For presentation by MHC molecules, antigens must be processed into peptides.

An **antigenic peptide** is a peptide that is able to induce an immune response and one that complexes with major histocompatibility complex (MHC), thereby permitting its recognition by a T cell receptor.

The **peptide-binding cleft** is that part of a major histocompatibility molecule that binds peptides for display to T lymphocytes. Paired α-helices on a floor of an eight-stranded β-pleated sheet comprise the cleft. Situated in and around this cleft are polymorphid residues that are the amino acids which differ among various MHC alleles.

Anchor residues are amino acid side chains of the peptide whose side chains fit into pockets in the peptide-binding cleft of the MHC molecule. The side chains anchor the peptide in the cleft of the MHC molecule by binding two complementary amino acids in the MHC molecule.

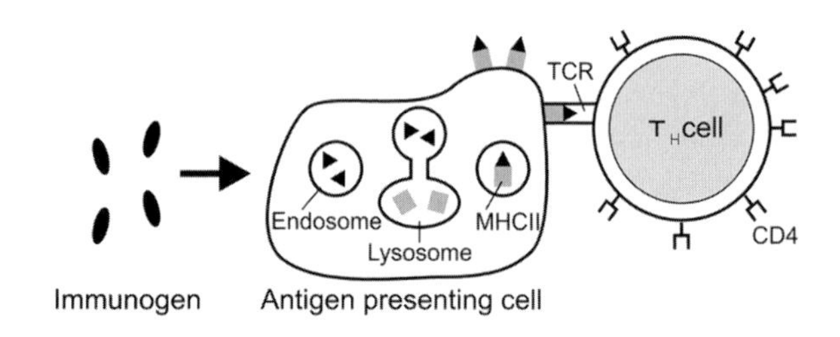

FIGURE 5.1 Capture, processing, and presentation of antigen by an antigen-presenting cell.

CD1 is an antigen that is a cortical thymocyte marker, which disappears at later stages of T cell maturation. The antigen is also found on interdigitating cells, fetal B cells, and Langerhans cells. These chains are associated with β_2-microglobulin and the antigen is thus analogous to classical histocompatibility antigens, but coded for by a different chromosome. More recent studies have shown that the molecule is coded for by at least five genes on chromosome 1, of which three produce recognized polypeptide products. CD1 may participate in antigen presentation.

Direct antigen presentation refers to cell surface allogeneic MHC molecule presentation by graft cells to T lymphocytes of the graft recipient, leading to T lymphocyte activation. This process does not require processing. Direct recognition of foreign MHC molecules is a cross-reaction between a normal TCR that recognizes self-MHC molecules plus foreign antigen and allogeneic MHC molecule–peptide complex. The powerful T cell response to allografts is due in part to direct presentation.

A **proteasome** (Figure 5.6) is a 650-kDa organelle in the cytoplasm termed the low molecular mass polypeptide complex. The proteasome is believed to generate peptides by degradation of proteins in the cytosol. It is a cylindrical structure comprised of as many as 24 protein subunits. The proteasome participates in degradation of proteins in the cytosol that are covalently linked, ubiquinated, prior to presentation to MHC class I-restricted T lymphocytes.

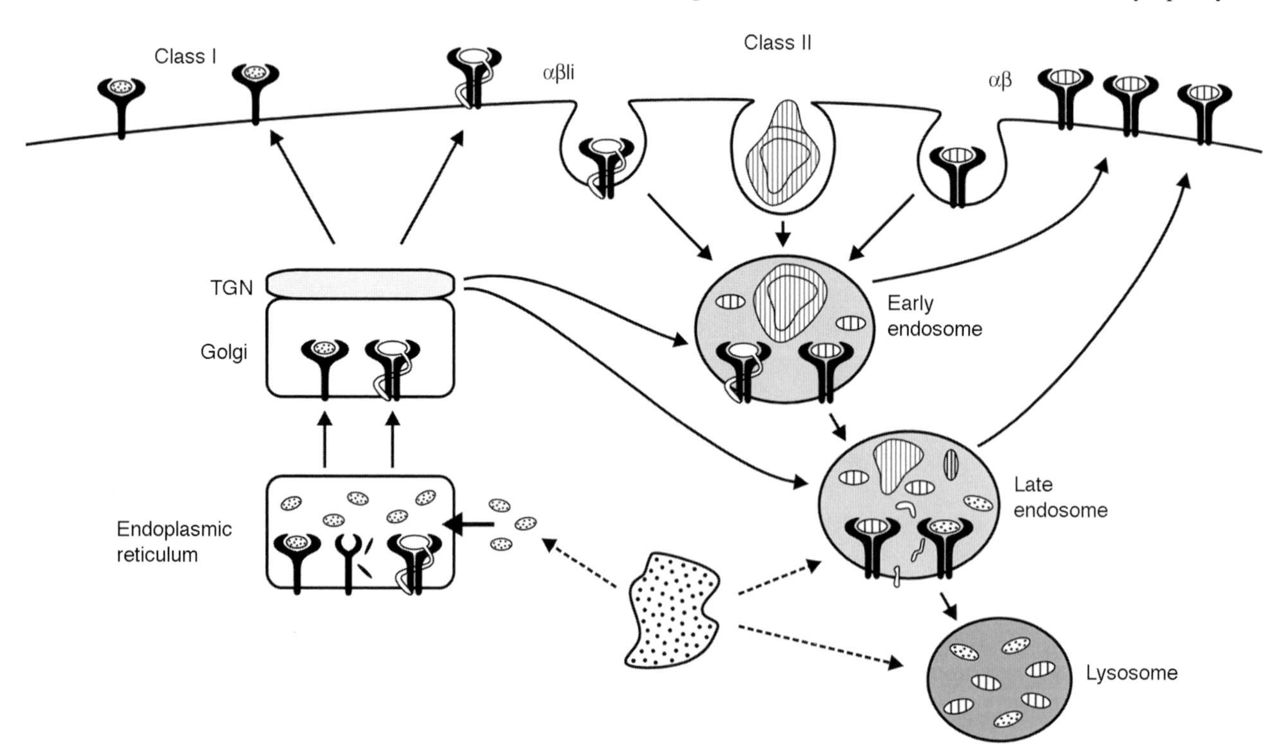

FIGURE 5.2 Processing pathways for class II-restricted antigen presentation.

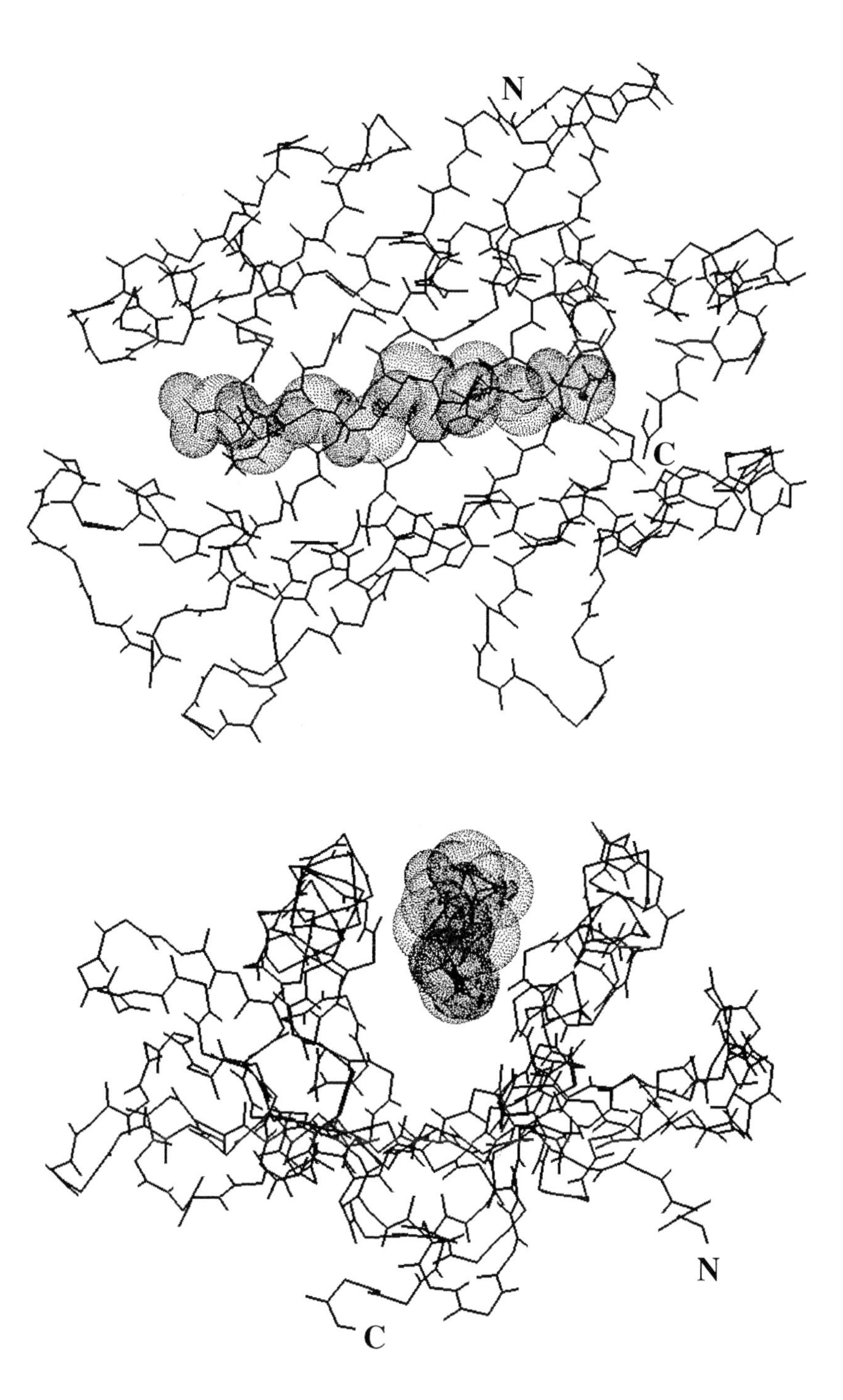

FIGURE 5.3 Presentation of MHC histocompatibility antigen HLA-B 270S complexed with nonapeptide ARG-ARG-ILE-LYS0ALA-ILE-THR-LEU-LYS. The C-terminal amino acid of the antigen-binding domain is protected by a N-methyl group. Three water molecules bridge the binding of the peptide to the histocompatibility protein.

Proteasomes that include MHC gene encoded subunits are especially adept at forming peptides that bind MHC class I molecules.

Indirect antigen presentation: In organ or tissue transplantation, the mechanism whereby donor allogeneic MHC molecules are present in microbial proteins. The recipient professional antigen-presenting cells process allogeneic MHC proteins. The resulting allogeneic MHC peptides are presented in association with recipient self-MHC molecules to host T lymphocytes. By contrast, recipient T cells recognize unprocessed allogeneic MHC molecules on the surface of the graft cells in direct antigen presentation.

The **immunological synapse** is the nanometer scale gap between a T cell and an antigen-presenting cell, which is

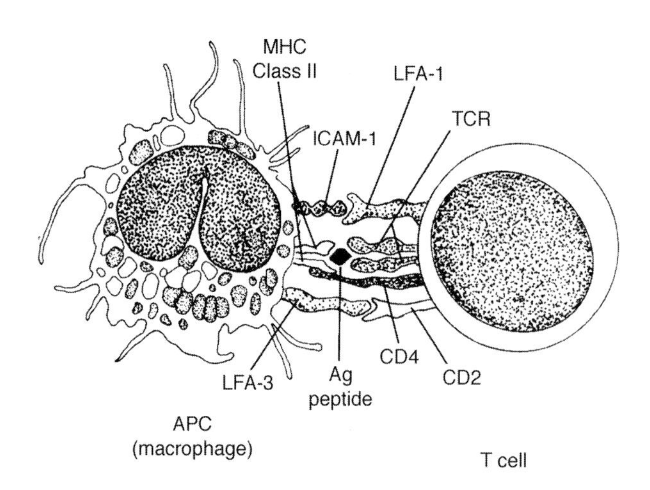

FIGURE 5.4 Antigen presentation.

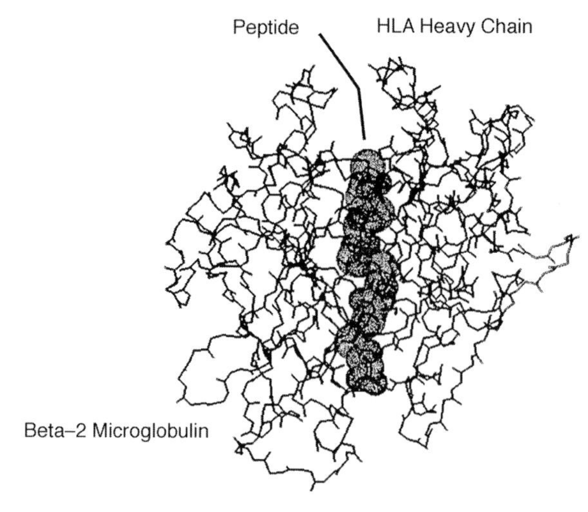

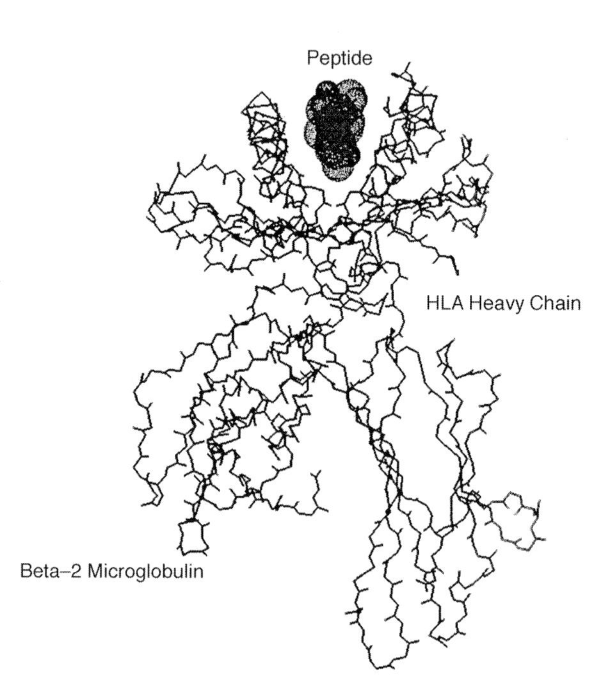

FIGURE 5.5 A schematic backbone structure of human class I histocompatibility antigen (HLA-A 0201) complexed with a decameric peptide from hepatitis B nucleocapsid protein (residues 18 to 27). Determined by x-ray crystallography.

the site of interaction between T cell antigen receptors and major histocompatibility complex molecule–peptide complexes that initiate adaptive immune responses.

Erp57 is a chaperone molecule that participates in the loading of peptide onto MHC class I molecules in the endoplasmic reticulum.

Transporter associated with antigen processing (TAP) refers to a TAP-binding heterodimeric protein in the rough endoplasmic reticulum membrane that transports peptides from the cytosol to the endoplasmic reticulum lumen. It is comprised of TAP 1 and TAP 2 subunits that bind peptides to class I MHC molecules.

Agrin is an aggregating protein crucial for formation of the neuromuscular junction. It is also expressed in lymphocytes and is important in reorganization of membrane lipid microdomains in setting the threshold for T cell signaling. T cell activation depends on a primary signal delivered through the T cell receptor and a secondary costimulatory signal mediated by coreceptors. Costimulation is believed to act through the specific redistribution and clustering of membrane and intracellular kinase-ridge lipid raft microdomains at the contact site between T cells and antigen-presenting cells. This site is known as the immunological synapse. Endogenous mediators of raft clustering in lymphocytes are essential for T cell activation. Agrin induces the aggregation of signaling proteins and the creation of signaling domains in both immune and nervous systems through a common lipid raft pathway.

A **lipid raft** is a membrane subdomain rich in cholesterol and glycosphingolipid-rich where cellular activation molecules are concentrated.

Proteasome genes are two genes in the MHC class II region that encode two proteasome subunits. The proteasome is a protease complex in the cytosol that may participate in the generation of peptides from proteins in the cytosol.

***LMP* genes** are two genes located in the MHC class II region in humans and mice that code for proteasome subunits. They are closely associated with the two *TAP* genes.

LMP-2 and LMP-7 are catalytic subunits of the organelle (proteasome) that degrades cytosolic proteins into peptides in the class I MHC pathway of antigen presentation. MHC genes encode these two subunits which are upregulated by IFN-γ and are especially significant in the generation of class I MHC-binding peptides.

Tapasin: TAP-associated protein is a chaperone molecule that participates in the assembly of peptide-MHC class I

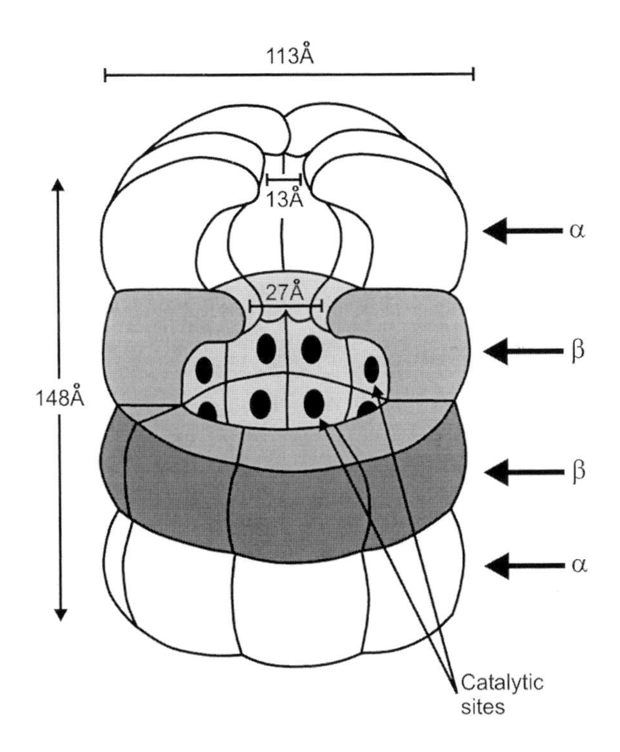

FIGURE 5.6 Longitudinal and transverse section through the 20S proteasome. The 20S proteasome is composed of two outer and two inner rings. The two outer rings each comprise seven copies of the 25.9-kDa α subunit.

molecule complexes in the endoplasmic reticulum. Cells deficient in this protein have unstable MHC class I cell surface molecules.

Tapasin (TAP-associated protein) is a molecule that is critical in MHC class I molecule assembly. Without this protein, MHC class I molecules are unstable on the cell surface.

The **transporter in antigen processing (TAP) 1 and 2 genes** (Figure 5.7) are in the MHC class II region that must be expressed for MHC class I molecules to be assembled efficiently. TAP 1 and TAP 2 are postulated to encode components of a heterodimeric protein pump that conveys cytosolic peptides to the endoplasmic reticulum. Here they associate with MHC class I heavy chains.

TAP 1 and TAP 2 genes: See transporter in antigen processing 1 and 2 genes.

An **antigen-presenting cell (APC)** is a cell that can process a protein antigen, break it into peptides, and present it in conjunction with class II MHC molecules on the cell surface where it may interact with appropriate T cell receptors (Figure 5.8). Dendritic cells, macrophages, Langerhans cells, and B cells process and present antigen to immunoreactive lymphocytes such as CD4$^+$, helper/inducer T cells (Figure 5.8). A MHC transporter gene-encoded peptide supply factor may mediate peptide

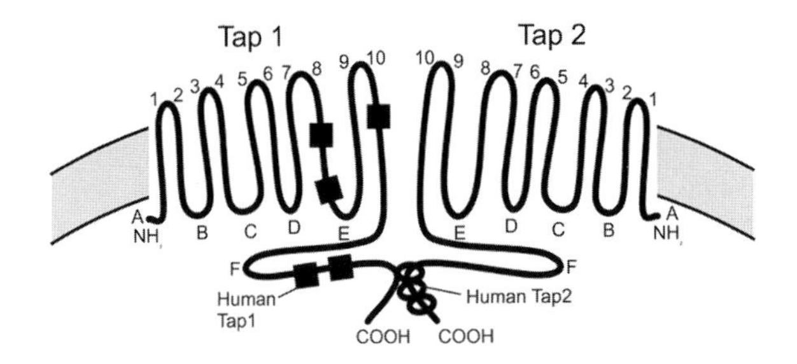

FIGURE 5.7 Topology of Tap 1 and Tap 2 proteins.

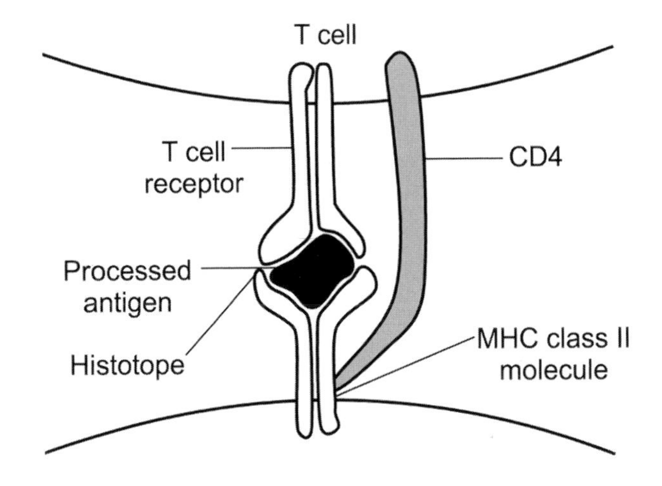

FIGURE 5.8 Antigen-presenting cell.

Cell type	Class II	Costimulators	Principal functions
Dendritic cells (Langerhans cells, lymphoid dendritic cells)	Constitutive	Constitutive	Inflammation of CD4$^+$ T cell response; allograft rejection
Macrophages	Inducible by IFN	Inducible by LPS	Development of CD4$^+$ effector T cells
B lymphocytes	Constitutive	Constitutive	Stimulation by CD4$^+$ helper T cells in humoral immune responses
Vascular endothelial cells	Inducible by IFN	Constitutive	Recruitment of antigen-specific T cells to site of antigen exposure or inflammation

FIGURE 5.9 Properties and functions of antigen-presenting cells.

antigen presentation. Other antigen-presenting cells that serve mainly as passive antigen transporters include B cells, endothelial cells, keratinocytes, and Kupffer cells. APCs include cells that present exogenous antigen processed in their endosomal compartment and presented together with MHC class II molecules. Other APCs present antigen that has been endogenously produced by the body's own cells with processing in an intracellular compartment and presentation together with class I MHC molecules. A third group of APCs present exogenous antigen that is taken into the cell and processed, followed by presentation together with MHC class I molecules. In addition to processing and presenting antigenic peptides in association with MHC class II molecules, an antigen-presenting cell must also deliver a costimulatory signal that is necessary for T cell activation. Professional APCs include dendritic cells, macrophages, and B cells, whereas nonprofessional APCs that function in antigen presentation for only brief periods include thymic epithelial cells and vascular endothelial cells. Dendritic cells, macrophages, and B cells are the principal antigen-presenting cells for T cells, whereas follicular dendritic cells are the main antigen-presenting cells for B cells.

Professional antigen-presenting cells are dendritic cells, macrophages, and B cells that are capable of initiating T lymphocyte responsiveness to antigen. These cells display antigenic peptide fragments in association with the proper class of MHC molecules and also bear costimulatory surface molecules. Dendritic cells are the most important professional APCs for initiating primary T lymphocyte responses. Among the three major antigen-presenting cells, dendritic cells are the only ones that continuously express high levels of costimulatory B7 and can present antigen via both class I MHC molecules and class II MHC molecules. Thus, they can activate both CD8 and CD4 T cells directly.

APC is the abbreviation for antigen-presenting cell.

Cathepsins are thiol and aspartyl proteases that have broad substrate specificities. Cathepsins represent the most abundant proteases of endosomes in antigen-presenting cells. They are believed to serve an important function in the generation of peptide fragments from exogenous protein antigens that bind to class II MHC molecules.

CD2 is a T cell adhesion molecule that binds to the LFA-3 adhesion molecule of antigen-presenting cells. Also called LFA-2.

A **circulating dendritic cell** is one that has taken up antigen and is migrating to a secondary lymphoid tissue such as a lymph node.

CD8 is a cell surface glycoprotein on T cells that recognizes antigens presented by MHC class I molecules. It binds to MHC class I molecules on antigen-presenting cells and serves as a coreceptor to facilitate the T cell's response to antigen.

The **CD8** molecule is a heterodimer of an α and β chain that are covalently associated by disulfide bond. The two chains of a dimer have similar structures, each having a single domain resembling an immunoglobulin variable domain and a stretch of peptide believed to be in a relatively extended conformation.

CD8 T cells comprise the T cell subset that expresses CD8 coreceptor and recognizes peptide antigens presented by MHC class I molecules.

CD20 is a B cell marker with a molecular weight of 33, 35, and 37 kDa that appears relatively late in the B cell maturation (after the pro-B cell stage) and then persists for some time before the plasma cell stage. Its molecular structure resembles that of a transmembrane ion channel. The gene is on chromosome 11 at band q12-q13. It may be involved in regulating B cell activation.

CD21 is a 145-kDa glycoprotein. Component of the B cell receptor. CD21 is a membrane molecule that participates in transmitting growth-promoting signals to the interior of the B cell. It is the receptor for the C3d fragment of the third component of complement, CR2. The CD21 antigen is a restricted B cell antigen expressed on mature B cells. It is present at high density on follicular dendritic cells (FDC), the accessory cells of the B zones. Also called complement receptor 2 (CR2).

CD22 is a molecule with an α130- and β140-kDa mol wt that is expressed in the cytoplasm of B cells of the pro-B and pre-B cell stage and on the cell surface on mature B cells with surface Ig. The antigen is lost shortly before the terminal plasma cell phase. The molecule has five extracellular immunoglobulin domains and shows homology with myelin adhesion glycoprotein and with N-CAM (CD56). It participates in B cell adhesion to monocytes and T cells. Also called BL-CAM.

CD28 is a T cell low-affinity receptor that interacts with B7 costimulatory molecules to facilitate T lymphocyte activation. More specifically, B7-1 and B7-2 ligands are expressed on the surface of activated antigen-presenting cells (APCs). Signals from CD28 to the T cell elevate expression of high affinity IL-2 receptor and increase the synthesis of numerous cytokines, including IL-2. CD28 regulates the responsiveness of T cells to antigen when they are in contact with APCs. It serves as a costimulatory receptor because its signals are synergistic with those provided by the T cell antigen receptor (TCR-CD3) in promoting T cell activation and proliferation. Without the signal from TCR-CD3, CD28's signal is only able to stimulate minimal T cell proliferation and may even lead to T cell unresponsiveness.

Tp44 (CD28) is a T lymphocyte receptor that regulates cytokine synthesis, thereby controlling responsiveness to antigen. Its significance in regulating activation of T lymphocytes is demonstrated by the ability of monoclonal antibody against CD28 receptor to block T cell stimulation by specific antigen. During antigen-specific activation of T lymphocytes, stimulation of the CD28 receptor occurs when it combines with the B7/BB1 coreceptor during the interaction between T and B lymphocytes. CD28 is a T lymphocyte differentiation antigen that four fifths of CD3/Ti positive lymphocytes express. It is a member of the immunoglobulin superfamily. CD28 is found only on T lymphocytes and on plasma cells. There are 134 extracellular amino acids with a transmembrane domain and a brief cytoplasmic tail in each CD28 monomer.

A **costimulator** is an antigen-presenting cell surface molecule that supplies a stimulus, serving as a second signal required for activation of naïve T lymphocytes, in addition to antigen (the "first signal"). An example of a costimulator is the B7 molecule on professional antigen-presenting cells that binds to the CD28 molecule on T lymphocytes.

Costimulatory molecules are membrane bound or secreted products of accessory cells that activate signal transduction events in addition to those induced by MHC/TCR interactions. They are required for full activation of T cells, and it is thought that adjuvants may work by enhancing the expression of costimulator molecules by accessory cells. The interaction of CD28/CTLA-4 with B7 to induce full transcription of IL-2 mRNA is an example of costimulator mechanisms.

A **costimulatory signal** is an extra signal requisite to induce proliferation of antigen-primed T lymphocytes. It is generated by the interaction of CD28 on T cells with B7 on antigen-presenting cells or altered self cells. In B cell activation an analogous second signal is illustrated by the interaction of CD40 on B cells with CD40L on activated T_H cells.

Cross-priming is the activation or priming of a naïve $CD4^+$ cytotoxic T lymphocyte specific for antigens of a third cell such as a virus-infected cell or tumor cell by a professional antigen-presenting cell. Cross-priming takes place when a professional antigen-presenting cell ingests an infected cell and the microbial antigens are processed and presented in association with class I MHC molecules. The professional antigen-presenting cell also costimulates the T cells. Also referred to as cross-presentation.

A **coreceptor** is a cell surface protein that increases the sensitivity of an antigen receptor to antigen by binding to associated ligands and facilitating in signaling for activation. CD4 and CD8 are T cell coreceptors that bind nonpolymorphic parts of a MHC molecule concurrently with the TCR binding to polymorphic residues and the bound peptide. It is a structure on the surface of a lymphocyte that binds to a part of an antigen simultaneously with membrane immunoglobulin (Ig) or T cell receptor (TCR) binding of antigen and which transmits signals required for optimal lymphocyte activation. CD4 and CD8 represent T cell coreceptors that bind nonpolymorphic regions of a major histocompatibility complex (MHC) molecule simultaneously with the binding of the T cell receptor to polymorphic residues and the exhibited peptide.

MHC restriction is the recognition of an antigen in the context of either class I or class II molecules by the T cell receptor for antigen. In the afferent limb of the immune response, when antigen is being presented at the surface of a macrophage or other antigen-presenting cell to $CD4^+$ T lymphocytes, this presentation must be in the context of MHC class II molecules for the $CD4^+$ lymphocyte to recognize the antigen and proliferate in response to it. By contrast, cytotoxic ($CD8^+$) T lymphocytes recognize foreign antigen, such as viral antigens on infected target cells, only in the context of class I MHC molecules. Once this recognition system is in place, the cytotoxic T cell can fatally injure the target cell through release of perforin molecules that penetrate the target cell surface. T cells recognize a firm peptide antigen only in the context of a specific allelic form of an MHC molecule to which it is bound.

CTLA-4 is a molecule that is homologous to CD28 and expressed on activated T cells (Figure 5.10). The genes for CD28 and CTLA-4 are closely linked on chromosome 2. The binding of CTLA-4 to its ligand B7 is an important costimulatory mechanism (see CD28 and costimulatory molecules). CTLA-4 is a high affinity receptor for B7 costimulatory molecules on T lymphocytes. CTLA4-Ig is a soluble protein composed of the CD28 homolog CTLA and the constant region of an IgG1 molecule. It is used experimentally to inhibit the immune response by blocking CD28-B7 interaction.

CTLA4-Ig is a soluble protein composed of the CD28 homolog CTLA and the constant region of an IgG1 molecule. It is used experimentally to inhibit the immune response by blocking CD28–B7 interaction.

Immune costimulatory molecules B7-1 and B7-2, together with their receptors CD28 and CTLA-4, constitute one of the dominant costimulatory pathways that regulate T- and B-cell responses. Although both CTLA-4 and CD28 can bind to the same ligands, CTLA-4 binds to B7-1 and B7-2 with a 20- to 100-fold higher affinity than CD28 and is involved in the downregulation of the immune response. B7-1 is expressed on activated B and activated T cells and macrophages. B7-2 is constitutively expressed on interdigitating dendritic cells, Langerhans cells, peripheral blood dendritic cells, memory B cells, and germical center B cells. It has been observed that both human and mouse B7-1 and B7-2 can bind to either human or mouse CD28 and CTLA-4, suggesting that there are conserved amino acids which form the B7-1/B7-2/CD28/CTLA-4 critical binding sites.

B7, B7-2: B7 is the ligand for CD28. B7 is expressed by accessory cells and is important in costimulataory mechanisms (Figure 5.11). Some APCs may upregulate expression of B7 following activation by various stimuli including

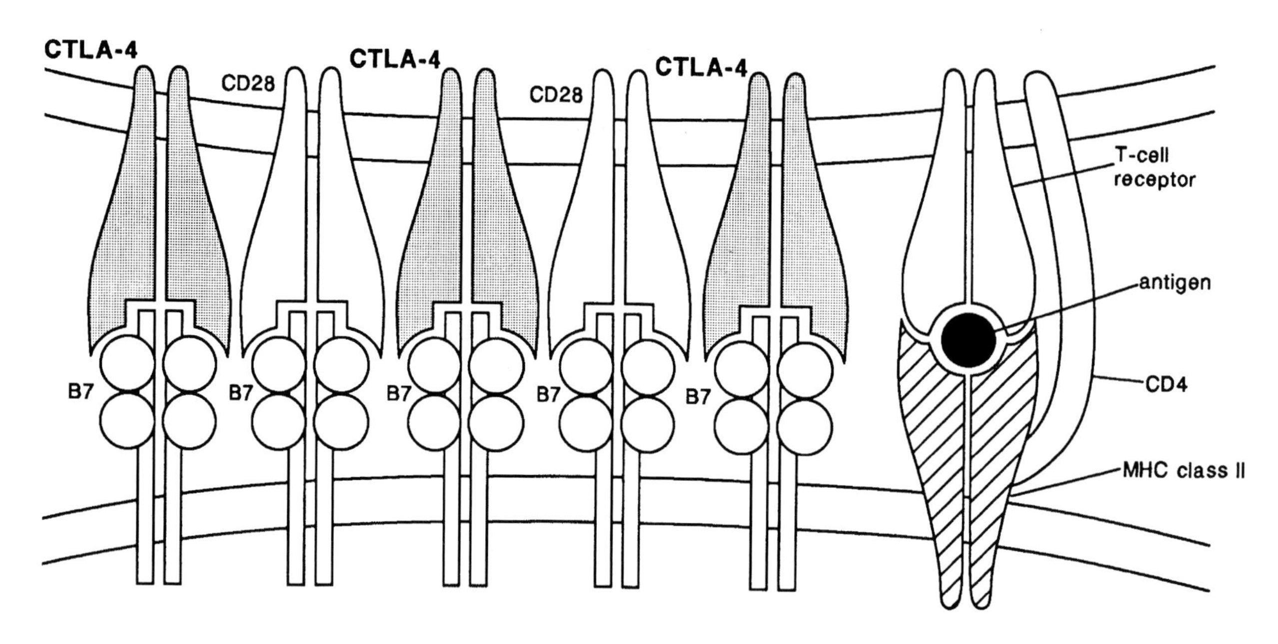

FIGURE 5.10 Participation of CTLA-4 molecules during antigen presentation.

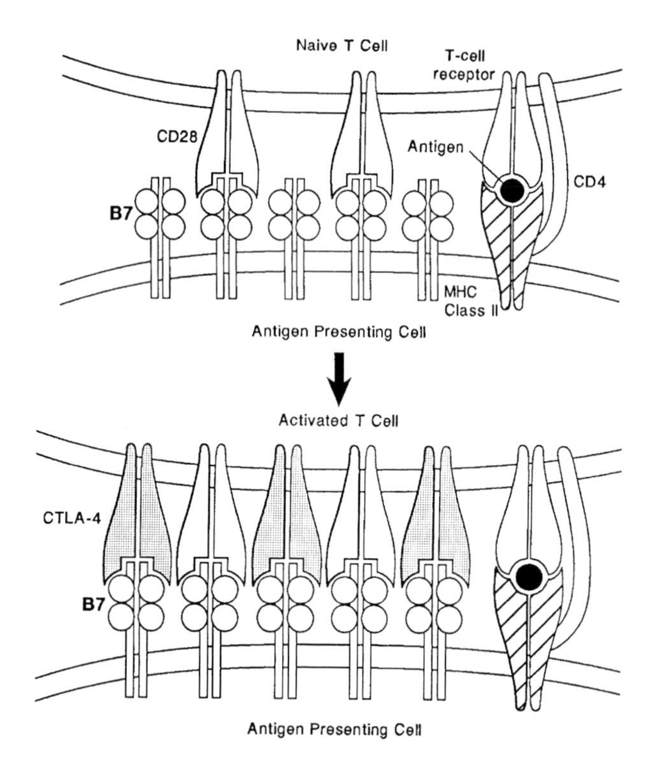

FIGURE 5.11 B7.

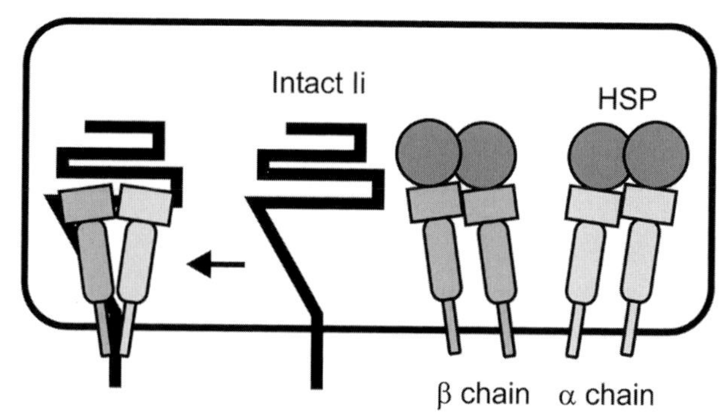

FIGURE 5.12 Invariant chain promotes assembly of class II heterodimers from free chains.

IFN-a, endotoxin, and MHC class II binding. B7 is also termed CD80. B7-2 is a costimulatory molecule whose sequence resembles that of B7. Dendritic cells, monocytes, activated T cells, and activated B lymphocytes may express B7-2.

An **invariant (Ii) chain** is a nonpolymorphic, 31-kDa glycoprotein that associates with class II histocompatibility molecules in the endoplasmic reticulum (Figure 5.12). It inhibits the linking of endogenous peptides with the class II molecule, conveying it to appropriate intracellular compartments. Truncation of the invariant chain stimulates a second signal that may function in the trans-Golgi network, prior to the conveyance of MHC class II molecules to the cell surface.

A **class II vesicle (CIIV)** is a murine B cell membrane-bound organelle that is critical in the class II MHC pathway of antigen presentation. It contains all constituents requisite for the formation of peptide antigen and class II MHC molecular complexes, including the enzymes that

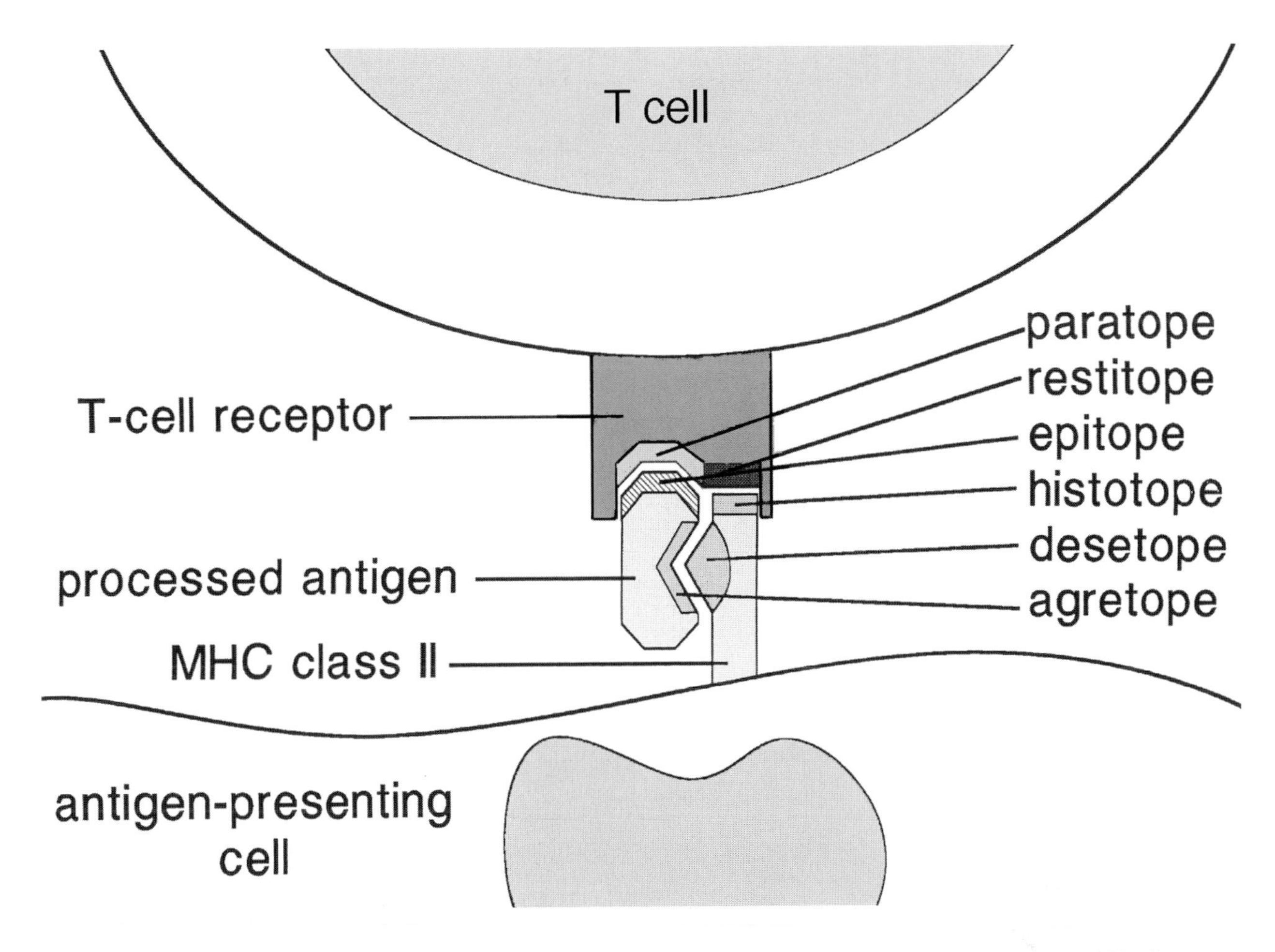

FIGURE 5.13 The schematic representation of the interaction of the class II MHC, processed peptide antigen, and T cell receptor molecules during antigen presentation.

degrade protein antigens, class II molecules, invariant chain, and HLA-DM.

I invariant (Ii): See invariant chain.

Desetope is a term derived from "determinant selection." It describes that region of class II histocompatibility molecules that reacts with the antigen during antigen presentation (Figure 5.13 and Figure 5.14). Allelic variation permits these contact residues to vary, which is one of the factors in histocompatibility molecule selection of a particular epitope that is being presented.

An **agretope** refers to the region of a protein antigen that combines with an **MHC** class II molecule during antigen presentation. This is then recognized by the T cell receptor **MHC** class II complex. Amino acid sequences differ in their reactivity with MHC class II molecules. A **histotope** is the portion of an MHC class II histocompatibility molecule that reacts with a T lymphocyte receptor. A **restitope** is that segment of a T cell receptor that makes contact and interacts with a class II histocompatibility antigen molecule during antigen presentation.

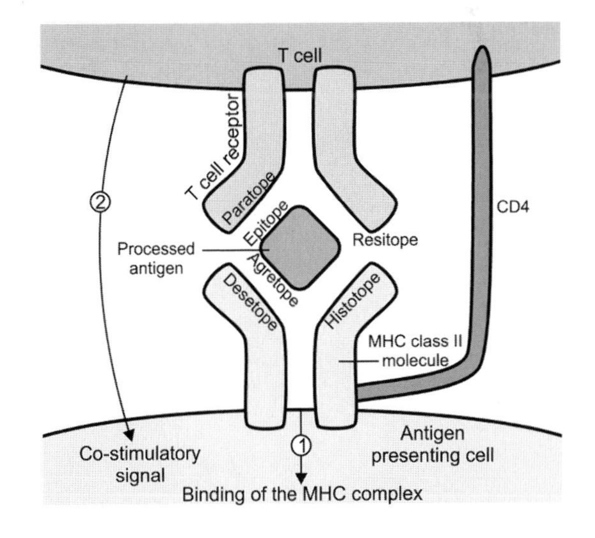

FIGURE 5.14 Costimulator for the activation of T cells.

CLIP is the processed fragment of invariant chain. In the MHC class II transport pathway, the peptide binding groove must be kept free of endogenous peptides. The cell uses one protein, called invariant chain (and its processed fragment CLIP), to block the binding site until needed. HLA-DM facilitates release of CLIP peptides and their exchange for antigenic peptides as they become available. As long as CLIP remains in the binding groove, antigenic peptides cannot bind.

A **superantigen** is an antigen such as a bacterial toxin that is capable of stimulating multiple T lymphocytes, especially $CD4^+$ T cells, leading to the release of relatively large quantities of cytokines. Selected bacterial toxins may stimulate all T lymphocytes in the body that contain a certain family of V β T cell receptor genes. Superantigens may induce proliferation of 10% of $CD4^+$ T cells by combining with the T cell receptor V β and to the MHC HLA-DR α-1 domain. Superantigens are thymus-dependent (TD) antigens that do not require phagocytic processing. Instead of fitting into the TCR internal groove where a typical processed peptide antigen fits, superantigens bind to the external region of the αβ TCR and simultaneously link to DP, DQ, or DR molecules on antigen-presenting cells (Figure 5.15). Superantigens react with multiple TCR molecules whose peripheral structure is similar. Thus, they stimulate multiple T cells that augment a protective T and B cell antibody response. This enhanced responsiveness to antigens such as toxins produced by staphylococci and streptococci is an important protective mechanism in the infected individual (Figure 5.16). Several staphylococcal enterotoxins are superantigens and may activate many T cells resulting in the release of large quantities of cytokines and producing a clinical syndrome resembling septic shock.

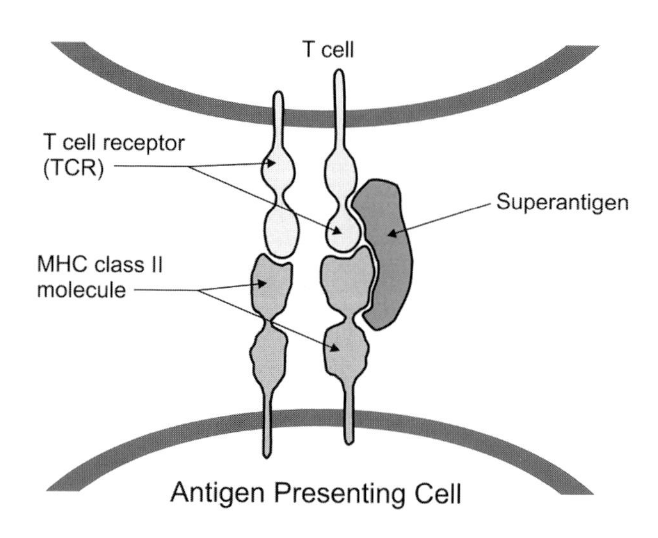

FIGURE 5.15 Superantigen.

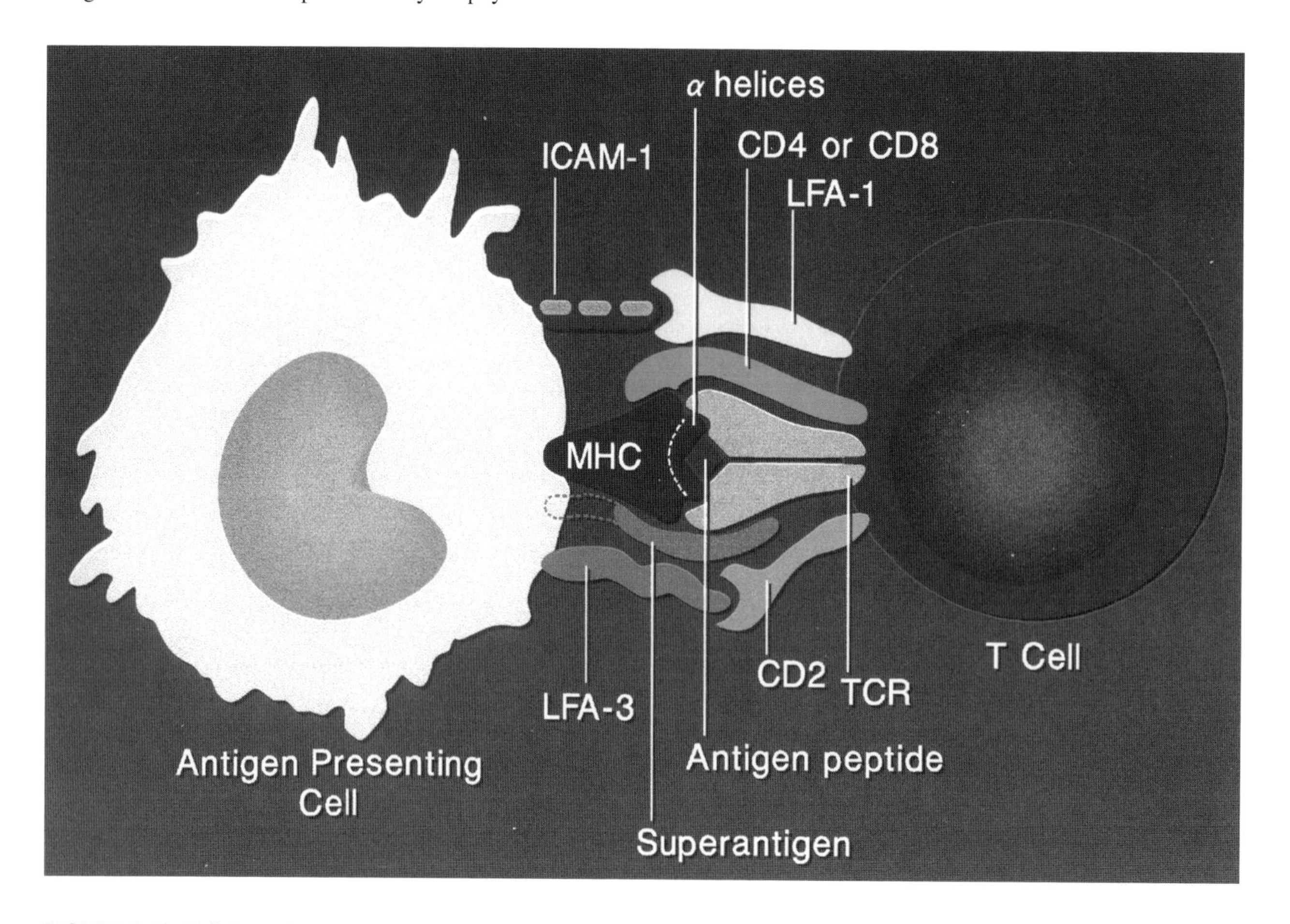

FIGURE 5.16 Cellular and molecular interactions in antigen presentation.

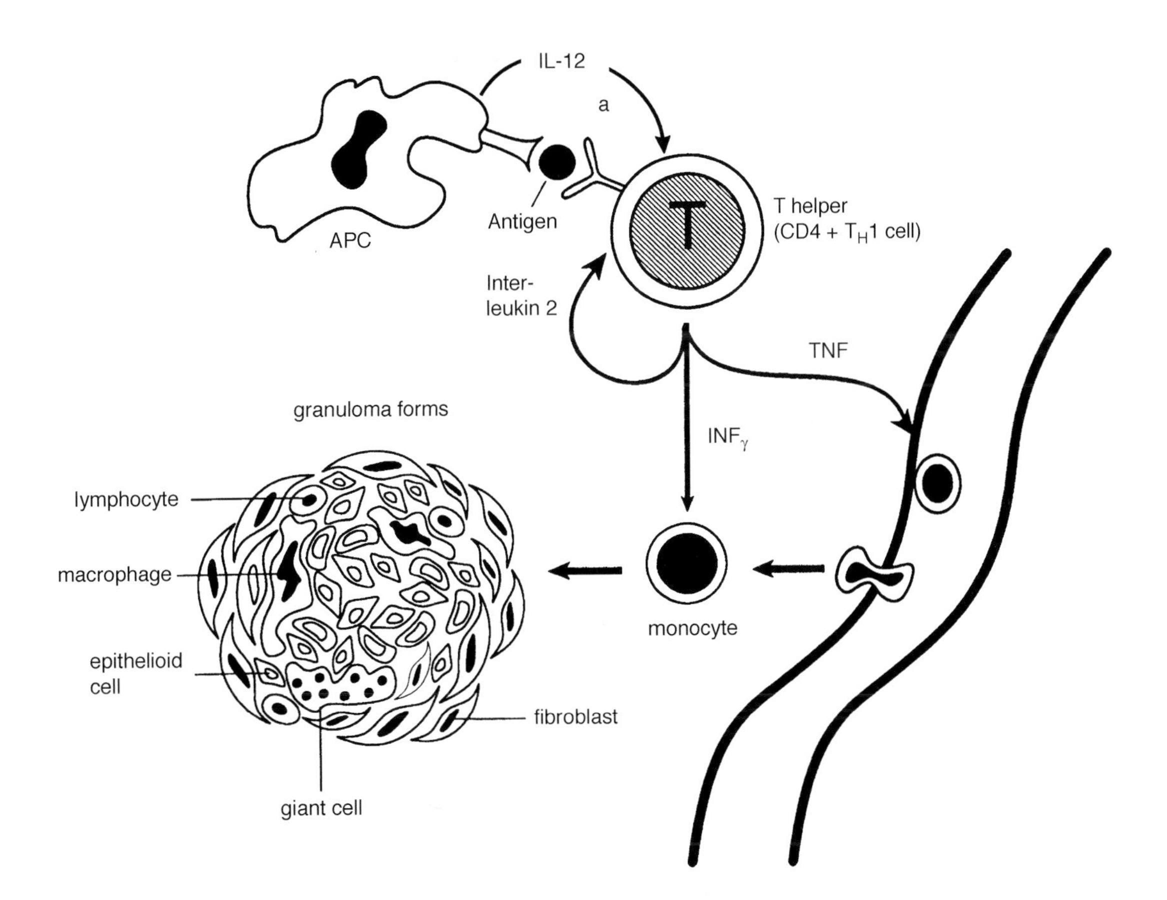

FIGURE 5.17 Granuloma.

MHC-I antigen presentation: Proteins in the cytosol, such as those derived from viruses, may be processed through the class I MHC route of antigen presentation. The multiprotein complex in the cytoplasm, known as the proteasome effects, involve proteolytic degradation of proteins in the cytoplasm to yield many of the peptides that are presented by class I MHC molecules. TAP molecules transport peptides from the cytoplasm to the endoplasmic reticulum, where they interact and bind to class I MHC dimeric molecules. Once the class I MHC molecules have become stabilized through peptide binding, the complex leaves the endoplasmic reticulum, entering the Golgi apparatus en route to the surface of the cell. Thus, mechanisms are provided through MHC-restricted antigen presentation to guarantee that peptides derived from extracellular microbial proteins can be presented by class II MHC molecules to $CD4^+$ helper T cells and that peptides derived from intracellular microbes can be presented by class I MHC molecules to $CD8^+$ cytotoxic T lymphocytes. The generation of microbial peptides produced through antigen processing to combine with self MHC molecules is critical to the development of an appropriate immune response.

A **granuloma** is a tissue reaction characterized by altered macrophages (epithelioid cells), lymphocytes, and fibroblasts (Figure 5.17). These cells form microscopic masses of mononuclear cells. Giant cells form from some of these fused cells. Granulomas may be of the foreign body type, such as those surrounding silica or carbon particles, or of the immune type that encircle particulate antigens derived from microorganisms. Activated macrophages trap antigen which may cause T cells to release lymphokines causing more macrophages to accumulate. This process isolates the microorganism. Granulomas appear in cases of tuberculosis and develop under the influence of helper T cells that react against *Mycobacterium tuberculosis*. Some macrophages and epithelioid cells fuse to form multinucleated giant cells in immune granulomas. There may also be occasional neutrophils and eosinophils. Necrosis may develop. It is a delayed type of hypersensitivity reaction that persists as a consequence of the continuous presence of foreign body or infection.

HAM-1 and HAM-2 (histocompatibility antigen modifier): These are two murine genes that determine formation of permeases that are antigen transporters (oligopeptides) from the cytoplasm to a membrane-bound compartment where antigen complexes with MHC class I and class II molecules. In man, the equivalents of HAM-1 and HAM-2 are termed ATP-binding cassette transporters.

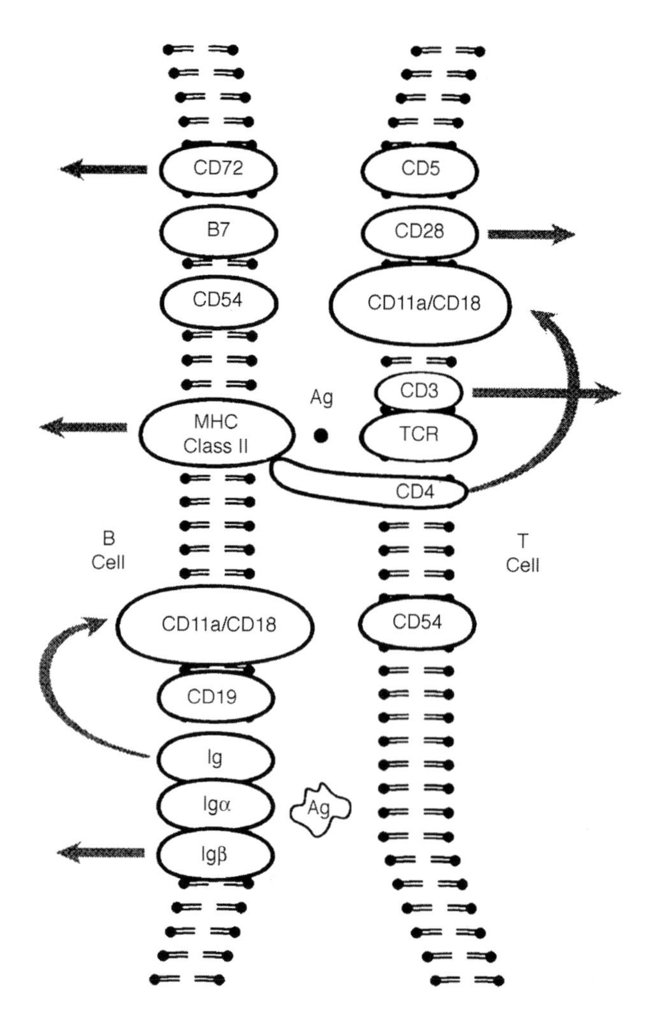

FIGURE 5.18 T–B cell interactions.

T lymphocyte–B lymphocyte cooperation refers to the association of B cell and helper T cell through a number of receptor–ligand interactions at the surfaces of both cell types which leads to B cell proliferation and differentiation into plasma cells that synthesize and secrete specific antibody specific for thymus-dependent antigen (Figure 5.15). B cell immunoglobulin receptors react with protein antigens. This is followed by endocytosis, antigen processing, and presentation to helper T lymphocytes. Their antigen-specific T cell receptors recognize processed antigens only in the context of MHC class II molecules on the B cell surface during antigen presentation. $CD4^+$ helper T cells secrete lymphokines, including IL-2, which promote B cell growth and differentiation into plasma cells that secrete specific antibodies. T cells are required for B cells to be able to switch from forming IgM to synthesizing IgG or IgA. B and T lymphocytes recognize different antigens. B cells may recognize peptides, native proteins, or denatured proteins. T cells are more complex in their recognition system in that a peptide antigen can be presented to them only in the context of MHC class II or class I histocompatibility molecules. Hapten–carrier complexes have been successfully used in delineating the different responses of B and T cells to each part of this complex. Immunization of a rabbit or other animal with a particular hapten–carrier complex will induce a primary immune response, and a second injection of the same hapten–carrier conjugate will induce a secondary immune response. However, linkage of the same hapten to a different carrier elicits a much weaker secondary response in an animal primed with the original hapten–carrier complex. This is termed the carrier effect. B lymphocytes recognize the hapten and T lymphocytes the carrier.

By light microscopy, **resting lymphocytes** appear as a distinct and homogeneous population of round cells, each with a large, spherical or slightly kidney-shaped nucleus which occupies most of the cell and is surrounded by a narrow rim of basophilic cytoplasm with occasional vacuoles. The nucleus usually has a poorly visible single indentation and contains densely packed chromatin. Occasionally, nucleoli can be distinguished. The small lymphocyte variant, which is the predominant morphologic form, is slightly larger than an erythrocyte. Larger lymphocytes, ranging between 10 and 20 μm in diameter, are difficult to differentiate from monocytes. They have more cytoplasm and may show azurophilic granules. Intermediate-size forms between the two are described. By phase contrast microscopy, living lymphocytes show a feeble motility with ameboid movements that give the cells a hand-mirror shape. The mirror handle is called a uropod. In large lymphocytes, mitochondria and lysosomes are better visualized, and some cells show a spherical, birefringent, 0.5-μm diameter inclusion, called a gall body. Lymphocytes do not spread on surfaces. The different classes of lymphocytes cannot be distinguished by light microscopy. By scanning electron microscopy, B lymphocytes sometimes show a hairy (rough) surface, but this is apparently an artifact. Electron microscopy does not provide additional information except for visualization of the cellular organelles which are not abundant. This suggests that the small, resting lymphocytes are end-stage cells. However, under appropriate stimulation, they are capable of considerable morphologic changes.

Cooperation refers to T lymphocyte–B lymphocyte cooperation.

Cooperativity refers to the effect observed when two binding sites are linked to their ligand to yield an effect of binding to both that is greater than the sum of each binding site acting independently.

Cognate interaction refers to the interaction of processed antigen on a B cell surface interacting with a T cell receptor for antigen resulting in B cell differentiation into an antibody-producing cell.

Cognate recognition refers to cognate interaction.

6 B Lymphocyte Development and Immunoglobulin Genes

The **bursa of Fabricius** (Figure 6.1) is located near the terminal portion of the cloaca and, like the thymus, is a lympho-epithelial organ. The bursa begins to develop after the 5th day of incubation and becomes functional around the 10th to 12th day. It has an asymmetric sac-like shape and a star-like lumen, which is continuous with the cloacal cavity. The epithelium of the intestine covers the bursal lumen but lacks mucous cells. The bursa contains abundant lymphoid tissue, forming nodules beneath the epithelium. The nodules show a central medullary region containing epithelial cells and project into the epithelial coating. The center of the medullary region is less structured and also contains macrophages, large lymphocytes, plasma cells, and granulocytes. A basement membrane separates the medulla from the cortex; the latter comprises mostly small lymphocytes and plasma cells. The bursa is well developed at birth but begins to involute around the 4th month; it is vestigial at the end of the first year. There is a direct relationship between the hormonal status of the bird and involution of the bursa. Injections of testosterone may lead to premature regression or even lack of development, depending on the time of hormone administration. The lymphocytes in the bursa originate from the yolk sac and migrate there via the blood stream. They comprise B cells which undergo maturation to immunocompetent cells capable of antibody synthesis. Bursectomy at the 17th day of incubation induces agammaglobulinemia, with the absence of germinal centers and plasma cells in peripheral lymphoid organs.

A **bursacyte** is a lymphocyte that undergoes maturation and differentiation under the influence of the bursa of Fabricius in avian species. This cell synthesizes the antibody which provides humoral immunity in this species. A bursacyte is a B lymphocyte. The anatomical site in mammals and other nonavian species that resembles the bursa of Fabricius in controlling B cell ontogeny is termed a **bursa equivalent**. Mammals do not have a specialized lymphoid organ for maturation of B lymphocytes. Although lymphoid nodules are present along the gut, forming distinct structures called Peyer's patches, their role in B cell maturation is no different from that of lymphoid structures in other organs. After commitment to B cell lineage, the B cells of mammals leave the bone marrow in a relatively immature stage. Likewise, after education in the thymus, T cells migrate from the thymus also in a relatively immature stage. Both populations continue their maturation process away from the site of origin and are subject to influences originating in the environment in which they reside.

Bursectomy refers to the surgical removal or ablation of the bursa of Fabricius, an outpouching of the hindgut near the cloaca in birds. Surgical removal of the bursa prior to hatching or shortly thereafter, followed by treatment with testosterone *in vivo*, leads to failure of the B cell limb of the immune response responsible for antibody production.

Hematogones are early precursor B cells that express immature cell surface antigens such as HLA-DR, CD10, CD19, and CD20. Morphologically they appear as small compact lymphocytes.

Large pre-B cells are cells with a surface pre-B cell receptor which is deleted on transition to small pre-B cells, which undergo light-chain gene rearrangement.

Pre-B cells develop (Figure 6.2 and Figure 6.3) from lymphoid stem cells in the bone marrow. These are large, immature lymphoid cells that express cytoplasmic T chains but no light chains or surface immunoglobulin, and are found in fetal liver and adult bone marrow. They are the earliest cells of the B cell lineage. Antigen is not required for early differentiation of the B cell series. Pre-B cells differentiate into immature B cells, followed by mature B cells that express surface immunoglobulin. Pre-B cell immunoglobulin genes contain heavy chain *V*, *D*, and *J* gene segments that are contiguous. No rearrangement of light-chain gene segments has yet occurred. In addition to their cytoplasmic IgM, pre-B cells are positive for CD10, CD19, and HLA-DR markers. Pre-B cells have rearranged heavy- but not light-chain genes. They express cytoplasmic Ig μ heavy chains and surrogate light chains but not Ig light chains. The pre-B cell receptor consists of μ chains and surrogate light chains. These receptors transmit signals that induce further pre-B cell maturation into immature B cells.

A **pre-B cell receptor** is a maturing B cell lymphocyte receptor comprised of a μ heavy chain and an invariant surrogate light chain. It is expressed at the pre-B cell stage. The two proteins comprising the surrogate light chain include the λ5 protein, which is homologous to the λ light

FIGURE 6.1 The bursa of Fabricius is an outpouching of the hindgut located near the cloaca in avian species that governs B cell ontogeny. This specific lymphoid organ is the site of migration and maturation of B lymphocytes.

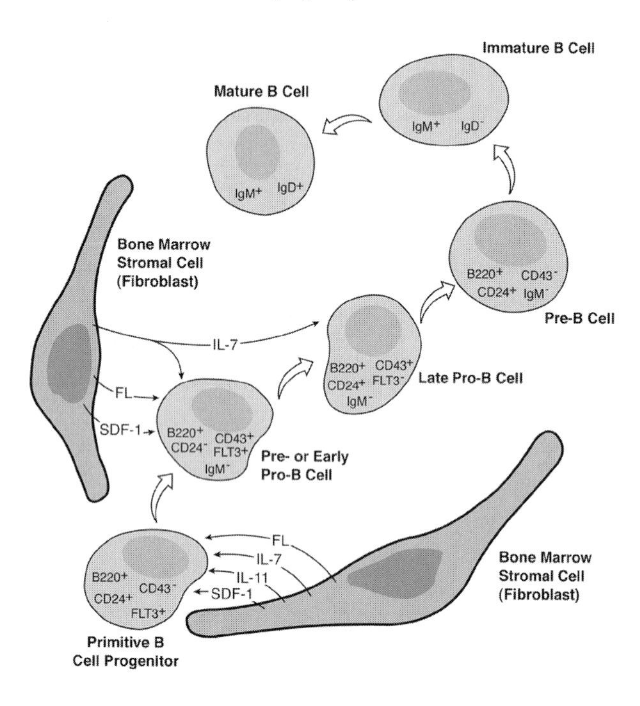

FIGURE 6.2 FLT-3 and bone marrow B cell lymphopoiesis.

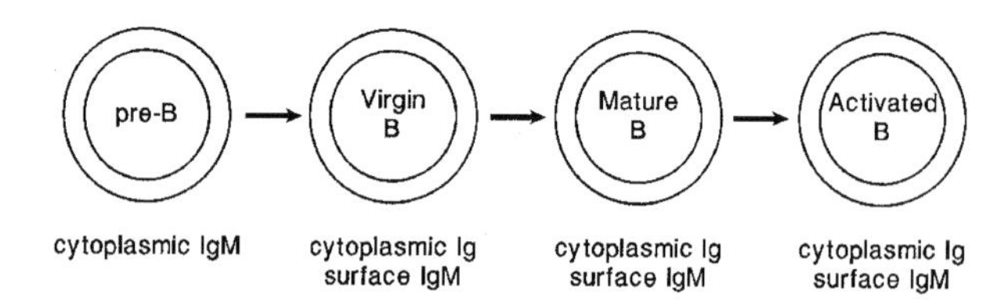

FIGURE 6.3 Pre-B cell development.

chain C domain, and the Vpre-B protein which is homologous to a Vdomain. Association of the pre-B cell receptor with the Igα and the Igβ signal transduction proteins forms the pre-B cell receptor complex. Stimulation of proliferation and continued maturation of developing B cells require pre-B cell receptors.

λ5: See pre-B cell receptor.

Lymphocyte maturation is the development of pluripotent bone marrow precursor cells into T or B lymphocytes that express antigen receptors that are present in peripheral lymphoid tissue. B cell maturation takes place in the bone marrow and T cell maturation is governed by the thymus.

λ5 B cell development: See Vpre-B protein.

Terminal deoxynucleotidyl transferase (TdT) is an enzyme catalyzing the attachment of mononucleotides to the 3′ terminus of DNA. It thus acts as a DNA polymerase. Tdt is an enzyme present in immature B and T lymphocytes, but not demonstrable in mature lymphocytes. TdT is present both in the nuclear and soluble fractions of thymus and bone marrow. The nuclear enzyme is also able to incorporate ribonucleotides into DNA. In mice, two forms of TdT can be separated from a preparation of thymocytes. They are designated peak I and peak II. They have similar enzymatic activities and appear to be serologically related but display significant differences in their biologic properties. Peak I appears constant in various strains of mice and at various ages. Peak II varies greatly. In some strains, peak II remains constant up to 6 to 8 months of age; in others, it declines immediately after birth. A total of 80% of bone marrow TdT is associated with a particular fraction of bone marrow cells separated on a discontinuous BSA gradient. This fraction represents 1 to 5% of the total marrow cells, but is O antigen negative. These cells become O positive after treatment with a thymic hormone, thymopoietin, suggesting that they are precursors of thymocytes. Thymectomy is associated with rapid loss of peak II and a slower loss of peak I in this bone marrow cell fraction. TdT is detectable in T cell leukemia, 90% of common acute lymphoblastic leukemia cases, and half of acute undifferentiated leukemia cells. Approximately one third of chronic myeloid leukemia cells in blast crisis and a few cases of pre-B cell acute lymphoblastic leukemia cases show cells that are positive for TdT. This marker is very infrequently seen in cases of chronic lymphocytic leukemia. In blast crisis, some cells may simultaneously express lymphoid and myeloid markers. Indirect immunofluorescence procedures can demonstrate TdT in immature B and T lymphocytes. It inserts nontemplated nucleotides (N-nucleotides) into the junctions between gene segments during T cell receptor and immunoglobulin heavy chain gene rearrangement.

Vpre-B: See pre-B cell receptor.

Vpre-B and **λ5** are proteins produced at the early pro-B cell stage of B cell development that are required for regulation of μ chain expression on the cell surface. Vpre-B

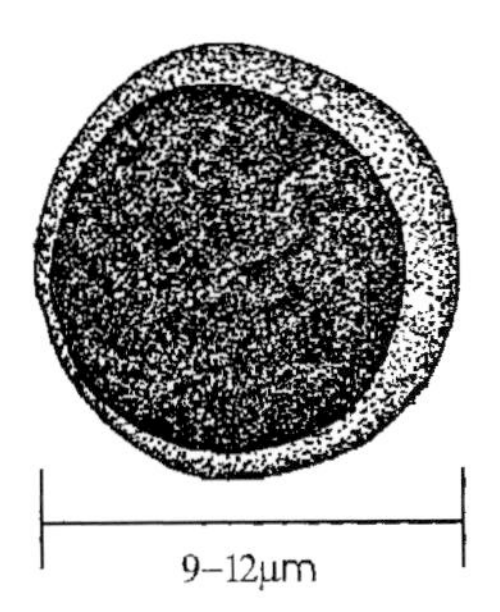

FIGURE 6.4 B cell.

and λ5 take the place of immunoglobulin light chains in pre-B cells. These proteins have an important role in B cell differentiation.

B cells (Figure 6.4) are lymphocytes that are derived from the fetal liver in the early embryonal stages of development and from the bone marrow thereafter. Plasma cells that synthesize antibody develop from precursor B cells.

A **mature B cell** is a B lymphocyte that expresses IgM and IgD and has become functionally capable of responding to antigen. They constitute the final step in B cell maturation of the bone marrow and reside in peripheral lymphoid organs.

B-1 cells are B lymphocytes that express the CD5 glycoprotein and synthesize antibodies of broad specificities. They comprise a minor population of B cells. Also called CD5 B cells.

CD5 was initially described as an alloantigen termed Ly-1 on murine T cells. Subsequently a pan T cell marker of similar molecular mass was found on human lymphocytes using monoclonal antibodies. This was named CD5. Thus, CD5 is homologous at the DNA level with Ly-1. CD5 is a 67-kDa type I transmembrane glycoprotein comprised of a single polypeptide of approximately 470 amino acids. The signal peptide is formed by the first 25 amino acids. CD5 is expressed on the surface of all αβ T cells but is absent or of low density on γδ T cells. It has been discovered on many murine B cell lymphomas as well as on endothelial cells of blood vessels in the pregnant sheep uterus. CD72 on B cells is one of its three ligands. CD5 is present on thymocytes and most peripheral T cells. It is believed to be significant for the activation of T cells and possibly B1 cells. CD5 is also present on a subpopulation of B cells termed B1 cells that synthesize polyreactive and autoreactive antibodies as well as the "natural antibodies" present in normal serum. Human chronic lymphocytic leukemia cells express CD5, which points to their derivation from this particular B cell subpopulation.

CD5 B cells constitute an atypical, self-renewing class of B lymphocytes that reside mainly in the peritoneal and pleural cavities in adults and which have a far less diverse receptor repertoire than do conventional B cells.

Preprogenitor cells comprise a pool of cells that represents a second step in the maturation of B cells and is induced by nonspecific environmental stimuli. They are the immature cells that by themselves are unable to mount an immune response but are the pool from which the specific responsive clones will be selected by the specific antigen. They are present both in the bone marrow and peripheral lymphoid organs such as the spleen, but in the latter they form a minor population. They are characterized by the presence of some surface markers, frequently doublet or triplet surface immunoglobulins, and are capable of being stimulated by selected activators. They are sometimes termed B1 cells.

B lymphocytes are lymphocytes of the B cell lineage that mature under the influence of the bursa of Fabricius in birds and in the bursa equivalent (bone marrow) in mammals. B cells occupy follicular areas in lymphoid tissues and account for 5 to 25% of all human blood lymphocytes that number 1000 to 2000 cells/mm^3. They comprise most of the bone marrow lymphocytes, and one-third to one-half of the lymph node and spleen lymphocytes, but less than 1% of those in the thymus. Nonactivated B cells circulate through lymph nodes and spleen. They are concentrated in follicles and marginal zones around the follicles. Circulating B cells may interact and be activated by T cells at extrafollicular sites where the T cells are present in association with antigen-presenting dendritic cells. Activated B cells enter the follicles, proliferate, and displace resting cells. They form germinal centers and differentiate into both plasma cells that form antibody and long-lived memory B cells. Those B cells synthesizing antibodies provide defense against microorganisms including bacteria and viruses. Surface and cytoplasmic markers reveal the stage of development and function of lymphocytes in the B cell lineage. Pre-B cells contain cytoplasmic immunoglobulins, whereas mature B cells express surface immunoglobulin and complement receptors. B lymphocyte markers include CD9, CD19, CD20, CD24, Fc receptors, B1, BA-1, B4, and Ia.

B1a B-cells (CD5) are a small population of B cells that express CD5 but to a lesser degree than CD5 expression on T cells. CD5 is a negative regulator of T cell receptor signaling. CD5 participates in B cell receptor-induced apoptosis of B1a cells.

B-2 cells are B lymphocytes that fail to express the CD5 glycoprotein. They synthesize antibodies of narrow specificities and comprise most of the B cell population.

Unprimed refers to animals or cells that have not come into contact previously with a particular antigen.

Tec kinase is a family of src-like tyrosine kinases that have a role in activation of lymphocyte antigen receptors through activation of PLC-γ. Btk in B lymphocytes, which is mutated in X-linked agammaglobulinemia (XLA), human immunodeficiency disease, and ltk in T lymphocytes, are examples of other Tec kinases.

Syk PTK is a 72-kDa phosphotyrosine kinase found on B cells and myeloid cells that is homologous to the ZAP-70 PTK found on T cells and NK cells. Both Syk and ZAP-70 play roles in the functions of distinct antigen receptors.

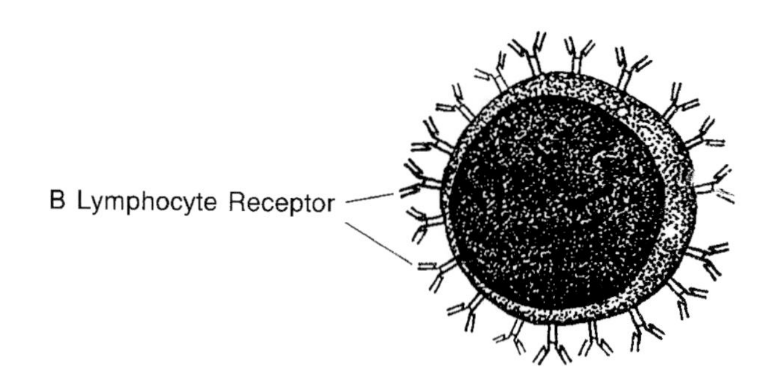

FIGURE 6.5 B lymphocyte receptor.

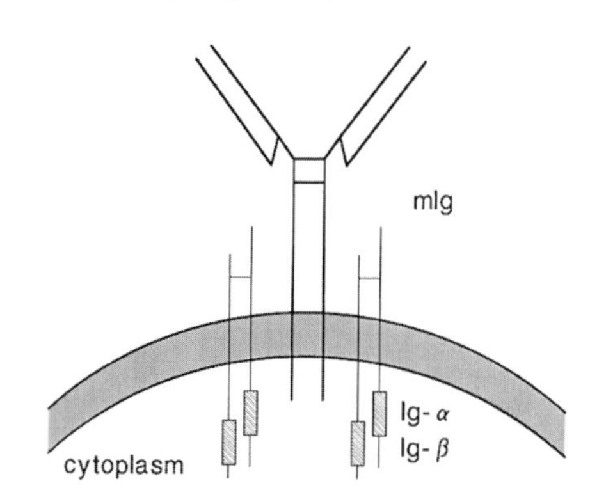

FIGURE 6.6 Schematic representation of Ig on a cell membrane.

B lymphocyte tolerance refers to the immunologic nonreactivity of B lymphocytes induced by relatively large doses of antigen. It is of relatively short duration. By contrast, T cell tolerance requires less antigen and is of longer duration. Exclusive B cell tolerance leaves T cells immunoreactive and unaffected.

B cell tolerance is manifested as a decreased number of antibody-secreting cells following antigenic stimulation, compared with a normal response. Hapten-specific tolerance can be induced by inoculation of deaggregated haptenated-gammaglobulins (Ig). Induction of tolerance requires membrane Ig crosslinking. Tolerance may have a duration of 2 months in B cells of the bone marrow and 6 to 8 months in T cells. Whereas prostaglandin E enhances tolerance induction, IL-1, LPS, or 8-bromoguanosine block tolerance instead of an immunogenic signal. Tolerant mice carry a normal complement of hapten-specific B cells. Tolerance is not attributable to a diminished number or isotype of antigen receptor. It has also been shown that the six normal activation events related to membrane Ig turnover and expression do not occur in tolerant B cells. Whereas tolerant B cells possess a limited capacity to proliferate, they fail to do so in response to antigen. Antigenic challenge of tolerant B cells induces them to enlarge and increase expression, yet they are apparently deficient in a physiologic signal required for progression into a proliferative stage.

A **B lymphocyte hybridoma** is a clone formed by the fusion of a B lymphocyte with a myeloma cell. Activated splenic B lymphocytes from a specifically immune mouse are fused with myeloma cells by polyethylene glycol. Thereafter, the cells are plated in multi-well tissue culture plates containing HAT medium. The only surviving cells are the hybrids, since the myeloma cells employed are deficient in hypoxanthine-guanine phosphoribosyl transferse and fail to grow in HAT medium. Wells with hybridomas are screened for antibody synthesis. This is followed by cloning, which is carried out by limiting dilution or in soft agar. The hybridomas are maintained either in tissue culture or through inoculation into the peritoneal cavity of a mouse that corresponds genetically to the cell strain. The antibody-producing B lymphocyte confers specificity and the myeloma cell confers immortality upon the hybridoma. B lymphocyte hybridomas produce monoclonal antibodies.

The **B lymphocyte receptor** (Figure 6.5) is an immunoglobulin anchored to the B lymphocyte surface. Its combination with antigen leads to B lymphocyte division and differentiation into memory cells, lymphoblasts, and plasma cells. The original antigen specificity of the immunoglobulin is maintained in the antibody molecules subsequently produced. B lymphocyte receptor immunoglobulins (Figure 6.6) are to be distinguished from those in the surrounding medium that adhere to the B cell surface through Fc receptors. See membrane immunoglobulin.

A **B cell antigen receptor** (Figure 6.7) is an antibody expressed on antigen reactive B cells that is similar to secreted antibody but is membrane-bound due to an extra domain at the Fc portion of the molecule. Upon antigen recognition by the membrane-bound immunoglobulin, noncovalently associated accessory molecules mediate transmembrane signaling to the B cell nucleus. The immunoglobulin and accessory molecule complex is similar in structure to the antigen receptor–CD3 complex of T lymphocytes. The cell surface membrane-bound immunoglobulin molecule serves as a receptor for antigen, together with two associated signal-transducing Igα/Igβ molecules.

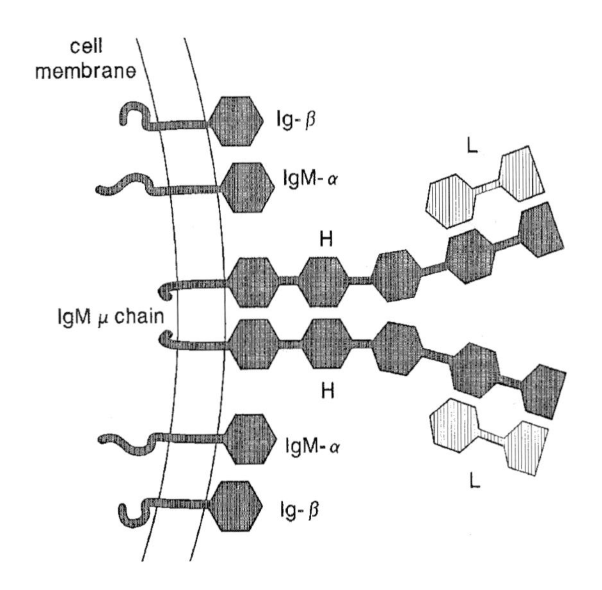

FIGURE 6.7 B cell antigen receptor.

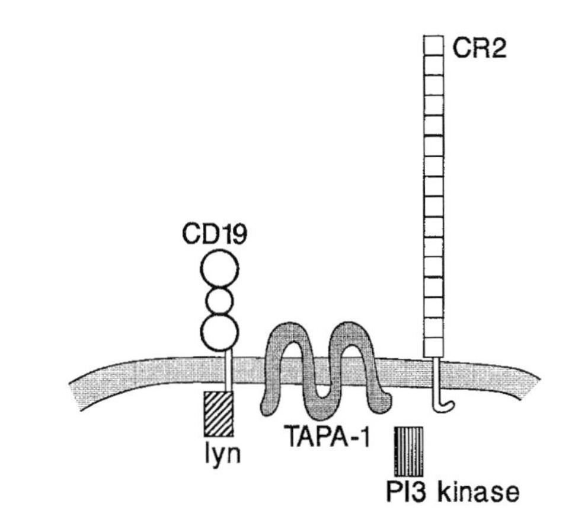

FIGURE 6.8 B cell coreceptor.

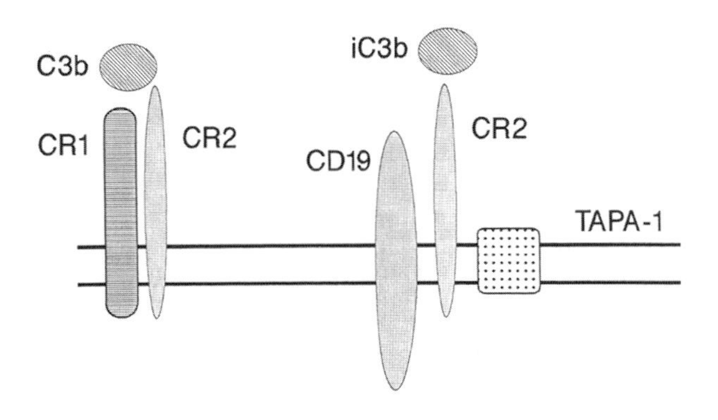

FIGURE 6.9 Complement receptor complexes on the surface of B cells include CR1, C3b, CR2, CD19, iC3b, CR2 (CD21), and TAPA-1. B cell markers that are used routinely for immunophenotyping by flow cytometry include CD19, CD20, and CD21.

Igα and Igβ are proteins on the B cell surface that are noncovalently associated with cell surface IgM and IgD. They link the B cell antigen-receptor complex to intracellular tyrosine kinases. Anti-Ig binding leads to their phosphorylation. Igα and Igβ are required for expression of IgM and IgD on the B cell surface. Disulfide bonds link Igα and Igβ pairs that are associated noncovalently with the membrane Ig cytoplasmic tail to form the B cell receptor complex (BRC). The Igα and Igβ cytoplasmic domains bear immunoreceptor tyrosine based activation motifs (ITAMs) that participate in early signaling when antigens activate B cells.

Igα/Igβ (CD79a/CD79b): The Igα/Igβ heterodimer interacts with immunoglobulin heavy chains for signal transduction. In the pro-B cell stage, rearrangement of the immunoglobulin heavy chain gene leads to expression of surface membrane immunoglobulin (mIgμ). mIgμ associates with Igα/Igβ and surrogate light chain in pre-B cells or ordinary light chains in B cells to form the precursor B cell receptor and B cell receptor, respectively. Igα and Igβ are expressed before immunoglobulin heavy chain gene rearrangement. They are products of mb-1 and B29 genes, respectively. Allelic exclusion is mediated through signal transduction via Igα and Igβ and depends on intact tyrosine residues.

Specificity refers to the recognition by an antibody or a lymphocyte receptor of a specific epitope in the presence of other epitopes for which the antigen-binding site of the antibody or of the lymphoid cell receptor is specific.

A **B cell coreceptor** (Figure 6.8) is a three-protein complex that consists of CR2, TAPA-1 (Figure 6.9), and CD19. CR2 unites not only with an activated component of complement, but also with CD23. TAPA-1 is a serpentine membrane protein. The cytoplasmic tail of CD 19 is the mechanism through which the complex interacts with lyn, a tyrosine kinase. Activation of the coreceptor by ligand binding leads to union of phosphatidyl inositol-3′ kinase with CD19 resulting in activation. This produces intracellular signals that facilitate B cell receptor signal transduction.

TAPA-1 is a serpentine membrane protein that crosses the cell membrane four times. It is one of three proteins comprising the B cell coreceptor. It is also call CD81.

CD19 (Figure 6.10) is an antigen with a 90-kDa mol wt that has been shown to be a transmembrane polypeptide with at least two immunoglobulin-like domains. The CD19 antigen is the most broadly expressed surface marker for B cells, appearing at the earliest stages of B cell differentiation. The CD19 antigen is expressed at all stages of B cell maturation, from the pro-B cell stage until just before the terminal differentiation to plasma cells. CD19 complexes with CD21 (CR2) and CD81 (TAPA-1). It is a coreceptor for B lymphocytes.

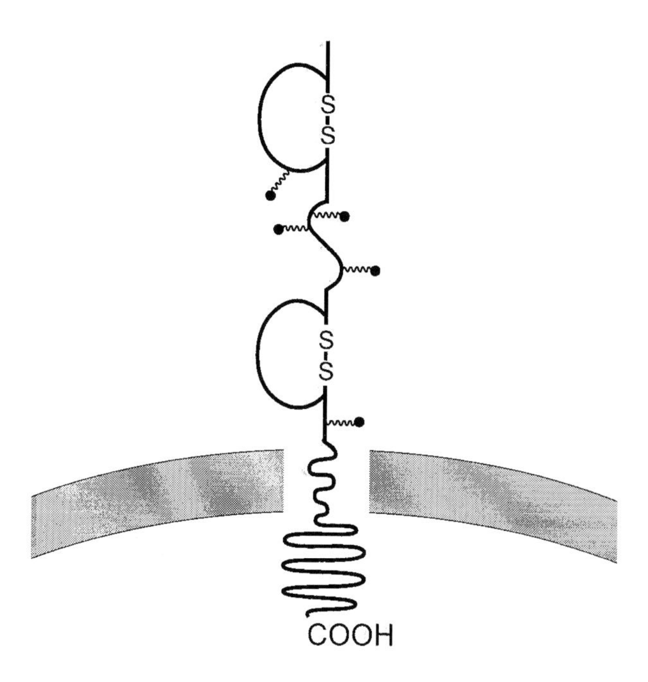

FIGURE 6.10 CD19.

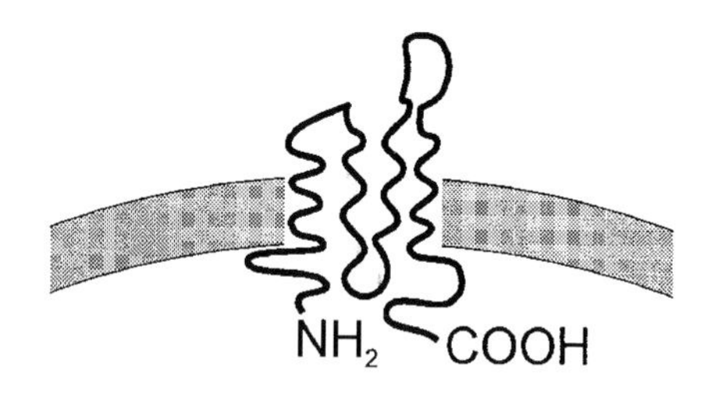

FIGURE 6.11 CD20.

CD20 (Figure 6.11) is a B cell marker with a 33-, 35-, and 37-kDa mol wt that appears relatively late in the B cell maturation (after the pro-B cell stage) and then persists for some time before the plasma cell stage. Its molecular structure resembles that of a transmembrane ion channel. The gene is on chromosome 11 at band q12-q13. It may be involved in regulating B cell activation.

CD21 (Figure 6.12) is an antigen with a 145-kDa mol wt, that is expressed on B cells and, even more strongly, on follicular dendritic cells. It appears when surface Ig is expressed after the pre-B cell stage and is lost during early stages of terminal B cell differentiation to the final plasma cell stage. CD21 is coded for by a gene found on chromosome 1 at band q32. The antigen functions as a receptor for the C3d complement component and also for Epstein–Barr virus. CD21, together with CD19 and CD81, constitutes the coreceptor for B cells. It is also termed CR2.

CD21 is a 145-kDa glycoprotein component of the B cell receptor. CD21 is a membrane molecule that participates in transmitting growth-promoting signals to the interior of the B cell. It is the receptor for the C3d fragment of the third component of complement, CR2. The CD21 antigen is a restricted B cell antigen expressed on mature B cells. It is present at high density on follicular dendritic cells (FDC), the accessory cells of the B zones. Also called complement receptor 2 (CR2).

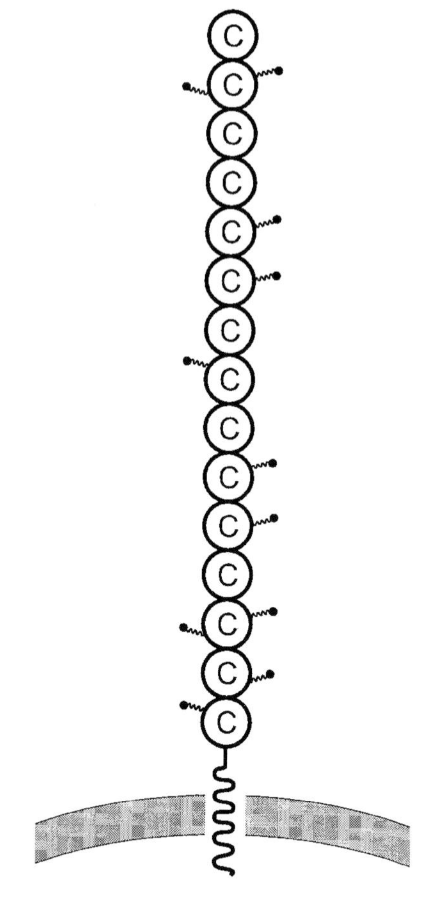

FIGURE 6.12 CD21.

CD22 is a molecule with an α130- and β140-kDa mol wt that is expressed in the cytoplasm of B cells of the pro-B and pre-B cell stage and on the cell surface on mature B cells with surface Ig. The antigen is lost shortly before the terminal plasma cell phase. The molecule has five extracellular immunoglobulin domains and shows homology with myelin adhesion glycoprotein and with N-CAM (CD56). It participates in B cell adhesion to monocytes and T cells. Also called BL-CAM.

A **plasmablast** is an immature cell of the plasma cell lineage that reveals distinctive, clumped nuclear chromatin developing endoplasmic reticulum and a Golgi apparatus. It is a B lymphocyte in a lymph node that is beginning to reveal plasma cell features.

A **plasmacyte** is a plasma cell.

Plasma cells (Figure 6.13 to Figure 6.15) are antibody-producing cells. Immunoglobulins are present in their cytoplasm and secretion of immunoglobulin by plasma cells has been directly demonstrated *in vitro*. Increased levels of immunoglobulins in some pathologic conditions are associated with increased numbers of plasma cells, and conversely, their number at antibody-producing sites increases following immunization. Plasma cells develop from B cells and are large spherical or ellipsoidal cells, 10 to 20 μm in size. Mature plasma cells have abundant cytoplasm, staining deep blue with Wright's stain, and have an eccentrically located round or oval nucleus, usually surrounded by a well-defined perinuclear clear zone. The nucleus contains coarse and clumped masses of chromatin, often arranged in a cartwheel fashion. The nuclei of normal, mature plasma cells have no nucleoli, but those of neoplastic plasma cells, such as those seen in multiple myeloma, have conspicuous nucleoli. The cytoplasm of normal plasma cells has conspicuous Golgi complex and rough endoplasmic reticulum and frequently contains vacuoles. The nuclear to cytoplasmic ratio is 1:2. By electron microscopy, plasma cells show very abundant endoplasmic reticulum, indicating extensive and active protein synthesis. Plasma cells do not express surface immunoglobulin or complement receptors, which distinguishes them from B lymphocytes.

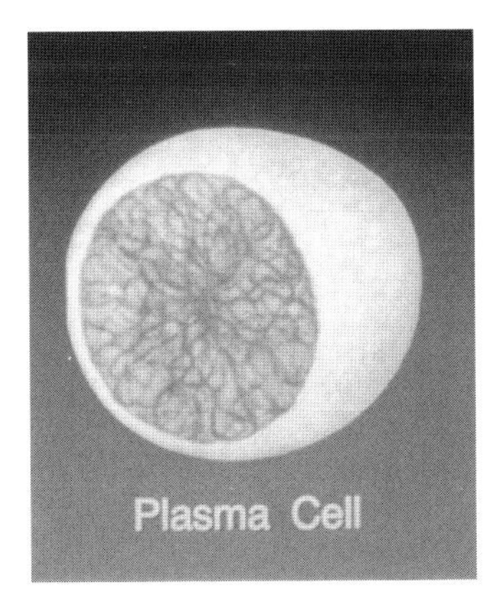

FIGURE 6.13 Plasma cell diagram.

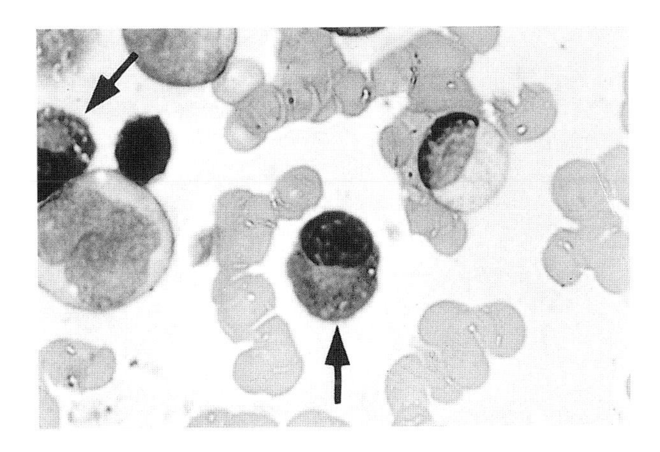

FIGURE 6.14 Plasma cell in peripheral blood smear.

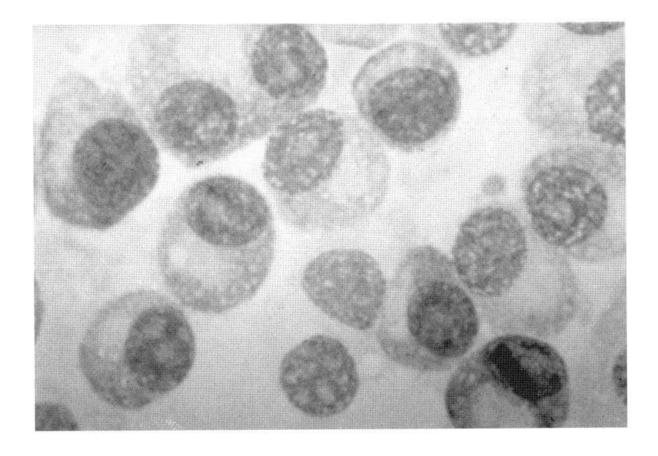

FIGURE 6.15 Plasma cell cluster in peripheral blood smear.

An example of **plasma cell antigen** is a murine plasmacyte membrane alloantigen. It may be designated PC-1, PC-2.

Pyroninophilic cells are cells whose cytoplasm stains red with methyl green pyronin stain. This signifies large quantities of RNA in the cytoplasm, indicating active protein synthesis. For example, plasma cells or other protein-producing cells are pyroninophilic.

Antibody-secreting cells are differentiated B lymphocytes that synthesize the secretory form of immunoglobulin. Antibody-secreting cells result from antigen stimulation. They may be found in the lymph nodes, spleen, and bone marrow.

A **signal peptide** is the leader sequence, a small sequence of amino acids, that shepherds the heavy or light chain through the endoplasmic reticulum and is cleaved from the nascent chains prior to assembly of a completed immunoglobulin molecule.

B cell activation (Figure 6.16) follows antigen binding to membrane immunoglobulin molecules on B lymphocyte surfaces. This interaction of antigen and membrane immunoglobulin may lead to two types of response. Either biochemical signals are conveyed to the cells via the B lymphocyte antigen receptor leading to lymphocyte activation, or antigen is taken into endosomal vesicles where protein antigens are processed and resulting peptides presented at the B lymphocyte surface to helper T cells. With respect to B lymphocyte antigen receptor signaling, the relatively short cytoplasmic tails of membrane IgM and IgD are unable to transduce signals caused by Ig clustering. Therefore, Igα and Igβ that are expressed on mature B cells is a noncovalent association with membrane Ig actually transduce signals (Figure 6.17).

B cell activation: B cell responses to antigen, whether T-independent or T-dependent, result in the conversion of small resting B cells to large lymphoblasts and then either into plasma cells that form specific antibody or into long-lasting memory B cells. Thymic independent antigens such as bacterial polysaccharides can activate B cells independently of T cells by crosslinking of the B cell receptor. By contrast, protein antigens usually require the intimate interaction of B cells with helper T cells. Antigen stimulation of the B cell receptor leads to endocytosis and degradation of the antigen captured by the B cell receptor. Peptides that result from degraded antigen are bound to MHC class II molecules and transported to the cell surface for presentation to T lymphocytes. T cells bearing a specific T cell receptor that recognizes the peptide-MHC complex presented on the B cell surface are activated. Activated T cells help B cells by either soluble mediators such as cytokines, i.e., IL-4, IL-5, and IL-6, or as membrane-bound stimulatory molecules such as CD40 ligand. In germinal centers, B cells are converted to large replicating centroblasts and then to nonreplicating centroblasts. In germinal centers, frequent Ig region mutations and the switch from IgM to IgG, IgA, or IgE production occur. Mutation increases the diversity of antigen-binding sites. Mutations that lead to loss of antigen binding cause the cells to die by apoptosis. The few cells to which mutation gives an immunoglobulin product that has high affinity for antigen are selected for survival. These antigen selected cells differentiate into plasma cells that produce antibody or into small long-lived memory B cells that enter the blood and lymphoid tissues.

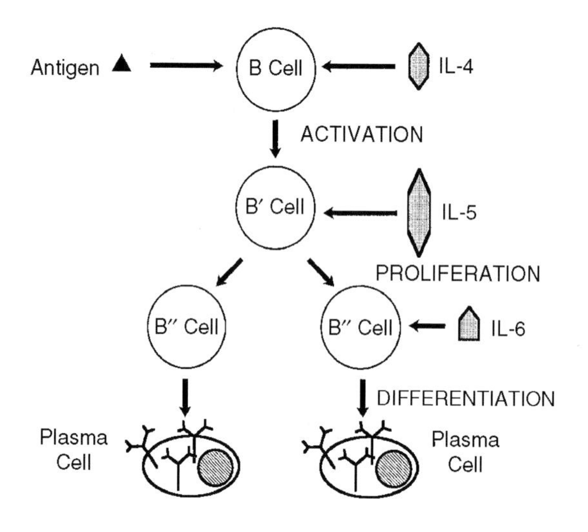

FIGURE 6.16 B cell activation.

Virgin B cells that have never interacted with antigen must have two separate types of signals to proliferate and differentiate. The antigen provides the first signal through interaction with surface membrane Ig molecules on specific B lymphocytes. Helper T cells and their lymphokines provide the second type of signal needed. Whereas polysaccharides and lipids, as nonprotein antigens, induce IgM antibody responses without antigen-specific T cell help, protein antigens, which are helper T cell-dependent, lead to the production of immunoglobulin of more than one isotype and of high affinity in addition to immunologic

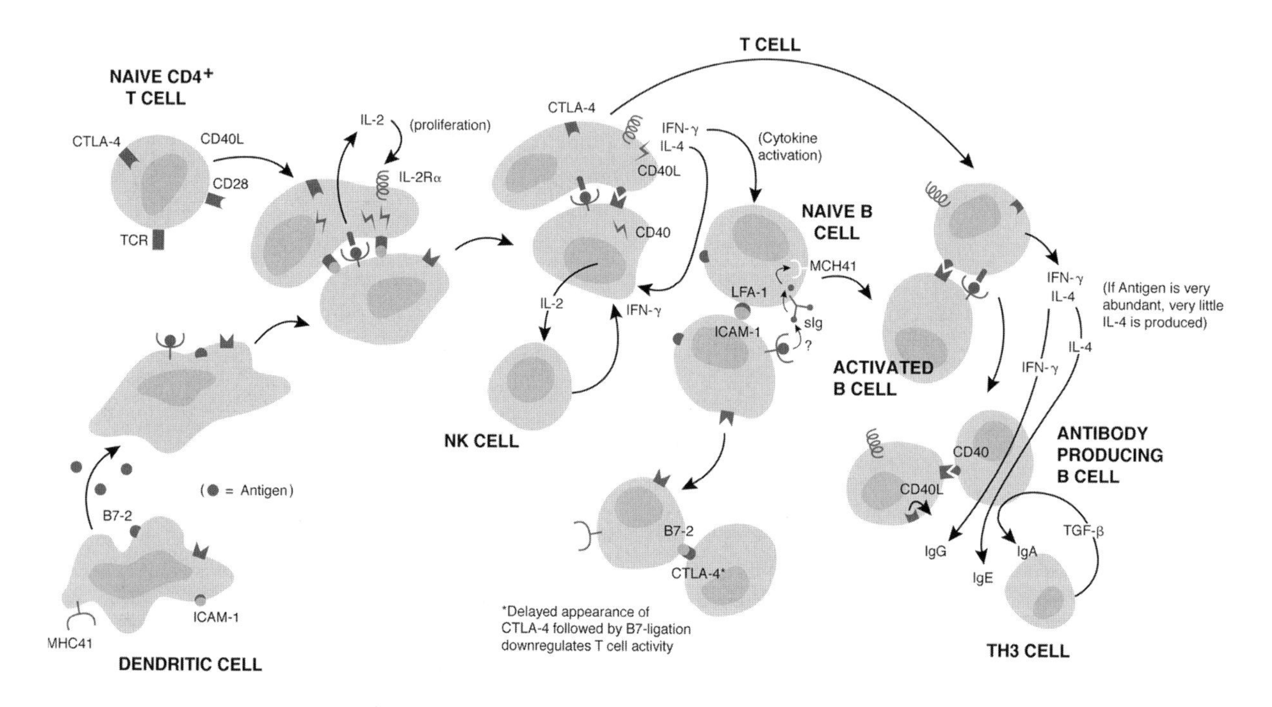

FIGURE 6.17 Hypothetical B7/CD40 pathway for B cell activation.

memory. Igα and Igβ together with membrane Ig molecules constitute the **B lymphocyte antigen receptor (BCR) complex**. The BCR complex is a B lymphocyte surface multiprotein complex that identifies antigen and transduces activating signals into the cell. The BCR comprises membrane immunoglobulin, which binds antigen, and Igα and Igβ proteins that initiate signals.

Lymphocyte antigen receptor complex: Igα and Igβ function in B lymphocytes as CD3 and proteins do in T cells. Required for signal transduction are immunoreceptor tyrosine-based activation motifs (ITAMs) in the cytoplasmic domains of Igα and Igβ. Crosslinking of the B cell receptor complex by antigen leads to increased cell size and cytoplasmic ribonucleic acid with increased biosynthetic organelles including ribosomes as resting cells enter G1 stage in the cell cycle. Class II molecules and B7-2 and B7-1 costimulators show increased expression. B cells stimulated by antigen are then able to activate helper T cells. Expression of receptors for T cell cytokines increases, thereby facilitating the ability of antigen-specific B cells to receive T cell help. The effect of B cell receptor complex signaling on proliferation and differentiation depends in part on the type of antigen. Following activation as a result of combination with antigen, cell proliferation and differentiation are facilitated by interaction with helper T lymphocytes. Helper T cells must recognize antigen and there must be interaction between protein antigen-specific B cells and T lymphocytes for antibody to be formed. When B cells, acting as antigen-presenting cells, interact with helper T lymphocytes that are specific for the peptide being presented, there are numerous ligand–receptor interactions that facilitate transmission of signals to B cells that are required to generate a humoral immune response. Among these are B7 molecules: CD28 and CD40:CD40 ligand interactions. Cytokines play important roles in antibody production by switching from one heavy chain isotype to another and by providing amplification mechanisms through augmentation of B lymphocyte proliferation and differentiation. Germinal centers are the sites of synthesis of antibodies of high affinity and of memory B cells.

Clonotypic is an adjective that defines the features of a specific B cell population's receptors for antigens that are products of a single B lymphocyte clone. Following release from the B cells, these antibodies should be very specific for antigen, have a restricted spectrotype, and should possess at least one unique private idiotypic determinant. Clonotypic may also be used to describe the features of a particular clone of T lymphocytes' specific receptor for antigen with respect to idiotypic determinants, specificity for antigen, and receptor similarity from one daughter cell of the clone to another.

Anti-B cell receptor idiotype antibodies interact with antigenic determinants (idiotopes) at the variable N-terminus of the heavy and light chains comprising the paratope region of an antibody molecule where the antigen-binding site is located. The idiotope antigenic determinants may be situated either within the cleft of the antigen-binding region or located on the periphery or outer edge of the variable region of heavy and light chain components. Anti-idiotypic antibodies also block T cell receptors for antigens for which they are specific.

Patching (Figure 6.18) describes the accumulation of membrane receptor proteins crosslinked by antibodies or lectins on a lymphocyte surface prior to capping. The antigen–antibody complexes are internalized following capping, which permits antigen processing and presentation in the context of MHC molecules. Membrane protein redistribution into patches is passive, not requiring energy. The process depends on the lateral diffusion of membrane constituents in the plane of the membrane.

Capping (Figure 6.19) refers to the migration of antigens on the cell surface to a cell pole following crosslinking of antigens by a specific antibody. These antigen–antibody complexes coalesce or aggregate into a "cap" produced by interaction of antigen with cell surface IgM and IgD molecules at sites distant from each other, as revealed by immunofluorescence. Capping is followed by interiorization of

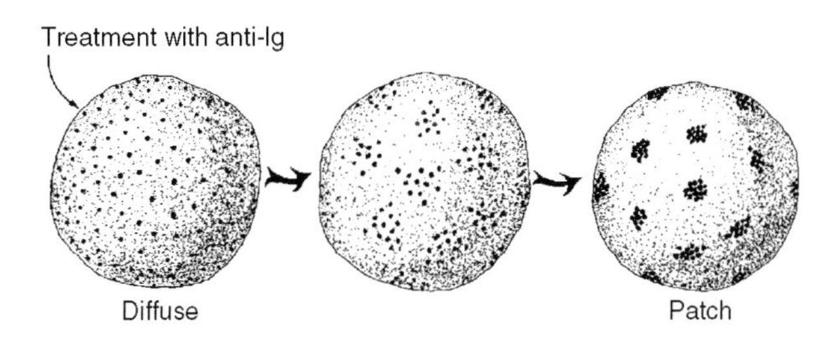

FIGURE 6.18 Patching.

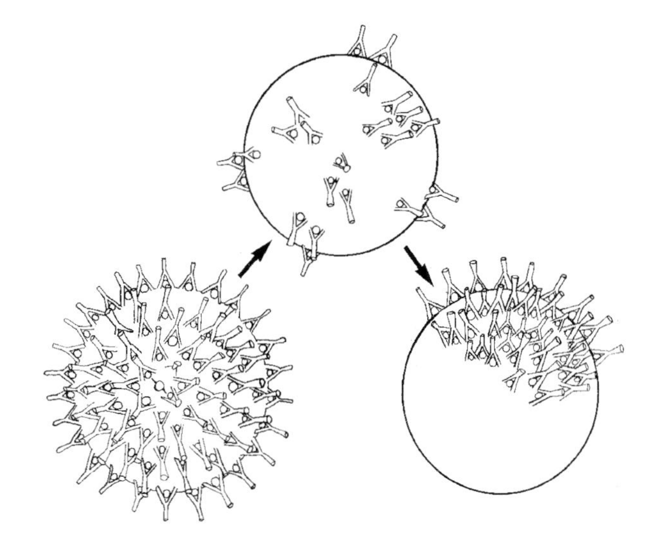

FIGURE 6.19 Capping.

the antigen. Following internalization, the cell surface is left bereft of immunoglobulin receptors until they are reexpressed.

The **capping phenomenon** is the migration of surface membrane proteins toward one pole of a cell following crosslinking by a specific antibody, antigen, or mitogen. Bivalent or polyvalent ligands cause the surface molecules to aggregate into patches. This passive process is referred to as patching. The ligand–surface molecule aggregates, in patches, move to a pole of the cell where they form a cap. If a cell with patches becomes motile, the patches move to the rear, forming a cluster of surface molecule–ligand aggregates that constitutes a cap. The process of capping requires energy and may involve interaction with microfilaments of the cytoskeleton. In addition to capping in lymphocytes, the process occurs in numerous other cells.

Cocapping: If two molecules are associated in a membrane, capping of one induced by its ligand may lead also to capping of the associated molecule. Antibodies to membrane molecule x may induce capping of membrane molecule y as well as of x if x and y are associated in the membrane. In this example, the capping of the associated y molecule is termed cocapping.

Colocalization is a mechanism of *differential redistribution* of membrane components into patches and caps, which has been employed to investigate possible interactions between various plasma membrane components or between the cell membrane and cytoplasmic structures.

Clustering: Monomeric antigens, monovalent lectins, and monovalent antibody to mIg or any other membrane component neither cap on the cell surface nor do they produce large clusters. Multivalent ligand binding is necessary for clustering as well as for capping. Clustering, unlike capping, is a passive redistribution process. Spotting or patching neither requires the cell to be living nor metabolically active. Clustering is affected not only by the factors that control phenomena occurring in a three-dimensional fluid aqueous phase but also by physicochemical properties of the plasma membrane. The outcome of cluster formation is influenced by physiological interactions between the membrane proteins themselves.

Receptor-mediated endocytosis is internalization into endosomes of cell surface receptor bound molecules such as B cell receptors to which antigens are bound and internalized.

Immune response (*Ir*) genes regulate immune responsiveness to synthetic polypeptide and protein antigens as demonstrated in guinea pigs and mice. This property is transmitted as an autosomal dominant trait that maps to the major histocompatibility complex (MHC) region.

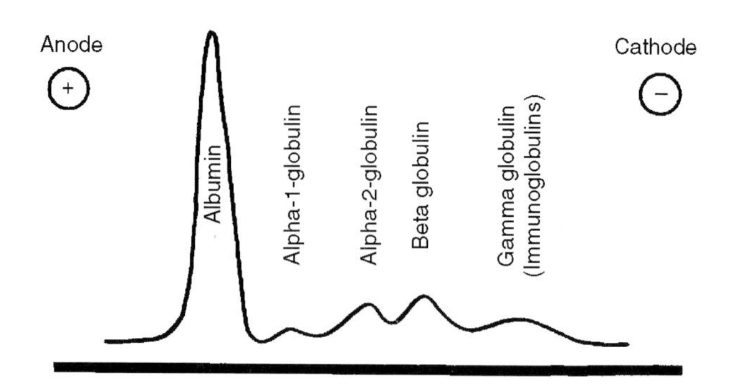

FIGURE 6.20 Electrophoresis of serum protein showing the gammaglobulin region that contains the immunoglobulins.

Ir genes control helper T lymphocyte activation, which is required for the generation of antibodies against protein antigens. T lymphocytes with specific receptors for antigen fail to recognize free or soluble antigens, but they do recognize protein antigens noncovalently linked to products of MHC genes termed class I and class II MHC molecules. The failure of certain animal strains to respond may be due to ineffective antigen presentation in which processed antigen fails to bind properly to class II MHC molecules or due to an ineffective interaction between the T cell receptor and the MHC class II-antigen complex.

Immunoglobulin genes (Figure 6.20) encode heavy and light polypeptide chains of antibody molecules and are found on different chromosomes, i.e., chromosome 14 for heavy-chain, chromosome 2 for κ light chain, and chromosome 22 for λ light chain. The DNA of the majority of cells does not contain one gene that encodes a complete immunoglobulin heavy or light polypeptide chain. Separate gene segments that are widely distributed in somatic cells and germ cells come together to form these genes. In B cells, gene rearrangement leads to the creation of an antibody gene that codes for a specific protein. Somatic gene rearrangement also occurs with the genes that encode T cell antigen receptors. Gene rearrangement of this type permits the great versatility of the immune system in recognizing a vast array of epitopes. Three forms of gene segments join to form an immunoglobulin light-chain gene. The three types include light-chain variable region (V_L), joining (J_L), and constant region (C_L) gene segments. V_H, J_H, and C_H as well as D (diversity) gene segments assemble to encode the heavy chain. Heavy and light chain genes have a closely similar organizational structure. There are 100 to 300 $V\kappa$ genes, five $J\kappa$ genes, and one $C\kappa$ gene on the κ locus of chromosome 2. There are 100 V_H genes, 30 D genes, 6 J_H genes, and 11 C_H genes on the heavy-chain locus of chromosome 14. Several V_λ, six J_λ, and six C_λ genes are present on the λ locus of chromosome 22 in humans. V_H and V_L genes are classified as V gene families, depending on the sequence homology of their nucleotides or amino acids.

Immunoglobulin gene superfamily: See immunoglobulin superfamily.

***J* exon** is a DNA sequence that encodes part of the third hypervariable region of a light or heavy chain located near the 5′ end of the κ, λ, and γ constant region genes. An intron separates the *J* exon from them. The *J* exon should not be confused with the J chain. The H constant region gene is associated with several *J* exons. The V region gene is translocated to a site just 5′ to one of the *J* exons during stem cell differentiation to a lymphocyte.

A **J gene segment** is a DNA sequence that codes for the carboxy terminal 12 to 21 residues of T lymphocyte receptor or immunoglobulin polypeptide chain variable regions. Through gene rearrangement, a J gene segment unites either a V or a D gene segment to intron 5′ of the C gene segment.

The **J region** is the variable part of a polypeptide chain, comprising a T lymphocyte receptor or immunoglobulin that a J gene segment encodes. The J region of an immunoglobulin light chain is comprised of the third hypervariable region carboxy terminal (1 or 2 residues) and the fourth framework region (12 to 13 residues). The J region of an immunoglobulin heavy chain is comprised of the third hypervariable region carboxy terminal portion and the fourth framework region (15 to 20 residues). The heavy chain's J region is slightly longer than that of the light chain. The variable region carboxy terminal portion represents the J region of the T cell receptor.

Junctional diversity: When gene segments join imprecisely, the amino acid sequence may vary and affect variable region expression. This can alter codons at gene segment junctions. These include the V-J junction of the genes encoding immunoglobulin κ and λ light chains and the V-D, D-J, and D-D junctions of genes encoding immunoglobulin heavy chains, or the genes encoding T cell receptor β and δ chains.

P-nucleotides: Palindromic or p-nucleotides are short inverted repeat nucleotide sequences in VDJ junctions of rearranged immunoglobulin and T cell receptor genes. They are generated from a hairpin intermediate during recombination and contribute to junctional diversity of antigen receptors.

The **12/23 rule:** Immunoglobulin or T cell receptor gene segments can be joined only if one has a recognition signal sequence with a 12-bp spacer and the other has a 23-bp spacer.

V(D)J recombination class switching is a mechanism to generate multiple-binding specificities by developing lymphocytes through exon recombination from a conservative number of gene segments known as variable (V), joining (J), and diversity (D) gene segments at seven different loci that include μ, κ, and λ for B cell immunoglobulin genes, and α, β, γ, and δ genes for T cell receptors.

RAG-1 and RAG-2 refer to recombination activating genes 1 and 2.

V(J) recombination: See class switching.

Recombination activating genes 1 and 2 (RAG-1 and RAG-2) are genes that activate Ig gene recombination. Pre-B cells and immature T cells contain them. It remains to be determined whether RAG-1 and RAG-2 encode the recombinases or the regulatory proteins that control recombinase function. RAG-1 and RAG-2 gene products are requisite for rearrangements of both Ig and TCR genes. In the absence of these genes neither Ig nor T cell receptor proteins are produced. This blocks the production of mature T and B cells.

Recombination recognition sequences are DNA sequences situated adjacent to the V, D, and J segments in antigen receptor loci that are recognized by the RAG-1/RAG-2 component of V(D)J recombinase. The recognition sequences are comprised of a highly conserved seven nucleotide heptamer situated adjacent to the V, D, or J coding sequence, followed by a 12 or 23 nonconserved nucleotide spacer and a highly conserved nonnucleotide segment termed the *nonamer.*

A **signal joint** is a structure produced by the precise joining of recognition signal sequences during somatic recombination that produces T cell receptor and immunoglobulin genes.

Somatic recombination refers to DNA recombination whereby functional genes encoding variable regions of antigen receptors are produced during lymphocyte development. A limited number of inherited or germ-line DNA sequences that are first separated from each other are assembled together by enzymatic deletion of intervening sequences and religation. This takes place only in developing B or T cells. Also called somatic rearrangement.

Unproductive rearrangements are DNA rearrangements of T cell receptor and immunoglobulin genes that produce a gene incapable of encoding a functional polypeptide chain.

A ***V* gene** encodes the variable region of immunoglobulin light or heavy chains. Although it is not in proximity to the *C* gene in germ-line DNA, in lymphocyte and plasma cell DNA the V gene lies near the 5′ end of the *C* gene from which it is separated by a single intron.

A **V gene segment** is a DNA segment encoding the first 95 to 100 amino acid residues of immunoglobulin and T cell polypeptide chain variable regions. There are two coding regions in the V gene segment which are separated by a 100- to 400-bp intron. The first 5′ coding region is an exon that codes for a brief untranslated mRNA region and for the first 15 to 18 signal peptide residues. The second 3′ coding region is part of an exon that codes for the terminal 4 signal peptide residues and 95 to 100 variable region residues. A J gene segment encodes the rest of the variable region. A D gene segment is involved in the encoding of immunoglobulin heavy chain and T cell receptor β and δ chains.

V(D)J recombinase is an enzyme that is able to identify and splice the V (variable), J (joining), and in some cases D (diverse) gene segments that confer antibody diversity. This collection of enzymes makes possible the somatic recombination events that produce functional antigen receptor genes in developing T and B lymphocytes. Some that include RAG-1 and RAG-2 are present only in developing lymphocytes, whereas others are ubiquitous DNA repair enzymes.

A **vector** is a DNA segment employed for cloning a foreign DNA fragment. A vector should be able to reproduce autonomously in the host cell, possess one or more selectable markers, and have sites for restriction endonucleases in nonessential regions that permit DNA insertion into the vector or replacement of a segment of the vector. Plasmids and bacteriophages may serve as cloning vectors.

In **allelic exclusion**, only one of two genes for which the animal is heterozygous is expressed, whereas the remaining gene is not. Immunoglobulin genes manifest this phenomenon. Allelic exclusion accounts for the ability of a B cell to express only one immunoglobulin or the capacity of a T cell to express a T cell receptor of a single specificity. Investigations of allotypes in rabbits established that individual immunoglobulin molecules have identical heavy chains and light chains. Immunoglobulin-synthesizing cells produce only a single class of H chain and one type of light chain at a time. Thus, by allelic exclusion, a cell that is synthesizing antibody expresses just one of two alleles encoding an immunoglobulin chain at a particular locus.

An **enhancer** is a segment of DNA that can elevate the amount of RNA a cell produces. This DNA sequence activates the beginning of RNA polymerase II transcription from a promoter. Although initially described in the DNA tumor virus SV40, enhancers have now been demonstrated in immunoglobulin μ and κ genes' J-C intron. Immunoglobulin enhancers function well in B cells, presumably due to precise regulatory proteins that communicate with the enhancer region.

Gene rearrangement refers to genetic shuffling that results in elimination of introns and the joining of exons to produce mRNA. Gene rearrangement within a lymphocyte signifies its dedication to the formation of a single cell type, which may be immunoglobulin synthesis by B lymphocytes or production of a β-chain receptor by T lymphocytes. Neoplastic transformation of lymphocytes may be followed by the expansion of a single clone of cells, which is detectable by Southern blotting.

Gene segments are multiple short DNA sequences in immunoglobulin and T cell receptor genes, which can undergo rearrangements in many different combinations to yield a vast diversity of immunoglobulin or T cell receptor polypeptide chains.

One-turn recombination signal sequences are immunoglobulin gene-recombination signal sequences separated by intervening sequence of 12 bp.

The **Pax-5 gene** is DNA that encodes B-cell specific activator protein (BSAP), which is required as a transcription factor for B lymphocyte development.

The **one gene, one enzyme theory (historical)** was an earlier hypothesis which proposed that one gene encodes one enzyme or other protein. Although basically true, it is now known that one gene encodes a single polypeptide chain, and it is necessary to splice out mRNA introns comprised of junk DNA before mRNA can be translated into a protein.

Receptor editing is the replacement of a light chain or a heavy chain of a self-reactive antigen receptor on an immature B cell with a light chain or heavy chain that does not confer autoreactivity. It is a mechanism whereby rearranged genes in B lineage cells may undergo secondary rearrangement forming different antigenic specificities. This process involves RAG gene reactivation, additional light chain VJ recombinations, and new Ig light chain synthesis, which permits the cell to express a different immunoglobulin receptor that does not react with self.

Phage antibody library: This is a library of cloned antibody variable region gene sequences that may be expressed as Fab or svFv fusion proteins with bacteriophage coat proteins. These can be exhibited on the phage surface. The phage particle contains the gene encoding a monoclonal recombinant antibody and can be selected from the library by binding of the phage to specific antigen.

P-addition is the appending of nucleotides from cleaved hairpin loops produced by the junction of V-D or D-J gene segments during rearrangements of immunoglobulin or T cell receptor genes.

D gene region is the diversity region of the genome that encodes heavy chain sequences in the immunoglobulin H chain hypervariable region.

D gene is a small segment of immunoglubulin heavy-chain and T-cell receptor DNA that encodes the third hypervariable region of most receptors.

The **D region** is a segment of an immunoglobulin heavy chain variable region or the β or δ chain of the T lymphocyte receptor coded for by a D gene segment. A few residues constitute the D region in the third hypervariable region in most heavy chains of immunoglobulins. The D or diversity region governs antibody specificity and probably T cell receptor specificity as well.

A **D gene segment** is the DNA region that codes for the D or diversity portion of an immunoglobulin heavy chain or a T lymphocyte receptor β or δ chain. It is the segment that encodes the third hypervariable region situated between the chain regions which the V gene segment and J gene segment encode. This part of the heavy-chain variable region is frequently significant in determining antibody specificity.

Diversity (D) segments are abbreviated coding sequences between the (V) and constant gene segments in the immunoglubulin heavy chain and T cell receptor β and δ loci. Together with J segments they are recombined somatically with V segments during lymphocyte development. The recombined VDJ DNA encodes the carboxy terminal ends of the antigen receptor V regions, including the third hypervariable (CDR) regions. D segments used randomly contribute to antigen receptor repertoire diversity.

***D* exon** is a DNA sequence that encodes a portion of the immunoglobulin heavy chain's third hypervariable region. It is situated on the 5′ side of *J* exons. An intron lies between them. During lymphocyte differentiation, V-D-J sequences are produced that encode the complete variable region of the heavy chain.

E2A is a transcription factor, critical for B lymphocyte development, that is necessary for recombinase-activating gene (RAG) expression and for lambda 5 pre-B cell component expression in B cell development. Also called CD62E.

Combinatorial joining is a mechanism for one exon to unite alternatively with several other gene regions, increasing the diversity of products encoded by the gene.

Combinatorial diversity refers to the numerous different combinations of variable, diversity, and joining segments that are possible as a consequence of somatic recombination of DNA in the immunoglubulin and TCR loci during B cell or T cell development. It serves as a mechanism for generating large numbers of different antigen receptor genes from a limited number of DNA gene segments.

A **coding joint** is a structure formed when a V gene segment joins imprecisely to a (D)J gene segment in immunoglobulin or T cell receptor genes.

A **mutation** is a structural change in a gene that leads to a sudden and stable alteration in the genotype of a cell, virus, or organism. It is a heritable change in the genome of a cell, a virus, or an organism apart from that induced through the incorporation of "foreign" DNA. It represents an alteration in DNA's base sequence. Germ cell mutations may be inherited by future generations, whereas somatic cell mutations are inherited only by the progeny of that cell produced through mitotic division. A point mutation is an alteration in a single base pair. Mutations in chromosomes may be expressed as translocation, deletion, inversion, or duplication.

Mutant is an adjective that describes a mutation that may have occurred in a gene, protein, or cell.

The **N region** is a brief segment of an immunoglobulin molecule's or T cell receptor chain's variable region that is not encoded by germ-line genes but instead by brief nucleotide (N) insertions at recombinational junctions. These N-nucleotides may be present both 3′ and 5′ to the rearranged immunoglobulin heavy chain's D gene segment as well as at the V-J, V-D-J, and D-D junctions of the variable region genes of the T lymphocyte receptor.

N-region diversification: In junction diversity, this is the addition at random of nucleotides that are not present in the genomic sequence at VD, DJ, and VJ junctions. TdT catalyzes N-region diversification, which takes place in TCR α and β genes and in Ig heavy chain genes but not in Ig light chain genes.

N-addition: To append nucleotides by TdT during D-J joining or V to DJ joining.

N-nucleotides are nucleotides that TdT adds to the 3′ cut ends of V, D, and J coding segments during rearrangement. They are added to junctions between V, D, and J gene segments in immunoglobulin or T cell receptor genes during lymphocyte development. When as many as 20 of these nucleotides are added, the diversity of the antibody and T cell receptor repertoires is expanded.

Immunoglobulin is the product of a mature B cell product synthesized in response to stimulation by an antigen. Antibody molecules are immunoglobulins of defined specificity produced by plasma cells. The immunoglobulin molecule consists of heavy (H) and light (L) chains fastened together by disulfide bonds. The molecules are subdivided into classes and subclasses based on the antigenic specificity of the heavy chains. Heavy chains are designated by lower case Greek letters (μ, γ, α, δ, and ε), and the immunoglobulins are designated IgM, IgG,

IgA, IgD, and IgE, respectively. The three major classes are IgG, IgM, and IgA, and the two minor classes are IgD and IgE, which together comprise less than 1% of the total immunoglobulins. The two types of light chains (termed κ and λ) are present in all five immunoglobulin classes, although only one type is present in an individual molecule.

IgG, IgD, and IgE have two H and two L polypeptide chains, whereas IgM and IgA consist of multimers of this basic chain structure. Disulfide bridges and noncovalent forces stabilize immunoglobulin structure. The basic monomeric unit is Y shaped, with a hinge region rich in proline and susceptible to cleavage by proteolytic enzymes. Both H and L chains have a constant region at the carboxy terminus and a variable region at the amino terminus. The two heavy chains are alike, as are the two light chains in any individual immunoglobulin molecule. Approximately 60% of human immunoglobulin molecules have κ light chains and 40% have λ light chains. The five immunoglobulin classes are termed isotypes based on the heavy chain specificity of each immunoglobulin class. Two immunoglobulin classes, IgA and IgG, have been further subdivided into subclasses based on H-chain differences. There are four IgG subclasses, designated IgG1 through IgG4, and two IgA subclasses, designated IgA1 and IgA2.

Digestion of IgG molecules with papain yields two Fab and one Fc fragments. Each Fab fragment has one antigen-binding site. By contrast, the Fc fragment has no antigen-binding site but is responsible for fixation of complement and attachment of the molecule to a cell surface. Pepsin cleaves the molecule toward the carboxy terminal end of the central disulfide bond, yielding an $F(ab')_2$ fragment and a pFc′ fragment. $F(ab')_2$ fragments have two antigen-binding sites. L chains have a single variable and constant domain, whereas H chains possess one variable and three to four constant domains.

Secretory IgA is found in body secretions such as saliva, milk, and intestinal and bronchial secretions. IgD and IgM are present as membrane-bound immunoglobulins on B cells, where they interact with antigen to activate B cells. IgE is associated with anaphylaxis, and IgG, which is the only immunoglobulin capable of crossing the placenta, is the major human immunoglobulin.

***C* gene:** DNA encodes the constant region of immunoglobulin heavy and light polypeptide chains. The heavy chain *C* gene is comprised of exons that encode the heavy chain's different homology regions.

***C* gene segment:** DNA encodes for a T cell receptor or an immunoglobulin polypeptide chain constant region. One or more exons may be involved. Constant region gene segments comprise immunoglobulin and T cell receptor gene loci DNA sequences that encode TCR α and β chains and nonvariable regions of immunoglobulin heavy and light polypeptide chains.

C segment is an exon that encodes an immunoglobulin molecule's constant region domain.

***V* gene:** Gene encoding the variable region of immunoglobulin light or heavy chains. Although it is not in proximity to the *C* gene in germ-line DNA, in lymphocyte and plasma-cell DNA the *V* gene lies near the 5′ end of the *C* gene from which it is separated by a single intron.

***V* gene segment:** DNA segment encoding the first 95 to 100 amino acid residues of immunoglobulin and T cell polypeptide chain variable regions. The two coding regions in the *V* gene segment are separated by a 100- to 400-bp intron. The first 5′ coding region is an exon that codes for a brief untranslated mRNA region and for the first 15 to 18 signal peptide residues. The second 3′ coding region is part of an exon that codes for the terminal 4 signal peptide residues and 95 to 100 variable region residues. A *J* gene segment encodes the rest of the variable region. A *D* gene segment is involved in the encoding of immunoglobulin heavy chain and T cell receptor β and δ chains.

Membrane immunoglobulin is cell surface immunoglobulin that serves as an antigen receptor. Virgin B cells contain surface membrane IgM and IgD molecules. Following activation by antigen, the B cell differentiates into a plasma cell that secretes IgM molecules. Whereas membrane-bound IgM is a four-polypeptide chain monomer, the secreted IgM is a pentameric molecule containing five four-chain unit monomers and one J chain. Other immunoglobulin classes have membrane and secreted types. IgG and IgA membrane immunoglobulins probably serve as memory B cell antigen receptors. That segment of the immunoglobulin introduced into the cell membrane is a hydrophobic heavy chain region in the vicinity of the carboxy terminus. Within a particular isotype, the heavy chain is of greater length in the membrane form than in the secreted molecule. This greater length is at the carboxy terminal end of the membrane form. Separate mRNA molecules from one gene encode the membrane and secreted forms of heavy chain.

Surface immunoglobulin: All immunoglobulin isotypes may be expressed on the surface of individual B cells, but only one isotype is expressed at any one time with the exception of unstimulated, mature B lymphocytes which coexpress surface IgM (sIgM) and surface IgD (sIgD). See B lymphocyte receptor.

Surrogate light chains are invariant light chains that are structurally homologous to kappa and lambda light chains and associate with pre-B cell μ heavy chains. They are the same in all B cells. V regions are absent in surrogate light chains. Low levels of cell surface μ-chain and surrogate light-chain complexes are believed to participate in stimulation of kappa or lambda light-chain synthesis and maturation of B cells. This complex of two nonrearranging polypeptide chains (V pre-B and λ5) synthesized by pro-B cells associate with Igμ heavy chain to form the pre-B cell receptor.

B cell mitogens are substances that induce B cell division and proliferation.

Protein A is a *Staphylococcus aureus* bacterial cell wall protein comprised of a solitary polypeptide chain whose binding sites manifest affinity for the Fc region of IgG. It combines with IgG1, IgG2, and IgG4, but not with the IgG3 subclasses in humans, the IgG subclasses of mice, or the IgG of rabbits and certain other species. It has been used extensively for the isolation of IgG during protein purification and for the protection of IgG in immune (antigen–antibody) complexes. Protein A is antiphagocytic, a property that may be linked to its ability to bind an opsonizing antibody's Fc region. Protein A has been postulated to facilitate escape from the immune response by masking its epitopes with immunoglobulins. It is used in immunology for mitogenic stimulation of human B lymphocytes, investigation of lymphocyte Fc receptors, and agglutination tests, as well as in detection and purification of immunoglobulins by the ELISA technique.

B cell-specific activator protein (BSAP) is a transcription factor that has an essential role in early and later stages of B cell development. It is encoded by the gene *Pax-5.*

B lymphocyte stimulatory factors include interleukins 4, 5, and 6.

B cell differentiation and growth factors are T lymphocyte-derived substances that promote differentiation of B lymphocytes into antibody producing cells. They can facilitate the growth and differentiation of B cells *in vitro*. Interleukins 4, 5, and 6 belong in this category of factors.

B cell tyrosine kinase (Btk) is an src-family tyrosine kinase that plays a critical role in the maturation of B lymphocytes. Btk gene mutations lead to crosslinked agammaglobulinemia in which B cell maturation is halted at the pre-B stage.

Early B cell factor (EBF) is a transcription factor required for early B cell development and for RAG expression.

EBF (early B-cell factor) is a transcription factor required for early B cell development and for RAG expression.

Inositol 1, 4, 5-triphosphate (IP_3) is a signaling molecule in the cytoplasm of lymphocytes activated by antigen that is formed by phospholipase C (PLC γ)-mediated hydrolysis of the plasma membrane. Phospholipid PIP_2. IP_3's principle function is to induce the release of intracellular Ca^{2+} from membrane-bound compartments such as the endoplasmic reticulum (ER).

LPS is the abbreviation for lipopolysaccharide, which may serve as an endotoxin, an oligomer of lipid and carbohydrate that comprises the Gram-negative bacterial endotoxin. LPS is used as a polyclonal activator of murine B cells, causing them to divide and differentiate into antibody-forming plasma cells.

LPS-binding protein (LBP) is a substance that can bind a bacterial lypopolysaccharide (LPS) molecule which enables it to interact with CD14 and LPS:LBP-binding protein on macrophages and selected other cells.

BCDF is B cell differentiation factors.

BCGF (B cell growth factors) include interleukins 4, 5, and 6.

AtxBm is the abbreviation for a so-called B cell mouse, which refers to a thymectomized irradiated adult mouse that has received a bone marrow transplant.

Ly antigen is a murine lymphocyte alloantigen that is expressed to different degrees on mouse T and B lymphocytes and thymocytes. Also referred to as Lyt antigen.

Lyb is a murine B lymphocyte surface alloantigen.

Ly1 B cell is a murine B lymphocyte that expresses CD5 (Ly1) epitope on its surface. This cell population is increased in inbred strains of mice susceptible to autoimmune diseases, such as the New Zealand mouse strain.

Lyb-3 antigen: Mature murine B cells express a surface marker designated Lyb-3. It is a single membrane-bound 68-kDa polypeptide. On sodium dodecyl sulfate polyacrylamide gel electrophoresis (SDS-PAGE), it appears distinct from the SIg chains δ and μ. It does not contain disulfide bridges. The gene coding for Lyb-3 appears crosslinked and recessive, and mutant mice lacking Lyb-3 antigens are known. Lyb-3 is involved in the cooperation between T and B cells in response to thymus-dependent antigens and seems to be manifested particularly when the amount of antigen used for immunization is suboptimal. The number of cells carrying Lyb-3 increases with the age of the animal.

Ly6 are GPI linked murine cell surface alloantigens found most often on T and B cells but found also on nonlymphoid tissues such as brain, kidney, and heart. Monoclonal antibodies to these antigens indicate TCR dependence.

Abelson murine leukemia virus (A-MuLV) is a B cell murine leukemia-inducing retrovirus that bears the v-*abl* oncogene. The virus has been used to immortalize immature B lymphocytes to produce pre-B cell or less differentiated B cell lines in culture. These have been useful in unraveling the nature of immunoglobulin differentiation such as H and L chain immunoglobulin gene assembly, as well as class switching of immunoglobulin.

The **Bcl-2** family of proteins consists of proteins which share homology to Bcl-2 in one or more of the Bcl-2 homology regions designated BH1, BH2, BH3, and BH4. Many of the family members have a carboxyl-terminal mitochondrial membrane targeting sequence. All have two central membrane-spanning helices, which are surrounded by additional amphipathic helices. X-ray crystallographic studies have shown that Bcl-X_L has structural similarity to diphtheria toxin and colicins. Diphtheria toxin is endocytosed by cells. Acidification of the endosome induces a conformational change in diphtheria toxin that triggers membrane insertion of the membrane spanning domains. Pore formation by dimerization is thought to occur and the toxic subunit diphtheria toxin is translocated from the endosomal lumen into the cytosol. It is clear from reconstitution assays that many members of the Bcl-2 family can form a pore which allows passage of ions or elicit the release of cytochrome c from isolated mitochondria. Passage of ions is more dramatic at low pH. The carboxylmitochondria targeting sequence is not required for *in vitro* pore formation.

Bcl-2 is a 25-kDa human oncoprotein that is believed to play a regulatory role in tissue development and maintenance in higher organisms by preventing the apoptosis of specific cell types. Bcl-2's inhibitory effect is influenced by the expression of other gene products such as *bax, bcl-xs bak*, and *bad* that promote apoptosis. Bcl-2 is situated at the outer membrane of mitochondria, the endoplasmic reticulum, and the nuclear membrane, and may prevent apoptosis either by acting at these locations or as an antioxidant that neutralizes the effects of reactive oxygen species that promote apoptosis or by obstructing mitochondrial channel openings, thereby preventing the release of factors that promote apoptosis. Bcl-2 is involved in the development of the adult immune system as demonstrated by studies with Bcl-2 knockout mice. Failure to induce normal levels of apoptosis due to overexpression of Bcl-2 may contribute to the development of lymphoproliferative disorders and acceleration of autoimmunity under the appropriate genetic background. The role of Bcl-2 in human SLE and PSS has not yet been fully defined.

Bcl-2 proteins regulate the rate at which apoptotic signaling events initiate or amplify caspase activity. Bcl-2 alters the apoptotic threshold of a cell rather than inhibiting a specific step in programmed cell death. Cells that overexpress Bcl-2 still carry out programmed cell death in response to a wide spectrum of apoptotic initiators. The dose of the initiator is greater than in the absence of Bcl-2. Several Bcl-2-related proteins have been identified. Whereas some Bcl-2 family members promote cell survival, others enhance the sensitivity of a cell to programmed cell death. Five homologs of Bcl-2 with anti-apoptotic properties include Bcl-X_L, Bcl-w, Mcl-l, NR-13, and A-1. By contrast, two proapoptotic members of the Bcl-2 family include Bax and Bak. Bcl-2 related proteins are present in the intracellular membranes including the endoplasmic reticulum, outer mitochondrial membrane, and outer nuclear membrane. Bcl-2 proteins are hypothesized to regulate membrane permeability.

Bcl-X_L is a Bcl-2 related protein. It is upregulated through the action of the costimulatory molecules CD28 and CD40. Bcl-X_L expression is claimed to prevent cell death in response to growth factor limitation or Fas signal transduction. Signal transduction through CD28 or CD40 induces Bcl-X_L expression only in antigen-activated cells. Without antigen receptor engagement, CD40 engagement promotes cell death. Bcl-X_L induction by either CD28 or CD40 is transient, persisting only 3 to 4 d. Protection of cells from death lasts only as long as Bcl-X_L is expressed.

E32 is a protein formed early in development of B lymphocytes that has a role in immunoglobulin heavy-chain transcription.

Oct-2 is a protein formed early in the development of B lymphocytes that has a role in immunoglobulin heavy-chain transcription.

Staphylococcal protein A is a substance derived from the cell wall of *Staphylococcus aureus* that interacts with IgG1, IgG2, and IgG4 subclasses. It stimulates human B cell activation.

7 Immunoglobulin Synthesis, Properties, Structure, and Function

Following enunciation of the clonal selection theory of antibody formation by Burnet in 1957, experimental evidence confirms the validity of this selective theory as opposed to the instructive theory of antibody formation that prevailed during the first half of the 20th century. As immunogeneticists attempted to explain the great diversity of antibodies encoded by finite quantities of DNA, Tonegawa offered a plausible explanation for the generation of antibody diversity in his studies of immunoglobulin gene C, V, J, and D regions and their rearrangement. It is necessary for those segments that encode genes and determine immunoglobulin H and L chains to undergo rearrangement prior to gene transcription and translation. Newly synthesized immunoglobulin molecules have different properties based on their immunoglobulin class or isotype. Nevertheless, antigen-binding specificities reside in the Fab regions of antibody molecules, which governs their interactions with antigens *in vitro* and *in vivo*. By contrast, complement binding and activation capabilities, binding to cell surface, and transport through cells reside in the Fc region of the molecule. The fate of immunoglobulin molecules also differs according to the immunoglobulin class, each with its own characteristic half-life. Only IgG is protected from catabolism by binding to a specific receptor. Some antibodies are protective, others cross the placenta from mother to fetus, whereas others participate in hypersensitivity reactions that lead to adverse effects in target tissues. Antibodies are a diverse and unique category of proteins whose antigen-binding diversity is expressed in the 1020 antibody molecules synthesized from the 1012 B lymphocytes found in the human body.

Antibodies are glycoprotein substances produced by B lymphoid cells in response to stimulation with an immunogen. They possess the ability to react *in vitro* and *in vivo* specifically and selectively with the antigenic determinants or epitopes eliciting their production or with an antigenic determinant closely related to the homologous antigen. Antibody molecules are immunoglobulins found in the blood and body fluids. Thus, all antibodies are immunoglobulins formed in response to immunogens. Antibodies may be produced by hybridoma technology in which antibody-secreting cells are fused by polyethylene glycol (PEG) treatment with a mutant myeloma cell line. Monoclonal antibodies are widely used in research and diagnostic medicine and have potential in therapy. Antibodies in the blood serum of any given animal species may be grouped according to their physicochemical properties and antigenic characteristics. Immunoglobulins are not restricted to the plasma but may be found in other body fluids or tissues, such as urine, spinal fluid, lymph nodes, and spleen. Immunoglobulins do not include the components of the complement system. Immunoglobulins (antibodies) constitute approximately 1 to 2% of the total serum proteins in health. γ Globulins are serum proteins that show the lowest mobility toward the anode during electrophoresis when the pH is neutral. γ Globulins comprise 11.2 to 20.1% of the total serum content in man. Antibodies are in the γ globulin fraction of serum. Electrophoretically they are the slowest migrating fraction.

Heteroclitic antibody is an antibody with greater affinity for a heterologous epitope than for the homologous one that stimulated its synthesis.

Heterocytotropic antibody is an antibody that has a greater affinity when fixed to mast cells of a species other than the one in which the antibody is produced. Frequently assayed by skin-fixing ability, as revealed through the passive cutaneous anaphylaxis test. Interaction with the antigen for which these "fixed" antibodies are specific may lead to local heterocytotrophic anaphylaxis.

Heterogenetic antibody: See heterophile antibody.

Heterophile antibody is an antibody found in an animal of one species that can react with erythrocytes of a different and phylogenetically unrelated species. These are often IgM agglutinins. Heterophile antibodies are detected in infectious mononucleosis patients who demonstrate antibodies reactive with sheep erythrocytes. To differentiate this condition from serum sickness, which also is associated with a high titer of heterophile antibodies, the serum sample is absorbed with beef erythrocytes which contain Forssmann antigen. This treatment removes the heterophile antibody reactivity from the serum of infectious mononucleosis patients.

High-titer, low-avidity antibodies (HTLA) are antibodies that induce erythrocyte agglutination at high dilutions in the Coombs' antiglobulin test. These antibodies cause only weak agglutination and are almost never linked to hemolysis of clinical importance. Examples of HTLA

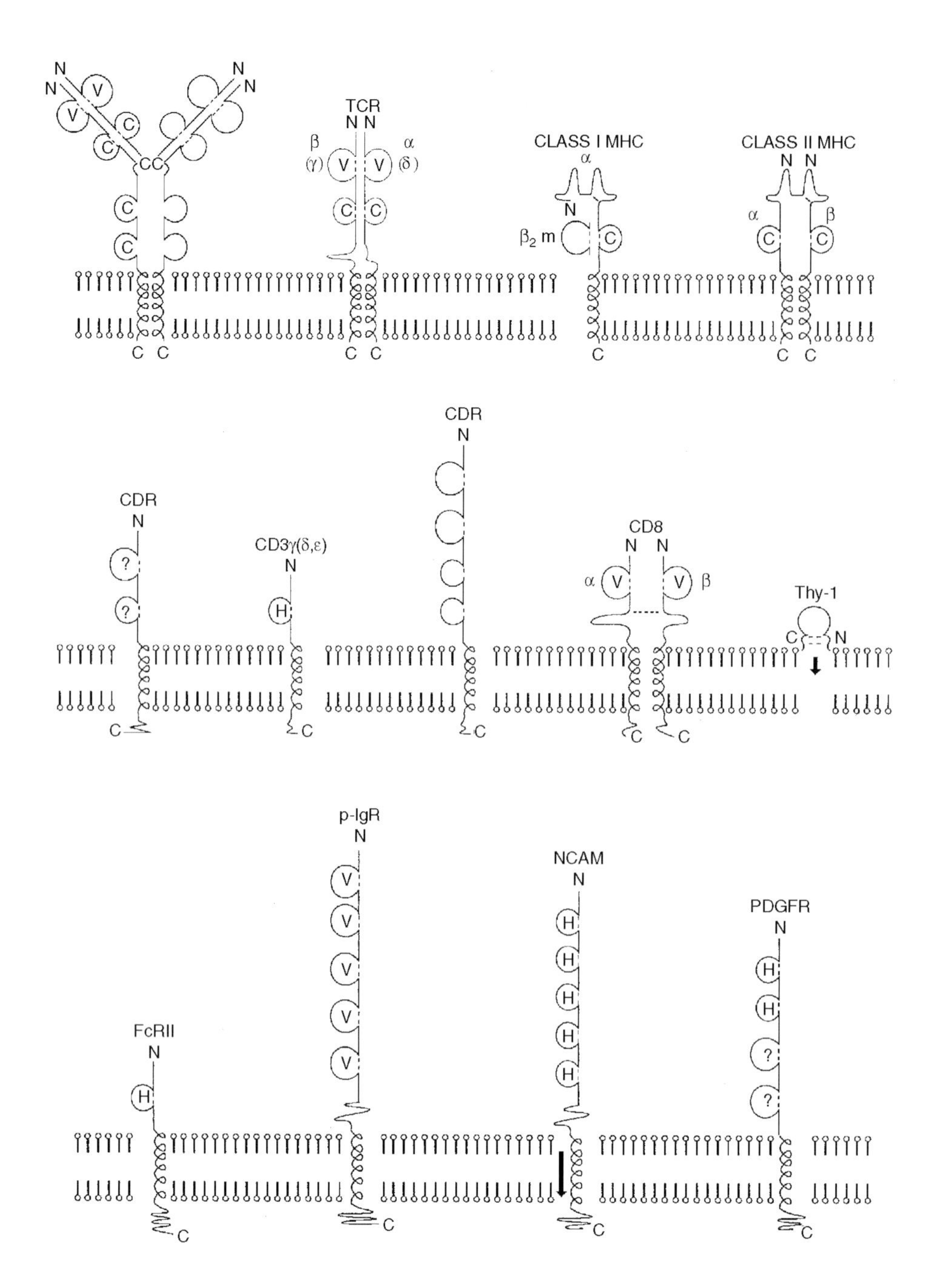

FIGURE 7.1 Immunoglobulin superfamily.

antibodies are anti-Bg[a], -Cs, -Ch, -Kna, -JMH, -Rg, and -Yk, among others.

Homocytotrophic antibody is an antibody that attaches better to animal cells of the same species in which it is produced than it does to animal cells of a different species. The term usually refers to an antibody that becomes fixed to mast cells of an animal of the same species, which results in anaphylaxis with the release of pharmacological mediators of immediate hypersensitivity. These include histamines and other vasoactive amines when the mast cells degranulate.

The **immunoglobulin superfamily** (Figure 7.1) is comprised of several molecules that participate in the immune response and show similarities in structure, causing them to be named the immunoglobulin supergene family. Included are CD2, CD3, CD4, CD7, CD8, CD28, T cell receptor (TCR), MHC class I and MHC class II molecules, leukocyte function associated antigen 3 (LFA-3), the IgG receptor, and a dozen other proteins. These molecules share in common with each other an immunoglobulin-like domain with a length of approximately 100-amino acid residues and a central disulfide bond that anchors and stabilizes antiparallel β strands into a folded structure resembling immunoglobulin. Immunoglobulin superfamily members may share homology with constant or variable immunoglobulin domain regions. Various molecules of the cell surface with polypeptide chains whose folded structures are involved in cell-to-cell interactions belong in this category. Single gene and multigene members are included.

Globulins are serum proteins comprised of α, β, and γ globulins and classified on the basis of their electrophoretic mobility. All three globulin fractions demonstrate anodic mobility that is less than that of albumin. α globulins have the greatest negative charge, whereas γ globulins have the least negative charge. Originally, globulins were characterized based on their insolubility in water, i.e., the euglobulins, or sparing solubility in water, i.e., the pseudoglobulins. Globulins are precipitated in half-saturated ammonium sulfate solution.

γ globulin is an obsolete designation for immunoglobulin. These serum proteins show the lowest mobility toward the anode during electrophoresis when the pH is neutral. The γ globulin fraction contains immunoglobulins. It is the most cationic of the serum globulins.

γ globulin fraction is the electrophonic fraction of serum in which most of the immunoglobulin classes are found.

Cohn fraction II is, principally, gammaglobulin isolated by ethanol fractionation of serum by the method of Cohn.

Antibody-binding site is the antigen-binding site of an antibody molecule, known as a paratope, which is comprised of heavy chain and light chain variable regions. The paratope represents the site of attachment of an epitope to the antibody molecule. The complementarity-determining hypervariable regions play a significant role in dictating the combining site structure together with the participation of framework region residues. The T cell receptor also has an antigen-binding site in the variable regions of its α and β (or γ and δ) chains.

Pyroglobulins are monoclonal immunoglobulins that undergo irreversible precipitation upon heating to 56°C. These monoclonal immunoglobulins are usually detected during routine inactivation of complement in serum by heating to 56°C in a water bath. Whereas most immunoglobulins are unharmed at this temperature, pyroglobulins precipitate. This may be attributable to formation of hydrophobic bonds linking immunoglobulin molecules as a consequence of diminished heavy chain polarity. Half of the pyroglobulin positive subjects have multiple myeloma, and the remaining half have a lymphoproliferative disorder such as macroglobulinemia, carcinoma, or systemic lupus erythematosus. Their relevance to disease is unknown.

Euglobulin is a type of globulin that is insoluble in water but dissolves in salt solutions. In the past, it was used to designate that part of the serum proteins that could be precipitated by 33% saturated ammonium sulfate at 4°C or by 14.2% sodium sulfate at room temperature. Euglobulin is precipitated from the serum proteins at low ionic strength.

Antibody specificity is a property of antibodies determined by their relative binding affinities, the intrinsic capacity of each antibody combining site, expressed as equilibrium dissociation (K_d) or association (K_a) for their interactions with different antigens.

Antiserum is a preparation of serum containing antibodies specific for a particular antigen, i.e., immunogen. A therapeutic antiserum may contain antitoxin, antilymphocyte antibodies, etc. An antiserum contains a heterogenous collection of antibodies that bind the antigen used for immunization. Each antibody has a specific structure, antigenic specificity, and cross-reactivity contributing to the heterogeneity that renders an antiserum unique.

Antiagglutinin is a specific antibody that interferes with the action of an agglutinin.

Antitoxin is an antibody specific for exotoxins produced by certain microorganisms such as the causative agents of diphtheria and tetanus. Prior to the antibiotic era, antitoxins were the treatment of choice for diseases produced by the soluble toxic products of microorganisms, such as those from *Corynebacteriumdiphtheriae* and *Clostridiumtetani* .

Antibody repertoire refers to all of the antibody specificities that an individual can synthesize.

Amboceptor (historical): Paul Ehrlich (circa 1900) considered antisheep red blood cell antibodies, known as amboceptors, to have one receptor for sheep erythrocytes and another receptor for complement. The term gained worldwide acceptance with the popularity of complement fixation tests for syphilis, such as the Wasserman reaction. The term is still used by some when discussing complement fixation.

MAC-1 is a monoclonal antibody specific for macrophages.

Humoral immunity is immunity attributable to specific immunoglobulin antibody and present in the blood plasma, lymph, other body fluids, or tissues. The antibody may also adhere to cells in the form of cytophilic antibody. Antibody or immunoglobulin-mediated immunity acts in conjunction with complement proteins to produce either beneficial (protective) or pathogenic (hypersensitivity tissue-injuring) reactions. Antibodies that are the messengers of humoral immunity are derived from B cells. For purposes of discussion, it is separated from so-called cellular or T cell-mediated immunity, though the two cannot be clearly distinguished since antibodies and T cells often participate in immune reactions together. However, the classification of humoral separate from cellular immunity is useful in understanding and explaining biological mechanisms.

Antibody detection: Techniques employed to detect antibodies include immunoprecipitation, agglutination, complement-dependent assays, labeled antiimmunoglobulin reagents, blotting techniques, and immunohistochemistry. Enzyme-based immunoassays, blotting methods, and immunohistochemistry are routine procedures to detect antibodies and to characterize their specificity.

Anti-DEX antibodies are murine α1-3 dextran-specific antibodies.

Antibody titer is the amount or level of circulating antibody in a patient with an infectious disease. For example, the reciprocal of the highest dilution of serum (containing antibodies) that reacts with antigen, e.g., agglutination, is the titer. Two separate titer determinations are required to reflect an individual's exposure to an infectious agent.

Antitoxin assay (historical): Antitoxins are assayed biologically by their capacity to neutralize homologous toxins as demonstrated by production of no toxic manifestations following inoculation of the mixture into experimental animals, e.g., guinea pigs. They may be tested serologically by their ability to flocculate (precipitate) toxin *in vitro*.

Univalent is a single binding site.

Univalent antibody is an antibody molecule with one antigen-binding site. Although incapable of leading to precipitation or agglutination, univalent antibodies or Fab fragments resulting from papain digestion of an IgG molecule might block precipitation of antigen by a typical bivalent antibody.

Antibody units: See titer.

Antitoxin unit: A unit of antitoxin is that amount of antitoxin present in 1/6000 g of a certain dried unconcentrated horse serum antitoxin which has been maintained since 1905 at the National Institutes of Health in Bethesda, MD. The standard antitoxin unit contained sufficient antitoxin to neutralize 100 MLD of the special toxin prepared by Ehrlich and used by him in titration of standard antitoxin. Both the American and international units of antitoxin are the same.

Combining site: See antigen-binding site.

An **antigen-binding site** is the location on an antibody molecule where an antigenic determinant or epitope combines with it. The antigen-binding site is located in a cleft bordered by the N-terminal variable regions of heavy and light chain parts of the Fab region. Also called paratope. Also refers to that part of a T cell receptor that binds antigen specifically.

An **antivenom** is antitoxin prepared specifically for the treatment of bite or sting victims of poisonous snakes or arthropods. Antibodies in this immune serum preparation neutralize the snake or arthropod venom. Also called antivenin or antivenene.

Titer is an approximation of the antibody activity in each unit volume of a serum sample. The term is used in serological reactions and is determined by preparing serial dilutions of antibody to which a constant amount of antigen is added. The end point is the highest dilution of antiserum in which a visible reaction with antigen, e.g., agglutination, can be detected. The titer is expressed as the reciprocal of the serum dilution which defines the end point. If agglutination occurs in the tube containing a 1:240 dilution, the antibody titer is said to be 240. Thus, the serum would contain approximately 240 units of antibody per milliliter of antiserum. The titer only provides an estimate of antibody activity. For absolute amounts of antibody, quantitative precipitation or other methods must be employed.

A **precipitating antibody** is a precipitin.

A **precipitin** is an antibody that interacts with a soluble antigen to yield an aggregate of antigen and antibody molecules in a lattice framework called a precipitate. Under appropriate conditions, the majority of antibodies can act as precipitins.

Antibody synthesis: The 10^{12} B lymphocytes that comprise the human immune system synthesize 10^{20} antibody (immunoglobulin) molecules present inside and on the surface of these cells and most of all in the serum. Other species have B cell and immunoglobulin molecule numbers relative to their body weight. B cells and immunoglobulin molecules are formed and degraded throughout the human lifespan.

A **paratope** is the antigen-binding site of an antibody molecule, the variable (V) domain, or T cell receptor that binds to an epitope on an antigen. It is the variable or Fv region of an antibody molecule and is the site for interaction with an epitope of an antigen molecule. It is complementary for the epitope for which it is specific.

A paratope is the portion of an antibody molecule where the hypervariable regions are located. There is less than 10% variability in the light and heavy chain amino acid positions in the variable regions. However, there is 20 to 60% variability in amino acid sequence in the so-called "hot spots" located at light chain amino acid positions 29 to 34, 49 to 52, and 91 to 95, and at heavy chain positions 30 to 34, 51 to 63, 84 to 90, and 101 to 110. Great specificity is associated with this variability and is the basis of an idiotype. This variability permits recognition of multiple antigenic determinants.

Intrabody is intracellular antibody that binds key targets to inhibit tumor growth. It is postulated that this might be accomplished by gene therapy. Intrabodies can be expressed within the cell in precise locations within the mammalian cells by modifying intrabody genes (in scFv or Fab format) with sequence-encoding classical intracellular trafficking signals.

Cross-reacting antibody reacts with epitopes on an antigen molecule different from the one that stimulated its synthesis. The effect is attributable to shared epitopes on the two antigen molecules.

Natural autoantibodies are polyreactive antibodies of low affinity that are synthesized by $CD5^+$ B cells that comprise 10 to 25% of circulating B lymphocytes in normal individuals, 27 to 52% in those with rheumatoid arthritis, and less than 25% in systemic lupus erythematosus patients. Natural autoantibodies, they may appear in first degree relatives of autoimmune disease patients as well as in older individuals. They may be predictive of disease in healthy subjects. They are often present in patients with bacterial, viral, or parasitic infections and may have a protective effect. In contrast to natural autoantibodies, they may increase in disease and may lead to tissue injury. The blood group isohemagglutins are also termed natural antibodies even though they are believed to be of heterogenetic immune origin as a consequence of stimulation by microbial antigens.

Cytophilic antibody: (1) An antibody that attaches to a cell surface through its Fc region. It binds to Fc receptors on the cell surface. For example, IgE molecules bind to the surface of mast cells and basophils in this manner. Murine IgG1, IgG2a, and IgG3 bind to mononuclear phagocytic cell surface Fc receptors through their Fc regions. IgG1 and IgG3 may also attach through their Fc regions to mononuclear phagocytic cell Fc receptors in humans. Immunoglobulin molecules that bind to macrophage surfaces through their Fc regions represent a type of cytophilic antibody. (2) Described in the 1960s as a globulin fraction of serum which is adsorbed to certain cells *in vitro* in a manner that allows them to specifically adsorb antigen. Sorkin, in 1963, suggested the possible significance of cytophilic antibody in anaphylaxis and other immunologic and/or hypersensitivity reactions.

Neonatal immunity: The quality and quantity of both humoral and cellular immune responses of neonates differ from those of adults. These differences may be consequences of the lower incidences and decreased functions of immunocompetent cells, such as B cells, T cells, and antigen-presenting cells early in ontogeny. Restriction of T and B cell function early in ontogeny compared with their function in the adult may be due partially to limitations on the diversity of antigen-specific repertoires. Differences in responses to various types of activation by neonatal and adult lymphocytes also affect immunity.

A **neonatal Fc receptor (FcRn)** is an Fc receptor specific for IgG that facilitates the transport of maternal IgG across the placenta and the neonatal intestinal epithelium. FcRn is similar to a class I molecule. An adult variety of this receptor protects plasma IgG antibodies from catabolism.

Cytotoxic antibody is an antibody that combines with cell surface epitopes followed by complement fixation that leads to cell lysis or cell membrane injury without lysis.

Cytotoxicity is the fatal injury of target cells by either specific antibody and complement or specifically sensitized cytotoxic T cells, activated macrophages, or natural killer (NK) cells. Dye exclusion tests are used to assay cytotoxicity produced by specific antibody and complement. Measurement of the release of radiolabel or other cellular constituents in the supernatant of the reacting medium is used to determine effector cell-mediated cytotoxicity.

Antibody humanization is the transference of the antigen-binding part of a murine monoclonal antibody to a human antibody.

Antibody-mediated suppression is the feedback inhibition that antibody molecules exert on their own further synthesis.

Cytotoxicity tests: (1) Assays for the ability of specific antibody and complement to interrupt the integrity of a cell membrane, which permits a dye to enter and stain the cell. The relative proportion of cells stained, representing dead cells, is the basis for dye exclusion tests. See microlymphocytotoxicity. (2) Assays for the ability of specifically sensitized T lymphocytes to kill target cells whose surface epitopes are the targets of their receptors. Loss of the structural integrity of the cell membrane is signified by the release of a radioisotope such as ^{51}Cr, which was taken up by the target cells prior to the test. The amount of isotope released into the supernatant reflects the extent of cellular injury mediated by the effector T lymphocytes.

Cytotrophic antibodies are IgE and IgG antibodies that sensitize cells by binding to Fc receptors on their surface, thereby sensitizing them for anaphylaxis. When the appropriate allergen crosslinks the Fab regions of the molecules, it leads to the degranulation of mast cells and basophils bearing IgE on their surface.

Nonprecipitating antibodies: The addition of antigen in increments to an optimal amount of antibody precipitates only approximately 78% of the amount of antibody that would be precipitated by one step addition to the antigen. This demonstrates the presence of both precipitating and nonprecipitating antibodies. Although the nonprecipitating variety cannot lead to the formation of insoluble antigen–antibody complexes, they can be assimilated into precipitates that correspond to their specificity. Rather than being univalent, as was once believed, they may merely have a relatively low affinity for the homologous antigen. Monogamous bivalency, which describes the combination of high-affinity antibody with two antigenic determinants on the same antigen particle, represents an alternative explanation for the failure of these molecules to precipitate with their homologous antigen. The formation of nonprecipitating antibodies, which usually represents 10 to 15% of the antibody population produced, is dependent upon such variables as heterogeneity of the antigen, characteristics of the antibody, and animal species.

The equivalence zone is narrower with native proteins of 40 to 60 kDa and their homologous antibodies than with polysaccharide antigens or aggregated denatured proteins and their specific antibody. The equivalence zone with synthetic polypeptide antigens varies with the individual compound used. The solubility of antibody–antigen complexes and the nature of the antigen are related to these variations at the equivalence zone. The extent of precipitation is dependent upon characteristics of both the antigen and antibody. At the equivalence zone, not all antigen and antibody molecules are present in the complexes. For example, rabbit anti-BSA (bovine serum albumin) precipitates only 46% of BSA at equivalence.

"O" phage antibody library refers to cloned antibody variable region gene sequences that may be expressed as Fab or svFv fusion proteins with bacteriophage coat proteins. These can be exhibited on the phage surface. The phage particle contains the gene encoding a monoclonal recombinant antibody and can be selected from the library by binding of the phage to specific antigen.

OKT monoclonal antibodies are commercially available preparations used to enumerate human T cells according to their surface antigens to determine the immunophenotype. OKT designations have been replaced by CD designations.

Phage display is a technique that permits expression of the humoral immune system *in vitro* by phage display technology and antibody engineering. Large libraries of antibody fragments are displayed on the surface of bacteriophage particles. Phages expressing desirable antibody specificity must be selected and expanded. Favorable mutations in the genes encoding a selected antibodies specificity must be selected. Antibody fragments are selected from large libraries constructed from B cells or assembled *in vitro* from the genetic elements encoding antibodies. This technique is rapid and unaffected by the immunogenecity of the target antigen. Selection procedures permit the isolation of antibodies specific for membrane molecules and epitopes. Antibody fragments can be tailored to have the desired avidity, pharmacokinetic properties, and biological effector functions. Monoclonal antibodies prepared from phage display libraries formed from human V regions constitute a molecule especially amenable for immunotherapy in humans.

Phage display library: Antibody-like phage produced by cloning immunoglobulin V region genes and filamentous phage which results in their expressing antigen-binding domains on their surfaces. Antigen-binding phage can be replicated in bacteria and used like antibodies. This method can be employed to develop antibodies of any specificity.

OKT4: See CD4.

A **polyvalent antiserum** is an antiserum comprised of antibodies specific for multiple antigens.

Antitoxin is an antibody specific for exotoxins produced by certain microorganisms such as the causative agents of diphtheria and tetanus. Prior to the antibiotic era, antitoxins were the treatment of choice for diseases produced by the soluble toxic products of microorganisms, such as those from *Corynebacterium diphtheriae* and *Clostridium tetani.*

OKT8: See CD8.

Antiantibody: In addition to their antibody function, immunoglobulin molecules serve as excellent protein immunogens when inoculated into another species or they may become autoantigenic even in their own host. The Gm antigenic determinants in the Fc region of an IgG molecule may elicit autoantibodies, principally of the IgM class, known as rheumatoid factor in individuals with rheumatoid arthritis. Antiidiotypic antibodies, directed against the antigen-binding N-terminal variable regions of antibody molecules, represent another type of antiantibody. Rabbit antihuman IgG (the Coombs' test reagent) is an antiantibody used extensively in clinical immunology to reveal autoantibodies on erythrocytes.

Antiimmunoglobulin antibodies are antibodies specific for immunoglobulin constant domains which render them useful for detection of bound antibody molecules in immunoassays. Antiisotype antibodies are synthesized in a different species; antiallotype antibodies are made in the same species against allotypic variants; and antiidiotype antibodies are induced against a single antibody molecule's unique determinants.

Antiimmunoglobulin antibodies are produced by immunizing one species with immunoglobulin antibodies derived from another.

Skin-fixing antibody is an antibody such as IgE that is retained in the skin following local injection, as in passive cutaneous anaphylaxis. Antibody with this property was referred to previously as reagin before IgE was described.

Catalytic antibodies are not only exclusively specific for a particular ligand but are also catalytic. Approximately 100 reactions have been catalyzed by antibodies. Among these are pericyclic processes, elimination reactions, bond-forming reactions, and redox processes. Most antibody-catalyzed reactions are highly stereospecific. For efficient catalysis, it is necessary to introduce catalytic functions within the antibody-combining site properly juxtaposed to the substrate. Catalytic antibodies resemble enzymes in processing their substrates through a Michaelis M complex in which the chemical transformation takes place followed by a product dissociation. See also abzyme.

A **catalytic antibody** is a monoclonal antibody into whose antigen-binding site the catalytic activity of a specific biological enzyme has been introduced. This permits enzymatic catalysis of previously arranged specificity to take place. Site-directed mutagenesis, in which a catalytic residue is added to a combining site by amino acid substitution, is used to attain the specificity. Specific catalysts can be generated by other mechanisms such as alternation of enzyme sites genetically or chemical alternation of receptors with catalytic properties.

Chimeric antibodies are antibodies that have, for example, mouse Fv fragments for the Ag-binding portion of the molecule but Fc regions of human Ig which convey effector functions.

Lysins are factors such as antibodies and complement or microbial toxins that induce cell lysis. For an antibody to demonstrate this capacity, it must be able to fix complement.

C_κ is an immunoglobulin κ light chain constant region. The corresponding exon is designated as C_κ.

C_L is an immunoglobulin light chain constant domain. The corresponding exon is designated C_L.

C_λ is an immunoglobulin λ light chain constant region. The corresponding exon is designated C_λ. There is more than one isotype in mouse and man.

C_μ is an immunoglobulin μ-chain constant region. The corresponding exon is designated as C_μ.

C_γ is an immunoglobulin γ chain constant region that is further subdivided into four isotypes in man that are indicated as $C_\gamma1$, $C_\gamma2$, $C_\gamma3$, and $C_\gamma4$. The corresponding exons are expressed by the same designations in italics.

C_H is an immunoglobulin heavy chain's constant region encoded by the C_H gene.

C_H1 is an immunoglobulin heavy chain's first constant domain encoded by the C_H1 exon.

C_H2 is an immunoglobulin heavy chain's second constant domain encoded by the C_H2 exon.

C_H3 is an immunoglobulin heavy chain's third constant domain encoded by the C_H3 exon.

C_H4 is an immunoglobulin heavy chain's fourth constant domain encoded by the C_H4 exon. Of the five immunoglobulin classes in man, only the μ heavy chain of IgM and the ε heavy chain of IgE possess a fourth domain.

Complementarity-determining region (CDR) refers to the hypervariable regions in an immunoglobulin molecule that form the three-dimensional cavity where an epitope binds to the antibody molecule. The heavy and light polypeptide chains each contribute three hypervariable regions to the antigen binding region of the antibody molecule. Together, they form the site for antigen binding. Likewise, the T cell receptor α and β chains each have three regions with great diversity that are analogous to the immunoglobulin's CDRs. These hypervariable areas are sites of binding for foreign antigen and self MHC molecular complexes.

Constant domain refers to the immunoglobulin C_H and C_L regions. A globular compact structure that consists of two antiparallel twisted β sheets. There are differences in the number and the irregularity of the β strands and bilayers in variable (V) and constant (C) subunits of immunoglobulins. C domains have a tertiary structure that closely resembles that of the domains, which are comprised of a five-strand β sheet and a four-strand β sheet packed facing one another. However, the C domain does not have a hairpin loop at the edge of one of the sheets. Thus, the C domain has seven or eight β strands rather than the nine that are found in the V domains. Refers also to the constituent domains of constant regions of T cell receptor polypeptides.

Constant region is that part of an immunoglobulin polypeptide chain that has an invariant amino acid sequence among immunoglobulin chains belonging to the same isotype and allotype. There is a minimum of two and often three to four domains in the constant region of immunoglobulin heavy polypeptide chains. The hinge region "tail end piece" (a carboxy terminal region) constitutes part of the constant region in selected classes of immunoglobulin. A few exons encode the constant region of an immunoglobulin heavy chain, and one exon encodes the constant region of an immunoglobulin light chain. The constant region is the location for the majority of isotypic and allotypic determinants. It is associated with a number of antibody functions. T cell receptor α, β, γ, and δ chains have constant regions coded for by three to four exons. MHC class I and II molecules also have segments that are constant regions in that they show little sequence variation from one allele to another. Refers also to the part of a T cell receptor (TCR) polypeptide chain that does not vary in sequence among different clones and is not involved in antigen binding.

Cyanogen bromide is a chemical that specifically breaks methionyl bonds. Approximately one half of the methionine residues in an IgG molecule, e.g., those in the Fc region, are cleaved by treatment with cyanogen bromide.

Distribution ratio is the plasma immunoglobulin to whole body immunoglobulin ratio.

Effector function refers to the nonantigen-binding functions of an antibody molecule that are mediated by the constant region of heavy chains. These include Fc receptor binding, complement fixation, binding to mast cells, etc. Effector function generally results in removal of antigen from the body, such as in phagocytosis or complement-mediated lysis.

The **hinge region** is an area of an immunoglobulin heavy chain situated between the first constant domain and the second constant domain (C_H1 and C_H2) in an immunoglobulin polypeptide chain. The high content of proline residues in this region provides considerable flexibility to this area, which enables the Fab region of an immunoglobulin molecule to combine with cell surface epitopes that it might not otherwise reach. Fab regions of an Ig molecule can rotate on the hinge region. There can be an angle up to 180 between the two Fab regions of an IgG molecule. In addition to the proline residues, there may be one or several half cysteines associated with the interchain disulfide bonds. Enzyme action by papain or pepsin occurs near the hinge region. Whereas, γ, α, and δ chains each contain a hinge region, μ and ε chains do not. The 5′ part of the C_H2 exon encodes the human and mouse a-chain hinge region. Four exons encode the γ-3 chains of humans and two exons encode human δ chains.

Homology unit: A structural feature of an immunoglobulin domain.

A **hot spot** is a hypervariable region in DNA that encodes the variable region of an immunoglobulin molecule's heavy (V_H) and light (V_L) polypeptide chains. These are also designated complementarity-determining regions (CDR). These are the areas for specific antigen binding, and they also determine the idiotype of an immunoglobulin molecule. The remaining background support structures of the heavy and light polypeptide chains are termed framework regions (FR). The κ and λ light chain hot spots are situated near amino acid residues 30, 50, and 95. Also called hypervariable regions.

Humanization refers to the genetic engineering of murine hypervariable loop specificity into human antibodies. The DNA encoding hypervariable loops of murine monoclonal antibodies or V regions selected in phage display libraries is inserted into the framework regions of human immunoglobulin genes. This technique permits the synthesis of antibodies of a particular specificity without inducing an immune response in the human subject treated with them.

A **humanized antibody** is an engineered antibody produced through recombinant DNA technology. A humanized antibody contains the antigen-binding specificity of an antibody developed in a mouse, whereas the remainder of the molecule is of human origin. To accomplish this, hypervariable genes that encode the antigen-binding regions of a mouse antibody are transferred to the normal human gene which encodes an immunoglobulin molecule that is mostly human but expresses the antigen-binding specificity of the mouse antibody in the variable region of the molecule. This greatly diminishes any immune response to the antibody molecule itself as a foreign protein by the human host, while retaining the desired functional capacity of reacting with the specific antigen.

An **immunoglobulin** is a mature B cell product synthesized in response to stimulation by an antigen. Antibody molecules are immunoglobulins of defined specificity produced by plasma cells. The immunoglobulin molecule consists of heavy (H) and light (L) chains fastened together by disulfide bonds. The molecules are subdivided into classes and subclasses based on the antigenic specificity of the heavy chains. Heavy chains are designated by lower case Greek letters (μ, γ, α, δ, and ε), and immunoglobulins are designated IgM, IgG, IgA, IgD, and IgE, respectively. The three major classes are IgG, IgM, and IgA, and the two minor classes are IgD and IgE, which together comprise less than 1% of the total immunoglobulins. The two types of light chains termed (κ and λ) are present in all five immunoglobulin classes, although only one type is present in an individual molecule. IgG, IgD, and IgE have two H and two L polypeptide chains, whereas IgM and IgA consists of multimers of this basic chain structure. Disulfide bridges and noncovalent forces stabilize immunoglobulin structure. The basic monomeric unit is Y shaped, with a hinge region rich in proline and susceptible to cleavage by proteolytic enzymes. Both H and L chains have a constant region at the carboxyl terminus and a variable region at the amino terminus. The two

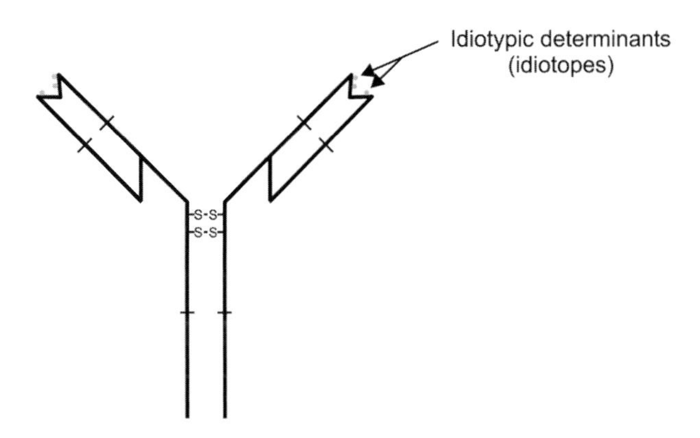

FIGURE 7.2 Schematic representation of idiotypes present on an immunoglobulin molecule.

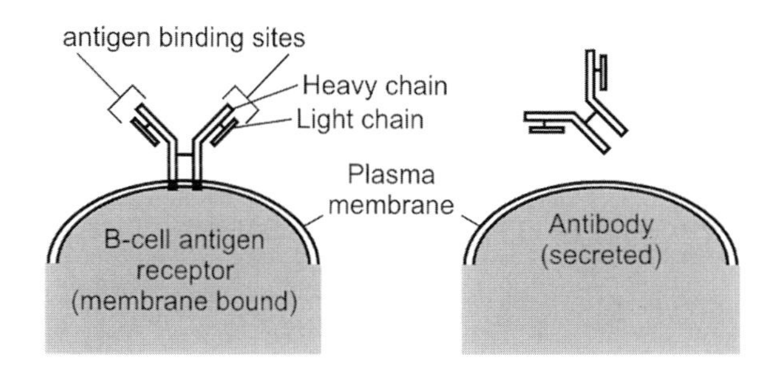

FIGURE 7.3 Schematic representation of an antigen receptor on the plasma membrane of a B cell.

heavy chains are alike, as are the two light chains, in any individual immunoglobulin molecule. Approximately 60% of human immunoglobulin molecules have κ light chains, and 40% have λ light chains. The five immunoglobulin classes are termed isotypes based on the heavy-chain specificity of each immunoglobulin class. Two immunoglobulin classes, IgA and IgG, have been further subdivided into subclasses based on H chain differences. The four IgG subclasses are designated as IgG_1 through IgG_4, and the two IgA subclasses are designated IgA_1 and IgA_2. Digestion of IgG molecules with papain yields two Fab and one Fc fragments. Each Fab fragment has one antigen-binding site but is responsible for fixation of complement and attachment of the molecule to a cell surface. Pepsin cleaves the molecule toward the carboxyl-terminal end of the central disulfide bone, yielding an $F(ab')_2$ fragment and a pFc′ fragment. $F(ab')_2$ fragments have two antigen-binding sites. L chains have a single variable and constant domain, whereas H chains possess one variable and three to four constant domains. Secretory IgA is found in body secretions such as saliva, milk, and intestinal and bronchial secretions. IgD and IgM are present as membrane-bound immunoglobulins on B cells, where they interact with antigen to activate B cells. IgE, associated with anaphylaxis and IgG, which is the only immunoglobulin capable of crossing the placenta, is the major human immunoglobulin.

Immunoglobulin structure: See immunoglobulin.

The **homology region** is a 105- to 115-amino acid residue sequence of heavy or light chains of immunoglobulins which have a primary structure that resembles other corresponding sequences of the same size. A **homology region** has a globular shape and an intrachain disulfide bond. The exons that encode homology regions are separated by introns. Light polypeptide chain homology regions are termed V_L and C_L. Heavy chain homology regions are designated V_H, C_H1, C_H2, and C_H3.

N-terminus is the amino end of a polypeptide chain bearing a free amino-NH_2 group.

Reagin (historical): (1) Obsolete term for a complement-fixing IgM antibody reacting in the Wassermann test for syphilis. (2) A name used previously for immunoglobulin E (IgE), the anaphylactic antibody in humans that fixes to tissue mast cells leading to release of histamine and vasoactive amines following interaction with specific antigen (allergen).

SCAB (single chain antigen-binding proteins) are polypeptides that join the light chain variable sequence of an antibody to the antibody heavy chain variable sequence. All monoclonal antibodies are potential sources of SCABs. They are smaller and less immunogenic than the intact heavy chains with immunogenic constant regions. Their many possible uses are in imaging and treatment of cancer, in cardiovascular disease, as biosensors, and for chemical separations.

ScFv is a single-chain molecule comprised of both heavy and light chain variable regions fastened together by a flexible linker.

Tail peptide is an immunoglobulin heavy polypeptide chain carboxy terminus that is separate from the carboxy terminal domain. This structure is present in membrane-anchored immunoglobulins. Whereas tail peptides of 20 amino acids each are present in IgM and IgA molecules that have been secreted, IgG and IgE molecules do not contain tail peptides.

Telencephalin is an immunoglobulin superfamily member with nine immunoglobulin-like domains that is expressed in the central nervous system. It is a human $\alpha_1\beta_2$ ligand that has a high homology with ICAM-1 (50%) and ICAM-3 (55%) in the amino-terminal first five domains. The clustered chromosomal location for ICAM-1, ICAM-3, and telencephalin is on human chromosome 19. Their corresponding α chain receptors are located on chromosome 16.

Ig is the abbreviation for immunoglobulin.

Immune serum is an antiserum containing antibodies specific for a particular antigen or immunogen. Such antibodies may confer protective immunity.

Immune serum globulin is an injectable immunoglobulin that consists mainly of IgG extracted by cold ethanol fractionation from pooled plasma of up to 1000 human donors. It is administered as a sterile 16.5 ± 1.5% solution to patients with immunodeficiencies and as a preventive against certain viral infections including measles and hepatitis A.

Immunoglobulin function is to link an antigen to its elimination mechanism (effector system). Antibodies induce complement activation and cellular elimination mechanisms that include phagocytosis and antibody-dependent cell-mediated cytotoxicity (ADCC). This type of activation usually requires antibody molecules clustered together on a cell surface rather than as free unliganded antibody. Antibodies can combine with virus particles to render them noninfectious *in vitro* through neutralization. IgG catabolism is regulated by the IgG concentration. All immunoglobulin classes can be expressed on B cell surfaces where they act as antigen receptors although this is mainly a function of IgM and IgD. Surface immunoglobulin has an extra C-terminal sequence compared to secreted immunoglobulin containing linker, transmembrane, and cytoplasmic segments.

Cerebrospinal fluid (CSF) immunoglobulins: In normal individuals, CSF immunoglobulins are derived from plasma by diffusion across the blood–brain barrier. The amount present is dependent on the immunoglobulin concentration in the serum, the molecular size of the immunoglobulin, and the permeability of the blood–brain barrier. IgM is normally excluded by virtue of its relatively large molecular size and low plasma concentration. However, in certain disease states, such as demyelinating diseases and infections of the central nervous system, immunoglobulins may be produced locally. The permeability of the blood–brain barrier is accurately reflected by the CSF total protein or albumin levels relative to those in the serum. By comparing these data, it is possible to derive information about deviation from normal. The comparative method is called Ig quotient and is calculated in various ways:

1. CSF-IgG/albumin (normal 13.9 + 14%)
2. CSF-IgG/total protein
3. CSF-IgA/albumin
4. CSF-κ/λ (ratio)

To correct for variations in the blood–brain barrier, the calculation can be modified to give a more sensitive quotient, which is represented as

$$\frac{\text{CSF IgG/serum IgG}}{\text{CSF albumin/serum albumin}}$$

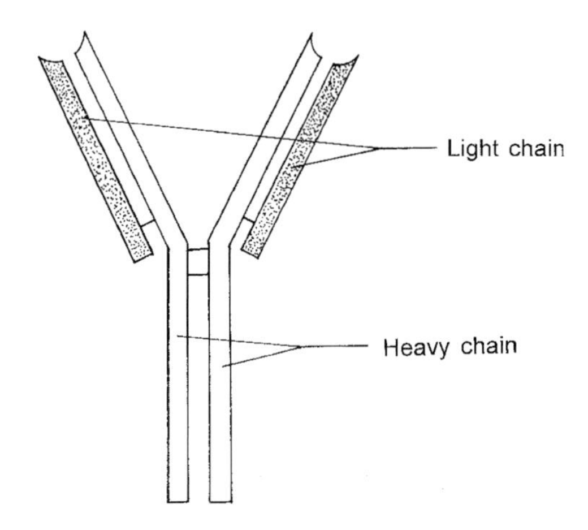

FIGURE 7.4 Immunoglobulin heavy chains that are fastened to each other or to light polypeptide chains by disulfide bonds.

The ratio of κ to λ light chains in CSF in comparison with that of these light chains in serum is significant in that some patients with local immunoglobulin production show a change in the ratio. An increase in the IgA present in CSF appears in some viral infections of the CNS in which antiviral antibodies are also detectable.

An **immunoglobulin heavy chain** (Figure 7.4) is a 5-kDa to 71-kDa polypeptide chain present in immunoglobulin molecules that serves as the basis for dividing immunoglobulins into classes. The heavy chain is comprised of three to four constant domains, depending on class, and one variable domain. In addition, a hinge region is present in some chains. There is approximately 30% homology with respect to amino acid sequence among the five classes of immunoglobulin heavy chain in humans. The heavy chain of IgM is μ, of IgG is γ, of IgA is α, of IgD is δ, and of IgE is ε. A heavy chain is a principal constituent of immunoglobulin molecules. Each immunoglobulin is comprised of at least one four-polypeptide chain monomer which consists of two heavy and two light polypeptide chains. The two heavy chains are identical in any one molecule as are the two light chains.

A **heavy chain** is an immunoglobulin polypeptide chain that designates the class of immunoglobulin. The five immunoglobulin classes are based on the heavy chains they possess and are IgM, IgG, IgA, IgD, and IgE. Each four-chain immunoglobulin molecule or each four-chain monomeric unit of IgM contains two heavy chains and two light chains. These are fastened together by disulfide bonds. At the amino terminus is the variable region of the heavy chain, designated V_H. Adjacent to this is the first constant region, designated C_H1 through C_H3 or C_H4 domains, based on immunoglobulin class. Heavy-chain antigenic determinants determine not only the immunoglobulin class but the subclass as well.

Heavy chain class refers to the immunoglobulin heavy polypeptide chain primary (antigenic) structure present in all members of a species that is different from the other heavy chain classes. Primary structural features governing immunoglobulin heavy chain class are located in the constant region. Lowercase Greek letters such as μ, γ, α, δ, and ε designate heavy chain class.

H chain (heavy chain) is a principal constituent of immunoglobulin molecules. Each immunoglobulin is comprised of at least one four-polypeptide chain monomer, which consists of two heavy and two light polypeptide chains. The two heavy chains are identical in any one molecule as are the two light chains.

Heavy chain subclass: Within an immunoglobulin heavy chain class, differences in primary structure associated with the constant region that can further distinguish these heavy chains of the same class are designated as subclasses. These differences are based on primary or antigenic structure. Heavy chain subclasses are designated as γ1, γ2, γ3, etc.

Immunoglobulin heavy chain binding protein (BiP) is a 77-kDa protein that combines with selected membrane and secretory proteins. It is believed to facilitate their passage through the endoplasmic reticulum.

A **light chain** (Figure 7.5) is a 22-kDa polypeptide chain found in all immunoglobulin molecules. Each four-chain immunoglobulin monomer contains two identical light polypeptide chains. They are joined to two like heavy chains by disulfide bonds. There are two types of light chains designated κ and λ. An individual immunoglobulin molecule possesses two light chains that are either κ or λ but never a mixture of the two. The types of light polypeptide chains occur in all five of the immunoglobulin classes. Each light chain has an N-terminal V region which constitutes part of the antigen-binding site of the antibody molecule. The C region or constant terminal reveals no variation except for the Km and Oz allotype markers in humans. **KM (formerly Inv)** is the designation for the κ light chain allotype genetic markers.

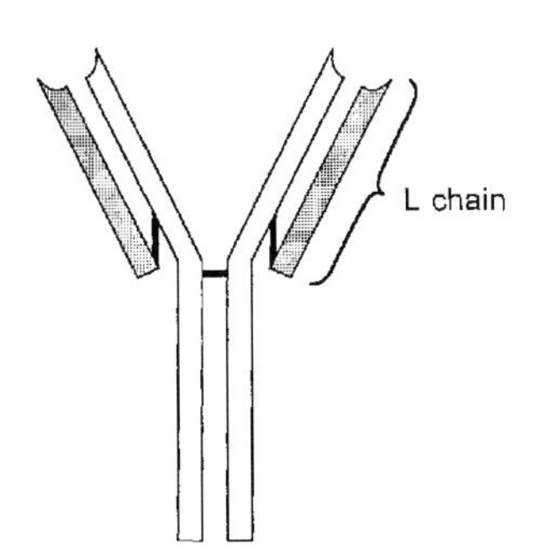

FIGURE 7.5 Light polypeptide chains of immunoglobulins that are fastened to heavy chains through disulfide bonds and are found in all classes of immunoglobulin.

κ chain is one of two types of light polypeptide chains present in immunoglobulin molecules of human and other species. κ light chains are found in approximately 60% of human immunoglobulin, whereas λ light chains are present in approximately 40% of human immunoglobulin molecules. A single immunoglobulin molecule contains either κ or λ light chains, not one of each.

Light chain type is a term for the classification of immunoglobulin light chains based on their primary or antigenic structure. Two types of light chains have been described and are designated as κ and λ. Two κ chains or two λ chains, never one of each, are present in each monomeric immunoglobulin subunit of vertebrate species.

Kappa (κ) is the designation for one of the two types of immunoglobulin light chain, with the other designated as lambda (λ).

Light chain subtype refers to the subdivision of a type of light polypeptide chain based on its primary or antigenic structure that appears in all members of an individual species. Subtype differences distinguish light chains that share a common type. These relatively minor structural differences are located in the light chain constant region. Oz^+, Oz^-, $Kern^+$, and $Kern^-$ markers represent subtypes of λ light chains in humans.

Km allotypes: Three Km allotypes have been described in human immunoglobulin κ light chains. They are designated Km1, Km1,2, and Km3. They are encoded by alleles of the gene that codes for the human κ light chain constant regions. Allotype differences are based on the amino acid residue at positions 153 and 191, which are in proximity to one another in a folded immunoglobulin Cκ domain. One person may have a maximum of two out of the three Km allotypes on their light chains. To fully express Km determinants, the heavy immunoglobulin chains should be present, probably to maintain appropriate three-dimensional configuration.

L chain is a 22-kDa polypeptide chain found in all immunoglobulin molecules. There are two types designated κ or λ. Each four-chain immunoglobulin monomer contains either two κ or two λ light chains. The two types of light chain never occur in one molecule under natural conditions.

Lambda (λ) chain is one of the two light polypeptide chain types found in immunoglobulin molecules. The κ light chain is the other type. Each immunoglobulin molecule contains either two λ or two κ light chains. The κ to λ light chains ratio differs among species. Approximately 60% of IgG molecules in humans are λ and 40% are λ.

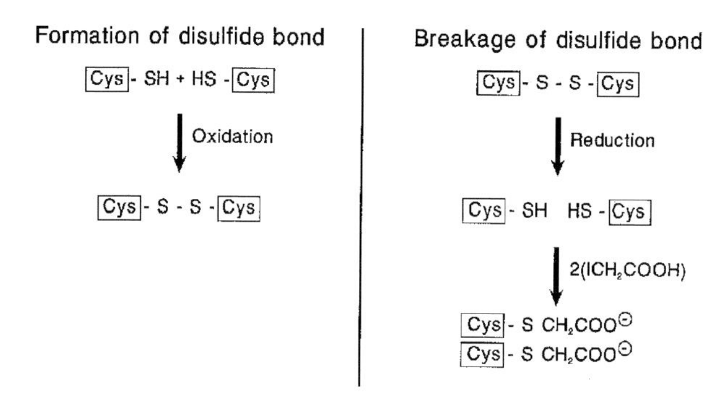

FIGURE 7.6 Depiction of the formation of disulfide bonds from the oxidation of two sulfhydryl groups as well as the breaking of disulfide bonds through reduction leading to sulfhydryl formation.

Mcg isotypic determinant is a human immunoglobulin λ chain epitope that occurs on some of every person's λ light polypeptide chains in immunoglobulin molecules. The Mcg isotypic determinant is characterized by asparagine at position 112, threonine at position 114, and lysine at position 163.

Inv is the former designation for human κ light polypeptide chain allotype. Allotypic epitopes in the immunoglobulin κ light chain constant region. Km replaced Inv.

Inv allotypes: Original terminology for Km allotypes, which is now the preferred nomenclature.

Inv allotypic determinant: See Km allotypic determinant.

Inv marker: See Km allotypic determinant.

Disulfide bonds (Figure 7.6) are the –S–S– chemical bonds between amino acids that link polypeptide chains together. Chemical reduction may break these bonds. Disulfide bonds in immunoglobulin molecules are either intrachain or interchain. The interchain disulfide bonds include linking heavy to heavy and heavy to light. The different types of bonds in immunoglobulin molecules differ in their ease of chemical reduction.

Domain is a region of a protein or polypeptide chain that is globular and folded with 40 to 400 amino acid residues. The domain may have a spatially distinct "signature" which permits it to interact specifically with receptors or other proteins. In immunology, it refers to the loops in polypeptide chains that are linked by disulfide bonds on constant and variable regions of immunoglobulin molecule light and heavy polypeptide chains or a compact TCR chain segment comprised of amino acids around an S–S bond.

An **immunoglobulin domain** (Figure 7.7) is an immunoglobulin heavy or light polypeptide chain structural

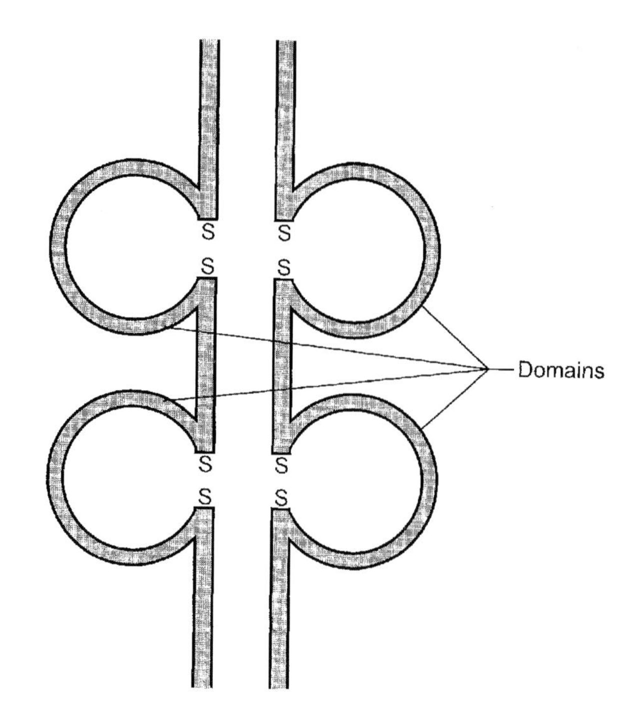

FIGURE 7.7 Domain structure of light or heavy polypeptide chains, the subunits of immunoglobulin molecules.

unit that is comprised of approximately 110 amino acid residues. Domains are loops that are linked by disulfide bonds on constant and variable regions of heavy and light chains. Immunoglobulin functions may be linked to certain domains. There is much primary and three-dimensional structural homology among immunoglobulin domains. A particular exon may encode an immunoglobulin domain.

S antibody is the sedimentation coefficient of immunoglobulin molecules such as IgG. 6.6 S immunoglobulins are usually referred to as 7 S immunoglobulins.

C region (constant region) is the abbreviation for the constant region carboxy terminal portion of immunoglobulin heavy or light polypeptide chain that is identical in a particular class or subclass of immunoglobulin molecules. C_H designates the constant region of the heavy chain of immunoglobulin, and C_L designates the constant region of the light chain of immunoglobulin.

Immunoglobulin-like domain is the 100-amino acid residue structure found in selected β sheet-rich proteins with intrachain disulfide bonds. It is found in immunoglobulins, interleukins 1 and 6, the T cell receptor, and platelet-derived growth factor.

C-terminus is the carboxy terminal end of a polypeptide chain containing a free –COOH group.

An **immunoglobulin fold** is an immunoglobulin domain's three-dimensional configuration. An immunoglobulin fold

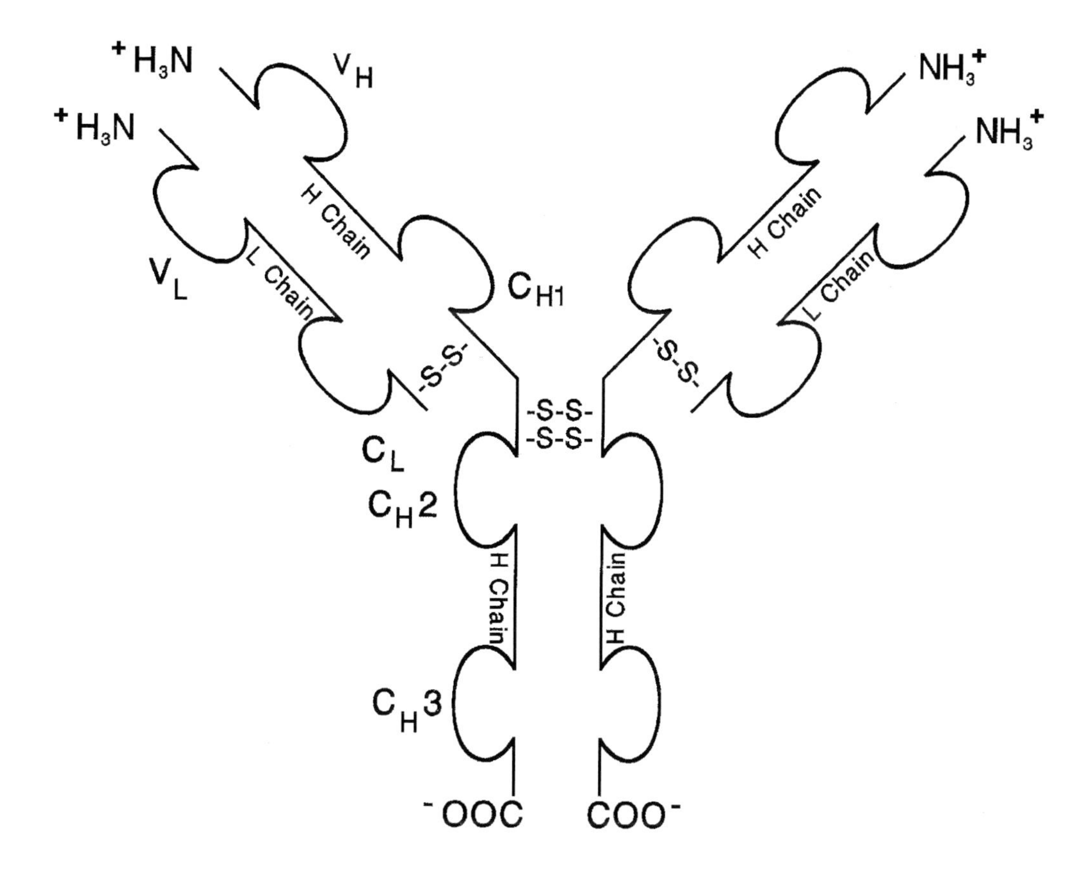

FIGURE 7.8 V_H and V_L regions on an antibody.

has a sandwich-like structure comprised of two β-pleated sheets that are nearly parallel. There are four antiparallel chain segments in one sheet and three in the other. Approximately 50% of the domain's amino acid residues are in the β-pleated sheets. The other 50% of the amino acid residues are situated in polypeptide chain loops and in terminal segments. The turns are sites of invariant glycine residues. Hydrophobic amino acid side chains are situated between the sheets.

The **V_H region** refers to the variable part of immunoglobulin heavy chain which is the part of a variable region encoded for by the V_H gene segment (Figure 7.8). The **V_L region** describes the variable portion of an immunoglobulin light chain. The symbol may be used to designate the V_L gene encoded segment. V_κ is a variable region of an immunoglobulin κ light chain. This symbol may also be used to signify that part of a variable region encoded by the V_κ gene segment.

Vλ is the variable region of an immunoglobulin λ light chain. The symbol may designate that part of a variable region which the Vλ gene segment encodes.

V_L region is the variable region of an immunoglobulin light chain. The symbol may be used to designate the V_L gene encoded segment.

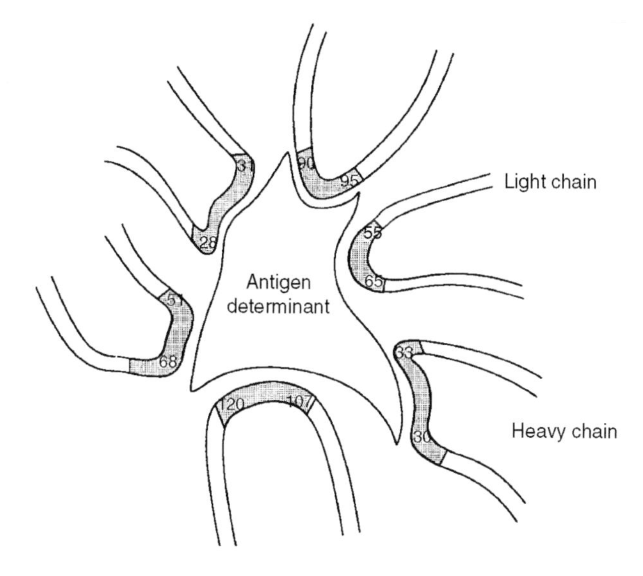

FIGURE 7.9 Depiction of the structure of the six hypervariable regions of an antibody.

Hypervariable regions (Figure 7.9) constitute a minimum of four sites of great variability which are present throughout the H and L chain V regions. They govern the antigen-binding site of an antibody molecule. Thus, grouping of these hypervariable residues into areas govern both confor-

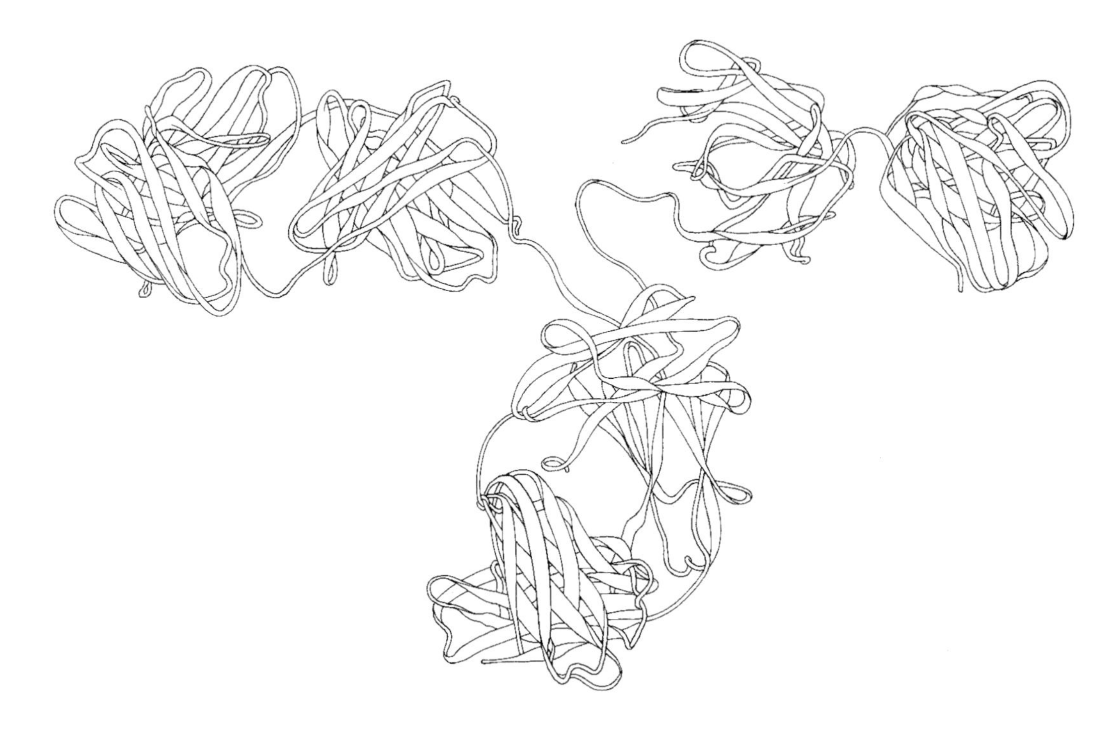

FIGURE 7.10 Illustration of intact monoclonal antibody for canine lymphoma.

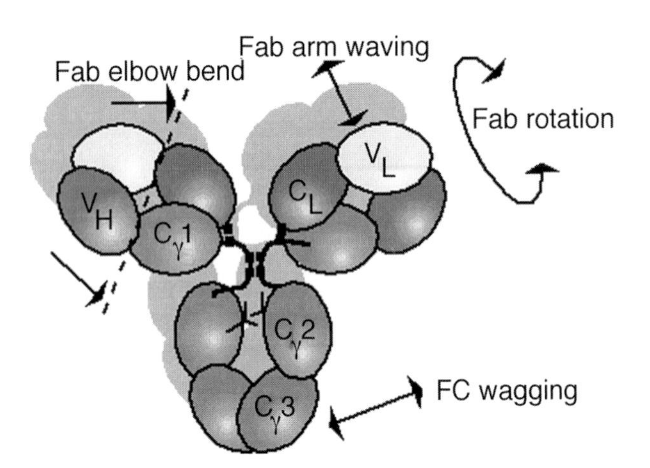

FIGURE 7.11 Cartoon of an IgG molecule.

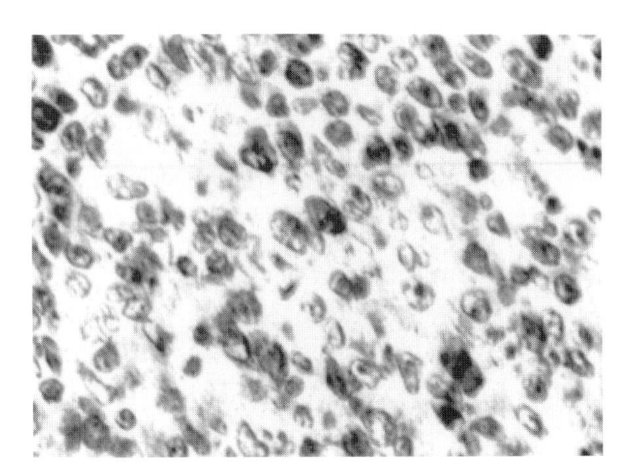

FIGURE 7.12 Photomicrograph of immunoglobulin 1 light "staining" by immunoperoxidase.

mation and specificity of the antigen-binding site upon folding of the protein molecule. Hypervariable residues are also responsible for variations in idiotypes between immunoglobulins produced by separate cell clones. Those parts of the variable region that are not hypervariable are termed the framework regions. Hypervariable regions are also called complementarity-determining regions (Figure 7.10 and Figure 7.11). See hot spot. The term refers also to those portions of the T cell receptor that constitute the antigen-binding site. Each antibody heavy chain and light chain and each TCR α chain and β chain possess three hypervariable loops, also called CDRs. Most of the variability between different antibodies or TCRs is present within these loops.

An **immunoglobulin light chain** is a 23-kDa, 214-amino acid polypeptide chain comprised of a single constant region and a single variable region that is present in all five classes of immunoglobulin molecules. The two types of light chains are designated κ and λ. They are found in association with heavy polypeptide chains and immunoglobulin molecules and are fastened to these structures through disulfide bonds.

An **immunoglobulin λ chain** (Figure 7.12) is a 23-kDa 214-amino acid residue polypeptide chain with a single variable region and a single constant region (Figure 7.13). The λ chains represent one of two light polypeptide chains

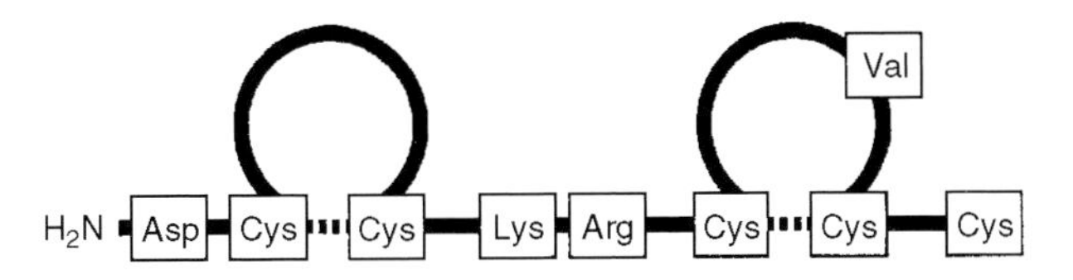

FIGURE 7.13 Lambda light chain showing domain structure.

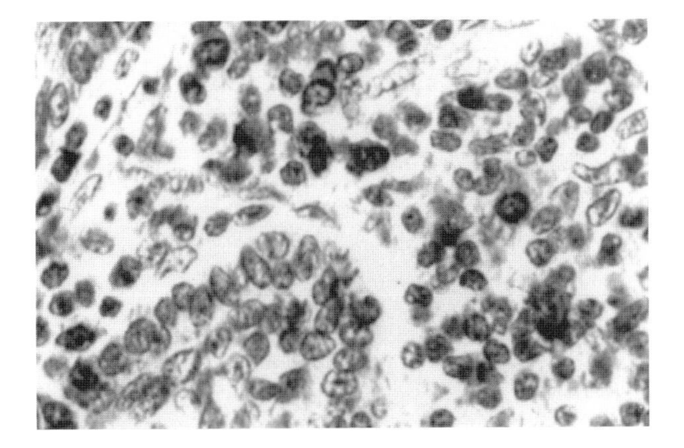

FIGURE 7.14 Photomicrograph of immunoglobulin κ light chain "staining" by immunoperoxidase.

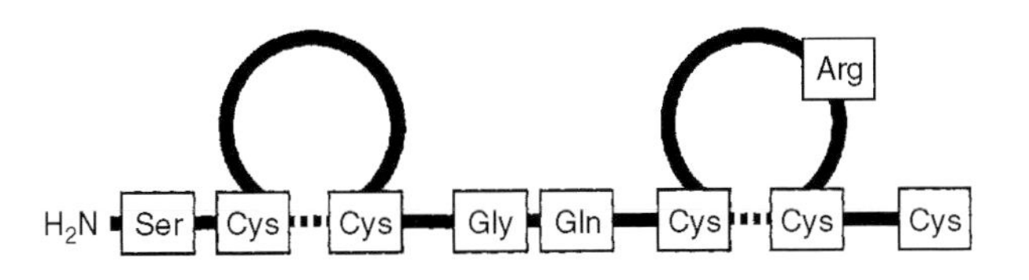

FIGURE 7.15 The κ light chain showing domain structure.

comprising all five classes of immunoglobulin molecules. Approximately 40% of immunoglobulin light chains in humans are λ. Wide variations in percentages are observed in other species. For example, the great majority of immunoglobulin light chains in horses and dogs are λ, whereas they constitute only 5% of murine light chains. Constant region differences among λ light chains of mice and humans distinguish the molecules into four isotypes in humans. A different C gene segment encodes the separate constant regions defining each λ light chain isotype. The human λ light chain isotypes are designated Kern-Oz$^+$, Kern$^+$Oz$^-$, and Mcg.

Immunoglobulin heavy chain-binding protein (BiP) is a 77-kDa protein that combines with selected membrane and secretory proteins. It is believed to facilitate their passage through the endoplasmic reticulum.

An **immunoglobulin κ chain** (Figure 7.14 and Figure 7.15) is a 23-kDa 214-amino acid residue polypeptide chain that is comprised of a single variable region and a single constant region. It is one of the two types of light polypeptide chain present in all five immunoglobulin classes. Approximately

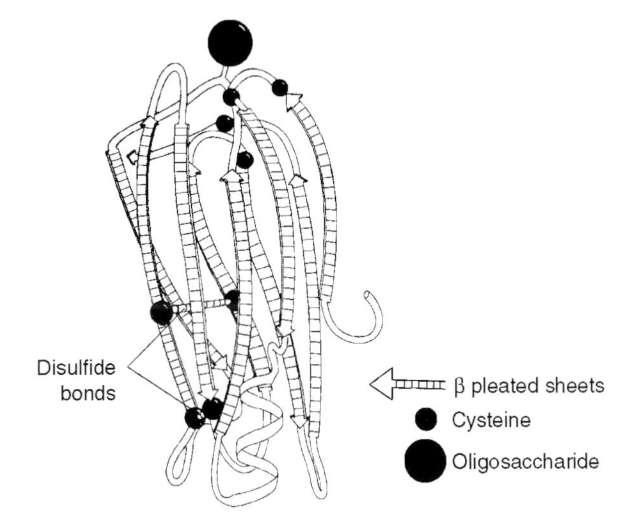

FIGURE 7.16 Structure of J chain that occurs in secretory IgA and IgM molecules and facilitates polymerization.

60% of light immunoglobulin chains in humans are κ with wide variations of their percentages in other species. Whereas κ chains are virtually absent in immunoglobulins of dogs, they comprise the vast majority of murine immunoglobulin light chains. κ light chain allotypes in man are termed Km1, Km1,2, and Km3.

A **J chain** (Figure 7.16) is a 17.6-kDa polypeptide chain present in polymeric immunoglobulins that include both IgM and IgA. It links four-chain immunoglobulin monomers to produce the polymeric immunoglobulin structure. J chains are produced in plasma cells and are incorporated into IgM or IgA molecules prior to their secretion. Incorporation of the J chain appears essential for transcytosis of these immunoglobulin molecules to external secretions. The J chain comprises 2 to 4% of an IgM pentamer or a secretory IgA dimer. Tryptophan is absent from both mouse and human J chains. J chains are comprised of 137-amino acid residues and a single complex N-linked oligosaccharide on asparagine. Human J chain contains three forms of the oligosaccharide which differ in sialic acid content. The J chain is fastened through disulfide bonds to penultimate cysteine residues of μ or α heavy chains. The human J chain gene is located on chromosome 4q21, whereas the mouse J chain gene is located on chromosome 5.

An **immunoglobulin class** (Figure 7.17) is a subdivision of immunoglobulin molecules based on antigenic and structural differences in the Fc regions of their heavy polypeptide chains. Immunoglobulin molecules belonging to a particular class have at least one constant region isotypic determinant in common. The different classes such as IgG, IgM, and IgA designate separate isotypes. Since the light chains of immunoglobulin molecules are one of two types, the heavy chains determine immunoglobulin class. There is

Ig	IgG	IgM	IgA	IgD	IgE
Serum concentration (mg/dl)	800-1700	50-190	140-420	0.3-0.40	<0.001
Total Ig (%)	85	5-10	5-15	<1	<1
Complement fixation	+	++++	-	-	-
Principal biological effect	Resistance-opsonin; secondary response	Resistance-prepcipitin; primary response	Resistance prevents movement across mucous membranes	?	Anaphylaxis
Principal site of action	Serum	Serum	Secretions	?; receptor for B cells	Mast cells
Molecular weight (kd)	154	900	160 (+ dimer)	185	190
Serum half-life (days)	23	5	6	2-3	2-3
Antibacterial lysis	+	+++	+	?	?
Antiviral lysis	+	+	+++	?	?
H-chain class	γ	μ	α	δ	ε
Subclass	$\gamma_1 \gamma_2 \gamma_3 \gamma_4$		$\alpha_1 \alpha_2$		

FIGURE 7.17 Table of human immunoglobulins and their properties.

Ig	IgG	IgM	IgA	IgE	IgD
H Chain Class	γ	μ	α	ε	δ
Subclass	γ_1 γ_2 γ_3 γ_4		α_1 α_2		

FIGURE 7.18 Summary of the heavy chain designations of immunoglobulins that determine class and of their subdivisions that determine subclass.

about 30% homology of amino acid sequence among the five immunoglobulin heavy chain constant regions in man. Heavy chains (or isotypes) also differ in carbohydrate content. Immunization of a nonhuman species with human immunoglobulin provides antisera that may be used for class or isotype determination. Ig G is divided into four subclasses and IgA is divided into two subclasses.

An **immunoglobulin subclass** (Figure 7.18) is a subdivision of immunoglobulin classes according to structural and antigenic differences in the constant regions of their heavy polypeptide chains. All molecules in an immunoglobulin subclass must express the isotypic antigenic determinants unique to that class, but they also express other epitopes that render that subclass different from others. IgG has four subclasses designated as IgG1, IgG2, IgG3, and IgG4. Whereas there is only 30% identity among the five immunoglobulin classes, there is three times that similarity among IgG subclasses. The IgA class is divisible into two subclasses, whereas the remaining three immunoglobulin classes have not been further subdivided into subclasses. The structural differences in subclasses are exemplified by the variations and number of inter-heavy chain disulfide bonds which the four IgG subclasses possess. The function of immunoglobulin molecules differs from one subclass to another, as exemplified by the inability of IgG4 to fix complement.

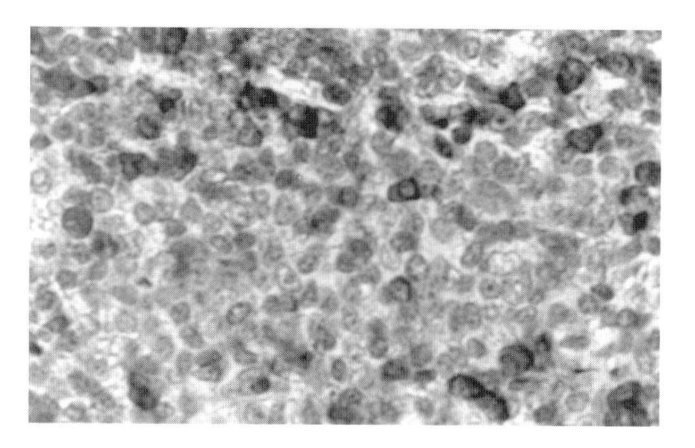

FIGURE 7.19 Photomicrograph of immunoglobulin G producing cells.

Immunoglobulin G (IgG) (Figure 7.19) comprises approximately 85% of the immunoglobulins in adults. It has a molecular weight of 154 kDa based on two L chains of 22,000 Da each and two H chains of 55,000 Da each. It has the longest half-life (23 d) of the five immunoglobulin classes, crosses the placenta, and is the principal antibody in the anamnestic or booster response. IgG shows high avidity or binding capacity for antigen, fixes complement, stimulates chemotaxis, and acts as an opsonin to facilitate phagocytosis (Figure 7.23 and Figure 7.24).

IgG: See immunoglobulin G.

The **IgG index** is the ratio of IgG and albumin synthesis in the brain and in peripheral tissues. It is increased in multiple central nervous system infections, inflammatory disorders, and neoplasms.

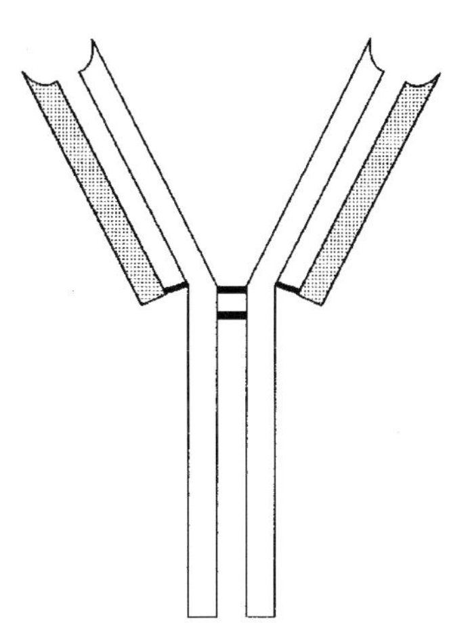

FIGURE 7.20 IgG1.

The **immunoglobulin gamma (γ) chain** is a 51-kDa, 450-amino acid residue heavy polypeptide chain comprised of one variable V_H domain and a constant region with three domains designated C_H1, C_H2, and C_H3. The hinge region is situated between C_H1 and C_H2. There are four subclasses of IgG in humans with four corresponding γ chain isotypes designated γ-1, γ-2, γ-3, and γ-4. IgG1, IgG2, IgG3, and IgG4 have differences in their hinge regions and differ in the number and position of disulfide bonds that link the two γ chains in each IgG molecule. There is only a 5% difference in amino acid sequence among human γ chain isotypes, exclusive of the hinge region. Cysteine residues, which make it possible for inter-heavy (γ) chain disulfide bonds to form, are found in the hinge area. IgG1 and IgG4 have 2 inter-heavy chain disulfide bonds, IgG2 has 4, and IgG3 has 11. Proteolytic enzymes, such as papain and pepsin, cleave an IgG molecule in the hinge region to produce Fab and $F(ab')_2$ and Fc fragments. Four murine isotypes have also been described. Two exons encode the carboxy terminal region of membrane γ chain. Two γ chains, together with two κ or γ light chains, fastened together by disulfide bonds, comprise an IgG molecule.

Immunoglobulin M (IgM) (Figure 7.25) comprises 5 to 10% of the total immunoglobulins in adults and has a half-life of 5 d. It is a pentameric molecule with 5 four-chain monomers joined by disulfide bonds and the J chain, with a total molecular weight of 900 kDa. Theoretically, this immunoglobulin has 10 antigen-binding sites. IgM is the most efficient immunoglobulin in fixing complement. A single IgM pentamer can activate the classic pathway. Monomeric IgM is found with IgD on the B lymphocyte cell surface, where it serves as the receptor for antigen. Because IgM is relatively large, it is confined to intravascular locations. IgM is particularly important for immunity against polysaccharide antigens on the exterior of pathogenic microorganisms. It also promotes phagocytosis and bacteriolysis through its complement activation activity.

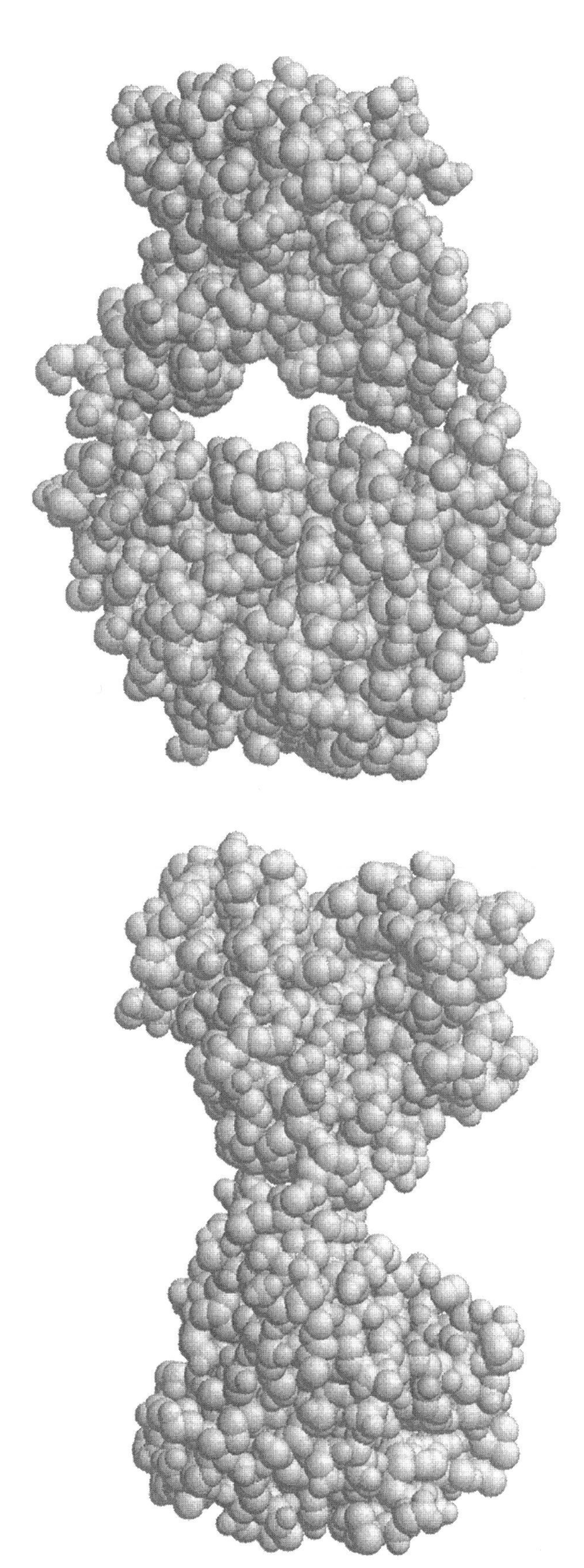

FIGURE 7.21 IgG1 Fab fragment.

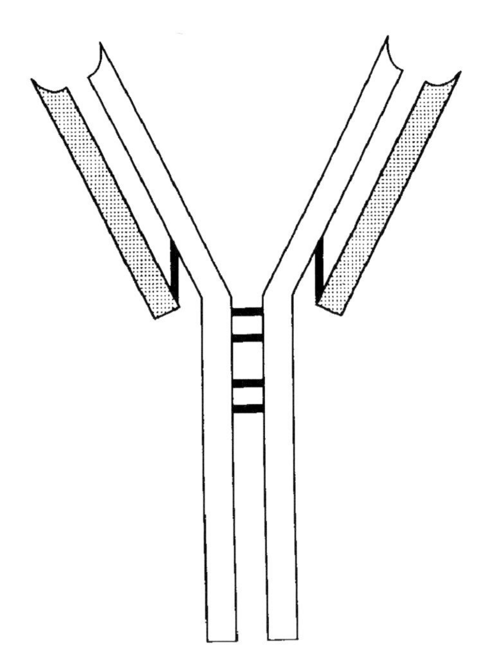

FIGURE 7.22 IgG2.

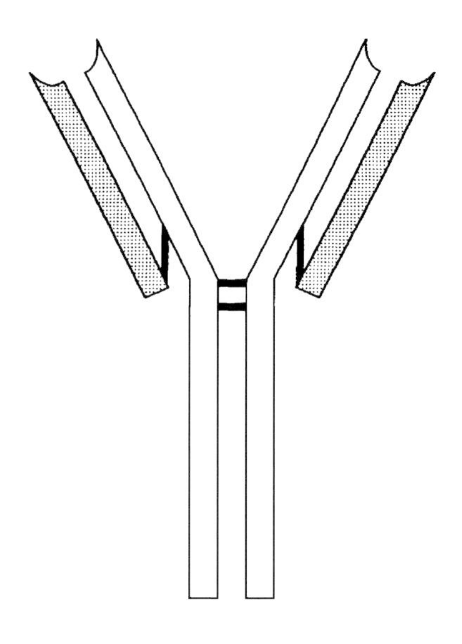

FIGURE 7.24 IgG4.

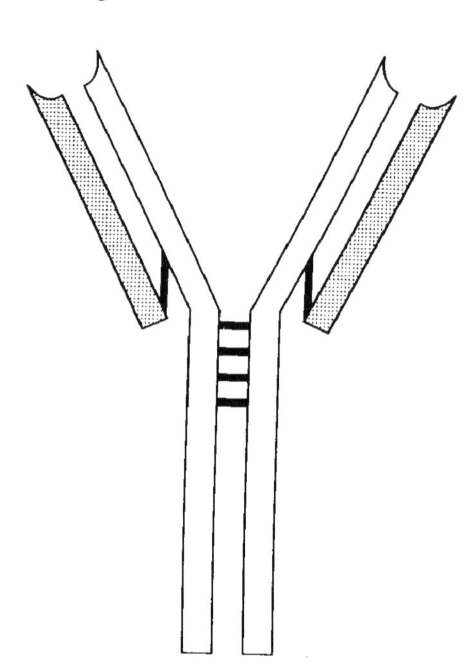

FIGURE 7.23 IgG3.

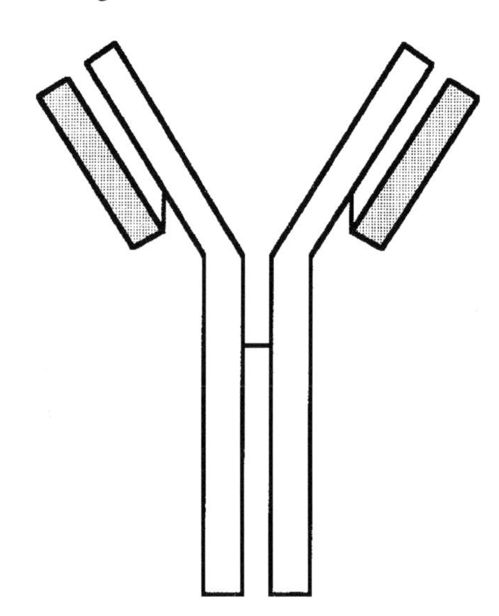

FIGURE 7.25 Monomeric IgM that contains two H chains and two κ or two l light chains.

Cell-surface immunoglobulin is the B cell receptor for antigen.

γM globulin is an obsolete term for IgM.

γ Macroglobulin is an obsolete term for IgM.

IgM: See immunoglobulin M.

19 S antibody refers to the sedimentation coefficient of the IgM class of immunoglobulin.

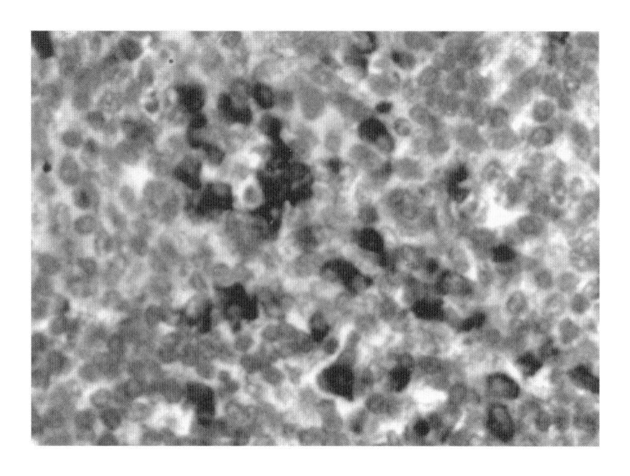

FIGURE 7.26 Photomicrograph of IgM producing cells (immunoperoxidase stain).

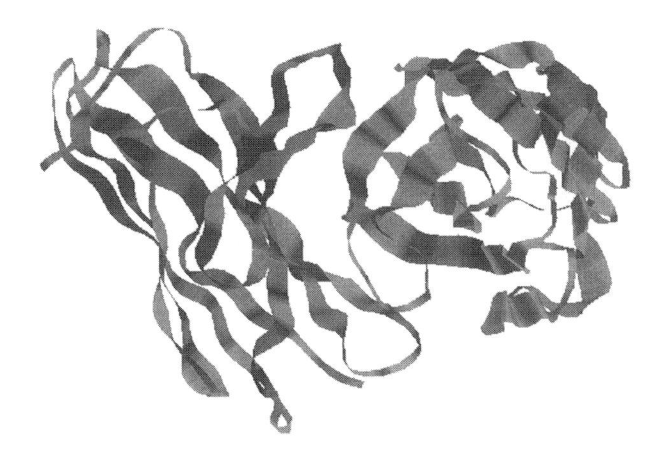

FIGURE 7.27 IgM Fv fragment.

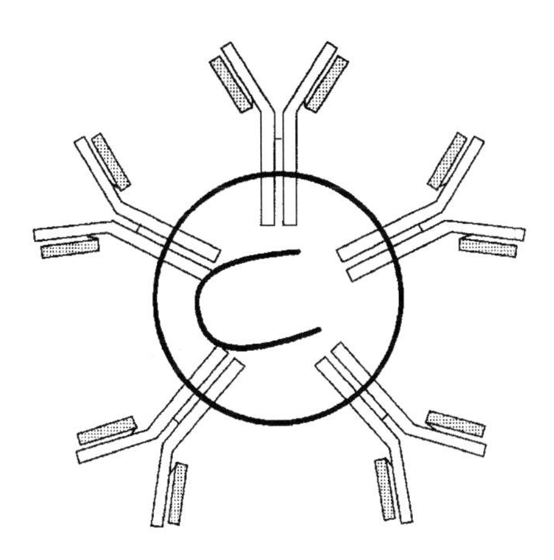

FIGURE 7.28 Pentameric IgM that consists of five 7S monomers comprised of two heavy and two light polypeptide chains each as well as one J chain per molecule.

Cell-bound antibody (cell-fixed antibody) is an antibody anchored to the cell surface either through its paratopes binding to cell epitopes or attachment of its Fc region to Fc receptors. An example is cytophilic antibody or IgE which may then react with antigen as their Fab regions are available.

The **IgM index** reflects total IgM formation in the blood–brain barrier. This is elevated in infectious meningoencephalomyelitis and may also be increased in central nervous system lupus erythematosus and multiple sclerosis.

μ chain is the IgM heavy polypeptide chain. Membrane μ chain is designated μ_m. Secreted μ chain is designated μ_s.

The **immunoglobulin mu (μ) chain** is a 72-kDa, 570-amino acid heavy polypeptide chain comprised of one variable region, designated V_H, and a four-domain constant region, designated C_H1, C_H2, C_H3, and C_H4. The μ chain does not have a hinge region. A "tail piece" is located at the carboxy terminal end of the chain. It is comprised of 18-amino acid residues. A cysteine residue at the penultimate position of a carboxy terminal region of the μ chain forms a disulfide bond that joins to the J chain. There are five N-linked oligosaccharides in the μ chain of humans. Secreted IgM (μ_s) and membrane IgM (μ_m) and μ chain differ only in the final 20-amino acid residues at the carboxy terminal end. The membrane form of IgM has 41 different residues substituted for the final 20 residues in the secreted form. A 26-residue region of this carboxy terminal section in the membrane form of IgM apparently represents the hydrophobic transmembrane part of the chain.

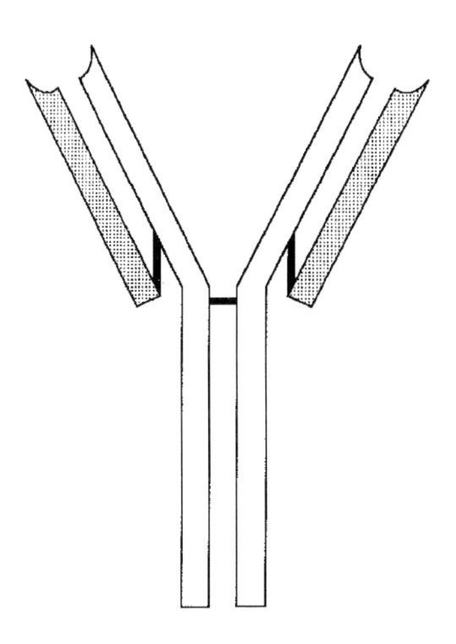

FIGURE 7.29 IgA1.

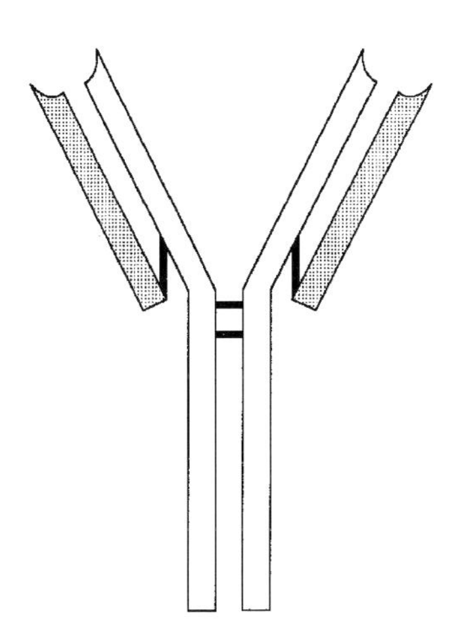

FIGURE 7.30 IgA2.

Immunoglobulin A (IgA) (Figure 7.29) comprises 5 to 15% of the serum immunoglobulins and has a half-life of 6 d. It has a mol wt of 160 kDa and a basic four-chain monomeric structure. However, it can occur as monomers, dimers, trimers, and multimers. It contains α heavy chains and κ or λ light chains. There are two subclasses of IgA designated as IgA1 and IgA2 (Figure 7.30). In addition to serum IgA, a secretory or exocrine variety appears in body secretions and provides local immunity. For example, the

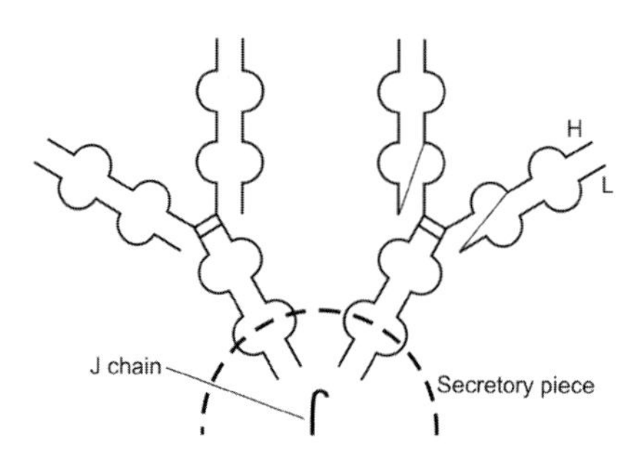

FIGURE 7.31 Secretory IgA that consists of two IgA monomers, a J chain, and secretory piece that is believed to protect the molecule from enzymatic digestion in the gut.

Sabin oral polio vaccine stimulates secretory IgA antibodies in the gut, which provides effective immunity against poliomyelitis (Figure 7.31). IgA-deficient individuals have an increased incidence of respiratory infections associated with a lack of secretory IgA in the respiratory system. Secretory or exocrine IgA appears in colostrum, intestinal, and respiratory secretions, saliva, tears, and other secretions.

IgA: See immunoglobulin A.

Secreted immunoglobulin (sIg) is a product of plasma cells that is secreted as free immunoglobulin, where it may circulate as a component of blood plasma or make up part of the protein content of other body fluids. This form of immunoglobulin does not possess a transmembrane domain.

T piece: See secretory piece.

Secretory piece is a 75-kDa polypeptide chain synthesized by epithelial cells of the gut for linkage to immunoglobulin A (IgA) dimers present in body secretions. Secretory component facilitates IgA transport across epithelial cells and protects secretory IgA released into the lumen of the gut from proteolytic digestion by enzymes in the secretions. It is not formed by plasma cells in the lamina propria of the gut that synthesize the IgA molecules with which it combines. Secretory component has a special affinity for mucous, thereby facilitating the attachment of IgA to the mucous membranes. Also called secretory component.

Transport piece: See secretory piece.

Immunoglobulin alpha (α) chain is a 58-kDa, 470-amino acid residue heavy polypeptide chain that confers class specificity on immunoglobulin A molecules. The chain is divisible into three constant domains, designated C_H1, C_H2, and C_H3, and one variable domain, designated V_H. A hinge region is situated between C_H1 and C_H2 domains. An additional segment of 18-amino acid residues at the penultimate position of the chain contains a cysteine residue where the J chain can be linked through a disulfide bond. The IgA subclass is divisible into IgA1 and IgA2

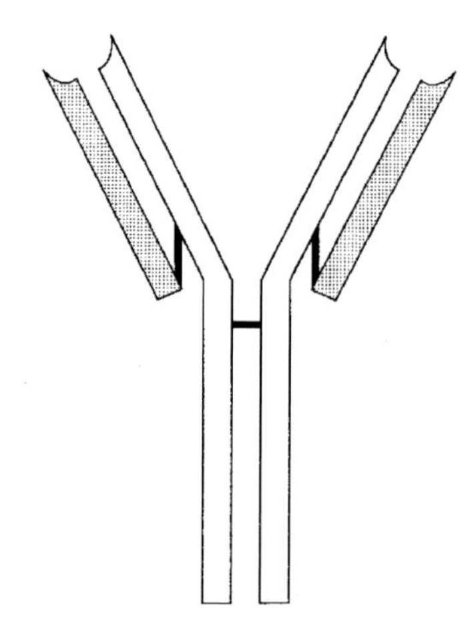

FIGURE 7.32 IgD structure showing a four-chain monomeric unit that consists of two δ heavy chains and either two κ or two λ light chains per molecule.

subclasses, reflecting two separate α chain isotypes. The α-2 chain has two allotypes designated A2m(1) and A2m(2) and does not have disulfide bonds linking H to L chains. Residues that are subclass specific are found in a number of positions in C_H1, the hinge region, and C_H2, where α-1 and α-2 chains differ, but α-2 chains are the same. Differences in the two α chains are found in two C_H1 and five C_H3 positions. Thus, there are three varieties of α heavy chains in humans.

α chain is the immunoglobulin (Ig) class-determining heavy chain found in IgA molecules.

Immunoglobulin class switching: See isotype switching, switch, switch cells, switch region, and switch site.

Immunoglobulin class switching is the mechanism whereby an IgM producing B cell switches isotype to begin producing IgG molecules instead. Further differentiation may lead to a B cell producing IgA. However, the antigen-binding specificity of the antibody molecules with a different isotype remains unchanged.

Immunoglobulin D (IgD) (Figure 7.32), which has a mol wt of 185 kDa, comprises less than 1% of serum immunoglobulins. It has the basic four-chain monomeric structure with two δ heavy chains (mol wt 63,000 Da each) and either two κ or two λ light chains (mol wt 22,000 Da each). The half-life of IgD is only 2 to 3 d, and the role of IgD in immunity remains elusive. Membrane IgD serves with IgM as an antigen receptor on B cell membranes.

IgD: See immunoglobulin D.

Immunoglobulin δ chain is a 64-kDa, 500-amino acid residue heavy polypeptide chain consisting of one variable region, designated V_H, and a three-domain constant region, designated C_H1, C_H2, and C_H3. There is also a 58-residue amino acid residue hinge region in human δ chains. Two exons encode the hinge region. IgD is very susceptible to the action of proteolytic enzymes at its hinge region. Two separate exons encode the membrane component of δ chain. A distinct exon encodes the carboxy terminal portion of the human δ chain that is secreted. The human δ chain contains three N-linked oligosaccharides. Two δ chains and two light chains, either κ or λ fastened together by disulfide bonds, constitute an IgD molecule.

δ chain is the heavy chain of immunoglobulin D (IgD).

Immunoglobulin E (IgE) (Figure 7.33) constitutes less than 1% of the total immunoglobulins and has a half-life of approximately 2.5 d. This antibody has a four-chain unit structure with two ε heavy chains (mol wt 75,000 Da each) and either two κ or two λ light chains per molecule (total mol wt 190 kDa). IgE does not precipitate with antigen *in vitro* and is heat labile. IgE is responsible for anaphylactic hypersensitivity in humans.

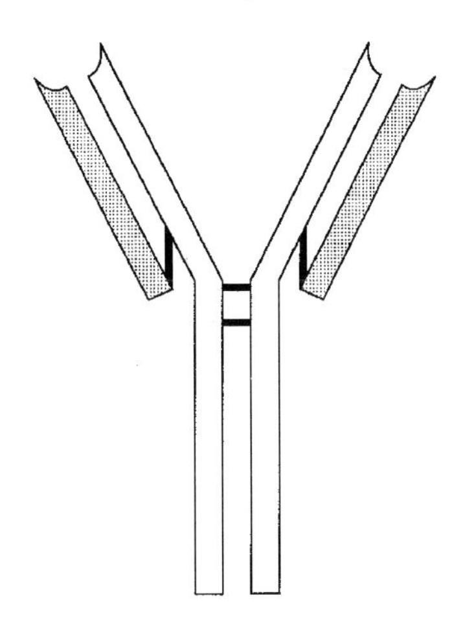

FIGURE 7.33 IgE molecule.

IgE: See immunoglobulin E.

The **immunoglobulin epsilon (ε) chain** 72-kDa, 550-amino acid residue heavy polypeptide chain comprised of one variable region, designated V_H, and a four-domain constant region, designated C_H1, C_H2, C_H3, and C_H4. This heavy chain does not possess a hinge region. In humans the ε heavy chain has 428 amino acid residues in the constant region. There is no carboxy terminal portion of the ε chains. Two ε heavy polypeptide chains, together with two κ or two λ light chains, fastened together by disulfide bonds, comprise an IgE molecule.

In **pepsin digestion** (Figure 7.34) a proteolytic enzyme is used to hydrolyze immunoglobulin molecules into $F(ab')_2$ fragments together with small peptides that represent what remains of the Fc fragment. Each immunoglobulin molecule yields only one $F(ab')_2$ fragment which is bivalent and may manifest many of the same antibody characteristics as intact IgG molecules sugh as antitoxic activity in neutralizing bacterial toxins. Cleaving the Fc region from an IgG molecule deprives it of its ability to fix complement and bind to Fc receptors on cell surfaces. Pepsin digestion is useful in diminishing the immunogenicity of antitoxins.

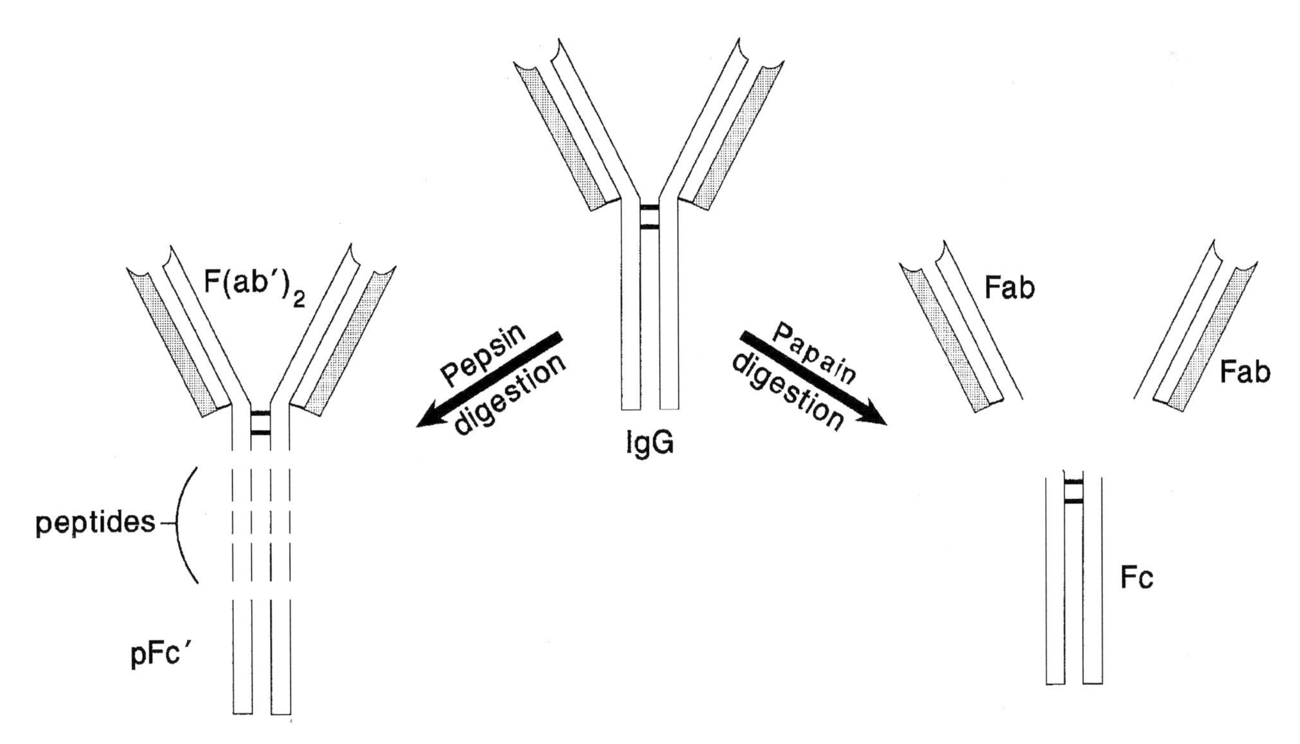

FIGURE 7.34 Pepsin digestion.

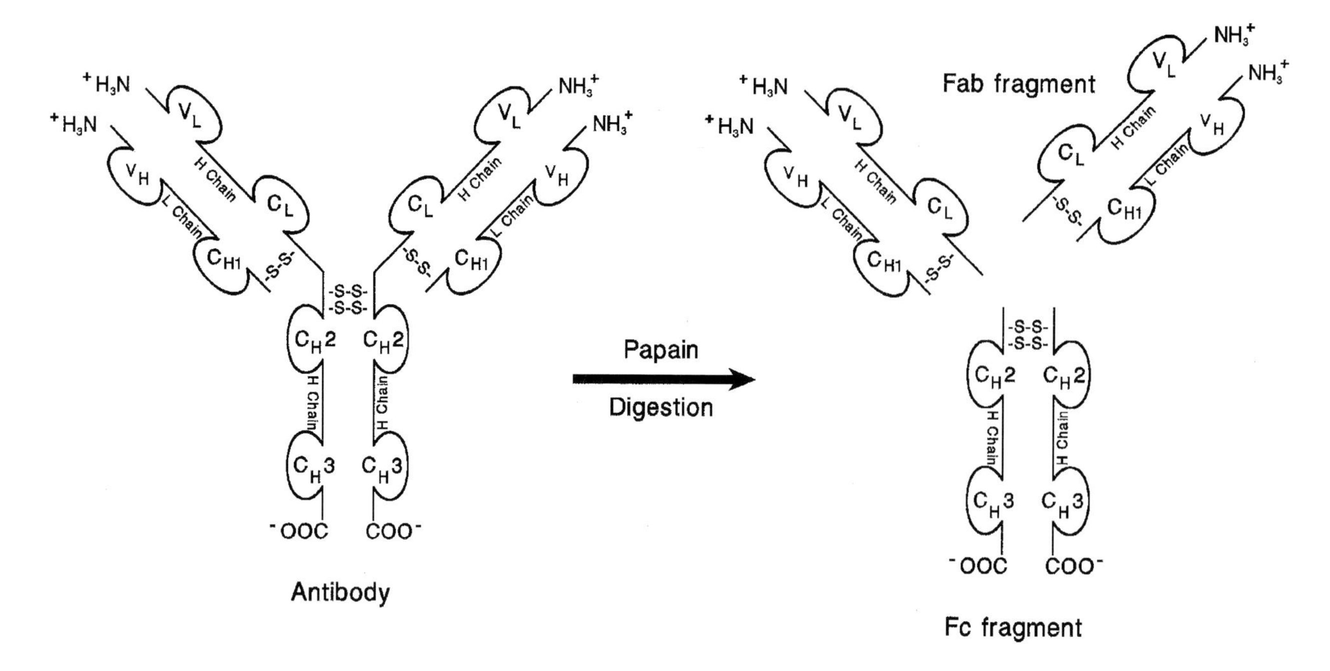

FIGURE 7.35 Papain digestion of IgG molecules yielding two Fab and one Fc fragments per molecule.

It converts them to $F(ab')_2$ fragments which retain antitoxin activity.

Pepsin digestion refers to a proteolytic enzyme used to cleave immunoglobulin molecules into $F(ab')_2$ fragments together with fragments of small peptides that represent what remains of the Fc fragment. Each immunoglobulin molecule yields only one $F(ab')_2$ fragment, which is bivalent and may manifest many of the same antibody characteristics as intact IgG molecules, such as antitoxic activity in neutralizing bacterial toxins. Cleaving the Fc region from an IgG molecule deprives it of its ability to fix complement and bind to Fc receptors on cell surfaces. Pepsin digestion is useful in diminishing the immunogenicity of antitoxins. It converts them to $F(ab')_2$ fragments which retain antitoxin activity.

pFc′ fragment is a fragment of pepsin digestion of IgG or of the Fc fragment. Pepsin digestion of IgG or of the Fc fragment yields low molecular weight peptides and a pFc′ fragment that is still capable of binding to an Fc receptor on a macrophage or monocyte. It is a 27-kDa dimer without a covalent bond comprised of two C_H3 domains, the carboxyl terminal 116 residues of each chain. Unlike the Fc′ fragment, it has the basic N-terminal and C-terminal peptides of this immunoglobulin domain.

Polymers are molecules comprised of more than one repeating unit. In immunology, immunoglobulins are comprised of more than one basic monomeric four-polypeptide chain unit. IgA may exist as dimers with two units or as multimers. The IgM molecule is pentameric, containing five monomeric units.

Papain (Figure 7.35) is a proteolytic enzyme extracted from *Carica papaya* that is used to digest each IgG immunoglobulin molecule into two Fab fragments and one crystallizable Fc fragment. This aids efforts to reveal the molecular structure of immunoglobulins. Papain cleaves the immunoglobulin G molecule on the opposite side of the central disulfide bond from pepsin, which cleaves the molecule to the C-terminus side, leading to the formation of 1 $F(ab')_2$ fragment, which is bivalent in contrast to the Fab fragments which are univalent. The Fc fragment of papain digestion has no antigen-binding capacity, although it does have complement-fixing functions and attaches immunoglobulin molecules to Fc receptors on a cell membrane. The enzyme has also been used to render red blood cell surfaces susceptible to agglutination by incomplete antibody.

Papain hydrolysis refers to cleavage of IgG molecules into two Fab fragments and one Fc fragment. When the immunoglobulin is exposed to papain with cysteine present, papain cleaves a histidyl-threonine peptide bond of the heavy chain.

Immunoglobulin fragment is a term reserved for products that result from the action of proteolytic enzymes on immunoglobulin molecules. Intrachain disulfide bonds can be severed by reduction in the presence of denaturing agents such as urea, guanidine, or detergents. Peptide bonds in intact domains are not easily split by proteolytic enzymes. Light chains can be cleaved at the V-C junction, giving rise to large segments that correspond to the V_L and C_L domains. Similar cleavage of the heavy chain is more difficult to achieve. Papain cleaves H chains at the

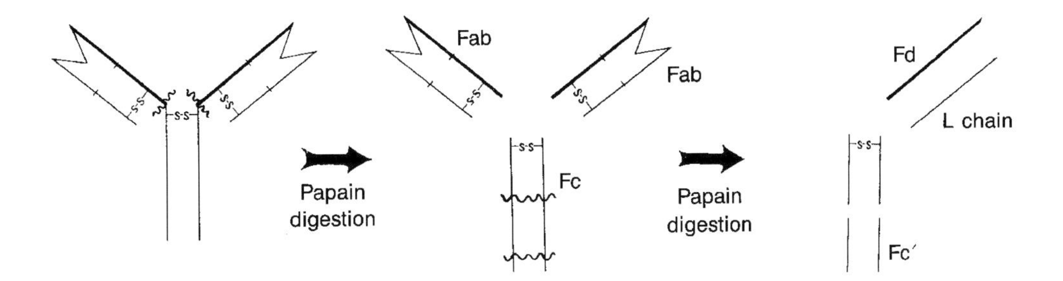

FIGURE 7.36 The Fab fragment is comprised of one light chain and the variable and C_H1 regions of a heavy chain. They are united by disulfide bonds and have a single binding site for antigen. The heavy chain part of a Fab fragment is referred to as Fd. Further digestion with papain yields an Fc′.

N-terminus of the H–H disulfide bonds, giving two individual portions of the terminus of the molecule, called Fab, and the fragment of the C-terminus region, Fc, which is crystallizable. In contrast, pepsin cleaves H chains at the C-terminus of the H–H disulfide bonds. Thus, the two Fab fragments will remain joined and are called $F(ab')_2$. It degrades the C_H2 domains, but splits the C_H3 domains, which remain noncovalently bonded in dimeric form and are called pFc′. Further digestion of the pFc′ with papain results in smaller dimeric fragments called Fc′Plasmin, which have been found to cleave the immunoglobulin molecule between C_H2 and C_H3, giving rise to a fragment designated Facb. The heavy chain portion of the Fab, designated Fd, and the heavy chain portion of the Fab′ fragment, designated Fd′, results from the breakdown of an $F(ab')_2$ fragment produced by pepsin digestion of the IgG molecule. The Fv fragment consists of the variable domain of heavy and light chains on an immunoglobulin molecule where antigen binding occurs.

Plasmin is a serine protease proteolytic enzyme in plasma that is generated from its inactive precursor plasminogen. It is a 90-kDa enzyme that derives from cleavage of a single arginyl–valyl bond in the C-terminal region of plasminogen. It consists of two unequal chains, termed heavy (A) and light (B) chains, linked by a single disulfide bond. The A chain derives from the N-terminal region plasminogen. The B chain carries the serine active site. Plasmin catalyzes the hydrolysis of fibrin. Thus, it facilitates the dissolution of intravascular blood clots. In addition to its fibrinolytic activity, plasmin has numerous other functions associated with coagulation, fibrinolysis, and inflammation that include (1) enhancement of antibody responses to both thymus-dependent and thymus-independent antigens, (2) augmentation of agglutination by lectins, (3) facilitation of the escape of cells from contact inhibition and culture, (4) enhancement of cytotoxicity with or without participation of antibodies, and (5) stimulation of B cell proliferation.

An **antibody fragment** is a product of enzymatic treatment of an antibody immunoglobulin molecule with an enzyme such as papain or pepsin. For example, papain treatment leads to the production of two Fab and one Fc fragments, whereas the use of pepsin yields the $F(ab')_2$ fragment. See the individual fragments for further information.

Active site is a crevice formed by the V_L and V_H regions of an immunoglobulin's Fv region. It may differ in size or shape from one antibody molecule to another. Its activity is governed by the amino acid sequence in this variable region and differences in the manner in which V_H and V_L regions relate to one another. Antibody molecule specificity is dependent on the complementary relationship between epitopes on antigen molecules and amino acid residues in the recess comprising the antibody active site. The V_L and V_H regions contain hypervariable areas that permit great diversity in the antigen-binding capacity of antibody molecules.

Binding site: In immunology, the paratope area of an antibody molecule that binds antigen or that part of the T cell receptor which is antigen binding.

A **Fab fragment** (Figure 7.36 and Figure 7.37) is a product of papain digestion of an IgG molecule. It is comprised of one light chain and the segment of heavy chain on the N-terminal side of the central disulfide bond. The light and heavy chain segments are linked by interchain disulfide bonds. It is 47 kDa and has a sedimentation coefficient of 3.5 S. The Fab fragment has a single antigen-binding site. There are two Fab regions in each IgG molecule.

Fd fragment is the heavy chain portion of a Fab fragment produced by papain digestion of an IgG molecule. It is on the N-terminal side of the papain digestion site.

Fd piece: See Fd fragment.

The **Fv region** (Figure 7.38) consists of the N-terminal variable segments of both the heavy and light chains in each Fab region of an immunoglobulin molecule with a four-chain unit structure.

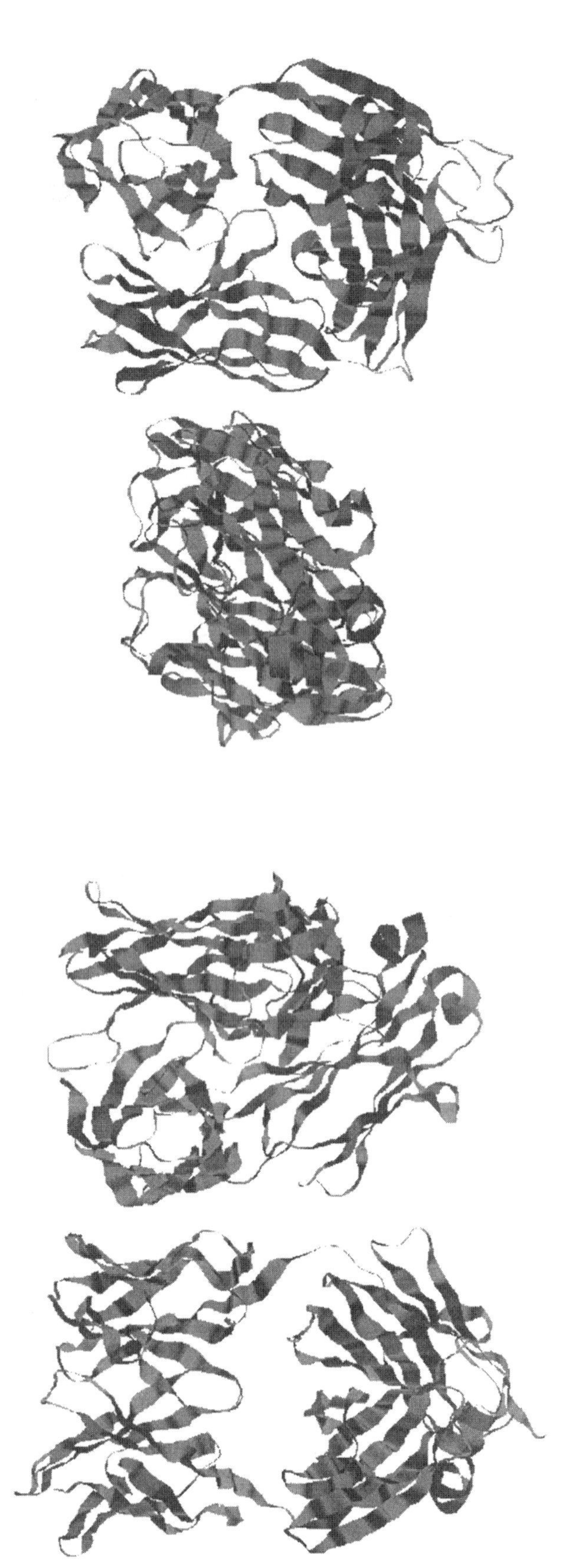

FIGURE 7.37 Fab fragment.

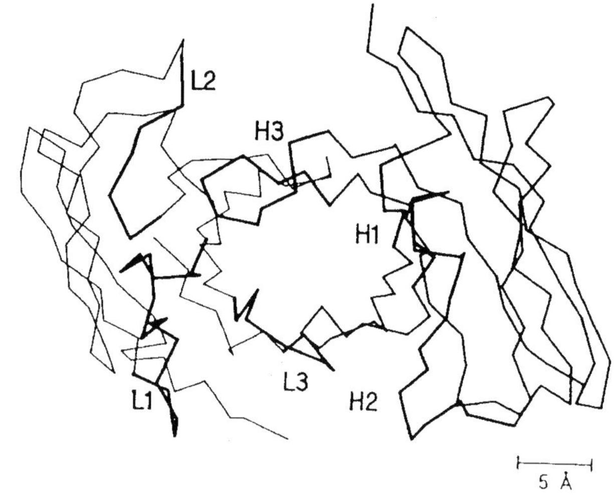

FIGURE 7.38 Fv region.

The **Wu-Kabat plot** (Figure 7.39) is a graph that demonstrates the extent of variability at individual amino acid residue positions in immunoglobulin and T cell receptor variable regions. Division of the different amino acid number at a given position by the frequency of the amino acid which occurs most commonly at that position gives the index of variability. The index varies between 1 and 400. To show the variability graphically, a biograph is prepared where the index is plotted at each residue position. This plot indicates the extent of variability at each position and is useful in localizing immunoglobulin and T cell receptor hypervariable regions. This analytic method revealed that in immunoglobulin heavy or light chains most variable residues are clustered in three hypervariable regions.

Kabat-Wu plot: See Wu-Kabat plot.

Variability plot: See Wu-Kabat plot.

Fc piece: See Fc fragment.

The **Fc fragment (fragment crystallizable)** (Figure 7.40) is a product of papain digestion of an IgG molecule. It is comprised of two C-terminal heavy chain segments (C_H2, C_H3) and a portion of the hinge region linked by the central disulfide bond and noncovalent forces. This 50-kDa fragment is unable to bind antigen, but it has multiple other biological functions, including complement fixation, interaction with Fc receptors on the cell surfaces and placental transmission of IgG. One Fc fragment is produced by papain digestion of each IgG molecule. The Fc region of an intact IgG molecule mediates effector functions by binding to cell surface receptors or C1q complement protein.

Gm allotype is a genetic variant determinant of the human IgG heavy chain. Allelic genes that encode the γ1, γ2, and γ3 heavy chain constant regions encode the Gm allotypes. They were recognized by the ability of blood sera from

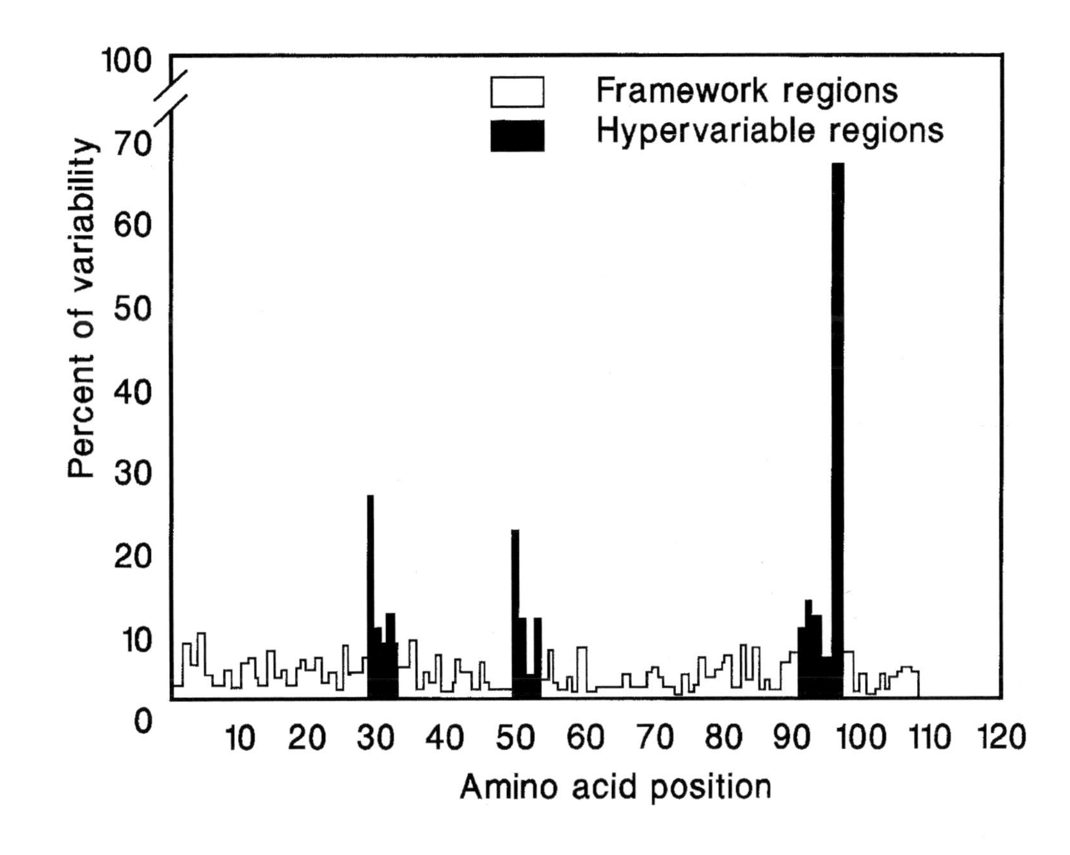

FIGURE 7.39 Wu-Kabat plot.

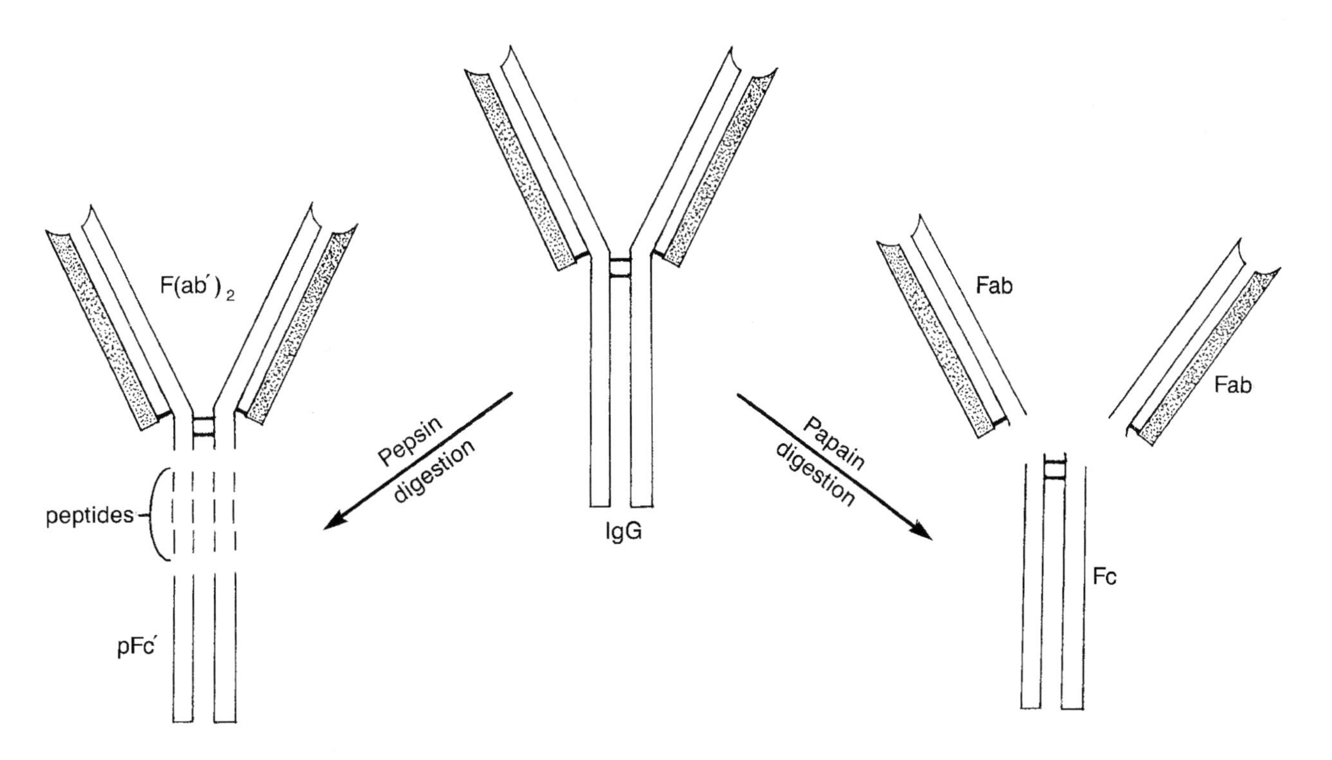

FIGURE 7.40 Fc fragment.

rheumatoid arthritis patients, which contain anti-IgG rheumatoid factor, to react with them. Gm allotypic determinants are associated with specific amino acid substitution in different γ chain constant regions in man. IgG subclasses are associated with certain Gm determinants. For example, IgG1 is associated with G1m(1) and G1m(4), and IgG3 is associated with G3m(5). Although the great majority of Gm allotypes are restricted to the IgG-γ chain Fc region, a substitution at position 214 of C_H1 of arginine yields the G1m(4) allotype, and a substitution at this same site of lysine yields G1m(17). For Gm expression, the light chain part of the molecule must be intact.

The **Fc′ fragment** is a product of papain digestion of IgG. It is comprised of two noncovalently bonded C_H3 domains that lack the terminal 13 amino acids. This 24-kDa dimer consists of the region between the heavy chain amino acid residues 14 through 105 from the carboxy terminal end. Normal human urine contains minute quantities of Fc′ fragment.

An **Fabc fragment** (Figure 7.41) is a 5-S intermediate fragment produced by partial digestion of IgG by papain in which only one Fab fragment is cleaved from the parent molecule in the hinge region. This leaves the Fabc fragment, which is comprised of a Fab region bound covalently to an Fc region and is functionally univalent.

Fab′ fragment (Figure 7.42) is a product of reduction of an F(ab′)$_2$ fragment that results from pepsin digestion of IgG. It is comprised of one light chain linked by disulfide bonds to the N-terminal segment of heavy chain. The Fab′ fragment has a single antigen-binding site. There are two Fab′ fragments in each F(ab′)$_2$ fragment.

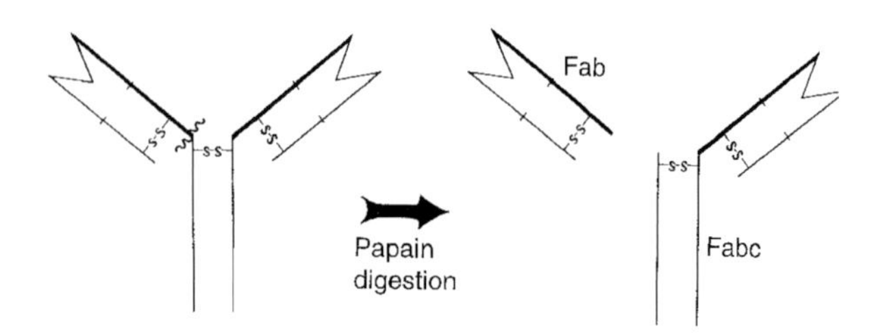

FIGURE 7.41 Generation of an Fabc fragment by papain digestion of IgG in which one Fab region is cleaved, leaving an Fabc fragment consisting of the Fc region and one Fab region of the molecule bearing a single antigen-binding site.

F(ab′)$_2$ fragment is a product of pepsin digestion of an IgG molecule. This 95-kDa immunoglobulin fragment has a valence or antigen-binding capacity of two, which renders it capable of inducing aggulination or precipitation with homologous antigen. However, the functions associated with the intact IgG molecule's Fc region, such as complement fixation and attachment to Fc receptors on cell surfaces, are missing. Pepsin digestion occurs on the carboxy terminal side of the central disulfide bond at the hinge region of the molecule, which leaves the central disulfide bond intact. The C_H2 domain is converted to minute peptides, yet the C_H3 domain is left whole, and the two C_H3 domains comprise the pFc′ fragment.

Fab″ fragment: See F(ab′)$_2$ fragment.

An **Fd′ fragment** (Figure 7.43) is the heavy chain portion of an Fab′ fragment produced by reduction of the F(ab′)$_2$ fragment that results from pepsin digestion of IgG. It is comprised of V_H1, C_H1, and the heavy chain hinge region. Fd′ contains 235-amino acid residues.

13-26 Fd′ piece: See Fd′ fragment.

An **Facb fragment** (Figure 7.44) is a fragment antigen and is complement binding. The action of plasmin on IgG molecules denatured by acid cleaves C_H3 domains from both heavy chain constituents of the Fc region. This yields a bivalent fragment functionally capable of precipitation

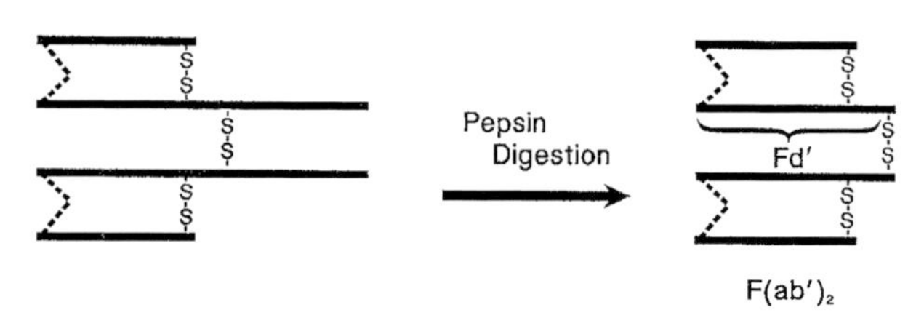

FIGURE 7.43 F(ab′)$_2$ fragment containing 2 Fd′ heavy chain portions.

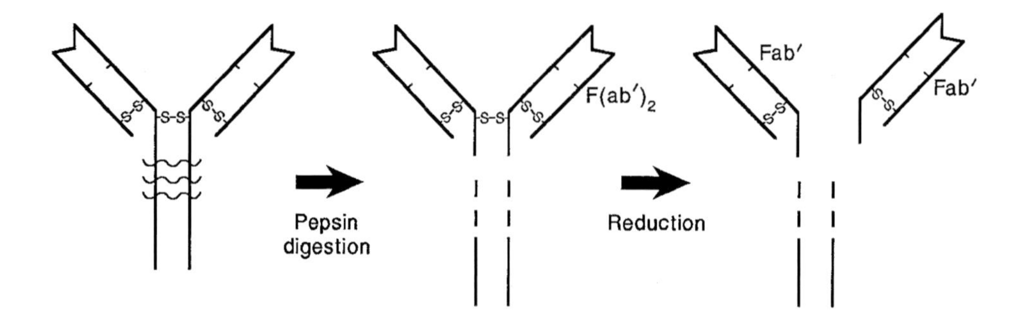

FIGURE 7.42 Pepsin digestion of an IgG molecule leading to the formation of F(ab′)$_2$ intermediate fragments and with reduction to the formation of Fab′ fragments.

and agglutination with an Fc remnant still capable of fixing complement.

Fb fragment (Figure 7.45) is the product of subtilisin digestion. It is comprised of the Fab fragments C_H1 and C_L (constant) domains.

An **Fv fragment** (Figure 7.46) consists of the N-terminal variable segments of both heavy (V_H) and light (V_L) chain domains that are joined by noncovalent forces. The fragment has one antigen-binding site.

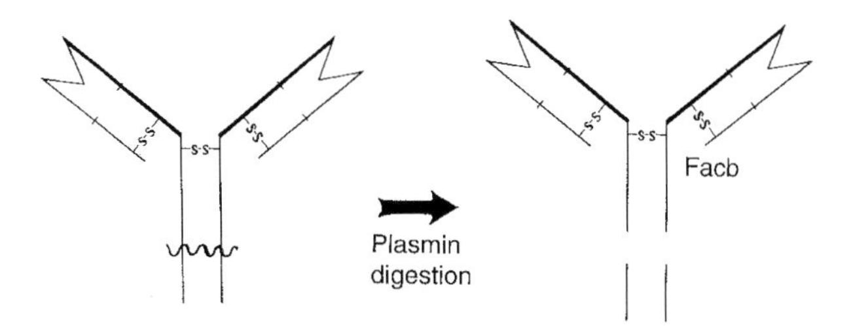

FIGURE 7.44 Generation of Facb fragment through plasmin digestion of an IgG molecule.

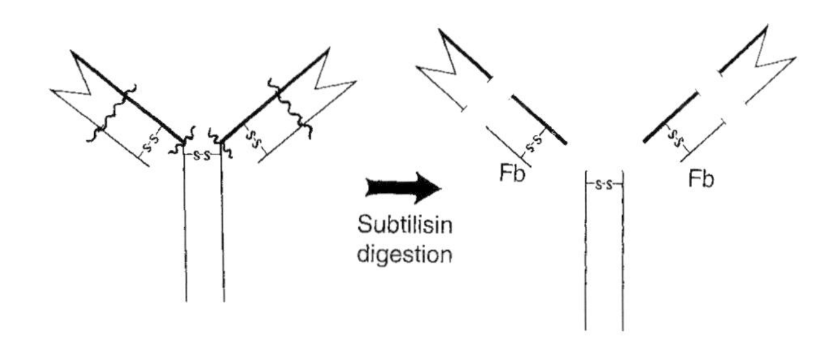

FIGURE 7.45 Formation of Fb fragments by digestion of IgG molecules with subtilisin.

Single-chain Fv fragment is a genetically engineered structure consisting of a heavy chain V region linked by a stretch of synthetic peptide to a light chain V region.

Single domain antibodies are antibodies capable of binding epitopes with high affinity even though they do not possess light chains. They are cloned from heavy chain variable regions and can be produced in days to weeks in contrast to monoclonal antibodies which require weeks to months to develop. Their relatively small size is a further advantage. Single domain antibodies with antigen-specific V_H domains are expected to find wide application in the future.

Isotype (Figure 7.47) refers to the antigens that determine the class or subclass of heavy chains or the type and subtype of light chains of immunoglobulin molecules. Every normal member of a species expresses each isotype. An immuno-

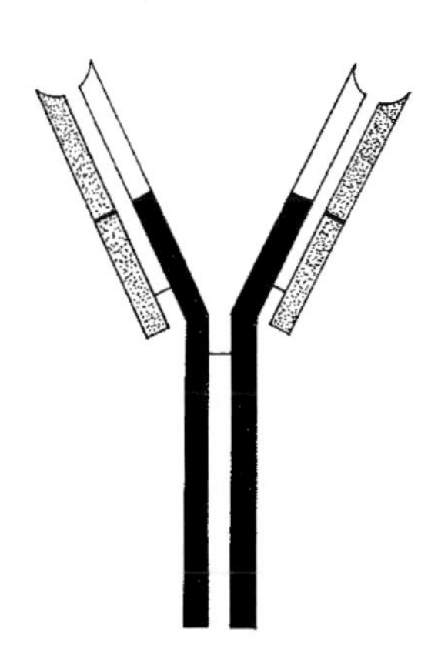

FIGURE 7.47 IgG showing that the isotype, designated in black, is determined by the heavy chain.

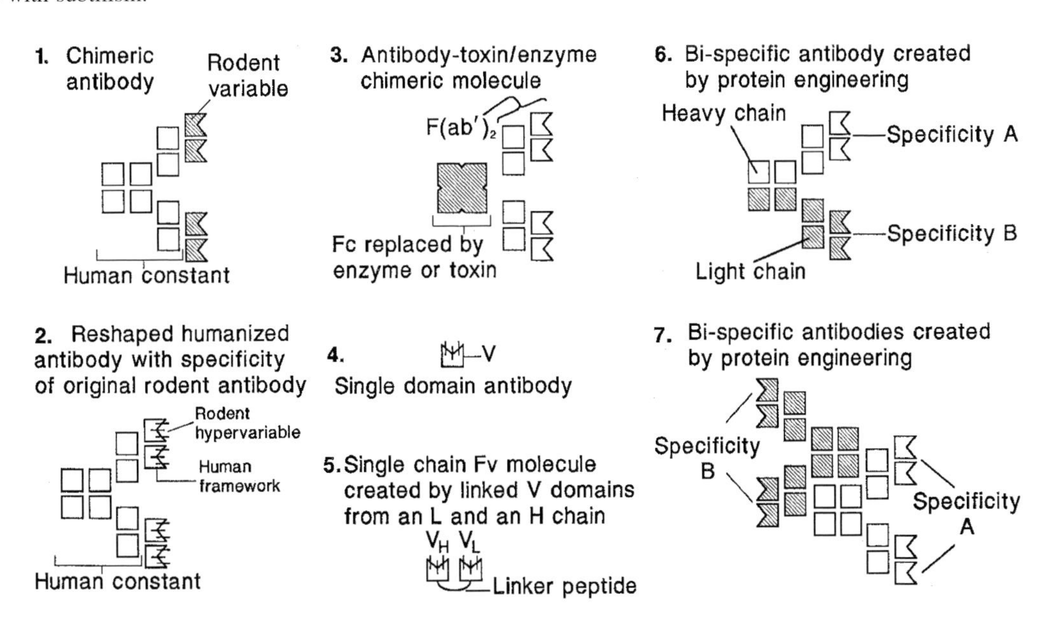

FIGURE 7.46 Genetically engineered antibodies.

globulin subtype is found in all normal individuals. Among the immunoglobulin classes, IgG and IgA have subclasses that are designated with Arabic numerals. They are distinguished according to domain number and size, as well as the number of both intrachain and interchain disulfide bonds of the constant region. The four isotypes of IgG are designated IgG1, IgG2, IgG3, and IgG4. The two IgA isotypes are designated IgA1 and IgA2. The μ, δ, and ε heavy chains and the κ and λ light chains each have one isotype. Immunoglobulin isotypes are responsible for the biological effector functions of an antibody molecule.

An **isotypic determinant** is an immunoglobulin epitope present in all normal individuals of a species. Isotypic determinants of immunoglobulin heavy and light chains determine immunoglobulin class and subclass and light chain type.

Oz isotypic determinant: Oz represents an isotypic marker. λ Light chains of human immunoglobulin that are Oz^+ contain lysine at position 190, whereas those that are Oz^- contain arginine at this position. A fraction of each person's λ light chains express Oz determinants.

Isotypic specificities refer to species-specific variability of antibody (human, mouse, rabbit, etc.). Examples of isotypes include IgG, IgM, and κ light chains.

An **isotope** is an isotypic determinant or epitope of an isotype.

Isotypic variation refers to differences among antigens found in members of a species such as the epitopes that differentiate immunoglobulin classes and subclasses and light chain types among immunoglobulin chains.

An **isoallotypic determinant** is an antigenic determinant present as an allelic variant on one immunoglobulin class or immunoglobulin subclass heavy chain that occurs on every molecule of a different immunoglobulin class or subclass heavy chain.

An **isoagglutinin** is an alloantibody present in some individuals of a species that is capable of agglutinating cells of other members of the same species.

An **allotype** (Figure 7.48) is a distinct antigenic form of a serum protein that results from allelic variations present on the immunoglobulin heavy-chain constant region. Allotypes were originally defined by antisera which differentiated allelic variants of Ig subclasses. The allotype is due to the existence of different alleles at the genetic locus, which determines the expression of a given determinant. Immunoglobulin allotypes have been extensively investigated in inbred rabbits. Currently, allotypes are usually defined by DNA techniques. To be designated as an official allotype, the polymorphism must be present in a reasonable subset of the population (approximately 1%) and follow Mendelian genetics. Allotype examples include the IgG3 Caucasian allotypes $G3m^b$ and $G3m^g$. These two alleles vary at positions 291, 296, and 384. Another example is the allotype at the IgA2 locus. The IgA2m(1) allele is European/Near Eastern, while IgA2m(2) is African/East Asian. The allotypic differences are in Ca_1 and Ca_3, and the IgA2m(2) allele has a shorter hinge than the IgA2m(1) allele. An **allotope** is an allotype's antigenic determinant Allotypic differences in immunoglobulin molecules have been important in solving the genetics of antibodies.

Latent allotype refers to the detection of an unexpected allotype in an animal's genetic constitution. This latent allotype is expressed as a temporary replacement of the nominal heterozygous allotype of the animal. This has been described among rabbit immunoglobulin allotypes. Whereas an F_1 rabbit would be expected to synthesize a1 or a2 heavy chain allotypes if one of its parents was homozygous for a1 and its other parent was homozygous for a2, the F_1 rabbit might, under certain circumstances of stimulation, produce immunoglobulin molecules of the a3 allotype, which would represent a latent allotype.

A 20-amino acid sequence situated at the N-terminus of free heavy and light polypeptide chains but not present in secreted immunoglobulins. Once the light and heavy polypeptide chains reach the cisternal space of the endoplasmic reticulum, the peptide is split from the polypeptide chains. It is thought to facilitate vectorial release of the chains and their secretion.

A **simple allotype** is an allotype that is different from another allotype in the sequence of amino acids at a single position or several positions. Alleles at one genetic locus often encode simple allotypes.

An **alloantibody** is an antibody that interacts with an alloantigen, such as the antibodies generated in the recipient of an organ allotransplant (such as kidney or heart), which then may react with the homologous alloantigen of the allograft. Alloantibodies interact with antigens that result from allelic variation at polymorphic genes. Examples include those that recognize blood group antigens and HLA class I and class II molecules.

Allogroup refers to several allotypes representing various immunoglobulin classes and subclasses inherited as a unit. Alleles that are closely linked encode the immunoglobulin heavy chains in an allogroup. An allogroup is a form of a haplotype.

B allotype is a rabbit immunoglobulin κ light chain allotype encoded by alleles at the K1 locus.

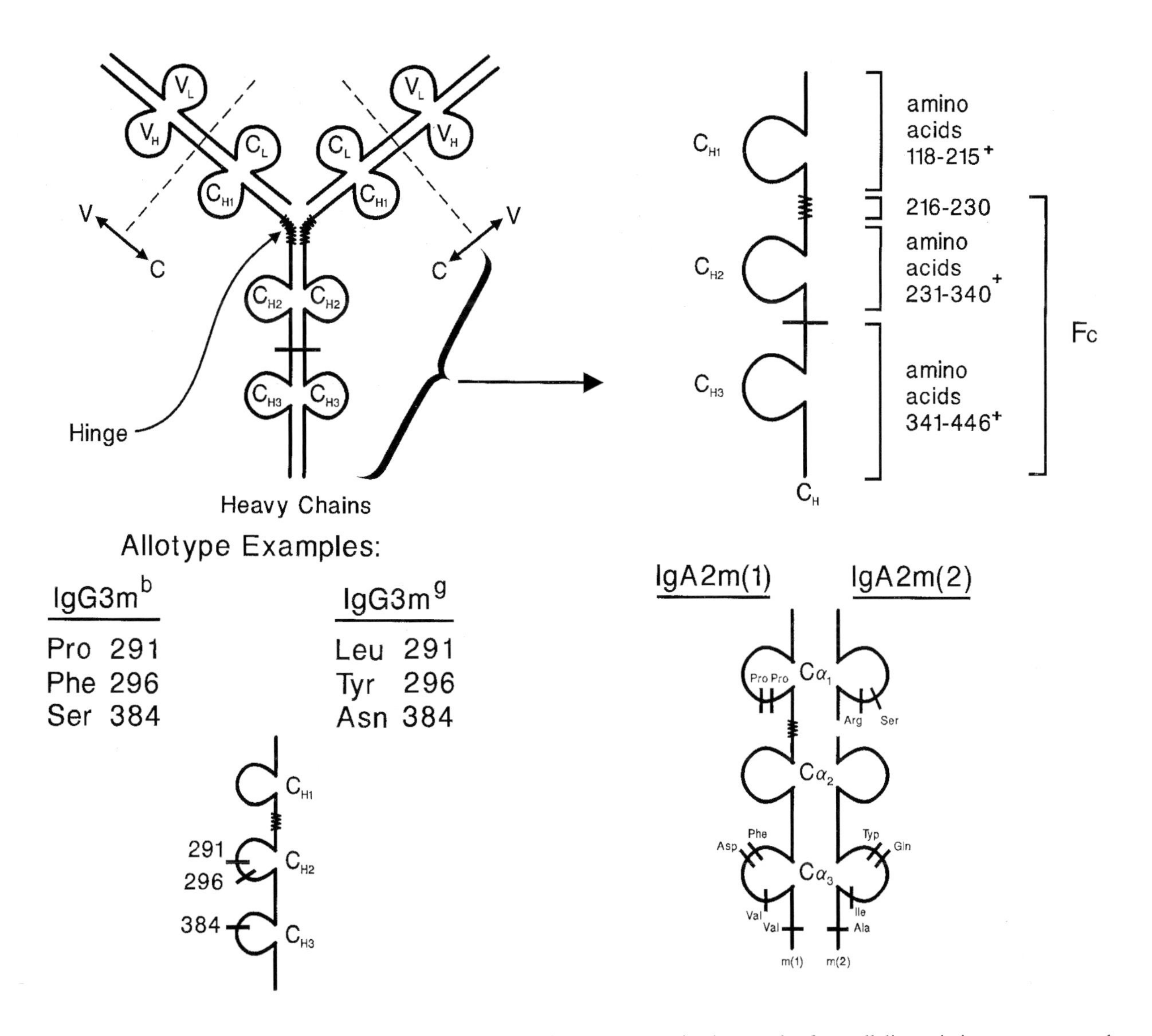

FIGURE 7.48 Allotype describes a distinct antigenic form of a serum protein that results from allelic variations present on the immunoglobulin heavy chain constant region.

A **complex allotype** is an allotype with multiple amino acid residue positions that are not the same as those of a different allotype at that same locus.

Allotype suppression is the failure to produce antibodies of a given allotype by offspring of mothers who have been immunized against the paternal Ig allotype before mating. It may also occur after administration of antibodies specific for the paternal allotype to heterozygous newborn rabbits. Animals conditioned in this way remain deficient for the suppressed allotype for months and possibly years thereafter. The animal switches to a compensatory increase in the production of the nonsuppressed allotype. In homozygotes, suppression involves heavy chain switching from one Ig class to another.

An **allotypic determinant** is an epitope on serum immunoglobulin molecules present in selected members of a species. Allotypic determinants are present in addition to other markers characteristic of the molecule as a whole for the respective species. The allotypic determinants or allotopes are characteristic of a given class and subclass of immunoglobulin and have been demonstrated in several species. The inheritance of genes controlling these determinants is strictly Mendelian and is not sex linked. Both H and κ L chains contain such determinants, and the encoding genes are not linked and are codominant in an individual. This means that markers present both in the father and mother are expressed phenotypically in a heterozygote. However, an individual immunoglobulin-producing cell expresses only one of a pair of allelic genes, since a single cell can utilize only one parental chromosome. In an individual, some cells use the information encoded on the chromosome derived paternally and other cells use that derived from the maternal contribution.

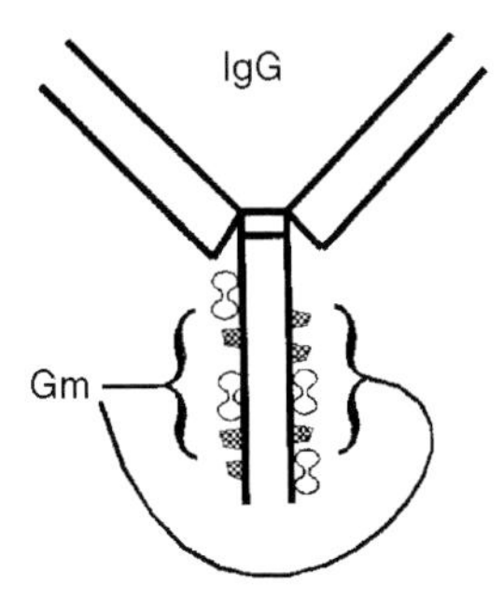

FIGURE 7.49 Illustration of the location of Gm marker specificities on the Fc region of an IgG molecule.

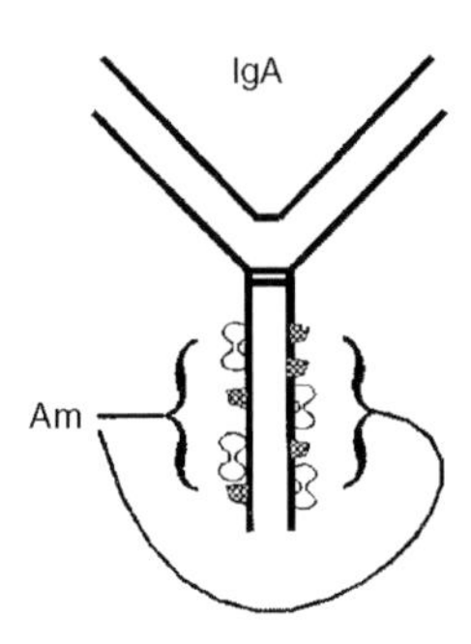

FIGURE 7.50 Illustration of the location of Am marker specificities on the Fc region of a serum IgA molecule.

So the individual is heterozygous, but a single cell only secretes products of one allele because of allelic exclusion. The allotypic markers of human IgG are designated Gm determinants. The Km(Inv) markers are characteristic for the C region of κ light chains.

An **allotope** is an epitope or antigenic determinant encoded by a single allelic form of a polymorphic gene within species. An allotope often designates an antigenic determinant on an antibody molecule that is encoded by C gene alleles or framework regions of V genes. The antigenic determinant of an allotype, also called allotypic determinant.

Allotypic marker: See allotypic determinant or allotope.

Allotypic specificities are genetically different antibody classes and subclasses produced within individuals of the same species. They are detected as changes in the amino acid residues present in specific positions in various polypeptide chains. The allotypic specificities are also called genetic markers. Gm and Km(Inv) markers are examples of allotypes of human IgG H chains and κ light chains, respectively.

Allotypy is a term that describes the various allelic types or allotypes of immunoglobulin molecules.

Gm allotype (Figure 7.49) refers to a genetic variant determinant of the human IgG heavy chain. Allelic genes that encode the γ-1, γ-2, and γ-3 heavy chain constant regions encode the Gm allotypes. They were recognized by the ability of blood sera from rheumatoid arthritis patients, which contain anti-IgG rheumatoid factor, to react with them. Gm allotypic determinants are associated with specific amino acid substitution in different γ chain constant regions in man. IgG subclasses are associated with certain Gm determinants. For example, IgG is associated with Glm(1) and G1m(4), and IgG3 is associated with G3m(5). Although the great majority of Gm allotypes are restricted to the IgG γ chain Fc region, a substitution at position 214 of CH1 of the arginine yields the G1m(4) allotype and a substitution at this same site of lysine yields

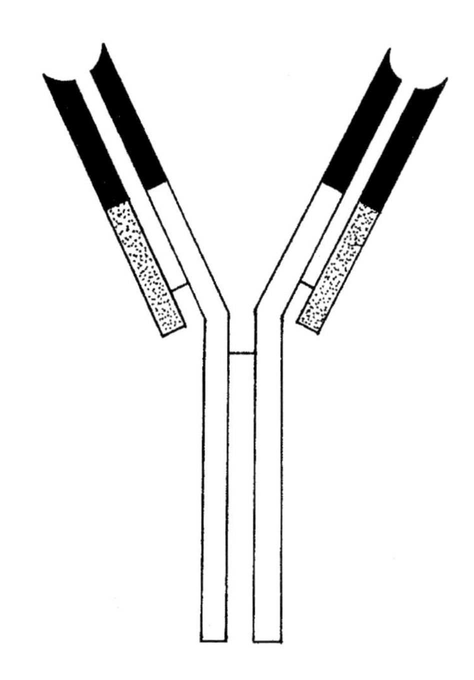

FIGURE 7.51 Idiotype which is determined by the variable regions of heavy and light chains of an immunoglobulin molecule.

G1m(17). For Gm expression, the light chain part of the molecule must be intact.

Gm marker: See Gm allotype.

An **Am allotypic marker** (Figure 7.50) is an allotypic antigenic determinant located on a heavy chain of the IgA molecule in man. Of the two IgA subclasses, the IgA1 subclass has no known allotypic determinant. The IgA2 subclass has two allotypic determinants designated A2m(1) and A2m(2) based on differences in α-2 heavy chain primary structures. Allelic genes at the A2m locus encode these allotypes which are expressed on the α-2 heavy chain constant regions.

Idiotype (Figure 7.51) refers to that segment of an immunoglobulin or antibody molecule that determines its specificity for antigen and is based on the multiple combinations of variable (V), diversity (D), and joining (J)

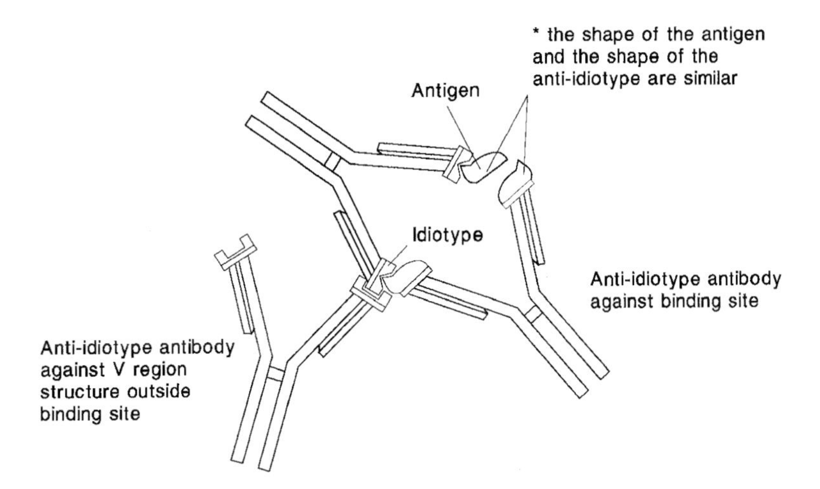

FIGURE 7.52 Idiotype network.

exons. The idiotype is located in the Fab region, and its expression usually requires participation of the variable regions of both heavy and light chains, namely the Fv fragment which contains the antigen-combining site. The antigen-binding specificity of the combining site may imply that all antibodies produced by an animal in response to a given immunogen have the same idiotype. This is not true since the antibody response is heterogeneous. There will usually be a major idiotype representing 20 to 70% of the specific antibody response. The remainder carry different idiotypes that may cross-react with the major idiotype. Cross-reacting idiotypy represents the extent of heterogeneity among the antibodies of a given specificity.

The unique antigenic determinants that govern the idiotype (Id) of an immunoglobulin molecule occur on the products of either a single or several clones of cells synthesizing immunoglobulins (Figure 7.52). This unique idiotypic determinant is sometimes called a private idiotype that appears in all V regions of immunoglobulin molecules whose amino acid sequences are the same. Shared idiotypes are also known as public idiotype determinants. These may appear in a relatively large number of immunoglobulin molecules produced by inbred strains of mice or other genetically identical animals in response to a specific antigen. The localization of idiotypes in the antigen-binding site of the V region of the molecule is illustrated by the ability of haptens to block or inhibit the interaction of antiidiotypic antibodies with their homologous antigenic markers or determinants in the antigen-binding region of antibody molecules.

Idiotype network is the interaction of idiotypes and anti-idiotypes that lead to the regulation of antibody synthesis or of lymphoid cells bearing receptors that express these idiotypic specificities. Response of one or a few lymphocyte clones to antigen leads to idiotype expansion and antiidiotype responses that downregulate the antigen-specific response.

Idiotype network theory: See network theory.

Idiotypic determinant: See epitope.

Private idiotypic determinant is a determinant produced by a particular amino acid sequence in the immunoglobulin heavy or light chain hypervariable region of an antibody synthesized by only one individual.

Public idiotypic determinant (IdX or CRI) is an idiotypic determinant present on antibody molecules in numerous individuals of one species. It may occur on antibodies with a single antigen specificity, but may be present on antibodies with separate specificities as well. The terms IdX or CRI signify that these are cross-reacting idiotypic determinants. These determinants are manifestations of immunoglobulin heavy and light chain amino acid sequence similarities.

An **idiotope** is an epitope or antigenic determinant in the hypervariable region of the N-terminus of an immunoglobulin molecule or T cell receptor molecule. An idiotope is an epitope or antigenic determinant of an idiotype on an antibody molecule's V region. This type of antigenic determinant is present on immunoglobulin molecules synthesized by one clone or a few clones of antibody-producing cells.

Idiotype suppression is the inhibition of idiotype antibody production by suppressor T lymphocytes activated by antiidiotype antibodies.

Idiotypic specificity refers to characteristic folding of the antigen-binding site, thereby exposing various groups which by their arrangement confer antigenic properties to the immunoglobulin V region itself. Antibodies against a specific antigen usually carry both a few predominant idiotypes and other similar, but not identical, cross-reacting

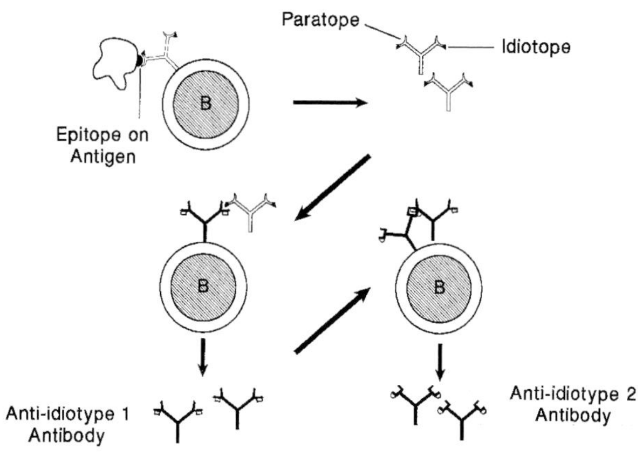

FIGURE 7.53 Jerne network theory.

idiotypes. The proportion between the two indicates the degree of heterogeneity of the antibody response in a given individual.

Jerne network theory (Figure 7.53) refers to Niels Jerne's hypothesis that antibodies produced in response to a specific antigen would themselves induce a second group of antibodies which would in turn downregulate the original antibody-producing cells. The second antigen (Ab-2) would recognize epitopes of the antibody-binding region of antibody. These would be antiidiotypic antibodies. Such antiidiotypic antibodies would also be reactive with the antigen-binding region of T cell receptors for which they were specific. Thus, a network of antiantibodies would produce a homeostatic effect on the immune response to a particular antigen. This theory was subsequently proven and confirmed by numerous investigators.

The **network theory** is a hypothesis proposed by Niels Jerne which explains immunoregulation through a network of idiotype–antiidiotype reactions involving T cell receptors and the antigen-binding regions of antibody molecules, i.e., the paratope regions. Exposure to antigens interrupts the delicate balance of the idiotype–antiidiotype network, leading to the increased synthesis of some idiotypes as well as of the corresponding antiidiotypes, leading to modulation of the response. Antiidiotypes occur following immunization against selected antigens and may prevent the response to the antigen. Selected antiidiotypic antibodies have a binding site that is closely similar to the immunizing epitope. It is referred to as the internal image of the epitope. Other antiidiotypic antibodies are directed to idiotopes of the antigen-binding region and are not internal images. Antiidiotypic antibodies with an internal image may be substituted for an antigen, leading to specific antigen-binding antibodies. These are the basis for so-called idiotypic vaccines in which the individual never has to be exposed to the infecting agent. Antiidiotypes may also block T cell receptors for the corresponding antigen.

The **network hypothesis** is Niels Jerne's theory that antiidiotypic antibodies form in response to the antigen-binding regions of antibody molecules or of lymphocyte surface receptors. These in turn elicit anti-antiidiotypic antibodies, etc. Each new immune response stimulated in this network interrupts the finely tuned immune network balance as antiidiotype antibodies are produced, eventually downregulating the response and bringing it back to homeostasis.

The **immune network hypothesis of Jerne:** The antigen-binding sites of antibody molecules (paratopes), which are encoded by variable region genes, have idiotopes as phenotypic markers. Each paratope recognizes idiotopes on a different antibody molecule. Interaction of idiotypes with antiidiotypes is physiologic idiotypy and is shared among Ig classes. This comprises antibodies produced in response to the same antigen. The idiotypic network consists of the interaction of idiotypes involving free molecules as well as B and T lymphocyte receptors. Idiotypes are considered central in immunoregulation involving autoantigens.

Internal image: According to the Jerne network theory, antibodies are produced against the antibodies induced by an external antigen. Some of the antiantibodies produced will bear idiotopes that precisely fit the paratope or antigen-binding site of the original antibody against the external antigen. Since they bear close structural similarity to the epitope on the antigen molecule that was originally administered, they are termed the internal image of the antigen.

Autobody (Figure 7.54) refers to an antibody that exhibits the internal image of an antigen as well as a binding site for an antigen. It manifests dual binding to both idiotope and epitope. It bears an idiotope that is complementary to its own antigen-binding site or paratope. Thus, it has self-binding potential. This type of antiidiotypic antibody has features of Ab1 and Ab2 on the same molecule, causing it to be designated Ab1-2 or *autobody*. The name points to the potential for self-aggregation of the molecules and the potential participation of autobodies in autoimmune

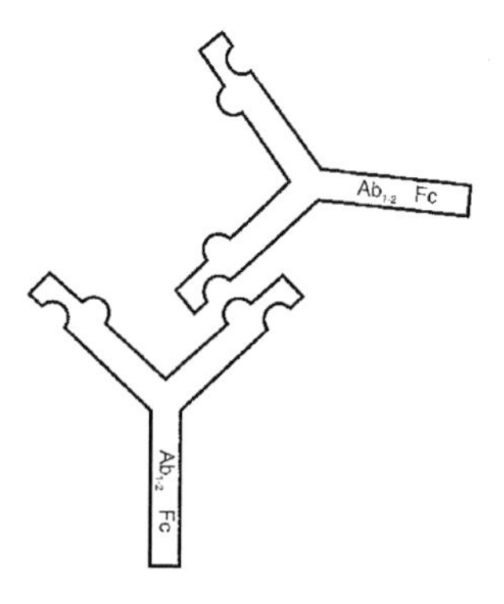

FIGURE 7.54 Autobody.

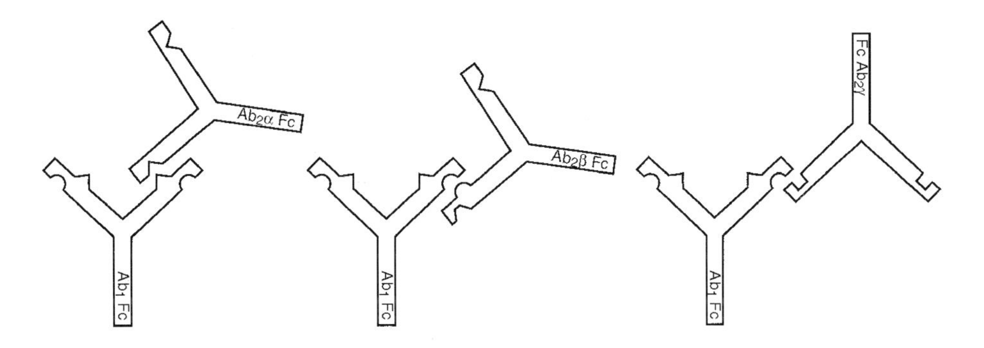

FIGURE 7.55 Homobody

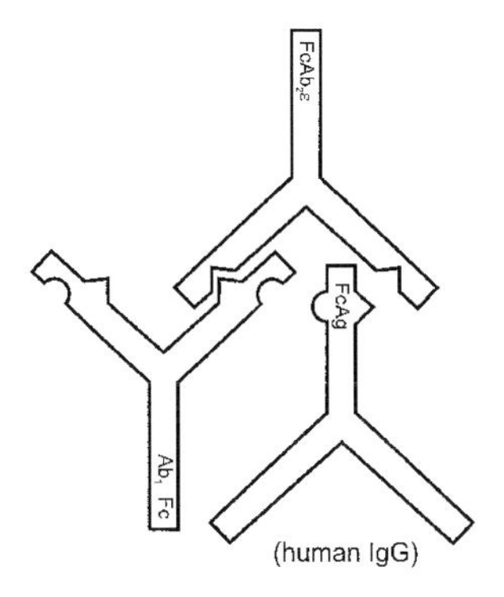

FIGURE 7.56 Epibody.

phenomena. Antibodies to phosphorylcholine (PC) epitope raised in Balb/c mice expressing the T15 idiotype self aggregate bind to one another.

Homobody (Figure 7.55) designates an idiotypic determinant of an antibody molecule whose three-dimensional structure resembles antigen. Also called internal image of antigen. For example, antiidiotypic antibodies of this type to insulin receptor may partially mimic the action of insulin.

Epibody (Figure 7.56) refers to an antiidiotypic antibody reactive with an idiotype of a monoclonal, human anti-IgG autoantibody as well as with human IgG Fc region. These antibodies identify an antigenic determinant associated with the sequence Ser-Ser-Ser. The ability of an epibody to identify an epitope shared by a rheumatoid factor idiotope and an Fcγ epitope demonstrates that this variety of antiidiotypic antibody may function as a rheumatoid factor.

The **Ehrlich side chain theory (historical)** was the first selective theory of antibody synthesis developed by Paul Ehrlich in 1900. Although elaborate in detail, the essential feature of the theory was that cells of the immune system possess the genetic capability to react to all known antigens and that each cell on the surface bears receptors with surface haptophore side chains. On combination with antigen, the side chains would be cast off into the circulation and new receptors would replace the old ones. These cast-off receptors represented antibody molecules in the circulation. Although far more complex than this explanation, the importance of the theory was in the amount of research stimulated to try to disprove it. Nevertheless, it was the first effort to account for the importance of genetics in immune responsiveness at a time when Mendel's basic studies had not even yet been "rediscovered" by De Vries.

The **side chain theory** is a concept proposed by Paul Ehrlich in 1899 which postulated that a cell possessed highly complex chemical aggregates with attached groupings, or "side chains," whose normal function was to anchor nutrient substances to the cell prior to internalization. These side chains, or receptors, were considered to permit cellular interaction with substances in the extracellular environment. Antigens were postulated to stimulate the cell by attachment to these receptors. Since antigens played no part in the normal economy of the cell, the receptors were diverted from their normal function. Stimulated by this derangement of its normal mechanism, the cell produced excessive new receptors, of the same type as those thrown out of action. The superfluous receptors were shed into the extracellular fluids and constituted specific antibodies with the capacity to bind homologous antigens. Ehrlich proposed a haptophore group that reacted with a corresponding group of an antigen, as in neutralization of toxin by antitoxin. For reactions such as agglutination or precipitation, he postulated another group termed the ergophobe group, which determined the change in antigen after the antibody was anchored by its haptophore group. Ehrlich proposed receptors with two haptophore groups: one that attached to antigen and the other to complement. The group that combined with the cell or other antigen was called the cytophilic group, and the group that combined with complement was the complementophilic group. Ehrlich named this type of receptor an amboceptor because both groups were supposed to be of the haptophore type. He considered toxins to have a haptophore group and a toxophore group. Detoxification without loss of antitoxin-binding capacity led Ehrlich to believe that a toxophore group had been altered while the haptophore group remained intact. Ehrlich's theory assumed that each antibody-forming cell had the ability to react to every antigen in nature. The demonstration by Landsteiner that antibodies could be formed against substances manufactured in the laboratory that had never existed before in nature led to abandonment of the side chain theory. Yet its basic premise as a selective hypothesis rather than an instructive theory was ultimately proven correct.

The **indirect template theory (historical)** was a variation of the template hypothesis which postulated that instructions for antibody synthesis were copied from the antigen configuration into the DNA encoding the specific antibody. This was later shown to be untenable and is of historical interest only.

The **self marker hypothesis (historical)** was a concept suggested by Burnet and Fenner in 1949 in an attempt to account for the failure of the body to react against its own antigens. They proposed that cells of the body contained a marker that identified them to the immunologically competent cells of the host as self. This recognition system was supposed to prevent the immune cells of the host from rejecting its own tissue cells. This hypothesis was later abandoned by the authors and replaced by the clonal selection theory of acquired immunity which Burnet proposed in 1957.

The **template theory (historical)** was an instructive theory of antibody formation which requires that the antigen must be present during the process of antibody synthesis. According to the refolding template theory, uncommitted and specific globulins could become refolded on the antigen, serving as a template for it. The cell thereupon releases the complementary antibodies, which thenceforth rigidly retain their shape through disulfide bonding. This theory had to be abandoned when it became clear that the specificity of antibodies in all cases is due to the particular arrangement of their primary amino acid sequence. The template theory could not explain immunological tolerance or the anamnestic (memory) immune response.

X cell: See XYZ cell theory.

The **XYZ cell theory (historical)** was an earlier concept of antibody synthesis which proposed (1) an immunocompetent "X cell" that had not participated previously in a specific immune response, (2) a "Y cell" activated immunologically by "X cell" interaction with antigen, and (3) a "Z cell" that synthesized antibodies following second contact with antigen.

Y cell: See XYZ cell theory.

Z cell: See XYZ cell theory.

The **instructional model** was a theory of antibody diversity that postulates antigen to serve as a template for the antibody, which assumes a complementary shape.

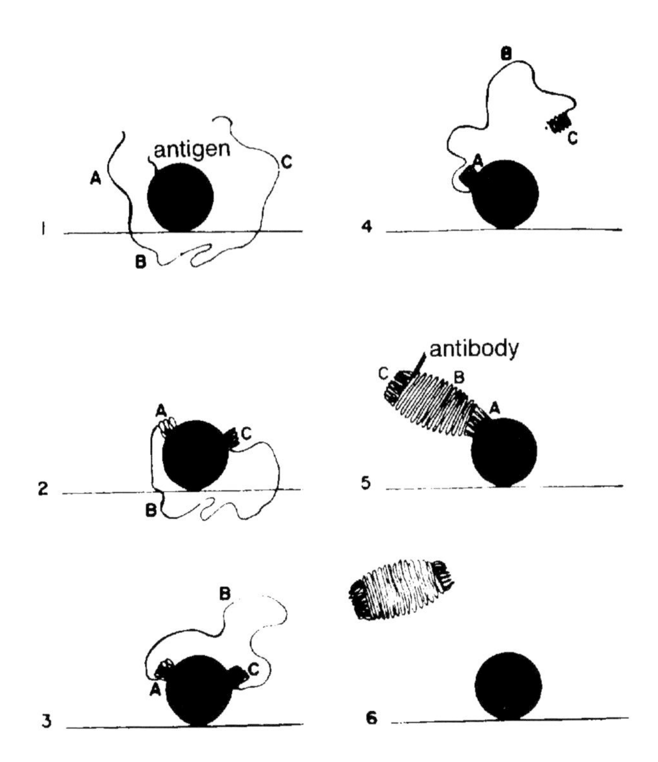

FIGURE 7.57 Instructive theory of antibody formation.

The **instructive theory (of antibody formation)** was a hypothesis that postulated acquisition of antibody specificity after contact with a specific antigen (Figure 7.57). According to one template theory of antibody formation, it was necessary that the antigen be present during the process of antibody synthesis. According to the refolding template theory, uncommitted and specific globulins could become refolded upon the antigen, serving as a template for it. The cell released the complementary antibodies, which rigidly retained their shape through disulfide bonding. This theory had to be abandoned when it was shown that the specificity of antibodies in all cases is due to the particular arrangement of their primary amino acid sequence. *De novo* synthesis template theories that recognized the necessity for antibodies to be synthesized by amino acids, in the proper and predetermined order, still had to contend with the serious objection that proteins cannot serve as informational models for the synthesis of proteins. Instructive theories were abandoned when immunologic tolerance was demonstrated and when antigen was shown not to be necessary for antibody synthesis to occur. The template theories had never explained the anamnestic (memory) immune response. Antibody specificity depends on the variable region amino acid sequence, especially the complementarity-determining or hypervariable regions.

The **selective theory** is a hypothesis that describes antibody synthesis as a process in which antigen selects cells expressing receptors specific for that antigen. The antigen–cell

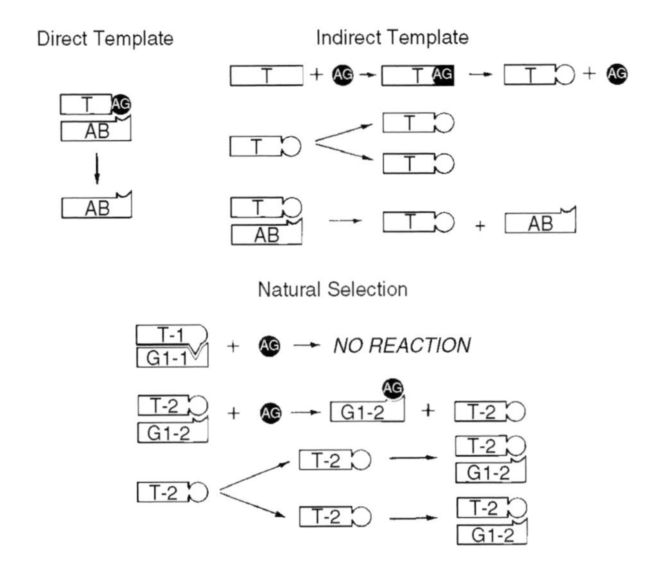

FIGURE 7.58 Comparison of template with natural selection theories of antibody synthesis.

receptor interaction leads to proliferation and differentiation of that clone of cells which synthesizes significant quantities of antibodies of a single specificity. Selective theories included the side chain theory of Paul Ehrlich proposed in 1899, the natural selection theory proposed by Niels Jerne in 1955, and the cell selection theory proposed by Talmage and by Burnet in 1957. Burnet termed his version of the theory the clonal selection theory of acquired immunity. The basic tenets of the clonal selection theory have been substantiated by the scientific evidence. The selective theories maintained that cells are genetically programmed to react to certain antigenic specificities prior to antigen exposure. They are in sharp contrast to the instructive theories which postulated that antigen was necessary to serve as a template around which polypeptide chains were folded to yield specific antibodies. This template theory was abandoned when antibody was demonstrated in the absence of antigen (Figure 7.58).

Endoplasmic reticulum is a structure in the cytoplasm comprised of parallel membranes that are connected to the nuclear membranes. Lipids and selected proteins are synthesized in this organelle. The membrane is continuous and convoluted. Electron microscopy reveals rough endoplasmic reticulum, which contains ribosomes on the side exposed to the cytoplasm, and smooth endoplasmic reticulum without ribosomes. Fatty acids and phospholipids are synthesized and metabolized in smooth endoplasmic reticulum. Selected membrane and organelle proteins, as well as secreted proteins, are synthesized in the rough endoplasmic reticulum. Cells such as plasma cells that produce antibodies or other specialized secretory proteins have abundant rough endoplasmic reticulum in the cytoplasm. Following formation, proteins move from the rough endoplasmic reticulum to the Golgi complex. They may be transported in vesicles that form from the endoplasmic reticulum and fuse with Golgi complex membranes. Once secreted proteins reach the endoplasmic reticulum lumen, they do not have to cross any further barriers prior to exit from the cell.

The **recombinatorial germ-line theory** was a hypothesis proposed by Dreyer and Bennett which postulated that variable-region and constant-region immunoglobulin genes were separated and rejoined in DNA levels. This concept was an important step toward understanding the generation of diversity in the production of antibody molecules, a puzzle finally solved by Tonegawa.

The **unitarian hypothesis** was the view that one type of antibody produced in response to an injection of antigen could induce agglutination, complement fixation, precipitation, and lysis based on the type of ligand with which it interacted. This view was in contrast to the earlier belief that separate antibodies accounted for every type of serological reactivity described above. Usually, more than one class of immunoglobulin may manifest a particular serological reactivity such as precipitation.

BiP is a chaperonin that binds unassembled heavy and light chains after they are synthesized in the endoplasmic reticulum. Chains that are malformed are not allowed to leave the ER and are thus not used in immunoglobulin assembly.

The **signal hypothesis** is a proposed mechanism for selection of secretory proteins by and for transport through the rough endoplasmic reticulum. The free heavy and light chain leader peptide is postulated to facilitate the joining of polyribosomes forming these molecules to the endoplasmic reticulum. It also refers to the release of heavy and light polypeptide chains through the endoplasmic reticulum membrane into the cisternal space followed by immunoglobulin secretion once the immunoglobulin molecule has been assembled. This refers to the pathways whereby proteins reach their proper cellular destinations. Important to protein targeting is the signal sequence, which is a short amino acid sequence at the amino terminus of a polypeptide chain. This signal sequence directs the protein to the proper destination in the cell and is removed either during passage or when the protein arrives at its final location.

Signal sequence: See signal hypothesis.

Chaperones are a group of proteins that includes BiP, a protein that binds the immunoglobulin heavy chain. Chaperones aid the proper folding of oligomeric protein complexes. They prevent incorrect conformations or enhance correct ones. Chaperones are believed to combine with the surfaces of proteins exposed during intermediate folding and to restrict further folding to the correct conformations. They take part in transmembrane targeting of selected proteins. Chaperones hold some proteins that are to be inserted into membranes in intermediate conformation in the cytoplasm until they interact with the target membrane. Besides BiP, they include heat shock protein 70 and 90 and nucleoplasmins.

Somatic hypermutation is the induced increase in frequency of mutation in rearranged variable-region DNA of immunoglobulin genes in activated B cells. This leads to the synthesis of variant antibodies some of which have a higher affinity for antigen. Somatic mutation may occur in germinal centers. Only somatic cells are affected. Somatic hypermutation is not inherited through the germline transmission.

Somatic mutation is a genetic variation in a somatic cell which is heritable by its progeny. Increased somatic mutation enhances diversity of an antibody molecule's light and heavy chain variable regions. IgG and IgA antibodies reveal somatic mutations more often than do IgM antibodies. It is a mechanism whereby point mutations are introduced into the rearranged immunoglobulin variable-region genes during B lymphocytes activation and proliferation.

Switch refers to the change within an immunologically competent B lymphocyte from synthesizing one isotype of heavy polypeptide chain, such as μ, to another isotype, such as γ, during differentiation. The switch signal comes from T cells. Isotype switching does not alter the antigen-binding variable region of the chain at the N-terminus.

Switch cells are a subset of T lymphocytes that governs isotype differentiation of B lymphocytes exiting the Peyer's patches to ensure that they become IgA-producing plasma cells when they home back to the lamina propria of the intestine from the systemic circulation.

Leader sequence: See leader peptide.

Switch defect disease: See hyperimmunoglobulin M syndrome.

A **transfectoma** is comprised of antibody-synthesizing cells that are generated by introducing antibody genes that have been genetically engineered into myeloma cells using genomic DNA.

The **switch region** is the amino acid sequence between the constant and variable portions of light and heavy polypeptide immunoglobulin chains. This amino acid segment is encoded by *D* and *J* genes. This segment of DNA controls recombination associated with immunoglobulin class switching. Specific switch region sequences are critical for switching from one immunoglobulin isotype to another and from one class to another. During isotype switching, the active heavy chain V-region exon undergoes somatic recombination with a 3′ constant-region gene at a switch region of DNA.

Switch site refers to breakage points on a chromosome where gene segments unite during gene rearrangement. In immunology, it often refers to an abbreviated DNA sequence 5′ to each gene encoding a heavy chain C region. It serves as an identification site for V region gene translocation in the process of switching gene expression from one immunoglobulin heavy chain class to another. There are numerous switch sites for each gene encoding the C region.

Clonal selection refers to antigen-mediated activation and proliferation of members of a clone of B lymphocytes bearing receptors for the antigen or for complexes of MHC and peptides derived from the antigen in the case of T lymphocytes.

The **clonal selection theory** is a selective theory of antibody formation proposed by F.M. Burnet who postulated the presence of numerous antibody-forming cells, each capable of synthesizing its own predetermined antibody (Figure 7.59). One of the cells, after having been selected by the best-fitting antigen, multiplies and forms a clone of cells which continue to synthesize the same antibody. Considering the existence of many different cells, each capable of synthesizing an antibody of a different specificity, all known facts of antibody formation are easily accounted for. An important element of the clonal selection theory was the hypothesis that many cells with different antibody specificities arise through random somatic mutations during a period of hypermutability early in the animal's life. Also early in life, the "forbidden" clones of antibody-forming cells (i.e., the cells that make antibody to the animal's own antigen) are still destroyed after encountering these autoantigens. This process accounted for an animal's tolerance of its own antigens. Antigen would have no effect on most lymphoid cells, but it would selectively stimulate those cells already synthesizing the corresponding antibody at a low rate. The cell surface antibody would serve as receptor for antigen and proliferate into a clone of cells, producing antibody of that specificity. Burnet introduced the *forbidden clone* concept to explain autoimmunity. Cells capable of forming antibody against a normal self antigen were *forbidden* and eliminated during embryonic life. During fetal development, clones that react with self antigens are destroyed or suppressed. The subsequent activation of suppressed clones reactive with self antigens in later life may induce autoimmune disease. D.W. Talmage proposed a cell selection theory of antibody formation, which was the basis for Burnet's clonal selection theory.

Monoclonal antibody (MAb) is an antibody synthesized by a single clone of B lymphocytes or plasma cells

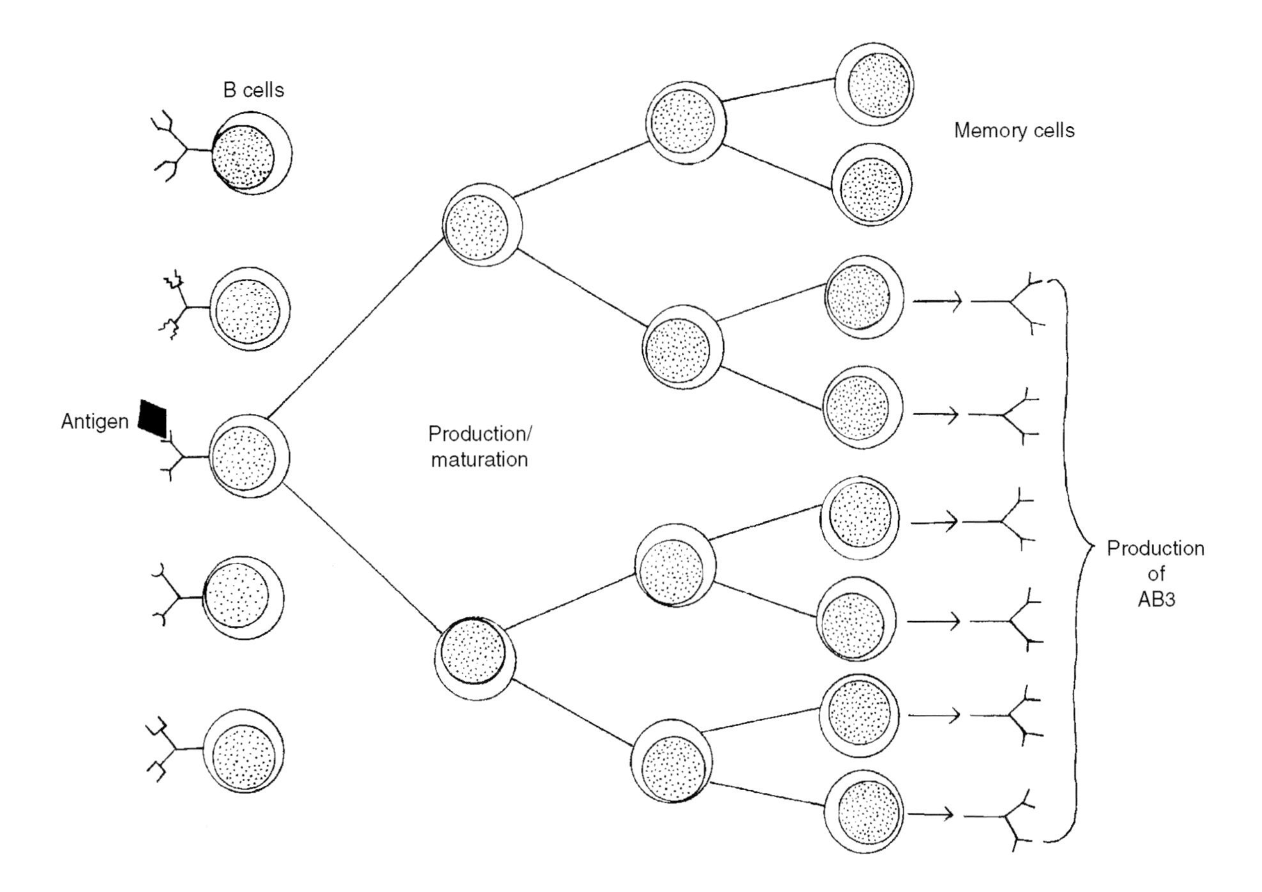

FIGURE 7.59 Clonal selection theory.

(Figure 7.60). The first to be observed were produced by malignant plasma cells in patients with multiple myeloma and associated gammopathies. The identical copies of the antibody molecules produced contain only one class of heavy chain and one type of light chain. Kohler and Millstein in the mid-1970s developed B lymphocyte hybridomas by fusing an antibody-producing B lymphocyte with a mutant myeloma cell that was not secreting antibody. The B lymphocyte product provided the specificity, whereas the myeloma cell conferred immortality on the hybridoma clone. Today monoclonal antibodies (MAb) are produced in large quantities against a plethora of antigens for use in diagnosis and sometimes in treatment. MAb are homogeneous and are widely employed in immunoassays, single antigen identification in mixtures, delineation of cell surface molecules, and assay of hormones and drugs in serum, among many other uses. Since the response to some immunogens is inadequate in mice, monoclonal antibodies have also been generated using rabbit cells. Monoclonal antibodies have been radioactively labeled and used to detect tumor metastases; to differentiate subtypes of tumors, with monoclonal antibodies against membrane antigens or intermediate filaments; to identify microbes in body fluids; and for circulating hormone assays. MAb may be used to direct immunotoxins or radioisotopes to tumor targets with potential for tumor therapy.

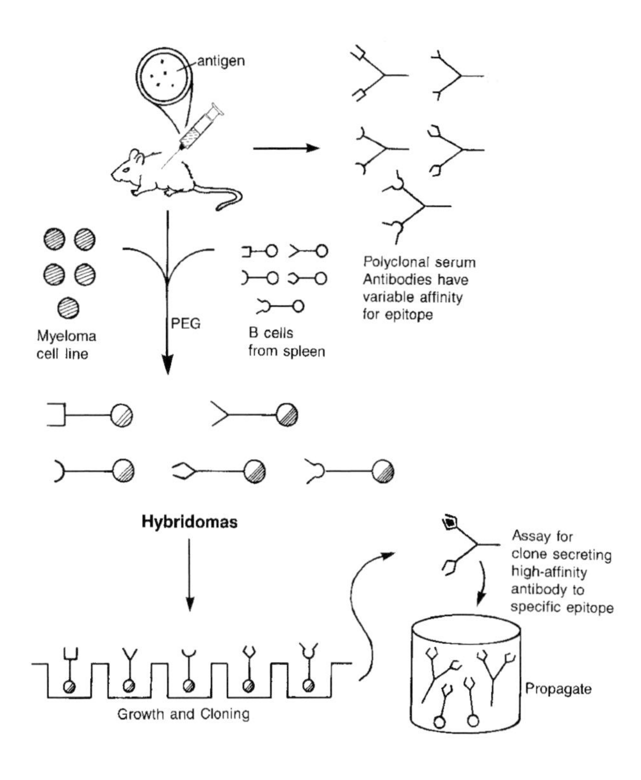

FIGURE 7.60 Monoclonal antibody.

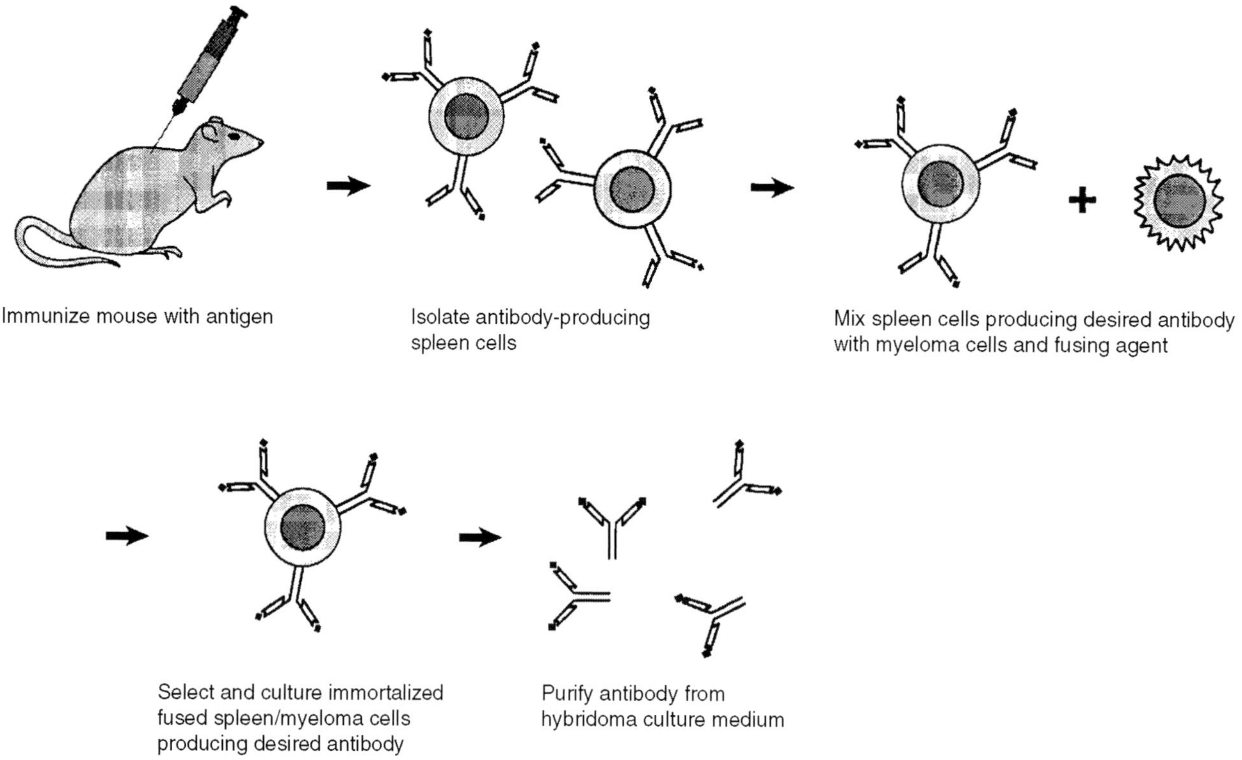

Immunize mouse with antigen

Isolate antibody-producing spleen cells

Mix spleen cells producing desired antibody with myeloma cells and fusing agent

Select and culture immortalized fused spleen/myeloma cells producing desired antibody

Purify antibody from hybridoma culture medium

FIGURE 7.61 Development of a hybridoma for monoclonal antibody production.

MAb is an abbreviation for monoclonal antibody.

Monoclonal is an adjective that refers to derivation from a single clone.

A **B lymphocyte hybridoma** is a hybrid cell produced by the fusion of a splenic antibody-secreting cell immunized against that particular antigen with a mutant myeloma cell from the same species that no longer secretes its own protein product (Figure 7.61 and Figure 7.62). Polyethylene glycol is used to effect cell fusion. Antibody-synthesizing cells do not secrete hypoxanthine guanine phosphoribosyl transferase (HGPRT), an enzyme needed for DNA nucleotide synthesis, but do provide the ability to produce a specific monoclonal antibody. The mutant myeloma cell line confers immortality upon the hybridoma. If the nucleotide synthesis pathway is inhibited, the myeloma cells become HGPRT-dependent. The antibody synthesizing cells provide the HGPRT and the mutant myeloma cell enables endless reproduction. Once isolated through use of a selective medium such as HAT, hybridoma cell lines can be maintained for relatively long periods. Hybridomas produce specific monoclonal antibodies that may be collected in great quantities for use in diagnosis and selected types of therapy (Figure 7.63 and Figure 7.64).

Polyclonal antibodies are multiple immunoglobulins responding to different epitopes on an antigen molecule. This multiple stimulation leads to the expansion of several antibody-forming clones whose products represent a mixture of immunoglobulins in contrast to proliferation of a single clone which would yield a homogeneous monoclonal antibody product. Thus, polyclonal antibodies represent the natural consequence of an immune response in contrast to monoclonal antibodies, which occur *in vivo* in pathologic conditions such as multiple myeloma or are produced artificially by hybridoma technology against one of a variety of antigens.

Doctrine of original antigenic sin: (Figure 7.65) The immune response against a virus, such as a parental strain, to which an individual was previously exposed may be greater than it is against the immunizing agent, such as type A influenza virus variant. This concept is known as the **doctrine of original antigenic sin.**

When an individual is exposed to an antigen that is similar but not identical to an antigen to which he was previously exposed by either infection or immunization, the immune response to the second exposure is still directed against the first antigen. This was first noticed in influenza virus infection. Due to antigenic drift and antigenic shift in influenza virus, reinfection with an antigenically altered strain generates a secondary immune response that is specific for the influenza virus strain that produced an earlier infection. Other highly immunogenic epitopes on a second and subsequent viruses are ignored.

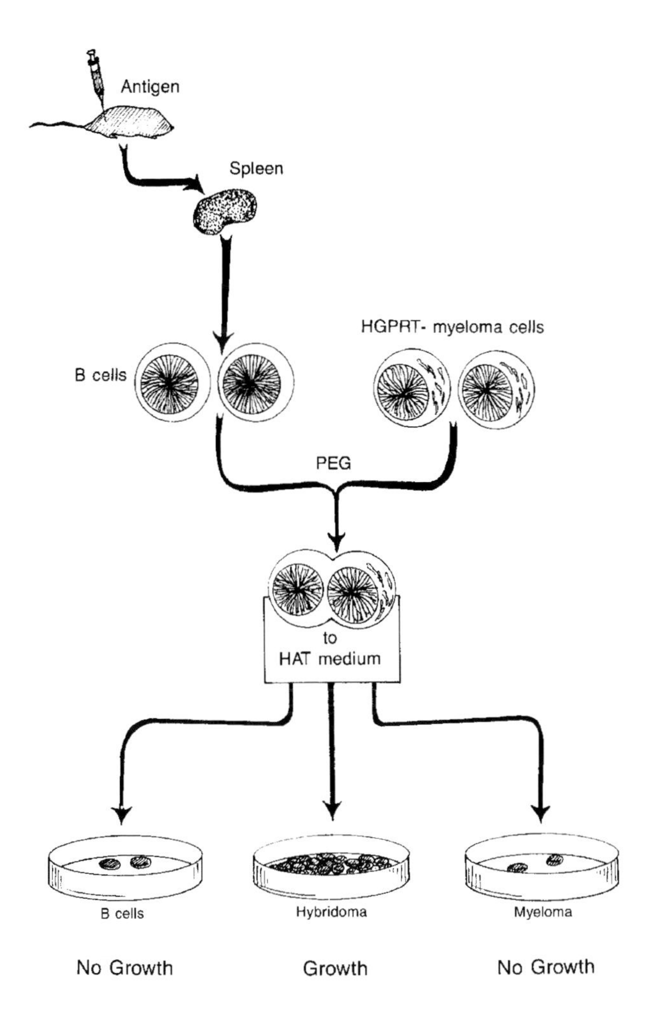

FIGURE 7.62 B lymphocyte hybridoma.

Maternal immunoglobulins are passed to the offspring both *in utero* and in the neonatal period. In humans, immunoglobulin G is transferred by way of the placenta as early as at 38 days gestation. Transfer remains stable until the 17th week and then increases until term. Only the IgG class of antibodies crosses the placenta. The SC portion of the immunoglobulin is bound to specific placental membrane receptors of the Sc γ type III variety (CD16) that possess binding sites for the Foc region of IgG in addition to epitopes that can be recognized by the antigen-binding site of IgG. The receptors have low affinity for aggregated or complex IgG and is strongest for IgG1 and IgG3, less for IgG 4, least for IgG2, and no affinity for IgM and IgA. The receptors are concentrated on the syncytiotrophoblast that makes direct contact with the mother's circulation. (See also Brambell receptor in the following text.) Human colostrum and fresh breast milk contain numerous host-resistance factors that are both specific and nonspecific. Secretory IgA is present in these fluids with highest concentration in colostrom that peaks during the first 34 d post partum. Breast-fed children receive 0.5 grams of secretory IgA per day. Human milk also contains IgG and IgM as well as secretory IgA but in lower concentrations. Factors other than immunoglobulins that protect against infections include lactoferrin that serves as a bacterial static agent depriving microbes of iron. Lysozyme is also present in milk. Breast milk also contains macrophages, T and B cells, neutrophils and epithelial cells, and immunomodulators. Diseases caused by maternal antibodies are listed in the separate diseases.

The **Brambell receptor (FcRB)**, named for F.W.R. Brambell, a pioneer in transmission of immunity, conserves IgG by binding to its Fc region and protects it from catabolism by lysosomes (Figure 7.66). It is an Fc receptor that transports IgG across epithelial surfaces. Its structure resembles that of an MHC class I molecule. The mechanism of binding is pH dependent. In the low pH of endosomes, the receptor FcRB binds to IgG. IgG is then transported to the luminal surface of the catabolic cell, where the neutral pH mediates release of the bound IgG (Figure 7.67 and Figure 7.68).

Half-life ($T_{1/2}$) is the time required for a substance to be diminished to one half of its earlier level by degradation or decay; by catabolism, as in biological half-life; or by elimination. In immunology, $T_{1/2}$ refers to the time in which an immunoglobulin remains in the blood circulation. For IgG, the half-life is 20 to 25 d; for IgA, 6 d; for IgM, 5 d; for IgD, 2 to 8 d; and for IgE, 1 to 5 d.

Antibody half-life is the mean survival time of any particular antibody molecule after its formation. It refers to the time required to rid the animal body of one half of a known amount of antibody. Thus, antibody half-life differs according to the immunoglobulin class to which the antibody belongs.

The **fractional catabolic rate** is the total plasma immunoglobulin percentage that is catabolized each day. Predicted from the half-life of plasma or from the excretion rate of catabolized immunoglobulin products in urine.

Avidity refers to the strength of binding between an antibody and its specific antigen. The stability of this union is a reflection of the number of binding sites that they share. **Avidity** is the binding force or intensity between multivalent antigen and multivalent antibody (Figure 7.69). Multiple binding sites on both the antigen and the antibody (e.g., IgM or multiple antibodies interacting with various epitopes on the antigen) and reactions of high affinity between each of the antigens and its homologous antibody all increase the avidity. Such nonspecific factors as ionic and hydrophobic interactions also increase avidity. Whereas affinity is described in thermodynamic terms, avidity is not, as it is described according to the assay procedure employed. The sum of the forces contributing to the avidity of an antigen and antibody

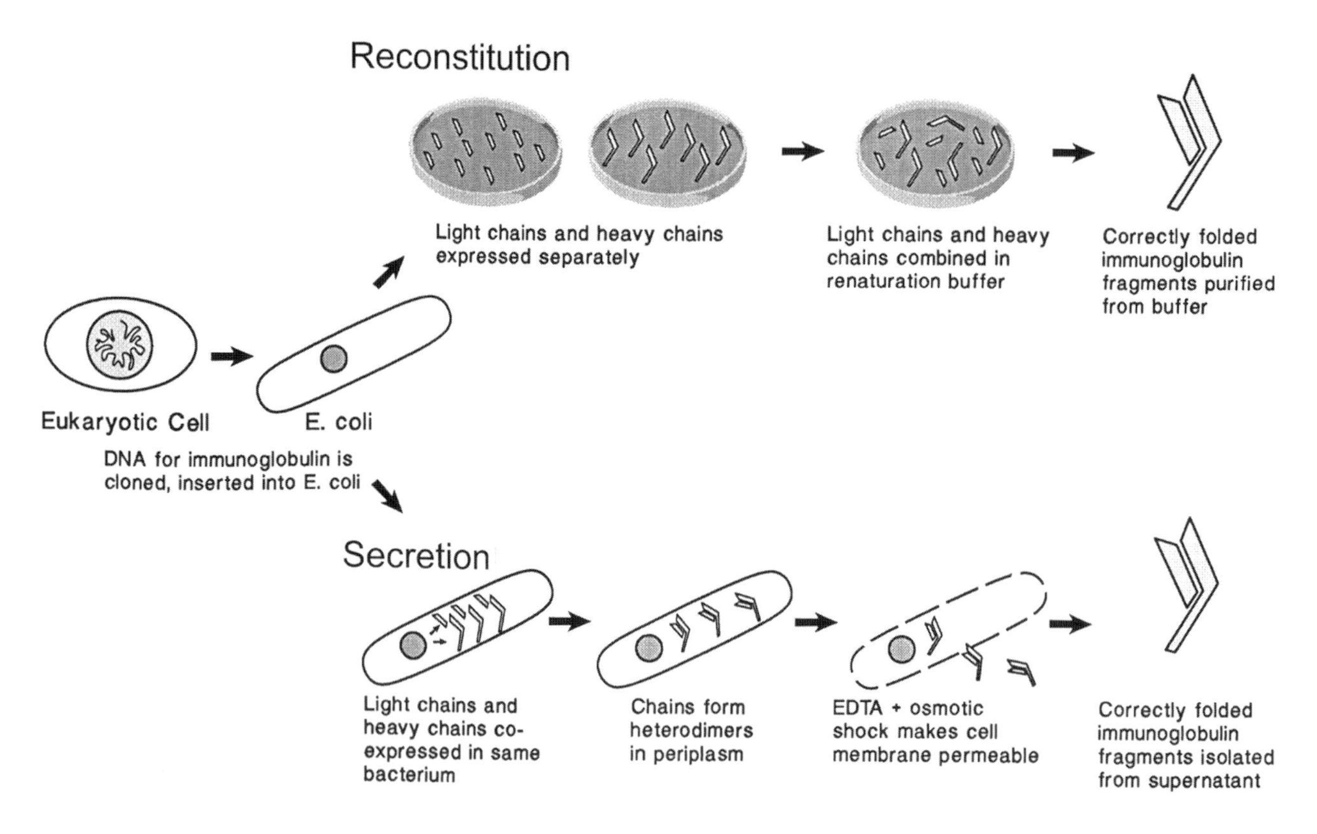

FIGURE 7.63 Beyond hybridoma technology, immunoglobulin subunits and fragments have been produced using cloned DNA expressed in bacteria.

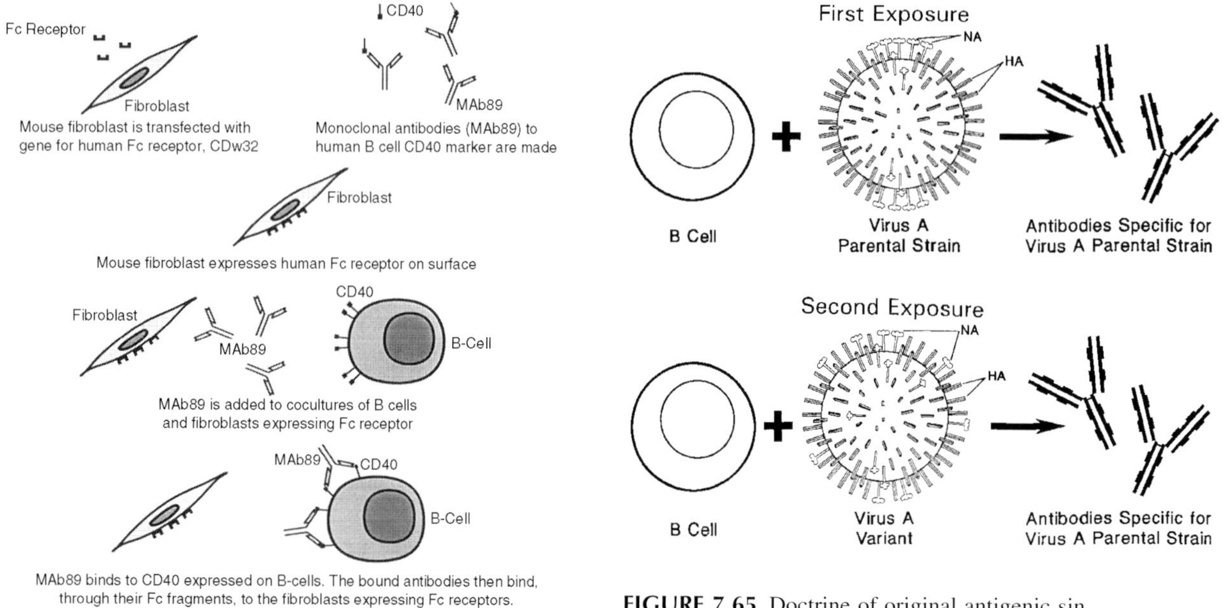

FIGURE 7.64 System for producing factor-dependent B cell lines.

FIGURE 7.65 Doctrine of original antigenic sin.

interaction may be greater than the strength of binding of the individual antibody–antigen combinations contributing to the overall avidity of a particular interaction. K_A, the association constant for the Ab + Ag = AbAg interaction, is frequently used to indicate avidity. Avidity may also describe the strength of cell-to-cell interactions which are mediated by numerous binding interactions between

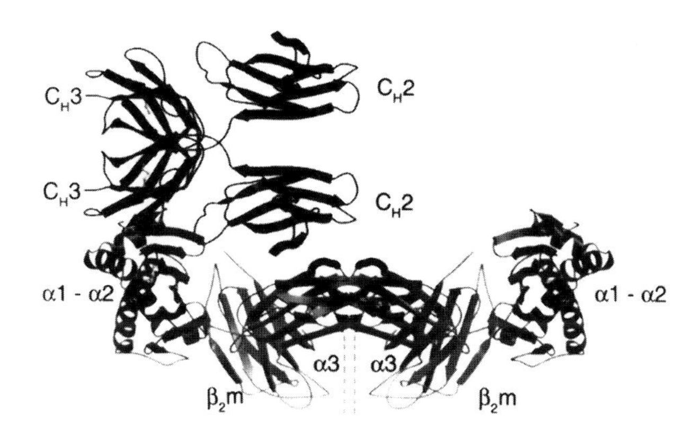

FIGURE 7.66 Brambell receptor.

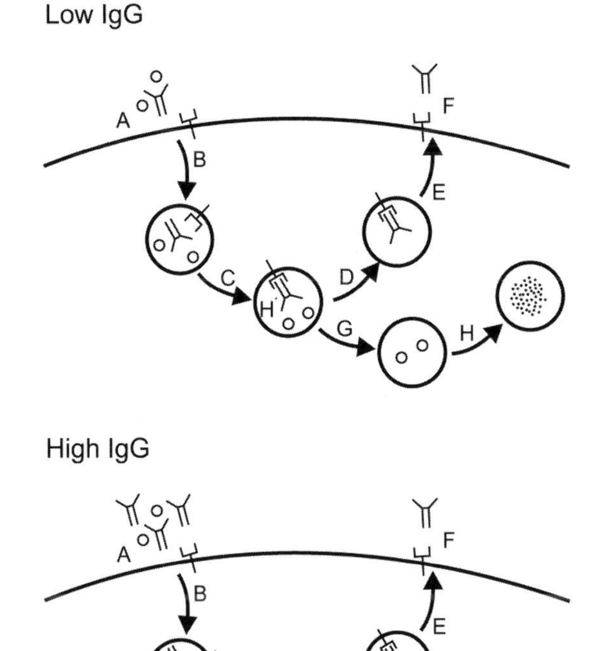

FIGURE 7.67 Mechanism of γ globulin protection from catabolism. IgG (Y) and plasma proteins (o) (A) are internalized into endosomes of endothelium (B), without prior binding. In the low pH (H^+) of the endosome (C), binding of IgG is promoted (D, E, F) IgG retained by receptor recycles to the cell surface and dissociates in the neutral pH of the extracellular fluid, returning to circulation (G, H). Unbound proteins are shunted to the lysosomes for degradation. With "low IgG," receptor efficiently "rescues" IgG from catabolism. With "high IgG," receptor is saturated and excess IgG passes to catabolism for a net acceleration of IgG catabolism.

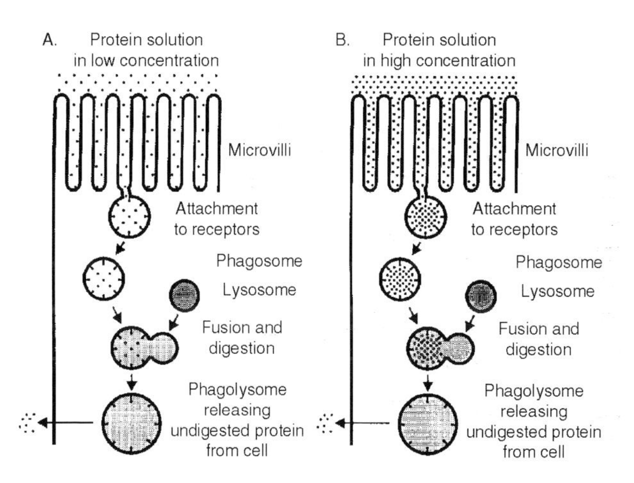

FIGURE 7.68 Mechanism of γ globulin transmission by the cell. **(A)** Concentration of γ globulin is only a little more than sufficient to saturate the receptors and the proportion degraded is less than 40%. **(B)** Concentration is about four times that in (A) and hence over 80% is degraded. The amount released from the cell remains constant, irrespective of concentration.

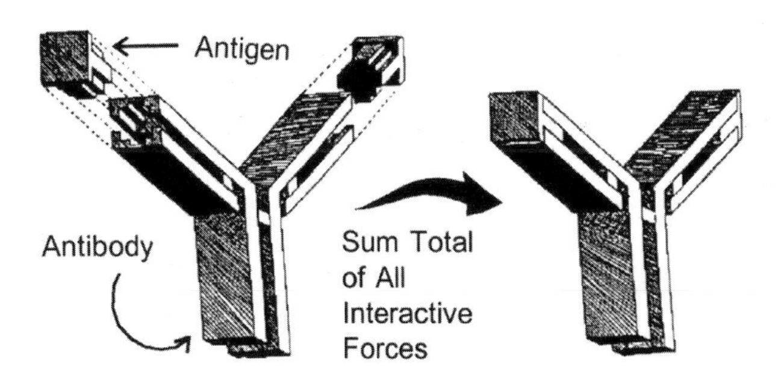

FIGURE 7.69 Avidity.

cell surface molecules. Avidity is distinct from affinity, which describes the strength of binding between a single molecular site and its ligand.

The **avidity hypothesis** was previously termed the affinity hypothesis of T cell selection in the thymus. The avidity hypothesis is based on the concept that T cells must have a measurable affinity for self MHC molecules to mature but not an affinity sufficient to cause activation of the cell when it matures, as this would necessitate deletion of the cell to maintain self tolerance.

Antibody affinity is the force of binding of one antibody molecule's paratope with its homologous epitope on the antigen molecule (Figure 7.70). It is a consequence of positive and negative portions affecting these molecular interactions.

Functional affinity is the association constant for a bivalent or multivalent antibody's interaction with a bivalent or multivalent ligand. The multivalent reactivity may enhance the affinity of multiple antigen–antibody reactions. Avidity has

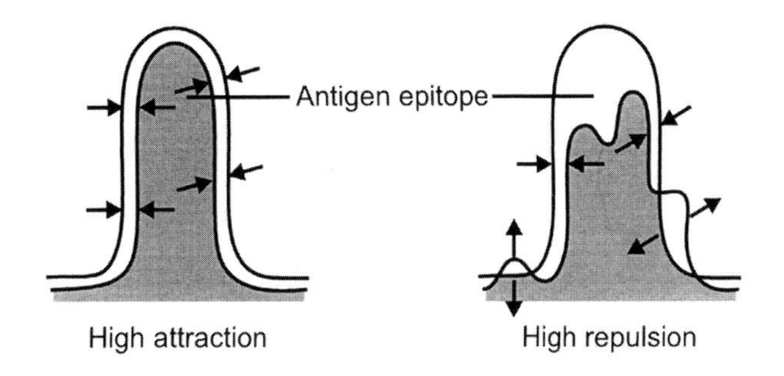

FIGURE 7.70 Antibody affinity.

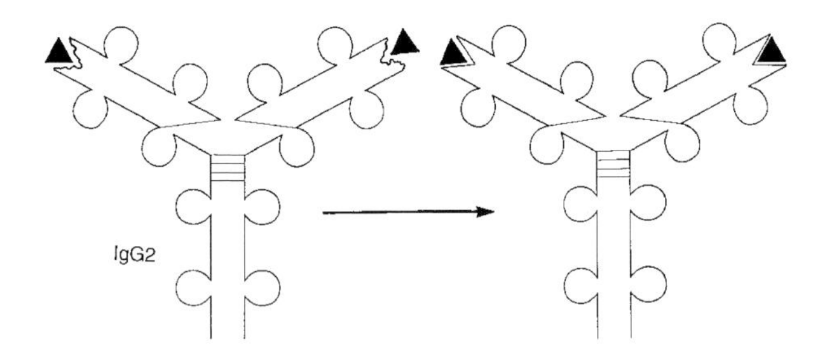

FIGURE 7.71 Affinity maturation.

a similar connotation, but it is a less precise term than is functional affinity.

Affinity maturation refers to the sustained increase in affinity of antibodies for an antigen with time following immunization (Figure 7.71). The genes encoding the antibody variable regions undergo somatic hypermutation with the selection of B lymphocytes whose receptors express high affinity for the antigen. The IgG antibodies that form following the early, heterogeneous IgM response manifest greater specificity and less heterogeneity than do the IgM molecules.

Hybrid antibody is an immunoglobulin molecule that may be prepared artificially but may never occur in nature (Figure 7.72). Each of two antigen-binding sites is of a different specificity. If IgG whole molecules or $F(ab')_2$ fragments prepared from them are subjected to mild reduction, the central disulfide bond is converted to sulfhydryl groups. A preparation of half molecules or F(ab′) fragments is produced. If either of these is reoxidized, hybrid molecules will constitute some of the products. These can be purified by passing over immunoadsorbents containing bound antigen of the appropriate specificity. Hybrid antibodies are monovalent and do not induce precipitation. They may be used in labeling cell surface antigens in which one antigen-binding site of the molecule is specific for cell surface epitopes and the other combines with a marker that renders the reaction product visible.

Bispecific antibody is a molecule that has two separate antigen-binding specificities. It may be produced by either cell fusion or chemical techniques. A bispecific antibody

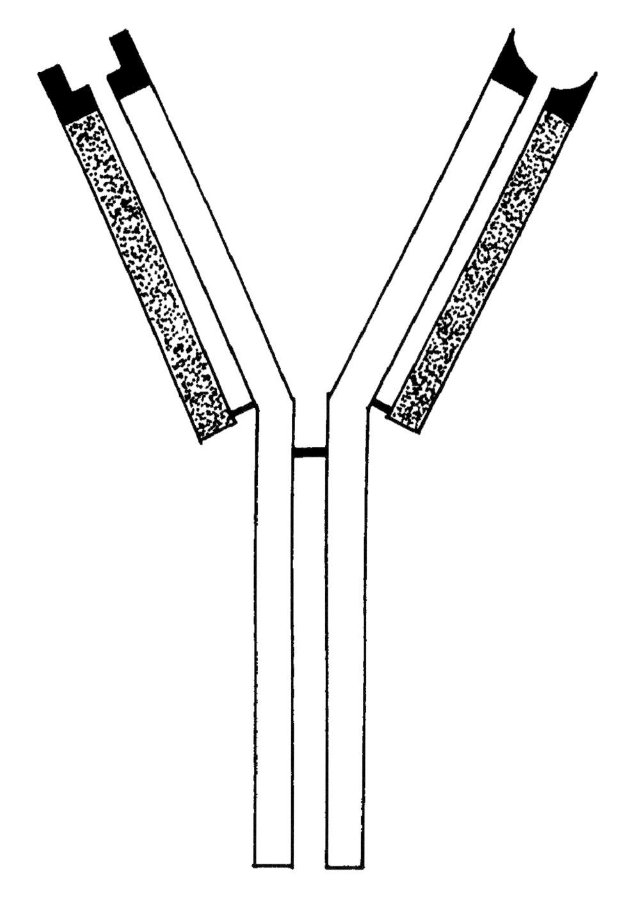

FIGURE 7.72 Hybrid antibody.

is an immunoglobulin molecule in which one of two antigen-binding sites is specific for one antigen-binding specificity, whereas the other antigen-binding site is specific for a different antigen specificity. This never occurs in nature, but it can be produced *in vitro* by treating two separate antibody specificities with mild reducing agents, converting the central disulfide bonds of both antibody molecules to sulfhydryl groups, mixing the two specificities of half molecules together, and allowing them to reoxidize to form whole molecules, some of which will be bispecific. They may be produced also by fusing hybridomas that synthesize different monoclonal antibodies. Bispecific antibodies were developed to redirect or enhance immune effector responses toward tumors and to induce killing of target cells in a non-MHC-restricted manner. One limb of the bispecific antibody recognizes a cell-surface antigen on the cytotoxic effector cell, while the other limb is specific for a tumor antigen. Recombinant DNA technology permitted the generation of bispecific scFv, minibodies, diabodies, and multivalent bispecific antibodies. Single-chain bispecific antibodies are composed of linked variable domains fused to human Fc domains. Miniantibodies comprise an scFv joined by a linker to a dimerization domain, while diabodies exploit the intrinsic nature of VH and VL within an Fv to pair.

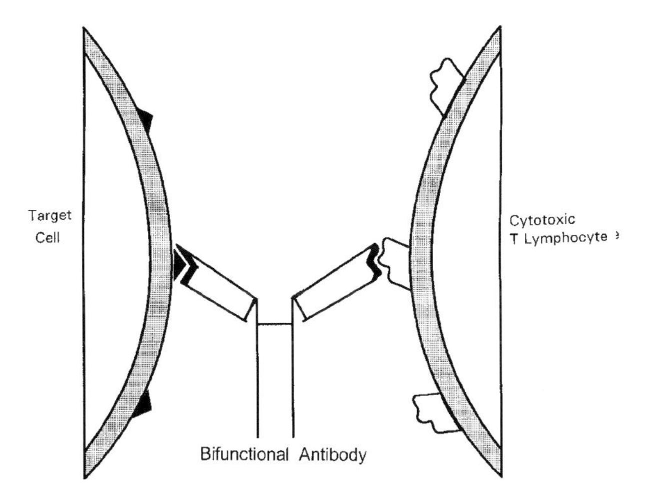

FIGURE 7.73 Bifunctional antibody.

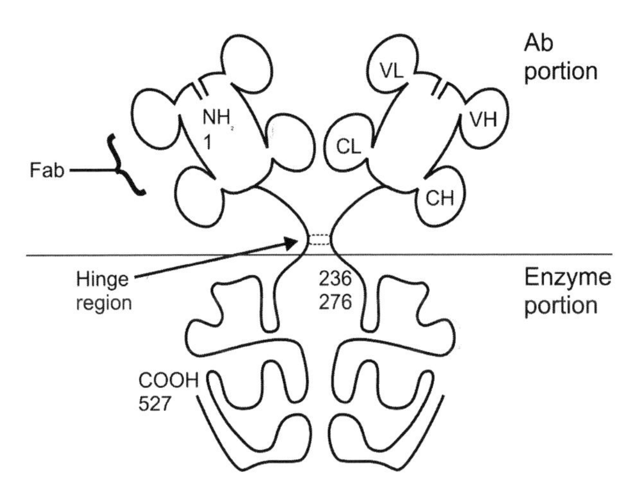

FIGURE 7.74 Abzyme (59D8-tPA).

A **bifunctional antibody** is an immunoglubulin molecule in which the Fab variable regions have different antigen binding specificities (Figure 7.73). Bifunctional antibodies are artificially formed immunoglobulins that manifest double, well-defined antigen specificity. They are used to focus the activity of an effector cell on a target cell by their ability to bind with one combining site on the effector cell and with another on the target cell. Antibodies to helper and cytotoxic effector T cell receptors can mimmic antigen and activate the T cells to proliferate, release lymphokines, or to become cytotoxic. Bispecific antibodies that bind to both effector cell and to target cell can activate the effector cell and guarantee intimate contact between effector and target cells. There are three basic techniques to prepare bispecific molecules: (1) heteroconjugate bispecific antibodies are produced by chemical linkage of two immunoglobulin molecules with different binding specificities; (2) hybridoma-specific antibodies are formed by hybridomas produced by fusing two lymphocytes that synthesize antibodies of different specificity; and (3) bispecific molecules are produced by genetic engineering which permits the insertion, within or adjacent to the genes encoding an immunoglobulin molecule, of oligonucleotides encoding another immunoglobulin, a desired immunogenic epitope, or an epitope responsible for interaction with a viral antigen. Bispecific antibodies were developed to redirect or enhance immune effector responses toward tumors and to induce killing of target cells in a non-MHC-restricted manner. One limb of the bispecific antibody recognizes a cell surface antigen on the cytotoxic effector cell, while the other limb is specific for a tumor antigen. Recombinant DNA technology has permitted the generation of bispecific scFv, minibodies, diabodies, and multivalent bispecific antibodies. Single-chain bispecific antibodies are composed of linked variable domains fused to human Fc domains. Miniantibodies comprise an scFv joined by a linker to a dimerization domain, while diabodies exploit the intrinsic nature of V_H and V_L within an Fv to pair.

An **abzyme** is the union of antibody and enzyme molecules to form a hybrid catalytic molecule (Figure 7.74). Specificity for a target antigen is provided through the antibody portion and for a catalytic function through the enzyme portion. Thus, these molecules have numerous potential uses. These molecules are capable of catalyzing various chemical reactions and show great promise as protein-clearing antibodies, as in the dissolution of fibrin clots from occluded coronary arteries in myocardial infarction.

Designer antibody is a genetically engineered immunoglobulin needed for a specific purpose (Figure 7.75). The term has been used to refer to chimeric antibodies produced by linking mouse gene segments that encode the variable region of immunoglobulin with those that encode the constant region of a human immunoglobulin. This technique provides the antigen specificity obtained from the mouse antibody, while substituting the less immunogenic Fc region of the molecule from a human source. This greatly diminishes the likelihood of an immune response in humans receiving the hybrid immunoglobulin molecules, since most of the mouse immunoglobulin Fc region epitopes have been eliminated through the human Fc substitution.

Passive immunization describes the transfer of a specific antibody or of sensitized lymphoid cells from an immune to a previously nonimmune recipient host. Unlike active immunity, which may be of a relatively long duration, passive immunity is relatively brief, lasting only until the injected immunoglobulin or lymphoid cells have disappeared. Examples of passive immunization include (1) the administration of γ globulin to immunodeficient individuals; and (2) the transfer of immunity from mother to

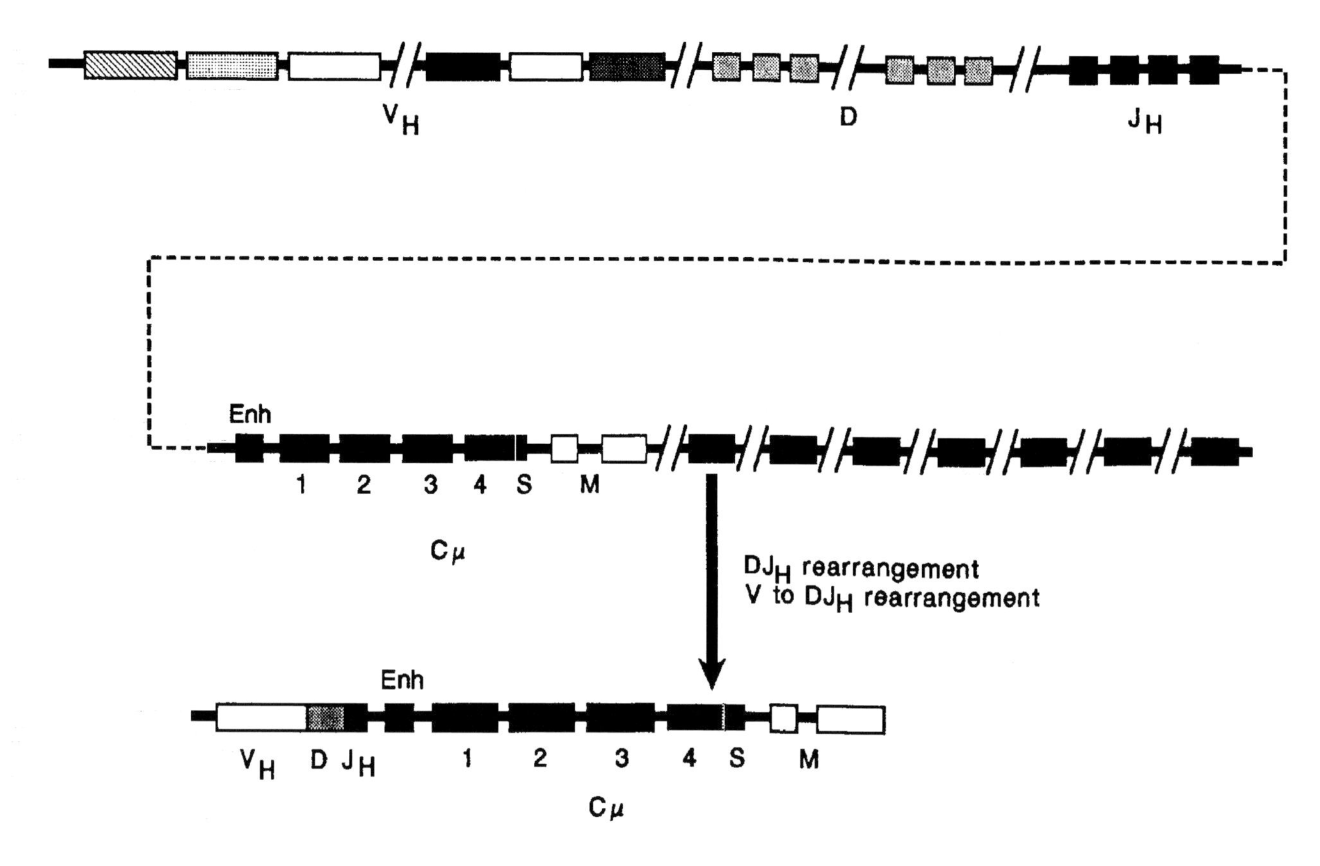

FIGURE 7.75 Designer antibody.

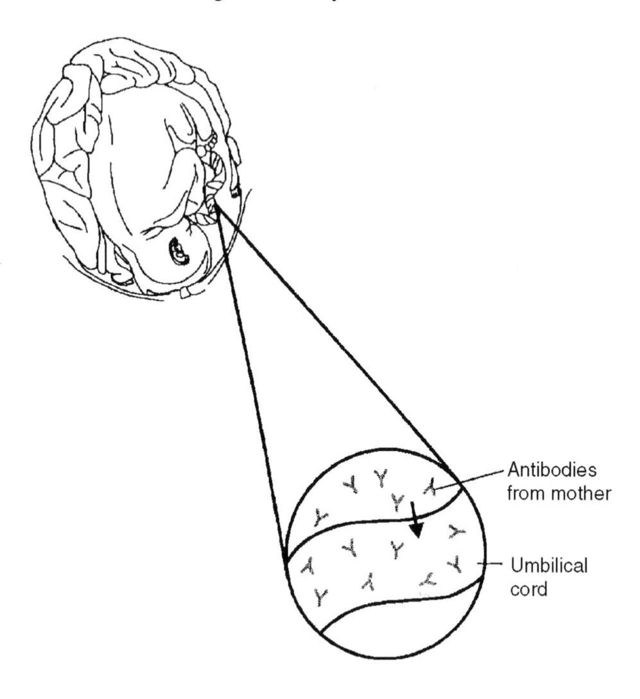

FIGURE 7.76 Passive immunization.

young (antibodies across the placenta or the ingestion of colostrum containing antibodies).

Antibody feedback (Figure 7.76) is the negative feedback system whereby antigen-specific antibodies downregulate further immune responses to that antigen. Several mechanisms may be responsible for this including (1) removal of the initiating stimulus by the antibody, (2) binding of antigen/IgG antibody immune complexes to the Fcγ receptor of B cells, and (3) inhibition of T cell responses by antigen/antibody complexes.

The use of Rh immune globulin to prevent erythroblastosis fetalis in the infants of Rh negative mothers is an example of antibody feedback. Secreted IgG antibodies may downregulate antibody production when antigen–antibody complexes simultaneously engage B cell membrane immunoglobulin and Fcγ receptors (FcγRII). The cytoplasmic tails of the Fcγ receptor transduce inhibitory signals inside the B cells.

An **antiidiotypic antibody** is an antibody that interacts with antigenic determinants (idiotopes) at the variable N-terminus of the heavy and light chains comprising the paratope region of an antibody molecule where the antigen-binding site is located. The idiotope antigenic determinants may be situated either within the cleft of the antigen-binding region or on the periphery or outer edge of the variable region of heavy and light chain components.

An antiidiotypic vaccine is an immunizing preparation of antiidiotypic antibodies that are internal images of certain exogenous antigens. To develop an effective antiidiotypic

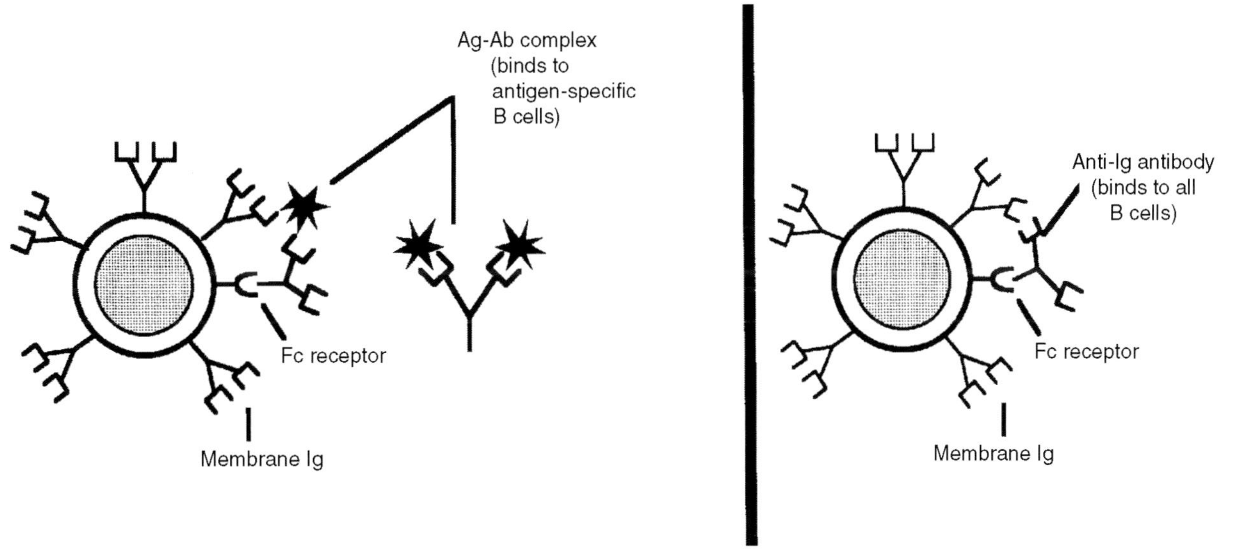

FIGURE 7.77 Antibody feedback.

vaccine, epitopes of an infectious agent that induce protective immunity must be identified. Antibodies must be identified that confer passive immunity to this agent. An antiidiotypic antibody prepared using these protective antibodies as the immunogen, in some instances, can be used as an effective vaccine. Antiidiotypic vaccines have effectively induced protective immunity against such viruses as rabies, coronavirus, cytomegalovirus, and hepatitis B; such bacteria as *Listeria monocytogenes*, *Escherichia coli*, and *Streptococcus pneumoniae*; and such parasites as *Schistosoma mansoni*. Antiidiotypic vaccination is especially desirable when a recombinant vaccine is not feasible. Monoclonal antiidiotypic vaccines represent a uniform and reproducible source for an immunizing preparation.

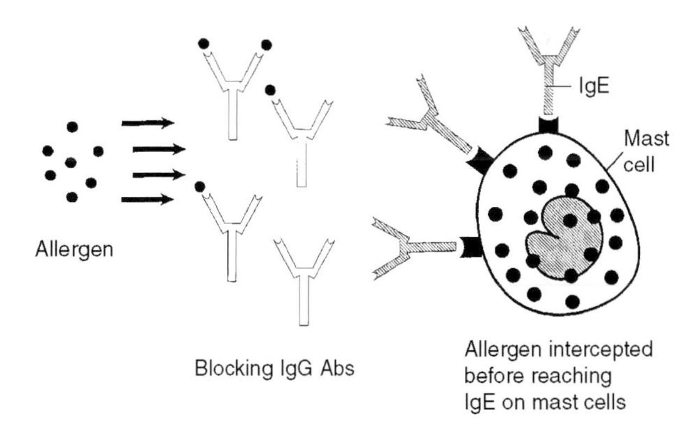

FIGURE 7.78 Blocking antibody.

Blocking is the prevention of nonspecific interaction of an antibody with a certain antigenic determinant, whose identification is sought, by washing with mammalian serum other than that being used in the test system. For example, enzyme-linked immunosorbent assays (ELISA) employ blocking.

Blocking antibody (Figure 7.78) is:

1. An incomplete IgG antibody that, when diluted, may combine with red blood cell surface antigens and inhibit agglutination reactions used for erythrocyte antigen identification. This can lead to errors in blood grouping for Rh, K, and k blood types. Pretreatment of red cells with enzymes may correct the problem.
2. An IgG antibody specifically induced by exposure of allergic subjects to specific allergens, to which they are sensitive, in a form that favors IgG rather than IgE production. The IgG, specific for the allergens to which they are sensitized, competes within IgE molecules bound to mast cell surfaces, thereby preventing their degranulation and inhibiting a type I hypersensitivity response.
3. A specific immunoglobulin molecule that may inhibit the combination of a competing antibody molecule with a particular epitope. Blocking antibodies may also interfere with the union of T cell receptors with an epitope for which they are specific, as occurs in some tumor-bearing patients with blocking antibodies which may inhibit the tumoricidal action of cytotoxic T lymphocytes.

Monogamous bivalency is the binding of a bivalent antibody molecule, such as IgG, with two identical antigenic determinants or epitopes on the same antigen molecule, in contrast to each Fab region of the IgG molecule uniting

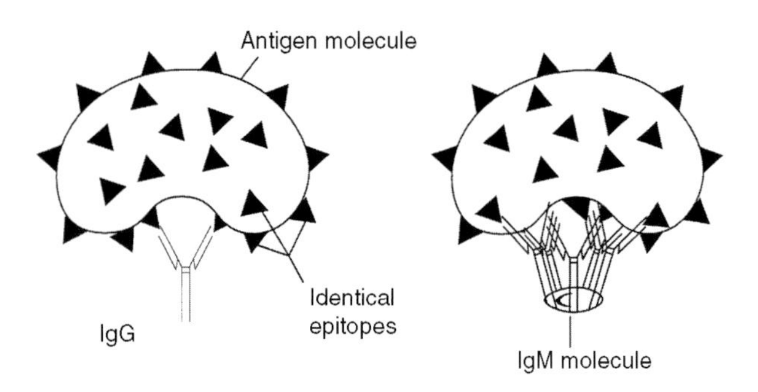

FIGURE 7.79 Monogamous bivalency and monogamous multivalency.

with an identical antigenic determinant on two separate antigen molecules (Figure 7.79). For this monogamous binding to take place, the epitopes must be positioned on the surface of the antigen molecule in such a manner that the binding of one Fab region to an epitope can position the remaining Fab of the IgG molecule for easy interaction with an adjacent identical epitope. Interaction of this type represents high affinity of binding, which lends a stability to the antigen–antibody complex. The combination of one IgM molecule to multiple epitopes on a single molecule of antigen represents **monogamous multivalency**.

Fc receptor is a structure on the surface of some lymphocytes, macrophages, or mast cells that specifically binds the Fc region of immunoglobulin, often when the Fc is aggregated. The Fc receptors for IgG are designated FcγR (Figure 7.80). Those for IgE are designated FcεR (Figure 7.81). IgM, IgD, and IgA Fc receptors have yet to be defined. Neutrophils, eosinophils, mononuclear phagocytes, B lymphocytes, selected T lymphocytes, and accessory cells bear Fc receptors for IgG on their surfaces. When the Fc region of immunoglobulin binds to the cation permease Fc receptor, there is an influx of Na^+ or K^+ that activates phagocytosis, H_2O_2 formation, and cell movement by macrophages. **Fc receptors** are found on 95% of human peripheral blood T lymphocytes. On about 75% of the cells, the FcR are specific for IgM; the remaining 20% are specific for IgG. The FcR-bearing T cells are also designated TM and the B cell responses to pokeweed mitogen (PWM). In cultures of B cells with PWM, the TM cells also proliferate, supporting the views on their helper effects. Binding of IgM to FcR of TM cells is not a prerequisite for helper activity. In contrast to TM cells, the TG cells effectively suppress B cell differentiation. They act on the TM cells, and their complexes. There are a number of other differences between TM and TG cells. Circulating TG cells may be present in increased number, often accompanied by a reduction in the circulating number of TM cells. Increased numbers of TG cells are seen in cord blood and in some patients with hypogammaglobulinemia, sex-linked agammaglobulinemia, IgA deficiency, Hodgkin's disease and thymoma, to mention only a few.

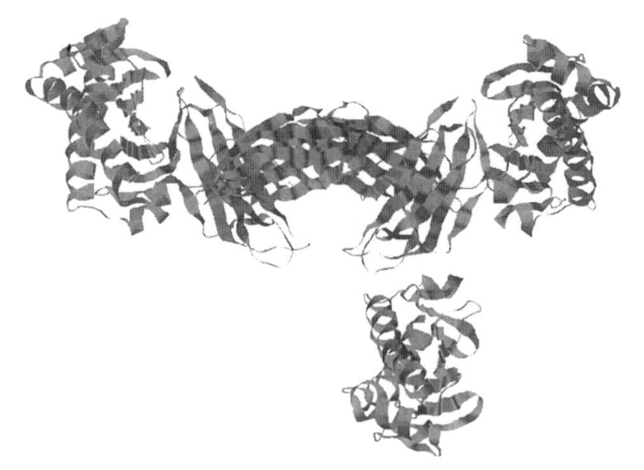

FIGURE 7.80 Fc (IgG) receptor (neonatal).

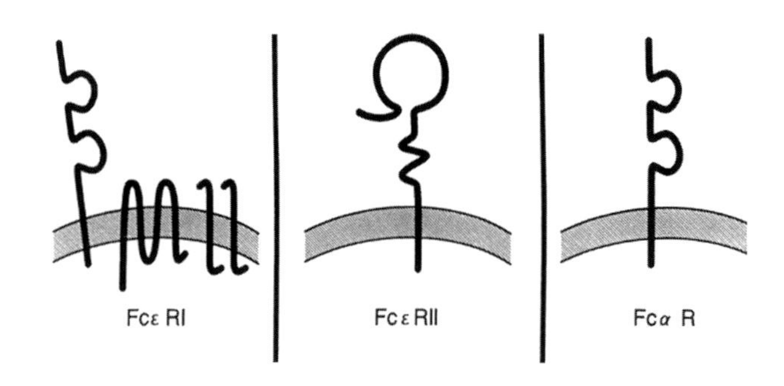

FIGURE 7.81 Fcα and Fcε receptors.

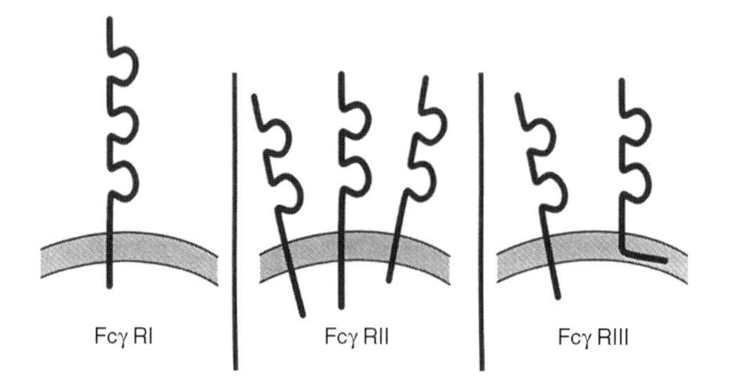

FIGURE 7.82 Fcγ receptors.

Fcγ receptors (FcγR) are receptors for the Fc region of IgG (Figure 7.82). B and T lymphocytes, natural killer cells, polymorphonuclear leukocytes, mononuclear phagocytes, and platelets contain FcγR. When these receptors bind immune complexes, the cell may produce leukotrienes, prostaglandins, modulate antibody synthesis, increase consumption of oxygen, activate oxygen metabolites, and become phagocytic. The three types of Fcγ receptors include FcγRI, or FcγRII (CD32), and FcγRIII (CD16). FcγRI represents a high-affinity receptor found on mononuclear phagocytes. In humans, it binds IgG1 and IgG3.

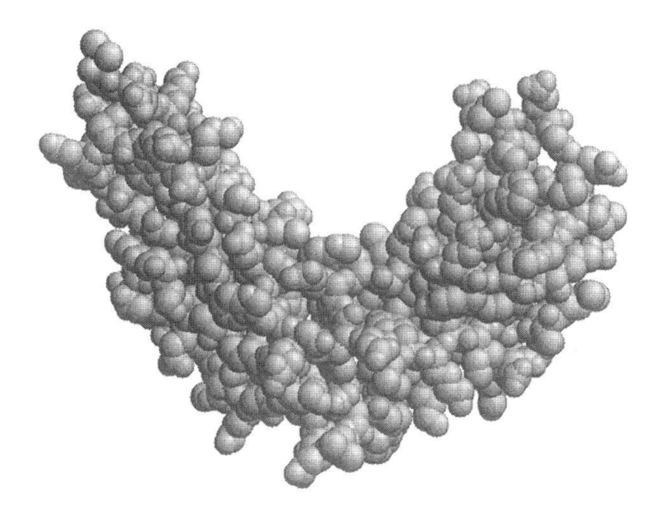

FIGURE 7.83 FcεRI.

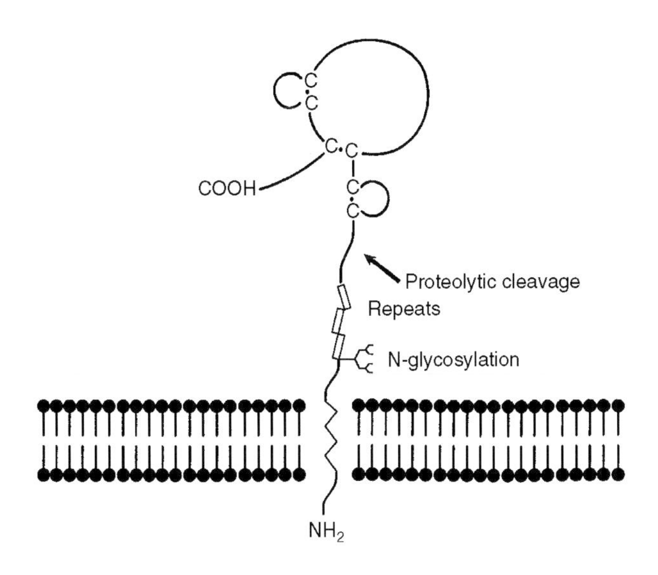

FIGURE 7.84 FcεRII.

FcγRII and FcγRIII represent low-affinity IgG receptors. Neutrophils, monocytes, eosinophils, platelets, and B lymphocytes express FcγRII on their membranes. Neutrophils, natural killer cells, eosinophils, macrophages, and selected T lymphocytes express FcγRIII on their membranes in humans and bind IgG1 and IgG3. Paroxysmal nocturnal hemoglobulinuria patients have deficient FcγRIII on their neutrophil membranes.

FcγRI (Figure 7.83) represents a high-affinity receptor found on mononuclear phagocytes. In humans it binds IgG1 and IgG3 .

FcγRII (Figure 7.84) is a membrane protein designated as CD32 that serves as a receptor for the Fc region of the IgG molecule.

FcγRII and FcγRIII represent low affinity IgG receptors. In humans, neutrophils, monocytes, eosinophils, platelets, and B lymphocytes express FcγRII on their membranes. Neutrophils, natural killer cells, eosinophils, macrophages, and selected T lymphocytes express FcγRIII on their membranes and bind IgGl and IgG3. Paroxysmal nocturnal hemoglobulinuria patients have deficient FcγRIII on their neutrophil membranes.

Fcε receptor (FcεR): Mast cell and leukocyte high affinity receptor for the Fc region of IgE. When immune complexes bind to Fcε receptors, the cell may respond by releasing the mediators of immediate hypersensitivity, such as histamine and serotonin. Modulation of antibody synthesis may also occur. There are two varieties of Fcε receptors, designated FcεRI and FcεRII (CD23). FcεRI represents a high-affinity receptor found on mast cells and basophils. It anchors monomeric IgE to the cell surface. It possesses 1α, 1β, and 2γ chains. FcεRII represents a low-affinity receptor. It is found on mononuclear phagocytes, B lymphocytes, eosinophils, and platelets. Subjects with increased IgE in the serum have elevated numbers of FcεRII on their cells. It is a 321-amino acid single polypeptide chain that is homologous with a sialoglycoprotein receptor.

The **polyimmunoglobulin receptor** is an attachment site for polymeric immunoglobulins located on epithelial cell and hepatocyte surfaces that facilitate polymeric IgA and IgM transcytosis to the secretions. After binding, the receptor–immunoglobulin complex is endocytosed and enclosed within vesicles for transport. Exocytosis takes place at the cell surface where the immunoglobulin is discharged into the intestinal lumen. A similar mechanism in the liver facilitates IgA transport into the bile. The receptor–polymeric immunoglobulin complex is released from the cell following cleavage near the cell membrane. The receptor segment that is bound to the polymeric immunoglobulin is known as the secretory component which can only be used once in the transport process.

A **genome** consists of all genetic information that is contained in a cell or in a gamete.

Genomic DNA is the DNA found in the chromosomes.

The **genetic code** includes the codons and nucleotide triplets correlating with amino acid residues in protein synthesis. The nucleotide linear sequence in mRNA is translated into the amino acid residue sequence.

Immunoglobulin genes encode heavy and light polypeptide chains of antibody molecules and are found on different chromosomes (i.e., chromosome 14 for heavy chain, chromosome 2 for κ light chain, and chromosome 22 for λ light chain (Figure 7.85 and Figure 7.86). The DNA of the majority of cells does not contain one gene that encodes a complete immunoglobulin heavy or light polypeptide chain. Separate gene segments that are widely distributed in somatic cells and germ cells come

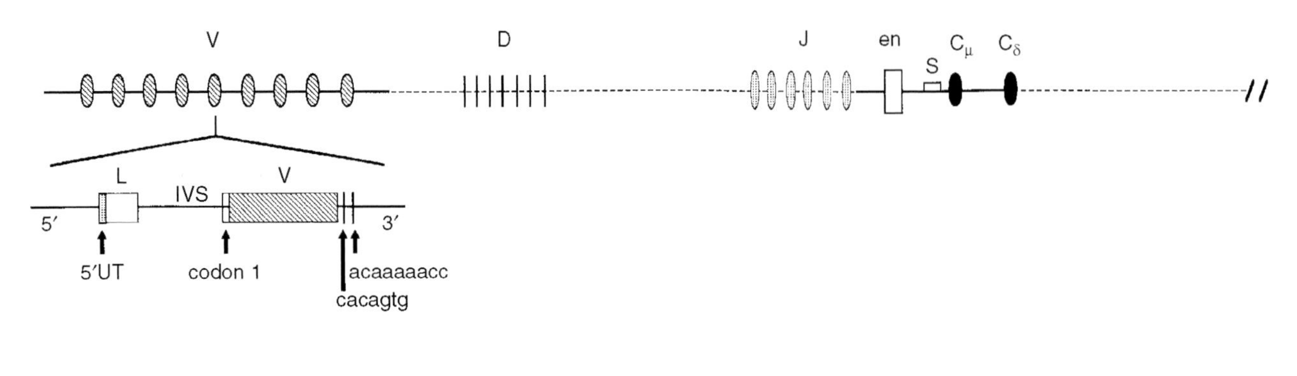

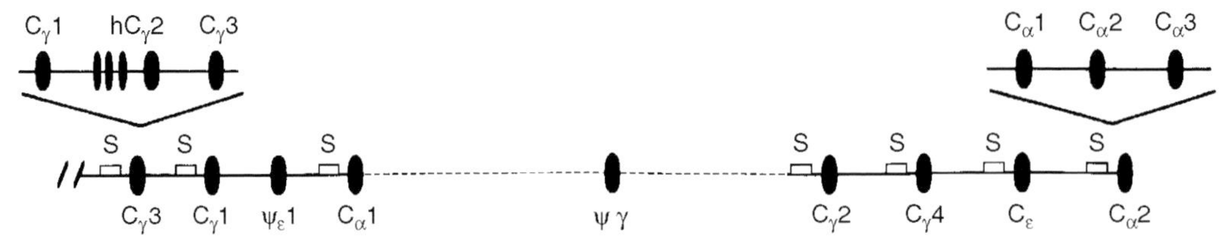

FIGURE 7.85 Immunoglobulin gene.

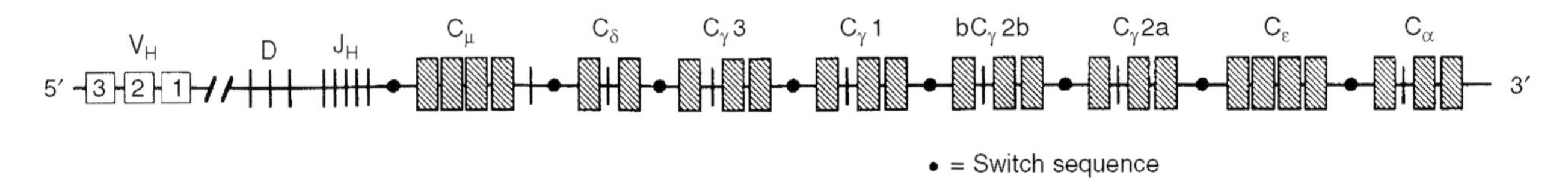

FIGURE 7.86 Mouse immunoglobulin gene.

together to form these genes. In B cells, gene rearrangement leads to the creation of an antibody gene that codes for a specific protein. Somatic gene rearrangement also occurs with the genes that encode T cell antigen receptors. Gene rearrangement of this type permits the great versatility of the immune system in recognizing a vast array of epitopes. Three forms of gene segments join to form an immunoglobulin light-chain gene. The three types include light-chain variable region (V_L), joining (J_L), and constant region (C_L) gene segments. V_H, J_H, and C_H as well as D (diversity) gene segments assemble to encode the heavy chain. Heavy and light chain genes have a closely similar organizational structure. There are 100 to 300 Vκ genes, five Jκ genes, and one C gene on the κ locus of chromosome 2. There are 100 V_H genes, 30 D genes, six J_H genes, and 11 C_H genes on the heavy chain locus of chromosome 14. Several Vλ, six Jλ, and six Cλ genes are present on the λ locus of chromosome 22 in humans. V_H and V_L genes are classified as V gene families, depending on the sequence homology of their nucleotides or amino acids.

Gene diversity is the determination of the extent of an immune response to a particular antigen or immunogen as determined by mixing and matching of exons from variable, joining, diversity, and constant region gene segments. Tonegawa received the Nobel Prize for revealing the mechanism of the generation of diversity in antibody formation.

Restriction fragment length polymorphism (RFLP) is the genome diversity in DNA from different subjects revealed by restriction map comparison. This diversity is based on differences in restriction fragment lengths, which are determined by sites of restriction, endonuclease cleavage of the DNA molecules, and revealed by preparing Southern blots using appropriate molecular hybridization probes. Polymorphisms may be demonstrated in exons, introns, flanking sequences, or any DNA sequence. Variations in DNA sequence show Mendelian inheritance. Results are useful in linkage studies and can help to identify defective genes associated with inherited disease.

The **gene conversion hypothesis** describes the method by which alternative sequences can be introduced into the MHC genes without reciprocal crossover events. This mechanism could account for the incredible polymorphism of alleles. The alternative sequences may include those found in the class I-like genes and pseudogenes present on chromosome 6. Hypothetically, gene conversion was an evolutionary event as well as an ongoing one, giving rise to new mutations; therefore new alleles, within a population.

Gene cloning is the use of recombinant DNA technology to replicate genes or their fragments.

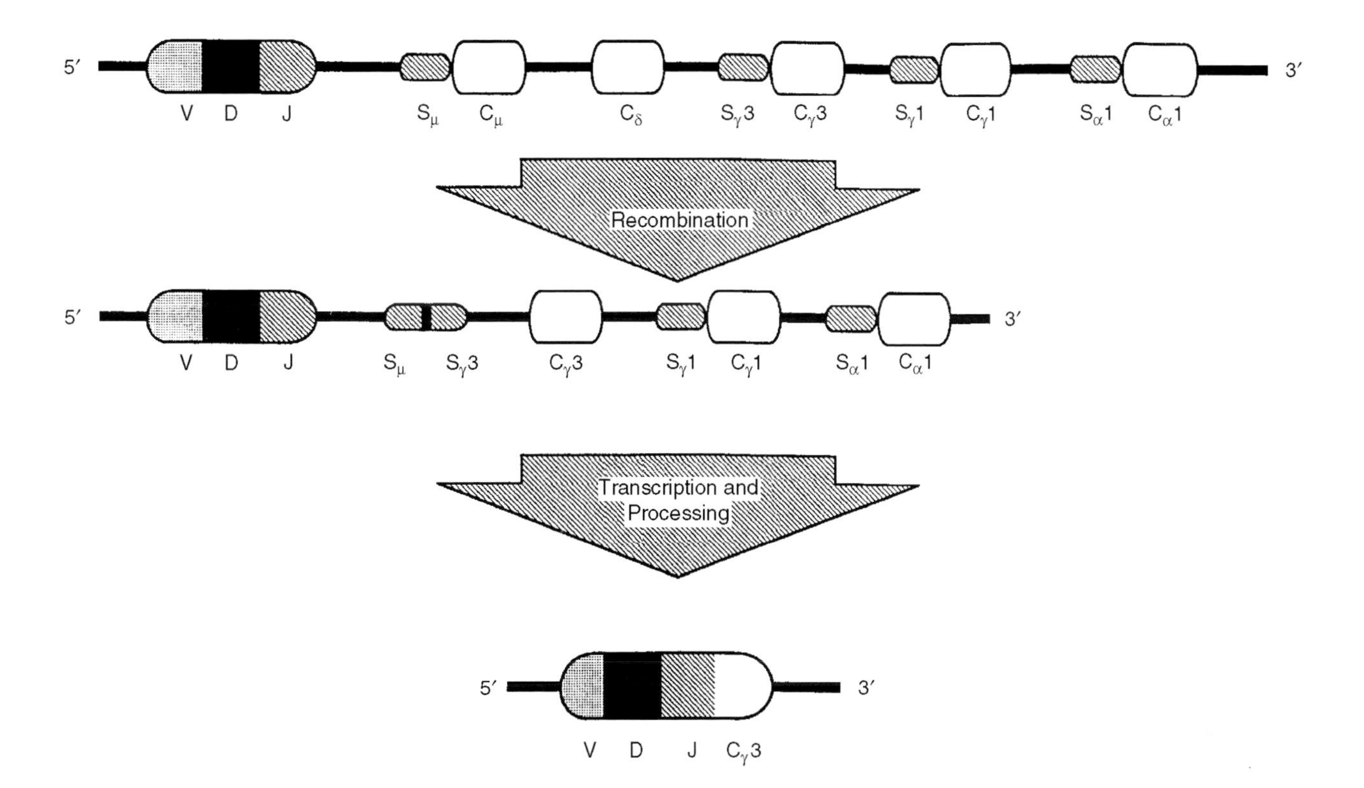

FIGURE 7.87 Isotype switching (from IgM to IgG).

Gene bank is a synonym for DNA library.

Isotype switching refers to the mechanism whereby a cell changes from synthesizing a heavy polypeptide chain of one isotype to that of another, as from μ chain to γ chain formation in B cells that have received a switch signal from a T cell (Figure 7.87).

Class switching (isotype switching) is a change in the isotype or class of an immunoglobulin synthesized by a B lymphocyte undergoing differentiation. IgM is the main antibody produced first in a primary humoral response to thymus-dependent antigens, with IgG being produced later in the response. A secondary antibody response to the same antigen results in the production of only small amounts of IgM but much larger quantities of IgG, IgA, or IgE antibodies. T_H cell lymphokines have a significant role in controlling class switching. Only heavy chain constant regions are involved in switching, with the light chain type and heavy chain variable region remaining the same. The specificity of the antigen-binding region is not altered. Mechanisms of class switching during B cell differentiation include the generation of transcripts processed to separate mRNAs and the rearrangements of immunoglobulin genes that lead to constant region gene segment transposition. Membrane IgM appears first on immature B cells, followed by membrane IgD as cell maturation proceeds. A primary transcript bearing heavy chain variable region, μ chain constant region, and δ chain constant region may be spliced to form mRNA that codes for each heavy chain. Following stimulation of B cells by antigen and T lymphocytes, class switching is probably attributable to immunoglobulin gene rearrangements. During switching, B cells may temporarily express more than one class of immunoglobulin. Class switching in B cells mediated by IL-4 is sequential, proceeding from Cμ to Cγ-1 to Cε. IgG1 expression replaces IgM expression as a consequence of the first switch. IgE expression replaces IgG1 expression as a result of the second switch. TGF-β and IL-5 have been linked to the secretion of IgA.

Heavy chain class (isotype) switching is the mechanism whereby a B cell changes the class (or isotype) of antibodies it synthesizes from IgM to IgG, IgE, or IgA without altering the antigen-binding specificity of the antibody. Helper T lymphocyte cytokines and CD40 ligand regulate heavy chain class switching which involves B cell VDJ segment recombination with downstream heavy chain gene segments.

The **genetic switch hypothesis** is a concept that predicts a switch in the gene governing heavy chain synthesis by plasma cells during immune response ontogeny.

Gene rearrangement refers to genetic shuffling that results in elimination of introns and the joining of exons to produce mRNA (Figure 7.88, Figure 7.89, and

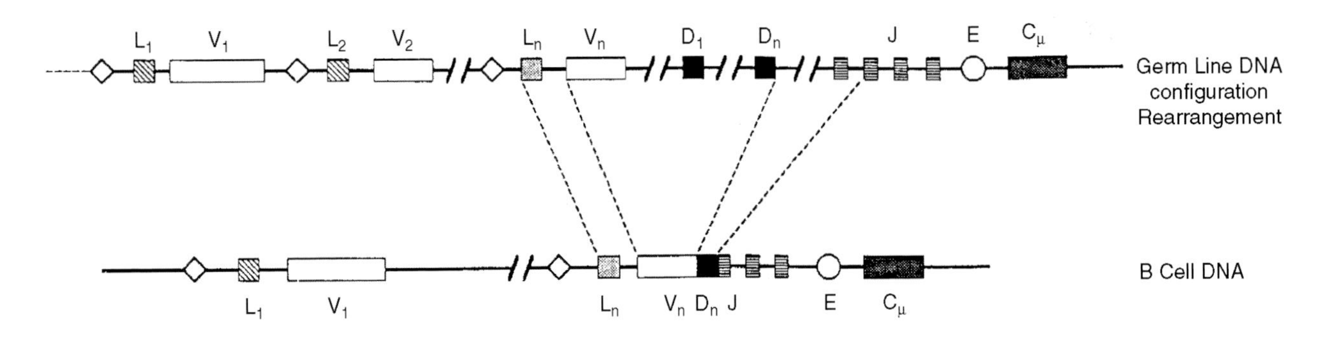

FIGURE 7.88 Immunoglobulin gene rearrangement.

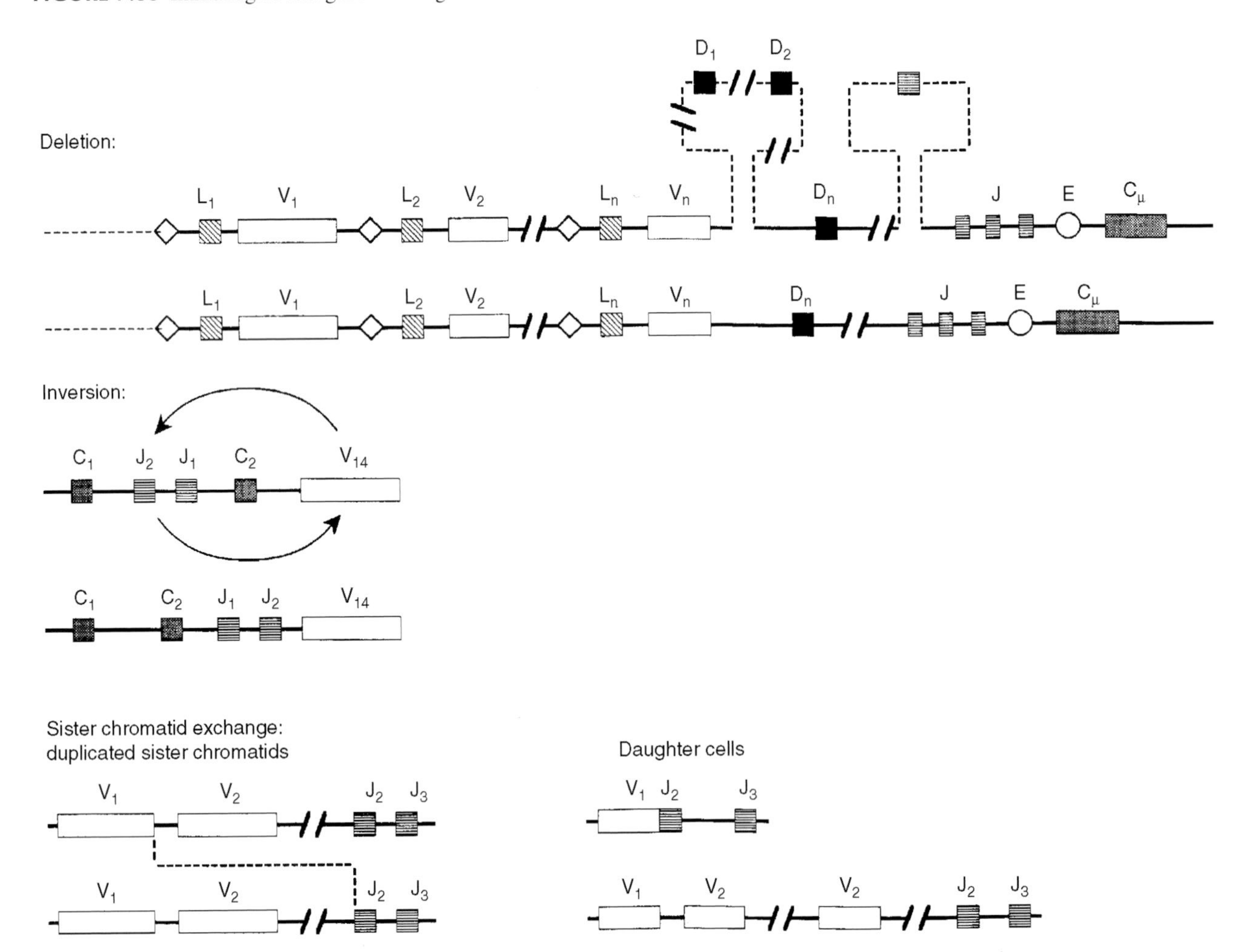

FIGURE 7.89 Immunoglobulin gene rearrangement.

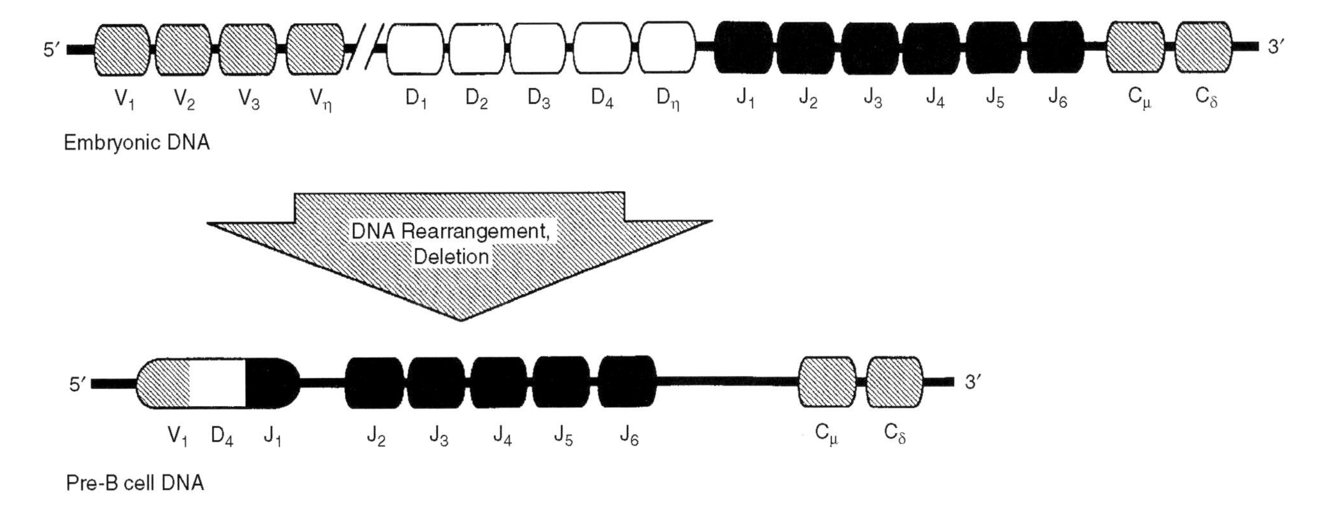

FIGURE 7.90 Gene rearrangement.

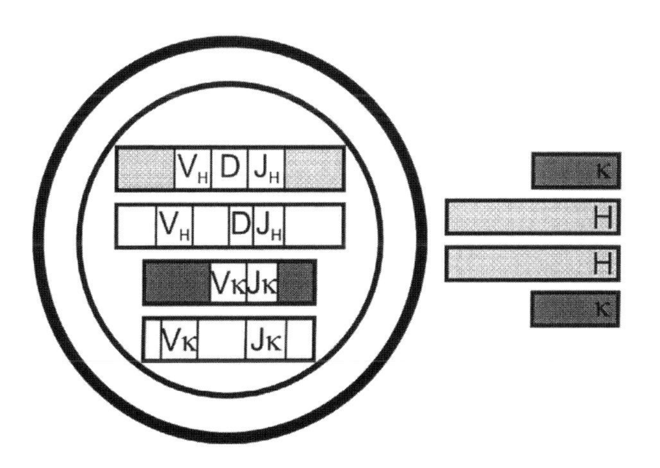

FIGURE 7.91 Allelic exclusion.

Figure 7.90). Gene rearrangement within a lymphocyte signifies its dedication to the formation of a single cell type, which may be immunoglobulin synthesis by B lymphocytes or production of a β chain receptor by T lymphocytes. Neoplastic transformation of lymphocytes may be followed by the expansion of a single clone of cells, which is detectable by Southern blotting. RNA splicing is the method whereby RNA sequences that are nontranslatable (known as introns) are excised from the primary transcript of a split gene. The translatable sequences (known as exons) are united to produce a functional gene product.

Allelic exclusion (Figure 7.91) occurs when only one of two genes for which the animal is heterozygous is expressed, whereas the remaining gene is not. Immunoglobulin genes manifest this phenomenon. Allelic exclusion accounts for the ability of a B cell to express only one immunoglobulin or the capacity of a T cell to express a T cell receptor of a single specificity. Investigations of allotypes in rabbits established that individual immunoglobulin molecules have identical heavy chains and light chains. Immunoglobulin-synthesizing cells produce only a single class of heavy chain and one type of light chain at a time. Thus, by allelic exclusion, a cell that is synthesizing antibody expresses just one of two alleles encoding an immunoglobulin chain at a particular locus.

Isotypic exclusion refers to the use by a B cell or antibody of one or the other of the light polypeptide chain isotopes, κ or λ. It is the productive rearrangement of light-chain genes such as the rearrangement of the κ gene which occurs when both κ gene alleles are rearranged.

Junctional diversity occurs when gene segments join imprecisely, and the amino acid sequence may vary and affect variable region expression, which can alter codons at gene segment junctions (Figure 7.92). These include the V-J junction of the genes encoding immunoglobulin κ and λ light chains and the V-D, D-J, and D-D junctions of genes encoding immunoglobulin heavy chains or the genes encoding T cell receptor and β and δ chains.

***D* exon** is a DNA sequence that encodes a portion of the immunoglobulin heavy chain's third hypervariable region (Figure 7.93). It is situated on the 59 side of J exons. An intron lies between them. During lymphocyte differentiation, V-D-J sequences are produced that encode the complete variable region of the heavy chain.

A ***V* gene** is a gene encoding the variable region of immunoglobulin light or heavy chains. Although it is not in proximity to the *C* gene in germ-line DNA, in lymphocyte and plasma-cell DNA the V gene lies near the 5′ end of the *C* gene from which it is separated by a single intron.

***V* gene segment** is a DNA segment encoding the first 95 to 100 amino acid residues of immunoglobulin and T cell polypeptide chain variable regions. The two coding

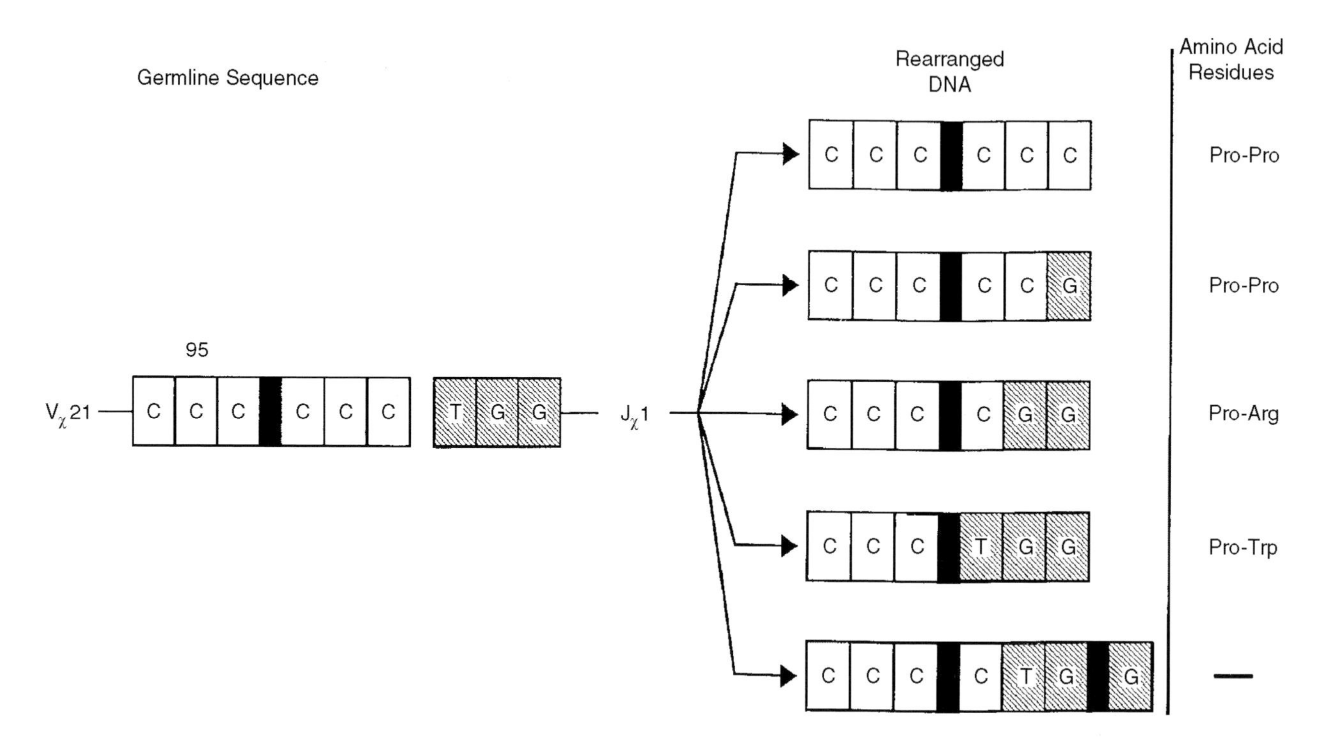

FIGURE 7.92 Junctional diversity.

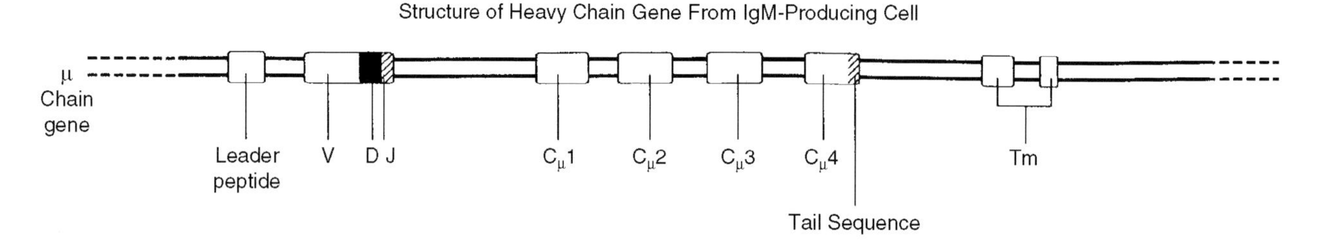

FIGURE 7.93 *D* exon and V segment.

regions in the V gene segment are separated by a 100- to 400-bp intron. The first 5′ coding region is an exon that codes for a brief untranslated mRNA region and for the first 15 to 18 signal peptide residues. The second 3′ coding region is part of an exon that codes for the terminal 4 signal peptide residues and 95 to 100 variable region residues. A J gene segment encodes the rest of the variable region. A D gene segment is involved in the encoding of immunoglobulin heavy chain and T cell receptor β and δ chains.

A **variable region** is that segment of an immunoglobulin molecule or antibody formed by the variable domain of a light polypeptide chain (κ or λ) or of a heavy polypeptide chain (α, γ, μ, δ, or ε). This is sometimes referred to as the Fv region and is encoded by the V gene. This antigen-binding region of the molecule is responsible for the specificity of the antigen bound. The antigen-binding variable sequences are present in the extended loop structures or hypervariable segment. Refers also to the variable region (V) region of T cell receptor α, β, γ, δ chains that contain variable amino acid sequences.

V region subgroups are individual chain V region subdivisions based on significant homology in amino acid sequence.

Framework regions (FR): Relatively invariant regions within variable domains of immunoglobulins and T cell receptors that constitute a protein scaffold for hypervariable regions interacting with antigen. Amino acid sequences in variable regions of heavy or light immunoglobulin chains other than the hypervariable sequences. There is much less variability in the framework region than in the hypervariable region. Two β-pleated sheets opposing one another comprise the structural features of an antibody domain's framework regions. Polypeptide chain loops join the β-pleated sheet strands. The framework regions contribute to the secondary and tertiary structure of the variable region domain, although they are less significant than the hypervariable regions for the antigen-binding site. The framework region forms the folding part of the immunoglobulin molecule. Light chain FRs are found at amino acid residues 1 to 28, 38 to 50, 56 to 89, and 97 to 107. Heavy chain FRs are present at amino acid residues 1 to 31, 35 to 49, 66 to 101, and 110 to 117.

Recombination activating genes (RAG-1 and RAG-2) are genes that activate immunoglobulin (Ig) gene recombination. Pre-B cells and immature T cells contain them. It remains to be determined whether RAG-1 and RAG-2 gene products are requisite for rearrangements of both Ig and TCR genes. In the absence of these genes, no Ig or TCR proteins are produced, which blocks the production of mature T and B cells.

A **genotype** is an organism's genetic makeup.

A **haplotype** consists of those phenotypic characteristics encoded by closely linked genes on one chromosome inherited from one parent. It frequently describes several MHC alleles on a single chromosome. Selected haplotypes are in strong linkage disequilibrium between alleles of different loci. According to Mendelian genetics, 25% of siblings will share both haplotypes.

b4, b5, b6, and b9 are the four alleles whose κ chains vary in multiple constant region amino acid residues.

Bystander B cells are nonantigen-specific B cells in the area of B cells specific for antigen. Released cytokines activate bystander B cells that synthesize nonspecific antibody following immunogenic challenge.

Forbidden clone theory: According to this hypothesis, self-reactive lymphocyte clones are eliminated in the thymus during embryonic life, but mutation later on may permit the reappearance of self-reactive clones of lymphocytes that induce autoimmunity.

Heat-labile antibody is an antigen-specific immunoglobulin that loses its ability to interact with antigen following exposure to heating at 56° C for 30 min.

Plasma half-life ($T_{1/2}$) refers to determination of the catabolic rate of any component of the blood plasma. With respect to immunoglobulins, it is the time required for one half of the plasma immunoglobulins to be catabolized.

Plasma pool is the amount of plasma immunoglobulin per unit of body weight. This may be designated as milligrams of immunoglobulin per kilogram of body weight.

Polyclone proteins are protein molecules from multiple cell clones.

Recombination signal sequences (RSSs) are abbreviated stretches of DNA that flank the gene segments that are rearranged to generate a V-region exon. They consist of a conserved heptamer and nonamer separated by 12 or 23 bp. Gene segments are joined only if one is flanked by an RSS containing a 12-bp spacer and the other is flanked by an RSS containing a 23-bp spacer. This is referred to as the 12-23 rule of gene segment joining.

Tryptic peptides are peptides formed by tryptic digestion of protein.

Heterocytotropic antibody is an antibody that has a greater affinity when fixed to mast cells of a species other than the one in which the antibody is produced. It is frequently assayed by skin-fixing ability, as revealed through the passive cutaneous anaphylaxis test. Interaction with the antigen for which these "fixed" antibodies are specific may lead to local heterocytotrophic anaphylaxis.

Binding protein is also called immunoglobulin heavy-chain binding protein.

T lymphocyte–T lymphocyte cooperation refers to signals from one T lymphocyte subpopulation to another, such as isotype class switching in the regulation of immunologic responsiveness.

V region subgroups are individual chain V region subdivisions based on significant homology in amino acid sequence.

8 Antigen–Antibody Interactions

Serology is the study of the *in vitro* reaction of antibodies in blood serum with antigens, i.e., usually those of microorganisms inducing infectious disease. Precipitation, agglutination, and complement fixation are serological methods used in diagnosis and research.

Stimulation of B lymphoid cells by antigen leads to the formation of immunoglobulin molecules (antibodies), which may enter into a number of different types of immunological and chemical reactions. These have been classified into (1) primary, (2) secondary, and (3) tertiary reactions.

The **primary reaction** is the actual binding of antibody, via its Fab or antigen binding fragment, to its homologous antigen forming an antibody–antigen complex (Figure 8.1). After the two substances are brought into contact, their initial union takes place almost instantaneously (within milliseconds).

Primary interaction refers to antigen and antibody binding that may or may not lead to a secondary visible reaction such as precipitation. Primary antigen–antibody interaction may be measured by equilibrium dialysis, the Farr assay, fluorescence polarization, fluorescence quenching, and selected radioimmunoassays such as radioimmunoelectrophoresis.

Antigen-binding capacity is determined by assay of the total capacity of antibody of all immunoglobulin classes to bind antigen. This refers to primary as opposed to secondary or tertiary manifestations of the antigen–antibody interaction. Equilibrium dialysis measures the antigen-binding capacity of antibodies with the homologous hapten, and the Farr test measures primary binding of protein antigens with the homologous antibody.

Valence is the number of antigen-binding sites on an antibody molecule or the number of antibody-binding sites on an antigen molecule. The valence is equal to the maximum number of antigen molecules that can combine with a single antibody molecule. Whereas antigens are usually multivalent, most antibodies are bivalent. IgM has a valence as high as 10, and IgA has a valence that differs depending on its level of polymerization. Antibodies of the IgG, IgD, and IgE classes have a valence of 2. Due to steric hindrance, antibodies usually bind less antigen than would be expected from their valence. Whereas antibody molecules are bivalent or multivalent, T lymphocyte receptors are univalent.

Polyvalent is a synonym for multivalent.

Heterocliticity is the preferential binding by an antibody to an epitope other than the one that generates synthesis of the antibody.

In immunology, **multivalent** refers to antibody or antigen molecules with a combining power greater than two.

Polyspecificity is the capacity to bind many different antigens. Also termed polyreactivity.

Multivalent antiserum is an immune serum preparation containing antibodies specific for more than two antigens. Multivalent means possessing more than two binding sites.

Secondary reactions are those visible effects resulting from antibody–antigen binding such as precipitation, agglutination, flocculation, complement fixation, and so on.

Tertiary reactions, which may result from either primary or secondary interactions of antibody with antigen, include those *in vivo* biological manifestations of antibody reactivity. Some *in vitro* secondary interactions, such as cytophilic reactions (adherence of the antibody via its Fc to a cell surface) may, when occurring *in vivo,* give rise to tertiary manifestations. Because reactions occur *in vivo*, they tend to be very complex and are subject to many variables.

In an immune response, antibodies are directed against specific conformational areas on the antigen molecule referred to as antigenic determinants. Antigens are macromolecules which stimulate antibody production. Antibody populations directed against these macromolecules are notoriously heterogeneous with respect to their antibody specificity and affinity since antibodies to different antigenic determinants may be present simultaneously in the sera. Bivalent and multivalent antibodies directed against multideterminant antigens result in the formation of large antibody–antigen aggregates of the type $(Ab)_x(Ag)_y$ varying in size, complexity, and solubility.

Landsteiner devised a method whereby an immune response could be directed against small molecules of known structure. He referred to these substances as haptens, which by themselves were too small to initiate an immune response, but were capable of reacting with the products of an immune response. He chemically coupled

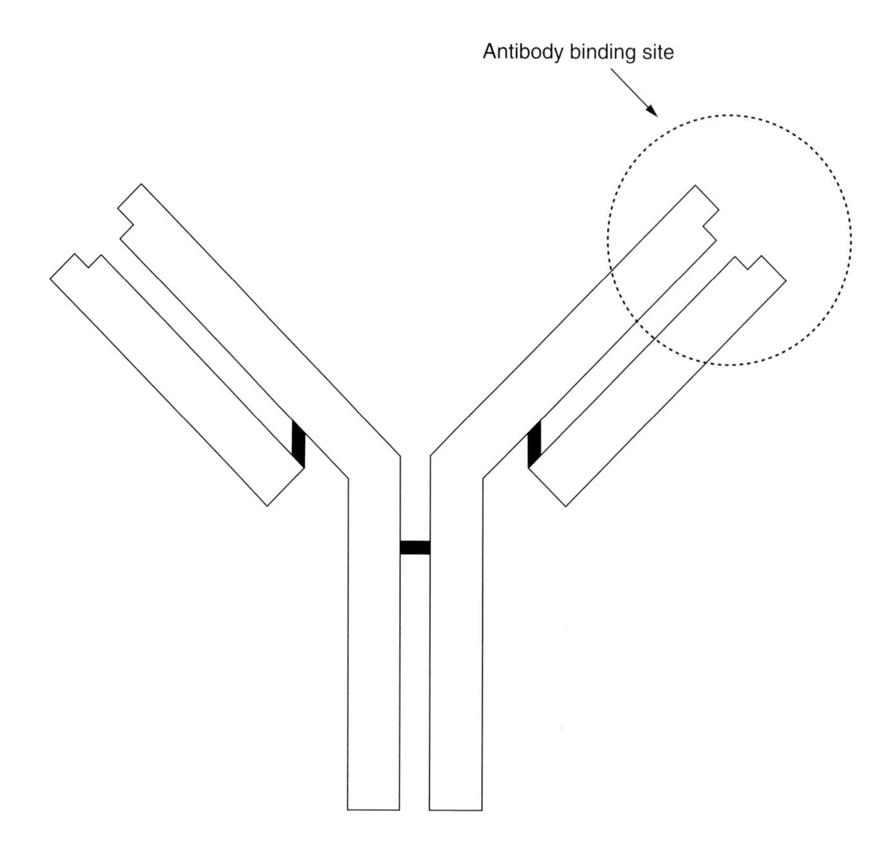

FIGURE 8.1 Antibody binding site.

these haptens to large biological macromolecules, such as ovalbumin, which he termed *carriers*, producing conjugated antigens capable of stimulating an immune response. The nonvalent hapten in pure form, together with serum antibodies, could then be used to study antibody–hapten interactions without the complications of multideterminant macromolecular antigens. These antibody populations were still heterogeneous with respect to structure, class, subclass, and so on. Isolation of monoclonal antibodies derived from the blood sera of multiple myeloma patients led to sequencing and x-ray diffraction studies on homogenous antibody populations. Recently, hybridoma technology has provided a mechanism to produce monoclonal homogenous antibodies *in vitro*.

The **noncovalent forces** of the antibody–antigen complex include hydrogen bonding, ionic or Coulombic bonding, Van der Waals interactions, hydrophobic bonding, and steric repulsion forces which are extremely sensitive to the distance between the interacting groups. Although charge on the antigen is not required for antigen–antibody binding, it may play a very important role in determining the stability of the antigen–antibody complex. There are discrepancies in the literature as to whether charged antigens elicit antibodies of reciprocal charge. It appears that the charge effect *per se* is exerted by microenvironment of the antigen and antibody molecules (such as pH, ionic

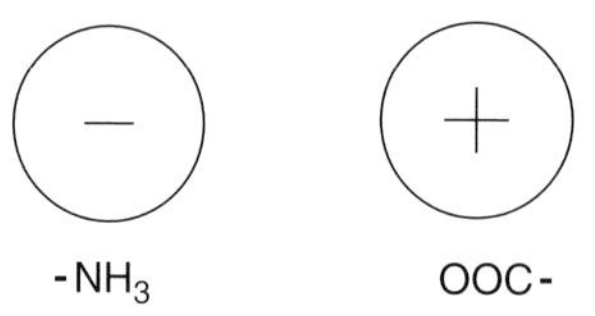

FIGURE 8.2 Electrostatic forces.

strength of solution, etc.) and not such so much by the net of the molecules as a whole.

Ionic or **Coulombic forces** of attraction result from the interaction between oppositely charged ionic groups on the antigen and antibody molecules (Figure 8.2). As can be seen from the equation of Coulomb:

$$F = \frac{Q^+Q^-}{\varepsilon r^2}$$

where ε is the dielectric constant of the medium, Q^+ and Q^- are the positive and negative charges in electrostatic units, respectively, and r is the distance between the centers of the charged sites. The Coulombic force of attraction is inversely proportional to the square of the distances between antigen and antibody.

Apolar or hydrophobic bonding is of considerable importance in the antigen–antibody complex when it is in an aqueous environment and may contribute greatly to its stabilization. A net attractive force resulting from a decrease in energy is that obtained from the preference of apolar or hydrophobic regions of the interacting molecules to associate with themselves rather than with solvent molecules (H_2O). The reaction is endothermic ($\Delta H > 0$); in order for it to be spontaneous ($\Delta H < 0$) it must occur through a concomitant increase in entropy through an entropy-driven reaction. This can be reviewed from the following thermodynamic relationship:

$$\Delta G = \Delta H - T\Delta S$$

where ΔG is the free energy change, ΔH is the enthalpy charge, ΔS is the entropy change, and T is the absolute temperature. When $\Delta H > 0$ (endothermic reaction), a positive ΔS is needed for the overall energy decrease ($\Delta G < 0$), resulting in the attractive force. The binding of this attractive force increases (ΔG becomes more negative) as ΔH decreases and as the temperature T increases.

Hydrogen bonds are formed between hydrogen atoms covalently linked to an electronegative atom and a second electronegative atom containing an unshared pair of electrons (Figure 8.3). The hydrogen atom becomes electron deficient through polarization of its electron cloud towards the electronegative atom covalently bonded to it, allowing for an electrostactic attraction to a relatively negative second electronegative atom. The contribution of hydrogen bonding to the stability of the complex is minor compared to the other forces involved and decreases with the sixth power of the distance between interaction groups.

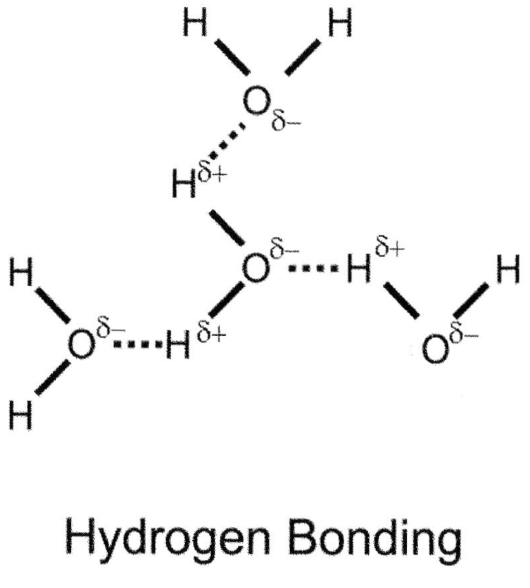

FIGURE 8.3 The hydrogen bonds are shown in dotted lines.

Hydrophilic is an adjective that describes a water-soluble substance. A cell membrane or protein that contains hydrophilic groups on its surface which attract water molecules.

Hydrophobic is a descriptor for a substance that is insoluble in water. Protein or membrane hydrophobic groups are situated inside these structures away from water.

Avidity is the strength of binding between an antibody and its specific antigen. The stability of this union is a reflection of the number of binding sites which they share. Avidity is the binding force or intensity between multivalent antigen and a multivalent antibody. Multiple binding sites on both the antigen and the antibody, e.g., IgM or multiple antibodies interacting with various epitopes on the antigen, and reactions of high affinity between each of the antigens and its homologous antibody all increase the avidity. Nonspecific factors such as ionic and hydrophobic interactions also increase avidity. Whereas affinity is described in thermodynamic terms, avidity is not, since it is described according to the assay procedure employed. The sum of the forces contributing to the avidity of an antigen–antibody interaction may be greater than the strength of binding of the individual antibody–antigen combinations contributing to the overall avidity of a particular interaction. Ka, the association constant for Ab + Ag = AbAg interaction, is frequently used to indicate avidity. Avidity may also describe the strength of cell-to-cell interactions, which are mediated by numerous binding interactions between cell surface molecules. Avidity is distinct from affinity, which describes the strength of binding between a single molecular site and its ligand.

Previously termed the affinity hypothesis of T cell selection in the thymus, the **avidity hypothesis** is based on the concept that T cells must have a measurable affinity for self major histocompatibility complex (MHC) molecules to mature but not an affinity sufficient to cause activation of the cell when it matures, since this would necessitate deletion of the cell to maintain self tolerance.

Affinity is the strength of binding between antigen and antibody molecules. It increases with linkage stability. The affinity constant reflects the strength of binding. The paratope of an antibody molecule views the epitope as a three-dimensional structure. Affinity (K_A) is a thermodynamic parameter that quantifies the strength of the association between antigen and antibody molecules in solution. It applies to a single species of antibody-combining sites interacting with a single species of antigen-binding sites. Affinity describes the strength of binding between a single binding site of an antibody molecule and its ligand or antigen. The dissociation constant (K_d) represents the affinity of molecule A for ligand B. This is the concentration of B required to occupy the combining sites of half the A molecules in solution. A smaller K_d signifies a stronger or higher affinity interaction and a lower concentration of ligand is required to bind with the sites.

Antibody affinity is the force of binding of one antibody molecule's paratope with its homologous epitope on the antigen molecule. It is a consequence of positive and negative portions affecting these molecular interactions.

Affinity constant: Determination of the equilibrium constant through application of the Law of Mass Action to interaction of an epitope or hapten with its homologous antibody. The lack of covalent bonds between the interacting antigen and antibody permits a reversible reaction.

A **Scatchard plot** is a graphic representation of binding data obtained by plotting r/c against r (see the Scatchard equation). The purpose of this plot is to determine intrinsic association constants and to ascertain how many noninteracting binding sites each molecule contains. A straight line with a slope of –K indicates that all the binding sites are the same and are independent. The plot should also intercept on the *r* axis of *n*. A nonlinear plot signifies that the binding sites are not the same and are not independent. The degree to which the sites are occupied is reflected by the slope (–K). An average association constant for ligand binding to heterogeneous antibodies is the reciprocal of the amount of free ligand needed for half saturation of antibody sites.

Scatchard equation: In immunology, an expression of the union of a univalent ligand with an antibody molecule. r/c = K – Kr. To obtain the average number of ligand molecules which an antibody molecule may bind at equilibrium, the bound ligand molar concentration is divided by the antibody molar concentration. This is designated as r. The free ligand molar concentration is represented by c. Antibody valence is designated by n, and the association constant is represented by K.

Scatchard analysis is a mathematical analytical method to determine the affinity and valence of a receptor–ligand interaction in equilibrium binding.

Antibody–antigen intermolecular forces: The various types of bonds that participate in the specific interaction between antibody and antigen molecules are relatively weak physical forces. They fall into three classes that include: (1) van der Waals or electrodynamic forces, (2) hydrogen bonding or polar forces, and (3) electrostatic forces. Covalent bonds are not involved in antibody–antigen interactions.

van der Waals forces (London forces) are weak forces of attraction between atoms, ions, and molecules. It is active only at short distances since this force varies inversely to the seventh power of the distance between ions or molecules. Thus, van der Waals forces may be important in antigen–antibody binding.

The van der Waals forces (or London forces) contribute somewhat to the stabilization of the antigen–antibody complex that results from attraction of oscillating dipoles of atoms and molecules moving their electrons from one side to another (Figure 8.4). Dispersion forces occur only when the two molecules are very close together and decrease with the sixth power of the distance between interaction sites.

The major attractive forces involved in the antigen–antibody complex have been described and are all inversely proportional to the distance between interacting groups. **Steric repulsion**, on the other hand, results in a repulsive force and tends to decrease the stability of the complex. This repulsive force results from the interpenetration of the electron clouds of the antigenic determinant and the antigen-binding region of the antibody. When the electron clouds of the antigenic determinant and the homologous antigen-binding site on the antibody molecule are not complementary, the steric repulsion between the two becomes great and decreases the stability of the complex. By contrast, when complementary electron clouds come together, steric repulsion forces are minimized. This permits a closer association between the two interacting molecules and increases the attractive forces described above, leading to formation of a stable complex.

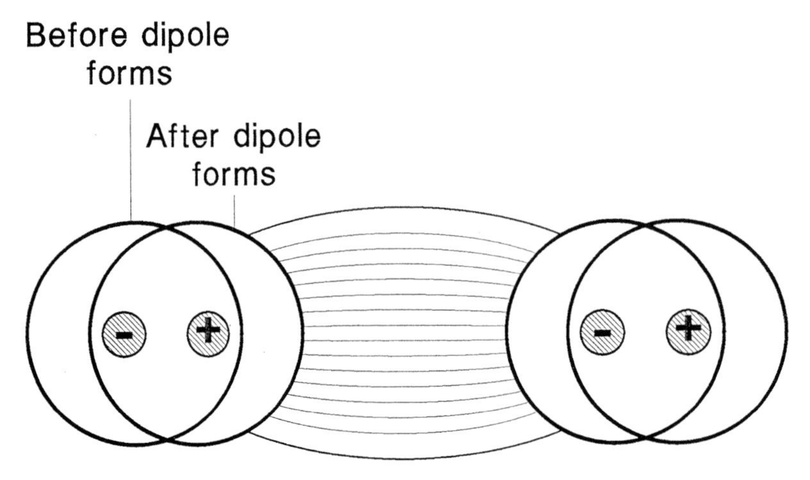

FIGURE 8.4 van der Waals forces.

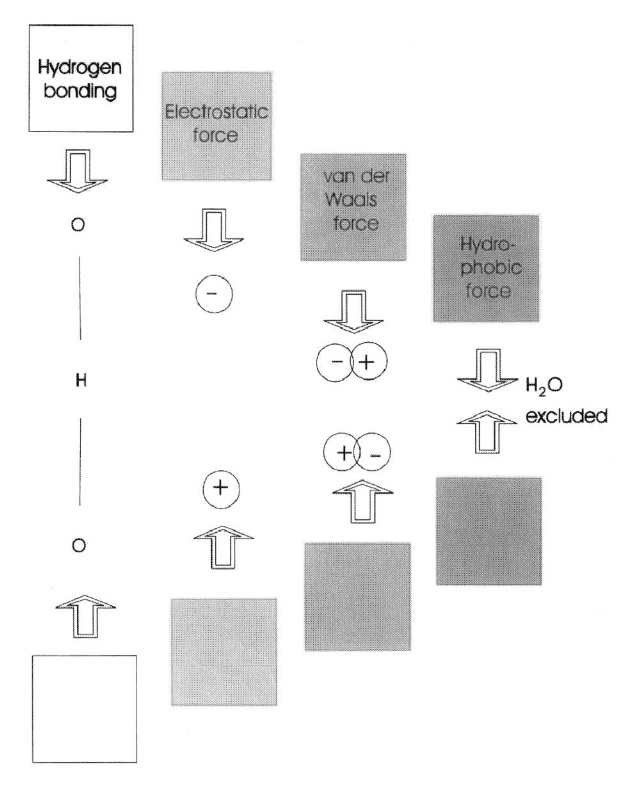

FIGURE 8.5 Attractive forces binding antigen to antibody.

Thus, steric repulsion provides the basis for the antigenic specificity of the antigen–antibody reaction.

In summary, the forces that account for the stability of an antigen–antibody complex are the following: (1) attractive forces resulting in increased binding or stability, and (2) repulsive forces resulting in decreased binding or stability (Figure 8.5).

The stability of the antigen–antibody complex, expressed as antibody affinity, is actually a sum of all attractive and repulsive forces acting at a given time (Figure 8.6).

When attractive forces exceed repulsive forces, an antigen–antibody complex may result if given sufficient time for interaction. Measurement of the quantity of complexes formed with respect to time (kinetically derived quantity) is referred to as **avidity**. Avidity may be measured by determining the time it takes for a given amount of radiolabeled antigen to dissociate from antibody. Differences in avidity of various antisera can be seen from a comparison of their precipitation curves (Figure 8.7).

In order for the classical laws of thermodynamics to apply, the following two assumptions must be made: (1) the reactants, both antigen and antibody, must be pure and in solution, and (2) the reactants must be homogeneous with

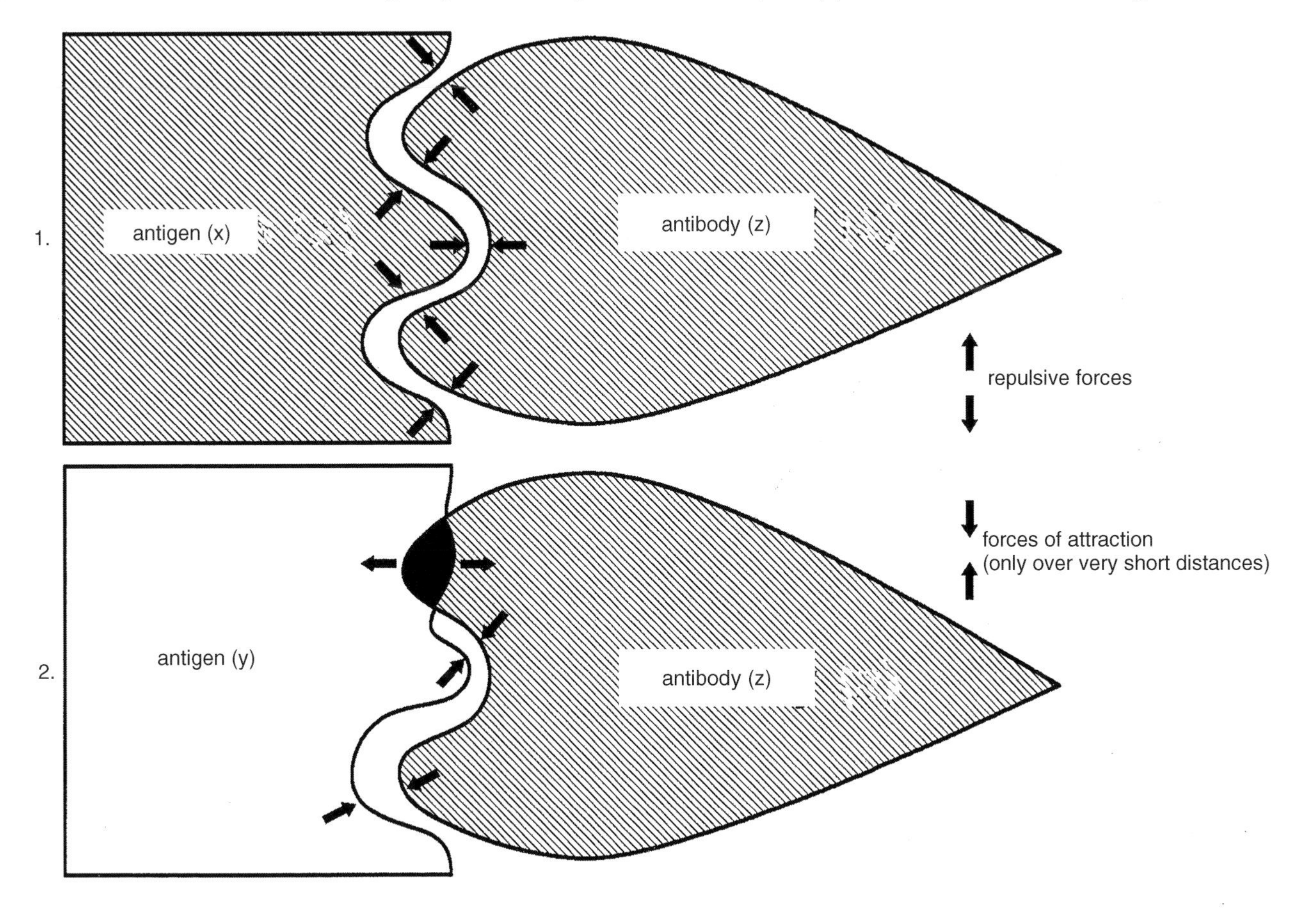

FIGURE 8.6 Antibody affinity.

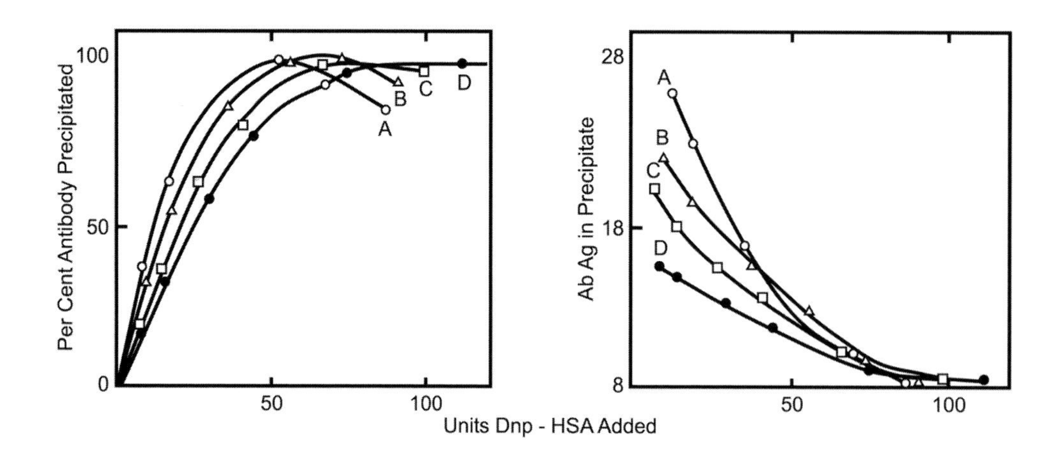

FIGURE 8.7 Precipitation curves showing differences in avidity of four antisera for the same antigen. The order of avidity of the sera is A>B>C>D.

regard to antigen binding sites and antigenic determinants. Though antibodies are extremely heterogeneous with respect to structure as well as multivalent with respect to antigen binding sites, measurements can be performed when purified antibodies are used to define monovalent haptens .

Using univalent haptens (H) and antibodies (Ab), the elementary reversible antibody–hapten reaction that gives rise to complexes of the AbH can be written as follows:

$$Ab + H \underset{k_r}{\overset{k_f}{\rightleftharpoons}} AbH$$

where k_f is the rate constant for the association reaction and k_r is the rate constant for the dissociation reaction, assuming that all antibody binding sites are identical and independent of each other. By applying the law of mass action, the equilibrium constant K, also called antibody affinity for reaction, can be derived as follows:

$$k_f[Ab][H] = k_r[AbH]$$

$$K = \frac{k_f}{k_r} = \frac{[AbH]}{[Ab][H]}$$

where [Ab], and [H], [AbH] are the free antibody concentration, free hapten concentration, and the antibody–hapten complex concentration (or the concentration of bound hapten or antibody sites occupied), respectively. Standard free energy change $\Delta G°$ for the association reaction of hapten and antibody is related to the equilibrium constant K by the following:

$$\Delta G^o = -RT \ln K$$

where R is the gas constant and T the temperature in degrees Kelvin. Affinity is a thermodynamically derived quantity which may be expressed by either K, which becomes more positive as affinity increases, or $\Delta G°$, which becomes more negative as affinity increases. Antibody affinity may be interpreted as measuring either (1) the strength of binding of the antibody to its homologous antigenic determinant or (2) the stability of the hapten–antibody complex. Avidity is really a measure of the "stickiness" of antibody toward hapten and is not, in actuality, a thermodynamically derived quantity as is affinity.

Having calculated the equilibrium constant, or antibody affinity K, standard enthalpy change $\Delta H°$ for the reaction can be calculated from the Van't Hoff equation which expresses the change in K as a function of T.

$$\frac{d(\ln K)}{dT} = \frac{-\Delta H°}{RT^2}$$

Integration of Equation 4 from T_1 to T_2 gives

$$\int_{T_1}^{T_2} d(\ln K) = \int_{T_1}^{T_2} \frac{-\Delta H°}{RT^2} = dT$$

which yields

$$\ln \frac{K_2}{K_1} = \frac{\Delta H°(T_2 - T_1)}{R(T_2 T_1)}$$

where K_2 and K_1 are the equilibrium constants at T_2 and T_1, respectively. Algebraic rearrangement gives the following expression for $\Delta H°$:

$$\Delta H° = \frac{\ln K_2 - \ln K_1}{\frac{1}{T_1} - \frac{1}{T_2}}$$

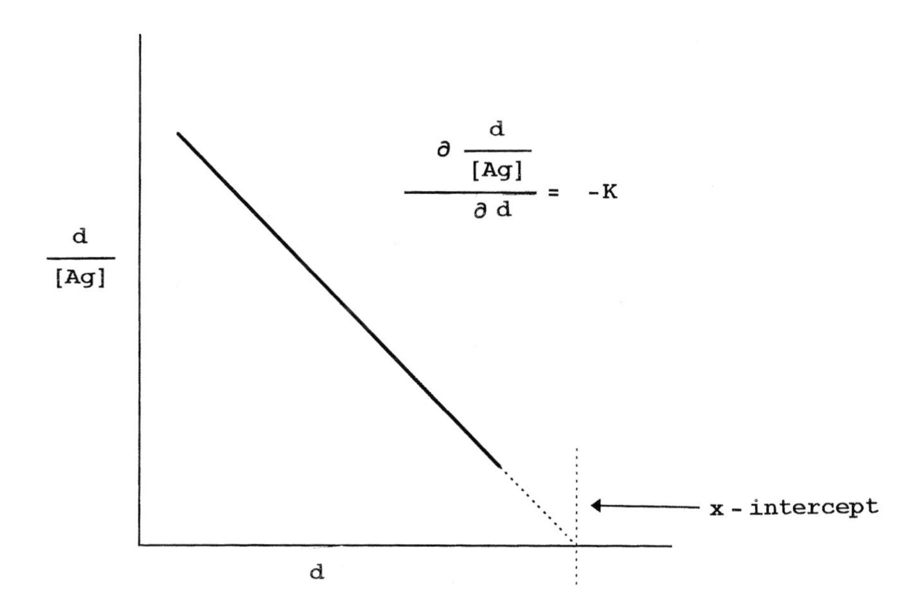

FIGURE 8.8 Scatchard plot.

Application of the Langmuir adsorption isotherm equation to antibody–hapten reactions has been shown to be useful in affinity measurements as well as in calculations of antibody valency. In an antibody–hapten solution where [H] is the concentration of free hapten, let d equal the fraction of antigen binding sites occupied and (1 – d) equal the fraction of available antigen binding sites.

Since the rate of reaction is directly proportional to the number of available antigen binding sites, the rate of the forward reaction, rate f, and the rate of the reverse reaction, rate r, at a given temperature T may be written as follows:

$$\text{rate f} = K_f\,(1 - d)\,[H]$$

$$\text{rate r} = K_r\,(d)$$

At equilibrium the forward and reverse rates are equal; therefore,

$$\text{rate f} = \text{rate r} \quad (1)$$

$$K_f\,(1 - d)\,[H] = K_r(d)$$

Solving for d gives the Langmuir adsorption isotherm equation as applied to the antigen–hapten reaction for univalent antibody and hapten:

$$d = \frac{K[H]}{1 + K[H]} \quad (2)$$

For antibody of valency n, the following also can be shown:

$$\frac{d}{n} = \frac{K[H]}{1 + K[H]} \quad (3)$$

Algebraic rearrangement of the Langmuir adsorption isotherm yields the Scatchard Equation

$$\frac{d}{[Ag]} = nK - dK \quad (4)$$

from which a plot of d/[Ag] vs. d for the ideal antibody–hapten system over a range of free hapten concentrations [H] gives a straight line with slope –K (Figure 8.8).

Extrapolation to the x-axis gives antibody valency n. The Scatchard plot allows for the calculation of antibody affinity and valency from the concentration of antigen and antigen–antibody complex. It can be shown from the Scatchard equation that when half the antigen binding sites in bivalent antibody (n = 2) are hapten bound (d = 1), K is equal to 1/[*Ag*].

$$K_0 = 2K - K = \frac{1}{[H]} \quad (5)$$

The average intrinsic association constant K_0 is thus defined as the reciprocal of free hapten concentration at equilibrium when half of the antigen binding sites on antibody molecules are bound by hapten.

Another method for obtaining antibody affinity and valency is the Langmuir plot using the following rearrangement of Equation 3.

$$\frac{1}{d} = \frac{1}{n}\cdot\frac{1}{[H]}\cdot\frac{1}{K} + \frac{1}{n} \quad (6)$$

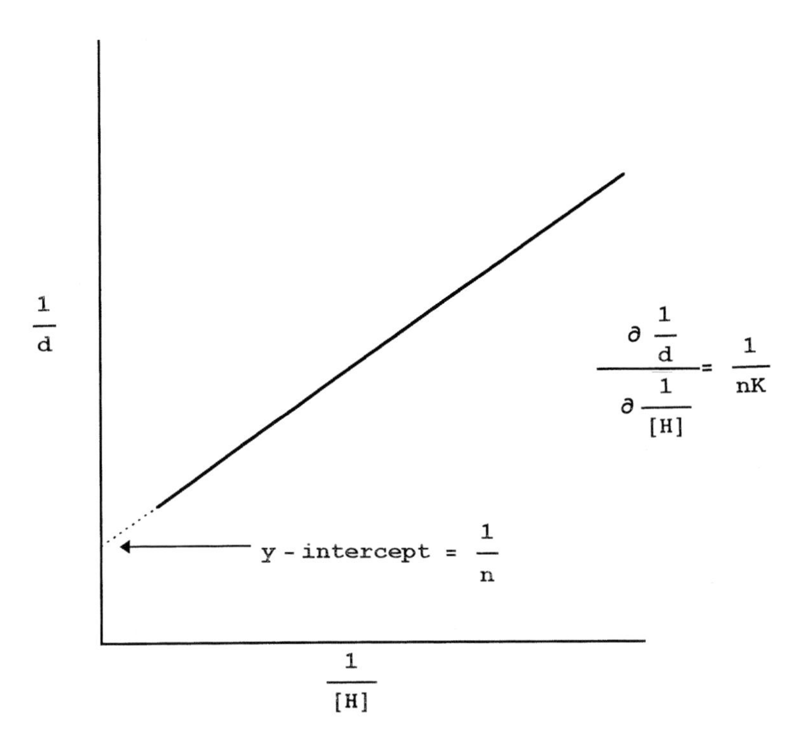

FIGURE 8.9 Langmuir plot.

A plot of l/d vs. reciprocal hapten concentration, the Langmuir plot, results in a slope of l/nK and y-intercept of l/n (Figure 8.9).

The following method may also be employed for calculation of antibody affinity which only requires the measurement of free and bound hapten concentrations. This is done by solving Equation 3 with respect to the antibody–hapten complex. In the following derivation, [AbH] is the hapten–antibody complex concentration or the concentration of bound antibody combining sites and $[Ab_t]$ is the concentration of total antibody combining sites.

$$\frac{d}{n} = \frac{K[H]}{1+K[H]} \qquad (7)$$

then becomes

$$\frac{[AbH]}{[Ab_t]} = \frac{K[H]}{1+K[H]} \qquad (8)$$

Rearrangement to a more suitable form for plotting gives

$$\frac{1}{AbH} = \frac{1}{[Ab_t]K[H]} + \frac{1}{[Ab_t]} \qquad (9)$$

A plot of 1/[AbH] vs. 1/[H] gives a line with a y-intercept of $1[Ab_t]$ and slope for $1/K[Ab_t]$. Thus antibody affinity is equal to the product of the y-intercept and the reciprocal of the slope (Figure 8.10).

Both the Scatchard and Langmuir plots theoretically give rise to linear relationships when applied to ideal antigen–antibody reactions. In reality, however, the plots may deviate considerably from linearity due to the heterogeneity of antibody affinities within an antibody population. Antibody heterogeneity has long been known and was originally described in terms of the Gaussian distribution function by Heidelberger and Kendall in 1935. A more modern approach to quantification of this affinity distribution uses the logarithmic transformation of the Sipsian distribution function,

$$\log\frac{d}{n-d} = a\log K_0 + a\log[Ag]$$

where d represents the moles of antigen or hapten bound per mole of antibody, k_0 the average intrinsic association constant, and a the index of heterogeneity. A plot of log d/(n – d) vs. log [Ag] yields a line of slope a (Figure 8.11).

The antibody population approaches homogeneity with respect to K as the heterogeneity index approaches unity. In addition K_0 can be obtained from the graph by extrapolation since K_0 is equal to 1/[Ag] when the log d/(n – d) is equal to 0 and represents the peak of the distribution.

Actual data can be obtained in the lab. An adequate means of measuring such quantities as bound and unbound hapten

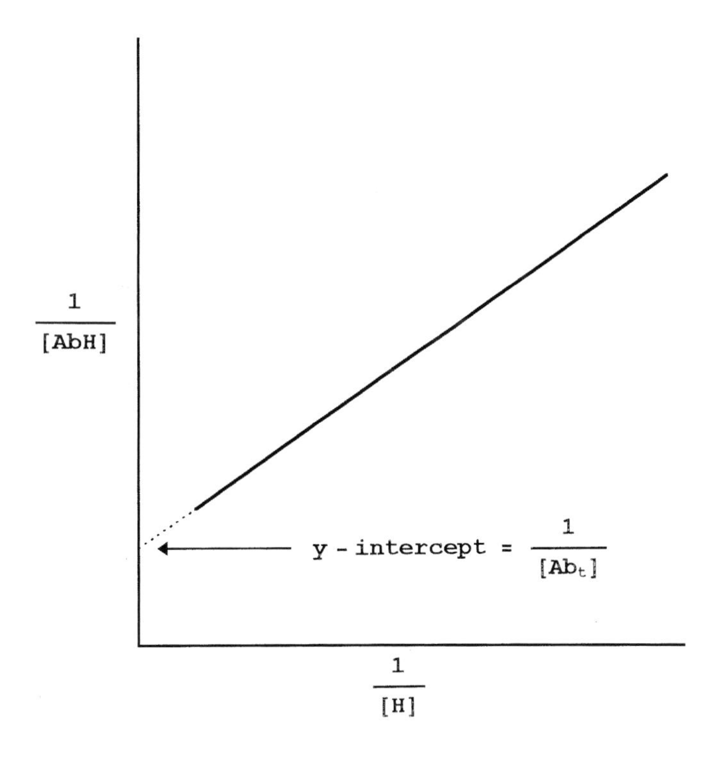

FIGURE 8.10 Modification of Langmuir plot.

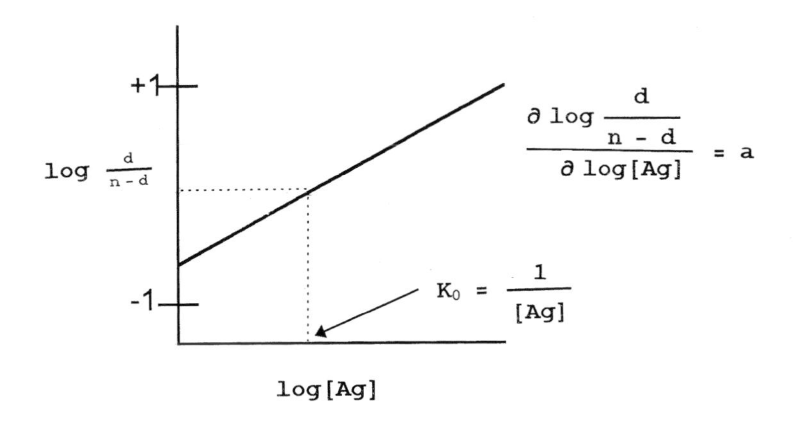

FIGURE 8.11 Plot of Sipsian distribution function.

concentrations must be available for calculations. The following three methods have been popular in the study of primary interaction between antibody and antigen.

Equilibrium dialysis was developed for the study of primary antibody–hapten interactions (Figure 8.12). The basis for the technique is as follows: Two cells are separated by a semipermeable membrane allowing free passage of hapten molecules but not larger antibody molecules. At time zero (t_0), there is a known concentration of hapten in cell A and antibody in cell B. Hapten from cell A diffuses across the membrane into cell B until, at equilibrium, the concentration of free hapten is the same in both cells A and B; that is, the rate of diffusion of hapten from cell A to B is the same as that from cell B to A. Though the concentrations of free hapten are the same in both cells, the total amount of hapten in cell B is greater because some of the hapten is bound to the antibody molecules. In order to obtain d and [H], a series of experiments are performed which varies the starting amount of hapten concentration while keeping antibody concentration constant.

Association constant (K_A) is a mathematical measurement of the reversible interaction between two molecular forms at equilibrium. The AB complex, free A and B concentrations at equilibrium, are expressed in K_A l/mol by [AB], [A], and [B]. The s (molecules of substance A) interact reversibly with the t (molecules of substance B); i.e., sA + tB1 A_sB_t; the association constant is $[A_sB_t]/[A]^s[B]^t$.

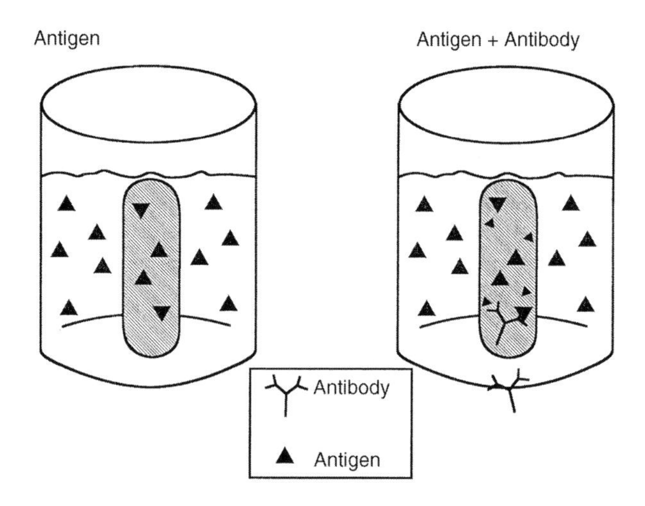

FIGURE 8.12 Equilibrium dialysis.

Molar concentrations at equilibrium are indicated by the symbols in brackets.

Surface plasmon resonance (SRP) is a phenomenon that is the basis for instruments designed to measure macromolecular interactions in real-time. It measures alterations in the refractive index of the medium surrounding a receptor immobilized on a solid support, which take place when a ligand binds. SRP is useful in the analysis of antigen–antibody interactions and can be employed to determine kinetic parameters (association and dissociation rate constants) and equilibrium binding constants, as well as measure concentrations, perform epitope mapping, and determine antibody isotypes.

In immunology, **steric hindrance** refers to interference between the interaction of the paratope of an antibody molecule with the homologous epitope on antigen molecules of varying sizes based on the shapes of the two reactants. Whereas IgM molecules potentially have an antigen-binding capacity of ten, only some of these may be able to interact with relatively large antigen molecules bearing epitopes because of the shapes of the two. By contrast, relatively small antigen molecules would be able to permit their epitopes to bind with more paratopes on the IgM molecule. Steric hindrance also refers to the blocking of ligand binding when a receptor site is already occupied by another ligand.

Dissociation constant refers to the equilibrium constant for dissociation. This is usually described in enzyme–substrate interactions. If the interaction of enzyme with substrate attains equilibrium prior to catalysis, the Michaelis constant (K_M) represents the dissociation constant. The Michaelis constant is equivalent to the concentration substrate when the reaction velocity is half maximal.

Binding constant: See association constant.

Intrinsic association constant describes univalent ligand binding to a special site on a protein macromolecule if all sites of this type are identical and noninteracting when found on the same molecule. Equilibrium dialysis and other techniques that determine bound and free ligand concentrations evaluate the intrinsic association constant.

Intrinsic affinity is a synonym for intrinsic association constant.

Sips distribution is the frequency distribution of antibody association constants in a heterogeneous mixture. The Sips distribution is very similar to the Gaussian (normal) distribution. It is employed to analyze data from antigen–antibody reactions measured by equilibrium dialysis.

A **Sips plot** is data representation produced in assaying ligand binding to antibodies by plotting log r/n-r against log c. A straight line signifies that the data are in agreement with the Sips equation. In this instance, the slope signifies heterogeneity of antibody affinity.

The **ammonium sulfate method** is a means of measuring the primary antigen-binding capacity of antisera and detects both precipitating and nonprecipitating antibodies. It offers an advantage over equilibrium dialysis in that large, nondializable protein antigens may be used. This assay is based on the principle that certain proteins are soluble in 50% saturated ammonium sulfate, whereas antigen–antibody complexes are not. Complexes may be separated from unbound antigen. Spontaneous precipitation will occur if precipitating-type antibody is used, until a point of antigen excess is reached where complex aggregation no longer occurs and soluble complexes form. Upon the addition of an equal volume of saturated ammonium sulfate solution (SAS), these complexes become insoluble, leaving radiolabeled antigen in solution. SAS fractionation does not significantly alter the stoichiometry of the antibody–antigen reaction and inhibits the release or exchange of bound antigen. The radioactivity of this "induced" precipitate is a measure of the antigen-binding capacity of the antisera as opposed to a measure of the amount of antigen or antibody spontaneously precipitated.

Fluorescence quenching is a method used to ascertain association constants of antibody molecules interacting with ligands. Fluorescence quenching results from the excitation energy transfer where certain electronically excited residues in protein molecules, such as tryptophan and tyrosine, transfer energy to a second molecule that is bound to the protein. Maximum emission is a wavelength of approximately 345 nm. The attachment of the acceptor molecule need not be covalent. This transfer of energy occurs when the absorbance spectrum of the acceptor molecule overlaps with that of the emission spectrum of the donor and takes place via resonance interaction (Figure 8.13).

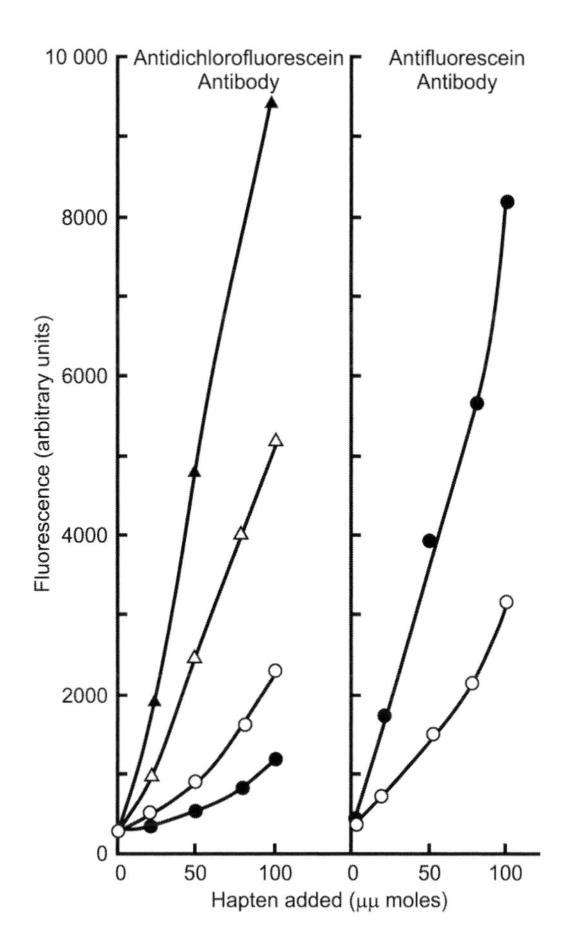

FIGURE 8.13 Titration curves using fluorescence quenching.

There is no need for direct contact between the two molecules for energy transfer. If the acceptor molecule is nonfluorescent, diminution of energy occurs through nonradiation processes. On the other hand, if the acceptor molecule is fluorescent, the transfer of radiation results in its own fluorescence (sensitized fluorescence). Fluorescence quenching techniques can provide very sensitive quantitative data on antibody–hapten interactions.

An **antigen–antibody complex** is a product of the union of antibody with soluble antigen in solution containing electrolyte. When the interaction takes place *in vitro*, it is called the precipitin reaction, but it may also take place *in vivo*. The relative proportion in which antigen and antibody combine varies their molar ratio. Excess antigen may lead to soluble complexes, whereas excess antibody may lead to insoluble complexes. *In vivo*, soluble complexes are more likely to produce tissue injury, whereas larger insoluble complexes are often removed by reticuloendothelial system cells. Also called immune complex.

Immune elimination is the accelerated removal of an antigen from the blood circulation following its interaction with specific antibody and elimination of the antigen–antibody complexes through the mononuclear phagocyte system. A few days following antigen administration, antibodies appear in the circulation and eliminate the antigen at a much more rapid rate than that occuring in nonimmune individuals. Splenic and liver macrophages express Fc receptors that bind antigen–antibody complexes and also complement receptors which bind those immune complexes that have already fixed complement. This is followed by removal of immune complexes through the phagocytic action of mononuclear phagocytes. Immune elimination also describes an assay to evaluate the antibody response by monitoring the rate at which a radiolabeled antigen is eliminated in an animal with specific (homologous) antibodies in the circulation.

Following the union of soluble macromolecular antigen with the homologous antibody in the presence of electrolytes *in vitro* or *in vivo*, complexes of increasing density form within seconds after contact in a lattice arrangement and settle out of solution, as in the **precipitation** or **precipitin reaction.** The materials needed for a precipitation reaction include antigen, antibody, and electrolyte. The reaction of soluble antigen and antibody in the precipitation test may be observed in liquid or in gel media The reaction in liquid media may be qualitative or quantitative. Following discovery of the precipitation reaction by Kraus in 1897, only quantitative and semiquantitative measurements of antibody could be made. The term precipitinogen is sometimes employed to designate the antigen, and precipitin is the antibody in a precipitation reaction.

Precipitation reaction: See precipitin test, precipitation, precipitation curve, and precipitation in gel media.

Immunoreaction refers to the interaction of antibody with antigen.

An **immunoreactant** is any substance, including immunoglobulins, complement components, and antigens, involved in immune reactions.

A **precipitin test** is an assay in which antibody interacts with soluble antigen in the presence of electrolyte to produce a precipitate. Both qualitative and quantitative precipitin reactions have been described.

Precipitin reaction: See precipitin test, precipitation, precipitation curve, and precipitation in gel media.

The **zone of equivalence** is that point in a precipitin antigen–antibody reaction *in vitro* where the ratio of antigen to antibody is equivalent. The supernatant contains neither free antigen nor antibody. All molecules of each have reacted to produce antigen–antibody precipitate. When a similar reaction occurs *in vivo*, immune complexes are deposited in the microvasculature, and serum sickness develops. Also called equivalence zone.

The **Dean and Webb titration** was a historically important assay for the measurement of antibody. While the quantity of antiserum is held constant, varying dilutions of antigen are added and the tube contents are mixed. The tube in which flocculation occurs first represents the endpoint. It is in this tube that the ratio of antigen to antibody is in optimal proportions.

Coprecipitation is the addition of an antibody specific for either the antigen portion or the antibody portion of immune complexes to effect their precipitation. Protein A may be added instead to precipitate soluble immune complexes. The procedure may be employed to quantify low concentrations of radiolabeled antigen that are combined with excess antibody. After soluble complexes have formed, antiimmunoglobulin or protein A is added to induce coprecipitation.

Ring precipitation test: (Figure 8.14) See ring test.

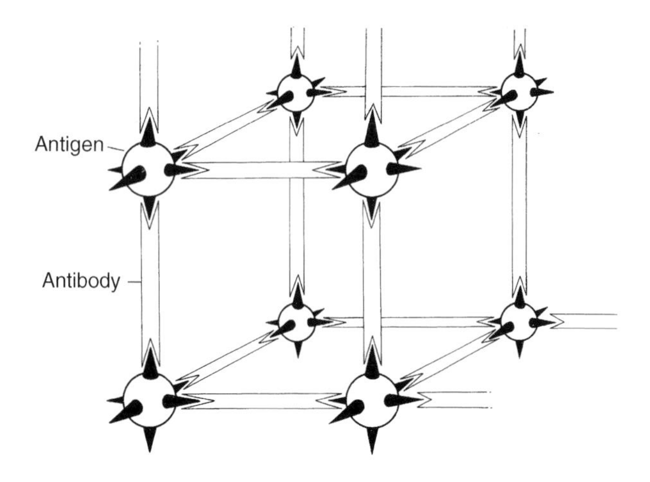

FIGURE 8.14 Precipitation test.

A **ring test** is a qualitative precipitin test used for more than a century, in which soluble antigen (or antibody) is layered onto an antibody (or antigen) solution in a serological tube or a capillary tube without agitating or mixing the two layers. If the antigen and antibody are specific for each other, a ring of precipitate will form at the interface. This simple technique was among the first antigen–antibody tests performed.

Quantitative precipitin reaction is an immunochemical assay based on the formation of an antigen–antibody precipitate in serial dilutions of the reactants, permitting combination of antigen and antibody in various proportions. The ratio of antibody to antigen is graded sequentially from one tube to the next. The optimal proportion of antigen and antibody is present in the tube that shows the most rapid flocculation and yields the greatest amount of precipitate. After washing, the precipitate can be analyzed for protein content through procedures such as the micro-Kjeldahl analysis to ascertain nitrogen content, spectrophotometric assay, or other techniques. Heidelberger and Kendall used the technique extensively, employing pneumococcus polysaccharide antigen and precipitating antibody in which nitrogen determinations reflected a quantitative measure of antibody content. The classic precipitin reaction may be illustrated using the serum of a rabbit immunized with egg albumin (Figure 8.15).

In this technique, a constant volume and concentration of rabbit antibody is placed in a row of serological tubes. Varying amounts of the egg albumin antigen are added and the tubes incubated. Let us say there is no precipitate in tube 1, a slight quantity in tube 2, a heavy amount in tubes 3, 4, and 5, a slight quantity in tube 6, and none in tube 7. All tubes are centrifuged and the supernatants tested for both unreacted antigen and antibody. There is excess antigen but no free antibody in tubes 1, 2, and 3. In the supernatant of tube 4 there is neither antigen nor antibody, therefore this tube is called the equivalence tube, where antigen and antibody are in identical proportions and completely reacted. In the supernatants of tubes 5, 6, and 7 there is antibody but no antigen. Why was there no

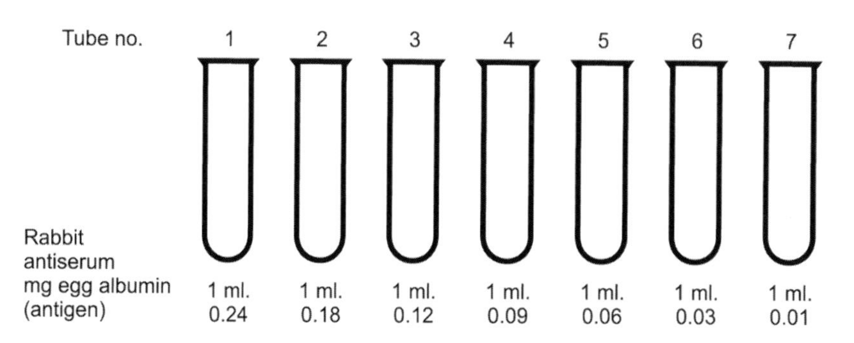

FIGURE 8.15 Precipitation reaction in liquid media.

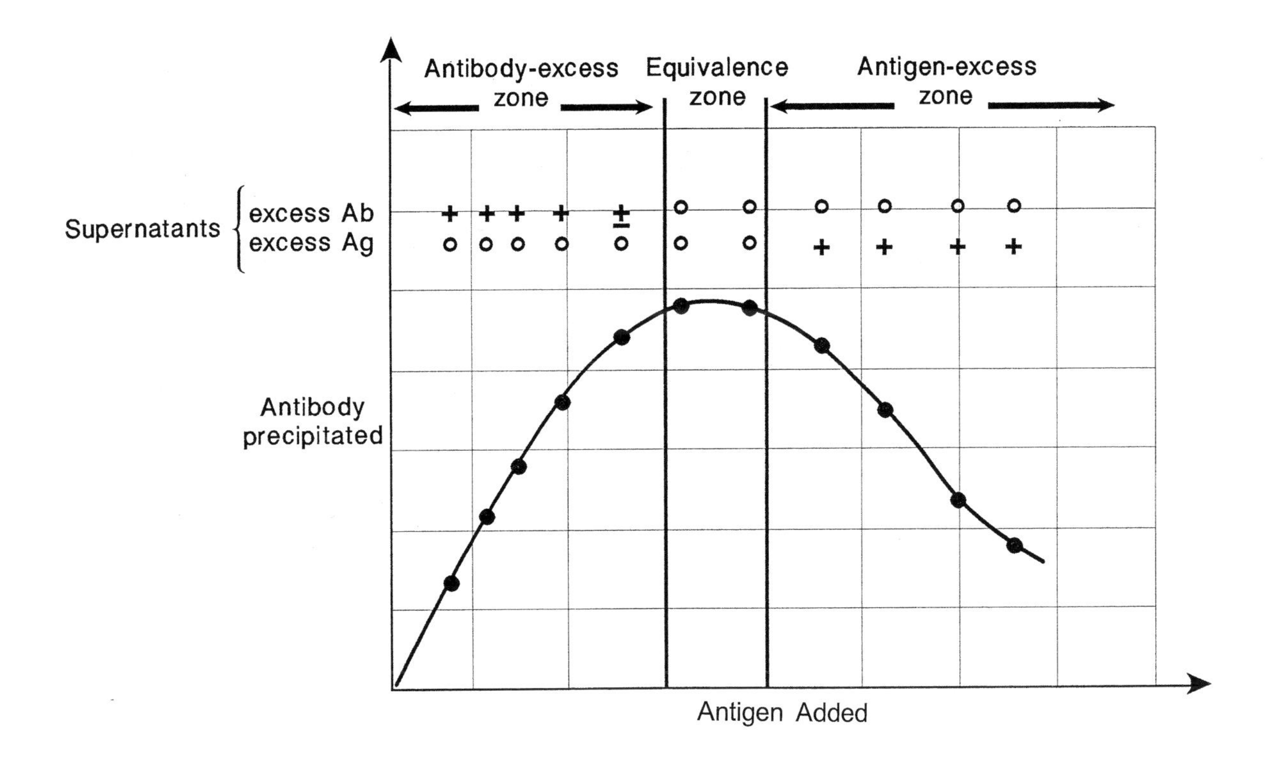

FIGURE 8.16 Precipitation curve.

precipitate in tubes 1 and 7? Both tubes contained antigen as well as antibody which reacted but did not form aggregates large enough to precipitate out of solution. Therefore, an excess of either antigen or antibody may inhibit precipitation, particularly an excess of antigen.

Quantitative gel diffusion test refers to estimation of the amount of antibody or antigen by a gel diffusion method such as single radial diffusion or Laurell rocket assay.

Milligrams of antibody in the precipitate are plotted on the ordinate and the milligrams of antigen added are plotted on the abscissa of a graph (Figure 8.16). The **precipitin curve** contains an ascending and a descending limb and zones of antibody excess, equivalence, and antigen excess. By testing with the homologous reagents, unreacted antibodies and antigens can be detected in the supernatants. If antigen is homogeneous or if antibodies specific for only one of a mixture of antigens are studied by the precipitin reaction, none of the supernatants contain both unreacted antibodies and unreacted antigens that can be detected.

The ascending limb of the precipitation curve represents the zone of antibody excess where free antibody molecules are present in the supernatants. The descending limb represents the zone of antigen excess where free antigen is present in the supernatants. Precipitation is maximum in the zone of equivalence (or equivalence point) where neither antigen nor antibody can be detected in the supernatants (Figure 8.17). In contrast to the nonspecific system described above, the presence of more than one antigen–antibody system in the reaction medium may be revealed by the demonstration of unreacted antibody and antigen in certain supernatants. This occurs when there is an overlap between the zone of antigen excess in one antigen–antibody combination with the zone of antibody excess of a separate antigen–antibody system (Figure 8.18).

The **lattice theory** (Figure 8.19) proposed by Marrack explains how multivalent antigen molecules and bivalent antibodies can combine to yield antigen to antibody ratios that differ from one precipitate to another, depending on the zone of the precipitin reaction in which they are formed. When the ratio of antibody to antigen is above 1, a visible precipitate forms. However, when the ratio is less than 1, soluble complexes result and remain in the supernatant. The soluble complexes are associated with the precipitin curve's descending limb. Also termed precipitin curve.

The **lattice theory** is the concept that soluble antigen and antibody combine with each other in the precipitation reaction to produce an interconnecting structure of molecules. This structure has been likened to a criss-cross pattern of wooden strips fastened together to reveal a series of diamond-shaped structures. Lattice formation requires interaction of bivalent antibodies with multivalent antigens to produce a connecting linkage of many molecules to produce a complex whose density becomes sufficient to settle out of solution. The more epitopes recognized by the antibody molecules present, the more extensive is the complex formation.

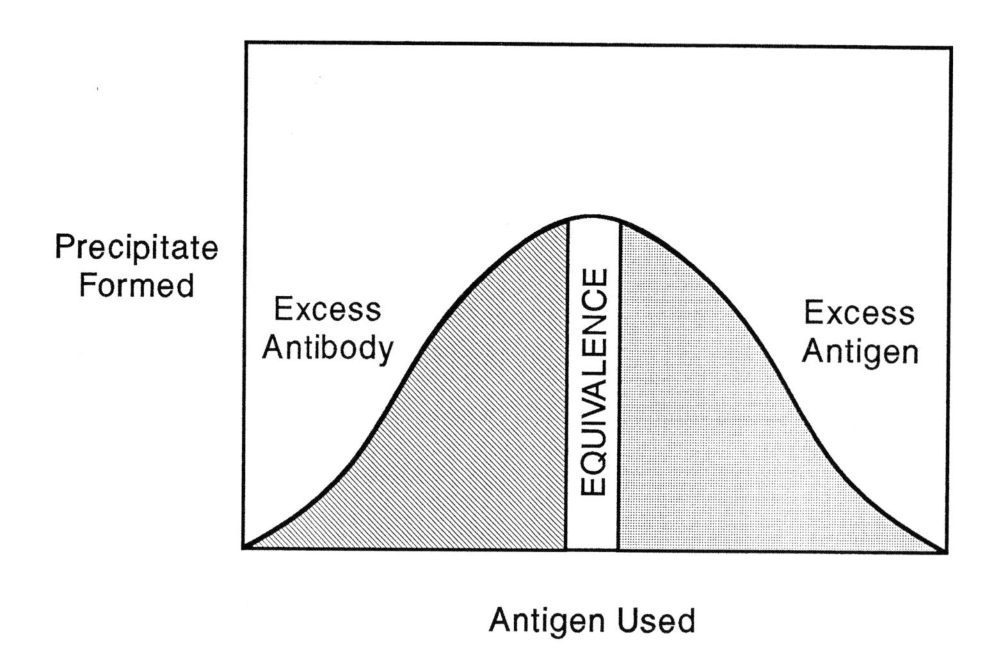

FIGURE 8.17 Precipitate formed vs. antigen used.

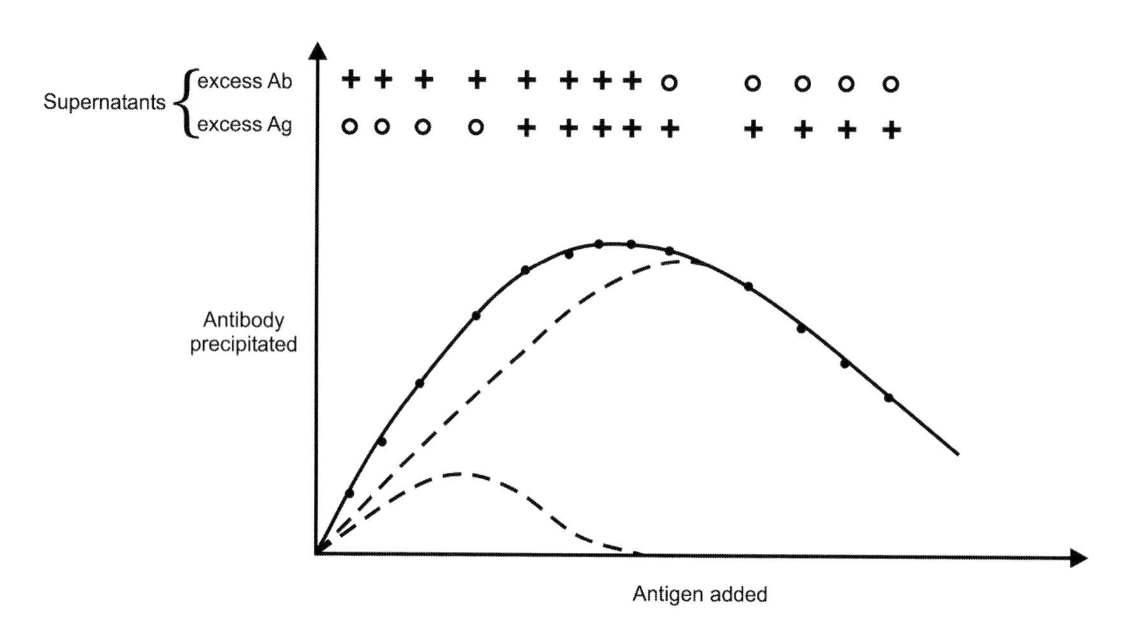

FIGURE 8.18 Precipitin curve for a multispecific system.

The incremental addition of antigen to an optimal amount of antibody precipitates only 78% of the antibody amount precipitated by one-step addition to the antigen. This demonstrates the presence of both precipitating and nonprecipitating antibodies. Although the nonprecipitating variety cannot lead to the formation of insoluble antigen–antibody complexes, they can be assimilated into precipitates that correspond to their specificity. Rather then being univalent, as was once believed, they may merely have a relatively low affinity for the homologous antigen. Monogamous bivalency, which describes the combination of high affinity antibody with two antigenic determinants on the same antigen particle, represents an alternative explanation of the failure of these molecules to participate with their homologous antigen. The formation of nonprecipitating antibodies, which usually represents 10 to 15% of the antibody population produced, is dependent on such variables as heterogeneity of the antigen, characteristics of the antibody, and animal species.

Factors that affect the precipitin reaction are pH, salt concentration (ionic strength), temperature, the presence of complement in the sera used, and time curve. The precipitin reaction usually remains unaffected by changes of pH in the range of 6.6 to 8.5. Synthetic antigens are the most sensitive to changes in pH and the greater the number of charged amino acids, the greater their sensitivity.

Electrolytes contribute to the stabilization of the reaction. The overall charge of the molecules used as antigens should be considered. Avian antisera show an anomalous behavior in that more precipitation is obtained in high NaCl concentration than occurs in physiological saline.

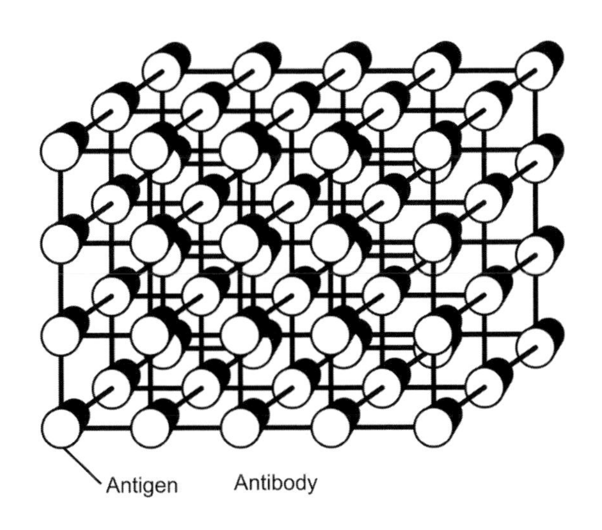

FIGURE 8.19 Antigen–antibody lattice formation.

Generally, the amount of antibody precipitated at 4°C is greater than at 37°C. Here again, synthetic polypeptides are the most sensitive to the temperature changes, but the individual variations should be determined experimentally. Small amounts of complement may be present and may persist for many months at 4°C. By binding to immune complexes, complement augments the specific precipitates and shifts the solubility equilibrium. It precipitates antigen–antibody complexes in the antigen excess zone. The error is greater with rabbit and guinea pig sera than with human sera.

The time required for the formation of a precipitate varies with the system and contrasts with the rapidity of antigen and antibody interaction. Generally, it depends on the ratio of antigen to antibody and is more rapid at the equivalence zone. The precipitin reaction should be viewed as a series of competing biomolecular reactions. Some antigens and antibodies require longer times for precipitation (for example, gelatin–antigelatin system requires 10 d).

Other factors such as the storage or serum, volumes, washing of the complexes, use of diluents, presence of active enzymes in serum, and state of aggregation of antigen may affect the course of the precipitation reaction.

Flocculation is a variant of the preciptin reaction in which soluble antigens interact with antibody to produce precipitation over a relatively narrow range of antigen to antibody ratios. **Flocculation** differs from the classic precipitin reaction in that insoluble aggregates are not formed until a greater amount of antigen is added than would be required in a typical precipitin reaction (Figure 8.20). If the antibody (or total protein) precipitated vs. antigen added is plotted, the plot does not extrapolate to the origin. In flocculation reactions, excess antibody as well as excess antigen inhibits precipitation. Precipitation occurs only over a narrow range of antibody to antigen ratios. Soluble antigen–antibody complexes are formed in both antigen and antibody excess.

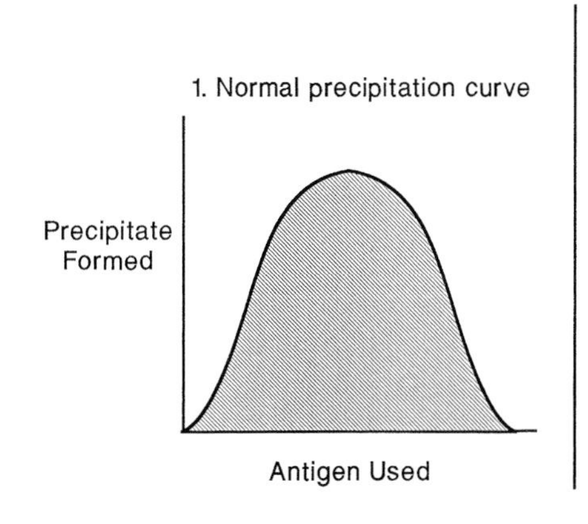

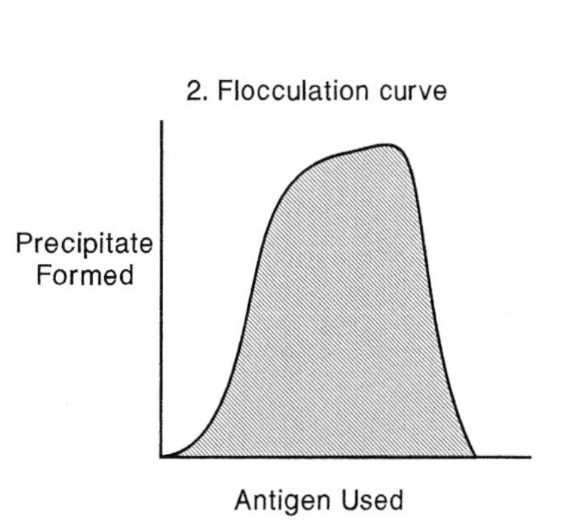

FIGURE 8.20 Flocculation.

Horse antisera commonly give flocculation reactions (for example, antisera to diphtheria and certain streptococcal toxins). The peculiar aspects of the flocculation reaction must be attributed to the reacting antibodies as opposed to the antigen, which gives a typical precipitin reaction with rabbit antisera. For many years this reaction was known as the toxin–antitoxin type of curve because it was observed with horse antibodies against diphtheria and tetanus toxins. In recent years it has been observed with blood sera from some patients with Hashimoto's thyroiditis. These patients develop autoantibodies against human thyroglobulin. This antithyroglobulin antibody may give a classic precipitin curve, but some individuals develop a flocculation type of antibody response against the antigen.

How do the flocculation and precipitin curves differ? In flocculation, soluble antigen–antibody complexes form in antigen as well as in antibody excess regions. In the precipitin reaction, precipitate is developed with even minute quantities of antigen, causing the curve to pass through the origin (Figure 8.21). The graph is a classic flocculation curve based on the data of Pappenheimer and Robinson. Roitt and associates demonstrated that one human antiserum to thyroglobulin gave a precipitin curve (Figure 8.22). They also showed a flocculation curve (Figure 8.23).

Although no satisfactory explanation has yet been offered for the flocculation curve, it may be attributable in part to such variables as antibody heterogeneity or the relative binding affinities of flocculating antibodies compared with those of precipitins. Different antigenic determinants may be involved in flocculation and precipitation.

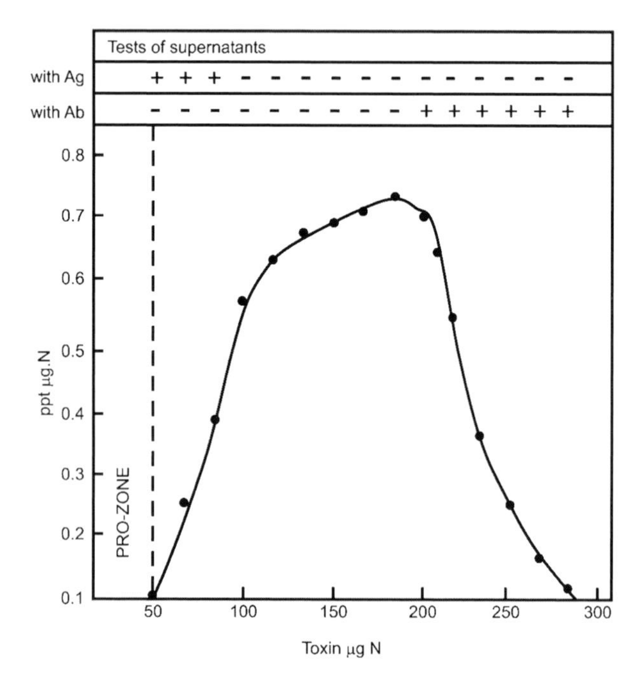

FIGURE 8.21 Flocculation curve of Pappenheimer and Robinson.

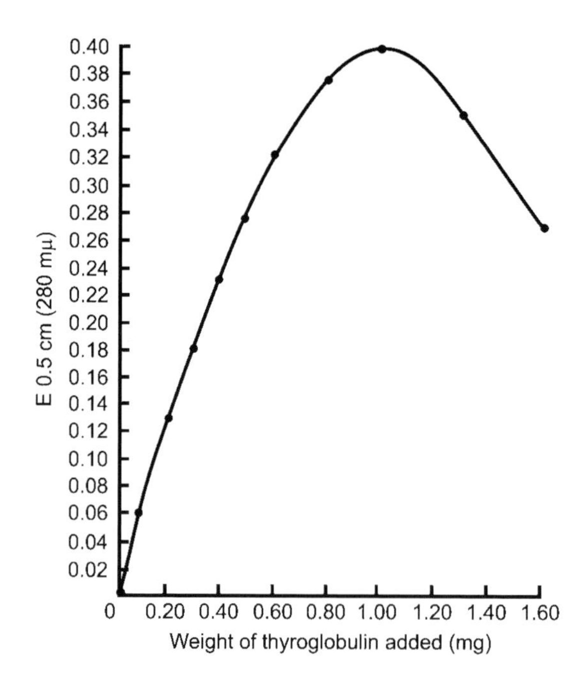

FIGURE 8.22 Precipitation curve with human thyroglobulin and homologous antibody.

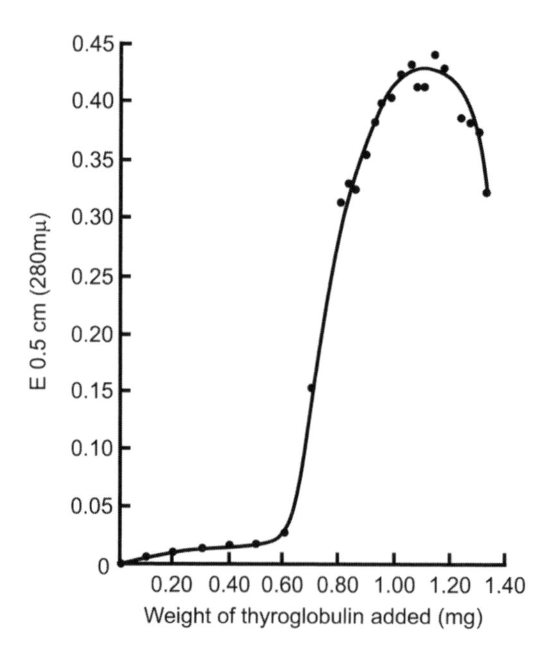

FIGURE 8.23 Flocculation curve with human thyroglobulin and homologous antibody.

The addition of toxin to the homologous antitoxin in several fractions with appropriate time intervals between them results in a greater toxicity of the mixture than would occur if all the samples of toxin were added at once. Therefore, a greater amount of antitoxin is required for neutralization if the toxin is added in divided doses than if all toxin is added at one time, or less toxin is required to neutralize the given quantity of antitoxin if all toxin is added at one time than if it is added in divided doses with time intervals between. This form of reaction has been called the **Danysz phenomenon** or **Danysz effect**. Neutralization in the above instances is tested by injection of the toxin–antitoxin mixture into experimental animals. This phenomenon is attributed to the combination of toxin and antitoxin in multiple proportions. The addition of one fraction of toxin to excess antitoxin leads to maximal binding of antitoxin by toxin molecules. When a second fraction of toxin is added, insufficient antitoxin is available to bring about neutralization. Therefore, the mixture is toxic due to uncombined excess toxin. Equilibrium is reached after an appropriate time interval. The interaction between toxin and antitoxin is considered to occur in two steps: (1) rapid combination of toxin and antitoxin and (2) slower aggregation of the molecules. These reactions are outlined in the steps shown above.

Nephelometry is a technique used to assay proteins and other biological materials through the formation of a precipitate of antigen and homologous antibody (Figure 8.24). The assay depends on the turbidity or cloudiness of a suspension. It is based on determination of the degree to which light is scattered when a helium–neon laser beam is directed through the suspension. Antigen concentration is ascertained using a standard curve devised from the light scatter produced by solutions of known antigen concentration. This method is used by many clinical immunology laboratories for the quantification of complement components and immunoglobulins in patients' sera or other body fluids.

Oudin in 1946 overlaid antibody incorporated in agar in a test tube with the homologous antigen. A band of precipitation appeared in the gel where the antigen–antibody interaction occurred. Mixtures of antibodies of several specificities were overlaid with a mixture of the homologous antigens, and a distinct band for each resulted. Oudin's technique involves simple (or single) diffusion in one dimension (Figure 8.25). Oakley and Fulthorpe (1953) placed antiserum incorporated into agar in the bottom of a tube, covered this with a layer of plain agar which was permitted to solidify, and then added antigen. This was double diffusion in one dimension. Double diffusion in two dimensions was developed by Ouchterlony and independently by Elek in 1948. Agar is poured on a flat glass surface such as a microscope slice, glass plate, or Petri dish. Wells or troughs are cut in the agar and these are filled with antigen and antibody solutions under study. Multiple component systems may be analyzed by use of this method and cross-reactivities detected. Double diffusion in

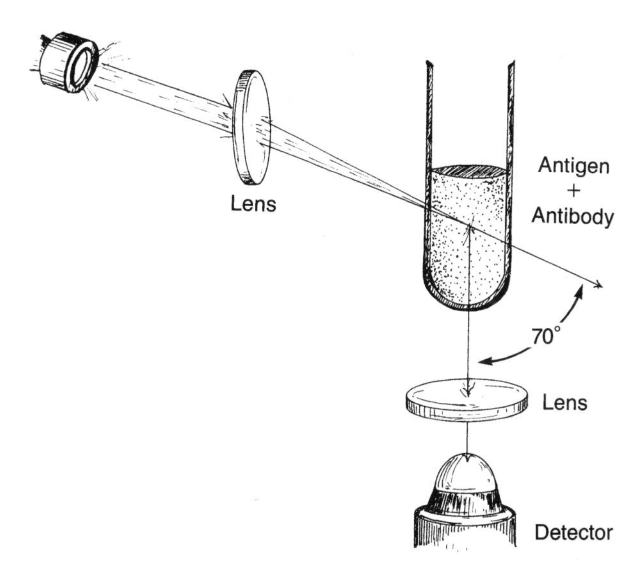

FIGURE 8.24 Nephelometry.

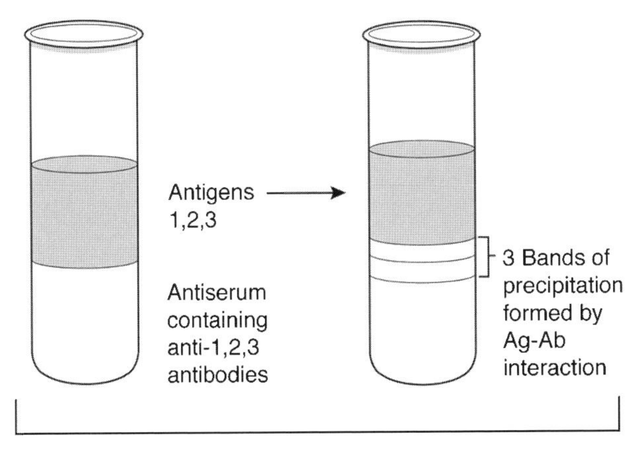

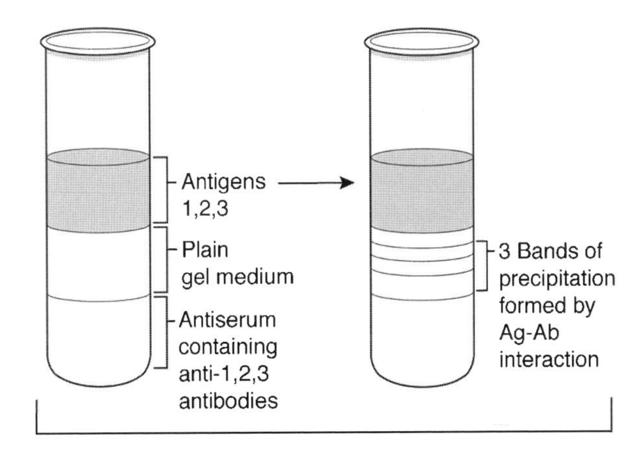

FIGURE 8.25 Precipitation in gel media.

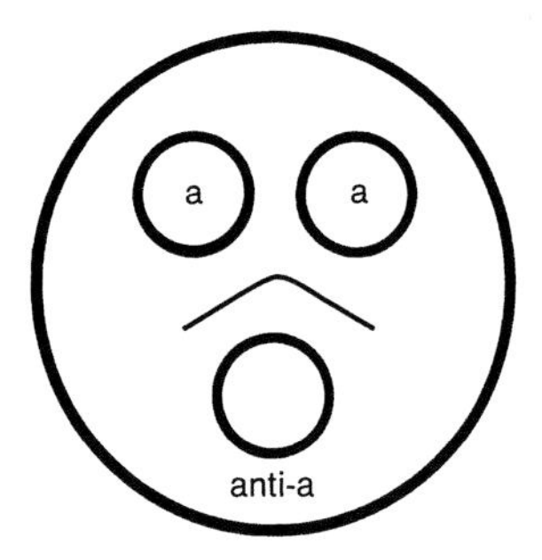

FIGURE 8.26 Reaction of identity.

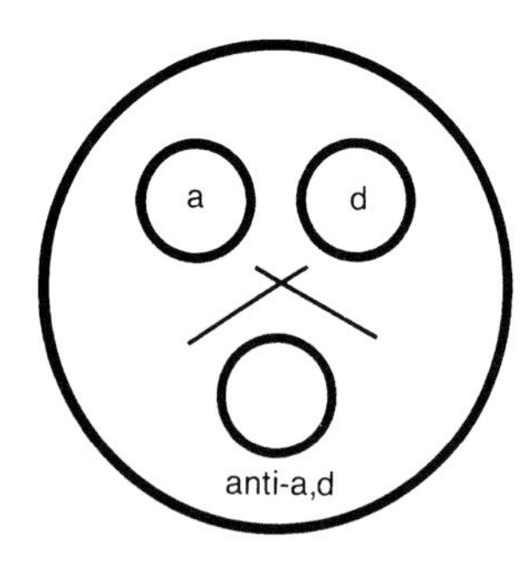

FIGURE 8.27 Reaction of nonidentity.

agar is a useful method to demonstrate similarity among structurally related antigens. Equidistant holes are punched in agar gel containing electrolytes. Antigen is placed in one well, antiserum in an adjacent well, and the plates are observed the following day for a precipitation line where antigen and antibody have migrated toward one another and reached equivalent concentrations. A single line implies a single antigen–antibody system. If agar plates containing one central well with others cut equidistant from it at the periphery are employed, a reaction of identity may be demonstrated by placing antibody in the central well and the homologous antigen in the adjacent peripheral wells.

A confluent line of precipitate is produced in the shape of an arc. This implies that the antigen preparations in adjacent peripheral wells are identical (**reaction of identity**); they have the same antigenic determinants (Figure 8.26). If antibodies against two unrelated antigen preparations are combined and placed in the central well and their homologous antigens placed in separate adjacent peripheral wells, a line of precipitation is produced by each antigen–antibody reaction to give the appearance of crossed sword points. This constitutes a **reaction of nonidentity** (Figure 8.27).

It implies that the antigenic determinants are different in each of the two sample of antigen. A third pattern known as a **reaction of partial identity** occurs when two antigen preparations that are related but not the same are placed in separate adjacent wells with an antibody preparation that crossreacts with both of them placed in a central well (Figure 8.28).

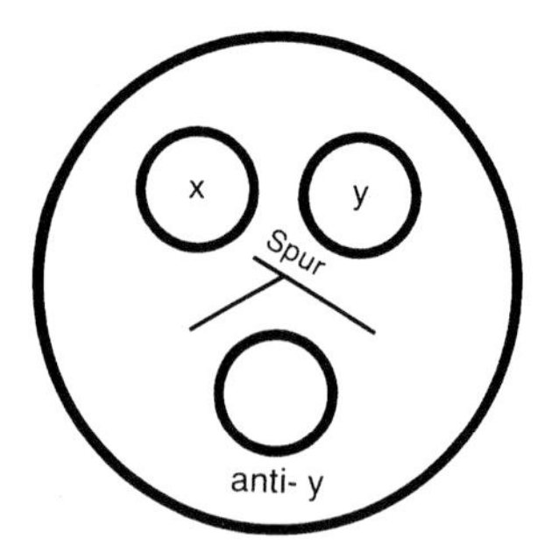

FIGURE 8.28 Reaction of partial identity.

The precipitation lines between each antigen–antibody system converge, but a spur or extension of one of the precipitation lines occurs. This reaction of partial identity with spur formation implies that the antigen preparations are similar but that one has an antigenic determinant not present in the other. A reaction of identity and nonidentity may be observed simultaneously, implying that two separate antigen preparations have both common and different antigenic determinants.

Reaction of identity: Double immunodiffusion in two dimensions in gel can reveal that two antigen solutions are identical. If two antigens are deposited into separate but adjacent wells and permitted to diffuse toward a specific antibody diffusing from a third well that forms a triangle with the other two, a continuous arc of precipitation is formed. This reveals the identity of the two antigens.

Reaction of nonidentity: Double immunodiffusion in two dimensions in gel can show that two antigen solutions are different, i.e., nonidentical. If each antigen solution is deposited into separate but adjacent wells and permitted to diffuse toward a combination of antibodies specific for each antigen diffusing from a third well that forms a triangle with the other two, the lines of precipitation form independently of one another and intersect, resembling crossed swords. This reaction reveals a lack of identity with no epitopes shared between the antigens detectable by these antibodies.

Reaction of partial identity: Double immunodiffusion in two dimensions in gel can show that two antigen solutions share epitopes but are not identical. If each antigen is deposited into separate but adjacent wells and permitted to diffuse toward specific antibodies diffusing from a third well that forms a triangle with the other two, a continuous arc of precipitation manifesting a spur is formed. This demonstrates that the two antigens share some epitopes, shown by the continuous arc, but not others, demonstrated by the spur.

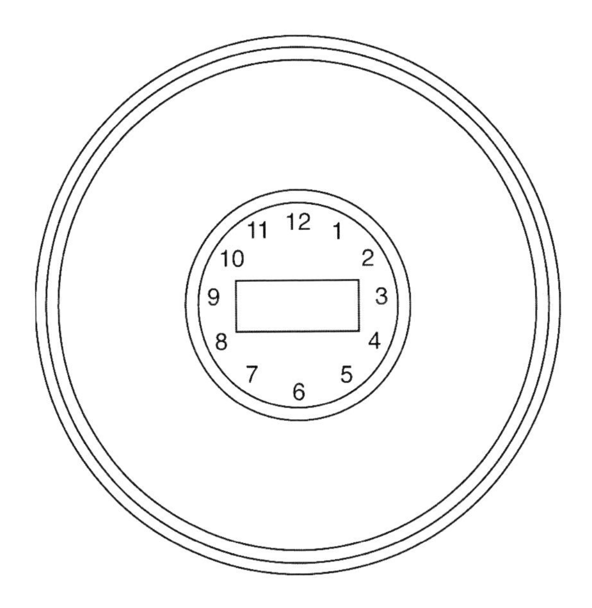

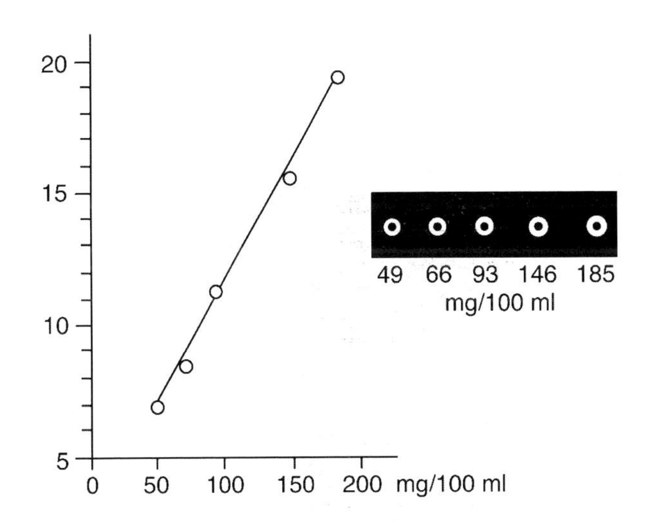

FIGURE 8.29 Radial immunodiffusion.

Partial identity: See reaction of partial identity.

A **spur** is an extension of a precipitation line observed in a two-dimensional double-immunodiffusion assay such as the Ouchterlony test. It represents a reaction of partial identity between two antigens that crossreact with the antibody.

Mancini, in 1965, developed a quantitative technique employing single radial diffusion to quantify antigens. Plates are poured in which specific antibody is incorporated into agar. Wells are cut and precise quantities of antigen are placed over time. The antigen is permitted to diffuse into the agar containing antibody and produce a ring of precipitation where they interact (Figure 8.29). The precipitation ring encloses an area proportional to the concentration of antigen measured 48 to 72 h following diffusion. Standard curves are employed using known antigen standards, and the antigen concentration as reflected by the diameter of the ring ascertained. The Mancini technique can detect as little as 1 to 3 μm/ml of antigen.

Radial immunodiffusion is a technique used to ascertain the relative concentration of an antigen. Antigen is placed in a well and permitted to diffuse into agar containing an appropriate dilution of an antibody. The area of the precipitin ring that encircles the well in the equivalence region is proportional to the antigen concentration.

Diffusion coefficient is a mathematical representation of a protein's diffusion rate in gel. The diffusion coefficient is useful in determining antigen molecular weight. It is the diffusion rate to concentration gradient ratio.

Agglutination is the combination of soluble antibody with particulate antigens in an aqueous medium containing electrolyte, such as erythrocytes, latex particles bearing antigen, or bacterial cells to form an aggregate which may be viewed either microscopically or macroscopically (Figure 8.30). If antibody is linked to insoluble beads or particles, they may be agglutinated by soluble antigen by reverse agglutination. Agglutination is the basis for multiple serological reactions including blood grouping, diagnosis of infectious diseases, rheumatoid arthritis (RA) test, etc. To carry out an agglutination reaction, serial dilutions of antibody are prepared and a constant quantity of particulate antigen is added to each antibody dilution. Red blood cells may serve as carriers for adsorbed antigen, e.g., tanned red cell or bis-diazotized red cell technique. Like precipitation, agglutination is a secondary manifestation of antigen–antibody interaction. As specific antibody crosslinks particulate antigens, aggregates form that become macroscopically visible and settle out of suspension. Thus, the agglutination reaction has a sensitivity 10 to 500 times greater than that of the precipitin test with respect to antibody detection. Agglutination permits phagocytic cells to engulf invading macroorganisms. This is a major role of agglutinin in the immune reaction (Figure 8.31).

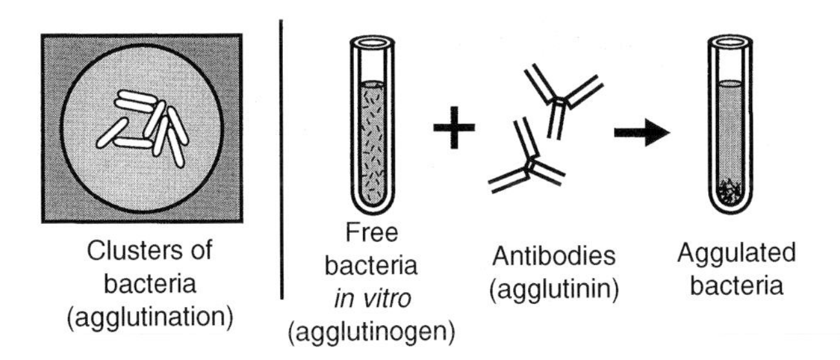

FIGURE 8.30 Agglutination.

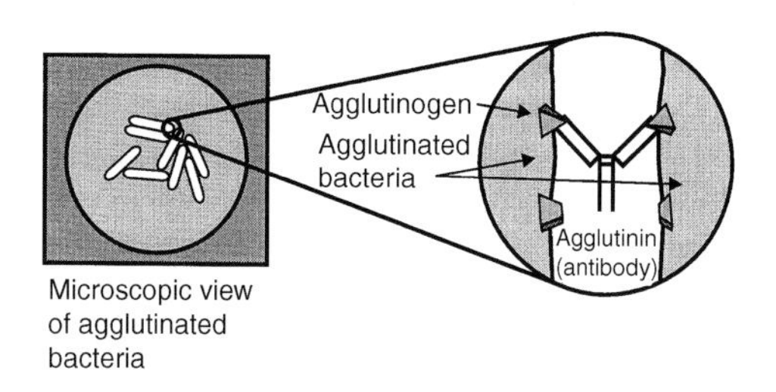

FIGURE 8.31 Bacterial agglutination.

However, massive agglutination may occur without immunization. Antibodies alone are not enough and phagocytes must be present to remove the precipitate in the circumstances. **Titer** is the quantity of a substance required to produce a reaction with a given volume of another substance.

An **agglutination titer** is the highest dilution of a serum which causes clumping of particles such as bacteria. Titer is an approximation of the antibody activity in each unit volume of a serum sample. The term is used in serological reactions and is determined by preparing serial dilutions of antibody to which a constant amount of antigen is added. The end point is the highest dilution of antiserum in which a visible reaction with antigen, e.g., agglutination, can be detected. The titer is expressed as the reciprocal of the serum dilution which defines the endpoint. If agglutination occurs in the tube containing a 1:240 dilution, the antibody titer is said to be 240. Thus, the serum would contain approximately 240 units of antibody per milliliter of antiserum. The titer only provides an estimate of antibody activity. For absolute amounts of antibody, quantitative precipitation or other methods must be employed.

Agglutination inhibition is diminished clumping of particles bearing antigen on their surface by the addition of soluble antigen that interacts with and blocks the agglutinating antibody.

Bacterial agglutination is antibody-mediated aggregation of bacteria. This technique has been used for a century in the diagnosis of bacterial diseases through the detection of an antibody specific for a particular microorganism or for the identification of a microorganism isolated from a patient.

An **agglutinin** is an antibody that interacts with antigen on the surface of particles such as erythrocytes, bacteria, or latex cubes to cause their aggregation or agglutination in an aqueous environment containing electrolyte. Substances other than agglutinin antibody that cause agglutination or aggregation of certain specificities of red blood cells include hemagglutinating viruses and lectins.

An **agglutinogen** is an antigen on the surface of particles such as red blood cells that react with the antibody known as agglutinin to produce aggregation or agglutination. The most widely known agglutinogens are those of the ABO and related blood group systems.

The **bentonite flocculation test** is an assay in which bentonite particles were used as carriers to adsorb antigens. These antigen-coated bentonite particles were then agglutinated by the addition of a specific antibody.

Passive hemagglutination is the aggregation by antibodies of erythrocytes bearing adsorbed or covalently bound soluble antigen on their surface.

Passive hemolysis is the lysis of erythrocytes used as carriers for soluble antigen bound to their surface. Following interaction with antibody, complement induces cell lysis. In passive hemolysis, the antigen is not a part of the cell surface structure but is only attached to it.

A **cross-reaction** is the reaction of an antigenic determinant with an antibody formed against another antigen. It is a laboratory technique used for matching blood for transfusion and organs for transplantation. In blood transfusion, donor erythrocytes are combined with the recipient's serum. If there is antibody in the recipient's serum that is specific for donor red cells, agglutination occurs. This represents the major part of the cross-match. In a negative test, the presence of incomplete antibodies may be detected by washing the red blood cells and adding rabbit antihuman globulin. Agglutination signifies the presence of incomplete antibodies. The minor part of the cross-match test consists of mixing recipient's red blood cells with donor serum. It is of less importance than the major cross-match because of the limited amount of serum in a unit of blood compared to the red cell volume of the recipient.

A **cross-reacting antibody** is an antibody that reacts with epitopes on an antigen molecule different from the one that stimulated its synthesis. The effect is attributable to shared epitopes on the two antigen molecules.

A **cross-reacting antigen** is an antigen that interacts with an antibody synthesized following immunogenic challenge with a different antigen. Epitopes shared between these two antigens or epitopes with a similar stereochemical configuration may account for this type of cross-reactivity. The presence of the same or of a related epitope between bacterial cells, red blood cells, or other types of cells may cross-react with an antibody produced against either of them.

Cross-reactivity is the ability of an antibody or T cell receptor to react with two or more antigens that share an epitope in common.

Cross-absorption is the use of cross-reacting antigens or cross-reacting antibodies to absorb antibodies or antigens, respectively.

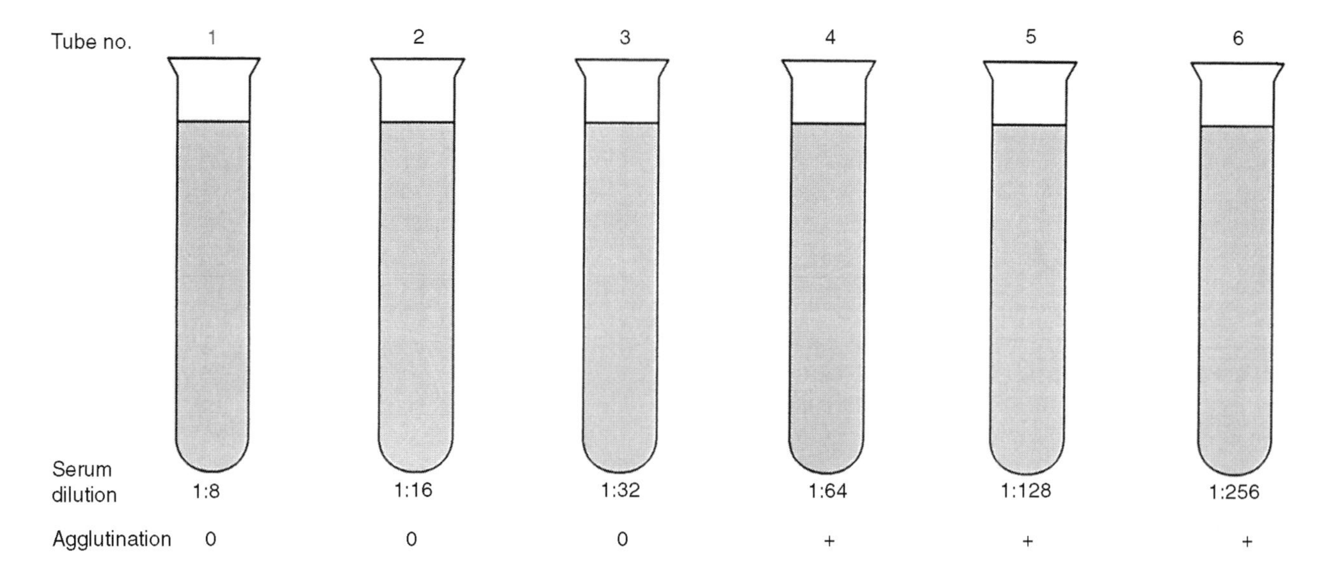

FIGURE 8.32 Prozone effect.

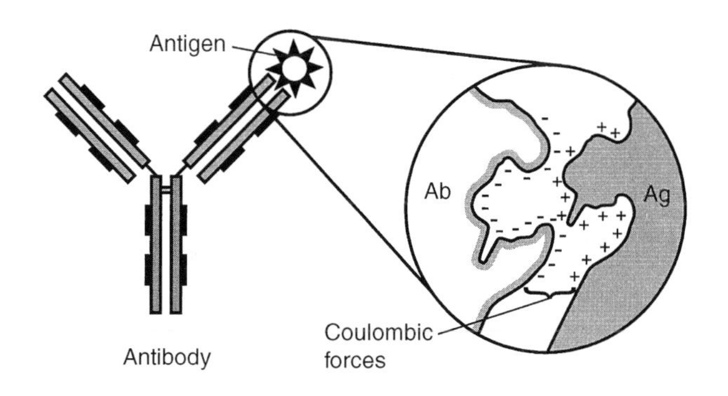

FIGURE 8.33 Antigen–antibody complex. Role of positive and negative charges in agglutination of antigen by antibody.

Dilution end point is a value expressed as the titer that reflects the lowest amount of an antibody giving a reaction. It is determined by serial dilution of the antibody in serum or other body fluid while maintaining a constant amount of antigen.

Final serum dilution is a serological term to designate the titration end point. It is the precise dilution of serum reached following combination with all components needed for the reaction, i.e., the addition of antigen and complement to the diluted serum.

End point is the greatest dilution of an antibody in solution that will still yield an identifiable reaction when combined with antigen. The reciprocal of this dilution represents the titer.

End-point immunoassay is a test in which the signal is measured as the antigen–antibody complex reaches equilibrium.

In agglutination and precipitation reactions, the lack of agglutination or precipitation in tubes where the antibody concentration is greatest is known as a **prozone** or **prozone phenomenon**. This is attributable to suboptimal agglutination or precipitation in the region of antibody excess. Agglutination or precipitation becomes readily apparent in the tubes where the same antibody is more dilute. The **prozone phenomenon** is attributable to either blocking antibody or antibody combining with only individual cells or molecules, as in antibody excess, or to some serum lipid or protein-induced nonspecific inhibition reaction. Soluble complexes of antigen and antibody may be present in the antibody excess zone of certain precipitation reactions. Excess antibody coating the surfaces of cells in an agglutination reaction in the antibody excess zone may prevent cross-linking, which is requisite for agglutination to become manifest. The prozone represents a false-negative reaction. When a serum sample is believed to contain a certain antibody that is being masked or is demonstrating a prozone phenomenon, the sample must be diluted serially to demonstrate reactivity in more dilute tubes to avoid reporting a false-negative result.

Inhibition zone: See prozone.

The **prozone** is that portion of the dilution range in which an immune serum of high agglutinin titer fails to agglutinate the homologues (Figure 8.32).

Antigens and antibodies agglutinate because of plus and minus charges (Figure 8.32). Antigen–antibody reactions are therefore surface phenomena. The antibody is the mirror image of the antigen.

Factors that affect the agglutination test are electrolytes, pH, and temperature. Salt decreases the potential difference between antigen particles and the surrounding liquid medium in which they are suspended. It also decreases cohesive forces between antigen particles. Agglutination occurs when the potential drops below 15 mV. As the salt concentration is increased the potential drops, favoring agglutination; but the cohesive forces between particles also drops, which is unfavorable to agglutination. Finally, a salt concentration is reached where no agglutination occurs. A concentration near 0.15 *M* is ideal.

With sera of high antibody titer, complete agglutination occurs over a wide pH range. Upon lowering the titer, the optimal pH also drops until acid agglutination occurs around a pH of 4. This acid agglutination is nonspecific, since even normal serum will agglutinate cells at this low pH. This type of nonspecific agglutination is not antibody dependent.

The rate of agglutination increases rapidly from 0 to 30°C. Above that temperature it is less rapid, and above 56°C the antibody molecules are injured by heat. Shaking or stirring accelerates agglutination.

Certain conditions must be established in the reacting medium for agglutination to take place. These include ionic strength and pH. Optimally, agglutination reactions are carried out in neutral dilute salt solutions such as 0.15 *M* sodium chloride. The significance of ionic strength is demonstrated by agglutination without antibodies at neutral pH or by bacteria bearing a negative surface charge following the addition of enough salt to induce damping of these charges. Low salt concentration below $10^{-3}M$ NaCl, may prevent agglutination of bacteria or other particulate antigens with antibody already attached to their surfaces. It appears necessary to dampen the highly negative surface charge of cells by counter ions to permit close enough contact between cells for bivalent antibody molecules to form specific connecting bridges between them.

Postzone refers to the lack of serologic reactivity as a consequence of high dilution of antibody, as in a serial dilution procedure. This is a zone of relative antigen excess.

A **soluble complex** is an immune (antigen–antibody) complex formed in excess antigen and rendered soluble. Antigen excess prevents lattice formation either *in vitro* or *in vivo*. Soluble complexes may produce tissue injury *in vivo*, which is more severe if complement has been fixed. C5a attracts neutrophils, and there is increased capillary permeability. PMNs, platelets, and fibrin are deposited on the endothelium. This is followed by thrombosis and necrosis. Immune complexes induce type III hypersensitivity reactions.

Antigen excess: The interaction of soluble antigen and antibody in the precipitin reaction leads to occupation of the antigen-binding sites of all the antibody molecules and leaves additional antigenic determinants free to combine with more antibody molecules if excess antigen is added to the mixture. This leads to the formation of soluble antigen–antibody complexes *in vitro*, i.e., the postzone in the precipitin reaction. A similar phenomenon may take place *in vivo* when immune complexes form in the presence of excess antigen. These are of clinical significance in that soluble immune complexes may induce tissue injury, leading to immunopathologic sequelae.

TAF is the abbreviation for toxoid antitoxin floccules.

Toxin neutralization (by antitoxin): Toxicity is titrated by injection of laboratory animals and the activity of antitoxins is evaluated by comparison with standard antitoxins of known protective ability. Antitoxin combines with toxin in varying proportions, depending on the ratio in which they are combined, to form complexes which prove nontoxic when injected into experimental animals. Mixing of antitoxin and toxin in optimal proportions may result in flocculation. If toxin is added to antitoxin in several fractions with time intervals between them instead of all at

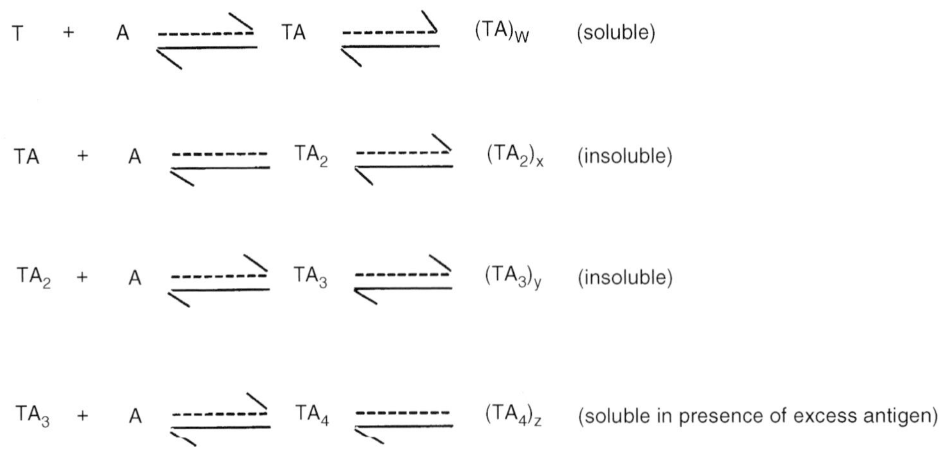

$$T + A \rightleftharpoons TA \rightleftharpoons (TA)_W \quad \text{(soluble)}$$

$$TA + A \rightleftharpoons TA_2 \rightleftharpoons (TA_2)_x \quad \text{(insoluble)}$$

$$TA_2 + A \rightleftharpoons TA_3 \rightleftharpoons (TA_3)_y \quad \text{(insoluble)}$$

$$TA_3 + A \rightleftharpoons TA_4 \rightleftharpoons (TA_4)_z \quad \text{(soluble in presence of excess antigen)}$$

FIGURE 8.34 Toxin neutralization (by antitoxin).

once, more antitoxin is required for neutralization to occur than would be necessary if all toxin had been added at once. This means that toxins are polyvalent. This phenomenon is explained by the ability of toxin to combine with antitoxin in multiple proportions. Neutralization does not destroy the reacting toxin. In many instances, toxin may be recovered by dilution of the toxin–antitoxin mixture. The effect of heat on a zootoxin is illustrated by the destruction of cobra venom antitoxin if cobra venom (toxin)–antivenom (antitoxin) mixtures are subjected to boiling. The venom or toxin remains intact. Since toxins have specific affinities for certain tissues of the animal body, such as the high affinity of tetanus toxin for nervous tissue, antitoxins are believed to act by binding toxins before they have the opportunity to combine with specific tissue cell receptors.

L+ dose (historical): The smallest amount of toxin which, when mixed with one unit of antitoxin and injected subcutaneously into a 250-g guinea pig, will kill the animal within 4 days. This is the unit used for standardization of antitoxin.

L_f dose (historical): The flocculating unit of diphtheria toxin is that amount of toxin which flocculates most rapidly with one unit of antitoxin in a series of mixtures containing constant amounts of toxin and varying amounts of antitoxin. This unit must be calculated.

L_f flocculating unit (historical): The flocculating unit of diphtheria toxin is that amount of toxin which flocculates most rapidly with one unit of antitoxin in a series of mixtures containing constant amounts of toxin and varying amounts of antitoxin. Historically, a unit of antitoxin was considered as the least quantity that would neutralize 100 minimal lethal doses of toxin administered to a guinea pig. Modern usage relates antitoxic activity to an international standard antitoxin.

L_0 dose (historical): This is the largest amount of toxin which, when mixed with one unit of antitoxin and injected subcutaneously into a 250-g guinea pig, will produce no toxic reaction.

L_r dose (historical): This is the least amount of toxin which, after combining with one unit of antitoxin, will produce a minimal skin lesion when injected intracutaneously into a guinea pig.

Antibody excess immune complexes (ABICs) may also result from the alteration of the immunoglobulin molecules, such as that seen in rheumatoid arthritis, or may be produced locally, such as in type B hepatitis. ABICs have a short intravascular life and, in contrast to antigen–excess immune complexes, adhere to platelets and cause platelet aggregation. The aggregation of platelets is not due to crosslinking of ABIC to platelets but to changes in the adhesive properties of the latter. ABIC also bind to neutrophils and induce the release of lysosomal enzymes without prior phagocytic activity.

Hemolysis is caused by interruption of the cellular integrity of red blood cells that may be either immune or nonimmune mediated. Clinically, immune hemolysis may be IgM mediated when immunoglobulins combine with red blood cell surfaces for which they are specific, such as the ABO blood groups, and activate complement to produce lysis. This results in the release of free hemoglobin in the intravascular space with serious consequences. By contrast, hemolysis mediated by IgG in the extravascular space may be less severe. There is an elevation of indirect bilirubin, since the liver may not be able to conjugate the bilirubin in case of massive hemolysis. There is an elevation of lactate dehydrogenase, and hemoglobin appears in the blood and urine. There is elevated urobilinogen in both urine and feces. Hemolysis may be also attributable to the action of enzymes or other chemicals acting on the cell membrane. It can also be induced by such mechanisms as placing the red cells in a hypotonic solution.

Lysis is disruption of cells due to interruption of their cell membrane integrity. This may be accomplished nonspecifically, as with hypotonic salt solution, or through the interaction of surface membrane epitopes with specific antibody and complement or with cytotoxic T lymphocytes.

The **Pfeiffer phenomenon (historical)** is the rapid lysis of *Vibrio cholerae* microorganisms that have been injected into the peritoneal cavity of guinea pigs immunized against them. The microorganisms are first rendered nonmotile, followed by complement-induced lysis in the presence of antibody. Immune bacteriolysis *in vivo* involving the cholera vibrio became known as the Pfeiffer phenomenon.

The **unitarian hypothesis** is the view that one type of antibody produced in response to an injection of antigen could induce agglutination, complement fixation, precipitation, and lysis based on the type of ligand with which it interacted. This view was in contrast to the earlier belief that separate antibodies accounted for every type of serological reactivity described above. Usually, more than one class of immunoglobulin may manifest a particular serological reactivity such as precipitation.

9 The Thymus and T Lymphocytes

The **thymus** is a triangular bilobed structure enclosed in a thin fibrous capsule and located retrosternally (Figure 9.1 and Figure 9.2). Each lobe is subdivided by prominent trabeculae into interconnecting lobules, and each lobule comprises two histologically and functionally distinct areas, cortex and medulla (Figure 9.3). The cortex consists of a mesh of epithelial reticular cells enclosing densely packed large lymphocytes (called thymocytes). It has no germinal centers.

The lymphoid cells are of mesenchymal origin; the epithelial component is of endodermal origin. The thymic cortex contains three types of epithelial reticular cells: type I, type II, and type III. Type I epithelial reticular cells are found at the outer part of the cortex isolating the thymus from the body. The middle of the cortex houses the type II epithelial reticular cells. They compartmentalize the cortex into small areas of lymphocytes. Deep in the cortex and at the corticomedullary junction are the type III epithelial reticular cells. Like type II cells these also compartmentalize the cortex. They also isolate the cortex from the medulla. The isolation provided by the type I and III epithelial reticular cells keep the thymocytes from coming into contact with foreign antigen.

The prothymocytes, which migrate from the bone marrow to the subcapsular regions of the cortex, are influenced by this microenvironment which directs their further development. The process of education is exerted by hormonal substances produced by the thymic epithelial cells. The cortical cells proliferate extensively. Some of these cells are short lived and die. The surviving cells acquire characteristics of thymocytes (Figure 9.4). The cortical cells migrate to the medulla (Figure 9.5) and from there to the peripheral lymphoid organs, sites of their main residence. The medullary areas of the thymus contain loosely packed thymocytes and many epithelial reticular cells (Figure 9.6).

Like the cortex, the medulla also consists of three types of epithelial reticular cells: type IV, type V, and type VI. Type IV cells are found at the corticomedullary junction and are associated with type III cells. The middle section of the medulla is composed of type V epithelial reticular cells. The most characteristic feature of the medulla, the Hassall's corpuscles (Figure 9.7 and Figure 9.8), are formed by the type VI epithelial reticular cells. These remnants of epithelial islands are histologically identifiable and are markers for thymic tissue. The thymocytes are small cells ready to exit the thymus. The blood supply to the cortex comes from capillaries that form anastomosing arcades. Drainage is mainly through veins; the thymus has no lymphatic vessels.

The thymus develops from the branchial pouches of the pharynx at about 6 weeks of embryonal age. In most species it is fully developed at birth. In humans, the weight of the thymus at birth is 10 to 15 g. It continues to increase in size, reaching a maximum (30 to 40 g) at puberty. It then begins to involute with increasing age, but the adult gland is still functional. The medulla involutes first with pyknosis and beading of the nuclei of small lymphocytes, giving a false impression of an increased number of Hassall's corpuscles. The cortex atrophies progressively. The blood–thymus barrier protects thymocytes from contact with antigen. Lymphocytes reaching the thymus are prevented from contact with antigen by a physical barrier. The first level is represented by the capillary wall with endothelial cells inside the pericytes outside of the lumen. Potential antigenic molecules, which escape the first level of control, are taken over by macrophages present in the pericapillary space. Further protection is provided by a third level, represented by the mesh of interconnecting epithelial cells, which enclose the thymocyte population. The effects of thymus and thymic hormones on the differentiation of T cells is demonstrable in animals congenitally lacking the thymus gland (nu/nu animals), neonatally or adult thymectomized animals, and in subjects with immunodeficiencies involving T cell function. Differentiation is associated with surface markers whose presence or disappearance characterizes the different stages of cell differentiation. There is extensive proliferation of the subcapsular thymocytes. The largest proportion of these cells die, but the remaining cells continue to differentiate. The differentiating cells become smaller in size and move through interstices in the thymic medulla. The fully developed thymocytes pass through the walls of the postcapillary venules to reach the systemic circulation and seed in the peripheral lymphoid organs. Part of them recirculate, but do not return to the thymus.

A **pharyngeal pouch** is an embryonic structure in the neck that provides the thymus, parathyroids, and other tissues with epithelial cells.

Pharyngeal pouch syndrome is thymic hypoplasia.

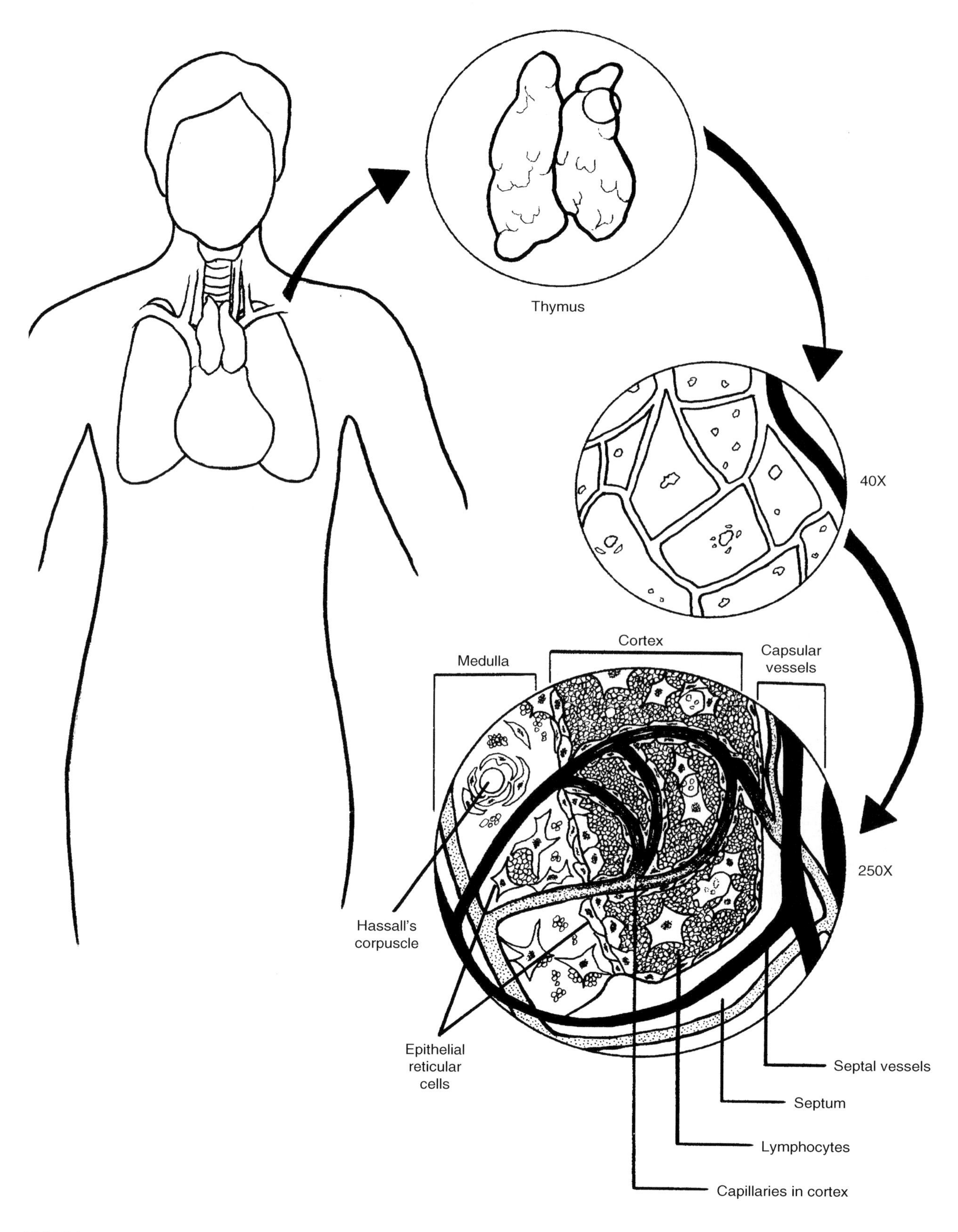

FIGURE 9.1 Diagram showing the location of the thymus in the human body with an enlarged view of its histology.

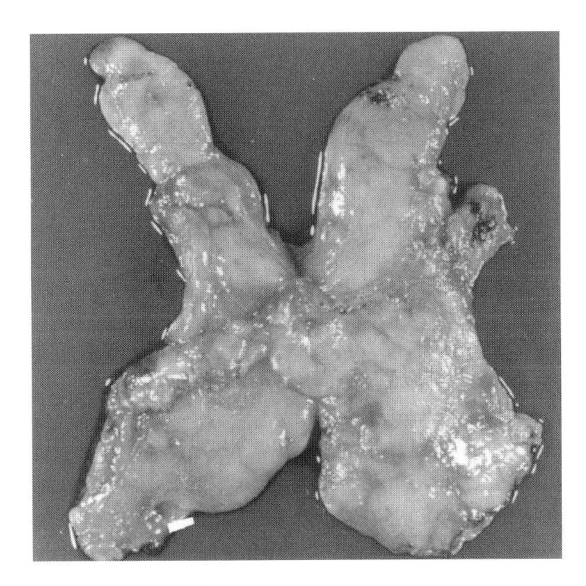

FIGURE 9.2 Normal adult thymus. The thymus often shows an X- or H-shaped configuration.

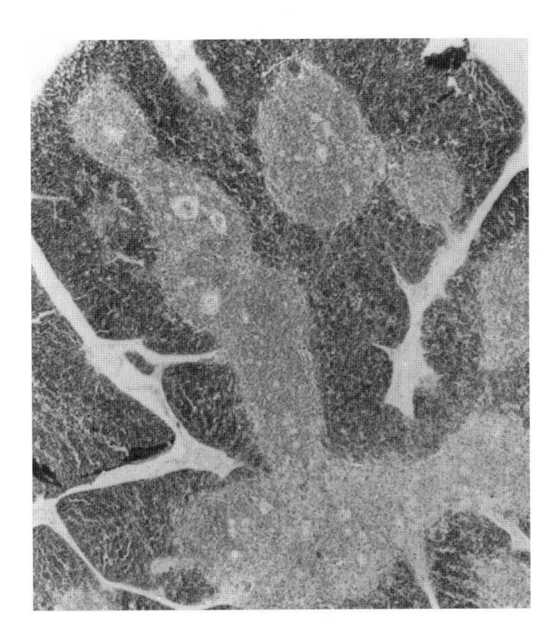

FIGURE 9.3 Normal thymus. The normal thymus in this infant shows the lobulation and sharp separation of cortex from the stalk-like medulla (× 40).

Neonatal thymectomy syndrome: See wasting disease.

Stromal cells are sessile cells that form an interconnected network which gives an organ structural integrity but also provides a specific inductive microenvironment that facilitates differentiation and maturation of incoming precursor cells. Stromal cells and their organization are fully as complex as the cells whose development they regulate. For example, stromal cells of the thymus have been the best characterized with respect to their role in T lymphocyte maturation.

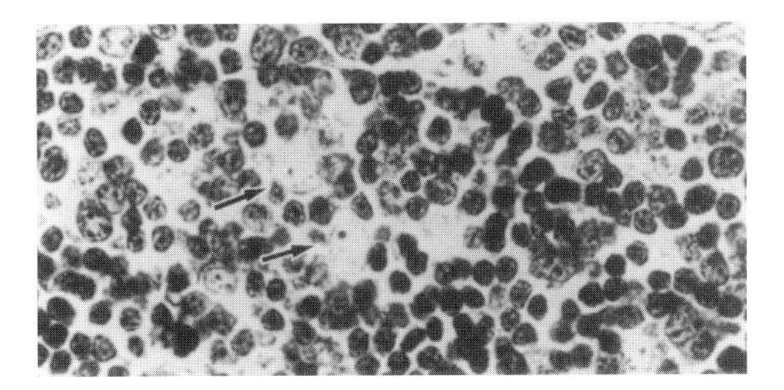

FIGURE 9.4 Thymic epithelial cells: cortex. Epithelial cells of the cortical type have large, round to oval, clear nuclei and conspicuous nucleoli.

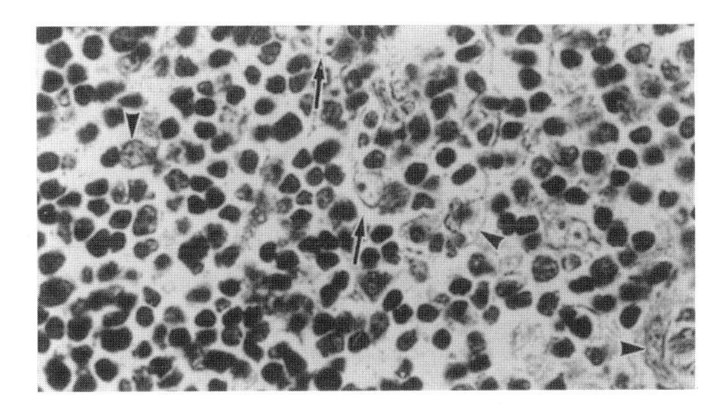

FIGURE 9.5 Thymic epithelial cells: cortico–medullary junction. Epithelial cells of cortical type (arrows) and medullary type (arrowheads) are intermingled.

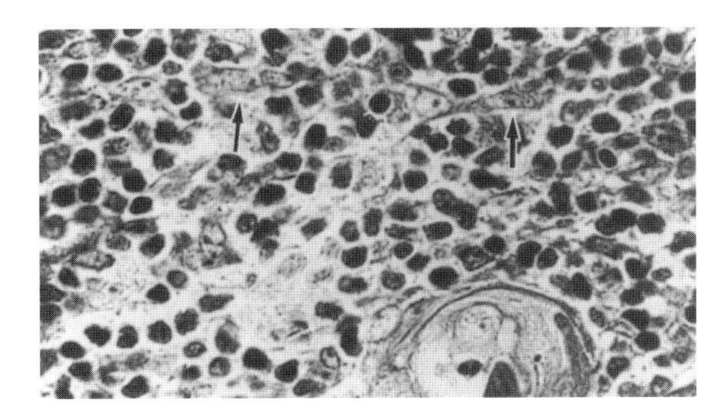

FIGURE 9.6 Thymic epithelial cells: medulla. Fusiform epithelial cells have spindle-shaped nucleus, coarse chromatin structure, and interconnecting cell process. A Hassall's corpuscle is seen in the right lower part.

Thymoma is a rare neoplasm of epithelial cells often with an associated thymic lymphoproliferation that is benign. Half of these tumors occur in patients with myasthenia gravis. They may also be associated with immunodeficiency.

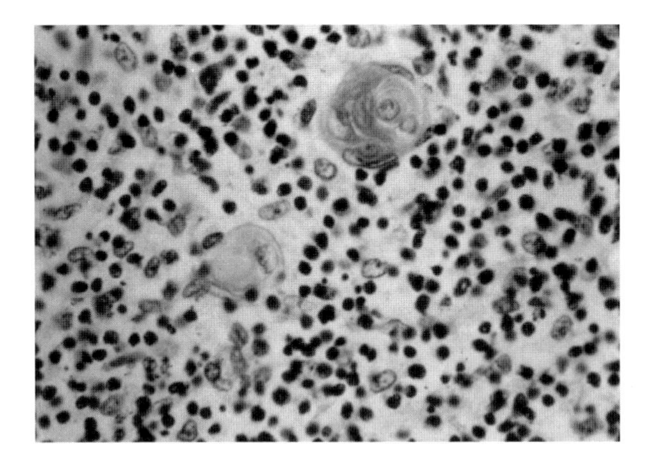

FIGURE 9.7 Hassall's corpuscle.

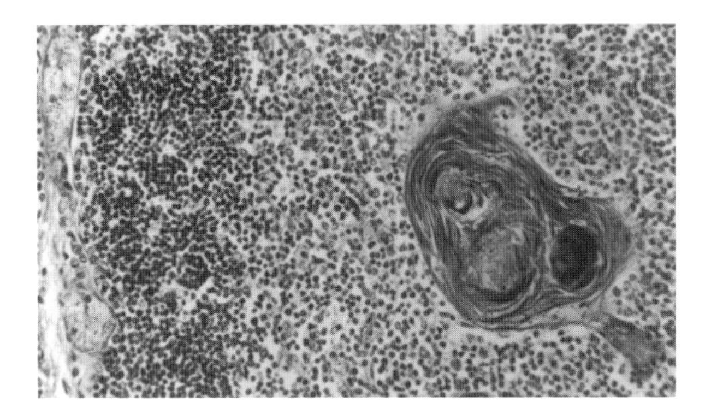

FIGURE 9.8 Normal thymus. This child's thymus shows the dense cortex composed predominantly of lymphocytes and the less dense medulla with fewer lymphocytes. Note the Hassall's corpuscle.

Myoid cells are present in the neonatal thymus of humans and other species. They contain skeletal myofibrils.

Status thymolymphaticus (historical) was **a** clinical condition described a half century ago as pathological enlargement of the thymus gland. Regrettably, it was treated with radiotherapy. Physicians of the time did not realize that the thymus enlarges under physiologic conditions, attaining a weight of 15 to 25 g at puberty followed by subsequent involution of the gland. Individuals subjected to radiation therapy were at increased risk of developing thyroid and breast cancer.

Hassall's corpuscles are epithelial cell whorls in the medulla of the thymus. They are thought to produce thymic hormones. Whereas the center may exhibit degeneration, cells at the periphery may reveal endocrine secretion granules.

Sialophorin (CD43) is a principal glycoprotein present on the surface of thymocytes, T cells, selected B lymphocytes, neutrophils, platelets, and monocytes. Monocyte and lymphocyte sialophorin is a 115-kDa polypeptide chain, whereas the platelet and neutrophil sialophorin is a 135-kDa polypeptide chain that has only a different content of carbohydrate from the first form. Galactose β1–3 galactosamine in O-linked saccharides bound to threonine or serine amino acid residues represents a site of attachment for sialic acid in thymocytes in the medulla and in mature T cells. Incomplete sialylation of thymocytes in the thymic cortex accounts for their binding to peanut lectin, whereas the more thoroughly sialylated structure on T cells and thymocytes in the medulla fails to bind the lectin. In humans, the sialophorin molecule is comprised of 400 amino acid residues. There is a 235-residue extracellular domain, a 23-residue transmembrane portion, and a 123-residue domain in the cytoplasm. Antibodies specific for CD43 can activate T cells. Wiskott–Aldrich syndrome patients have T cells with defective sialophorin.

LGSP (leukocyte sialoglycoprotein) is a richly glycosylated protein present on thymocytes and T lymphocytes. B lymphocytes are devoid of leukocyte sialoglycoprotein.

The ***c-myb*** **gene** encodes formation of a DNA-binding protein that acts during early growth and differentiation stages of normal cells. A c-*myb* gene is expressed mainly in hematopoietic cells, especially bone marrow hematopoietic precursor cells, but it is greatest in the normal murine thymus. The highest c-*myb* expression is in the double-negative thymocyte subpopulation.

The **blood–thymus barrier** protects thymocytes from contact with antigen (Figure 9.9). Lymphocytes reaching the thymus are prevented from contact with antigen by a physical barrier. The first level is represented by the capillary wall endothelial cells inside the pericytes outside of the lumen. Potential antigenic molecules, which escape the first level of control, are taken over by macrophages present in the pericapillary space. Further protection is provided by a third level, represented by the mesh of interconnecting epithelial cells, which enclose the thymocyte population. The effects of thymus and thymic hormones on the differentiation of T cells is demonstrable in animals congenitally lacking the thymus gland (nu/nu animals), neonatally or adult thymectomized animals, and in subjects with immunodeficiencies involving T cell function. Differentiation is associated with surface markers whose presence or disappearance characterizes the different stages of cell differentiation. There is extensive proliferation of the subcapsular thymocytes. The largest proportion of these cells die, but remaining cells continue to differentiate. The differentiating cells become smaller in size and move through interstices in the thymic medulla. The fully developed thymocytes pass through the walls of the postcapillary venules to reach the systemic circulation

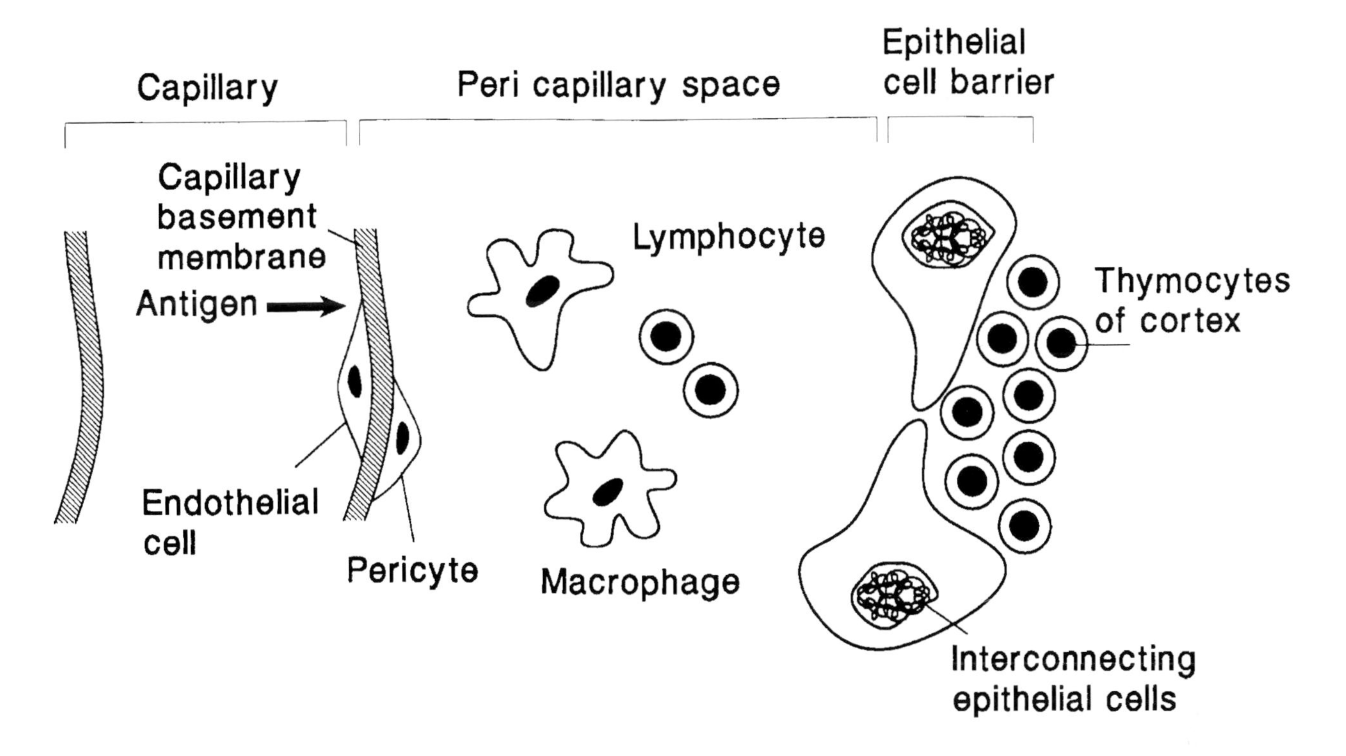

FIGURE 9.9 Three levels of lymphocyte protection which form the blood–thymus barrier: capillary wall, macrophages in pericapillary space, and a wall of epithelial cells.

and seed in the peripheral lymphoid organs. Some of them recirculate but do not return to the thymus.

Thymic epithelial cells are present in the cortex (Figure 9.10) and in the medulla of the thymus (Figure 9.11), and are derived from the third and fourth pharyngeal pouches. They affect maturation and differentiation of thymocytes through the secretion of thymopoietin, thymosins, and serum thymic factors. Thymic epithelial cells express both MHC class I and class II molecules. They secrete IL-7 which is required for early T lymphocyte development during positive selection. Maturing T lymphocytes must recognize self-peptides bound to thymic epithelial cell surface MHC molecule to avoid programmed cell death.

Thymic nurse cells are relatively large epithelial cells that are very near thymic lymphocytes and are believed to have a significant role in T lymphocyte maturation and differentiation.

A **thymocyte** is a developing T lymphocyte present in the thymus gland.

Adenosine deaminase (ADA) is a 38-kDa deaminating enzyme that prevents increased levels of adenosine, adenosine trisphosphate (ATP), deoxyadenosine, deoxy-ATP, and *S*-adenosyl homocysteine. It is encoded by the

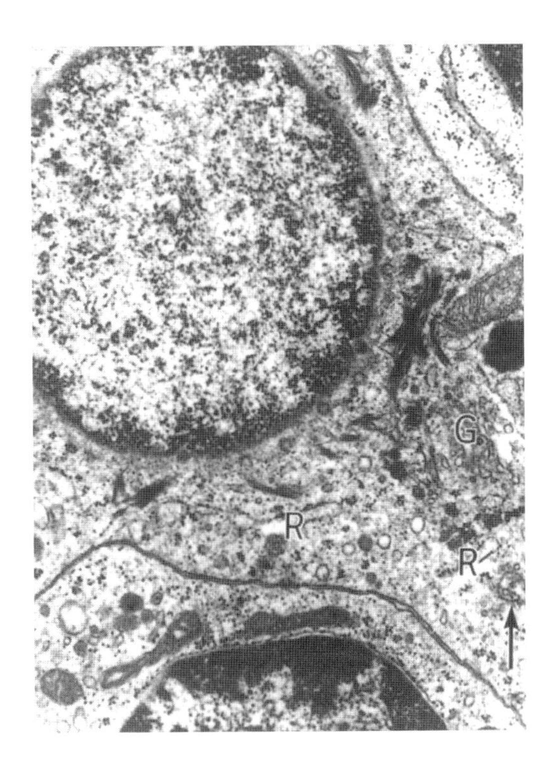

FIGURE 9.10 Type II "pale" epithelial cell in outer cortex. R: profiles of RER; G: a Golgi complex; arrow: multivesicular body (× 9000).

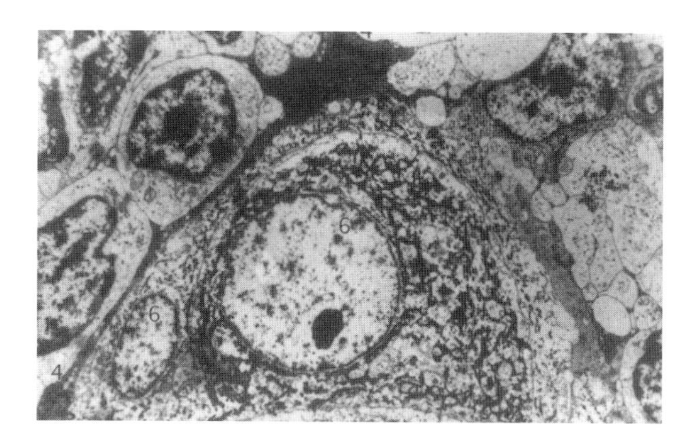

FIGURE 9.11 Two type VI "large medullary" epithelial cells (6) adjacent to a Hassall's corpuscle and two type IV epithelial cells (4) in the medulla (× 7000).

chromosome 20 q13-ter gene. Elevated adenosine levels block DNA methylation within cells, leading to their death. Increased levels of deoxy-ATP block ribonucleoside-diphosphate reductase, which participates in the synthesis of purines. The absence of adenosine deaminase (ADA) enzyme that participates in purine metabolism causes toxic purine nucleosides and nucleotides to accumulate, leading to death of thymic developing lymphocytes.

Single-positive thymocytes represent a late stage of T cell development in the thymus in which maturing T cell precursors express either CD4 or CD8 molecules but not both. They have matured from the double-positive stage and are present principally in the medulla.

Thymic leukemia antigen (TL) is an epitope on thymocyte membrane of TL^+ mice. As the T lymphocytes mature, antigen disappears but resurfaces if leukemia develops. TL antigens are specific and are normally present on the cell surface of thymocytes of certain mouse strains. They are encoded by a group of structural genes located at Tla locus in the linkage group IX, very close to the D pole of the H-2 locus on chromosome 17. There are three structural TL genes, one of which has two alleles. The TL antigens are numbered from 1 to 4, specifying four antigens: TL.1, TL.2, TL.3, and TL.4. TL.3 and TL.4 are mutually exclusive. Their expression is under the control of regulatory genes, apparently located at the same Tla locus. Normal mouse thymocytes belong to three phenotypic groups: Tl^-, Tl.2, and TL.1, 2, 3. Development of leukemia in the mouse induces a restructuring of the Tl surface antigens of thymocytes with expression of TL.1 and TL.2 in TL^- cells, expression of TL.1 in TL.2 cells, and expression of TL.4 in both TL^- and TL.2 cells. When normal thymic cells leave the thymus, the expression of TL antigen ceases. Thus thymocytes are TL^+ (except the TL^- strains) and the peripheral T cells undergo antigenic modulation. In transplantation experiments TL^+ tumor cells undergo antigenic modulation. Tumor cells exposed to homologous antibody stop expressing the antigen and thus escape lysis when subsequently exposed to the same antibody plus complement.

T cell development: Stem cells in the bone marrow that are destined to develop into T cells migrate to the thymus where they undergo maturation and development. These precursors of T cells possess unrearranged TCR genes and express neither CD4 nor CD8 markers. Thymocytes, the developing T cells, are found first in the outer cortex where their numbers increase. The TCR genes are rearranged, and CD3, CD4, CD8, and T cell receptor molecules are expressed on the surface. During maturation, these cells pass from the cortex to the medulla. As maturation proceeds, $CD4^-CD8^-$ (double negative) T cells develop into $CD4^+CD8^+$ double positive cells that then become either $CD4^+$ $CD8^-$ or $CD4^-CD8^+$ single positive T cells. Somatic rearrangement of variable, diversity (β) and joining gene segments in the area of C gene segments lead to the production of functional genes that encode TCR α and β polypeptides. Many T cell specificities result from the numerous combinations possible for joining of separate gene segments in addition to various mechanisms for junction diversity. Somatic rearrangement of germ-line genes is also responsible for the functional genes that encode TCR γ and δ polypeptides. Even though there are fewer V genes in the γ and δ loci and greater junction diversity, the mechanisms to produce γδ diversity resemble those for the αβ receptor. A few cortical thymocytes express γδ receptors. Thereafter, a line of developing T lymphocytes express numerous αβ TCR receptors. Beta chains appear first, followed by α chains of the TCR. The β chain associates itself with an invariant pre-T α surrogate alpha chain. Signals transduced by the pTαβ receptor facilitate expression of CD4 and CD8 and expansion of immature thymocytes. $CD4^+CD8^+$ cortical thymocytes first express αβ receptors. Self-HC restriction and self-tolerance develop as a consequence of the interaction between cortical epithelial cells and nonlymphoid cells derived from the bone marrow that both express MHC. This leads to selection of those T cells that are to be saved. During positive selection, $CD4^+CD8^+$ TCR αβ thymocytes recognize peptide–MHC complexes on thymic epithelial cells with low avidity. This saves them from programmed cell death or apoptosis. Recognition of self-peptide-MHC complexes on thymic antigen presenting cells with avidity by $CD4^+CD8^+$TCR αβ thymocytes leads to apoptosis. The majority of cortical thymocyes are killed during selection processes. Those αβ thymocytes that remain undergo maturation and proceed to the medulla where they become single positive cells that are either $CD4^+CD8^-$, or $CD4^-CD8^+$. During residence in the medulla, these cells become either helper or cytolytic cells prior to their journey to the peripheral

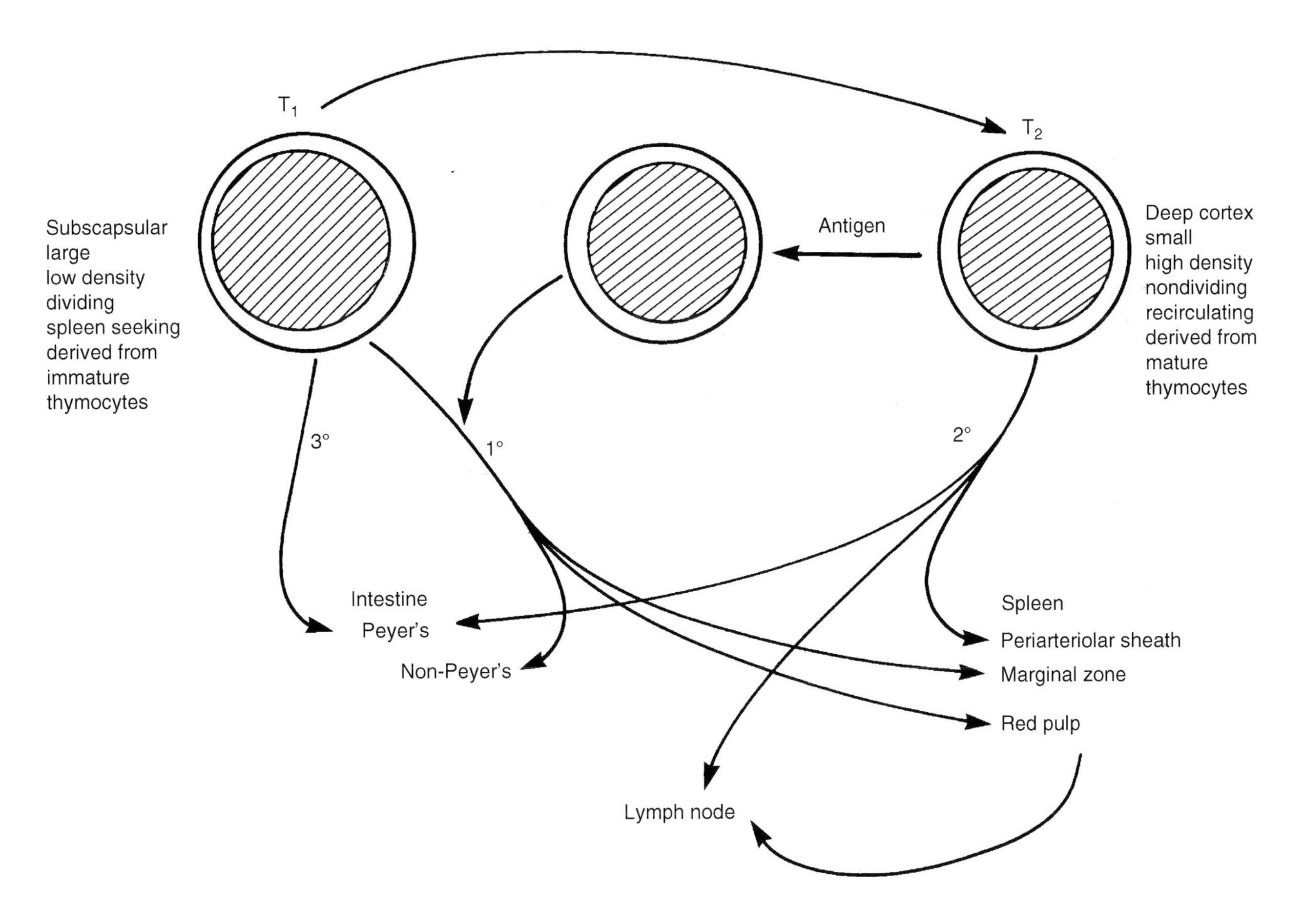

FIGURE 9.12 Migration patterns of thymus cells.

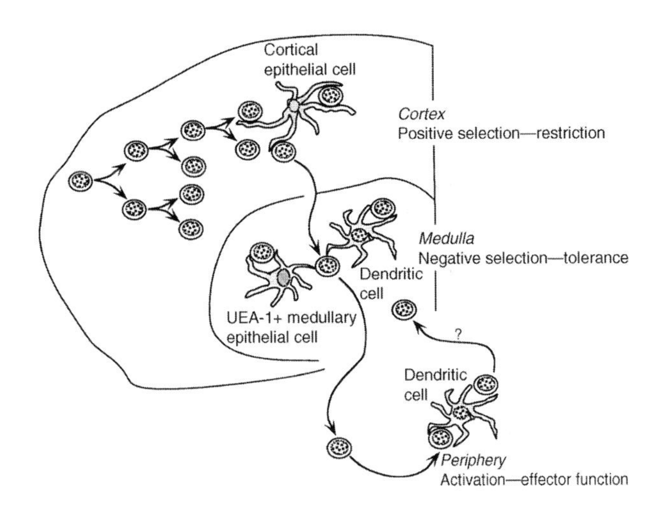

FIGURE 9.13 A model for the thymic compartment specialization.

lymphoid tissues where they function as self-MHC-restricted helper T cells or precytotoxic T lymphocytes capable of responding to foreign antigen.

Adaptive differentiation refers to acquisition of the ability to identify MHC class II antigens by thymocytes undergoing differentiation and maturation to $CD4^+$ T helper/inducer cells in the thymus.

Lymphocyte maturation refers to the development of pluripotent bone marrow precursor cells into T or B lymphocytes that express antigen receptors that are present in peripheral lymphoid tissue. B cell maturation takes place in the bone marrow and T cell maturation is governed by the thymus.

A **pro-T cell** is the earliest identifiable thymocyte and is recognized by expression of cell surface antigens such as CD2, CD7, or CD3 M protein in the cytoplasm. Rearrangements of δ, γ, and β TCR genes accompanies differentiation of pro-T cells into pre-T cells.

T cell maturation: See thymus cell differentiation.

T cell migration: Cells leaving the thymus migrate to all peripheral lymphoid organs seeding in the T-dependent regions of the lymph nodes and spleen and at the periphery of the lymphoid follicles. The rate of release of thymocytes from the thymus is markedly increased following antigenic stimulation. The patterns of migration of T cells (as well as of B cells) have been studied by adoptive transfer of labeled purified cells into irradiated syngeneic mice matched for age and sex.

Somatic recombination is DNA recombination whereby functional genes encoding variable regions of antigen receptors are produced during lymphocyte development. A limited number of inherited, or germline DNA sequences that are first separated from each other are assembled together by enzymatic deletion of intervening sequences and religation. This takes place only in developing B or T cells. Also called somatic rearrangement.

Double-negative (DN) cell is a stage in the development of T lymphocytes in which differentiating α/β T lymphocytes do not possess a T cell receptor and do not manifest either the CD4 or the CD8 coreceptor.

Double-negative thymocytes are CD4⁻CD8⁻ thymocytes that are few in number and serve as progenitors for all other thymocytes. They represent an intermediate step between pluripotent bone marrow stem cells and immature cells destined to follow T cell development. Significant heterogeneity is present in this cell population. Peripheral extrathymic CD4⁻CD8⁻ T cells have been examined in both skin and spleens of mice, and like their corresponding cells in the thymus, CD4⁻CD8⁻ T cells express T cell receptor (TCR) γδ proteins. These double-negative cell populations are greatly expanded in certain autoimmune mouse strains such as those expressing the *lpr* or *gld* genes. Available evidence reveals two thymic populations of CD4⁻CD8⁻ cells. Immature double-negative bone marrow graft cells have stem cell features. However, double-negative cells of greater maturity and without stem cell functions quickly repopulate the thymus. Most double-negative thymocytes are in an early stage of development and fail to express antigen receptors.

A **double-positive (DP) cell** is a T lymphocyte developmental stage in which differentiating α/β T cells manifest the pre-T cell receptor and both the CD4 and CD8 coreceptors.

Double-positive thymocytes are cells at an intermediate stage of T lymphocyte development in the thymus that express both the CD4 and CD8 coreceptor proteins. They also express T cell receptors and are exposed to selection processes that culminate in mature single-positive T lymphocytes that express only CD4 or CD8.

Epithelial thymic-activating factor (ETAF) is an epithelial cell-culture product capable of facilitating thymocyte growth. The activity is apparently attributable to interleukin-1.

Pre-T cells are developed from pro-T cells through gene rearrangement (Figure 9.14). They give rise to γδ TCR-bearing cells through rearrangement and expression of γ and δ TCR genes. Pre-T cells that give rise to T lymphocytes expressing the αβ T cell receptor rearrange TRC genes and delete the δ TCR genes situated on the chromosome between the Vd and Cd genes.

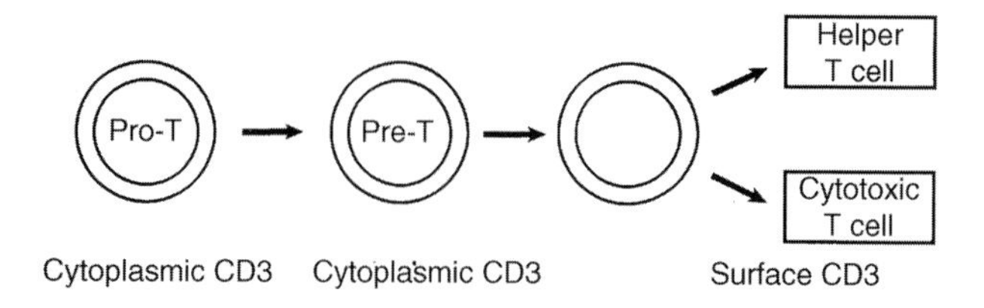

FIGURE 9.14 Schematic of pre-T cells.

Pre-T cells represent a stage in the maturation of T lymphocytes in the thymus in which cells express the T cell receptor β (TCR) chain but not the alpha chain or the CD4 or CD8 markers. The pre-T cell expresses TCR β polypeptide chain on the cell surface together with the pTα (gp33) molecule to form the pre-T cell receptor.

Pre-T cell receptor (Pre-TCR) is a pre-T lymphocyte cell surface receptor comprised of the TCR β chain and an invariant pre-T α protein. This configuration associates with CD3 and ζ (zeta) molecules to comprise the pre-T cell receptor complex. This receptor functions similar to the pre-B cell receptor in B cell development by delivering signals that induce further proliferation, antigen receptor gene rearrangements, and other maturational steps. No ligand has been described that binds specifically to a pre-T cell receptor.

pTα refers to pre-T-cell receptor.

A **prothymocyte** is a hematopoietic stem cell from the bone marrow which migrates to the thymus by the blood circulation and enters through the epithelial cell lining of the cortex. Prothymocytes (pre-T cells) differentiate in the thymus microenvironment. The prothymocytes are educated in the thymus to function as T cells. There are four thymic peptide hormones termed thymulin, thymosin α, thymosin β, and thymopoietin. These hormones are significant in T lymphocyte proliferation and differentiation. Direct interaction with the thymus epithelium, which expresses HLA antigens, is necessary for forming functional T lymphocytes and for learning to recognize major histocompatibility complex (MHC) antigens. Prothymocytes proliferate and migrate from the cortex to the medulla. Some of them are short-lived and die. The long-lived cells acquire new characteristics and are called thymocytes. They exit the thymus as immature cells and seed specific areas of the peripheral lymphoid organs where they continue to differentiate through a process driven by an antigen. From these areas, they recirculate throughout the body.

Pre-T lymphocyte (see prothymocyte) is the earliest identifiable thymocyte that is recognized by expression of cell surface antigens such as CD2 and CD7 and by CD3 ε

FIGURE 9.15 Activity of terminal deoxynucleotidyl transferase (TdT).

protein in the cytoplasm, but is absent from the cell surface. Rearrangement of δ, γ, and β TCR genes accompanies differentiation of pro-T cells into pre-T cells.

Negative selection is the process whereby those thymocytes that recognize *self antigens* in the context of self-MHC undergo clonal deletion (apoptosis) or clonal anergy (inactivation). The resulting cell population is self-MHC restricted and self antigen tolerant. (See positive selection.) Autoreactive B cells undergo a similar process in the bone marrow.

Positive selection is the survival of those thymocytes that recognize self-MHC as well as self or foreign antigen, and the death of those that do not recognize self-MHC. The resulting cell population is self-MHC restricted and capable of interacting with both self and foreign antigens. (See negative selection.)

The **differential signaling hypothesis** is a proposal which considers that antigens that differ qualitatively might mediate positive and negative selection of thymic T lymphocytes.

DNA nucleotidyltransferase (terminal deoxynucleotidyl transferase [TdT]) is DNA polymerase that randomly catalyzes deoxynucleotide addition to the 3′-OH end of a DNA strand in the absence of a template (Figure 9.15). It can also be employed to add homopolymer tails. Immature T and B lymphocytes contain TdT. The thymus is rich in TdT, which is also present in the bone marrow. TdT inserts a few nucleotides in T cell receptor genes and immunoglobulin gene segments at the V-D, D-J, and V-J junctions. This enhances sequence diversity.

Terminal deoxynucleotidyl transferase (TdT) is an enzyme catalyzing the attachment of mononucleotides to the 3′ terminus of DNA. It thus acts as a DNA polymerase. It is an enzyme present in immature B and T lymphocytes, but not demonstrable in mature lymphocytes. TdT is present both in the nuclear and soluble fractions of thymus and bone marrow. The nuclear enzyme is also able to incorporate ribonucleotides into DNA. In mice, two forms of TdT can be separated from a preparation of thymocytes. They are designated peak I and peak II. They have similar enzymatic activities and appear to be serologically related, but display significant differences in their biologic properties. Peak I appears constant in various strains of mice and at various ages. Peak II varies greatly. In some strains, peak II remains constant up to 6 to 8 months of age; in others, it declines immediately after birth. A total of 80% of bone marrow TdT is associated with a particular fraction of bone marrow cells separated on a discontinuous BSA gradient. This fraction represents 1 to 5% of the total marrow cells, but is O antigen negative. These cells become O positive after treatment with a thymic hormone, thymopoietin, suggesting that they are precursors of thymocytes. Thymectomy is associated with rapid loss of peak II and a slower loss of peak I in this bone marrow cell fraction. TdT is detectable in T cell leukemia, 90% of common acute lymphoblastic leukemia cases, and half of acute undifferentiated leukemia cells. Approximately one third of chronic myeloid leukemia cells in blast crisis and a few cases of pre-B cell acute lymphoblastic leukemia cases show cells that are positive for TdT. This marker is very infrequently seen in cases of chronic lymphocytic leukemia. In blast crisis, some cells may simultaneously express lymphoid and myeloid markers. Indirect immunofluorescence procedures can demonstrate TdT in immature B and T lymphocytes. It inserts nontemplated nucleotides (N-nucleotides) into the junctions between gene segments during T cell receptor and immunoglobulin heavy chain gene rearrangement.

Thymus cell differentiation is stem cell maturation and differentiation into mature T lymphocytes (Figure 9.16

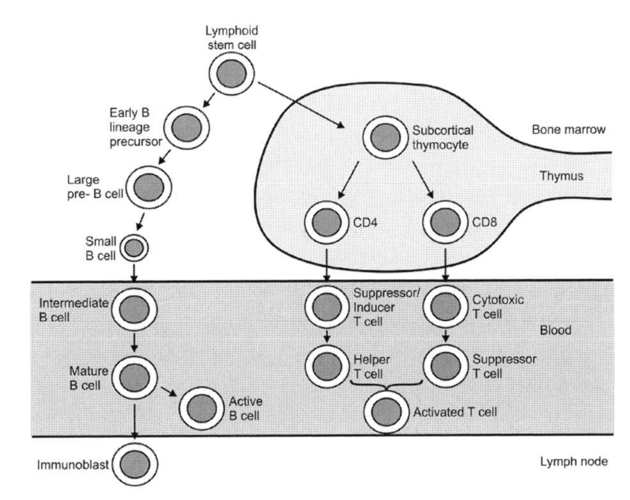

FIGURE 9.16 T cell maturation.

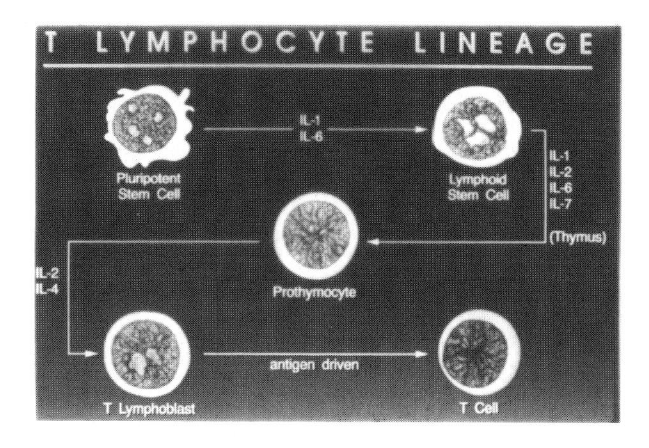

FIGURE 9.17 Differentiation of a stem cell into a mature T cell.

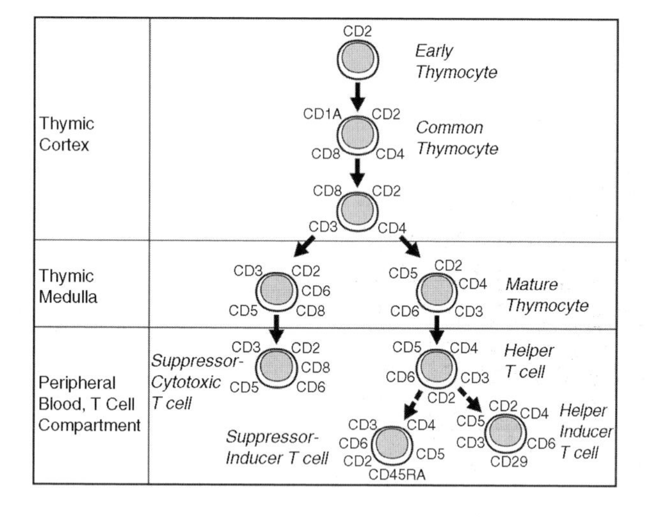

FIGURE 9.18 Diagram of T cell maturation showing addition of CD (cluster of differentiation) markers and the position of the T cell (in thymus or periphery) when the marker is added.

and Figure 9.17). This is accompanied by the appearance and disappearance of specific surface CD antigens (Figure 9.18). In humans, the differentiation of CD38 positive stem cells into early thymocytes is signaled by the appearance of CD2 and CD7, followed by the transferrin receptor marker. This is followed by expression of CD1, which identifies thymocytes in the mid-stage of differentiation when T cell receptor genes γ and δ and later α and β rearrange. This is followed by the expression of CD3, CD4, and CD8 surface antigens by thymocytes, yet CD1 usually disappears at this time. Ultimately, the $CD4^+$, $CD8^-$, and the $CD4^-CD8^+$ subpopulations which both express the CD3 pan-T cell marker appear. An analogous maturation of T cells takes place in mice. Thymus cell differentiation is also called thymus cell education.

Thymus cell education refers to thymus cell differentiation.

A **T cell-dependent (TD) antigen** is an immunogen that is much more complex than the T cell-independent (TI) antigens. They are usually proteins, protein-nuclear protein conjugates, glycoproteins, or lipoproteins. They stimulate all five classes of immunoglobulin, elicit an anamnestic or memory response, and are present in most pathogenic microorganisms. These properties ensure that an effective immune response can be generated in a host infected with these pathogens.

A **thymus-dependent (TD) antigen** is an immunogen that requires T lymphocyte cooperation for B cells to synthesize specific antibodies. Presentation of thymus-dependent antigen to T cells must be in the context of MHC class II molecules. Thymus-dependent antigens include proteins, polypeptides, hapten–carrier complexes, erythrocytes, and many other antigens that have diverse epitopes. T dependent antigens contain some epitopes that T cells recognize and others that B cells identify. T cells produce cytokines and cell surface molecules that induce B cell growth and differentiation into antibody-secreting cells. Humoral immune responses to T-dependent antigens are associated with isotype switching, affinity maturation, and memory. The response to thymus-dependent antigens shows only minor heavy chain isotype switching or affinity maturation, both of which require helper T cells signals.

T-dependent antigen is a thymus-dependent antigen.

TD antigen is a thymus-dependent antigen.

Inositol 1, 4, 5-triphosphate (IP_3) is a signaling molecule in the cytoplasm of lymphocytes activated by antigen that is formed by phospholipase C (PLC γ_1)-mediated hydrolysis of the plasma membrane. Phospholipid PIP_2. IP_3's principal function is to induce the release of intracellular Ca^{2+} from membrane-bound compartments such as the endoplasmic reticulum (ER).

Linked recognition is the need for helper T cell and B cells participating in the antibody response to a thymus-dependent antigen to interact with different epitopes linked physically in the same antigen. Epitopes that B cell and helper T cells recognize must be linked physically for the helper T cell to activate the B cell, which constitutes linked recognition.

Thymus-dependent areas are regions of peripheral lymphoid tissues occupied by T lymphocytes. Specifically, these include the paracortical areas of lymph nodes, the zone between nodules and Peyer's patches, and the center of splenic Malpighian corpuscles. These regions contain small lymphocytes derived from the circulating cells that reach these areas by passage through high endothelial venules. Proof that these anatomical sites are thymus-dependent areas is provided by the demonstration that animals thymectomized as neonates do not have lymphocytes in these areas. Likewise, humans or animals with thymic hypoplasia or congenital aplasia of the thymus reveal no T cells in these areas.

Traffic area refers to thymus-dependent area.

Thymus-dependent cells are lymphoid cells that mature only under the influence of the thymus.

A **T cell-independent (TI) antigen** is an immunogen that is simple in structure, often a polysaccharide such as the polysaccharide of the pneumococcus, a dextran polyvinyl hooter, or a bacterial lipopolysaccharide. They elicit an IgM response only and fail to stimulate an anamnestic response. They are not found in most pathogenic microbes.

T-independent antigen is a thymus-independent antigen. Humoral immune responses to T-independent antigens show only minor heavy chain isotype switching or affinity maturation, both of which require helper T cell signals.

A **thymus-independent (TI) antigen** is an immunogen that can stimulate B cells to synthesize antibodies without participation by T cells. These antigens are less complex than are thymus-dependent antigens. They are often polysaccharides that contain repeating epitopes or lipopolysaccharides (LPS) derived from Gram-negative microorganisms. Thymus-independent antigens induce IgM synthesis by B lymphocytes without cooperation by T cells. They also do not stimulate immunological memory. Murine TI antigens are classified as either TI-1 or TI-2 antigens. LPS, which activate murine B cells without participation by T or other cells, are typical TI-1 antigen. Low concentrations of LPS stimulate synthesis of specific antigen, whereas high concentrations activate essentially all B cells to grow and differentiate. TI-2 antigens include polysaccharides, glycolipids, and nucleic acids. When T lymphocytes and macrophages are depleted, no antibody response develops against them.

T cell domains are specific areas in lymph nodes and other lymphoid organs where T lymphocytes localize preferentially.

Cluster of differentiation (CD) is the term given to cell surface molecules comprising epitopes, identifiable by monoclonal antibodies, on the surfaces of hematopoietic (blood) cells in man, as well as in mice and other animals. CD markers are given numerical designations in man, but separate designations equivalent to the human CD numbers are given to animal determinants. Some individuals use the CD designation to see the antibodies which identify a particular antigen.

Pan-T cell markers are surface epitopes found on all normal T lymphocytes. These include the 50-kDa CD2 molecule that is the sheep erythrocyte rosette marker and is found exclusively on T lymphocytes, the 41-kDa CD7 molecule, CD1 present on peripheral T lymphocytes and cortical thymocytes, the mature T lymphocyte marker CD3, and CD5.

CD2 (Figure 9.19) is a T cell antigen that is the receptor molecule for sheep red cells and is also referred to as the T11 antigen or the leukocyte function associated antigen-2 (LFA-2). The molecule has a MW of 50 kDa. The antigen also seems to be involved in cell adherence, probably binding LFA-3 as its ligand. CD2 can activate T lymphocytes.

A **rosette** consists of cells of one type surrounding a single cell of another type. In immunology, it was used as an early method to enumerate T cells. E rosettes form when CD2 markers (LFA-2) on human T lymphocytes adhere to LFA-3 molecules on sheep red cells surrounding them to give a rosette arrangement (Figure 9.20 and Figure 9.21). This method is useful because sheep red cells do not form spontaneous rosettes with human B lymphocytes. $CD2^-$ T cells are now enumerated by the use of monoclonal

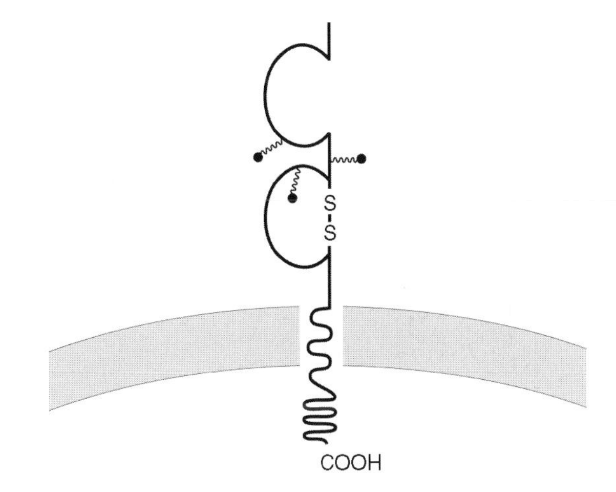

FIGURE 9.19 Structure of CD2.

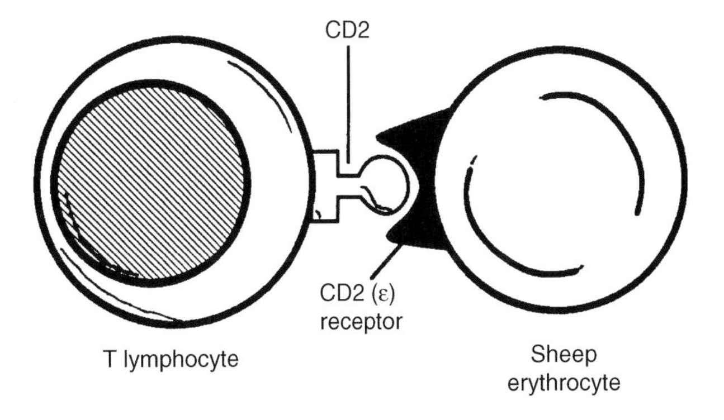

FIGURE 9.20 Adhesion of a T lymphocyte and sheep red blood cell.

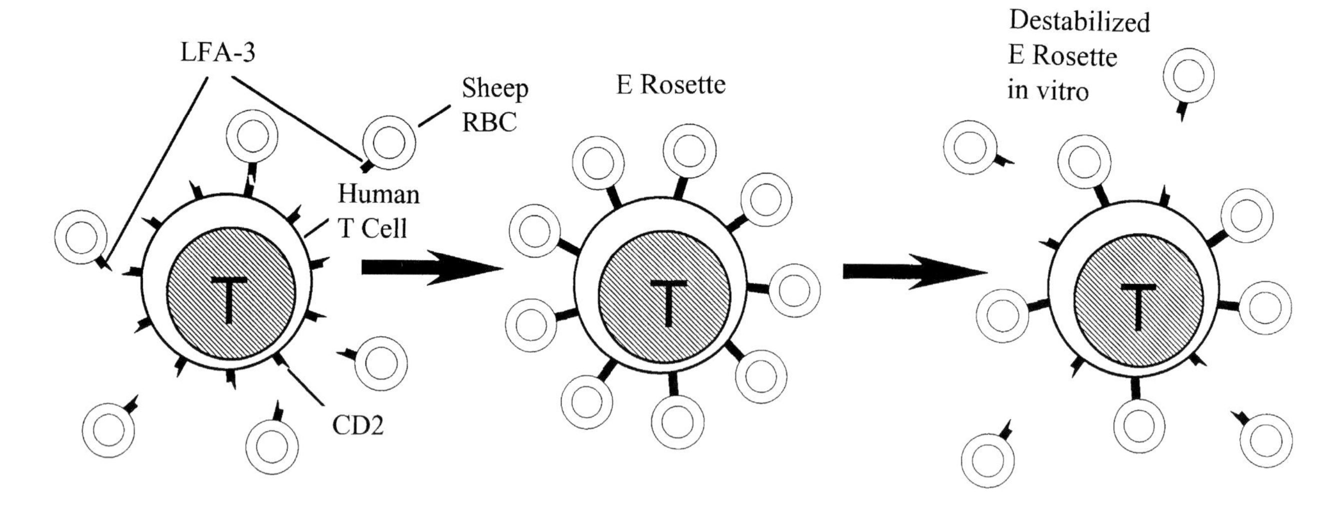

FIGURE 9.21 Formation of a rosette.

antibodies and flow cytometry. Another example was the use of the EAC rosette, consisting of erythrocytes coated with antibody and complement which surrounded a B cell bearing Fc receptors or complement receptors on its surface.

E rosette refers to a human T lymphocyte encircled by rings of sheep red blood cells. It was used previously as a method to enumerate T lymphocytes. See also E rosette-forming cell.

SRBC is the abbreviation for sheep red blood cells.

E rosette-forming cell refers to the formation of a complex of sheep red cells encircling a T lymphocyte to form a rosette. This was one of the first methods to enumerate human T cells, as the sheep red cells did not form spontaneous rosettes with human B lymphocytes. This is now known to be due to the presence of the CD2 marker, which is a cell adhesion (LFA-2) molecule, on T cells. CD2 positive T cells are now enumerated by the use of monoclonal antibodies and flow cytometry.

CD2R is a molecule restricted to activated T cells and some NK cells that has a molecular weight of 50 kDa. The CD2R epitope is unrelated to LFA-3 sites. The antigen is of importance in T cell maturation since certain CD2 antibodies in combination with CD2R antibodies or LFA-3 may induce T cell proliferation. CD2R is a conformational form of CD2 that is activation dependent.

CD1 (Figure 9.22) is an antigen that is a cortical thymocyte marker, which disappears at later stages of T cell maturation. The antigen is also found on interdigitating cells, fetal B cells, and Langerhans cells. These chains are associated with β_2-microglobulin and the antigen is thus analogous to classical histocompatibility antigens but coded for by a different chromosome. More recent studies

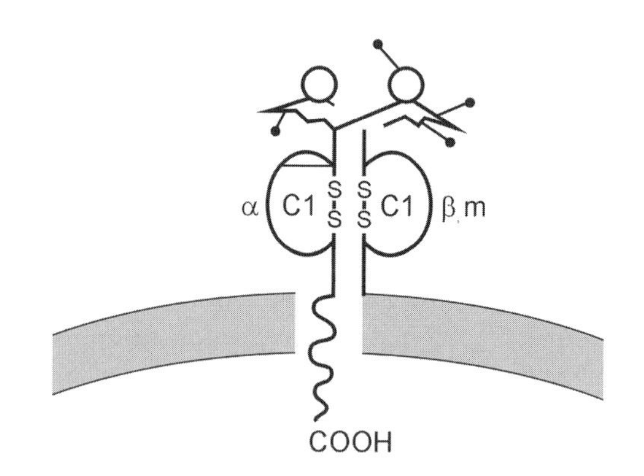

FIGURE 9.22 Structure of CD1.

have shown that the molecule is coded for by at least five genes on chromosome 1, three of which produce recognized polypeptide products. CD1 may participate in antigen presentation.

CD1a is an antigen that is a membrane glycoprotein with a molecular weight of 49,12 kDa. It is expressed strongly on cortical thymocytes.

CD1b is an antigen that is a membrane glycoprotein with a molecular weight of 45,12 kDa. It is expressed moderately on thymocytes.

CD1c is an antigen that is a membrane glycoprotein with a molecular weight of 43,12 kDa. It is expressed weakly on cortical thymocytes.

T4 antigen: See CD4.

CD4 (Figure 9.23 to Figure 9.26) is a single chain glycoprotein, also referred to as the T4 antigen, that has a molecular weight of 56 kDa and is present on approximately two-thirds of circulating human T cells, including most T cells of the helper/inducer type. The antigen is also

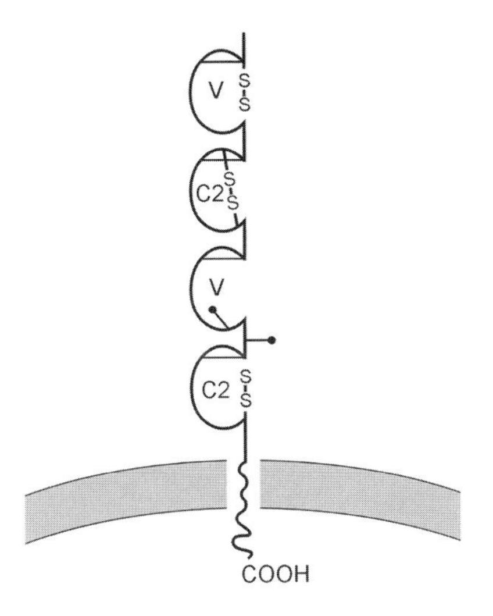

FIGURE 9.23 Structure of CD4.

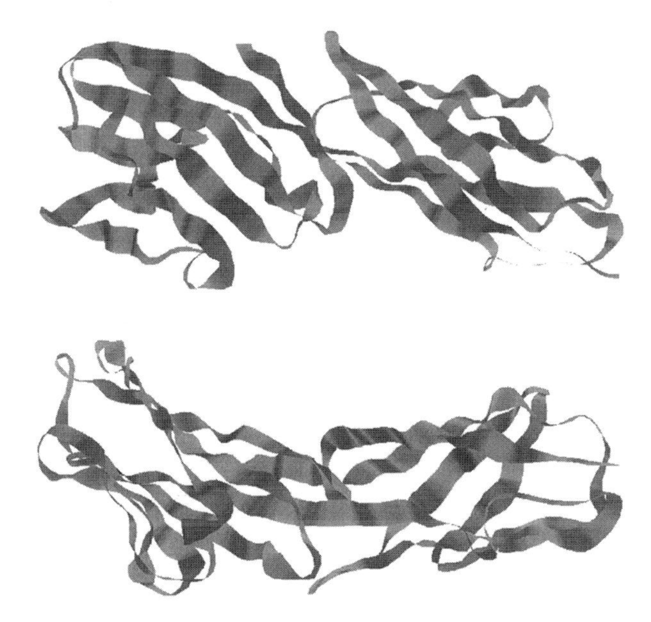

FIGURE 9.24 Ribbon structure of T cell surface glycoprotein CD4.

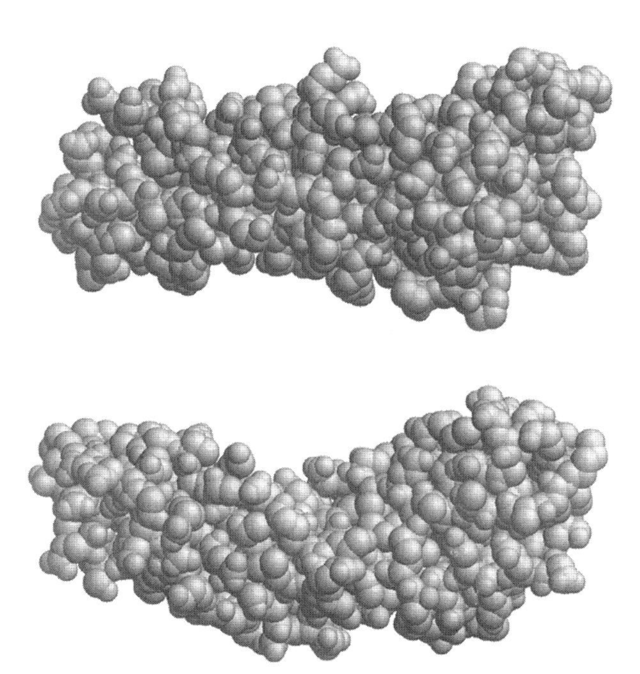

FIGURE 9.25 Space-filling model of CD4 type I crystal form. Human recombinant form expressed in Chinese hamster ovary cells.

FIGURE 9.26 Ribbon structure of CD4 domains 3 and 4. Rat recombinant form expressed in Chinese hamster ovary cells.

found on human monocytes and macrophages. The molecule is a receptor for gp120 of HIV-1 and HIV-2 (AIDS viruses). This antigen binds to class II MHC molecules on antigen-presenting cells (APC) and may stabilize APC/T cell interactions. It is physically associated with the intracellular tyrosine protein kinase known as $p56^{lck}$, which phosphorylates nearby proteins. This antigen is thereby relaying a signal to the cells. Crosslinking of CD4 may induce activation of this enzyme and phosphorylation of CD3.

The **CD4 molecule** exists as a monomer and contains four immunogloblin-like domains. The first domains of CD4 form a rigid rod-like structure that is linked to the two carboxyl-terminal domains by a flexible link. The binding site for MHC class II molecules is believed to involve the D_1 and D_2 domains of CD4.

CD4 T cells comprise the T cell subset that expresses the CD4 coreceptor and recognizes peptide antigens presented by MHC class II molecules.

Helper CD4⁺ T cells are CD4⁺ T lymphocytes that facilitate antibody formation by B cells following antigenic challenge. T_H2 T cells that synthesize cytokines IL-4 and IL-5 represent the most efficient helper T cells in contrast to helper CD4⁺ T cells. CD4⁺ T lymphocytes kill the cells with which they interact.

Inflammatory CD4 T cells are armed effector T_H1 cells that synthesize IFN-γ and TNF cytokines when they recognize antigen. Their principle function is to activate macrophages. Selected T_H1 cells may also mediate cytotoxicity.

CD29 is an antigen with a mol wt of 130 kDa, that is very widely distributed on human tissues (e.g., nerve, connective tissue, endothelium), as well as on many white cell populations. The structure recognized is now known to be the platelet gpIIa, the integrin β-1 chain, the common β-subunit of very late antigens (VLA) 1 to 6, and the fibronectin receptor. The CD29 antigen is termed gpIIa in contrast to CD31 which is called gpIIa′. Together with α1 chain (CDw49), it forms a heterodimeric complex. CD29 associates with CD49a in VLA-1 integrin. Antibodies against CD29 appear to define a subset of cells among CD4 positive T cells which provide help for antibody production. Like the antibody belonging to CD45RO (UCHL1), CD29 antibodies appear to be reciprocal (when T4 positive cells are analyzed) to those of CD45RA antibodies against the B cell restricted form of LCA. This form recognizes a T cell subset of CD4⁺ cells suppressing antibody production. The molecule is involved in the mediation of cell adhesion to cells or matrix, especially in conjunction with CD49.

Helper T cells are CD4⁺ helper/inducer T lymphocytes. They represent a subset of T cells that are critical for induction of an immune response to a foreign antigen. Antigen is presented by an antigen-presenting cell such as the macrophage in the context of self-MHC class II antigen and IL-1. Once activated, the CD4⁻ T cells express IL-2 receptors and produce IL-2 molecules which can act in an autocrine fashion by combining with the IL-2 receptors and stimulating the CD4⁻ cells to proliferate. Differentiated CD4⁻ lymphocytes synthesize and secrete lymphokines that affect the function of other cells of the immune system such as CD8⁻ cells, B cells, and NK cells. B cells differentiate into plasma cells that synthesize antibody. Activated macrophages participate in delayed-type hypersensitivity (type IV) reactions. Cytotoxic T cells also develop. Murine monoclonal antibodies are used to enumerate CD4⁻ T lymphocytes by flow cytometry.

A **nonspecific T lymphocyte helper factor** is a soluble factor released by CD4⁻ helper T lymphocytes, which nonspecifically activates other lymphocytes.

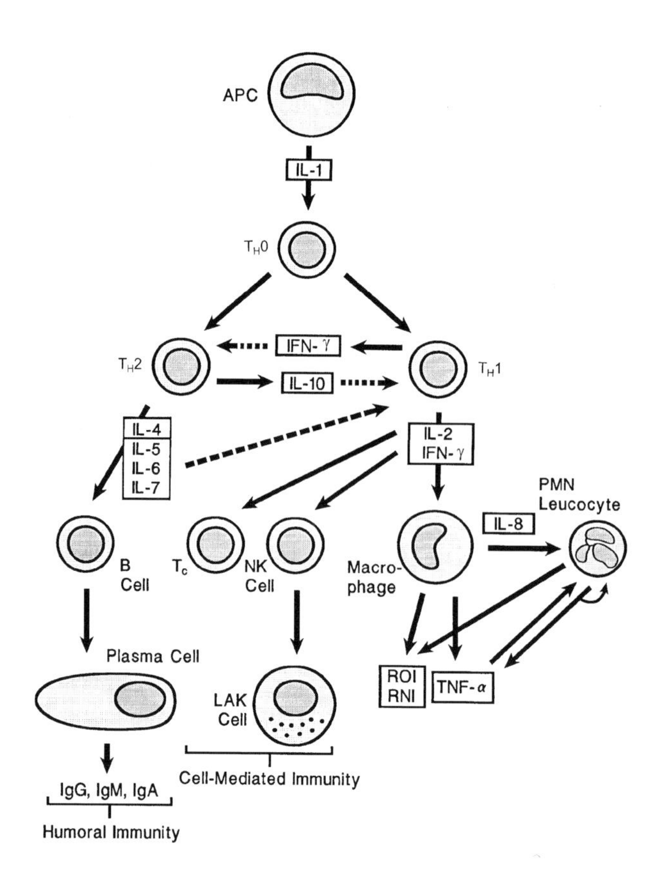

FIGURE 9.27 Functions of T_H1 and T_H2 cells in the immune response via the release of cytokines.

T cell nonantigen-specific helper factor is a substance that provides nonspecific help to T lymphocytes.

T_H0 cells are a subset of CD4⁻ cells in both humans and mice based on cytokine production and effector functions. T_H0 cells synthesize multiple cytokines. They are responsible for effects intermediate between those of T_H1 and T_H2 cells, based on the cytokines synthesized and the responding cells. T_H0 cells may be precursors of T_H1 and T_H2 cells (Figure 9.27).

T_H1 cells are a subset of CD4⁻ cells which synthesize interferon-gamma (IFN-γ), IL-2, and tumor necrosis factor (TNF)-β. They are mainly responsible for cellular immunity against intracellular microorganisms and for delayed-type hypersensitivity reactions. They affect IgG2a antibody synthesis and antibody-dependent cell-mediated cytotoxicity. T_H1 cells activate host defense mediated by phagocytes. Intracellular microbial infections induce T_H1 cell development which facilitates elimination of the microorganisms by phagocytosis. T_H1 cells induce synthesis of antibody that activates complement and serves as an opsonin that facilitates phagocytosis. The IFN-γ they produce enhances macrophage activation (Figure 9.27). The cytokines released by T_H1 cells activate NK cells, macrophages, and CD8⁺ T cells. Their main function is to induce phagocyte-mediated defense against infections, particularly by intracellular microorganisms.

T_H2 cells are a subset of CD4⁻ cells which synthesize IL-4, IL-5, IL-6, IL-9, IL-10, and IL-13. They greatly facilitate IgE and IgG1 antibody responses, and mucosal immunity, by synthesis of mast cell and eosinophil growth and differentiation factors and facilitation of IgA synthesis. IL-4 facilitates IgE antibody synthesis. IL-5 is an eosinophil activating substance. IL-10, IL-13, and IL-4 suppress cell-mediated immunity. T_H2 cells are principally responsible for host defense exclusive of phagocytes. They are crucial for the IgE and eosinophil response to helminths and for allergy attributable to activation of basophils and mast cells through IgE (Figure 9.27).

An **inducer T lymphocyte** is a cell required for the initiation of an immune response. The inducer T lymphocyte recognizes antigens in the context of MHC class II histocompatibility molecules. It stimulates helper, cytotoxic, and suppressor T lymphocytes, whereas helper T cells activate B cells. The human leukocyte common antigen variant, termed 2H4, occurs on the inducer T cell surface, and 4B4 surface molecules are present on $CD4^-$ helper T cells. $CD4^-$ T lymphocytes must be positive for either 4B4 or for 2H4.

L3T4 is a CD4 marker on mouse lymphocytes that signifies the T helper/inducer cell. It is detectable by specific monoclonal antibodies and is equivalent to the $CD4^-$ lymphocyte in man. $L3T4^-$ T lymphocytes are murine $CD4^-$ T cells.

$L3T4^+$ T lymphocytes are murine $CD4^+$ T cells.

CD8 (Figure 9.28 to Figure 9.30) is a cell surface glycoprotein antigen, also referred to as the T8 antigen, that has

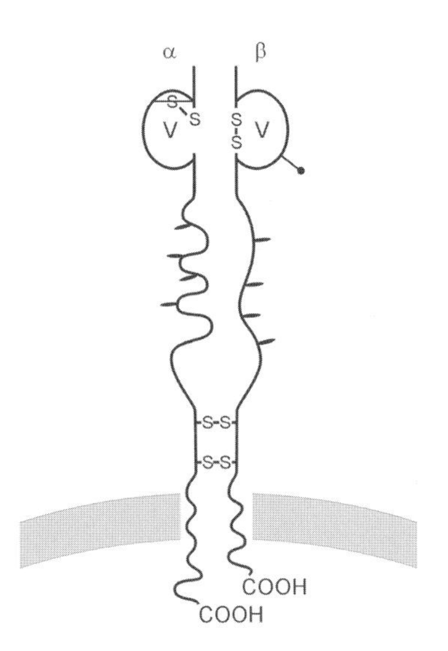

FIGURE 9.28 Structure of CD8.

FIGURE 9.29 Ribbon structure of a human CD8 T cell receptor.

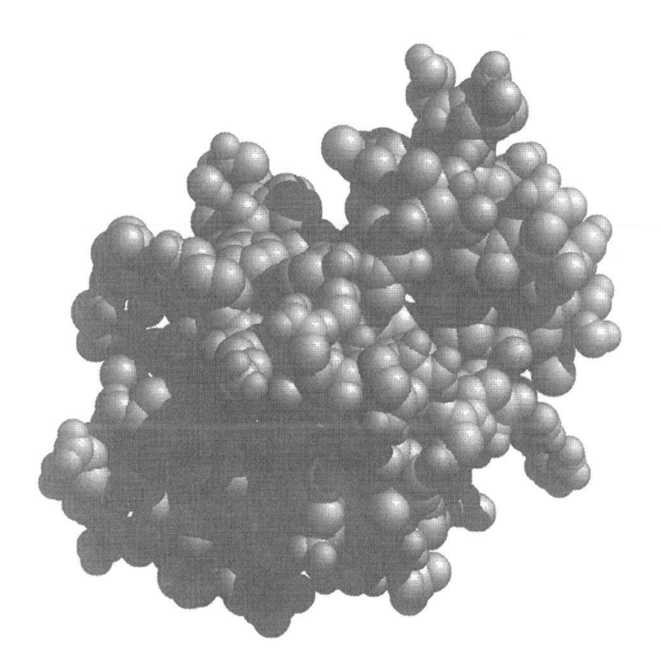

FIGURE 9.30 Space-filling model of human CD8 T cell receptor.

a molecular weight of 32 to 34 kDa. The CD8 antigen consists of two polypeptide chains, α and β, which may exist in the combination α/α homodimer of α/β heterodimer. Most antibodies are against the α-chain. This antigen binds to class I MHC molecules on antigen presenting cells (APC), and may stabilize APC/class I cell interactions. It serves as a coreceptor to facilitate the response to the T cell antigen. Like CD4, it appears to be physically associated with the $p56^{lck}$ protein tyrosine kinase, which phosphorylates nearby proteins. This antigen thereby relays a signal to the cell. It is widely used as a marker for the subpopulation of T cells that includes suppressor/cytotoxic cells. The antigen is also present on splenic sinusoidal lining cells.

T8 antigen: See CD8.

The **CD8 molecule** is a heterodimer of an α and β chain that are covalently associated by disulfide bond. The two chains of a dimer have similar structures, each having a single domain resembling an immunoglobulin variable domain and a stretch of peptide believed to be in a relatively extended confirmation.

CD8 T cells comprise the T cell subset that expresses CD8 coreceptor and recognizes peptide antigens presented by MHC Class I molecules.

Cytotoxic T lymphocyte precursor (CTLp) is a progenitor that develops into a cytotoxic T lymphocyte after it has reacted with antigen and inducer T cells.

Cytotoxicity refers to the fatal injury of target cells by either specific antibody and complement or specifically sensitized cytotoxic T cells, activated macrophages, or natural killer (NK) cells. Dye exclusion tests are used to assay cytotoxicity produced by specific antibody and complement. Measurement of the release of radiolabel or other cellular constituents in the supernatant of the reacting medium is used to determine effector cell-mediated cytotoxicity.

Cytotoxic is the ability to kill cells.

Cytotoxic T lymphocytes (CTLs) are a subset of antigen-specific effector T cells that have a principal role in protection and recovery from viral infection, that mediate allograft rejection, participate in selected autoimmune diseases, participate in protection and recovery from selected bacterial and parasitic infections, and are active in tumor immunity. They are $CD8^+$, Class I MHC–restricted, nonproliferating endstage effector cells. However, this classification also includes T cells that invoke one or several mechanisms to produce cytolysis, including perforin/granzyme, FasL/Fas, tumor necrosis factor α (TNF α), synthesize various lymphokines by T_H1 and T_H2 lymphocytes, and recognize foreign antigen in the context of either class I or class II MHC molecules.

Cytotoxic T lymphocytes (CTLs) are specifically sensitized T lymphocytes that are usually $CD8^+$ and recognize antigens in the context of MHC class I histocompatibility molecules through the T cell receptor on cells of the host infected by viruses or that have become neoplastic. Following recognition and binding, death of the target cell occurs a few hours later. CTLs secrete lymphokines that attract other lymphocytes to the area and release serine proteases and perforins that produce ion channels in the membrane of the target leading to cell lysis (Figure 9.31). Interleukin-2, produced by $CD4^+$ cells, activates cytotoxic T cell precursors. Interferon-γ generated from CTLs activates macrophages. CTLs have a significant role in the rejection of allografts and in tumor immunity. A minor population of $CD4^+$ lymphocytes may also be cytotoxic, but they recognize target cell antigens in the context of MHC class II molecules.

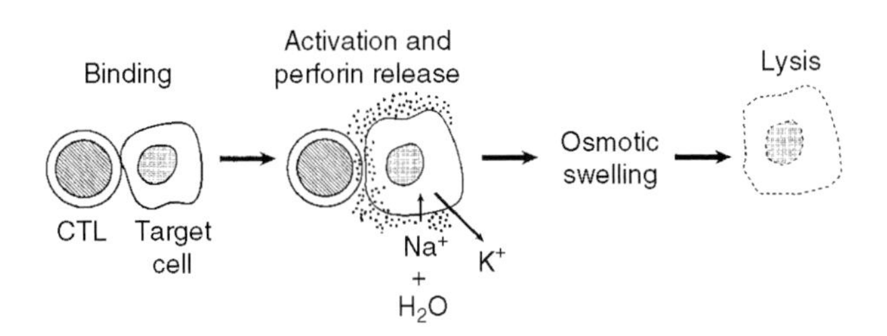

FIGURE 9.31 CTL-mediated target cell lysis.

Tc lymphocyte: See cytotoxic T lymphocyte.

Cytotoxic T cells are T lymphocytes that fatally injure other cells. Most are MHC Class I–restricted $CD8^+$ T lymphocytes even though $CD4^+$ T cells may serve as killer cells in some instances. Cytotoxic T cells are significant in host resistance against viruses and other cytosolic pathogens.

A **killer T cell** is a T lymphocyte that mediates a lethal effect on a target cell expressing a foreign antigen bound to MHC molecules on a target cell surface. A synonym for a cytotoxic T cell.

Leukophysin is an RNA helicase A-related molecule identified in cytotoxic T cell granules and vesicles. It is a 28-kDa protein of cytotoxic T lymphocytes and U937 monocytic cells that is located in the membrane of high density granules as well as lighter cytoplasmic granules or vesicles. Based on the expression of the C-terminal LKP epitope, vesicular structures and granules have been detected in CTL that are distinct from classical granzyme-containing cytolytic granules.

LTα is a mediator of killing by cytolytic T cells, "helper-killer" T cells, natural killer cells, and lymphokine-activated killer cells.

Lytic granules include perforin and granzyme-containing intracellular storage granules of cytotoxic T cells and NK cells.

They are characteristic of armude effector cytotoxic cells.

CTL is the abbreviation for cytotoxic T lymphocyte.

Cytotoxic CD8 T cells comprise a T lymphocyte subset that expresses the CD8 coreceptor and recognizes peptide antigen presented in the context of MHC Class I molecules.

Cytotoxicity assays are techniques to quantify the action of immunological effector cells in inducing cytolysis of target cells. The cell death induced is either programmed cell death (apoptosis) in which the dying cell's nuclear DNA disintegrates and the cell membrane increases in permeability, or it leads to necrosis that does not involve active metabolic processes and to increased membrane permeability without immediate nuclear disintegration. Cell-mediated cell lysis usually induces apoptosis, whereas antibody and complement usually induce necrosis. Cell death is determined by measurement of increased membrane permeability and by detecting DNA disintegration. The two methods used to determine membrane permeability of cells include dye exclusion in which trypan blue is used to stain dead cells but not viable ones, and the other is the chromium-release assay in which target cells are labeled with radioactive ^{51}Cr, which is released from cells that develop increased membrane permeability as a consequence of immune attack. To quantitatively measure either DNA disintegration, ^{125}IUdR or ^{3}H-thymidine can be used to label nuclear DNA.

A **lethal hit** refers to the induction of irreversible injury to a target cell following binding with a cytotoxic T lymphocyte. There is exocytosis, cytotoxic prelymphocye granules, polymerization of perforin in target cell membranes, and the passage of calcium ions and granzymes (apoptosis-inducing enzymes) into the cytoplasm of the target cell.

Cytotoxicity tests: (1) Assays for the ability of specific antibody and complement to interrupt the integrity of a cell membrane, which permits a dye to enter and stain the cell. The relative proportion of cells stained, representing dead cells, is the basis for dye exclusion tests. (See microlymphocytotoxicity.) (2) Assays for the ability of specifically sensitized T lymphocytes to kill target cells whose surface epitopes are the targets of their receptors. Loss of the structural integrity of the cell membrane is signified by the release of a radioisotope such as ^{51}Cr, which was taken up by the target cells prior to the test. The amount of isotope released into the supernatant reflects the extent of cellular injury mediated by the effector T lymphocytes.

Perforin from cytolytic T lymphocyte and NK cell granules produces target cell lysis. Perforin isolation and sequence determination through cDNA cloning made possible studies on perforin structure and function. This is a 70-kDa glycoprotein. Granules containing perforin mediate lysis in a medium with calcium. A transmembrane polyperforin tubular channel is formed in the membrane's lipid bilayer. Perforin that has been released inserts into the membrane of the target cell to produce a 5- to 20-nm doughnut-shaped polymeric structure that serves as a stable conduit for intracellular ions to escape to the outside extracellular environment, promoting cell death. Both murine and human perforin cDNA have now been cloned and sequenced. They share 67% homology, and each is 534 amino acids long. Perforin and C9 molecules appear to be related in a number of aspects. This is confirmed by sequence comparison, functional studies, and morphologic and immunologic comparisons. Whereas lysis of host cells by complement is carefully regulated, cytotoxic T lymphocytes lyse virus-infected host cells. Decay-accelerating factor (DAF) and HRF guard host cells against lysis by complement. By contrast, perforin is able to mediate lysis without such control factors. Using perforin cDNA as the hybridization probe, mRNA levels have been assayed for perforin expression. Thus far, all cytotoxic T lymphocytes have been shown to express the perforin message. Perforin expression is an *in vitro* phenomenon, but perforin expression has been shown *in vivo*. This was based upon cytotoxic CD8$^+$ T cells containing perforin mRNA.

Cytotoxins are proteins synthesized by cytotoxic T lymphocytes that facilitate target cell destruction. Cytotoxins include perforins and granzymes or fragmentins.

Cytolysin is a substance such as perforin that lyses cells.

Cytolytic is an adjective describing the property of disrupting a cell.

Cytolytic reaction refers to cell destruction produced by antibody and complement or perforin released from cytotoxic T lymphocytes.

A **suppressor cell** is part of a lymphoid cell subpopulation that is able to diminish or suppress the immune reactivity of other cells. An example is the CD8$^+$ suppressor T lymphocyte subpopulation detectable by monoclonal antibodies and flow cytometry in peripheral blood lymphocytes.

Suppressor/inducer T lymphocyte are a subpopulation of T lymphocytes which fail to induce immunosuppression themselves but are claimed to activate suppressor T lymphocytes.

Ts: (1) Suppressor T lymphocyte. (2) Secreted immunoglobulin heavy chain tail (C-terminal) polypeptide.

Ts1, Ts3 lymphocytes are suppressor T lymphocyte subpopulations.

TsF is an abbreviation for suppressor T cell factor.

Suppressin is a 63-kDa single polypeptide chain molecule with multiple disulfide linkages and a p.i. of 8.1. It is produced by the pituitary and by lymphocytes and is a negative regulator of cell growth. It inhibits lymphocyte proliferation and is more effective on T cells than on B cells. Suppressin has properties similar to TGF-β, although it is structurally different. Antisuppressin antibody leads to T or B cell proliferation.

Suppressor T cells (Ts cells) comprise a T lymphocyte subpopulation that diminishes or suppresses antibody formation by B cells or downregulates the ability of T lymphocytes to mount a cellular immune response. Ts cells may induce suppression that is specific for antigen, idiotype, or nonspecific suppression. Some $CD8^+$ T lymphocytes diminish T helper $CD4^+$ lymphocyte responsiveness to both endogenous and exogenous antigens. This leads to suppression of the immune response. An overall immune response may be a consequence of the balance between helper T lymphocyte and suppressor T lymphocyte stimulation. Suppressor T cells are also significant in the establishment of immunologic tolerance and are particularly active in response to unprocessed antigen. The inability to confirm the presence of receptor molecules on suppressor cells has cast a cloud over the suppressor cell; however, functional suppressor cell effects are indisputable. Some suppressor T lymphocytes are antigen specific and are important in the regulation of T helper cell function. Like cytotoxic T cells, T suppressor cells are MHC class I restricted. T cells may act as suppressors of various immune responses by forming inhibitory cytokines.

The **suppressor T cell factor (TsF)** is a soluble substance synthesized by suppressor T lymphocytes, which diminishes or suppresses the function of other lymphoid cells. The suppressor factor downregulates immune reactivity. TsF is also used as an abbreviation for suppressor T cell factor.

T cell antigen-specific suppressor factor: A soluble substance that is produced by a suppressor T cell after it has been activated. This suppressor factor has been claimed to bind antigen and cause the immune response to be suppressed in a manner that is antigen specific.

A **nonspecific T cell suppressor factor** is a $CD8^+$ suppressor T lymphocyte soluble substance that nonspecifically suppresses the immune response.

Lyt antigens are murine T cell surface alloantigens that distinguish T lymphocyte subpopulations designated as helper (Lyt1) and suppressor (Lyt2 and Lyt3) antigens. Corresponding epitopes on B cells are termed Ly.

Lyt 1,2,3 is a category of murine T lymphocyte surface antigens that is used to subdivide them into helper T cells (Lyt1) and suppressor T cells (Lyt2 and Lyt3).

Thy (θ) are epitopes found on murine thymocytes and the majority of murine T lymphocytes.

θ antigen: See Thy 1 antigen.

Thy-1 is a murine and rat thymocyte surface glycoprotein that is also found in neuron membranes of several species. Thy-1 was originally termed the θ alloantigen. Thy.1 and Thy.2 are the two allelic forms. A substitution of one amino acid, arginine or glutamine, at position 89 represents the difference between Thy1.1 and Thy1.2. Mature T cells and thymocytes in the mouse express Thy-1. The genes encoding Thy-1 are present on chromosome 9 in the mouse. Whereas few human lymphocytes express Thy-1, it is present on the surfaces of neurons and fibroblasts.

Thy 1 antigen is a murine isoantigen present on the surface of thymic lymphocytes and on thymus-derived lymphocytes found in peripheral lymphoid tissues. Central nervous system tissues may also express Thy 1 antigen.

$Thy\text{-}1^+$ dendritic cells are cells derived from the T lymphocyte linege found within the epithelium of the mouse epidermis.

Thymic stromal-derived lymphopoietin (TSLP) is a cytokine isolated from a murine thymic stromal cell line that possesses a primary sequence distinct from other known cytokines. The cDNA encodes a 140-amino acid protein that includes a 19-amino acid signal sequence. TSLP stimulates $B220^+$ bone marrow cells to proliferate and express surface μ. TSLP synergizes with other signals to induce thymocyte and peripheral T cell proliferation but is not mitogenic for T cells alone.

Thymulin is a nonapeptide (Glu-Ala-Lys-Ser-Gln-Gly-Ser-Asn) extracted from blood sera of humans, pigs, and calf thymus. Thymulin shows a strong binding affinity for the T cell receptor on the lymphocyte membrane. Its zinc-binding property is associated with biological activity. Thymulin's enhancing action is reserved exclusively for T lymphocytes. It facilitates the function of several T lymphocyte subpopulations, but mainly enhances T suppressor lymphocyte activity. Formerly called FTS.

Contrasuppression is a part of the immunoregulatory circuit that prevents suppressor effects in a feedback loop (Figure 9.32). This is a postulated mechanism to counteract the function of suppressor cells in a feedback-type mechanism. Proof of contrasuppressor and suppressor cell circuits awaits confirmation by molecular biologic techniques. A contrasuppressor cell is a T cell that opposes the action of a suppressor T lymphocyte.

CD3 is a molecule, also referred to as the T3 antigen, that consists of five different polypeptide chains with molecular weight ranging from 16 to 28 kDa. The five chains are designated γ, δ, ε, ζ, and η, with most CD3 antibodies being against the 20 kDa ε-chain. Physically, they are closely associated with each other and also with the T cell antigen receptor in the T cell membrane (Figure 9.33).

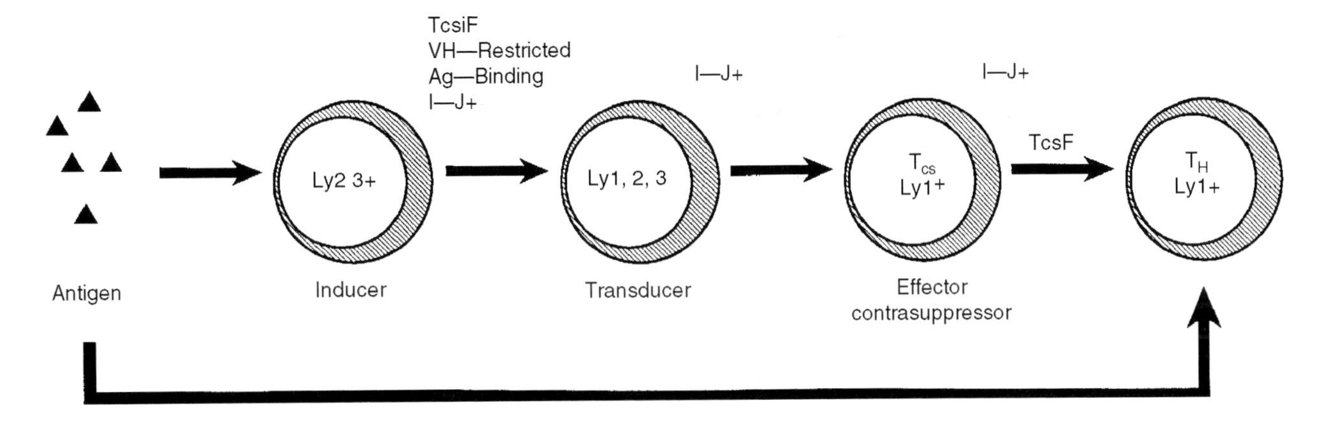

FIGURE 9.32 Contrasuppression.

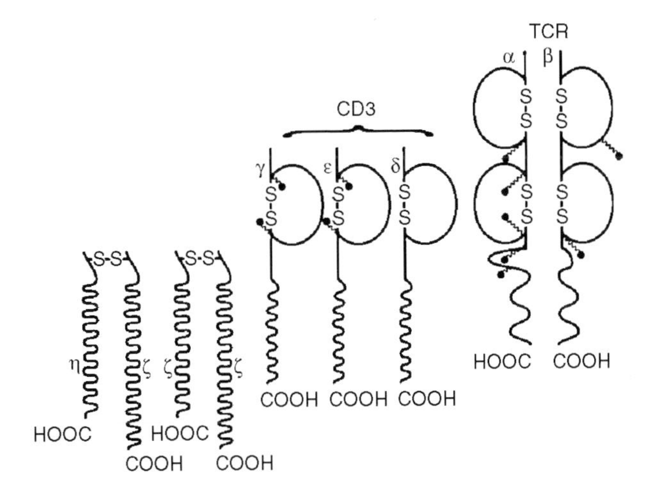

FIGURE 9.33 Structure of CD3/TCR complex.

Incubation of T cells with CD3 antibodies induces calcium flux and proliferation. This group of molecules may therefore transmit a signal to the cell interior following binding of the antigen to the antigen receptor.

The **CD3 complex** is a combination of α:β or γ:δ T cell receptor chains with the invariant subunits CD3γ, δ, and ε, and the dimeric ζ chains.

T3 antigen: See CD3.

Lck, fyn, ZAP (phosphotyrosine kinases in T cells) are the PTKs associated with early signal transduction in T cell activation (Figure 9.34). Lck is an src type PTK. It is found on T cells in physical association with CD4 and CD8 cytoplasmic regions. Deficiency of lck results in decreased stimulation of T cells and decreased T cell growth. Fyn is also an src PTK; however, it is found on hematopoietic cells. Increased fyn results in enhanced T cell activation, but a deficiency of fyn has not been shown to decrease T cell growth. Fyn deficiency inhibits T cell activation in only some T cell subsets. ZAP or ζ-associated protein kinase is like the syk PTK in B cells. It is found

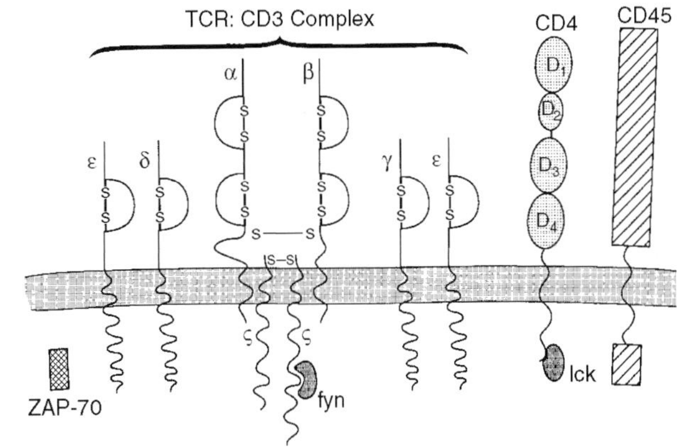

FIGURE 9.34 Structure of TCR/CD3 complex showing the lck, fyn, and ZAP phosphotyrosine kinases.

only on T cells and NK cells. Activity of the enzyme is dependent on association of ZAP with the ζ chain following TCR activation.

Zeta-associated protein of 70 kDa (ZAP-70) is a 70-kDa tyrosine kinase present in the cytosol that is believed to participate in maintaining T lymphocyte receptor signaling. It is similar to syk in B lymphocytes. It belongs to the src family of cytoplasmic protein tyrosine kinases and is required for early signaling in antigen-induced T lymphocyte activation. It binds to cytoplasmic tail phosphorylated tyrosines of the ζ chain of the T cell receptor complex. This is followed by phosphorylation of adapter proteins that recruit other signaling cascade components. It is similar to syk in B lymphocytes. See also lck, fyn, ZAP (phosphotyrosine kinases in T cells).

A **C gene segment** is DNA encoding for a T cell receptor or an immunoglobulin polypeptide chain constant region. One or more exons may be involved. Constant region gene segments comprise immunoglobulin and T cell receptor gene loci DNA sequences that encode TCR α and β chains and nonvariable regions of immunoglobulin heavy and light polypeptide chains.

A **coding joint** is a structure formed when a V gene segment joins imprecisely to a (D)J gene segment in immunoglobulin or T-cell receptor genes.

Combinatorial diversity refers to the numerous different combinations of variable, diversity, and joining segments that are possible as a consequence of somatic recombination of DNA in the immunoglubulin and TCR loci during B cell or T cell development. It serves as a mechanism for generating large numbers of different antigen receptor genes from a limited number of DNA gene segments.

Diversity (D) segments are abbreviated coding sequences between the (V) and constant gene segments in the immunoglubulin heavy chain and T cell receptor β and δ loci. Together with J segments they are recombined somatically with V segments during lymphocyte development. The recombined VDJ DNA encodes the carboxy terminal ends of the antigen receptor V regions, including the third hypervariable (CDR) regions. D segments used randomly contribute to antigen receptor repertoire diversity.

Gene rearrangement refers to genetic shuffling that results in elimination of introns and the joining of exons to produce mRNA. Gene rearrangement within a lymphocyte signifies its dedication to the formation of a single cell type, which may be immunoglobulin synthesis by B lymphocytes or production of a β chain receptor by T lymphocytes. Neoplastic transformation of lymphocytes may be followed by the expansion of a single clone of cells, which is detectable by Southern blotting.

Gene segments are multiple short DNA sequences in immunoglobulin and T cell receptor genes, which can undergo rearrangements in many different combinations to yield a vast diversity of immunoglobulin or T cell receptor polypeptide chains.

A **J gene segment** is a DNA sequence that codes for the carboxy terminal 12 to 21 residues of T lymphocyte receptor or immunoglobulin polypeptide chain variable regions. Through gene rearrangement, a J gene segment unites either a V or a D gene segment to intron 5′ of the C gene segment.

The **J region** is the variable part of a polypeptide chain comprising a T lymphocyte receptor or immunoglobulin that a J gene segment encodes. The J region of an immunoglobulin light chain is comprised of the third hypervariable region carboxy terminal (1 or 2 residues) and the fourth framework region (12 to 13 residues). The J region of an immunoglobulin heavy chain is comprised of the third hypervariable region carboxy terminal portion and the fourth framework region (15 to 20 residues). The heavy chain's J region is slightly longer than that of the light chain. The variable region carboxy terminal portion represents the J region of the T cell receptor.

A **signal joint** is a structure produced by the precise joining of recognition signal sequences during somatic recombination that produces T cell receptor and immunoglobulin genes.

Specificity refers to recognition by an antibody or a lymphocyte receptor of a specific epitope in the presence of other epitopes for which the antigen-binding site of the antibody or of the lymphoid cell receptor is specific.

There are four separate sets of **T cell receptor genes** that encode the antigen–MHC binding region of the T cell receptor. Most (approximately 95%) peripheral T lymphocytes express α and β gene sets. Only approximately 5% of circulating peripheral blood T cells and a subset of T lymphocytes in the thymus express γ and δ genes. The αβ chains or the γδ chains, encoded by their respective genes, form an intact T cell receptor and are associated with γ, δ, ε, ζ, and η chains that comprise the CD3 molecular complex. The arrangement of TCR genes resembles that of genes that encode immunoglobulin heavy chains. The TCR δ genes are located in the center of the α genes. V, D, and J segment recombination permits TCR gene diversity. Rearrangement of a V α segment to a J α segment yields an intact variable region. There are two sets of D, J, and C genes at the β locus. During joining, marked diversity is achieved by a V-J and V-D-D-J as well as V-D-J rearrangements. Humans have eight V γ, three J γ, and the initial C γ gene. Before reaching Cγ2, there are two more J γ genes. The δ locus contains five V δ, two D δ, and six J δ genes. TCR gene recombination takes place by mechanisms that resemble those of B cells genes. B and T lymphocytes have essentially the same rearrangement enzymes. TCR genes do not undergo somatic mutation, which is essential to immunoglobulin diversity.

T cell receptor (TCR) is a T cell surface structure that is comprised of a disulfide-linked heterodimer of highly variable α and β chains expressed at the cell membrane as a complex with the invariant CD3 chains. Most T cells that bear this type of receptor are termed αβ T cells. A second receptor, the γδ TCR, is comprised of variable γ and δ chains expressed with CD3 on a smaller subset of T lymphocytes that recognize different types of antigens. Both of these types of receptors are expressed with a disulfide-linked homodimer of ξ chains. The TCR is a receptor for antigen on $CD4^+$ and $CD8^+$ T lymphocytes that recognizes foreign peptide–self-MHC molecular complexes on the surface of antigen presenting cells. In the predominant αβ TCR, the two disulfide-linked transmembrane α and β polypeptide chains each bear one N-terminal Ig-like variable (V) domain, one Ig-like constant (C) domain, a hydrophobic transmembrane region, and a short cytoplasmic region. See also T lymphocyte antigen receptor, T cell receptor genes, and T cell receptor γδ.

There are two types of **T cell antigen receptors**: TCR1, which appears first in ontogeny, and TCR2 (Figure 9.35). TCR2 is a heterodimer of two polypeptides (α and β); TCR1 consists of γ and δ polypeptides (Figure 9.36). Each of the two polypeptides comprising each receptor has a constant and a variable region (similar to immunoglobulin). Reminiscent of the diversity of antibody molecules, T cell antigen receptors can likewise identify a tremendous number of antigenic specificities (estimated to be able to recognize 10^{15} epitopes).

The TCR is a structure comprised of a minimum of seven receptor subunits whose production is encoded by six separate genes. Following transcription, these subunits are assembled precisely. Assimilation of the complete receptor complex is required for surface expression of TCR subunits. Numerous biochemical events are associated with activation of a cell through the TCR receptor. These events ultimately lead to receptor subunit phosphorylation.

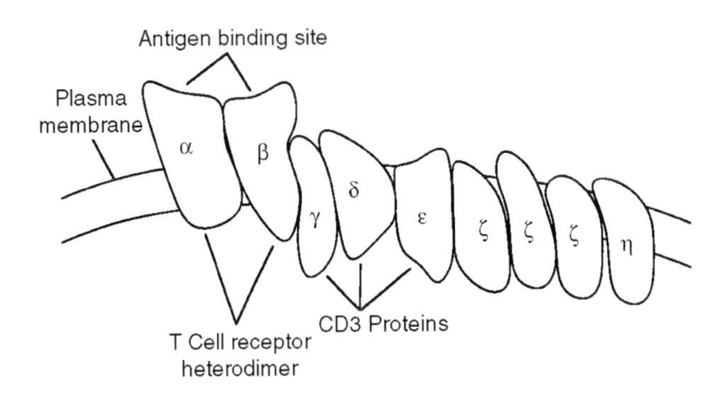

FIGURE 9.35 T lymphocyte receptor.

T cells may be activated by the interaction of antigen, in the context of MHC, with the T cell receptor. This involves transmission of a signal to the interior through the CD3 protein to activate the cell.

T lymphocyte receptor: See T lymphocyte antigen receptor.

V_T region is the T lymphocyte antigen receptor variable region.

T cell receptor complex refers to the combination of T cell receptor α and β chains and the invariant signaling proteins CD3 γ, δ, and ε, and the ζ chain.

The **antigen recognition activation motif** is a conserved sequence of 17–amino acid residues which contains two tyrosine-X-X-leucine regions. This motif is found in the cytoplasmic tails of the FceRI-b and FceRI-a chains, the z and h chains of the TCR complex, the IgG and IgA proteins of membrane IgD and IgM, and the α, γ, and ε chains of CD3. The antigen recognition activation motif is thought to be involved in signal transduction.

fyn: See lck, fyn, ZAP (phosphotyrosine kinases in T cells).

γδ T cell receptor (TCR): Double-negative CD4-CD8- T cells may express CD3-associated γδ TCRs. γδ TCRs may be capable of reacting with polymorphic ligand(s). The initial receptor to be expressed during thymic ontogeny is the γδ TCR. Cells lacking γδ TCR expression may subsequently rearrange α and β chains, resulting in αβ TCR expression. Thus, a single cell never normally expresses both receptors. Cells expressing the αβ TCR mediate

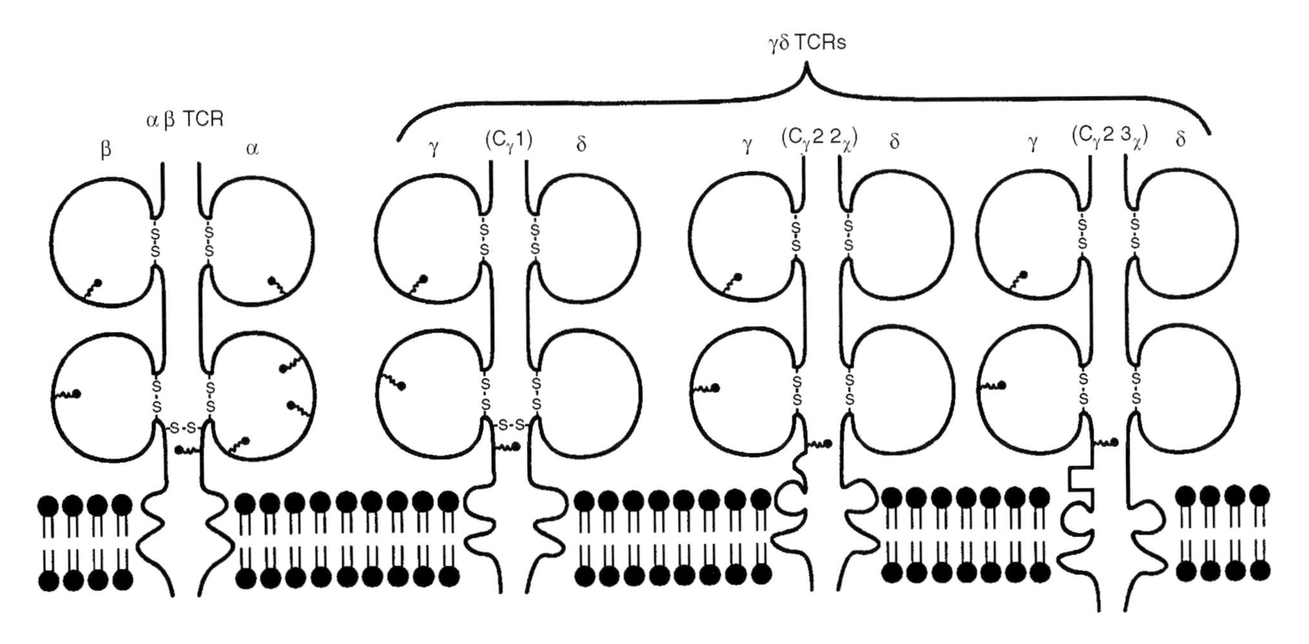

FIGURE 9.36 Structure of T cell receptors showing αβ and γδ receptors.

helper T cell and cytotoxic T cell functions. Whereas γδ TCRs are expressed on double-negative cells only, αβ TCRs may be present on either $CD4^+$ or $CD8^+$. The γδ TCR apparently can function in the absence of MHC molecules, in contrast to the association of $CD4^+$ with MHC class II or $CD8^+$ with MHC class I recognition by cells expressing αβ TCRs.

The **γδ T cell receptor** is a far less common receptor than the αβ TCR. It is comprised of γ and δ chains and occurs on the surface of early thymocytes and a small number of peripheral blood lymphocytes. The γδ TCR appears on double-negative $CD4^-CD8^-$ cells. Thus, the γδ heterodimer resembles its αβ counterpart in possessing both V and C regions, but has less diversity. TCR specificity and diversity are attributable to the multiplicity of germline V gene segments subjected to somatic recombination in T cell ontogeny, leading to a complete TCR gene. Cells bearing the γδ receptor often manifest target cell killing that is not MHC restricted. Monoclonal antibodies to specific TCR V regions are being investigated for possible use in the future treatment of autoimmune diseases. γδ T cells are sometimes found associated with selected epithelial surfaces, especially in the gut. The TCR complex is made up of the antigen-binding chains associated at the cell level with the signal transduction molecules CD3 together with zeta (ζ) and eta (η).

γδ TCR-expressing cells might protect against microorganisms entering through the epithelium in the skin, lung, intestines, etc. γδ TCR-bearing cells constitute a majority of T cells during thymic ontogeny and in the epidermis of the mouse. It is not known whether the epidermis/epithelium in the skin can function as a site for T cell education and maturation. γδ TCR represents an evolutionary precursor of the αβ TCR, as reflected by the relatively low percentages of cells expressing the γδ TCRs in adults and the fact that cells expressing the αβ TCR carry out the principal immunologic functions. Diversity of the γδ TCR and lymphokine synthesis by cells expressing the γδ TCR attest to the significance of the cells in the immune system. Little is known concerning the types of antigen recognized by gamma delta TCRs. They fail to recognize peptide complexes bound to polymorphic MHC molecules.

Clonotypic is an adjective that defines the features of a specific B cell population's receptors for antigen that are products of a single B lymphocyte clone. Following release from the B cells, these antibodies should be very specific for antigen, have a restricted spectrotype, and should possess at least one unique private idiotypic determinant. Clonotypic may also be used to describe the features of a particular clone of T lymphocytes' specific receptor for antigen with respect to idiotypic determinants, specificity for antigen, and receptor similarity from one daughter cell of the clone to another.

ζ (zeta) chain is a T cell receptor complex expressed as a transmembrane protein in T lymphocytes that contains ITAMs in its cytoplasmic tail and binds ZAP-70 protein tyrosine kinase during T lymphocyte activation.

A **homodimer** is a protein comprised of dual peptide chains that are identical.

Heterodimer is a molecule comprised of two components that are different but closely joined structures, such as a protein comprised of two separate chains. Examples include the T cell receptor comprised of either α and β chains or of γ and δ chains, and class I as well as class II histocompatibility molecules.

Intraepithelial T lymphocytes are T cells present in the epidermis and in the mucosal epithelial layer. They characteristically manifest a limited diversity of antigen receptors. They may identify glycolipids from microbes associated with nonpolymorphic major histocompatibility complex (MHC) class I–like molecules. They can function as effector cells of innate immunity and facilitate host defense by secreting cytokines, activating phagocytes, and destroying infected cells.

Self-MHC restriction refers to the confinement of antigens that an individual's T cells can recognize to complexes of peptides bound to major histocompatibility complex (MHC) molecules that existed in the thymus during T lymphocyte maturation, i.e., self-MHC molecules. Positive selection leads to self-MHC restriction of the T cell repertoire.

Self-restriction: See MHC restriction.

Anti-T cell receptor idiotype antibodies interact with antigenic determinants (idiotypes) at the variable N terminus of the heavy and light chains comprising the paratope region of an antibody molecule where the antigen-binding site is located. The idiotope antigenic determinants may be situated either within the cleft of the antigen-binding region or located on the periphery or outer edge of the variable region of heavy and light chain components. Antiidiotypic antibodies also block T cell receptors for antigen for which they are specific.

Second messengers (IP_3 and DAG): Upon stimulation of the T cell receptor, protein tyrosine kinase (PTK) becomes activated. PTK then phosphorylates phospholipase C (PLC), which in turn hydrolyzes phosphatidylinositol 4,5-bisphosphate (PIP_2). The products of PIP_2 hydrolysis are the intracellular second messengers inositol 1,4,5-triphosphate (IP_3) and 1,2-diacylglycerol (DAG). IP_3 leads to increased Ca^{++} release from intracellular stores and

DAG leads to increased levels of protein kinase C. Protein kinase C and Ca^{++} signals, such as the interaction of Ca^{++}/calmodulin to activate calcineurin, are associated with gene transcription and T cell activation. Second messenger systems are also important in B cell activation as a means of stimulating the resting B cell to enter the cell cycle. Through these and other second messenger systems, extracellular signals are received at the cell membrane and relayed to the nucleus to induce responses at the genetic level.

Lymphoid enhancer factor-1 (LEF-1) is a cell-type specific transcription factor and a member of the family of High Motility Group (HMG) domain proteins that recognizes a specific nucleotide sequence in the T cell receptor (TCR) alpha enhancer. The function of LEF-1 is dependent, in part, on the HMG domain that induces a sharp bend in the DNA helix and on an activation domain that stimulates transcription only in the specific context of other enhancer-binding proteins.

αβ T cells are T lymphocytes that express αβ chain heterodimers on their surface. The vast majority of T cells are of the αβ variety (Figure 9.37). T lymphocytes that express an antigen receptor are composed of α and β polypeptide chains. This population, to which most T cells belong, includes all those that recognize peptide antigen presented by major histocompatibility complex (MHC) class I and class II molecules.

αβ T cell receptor (αβ TCR): The structure on both $CD4^+$ and $CD8^+$ T lymphocytes that recognizes peptide antigen presented in the context of a major histocompatibility complex (MHC) molecule. The variable (V) regions of both α and β chains comprise the antigen-binding site. The αβ TCR also contains constant (C) regions. There is structural homology between TCR (V) and (C) regions and the corresponding regions of an immunoglobulin molecule. The αβ TCR is the most common form of T cell receptor, accounting for approximately 95% of T cells in human blood with γδ TCR comprising the remaining 5%.

A **silencer sequence** blocks transcription of the T cell receptor α chain. This sequence is found 5′ to the α chain enhancer in non-T cells and in those with γδ receptors.

Silencers are nucleotide sequences that downregulate transcription, functioning in both directions over a distance.

γδ T cells are early T lymphocytes that express γ and δ chains comprising the T cell receptor of the cell surface. They comprise only 5% of the normal circulating T cells in healthy adults. γδ T cells "home" to the lamina propria of the gut. Their function is not fully understood.

Tp44 (CD28) is a T lymphocyte receptor that regulates cytokine synthesis, thereby controlling responsiveness to antigen. Its significance in regulating activation of T lymphocytes is demonstrated by the ability of monoclonal antibody against CD28 receptor to block T cell stimulation by specific antigen. During antigen-specific activation of T lymphocytes, stimulation of the CD28 receptor occurs when it combines with the B7/BB1 coreceptor during the interaction between T and B lymphocytes. CD28 is a T lymphocyte differentiation antigen that four fifths of CD3/Ti positive lymphocytes express. It is a member of the immunoglobulin superfamily. CD28 is found only on T lymphocytes and on plasma cells. There are 134 extracellular amino acids with a transmembrane domain and a brief cytoplasmic tail in each CD28 monomer.

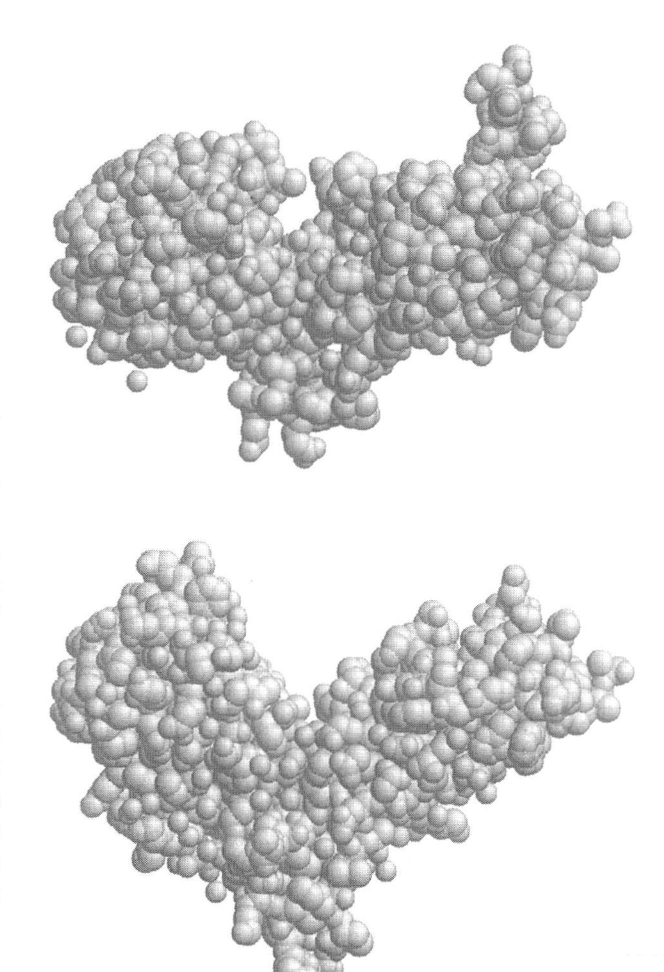

FIGURE 9.37 β chain of a T cell antigen receptor of a mouse.

CD5 (Figure 9.38), also referred to as the T1 antigen, was initially described as an alloantigen termed Ly-1 on murine T cells. Subsequently a pan-T cell marker of similar molecular mass was found on the majority of human T lymphocytes using monoclonal antibodies. This was named CD5. Thus, CD5 is homologous at the DNA level with Ly-1. CD5 is a 67-kDa type I transmembrane glycoprotein comprised

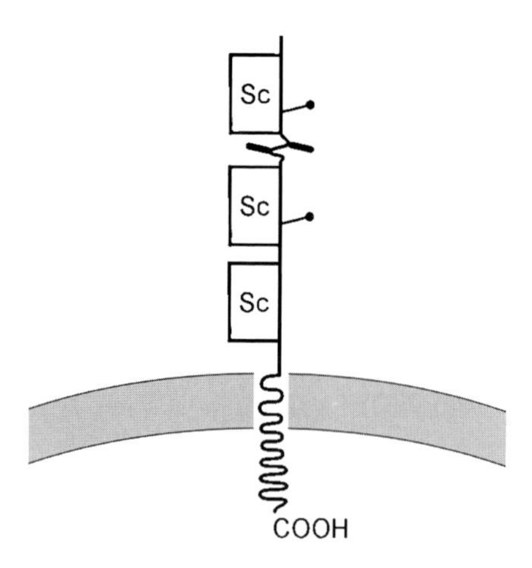

FIGURE 9.38 Structure of CD5.

of a single chain glycopolypeptide of approximately 470 amino acids. The signal peptide is formed by the first 25 amino acids. CD5 is expressed on the surface of all αβ T cells but is absent or of low density on γδ T cells. Its density increases with maturation of T cells. It is also found on a subpopulation of B cells. CD5 has been discovered on many murine B cell lymphomas as well as on endothelial cells of blood vessels in the pregnant sheep uterus. Sensitive immunohistochemical techniques have demonstrated its presence on immature B cells in the fetus and at low levels on mantle zone B cells in adult human lymphoid tissue. CD72 on B cells is one of its three ligands. CD5 is present on thymocytes and most peripheral T cells. It is believed to be significant for the activation of T cells and possibly B1 cells. CD5 is also present on a subpopulation of B cells termed B1 cells that synthesize polyreactive and autoreactive antibodies as well as the natural antibodies present in normal serum. Human chronic lymphocytic leukemia cells express CD5, which points to their derivation from this particular B cell subpopulation. CD5 bonds to CD72.

T1 antigen: See CD5.

CD6 is a molecule sometimes referred to as the T12 antigen. It is a single chain glycopolypeptide with a molecular weight of 105 kDa, and is present on the majority of human T cells (similar in distribution to CD3). It stains some B cells weakly.

CD7 is an antigen, with a molecular weight of 40 kDa that is present on the majority of T cells. It is useful as a marker for T cell neoplasms when other T cell antigens are absent. The CD7 antigen is probably an Fc receptor for IgM.

CD45 (Figure 9.39) is an antigen that is a single chain glycoprotein referred to as the leukocyte common antigen (or T-200). It consists of at least five high molecular weight

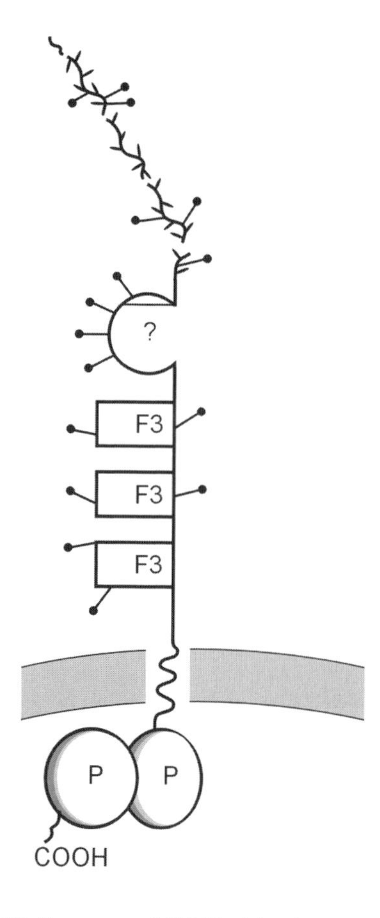

FIGURE 9-39 Structure of CD45. Also known as leukocyte common antigen (LCA).

glycoproteins present on the surface of the majority of human leukocytes (molecular weights 180, 190, 205, and 220 kDa). The different isoforms arise from a single gene via alternative mRNA splicing. The variation between the isoforms is all in the extracellular region. The larger (700–amino acid) intracellular portion is identical in all isoforms and has protein tyrosine phosphatase activity. It can potentially interact with intracellular protein kinases such as $p56^{lck}$, which may be involved in triggering cell activation. By dephosphorylating proteins, CD45 would act in an opposing fashion to a protein kinase. It facilitates signaling through B and T cell antigen receptors. Leukocyte common antigen (LCA) is present on all leukocyte surfaces. It is a transmembrane tyrosine phosphatase that is expressed in various isoforms on different types of cells, including the different subtypes of T cells. The isoforms are designated by CD45R followed by the exon whose presence gives rise to distinctive antibody-binding patterns.

CD40 ligand is a molecule on T cells that interacts with CD40 on B cell proliferation.

T-200 is an obsolete term for leukocyte common antigen (CD45).

Leukocyte common antigen (LCA, CD45) is a family of high molecular weight glycoproteins (180 to 220 kDa) densely expressed on lymphoid and myeloid cells including lymphocytes, monocytes, and granulocytes. Expression of LCA on leukocytes, but not on other cells, makes this a valuable marker in immunophenotyping of tumors with respect to determination of histogenetic origin. LCA antigen function is unknown, but it has a high carbohydrate content and is believed to be associated with the cytoskeleton. LCA molecules are heterogeneous and appear on T and B lymphocytes as well as selected other hematopoietic cells. Some LCA epitopes are present in all LCA molecules, while others are confined to B lymphocyte LCA and still other epitopes are associated with B cell, $CD8^+$ T cell, and most $CD4^+$ T cell LCA molecules. About 30 exons are present in the gene that encodes LCA molecules. It is designated as CD45, a protein tyrosine phosphatase that is found on all leukocytes and exists in several forms. By immunoperoxidase staining, it can be demonstrated in sections of paraffin-embedded tissues containing these cell types. It is a valuable marker to distinguish lymphoreticular neoplasms from carcinomas and sarcomas. T200 is an obsolete term for leukocyte common antigen (LCA).

A **T lymphocyte (T cell)** (Figure 9.40) is a thymus-derived lymphocyte that confers cell-mediated immunity. They cooperate with B lymphocytes, enabling them to synthesize antibody specific for thymus-dependent antigens, including switching from IgM to IgG and/or IgA production. T lymphocytes exiting the thymus recirculate in the blood and lymph and in the peripheral lymphoid organs. They migrate to the deep cortex of lymph nodes. Those in the blood may attach to postcapillary venule endothelial

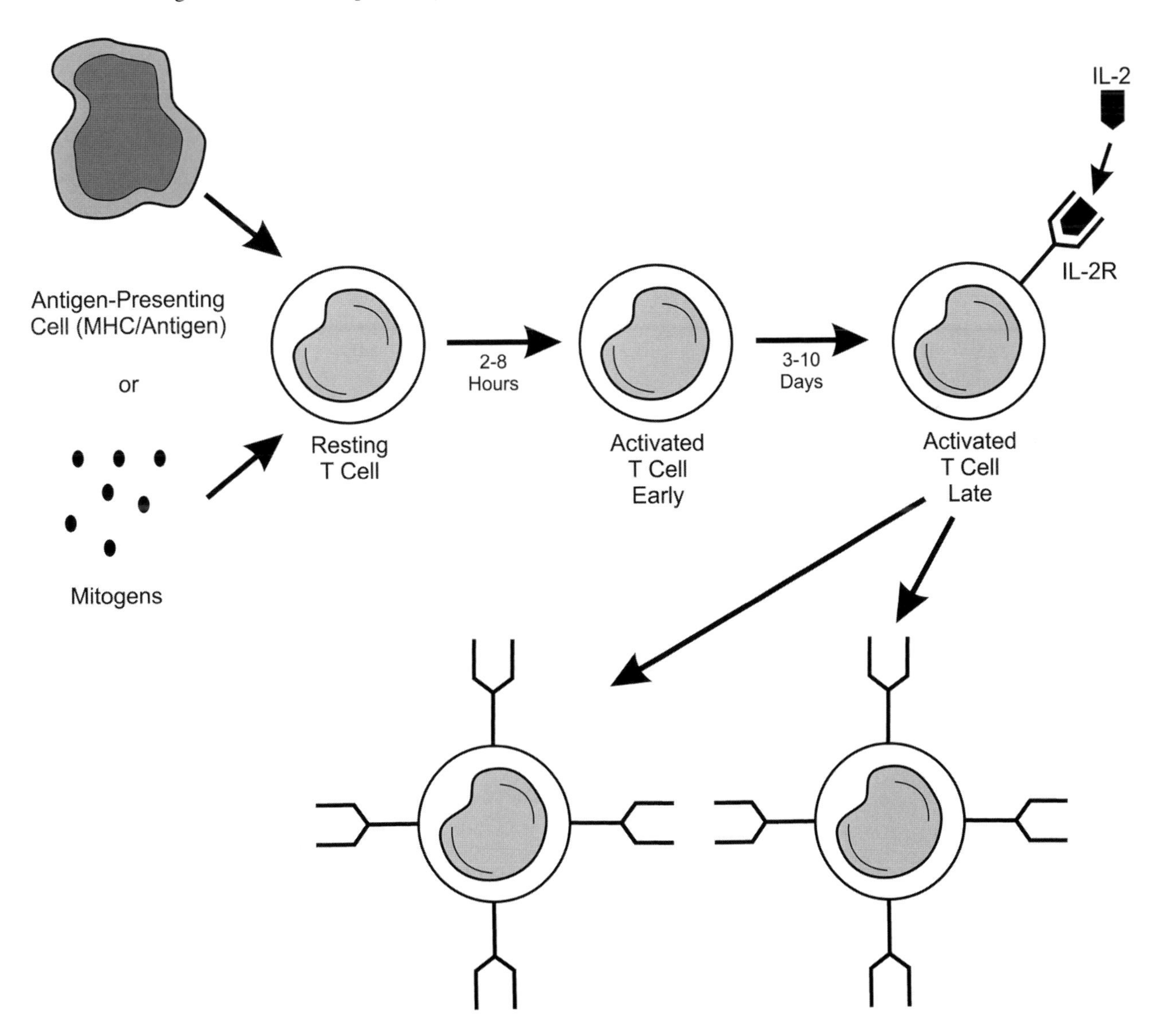

FIGURE 9.40 T cell activation.

cells of lymph nodes and to the marginal sinus in the spleen. After passing across the venules into the splenic white pulp or lymph node cortex, they reside there for 12 to 24 h and exit by the efferent lymphatics. They proceed to the thoracic duct and from there proceed to the left subclavian vein where they enter the blood circulation. Mature T cells are classified on the basis of their surface markers such as CD4 and CD8. $CD4^+$ T lymphocytes recognize antigens in the context of MHC class II histocompatibility molecules, whereas $CD8^+$ T lymphocytes recognize antigen in the context of class I MHC histocompatibility molecules. The $CD4^+$ T cells participate in the afferent limb of the immune response to exogenous antigen, which is presented to them by antigen-presenting cells. This stimulates the synthesis of IL-2, which activates $CD8^+$ T cells, NK cells, and B cells, thereby orchestrating an immune response to the antigen. Thus, they are termed helper T lymphocytes. They also mediate delayed-type hypersensitivity reactions. $CD8^+$ T lymphocytes include cytotoxic and suppressor cell populations. They react to endogenous antigen and often express their effector function by a cytotoxic mechanism, e.g., against a virus-infected cell. Other molecules on mature T cells in humans include the E rosette receptor CD2 molecule, the T cell receptor, the pan-T cell marker termed CD3, and transferrin receptors.

T lymphocytes are T cells. T lymphocyte precursors are detectable in the human fetus at 7 weeks of gestation. Between 7 and 14 weeks of gestation, thymic changes begin to imprint thymic lymphocytes as T cells. The maturation (mediated by hormones such as thymosin, thymulin, and thymopoietin II) can be followed by identification of surface (cluster of differentiation [CD]) markers detectable by immunophenotyping methods. CD3, a widespread T cell marker, serves as a signal transducer from the antigen receptor to the cell interior. Thus, the CD3 molecule is intermittently associated with the T cell receptor for antigen. T lymphocytes in the medulla initially express both CD4 and CD8 class markers; however, these cells will later differentiate into either $CD4^+$ helper cells or $CD8^+$ suppressor cells. The $CD4^+$ cells, characterized by a 55-kDa surface marker, communicate with macrophages and B cells bearing MHC class II molecules during antigen presentation. The $CD8^+$ suppressor/cytotoxic cells interact with antigen-presenting cells bearing MHC class I molecules.

Ly antigen is a murine lymphocyte alloantigen that is expressed to different degrees on mouse T and B lymphocytes and thymocytes. Also referred to as Lyt antigen.

T cells are derived from hematopoietic precursors that migrate to the thymus where they undergo differentiation which continues thereafter to completion in the various lymphoid tissues throughout the body or during their circulation to and from these sites. The T cells are primarily involved in the control of the immune responses by providing specific cells capable of helping or suppressing these responses. They also have a number of other functions related to cell-mediated immune phenomena.

T cell clonal expansion following the initial encounter with antigen is a critical mechanism of the effector limb of the immune response. This permits the immune system to respond to a broad spectrum of pathogens. Clonal expansion is closely regulated to prevent inappropriate responses. Increased clonal expansion would be desirable during the early course of an infection and in the prevention of growth and survival of T cells mediating autoimmune disease.

A **cloned T cell line** is a lineage of T lymphocytes that grow continuously from a single progenitor cell. Stimulation with antigen from time to time is necessary to maintain their growth. They are used in experimental immunology to investigate T cell function and specificity.

T cell specificity: See MHC restriction.

Fc receptors on human T cells: Fc receptors are carried on 95% of human peripheral blood T lymphocytes. On about 75% of the cells, the FcR are specific for IgM; the remaining 20% are specific for IgG. The FcR-bearing T cells are also designated T_M and T_G or T_μ and T_γ. The T_M cells act as helpers in B cell function. They are required for the B cell responses to pokeweed mitogen (PWM). In cultures of B cells with PWM, the T_M cells also proliferate, supporting the views on their helper effects. Binding of IgM to FcR of T_M cells is not a prerequired for helper activity. In contrast to the T_M cells, the T_G cells effectively suppress B cell differentiation. They act on the T_M cells, and their suppressive effect requires prior binding of their Fc-IgG receptors by IgG immune complexes. There are a number of other differences between T_M and T_G cells. Circulating T_G cells may be present in increased numbers, often accompanied by a reduction in the circulating number of T_M cells. Increased numbers of T_G cells are seen in cord blood and in some patients with hypogammaglobulinemia, sex-linked agammaglobulinemia, IgA deficiency, Hodgkin's disease, and thymoma, to mention only a few.

T cell rosette: See E rosette.

T_R1: See regulatory T cells.

Armed effector T cells are activated effector T cells that are induced to mediate their effector functions as soon as they contact cells expressing the peptide:MHC complex for which they are specific. By contrast, memory T lymphocytes require activation by antigen-presenting cells to empower them to carry out effector functions.

Subset refers to a subpopulation of cells such as T lymphocytes in samples of peripheral blood. Subsets are identified by immunophenotyping through the use of monoclonal antibodies and by flow cytometry. The cells are separated based on their surface CD (cluster of differentiation) determinants such as CD4 that identifies helper/inducer T lymphocytes and CD8 that identifies suppressor/cytotoxic T lymphocytes.

Ly6 is a GPI-linked murine cell surface alloantigens found most often on T and B cells but found also on nonlymphoid tissues such as brain, kidney, and heart. Monoclonal antibodies to these antigens indicate T cell receptor (TCR) dependence.

Helper/suppressor ratio: The ratio of $CD4^+$ helper/inducer T lymphocytes to $CD8^+$ suppressor/cytotoxic T lymphocytes. This value normally ranges between 1.5 and 2.0. However, in certain virus infections, notably AIDS, the ratio becomes inverted as a consequence of greatly diminished $CD4^+$ lymphocytes and either stationary or elevated levels of $CD8^+$ lymphocytes. Inversion of the ratio continues as the clinical situation deteriorates. Eventually, the $CD4^+$ cells may completely disappear in the AIDS patient. Other conditions in which the helper/suppressor (CD4/CD8) ratio is decreased include other viral infections such as herpes, cytomegalovirus, Epstein–Barr virus infection, and measles, as well as in graft-vs.-host disease, lupus erythematosus with renal involvement, severe burns, exercise, myelodysplasia, acute lymphocytic leukemia in remission, severe sunburn, exercise, and loss of sleep. Elevated ratios may occur in such conditions as atopic dermatitis, Sezary syndrome, psoriasis, rheumatoid arthritis, primary biliary cirrhosis, lupus erythematosus without renal involvement, chronic autoimmune hepatitis, and type I insulin-dependent diabetes mellitus.

T cell replacing factor (TRF) is an earlier term for B cell differentiation factor derived from $CD4^+$ helper T lymphocytes which permits B lymphocytes to synthesize antibody without the presence of T lymphocytes.

Concanavalin A (con A) is a jack bean *(Canavalia ensiformis)* lectin that induces erythrocyte agglutination and stimulates T lymphocytes to undergo mitosis and proliferate. Con A interacts with carbohydrate residues rich in mannose. Macrophages must be present for T lymphocytes to proliferate in response to con A stimulation. There are four 237-amino acid residue subunits in con A. There is one binding site for saccharide, one for Ca^{2+}, and one for a metallic ion such as Mn^{2+} in each con A subunit. T lymphocytes stimulated by con A release interleukin-2. Cytotoxic T lymphocytes stimulated by con A induce lysis of target cells without regard to the antigen specificity of either the effector or target cell. This could be induced by crosslinking of the effector and target cells by con A, which is capable of linking to high-mannose oligosaccharides on target cell surfaces as well as to high-mannose sugars on the T cell receptor. Con A binds readily to ordinary cell membrane glycoproteins such as glucopyranosides, fructofuranosides, and mannopyranosides.

PHA is the abbreviation for phytohemagglutinin. PHA is principally a T lymphocyte mitogen, producing a greater stimulatory effect on $CD4^+$ helper/inducer T lymphocytes than on CD8 suppressor/cytotoxic T cells. It has a weaker mitogenic effect on B lymphocytes.

Mature T cells are classified on the basis of their surface markers such as CD4 and CD8. $CD4^+$T lymphocytes recognize antigens in the context of MHC class II histocompatibility molecules, whereas $CD8^+$T lymphocytes recognize antigen in the context of class I MHC histocompatibility molecules. The $CD4^+$T cells participate in the afferent limb of the immune response to exogenous antigen, which is presented to them by antigen-presenting cells. This stimulates the synthesis of IL-2, which activates $CD8^+$T cells, NK cells, and B cells, thereby orchestrating an immune response to the antigen. Thus, they are termed helper T lymphocytes. They also mediate delayed-type hypersensitivity reactions. $CD8^+$T lymphocytes include cytotoxic and suppressor cell populations. They react to endogenous antigen and often express their effector function by a cytotoxic mechanism, e.g., against a virus-infected cell. Other molecules on mature T cells in humans include the E rosette receptor CD2 molecule, the T cell receptor, the pan-T cell marker termed CD3, and transferrin receptor.

A **T lymphocyte clone** is a daughter cell of one T lymphocyte derived from the blood or spleen that is added to culture medium and activated by antigen. Those T cells that are stimulated by the antigen form blasts which can be separated from the remaining T cells by density gradient centrifugation. The T cells that have responded to antigen are diluted, and aliquots are dispensed into tissue culture plates to which antigen and interleukin-2 are added. Each well contains a single lymphocyte. This method provides individual T cell clones.

A **T lymphocyte hybridoma** is produced by fusing a murine splenic T cell and an AKR strain BW5147 thymoma cell using polyethylene glycol (PEG). The hybrid cell clone is immortal and releases interleukin-2 when activated by antigen provided by an antigen-presenting cell.

A **T lymphocyte subpopulation** is a subset of T cells that have a specific function and express a specific cluster of differentiation (CD) markers or other antigens on their surface. Examples include the CD4+ helper T lymphocyte subset and the CD8+ suppressor/cytotoxic T lymphocyte subset.

Calcineurin is a protein phosphatase that is serine/threonine specific. Activation of T cells apparently requires deletion of phosphates from serine or threonine residues. Its action is inhibited by the immunosuppressive drugs, cyclosporin-A and FK-506. Cyclosporin-A and FK-506 combine with immunophilin intracellular molecules to form a complex that combines with calcineurin and inhibits its activity. (Figure 9.41 and Figure 9.42)

Phorbol ester(s) (Figure 9.43) are esters of phorbol alcohol (4,9,12-β,13,20-pentahydroxy-1, 6-tigliadien-3-on) found in croton oil and myristic acid. Phorbol myristate acetate (PMA), which is of interest to immunologists, is a phorbol ester that is 12-*O*-tetradecanoylphorbol-13-acetate (TPA). This is a powerful tumor promoter that also exerts pleotrophic effects on cells in culture, such as stimulation of macromolecular synthesis and cell proliferation, induction of prostaglandin formation, alteration in the morphology and permeability of cells, and disappearance of surface fibronectin. PMA also acts on leukocytes. It links to and stimulates protein kinase C, leading to threonine and serine residue phosphorylation in the transmembrane protein cytoplasmic domains such as in the CD2 and CD3 molecules. These events enhance interleukin-2 receptor expression on T cells and facilitate their proliferation in the presence of interleukin-1 as well as TPA. Mast cells, polymorphonuclear leukocytes, and platelets may all degranulate in the presence of TPA.

Protein kinase C (Figure 9.44) is a serine/threonine kinase that Ca2+ activates in the cytoplasm of cells. It participates in T and B cell activation and is a receptor

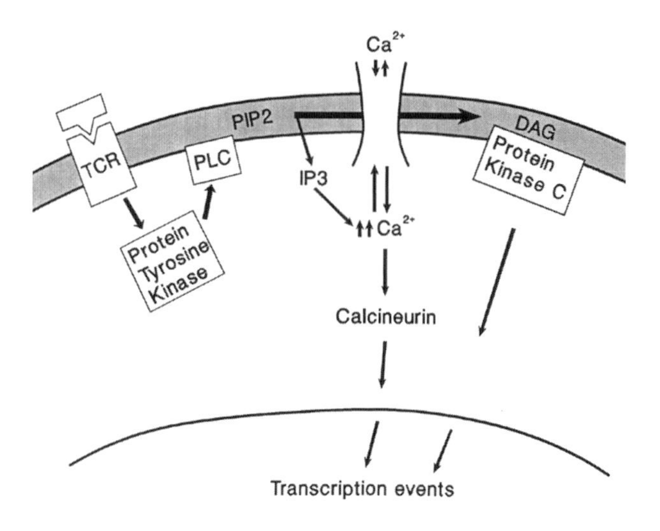

FIGURE 9.41 Schematic representation of cellular events upon binding of an activated T cell to the T cell receptor.

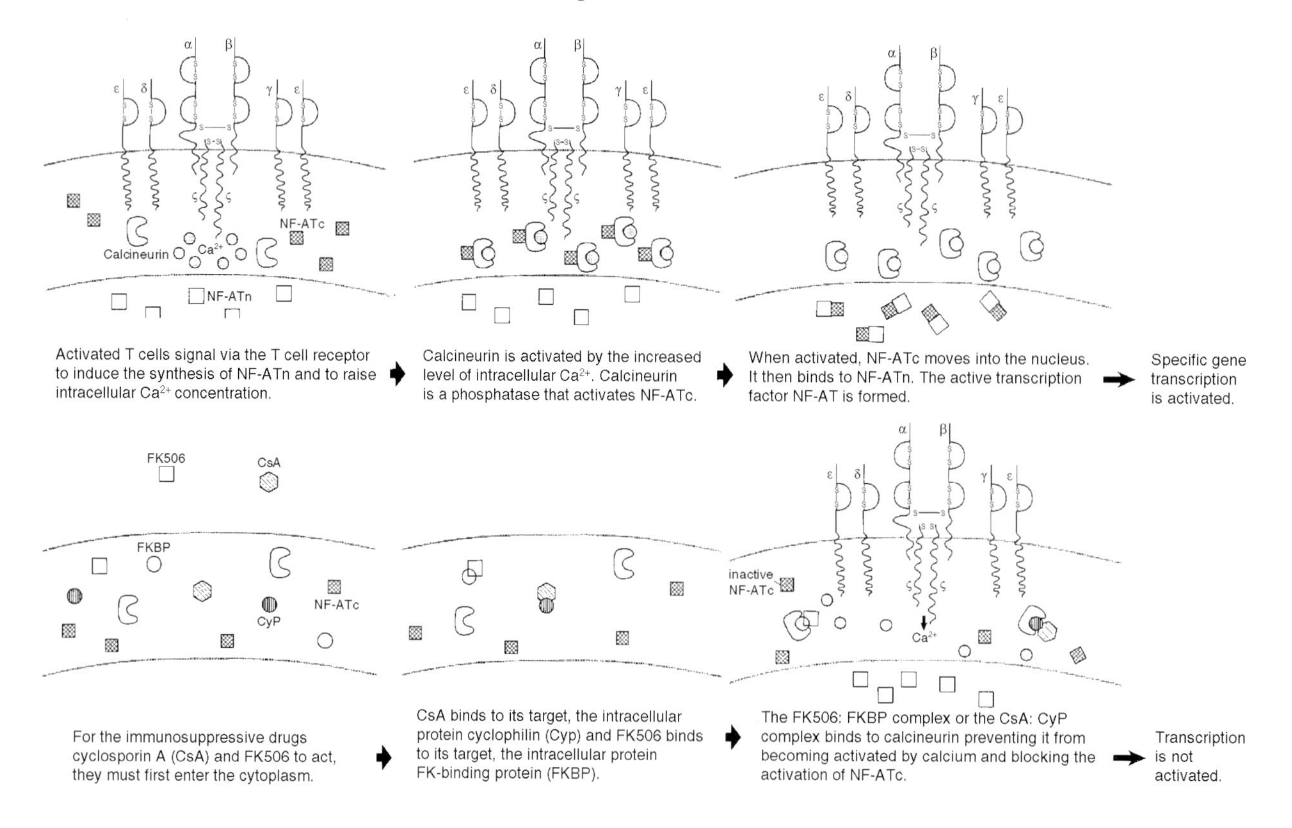

FIGURE 9.42 Calcineurin.

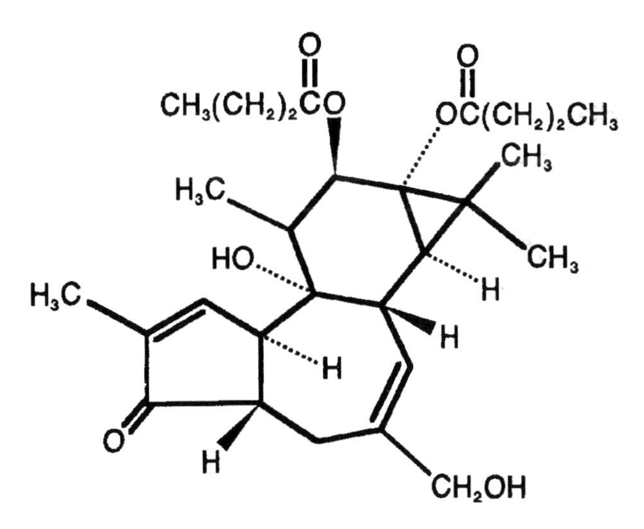

FIGURE 9.43 Structure of phorbol ester.

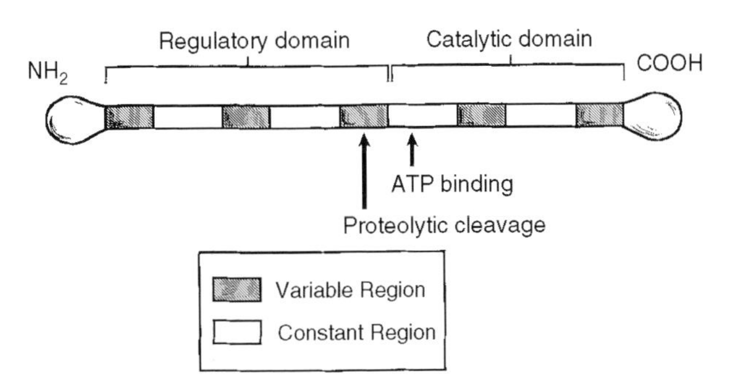

FIGURE 9.44 Protein kinase C.

for phorbol ester that acts by signal transduction, leading to hormone secretion, enzyme secretion, neurotransmitter release, and mediation of inflammation. It is also involved in lipogenesis and gluconeogenesis. PKC participates also in differentiation of cells and in tumor promotion.

By light microscopy, **resting lymphotytes** (Figure 9.45) appear as a distinct and homogeneous population of round cells, each with a large, spherical or slightly kidney-shaped nucleus which occupies most of the cell and is surrounded by a narrow rim of basophilic cytoplasm with occasional vacuoles. The nucleus usually has a poorly visible single indentation and contains densely packed chromatin. Occasionally, nucleoli can be distinguished. The small lymphocyte variant, which is the predominant morphologic form, is slightly larger than an erythrocyte. Larger lymphocytes, ranging between 10 to 20 μm in diameter, are difficult to differentiate from monocytes. They have more cytoplasm and may show azurophilic

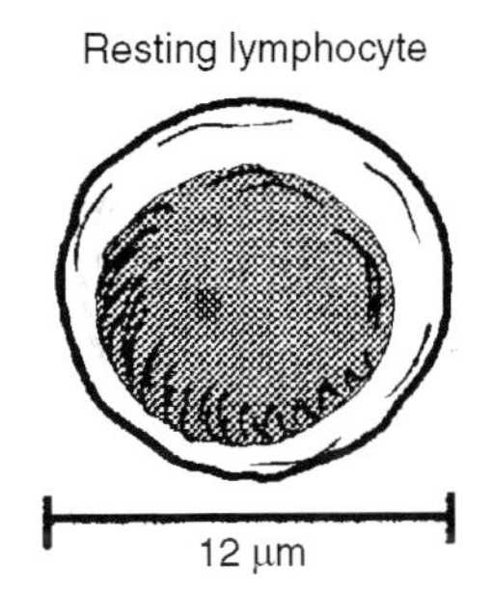

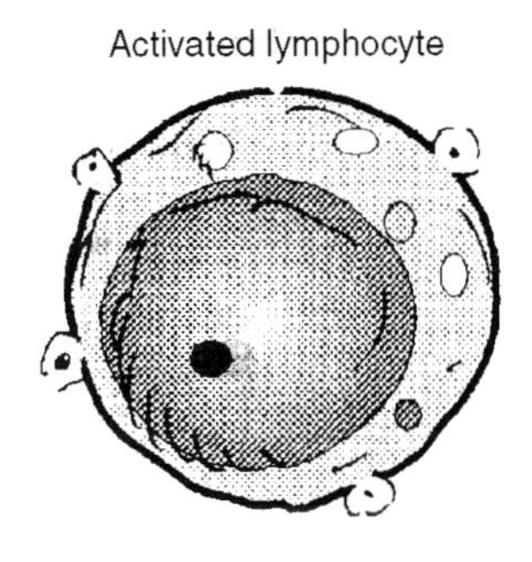

FIGURE 9.45 Comparison of a resting and an activated lymphocyte.

granules. Intermediate-size forms between the two are described. By phase contrast microscopy, living lymphocytes show a feeble motility with amoeboid movements that give the cells a hand-mirror shape. The mirror handle is called a uropod. In large lymphocytes, mitochondria and lysosomes are better visualized, and some cells show a spherical, birefringent, 0.5-μ diameter inclusion, called a Gall body. Lymphocytes do not spread on surfaces. The different classes of lymphocytes cannot be distinguished by light microscopy. By scanning electron microscopy, B lymphocytes sometimes show a hairy (rough) surface, but this is apparently an artifact. Electron microscopy does not provide additional information except for visualization of the cellular organelles which are not abundant. This suggests that the small, resting lymphocytes are end-stage cells. However, under appropriate stimulation, they are capable of considerable morphologic changes.

Activated lymphocytes are cells with surface receptors that interact with specific antigens or mitogens such as phytohemmagglutinin, concanavalin A, or staphylococcal protein A. The morphologic appearance of activated (or stimulated) lymphocytes is characteristic, and in this form the cells are called immunoblasts. These cells increase in size from 15 to 30 μm in diameter, show increased cytoplasmic basophilia, and develop vacuoles, lysosomes, and ribosomal aggregates. Pinocytotic vesicles are present on the cell membrane. The nucleus contains very little chromatin, which is limited to a thin marginal layer, and the nucleolus becomes conspicuous. The array of changes that follows stimulation is called transformation. Such cells are called transformed cells. An activated B lymphocyte may synthesize antibody molecules, whereas an activated T cell may mediate a cellular immune reaction.

Tec kinase belongs to a family of src-like tyrosine kinases that have a role in activation of lymphocyte antigen receptors through activation of PLC-γ. Btk in B lymphocytes, which is mutated in X-linked agammaglobulinemia (XLA), human immunodeficiency disease, and ltk in T lymphocytes, are examples of other Tec kinases.

Adaptor proteins are critical linkers between receptors and downstream signaling pathways that serve as bridges or scaffolds for recruitment of other signaling molecules. They are functionally heterogeneous, yet share an SH domain that permits interaction with phosphotyrosine residues formed by receptor-associated tyrosine kinases. During lymphocyte activation, they may be phosphorylated on tyrosine residues which enables them to combine with other homology-2 (SH2) domain-containing proteins. LAT, SLP-76, and Grb-2 are examples of adaptor molecules that participate in T cell activation.

Linker of activation in T cells (LAT) is an adaptor cytoplasmic protein with several tyrosines that are phosphorylated by ZAP-70, a tyrosine kinase. It associates with membrane lipid rafts and coordinates downstream signaling events in T lymphocyte activation.

LAT: See linker of activation in T cells.

NF-AT: See nuclear factor of activated T cells.

H7 is a pharmacological agent used to study T cell activation. Its target is protein kinase C.

Herbimycin A is an inhibitor of T cell activation. Its target is the src family of protein tyrosine kinases.

Transferrin receptor (T9) is the receptor on the membranes of cells for the 76-kDa protein transferrin in serum that serves as a conveyer for ferric ($Fe3^+$) iron. Monoclonal antibody can detect the transferrin receptor on activated T lymphocytes, even though resting T cells are essentially bereft of this receptor. Primitive thymocytes also express it. Transferrin receptor is comprised of two 100-kDa polypeptide chains that are alike and fastened together by disulfide bonds. The transferrin receptor gene is located on chromosome 3p12-ter in man.

Rac is a small guanine nucleotide-binding protein that the GDP-GTP exchange factor Vav activates during the early stages of T lymphocyte activation. GTP-Rac activates a protein kinase cascade that leads to activation of the stress-activated protein (SAP) kinase, c-Jun N-terminal kinase (JNK), and p38 kinase, which are similar to the MAP kinases.

ICOS is a CD28-related protein that develops on activated T lymphocytes and is able to enhance T cell responses. Licos, the ligand to which it binds, is distinct from B7 molecules.

A **sensitized lymphocyte** is a primed lymphocyte that has been previously exposed to a specific antigen.

Tac is a cell surface protein on T lymphocytes that binds IL-2. It is a 55-kDa polypeptide (p55) that is expressed on activated T lymphocytes. Tac is an abbreviation for T activation. The p55 Tac polypeptide combines with IL-2 with a kDa of about 10^{-8} *M*. Interaction of IL-2 with p55 alone does not lead to activation. IL-2 binds to a second protein termed p70 or p75 that has a higher affinity of binding to p55. T cells expressing p70 or p75 alone are stimulated by IL-2. However, cells that express both receptor molecules bind IL-2 more securely and can be stimulated with a relatively lower IL-2 concentration compared with stimulation when p70 or p75 alone interacts with IL-2. Anti-Tac monoclonal antibody can inhibit T cell proliferation.

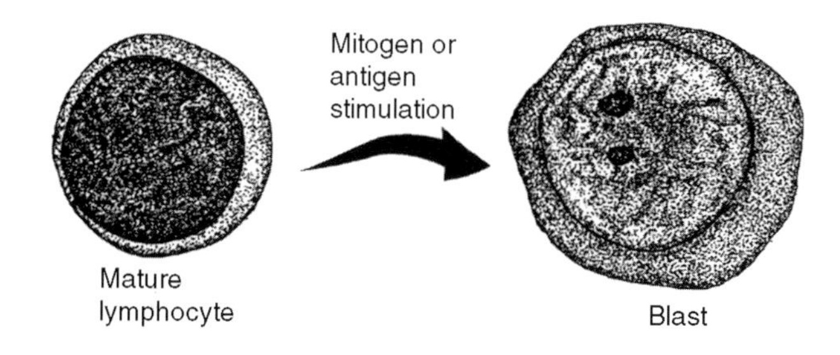

FIGURE 9.46 Blastogenesis.

Tac antigen: See CD25.

Blastogenesis (Figure 9.46), or blast transformation, is the activation of small lymphocytes to form blast cells. A blast cell is a relatively large cell that is greater than 8 μm in diameter with abundant RNA in the cytoplasm, a nucleus containing loosely arranged chromatin, and a prominent nucleolus. Blast cells are active in synthesizing DNA and contain numerous polyribosomes in the cytoplasm.

Thymic hormones are soluble substances synthesized by thymic epithelial cells that promote thymocyte differentiation. They include thymopoietin and thymosins, peptides that help to regulate differentiation of T lymphocytes.

Thymic hormones and peptides are polypeptides that have been extracted from thymus glands include thymulin, thymopoietin, thymic humoral factor (THF), and thymosins. Thus far, only thymulin is characterized as a hormone that fits the accepted classical endocrine and physiological criteria, i.e., it is thymus restricted and regulated in its secretions.

Thymic humoral factor(s) (THFs) are soluble substances such as thymosins, thymopoietin, serum thymic factor, etc., which are synthesized by the thymus and govern differentiation and function of lymphocytes.

Thymosine (Figure 9.47) is a 12-kDa protein hormone produced by the thymus gland that can provide T lymphocyte immune function in animals that have been thymectomized.

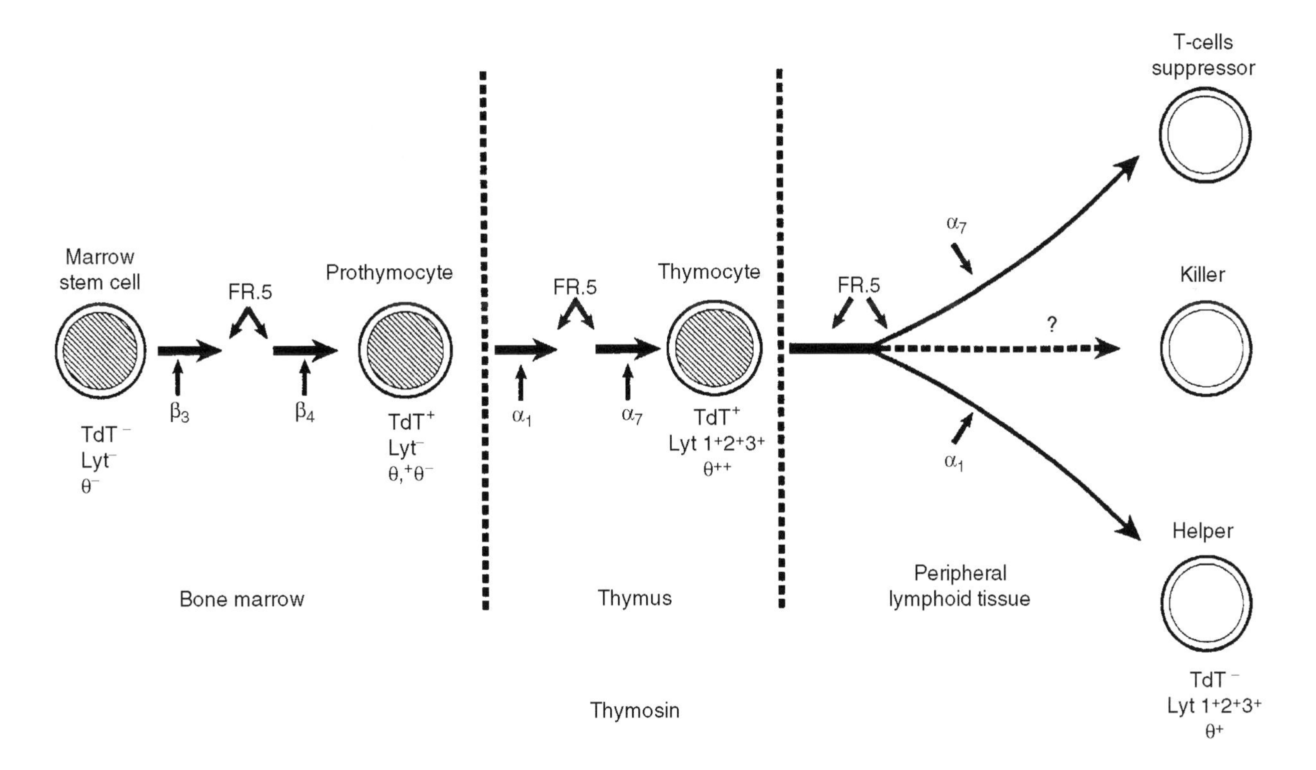

FIGURE 9.47 Proposed site of action of thymosin polypeptides on maturation of T cell subpopulations.

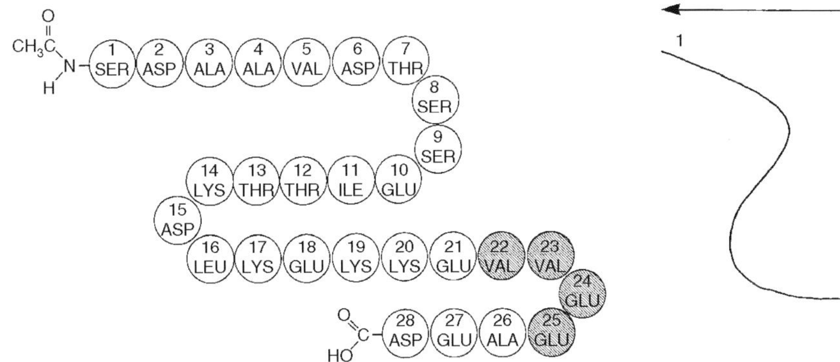

FIGURE 9.48 Structure of thymosin α-1.

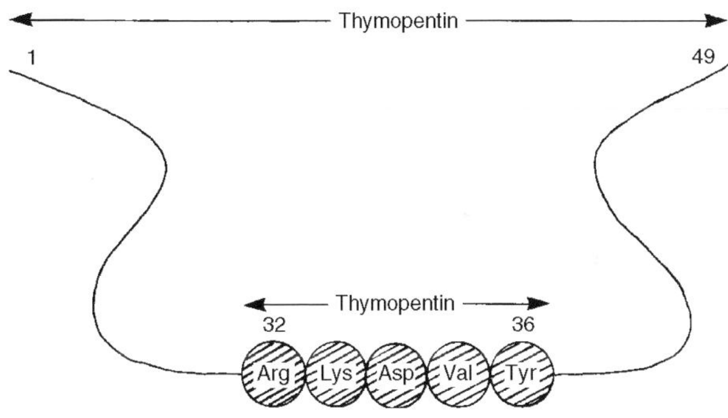

FIGURE 9.49 Structure of thymopoietin and thymopentin.

Thymosin α-1 (thymopoietin) is a hormone produced by the thymus that stimulates T lymphocyte helper activity (Figure 9.48). It induces production of lymphokines such as interferon and macrophage-inhibiting factor. It also enhances Thy-1.2 and Lyt-1-2-3 antigens of T lymphocytes. It may also alter thymocyte TdT concentrations.

Thymopoietin is a 49–amino acid polypeptide thymic hormone secreted by epithelial cells in the thymus. It affects neuromuscular transmission, induces early T lymphocyte differentiation, and affects immune regulation. Thymopoietin functions biologically to normalize immune imbalances related to either hypo- or hyperresponsiveness. These could be related to thymic involution with age, thymectomy, or other factors that result in immunologic imbalance. It is a 7-kDa protein that facilitates the expression of Thy-1 antigen on T lymphocytes, possibly interfering with neuromuscular transmission, which has often been implicated in myasthenia gravis patients who develop thymoma.

Thymopentin (TP5) is a synthetic pentapeptide, Arg-Lys-Asp-Val-Tyr, which corresponds to amino acid residues 32 to 36 of the thymic hormone thymopoietin. Thymopentin is the minimal fragment that can produce the biological activities of thymopoietin, i.e., thymopentin is the active site of thymopoietin (Figure 9.49).

The **cell-mediated immune response** is a host defense mediated by antigen-specific T lymphocytes together with nonspecific cells of the immune system. It offers protection against intracellular bacteria, viruses, and neoplasms, and mediates graft rejection. It may be transferred passively with primed T lymphocytes.

Cell-mediated immunity (CMI) is the limb of the immune response that is mediated by specifically sensitized T lymphocytes that produce their effect through direct reaction in contrast to the indirect effect mediated by antibodies of the humoral limb produced by B lymphocytes. The development of cell-mediated immunity to an exogenous antigen first involves processing of the antigen by an antigen-presenting cell such as a macrophage. Processed antigen is presented in the context of MHC class II molecules to a $CD4^+$ T lymphocyte. IL-1 β is also released from the macrophage to induce IL-2 synthesis in $CD4^+$ lymphocytes. The IL-2 has an autocrine effect on the cells producing it, causing their proliferation and also causing proliferation of other lymphocyte subsets including $CD8^+$ suppressor/cytotoxic T cells, B lymphocytes that form antibody, and natural killer (NK) cells. Cell-mediated immunity is of critical importance in the defense against mycobacterial and fungal infections, resistance to tumors, and for the significance of its role in allograft rejection.

Thymin is a hormone extracted from the thymus that has an activity resembling that of thymopoietin.

Athymic nude mice are a strain which lack a thymus and have no hair (Figure 9.50). T lymphocytes are absent; therefore, no manifestations of T cell immunity are present, i.e., they do not produce antibodies against thymus-dependent antigens and fail to reject allografts. They possess a normal complement of B and NK cells. These nude or nu/nu mice are homozygous for a mutation, v on chromosome 11, which is inherited as an autosomal recessive trait. These features make the strain useful in studies evaluating thymic-independent immune responses.

FIGURE 9.50 Athymic (nude) mouse.

A **nude mouse** is one that is hairless and has a congenital absence of the thymus and of T lymphocyte function. They serve as highly effective animal models to investigate immunologic consequences of not having a thymus. They fail to develop cell-mediated (T lymphocyte-mediated) immunity, are unable to reject allografts, and are unable to synthesize antibodies against the majority of antigens. Their B lymphocytes and natural killer cells are normal even though T lymphocytes are missing. Also called nu/nu mice. Valuable in the investigation of graft-vs.-host disease.

A **thymectomy** is the surgical removal of the thymus. Mice thymectomized as neonates fail to develop cell-mediated immunity and humoral immunity to thymus-dependent antigens. Neonatal thymectomy in mice also leads to a chronic and eventually fatal disease called runt disease or wasting disease, or runting syndrome, which is characterized by lymphoid atrophy and weight loss. Animals may develop ruffled fur, diarrhea, and a hunched appearance. Gnotobiotic (germ-free) animals fail to develop wasting disease following neonatal thymectomy. Thus, thymectomy of animals that are not germ free may lead to fatal infection as a consequence of greatly decreased cell-mediated immunity. Wasting may appear in immunodeficiencies such as AIDS, as well as in graft-vs.-host (GVH) reactions.

Runt disease also results when neonatal mice of one strain are injected with lymph node or splenic lymphocytes from a different strain. It is accompanied by weight loss, failure to thrive, diarrhea, splenomegaly, and even death. The immune system of the neonatal animal is immature and reactivity against donor cells is weak or absent. Runt disease is an example of a graft-vs.-host reaction.

Runting syndrome is characterized by wasting, ruffled fur, diarrhea, lethargy, and debilitation. This is a consequence of thymectomy in neonatal mice that develop lymphopenia, lose lymphocytes from their lymphoid tissues, and become immunologically incompetent.

Apoptosis (Figure 9.51 and Figure 9.52) is programmed cell death in which the chromatin becomes condensed and the DNA is degraded. The immune system employs apoptosis for clonal deletion of cortical thymocytes by antigen in immunologic tolerance. Contrary to necrosis, programmed cell death occurs to regulate steady-state levels of hematopoietic cells that undergo cellular division and differentiation. No local inflammatory response is induced with apoptosis because cellular contents, which stimulate inflammation, are not released. Regulatory genes that control apoptosis have been identified. These include:

bcl-2 prevents apoptosis
bax promotes apoptosis by opposing *bcl-2*

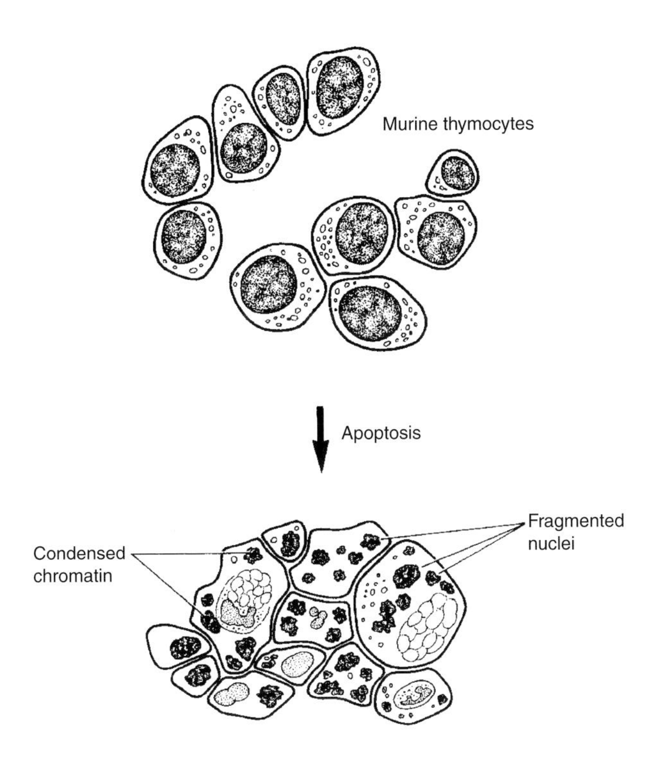

FIGURE 9.51 Apoptosis with phagocytosis.

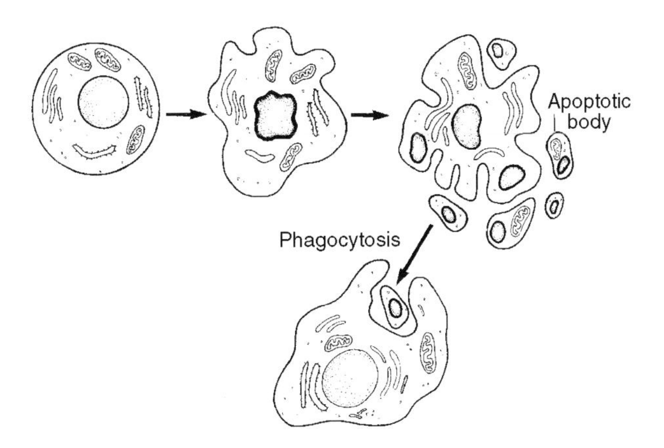

FIGURE 9.52 Apoptosis.

bcl-XI prevents apoptosis
bcl-Xs promotes apoptosis by opposing *bcl-XI*
ICE promotes apoptosis by encoding enzyme IL-lb convertase
fas/apo-1 promotes apoptosis

Apoptosis regulates the number of active immune cells in circulation so that unnecessarily activated lymphocytes are eliminated once an antigenic response has ended. As the response subsides, signals promoting apoptosis appear. For instance, B-cell activation induces increased cytokine receptors and decreased *bcl-2* expression. This makes the cell more susceptible to apoptosis than naïve B-cells or memory cells.

A healthy organism is an exquisitely integrated collection of differentiated cells, which maintain a balance between life and death. Some cells are irreplaceable; some cells complete their functions and are then sacrificed; and some cells live a finite lifetime to be replaced by yet another generation. A failure of cells to fulfill their destiny has catastrophic consequences for the organism. Apoptosis is the last phase of a cell's destiny. It is the controlled disassembly of a cell. When apoptosis occurs on schedule, neighboring cells and, more importantly, the organism itself, are not adversely affected. Apoptosis gone awry, however, has dire consequences. When apoptosis occurs in irreplaceable cells, as in some neurodegenerative disorders, functions critical to the organism are lost. When cells fail to undergo apoptosis after serving their purpose, as in some autoimmune disorders, escaped cells adversely affect the organism. When cells become renegade and resist apoptosis, as in cancer, the outlaw cells create a dire situation for the organism. Mistiming of, or errors in, apoptosis can have devastating consequences on development. Apoptotic fidelity is, therefore, critical to the well-being of an organism. The process of apoptosis (programmed cell death) is regulated by signals generated when cytokines bind to their receptors. There are two types of cytokine-induced signals. The first initiates apoptosis. Cytokines producing an inductive signal include TNFα, FAS/APO-1 ligand, and TRAIL/APO-2 ligand. The second is an inhibitory signal that suppresses apoptosis. Cytokines producing inhibitory signals include those required for cell survival.

Apoptosis proceeds through cleavage of vital intracellular proteins. Caspases are inactive until a signal initiates activation of one, starting a cascade in which a series of other caspases are proteolytically activated. Although both signaling processes affect caspase activation, the mechanism differs. Apoptosis is characterized by degradation of nuclear DNA, degeneration and condensation of nuclei, and phagocytosis of cell residue. Proliferating cells often undergo apoptosis as a natural process, and proliferating lymphocytes manifest rapid apoptosis during development and during immune responses. In contrast to the internal death program of apoptosis, necrosis describes death from without.

Snell-Bagg mice comprise a mutant strain (dw/dw) of inbred mice with pituitary dwarfism that have a diminutive thymus, deficient lymphoid tissue, and decreased cell-mediated immune responsiveness.

10 Cytokines and Chemokines

Leukocytes or other types of cells produce soluble proteins or glycoproteins termed *cytokines* that serve as chemical communicators from one cell to another. Cytokines are usually secreted, although some may be expressed on a cell membrane or maintained in reservoirs in the extracellular matrix. Cytokines combine with surface receptors on target cells that are linked to intracellular signal transduction and second messenger pathways. Their effects may be autocrine, acting on cells that produce them, or paracrine, acting on neighboring cells, and only rarely endocrine, acting on cells at distant sites.

Cytokines are immune system proteins that are biological response modifiers. They coordinate antibody and T cell immune system interactions and amplify immune reactivity. Cytokines include monokines synthesized by macrophages and lymphokines produced by activated T lymphocytes and natural killer cells. Monokines include interleukin-1, tumor necrosis factor, α and β interferons, and colony-stimulating factors. Lymphokines include interleukins-2 to 6, γ interferon, granulocyte-macrophage colony-stimulating factor, and lymphotoxin. Endothelial cells, fibroblasts, and selected other cell types may also synthesize cytokines (Figure 10.1).

Lymphokine research can be traced to the 1960s when macrophage migration inhibitory factor was described. It is believed to be due to more than one cytokine in lymphocyte supernatants. Lymphotoxin was described in activated lymphocyte culture supernatants in the late 1960s and lymphokines were recognized as cell-free soluble factors formed when sensitized lymphocytes respond to specific antigen. These substances were considered responsible for cell-mediated immune reactions. Interleukin-2 was described as T cell growth factor. Tumor necrosis factor (TNF) was the first monocyte/macrophage-derived cytokine or monokine to be recognized. Other cytokines derived from monocytes include lymphocyte activation factor (LAF), later named interleukin-1. It was found to be mitogenic for thymocytes. Whereas immunologists described lymphokines and monokines, virologists described interferons. Interferon is a factor formed by virus-infected cells that are able to induce resistance of cells to infection with homologous or heterologous viruses. Subsequently, interferon-γ, synthesized by T lymphocytes activated by mitogen, was found to be distinct from interferons-α and -β and to be formed by a variety of cell types. Colony-stimulating factors (CSFs) were described as proteins capable of promoting proliferation and differentiation of hematopoietic cells. CSFs promote granulocyte or monocyte colony formation in semisolid media. Proteins that facilitate the growth of nonhematopoietic cells are not usually included with the cytokines. Transforming growth factor β has an important role in inflammation and immunoregulation as well as immunosuppressive actions on T cells.

Cytokine receptors are sites on cell surfaces where cytokines bind, leading to new activities in the cell that include growth, differentiation, or death. Cytokines bind to these high-affinity cell surface receptors that share some features in common. They have a high affinity for ligand. Typically, a hundred to a few thousand receptors are present per cell. Most cytokine receptors are glycosylated integral type I membrane proteins. Functional cytokine receptors are usually complex structures requiring the formation of homologous or heterologous associations between receptor chains. By contrast, receptors of the chemokine family belong to the serpentine superfamily, which span the membrane seven times and do not appear to form multimeric complexes.

Classes of cytokine receptors include the immunoglobulin receptor superfamily, the hematopoietic/cytokine receptor superfamily, the nerve-growth factor receptor superfamily, the G-protein-coupled receptor superfamily, and the receptor tyrosine or serine kinases plus an unclassified group.

Cytokine receptor families is a classification system of cytokine receptors according to conserved sequences or folding motifs. Type I receptors share a trytophan-serine-X-trytophan-serine or WSXWS sequence on the proximal extracellular domains. Type I receptors recognize cytokines with a structure of four α-helical strands, including IL-2 and G-CSF. Type II receptors are defined by the sequence pattern of Type I and Type II interferon receptors. Type III receptors are those found as receptors for TNF. CD40, nerve growth factor receptor, and Fas protein have sequences homologous to those of Type III receptors. A fourth family of receptors has extracellular domains of the Ig superfamily. IL-1 receptors as well as some growth factors and colony-stimulating factors have Ig domains. The fifth superfamily of receptors displays a seven-transmembrane α-helical structure. This motif is shared by many of the receptors linked to GTP-binding proteins.

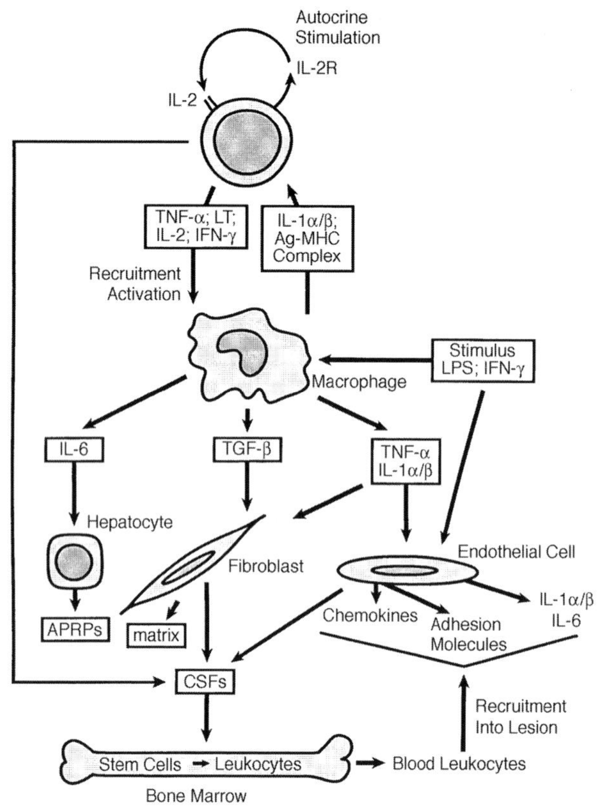

FIGURE 10.1 Generation of cytokines by endothelial cells, fibroblasts, T-helper lymphocytes, and monocyte/macrophages.

Type I cytokine receptors are cytokine receptors that possess conserved structural motifs in their extracellular domains and bind cytokines that fold into four α-helical strands. They include growth hormone, IL-2, IL-3, IL-4, IL-5, IL-6, IL-7, IL-9, IL-11, IL-13, IL-15, GM-CSF, and G-CSF. There is a ligand-binding chain and one or more signal-transducing chains in some of these receptors. The same structural motifs are found in all of these chains. When bound to their cytokine ligands, Type I cytokine receptors become dimerized. They signal through JAK-STAT pathways.

A principal feature of cytokines is that their effects are pleiotropic and redundant. Cytokines do not exert a specific effect on one type of target cell. Most of them have a broad spectrum of biological effects on more than one type of tissue and cell. Various cytokines may interact with the same cell type to produce similar effects. Cytokine receptors are grouped into two classes. Class I receptors include those that bind a number of interleukins, including IL-2, IL-3, IL-4, IL-6, IL-7, IL-9, IL-11, IL-12, and IL-15. Other type I receptors are those for erythropoietin (EPO), growth hormone (GH), granulocyte-colony-stimulating factor (G-CSF), granulocyte-macrophage colony-stimulating factor (GM-CSF), leukemia inhibitory factor (LIF), and ciliary neurotrophic factor (CNTF). Class II receptors include those that bind interferons IFN-α/β, IFNγ, and IL-10. Cytokine receptors are usually comprised of multipeptide complexes with a specific ligand-binding subunit and a signal-transducer subunit that is class specific. Receptors for IL-3, IL-5, and GM-CSF possess a unique a component and a common b signal-transducer subunit, which accounts for the redundant effects these molecules have on hematopoietic cells. IL-6 CNTF, LIF, Oncostatin M (OM), and interleukin-11 (IL-11) belong to a family of receptors that shares a gp130 signal-transducer subunit in common, which explains some of the similar functions these molecules have on various tissues. In the IL-2 receptor system, which is comprised of three peptide chains α, β, and γ, a β and γ gene heterodimer accounts for IL-2 signal transduction with respect to T cell growth. IL-4, IL-7, and IL-15 also promote growth of T cells. IL-4, IL-7, and IL-15 use the IL-2 receptor γ chain, which accounts for the redundant functions of these cytokines as T cell growth factors.

Cytokines have been named on the basis of the cell of the origin or of the bioassay used to define them. This system has lead to misnomers such as tumor necrosis factor which is more appropriately termed an immunomodulator and pro-inflammatory cytokine. The interleukins now number 17, even though IL-8 is in fact a member of the chemokine cytokine family. The chemokines share a minimum of 25% amino acid homology, are similar in structure, and bind to 7 rhodopsin superfamily transmembrane-spanning receptors. The chemokines are comprised of the CSC or α chemokines encoded by genes on chromosome 4 in humans and the CC or β chemokines encoded by genes on chromosome 17. The CXC chemokines attract neutrophils whereas CC chemokines attract monocytes.

Autocrine describes the action of a hormone, cytokine, or other secreted molecule on the same cell that synthesized it.

An **autocrine factor** is a molecule that acts on the same cell that synthesized it, such as IL-2 stimulating the T lymphocyte that produced it to undergo mitosis.

Paracrine describes local effects of a hormone acting on cells in its immediate vicinity.

Soluble cytokine receptors include IL-1, type I-R, IL-1, type II-R, IL-2Rα, IL-2Rβ, IL-4R, IL-5Rα, IL-6Rα, gp130 IL-6Rβ, ciliary neurotrophic factor (CNF)-Rα, and growth hormone R. Most of these function by blocking ligand binding.

A **cytokine-specific subunit** is a multimeric cytokine receptor polypeptide chain that interacts stereospecifically with the target cytokine. It is responsible for the specificity of a receptor for a particular cytokine.

Growth factors refer to cytokines that facilitate the growth and proliferation of cells. Examples include platelet-derived growth factor, erythropoietin, interleukin-2 (T cell growth factor), and many others.

A **paracrine factor** is a molecule that produces its effect on cells in close proximity to those that synthesize the factor. Powerful cytokine molecules act in this way.

T lymphocyte-conditioned medium is a cell culture medium containing multiple lymphokines that have been released from T cells stimulated by antigens or lectins.

Signal transducers and activators of transcription (STATs): See Janus kinases.

Synergism: A mutual assistance such as the finding that the combined effect of two different cytokines is greater than the action of either alone. It may be a consequence when the combined effect of two different cytokines is greater than the action of either alone.

A **signal transducer and activator of transcription (STAT)** is a protein-signaling molecule and transcription factor in response to cytokines binding to type 1 and type II cytokine receptors. STATs exist as inactive monomers in the cell cytoplasm and are transported to cytoplasmic tails of crosslinked cytokine receptors where they are tyrosine-phosphorylated by Janus kinases (Jaks). Phosphorylated STAT proteins dimerize and migrate to the nucleus where they bind to specific sequences in the promoter regions of various genes and activate their transcription. Different cytokines activate different STATs.

Skin-reactive factor (SRF) is a lymphokine that induces vasodilation and increased vascular permeability when injected into the skin.

Permeability factors: See vascular permeability factors.

Permeability-increasing factor is a lymphokine that enhances the permeability of vessels.

Janus kinases (Jaks) are tyrosine kinases that associate with cytoplasmic tails of various cytokine receptors such as receptors for IL-2, IL-4, IFN-γ, and IL-12, among others. They are activated by the aggregation of cytokine receptors. Jaks phosphorylate cytokine receptors that facilitate the binding of STATs. This is followed by phosphorylation and activation of STATs by Jaks. Jak kinases associate selectively with different cytokine receptors. STATs in the cytosol transfer to the nucleus following phosphorylation, where they activate various genes. There are four known Jak kinases — Jak1, Jak2, Jak3, and Tyk2.

The **JAK-STAT signaling pathway** is a signaling mechanism induced by the binding of cytokine to type I and type II cytokine receptors. There is sequential activation of receptor-associated Janus-kinase (JAK) tyrosine kinases, JAK-mediated tyrosine phosphorylation of cytokine receptor cytoplasmic tails, docking of signal transducer and activator of transcription (STATs) to phosphorylated receptor chains, JAK-mediated tyrosine phosphorylation of the associated STATs, dimerization and nuclear translocation of the STATs, and binding STAT to regulatory regions of target genes leading to transcriptional activation of those genes.

Cytokine inhibitors are prostaglandins, especially PGE_2, synthesized by monocytes and macrophages, and have been suggested to provide feedback inhibition of cytokine synthesis *in vivo*. Cytokine production inhibition is a possible mechanism whereby PGE exerts its antiinflammatory effect. Rats have been treated with a monoclonal antibody specific for murine IL-6, which mediates the acute phase response. This has led to increased survival in *E. coli* shock. β-adrenergic agonists increase cAMP levels and THP-1 cells and suppress synthesis of TNF α. A β-adrenergic antagonist can abolish inhibition of TNF-α synthesis. Isoproterenol, a β-adrenergic agonist, inhibits TNFα synthesis. Dexamethasone and cAMP phosphodiesterase (PDE-IV) inhibitors suppress TNF-α secretion by LPS-activated human monocytes. IL-1 receptor antagonist (IL-1ra) is the only known natural cytokine antagonist. It inhibits the binding of IL-1α and IL-1β to their cell surface receptors. Selected antibiotics such as fluroquinolones, clarithromycin, and tetrandrine inhibit IL-1 and TNF-α synthesis. Some antiinflammatory drugs also have cytokine inhibitor capacity.

Cytokine autoantibodies: Autoantibodies that may inhibit cytokine functions and lead to cytokine deficiency. Autoimmune disease may occur and the action of the cytokine may be inhibited. By contrast, these autoantibodies may serve as cytokine-specific carriers in the circulation. For example, insulin autoantibodies may prolong the release of active insulin to the tissues leading to hypoglycemia in nondiabetics and a significant decrease in the exogenous insulin requirement in diabetic patients. AIDS patients may develop autoantibodies against IL-2, and antibodies against TNF-α have been used successfully to treat rheumatoid arthritis. Both normal and inflammatory disease patients may develop autoantibodies against IL-1α. Cytokine activity is enhanced even in the presence of cytokine autoantibodies *in vivo* by a mechanism that delays rapid catabolism of cytokines from the circulation. The clinical relevance of cytokine autoantibodies *in vivo* remains to be determined. However, these autoantibodies portend a poor prognosis in any disease. Methods for cytokine autoantibody detection include bioassays, immunometric assays, and blotting techniques.

Cytokine assays are tests based on the biological properties, immunological recognition (ELISA or radio-immunoassay), competitive binding to the receptor molecule and inference from the transcription of the mRNA. Each method has its own advantages and disadvantages. Bioassays are quite sensitive and verify biological activity of the cytokine but are not always reproducible or specific. By contrast, immunoassays are reproducible and specific but not nearly as sensitive as bioassays.

Intracellular cytokine staining refers to the use of fluorescent-labeled anticytokine antibodies to "stain" permeabilized cells that synthesize the cytokine in question.

A **lymphokine** is a nonimmunoglobulin polypeptide substance synthesized mainly by T lymphocytes that affects the function of other cells. It may either enhance or suppress an immune response. Lymphokines may facilitate cell proliferation, growth, and differentiation, and they may act on gene transcription to regulate cell function. Lymphokines have either a paracrine or autocrine effect. Many lymphokines have now been described. Well-known examples of lymphokines include interleukin-2, interleukin-3, migration inhibitory factor (MIF), and gamma interferon. The term cytokine includes lymphokines, soluble products produced by lymphocytes, as well as monokines and soluble products produced by monocytes. Lymphokines are more frequently referred to currently as cytokines formed by lymphocytes, soluble protein mediators of immune responses.

A **monokine** is a cytokine produced by monocytes. Any one of a group of biologically active factors secreted by monocytes and macrophages, which has a regulatory effect on the function of other cells such as lymphocytes. Examples include interleukin-1 and tumor necrosis factor.

MIF (macrophage/monocyte migration inhibitory factor) is a substance synthesized by T lymphocytes in response to immunogenic challenge that inhibits the migration of macrophages. MIF is a 25-kDa lymphokine. Its mechanism of action is by elevating intracellular cAMP, polymerizing microtubules, and stopping macrophage migration. MIF may increase the adhesive properties of macrophages, thereby inhibiting their migration. The two types of the protein MIF include one that is 65 kDa with a pI of 3 to 4 and another that is 25 kDa with a pI of approximately 5.

Recombinant DNA technology has permitted the preparation of relatively large quantities of cytokines, enabling investigators to perform x-ray and NMR studies to determine molecular structure. Other investigations of gene organization, chromosomal location, and receptor usage have permitted the classification of cytokines into six separate families.

Macrophage migration inhibitory factor is also termed migration inhibitory factor.

Chemokines are a family of 8- to 10-kDa chemotactic cytokines that share structural homology. They are chemokinetic and chemotactic, stimulating leukocyte movement and directed movement. Two internal disulfide loops are present in chemokine molecules that may be subdivided according to the position of the two amino terminal cysteine residues, i.e., adjacent (cys-cys) or separated by a single amino acid (cys-X-cys). Activated mononuclear phagocytes as well as fibroblasts, endothelium, and megakaryocytes synthesize cys-X-cys chemokines, including interleukin-8, which act mainly on polymorphonuclear neutrophils as acute inflammatory mediators. Activated T lymphocytes synthesize cys-cys chemokines that act principally on mononuclear inflammatory cell subpopulations. Cys-X-cys and cys-cys chemokines combine with heparan sulfate proteoglycans on endothelial cell surfaces. They may activate chemokinesis of leukocytes that adhere to endothelium via adhesion molecules. Chemokine receptors are being characterized, and selected ones interact with more than one chemokine.

Chemokines are molecules that recruit and activate leukocytes and other cells at sites of inflammation. They exhibit both chemoattractant and cytokine properties. There are two groups of chemokines. Those that mainly activate neutrophils are the α-chemokines (C-X-C chemokines). By contrast, those that activate monocytes, lymphocytes, basophils, and eosinophils are designated β-chemokines (C-C chemokines). Blocking chemokine function can exert a major effect on inflammatory responses.

A **chemokine receptor** is a cell surface molecule that transduces signals stimulating leukocyte migration following the binding of the homologous chemokine. These receptors belong to the seven-transmembrane and α-helical G-protein-linked family.

Chemokine β receptor-like 1 is a member of the G-protein-coupled receptor family, the chemokine receptor branch of the rhodopsin family. It is expressed on neutrophils and monocytes but not on eosinophils. It may be found in the brain, placenta, lung, liver, and pancreas.

The **C-C subgroup** is a chemokine subgroup in which adjacent cysteines are linked by disulfide bonds.

Eotaxin is a β chemokine family (CC). It was purified from bronchoalveolar lavage fluid proteins. It has 53% homology with human MCP-1, 44% with guinea pig MCP-1, 31% with human MIP-1α, and 26% with human RANTES. It is induced locally following transplantation of IL-4-secreting tumor cells suggesting that it may contribute to eosinophil recruitment and antitumor activity of IL-4. There is a known human homolog of eotaxin. It is expressed by mouse endothelial cells, alveolar macrophages, lung, intestine, stomach, heart, thymus, spleen, liver, testes, and kidney. The target cell is eosinophils. It is constitutively expressed in guinea pig lungs. Its level is increased within 30 min after challenge of sensitized guinea pigs with aerosolized antigen in association with eosinophil infiltration. Eotaxin protein can induce accumulation of eosinophils but not neutrophils following intradermal injection. Eotaxin protein administered to guinea pigs in aerosolized form induces eosinophil but not neutrophil accumulation in bronchalveolar fluid. It facilitates chemotaxins eosinophils.

Eotaxin-1 and **eotaxin-2** are CC-chemokines that have a specific action on eosinophils.

C10 is a chemokine of the β family (CC family) that has been found in mice but not humans. Its biological significance is unknown. It is expressed on bone marrow cells, myeloid cell lines, macrophages, and T lymphocytes.

The **C-X-C subgroup** is a chemokine family in which a disulfide bridge between cysteines is separated by a different amino acid residue (X).

Granulocyte chemotactic protein-2 (GCP-2) is a chemokine of the α family (CXC family). Osteosarcoma cells can produce both human GCP-2 and interleukin-8 (IL-8). The bovine homolog of human CP-2 has been demonstrated in kidney tumor cells. Human as well as bovine GCP-2 are chemotactic for human granulocytes and activate postreceptor mechanisms that cause the release of gelatinase B, which portends a possible role in inflammation and tumor cell invasion. Tissue sources include osteosarcoma cells and kidney neoplastic cells. Granulocytes are the target cells.

HCC-1 is a β chemokine (CC family) isolated from the hemofiltrate of chronic renal failure patients. HCC-1 has 46% sequence identity with MIP-1α and MIP-1β, and 29 to 37% with the remaining human β chemokines. In contrast to other β chemokines, HCC-1 is expressed constitutively in normal spleen, liver, skeletal and heart muscle, gut, and bone marrow. It reaches significant concentrations (1 to 80 n*M*) in plasma. HCC-1 is expressed by monocytes and hematopoietic progenitor cells.

CXCR-4 is a chemokine receptor. CXCR-4 and its ligand, PBSF/SDF-1, are required for later stages of development of the vascularization of the gastrointestinal tract.

CC chemokine receptor 1 (CC CKR-1) belongs to the G-protein coupled receptor family, the chemokine receptor branch of the rhodopsin family. It is expressed on neutrophils, monocytes, eosinophils, B cells, and T cells. Tissue sources include the placenta, lung, and liver. CC CKR-1 RNA is present in peripheral blood and synovial fluid of RA patients but not osteoarthritis patients. Also termed MIP-1 α receptor, RANTES receptor and CCR-1.

CC Chemokine receptor 2 (CC CKR-2) belongs to the G-protein coupled receptor family, the chemokine receptor branch of the rhodopsin family. Ligands include MCP-1 and MCP-3. It is expressed on monocytes. Tissue sources include kidney, heart, bone marrow, lung, liver, and pancreas. Also called MCP-1 receptor, CCR2 A, and CCR2 B.

CC chemokine receptor 3 (CC CKR-3) is a member of the G-protein-coupled receptor family, the chemokine receptor branch of the rhodopsin family. Tissue sources include the human monocyte cDNA library. Also termed eosinophil chemokine receptor, RANTES receptor, CCR3, and eotaxin receptor.

CC chemokine receptor 4 (CC CKR-4) mRNA is present in leukocyte-rich tissue. It is found in IL-5-treated basophils. Ligands include MIP-1α, RANTES, and MCP-1. Tissue sources include the immature basophilic cell line KU-812. Also termed CCR4.

Lymphocyte chemokine (BLC) is a CSC chemokine that induces B lymphocytes and activated T cells to enter peripheral lymphoid tissue follicles by binding to the CXCR5 receptor.

Duffy antigen/chemokine receptor (DARC) is a member of the G-protein-coupled receptor family, the chemokine receptor branch of the rhodopsin family. It is expressed in erythroid cells in bone marrow. Ligands include the human malarial parasite *Plasmodium vivax*. β chemokine ligands include RANTES and MCP-1, and α chemokine ligands include IL-8 and MGSA/GRO. DARC is also expressed in endothelial cells lining postcapillary venules and splenic sinusoids in individuals who are Duffy-negative. It is also found in adult spleen, kidney, brain, and fetal liver. This receptor is also expressed in K562 and HEL cell lines. Also called erythrocyte chemokine receptor (erythrocyte CKR), RBC chemokine receptor, gpFy, and gpD.

EBI1 is an orphan chemokine receptor expressed on normal lymphoid tissues as well as several B and T lymphocyte cell lines. EBI1 mRNA is detected in EBV-positive B cell lines. The tissue source is Epstein–Barr-induced cDNA. Also termed BLR2.

ECRF3 is a member of the G-protein coupled receptor family, the chemokine receptor branch of the rhodopsin family expressed on T cells. The tissue source is herpesvirus saimiri of the γ-herpesvirus family. *In vitro*, it facilitates calcium efflux. Ligands include MGSA/GRO-α, NAP-2, and IL-8.

***N*-formyl peptide receptor (FPR)** is a member of the G-protein-coupled receptor family, the chemokine receptor branch of the rhodopsin family. Tissue sources include HL-60 cells differentiated with Bt2 cAMP. Undifferentiated HL-60 cells transfected with FPR-bound FMLP with two affinities. COS7 cells transfected with FPR-bound FMLPK-Pep12 with low and high affinity. The receptor is expressed in neutrophils. Dibutyryl cAMP includes FPR transcription in HL-60 cells.

Lymphotactin (Ltn) is a member of the γ family (C family) of chemokines. Human lymphotactin (Ltn) resembles some β-chemokines but is lacking the first and third cysteine residues that are characteristic of the α and β chemokines. Ltn is chemotactic for lymphocytes but not for monocytes. Ltn employs a unique receptor. An ATAC cDNA clone derived from human T cell activation genes encodes a protein 73.8% identical to mouse lymphotactin. Tissue sources include thymocytes and activated T cells. T lymphocytes are the target cells.

Macrophage chemotactic and activating factor (MCAF) (MCP-1) is a chemoattractant and activator of macrophages produced by fibroblasts, monocytes, and endothelial cells as a result of exogenous stimuli and endogenous cytokines such as tumor necrosis factor (TNF), interleukin-1 (IL-1), and PDGF. It also has a role in activating monocytes to release an enzyme that is cytostatic for some tumor cells. MCAF also plays a role in endothelial leukocyte adhesion moledule 1 (ELAM-1) and CD11a and b surface expression in monocytes, and is a potent degranulator of basophils. MCP-1 is a member of the chemokine β family and has a CC (cysteine-cysteine) amino acid sequence. It is chemotactic toward monocytes *in vivo* and *in vitro* , and it activates monocytes. MCP-1 shares 21% amino-acid sequence homology with IL-8. The 76-amino acid mature form is derived from a 99-amino acid precursor. The MCP-1 receptor (CCR2A CCR2B) is a seven-transmembrane-spanning, G-protein-coupled molecule of 39 kDa with homology to other cyokine receptors. Various normal and malignant cell types synthesize MCP-1, which can induce not only chemotaxis but also enzyme release and increased β_2-integrin cell adhesion molecule expression in monocytes and facilitate monocyte cytostatic activity against tumor cells that have been activated by oxidized low-density lipoprotein (LDL).

Monocyte chemoattractant protein-1 (MCP-1) is a chemokine of the β family (CC family). MCP-1 is a prototypic β-family chemokine that was first isolated as a product of the immediate early gene, *JE,* induced by PDGF. Cloning of the human homologue of *JE* reveals an encoded protein that is identical to an authentic chemokine MCP-1, which is believed to be one of the most significant chemokines in chronic inflammatory diseases controlled by mononuclear leukocytes. Tissue sources include fibroblasts, monocytes, macrophages, mouse spleen lymphocytes, and endothelial cells among others. Target cells include monocytes, hematopoietic precursors, T lymphocytes, basophils, eosinophils, mast cell, NK cells, and dendritic cells.

Monocyte chemoattractant of a protein-2 (MCP-2) is a chemokine of the β family (CC family). It is a variant of MCP-1, but MCP-2 is an independent chemokine with several distinct biological properties that include eosinophil and basophil activation. The principal differences between MCP-1 and MCP-2 are in the N-terminal molecular region. Tissue sources include fibroblasts, peripheral blood mononuclear cells, leteal cells, and osteosarcoma cell line MG 63. Target cells include T lymhocytes, monocytes, eosinophils, basophils, and NK cells.

RANTES is an 8-kDa protein comprised of 68-amino acid residues. It belongs to the PF4 superfamily of chemoattractant proteins. RANTES chemoattracts blood monocytes as well as $CD4^+/CD45RO^+$ T cells *in vitro* and is useful in research on inflammation. A chemokine of the β family (CC family), RANTES was first identified by molecular cloning as a transcript expressed in T cells but not B cells. It is the only β chemokine in platelets and demonstrates powerful chemotactic and activating properties for basophils, eosinophils, and NK cells. It has an HIV-suppressive effect and acts synergistically with MIP-1α and MIP-1β in the suppression of HIV. Tissue sources include T lymphocytes, monocytes, epithelial cells, mesangial cells, platelets, and eosinophils among many other cell types. Monocytes, T lymphocytes, basophils, eosinophils, NK cells, dendritic cells, and mast cells are target cells.

Monocyte chemoattractant protein-3 (MCP-3) is a chemokine of the β family (CC family) that has very different binding characteristics and actions from MCP-1. MCP-3 binds to unique monocyte receptors shared by MCP-1. In contrast to MCP-1, MCP-3 is chemotactic to eosinophils and more potent toward basophils. It is the only β chemokine that fails to form dimers at elevated concentrations. Tissue sources include fibroblasts, platelets, mast cells, monocytes, and osteosarcoma MG 63. Target cells include monocytes, T cells, basophils, eosinophils, activated NK cells, dendritic cells, and neutrophils.

Melanoma growth stimulatory activity (MGSA) is a chemokine of the α family (CXC family). It is a mitogenic polypeptide secreted by human melanoma cells. The MGSA/GRO-α gene product has powerful chemotactic, growth regulatory, and transforming functions. There are many tissue sources that include monocytes, neutrophils, bronchoalveolar macrophages, and endothelial cells, among other cell types. Neutrophils, lymphocytes, monocytes, and epidermal melanocytes are target cells.

Connective tissue-activating peptide-III (CTAP-III) is a platelet granule peptide derived by proteolytic processing from its CXC chemokine precursor, leukocyte-derived growth factor. It has a powerful effect on fibroblast growth, wound repair, inflammation, and neoplasia. CTAP-III and neutrophil activating peptide-2 (NAP-2) are heparanases whose growth-promoting activities could be a consequence of heparan sulfate solubilization and bound growth factors in extracellular matrix. CTAP-III is a member of the family of molecules previously referred to as *histamine release factors*. In allergic state late-phase reactions, CTAP-III may either stimulate or inhibit histamine release based on the relative concentration, pattern of release, and responsiveness of basophils and mast cells. The ability of CTAP-III to activate neutrophils is a critical link between platelet activation and neutrophil recruitment in stimulation, which are key features of inflammatory reactions in allergic conditions.

MIG, a monokine induced by interferon gamma, is a chemokine of the α family (CXC family). MIG was derived from a cDNA library from lymphokine-activated macrophages. IFN-γ can induce macrophages to express the *MIG* gene. Tissue sources include lymphokine-activated macrophages, IFN-γ treated human peripheral blood monocytes, and IFN-γ treated human monocytic cell line THP-1. Target cells include human tumor-infiltrating lymphocytes (TILs) and monocytes.

Neutrophil-activating protein 2 (NAP-2) is a chemokine of the α family (CXC family). NAP-2 is a proteolytic fragment of PBP corresponding to amino acids 25 to 94. CTAP-III and LA-PF4 or β-TG, released from activated platelets, are inactive NAP-2 precursors. Leukocytes and leukocyte-derived proteases convert the inactive precursors into NAP-2 by proteolytic cleavage at the N-terminus. Platelets represent the tissue source, whereas neutrophils, basophils, eosinophils, fibroblasts, NK cells, megakaryocytes, and endothelial cells are target cells for NAP-2.

US28 is a member of the G-protein-coupled receptor family, the chemokine receptor branch of the rhodopsin family. It is expressed in the late phase of lytic infection of leukocytes. Tissues sources include human CMV DNA and CMV-infected human fibroblasts. Ligands include COS-7 cells transfected with US28-bound ^{125}I-MCP-1 and ^{125}I-RANTES. Also called human cytomegalovirus (CMV) G-protein-coupled receptor.

V28 is an orphan chemokine receptor expressed in neural and lymphoid tissue and on THP-1 cell line. The tissue source is peripheral blood mononuclear cells.

MCP-1 in atherosclerosis: Chemokines are involved in the pathogenesis of atherosclerosis by promoting directed migration of inflammatory cells. Monocyte chemoattractant protein-1 (MCP-1), a CC chemokine, has been detected in atherosclerotic lesions by anti-MCP-1 antibody detection and *in situ* hybridization. MCP-1 mRNA expression has been detected in endothelial cells, macrophages, and vascular smooth muscle cells in atherosclerotic arteries of patients undergoing bypass revascularization. MCP-1 functions in the development of atherosclerosis by recruiting monocytes into the subendothelial cell layer. MCP-1 is critical for the initiation and development of atherosclerotic lesions. During the progression of atherosclerosis, there is an accumulation of low-density lipoprotein (LDL) within macrophages and monocytes present in the intimal layer. Deposition of lipids within these cells leads to the formation and eventual enlargement of atherosclerotic lesions. Studies suggest a noncholesterol-mediated effect of MCP-1 in the development of atherosclerotic lesions. MCP-1 plays a crucial role in initiating atherosclerosis by recruiting macrophages and monocytes to the vessel wall.

LESTR is a leukocyte-derived seven-transmembrane domain receptor. It is a member of the G-protein-coupled receptor family, the chemokine receptor branch of the rhodopsin family. Tissue sources include human blood monocyte cDNA library and human fetal spleen. It is expressed on monocytes, neutrophils, lymphocytes, and PHA-activated T cell blasts. CHO and COS cells transfected with LESTR fail to bind IL-8, NAP-2, GRO-α, MCP-1, MCP-3, MIP-1α, and RANTES. Also called fusin and CXCR4.

Intercrine cytokines is a family comprised of a minimum of 8- to 10-kDa cytokines that share 20 to 45% amino acid sequence homology. All are believed to be basic heparin-binding polypeptides with proinflammatory and reparative properties. Their cDNA has conserved single open reading frames, 5′ region typical signal sequences, and 3′ untranslated regions that are rich in AP sequences. Human cytokines that include interleukin-8, platelet factor 4, β thromboglobulin, IP-10, and melanoma growth-stimulating factor or GRO comprise a subfamily encoded by genes on chromosome 4. They possess a unique structure. LD78, ACT-2, I-309, RANTES, and macrophage chemotactic and activating factor (MCAF) comprise a second subset and are encoded by genes on chromosome 17 of man. Human chromosome 4 bears the intercrine α genes and human chromosome 17 bears the intercrine β genes. Four cysteines are found in the intercrine family. Adjacent cysteines are present in the intercrine β subfamily which includes huMCAF, huBLD-78, huACT-II, huRANTES, muTCA-III, muJE, muMIP-1 α, and muMIP-1 β. One amino acid separates cysteines of the intercrines alpha-subfamily, which is comprised of huPF-4, hubetaTG, huIL-8, ch9E3, huGRO, huIP-10, and muMIP-2. The cysteines are significant for tertiary structure and for intercrine binding to receptors.

Interleukin(s) (IL) are a group of cytokines synthesized by lymphocytes, monocytes, and selected other cells that promote growth of T cells, B cells, and hematopoietic stem cells and have various other biological functions. See IL-1 to IL-18.

IL is the abbreviation for interleukins.

Interleukin-1 (IL-1) is a cytokine synthesized by activated mononuclear phagocytes that have been stimulated by ribopolysaccharide or by interaction with CD4+ T lymphocytes. It is a monokine and is a mediator of inflammation, sharing many properties in common with tumor necrosis factors (TNF). IL-1 is comprised of two principal polypeptides of 17 kDa, each with isoelectric points of 5 and 7. They are designated IL-1α and IL-1β, respectively. Genes found on chromosome 2 encode these two molecular species. They have the same biological activities and bind to the same receptor on cell surfaces. Both IL-1α and IL-1β are derived by proteolytic cleavage of 33-kDa precursor molecules. IL-1α acts as a membrane-associated substance, whereas IL-1β is found free in the circulation. IL-1 receptors are present on numerous cell types. IL-1 may either activate adenylate cyclase, elevating cAMP levels and then activating protein kinase A, or it may induce nuclear factors that serve as cellular gene transcriptional activators. IL-1 may induce synthesis of enzymes that generate prostaglandins, which may in turn induce fever, a well-known action of IL-1. IL-1's actions differ according to whether it is produced in lower or in higher concentrations. At low concentrations, the effects are mainly immunoregulatory. IL-1 acts with polyclonal activators to facilitate CD4+ T lymphocyte proliferation, as well as B lymphocyte growth and differentiation. IL-1 stimulates multiple cells to act as immune or inflammatory response effector cells. It also induces further synthesis of itself, as well as of IL-6, by mononuclear phagocytes and vascular endothelium. It resembles tumor necrosis factor (TNF) in inflammatory properties. IL-1 secreted in greater amounts produces endocrine effects as it courses through the peripheral blood circulation. For example, it produces fever and promotes the formation of acute-phase plasma proteins in the liver. It also induces cachexia. Natural inhibitors of IL-1 may be produced by mononuclear phagocytes activated by immune complexes in humans. The inhibitor is biologically inactive and prevents the action of IL-1 by binding with its receptor, serving as a competitive inhibitor. Corticosteroids and prostaglandins suppress IL-1 secretion. IL-1 was formerly called lymphocyte-activating factor (LAF).

LAF (lymphocyte-activating factor): See interleukin-1.

IL-1 is the abbreviation for interleukin-1.

Lymphocyte activation factor (LAF): See interleukin-1.

A **pyrogen** is a substance that induces fever. It may be either endogenously produced, such as interleukin-1 released from macrophages and monocytes, or it may be an endotoxin associated with Gram-negative bacteria produced exogenously that induces fever.

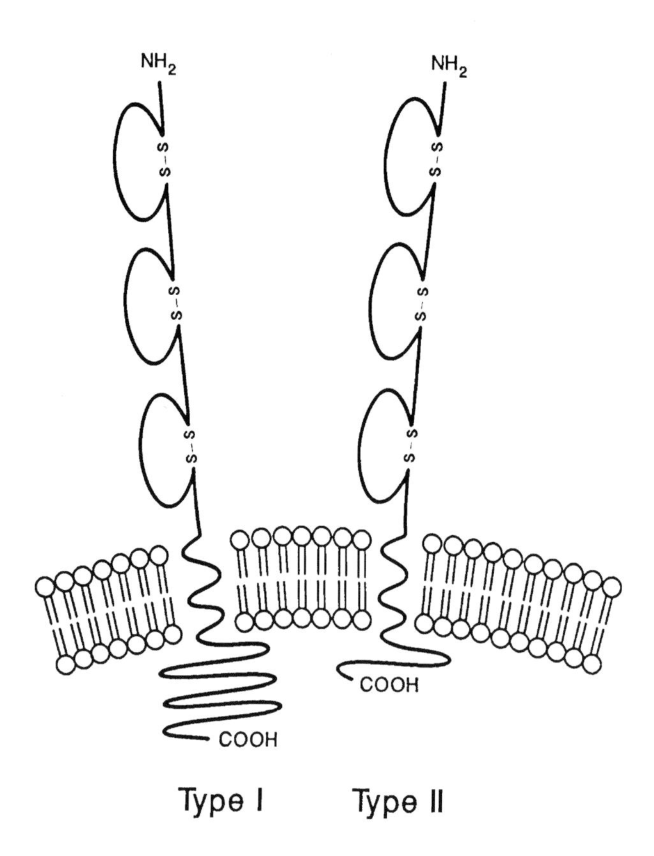

FIGURE 10.2 IL-1 receptors, type I and type II.

An **endogenous pyrogen** is a cytokine that induces an increase in body temperature, in contrast to exogenous pyrogens such as endotoxin from Gram-negative bacteria that elevate body temperature by activating endogenous pyrogen synthesis and release.

Interleukin-1 (IL-1) receptor (Figure 10.2) is an 80-kDa receptor on T lymphocytes, chondrocytes, osteoblasts, and fibroblasts, which binds IL-1α and IL-1β (Figure 10.3). Helper/inducer CD4+ T lymphocytes are richer in IL-1R than are suppressor/cytotoxic (CD8+) T cells. The IL-1R has an extracellular portion that binds ligand and contains all N-linked glycosylation sites. A 217-amino acid segment, apparently confined to the cytoplasm, could be involved in signal transduction. Further studies of ligand-binding have been facilitated through the development of a soluble form of the cloned IL-1 receptor molecule which contains the extracellular part but not the transmembrane cytoplasmic region of the molecule. IL-1α and IL-1β molecules bind with equivalent affinities. The IL-1 receptor has been claimed to have more than one subunit. This is based on the demonstration of bands such as a 100-kDa band in addition to that characteristic of the receptor, which is 80 kDa. The recombinant IL-1R functions in signal transduction. When the cytoplasmic part of the IL-1R is depleted, the molecule does not function. The human T cell IL-1R has now been cloned and found to be quite similar to its

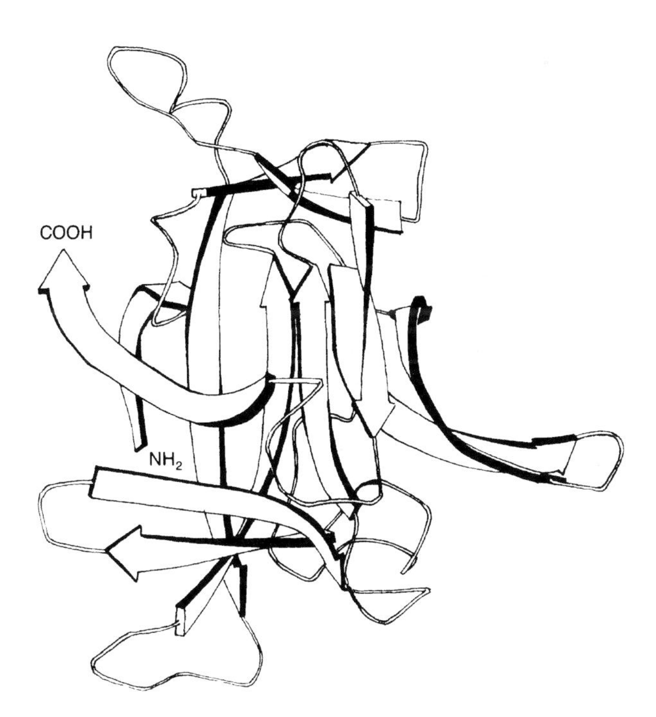

FIGURE 10.3 Tube structure of backbone of interleukin-1α. Resolution 2.3 Å. This entry contains only α carbon coordinates. Interleukin-2 complex (human). Theoretical model. This interleukin-2 complex was generated by homology modeling. This is a shape-filling diagram.

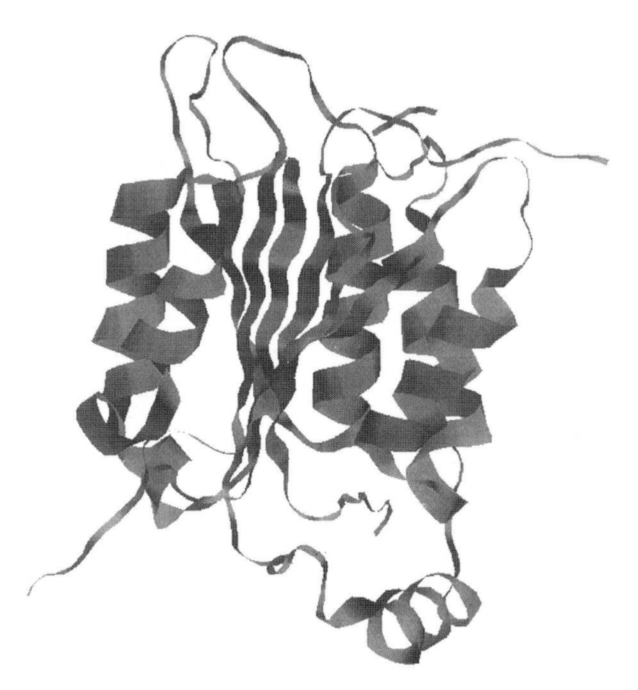

FIGURE 10.4 Human IL-1β converting enzyme. X-ray diffraction. Resolution 2.6 Å.

murine counterpart. Two affinity classes of binding sites for IL-1 have been described (Figure 10.4 and Figure 10.5).

IL-1 receptor antagonist (IL-1ra) is an IL-1 inhibitor that is structurally homologous to IL-1 and is synthesized by mononuclear phagocytes. This biologically inactive molecule blocks IL-1 activity by binding to its receptor. It plays a central role in acute and chronic inflammation, both locally and systemically. It is produced primarily by monocytes and macrophages and is synthesized as a precursor that is cleaved by IL-1β-converting enzyme (ICE) to mature IL-1β plus a prosegment. Some combination of mature IL-1β, IL-1β precursor, and the prosegment are released from the cell.

Interleukin-1 receptor antagonist protein (IRAP) (Figure 10.6) is a substance on T lymphocytes and endothelial cells that inhibits IL-1 activity.

Interleukin-1 receptor deficiency refers to $CD4^+$ T cells deficient in IL-1 receptors in affected individuals, which fail to undergo mitosis when stimulated and fail to generate IL-2. This leads to a lack of immune responsiveness and constitutes a type of combined immunodeficiency. Opportunistic infections are increased in affected children who have inherited the condition as an autosomal recessive trait.

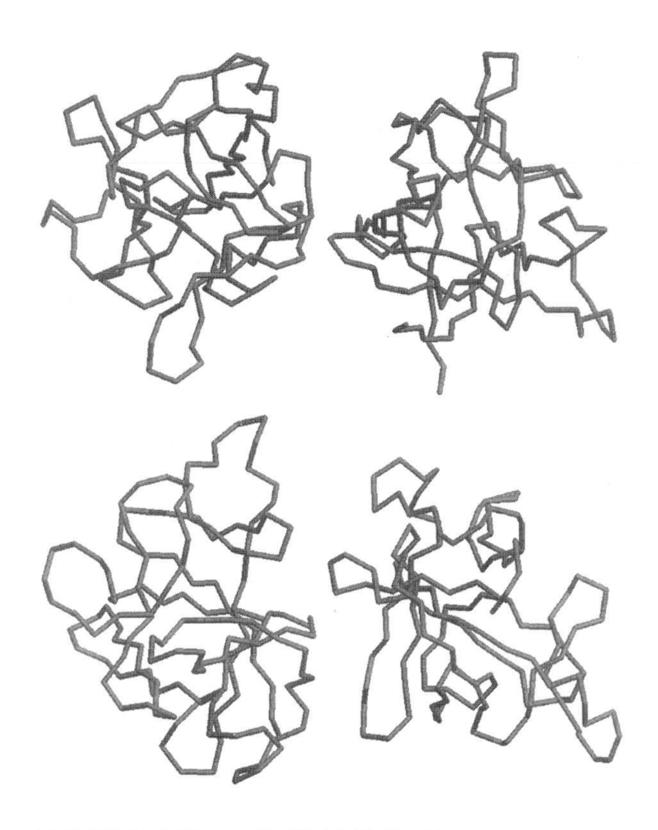

FIGURE 10.5 Human IL-1β. NMR. Backbone structure.

Interleukin-2 (IL-2) (Figure 10.7) is a 15.5-kDa glycoprotein synthesized by $CD4^+$ T helper lymphocytes. It was formerly called T-cell growth factor. IL-2 has an autocrine effect acting on the $CD4^+$ T cells that produce it.

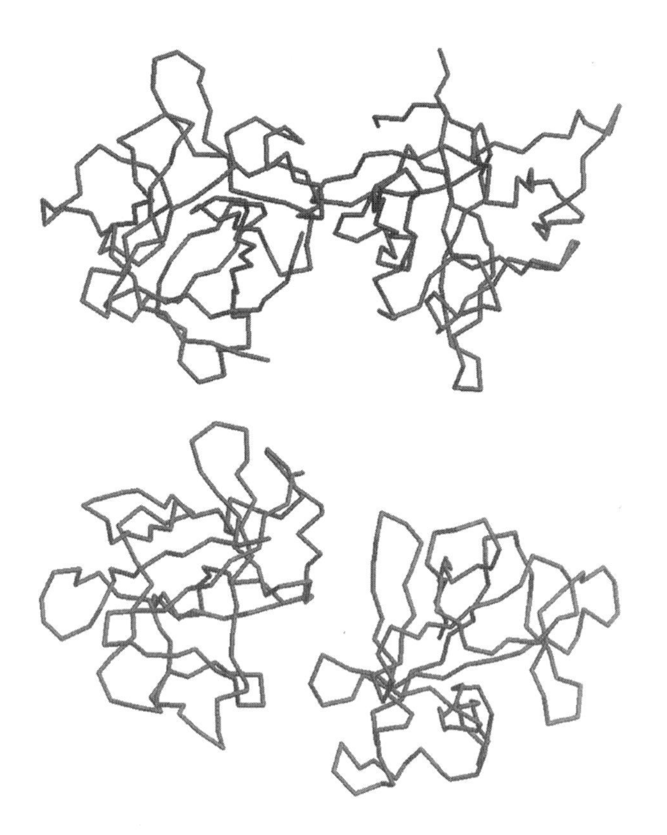

FIGURE 10.6 IL-1 receptor antagonist protein. Resolution 3.2 Å.

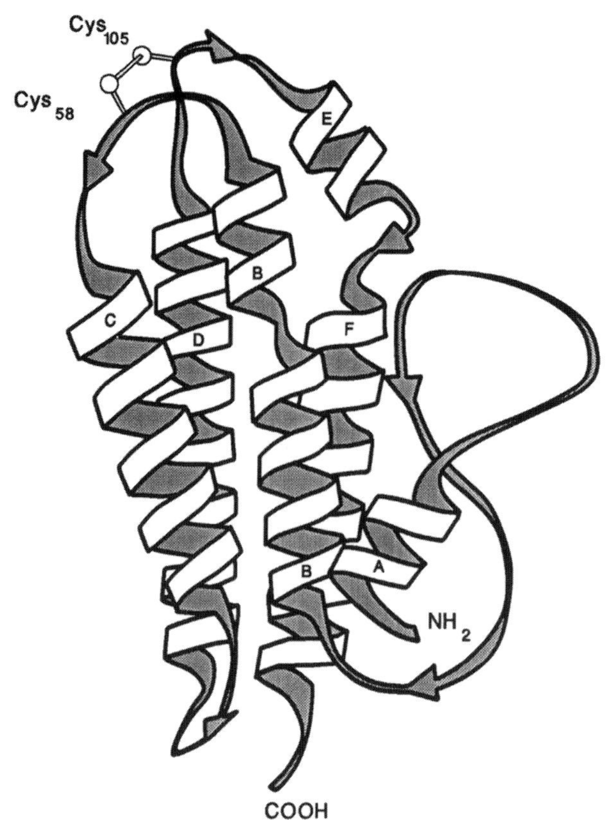

FIGURE 10.7 Schematic representation of IL-2.

Although mainly produced by $CD4^+$ T cells, a small amount is produced by $CD8^+$ T cells. Physiologic amounts of IL-2 do not have an endocrine effect, since IL-2 acts on the cells producing it or on those nearby, acting as a paracrine growth factor. IL-2's main effects are on lymphocytes. The amount of IL-2 which $CD4^+$ T lymphocytes synthesize is a principal factor in determining the strength of an immune response. IL-2 also facilitates formation of other cytokines produced by T lymphocytes, including interferon γ and lymphotoxins. Inadequate IL-2 synthesis can lead to antigen-specific T lymphocyte anergy. It interacts with T lymphocytes by reacting with IL-2 receptors. IL-2 also promotes NK cell growth and potentiates the cytolytic action of NK cells through generation of lymphokine-activated killer (LAK) cells. Although NK cells do not have the p55 lower-affinity receptor, they do express the high-affinity p70 receptor and thus require high IL-2 concentrations for their activation. IL-2 is a human B cell growth factor and promotes synthesis of antibody by these cells. However, it does not induce isotype switching. IL-2 promotes the improved responsiveness of immature bone marrow cells to other cytokines. In the thymus, it may promote immature T cell growth. The IL-2 gene is located on chromosome 4 in man. Corticosteroids, cyclosporin A, and prostaglandins inhibit IL-2 synthesis and secretions.

IL-2 is the abbreviation for interleukin-2.

Capillary leak syndrome: The therapeutic administration of GM-CSF (granulocyte-macrophage colony-stimulating factor) may lead to progressive dyspnea and pericarditis in the treatment of patients bearing metatastic solid tumors. Interleukin-2 (IL-2) may likewise induce the effect when it is used to treat tumor patients. IL-2 treatment may lead to the accumulation of 10 to 20 l of fluid in peripheral tissues with resultant disorientation, confusion, and pronounced fever.

T-cell growth factor (TCGF): See interleukin-2.

T-cell growth factor 1: Interleukin-2 (IL-2).

TCGF (T-cell growth factor): Interleukin-2.

Interleukin-2 receptor (IL-2R) is also known as CD25. IL-2R is a structure on the surface of T lymphocytes, natural killer, and B lymphocytes, characterized by the presence of a 55-kDa polypeptide, p55, and a 70-kDa polypeptide termed p70, which interacts with IL-2 molecules at the cell surface. The p55 polypeptide chain is referred to as Tac antigen, an abbreviation for T activation. The expression of both p55 and p70 permits a cell to bind

IL-2 securely with a K_d of about 10^{-8}. P55, the low-affinity receptor, apparently complexes with p70, the high-affinity receptor, to accentuate the p70 receptor's affinity for IL-2. This permits increased binding in cells expressing both receptors. In addition, lesser quantities of IL-2 than would otherwise be required for stimulation are effective when both receptors are present on the cell surface. Antibodies against p55 or p70 can block IL-2 binding. Powerful antigenic stimulation such as in transplant rejection may lead to the shedding of p55 IL-2 receptors into the serum. The gene encoding the p55 chain is located on chromosome 10p14 in man. IL-2, IL-1, IL-6, IL-4, and TNF may induce IL-2 receptor expression.

IL-2 receptor (CD25): See interleukin-2 receptor (IL-2R).

CD25 is a single-chain glycoprotein, often referred to as the α-chain of the interleukin-2 receptor (IL-2R) or the Tac antigen, that has a mol wt of 55 kDa and is present on activated T and B cells and activated macrophages. It functions as a receptor for IL-2. Together with the β-chain of the IL-2R (p75, mol wt 75 kDa), the CD25 antigen forms a high-affinity receptor complex for IL-2. The gene for IL-2R has been located as a single gene on chromosome 10. It associates with CD122 and complexes with the IL-2Rβγ high-affinity receptor. It facilitates T cell growth.

Interleukin-2 receptor α subunit (IL-2Rα) (Figure 10.8) is a 55-kDa polypeptide subunit of the IL-2R with a K_d of 10^{-8} *M*. The α subunit is responsible for increasing the affinity between cytokine and receptor; however, it has no role in signal transduction. It is expressed only on antigen stimulation of T cells, usually within 2 h. Following long-term T cell activation, the α subunit is shed, making it a potential candidate for a serum marker of strong or prolonged antigen stimulation. The gene encoding IL-2Rα is located on chromosome 10p14 in man.

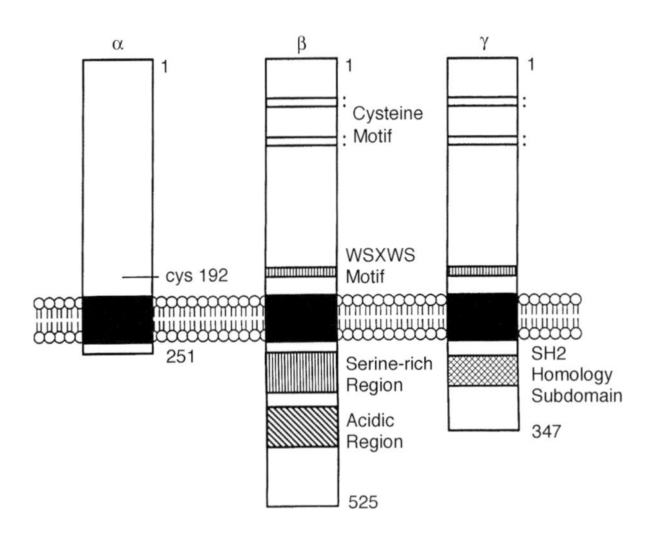

FIGURE 10.8 Schematic representation of interleukin-2 receptor αβγ subunit (IL2Rαβγ).

Interleukin-2 receptor β subunit (IL-2Rβ) (Figure 10.8) is a 70- to 74-kDa subunit of IL-2R with a K_d of 10^{-9} *M*. The β subunit is a member of the cytokine receptor family type I due to its tryptophan serine-X-tryptophan-serine (WSXWS) domain. It is a constitutive membrane protein coordinately expressed with IL-2Rγ.

Interleukin-2 receptor βγ subunit (IL-2Rβγ) is a heterodimer found on resting T cells. Only those T cells expressing IL-2Rβγ are capable of growth in response to IL-2, as this is the portion of the receptor responsible for signal transduction.

Interleukin-2 receptor γ subunit (IL-2Rγ) (Figure 10.8) is also a type I (WSXWS) receptor that is associated with IL-4 and IL-7 receptors as well as IL-2Rγ. Mutations in the γ subunit have been found in some SCIDS cases with X-linked inheritance, resulting in decreased proliferation of B and T cells.

Interleukin-2 receptor αβγ subunit (IL-α2Rβγ) is a complete IL-2 receptor. It consists of two distinct polypeptides —IL-2α, which is induced upon activation, and IL-2βγ, which is present on resting T cells. Upon expression of all three proteins, affinity increases to 10^{-11} *M*, and very low (physiologic) levels of IL-2 are capable of stimulating the cell. IL-2R is found on T lymphocytes, natural killer cells, and B lymphocytes, although natural killer cells do not express IL-2Rα. IL-2, IL-1, IL-6, IL-4, and TNF may induce IL-2R expression.

Differentiation factors are substances that facilitate maturation of cells, such as the ability of interleukin-2 to promote the growth of T cells.

Interleukin-3 (IL-3) is a 20-kDa lymphokine synthesized by activated CD4+ T helper lymphocytes, which acts as a colony-stimulating factor by facilitating proliferation of some hematopoietic cells and promoting proliferation and differentiation of other lymphocytes. It acts by binding to high- and low-affinity receptors and by inducing tyrosine phosphorylation and colony formation of erythroid, myeloid, megakaryocytic, and lymphoid hematopoietic cells. It also facilitates mast cell proliferation and the release of histamine. It facilitates T lymphocyte maturation through induction of 20α-hydroxysteroid dehydrogenase. The gene encoding IL-3 is situated on the long arm of chromosome 5.

IL-3 is the abbreviation for interleukin-3.

Mast cell growth factor-1 is a synonym for interleukin-3.

Interleukin-3 receptor (IL-3R) (Figure 10.9) is a low affinity IL-3 binding α subunit (IL-3α) (CD123) that associates

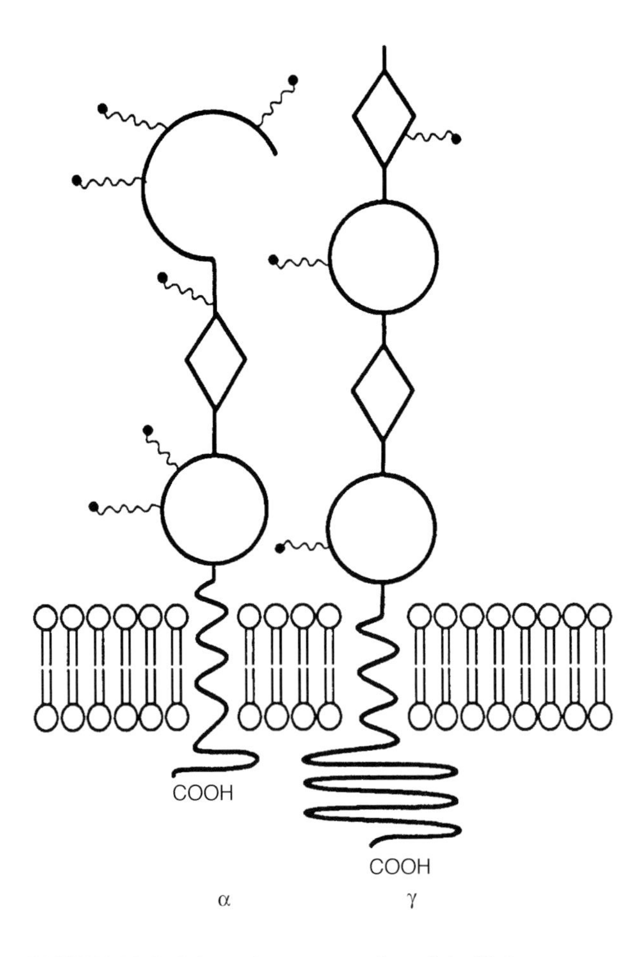

FIGURE 10.9 Schematic representation of the IL-3 receptor.

with a β subunit to produce a high-affinity IL-3 receptor. The IL-3α subunit has an N-terminal region consisting of a 100-amino acid residue, a cytokine receptor domain, and a fibronectin III domain containing the WSXWS motif. The truncated cytoplasmic domain is associated with the inability to signal. There are two homologous segments in the β subunit's extracellular region. A cytokine receptor domain followed by a fibronectin domain is present in each of these segments. The human alpha chain contains six potential N-linked glycosylation sites whereas the β chain has three. Tyrosine and serine/threonine phosphorylation of numerous cellular proteins occurs rapidly following union of IL-3 with its receptor. The β subunit is required for signal transduction.

Interleukin-4 (IL-4) (B cell growth factor) (Figure 10.10 and Figure 10.11) is a 20-kDa cytokine produced by CD4+ T lymphocytes, and also by activated mast cells. Most studies of IL-4 have been in mice, where it serves as a growth and differentiation factor for B cells and as a switch factor for synthesis of IgE. It also promotes growth of a cloned CD4+ T cell subset. Further properties of murine IL-4 include its function as a growth factor for mast cells and as an activation factor for macrophages. It also causes resting B lymphocytes to enlarge, and it enhances class II MHC molecule expression. IL-4 was previously termed B-cell growth factor I (BCGF-1) and also termed B-cell-stimulating factor 1 (BSF-1). In man, CD4+ T lymphocytes also produce IL-4, but the human variety has not been shown to serve as a B cell or mast cell growth factor. Human IL-4 also fails to activate macrophages. Both murine and human IL-4 induces switching of B lymphocytes to synthesize IgE. Thus, IL-4 may be significant in allergies. Human IL-4 also induces CD23 expression by B lymphocytes and macrophages in man. IL-4 may have some role in cell-mediated immunity. IL-4 induces differentiation of T_h2 cells from naïve CD4+ precursor cells, stimulates IgE formation by B cells, and suppresses IFN-γ-dependent macrophage functions.

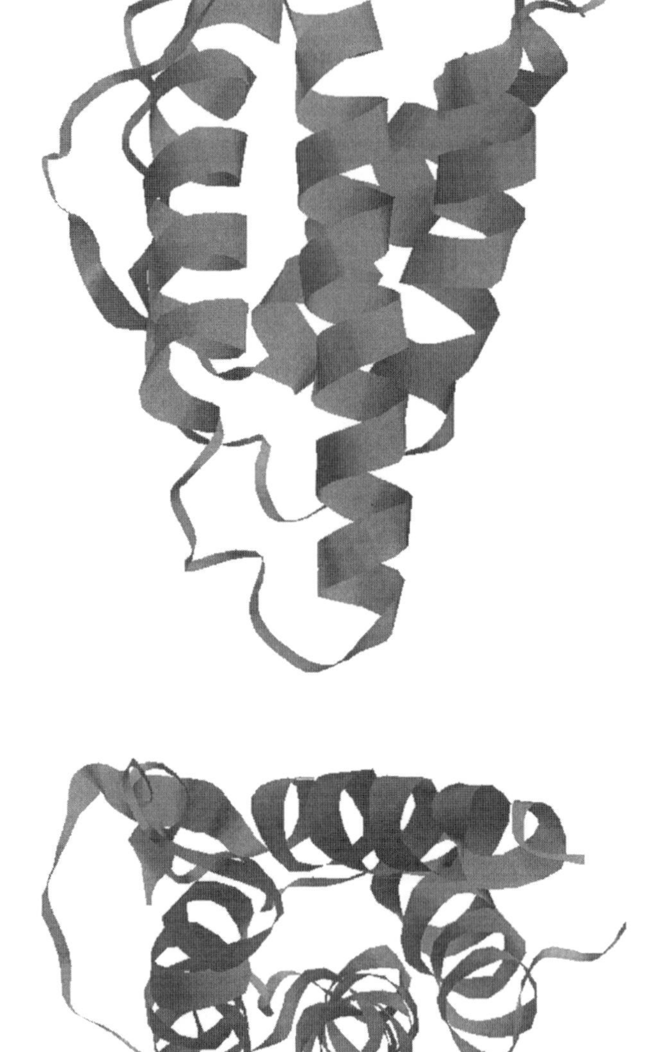

FIGURE 10.10 Human IL-4 NMR.

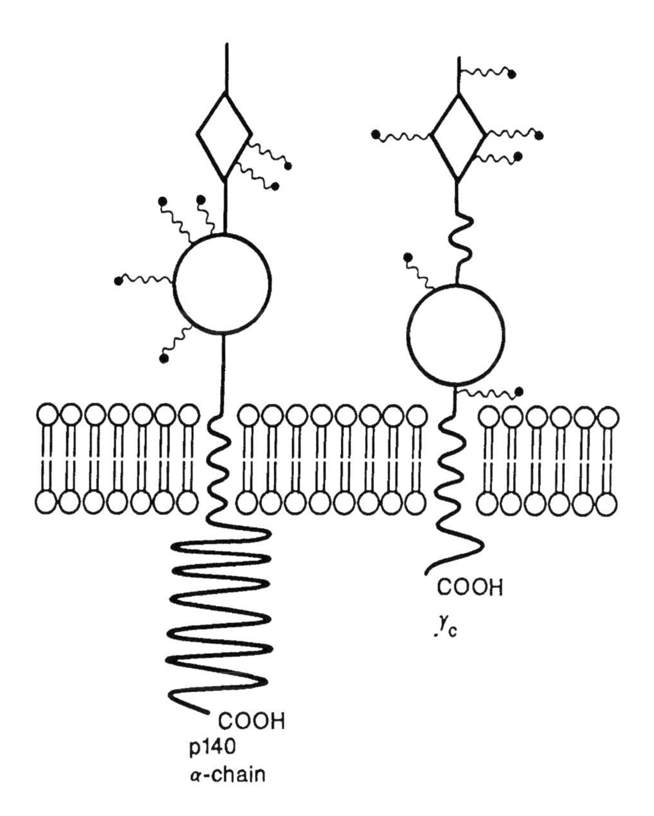

FIGURE 10.11 Schematic representation of the IL-4 receptor.

B cell-stimulating factor 1 (BSF-1) is an earlier term for interleukin-4.

IL-4 is the abbreviation for interleukin-4.

Mast cell growth factor-2 is a synonym for interleukin-4.

T cell growth factor 2 is interleukin-4 (IL-4).

B cell growth factor I (BCGF-1) is an earlier term for interleukin-4.

Interleukin-4 receptor (IL-4R) is a structure comprised of two major complexes — one of which, a 140-kDa single-chain, constitutes a high affinity IL-4 binding site (IL-4 Rα). The IL-4 Rα possesses pairs of cysteine residues and the WSXWS motif present in other members of the hematopoietin receptor superfamily, a classic transmembrane domain, and a 500-amino acid cytoplasmic domain. The IL-4 receptor complex also contains other polypeptide chains. Following the binding of IL-4, the IL-2 receptor γ chain associates with the IL-Rα chain. This complex comprises the Type I IL-4 receptor. The IL-13 low-affinity binding chain (IL-13Rα) can also associate with IL-4Rα. The IL-4 receptor containing the IL-13Rα is termed Type II.

B-cell growth factor (BCGF): See interleukins 4, 5, and 6.

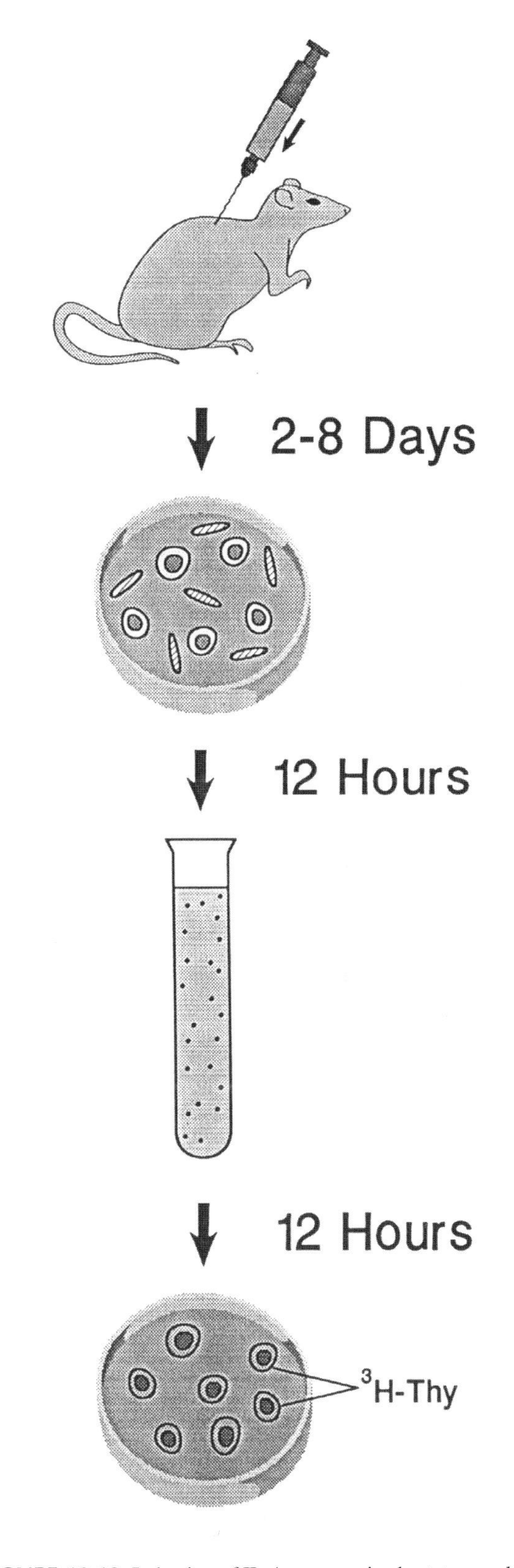

FIGURE 10.12 Induction of IL-4 response in short-term culture.

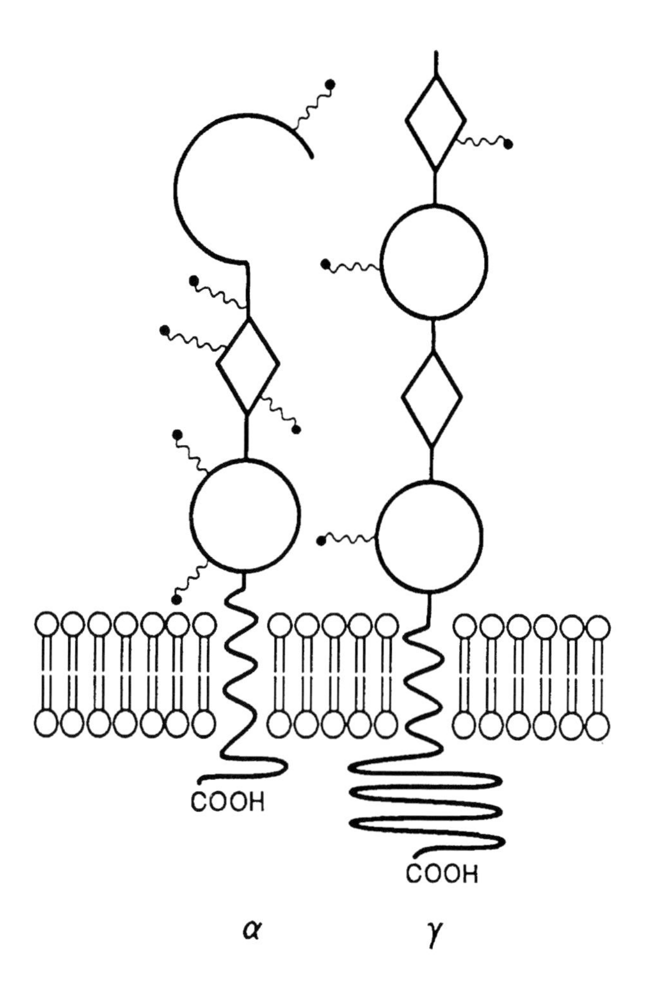

FIGURE 10.13 Schematic representation of the IL-5 receptor.

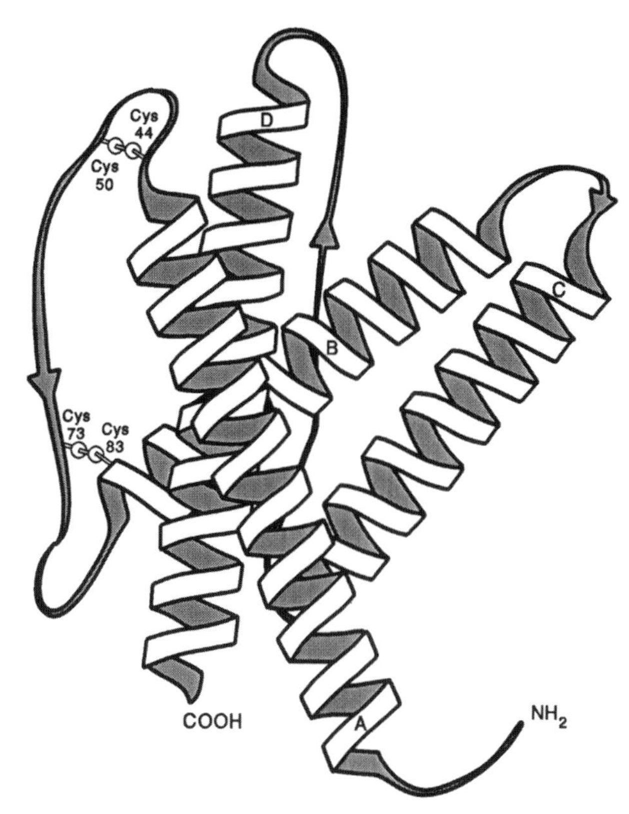

FIGURE 10.14 Schematic representation of IL-6.

Interleukin-5 (IL-5) (eosinophil differentiation factor) (Figure 10.13) is a 20-kDa cytokine synthesized by some activated CD4+ lymphocytes and by activated mast cells. Formerly it was called T-cell replacing factor or B-cell growth factor II. It facilitates B-cell growth and differentiation into cells that secrete IgA. It is a costimulator with IL-2 and IL-4 of B-cell growth and differentiation. IL-5 also stimulates eosinophil growth and differentiation. It activates mature eosinophils to render them capable of killing helminths. Through IL-5, T lymphocytes exert a regulatory effect on inflammation mediated by eosinophils. Because of its action in promoting eosinophil differentiation, it has been called eosinophil differentiation factor (EDF). IL-5 can facilitate B-cell differentiation into plaque-forming cells of the IgM and IgG classes. In parasitic diseases, IL-5 leads to eosinophilia.

Interleukin-5 receptor complex is a structure comprised of an α chain and a β chain, both of which resemble members of the hematopoietin receptor superfamily. The α chain is a 415-amino acid glycoprotein and binds with low affinity to IL-5. Even though the β chain does not bind to IL-5, it associates with the α chain to form the high-affinity receptor. The β chain is common to IL-3 and GM-CSF receptors.

IL-5 is an abbreviation for interleukin-5.

Thymus-replacing factor (TRF) is interleukin-5.

TRF is an abbreviation for T-cell replacing factor.

B cell growth factor II (BCGF-2) is an earlier term for interleukin-5.

Thymic stromal-derived lymphopoietin (TSLP) is a cytokine isolated from a murine thymic stromal cell line that possesses a primary sequence distinct from other known cytokines. The cDNA encodes a 140-amino acid protein that includes a 19-amino acid signal sequence. TSLP stimulates B220+ bone marrow cells to proliferate and express surface μ. TSLP synergizes with other signals to induce thymocyte and peripheral T cell proliferation but is not mitogenic for T cells alone.

Interleukin-6 (IL-6) (Figure 10.14 to Figure 10.16) is a 26-kDa cytokine produced by vascular endothelial cells, mononuclear phagocytes, fibroblasts, activated T lymphocytes, and various neoplasms such as cardiac myxomas, bladder cancer, and cervical cancer. It is secreted in response to IL-1 or TNF. Its main actions are on hepatocytes

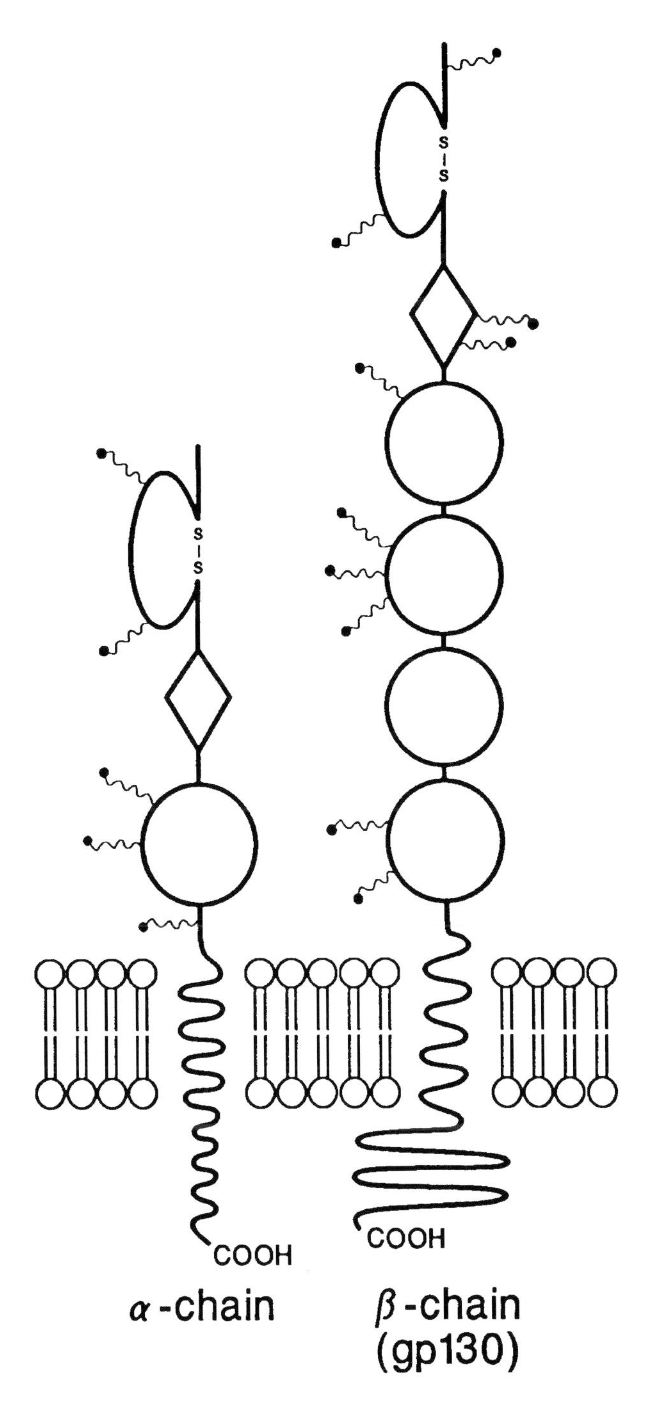

FIGURE 10.15 Schematic representation of the IL-6 receptors: IL-6 receptors of high affinity are comprised of two subunits that are noncovalently associated. There is low affinity binding of IL-6 to the IL-6Rα chain but no signal is produced. The gp130 extracellular domain is comprised of an IgSF C to set domain at the N-terminus. The cytokine receptor-SF domain and four fibronectin III domain follow. The WSXWS motif is present only in the first of the fibronectin III domain. There are 5 potential N-linked glycosylation sites in the human IL-6Fα chain and 10 in the gp130 IL-6 receptor.

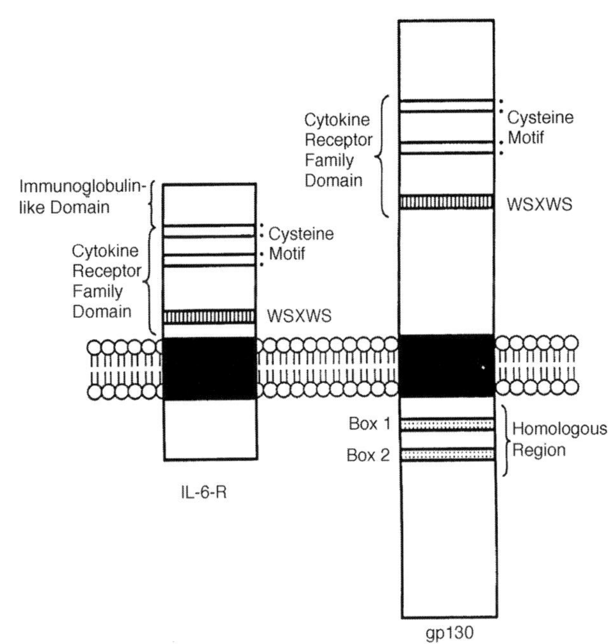

FIGURE 10.16 Schematic representation of the IL-6 receptor.

and B cells. Although it acts on many types of cells, a significant function is its ability to cause B lymphocytes to differentiate into cells that synthesize antibodies. IL-6 induces hepatocytes to form acute-phase proteins that include fibrinogen. It is the main growth factor for activated B lymphocytes late in B-cell differentiation. IL-6 is a growth factor for plasmacytoma cells which produce it. IL-6 also acts as a costimulator of T lymphocytes and thymocytes. It acts in concert with other cytokines that promote the growth of early bone marrow hematopoietic stem cells, and together with IL-1 to costimulate activation of T_H cells. IL-6 was formerly termed B-cell differentiation factor (BCDF) and B-cell stimulating factor 2 (BSF-2). It has also been implicated in the pathogenesis of plaques in psoriasis (Figure 10.17). Together with IL-2 and TNF-α, IL-6 produces a broad spectrum of responses early in infection.

IL-6 is an abbreviation for interleukin-6.

Ciliary neurotrophic factor (CNTF) is a protein hormone related to the IL-6 family. It has several functions in the nervous system which are similar to those of leukemia inhibitory factor.

Hepatocyte-stimulating factor is a substance indistinguishable from interleukin-6, classified as a monokine, that stimulates hepatocytes to produce acute-phase reactants.

IL-6 receptor is a 468-amino acid residue structure on the membrane that shows homology with an immunoglobulin domain.

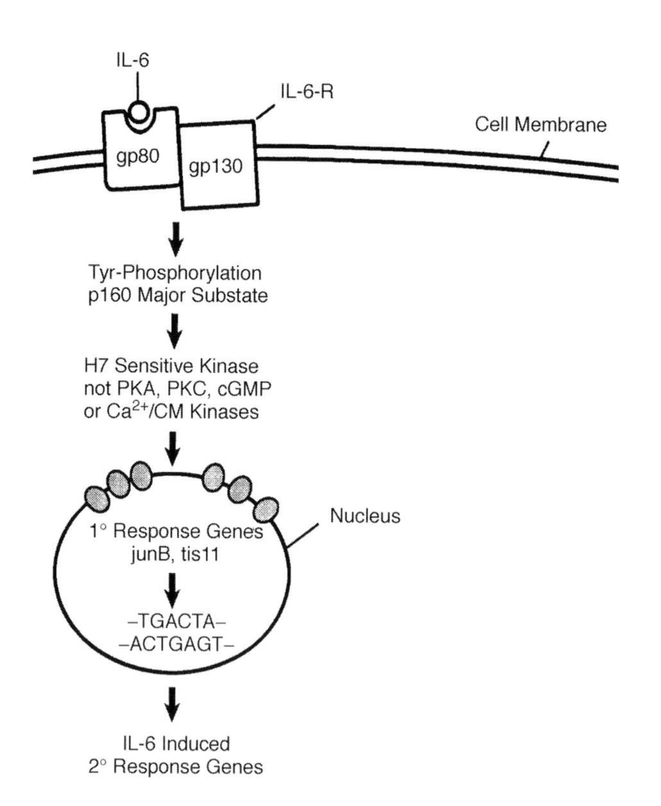

FIGURE 10.17 Pathway of IL-6 induction and signaling.

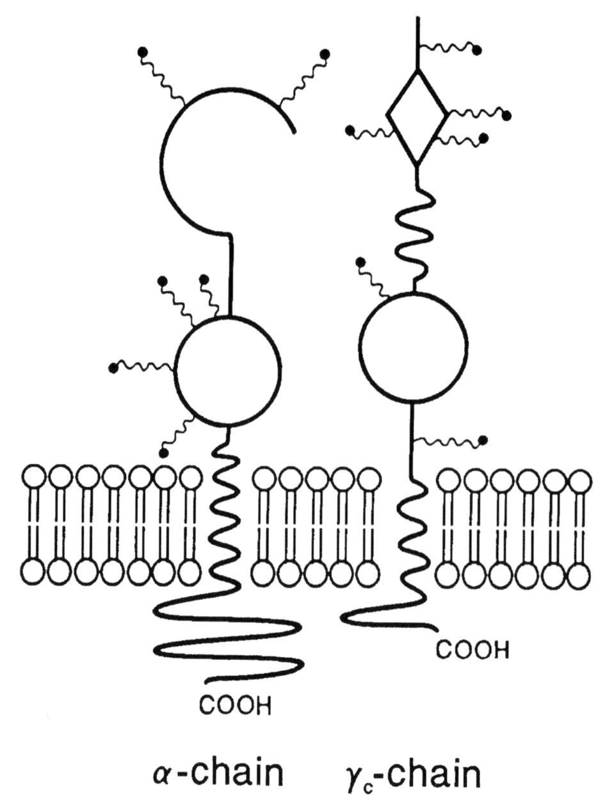

FIGURE 10.18 Schematic representation of the IL-7 receptor that is comprised of an IL-7 binding chain designated CD127 and the γ chain of the IL-2R. Following a 100-amino acid N-terminal region, there is a fibronectin III domain with a WSXWS motif in the IL-7R γ chain that augments IL-7 binding. In the human form of the IL-7 receptor, there are five potential N-linked glycosylation sites.

B cell-stimulating factor 2 (BSF-2) is an earlier term for interleukin-6.

Interleukin-6 receptor is a structure comprised of 468 amino acids, 19-amino acid signal peptide, 339-amino acid extracellular portion, 28-amino acid transmembrane domain, and 82-amino acid intracytoplasmic portion. There are 6 potential *N*-linked glycosylation sites and 11 cysteine residues. The mature receptor has a mol wt of 80 kDa and has *O*- and *N*-glycosylation. The IL-6 receptor system is comprised of two functional chains, the ligand-binding 80-kDa, IL-6 receptor and the nonligand-binding gp130. This receptor is expressed on lymphoid as well as nonlymphoid cells.

Interleukin-7 (IL-7) (Figure 10.18) facilitates lymphoid stem cell differentiation into progenitor B cell. Principally a T lymphocyte growth factor synthesized by bone-marrow stromal cells, it promotes lymphopoiesis, governing stem-cell differentiation into early pre-T and B cells. It is also formed by thymic stroma and promotes the growth and activation of T cells and macrophages. It also enhances fetal and adult thymocyte proliferation.

IL-7 is an abbreviation for interleukin-7. cDNA that encodes the receptor for IL-7, known as IL-7Rα, has been cloned from both humans and mouse.

Interleukin-8 (IL-8) (neutrophil-activating protein 1) is an 8-kDa protein of 72 residues produced by macrophages and endothelial cells. It has a powerful chemotactic effect on T lymphocytes and neutrophils and upregulates the binding properties of leukocyte adhesion receptor CD11b/CD18. IL-8 regulates expression of its own receptor on neutrophils and has antiviral, immunomodulatory, and antiproliferative properties. It prevents adhesion of neutrophils to endothelial cells activated by cytokines, thereby blocking neutrophil-mediated injury. IL-8 participates in inflammation and the migration of cells. It facilitates neutrophil adherence to endothelial cells, accomplishing this through the induction of β_2 integrins by neutrophils (Figure 10.19 to Figure 10.21).

Interleukin-8 (IL-8) is a chemokine of the α family (CXC family). It is the prototypic and most widely investigated chemokine. Its powerful neutrophil chemotactic action makes IL-8 a primary regulatory molecule of acute inflammation. High-affinity neutralizing anti-IL-8 antibodies have been used to reveal the regulatory action of IL-8 on neutrophil infiltration into tissues. IL-8 also has a second significant effect in promoting angiogenesis. IL-8 contains the

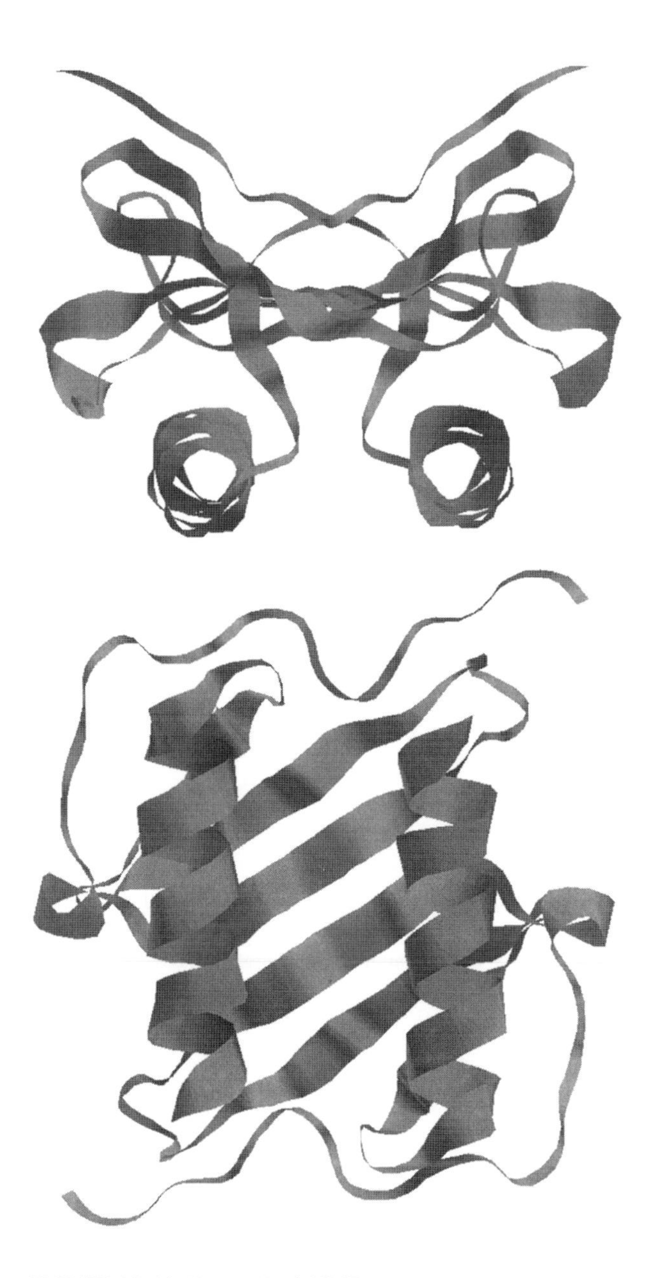

FIGURE 10.19 Human IL-8 NMR.

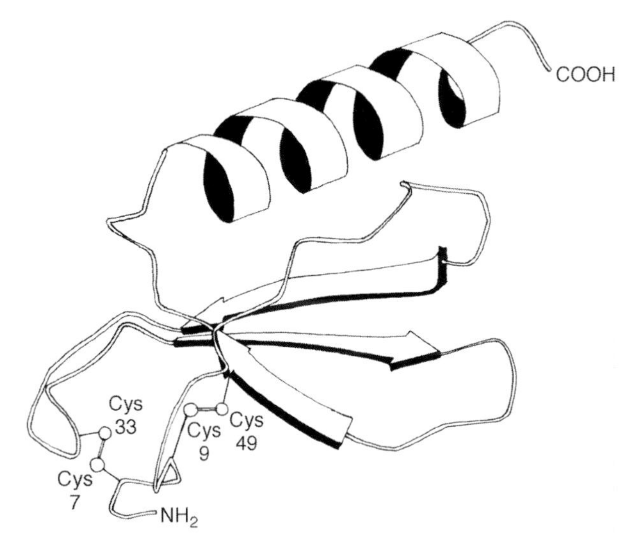

FIGURE 10.20 A three-dimensional minimized mean structure, derived by NMR, of the polypeptide backbone structure of human recombinant interleukin-8.

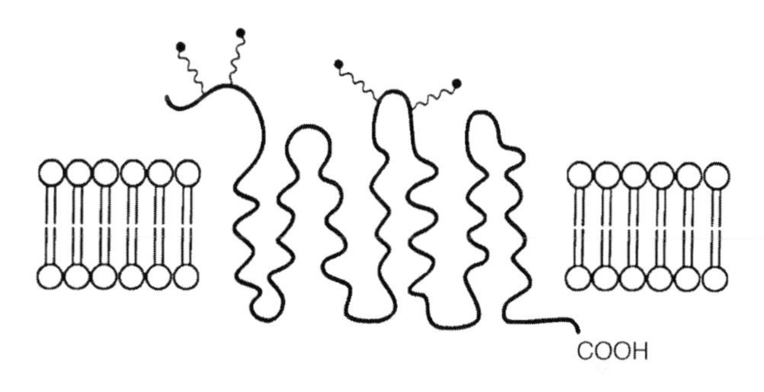

FIGURE 10.21 Schematic representation of the IL-8 receptor, which may be of either high or low affinity. These receptors are seven-transmembrane spanning and are G-protein linked. They belong to the rhodopsin superfamily. Whereas the high-affinity receptor binds IL-8 exclusively, the low-affinity receptor also shows specificity for NAP-2 and GRO/MPSA.

ELR motif that is crucial to the chemotactic action of several α-chemokine family members. Monocytes, macrophages, $CD4^+$, $CD8^+$, and $CD45RA^+$ lymphocytes, among many other cell types, serve as tissue sources. Neutrophils, T lymphocytes, basophils, eosinophils, keratinocytes, and HUVECs cells serve as target cells. There are two types of human IL-8 receptor, type A and type B, which consist of 351 and 360 amino acids, respectively. IL-8 belong to a family of G-protein-coupled receptors with seven transmembrane domains and have high sequence homology with each other at the amino acid level.

Monocyte-derived neutrophil chemotactic factor: See interleukin-8.

NAP-1 is neutrophil attractant or activation protein-1. See interleukin-8.

LDCF (lymphocyte-derived chemotactic factor) is chemotactic especially for mononuclear phagocytes.

Interleukin-8 receptor, type A (IL-8RA) is the high-affinity IL-8 receptor, class I. IL-8R belongs to the G-protein-coupled receptor family, the chemokine receptor branch of the rhodopsin family. The tissue source is human neutrophils. Murine IL-8 receptor knockout mice develop splenomegaly, enlarged cervical lymph nodes, and extramedullary myelopoiesis, and impaired neutrophil acute migration is expressed on hematopoietic cells. IL-8RA mRNA is present in neutrophils, monocytes, basophils, and freshly isolated T cells.

Neutrophil-activating protein-1 (NAP-1) is the former term for IL-8.

Interleukin-8 receptor, type B (IL-8RB) is a low-affinity IL-8R, class II, IL-8 receptor, CXCR2, F3R (rabbit). IL-8RB mRNA is present in neutrophils, monocytes, basophils, T cells, and primary keratinocytes. It is expressed in the lesional skin of psoriasis patients. G-CSF upregulates IL-8RB mRNA expression, whereas IL-8 and MGSA down-regulate IL-8RB surface expression on human neutrophils.

Neutrophil-activating factor-1 is interleukin-8 (IL-8).

Neutrophil-activating peptide 2 is a 75-amino aide chemokine produced by the proteolysis of beta-thromboglobulin (β-TG) by neutrophil cathepsin-G. It can activate neutrophils and monocytes by binding to their interleukin-8 (IL-8) type B receptors. NAP-2 has about 60% amino acid sequence similarity with platelet factor 4 (PF-4). NAP-2 is believed to augment inflammation by cooperative interactions between platelets and neutrophils. Autoantibodies against NAP-2 and IL-8 have been identified in heparin-associated thrombocytopenia, but their clinical significance remains to be determined.

Neutrophil-attracting peptide (NAP-2) is a chemoattractant of neutrophils to sites of platelet aggregation. NAP-2 competes weakly with IL-8 for the IL-8 type II receptor. However, since it can be found at much higher concentrations than IL-8 at platelet aggregation sites, NAP-2 is considered an active participant in the inflammatory process.

Interleukin-9 (IL-9) (Figure 10.22 and Figure 10.23) is a cytokine that facilitates the growth of some T helper cell clones but not of clones of cytolytic T lymphocytes. It is encoded by genes comprised of 5 exons in the 4 kb segment of DNA in both mice and humans. In the presence of erythropoietin, IL-9 supports erythroid colony formation. In conjunction with IL-2, IL-3, IL-4, and erythropoietin, IL-9 may enhance hematopoiesis *in vivo*. It may facilitate bone marrow–derived mast cell growth stimulated by IL-3 and fetal thymocyte growth in response to IL-2. T_H2 cells preferentially express IL-9 following stimulation with ConA or by antigen presented on syngeneic antigen presenting cells.

IL-9 is the abbreviation for interleukin-9. The murine IL-9R encodes a 468-amino acid polypeptide with two potential N-linked glycosylation sites and six cysteine residues in the extracellular domain. Contemporary research suggests that RL-9R is a member of the hematopoietin receptor superfamily. The human IL-9R encodes a 522-amino acid polypeptide with 53% homology with the mIL-9R.

Interleukin-9 (murine growth factor P40, T cell growth factor III) is a hematopoietic growth factor glycoprotein derived from a megakaryoblastic leukemia. Selected human T lymphocyte lines and peripheral lymphocytes activated by mitogen express it. IL-9 is related to mast cell growth-enhancing activity both structurally and functionally. The genes encoding IL-9 are located on chromosomes 5 and 13.

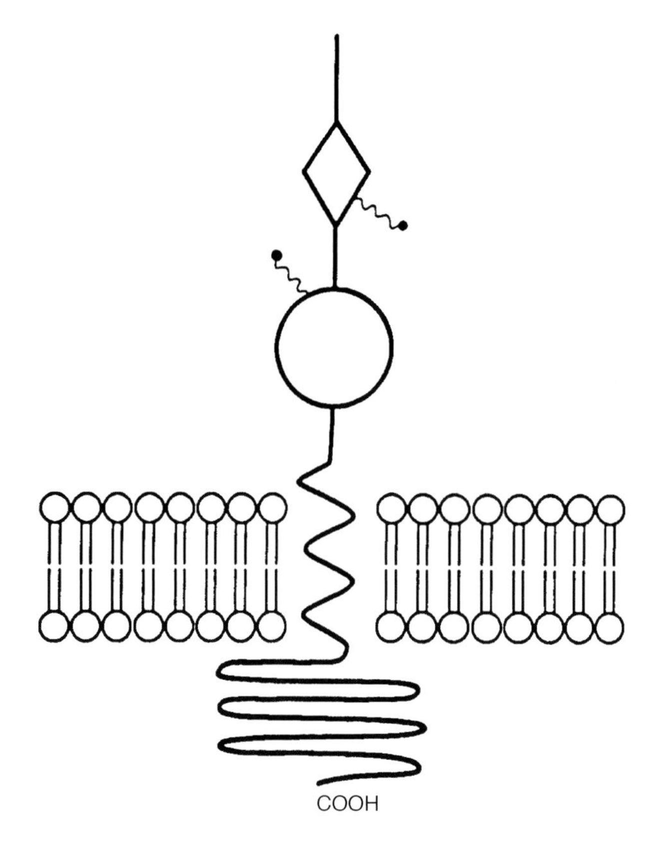

FIGURE 10.22 Schematic representation of the IL-9 receptor, only one type of which has been found on murine T cell lines. Investigation of the recombinant murine receptor has demonstrated association between the IL-2R gamma chain and the IL-9R. Macrophages, some T cell tumors, and mast cell lines have been shown to express IL-9 receptors.

Interleukin-10 (IL-10) (cytokine synthesis inhibitory factor) (Figure 10.24 and Figure 10.25) is an 18-kDa polypeptide devoid of carbohydrate in humans that acts as a cytokine synthesis inhibitory factor. It is expressed by $CD4^+$ and $CD8^+$ T lymphocytes, monocytes, and macrophages, activated B lymphocytes, B lymphoma cells, and keratinocytes. It inhibits some immune responses and facilitates others. It inhibits cytokine synthesis by T_H1 cells and blocks antigen presentation and the formation of interferon γ. It also inhibits the macrophage's ability to present antigen and to form IL-1, IL-6, and TNFα. It also participates in IgE regulation. Although IL-10 suppresses cell-mediated immunity, it stimulates B lymphocytes, IL-2 and IL-4 T lymphocyte responsiveness *in vitro,* and murine mast cells exposed to IL-3 and IL-4. IL-10 might have future value

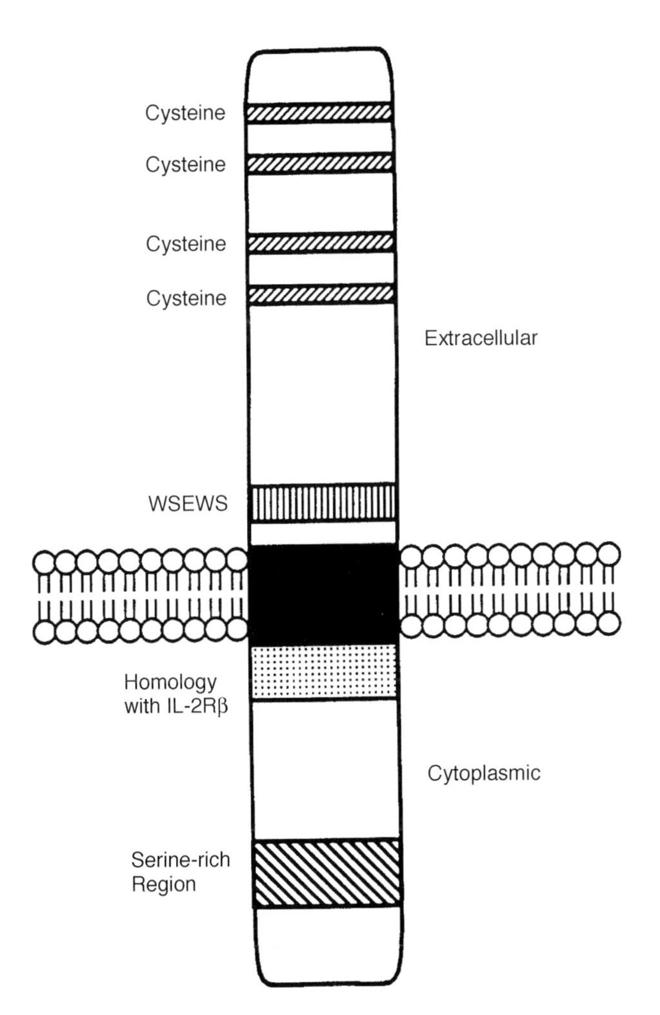

FIGURE 10.23 Schematic representation of the IL-9 receptor.

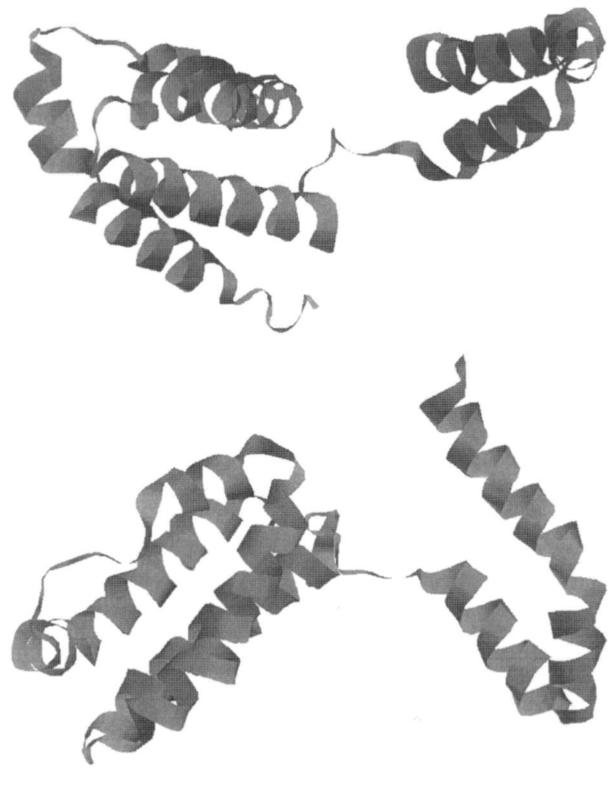

FIGURE 10.24 IL-10. By x-ray diffraction. Resolution 1.8 Å.

in suppressing T-lymphocyte autoimmunity in multiple sclerosis and type I diabetes mellitus and in facilitating allograft survival. T_H2 cells of murine T-clones secrete IL-10, which suppresses synthesis of cytokines by T_H1 cells. IL-10's principal function is to inhibit activated macrophages, thereby maintaining homeostatic control of innate and cell-mediated immune reactions.

Cytokine synthesis inhibitory factor: See interleukin-10.

Interleukin-10 (IL-10) is a multifunctional cytokine that inhibits activation and affects function of T cells, monocytes, and macrophages. It has diverse effects on hemopoietic cell types. It emits and ultimately terminates inflammatory responses. IL-10 regulates growth and/or differentiation of B cell, NK cells, cytoxic and helper T cells, mast cells, granulocytes, dendritic cells, keratinocytes, and endothelial cells. It has a key role in differentiation and function of the T regulatory cell which is critical in the control of immune responses and tolerance *in vivo.*

CSIF is the abbreviation for cytokine synthesis inhibitory factor. See interleukin-10.

IL-10 is the abbreviation for interleukin-10. Both the mouse and human IL-10 receptors have been identified, characterized, and cloned. There is 75% DNA sequence homology between the mouse IL-10R and the human IL-10R. The human IL-10 receptor is a 110-kDa glycoprotein.

Interleukin-11 (IL-11) is a cytokine produced by stromal cells derived from the bone marrow of primates. It is a growth factor that induces IL-6-dependent murine plasmacytoma cells to proliferate. IL-11 has several biological actions that include its hematopoietic effect. In man, the genomic sequence and gene encoding IL-11 is comprised of five exons and four introns. The gene is located at band 19q13.3-13.4 on the long arm of chromosome 19. It may facilitate plasmacytoma establishment, possibly representing an important role for IL-11 in tumorigenesis. In combination with IL-3, IL-11 can potentiate megakaryocyte growth, producing increased numbers, size, and ploidy values. It may be important in the formation of platelets. In the presence of functional T lymphocytes, IL-11 can stimulate the production of B cells that secrete IgG. It has a synergistic effect in primitive hematopoietic cell proliferation that is IL-3 dependent. The **interleukin-11 receptor** designated IL-11R is comprised of at least one ligand-binding subunit (IL-11Rα) and a signal transduction subunit (gp130). The murine and human IL-11R cDNAs encode a 432- and 422-amino acid polypeptide, respectively, with two potential N-linked glycosylation sites and

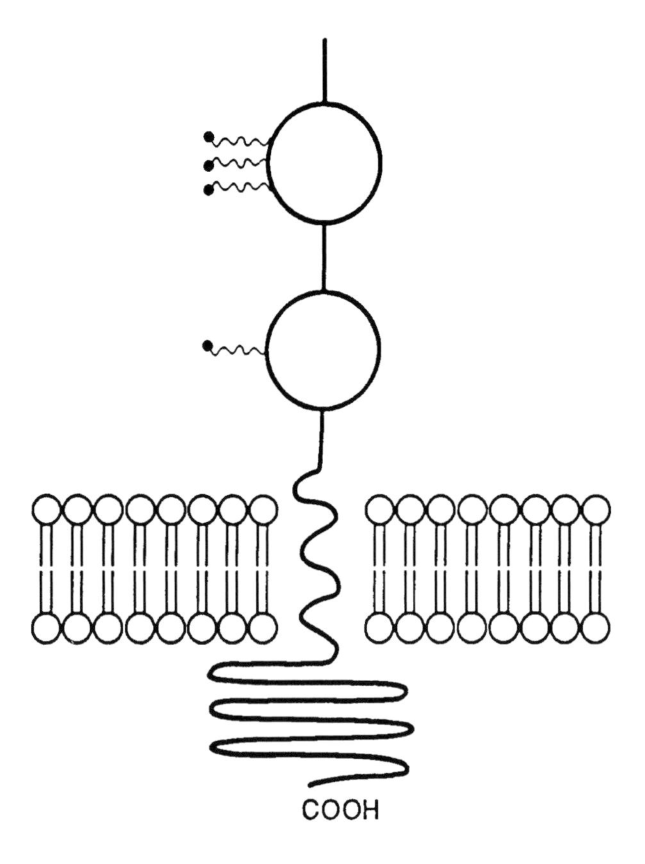

FIGURE 10.25 Schematic representation of the IL-10 receptor that is a class II cytokine receptor family molecule. There are two homologous fibronectin III domains in the 220-amino acid extracellular region. There are two conserved tryptophans and one pair of conserved cysteines in the first fibronectin domain and a disulfide loop produced from a second pair of conserved cysteines. However, there is no WSXWS motif, which is found in class I cytokine receptors. The human IL-10 receptor has six potential N-linked glycosylation sites. Signal transduction is believed to involve the JAK family of kinases, but this remains to be elucidated.

four cysteine residues in the extracellular domain. IL-11Rα is believed to be a member of the hematopoietin receptor superfamily. Murine and human IL-11Rα have 84% homology at the amino acid level and 85% homology at the nucleotide level.

IL-11 is the abbreviation for interleukin-11.

Interleukin-12 (IL-12) is a heterodimeric molecule comprised of 35-kDa and 40-kDa chains linked by disulfide bonds. It acts on T cells as a cytotoxic lymphocyte maturation factor (CLMF). It also serves as a natural killer (NK) cell stimulatory factor (NKSF). IL-12 is a growth factor for activated $CD4^+$ and $CD8^+$ T lymphocytes and for NK cells. It facilitates NK cell and LAK cell lytic action exclusive of IL-2. It can induce resting peripheral blood mononuclear cells to form interferon gamma *in vitro*. IL-12 may act synergistically with IL-2 to increase responses by cytotoxic lymphocytes. It may have potential as a therapeutic agent in the treatment of tumors or infections, especially if used in combination with IL-2. IL-12 promotes IFN-γ formation by NK cells and T cells, enhances the cytolytic activity NK cells in CTLs, and facilitates the development of T_H2 cells. **Interleukin-12 receptor** is designated as **IL-12R**. Two receptor chains, IL-12Rβ_1 and IL-12Rβ_2, have been identified and cloned. Both β_1 and β_2 chains are members of the cytokine receptor superfamily. They are related to gp130 of the IL-6R and to receptors of LIF and G-CSF. Coexpression of both these receptors confers IL-12 responsiveness. Both human- and mouse-activated T cells have IL-12 binding sites of high, intermediate, and low affinities.

IL-12 is the abbreviation for interleukin-12.

Interleukin-13 (IL-13) (Figure 10.26) is a cytokine expressed by activated T cells that inhibits inflammatory cytokine production induced by lipopolysaccharide in human peripheral blood monocytes. It synergizes with interleukin-2 to regulate the synthesis of interferon-γ in large granular lymphocytes. Mapping reveals that the IL-13 gene is linked closely to the IL-4 gene. IL-13 could be considered a modulator of B cell responses. It stimulates B-cell proliferation with anti-Ig and anti-CD40 antibodies and promotes IgE synthesis. It also induces resting B cells to express CD23/FcεRII and class II MHC antigens. The biological activities of IL-13 are similar to those of IL-4. Therefore, both IL-13 and IL-4 may contribute much to the development of allergies. Although IL-13 and IL-4 affect B cells similarly, they have different functions. IL-13 induces less IgE and IgG4 production than does IL-4. IL-13 does not act on T cells or T cell clones; it also fails to induce expression of CD8α on $CD4^+$ T cell clones and has no T-cell growth-promoting activity.

IL-13 suppresses cell-mediated immunity. It suppresses the cytotoxic functions of monocytes/macrophages and the generation of proinflammatory cytokines. IL-13 has a pleiotrophic action on monocytes/macrophages, neutrophils, and B cells. It usually produces an inhibitory effect on monocytes/macrophages and downregulates Fcγ receptors and secretion of inflammatory cytokines and nitric oxide induced by lipopolysaccharide. IL-13 inhibits macrophage function in cell-mediated immunity. It induces neutrophils to form IL-1 receptor antagonist and IL-1 receptor II. This leads to inhibition of IL-1, a central inflammatory mediator, which is in accord with the antiinflammatory function of IL-13. The high-affinity **IL-13 receptor complex** consists of the IL-4Rα chain and an IL-13 binding protein designated the IL-13Rα chain. The **IL-13 receptor** is expressed on B cells, monocytes/macrophages, basophils, mast cells, and endothelial cells but not on T cells. The IL-13R complex also acts as a second receptor for IL-4. The IL-13Rα chain is a specific binding protein for IL-13. IL-4 can signal through both

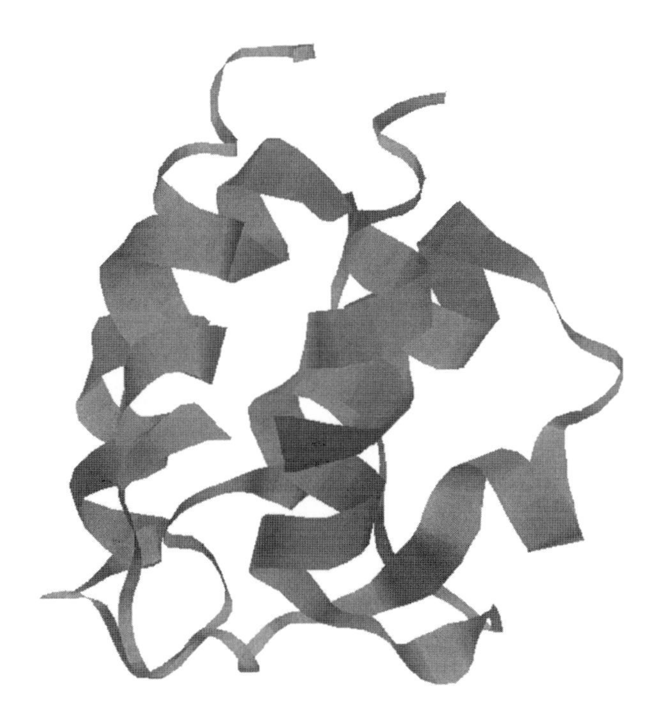

FIGURE 10.26 A ribbon diagram of IL-13 that constitutes a theoretical model of human IL-13. This represents the first of two alternative structures of IL-13 generated by homology modeling.

IL-13R and IL-4R complexes. Both IL-4 and IL-13 can mediate their biological actions through the IL-13R complex.

IL-13 is the abbreviation for interleukin-13.

Interleukin-14 (IL-14) was formerly known as high-molecular-weight B cell growth factor (HMW-BCGF). IL-14 is a cytokine produced by follicular dendritic cells, germinal center T cells, and some malignant B cells. Normal and malignant B cells, notably germinal center B cells and NHL-B cells, respectively, express receptors for IL-14. Its predominant activity is to enhance the proliferation of B cells and to induce memory B-cell production and maintenance. Work with NHL-B cell lines has shown that inhibition of the expression of the IL-14 gene results in diminished cell growth and eventual cell death.

IL-14 is the abbreviation for interleukin-14.

Interleukin-15 (IL-15) is a T-cell growth factor synthesized by mononuclear phagocytes and other cells in response to viral infections. It has many of the biological properties of IL-2. Its principal function is to stimulate the proliferation of natural killer (NK) cells. IL-15 enhances peripheral blood T cell proliferation, and *in vitro* studies demonstrate its ability to induce cytotoxic T cells. IL-15 mRNA has not been found in activated peripheral T cells, but monocyte-enriched peripheral blood cell lines as well as placental and skeletal muscle tissues do express IL-15. The **IL-2 receptor** and **IL-15 receptor** also share a common component for successful signal transduction. IL-15 receptors are widely expressed. IL-15 plays a major role related to NK cell development and cytolytic activity. The receptor for IL-15 on T cells contains IL-2Rβ,γ_c, and one unique protien, IL-15Rα; alternate IL-15 receptor, designated IL-15RX, has been detected on mast cells. IL-15 mediates its action by combination with a heterotrimeric receptor that is comprised of both β and the γ_c chains of the IL-2R as well as a unique IL-15 binding subunit. This subunit is termed IL-15α. The IL-15Rα chain is required for high affinity binding but not signaling by IL-15. The signaling pathways for IL-2 and IL-15 are the same.

IL-15 is the abbreviation for interleukin-15.

Interleukin-16 (IL-16) is a 13.2-kDa protein containing 130 amino acid residues. It is also called lymphocyte chemoattractant (LCA). IL-16 activates at migratory response in CD4$^+$ T cells and CD4$^+$ monocytes and eosinophils. Human IL-16 also induces IL-2 receptor expression by T lymphocytes. r IL-16 has been found to suppress T-cell proliferation and mixed lymphocyte reactions. IL-16 is structurally distinct from chemokines and in the active form is a homotetramer comprised of four 16-kDa chains. It is synthesized by CD4$^+$ T cells, mass cells, and eosinophils. Its gene is located on human chromosome 15q26.1. CD4 may be a receptor for IL-16. IL-16 is chemotactic for resting as well as activated T cells, eosinophils, and monocytes. It facilitates CD4$^+$ T cell adhesion, expression of IL-2Rα (CD25), and HLA-DR, as well as cytokine synthesis. It suppresses antigen and alloantigen (MLR)-induced lymphocyte proliferation. IL-16 does not competitively inhibit HIV-gp120 binding. It is a potent survival factor and counters apoptotic effects of growth-inducing cytokines such as IL-2. It has been found in bronchial airway epithelial cells and fluid of asthmatics. It may exacerbate allergic reactions. IL-16 is a proinflammatory and immunomodulatory cytokine.

IL-16 is the abbreviation for interleukin-16.

Interleukin-17 (IL-17) is a glycoprotein of 155 amino acids secreted as a homodimer by activated memory $CD4^+$ T cells. It is also called cytotoxic T lymphocyte associated antigen 8 (CTLA). It is believed to share 57% amino acid identity with the protein predicted from ORF13, an open reading frame of *Herpesvirus saimiri* . hIL-17 has no direct effect on cells of hematopoietic origin but does stimulate epithelial, endothelial, and fibroblastic cells to secrete cytokines such as IL-6, IL-8, and granulocyte-colony-stimulating factor, as well as prostaglandin E2. hIL-17 may be an early initiator of T cell dependent inflammatory reactions and part of the cytokine network linking the immune system to hematopoiesis. Human CTLA-8 expression is restricted to memory $CD4^+$ lymphocytes. It induces epithelial, endothelial, and fibroblastic cells to activate NKFB and to synthesize IL-6, IL-8, G-CSF, and PGE_2 but not IL-1 or TNF. IL-17 is located on murine chromosome 1A and human chromosome 2q31. IL-17 induces fibroblasts to better support the growth and differentiation of $CD34^+$ progenitor cells. IL-17 is an inducer of stromal cell proinflammatory and hematopoietic cytokines.

IL-17 is the abbreviation for interleukin-17.

Interleukin-18 (IL-18) is a cytokine produced not only by immune cells but also by nonimmune cells. It is structurally homologous to IL-1 and its receptor belongs to the IL-1/Toll-like receptor (TLR) superfamily, but its function differs from that of IL-1. Together with IL-12, IL-18 stimulates Th1-mediated immune responses that are critical for host defense against infection with intracellular microbes through the induction of IFN-γ. IL-18 is believed to be a potent proinflammatory cytokine, since overproduction of IL-12 and IL-18 induces severe inflammatory disorders. IL-18 mRNA is expressed in a broad range of cell types including Kupffer cells, macrophages, T cells, B cells, dendritic cells, osteoblasts, keratinocytes, astrocytes, and microglias. IL-18 enhances IL-12-driven Th1 immune responses, but it can also stimulate TH2 immune responses in the absence of IL-12.

IL-18 is the abbreviation for interleukin-18.

Interleukin-19 (IL-19) is a novel homologue of IL-10 designated as IL-19 that shares 21% amino acid identity with IL-10. The expression of IL-19 mRNA can be induced in monocytes by LPS treatment. GM-CSF can directly induce IL-19 gene expression in monocytes. IL-19 does not bind or signal through the canonical IL-10 receptor complex, suggesting the existence of an IL-19-specific receptor complex which remains to be demonstrated.

Interleukin-20 (IL-20) is a novel homologue, overexpression of which in transgenic mice causes neonatal lethality with skin abnormalities that include aberrant epidermal differentiation. Recombinant IL-20 protein stimulates a signal transduction pathway through STAT3 in a keratinocyte line. An IL-20 receptor was identified as a heterodimer to orphan class II cytokine receptor subunits. Both receptor subunits are expressed in skin and are significantly upregulated in psoriatic skin.

Interleukin-21 (IL-21) is a cytokine that is most closely related to IL-2 and IL-15. Its receptor has been designated as IL-21R. *In vitro* assays suggest that IL-21 has a role in the proliferation and maturation of natural killer (NK) cell population from bone marrow, in the proliferation of mature B cell populations costimulating with anti-CD40, and in the proliferation of T cells costimuating with anti-CD3.

Interleukin-22 (IL-22) is a human cytokine distantly related to interleukin-10. IL-22 is produced by activated T cells. It is a ligand for CRF2-4. IL-22 is a member of the class II cytokine receptor family. No high-affinity ligand is known for this receptor although it has been reported to serve as a second component in IL-10 signaling. A new member of the interferon receptor family, termed IL-22R, serves as a second component together with CRF2-4 to facilitate IL-22 signaling. IL-22 does not bind the IL-10 receptor. In contrast to IL-10, IL-22 does not inhibit the synthesis of proinflammatory cytokines by monocytes in response to LPS. Nor does it affect IL-10 function on monocytes. IL-22 does have a modest inhibitory effect on IL-4 synthesis by TH2 T cells.

Interleukin-23 (IL-23) is a composite cytokine that binds to IL-12R β1 but fails to engage IL-12R β2. IL-23 activates Stat 4 in PHA blast T cells. It induces strong proliferation of mouse memory [$CD4^+$ CD45Rb (low)] T cells, a unique activity of IL-23, whereas IL-12 has no effect on this cell population. Similar to IL-12, human IL-23 stimulates IFN-γ production and proliferation in PHA blast T cells as well as in CD45RO (memory) T cells.

Interferons (IFNs) comprise a group of immunoregulatory proteins synthesized by T lymphocytes, fibroblasts, and other types of cells following stimulation with viruses, antigens, mitogens, double-stranded DNA, or lectins. Interferons are classified as α and β, which have antiviral properties, and as γ, which is known as immune interferon. Interferons α and β share a common receptor, but γ has its own. Interferons have immunomodulatory functions. They enhance the ability of macrophages to destroy tumor cells, viruses, and bacteria. Interferons α and β were formerly classified as type I interferons. They are acid stable and synthesized mainly by leukocytes and fibroblasts. Interferon γ is acid-labile and is formed mainly by T lymphocytes stimulated by antigen or mitogen. This immune interferon has been termed type II interferon in the past. Whereas the ability of interferon to prevent infection of noninfected cells is species specific, it is not virus specific. Essentially, all viruses are subject to its inhibitory action. Interferons induce formation of a second inhibitory protein that prevents viral messenger RNA translation. In addition to gamma interferon formation by T cells activated with mitogen, natural killer cells also secrete it. Interferons are not themselves viricidal.

IFN is an abbreviation for interferon.

Type I interferons (IFN-α, IFN-β) are a cytokine family of proteins comprised of several of the IFN-α variety that are structurally similar, together with a single IFN-β protein. All of the type I interferons manifest potent antiviral activity. Mononuclear phagocytes represent the principal source of IFN-α, whereas IFN-β is synthesized by numerous cell types that include fibroblasts. The same cell surface receptor binds both IFN-α and IFN-β, both of which have similar biologic effects. Type I interferons block viral replication, potentiate the lytic activity of NK cells, enhance Class I MHC expression on virus-infected cells, and induce the development of human T_H1 cells.

Type II interferon is a synonym for interferon γ.

Interferon regulatory factors (IRF) comprise a family of transcription factors that have a novel helix-turn-helix DNA-binding motif. In addition to two structurally related members, IRF-1 and IRF-2, seven additional members have been described. Virally encoded IRFs that may interfere with cellular IRFs have also been identified. They have a functional role in the regulation of host defense, such as innate and adaptive immune responses in oncogenesis.

Adrenergic receptor agonists: The β2-adrenergic receptor is expressed on Th1 but not on Th2 clones. An agonist for this receptor selectively suppresses IFN-γ synthesis *in vitro*; increases IL-4, IL-5, and IL-10 synthesis in spleen cells from treated mice; and increases IgE levels *in vivo*. Epinephrine and norepinephrine inhibit IL-12 synthesis but enhance IL-10 formation. β2-adrenergic receptor stimulation favors Th2 immune responses.

Interferon γ (IFN-γ) inducible protein-10 (IP-10) is a chemokine of the α family (CXC family). IP-10 is a gene product following stimulation of cells with IFN-γ. IP-10 does not posses the ELR motif that determines the biological significance of a particular α chemokine. IP-10 has an angiostatic potential, which is opposite to the angiogenic effect of other ELR-containing α chemokines that renders IP-10 a unique α chemokine family member. Tissue sources include endothelial cells, monocytes, fibroblasts, and keratinocytes. High levels of IP-10 transcripts are present in lymphoid organs. Monocytes, progenitor cells, and NK cells serve as target cells.

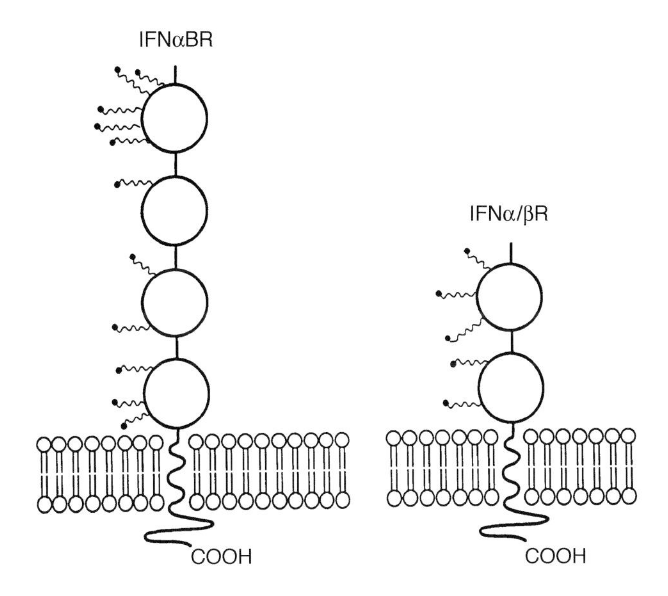

FIGURE 10.27 Three-dimensional crystal structure of recombinant murine interferon α. There is overall similarity of the basic polypeptide chain folding among all the interferon α and interferon β molecules from various sources.

Interferon(s) α (IFN-α) (Figure 10.27) contains at least 13 immunomodulatory 189-amino acid residue glycoproteins synthesized by macrophages and B cells that are able to prevent the replication of viruses, are antiproliferative, and are pyrogenic, inducing fever. IFN-α stimulates natural killer cells and induces expression of class I MHC antigens. It also has an immunoregulatory effect through alteration of antibody responsiveness. The 14 genes that encode IFN-α are positioned on the short arm of chromosome 9 in man. Polyribonucleotides as well as RNA or DNA viruses may induce IFN-α secretion. Recombinant IFN-α has been prepared and used in the treatment of hairy cell leukemia, Kaposi's sarcoma, chronic myeloid leukemia, human papilloma virus-related lesions, renal cell carcinoma, chronic hepatitis, and selected other conditions. Patients may experience severe flu-like symptoms as long as the drug is administered. They also have malaise, headache, depression, and supraventricular tachycardia and may possibly develop congestive heart failure. Bone marrow suppression has been reported in some patients.

Interferon β (IFN-β) is an antiviral, 20-kDa protein comprised of 187 amino acid residues. It is produced by fibroblasts and prevents replication of viruses. It has 30% amino acid sequence homology with interferon α. RNA or DNA viruses or polyribonucleotides can induce its secretion. The gene encoding it is located on chromosome 9 in man.

Interferon γ (IFN-γ) is a glycoprotein that is a 21- to 24-kDa homodimer synthesized by activated T lymphocytes and natural killer (NK) cells, causing it to be classified as a lymphokine. IFN-γ has antiproliferative and antiviral properties. It is a powerful activator of mononuclear phagocytes, increasing their ability to destroy intracellular microorganisms and tumor cells. It causes many types of cells to express class II MHC molecules and can also increase expression of class I. It facilitates differentiation of both B and T lymphocytes. IFN-γ is a powerful activator of NK cells and also activates neutrophils and vascular endothelial cells. It is decreased in chronic lymphocytic leukemia, lymphoma, and IgA deficiency, as well as in those infected with rubella, Epstein–Barr virus, and cytomegalovirus. Recombinant IFN-γ has been used for treatment of a variety of conditions including chronic lymphocytic leukemia, mycosis fungoides, Hodgkin's disease, and various other disorders. It has been found effective in decreasing synthesis of collagen by fibroblasts and might have potential in the treatment of connective tissue diseases. People receiving it may develop headache, chills, rash, or even acute renal failure. The one gene that encodes IFN-γ in man is found on the long arm of chromosome 12. Also termed immune or type II interferon.

γ interferon: See interferon γ.

Immune interferon is a synonym for interferon γ.

Macrophage-activating factor (MAF) is a lymphokine such as γ interferon that accentuates the ability of macrophages to kill microbes and tumor cells. A lymphokine that enhances a macrophage's phagocytic activity and bactericidal and tumoricidal properties.

MAF is the abbreviation for macrophage-activating factor.

Macrophage chemotactic factors (MCF) are cytokines that act together with macrophages to facilitate migration. Among these substances are interleukins and interferons.

Interferon γ receptor (Figure 10.28) is a 90-kDa glycoprotein receptor comprised of one polypeptide chain. The only cells found lacking this receptor are erythrocytes. It is encoded by a gene on chromosome 6q in man.

Tumor necrosis factor (TNG) family: Tumor necrosis factor (TNF) is responsible for lypopolysaccharide (LPS)-induced hemorrhagic necrosis of tumors in animals. It was subsequently identified as "cachectin," a factor responsible for wasting during parasitic infections or neoplasia. TNF cDNA cloning and purification of TNF protein showed that TNF is related structurally to lymphotoxin, a product of activated T lymphocytes, which is termed LT-α (formerly termed TNF-β). The designation TNF now refers to is what was formerly termed TNF-α. Both TNF and LT-α are synthesized by different cells. They occupy the same receptors and can produce similar biological activities. LT-α forms a heterocomplex with LT-β. This complex requires a different receptor (LT-β receptor) and performs biologic functions different from those of LT-α alone. TNF and LT-α are the first described members of the large family of ligands and receptors that also includes LT-β, the Fas/Apol receptor and its ligand, TRAIL/Apo2L and its receptor, CD40, and many others. They are all members of the tumor necrosis factor family.

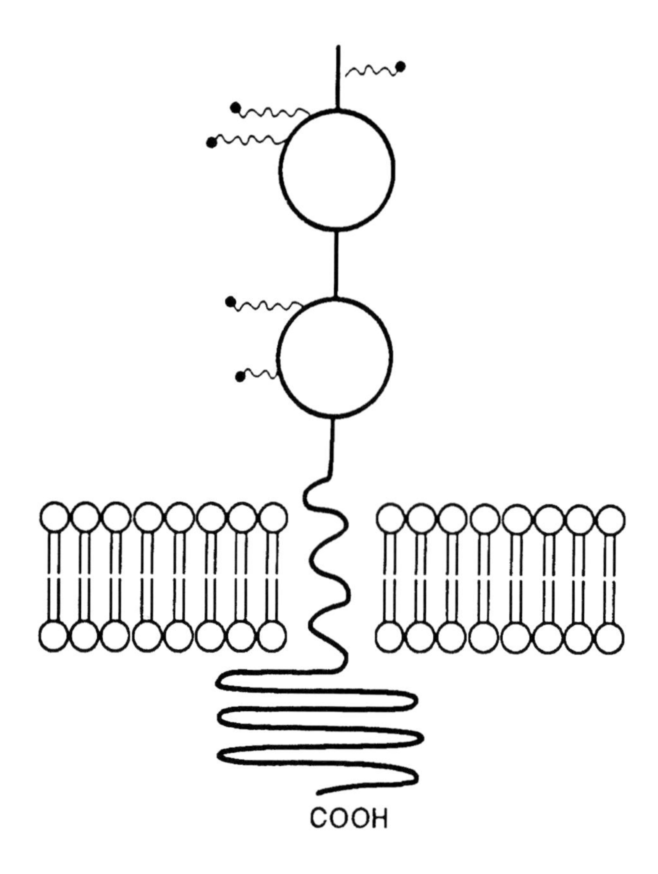

FIGURE 10.28 A preliminary three-dimensional structure of human interferon γ (recombinant form). Resolution 3.5 Å.

TNF is the abbreviation for tumor necrosis factor.

Cachectin is an earlier name for tumor necrosis factor α found in the blood serum and associated with wasting in these individuals. See tumor necrosis factor α (TNF-α).

Tumor necrosis factor α (TNF-α) (Figure 10.29) is a cytotoxic monokine produced by macrophages stimulated with bacterial endotoxin. TNF-α participates in inflammation, wound healing, and remodeling of tissue. TNF-α, which is also called cachectin, can induce septic shock and cachexia. It is a cytokine comprised of 157 amino acid residues. It is produced by numerous types of cells including monocytes, macrophages, T lymphocytes, B lymphocytes, NK cells, and other types of cells stimulated by endotoxin or other microbial products. The genes encoding TNF-α and TNF-β (lymphotoxin) are located on the short arm of chromosome 6 in man in the MHC region. High levels of TNF-α are detectable in the blood

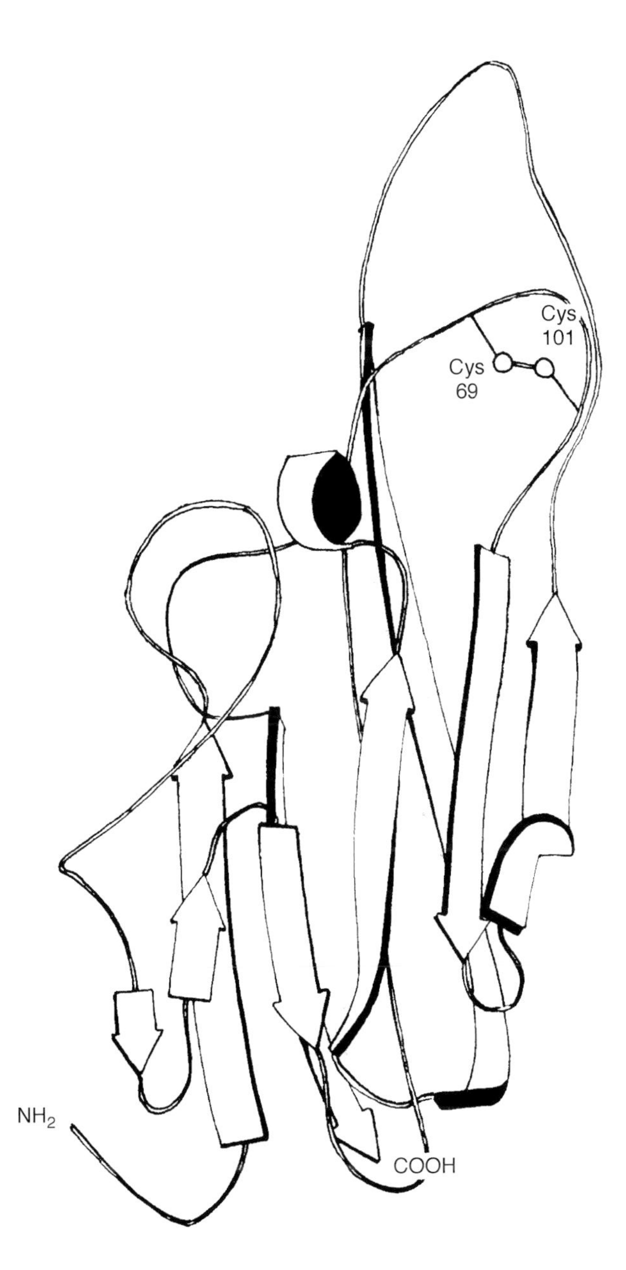

FIGURE 10.29 Molecular model of tumor necrosis factor-α (cachectin)(human recombinant form). Resolution 2.6 Å. The molecule exists as a trimer in which the three subunits are related by approximately three-fold symmetry.

circulation very soon following administration of endotoxin or microorganisms. The administration of recombinant TNF-α induces shock, organ failure, and hemorrhagic necrosis of tissues in experimental animals including rodents, dogs, sheep, and rabbits, closely resembling the effects of lethal endotoxemia. TNF-α is produced during the first 3 days of wound healing. It facilitates leukocyte recruitment, induces angiogenesis, and promotes fibroblast proliferation. It can combine with receptors on selected tumor cells and induce their lysis. TNF mediates the antitumor action of murine natural cytotoxic (NC) cells, which distinguishes their function from that of natural killer (NK) and cytotoxic T cells. TNF-α was termed cachectin because of its ability to induce wasting and anemia when administered on a chronic basis to experimental animals. Thus, it mimics the action in cancer patients and those with chronic infection with human immunodeficiency virus or other pathogenic microorganisms. It can induce anorexia, which may lead to death from malnutrition. Both TNF-α and TNF-β are cytotoxic for tumor cells but not for normal cells. TNF induces vascular endothelial cells to express new adhesion molecules, induces macrophages and endothelial cells to secrete chemokines, and facilitates apoptosis of target cells. Large amounts of TNF produced during severe infections may lead to systemic effects that include fever, synthesis of acute phase proteins by the liver, and cachexia. Very large quantities of TNF may induce intravascular thrombosis and shock. Of all its biological effects, the cytotoxic effects of TNF-α and its induction of apoptosis are the most important.

SODD (silencer of death domains): The tumor necrosis factor (TNF) receptor superfamily contains several members with homologous cytoplasmic domains known as death domains (DD). The intercellular DD are important in initiating apoptosis and other signaling pathways following ligand binding by the receptors. In the absence of ligand, DD-containing receptors are maintained in an inactive state. TNF Ri contains a cytoplasmic DD required for signaling pathways associated with apoptosis and NF-κB activation. Jiang et al. identified a widely expressed 60-kDa protein named SODD (silencer of death domains) associated with the DD of TNF RI and DR3. Overexpression of SODD suppresses TNF-induced cell death and NF-κB activation, demonstrating its role as a negative regulatory protein for these signaling pathways. TNF-induced receptor trimerization aggregates the DD of TNF RI and recruits the adapter protein TRADD. This in turn promotes the recruitment of the DD-containing cytoplasmic proteins FADD, TRAF2 and RIP to form an active TNF RI signaling complex. In contrast, SODD acts as a silencer of TNF RI signaling and does not interact with TRADD, FADD, or RIP. It is associated with the DD of TNF RI and maintains TNF RI in an inactive, monomeric state. TNF-induced aggregation of TNF RI promotes the disruption of the SODD–TNF RI complex. SODD does not interact with the DDs of other TNF receptor superfamily members such as Fas, DR4, DR5, or TNF RII. SODD association with TNF-RI may represent a general model for the prevention of spontaneous TNF signaling by other DD-containing receptors.

TACI (transmembrane activator and CAML-interactor) is a TNF-receptor family member that is one of the two major receptors for BLyS. Dendritic cells, B cells, and T cells all express this receptor which is believed to be important for receiving signals from BLyS.

TNF receptor-associated factors (TRAFs) are adapter molecules that interact with cytoplasmic domains in TNF receptor molecules. This family includes tumor necrosis factor (TNF) RII, lymphotoxin (LT)-β receptor, and CD40. A cytoplasmic motif in each of these receptors binds separate TRAFs that interact with other signaling molecules, resulting in transcription factor AP-1 and NF-$_{\kappa}$B transcription factors.

TRAFs are signal transducing molecules comprised of at least six members that bind to various TNF family receptors or TNFRs. They all share a TRAF domain in common and play a critical role as signal transducers between upstream members of the TNFR family and downstream transcription factors.

TRAIL (TNF-related apoptosis-inducing ligand) is a member of the TNF family of cytokines. Several TRAIL receptors have been identified, including decoy receptors that function to antagonize TRAIL-induced apoptosis.

BlyS is a TNF cytokine family molecule that is secreted by T lymphocytes and is critical in the formation of germinal centers and plasma cells and may be significant in maturation of dendritic cells.

TNF-related activation-induced cytokine (TRANCE) (RANK Ligand) is a member of the TNF family and is involved in regulating the function of dendritic cells and osteoclasts. RANK is the cell surface signaling receptor for TRANCE. Osteoprotegerin also binds TRANCE and serves as a decay receptor that counterbalances the effects of TRANCE.

Tumor necrosis factor β (TNF-β) is a 25-kDa protein synthesized by activated lymphocytes. It can kill tumor cells in culture, induce expression of genes, stimulate proliferation of fibroblasts, and mimic most of the actions of tumor necrosis factor α (cachectin). It participates in inflammation and graft rejection and was previously termed lymphotoxin. TNF-β and TNF-α have approximately equivalent affinity for TNF receptors. Both 55- and 80-kDa TNF receptors bind TNF-β. TNF-β has diverse effects that include killing of some cells and causing proliferation of others. It is the mediator whereby cytolytic T cells, natural killer cells, lymphokine-activated killer cells, and "helper-killer" T cells induce fatal injury to their targets. TNF-β and TNF-α have been suggested to play a role in AIDS, possibly contributing to its pathogenesis.

Tumor necrosis factor receptor (Figure 10.30 and Figure 10.31) is a receptor for tumor necrosis factor that is comprised of 461 amino acid residues and possesses an extracellular domain that is rich in cysteine. Cell surface receptors for tumor necrosis factor TNF α and tumor necrosis factor β TNF-β (LT) are expressed by most cell types. The two TNF receptors are designated TNF-R1, which mediates most biologic effects of tumor necrosis factor N (TNF RII). TNF receptors possess cysteine-rich extracellular motifs that include Fas and CD40.

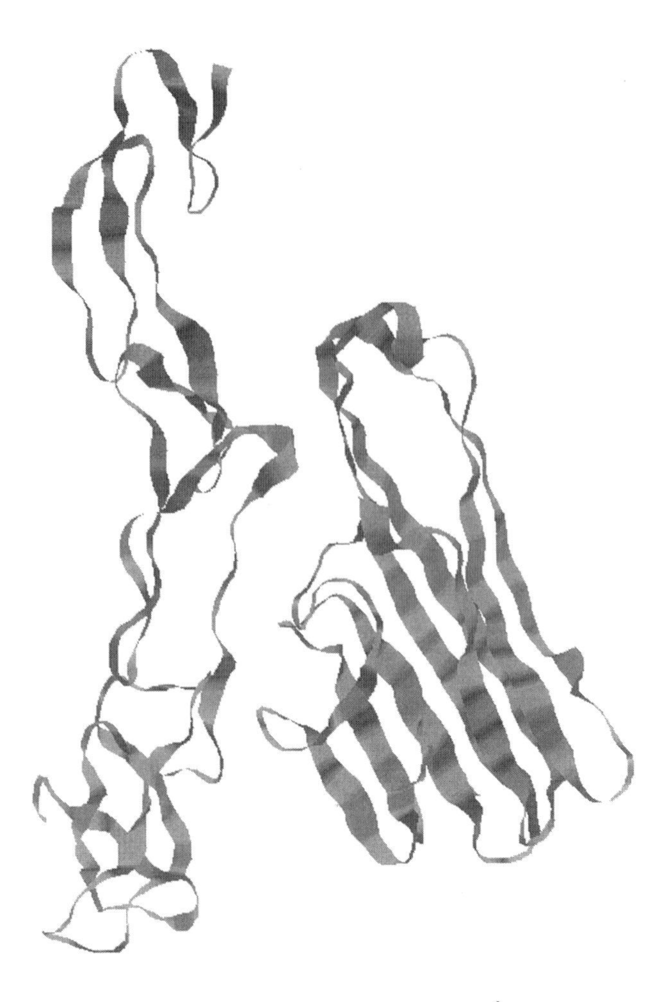

FIGURE 10.30 TNF receptor. Resolution 2.85 Å.

Tumor necrosis factor receptors (TNF receptors) interact with three structurally related ligands: TNF-α, lymphotoxin α (LTα), or TNF-β and LTβ. Two of the three receptors have been given the designation CD120a (type I, p55 or p60 receptor) and the other CD120b (type II, p75 or p80 receptor). The third one is designated the LTβ receptor (LTβ-R), the type III TNF receptor, or TNFR-RP. TNF receptors may be cell-bound or soluble. Both CD120a and CD120b may bind cell-bound and soluble forms of homotrimers of TNF-α and LTα. CD120b binds more effectively to cell bound TNF-α. LTβ-R binds only to heterotrimers of LTα and LTβ. CD120a, CD120b, and the LTβ-R represent single-tranmembrane type I proteins. Activation of CD120a in tissue cultured cells by antibody induces numerous TNF effects. CD120b is believed to be important in the immunoregulatory activities of the TNFs.

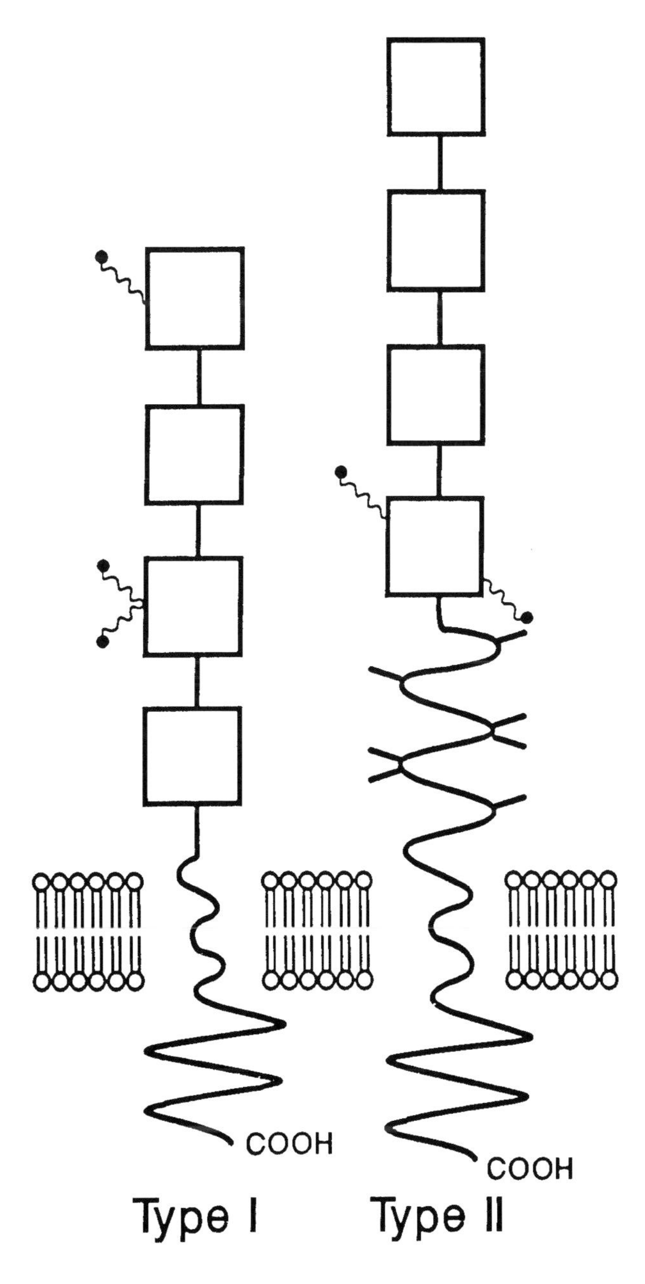

FIGURE 10.31 Schematic representation of TNF receptors. The two receptors for TNF are designated type I (CD120a) and type II (CD120b). The 55-kDa type I receptor and the 75-kDa type II receptor both bind TNF-α and TNF-β (lymphotoxin). These two receptors belong to the NGRF/TNFR superfamily. The extracellular domain contains four cys-rich repeat. The two TNF receptors have essentially no homology in their intracellular domains. Soluble forms of both receptors have been described in urine and serum. With the exception of erythrocytes and resting T cells, TNF receptors are found on most all types of cells. Whereas, the type Ip55 receptor is widely distributed on various cell types, the type Iip75 receptor appears confined to hematopoietic cells. Whereas there are three potential N-linked glycosylation sites in human P55 type I TNF receptors, there are two in the p75 type II form.

Lymphotoxin (LT) is a T lymphocyte lymphokine that is a heterodimeric glycoprotein comprised of a 5- and a 15-kDa protein fragment. This cytokine is inhibitory to the growth of tumors either *in vitro* or *in vivo,* and it also blocks chemical-, carcinogen-, or ultraviolet light-induced transformation of cells. Lymphotoxin has cytolytic or cytostatic properties for tumor cells that are sensitive to it. Approximately three quarters of the amino acid sequence is identical between human and mouse lymphotoxin. Human lymphotoxin has 205 amino acid residues, whereas the mouse variety has 202 amino residues. Lymphotoxin does not produce membrane pores in its target cells, such as those produced by perforin or complement, but it is taken into cells after it is bound to their surface and it subsequently interferes with metabolism. Lymphotoxin is also called tumor necrosis factor β (TNF-β). This T-cell cytokine is homologous to and binds to the same receptors as TNF. It is pro-inflammatory, activating both endothelial cells and neutrophils. I t is necessary for normal lymphoid organ development. A surface form of lymphotoxin on T cells is mainly a heterotrimer of one LTα subunit with two LTβ molecules (LT$\alpha_1\beta_2$). LTα and LTβ are related to tumor necrosis factor α (TNF-α), sharing sequence and structural characteristics in addition to a tight genetic linkage. TNF-α, LTα, and LT$\alpha_1\beta_2$ all bind to TNF receptor family molecules.

Lymphocytotoxin: See lymphotoxin.

Colony-stimulating factors (CSF) (Figure 10.32 and Figure 10.33) are glycoproteins that govern the formation, differentiation, and function of granulocytes and monomacrophage system cells. CSF promotes the growth, maturation, and differentiation of stem cells to produce progenitor cell colonies. They facilitate the development of functional end-stage cells. They act on cells through specific receptors on the target cell surface. T cells, fibroblasts, and endothelial cells produce CSF factors. Different colony-stimulating factors act on cell-line progenitors that include CFU-E (red blood cell precursors), GM-CFC (granulocyte-macrophage colony forming cells) (Figure 10.32), MEG-CFC (megakaryocyte-colony forming cells), EO-CFC (eosinophil-leukocyte colony forming cells), T cells, and B cells. Colony-stimulating factors promote the clonal growth of cells. They include granulocyte CSF, which is synthesized by endothelial cells, macrophages, and fibroblasts. It activates the formation of granulocytes and is synergistic with IL-3 in the generation of megakaryocytes and granulocytes-macrophages. Endothelial cells, T lymphocytes, and fibroblasts form granulocyte-macrophage CSF, which stimulates granulocyte and macrophage colony formation. It also stimulates megakaryocyte blast cells. Colony-stimulating factor-1 is produced by endothelial cells, macrophages, and fibroblasts, and induces the generation of macrophage colonies. Multi-CSF (interleukin-3) is produced by T lymphocytes

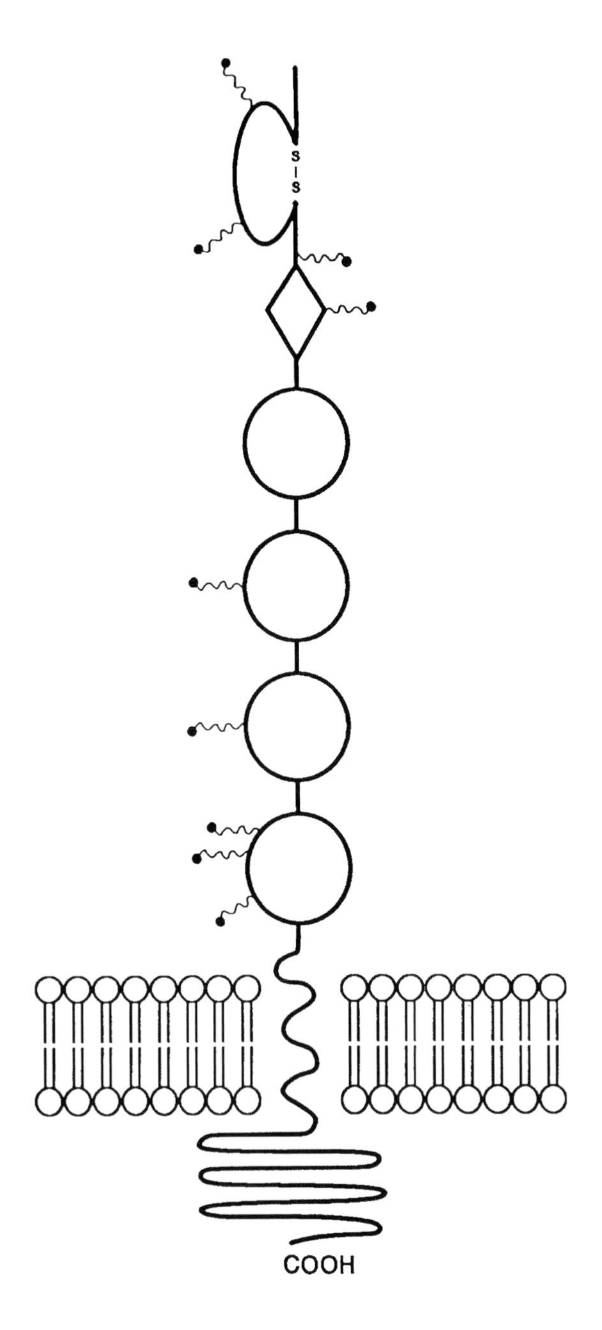

FIGURE 10.32 Schematic representation of the G-CSF receptor, which is comprised of an immunoglobulin domain, a hematopoietin domain, and three fibronectin III domains. There are two forms of the human receptor, i.e., the 25.1 form that has a C kinase phosphorylation site, and a second form in which the transmembrane region has been deleted. The mouse receptor, which shares 62.5% homology with the human receptor, is similar to the 25.1 form. The human G-CSF receptor has 46.3% sequence homology with IL-5 receptor's gp130 chain. The hematopoietin domain contains the binding site for G-CSF, yet proliferative signal transduction requires the membrane proximal 57 amino acids. Acute-phase protein induction mediated by G-CSF involves residues 57 to 96. G-CSF receptors are found on neutrophils, platelets, myeloid leukemia cells, endothelium, and placenta. The human form contains 9 potential N-linked glycosylation sites. It is believed that the receptor binds and mediates autophosphorylation of JAK-2 kinase.

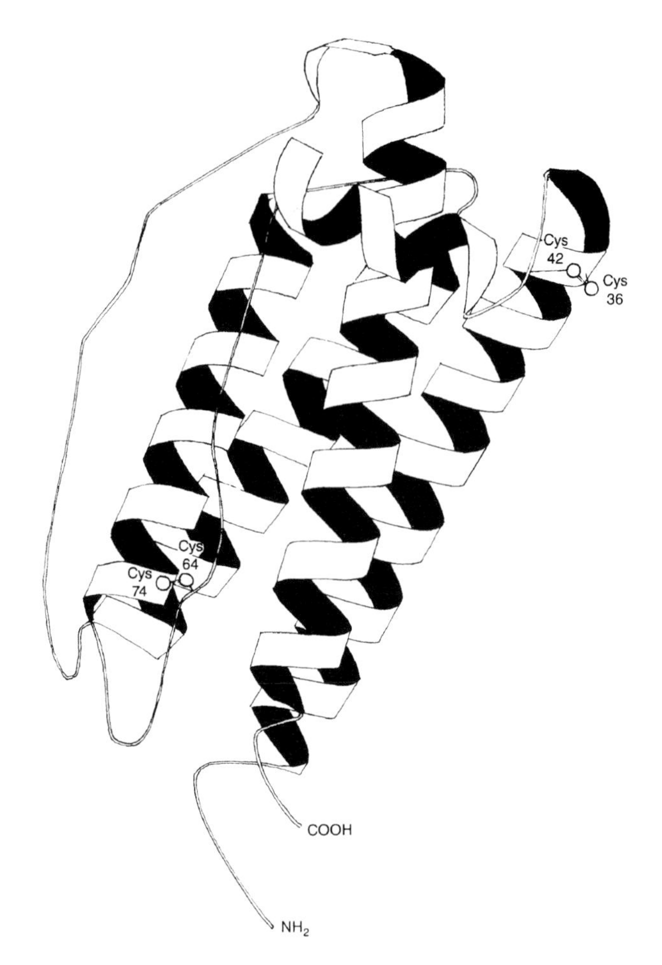

FIGURE 10.33 Granulocyte-stimulating factor (G-CSF).

and activates the generation of granulocytes, macrophages, eosinophils, and mast cell colonies. It is synergistic with other factors in activating hematopoietic precursor cells. Renal interstitial cells synthesize erythropoietin that activates erythroid colony formation.

Colony-stimulating factors (CSFs) are cytokines that facilitate the expansion and differentiation of progenitor cells in the bone marrow. They are critical for erythocyte, granulocyte, monocyte, and lymphocyte maturation. GM-CSF, c-kit ligand, IL-3, and IL-7 are examples of CSFs.

Granulopoietin is a 45-kDa glycoprotein produced by monocytes, which governs granulocyte formation in the bone marrow. Also called colony-stimulating factor.

An **inducer** is a substance that promotes cellular differentiation, such as colony-stimulating factors.

The **c-kit ligand** is a cytokine, also termed stem cell factor, that interacts with a tyrosine-kinase membrane receptor of pluripotent stem cells. The receptor, which contains a five-Ig domain extracellular structure, is encoded by the cellular oncogene, *c-kit*. Bone-marrow stromal cells such as fibroblasts, endothelial cells, and adipocytes produce a 27-kDa transmembrane form and a 24-kDa secreted form of a c-kit ligand. This cytokine alone apparently does not induce colony formation but is postulated to render stem cells reactive with other colony-stimulating factors.

CSF is an abbreviation for (1) colony-stimulating factor or (2) cerebrospinal fluid.

Granulocyte colony-stimulating factor (G-CSF) is a cytokine synthesized by activated T lymphocytes, macrophages, and endothelial cells at infection sites that causes bone marrow to increase the production and mobilization of polymorphonuclear neutrophils to replace those spent by inflammatory processes.

G-CSF is a biological response modifier that facilitates formation of granulocytes in the bone marrow. It was first licensed by the FDA in 1991 and may be useful to reactivate granulocyte production in the marrow of irradiated or chemotherapy-treated patients. The genes for G-CSF are found on chromosome 17. Endothelial cells, macrophages, and fibroblasts produce G-CSF, which functions synergistically with IL-3 in stimulating bone marrow cells. G-CSF induces differentiation and clonal extinction in certain myeloid leukemia cell lines. It promotes almost exclusively the development of neutrophils from normal hepatopoietic progenitor cells.

Granulocyte–macrophage colony-stimulating factor (GM-CSF) is a cytokine that participates in the growth and differentiation of myeloid and monocytic lineage cells which include dendritic cells, monocytes, macrophages, and granulocyte lineage cells. **GM-CSF** is a growth factor for hemopoietic cells that is synthesized by lymphocytes, monocytes, fibroblasts, and endothelial cells. It has been prepared in recombinant form to stimulate production of leukocytes in AIDS patients and to initiate hematopoiesis following chemotherapy of bone marrow recipients. Patients with anemia and malignant neoplasms might also derive benefit from GM-CSF administration. It is a cytokine that induces the proliferation, differentiation, and functional activation of hematopoietic cells. It has an established role in clinical medicine. GM-CSF is produced by T lymphocytes, macrophages, fibroblasts, and endothelial cells that have been stimulated by antigen, lectin, interleukin-1 (IL-1), lipopolysaccharide, tumor necrosis factor α, and phorbol ester. Human and mouse GM-CSFs contain 124 and 127 amino acids, respectively. The mature protein is preceded by a 17-amino acid leader sequence. GM-CSF has been shown by x-ray crystallography to combine a two-stranded antiparallel β sheet with an open bundle of four α helices. A single copy of 2.5 kb contains 4 exons and 3 introns. In the mouse, it has been mapped to chromosome 11 and in the human to the long arm of chromosome 5. It induces proliferation and differentiation for granulocyte, monocyte, and eosinophil progenitors and as an enhancer of the function of mature effector cells. GM-CSF enhances the function of granulocytes and macrophages in immune responses. Its biological effects are mediated through binding to specific cell surface receptors which may be either of low or high affinity on hematopoietic cells. *In vivo*, GM-CSF is a potent stimulator of hematopoiesis. Its administration is usually well tolerated but may be associated with bone pain and influenza-like symptoms including fever, flushing, malaise, myalgia, arthralgia, anorexia, and headache, but usually resolve with continued administration. Higher doses of GM-CSF may lead to capillary leak syndrome. GM-CSF has been used to support chemotherapy and to manage cytopenias associated with HIV infection. Low doses can elevate neutrophil counts. It has been used as a supportive agent to manage secondary infections in following chemotherapy for AIDS-related non-Hodgkin's lymphoma. GM-CSF has been used following autologous bone marrow transplantation to reduce neutropenia and decrease serious infection and the use of antibiotics. It has also been used to mobilize peripheral blood progenitor cells for collection by apheresis and subsequent transplantation.

CFU is an abbreviation for colony-forming unit.

CFU-GEMM is a colony-stimulating factor that acts on multiple cell lines which include erythroid cells, granulocytes, megakaryocytes, and macrophages. The pancytopenia observed in myelodysplasia and Fanconi's disease has been attributable to the total lack of CFU-GEMM.

Granulocyte–monocyte colony-stimulating factor is a cytokine synthesised by activated T lymphocytes, macrophages, stromal fibroblasts, and endothelial cells that leads to increased production of neutrophils and monocytes in the bone marrow. GM-CSF also activates macrophages and facilitates the differentiation of Langerhans cells into mature dendritic cells.

Macrophage colony-stimulating factor (M-CSF) (Figure 10.34 to Figure 10.36) facilitates growth, differentiation, and survival and serves as an activating mechanism for macrophages and their precursors. It is derived from numerous sources such as lymphocytes, monocytes, endothelial cells, fibroblasts, epithelial cells, osteoblasts, and myoblasts. X-ray crystallography of recombinant CSF reveals a structure in which four α helices are placed end to end in two bundles. Human and mouse M-CSF share 82% homology in the N-terminal 227 amino acids of the mature sequence, but there is only 47% homology in the remainder of the molecular structure. M-CSF derived from humans and from mice has been previously termed colony-stimulating factor (CSF-1). M-CSF is homodimeric

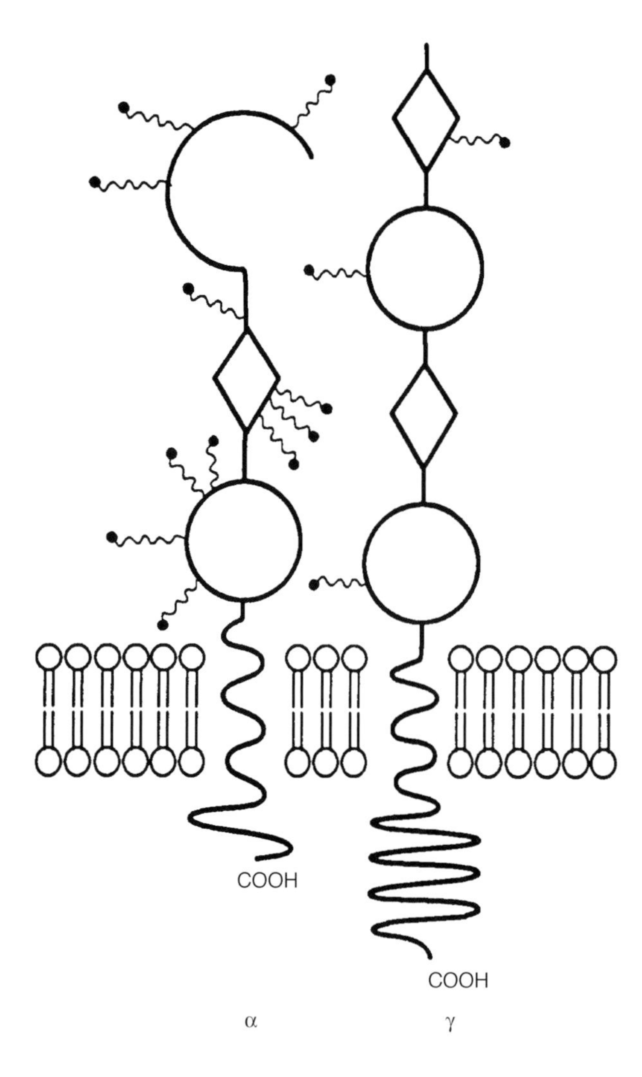

FIGURE 10.34 Schematic representation of a macrophage colony-stimulating factor.

and is secreted as an 80- to 100-kDa glycoprotein, or a 130- to 160-kDa chondroitin sulfate proteoglycan, or is expressed as a biologically active cell surface membrane 68- to 86-kDa glycoprotein. Both forms are present in the blood circulation. The cell surface form participates in local regulation, whereas the proteoglycan form may be sequestered to specific sites. The CSF-1 receptor mediates the effects of CSF-1. This receptor is expressed on osteoclasts and tissue macrophages and their precursors, as well as on embryonic cells, decidual cells, and trophoblasts. Circulating CSF-1 is postulated to be derived from endothelial cells that line small blood vessels.

Monocyte colony-stimulating factor (MCSF) is a cytokine that induces the production of monocytes from bone marrow precursor cells. It is synthesized by T lymphocytes, macrophages, endothelial cells, and stromal fibroblasts.

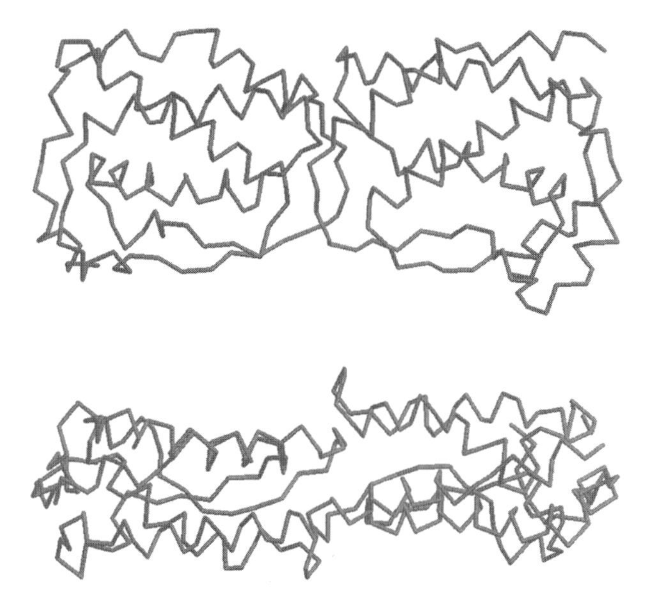

FIGURE 10.35 Human macrophage colony-stimulating factor. Resolution 2.5 Å.

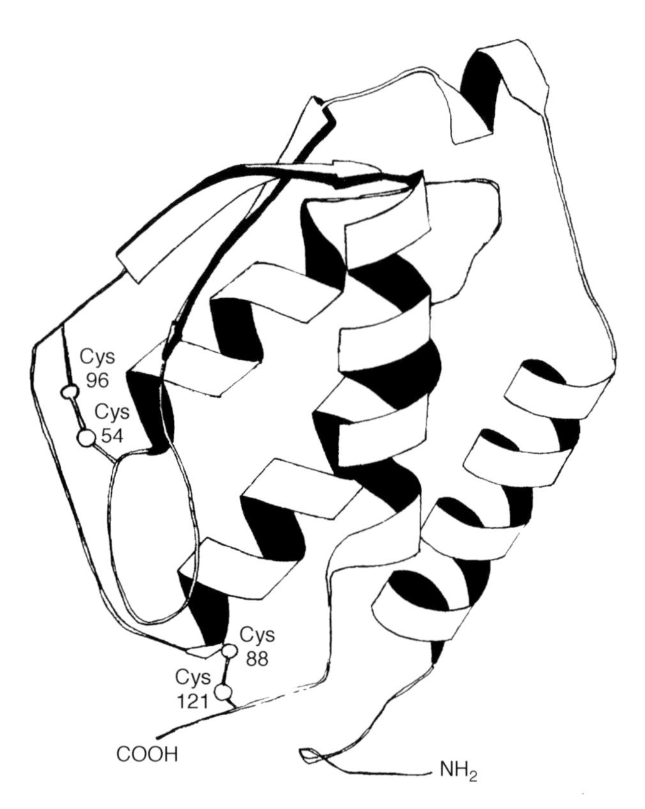

FIGURE 10.36 Human macrophage colony-stimulating factor (α form, soluble) (human protein recombinant form). Three-dimensional structure of dimeric human recombinant macrophage colony-stimulating factor. Resolution 2.5 Å.

L cell conditioned medium is a powerful growth factor for macrophages, termed macrophage colony-stimulating factor (MCSF), present in L-cell cultures.

Transfer factor (TF) is a substance in extracts of leukocytes that were as effective as viable lymphoid cells in transferring delayed-type hypersensitivity. The active principle is not destroyed by treatment with DNase or RNase. Originally described by H. S. Lawrence, transfer factor has been the subject of numerous investigations. Transfer factor is dialyzable and is less than 10 kDa. It has been separated on Sephadex®. Following demonstration of its role in humans, transfer factor was later shown to transfer delayed-type hypersensitivity in laboratory animals. It was shown to be capable of also transferring cell-mediated immunity as well as delayed-type hypersensitivity between members of numerous animal species. It also became possible to transfer delayed-type hypersensitivity across species barriers using transfer factor. Attempts to identify purified transfer factor have remained a major challenge. However, it has now been shown that transfer factor combines with specific antigen. Urea treatment of a solid-phase immunosorbent permits its recovery. Thus, specific transfer factor is generated in an animal that has been immunized with a specific antigen. T helper lymphocytes produce transfer factor. TF combines with T suppressor cells as well as with Ia antigen on B lymphocytes and macrophages. It also interacts with antibody specific for V-region antigenic determinants. It may be a fragment of the T-cell receptor for antigen. Transfer factor has been used as an immunotherapeutic agent for many years to treat patients with immunodeficiencies of various types. It produces clinical improvement in numerous infectious diseases caused by viruses and fungi. Transfer factor improves cell-mediated immunity and delayed-type hypersensitivity response, i.e., it restores decreased cellular immunity to some degree.

Insulin-like growth factors (IGFs) (Figure 10.37) consists of IGF-I and IGF-II which are prohormones with M_T of 9K and 14K, respectively. IGF-I is a 7.6-kDa side-chain polypeptide hormone that resembles proinsulin structurally. It is formed by the liver and by fibroblasts. IGF-I is the sole effector of growth hormone activity. It is a primary growth regulator that is age dependent. IGF is expressed in juvenile life and is detectable in the circulating blood plasma. Levels of IGF-I increase in the circulation during juvenile life but decline after puberty. Circulating IGFs are not free in the plasma but are associated with binding proteins that may have the function of limiting the bioavailability of circulating IGFs, which may be a means of controlling growth factor activity. IGF-II is present mainly during the embryonic and fetal stages of mammalian development in various tissues. It is also present in the circulating plasma in association with binding proteins, reaching its highest level in the fetal circulation and declining following birth. IGF-II is important for growth of the whole organism.

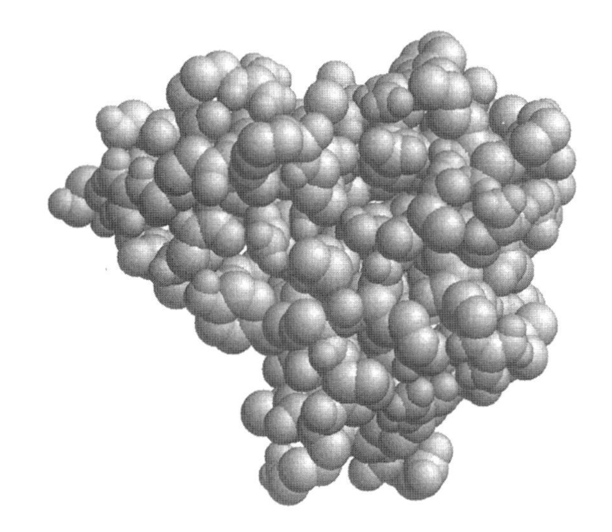

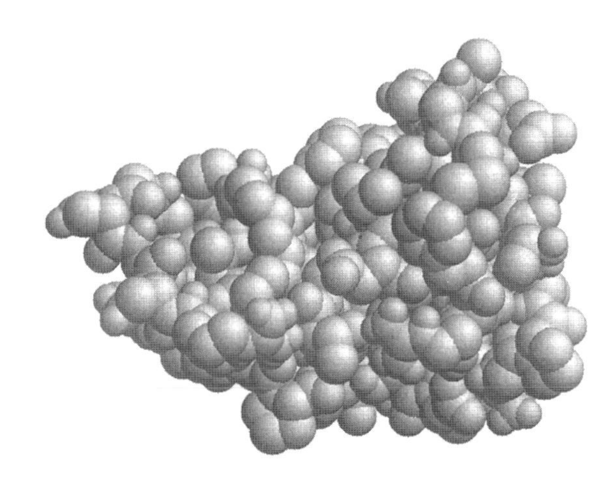

FIGURE 10.37 Insulin-like growth factors (IGFs).

TGF (transforming growth factor[s]) are polypeptides produced by virus-transformed 3T3 cells that induce various cells to alter their phenotype.

Transforming growth factor-α (TGF-α) is a polypeptide produced by transformed cells (Figure 10.38). It shares approximately one third of its 50-amino acid sequence with epidermal growth factor (EGF). TGF-α has a powerful stimulatory effect on cell growth and promotes capillary formation. It is a 5.5-kDa protein comprised of 50 amino acid residues (human form). This polypeptide growth factor induces proliferation of many types of epidermal and epithelial cells. It facilitates growth of selected transformed cells.

There are five **TGF-βs (transforming growth factor-βs)** that are structurally similar in the C-terminal region of the

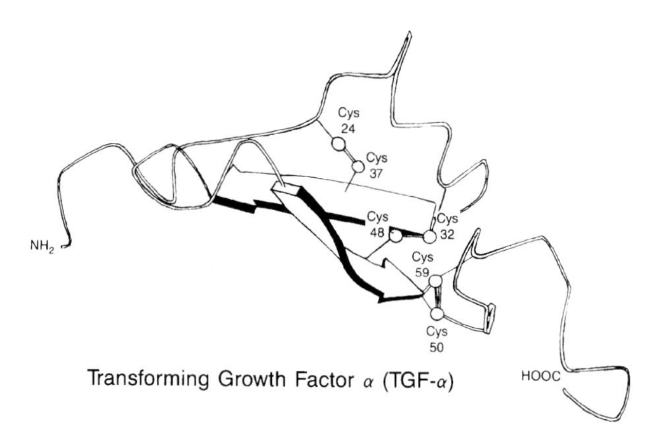

FIGURE 10.38 Transforming growth factor-α (NMR minimized average structure). The structure was determined by a combination of two-dimensional NMR distance geometry and restrained molecular dynamics using the program (GROMOS). United atoms were used for all nonpolar hydrogen atoms and so are not included in this coordinate entry.

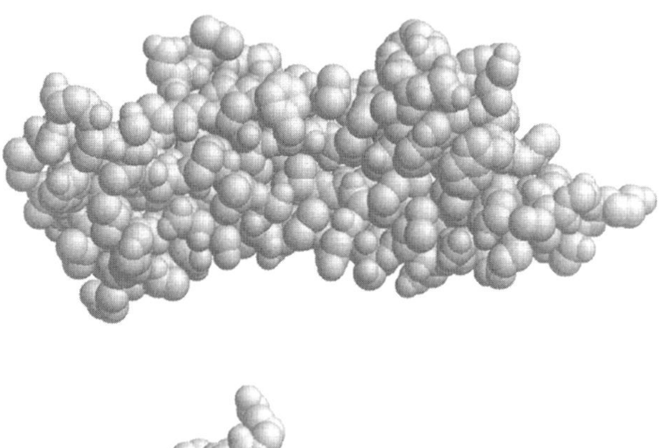

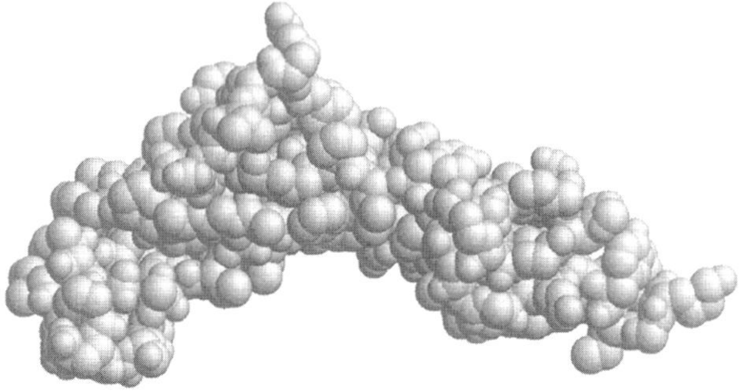

FIGURE 10.39 Human TGF-β2. Resolution 1.8 Å.

protein. They are designated TGF-β1 through TGF-β5 and have similar functions with respect to their regulation of cellular growth and differentiation. After being formed as secretory precursor polypeptides, TGF-β1, TGF-β2 (Figure 10.39), and TGF-β3 molecules are altered to form a 25-kDa homodimeric peptide. The ability of TGF-β to regulate growth depends on the type of cell and whether or not other growth factors are also present. It also regulates deposition of extracellular matrix and cell attachment to it. It induces fibronectin, chondroitin/dermatin sulfate proteoglycans, collagen, and glycosaminoglycans. TGF-β also promotes the formation and secretion of protease inhibitors. It has been shown to increase the rate of wound healing and induce granulation tissue. It also stimulates proliferation of osteoblasts and chondrocytes. TGF-β inhibits bone marrow cell proliferation and also blocks interferon α-induced activation of natural killer (NK) cells. It diminishes IL-2 activation of lymphokine activated killer cells. TGF-β decreases cytokine-induced proliferation of thymocytes and also decreases IL-2-induced proliferation and activation of mature T lymphocytes. It inhibits T cell precursor differentiation into cytotoxic T lymphocytes. TGF-β may reverse the activation of macrophages by preventing the development of cytotoxic activity and superoxide anion formation that is needed for antimicrobial effects. In addition to suppressing macrophage activation, TGF may diminish MHC class II molecule expression. It also decreases Fcε receptor expression in allergic reactions. TGF-β has potential value as an immunosuppressant in tissue and organ transplantation. It may protect bone marrow stem cells from the injurious effects of chemotherapy. It may also have use as an antiinflammatory agent based on its ability to inhibit the growth of both T and B cells. TGF-β has potential as a possible treatment for selected autoimmune diseases. It diminishes myocardial damage associated with coronary occlusion, promotes wound healing, and may be of value in restoring collagen and promoting formation of bone in osteoporosis patients. TGF-β is produced by activated T lymphocytes, mononuclear phagocytes, as well as other cells. Its main effect is to inhibit proliferation and differentiation of T cells, inhibit macrophage activation, and counteract proinflammatory cytokines.

Transforming growth factor β (TGF-β): See TGF-β.

Leukocyte inhibitory factor (LIF) is a lymphokine that prevents polymorphonuclear leukocyte migration. T lymphocytes activated *in vitro* may produce this lymphokine, which can interfere with the migration of polymorphonuclear neutrophils from a capillary tube, as observed in a special chamber devised for the laboratory demonstration of this substance. Serine esterase inhibitors inhibit LIF activity, although they do not have this effect on macrophage migration inhibitory factor (MIF). This inhibitor is released by normal lymphocytes stimulated with the lectin concanavalin A or by sensitized lymphocytes challenged with the specific antigen. LIF is a 65- to 70-kDa protein.

Leukocyte migration inhibitory factor: See leukocyte inhibitory factor.

Leukemia inhibitory factor (LIF) (Figure 10.40 and Figure 10.41) is a lymphoid factor that facilitates maintenance of embryonic stem cells through suppression of spontaneous generation. It also induces mitogenesis of selected cell

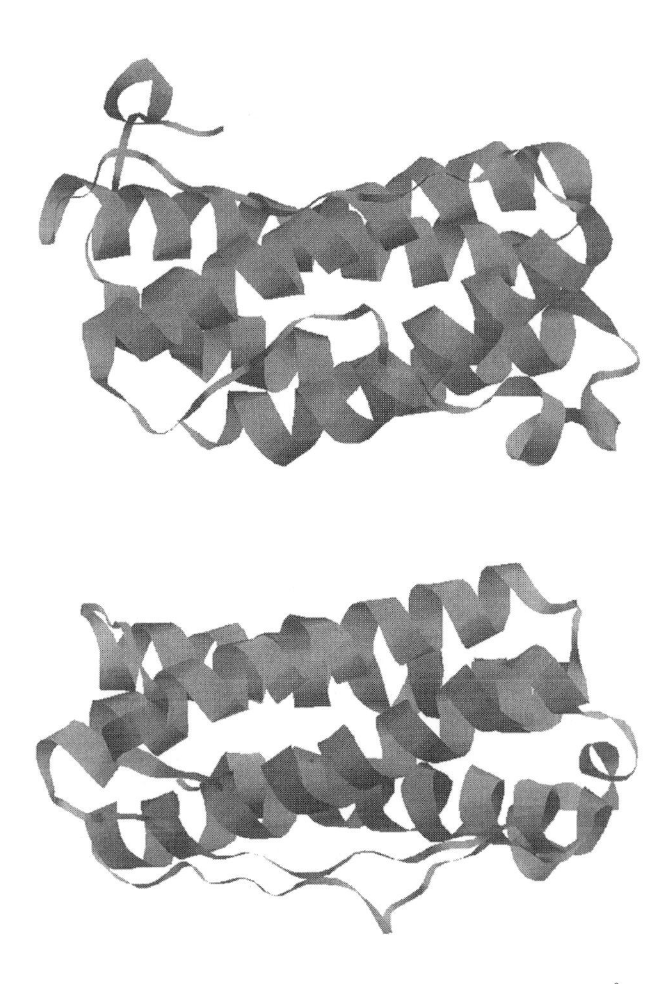

FIGURE 10.40 Leukemia inhibitory factor. Resolution 2.0 Å.

lines, stimulation of bone remodeling, facilitation of megakaryocyte formation *in vivo,* and suppression of cellular differentiation in culture. The recombinant form is a 20-kDa protein comprised of 180 amino acid residues. This cytokine has pleitropic activities. LIF is a member of the IL-6 family of cytokines which include oncostatin m, ciliary neurtrophic factor (CNTF), interleukin-11, interleukin-6, and cardiotrophin 1. The receptors for these cytokines consist of a cytokine-specific ligand-binding chain in the shared gp 130 transducer chain. Two signal-transducing pathways downstream of gp130 include the janus kinase (JA)/signal transducer and activator of transcription (STAT) and the Ras/mitogen-activated protein kinase (MAPK) pathway. LIF can induce the same acute-phase proteins has IL-6. It is released at local injury cite where monocytes and polymorphic nuclear cells are recruited. LIF can induce stimulation of proliferation of cancer cells. It is also modulates tumor cell capacity to adhere to matrix components.

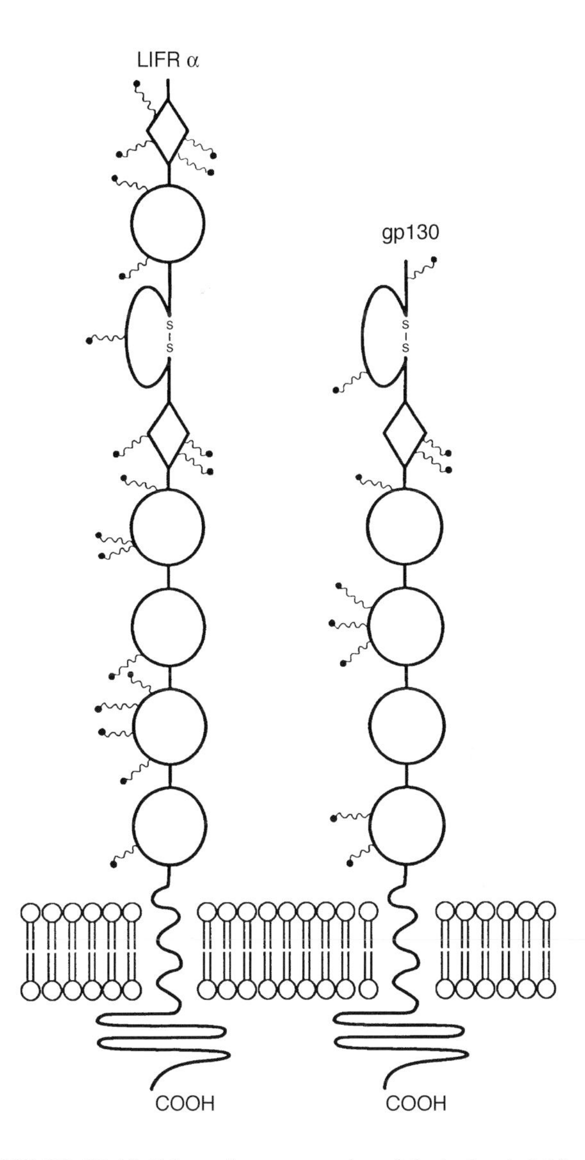

FIGURE 10.41 Schematic representation of the leukemia inhibitory factor.

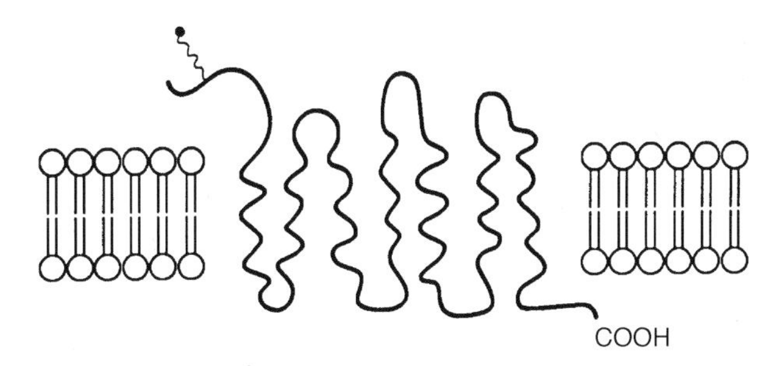

FIGURE 10.42 Schematic representation of MIP-1α.

Macrophage inflammatory protein-1α (MIP-1α) (Figure 10.42) is an endogenous fever-inducing substance that binds heparin and is resistant to cyclooxygenase inhibition. Macrophages stimulated by endotoxin may secrete this protein, termed MKP-1, which differs from tumor necrosis factor (TNF) and IL-1 as well as other endogenous pyrogens because its action is not associated with prostaglandin synthesis. It appears indistinguishable from hematopoietic stem cell inhibitor and may function in growth regulation of hematopoietic cells.

Macrophage inflammatory protein 1α (MIP-1α) is a chemokine of the β family (CC family). It has a broad spectrum of biological activities that include prostaglandin-independent pyrogenic activity, possible participation in wound healing, monocyte chemotaxis, and suppression of immature bone marrow stem and progenitor cells. Significantly, MIP-1α has an HIV-suppressive effect. Tissue sources include fibroblasts, monocytes, lymphocytes, neutrophils, eosinophils, smooth muscle cells, mast cells, platelets, and bone marrow stromal cells, among many other cell types. T lymphocytes, basophils, hematopoietic precursor cells, monocytes, eosinophils, neutrophils, mast cells, natural killer (NK) cells, and dendritic cells, among many other cell types, are target cells.

MIP-1 (macrophage inflammatory protein-1-α): See macrophage inflammatory protein.

MIP-1α receptor is a seven-membrane-spanning structure. There is 32% homology between the receptor for MIP-1α and those for IL-8. A calcium flux occurs when MIP-1α binds to its receptor. Murine T cells, macrophages, and eosinophils have been shown to express a receptor for MIP-1α. The human MIP-1α receptor has one potential N-linked glycosylation site. The gene for the MIP-1α receptors in humans is found on chromosome 3p21.

MIP-1β is a chemokine produced by monocytes and lymphocytes and stimulates growth of myelopoietic cells and promotes leukocyte chemoattraction. This molecule belongs to the CC family of chemokine–intercrine cytokines and is related to MIP-1α. Human and murine forms of MIP-1β share 70% homology. T lymphocytes, B cells, and macrophages represent sources of this chemokine. Human MIP-1β chemoattracts memory T lymphocytes. The major classes of cytokine receptors are arranged according to their corresponding superfamilies.

Macrophage inflammatory protein 1β (MIP-1β) is a chemokine of the β family (CC family). It shares 70% homology with MIP-1α. Although the two molecules resemble one another structurally, they are significantly different in functions. Unlike MIP-1α, MIP-1β does not activate neutrophils. Also unlike MIP-1α, which inhibits early hematopoietic progenitor growth, MIP-1β potentiates it. Both MIP-1α and MIP-1β have a synergistic HIV-suppressive effect. Tissue sources include monocytes, fibroblasts, T lymphocytes, B lymphocytes, neutrophils, smooth muscle cells, mast cells, and selected tumor cell lines. Monocytes, T lymphocytes, hematopoietic precursor cells, and basophils are target cells.

Macrophage inflammatory protein-2 (MIP-2) is a chemokine of the α family (CXC family). The MIP-2 class is comprised of MIP-2α, also termed GRO-β gene product, and MIP-2β, the GRO-γ gene product. MIP-2 has a role in anti-GBM antibody-induced glomerulonephritis in mice. Anti-MIP-2 antibody injection 30 min before anti-GBM antibody effectively decreases neutrophil influx and PAS positive deposits containing fibrin. Tissue sources include mast cells, cardiac myocytes, mesangial cells, alveolar macrophages, epidermal cells, and human nasal and bronchial epithelium. Neutrophils, basophils, and epithelial cells are the target cells.

Cytokines play a significant role in the homeostasis of normal tissues and cells as well as in the induction and perpetuation of pathologic processes. Thus, much pharmaceutical industry research has been aimed at blocking the synthesis or action of one or more cytokines. Various methods have been tested to block cytokines. Studies have revealed that suppression of transcription of some cytokines can render experimental animals susceptible to life-threatening inflammatory conditions. Thus, prolonged treatment with a cytokine-blocking agent might produce pathologic changes. Modulating cytokines as therapeutic procedures must be approached with caution and require further study. Nevertheless, cytokines are major therapeutic targets in many human diseases. Thus, ways are being sought to modulate them to produce a desired therapeutic effect. The four major approaches to modulation of cytokines include (1) inhibition cytokine synthesis, (2) inhibition of cytokine release, (3) inhibition of cytokine action, or (4) inhibition of cytokine intracellular signaling pathways.

Therapeutic cytokines that include γ interferon, α interferon, erythropoietin, and colony-stimulating factor are already being used clinically. In the future, cytokine modulators will hopefully be used to downregulate the action of selected cytokines. One such approach would be the therapy of septic shock in which antiendotoxin or anticytokine agents might be used for therapeutic purposes. Rheumatoid arthritis and other chronic diseases may also be treated by anticytokine therapies. Approaches will probably include blocking single cytokines by cloned chimerized antibodies or soluble receptor–antibody complexes to diminish tissue pathology. It remains to be determined whether pathologic changes induced by cytokines can be diminished by the use of agents that neutralize or inhibit synthesis of individual cytokines. Nevertheless, blocking the pathology of many diseases may depend upon targeting specific cytokines in the future.

Platelet factor 4 (PF4) is a chemokine of the α family (CXC family). It is a heat-stable heparin-binding protein that is stored together with fibronectin, fibrinogen, thrombospondin, von Willebrand factor, and β-TG in the α-granules of platelets. PF4 is secreted from stimulated and aggregated platelets. It is a megakaryocyte differentiation marker and is a negative autocrine regulator of human megakaryocytopoiesis. A related protein termed low-affinity platelet factor 4 (LA-PF4) shares 50% homology with PF4, but unlike PF4 is an active mitogenic and chemotactic agent that fails to bind heparin. PF4 interacts with a carrier molecule, a 53-kDa proteoglycan, *in vivo*. Platelets and megakaryocytes represent the tissue source. Fibroblasts, platelets, adrenal microvascular pericytes, mast cells, basophils, and megakaryocytes are the target cells. PF4 levels increase during cardiopulmonary bypass surgery, in arterial thrombosis, following surgery, in acute myocardial infarction, during acute infections, and during inflammation. Its sequence resembles that of β-thromboglobulin (βTG) and neutrophil activating peptide-2 (NAP-2). PF4 activities include heparin binding, inhibition of angiogenesis, induction of ICAM-1 on endothelial cells, promotion of neutrophil adhesion to endothelium, increased fibrin, fiber formation, inhibition of other chemotactic factor effects, *in vivo* recruitment of neutrophils, fibroblasts migration during wound repair, and reversal of Con-A-induced suppression of lymphocyte activity.

Platelet-activating factor receptor (PAFR) is a member of the G-protein-coupled receptor family, the chemokine receptor branch of the rhodopsin family. Tissue sources include HL-60 granulocyte cDNA library. PAFR's classic ligand is the proinflammatory lipid PAF. COS-7 cells expressing PAFR possess a single high-affinity binding site for PAF. Transcript #1 is ubiquitous and is expressed on peripheral leukocytes, whereas transcript #2 is expressed in placenta, heart, and lung.

Single cysteine motif-1 (SCM-1) is a member of the γ family of chemokines (C family). cDNA clones derived from human peripheral blood mononuclear cells stimulated with PHA encode SCM-1, which is significantly related to the α and β chemokine. It has only the second and fourth of four cysteines conserved in these proteins. It is 60.5% identical to lymphotactin. SCM-1 and lymphotactin are believed to represent human and mouse prototypes of a γ (C) chemokine family. It is expressed by human T cells and spleen.

Stromal cell-derived factor-1 (SDF-1) is an α family chemokine (CXC family). The stromal cell line PA6 synthesizes SDF-1. It promotes proliferation of stromal cell–dependent pre-B-cell of the stromal cell–dependent pre-B cell clone, DW34. Alternative splicing of the *SDF-1* gene yields SDF-1α and SDF-1β. SDF-1 is believed to be the natural ligand for the LESTR receptor. It is expressed by stromal cells, bone marrow, liver tissue, and muscle.

Osteoclast-activating factor (OAF) is a lymphokine produced by antigen-activated lymphocytes that promotes bone resorption through activation of osteoclasts. Besides lymphotoxin produced by T cells, interleukin-1, tumor necrosis factor, and prostaglandins synthesized by macrophages also have OAF activity. Osteoclast-activating factors might be responsible for the bone resorption observed in multiple myeloma and T cell neoplasms.

Biological response modifiers (BRM) include a broad spectrum of molecules, such as cytokines, that alter the immune response. They include substances such as interleukins, interferons, hematopoietic colony-stimulating factors, tumor necrosis factor, B-lymphocyte growth and differentiating factors, lymphotoxins, and macrophage-activating and chemotactic factors, as well as macrophage inhibitory factor, eosinophil chemotactic factor, osteoclast activating factor, etc. BRM may modulate the immune system of the host to augment antitumor defense mechanisms. Some have been produced by recombinant DNA technology and are available commercially. An example is α interferon used in the therapy of hairy cell leukemia.

BRMs is an abbreviation for biological response modifiers.

Keratinocyte growth factor (KGF) is a 19-kDa protein comprised of 163 amino acid residues that binds heparin and facilitates the growth of keratinocytes and other epithelial cells that comprise 95% of the epidermis.

11 The Complement System

Throughout the ages man has been fascinated and, at times, obsessed by the marvelous, mysterious, and even baffling qualities of the blood. In 1889, Hans Buchner described a heat-labile bactericidal principle in the blood which was later identified as the complement system. In 1894, Jules Bordet working at the Pasteur Institute in Metchnikoff's laboratory discovered that the lytic or bactericidal action of freshly drawn blood, which has been destroyed by heating, was promptly restored by the addition of fresh, normal, unheated serum. Paul Ehrlich called Bordet's alexine "*das Komplement*." In 1091, Bordet and Gengou developed the complement fixation test to measure antigen–antibody reactions. Ferrata, in 1907, recognized complement to be a multiple component system, a complex of protein substances of mixed globulin composition present in normal sera of many animal species.

The classical pathway of complement activation was described first by investigators using sheep red blood cells sensitized with specific antibody and lysed with guinea pig or human complement. In addition to immune lysis, complement has many other functions and is important in the biological amplification mechanism that is significant in resistance against infectious disease agents. The mechanism of action of complement in the various biological reactions in which it participates has occupied the attention of a host of investigators. In 1954, Pillemer et al. suggested the existence of a nonantibody-dependent protein in the serum which is significant for early defense of the host against bacteria and viruses. This protein was named properdin. It acts in combination with certain inorganic ions and complement components that make up the so-called properdin system, which constitutes a part of the natural defense mechanism of the blood.

Multiple plasma proteins may be activated during inflammation. Immune complexes activate the classic pathway of complement whereas bacterial products activate the alternative pathway without participation by specific antibody. Many antimicrobial effects are produced by complement. C5a, C5b67, and C3a induce chemotaxis of leukocytes. C3b has opsonic properties. The membrane attack complex leads to lysis of bacterial cells. Complement also facilitates the antimicrobial effects of PMNs and macrophages through the alternative pathway (Figure 11.1).

Complement (C) is a system of 20 soluble plasma and other body fluid proteins, together with cellular receptors for many of them and regulatory proteins found on blood and other tissue cells. These proteins play a critical role in aiding phagocytosis of immune complexes, which activate the complement system. These molecules and their fragments, resulting from the activation process, are significant in the regulation of cellular immune responsiveness. Once complement proteins identify and combine with target substance, serine proteases are activated. This leads ultimately to the assembly of C3 convertase, a protease on the surface of the target substance. The enzyme cleaves C3, yielding a C3b fragment that is bound to the target through a covalent linkage. C3b or C3bi bound to phagocytic cell surfaces become ligands for C3 receptors, as well as binding sites for C5. The union of C5b with C6, C7, C8, and C9 generates the membrane attack complex (MAC) which may associate with the lipid bilayer membrane of the cell to produce lysis, which is critical in resistance against certain species of bacteria. The complement proteins are significant, nonspecific mediators of humoral immunity. Multiple substances may trigger the complement system. There are two pathways of complement activation, one designated the classical pathway, in which an antigen, e.g., red blood cell, and antibody combine and fix the first subcomponent designated C1q. This is followed in sequence: C1qrs, 4,2,3,5,6,7,8,9, to produce lysis. The alternative pathway does not utilize C1, 4, and 2 components. Bacterial products such as endotoxin and other agents may activate this pathway through C3. There are numerous biological activities associated with complement besides immune lysis. These include the formation of anaphylatoxin, chemotaxis, opsonization, phagocytosis, bacteriolysis, hemolysis, and other amplification mechanisms.

The **classic pathway of complement** (Figure 11.2) is a mechanism to activate C3 through participation by the serum proteins C1, C4, and C2. Either IgM or a doublet of IgG may bind the C1 subcomponent C1q. Following subsequent activation of C1r and C1s, the two C1s substrates, C4 and C2, are cleaved. This yields C4b and C2a fragments that produce C4b2a, known as C3 convertase. It activates opsonization, chemotaxis of leukocytes, increased permeability of vessels, and cell lysis. Activators of the classic pathway include IgM, IgG, staphylococcal protein A, C-reactive protein, and DNA. C1 inhibitor blocks the classical pathway by separating C1r and C1s from C1q. C4-binding protein also blocks the classical

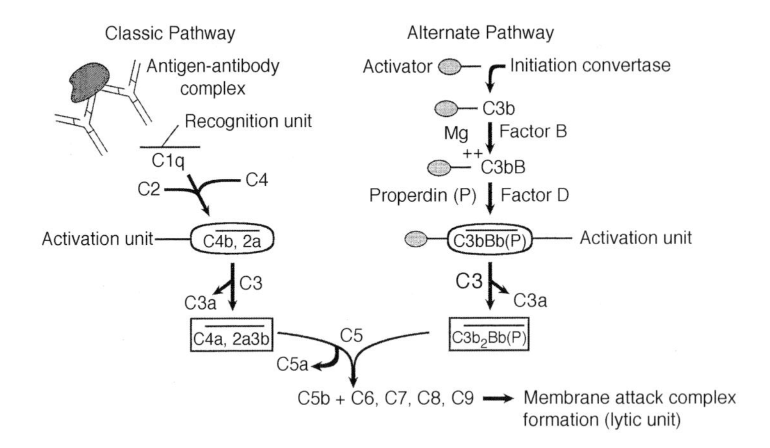

FIGURE 11.1 The classical and alternative pathways of the complement system.

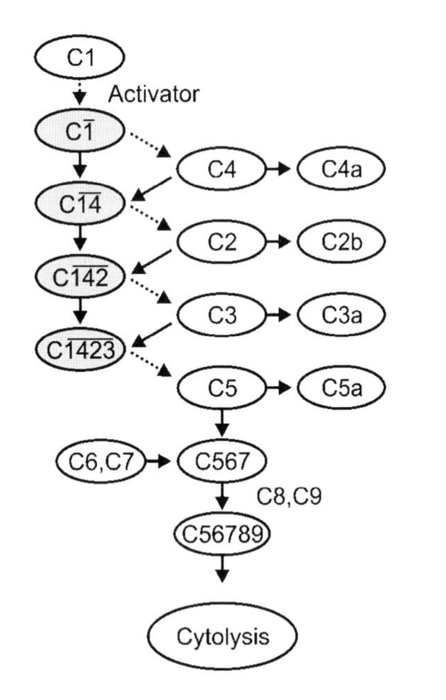

FIGURE 11.2 The classical pathway of complement activation.

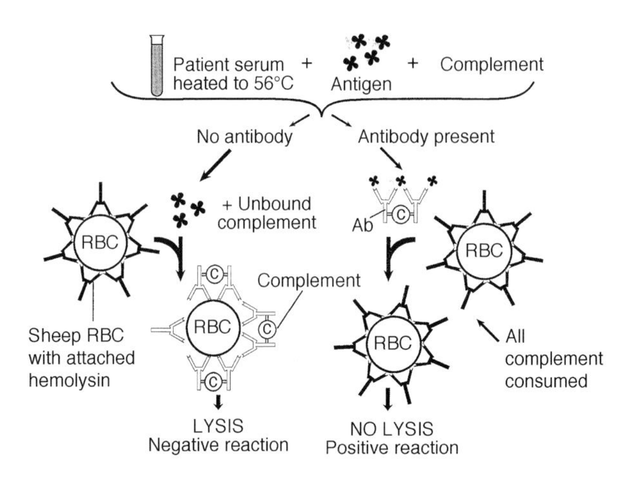

FIGURE 11.3 Complement split products in an inflammatory response.

pathway by linking to C4b, separating it from C2a, and permitting factor I to split the C4b heavy chain to yield C4bi, which is unable to unite with C2a, thereby inhibiting the classical pathway.

The first complement component designated C1 consists of a complex of three separate proteins designated C1q, C1r, and C1s. The formation of an active fragment following proteolysis of a component such as C3 is designated by a lowercase letter such as C3a and C3b (Figure 11.3). Active fragments degrade either spontaneously or by the action of serum proteases. Inactive fragments are designated by i, such as iC3b. The interaction of antibody with antigen forms antigen–antibody complexes that initiate the classical pathway. One molecule each of C1q and of C1s together with two molecules of C1r comprise the C1 complex or **recognition unit**. C1q facilitates recognition unit binding to cell surface antigen–antibody complexes. To launch the classical complement cascade, C1q must link to two IgG antibodies through their Fc regions. By contrast, one pentameric IgM molecule attached to a cell surface may interact with C1q to initiate the classical pathway. Binding of C1q activates C1r to become activated $\overline{C1r}$. This in turn activates C1s.

To create the **activation unit**, $\overline{C1s}$ splits C4 to produce C4a and C4b and C2 to form C2a and C2b. The ability of a single recognition unit to split numerous C2 and C4 molecules represents an amplification mechanism in the cascade. The union of C4b and C2a produces C4b2a which is known as **C3 convertase**. C4b2a binds to the cell membrane and splits C3 into C3a and C3b fragments. The ability of C4b2a to cleave multiple C3 molecules represents another amplification mechanism. The interaction of C3b with C4b2a bound to the cell membrane produces a complete activation unit, C4b2a3b, which is termed C5 convertase. It splits C5 into C5a and C5b fragments and represents yet another amplification mechanism because a single molecule of C5 convertase may cleave multiple C5 molecules.

Classical pathway: See classical complement pathway.

Alexine (or alexin) is a historical synonym for complement. Hans Buchner (1850–1902) found that the bactericidal property of cell-free serum was destroyed by heat. He named the active principle alexine (which in Greek means to ward off or protect). Jules Bordet (1870–1961) studied this thermolabile and nonspecific principle in blood serum that induces the lysis of cells, e.g., bacterial cells, sensitized with specific antibody. Bordet and Gengou went on to discover the complement fixation reaction. (Bordet received the Nobel Prize for Medicine in 1919.)

The **complement system** is a group of more than 30 plasma and cell surface proteins that faciliate the destruction of pathogenic microorganisms by phagocytes or by lysis.

Complement activation is the initiation of a series of reactions involving the complement proteins of plasma that may result in either the death and elimination of a pathogenic microorganism or the amplification of the complement cascade effect.

In the early complement literature, complement activity in the pseudoglobulin fraction of serum is called the **end piece**, in contrast to the activity in the euglobulin fraction which is called the **mid-piece** of complement. Current information reveals that what was referred to as the **end piece** did not contain C1 but did contain all of the C2 and some other complement components.

Hypocomplementemia is, literally, diminished complement in the blood. This can occur in a number of diseases in which immune complexes fix complement *in vivo*, leading to a decrease in complement protein. Examples include active cases of systemic lupus erythematosus, proliferative glomerulonephritis, and serum sickness. Protein deficient patients may also have diminished plasma complement protein levels.

Procomplementary factors are serum components of selected species such as swine that resemble complement in their action and confuse evaluation of complement fixation tests.

The **RCA locus (regulator of complement activation)** is a locus on the long arm of chromosome 1 with a 750-kb DNA segment containing genes that encode complement receptor 1, complement receptor 2, C4-binding protein, and decay-accelerating factor. These substances regulate the activation of complement through combination with C4b, C3b, or C3dg. A separate chromosome 1 gene, not in the RCA locus, encodes factor H.

RCA is an abbreviation for regulators of complement activity.

The **lectin pathway of complement activation** is a complement activation pathway, not involving antibody, that is initiated by the binding of microbial polysaccharide to circulating lectins such as MBL, which structurally resembles C1q. Similar to C1q, it activates the C1r–C1s enzyme complex or another serine esterase, termed mannose-binding protein-associated serine esterase. Thereafter, all steps of the lectin pathway are the same as in the classic pathway following cleavage of C4. The activation of complement without antibody in which microbial polysaccharide are bound to circulating lectins such as plasma mannose-binding lectin (MBL). MBL resembles C1q structurally and activates the C1r–C1s enzyme complex or activates another serine esterase termed mannose-binding protein-associated serine esterase. Beginning with C4 cleavage, the lectin pathway is the same as the classical pathway.

Mannan-binding protein (Man-BP) is a substance that induces carbohydrate-mediated activation of complement, in contrast to complement activation mediated by immune complexes, which is initiated by C1q. For example, the inability to opsonize *Saccharomyces cerevisiae* may be attributable to Man-BP deficiency. Alveolar macrophage mannose receptors facilitate ingestion of *Pneumocystis carinii*.

Convertase is an enzyme that transforms a complement protein into its active form by cleaving it. The generation of C3/C5 convertase is a critical event in complement activation.

C3 convertase is an enzyme that splits C3 into C3b and C3a. There are two types: one in the classical pathway designated C4b2a and one in alternative pathway of complement activation termed C3bBb. An amplification loop with a positive feedback is stimulated by alternative pathway C3 convertase. Each of the two types of C3 convertase lacks stability, leading to the ready disassociation of their constituents. However, C3 nephritic factor can stabilize both classical and alternative pathway C3 convertases. Properdin may stabilize alternative pathway C3 convertase. C2a and Bb contain the catalytic sites.

The terminal stage of the classical pathway involves creation of the **membrane attack unit**, which is also called the lytic unit exclusive of the recognition and activation units. C5b binds to the cell membrane. This is followed by the successive interaction of single molecules of C6, C7, and C8 with the membrane-bound C5b. Finally, further interaction with several C9 molecules finishes formation of the lytic unit through noncovalent interactions without enzymatic alteration. Formation of a membrane attack unit or membrane attack complex (MAC) leads to a cell-membrane lesion that permits loss of K^+ and ingress of Na^+ and water, leading to hypotonic lysis of cells.

Not all C3b produced in classic complement activation unites with C4b2a to produce C5 convertase. Some of it binds directly to the cell membrane and acts as an opsonin, which makes the cell especially delectable to phagocytes such as neutrophils and macrophages that have receptors for C3b.

Complement fragments C3a and C5a serve as powerful anaphylatoxins that stimulate mast cells to release histamine, which enhances vascular permeability and smooth muscle contraction. Neutrophils release hydrolytic enzymes and platelet aggregate leading to microthrombosis, blood stasis, edema formation, and local tissue injury/destruction. C5a not only acts as an anaphylatoxin but is also chemotactic for PMNs and macrophages.

The **single hit theory** is based on the hypothesis that hemolysis of red blood cells sensitized with specific antibody is induced by perforation of the membrane by complement at only one site rather than at multiple sites on the membrane.

Nonimmunologic classic pathway activators are selected microorganisms such as *Escherichia coli* and low-virulence *Salmonella* strains, as well as certain viruses such as parainfluenza that react with C1q, leading to C1 activation without antibody. Thus, this represents classic pathway activation, which facilitates defense mechanisms. Various other substances such as myelin basic protein, denatured bacterial endotoxin, heparin, and urate crystal surfaces may also directly activate the classic complement pathway.

C1 (Figure 11.4) is a 750-kDa multimeric molecule comprised of one C1q subcomponent, two C1r, and two C1s subcomponents. The classical pathway of complement activation begins with the binding of C1q to IgM or IgG

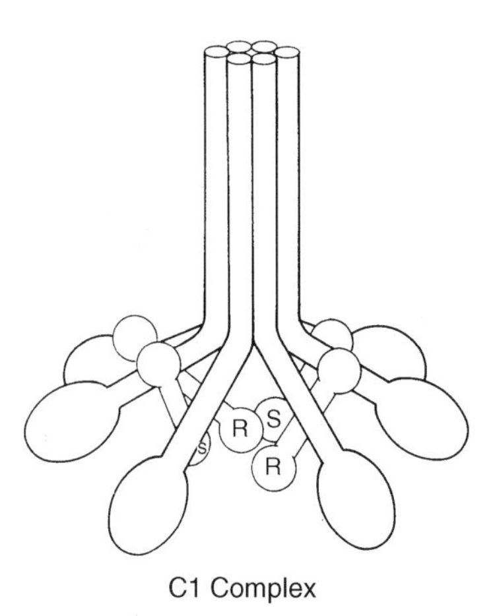

FIGURE 11.4 The subcomponents of C1 designated C1qrs.

molecules. C1q, C1r, and C1s form a macromolecular complex in a Ca^{2+}-dependent manner. The 400-kDa C1q molecule possesses three separate polypeptide chains that unite into a heterotrimeric structure resembling stems which contain an amino terminal in triple helix and a globular structure at the carboxy terminus that resembles a tulip. Six of these tulip-like structures with globular heads and stems form a circular and symmetric molecular complex in the C1q molecule. There is a central core. The serine esterase molecules designated C1r and C1s are needed for the complement cascade to progress. These 85-kDa proteins that are single-chains unite in the presence of calcium to produce a tetramer comprised of two C1r and two C1s subcomponents to form a structure that is flexible and has a C1s-C1r-C1r-C1s sequence. When at least two C1q globular regions bind to IgM or IgG molecules, the C1r in a tetramer associated with the C1q molecule becomes activated, leading to splitting of the C1r molecules in the tetramer with the formation of a 57-kDa and a 28-kDa chain. The latter, termed C1r, functions as a serine esterase, splitting the C1s molecules into 57-kDa and 28-kDa chains. The 28-kDa chain derived from the cleavage of $C\overline{1s}$ molecules designated C1s also functions as a serine esterase cleaving C4 and C2 and causing progression of the classical complement pathway cascade.

C1 esterase inhibitor is a serum protein that counteracts activated C1. This diminishes the generation of C2b which facilitates development of edema. C1 inhibitor (C1 INH) is a 478-amino acid residue single polypeptide chain protein in the serum. It blocks C1r activation, prevents $C\overline{1r}$ cleavage of $C\overline{1s}$, and inhibits $C\overline{1s}$ splitting of C4 and C2. The molecule is highly glycosylated with carbohydrates making up approximately one half of its content. It contains seven O-linked oligosaccharides linked to serine and six N-linked oligosaccharides tethered to an asparagine residue. Besides its effects on the complement system, C1 INH blocks factors in the blood clotting system that include kallikrein, plasmin, factor XIa, and factor XIIa. C1 INH is an α_2 globulin and is a normal serum constituent which inhibits serine protease. The 104-kDa C1 INH interacts with activated C1r or C1s to produce a stable complex. This prevents these serine protease molecules from splitting their usual substrates. Either $C\overline{1s}$ or $C\overline{1r}$ can split C1 INH to uncover an active site in the inhibitor, which becomes bound to the proteases through a covalent ester bond. By binding to most of the C1 in the blood, C1 INH blocks the spontaneous activation of C1. C1 INH binding blocks conformational alterations that would lead to spontaneous activation of C1. When an antigen–antibody complex binds C1, C1 INH's inhibitory influence on C1 is relinquished. Genes on chromosome 11 in man encode C1 INH. C1r and C1s subcomponents disengage from C1q following their interaction with C1 INH. In hereditary angioneurotic edema, C1 INH formation is defective.

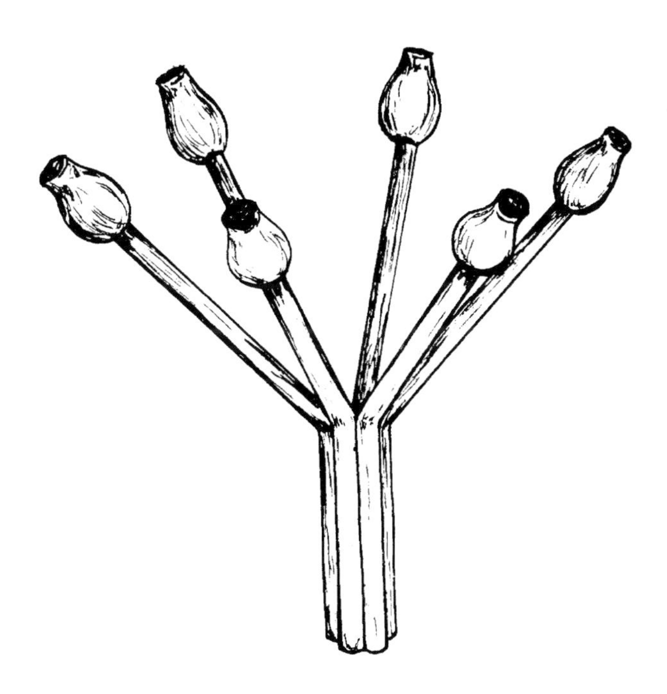

FIGURE 11.5 Cl.

In acquired C1 INH deficiency, there is elevated catabolism of C1 INH.

C1q (Figure 11.5) is an 8-polypeptide chain subcomponent of C1, the first component of complement. It commences the classical complement pathway. The three types of polypeptide chains are designated A chain, B chain, and C chain. Disulfide bonds link these chains. The C1q molecule's triple helix structures are parallel and resemble the stems of six tulips in the amino terminal half of their structure. They then separate into six globular regions, which resemble the head of a tulip. The molecule is arranged in a heterotrimeric rod-like configuration, bearing a collagen-like triple helix at its amino terminus and a tulip-like globular region at its carboxy terminus. The combination of six of the rod-shaped structures leads to a symmetric molecular arrangement comprised of three helices at one terminus and the globular (tulip-like) heads at the other terminus. The binding of antibody to C1q initiates the classic complement pathway. It is the globular C-terminal region of the molecule that binds to either IgM or IgG molecules. A tetramer comprised of two molecules of C1r and two molecules of C1s bind by Ca^{2+} to the collagen-like part of the stem. C1q A chain and C1q B chain are coded for by genes on chromosome 1p in man. The interaction of C1q with antigen–antibody complexes represents the basis for assays for immune complexes in patients' serum. IgM, IgG1, IgG2, and IgG3 bind C1q, whereas IgG4, IgE, IgA, and IgD do not.

Anti-Clq antibody is present in the majority of patients with hypocomplementemic urticarial vasculitis syndrome (HUVS) and in 30 to 60% of systemic lupus erythematosus patients. Clq is strikingly decreased in the blood sera of HUVS patients, even though their Clr and Cls levels are within normal limits and C5 to C9 are slightly activated.

C1q autoantibodies are detectable in 14 to 52% of SLE patients, 100% of hypocomplementemic urticarial vasculitis syndrome (HUVS), rheumatoid arthritis (RA) (5% in uncomplicated RA and 77% of RA patients with Felty syndrome), 73% of membranoproliferative glomerulonephritis type I, 45% of membranoproliferative glomerulonephritis type II and III, 94% of mixed connective tissue disease, and 42% of polyarthritis nodosa. Lupus nephritis patients usually reveal the IgG isotype. Rising levels of C1q autoantibodies portend renal flares in systemic lupus erythematosus patients. The rare condition HUVS can occur together with SLE and is marked by diminished serum C1q and recurrent idiopathic urticaria with leukocytoclastic vasculitis.

C1q binding assay for circulating immune complexes(CIC): There are two categories of methods to assay circulating immune complexes: (1) The specific binding of CIC to complement components such as C1q or the binding of complement activation fragments within the CIC to complement receptors, as in the Raji cell assay. (2) Precipitation of large and small CIC by polyethylene glycol. The C1q binding assay measures those CIC capable of binding C1q, a subcomponent of the C1 component of complement and capable of activating the classical complement pathway.

Clq facilitates **recognition unit** binding to cell surface antigen–antibody complexes. To launch the classical complement cascade, Clq must link to two IgG antibodies through their Fc regions. By contrast, one pentameric IgM molecule attached to a cell surface may interact with Clq to initiate the classical pathway. Binding of Clq activates Clr to become activated $\overline{\text{Clr}}$. This, in turn, activates Cls.

C1q receptors (C1q-R) binds the collagen segment of C1q fixed to antigen–antibody complexes. The C1q globular head is the site of binding of the Fc region of an immunoglobin. Thus, the C1q-R can facilitate the attachment of antigen–antibody complexes to cells expressing C1q-R and Fc receptors. Neutrophils, B cells, monocytes, macrophages, NK cells, endothelial cells, platelets, and fibroblasts all express C1q-R. C1q-R stimulation on neutrophils may lead to a respiratory burst.

The **collectin receptor** is the receptor of C1q, a subcomponent of the complement component C1.

C1r is a subcomponent of C1, the first component of complement in the classical activation pathway. It is a serine esterase. Ca^{2+} binds C1r molecules to the stem of a C1q molecule. Following binding of at least two globular regions of C1q with IgM or IgG, C1r is split into a 463-amino acid residue α chain, the N-terminal fragment, and a 243-amino acid residue carboxy-terminal β-chain fragment where the active site is situated. C1s becomes activated when C1r splits its arginine–isoleucine bond.

C1s is a serine esterase that is a subcomponent of C1, the first component of complement in the classical activation pathway. Ca^{2+} binds two C1s molecules to the C1q stalk. Following activation, $C\overline{1r}$ splits the single-chain 85-kDa C1s molecule into a 431-amino acid residue A chain and a 243-amino acid residue B chain where the active site is located. $C\overline{1s}$ splits a C4 arginine–alanine bond and a C2 arginine–lysine bond.

C2 (complement component 2) is the third complement protein to participate in the classical complement pathway activation. C2 is a 110-kDa single-polypeptide chain that unites with C4b molecules on the cell surface in the presence of Mg^{2+}. $C\overline{1s}$ splits C2 following its combination with C4b at the cell surface. This yields a 35-kDa C2b molecule and a 75-kDa C2a fragment. Whereas C2b may leave the cell surface, C2a continues to be associated with surface C4b. The complex of $C\overline{4b2a}$ constitutes classical pathway C3 convertase. This enzyme is able to bind and split C3. C4b facilitates combination with C3. C2b catalyzes the enzymatic cleavage. C2a contains the active site of classical pathway C3 convertase ($C\overline{4b2a}$). C2 is encoded by genes on the short arm of chromosome 6 in man. ***C2A***, ***C2B,*** and ***C2C*** alleles encode human C2. Murine C2 is encoded by genes at the S region of chromosome 17.

C2a is principal substance produced by $C\overline{1s}$ cleavage of C2. N-linked oligosaccharides may combine with C2a at six sites. The 509-carboxy-terminal amino acid residues of C2 constitute C2a. The catalytic site for C3 and C5 cleavage is located in the 287-residue carboxy-terminal sequence. The association of C2a with C4b yields the C3 convertase ($C\overline{4b2a}$) of the classical pathway.

C2b is a 223-amino acid terminal residue of C2 that represents a lesser product of $C\overline{1s}$ cleavage of C2. There are three abbreviated 68-amino acid residue homologous repeats in C2b that are present in C3- or C4-binding proteins. N-linked oligosaccharides combine with C2b at two sites. A peptide split from the carboxy terminus of C2b by plasmin has been implicated in the formation of edema in hereditary angioneurotic edema patients.

C2 and B (Figure 11.6 and Figure 11.7) genes are situated within the MHC locus on the short arm of chromosome 6. They are termed MHC class III genes. *TNF*-α and *TNF*-β genes are situated between the *C2* and *HLA-B* genes. Another gene designated *FD* lies between the *Bf* and *C4a* genes. *C2* and *B* complete primary structures have been deduced from cDNA and protein sequences. *C2* is comprised of 732 residues and is an 81-kDa molecule, whereas *B* contains 739 residues and is an 83-kDa molecule. Both proteins have a three-domain globular structure.

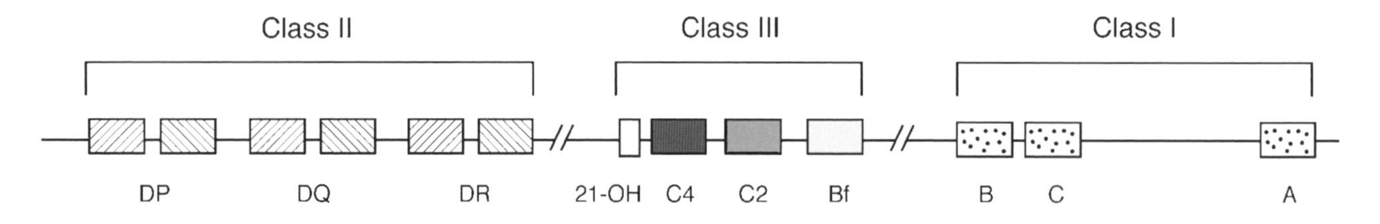

FIGURE 11.6 *C2* and *B* genes are situated within the MHC locus on the short arm of chromosome 6.

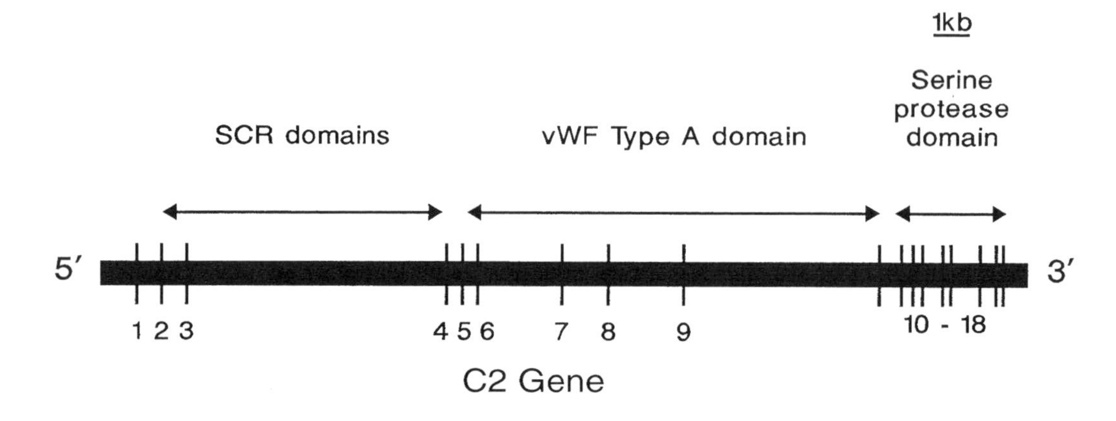

FIGURE 11.7 Schematic representation of the exon/intron organization of the human *C2* gene. Exons, represented by vertical bars, are numbered.

During C3 convertase formation, the amino terminal domains C2b or Ba are split off. They contain consensus repeats that are present in CR1, CR2, H, DAF, and C4bp, which all combine with C3 and/or C4 fragments and regulate C3 convertases. The amino acid sequences of the C2 and B consensus repeats are known. C2b contains site(s) significant for C2 binding to C4b. Ba, resembling C2b, manifests binding site(s) significant in C3 convertase assembly. Available evidence indicates that C2b possesses a C4b-binding site and that Ba contains a corresponding C3b-binding site.

In considering assembly and decay of C3 convertases, initial binding of the three-domain structures *C2* or *B* to activator-bound C4b or C3b, respectively, requires one affinity site on the C2b/Ba domain and another on one of the remaining two domains. A transient change in C2a and Bb conformation results from *C2* or *B* cleavage by C1s or D. This leads to greater binding affinity, Mg^{+2} sequestration, and acquisition of proteolytic activity for C3. C2a or Bb dissociation leads to C3 convertase decay. Numerous serum-soluble and membrane-associated regulatory proteins control the rate of formation and association of C3 convertases.

Complotype is an MHC class III haplotype. It is a precise arrangement of linked alleles of MHC class III genes that encode C2, factor B, C4A, and C4B MHC class III molecules in man. Caucasians have 12 ordinary complotypes that may be in positive linkage disequilibrium.

Regulation of complement activation (RCA) cluster is a gene cluster on a 950-kb DNA segment of chromosome 1's long arm that encodes the homologous proteins CR1, CR2, C4bp, and DAF.

Regulators of complement activity (RCA): A group of proteins including C4bP, decay-accelerating factor (DAF), membrane cofactor protein (MCP), factor H, CR1 gel caps, and CR2 gel caps. They are comprised of 60- to 70-amino acid consensus repeat sequences.

Ss protein is a hemolytically active substance encoded by the murine complement locus C4B.

slp: Abbreviation for **sex-limited protein** that is encoded by genes at the C4A complement locus in mice. It is restricted to males but may be induced in females by androgen administration.

C3 (complement component 3) is a 195-kDa glycoprotein heterodimer that is linked by disulfide bonds. It is the fourth complement component to react in the classical pathway, and it is also a reactant in the alternative complement pathway. C3 contains α and β polypeptide chains and has an internal thioester bond which permits it to link covalently with surfaces of cells and proteins. Much of the C3 gene structure has now been elucidated. It is believed to contain approximately 41 exons. Eighteen of 36 introns have now been sequenced. The C3 gene of man is located on chromosome 19. Hepatocytes, monocytes, fibroblasts, and endothelial cells can synthesize C3. More than 90% of serum C3 is synthesized in the liver. The concentration of C3 in serum exceeds that of any other complement component. Human C3 is generated as a single chain precursor which is cleaved into the two-chain mature state. C3 molecules are identical antigenically, structurally, and functionally, regardless of cell source. Hepatocytes and monocytes synthesize greater quantities of C3 than do epithelial and endothelial cells. C3 convertases split a 9-kDa C3a fragment from the α chain of C3. The other product of the reaction is C3b which is referred to as metastable C3b and has an exposed thioester bond. Approximately 90% of the metastable C3b thioester bonds interact with H_2O to form inactive C3b byproducts that have no role in the complement sequence. 10% of C3b molecules may bind to cell substances through covalent bonds or with the immunoglobulin bound to C4b2a. This interaction leads to the formation of $\overline{\text{C4b2a3b}}$ which is classical pathway C5 convertase, and serves as a catalyst in the enzymatic splitting of C5 which initiates membrane attack complex (MAC) formation. When C3b, in the classical complement pathway, interacts with E (erythocyte), A (antibody), C1 (complement 1), and 4b2a, EAC14b2a3b is produced. As many as 500 C3b molecules may be deposited at a single EAC14b2a complex on an erythrocyte surface. C3S (slow electrophoretic mobility) and C3F (fast electrophoretic mobility) alleles on chromosome 19 in man encode 99% of C3 in man, with rare alleles accounting for the remainder. C3 has the highest concentration in serum of any complement system protein with a range of 0.552 to 1.2 mg/ml. Following splitting of the internal thioester bond, it can form a covalent link to amino or hydroxyl groups on erythrocytes, microorganisms, or other substances. C3 is an excellent opsonin. C3 was known in the past as β_1C globulin.

Pro-C3 is a polypeptide single chain that is split into C3 α and β (amino terminal) chains.

β_1C globulin is the globulin fraction of serum that contains complement component C3. On storage of serum, β_1C dissociates into β_1A globulin, which is inactive.

β_1A globulin is a breakdown product of β_1C globulin. It has a molecular weight less than that of β_1C globulin, and its electrophoretic mobility is more rapid than that of the β_1C globulin. β_1A degradation is linked to the disappearance of C3 activity.

C3 convertase (Figure 11.8) is an enzyme that splits C3 into C3b and C3a. There are two types, one in the classical

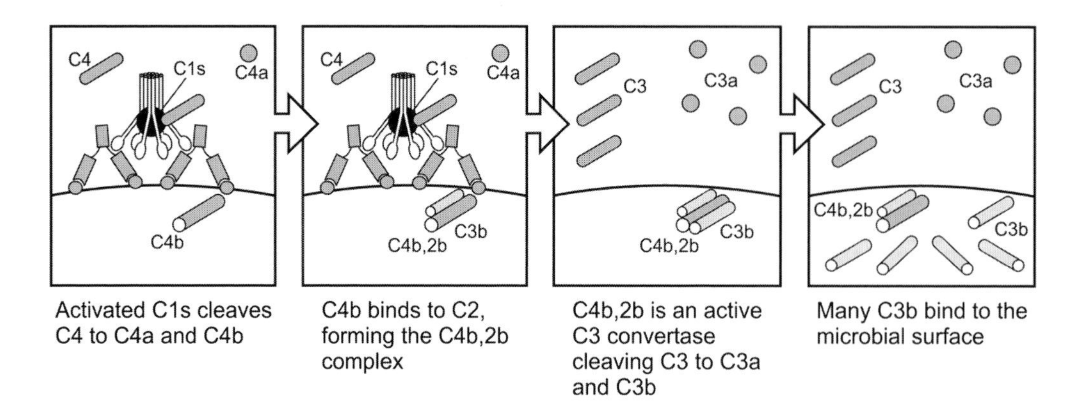

FIGURE 11.8 The classical pathway of complement activation generates a C3 convertase.

pathway designated $\overline{C4b2a}$ and one in the alternative pathway of complement activation termed $\overline{C3bBb}$. An amplification loop with a positive feedback is stimulated by alternative pathway C3 convertase. Each of the two types of C3 convertase lacks stability leading to ready disassociation of their constituents. However, C3 nephritic factor can stabilize both classical and alternative pathway C3 convertases. Properdin may stabilize alternative pathway C3 convertase. C2a and Bb contain the catalytic sites.

C3a is a low-molecular-weight (9 kDa) peptide fragment of complement component C3. It is comprised of the 77 N-terminal end residues of C3 α chain. This biologically active anaphylatoxin, which induces histamine release from mast cells and causes smooth muscle contraction, is produced by the cleavage of C3 by either classical pathway C3 convertase, i.e., $\overline{C4b2a}$, or alternative complement pathway C3 convertase, i.e., $\overline{C3bBb}$. Anaphylatoxin inactivator, a carboxy peptidase N, can inactivate C3a by digesting the C-terminal arginine of C3a.

The **C3a receptor (C3a-R)** is a protein on the surface membrane of mast cells and basophils. It serves as a C3a anaphylatoxin receptor.

C3a/C4a receptor (C3a/C4a-R): C3a and C4a share a common receptor on mast cells. When a C-terminal arginine is removed from C3a and C4a by serum carboxy peptidase N (SCPN), these anaphylatoxins lose their ability to activate cellular responses. Thus, $C3a_{des\ Arg}$ and $C4a_{des\ Arg}$ lose their ability to induce spasmogenic responses. C3a-R has been demonstrated on guinea pig platelets. Eosinophils have been found to bind C3a.

C3b is a principal fragment produced when complement component C3 is split by either classical or alternative pathway convertases, i.e., $\overline{C4b2a}$ or $\overline{C3bBb}$, respectively. It results from C3 convertase digestion of C3's α chain. C3b is an active fragment as revealed by its combination with factor B to produce $\overline{C3bBb}$, which is the alternative pathway C3 convertase. Classical complement pathway C5 convertase is produced when C3b combines with $\overline{C4b2a}$ to yield $\overline{C4b2a3b}$. Factor I splits the arginine-serine bonds in C3b, if factor H is present, to yield C3bi. This produces the C3f peptide. Particle-bound C3b interacts with complement receptor 1. C3b interacts with C3b receptors on macrophages, B lymphocytes, polymorphonuclear neutrophils (PMNs), and possibly T cells. It promotes phagocytosis and immune adherence and may function as an opsonin.

Cobra venom factor (CVF) is a protein in the venom of the Indian cobra, *Naja naja*. It is the equivalent of mammalian C3b, which means that it can activate the alternative pathway of complement. Mammalian factor I does not inactivate cobra venom, which leads to the production of a stable alternative pathway, C3 convertase, if CVF is injected intravenously into a mammal. Thus, the injection of cobra venom factor into mammals has been used to destroy complement activity for experimental purposes. A component of cobra venom interacts specifically with the serum complement system, leading to its continous activation. CVF is a structural and functional analog of complement component C3. It has been employed to decomplement laboratory animals to investigate complement's biological functions. It has been used in animal experiments involving xenotransplantation to show that complement is a principal contributor to hyperacute rejection of a transplanted organ. It has also been used in antibody conjugates to render monoclonal antibodies cytotoxic.

An **activation unit** is generated by the interaction of C3b with C4b2a bound to the cell membrane.

C4 (complement component 4) is a 210-kDa molecule comprised of α, β, and δ chains. The α chain has an internal thioester bond linking a cysteine residue and adjacent glutamate residue. C4 reacts immediately following C1 in the classical pathway of complement activation. $\overline{C1s}$

splits the α chain of C4 at position 76–77, where an arginine–alanine bond is located. This yields a 8.6-kDa C4a fragment, an anaphylatoxin, and C4b, which is a larger molecule. C4b remains linked to C1. Many C4b molecules can be formed through the action of a single $\overline{C1s}$ molecule. Enzymatic cleavage renders the α chain thioester bond of the C4b fragment very unstable. The molecule's chemically active form is termed metastable C4b. C4bi intermediates form when C4b thioester bonds and water molecules react. C4b molecules may become bound covalently to cell surfaces when selected C4b thioester bonds undergo transesterification, producing covalent amide or ester bonds with proteins or carbohydrates on the cell surface. This enables complement activation to take place on the surfaces of cells where antibodies bind. C4b may also link covalently with antibody. C4 is first formed as a 1700-amino acid residue chain which contains β, α, and γ chain components joined through connecting peptides. *C4A* and *C4B* genes located at the major histocompatibility complex on the short arm of chromosome 6 in man encode C4. *Slp* and *Ss* genes located on chromosome 17 in the mouse encode murine C4.

β_1E globulin is the globulin fraction of serum that contains complement component C4 activity.

C4 allotypes: Complement component 4 (C4) is encoded by two genes designated C4A and C4B, located within the MHC class III region on the short arm of chromosome 6 in humans. Each locus has numerous allelic forms, including nonexpressed or null alleles. C4A is usually Rodgers-negative, Chido-negative, whereas C4B is usually Rogers-negative and Chido-positive. C4 protein is highly polymorphic with more than 40 variants that include null alleles (C4Q0) at both loci. Null alleles are defined by the absence of C4 protein in plasma and exist in normal populations at frequencies of 0.1 to 0.3%. The presence and number of null alleles determine the expected reference range of serum C4 for a given person.

Pro-C4 is a polypeptide single chain that is split into C4 α, β (amino terminal), and γ (carboxy terminal) chains.

C4a is a 76-amino acid terminal residue peptide produced by CTS cleavage of C4. Together with C3a and C5a, C4a is an anaphylatoxin which induces degranulation of mast cells and basophils associated with histamine release and the features of anaphylaxis. However, the anaphylatoxin activity of C4a is 100 times weaker than is that of the other two anaphylatoxin molecules.

C4A is a very polymorphic molecule expressing the Rodgers epitope that is encoded by the *C4A* gene. The equivalent murine gene encodes a sex-limited protein (SLP). It has less hemolytic activity than does C4B. C4A and C4B differ in only four amino acid residues in the α chain's C4d region. C4A is Pro-Cys-Pro-Bal-Leu-Asp, whereas C4B is Leu-Ser-Pro-Bal-Ile-His.

C4B is a polymorphic molecule that usually expresses the Chido epitope and is encoded by the *C4B* gene. The murine equivalent gene encodes an Ss protein. It shows greater hemolytic activity than does C4A.

C4b is the principal molecule produced when $\overline{C1s}$ splits C4. C4b is that part of the C4 molecule that remains after C4a has been split off by enzymatic digestion. C4b unites with C2a to produce $\overline{C4b2a}$, an enzyme which is known as the classical pathway C3 convertase. Factor I splits the arginine–asparagine bond of C4b at position 1318-1319 to yield C4bi, if C4b binding protein is present. C4b linked to particulate substances reacts with complement receptor 1.

C3a/C4a receptor (C3a/C4a-R) is a common receptor on mast cells. When a C-terminal arginine is removed from C3a and C4a by SCPN, these anaphylatoxins lose their ability to activate cellular responses. Thus, $C3a_{des\ Arg}$ and $C4a_{des\ Arg}$ lose their ability to induce spasmogenic responses. C3a-R has been demonstrated on guinea pig platelets. Eosinophils have been found to bind C3a.

C5 (complement component 5) is a component comprised of α and β polypeptide chains linked by disulfide bonds that react in the complement cascade following C1, C4b, C2a, and C3b fixation to complexes of antibody and antigen. The 190-kDa dimeric C5 molecule shares homology with C3 and C4, but does not possess an internal thioester bond. C5 combines with C3b of C5 convertase of either the classical or the alternative pathway. C5 convertases split the α chain at an arginine–leucine bond at position 74-75, producing an 11-kDa C5a fragment, which has both chemotactic action for neutrophils and anaphylatoxin activity. It also produces a 180-kDa C5b fragment that remains anchored to the cell surface. C5b maintains a structure that is able to bind with C6. C5 is a β1F globulin in man. C5b complexes with C6, C7, C8, and C9 to form the membrane attack complex (MAC) which mediates immune lysis of cells. Murine C5 is encoded by genes on chromosome 2.

β_1F globulin is the globulin fraction of serum that contains complement component C5 activity.

Pro-C5 is a polypeptide single chain that is split into C5 α and β (amino terminal) chains.

C5a is a peptide split from C5 through the action of C5 convertases, $\overline{C4b2a3b}$ or $\overline{C3bBb3b}$. It is comprised of the C5 α chain's 74-amino terminal residues. C5a is a powerful chemotactic factor and is an anaphylatoxin, inducing mast cells and basophils to release histamine. It also causes smooth muscle contraction. C5a promotes the production of superoxide in polymorphonuclear neutrophils (PMNs) and accentuates CR3 and Tp150,95 expression in their membranes. In addition to chemotaxis, it may facilitate PMN degranulation. Human serum contains anaphylatoxin inactivator that has carboxypeptidase N properties. It deletes C5a's C-terminal arginine which yields $C5a_{des\ Arg}$. Although deprived of anaphylatoxin properties, $C5a_{des\ Arg}$ demonstrates limited chemotactic properties.

$C5a_{74des\ Arg}$ is that part of C5a that remains following deletion of the carboxy terminal arginine through the action of anaphylatoxin inactivator. Although deprived of C5a's anaphylatoxin function, $C5a_{74des\ Arg}$ demonstrates limited chemotactic properties. This very uncommon deficiency of C2 protein in the serum has an autosomal recessive mode of inheritance. Affected persons have an increased likelihood of developing type III hypersensitivity disorders mediated by immune complexes, such as systemic lupus erythematosus. Whereas affected individuals possess the C2 gene, mRNA for C2 is apparently absent. Individuals who are heterozygous possess 50% of the normal serum levels of C2 and manifest no associated clinical illness.

C5a receptor (C5a-R) is a receptor found on phagocytes and mast cells that binds the anaphylatoxin C5a, which plays an important role in inflammation. SCPN controls C5a function by eliminating the C-terminal arginine. This produces $C5a_{des\ Arg}$. Neutrophils are sites of C5a catabolism. C5a-R is a 150- to 200-kDa oligomer comprised of multiple 40- to 47-kDa C5a-binding components. C5a-R mediates chemotaxis and other leukocyte reactions.

C5b is the principal molecular product that remains after C5a has been split off by the action of C5 convertase on C5. It has a binding site for C6 and complexes with it to begin generation of the MAC of complement, which leads to cell membrane injury and lysis.

C5 convertase is a molecular complex that splits C5 into C5a and C5b in both the classical and the alternative pathway of complement activation. Classical pathways C5 convertase is comprised of $C\overline{4b2a3b}$ whereas alternative pathway C5 convertase is comprised of $C\overline{3bBb3b}$. C2a and Bb contain the catalytic sites.

C6 (complement component 6) is a 128-kDa single polypeptide chain that participates in the MAC. It is encoded by *C6A* and *C6B* alleles. It is a β2 globulin.

C7 (complement component 7) is an 843-amino acid residue polypeptide chain that is a β_2 globulin. C5b67 is formed when C7 binds to C5b and C6. The complex has the appearance of a stalk with a leaf type of structure. C5b constitutes the leaf, and the stalk consists of C6 and C7. The stalk facilitates introduction of the C5b67 complex into the cell membrane, although no transmembrane perforation is produced. C5b67 anchored to the cell membrane provides a binding site for C8 and C9 in formation of the MAC. N-linked oligosaccharides bind to asparagine at positions 180 and 732 in C7.

C8 (complement component 8) is a 155-kDa molecule comprised of a 64-kDa α chain, a 64-kDa β chain, and a 22-kDa γ chain. The α and γ chain are joined by disulfide bonds. Noncovalent bonds link α and γ chains to the β chain. The C5b678 complex becomes anchored to the cell surface when the γ chain inserts into the membrane's lipid bilayer. When the C8 β chain combines with C5b in C5b67 complexes, the α chain regions change in conformation from β-pleated sheets to α helixes. The C5b678 complex has a limited capacity to lyse the cell to which it is anchored, since the complex can produce a transmembrane channel. The α chain of C8 combines with a single molecule of C9, thereby inducing C9 polymerization in the MAC. Genes at three different loci encode C8 α, γ, and β chains. One third of the amino acid sequences are identical between C8 α and β chains. These chains share the identity of one-quarter of their amino-acid sequences with C7 and C9. C8 is a β_1 globulin. In humans, the C8 concentration is 10 to 20 μ/ml.

C9 (complement component 9) is a 535-amino acid residue single-chain protein that binds to the C5b678 complex on the cell surface. It links to this complex through the α chain of C8, changes in conformation, significantly increases its length, and reveals hydrophobic regions that can react with the cell membrane lipid bilayer. With Zn^{2+} present, a dozen C9 molecules polymerize to produce 100nm diameter hollow tubes that are positioned in the cell membrane to produce transmembrane channels. A total of 12 to 15 C9 molecules interact with one C5b678 complex to produce the membrane attack complex (MAC). When viewed by an electron microscope, the pores in the plasma membrane produced by the poly-C9 have a 110-Å internal diameter, a 115-Å stalk anchored in the membrane's lipid bilayer, and a 100-Å structure above the membrane that gives an appearance of a doughnut when viewed from above. Similar pores are produced by proteins released from cytotoxic T lymphocytes and natural killer cells called perforin or cytolysin. Sodium and water quickly enter the cells through these pores, leading to cell swelling and lysis. C9 shares one fourth the amino acid sequence identity with the α and β chains of C7 and C8. It resembles perforin structurally. No polymorphism is found in C9, which is encoded by genes on chromosome 5 in man.

Doughnut structure is the assembly and insertion of complement C9 protein monomers into a cell membrane to produce a transmembrane pore which leads to cell destruction through the cytolytic action of the complement cascade.

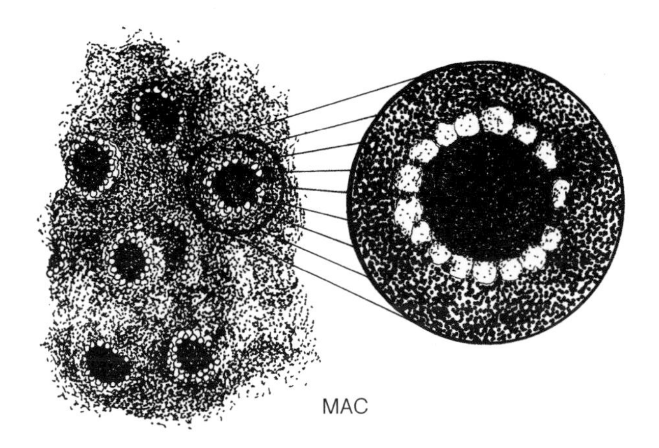

FIGURE 11.9 Membrane perforations resulting from membrane attack complex (MAC) action.

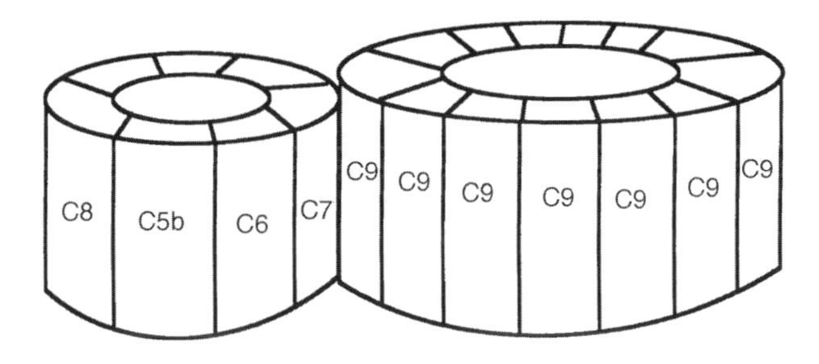

FIGURE 11.10 Membrane attack complex.

Complement multimer is a doughnut-shaped configuration as a part of the complement reaction sequence.

The **membrane attack complex (MAC)** (Figure 11.9 and Figure 11.10) consists of the five terminal proteins C5, C6, C7, C8, and C9 associate into MAC on a target-cell membrane to mediate injury. Initiation of MAC assembly begins with C5 cleavage into C5a and C5b fragments. A $(C5b678)_1(C9)_n$ complex then forms either on natural membranes or, in their absence, may combine with such plasma inhibitors as lipoproteins, antithrombin III, and S protein. cDNA sequencing reveals the primary structure of all five related complement proteins. C9 is believed to have a five-domain structure. C9 and C8 alpha proteins resemble each other not only structurally but also in sequence homologies. Both bind calcium and furnish domains that bind lipid, enabling MAC to attach to the membrane.

Mechanisms proposed for complement-mediated cytolysis include extrinsic protein channel incorporation into the plasma membrane or membrane deformation and destruction. Central regions of C6, C7, C8α, C8β, and C9 have been postulated to contain amphiphilic structures which may be membrane anchors.

A single C9 molecule per C5b-8 (Figure 11.11) leads to erythrocyte lysis. C8 polymerization is not required.

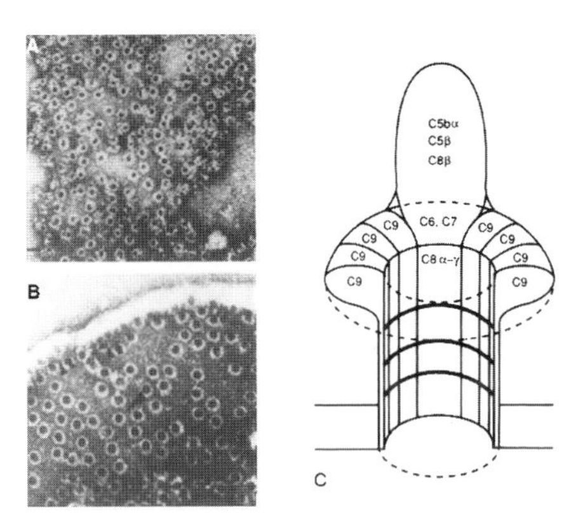

FIGURE 11.11 **A**. Electron micrograph of complement lesions (approximately 100 C) in erythrocyte membranes formed by poly C9 tubular complexes. **B**. Electron micrograph of complement lesions (approximately 160 C) induced on a target cell by a clone cytolytic T lymphocyte (CTL) line. CTL- and natural killer (NK)-induced membrane lesions are formed by tubular complexes of perforin which is homologous to C9. Therefore, except for the larger internal diameter, the morphology of the lesions is similar to that of complement mediated lesions. **C**. Model of the MAC subunit arrangement.

Gram-negative bacteria, which have both outer and inner membranes, resist complement action by lengthening surface carbohydrate content, which interferes with MAC binding. MAC assembly and insertion into the outer membrane is requisite for lysis of bacteria. Nucleated cells may rid their surfaces of MAC through endocytosis or exocytosis. Platelets have provided much data concerning sublytic actions of C5b-9 proteins.

Control proteins acting at different levels may inhibit killing of homologous cells mediated by MAC. Besides C8-binding protein or homologous restriction factor (HRF) found on human erythrocyte membranes, the functionally similar but smaller phosphatidylonositol glycan (PIG)-tailed membrane protein harnesses complement-induced cell lysis. Sublytic actions of MAC may be of greater consequence for host cells than are its cytotoxic effects.

MIRL (membrane inhibitor of reactive lysis) is an inhibitor of membrane attack complexes on self-tissue and is also known as CD59.

CD59 (Figure 11.12) is an 18-kDa membrane glycoprotein that regulates the activity of the complement MAC. A glycosylphosphatidylinositol anchor attaches CD59 to the outer layer of the membrane. It is shed from cell surfaces into body fluid. There is no evidence for a secreted form. CD59 inhibits MAC formation by binding to sites on C8 and C9 which blocks the uptake and incorporation of multiple C9 molecules into the complex. CD59 can interact with C8 and C9 from heterologous species but works best with homologous proteins, which is the

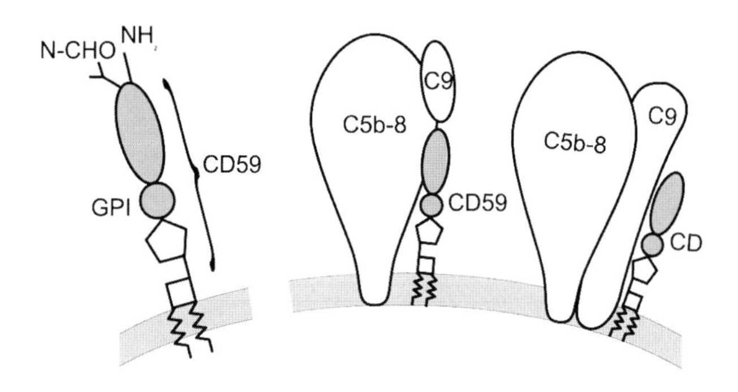

FIGURE 11.12 CD59.

basis for homologous restriction (of lysis). Paroxysmal nocturnal hemoglobinuria (PNH)–affected cells deficient in C59 and other complement regulatory proteins are very sensitive to lysis. Also called protective membrane inhibitor of reactive lysis (MIRL), homologous restriction factor 20 (HRF 20), and membrane attack complex inhibitory factor (MACIF).

Complement membrane attack complex: See membrane attack complex.

MAC is an abbreviation for membrane attack complex of the complement system.

Immune cytolysis is complement-induced destruction or lysis of antibody-coated target cells such as tumor cells. Immune cytolysis of red blood cells is referred to as immune hemolysis. Immune cytolysis refers also to the destruction of target cells by cytotoxic T lymphocytes or natural killer (NK) cells (including K cells) through the release of perforins.

Immune hemolysis is the lysis of erythrocytes through the action of specific antibody and complement.

Clusterin (serum protein SP-40,40) is a complement regulatory protein that inhibits membrane attack complex (MAC) formation by blocking the fluid-phase of the MAC. Its substrate is C5b67.

S protein is an 83-kDa serum protein in man that prevents generation of the MAC of complement. The S protein molecule is comprised of one 478-amino acid residue polypeptide chain. Its mechanism of action is to inhibit insertion of the C5b67 complex into the membrane of a cell by first linking three of its molecules to each free C5b67 complex. It also inhibits C9 from polymerizing on C5b678 complexes. See vitronectin.

SP-40,40 is a heterodimeric serum protein derived from soluble C5-9 complexes that may modulate the membrane attack complex's cell lysing action.

Protectin (CD59) is a protein in the cell surface that guards host cells from complement-mediated injury. It prevents the formation of the membrane attack complex by inhibiting C8 and C9 binding to the C5b,6,7 complex.

Suramin {Antrypol, 8,8′-(carbonyl-bis-[imino-3,1-phenylenecarbonylimino])-bis-1,3,5-naphthalene trisulfonic acid} is a therapeutic agent for African sleeping sickness produced by trypanosomes. Of immunologic interest is its ability to combine with C3b, thereby blocking factor H and factor I binding. The drug also blocks lysis mediated by complement by preventing attachment of the MAC of complement to the membranes of cells.

The **terminal complement complex (TCC)** is a sequence that is the same whether activation is initiated via the classical, alternative, or lectin pathway. Following cleavage of C5 by either the classical or the alternative C5 convertase, the terminal complement components C6, C7, C8, and C9 are sequentially but nonenzymatically activated, resulting in the formation of TCC. TCC can be generated on a biologic target membrane as potentially membranolytic MAC or in extracellular fluids as nonlytic SC5b-9 in the presence of S protein (also called vitronectin). Both forms consist of C5b and the complement proteins C6, C7, C8, and C9. Although some lytic activity is expressed by the C5b-8 complex, efficient lysis is dependent on an interaction with C9, facilitated by the α-moiety of C8.

Terminal complement complex deficiency is the hereditary deficiency of a terminal complement component that leads to an inability to form a functional TCC with absence of hemolysis and bactericidal activity. Frequently, there are terminal complement deficiencies in patients with meningococcal infections, suggesting that the cytolytic activity of the complement system is critical in resistance to *Nisseria meningitidis*.

Terminal complement components are those constituents of the complement system that assemble to produce the MAC.

Anaphylatoxins (Figure 11.13) are substances generated by the activation of complement that lead to increased vascular permeability as a consequence of the degranulation of mast cells with the release of pharmacologically active mediators of immediate hypersensitivity. These biologically active peptides of low molecular weight are derived from C3, C4, and C5. They are generated in serum during fixation of complement by Ag-Ab complexes, immunoglobulin aggregates, etc. Small blood vessels, mast cells, smooth muscle, and leukocytes in peripheral blood are targets of their action. Much is known about their primary structures. These complement fragments are designated C3a, C4a, and C5a. They cause smooth muscle contraction, mast cell degranulation with histamine release, increased vascular permeability, and the triple response in skin. They induce anaphylactic-like symptoms upon parenteral inoculation.

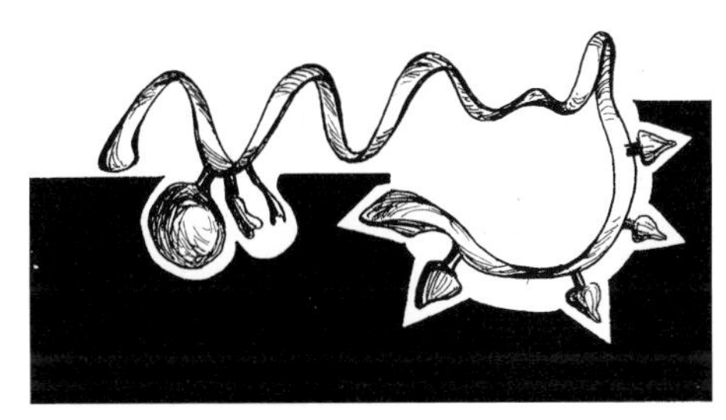

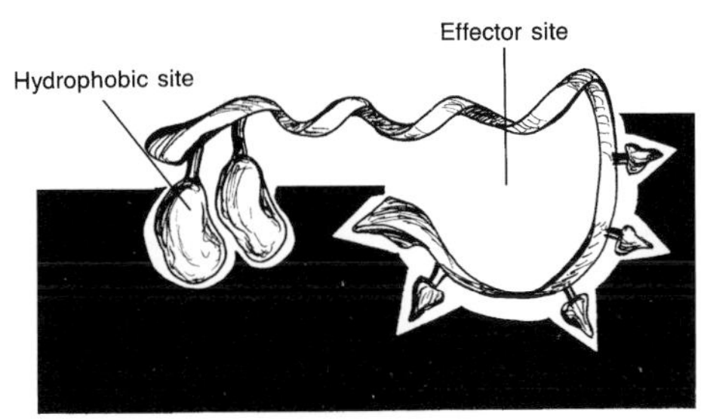

FIGURE 11.13 Anaphylatoxin–receptor interactions.

C5aR (C5 anaphylatoxin receptor) belongs to the G-protein-coupled receptor family, the chemokine receptor branch of the rhodopsin family. COS-7 cells and HEK 293 cells expressing C5aR bind C5a with high affinity. C5aR mRNA is present in peripheral blood monocytes, granulocytes, and in the myeloid fraction of bone marrow. C5aR message is present in the KG-1 monocytic cell line.

Anaphylatoxin inhibitor (Ana INH) is a 300-kDa α globulin. It is a carboxy peptidase that cleaves anaphylatoxin's carboxy terminal arginine. The enzyme acts on all three forms including C3a, C4a, and C5a, inactivating rather than inhibiting them.

The **alternative complement pathway** (Figure 11.14) is a nonantibody-dependent pathway for complement activation in which the early components C1, C2, and C4 are not required. It involves the protein properdin factor D, properdin factor B, and C3b, leading to C3 activation and continuing to C9 in a manner identical to that which takes place in the activation of complement by the classical pathway. Substances such as endotoxin, human IgA, microbial polysaccharides, and other agents may activate complement by the alternative pathway. The C3bB complex forms as C3b combines with factor B. Factor D splits factor B in the complex to yield the Bb active fragment

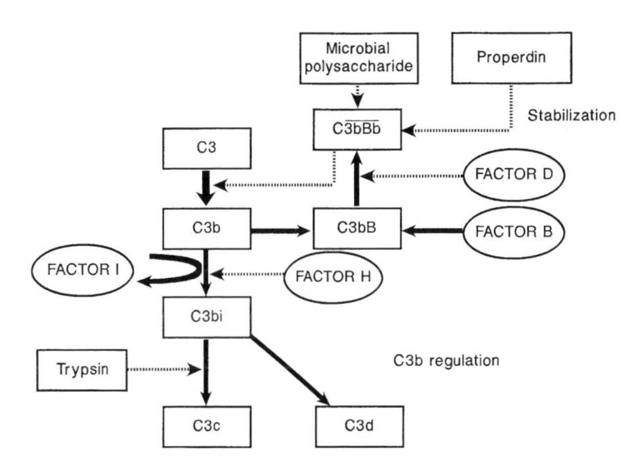

FIGURE 11.14 Alternative complement pathway.

that remains linked to C3b and Ba, which is inactive and is split off. C3bBb, the alternative pathway C3 convertase, splits C3 into C3b and C3a, thereby producing more C3bBb, which represents a positive feedback loop (Figure 11.15). Factor I, when accompanied by factor H, splits C3b's heavy chain to yield C3bi, which is unable to anchor Bb, thereby inhibiting the alternative pathway. Properdin and C3 nephritic factor stabilize C3bBb. C3 convertase stabilized by properdin activates complement's late components, resulting in opsonization, chemotaxis of leukocytes, enhanced permeability, and cytolysis. Properdin, IgA, IgG, lipopolysaccharide, and snake venom can initiate the alternate pathway. Trypsin-related enzymes can activate both pathways.

Alternative pathway is the activation of complement not by antibody but through the binding of complement protein C3b to the surface of a pathogenic microorganism. This innate immune mechanism amplifies complement activation through the classic pathway.

Inulin is a homopolysaccharide of D-fructose that is found in selected plants such as dahlias. Although it is used to measure renal clearance, soluble inulin is known to activate the alternative complement pathway.

Zymosan is a complex polysaccharide derived from dried cell walls of the yeast *Saccharomyces cerevisiae,* which activates the alternative complement pathway. It binds to C3b and is useful in investigations of opsonic phagocytosis.

Properdin (factor P) is a globulin in normal serum that has a central role in activation of the alternative complement pathway activation. Additional factors such as magnesium ions are required for properdin activity. It is an alternative complement pathway protein that has a significant role in resistance against infection. It combines with and stabilizes the C3 convertase of the alternate pathway,

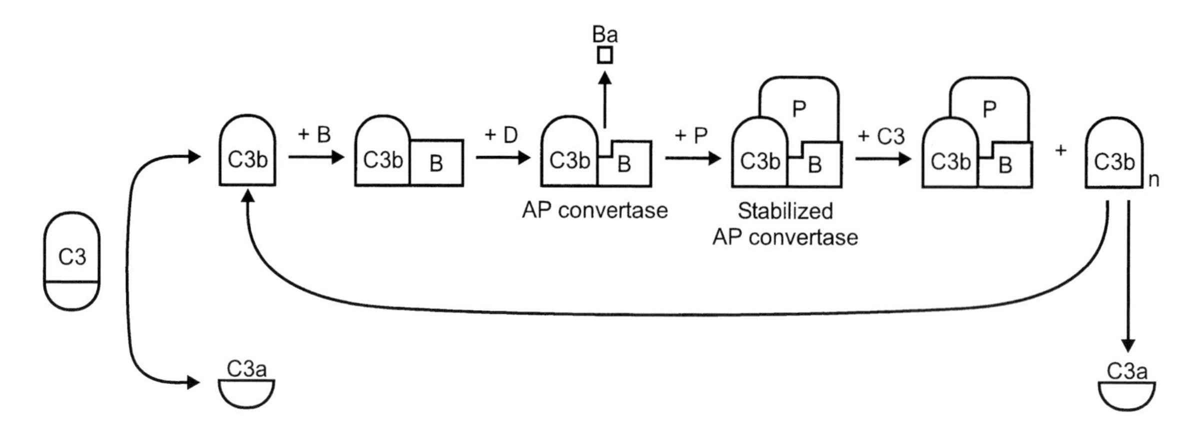

FIGURE 11.15 The feedback loop of the alternative pathway.

which is designated C3bBb. Properdin is a 441-amino acid residue polypeptide chain with two points where N-linked oligosaccharides may become attached. Electron microscopy reveals it to have a cyclic oligomer conformation. Molecules are comprised of six repeating 60-residue motifs which are homologous to 60 amino acids at C7, C8 α, and C8 β amino carboxy terminal ends and the C9 amino-terminal end.

The **properdin system** is an older term for alternative complement pathway. It consists of several proteins that are significant in resistance against infection. The first protein to be discovered was properdin. The properdin system also consists of factor B, a 95-kDa β_2 globulin which is also termed C3 proactivator, glycine-rich β glycoprotein, or β_2 glycoprotein II. Properdin factor D is a 25-kDa, α globulin which is also termed C3 proactivator convertase or glycine-rich β glycoproteinase. The properdin system, or alternative pathway, does not require the participation of antibody to activate complement. It may be activated by endotoxin or other substances. See alternative complement pathway.

P is the abbreviation for properdin. Also called factor P.

The **properdin pathway** is an earlier synonym for alternative pathway of complement activation.

Properdin deficiency is an X-linked recessive disorder in which there is an increased susceptibility to infections by *Neisseria* microorganisms. Males with the deficiency reveal 2% or less of normal serum levels of properdin. In some heterozygous females, the serum properdin level may be only 50%, while in others serum properdin levels are normal.

Alternative pathway C3 convertase (Figure 11.16) is an alternate unstable C3bB complex that splits C3 into C3a and C3b. Factor P, also known as properdin, stabilizes C3bB to yield C3bBbP. C3 nephritic factor can also stabilize C3bB.

C3 tickover: Alternative pathway C3 convertase perpetually generates C3b. C3 internal thioester bond hydrolysis is the initiating event.

C3bi (iC3b) (Figure 11.17) is the principal molecular product when factor I cleaves C3b. If complement receptor 1 or factor H is present, factor I can split C3bi's arginine–glutamic acid bond at position 954-955 to yield C3c and C3dg. C3bi attached to particles promotes phagocytosis when combined with complement receptor 3 on the surface of polymorphonuclear neutrophils (PMNs) and monocytes. It also promotes phagocytosis by binding to conglutinin in the serum of cows.

C3b (inactivated C3b): Also designated as C3bi.

Mo1 is an adhesive glycoprotein present on neutrophils and monocytes. It is termed Leu-CAM, which is an iC3b receptor, and thus helps mediate monocyte functions that are complement dependent.

CD11b: α chain of Mac-1 (C3bi receptor). It has a mol wt of 170 kDa and is present on granulocytes, monocytes, and NK cells.

iC3b-Neo: Complement component C3 expresses a neoepitope that is not present on native C3 or C3 degradation products following activation or cleavage by C3 convertases of either the classical or alternative pathway. Increased iC3b concentrations have been found in septic shock, transplant rejection, and in some patients with SLE but not others. Localized elevations have been found in atherosclerotic lesions, suggesting a role for complement in atherosclerosis. The clinical significance of iC3b determinations has not yet been determined.

C3c is the principal molecule that results from factor I cleavage of C3bi when factor H or complement receptor 1 is present. C3c is comprised of 27- and 43-kDa α chain fragments linked through disulfide bonds to a whole β chain.

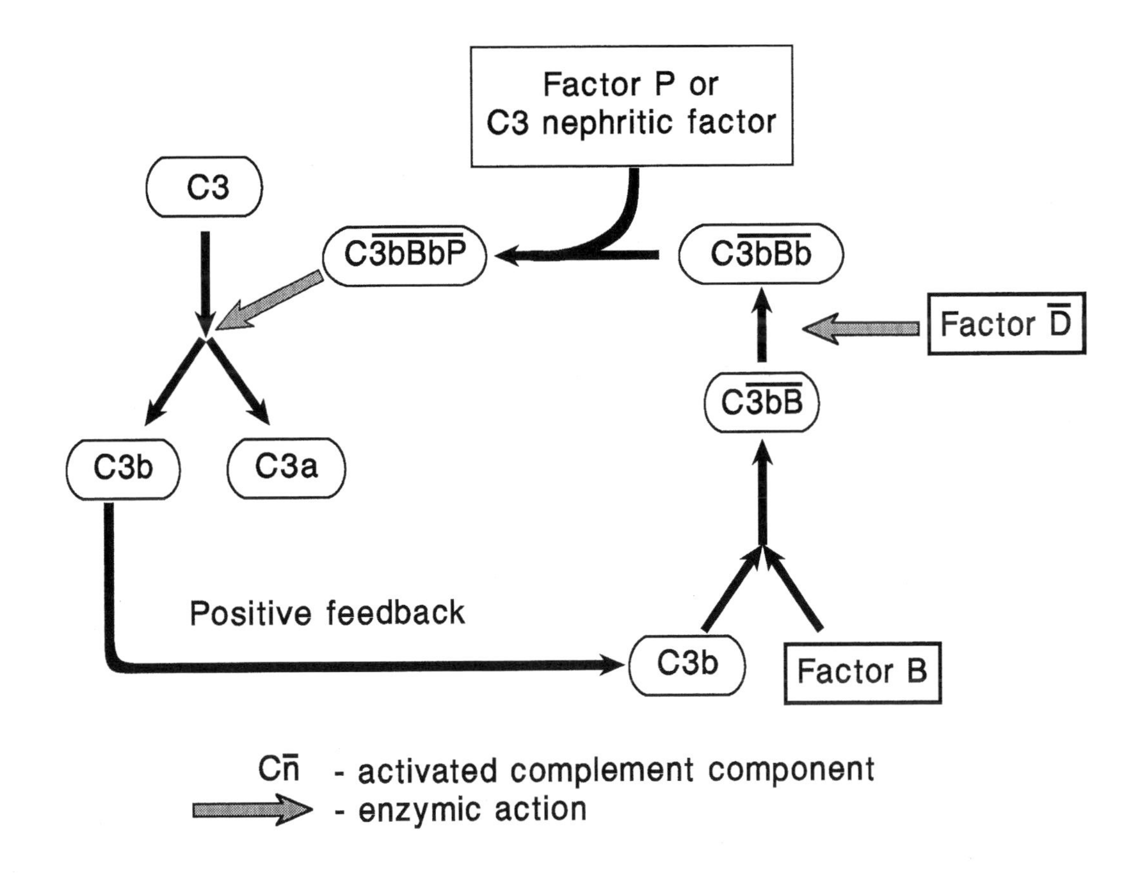

FIGURE 11.16 Alternative pathway C3 convertase.

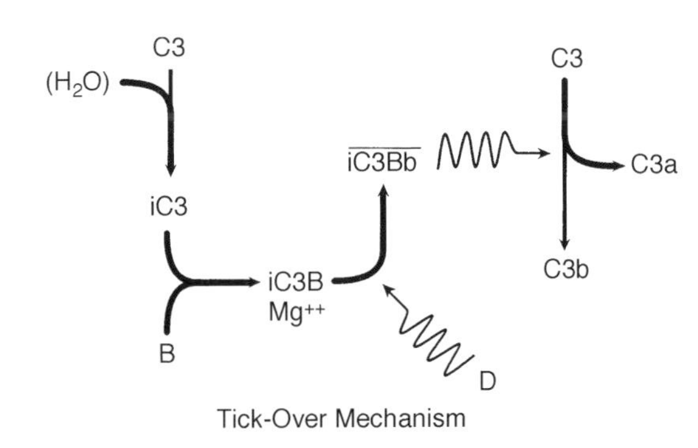

FIGURE 11.17 Through the C3 tickover mechanism alternative pathway, C3 convertase perpetually generates C3b. C3 internal thioester bond hydrolysis is the initiating event.

C3d is a 33-kDa B-cell growth factor that is formed by proteolytic enzyme splitting of a lysine–histidine bond in C3dg at position 1001-1002. C3d is comprised of the carboxy terminal 301-amino acid residues of C3dg. It interacts with complement receptor 2 on the surface of B cells. C3d contains the C3 α chain's thioester.

C3dg is a 41-kDa, 349-amino acid residue molecule formed by the cleavage of C3bi with factor H or complement receptor 1 present. Polymorphonuclear neutrophil leukocytes express complement receptor 4, which is reactive with C3dg. Complement receptor 2 on B cells is also a C3dg receptor.

C3e is a C3c α chain nonapeptide that causes leukocytosis. The peptide is comprised of Thr-Leu-Asp-Pro-Glu-Arg-Leu-Gly-Arg.

C3f is a 17-amino acid residue peptide split from the α chain of C3b by factor I with factor H or complement receptor 1 present.

C3g is an 8-kDa molecule comprised of C3dg's amino-terminal 47-amino acid residues. Trypsin digestion of C3dg yields C3g, whose function is unknown.

C3 nephritic factor (C3NeF) is an IgG autoantibody to the alternate complement pathway C3 convertase that mimics the action of properdin. C3NeF is present in the serum of patients with membranoproliferative glomerulonephritis type II (dense deposit disease). It stabilizes the alternate pathway C3 convertase, thereby enhancing the breakdown of C3, and produces hypocomplemententemia. Rarely, C3NeF may be IgG autoantibodies to C3 convertase $\overline{C4b2a}$ of the classical pathway. Systemic lupus erythematosus patients may contain antibodies against $\overline{C4b2a}$ that stabilize the classical pathway C3 convertase, leading to increased *in vivo* cleavage of C3.

NFt (C3bBb-P stabilizing factor) is another autoantibody to an alternate pathway C3 convertase present in patients with MPGN types 1 and 3. In patients with MPGN, NFt activity is inversely correlated with serum C3 concentrations.

NFc (nephritic factor of the classical pathway) binds to a neoantigen of the C4b2a complex and is associated with low C4 and C3 concentrations. It is found in patients with MPGN type I, acute postinfectious glomerulonephritis, SLE, and chronic glomerlunephritis.

Factor P (properdin) is a key participant in the alternative pathway of complement activation. It is a gammaglobulin, but not an immunoglobulin, that combines with C3b and stabilizes alternative pathway C3 convertase (C3bB) to produce C3bBbP. Factor P is a 3- or 4-polypeptide chain structure.

Factor B is an alternative complement pathway component. It is a 739-amino acid residue single-polypeptide chain that combines with C3b and is cleaved by factor D to produce alternative pathway C3 convertase. Cleavage by factor D is at an arginine–lysine bond at position 234 to 235 to yield an amino-terminal fragment Ba. The carboxy terminal fragment termed Bb remains attached to C3b. C3bBb is C3 convertase, and C3bBb3b is C5 convertase of the alternative complement pathway. The Bb fragment is the enzyme's active site. There are three short homologous, 60-amino acid residue repeats in factor B, and it possesses four attachment sites for N-linked oligosaccharides. Alleles for human factor B include *BfS* and *BfF.* The factor B gene is located in the major histocompatibility complex situated on the short arm of chromosome 6 in man and on chromosome 17 in mice. Also called C3 proactivator.

C3 PA (C3 proactivator) is an earlier designation for factor B.

Factor D is a serine esterase of the alternative pathway of complement activation. It splits factor B to produce Ba and Bb fragments. It is also called C3 activator convertase.

Factor D deficiency is an extremely rare genetic deficiency of factor D which has an X-linked or autosomal recessive pattern of inheritance. There is only 1% of physiologic amounts of factor D in the serum of affected patients, which renders them susceptible to repeated infection by *Neisseria* microorganisms. There are half the physiologic levels of factor D in the serum of heterozygotes which have no clinical symptoms related to this deficiency.

Factor H is a regulator of complement in the blood under physiologic conditions. Factor H is a glycoprotein in serum that unites with C3b and facilitates dissociation of alternative complement pathway C3 convertase, designated C3bBb, into C3b and Bb. Factor I splits C3b if factor H is present. In man, factor H is a 1231-amino acid residue single-polypeptide chain. It is comprised of 20 short homologous repeats that are comprised of about 60 residues present in proteins, which interact with C3 or C4. Factor H is an inhibitor of the alternative complement pathway. Previously called β-1H globulin.

β_1H: See factor H.

Factor H receptor (fH-R) is a receptor that initial studies have shown to be comprised of a 170-kDa protein expressed by Raji cells and tonsil B cells. Neutrophils, B lymphocytes, and monocytes express fH-R activity.

Factor H deficiency is an extremely rare genetic deficiency of Factor H which has an autosomal recessive mode of inheritance. Only 1% of the physiologic level of Factor H is present in the serum of affected individuals, which renders them susceptible to recurrent infections by pyogenic microorganisms. People who are heterozygotes contain 50% of normal levels of factor H in their serum and show no clinical effects.

Factor I (Figure 11.18) is a serine protease that splits the α chain of C3b to produce C3bi and the α chain of C4b to yield C4bi. Factor I splits a 17-amino acid residue peptide termed C3f, if factor H or complement receptor 1 are present, from the C3b α chain to yield C3bi. Factor I splits the C3bi, if complement receptor I or factor H are present, to yield C3c and C3dg. Factor I splits the C4b α chain, if C4-binding protein is present, to yield C4bi. C4c and C4d are produced by a second splitting of the α chain of C4bi. Factor I is a heterodimeric molecule. It is also called C3b/C4b inactivator.

C3b inactivator: See factor I.

C4b inactivator: See factor I.

Membrane cofactor of proteolysis (MCP or CD46) is a host-cell membrane protein that functions in association with factor I to cleave C3b to its inactive derivative iC3b, thereby blocking formation of convertase.

Membrane cofactor protein (MCP) is a regulatory protein of the complement system that is a type I transmembrane protein which is a cofactor for the factor I-mediated proteolytic cleavage and inactivation of C3b and C4b deposited on self tissue. MCP protects the cell on which

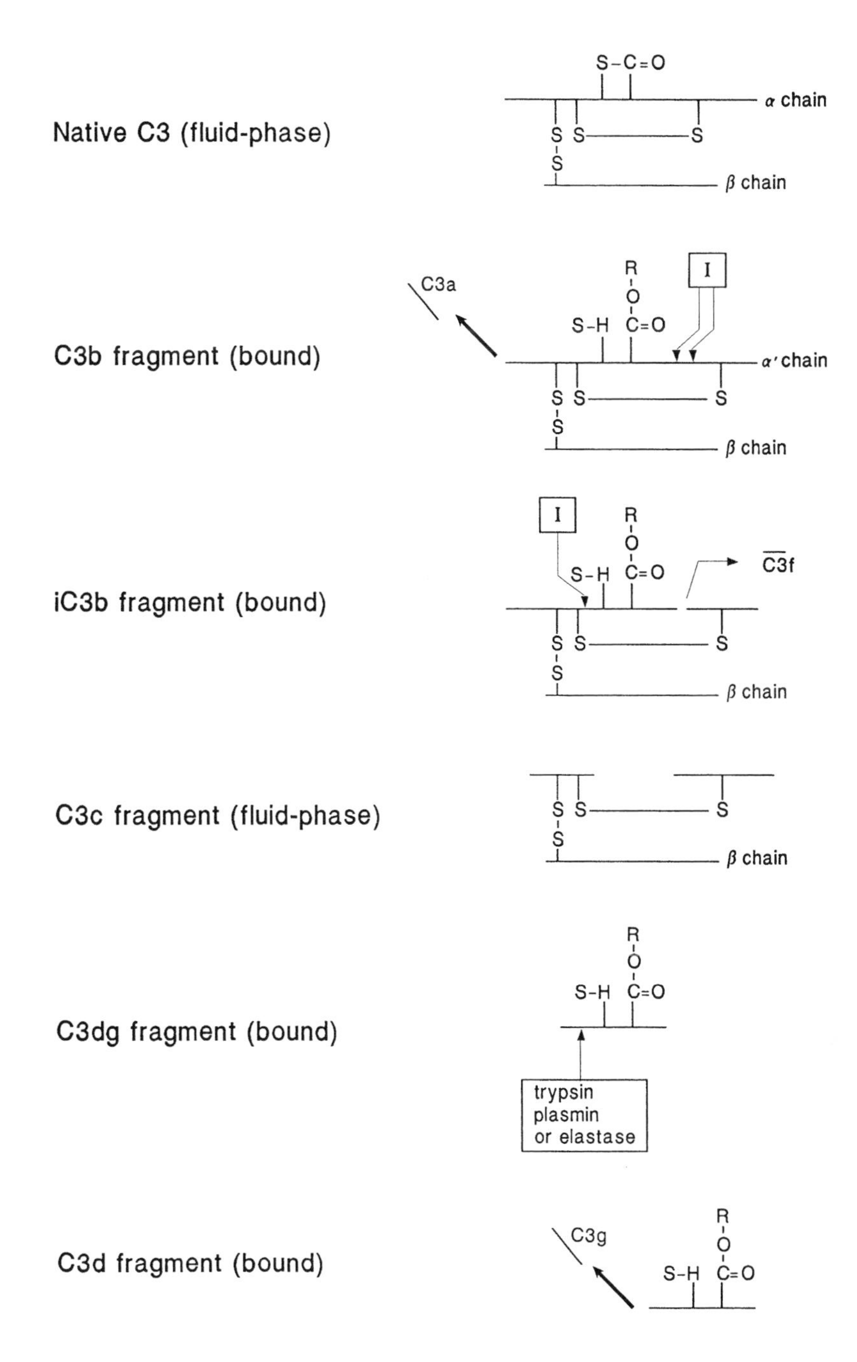

FIGURE 11.18 The binding of Factor I to C3b.

it is located rather than neighboring cells. It also serves as the measles virus receptor and participates in the adherence of *Streptococcus pyogenes* (group A streptococcus) to endothelial cells. MCP is a member of a group of proteins that are structurally, functionally, and genetically related. They are termed regulators of complement activation (RCA) and include decay-accelerating factor DAF (CD55), complement receptors 1 and 2 and factor H, and C4-binding protein in plasma. Also termed CD46.

Factor I deficiency is a very uncommon genetic deficiency of C3b inactivator. It has an autosomal recessive pattern of inheritance. There is less than 1% of the physiologic level of factor I in the serum of affected subjects, which renders them susceptible to repeated infections by pyogenic microorganisms. These individuals also reveal deficiencies of factor B and C3 in their serum since these components are normally split *in vivo* by alternative pathway C3 convertase (C3bBb), which factors I and H inhibit under physiologic conditions. These patients may develop urticaria because of the formation of C3a which induces the release of histamine.

C4b-binding protein (C4bp) is a 600-kDa protein in serum capable of binding six C4b molecules at once by means of seven spokes extending from a core at the center. C4b halts progression of complement activation. Factor I splits C4b molecules captured by C4bp. C4bp belongs to the regulators of complement activity molecules. C4bp interferes with C2a association with C4b. It also promotes C4b2a dissociation into C4b and C2a. It is also needed for the action of factor I in splitting C4b to C4bi and of C4bi into C4c and C4d. The *C4bp* gene is located on chromosome 1q3.2.

Long homologous repeat: See consensus sequence of C3/C4-binding proteins.

C4bi (iC4b) is the principal product of the reaction when factor I splits C4b. When C4b-binding protein is present, C4bi splits an α-chain arginine–threonine bond to yield C4c and C4d.

iC4b is a synonym for C4bi.

C4c is the principal product of Factor I cleavage of C4bi when C4b-binding protein is present. This 145-kDa molecule, of unknown function, is comprised of β and γ chains of C4 and two α-chain fragments.

C4d is a 45-kDa molecule produced by factor I cleavage of C4bi when C4b-binding protein is present. C4d is the molecule where Chido and Rodgers epitopes are located. It is also the location of the C4 α chain's internal thioester bond.

Complement inhibitors are protein inhibitors that occur naturally and block the action of complement components, including factor H, factor I, C1 inhibitor, and C4-binding protein (C4BP). Also included among complement inhibitors are heating to 56°C to inactivate C1 and C2, combination with hydrazine and ammonia to block the action of C3 and C4, and the addition of zymosan or cobra venom factor to induce alternate pathway activation of C3, which consumes C3 in the plasma.

Complement receptors (Figure 11.19) are receptors for products of complement reactions. Proteolytic cleavage of human complement component C3 takes place following activation of either the classical or the alternative complement pathway. Following the generation of C3a and C3b, the C3b covalently binds to bacteria, immune complexes, or some other target and then unites with a high-affinity receptor termed the C3b/C4b receptor currently known as CR1. Subsequent proteolytic cleavage of the bound C3b is attributable to factor I and a cofactor. This action yields C3bi, C3d, g, and C3c, which interact with specific receptors. CR2 is the C3dg receptor, and CR3 is the C3bi receptor.

Membrane complement receptors are receptors expressed on blood cells and tissue macrophages of man. These include Clq-R (Clq receptor), CR1 (C3b/C4b receptor; CD35), CR2 (C3d/Epstein-Barr virus (EBV) receptor; CD21), CR3 (iC3b receptor; CD11b/CD18), CR4 (C3bi receptor; CD11c/CD18), CR5 (C3dg-dimer receptor), fH-R (factor H receptor), C5a-R (C5a receptor) and C3a/C4a-R (C3a/C4a receptor). Ligands for C receptors generated by either the classic or alternate pathways include fluid-phase activation peptides of C3, C4, and C5 designated C3a, C4a, and C5a, which are anaphylatoxins that interact with either C3a/C4a-R or C5a-R, and participate in inflammation. Other ligands for C receptors include complement proteins deposited on immune complexes that are either soluble or particulate. Fixed C4 and C3 fragments (C4b,

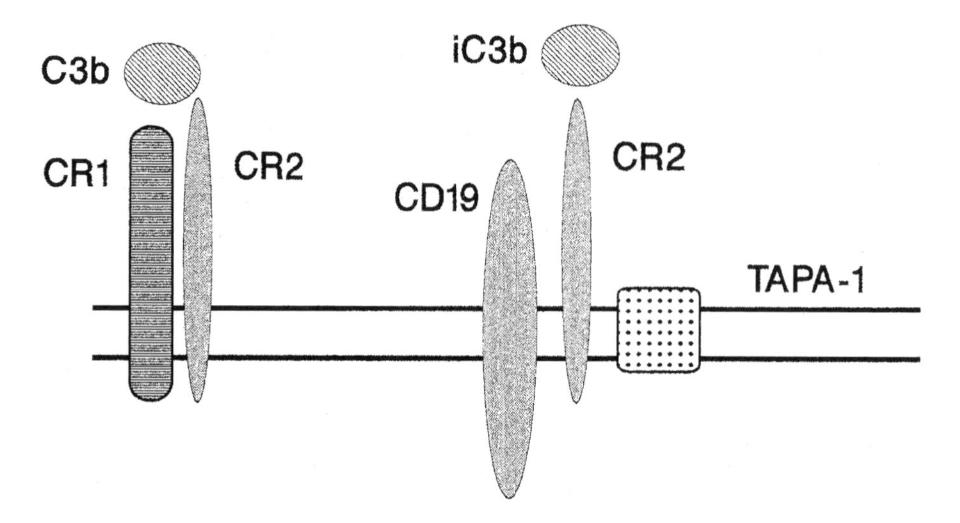

FIGURE 11.19 Complement receptor complexes on the surface of B cells.

C3b, C3bi, C3dg, and C3d), Clq, and factor H constitute these ligands. These receptors play a major role in facilitating improved recognition of pathogenic substances. They aid the elimination of bacteria and soluble immune complexes. Neutrophils, monocytes, and macrophages express C3 receptors on their surface. Neutrophils and erythrocytes express immune adherence receptors, termed CR1, on their surface. Four other receptors for C3 are designated CR2, CR3, CR4, and CR5. Additional receptors for complement components other than C3 and a receptor for C3a are termed Clq-R (Clq receptor), C5a-R (C5a receptor), C3a-R (C3a receptor), and fH-R (factor H receptor).

Complement receptor 1 (CR1) is a CR1 membrane glycoprotein found on human erythrocytes, monocytes, polymorphonuclear leukocytes, B cells, a T cell subset, mast cells, and glomerular podocytes. On red cells, CR1 binds C3b or C4b components of immune complexes, facilitating their transport to the mononuclear phagocyte system. CR1 facilitates attachment, endocytosis, and phagocytosis of C3b/C4b-containing complexes to macrophages or neutrophils and may serve as a cofactor for factor I-mediated C3 cleavage. The identification of CR1 cDNA has made possible the molecular analysis of CR1 biological properties.

C3b receptors: See complement receptor 1 (CR1).

CR1: See complement receptor 1.

Immune adherence receptor is a synonym for complement receptor 1.

Immune adherence is the attachment of antigen–antibody complexes or of antibody-coated bacteria or other particles carrying C3b or C4b to complement receptor 1 (CR1)-expressing cells. Erythrocytes of primates, B cells, T cells, phagocytic cells, and glomerular epithelial cells all express CR1. It is absent on the erythrocytes of other mammals but is found on their thrombocytes. Immune adherence facilitates the elimination of antigen–antibody complexes from the blood circulation, especially through their attachment to red blood cells and platelets followed by uptake by phagocytic cells through complement receptor 3.

Liberated CR1 is a truncated complement receptor-1 (CR1) without transmembrane and intracytoplasmic domain that may help to limit the size of myocardial infarcts by diminishing complement activation. Liberated CR1 may also play a therapeutic role in other types of ischemia, burns, autoimmunity, and inflammation because it is a natural inhibitor of complement activation.

CD35 is an antigen that is known as CR1 that binds C3b and C4b. The CD35 antigen includes four different allotypes termed C, A, B, and D, with mol wts of 160 (C), 190 (A), 220 (B), and 250 kDa (D), respectively. The antigen is widely distributed on various cell types, including erythrocytes, B cells, monocytes, granulocytes (negative on basophils), some NK cells, and follicular dendritic cells. The functions of CR1 are processing of immune complexes and promotion of binding and phagocytosis of C3b-coated particles/cells.

sCR1 (soluble complement receptor type 1) is a substance prepared by recombinant DNA technology that combines with activated C3b and C4. This facilitates complement factor I's inactivation of them. sCR1 significantly diminishes myocardial injury induced by hypoxia in rats.

Complement receptor 2 (CR2) is a receptor for C3 fragments that also serves as a binding site for Epstein–Barr virus (EBV). It is a receptor for C3bi, C3dg, and C3d, based on its specificity for their C3d structure. B cells, follicular dendritic cells of lymph nodes, thymocytes, and pharyngeal epithelial cells, but not T cells, express CR2. EBV enters B lymphocytes by way of CR2. The gene encoding CR2 is linked closely with that of CR1. A 140-kDa single-polypeptide chain makes up CR2, which has a short consensus repeat (SCR) structure similar to that of CR1. CR2 may be active in B cell activation. Its expression appears restricted to late pre-B and to mature B cells. CR2 function is associated with membrane IgM. Analysis of cDNA clones has provided CR2's primary structure.

CR2: See complement receptor 2.

CR2, Type II complement receptor is a coreceptor on B lymphocytes that binds to antigens coated with complement at the same time that membrane immunoglobulin binds an epitope of the antigen.

Complement receptor 3 (CR3) is a principal opsonin receptor expressed by monocytes, macrophages, and neutrophils. It plays an important role in the removal of bacteria. CR3 binds fixed C3bi in the presence of divalent cations. It also binds bacterial lipopolysaccharides and β glucans of yeast cell walls. The latter are significant in the ability of granulocytes to identify bacteria and yeast cells. CR3 is an integrin type of adhesion molecule that facilitates the binding of neutrophils to endothelial cells in inflammation. CR3 enables phagocytic cells to attach to bacteria or yeast cells with fixed C3bi, β glucans, or lipopolysaccharide on their surface. This facilitates phagocytosis and the respiratory burst. CR3 is comprised of a 165-kDa α-glycoprotein chain and a 95-kDa β-glycoprotein chain. CR3 appears related to LFA-1 and P150, has 95 molecules, and shares a β chain with them. All three of these molecules are of critical significance in antigen-independent cellular adhesion, which confines leukocytes to inflammatory areas, among other functions. Deficient surface expression of these molecules occurs in leukocyte adhesion deficiency (LAD), in which patients experience repeated bacterial infections. The primary defect appears associated with the common β chain. Besides C3bi binding, CR3 is of critical significance in IgG- and CR1-facilitated phagocytosis by neutrophils and monocytes. CR3 has a more diverse function than does either CR1 or CR2.

CR3 deficiency syndrome: See leukocyte adhesion deficiency.

Complement receptor 4 (CR4) is a glycoprotein membrane receptor for C3dg on polymorphonuclear neutrophils (PMNs), monocytes, and platelets. CR4 facilitates Fc receptor-mediated phagocytosis and mediates Fc-independent phagocytosis. It consists of a 150-kDa α chain and a 95-kDa β chain. Chromosome 16 is the site of genes that encode the α chain, whereas chromosome 21 is the site of genes that encode the β chain. Tissue macrophages express CR4. It is an integrin with a β chain in common with CR3 and LFA-1.

Complement receptor 5 (CR5) is a receptor that binds C3bi, C3dg, and C3d fragments based on its specificity for their C3d component. Reactivity is only in the fluid phase and not when the fragments are fixed. CR5 is the C3dg-dimer receptor. Neutrophils and platelets manifest CR5 activity.

Decay-accelerating factor (DAF) (Figure 11.20) is a 70-kDa membrane glycoprotein of normal human erythrocytes, leukocytes, and platelets, but it is absent from the red blood cells of paroxysmal nocturnal hemaglobulinuria patients. It facilitates dissociation of classical complement pathway C3 convertase (C4b2a) into C4b and C2a. It also promotes the dissociation of alternative complement pathway C3 convertase (C3bBb) into C3b and Bb. DAF inhibits C5 convertases (C4b2a3b and C3bBb3b) on the cell surface. It reacts with the convertases and destabilizes them by inducing rapid dissociation of a catalytic subunit C2a or Bb. The inhibitory effect of DAF is restricted to C3/C5 convertases bound to host cells but it does not inhibit normal complement activation on microbial or immune complex targets. DAF is found on selected mucosal epithelial cells and endothelial cells. It prevents complement cascade amplification on the surfaces of cells to protect them from injury by autologous complement. The physiologic function of DAF may be to protect cells from lysis by serum. DAF competes with C2 for linkage with C4b to block C3 convertase synthesis in the classical pathway. The DAF molecule consists of a single chain bound to the cell membrane by phosphatidyl inositol. Paroxysmal nocturnal hemoglobulinuria develops as a consequence of DAF deficiency.

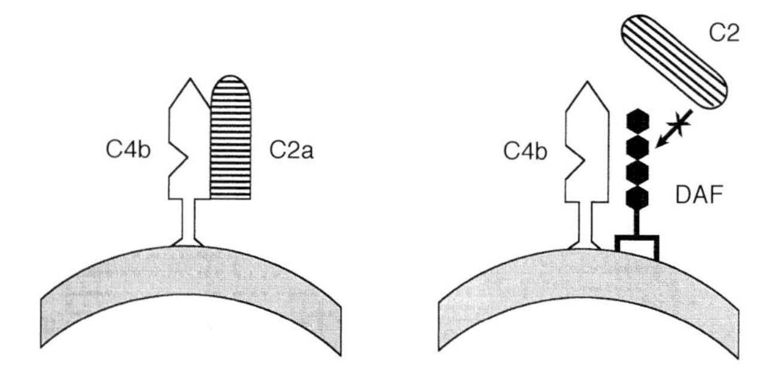

FIGURE 11.20 Decay acceleration factor.

DAF: See decay-accelerating factor.

Homologous restriction factor (HRF) is an erythrocyte surface protein that prevents cell lysis by homologous complement on its surface. It bears a structural resemblance to C8 and C9.

Paroxysmal nocturnal hemoglobinuria (PNH) is a rare form of hemolytic anemia in which the red blood cells, as well as neutrophils and platelets, manifest strikingly increased sensitivity to complement lysis. PNH red blood cell membranes are deficient in decay-accelerating factor (DAF), LFA-3, and Fc γ RIII. Without DAF, which protects the cell membranes from complement lysis by classic pathway C5 convertase and decreases membrane attack complex formation, the erythrocytes and lymphocytes are highly susceptible to lysis by complement. Interaction of these PNH erythrocytes with activated complement results in excessive C3b binding that leads to the formation of more C3b through the alternate complement pathway by way of factors B and D. Intravascular hemolysis follows activation of C5 convertase in the C5-9 membrane attack complex (MAC). The blood platelets and myelocytes in affected subjects are also DAF deficient and are readily lysed by complement. There is leukopenia, thrombocytopenia, iron deficiency, and diminished leukocyte alkaline phosphatase. The Coombs' test is negative, and there is also very low acetylcholine esterase activity in the red cell membrane. No antibody participating in this process has been found in either the serum or on the erythrocytes. The disease is suggested by episodes of intravascular hemolysis, iron deficiency, and hemosiderin in the urine. It is confirmed by hemolysis in acid medium, termed the HAM test.

Complement fixation (Figure 11.21) is a primary union of an antigen with an antibody in the complement fixation reaction that takes place almost instantaneously and is invisible. A measured amount of complement present in the reaction mixture is taken up by complexes of antigen and antibody. Complement fixation is the consumption or binding of complement by antigen–antibody complexes. This serves as the basis for a serologic assay in which antigen is combined with a serum specimen suspected of containing the homologous antibody. Following the addition of a measured amount of complement, which is fixed

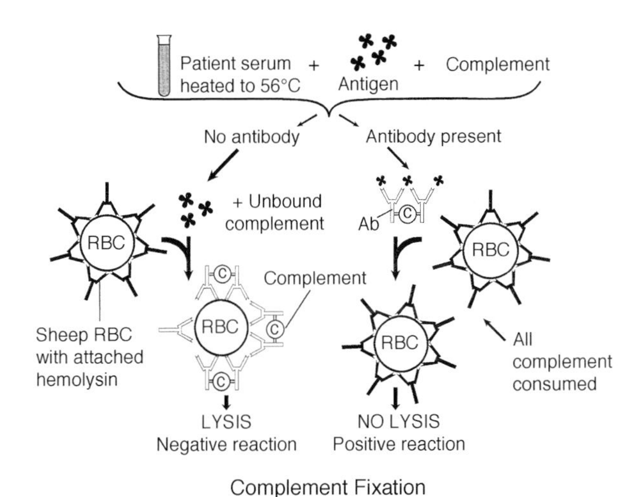

FIGURE 11.21 Complement fixation.

or consumed only if antibody was present in the serum and has formed a complex with the antigen, sheep red blood cells sensitized (coated) with specific antibody are added to determine whether or not complement has been fixed in the first phase of the reaction. Failure of the sensitized sheep red blood cells to lyse constitutes a positive test since no complement is available. However, sheep red blood cell lysis indicates that complement was not consumed during the first phase of the reaction, implying that homologous antibody was not present in the serum, and complement remains free to lyse the sheep red blood cells sensitized with antibody. Hemolysis constitutes a negative reaction. The sensitivity of the complement fixation test falls between that of agglutination and precipitation. Complement fixation tests may be carried out in microtiter plates, which are designed for the use of relatively small volumes of reagents. The lysis of sheep red blood cells sensitized with rabbit antibody is measured either in a spectrophotometer at 413 nm or by the release of ^{51}Cr from red cells that have been previously labeled with the isotope. Complement fixation can detect either soluble or insoluble antigen. Its ability to detect virus antigens in impure tissue preparations makes the test still useful in diagnosis of virus infections.

Complement fixing antibody is an antibody of the IgG or IgM class that binds complement after reacting with its homologous antigen. This represents complement fixation by the classic pathway. Nonantibody mechanisms, as well as IgA, fix complement by the alternative pathway.

The **complement fixation assay** is a serologic test based on the fixation of complement by antigen–antibody complexes. It has been applied to many antigen–antibody systems and was widely used earlier in the century as a serologic test for syphilis.

The **complement fixation inhibition test** is an assay in which a known substance prevents interaction of antibody and antigen, thereby preventing complement uptake and fixation.

The **CH_{50} unit** is the amount of complement (serum dilution) that induces lysis of 50% of erythrocytes sensitized (coated) with specific antibody. More specifically, the 50% lysis should be of 5×10^8 sheep erythrocytes sensitized with specific antibody during 60 min of incubation at 37°C. To obtain the complement titer, i.e., the number CH50 present in 1 ml of serum that has not been diluted, the log $y/1 - y$ (y = % lysis) is plotted against the log of the quantity of serum. At 50% lysis, the plot approaches linearity near $y/1 - y$.

The **indirect complement fixation test** is a complement fixation assay for antibodies that are unable to fix guinea pig complement. It involves the addition of a rabbit antibody of established guinea pig complement-fixing capacity to an antigen–(avian) antibody–guinea pig complement complex. This is followed by the addition of a visible hemolytic system. Cell lysis indicates the initial presence of avian antibody.

AH50 is an assay to measure the functional activity of the alternative complement pathway. This alternative pathway (AP) hemolytic assay (AH50) employs complement-mediated lysis of rabbit red blood cells in Mg^{2+}-EGTA buffer. The EGTA chelates Ca^{2+} needed for complement C1 macromolecular assembly, thereby blocking classical complement pathway activity.

The AH50 assay is valuable to screen for homozygous deficiencies of AP constituents including C3, factor I, factor B, properdin, factor H, and factor D. Deficiencies of the fluid phase regulatory proteins of the AP, factor I, or factor H induce a secondary deficiency of C3, properdin, and factor B as a result of uncontrolled consumption of these proteins. The lack of hemolysis in both the AH50 and CH50 assays strongly suggests a deficiency in a complement terminal pathway component (C5–C9). When the AH50 values are normal and there is no lysis in CH50, there is a classical pathway component deficiency (C1, C2, or C4). A normal CH50 and a lack of hemolysis in AH50 firmly suggests an AP component deficiency. A lack of lysis strongly points to homozygous deficiency. Low levels of lysis in AH50 or CH50 assays suggest either heterozygous deficiency or complement activation. Hemolytic functional assays for individual complement components facilitate the detection of homozygous and some heterozygous deficiencies in addition to mutations that lead to inactive complement protein. Usually, AP deficiency is associated with recurrent bacterial infections. Homozygous factor H deficiency has been observed in glomerulonephritis, recurrent pyogenic infection, hemolytic–uremic syndrome, systemic lupus erythematosus, or in healthy subjects. Factor D deficiency has been linked to recurrent upper respiratory infections with *Neisseria* species.

Minimal hemolytic dose (MHD) refers to an amount of complement used in a complement fixation test. (MHD is also an abbreviation for minimal hemagglutinating dose of a virus, as in a hemagglutination inhibition test.) MHD is the smallest amount of complement that can completely lyse a defined volume of a standardized suspension of red blood cells sensitized with antibody.

The **Neisser-Wechsberg phenomenon** is the deviation of complement by antibody. Although complement does not react with antigen alone, it demonstrates weak affinity for unreacted antibody.

HD_{50} is an uncommon designation for CH_{50}, which refers to hemolytic complement activity of a serum sample.

Complement deviation (Neisser-Wechsberg phenomenon) is the blocking of complement fixation or of complement-induced lysis when excess antibody is present. There is deviation of the complement by the antibody.

Anticomplementary is the action of any agent or treatment that interferes with complement fixation through removal or inactivation of complement components or cascade reactants. Multiple substances may exhibit anticomplementary activity. These are especially significant in complement-fixation serology since anticomplementary agents may impair the evaluation of test results.

Cytolytic is an adjective describing the property of disrupting a cell.

Cascade reaction is a sequential event in such enzymatic reactions as complement fixation and blood coagulation in which each stage of the reaction triggers the next appropriate step.

Decomplementation is the deliberate inactivation of complement either *in vitro* or *in vivo*. To decomplement serum to remove hemolytic action, the specimen may be heated to 56°C for 30 min. Other methods for inactivation of complement include the addition of cobra venom factor, zymosan, or other substances which take up complement from the medium in which they are placed. Removal of complement activity in living animals may be accomplished through the injection of cobra venom factor or other substances to use up or inactivate the complement system.

Hemolysin is a complement-fixing antibody that lyses erythrocytes in the presence of complement. It is an antibody which acts together with complement to produce an interruption in the membrane integrity of red blood cells, causing disruption. Historically, the term refers to rabbit antisheep erythrocyte antibody used in the visible part of a complement fixation test. Microbial products such as streptolysin O may disrupt (lyse) red blood cells in agar medium.

Heat inactivation is the loss of biological activity through heating. Heating a serum sample at 56°C in a water bath for 30 min destroys complement activity through inactivation of C1, C2, and factor B. By diminishing the heat to 52°C for 30 min, only factor B is destroyed, whereas C1 and C2 remain intact. This inactivates the alternative complement pathway but not the classical pathway. IgG, IgM, and IgA are resistant to incubation in a 56°C water bath for 30 min, whereas IgE antibodies are destroyed by this temperature.

The **hemolytic system** is the visible phase of a complement fixation test in which sheep red blood cells combine with their homologous antibody and are added to a system where antigen is mixed with patient serum, presumed to contain specific antibody, followed by complement. If antibody is present in the patient's serum and combines with the antigen, the added complement is fixed and is no longer available to react with antibody-coated sheep red cells, which are added as the second phase or visible reaction. Lysis indicates the presence of free complement from the first phase of the reaction, indicating that antibody against the antigen in question was not present in the patient's serum. By contrast, no hemolysis indicates the presence of the antibody in the patient's serum.

The hemolytic system may also be used to describe a reaction in which erythrocytes are combined with their homologous antibody. The addition of all nine components of complement together with calcium and magnesium ions followed by incubation at 37°C results in immune lysis (hemolysis).

EAC is an abbreviation for erythrocyte (E), antibody (A), and complement (C), as designated in studies involving the complement cascade. Historically, sheep red blood cells have been combined with antibody specific for them, and lysis is induced only after the addition of complement. This interaction of the three reagents has served as a useful mechanism to study the reaction sequence of the multiple component complement system, which is often written as EAC1423, etc. EAC may also be employed to reveal the presence of complement receptors 1 through 4. The combination of IgM-coated sheep red blood cells with sublytic quantities of complement to produce EAC provides a mechanism whereby rosettes are formed by cells bearing complement receptors once the EAC are layered onto test cells at room temperature. EAC rosette erythrocytes such as sheep red blood cells, coated or synthesized with antibody and complement, encircle human B lymphocytes to form a rosette. This is based on erythrocyte binding to C3b receptors. This technique has been used in the past to identify B lymphocytes. However, phagocytic cells bearing C3b receptors may also form EAC rosettes. This type of rosette is in contrast to the E rosette, which identifies T lymphocytes. B lymphocytes are now enumerated by monoclonal antibodies against the B lymphocyte CD markers using flow cytometry.

A **target cell** is a cell that is the object of an immune attack mediated either by antibodies and complement or by specifically immune lymphoid cells. The target cell must bear an antigen for which the antigen-binding regions of either antibody molecules or of the T cell receptors are specific.

A **cytolytic reaction** is cell destruction produced by antibody and complement or perforin released from cytotoxic T lymphocytes. Antibody that combines with cell surface epitopes followed by complement fixation that leads to cell lysis or cell membrane injury without lysis.

Cytotoxicity is the fatal injury of target cells by either specific antibody and complement or specifically sensitized cytotoxic T cells, activated macrophages, or natural killer (NK) cells. Dye exclusion tests are used to assay cytotoxicity produced by specific antibody and complement. Measurement of the release of radiolabel or other cellular constituents in the supernatant of the reacting medium is used to determine effector cell-mediated cytotoxicity.

A **one-hit theory** is a concept related to complement activation which states that only a single site of preparation is necessary for red cell lysis during complement-antibody-mediated injury to the cell membrane.

Inactivation is a term used mostly by immunologists to signify loss of complement activity in a serum sample that has been heated to 56°C for 30 min or by hydrazine treatment. Inactivation also applies to chemical or heat treatment of pathogenic microorganisms in a manner that preserves their antigenicity for use in inactivated vaccines.

Reactive lysis is dissolution of red blood cells not sensitized with antibody. Initiated by C5b and C6 complexes in the presence of C7, C8, and C9. The activation of complement leads to lysis as a "bystander phenomenon."

The **von Krough equation** is a mathematical equation to ascertain serum hemolytic complement titer. It correlates complement with the extent of lysis of red blood cells coated with antierythrocyte antibodies under standard conditions.

Complement deficiency conditions are rare inherited deficiencies. In healthy Japanese blood donors, only 1 in 100,000 persons had no C5, C6, C7, and C8. No C9 was contained in 3 of 1000 individuals. Most individuals with missing complement components do not manifest clinical symptoms. Additional pathways provide complement-dependent functions that are necessary to preserve life. If C3, factor I, or any segment of the alternative pathway is missing, the condition may be life-threatening, with markedly decreased opsonization in phagocytosis. C3 is depleted when factor I is absent. C5, C6, C7, or C8 deficiencies are linked with infections, mainly meningococcal or gonococcal, which usually succumb to the bactericidal action of complement. Deficiencies in classical complement pathway activation are often associated with connective tissue or immune complex diseases. Systemic lupus erythematosus may be associated with C1qrs, C2, or C4 deficiencies. Hereditary angioedema (HAE) patients have a deficiency of C1 inactivator. A number of experimental animals with specific complement deficiencies has been described, such as C6 deficiency in rabbits and C5 deficiency in mice. Acquired complement deficiencies may be caused by either accelerated complement consumption in immune complex diseases with a type III mechanism or by diminished formation of complement proteins as in acute necrosis of the liver.

C1 deficiencies are also very rare. Only a few cases of C1q and C1r, or C1r and C1s deficiencies have been reported. These have an autosomal recessive mode of inheritance. Patients with these defects may manifest systemic lupus erythematosus, glomerulonephritis, or pyogenic infections. They have an increased incidence of type III (immune complex) hypersensitivity diseases. Half of C1q-deficient persons may contain physiologic levels of mutant C1q that are not functional.

C1 inhibitor (C1 INH) deficiencies are the most frequently found deficiency of the classic complement pathway and may be seen in patients with hereditary angioneurotic edema. This syndrome may be expressed as either a lack of the inhibitor substance a functionally inactive C1 INH. The patient develops edema of the face, respiratory tract, including the glottis and bronchi, and the extremities. Severe abdominal pain may occur with intestinal involvement. Since C1 INH can block Hagemann factor (Factor XII) in the blood-clotting mechanism, its absence can lead to the liberation of kinin and fibrinolysis, which results from the activation of plasmin. The disease is inherited as an autosomal dominant trait. When edema of the larynx occurs, the patient may die of asphyxiation. When abdominal attacks occur, there may be watery diarrhea and vomiting. These bouts usually span 48 h and are followed by a rapid recovery. During an attack of angioedema, C1r is activated to produce C1s, which depletes its substrates C4 and C2. The action of activated C1s on C4 and C2 leads to the production of a substance that increases vascular permeability, especially that of postcapillary venules. C1 and C4 cooperate with plasmin to split this active peptide from C2. Of the families of patients with hereditary angioneurotic edema, 85% do not contain C1 INH. Treatment is by preventive maintenance. Patients are given inhibitors of plasmin such as aminocaproic acid and tranexamic acid. Methyl testosterone, which causes synthesis of normal C1 INH in angioneurotic edema patients, is effective by an unknown mechanism.

Hereditary angioneurotic edema (HANE) is a disorder in which recurrent attacks of edema, persisting for 48 to 72 h, occur in the skin and gastrointestinal and respiratory tracts. It is nonpitting and life threatening if laryngeal edema becomes severe enough to obstruct the airway. Edema in the jejunum may be associated with abdominal cramps and bilious vomiting. Edema of the colon may lead to watery diarrhea. There is no redness or itching associated with edema of the skin. Tissue trauma or no apparent initiating cause may induce an attack. HANE is a condition induced by C1q esterase inhibitor (Clq-INH) deficiency in which immune complexes induce uptake of activated C1, C4, and C2. This may be activated by trauma, cold, vibration or other physical stimuli, histamine release, or menstruation. HANE induces nonpitting swelling that is not pruritic and not urticarial that reaches a peak within 12 to 18 h. It occurs on the face, extremities, toes, fingers, elbows, knees, gastrointestinal mucosa, and oral pharynx. It can produce edema of the epiglottis, which is fatal in approximately one third of the cases. Patients may have nausea and abdominal pain with vomiting. HANE is inherited in an autosomal dominant fashion. Heterozygotes for the defect develop the disorder. Greatly diminished C1 INH levels (5 to 30% of normal) are found in affected individuals. Activation of C1 leads to increased cleavage of C4 and C2, decreasing their serum levels during an attack. C1 INH is also a kinin system inactivator. The C1 INH deficiency in HANE permits a kinin-like peptide produced from C2b to increase vascular permeability, leading to manifestations of HANE. Some have proposed that bradykinin may represent the vasopermeability factor. Four types have been described. Two are congenital. In type I, which accounts for 85% of cases, Clq-INH is diminished to 30% of normal. In type II variant, the product of the gene is present but does not function properly. Type II may be acquired or autoimmune. The acquired type is associated with certain lymphoproliferative disorders such as Waldenström's macroglobulinemia, IgA myeloma, chronic lymphocytic leukemia, or other types of B-cell proliferation. The autoimmune type is linked to IgG1 autoantibodies. There is unregulated Cls activation. Hereditary angioneurotic edema has been treated with ε aminocaproic acid and transexamic acid, but they do not elevate C1 INH or C4 levels. Anabolic steroids such as danazol and stanozolol, which activate C1 INH synthesis in affected individuals, represent the treatment of choice.

Angioedema refers to the significant localized swelling of tissues as a consequence of complement activation which takes place when C1 esterase inhibitor is lacking. Angioedema may also describe skin swelling following IgE-mediated allergic reactions that cause increased permeability of subcutaneous blood vessels.

Angiogenesis factor is a macrophage-derived protein that facilitates neovascularization through stimulation of vascular endothelial cell growth. Among the five angiogenesis factors known, basic fibroblast growth factor may facilitate neovascularization in type IV delayed hypersensitivity responses.

C1q deficiency may be found in association with lupus-like syndromes. C1r deficiency, which is inherited as an autosomal recessive trait, may be associated with respiratory tract infections, glomerulonephritis, and skin manifestations which resemble a systemic lupus erythematosus (SLE)-like disease. C1s deficiency is transmitted as an autosomal dominant trait, and patients may again show SLE-like signs and symptoms. Their antigen–antibody complexes can persist without resolution.

C2 deficiency is rarely found in individuals. Although no symptoms are normally associated with this trait, which has an autosomal recessive mode of inheritance, autoimmune-like manifestations that resemble features of certain collagen-vascular diseases, such as systemic lupus erythematosus, may appear. Thus, many genetically determined complement deficiencies are not associated with signs and symptoms of disease. When they do occur, it is usually manifested as an increased incidence of infectious diseases which affect the kidneys, respiratory tract, skin, and joints.

C3 deficiency is an extremely uncommon genetic disorder that may be associated with repeated serious pyogenic bacterial infections and that may lead to death. The C3 deficient individuals are deprived of appropriate opsonization, prompt phagocytosis, and the ability to kill infecting microorganisms. There is defective classical and alternative pathway activation. Besides infections, these individuals may also develop an immune complex disease such as glomerulonephritis. C3 levels that are one half normal in heterozygotes are apparently sufficient to avoid the clinical consequences induced by a lack of C3 in the serum.

C4 deficiency is an uncommon genetic defect with an autosomal recessive mode of inheritance. Affected individuals have defective classical complement pathway activation. Those who manifest clinical consequences of the defect may develop SLE or glomeruloneophritis. Half of the patients with C4 and C2 deficiencies develop SLE, but deficiencies in these two complement components are not usually linked to increased infections.

C5 deficiency is a very uncommon genetic disorder that has an autosomal recessive mode of inheritance. Affected individuals have only trace amounts of C5 in their plasma. They have a defective ability to form the MAC, which is necessary for the efficient lysis of invading microorganisms. They have an increased susceptibility to disseminated infections by *Neisseria* microorganisms such as *N. meningitidis* and *N. gonorrhoeae*. Heterozygotes may manifest 13 to 65% of C5 activity in their plasma and usually show no clinical effects of their partial deficiency. C5 deficient mice also have been described.

C6 deficiency is a highly uncommon genetic defect with an autosomal recessive mode of inheritance in which affected individuals have only trace amounts of C6 in their plasma. They are defective in the ability to form an MAC and have increased susceptibility to disseminated infections by *Neisseria* microorganisms, which include gonococci and meningococci. C6-deficient rabbits have been described.

C7 deficiency is a highly uncommon genetic disorder with an autosomal recessive mode of inheritance in which the serum of affected persons contains only trace amounts of C7 in the plasma. They have a defective ability to form an MAC and show an increased incidence of disseminated infections caused by *Neisseria* microorganisms. Some may manifest an increased propensity to develop immune complex (type III hypersensitivity) diseases such as glomerulonephritis or systemic lupus erythematosus.

C8 deficiency is a highly uncommon genetic disorder with an autosomal recessive mode of inheritance in which affected individuals are missing C8 α, γ, or β chains. This is associated with a defective ability to form an MAC. Individuals may have an increased propensity to develop disseminated infections caused by *Neisseria* microorganisms such as meningococci.

C9 deficiency is a highly uncommon genetic disorder with an autosomal recessive mode of inheritance in which only trace amounts of C9 are present in the plasma of affected persons. There is a defective ability to form the MAC. The serum of C9-deficient subjects retains its lytic and bactericidal activity, even though the rate of lysis is decreased compared to that induced in the presence of C9. There are usually no clinical consequences associated with this condition. The disorder is more common in the Japanese than in most other populations.

Hereditary complement deficiencies are associated with defects in activation of the classical pathway that lead to increased susceptibility to pyogenic infections. A deficiency of C3 leads to a defect in activation of both the classical and alternative pathways that lead to an increased frequency of pyogenic infections that may prove fatal. Such individuals also have defective opsonization and phagocytosis. Defects of alternate pathway factors D and P lead to impaired activation of the alternative pathway with increased susceptibility to pyogenic infections. Deficiencies of C5 through C9 are associated with defective MAC formation and lysis of cells, including bacteria. This produces increased susceptibility to disseminated *Neisseria* infection.

Conglutinin is a bovine serum protein that reacts with fixed C3. It causes the tight aggregation, i.e., conglutination, of red blood cells coated with complement. Conglutinin, which is confined to sera of Bovidae, is not to be confused with immunoconglutinin, which has a similar activity, but is produced in other species by immunization with complement-coated substances or may develop spontaneously following activation of complement *in vivo*. Conglutinin reacts with antigen–antibody–complement complexes in a medium containing Ca^{2+}. The N-linked oligosaccharide of the C3bi α chain is its ligand. The phagocytosis of C3bi-containing immune complexes is increased by interaction with conglutinin. Conglutinin contains 12 33-kDa polypeptide chains that are indistinguishable and are grouped into four subunits. Following a 25-residue amino terminal sequence, there is a 13-kDa sequence that resembles collagen in each chain. The 20-kDa carboxy terminal segments form globular structures that contain disulfide-linked chains.

Conglutination is the strong agglutination of antigen–antibody–complement complexes by conglutinin, a factor present in normal sera of cows and other ruminants. The complexes are similar to EAC1423 and are aggregated by conglutinin in the presence of Ca^{2+}, which is a required cation. Conglutination is a sensitive technique for detecting complement-fixing antibodies.

Conglutinin solid phase assay is a test that quantifies C3bi-containing complexes that may activate complement by either the classical or the alternate pathways.

Conglutinating complement absorption test is an assay based on the removal of complement from the reaction medium if an antigen–antibody complex develops. This is a test for antibody. As in the complement fixation test, a visible or indicator combination must be added to determine whether any unbound complement is present. This is accomplished by adding sensitized erythrocytes and conglutinin, which is prepared by combining sheep erythrocytes with bovine serum that contains natural antibody against sheep erythrocytes as well as conglutinin. Horse serum may be used as a source of nonhemolytic complement for the reaction. Aggregation of the erythrocytes constitutes a negative test.

Immunoconglutinin is an autoantibody, usually of the IgM class, that is specific for neoantigens in C3bi or C4bi. It may be stimulated during acute and chronic infections caused by bacteria, viruses, or parasites and in chronic inflammatory disorders. The level is also increased in many autoimmune diseases, as well as following immunization with various immunogens. C3 nephritic factor is an example of an immunoconglutinin. Also called *immune conglutinin,* but it should not be confused with *conglutinin.* See immunoconglutination.

Immunoconglutination is the aggregation of C3bi- or C4bi-coated erythrocytes or bacteria by antibodies to C3bi or C4bi, produced by immunizing animals with erythrocytes or bacteria that have interacted with antibody and complement. The antibody is termed immunoconglutinin, which resembles conglutinin in its activity, but should not be confused with it. Autoantibodies specific for C3bi or C4bi may develop as a result of an infection and produce immunoconglutination.

12 Types I, II, III, and IV Hypersensitivity

Hypersensitivity is increased reactivity or increased sensitivity by the animal body to an antigen to which it has been previously exposed. The term is often used as a synonym for allergy, which describes a state of altered reactivity to an antigen. Hypersensitivity has been divided into categories based upon whether it can be passively transferred by antibodies or by specifically immune lymphoid cells. The most widely adopted current classification is that of Coombs and Gell that designates immunoglobulin-mediated (immediate) hypersensitivity reactions as types I, II, and III, and lymphoid cell-mediated (delayed-type) hypersensitivity/cell-mediated immunity as a type IV reaction. "Hypersensitivity" generally represents the "dark side," signifying the undesirable aspects of an immune reaction, whereas the term "immunity" implies a desirable effect.

Sensitization refers to exposure of an animal to an antigen for the first time in order that subsequent or secondary exposure to the same antigen will lead to a greater response. The term has been used especially when the reaction induced is more of a hypersensitive or allergic nature than of an immune-protective type of response. Thus, an allergic response may be induced in a host sensitized by prior exposure to the same allergen.

Passive transfer is the transfer of immunity or hypersensitivity from an immune or sensitized animal to a previously nonimmune or unsensitized (and preferably syngeneic) recipient animal by serum containing specific antibodies or by specifically immune lymphoid cells. The transfer of immunity by lymphoid cells is referred to as adoptive immunization. Humoral immunity and antibody-mediated hypersensitivity reactions are transferred with serum, whereas delayed-type hypersensitivity, including contact hypersensitivity, is transferred with lymphoid cells. Passive transfer was used to help delineate which immune and hypersensitivity reactions were mediated by cells and which were mediated by serum.

Passive sensitization refers to the transfer of antibodies or primed lymphocytes from a donor previously exposed to antigen to a normal recipient for the purpose of conveying hypersensitivity from a sensitized to a nonsensitized individual. The Prausnitz-Küstner reaction is an example.

Hypersensitivity diseases are disorders mediated at least in part by immune mechanisms such as autoimmune diseases in which autoantibodies react with basement membranes such as the kidney, lung, and skin, or in which autoantibodies react against cell constituents such as DNA as in systemic lupus erythematosus. Any of the four mechanism of hypersensitivity reaction may participate in the production of a hypersensitivity disease.

Allergic alveolitis: See farmer's lung.

Allergic granulomatosis is a type of pulmonary necrotizing vasculitis with granulomas in the lung and pulmonary vessel walls. There may be infiltrates of eosinophils in the tissues and asthma. Also called Churg-Strauss syndrome.

Churg-Strauss syndrome (allergic granulomatosis) is a combination of asthma associated with necrotizing vasculitis, eosinophilic tissue infiltrates, and extravascular granulomas.

Allergic orchitis: The immunization of guinea pigs with autologous extracts of the testes incorporated into Freund's complete adjuvant leads to lymphocytic infiltrate in the testis and antisperm cytotoxic antibodies in the serum 2 to 8 weeks after inoculation. Human males who have been vasectomized may also develop allergic orchitis.

Hypersensitivity pneumonitis: Also called pigeon fancier's lung.

Pigeon breeder's lung: Hypersensitivity pneumonitis. Also called pigeon fancier's lung.

***Phoma* species** are aeroallergenic fungi that can induce hypersensitivity pneumonitis (HP). These fungi can induce a type of hypersensitivity pneumonitis known as shower curtain disease.

Alternaria species are aeroallergenic fungi which can induce hypersensitivity pneumonitis (HP). These fungi cause a form of HP known as woodworker's lung disease, as well as immunoglobulin E (IgE)-mediated allergic disease.

Immediate hypersensitivity is an antibody-mediated hypersensitivity. In humans, this homocytotrophic antibody is IgE and in certain other species IgG1. The IgE antibodies are attached to mast cells through their Fc receptors. Once the Fab regions of the mast cell-bound IgE molecules interact with specific antigen, vasoactive amines are released from the cytoplasmic granules as described under type I hypersensitivity reactions. The term "immediate" is used to indicate that this type of reaction occurs within seconds to minutes following contact of a cell-fixed IgE antibody with antigen. Skin tests and RAST are useful to detect immediate hypersensitivity in humans, and passive cutaneous anaphylaxis reveals immediate hypersensitivity in selected other species. Examples of immediate hypersensitivity in humans include the classic anaphylactic reaction to penicillin administration, hay fever, and environmental allergens such as tree and grass pollens, bee stings, etc.

The **triple response of Lewis** refers to skin changes in immediate hypersensitivity illustrated by striking the skin with a sharp object such as the side of a ruler. The first response termed the "stroke response" is caused by the production of histamine and related mediators at the point of contact with the skin. The second response is a flare produced by vasodilation and resembles a red halo. The third response is a wheal characterized by swelling and blanching induced by histamine from mast cell degranulation. The swelling is attributable to edema between the junctions of cells that become rich in protein and fluid.

Type I anaphylactic hypersensitivity (Figure 12.1) is a reaction mediated by IgE antibodies reactive with specific allergens (antigens that induce allergy) attached to basophil or mast cell Fc receptors. Crosslinking of the cell-bound IgE antibodies by antigen is followed by mast cell or basophil degranulation, with release of pharmacological mediators. These mediators include vasoactive amines such as histamine, which causes increased vascular permeability, vasodilation, bronchial spasm, and mucous secretion (Figure 12.2). Secondary mediators of type I hypersensitivity include leukotrienes, prostaglandin D_2, platelet-activating factor, and various cytokines. Systemic anaphylaxis is a serious clinical problem and can follow the injection of protein antigens such as antitoxin, or of drugs such as penicillin.

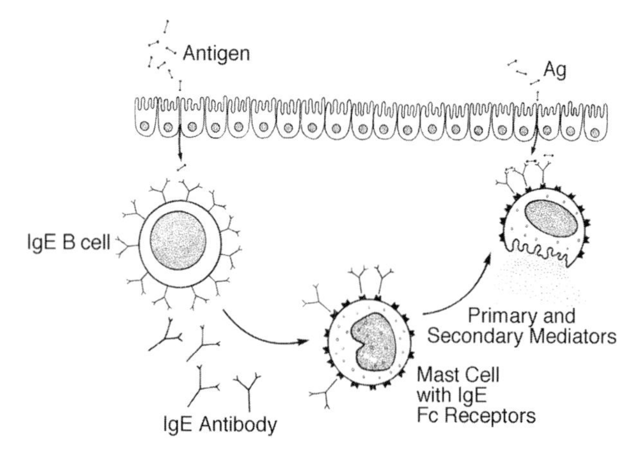

FIGURE 12.1 A type I hypersensitivity reaction in which antigen molecules cross-link IgE molecules on the surface of mast cells, resulting in their degranulation with the relase of both primary and secondary mediators of anaphylaxis.

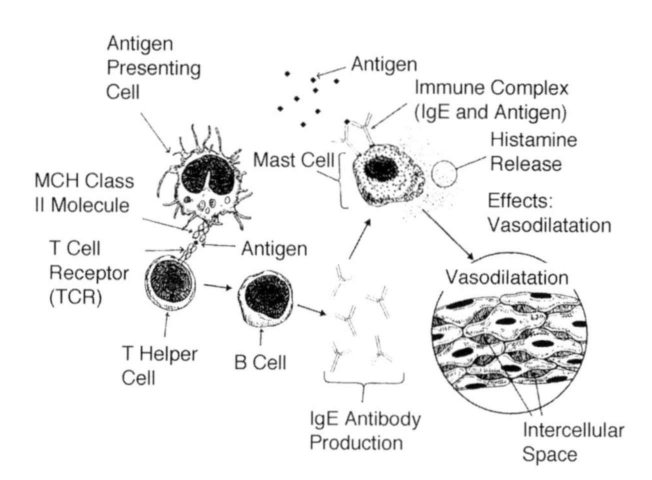

FIGURE 12.2 Schematic representation of events that follow degranulation of mast cells in tissues, resulting in vasodilatation of capillaries, leading to the changes associated with type I hypersensitivity reactions in tissues.

Skin-sensitizing antibody is an antibody, usually of the IgE class, that binds to the Fc receptors of mast cells in the skin, thereby conditioning this area of skin for a type I immediate hypersensitivity reaction following crosslinking of the IgE Fab regions by a specific allergen (antigen). In guinea pig skin, human IgG1 antibodies may be used to induce passive cutaneous anaphylaxis.

Dander antigen is a combination of debris such as desquamated epithelial cells, microorganisms, hair, and other materials trapped in perspiration and sebum that are constantly deleted from the skin. Dander antigens may induce immediate, IgE-mediated, type I hypersensitivity reactions in atopic individuals.

Anaphylaxis is a shock reaction that occurs within seconds following the injection of an antigen or drug or after a bee sting to which the susceptible subject has IgE-specific antibodies. There is embarrassed respiration due to laryngeal and bronchial constriction and shock associated with decreased blood pressure. Signs and symptoms differ among species based on the primary target organs or tissues. Whereas IgE is the anaphylactic antibody in man, IgG1 may mediate anaphylaxis in selected other species. Type I hypersensitivity occurs following the cross linking of IgE antibodies by a specific antigen or allergen on the surfaces of basophils in the blood or mast cells in the tissues. This causes the release of the pharmacological mediators of immediate hypersensitivity, with a reaction occurring within seconds of contact with antigen or allergen (Figure 12.3). Eosinophils, chemotactic factor,

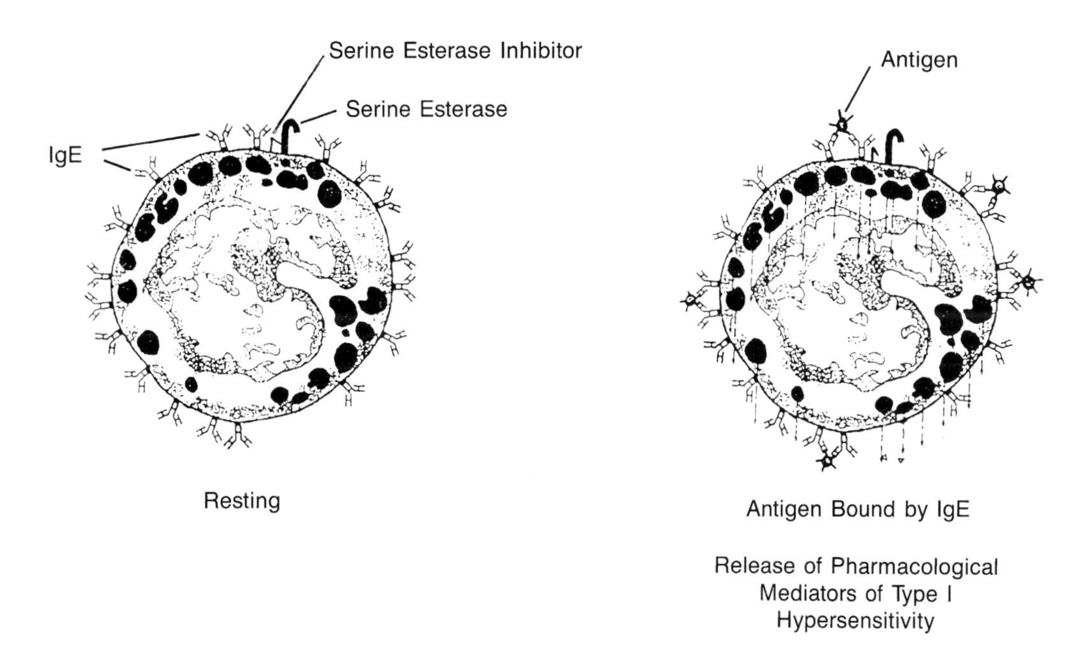

FIGURE 12.3 Comparison of a resting basophil or mast cell in which serine esterase activity is blocked by serine esterase inhibitor with a mast cell or basophil undergoing degranulation as a consequence of antigen interaction with Fab regions of cell-surface IgE molecules in which the serine esterase inhibitor activity is removed.

heparin, histamine, and serotonin, together with selected other substances, are released during the primary response. Acute-phase reactants are formed and released in the secondary response. Secondary mediators include slow-reacting substance of anaphylaxis (SRS-A), bradykinin, and platelet-activating factor. In addition to systemic anaphylaxis described above, local anaphylaxis may occur in the skin, gut, or nasal mucosa following contact with the antigen. The skin reaction, called urticaria, consists of a raised wheal surrounded by an area of erythema. Cytotoxic anaphylaxis follows the interaction of antibodies with cell surface antigens. See also aggregate anaphylaxis.

Anaphylactic shock is cardiovascular collapse and suffocation attributable to tracheal swelling that results from a systemic anaphylactic (immediate hypersensitivity) reaction following an antigen that has been systemically administered. This allergic reaction is a consequence of the binding of antigen to IgE antibodies on mast cells in the connective tissue throughout the body, resulting in a disseminated release of inflammatory mediators.

Arachidonic acid (AA) and leukotrienes are critical constituents of mammalian cell membranes. AA is released from membrane phospholipids as a consequence of membrane alterations or receptor-mediated signaling that leads to activation of phospholipase A_2, PLA_2. Oxidative metabolism of AA through cyclooxygenase leads to the formation of various prostaglandins through the cytochrome P450 system or through one of the lipoxygenases. Molecules derived from the lipoxygenase pathway of arachidonic acid are termed leukotrienes (LTs). They are derived from the combined actions of 5-lipoxygenase (5-LOX) and 5-LOX-activating protein (FLAP). Initially 5-hydroperoxyeicosatetraenoic (5-HPETE) acid is formed, followed by LTA_4. LTB_4 is an enzymatic, hydrolytic product of LTA_4. It differs from the other leukotrienes, LTC_4, LTD_4, and LTE_4, in that it does not have a peptide component. LTB_4 has powerful leukocytotropic effects. It is a powerful chemokinetic and chemotactic agent with the ability to induce neutrophil aggregatoin, degranulation, hexose uptake, and enhanced binding to endothelial cells. It can cause cation fluxes, augment cytoplasmic calcium concentrations form intracellular pools, and activate phosphatidylinositol hydrolysis. LTB_4 can act synergistically with prostaglandins E_1 (PGE_1) and E_2 (PGE_2) to induce macromolecule leakage in the skin through increased vascular permeability. When injected into the skin of a guinea pig, it can induce leukocytoclastic vasculitis. Human polymorphonuclear neutrophils have two sets of plasma membranes receptors. A high-affinity receptor set mediates aggregation, chemokinesis, and increased surface adherence, whereas the low-affinity receptor set mediates degranulation and increased oxidative metabolism. A fraction of $CD4^+$ and $CD8^+$ T lymphocytes bind LTB_4 but LTB_4 receptors on lymphocytes remain to be characterized.

The **cyclooxygenase pathway** is the method whereby prostaglandins are produced by enzymatic metabolism of arachidonic acid derived from the cell membrane, as in type I (anaphylactic) hypersensitivity reactions.

Platelet-activating factor (PAF) is a phospholipid with a mol wt of about 300 to 500 Da formed by leukocytes, macrophages, mast cells, and endothelial cells that induces aggregation of platelets and promotes amine secretion, aggregation of neurophils, release of enzymes, and an increase in vascular permeability. Its effect resembles that of IgE-mediated changes in anaphylaxis and cold urticaria. It may also participate in endotoxin shock and is derived from phosphorylcholine. The combination of antigens with the Fab regions of antibody molecules bound through Fc receptors to mast cells, polymorphonuclear leukocytes, and macrophages results in platelet-activating factor release. PAF release accompanies anaphylactic shock. It apparently mediates inflammation and allergic reactions. PAF induces a transient reduction in blood platelets, causes hypotension, and facilitates vascular permeability, but it has no effects on contracting smooth muscle and has no chemotactic activity. There is probably more than a single compound with this activity. PAF is resistant to arylsulfatase B, but is sensitive to phospholipases. PAF may induce bronchoconstriction and vascular dilation and leak, and may serve as a significant mediator of asthma.

Biogenic amines are nonlipid substances of low molecular weight, such as histamine, that have an amine group in common. Biogenic amines are stored in the cytoplasmic granules of mast cells, from which they may be released to mediate the biological consequences of immediate hypersensitivity reactions. Also called vasoactive amines.

Active kinins are characterized by a nonapeptide amino acid sequence whose prototype is bradykinin, the active kinin generated from plasma kininogen. Generation of the other forms depends on the enzyme and substrate used, but they differ in length and the additional residues. Tissue kallikreins are best activated by enzymes like trypsin and hydrolyze both low-molecular-weight kininogen (LMK) and high-molecular-weight kininogen (HMK) to give kallidin, a tissue form of kinin. Bradykinin is formed by the action of plasma kallikrein on HMK or by the action of trypsin on both LMK and HMK. Met-lys-bradykinin, another kinin, results from hydrolysis by plasma kallikrein activated by acidification. Other active kinins have also been described. Besides being the precursors for the generation of kinins, kininogens also affect the coagulation system, with the activation of the Hageman factor (HF) being the link between the two systems.

The **lipoxygenase pathway** refers to the enzymatic metabolism of arachidonic acid derived from the cell membrane, which is the source of leukotrienes.

Mast cell activation: Mast cells may be activated immunologically through crosslinking by antigen of surface IgE attached to FcεRI. They may be activated also by anti-IgE antibody, C5a, substance P, or local trauma.

Substance P is a tachykinin which may induce joint inflammation when released at local sites. It facilitates synthesis by monocytes of IL-1, IL-6, and TNF-α, and stimulates synovial cells to produce prostaglandins. Its receptor is designated NK1. Substance P is a principal component of the neuroimmune axis and axon reflex. It is a mediator of neurogenic inflammation, causing vasodilatation and plasma extravasation. It induces histamine release from MAST cells, lymphocyte proliferation, and immunoglobulin and cytokine secretion from B-lymphocytes and monocytes, respectively, macrophage stimulation, immune complex formation, eosinophil peroxidase secretion, and chemotaxis in response to platelet activating factor (PAF). SP induces polymorphonuclear leukocyte (PMNL) chemotaxis, phagocytosis, respiratory burst activity, exocytosis, and antibody-mediated cell cytotoxicity. It elevates superoxide production in PMNL and facilitates tumor necrosis factor-dependent interleukin-8 secretion. SP derived from primary afferant nerves has a proinflammatory effect on neutrophils, leading to increased adhesion to bronchial epithelial cells in acute and chronic bronchitis. SP levels are increased in the sputum of asthmatic and chronic bronchitis patients as well as those with nasal allergy. SP increases PMNL infiltration into the skin in allergic contact dermatitis. It enhances prostaglandin D2 and leukotriene C4 release from human nasal mucosa and skin mast cells. In vessels, SP induces intercellular adhesion molecule-1 (ICAM-1) expression on vascular endothelial cells. It induces degranulation of mast cells and facilitates transendothelial migration.

SRS-A is an abbreviation for slow-reacting substance of anaphylaxis.

EIA is an abbreviation for exercise-induced anaphylaxis.

Mast cell tryptase is a serine protease present in secretory granules of mast cells and released together with histamine during mast cell activation. Serum tryptase is a clinical indicator of diseases of mast–cell activation such as systemic anaphylaxis or mastocytosis. It is a better *in vitro* marker of anaphylaxis than histamine because tryptase has a slower release and is more stable. The half-life of tryptase is 90 min. It takes at least 15 min to reach detectable concentrations. The best time to measure tryptase is 1 to 2 h but not more than 6 h after the reaction. In anaphylaxis, both α and β tryptase are elevated. Most studies indicate that tryptase is increased in postmortem blood following severe anaphylaxis and is a reliable postmortem indicator of fatal anaphylaxis.

Vasoactive amines are amino group-containing substances that include histamine and serotonin, which cause dilatation of the peripheral vasculature and increase the permeability of capillaries and small vessels.

ECF-A (eosinophil chemotactic factor of anaphylaxis) is a 500-Da acidic polypeptide that attracts eosinophils. Interaction of antigen with IgE antibody molecules on the surface of mast cells causes ECF-A to be released from the mast cells.

H1 receptors are histamine receptors on vascular smooth-muscle cells through which histamine mediates vasodilation.

Vasoconstriction leads to diminished blood flow as a consequence of contraction of precapillary arterioles.

Vasodilatation leads to increased blood flow through capillaries as a consequence of precapillary arteriolar dilatation.

H2 receptors are histamine receptors on different types of tissue cells through which histamine mediates bronchial constriction in asthma, gastrointestinal constriction in diarrhea, and endothelial constriction resulting in edema.

Paradoxical reaction (historical) refers to death of experimental animals from anaphylaxis when administered a second injection of an antigen to which they had been previously immunized. Early workers administering repeated injections of tetanus toxoid observed the phenomenon. This term is no longer in use.

Sulfite sensitivity: Sulfites or sulfating agents, such as surfur dioxide, bisulfite salt, and metabisulfite salt, which are in broad use as food additives, can induce reactions marked by angioedema, laryngeal edema, asthma, and anaphylaxis. Sulfite reactions occur in atopic and selected nonatopic patients but are more commonly observed in chronic asthma. The hyperreactivity to sulfur dioxide generated from sulfites is believed to involve afferent cholinergic receptors in the tracheobronchial tree, IgE-mediated reactions in a few patients, and in sulfite oxidase deficiency. Kinins are believed to play a role in mediation of bronchial constriction. Treatment includes cromolyn to stabilize mast cells, atropine to block cholinergic sensitivity, and cyanocobalamin to assist sulfite oxidation in sulfite-oxidase deficient patients. Diagnosing sulfite sensitivity is based on metabisulfite challenge.

H1, H2 blocking agents: See antihistamine.

Mast cell–eosinophil axis is a term that characterizes the interactions between mast cells and eosinophils during inflammatory reactions recognized as immediate hypersensitivity. This involves the attraction of eosinophils and their activation by mast cell-derived ECF-A, as well as a dampening effect exerted by eosinophils upon mast cells. During this process, the released mediators influence the reactions in the microenvironment. When the causative agent is a parasite, the antiparasitic cytotoxic mechanisms of eosinophils reinforce the defense. The other effector cells attracted to the involved sites join forces in the defense activities. The inhibitory effects of eosinophils upon mast cells are exerted through a number of enzymes which inactivate or destroy some of the mast cell–derived mediators. Intact granules released from mast cells in the microenvironment are phagocytized by eosinophils and can be demonstrated in these cells by metachromatic staining. This represents an important detoxification mechanism since even intact granules have been shown to exert proteolytic activity.

Hives is a wheal and flare reaction of the anaphylactic type produced in the skin as a consequence of histamine produced by activated mast cells. There is edema, erythema, and pruritus. "Hives" is a synonym for urticaria.

Schultz-Dale test (historical): Strong contraction of the isolated uterine horn muscle of a virgin guinea pig that has been either actively or passively sensitized occurs following the addition of specific antigen to the 37°C tissue bath in which it is suspended. This reaction is the basis for an *in vitro* assay of anaphylaxis termed the Schultz-Dale test. Muscle contraction is caused by the release of histamine and other pharmacological mediators of immediate hypersensitivity, following antigen interaction with antibody fixed to tissue cells.

A **scratch test** is a skin test for the detection of IgE antibodies against a particular allergen that are anchored to mast cells in the skin. After scratching the skin with a needle, a minute amount of aqueous allergen is applied to the scratch site and the area is observed for the development of urticaria manifested as a wheal and flare reaction. This signifies that the IgE antibodies are specific for the applied allergen and lead to the degranulation of mast cells with release of pharmacologic mediators of hypersensitivity, such as histamine.

The **kallikrein–kinin system** is comprised of vasopressive peptides that control blood pressure through maintenance of regional blood flow and the excretion of water and electrolytes. Kallikrein causes the release of renin and the synthesis of kinins, which interact with the immune system and increase urinary sodium excretion, as well as acting as powerful vasodilators.

A **shock organ** is a particular organ involved in a specific reaction such as an anaphylactic reaction.

Shocking dose is the amount of antigen required to elicit a particular clinical response or syndrome.

PK test is an abbreviation for Prausnitz-Küstner reaction.

P-K reaction: See Prausnitz-Küstner reaction.

The **prick test** is an assay for immediate (IgE-mediated) hypersensitivity in humans. The epidermal surface of the skin on which drops of diluted antigen (allergen) are placed is pricked by a sterile needle passed through the allergen. The reaction produced is compared with one induced by histamine or another mast-cell secretogogue. This test is convenient, simple, rapid, and produces little discomfort for the patient in comparison with the intradermal test. It even may be used for infants.

Disodium cromoglycate is a drug that is valuable in the treatment of immediate (type I) hypersensitivity reactions, especially allergic asthma. Although commonly used as an inhalant, the drug may also be administered orally or applied topically to the nose and eyes. Mechanisms of action that have been postulated include mast cell membrane stabilization and bridging of IgE on mast cell surfaces, thereby blocking bridging by antigen.

Cromolyn sodium is a drug that blocks the release of pharmacological mediators from mast cells and diminishes the symptoms and tissue reactions of type I hypersensitivity (i.e., anaphylaxis) mediated by IgE. Although cromolyn sodium's mechanism of action remains to be determined, it apparently inhibits the passage of calcium through the cell membrane. The drug inhibits mast-cell degranulation but has no adverse effect on the linkage of IgE to the mast cell surface or to its interaction with antigen. Cromolyn sodium is inhaled as a powder or applied topically to mucous membranes. This is usually the treatment for asthma, allergic rhinitis, and allergic conjunctivitis, and has low cytotoxicity.

Intal® is a commercial preparation of disodium cromoglycate.

Kallikrein is an enzyme that splits kininogens to generate bradykinin, which has an effect on pain receptors and smooth muscle and exerts a chemotactic effect on neutrophils. Bradykinin is a nonapeptide that induces vasodilation and increases capillary permeability. Kallikreins, also known as kininogenases, are present both in the plasma and in tissue and also in glandular secretions such as saliva, pancreatic juice, tears, and urine. Trypsin, pepsin, proteases of snake venoms, and bacterial products are also able to hydrolyze kininogens, but the substrate specificity and potency of kallikreins is greater. Plasma and tissue kallikreins are physically and immunologically different. It is not known whether plasma contains more than one form of kallikrein.

Kallikrein inhibitors are natural inhibitors of kallikreins and belong to the group of natural inhibitors of proteolysis. They also inhibit other proteolytic enzymes, but each has its own preference for one protease or another. Apronitin, also known by its registered name of Trasylol®, is particularly active on tissue kallikreins.

Kininases are enzymes in the blood that degrades kinins to inactive peptides. Inactivation occurs when any of the eight bonds in the kinin is cleaved. There are two kininases of plasma. Kininase I, or carboxy peptidase N, cleaves the C-terminal arginine of kinins and of anaphylatoxins. It differs from pancreatic carboxy peptidase B with respect to molecular weight, subunit structure, carbohydrate content, antigenic properties, substrate specificity, and inhibition pattern. The purified enzyme has a mol wt of 280 kDa. The site of synthesis of carboxy peptidase N is believed to be the liver. Kininase II, a peptidyl dipeptidase, cleaves the C-terminal Phe-Arg of kinins and also liberates angiotensin II from angiotensin I. (Angiotensin is another vasoactive substance.) The vascular endothelium of both lung and peripheral vascular bed are rich in this enzyme. Peptidyl dipeptidase is present also outside the circulatory system and has different functions at these sites. Other enzymes with kininase activity are present in the spleen and kidney (cathepsin) and the endothelial cells of the gastrointestinal tract.

Latex allergy is hypersensitivity to natural latex, which is the milky sap from the rubber tree *Hevea brasiliensis*. The symptoms may range from hand dermatitis to life-threatening anaphylaxis. Of the general population, 3% shows sensitivity to latex but 2 to 25% of healthcare workers manifest this type of hypersensitivity. Among nurses, 9% have natural latex-specific IgE antibodies. Patients with increased frequency of operations have an increased prevalence of latex allergy (29%). The number of operations and total serum IgE are the most important factors in predicting a latex allergy. Natural latex-specific IgE is detectable by RAST, EIA, flow cytometry, and electrochemiluminescence. Patients with latex allergy have an increased frequency of allergy to foods such as avocado, potato, banana, tomato, chestnut, and kiwi.

Kininogens are the precursors of kinins, and are glycoproteins synthesized in the liver. Plasma kininogens comprise two, possibly three, classes of compounds with species variation: (1) Low mol-wt kininogens (LMK) are acidic proteins with a molecular weight of about 57 kDa and are susceptible to conversion into kinins by kininogen-converting enzymes (kallikreins) of tissue origin. LMK has two forms, I and II, which represent the main plasma kininogens. (2) High mol-wt kininogens (HMK) are α glycoproteins with a mol wt of 97 kDa, which can be converted into kinins both by plasma and tissue kallikreins. HMK exists in two forms, a and b, which differ both in enzyme sensitivity and generated kinin.

Kinins are a family of straight-chain polypeptides generated by enzymatic hydrolysis of plasma α-2 globulin precursors, collectively called kininogens. They exert potent vasomotor effects, causing vasodilatation of most vessels in the body but vasoconstriction of the pulmonary bed.

They also increase vascular permeability and promote the diapedesis of leukocytes. Peptide kinins are released during an inflammatory response.

Active anaphylaxis is an anaphylactic state induced by natural or experimental sensitization in atopic subjects or experimental animals. See also anaphylaxis.

Cytotropic anaphylaxis is a form of anaphylaxis caused by antigen-binding to reaginic IgE antibodies. The latter are cytotropic, that is, they bind to cells. Cytotropic antibodies (IgE) bind to specific receptors on the mast-cell surface. The receptors are in close proximity to a serine esterase enzyme causing the release of mast cell granules and to a natural inhibitor of this enzyme. As long as the surface IgE has no bound antigen, the status quo of the cell is maintained. It is believed that binding of the antigen induces a conformational alteration in the IgE with displacement of the inhibitor from its steric relationship to the enzyme. The inhibition-free enzyme mediates the release. The process requires energy and is Ca^{2+} dependent.

Passive anaphylaxis is an anaphylactic reaction in an animal that has been administered an antigen after it has been conditioned by an inoculation of antibodies derived from an animal immunized against the antigen of interest.

Antianaphylaxis is the inhibition of anaphylaxis through desensitization. This is accomplished by repeated injections of the sensitizing agent too minute to produce an anaphylactic reaction.

Desensitization is a method of treatment used by allergists to diminish the effects of IgE-mediated type I hypersensitivity (Figure 12.4). The allergen to which an individual has been sensitized is repeatedly injected in a form that favors the generation of IgG-(blocking) antibodies rather than IgE antibodies that mediate type I hypersensitivity in humans. This method has been used for many years to diminish the symptoms of atopy, such as asthma and allergic rhinitis, and to prevent anaphylaxis produced by bee venom. IgG antibodies are believed to prevent antigen interaction with IgE antibodies anchored to mast cell surfaces by intercepting the antigen molecules before they reach the cell-bound IgE. Thus, a type I hypersensitivity reaction of the anaphylactic type is prevented.

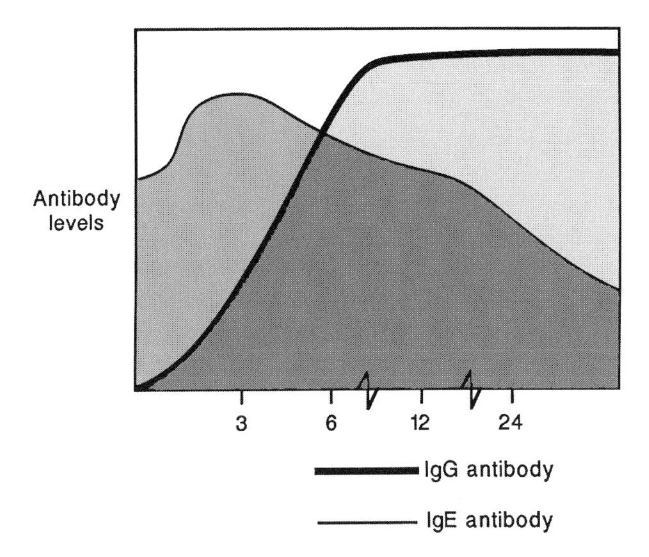

FIGURE 12.4 Desensitization.

Allergen immunotherapy: Desensitization treatment. See desensitization.

Allergic disease immunotherapy refers to the principal immunologic, as opposed to environmental or pharmacologic, treatment of allergic diseases that include allergic rhinitis, conjunctivitis, and asthma. The allergen to which a patient exhibits immunoglobulin E (IgE)-dependent sensitivity is first identified, followed by subcutaneous administration of minute quantities of natural extracts containing these allergens. The aim is to modify the immune response responsible for maintaining atopic symptoms.

Hyposensitization is a technique to decrease responsiveness to antigens acting as allergens in individuals with immediate (type I) hypersensitivity to them. Since the reaction is mediated by IgE antibodies specific for the allergen becoming fixed to the surface of the patient's mast cells, the aim of this mode of therapy is to stimulate the production of IgG-blocking antibodies that will combine with the allergen before it reaches the mast cell–bound IgE antibodies. This is accomplished by graded administration of an altered form of the allergen that favors the production of IgG rather than IgE antibodies. In addition to intercepting allergen prior to its interaction with IgE, the IgG antibodies may also inhibit further production of IgE.

Bradykinin is a 9-amino acid peptide split by plasma kallikrein from plasma kininogens. It produces slow, sustained, smooth muscle contraction. Its action is slower than is that of histamine. Bradykinin is produced in experimental anaphylaxis in animal tissues. Its sequence is Arg-Pro-Pro-Gly-Phe-Ser-Pro-Phe-Arg (Figure 12.5). Besides anaphylaxis, bradykinin is also increased in endotoxin shock. Lysyl-bradykinin (kallidin), which is split from

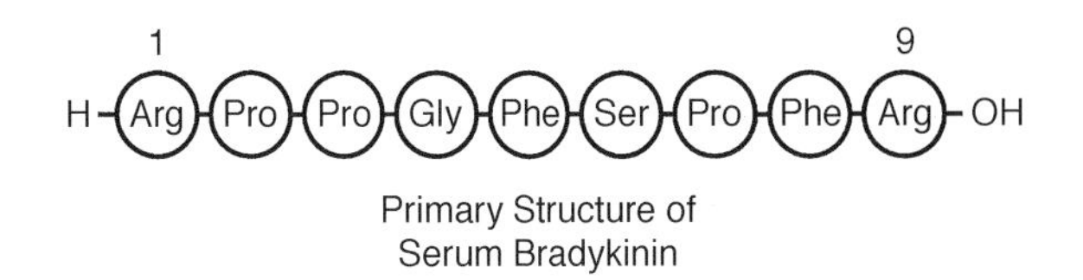

FIGURE 12.5 Active kinins.

kininogens by tissue kallikreins, also has a lysine residue at the amino terminus.

Systemic anaphylaxis is a type I, immediate, anaphylactic type of hypersensitivity mediated by IgE antibodies anchored to mast cells that become crosslinked by homologous antigen (allergen) causing release of the pharmacological mediators of immediate hypersensitivity, producing lesions in multiple organs and tissue sites. This is in contrast to local anaphylaxis, where the effects are produced in isolated anatomical location. The intravenous administration of a serum product, antibiotic, or other substance against which the patient has anaphylactic IgE-type hypersensitivity may lead to the symptoms of systemic anaphylaxis within seconds and may prove lethal.

Passive systemic anaphylaxis renders a normal, previously unsensitized animal susceptible to anaphylaxis by a passive injection, often intravenously, of homocytotrophic antibody derived from a sensitized animal, followed by antigen administration. Anaphylactic shock occurs soon after the passively transferred antibody and antigen interact *in vivo*, releasing the mediators of immediate hypersensitivity from mast cells of the host.

Generalized anaphylaxis occurs when the signs and symptoms of anaphylactic shock manifest within seconds to minutes following the administration of an antigen or allergen that interacts with specific IgE antibodies bound to mast cell or basophil surfaces, causing the release of pharmacologically active mediators that include vasoactive amines from their granules. Symptoms may vary from transient respiratory difficulties (due to contraction of the smooth muscle and terminal bronchioles) to even death.

Local anaphylaxis is a relatively common type-I immediate hypersensitivity reaction. Local anaphylaxis is mediated by IgE crosslinked by allergen molecules at the surface of mast cells, which then release histamines and other pharmacological mediators that produce signs and symptoms. The reaction occurs in a particular target organ such as the gastrointestinal tract, skin, or nasal mucosa. Hay fever and asthma represent examples.

Basophils are polymorphonuclear leukocytes of the myeloid lineage with distinctive basophilic secondary granules in the cytoplasm that frequently overlie the nucleus. These granules are storage depots for heparin, histamine, platelet-activating factor, and other pharmacological mediators of immediate hypersensitivity (Figure 12.6). Degranulation of the cells with release of these pharmacological mediators takes place following crosslinking by allergen or antigen of Fab regions of IgE receptor molecules bound through Fc receptors to the cell surface. They comprise less than 0.5% of peripheral blood leukocytes. Following crosslinking of surface-bound IgE molecules by specific allergen or antigen, granules are released by exocytosis.

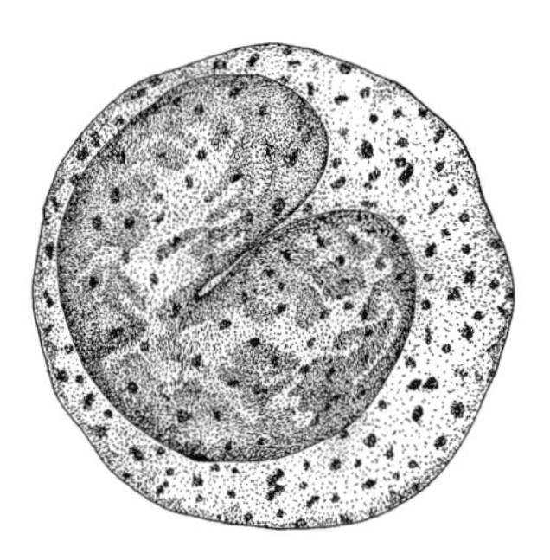

FIGURE 12.6 A peripheral–blood basophil showing a bilobed nucleus and cytoplasmic granules that represent storagae sites of pharmacological mediators of immediate type I hypersensitivity.

HOOC ... O ... COOH; O ... OCH_2CHCH_2O ... O; OH

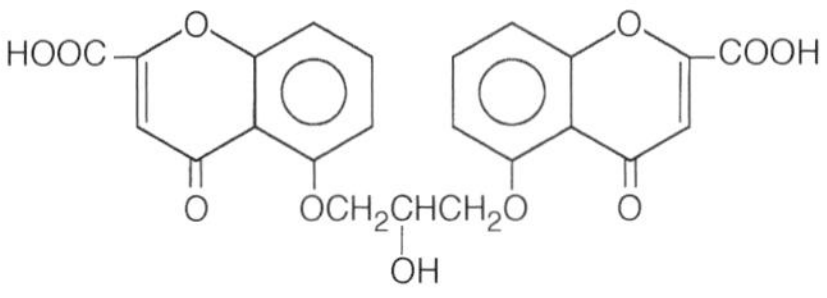

FIGURE 12.7 Structural formula for cromolyn.

Substances liberated from the granules are pharmacological mediators of immediate (type I) anaphylactic hypersensitivity.

Degranulation is a mechanism whereby cytoplasmic granules in cells fuse with the cell membrane to discharge the contents from the cell. A classic example is degranulation of the mast cell or basophil in immediate (type I) hypersensitivity. In phagocytic cells, cytoplasmic granules combine with phagosomes and release their contents into the phagolysosome formed by their union.

Cromolyn (1,3-*bis* [2-carboxychromon-5-yloxy-2-hydroxypropane]) is a therapeutic agent that prevents mast-cell degranulation (Figure 12.7). It has proven effective in the therapy of selected allergies that include allergic rhinitis and asthma.

Aggregate anaphylaxis is a form of anaphylaxis caused by aggregates of antigen and antibody in the fluid phase. The aggregates bind complement-liberating complement fragments C3a, C5a, and C4a, also called anaphylatoxins, which induce the release of mediators. Preformed aggregates of antigen-antibody complexes in the fluid phase fix complement. Fragments of complement components, the anaphylatoxins, may induce experimentally the release of mediators from mast cells. There is no evidence that these components play a role in anaphylactic reactions *in vivo*. Aggregates of antigen–IgG antibody, however, may induce anaphylaxis, whose manifestations are different in the various species.

Histamine (Figure 12.8) is a biologically active amine, i.e., β-aminoethylimidazole, of 111-kDa mol wt that

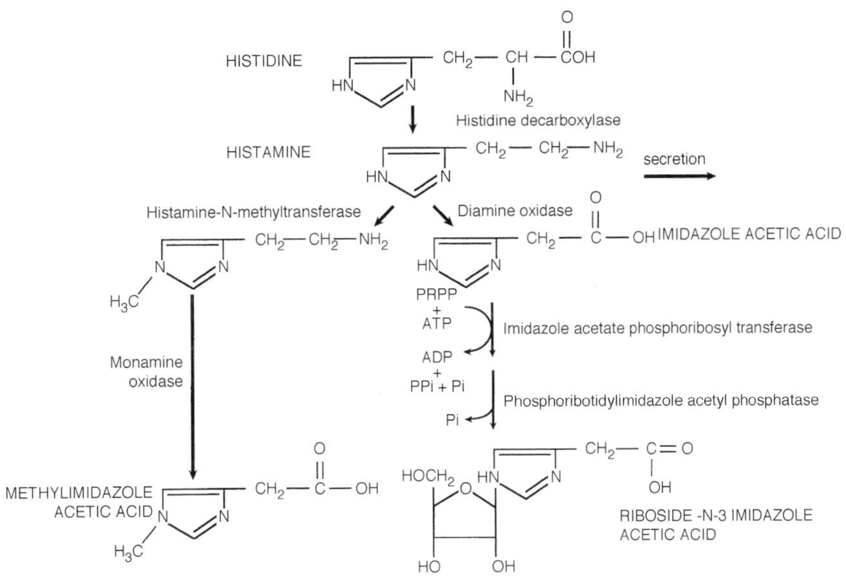

FIGURE 12.8 Schematic representation of the synthesis of histamine, a principal mediator of immediate (type I) hypersensitivity reactions.

induces contraction of the smooth muscle of human bronchioles and small blood vessels, increased capillary permeability, and increased secretion by the mucous glands of the nose and bronchial tree. It is a principal pharmacological mediator of immediate (type I) hypersensitivity (anaphylaxis) in man and guinea pigs. Although found in many tissues, it is especially concentrated in mast cells of the tissues and basophils of the blood. It is stored in their cytoplasmic granules and is released following crosslinking of IgE antibodies by a specific antigen on their surfaces. It is produced by the decarboxylation of histidine through the action of histidine decarboxylase. When histamine combines with H_1 receptors, smooth muscle contraction and increased vascular permeability may result. Combination with H_2 receptors induces gastric secretion and blocks mediator release from mast cells and basophils. It may interfere with suppressor T cell function. Histamine attracts eosinophils that produce histaminase, which degrades histamine.

Plasma histamine: Even though histamine released from basophils and mast cells is a critical step in immediate hypersensitivity reactions, plasma histamine levels are rarely determined due to the short half-life of histamine in plasma. Thus, mast cell tryptase and thromboxane A2 are the preferred analytes due to their longer half-life.

Histaminase is a common tissue enzyme, termed diamine oxidase, which transforms histamine into imidazoleacetic acid, an inactive substance.

Histamine-releasing factors (HRF) are protein cytokines, i.e., lymphokines produced from antigen-stimulated lymphocytes, which induce the release of histamine from basophils and mast cells. The 12-kDa HRF is similar to connective tissue–activating peptide-III (CTAP-III) and its degradation product, neutrophil-activating peptide 2 (NAP-2). The sera of patients with chronic idiopathic urticaria (CIU) contain a histamine-releasing factor in the IgG fraction of serum, which correlates with disease activity.

Antihistamine is a substance that links to histamine receptors, thereby inhibiting histamine action. Antihistamine drugs derived from ethylamine block H1 histamine receptors, whereas those derived from thiourea block the H2 variety.

Anaphylactoid reaction is a response resembling anaphylaxis, except that it is not attributable to an allergic reaction mediated by IgE antibody. It is due to the nonimmunologic degranulation of mast cells such as that caused by drugs or chemical compounds like aspirin, radiocontrast media, chymopapain, bee or snake venom, and gum acacia which trigger release of the pharmacological mediators of immediate hypersensitivity, including histamine and other vasoactive molecules.

Dermatographism is a wheal and flare reaction of the immediate hypersensitivity type induced by scratching the skin. Thus, minor physical trauma induces degranulation of mast cells with the release of the pharmacological mediators of immediate hypersensitivity through physical stimulation. It is an example of an anaphylactoid reaction.

First-use syndrome is an anaphylactoid reaction that may occur in some hemodialysis patients during initial use of a dialyzer. It may be produced by dialyzer material or by residual ethylene oxide used for sterilization.

A **pseudoallergic reaction** is a nonimmunological clinical syndrome characterized by signs and symptoms that mimic or resemble immune-based allergic or immediate hypersensitivity reactions. However, pseudoallergic reactions are not mediated by specific antibodies or immune lymphoid cells.

Pseudoallergy is an anaphylaxis-like reaction in some individuals that occurs suddenly, frequently following food ingestion, and represents an anaphylactoid reaction. It may be induced by a psychogenic factor, a metabolic defect, or other nonimmunological cause. This is not an immune reaction and is classified as an anaphylactoid reaction.

Photoallergy is an anaphylactoid reaction induced by exposing an individual to light.

Serotonin (5-hydroxytryptamine) [5-HT] is a 176-M_w catecholamine found in mouse and rat mast cells and in human platelets, which participates in anaphylaxis in several species such as the rabbit but not in humans (Figure 12.9). 5-HT induces contraction of smooth muscle, enhances vascular permeability of small blood vessels, and induces large blood vessel vasoconstriction. It is derived from tryptophan by hydroxylation to 5-hydroxytryptophan and decarboxylation to 5-hydroxytryptamine. In humans, gut enterochromaffin cells contain 90% of 5-HT, with the remainder accruing in blood platelets and the brain. 5-HT is a potent biogenic amine with wide species distribution. 5-HT may stimulate phagocytosis by leukocytes and interfere with the clearance of particles by the mononuclear phagocyte system. Immunoperoxidase staining for 5-HT, which is synthesized by various neoplasms, especially carcinoid tumors, is a valuable aid in surgical pathologic diagnosis of tumors producing it.

Vascular permeability factors are substances such as serotonins, histamines, kinins, and leukotrienes that increase the spaces between cells of capillaries and small vessels, facilitating the loss of protein and cells into the extravascular fluid.

An **eicosanoid** is an arachidonic acid-derived 20-carbon cyclic fatty acid (Figure 12.10). It is produced from membrane phospholipids. Eicosanoids, as well as other arachidonic acid metabolites, are elevated during shock and following injury and are site specific. They produce various effects, including bronchodilation, bronchoconstriction, vasoconstriction, and vasodilation. Eicosanoids include leukotrienes, prostaglandins, thromboxanes, and prostacyclin.

HO CH₂CH₂NH₂ H N

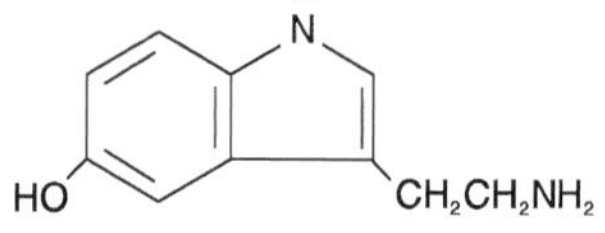

FIGURE 12.9 Structural formula of serotonin (5-hydroxytryptamine) that is released and participates in the mediation of anaphylactic reactions in some species and not others.

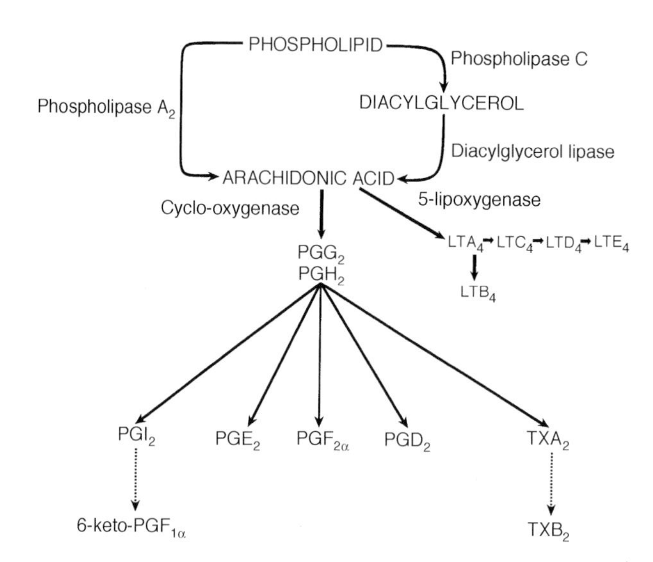

FIGURE 12.10 Schematic diagram of eicosanoids.

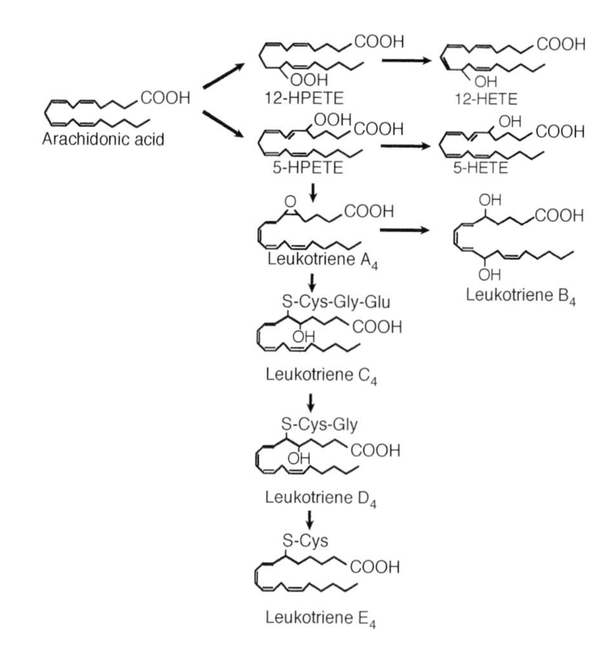

FIGURE 12.11 The lipoxygenase pathway of arachidonic acid metabolism that participates in the mediation of type I hypersensitivity reactions.

A **leukotriene** is a product of the enzymatic metabolism of arachidonic acid derived from the cell membrane. They are generated during an anaphylactic reaction by way of the lipoxygenase pathway (Figure 12.11). A leukotriene may include lipid mediators of inflammation and type I hypersensitivity. In the past they were referred to as slow-reacting substances of anaphylaxis (SRS-A). These lipid-inflammatory mediators are produced by the lipoxygenase pathway in various types of cells. Most cells synthesize plentiful amounts of leukotriene C and (LTC) and its breakdown products LTD and LTE that bind to smooth muscle cell receptors to produce prolonged bronchoconstriction. Leukotrienes play a significant role in the pathogenesis of bronchial asthma. Slow-reacting substance of anaphylaxis, used previously,represented the collective action of LTD and LTE. See also arachidonic acid (AA) and leukotrienes.

Slow-reacting substance of anaphylaxis (SRS-A) is a 400-kDa mol wt acidic lipoprotein derived from arachidonic acid that induces the slow contraction of bronchial smooth muscle and is produced following exposure to certain antigens. It is comprised of leukotrienes LTC_4, LTD_4, and LTE_4, which produce the effects observed in anaphylactic reactions. It is released, *in vitro*, in effluents from synthesized lung tissue of guinea pig, rabbit, and rat profused with antigen. It has also been demonstrated in human lung tissue and nasal polyps. It contracts smooth muscle of guinea pig ileum. *In vitro*, it also increases vascular permeability upon intracutaneous injection and decreases pulmonary compliance by a mechanism independent of vagal reflexes. It also enhances some of the smooth muscle effects of histamine. The source of SRS-A is mast cells and certain other cells. It is found in immediate (type I) hypersensitivity reactions. SRS-A is not stored in a preformed state and is sequentially synthesized and released. The effects have a latent period before becoming manifest. Antihistamines do not neutralize the effects of SRS-A.

Prostaglandins (PG) are a family of biologically active lipids derived from arachidonic acid through the effects of the enzyme cyclooxygenase (Figure 12.12 and Figure 12.13). Although first described in the prostate gland, they are now recognized in practically all tissues of mammals. The hormonal effects of prostaglandin include decreasing blood pressure, stimulating contraction of smooth muscle, and regulation of inflammation, blood clotting, and the immune response. Prostaglandins are grouped on the basis of their substituted five-membered ring structure. During anaphylactic reactions mediated by IgE on mast cells, PGD_2 is released, producing small blood-vessel dilation and constriction of bronchial and pulmonary blood vessels. Mononuclear phagocytes may release PGE_2 after binding of immune complexes to Fc′ receptors. Other effects of PGE_2 include blocking of MHC class II molecule expression in T cells and macrophages and inhibition of T-cell growth. PGD_2 and PGE_2 both prevent aggregation of platelets. Antiinflammatory agents such as aspirin block prostaglandin synthesis.

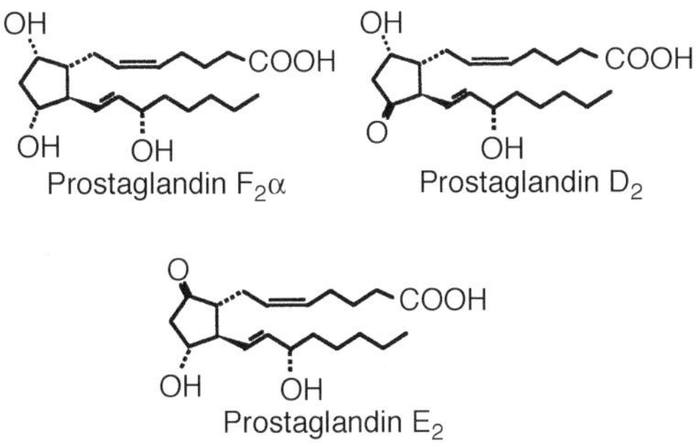

Prostaglandin $F_2\alpha$ Prostaglandin D_2 Prostaglandin E_2

FIGURE 12.12 Structural formulae of prostaglandins that are released and facilitate mediation of type I hypersensitivity reactions.

Thromboxanes comprise a group of biologically active compounds with a physiological role in homeostasis and a pathophysiological role in thromboembolic disease and anaphylactic reactions. They are cyclopentane derivatives of polyunsaturated fatty acids and are derived by isomerization from prostaglandin endoperoxide PGH_2, the immediate precursor. The isomerizing enzyme is called thromboxane synthetase. The active compound, thromboxane A_2, is unstable (half-life about 35 sec at pH 7.4), being degraded to thromboxane B_2 that is stable but inactive on blood vessels; it has, however, polymorphonuclear cell chemotactic activity. The short notation is TXA_2 and TXB_2. TXA_2 and TXB_2 represent the major pathway of conversion of prostaglandin endoperoxide precursors. TXA_2, derived from prostaglandin G_2 generated from arachidonic acid by cyclooxygenase, increases following injury to vessels. It stimulates a primary hemostatic response. TXA_2 is a potent inducer of platelet aggregation, smooth muscle contraction, and vasoconstriction. TXA was previously called rabbit aorta contacting substance (RACS) and is isolated from lung perfusates during anaphylaxis. It appears to be a peptide containing less than ten amino acid residues. Thromboxane formation in platelets is associated with the dense tubular system. PMNs and spleen, brain, and inflammatory granulomas have been demonstrated to produce thromboxanes.

Nonsteroidal antiinflammatory drugs (NSAIDS) are used in the treatment of arthritis. A group of drugs used in the treatment of rheumatoid arthritis, gouty arthritis, ankylosing spondylitis, and osteoarthritis, the drugs are weak organic acids. They block prostaglandin synthesis by inhibiting cyclooxygenase and lipoxygenase. They interrupt membrane-bound reactions such as NADPH oxidase in neutrophils, monocyte phospholipase C, and processes regulated by G protein. They also exert a number of other possible activities such as diminished generation of free radicals and superoxides, which may alter intracellular cAMP levels, diminishing vasoactive mediator release from granulocytes, basophils, and mast cells. Prostacyclin (PC) is a derivative of arachidonic acid that is

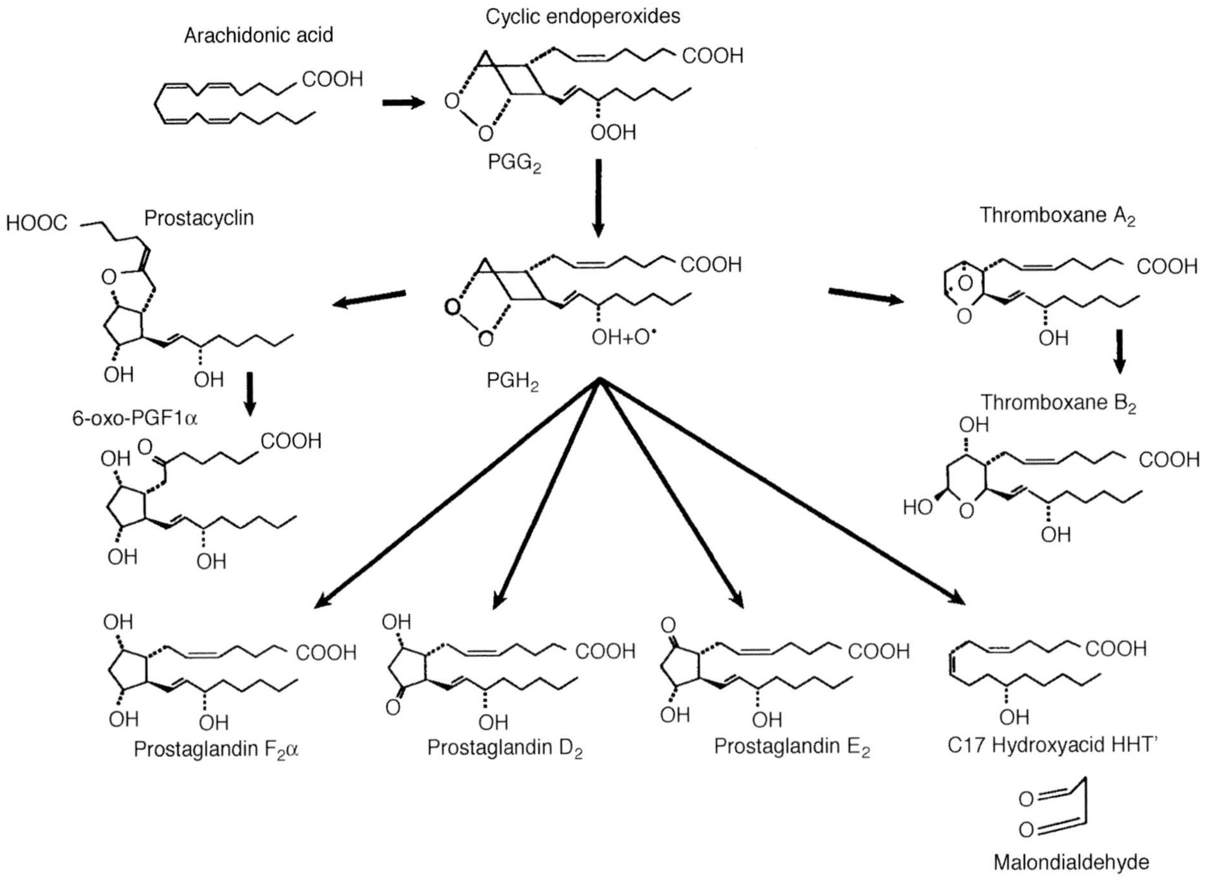

FIGURE 12.13 Cyclooxygenase pathway of arachidonic acid metabolism that participates in the mediation of type I hypersensitivity reactions.

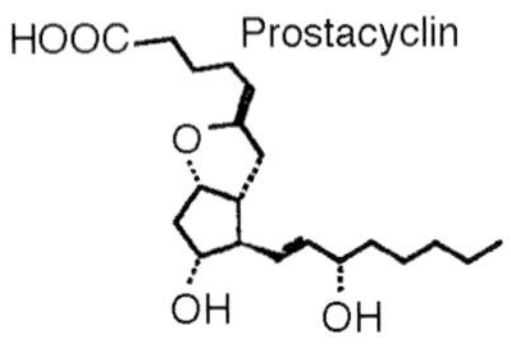

FIGURE 12.14 Structural formula for prostacyclin, which is released and which participates in mediation of the effects on tissues of type I immediate hypersensitivity.

related to prostaglandins. It has a second five-membered ring (Figure 12.14). It inhibits aggregation of platelets and is a potent vasodilator. Prostacyclin's actions are the opposite of the actions of thromboxanes.

Anaphylatoxins are substances generated by the activation of complement that lead to increased vascular permeability as a consequence of the degranulation of mast cells with the release of pharmacologically active mediators of immediate hypersensitivity. These biologically active peptides of low molecular weight are derived from C3, C4, and C5. They are generated in serum during fixation of complement by Ag-Ab complexes, immunoglobulin aggregates, etc. Small blood vessels, mast cells, smooth muscle, and leukocytes in peripheral blood are targets of their action. Much is known about their primary structures. These complement fragments are designated C3a, C4a, and C5a. They cause smooth muscle contraction, mast-cell degranulation with histamine release, increased vascular permeability, and the triple response in skin. They induce anaphylactic-like symptoms upon parenteral inoculation.

Anaphylatoxin inactivator is a 300-kDa α globulin carboxy peptidase in serum that destroys the anaphylatoxin activity of C5a, C3a, and C4a by cleaving their carboxy terminal arginine residues.

AnaINH is an abbreviation for anaphylatoxin inhibitor.

An **anaphylatoxin inhibitor (AnaINH)** is a A 300-kDa globulin carboxy peptidase that cleaves anaphylatoxin's carboxy terminal arginine. The enzyme acts on all three forms including C3a, C4a, and C5a, inactivating rather than inhibiting them.

Atopy is a type of immediate (type I) hypersensitivity to common environmental allergens in humans mediated by humoral antibodies of the IgE class formerly termed reagins, which are able to passively transfer the effect. Atopic hypersensitivity states include hay fever, asthma, eczema, urticaria, and certain gastrointestinal disorders. There is a genetic predisposition to atopic hypersensitivities, which affect more than 10% of the human population. Antigens that sensitize atopic individuals are termed allergens. They include (1) grass and tree pollens; (2) dander, feathers, and hair; (3) eggs, milk, and chocolate; and (4) house dust, bacteria, and fungi. IgE antibody is a skin-sensitizing homocytotropic antibody that occurs spontaneously in the sera of human subjects with atopic hypersensitivity. IgE antibodies are nonprecipitating (*in vitro*), are heat sensitive (destroyed by heating to 60°C for 30 to 60 min), are not able to pass across the placenta, remain attached to local skin sites for weeks after injection, and fail to induce passive cutaneous anaphylaxis (PCA) in guinea pigs.

Atopic is an adjective referring to clinical manifestations of type I (IgE-mediating) types of hypersensitivity to environmental antigens such as pollen or house dust resulting in allergic rhinitis (hay fever), asthma, eczema, or food allergies. Individuals with such allergies are described as atopic.

Atopic hypersensitivity: See atopy.

Atopic dermatitis is a chronic eczematous skin reaction marked by hyperkeratosis and spongiosis, especially in children with a genetic predisposition to allergy. These are often accompanied by elevated serum IgE levels, which are not proved to produce the skin lesions.

***Cladosporium* species** are aeroallergenic fungi that can induce hypersensitivity pneumonitis (HP) of the type known as hot tub lung disease, as well as IgE-mediated allergic disease which is frequent in persons with atopic disease. Symptoms of asthma that follow seasonal variations in *Cladosporium* spore counts and the demonstration of IgE reactivity aid diagnosis of allergy attributable to this microorganism.

Hay fever: Allergic rhinitis, recurrent asthma, rhinitis, and conjunctivitis in atopic individuals who inhale allergenic (antigenic) materials such as pollens, animal dander, house dust, etc. These substances do not induce allergic reactions in nonatopic (normal) individuals. Hay fever is a type I immediate hypersensitivity reaction mediated by homocytotrophic IgE antibodies specific for the allergen for which the individual is hypersensitive. Hay fever is worse during seasons when airborne environmental allergens are most concentrated.

Atopic allergy or atopy is an increased tendency of some members of the population to develop immediate hypersensitivity reactions, often mediated by IgE antibodies, against innocuous substances.

Allergic rhinitis is a condition characterized by a pale and wet nasal mucosa with swollen nasal turbinates. When hayfever is seasonal or environmental, it is allergic. One must first distinguish between infectious and noninfectious rhinitis. Both types of patients may have nasal polyps, and the patient is often anosmic. Noninfectious rhinitis that is perennial is difficult to establish as allergic.

Allergic conjunctivitis is a hypersensitivity reaction in the conjunctiva of the eye, often induced by airborne antigens.

Dermatophagoides is the house mite genus. Constituents of house mites represent the main allergen in house dust.

Dermatophagoides pteronyssinus is a house dust mite whose antigens may be responsible for house dust allergy in atopic individuals. It is a common cause of asthma.

Eczema is a skin lesion that is characterized as a weeping eruption consisting of erythema, pruritus with edema, papules, vesicles, pustules, scaling, and possible exudation. It occurs in individuals who are atopic, such as those with atopic dermatitis. Application to the skin or the ingestion of drugs that may themselves act as haptens may induce this type of hypersensitivity. It may be seen in young children who subsequently develop asthma in later life.

An **eczematoid skin reaction** is characterized by the appearance of erythematous, vesicular, and pruritic lesions on the skin that resemble eczema but are not due to atopy.

Allergen is an antigen that induces an allergic or hypersensitivity response in contrast to a classic immune response produced by the recipient host in response to most immunogens. Allergens include such environmental substances as pollens, i.e., their globular proteins, from trees, grasses, and ragweed, as well as certain food substances, animal danders, and insect venom. Selected subjects are predisposed to synthesizing IgE antibodies in response to allergens and are said to be atopic. The crosslinking of IgE molecules anchored to the surfaces of mast cells or basophils through their Fc regions results in the release of histamine and other pharmacological mediators of immediate hypersensitivity from mast cells/basophils.

Drug allergy is an immunopathologic or hypersensitivity reaction to a drug. Some drugs are notorious for acting as haptens that bind to proteins in the skin or other tissues that act as carriers. This hapten–carrier complex elicits an immune response manifested as either antibodies or T lymphocytes. Any of the four types of hypersensitivity in the Coombs' and Gell classification may be mediated by drug allergy. One of the best known allergies is hypersensitivity to penicillin. Antibodies to a drug linked to carrier molecules in the host may occur in autoimmune hemolytic anemia or thrombocytopenia, anaphylaxis, urticaria, and serum sickness. Skin eruptions are frequent manifestations of a T cell-mediated drug allergy.

Isoallergens are allergenic determinants with similar size, amino acid composition, peptide fingerprint, and other characteristics. They are present in a given material, but each is able individually to sensitize a susceptible subject. They are molecular variants of the same allergen.

Aspirin (ASA) acetyl salicylic acid is an antiinflammatory, analgesic, and antipyretic drug that blocks the synthesis of prostaglandin. It may induce atopic reactions such as asthma and rhinitis due to intolerance and idiosyncratic reactivity against the drug.

Aspirin sensitivity reactions: A hypersensitivity response to aspirin (acetylsalicylic acid ASA) as a manifestation of allergy and asthma. This allergic reaction may be manifested as either urticaria and angioedema or rhinoconjunctivitis with bronchospasm. Aspirin sensitivity is sometimes termed a "pseudoallergic reaction." Possible mechanisms are: cyclooxygenase blockade resulting in an altered PGE_2 (bronchodilator) to PHF_2 (bronchoconstrictor) ratio: a shunting of arachidonate from the cyclooxygenase to the 5-lipoxygenase pathway with eosinophilia and viral infection, in which altered prostaglandin regulation of cytotoxic lymphocytes is postulated. Mast cells and eosinophils in the nasal mucosa and periphery point to a possible role of these cells in aspirin hypersensitivity. Patients with this type of allergic reaction present clinically with such features as rhinitis, nasal polyps, eosinophilia in nasal smears, abnormal sinus radiographs, and chronic asthma. ASA-sensitivity is associated with various NSAIDs.

A **primary allergen** is the antigenic material that sensitized a patient who subsequently shows cross-sensitivity to a related allergen.

Secondary allergen is an agent that induces allergic symptoms because of cross-reactivity with an allergen to which the individual is hypersensitive.

An **allergic response** is a response to antigen (allergen) that leads to a state of increased reactivity or hypersensitivity rather than a protective immune response.

Allergoids are allergens that have been chemically altered to favor the induction of IgG rather than IgE antibodies to diminish allergic manifestations in the hypersensitive individual. These formaldehyde-modified allergens are analogous to toxoids prepared from bacterial exotoxins. Some of the physical and chemical characteristics of allergens are similar to those of other antigens. However, the molecular weight of allergens is lower.

Allergy is a term coined by Clemens von Pirquet in 1906 to describe the altered reactivity of the animal body to antigen. Presently, the term allergy refers to altered immune reactivity to a spectrum of environmental antigens, including pollen, insect venom, and food. Allergy is also referred to as hypersensitivity and usually describes type I immediate hypersensitivity of the atopic/anaphylactic type. Some allergies, especially the delayed T cell type, develop in subjects infected with certain microorganisms such as *Mycobacterium tuberculosis* or certain pathogenic fungi. Allergy is a consequence of the interaction between antigen (or allergen) and antibody or T lymphocytes produced by previous exposure to the same antigen or allergen.

An **allergic reaction** is a response to antigens or allergens in the environment as a consequence of either preformed antibodies or effector T cells. Allergic reactions are mediated by a number of immune mechanisms, the most common of which is type I hypersensitivity, in which IgE antibody specific for an allergen is carried on the surface of mast cells. Combining with specific antigen leads to the release of pharmacological mediators, resulting in clinical symptoms of asthma, hayfever, and other manifestations of allergic reactions.

A **food allergy** is a type I (anaphylactic) or type III (antigen–antibody complex) hypersensitivity mechanism response to allergens or antigens in foods that have been ingested, which may lead to intestinal distress, producing nausea, vomiting, and diarrhea. There may be edema of the buccal mucosa, generalized urticaria, or eczema. Food categories associated with food allergy in some individuals include eggs, fish, or nuts. Atopic sensitization to cow's milk is the most common food allergy; casein is the major allergenic and antigenic protein in cow's milk. Both skin tests and RAST tests using the appropriate allergen or antigen may identify individuals with a particular food allergy.

House dust allergy is a type-I immediate hypersensitivity reaction in atopic individuals exposed to house dust in which the principal allergen is *Dermatophagoides pteronyssinus*, the house dust mite. The condition is expressed as a respiratory allergy with the atopic subject manifesting either asthma or allergic rhinitis.

Horse serum sensitivity is an allergic or hypersensitive reaction in a human or other animal receiving antitoxin or antithymocyte globulin generated by immunization of horses whose immune serum is used for therapeutic purposes. Classic serum sickness is an example of this type of hypersensitivity, which first appeared in children receiving diphtheria antitoxin early in the 20th century.

A **pollen hypersensitivity** is an immediate (type I) hypersensitivity which atopic individuals experience following inhalation of pollens such as ragweed in the U.S. This is an IgE-mediated reaction that results in respiratory symptoms expressed as hay fever or asthma. Sensitivity to certain pollens can be detected through skin tests with pollen extracts.

Wheal and flare reaction is an immediate hypersensitivity, IgE-mediated (in humans) reaction to an antigen. Application of antigen by a scratch test in a hypersensitive individual may be followed by erythema, which is the red flare, and edema, which is the wheal. Atopic subjects who have a hereditary component to their allergy experience the effects of histamine and other vasoactive amine released from mast-cell granules following crosslinking of surface IgE molecules by antigen or allergen.

Venom is a poisonous or toxic substance which is produced by species such as snakes, arthropods, and bees. The poison is transmitted through a bite or sting.

Hypersensitivity angiitis is a small vessel inflammation most frequently induced by drugs.

Hypersensitivity pneumonitis is a lung inflammation induced by antibodies specific for substances that have been inhaled. Within hours of inhaling the causative agent dyspnea, chills, fever, and coughing occur. Histopathology of the lung reveals inflammation of alveoli in the interstitium, with obliterating bronchiolitis. Immunofluorescence examination reveals deposits of C3. Hyperactivity of the lungs to airborne immunogens or allergens may ultimately lead to interstitial lung disease. An example is farmer's lung, which is characterized by malaise, coughing, fever, tightness in the chest, and myalgias. Of the numerous syndromes and associated antigens that may induce hypersensitivity pneumonitis, humidifier lung (thermophilic actinomycetes), bagassosis (***Thermoactinomyces vulgaris***), and bird fancier's lung (bird droppings) are well known.

Hypersensitivity vasculitis is an allergic response to drugs, microbial antigens, or antigens from other sources, leading to an inflammatory reaction involving small arterioles, venules, and capillaries.

Ragweed (*Ambrosia)* is distributed throughout the warmer parts of the Western Hemisphere but poses the greatest clinical problem in North America (Figure 12.15). Airborne ragweed pollen constitutes a troublesome respiratory allergen. Ragweed pollen appears in the air in northern states of the U.S. and adjacent eastern Canada in the latter days of July, when it reaches levels of up to thousands of grains per cubic meter by early September and then declines. Pure ragweed pollinosis has usually subsided by mid-October in the north. Short ragweed (*A. artemisiifolia)* and giant ragweed *(A. trifida)* are found widely from the Atlantic coast to the Midwest, Ozark plateau, and Gulf states. Short ragweed reaches to northern Mexico and is only sparsely found in the Pacific Northwest. Giant ragweed is abundant in the Mississippi Delta and along the flood plains of southeastern rivers. Pollen allergens of ragweed that have been characterized include the following:

FIGURE 12.15 *Ambrosia artemisiifolia*, commonly termed ragweed.

Allergen Source	**Allergen**
Ambrosiae artemisiifolia	Mol wt (kDA)
(ragweed)	SDS-page
AMB alpha I (AGE)	38
AMB alpha II (AGK)	38
AMB alpha III (Ia3)	12
AMB alpha IV (Ra4)	23
AMB alpha V (Ra5)	5
AMB alpha VI (Ra6)	8
AMB alpha VII	—
Ambrosiae trifida (giant ragweed) AMB T V (Ra6)	4.4

Urticaria is a pruritic skin rash identified by localized elevated, edematous, erythematous, and itching wheals with a pale center encircled by a red flare. It is caused by the release of histamine and other vasoactive substances from mast-cell cytoplasmic granules based upon immunologic sensitization or due to physical or chemical substances. It is a form of type-I immediate hypersensitivity. Urticaria is mediated by IgE antibodies in man. The action of allergen or antigen with IgE antibodies anchored to mast cells can lead to this form of cutaneous anaphylaxis or hives. The wheal is due to leakage of plasma from venules, and the flare is caused by neurotransmitters. Also called hives.

In **food and drug additive reactions**, urticaria and angioneurotic edema are the main symptoms.

Cold hypersensitivity is a localized wheal and flare reaction and bradycardia that follow exposure to low temperatures that induce overstimulation of the autonomic nervous system.

Cold urticaria occurs soon after exposure to cold. The lesions are usually confined to the exposed areas. The condition has been observed in patients with underlying conditions that include cryoglobulinemia, cryofibrinogenemia, cold agglutinin disease, and paroxysmal cold hemoglobinuria. Although the mechanism is unknown, cold exposure has been shown to cause the release of histamine and other mediators. Cold sensitivity has been passively transferred in individuals with abnormal proteins. Cryoprecipitates may fix complement and lead to the generation of anaphylatoxin. The condition can be diagnosed by placing ice cubes on the forearm for 4 min and observing, for the following 10 min, for the appearance of urticaria when the area is rewarmed. Treatment consists of limiting exposure to cold and the administration of antihistamines such as oral cyproheptadine. Individuals with an abnormal protein should have the underlying disease treated.

Cholinergic urticaria refers to skin edema induced by an aberrant response to acetylcholine that occurs following diminished cholinesterase activity.

Phacoanaphylaxis is a hypersensitivity to lens protein of the eye following an injury that introduces lens protein, normally a sequestered antigen, into the circulation. The immune system does not recognize it as self and responds to it as it would any other foreign antigen.

The **Prausnitz-Küstner (P-K) reaction (historical)** is a skin test for hypersensitivity in which serum containing IgE antibodies specific for a particular allergen is transferred from an allergic individual to a nonallergic recipient by intradermal injection. This is followed by injection of the antigen or allergen in question into the same site as the serum injection. Fixation of the IgE antibodies in the "allergic" serum to mast cells in the recipient results in local release of the pharmacological mediators of immediate hypersensitivity, which includes histamine. It results in a local anaphylactic reaction with a wheal and flare response. This reaction is no longer used because of the risk of transmitting hepatitis or AIDS.

Cutaneous anaphylaxis is a local reaction specifically elicited in the skin of an actively or passively sensitized animal. Causes of cutaneous anaphylaxis include immediate wheal and flare response following prick tests with drugs or other substances; insect stings or bites; and contact urticaria in response to food substances such as nuts, fish, or eggs, or other substances such as rubber, dander, or other environmental agents. The signs and symptoms of anaphylaxis are associated with the release of chemical mediators that include histamine and other substances from mast cell or basophil granules, following crosslinking of surface IgE by antigen or nonimmunological degranulation of these cells. The pharmacological mediators act principally on the blood vessels and smooth muscle. The skin may be the site where an anaphylactic reaction is induced or it can be the target of a systemic anaphylactic reaction resulting in itching (pruritus), urticaria, and angioedema.

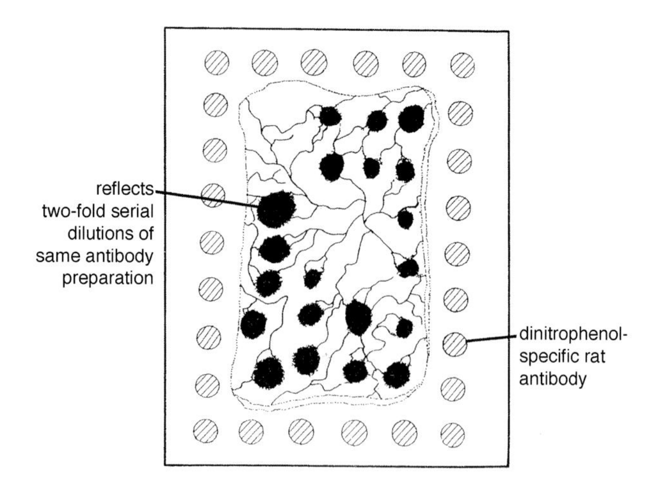

FIGURE 12.16 Passive cutaneous anaphylaxis in rats to dinitrophenol-specific rat reagin antibody. The diminished size of areas of increased capillary permeability is a consequence of twofold serial dilutions of the antibody.

Passive cutaneous anaphylaxis (PCA) is a skin test that involves the *in vivo* passive transfer of homocytotropic antibodies that mediate type I immediate hypersensitivity (e.g., IgE in humans) from a sensitized to a previously nonsensitized individual by injecting intradermally the antibodies, which become anchored to mast cells through their Fc receptors (Figure 12.16). This is followed hours or even days later by intravenous injection of antigen mixed with a dye such as Evans Blue. Crosslinking of the cell-fixed (e.g., IgE) antibody receptors by the injected antigen induces a type-I immediate hypersensitivity reaction in which histamine and other pharmacological mediators of immediate hypersensitivity are released. Vascular permeability factors act on the vessels to permit plasma and dye to leak into the extravascular space, forming a blue area which can be measured with calipers. In humans, this is called the Prausnitz-Küstner (PK) reaction.

PCA is an abbreviation for passive cutaneous anaphylaxis.

Reverse passive cutaneous anaphylaxis (RPCA) is a passive cutaneous anaphylaxis assay in which the order of antigen and antibody administration is reversed, i.e., the antigen is first injected, followed by the antibody. In this case, the antigen must be an immunoglobulin that can fix to tissue cells.

Asthma is a disease of the lungs characterized by reversible airway obstruction (in most cases), inflammation of the airway with prominent eosinophil participation, and increased responsiveness by the airway to various stimuli (Figure 12.17). There is bronchospasm associated with recurrent paroxysmal dyspnea and wheezing. Some cases

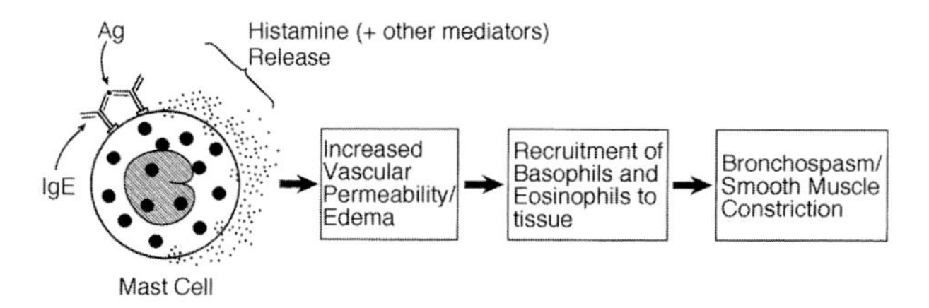

FIGURE 12.17 Events occurring in asthma.

of asthma are allergic, i.e., bronchial allergy, mediated by IgE antibody to environmental allergens. Other cases are provoked by nonallergic factors that are not discussed here.

Charcot-Leyden crystals are crystals present in asthmatic patients' sputum that are hexagonal and bipyramidal. They contain a 13-kDa lysophospholipase derived from the eosinophil cell membrane.

Major basic protein (MBP) is a 10- to 15-kDa protein present in eosinophilic granules. It has an isoelectric point that exceeds pH 10; thus the descriptor "basic." MBP induces injury to the bronchial epithelium and is linked to asthma. When inoculated intracutaneously, it can induce a wheal and flare response. Thus, this substance induces tissue injury in allergic and inflammatory diseases.

Allergic asthma is bronchial constriction that is a consequence of an allergic reaction to an inhaled antigen.

Bronchial asthma is intermittent and reversible airway obstruction that results from repeated hypersensitivity reactions, leading to inflammatory disease in the lung. Chronic inflammation of the bronchi with eosinophils, hypertrophy, and hypereactivity of bronchial smooth muscle cells occur.

Theophylline (1,3,dimethylxanthine) is a compound used to treat acute bronchial asthma because of its powerful smooth-muscle relaxing activity. Aminophylline, a salt of theophylline, is also a smooth-muscle relaxant used to induce bronchodilation in the treatment of asthma.

Isoproterenol (*dl*-β-[3,4-dihydroxyphenyl]-α-isopropylaminoethanol): A β-adrenergic amine that is used to treat patients with asthma. It is able to relax the bronchial smooth muscle constriction that occurs in asthmatics.

Status asthmaticus is a clinical syndrome characterized by diminished responsiveness of asthmatic patients to drugs to which they were formerly sensitive. Patients may not respond to adrenergic bronchodilators and are hypoxemic. Treatment is with oxygen, aminophylline, and methylprednisone.

Exercise-induced asthma refers to an attack of asthma brought on by exercise.

MK-571 is a powerful synthetic antagonist of leukotriene D_4 receptor that prevents bronchoconstriction induced by exercise in patients with asthma.

Metaproterenol (*dl*-β-[3,5-dihydroxyphenyl]-α-isopropylamino-ethanol) is a β-adrenergic amine that induces smooth muscle relaxation, especially in the bronchi. This substance has been used to treat asthma.

Extrinsic asthma is caused by antigen–antibody reactions, with two mutually nonexclusive forms: (1) atopic, involving IgE antibodies, and (2) nonatopic, involving either antibodies other than IgE or immune complexes. The atopic form usually begins in childhood and is preceded by other atopic manifestations such as paroxysmal rhinitis, seasonal hay fever, or infantile eczema. A familial history is usually obtainable. Other environmental agents may sometimes be recognized in the nonatopic form. Patients with extrinsic asthma respond favorably to bronchodilators.

Type II antibody-mediated hypersensitivity is induced by antibodies and has three forms (Figure 12.18).

The classic type of hypersensitivity involves the interaction of antibody with cell membrane antigens followed by complement lysis (Figure 12.19). These antibodies are directed against antigens intrinsic to specific target tissues. Antibody-coated cells also have increased susceptibility to phagocytosis. Examples of type II hypersensitivity include the antiglomerular basement membrane antibody that develops in Goodpasture's syndrome and antibodies that develop against erythrocytes in Rh incompatibility, leading to erythroblastosis fetalis or autoimmune hemolytic anemia.

A second variety of type II hypersensitivity is antibody-dependent cell-mediated cytotoxicity (ADCC). Killer (K) cells or NK cells, which have Fc receptors on their surfaces, may bind to the Fc region of IgG molecules. They may react with surface antigens on target cells to produce lysis of the antibody-coated cell. Complement fixation is not required and does not participate in this reaction. In addition to K and NK cells, neutrophils, eosinophils, and macrophages may participate in ADCC.

A third form of type II hypersensitivity is antibody against cell surface receptors that interfere with function, as in the case of antibodies against acetylcholine receptors in motor endplates of skeletal muscle in myasthenia gravis (Figure 12.20). This interference with neuromuscular transmission results in muscular weakness, ultimately affecting the muscles of respiration and producing death. By contrast, stimulatory antibodies develop in hyperthyroidism (Graves' disease). They react with thyroid-stimulating hormone receptors on thyroid epithelial cells to produce hyperthyroidism.

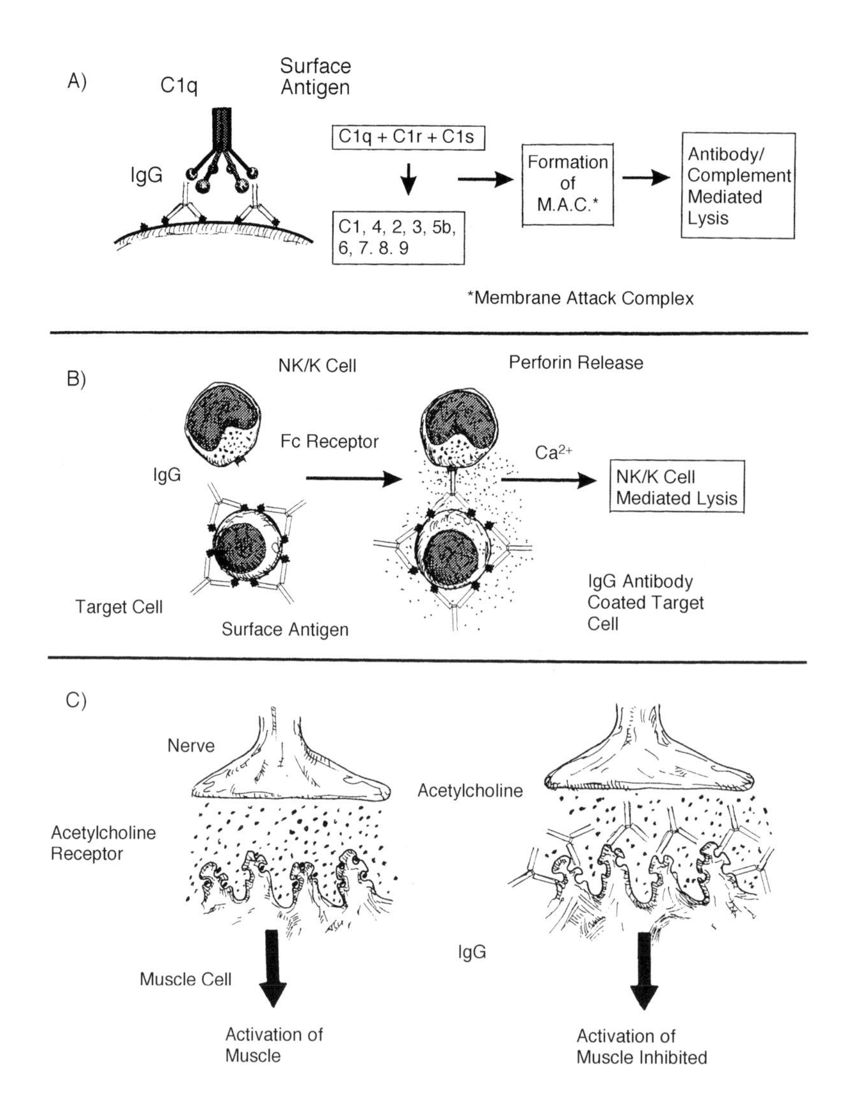

FIGURE 12.18 Three separate forms of type II hypersensitivity. The uppermost diagram depicts antibody- and complement-mediated lysis of a nucleated cell as a consequence of formation of the membrane attack complex. The middle diagram shows antibody-dependent cell-mediated cytotoxicity through the action of either an NK or a K cell with surface antibody specific for a target cell. The bottom figure illustrates inhibition of transmission of the nerve impulse by antibodies against acetycholine receptors as occurs in myasthenia gravis.

Antibody-dependent cell-mediated cytotoxicity (ADCC) is a reaction in which T lymphocytes; NK cells, including large granular lymphocytes; neutrophils; and macrophages may lyse tumor cells, infectious agents, and allogeneic cells by combining through their Fc receptors with the Fc region of IgG antibodies bound through their Fab regions to target cell surface antigens. Following linkage of Fc receptors with Fc regions, destruction of the target is accomplished through released cytokines. It represents an example of participation between antibody molecules and immune system cells to produce an effector function. NK cells mediate most ADCC through the Fc receptor FcγRIII or CD 16 on their surface.

Long-acting thyroid stimulator (LATS) is an IgG autoantibody that mimics the action of thyroid-stimulating hormone in its effect on the thyroid (Figure 12.21). The majority of patients with Graves' disease, i.e., hyperthyroidism, produce LATS. This IgG autoantibody reacts with the receptors on thyroid cells that respond to thyroid-stimulating hormone. Thus, the antibody–receptor interaction results in the same biological consequence as does hormone interaction with the receptor. This represents a stimulatory type of hypersensitivity and is classified in the Gell and Coombs classification as one of the forms of type II hypersensitivity.

Immune complex reactions are type III hypersensitivity in which antigen–antibody complexes fix complement and induce inflammation in tissues such as capillary walls.

Type III immune complex-mediated hypersensitivity is a type of hypersensitivity mediated by antigen–antibody–complement complexes (Figure 12.22). Antigen–antibody

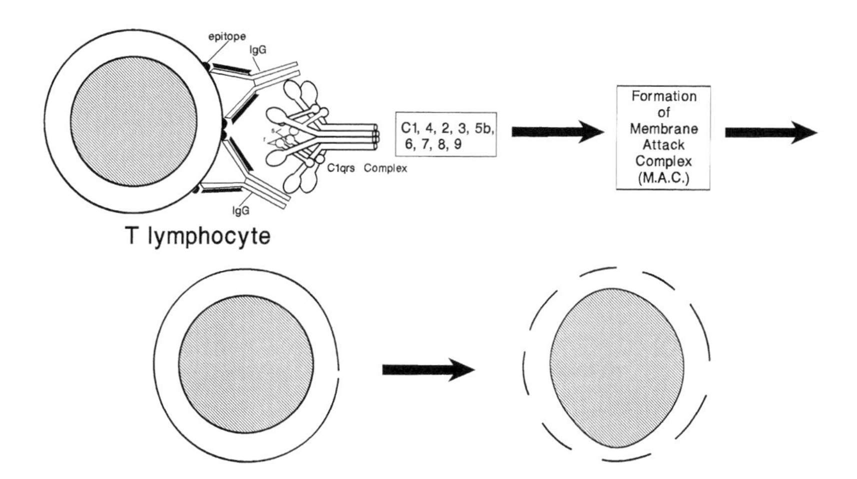

FIGURE 12.19 Schematic representation of the action of specific IgG antibody on surface epitopes of a T lymphocyte leading to antibody–complement mediated lysis of that cell.

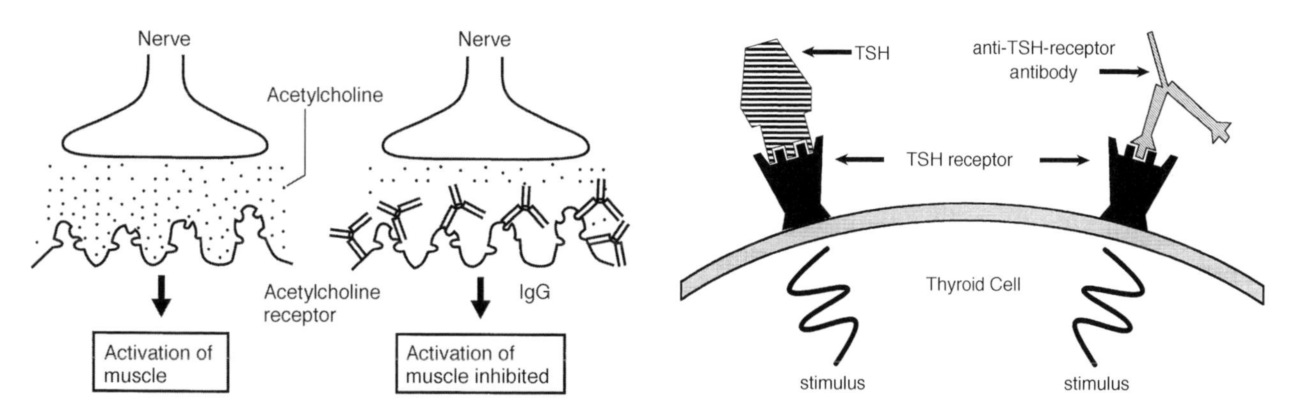

FIGURE 12.20 Schematic representation of the interference by acetylcholine receptor antibodies of chemical transmission of the nerve impulse. Acetylcholine receptor (AchR) antibodies are IgG autoantibodies that cause loss of function of acetylcholine receptors, which are critical to chemical transmission of the nerve impulse at the neuromuscular junction. This represents a type II mechanism of hypersensitivity according to the Coombs and Gell classification. AchR antibodies are heterogeneous with some showing specificity for antigenic determinants other than those that serve as acetylcholine or alpha–bungarotoxin binding sites. As many as 85 to 95% of myasthenia gravis patients may manifest acetylcholine receptor antibodies.

FIGURE 12.21 Schematic representation of the third form of type II hypersensitivity in which long-acting thyroid stimulator (LATS), an IgG antibody specific for the TSH receptor, leads to continuous stimulation of thyroid parenchymal cells, causing hyperthyroidism. The IgG antibody mimics the action of TSH.

complexes can stimulate an acute inflammatory response that leads to complement activation and PMN leukocyte infiltration. The immune complexes are formed either by exogenous antigens such as those from microbes or by endogenous antigens such as DNA, a target for antibodies produced in systemic lupus erythematosus. Immune complex–mediated injury may be either systemic or localized. In the systemic variety, antigen–antibody complexes are produced in the circulation, deposited in the tissues, and initiate inflammation. Acute serum sickness occurred in children treated with diphtheria antitoxin earlier in this century as a consequence of antibody produced against the horse serum protein. When immune complexes are deposited in tissues, complement is fixed, and PMNs are attracted to the site. Their lysosomal enzymes are released, resulting in tissue injury. Localized immune complex disease, sometimes called the Arthus reaction, is characterized by an acute immune complex vasculitis with fibrinoid necrosis occurring in the walls of small vessels.

Auer's colitis is an Arthus reaction in the intestine produced by the inoculation of albumin serving as antigen into the colon of rabbits that have developed antialbumin antibodies. An inflammatory lesion is produced in the colon and is marked by hemorrhage and necrosis.

Vasculitis is inflammation of the wall of any size blood vessel that may be accompanied by necrosis. Vasculitis with an immunologic basis is often associated with immune complex disease, in which deposition of complement-fixing

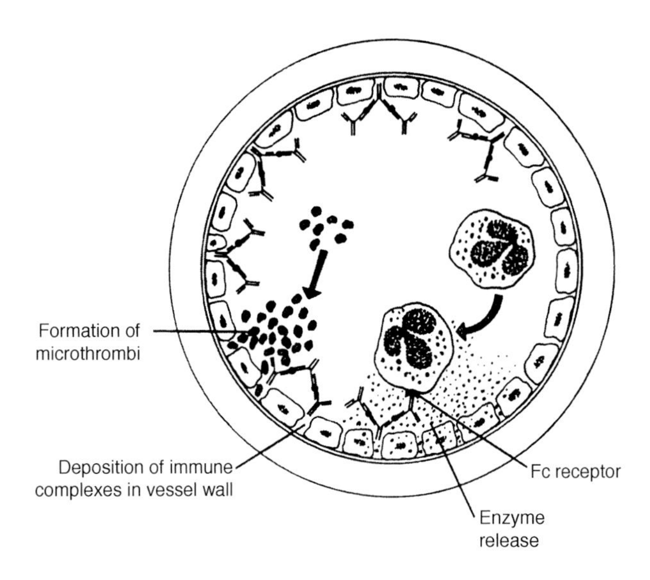

FIGURE 12.22 Schematic representation of the formation and deposition of immune complexes in vessel walls in type III hypersensitivity.

microprecipitates in the vessel wall may attract polymorphonuclear neutrophils and lead to tissue injury associated with acute inflammation.

Toxic complexes refers to increased levels of circulating immune complexes that may be harmful and that trigger type III hypersensitivity reactions. The soluble complexes are pathogenic. The classic description regards such complexes as "toxic," and the term "toxic complexes" was frequently used in the literature. Such complexes are characterized by (1) formation in a zone of moderate antigen excess, (2) the antibody in the complex has no cytotropic affinity for tissues, and (3) the complex is able to activate the complement system. Complex formation is associated with conformational changes in the antibody molecule, and the activity of the complex depends on the antibody and not on the antigen. Antibodies produced in some species such as rabbit, human, and guinea pig have the above properties. Those produced in other species such as bovine, chicken, and horse are inactive in this respect. Fixation of the complexes occurs by the Fc portion of the antibody in the complex. They stick to cells and basement membranes, causing injury to the endothelium of small vessels. The injury may occur at the local site of antigen injection or may be systemic when antigen is injected intravenously. The chain of events characteristic for inflammation is set in motion with liberation of vasoactive amines and involvement of polymorphonuclear leukocytes.

Complex release activity refers to the binding of injected preformed complexes to the endothelial cell membranes immediately after their injection into experimental animals. The amount of such binding decreases with age, favoring deposition of such complexes within tissues.

CIC is the abbreviation for circulating immune complexes.

Serum sickness is a systemic reaction that follows the injection of a relatively large single dose of serum (e.g., antitoxin) into humans or other animals. It is characterized by systemic vasculitis (arteritis), glomerulonephritis, and arthritis. The lesions follow the deposition in tissues, such as the microvasculature, of immune complexes that form after antibody appears in the circulation between the 5th and 14th d following antigen administration. The antigen–antibody complexes fix complement and initiate a classic type III hypersensitivity reaction, resulting in immune-mediated tissue injury. Patients may develop fever, lymphadenopathy, urticaria, and sometimes arthritis. The pathogenesis of serum sickness is that of a classic type III reaction. Antigen escaping into circulation from the site of injection forms immune complexes that damage the small vessels. The antibodies involved in the classic type of serum sickness are of the precipitating variety, usually IgG. They may be detected by passive hemagglutination. Pathologically, serum sickness is a systemic immune complex disease characterized by vasculitis, glomerulonephritis, and arthritis due to the intravascular formation and deposition of immune complexes that subsequently fix complement and attract polymorphonuclear neutrophils to the site through the chemotactic effects of C5a, thereby initiating inflammation. The classic reaction, which occurs 7 to 15 d after the triggering injection, is called the primary form of serum sickness. Similar manifestations appearing only 1 to 3 d following the injection represent the accelerated form of serum sickness and occur in subjects presumably already sensitized. A third form, called the anaphylactic form, develops immediately after injection. This latter form is apparently due to reaginic IgE antibodies and usually occurs in atopic subjects sensitized by horse dander or by previous exposure to serum treatment. The serum sickness-like syndromes seen in drug allergy have a similar clinical picture and similar pathogenesis (Figure 12.23). See also type III immune complex-mediated hypersensitivity.

Herxheimer reaction is a serum sickness (type III) form of hypersensitivity that occurs following the treatment of

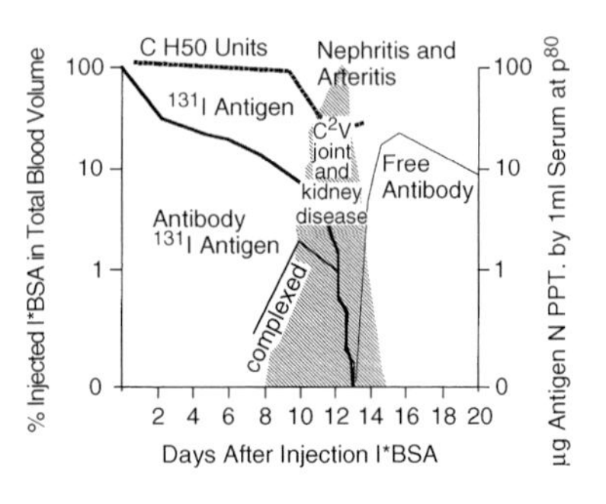

FIGURE 12.23 Serum sickness.

selected chronic infectious diseases with an effective drug. When the microorganisms are destroyed in large numbers in the blood circulation, a significant amount of antigen is released from the disrupted microbes, which tend to react with preformed antibodies in the circulation. This type of reaction has been described following the use of effective drugs to treat syphilis, trypanosomiasis, and brucellosis.

Arthus reaction is induced by repeated intradermal injections of antigen into the same skin site. It is dependent upon the development of humoral antibodies of the precipitin type, which react *in vivo* with specific antigen at a local site. It may also be induced by the inoculation of antigen into a local skin site of an animal possessing preformed IgG antibodies specific for the antigen. Immune complexes are comprised of antigen, antibody, and complement formed in vessels. The chemotactic complement fragment C5a and other chemotactic peptides produced attract neutrophils to antigen–antibody–complement complexes. This is followed by lysosomal enzyme release, which induces injury to vessel walls with the development of thrombi, hemorrhage, edema, and necrosis. Events leading to vascular necrosis include blood stasis; thrombosis; capillary compression in vascular injury, which causes extravasation; venule rupture; hemorrhage, and local ischemia. There is extensive infiltration of polymorphonuclear cells, especially neutrophils, into the connective tissue. Grossly, edema, erythema, central blanching, induration, and petechiae appear. Petechiae develop within 2 h, reach a maximum between 4 and 6 h, and then may diminish or persist for 24 h or longer with associated central necrosis, depending on the severity of the reaction. If the reaction is more prolonged, macrophages replace neutrophils; histiocytes and plasma cells may also be demonstrated (Figure 12.24). The Arthus reaction is considered a form of immediate-type hypersensitivity, but it does not occur as rapidly as does anaphylaxis. It takes place during a 4-h period and diminishes after 12 h. Thereafter, the area is cleared by mononuclear phagocytes. The passive cutaneous Arthus reaction consists of the inoculation of antibodies intravenously into a nonimmune host, followed by local cutaneous injection of antigen. The reverse passive cutaneous Arthus reaction requires the intracutaneous injection of antibodies, followed by the intravenous or incutaneous (at the same site) administration of antigen. The Arthus reaction is a form of type III hypersensitivity since it is based upon the formation of immune complexes with complement fixation. Clinical situations for which it serves as an animal model include serum sickness, glomerulonephritis, and farmer's lung.

Fibrinoid necrosis refers to tissue death in which there is a smudgy eosinophilic deposit that resembles fibrin microscopically and camouflages cellular detail. It is induced by proteases released from neutrophils that digest

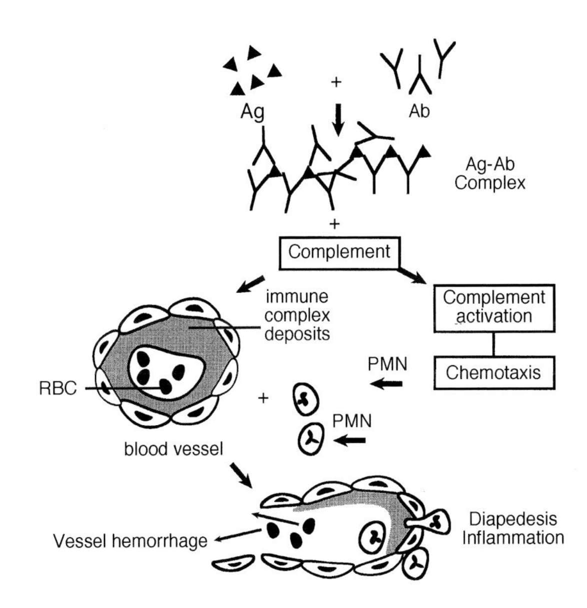

FIGURE 12.24 Schematic representation of molecular, cellular, and tissue interactions in the Arthus reaction.

the tissue and cause fibrin deposition. Fibrinoid necrosis is seen in tissues in a number of connective tissue diseases with immune mechanisms. An example is systemic lupus erythematosus. Fibrinoid necrosis is classically seen in the walls of small vessels in immune complex vasculitis such as occurs in the Arthus reaction.

A **passive Arthus reaction** is an inflammatory vasculitis produced in experimental animals by the passive intravenous injection of significant amounts of precipitating IgG antibody, followed by the intracutaneous or subcutaneous injection of the homologous antigen for which the antibodies are specific. This permits microprecipitates to occur in the intercellular spaces between the intravascular precipitating antibody and antigen in the extravascular space. This is followed by interaction with complement, attraction of polymorphonuclear leukocytes, and an inflammatory response as described under Arthus reaction.

Bagassosis is a hypersensitivity among sugarcane workers to a fungus, *Thermoactinomyces saccharic*, that thrives in the pressing from sugar cane. The condition is expressed as a hypersensitivity pneumonitis. Subjects develop type III (Arthur reaction) hypersensitivity following inhalation of dust from molding hot sugarcane bagasse.

Reverse anaphylaxis is anaphylaxis produced by the passive transfer of serum antibody from a sensitized animal to a normal untreated recipient only after the recipient had been first injected with the antigen. Thus, the usual order of administration of antigen and antibody are reversed compared to classic anaphylaxis.

Reverse passive Arthus reaction is a reaction that differs from a classic Arthus reaction only in that the precipitating antibody is injected into an animal intracutaneously, and after an interval of $^1/_2$ h to 2 h, the antigen is administered intravenously. In this situation, antigen, rather than antibody, diffuses from the blood into the tissues, and antibody, rather than antigen, diffuses into the tissue, where it encounters and interacts with antigen with the consequent typical changes in the microvasculature and tissues associated with the Arthus reaction.

Shwartzman (or Shwartzman-Sanarelli) reaction is a nonimmunologic phenomenon in which endotoxin (lipopolysaccharides) induces local and systemic reactions. Following the initial or preparatory injection of endotoxin into the skin, polymorphonuclear leukocytes accumulate and are then thought to release lysosomal acid hydrolases that injure the walls of small vessels, preparing them for the second provocative injection of endotoxin. The intradermal injection of endotoxin into the skin of a rabbit, followed within 24 h by the intravenous injection of the same or a different endotoxin, leads to hemorrhage at the local site of the initial injection (Figure 12.25 and Figure 12.26). Although the local Shwartzman reaction may resemble an Arthus reaction in appearance, the Arthus reaction is immunological, whereas the Shwartzman reaction is not. In the Shwartzman reaction, there is insufficient time between the first and second injections to induce an immune reaction in a previously unsensitized host. There is also a lack of specificity since even a different endotoxin may be used for first and second injections.

The generalized or systemic Shwartzman reaction again involves two injections of endotoxin. However, both are administered intravenously, one 24 h following the first. The generalized Shwartzman reaction is the experimental equivalent of disseminated intravascular coagulation that occurs in a number of human diseases. Following the first injection, sparse fibrin thrombi (Figure 12.27) are formed in the vasculature of the lungs, kidney, liver, and capillaries of the spleen. There is blockade of the reticuloendothelial system as its mononuclear phagocytes proceed to clear thromboplastin and fibrin. Administration of the second dose of endotoxin while the reticuloendothelial system is blocked leads to profound intravascular coagulation since the mononuclear phagocytes are unable to remove the thromboplastin and fibrin. There is bilateral cortical necrosis of the kidneys and splenic hemorrhage and necrosis. Neither platelets nor leukocytes are present in the fibrin thrombi that are formed.

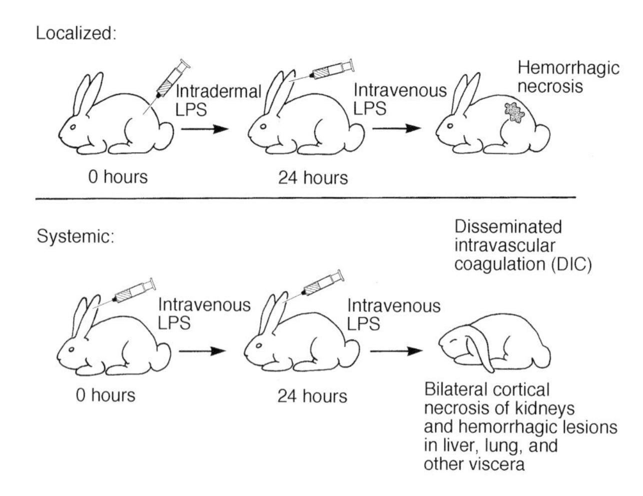

FIGURE 12.25 Schematic representation of the localized and systemic Shwartzman reaction.

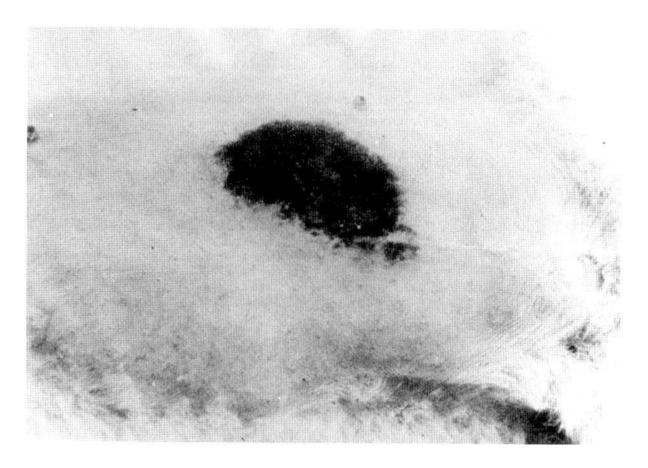

FIGURE 12.26 The ventral surface of a rabbit in which the localized Shwartzman reaction has been induced with endotoxin showing hemorrhage and necrosis.

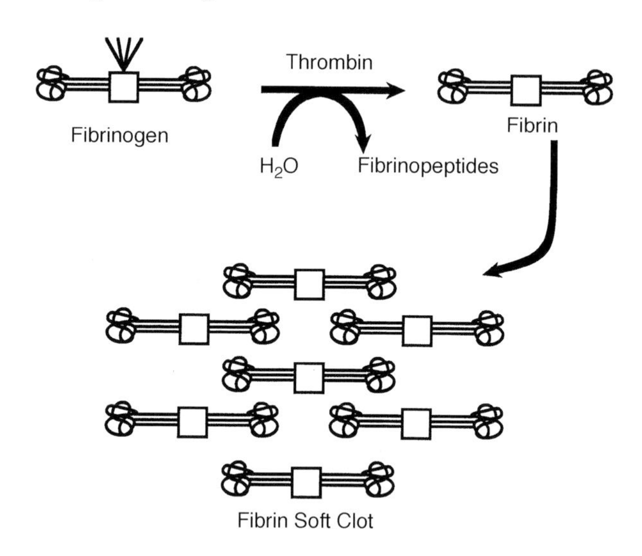

FIGURE 12.27 Fibrin.

Sanarelli-Shwartzman reaction: See Shwartzman reaction.

DIC is an abbreviation for disseminated intravascular coagulation.

Disseminated intravascular coagulation (DIC) is a tendency to favor coagulation over fibrinolysis in the blood circulation as a consequence of various factors. In DIC, 30 to 65% of the cases are due to infection. It may occur as fast DIC or slow DIC. The fast variety is characterized by acute, fulminant, consumptive coagulopathy with bleeding as a result of Gram-negative substances, massive tissue injury, burns, etc. Deficient or consumed coagulation factors must be replaced in fast DIC. Slow DIC accompanies chronic diseases characterized by thrombosis, microcirculatory ischemia, and end-organ function. Examples include acute promyelocytic leukemia, neoplasia, vasculitis, and other conditions. The pathogenesis includes endothelial cell damage produced by endotoxins or other agents, activation of platelets, and activation of the intrinsic coagulation pathway. Tissue thromboplastin may be released by trauma or neoplasia, leading to the activation of the extrinsic pathway. Patients with DIC have elevated partial thromboplastin time, prothrombin time, fibrinopeptide A, and fibrinogen degradation products and have decreased fibrinogen, factor V, platelets, antithrombin III, factor VIII, and plasminogen. There is diffuse cortical necrosis of the kidneys.

Type IV cell-mediated hypersensitivity is a form of hypersensitivity mediated by specifically sensitized cells (Figure 12.28). Whereas antibodies participate in type I, II, and III reactions, T lymphocytes mediate type IV hypersensitivity. Two types of reactions, mediated by separate T cell subsets, are observed. Delayed-type hypersensitivity (DTH) is mediated by $CD4^+$ T cells, and cellular cytotoxicity is mediated principally by $CD8^+$ T cells.

A classic delayed hypersensitivity reaction is the tuberculin or Mantoux reaction. Following exposure to *Mycobacterium tuberculosis*, $CD4^+$ lymphocytes recognize the microbe's antigens complexed with class II MHC molecules on the surface of antigen-presenting cells that process the mycobacterial antigens. Memory T cells develop and remain in the circulation for prolonged periods. When tuberculin antigen is injected intradermally, sensitized T cells react with the antigen on the antigen-presenting cell's surface, undergo transformation, and secrete lymphokines that lead to the manifestations of hypersensitivity. Unlike an antibody-mediated hypersensitivity, lymphokines are not antigen specific.

In T-cell-mediated cytotoxicity, $CD8^+$ T lymphocytes kill antigen-bearing target cells. The cytotoxic T lymphocytes play a significant role in resistance to viral infections. Class I MHC molecules present viral antigens to $CD8^+$ T-lymphocytes as a viral peptide–class I molecular complex, which is transported to the infected cell's surface. Cytotoxic $CD8^+$ cells recognize this and lyse the target before the virus can replicate, thereby stopping the infection.

Chlorodinitrobenzene (1-chlor-2,4-dinitrobenzene) (More often termed dinitrochlorobenzene [DNCB]) it is a chemical substance used to test for a patient's ability to develop the type of delayed-type hypersensitivity referred to as contact hypersensitivity. This is a type IV hypersensitivity reaction. The chemical is applied to a patient's forearm. Following sufficient time for sensitization to develop, the patient's other forearm is exposed to a second (test) dose of the same chemical. In an individual with an intact cell-mediated limb of the immune response, a positive reaction develops at the second challenge site within 48 to 72 h. Individuals with cell-mediated immune deficiency disorders fail to develop a positive delayed-type hypersensitivity reaction.

A **zirconium granuloma** is a tissue reaction in axillary regions of subjects who use solid antiperspirants containing zirconium. The granuloma develops as a consequence of sensitization to zirconium.

Cell-mediated hypersensitivity refers to delayed-type hypersensitivity and to type IV cell-mediated hypersensitivity.

Cellular allergy refers to delayed-type hypersensitivity, type IV cell-mediated hypersensitivity, and cell-mediated immunity.

Purified protein derivative (PPD) is a derivative of the growth medium in which *Mycobacterium tuberculosis* has been cultured. It is a soluble protein that is precipitated from the culture medium by trichloroacetic acid. It is used for tuberculin skin tests.

Fluorodinitrobenzene: See dinitrofluorobenzene (DNFB).

Cellular hypersensitivity refers to delayed-type hypersensitivity, type IV cell-mediated hypersensitivity, and cell-mediated immunity.

Corneal test refers to corneal response.

Corneal response: In an animal that has been previously sensitized to an antigen, the cornea of the eye may become clouded (or develop opacities) after injection of the same antigen into it. There is edema and lymphocytic and macrophage infiltration into the area. The response has been suggested to represent cell-mediated immunity.

A **patch test** is an assay to determine the cause of skin allergy, especially contact allergic (type IV) hypersensitivity. A small square of cotton, linen, or paper impregnated with the suspected allergen is applied to the skin for 24 to 48 h. The test is read by examining the site 1 to 2 d after applying the patch. The development of redness (erythema), edema, and formation of vesicles constitute a positive test. The impregnation of tuberculin into a patch was used by Vollmer for a modified tuberculin test. There are multiple chemicals, toxins, and other allergens that

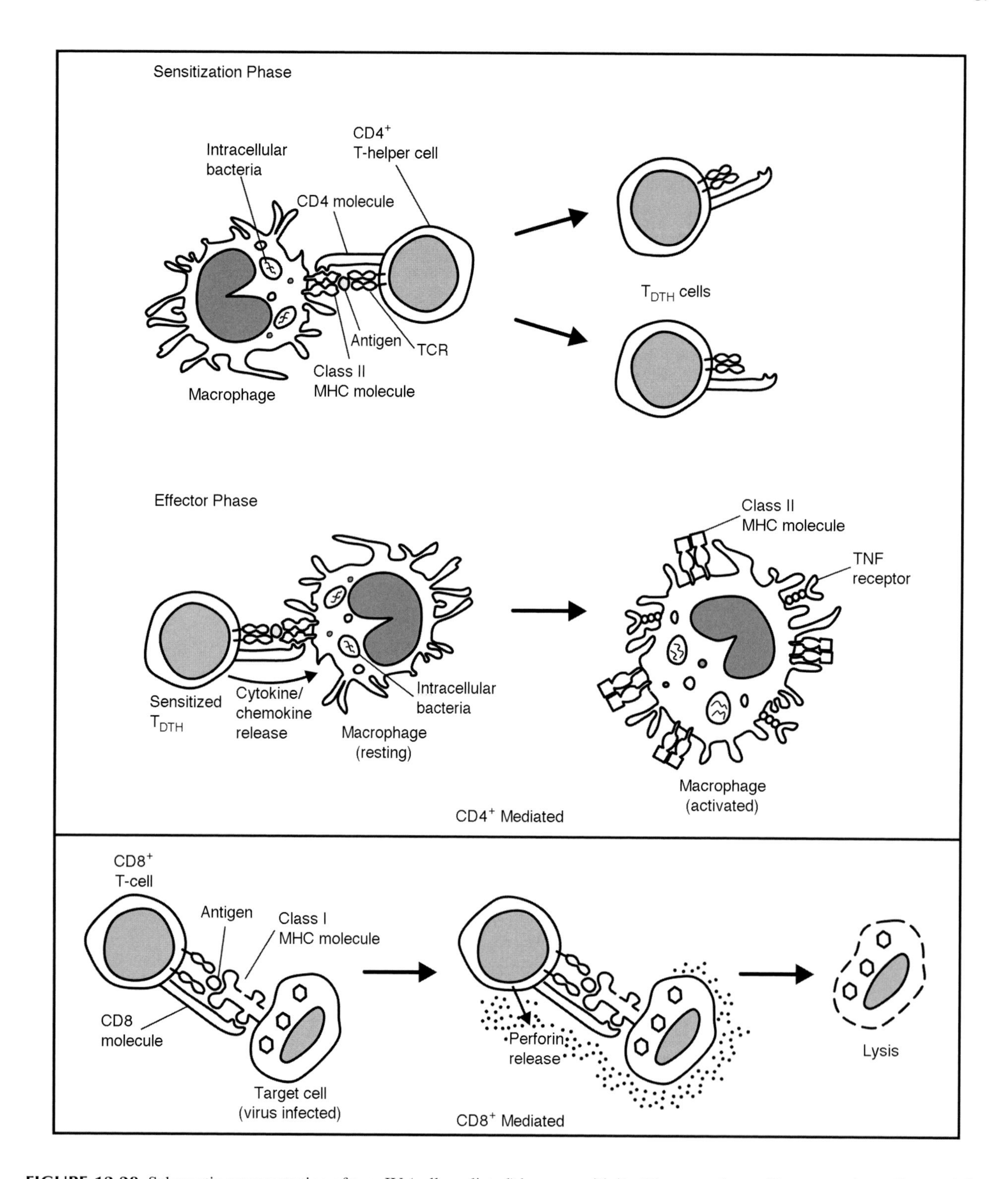

FIGURE 12.28 Schematic representation of type IV (cell-mediated) hypersensitivity. The upper frame illustrates tuberculin reactivity in the skin that is mediated by CD4+ helper/inducer T cells and represents a form of bacterial allergy. The lower frame illustrates the cytotoxic action of CD8+ T cells against a virus-infected target cell that presents antigen via class I MHC molecules to its TCR, resulting in the release of perforin molecules that lead to target cell lysis.

may induce allergic contact dermatitis in exposed members of the population.

Delayed-type hypersensitivity (DTH) is a cell-mediated immunity, or hypersensitivity mediated by sensitized T lymphocytes (Figure 12.29). Although originally described as a skin reaction that requires 24 to 48 h to develop following challenge with antigen, current usage emphasizes the mechanism, which is T-cell mediated, as opposed to emphasis on the temporal relationship of antigen

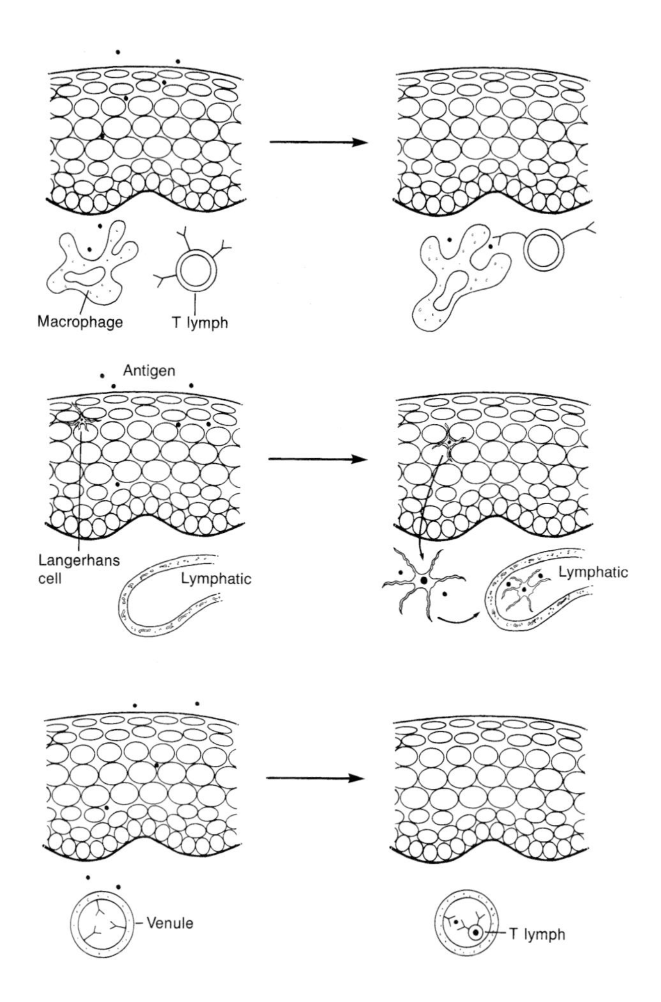

FIGURE 12.29 Diagram of delayed-type hypersensitivity.

injection and host response. The $CD4^+$ T lymphocyte is the principal cell that mediates delayed-type hypersensitivity reactions. To induce a DTH reaction, antigen is injected intradermally in a primed individual. If the reaction is positive, an area of erythema and induration develops 24 to 48 h following antigen challenge. Edema and infiltration by lymphocytes and macrophages occur at the local site. The $CD4^+$ T lymphocytes identify antigen on Ia-positive macrophages and release lymphokines, which entice more macrophages to enter the area where they become activated.

Skin tests are used clinically to reveal delayed-type hypersensitivity to infectious disease agents. Skin test antigens include such substances as tuberculin, histoplasmin, and candidin. Tuberculin or purified protein derivative (PPD), which are extracts of the tubercle bacillus, have long been used to determine whether or not a patient has had previous contact with the organism from which the test antigen was derived. Delayed-type hypersensitivity reactions are always cell mediated. Thus, they have a mechanism strikingly different from anaphylaxis or the Arthus reaction, which occur within minutes to hours following exposure of the host to antigen and are examples of antibody-mediated reactions. DTH is classified as type IV hypersensitivity (Coombs and Gell classification). A T_{DTH} lymphocyte is a delayed-type hypersensitivity T lymphocyte. Delayed-type hypersensitivity may be either permanent, persisting from month to years after sensitization, as occurs in classic tuberculin-type hypersensitivity, or transient, which resembles the permanent type morphologically but disappears 1 to 2 weeks following induction of sensitization. In the permanent type the inflammatory reaction remains prominent 72 to 96 h following intradermal injection of antigen, but the inflammatory reaction disappears 1 to 2 weeks after induction of sensitization in the transient type in which the inflammatory lesion peaks at 24 h but disappears by 48 to 72 h (Jones-Mote hypersensitivity). The activation of T cells in delayed-type hypersensitivity is associated with the secretion of cytokines. Activation of different subsets of T helper (T_H) cells leads to secretion of different types of cytokines. Those associated with a T_H1 $CD4^+$ cellular response profile include IFN-γ, IL-2, TNF-β, TNF-α, GM-CSF, and IL-3. Cytokines involved in a T_H2 $CD4^+$ cellular response include IL-3, IL-4, IL-5, IL-6, IL-10, IL-13, TNF-α, and GM-CSF. T_H2-type immediate IgE-mediated hypersensitivity immune responses are induced by allergens such as animal dander, dust mites, and pollens which underly asthma. Positive tuberculin reactions induce T_H1-type delayed-type hypersensitivity responses associated with T_H1 cytokines. T_H1 immunity might possibly inhibit atopic allergies by repressing T_H2 immune responses.

Cutaneous sensitization refers to the application of antigen to the skin to induce hypersensitivity.

DTH is an abbreviation for delayed-type hypersensitivity.

A **DTH T cell** is a $CD4^+$ T lymphocyte sensitized against a delayed-type hypersensitivity antigen.

Cellular and humoral metal hypersensitivity: Metal ions interact with proteins in several ways. Mercury and gold form metal–protein complexes by binding with high affinity to thiol groups of cysteine. These metal–protein complexes are able to activate T lymphocytes either with or without antigen or by superantigen stimulation. Mercury is used as a preservative and as a form of dental amalgam. Mercury compounds can induce contact sensitivity and glomerulonephritis in humans. Gold salts are used as antirheumatic drugs and can cause contact dermatitis, stomatitis, penumonitis, glomerulonephritis, increased levels of serum immunoglobulins, antinuclear autoantibodies, thrombocytopenia, and asthma in gold miners. Cadmium chloride can induce renal tubular damage in mice via interaction with T cells with specific Hsp 70 on tubular cells. Occupational exposure to beryllium salts may lead to chronic interstitial granulomatous lung disease.

$CD4^+$ T lymphocytes from berylliosis patients react to beryllium salts in an MHC class II–restricted manner. Silica, silicone, and sodium silicate activate $CD4^+$ memory T lymphocytes in women with silicone breast implants. Silicone hypersensitivity results in high levels of IL-1 and IL-1RA in the circulation. Silicone immune disease reaction is associated with the synthesis of autoantibodies to multiple endocrine organs, which is compatible with an immune-mediated endocrinopathy. Nickel sulfate, potassium dichromate, cobalt chloride, palladium chloride, and gold sodium thiosulfate represent metal allergy in patients with symptoms resulting from dental restoration. Lead and cadmium can lead to suppression of cell-mediated immunity. Laboratory methods of assessment include EIA, memory lymphocyte immunostimulation assay (MELISA), and lymphocyte proliferation tests.

Dhobi itch is a contact hypersensitivity (type IV hypersensitivity) induced by using a laundry marking ink made from Indian ral tree nuts. It occurs in subjects sensitized by wearing garments marked with such ink. It induces a dermatitis at sites of contact with the laundry marking ink.

Infection or **bacterial allergy** is a hypersensitivity, especially of the delayed T-cell type, that develops in subjects infected with certain microorganisms such as *Mycobacterium tuberculosis* or certain pathogenic fungi.

An **infection allergy** is a T cell-mediated delayed-type hypersensitivity associated with infection by selected microorganisms such as *Mycobacterium tuberculosis.* This represents type IV hypersensitivity to antigenic products of microorganisms inducing a particular infection. It also develops in *Brucellosis* , *Lymphogranuloma venereum*, mumps, and vaccinia. Also called infection hypersensitivity.

Bacterial allergy is delayed-type hypersensitivity of infection such as in tuberculosis.

Infection hypersensitivity is a tuberculin-type sensitivity that is more evident in some infections than others. It develops with great facility in tuberculosis, *Brucellosis* , *Lymphogranuloma venereum* , mumps, and vaccinia. The sensitizing component of the antigen molecule is usually protein, although polysaccharides may induce delayed reactivity in cases of systemic fungal infections such as those caused by *Blastomyces* , *Histoplasma* , and *Coccidioides* .

Bacterial hypersensitivity: See delayed-type hypersensitivity (type IV).

The **Koch phenomenon** is a delayed hypersensitivity reaction in the skin of a guinea pig after it has been infected with *Mycobacterium tuberculosis* (Figure 12.30). Robert Koch described the phenomenon in 1891 following the injection of either living or dead *M. tuberculosis* microorganisms into guinea pigs previously infected with the same microbes. He observed a severe necrotic reaction at the site of inoculation, which occasionally became generalized and induced death. The injection of killed *M. tuberculosis* microorganisms into healthy guinea pigs caused no ill effects. This is a demonstration of cell-mediated immunity and is the basis for the tuberculin test.

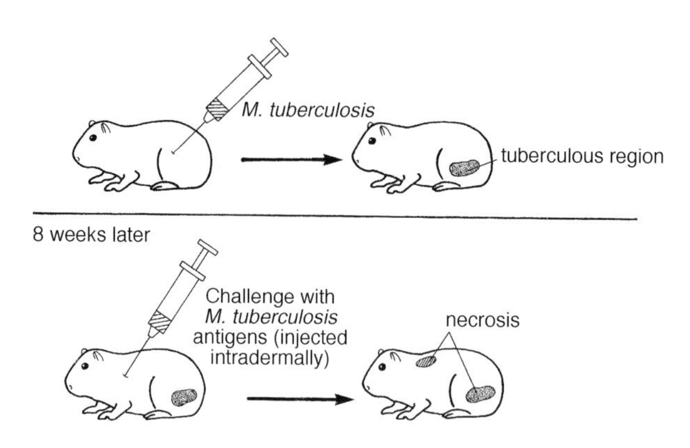

FIGURE 12.30 Illustration of the Koch phenomenon.

Tuberculid is a hypersensitivity skin reaction to mycobacteria. The lesion may be either a papulonecrotic tuberculid, with sterile papules ulcerated in the center and obliterative vasculitis, or crops of small red papules, with a sarcoid-like appearance that represent lichen scrofulosorum.

Tuberculin is a sterile solution containing a group of proteins derived from culture medium where *Mycobacterium tuberculosis* microorganisms have been grown. It has been used for almost a century as a skin test preparation to detect delayed-type (type IV) hypersensitivity to infection with *M. tuberculosis.* Many tuberculin preparations have been used in the past, but only old tuberculin (OT) and purified protein derivative (PPD) are still used. Whereas OT is a heat-concentrated filtrate of the culture medium in which *M. tuberculosis* was grown, PPD of tuberculin is a trichloroacetic acid precipitate of the growth medium. Tuberculin is a mitogen for murine B lymphocytes, as well as a T-lymphocyte mitogen.

Tuberculin hypersensitivity is a form of bacterial allergy specific for a product in culture filtrates of *Mycobacterium tuberculosis* which, when injected into the skin, elicits a cell-mediated delayed-type hypersensitivity (type IV) response. Tuberculin-type hypersensitivity is mediated by $CD4^+$ T lymphocytes. Following the intracutaneous inoculation of tuberculin extract or purified protein derivative (PPD), an area of redness and induration develops at the site within 24 to 48 h in individuals who have present or past interaction with *M. tuberculosis.*

PPD is an abbreviation for purified protein derivative of tuberculin.

Tuberculin reaction is a test of *in vivo* cell-mediated immunity. Robert Koch observed a localized lesion in the skin of tuberculous guinea pigs inoculated intradermally with broth from a culture of tubercle bacilli. The body's immune response to infection with the tubercle bacillus is signaled by the appearance of agglutinins, precipitins, opsonins, and complement-fixing antibodies in the serum. This humoral response is, however, not marked, and such antibodies are present in low titer. The most striking response is the development of delayed-type hypersensitivity (DTH), which has a protective role in preventing reinfection with the same organism. Subcutaneous inoculation of tubercle bacilli in a normal animal produces no immediate response, but in 10 to 14 d a nodule develops at the site of inoculation. The nodule then becomes a typical tuberculous ulcer. The regional lymph nodes become swollen and caseous. In contrast, a similar inoculation in a tuberculous animal induces an indurated area at the site of injection within 1 to 2 d. This becomes a shallow ulcer which heals promptly. No swelling of the adjacent lymphatics is noted. The tubercle bacillus antigen responsible for DTH is wax D, a lipopolysaccharide–protein complex of the bacterial cell wall. The active peptide comprises diaminopimelic acid, glutamic acid, and alanine. Testing for DTH to the tubercle bacillus is done with tuberculin, a heat-inactivated culture extract containing a mixture of bacterial proteins, or with PPD, a purified protein derivative of culture in nonproteinaceous media. Both these compounds are capable of sensitizing the recipient themselves. The protective role of DTH is supported by the observation that in positive reactors living cells are usually free of tubercle bacilli and the bacteria are present in necrotic areas, separated by an avascular barrier. By contrast, in infected individuals giving a negative reaction, the tubercle bacilli are found in great numbers in living tissues. The reaction is permanently or transiently negative in individuals whose cell-mediated immune responses are transiently or permanently impaired.

Tuberculin test is the 24- to 48-h response to intradermal injection of tuberculin. If positive, it signifies delayed-type hypersensitivity (type IV) to tuberculin and implies cell-mediated immunity to *Mycobacterium tuberculosis*. The intradermal inoculation of tuberculin or of PPD leads to an area of erythema and induration within 24 to 48 h in positive individuals. A positive reaction signifies the presence of cell-mediated immunity to *M. tuberculosis* as a consequence of past or current exposure to this microorganism. However, it is not a test for the diagnosis of active tuberculosis.

A **tuberculin-type reaction** is a cell-mediated delayed-type hypersensitivity skin response to an extract such as candidin, brucellin, or histoplasmin. Individuals who have positive reactions have developed delayed-type hypersensitivity or cell-mediated immunity mediated by T lymphocytes following contact with the microorganism in question.

A **tuberculosis immunization** is the induction of protective immunity through injection of an attenuated vaccine containing Bacille-Calmette-Guerin (BCG). This vaccine was more widely used in Europe than in the U.S. in an attempt to provide protection against development of tuberculosis. A local papule develops several weeks after injection in individuals who were previously tuberculin negative, as it is not administered to positive individuals. It is claimed to protect against development of tuberculosis, although not all authorities agree on its efficacy for this purpose. In recent years, oncologists have used BCG vaccine to reactivate the cellular immune system of patients bearing neoplasms in the hope of facilitating antitumor immunity.

Anergy is a diminished or absent delayed-type hypersensitivity, i.e., type IV hypersensitivity, as revealed by lack of responsiveness to commonly used skin test antigens, including PPD, histoplasmin, candidin, etc. Decreased skin test reactivity may be associated with uncontrolled infection, tumor, Hodgkin's disease, sarcoidosis, etc. (Figure 12.31). There is decreased capacity of T lymphocytes to secrete lymphokines when their T cell receptors interact with specific antigen.

Contact hypersensitivity is a type IV delayed-type hypersensitivity reaction in the skin characterized by a delayed-type hypersensitivity (cell-mediated) immune reaction produced by cytotoxic T lymphocytes invading the epidermis (Figure 12.32). It is often induced by applying a skin-sensitizing simple chemical such as dinitrochlorobenzene (DNCB) that acts as a hapten uniting with proteins in the skin, leading to the delayed-type hypersensitivity response mediated by $CD4^+$ T cells. Although substances such as DNCB alone are not antigenic, they may combine with epidermal proteins which serve as carriers for these simple chemicals acting as haptens. Contact

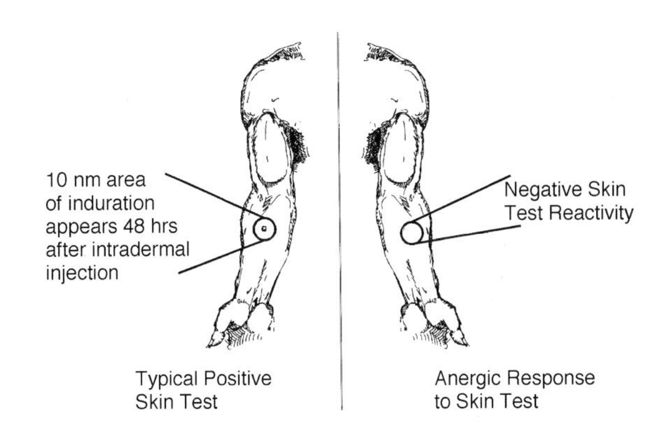

FIGURE 12.31 Representation of anergy.

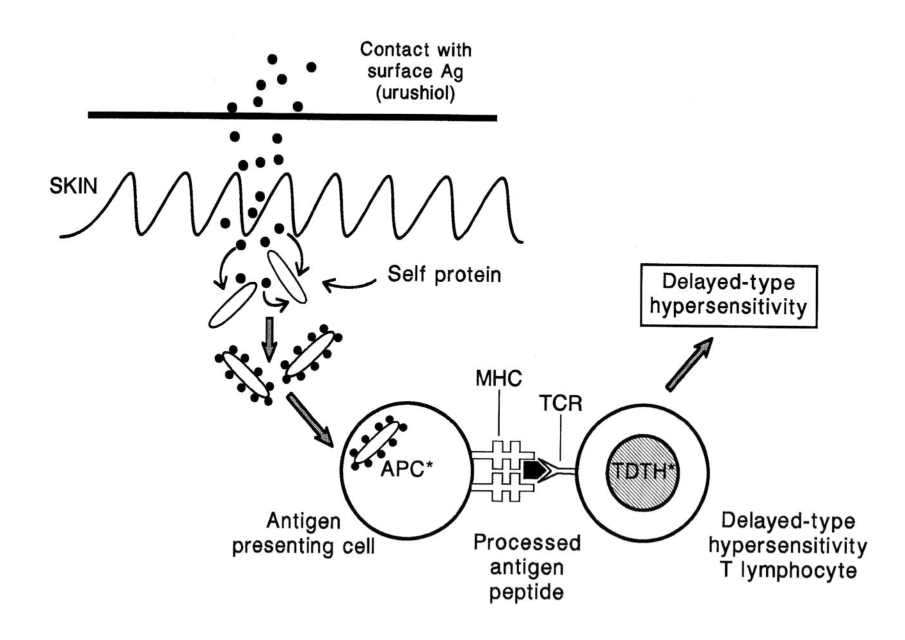

FIGURE 12.32 Contact sensitivity to poison ivy plants containing the chemical urushiol that induces delayed-type hypersensitivity mediated by $CD4^+$ T cells with skin lesions.

hypersensitivity may follow sensitization by topical drugs, cosmetics, or other types of contact chemicals. The causative agents, usually simple, low-molecular-weight compounds (mostly aromatic molecules), may also behave as haptens. The development of sensitization depends on the penetrability of the agent and its ability to form covalent bonds with protein. Part of the sensitizing antigen molecule is thus represented by protein, usually the fibrous protein of the skin. Local skin conditions that alter local proteins, such as inflammation, stasis, and others, facilitate the development of contact hypersensitivity, but some chemicals such as penicillin, picric acid, or sulfonamides are unable to conjugate to proteins, though their degradation products may have this property.

Allergic contact dermatitis is delayed-type hypersensitivity mediated by specifically sensitized T lymphocytes (type IV hypersensitivity) in response to the covalent linkage of low-molecular-weight chemicals, often of less than 1000 M_r to proteins in the skin. The inflammation induced by these agents is manifested as erythema and swelling at approximately 12 h after contact and is maximal at 24 to 48 h. Blisters form that are filled with serum, neutrophils, and mononuclear cells. There is perivascular cuffing with lymphocytes, vesiculation, and necrosis of epidermal cells. Basophils, eosinophils, and fibrin deposition appear together with edema of the epidermis and dermis. Langerhans cells in the skin serve as antigen-processing cells where the allergen has penetrated. Sensitization lasts for many years and becomes generalized in the skin. Chemicals become conjugated to skin proteins and serve as haptens. Therefore, the hapten alone can elicit the hypersensitivity once sensitization is established. After blistering, there is crust formation and weeping of the lesion. It is intensely pruritic and painful. Metal dermatitis, such as that caused by nickel, occurs as a patch, which corresponds to the area of contact with the metal or jewelry. Dyes in clothing may produce skin lesions at points of contact with the skin. The patch test is used to detect sensitivity to contact allergens. Rhus dermatitis represents a reaction to urushiols in poison oak or ivy which elicit vesicles and bullae on affected areas. Treatment is with systemic corticosteroids or the application of topical steroid cream to localized areas. Dinitrochlorobenzene (DNCB) and dinitrofluorobenzene (DNFB) are chemicals that have been used to induce allergic contact dermatitis in both experimental animals and in man.

Oxazolone (4-ethoxymethylene-2-phenyloxazol-5-one) is a substance used in experimental immunology to induce contact hypersensitivity in laboratory animals.

Picryl chloride (1-chloro-2,4,6-trinitrobenzene) is a substance used to add picryl groups to proteins. When applied to the skin of an experimental animal such as a guinea pig, a solution of picryl chloride may conjugate with skin proteins, where it acts as a hapten and may induce contact (type IV) hypersensitivity.

Contact sensitivity (CS) (or **allergic contact dermatitis**) is a form of DTH reaction limited to the skin and consisting of eczematous changes. It follows sensitization by topical drugs, cosmetics, or other types of contact chemicals. The causative agents, usually simple, low-molecular-weight compounds (mostly aromatic molecules), behave as haptens. The development of sensitization depends on the penetrability of the agent and its ability to form covalent bonds with protein. Part of the sensitizing antigen molecule is thus represented by protein, usually the fibrous protein of the skin. Local skin conditions that alter local proteins, such as inflammation, stasis, and others, facilitate the development of CS, but some chemicals such as penicillin, picric acid, or sulfonamides are unable to conjugate to proteins. It is believed that in this case the degradation products of such chemicals have this property. CS may also be induced by hapten conjugates given by other routes in adjuvant. The actual immunogen in CS remains unidentified. CS may also have a toxic, nonimmunologic component, and frequently both toxic and sensitizing effects can be produced by the same compound. With exposure to industrial compounds, an initial period of increased sensitivity is followed by a gradual decrease in reactivity. This phenomenon is called hardening and could represent a process of spontaneous desensitization. The histologic changes in CS are characteristic. Vascular endothelial cells in skin lesions produce cytokine-regulated surface molecules such as IL-2. IL-4 mRNA is strongly expressed in allergic contact dermatitis lesions. IFN-γ mRNA is the predominant cytokine in tuberculin reactions. IL-10 mRNA overexpression in atopic dermatitis might facilitate upregulation of humoral responses and downregulation of TH1 responses. Selected allergic subjects manifest several types of autoantibodies including IgE and β-adrenergic receptor autoantibodies.

Contact dermatitis is a type IV, T lymphocyte-mediated hypersensitivity reaction of the delayed type that develops in response to an allergen applied to the skin.

Vesiculation refers to the development of minute intraepidermal fluid-filled spaces, i.e., vesicles seen in contact dermatitis.

Dermatitis venenata: See contact dermatitis.

An **id reaction** is a dermatophytid reaction. It is a sudden rash linked to, but anatomically separated from, an inflammatory reaction of the skin in a sensitized individual with the same types of lesions elsewhere. The hands and arms are usual sites of id reactions that are expressed as sterile papulovesicular pustules. They may be linked with dermatophytosis such as tinea capitis or tinea pedis. They may also be associated with stasis dermatitis, contact dermatitis, and eczema.

Dermatophytid reaction: See id reaction.

Poison ivy is a plant containing the chemical urushiol, which may induce severe contact hypersensitivity of the skin in individuals who have come into contact with it (Figure 12.33). Urushiol is also found in mango trees, Japanese lacquer trees, and cashew plants. Urushiol is present not only in *Toxicodendron radicans* (poison ivy) found in the Eastern U.S., but also in *T. diversilobium* (poison oak) found in the Western U.S. and *T. vernix*

FIGURE 12.33 Poison ivy.

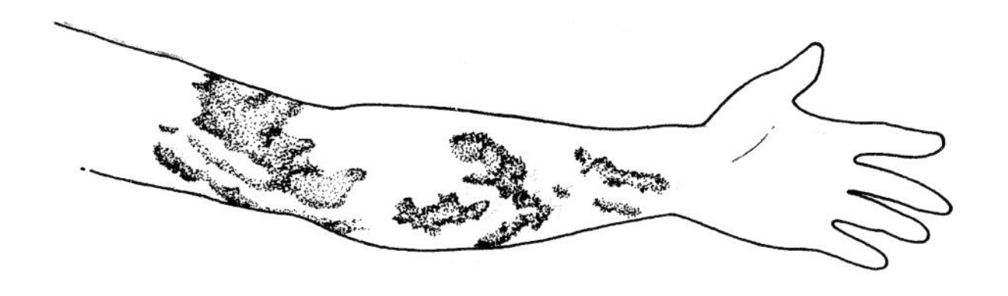

FIGURE 12.34 Contact hypersensitivity induced by exposure to poison ivy.

(poison sumac) found in the Southern U.S. Setting fire to these plants is also hazardous in that the smoke containing the chemical may induce tracheitis and pulmonary edema in allergic individuals. The chemical may remain impregnated in unwashed clothing for long periods of time and cause reactions in people who come into contact with it. Pentadecacatechol, a potent contact-sensitizing agent in leaves of the poison ivy plant, is a frequent cause of the contact hypersensitivity.

Poison ivy hypersensitivity is principally type IV contact hypersensitivity induced by urushiols, which are chemical constituents of poison ivy (*Rhus toxicodendron*), when poison ivy plants containing this chemical come into contact with the skin (Figure 12.34). The urushiol acts as a hapten by complexing with skin proteins to induce cellular (type IV) hypersensitivity on contact. Also called delayed-type hypersensitivity.

Urushiols are catechols in poisoning (*Rhus toxicodendron)* plants that act as allergens to produce contact hypersensitivity, i.e., contact dermatitis at skin sites touched by the urushiol-bearing plant. The cutaneous lesion is T-cell mediated and is classified as type IV hypersensitivity. There are four *Rhus* catechols that differ according to pentadecyl side-chain saturation. They induce type IV delayed hypersensitivity. These substances present in such plants as poison oak, poison sumac, and poison ivy (Figure 12.35).

Pentadecacatechol is the chemical constituent of the leaves of poison ivy plants that induces cell-mediated immunity associated with hypersensitivity to poison ivy.

FIGURE 12.35 *Rhus toxicodendron* is now commonly called *Toxicodendron radicans*.

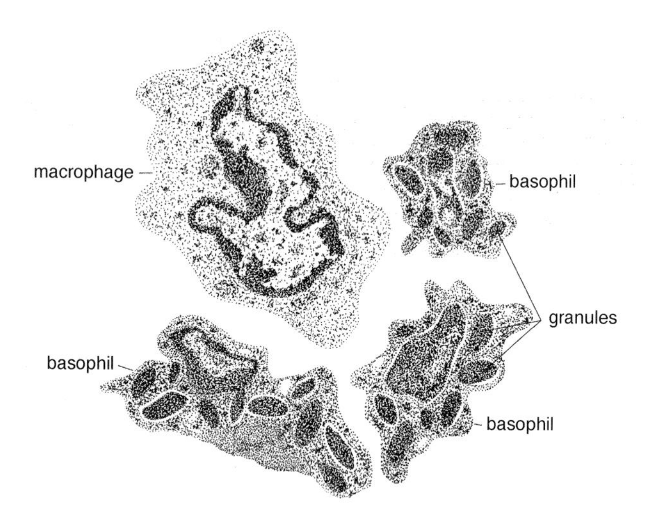

FIGURE 12.36 Cutaneous basophil hypersensitivity.

Cutaneous basophil hypersensitivity (Jones-Mote hypersensitivity) is a type of delayed (type IV) hypersensitivity in which there is prominent basophil infiltration of the skin immediately beneath the epidermis (Figure 12.36). It can be induced by the intradermal injection of a soluble antigen such as ovalbumin incorporated into Freund's incomplete adjuvant. Swelling of the skin reaches a maximum within 24 h. The hypersensitivity reaction is maximal between 7 and 10 d following induction and vanishes when antibody is formed. Histologically, basophils predominate, but lymphocytes and mononuclear cells are also present. Jones-Mote hypersensitivity is greatly influenced by lymphocytes that are sensitive to cyclophosphamide (suppressor lymphocytes).

Jones-Mote reaction is a delayed-type (type IV) hypersensitivity to protein antigens associated with prominent basophil infiltration of the skin immediately beneath the dermis, which gives the reaction the additional name "cutaneous basophil hypersensitivity." Compared with the other forms of delayed-type hypersensitivity, it is relatively weak and appears on challenge several days following sensitization with minute quantities of protein antigens in aqueous medium or in incomplete Freund's adjuvant. No necrosis is produced. Jones-Mote hypersensitivity can be produced in laboratory animals such as guinea pigs appropriately exposed to protein antigens in aqueous media or in incomplete Freund's adjuvant. It can be induced by the intradermal injection of a soluble antigen such as ovalbumin incorporated into Freund's incomplete adjuvant. Swelling of the skin reaches a maximal between 7 and 10 d following induction and vanishes when antibody is formed. Histologically, basophils predominate, but lymphocytes and mononuclear cells are also present. Jones-Mote hypersensitivity is greatly influenced by lymphocytes that are sensitive to cyclophosphamide (suppressor lymphocytes). It can be passively transferred by T lymphocytes.

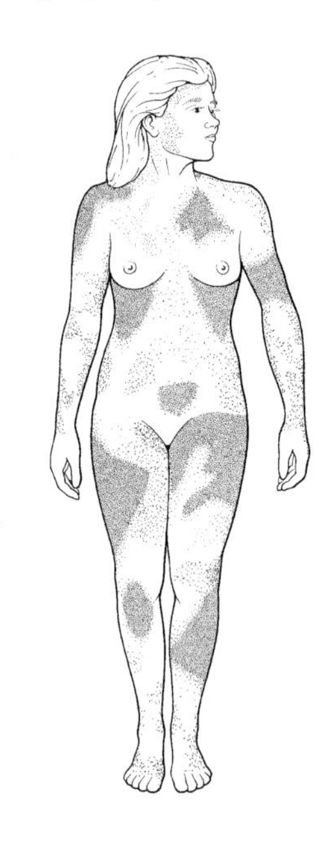

FIGURE 12.37 Penicillin hypersensitivity.

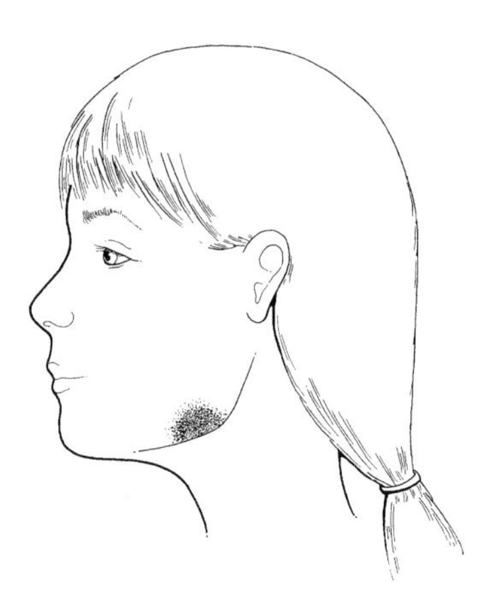

FIGURE 12.38 Fixed drug eruption.

Penicillin hypersensitivity is an allergic reaction to penicillin or its degradation products such as penicillinic acid and may be either antibody mediated or cell mediated (Figure 12.37). Penicillin derivatives may act as haptens by conjugating to tissue proteins to yield penicilloyl derivatives. These conjugates may induce antibody-mediated hypersensitivity manifested as an anaphylactic reaction when the patient is subsequently exposed to penicillin, or it may be manifested as a serum sickness-type reaction with fever, urticaria, and joint pains. Penicillin hypersensitivity may also be manifested as hemolytic anemia in which the penicillin derivatives have become conjugated to the patient's red blood cells or as allergic contact dermatitis, especially in pharmacists or nurses who come into contact with penicillin on a regular basis. Whereas the patch test using material impregnated with penicillin may be applied to the skin to detect cell-mediated (delayed-type, type IV) hypersensitivity, individuals who have developed anaphylactic hypersensitivity with IgE antibodies specific for penicilloyl–protein conjugates may be identified by injecting penicilloyl–polylysine into their skin. The development of a wheal and flare response signifies the presence of IgE antibodies, which mediate anaphylactic reactivity in man.

Fixed drug eruption is a hypersensitivity reaction to a drug that appears at the same local site on the body surface regardless of the route by which the drug is administered. The lesion is a clearly circumscribed plaque that is reddish-brown or purple and edematous (Figure 12.38). It may be covered by a bulla. Common sites of occurrence include the extremities, hands, and glans penis. Drugs that may induce this reaction include suflonamides, barbiturates, quinine, and tetracycline. There is hydropic degeneration of the basal layer.

13 Immunoregulation and Immunologic Tolerance

Immunoregulation refers to control of the immune response usually by its own products such as: the idiotypic network of antibody regulation described by Niels Jerne; feedback inhibition of antibody formation by antibody molecules; T cell receptor interaction with antibodies specific for them; and the effect of immunosuppressive and immunoenhancing cytokines on the immune response; in addition to other mechanisms. Refers to control of both humoral and cellular limbs of the immune response by mechanisms such as antibody feedback inhibition, the immunoglobulin idiotype and antiidiotype network, helper and suppressor T cells, and cytokines. Results of these immunoregulatory interactions may lead to either suppression or potentiation of one or the other limb of the immune response.

Unresponsiveness describes the failure to respond to an immunogenic (antigenic) stimulus. Unresponsiveness may be antigen-specific as in immunological tolerance or nonspecific as a consequence of suppression of the immune system in general by whole-body irradiation or by immunosuppressive drugs such as cyclosporine.

Immunological unresponsiveness is characterized by failure to form antibodies or develop a lymphoid cell-mediated response following exposure to immunogen (antigen). Immunosuppression that is specific for only one antigen with no interference with the response to all other antigens is termed immunological tolerance. By contrast, the administration of powerful immunosuppressive agents such as azathioprine, cyclosporine, or total body irradiation causes generalized immunological unresponsiveness to essentially all immunogens to which the host is exposed.

Immunological inertia refers to specific immunosuppression related to paternal histocompatibility antigens during pregnancy, such as suppression of maternal immune reactivity against fetal histocompatibility antigens.

Clonal restriction describes an immune response that is limited to the expression of a few lymphoid cell clones.

Immunologic competence is the capability to mount an immune response.

A **contrasuppressor cell** is a T cell that opposes the action of a suppressor T lymphocyte.

Contrasuppression is a part of the immunoregulatory circuit that prevents suppressor effects in a feedback loop. This is a postulated mechanism to counteract the function of suppressor cells in a feedback-type mechanism. Proof of contrasuppressor and suppressor cell circuits awaits confirmation by molecular biologic techniques.

Tolerance is an active state of unresponsiveness by lymphoid cells to a particular antigen (tolerogen) as a result of the cells' interaction with that antigen. The immune response to all other immunogens is unaffected. Thus, this is an acquired nonresponsiveness to a specific antigen. When inoculated into a fetus or a newborn, an antigenic substance will be tolerated by the recipient in a manner that will prevent manifestations of immunity when the same individual is challenged with this antigen as an adult. This treatment has no suppressive effect on the response to other unrelated antigens. Immunologic tolerance is much more difficult to induce in an adult whose immune system is fully developed. However, it can be accomplished by administering repetitive minute doses of protein antigens or by administering them in large quantities. Mechanisms of tolerance induction have been the subject of numerous investigations, and clonal deletion is one of these mechanisms. Either helper T or B lymphocytes may be inactivated or suppressor T lymphocytes may be activated in the process of tolerance induction. In addition to clonal deletion, clonal anergy and clonal balance are among the complex mechanisms proposed to account for self-tolerance in which the animal body accepts its own tissue antigens as self and does not reject them. Nevertheless, certain autoantibodies form under physiologic conditions and are not pathogenic. However, autoimmune phenomena may form under disease conditions and play a significant role in the pathogenesis of autoimmune disease.

An immunological adaptation to a specific antigen is distinct from unresponsiveness, which is the genetic or pathologic inability to mount a measurable immune response. Tolerance involves lymphocytes as individual cells, whereas unresponsiveness is an attribute of the whole organism. The humoral or cell-mediated response may be affected individually or at the same time. The genetic form of unresponsiveness has been demonstrated with the immune response to synthetic antigens and has led to characterization of the immune response (Ir) locus of the major histocompatibility complex. The immune response of experimental animals, which are classified as high, intermediate, or as nonresponders, is not defective, but is not reactive to the particular antigen. In some cases, suppressor cells prevent the development of an appropriate response. Unresponsiveness may also be the result of immunodeficiency states, some with clinical expression, or may be induced by immunosuppressive therapy such as that following X-irradiation, chemotherapeutic agents, or antilymphocyte sera. Tolerance, as the term is currently used, has a broader connotation and is intended to represent all instances in which an immune response to a given antigen is not demonstrable. Immunologic tolerance refers to a lack of response as a result of prior exposure to antigen.

A **freemartin** is the female member of dizygotic cattle twins where the other twin is a male. Their placentas are fused *in utero*, causing them to be exposed to each other's cells *in utero* prior to the development of immunologic maturity. This renders the animals immunologically tolerant of each other's cells and prevents them from rejecting grafts from the other twin. The female twin has reproductive abnormalities and is sterile.

Tolerogenic refers to the capacity of a substance such as an antigen to induce immunologic tolerance.

The **Brester-Cohn theory** maintains that the self/not-self discrimination occurs at any stage of lymphoreticular development. The concept is based on three principles: (1) engagement of the lymphocyte receptor by antigen provides signal (1), and signal (1) alone is a tolerogenic signal for the lymphocyte; (2) provision of signal (1) in conjunction with signal (2), a costimulatory signal, results in lymphocyte induction; (3) delivery of signal (2) requires associative recognition of two distinct epitopes on the antigen molecule. The requirement for associative recognition blocks the development of autoimmunity in an immune system where diversity is generated randomly throughout an individual's lifetime.

Cross-sensitivity follows induction of hypersensitivity to a substance by exposure to another substance containing cross-reacting antigens.

Immunologic tolerance is an active but carefully regulated response of lymphocytes to self antigens. Autoantibodies are formed against a variety of self antigens. Maintenance of self-tolerance is a quantitative process. When comparing the case with which T and B cell tolerance may be induced, it was found that T cell tolerance is induced more rapidly and is longer lasting than B cell tolerance. For example, T cell tolerance may be induced in a single day, whereas B cells may require 10 d for induction. In addition, 100 times more tolerogen may be required for B cell tolerance than for T cell tolerance. The duration of tolerance is much greater in T cells, which is 150 d, compared with that in B cells, which is only 50 to 60 d. T suppressor cells are also very important in maintaining natural tolerance to self antigens. For example, they may suppress T helper cell activity. Maintenance of tolerance is considered to require the continued presence of specific antigens. Low antigen doses may be effective in inducing tolerance in immature B cells leading to clonal abortion, whereas T cell tolerance does not depend upon the level of maturation. Another mechanism of B cell tolerance is cloning exhaustion, in which the immunogen activates all of the B lymphocytes specific for it. This leads to maturation of cells and transient antibody synthesis, thereby exhausting and diluting the B cell response. Antibody-forming cell blockade is another mechanism of B cell tolerance. Antibody-expressing B cells are coated with excess antigen, rendering them unresponsive to the antigen.

Adoptive tolerance is the passive transfer of immunologic tolerance with lymphoid cells from an animal tolerant to that antigen to a previously nontolerant and irradiated recipient host animal.

Cross-tolerance is the induction of immunologic tolerance to an antigen by exposure of the host to a separate antigen containing cross-reacting epitopes under conditions that favor tolerance induction.

Immune tolerance is synonymous with immunologic tolerance.

Termination of tolerance: In several forms of tolerance, the unresponsive state can be terminated by appropriate experimental manipulation. There are several methods for the termination of tolerance: (1) Injection of normal T cells. Tolerance to heterologous γ globulin can be terminated by normal thymus cells. It is, however, possible only in adoptive transfer experiments with cells of tolerant animals at 81 d after the induction of tolerance and after supplementation with normal thymus cells. By this time, B cell tolerance vanishes, and only the T cells remain tolerant. Similar experiments at an earlier date do not terminate tolerance. (2) Allogeneic cells, injected at the time when B cell tolerance has vanished or has not yet been induced, can also terminate or prevent tolerance. The mechanism is not specific and involves the allogeneic effect factor with activation of the unresponsive T cell population. (3) Injection of lipopolysaccharide (LPS). This polyclonal B cell activator is also capable of terminating tolerance if the B cells are competent. It has the ability to bypass the requirements for T cells in the response to the immunogen by providing the second (mitogenic) signal required for a response. The termination of tolerance by LPS does not involve T cells at all. LPS may also circumvent tolerance to self by a similar mechanism. (4) Crossreacting immunogens. Crossreacting immunogens (some heterologous protein in aggregated form or a different heterologous protein) also are capable of terminating tolerance to the soluble form of the protein. Termination also occurs by a mechanism that bypasses the unresponsive T cells and is obtainable at time intervals after tolerization when the responsiveness of B cells is restored. The antibody produced to the crossreacting antigen also reacts with the tolerogenic protein and is indistinguishable from the specificity produced by this protein in the absence of tolerance.

A **tetraparental chimera** is produced by the deliberate fusion of two-, four-, or eight-cell stage murine blastocyst ultimately yielding a mouse that is a chimera with contributions from four parents. These animals are of great value in studies on immunological tolerance.

A **tetraparental mouse** is an allophenic mouse.

An **immunomodulator** is an agent that alters the level of an immune response.

Tolerogen is an antigen that is able to induce immunologic tolerance. The production of tolerance rather than immunity in response to antigen depends on such variables as physical state of the antigen, i.e., soluble or particulate, route of administration, level of maturation of the recipient's immune system, or immunologic competence. For example, soluble antigens administered intravenously will favor tolerance in many situations as opposed to particulate antigens injected into the skin that might favor immunity. Immunologic tolerance with cells is easier to induce in the fetus or neonate than it is in adult animals that would be more likely to develop immunity rather than tolerance.

Self-peptides are formed from body proteins. When there is no infection, these peptides occupy the peptide binding sites of MHC molecules on cell surfaces.

Self-tolerance is a term used to describe the body's acceptance of its own epitopes as self antigens (Figure 13.1). The body is tolerant to these autoantigens, which are exposed to the lymphoid cells of the host immune system. Tolerance to self antigens is developed during fetal life. Thus, the host is immunologically tolerant to self or autoantigens. Self-tolerance is due mainly to inactivation or killing of self-reactive lymphocytes induced by exposure to self antigens. Failure of self-tolerance in the normal immune system may lead to autoimmune diseases. See also tolerance, and immunologic tolerance.

Acquired tolerance is induced by the inoculation of a neonate or fetus *in utero* with allogeneic cells prior to maturation of the recipient's immune response (Figure 13.2). The inoculated antigens are accepted as self. Immunologic tolerance may be induced to some soluble antigens by low-dose injections of neonates with the antigen or to older animals by larger doses, the so-called low-dose and high-dose tolerance, respectively.

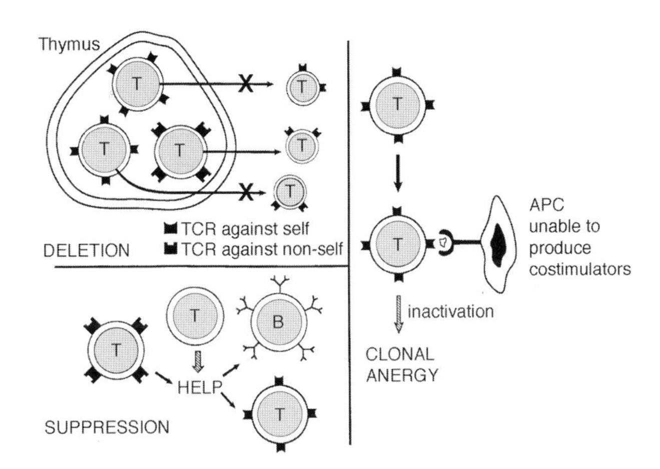

FIGURE 13.1 A schematic representation of the mechanism of self-tolerance.

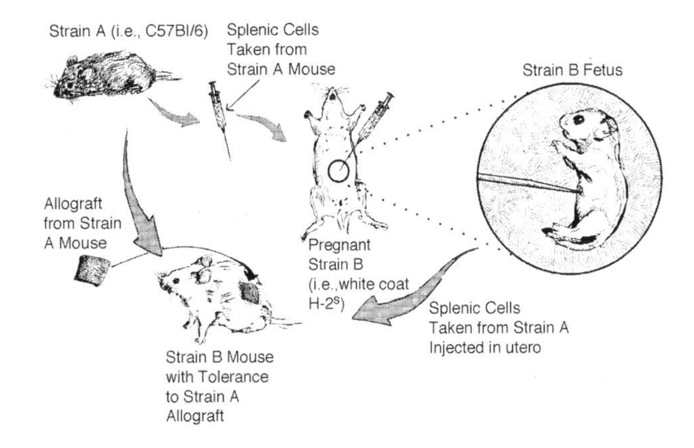

FIGURE 13.2 Acquired immunologic tolerance.

High-dose tolerance is specific immunologic unresponsiveness induced in immunocompetent adult animals by the repeated administration of large doses of antigen (tolerogen), if the substance is a protein. A massive single dose is administered if the substance is a polysaccharide. Although no precise inducing dose of antigen can be defined, in high-dose tolerance the antigen level usually exceeds 10^{-4} mol Ag per kilogram of body weight. This is also called high-zone tolerance.

High-zone tolerance is antigen-induced specific immunosuppression with relatively large doses of protein antigens (tolerogens). B-cell tolerance usually requires high antigen doses. High-zone tolerance is relatively short-lived. Called also high-dose tolerance. See high-dose tolerance.

Low-dose (or **low-zone**) **tolerance** is an antigen-specific immunosuppression induced by the administration of antigen in a suboptimal dose. Low-dose tolerance is achieved easily in the neonatal period, in which the lymphoid cells of the animal are not sufficiently mature to mount an antibody or cell-mediated immune response. This renders helper T lymphocytes tolerant, thereby inhibiting them from signaling B lymphocytes to respond to immunogenic challenge. Although no precise inducing dose of antigen can be defined, in low-dose tolerance 10^{-8} mol Ag per kilogram of body weight is usually effective. Low-dose tolerance is relatively long-lasting. This is also called low-zone tolerance.

Central tolerance is the mechanism involved in the functional inactivation of cells required for the initiation of an immune response. Central tolerance affects the afferent limb of the immune response, which is concerned with sensitization and cell proliferation. It is established in lymphocytes developing in central lymphoid organs and prevents the emergence of lymphocytes with high-affinity receptors for self antigens present in bone marrow or thymus.

Peripheral tolerance is involved in the inhibition of expression of the immune response. The cells delivering the actual response are functionally impaired but not defective. Peripheral tolerance affects the efferent limb of the immune response which is concerned with the generation of effector cells.

Mechanisms that interfere with maturation or stimulation of lymphocytes with the potential for reacting with self include self-tolerance, which is acquired and not inherited. Lymphocytes reactive with self may either be inhibited from responding to self or inactivated upon combination with self antigens. Self-tolerance may involve both central and peripheral tolerance. In central tolerance, immature lymphocytes capable of reacting with self encounter self-antigen producing tolerance instead of activation. By contrast, peripheral tolerance involves the interaction of mature self-reactive lymphocytes with self antigens in peripheral tissues if the lymphocytes are under conditions that promote tolerance instead of activation. Clonal deletion and clonal anergy are also principal mechanisms of tolerance in clones of lymphocytes reactive with self antigens.

Control tolerance refers to the mechanism that involves the absence or functional inactivation of cells requisite for the initiation of an immune response. These cells are defective or inactivated. Control tolerance affects the afferent limb of the immune response, which is concerned with sensitization and cell proliferation.

Clonal ignorance also may play a role that is yet ill-defined. Tolerance of T or B lymphocytes reactive with self antigen may also contribute to tolerance to self proteins. Properties of self antigens that render them tolerogenic and govern their ability to induce central or peripheral tolerance include self-antigen concentration and persistence in the formative lymphoid organs, the type and strength of signals that self antigens activate in lymphocytes, and the ability to recognize antigens in the absence of costimulators.

T cell tolerance to self antigens involves the processing and presentation of self proteins complexed with MHC molecules on antigen-presenting cells of the thymus. The interaction of immature T cells in the thymus with self peptide–MHC molecules leads to either clonal deletion or clonal anergy. Through this mechanism of negative selection, the T lymphocytes exiting the thymus are tolerant to self antigens. Tolerance of T cells to tissue antigens not represented in the thymus is maintained by peripheral tolerance. It is attributable to clonal anergy in which antigen-presenting cells recognize antigen in the absence of costimulation. Thus, cytokines are not activated to stimulate a T cell response. The activation of T lymphocyte by high antigen concentration may lead to their death through Fas-mediated apoptosis. Regulatory T lymphocytes may also suppress the reactivity of T lymphocytes specific for self antigens. IL-10 or TGF-β or some other immunosuppressive cytokine produced by T lymphocytes reactive with self may facilitate tolerance. Clonal ignorance may also be important in preventing autoimmune reactivity to self.

Anergic B cells: Lymphocyte anergy, also termed clonal anergy, is the failure of B cell (or T cell) clones to react against antigen and may represent a mechanism to maintain immunologic tolerance to self. Anergic B cells express IgD at levels equal to that of normal B cells but they downregulate IgM 5- to 50-fold. This is associated with inhibition in signaling, resulting in diminished phosphorylation of critical signal transduction molecules associated with surface immunoglobulin. Receptor stimulation of anergic B cells fails to release intracellular calcium, a critical step in B cell activation. Anergic B cells are unable to respond to subsequent exposure to cognate antigen. Anergy may be a means whereby the immune system silences potentially harmful B cell clones, yet permitting B cells to live long enough to be exported to peripheral lymphoid organs where anergic B cells may encounter a foreign antigen to which they have a higher affinity than their affinity for self antigen. If this were so, anergic B cells would be activated and contribute to a protective immune response.

B cell tolerance is manifested as a decreased number of antibody-secreting cells following antigenic stimulation, compared with a normal response. Hapten-specific tolerance can be induced by inoculation of deaggregated haptenated gammaglobulins (Ig). Induction of tolerance requires membrane Ig crosslinking. Tolerance may have a duration of 2 months in B cells of the bone marrow and 6 to 8 months in T cells. Whereas prostaglandin E enhances tolerance induction, IL-1, LPS, or 8-bromoguanosine block tolerance instead of an immunogenic signal. Tolerant mice carry a normal complement of hapten-specific B cells. Tolerance is not attributable to a diminished number or isotype of antigen receptors. It has also been shown that the six normal activation events related to membrane-Ig turnover and expression do not occur in tolerant B cells. Whereas tolerant B cells possess a limited capacity to proliferate, they fail to do so in response to antigen. Antigenic challenge of tolerant B cells induces them to enlarge and increase expression, yet they are apparently deficient in a physiologic signal required for progression into a proliferative stage.

Much of the understanding of B cell tolerance to self has been developed through models that permit the investigation of B cell development and function following exposure to self antigen in the absence of T cell help. Both central and peripheral mechanisms may be involved in B-lymphocyte tolerance to self antigens. The amount and valence of antigens in the bone marrow control the fate of immature B lymphocytes specific for these self antigens. Concentrated antigens that are multivalent may cause death of B lymphocytes. Other B cells specific for self antigens may survive to maturity following interaction with specific antigen but are permanently barred from migration to lymphoid follicles in the peripheral lymphoid tissues. Therefore, they do not respond to antigen in peripheral lymphoid sites. Soluble self antigens in lower concentrations may interact with B cells to produce anergy, which could be attributable to diminished membrane-Ig receptor expression on B lymphocytes or a failure in transmission of activation signals following interaction of antigen with its receptor. Thus, the power of the signal induced by antigen may determine the fate. In brief, larger concentrations of multivalent antigen may lead to unresponsiveness.

If specific helper T lymphocytes are absent in peripheral lymphoid tissues, mature B cells may interact with self antigen there to become tolerant. Following interaction with antigen, some B lymphocytes may be unable to activate tyrosine kinases and others may diminish their antigen receptor expression after they have interacted with self antigen. Exposure to self antigen fails to cause self-reactive B lymphocyte proliferation or increased expression of costimulators. These B cells also fail to become activated when aided by T cells. Other B cells may become blocked from terminal differentiation into antibody-forming cells following reaction with self antigens. B cells capable of reacting with self may remain inactive in the absence of helper T cell activity.

B lymphocyte tolerance refers to immunologic nonreactivity of B lymphocytes induced by relatively large doses of antigen. It is of relatively short duration. By contrast, T cell tolerance requires less antigen and is of a longer duration. Exclusive B cell tolerance leaves T cells immunoreactive and unaffected.

Lymphocyte anergy refers to the failure of clones of T or B lymphocytes to react to antigen and may represent a mechanism to maintain immunologic tolerance to self. Also called clonal anargy. Antigen stimulation of a lymphocyte without costimulation leads to tolerance.

Clonal deletion (negative selection) is the elimination of self-reactive T lymphocyte in the thymus during the development of natural self-tolerance. T cells recognize self antigens only in the context of MHC molecules. Autoreactive thymocytes are eliminated following contact with self antigens expressed in the thymus before maturation is completed. The majority of CD4$^+$ T lymphocytes in the blood circulation that survived clonal deletion in the thymus failed to respond to any stimulus. This reveals that clonal anergy participates in suppression of autoimmunity. Clonal deletion represents a critical mechanism to rid the body of autoreactive T lymphocytes. This is brought about by minor lymphocyte stimulation (MLS) antigens that interact with the T cell receptor's Vβ region of the T lymphocyte receptor, thereby mimicking the action of bacterial super antigen. Intrathymic and peripheral tolerance in T lymphocyte can be accounted for by clonal deletion and functional inactivation of T cells reactive against self.

Clonal anergy is the interaction of immune system cells with an antigen, without a second antigen signal, of the type usually needed for a response to an immunogen. This leads to functional inactivation of the immune system cells in contrast to the development of antibody formation or cell-mediated immunity. Clonal ignorance refers to lymphocytes that survive the principal mechanisms of self-tolerance and remain functionally competent but are unresponsive to self antigens and do not cause autoimmune reactions.

It is convenient to describe **clonal balance** as an alteration in the helper/suppressor ratio with a slight predominance of helper activity. Factors that influence the balance of helper/suppressor cells include aging, steroid hormones, viruses, and chemicals. The genetic constitution of the host and the mechanism of antigen presentation are the two most significant factors that govern clonal balance. Immune response genes associated with MHC determine class II MHC antigen expression on cells presenting antigen to helper CD4$^+$ lymphocytes. Thus, the MHC class II genotype may affect susceptibility to autoimmune disease. Other genes may be active as well. Antigen presentation exerts a major influence on the generation of an autoimmune response. Whereas a soluble antigen administered intravenously with an appropriate immunologic adjuvant may induce an autoimmune response, leading to immunopathologic injury, the same antigen administered intravenously without the adjuvant may induce no detectable response. Animals rendered tolerant to foreign antigens possess suppressor T lymphocytes associated with the induced unresponsiveness. Thus, self-tolerance could be due, in part, to the induction of suppressor T cells. This concept is called clonal balance rather than clonal deletion. Self antigens are considered to normally induce mostly suppressor rather than helper T cells, leading to a negative suppressor balance in the animal body. Three factors with the potential to suppress immune reactivity against self include nonantigen-specific suppressor T cells, antigen-specific suppressor T cells, and antiidiotypic antibodies. Suppressor T lymphocytes may leave the thymus slightly before the corresponding helper T cells. Suppressor T cells specific for self antigens are postulated to be continuously stimulated and usually in greater numbers than the corresponding helper T cells.

Clonal expansion refers to antigen-specific lymphocyte proliferation in response to antigenic stimulation that precedes differentiation into effector cells. It is a critical mechanism of adaptive immunity that enables rare antigen-specific cells to proliferate sufficiently to combat the pathogenic microorganisms that provoked the response.

Psychoneuroimmunology is the study of central nervous system and immune system interactions in which neuroendocrine factors modulate immune system function. For example, psychological stress as in bereavement or other causes may lead to depressed immune function through neuroendocrine immune system interaction. There are multiple bidirectional interactions among the nervous, immune, and endocrine systems. This represents an emerging field of contemporary immunological research.

Immune–neuroendocrine axis describes the bidirectional regulatory circuit between the **immune and neuroendocrine systems** (Figure 13.3 and Figure 13.4). The neuroendocrine and immune systems affect each other. Receptors for neurally active polypeptides, neurotransmitters, and hormones are present on immune system cells, whereas receptors reactive with products of the immune system may be identified on nervous system cells. Neuroendocrine hormones have variable immunoregulatory effects mediated through specific receptors. Neurotransmitter influence on cell function is determined in part by the receptor-linked signal amplification associated with second messenger systems. Neuroimmunoregulation is mediated either through a neural pathway involving pituitary peptides and adrenal steroid hormones or through a second pathway consisting of direct innervation of immune system tissues. The thymus, spleen, bone marrow, and perhaps other lymphoid organs contain afferent and efferent nerve fibers. ACTH, endorphins, enkephalins, and adrenal cortical steroids derived from the pituitary represent

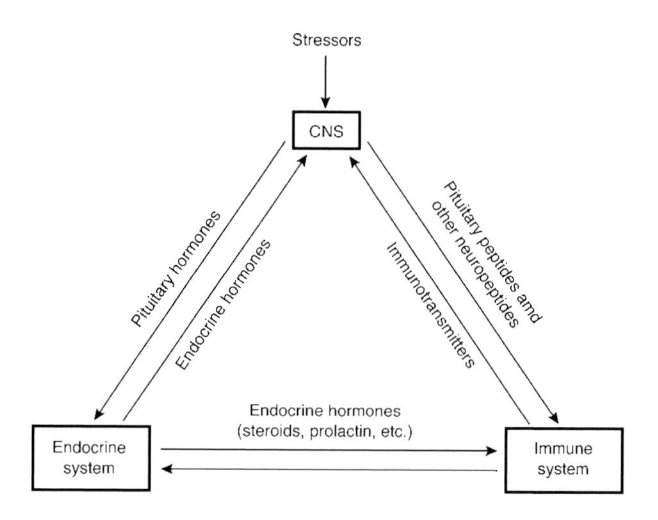

FIGURE 13.3 Interactions among the endocrine, central nervous, and immune systems.

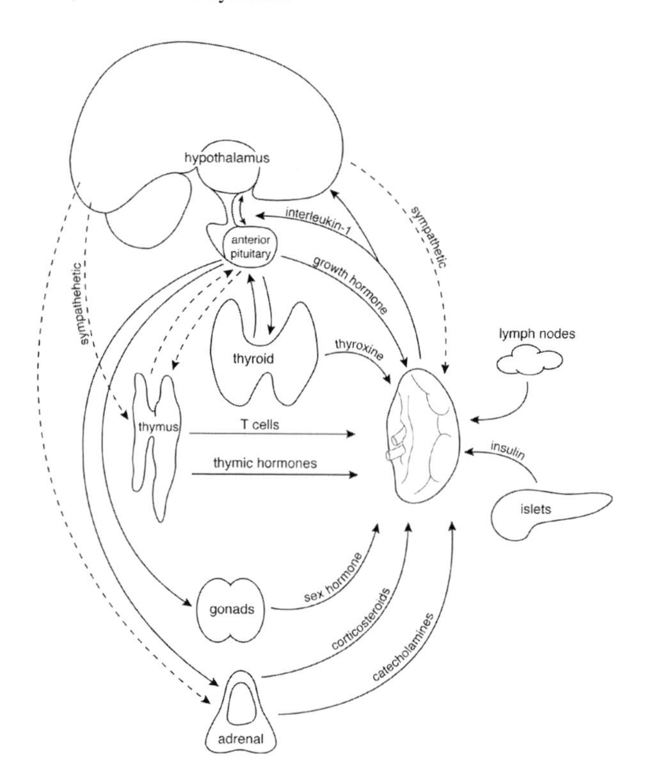

FIGURE 13.4 Immune–neuroendocrine axis.

one direction for modulating the immune response. For example, prolactin (PRL) regulates lymphocyte function. Stimulated lymphocytes or nonstimulated macrophages may produce neuroendocrine hormone-related peptides, ACTH, and endorphins. The participation of these substances in a stress response represents the opposite direction of regulation. Thymic hormones may also induce an endocrine response.

Neuroendocrine hormones may exert either a positive or negative regulatory effect on the macrophage, which plays a key role in both inflammation and immune responsiveness. Leukocyte mediators are known to alter both central nervous system and immune system functions. IL-1 acts on the hypothalamus to produce fever and participates in antigen-induced activation. IL-1 is synthesized by macrophages and has a major role in inflammation and immune responsiveness.

Substance P (SP) has been postulated to have an effect in hypersensitivity diseases including arthritis and asthma. The nervous system may release SP into joints in arthritis and into the respiratory tract in asthma, perpetuating inflammation. SP has also been found to participate in immune system functions such as the induction of monocyte chemotaxis in *in vitro* experiments. Through their effect on mast cells, enkephalins influence hypersensitivity reactions. Enkephalins have been shown to diminish antibody formation against cellular and soluble immunogens and to diminish passive cutaneous anaphylaxis. Mast cells, which can be stimulated by either immunologic or nonimmunologic mechanisms, may be significant in immune regulation of neural function.

Stress and immunity: Stressors can alter many facets of the immune response. There are numerous bidirectional pathways of communication between the immune system and the brain. Stress may function through this neuroimmune network to influence immune responses. However, there are different types of stress and different means of stress perception leading to the production of different combinations of autonomic activation and hormones. Immune system-derived information may act on the nervous system, such as during an infection, which might also trigger changes in behavior patterns that resemble stress-associated behavior. The immune system may serve as a sensory organ, transmitting to the brain information about antigens through lymphocyte-derived hormones, while being modulated by neural factors (stress related or otherwise). Immune, neural, and psychosocial realms are coordinated as coherent processes within an individual's life in a manner that permits the context, interpretation, and meaning of stress for that individual to determine the effects on immunity.

Sulzberger-Chase phenomenon refers to the induction of immunological unresponsiveness to skin-sensitizing chemicals such as picryl chloride by feeding an animal (e.g., guinea pig) the chemical in question prior to application to the skin. Intravenous administration of the chemical may also block the development of delayed-type hypersensitivity when the same chemical is later applied to the skin. Simple chemicals such as picryl chloride may induce contact hypersensitivity when applied to the skin of guinea pigs. The unresponsiveness may be abrogated by adoptive immunization of a tolerant guinea pig with lymphocytes from one that has been sensitized by application of the chemical to the skin without prior oral feeding.

Oral tolerance is discussed in Chapter 15.

Split tolerance includes several mechanisms: (1) Specific immunological unresponsiveness (tolerance) affecting either the B cell (antibody) limb or the T lymphocyte (cell-mediated) limb of the immune response. The unaffected limb is left intact to produce antibody or respond with cell-mediated immunity, depending on which limb has been rendered specifically unresponsive to the antigen in question. (2) The induction of immunologic tolerance to some epitopes of allogeneic cells, while leaving the remaining epitopes capable of inducing an immune response characterized by antibody production and/or cell-mediated immunity.

In **immune deviation**, antigen-mediated suppression of the immune response may selectively affect delayed-type hypersensitivity, leaving certain types of immunoglobulin responses relatively intact and unaltered. This selective suppression of certain phases of the immune response to an antigen without alteration of others has been termed "immune deviation." Thus, "split tolerance" or immune deviation offers an experimental model for dissection of the immune response into its component parts. It is necessary to use an antigen capable of inducing formation of humoral antibody and development of delayed-type hypersensitivity to induce immune deviation. Since it is essential that both humoral and cellular phases of the immune response be directed to the same antigenic determinant group, defined antigens are required. Immune deviation selectively suppresses delayed-type hypersensitivity and IgG_2 antibody production. By contrast, immunologic tolerance affects both IgG_1 and IgG_2 antibody production and delayed-type hypersensitivity. For example, prior administration of certain protein antigens to guinea pigs may lead to antibody production. However, the subsequent injection of antigen incorporated into Freund's complete adjuvant leads to deviation from the expected heightened delayed-type hypersensitivity and formation of IgG_2 antibodies to result in little of either, i.e., negligible delayed-type hypersensitivity and suppression of IgG_2 formation. Recent advances in understanding of cytokines and T cell subsets has rekindled interest in immune deviation, with the delineation of the Th1 and Th2 subsets of $CD4^+$ effector T lymphocytes, a cellular framework for immune deviation being available. Powerful cell-mediated (DTH) responses occur when Th1 cells secreting IL-2 and IFN-γ are preferentially activated under the influence of macrophage-derived IL-12. By contrast, synthesis of most antibody classes is favored by stimulation of IL-4 secreting Th2 cells. Cross-inhibition of Th2 cells by IFN-γ and Th1 cells by IL-4 and IL-10 reinforces deviation down one rather than the other T-cell pathway.

Liacopoulos phenomenon (nonspecific tolerance): The daily administration of 0.5 to 1.0 g of bovine γ globulin and bovine serum albumin to guinea pigs for at least 8 d suppresses their immune response to these antigens. If an unrelated antigen is injected and then continued for several days thereafter, the response to the unrelated antigen is reduced. This phenomenon has been demonstrated for circulating antibody, delayed hypersensitivity, and graft-vs.-host reaction. It describes the induction of nonspecific immunosuppression for one antigen by the administration of relatively large quantities of an unrelated antigen.

Anergy refers to diminished or absent delayed-type hypersensitivity, i.e., type IV hypersensitivity, as revealed by lack of responsiveness to commonly used skin-test antigens including PPD, histoplasmin, and candidin. Decreased skin-test reactivity may be associated with uncontrolled infection, tumor, Hodgkin's disease, or sarcoidosis. There is decreased capacity of T lymphocytes to secrete lymphokines when their T cell receptors interact with a specific antigen. Anergy describes nonresponsiveness to antigen. Individuals are anergic when they cannot develop a delayed-type hypersensitivity reaction following challenge with an antigen. T and B lymphocytes are anergic when they cannot respond to their specific antigen.

Immunological ignorance is a type of tolerance to self in which a target antigen and lymphocytes capable of reacting with it are both present simultaneously in an individual without an autoimmune reaction occurring. The abrogation of immunologic ignorance may lead to autoimmune disease.

Infectious tolerance was described in the 1970s. Animals rendered tolerant to foreign antigens were found to possess suppressor T lymphocytes associated with the induced unresponsiveness. Thus, self-tolerance was postulated to be based on the induction of suppressor T cells. Rose has referred to this concept as clonal balance rather than clonal deletion. Self antigens are considered to normally induce mostly suppressor rather than helper T cells, leading to a negative suppressor balance in the animal body. Three factors with the potential to suppress immune reactivity against self include nonantigen-specific suppressor T cells, antigen-specific suppressor T cells, and antiidiotypic antibodies. Rose suggested that suppressor T lymphocytes leave the thymus slightly before the corresponding helper T cells. Suppressor T cells specific for self antigens are postulated to be continuously stimulated and usually in greater numbers than the corresponding helper T cells.

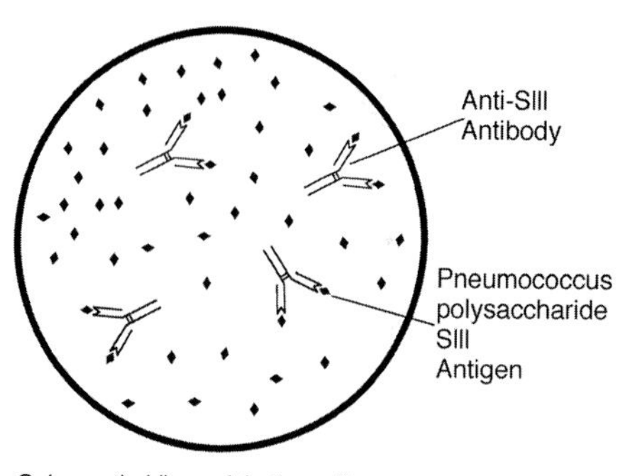

FIGURE 13.5 Immunologic paralysis (schematic view of antigen excess).

Paralysis is the masking of an immune response by the presence of excessive quantities of antigen. This mimics acquired immunologic tolerance and is considered a false tolerance state.

Immunologic (or **immune**) **paralysis** is an immunologic unresponsiveness induced by the injection of large doses of pneumococcal polysaccharide into mice where it is metabolized slowly (Figure 13.5). Any antibody that is formed is consumed and not detectable. The pneumococcal polysaccharide antigen remained in tissues of the recipient for months, during which time the animals produced no immune response to the antigen. Immunologic paralysis is much easier to induce with polysaccharide than with protein antigens. It is highly specific for the antigen used for its induction. Felton's first observation of immunologic paralysis preceded the demonstration of acquired immunologic tolerance by Medawar et al.

The **Felton phenomenon** is specific immunologic unresponsiveness or paralysis induced by the inoculation of relatively large quantities of pneumococcal polysaccharide into mice.

Immunological suicide describes the use of an antigen deliberately labeled with high-dose radioisotope to kill a subpopulation of lymphocytes with receptors specific for that antigen following antigen binding.

Immunologic enhancement refers to the prolonged survival, conversely the delayed rejection, of a tumor allograft in a host as a consequence of contact with specific antibody. Both the peripheral and central mechanisms have been postulated. Coating of tumor cells with antibody was presumed, in the past, to interfere with the ability of specifically reactive lymphocytes to destroy them, but a central effect in suppressing cell-mediated immunity, perhaps through suppressor T lymphocytes, is also possible.

14 Autoimmunity

Autoimmunity is an immune reactivity involving either antibody-mediated (humoral) or cell-mediated limbs of the immune response against the body's own (self) constituents, i.e., autoantigens. When autoantibodies or autoreactive T lymphocytes interact with self-epitopes, tissue injury may occur; for example, in rheumatic fever the autoimmune reactivity against heart muscle sacrolemmal membranes occurs as a result of cross-reactivity with antibodies against streptococcal antigens (molecular mimicry). Thus, the immune response can be a two-edged sword, producing both beneficial (protective) effects and leading to severe injury to host tissues. Reactions of this deleterious nature are referred to as hypersensitivity reactions and are subgrouped into four types.

Abrogation of tolerance to self antigens often leads to autoimmunity. This may result from altered regulation of lymphocytes reactive with self or in aberrations in self-antigen presentation. Many factors participate in the generation of autoimmunity. Autoimmune reactants may be a consequence, and not a cause, of a disease process. Autoimmune diseases may be organ-specific, such as autoimmune thyroiditis, or systemic, such as systemic lupus erythematosus. Different hypersensitivity mechanisms, classified from I to IV, may represent mechanisms by which autoimmune diseases are produced. Thus, antibodies or T cells may be the effector mechanisms mediating tissue injury in autoimmunity. Helper T cells control the immune response to protein antigens. Therefore, defects in this cell population may lead to high-affinity autoantibody production specific for self antigens. MHC molecules, which are often linked genetically to the production of autoimmune disease, present peptide antigens to T lymphocytes. Various immunologic alterations may lead to autoimmunity. Experimental evidence supports the concept that autoimmunity may result from a failure of peripheral T lymphocyte tolerance, but little is known about whether or not loss of peripheral B-cell tolerance is a contributory factor in autoimmunity. Processes that activate antigen-presenting cells in tissues, thereby upregulating their expression of costimulators and leading to the formation of cytokines, may abrogate T-lymphocyte anergy. The mouse model of human systemic lupus erythematosus involves *lpr/lpr* and *gld/gld* mice that succumb at 6 months of age from profound systemic autoimmune disease with nephritis and autoantibodies. The *lpr/lpr* is associated with a defect in the gene that encodes **Fas**, which determines the molecule that induces cell death. The *gld/gld* is attributable to a point mutation in the **Fas ligand**, which renders the molecule unable to signal. Thus, abnormalities in the Fas and Fas ligand prolong the survival of helper T cells specific for self antigens since they fail to undergo activation-induced cell death. Thus, this deletion failure mechanism involves peripheral tolerance rather than central tolerance. A decrease of regulatory T cells which synthesize lymphokines that mediate immunosuppression and maintain self-tolerance might lead to autoimmunity even though no such condition has yet been described.

Autoantigens are normal body constituents recognized by autoantibodies specific for them. T cell receptors may also identify autoantigen (self antigen) when the immune reactivity has induced a cell-mediated T lymphocyte response.

An **autoantibody** recognizes and interacts with an antigen present as a natural component of the individual synthesizing the autoantibody (Figure 14.1 through Figure 14.3). The ability of these autoantibodies to "crossreact" with corresponding antigens from other members of the same species provides a method for *in vitro* detection of such autoantibodies. **Autoallergy** is a tissue injury or disease induced by immune reactivity against self antigens.

Horror autotoxicus (historical) is the term coined by Paul Ehrlich *(ca.* 1900) to account for an individual's failure to produce autoantibodies against his own self-constituents even though they are excellent antigens or immunogens in other species. This lack of immune reactivity against self was believed to protect against autoimmune disease. This lack of self-reactivity was postulated to be a fear of poisoning or destroying one's self. Abrogation of horror autotoxicus leads to autoimmune disease. Horror autotoxicus was later (1959) referred to as self-tolerance by F.M. Burnet.

Autoallergy is tissue injury or disease induced by immune reactivity against self antigens.

Autoagglutination is the spontaneous aggregation of erythrocytes, microorganisms, or other particulate antigens in a saline suspension, thereby confusing interpretation of bacterial agglutination assays. The term refers also to the aggregation of an individual's cells by their own antibody.

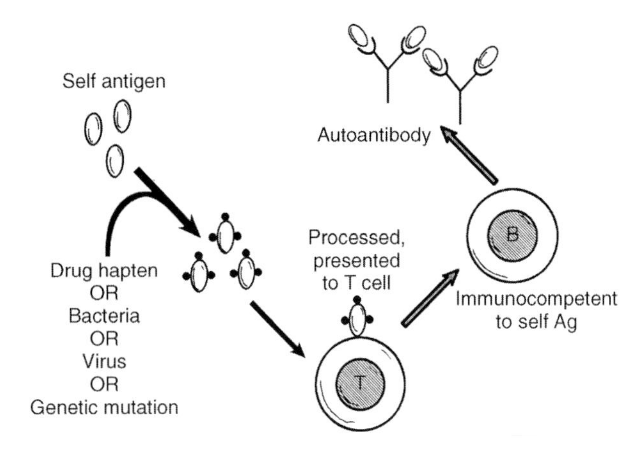

FIGURE 14.1 Autoantibody formation.

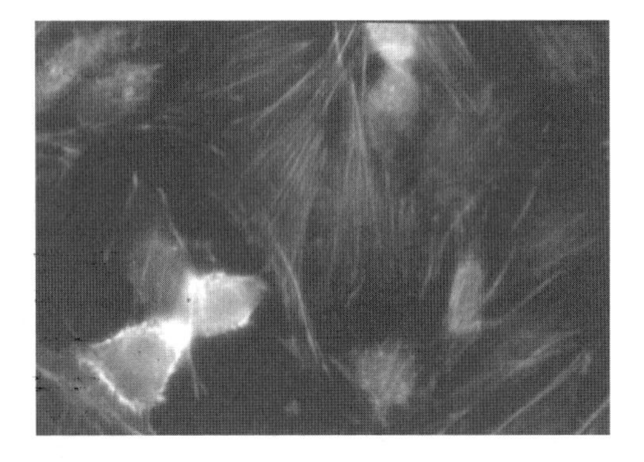

FIGURE 14.2 Antiactin autoantibody.

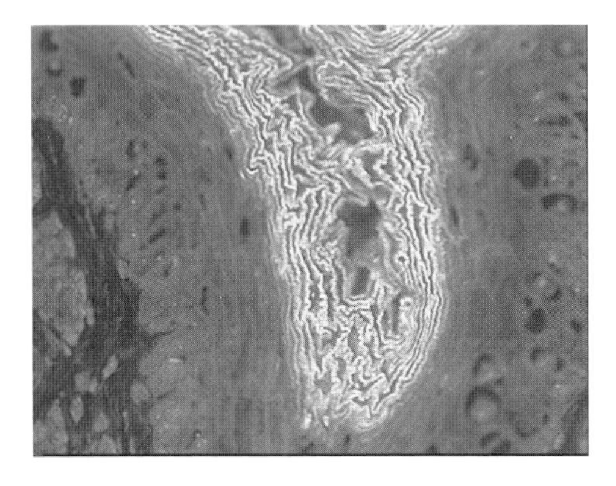

FIGURE 14.3 Antikeratin autoantibody.

Autoimmune complement fixation reaction is the ability of human blood serum from patients with certain autoimmune diseases such as systemic lupus erythematosus, chronic active hepatitis, etc., to fix complement when combined with kidney, liver, or other tissue suspensions in saline.

Autoreactivity is an immune response against self antigens.

Autosensitization is the development of reactivity against one's own antigens, i.e., autoantigens, which occurs in autoimmunity or in autoimmune disease.

Witebsky's criteria: According to criteria suggested by Ernest Witebsky, an autoimmune response should be considered as the cause of a human disease if (1) it is regularly associated with that disease, (2) immunization of an experimental animal with the antigen from the appropriate tissue causes it to make an immune response (form antibodies or develop allergy), (3) associated with this response the animal develops pathological changes that are basically similar to those of humans, and (4) the experimental disease can be transferred to a nonimmunized animal by serum or by lymphoid cells.

Altered self refers to the concept that the linkage of nonself peptide to MHC yields a peptide–MHC structure different from any found in normal cells of the individual.

An **autoimmune response** is an antibody or T-cell immune response to self antigens.

Autoreactive T lymphocytes in selected diseases may represent a failure of normal regulation, or autoreactive T cells present in a normal healthy individual may be a necessary aspect of the immune system. Autoreactive T cells develop from mature antigen-dependent precursor cells that have changed physiologically in a manner that restores their thymic selected ability to respond to self. Any mechanism that returns T cells to a resting state would halt autoreactive expansion until stimulation is induced once again by a specific foreign antigen. Chronic stimulation in the presence of continuous expression of high levels of MHC class II molecules together with abnormal immune regulation can lead to severe and persistent inflammation.

Cytokine autoantibodies are autoantibodies that may inhibit cytokine functions and lead to cytokine deficiency. Autoimmune disease may occur and the action of the cytokine may be inhibited. By contrast, these autoantibodies may serve as cytokine-specific carriers in the circulation. For example, insulin autoantibodies may prolong the release of active insulin to the tissues, leading to hypoglycemia in nondiabetics and a significant decrease in the exogenous insulin requirement in diabetic patients. AIDS patients may develop autoantibodies against IL-2, and antibodies against TNF-α have been used successfully to treat rheumatoid arthritis. Both normal and inflammatory disease patients may develop autoantibodies against IL-1α. Cytokine activity is enhanced even in the presence of cytokine autoantibodies *in vivo* by a mechanism that delays rapid catabolism of cytokines from the circulation. The clinical relevance of cytokine autoantibodies *in vivo* remains to be determined. However, these autoantibodies portend a poor prognosis in any disease. Methods for cytokine autoantibody detection include bioassays, immunometric assays, and blotting techniques.

Cytoskeletal autoantibodies are specific for cytoskeletal proteins that include microfilaments (actin), microtubules (tubulin), and intermediate filaments (acidic and basic keratins, vimentin, desmin, glial fibrillary acidic protein, peripherin, neurofilaments, α-internexin, nuclear lamins), and are present in low titers in a broad spectrum of diseases that include infection, autoimmune diseases, selected chronic liver diseases, biliary cirrhosis, Crohn's disease, myasthenia gravis, and angioimmunoblastic lymphadenopathy. These antibodies are not useful for diagnosis.

Cytoskeletal antibodies are specific for cytoskeletal proteins that include cytokeratins, desmin, actin, titin, vimentin, and tropomyosin. They have been demonstrated in some patients with various diseases that include autoimmune diseases, chronic active hepatitis and other liver disease, infection, myasthenia gravis, and Crohn's disease. These antibodies are not helpful in diagnosis.

Determinant spreading is an amplification mechanism in inflammatory autoimmune disease, in which the initial T lymphocyte response diversifies through induction of T cells against additional autoantigenic determinants. The response to the original epitope is followed by intramolecular spreading, activation of T lymphocytes for other cryptic or subdominant self-determinants of the same antigen during chronic and progressive disease; intermolecular spreading involves epitopes on other unrelated self antigens. A multideterminant protein antigen has dominant, subdominant, and cryptic T-cell epitopes. The dominant epitopes are the ones most efficiently processed and presented from native antigen. By contrast, cryptic determinants are inefficiently processed and/or presented. Only by immunization with a peptide that often requires no additional processing can a response to a cryptic determinant be mounted. Subdominant determinants fall between these two types.

Some mechanisms in **drug-induced autoimmunity** are similar to those induced by viruses (Figure 14.4). Autoantibodies may appear as a result of the helper determinant effect. With some drugs such as hydantoin, the mechanism resembles that of the Epstein–Barr virus (EBV). The generalized lymphoid hyperplasia also involves clones specific for autoantigens. A third form is that seen with α-methyldopa. This drug induces the production of specific antibodies. The drug attaches to cells *in vivo* without changing the surface antigenic makeup. The antibodies, which often have anti-e (Rh series) specificity, combine with the drug on cells, fix complement, and induce a bystander type of complement-mediated lysis. Another form of drug-induced autoimmunity is seen with nitrofurantoin, in which the autoimmunity involves cell-mediated phenomena without evidence of autoantibodies.

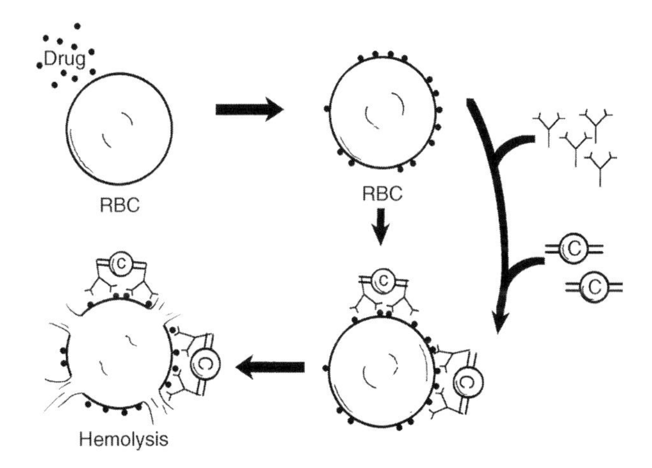

FIGURE 14.4 Drug-induced hemolysis.

Natural autoantibodies are polyreactive antibodies of low affinity that are synthesized by $CD5^+$ B cells, which comprise 10 to 25% of circulating B lymphocytes in normal individuals, 27 to 52% in those with rheumatoid arthritis, and less than 25% in systemic lupus erythematosus patients. Natural autoantibodies may appear in first-degree relatives of autoimmune disease patients as well as in older individuals. They may be predictive of disease in healthy subjects. They are often present in patients with bacterial, viral, or parasitic infections and may have a protective effect. In contrast to natural antibodies, autoantibodies may increase in disease and may lead to tissue injury. The blood group isohemagglutinins are also termed natural antibodies even though they are believed to be of heterogenetic immune origin as a consequence of stimulation by microbial antigens.

Pathologic autoantibodies are autoantibodies generated against self antigens that induce cell and tissue injury following interaction with the cells bearing epitopes for which they are specific. Many autoantibodies are physiologic, representing an epiphenomenon during autoimmune stimulation, whereas others contribute to the pathogenesis of tissue injury. Autoantibodies that lead to red blood-cell destruction in autoimmune hemolytic anemia represent pathogenic autoantibodies, whereas rheumatoid factors such as IgM anti-IgG autoantibodies have no proven pathogenic role in rheumatoid arthritis.

Sequestered antigen is anatomically isolated and not in contact with the immunocompetent T and B lymphoid cells of the immune system (Figure14.5). Examples include myelin basic protein, sperm antigens, and lens protein antigens. When a sequestered antigen such as myelin basic protein is released by one or several mechanisms including viral inflammation, it can activate both immunocompetent T and B cells. An example of the sequestered antigen release mechanism of autoimmunity is found in

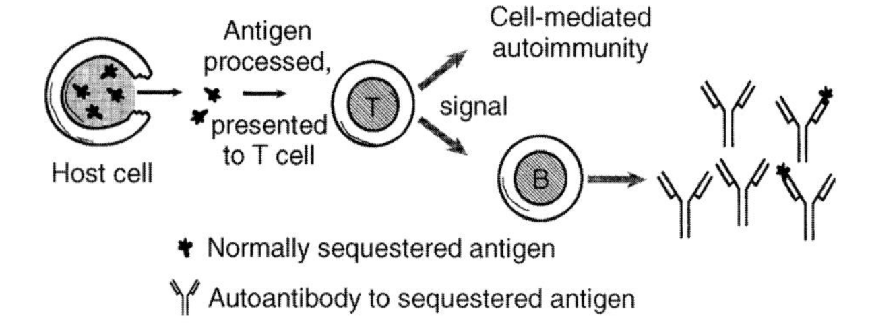

FIGURE 14.5 Release of sequestered antigen.

experimental and postinfectious encephalomyelitis. Cell-mediated injury represents the principal mechanism in experimental and postviral encephalitis. In vasectomized males, antisperm antibodies are known to develop when sperm antigens become exposed to immunocompetent lymphoid cells. Likewise, lens protein of the eye that enters the circulation as a consequence of either crushing injury to an eye or exposure of lens protein to immunocompetent cells inadvertently through surgical manipulation may lead to an antilens protein immune response. Autoimmunity induced by sequestered antigens is relatively infrequent and is a relatively rare cause of autoimmune disease.

Sex hormones and immunity: Females have been recognized as more susceptible to certain autoimmune diseases than males, which immediately led to suspicion that sex hormones might play a role. The exact mechanisms through which sex steroids interact with the immune system remain to be determined, but it is known that these hormones have a direct effect on immune system cells or indirectly through cells that control growth and development of the immune system or through organs that are ultimately destroyed by autoimmune reactions. Female mice synthesize more antibody in response to certain antigens than do males, but humans injected with vaccines show essentially identical antibody responses regardless of sex. Murine cell-mediated responses to selected antigens were stronger in females than in males. Sex steroids have a profound effect on the thymus. Androgen or estrogen administration to experimental animals led to thymic involution, whereas castration led to thymic enlargement. The thymus is markedly involuted during pregnancy, and achieves its normal size and shape following parturition. Sex steroids have numerous targets that include the bone marrow and thymus where the precursors of immunity originate and differentiate. Further studies have investigated how sex hormones control cytokine genes. Thus, sex steroid hormones affect the development and function of various immune system cells.

In an **autoimmune disease**, pathogenic consequences including tissue injury may be produced by autoantibodies or autoreactive T lymphocytes interacting with self epitopes, i.e., autoantigens. The mere presence of autoantibodies or autoreactive T lymphocytes does not prove that there is any cause-and-effect relationship between these components and a patient's disease. To show that autoimmune phenomena are involved in the etiology and pathogenesis of human disease, Witebsky suggested that certain criteria be fulfilled (See Witebsky's criteria). In addition to autoimmune reactivity against self-constituents, tissue injury in the presence of immunocompetent cells corresponding to the tissue distribution of the autoantigen, duplication of the disease features in experimental animals injected with the appropriate autoantigen, and passive transfer with either autoantibody or autoreactive T lymphocytes to normal animals offer evidence in support of an autoimmune pathogenesis of a disease. Individual autoimmune diseases are discussed under their own headings, such as systemic lupus erythematosus, autoimmune thyroiditis, etc. Autoimmune diseases may be either organ specific, such as thyroiditis diabetes, or systemic, such as systemic lupus erythematosus.

Autoimmune disease animal models: Studies of human autoimmune disease have always been confronted with the question of whether immune phenomena, including the production of autoantibodies, represent the cause or a consequence of the disease. The use of animal models has helped to answer many of these questions. By using the rat, mouse, guinea pig, rabbit, monkey, chicken, and dog among other species, many of these questions have been answered. A broad spectrum of human autoimmune diseases has been clarified through the use of animal models that differ in detail but have nevertheless provided insight into pathogenic mechanisms, converging pathways and disturbances of normal regulatory function related to the development of autoimmunity.

Autoimmune disease spontaneous animal models: Animal strains based on years of selective breeding develop certain organ-specific or systemic autoimmune diseases spontaneously without any experimental manipulation. These models resemble the human condition to a remarkable degree in many cases and serve as valuable models to investigate pathogenetic mechanisms underlying disease development. Spontaneous animal models for organ-specific autoimmune diseases include the obese strain of chickens that are an animal model for Hashimoto's thyroditis. Animal strains with spontaneous insulin-dependent diabetes mellitus (IDDM) include the NOD mouse and the DP-BB rat that both develop humoral and cellular autoimmune responses against islet β cells of the pancreas. Spontaneous animal models of systemic autoimmune diseases include the University of California at Davis (UCD) 200 strain as an animal model for progressive systemic sclerosis (SSc)-scleroderma. Several mouse strains have been developed that develop systemic lupus erythematosus-like autoimmunity. These include the New Zealand black (NZB), (NZB × New Zealand White (NZW)F1, (NZB × SWR)F1, BXSB, and MRL mice. These animal models have contributed greatly to knowledge of the etiopathogenesis, genetics, and molecular defects responsible for autoimmunity. Mechanisms include defects in lymphoid lineage, endocrine alterations, target organ defects, endogenous viruses, and/or mutations in immunologically relevant molecules such as MHC and cell receptor genes. Many of these have been implicated in animal autoimmune diseases and in some human disease cases.

In **organ-specific autoimmune diseases,** immune reactivity against specific organs, such as the thyroid in Hashimoto's thyroiditis, leads to cell and tissue damage to specific organs. By contrast, systemic lupus erythematosus affects a wide variety of tissues and organs of the body.

An **organ-specific antigen** is an antigen unique to a particular organ even though it may be found in more than one species.

A **tissue-specific antigen** is an antigen restricted to cells of one type of tissue. Tissue-specific or organ-specific autoantibodies occur in certain types of autoimmune diseases. Organ-specific or tissue-specific antibodies are often not species specific. For example, autoantibodies against human thyroglobulin may cross-react with the corresponding molecules of other species.

Lymphadenoid goiter: See Hashimoto's thyroiditis.

Autoimmune thyroiditis: Hashimoto's disease (chronic thyroiditis) is an inflammatory disease of the thyroid found most frequently in middle-aged to older women. There is extensive infiltration of the thyroid by lymphocytes that completely replace the normal glandular structure of the organ. There are numerous plasma cells, macrophages, and germinal centers which give the appearance of node structure within the thyroid gland. Both B cells and $CD4^+$ T lymphocytes comprise the principal infiltrating lymphocytes. Thyroid function is first increased as the inflammatory reaction injures thyroid follicles, causing them to release thyroid hormones. However, this is soon replaced by hypothyroidism in the later stages of Hashimoto's thyroiditis. Patients with this disease have an enlarged thyroid gland. There are circulating autoantibodies against thyroglobulin and thyroid microsomal antigen (thyroid peroxidase). Cellular sensitization to thyroid antigens may also be detected. Thyroid hormone replacement therapy is given for the hypothyroidism that develops.

Parathyroid hormone autoantibodies to parathyroid hormone have been observed in unexplained hypocalcemia and hypoparathyroidism associated with normal or increased concentrations of immunoreactive parathyroid hormone.

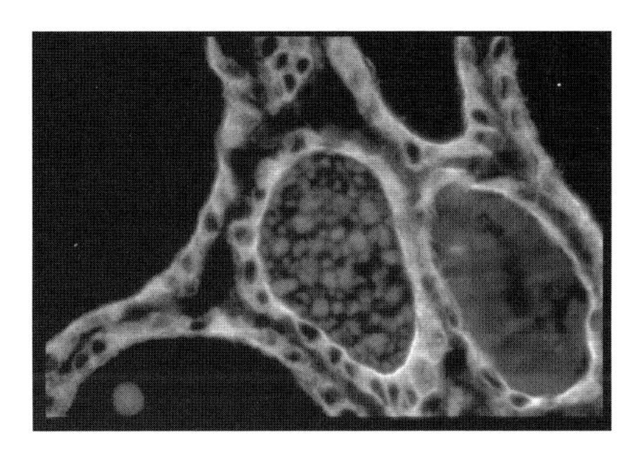

FIGURE 14.6 Thyroid autoantibody.

Thyroid antibodies include autoantibodies present in patients with Hashimoto's thyroiditis or those with thyrotoxicosis (Graves' disease) that are organ specific for the thyroid (Figure 14.6). Antibodies against thyroglobulin and antibodies against the microsomal antigen of thyroid acinar cells may appear in patients with autoimmune thyroiditis. Antibodies against TSH receptors appear in Graves' disease patients and cause stimulatory hypersensitivity. They mimic the action of TSH. This is an IgG molecule termed long-acting thyroid stimulator (LATS). LATS levels are increased in many patients with thyrotoxicosis or Graves' disease.

Thyrotoxicosis is a disease of the thyroid, in which there is hyperthyroidism with elevated levels of thyroid hormones in the blood and thyroid gland hyperplasia or hypertrophy. Thyrotoxicosis may be autoimmune, as in Graves' disease, in which there may be diffuse goiter. Autoantibodies specific for thyroid antigens mimic thyroid-stimulating hormones (TSH) by stimulating thyroid cell function. In addition, patients with Graves' disease develop ophthalmopathy and proliferative dermopathy. The disease occurs predominantly in females (70 females to 1 male) and appears usually in the 30- to 40-year-old age group. In Caucasians, it is a disease associated with DR3. Patients may develop nervousness, tachycardia, and numerous other symptoms of hyperthyroidism. They also have increased levels of total and free T3 and T4. There is a diffuse and homogeneous uptake of radioactive iodine in these patients. Three types of antithyroid antibodies occur: (1) thyroid-stimulating immunoglobulin, (2) thyroid growth-stimulating immunoglobulin, and (3) thyroid binding-inhibitory immunoglobulin. Their presence confirms a diagnosis of Graves' disease. The thyroid gland may be infiltrated with lymphocytes. Long-acting thyroid stimulator (LATS) is classically associated with thyrotoxicosis. It is an IgG antibody specific for thyroid hormone receptors. It induces thyroid hyperactivity by combining with TSH receptors.

Thyroglobulin is a thyroid protein demonstrable by immunoperoxidase staining that serves as a marker for papillary and/or follicular thyroid carcinomas.

Thyroglobulin autoantibodies are found in the blood sera of patients with such thyroid disorders as chronic lymphocytic (Hashimoto's) thyroiditis (76 to 100%), primary myxedema (72%), hyperthyroidism (33%), colloid goiter (7%), adenoma (28%), thyroid cancer based on type (13 to 65%), and pernicious anemia (27%); in patients with autoimmune diseases; Addison's disease (28%); and diabetes mellitus (20%). Thyroglobulin autoantibodies can be detected in normal subjects, which diminishes useful clinical usefulness. 18% of women have these antibodies which increase in frequency up to 30% as they age. They are much more prevalent in men (approximately 3 to 6%) and also increase with age. EIA, immunoradiometric and chemiluminescence are the assay methods of choice for these autoantibodies.

Thyrotropin is thyroid-stimulating hormone (TSH).

Thyrotropin receptor autoantibodies are autoantibodies specific for thyrotopin receptor (TSHR) and are very heterogeneous. Some TSHR antibodies stimulate the receptor and are associated with Graves' disease (hyperthyroidism). Others are inhibitory. Thyrotropin-binding inhibitory immunoglobulin (TBII) antibodies bind to TSHR and inhibit binding of TSH to the receptor. Thyroid-stimulating antibodies (TSAb) mimic TSH and induce thyroid cells to increase cAMP production in a bioassay. Long-acting thyroid stimulator (LATS), a growth-promoting IgG immunoglobulin found in Graves' disease, interacts with the TSHR. TSHR antibodies are measured by receptor assays. TSAb are 95% in Graves' disease. They correlate closely with disease activity. TSBAb are associated closely with atrophic gastritis and severe hypothyroidism, yet atrophic thyroiditis may occur in the absence of TSBAb. TBII and TSBAb occur in nongoitrous autoimmune thyroiditis. Placental passage of TSHR antibodies can induce transient hypo- or hyperthyroidism.

Autoimmune thyroiditis: See thyroiditis, autoimmune.

Thyroid autoimmunity animal models: Spontaneous and experimentally induced thyroiditis are the two models used in research. Spontaneous models of thyroiditis include OS chickens, BUF and BB rat strains, NOD mice, and a special colony of beagle dogs. Genetic as well as environmental factors contribute to autoimmune thyroid disease. Genetic susceptibility to thyroid autoimmunity is multifaceted and diverse. MHC class II genes are the principal genetic determinants of susceptibility in most species. Autoimmunity commences as a response to a restricted region of the thyroglobulin molecule, but a number of different epitopes is present in that region. Autoimmunity differs from one species to another. Experimental thyroiditis in mice is produced mainly by cytotoxic T cells, whereas antibody has a significant pathogenetic role in spontaneous thyroiditis of the OS chicken.

Spontaneous autoimmune thyroiditis (SAT) occurs in the obese strain (OS) of chickens. It is an animal model of Hashimoto's thyroiditis in man. Many mononuclear cells infiltrate the thyroid gland, leading to disruption of the follicular architecture. Immunodysregulation is critical in the etiopathogenesis of SAT. Endocrine abnormalities play a role in the pathogenesis, and disturbances in communication between the immune and endocrine systems occur in the disease. Dysregulation leads to hyperreactivity of the immune system, which, combined with a primary genetic defect-induced alteration of the thyroid gland, leads to autoimmune thyroiditis. Both T and B lymphocytes are significant in the pathogenesis of SAT. T-cell effector mechanisms have greater influence than do humoral factors in initiation of the disease, and most lymphocytes infiltrating the autoimmune thyroid are mature cells. Not only do autoantibodies have a minor role in the pathogenesis, but T cells, rather than B cells, of OS chickens are defective.

LATS (long-acting thyroid stimulator): See long-acting thyroid stimulator.

Long-acting thyroid stimulator (LATS) is an IgG autoantibody that mimics the action of thyroid-stimulating hormone in its effect on the thyroid. The majority of patients with Graves' disease, i.e., hyperthyroidism, produces LATS. This IgG autoantibody reacts with the receptors on thyroid cells that respond to thyroid-stimulating hormone. Thus, the antibody–receptor interaction results in the same biological consequence as does hormone interaction with the receptor. This represents a stimulatory type of hypersensitivity and is classified in the Gell and Coombs' classification as one of the forms of type II hypersensitivity.

LATS protector is an antibody found in Graves' disease patients that inhibits LATS neutralization *in vitro*. This forms the basis for a LATS protector assay in which serum from Graves' patients is tested for the ability to "protect" a known LATS serum from being neutralized by binding to human thyroid antigen.

Drug-induced immune hemolytic anemia: Acquired hemolytic anemia that develops as a consequence of immunological reactions following the administration of certain drugs. Clinically, they resemble autoimmune hemolytic anemia of idiopathic origin. A particular drug may induce hemolysis in one patient, thrombocytopenia in another, neutropenia in a different patient, and occasionally combinations of these in a single patient. The drug-induced antibodies that produce these immune cytopenias are cell-specific. Drugs that cause hemolysis by complement-mediated lysis include quinine, quinidine, and rifampicin, as well as chlorpropamide, hydrochlorothiazide, nomifensine phenacetin, salicylazosulfapyridine, the sodium salt of *p*-aminosalicylic acid, and stibophen. Drug-dependent immune hemolytic anemia, in which the mechanism is extravascular hemolysis, may occur with prolonged high-dose penicillin therapy or other penicillin derivatives as well as cephalosporins and tetracycline.

Autoimmune and lymphoproliferative syndrome is a rare disease that leads to lymphoid enlargement and immune cytopenias. In kindred, the disorder results from a dominant nonfunctional Fas molecule. Like *lpr* and *gld* mice, children develop "double-negative" T cells and hypergammaglobulinemia. Unlike the mouse mutations, affected patients rarely develop antinuclear antibodies or lupus-like renal pathology. There may be an increased susceptibility, based on long-term observations of a few patients.

Although both warm-antibody and cold-antibody types of **autoimmune hemolytic anemia** are known, the warm-antibody type is the most common and is characterized by a positive direct antiglobulin (Coombs' test) associated with lymphoreticular cancer or autoimmune disease and splenomegaly. Patients may have anemia, hemolysis, lymphadenopathy, hepatosplenomegaly, or features of autoimmune disease. They commonly have a normochromic, normocytic anemia with spherocytosis and nucleated red blood cells in the peripheral blood. Leukocytosis and thrombocytosis may also occur. There is a significant reticulocytosis and an elevated serum indirect (unconjugated) bilirubin. IgG and complement adhere to red blood cells. Antibodies are directed principally against Rh antigens. There is a positive indirect antiglobulin test in 50% of cases and agglutination of enzyme-treated red blood cells in 90% of cases. In the cold agglutinin syndrome, IgM antibodies with an anti-I specificity are involved. Warm-autoantibody autoimmune hemolytic anemia has a fairly good prognosis.

Autoimmune hemophilia is an acquired disorder that resembles the inborn disease of coagulation due to a deficiency or dysfunction of Factor VIII. Patients develop an autoantibody that is able to inactivate Factor VIII. This autoanti-Factor VIII antibody leads to an acquired hemophilia that resembles inherited hemophilia. This is a rare disorder that presents with spontaneous bleeding that can be life threatening. It may occur spontaneously or may be associated with other autoimmune disorders. It may result from the treatment of inherited hemophilia with preparations of Factor VIII. Soft tissues and muscles, the gut, the postpartum uterus, and retroperitoneum all represent sites of hemorrhage. Bleeding may also occur following surgery. The acquired antibody to Factor VIII is usually IgG, of the IgG4 subclass. Antiidiotypic antibodies may also be formed against anti-Factor VIII antibodies.

Autoimmune lymphoproliferative syndrome (ALPS) is a rare disorder of children characterized by lymphocytosis, hypergammaglobulinemia, hepatosplenomegaly, prominent lymphadenopathy, antinuclear and anti-red blood cell antibodies, and accumulation of $CD4^-CD8^-$ T cells. It is believed to be associated with a defect in the Fas-Fas L apoptosis signaling system. Some children with the disorder have autoimmune hemolytic anemia, neutropenia, and thrombocytopenia. ALPS is associated with inherited genes that encode defective versions of the Fas protein. *Fas* alleles in children with ALPS have been shown to be heterozygous and to manifest interference with T-cell apoptosis. *Fas* mutations in these children included single- or multiple-based lesions, substitutions or duplications leading to premature termination of transcription, or aberrant *Fas* mRNA splicing. These alterations produce truncated or elongated forms of *Fas* and are not detected in normal persons. Murine LPR and GLD disease models are being used to investigate how *Fas* molecule mutations might interfere at the molecular level with cell death signaling and contribute to human lymphoproliferative and autoimmune disorders.

Autoimmune neutropenia can be either an isolated condition or secondary to autoimmune disease. Patients may have either recurrent infections or remain asymptomatic. Antigranulocyte antibodies may be demonstrated. There is normal bone-marrow function with myeloid hyperplasia and a shift to the left as a result of increased granulocyte destruction. The autoantibody may suppress myeloid cell growth. The condition is treated by immunosuppressive drugs, corticosteroids, or splenectomy. Patients with systemic lupus erythematosus and Felty's syndrome (rheumatoid arthritis, splenomegaly, and severe neutropenia), as well as other autoimmune diseases, may manifest autoimmune neutropenia.

Autoimmune thrombocytopenic purpura is a disease in which patients synthesize antibodies specific for their own blood platelets. The antibody–platelet reactants become bound to cells bearing Fc receptors and complement receptors, leading to decreased blood platelets followed by purpura or bleeding.

Canale-Smith syndrome is an autoimmune and lymphoproliferative syndrome (ALPS). It is a rare disease that leads to lymphoid enlargement and immune cytopenia. It is a consequence of a dominant nonfunctional *Fas* molecule. Children develop "double-negative" T lymphocytes and hypergammaglobulinemia. Unlike *lpr* and *gld* mice, human patients rarely develop antinuclear antibodies or lupus-like renal pathology. However, in the few cases studied, they do have an increased susceptibility to malignancy.

Cold-reacting autoantibodies represent a special group of both naturally occurring and pathologic antibodies characterized by the unusual property of reacting with the corresponding antigen at low temperature. Those reacting with red blood cells are also called cold agglutinins, although they may also react with other cells. Those with a more restricted range of targets are called cytotoxins.

Eosinophilia refers to an elevated number of eosinophils in the blood. It occurs in immediate, type I hypersensitivity reactions, including anaphylaxis and atopy, and is observed in patients with parasitic infestations, especially by nematodes.

Granulocyte autoantibodies are the principal cause of autoimmune neutropenia (AIN). Autoantibodies to the neutrophil-specific antigens of the NA system are associated with this condition. Fifty percent of SLE patients have granulocyte autoantibodies which are also found in febrile nonhemolytic transfusion reactions, transfusion associated acute lung injury, and alloimmune neonatal neutropenia (ANN). Granulocyte autoantibodies on the neutrophil surface can be detected by immunofluorescence (GIFT), EIA, RIA, flow cytometry (FC), or indirectly by detection of autoantibody effects. Granulocyte autoantibodies are mainly of the IgG isotype. The preferred method of choice for assay of these autoantibodies is by flow cytometry.

Granulocytopenia is an abnormally low number of blood granulocytes.

HLA-B is a class I histocompatibility antigen in humans. It is expressed on the nucleated cells of the body. Tissue typing to define an individual's HLA-B antigens employs lymphocytes.

HLA-C is a class I histocompatibility antigen in humans. It is expressed on nucleated cells of the body. Lymphocytes are employed for tissue typing to determine HLA-C antigens. HLA-C antigens play little or no role in graft rejection.

HLA-A, HLA-B, and **HLA-C** are highly polymorphic human MHC class I genes.

HLA Class I: See MHC genes and Class I MHC molecules.

HLA Class II: See MHC genes and Class II MHC molecules.

Immune neutropenia is neutrophil degradation by antibodies termed leukoagglutinins, which are specific for neutrophil epitopes. Penicillin or other drugs, as well as blood transfusions, may induce immune neutropenia. The condition may be associated with such autoimmune disorders as systemic lupus erythematosus and may occur in neonates through passage of leukoagglutinins from mother to young.

Innocent bystander hemolysis: Drugs acting as haptens induce immune hemolysis of "innocent" red blood cells. The reaction is drug-specific and involves complement through activation of the alternate complement pathway. The direct antiglobulin (Coombs') test identifies split products that are membrane associated, yet the indirect Coombs' test remains negative.

Lymphocytotoxic autoantibodies are a heterogenous group of autoantibodies that include (1) those that occur naturally and in selected diseases, are of the IgM isotype, and are cold-reactive at temperatures near 15°C, and (2) lymphocytotoxic autoantibodies associated with infectious mumps, herpes, mycoplasma and chronic parasitic infections, as well as the autoimmune disease SLE and RA. Most SLE patient sera react with CD45 on mitogen-stimulated T lymphocytes, and some interact with T cell receptor, β_2 microglobulin, as well as selected B cell antigens. Alloantibodies against lymphocytes usually result form immunization, blood transfusion, transplantation, and pregnancy. IgG alloantibodies are warm-reactive, i.e., either 37°C or room temperature. Class I and Class II MHC antigen on the lymphocyte surface are the targets of lymphocyte autoantibodies. The microlymphocytotoxic text is the method of choice for lymphocytotoxic autoantibody detection.

Metatype autoantibodies are anti-immunoglobulin antibodies that recognize an antibody-liganded active site but are not specific for either the ligand or the idiotype alone. Metatype antibody interactions with anti-metatype-metatype immunoglobulins serve as a model for T-cell receptor and antigen-MHC complex interaction which could be considered as metatype immunoglobulins and liganded antibody, respectively.

A **neutrophil leukocyte** is a peripheral blood polymorphonuclear leukocyte derived from the myeloid lineage. Neutrophils comprise 40–75% of the total white blood count numbering 2500–7500 cells per cubic millimeter. They are phagocytic cells and have a multilobed nucleus and azurophilic and specific granules that appear lilac following staining with Wright's or Giemsa stains. They may be attracted to a local site by such chemotactic factors as C5a. They are the principal cells of acute inflammation and actively phagocytize invading microorganisms. Besides serving as the first line of cellular defense in infection, they participate in such reactions as the uptake of antigen–antibody complexes in the Arthus reaction.

Neutrophil cytoplasmic antibodies are autoantibodies specific for myeloid-specific lysosomal enzymes. Anti-neutrophil cytoplasmic antibodies (ANCA) are of two types. The C-ANCA variety stains the cytoplasm where it reacts with alpha granule proteinase-3 (PR-3). By contrast, P-ANCA stains the perinuclear zone through its reaction with myeloperoxidase (MPO). Of Wegener's granulomatosis patients with generalized active disease, 84 to 100% develop C-ANCA, although fewer individuals with the limited form of Wegener's granulomatosis develop these antibodies. Organ involvement cannot be predicted from the identification of either C-ANCA or P-ANCA antibodies. P-ANCA antibodies reactive with myeloperoxidase may be found in patients with certain vasculitides that include Churg-Strauss syndrome, polyarteritis nodosa, microscopic polyarteritis, and polyangiitis. A P-ANCA staining pattern unrelated to antibodies against PR-3 or to MPO has been described in inflammatory bowel disease (IBD). A total of 59–84% of ulcerative colitis patients and 65–84% of primary sclerosing cholangitis patients are positive for P-ANCA. By contrast, only 10 to 20% of Crohn's disease patients are positive for it. These antibodies are classified as neutrophil nuclear antibodies (ANAs). Hep-2 cells are used to differentiate granulocyte-specific ANAs from ANAs that are not tissue-specific (Figures 14.7–14.10).

New Zealand black (NZB) mice are an inbred strain of mice that serves as an animal model of autoimmune hemolytic anemia. They develop antinuclear antibodies in low titer, have defective T lymphocytes, defects in DNA repair, and have B cells that are spontaneously activated.

Paroxysmal cold hemoglobinuria (PCH) is a rare type of disease that accounts for 10% of cold autoimmune hemolytic anemias. It may be either a primary idiopathic disease or secondary to syphilis or viral infection and is characterized by the passage of hemoglobin in the urine after exposure to cold. In addition to passing dark brown urine, the patient may experience chills, fever, and pain in the back, legs, or abdomen. The disease is associated with a hemolysin termed the Donath-Landsteiner antibody which is a polyclonal IgG antibody. It sensitizes the patient's red blood cells in the cold, complement attaches to the erythrocyte surface, and hemolysis occurs on

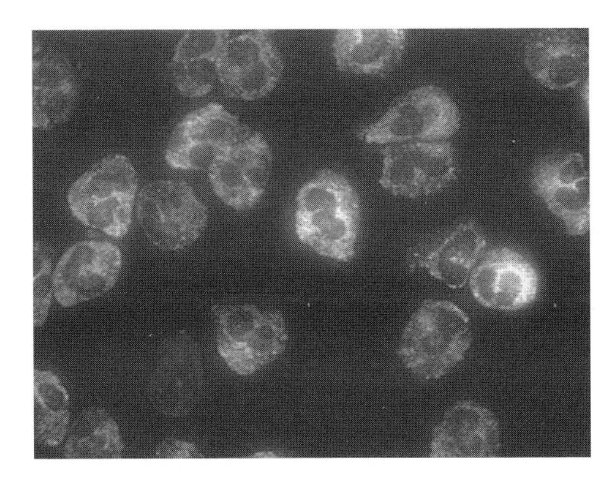

FIGURE 14.7 Cytoplasmic antineutrophil cytoplasmic antibodies C-ANCA (ethanol fixation).

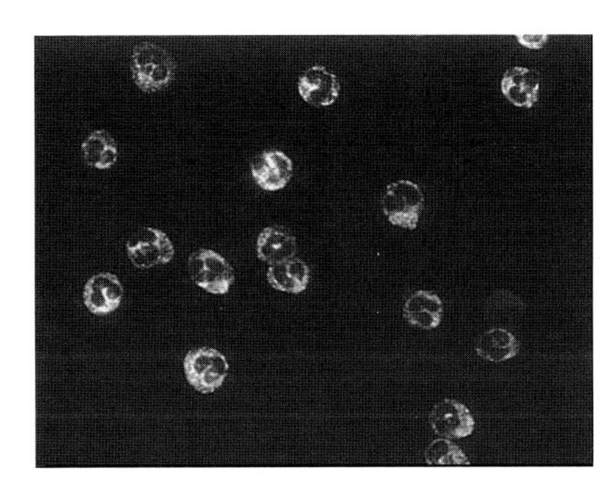

FIGURE 14.8 Neutrophil cytoplasmic antibody C-ANCA (formalin fixation).

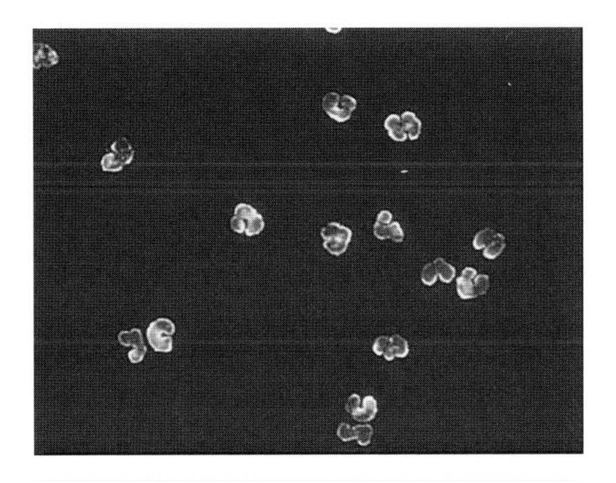

FIGURE 14.9 Neutrophil cytoplasmic antibody P-ANCA (ethanol fixation).

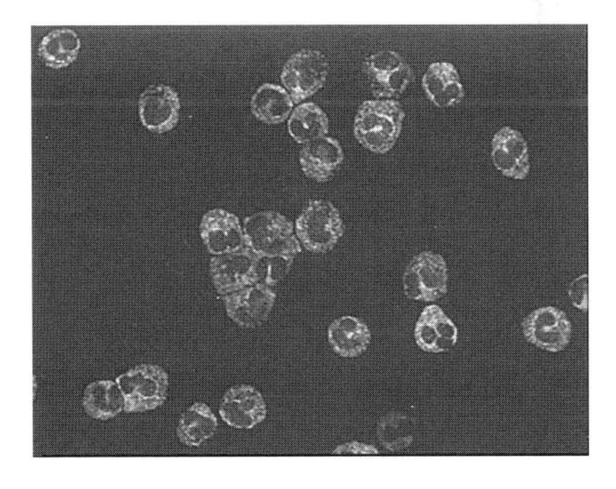

FIGURE 14.10 Neutrophil cytoplasmic antibody P-ANCA (formalin fixation).

warming to 37°C. The antibody's specificity is for the P antigen of red blood cells.

Pernicious anemia (PA) is an autoimmune disease characterized by the development of atrophic gastritis, achlorhydria, decreased synthesis of intrinsic factor, and malabsorption of vitamin B_{12}. Patients present with megaloblastic anemia caused by the vitamin B_{12} deficiency that develops. The majority of pernicious anemia patients develop antiparietal cell antibodies, and at least half of them also develop antibodies against intrinsic factor, which is necessary for the absorption of B_{12}. The antiparietal cell antibodies are against a microsomal antigen found in gastric parietal cells. Intrinsic factor is a 60-kDa substance that links to vitamin B_{12} and aids its uptake in the small intestine. PA may be a complication of common variable immunodeficiency or may be associated with autoimmune thyroiditis. PA is caused principally by injury to the stomach mediated by T lymphocytes. Patients may manifest megaloblastic anemia, deficiency of vitamin B_{12}, and increased gastrin in serum.

Prothrombin antibodies are antibodies against prothrombin or to prothrombin–phospholipid complex, which have been associated with antiphospholipid syndrome (APF). The test is based on the ability of patients' sera to inhibit prothrombin activation to thrombin. Either factor Xa or purified snake venom can cleave prothrombin to produce thrombin. Prothrombin antibodies in patient sera inhibit the capacity of these proteins to activate prothrombin. Prothrombin antibodies prevent prothrombin activation but do not interfere with prothrombin activity on fibrin and are therefore associated with thrombosis, not bleeding. Increased concentrations of antibodies against prothrombin increase the risk of deep venous thrombosis and pulmonary embolism in middle-aged men.

Platelet autoantibodies: Platelets possess surface FcRII that combine to IgG or immune complexes. The platelet surfaces can become saturated with immune complexes, as in autoimmune (or idiopathic) thrombocytopenic purpura (ITP) or AITP. Fab-mediated antibody binding to platelet antigens may be difficult to distinguish from Fc-mediated binding of immune complexes to the surface.

Idiopathic thrombocytopenic purpura is an autoimmune disease in which antiplatelet autoantibodies destroy platelets. Splenic macrophages remove circulating platelets coated with IgG autoantibodies at an accelerated rate. Thrombocytopenia occurs even though the bone marrow increases platelet production. This can lead to purpura and bleeding. The platelet count may fall below 20,000 to 30,000 per microliter. Antiplatelet antibodies are detectable in the serum and on platelets. Platelet survival is decreased. Splenectomy is recommended in adults. Corticosteroids facilitate a temporary elevation in the platelet count. This disease is characterized by decreased blood platelets, hemorrhage, and extensive thrombotic lesions.

Anti-intrinsic factor autoantibodies are antibodies against the glycoprotein, intrinsic factor, with a molecular weight of 44-kDa. Intrinsic factor binds vitamin B_{12}. Radio immunoassay can detect two separate antibodies: one that reacts with the binding site for vitamin B_{12}, thereby blocking the subsequent binding of intrinsic factor with the vitamin, whereas the other reacts with an antigenic determinant remote from this site.

Autoantibodies to gastric parietal cells are specific for both the catalytic 100-kDa glycoprotein (α) and the 60 to 90-kDa glycoprotein (β) subunit of the gastric H^+/K^+ ATPase, the enzyme that acidifies gastric juice. Detection is by immunofluorescence.

Autoimmune gastritis is an organ-specific autoimmune disease in which autoantibodies are formed against gastric antigens, mononuclear cells infiltrate through target organs with destruction; it is a regenerative response of the affected tissue to corticosteroid and immunosuppressive drugs, and has familial predisposition and association with other autoimmune diseases. The molecular target of parietal cell autoantibodies is the gastric H^+/K^+ ATPase located on secretory membranes of gastric parietal cells. See also pernicious anemia.

Colon autoantibodies: Seventy-one percent of ulcerative colitis (UC) patients have IgG autoantibodies reactive with rat epithelial cells. A monoclonal antibody against colonic epithelium cross-reacted with Barrett epithelium adenocarcinoma and Barrett esophagus epithelium. Other studies have revealed shared phenotypic expression of colonic, biliary, and Barrett epithelium. Other studies have revealed shared phenotypic expression of colonic, biliary, and Barrett epithelium. Seventy percent of patients with colitis manifest antibodies reactive with neutrophil cytoplasm distinct from cANCA and pANCA. Of the IgG subclasses of antibody against colonic antigens, IgG 4 and IgG1 have shown the greatest reactivity in separate studies and have differentiated ulcerative colitis patients from Crohn's disease.

Endomysial autoantibodies are IgA autoantibodies that react with the reticulin constituent of endomysium in primate smooth muscle. These antibodies are present in 70 to 80% of dermatitis herpetiformis (DH) patients on a regular gluten-containing diet and in all patients with celiac disease (CD), a gluten-sensitive enteropathy, with severe villous atrophy. IgA EmA specificity for active gluten-sensitive enteropathy is >98%. IgA EmA are better

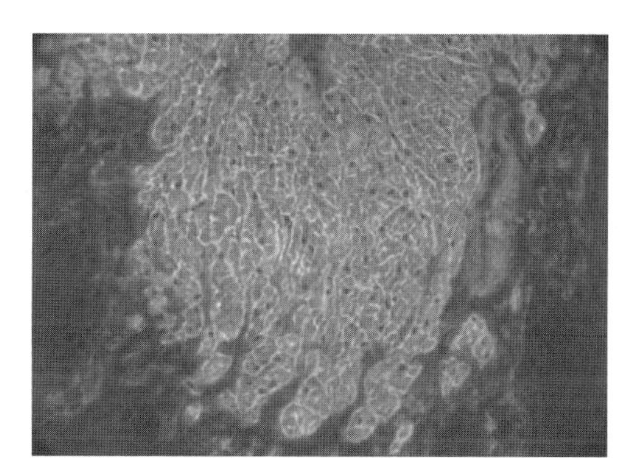

FIGURE 14.11 Endomysial autoantibody.

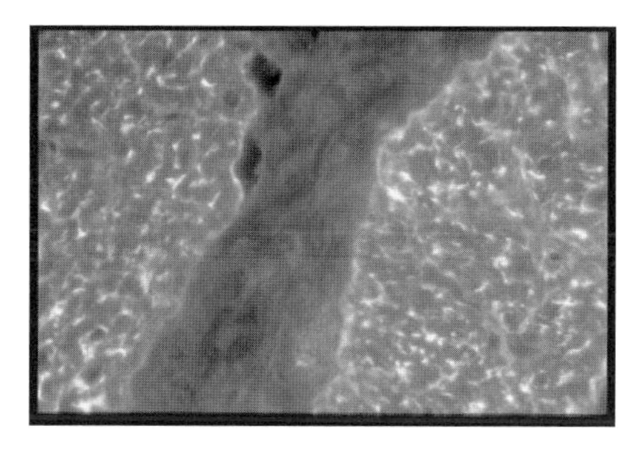

FIGURE 14.12 Endomysial autoantibody. (400x)

predictors for CD than are IgA gliadin autoantibodies (AGA). A positive IgA EmA is confirmatory evidence and indicates the need for intestinal biopsy. IgA AGA and IgA-R1-reticulin autoantibodies (IgA ARA) are each present in approximately one quarter of DH patients and in 93% and 44%, respectively, of patients with active CD. The incidence of CD is 10 times greater (with selective IgA deficiency) compared with subjects without selective IgA disease.

Gastric cell cAMP stimulating autoantibodies are detected in young males with long-standing duodenal ulcer disease, a family history of this condition, and poor responsiveness to H2-receptor antagonists, but other investigators have failed to confirm these results.

Gastrin-producing cell autoantibodies (GPCA) are found in 8 to 16% of patients with antral (type B) chronic atrophic gastritis, which primarily affects the antral mucosa. There are no parietal cell autoantibodies (PCA) and no association with pernicious anemia or with polyendocrinopathy. The GPCA are believed to be responsible for the diminished gastrin secretion and fewer gastrin-producing cells in antral gastritis. PCA are only found in patients with normal antral mucosa. By contrast, GPCA are present only in patients with normal antral mucosa or

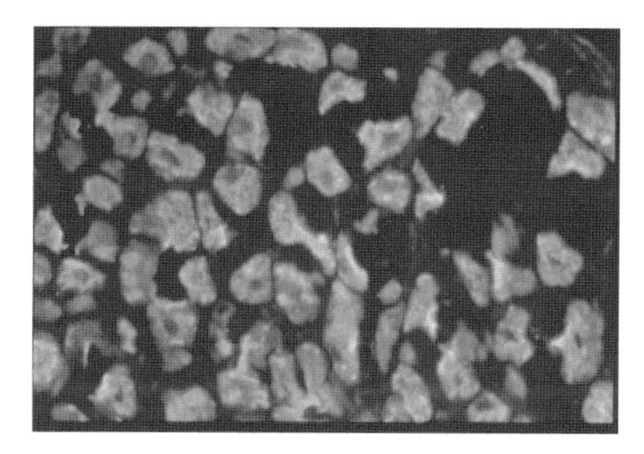

FIGURE 14.13 Gastric parietal cell autoantibody.

mild antral gastritis but not in individuals with moderate or severe antral gastritis. *Helicobacter pylori* has a world-established association with antral (type B) gastritis and peptic ulcer disease, but antibodies against *H. pylori* are not helpful for monitoring the disease.

Gliadin autoantibodies are used to screen persons at risk for developing celiac disease and other gluten-sensitive enteropathies including dermatitis herpetiformis (DH) and in monitoring patient adherence to a gluten-free diet (GFD) (Figure 14.13). Gliadins are a type of protein present in the gluten of wheat and rye grains. In genetically susceptible subjects, α-gliadins activate celiac disease (CD), a gastrointestinal disorder in which jejunal mucosa is flattened. IgG AGA are more sensitive (~100) than IgA AGA (~50%); IgA AGA are more specific (~95%) than IgG AGA (~60%). Both isotypes of the autoantibodies increase significantly during a challenge with gluten. Perhaps even months prior to clinical relapse, CD is 10 times more frequent in subjects with selective IgA deficiency. EIA is the method of choice to detect AGA in CD.

Intrinsic factor is a glycoprotein produced by parietal cells of the gastric mucosa, which is necessary for vitamin B_{12} absorption. A lack of intrinsic factor leads to vitamin B_{12} deficiency and pernicious anemia.

Intrinsic factor antibodies are found in three fourths of pernicious anemia patients that are specific for either the binding site (type I or "blocking antibodies") or some other intrinsic factor antigenic determinant (type II antibodies).

Parietal cell antibodies are present in 50 to 100% of pernicious anemia (PA) patients. They are also found in 2% of normal individuals. Their frequency increases with aging and in subjects with insulin-dependent diabetes mellitus. The frequency of parietal cell antibodies diminishes with disease duration in pernicious anemia. Parietal cell autoantibodies in pernicious anemia and in autoimmune

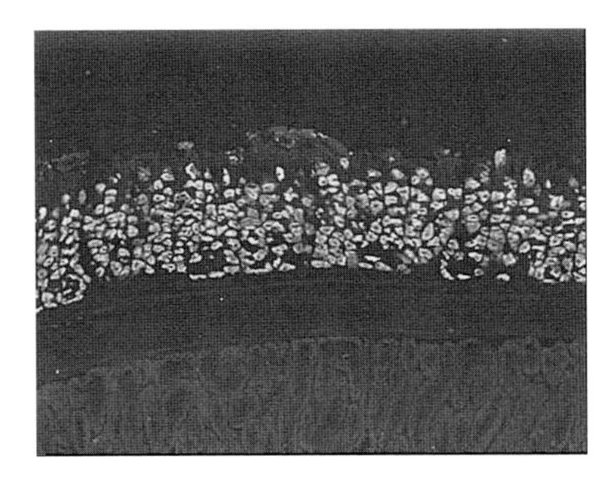

FIGURE 14.14 Autoantibody: Gastric parietal cell + kidney.

gastritis recognize the α and β subunits of the gastric proton pump (H^+/K^+ ATPase), which are the principal target antigens. Parietal cell antibodies react with the α and β subunits of the gastric proton pump and inhibit the gastric mucosa's acid-producing H^+/K^+ adenosine triphosphatase. Parietal cell antibodies relate to type A gastritis for which fundal mucosal atrophy, achlorhydria, development of pernicious anemia, and autoimmune endocrine disease are characteristic.

Parietal cell autoantibodies are specific for the α and β subunits of the gastric H^+/K^+ -ATPase (gastric proton pump) which secretes acid into the stomach lumen and is expressed specifically in gastric parietal cells (Figure 14.14). Although parietal cell antibodies can occur in 5% of healthy subjects, they occur in high frequency in pernicious anemia. They are related to type A gastritis in which there is atrophy of the fundal mucosa, achlorhydria, tendency to evolve into pernicious anemia, and association with autoimmune endocrine disease. PCA can be detected by IFA as reticular cytoplasmic staining using mouse stomach frozen sections. PCA have been found in 100% of primary biliary cirrhosis patients, in three fourths of autoimmune hepatitis cases, and in 29% of chronic liver disease patients.

Autoantibodies against pepsinogen develop in autoimmune atrophic gastritis patients with pernicious anemia. Three fourths of peptic ulcer patients have pepsinogen antibodies.

Reticulin autoantibodies include R1-ARA, the only one of five separate reticulin autoantibodies that is of diagnostic and pathogenic significance (Figure 14.15). IgA R1-ARA are highly specific for untreated CD. Approximately 2.6% of healthy children possess reticulin autoantibodies. In IgA-deficient subjects under evaluation for CD, the detection of IgG-ARA, IgG gliadin autoantibodies (AGA), and IgG endomysial autoantibodies (EmA) may be helpful.

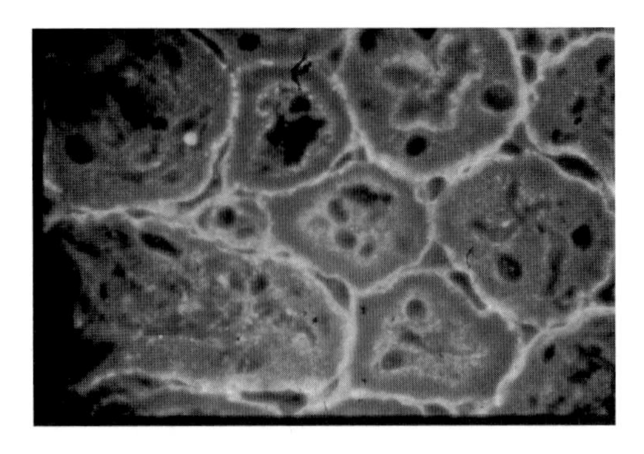

FIGURE 14.15 Reticulin autoantibody.

R1-ARA classically disappear 3 to 12 months following the maintenance of a strict gluten-free diet in both adults and children with CD. IgA, ARA, as well as IgA AGA and IgA EmA can be detected in dermatitis herpetiformis (DH).

Tissue transglutaminase autoantibodies are IgA autoantibodies against tissue transglutaminase (tTG) that have a close correlation with active CD. Tissue transglutaminase is an intracellular enzyme found in the endomysium and is released following cellular injury or wounding. The enzyme facilitates deamidation of extracellular proteins, including gluten peptides which leads to enhanced T cell-stimulatory activity. tTG antibodies are transient, disappearing when gluten is eliminated from the diet and mucosal healing has taken place. Available data suggest that tTG autoantibodies are very specific for celiac disease.

Acetaldehyde adduct autoantibodies are found in approximately 73% of alcoholics and in 39% of nonalcoholic liver disease (e.g., primary biliary cirrhosis, chronic active hepatitis, and acute viral and drug-induced hepatitis). The highest titers of these autoantibodies are found in advanced stages of alcoholic and nonalcoholic liver disease. Alcoholics develop aceteldehyde adduct autoantibodies to apo B-containing lipoproteins, particularly very low density lipoproteins (VLDL). A total of 33% of alcoholic heart muscle disease patients develop cardiac–protein–acetaldehyde adduct autoantibodies which could be a potential marker for this heart condition.

Asialoglycoprotein receptor (ASGP R) autoantibodies against this liver-specific membrane receptor occur at high frequency in autoimmune liver diseases, especially autoimmune hepatitis, and may occur also in primary biliary cirrhosis (PBC), viral hepatitis, and other liver diseases but at a lower frequency. Anti-ASGPR antibodies correlate with disease activity. Anti-ASGP R antibodies against human-specific epitopes are closely linked to autoimmune hepatitis. T lymphocytes specific for the ASG R have been isolated from the liver of autoimmune hepatitis type I patients. Tissue expression of ASGP R is most prominent in periportal areas where piecemeal necrosis is observed as a marker of severe inflammatory activity. ASGP R antibodies may be used as a diagnostic marker for autoimmune hepatitis if other markers are negative and autoimmune liver disease is suspected.

Approximately 10 to 20% of chronic hepatitis cases are attributable to **autoimmune hepatitis**. The diagnosis of autoimmune hepatitis is based on clinical and laboratory criteria defined by the International Autoimmune Hepatitis Group (IAHG). Circulating autoantibodies are a hallmark of this syndrome. Whereas immune defects occur during the chronic course of viral hepatitis, loss of tolerance to autologous liver tissue is the principal pathogenic mechanism in autoimmune hepatitis. It is important to distinguish between autoimmune and viral hepatitis since interferons administered for viral hepatitis are contraindicated in autoimmune liver disease. Immunosuppression prolongs survival in autoimmune hepatitis but favors viral replication. Drug-induced liver injury may also be immune-mediated.

Chronic active hepatitis, autoimmune is a disease that occurs in young females who may develop fever, arthralgias, and skin rashes. They may be of the HLA-B8 and DR3 haplotype and suffer other autoimmune disorders. Most develop antibodies to smooth muscle, principally against actin, and autoantibodies to liver membranes. They also have other organ- and nonorgan-specific autoantibodies. A polyclonal hypergammaglobulinemia may be present. Lymphocytes infiltrating portal areas destroy hepatocytes. Injury to liver cells produced by these infiltrating lymphocytes produces piecemeal necrosis. The inflammation and necrosis are followed by fibrosis and cirrhosis. The T cells infiltrating the liver are $CD4^+$. Plasma cells are also present, and immunoglobulins may be deposited on hepatocytes. The autoantibodies against liver cells do not play a pathogenetic role in liver injury. There are no serologic findings that are diagnostic. Corticosteroids are useful in treatment. The immunopathogenesis of autoimmune chronic active hepatitis involves antibody, K-cell cytotoxicity, and T-cell reactivity against liver membrane antigens. Antibodies and specific T suppressor cells reactive with LSP are found in chronic active hepatitis patients, all of whom develop T cell sensitization against asialoglycoprotein (AGR) antigen. Chronic active hepatitis has a familial predisposition.

Smooth muscle antibodies are autoantibodies belonging to the IgM or IgG class that are found in the blood sera of 60% of chronic active hepatitis patients (Figure 14.16 and Figure 14.17). In biliary cirrhosis patients, 30% may also be positive for these antibodies. Low titers of smooth muscle antibodies may be found in certain viral infections of the liver. The anti-actin variety of smooth muscle antibodies is especially associated with autoimmune liver disease, whereas SMA directed to intermediate filaments can be present in virus-induced liver disease. The presence of SMA is not predictive of the development of liver disease and is not helpful for prognosis in autoimmune chronic active hepatitis patients.

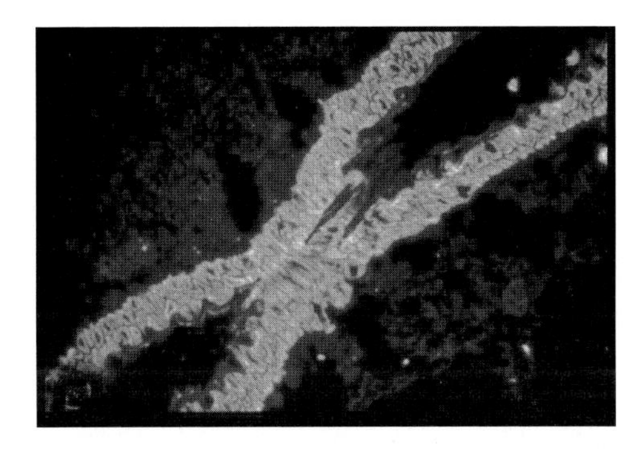

FIGURE 14.16 Smooth muscle autoantibody.

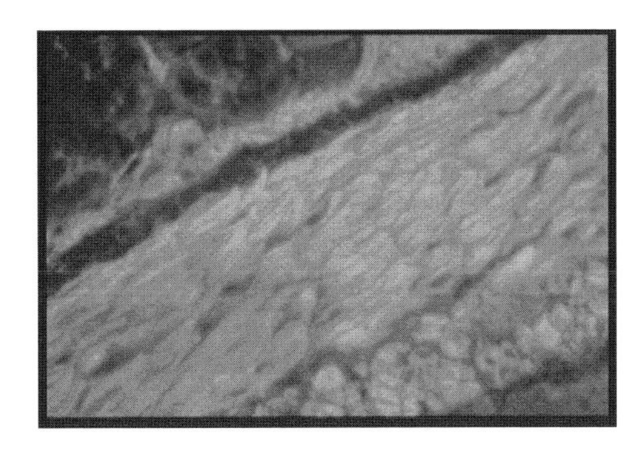

FIGURE 14.17 Smooth muscle autoantibody. (400x)

Endoplasmic reticulum autoantibodies are autoantibodies against trifluoroacetylated hepatic proteins (TFA-proteins) in patients who have developed a severe form of halothane-induced hepatitis alluded to be induced by an immune response. The sera of these patients also contain antibodies that recognize various non-TFA-modified proteins which may be present in patients exposed to halothane but do not develop hepatitis. Halothane anesthetic can lead to either mild halothane-induced liver damage, which is of no clinical significance, or the severe form. The clinical significance of antibodies to TNA and non-TFA proteins remains to be determined.

GOR autoantibodies are detected in approximately 80% of HCV-RNA-positive chronic hepatitis C infection, less than 10% of chronic HBV or alchol misuse, and in approximately 2% of normal subjects. These autoantibodies recognize an epitope of the sequence GRRGQKAKSNPN-RPL that is common to a presumed core gene product of HCV and a host nuclear component. GOR autoantibodies

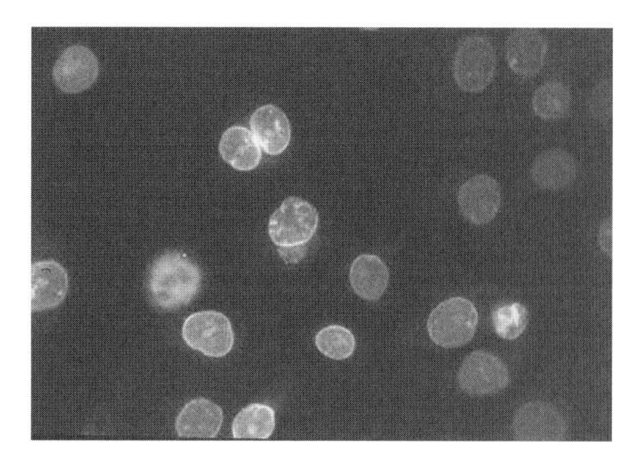

FIGURE 14.18 Lamin autoantibody.

are detected also in most HCV-antibody-positive patients with LKM-1-autoantibody-positive autoimmune hepatitis type 2b.

Autoantibodies against lamin, a nuclear antigen, are present in the sera of chronic autoimmune disease patients manifesting hepatitis, leukocytoclastic angiitis or brain vasculitis, cytopenia, and circulating anticoagulant or cardiolipin antibodies (Figure 14.18). They form a rim-type antinuclear staining pattern in immunofluorescence assays. A minority of systemic lupus erythematosus patients develop antibodies to lamin. They are found in selected patients with autoimmune and inflammatory diseases. IFA is the method of choice for their detection. Lamin autoantibodies occur in selected patients with chronic autoimmune disease marked by delta hepatitis, cytopenia with circulating anticoagulants or cardiolipin antibodies, and cutaneous leukocytoclastic angiitis or possibly brain vasculitis. They may occur naturally be cross-reacting or formed in response to antigen. They have no known clinical significance.

Liver cytosol autoantibodies are autoantibodies against liver cystolic antigen type I (LC1). They are present in one fifth of liver–kidney microsome-1 (LKM-1) autoantibody-positive patients, in less than 1% of chronic hepatitis C virus infected, and in some AIH patients.

Liver membrane antibodies are antibodies specific for the 26-kDa LM protein target antigen in the sera of 70% of autoimmune chronic active hepatitis patients who are HBsAg-negative. These antibodies are demonstrable by immunofluorescence. They may be demonstrated also in primary biliary cirrhosis, chronic hepatitis B, alcoholic liver disease, and sometimes in Sjögren's syndrome patients. "Lupoid" autoimmune chronic active hepatitis patients may develop antibodies not only against liver membrane but also against smooth muscle and nuclear constituents. Liver membrane antibodies are not useful for either diagnosis or prognosis.

Liver membrane autoantibodies are a heterogeneous group of autoantibodies that include liver-specific membrane lipoprotein (LSMP) autoantibodies, which are linked to chronic autoimmune hepatitis (AIH) that lack disease specificity. They are useful prognostically for treatment withdrawal. Characteristic autoantibodies usually aid in the diagnosis of autoimmune hepatitis. Persistence or reappearance of liver-specific protein antibodies in chronic AIH patients in remission during treatment withdrawal may signify reactivation of the disease.

LM autoantibodies are antimicrosomal antibodies that react exclusively with liver tissue. They have been found in drug-induced hepatitis caused by dihydralazine and interact with cytochrome P450-1 A2. LM antibodies against cytochrome P450-1A2 in nondrug-induced autoimmune liver disease suggested that the liver disease was part of the autoimmune polyendocrine syndrome type II (APS-1).

Lupoid hepatitis is an autoimmune hepatitis that appears usually in young females who may produce antinuclear, antimitochondria, and antismooth muscle antibodies. Of these patients, 15% may show LE cells in the blood. This form of hepatitis has the histologic appearance of chronic active hepatitis, which generally responds well to corticosteroids.

Mitochondrial antibodies are IgG antibodies present in 90 to 95% of primary biliary cirrhosis (PBC) patients (Figure 14.19 through Figure 14.21). These antibodies, which are of doubtful pathogenic significance, are specific for the pyruvate dehydrogenase enzyme complex E2 component situated at the inner mitochondrial membrane (M2). These antibodies are also specific for another E2-associated protein. Other antimitochondrial antibodies include the M1 antibodies of syphilis, the M3 antibodies found in pseudolupus, the M5 antibodies in collagen diseases, the M6 antibodies in hepatitis induced by iproniazid, the M7 antibodies associated with myocarditis and

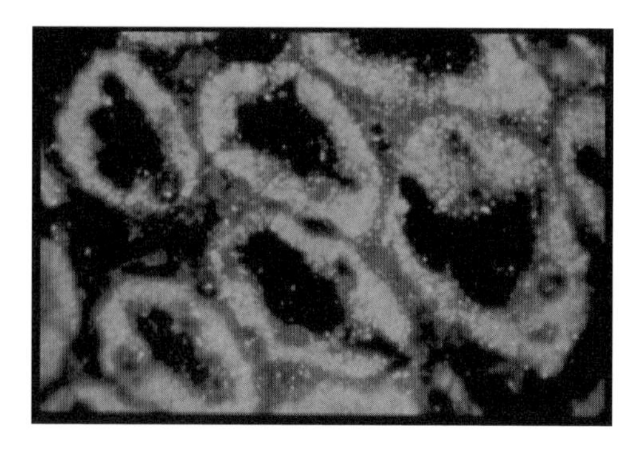

FIGURE 14.19 Mitochondrial antibody. (Mouse kidney, 500 ×)

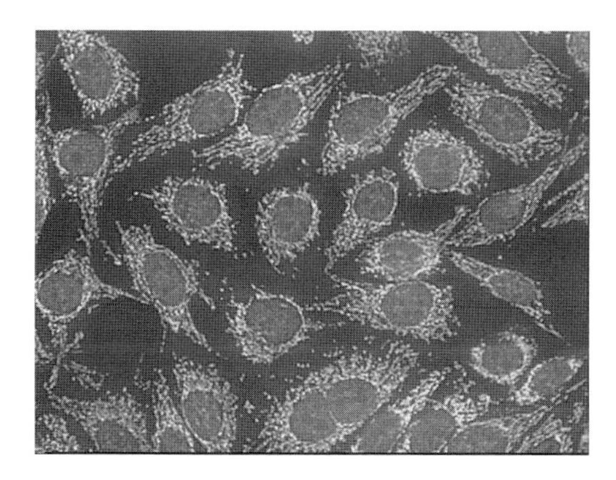

FIGURE 14.20 Mitochondrial autoantibodies. (HEP-2 cells)

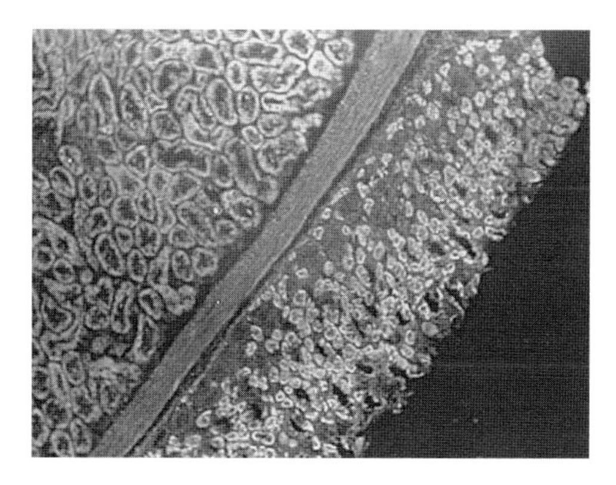

FIGURE 14.21 Mitochondrial autoantibodies. (Mouse kidney/stomach)

cardiomyopathy, the M8 antibodies that may be a marker for prognosis, and the M9 antibodies that serve as a marker for beginning primary biliary cirrhosis. Mitochondrial antibody titer may be an indicator of primary biliary cirrhosis progression. A patient with a titer of 1 to 40 or more should be suspected of having primary biliary cirrhosis, whether or not symptoms are present and even if the alkaline phosphatase is normal.

Mitochondrial autoantibodies (MA) are IgG autoantibodies specific for the E2 component (lipoate acetyltransferase) of the pyruvate dehydrogenase enzyme complex located at the inner mitochondrial membrane (M2) as well as against another protein in E2 preparations. They are present in 90 to 95% of patients with primary biliary cirrhosis (PBC). However, they are not believed to be pathogenic. MA that do not react with M2 are found in other conditions such as M1 autoantibodies in syphillis, M3 autoantibodies in pseudolupus, M5 autoantibodies in poorly defined collagen diseases, M6 autoantibodies in iproniaizid-induced hepatitis, M7 autoantibodies in cardiomyopathy and myocarditis, and M8 autoantibodies that serve as indicator of early PBC. The titer of MA in PBC correlates with disease progression, i.e., a titer of 1:40 or grater suggests PBC in the absence of symptoms and in the presence of a normal alkaline phosphatase. Nuclear autoantibody (ANA) patterns in PBC include (1) multiple nuclear dots (MND-ANA) present in 10 to 15% of PBC and mainly associated with sicca syndrome, and (2) membrane-associated ANA (MANA) present in 25 to 50% of PBC.

Primary sclerosing cholangitis (PSC) is a possible autoimmune liver disease characterized by inflammation of the large intra- and extrahepatic bile ducts, which eventually leads to strictures and dilatations, with the bile ducts eventually becoming stenotic and small intrahepatic bile ducts disappearing. Even though the etiology is unknown, chronic portal bacteremia, toxic bile acids, chronic viral infections, ischemic vascular injury, and immunoregulatory disorders have been implicated. Antineutrophil cytoplasmic antibodies (ANCA) together with antinuclear antibodies (ANA) are principal features of PSC. Of lesser importance are anti-colon epithelial autoantibodies that have been described. There is a male predominance, in contrast to most autoimmune diseases. Lymphocytes infiltrate and destroy bile ducts, there is hypergammaglobulinemia, circulary immune complexes, increased metabolism of complement component C3, and classical pathway activation of complement.

Liver–kidney microsomal antibodies are present in a subset of individuals with autoimmune chronic active hepatitis who are ANA-negative. By immunofluorescence, these antibodies can be shown to interact with hepatocyte and proximal renal tubule cell cytoplasm. These antibodies are not demonstrable in the sera of non-A, non-B chronic active hepatitis patients.

Liver–kidney microsomal autoantibodies are associated with autoimmune hepatitis type II. Their principal feature is the exclusive staining of the P3 portion of the proximal renal tubules. A total of 67% of patients with this disease have antibodies to liver cytosol-type I (anti-LC-1). Cytochrome P450 2D6 is the major antigen for LKM-1 autoantibodies.

Liver–kidney microsomal autoantibodies are specific for cytoplasmic constituents of hepatocytes and proximal renal tubule cells. They are present in selected patients with ANA-negative autoimmune chronic active hepatitis (CAH) (HbsAg negative). They are not found in CAH non-A, non-B. LKM-1 autoantibody-positive CAH (autoimmune chronic active hepatitis type II) is the most frequent autoimmune liver disease in children, with a relatively poor prognosis. LKM-1 autoantibodies are not believed to be pathogenic in autoimmune CAH of children. They show specificity for cytochrome P450. LKM-2 autoantibodies are present in tienilic acid (Ticrynafen)–induced hepatitis. LKM-3 autoantibodies are present in 10% of chronic delta virus hepatitis cases. The LKM-1 antigen is cytochrome P450IID6. The LKM-2 antigen is cytochrome P450IIC8/9/10 (Figure 14.22).

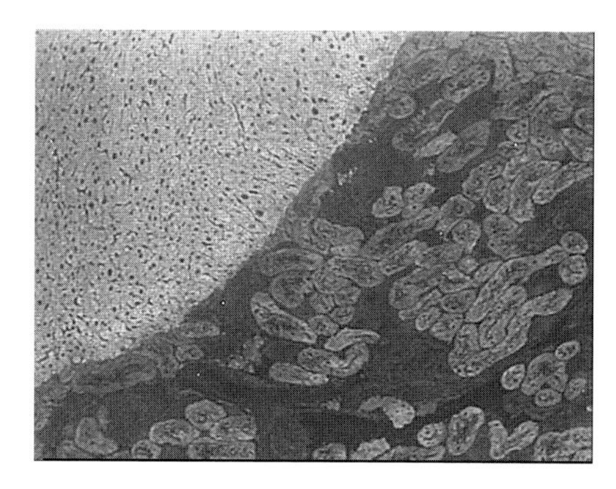

FIGURE 14.22 Liver–kidney microsome (LKM-L) autoantibodies.

Liver–kidney microsome (LKM-1) autoantibodies are associated with autoimmune hepatitis type II. Their principal feature is the exclusive staining of the P3 portion of the proximal renal tubules. A total of 67% of patients with this disease have antibodies to liver cytosol type I (anti-LC1). Cytochrome P450 2D6 is the major antigen for LKM-1 autoantibodies, which are specific for cytoplasmic constituents of hepatocytes and proximal renal tubule cells. They are present in selected patients with antinuclear antibody (ANA)-negative autoimmune chronic active hepatitis (CAH) (HbsAg negative). They are not found in CAH non-A, non-B. LKM-1 autoantibody-positive CAH type II is the most frequent autoimmune liver disease in children, who have a relatively poor prognosis. LKM-1 autoantibodies are not believed to be pathogenic in autoimmune CAH of children. They show specificity for cytochrome P450. LKM-2 autoantibodies are present in tienilic acid (Ticrynafen)-induced hepatitis. LKM-3 autoantibodies are present in 10% of chronic delta virus hepatitis cases. The LKM-1 antigen is cytochrome P450 IID6. The LKM-2 antigen is cytochrome P450IIC8/9/10.

Liver–kidney microsome 2 (LKM-2) autoantibodies are associated only with drug-induced hepatitis caused by tienilic acid and not an autoimmune hepatitis. These antibodies are specific for cytochrome P450 2C9.

Liver–kidney microsome 3 (LKM-3) autoantibodies detect a protein band of 55 kDa in approximately 10% of autoimmune chronic hepatitis type II.

Autoimmune tubulointerstitial nephritis is a renal disease believed to be mediated by antitubular basement membrane (TBM) antibody. It is characterized by linear deposition of IgG and occasionally complement along tubular basement membranes. It occurs together with serum anti-TBM antibodies in 10 to 20% of cases. Symptoms include tubular dysfunction with polyuria, proteinuria, aminoaciduria, and glucosuria. TBM antigens that have been identified include a 58-kDa "TIN-antigen."

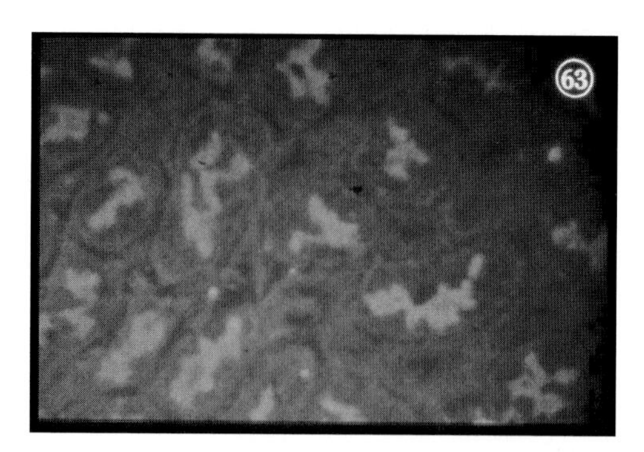

FIGURE 14.23 Brush border autoantibody (not kidney).

Other nephritogenic tubular interstitial antigens have also been identified in both animal models of the disease and in humans.

Brush border autoantibodies (Figure 14.23) are associated with Heymann nephritis in rats as well as half of patients with ulcerative proctocolitis, 20% of patients with antibodies to *Yersinia enterocolitica* 0:3, and extensively burned patients.

Entactin/nidogen autoantibodies are autoantibodies against entactin, a 150-kDa sulphated glycoprotein present in the glomerular plasma membrane. It is believed to be synthesized by endothelial and epithelial cells in the developing kidney but in the mature tissue. It is believed to have role in cellular adhesion to extracellular material. Laminin and entactin form a complex *in vivo*. Nidogen, another sulfated glycoprotein, was identified in basement membrane matrix of EHS sarcoma. This is a 150-kDa glycoprotein and, similar to entactin, reveals an *in vitro* binding affinity for laminin. Subsequent investigations have shown nidogen and entactin to be similar or identical. More than 40% of 206 patients with glomerulonephritis develop IgG, IgM, and IgA autoantibodies against entactin/nidogen. Entactin autoantibodies of IgG, IgA, and IgM classes have been found in the sera of systemic lupus erythematosus patients, but IgA entactin autoantibodies are found most frequently in patients with IgA nephropathy.

Intercalated cell autoantibodies are autoantibodies against renal tubular cells (intercalated cells) detected from time to time by using immunofluorescence with frozen sections of normal human or rabbit kidney. The clinical significance with these antibodies remains to be determined.

Glomerular basement membrane autoantibodies Figure 14.24: Autoantibodies against the noncollagenous portion (NC1) of the $\alpha3$ chain of type IV collagen present in both GBM and alveolar basement membranes are revealed by EIA in Goodpasture's syndrome. NC1 antibodies comprise 1% of the total IgM in Goodpasture's syndrome patients, and 90% of these autoantibodies are

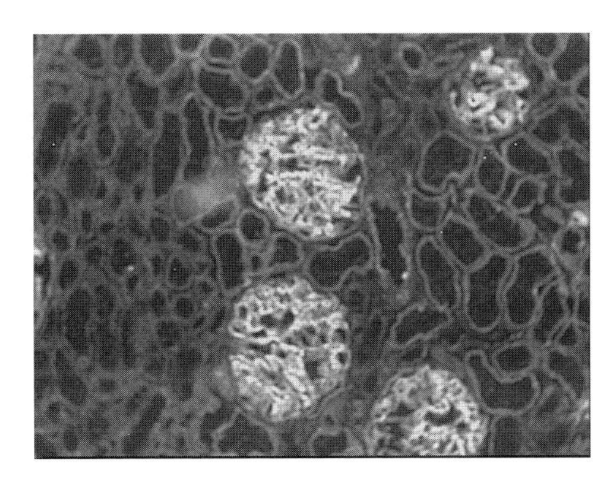

FIGURE 14.24 Glomerular basement membrane autoantibodies.

specific for α3 (IV)-chain. Antibodies to the other α (IV)-chains occur in 80% of the patients. Both C-X-C and C-C chemokines are associated with glomerular polymorphonuclear neutrophil and monocyte/macrophage infiltration in glomerular nephritis induced by GBM antibody. A total of 60% of human anti-GBM nephritis patients reveal antibody deposits along tubule basement membranes and manifest tubulointerstitial injury. Of Goodpasture's syndrome patients, 10 to 35% develop P-ANCA with myeloperoxidase antibody activity. ANCA and GBM antibodies may appear together in the sera of patients with RPGN associated with GBM disease and systemic vasculitis.

Nephritic factor: C3 nephritic factor (C3NeF) is a substance found in the serum of patients with type II membranoproliferative glomerulonephritis, i.e., dense deposit disease. This factor is able to activate the alternate complement pathway. It is an immunoglobulin that interacts with alternate complement pathway C3 convertase and stabilizes it. Thus, it activates the pathway and leads to the generation of complement fragments that are biologically active. C3NeF is an autoantibody against alternate pathway C3 convertase. Membranoproliferative glomerulonephritis patients have a genetic predisposition to developing the disease. The excessive C3 consumption leads to hypocomplementemia.

Nephritic factor autoantibodies: The most frequently encountered of the four types of autoantibodies reactive with the complement system in membranoproliferative glomerulonephritis (MPGN) are antibodies to a neoepitope on Bb moiety of C3bBb (properdin-independent C3NeF), the alternate pathway' amplification C3 convertase. These autoantibodies are designated C3NeF (or as Nfa, NFII, or C3bBb stabilizing factor). They belong to either thr IgG or IgM class and are found in Type II (dense deposit) MPGN. They have also been found in partial lipodystrophy (PLD), poststreptococcal glomerulonephritis, systemic lupus erythmatosus, idiopathic rapidly progressive glomerulonephritis, and rarely in MPGN Types I and III. MPGN patients' serum contain antiidiotype antibodies. C3NeF (Nfa) interaction with insomologous antigen induces prolonged *in vitro* C3bBb C3-cleaving activity half-life and decreased serum C3 concentrations. The presence of these antibodies is usual to distinguish types of MPGN and to monitor treatment.

Tubular basement membrane autoantibodies are specific for tubular basement membranes (TBM) and are detectable by immunofluorescence in some renal allotransplant patients, following certain types of drug therapy, and occasionally without a known cause. In human tubulointerstitial nephritis, TBM autoantibodies are generated that react with a major 58-kDa antigen (TIN-antigen) and minor 160-, 175-, and 300-kDa antigens that are related to laminin and entactin/nidogen. A total of 22% of patients with various types of interstitial nephritis manifest autoantibodies against TIN antigen. The normal urine may reveal human TBM antigens, which are capable of inducing tubulointerstitial nephritis in rats. TBM autoantibodies are only rarely detected in human cases of tubulointerstitial nephritis, but they are often prominent as GBM autoantibodies in Goodpasture's syndrome patients. TBM autoantibodies are of unknown clinical significance.

Mercury and immunity: Mercury does not have any known physiologic function in humans. This highly toxic metal affects enzyme function and calcium ion channels. It is especially toxic for the kidneys and central nervous system and may affect the reproductive system. It has been postulated to have a carcinogenic effect. Mercury is allergenic, affects immunoglobulin synthesis, and induces multispecific antibodies and autoimmune reactions. Mercuric chloride serves as a polyclonal activator of both B and T lymphocytes. Mercury's immunological effects are under genetic influence.

Alveolar basement membrane autoantibodies (ABM autoantibodies) are present in the blood sera and as linear deposits on ABM of patients with rapidly progressive glomerulonephritis (RPGN) and pulmonary hemorrhage (Goodpasture's syndrome). They are also present in the blood sera of patients with glomerular basement membrane (GBM) nephritis alone. Certain patients with GBM nephritis have pulmonary involvement, whereas others do not. This might be explained based on different reactivities with ABM. Linear immunoglobulin staining of ABM in ABM disease limited to the lungs is highly specific. Endothelial cell injury, such as that induced by infection, is considered significant in the pathogenesis of Goodpasture's hemorrhagic pneumonitis. Pulmonary hemorrhage has been associated with antineutrophil cytoplasmic antibodies (ANCA), which may sometimes be IgM isotype restricted. In the absence of GBM autoantibodies and ANCA, intra-alveolar hemorrhage may be associated with cardiolipin antibodies. Early diagnosis of patients with these types of clinical conditions is critical since mortality rates exceed 75%.

Goodpasture's antigen is found in the noncollagenous part of type IV collagen. It is present in human glomerular and alveolar basement membranes, making them a target for injury-inducing anti-GBM antibodies in blood sera of Goodpasture's syndrome patients. Interestingly, individuals with Alport's (hereditary) nephritis *do not* have the Goodpasture antigen in their basement membranes. Thus, renal transplants stimulate anti-GBM antibodies in Alport's patients.

Goodpasture's syndrome is a disease with pulmonary hemorrhage (with coughing up blood) and glomerulonephritis (with blood in the urine). It is induced by antiglomerular basement membrane autoantibodies that also interact with alveolar basement membrane antigens. A linear pattern of immunofluorescent staining confirms interaction of the IgG antibodies with basement membrane antigens in the kidney and lung, leading to membrane injury with pulmonary hemorrhage and acute (rapidly progressive or crescentic) proliferative glomerulonephritis. Pulmonary hemorrhage may precede hematuria. In addition to linear IgG, membranes may reveal linear staining for C3.

Lung autoantibodies have been described in the sera of some patients with farmer's lung. Cytotoxic autoantibodies against lung tissue have been reported in sarcoidosis and in extrinsic asthma but are not well defined. Antibasement membrane antibodies specific for lung and kidney antigens are recognized in Goodpasture's syndrome.

Adrenal autoantibody (AA) is found in two thirds of idiopathic (autoimmune) Addison disease patients (Figure 14.25). Although once the most common cause of Addison's disease, tuberculosis is now only infrequently associated with it. Adrenal autoantibodies react with the zona glomerulosa of the adrenal cortex. The autoantigens linked to this disease are cytochrome P450 enzyme family members that participate in steroidogenesis in the adrenal and other steroid-producing organs. The enzyme autoantigens include P450 steroid 21 hydroxylase (21 OH), P450 side-chain cleavage enzyme (P450 scc), and 17 α-hydroxylase (17-αRH). Autoantibodies in the adult autoimmune Addison's disease may be directed against cytochrome P450 steroid 21 OH, the only adrenal-specific enzyme. Other autoantibodies such as those against P450 scc present in the adrenals, gonads, and placenta, and 17-α-OH present in the adrenals and gonads, are found in autoimmune polyglandular syndromes (APGS). APGS type I autoimmune polyendocrinopathy–candidiasis–ectodermaldystrophy ({APECED}) begins in early childhood with mucocutaneous candidiasis followed by Addison's disease and often gonadal failure in hypoparathyroidism. APS type II (Schmidt syndrome) is mostly an adult disease inherited in an autosomal dominant manner and is marked by a variable combination of Addison's disease, insulin dependent type I diabetes mellitus (IDDM), and autoimmune thyroid disorders. It is liked to HLA-DR3. Steroid cell antibodies have been associated with autoimmune gonadal failure. They are present in 18% of patients with Addison's disease alone or in association with APGS. The most frequently encountered autoantigen in Addison's disease is the adrenal cytochrome P450 enzyme 21 OH. Autoantibodies to 17-α-OH are found in 55% of APTS type I patients and in 33% of APGS type II patients. Autoantibodies against P450 scc are found 45% of APGS type I patients and 42% of APGS type II patients.

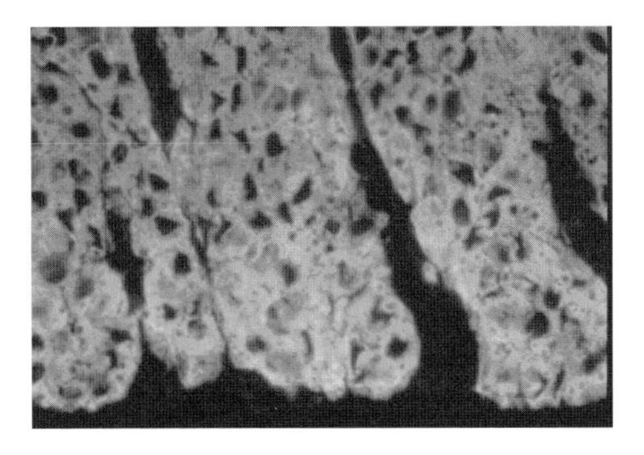

FIGURE 14.25 Adrenal autoantibody.

Autoimmune adrenal failure: Autoimmune destruction of the adrenal gland or idiopathic adrenal atrophy is the principal cause of primary adrenal failure in North America. The eponym for this condition is Addison's disease. It typically presents in females in their 20s to 30s. It may be part of a polyglandular endocrine disorder but in half of the cases it is an isolated autoimmune endocrine disorder. One study recorded 93 cases per 1,000,000 population with 40% manifesting autoimmune disease and 27% remaining unclassified. Autoimmune adrenal disease is believed to comprise approximately two thirds of cases of primary adrenal disease in North America. The pathologic findings in the adrenal gland consist of inflammation with lymphocyte and plasma cell infiltration. By the time of diagnosis, fibrosis is present. Germinal centers are rare and antibody binding to cortical cells can be detected by immunohistochemistry. In the advanced stages of the disease, the adrenal cortex is completely destroyed. Autoantibodies are present in 60 to 75% of patients with autoimmune adrenal disease and are more common with polyglandular syndrome. These antibodies are directed against specific cytochrome P450 enzymes involved in steroidogenesis. Antibodies to 17 α-hydroxylase (P450 C17), 21 α- hydroxylase (P450 C21), and the side-chain cleavage enzyme (P450 scc) have all been identified. The major symptom of acute insufficiency is postural hypotension.

Autoimmune polyglandular syndromes is a designation for Addison's disease and associated diseases based on classification into Types I, II, and III APSs. Type I APS patients have chronic mucocutaneous candidiasis, adrenal insufficiency, and hypoparathyrodism. It also may include gonadal failure, alopecia, malabsorption, pernicious anemia, and chronic active hepatitis. Candidiasis and malabsorption may also be present. Type II APS consists of three primary disorders: Addison's disease, insulin-dependent diabetes mellitus, and autoimmune thyroid disease. Type III APS consists of a subgroup of Type II without Addison's disease but who have thyroid autoimmunity and pernicious anemia as well as vitiligo.

Corticotrophin receptor autoantibodies (CRA) are adrenal antibodies that have a role in the pathogenesis of Addison's disease. Corticotrophin receptor antibodies (CRA) may block ACTH binding to specific receptors on cells of the adrenal cortex. Corticotrophin receptor antibodies of the stimulatory type may be found in Cushing's syndrome attributable to primary pigmented nodular adrenocortical disease. CRA have high sensitivity specificity and predictive value for idiopathic Addison's disease.

Polyendocrine deficiency syndrome (polyglandular autoimmune syndrome) refers to two related endocrinopathies with gonadal failure that may be due to defects in the hypothalamus. There is vitiligo and autoimmune adrenal insufficiency. Four fifths of the patients have autoantibodies. Type I occurs in late childhood and is characterized by hypoparathyroidism, alopecia, mucocutaneous candidiasis, malabsorption, pernicious anemia, and chronic active hepatitis. It has an autosomal recessive mode of inheritance. Type II occurs in adults with Addison's disease and autoimmune thyroiditis or insulin-dependent diabetes mellitus.

Steroid cell antibodies are IgG antibodies that interact with antigens in the cytoplasm of cells producing steroids in the ovary, testes, placenta, and adrenal cortex. Patients with Addison's disease with ovarian failure or hypoparathyroidism develop these antibodies which are rarely associated with primary ovarian failure in which organ-specific and nonorgan-specific autoantibodies are prominent.

Acanthosis nigricans is a condition in which the afflicted subject develops insulin receptor autoantibodies associated with insulin-resistant diabetes mellitus, as well as thickened and pigmented skin.

Antibodies to Mi-1 and Mi-2 have been found exclusively in dermatomyositis (DM) patients (15 to 35%). Anti-Mi-1 has been found in a small percentage of DM but also in 5% of patients with SLE, including 7% of those with anti-RNP and 9% of those with anti-Sm. It was also shown to bind bovine IgG and is not helpful in diagnosis.

GAD-65 is a major autoantigen in insulin-dependent diabetes mellitus. It is expressed primarily in human β cells.

ICA512 (IA-2) is a protein tyrosine phosphatase-like molecule that belongs to a group of transmembrane molecules. It was discovered by screening islet expression libraries with human insulin-dependent diabetes mellitus sera. IA-2 and phogrin are the two homologues. The IA-2 antigen is expressed in islets as well as in the brain.

Insulin-dependent (type I) diabetes mellitus (IDDM) refers to juvenile-onset diabetes caused by diminished capacity to produce insulin. Genetic factors play a major role, as the disease is more common in HLA-DR3 and HLA-DR4 positive individuals. There are significant autoimmune features that include IgG autoantibodies against glucose transport proteins and anticytoplasmic and antimembrane antibodies directed to antigens in the pancreatic islets of Langerhans. β Cells are destroyed, and the pancreatic islets become infiltrated by T lymphocytes and monocytes in the initial period of the disease. Experimental models of IDDM include the NOD mouse and the BB rat.

Insulin receptor autoantibodies induce insulin resistance (type B insulin resistance). This syndrome may be accompanied by acanthosis nigricans and autoimmune disease. Insulin receptor antibodies may lead to hypoglycemia. Selected patients with insulin-dependent diabetes mellitus and noninsulin-dependent diabetes mellitus have antibodies that react with the insulin receptor at its insulin-binding site or other binding sites. Insulin autoantibodies and islet cell antibodies in first degree relatives of patients with IDDM have a predictive of 60 to 77% for the development of IDDM within 5 to 10 years.

Insulin resistance is the diminished responsiveness to insulin as revealed by its decreased capacity to induce hypoglycemia, which may or may not have an immunologic basis. The administration of bovine or porcine insulin to humans can lead to antibody production that may contribute to insulin resistance.

Insulitis is the lymphocytic or other leukocytic infiltration of the islets of Langerhans in the pancreas, which may accompany the development of diabetes mellitus.

Islet cell autoantibodies (ICA) are autoantibodies specific for the β cells of the pancreatic islets (Figure 14.26). They are found in 80% of newly diagnosed insulin-dependent diabetes mellitus (IDDM) by indirect immunofluorescence on frozen sections of human pancreas. Individuals with high titers of persistent ICA are more likely to have IDDM than those with fluctuating ICA. Half of relatives

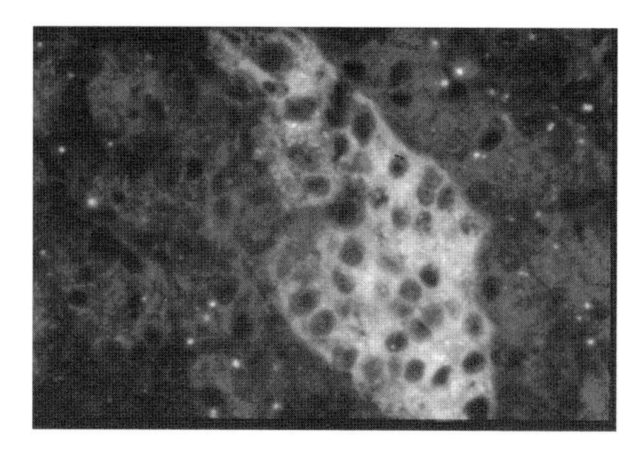

FIGURE 14.26 Islet cell autoantibodies (ICA).

with a single positive ICA assay and 60 to 80% of relatives with both ICA and insulin autoantibodies (IAA) will develop IDDM within 10 years. Strong persistently positive ICA (greater than 40 JDF-U) are the best predictors of subsequent onset of IDDM, especially if there is greatly diminished insulin secretion. ICA are found in 31% of women who develop gestational diabetes mellitus.

Juvenile onset diabetes is a synonym for type I insulin-dependent diabetes mellitus.

A **NOD (nonobese diabetic) mouse** is one of a mutant mouse strain that spontaneously develops type I, insulin-dependent diabetes mellitus, an autoimmune disease. There is an autosomal recessive pattern of inheritance for the NOD mutation. Lymphocytes infiltrate NOD mouse islets of Langerhans in the pancreas and kill β cells. There is a defect in the HLA-DQ part of the MHC class II region in humans with insulin-dependent diabetes mellitus and in the class II IA region of the mouse MHC class II. A major DNA segment is missing from the NOD mouse MHC IE region. When the IE segment is inserted or the IA defect is corrected in transgenic NOD mice, disease progression is halted.

A **NON mouse** is the normal control mouse for use in studies involving the NOD mouse strain that spontaneously develops type I (insulin-dependent) diabetes mellitus. The two strains differ only in genes associated with the development of diabetes.

Antiplacental alkaline phosphatase (PLAP) antibody is normally produced by syncytiotrophoblasts after the 12th week of pregnancy. Human placental alkaline phosphatase is a member of a family of membrane-bound alkaline phosphatase enzymes and isoenzymes. It is expressed by both malignant somatic and germ cell tumors. PLAP immunoreactivity can be used in conjunction with epithelial membrane antigen (EMA) and keratin to differentiate between germ cell and somatic tumor metastases. Germ cell tumors appear to be universally reactive for PLAP, whereas somatic tumors show only 15 to 20% reactivity.

More than 95% of female **D3TX mice** develop autoimmune oophoritis, whereas only 30% of D3TX male mice develop autoimmune orchitis. Following vasectomy of the D3TX mice, the orchitis increased to over 90%. Since vasectomy is associated with leakage of sperm antigens, the different incidence of orchitis and oophoritis in D3TX mice is believed to indicate a difference in the sequestration between testicular and ovarian antigens.

Endometrial antibodies are IgG autoantibodies present in two thirds of women with endometriosis. These antibodies react with the epithelial glandular portion but not the stromal component of endometrium. They have been suggested as a possible cause for infertility that occurs in approximately 30 to 40% of women with endometriosis.

Ovary autoantibodies (Figure 14.27): Autoantibodies to adrenal glands, ovary, placenta, and testes are known as steroid cell autoantibodies. Adrenal gland steroid autoantibodies commonly react with ovary and testes antigens. The majority of females with Addison's disease and amenorrhea with autoimmune gonadal failure develop serum steroid cell autoantibodies. These autoantibodies occur in 78% of premature ovarian failure and Addison's disease patients.

Ovary antibodies (OA) are present in 15 to 50% of premature ovarian failure patients in whom ovarian function failed after puberty but prior to 40 years of age. These individuals have elevated levels of gonadotropins in serum but diminished serum estradiol levels. Premature ovarian failure is associated with an increased incidence of autoimmune disease, especially of the thyroid. They also develop autoantibodies of both the organ specific and non-organ-specific types. These women are often HLA-DR3 positive and have elevated CD4/CD8 ratios. Ovarian antibodies are specific for steroid cells, which causes them to

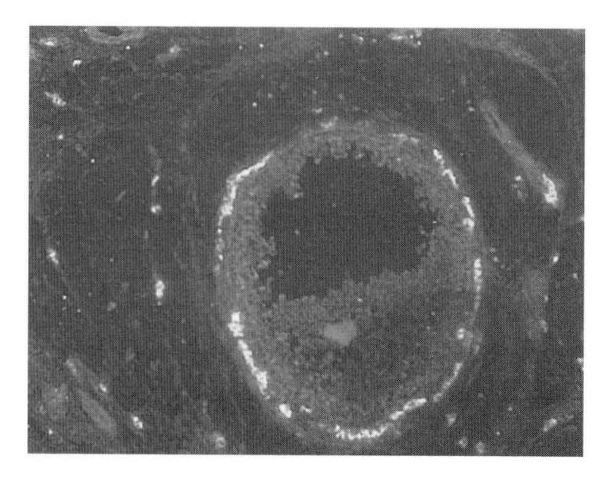

FIGURE 14.27 Ovary autoantibodies.

cross-react with steroid-synthesizing cells in the placenta, adrenals, and testes.

Endometrial autoantibodies are IgG autoantibodies present in 50 to 74% of endometriosis patients. They react with the epithelial (glandular) component and not the stromal component of normal endometrium regardless of the menstrual cycle phase. These autoantibodies have teen suggested as a possible cause of infertility observed in 30 to 40% of endometriosis patients. The blood sera of endometriosis patients react with both normal and endometriotic tissue. Antibody titer is not correlated with stage of endometriosis or the menstrual cycle phase. Endometriosis patients often manifest lupus anticoagulant (45%) antinuclear antigen reactivity (10 to 25%) and elevated IgG concentrations (95%). They also form antibodies to phospholipids and histones.

Testicular autoimmunity: Natural animal models of infertility include beagles and mink. A colony of mink has been developed with approximately 20 to 30% infertile males. They develop circulating antibodies to acrosomal antigens of spermatozoa that may be detected in their blood sera. Heavy granular deposits of IgG and/or C3 have been found in the basal lamina of the seminiferous tubules in 71% of mink with late infertility. Immunoglobulins eluted from these deposits contained antibodies to the acrosome of mink spermatozoa. Spermatozoal antigens were not detected in the immune deposits.

Vasectomy causes autoimmunity to spermatozoa antigen in various animal species that include monkeys, guinea pigs, rabbits, and mice. Autoimmune responses to spermatozoa in vasectomized men led to concerns about the safety of vasectomy. Cohort investigations of vasectomized men have established that vasectomy is a safe contraceptive procedure. Even though the mechanisms of vasectomy-induced autoimmunity remain elusive, animal studies have revealed that autosensitization to spermatozoal autoantigens can occur without the use of adjuvants or other agents or processes that are designated as "dangerous" agents.

Immunological infertility: Infertility in 12 to 25% of cases in which infertile couples experience infertility even though they manifest no significant abnormalities upon physical examination. These cases of unexplained infertility may be caused by autoimmune responses to organ-specific antigens of the reproductive tract of both males and females and isoimmune reactions of females against semen components. Immune responses may cause or contribute to infertility in approximately 10% of these couples.

Sperm antibodies are antibodies specific for the head or tail of sperm. These antibodies are synthesized by 3% of infertile males and 2 to 9% of infertile females. There is a positive correlation between the titer of antibody against sperm and the couple's infertility. Treatment includes corticosteroid therapy or use of a condom to permit waning of immunologic memory in the female or washing of sperm prior to insemination.

Sperm autoantibodies can be formed in either the male or the female and play a role in infertility. Not all antibodies against spermatozoa interfere with sperm function and fertilization. Antibodies in cervical mucous or on sperm are associated with a reduction in cervical mucous penetration. Vasectomy causes the production of sperm autoantibodies. In vasectomized men, the presence of both IgA sperm antibodies on all sperm and a heightened immune response (antibody titer greater than 1:256) is associated with conception failure. Assays for circulating sperm antibodies vary in sensitivity and specificity. There are many variables to be considered such as class of immunoglobulin, different sperm antigen, etc. Of the various methods to measure sperm antibodies, the direct immunobead test (dIBT) is the method of choice. This is based on rosette formation between viable sperm and plastic beads coated with antiserum to human immunoglobulins. It allows measurement of class-specific antibodies (IgG, IgM, or IgA) as well as the antibody attachment site, i.e., head or tail. Antibodies can be measured in cervical mucus, seminal plasma, or serum, as well as an antibody complex on the surface of donor sperm. The measurement of sperm antibodies on sperm is the preferred technique. Three major sperm surface glycoproteins have been identified. Sperm antibodies include those reactive with galactosyl transferase.

Acetylcholine receptor (AChR) antibodies are IgG autoantibodies that cause loss of function of AChRs that are critical to chemical transmission of the nerve impulse at the neuromuscular junction. This represents a type II mechanism of hypersensitivity, according to the Coombs and Gell classification. AChR antibodies are heterogeneous with some showing specificity for antigenic determinants other than those that serve as acetylcholine or α-bungarotoxin binding sites. As many as 85 to 95% of myasthenia gravis patients may manifest AChR antibodies.

Acetylcholine receptor (AChR) binding autoantibodies are those that react with several epitopes other than the binding site for ACh or α-bungarotoxin. They are found in 88% of patients with generalized myasthenia gravis (MG), 70% of ocular myasthenia, and in 80% of MG in remission. They decrease in titer as weakness improves with immunosuppressive therapy. AChR blocking autoantibodies are those that react with the AChR binding site. They are found in 50% of MG patients, in 30% with ocular MG, and in 20% of MG in remission. AChR modulating autoantibodies are those that crosslink AchRs and induce their removal from muscle membrane surfaces. They are present in more than 90% of MG. AChR autoantibodies of one or more types are present in at least 80% of ocular MG.

Bungarotoxin is an anticholinergic neurotoxin extracted from the venom of Australian snakes belonging to the genus *Bungarus*. It binds to acetylcholine receptors of the nicotinic type and inhibits depolarization of the neuromuscular junction's postsynaptic membrane, leading to muscular weakness.

EAMG (experimental autoimmune myasthenia gravis) can be induced in more than one species of animals by immunizing them with purified AchR from the electric ray (*Torpedo californica*). The autoantigen, nicotinic AchR, is T cell-dependent. The *in vivo* synthesis of anti-AchR antibodies requires helper T cell activity. Antibodies specific for the nicotinic acetylcholine receptors (AchR) of skeletal muscle react with the postsynaptic membrane at the neuromuscular junction.

Neuromuscular junction autoimmunity: The three autoantibody-mediated disorders of the neuromuscular junction include: (1) myasthenia gravis (MG) in which antibodies lead to loss of the muscle acetylcholine receptor (AChR). This disease satisfies all Witebsky's criteria for antibody-mediated autoimmune disease, listed elsewhere in this volume. (2) The Lambert-Eaton myasthenia syndrome (LEMS) is an antibody-mediated presynaptic disease in which the target is the nerve terminal voltage-gated calcium channel. It meets two of the Witebsky's criteria. (3) Acquired Neuromyotonia (Isaacs' syndrome), which is induced by autoantibodies against voltage-gated potassium channels in peripheral motor nerves. This leads to increased excitability and continuous muscle fiber activity. This disorder satisfies two of Witebsky's criteria.

Stiff man syndrome (SMS) refers to a progressive involuntary axial and lower limb rigidity which disappears during sleep. It results from an imbalance between descending inhibitory gamma amino butyric acid (GABA) ergic and excitatory aminergic pathways. 60% of SMS patients have circulating antibody against glutamic acid decarboxylase (GAD), an enzyme that has 65-kDa and 67-kDa isoforms. GAD is a major antigenic target in insulin-dependent diabetes mellitus (IDDM). Antibody levels to GAD are 100- to 500-fold higher in SMS than in IDDM. SMS anti-GAD bind to both linear epitopes and epitopes that depend on protein conformation. SMS is associated with other autoimmune diseases that include myasthenia gravis, pernicious anemia, vitiligo, autoimmune adrenal failure, ovarian failure, thyroid disease, and IDDM.

Striational antibodies are demonstrable in 80 to 100% of MG patients with thymoma. They are not present in 82 to 100% of MG patients who do not have thymoma. If striational antibodies are not demonstrable in an MG patient, the individual probably does not have a thymoma. One quarter of rheumatoid arthritis patients receiving penicillamine therapy develop IgM striational antibodies. Patients receiving immunosuppressive therapy may also be monitored for striational antibodies to signify the development of autoimmune reactions following bone marrow transplantation.

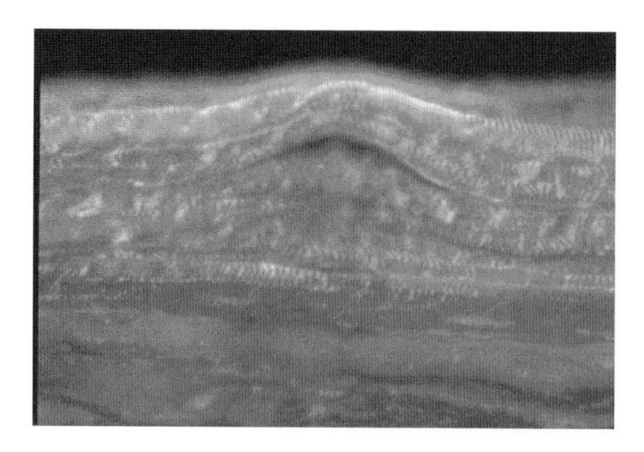

FIGURE 14.28 Striational autoantibodies.

Striational autoantibodies (StrAb) are associated with the presence of thymoma in patients with MG. Thymoma occurs in 10% of MG patients and MG occurs in 15 to 80% of thymoma patients. Striational antibodies react with proteins in the contractile constituents of skeletal muscle. The relevant epitopes remain to be demonstrated. Autoantigens reactive with striational autoantibodies include actin, α-actinin, myosin, and titin (connectin). The ryanodine receptor (sarcoplasmic reticulum calcium release channel protein). Striational autoantibodies can be detected by immunofluorescence using cryostat sections of skeletal muscle, which is the method of choice for striational autoantibody screening. These autoantibodies are most often found in MG patients over 60 years of age. They are present in 80 to 90% of patients with both MG and thymoma, in approximately 30% of patients with acquired (adult onset) MG, and 24% of thymoma patients without clinical signs of MG. Striational autoantibodies are absent in more than 70% of MG patients without thymoma. Thus, they are very sensitive and specific for thymoma in MG patients and their absence rules out a diagnosis of thymoma in MG.

Voltage-gated-calcium channel autoantibodies: Autoantibodies specific for voltage-gated calcium channels (VGCCs) located on the presynaptic nerve terminal are generated during an autoimmune disease associated with small-cell lung cancer (SCLC), known as Lambert-Eaton myasthenic syndrome. Although the exact role of N-type VGCC autoantibodies remains to be determined, they may contribute to disturbance of autonomic neurotransmitter release associated with LEMS.

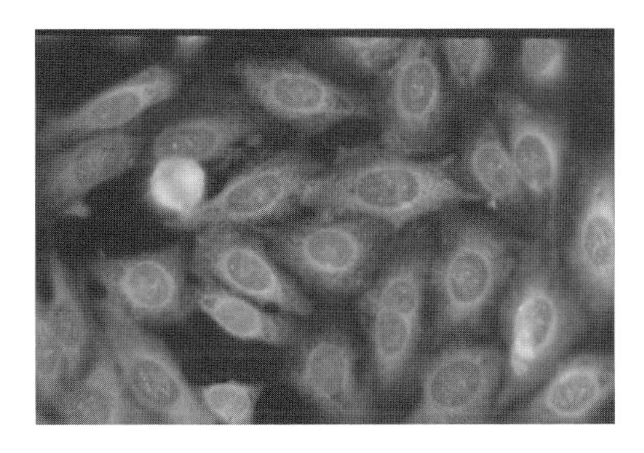

FIGURE 14.29 Alanyl-tRNA synthetase autoantibodies.

Alanyl-tRNA synthetase autoantibodies (Figure 14.29) against the synthetases (histidyl-, threonyl-, alanyl-, isoleucyl- and glycyl-tRNA synthetases) have been detected in some patients with polymyositis (PM) and dermatomyositis (DM). Histidyl-tRNA synthetase (Jo-1) antibodies are the most common (20 to 30% of the patients with PM/DM).

Amino acyl tRNA synthetases refer to myositis-specific autoantibodies that include antibodies against histidyl tRNA synthetase (HRS), which was first known as Jo-1. They are found in 20% of patients with polymyositis mainly or in dermatomyositis. Autoantibodies to analyl (PL-12), threonyl (PL-7), glycyl (EJ), and isoleucyl (OJ) tRNA synthetases have also been reported. These aminoacyl synthetases bind their corresponding amino acids to the tails of their respective tRNAs. Essentially all Jo-1 positive sera possess antibodies that react with the amino terminal (amino acids 1 to 44) of the protein but most also have antibodies against epitopes positioned further toward the carboxyl end of the sequence. Anti-HRS antibodies are principally IgG_1 that persist throughout the course of the disease. Patients with autoantibodies against tRNA synthetases have several clinical features in common that include interstitial lung disease and pulmonary fibrosis (50 to 100%), arthritis (60 to 100%), Raynaud's phenomenon (60 to 93%), fever, and "mechanic's hands." In patients with antisynthetase syndrome, anti-Jo-1 antibodies are most common in polymyositis, whereas autoantibodies against other aminoacyl-tRNA synthetases are more often found in dermatomyositis. Antibodies to HRS show a strong association with HLA-DR3.

Anti-Ku autoantibodies are myositis-associated autoantibodies that occur in a few (5 to 12%) Japanese patients with overlap syndrome. These antibodies are specific for 70-kDa anti 80-kDa DNA-binding proteins that represent the Ku antigen. These antibodies have also been found in patients with systemic lupus erythematosus, pure dermatomyositis, pure scleroderma, thyroid disease, or Sjögren's syndrome.

Ku autoantibodies against the Ku epitope reveal a strong association with systemic autoimmunity in Japanese patients in contrast to SLE and overlap sydromes in Americans. These antibodies are also present in selected patients with mixed connective tissue disease, scleroderma, polymyositis, Graves' disease, and primary pulmonary hypertension, making them nondiagnostic for any specific autoimmune disease. The Ku autoantigen is a heterodimeric nucleolar protein comprised of 70- and 80- to 86-kDa subunits which, in association with a 350-kDa catalytic subunit, is the DNA-binding component of a DNA-dependent protein kinase involved in double-stranded DNA repair and V(D)J recombination. Ku autoantibodies are of unknown pathogenicity and clinical significance.

Anti-PM/Scl autoantibodies (Figure 14.30) are myositis-associated autoantibodies specific for the nucleolar antigen Pm-Scl, which are found among many patients with an overlap syndrome with features such as scleroderma and polymyositis/dermatomyositis (25%). This autoantibody is found in a few patients with pure polymyositis or scleroderma. It is specific for a complex of nucleolar proteins, the principal antigen having a mass of 100 kDa, and is found mainly in Caucasian patients with the overlap syndrome.

Autoimmune cardiac disease: Rheumatic fever is a classic example of microbial-induced autoimmune heart disease. The immune response against the M protein of group A streptococci crossreact with cardiac proteins such as tropomyosin and myosin. The M protein contains numerous epitopes that participate in these cross-reactions. A second cross-reactive protein in the streptococcal membrane has been purified to a series of four peptides ranging

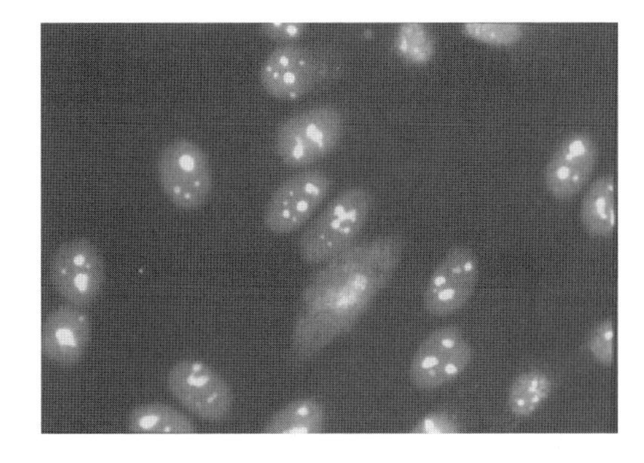

FIGURE 14.30 Anti PM/Scl autoantibodies.

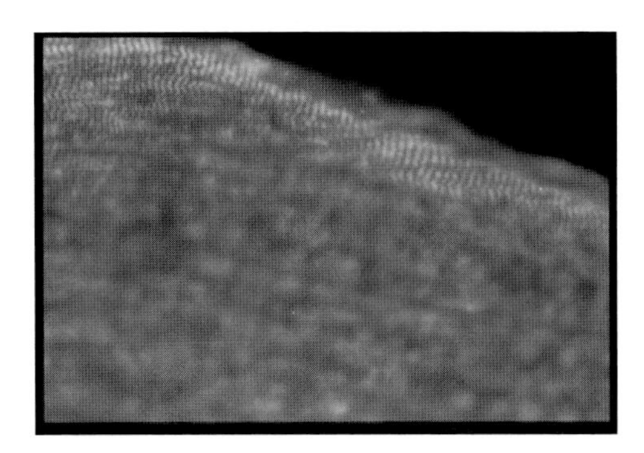

FIGURE 14.31 Autoimmune cardiac disease.

in molecular weight from 22 to 23 kDa. Patients with rheumatic fever may develop antibody that binds to the cytoplasm of cells of the caudate nuclei with specificity for its cells. Autoantibodies against microorganisms cross-react with cardiac tissue. A monoclonal antibody termed D8/17 identifies all rheumatics. This antibody is not related to the MHC system and serves as a B cell marker associated with rheumatic fever, although no specific role for the antigen in the disease has been demonstrated. Numerous microbes and viruses can induce acute myocarditis. These cases are characterized by the presence of lymphocytic infiltrates and increased titers of heart-reactive antibodies. Important causative agents include group B coxsackieviruses that cause acute cardiac inflammation in humans. Rose et al., using an experimental mouse model, showed that only those mice with heart-reactive antibodies in their sera went on to develop chronic cardiomyopathy with antibodies primarily against the cardiac isoform of myosin in their model of acute myocarditis. Postpericardiotomy syndrome occurs in both adults and children 10 to 14 days after surgery and is characterized by fever, chest pain, and pericardial and pleural effusions. This condition is associated with the presence of high-titer, heart-reactive antibodies in the blood sera. The heart-specific antibodies are believed to play a role in the disease pathogenesis. Since many microbes share epitopes with human tissues, cross-reactions between antibodies against the microbe and human tissues may be harmless or they may lead to serious autoimmune consequences in genetically susceptible hosts. As much attention has been given to antibodies, cell-mediated immunity may play a larger role than previously thought in both rheumatic fever and Chagas' disease, in which T cells are specifically cytotoxic for the target organ. Cytotoxic T cells specific for cardiac myofibers appear in both rheumatic fever and Chagas' disease. Only selected individuals with cardiomyopathy develop progressive autoimmune disease after active infection.

Multicatalytic proteinase autoantibodies: A total of 35% of systemic lupus erythematosus patients manifest autoantibodies to multicatalytic proteinases (proteasomes), which are macromolecular structures comprised of at least 14 subunits that are involved in the intracellular degradation of proteins. They are not usually found in myositis, scleroderma, or Sjögren's syndrome.

Myocardial autoantibodies (MyA) have been demonstrated in two thirds of coronary artery bypass patients and in Dressler syndrome, but this is not necessarily related to postcardiotomy syndrome. A majority of acute rheumatic fever patients manifest myocardial antibodies reactive with sarcolemmal, myofibrillar, or intermyofibrillar targets. Dilated cardiomyopathy patients and patients with systemic hypertension/autoimmune polyendocrinopathy may develop autoantibodies against myocardium. MyA with sarcolemmal, intermyofibrillar patterns are demonstrable in most acute rheumatic fever patients. Their molecular mimicry is believed to have a role in pathogenesis. Idiopathic dilated cardiomyopathy patients'sera react with the adenine nuclear translocator protein, a mitochondrial branched chain α-ketoacid dehyrogenase, cardiac β adrenoreceptor protein, and heat shock protein (HSP)-60. The autoantigen in Dressler syndrome has not yet been identified.

Postcardiotomy syndrome is a condition that follows heart surgery or traumatic injury. Autoantibodies against heart antigens may be demonstrated by immunofluorescence within weeks of the surgery or trauma. Corticosteroid therapy represents an effective treatment. Myocardial infarct patients may also develop a similar condition.

Threonyl-transfer RNA synthetase autoantibodies: (Figure 14.32) Autoantibodies (PL-7) to the 80-kDa threonyl-tRNA synthetase protein are highly specific for myositis. PL-7 autoantibodies have been found together with autoantibodies against aminoacyl tRNA synthetases, including histidyl, alanyl, isoleucyl, and glycyl tRNA synthetases, associated with the antisynthetase syndrome which is characterized by acute-onset steroid responsive

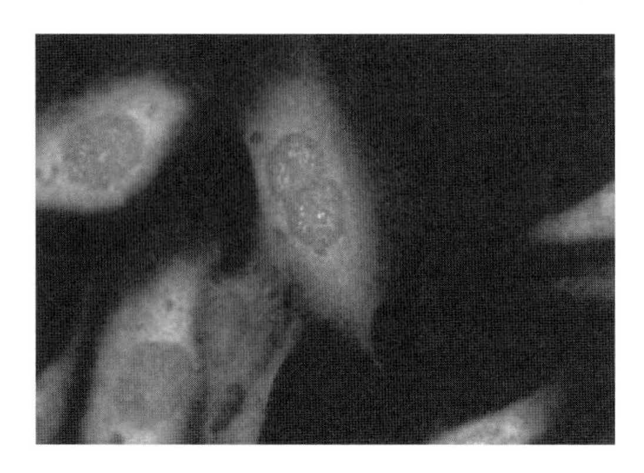

FIGURE 14.32 Threonyl-transfer RNA synthetase autoantibodies.

myositis with interstitial lung disease, fever, symmetrical arthritis, Raynaud phenomenon, and mechanic's hands.

U2 snRNP autoantibodies are antibodies against U2 snRNP (small nuclear ribonucleoparticles) comprised of U2 snRNA, and eight associated polypeptides. A′ and B′ are unique to U2 snRNP and six (B′, B, D, E, F, and G) are shared with U1 and other snRNP. Anti-U2 sera that react with β polypeptide are often present in patients with overlap syndromes with myositis and may be associated with antibodies against U1 snRNP polypeptides (70 kDa, A and C). These patients may also have antibodies against U1 snRNP polypeptide (70 kDa, A and C). U1 snRNP antibodies that interact with 70-kDa polypeptide were previously designated RNP or nRNP and are a principal feature of mixed connective tissue disease.

Amphiphysin autoantibodies: Autoantibodies against amphiphysin, which is one of two known target autoantigens in stiff man syndrome (SMS), a central nervous system disease characterized by progressive muscle rigidity and painful spasms. Amphiphysin is a 125- to 128-kDa cytoplasmic synaptic vesicle-associated protein expressed in neurons, certain types of endocrine cells, and spermatocytes. Its biological function may involve synaptic vesicle endocytosis. Autoantibodies to glutamic acid decarboxylase (GAD) are detected in 60% of SMS. Amphiphysin autoantibodies are present in SMS patients negative for GAD autoantibody, all of whom are female breast cancer patients. The amphiphysin autoepitope is present in the C-terminal region of the protein. Small-cell lung carcinoma patients who present with paraneoplastic encephalomyelitis have been reported to manifest IgG autoantibodies against amphiphysin.

Anti-Hu antibodies are autoantibodies associated with paraneoplastic encephalomyeloneuronitis (PEMN). They react with neuronal nuclei, recognizing 35- to 40-kDa antigens present in neuronal nuclei and also detectable in the nuclei of cells from small-cell carcinoma of the lung. The antibody may be responsible for death of neurons.

Carcinomatous neuropathy refers to neurological findings in tumor-bearing patients who have no nervous system metastasis. Sensory carcinomatous neuropathy, in which patients develop autoantibodies specific for neurone cytoplasm RNA-protein, is an example of a carcinomatous neuropathy.

Diabetes insipidus is a chronic idiopathic disease due to deficient secretion of vasopressin from the posterior lobe of the pituitary gland. Most cases of diabetes insipidus have a genetic basis or are secondary to recognized conditions that injure or destroy the hypothalamic neurohypophyseal complex, but there are cases in which an autoimmune response may be the cause. Antibodies have been demonstrated against cells that secrete vasopressin. The biopsy samples from rare cases have been consistent with autoimmune injury as reflected by a lymphocyte-plasma cell infiltrate in the neurohypophysis. Most serological studies for antibodies to vasopressin secreting cells have been reported.

Dopamine neuron autoantibodies against dopamine neurons in the substantia nigra are found in approximately 78% of CSF from Parkinson disease patients and in 3% of normal CSF from subjects with other neurological diseases. Transplantation of adrenal medulla leads to disappearance of the autoantibodies.

EAE is an abbreviation for experimental allergic encephalomyelitis.

Encephalitogenic factors are myelin basic protein or related molecules found in the brain that can induce experimental allergic encephalomyelitis if administered to experimental animals together with Freund's complete adjuvant. The smallest constituent of myelin that is capable of inducing experimental encephalomyelitis is a nonapeptide (Phe-Leu-Trp-Ala-Glu-Gly-Gln-Lys).

Enteric neuronal autoantibodies are IgG autoantibodies that react strongly with nuclei and weakly with cytoplasm of enteric nuclei of myenteric and submucosal plexuses. They develop in selected patients with the paraneoplastic syndrome marked by intestinal pseudo-obstruction in association with small-cell lung carcinoma. These autoantibodies are nonreactive with nuclei of non-neuronal tissue. Myenteric neuronal autoantibodies are specific for enteric neuronal neurofilaments and can be detected in scleroderma and visceral neuropathy patients.

Ergotype refers to a T lymphocyte being activated. The injection of antiergotype T cells blocks full-scale activation of T lymphocytes and may prevent development of experimental autoimmune disease in animal models. An example is experimental allergic encephalomyelitis (EAE), in which antiergotype T lymphocytes may prevent full T-lymphocyte activation.

Ganglioside autoantibodies are autoantibodies to the monosialogangliosides GM1, GM2, and GM3, and are present in Guillain–Barré syndrome (GBS), multiple sclerosis (MS), and rheumatic disorders with neurologic involvement, motor neuron diseases (MNDs), and various neoplasms. These autoantibodies cross-react with one or more polysialogangliosides that include GD1a, GD1b, GD2, GD3, GT1a, GT1b, and GQ1b. Cross- or polyreactivity patterns lend some clinical specificity to the anti-ganglioside response in GBS, Miller Fisher syndrome (MFS), MNDs, and other neurological diseases that include schizophrenia, AIDS dementia complex, and neuropsychiatric SLE.

Glutamic acid decarboxylase autoantibodies are associated with a rare neurological disorder termed stiff man syndrome (SMS) and with IDDM. Autoantibodies in the sera of SMS patients react with denatured GAD on immunoblots. By contrast, antibodies in IDDM patients react only with native (undenatured) GAD by immunoprecipitation. In SMS, the frequency of GAD autoantibodies ranges from 60 to 100%; in IDDM the frequency ranges from 25 to 79%. GAD autoantibodies are almost always present with islet cell antibodies (ICA) and may represent a constituent of ICA.

Guillain-Barré syndrome is a type of idiopathic polyneuritis in which autoimmunity to peripheral nerve myelin leads to a condition characterized by chronic demyelination of the spinal cord and peripheral nerves.

Marek's disease is a lymphoproliferative disease of chickens induced by a herpes virus. Demyelination may occur as a consequence of autoimmune lymphocyte reactivity.

MBP is an abbreviation for myelin basic protein or major basic protein.

Myelin autoantibodies: The detection of autoantibodies to myelin used previously to screen for autoantibodies in idiopathic and paraproteinemic neuropathy patients leads to inconsistent results due to imprecise myelin preparations, leading to poor clinical specificity. Autoantibodies against myelin in Guillain-Barré (GBS) and sensory polyneuropathy with monoclonal gammopathy of unknown significance involve autoantibodies against glycolipids, including sulfatide, Forssmann antigen, galactosyl-cerebroside (Gal-Cer) and gangliosides.

Myelin-associated glycoprotein (MAG) autoantibodies are specific for the glycoprotein constituent of myelin that are found in the periaxonal region, Schmidt-Lanterman incisures, lateral loops, and outer mesaxon of the myelin sheath, belong to the immunoglobulin superfamily, and act as adhesion molecules that facilitate myelination. MAG autoantibodies recognize the N-linked L1 and J1 carbohydrate moeity, which is also found on PO and P2 glycoproteins, sulfate-3-glucuronyl paragloboside (SGPG), and sulfate-3-glucuronyl lactosaminyl paragloboside (SGLPG) glycolipids of peripheral nerves, neural cell adhesion molecule and Li and J1 of lymphocytes and human natural killer cells. MAG autoantibodies are present in half of polyneuropathy patients with monoclonal gammopathy, which may be associated with other lymphoid proliferative diseases. T lymphocyte responses to MAG have been found in multiple sclerosis.

Myelin basic protein (MBP) is a principal constituent of the lipoprotein myelin that first appears during late embroygenesis. It is a 19-kDa protein that is increased in multiple sclerosis patients who may generate T lymphocyte reactivity against MBP. T lymphocytes with the V β-17 variant of the T cell receptor are especially prone to react to MBP.

Myelin basic protein (MBP) antibodies are antibodies against myelin proteins and have been investigated for their possible role in the demyelination that accompanies multiple sclerosis, acute idiopathic optic neuritis, Guillain-Barré syndrome, chronic relapsing polyradiculoneuritis, carcinomatous polyneuropathy, and subacute sclerosing panencephalitis. Antibodies against myelin basic protein are well known to play a role in experimental allergic encephalomyelitis in laboratory animals, but the role they play in patients with multiple sclerosis or other neurological diseases has yet to be established. MBP autoantibody titers are closely correlated with exacerbations and relapses in MS and are present in 75% of MS patients and 89% of optic neuritis subjects. They have also been detected in 64% of HIV-associated neurological syndrome, in 58% of autistic children, in 30% of Japanese encephalitis patients, and 30% of subacute sclerosing panencephalitis (SSPE) patients. Humoral as well as cellular immune responses participate in inflammatory demyelination.

Neurological autoimmune diseases are disorders that can alter function of the nerve–muscle junction or the peripheral nerves or the central nervous system. Nerve–muscle junction disorders include myasthenia gravis and Lambert-Eaton syndrome. Guillain-Barré syndrome is an example of the peripheral nerve disorder. Multiple sclerosis and post infectious vaccination encephalomyelitis are examples of central nervous system disorders.

Neuronal autoantibodies are present in the cerebrospinal fluid of approximately three fourths of systemic lupus erythematosus (SLE) patients with neuropsychiatric manifestations. 11% of SLE patients who do not manifest neuropsychiatric disease also develop them. Neuronal antibodies are identified by their interaction with human neuroblastoma cell surface antigens. The presence of neuronal antibodies generally indicates CNS-SLE. The lack of neuronal antibodies in the serum and cerebrospinal fluid mitigates against CNS involvement in SLE patients. Neuronal autoantibodies may be present in Raynaud's phenomenon and antiphospholipid syndromes. They may be detected by flow cytometry, but their clinical usefulness remains to be determined.

Paraneoplastic autoantibodies cross-react with tumor and normal tissue in the same patient. Examples are Yo antibodies against cerebellar Purkinje cells that occur in paraneoplastic cerebellar degeneration, neuronal nuclear (Hu) antibodies in paraneoplastic subacute sensory neuronapathy and sensory neuropathies, antikeratinocyte polypeptides in paraneoplastic pemphigus, antibodies against voltage-gated calcium channels in Lambert-Eaton syndrome, antibodies against retina in retinopathy associated with cancer, and antibodies against myenteric and submucosal plexuses in pseudo-obstruction of the intestine.

Postinfectious encephalomyelitis is a demyelinating disease following a virus infection that is mediated by autoimmune delayed-type (type IV) hypersensitivity to myelin.

Proteolipid protein autoantibodies: Proteolipid protein (PLP) is a principal structural protein of the myelin sheath of the mammalian central nervous system and plays a critical role in myelination. It has been used to induce experimental allergic encephalomyelitis (EAE) in rabbits, producing legions that resemble those formed in other species following immunization with whole CNS tissue or myelin basic protein. The PLP antibody response in rabbits correlates with the clinical and histopathologic severity of exerimental and allergic enchephalomyelitis. PLP autoantibodies are correlated to a specific subset of multiple sclerosis patients.

Pituitary autoantibodies are autoantibodies most often found in the empty sella syndrome. They are less often found in the blood sera of patients with pituitary adenomas, prolactinomas, acromegaly, idiopathic diabetes insipidus, Hashimoto thyroiditis, Graves' disease, and POEMS (polyneuropathy, organomegaly, endocrinopathy, M protein, and skin changes) syndrome. They are also associated with IDDM and ACTH deficiency. They may portend an unfavorable consequence of pituitary microsurgery for Cushing's disease. A total of 48% of children with cryptorchidism may develop autoantibodies against FSH-secreting and LH-secreting pituitary cells. Autoantibodies against pituitary hormones have been discovered in 45% of patients with either pituitary tumor or empty sella syndrome but not in normal individuals. A full 100% develop ACTH autoantibodies, 20% form TSH and GH autoantibodies. Empty sella syndrome may be induced by hypophysitis secondary to pituitary autoantibodies in adults. Autoantibodies against prolactin develop in 16% of patients with idiopathic hyperprolactinemia. Patients with these autoantibodies do not usually have clinical symptoms of hyperprolactinemia.

Paraneoplastic pemphigus is a rare autoimmune condition that may occasionally be seen in lymphoproliferative disorders. It is caused by autoantibodies against desmoplakin I and bullous pemphigoid antigen, as well as other epithelial antigens. Clinically, there is erosion of the oropharynx and vermilion border, as well as pseudomembranous conjunctivitis and erythema of the upper trunk skin.

Pemphigoid is a blistering disease of the skin in which bullae form at the dermal–epidermal junction, in contrast to the intraepidermal bullae of pemphigus vulgaris.

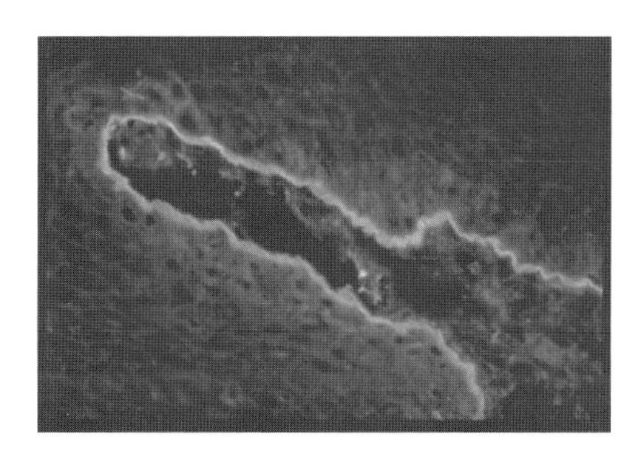

FIGURE 14.33 Pemphigoid-autoantibody to dermal basement membrane.

Autoantibodies develop against the dermal basement membrane (Figure 14.33). By using fluorochrome-labeled goat or rabbit anti-human IgG, linear fluorescence can be demonstrated at the base of the subepidermal bullae by immunofluorescence microscopy. Dermal basement membrane IgG autoantibodies can also be demonstrated in the patient's serum. This disease occurs principally in elderly individuals. C3 linear fluorescence at the dermal–epidermal junction is often demonstrable as well.

Pemphigus erythematosus (Senear-Usher syndrome) is a clinical condition with immunopathologic characteristics of both pemphigus and lupus erythematosus. Skin lesions may be on the seborrheic regions of the head and upper trunk, as seen in *pemphigus foliaceous*. However, immune deposits are also demonstrable at the dermal–epidermal junction and in skin biopsy specimens obtained from areas exposed to sunlight, reminiscent of lupus erythematosus. Light microscopic examination may reveal an intraepidermal bulla of the type seen in pemphigus foliaceous. Facial skin lesions may even include the "butterfly rash" seen in lupus. Immunofluorescence staining may reveal intercellular IgG and C3 in a "chickenwire" pattern in the epidermis with concomitant granular immune deposits containing immunoglobulins and complement at the dermal–epidermal junction. The serum may reveal both antinuclear antibodies and pemphigus antibodies. Pemphigus erythematosus has been reported in patients with neoplasms of internal organs, and in drug addicts, among other conditions. Indirect immunofluorescence using serum with both antibodies may reveal simultaneous staining for intercellular antibodies and peripheral (rim) nuclear fluorescence in the same specimen of monkey esophagus used as a substrate.

Pemphigus foliaceus is a type of pemphigus characterized by subcorneal blisters and anti-dsg1 autoantibodies. Patients develop fragile blisters that rupture early, leaving areas of denuded skin. One form of the disease affects individuals of all ethnic backgrounds whereas a second is

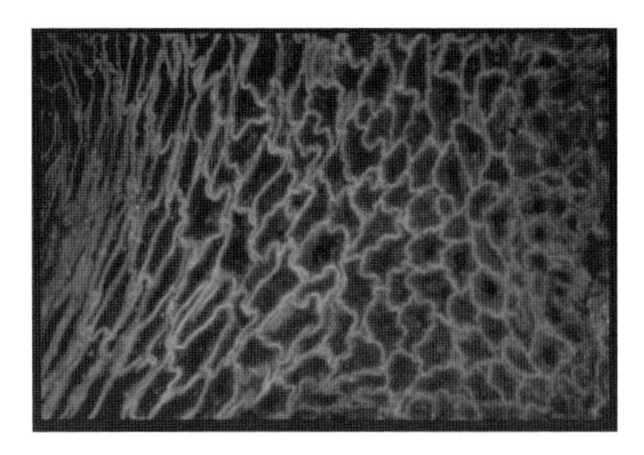

FIGURE 14.34 Pemphigus (esophagus substrate) — autoantibody to intercellular substance.

endemic to certain regions of Brazil and is known as fogo selvagem. This disease rarely involves mucosal surfaces. Histologic studies reveal acantholysis in the subcorneal layers of the epidermis.

Pemphigus vulgaris is a blistering lesion of the skin and mucous membranes (Figure 14.34). The bullae develop on normal-appearing skin and rupture easily. The blisters are prominent on both the oral mucosa and anal/genital mucous membranes. The disorder may have an insidious onset appearing in middle-aged individuals, and tends to be chronic. It may be associated with autoimmune diseases, thymoma, and myasthenia gravis. Certain drugs may induce a pemphigus-like condition. By light microscopy, intraepidermal bullae are present. There is suprabasal epidermal acantholysis with only mild inflammatory reactivity in early pemphigus. Suprabasal unilocular bullae develop, and there are autoantibodies to intercellular substance with activation of classic pathway-mediated immunologic injury. Acantholysis results as the epidermal cells become disengaged from one another as the bulla develops. Epidermal proteases activated by autoantibodies may actually cause the loss of intercellular bridges. Immunofluorescence staining reveals IgG, Clq, and C3 in the intercellular substance between epidermal cells. In pemphigus vulgaris patients, 80 to 90% have circulating pemphigus antibodies. Their titer usually correlates positively with clinical manifestations. Corticosteroids and immunosuppressive therapy, as well as plasmapheresis have been used with some success.

Sympathetic nervous system autoantibodies: Complement-fixing sympathetic ganglia (CF-SG) autoantibodies and complement-fixing adrenal medullary (CF-ADM) autoantibodies present in some prediabetic and IDDM patients are associated with decreased catecholamine response to posture. IDDM patients may manifest cf. autoantibodies against vagus nerve that correlate with the presence of CF-SG and CF-ADM autoantibodies. CF-V autoantibodies (parasympathetic nervous system autoantibodies) may also be found in IDDM, but their clinical significance remains to be determined.

Desmoglein is a transmembrane glycoprotein that is one of the three components that make up a complex of epidermal polypeptides formed from the immunoprecipitate of pemphigus foliaceous autoantibodies.

Autoimmune skin diseases include pemphigus, pemphigoid, epidermolysis bullosa acquisita, linear bullous IgA disease, Herpes gestationes, and cutaneous lupus erythematosus. Probable autoimmune skin diseases include psoriasis, lichen ruber planus, alopecia areata, actinic reticuloid, morphea, and scleroderma. Possible autoimmune diseases include pyoderma gangrenosum, parapsoriasis, and sarcoidosis.

Chief cell autoantibodies: Antibodies reactive with chief cells that are a principal source of PG1 (the main form of serum pepsinogen). They are significantly diminished in atrophic gastritis type A, which is associated with pernicious anemia. Autoantibodies against pepsinogen, a 41-kDa antigen, are more frequent in pernicious anemia than are intrinsic factor antibodies. Pepsinogen autoantibodies are also present in approximately 50% of active duodenal ulcer patients, a quarter of whom may have autoantibodies against H^+/K^+ ATPase.

Herpes gestationis (HG) autoantibodies: Autoantibodies from patients with herpes gestationis (HG) and bullous pemphigoid (BP) react with epidermal hemidesmosome (230-kDa and 180-kDa) proteins, respectively, which helps to anchor basal keratinocytes to the lamina densa of the basement membrane. Although not present in normal pregnancy, HG autoantibodies are found in 71 to 89% of positive HG sera and in 47 to 53% of BP sera. IgG autoantibodies to the basement membrane zone (BMZ) are revealed by direct immunofluorescence in 30 to 50% of HG cases, whereas C3 is detected at the basement membrane zone of perilesional skin in approximately 100% of HG. With immunoblotting, serum antibodies to heterogeneous hemidesmosomal components of the BMZ can be detected in about 90% of HG patients. The MCW-1 antigenic site that comprises a part of the noncollagenous domain is believed to be involved in subepidermal blister formation. There is an increased risk for development of other autoimmune diseases, such as Graves' disease, in herpes gestationis patients and their relatives.

Skin autoantibodies detectable by immunofluorescence that are associated with selected bullous skin diseases and are useful in categorizing them. The three principal categories of bullous skin diseases include intraepidermal bullae with an immunological etiology, subepidermal bullae with an immunological pathogenesis, and nonimmune bullous disorders.

Vitiligo is loss of skin pigmentation as a consequence of autoantibodies against melanocytes. The Smyth chicken is a partially inbred line of birds that exhibits a posthatching depigmentation of the feathers as a consequence of an autoimmune process. A total of 95% of the depigmented chicks have autoantibodies that are detectable several weeks prior to the appearance of depigmentation. The autoantigen is a tyrosinase-related protein. Smyth chicken amelanosis and human vitiligo are similar in that their onset is in early adulthood and is often associated with other autoimmune diseases, especially those of the thyroid gland. It may also result from a polygenic disorder or sporadically. Vitiligo is a syndrome marked by acquired loss of pigmentation in a usually symmetrical but spotty distribution, usually involving the central face and lips, genitalia, hands, and extremities. Skin pigmentation is produced by melanin, which is contained in melanosomes that are transferred to keratinocytes to protect the skin against light. There are two stages in the production of vitiligo that may be sequential. Type I vitiligo is marked by decreased melanocyte tyrosinase activity and type II vitiligo is characterized by destruction of melanocytes. Melanocytes and keratinocytes in the border of vitiligo lesions exhibit increased ICAM-1 expression.

Autoimmune uveoretinitis is an ocular inflammation, which is the leading cause of visual impairment in a significant segment of the population. T-cell autoimmunity is postulated to have a significant role in the pathogenesis of at least some of these conditions.

Dalen-Fuchs nodule is a hemispherical granulomatous nodule composed of epithelioid cells and retinal epithelial cells in the choroid of the eye in sympathetic ophthalmia patients and in some other diseases.

Endophthalmitis phacoanaphylactica is a clinical condition that is a consequence of autosensitization to one's own lens antigens. It results from the accidental release of lens protein into the blood circulation during cataract removal in humans. This interaction of a normally sequestered antigen, i.e., lens protein, with the host immune system activates an autoimmune response that results in inflammation of the eye concerned.

Keratoconjunctivitis sicca is a condition characterized by hyperemia of the conjunctiva, lacrimal deficiency, corneal epithelium thickening, itching and burning of the eye, and often reduced visual acuity.

Lens-induced uveitis is a term for inflammatory reactions in the eye related to sensitization or toxicity to lens material. It also includes inflammation that occurs following dislocation and breakdown of the lens. The lens contains very strong organ-specific antigen that can stimulate the formation of autoantibodies. Lens antigens are normally sequestered and do not induce an antibody response until exposed to the immune system of the host. The principal lens antigen is α crystallin. Evidence that lens-induced uveitis is an immunological disease is based mainly on animal studies in which its potent immunogenic properties and capacity to produce autoimmune disease have been revealed. The condition is treated by surgically removing the lens or its remnants soon after diagnosis.

Phacoanaphylactic endophthalmitis: Introduction of lens protein into the circulation following an acute injury of the eye involving the lens may result in chronic inflammation of the lens as a consequence of autoimmunity to lens protein.

Postinfectious iridocyclitis refers to an inflammation of the iris and ciliary body of the eye. It may occur after a virus or bacterial infection and is postulated to result from an autoimmune reaction.

Retina autoantibodies are found in cancer-associated retinopathy (CAR syndrome) patients who are ANA-negative. These antibodies are specific for five antigens: retinal S-antigen, which is associated with ocular inflammation of uveitis; rhodopsin; interphotoreceptor in retinol-binding protein (IRBP); phosducin; and CAR autoantigen (recoverin), a 23-kDa protein which is closely associated with CAR syndrome. Retina antibodies are assayed using EIA, immunoblotting, and dot blotting. There is no known clinical correlation for retina autoantibodies.

Sympathetic ophthalmia is the uveal inflammation of a healthy uninjured eye in an individual who has sustained a perforating injury to the other eye. The uveal tract reveals an infiltrate of lymphocytes and epithelioid cells, and there is granuloma formation. The mechanism has been suggested to be autoimmunity expressed as T lymphocyte-mediated immune reactivity against previously sequestered antigens released from the patient's other injured eye.

EVI antibodies are autoantibodies found in Chaga's disease to endocardium, vascular structures, and interstitium of striated muscle. The target is laminin, but the relevance is in doubt because other diseases produce antilaminin antibodies that do not produce the unique pathology seen in Chagas' disease.

β-adrenergic receptor antibodies: β_2 Receptor autoantibodies may have a significant role in selected diseases of humans that include Chagas' disease, asthma, and dilated cardiomyopathies.

Aging and immunity: Aging is accompanied by many changes in the immune system. The involution of the thymus gland sets the pace for immune senescence. There are changes in the distribution and function of both T and B lymphocytes. A striking feature of immune senescence is the increased frequency of autoimmune reactions in both humans and experimental animals. There is also an increased instance of neoplasia with increasing age as immune surveillance mechanisms falter.

α2-plasmin inhibitor-plasmin complexes (α2PIPC) are complexes formed by the combination of α_2-PI or α_2-macroglobulin with plasmin, the active principle in fibrinolysis. These complexes are found in elevated quantities in plasma of systemic lupus erythematosus (SLE) patients with vasculitis compared to plasma of SLE patients without vasculitis.

Alopecia areata is a partial or patchy loss of hair from the scalp or other hair sites, which is believed to have an autoimmune basis, even though no autoantigenic molecule in hair follicles has been identified. The inheritance is polygenic with a contribution from MHC genes. Other autoimmune diseases associated with alopecia areata include the thyrogastric group, pernicious anemia, Addison's disease, and diabetes, as well as vitiligo and systemic lupus erythematosus (SLE), among others. There is supportive serological evidence for autoimmunity in alopecia areata. Organ-specific autoantibodies are increased in frequency especially to thyroid micosomal antigen and to other tissue antigens as well. Affected hair follicles are encircled by dense lymphocytic infiltrates that consist of T cells with $CD8^+$ T cell subset predominance.

Anticardiolipin antibody syndrome: Circulating lupus anticoagulant syndrome (CLAS). A clinical situation in which circulating anticardiolipin antibodies may occur in patients with lupus erythematosus in conjunction with thromboembolic events linked to repeated abortions caused by placenta vasculothrombosis, repeated myocardial infarction, pulmonary hypertension, and possibly renal and cerebral infarction. There is neurologic dysfunction, including a variety of manifestations such as myelopathy, transient ischemic attacks, chorea, epilepsy, etc. There may be hemolytic anemia, thrombocytopenia, and Coombs' positive reactivity. Immunoglobulin G (IgG) anticardiolipin antibodies manifest 80% specificity for the anticardiolipin antibody syndrome. Anticardiolipin antibodies and DNA show cross-reactivity.

Systemic autoimmunity refers to the formation of antibodies specific for self-constituents leading to type III hypersensitivity, in which immune complexes are deposited in the tissues leading to pathological sequelae. The prototype for systemic autoimmune disease is SLE, in which autoantibodies specific for DNA, RNA, and proteins, associated with nucleic acids, form immune complexes that are deposited in small blood vessels, fix complement, and incite inflammation leading to vascular injury.

Systemic lupus erythematosus (SLE) is the prototype of connective tissue diseases that involves multiple systems and has an autoimmune etiology. It is a disease with an acute or insidious onset. Patients may experience fever, malaise, loss of weight, and lethargy. All organ systems may be involved. Patients form a plethora of autoantibodies, especially antinuclear autoantibodies. SLE is characterized by exacerbations and remissions. Patients often have injury to the skin, kidneys, joints, and serosal membranes. SLE occurs in 1 in 2500 people in certain populations. It has a 9:1 female to male predominance. Its cause remains unknown. Antinuclear antibodies produced in SLE fall into four categories that include (1) antibodies against DNA, (2) antibodies against histones, (3) antibodies to nonhistone proteins bound to RNA, and (4) antibodies against nucleolar antigens. Indirect immunofluorescence is used to detect nuclear fluorescence patterns that are characteristic for certain antibodies. These include homogeneous or diffuse staining, which reveals antibodies to histones and deoxyribonucleoprotein; rim or peripheral staining, which signifies antibodies against double-stranded DNA; speckled pattern, which indicates antibodies to non-DNA nuclear components, including histones and ribonucleoproteins; and the nucleolar pattern in which fluorescent spots are observed in the nucleus and reveal antibodies to nucleolar RNA. Antinuclear antibodies most closely associated with SLE are anti-double-stranded DNA and anti-Sm (Smith) antibodies. There appears to be genetic predisposition to the disease, which is associated with DR2 and DR3 genes of the MHC in Caucasians of North America. Genes other than HLA genes are also important. In addition to the anti-double-stranded DNA and anti-Sm antibodies, other immunologic features of the disease include depressed serum complement levels, immune deposits in glomerular basement membranes and at the dermal–epidermal junction, and the presence of multiple other autoantibodies. Of all the immunologic abnormalities, the hyperactivity of B cells is critical to the pathogenesis of SLE. B cell activation is polyclonal, leading to the formation of antibodies against self and nonself antigens. In SLE, there is a loss of tolerance to self-constituents, leading to the formation of antinuclear antibodies. The polyclonal activation leads to antibodies of essentially all classes in immune deposits found in renal biopsy specimens by immunofluorescence. In addition to genetic factors, hormonal and environmental factors are important in producing the B cell activation.

Nuclei of injured cells react with antinuclear antibodies, forming a homogeneous structure called an LE body, or a hematoxylin body which is usually found in a neutrophil that has phagocytized the injured cell's denatured nucleus. Tissue injury in lupus is mediated mostly by an immune complex (type III hypersensitivity). There are also autoantibodies specific for erythrocytes, leukocytes, and platelets that induce injury through a type II hypersensitivity mechanism. There is an acute necrotizing vasculitis involving small arteries and arterioles present in tissues in lupus. Fibrinoid necrosis is classically produced. Most SLE patients have renal involvement that may take several forms, with diffuse proliferative glomerulonephritis being the most serious. Subendothelial immune deposits in the kidneys of lupus patients are typical and may give a "wire loop" appearance to a thickened basement membrane. In the skin, immunofluorescence can demonstrate deposition of immune complexes and complement at the dermal–epidermal junction. Immune deposits in the skin are especially prominent in sun-exposed areas of the skin. Joints may be involved, but the synovitis is nonerosive. Typical female patients with lupus have a butterfly rash over the bridge of the nose in addition to fever and pain in the peripheral joints. However, the presenting complaints in SLE vary widely. Patients may have central nervous system involvement, pericarditis, or other serosal cavity inflammation. There may be pericarditis as well as involvement of the myocardium or of the cardiac valves to produce Libman-Sacks endocarditis. There may be splenic enlargement, pleuritis, and pleural effusion or interstitial pneumonitis, as well as other organ or system involvement. Patients may also develop antiphospholipid antibodies called lupus anticoagulants. They may be associated with a false-positive VDRL test for syphilis. A drug such as hydrazaline may induce a lupus-like syndrome. However, the antinuclear antibodies produced in drug-induced lupus are often specific for histones, a finding not commonly found in classic SLE. Lupus erythematosus induced by drugs remits when the drug is removed. Discoid lupus refers to a form of the disease limited to the skin. Corticosteroids have proven very effective in suppressing immune reactivity in SLE. In more severe cases, cytotoxic agents such as cyclophosphamide, chlorambucil, and azathioprine have been used. See LE cell.

Antinuclear antibodies (ANA) are autoantibodies that react not only with cell nuclei from many sources but are also directed against antigens in the nuclei of the host, leading to their classification as autoantibodies (Figure 14.35 through Figure 14.38). Although low levels of antibodies against cell nuclei may be found in some healthy humans, increased titers above the normal serum level are considered to constitute a positive ANA test. Autoimmune diseases associated with increased titers of autoantibodies include systemic lupus erythematosus (SLE), rheumatoid arthritis (RA), Sjögren's syndrome, scleroderma, polymyositis, mixed connective tissue disease (MCTD), myasthenia gravis, and chronic active hepatitis. Most ANA are specific for nucleic acids or proteins associated with nucleic acids. Only nucleoli and centromeres of chromosomes can be distinguished by the indirect immunofluorescence technique that is the method of choice as a screening procedure before more specific techniques are employed.

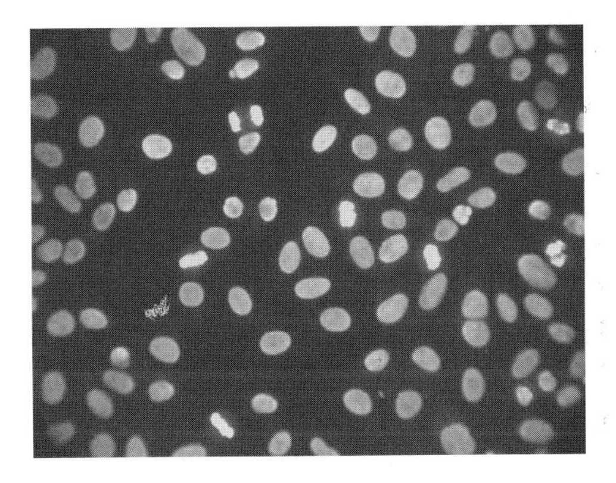

FIGURE 14.35 Antinuclear antibodies.

Antinuclear antibodies (ANA) are found in the circulation of patients with various connective tissue disorders. They may show specificity for various nuclear antigens, including single- and double-stranded DNA, histones, and ribonucleoprotein. To detect antinuclear antibodies, the patient's serum is incubated with Hep-2 cells, and the pattern of nuclear staining is determined by fluorescence microscopy. The homogeneous pattern of staining represents the morphologic expression of antinuclear antibodies specific for ribonucleoprotein which is positive in more than 95% of cases of systemic lupus erythematosus and drug-induced lupus erythematosus, and in 70 to 90% in diffuse systemic sclerosis and limited scleroderma (CREST) cases. It is also positive in 50 to 80% of Sjögren's syndrome and in 40 to 60% of inflammatory myopathies. The test is also positive in progressive systemic sclerosis, rheumatoid arthritis, and other connective tissue disorders. Peripheral nuclear staining represents the morphologic expression of DNA antibodies associated with systemic lupus erythematosus. Nucleolar fluorescence signifies anti-RNA antibodies of the type that occurs in progressive systemic sclerosis (scleroderma). The speckled pattern of staining is seen in several connective tissue diseases. When antinuclear antibodies (ANAs) reach elevated titers significantly above the normal serum level, ANA tests are considered positive. The indirect immunofluorescence technique (IFT) is used as a screening technique before more specific methods are used. Most ANAs are specific for nucleic acids or proteins associated with nucleic acids. Only nucleoli in centromeres of chromosomes can be distinguished by IFT as separate antigens. Nucleosomes are irrelevant autoantigens for the formation of antibodies against nucleosomes, histones, and DNA. Antibodies against specific nuclear antigens are listed under each individually.

ANA is an abbreviation for antinuclear antibodies.

Anti-double-stranded DNA (Figure 14.36 and Figure 14.38) are antibodies present in the blood sers of systemic lupus erythematosus (SLE) patients. Among the detection methods is an immunofluorescence technique (IFT) using *Crithidia luciliae* as the substrate. In this method, fluorescence of the kinetoplast, which contains mitochondrial DNA, signals the presence of anti-dsDNA antibodies. This technique is useful for assaying SLE serum, which is usually positive in patients with active disease. A rim or peripheral pattern of nuclear staining of cells interacting with antinuclear antibody represents morphologic expression of anti-double-stranded DNA antibody.

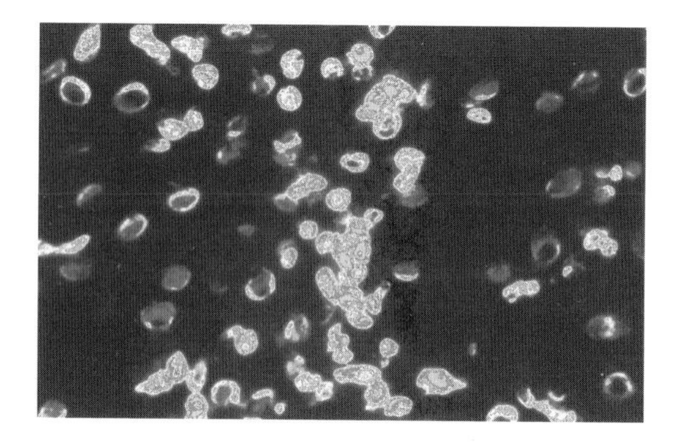

FIGURE 14.36 Anti-double stranded DNA autoantibodies.

FIGURE 14.37 Antihistone (H2A-H2B) DNA complex autoantibodies.

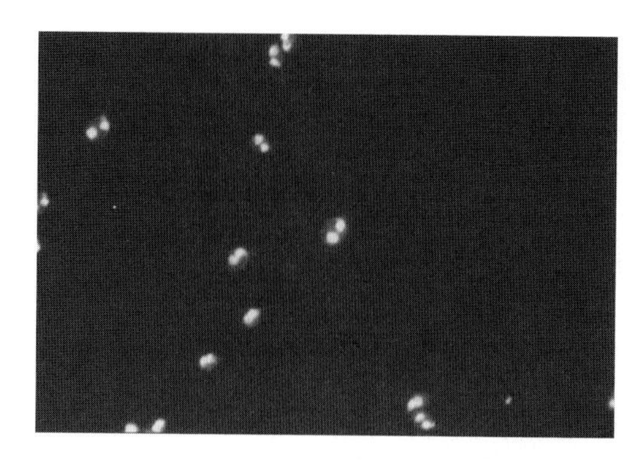

FIGURE 14.38 Anti double stranded DNA autoantibodies (*Crithidia luciliae*).

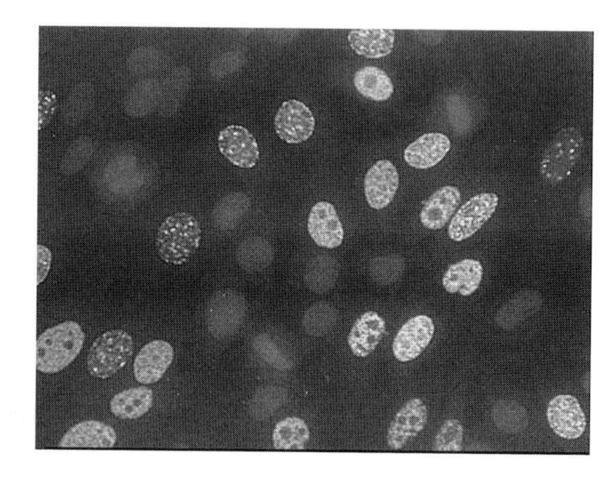

FIGURE 14.39 Anti-PCNA antibodies.

Anti-PCNA antibodies are present in the sera of 3% of SLE patients. Epitope mapping reveals that the antibodies bind conformational epitopes of this antigen (Figure 14.39).

Anti-Sm autoantibodies are found in the sera of patients with systemic lupus erythematosus. They are usually accompanied by antinuclear ribonucleoprotein (nRNP) antibodies. The U1RNP particle has both Sm and RNP binding specificities. The difference is that the RNP particles bound by U2, U4/6, and U5 are bound by anti-Sm autoantibodies but not by anti-nRNP autoantibodies. Sm antigen is a nonhistone nucleoprotein composed of several polypeptides of differing molecular weights. Autoantibodies against Sm antigen precipitate the U1, U2, U4/6, and U5 small nuclear RNAs. The Sm antigen is involved in normal posttranscriptional, premessenger RNA processing to excise introns. Autoantibodies to Sm antigen have been observed in 15 to 30% of SLE sera as a diagnostic marker. It is believed that IgG anti-Sm correlates with lupus disease activity and is a useful variable in predicting exacerbation and prognosis of SLE. IgG anti-Sm is specifically detected in patients with SLE. IgG anti-Sm has rarely been detected in SLE. ELISA methodology is used to quantitate this antibody.

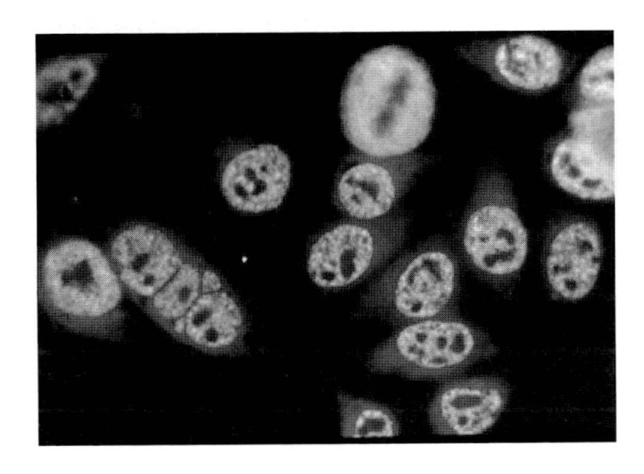

FIGURE 14.40 Anti-SM/RNP autoantibodies.

Anti-snRNP (Sm, U1-RNP, U2-RNP) (Figure 14.40): Sm antibodies are specific for the D proteins and the BB[1] doublet of the Usn RNPs. Sm-specific antibodies precipitate U1, U2, U4, U5, and U6 smRNPs. Anti-Sm autoantibodies are found exclusively in SLE patients and are considered disease specific. Antibodies to U1-RNP are a marker for mixed connective tissue disease (MCTD).

Anti-Sm antibodies do not show the increase in titer observed with anti-double-stranded DNA antibodies during exacerbations of systemic lupus erythematosus.

Anti-Sm (Smith) autoantibodies, which recognize a subset of the protein components of U1 SnRNP, occur in approximately 29% of SLE patients and are quite specific for this disease. They are specific for polypeptides of U1, U2, and U4-6 snRNPs, which are significant in pre-mRNA splicing. Anti-Sm antibodies recognize a group of proteins (B′/B, D, E, F, and G) that forms a stable subparticle termed the *Sm core particle*, which is associated with the U1 snRNP, as well as a series of other snRNPs, the most abundant of which are the U2, U5, and U4/U5 snRNPs. The assembly of these proteins and RNAs into macromolecular complexes is responsible for the patterns found on protein and RNA immunoprecipitation. Anti-Sm antibodies do not show the increase in titer observed with anti-double-stranded DNA antibodies during exacerbations of the disease.

Spliceosomal snRNP autoantibodies: Sm autoantibodies are those reactive with the Sm core protein (B, V, E) that are common to the U1 snRNP, U2 snRNP, and the other uridine-rich snRNPs (U4/U6, U5, U7, and U11/U12). The term U1 snRNP autoantibodies should be restricted to autoantibodies reactive with non-Sm protein epitopes characteristic of individual snRNPs, e.g., 70K, A, and C for U1; AN and BO for U2; and 120 kDa/150 kDa for U4/U6. Sm autoantibodies occur in 20 to 30% of Caucasian adults and children with SLE and in 30 to 40% of Asians and African Americans with SLE. MCTD is characterized by U1 snRNP (100%) and is occasionally accompanied by U2 snRNP (approximately 15%) autoantibodies.

ENA antibodies are specific for extractable nuclear antigens. This category includes antibodies to ribonucleoprotein (RNP), presently termed U1 snRNP or U1 RNP in addition to Sm antibodies that have specificity for Smith antigen. Sm antibodies are associated with SLE, whereas U1 snRNP antibodies in high titer are detected in patients with mixed connective tissue disease. U4/U6 snRNP antibodies are detectable in patients with systemic sclerosis.

Ribosomal P protein autoantibodies (RPP) are found in 45 to 90% of SLE patients with severe depression or psychosis, but are also present in 7 to 20% of SLE patients who are not psychotic. P (phospho) proteins (P0, P1, and P2) are the most common antigens. These phosphoproteins are also known as A (alanine-rich) proteins. Antibodies to P proteins, which are specific for a common carboxy terminus epitope, occur in 12 to 19% of SLE patients. Other ribosomal antigens include L12 protein, L5/5s, S10, Ja, L7, and ribosomal ribonucleoprotein (rRNA). Anti-L7, anti-rRNA, anti-S10, and anti-Ja autoantibodies are more common than those against L12 and L5/5S, which are rare. A positive RPP autoantibody test is not diagnostic of lupus psychosis since approximately 50% of patients with RPP autoantibodies do not have severe behavioral problems. These antibodies are found in Sjögren's syndrome patients with CNS abnormalities. Immunoblotting techniques reveal autoantibodies to ribosomal proteins in about 42% of SLE and 55% of RA sera. RPP autoantibodies are rare in systemic sclerosis and, when present, indicate an overlap with SLE. The clinical significance of these autoantibodies is doubtful and remains to be determined. Firm data suggest that RPP autoantibodies may play a role in the pathogenesis of lupus hepatitis.

Antinucleosome antibodies are antibodies specific for nucleosomes or subnucleosomal structures that consist of DNA plus core histones. These autoantibodies were the first to be associated with SLE and were formerly referred to as LE factors responsible for the so-called "LE cell phenomenon."

Crithidia luciliae is a hemoflagellate possessing a large mitochondrion that contains concentrated mitochondrial DNA in a single large network called the kinetoplast. It is used in immunofluorescence assays for the presence of anti-dsDNA antibodies in the blood sera of SLE patients.

***Crithidia* assay** refers to the use of a hemoflagellate termed *Crithidia luciliae* to measure anti-dsDNA antibodies in the serum of SLE patients by immunofluorescence methods. The kinetoplast of this organism is an altered mitochondrion that is rich in double-stranded DNA.

Double-stranded DNA autoantibodies: Increased concentrations of high affinity anti-dsDNA autoantibody (double-stranded DNA) autoantibodies are detected by the Farr (ammonium sulfate precipitation) assay and serve as a reliable predictor of SLE and to monitor treatment. dsDNA antibody levels are interpreted in conjunction with serum C3 or C4 concentrations. When detected, these antibodies can foretell the development of SLE within a year in approximately two thirds of patients without clinical evidence of SLE. Currently used tests often measure dsDNA autoantibodies of varying affinities that give positive and confusing results in non-SLE patients. Autoantibodies against ssDNA have no clinical significance except as a general screen for autoantibodies. If the dsDNA antibody level doubles or exceeds 30 IU/ml earlier than 10 weeks, an exacerbation of SLE is likely, especially if there is an associated decrease in the serum C4 concentration. This reflects selective stimulation of B-cell stimulation known to occur in SLE patients. The Farr assay is more reliable than is the EIA method since the Farr assay measures high-avidity antibodies to dsDNA. In SLE patients with CNS involvement, low avidity anti-dsDNA autoantibodies are more common than high-avidity anti-dsDNA autoantibodies. Blood sera from patients suspected of having SLE should be assayed for autoantibodies to dsDNA (Farr technique) and to Sm (EIA) with IB confirmation), since either or both of these autoantibodies are essentially diagnostic of SLE.

ENA autoantibodies are specific for extractable nuclear antigens. They are small nuclear ribonucleoproteins (snRNPs) and small cellular ribonucleoproteins (scRNPs). This category includes antibodies to RNP, presently termed U1 snRNP or U1 RNP in addition to Sm antibodies which have specificity for Smith antigen. Sm antibodies are associated with SLE, whereas U1 snRNP antibodies in high titer are detected in patients with mixed connective tissue disease. U4/U6 snRNP antibodies are detectable in patients with systemic sclerosis. U1 snRNP autoantibodies bind to RNP proteins A and C. Sm autoantibodies bind to RNP proteins B′/B, D, and E. scRNP autoantibodies include SS-A/Ro with specificity for 60-kDa and 52-kDa proteins and SS-B/La with specificity for a 48-kDa protein.

FANA is an abbreviation for fluorescent antinuclear antibody.

Hematoxylin bodies are nuclear aggregates of irregular shape found in areas of fibrinoid change or fibrinoid necrosis in SLE patients. These homogeneous-staining nuclear masses contain nuclear protein and DNA as well as anti-DNA. They are probably formed from injured cell nuclei that have interacted with antinuclear antibodies *in vivo*. Hematoxylin staining imparts a bluish-purple color to hematoxylin bodies. They may be viewed in the kidney, lymph nodes, spleen, lungs, atrial endocardium, synovium, and serous membranes. Hematoxylin bodies of the tissue correspond to LE cells of the peripheral blood.

Histone antibodies of the IgG class against H2A–H2B histones are detectable in essentially all procainamide-induced lupus patients manifesting symptoms. They are also present in approximately one fifth of SLE patients and in procainamide-treated persons who do not manifest symptoms. Antibodies against H2A, H2B, and H2A–H2B complex react well with histone fragments resistant to trypsin. By contrast, antibodies in the sera of SLE patients manifest reactivity for intact histones but not with their fragments. In lupus induced by hydralazine, antihistone antibodies react mainly with H3 and H4 and their fragments that are resistant to trypsin.

Histone (H2A–H2B)–DNA complex autoantibodies (IgG) may develop in drug-induced lupus (DIL) in the absence of other autoantibodies. They show high reactivity with histone H2A–H2B dimers when induced by procainamide. The autoantibodies diminish when the drug is discontinued. IgM autoantibodies against histone H1–H4 autoantibodies are found frequently in SLE and in persons receiving a variety of medications. Autoantibodies against histone (H2A–H2B)–DNA complexes are generated in DIL attributable to procainamide, quinidine, acebutolol, penicillamine, and isoniazid, but not methyldopa. H2A and H2B histone linear epitopes are identified by separate histone autoantibody populations associated with SLE, DIL, juvenile rheumatoid arthritis, and scleroderma.

Histone autoantibodies (non-H2A–H2B)–DNA): H1–H4 autoantibodies of the IgM isotype are broadly reactive and are frequently found in patients with SLE as well as in normal persons taking various medications. Patients with localized scleroderma (40 to 60%) and those with generalized morphea (80%) demonstrate autoantibodies against histones H1, H2A, and H2B. Systemic sclerosis patients (29%) and diffuse cutaneous systemic sclerosis patients (44%) reveal histone H1 autoantibodies, which indicate the severity of pulmonary fibrosis in systemic sclerosis. Autoantibodies against histone (H2A–H2B)–DNA complexes are seen more frequently in scleroderma-related disorders than in SLE. Elevated IgA antibodies against all H1, H2A, H2B, H3, and H4 are all detectable in IgA nephropathy, primary glomerulonephritis, membranous glomerulonephritis, and idiopathic nephrotic syndrome. Serum histone autoantibody titer is also positively correlated with the extent of dementia in Alzheimer's disease.

Ki autoantibodies are specific for a 32kD nonhistone nuclear protein from rabbit thymus that are present in 19 to 21% of SLE patients by EIA and 8% by DD. They are present in smaller quantities in mixed connective tissue disease, systemic sclerosis, and RA. At present this

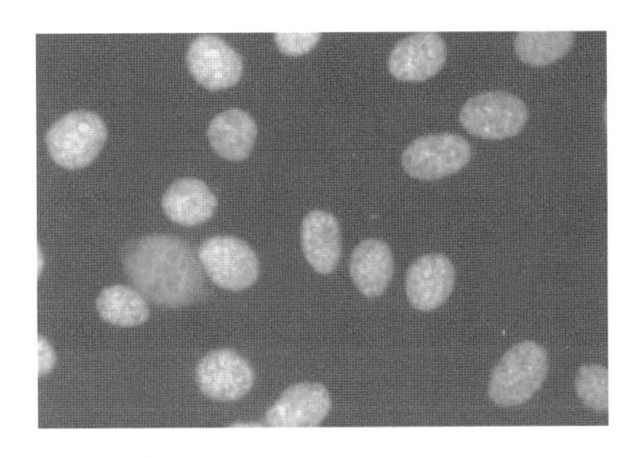

FIGURE 14.41 Ku antibodies.

autoantibody is not clinically useful. The Ki protein contains an antigenic determinant that is homologous to Sv40 large T antigen nuclear localization signal.

Ku is a 70-kDa to 80-kDa DNA-binding protein that is the target of autoantibodies in a few Japanese myositis patients manifesting the overlap syndrome. Nuclear factor IV (NF IV) and Ku antigen are the same. Anti-Ku antibodies recognize conformational epitopes.

Ku antibodies are detectable in 15 to 50% of individuals with Sjögren's syndrome, SLE, mixed connective tissue disease, and scleroderma. They are found in 5 to 15% of myositis patients in the U.S.A. (Figure 14.41).

An **LE cell** is a neutrophil (PMN) in the peripheral blood or synovium of lupus erythematosus patients produced when the PMN phagocytizes a reddish-purple staining homogeneous lymphocyte nucleus with Wright's stain that has been coated with antinuclear antibody. In addition to lupus erythematosus, LE cells are seen also in scleroderma, drug-induced lupus erythematosus, and lupoid hepatitis.

In **LE cell "prep,"** glass beads are added to heparinized blood samples, causing nuclei to be released from some blood cells which become coated with antinuclear antibody present in the serum. These opsonized nuclei are then phagocytized by polymorphonuclear neutrophils to produce LE cells. The LE cells produce homogeneous chromatin, which imparts a glassy appearance to the phagocytized nuclear material.

LE cell test refers to a diagnostic test (no longer used) which detects antinuclear antibodies in the blood sera of SLE patients. Antinuclear antibodies present in the serum react with nascent lymphocyte nuclei and serve as opsonins, enhancing phagocytosis of the nucleus–antibody complex by polymorphonuclear neutrophils. Thus, the appearance of a polymorphonuclear neutrophil with its own nucleus displaced to the periphery by an ingested lymphocyte nucleus, which appears as a homogeneous mass and is coated with antinuclear antibody, represents the so-called LE cell. These cells develop following incubation of blood containing the appropriate antibody for 1 h at 37°C. This earlier diagnostic test for the presence of antinuclear antibody in SLE has been replaced by a more sophisticated antinuclear antibody test. LE cells are also present in other connective tissue diseases in addition to SLE.

LE factor refers to the antinuclear antibodies present in the blood sera of SLE patients. LE factor facilitates LE cell formation

Antiphospholipid syndrome (APS) describes a thrombotic event together with antiphospholipid antibodies. Antiphospholipid antibodies are not always associated with thrombosis. Multiple factors cause thrombosis, including the antiphospholipid antibodies. There may be recurrent fetal loss three times greater than expected or thrombocytopenia. Antiphospholipid syndrome is a condition with four common clinical features: venous thrombosis; arterial thrombosis; pregnancy loss; or thrombocytopenia. Fifty percent of patients present with primary APS (not associated to systemic disease) and the remaining 50% have secondary APS associated with such diseases as SLE, other connective tissue diseases, HIV or other infections, malignancy, and drug use. APS is a leading cause of hypercoagulability and thrombosis. Thrombosis is the most common presentation of APS. Anticardiolipin (ACA)-positive APS and lupus anticoagulant (LA)-positive APS are the two syndromes. ACA-APS is more common than LA-APS with a relative prevalence of 5 to 2. Both are associated with thrombosis, fetal wastage, and thrombocytopenia. ACA-APC is associated with both arterial and venous thrombosis including deep vein thrombosis and pulmonary embolism, premature coronary artery disease, and premature cerebrovascular disease. By contrast, LA-APS is more commonly associated with venous thrombosis involving mesenteric, renal, hepatic, and portal veins and vena cava. Antiphospholipid antibodies (aPL) are demonstrated by either prolongation of a phospholipid-dependent coagulation test (lupus) or by their reactivity to anticoagulant (LA anionic phospholipids [PL]) in a solid-phase immunoassay. To diagnose APS, at least one of the above clinical features and a positive test are required.

Antiphospholipid antibodies are a heterogeneous group of immunoglobulins directed against negatively charged phospholipids, protein–phospholipid complexes, and plasma proteins. Antiphospholipid antibodies occur in autoimmune disease, bacterial fungal and viral infections, malignant tumors, and specific medications, or may appear in normal healthy people. They are associated with both venous and arterial thrombosis. They occur in systemic lupus erythematosus, rheumatoid arthritis, Sjögren's syndrome, Reiter's syndrome, and possibly scleroderma and polyarteritis. These antibodies may appear in HIV, Lyme disease, ornithosis, adenovirus, rubella and chicken pox, smallpox vaccination, and syphilis. See also anticardiolipin antibody syndrome and lupus anticoagulant.

Phospholipid autoantibodies include those specific for cardiolipin, phosphatidylserine, and lupus present in individuals with antiphospholipid antibody syndrome and SLE in addition to drug-induced disorders and infectious and neurological diseases. Phospholipid autoantibodies, a primary marker for antiphospholipid antibody syndrome, are linked to increased risk of thrombosis, thrombocytopenia, and recurrent abortion.

Beta-2 glycoprotein-I autoantibodies: Autoantibodies against beta-2 glycoprotein-1 (b-2-GPI), a natural serum coagulation inhibitor. This represents one of the interactions of antiphospholipid antibodies with components of the clotting system. Antiphospholipid autoantibodies occur in antiphospholipid syndrome (APS), which is associated with recurrent thrombotic events, repeated fetal loss and/or thrombocytopenia, usually associated with systemic lupus erythematosus, as well as in patients without autoimmune disease. b-2-GPI, a 50-kDa plasma protein, binds to phospholipids (PL) and lipoproteins. Binding to PL inhibits the intrinsic blood coagulation pathway, prothrombinase activity, and ADP-dependent platelet aggregation, and induces a conformational change in b2-GPI causing b2-GPI autoantibodies to bind only to b2-GPI when it is complexed to PL.

C1q autoantibodies are detectable in 14 to 52% of SLE patients, 100% of HUVS, rheumatoid arthritis (RA) (5% in uncomplicated RA and 77% of RA patients with Felty syndrome), 73% of membranoproliferative glomerulonephritis type I, 45% of membranoproliferative glomerulonephritis type II and III, 94% of mixed connective tissue disease, and 42% of polyarteritis nodosa. Lupus nephritis patients usually reveal the IgG isotype. Rising levels of C1q autoantibodies portend renal flares in systemic lupus erythematosus patients. The rare condition HUVS can occur together with SLE and is marked by diminished serum C1q and recurrent idiopathic urticaria with leukocytoclastic vasculitis.

Cardiolipin autoantibodies: Anticardiolipin antibodies (ACA) may be linked to thrombocytopenia, thrombotic events, and repeated fetal loss in systemic lupus erythematosus patients. Caucasian, but not Chinese, patients with SLE may have a relatively high incidence of valve defects associated with the presence of these antibodies. Other conditions associated with ACA include adrenal hemorrhage and Addison's disease, livedo reticularis (LR), livedo reticularis with cardiovascular disease (Sneddon syndrome), possibly polymyalgia rheumatica/giant cell arteritis, and possibly focal CNS lupus. IgG, IgM, and IgA cardiolipin autoantibodies are assayed by EIA for the detection of antiphospholipid antibodies (aPL) in patients believed to have APS. Combined testing for phosphatidylserine autoantibodies and lupus anticoagulant in additioin to anticardiolipin autoantibodies improves the sensitivity for the detection of antiphospholipid antibodies. High anticardiolipin autoantibody concentrations are associated with an increased risk of venous and arterial thrombosis, recurrent pregnancy loss, and thrombocytopenia.

Circulating anticoagulant refers to antibodies specific for one of the blood coagulation factors. They may be detected in the blood serum of patients treated with penicillin, streptomycin, or isoniazid; in systemic lupus erythematosus patients; or following treatment of hemophilia A or B patients with factor VIII or factor IX. These are often IgG4 antibodies.

Circulating lupus anticoagulant syndrome (CLAS) refers to the occurrence in lupus patients, who are often ANA negative, of thromboses that are recurrent, kidney disease, and repeated spontaneous abortions. There is an IgM gammopathy and fetal wastage that occurs repeatedly.

Collagen (types I, II, and III) autoantibodies are autoantibodies found in various autoimmune diseases such as mixed connective tissue disease (MCTD), SLE, progressive systemic sclerosis (PSS), RA, and vasculitis. Approximately 85% of SLE patients have antibodies against type I collagen. Type II collagen autoantibodies in rodents and monkeys have led to arthritis. Autoantibodies to types I, II, and III collagen have been identified in adult and juvenile RA and relapsing polychondritis. Individuals with hypersensitivity reactions to collagen implants may manifest immunity to native and denatured collagens. Autoantibodies to type II collagen have been associated with RA and relapsing polychondritis patients. Autoimmunity to type II collagen has been hypothesized to have a role in the pathogenesis of RA. Autoantibodies to type II collagen have also been reported in scleroderma and SLE as well as possibly in inner ear diseases. Collagen type III antibodies have been found in 44% of SLE patients and 85% of RA patients.

Collagen disease and arthritis panel: A cost effective battery of tests to diagnose rheumatic disease that includes the erythrocyte sedimentation rate and assays for rheumatoid factor (RA test), antinuclear antibody, uric acid levels, and C-reactive protein.

Collagen type IV autoantibodies are collagen type IV that is present in all human basement membranes including those of the kidney, eye, cochlea, lung, placenta, and brain. Triple-helical molecules comprised of two $\alpha 1$ and one α2 chains in a chickenwire-like network. Four other type IV collagen chains (α3 to α6) from a similar network. Autoantibodies against type IV collagen occur in progressive systemic sclerosis, Raynaud's phenomenon, and scleroderma. 70% of SLE, Raynaud's phenomenon, polyarteritis nodosa, and vasculitides have collagen type IV autoantobodies. 35 to 44% of thromboangiitis obliterans (Burger's disease) have autoantibodies against types I and IV collagen. Three fourths of patients develop cell-mediated immunity to these two collagens. Autoantibodies to basement membrane and interstitial collagens play a role in the pathogenesis of scleroderma. Autoantibodies to type IV collagen are found also in post streptococcal glomerulonephritis. Circulating antibodies specific for the NC1 domain of type IV collagen are present in Goodpasture's syndrome and lead to rapidly progressive glomerulonephritis in these subjects. Most Alport syndrome patients lack α3, α4, and α5 chains in their golmerular basement membrane.

Collagen vascular diseases are a category of connective tissue diseases in which type III hypersensitivity mechanisms with immune complex deposition play a major role. These diseases are characterized by inflammation and fibrinoid necrosis in tissues. Patients may manifest involvement of multiple systems, including the vasculature, joints, skin, kidneys, and other tissues. These are classic, systemic autoimmune diseases in most cases. The prototype of this category is systemic lupus erythematosus. Of the multiple disorders included in this category are dermatomyositis, polyarteritis nodosa, progressive systemic sclerosis (scleroderma), rheumatoid arthritis, and mixed connective tissue disease. They are treated with immunosuppressive drugs, especially corticosteroids.

Connective tissue disease is one of a group of diseases, formerly known as collagen vascular diseases, that affect blood vessels producing fibrinoid necrosis in connective tissues. The prototype of a systemic connective tissue disease is systemic lupus erythematosus. Also included in this classification are systemic sclerosis (scleroderma), rheumatoid arthritis, dermatomyositis, polymyositis, Sjögren's syndrome, polyarteritis nodosa, and a number of other disorders that are believed to have an immunological etiology and pathogenesis. They are often accompanied by the development of autoantibodies such as antinuclear antibodies, or antiimmunoglobulin antibodies such as rheumatoid factors.

α2-plasmin inhibitor-plasmin complexes (α2PIPC) are formed by the combination of α_2-PI or α_2-macroglobulin with plasmin, the active principle in fibrinolysis. These complexes are found in elevated quantities in plasma of SLE patients with vasculitis compared to plasma of SLE patients without vasculitis.

Lupus anticoagulant is IgG or IgM antibody that develops in lupus erythematosus patients, in certain individuals with neoplasia or drug reactions, in some normal persons, and was recently reported in AIDS patients who have active opportunistic infections. These antibodies are specific for phospholipoproteins or phospholipid constituents of coagulation factors. *In vitro,* these antibodies inhibit coagulation dependent upon phospholipids. Members of a family of acquired circulating anticoagulants, lupus anticoagulants are immunoglobulins that lead to prolonged coagulation screening tests that include activated partial thromboplastin time (APTT) and prothrombin time (PT). LAC shows species specificity for a prothrombin–anionic phospholipid neoepitope in primary antiphospholipid syndrome (PAPS). LAC assays may be useful to predict thrombosis.

Lupus erythematosus is a connective tissue disease associated with the development of autoantibodies against DNA, RNA, and nucleoproteins. It is believed to be due to hyperactivity of the B cell limb of the immune response. Clinical manifestations include skin lesions (including the so-called butterfly rash on the cheeks and across the bridge of the nose) that are light sensitive. Patients may develop vasculitis, arthritis, and glomerulonephritis. When the disease is confined to the skin, it is referred to as discoid LE or cutaneous LE. Approximately 75% of lupus patients have renal involvement.

Lupus erythematosus and pregnancy: Pregnant lupus erythematosus patients may experience fetal wasting caused by thromboses. Spontaneous abortion together with anticardiolipin antibody and anti-Rh antibodies may be linked to fertility failure and death of the fetus.

Drug-induced lupus erythematosus: Certain drugs such as procainamide, hydralazine, D-penicillamine, phenytoin, isoniazid, and ergot substances may produce a condition that resembles lupus in patients receiving these substances. Most of these cases develop antinuclear and antihistone antibodies, with only approximately one-third of them developing clinical signs and symptoms of lupus, such as arthralgia, serositis, and fever. These cases do not usually develop the renal and CNS lesions seen in classic lupus. Nonacetylated metabolites accumulate in many of these individuals who are described as "slow acetylators." These nonacetylated metabolites act as haptens by combining with macromolecules. This may lead to an autoimmune response due to metabolic abnormality.

Lupus nephritis refers to renal involvement in cases of SLE that can be classified into five patterns according to morphologic criteria developed by WHO. These include (1) normal by light electron and immunofluorescent microscopy (Class I), (2) mesangial lupus glomerulonephritis (Class II), (3) focal proliferative glomerulonephritis

(Class III), (4) diffuse proliferative glomerulonephritis (Class IV), and (5) membranous glomerulonephritis (Class V). None of these patterns is specific for lupus.

Systemic lupus erythematosus, animal models: The main lupus models include inbred murine mouse strains that develop lupus. They are best suited for investigation because their serologic, cellular, and histopathologic characteristics closely resemble those of the corresponding human disease. There are three principal types of lupus-prone mice. They include the New Zealand (NZ) mice (NZB, NZW, and their F1 hybrid), the BXSB mouse, and MRL mouse. In addition, single autosomal recessive mutations termed *gld* (generalized lymphoproliferative disease) and lpr^{cg} (*lpr*-complementing gene) have been described.

An **MRL-*lpr/lpr* mouse** is from a strain genetically prone to develop lupus erythematosus-like disease spontaneously. Its congenic subline is MRL-+/+. The lymphoproliferation (*lpr*) gene in the former strain is associated with development of autoimmune disease, i.e., murine lupus. Although the MRL-+/+ mice are not normal immunologically, they develop autoimmune disease only late in life and without lymphadenopathy. MRL-*lpr/lpr* mice differ from New Zealand mice mainly in the development of striking lymphadenopathy in both males and females of the MRL-*lpr/lpr* strain between 8 and 16 weeks of age with a 100-fold increase in lymph-node weight. Numerous Thy-1^+, Ly-1^+, Ly-2^-, and L3T4-lymphocytes express and rearrange α and β genes of the T-cell receptor but fail to rearrange immunoglobulin genes that are present in the lymph nodes. Multiple antinuclear antibodies, including anti-Sm, are among serological features of murine lupus in the MRL-*lpr/lpr* mouse model. These are associated with the development of immune complexes that mediate glomerulonephritis. Although the *lpr* gene is clearly significant in the pathogenesis of autoimmunity, the development of anti-DNA and anti-Sm even in low titers and late in life in the MRL-+/+ congenic line points to the role of factors other than the *lpr* gene in the development of autoimmunity in this strain.

BXSB mice are genetically prone to developing lupus erythematosus–like disease spontaneously. The BXSB strain manifests serologic aberrations and immune complex glomerulonephritis, but demonstrates a distinct and significant acceleration of the disease in males. Among other features, BXSB strains develop moderate lymphadenopathy, which reaches 10 to 20 times greater than normal. The B cell content of these proliferating male lymph nodes may reach 70%. B cell content also develops significant levels of antinuclear antibodies, including anti-DNA, diminished complement, and immune complex-mediated renal injury. Acceleration of this autoimmune disease in the male rather than in the female has been shown not to be hormone mediated.

The ***gld* gene** is a murine mutant gene on chromosome 1. When homozygous, the *gld* gene produces progressive lymphadenopathy and lupus-like immunopathology.

New Zealand white (NZW) mice are an inbred strain of white mice that, when mated with NZB strain that develop autoimmunity, produce an F_1 generation of NZB/NZW mice that represent an animal model of autoimmune disease and especially of an LE-like condition.

NZB/NZW F_1 hybrid mice are a strain genetically prone to develop LE-like disease spontaneously. This was the first murine lupus model developed from the NZB/B_1 mouse model of autoimmune hemolytic anemia, mated with the NZW mouse, that develops a positive Coombs test, antinuclear antibodies, and glomerulonephritis. This F_1 hybrid develops various antinuclear autoantibodies and forms immune complexes that subsequently induce glomerulonephritis and shorten the life span due to renal disease. Hemolytic anemia is minimal in the F_1 strain compared to the NZB parent. NZB mice show greater lymphoid hyperplasia than do the F_1 hybrids, yet the latter show features that are remarkably similar to those observed in human lupus, such as major sex differences.

Ubiquitin autoantibodies are specific for ubiquitin, a highly conserved protein of 76 amino acids found universally in eukaryotic cells present among antibodies of 29 to 79% of SLE patients. Ubiquitin antibodies may be inversely related to dsDNA antibodies and disease activity. EIA and immunoblotting are the methods of choice to detect ubiquitin autoantibodies. Renal biopsies in 29% of lupus nephritis patients reveal ubiquitin autoantibodies, which points to the possible role in lupus nephritis. Immunohistochemical staining has revealed the presence of intracellular ubiquitinated filamentous inclusions in human chronic neurodegenerative diseases, e.g., in the motor cortex and in the spine in motor neurone disease, and in brain cortical regions in one of the major newly recognized forms of dementia, diffuse Lewy body disease. Ubiquitinated filamentous inclusions have also been observed in some liver and viral diseases (e.g., I hepatocytes in alcoholic liver disease and in Epstein-Barr virus–transformed human lymphocytes). Ubiquitin–protein conjungates have been found in the primary (azurophilic) lysosome-related granules in mature polymorphonuclear neutrophils.

Anti-RA-33 are antibodies considered specific for rheumatoid arthritis that have now been found in other autoimmune diseases and are detected by immunoblotting.

Gold therapy refers to a treatment for arthritis that inhibits oxidative degradation of membrane proteins and lipids and counteracts singlet oxygen produced as free radicals.

It is administered in such forms as aurothioglucose and gold sodium thiomalate. Gold may induce gastrointestinal symptoms such as nausea and vomiting, diarrhea, and abdominal pain, and renal symptoms such as nephrotic syndrome and proteinuria, as well as skin rashes, hepatitis, or blood dyscrasias. Renal biopsy may reveal IgG and C3 in a "motheaten" pattern in the glomerular basement membrane, as well as "feathery crystals" in the renal tubules.

Granulocyte-specific antinuclear autoantibodies (GS-ANA) are specific antinuclear autoantibodies that react with neutrophil nuclei to produce a homogeneous staining pattern by immunofluorescence but not with other substrates used for ANA detection. They do not react with Hep-2 cells or with liver substrates. The exact histone target antigens of GS-ANA remain to be determined. GS-ANA are found in the sera of patients with active RA, with vasculitis, and/or neutropenia in frequencies approaching 75% in 90% of Felty's syndrome. They are found in only 41% of effectively managed RA patients and in 17% of juvenile rheumatoid arthritis. GS-ANA correlate well with erosive joint disease. These antibodies can be demonstrated in five immunoglobulin isotypes with mostly IgG and IgM in RA and Felty's syndrome. Indirect immunofluorescence is the preferred method of detection.

Heterogeneous nuclear ribonucleoprotein (RA-33) autoantibodies: hnRNP autoantibodies are associated mainly with RA, SLE, MCTD, and less often with other connective tissue diseases. The role of these autoantibodies in the pathogenesis of the diseases cited is unknown. RA-33 autoantibodies have been correlated with severe erosive arthritis in lupus patients. These autoantibodies appear early in the course of the disease, which renders them valuable in the diagnosis of early rheumatoid arthritis, especially rheumatoid factor (RF-negative RA). Both EIA and immuno-blotting techniques have been used to detect these antibodies.

HLA-B27-related arthropathies are joint diseases that occur with increased frequency in individuals who are HLA-B27 antigen positive. Juvenile rheumatoid arthritis, ankylosing spondylitis, Reiter's syndrome, *Salmonella*-related arthritis, psoriatic arthritis, and *Yersinia* arthritis belong to this group.

Ibuprofen, ($^{+}$)-2-(*p*-isobutylphenyl) propionic acid, is an antiinflammatory drug used in the therapy of patients with rheumatoid arthritis, ankylosing spondylitis, and juvenile rheumatoid arthritis.

Indomethacin, 1-(4-chlorobenzylyl)-5-methoxy-2-methyl-1H-indole-3-acetic acid, is a drug that blocks synthesis of prostaglandin. It is used for therapy of rheumatoid arthritis and ankylosing spondylitis. It may counter the effects of suppressor macrophages.

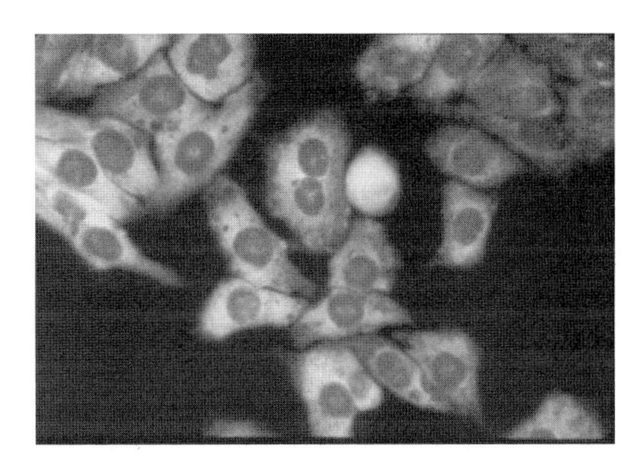

FIGURE 14.42 Antiribosomal RNP autoantibodies.

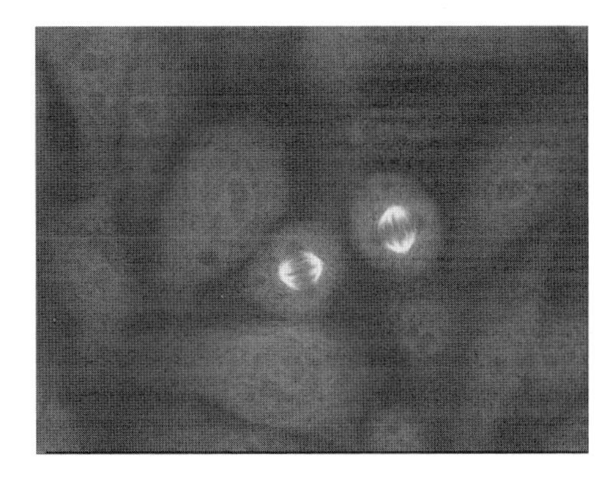

FIGURE 14.43 Mitotic spindle apparatus autoantibodies.

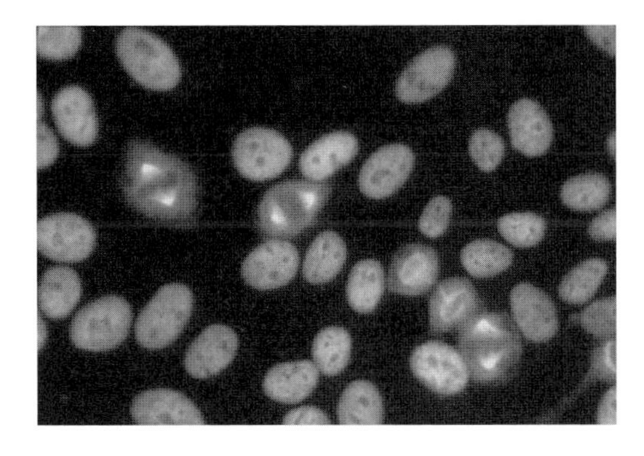

FIGURE 14.44 Mitotic spindle apparatus (NUMA) autoantibodies.

Mitotic spindle apparatus autoantibodies (Figure 14.43 and Figure 14.44) are present in the sera of individuals with various autoimmune diseases that include rheumatoid arthritis, Sjögren's syndrome, Hashimoto thyroiditis, and localized scleroderma, in addition to respiratory infection, dilated cardiomyopathy, and melanoma. The clinical usefulness of autoantibodies to the mitotic spindle apparatus (NuMA) protein is unknown.

Perinuclear factor (profillagrin) autoantibodies are autoantibodies against perinuclear granules of human buccal mucosal cells. They are positive in 78% of classical RA patients and in 40% of IgM RF-negative RA patients. These antibodies signal poor prognosis when found in RF-negative individuals. These autoantibodies are present in much higher titers in RA patients than in individuals with other rheumatic diseases. The principal target antigen is profillagrin, an insoluble protein rich in histidine.

Rheumatoid arthritis is an autoimmune inflammatory disease of the joints which is defined according to special criteria designated 1 through 7. Criteria 1 through 4 must be present for more than 6 weeks. The "revised criteria" for rheumatoid arthritis are as follows: (1) morning stiffness in and around joints lasting at least 1 h before maximum improvement; (2) soft tissue swelling (arthritis) of three or more joints observed by a physician; (3) swelling (arthritis) of the proximal interphalangeal, metacarpal phalangeal, or wrist joints; (4) symmetric swelling (arthritis); (5) rheumatoid nodules; (6) presence of rheumatoid factors; and (7) roentgenographic erosions. $CD4^+$ T cells, activated B cells, and plasma cells are present in the inflamed joint lining termed the synovium, and multiple proinflammatory cytokines such as IL-1 and TNF are found in synovial joint fluid. The disease is accompanied by the production of rheumatoid factor, usually an IgM anti-IgG antibody.

Rheumatoid arthritis cell (RA cell) is an irregular neutrophil that contains a variable number of black-staining cytoplasmic inclusions that are 0.2 to 2.0 µm in diameter. These cells contain IgM rheumatoid factor, complement, IgG, and fibrin. They are found in synovial fluid of RA patients. Although RA cells may constitute 5 to 100% of a RA patient's neutrophils, they may also be present in patients with other connective tissue diseases.

RA cell is an abbreviation for rheumatoid arthritis cell.

Rheumatoid factor (RF) is an autoantibody present in the serum of patients with rheumatoid arthritis, but also found with varying frequency in other diseases such as subacute bacterial endocarditis, tuberculosis, syphilis, sarcoidosis, hepatic diseases, and others, as well as in the sera of some human allograft recipients and apparently healthy persons. RF are immunoglobulins, usually of the IgM class and to a lesser degree of the IgG or IgA classes, with reactive specificity for the Fc region of IgG. This antiimmunoglobulin antibody, which may be either monoclonal or polyclonal, reacts with the Fc region epitopes of denatured IgG, including the Gm markers. Most RF are isotype specific, manifesting reactivity mainly for IgG1, IgG2, and IgG4, but only weakly reactive with IgG3. Antigenic determinants of IgG that are potentially reactive with RF include (1) subclass-specific or genetically defined determinants of native IgG (IgG1, IgG2, IgG4, and Gm determinants); (2) determinants present on complexed IgG, but absent on native IgG; and (3) determinants exposed after enzymatic cleavage of IgG. The Gm determinants are allotypic markers of the human IgG subclasses. They are located in the IgG molecule as follows: in the C_H1 domain in IgG1, in the C_H2 domain in IgG2, and in C_H2 and C_H3 domains in IgG4. Although rheumatoid factor titers may not be clearly correlated with disease activity, they may help perpetuate chronic inflammatory synovitis. When IgM rheumatoid factors and IgG target molecules react to form immune complexes, complement is activated, leading to inflammation and immune injury. IgG rheumatoid factors may self-associate to form IgG–IgG immune complexes that help perpetuate chronic synovitis and vasculitis. IgG RF synthesized by plasma cells in the rheumatoid synovium fix complement and perpetuate inflammation. IgG RF has been shown in microbial infections, B lymphocyte proliferative disorders and malignancies, non-RA patients, and aging individuals. RF might have a physiologic role in removal of immune complexes from the circulation. Rheumatoid factors (RF) were demonstrated earlier by the Rose-Waaler test, but are now detected by the latex agglutination (or RA) test, which employs latex particles coated with IgG.

A **rheumatoid nodule** is a granulomatous lesion characterized by central necrosis encircled by a palisade of mononuclear cells and an exterior mantle of lymphocytic infiltrate. The lesions occur as subcutaneous nodules, especially at pressure points such as the elbow, and in individuals with rheumatoid arthritis or other rheumatoid diseases.

Rheumatoid pneumonitis is a diffuse interstitial pulmonary fibrosis that causes varying degrees of pulmonary impairment in 2% of rheumatoid arthritis patients. It may result from the coincidental occurrence of rheumatoid arthritis and interstitial pneumonia, which is a rare situation. Gold therapy used for RA, smoking, or contact with an environmental toxin may induce interstitial pneumonia.

SNagg refers to agglutinating activity by rheumatoid factor in certain normal sera, as revealed by a positive RA test.

Juvenile rheumatoid arthritis (JRA): When a child has inflammation in one or more joints that persists for a minimum of 3 months with no other explanation for arthritis, JRA should be suspected. It may be associated with uveitis, pericarditis, rheumatoid nodules, fever, and rash; however, RF is found in the serum in less than 10% of these patients. Subgroups include (1) polyarticular disease divided into group 1, which is HLA-DR4 associated, and group 2, which is HLA-DR5 and DR8 associated; (2) systemic Still's disease with hepatosplenomegaly, enlarged lymph nodes, pericardial inflammation, rash, and fever; and (3) pauciarticular disease in which one to nine joints are affected and associated with uveitis and anti-DNA antibodies, principally in young women.

RANA autoantibodies are RA-associated nuclear antigen autoantibodies. RA patients develop antibodies to Epstein–Barr related antigens that include viral capsid antigen, early antigen, Epstein–Barr nuclear antigen, and RA nuclear antigen. RANA antibodies are present in approximately 60% of RA patients as revealed by IFA or EIA techniques using synthetic peptide P62 that is the equivalent of the principal epitope of RANA. The clinical significance of RANA antibodies remains to be determined.

Anticentriole antibodies may occur in sera specific for the mitotic spindle apparatus (MSA). They are found rarely in the sera of subjects developing connective tissue diseases such as scleroderma.

Antifibrillarin antibodies (Figure 14.44) are specific for a 34-kDa protein constituent of U3 ribonucleoproteins (RNP) and are detectable in the serum of about 8% of patients with diffuse and limited scleroderma. Also called anti-U3 RNP antibodies.

Centriole antibodies are sometimes detected in blood sera also containing antibodies against mitotic spindle apparatus (MSA). They are only rarely found in patients developing connective tissue disease of the scleroderma category. Selected centriole antibodies may be directed against the glycolytic enzyme enolase.

Fibrillarin autoantibodies are discovered in 4 to 9% of scleroderma patients (Figure 14.45). Fibrillarin is the main 34-kDa component of U3-RNP. 56% of Black patients with diffuse cutaneous scleroderma compared with only 14% of Black patients with limited cutaneous scleroderma or white patients with diffuse and limited cutaneous scleroderma (5% and 4%, respectively), had antifibrillarin autoantibodies in their sera. Mercuric chloride can induce fibrillarin antibodies in mouse strains that recognize the same epitopes as fibrillarin antibodies found in scleroderma patients and are mainly of the IgG class. The demonstration that selected HLA haplotypes occur at a greater frequency in patients with fibrillarin antibodies suggests a genetic predisposition to developing these autoantibodies. Indirect immunofluorescence and immunoblotting are the methods of choice for antifibrillarin antibody detection.

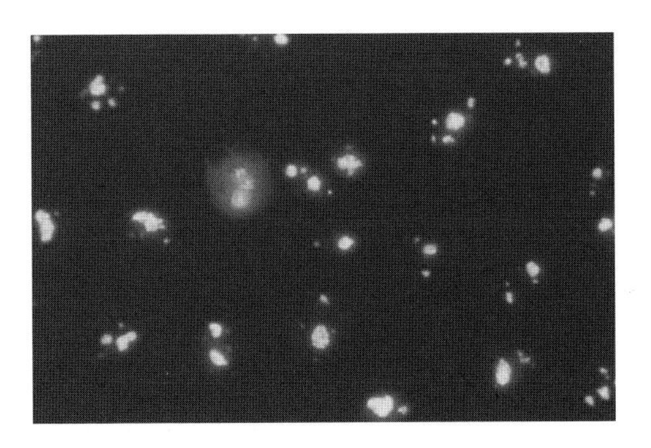

FIGURE 14.45 Antifibrillarin antibodies.

Nucleolar autoantibodies are associated with systemic sclerosis (SSc). They include autoanitbodies to nucleolar 7 to 2 RNA (present in only a small percentage of SSc), RNA polymerase I in 4% of SSc, the fibrillarin component of U3 RNP in 6% of SSc patients, and PM-Scl autoantibodies in 3% of polymyositis–scleroderma overlap syndromes. Nucleolar autoantibodies have also been observed in chronic graft-vs.-host disease following bone marrow transplantation. Indirect immunofluorescence is used to demonstrate nucleolar staining patterns that include speckled or punctuate (RNA polymerase I specificity), homogenous (PM-Scl specificity), and clumpy (U3 RNP specificity). Hep-2 cells are useful as a substrate to detect nucleolar autoantibodies.

PM-Scl autoantibodies react with nucleoli and with a complex of 16 (2 to 110-kDa) proteins. They are detected in subjects with homogenous overlap connective tissue disease marked by Reynaud's phenomenon, scleroderma, myositis arthritis, and pulmonary restriction. These autoantibodies are closely linked to the MHC class II antigen, HLA-DR. PM-Scl autoantibodies represent a good prognostic sign.

RNA polymerase represents a multiprotein complex comprised of eight to fourteen polypeptides involved in the transcription of different sets of genes into RNA. RNAP I, II, and III direct the synthesis of ribosomal RNA, messenger RNA, and selected small nuclear or cytoplasmic RNAs, including transfer RNA, respectively. Each human RNAP is comprised of two large subunits (126 to 192 kDa) and at least six small subunits (14 to 18 kDa), three of which are shared by all three RNAPs classes. Autoantibodies against RNAP I and III are very specific for systemic sclerosis and may portend a poor prognosis. Autoantibodies to RNAP II are also associated with systemic sclerosis. Although first believed to be diagnostic for systemic sclerosis, autoantibodies against RNAP II have also been detected in patients with SLE and overlap syndromes.

RNA polymerases I, II, and III autoantibodies: (Figure 14.46) Polymerase I antibodies that are found in 4% of systemic sclerosis patients give a speckled or punctate nucleolar staining pattern. Autoantibodies are also found in the urine of 46% of SLE and 19% of RA patients. Autoantibodies against RNA polymerase I, II, and III are specific for systemic sclerosis (SSc), especially diffuse cutaneous SSc. Clinically, these autoantibodies are associated with an increased frequency of heart and kidney involvement and poor 5-year survival. Of the SSc sera, 23% contain antibodies against RNA polymerase III,

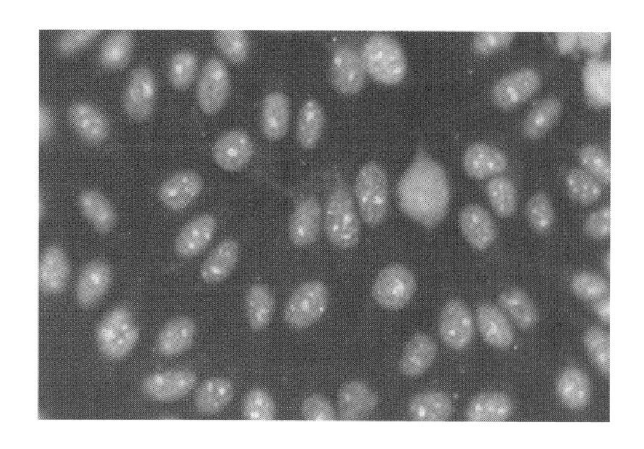

FIGURE 14.46 Anti-RNA polymerase I antibody.

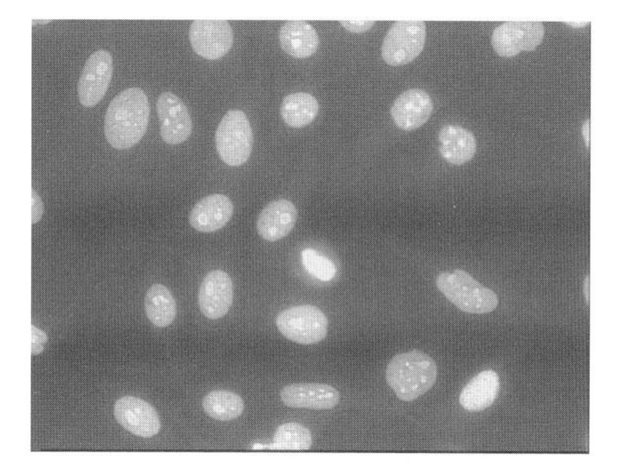

FIGURE 14.47 Scl-70 (topoisomerase I).

which is a marker for SSc with diffuse or extensive cutaneous involvement. Autoantibodies against RNA polymerase I are present in 4 to 33% of SSc patients and in subjects with SLE or overlap syndromes. Autoantibodies against RNA polymerase II are frequently accompanied by autoantibodies against RNA polymerase I and III.

Topoisomerase I is a 100-kDa nuclear enzyme that induces autoantibodies. This enzyme is concerned with the relaxation of supercoiled DNA by nicking and releasing one strand of the DNA duplex.

Scl-70 (topoisomerase I) autoantibody (Figure 14.47) is an antinuclear antibody found in 20 to 40% of diffuse scleroderma (progressive systemic sclerosis) patients who experience extensive and rapid skin involvement as well as early visceral manifestations. Two thirds of patients with these autoantibodies have diffuse scleroderma. Of limited scleroderma patients, 20% manifest topoisomerase I autoantibodies. Scl-70 autoantibodies or a DR3-DRw52a tissue type increase the risk for development of pulmonary fibrosis in scleroderma 17-fold. These autoantibodies are occasionally found in classical SLE without manifestations of scleroderma. Topoisomerase I autoantibodies portend a poor prognosis in Raynaud's phenomenon. Silica can induce a scleroderma-like condition. Some of these patients develop Scl-70 autoantibodies. The preferred technique for detection of IgG autoantibodies against topoisomerase I is EIA with immunoblot confirmation. The relevance of an association of topoisomerase I autoantibodies with neoplasms remains to be determined.

Scleroderma: See progressive systemic sclerosis.

Systemic sclerosis: See progressive systemic sclerosis.

Gastrin receptor antibodies are claimed to be present in 30% of pernicious anemia patients by some investigators but this is not confirmed by others. Gastric receptor antibodies may inhibit gastrin binding and have been demonstrated to bind gastric parietal cells.

Anticentromere autoantibody (Figure 14.47): The centromere antigen complex consists of the antigenic proteins CENP-A (18 kDa), CENP-B (80 kDa), and CDNP-C (140 kDa). Antibodies against the centromere antigens CENP-A and CENP-B are present in approximately 95% of CREST syndrome patients. They are much less frequent in cases of diffuse scleroderma and are considered a significant diagnostic marker for scleroderma with limited skin involvement, and also have prognostic significance. Indirect immunofluorescence and immunoblotting are the suggested methods for assay. Of primary biliary cirrhosis patients, half of whom also have manifestations of systemic sclerosis, 12% have anticentromere antibodies. Anticentromere antibodies do not affect survival and pulmonary hypertension in patients with limited scleroderma. However, survival is much longer in anticentromere positive patients with limited scleroderma than in anti-Scl-70 positive diffuse scleroderma patients

Centromere autoantibodies against centromeres (CENA) occur in 22% of systemic sclerosis patients, most of whom have CREST syndrome rather than diffuse scleroderma. They occur in 12% of primary biliary cirrhosis (PBC) cases, half of whom develop scleroderma. Limited scleroderma, known as CREST, occurs in 4 to 18% of PBC cases. CENA-positive limited scleroderma patients are more likely to have calcinosis and telangiectasia and less often pulmonary interstitial fibrosis compared with CENA-negative, limited scleroderma. When patients with Raynaud's syndrome become CENA-positive, they may be developing limited scleroderma. CENA may occur in silicone breast implant patients who develop symptoms associated with connective tissue type diseases. Antinuclear antibodies develop in 58% (470/813) of these patients. Antibody levels and clinical symptoms diminished when the implants were removed.

CREST syndrome is a relatively mild clinical form of scleroderma (progressive systemic sclerosis). CREST is an acronym for calcinosis, Raynaud's phenomenon, esophageal dysmotility, sclerodactyly, and telangiectasia. Skin lesions are usually limited to the face and fingers,

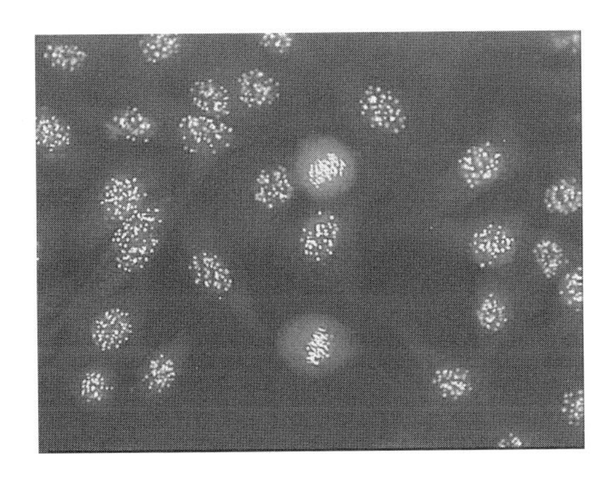

FIGURE 14.48 Anticentromere autoantibody.

with only later visceral manifestations. Most (80 to 90%) of CREST patients have anticentromere antibodies. The prognosis of CREST is slightly better than that of other connective tissue diseases, but biliary cirrhosis and pulmonary hypertension are complications. CREST patients may develop anticentromere antibodies, which may also occur in progressive systemic sclerosis patients, in aged females, or in individuals with HLA-DR1 (Figure 14.48).

Kinetochore autoantibodies are specific for the mitotic spindle apparatus that are also termed centromere autoantibodies. They have been found in the cutaneous form of systemic sclerosis, Raynaud's phenomenon, and primary biliary cirrhosis patients with features of systemic sclerosis.

Sjögren's syndrome is a condition in which immunologic injury to the lacrimal and salivary glands leads to dry eyes (keratoconjunctivitis sicca) and dry mouth (xerostomia). It may occur alone as sicca syndrome (primary form) or together with an autoimmune disease such as RA (secondary form). The lacrimal and salivary glands show extensive lymphocytic infiltration and fibrosis. Most of the infiltrating cells are $CD4^+$ T cells, but there are some B cells (plasma cells) that form antibody. Approximately 75% of the patients form rheumatoid factor. The LE cell test is positive in 25% of the patients. Numerous antibodies are produced, including autoantibodies against salivary duct cells, gastric parietal cells, thyroid antigens, and smooth muscle mitochondria. Antibodies against ribonucleoprotein antigens are termed SS-A (Ro) and SS-B (La). Approximately 90% of the patients have these antibodies. Anti-SS-B shows greater specificity for Sjögren's syndrome than do anti-SS-A antibodies, which also occur in SLE. Sjögren's syndrome patients who also have RA may have antibodies to rheumatoid-associated nuclear antigen (RANA). There is a positive correlation between HLA-DR3 and primary Sjögren's syndrome, and between HLA-DR4 and secondary Sjögren's syndrome associated with RA. Genetic predisposition, viruses, and disordered immunoregulation may play a role in the pathogenesis. About 90% of the patients are 40- to 60-year-old females. In addition to dry eyes and dry mouth, with associated visual or swallowing difficulties, 50% of the patients show parotid gland enlargement. There is drying of the nasal mucosa with bleeding, pneumonitis, and bronchitis. A lip biopsy to examine minor salivary glands is needed to diagnose Sjögren's syndrome. Inflammation of the salivary and lacrimal glands was previously called Mikulicz's disease. Mikulicz's syndrome refers to enlargement of the salivary and lacrimal glands due to any cause. Enlarged lymph nodes that reveal a pleomorphic cellular infiltrate with many mitoses are typical of Sjögren's syndrome and have been referred to as "pseudolymphoma." These patients have a 40-fold greater risk than do others of lymphoid neoplasms.

Anti-La/SS-B autoantibodies are found in Sjögren's syndrome and SLE. The La/SS-B antigen is a 47-kDa ribonucleic protein that is associated with a spectrum of small RNAs. La /SS-B is readily susceptible to proteolysis, resulting in many smaller (42 kDa, 32 kDa, and 27 kDa) but still immunoreactive polypeptides. The La/SS-B antigen primarily resides in the nucleus and is associated with RNA polymerase III transcripts. The La/SS-B antigen shows strong conservation across species. The presence of autoantibodies against the La/SS-B antigen has been advocated as a diagnostic aid in Sjögren's syndrome patients. Autoantibodies against La/SS-B are also commonly found in SLE and subacute cutaneous lupus. There is a correlation between anti-La/SS-B and the absence of nephritis in SLE patients. SLE patients with precipitating anti-Ro/SS-A antibodies have a high incidence of serious nephritis (53%), while those with both anti Ro/SS-A and anti-La/SS-B have a low (9%) frequency of nephritis.

Ro/SS-A and La/SS-B: Heterogeneous ribonucleoprotein complexes that consist of antigenic proteins (two principal proteins of 52 kDa [Ro52] and 60 kDa [Ro60] for Ro [SSA] and one protein of 48 kDa for La[SSB]) are associated with small cytoplasmic RNAs (hY-RNAs). Antibodies to Ro(SSA) and La(SSB) antigens are found frequently in the blood sera of patients with primary Sjögren's syndrome.

Anti-SS-A (Figure 14.49) is the anti-RNA antibody that occurs in Sjögren's syndrome patients. The antibody may pass across the placenta in pregnant females and be associated with heart block in their infants.

SS-A Ro is an antigen in the cytoplasm to which 25% of lupus erythematosus patients and 40% of Sjögren's syndrome patients synthesize antibodies.

SS-A/Ro antibodies are specific for SS-A/Ro antigen, which consists of 60-kDa and 52-kDa polypeptides associated with Ro RNAs. They may be demonstrated by

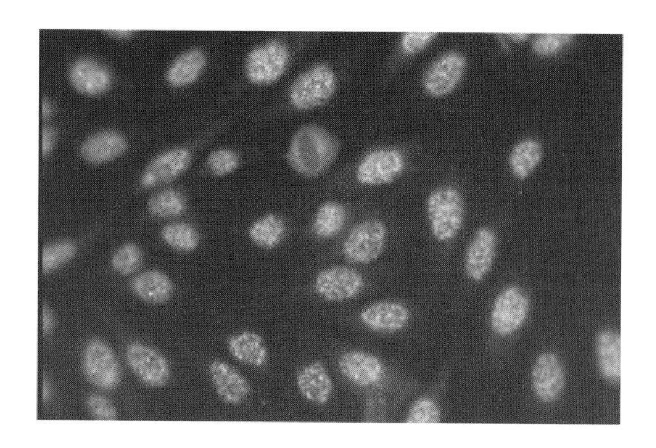

FIGURE 14.49 Anti-SS-A autoantibody.

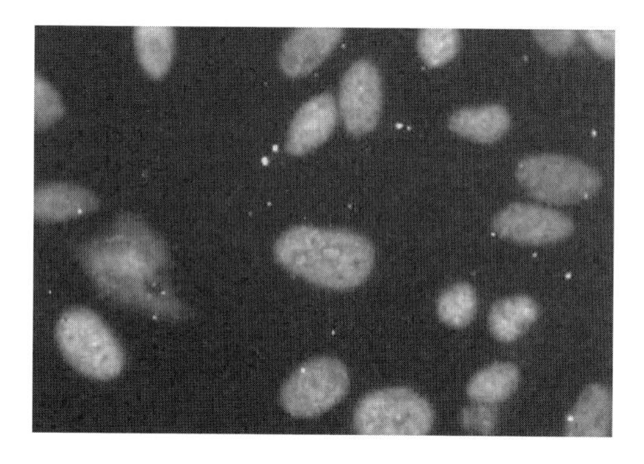

FIGURE 14.50 Anti-SS-B autoantibody.

immunodiffusion in the sera of 35% of SLE patients and 60% of Sjögren's syndrome patients.

Anti-SS-B (Figure 14.50) is an anti-RNA antibody detectable in patients with Sjögren's syndrome as well as other connective tissue (rheumatic) diseases.

SS-B La is an antigen in the cytoplasm to which Sjögren's syndrome and lupus erythematosus patients form antibodies. Anti-SS-B antibodies may portend a better prognosis in patients with lupus erythematosus.

Anti-scRNP (Ro/SS-A, La/SS-B): Anti-La/SS-B antibodies are found mainly in patients with Sjögren's syndrome but also in those with SLE and RA. These antibodies are usually found together with anti-Ro/SS-A. Anti-Ro/SS-A frequently occur alone in SLE and systemic sclerosis. There is no correlation between anti-Ro/SS-A and anti-La/SS-B antibody in patients with SLE and disease activity. Maternal anti-Ro/SS-A is the greatest single risk factor for the development of intrauterine or neonatal complete heart block in a child.

Benign lymphoepithelial lesion is an autoimmune lesion in lacrimal and salivary glands associated with Sjögren's syndrome. There are myoepithelial cell aggregates together with extensive lymphocyte infiltration.

Sicca complex is a condition characterized by dryness of mucus membranes, especially of the eyes, producing keratoconjunctivitis attributable to decreased tearing that results from lympocytic infiltration of the lacrimal glands, and by dry mouth (xerostomia), associated with decreased formation of saliva as a result of lymphocytic infiltration of the salivary glands, producing obstruction of the duct. It is most frequently seen in patients with Sjögren's syndrome, but it also may be seen in cases of sarcoidosis, amyloidosis, hemochromatosis, deficiencies of vitamins A and C, scleroderma, and hyperlipoproteinemia types IV and V.

SS-B/La antibodies are specific for SS-B/La antigen, which is a 48-kDa nucleopolasmic phosphoprotein associated with selected Ro small RNA (Ro hY1–hY5). These antibodies may be demonstrated by EIA in Sjögren's syndrome that is either primary or secondary to RA or in SLE.

Antibodies to histidyl t-RNA synthetase (anti-HRS) are antibodies whose formation is closely associated with DR3 in Caucasians. In African Americans, DR6 may be increased in patients with this autoantibody. The supertypic DR specificity DR52 is found among essentially all individuals with anti-HRS. This specificity is encoded by the *DRB3* gene, which is present on haplotypes bearing DR3, DR5, or DR6, and is also represented on the DR8 molecule. Essentially all patients with anti-HRS had one or two of these DR alleles. All were DR52 positive. The first hypervariable regions of the *DRB1* genes encoding DR3, DR5, DR6, and DR8 have homologous sequences. This homology is believed to predispose to the development of immune reactivity against HRS.

Dermatomyositis is a connective tissue or collagen disease characterized by skin rash and muscle inflammation. It is a type of polymyositis presenting with a purple-tinged skin rash that is prominent on the superior eyelids, extensor joints surfaces, and base of the neck. There is weakness, muscle pain, edema, and calcium deposits in the subcutaneous tissue, especially prominent late in the disease. Blood vessels reveal lymphocyte cuffing, and autoantibodies against tRNA synthetases appear in the serum.

Jo-1 autoantibodies (Figure 14.51): A total of 30% of adult patients with myositis (including polymositis, dermatomyositis, and combination syndromes) form autoantibodies against the Jo-1 antigen (54-kDa histidyl-tRNA synthetase). These antibodies against Jo-1 occur in approximately 60% of patients with both myositis and interstitial lung disease. Jo-1 antibodies are most often found, in addition to other aminoacyl sythetases, in individuals with antisynthetase syndrome, which is marked by acute onset, steroid-responsive myositis with interstitial lung disease, fever, symmetrical arthritis, Raynaud's phenomenon, and mechanic's hands. Jo-1 autoantibodies

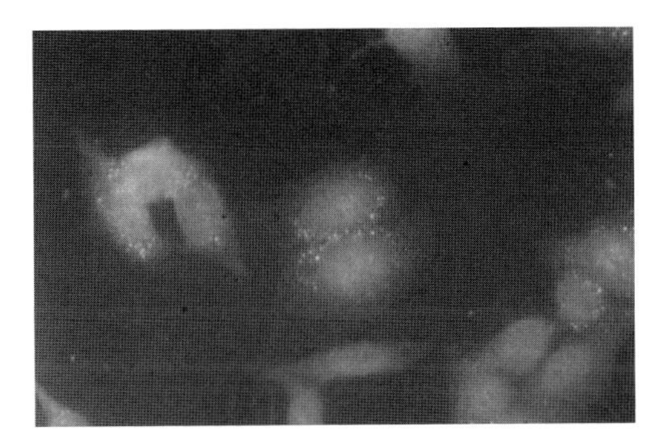

FIGURE 14.51 Jo-1 autoantibodies.

in patients with idiopathic polymyositis are usually associated with severe relapse and poor prognosis.

Jo1 syndrome is a clinical condition in which anti-Jo1 antigen (histidyl-tRNA synthetase) antibodies are produced. Arthritis and myositis, as well as interstitial lung disease, may be present. One quarter of myositis patients may manifest anti-Jo1 antibodies.

Myositis-associated autoantibodies specific for nucleolar antigen PM-Scl are found in patients with features of scleroderma and polymyositis/dematomyositis. This antibody is specific for a complex of nucleolar proteins. The principal antigen, having a molecular mass of 100 kDa is found mainly in Caucasians with the overlap syndrome. Antibodies against the Ku antigen are found in a few Japanese patients with overlap syndrome. The Ku antibody is specific for a 70-kDa and an 80-kDa DNA-binding proteins. Anti-Ku autoantibodies are also found in SLE, dermatomyositis, scleroderma, thyroid disease, or Sjögren's syndrome. Other autoantibodies found in inflammatory myopathies include anti-U_1 RNP that helps define mixed connective tissue disease. High titers of anti-RNP may also be found in inflammatory muscle disease. Autoantibodies against SS-A (Ro), antithyroid microsome antibodies, and rheumatoid factor have also been found in myositis syndromes.

Myositis-specific autoantibodies: Many patients with idiopathic inflammatory myopathies (IIMs) generate autoantibodies against aminoacyl-transfer (tRNA) synthetases (myositis-specific antibodies), which represent a group of cellular enzymes that catalyze binding of one amino acid to its tRNA. Autoantibodies against eight of these synthetases have been found in IIMs. J o-1 antibodies specific for histidyl-tRNA synthetase are most common, i.e., found in 20 to 30% of IIMs. Aminoacyl-tRNA synthetase autoantibodies are closely associated with interstitial lung disease, arthritis, and Raynaud's phenomenon. Other myositis-specific autoantibodies include those that react with signal recognition particles (SRP) in acute onset severe myalgic IIM, Mi-2 antibodies reactive with 235 to 240 kDa nuclear antigen in 15 to 25% of dermatomyositis patients, antielongation factor Iα autoantibodies, and RNA-reactive autoantibodies.

Signal recognition particle autoantibodies against SRP are mitochondrial, ribosomal, and certain cytoplasmic autoantibodies that stain Hep-2 cell cytoplasm and can be assayed by RIP, EIA, and cytoplasmic (non-Jo-1) fluorescence in approximately 4% of polymyositis/dermatomyositis (PM/DM) patient sera. SRP autoantibodies characterize a homogeneous group of patients with similar clinical features (typically Black females with acute onset of severe polymyositis in the fall, together with cardiac involvement and resistance to therapy, including corticosteroids, and with a mortality of 75% at 5 years). SRP autoantibodies are not usually detected in patients with arthritis, dermatomyositis, pulmonary fibrosis, or Raynaud's phenomenon.

Threonyl-transfer RNA synthetase antibodies (Figure 14.51) against threonyl-tRNA synthetase (threonyl RS) protein that shows high specificity for myositis. These antibodies were demonstrated in 4% of polymyositis/dermatomyositis patients. Antithreonyl RS antibodies have also been linked to the antisynthetase syndrome, in which there is fever, Raynaud's phenomenon, symmetrical arthritis, interstitial lung disease, myositis, and mechanic's hands.

Antiendothelial cell autoantibodies may cause vasculitis as part of an autoimmune response.

Cytoplasmic antigens are immunogenic constituents of cell cytoplasm that induce autoantibody formation in patients with generalized autoimmune diseases who also manifest antinuclear antibodies. Thus, anticytoplasmic antibodies are included under the term "ANA."

Endothelial cell autoantibodies (ECA) are a heterogeneous family of antibodies that react with various antigens expressed on resting and activated endothelial cells. They are IgG antibodies present in the blood sera of SLE patients that may mediate immunologic injury to blood-vessel walls. Some are associated with nephritis in SLE, but ECA are not SLE restricted. Cytotoxic antibodies against vascular endothelial cells have been demonstrated in subjects with hyperacute rejection of cardiac allografts who had a compatible direct lymphocytotoxic crossmatch, in patients who had rejected renal allografts, in hemolytic–uremic syndrome, and in Kawasaki syndrome. ECA that are not cytolytic have been demonstrated in patients with micropolyarteritis and Wegener's granulomatosis, as well as in approximately 44% of dermatomyositis patients, particularly those who also have interstitial lung disease. Serum antibodies that bind to vascular endothelial cells, some with anti-HLA class I specificity, have been demonstrated in about one third of patients with IgA nephropathy. Sera from scleroderma, Wegener's granulomatosis, and microscopic polyarteritis patients mediate antibody-dependent cellular cytotoxicity. Cytokine-mediated activation of vascular endothelium is believed to be significant in the pathogenesis of antibody-mediated and cell-mediated entry to blood vessels. They may be involved in the pathogenesis of rheumatoid vasculitis.

Giant cell arteritis is an inflammation of the temporal artery in middle-age subjects that may become a systemic arteritis in 10 to 15% of affected subjects. Blindness may occur eventually in some of them. There is nodular swelling of the entire artery wall. Neutrophils, mononuclear cells, and eosinophils may infiltrate the wall with production of giant-cell granulomas.

Anti-U1 RNP autoantibodies to ribonucleoprotein (U1 RNP) have been used to partially define MCTD. High titer anti-RNP occurs also in inflammatory muscle disease. Autoantibodies against U1 RNP in Caucasians with or without MCTD are associated with DR4 anti-U1Sn RNP. Autoantibodies in Japanese demonstrated an increased frequency of DQ3, only a few of whom had MCTD. Further studies demonstrated an association between DRB1 *0401/DRB4 *0101/DQA1*03/DQB1*0301, and MCTD.

Speckled pattern is a type of immunofluorescence produced when the serum from individuals with one of several connective tissue diseases is placed in contact with the human epithelial cell line HEp-2 and "stained" with fluorochrome-labeled goat or rabbit antisera against human immunoglobulin. The speckled pattern of fluorescence occurs in mixed connective tissue disease, lupus erythematosus, polymyositis, sicca syndrome, Sjögren's syndrome, drug-induced immune reactions, and rheumatoid arthritis. It is the most frequent and shows the greatest variations of immunofluorescent nuclear staining patterns. It is classified as (1) fine speckles associated with anticentromere antibody; (2) coarse speckles associated with antibodies against the nonhistone nuclear proteins Scl-70, nRNP, SS-B/La, and Sm; (3) large speckles that may be limited to 3 to 10 per nucleus, which are seen in undifferentiated connective tissue disease and represent IgM antibody against class H3 histones.

U1 snRNP autoantibodies against 70-kDa, A and C protein constituents of U1 snRNPs are often found only in SLE and SLE-overlap syndrome which include MCTD, and less frequently in SLE. U1 snRNP autoantibodies were formerly termed RNP (ribonucleoprotein) or nRNP (nuclear RNP) antibodies. A total of 38% of sera that react with protein constituents of U1 snRNPs also interact with the RNA portion of the snRNP particle. Unlike dsDNA antibodies, titers of antibodies against 70-kDa or A constituents of U1 snRNP are not helpful in monitoring disease activity or predicting SLE flares. High-titer RNP antibodies are not linked to an increased risk of fetal loss in SLE. Sm antibody sera do not contain U1 RNA antibodies. A total of 60% of sera with U1 smRNP antibodies have fall quantities of Sm antibodies and may be designated as anti-RNP/Sm sera. Besides Ul smRNP and Sm antibodies, a less common U1/U2 snRNP antibody reactive with Ul-A, U2-B′ proteins is present in SLE-overlap syndromes and in SLE. Specific T-lymphocyte and antibody epitopes are present in the 70-kDa protein of U1 snRNP. These epitopes are recognized in MCTD and to a lesser degree in other systemic rheumatic diseases.

Silicate autoantibodies are stimulated by silicone plastic polymers used for breast implants in cosmetic and reconstructive surgery, vascular prostheses, and joint repair and replacement. Silicone-induced human adjuvant disease characterized by autoimmune disease-like symptoms, granulomas, and serological abnormalities have been reported since 1964. Some individuals with breast and joint implants have been reported to develop delayed-type hypersensitivity to silicone plastic as interpreted by refractile particles in phagocytes and passage between lymphocytes and macrophages. Women who develop either implant ruptures or leakage of their silicone gel implant may develop high titers of silicone autoantibodies as measured using silicate-coated plates in EIA. Only 4% of 249 patients with silicone breast implants developed IgG binding to fibronectin-laminin adsorbed to silicone, and only 2% showed IgG binding to silicone film alone. Thus, patients occasionally develop antibodies against fibronectin and laminin denatured by silicone. Whereas 30% of 40 symptomatic women with silicone breast implants formed IgG antibodies that reacted with BSA-bound silicate, 9% (8/91) of asymptomatic silicone breast implant patients also formed these antibodies in low concentrations.

T cell vaccination (TCV) is a technique to modulate an immune response in which T lymphocytes are administered as immunogens. The vaccine is composed of T cells specific for the target auto-antigen in an autoimmune response to be modulated. For example, antimyelin basic protein (MBP) $CD4^+$ 8-T cells, serving as a vaccine, were irradiated (1500 R) and injected into Lewis rats. Vaccination with the attenuated anti-MBP T cells induced resistance against subsequent efforts to induce experimental allergic encephalomyelitis (EAE) by active immunization to myelin basic protein in adjuvant. This technique could even induce resistance against EAE adaptively transferred by active anti-MBP T cells. This method was later successfully applied to other disease models of adjuvant arthritis, collagen-induced arthritis, experimental autoimmune neuritis, experimental autoimmune thyroiditis, and IDDM. TCV was demonstrated to successfully abort established autoimmune disease and spontaneous autoimmune disease in the case of IDDM.

Chemokine autoantibodies are autoantibodies against members of the chemokine family that include macrophage inflammatory proteins (MIP-1-α), i.e., stem-cell inhibitor), macrophage inflammatory protein (MIP-1-β, macrophage inflammatory proteins (MIP)-2-α (i.e., GRO-β), GRO-α, MIP-2-α (i.e., GRO-γ), platelet factor-4 (PF4), IL-8, macrophage chemotactic and activating factor, IP-10, monocyte chemoattractant protein (MCP)-1, and RANTES. MCP-1, MIP-1, and RANTES are the mononuclear cell chemoattractant equivalent of IL-8. All these molecules are able to stimulate leukocyte movement (chemokinesis) and directed movement (chemotaxis). IL-8, the best-known member of this subfamily, is a proinflammatory cytokine synthesized by various cells that act as a neutrophil activator and chemotactic factor. It is believed to be responsible for the induction and maintenance of localized inflammation. Monoclonal antibodies against IL-8 and MCP-1 receptors can distinguish between MCP-1- and IL-8-responsive T lymphocyte subsets. The EIA technique is the preferred method to assay chemokines and chemokine autoantibodies.

Golgi autoantibodies (Figure 14.52) against golgi apparatus cisternal and vesicular membranes are very rare. A total of 10% of normal controls have no titer Golgi autoantibodies. The Golgi apparatus is a target of heterogeneous autoantibodies with specificity for five autoantigens that include golgin 95, golgin 160, golgin 160, golgin 97, golgin 180, macrogolgin, and GCp372. Immunofluorescence using Hep-2 cells is the technique employed to detect the golgi autoantibodies but no diagnostic significance has yet been correlated with disease activity.

Heteroantibody refers to an autoantibody with the ability to cross-react with an antigen of a different species.

I-K is the abbreviation for immunoconglutinin.

Paraneoplastic autoimmune syndromes are diseases that affect specific organ systems and are induced by a tumor but are caused by remote effects of the neoplasm and not by direct infiltration or by tumor metastases.

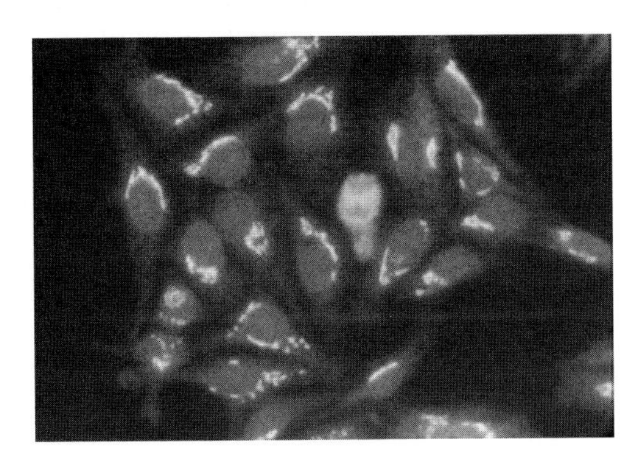

FIGURE 14.52 Golgi autoantibodies.

Neoplasms may induce perturbations of the immune system with injurious effects of various organ systems such as the central nervous system, the eye, and the skin. Tumor-immune system interaction induces tissue injury in distant organs by various mechanisms such as autoantibody-induced tissue injury.

Exposure of humans to **xenobiotics**, including chemicals and drugs, may lead to autoimmune responses resembling the response to infection with microorganisms, but actual autoimmune disease is observed less frequently. Various animal models of autoimmunity have been induced by chemicals such as drug-induced lupus, which nevertheless does not correspond exactly to the human disease. Yet, renal autoimmune disease caused by metals shows close similarity in animals and humans. Isoniazid, hydralazine, and procainamide have been implicated in drug-induced lupus. Iodine administration to experimental animals prone to developing spontaneous autoimmune thyroiditis favors the development of this disease. Toxic oil syndrome has been induced in B10.S mice by the intraperitoneal administration of oleic acid anilide. $HgCl_2$ and administration to rats, mice, and rabbits leads to the production of autoimmune kidney disease. Streptozotocin administration induces an experimental model of insulin-dependent diabetes mellitus in animal models. The anesthetic halothane induces liver changes in rabbits and rats that lead to the formation of halothane antigens, and complete Freund adjuvant can induce adjuvant arthritis in rats.

Addison's disease: See autoimmune adrenal disease.

Chrysotherapy: See gold therapy.

Undifferentiated connective tissue disease refers to a prodromal phase of a collagen vascular or connective tissue disease. At this stage, the principal clinical manifestations have not yet become apparent.

Molecular mimicry is the sharing of antigenic determinants or epitopes between cells of an immunocompetent host and a microorganism. This may lead to pathologic sequelae if antibodies produced against the microorganism combine with antigens of self and lead to immunologic injury. Ankylosing spondylitis and rheumatic fever are examples. Immunologic cross-reactivity between a viral antigen and a self antigen or between a bacterial antigen such as streptococcal M protein and human myocardial sarcolemmal membranes may lead to tissue injury.

Genetics has long been recognized to play an important role in autoimmunity. Susceptibility genes increase probability but do not determine by themselves development of an autoimmune disease. MHC genes have been shown to have a strong association with selected autoimmune diseases. This is especially true of the class II MHC genes. In addition, some non-MHC genes may also play a role in the development of autoimmunity. There is a higher frequency of certain HLA alleles in persons with selected autoimmune diseases than in members of the general population.

Cross reactive epitopes involving molecular mimicry may be either sequential (common amino-acid determinants between host and pathogen) or conformational (cross-reactivity to chemical moieties such as glycans, phosphates, or sulfates that reveal a specific tertiary configuration and electrostatic potential). Rothbard epitopes, which are common cognate T cell epitopes, may also induce cross reactive T lymphocyte responses. Molecular mimicry may also be observed in the idiotype network in which anti-id antibodies can generate a mirror image of the original antigen.

Anti-phospholipid antibodies: See lupus anticoagulant.

Anti-rRNP: Antibodies that react with ribosomal phosphoproteins P1 and P2. Anti-P is present in 10% of SLE patients.

Virus infection associated autoantibodies: Viral infections may stimulate the production of autoantibodies in three ways:

1. By complexing with cell surface histocompatibility antigens to form new immunogenic units
2. By nonspecifically stimulating the proliferation of lymphocytes, e.g., after infection with Epstein-Barr virus; the nonspecific response includes clones of cells specific for autoantigens
3. By inducing the expression of antigens normally repressed in the host cells' autoantigens

Normal body constituents are recognized by autoantibodies specific for them. T cell receptors may also identify autoantigens (self antigen) when the immune reactivity has induced a cell-mediated T lymphocyte response.

APECED (autoimmune polyendocrinopathy-candidiasis-ectodermal dystrophy) is an autosomal recessive disorder characterized by hypoparathyroidism, adrenal cortical failure, gonadal failure, candidiasis, and malabsorption.

Beta-adrenergic receptor autoantibodies against β1-adrenoreceptors are detected in a significant proportion of individuals with dilated cardiomyopathy. The human heart contains both β1-adrenoreceptors and a substantial number of β2-adrenoreceptors. Both types of receptors are diminished in chronic heart failure. β1-adrenoreceptors are decreased in all types of heart failure, whereas β2-adrenoreceptors are diminished in mitral valve disease, tetralogy of Fallot, and end stage ischemic cardiomyopathy. In total, 30 to 40% of dilated cardiomyopathy, 22% of ischemic cardiomyopathy, and 25% of alcoholic cardiomyopathy patients develop autoantibodies against β-adrenoreceptors. In familial idiopathic dilated cardiomyopathy, 62% of affected family members manifest the β-adrenoreceptor autoantibodies, compared to 29% in unaffected members. Dilated cardiomyopathy patients have an increased frequency of HLA-DR4 antigen compared to normal persons. 60 to 80% of patients with HLA-DR4 and dilated cardiomyopathy develop β-adrenoreceptor antibodies.

Deoxyribonucleoprotein antibodies are reactive with insoluble deoxyribonucleoprotein (DNP), which occur in 60 to 70% of patients with active systemic lupus erythematosus. IgG DNP antibodies that fix complement cause the LE cell phenomenon and yield a homogeneous staining pattern.

GM_1 autoantibodies are specific for a principal sialic acid-containing glycolipid enriched in peripheral and central nervous system myelin, i.e., Schwann cell and oligodendrocytes. It participates in the recognition of cells, compaction of myelin, signal transduction, and chemokine binding. Autoantibodies are specific for the terminal sialic acid and/or the galactosyl (β1-3) *N*-acetylgalactosamine epitope of GM_1. GM_1 antibodies are present in amyotrophic lateral sclerosis (ALS), Guillain-Barré syndrome (GBS), and other motor neuron diseases (MND), including chronic inflammatory demyelinating polyardiculoneuropathy (CIDP) and multifocal motor neuropathy (MMN). Sensorimotor neuropathies with plasma cell dyscrasia (monoclonal gammopathy) associated with IgM GM_1 paraproteins at frequencies of 50 to 60%. Acute peripheral neuropathy patients may demonstrate IgG GM_1 antibodies, especially following an infection with *Campylobacter jejuni*. There may be molecular mimicry between gangliosides and bacterial lipopolysaccharides (LPS). EIA detection of GM_1 antibody reveals 50% sensitivity and 80% specificity for motor neuron diseases. These antibodies may also be associated with systemic infections and autoimmune diseases with neurologic involvement.

Multiple autoimmune disorders (MAD): In type I MAD, a patient must manifest a minimum of two of the diseases designated Addison's disease, mucocutaneous candidiasis, or hypoparathyroidism. Type II MAD is known as Schmidt syndrome, in which patients manifest at least two conditions from a category that includes autoimmune thyroid disease, Addison's disease, mucocutaneous candaiasis, and insulin-dependent diabetes mellitus, with or without hypopituitarism.

Oxidized low-density lipoprotein autoantibodies against ox-LDL represent a good marker of LDL oxidation. Ox-LDL is immunogenic and leads to increased IgG but not IgM or IgA ox-LDL antibodies with carotid atherosclerosis. Ox-LDL antibodies are also characterized as antiphospholipid antibodies whereas most antibodies against ox-LDL are specific for apolipoprotein B epitopes, the major lipoprotein of LDL, and a considerable number are reactive against peroxidized phospholipid.

A **side effect** is an unwanted reaction to a drug administered for some desirable curative or other effect.

Polyendocrine autoimmunity: See polyendocrine deficiency syndrome (polygranular autoimmune syndrome).

Sedormid® purpura (historical): In the past, a form of thrombocytopenic purpura occurring in patients who received the drug Sedormid® (allyl-isopropyl-acetyl carbamide). The drug served as a hapten complexing with blood platelets. This platelet–drug complex was recognized by the immune system as foreign. Antibodies formed against it lysed the patient's blood platelets in the presence of complement, leading to thrombocytopenia. This was followed by bleeding, which caused the purpura manifested on the skin. This was a type II hypersensitivity reaction. Sedormid® is no longer used, but the principle of hypersensitivity it induced is useful for understanding autoimmunity induced by certain drugs.

IgG-induced autoimmune hemolysis, two-fifths of which are secondary to other diseases such as neoplasia, including chronic lymphocytic leukemia or ovarian tumors. It may also be secondary to connective tissue diseases such as lupus erythematosus, rheumatoid arthritis, and progressive systemic sclerosis. Patients experience hemolysis leading to anemia with fatigue, dizziness, palpitations, exertion dyspnea, mild jaundice, and splenomegaly. Patients manifest a positive Coomb's antiglobulin test. Their erythrocytes appear as spherocytes and schistocytes, and there is evidence of erythrophagocytosis. There is erythroid hyperplasia of the bone marrow, and lymphoproliferative disease may be present. Glucocorticoids and blood transfusions are used for treatment. Approximately 66% improve after splenomectomy, but often relapse. Three quarters of these individuals survive for a decade.

15 Mucosal Immunity

Local immunity refers to immunologic reactivity confined principally to a particular anatomic site such as the respiratory or gastrointestinal tract. Local antibodies as well as lymphoid cells present in the area may mediate a specific immunologic effect. For example, secretory IgA produced in the gut may react to food or other ingested antigens.

The **mucosal immune system** describes aggregates of lymphoid tissues or lymphocytes near mucosal surfaces of the respiratory, gastrointestinal, and urogenital tracts (Figure 15.1 to Figure 15.4). There is local synthesis of secretory IgA and T cell immunity at these sites. The mucosal epithelial layer serves as a mechanical barrier against foreign antigens and invading microorganisms. A specialized immune system, sometimes referred to as the common mucosal immune system (CMIS), located at epithelial surfaces, represents a critical defense mechanism. The mucosal immune system consists of secretory IgA molecules produced by plasma cells in the lamina propria and subsequently transported across epithelial cells with the aid of the polyimmunoglobulin receptor. Both $\alpha\beta$ and $\gamma\delta$ T lymphocytes are present in the mucosal epithelial layer as intraepithelial lymphocytes. They are also found in the lamina propria of the mucosa, where they serve as an integral component of the cellular immune system. These T lymphocytes function in the induction and regulation of responses by antigen-specific IgA B cells as well as effector T cells. The epithelial cells lining mucosal surfaces furnish signals that are significant for the initiation of the mucosal inflammatory response and critical communications between epithelial cells and mucosal lymphoid cells. The immune response to oral antigens differs from the response to parenterally administered immunogens. Oral tolerance may follow the ingestion of some protein antigens, but a vigorous local mucosal immune response with the production of high concentrations of IgA may follow oral immunization with selected vaccines such as the Sabin oral polio vaccine.

Mucosa-associated lymphoid tissue (MALT) includes extranodal lymphoid tissue associated with the mucosa at various anatomical sites, including the skin (SALT), bronchus (BALT), gut (GALT), nasal associated lymphoid tissue (NALT), breast, and uterine cervix. The mucosa-associated lymphoid tissues provide localized or regional immune defense since they are in immediate contact with foreign antigenic substances, thereby differing from the lymphoid tissues associated with lymph nodes, spleen, and thymus. Secretory or exocrine IgA is associated with the MALT system of immunity. The lymphoid tissues comprising MALT include intraepithelial lymphocytes, principally T lymphocytes, together with B cells beneath the mucosal epithelia and in the lamina propria.

Gut-associated lymphoid tissue (GALT) describes lymphoid tissue in the gastrointestinal mucosa and submucosa. It constitutes the gastrointestinal immune system (Figure 15.5). GALT is present in the appendix, in the tonsils, and in the Peyer's patches subjacent to the mucosa. GALT represents the counterpart of BALT and consists of radially arranged and closely packed lymphoid follicles which impinge upon the intestinal epithelium, forming dome-like structures. In GALT, specialized epithelial cells overlie the lymphoid follicles, forming a membrane between the lymphoid cells and the lumen. These cells are called M cells. They are believed to be "gatekeepers" for molecules passing across. Other GALT components include IgA-synthesizing B cells and intraepithelial lymphocytes such as $CD8^+$ T cells, as well as the lymphocytes in the lamina propria that include $CD4^+$ T lymphocytes, B lymphocytes which synthesize IgA, and null cells.

The **lamina propria** is the thin connective tissue layer that supports the epithelium of the gastrointestinal, respiratory, and genitourinary tracts (Figure 15.6). The epithelium and lamina propria form the mucous membrane. The lamina propria may be the site of immunologic reactivity in the gastrointestinal tract, representing an area where lymphocytes, plasma cells, and mast cells congregate.

A polyimmuoglobulin receptor is an attachment site for polymeric immunoglobulins located on epithelial cell and hepatocyte surfaces that facilitate polymeric IgA and IgM transcytosis to the secretions. After binding, the receptor immunoglobulin complex is endocytosed and enclosed within vesicles for transport. Exocytosis takes place at the cell surface where the immunoglobulin is discharged into the intestinal lumen. A similar mechanism in the liver facilitates IgA transport into the bile. The receptor–polymeric immunoglobulin complex is released from the cell following cleavage near the cell membrane. The receptor segment that is bound to the polymeric immunoglobulin is known as the secretory component, which can only be used once in the transport process.

Lymphocytes in the gastrointestinal mucosa may be present in the lamina propria, Peyer's patches, and the epithelial layer. In humans, the intraepithelial lymphocytes are mostly $CD8^+$ T cells. $\gamma\delta$ T cells make up these intraepithelial lymphocytes to varying degrees according to the species but constitute approximately 10% in humans. The $\gamma\delta$ as well as the $\alpha\beta$ intraepithelial T cells have restricted specificities probably corresponding to antigens frequently found in the gut. Numerous activated B cells, plasma cells, activated $CD4^+$ T cells, eosinophils, macrophages, and mast cells are present in the lamina propria of the intestine. T cells are believed to interact with antigen in regional mesenteric lymph nodes and then return to the intestinal lamina propria.

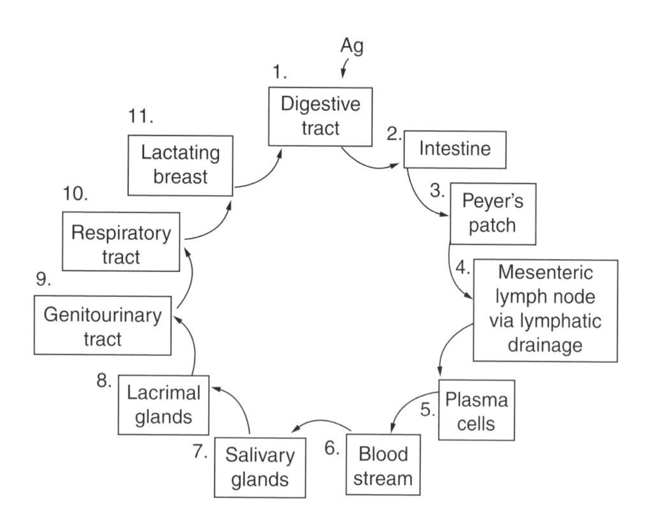

FIGURE 15.1 Secretory immune system.

Mucosa homing is a selective return of immunologically reactive lymphoid cells that originated in mucosal follicles, migrated to other anatomical locations, and then returned to their site of origin in mucosal areas. Receptors on lymphoid cells and ligands on endothelial cells are responsible for the cellular migration involving the mucosal immune system (Figure 15.7).

MadCAM-1 is a mucosal addressin cell adhesion molecule-1 which is an addressin in Peyer's patches of mice (Figure 15.8). This three-Ig domain structure with a polypeptide backbone binds the $\alpha 4\beta 7$ integrin. MadCAM-1 facilitates access of lymphocytes to the mucosal lymphoid tissues, as in the gastrointestinal tract. Addressin is a molecule such as a peptide or protein that serves as a homing device to direct a molecule to a specific location. Lymphocytes from Peyer's patches home to mucosal endothelial cells bearing ligands for the lymphocyte homing receptor.

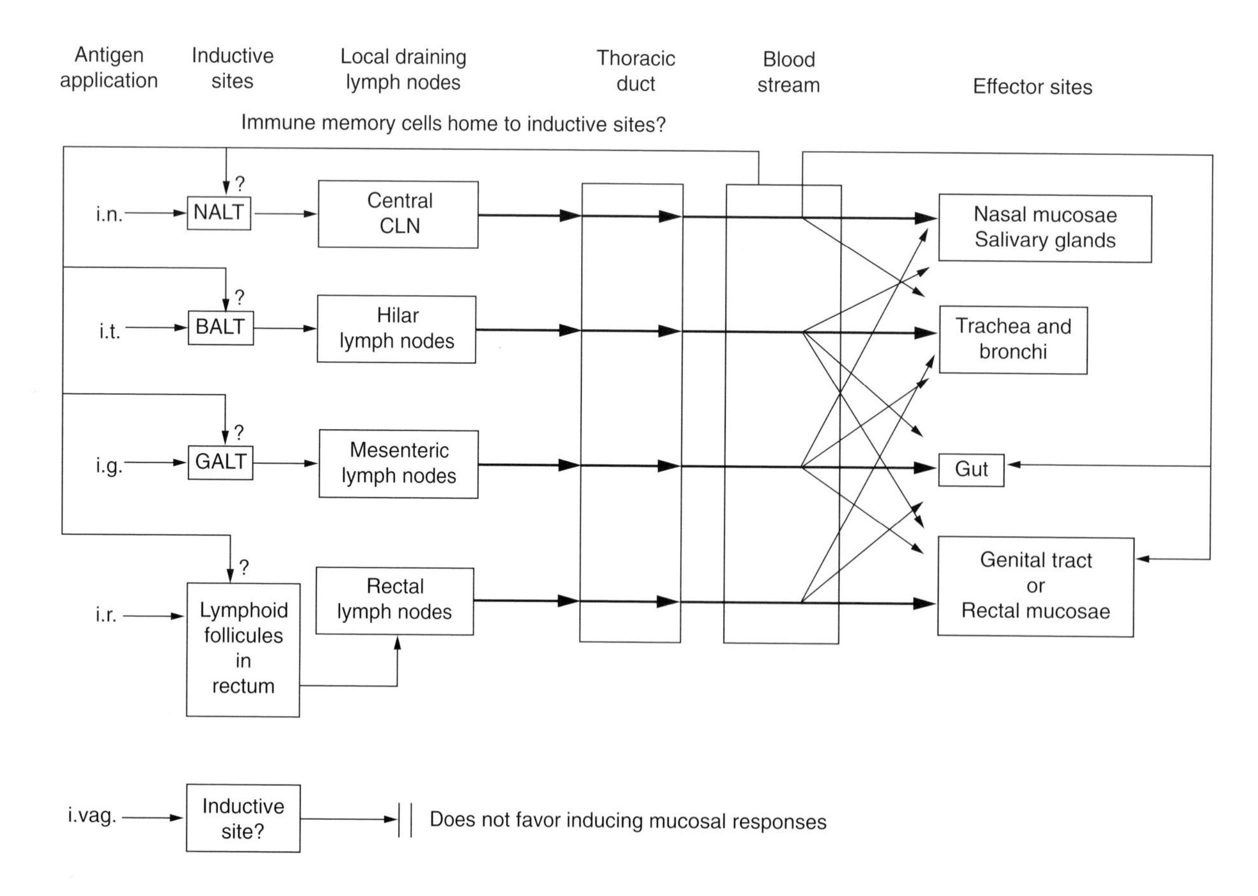

FIGURE 15.2 Compartmentalized common mucosal immune system.

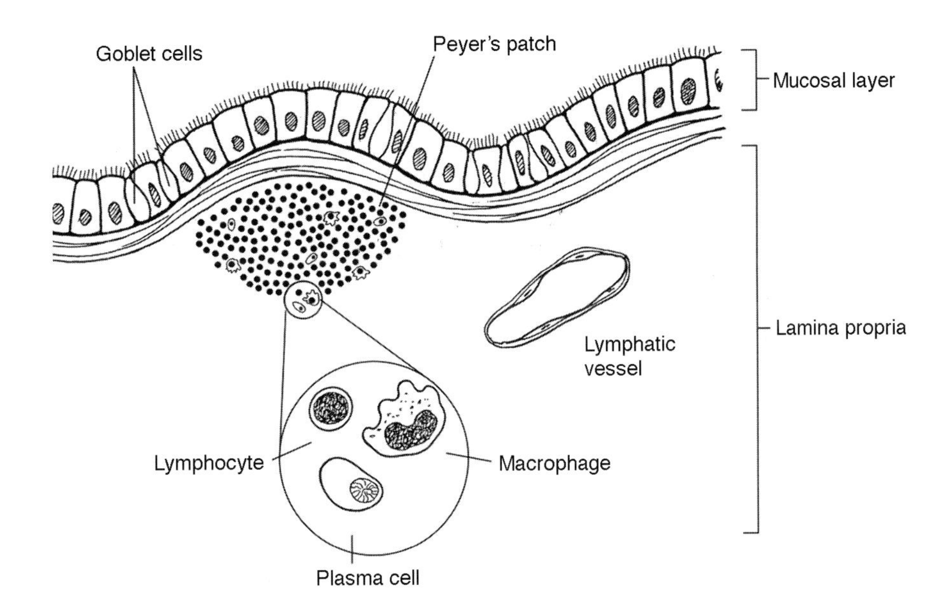

FIGURE 15.3 The mucosal system and its cellular components.

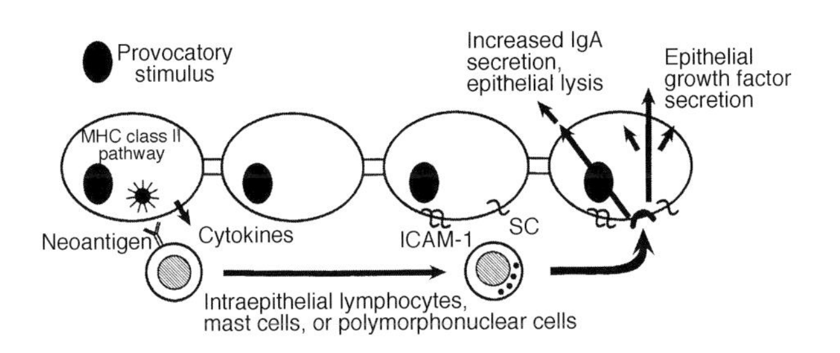

FIGURE 15.4 Common mucosal inflammatory pathway.

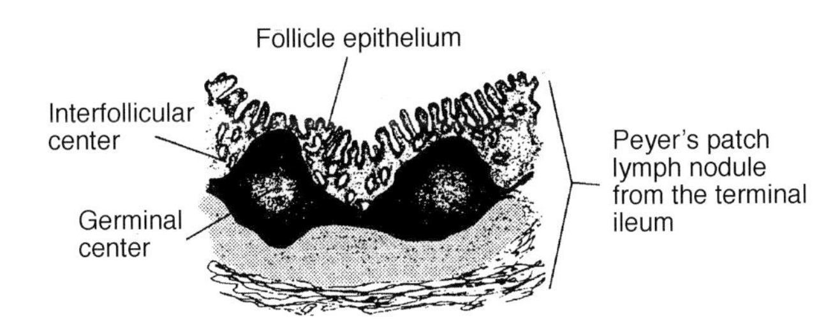

FIGURE 15.5 GALT (gut-associated lymphoid tissue).

L-Selectin is a molecule found on lymphocytes that is responsible for the homing of lymphocytes to lymph node high endothelial venules. L-Selectin is also found on neutrophils where it acts to bind the cells to activated endothelium early in the inflammatory process. L-Selectin is also called CD62L.

Mucosal lymphoid follicles include such structures as Peyer's patches in the small intestine and pharyngeal tonsils. The appendix and other areas of the gastrointestinal tract and respiratory tract contain similar aggregrates of lymphoid cells. Germinal centers at the center of lymphoid follicles have an abundance of B cells. $CD4^+$ T cells are present in interfollicular regions of Peyer's patches. One-half to three quarters of the lymphocytes in murine Peyer's patches are B cells, whereas 10 to 30% are T lymphocytes. M cells overlying Peyer's patches are membranous (M) cells devoid of microvilli, are pinocytic, and convey macromolecules to subepithelial tissues from the lumen of the intestine. Although M cells are believed to transport antigens to Peyer's patches, they do not act as antigen-presenting cells. Lymphoid cells in the blood migrate to the gut mucosa. The integrin α_4 associated β_7 is critical for endothelial binding of lymphocytes in the intestine and migration of cell into the mucosa.

The **M cell** is a gastrointestinal tract epithelial cell that conveys microorganisms and macromolecular substances from the gut lumen to Peyer's patches. M cells are nonantigen-presenting cells found in the epithelial layer of the Peyer's patches that, nevertheless, may have an important role in antigen delivery. They have relatively large surfaces with microfolds that attach to microorganisms and macromolecular surfaces. The M cell cytoplasmic processes extend to $CD4^+$ T cells underneath them. Materials attached to microvilli are conveyed to coated pits and moved to the basolateral surface, which has pronounced invaginations rich in leukocytes and mononuclear phagocytes. Thus, materials gaining access by way of M cells come into contact with lymphoid cells as they reach the basolateral surface. This is believed to facilitate induction of immune responsiveness.

The term "**microfold cells**" refers to M cells.

Lymphocytes present in the intestinal epithelium or other specialized epithelial layers are termed **intraepithelial lymphocytes**.

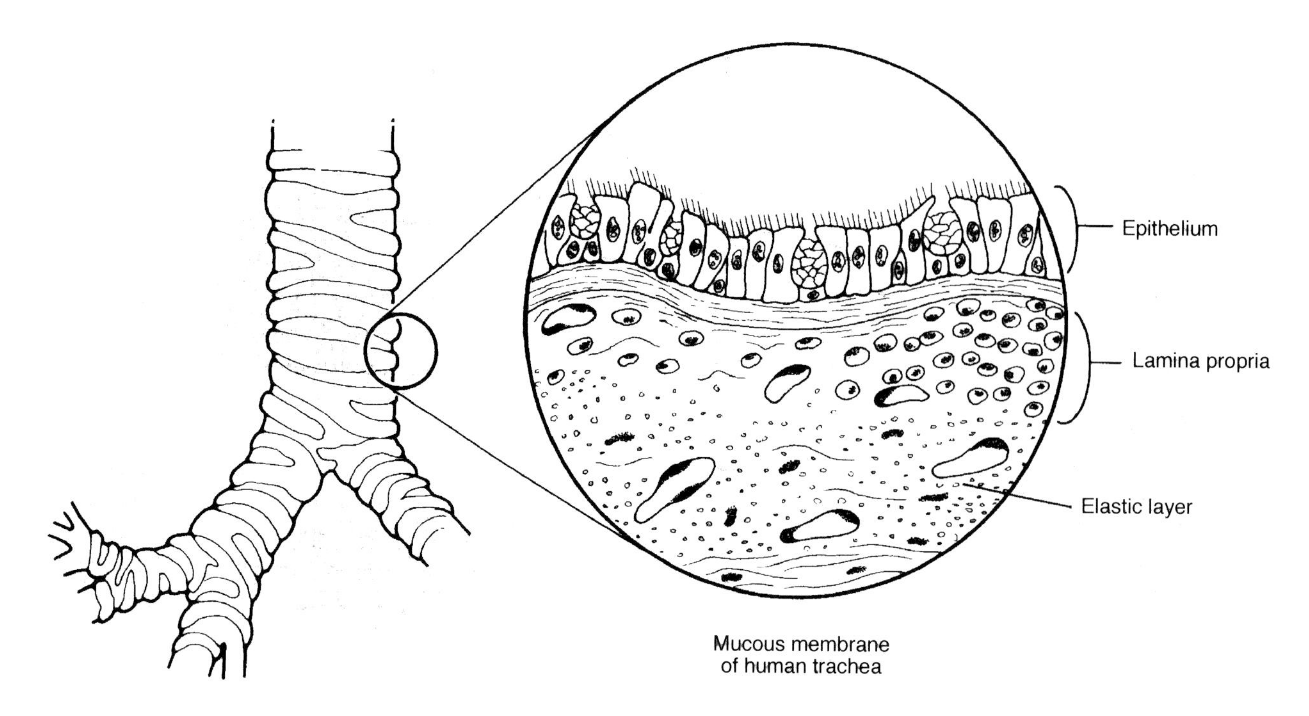

FIGURE 15.6 Lamina propria.

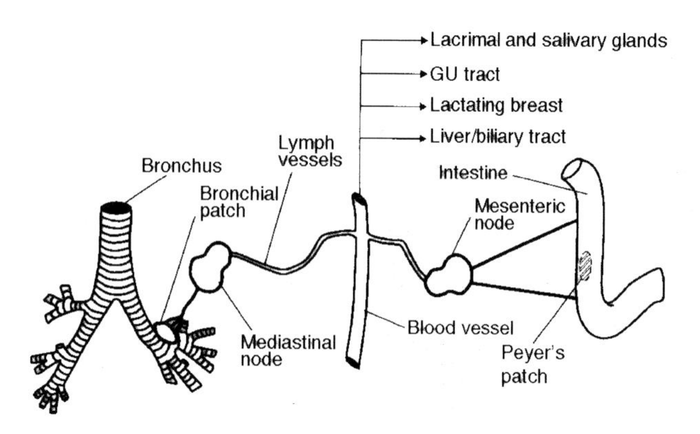

FIGURE 15.7 Cell traffic.

Inductive sites are small patches of mucous membrane that overlie organized lymphoid follicles and contain M cells.

MIC molecules are MHC class I-like molecules expressed in the gastrointestinal tract during stress. Genes within the class I region of the human MHC encode these molecules.

The **secretory immune system** is a major component of the immune system that provides protection from invading microorganisms at local sites. Much of the effect is mediated by secretory IgA molecules in the secretions at the mucosal surface. Immunoglobulins may also be in clotted fluids where they protect against microorganisms.

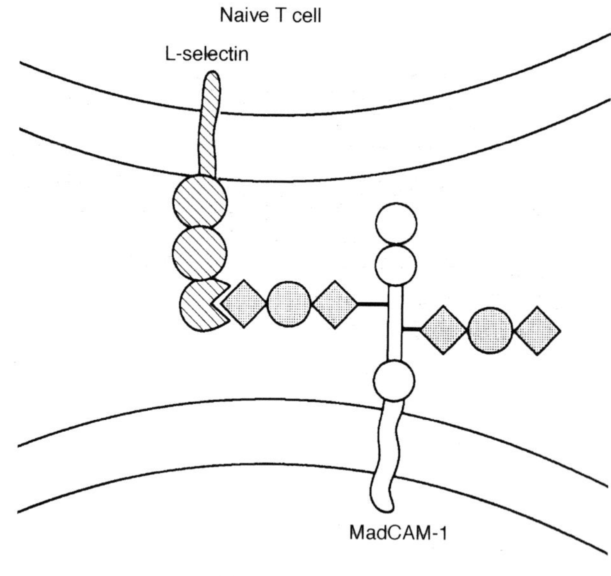

FIGURE 15.8 MadCAM-1.

Immune exclusion describes prevention of antigen entry into the body by the products of a specific immune response, such as the blocking of an antigen's access to the body by mucosal surfaces when secretory IgA specific for the antigen is present.

Secretory IgA (Figure 15.9) is a dimeric molecule comprised of two IgA monomers joined by a J polypeptide chain

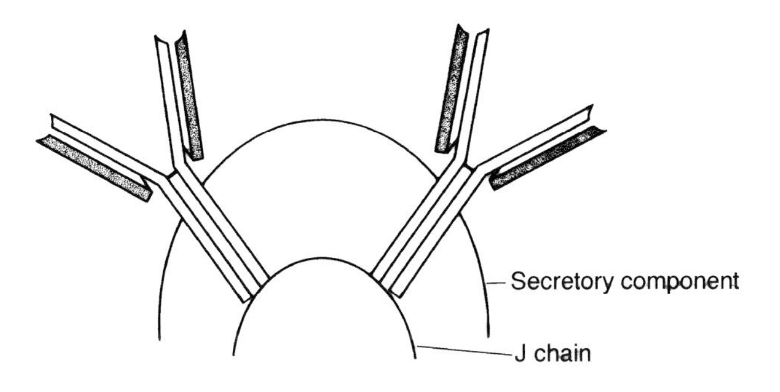

FIGURE 15.9 Secretory IgA.

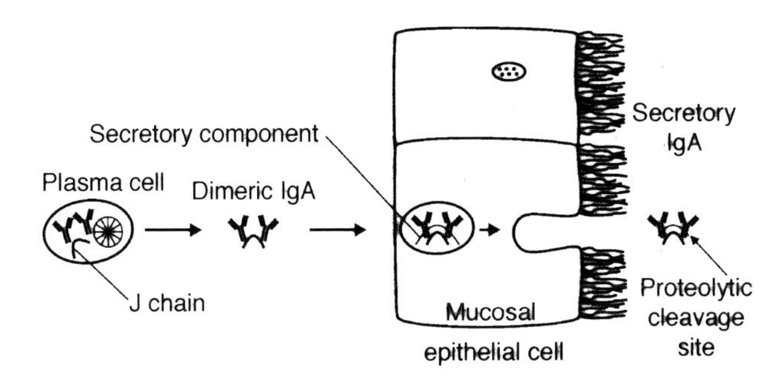

FIGURE 15.10 Secretory component.

and a glycopeptide secretory component (Figure 15.10). This is the principal molecule of mucosal immunity. IgA is the only immunoglobulin isotype that can be selectively passed across mucosal walls to reach the lumens of organs lined with mucosal cells. Specific FcαR that bind IgA molecular dimers are found on intestinal epithelial cells. Specific FαR that bind IfA molecular dimers are found on intestinal epithelial cells. The FcαR, also known as secretory (S protein), joins the antibody molecule to the epithelial cell's basal surface that is exposed to the blood. It is bound to the polyimmunoglobulin receptor on the epithelial cell's basolateral surface and facilitates vesicular transport of the anchored IgA across the cell to the surface of the mucosa. Once this complex reaches its destination, FcαR (S protein) is split in a manner that permits the dimeric IgA molecule to retain an attached secretory piece which has a strong affinity for mucous, thereby facilitating the maintenance of IgA molecules on mucosal surfaces. The secretory piece also has the important function of protecting the secreted IgA molecules from proteolytic digestion by enzymes of the gut. These latter two functions are in addition to its active role in transporting the IgA molecule through the epithelial cell. Secretory or exocrine IgA appears in the colostrum, intestinal and respiratory secretions, saliva, tears, and other secretions.

Transcytosis is the active transport of molecules across epithelial cells. IgA molecules are transported by transcytosis across intestinal epithelial cells in vesicles that are formed on the basolateral surface and fuse with the apical surface in contact with the intestinal lumen.

Antiseptic paint is a colloquial designation for the coating effect of secretory IgA, such as that produced locally in the gut, on mucosal surfaces, thereby barring antigen access.

Secretory component (T piece) or **secretory piece** is a 75-kDa molecule synthesized by epithelial cells in the lamina propria of the gut that becomes associated with IgA molecules produced by plasma cells in the lamina propria of the gut (Figure 15.10) as they move across the epithelial cell layer to reach the mucosal surface of intestine to provide local immunity. It can be found in three molecular forms: as an SIgM and SIgA stabilizing chain, as a transmembrane receptor protein, and as free secretory component in fluids. Secretory piece also has the important function of protecting the secreted IgA molecules from proteolytic digestion by enzymes of the gut. These latter two functions are in addition to its active role in transporting the IgA molecule through the epithelial cell. It is also called secretory component and is a fragment of the poly-Ig receptor that remains bound to Ig following transcytosis across the epithelium and cleavage. It is not formed by plasma cells in the lamina propria of the gut that synthesize the IgA molecules with which it combines. Secretory component has a special affinity for mucous, thereby facilitating IgA's attachment to the mucous membranes.

The **sacculus rotundus** describes the lymphoid tissue-rich terminal segments of the ileum in the rabbit. It is a part of the gut-associated lymphoid tissue (GALT).

Secretory component deficiency is a lack of IgA in secretions as a consequence of gastrointestinal tract epithelial cells' inability to produce secretory component to be linked to the IgA molecules synthesized in the lamina propria of the gut to prevent their destruction by the proteolytic enzymes in the gut lumen. The disorder is very infrequent but is characterized by the protracted diarrhea associated with gut infection.

***Helicobacter pylori* immunity:** Both circulating and local humoral antibody responses follow *H. pylori* colonization of the gastric mucosa. During a prolonged mucosal infection, IgG1 and IgG4, and often, IgG2 antibodies are detectable but IgG3 and IgM antibodies only rarely. IgA antibodies are usually also present. The initial IgM response is followed later by IgA with conversion to IgG 22 to 33 days after infection. IgA antibodies are found at the local mucosal level. They are secreted into the gastric juice. IgG produced locally is rapidly inactivated when it reaches the gastric juice. A systemic IgG response is present throughout the infection and diminishes, only prolonging successful therapy. If the infection reappears, the IgG antibody titer rises. There is great variability in the specificity of circulating host antibody against *H. pylori* . This is attributable in part to variations in host response and to a lesser degree to antigenic diversity of the microorganisms, such as variation in the Vac A and Cag A proteins. Most infected subjects synthesize antibodies against numerous antigens, including the urease subunits, the flagellins, and the 54-kDa hsp60 homolog. Antibodies usually develop to Vac A and Cag A polypeptides if they are present in the infecting strain. Although of variable complexity, the antigens all include urease. Plasma cells, lymphocytes, and monocytes infiltrate the superficial layers of the lamina propria in *H. pylori* -associated gastritis. Half of the mature B cells and infiltrate are B cells that are producing mostly IgA but also IgG and IgM. These cells produce antibodies specific for *H. pylori* and are mostly of the $CD8^+$ subset, although $CD4^+$ T cells are also increased as well as $\gamma\delta$ T cells. Gastric epithelial cells aberrantly express HLA-DR during *H. pylori* infection, w hich is also associated with elevated synthesis of IL-1, IL-6, and TNFα in the gastric mucosa. *H. pylori* induces Il-1, IL-6, and TNF-α in the gastric mucosa. It induces IL-8 expression in gastric epithelium, which induces neutrophil chemotaxis. Of the non-Hodgkins lymphoma cases affecting the stomach, 92% are associated with *H. pylori* infection. The bacteria may engage an immune avoidance by continually losing highly antigenic material such as urease and flagella sheath from the bacterial surface thereby diminishing the effectiveness of bound antibodies. The immune response may also be downregulated during infection as antigen-specific responsiveness of local and circulating T cells is diminished. The complexity of the disease makes development of effective vaccines difficult. An oral subunit vaccine used with a mucosal adjuvant has protected animal models. Thus immunotherapy might be a future option in prevention.

Bronchial-associated lymphoid tissue (BALT) is present in both mammals, including humans, and birds (Figure 15.11).

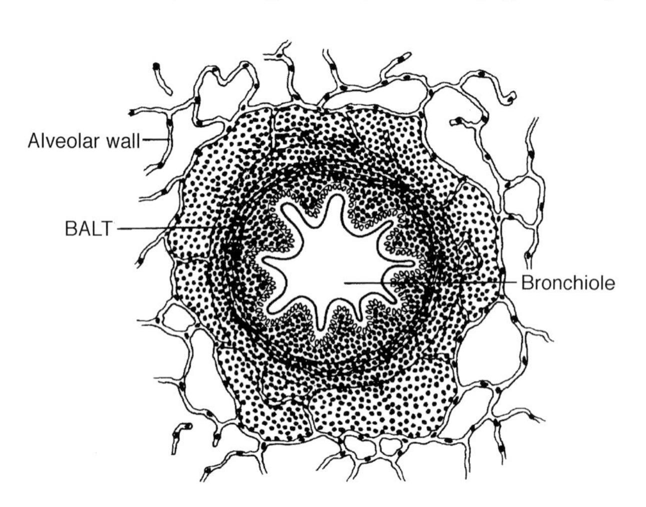

FIGURE 15.11 BALT (bronchial-associated lymphoid tissue).

In many areas it appears as a collar containing nodules located deep around the bronchus and connected with the epithelium by patches of loosely arranged lymphoid cells. Germinal centers are absent (except in the chicken), although cells in the center of nodules stain lighter than do those at the periphery. Plasma cells are present occasionally beneath the epithelium. The cells in BALT have a high turnover rate and apparently do not produce IgG. BALT development is independent of that of the peripheral lymphoid tissues or antigen exposure. The cells of BALT apparently migrate there from other lymphoid areas. This tissue plays an important role in mounting an immune response to inhaled antigens in respiratory infectious agents.

Oral immunology: Saliva not only rinses the oral cavity but contains numerous molecules such as lysozyme and secretory immunoglobulin A which as part of the mucosal immune system help to protect the oral cavity. Polymorphonuclear leukocytes are important in protection of gingival tissues and ultimately the periodontium. In addition to secretory IgA, the systemic vascular humoral immune response is significant in oral immunity. The relative contribution of T_H1 cells and T_H2 cells in an immune response to plaque bacterial pathogens is significant in periodontal disease. Individuals with immunodeficiencies often have increased mucosal infections by opportunistic microorganisms, such as by *Canda albicans.* Immunopathologic mechanism that involve types II, III, and IV hypersensitivity may be involved in the development and progression of chronic periodontitis. Vaccines may be used in the future to prevent or control dental caries and periodontal diseases. Oral or intranasal bouts of vaccine administration may prove useful to protect against oral infections.

Oral unresponsiveness is the mucosal immune system's selective ability to not react immunologically against antigens of food and intestinal microorganisms even though it responds vigorously to pathogenic microorganisms.

Oral feeding of a protein antigen may lead to profound systemic immunosuppression involving both the B cell (antibody-mediated) and T cell (cell-mediated) limbs of the immune response to that specific antigen (Figure 15.12). T cell clonal anergy is induced to some protein antigens administered by this route. Antigen presented by antigen-presenting cells deficient in costimulatory molecules may induce tolerance. Yet, possible nonprofessional antigen-presenting cells involved in the induction of oral tolerance have not been identified.

It has also been postulated that the immunosuppressive cytokine TGF-b may be released during the induction of **oral tolerance** through its ability to block lymphocyte proliferation and to induce B lymphocyte switching to IgA production. Paradoxically, it remains a mystery why large doses of soluble proteins induce systemic T lymphocyte tolerance,

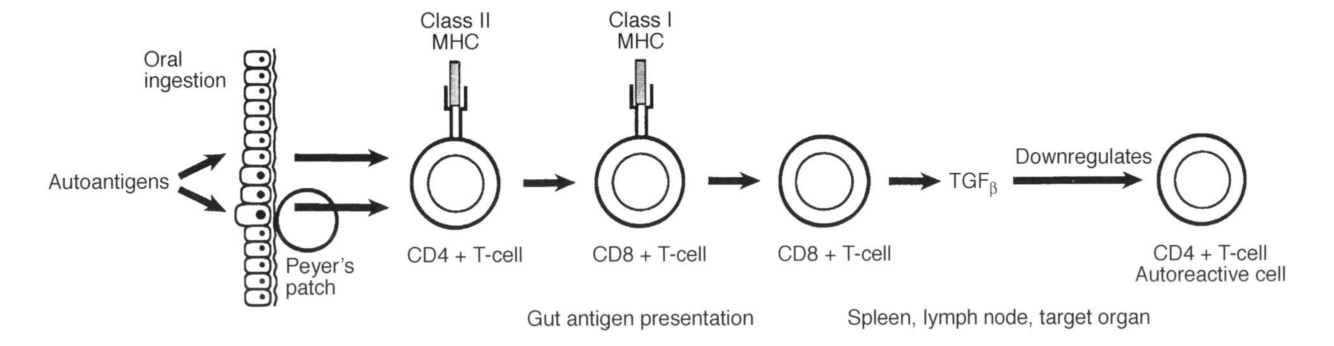

FIGURE 15.12 Antigen feeding.

whereas the components of vaccines such as the Sabin polio vaccine induce an effective local immune response. Specific local immunity may be attributable to activation or infection of antigen-presenting cells in the intestinal epithelium. Nevertheless, oral tolerance is beneficial through its prevention of the body's reaction to oral antigens or food or gastrointestinal bacteria. Feeding of autoantigen to induce oral tolerance has potential therapeutic value for the treatment of autoimmune disease.

Antigen fed orally soon reaches the lymphatics of the intestine and is transported to the mesenteric lymph nodes, where it stimulates an immune response. Antigen may also reach Peyer's patches through M cells and led to a T and B cell response. Activated lymphocytes in mesenteric lymph nodes may migrate to the lamina propria, whereas those in the Peyer's patches may reach either the lamina propria or mesenteric lymph nodes.

The epithelial layer is not only a mechanical barrier against pathogenic microorganisms but is the site in the gastrointestinal and respiratory tracts of secretory IgA antibodies. This class of immunoglobulin is also responsible for the passive transfer of immunity from mother to young through the milk and colostrum. Most of the antibodies produced in the normal adult are of the secretory IgA class. Antigens conveyed to Peyer's patches of the intestine activate T lymphocytes and follicular B cells. IgA-synthesizing lymphocytes enter the lamina propria. IL-5 and TGF-β facilitate IgA isotype switching. Antibody affinity maturation occurs in germinal centers of Peyer's patches where stimulated B lymphocytes have seeded and proliferated. IgA-synthesizing B lymphocytes populate the lamina propria or other mucosal tissues.

The relatively large quantity of secretory IgA synthesized in tissues of the mucosal immune system is attributable to the tendency of B cells that form IgA to populate the lamina propria and Peyer's patches. IgA-producing B lymphocytes in other parts of the body are responsible for serum IgA.

Oral tolerance is antigen-induced specific suppression of humoral and cell-mediated immunity to an antigen following oral administration of that antigen as a consequence of anergy of antigen-specific T lymphocytes for the formation of immunosuppressive cytokines such as transforming growth factor-β. Oral tolerance may inhibit immune responses against food antigens and bacteria in the intestine. Proteins passing through the gastrointestinal tract induce antigen-specific hyporesponsiveness. Oral tolerance is believed to have evolved to permit the gut-associated immune system to be exposed to external proteins without becoming sensitized. If proteins such as ovalbumin or myelin basic protein (MBP) are fed to animals which are then immunized, the immune response against the fed antigen, but not against the control antigen, is subsequently diminished. Based on the quantity of antigen fed, orally administered antigen may induce regulatory cells that suppress the antigen-specific response (low doses) or inhibit antigen-specific T cells by induction of clonal anergy (high doses). Antigens passing through the gastrointestinal tract preferentially induce T helper (Th2)-type T cells that secrete interleukin-4, IL-10, and transforming growth factor β (TGFβ). These cells migrate from the gut to organs that contain the fed antigen, where the Th2 cells are stimulated locally to release antiinflammatory cytokines.

Mucosal tolerance is synonymous with oral tolerance.

Maternal immunity describes passive immunity conferred on the neonate by its mother. This is accomplished prepartum by active immunoglobulin transport across the placenta from the maternal to the fetal circulation in primate animals including humans. Other species such as ungulates transfer immunity from mother to young by antibodies in the colostrum since the intestine can pass immunoglobulin molecules across its surface in the early neonatal period. The egg yolk of avian species is the mechanism through which immunity is passed from mother to young in birds.

Colostrum is immunoglobulin-rich first breast milk formed in mammals after parturition. The principal immunoglobulin is IgA with lesser amounts of IgG. It provides passive immune protection of the newborn prior to maturation of its own immune competence.

Coproantibody is a gastrointestinal tract antibody, commonly of the IgA class, which is present in the intestinal lumen or feces.

The **Sulzberger-Chase phenomenon** is the induction of immunological unresponsiveness to skin-sensitizing chemicals such as picryl chloride by feeding an animal (e.g., guinea pig) the chemical in question prior to application to the skin. Intravenous administration of the chemical may also block the development of delayed-type hypersensitivity when the same chemical is later applied to the skin. Simple chemicals such as picryl chloride may induce contact hypersensitivity when applied to the skin of guinea pigs. The unresponsiveness may be abrogated by adoptive immunization of a tolerant guinea pig with lymphocytes from one that has been sensitized by application of the chemical to the skin without prior oral feeding.

Chase-Sulzberger phenomenon: See Sulzberger-Chase phenomenon.

Skin immunity: The skin, the largest organ in the body, shields the body's interior environment from a hostile exterior (Figure 15.13). The skin defends the host through stimulation of inflammatory and local immune responses. Antigen applied to or injected into the skin drains to the regional lymph nodes through the skin's extensive lymphatic network. Cells of both the epidermis, papillary, and reticular dermis have critical roles in the skin's immune function. Keratinocytes, which are epidermal epithelial cells, secrete various cytokines such as granulocyte–macrophage colony-stimulating factor, interleukin-1, interleukin-3, interleukin-6, and tumor necrosis factor. T lymphocytes in the skin may secrete IFN-γ or other cytokines that cause these epithelial cells to synthesize chemokines that lead to leukocyte chemotaxis and activation. Stimulation by IFN-γ may also lead to their expression of class II MHC molecules. Other epidermal cells include Langherans cells, which form an extensive network in the epidermis that permits their interaction with any antigen entering the skin. The few lymphocytes in the epidermis are principally $CD8^+$ T cells often with restricted antigen receptors. In the mouse, these are mostly γδ T cells. Skin lymphocytes and macrophages are mostly in the dermis. The T cells are of $CD4^+$ and $CD8^+$ phenotypes and often perivascular.

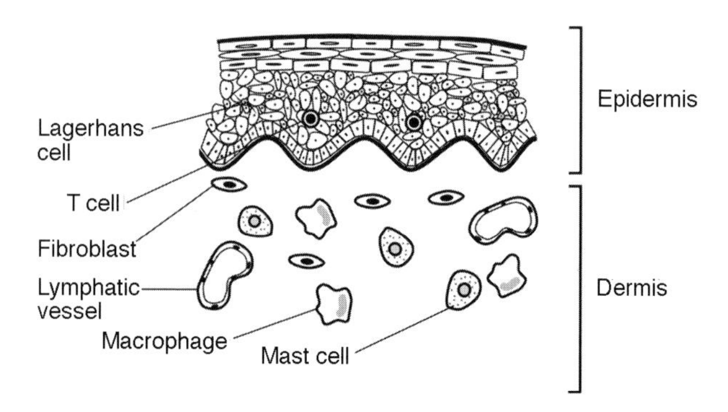

FIGURE 15.13 The cutaneous immune system and its cellular components.

The **cutaneous immune system** is comprised of adaptive and innate immune system constituents present in the skin that function in concert to detect and respond to environmental antigens. Among the cutaneous immune system constituents are keratinocytes, Langerhans cells, intraepithelial lymphocytes, and dermal lymphocytes.

Intraepidermal lymphocytes are primarily $CD8^+$ T cells within the dermis. Murine intraepidermal lymphocytes express mainly the γδ receptor. It is believed that these cells have a more restricted repertoire of antigen receptors than those homing to extracutaneous sites. Most skin-associated lymphocytes are found in the dermis with only about 2% in the epidermis. Intraepidermal lymphocytes express CLA-1, which may play a role in homing.

Langerhans cells are the principal antigen-presenting cells in the skin. They encounter antigen entering through the skin and process it. Langerhans cells may reach the parafollicular cortical areas of the regional lymph nodes via lymphatic vessels of the skin. These Langerhans cells develop into effective antigen-presenting cells as a result of upregulation of class II MHC molecules and of costimulatory molecules. In the lymph nodes they present antigen to $CD4^+$ T lymphocytes after becoming interdigitating dendritic cells. Dermal macrophages may present protein antigens to previously activated T cells. Delayed type hypersensitivity reactions are mediated by T cells in the skin reacting to soluble protein antigens or to chemicals acting as haptens that combine with self proteins producing new epitopes. This cell-mediated immune reaction is accompanied by the release of cytokines from the activated T cells. With respect to the antibody response of the skin, secretory IgA is found in sweat, which aids in defense against infection. IgE on the surface of mast cells in the dermis participates in the generation of immediate type I hypersensitivity reactions in the skin.

$Thy-1^+$ dendritic cells are found within the epithelium of the mouse epidermis.

The **cutaneous lymphocyte antigen** HECA-452 epitope is expressed on a skin-associated subset of memory T cells that are active in recirculation and homing to skin sites.

Cutaneous sensitization may be induced by application of antigen to the skin to induce hypersensitivity.

Cutaneous T-cell lymphoma describes a malignant growth of T lymphocytes that home to the skin, such as *mycosis fungoides*.

Photoimmunology is the investigation of the effects of photons on the immune system. The effects of photons on the immune system are initiated in the skin from interaction of ultraviolet radiation with immune cells. Ultraviolet-induced immune suppression may play a role in human skin cancer induction.

16 Immunohematology

Immunohematology is the study of blood group antigens and antibodies and their interactions in health and disease. Both the cellular elements and the serum constituents of the blood have distinct profiles of antigens. There are multiple systems of blood cell groups, all of which may stimulate antibodies and interact with them. These may be associated with erythrocytes, leukocytes, or platelets.

An article in the *Berliner klinische Wochenschrift* in 1900 by Ehrlich and Morgenroth, which described blood groups in goats based on antigens of their red cells, led Karl Landsteiner, a Viennese pathologist, to successfully identify the human ABO blood groups. He took samples of blood from his assistant Zaritsch and from colleagues Sturli and Ardhein, and Dr. Pietschnig. The small tables Landsteiner used to illustrate his reason bear the names of these colleagues. In 1902, Sturli and Alfred Descastello, under Landsteiner's direction, designated one more group which was actually not named "AB" until 10 years later, when von Dungen and Hirszfeld, studying the genetic inheritance of blood types, designated the fourth type and gave Landsteiner's "C" group the designated "O." It was for this discovery, rather than his elegant studies on immunochemical specificity, that Landsteiner won the Nobel Price in Medicine 30 years later. Among those who worked with him in the field of immunohematology were Alexander Weiner and Philip Levine. Landsteiner and Levine discovered the M and N blood group antigens by injecting human erythrocytes into rabbits. The antisera which they raised were able to divide human blood into three groups, M, N, and MN, based on their antigenic content. They showed that these antigens were under the genetic control of codominant alleles. Landsteiner and Weiner described the Rh factor in 1940. At first thought to be a simple system involving a single antigen, it was shown to be genetically, immunologically, and clinically complex. In studies of M-like factors on the erythrocytes of rhesus monkeys, antisera raised by injecting rabbits with rhesus red cells cross-reacted with human erythrocytes containing M antigen. It was subsequently demonstrated that the red cells of about 85% of the human population reacted with antisera against rhesus red cells. Thus, those individuals who shared an antigen with the rhesus monkeys red cells were termed Rh positive and those who did not were termed Rh negative. It was later shown that multiple pregnancies with Rh(D) positive fetuses in Rh negative mothers leads to stimulation of maternal anti-D antibodies of the IgG class which cross the placenta and cause lysis of fetal red cells. Unless the mother is treated with antibody against the D-antigen after parturition, hemolytic disease of the newborn (erythroblastosis fetalis) may result on subsequent births. Besides those mentioned above, other red blood cell antigens discovered in the intervening years included Kell, Diego, P, Duffy, and I blood group systems and soluble antigens such as the Lewis, Lutheran antigens that are in the secretions and are adsorbed to the red cell surface. Although most red cell groups are inherited as autosomal characteristics, the Xg blood group system is sex linked. Historically, new red blood cell antigens were discovered as a result of transfusion incompatibility reactions that could not be explained on the basis of existing or known antigens (Figure 16.1).

Blood grouping (Figure 16.2 and Figure 16.3) is the classification of erythrocytes based on their surface isoantigens. Among the well-known human blood groups are the ABO, Rh, and MNS systems.

Plasma is the transparent yellow fluid that constitutes 50 to 55% of the blood volume. It is 92% fluid and 7% protein. Inorganic salts, hormones, sugars, lipids, and gases make up the remaining 1%. Plasma from which fibrinogen and clotting factors have been removed is known as serum.

Blood group antigens are erythrocyte surface molecules that may be detected with antibodies from other individuals such as the ABO blood group antigens. Various blood group antigen systems, including Rh (Rhesus) may be typed in routine blood banking procedures. Other blood group antigen systems may be revealed through cross-matching.

ABO blood group substances are glycopeptides with oligosaccharide side chains manifesting ABO epitopes of the same specificity as those present on red blood cells of the individual in whom they are detected. Soluble ABO blood group substances may be found in mucous secretions of man such as saliva, gastric juice, ovarian cyst fluid, etc. Such persons are termed secretors, whereas those without the blood group substances in their secretions are nonsecretors.

FIGURE 16.1 Engraved title page from G. A. Mercklin, tractatio med. *Curiosa de Ortu et Sanguinis*, 1679. This is one of the best early pictures of blood transfusion. (From the Cruse Collection, Middleton Library, University of Wisconsin.)

Tabelle I, betreffend das Blut sechs anscheinend gesunder Männer.

Sera						
Dr. St.	−	+	+	+	+	−
Dr. Plecn.	−	−	+	+	−	−
Dr. Sturl.	−	+	−	−	+	−
Dr. Erdh.	−	+	−	−	+	−
Zar.	−	−	+	+	−	−
Landst.	−	+	+	+	+	−
Blutkörperchen von:	Dr. St.	Dr. Plecn.	Dr. Sturl.	Dr. Erdh.	Zar.	Landst.

Tabele II, betreffend das Blut von sechs anscheinend gesunden Puerperae.

Sera						
Seil.	−	−	+	−	−	+
Linsm.	+	−	+	+	+	+
Lust.	+	−	−	+	+	−
Mittelb.	−	−	+	−	−	+
Tomsch.	−	−	+	−	−	+
Graupn.	+	−	−	+	+	−
Blutkörperchen von:	Seil.	Linsm.	Lust.	Mittelb.	Tomsch.	Graupn.

FIGURE 16.2 Table illustrating ABO blood groups. (*Wien. klin. Wochenschr.* 14: 1132–1134, 1901.)

Blood Type	RBC Surface Antigen	Antibody in Serum
A	A antigen	Anti-B
B	B antigen	Anti-A
AB	AB antigens	No antibody
O	No A or B antigens	Both anti-A and anti-B

FIGURE 16.3 ABO blood group antigens and antibodies.

The ABO blood group system was the first described of the human blood groups based upon carbohydrate alloantigens present on red cell membranes. Anti-A or anti-B isoagglutinins (alloantibodies) are present only in the blood sera of individuals not possessing that specificity, i.e., anti-A is found in the serum of group B individuals and anti-B is found in the serum of group A individuals. This serves as the basis for grouping humans into phenotypes designated A, B, AB, and O. Type AB subjects possess neither anti-A nor anti-B antibodies, whereas group O persons have both anti-A and anti-B antibodies in their serum. Blood group methodology to determine the ABO blood type makes use of the agglutination reaction. The ABO system remains the most important in the transfusion of blood and is also critical in organ transplantation. Epitopes of the ABO system are found on oligosaccharide terminal sugars. The genes designated as A/B, Se, H, and Le govern the formation of these epitopes and of the Lewis (Le) antigens (Figure 16.4). The two precursor substances type I and type II differ only in that the terminal galactose is joined to the penultimate *N*-acetylglucosamine in the b 1–3 linkage in type I chains, but in the b 1–4 linkage in type II chains.

AB blood group: See ABO blood group system.

Landsteiner's rule (historical): Landsteiner (1900) discovered that human red blood cells could be separated into four groups based on their antigenic characteristics. These were designated as blood groups O, A, B, and AB. He found naturally occurring isohemagglutinins in the sera of individuals specific for the (ABO) blood group antigen which they did not possess (i.e., anti-A and anti-B isohemagglutinin in group O subjects; anti-B in group A individuals; and anti-A in group B persons; neither anti-A

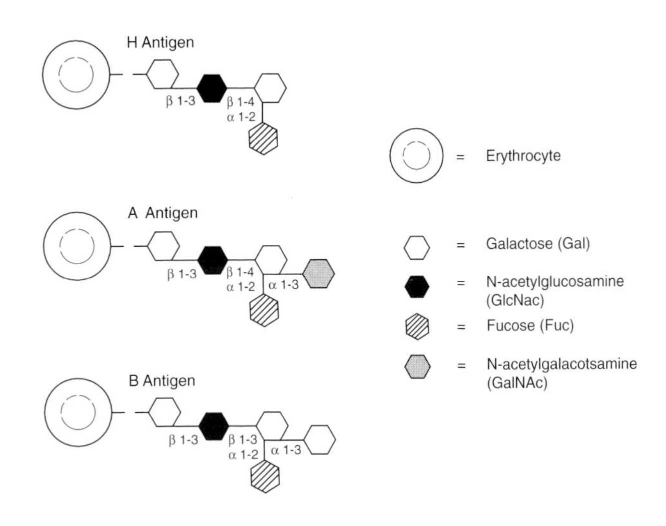

FIGURE 16.4 Chemical structure of A, B, and H antigens of the ABO blood group system.

nor anti-B in individuals of group AB). This principle became known as Landsteiner's rule.

ABO blood-group antigen: Glycosphingolipid epitopes on erythrocytes and numerous other types of cells. These antigens are governed by alleles that encode enzymes needed for their synthesis. They differ among individuals and may serve as alloantigens that lead to hyperacute rejection of allografts and to blood transfusion reactions.

Natural antibodies are those found in the serum of an individual who has no known previous contact with that antigen, such as by previous immunization or infection with a microorganism containing that antigen. The anti-A and anti-B antibodies related to the ABO blood group system are natural antibodies. Natural antibodies may be a consequence of exposure to cross-reacting antigen(s), e.g., ABO blood group antibodies resulting from exposure to bacterial antigens in the gut. The term also refers to IgM antibodies produced by B-1 cells specific for microorganism found in the environment and gastrointestinal tract. There are two kinds of natural antibodies in blood sera. These include (1) specific, antigen-induced antibodies whose synthesis depends on external antigenic stimuli and corresponds to acquired specificities; (2) a second type that expresses broad specificity, is genetically determined, and does not depend on a specific antigenic stimulus. Both kinds of natural antibodies of the IgM, IgG, and IgA isotopes specific for may antigens are present in normal sera of humans and other animals. Natural antibodies have a variety of biological functions ranging from physiological to pathological effects.

Null phenotype is the failure to express protein because the gene that encodes it is either defective or absent on both inherited haplotypes. Most of these involve erythrocytes, but are quite rare.

B blood group: See ABO blood group system.

H antigen is an ABO blood group system antigen that is also called H substance. It may also designate a histocompatibility antigen.

H substance is a basic carbohydrate of the ABO blood group system structure of man. Most people express this ABO-related antigen. "Secretors" have soluble H substance in their body fluids.

Acquired B antigen refers to the alteration of A1 erythrocyte membrane through the action of such bacteria as *Escherichia coli, Clostridium tertium,* and *Bacteroides flagilis* to make it react as if it were a group B antigen. The named microorganisms can be associated with gastrointestinal infection or carcinoma.

O antigen: In the ABO blood group system, O antigen is an oligosaccharide precursor form of A and B antigens: a fucose-galactose-*N*-acetylglucosamine-glucose.

The **O blood group** is one of the groups described by Landsteiner in the ABO blood group system. See ABO blood group system.

A blood group: See ABO blood group system.

Codominant describes the expression of both alleles of a pair in the heterozygote. The traits which they determine are codominant as in the expression of blood group A and B epitopes in type AB persons.

The **absorption elution test** is used for the identification of the ABO type in stains of semen or blood on clothing. ABO blood group antibodies are applied to the stain after it has been exposed to boiling water. Following washing to remove unfixed antibody, the preparation is heated to 56°C in physiological saline. An antibody that may be eluted from the stain is tested with erythrocytes of known ABO specificity to determine the ABO type. This is being replaced by DNA analysis of such specimens by forensics experts.

Immune antibody is a term used to distinguish an antibody induced by transfusion or other immunogenic challenge in contrast to a natural antibody such as the isohemagglutinins against ABO blood group substances found in humans.

In the ABO blood group system, O antigen is an oligosaccharide precursor form of A and B antigens; a fucose-galactose-*N*-acetylglucosamine-glucose. The O blood group is one of those described by Landsteiner (Figure 16.5 to Figure 16.7).

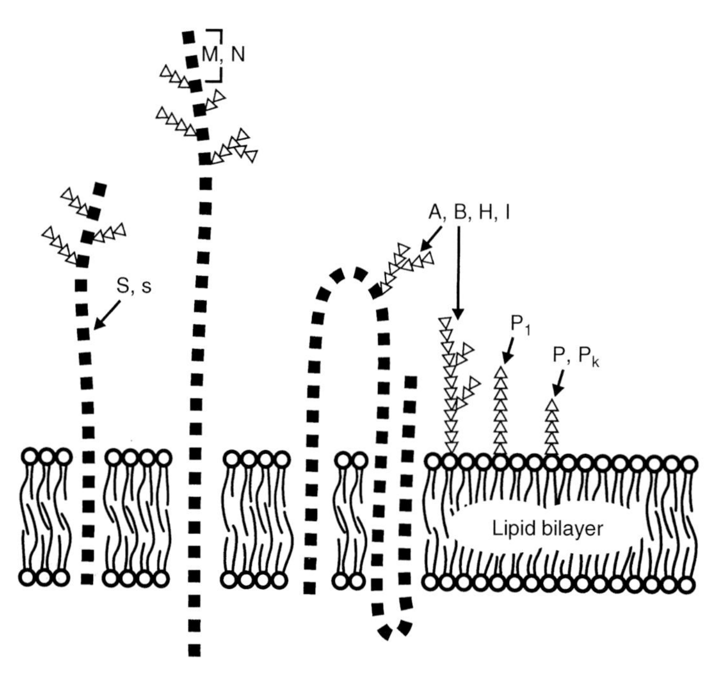

FIGURE 16.5 Schematic representation of membrane glycoproteins and glycosphingolipids that carry blood group antigens.

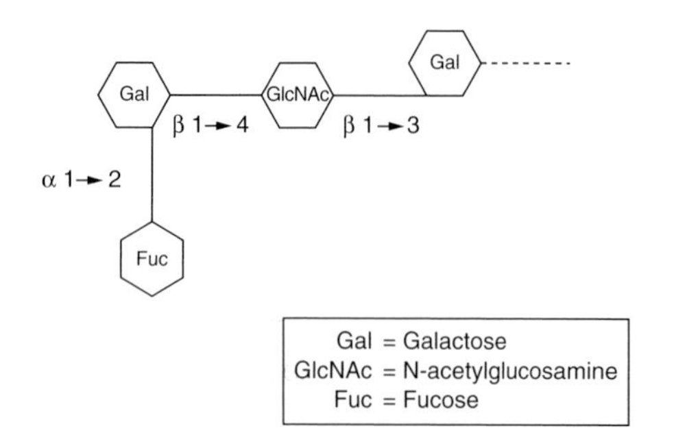

FIGURE 16.6 Chemical structure of H antigen, which is a specificity of the ABO blood group system.

The Bombay phenotype (O_h) is an ABO blood group antigen variant on human erythrocytes in rare subjects. These red blood cells do not possess A, B, or H antigens on their surfaces, even though the subject does have anti-A, anti-B, and anti-H antibodies in the serum. The Bombay phenotype may cause difficulties in crossmatching for transfusion.

The **para-Bombay phenotype** is a variant Bombay phenotype that is of the ABO blood group system. Individuals expressing it have an Se secretor gene that encodes synthesis of blood groups A and B which are detectable in secretions. However, these subjects do not produce A and B erythrocytes as the H gene is absent. By comparison, Bombay phenotype individuals do not have the H gene or the enzyme it produces, i.e., fucosyl transferase, and do not have A or B blood group substances on their erythrocytes or in their secretions.

A **hemagglutinin** is a red blood cell agglutinating substance. Antibodies, lectins, and some viral glycoproteins may induce erythrocyte agglutination. In immunology, hemagglutinin usually refers to an antibody that causes red blood cell aggregation in physiological salt solution either at 3°C, in which case they are termed warm hemagglutinins, or at 4°C, in which case they are referred to as cold hemagglutinins.

A **saline agglutinin** is an antibody that causes the aggregation or agglutination of cells such as red blood cells or bacterial cells or other particles in 0.15 *M* salt solutions without additives.

Hemagglutination is the aggregation of red blood cells by antibodies, viruses, lectin, or other substances.

Isohemagglutinin is an antibody in some members of a species, which recognizes erythrocyte isoantigens on the surfaces of red blood cells from other members of the same species. In the ABO blood group system, the anti-A antibodies in the blood sera of group B individuals and the anti-B antibodies in the blood sera of group A individuals are examples of isohemagglutinins.

The **hemagglutination test** is an assay based upon the aggregation of red blood cells into clusters either through the action of antibody specific for their surface epitopes or

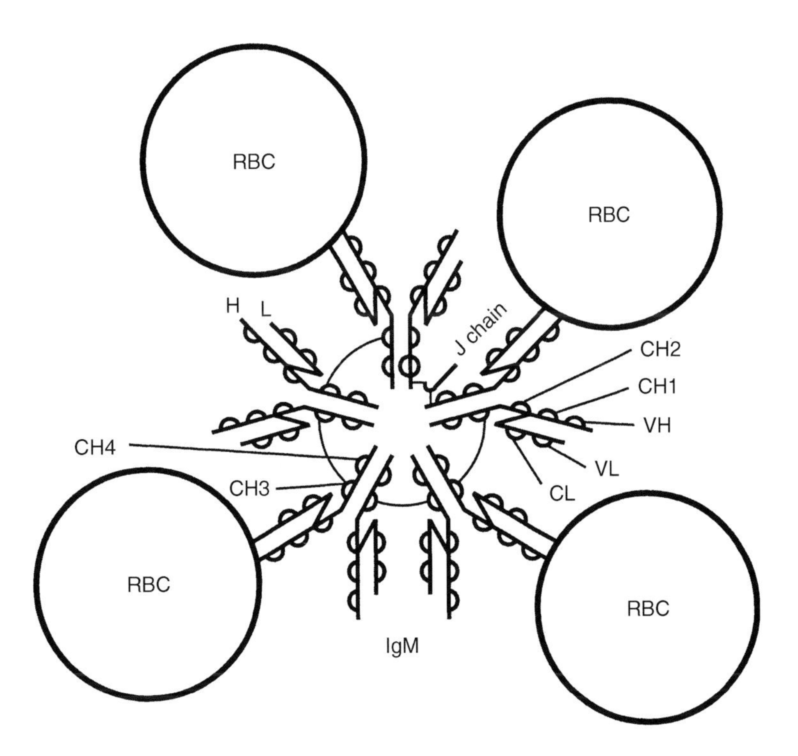

FIGURE 16.7 Schematic representation of the agglutination of human red cells by the natural isohemagglutinins, which are antibodies of the IgM class that constitute the natural isohemagglutinins in serum.

through the action of a virus that possesses a hemagglutinin as part of its structure and which does not involve antibody.

The **hemagglutination inhibition reaction** is a serological test based upon inhibiting the aggregation of erythrocytes bearing antigen. The technique may be employed for diagnosis of such viral infections as rubella, variola-vaccinia, rubeola, herpes zoster, herpes simplex types I and II, cytomegalovirus, and Epstein-Barr virus. It is also being used in the diagnosis of adenovirus, influenza, coronavirus, parainfluenza, mumps, and the viral diseases that include St. Louis, Eastern, Venezuelan, and Western equine encephalitides. It has also been used in diagnosing various bacterial and parasitic diseases.

The **hemagglutination inhibition test** is an assay for antibody or antigen based on the ability to interfere with red blood cell aggregation. Certain viruses are able to agglutinate red blood cells. In the presence of antiviral antibody, the ability to agglutinate erythrocytes is inhibited. Thus, this serves as a basis to assay the antibody.

Erythrocyte autoantibodies are autoantibodies against erythrocytes. They are of significance in the autoimmune form of hemolytic anemia and are usually classified into cold and warm varieties by the thermal range of their activity.

Direct agglutination is the aggregation of particulate antigens such as microorganisms, red cells, or antigen-coated latex particles when they react with specific antibody.

Polyagglutination is the aggregation of erythrocytes by antibodies, autoagglutinins, or alloagglutinins in blood serum. Polyagglutination also refers to aggregation of normal red blood cells treated with neuraminidase and also to red blood cells with altered membranes that are improperly aggregated by anti-A or anti-B antibodies. This is linked to altered glycoproteins such as Tn, T, and Cad. Acquired B antigens may also lead to polyagglutination, as can bacterial infection of the patient. Serum contaminants including detergents, microbes, metal cations, or silica may also cause polyagglutination. Also called panagglutination.

Panagglutination is the aggregation of cells with multiple antigenic specificities by certain blood sera, such as agglutination of normal red blood cells by a particular serum sample. It may also refer to an antibody that identifies an antigenic specificity held in common by a group of cells bearing a common antigenic specificity even though they differ in other antigenic specificities. Contamination of blood sera or of cells to be typed can result in aggregation of all the cells, leading to false-positive results as in blood grouping or cross-matching procedures.

Front typing (Figure 16.8) refers to blood typing for transfusion. Antibodies of known specificity are used to identify erythrocyte ABO antigens. Differences between front and back typing might be attributable to acquired group B or B subtypes, diminished immunoglobulins, anti-B and anti-A_1 antibody polyagglutination, rouleaux

Front Typing		Back Typing			
Reaction of Cells Tested with		Reaction of Serum Tested against			Interpretation
Anti-A	Anti-B	A Cells	B Cells	O Cells	ABO Group
O	O	+	+	O	O
+	O	O	+	O	A
O	+	+	O	O	B
+	+	O	O	O	AB

+ = agglutination
O = no agglutination

FIGURE16.8 ABO blood grouping by front and back typing.

Secretors in ABO Blood Group	Antigens in Saliva
O	A, H
A	B, H
B	A, B, H
AB	H

FIGURE 16.9 The secretor phenomenon in which AB and H substances are detectable in the saliva of individuals with ABO blood groups.

formation, cold agglutinins, Wharton's jelly, or two separate cell populations.

Back typing refers to the interaction of antibodies in an individual's serum with known antigens of an erythrocyte panel to ascertain whether or not the person's serum contains antierythrocyte antibodies. Also called reversed typing.

A **secretor** (Figure 16.9) is an individual who secretes ABH blood group substances into body fluids such as saliva, gastric juice, tears, ovarian cyst fluid, etc. At least 80% of the human population are secretors. The property is genetically determined and requires that the individual be either homozygous (*Se/Se*) or heterozygous (*Se/se*) for the *Se* gene.

A **nonsecretor** is an individual whose body secretions such as gastric juice, saliva, tears, and ovarian cyst mucin do not contain ABO blood group substances. Nonsecretors make up approximately one fifth of the population and are homozygous for the gene *se.*

The **Rhesus blood group system** (Figure 16.10) is comprised of Rhesus monkey erythrocyte antigens such as the D antigen, which are found on the red cells of most humans who are said to be Rh positive. This blood group system was discovered by Landsteiner et al. in the 1940s when they injected rhesus monkey erythrocytes into rabbits and guinea pigs. Subsequent studies showed the system to be quite complex, and the rare Rh alloantigens

Haplotype	Fisher-Race	Wiener	Frequency (%) Whites	African Am.
R^1	*CDe*	Rh_1	42	17
r	*cde*	*rh*	37	26
R^2	*cDE*	Rh_2	14	11
R^o	*cDe*	Rh_0	4	44
r'	*Cde*	rh'	2	2
r''	*cdE*	rh''	1	<1
R^z	*CDE*	Rh_z	Very rare	Very rare
R^y	*CdE*	rh^y	Very rare	Very rare

FIGURE 16.10 Principal Rh genes and their frequencies of occurrence among Whites and African Americans.

are still not characterized biochemically (Figure 16.11). Three closely linked pairs of alleles designated *Dd, Cc,* and *Ee* are postulated to be at the Rh locus, which is located on chromosome 1. There are several alloantigenic determinants within the Rh system. Clinically, the D antigen is the one of greatest concern, since RhD negative individuals who receive RhD positive erythrocytes by transfusion can develop alloantibodies that may lead to severe reactions with further transfusions of RhD positive blood. The D antigen also poses a problem in RhD negative mothers who bear a child with RhD positive red cells inherited from the father. The entrance of fetal erythrocytes into the maternal circulation at parturition or trauma during the pregnancy (such as in amniocentesis) can lead to alloimmunization against the RhD antigen, which may cause hemolytic disease of the newborn in subsequent pregnancies. This is now prevented by the administration of $Rh_o(D)$ immune globulin to these women within 72 h of parturition. Further confusion concerning this system has been caused by the use of separate designations by the Wiener and Fisher systems. Rh antigens are a group of 7- to 10-kDa, erythrocyte membrane-bound antigens that are independent of phosphatides and proteolipids. Antibodies against Rh antigens do not occur naturally in the serum.

Rhesus antigen refers to an erythrocyte antigen of man that shares epitopes in common with rhesus monkey red blood cells. Rhesus antigens are encoded by allelic genes. D antigen has the greatest clinical significance as it may stimulate antibodies in subjects not possessing the antigen and induce hemolytic disease of the newborn or cause transfusion incompatibility reactions. Rhesus antibody reacts with rhesus antigen, especially RhD (Figure 16.12).

Rh_{null} designates human erythrocytes that fail to express Rh antigens due to either the homozygous inheritance of the X^Or gene, which causes a regulator-type defect, or the inheritance of an amorphic gene (— / —). The Rh_{null} phenotype is associated with diminished erythrocyte survival.

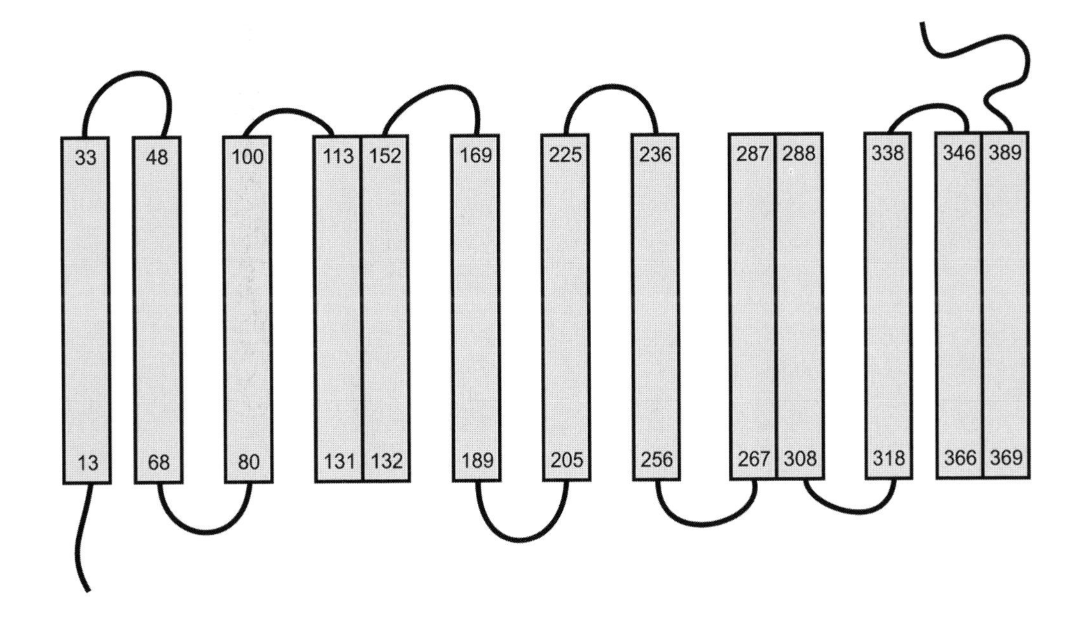

FIGURE 16.11 Schematic representation of the suggested molecular structure of the Rh polypeptide.

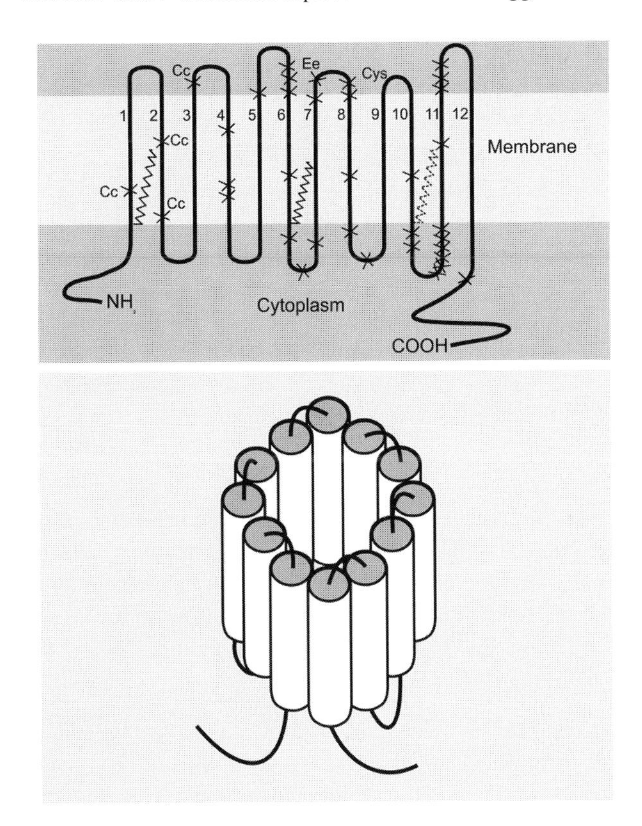

FIGURE 16.12 Schematic representation of CcEe and D polypeptide topology within the erythrocyte membrane. There are 12 membrane-spanning domains and cytoplasmic N- and C-termini. The linear diagram depicts probable sites of palmitoylation (Cys-Leu-Pro Motifs).

Rh_0D immune globulin is prepared from the serum of individuals hyperimmunized against Rh_0D antigen. It is used to prevent the immunization of Rh^- mothers by Rh_0D^+ erythrocytes of the baby, especially at parturition when the baby's red cells enter the maternal circulation in significant quantities, but also at any time during the pregnancy after trauma that might introduce fetal blood into the maternal circulation. This prevents hemolytic disease of the newborn in subsequent pregnancies. The dose used is effective in inhibiting immune reactivity against 15 ml of packed $Rh_0(D)^+$ red blood cells. It should be administered within 72 h of parturition. It may be used also following inadvertent or unavoidable transfusion of RhD^+ blood to RhD^- recipients, especially to a woman of childbearing years.

RhoGAM refers to $Rh_0(D)$ immune globulin.

Lw antibody is an antibody that was first believed to be an anti-Rh specificity, but was subsequently shown to be directed against a separate red-cell antigen closely linked to the *Rh* gene family. Its inheritance is separate from that of the Rh group. Lw is the designation given to recognize the research of Landsteiner and Wiener on the Rhesus system. The rare anti-Lw antibody reacts with Rh^+ or Rh^- erythrocytes and are nonreactive with Rh_{null} red cells.

Rhesus antibody is an antibody reactive with rhesus antigen, especially RhD.

There are 12 membrane-spanning domains and cytoplasmic N- and C-termini. The linear diagram depicts probable sites of palmitoylation (Cys-Leu-Pro Motifs).

Polymorphism refers to the occurrence of two or more forms, such as ABO and Rh blood groups, in individuals of the same species. This is due to two or more variants at a certain genetic locus occurring with considerable frequency

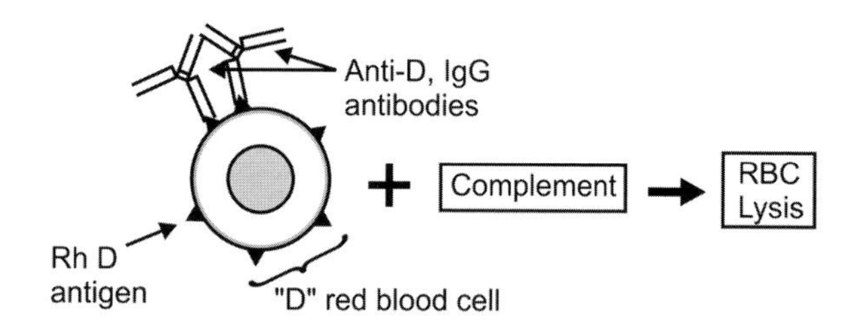

FIGURE 16.13 Complement-mediated lysis of RhD antigen-positive red blood cells through doublets of anti-D, IgG antibodies on the red cell surface.

in a population. Polymorphisms are also expressed in the HLA system of human leukocyte antigens as well as in the allotypes of immunoglobulin γ and κ chains.

Anti-D (Figure 16.13) is an antibody against the Rh blood group D antigen. This antibody is stimulated in RhD negative mothers by fetal RhD positive red blood cells that enter her circulation at parturition. Anti-D antibodies become a problem usually with the third pregnancy, resulting from the booster immune response against the D antigen to which the mother was previously exposed. IgG antibodies pass across the placenta, leading to hemolytic disease of the newborn (erythroblastosis fetalis). Anti-D antibody (Rhogam®) administered up to 72 h following parturition may combine with the RhD positive red blood cells in the mother's circulation, thereby facilitating their removal by the reticuloendothelial system. This prevents maternal immunization against the RhD antigen.

Rhesus incompatibility refers to the stimulation of anti-RhD antibodies in an Rh negative mother when challenged by RhD positive red cells of her baby (especially at parturition) that may lead to hemolytic disease of the newborn. The term also refers to the transfusion of RhD positive blood to an Rh negative individual who may form anti-D antibodies against the donor blood, leading to subsequent incompatibility reactions if given future RhD positive blood.

Dextrans (Figure 16.14) of relatively low molecular weight have been used as plasma expanders.

Coombs' test (Figure 16.15) is an antiglobulin assay that detects immunoglobulin on the surface of a patient's red blood cells. The test was developed in the 1940s by Robin Coombs to demonstrate autoantibodies on the surface of red blood cells that fail to cause agglutination of these red cells. In the direct Coombs' test, rabbit antihuman immunoglobulin is added to a suspension of patient's red cells, and if they are coated with autoantibody, agglutination results. In the indirect Coombs' test, the patient's serum can be used to coat erythrocytes, which are then washed and the antiimmunoglobulin reagent added to produce agglutination, if the

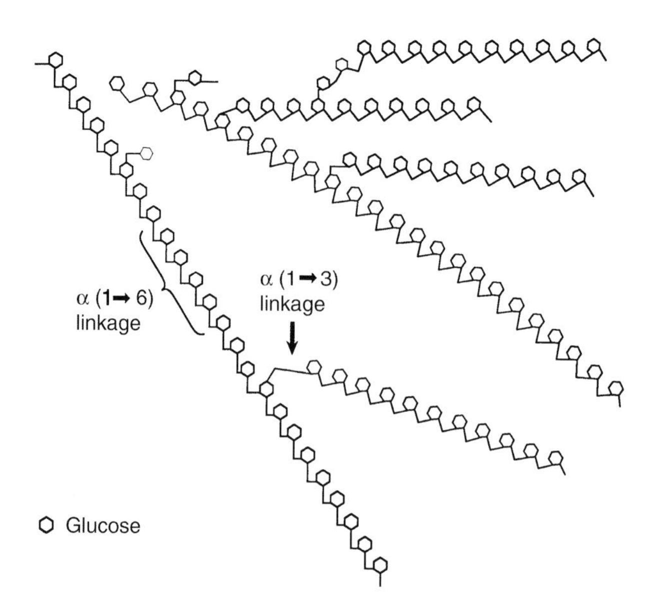

FIGURE 16.14 α (1–6) linkages and β (1–3) linkages are shown in the dextran molecule.

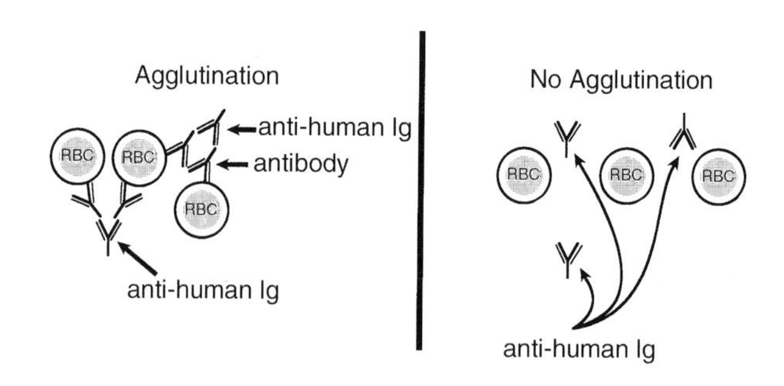

FIGURE 16.15 Schematic representation of the mechanism of the Coombs' test.

antibodies in question had been present in the serum sample. The Coombs' test has long been a part of an autoimmune disease evaluation of patients. An incomplete antibody is nonagglutinating and must have a linking agent such as anti-IgG to reveal its presence in an agglutination reaction.

The **dot DAT** is a variation of the Coombs' test known as a dot blot direct antiglobulin test. IgG is fixed on a solid phase support or nitrocellulose membrane. The patient's erythrocytes are incubated on the membrane. This technique eliminates subjective interpretation of results, which diminishes the number of false positives and false negatives.

DAT is an **a**bbreviation for direct antiglobulin test. See the direct Coombs' test.

The **direct Coombs' test:** See direct antiglobulin test.

The **indirect Coombs' test** is an indirect antiglobulin test.

Polyspecific antihuman globulin (AHG) is known as the Coombs' reagent, which consists of antibody against human IgG and C3d. It may also have anti-C3b, anti-C4b, and anti-C4d antibodies. Although it demonstrates only minimal reactivity with IgM and IgA heavy chains, it may interact with these molecules by reacting with their κ or λ light chains. It is used for the direct antiglobulin test.

Antiglobulin is an antibody raised by immunization of one species, such as a rabbit, with immunoglobulin from another species, such as man. Rabbit antihuman globulin has been used for many years in an antiglobulin test to detect incomplete antibodies coating red blood cells, as in erythroblastosis fetalis or autoimmune hemolytic anemia. Antiglobulin antibodies are specific for epitopes in the Fc region of immunoglobulin molecules used as immunogen, rendering them capable of agglutinating cells whose surface antigens are combined with the Fab regions of IgG molecules whose Fc regions are exposed.

The **antiglobulin test** (Figure 16.16) is an antibody raised by immunization of one species, such as a rabbit, with immunoglobulin from another species, such as man. Rabbit antihuman globulin has been used for many years in an antiglobulin test to detect incomplete antibodies coating red blood cells as in erythroblastosis fetalis or autoimmune hemolytic anemia.

Antiglobulin antibodies are specific for epitopes in the Fc region of immunoglobulin molecules used as immunogen, rendering them capable of agglutinating cells whose surface antigens are combined with Fab regions of IgG molecules whose Fc regions are exposed.

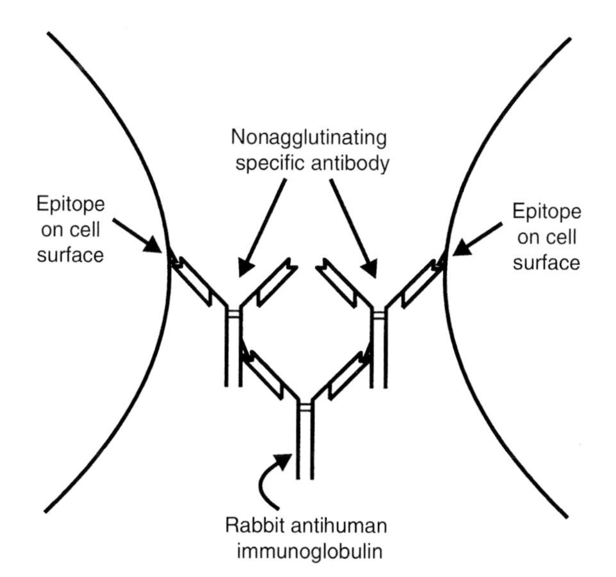

FIGURE 16.16 Schematic representation of the mechanism of the antiglobulin test to demonstrate nonagglutinating antibodies on red cell surfaces in autoimmune hemolytic anemia.

Antiglobulin test: When red blood cells are coated with antibodies that are not agglutinable in saline, such as those from an infant with erythroblastosis fetalis, a special antihuman immunoglobulin prepared by immunizing rabbits with human IgG may be employed to crosslink the antibody-coated red cells to produce agglutination. Although previously considered to be incomplete antibodies, they are known to be bivalent, but may be of a smaller size than saline agglutinable type antibodies. R.R.A. Coombs developed this test in England in the 1940s. In addition to its usefulness in hemolytic disease of the newborn, the Coombs' test detects incomplete antibody-coated erythrocytes from patients with autoimmune hemolytic anemia. In the direct Coombs' test, red blood cells linked to saline nonaggglutinable antibody are first washed, combined with rabbit antihuman immunoglobulin serum, and then observed for agglutination. In the indirect Coombs' test, serum containing the saline nonagglutinable antibodies is combined with red blood cells which are coated, but not agglutinated. The rabbit antihuman immunoglobulin is then added to these antibody-coated red cells, and agglutination is observed as in the direct Coombs' reaction. A third assay termed the "non-gamma" test requires the incubation of erythrocytes with anti-C3 or anti-C4 antibodies. Agglutination reflects the presence of these complement components on the red blood cell surface. This is an indirect technique to identify IgM antibodies that have fixed complement, such as those that are specific for Rh blood groups.

The **antiglobulin inhibition test** is an assay based upon interference with the antiglobulin test through reaction of the antiglobulin reagent with antibody against it prior to combination with incomplete antibody-coated erythrocytes. This is the basis for the so-called antiglobulin consumption test.

The **direct antiglobulin test** is an assay in which washed erythrocytes are combined with antiglobulin antibody. If the red cells had been coated with nonagglutinating (incomplete) antibody *in vivo,* agglutination would occur. Examples of this in humans include hemolytic disease of the newborn, in which maternal antibodies coat the infant's erythrocytes, and autoimmune hemolytic anemia, in which the subject's red cells are coated with autoantibodies. This is the basis of the direct Coombs' test.

The **indirect antiglobulin test** is a method to detect incomplete (nonagglutinating) antibody in a patient's serum. Following incubation of red blood cells or other cells possessing the antigen for which the incomplete antibodies of interest are specific, rabbit antihuman globulin is added to the antibody-coated cells which have been first washed. If agglutination results, incomplete agglutinating antibody is present in the serum with which the antigen-bearing red cells have been incubated.

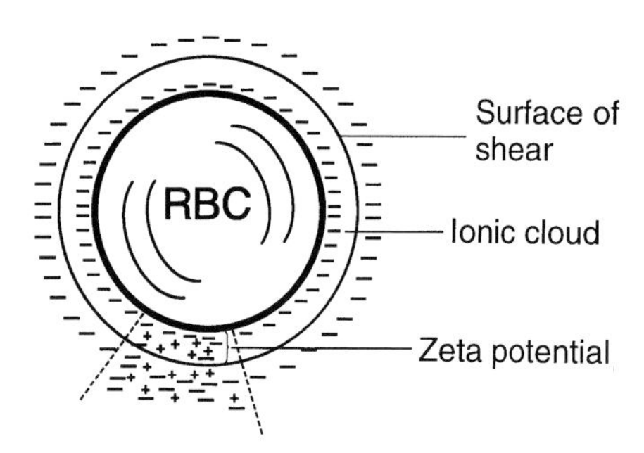

FIGURE 16.17 Schematic representation of the zeta potential surrounding red blood cells.

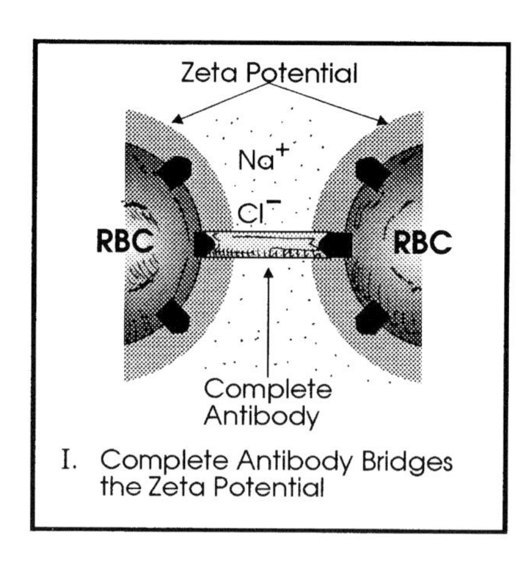

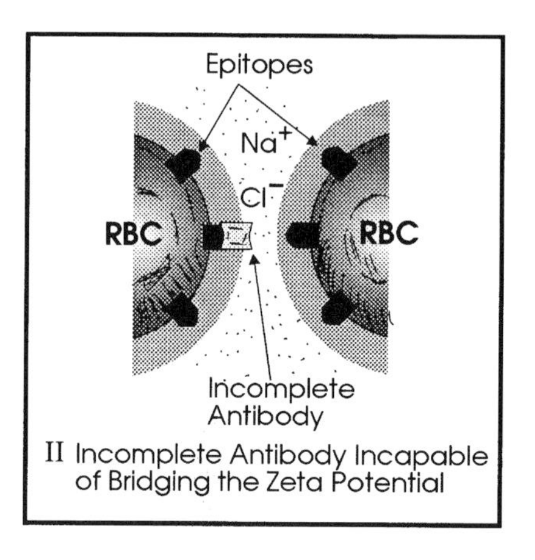

FIGURE 16.18 Comparison of the ability of complete antibody to bridge the zeta potential with the inability of incomplete antibody to do so.

A **public antigen (supratypic antigen)** is an epitope which several distinct or private antigens have in common. A public antigen such as a blood group antigen is one that is present in greater than 99.9% of a population. It is detected by the indirect antiglobulin (Coombs' test). Examples include Ve, Ge, Jr, Gy^a, and Ok^a. Antigens that occur frequently but are not public antigens include MNs, Lewis, Duffy, P, etc. In blood banking, there is a problem finding a suitable unit of blood for a transfusion to recipients who have developed antibodies against public antigens.

The **zeta potential** (Figure 16.17 and Figure 16.18) is a collective negative charge on erythrocyte surfaces that causes them to repulse one another in cationic medium. Some cations are red-cell surface bound, whereas others are free in the medium. The boundary of shear is between the two cation planes, where the zeta potential may be determined as $-mV$. IgM antibodies have an optimal zeta potential of −22 to −17 mV, and IgG antibodies have an optimum of −11 to −4.5 mV. The less the absolute mV, the less the space between cells in suspension. The addition of certain proteins, such as albumin, to the medium diminishes the zeta potential.

Serum albumin is the principal protein of human blood serum that is soluble in water and in 50% saturated sodium sulfate. At pH 7.0, it is negatively charged and migrates toward the anode during electrophoresis. It is important in regulating osmotic pressure and binding of anions. Serum albumin (e.g., bovine serum albumin, BSA) is commonly used as an immunogen in experimental immunology.

An **albumin agglutinating antibody** is an antibody that does not agglutinate erythrocytes in physiological saline solution but does cause their aggregation in 30% bovine serum albumin (BSA). Antibodies with this property have long been known as "incomplete antibodies" and are of interest in red blood cell typing.

Bromelin is an enzyme that has been used to render erythrocyte surfaces capable of being agglutinated by incomplete antibody.

Ficin is a substance employed to delete sialic acid from cell surfaces, which is especially useful in blood grouping to decrease the zeta potential and facilitate otherwise poorly agglutinating antibodies. Erythrocytes treated with ficin reveal enhanced expression of Kidd, Ii, Rh, and Lewis antigens. The treatment destroys MNSs, Lutheran, Duffy, Chido, Rogers, and Tn, among other antigens.

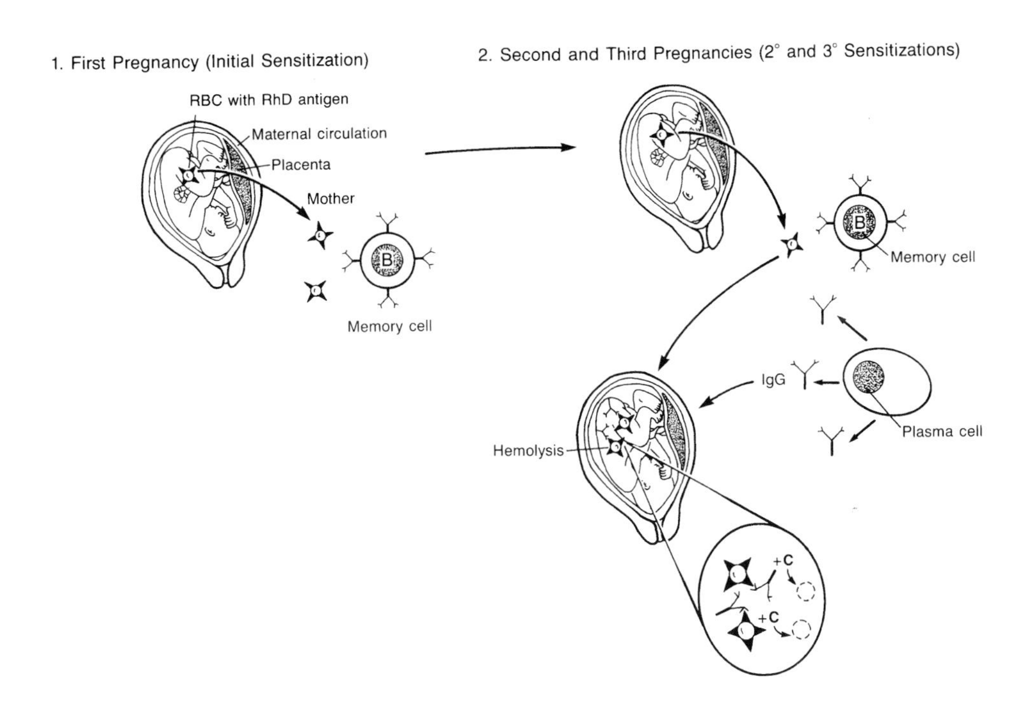

FIGURE 16.19 Representation of the mechanism of hemolytic disease of the newborn.

Incomplete antibody is a nonagglutinating antibody that must have a linking agent such as anti-IgG to reveal its presence in an agglutination reaction. See Coombs' test.

Hemolytic disease of the newborn (HDN) (Figure 16.19) is a condition in which a fetus with RhD positive red blood cells can stimulate an RhD negative mother to produce anti-RhD IgG antibodies that cross the placenta and destroy the fetal red blood cells when a sufficient titer is obtained. This is usually not until the third pregnancy with an Rh positive fetus. At parturition, the RhD positive red blood cells enter the maternal circulation, and subsequent pregnancies provide a booster to this response. With the third pregnancy, a sufficient quantity of high-titer antibody crosses the placenta to produce considerable lysis of fetal red blood cells. This may lead to erythroblastosis fetalis (hemolytic disease of the newborn). A total of 70% HDN are due to RhD incompatibility between the mother and fetus. Exchange transfusions may be required for treatment. Two other antibodies against erythrocytes that may likewise be a cause for transfusion exchange include anti-Fy^a and Kell. As bilirubin levels rise, the immature blood–brain barrier permits bilirubin to penetrate and deposit on the basal ganglia. Injection of the mother with anti-D antibody following parturition unites with the RhD positive red cells, leading to their elimination by the mononuclear phagocyte system.

Hemolytic anemia of the newborn: See hemolytic disease of the newborn.

HDN: Hemolytic disease of the newborn.

Erythroblastosis fetalis: A human fetal disease induced by IgG antibodies passed across the placenta from mother to fetus that are specific for fetal red blood cells, leading to their destruction. Although not often a serious problem until the third pregnancy, the escape of fetal red blood cells into the maternal circulation, especially at the time of parturition, produces a booster response in the mother of the IgG antibody that produces an even more severe reaction in the second and third fetus. The basis for this reaction is an isoantigen such as RhD antigen not present in the mother but present in the fetal red cells and inherited from the father. Clinical consequences of this maternal–fetal blood group incompatibility include anemia, jaundice, kernicterus, hydrops fetalis, and even stillbirth. Preventive therapy now includes administration of anti-D antiserum (RhIG) within 72 h following parturition. This antibody combines with the fetal red cells dumped into the mother's circulation at parturition and dampens production of a booster response. An antibody-mediated Type II hypersensitivity reaction.

Kernicterus describes deposition in the skin, leading to yellowish discoloration, as well as deposition in the central nervous system of erythrocyte breakdown products in the blood of infants with erythroblastosis fetalis. It may lead to neurologic dysfunction.

Hydrops fetalis is a hydropic condition that occurs in newborns who may appear puffy and plethoric, and that may be induced by either immune or nonimmune mechanisms. In the immune type, the mother synthesizes IgG antibodies specific for antigens of the offspring, such as anti-RhD erythrocyte antigen. These IgG antibodies pass across the placenta into the fetal circulation causing hemolysis. Nonimmune hydrops results from various etiologies not discussed here.

P antigen (Figure 16.20) is an ABH blood group-related antigen found on erythrocyte surfaces that is comprised of the three sugars galactose, *N*-isoacetyl-galactosamine and *n*-acetyl-glucosamine. The P antigens are designated P_1, P_2, P^k, and p. P_2 subjects rarely produce anti-P_1 antibody which may lead to hemolysis in clinical situations. Paroxysmal cold hemagglutinaria patients develop a biphasic autoanti-P antibody that fixes complement in the cold and lyses red blood cells at 37°C.

Donath-Landsteiner antibody is an immunoglobulin specific for P blood group antigens on human erythrocytes. This antibody binds to the patient's red blood cells at cold temperatures and induces hemolysis on warming. It occurs in subjects with paroxysmal cold hemaglobulinemia (PCH). Also called Donath-Landsteiner cold autoantibody.

The **MNSs blood group system** (Figure 16.21) refers to human erythrocyte glycophorin epitopes. There are four distinct sialoglycoproteins (SGP) on red cell membranes. These include α-SGP (glycophorin A, MN), β-SGP (glycophorin C), γ-SGP (glycophorin D), and δ-SGP (glycophorin B). MN antigens are present on α-SGP and δ-SGP. M and N antigens are present on α-SGP, with approximately one-half million copies detectable on each erythrocyte. This is a 31-kDa structure that is comprised of 131 amino acids, with about 60% of the total weight attributable to carbohydrate. This transmembrane molecule has a carboxyl terminus that stretches into the cytoplasm of the erythrocyte with a 23-amino acid hydrophobic segment embedded in the lipid bilayer. The amino terminal segment extends to the extracellular compartment. Blood group antigen activity is in the external segment. In α-SGP with M antigen activity, the first amino acid is serine, and the fifth is glycine. When it carries N antigen activity, leucine and glutamic acid replace serine and glycine at positions 1 and 5, respectively. The Ss antigens are encoded by allelic genes at a locus closely linked to the MN locus. The U antigen is also considered a part of the MNS system. Whereas anti-M and anti-N antibodies may occur without red cell stimulation, antibodies against Ss and U antigens generally follow erythrocyte stimulation. The MN and Ss alleles positioned on chromosome 4 are linked. Antigens of the MNSs system may provoke the formation of antibodies that can mediate hemolytic disease of the newborn (Figure 16.22).

Phenotype	Reactions with Anti- P_1	P	P^k	PP_1P^k	Phenotype Frequency Whites	African Am.
P_1	+	+	0	+	79	94
P_2	0	+	0	+	21	6
p	0	0	0	0	Very rare	
P_1^k	+	0	+	+	Very rare	
P_2^k	0	0	+	+	Very rare	

FIGURE16.20. P antigen.

Reactions Anti-M	Anti-N	Phenotype	Phenotype Frequency Whites	African Am.
+	0	M+N-	28	26
+	+	M+N+	50	45
0	+	M-N+	22	30
Anti-s	Anti-s			
+	0	S+s-	11	3
+	+	S+s+	43	28
0	+	S-s+	45	69
0	0	S-s-	0	<1

FIGURE 16.21 MNSs blood group system.

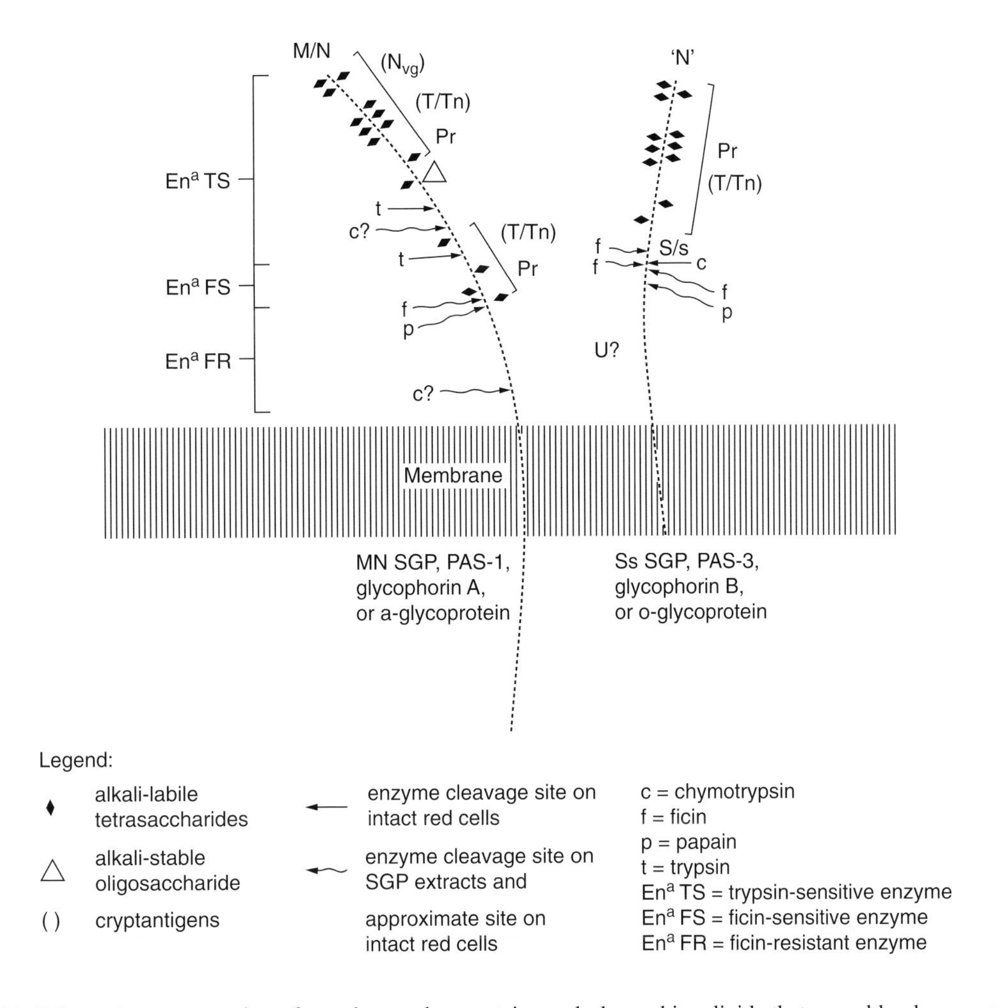

FIGURE 16.22 Schematic representation of membrane glycoproteins and glycosphingolipids that carry blood group antigens.

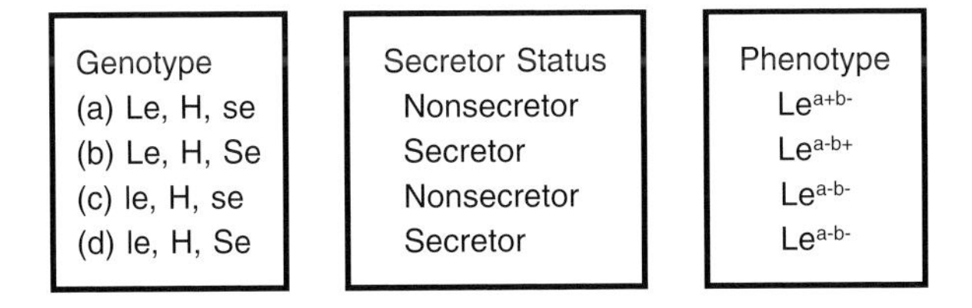

FIGURE 16.23 Lewis blood group system.

U antigen is a rare MNS erythrocyte antigen present in fewer than 1% of African Americans and absent from Caucasian red blood cells. When U antigen is not present, s antigen is not expressed. Membrane sialoglycoprotein, and glycophorins A and B, are requisite for U antigen expression.

The **Lewis blood group system** (Figure 16.23 and Figure 16.24) is an erythrocyte antigen system that differs from other red cell groups in that the antigen is present in soluble form in the blood and saliva. Lewis antigens are adsorbed from the plasma onto the red cell membrane. The Lewis phenotype expressed is based on whether the individual is a secretor or a nonsecretor of the Lewis gene product. Expression of the Lewis phenotype is dependent also on the ABO phenotype. Lewis antigens are carbohydrates chemically. Lewis blood secretors have an increased likelihood of urinary tract infections induced by *Escherichia coli* or other microbes because of the linkage of carbohydrate residues of glycolipids and glycoproteins on urothelial cells.

Lewis^x/Sialyl-Lewis^{x}CD15/CD15S: The blood group-related antigen Lewis^x (Le^x) and related oligosaccharide sequences on glycoproteins and glycolipids serve as ligands for the selectins, the leukocyte-endothelium adhesion molecules that are critical to the early stages of leukocyte recruitment in inflammation. Le^x and sialyl-Lewis^x are human granulocyte and monocyte markers and are designated CD15 and CD15s, respectively. Monocytes express mainly the sialyl-Le^x and the sialic acid masks expression of Le^x antigen. Le^x and sialyl-Le^x are tumor-associated antigens.

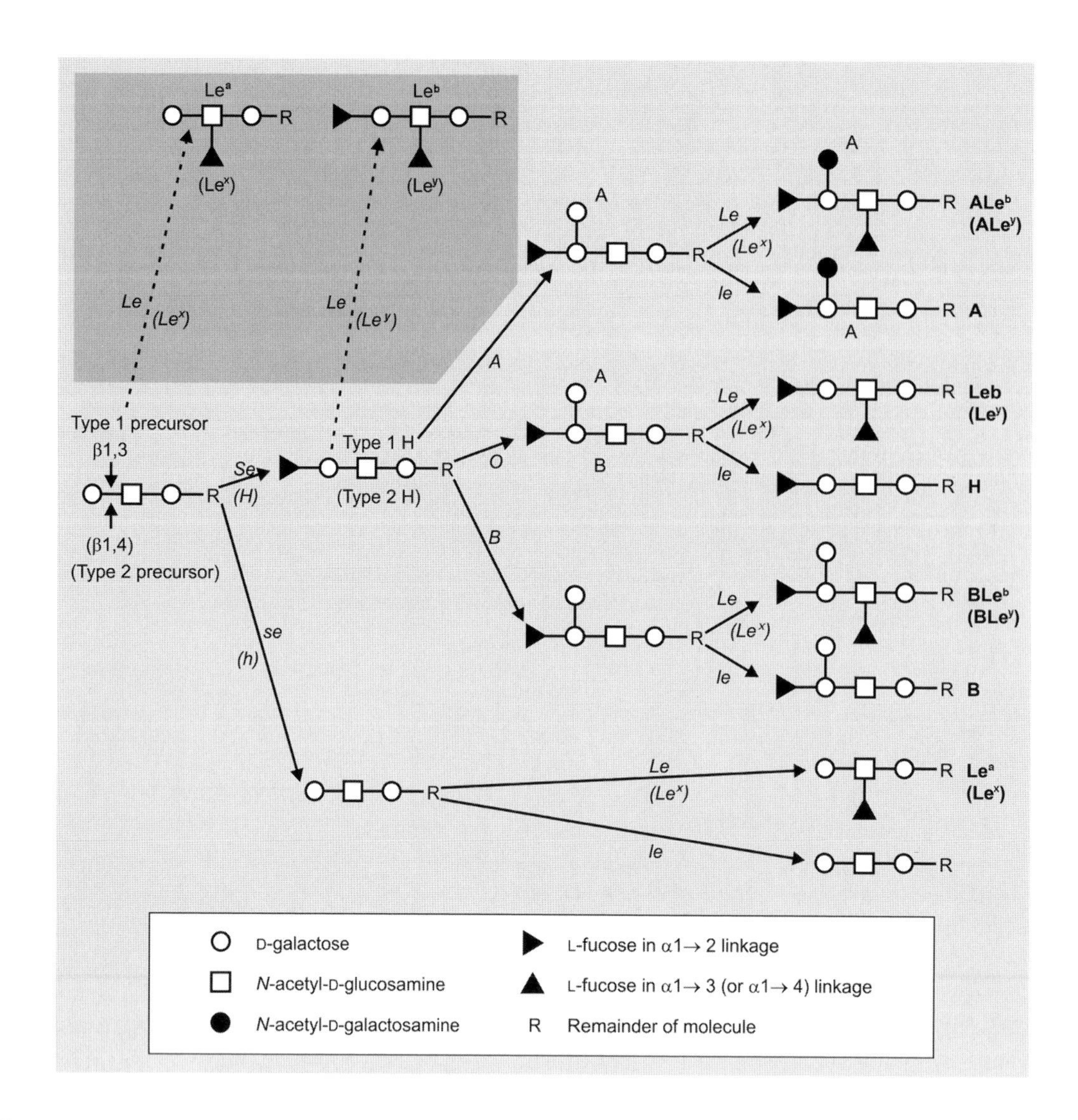

FIGURE 16.24 Schematic representation of the biosynthetic pathways of ABH, Lewis, and XY antigens derived from type I and type II core chains. Genes controlling steps in the pathway are shown in italics. Type I and type II precursors differ in the nature of the linkage between the nonreducing terminal galactolose and *N*-acetylglucosamine: Beta 1–3 in type I and Beta 1–4 in type II. Type II structures and the genes acting on them are shown in parenthesis. Dash lines show how Lea and (Lex) Led (Ley), produced from the precursor and H structures, respectively are not substrate for the H-, Se-, or ABO-transferases and remain unconverted.

The oligosaccharides may be inappropriately expressed on tumor cells but have been established as distinctive markers of myeloid cells in human peripheral blood. Le^x- and Le^a-related sequences serve as ligands for carbohydrate binding receptors, the selectins. All three selectins bind to sialyl-Le^x-related sequences when they are exhibited in the clustered state on protein or lipid. E-selectin also binds the asialo-Le^x sequence, but less avidly.

The **Kell blood group system** (Figure 16.25) named for an antibody that induces hemolytic disease of the newborn, described in 1946, showed specificity for the K(KEL1) antigen. A total of 9% of Caucasian and 2% of African Americans have the *K* gene that encodes this antigen. Subsequently, the *K* allele was identified. Anti-K(KEL2) antibodies reacted with the erythrocytes of more than 99% of the random population. Kell system antigens are present only in relatively low density on the erythrocyte membrane. The strong immunogenicity of the K antigen leads to the presence of anti-K antibodies in sera of transfused patients. Anti-K antibodies cause hemolytic transfusion reactions of both immediate and delayed varieties. A total of 90% of donors are K–, which considerably simplifies the task of finding compatible blood for patients with anti-K.

The **McCleod phenotype** reflects human erythrocytes without Kell or Cellano antigens. These red cells lack Kx, a precursor in the biosynthetic pathway of the Kell blood group system. Kx is encoded by a gene on the X chromosome termed X^lk and is normally found on granulocytes

Phenotype	Reactions with Anti- K	K	Kp^a	Kp^b	Js^a	Js^b	Phenotype Frequency Whites	African Am.
K+k-	+	0					0.2	Rare
K+k+	+	+					8.8	2
K-k+	0	+					91	98
Kp (a+b-)			+	0			Rare	0
Kp (a+b+)			+	+			2.3	Rare
Kp (a-b+)			0	+			97.7	100
Js (a+b-)					+	0	0	1
Js (a+b+)					+	+	Rare	19
Js (a-b+)					0	+	100	80
K_0	0	0	0	0	0	0	Very rare	Very rare

FIGURE 16.25 Kell blood group system.

Phenotype	Reactions with Anti- Fy^a	Fy^b	Phenotype Frequency Whites	African Am.
Fy (a+b-)	+	0	17	9
Fy (a+b+)	+	+	49	1
Fy (a-b+)	0	+	34	22
Fy (a-b-)	0	0	Very rare	68

FIGURE 16.26 Duffy blood group.

and fibroblasts. Red cells lacking Kx have decreased survival, diminished permeability to water, and are acanthocytic morphologically with spikes on their surface. They also have decreased expression of Kell system antigens. This group of erythrocyte abnormalities is termed the McCleod phenotype. Subjects with McCleod erythrocytes have a neuromuscular system abnormality characterized by elevated serum levels of creatine phosphokinase (CPK). Older individuals may have disordered muscular functions. The X^1k gene maps to the short arm of the X chromosome where it is linked to the chronic granulomatous disease gene.

The **Duffy blood group** (Figure 16.26) is comprised of human erythrocyte epitopes encoded by *Fya* and *Fyb* genes, located on chromosome 1. Since these epitopes are receptors for *Plasmodium vivax*, African Americans who often express the Fy(a–b–) phenotype are not susceptible to the type of malaria induced by this species. Mothers immunized through exposure to fetal red cells bearing the Duffy antigens, which she does not possess, may synthesize antibodies that cross the placenta and induce hemolytic disease of the newborn (Figure 16.27).

The **Kidd blood group system** (Figure 16.28) was named for anti-Jk^a antibodies which were originally detected in the blood serum of a woman giving birth to a baby with hemolytic disease of the newborn. Anti-Jk^b antibodies

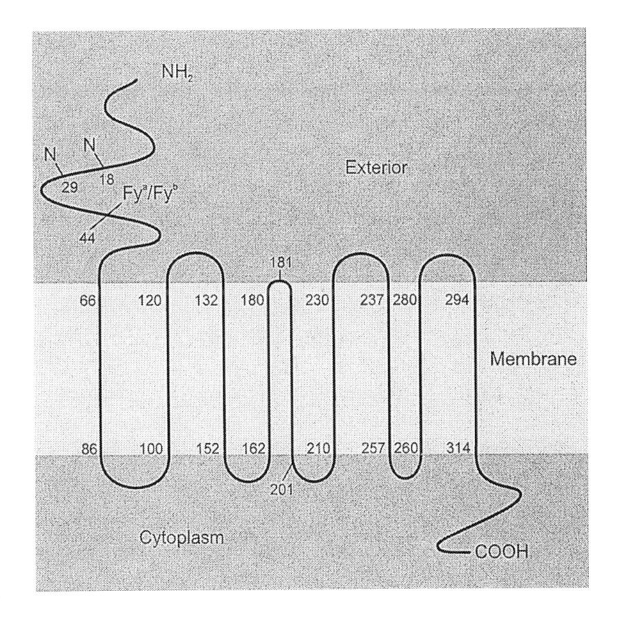

FIGURE 16.27 Schematic representation of the proposed topography of the Duffy glycoprotein within the red cell membrane. Numbers represent amino acid residues with the transcription-initiating methionine residue as 1. An extracellular N-terminal domain of 65 amino acids containing two *N*-glycosylation sites (N) and the site of the Fya/Fyb polymorphism is followed by nonmembrane-spanning domains, or alternatively, seven-membrane-spanning domains in common with other chemokine receptors.

Phenotype	Reactions with Anti- Jk[a]	Jk[b]	Phenotype Frequency Whites	African Am.
Jk (a+b-)	+	0	28	57
Jk (a+b+)	0	+	49	34
Jk (a-b+)	+	+	23	9
Jk (a-b-)	0	0	Very rare	Very rare

FIGURE 16.28 Kidd blood group system.

Phenotype	Antigen Expression I	i
I adult	Strong	Very weak
I cord	Weak	Strong
i adult	Very weak	Strong

FIGURE 16.29 Ii antigens.

were discovered in the serum of a patient following a transfusion reaction. Although Kidd system antibodies sometimes lead to HDN, it is not usually severe. However, the antibodies are problematic and cause severe hemolytic transfusion reactions, especially of the delayed type. These occur when antibodies developing quickly in a booster response to antigens on transfused erythrocytes destroy red cells in the circulation. As shown in the table, four phenotypes are revealed by the reactions of anti-Jk[a] and anti-Jk[b] antibodies. A dominant inhibitor gene (In[Jk]) may encode a null phenotype. Jk3 is believed to be present on both Jk(a$^+$) and Jk(b$^+$) red cells. Anti-Jk3 is frequently induced by red blood cell stimulation.

A **warm antibody** is an antibody that reacts best at 37°C. It is usually an IgG agglutinin and shows specificity for selected erythrocyte antigens that include KELL, DUFFY, KIDD, and Rh. It may be associated with immune hemolysis.

Ii antigens (Figure 16.29) are two nonallelic carbohydrate antigens (epitopes) on the surface membrane of human erythrocytes. They may also occur on some nonhematopoietic cells. The i epitope is found on fetal erythrocytes and red cell blood precursors. The I antigen is formed when aliphatic galactose-*N*-acetyl-glucosamine is converted to a complex branched structure. I represents the mature form and i the immature form. Mature erythrocytes express I. Antibodies against i antigen are hemolytic in cases of infectious mononucleosis.

Anti-I refers to antibodies against the I blood group antigen, which is present on the majority of adult red blood cells in man. The Ii antigens are present in the subterminal portions of the oligosaccharides which are ultimately converted to H and A or B antigens. I and i configurations are present on membrane-associated glycoproteins and glycosphingolipids. The heterogeneity observed with different anti-I antisera may reflect the recognition of different parts of the branched oligosaccharide chain. Fetal erythrocytes contain abundant i antigen but few branched oligosaccharides and little I antigen. The I antigen develops during the first 2 years of life with simultaneous loss of i. Anti-I is a common autoantibody that is frequently present as a cold-reacting agglutinin. Anti-I is of pathologic significance in many cases of CHD, when it acts as a complement-binding monoclonal antibody. Autoanti-I is of less significance in cold hemagglutinin disease than is anti-I. Thus, anti-I acting as a cold agglutinin may be detected as an autoantibody in a number of cases of cold antibody-type hemolytic anemia and in patients with *Mycoplasma pneumoniae* infection.

The **Lutheran blood group** (Figure 16.30) consists of human erythrocyte epitopes recognized by alloantibodies against Lua and Lub products. Antibodies developed against Lutheran antigens during pregnancy may induce hemolytic disease of the newborn.

The **Chido (Ch) and Rodgers (Rg) antigens** (Figure 16.31) are epitopes of C4d fragments of human complement component C4. They are not intrinsic to the erythrocyte membrane. The Chido epitope is found on C4d from C4B, whereas the Rodgers epitope is found on C4A derived from C4d. The Rodgers epitope is Val-Asp-Leu-Leu, and the Chido epitope is Ala-Asp-Leu-Arg. They are situated at residue positions 1188 to 1191 in the C4 α chain's C4d region. Antibodies against Ch and Rg antigenic determinants agglutinate saline suspensions of red blood cells coated with C4d. Since C4 is found in human serum, anti-Ch and anti-Rg are neutralized by sera of most individuals which contain the relevant antigens. Ficin and papain destroy these antigens.

The **Rodgers (Rg) antigens** are epitopes of C4d fragments of human complement component C4. They are not intrinsic to the erythrocyte membrane. The Chido epitope is found on C4d from C4B, whereas the Rodgers epitope

Phenotype	Reactions with Anti- Lu^a	Reactions with Anti- Lu^b	Phenotype Frequency
Lu (a+b-)	+	0	0.15
Lu (a+b+)	+	+	7.5
Lu (a-b+)	0	+	92.35
Lu (a-b-)	0	0	Very rare

FIGURE 16.30 Lutheran blood group.

Phenotype	C4d Component Present	Frequency (%) Whites
Ch(a+), Rg(a+)	C4dS, C4df	95
Ch(a-), Rg(a+)	C4df	2
Ch(a+), Rg(a-)	C4dS	3
Ch(a-), Rg(a-)	None	Very rare

FIGURE 16.31 Chido (Ch) ahd Rodgers (Rg) antigens.

Phenotype	Reactions with Anti- Xg^a	Phenotype Frequency Males	Phenotype Frequency Females
Xg (a+)	+	65.6	88.7
Xg (a-)	0	34.4	11.3

FIGURE 16.32 Xg^a, the sex-linked blood antigen.

is found on C4A derived from C4d. The Rodgers epitope is Val-Asp-Leu-Leu, and Chido epitope is Ala-Asp-Leu-Arg. They are situated at residue positions 1188 to 1191 in the C4 α chain's C4d region. Antibodies against Ch and Rg antigenic determinants agglutinate saline suspensions of red blood cells coated with C4d. Since C4 is found in human serum, anti-Ch and anti-Rg are neutralized by sera of most individuals which contain the relevant antigens. Ficin and papain destroy these antigens.

C4A is a very polymorphic molecule expressing the Rodgers epitope that is encoded by the C4A gene. The equivalent murine gene encodes sex-limited protein (SLP). It has less hemolytic activity than does C4B. C4A and C4B differ in only four amino acid residues in the α chain's C4d region. C4A is Pro-Cys-Pro-Bal-Leu-Asp, whereas C4B is Leu-Ser-Pro-Bal-Ile-His.

C4B is a polymorphic molecule that usually expresses the Chido epitope and is encoded by the C4B gene. The murine equivalent gene encodes an Ss protein. It shows greater hemolytic activity than does C4A.

Xg^a, the sex-linked blood antigen (Figure 16.32) is an antibody more common in women than in men. It is specific for the Xg^a antigen, in recognition of its X-born pattern of inheritance. This table gives phenotype frequencies in Caucasian males and females. The antibody is relatively uncommon and has not been implicated in hemolytic disease of the newborn or hemolytic transfusion reaction even though it can bind complement and may occasionally be an autoantibody. Anti-Xg^a antibodies might be of value in identifying genetic traits transmitted in association with the X chromosome.

Cryptantigens are surface antigens of red cells not normally detectable, but demonstrable by microbial enzyme action that leads to the modification of cell surface carbohydrates. Naturally occurring IgM antibodies in normal serum may agglutinate these exposed antigens.

Cold antibodies are antibodies that occur at higher titers at 4°C rather than at 37°C.

Cold agglutinin is an antibody that agglutinates particulate antigen, such as bacteria or red cells, optimally at temperatures less than 37°C. In clinical medicine, the term usually refers to antibodies against red blood cell antigens as in the cold agglutinin syndrome.

Cold hemagglutinin disease: See cold agglutinin syndrome.

PPLO (pleuropneumonia-like organisms): *Mycoplasma pneumoniae.* A microorganism that causes asymptomatic respiratory tract infection or upper respiratory tract inflammation. It spreads in the air. Patients develop headache, muscle pain, chest tenderness, and a low-grade fever. They may manifest cold agglutinin-induced hemolysis.

Cold agglutinin syndrome is an immune condition in which IgM autoantibodies agglutinate erythrocytes most effectively at 4°C. Normal individuals may have cold agglutinins in low titer (<1:32). Certain infections, such as cytomegalovirus, trypanosomias, mycoplasma, malaria, and Epstein-Barr virus infection, are followed by the development of polyclonal cold agglutinins. These antibodies are of concern only if they are hemolytic. Acquired hemolytic anemia patients with a positive direct Coombs' test should be tested for cold agglutinins. For example, they might have anti-Pr, anti-I, anti-i, anti-Sd^a, or anti-Gd. Aged individuals suffering from monoclonal κ proliferation or simultaneous large-cell lymphoma may develop cold agglutinin syndrome. It also occurs in the younger age group in whom anti-I antibodies have been synthesized following an infection with *Mycoplasma pneumoniae* or in whom anti-i antibodies associated with infectious mononucleosis have formed. C3d coats the cells. Agglutination and complement fixation may take place intravascularly in parts of the body exposed to the cold. When the red blood cells with attached complement are warmed to 37°C (normal body temperature), mild hemolysis occurs.

CHAD is an abbreviation for cold hemagglutinin disease.

Platelet antigens are surface epitopes on thrombocytes that may be immunogenic, leading to platelet antibody formation resulting in such conditions as neonatal alloimmune thrombocytopenia and posttransfusion purpura. The Pl^{A1} antigen may induce platelet antibody formation in Pl^{A1} antigen-negative individuals. Additional platelet antigens associated with purpura include Pl^{A2}, Bak^a, and HLA-A2. Anti-Bak^a is a normal human platelet (thrombocyte) antigen. Anti-Bak^a IgG antibody synthesized by a Bak^a-negative female may be passively transferred across the placenta to induce immune thrombocytopenia in the neonate.

Platelets possess surface FcγRII that combine to IgG or immune complexes (Figure 16.33 and 16.34). The platelet surfaces can become saturated with immune complexes as in autoimmune or idiopathic thrombocytopenic purpura (ITP) or AITP. Fab-mediated antibody binding to platelet antigens may be difficult to distinguish from Fc-mediated binding of immune complexes to the surface.

The administration of platelet concentrates prepared by centrifuging a unit of whole blood at low speed to provide 40 to 70 ml of plasma that contains 3–4 × 10^{11} platelets for platelet transfusion. This amount can increase an adult's platelet concentration by 10,000/mm^3. It is best to store platelets at 20 to 24°C, subjecting them to mild agitation. They must be used within 5 d of collection.

Bak^a is a normal human platelet (thrombocyte) antigen. Anti-Bak^a IgG antibody synthesized by a Bak^a-negative pregnant female may be passively transferred across the placenta to induce immune thrombocytopenia in the neonate.

P1^{A1} antibodies, which are specific for the P1^{A1} antigen, are responsible for three fourths of the cases of neonatal alloimmune thrombocytopenic purpura and posttransfusion purpura. Anti-P1^{A1} antibodies prevent clot retraction and platelet aggregation.

Drugs such as amphotericin B, cephalothin, methicillin, pentamidine, trimethoprim-sulfamethoxizole, and vancomycin may all induce the synthesis of **antiplatelet antibodies.**

Antigranulocyte antibodies play a significant role in the pathogenesis of febrile transfusion reactions; drug-induced neutropenia; isoimmune neonatal neutropenia; autoimmune neutropenia, including Felty's syndrome; Graves' disease; Evans syndrome; SLE; and primary autoimmune neutropenia of children. Antigranulocyte antibodies are best detected and quantitated by flow cytometry.

Thrombocytopenia describes diminished blood platelet numbers with values below 100,000 per cubic millimeter of blood compared to a normal value of 150,000 to 300,000 platelets per cubic millimeter of blood. This decrease in numbers of blood platelets can lead to bleeding.

Thrombocytosis refers to elevated blood platelet numbers with values exceeding 600,000 thrombocytes per cubic millimeter of blood compared to a normal value of 150,000 to 300,000 blood platelets per cubic millimeter of blood.

Platelet transfusion: The administration of platelet concentrates prepared by centrifuging a unit of whole blood at low speed to provide 40 to 70 ml of plasma that contains 3–4 × 10^{11} platelets. This amount can increase an adult's platelet concentration by 10,000 per cubic millimeter of blood. It is best to store platelets at 20 to 24°C, subjecting them to mild agitation. They must be used within 5 d of collection.

T antigen(s) is (1) an erythrocyte surface antigen that is shielded from interaction with the immune system by an *N*-acetyl-neuraminic acid residue. Thus, antibody is formed against this antigen once bacterial infection has diluted this neuraminic acid residue. Antibodies produced can cause polyagglutination of red cells bearing the newly revealed T antigen. (2) Several 90-kDa nuclear proteins that combine with DNA and are critical in transcription and replication of viral DNA in the lytic cycle. T antigen participates in the change from early to late stages of

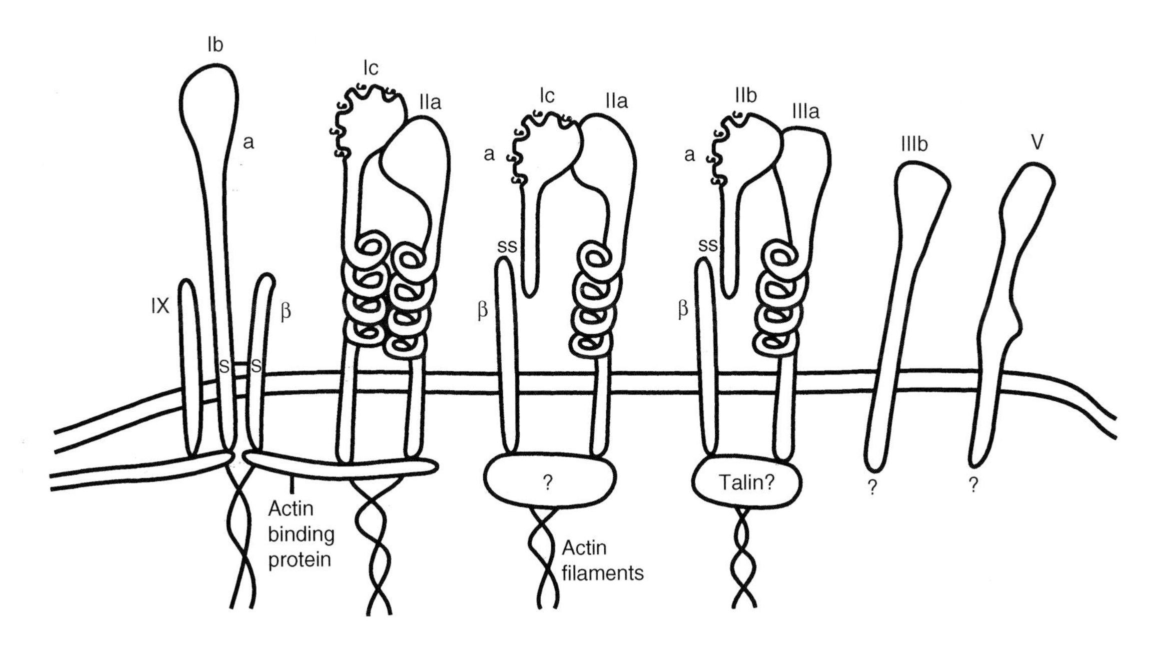

FIGURE 16.33 Schematic representation of the principal platelet membrane glycoproteins, indicating known or suspected complexes, disulfide bonds between chains, calcium-bonding domains, and interactions with cytoskeletal components.

Antigen System	Glycoprotein (GP) Location	Other Names	Antigens	Other Names	Phenotype Frequency (%)	
					White	Japanese
HPA-1	GPIIIa	Zw, Pl^{A}	HPA-1a	Zw^{a}, Pl^{A1}	97.9	99.9
			HPA-1b	Zw^{b}, Pl^{A2}	26.5	3.7
HPA-2	GPIb	Ko, Sib	HPA-2a	Ko^{b}	99.3	NT
			HPA-2b	Ko^{a}, Sib^{a}	14.6	25.4
HPA-3	GPIIb	Bak, Lek	HPA-3a	Bak^{a}, Lek^{a}	87.7	78.9
			HPA-3b	Bak^{b}	64.1	NT
HPA-4	GPIIIa	Pen, Yuk	HPA-4a	Pen^{a}, Yuk^{b}	99.9	99.9
			HPA-4b	Pen^{b}, Yuk^{a}	0.2	1.7
HPA-5	GPIa	Br, Hc, Zav	HPA-5a	Br^{b}, Zav	99.2	NT
			HPA-5b	Br^{a}, Zav^{a}, Hc^{a}	20.6	NT

FIGURE 16.34 Nomenclature and phenotype frequency of human platelet antigens.

transcription. (3) An epitope that shares homology at the N-terminal sequence with the SV40 virus T antigen.

T agglutinin is an antibody that occurs naturally in the blood serum of man, which agglutinates red blood cells expressing T antigen as a result of their exposure to bacteria or as a consequence of treatment with neuraminidase. This antibody is of interest in transfusion medicine as it may confuse blood grouping or cross-matching procedures by giving a false-positive reaction when red blood cell suspensions contaminated with microorganisms are used.

T activation: The use of bacterial neuraminidase to cleave *N*-acetyl (sialic acid) residue to uncover antigenic determinants (epitopes) which have been masked or hidden. This permits the treated cells to be agglutinated by natural antibodies in the blood of most individuals. Aged blood can be used to detect T activation.

Senescent cell antigen is a neoantigen appearing on old red blood cells that binds IgG autoantibodies. Senescent cell antigen is also found on lymphocytes, platelets, neutrophils, adult human liver cells (in culture), and human embryonic renal cells (in culture). Its appearance on aging somatic cells probably represents a physiologic process to remove senescent and injured cells that have fulfilled their function in the animal organism. Macrophages are able to identify and phagocytize dying and aging self cells that are no longer functional, without disturbing mature healthy cells.

Transfusion describes the transplantation of blood cells, platelets, and/or plasma from the blood circulation of one individual to another. Acute blood loss due to hemorrhage or the replacement of deficient cell types due to excess destruction or inadequate formation are indications for blood transfusion. With the description of human blood groups by Landsteiner in 1900, the transfusion of blood from one human being to another became possible. This ushered in the field of transfusion medicine, which relates to substitution therapies with human blood, protein deficiencies, and blood loss. Peripheral blood cell and plasma collection, processing, storage, compatibility matching, and transfusion are routine procedures in medical centers throughout the world. The description of multiple other blood group systems followed the initial description of the AB0 types. Modern-day blood group serology and immunohematology laboratories consider all aspects of allo- and auto-antibodies against red cells in clinical transfusion. Bloodborne viruses are recognized as critical risk factors in transfusion. In recent years, the rate of viral transmission through transfusions has been greatly diminished with the development of adequate methods of screening for HIV, hepatitis viruses, and other infectious agents.

A **universal recipient** is an ABO blood group individual whose cells express antigens A and B but whose serum does not contain anti-A and anti-B antibodies. Thus, red blood cells containing any of the ABO antigens may be transfused to them without inducing a hemolytic transfusion reaction, i.e., from an individual with type A, B, AB, or O. It is best if the universal recipient is Rh^+, i.e., has the Rh D antigen on his erythrocytes, to avoid developing a hemolytic transfusion reaction. However, blood group systems other than ABO may induce hemolytic reactions in a universal recipient. Thus, it is best to use type-specific blood for transfusions.

An **immediate-spin crossmatch** is a test for incompatibility between donor erythrocytes and the recipient patient's serum. This assay reveals ABO incompatibility in practically all cases, but is unable to identify IgG alloantibodies against erythrocyte antigens.

A **universal donor** is a blood group O RhD^- individual whose erythrocytes express neither A nor B surface antigens. This type of red blood cell fails to elicit a hemolytic transfusion reaction in recipients who are blood group A, B, AB, or O. However, group O individuals serving as universal donors may express other blood group antigens on their erythrocytes that will induce hemolysis. It is preferable to use type-specific blood for transfusions, except in cases of disaster or emergency.

Alloimmunization describes an immune response provoked in one member or strain of a species with an alloantigen derived from a different member or strain of the same species. Examples include the immune response in humans following transplantation of a solid organ graft such as a kidney or heart from one individual to another. Alloimmunization with red blood cell antigens in humans may lead to pathologic sequelae such as hemolytic disease of the newborn (erythroblastosis fetalis) in a third $Rh(D)^+$ baby born to an $Rh(D)^-$ mother.

Exchange transfusion is a method that involves replacing the entire blood volume of a patient with donor blood. This is done to remove toxic substances such as those formed in kernicterus in infants with erythroblastosis fetalis or may be employed to remove anti-Rh antibodies causing hemolytic disease of the newborn.

Isoimmunization describes an immune response induced in the recipient of a blood transfusion in which the donor red blood cells express isoantigens not present in the recipient. The term also refers to maternal immunization by fetal red blood cells bearing isoantigens the mother does not possess.

Incompatibility refers to dissimilarity between the antigens of a donor and recipient as in tissue allotransplantation or blood transfusions. The transplantation of a histoincompatible organ or the transfusion of incompatible blood into a recipient may induce an immune response against the antigens not shared by the recipient in injurious consequences.

Isophile antibody is an n antibody induced by and specifically reactive with erythrocytes, but it is not reactive with other species' red blood cells. These antibodies are against antigens of red blood cells unique to the species from which they were derived.

An **isophile antigen** is an antigen that is species specific; often refers to erythrocyte antigens.

Transfusion reaction(s) may be either immune or nonimmune reactions that follow the administration of blood. Transfusion reactions with immune causes are considered serious and occur in 1 in 3000 transfusions. Patients may develop urticaria, itching, fever, chills, chest pains, cyanosis, and hemorrhage. The appearance of these symptoms together with an increase in temperature by 1°C signals the need to halt the transfusion. Immune, noninfectious transfusion reactions include allergic urticaria (immediate hypersensitivity); anaphylaxis, as in the administration of blood to IgA-deficient subjects, some of whom develop anti-IgA antibodies of the IgE class; and serum sickness, in which the serum proteins such as immunoglobulins induce the formation of precipitating antibodies that lead to immune complex formation.

Infused immunocompetent T lymphocytes react against histoincompatible immune system cells of the recipient, leading to **transfusion-associated graft-vs.-host disease (TAGVHD)**. This is likely to occur in patients who have

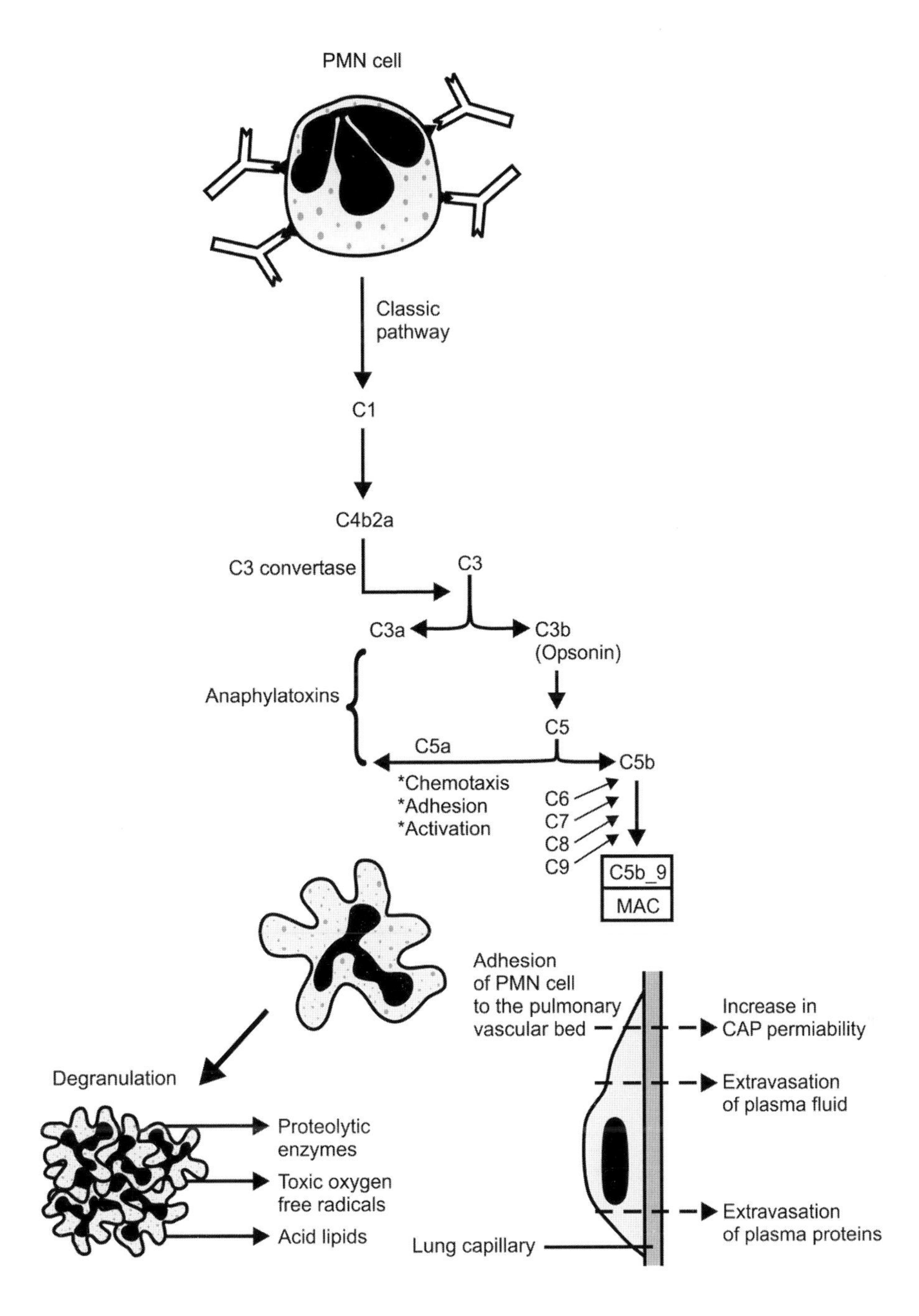

FIGURE 16.35 Schematic representation of molecular and cellular events that lead to the production of transfusion-related acute lung injury (TRALI), believed to be a form of adult respiratory distress syndrome (ARDS).

been either immunocompromised or treated with chemotherapy for tumors. The patient may develop a skin rash and have profound pancytopenia, as well as altered liver function tests. Three weeks following transfusion, 84% may die. To avoid graft-vs.-host reactivity induced by a blood transfusion, any blood product containing lymphocytes should be subjected to 1500 rad prior to administration.

TRALI (transfusion-related acute lung injury) (Figure 16.35) is a form of acute respiratory distress that often occurs within 4 h following a blood transfusion. It is attributable to leukocyte antibodies and is an acute pulmonary reaction leading to noncardiac pulmonary edema. It is a form of ARDS with a reasonably good prognosis. Mortality is 10% as opposed to 50 to 60% for other forms of ARDS. A total of 80% of TRALI patients experience rapid resolution of pulmonary infiltrates and restoration of arterial blood gas values to normal within 96 h. A total of 17% of TRALI patients retain pulmonary infiltrates for a week following the transfusion reactions. TRALI reactions have been reported in 1 in 5000 units of blood transfused. Leukoagglutinating antibodies as wall as some lymphocytotoxins have been implicated. The offending antibody is passively transfused in donor plasma, rather than donor's leukocytes reacting with recipient antibody. Both donor granulocyte antibodies and donor lymphocyte antibodies have been implicated in TRALI reactions. A total of 65% of cases revealed the presence of HLA-specific antibodies. However, HLA antibodies may be present in the plasma of donors but not cause TRALI reactions. Donor plasma implicated in TRALI reactions are often from multiparous females and individuals who have received multiple blood transfusions. There is difficulty in explaining the pathophysiology of a mechanism whereby such a small amount of antibody could induce a severe clinical reaction unless it initiates an amplification mechanism such as activation of complement. Such a mechanism could cause the formation of C5a that attaches to granulocytes, altering their membrane in such a way that they adhere nonspecifically to various surfaces. Once these cells are sequestered in the pulmonary vascular bed, they may become activated and release proteolytic enzymes in toxic oxygen metabolites, leading to acute lung injury. Pulmonary sequestration of granulocytes could lead to further endothelial injury and microvascular occlusion. The activation of complement, generation of C5a, and pulmonary sequestration of granulocytes which occur when blood comes into contact with hemodialysis membranes, further support a role for complement activation. In summary, TRALI depends on the simultaneous presence of antibody, complement and antigen positive cells leading to extensive capillary leakage.

17 Immunological Diseases and Immunopathology

Immunological diseases include those conditions in which there is either an aberration in the immune response or the immune response to the disease agent leads to pathological changes. This category includes diseases with an immunological etiology or pathogenesis, immunodeficiency, hyperactivity of the immune response, or autoimmunity that leads to pathological sequelae.

Immunopathology is the study of disease processes that have an immunological etiology or pathogenesis involving either humoral antibody (from B cells) and complement or T cell-mediated or cytokine mechanisms. Immunologic injury of tissues and cells may be mediated by any of the four types of hypersensitivity (described separately).

Immunopathic refers to injury to cells, tissues, or organs induced by either humoral (antibodies) or cellular products of an immune response.

Immunoparasitology is the study of immunologic aspects of the interaction between animal parasites and their hosts.

BLOOD

Autoimmune neutropenia can be either an isolated condition or secondary to autoimmune disease. Patients may either have recurrent infections or remain asymptomatic. Antigranulocyte antibodies may be demonstrated. There is normal bone marrow function with myeloid hyperplasia and a shift to the left as a result of increased granulocyte destruction. The autoantibody may suppress myeloid cell growth. The condition is treated by immunosuppressive drugs, corticosteroids, or splenectomy. Patients with systemic lupus erythematosus (SLE), or Felty's syndrome [rheumatoid arthritis (RA), splenomegaly, and severe neutropenia] as well as other autoimmune diseases may manifest autoimmune neutropenia.

Granulocyte antibodies: IgG and/or IgM antibodies are present in approximately 33% of adult patients with idiopathic neutropenia. They are also implicated in the pathogenesis of drug-induced neutropenia, febrile transfusion reactions, isoimmune neonatal neutropenia, Evans syndrome, primary autoimmune neutropenia of early childhood, SLE, Graves' disease, and the neutropenia of Felty syndrome and selected other autoimmune diseases.

Agranulocytosis A striking decrease in circulating granulocytes including neutrophils, eosinophils, and basophils as a consequence of suppressed myelopoiesis. The deficiency of polymorphonuclear leukocytes leads to decreased resistance and increased susceptibility to microbial infection. Patients may present with pharyngitis. The etiology may be either unknown or follow exposure to cytotoxic drugs such as nitrogen mustard or following administration of the antibiotic chloramphenicol.

Idiopathic thrombocytopenic purpura is an autoimmune disease in which antiplatelet autoantibodies destroy platelets. Splenic macrophages remove circulating platelets coated with IgG autoantibodies at an accelerated rate. Thrombocytopenia occurs even though the bone marrow increases platelet production. This can lead to purpura and bleeding. The platelet count may fall below 20,000 to 30,000/μl. Antiplatelet antibodies are detectable in the serum and on platelets. Platelet survival is decreased. Splenectomy is recommended in adults. Corticosteroids facilitate a temporary elevation in the platelet count. This disease is characterized by decreased blood platelets, hemorrhage, and extensive thrombotic lesions.

ITP is an abbreviation for idiopathic thrombocytopenic purpura.

Aplastic anemia: Bone marrow stem cell failure that leads to cessation of formation of the blood cellular components. Bone marrow transplantation is the recommended treatment.

Leukemia is characterized by clonal expansion of lymphohematopoietic cells. Acute leukemias are characterized by elevated numbers of immature lymphohematopoietic cells termed blasts in the blood and/or in the bone marrow. By contrast, chronic leukemias are marked by a neoplastic population of mature-appearing cells. With increasing age, there is an increased incidence of acute and chronic myeloid leukemias and of chronic lymphoid leukemias. By contrast, acute lymphoblastic leukemia occurs more frequently among children, reaching a sharp peak at 3 to 4 years of age. Leukemia cell biology in classification has been dramatically advanced through immunological characterization of normal and leukemic hematopoietic cells, the demonstration of immunoglobulin and T cell receptor gene rearrangement, and in the molecular elucidation of genetic abnormalities associated with leukemogenesis.

Leukemias are classified as lymphocytic–myelocytic or monocytic. Lymphocytic leukemias are derived from B or T lymphocyte precursors, myelogenous leukemias originate from granulocyte or monocyte precursors, and erythroid leukemias develop from red blood cell precursors.

Leukemia viruses: Leukemia is a neoplasm of hematopoietic cells. It may have a viral etiology in humans to produce adult T cell leukemia as well as in mice, cats, cattle, and birds. The leukemia-inducing viruses are members of the Oncovirinae subfamily of the Retroviridie family. Additional retroviruses comprise the lentiviruses, a subfamily of pathogenic slowviruses that include the HIV-viruses and the spumaviruses or formiviruses that induce persistent infections unaccompanied by clinical disease. Oncoviruses infect target cells and cause their transformation to produce infected cells with tumor-producing potential. Viral carcinogenesis involves (1) overexpression of a viral *onc* gene, (2) protooncogene (*c-onc*) capture in a retroviral vector and the occurrence of mutations in the captured gene that render it highly oncogenic (*v-onc*), (3) inclusion of two cooperation *v-oncs* in the same provirus, (4) viral protein and activation of a *c-onc* cooperation, and, (5) reorganization of cellular transcription processes by a viral transactivation protein.

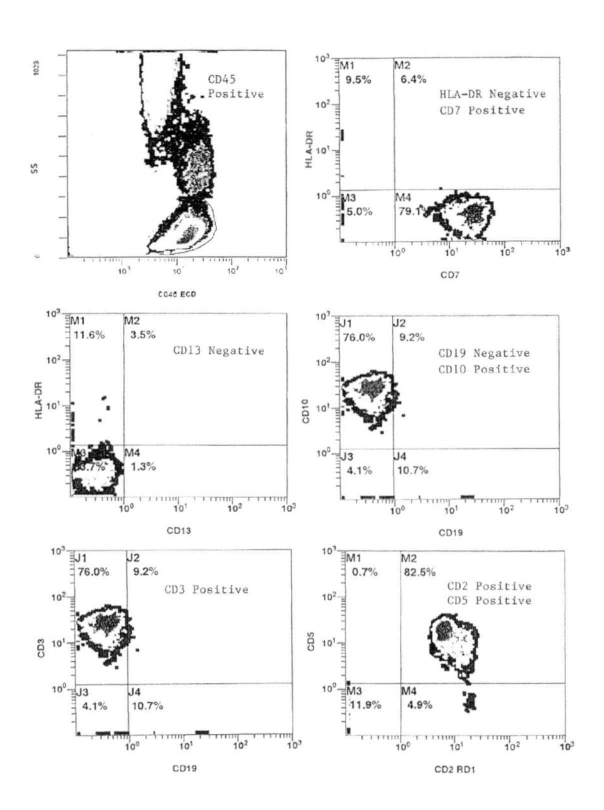

FIGURE 17.2 T cell acute lymphocytic leukemia (T-ALL).

Acute lymphoblastic leukemia (ALL) (Figure 17.1 to Figure 17.4) consists of a heterogeneous group of lymphopoietic stem cell disorders in which lymphoblasts accumulate in the bone marrow and suppress normal hemopoietic cells. A total of 80% of all ALL cases are of the B cell type whereas the remainder are T cell with rare cases of null cell origin. Of leukemias, 10% are ALL and 60% of ALL cases occur in children. Chromosomal abnormalities have been found in most cases of ALL. B-ALL (L3) is characterized by one of three chromosomal translocations. Most B-ALL cases have a translocation of *c-myc* protooncogene on chromosome 8 to the immunoglobulin coding gene region on chromosome 14. Lymphoblasts that are unable to differentiate and mature continue to accumulate in the bone marrow. ALL patients develop anemia, granulocytopenia, and thrombocytopenia. L1, L2, and L3 lymphoblast cytologic subtypes are recognized. ALL is diagnosed by the demonstration of lymphoblasts in the bone marrow. The total leukocyte count is normal or decreased in half of the cases with or without lymphoblasts in the peripheral blood. An elevated leukocyte count is usually accompanied by lymphoblasts in the peripheral blood. Patients develop a normochromic, normocytic anemia with thrombocytopenia and neutropenia. They may develop weakness, malaise, and pallor secondary to anemia. Half the individuals develop bleeding secondary to thrombocytopenia and many develop bacterial infections secondary to neutropenia. Patients also experience bone pain. There is generalized lymphadenopathy, especially affecting the cervical lymph nodes. Frequently, there is hepatosplenomegaly and leukemic meningitis. The age at

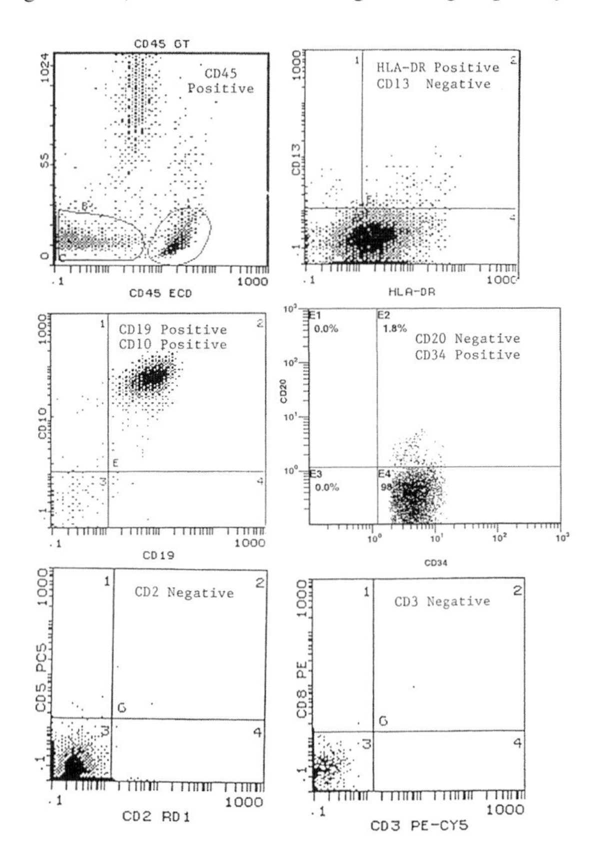

FIGURE 17.1 Pre-B cell acute lymphoblastic leukemia.

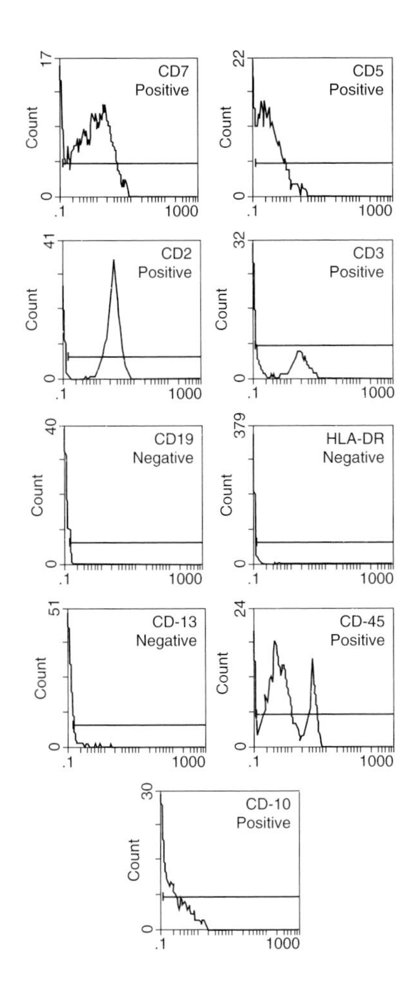

FIGURE 17.3 Acute lymphoblastic leukemia (ALL).

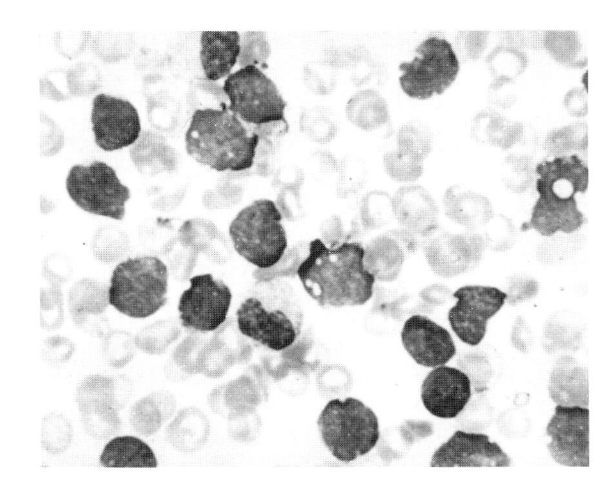

FIGURE 17.4 Hand mirror cell.

onset and initial total blood leukocyte count are valuable prognostic features. Of children with L1 morphology cells, 90% experience complete remission with chemotherapy. Patients may experience a profound reduction in the concentration of the serum immunoglobulins, possibly due to malignant expansion of suppressor T lymphocytes.

Common acute lymphoblastic leukemia antigen (CALLA) is a 100-kDa surface membrane glycoprotein present on human leukemia cells and, to a lesser degree, on other cells that include granulocytes and kidney cells. It is a zinc metalloproteinase and is classified as CD10/neutral endopeptidase 24.11. Four fifths of non-T cell leukemias express CALLA. Under physiologic conditions, 1% of cells in the bone marrow express CALLA. The presence of CALLA is revealed by monoclonal antibodies and flow cytometry using bone marrow or other cells. Bone marrow to be used for autologous transplants may be purged of CALLA positive lymphocytic leukemia cells by using anti-CALLA monoclonal antibodies and complement. It is a pre-B lymphoblast marker that is the most frequent type of cell in ALL. If Ia antigen is present together with CALLA, this portends a favorable prognosis. CALLA may also be positive in Burkitt's lymphoma, B cell lymphomas, and 40% of T cell lymphoblastic lymphomas. All blasts are usually positive not only for CALLA and the Ia antigen but also for TdT.

CALLA is an abbreviation for common acute lymphoblastic leukemia antigen. Also known as CD10.

HAM is an abbreviation for HTLV-1-associated myelopathy.

HTLV is an abbreviation for human T cell leukemia virus. A retrovirus that infects human $CD4^+$ T lymphocytes and produces adult T cell leukemia.

A **chromosomal translocation** is a DNA sequence rearrangement between chromosomes, which is frequently associated with neoplasia. Lymphocyte malignancies may be associated with chromosomal translocations involving an immunoglobulin or T cell receptor locus and chromosomal segment containing a cellular oncogene.

B cell chronic lymphocytic leukemia/small lymphocytic lymphoma (B-CLL/SLL) (Figure 17.5) is the most

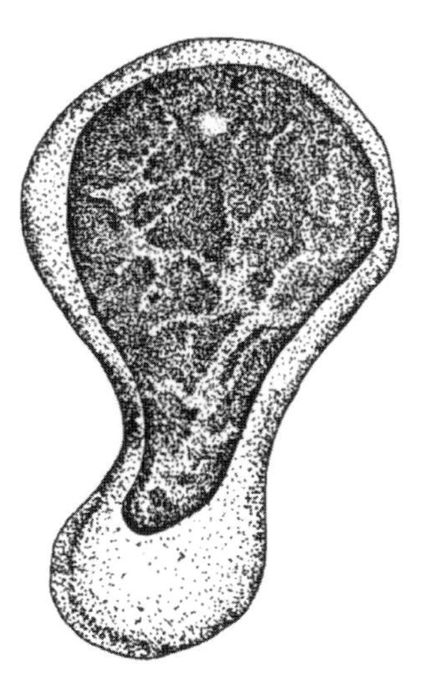

FIGURE 17.5 B cell chronic lymphocytic leukemia (B-CLL).

common leukemia in the Western world. The malignant cell is a small fragile B lymphocyte whose immunophenotype resembles that of lymphocytes in the mantle zone (MZ) of secondary lymphoid follicles. The most distinctive feature is the coexpression of CD5 with CD19 and CD20 with very faint amounts of monoclonal surface immunoglobulin. The finding of somatic mutation in half the cases indicates that a memory B cell has been exposed to antigen in germinal centers of secondary follicles. The remaining cases that show no mutations originate from naïve B cells that have not responded to antigen. The prognosis of cases arising from naïve B cells is worse than the prognosis from those arising from memory B cells. The disease has an insidious onset and is not curable. Patients are predisposed to repeated infections, have abdominal discomfort, and bleeding from mucosal surfaces. They may have general localized or generalized lymphadenopathy, spenomegaly, hepatomegaly, petechiae, and pallor that develop an absolute lymphocytosis with small lymphocytes expressing modest amounts of pale basophilic cytoplasm and nuclei with round contours and mature or "blocky" chromatin clumping with or without a small nucleolus. By immunophenotyping, the $CD5^+/CD10^-/CD23^+$ profile in a mature monoclonal B lymphocytosis is virtually diagnostic of B-CLL/SLL. A total of 80% of cases have clonal aberrations. The most common involves deletions at 13q.

B-CLL/SLL is a lymphoproliferative disorder characterized by sustained lymphocytosis of lymphocytes that are light-chain restricted. There is splenomegaly, lymphadenopathy, and hepatomegaly with lymphocytosis ranging from 4×10^9/l to lymphocyte counts exceeding 400×10^9/l. The lymphocytes are relatively small with condensed nuclear chromatin and sparse cytoplasm. They have a uniform appearance and injured cells are often present. Nucleoli are usually not visible. The mixed cell type may reveal both large and small lymphocytes. The more diffuse the pattern of involvement, the more aggressive the disease. B cell CLL lymphocytes express pan-B cell antigens and meager quantities of light-chain-restricted immunoglobulins on the cell surface. The lymphocytes usually express CD5, a pan-T cell antigen and rosette spontaneously with mouse red blood cells. One half to three fourths of B-CLL patients are hypogammaglobulinemic. Autoimmune hemolytic anemia, neutropenia, or thrombocytopenia develops in 15 to 30% of cases. Those cases of B cell CLL that become aggressive with pyrexia, weight loss, and fatigue and large cell lymphoma are referred to as having undergone Richter transformation, usually leading to death.

Chronic lymphocytic leukemia (CLL) is a peripheral B cell neoplasm in which the peripheral blood is flooded with small lymphocytes exhibiting condensed chromatin and scant cytoplasm. Disrupted tumor cells termed smudge cells represent a characteristic finding. Most patients have an absolute lymphocyte count of greater than 4000/mm^3 of blood. CLL is the most common leukemia of adults in the Western world. The tumor cells resemble a small subset of circulating B cells that express the surface marker CD45. The tumor cells express the pan-B cell markers CD19 and CD20. There is dim surface expression of immunoglobulin D heavy chain, and either kappa or lambda light-chain chromosomal anomalies include trisomy 12 deletions of 13q 12–14 and deletions of 11q. CLL may be asymptomatic and may disrupt normal immune function through unknown mechanics.

Follicular lymphoma is the most common form of non-Hodgkin's lymphoma in the U.S., comprising approximately 45% of adult lymphomas. The neoplastic cells closely resemble normal germinal center B cells. The neoplastic cells resemble normal follicular center B cells, expressing CD19, CD20, CD10, and monotypic surface immunoglobulin. CD5 is not expressed, which differentiates it from CLL, SLL, and mantle cell lymphoma. Follicular lymphoma cells express BCL2 protein in contrast to normal follicular center B cells, which are BCL2 negative. A (14:18) translocation is characteristic of follicular lymphoma in which there is juxtaposition of the IgH locus on chromosome 14 and the BCL2 locus on chromosome 18.

Hairy cell leukemia is a B lymphocyte leukemia in which the B cells have characteristic cytoplasmic filopodia.

Histiocytic lymphoma is a misnomer for large cell lymphomas, principally B cell tumors. The term histiocytic lymphoma more accurately describes a lymphoma of macrophage lineage.

Histiocytosis X is a descriptor for neoplasms of macrophage lineage. Included in this category are Letterer-Siwe disease, Hand-Schüller-Christian disease, and eosinophilic granuloma of bone.

Letterer-Siwe disease is a macrophage lineage tumor (histiocytosis X) that may appear in the skin, lymph nodes, and spleen.

White pulp disease is a lymphoproliferative disease that expresses major anatomical changes in the splenic white pulp. Diseases producing this effect include histiocytic lymphoma, lymphocytic leukemia, and Hodgkin's disease.

Eosinophilic granuloma is a subtype of a macrophage lineage (histiocytosis X) tumor that contains eosinophils, especially in bone.

T cell leukemia includes adult T cell leukemia/lymphoma.

T cell leukemia viruses are retroviruses such as HTLV-I that induce human T cell leukemia, and HTLV-II which has been associated with hairy cell leukemia.

Reticulum cell sarcoma is an obsolete term for large-cell lymphoma.

T cell lymphoma (TCL) is neoplastic proliferation of T lymphocytes. A condition that is diagnosed by determining whether or not there has been rearrangement of the genes encoding the T lymphocyte receptor β chain.

Mantle zone lymphoma is a follicular lymphoma of intermediate grade. Small lymphocyte proliferation in the mantle zone encircling benign germinal centers, splenomegaly, and generalized lymphadenopathy are present. Histopathologically, B cells vary in size from small to relatively large blasts containing clumped chromatin. Immunoglobulin M (IgM) is usually present.

Cutaneous T cell lymphoma is a malignant growth of T lymphocytes that home to the skin, such as mycosis fungoides.

Mycosis fungoides is a chronic disorder involving the lymphoreticular system. There is major involvement in which the skin appears scaly with eczematous areas that are erythematous. There is infiltration by lichenified plaques. Finally, ulcers and neoplasms of the internal organs and lymph nodes develop. Cells in the skin lesion reveal markers that identify them as T lymphocytes. This disease usually appears after 50 years of age and is more frequent in males than females and more frequent in African Americans than Caucasians. There is an increase in the number of null cells in the blood circulation, where there is a simultaneous decrease in the numbers of B and T lymphocytes. T cell immunity is diminished both *in vitro* and *in vivo*, as revealed by diminished lymphocyte unresponsiveness to mitogens and by a poor response and skin test. Immunoglobulin A and E levels may be elevated in the serum.

CD10 (CALLA) is an antigen, also referred to as common acute lymphoblastic leukemia antigen, that has a mol wt of 100 kDa. CD10 is now known to be a neutral endopeptidase (enkephalinase). It is present on many cell types, including stem cells, lymphoid progenitors of B and T cells, renal epithelium, fibroblasts, and bile canaliculi.

CD14 is an antigen that is a single-chain membrane glycoprotein with a mol wt of 55 kDa. It is found principally on monocytes, but also on granulocytes, dendritic reticulum cells, and some tissue macrophages. It serves as a receptor for lipopolysaccharide (LPS) and for lipopolysaccharide binding protein (LBP).

CD15 is a carbohydrate antigen that is often referred to as hapten X and consists of galactose, fucose, and *N*-acetyl-glucosamine linked in a specific sequence. The antigen appears to be particularly immunogenic in the mouse, and numerous monoclonal antibodies of this oligosaccharide specificity have been produced. CD15 is present in neutrophil secondary granules of granulocytic cells which express the antigen strongly late in their maturation. It is also present on eosinophils, monocytes, Reed–Sternberg and Hodgkin's cells, but can also be found on non-Hodgkin's cells. CD15 is also termed Lewis-x (Le^x).

CD23 is an antigen that is an integral membrane glycoprotein with a mol wt of 45 to 50 kDa. The CD23 antigen has been identified as the low-affinity Fc-IgE receptor (FceRII). Two species of FceRII/CD23 have been found. FceRIIa and FceRIIb differ only in the N-terminal cytoplasmic region, but share the same C-terminal extracellular region. FceRIIa is strongly expressed on IL-4-activated B cells and weakly on mature B cells. FceRIIa is not found on circulating B cells and bone marrow B cells positive for surface IgM, IgD. FceRIIb is expressed weakly on monocytes, eosinophils, and T cell lines. However, IL-4-treated monocytes show a stronger staining. The CD23 molecule may also be a receptor for B cell growth factor (BCGF).

CD30 is a molecule that is a single-chain glycoprotein, also referred to as the Ki-1 antigen. It has a mol wt of 105 kDa, and is present on activated T and B cells, embryonal carcinoma cells, and Hodgkin's and Reed-Sternberg cells. It is also found on a minority of non-Hodgkin's lymphomas (referred to as anaplastic large-cell lymphomas) with a characteristic anaplastic morphology.

Myeloid antigen is a surface epitope of myeloid leukocytes. Examples include CD13, CD14, and CD33. A poor prognosis is indicated when the leukocytes of a patient with acute lymphocytic leukemia express myeloid antigens. The expression of these antigens is a better indicator of decreased survival than are other features of the disease.

CD33 is an antigen that is a single-chain transmembrane glycoprotein with a mol wt of 67 kDa. It is restricted to myeloid cells and is found on early progenitor cells, monocytes, myeloid leukemias, and weakly on some granulocytes.

CD34 is a molecule (mol wt 105 to 120 kDa) that is a single-chain transmembrane glycoprotein present on immature haematopoietic cells and endothelial cells as well as bome marrow stromal cells. Three classes of CD34 epitopes have been defined by differential sensitivity to enzymatic cleavage with neuraminidase and with glycoprotease from *Pasteurella haemolytica*. Its gene is on chromosome 1. CD34 is the ligand for L-selectin (CD62L).

CD41 is an antigen, equivalent to glycoprotein IIb/IIIa, that is found on platelets and megakaryocytes. This structure is a Ca^{2+}-dependent complex between the 110-kDa gpIIIa (CD61) and the 135-kDa gpIIb molecules. The gpIIb molecule consists of two chains: an α-chain of 120 kDa and a β-chain of 23 kDa. The gpIIIa (CD61) is a single-chain protein, but both gpIIIa (CD61) and gpIIb contain transmembrane domains. The molecule is a receptor for fibrinogen, but it also binds von Willebrand factor, thrombospondin, and fibronectin. CD41 is absent or reduced in Glanzmann's thrombasthenia (hereditary gpIIb/IIIa deficiency). It has a role in platelet aggregation and activation.

CD44 is an ubiquitous multistructural and multifunctional cell-surface glycoprotein that participates in adhesive cell-to-cell and cell-to-matrix interactions. It also plays a role in cell migration and cell homing. Its main ligand is hyaluronic acid (HA), hyaluronate, and hyaluronan. It is expressed by numerous cell types of lymphohematopoietic origin, including erythrocytes, T and B lymphocytes, natural killer cells, macrophages, Kupffer cells, dendritic cells, and granulocytes. It is also expressed in other types of cells such as fibroblasts and CNS cells. In addition to hyaluronic acid, CD44 also interacts with other ECM ligands such as collagen, fibronectin, and laminin. In addition to its function stated above, CD44 facilitates lymph node homing via binding to high endothelial venules, presentation of chemokines or growth factors to migrating cells, and growth signal transmission. CD44 concentration may be observed in areas of intensive cell migration and proliferation as in wound healing, inflammation, and carcinogenesis. Many cancer cells and their metastases express high levels of CD44. It may be used as a diagnostic or prognostic marker for selected human malignant diseases.

CD56 is a 220/135 kDa molecule that is an isoform of the neural adhesion molecule (N-CAM). It is used as a marker of NK cells, but it is also present on neuroectodermal cells.

CD57 is a 110-kDa myeloid-associated glycoprotein that is recognized by the antibody HNK1. It is a marker for NK cells, but it is also found on some T and B cells. CD57 is an oligosaccharide present on multiple cell surface glycoproteins.

Acute myelogenous leukemias (AML) (Figure 17.6) consist of a heterogeneous group of disorders characterized by neoplastic transformation in a multipotential hematopoietic stem cell or in one of restricted lineage potential. Since multipotential hematopoietic stem cells are the precursors of granulocytes, monocytes, erythrocytes, and megakaryocytes, one or all of theses cell types may be affected. Differentiation usually ends at the blast stage, causing myeloblasts to accumulate in the bone marrow. AML is diagnosed by the discovery of 30% myelogenous blasts in the bone marrow, whether or not they are present in the peripheral blood. The total peripheral blood leukocyte count may be normal, low, or elevated. There may be thrombocytopenia, neutropenia, and anemia. The presenting symptoms and signs are nonspecific and may be secondary to anemia. Patients may experience fatigue, weakness, and pallor. One third of AML patients are found to have hepatosplenomegaly. One third of patients may have bleeding secondary to thrombocytopenia. GI tract or CNS hemorrhage may occur when platelet counts decrease below 20,000/µl. Decreased neutrophil counts may increase secondary infections.

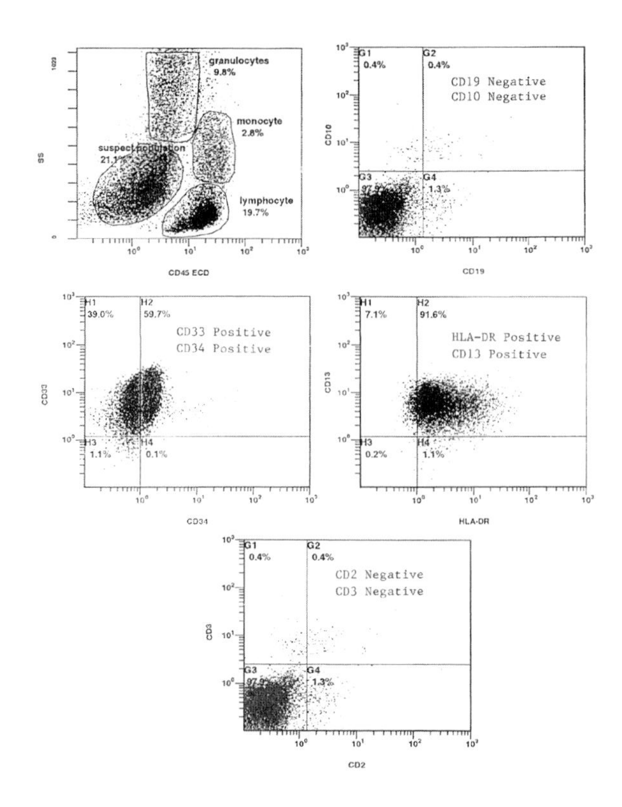

FIGURE 17.6 Acute myeloid leukemia (AML).

Unfavorable prognostic findings include (1) an age at diagnosis of greater than 60 years, (2) 5q- and 7q-chromosomal abnormalities, (3) a history of myelodysplastic syndrome, (4) a history of radiation or chemotherapy for cancer, and (5) a leukocytosis exceeding 100,000/µl. If untreated, AML leads to death in less than 2 months through hemorrhage or infection. Both chemotherapy and bone marrow transplantation are modes of therapy.

Lineage infidelity refers to cells that undergo neoplastic transformation and may express molecules on their surface that are alien to the cell's lineage.

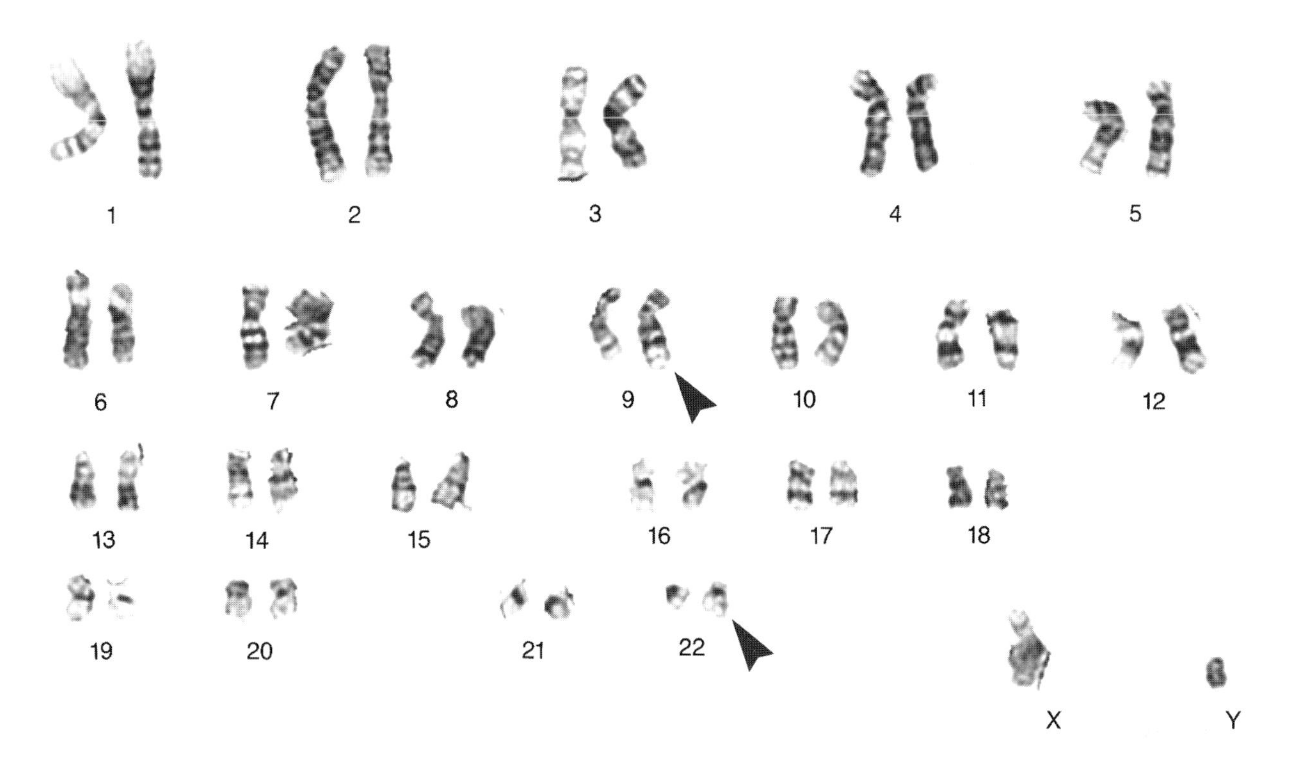

FIGURE 17.7 Philadelphia chromosome.

Chronic myelogenous leukemia (CML) is a disease affecting adults between the ages of 25 and 60 years that has a distinctive molecular abnormality consisting of a translocation involving the *Bcr* gene on chromosome 9 and the *Abl* gene on chromosome 22. The *Bcr/Abl* fusion gene that results controls the synthesis of a 210-kDa fusion protein that has tyrosine kinase activity. A total of 90% of cases have the Ph1 karyotype. There is a striking increase in neoplastic granulocytic precursors in the bone marrow. The target of transformation is a pluripotent stem cell. The *Bcr/cAbl* fusion gene can be detected by either chromosomal analysis or PCR-based molecular assays. There is also nearly total lack of leukocyte alkaline phosphatase.

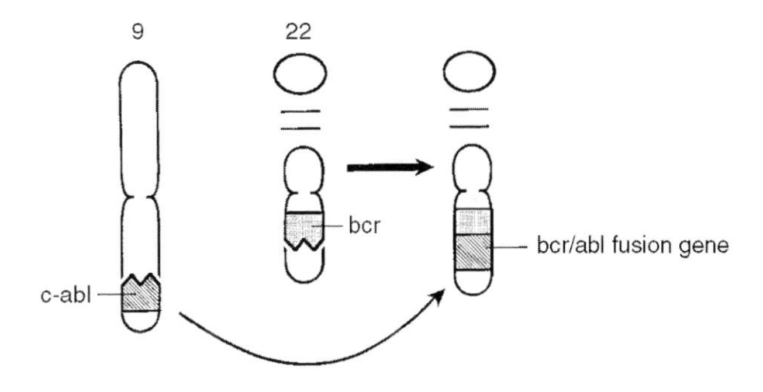

FIGURE 17.8 Philadelphia chromosome.

Philadelphia chromosome (Figure 17.7 and Figure 17.8) is the translocation of one arm of chromosome 22 to chromosome 9 or 6 in chronic granulocytic leukemia cells in humans.

Chronic myeloid leukemia is characterized by cell types in the circulation that are in the late stages of granulocyte maturation. These include mature granulocytes, myelocytes, and metamyelocytes.

Angioimmunoblastic lymphadenopathy (AILA) (Figure 17.9) is the proliferation of hyperimmune B lymphocytes. Immunoblasts, both large and small, form a pleomorphic infiltrate together with plasma cells in lymph nodes revealing architectural effacement. There is arborization of newly formed vessels and proliferating

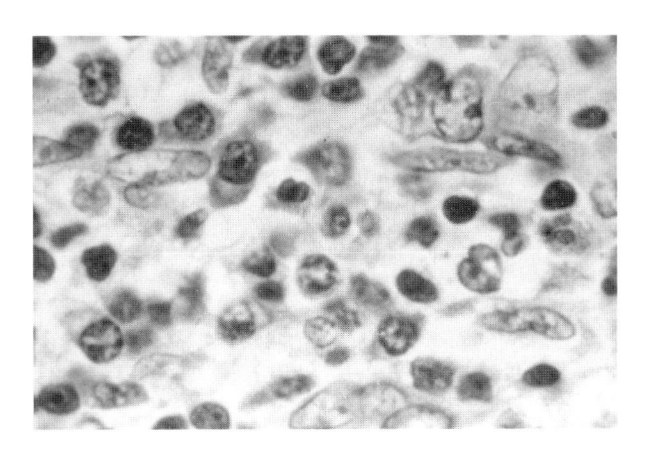

FIGURE 17.9 Angioimmunoblastic lymphadenopathy.

vessels with hyperplasia of endothelial cells. In the interstitium, amorphous eosinophilic PAS positive deposits, possibly representing debris from cells, are found. Fever, night sweats, hepatosplenomegaly, generalized lymphadenopathy, weight loss, hemolytic anemia, polyclonal gammopathy, and skin rashes may characterize the disease in middle-aged to older subjects. Patients live approximately 15 months, with some developing monoclonal gammopathy or immunoblastic lymphomas. AILA must be differentiated from AIDS, Hodgkin's disease, immunoblastic lymphoma, histocytosis X, and a variety of other conditions affecting the lymphoid tissues.

AILA is an abbreviation for angioimmunoblastic lymphadenopathy.

Systemic immunoblastic proliferation: A condition characterized by gene translocation and immature lymphocyte proliferation. Clinically, patients manifest rash, dyspnea, hepatosplenomegaly, and lymphadenopathy and show an increased incidence of immunoblastic lymphoma.

Immunoblastic lymphadenopathy: See angioimmunoblastic lymphadenopathy (AILA).

Immunoblastic sarcoma is a lymphoma comprised of immunoblast-like cells that are divided into B and T cell immunoblastic sarcomas. Both are malignant lymphomas. B cell immunoblastic sarcoma is characterized by large immunoblastic plasmacytoid cells and Reed-Sternberg cells. This is the most frequent lymphoma that occurs in individuals with natural immunodeficiency, with suppression of the immune system, or who manifest immunoproliferative disorders such as angioimmunoblastic lymphadenopathy, or autoimmune diseases such as Hashimoto's thyroiditis, α chain disease, etc. The disease has a poor prognosis. T cell immunoblastic sarcoma is less frequent than the B cell variety. The tumor cells are large and have a clear cytoplasm containing round to oval nuclei with fine chromatin. One to three nucleoli are present amidst lobulation and nuclear folding. Other cells present include plasma cells and histiocytes. Patients with *mycosis fungoides* who develop polyclonal gammopathy and general lymphadenopathy may develop a T cell immunoblastic sarcoma.

Plasmacytoma (Figure 17.10) is a plasma cell neoplasm. Also termed myeloma or multiple myeloma. To induce an experimental plasmacytoma in laboratory mice or rats, paraffin oil is injected into the peritoneum. Plasmacytomas may occur spontaneously. These tumors, comprised of neoplastic plasma cells, synthesize and secrete monoclonal immunoglobulins yielding a homogenous product that forms a spike in electrophoretic analysis of the serum. Plasmacytomas were used extensively to generate monoclonal immunoglobulins prior to the development of B cell hybridoma technology to induce monoclonal antibody synthesis at will to multiple antigens.

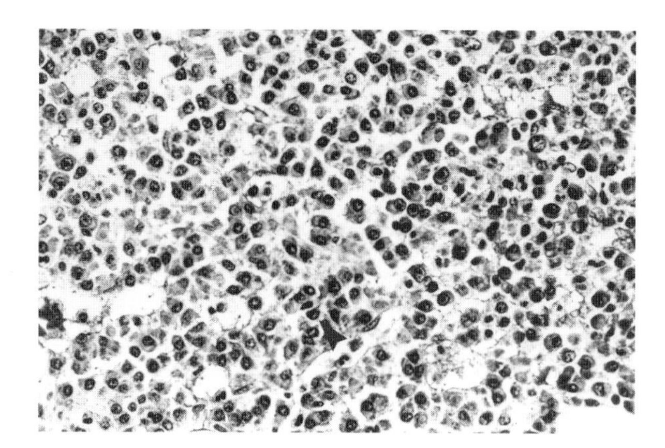

FIGURE 17.10 Bone marrow plasmacytoma with pure plasma cell infiltrate.

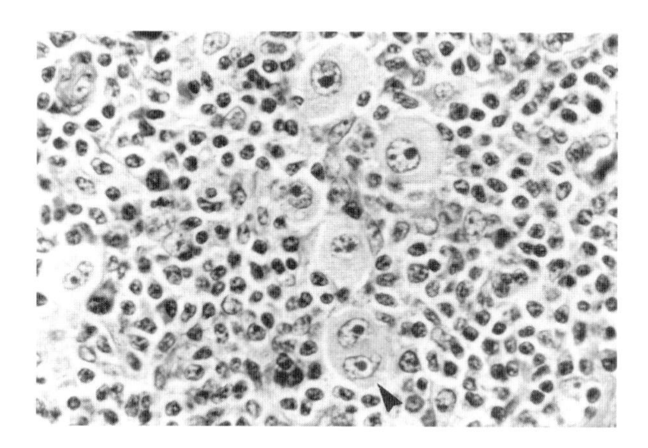

FIGURE 17.11 Hodgkin's disease. Hodgkin cell and Reed-Sternberg cell.

Hodgkin's disease (Figure 17.11) is a type of lymphoma that involves the lymph nodes and spleen, causing a replacement of the lymph node architecture with binucleated giant cells known as Reed-Sternberg cells, reticular cells, neutrophils, eosinophils, and lymphocytes. There is both lymphadenopathy and splenomegaly. Patients manifest a deficiency of cell-mediated immunity which causes skin tests of the tuberculin type to be negative. By contrast, there is no alteration in their B cell function. They may have an increase in suppressor cell activity. There is also increased susceptibility to opportunistic infections. Antigen-presenting cells resembling dendritic cells apparently represent the transformed cell type. Hodgkin's lymphoma, characterized by a predominance of lymphocytes, has a much better prognosis than does the nodular sclerosis variety of the disease in which the predominant cell type is nonlymphoid.

BLA-36 is an antigen demonstrable by immunoperoxidase staining in Reed-Sternberg cells of all types of Hodgkin's disease and in activated B lymphocytes and B cell lymphomas.

A **lymphoma** is a malignant neoplasm of lymphoid cells. Hodgkin's disease, non-Hodgkin's lymphoma, and Burkitt's lymphoma are examples. Lymphomas or lymphocyte tumors grow in lymphoid or other tissues but fail to enter the blood in large numbers. The numerous types of lymphomas are characterized by various classes of transformed lymphoid cells. Lymphomas often mainfest the phenotyped of the normal lymphocytes from which they arose.

Castleman's disease, also called giant lymph node hyperplasia, is a disease of unknown etiology which involves both lymph nodes and extranodal tissues. Two histopathologic subtypes have been described. The first, termed hyaline-vascular angiofollicular lymph node hyperplasia, accounts for 90% of the cases. It usually affects young men who present with an asymptomatic mass in the mediastinum. Histopathologically it reveals numerous small, follicle-like structures, frequently with radially penetrating, thick-walled, hyalinized vessels; concentrically arranged small lymphocytes around the follicular structures called "onion skinning" and extensive proliferation of capillaries in the interfollicular areas. The second type of Castleman's disease is plasma cell angiofollicular lymph node hyperplasia which comprises 10% of the cases. It is either a localized mass or a multicentered systemic disorder. The mass may consist of multiple matted lymph nodes with histopathologic features that include large hyperplastic follicles with less permanent penetrating vessels than in the hyaline-vascular type; pronounced interfollicular plasma cytosis and permanent vascularity. A multicentric type of the plasma cell variant of angiofollicular lymph node hyperplasia is more aggressive. Affected patients may have an increased risk for developing Kaposi's sarcoma or immunoblastic lymphoma. Clinical features of plasma cell angiofollicular hyperplasia include fever, polyclonal hypergammaglobulinemia, elevated sedimentation rate, and anemia.

Pseudolymphoma is hyperplasia of lymphoid tissue in which there is a uniform accumulation of lymphocytes. Unlike lymphomas, the cells are polyclonal. The architecture of the lymph node is well preserved with distinct cortical germinal centers with little or minimal capsular infiltration by lymphocytes. Inflammatory cells are detectable between germinal centers, but mitoses occur only within the germinal centers. The reticular framework remains intact. The lymphoid hyperplasia that characterizes pseudolymphoma may be found in various locations such as gastrointestinal tract, lung, breast, salivary gland, mediastinum, skin, soft tissue, and other areas. Pseudolymphomas may occur in individuals who later develop lymphomas.

Lymphomatosis refers to numerous lymphomas occurring in different parts of the body, such as those occurring in Hodgkin's disease.

Biclonality: In contrast to uncontrolled proliferation of a single clone of neoplastic cells which is usually associated with tumors, rarely two neoplastic cell clones may proliferate simultaneously, leading to a biclonality. For example, either neoplastic B or T lymphocytes could demonstrate this effect.

Progressive transformation of germinal centers (PTGC) refers to germinal center enlargement in the presence of follicular hyperplasia and loss of the distinct boundary between the mantle zone and the germinal center. Transformed germinal centers contain small lymphocytes with diffuse immunoblasts and histiocytes and often occur in one enlarged lymph node in young men. It is not believed to be neoplastic or to portend development of lymphoma in the future, although it may be observed in some patients with nodular lymphocyte-predominant Hodgkin's disease.

Reticulosis: See lymphoma.

Reed-Sternberg cells (Figure 17.12 and Figure 17.13) are binucleated giant cells that contain prominent nucleoli. They are classically associated with Hodgkin's disease.

Immunoproliferative small intestinal disease (IPSID) (Mediterranean lymphoma α heavy chain disease) A varied group of disorders in which there is monoclonal synthesis of immunoglobulin heavy chain (often α). Light chains are not produced. The variable region and often the C_H1 constant region may be missing. Whereas the monoclonal protein is usually elevated in the α chain, some cases may

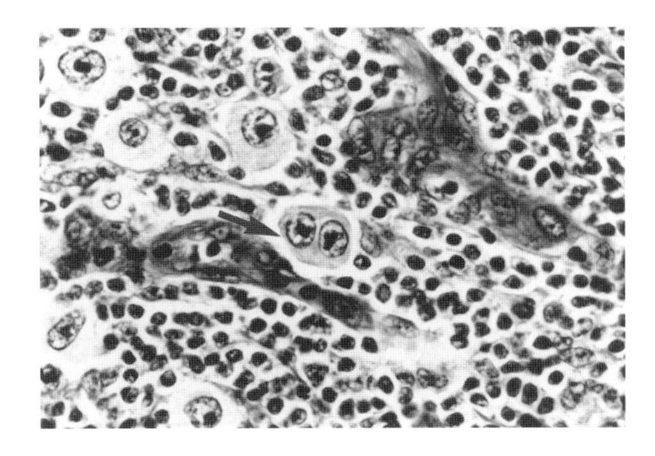

FIGURE 17.12 Reed-Sternberg cell.

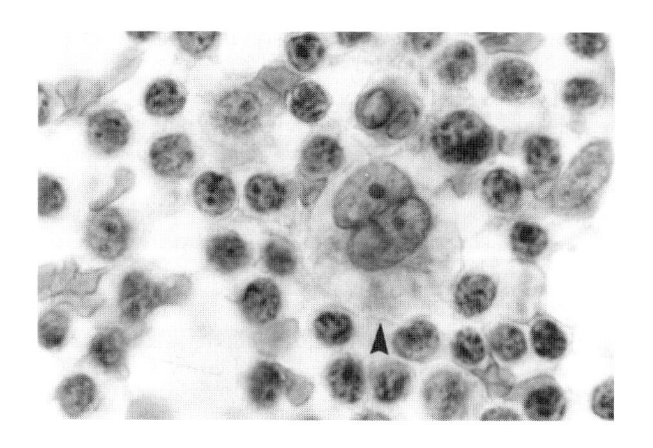

FIGURE 17.13 Touch preparation. Reed-Sternberg cell.

manifest γ or μ chain elevations. Patients experience weight loss, pain in the abdomen, diarrhea, and malabsorption. There is expansion of the mesenteric lymphoid tissue and of the proximal small intestine. There is also clubbing of the fingers. Tetracyclines are the recommended treatment.

SKIN

Allergic contact dermatitis is a delayed-type hypersensitivity mediated by specifically sensitized T lymphocytes (type IV hypersensitivity) in response to the covalent linkage of low-molecular-weight chemicals, often of less than 1,000M_r to proteins in the skin. The inflammation induced by these agents is manifested as erythema and swelling at approximately 12 h after contact and is maximal at 24 to 48 h. Blisters filled with serum, neutrophils, and mononuclear cells form. There is perivascular cuffing with lymphocytes, vesiculation, and necrosis of epidermal cells. Basophils, eosinophils, and fibrin deposition appear together with edema of the epidermis and dermis. Langerhans cells in the skin serve as antigen processing cells where the allergen has penetrated. Sensitization lasts for many years and becomes generalized in the skin. Chemicals become conjugated to skin proteins and serve as haptens. Therefore, the hapten alone can elicit the hypersensitivity once sensitization is established. After blistering, there is crust formation and weeping of the lesion. It is intensely pruritic and painful. Metal dermatitis, such as that caused by nickel, occurs as a patch that corresponds to the area of contact with the metal or jewelry. Dyes in clothing may produce skin lesions at points of contact with the skin. The patch test is used to detect sensitivity to contact allergens. Rhus dermatitis represents a reaction to urushiols in poison oak or ivy that elicit vesicles and bullae on affected areas. Treatment is with systemic corticosteroids or the application of topical steroid cream to localized areas. Dinitrochlorobenzene (DNCB) and dinitrofluorobenzene (DNFB) are chemicals that have been used to induce allergic contact dermatitis in both experimental animals and in man.

Atopic dermatitis (Figure 17.14) is a chronic eczematous skin reaction marked by hyperkeratosis and spongiosis especially in children with a genetic predisposition to allergy. These are often accompanied by elevated serum IgE levels, which are not proved to produce the skin lesions.

Bronchiectasis is chronic dilatation of the bronchi of the lungs associated with expectoration of mucopurulent material.

Bullous pemphigoid (Figure 17.15 and Figure 17.16) is a blistering skin disease with fluid-filled bullae developing at flexor surfaces of extremities, groin, axillae, and inferior

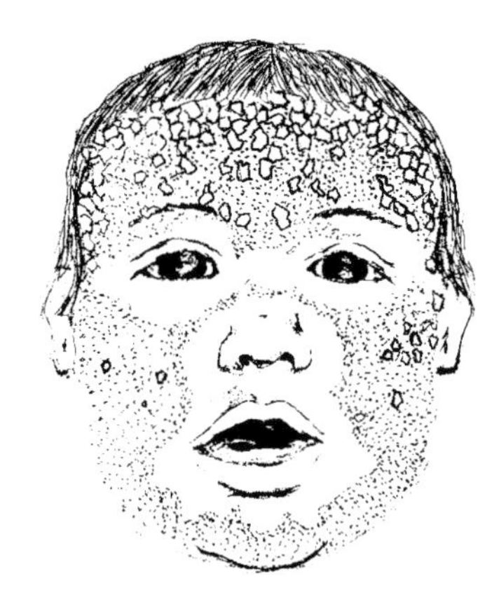

FIGURE 17.14 Atopic dermatitis.

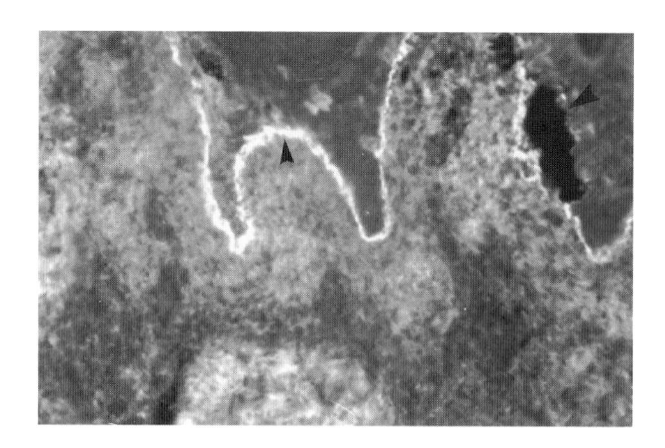

FIGURE 17.15 Bullous pemphigoid.

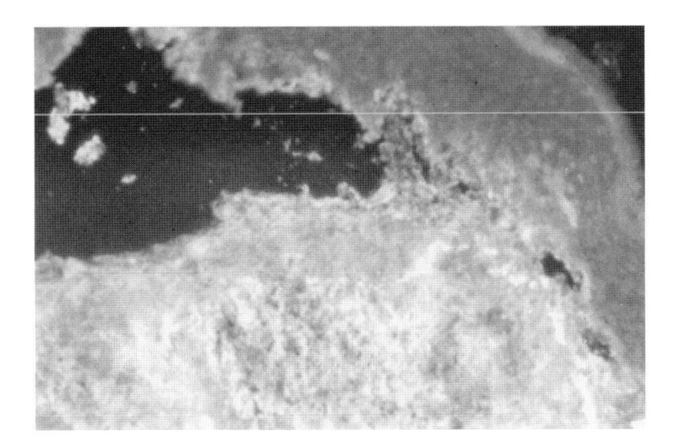

FIGURE 17.16 Bullous pemphigoid.

abdomen. IgG is deposited in a linear pattern at the lamina lucida of the dermal–epidermal junction in most (50 to 90%) patients and at linear C3 in nearly all cases. The blisters are subepidermal bullae filled with fluid containing fibrin, neutrophils, eosinophils, and lymphocytes. Antigen–antibody–complement interaction and mast cell degranulation release mediators that attract inflammatory cells and facilitate dermal–epidermal separation.

Bullous pemphigoid antigen: The principal antigen is a 230-kDa basic glycoprotein produced by keratinocytes in the epidermis. Autoantibody and complement react with this antigen to produce bullous pemphigoid skin lesions.

Dermatitis herpetiformis (DH) is a skin disease with grouped vesicles and urticaria that is related to celiac disease. Dietary gluten exacerbates the condition and should be avoided to help control it. Most patients (70%) manifest no bowel disease symptoms. IgA and C3 granular immune deposits occur along dermal papillae at the dermal–epidermal junction. Groups of papules, plaques, or vesicles appear in a symmetrical distribution on knees, elbows, buttocks, posterior scalp, neck, and superior back region. The disease is chronic, unless gluten is deleted from the diet. Neutrophils (PMNs) and fibrin collect at dermal papillae tips, producing microabscesses. Microscopic blisters, which ultimately develop into subepidermal blisters, may develop at the tips of these papillae. These lesions must be distinguished from those of bullous pemphigoid. There is an association of DH with HLA-B8, HLA-DR3 and HLA-B44, and HLA-DR7 haplotypes.

Pemphigus vulgaris (Figure 17.17 to Figure 17.19) is a blistering lesion of the skin and mucous membranes. The bullae develop on normal appearing skin and rupture easily. The blisters are prominent on both oral and anal genital mucous membranes. The disorder may have an insidious onset appearing in middle-aged individuals and tends to be chronic. It may be associated with autoimmune diseases, thymoma, and myasthenia gravis. Certain drugs may induce a pemphigus-like condition. By light microscopy, intraepidermal bullae are present. There is suprabasal epidermal acantholysis with only mild inflammatory reactivity in early pemphigus. Suprabasal unilocular bullae develop and there are autoantibodies to intercellular substance with activation of classic pathway-mediated immunologic injury. Acantholysis results as the epidermal cells become disengaged from one another as the bulla develops. Epidermal proteases activated by autoantibodies may actually cause the loss of intercellular bridges. Immunofluorescence staining reveals

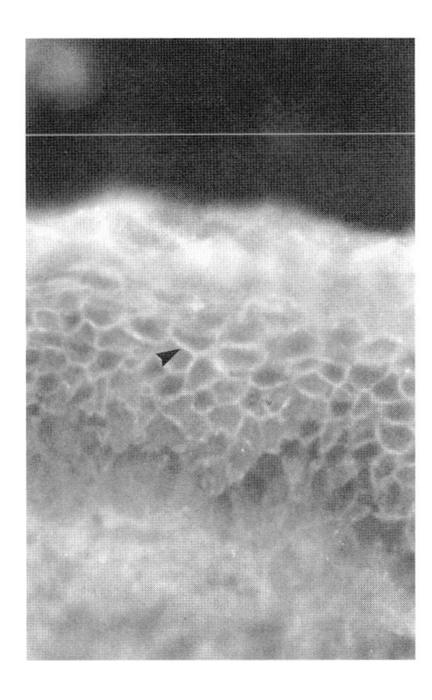

FIGURE 17.17 Pemphigus vulgaris. "Chickenwire staining" antibody to intercellular antigen.

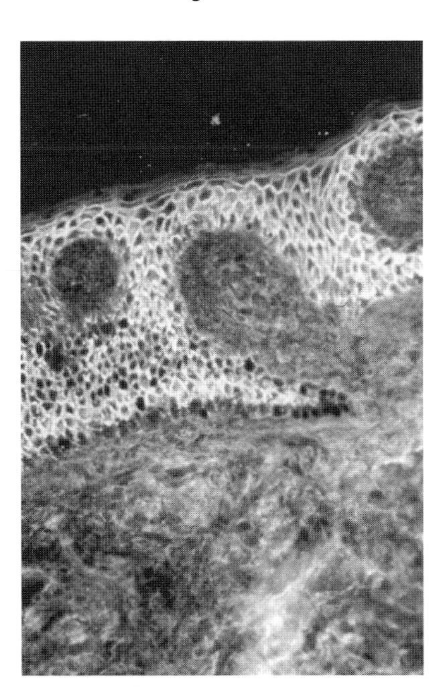

FIGURE 17.18 Pemphigus vulgaris. "Chickenwire staining" antibody to intercellular antigen.

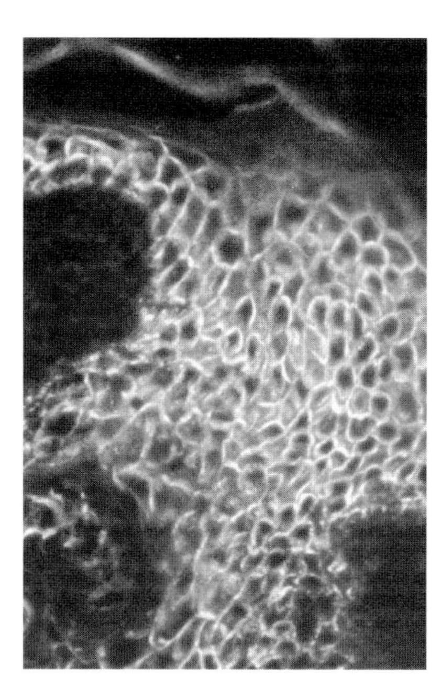

FIGURE 17.19 Pemphigus vulgaris. "Chickenwire staining" antibody to intercellular antigen, at higher magnification.

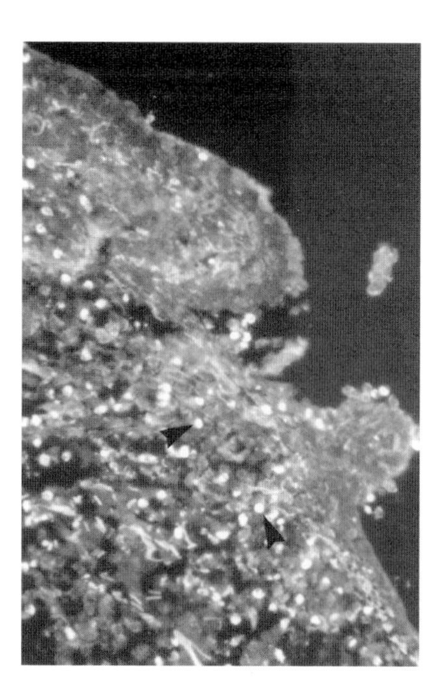

FIGURE 17.20 Erythema multiforme. Immunocytes in dermis.

IgG, Clq, and C3 in the intercellular substance between epidermal cells. Among pemphigus vulgaris patients, 80 to 90% have circulating pemphigus antibodies. Their titer usually correlates positively with clinical manifestations. Corticosteroids and immunosuppressive therapy, as well as plasmapheresis, have been used with some success.

Erythema multiforme (Figure 17.20) is a skin lesion resulting from subcutaneous vasculitis produced by immune complexes. They are frequently linked to drug reactions. The lesions are identified by a red center encircled by an area of pale edema encircled by a red or erythematous ring. This gives it a target appearance. Erythema multiforme usually signifies a drug allergy or may be linked to systemic infection. Lymphocytes and macrophages infiltrate the lesions. When there is involvement and sloughing of the mucous membranes, the lesion is considered quite severe and even life threatening. This form is called the Stevens-Johnson syndrome.

Immunocyte (Figure 17.21 and Figure 17.22) literally means "immune cell." The term is sometimes used by pathologists to describe plasma cells in stained tissue sections, e.g., in the papillary or reticular dermis in erythema multiforme.

Psoriasis vulgaris (Figure 17.23) is a chronic, recurrent, and papulosquamous disease. Clinical features include the appearance of a discrete, papulosquamous plaque on areas of trauma such as the elbow, knee, or scalp, although it may appear elsewhere on the skin. There is a relatively high instance of HLA-B13 and B17 antigen and decreased

FIGURE 17.21 Immunocytes in reticular dermis.

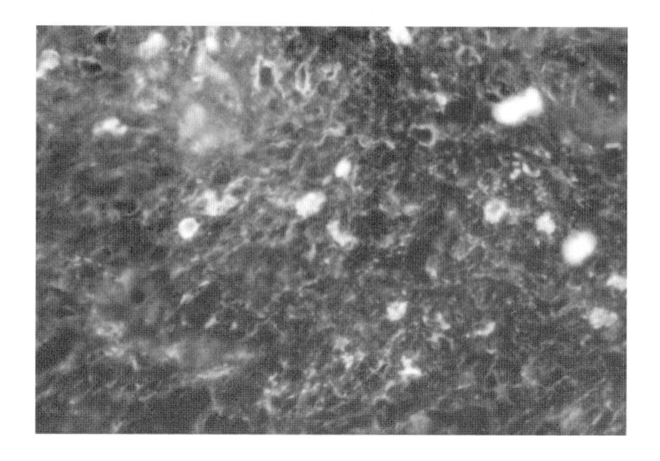

FIGURE 17.22 Immunocytes in dermis.

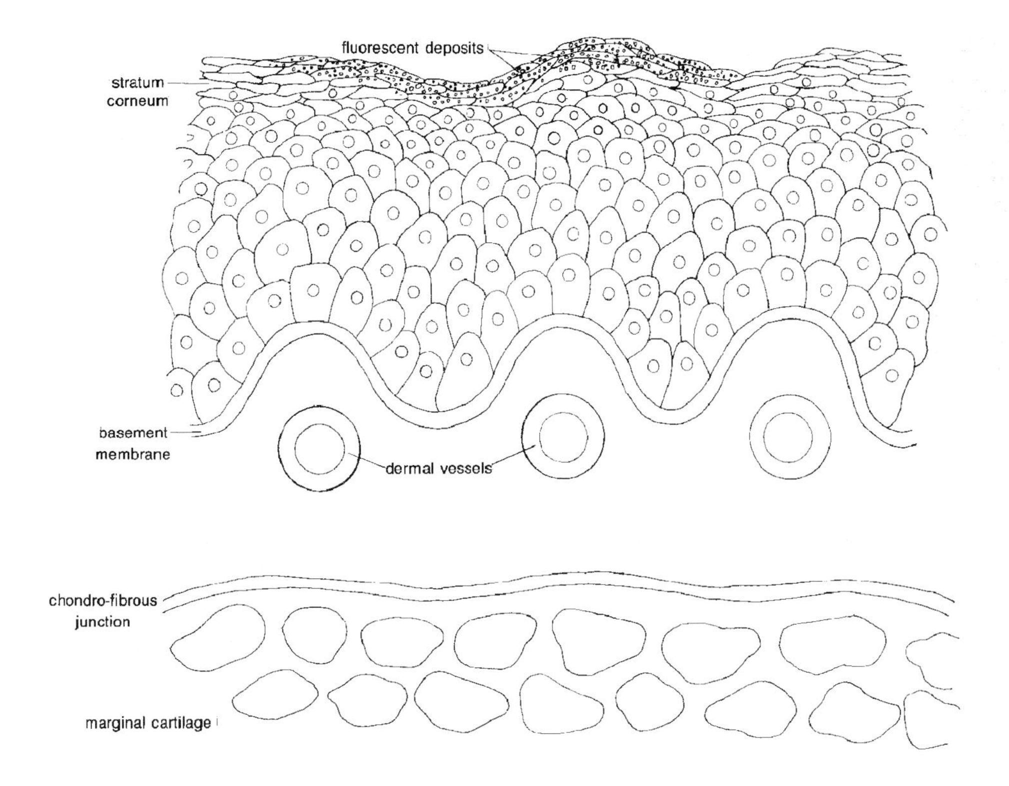

FIGURE 17.23 Psoriasis vulgaris.

T suppressor cell function in psoriasis patients. It may coexist with lupus erythematosus in some individuals. Peripheral blood helper/inducer CD4+ T lymphocytes are significantly decreased in psoriasis patients. It may be treated with psoralens and long-wave ultraviolet radiation. Psoriasis patients develop Monro microabscesses, hyperkeratosis, parakeratosis, irregular acanthosis, papillary edema, and mild chronic inflammation of the dermis. By immunofluorescence, there are focal granular or globular deposits of immunoglobulins and C3 in the stratum corneum. The finely granular deposits principally contain IgG, IgA, and C3. They are deposited in areas where stratum corneum antigens are located. C3 and properdin deposits suggest activation of the alternate complement pathway.

Sweet's syndrome (acute febrile neutrophilic dermatosis) A syndrome with neutrophilia, fever, and erythematous, painful skin plaques with pronounced dermal neutrophilic inflammation. About 10 to 15% of the cases may have an underlying malignant disease such as a myeloid proliferative disorder, acute myelogenous leukemia, or other tumor. The cutaneous lesions may become vesicular, and pustular skin lesions resemble those in bowel-bypass syndrome. Patients may develop arthritis, myalgia, conjunctivitis, and proteinuria. Immunofluorescence reveals IgG, IgM, and C3 in some lesions. Systemic steroid treatment has proven effective in improving skin lesions.

VASCULATURE

Leukocytoclastic vasculitis (Figure 17.24) is a type of vasculitis in which there is karyorrhexis of inflammatory cell nuclei. Fragments of neutrophil nuclei and immune complexes are deposited in vessels. Direct immunofluorescence reveals IgM, C3, and fibrin in vessel walls. There is nuclear dust, necrotic debris, and fibrin staining of the postcapillary venules. Leukocytoclastic vasculitis represents a type of allergic cutaneous arteriolitis or necrotizing angiitis. It is seen in a variety of diseases including Henoch-Schonlein purpura, rheumatoid arthritis, polyarteritis nodosa, and Wegener's granulomatosis.

Hypocomplementemic vasculitis urticarial syndrome is a type of systemic inflammation with leukocytoclastic vasculitis. There is diminished serum complement levels and urticaria.

Nuclear dust (leukocytoclasis) is extensive basophilic granular material representing karyolytic nuclear debris that accompanies areas of inflammation and necrosis, as in leukocytoclastic vasculitis.

Wegener's granulomatosis (Figure 17.25 to Figure 17.27) refers to necrotizing sinusitis with necrosis of both the upper and lower respiratory tract. The disease is characterized by granulomas, vasculitis, granulomatous arteritis, and glomerulonephritis. The condition is believed to have an immunological etiology, although this remains to be

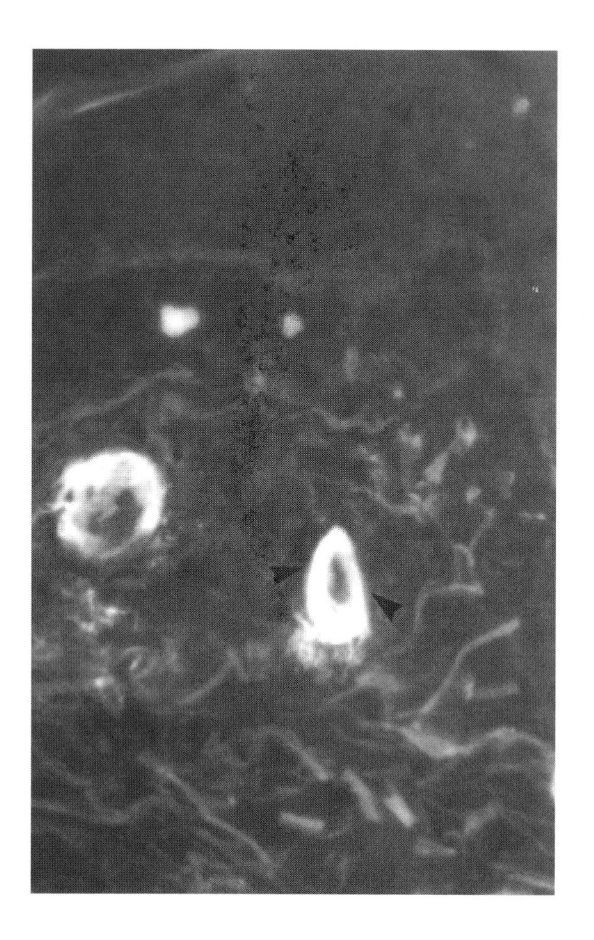

FIGURE 17.24 Leukocytoclastic vasculitis.

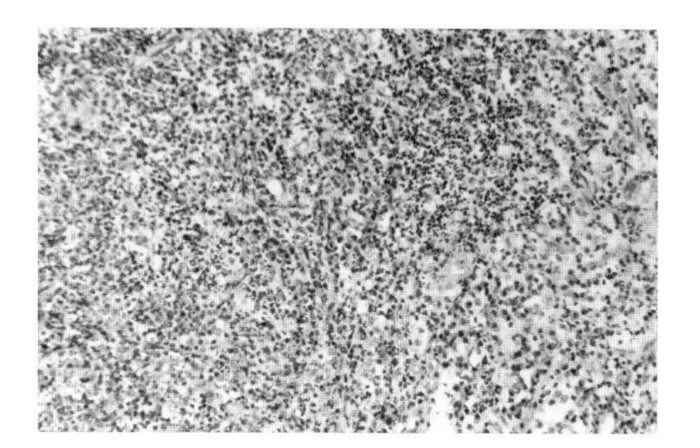

FIGURE 17.25 Inflammation of ethmoid and maxillary sinuses.

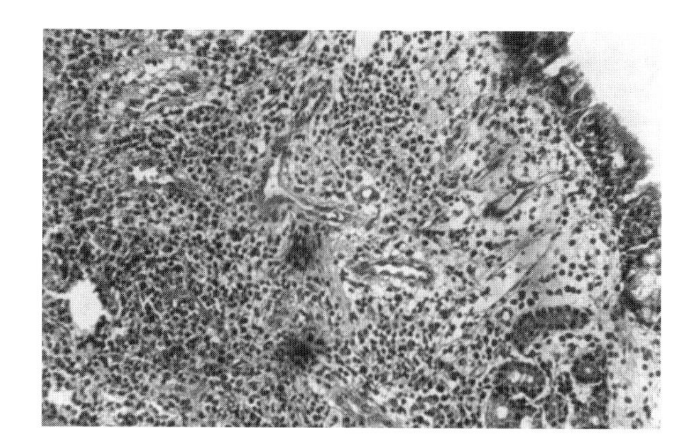

FIGURE 17.26 Necrosis in respiration epithelium.

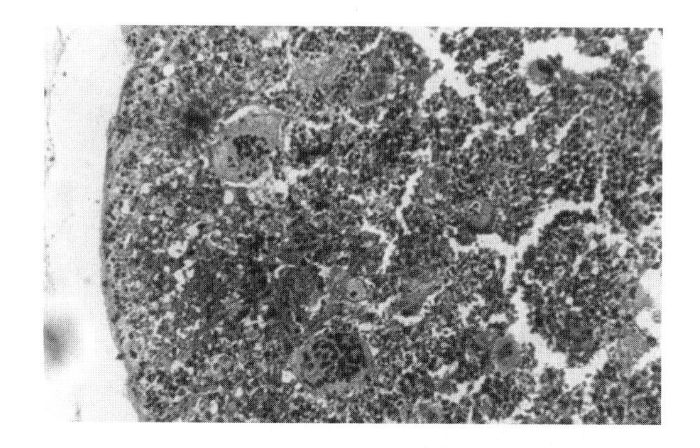

FIGURE 17.27 Giant cells in the respiratory epithelium.

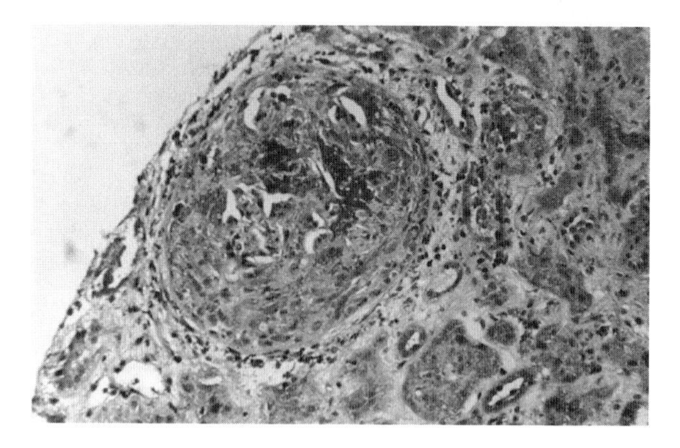

FIGURE 17.28 Obsolescent renal glomerulus in Wegener's granulomatosis.

proven. Patients develop antineutrophil cytoplasmic antibodies (c-ANCA). It may be treated successfully in many subjects by cyclophosphamide therapy (Figure 17.28 and Figure 17.29).

Antineutrophil cytoplasmic antibodies (ANCA) (Figure 17.30) are antibodies detected in 84 to 100% of generalized active Wegener's granulomatosis patients. These antibodies react with the cytoplasm of fixed neutrophils. They may also be detected in patients with microscopic polyarteritis. ANCA may be quantified by flow cytometry in conjunction with indirect immunofluorescence microscopy, which permits observation of antibody reactivity with the cytoplasm. There is a positive correlation between antibody levels and disease activity, with a decrease following therapy. The staining pattern is to be distinguished from that produced by antimyeloperoxidase antibodies that display perinuclear staining. One of the antibodies producing diffuse cytoplasmic fluorescence is against neutrophil proteinase-3.

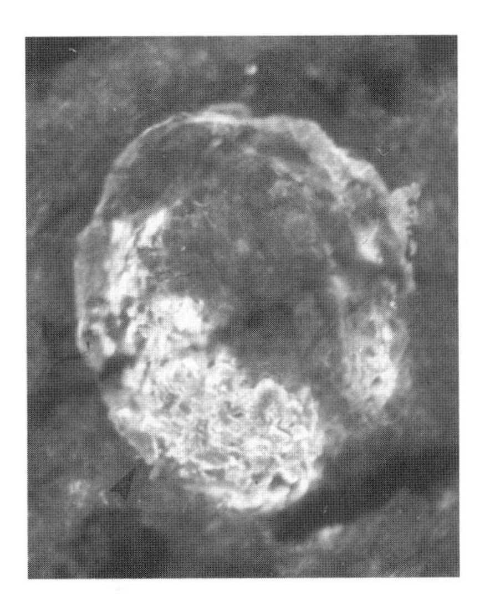

FIGURE 17.29 Immunofluorescence of renal glomerulus in Wegener's granulomatosis.

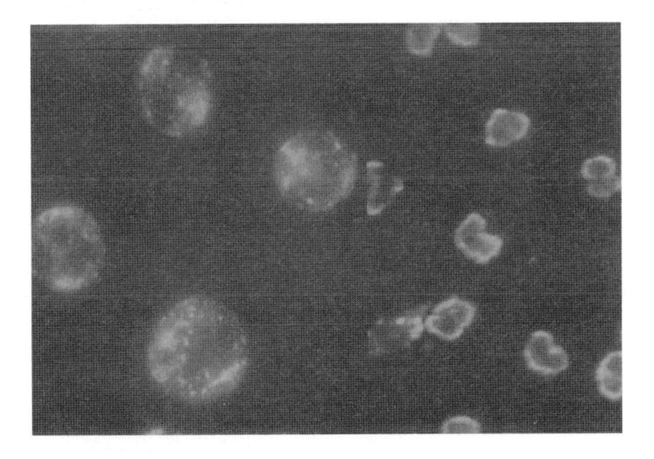

FIGURE 17.30 Antineutrophil cytoplasmic antibodies (ANCA).

Sera from some HIV-positive subjects may prove false positive for ANCA.

Antineutrophil cytoplasmic antibody occurs in 84 to 100% of active generalized Wegener's granulomatosis patients. This antibody is assayed by flow cytometry and indirect fluorescence microscopy. HIV-1 infected patients may be biologically false-positive for neutrophil cytoplasmic antibody.

Antineutrophil cytoplasmic autoantibodies (pANCA) recognize neutrophil myeloperoxidase. Of patients with ulcerative colitis, 50 to 80% express pANCA. It is unlikely that ANCA have a pathogenic role in inflammatory bowel disease since the presence of ANCA does not correlate with disease activity or extent. There is no defect in neutrophil function in the presence of ANCA.

Atypical antineutrophil cytoplasmic antibodies (atypical p-ANCA) are present in patients with ulcerative colitis (UC), primary sclerosing cholangitis (PSC), and autoimmune hepatitis (AIH). Atypical p-ANCA reacts with nuclear envelope proteins and neutrophils. Immunoblotting has revealed reactivity to a myeloid-specific 50-kDa nuclear protein with an isoelectric point of pH 6.0 found in 92% of patients with inflammatory bowel or hepatobiliary disease and atypical p-ANAC. Antibodies against the 50 kDa protein give a nuclear rim-like fluorescence on myeloid cells observed by immunofluorescence microscopy. Thus, the atypical p-ANCA in UC, PSC, or AIH recognize a 50-kDa myeloid-specific nuclear envelope protein.

ANNA is an abbreviation for antineutrophil nuclear antibodies.

Polyarteritis nodosa is a necrotizing vasculitis of small and medium-sized muscular arteries. It often involves renal and visceral arteries and spares the pulmonary circulation. The disease is often characterized by immune complex deposition in arteries, associated with chronic hepatitis B virus infection. Early lesions of the vessels often reveal hepatitis B surface antigen, IgM, and complement components. This uncommon disease has a male to female ratio of 2.5:1. The mean age at onset is 45 years. Characteristically the kidneys, heart, abdominal organs, and both peripheral and central nervous systems are involved. Lesions of vessels are segmental and show preference for branching and bifurcation in small and medium-sized muscular arteries. Usually, the venules and veins are unaffected and only rarely are granulomas formed. Characteristically, aneurysms form following destruction of the media and internal elastic lamina. There is proliferation of the endothelium with degeneration of the vessel wall and fibrinoid necrosis, thrombosis, ischemia, and infarction developing. Other than hepatitis B, polyarteritis nodosa is also associated with tuberculosis, streptococcal infections, and otitis media. Presenting signs and symptoms include weakness, abdominal pain, leg pain, fever, cough, and neurologic symptoms. There may be kidney involvement, arthritis, arthralgia, or myalgia, as well as hypertension. Up to 40% of patients may have skin involvement manifested as a maculopapular rash. Laboratory findings include elevated erythrocyte sedimentation rate, leukocytosis, anemia, thrombocytosis, and cellular casts in the urinary sediment signifying renal glomerular disease. Angiography is important in revealing the presence of aneurysm and changes in vessel caliber. There is no diagnostic immunologic test, but immune complexes, cryoglobulins, rheumatoid factor, and diminished complement component levels are often found. Biopsies may be taken from skeletal muscle or nerves for diagnostic purposes. Corticosteroids may be used, but cyclophosphamide is the treatment of choice in the severe progressive form. Hypersensitivity angiitis is a type of small vessel inflammation most frequently induced by drugs.

Periarteritis nodosa is a synonym for polyarteritis nodosa.

Henoch-Schoenlein purpura is a systemic form of small vessel vasculitis that is characterized by arthralgias, nonthrombocytopenic purpuric skin lesions, abdominal pain with bleeding, and renal disease. Immunologically, immune complexes containing IgA activate the alternate pathway of complement. Patients may present with upper respiratory infections preceding onset of the disease. Certain drugs, food, and immunizations have also been suspected as etiologic agents. The disease usually occurs in children 4 to 7 years of age, although it can occur in adults. Histopathologically, there is a diffuse leukocytoclastic vasculitis involving small vessels. The submucosa or subserosa of the bowel may be sites of hemorrhage. There may be focal or diffuse glomerulonephritis in the kidneys. Children may manifest lesions associated with the skin, gastrointestinal tract, or joints, whereas in adults the disease is usually associated with skin findings. The skin lesions begin as a pruritic urticarial lesion that develops into a pink maculopapular spot that matures into a raised and darkened lesion. The maculopapular lesion may ultimately resolve in 2 weeks without leaving a scar. Patients may also have arthralgias associated with the large joints of the lower extremities. Skin biopsy reveals the vasculitis, and immunofluorescence examination shows IgA deposits in vessel walls, which is in accord with a diagnosis of Henoch-Schoenlein purpura.

Purpura is characterized by purple areas on the skin caused by bleeding into the skin.

Hypergammaglobulinemic purpura is also called purpura hyperglobulinemia.

MUSCLE

Inflammatory myopathy (Figure 17.31) occurs when there is a necrosis and phagocytosis of muscle fibers with inflammatory cells.

Dermatomyositis (Figure 17.32 to Figure 17.34) is a connective tissue or collagen disease characterized by skin rash and muscle inflammation. It is a type of polymyositis presenting with a purple-tinged skin rash that is prominent on the superior eyelids, extensor joint surfaces, and base of the neck. There is weakness, muscle pain, edema, and calcium deposits in the subcutaneous tissue, especially prominent late in the disease. Blood vessels reveal lymphocyte cuffing, and autoantibodies against tRNA synthetases appear in the serum.

Polymyositis is an acute or chronic inflammatory disease of muscle that occurs in women twice as commonly as in men. Lymphocytes in polymyositis subjects produce a

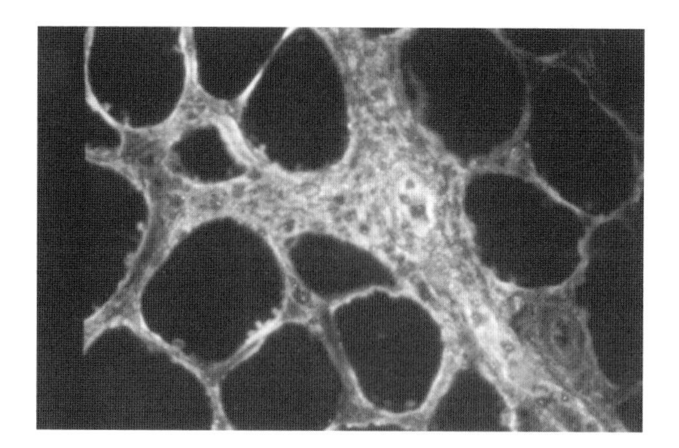

FIGURE 17.31 Inflammatory myopathy. Both regenerating and atrophic fibers are present with fibrosis.

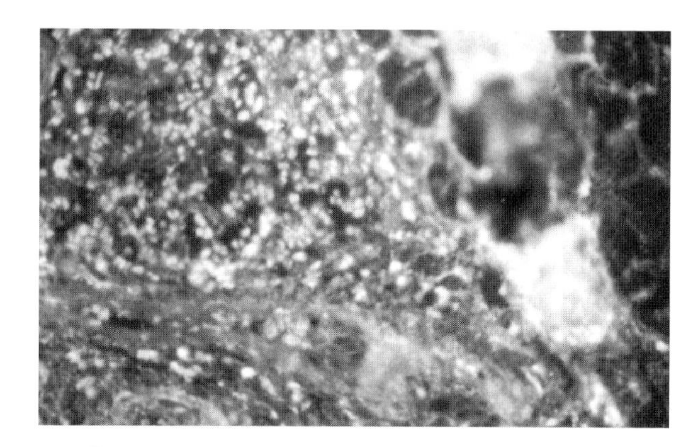

FIGURE 17.32 Dermatomyositis. Anti-IgG staining.

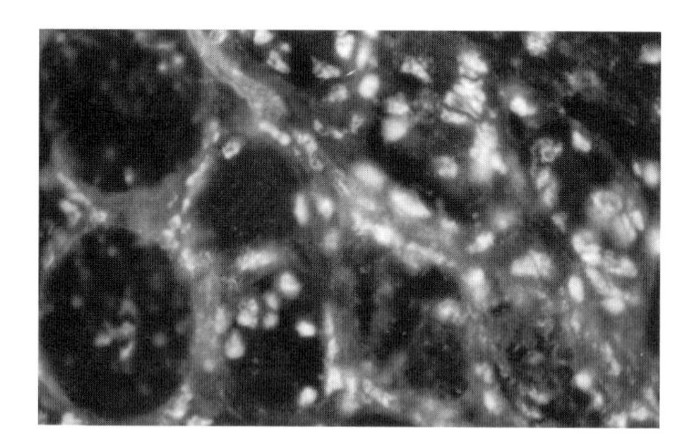

FIGURE 17.33 Dermatomyositis. Anti-IgG staining.

cytotoxin when incubated with autologous muscle. Biopsies of involved muscle reveal infiltration by lymphocytes and plasma cells. Antibodies can be demonstrated against the nuclear antigens Jo-1, PM-Scl, and RNP. Patients may develop polyclonal hypergammaglobulinemia. One fifth of the patients may develop rheumatoid factors and antinuclear antibodies. Cellular immunity appears important in the pathogenesis. This is exemplified by lymphocytes

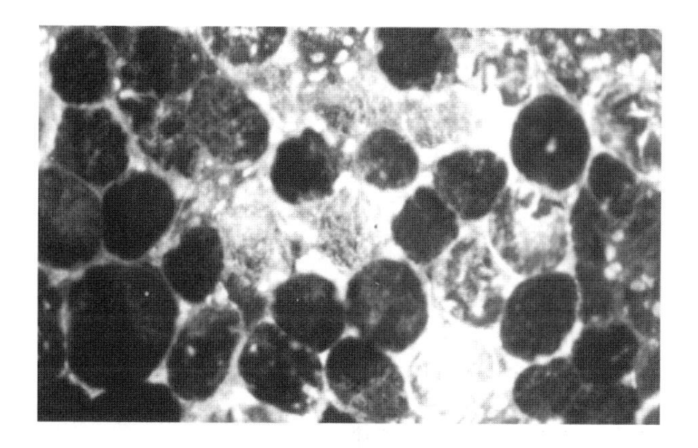

FIGURE 17.34 Dermatomyositis. Anti-IgG-staining.

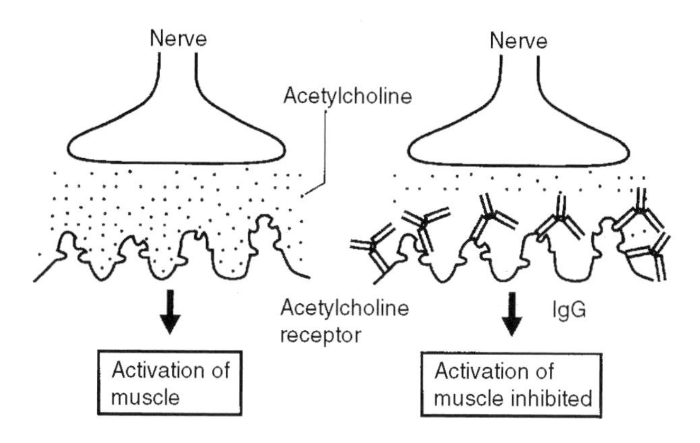

FIGURE 17.35 Myasthenia gravis.

of patients with polymyositis responding to their own muscle antigens as if they were alien. Patients often complain of muscle weakness, especially in the proximal muscles of extremities. To diagnose polymyositis, a minimum of three of the following must be present: (1) shoulder or pelvic girdle weakness, (2) myositis as revealed by biopsy, (3) increased levels of muscle enzymes, and (4) electromyographic findings of myopathy. Corticosteroids have been used to decrease muscle inflammation and increase strength. Methotrexate or other cytotoxic agents may be used when steroids prove ineffective.

NEUROMUSCULAR

Myasthenia gravis (MG) (Figure 17.35) is an autoantibody-mediated autoimmune disease. Antibodies specific for the nicotinic acetylcholine receptor (AChR) of skeletal muscle react with the postsynaptic membrane at the neuromuscular junction and diminish the number of functional receptors. Patients develop muscular weakness and some voluntary muscle fatigue. Thus, MG represents a receptor disease mediated by antibodies. The nicotinic AChR is the autoantigen. Contemporary research hopes to identify epitopes on the autoantigen(s) that interact with B and T cells in an autoimmune response. AChR is a four-subunit transmembrane protein. Most autoantibodies in humans are against the main immunogenic regions (MIR). Antibodies against the MIR crosslink AChR molecules, leading to their internalization and lysosomal degradation. This leads to a decreased number of postsynaptic membrane AChRs. Humans with MG and animals immunized against AChR develop circulating antibodies and clinical manifestations of MG. A subgroup of patients have seronegative MG; they resemble classic MG patients clinically but have no anti-AChR antibodies in their circulation. Since the IgG anti-AChR autoantibodies cross the placenta from mother to fetus, newborns of mothers with this disease may also manifest signs and symptoms of the disease. Neonatal MG establishes the antibody-mediated autoimmune nature of the disease. The thymus of an MG patient may reveal either lymphofollicular hyperplasia (70%) or thymoma (10%). Anti-AChR synthesizing B cells and T helper lymphocytes may be found in hyperplastic follicles. These are often encircled by myoid cells that express AChR. Interdigitating follicular dendritic cells closely associated with myoid cells have been suggested to present AChR autoantigen to autoreactive T helper lymphocytes. Antiidiotypic antibodies have been used to suppress or enhance experimental autoimmune myasthenia gravis (EAMG), depending on the antibody concentration employed. Conjugate immunotoxins to anti-Id antibodies have been able to suppress autoimmunity to AchR. Thymectomy and anticholinesterase drugs have proven useful in treatment.

Experimental autoimmune myasthenia gravis (EAMG): Myasthenia gravis (MG) is an autoantibody-mediated autoimmune disease. Experimental forms of MG were made possible through the ready availability of AChR from electric fish. Monoclonal antibodies were developed, followed by molecular cloning techniques that permitted definition of the AChR structure. EAMG can be induced in more than one species of animals by immunizing them with purified AChR from the electric ray (*Torpedo californica*). The autoantigen, nicotinic AChR, is T cell dependent. The *in vivo* synthesis of anti-AChR antibodies requires helper T cell activity. Antibodies specific for the nicotinic AChR of skeletal muscle react with the postsynaptic membrane at the neuromuscular junction.

Thymic medullary hyperplasia refers to the finding of germinal centers in the thymic medulla in myasthenia gravis patients. However, normal thymus glands may occasionally contain germinal centers, although the vast majority of normal thymus glands do not.

Acetycholine receptor (AChR) antibodies are IgG autoantibodies that cause loss of function of acetylcholine receptors that are critical to chemical transmission of the nerve impulse at the neuromuscular junction. This represents a type II mechanism of hypersensitivity according

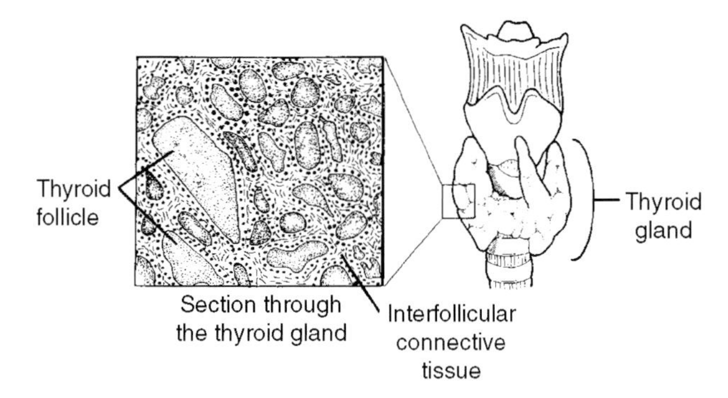

FIGURE 17.36 Thyroid gland.

to the Coombs and Gell classification. AchR antibodies are heterogeneous with some showing specificity for antigenic determinants other than those that serve as acetylcholine or alphabungaratoxin binding sites. As many as 85 to 95% of myasthenia gravis patients may manifest acetylcholine receptor antibodies.

THYROID

Chronic lymphocytic thyroiditis is characterized by profound infiltration of the thyroid by lymphocytes, leading to the extensive injury of thyroid follicular structure. Even though the gland becomes enlarged, its function diminishes, leading to hypothyroidism. Women are affected much more commonly than are men. Antibodies detectable in the serum are specific for the 107-kDa thyroid microsomal peroxidase, the thyrotropin receptor, and thyroglobulin. Also called Hashimoto's thyroiditis. Thyroid hormone replacement therapy is the usual approach to treatment.

Hashimoto's disease (chronic thyroiditis) is an inflammatory disease of the thyroid (Figure 17.36) found most frequently in middle-aged to older women. There is extensive infiltration of the thyroid by lymphocytes that completely replace the normal glandular structure of the organ. There are numerous plasma cells and macrophages and germinal centers which give the appearance of node structure within the thyroid gland. Both B cells and $CD4^+$ T lymphocytes comprise the principal infiltrating lymphocytes. Thyroid function is first increased as inflammatory reaction injures thyroid follicles, causing them to release thyroid hormones. However, this is soon replaced by hypothyroidism in the later stages of Hashimoto's thyroiditis. Patients with this disease have an enlarged thyroid gland. There are circulating autoantibodies against thyroglobulin and thyroid microsomal antigens (thyroid peroxidase). Cellular sensitization to thyroid antigens may also be detected. Thyroid hormone replacement therapy is given for the hypothyroidism that develops (Figure 17.37 to Figure 17.38A).

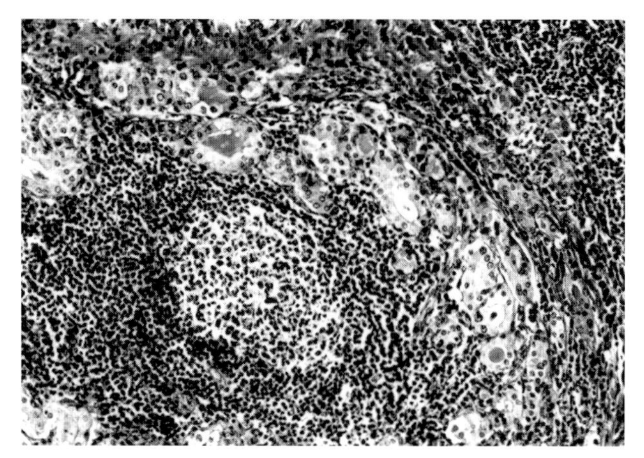

FIGURE 17.37 Hashimoto's thyroiditis.

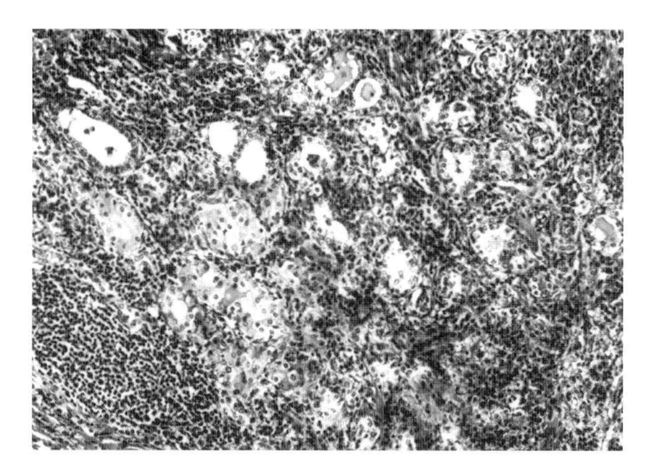

FIGURE 17.38 Hashimoto's thyroiditis.

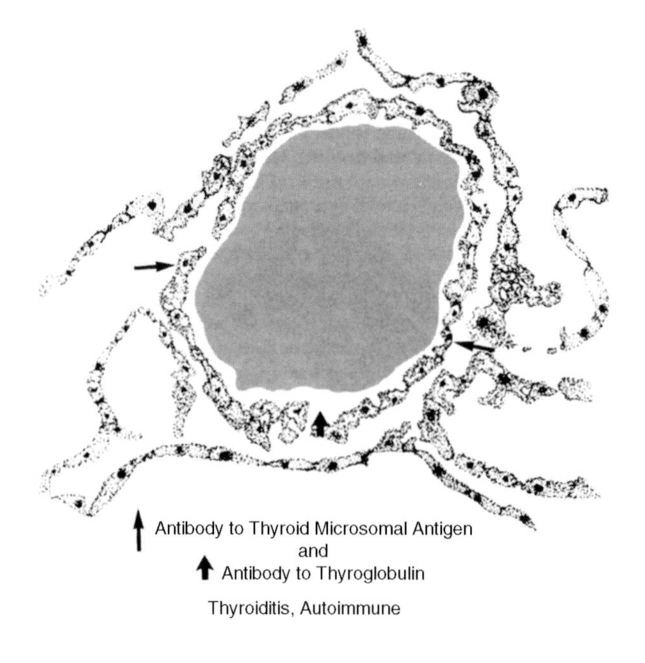

FIGURE 17.38A Thyroiditis, autoimmune.

Thyroid autoantibodies are found in patients with Hashimoto's thyroiditis or those with thyrotoxicosis (Graves' disease) that are organ-specific for the thyroid. Antibodies against thyroglobulin and antibodies against the microsomal antigen of thyroid acinar cells may appear in patients with autoimmune thyroiditis. Antibodies against TSH receptors appear in Graves' disease patients and cause stimulatory hypersensitivity. They mimic the action of TSH. This is an IgG molecule termed "long acting thyroid stimulator" (LATS). LATS levels are increased in many patients with thyrotoxicosis or Graves' disease.

Experimental autoimmune thyroiditis (EAT) is a murine model for Hashimoto's thyroiditis. There is a strong MHC genetic component in susceptibility to Hashimoto's thyroiditis, which has been shown to reside in the *IA* subregion of the murine MHC (H-2), governing the immune response (*Ir*) genes to mouse thyroglobulin (MTg). Following induction of EAT with MTg, autoantibodies against MTg appear, and mononuclear cells infiltrate the thyroid. Repeated administration of soluble, syngeneic MTg without adjuvant leads to thyroiditis only in the murine haplotype susceptible to EAT. Autoreactive T cells proliferate *in vitro* following stimulation with MTg. The disease can be passively transferred to naïve recipients by adoptive immunization and differentiate into cytotoxic T lymphocytes (Tc) *in vitro*. Thus, lymphoid cells rather than antibodies represent the primary mediator of the disease. *In vitro* proliferation of murine-autoreactive T cells was found to show a good correlation with susceptibility to EAT and was dependent on the presence of Thy-1^+, Lyt-1^+, Ia^+, and $L3T4^+$ lymphocytes. Effector T lymphocytes (T_E) in EAT comprise various T cell subsets and Lyt-1 (L3T4) and Lyt-2 phenotypes. T lymphocytes cloned from thyroid infiltrates of Hashimoto's thyroiditis patients reveal numerous cytotoxic T lymphocytes and clones synthesizing IL-2 and IFN-γ. While the T cell subsets participate in the pathogenesis of Hashimoto's thyroiditis, autoantibody synthesis appears to aid perpetuation of the disease or result from it.

EAT includes a murine model for Hashimoto's thyroiditis. There is a strong MHC genetic component in susceptibility to Hashimoto's thyroiditis, which has been shown to reside in the IA subregion of the murine MHC (H-2), governing the immune response(Ir) genes to mouse thyroglobulin (MTg). Following induction of EAT with MTg, autoantibodies against MTg appear and mononuclear cells infiltrate the thyroid. Repeated administration of soluble, syngeneic MTg without adjuvant leads to thyroiditis only in the murine haplotype susceptible to EAT. Autoreactive T cells proliferate *in vitro* following stimulation with MTg. The disease can be passively transferred to naïve recipients by adoptive immunization and differentiate into cytotoxic T lymphocytes (Tc) *in vitro*. Thus, lymphoid cells rather than antibodies represent the primary mediator of the disease. *In vitro* proliferation of murine-autoreactive T cells has been found to show a good correlation with susceptibility to EAT and to be dependent on the presence of Thy-1^+, Lyt-1^+, Ia^+, and $L3T4^+$ lymphocytes. Effector T lymphocytes (T_E) in EAT comprise various T cell subsets and Lyt-1 (L3T4) and Lyt-2 phenotypes. T lymphocytes cloned from thyroid infiltrates of Hashimoto's thyroiditis patients reveal numerous cytotoxic T lymphocytes and clones synthesizing IL-2 and IFN-γ. While the T cell subsets participate in pathogenesis of Hashimoto's thyroiditis, autoantibody synthesis appears to aid perpetuation of the disease or result from it.

Experimental allergic thyroiditis is an autoimmune disease produced by injecting experimental animals with thyroid tissue or extract or thyroglobulin incorporated into Freund's complete adjuvant. It represents an animal model of Hashimoto's thyroiditis in humans, with mixed extensive lymphocytic infiltrate. Another animal model is the spontaneous occurrence of this disease in the obese strain of chickens as well as in Buffalo rats.

Graves' disease (hyperthyroidism) is a thyroid gland hyperplasia with increased thyroid hormone secretion that produces signs and symptoms of hyperthyroidism in the patient. Patients may develop IgG autoantibodies against thyroid-stimulating hormone (TSH) receptors. This autoantibody is termed long-acting thyroid stimulator (LATS). When the LATS IgG antibody binds to the TSH receptor, it has a stimulatory effect on the thyroid promoting hyperactivity. This IgG autoantibody can cross the placenta and produce transient hyperthyroidism in a newborn infant. The disease has a female predominance.

Hyperthyroidism is a metabolic disorder attributable to thyroid hyperplasia with an elevation in thyroid hormone secretion. Also called Graves' disease.

Thyrotoxicosis is a disease of the thyroid in which there is hyperthyroidism with elevated levels of thyroid hormones in the blood and thyroid gland hyperplasia or hypertrophy. Thyrotoxicosis may be autoimmune, as in Graves' disease, in which there may be diffuse goiter. Autoantibodies specific for thyroid antigens mimic thyroid stimulating hormones (TSH) by stimulating thyroid cell function. In addition, patients with Graves' disease develop ophthalmopathy and proliferative dermopathy. It occurs predominantly in females (70 females to 1 male) and appears usually in the 30- to 40-year-old age group. In Caucasians, it is a disease associated with DR3. Patients may develop nervousness, tachycardia, and numerous other symptoms of hyperthyroidism. They also have increased levels of total and free T3 and T4. There is a diffuse and homogenous uptake of radioactive iodine in these patients. Three types of antithyroid antibodies occur: (1) thyroid-stimulating immunoglobulin, (2) thyroid growth-stimulating immunoglobulin, and (3) thyroid binding-inhibitory immunoglobulin. Their presence confirms a diagnosis of Grave's disease. The thyroid gland may be infiltrated with lymphocytes. Long-acting thyroid stimulator (LATS) is classically associated with thyrotoxicosis. It is an IgG antibody specific for thyroid hormone receptors. It induces thyroid hyperactivity by combining with TSH receptors.

Long-acting thyroid stimulator (LATS) is an IgG autoantibody that mimics the action of thyroid-stimulating hormone in its effect on the thyroid. The majority of patients with Graves' disease, i.e., hyperthyroidism, produce LATS. This IgG autoantibody reacts with the receptors on thyroid cells that respond to thyroid-stimulating hormone. Thus, the antibody–receptor interaction results in the same biological consequence as does hormone interaction with the receptor. This represents a stimulatory type of hypersensitivity and is classified in the Gell and Coombs classification as one of the forms of Type II hypersensitivity.

LUNG

Usual interstitial pneumonitis (Figure 17.39 to Figure 17.41) is a lung disease associated with interstitial inflammation and fibrosis that leads to progressive respiratory insufficiency. It is the most frequent type of idiopathic interstitial pneumonitis and has been referred to by various names including Hamman-Rich syndrome. It is believed to have an immunologic basis with 20% of cases associated with collagen vascular diseases such as rheumatoid arthritis (RA), progressive systemic sclerosis, and systemic lupus erythematosus (SLE). Autoantibodies in these patients include antinuclear antibodies and rheumatoid factor. Immune complexes may be found in the blood, alveolar walls, and bronchoalveolar lavage fluid, yet the antigen remains unknown. Alveolar macrophages are believed to become activated after phagocytizing immune complexes. This might be followed by the release of cytokines which attract neutrophils that cause injury of alveolar walls leading to interstitial fibrosis. Pathologically, this is chronic inflammation in the interstitial space with extensive alveolar damage and damage of normal lung parenchyma. Areas of diffuse alveolar damage contain infiltrates of lymphocytes and plasma cells in alveolar walls and hypoplasia of type II pneumocytes. Fibrosis varies from mild to severe, leading even to honeycomb lung in string cases. The distal acinus shrinks and proximal bronchioles dilate as fibrosis may be accompanied by pulmonary hypertension. UIP may occur over a 5- to 10-year period with development of dyspnea on exertion and dry cough. The disease may follow acute viral infection of the respiratory tract in one third of patients or in rare cases may have acute fulminating interstitial inflammation and fibrosis leading to Hamman-Rich syndrome that may lead to death without delay. Treatment modalities include

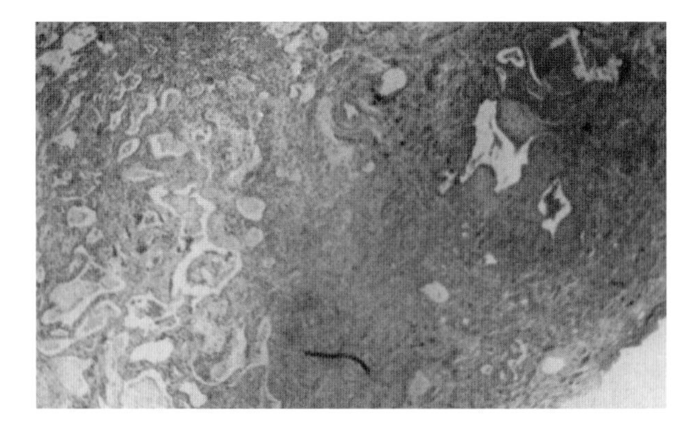

FIGURE 17.39 Usual interstitial pneumonitis. Biopsy reveals idiopathic pulmonary fibrosis. Histopathology reveals a bronchus that has divided chronic lymphocytic infiltrate, interstitial inflammation, and fibrosis edema. Only a few air spaces are irregularly located.

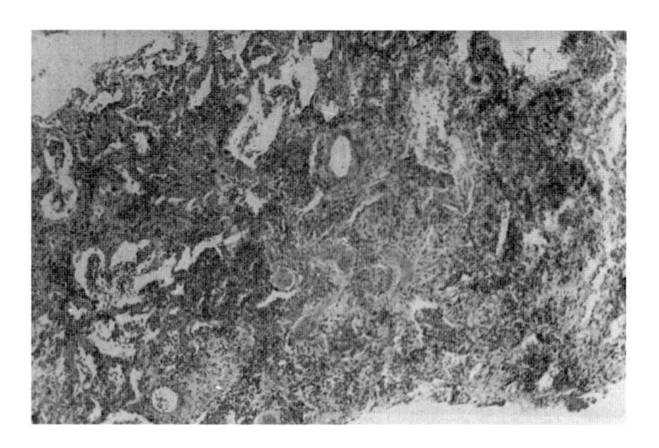

FIGURE 17.40 Usual interstitial pneumonitis. Histopathology reveals fibrosis and inflammation, numerous plasma cells, fibroblasts, histiocytes, and vascular destruction.

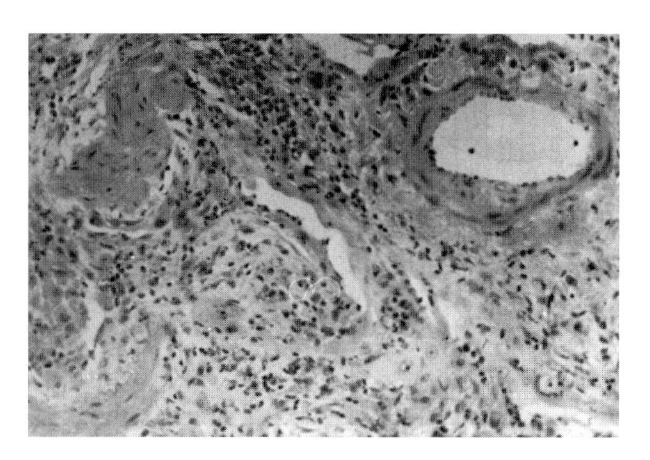

FIGURE 17.41 Usual interstitial pneumonitis.

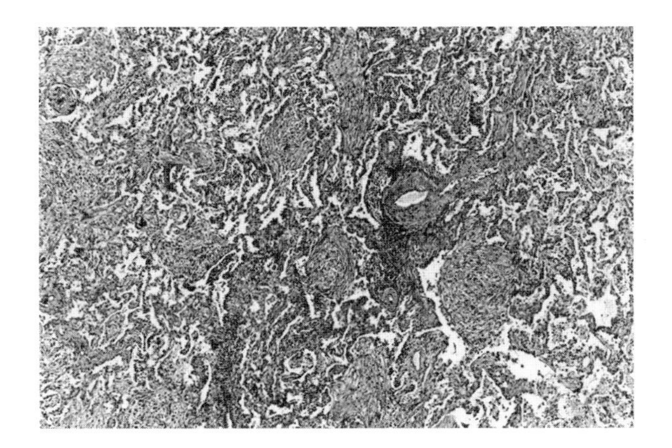

FIGURE 17.42 Lymphocytic interstitial pneumonia (LIP) with organization.

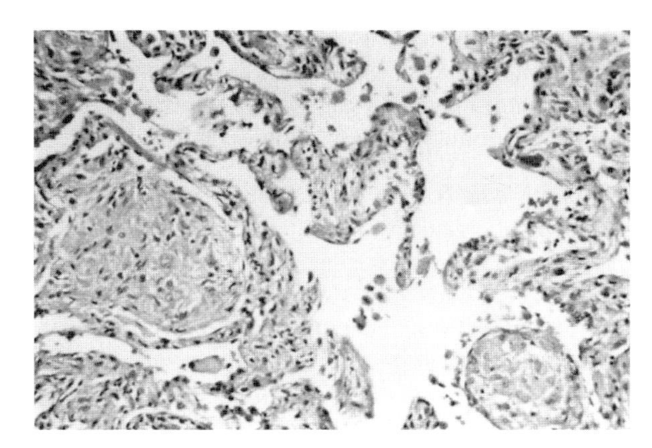

FIGURE 17.43 Farmer's lung.

corticosteroids or cytophosphamide in lung transplantation.

Lymphocytic interstitial pneumonia (LIP) (Figure 17.42) is a diffuse pulmonary disease of middle-aged females who may also have Sjögren's disease, hypergammaglobulinemia, or hypogammaglobulinemia. Patients develop shortness of breath, and reticulonodular infiltrates appear on chest films. Mature lymphocytes and plasma cells appear in the nodular interstitial changes in alveolar and interlobular septae with perivascular accumulation of round cells. LIP may resemble lymphoma based on the monotonous accumulation of small lymphocytes, and patients may ultimately develop end-stage lung disease or lymphoma.

Farmer's lung (Figure 17.43) is a pulmonary disease of farm workers who have been exposed repeatedly to organic dust and fungi such as the *Aspergillus* species. It occurs as an extrinsic allergic alveolitis or hypersensitivity pneumonitis in nonatopic subjects. It is mediated by IgG1. Antibodies specific for moldy hay, in which a number of fungi grow readily, are manifest in 90% of individuals. These include *Microspora vulgaris* and *Micropolyspora faeni*. The pathogenesis is believed to involve a type III hypersensitivity mechanism with the deposition of immune complexes in the lung. Patients become breathless within hours after inhaling the dust and may develop interstitial pneumonitis with cellular infiltration of the alveolar walls where monocytes and lymphocytes are prominent. This may lead to pulmonary fibrosis following chronic inflammation, peribronchiolar granulomatous reaction, and foreign body-type giant cell reactions. Corticosteroids are used for treatment. Interaction of the IgG antibodies with the inhaled allergen in the alveolar walls of the lung leads to inflammation and compromised gas exchange.

Adult respiratory distress syndrome (ARDS) is embarrassed respiratory function as a consequence of pulmonary edema caused by increased vascular permeability.

Adult T cell leukemia–lymphoma (ATLL) is a lymphoproliferative neoplasm of mature lymphocytes that progresses rapidly. This has been linked to the HTLV-1 retrovirus infection which has been observed in Japan, Africa, the Caribbean, and the southeastern U.S. Patients develop hypercalcemia, progressive skin changes, enlarged hilar, and retroperitoneal and peripheral lymph nodes without mediastinal node enlargement. There may be involvement of the lungs, gastrointestinal tract, and central nervous system, as well as opportunistic infections. The condition occurs in five clinical forms.

Faenia rectivirgula is the most frequently encountered inhalant in farmer's lung disease, the most common type of hypersensitivity pneumonitis in the U.S. This agent or its antigens induce IL-1 synthesis by alveolar macrophages and facilitate increased secretion of tumor necrosis factor-α (TNF-α) from alveolar macrophages and monocytes, revealing cytokine participation in the pathogenesis of hypersensitivity pneumonitis. Precipitin antibodies are formed in a number of the subjects.

Extrinsic allergic alveolitis is inflammation in the lung produced by immune reactivity, mainly of the granulomatous type, due to inhaled antigens such as dust, bacteria, mold, grains, or other substances. Also called farmer's lung.

Fog fever is an episode of acute respiratory distress in cows approximately 7 d after their removal to a pasture where hay has been recently cut. They may die within 1 d, developing pulmonary edema with extensive emphysema. This disease may present as an atopic allergy in sensitized animals who are exposed to grass proteins, pollen, and fungal spores. A nonimmunologic intoxication has also been suggested as a cause. Cattle may also have a similar reaction, which resembles farmer's lung in man, if they have been fed hay containing *Micropolyspora faeni* spores. Precipitating antibodies are present in their blood sera.

Byssinosis is a disease of people who work with cotton, flax, jute, and hemp, probably attributable to hypersensitivity to vegetable fiber dust. Patients develop tightness in the chest upon returning to work after several days' absence.

Thermoactinomyces species are Gram-positive, endospore-forming microorganisms, which together with *Aspergillus fumigatus* are the most common causes of the hypersensitivity pneumonitis (HP) termed farmer's lung (FL). Other thermophilic actinomycetes also play a role in the pathogensis of this disease. Antibodies reactive with a 55-kDa band are frequent in the sera of patients with FL but the significance is not known.

Bagassosis is hypersensitivity among sugarcane workers to a fungus, *Thermoactinomyces saccharic,* which thrives in the pressings from sugarcane. The condition is expressed as a hypersensitivity pneumonitis. Subjects develop type III (Arthus reaction) hypersensitivity following inhalation of dust from molding hot sugarcane bagasse.

Sugarcane worker's lung: See bagassosis and farmer's lung.

Sarcoidosis (Figure 17.44 and Figure 17.45) is a systemic granulomatous disease that involves lymph nodes, lungs, eyes, and skin. There is a granulomatous hypersensitivity reaction that resembles that of tuberculosis and fungus infections. Sarcoidosis has a higher incidence in African-Americans than in Caucasians and is prominent geographically in the southeastern U.S. It is of unknown etiology. Immunologically, there is a decrease in circulating T cells. There is decreased delayed-type hypersensitivity as manifested by anergy to common skin test antigens. Increased antibody formation leads to polyclonal hypergammaglobulinemia. There is a marked cellular immune response in local areas of disease activity. Tissue lesions consist of inflammatory cells and granulomas, comprised of activated mononuclear phagocytes such as epithelioid cells, multinucleated giant cells, and macrophages. Activated T cells are present at the periphery of the granuloma. $CD4^+$ T cells appear to be the immunoregulatory cells governing granuloma formation. Mediators released from T cells nonspecifically stimulate B cells, resulting in the polyclonal hypergammaglobulinemia. The granulomas are typically noncaseating, distinguishing them from those produced in tuberculosis. Patients may develop fever, polyarthritis, erythema nodosum, and iritis. They also may experience loss of weight, anorexia, weakness, fever, sweats, nonproductive cough, and increasing dyspnea on exertion. Pulmonary symptoms occur in more than 90% of the patients. Angiotensin-converting enzyme is increased in the serum of sarcoid patients. Disease activity is monitored by measuring the level of this enzyme in the serum. The subcutaneous inoculation of sarcoidosis lymph node extracts into patients diagnosed with sarcoidosis leads to a granulomatous reaction in the skin 3 to 4 weeks after inoculation. This was used in the past as a diagnostic test of questionable value termed the Kveim reaction. Sarcoidosis symptoms can be treated with corticosteroids, but only in patients where disease progression occurs. It is a relatively mild disease, with 80% resolving spontaneously and only 5% dying of complications. Evidence has been provided for oligoclonal expansion of $\alpha\beta$ T cell subsets and predominant expression of type 1 cytokines (interferon γ [IFN-γ] and interleukin 2 [IL-12]) at sites of inflammation, suggesting that sarcoidosis is an antigen-driven, T_H1-mediated immune disorder.

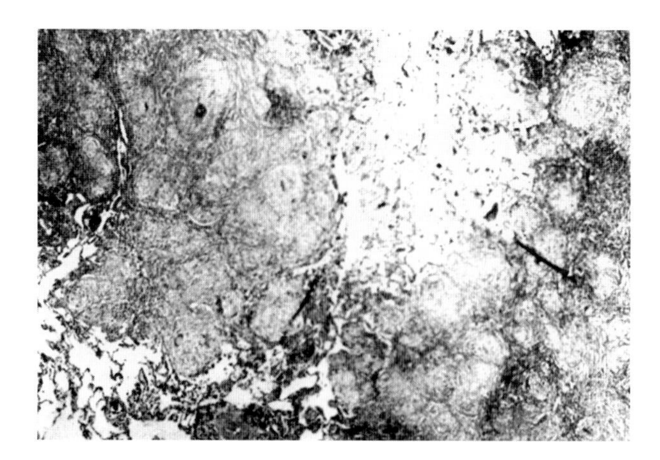

FIGURE 17.44 Open lung biopsy showing sarcoidosis.

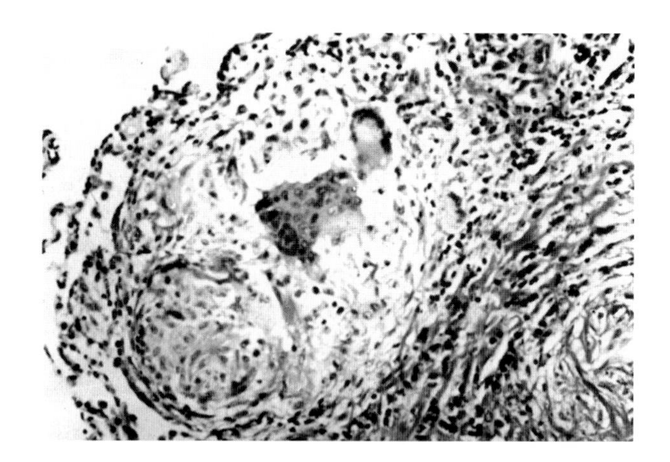

FIGURE 17.45 Open lung biopsy showing sarcoidosis.

The **Kveim reaction (historical)** is a skin reaction for the diagnosis of sarcoidosis in which ground lymph-node tissue of a known sarcoidosis patient is suspended in physiological salt solution and inoculated intracutaneously into a suspected sarcoidosis patient. A positive reaction, on histopathologic examination of an injection site biopsy

1 month to 6 weeks after inoculation, reveals a nodular epithelioid cell granuloma-like reaction. A positive Kveim test confirms the diagnosis of sarcoidosis. The danger of possibly transmitting hepatitis, AIDS, and other viruses precludes the use of this reaction.

Asthma is a disease of the lungs characterized by reversible airway obstruction (in most cases), inflammation of the airway with prominent eosinophil participation, and increased responsiveness by the airway to various stimuli. There is bronchospasm associated with recurrent paroxysmal dyspnea and wheezing. Some cases of asthma are allergic, i.e., bronchial allergy, mediated by IgE antibody to environmental allergens. Other cases are provoked by nonallergic factors that are not discussed here.

Charcot-Leyden crystals are hexagonal and bipyramidal crystals present in asthmatic patients' sputum. They contain a 13-kDa lysophospholipase derived from the eosinophil cell membrane.

Intrinsic asthma is nonallergic or idiopathic asthma that usually occurs first during adulthood and follows a respiratory infection. Patients experience chronic or recurrent obstruction of bronchi associated with exposure to pollen or other allergens. This is in marked contrast to patients with extrinsic (allergic) asthma mediated by immune (IgE) mechanisms in the bronchi. Intrinsic asthma patients have negative skin tests to ordinary allergens when the IgE content of their serum is normal. They do manifest eosinophilia. There is no family history of atopic diseases.

Churg-Strauss syndrome (allergic granulomatosis) is a combination of asthma associated with necrotizing vasculitis, eosinophilic tissue infiltrates, and extravascular granulomas.

Bird fancier's lung (Figure 17.46 and Figure 17.47) is a respiratory distress in subjects who are hypersensitive to plasma protein antigens of birds following exposure of the subject to bird feces or skin and feather dust. Hypersensitive subjects have an Arthus type of reactivity or type III hypersensitivity to the plasma albumin and globulin components. Precipitates may be demonstrated in the blood sera of hypersensitivity subjects.

Pulmonary vasculitis (Figure 17.48 to Figure 17.51) is a vasculitis characterized by chronic inflammation, necrotizing and nonnecrotizing granulomatous inflammation, fibrinoid necrosis, arterial wall medial thickening, and intimal proliferation.

Hypersensitivity pneumonitis is lung inflammation induced by antibodies specific for substances that have been inhaled. Within hours of inhaling the causative agent, dyspnea, chills, fever, and coughing occur. Histopathology of the lung reveals inflammation of alveoli in the interstitium

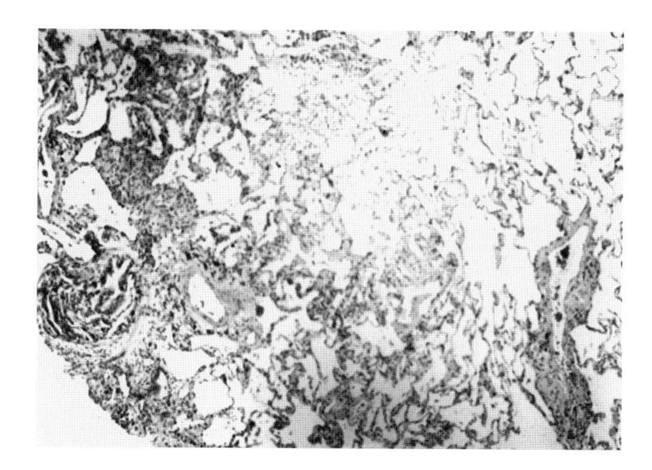

FIGURE 17.46 Bird fancier's lung. Lung biopsy showing an interstitial granulomatous pneumonitis consistent with bird fancier's lung.

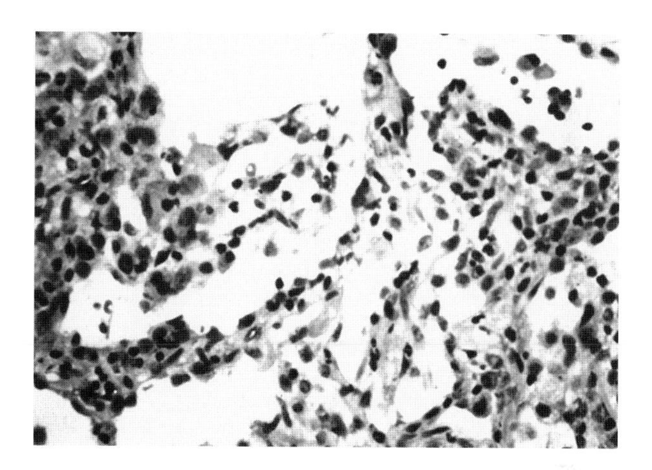

FIGURE 17.47 Bird fancier's lung. Lung biopsy showing an interstitial granulomatous pneumonitis consistent with bird fancier's lung.

with obliterating bronchiolitis. Immunofluorescence examination reveals deposits of C3. Hyperactivity of the lungs to airborne immunogens or allergens may ultimately lead to interstitial lung disease. An example is farmer's lung, which is characterized by malaise, coughing, fever, tightness in the chest, and myalgias. Of the numerous syndromes and associated antigens that may induce hypersensitivity pneumonitis, humidifier lung (thermophilic actinomycetes), bagassosis (*Thermoactinomyces vulgaris*), and bird fancier's lung (bird droppings) are well known. Types of hypersensitivity pneumonitis caused by *Penicillum* species include cheese workers disease, humidifier lung, woodman's disease, and cork worker's disease.

Aspergillus species are aeroallergenic fungi that may induce hypersensitivity pneumonitis (HP). *Aspergillus* species together with the thermophilic actinomyces are the most common causes of the hypersensitivity pneumonitis known as farmer's lung disease.

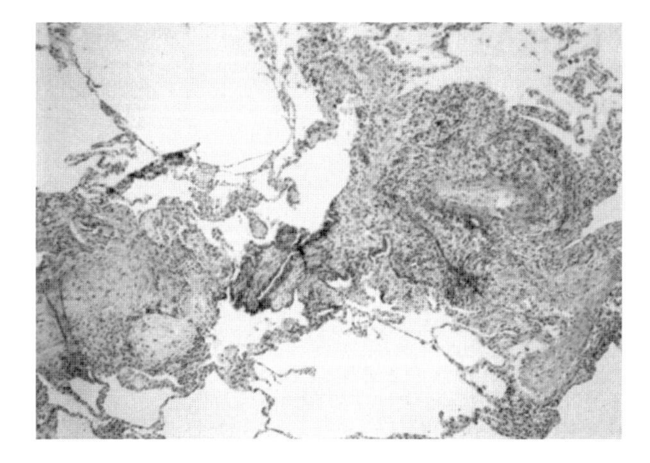

FIGURE 17.48 Pulmonary vasculitis. There is necrosis of endothelial cells and supporting stromal structures with acute chronic inflammation. There is exudation of polymorphonuclear leukocytes, eosinophils, and extravasated erythrocytes. Magnification: 35×.

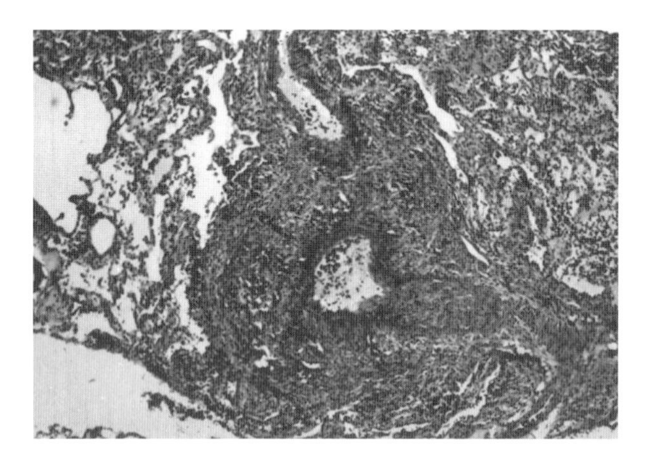

FIGURE 17.49 Pulmonary vasculitis. There is necrosis of endothelial cells and supporting stromal structures with acute chronic inflammation. There is exudation of polymorphonuclear leukocytes, eosinophils, and extravasated erythrocytes. Magnification: 50×.

MULTISYSTEM

Rheumatic fever (RF) is an acute, nonsuppurative, and inflammatory disease that is immune mediated and occurs mainly in children a few weeks following an infection of the pharynx with group A β hemolytic streptococci. M protein, a principal virulence factor associated with specific strains of the streptococci, induces antibodies that cross-react with epitopes of human cardiac muscle. These antibodies may not produce direct tissue injury, but together with other immune mechanisms evoke acute systemic disease characterized mainly by polyarthritis, skin lesions, and carditis. Whereas the arthritis and skin lesions resolve, the cardiac involvement may lead to permanent injury to the valves producing fibrocalcific deformity. Foci

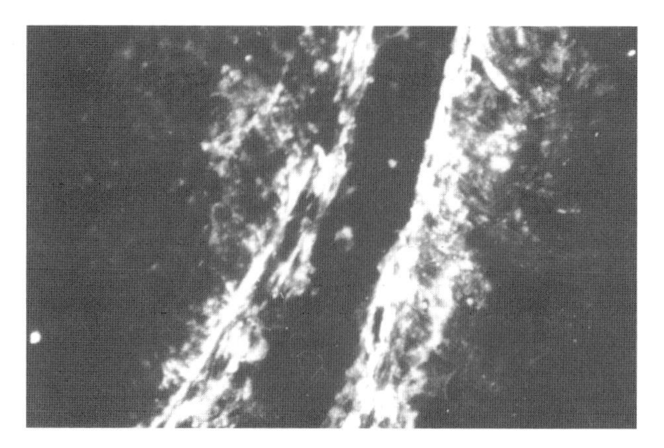

FIGURE 17.50 Pulmonary vasculitis. Direct immunofluorescence reveals coalescent and granular deposits of IgM and C3 in the walls of some muscular arteries and occasional large veins consistent with immune complex vasculitis involving large vessels. Magnification: 200×.

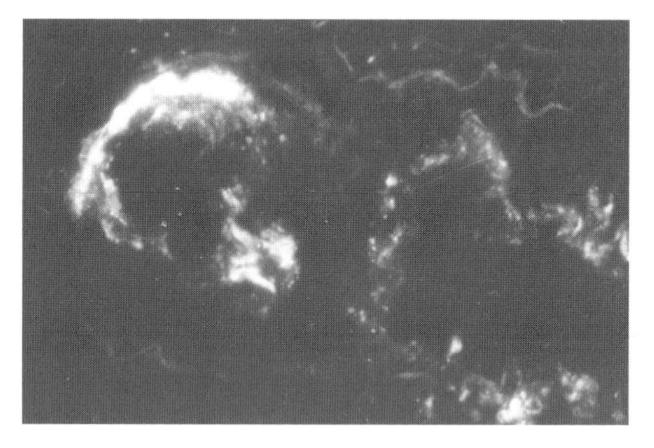

FIGURE 17.51 Pulmonary vasculitis. Direct immunofluorescence reveals coalescent and granulat deposits of IgM and C3 in the walls of some muscular arteries and occasional large veins consistent with immune complex vasculitis involving large vessels. Magnification: 500×.

of necrosis of collagen with fibrin deposition surrounded by lymphocytes, macrophages, and plump modified histiocytes are termed Aschoff bodies, which are ultimately replaced years later by fibrous scars. Aschoff bodies may be found in any of the three layers of the heart. In the pericardium, they are accompanied by serofibrinous (bread and butter) pericarditis. In the myocardium, they are scattered in the interstitial connective tissue often near blood vessels. There may be dilatation of the heart and mitral valve ring. There may be inflammation of the endocardium, mainly affecting the left-sided valves. Small vegetations may form along the lines of closure. Other tissues may be affected with the production of acute nonspecific arthritis affecting the larger joints. Fewer than half of the patients develop skin lesions such as subcutaneous nodules or erythema marginatum. Subcutaneous nodules that appear at pressure points overlying extensor tendons of extremities at the wrist, elbows, ankles, and knees consist of central fibrinoid necrosis enclosed by a palisade of fibroblasts and mononuclear inflammatory cells. Rheumatic arteritis has been described in coronary, renal, mesenteric, and cerebral arteries, as well as in the aorta and pulmonary vessels. Rheumatic interstitial pneumonitis is a rare complication of the disease. Antistreptolysin O (ASO) and antistreptokinase (ASK) antibodies are found in the serum of affected individuals. Myocarditis that develops during an acute attack may induce arrhythmias such as atrial fibrillation or cardiac dilatation with potential mitral valve insufficiency. Long-term antistreptococcal therapy must be given to any patient with a history of rheumatic fever, as subsequent streptococcal infections may worsen the carditis.

Erythema marginatum is immune complex-induced vasculitis in the subcutaneous tissues associated with rheumatic fever.

St. Vitus dance (chorea) is muscular twitching movements that are involuntary and may occur in acute rheumatic fever.

Chorea is involuntary muscle twitching that occurs in cases of acute rheumatic fever and is commonly known as St. Vitus dance.

Polyarthritis is multiple joint inflammation as occurs in rheumatic fever, SLE, and related diseases.

Jones criteria are signs and symptoms used in the diagnosis of acute rheumatic fever.

Aschoff bodies (Figure 17.52) are areas of fibrinoid necrosis encircled first by lymphocytes and macrophages with a rare plasma cell. The mature Aschoff body reveals prominent modified histiocytes termed Anitschkow cells or Aschoff cells in the inflammatory infiltrate. These cells have round to oval nuclei with wavy, ribbon-like chromatin and amphophilic cytoplasm. Aschoff bodies are pathognomonic of rheumatic fever. They may be found in any of the heart's three layers, i.e., pericardium, myocardium, or endocardium.

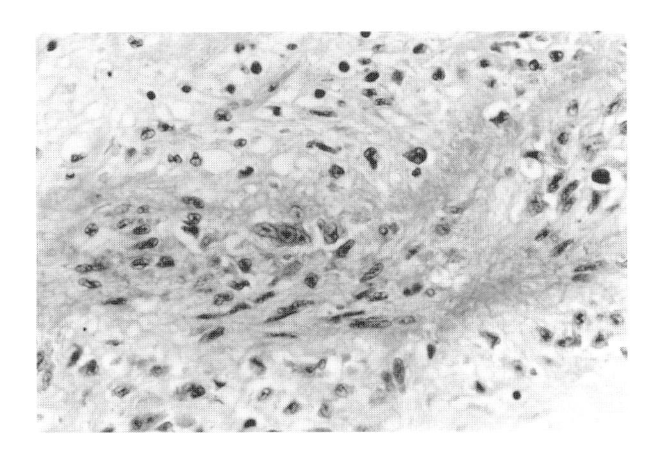

FIGURE 17.52 Aschoff body.

Takayasu's arteritis is inflammation and stenosis involving large and intermediate sized arteries, including the aortic arch. The disease occurs in the 15- to 20-year-old age group with a 9:1 female predominance. Mononuclear and giant cells infiltrate all layers of the wall of involved arteries, reflecting a true panarteritis. There may be intimal proliferation, fibrosis, elastic lamina disruption, and vascularization in the media. The disease begins with an inflammatory phase, followed within several weeks to several years (up to 8 years) by a chronic occlusive phase. In the initial inflammatory phase, patients develop fever, malaise, weakness, night sweats, arthralgias, and myalgias. Symptoms in the chronic phase are related to ischemia of involved organs. Vascular insufficiency is indicated by decreased or absent radial, ulnar, and carotid pulse. Approximately one third of the patients may have cardiac symptoms such as palpitations and congestive heart failure secondary to hypertension. IgG, IgA, and IgM may be elevated, and the erythrocyte sedimentation rate is usually increased. Corticosteroids may be helpful in controlling inflammation. Cyclophosphamide has been successfully used in those not responsive to corticosteroid therapy.

Kawasaki's disease is mucocutaneous lymph node syndrome that occurs in children under 5 years of age. The incubation period may be 1 to 2 weeks. It is an acute febrile disease characterized by erythema of the conjunctiva and oral cavity, skin rash, and swollen (especially cervical) lymph nodes. It occurs mostly in Japan, with some cases in the U.S. Cardiac lesions such as coronary artery aneurysms may be found in 70% of the patients. Coronary arteritis causes death in 1 to 2% of patients. There is necrosis and inflammation of the vessel wall. The etiology is unknown, but it has been suggested to be infection by a retrovirus. Associated immunoregulatory disorders include T and B lymphocyte activation, circulating immune complexes, and autoantibodies cytolytic for endothelial cells activated by cytokines.

Mucocutaneous lymph node syndrome: See Kawasaki's disease.

Acute rheumatic fever: Following a group A β hemolytic streptococcal infection of the throat, inflammation of the heart, joints, and connective tissue may follow. Arthritis may affect several joints in a migratory pattern. Carditis, inflammation of the heart, may be associated with the development of high-titer heart-reactive antibodies (HRA), which have been implicated in the pathogenesis of rheumatic fever. A patient may develop HRA, rheumatic fever, or both. HRA development in rheumatic fever represents molecular mimicry. Selected antistreptococcal cell wall M protein antibodies are cross-reactive through molecular mimicry with myocardial epitopes of the human heart. These serve as sites of attachment for IgG and complement molecules that are detectable by immunofluorescence examination. The antibodies on the cardiac muscle are found at sarcolemmal and subsarcolemmal sites and in the pericardium. The crossreactivity also involves heart valve glycoproteins and the myocardial conduction system. There may also be a cell-mediated immune attack, as revealed by the accumulation of CD4 T lymphocytes in valvular tissues. There appears to be a positive correlation between the development of rheumatic fever and HRA titers.

Lymphomatoid granulomatosis is vasculitis in the lung of unknown etiology with an ominous prognosis. Atypical lymphocytes and plasma cells extensively infiltrate the pulmonary vasculature. Many of these lymphocytes are undergoing mitosis. The lungs may develop cavities, and occasionally, the nervous system, skin, and kidneys may be sites of nodular vasculitis.

MLNS (mucocutaneous lymph node syndrome): See Kawasaki's disease.

Hypersensitivity angiitis is small vessel inflammation most frequently induced by drugs.

Hypersensitivity vasculitis is an allergic response to drugs, microbial antigens, or antigens from other sources, leading to an inflammatory reaction involving small arterioles, venules, and capillaries.

DIGESTIVE SYSTEM

Crohn's disease is a condition usually expressed as ileocolitis, but it can affect any segment of the gastrointestinal tract. Crohn's disease is associated with transmural granulomatous inflammation of the bowel wall characterized by lymphocyte, plasma cell, and eosinophil infiltration. Goblet cells and gland architecture are not usually affected. Granuloma formation is classically seen in Crohn's disease, appearing in 70% of patients. The etiology is unknown. *Mycobacterium paratuberculosis* has been found in a few patients with Crohn's disease, although no causal relationship has been established. An immune effector mechanism is believed to be responsible for maintaining chronic disease in these patients. Their serum immunoglobulins and peripheral blood lymphocyte counts are usually normal except for a few diminished T cell counts in selected Crohn's disease patients. Helper/suppressor ratios are also normal. Active disease has been associated with reduced suppressor T cell activity, which returns to normal during remission. Patients have complexes in their blood that are relatively small and contain IgG, although no antigen has been identified. The complexes may be merely aggregates of IgG. Complexes in Crohn's disease patients are associated with involvement of the colon and are seen less often in those with the disease confined to the ileum. During active disease, serum concentrations of C3, factor B, C1 inhibitor, and C3b inactivator are elevated but return to normal during remission. Patients with long-standing disease often develop high titers of immunoconglutinins which are antibodies to activated C3. High-titer antibodies against bacterial antigens such as those of *Escherichia coli* and *Bacteroides* cross-react with colonic goblet cell lipopolysaccharides. Patients' peripheral blood lymphocytes can kill colonic epithelial cells *in vitro*. Colonic mucosa lymphocytes in these patients are also cytotoxic for colonic epithelial cells.

Gluten-sensitive enteropathy (celiac sprue, nontropical sprue) is a disease that results from defective gastrointestinal absorption due to hypersensitivity to cereal grain storage proteins, including gluten or its product gliadin, present in wheat, barley, and oats. There is diarrhea, weight loss, and steatorrhea. It is characterized by villous atrophy and malabsorption in the small intestine. It occurs mostly in Caucasians and occasionally in African Americans, but not in Asians, and is associated with HLA-DR3 and DR7. The disease may be limited to the intestines or associated with dermatitis herpetiformis, a vesicular skin eruption. Antigliadin antibodies which are IgA are formed, and lymphocytes and plasma cells appear in the lamina propria in association with villous atrophy. Diagnosis is made by showing villous atrophy in a small intestinal biopsy. Administering a gluten-free diet leads to resolution of the disease.

Colon antibodies are IgG antibodies in the blood sera of 71% of ulcerative colitis patients and may be shown by flow cytometry to react with rat colon epithelial cells. Antibodies reactive with a 40-kDa constituent of normal colon extracts have been found in the blood sera of 79% of ulcerative colitis patients. Antineutrophil cytoplasmic antibodies, distinct from the P-ANCA of systemic vasculitis and the C-ANCA of Wegener's granulomatosis, are detectable in 70% of ulcerative colitis patients.

Colon autoantibodies: 71% of ulcerative colitis (UC) patients have IgG autoantibodies reactive with rat epithelial cells. A colon autoantibody is a monoclonal antibody against colonic epithelium crossreactive with Barrett epithelium adenocarcinoma and Barrett esophagus epithelium. Other studies have revealed shared phenotypic expression of colonic, biliary, and Barrett epithelium. 70% of the colitis patients manifest antibodies reactive with neutrophil cytoplasm distinct from c-ANCA and p-ANCA. Of the IgG subclasses of antibody against colonic antigens, IgG 4 and IgG1 have shown the greatest reactivity in separate studies and have differentiated ulcerative colitis patients from Crohn's disease.

Antigliadin antibodies (AGA) are specific for gliadin, a protein present in wheat and rye grain gluten. Antigliadins are required for the development of celiac disease with associated jejunal mucosal flattening in genetically prone subjects. Thus, antibodies against gliadin may be used for population screening for gluten-sensitive enteropathies such as celiac disease and dermatitis herpetiformis. Environmental agents, such as an adenovirus in the intestine, may induce an aberrant immune response to gluten in genetically susceptible subjects such as in HLA-DR3-DQw2 and HLA-B8 individuals.

Celiac sprue (gluten-sensitive enteropathy) results from hypersensitivity to cereal grain storage proteins, including gluten or its product gliadin, present in oats, wheat, and barley. It is characterized by villous atrophy and malabsorption in the small intestine. It occurs mostly in Caucausians and occasionally in African Americans, but not in Asians. Individual patients may have the disease limited to the intestines or associated with dermatitis herpetiformis, a vesicular eruption of the skin. The muscosa of the small intestine shows the greatest reactivity in areas in contact with gluten-containing food. Antigliadin antibodies are formed, and lymphocytes and plasma cells appear in the lamina propria in association with villous atrophy. Gluten-sensitive enteropathy is associated with HLA-DR3, -DR7, and -DQ2, as well as with HLA-B8. Diagnosis is made by showing villous atrophy in a biopsy of the small intestine. Administering a gluten-free diet leads to resolution of the disease. Antigliadin antibodies are of the IgA class. Both T and B cell limbs of the immune response participate in the pathogenesis of this disease. Patients may also develop IgA or IgG reticulin antibodies. An α gliadin amino acid sequence shares homology with adenovirus 12Elb early protein. Celiac disease patients develop immune reactivity for both gliadin and this viral peptide sequence. The disease occurs more frequently in individuals exposed earlier to adenovirus 12. Patients develop weight loss, diarrhea, anemia, petechiae, edema, and dermatitis herpetiformis, among other signs and symptoms. Lymphomas such as immunoblastic lymphoma may develop in 10 to 15% of untreated patients.

Celiac disease: See gluten-sensitive enteropathy.

Whipple's disease is a disorder characterized immunologically by massive infiltration of the lamina propria with PAS positive macrophages. There are secondary T cell abnormalities. This is a rare infectious disease produced by one or more microorganisms, which remain unidentified. Diagnosis is established by intestinal biopsy, showing microorganisms and numerous macrophages containing microbial cell wall debris in their cytoplasm. The cellular infiltration is associated with "clubbed" villi, lymphatic obstruction, malabsorption, and protein-losing enteropathy. Loss of lymphocytes into the gastrointestinal tract as a result of lymphatic obstruction may be associated with the development of lymphopenia and secondary T cell immune deficiency.

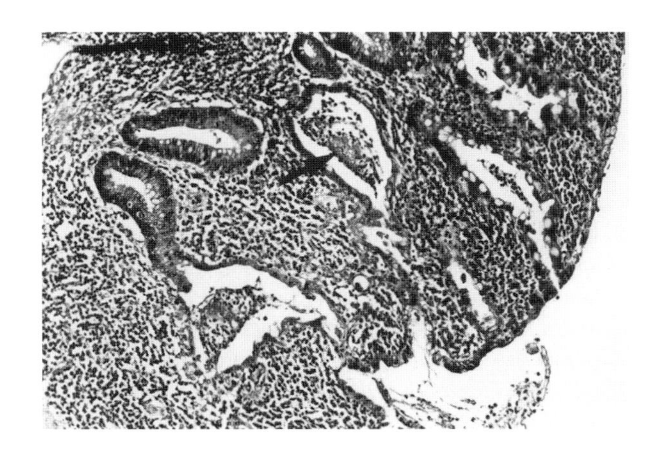

FIGURE 17.53 Ulcerative colitis. Crypt abscess.

Ulcerative colitis (immunologic colitis) (Figure 17.53) is an ulcerative condition that may involve the entire colon but does not significantly affect the small intestine. There is neutrophil, plasma cell, and eosinophil infiltration of the colonic mucosa. This is followed by ulceration of the surface epithelium, loss of globlet cells, and formation of crypt abscess. The etiology is unknown. An immune effector mechanism is believed to maintain chronic disease in these patients. Their serum immunoglobulins and peripheral blood lymphocytes count are usually normal. Complexes present in the blood are relatively small and contain IgG, although no antigen has been identified. The complexes could be merely aggregates of IgG. These patients have diarrhea with blood and mucus in the stool. The signs and symptoms are intermittent, and there is variation in the severity of colon lesions. The patient's lymphocytes are cytotoxic for colon epithelial cells. Antibodies against *Escherichia coli* may cross-react with colonic epithelium in these patients. However, whether or not such antibodies have a role in etiology and pathogenesis remains to be proven.

Inflammatory bowel disease (IBD) is a general term that applies to ulcerative colitis and Crohn's disease, as well as to idiopathic inflammatory bowel disease that resembles the other two. There is a hereditary predisposition to IBD. Intestinal epithelial cells express HLA-DR (MHC class II) antigens in Crohn's disease, ulcerative colitis, and infectious colitis patients that might render them capable of becoming autoantigen-presenting cells. Inflammatory bowel disease patients may become sensitized to cow's milk antigens, developing IgG and IgM antibodies against this protein. Leukotrienes have been shown to be of greater significance than prostaglandins in mediating inflammation in ulcerative colitis. Mucocutaneous conditions such as oral ulcers, epidermolysis bullosa acquisita, erythema nodosa, etc., as well as eye diseases such as uveitis and iridocyclitis, may be associated with IBD. Some IBD patients may also have cirrhosis, chronic active hepatitis, or joint involvement such as ankylosing spondylitis. IBD patients may have abdominal pain, fever, and diarrhea. Granulomas may develop in the gut wall, and lymphocytes that stain for IgA in the cytoplasm are often abundant. Autoantibodies reactive with fetal colon antigens may be present. IBD develops in gene knockout mice lacking IL-2, IL-10, or the TCRα chain. Inflammation is mediated in part through the action of cytokines that include IL-1α, IL-1β, IL-6, or tumor necrosis factor α (TNF-α). IL-1α and IL-1β induce fever, stimulate acute phase protein synthesis, and activate the immune system. TNF-α and IL-6 enhance cellular catabolism, induce acute phase proteins, and stimulate pyrogenic activity. TNF-α can induce IL-1 and IL-6. TNF-α and IL-2 stimulate prostaglandin I_2 ($PGIL_2$), PGE_2, and platelet-activating factor (PAF) secretion.

Immunologic colitis is an ulcerative condition that may involve the entire colon but does not significantly affect the small intestine. There is neutrophil and plasma cell eosinophil infiltration of the colonic mucosa. This is followed by ulceration of the surface epithelium, loss of goblet cells, and formation of crypt abscess. The etiology is unknown. An immune effector mechanism is believed to maintain chronic disease in these patients. Their serum immunoglobulins and peripheral blood lymphocyte counts usually are normal. Complexes present in the blood are relatively small and contain IgG, although no antigen has been identified. The complexes could be merely aggregates of IgG. These patients have diarrhea with blood and mucus in the stool. The signs and symptoms are intermittent, and there is variation in the severity of colon lesions. The patient's lymphocytes are cytotoxic for colon epithelial cells. Antibodies against *Escherichia coli* may cross-react with colonic epithelium in these patients. However, whether or not such antibodies have a role in etiology and pathogenesis remain to be proven.

Erythema nodosum is characterized by slightly elevated erythematous nodules that develop on the shins, and sometimes the forearm and head, and are quite painful. These nodular lesions represent subcutaneous vasculitis involving small arteries. The phenomenon is associated with infection and is produced by antigen–antibody complexes. Erythema nodosum may be an indicator of inflammatory bowel disease, histoplasmosis, tuberculosis, sarcoidosis, or leprosy. It can follow the use of certain drugs. Although claimed in the past to be due to antigen–antibody deposits in the walls of small venules, the immunologic mechanism may involve type IV (delayed-type) hypersensitivity in the small venules. Neutrophils, macrophages, and lymphocytes infiltrate the subcutaneous fat.

Regional enteritis is inflammatory bowel disease characterized by chronic segmental lymphocytic and granulomatous inflammation of the gastrointestinal tract.

IBD is an abbreviation for inflammatory bowel disease.

Sprue: See gluten-sensitive enteropathy.

Pernicious anemia (PA) (Figure 17.54 and Figure 17.55) is an autoimmune disease characterized by the development of atrophic gastritis, achlorhydria, decreased synthesis of intrinsic factor, and mild absorption of B_{12}. Patients present with megaloblastic anemia caused by the B_{12} deficiency that develops. The majority of pernicious anemia patients develop antiparietal cell antibodies and at least half of them also develop antibodies against intrinsic factor, which is necessary for the absorption of B_{12}. The antiparietal cell antibodies are against a microsomal antigen found in gastric parietal cells. Intrinsic factor is a 60-kDa substance that links to vitamin B_{12} and aids its uptake

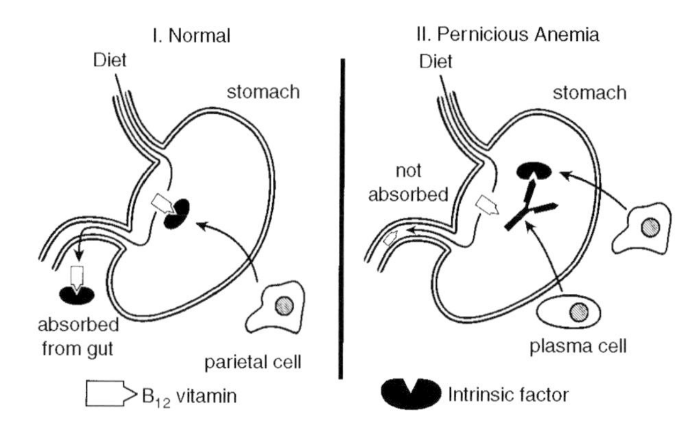

FIGURE 17.54 Pernicious anemia.

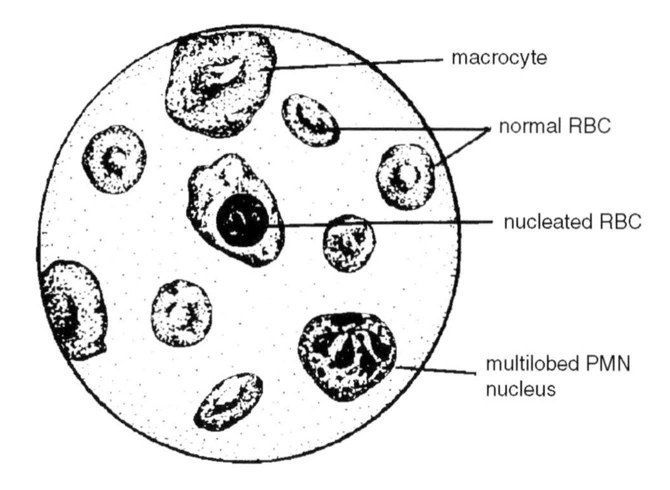

FIGURE 17.55 Macrocytic anemia.

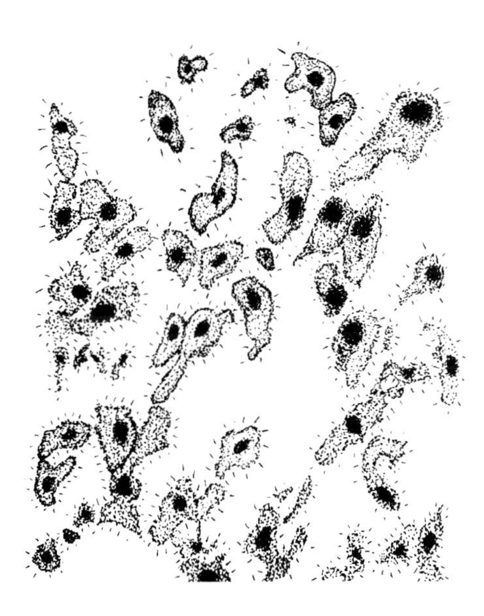

FIGURE 17.56 Pernicious anemia. Parietal cell antibody (FITC-labeled).

in the small intestine. PA may be a complication of common variable immunodeficiency or may be associated with autoimmune thyroiditis. Pernicious anemia is caused principally by injury to the stomach mediated by T lymphocytes. Patients may manifest megaloblastic anemia, deficiency of vitamin B_{12}, and increased gastrin in serum.

Parietal cell antibodies (Figure 17.56) are antibodies present in 50 to 100% of pernicious anemia (PA) patients. They are also found in 2% of normal individuals. Their frequency increases with aging and in subjects with insulin-dependent diabetes mellitus (IDDM). The frequency of parietal cell antibodies diminishes with disease duration in pernicious anemia. Parietal cell autoantibodies in pernicious anemia and in autoimmune gastritis recognize the α and β-subunits of the gastric protein proton pump (H^+/K^+ATPase) which are the principal target antigens. Parietal cell antibodies react with α and β subunits of the gastric proton pump and inhibit the gastric mucosa's acid-producing H^+/K^+ adenosine triphosphatase. Parietal cell antibodies relate to type A gastritis in which there is fundal mucosal atrophy, achlorhydria, development of pernicious anemia, and autoimmune endocrine disease.

LIVER

Chronic active hepatitis (autoimmune) (Figure 17.57) is a disease that occurs in young females who may develop fever, arthralgias, and skin rashes. They may be of the HLAB8, -DR3 haplotype and suffer other autoimmune disorders. Most develop antibodies to smooth muscle, principally against actin, and autoantibodies to liver membranes. They also have other organ- and nonorgan-specific autoantibodies. A polyclonal hypergammaglobulinemia

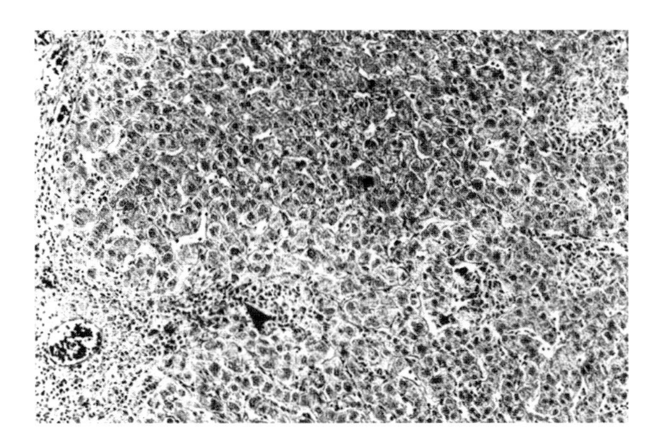

FIGURE 17.57 Chronic active hepatitis piecemeal necrosis.

may be present. Lymphocytes infiltrating portal areas destroy hepatocytes. Injury to liver cells produced by these infiltrating lymphocytes produces piecemeal necrosis. The inflammation and necrosis are followed by fibrosis and cirrhosis. The T cells infiltrating the liver are $CD4^+$. Plasma cells are also present and immunoglobulins may be deposited on hepatocytes. The autoantibodies against liver cells do not play a pathogenetic role in liver injury. There are no serologic findings that are diagnostic. Corticosteroids are useful in treatment. The immunopathogenesis of autoimmune chronic active hepatitis involves antibody, K cell cytotoxicity, and T cell reactivity against liver membrane antigens. Antibodies and specific T suppressor cells reactive with LSP are found in chronic active hepatitis patients, all of whom develop T cell sensitization against asialo-glycoprotein (AGR) antigen. Chronic active hepatitis has a familial predisposition.

Granulomatous hepatitis is granulomatous inflammation of the liver.

Piecemeal necrosis is death of individual liver cells which are encircled by lymphocytes in chronic active hepatitis.

Silicosis: The inhalation of silica particles over a prolonged period of time produces a chronic, nodular, and densely fibrosing pneumoconiosis that has an insidious onset and progresses even in the absence of continued exposure to silica dust. Lymphocytes and alveolar macrophages are quickly attracted to the particles that are phagocytized by the macrophages. Some macrophages remain in the interstitial tissue or in the pulmonary lymphatic channels. Interaction of macrophages and silica particles leads over time to collagenous fibrosis and fibrosing nodules. The silica dust–macrophage interaction may cause the secretion of IL-1 which recruits T helper cells that produce IL-2 which induces proliferation of T lymphocytes. They in turn produce a variety of lymphokines. Thus, immunocompetent cells mediate the collagenous reaction. Activated T cells interact with B lymphocytes, which synthesize increased amounts of IgG, IgM, rheumatoid factor, antinuclear antibodies, and circulating immune complexes. Collagenous silicotic nodules may coalesce, producing fibrous scars. The disease may be complicated by the development of RA, pulmonary tuberculosis, emphysema, or several other diseases, but there is no increased incidence of lung cancer in silicosis.

Berylliosis is a disease induced by the inhalation of beryllium in dust. Subjects may develop a delayed-type hypersensitivity (type IV) reaction to beryllium–macromolecular complexes. Either an acute chemical pneumonia or chronic pulmonary granulomatous disease that resembles sarcoidosis may develop. Granulomas may form and lead to pulmonary fibrosis. The skin, lymph nodes, or other anatomical sites may be affected. Subjects who develop the chronic progressive granulomatous pulmonary disease appear to be sensitized by beryllium.

Caplin's syndrome is characterized by rheumatoid nodules in the lungs of subjects such as hard coal miners in contact with silica.

Cellular interstitial pneumonia is inflammation of the lung in which the alveolar walls are infiltrated by mononuclear cells.

Smooth muscle antibodies are autoantibodies belonging to the IgM or IgG class that are found in the blood sera of 60% of chronic active hepatitis patients. 30% of biliary cirrhosis patients may also be positive for these antibodies. Low titers of smooch muscle antibodies may be found in certain viral infections of the liver.

Primary biliary cirrhosis (PBC) is a chronic liver disease of unknown cause that affects middle-aged women in 90% of the cases. There is chronic intrahepatic cholestasis caused by chronic inflammation and necrosis of intrahepatic bile ducts with progression to biliary cirrhosis. It is believed to be an autoimmune disease based on its association with autoimmune conditions and the presence of autoantibodies. Patients develop pruritis, fatigue, steatorrhea, renal tubular acidosis, hepatic osteodystrophy, and increased incidence of hepatocellular carcinoma and breast carcinoma. Four fifths of the patients also have a connective tissue or autoimmune disease such as RA, autoimmune thyroiditis, scleroderma, and Sjögren's syndrome. Most patients manifest a high titer of antimitochondrial antibodies. There is elevated IgM in the serum. Lymphocytic infiltration occurs together with intrahepatic bile duct destruction. Alkaline phosphatase is greatly increased, in addition to the elevation of IgM and antimitochondrial antibodies. The M2 antimitochondrial antibody is most frequently associated with PBC. In addition to hepatosplenomegaly and skin hyperpigmentation, patients may develop severe jaundice, petechiae, and purpura as the disease progresses. Liver transplantation is the only treatment for end-stage disease.

PBC is an abbreviation for primary biliary cirrhosis.

Hepatitis immunopathology panel is a profile of assays that are very useful to establish the clinical and immune status of a patient believed to have hepatitis. The panel for acute hepatitis may include hepatitis B surface antigen (HbsAg), antibody to hepatitis B core antigen (anti-HBc), antihepatitis B surface antigen (anti-Abs), antihepatitis A (IgM), anti-Hbe, and antihepatitis C. The panel for chronic hepatitis (carrier) includes all of these except for the antihepatitis A test.

Soluble liver antigen antibodies are autoantibodies that react with liver but not kidney as revealed by immunofluorescence. They are found in a subgroup of chronic autoimmune hepatitis (AIH) patients with negative antinuclear antibodies and negative liver–kidney microsome-1 (LKM-1) autoantibodies, and a low prevalence of autoantibodies against smooth muscle mitochondria and liver membrane antigens. These autoantibodies are believed to be identical to LP2 autoantibodies that are present in 45% of cryptogenic chronic active hepatitis patients.

Hepatitis B (Figure 17.58) is a DNA virus that is relatively small and has four open reading frames. The S gene codes for HbsAg. The P gene codes for a DNA polymerase. There is an X gene and core gene that code for HbcAg and the precore area that codes for Hbe-Ag.

Serum hepatitis (hepatitis B) is an infection with a relatively long incubation period lasting between 2 and 5 months caused by a double-stranded DNA virus. Transmission may be through the administration of serum or blood products from one person to another. Additional persons at high risk include drug addicts, dialysis patients,

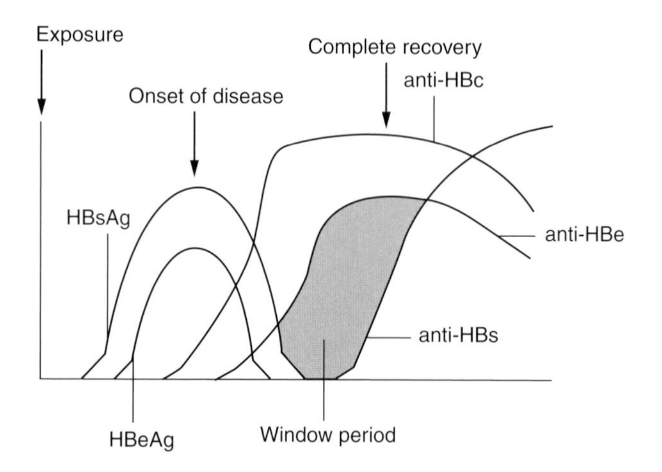

FIGURE 17.58 Hepatitis B.

health care workers, male homosexuals, and newborns with HIV-infected mothers. Chronic infection may occur in immunosuppressed individuals or those with lymphoid cancer. The infection may be acute or chronic, and hepatocellular carcinoma may be a complication. In acute HBV infection, the hepatocytes may undergo ballooning and eosinophilic degeneration. There may be focal necrosis of hepatocytes and lymphocytic infiltration of the portal areas and liver parenchyma. There may be central and mid-zonal necrosis of hepatocytes. Chronic type B hepatitis may progress to cirrhosis and hepatocellular carcinoma. Hepatocellular injury is induced by immunity to the virus. HBsAg, HBeAg, and IgM anti-HBc are present in the serum in acute hepatitis. In the convalescent phase of acute hepatitis, IgG anti-HBs appears in the serum. HBsAg, HBeAg, IgG anti-HBc in high titer, HBV DNA, and DNA polymerase appear in the serum in the active viral replication phase of chronic hepatitis. In the viral integration phase of chronic hepatitis, HBsAg, anti-HBc, and anti-HBe appear in the serum. Acute hepatitis B patients often recover completely. See hepatitis B.

Delta agent (hepatitis D virus [HDV]) is a viral etiologic agent of hepatitis that is a circular single-stranded incomplete RNA virus without an envelope. It is a 1.7-kb virus and consists of a small, highly conserved domain and a larger domain manifesting epitopes. HDV is a subviral satellite of the hepatitis B virus (HBV), on which it depends to fit its genome into viral particles. Thus the patient must first be infected with HBV to have HDV. Individuals with the delta agent in their blood are positive for HbsAg, anti-HBC, and often Hbe. This agent is frequently present in IV drug abusers and may appear in hemophiliacs and AIDS patients.

Hepatitis non-A, non-B (C) (NAN BH) is the principal cause of hepatitis that is transfusion-related. Risk factors include intravenous drug abuse (42%), unknown risk factors (40%), sexual contact (6%), blood transfusion (6%), household contact (3%), and health professionals (2%). There are 150,000 cases per year in the U.S. Of these, 30 to 50% become chronic carriers and one fifth develop cirrhosis. Parenteral NANBH is usually hepatitis C, and enteric NANBH is usually hepatitis E.

Non-A non-B hepatitis: See hepatitis non-A, non-B.

Hepatitis serology refers to hepatitis B antigens and antibodies against them. Core antigen is designated HBc. The HBc particle is comprised of double-stranded DNA and DNA polymerase. It has an association with Hbe antigens. Core antigen signifies persistence of replicating hepatitis B virus. Anti-HBc antibody is a serologic indicator of hepatitis B. It is an IgM antibody that increases early and is still detectable 20 years postinfection. The IgM anti-HBc antibody assay is the one best antibody assay for acute hepatitis B. The Hbe antigen (Hbe) follows the same pattern as HbsAg antigen. When found, it signifies a carrier state. The anti-Hbe increases as Hbe decreases. It appears in patients who are recovering and may last for years after the hepatitis has been resolved. The first antigen that is detectable following hepatitis B infection is surface antigen (HBs). It is detectable a few weeks before clinical disease and is highest with the first appearance of symptoms. This antigen disappears 6 months from infection. Antibody to HBs increases as the HbsAg levels diminish. Anti-HBs often is detectable for the lifetime of the individual.

PANCREAS

Insulin-dependent diabetes mellitus (IDDM), type 1 (Figure 17.59 and Figure 17.60) is a disease in which autoantibodies against islet cells (and insulin) may be identified. Among the three to six genes governing susceptibility to type I diabetes are those encoding the MHC. Understanding human diabetes has been greatly facilitated by both immunological and genetic studies in experimental animal models including nonobese diabetic mice (NOD mice) and biobreeding rats (BB rats). Human type I diabetes mellitus results from autoimmune injury of pancreatic β

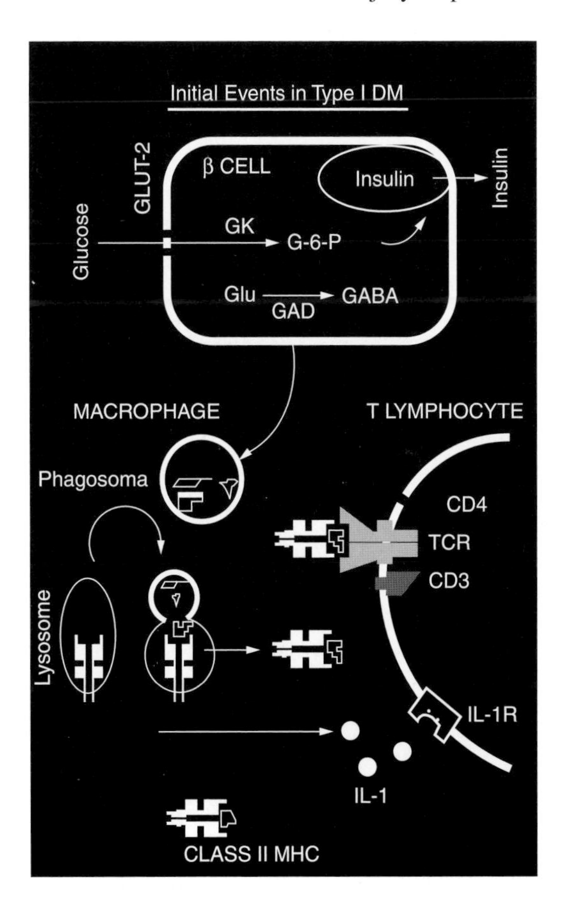

FIGURE 17.59 Diabetes mellitus, type I.

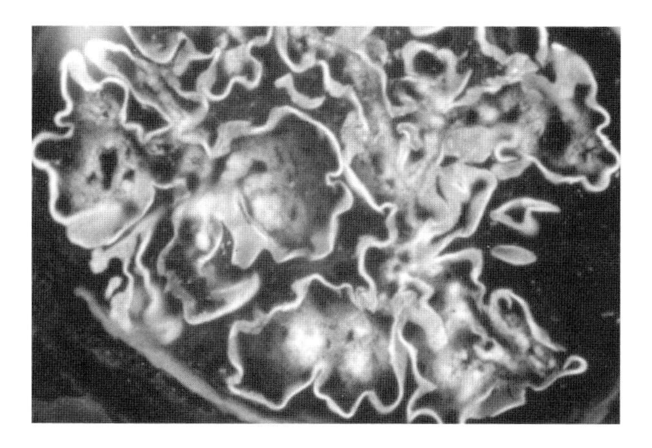

FIGURE 17.60 Diabetic glomerulopathy. Discontinuous linear fluorescence — anti-IgG.

cells. Specific autoantibodies signal pancreatic β cell destruction. The autoantibodies are against islet cell cytoplasmic or surface antigens or insulin. Antiidiotypic antibodies may also develop against antiinsulin antibodies, possibly leading to antibody blockade of insulin receptors and thereby inducing insulin resistance and β cell exhaustion. Autoantibodies have also been demonstrated against a 64-kDa third islet cell antigen which could represent a primary target of autoimmune reactivity in type I diabetes and have been found in the sera of diabetics before clinical onset of the disease. HLA typing is also useful. DNA sequence analysis has revealed that alleles of HLA-DQ β chain govern diabetes susceptibility and resistance. The amino acid at position 57 has a critical role in disease susceptibility and resistance. Although pancreatic β cells fail to express MHC class II antigens under normal circumstances, they become Ia MHC class II antigen positive following stimulation by INF-α and TNF or lymphotoxin. Class II positive pancreatic β cells may present islet cell autoantigens to T lymphocytes inaugurating an immune response.

Patients at risk for diabetes or prediabetes might benefit from immunosuppressive therapy such as cyclosporine, although most are still treated with insulin. Transplantation of pancreatic islet cells remains a bright possibility in future treatment strategies.

Diabetes mellitus, insulin dependent (type I): In type I (autoimmune) diabetes mellitus, autoantibodies against islet cells (and insulin) may be identified. Among the three to six genes governing susceptibility to type I diabetes are those encoding the MHC. Understanding human diabetes has been greatly facilitated by both immunologic and genetic studies in experimental animal models including nonobese diabetic mice (NOD mice) and biobreeding rats (BB rats).

Human type I diabetes mellitus results from autoimmune injury of pancreatic β cells. Specific autoantibodies signal pancreatic β cell destruction. The autoantibodies are against islet cell cytoplasmic or surface antigens or insulin. Antiidiotypic antibodies may also develop against antiinsulin antibodies, possibly leading to antibody blockade of insulin receptors and thereby inducing insulin resistance and β cell exhaustion. Autoantibodies have also been demonstrated against a 64-kDa third islet cell antigen which could represent a primary target of autoimmune reactivity in type I diabetes and have been found in the sera of the diabetic before clinical onset of the disease. HLA typing is also useful. DNA sequence analysis has revealed that alleles of HLA-DQ β chain govern diabetes susceptibility and resistance. The amino acid at position 57 has a critical role in disease susceptibility and resistance. Although pancreatic β cells fail to express MHC class II antigens under normal circumstances, they become Ia MHC class II antigen positive following stimulation by IFN-γ and TNF or lymphotoxin. Class II positive pancreatic β cells may present islet cell autoantigens to T lymphocytes inaugurating an autoimmune response. Patients at risk for diabetes or prediabetes might benefit from immunosuppressive therapy, such as by cyclosporine, although most are still treated with insulin. Transplantation of pancreatic islet cells remains a bright possibility in future treatment strategies.

Acanthosis nigricans (Figure 17.61) is a condition in which the afflicted subject develops insulin receptor autoantibodies associated with insulin-resistant diabetes mellitus, as well as thickened and pigmented skin.

Immune complex disease (ICD) (Figure 17.62 to Figure 17.64): As described under type III hypersensitivity reaction, antigen–antibody complexes have a fate that depends in part on the size. The larger insoluble immune complexes are removed by the mononuclear phagocyte system. The smaller immune complexes become lodged in the microvasculature such as the renal glomeruli. They may activate the complement system and attract polymorphonuclear

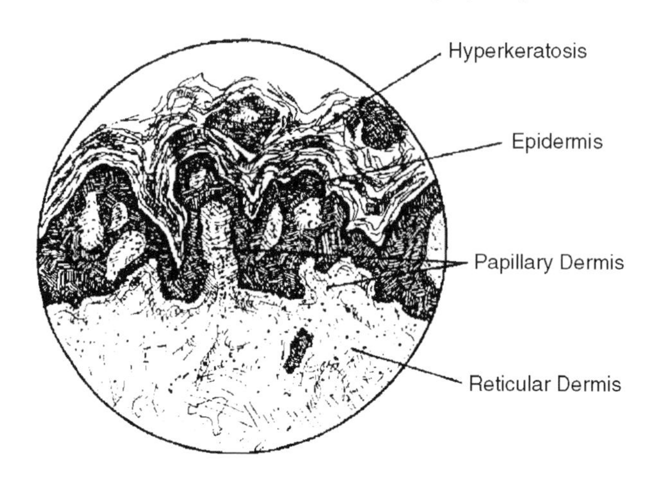

FIGURE 17.61 Acanthosis nigricans.

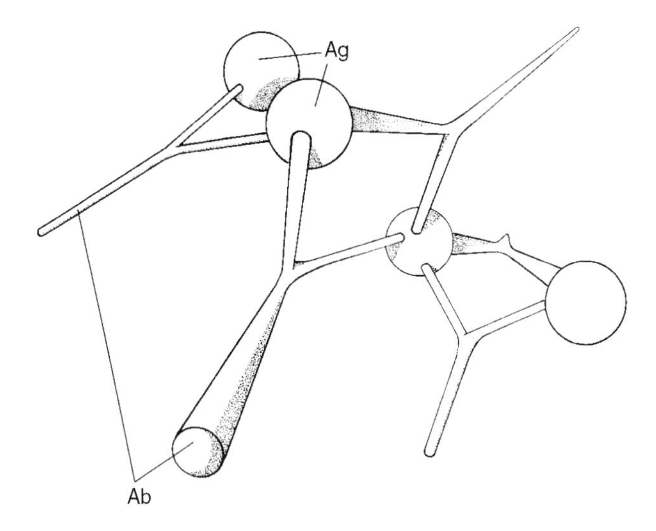

FIGURE 17.62 Immune complex.

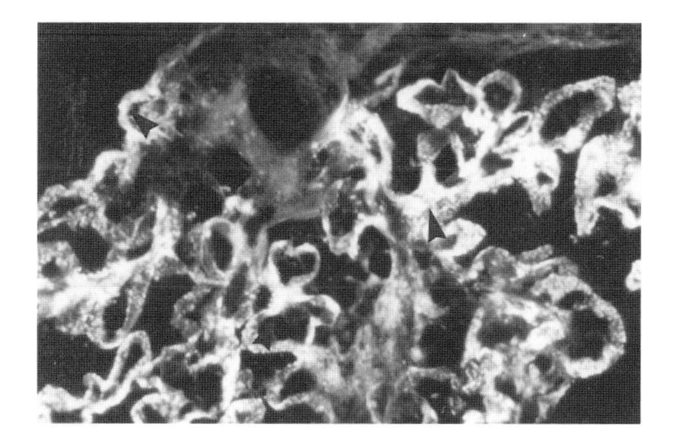

FIGURE 17.63 Immune complexes in renal glomerulus.

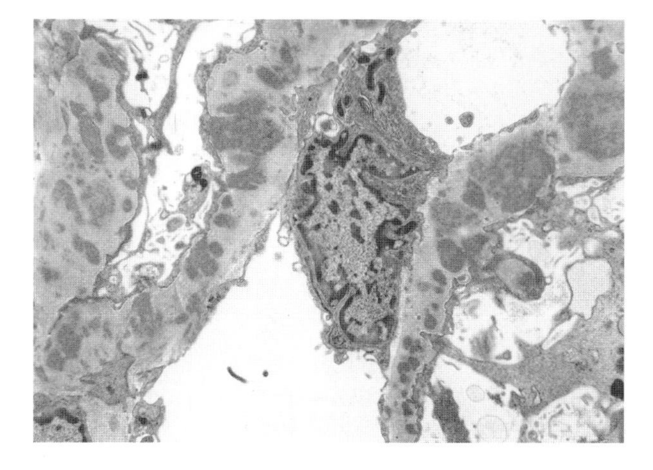

FIGURE 17.64 Electron dense deposits. Immune complex disease (ICD).

neutrophils, initiating an inflammatory reaction. The antigen of an immune complex may be from microorganisms such as streptococci leading to subepithelial deposits in renal glomeruli in poststreptococcal glomerulonephritis, or they may be endogenous such as DNA or nuclear antigens in SLE. Diphtheria antitoxin prepared in horses induced serum sickness when the foreign horse serum proteins stimulated antibodies in human recipients. Immune complex disease is characterized clinically by fever, joint pain, lymphadenopathy, eosinophilia, hypocomplementemia, proteinuria, purpura, and urticaria, among other features. Laboratory techniques to detect immune complexes include the solid-phase Clq assay, the Clq-binding assay, the Raji cell technique, and the staphylococcal protein assay, among other methods. Most autoimmune diseases have a type III (antigen–antibody complex)-mediated mechanism. Connective tissue diseases such as SLE, polyarteritis nodosa, progressive systemic sclerosis, dermatomyositis, RA, and others fall within this category. Viral infections such as hepatitis B, cytomegalovirus, infectious mononucleosis, and Dengue, as well as neoplasia such as carcinomas and melanomas, may be associated with immune complex formation. See type III hypersensitivity reaction.

Immune complex pneumonitis is a type III, immune complex Arthus-type reaction in the pulmonary alveoli.

Glomerulonephritis (GN) is a group of diseases characterized by glomerular injury. Immune mechanisms are responsible for most cases of primary glomerulonephritis and many of the secondary glomerulonephritis group. Over 70% of glomerulonephritis patients have glomerular deposits of immunoglobulins, frequently with complement components. Antibody-associated injury may result from the deposition of soluble circulating antigen–antibody complexes in the glomerulus or by antibodies reacting in the glomerulus either with antigens intrinsic to the glomerulus or with molecules planted within the glomerulus. Cytotoxic antibodies may also cause glomerular injury. Goodpasture's syndrome is an example of a disease in which antibodies react directly with the glomerular basement membrane, interrupting its integrity and permitting red blood cells to pass into the urine. Antigen–antibody complexes such as those produced in SLE may be deposited in the walls of the peripheral capillary loops, especially in a subendothelial location, leading to various manifestations of glomerulonephritis depending on the stage of the disease. In membranoproliferative glomerulonephritis type I, IgG and C3 are found deposited in the glomerulus, whereas in dense deposit disease (membranoproliferative glomerulonephritis type II), C3 alone is demonstrable in the dense deposit within the capillary wall. The types of primary glomerulonephritis (GN) include acute diffuse proliferative GN, rapidly progressive (crescentic) GN, membranous GN, lipoid nephrosis, focal segmental glomerulosclerosis, membranoproliferative GN, IgA nephropathy, and chronic GN. Secondary diseases affecting the glomeruli include SLE, diabetes mellitus, amyloidosis, Goodpasture's syndrome, polyarteritis nodosa, Wegener's granulomatosis, Henoch-Schönlein purpura, and bacterial endocarditis.

Membranous glomerulonephritis (Figure 17.65 to Figure 17.67) is a disease induced by deposition of electron-dense, immune (Ag-Ab) deposits in the glomerular basement membrane in a subepithelial location. This leads to progressive thickening of glomerular membranes. Most cases are idiopathic, but membranous glomerulonephritis may follow development of other diseases such as SLE; lung or colon carcinoma; and exposure to gold, mercury, penicillamine, or captopril. It can also be a sequela of certain infections, e.g., hepatitis B; or metabolic disorders, e.g., diabetes mellitus. Clinically, it is a principal cause of nephrotic syndrome in adults. The subepithelial immune deposits are shown by immunofluorescence to contain both immunoglobulins and complement. Proteinuria persists in 70 to 90% of cases, and half of the patients develop renal insufficiency over a period of years. A total of 10 to 30% of cases have a less severe course.

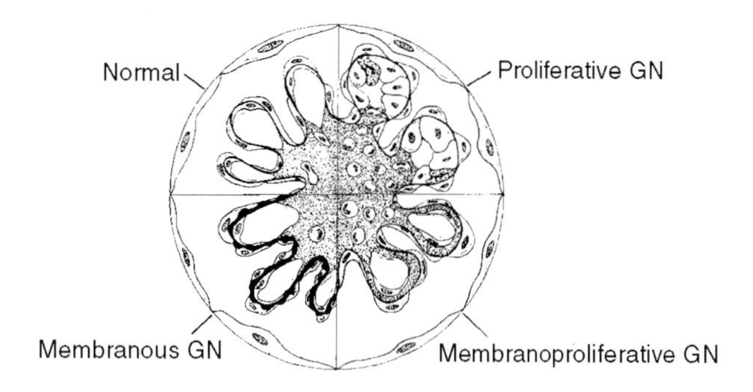

FIGURE 17.65 Types of glomerulonephritis.

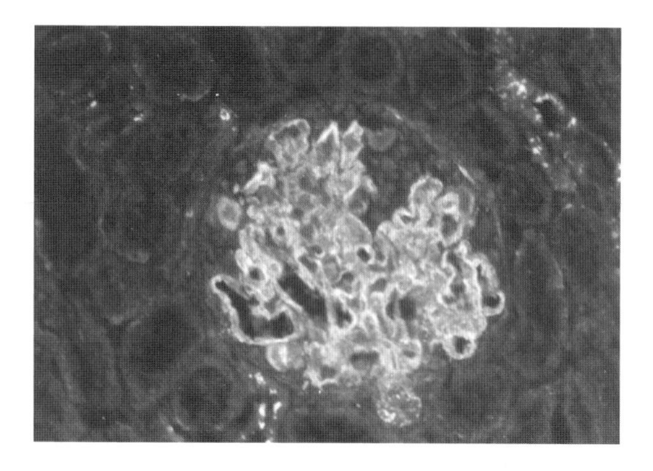

FIGURE 17.66 Glomerulonephritis.

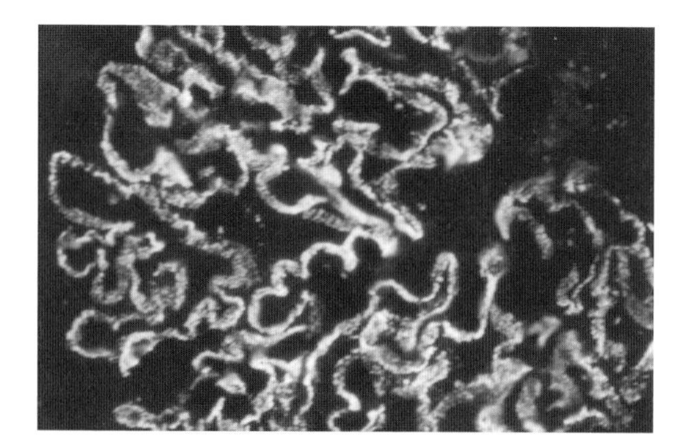

FIGURE 17.67 Membranous glomerulonephritis.

Membranoproliferative glomerulonephritis (MPGN) is a nephropathy in which the glomerular basement membrane is altered and glomerular cells proliferate, especially in mesangial areas, leading to the synonym mesangiocapillary GN. Patients may present with hematuria and/or proteinuria. In type I MPGN, IgG-, Clq-, C4-, and C3-containing subendothelial electron-dense deposits are present in approximately 66% of cases, as revealed by immunofluorescence and electron microscopy. Conventional light microscopy reveals splitting of basement membranes. In type II MPGN, also called dense deposit disease, the glomerular basement membrane's lamina densa appears as an electron-dense ribbon on either side of which C3 can be detected by immunofluorescence. Complement is fixed only by the alternate pathway. Type II patients have C3 nephritic factor (C3NeF) in their serum, which facilitates stabilization of alternate C3 convertase, thereby promoting C3 degradation and hypocomplementemia. Half of these patients develop chronic renal failure over a 10-year period.

Entactin and **nidogen** are glomerular basement membrane antigens that might have a possible role in glomerular basement membrane nephritis. Entactin is a 150-kDa glomerular membrane sulfated glycoprotein that is probably synthesized by endothelial and epithelial cells in the developing kidney. It has a role in cellular adhesion to extracellular material. It is complexed with laminin *in vivo*. Nidogen, also a basement membrane constitutent, is a 150-kDa glycoprotein that has binding affinity for laminin. Entactin and nidogen closely resemble one another and may be identical. Serum IgG, IgM, and/or IgA antibodies specific for entactin/nidogen have been found in over 40% of patients with glomerulonephritis in one study.

Immunotactoid glomerulopathy is a renal malady characterized by glomerular deposits of fibrillar material comprised of 10 to 48.9 nm microfibrils or microtubules as viewed by electron microscopy. These glomerular fibrillar deposits are not birefringent when stained with Congo red and examined by polarizing light microscopy. This differentiates them from amyloid. Usually, no extraglomerular fibrillar deposits are present which differentiate the condition from amyloid or light-chain deposition. It is not associated with concomitant systemic disease such as cryoglobulinemia or SLE. It typically affects middle-aged males who may manifest hypertension, nephrotic range proteinuria, and microscopic hematuria.

Acute poststreptococcal glomerulonephritis is a disease of the kidney with a good prognosis for most patients, which follows streptococcal infection of the skin or streptococcal pharyngitis by 10 d to 3 weeks in children. The subject presents with hematuria, fever, general malaise, facial swelling, and smokey urine. There is mild proteinuria and red blood cells in the urine. These children often have headaches, are hypertensive, and have an elevated blood pressure. Serum C3 levels are decreased. The erythrocyte sedimentation rate (ESR) is elevated, and an abdominal x-ray may reveal enlarged kidneys. Renal biopsy reveals infiltration of glomeruli by polymorphonuclear leukocytes and monocytes and a diffuse proliferative process. Immunofluorescence shows granular deposits of IgG and C3 on the epithelial side of peripheral capillary loops. Electron microscopy shows subepithelial "humps." The process usually resolves spontaneously within a week after onset of renal signs and symptoms, the patient becomes afebrile, and the malaise disappears. It is attributable to nephritogenic streptococci of types 1, 4, 12, and 49.

Interstitial nephritis is inflammation characterized by mononuclear cell (lymphocytic) infiltrate in the interstitium surrounding renal tubules. This occurs following autoantibody reaction with tubular basement membrane in the kidney. Other etiologies include analgesic abuse.

Nephritic syndrome is a clinical complex of acute onset characterized by hematuria with red cell and hemoglobin casts in the urine, oliguria, azotemia, and hypertension. There may be limited proteinuria. Inflammatory reactions within glomeruli cause injury to capillary walls, which permits the release of erythrocytes into the urine. Acute diffuse proliferative glomerulonephritis is an example of a primary glomerular disease associated with acute nephritic syndrome.

Nephrotic syndrome is a clinical complex that consists of massive proteinuria with the loss of greater than 3.5 g of protein per day; generalized edema; hypoalbuminemia, i.e., less than 3 g/dl; hyperlipidemia; and lipiduria. Nephrotic syndrome in children often follows lipoid nephrosis, whereas in adults it is frequently associated with membranous glomerulopathy. Both are primary glomerular diseases.

α1-microglobulin is a 30-kDa protein that belongs to the lipocalin family and possesses hydrophobic prosthetic groups. It is synthesized in the liver and is present in the urine and serum. α_1M may be complexed with monomeric IgA. It may be increased in IgA nephropathy. Elevated serum α_1M in patients with AIDS may signify renal pathology. α_1M blocks antigen stimulation and migration of granulocytes. It has a role in immunoregulation and functions as a mitogen.

α2 macroglobulin (α2M) is a 725-kDa plasma glycoprotein that plays a major role in inhibition of proteolytic activity generated during various extracellular processes. α_2M is synthesized in the liver and reticuloendothelial system. α_2M is produced by lymphocytes and is found associated with the surface membrane of a subpopulation of B cells. It has the unique property of binding all active endopeptidases. Other enzymes or even the inactive forms of proteinases are not bound. Complexes of α_2M with proteinase are rapidly cleared from circulation (minutes), in contrast to the turnover of α_2M which requires several days. Some of the roles of α_2M include (1) regulation of the extracellular proteolytic activity resulting from clotting, fibrinolysis, and proteinases of inflammation, and (2) specific activity against some proteinases of fungal or bacterial origin. It is elevated significantly in nephrotic syndrome. Increased levels have been reported also in atopic dermatitis and ataxia telangiectasia.

Proteinuria is protein in the urine.

Hematuria is either macroscopic or microscopic blood in the urine from any cause, for example from glomerular basement membrane injury or renal stones.

Hypocomplementemic glomerulonephritis is decreased complement in the blood during the course of chronic progressive glomerulonephritis, in which C3 is deposited in the glomerular basement membrane of the kidney.

Masugi nephritis (Figure 17.68) is an experimental model of human antiglomerular basement membrane (anti-GBM) nephritis. The disease is induced by the injection of rabbit anti-rat glomerular basement membrane antibody into rats. The antiserum for passive transfer is raised in rabbits immunized with rat kidney basement membranes. The passively administered antibodies become bound to the glomerular basement membrane, fix complement, and induce glomerular basement membrane injury with increased permeability. Neutrophils and monocytes may infiltrate the area. Masugi nephritis is an experimental model of Goodpasture's syndrome in man. This is also called nephrotoxic nephritis.

Heymann antigen: A 330-kDa glycoprotein (GP330) present on visceral epithelial cell basal surfaces in coated pits as well as on tubular brush borders. It participates in production of experimental nephritis in rats. See also Heymann glomerulonephritis.

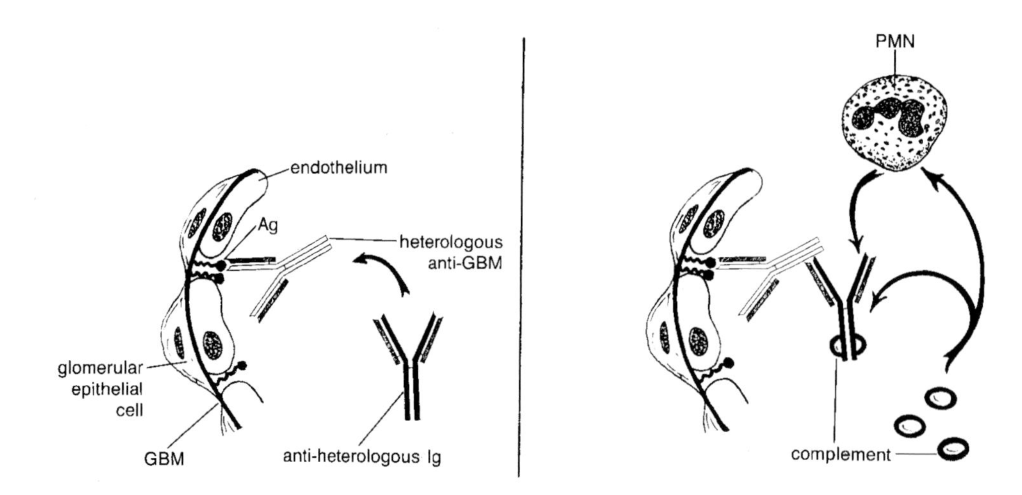

FIGURE 17.68 Masugi nephritis.

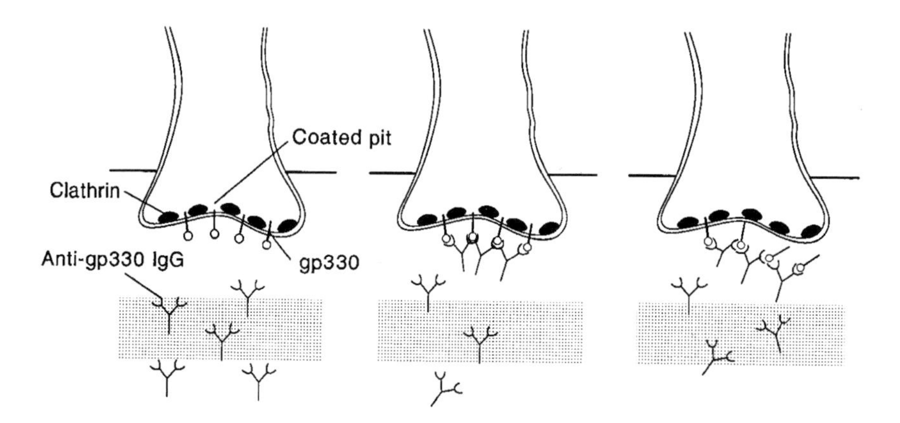

FIGURE 17.69 Heymann glomerulonephritis.

Heymann glomerulonephritis (Figure 17.69) is an experimental model of membranous glomerulonephritis induced by immunizing rats with proximal tubule brush border preparations containing subepithelial antigen or Heymann factor, a 330-kDa protein incorporated in Freund's adjuvant. The rats produce antibodies against brush border antigens, and membranous glomerulonephritis is induced. Autoantibodies combine with shed epithelial cell antigen. The union of antibody with Heymann antigen, distributed in an interrupted manner along visceral epithelial cell surfaces, leads to subepithelial electron-dense, granular deposits. Immunoglobulins and complement are deposited in a granular rather than linear pattern along the glomerular basement membrane, as revealed by immunofluorescence. The glomerulonephritis results from interaction of antibrush border antibody with the 330-kDa glycoprotein that is fixed, but discontinuously distributed on the base of visceral epithelial cells and is crossreactive with brush border antigens. Heymann nephritis closely resembles human membranous glomerulonephritis.

Heymann's nephritis: See Heymann glomerulonephritis.

Poststreptococcal glomerulonephritis is an acute proliferative glomerulonephritis that may follow a streptococcal infection of the throat or skin by 1 to 2 weeks. It is usually seen in 6 to 10 year old children, but it may occur in adults as well. The onset is heralded by evidence of acute nephritis. A total of 90% of patients have been infected with Group A β hemolytic streptococci that are nephritogenic, specifically types 12, 4, and 1 which are revealed by their cell wall M protein. Poststreptococcal glomerulonephritis is mediated by antibodies induced by the streptococcal infection. Most patients show elevated antistreptolysin-O (ASO) titers. Serum complement levels are decreased. Immunofluorescence of renal biopsies demonstrates granular immune deposits that contain immunoglobulin and complement in the glomeruli. This is confirmed by electron microscopy. The precise streptococcal antigen has never been identified; however, a cytoplasmic antigen termed endostreptosin together with some cationic streptococcal antigens are found in glomeruli. Subepithelial immune deposits appear as "humps." They are antigen–antibody complexes that may also appear in the mesangium or occasionally in a subendothelial or intramembranous position. These immune deposits stain positively for IgG and complement by immunofluorescence. Affected children develop fever, nausea, oliguria, and hematuria within 2 weeks following a streptococcal sore throat or skin infection. Erythrocyte casts and mild proteinuria may be identified. There may be periorbital edema and hypertension upon examination. The BUN and ASO titer may also be elevated. More than 95% of children with poststreptococcal glomerulonephritis recover, although a few, less than 1%, develop rapidly progressive glomerulonephritis and a few others develop chronic glomerulonephritis.

Humps (Figure 17.70) are immune deposits containing IgG and C3, as well as the alternate complement pathway components properdin and Factor B, that occur in post infectious glomerulonephritis on the subepithelial side of peripheral capillary basement membranes. They resolve within 4 to 8 weeks of the infection in most individuals. They may also occur in selected other nonstreptococcal postinfectious glomerulonephritides.

IgA nephropathy (Berger's disease) (Figure 17.71 to Figure 17.73) is a type of glomerulonephritis in which prominent IgA-containing immune deposits are present in mesangial areas. Patients usually present with gross or microscopic hematuria and often mild proteinuria. By light microscopy, mesangial widening or proliferation may be observed. However, immunofluorescence microscopy demonstrating IgA and C3, fixed by the alternative pathway, is a requisite for diagnosis. Electron microscopy confirms the presence of electron-dense deposits in mesangial areas. Half of the cases progress to chronic renal failure over a 20-year course.

Berger's disease is a type of glomerulonephritis in which prominent IgA-containing immune deposits are present in mesengial areas. Patients usually present with gross or microscopic hematuria and often mild proteinuria. By light microscopy, mesangial widening or proliferation may be observed. However, immunofluorescence microscopy demonstrating IgA and C3, fixed by the alternative

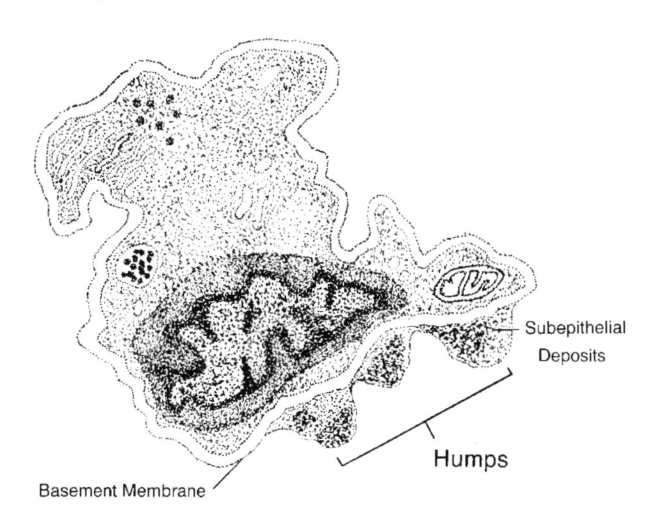

FIGURE 17.70 Hump.

FIGURE 17.71 IgA nephropathy. Granular immune deposits in mesangial areas.

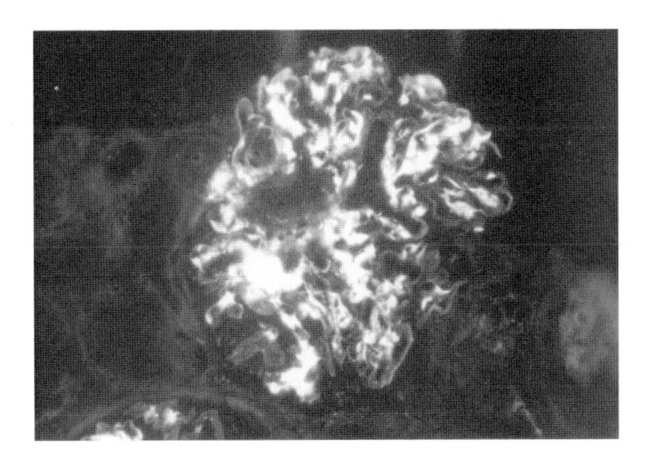

FIGURE 17.72 Berger's disease. Granular immune deposits containing IgG, IgA, and C3 in mesangial areas.

FIGURE 17.73 Berger's disease. Granular immune deposits containing IgG, IgA, and C3 in mesangial areas.

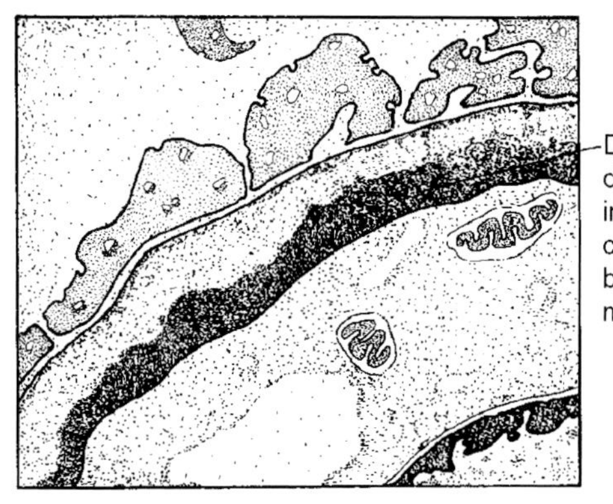

FIGURE 17.74 Dense deposit disease.

pathway, is required for diagnosis. Electron microscopy confirms the presence of electron-dense deposits in mesangial areas. Half of the cases progress to chronic renal failure over a 20-year course.

End-stage renal disease (ESRD) is chronic renal failure. Approximately one third of cases are linked to diabetes mellitus. Kidneys of patients on chronic dialysis ultimately develop ESRD. There is proliferation of intravascular smooth muscle induced by ischemia. There may also be venous thrombosis.

Dense-deposit disease (Figure 17.74) is a type II membranoproliferative glomerulonephritis characterized by the deposition of electron-dense material, often containing C3, in the peripheral capillary basement membrane of the glomerulus. C3 is decreased in the serum as a consequence of alternate complement pathway activation. C4 is normal. There is an increase in sialic acid-rich glomerular basement membrane glycoproteins. Patients may possess a serum factor termed nephritic factor that activates the alternate complement pathway. This factor is an immunoglobulin molecule that reacts with alternate complement pathway-activated components such as the bimolecular C3b and activated factor B complex. Nephritic factor stabilizes alternate pathway C3 convertase.

Basement membrane antibody refers to antibodies specific for the basement membrane of various tissues such as the lung basement membranes, the glomerular basement membrane, etc. This antibody is usually observed by immunofluorescence and less often by immunoperoxidase technology.

Goodpasture's syndrome (Figure 17.75 to Figure 17.77) is a disease with pulmonary hemorrhage (with coughing up blood) and glomerulonephritis (with blood in the urine), induced by antiglomerular basement membrane autoantibodies that also interact with alveolar basement membrane antigens. A linear pattern of immunofluorescent staining confirms interaction of the IgG antibodies with basement membrane antigens in the kidney and lung, leading to membrane injury with pulmonary hemorrhage and acute (rapidly progressive or crescentic) proliferative glomerulonephritis. Pulmonary hemorrhage may precede

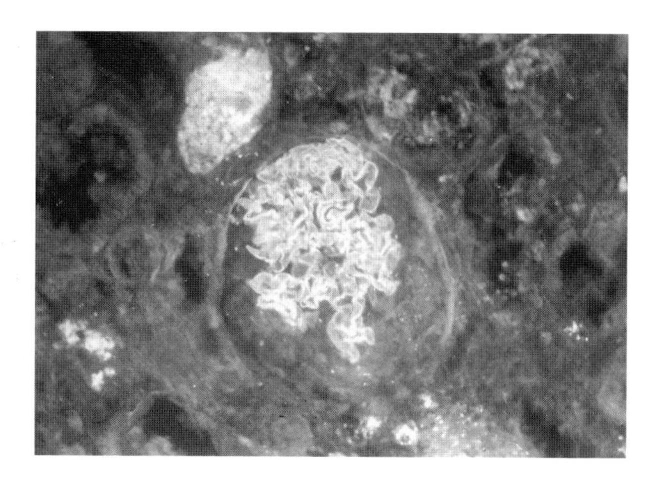

FIGURE 17.75 Goodpasture's syndrome. Kidney section. Stained with antiglomerular basement membrane (anti-GBM) antibody.

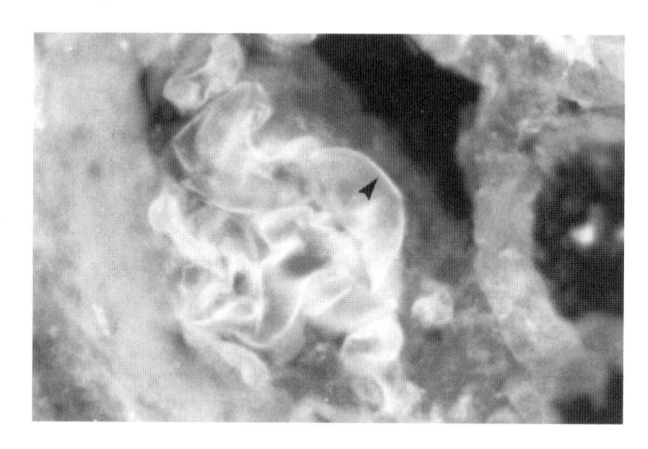

FIGURE 17.76 Goodpasture's syndrome. Kidney section. Stained with anti-GBM antibody.

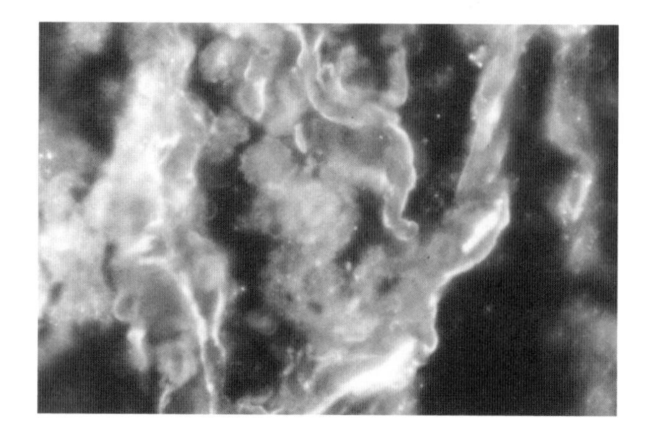

FIGURE 17.77 Goodpasture's syndrome. Liver section. Stained with antilung basement membrane antibody.

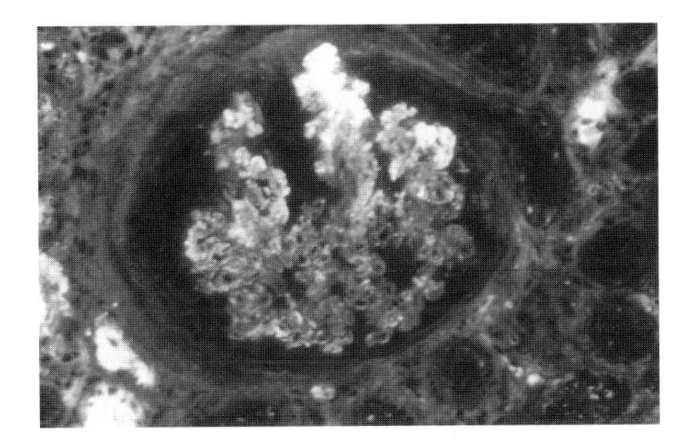

FIGURE 17.78 Focal segmental glomerulosclerosis.

hematuria. In addition to linear IgG, membranes may reveal linear staining for C3 (Figure 17.78).

Goodpasture's antigen is an antigen found in the noncollagenous part of type IV collagen. It is present in human glomerular and alveolar basement membranes, making them a target for injury-inducing anti-GBM antibodies in blood sera of Goodpasture's syndrome patients. Interestingly, individuals with Alport's (hereditary) nephritis do not have the Goodpasture antigen in their basement membranes. Thus, renal transplants stimulate anti-GBM antibodies in Alport's patients.

Linear staining is the interaction of IgG and possibly C3 on peripheral capillary loops of renal glomeruli in antiglomerular basement membrane diseases such as Goodpasture's syndrome. The use of fluorescein-labeled goat or rabbit antiimmunoglobulin preparations permits this smooth, thin, delicate, ribbon-like staining pattern to be recognized by immunofluorescence microscopy. It is in sharp contrast to the lumpy bumpy pattern of immunofluorescence staining seen in immune complex diseases.

NERVOUS SYSTEM

Multiple sclerosis (MS) (Figure 17.79) is a demyelinating nervous system disease of unknown cause. It is most frequent in young adult females and has an incidence of 1 in 2500 individuals in the U.S. MS shows a disease association with HLA-A3, B7, and Dw2 haplotypes. Patients express multiple neurological symptoms that are worse at some times than at others. They have paresthesias, muscle weakness, visual and gait disturbances, ataxia, and hyperactive tendon reflexes. There is infiltration of lymphocytes and macrophages in the nervous system, which facilitates demyelination. Autoimmune mechanisms mediated by T cells, which constitute the majority of infiltrating lymphocytes, are involved. At least 20 viruses have been suggested to play a role in the etiology of MS. Infected oligodendrocytes are destroyed by the immune mechanism, and there also may be "innocent bystander" demyelination. Antibodies against HTLV-I GAG (p24) protein have been identified in the cerebrospinal fluid of MS patients. HTLV-I gene sequences have been identified in monocytes of MS patients. An oligoclonal increase in CSF IgG occurs in 90% of MS patients. There is inflammation, demyelination, and glial scarring. Periventricular, frontal, and temporal areas of the brain are first involved, followed by regions of the brain stem, optic tracts, and white matter of the cortex with patchy lesions of the spinal cord. Attempts at treatment have included cop-1, a polypeptide mixture that resembles myelin basic protein, and numerous other agents.

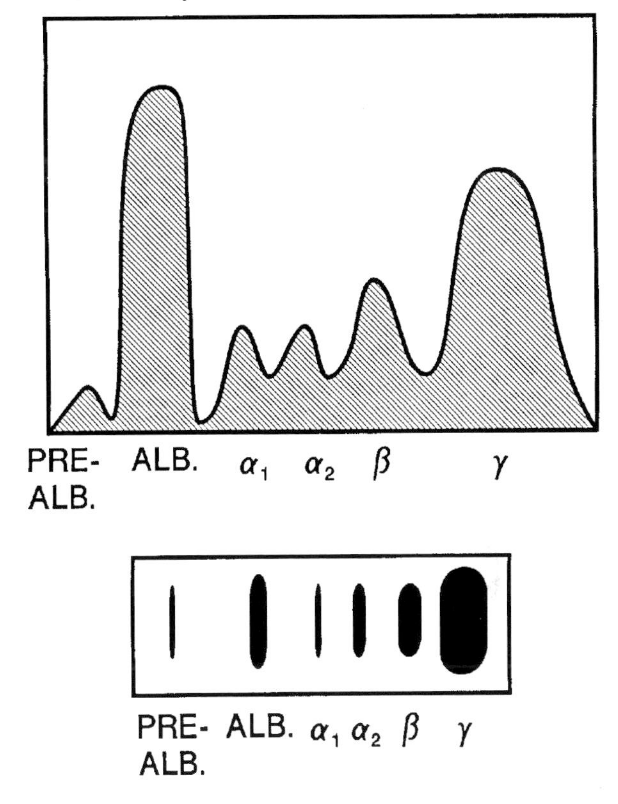

FIGURE 17.79 Multiple sclerosis (100× concentrate).

Miller-Fisher syndrome is a variant of Guillian-Barré (GBS) syndrome that has a subacute onset and follows infectious illnesses of which *Campylobacter jejuni* infection is the most common. Fewer than 5% of GBS cases have the MFS. Clinically, it presents as ophthalmoplegia, ataxia, and areflexia. There is little or no weakness of the extremity and trunk muscles. The spinal fluid protein is seldom increased and nerve conduction studies are usually normal. Almost all MFS patients have circulating polyclonal antibodies to GQb1 ganglioside, a minor ganglioside component of both the central and peripheral nervous systems. These antibodies are highest initially and fall with recovery.

Oligoclonal response is an immune response characterized by only a few separate clones of immunocompetent cells responding to yield a small number of immunoglobulin bands in agarose gel electrophoresis.

Oligoclonal bands: When cerebrospinal fluid of some multiple sclerosis patients is electrophoresed in agarose gel, immunoglobulins with restricted electrophoretic mobility may appear as multiple distinct bands in the γ region. Although nonspecific, 90 to 95% of multiple sclerosis patients show this. They may appear in selected other central nervous system diseases such as herpetic encephalitis, bacterial or viral meningitis, carcinomatosis, toxoplasmosis, neurosyphilis, progressive multifocal leukoencephalopathy, and subacute sclerosing panencephalitis and may appear briefly during the course of Guillian-Barré disease, lupus erythematosus vasculitis, spinal cord compression, diabetes, and amyotrophic lateral sclerosis.

Myelin basic protein (Figure 17.80) is a principal constituent of the lipoprotein myelin that first appears during late embryogenesis. It is a 19-kDa protein that is increased in multiple sclerosis patients who may generate T lymphocyte reactivity against MBP. T lymphocytes with the V-β 17 variant of the T cell receptor are especially prone to react to MBP.

When cerebrospinal fluid of some multiple sclerosis patients is electrophoresed in agarose gel, immunoglobulins with restricted electrophoretic mobility may appear as multiple distinct oligoclonal bands in the gamma region (Figure 17.81). Although nonspecific, 90 to 95% of multiple sclerosis patients show this. They may appear in selected other central nervous system diseases such as herpetic encephalitis, bacterial or viral meningitis, carcinomatosis, toxoplasmosis, neurosyphilis, progressive multifocal leukoencephalopathy, and subacute sclerosing panencephalitis, and may appear briefly during the course of Guillian-Barré disease, lupus erythematosus vasculitis, spinal cord compression, diabetes, or amyotrophic lateral sclerosis.

Experimental allergic encephalomyelitis is an autoimmune disease induced by immunization of experimental animals with preparations of brain or spinal cord incorporated into Freund's complete adjuvant. After 10 to 12 d, perivascular accumulations of lymphocytes and mononuclear phagocytes surround the vasculature of the brain and spinal cord white matter. Demyelination may also be present, worsening as the disease becomes chronic. The animals often develop paralysis. The disease can be passively transferred from a sick animal to a healthy one of the same strain with T lymphocytes, but not with antibodies. The mechanism involves T cell receptor interaction with an 18-kDa myelin basic protein molecule, which is an organ-specific antigen of nervous system tissue. The

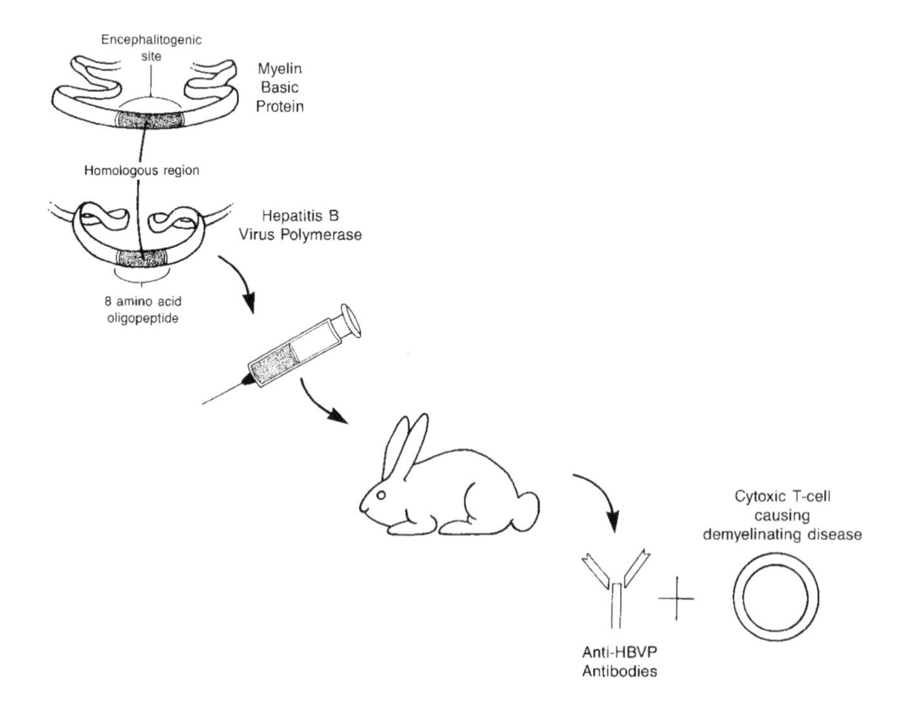

FIGURE 17.80 Myelin basic protein.

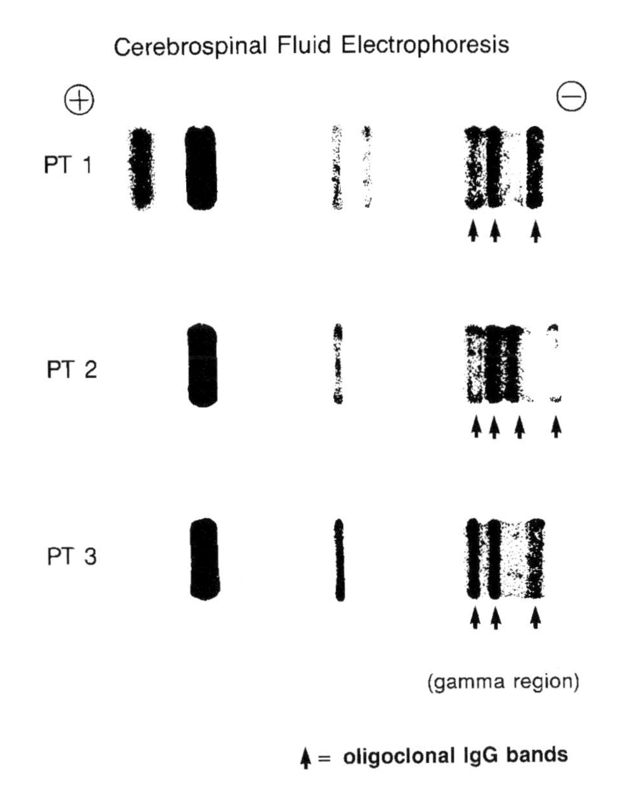

FIGURE 17.81 Oligoclonal bands.

CD4+ T lymphocyte represents the phenotype that is reactive with myelin basic protein. The immune reaction induces myelinolysis, wasting, and paralysis. Peptides derived from myelin basic protein (MBP) itself may be used to induce experimental allergic encephalomyelitis in animals. This experimental autoimmune disease is an animal model for multiple sclerosis and postvaccination encephalitis in man.

Experimental allergic neuritis is an experimental disease induced by injecting rats with peripheral nerve incorporated into Freund's complete adjuvant. P2 antigen is involved. Lymphocytes and macrophages infiltrate the sciatic nerve, and paralysis may develop.

Experimental allergic orchitis is an experimental autoimmune disease induced by injecting experimental animals with isogeneic or allogeneic testicular tissue incorporated into Freund's complete adjuvant.

Acute disseminated encephalomyelitis is brain inflammation that may be a sequela of certain acute viral infections such as measles in children or following vaccination. It was reported in some subjects following smallpox vaccination and in early recipients of rabies vaccine containing nervous system tissues. Symptoms include neck stiffness, headache, disorientation, and coma. Elevated quantities of protein and lymphocytes appear in the cerebrospinal fluid. Histopathologically, lymphocytes, plasma cells, and polymorphonuclear leukocytes may form perivascular infiltrates. The pathological changes are probably attributable to immune reactivity against the central nervous system constituent myelin basic protein and may represent the human equivalent of experimental allergic encephalomyelitis produced in animals. See also the animal model of this condition termed experimental allergic encephalomyelitis (EAE).

Experimental autoimmune encephalomyelitis is autoimmune demyelinating central nervous system disease in animal models induced by immunizing rats with myelin basic protein from the myelin sheath of nerves, incorporated into Freund's adjuvant. CD4+ T lymphocytes with specificity for myelin sheath proteins secrete cytokines that mediate the disease.

Anti myelin-associated glycoprotein (MAG) antibodies are associated with demyelinating neuropathy which is a slowly progressive distal and symmetrical sensory or sensorimotor neuropathy involving both arms and legs. Intention tremor may be present. Spinal fluid protein is often increased but cells are absent. There is demyelination and occasional axonal degeneration. Monoclonal anti-MAG and complement deposits have been found on the myelin sheaths. Selected nerves may show widening of the myelin lamellie. Most patients have a monoclonal gammopathy of the IgM type. The monoclonal IgM causes the neuropathy.

Antineutrophil cytoplasmic antibodies (ANCA): A heterogeneous group of autoantibodies specific for constituents of neutrophilic granulocytes. These autoantibodies are valuable serological markers for the diagnostic and therapeutic management of patients with systemic vasculitides such as Wegener's granulomatosis and microscopic polyangiitis, in which they recognize well defined cytoplasmic antigens such as proteinase III and myeloperoxidase. Two well-established ANCA staining patterns can be distinguished on ethanol-fixed neutrophils: a diffuse cytoplasmic fluorescent pattern (c-ANCA) and a fine homogeneous labeling of the perinuclear cytoplasm (pANCA). c-ANCA and classic pANCA in systemic vasculitides are autoantibodies specific for cytoplasmic antigens present in azurophil and specific granules of neutrophils. By contrast, atypical pANCA in inflammatory bowel disease and hepatobiliary disorders do not react with cytoplasmic structures. Their fluorescence pattern, revealed by indirect immunofluorescence microscopy, is characterized by a broad inhomogeneous labeling of the nuclear periphery together with multiple intranuclear fluorescent foci.

Experimental autoimmune neuritis (EAN) is a condition induced in mammalian species by immunization with peripheral nervous system (PNS) myelin, purified PNS myelin proteins, or synthetic peptides of PNS antigens. Basic antigens include myelin protein, the PO glycoprotein and the P2 protein. EAN develops 10 to 14 d after immunization with ascending paraparesis and paralysis. There is infiltration of PNS tissue by $CD4^+$ T cells and macrophages. Macrophages cause demyelination. Autoantibodies to myelin may also have a role in EAN pathogenesis.

Experimental autoimmune oophoritis is an ovarian autoimmune disease induced by immunization with synthetic peptides of mouse CP3, a glycoprotein with sperm receptor activity located in the zona pellucida of developing and mature oocytes. A single subcutaneous injection of ZP3 peptide incorporated into Freund's complete adjuvant can induce the disease in (C57B1/6 × A/J) S_1 hybrid mice. It is marked by ovarian inflammation that may be followed by ovarian follicle loss and ovarian atrophy. The infiltrating lymphocytes are comprised mainly of $CD4^+$ T lymphocytes. There are also many macrophages and IgG found in the jona pellucida.

Postinfectious encephalomyelitis is a demyelinating disease following a virus infection that is mediated by autoimmune delayed-type (type IV) hypersensitivity to myelin.

Anti-tau antibodies are rabbit antibodies against tau, one of the microtubule-associated proteins (MAPs) in the central nervous system. In the physphorylated form, tau is a major component of the paired helical filaments of the neurofibrillary tangles developed in Alzheimer's disease. One of the functions of tau is to stabilize microtubules. Phosphorylation of tau reduces the stabilizing effect. The C-terminal part of the tau protein shows a high degree of homology with other MAPs, such as MAP2, and it is suggested that it might serve as a microtubule-binding domain. The antibody reacts on immunoblots with tau protein and with tau from Alzheimer paired helical filaments. Brain tissue from patients with Alzheimer's disease is characterized by several histopathological lesions such as neurofibrillary tangles and senile plaques. For the study of lesions associated with Alzheimer's disease, the anti-tau antibody can be used in combination with mouse monoclonal anti-beta-amyloid and rabbit anti-ubiquitin. Neurofibrillary tangles are labeled by anti-tau whereas the senile plaques are labeled by the beta amyloid and the ubiquitin antibodies.

EYE

Sympathetic ophthalmia is uveal inflammation of an otherwise healthy, uninjured eye in an individual who has sustained a perforating injury to the other eye. The uveal tract reveals an infiltrate of lymphocytes and epithelioid cells, and there is granuloma formation. The mechanism has been suggested to be autoimmunity expressed as T lymphocyte-mediated immune reactivity against previously sequestered antigens released from the patient's other injured eye.

Vogt-Koyanagi-Harada (VKH) syndrome is uveal inflammation of the eye(s) with acute iridocyclitis, choroiditis, and retinal detachment (Figure 17.82 and Figure 17.83). Initial manifestations include headache, dysacusis, and sometimes vertigo. Scalp hair may show patchy loss or whitening. Vitiligo and poliosis often occur. The development of delayed-type hypersensitivity to melanin-containing tissue has been postulated. Apparently, pigmented constituents of the eye, hair, and skin are altered by some type of insult in a manner that leads to a delayed-type hypersensitivity response to them. Possible autoantigens are soluble substances from the retinal photoreceptor layer. There is a predisposition to VKH in Asians. Uveal tissue extracts have been used in delayed-hypersensitivity skin tests. Antibodies to uveal antigens may be present, but

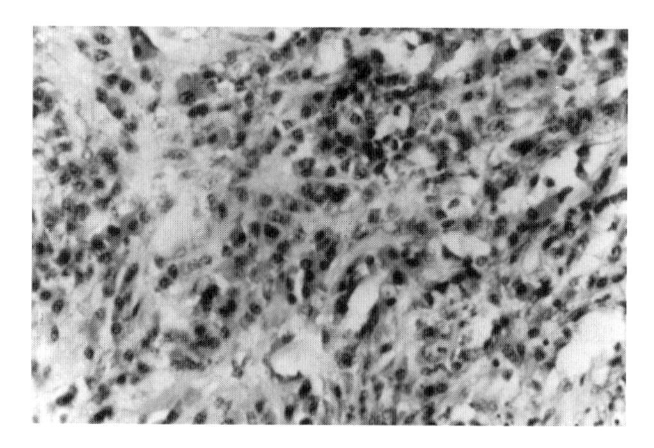

FIGURE 17.82 Hematoxylin and eosin (H&E) stained eye section showing extensive mononuclear cellular infiltrate of lymphocytes and plasma cells.

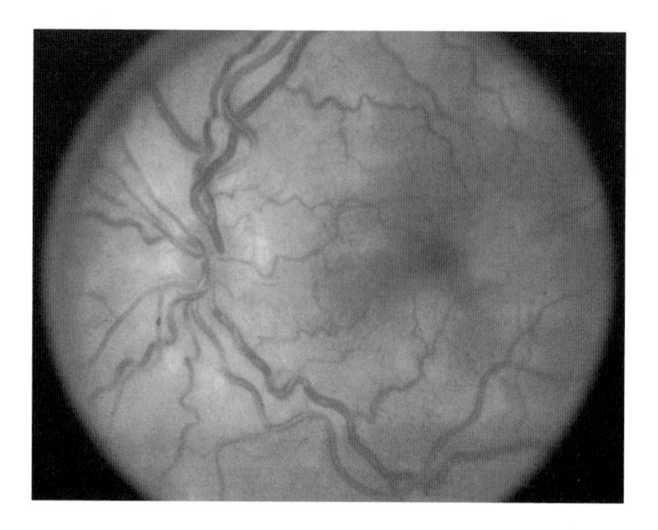

FIGURE 17.83 VKH funduscopic examination reveals venous engorgement and serous retinal detachment.

they are not specific for VKH. VKH is a common cause of uveitis in Japan. Hereditary factors play a predisposing role. VKH syndrome is closely associated with HLA-DR4. Corticosteroids, as well as chlorambucil, cyclophosphamide, or cyclosporin A have been used for treatment.

Cogan's syndrome is corneal inflammation (interstitial keratitis) and inflammation of the ear, leading to nausea, vomiting, vertigo, and ringing in the ears. This may be associated with connective tissue disease or occur following an infection.

Uveitis is uveal tract inflammation involving the uvea, iris, ciliary body, and choroid of the eye. It may be associated with Behcet's disease, sarcoidosis, and juvenile rheumatoid arthritis.

Experimental autoimmune uveitis (EAU) is a disease of the neural retina and related tissues induced by immunization with purified retinal antigens. It is a model for various human ocular inflammatory diseases, possibly autoimmune, that are accompanied by immunologic responses to ocular antigens. The most frequently used antigens are retinal soluble antigen (S-Ag), the interphotoreceptor retinoid-binding protein (IRBP), and rhodopsin. Immunization with these antigens incorporated into complete Freund's adjuvant leads to injury of the photoreceptor cell layer of the retina. Histopathologically, there is infiltration of the retina with inflammatory cells and injury to photoreceptors.

Mooren's ulcer is a chronic progressive marginal corneal degeneration of unknown cause. It has an increased incidence in older persons and is associated with marked pain and ceaseless melting of the peripheral cornea. Autoimmune phenomena and collagenolytic enzymes have been considered in the pathogenesis of this condition. Circulating antibodies to human corneal epithelium and serum antibodies to a bovine corneal antigen (CO-Ag) have been demonstrated. Immunoglobulins and complement components have been discovered in the affected epithelium and stroma in patterns not found in normal controls. It remains to be determined whether or not these immunological phenomena lead to corneal inflammation or are a consequence of it.

Cicatrical ocular pemphigoid (Figure 17.84 and Figure 17.85) is a rare blistering disorder of the conjunctival mucous membrane. This may lead to scarring that can result in blindness if untreated with corticosteroids. Histopatholocally, the subepithelial unilocular bulla is accompanied by a mild inflammatory reaction. Direct immunofluorescence reveals a diffuse, linear deposition

FIGURE 17.84 Cicatrical ocular pemphigoid. Stained with anti-IgG antibody. Linear fluorescence at the junction between the epithelium and the substantia propria.

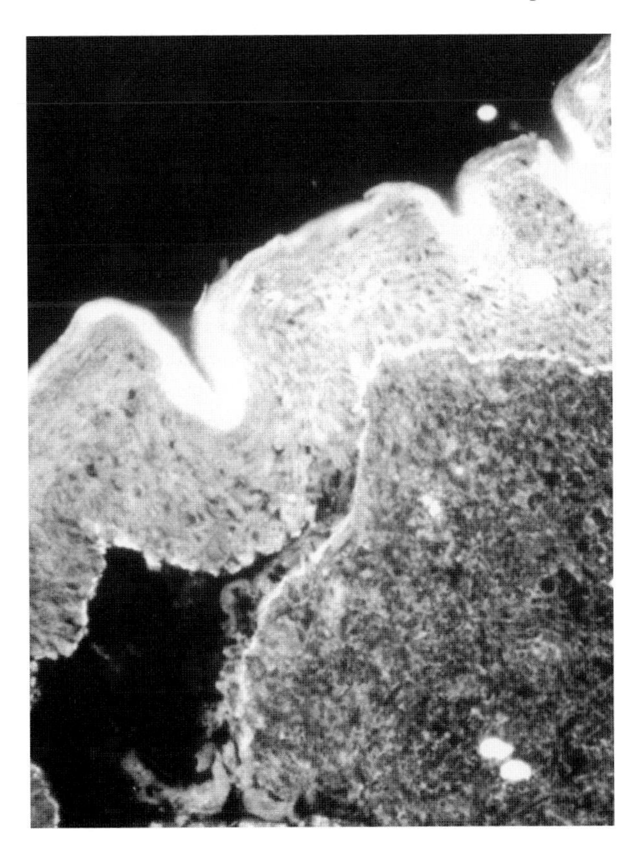

FIGURE 17.85 Cicatrical ocular pemphigoid. Stained with anti-IgG antibody. Linear fluorescence at the junction between the epithelium and the substantia propria.

of immunoglobulins and components, mainly IgG and C3, at the epithelial–subepithelial junction.

SPERMATOZOA

Antisperm antibody (Figure 17.86) is specific for any one of several sperm constituents. Antisperm agglutinating antibodies are detected in blood serum by the Kibrick sperm agglutination test that uses donor sperm. Sperm-immobilizing antibodies are detected by the Isojima test. The subject's serum is incubated with donor sperm and motility examined. Testing for antibodies is of interest to couples with infertility problems. Treatment with relatively small doses of prednisone is sometimes useful in improving the situation by diminishing antisperm antibody titers. One-half of infertile females manifest IgG or IgA sperm immobilizing antibodies which affect the tail of the spermatozoa. By contrast, IgM antisperm head-agglutinating antibodies may occur in homosexual males.

CARTILAGE

Relapsing polychondritis (Figure 17.87 to Figure 17.88) is an inflammation of the cartilage, especially that of the external pinnae of the ears, causing them to lose their structural integrity. It appears to have an immunological etiology, and anticollagen antibodies may be demonstrated in the serum of patients.

SYSTEMIC AUTOIMMUNE DISEASES

Systemic lupus erythematosus (SLE) (Figure 17.89 to Figure 17.94) is the prototype of connective tissue diseases that involves multiple systems and has an autoimmune etiology. It is a disease with an acute or insidious onset. Patients may experience fever, malaise, loss of weight, and lethargy. All organ systems may be involved. Patients form a plethora of autoantibodies, especially antinuclear autoantibodies. SLE is characterized by exacerbations and remissions. Patients often have injury to the skin, kidneys,

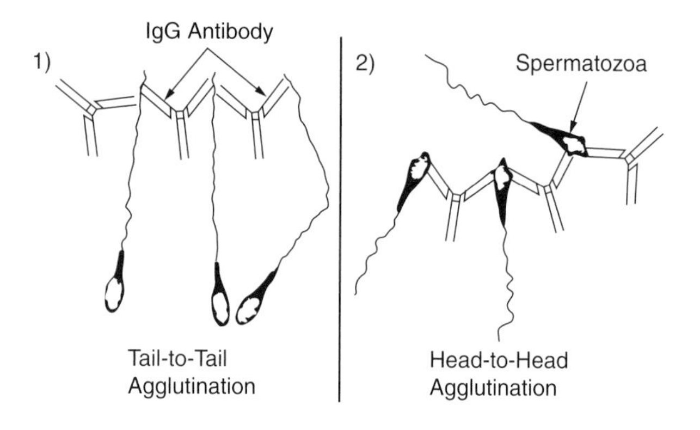

FIGURE 17.86 Antisperm antibody.

joints, and serosal membranes. SLE occurs in 1 in 2500 people in certain populations. It has a 9:1 female to male predominance. Its cause remains unknown. Antinuclear antibodies produced in SLE fall into four categories that include (1) antibodies against DNA, (2) antibodies against histones, (3) antibodies to nonhistone proteins bound to RNA, and (4) antibodies against nucleolar antigens.

Indirect immunofluorescence is used to detect nuclear fluorescence patterns that are characteristic for certain antibodies. These include homogeneous or diffuse staining, which reveals antibodies to histones and deoxyribonucleoprotein; rim or peripheral staining, which signifies antibodies against double-stranded DNA; speckled pattern, which indicates antibodies to non-DNA nuclear components, including histones and ribonucleoproteins; and the nucleolar pattern in which fluorescent spots are observed in the nucleus and reveal antibodies to nucleolar RNA. Antinuclear antibodies most closely associated with SLE are anti-double-stranded DNA and anti-Sm (Smith) antibodies.

There appears to be genetic predisposition to the disease, which is associated with DR2 and DR3 genes of the MHC in Caucasians of North America. Genes other than HLA genes are also important. In addition to the anti-double-stranded DNA and anti-Sm antibodies, other immunologic features of the disease include depressed serum complement levels, immune deposits in glomerular basement membranes and at the dermal–epidermal junction, and the presence of multiple other autoantibodies. Of all the immunologic abnormalities, the hyperactivity of B cells is critical to the pathogenesis of SLE. B cell activation is polyclonal, leading to the formation of antibodies against self and nonself antigens. In SLE, there is a loss of tolerance to self-constituents, leading to the formation of antinuclear antibodies. The polyclonal activation leads to antibodies of essentially all classes in immune deposits found in renal biopsy specimens by immunofluorescence. In addition to genetic factors, hormonal and environmental factors are important in producing the B cell activation. Nuclei of injured cells react with antinuclear antibodies, forming a homogeneous structure called an LE body or a hematoxylin body, which is usually found in a neutrophil that has phagocytized the injured cell's denatured nucleus. Tissue injury in lupus is mediated mostly by an immune complex (type III hypersensitivity). There are also autoantibodies specific for erythrocytes, leukocytes, and platelets that induce injury through a type II hypersensitivity mechanism. There is an acute necrotizing vasculitis involving small arteries and arterioles present in tissues in lupus. Fibrinoid necrosis is classically produced. Most SLE patients have renal involvement, which may take several forms, with diffuse proliferative glomerulonephritis being the most serious. Subendothelial immune deposits in the kidneys of lupus patients are typical and may give a "wire loop" appearance to a thickened basement membrane.

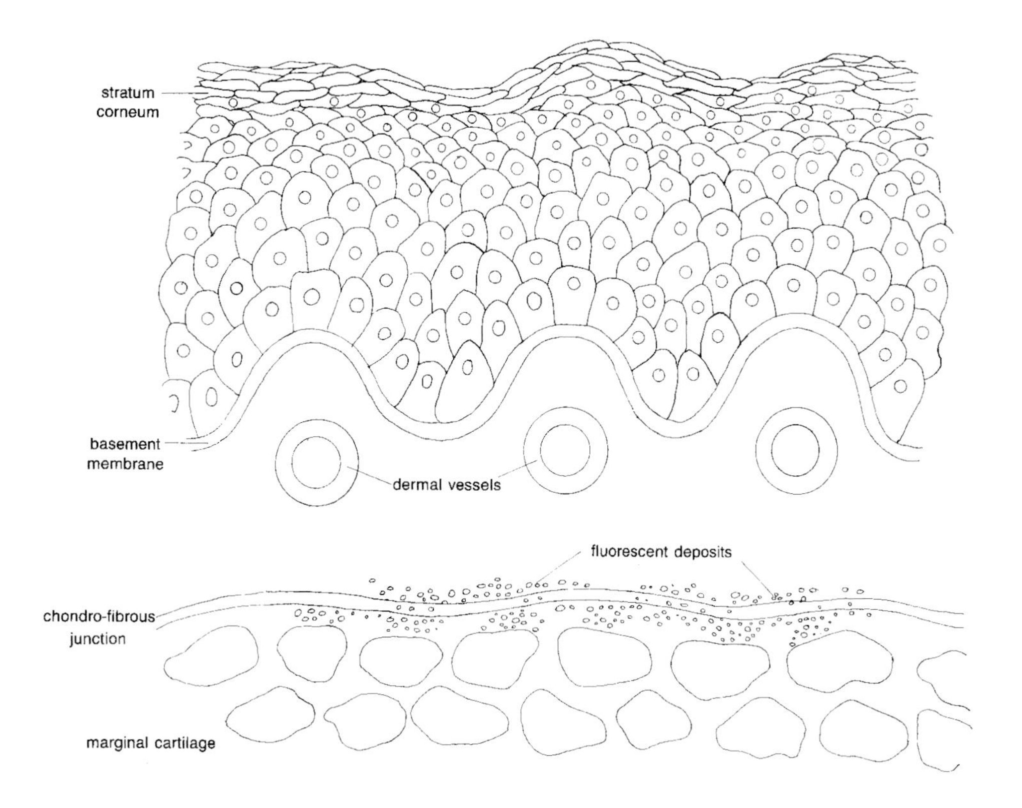

FIGURE 17.87 Relapsing polychondritis.

FIGURE 17.88 Relapsing polychondritis.

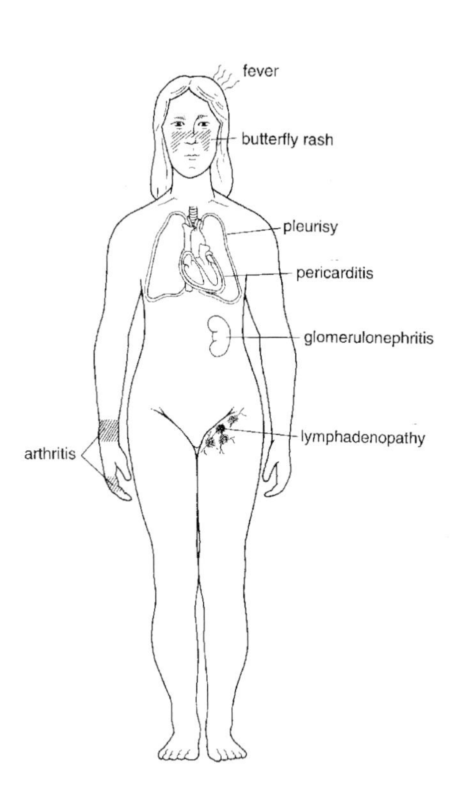

FIGURE 17.89 Systemic lupus erythematosus.

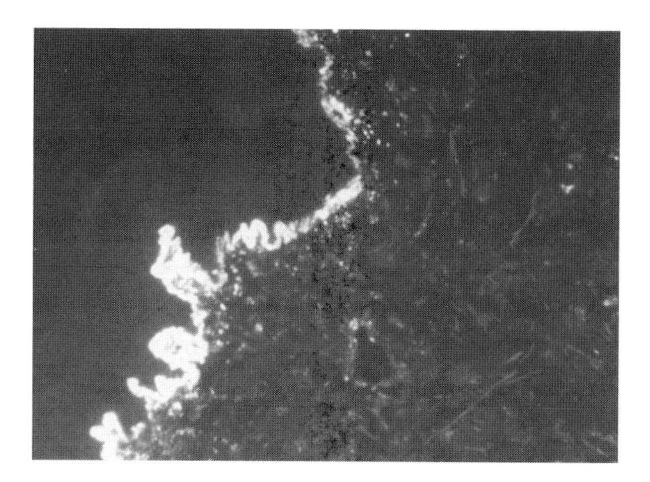

FIGURE 17.90 Lupus band test. (Skin)

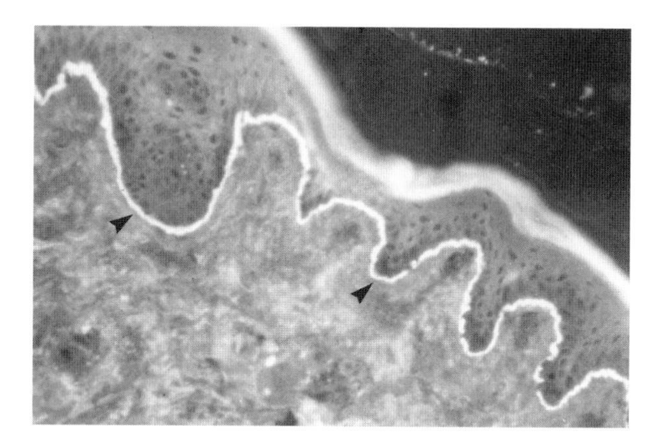

FIGURE 17.91 Lupus erythematosus. Skin. Immune deposits at dermal–epidermal junction.

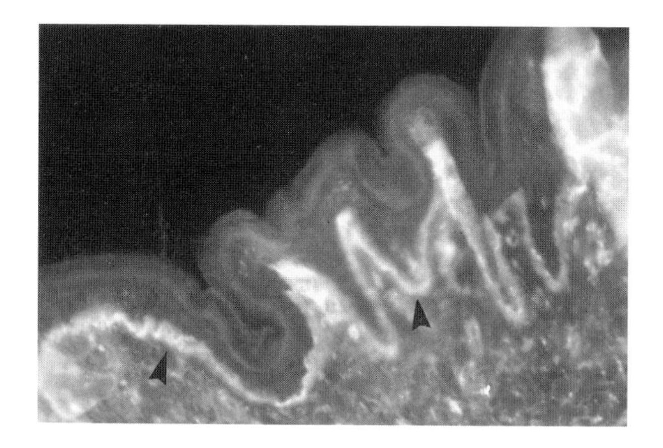

FIGURE 17.92 Systemic lupus erythematosus. Immune deposits at dermal–epidermal junction.

In the skin, immunofluorescence can demonstrate deposition of immune complexes and complement at the dermal–epidermal junction. Immune deposits in the skin are especially prominent in sun-exposed areas of the skin. Joints

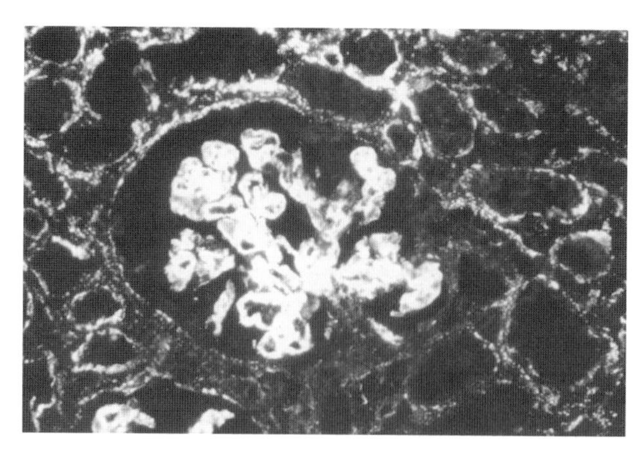

FIGURE 17.93 Diffuse proliferative lupus nephritis.

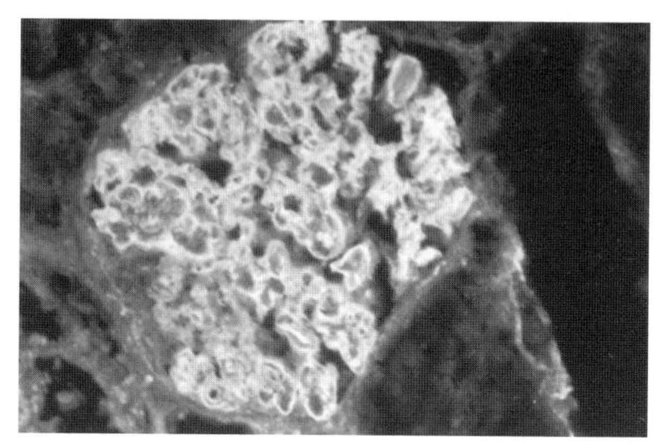

FIGURE 17.94 Systemic lupus erythematosus. Kidney. Diffuse immune deposits on peripheral capillary loops.

may be involved, but the synovitis is nonerosive. Typical female patients with lupus have a butterfly rash over the bridge of the nose in addition to fever and pain in the peripheral joints. However, the presenting complaints in SLE vary widely. Patients may have central nervous system involvement, pericarditis, or other serosal cavity inflammation. There may be pericarditis as well as involvement of the myocardium or of the cardiac valves to produce Libman-Sacks endocarditis. There may be splenic enlargement, pleuritis, and pleural effusion or interstitial pneumonitis, as well as other organ or system involvement. Patients may also develop antiphospholipid antibodies called lupus anticoagulants. They may be associated with a false-positive VDRL test for syphilis. A drug such as hydrazaline may induce a lupus-like syndrome. However, the antinuclear antibodies produced in drug-induced lupus are often specific for histones, a finding not commonly found in classic SLE. Lupus erythematosus induced by drugs remits when the drug is removed. Discoid lupus refers to a form of the disease limited to the skin. Corticosteroids have proven very effective in suppressing immune reactivity in SLE. In more severe cases, cytotoxic agents such as cyclophosphamide, chlorambucil, and azathioprine have been used. See LE cell.

Band test (see Figure 17.90) refers to antigen–antibody deposits at the dermal–epidermal junction in patients with lupus erythematosus. They may consist of IgG, IgM, IgA, and C3. Deposits are not found in uninvolved areas of the dermal–epidermal junction in discoid lupus erythematosus (DLE), but are present in both involved and uninvolved areas of the dermal–epidermal junction in SLE. Immune complexes at the dermal–epidermal junction appear in 90 to 95% of SLE patients. Of these, 90% reveal them in skin exposed to sunlight, and 50% have deposits in skin that is not exposed to the sun. Also called lupus band test. Immunofluorescence bands also occur in some cases of anaphylactoid purpura, atopic dermatitis, contact dermatitis, autoimmune thyroiditis, bullous pemphigoid, cold agglutinin syndrome, dermatomyositis, hypocomplementemic vasculitis, polymorphous light eruption, rheumatoid arthritis, scleroderma, and a number of other conditions.

Drug-induced lupus (DIL): Drugs which by themselves can induce a limited form of SLE include the aromatic amines or hydrazines, the two most common being procainamide and hydralazine. Other drugs that can induce DIL include isoniazid, methyldopa, quinidine, and chlorpromazine. This drug-induced disease remits on discontinuation of the drug. The autoimmune response is very restricted.

Liquefactive degeneration is dermal–epidermal interface liquefaction that is induced by immune mechanisms. This engages basal cells, leading to coalescing subepidermal vesicles in such skin diseases as dermatitis herpetiformis, erythema multiforme, fixed-drug reaction, lichen planus, lupus erythematosus, and many other skin conditions.

MCTD is an abbreviation for mixed-connective tissue disease.

Raynaud's phenomenon refers to episodes of vasospasm in the fingers when the hands are exposed to cold temperatures. The ischemia of the fingers is characterized by severe pallor and is often accompanied by paresthesias and pain. It is brought on by cold, emotional stress, or anatomic abnormality. When the condition is idiopathic or primary, it is called Raynaud's disease. Raynaud's phenomenon is seen in several connective tissue diseases including SLE and systemic sclerosis. Subjects with cryoglobulinemia may also manifest the phenomenon.

Mixed-connective tissue disease (MCTD) is a connective tissue disease that shares characteristics in common with SLE, RA, and dermatomyositis. By immunofluorescence, a speckled nuclear pattern attributable to antinuclear antibody in the circulation is revealed. There are high titers of antinuclear antibodies specific for nuclear ribonucleoproteins. Treatment with corticosteroids is quite effective. Also called Sharp syndrome or overlap syndrome.

Vinyl chloride (VC) is a synthetic resin that represents a toxin of immunologic significance in occupational exposure. It has been associated with mixed connective tissue disease. Autoantibodies are formed against ribonuclear proteins, RP, and arthritis. HLA-disease association studies have implicated HLA-DR5 of haplotype A1B8 and HLA-DR3 with these syndromes. Oxidized metabolites of VC are highly reactive and bind to sulfhydryl groups and amino radicals. It also binds to nucleotides. VC can induce oxidative cell injury and may enhance the immunogenicity of self molecules. Oxidizing agents may inactivate T suppressor cells. Cell-surface thiols influence the activities of lymphocytes. The reduction of disulfide bridges to thiols is linked to increased cellular activity. By contrast, thiol oxidation to disulfides decreases cell responsiveness. $CD8^+$ T cells are more sensitive to thiol blockers and oxidants than are $CD4^+$ T cells. Exposure to vinyl chloride, mercury, iodide, and toxic oil may pose interactions with cellular thiols, leading to preferential inactivation of $CD8^+$ T suppressor cells.

Wire loop lesion refers to thickening of capillary walls as a result of subendothelial immune complex deposits situated between the capillary endothelium and the glomerular basement membrane. Wire loop lesions are seen in diffuse proliferative lupus nephritis and are characteristic of class IV lupus erythematosus. They may be seen also in progressive systemic sclerosis and may appear together with crescent formation, necrosis, and scarring.

Butterfly rash (Figure 17.95) is a facial rash in the form of a butterfly across the bridge of the nose. Seen especially in patients with lupus erythematosus. These areas are photosensitive and consist of erythematous and scaly patches that may become bulbous or secondarily infected. The rash is not specific for lupus erythematosus, since butterfly-type

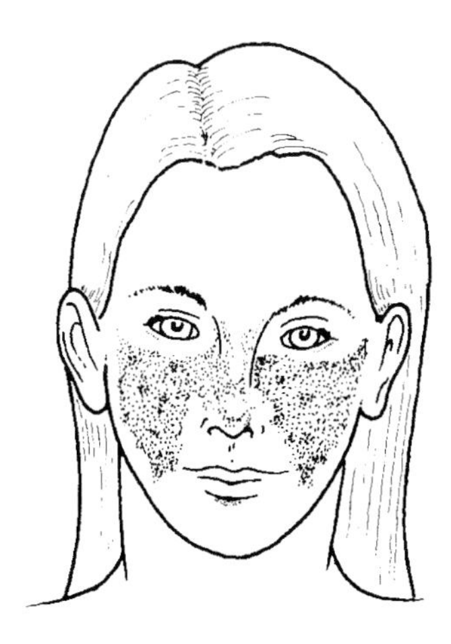

FIGURE 17.95 Butterfly rash.

rashes may also occur in various other conditions including AIDS, dermatomyositis, ataxia-telangiectasia, erysipelas, pemphigus erythematosus, and pemphigus foliaceous, etc.

An **LE cell** (see Figure 17.96 and Figure 17.97) is a neutrophil (PMN) in the peripheral blood or synovium of lupus erythematosus patients produced when the PMN phagocytizes a reddish-purple staining homogeneous lymphocyte nucleus that has been coated with antinuclear antibody. In addition to lupus erythematosus, LE cells are seen also in scleroderma, drug-induced lupus erythematosus, and lupoid hepatitis.

Antinuclear antibodies (ANA) (Figure 17.98 to Figure 17.104) are antibodies found in the circulation of patients with various connective tissue disorders. They may show specificity for various nuclear antigens including single- and double-stranded DNA, histones, and ribonucleoprotein. To detect antinuclear antibodies, the patient's serum is incubated with Hep-2 cells and the pattern of nuclear staining is determined by fluorescence microscopy. The homogeneous pattern of staining represents the morphologic expression of antinuclear antibodies specific for ribonucleoprotein which is positive in SLE, progressive systemic sclerosis, RA, and other connective tissue disorders. Peripheral nuclear staining represents the morphologic expression of DNA antibodies associated with SLE. Nucleolar fluorescence signifies anti-RNA antibodies of the type that occurs in progressive systemic sclerosis (scleroderma). The speckled pattern of staining is seen in several connective tissue diseases.

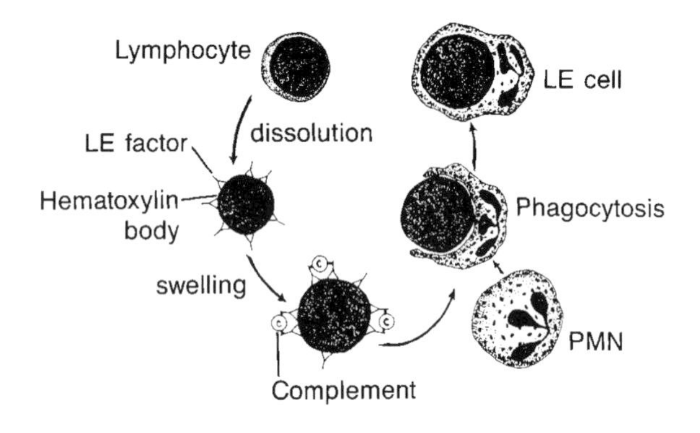

FIGURE 17.96 Formation of an LE cell.

FIGURE 17.97 Lupus erythematosus cell.

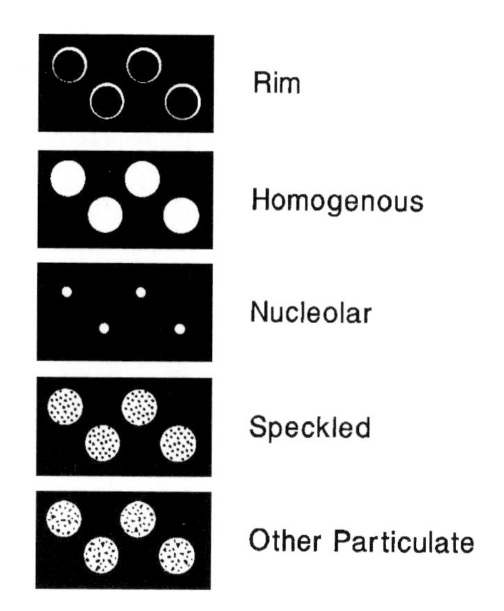

FIGURE 17.98 Antinuclear antibody.

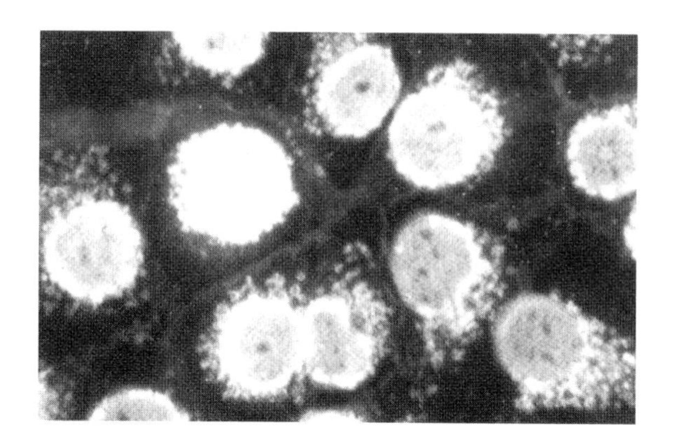

FIGURE 17.99 ANA/AMA mixed pattern.

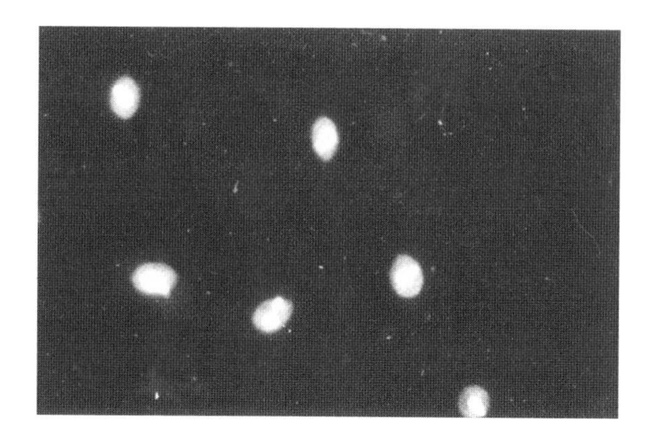

FIGURE 17.100 nDNA antibody. Positive reaction.

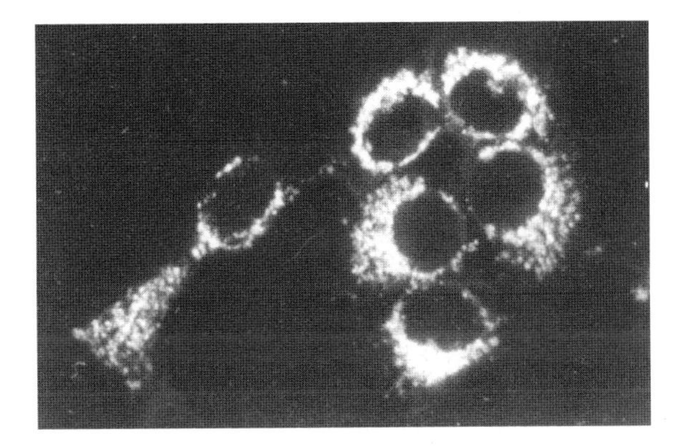

FIGURE 17.101 AMA positive pattern.

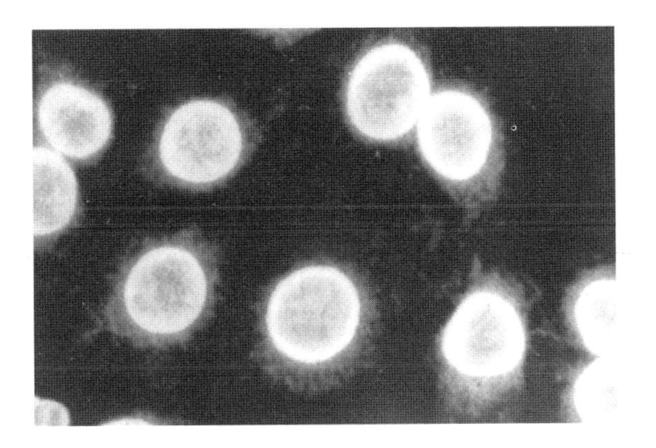

FIGURE 17.102 ANA peripheral pattern.

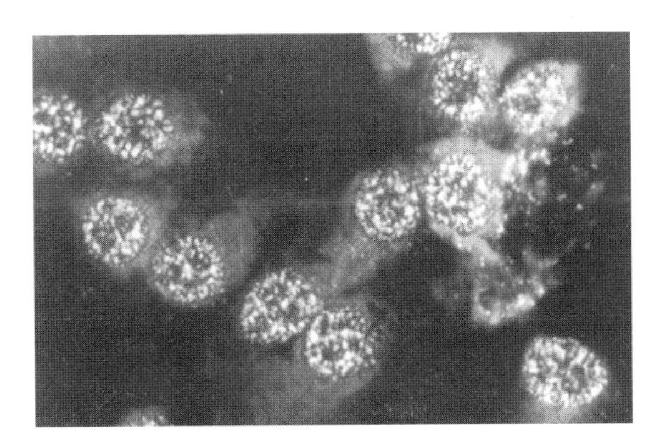

FIGURE 17.103 ANA speckled pattern.

Anti-doublestranded DNA (Anti-dsDNA) (Figure 17.105) are antibodies present in the blood sera of SLE patients. Among the detection methods is an immunofluorescence technique (IFT) using *Crithidia luciliae* as the substrate. In this method, fluorescence of the kinetoplast, which contains mitochondrial DNA, signals the presence of anti-dsDNA antibodies. This technique is useful for assaying SLE serum, which is usually positive in patients with active disease. A rim or peripheral pattern of nuclear staining of cells interacting with antinuclear antibody represents morphologic expression of anti-double-stranded DNA antibody.

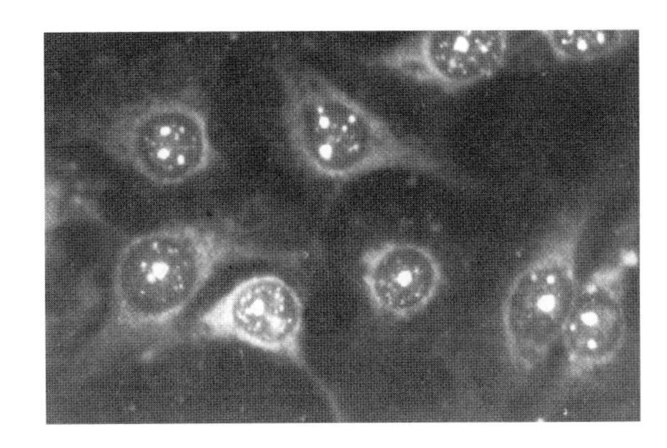

FIGURE 17.104 ANA nucleolar pattern.

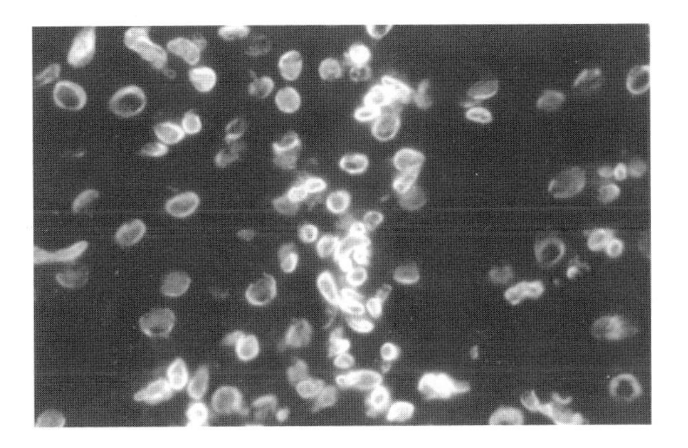

FIGURE 17.105 Anti-doublestranded DNA.

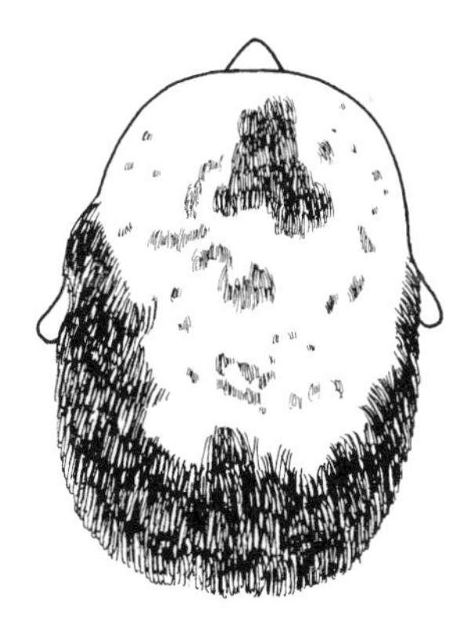

FIGURE 17.106 Alopecia areata.

Alopecia areata (Figure 17.106) describes hair loss in subjects who demonstrate autoantibodies against hair follicle capillaries.

Lupus anticoagulant is an IgG or IgM antibody that develops in lupus erythematosus patients as well as in certain individuals with neoplasia or drug reactions, in some normal persons, and recently reported in AIDS patients who have active opportunistic infections. These antibodies are specific for phospholipoproteins or phospholipid constituents of coagulation factors. *In vitro*, these

FIGURE 17.107 Discoid lupus erythematosus.

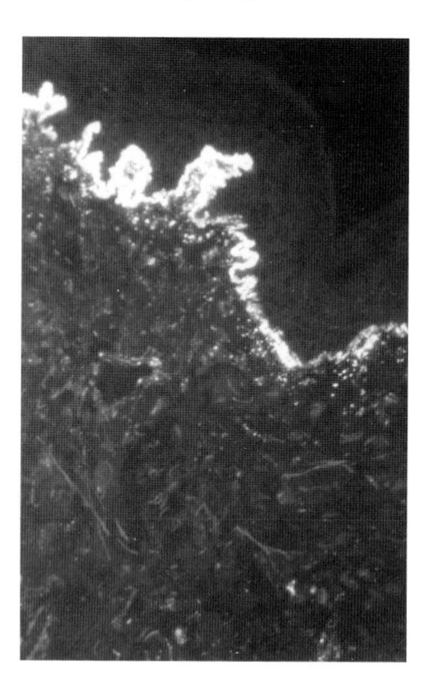

FIGURE 17.108 Discoid lupus erythematosus. Skin. Immune deposits at dermal–epidermal junction.

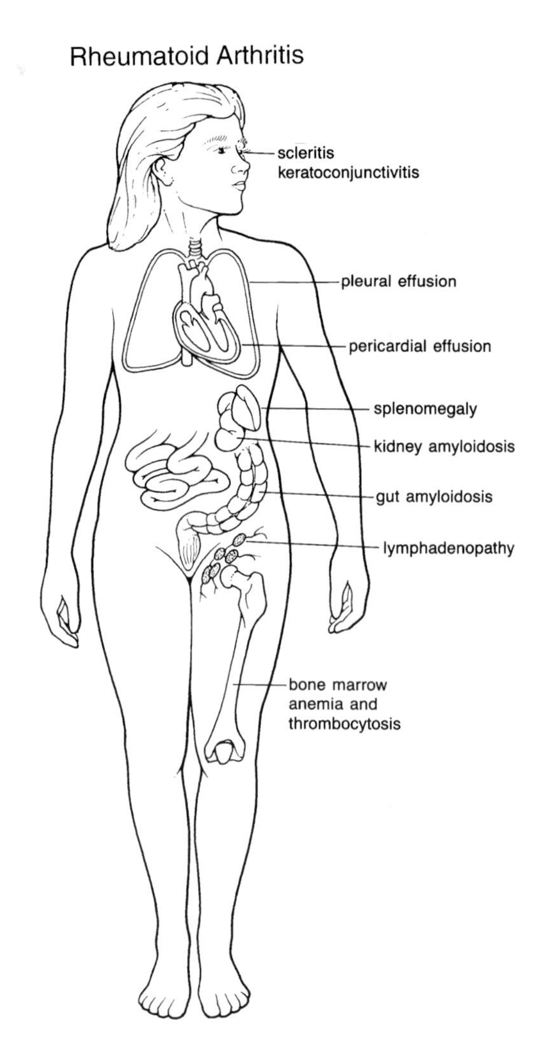

FIGURE 17.109 Rheumatoid arthritis.

antibodies inhibit coagulation dependent upon phospholipids. If there are no other platelet or coagulation defects, they do not cause coagulopathy.

Discoid lupus erythematosus (Figure 17.107 and Figure 17.108) is a type of lupus erythematosus that involves only the skin, which manifests a characteristic rash. The viscera are not involved, but the skin manifests erythematous plaques and telangiectasis with plugging of the follicles. Also called cutaneous lupus erythematosus.

Rheumatoid arthritis (RA) (Figure 17.109) is an autoimmune inflammatory disease of the joints which is defined according to special criteria designated 1 through 7. Criteria 1 through 4 must be present for more than 6 weeks. The "revised criteria" for rheumatoid arthritis are as follows: (1) morning stiffness in and around joints lasting at least 1 h before maximum improvement (2) soft tissue swelling (arthritis) of three or more joints observed by a physician, (3) swelling (arthritis) of the proximal interphalangeal, metacarpophalangeal, or wrist joints, (4) symmetric swelling (arthritis), (5) rheumatoid nodules, (6) presence of rheumatoid factors, and (7) roentgenographic erosions.

Pannus (Figure 17.110) refers to a granulation tissue reaction that is chronic and progressive and produces joint erosion in patients with rheumatoid arthritis. It is a structure that develops in synovial membranes during the chronic proliferative and destructive phase of rheumatoid arthritis. It is a membrane of granulation tissue induced by immune complexes that are deposited in the synovial membrane. They stimulate macrophages to release interleukin-1, fibroblast-activating factor, prostaglandins, substance P, and platelet-derived growth factor. This leads to extensive injury to chondroosseous tissues. The articular surface of the joint is covered by this synovitis. There is edema, swelling, and erythema in the joints. Palisades of histiocytes are present. This entire process can fill the joint space, leading to demineralization and cystic resorption.

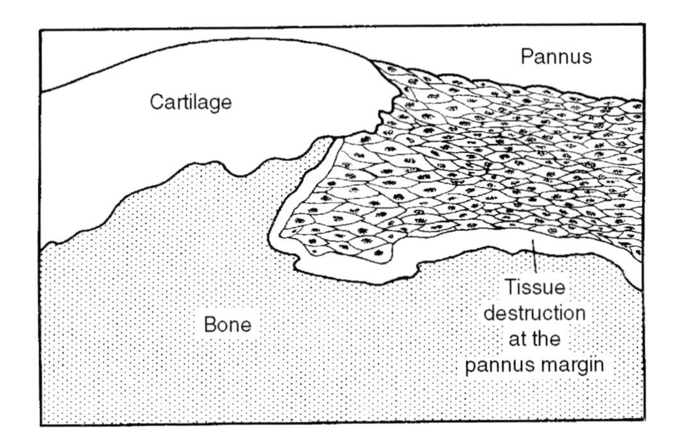

FIGURE 17.110 Pannus.

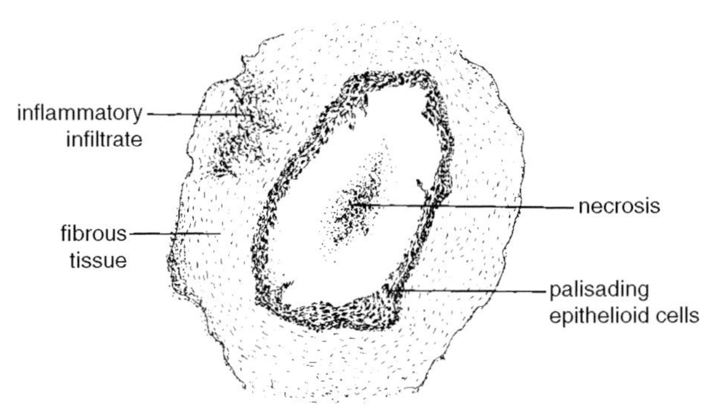

FIGURE 17.111 Rheumatoid nodule.

Rheumatoid factor (RF) is an autoantibody present in the serum of patients with RA, but also found with varying frequency in other diseases such as subacute bacterial endocarditis, tuberculosis, syphilis, sarcoidosis, hepatic diseases, and others, as well as in the sera of some human allograft recipients and apparently healthy persons. RF are immunoglobulins, usually of the IgM class, and to a lesser degree of the IgG or IgA classes, with reactive specificity for the Fc region of IgG. This antiimmunoglobulin antibody, which may be either monoclonal or polyclonal, reacts with the Fc region epitopes of denatured IgG, including the Gm markers. Most RF are isotype specific, manifesting reactivity mainly for IgG1, IgG2, and IgG4, but only weakly reactive with IgG3. Antigenic determinants of IgG that are potentially reactive with RF include (1) subclass-specific or genetically defined determinants of native IgG (IgG1, IgG2, IgG4, and Gm determinants), (2) determinants present on complexed IgG, but absent on native IgG, and (3) determinants exposed after enzymatic cleavage of IgG. The Gm determinants are allotypic markers of the human IgG subclasses. They are located in the IgG molecule as follows: in the C_H1 domain in IgG1, in the $C_H2$domain in IgG2, and in C_H2 and C_H3 domains in IgG4.

Although rheumatoid factor titers may not be clearly correlated with disease activity, they may help perpetuate chronic inflammatory synovitis. When IgM rheumatoid factors and IgG target molecules react to form immune complexes, complement is activated leading to inflammation and immune injury. IgG rheumatoid factors may self-associate to form IgG–IgG immune complexes that help perpetuate chronic synovitis and vasculitis. IgG RF synthesized by plasma cells in the rheumatoid synovium fix complement and perpetuate inflammation. IgG RF has been shown in microbial infections, B lymphocyte proliferative disorders and malignancies, non-RA patients, and aging individuals. RF might have a physiologic role in removal of immune complexes from the circulation. RFs were demonstrated earlier by the Rose-Waaler test but are now detected by the latex agglutination (or RA) test, which employs latex particles coated with IgG.

RF is an abbreviation for rheumatoid factor(s).

A **rheumatoid nodule** (Figure 17.111) is a granulomatous lesion characterized by central necrosis encircled by a palisade of mononuclear cells and an exterior mantle of lymphocytic infiltrate. The lesions occur as subcutaneous nodules, especially at pressure points such as the elbow, in individuals with RA or other rheumatoid diseases.

Rheumatoid arthritis cell (RA cell) is an irregular neutrophil that contains a variable number of black-staining cytoplasmic inclusions that are 0.2 to 2.0 μ in diameter. These cells contain IgM rheumatoid factor, complement, IgG, and fibrin. They are found in synovial fluid of RA patients. Although RA cells may constitute 5 to 1005 of an RA patient's neutrophils, they may also be present in patients with other connective tissue diseases.

RA is an abbreviation for rheumatoid arthritis.

RANA (rheumatoid arthritis-associated nuclear antigen) is an antigen of Epstein-Barr virus-immortalized lymphoid cell lines that reacts with blood sera from RA patients.

Ragocyte is a polymorphonuclear leukocyte containing IgG-rheumatoid factor–complement–fibrin conglomerates. These cells are found in the joints of RA patients. Also called RA cells.

Ragg refers to agglutinating activity by rheumatoid factor in RA patients' sera, as revealed by a positive RA rheumatoid arthritis test.

RA-33 is an RA-specific antigen that is identical to the heterogeneous ribonucleoprotein rhnRNP-A2.

Perinuclear antibodies are antibodies against perinuclear granules in buccal mucosal cells in man. They are present in about 78% of patients with classical RA and in 40% of

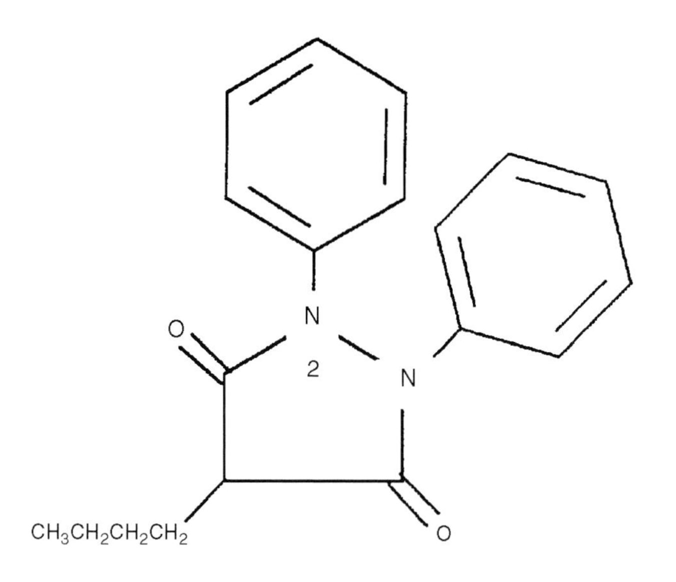

FIGURE 17.112 Phenylbutazone.

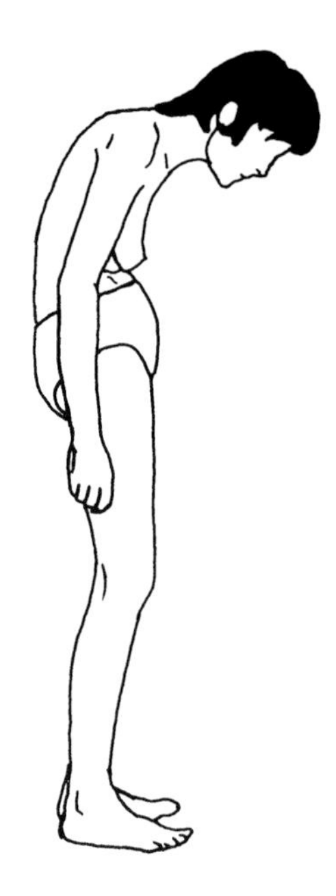

FIGURE 17.113 A patient with severe ankylosing spondylitis.

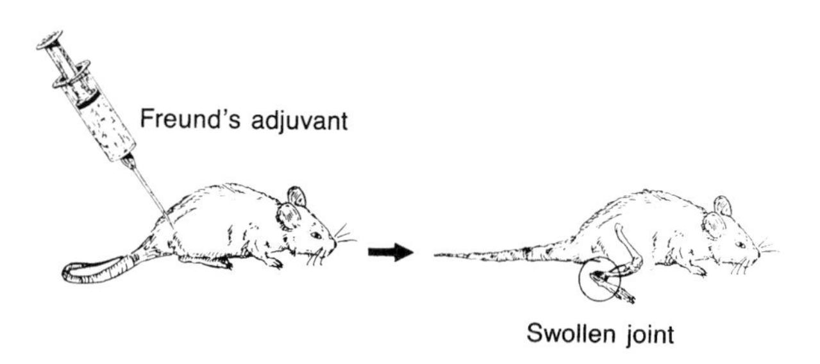

FIGURE 17.114 Adjuvant disease.

RA patients that are IgM rheumatoid factor negative. Their presence portends a poor prognosis in the rheumatoid factor negative group. Perinuclear antibodies may also be found in selected other rheumatic diseases and are often present in subjects infected with Epstein-Barr virus. They are also demonstrable in approximately one fourth of primary biliary cirrhosis patients.

Crystallographic antibodies are antibodies against antigenic crystals, such as those found in patients with gout who develop IgG antibodies against monosodium urate monohydrate (MSUM). These antibodies facilitate crystallization of more MSUM that can lead to reoccurrence of gout.

Phenylbutazone (Figure 17.112) is a drug that prevents synthesis of prostaglandins and serves as a powerful anti-inflammatory drug. It is used in therapy of RA and ankylosing spondylitis.

Ankylosing spondylitis (Figure 17.113) is a chronic inflammatory disease affecting the spine, sacroiliac joints, and large peripheral joints. There is a strong male predominance with onset in early adult life. The erythrocyte sedimentation rate is elevated, but subjects are negative for rheumatoid factor and antinuclear antibodies. Pathologically, there is chronic proliferative synovitis which resembles that seen in RA. The sacroiliac joints and interspinous and capsular ligaments ossify when the disease advances. There is a major genetic predisposition as revealed by increased incidence in selected families. Of ankylosing spondylitis patients, 90% are positive for HLA-B27, compared with 8% among Caucasians in the U.S. The HLA-B27 genes may be linked to genes that govern pathogenic autoimmunity. There may be increased susceptibility to infectious agents or molecular mimicry between HLA-B27 and an infectious agent such as *Klebsiella pneumoniae*, leading to the synthesis of a cross-reacting antibody. Treatment is aimed at diminishing inflammation and pain and providing physical therapy.

Adjuvant disease (Figure 17.114) may follow the injection of rats with Freund's complete adjuvant, a water-in-oil emulsion containing killed, dried mycobacteria, e.g., *Mycobacterium tuberculosis*, leads to the production of aseptic synovitis, which closely resembles RA in humans. Sterile inflammation occurs in the joints and lesions of the skin. In addition to swollen joints, inflammatory lesions of the tail may also result in animals developing adjuvant arthritis, which represents an animal model for RA.

Sjögren's syndrome (Figure 17.115 and Figure 17.116) is a condition in which immunologic injury to the lacrimal and salivary glands leads to dry eyes (keratoconjunctivitis sicca) and dry mouth (xerostomia). It may occur alone as sicca syndrome (primary form) or together with an autoimmune disease such as RA (secondary form). The lacrimal

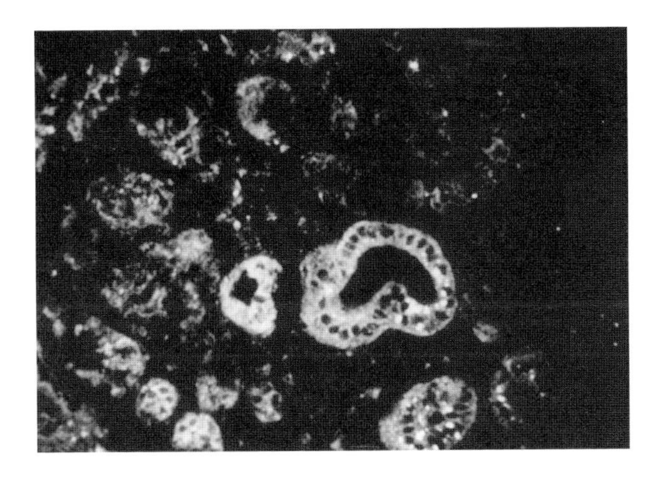

FIGURE 17.115 Sjögren's syndrome. Salivary gland. Antibody to ductal epithelium.

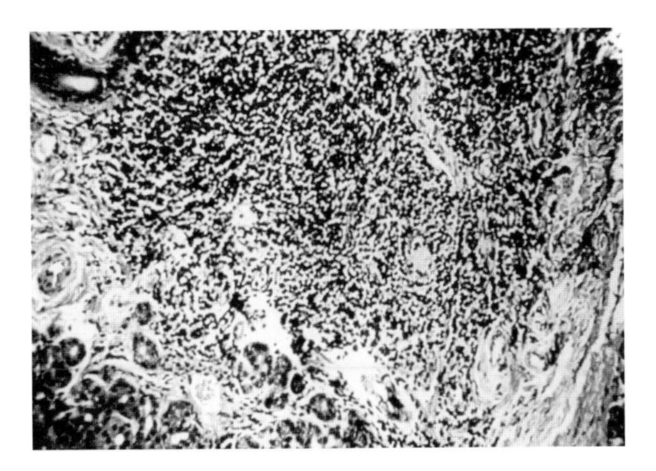

FIGURE 17.116 Sjögren's syndrome.

and salivary glands show extensive lymphocytic infiltration and fibrosis. Most of the infiltrating cells are $CD4^+$ T cells, but there are some B cells (plasma cells) that form antibody. Approximately 75% of the patients form rheumatoid factor. The LE cell test is positive in 25% of the patients. Numerous antibodies are produced, including autoantibodies against salivary duct cells, gastric parietal cells, thyroid antigens, and smooth muscle mitochondria. Antibodies against ribonucleoprotein antigens are termed SS-A (Ro) and SS-B (La). Approximately 90% of the patients have these antibodies. Anti-SS-B shows greater specificity for Sjögren's syndrome than do anti-SS-A antibodies which also occur in SLE. Sjögren's syndrome patients who also have RA may have antibodies to rheumatoid-associated nuclear antigen (RANA). There is a positive correlation between HLA-DR3 and primary Sjögren's syndrome and between HLA-DR4 and secondary Sjögren's syndrome associated with RA. Genetic predisposition, viruses, and disordered immunoregulation may play a role in the pathogenesis. About 90% of the patients are 40- to 60-year-old females. In addition to dry eyes and dry mouth, with associated visual or swallowing difficulties, 50% of the patients show parotid gland enlargement. There is drying of the nasal mucosa with bleeding, pneumonitis, and bronchitis. A lip biopsy to examine minor salivary glands is needed to diagnose Sjögren's syndrome. Inflammation of the salivary and lacrimal glands was previously called Mikulicz's disease. Mikulicz's syndrome refers to enlargement of the salivary and lacrimal glands due to any cause. Enlarged lymph nodes that reveal a pleomorphic cellular infiltrate with many mitoses are typical of Sjögren's syndrome and have been referred to as "pseudolymphoma." These patients have a 40-fold greater risk of lymphoid neoplasms than do others.

Experimental autoimmune sialoadenitis (EAS) is an animal model used to investigate the pathogenesis of salivary gland injury in Sjögren's syndrome. It may be induced in LEW rats by immunization with allogeneic or syngeneic submandibular gland tissue in complete Freund's adjuvant and B. *pertussis*. Autoantibodies are produced against antigens of salivary ducts as well as mononuclear cell infiltration of salivary tissues, comprised of $CD4^+$, $CD8^+$, and B lymphocytes. There is acute injury of glands, yet tissue damage resolves spontaneously 4 to 5 weeks after immunization. $CD4^+$ T lymphocytes have a critical role in EAS induction.

Mikulicz's syndrome is lymphocytic inflammation in the parotid gland. This condition represents a type of Sjögren's syndrome.

SSPE is an abbreviation for subacute sclerosing panencephalitis.

SS-A is anti-RNA antibody that occurs in Sjögren's syndrome patients. The antibody may pass across the placenta in pregnant females and be associated with heart block in these infants. SS-A Ro is an antigen in the cytoplasm to which one quarter of lupus erythematosus patients and 40% of Sjögren's syndrome synthesize antibodies.

A total of 60% of **Sjögren's syndrome (SS)** and 35% of **systemic lupus erythematosus (SLE)** patients have been shown by immunodiffusion to develop antibodies against SS-A/Ro antigen, (60-kDa and 52-kDa polypeptides complex with Ro RNAs). Upto 96% of primary Sjögren's syndrome patients and most individuals with SS secondary to RA or SLE have been shown by EIA to have SS-A antibodies. SS-A antibody levels are significantly higher in primary than in secondary (RA or SLE) Sjögren's syndrome. Cryostat sections of solid tissue and Hep-2 cell lines are used to detect SS-A/Ro and SS-B/La antibodies. There is a 5% risk of congenital heart block in babies born to mothers with SS-A/Ro antibodies. Passage of maternal SS-A antibodies across the placenta signifies neonatal lupus erythematosus. These antibodies from the mother may mediate congenital heart block. Mothers of babies with neonatal lupus erythematosus may range from asymptomatic to sicca syndrome or SLE. SS-A/Ro antibodies are demonstrable in cases of C2 and C4 deficiency, in subacute cutaneous lupus, and in vasculitis with Sjögren's syndrome. Fewer than 1% of normal subjects have low levels of SS/A/Ro antibodies, detectable by EIA. The 2 to 5% of lupus patients who are ANA negative may manifest subacute cutaneous lupus, positive IgM-RF, and SS-A antibodies. Approximately half of patients with Sjögren's syndrome and those with SLE react with the 60-kDa as well as the 52-kDa components of the SS-A/Ro particle that is comprised of these proteins and RNA. Up to 40% of patients with Sjögren's syndrome react with the 52-kDa protein alone, whereas 20% of SLE patients react only with the 60-kDa component. Most patients with neonatal lupus erythematosus and complete heart block reveal antibodies against both 52-kDa SS-A/Ro protein and 48-kDa SS-B/La protein. Fetal cardiac tissue of 18 to 24 week gestation contains these antigens in significant quantities. The synthesis of SS-A, but not SS-B, antibodies within the blood–brain barrier has been reported in cerebral vasculopathy.

SS-B is an anti-RNA antibody detectable in patients with Sjögren's syndrome as well as other connective tissue (rheumatic) diseases. SS-B/La is an antigen in the cytoplasm to which Sjögren's syndrome and lupus erythematosus patients form antibodies. Anti-SS-B antibodies may portend a better prognosis in patients with lupus erythematosus.

Progressive systemic sclerosis (scleroderma) (Figure 17.117) is a connective tissue or collagen–vascular disease in which the skin and submucosal connective tissue become thickened and scarred. There is increased collagen deposition in the skin. It is slowly progressive and chronic and may involve internal organs. The female to male ratio is 2:1. Although the etiology is unknown, patients demonstrate antinuclear antibodies, rheumatoid factor, and polyclonal hypergammaglobulinemia. There is no demonstrable immunoglobulin at the dermal–epidermal junction. These patients may have altered cellular immunity. The epidermis is thin, dermal appendages atrophy, and rete pegs are lost. There is also a marked increase in collagen deposition in the reticular dermis together with fibrosis and hyalinization of arterioles. The GI tract may also reveal increased collagen deposition in the lamina propria, submucosa, and muscularis layers. At the onset, 90% of affected individuals experience Raynaud's phenomenon. Skin changes are usually the initial manifestation with involvement of the hands, feet, forearms, and face, or possibly diffuse involvement of the trunk. A variation of the disease is called the CREST syndrome, which consists of calcinosis, Raynaud's, esophageal dysmotility, sclerodactyly, and telangiectasia. This form of the disease may become stabilized for a number of years. The skin may exhibit a tight, smooth, and waxy appearance in the sclerotic phase with no wrinkles or folds apparent. Ulcers may develop on the fingertips in many patients, with a mask-like appearance of the face with thin lips. The skin may either become atrophic or return to a normal soft structure. The lungs may be involved, leading to dyspnea on exertion. Pulmonary fibrosis may lead to cor pulmonale. The principal immunologic findings include antinuclear antibodies with a speckled or nucleolar pattern, anticentromere antibodies that are found in individuals with known CREST syndrome, and the development of antibodies specific for acid-extractable nuclear antigen. Approximately one third of the individuals with diffuse involvement of the trunk reveal antibodies specific for topoisomerase (anti-Scl-70 antibodies).

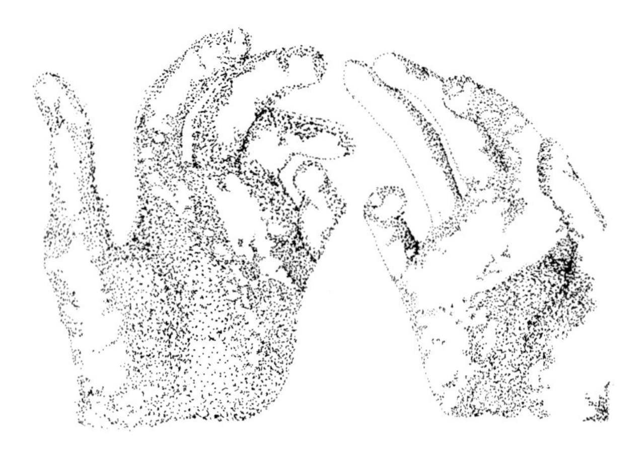

FIGURE 17.117 Systemic sclerosis.

Tight skin-1 mouse (Tsk1) is a mouse strain that developed from a spontaneous dominant mutation in the inbred stain B10.D2(58N)/Sn. It represents a genetically transmitted model of systemic sclerosis. The effects of the Tsk1 mutation include an excessive accumulation of collagen in the dermis and various internal organs, thereby mimicking major aspects of human systemic sclerosis. The principal visceral changes in $Tsk1/^{+}$ occur in the lungs and heart. Lungs are greatly distended and histologically resemble human emphysema with little fibrosis. Alveolar spaces are markedly dilated with thin disrupted walls and subpleural cysts and bullae. There is myocardial hypertrophy with increased collagen deposition.

Tight skin-2 mouse (Tsk2) is a mouse strain with features resembling those of both human systemic sclerosis and the original tight skin mutation of the mouse, Tsk1.

Anti-topoisomerase I (Scl 70) antibodies are associated with diffuse cutaneous systemic sclerosis and are specific to this disease. This renders Scl 70 as an important marker for systemic sclerosis.

Scl-70 antibody is an antibody found in as many as 70% of diffuse-type scleroderma (progressive systemic sclerosis)

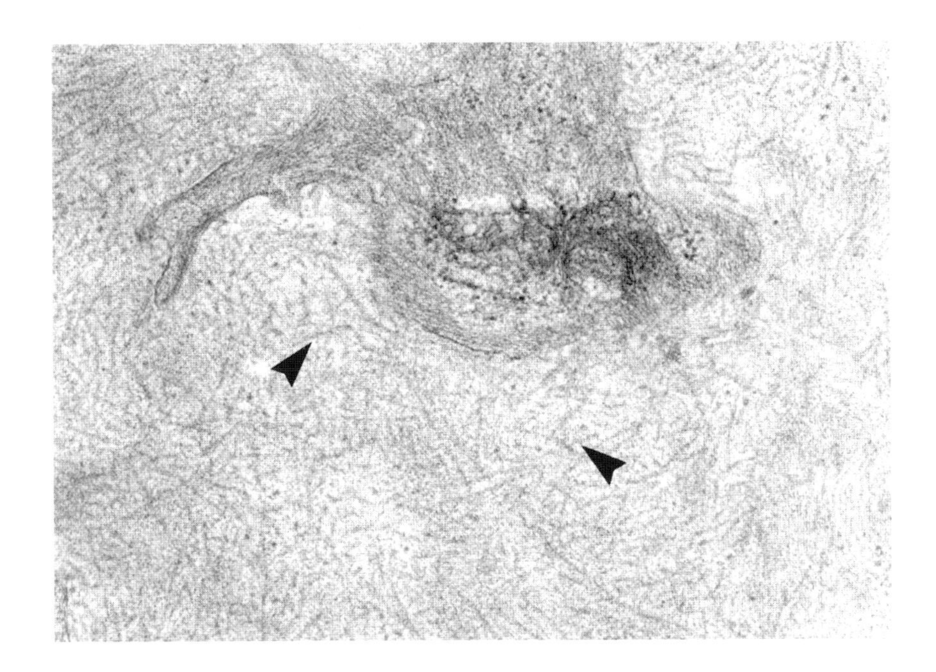

FIGURE 17.118 Amyloidosis. Amyloid fibrils.

patients, who experience extensive and rapid skin involvement as well as early visceral manifestations.

CREST complex includes calcinosis, Raynaud's phenomenon, esophageal dysmotility, sclerodactyly, and telangiectasia associated with mixed connective tissue disease. The prognosis of CREST is slightly better than that of other connective tissue diseases, but biliary cirrhosis and pulmonary hypertension are complications. CREST patients may develop anticentromere antibodies which may also occur in progressive systemic sclerosis patients, in aged females, or in individuals with HLA-DR1. CREST syndrome is a relatively mild clinical form of scleroderma (progressive systemic sclerosis. CREST is an acronym for calcinosis, Raynaud's phenomenon, esophageal dysmotility, sclerodactyly, and telangiectasia. Skin lesions are usually limited to the face and fingers with only lateral visceral manifestations.

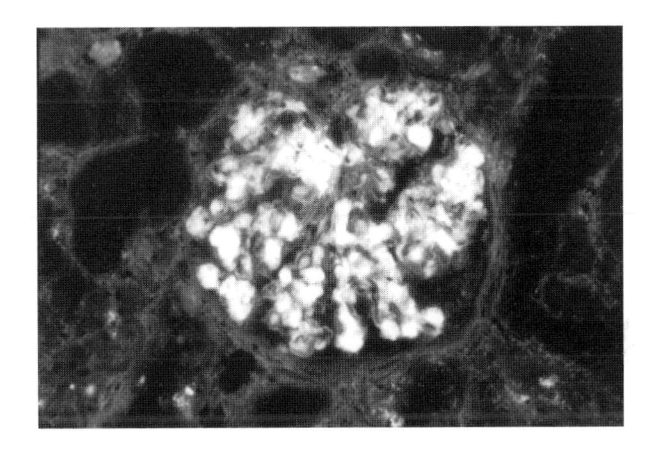

FIGURE 17.119 Amyloidosis. (Kidney)

GAMMAPATHIES

Amyloidosis (Figure 17.118 to Figure 17.120) is a constellation of diseases characterized by the extracellular deposition of fibrillar material that has a homogeneous and eosinophilic appearance in conventional staining methods. It may compromise the function of vital organs. Diseases with which it is associated may be inflammatory, hereditary, or neoplastic. All types of amyloid link to Congo red and manifest an apple-green birefringence when viewed by polarizing light microscopy after first staining with Congo red. By electron microscopy, amyloid has a major fibrillar component and a minor rod-like structure which is shaped like a pentagon with a hollow core

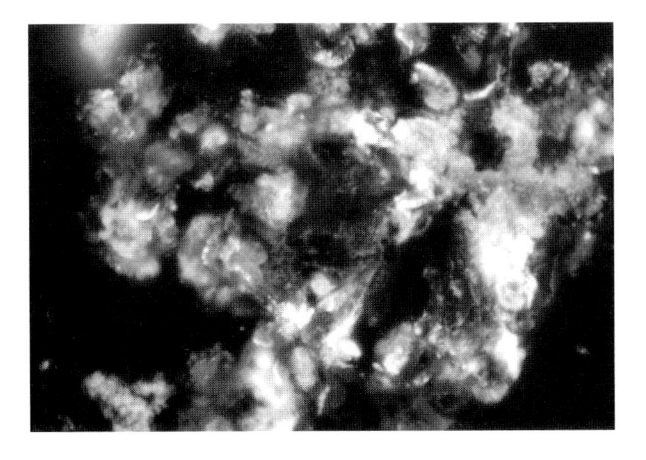

FIGURE 17.120 Amyloidosis. (Kidney)

when observed on end, i.e., the P component. All forms of amyloid share the P component in common. It is found as a soluble serum protein in the circulation (SAP). Amyloid has a b-pleated sheet structure; it is insoluble in physiologic saline but is soluble in distilled water.

The classification of amyloidosis depends upon the clinical presentation, anatomic distribution, and chemical content of the amyloid. In the U.S., AL amyloid is the most common type of amyloidosis which occurs in association with multiple myeloma and Waldenström's macroglobulinemia. These patients have free light-chain production in association with the development of Bence-Jones proteins in myeloma. The light-chain quality and degradation mechanisms are critical in determining whether or not Bence-Jones proteins will be deposited as amyloid. Chronic inflammation leads to increased levels of SAA, which are produced by the liver following IL-6 and IL-1 stimulation. Normally, SSA is degraded by the enzymes of monocytes. Thus, individuals with a defect in the degradation process could generate insoluble AA molecules. Likewise, there could be a defect in the degradation of immunoglobulin light chains in subjects who develop AL amyloidosis.

Amyloidosis secondary to chronic inflammation is severe with kidney, liver, spleen, lymph node, adrenal, and thyroid involvement. These secondary amyloidosis deposits consist of amyloid A protein (AA), which makes up 85 to 90% of the deposits, and serum amyloid P component, which accounts for the remainder of the deposit. The AL type of amyloidosis more often involves the heart, gastrointestinal tract, respiratory tract, peripheral nerves, and tongue. Amyloidosis may also be heredofamilial or associated with aging.

Sago spleen describes the replacement of lymphoid follicles by amyloid deposits that are circular and transparent in amyloidosis of the spleen.

Serum amyloid A component (SAA) is a 12-kDa protein in serum which is a precursor of the AA class of amyloid fibril protein. SAA is formed in the liver and associates with HDL3 lipoproteins in the circulation. It is a 114-amino acid residue polypeptide chain. Its conversion to AA involves splitting of peptides from both amino and carboxy terminals to yield an 8.5-kDa protein that forms fibrillar amyloid deposits. There may be a 1000-fold increase in SAA levels during inflammation.

Serum amyloid P component (SAP) is a protein in the serum that constitutes a minor second component in all amyloid deposits. By electron microscopy, it appears to be a doughnut-shaped pentagon with an external diameter of 9 nm and an internal diameter of 4 nm. There are five globular subunits in each pentagon. The amyloid P component in serum consists of two doughnut-shaped structures. Unlike serum amyloid A component, it does not increase during inflammation. The P component makes up 10% of amyloid deposits and is indistinguishable from normal α_1 serum glycoprotein. This 180- to 212-kDa substance shows close structural homology with C-reactive protein.

Monoclonal immunoglobulin deposition disease (MIDD) is a condition in which monotypic light or heavy chains are deposited in the tissues. The deposits may be fibrillar or nonfibrillar. The AL type of amyloidosis, characterized by light chains and amyloid P component, represents an example of fibrillar deposits, whereas light chain or light and heavy chain diseases represent nonfibrillar deposits. Patients may develop azotemia, albuminuria, hypogammaglobulinemia, cardiomyopathy, or nephropathy. Some of them may develop multiple myeloma or another plasma cell neoplasm. Another variety of MIDD is amyloid H, which is comprised of V_H, D_D, J_H, and C_H3 domains.

Amyloid (Figure 17.121 and Figure 17.122) is an extracellular, homogenous, and eosinophilic material deposited in various tissues in disease states designated as primary and secondary amyloidosis. It is composed chiefly of protein and shows a green birefringence when stained with

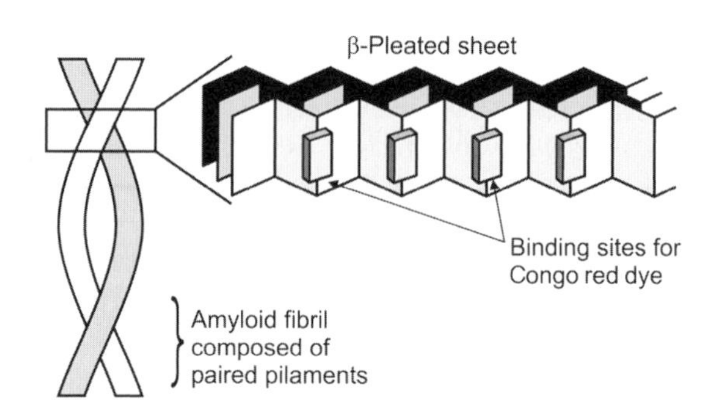

FIGURE 17.121 Amyloid.

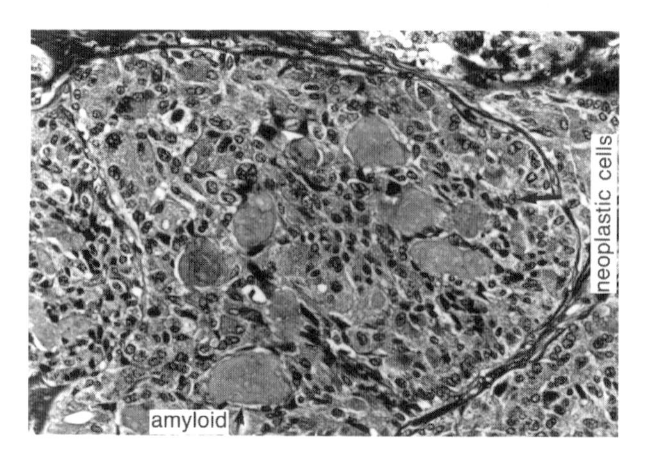

FIGURE 17.122 Medullary carcinoma of thyroid with amyloid deposition.

Congo red and observed by polarizing light microscopy. By electron microscopy, the fibrillar appearance is characteristic. By x-ray crystallography, it shows a b-pleated sheet structure arranged in an antiparallel fashion. The amino termini of the individual chains face opposite directions, and the chains are bound by hydroxyl bonds. Amyloid consists of two principal and several minor biochemical varieties. Pathogenetic mechanisms for its deposition differ, although the deposited protein appears similar from one form to another. Amyloid consists of nonbranching fibrils 7.5 to 10 nm wide and are of indefinite length. X-ray crystallography reveals a b-pleated sheet configuration which gives the protein its optical and staining properties. It also has a P component that is nonfibrillary, is pentagonal in structure, and constitutes a minor component of amyloid. Chemically, amyloid falls into two principal classes, i.e., AL, consisting of amyloid light chains, and AA (amyloid associated), comprising a nonimmunoglobulin protein called AA. These molecules are antigenically different and have dissimilar deposition patterns based on the clinical situation. AL amyloid is comprised of whole immunoglobulin light chains, their N-terminal fragments, or a combination of the two. λ Light chains rather than κ are the ones usually found in AL. Proliferating immunoglobulin-producing B cells, as in B cell dyscrasias, produce AL amyloid protein. AA amyloid fibroprotein is not an immunoglobulin and has a mol wt of 8.5 kDa. Serum amyloid-associated protein (SAA) is the serum precursor of AA amyloid. It constitutes the protein constituent of a high-density lipoprotein and acts as an acute-phase reactant. Thus, its level rises remarkably within hours of an acute inflammatory response. AA protein is the principal type of amyloid deposited in the tissues during chronic inflammatory diseases. Several other distinct amyloid proteins also exist.

Protein AA is an 8.5-kDa protein isolated by gel filtration and comprised of 76 amino acids. The sequence has no relationship to that of immunoglobulins or to other known sequences of human proteins. The sequence of protein AA extracted from various tissues appears identical, but several genetic variants are identified by sequence analysis. The molecule lacks cysteine and accordingly has no crosslinks. The N-terminal residue is arginine (in humans), and in some, the second amino acid is phenylalanine. It also lacks carbohydrate or other attached small molecules. Amyloid protein AA is insoluble in ordinary aqueous solvents. Protein AA is the predominant component in the secondary form of amyloidosis.

Protein P is a 23-kDa pentameric protein detectable in amyloid deposits.

Protein SAA is a soluble precursor of AA which is present in minor quantities in the serum. It has a mol wt of 100 kDa. SAA is present in the cord blood and during the first three decades of life and has an average concentration of 50 ± 40 ng/ml. In the following three decades of life it shows a slow but steady increase in concentration, with doubling of the level in the eighth decade. Levels of the serum precursor are elevated in almost all patients with amyloidosis except in cases of extreme protein dysfunction due to severe nephrosis. Levels reaching 700 ng/ml are detectable in patients with tuberculosis, lymphoma, carcinoma, and leukemia, but SAA levels have no value in the diagnosis of amyloidosis. High levels of SAA are seen in the secondary form of the disease. The level of SAA may increase transiently during various infections.

AA amyloid is a nonimmunoglobulin amyloid fibril of the type seen following chronic inflammatory diseases such as tuberculosis and osteomyelitis or, more recently, chronic noninfectious inflammatory disorders. Kidneys, liver, and spleen are the most significant areas of AA (amyloid-associated) deposition. The precursor for AA protein is apo-SAA (serum amyloid-associated), with a monomer mol wt of 12.5 kDa which is found in the circulation as a 220- to 235-kDa molecular complex because it is linked to high-density lipoproteins. IL-6 stimulates its synthesis. AA deposition is either associated with an amyloidogenic isotypical form of SAA or results from the inability to completely degrade SAA. Amyloid consists of nonbranching fibrils 7.5 to 10 nm in width and of indefinite length. Chemically, amyloid occurs in two classes. The AL (amyloid light) chain type consists of immunoglobulin light chains or parts of them. The AA-type is derived from the SAA protein in the serum. SAA acts like an acute-phase reactant increasing greatly during inflammation. Thus, AA protein is the principal type of amyloid deposited in chronic inflammatory diseases.

AL amyloid consists of either whole immunoglobulin light chains or their N-terminal fragments or a combination of the two. The λ light chain especially gives rise to AL. AL amyloid protein is often deposited following or during B cell disorders. Other biochemical forms of amyloid include transthyretin, β_2 microglobulin, β_2 amyloid protein, or additional forms. Amyloid filaments stained with Congo red exhibit green birefringence with polarized light.

Amyloid P component has a mol wt of 180 kDa. It migrates in electrophoresis with the α globulin fraction, and by electron microscopy has a pentagonal shape, suggesting that it consists of subunits linked by hydrogen bonds. It is a minor component of all amyloid deposits and is nonfibrillar. It is a normal a1 glycoprotein and has close structural homology with the C-reactive protein. It has an affinity for amyloid fibrils and accounts for their PAS positive staining quality.

Multiple myeloma (Figure 17.123 to Figure 17.125) is a clinical condition in which a plasmacytoma or plasma cell

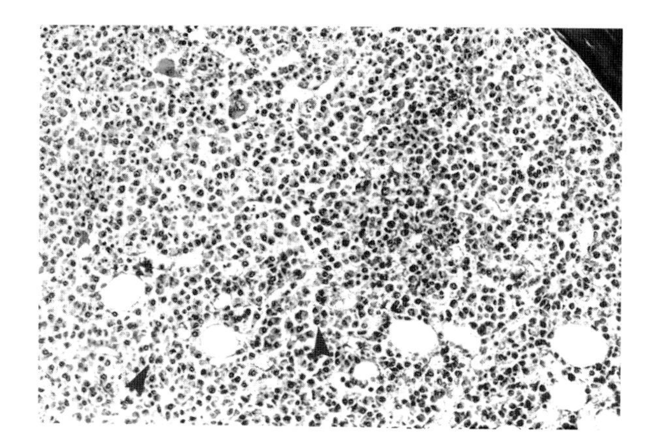

FIGURE 17.123 Multiple myeloma. Bone marrow plasma cell myeloma.

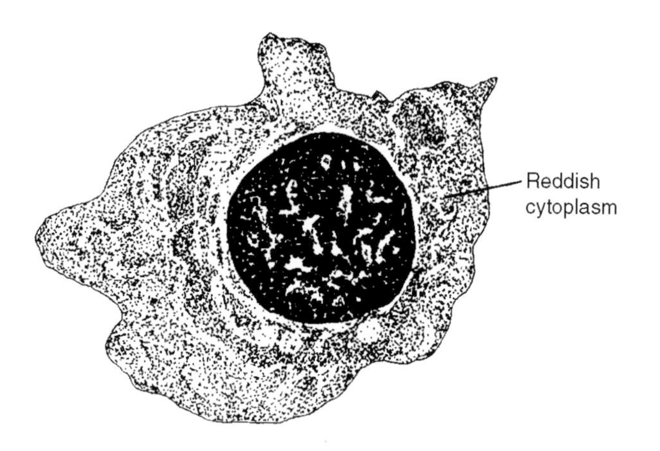

FIGURE 17.124 Flame cell.

neoplasm is associated with the production of a paraprotein that appears in the serum. The neoplastic plasma cells usually synthesize and secrete monoclonal, highly homogenous immunoglobulins. Serum electrophoresis reveals a narrow monoclonal band in 98 to 99% of patients. Up to 80% of myeloma patients manifest IgG paraimmunoglobulin, while 15% have monoclonal IgA. A few cases of the IgD and IgE types have been described. Homogeneous light chain dimers, which are identical to the corresponding light chain portion of immunoglobulin in the individual's blood, appear in the urine. These light chain dimers in the urine are called Bence-Jones proteins. These segments of light polypeptide chains do not represent degradation products of immunoglobulin, since they are synthesized separately from it. The disease affects 3 in 100,000 persons, usually men over 50 years of age. Patients develop anemia, anorexia, and weakness. The tumor infiltrates the bone marrow cavities, ultimately leading to erosion of the bone cortex. This may take years. Osteolytic lesions are the hallmark of multiple myeloma. The long bones, ribs, vertebrae, and skull manifest diffuse osteoporosis which leads to the appearance of punched-out

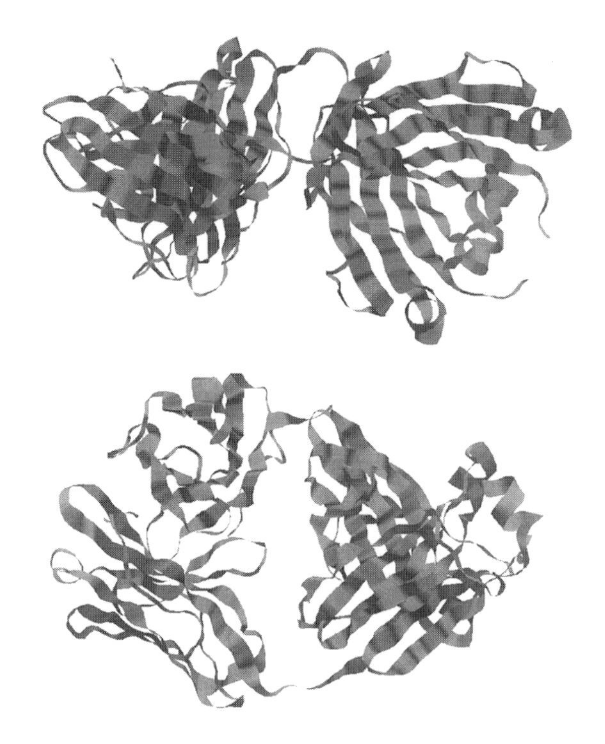

FIGURE 17.125 Immunoglobulin λ chain dimer (MCG) complex with *N*-acetyl-L-GLN-D-PHE-L-HIS-D-PRO-OH from a patient with multiple myeloma and amyloidosis.

areas and pathologic fractions. Tumor invasion of the marrow and erosion of the cortex, as well as osteoclast-activating substances, produce the bone lesions. Lung or renal infections may also occur. Hypogammaglobulinemia results from decreased functioning of normal plasma cells that leads to diminished antibody to combat infections. The malignant plasma cells produce an excess of nonsense paraimmunoglobulin, which does not protect against infection. There is also defective phagocytic activity. Patients may have altered B cell function and increased susceptibility to pyogenic infections. Some patients may develop myeloma kidney, signified by proteinuria, followed by oliguria, kidney failure, and possibly death.

Ig myeloma subclasses: The most frequent myeloma paraimmunoglobulin is IgG-κ, which belongs to the IgG1 subclass most commonly and to the IgG4 subclass only rarely. IgG3 paraimmunoglobulins are associated with increased serum viscosity, which leads to aggregates that are related to concentration and may be associated with serum cryoglobulin in an individual patient. In IgG paraimmunoglobulin myeloma, the serum concentration of IgG may reach 2000 mg/dl. Concomitantly there is a reduction in other immunoglobulins. Since the IgG myeloma paraimmunoglobulin belongs to one heavy-chain subtype, the Gm allotypes are restricted in a manner similar to the restriction of light chains to either the κ or λ type.

M protein refers to monoclonal immunoglobulin or immunoglobulin components such as a myeloma protein. The M protein represents 3 to 10% of the total serum proteins. This level remains constant throughout life or decreases with age. Group A β hemolytic streptococcal type-specific cell surface antigens, such as streptococcal M protein.

Myeloma is a plasmacytoma that represents proliferation of a neoplastic plasma cell clone in bone marrow with the production of a monoclonal immunoglobulin paraprotein. Besides myeloma that occurs in humans, the experimental variety can be induced in certain inbred mouse strains such as BALB/c by intraperitoneal injection of mineral oil.

IgA paraproteinemia is myeloma in which the paraimmunoglobulin belonging to the IgA class occurs in only about one fifth of affected patients. IgA myeloma patients are reputed to have lower survival rates than those patients affected with IgG myeloma, but this claim awaits confirmation. IgA myeloma may be associated with myeloma cell infiltration of the liver, leading to jaundice and altered liver function tests. It is frequently associated with hyperviscosity syndrome, which may be related to the ability of IgA molecules to polymerize as well as to form complexes with such substances as haptoglobin, α-1 antitrypsin, β lipoprotein, antihemophilic factor, and albumin.

Paraprotein refers to homogeneous, monoclonal immunoglobulin molecules synthesized by an expanding clone of plasma cells, as observed in patients with multiple myeloma or with Waldenström's macroglobulinemia. The homogeneity of the paraprotein is reflected by all molecules belonging to the same immunoglobulin class and subclass as well as the same light chain type. On electrophoresis, a serum paraprotein appears as a distinct band. This is a consequence of a biologic event such as neoplastic transformation rather than antigenic stimulation.

Myeloma protein is the immunoglobulin synthesized in excess in patients with multiple myeloma (plasmacytoma). Myeloma proteins are products of proliferating plasma cells of a malignant clone. The heavy and light chains are usually assembled to produce the homogeneous monoclonal paraimmunoglobulin, but if the synthesis of light chains is exclusive or exceeds that of heavy chains, a Bence-Jones protein may appear in addition to the paraimmunoglobulin or it may occur alone. A myeloma protein may be either a whole molecule of monoclonal immunoglobulin or part of the molecule synthesized by malignant plasma cells.

Myeloma, IgD is a myeloma in which the monoclonal immunoglobulin is IgD. It constitutes 1 to 2% of myelomas and usually occurs in older males. There is lymphadenopathy and hepatosplenomegaly. Approximately one half of the cases develop dissemination beyond the bones. Patients develop osteolysis, hypercalcemia, anemia, azotemia, and aberrant plasma cells and plasma blasts.

IgM paraproteinemia: Occasional cases of IgM myeloma occur and are characterized by infiltration of the bone marrow by plasmacytes, numerous osteolytic lesions, and occasionally bleeding diathesis. IgM myeloma is distinct from true Waldenström's macroglobulinemia, but is rare compared with the IgG and IgA types. Myelomas associated with the minor classes of immunoglobulin IgD and IgE occur with even less frequency than do the IgM myelomas.

Flame cells are plasma cells whose cytoplasm stains intensely eosinophilic and contains glycoprotein globules. They occur in IgA myelomas especially, but also in Waldenström's macroglobulinemia and in the leptomeninges of African trypanosomiasis patients near neutrophil aggregates.

H-chain disease is a variant of multiple myeloma in which immunoglobulin free heavy chains are demonstrable in the patient's serum. This condition, characterized by free H chains in the serum but not free light chains, is called heavy-chain disease.

Pseudolymphomatous lymphadenitis refers to hyperplasia of lymphoid organs that is similar to lymphoma except that the hyperplasia is reversible in this condition.

Myelomatosis is a condition in which bone marrow plasma cells undergo malignant transformation and produce excessive homogeneous monoclonal immunoglobulin molecules that represent a paraprotein of a specific immunoglobulin isotype such as IgG or IgA. IgD and IgE myelomas also occur from time to time. Serum electrophoresis reveals a clearly demarcated band. Isoelectric focusing shows a classic monoclonal banding pattern. Some of the patients also have Bence-Jones protein in the urine. Patients are often males past 50 years of age and commonly present with spontaneous bone fracture or anemia due to replacement of bone marrow.

Heavy-chain diseases: Monoclonal gammopathies or paraproteinemias associated with lymphoproliferative disease and characterized by excess synthesis of Fc fragments of immunoglobulins that appear in the serum and/or urine. The most common is the α heavy-chain disease (Seligmann's disease) in which patients produce excess incomplete α chains of IgA1 molecules. It mainly affects Sephardic Jews, Arabs, and other Mediterranean residents, as well as subjects in South America and Asia. It may appear in childhood or adolescence as a lymphoproliferative disorder associated with the respiratory tract or gastrointestinal tract. Patients may manifest malabsorption, diarrhea, steatorrhea, hepatic dysfunction, weight loss, lymphadenopathy, hypocalcemia, and extensive mononuclear infiltration.

α Heavy-chain disease may either spontaneously remit or respond favorably to treatment with antibiotics. It may require chemotherapy in some cases. δ Heavy-chain disease has been reported in an elderly male demonstrating abnormal plasma cell infiltration of the bone marrow together with osteolytic lesions. γ Heavy-chain disease (Franklin's disease) occurs in older men and may either induce death rapidly (within weeks) or last for years. Death usually takes place within 1 year because of infection. Patients develop lymphoproliferation with fever, anemia, fatigue, angioimmunoblastic lymphadenopathy, hepatosplenomegaly, eosinophilic infiltrates, leukopenia, lymphoma, autoimmune disease, or tuberculosis. There is elevated IgG1 in the serum. It has been treated with cyclophosphamide, vincristine, and prednisone. μ Chain disease is a rare condition of middle-aged to older individuals. Some patients may ultimately develop chronic lymphocytic leukemia. μ Chain disease is characterized by lymphadenopathy, hepatosplenomegaly, infiltration of the bone marrow by vacuolated plasma cells, and frequently by elevated synthesis of κ light chain.

Gammopathy is an abnormal increase in immunoglobulin synthesis. Gammopathies that are monoclonal usually signify malignancy such as multiple myeloma, Waldenström's disease, heavy-chain disease, or chronic lymphocytic leukemia. Benign gammopathies occur in amyloidosis and monoclonal gammopathy of undetermined etiology. Inflammatory disorders are often accompanied by benign polyclonal gammopathies. These include rheumatoid arthritis, lupus erythematosus, tuberculosis, cirrhosis, and angioimmunoblastic lymphadenopathy.

Hypergammaglobulinemia: Elevated serum γ globulin (immunoglobulin) levels. A polyclonal increase in immunoglobulins in the serum occurs in any condition where there is continuous stimulation of the immune system, such as chronic infection, autoimmune disease, SLE, etc. Hypergammaglobulinemia may also result from a monoclonal increase in immunoglobulin production, as in multiple myeloma, Waldenström's macroglobulinemia, or other conditions associated with the formation of monoclonal immunoglobulins. Repeated immunization may also induce hypergammaglobulinemia.

M component is a spike or defined peak observed on electrophoresis of serum proteins, which suggests monoclonal proliferation of mature B lymphocytes synthesizing immunoblogulin G (IgG), IgA, or IgM. M component can be seen in such diseases as multiple myeloma, heavy-chain disease, and Waldenström's macroglobulinemia.

Light-chain disease is a paraproteinemia termed Bence-Jones myeloma which makes up one fifth of all myelomas. Excess monoclonal light chains are produced. These are linked to renal amyloidosis and renal failure as a consequence of blockage of tubules by certain Bence-Jones proteins. Four fifths of patients have monoclonal light chains in the blood circulation, and 60% have diminished γ globulin and lytic lesions of the bone. The λ type is usually more severe than the κ type of light-chain disease. Light-chain disease patients have a worse course than do patients with IgA or IgG myelomas.

Monoclonal gammopathy is a pathologic state characterized by a monoclonal immunoglobulin (M component) in the patient's serum. Examples include multiple myeloma, benign monoclonal gammopathy, and Waldenström's macroglobulinemia.

Plasmacytoma is a plasma cell neoplasm. Also termed myeloma or multiple myeloma. To induce an experimental plasmacytoma in laboratory mice or rats, paraffin oil is injected into the peritoneum. Plasmacytomas may occur spontaneously. These tumors, comprised of neoplastic plasma cells, synthesize and secrete monoclonal immunoglobulins, yielding a homogeneous product that forms a spike in electrophoretic analysis of the serum. Plasmacytomas were used extensively to generate monoclonal immunoglobulins prior to the development of B cell hybridoma technology to induce monoclonal antibody synthesis at will to multiple antigens.

Monoclonal gammopathy of undetermined significance (MGUS) is a benign monoclonal gammopathy. It is a condition in which the serum contains an M component, serum albumin is less than 2 g/dl, and there is no Bence-Jones protein in the urine. There are no osteolytic lesions, and plasma cells comprise less than 5% of bone marrow constituents. Of these patients, 20 to 40% ultimately develop a monoclonal malignancy. When monoclonal spikes exceed 2 g/dl with other immunoglobulins decreased and no Bence-Jones protein in the urine, the condition is usually malignant. Only 1 to 2% of myelomas are nonsecretory monoclonal gammopathies. More than half of the patients with elevated monoclonal IgM continue as MGUS, with other cases evolving into Waldenström's macroglobulinemia and a few progressing to lymphoma or chronic lymphocytic leukemia.

Plasma cell dyscrasias are lymphoproliferative disorders in which there is monoclonal plasma cell proliferation, leading to such conditions as multiple myeloma or to the less ominous extramedullary plasmacytoma.

Plasma cell leukemia is a malignancy associated with plasma cells in the circulating blood that constitute greater than 20% of the leukocytes. The absolute plasma cell number is more than 2000 per cubic millimeter of blood. Advanced cases reveal extensive infiltration of the tissue with plasma cells and replacement of the marrow. Reactive plasmacytosis must be considered in the diagnosis.

MGUS: See monoclonal gammopathy of undetermined significance.

Benign monoclonal gammopathy is a paraproteinemia that occurs in normal healthy subjects who develop the serum changes characteristic of myeloma, i.e., a myeloma protein-type immunoglobulin spike on electrophoresis. They have none of the clinical signs and symptoms of multiple myeloma and have an excellent prognosis.

Heavy-chain disease is a type of myelomatosis in which aberrant monoclonal immunoglobulin μ chains are present in the serum but not in the urine, and Bence-Jones proteins are present in the urine. Although very uncommon, this condition may be associated with chronic lymphocytic leukemia or reticulum cell sarcoma. Vacuolated plasma cells have been demonstrated in the bone marrow and are very suggestive of the diagnosis of μ heavy-chain disease. The μ chains produced by bone marrow plasma cells have deletions in the variable region and involve the C_H1 domain, but have a normal sequence in the C_H2 domain. Light chains synthesized by these patients are not incorporated into molecular IgM; therefore, these individuals demonstrate a distinct failure in assembly of immunoglobulin molecules. Heavy and light chains have different electrophoretic mobilities, which becomes an important observation in establishing a diagnosis of μ heavy-chain disease by electrophoresis.

Cryofibrinogenemia is cryofibrinogen in the blood that is either primary or secondary to lymphoproliferative and autoimmune disorders, tumors, acute or chronic inflammation. Cryofibrinogenemia is often associated with IgA nephropathy.

Polyclonal hypergammaglobulinemia is an elevation in the blood plasma of γ globulin which is comprised of increased quantities of the different immunoglobulin classes rather than an increase of just one immunoglobulin class.

Cryoglobulin is a serum protein that precipitates or gels in the cold. It is an immunoglobulin that gels or precipitates when the temperature falls below 37°C. Cryoglobulins undergo reversible precipitation at cold temperatures. Most of them are complexes of immunoglobulin molecules, but nonimmunologic cryoprecipitate proteins such as cryofibrinogen or C-reactive protein–albumin complexes may also occur. When the concentration of cryoglobulins is relatively low, precipitation occurs near 4°C, but if the concentration is greater, precipitation occurs at a higher temperature. Cryoglobulins are usually associated with infectious, inflammatory, and neoplastic processes. They are found in different body fluids and they also appear in the urine. Cryoglobulins are divisible into three groups: type I are monoclonal immunoglobulins, usually IgM, associated with malignant B cell neoplasms; type II consists of mixed cryoglobulins with a monoclonal constituent specific for polyclonal IgG; and type III are mixed cryoglobulins comprised of polyclonal immunoglobulins as immunoglobulin–antiimmunoglobulin complexes. Cryoglobulin is not present in normal serum. See cryoglobulinemia.

Umbrella effect is the masking, by relatively large amounts of IgG, of low immunoglobulin light-chain concentrations in early IgM macroglobulinemia and IgA myeloma. This is observed in immunoelectrophoresis. Immunofixation electrophoresis, employing fluorochrome-labeled antibodies, can resolve this masking effect.

α Heavy-chain disease is a rare condition in individuals of Mediterranean extraction who may develop gastrointestinal lymphoma and malabsorption with loss of weight and diarrhea. The aberrant plasma cells infiltrating the lamina propria of the intestinal mucosa and the mesenteric lymph nodes synthesize α chains alone, usually α-1, with no production of light chains. Even though the end-terminal sequences are intact, a sequence stretching from the V region through much of the Ca-1 domain is deleted. Thus, there is no cysteine residue to crosslink light chains. α Heavy-chain disease is more frequent than the γ type. It has been described in North Africa, the Near East, the Mediterranean area, and some regions of southern Europe. Rare cases have been reported in the U.S. The condition may prove fatal, even though remissions may follow antibiotic therapy.

Cryoglobulinemia refers to cryoglobulins in the blood that are usually monoclonal immunoglobulins, i.e., IgG or IgM. Polymeric IgG3 may be associated with cryoglobulinemia in which the protein precipitates in those parts of the body exposed to cooling. Cryoglobulinemia patients develop embarrassed circulation following precipitation of the protein in peripheral blood vessels. This may lead to ulcers on the skin and to gangrene. More commonly, patients may manifest Raynaud's phenomenon following exposure of the hands or other parts of the anatomy to cold. Cryoglobulins may be detected in patients with Waldenström's macroglobulinemia, multiple myeloma, or SLE. Cryoglobulinemias are divisible into three types: type I monoclonal cryoglobulinemia, which is often associated with a malignant condition, i.e., IgG-multiple myeloma, IgM-macroglobulinemia, lymphoma or chronic lymphocytic leukemia, and benign monoclonal gammopathy; type II polymonoclonal cryoglobulinemia with mixed immunoglobulin complexes such as IgM–IgG, IgG–IgG, and IgA–IgG that may be linked to connective tissue disease such as RA, or Sjögren's syndrome, or lymphoreticular disease; and type III mixed polyclonal–polyclonal cryoglobulinemia with IgG and IgM mixtures, rarely including IgA, associated with infections, lupus erythematosus, rheumatoid arthritis, Epstein-Barr and cytomegalovirus inclusion virus, Sjögren's syndrome, crescentic and membranoproliferative glomerulonephritis, subacute bacterial infections, biliary cirrhosis, diabetes mellitus, and chronic active hepatitis.

Dysgammaglobulinemia: A pathologic condition associated with selective immunoglobulin deficiency, i.e., depression of one or two classes of serum immunoglobulins and normal or elevated levels of the other immunoglobulin classes. The term has also described antibody-deficient patients whose antibody response to immunogenic challenge is impaired even though their immunoglobulin levels are normal. Use of the term is discouraged, as it is imprecise. Also called dysimmunoglobulinemia.

γ Heavy-chain disease, also called Franklin's disease, is a very rare syndrome in which the myeloma cells synthesize γ heavy chains only. Clinically, this disease affects mostly older individuals who have either a gradual or sudden onset. The patients may be weak, have fever and malaise, and demonstrate lymphadenopathy over a period of time. Swelling of the uvula and edema of the palate may be a consequence of lymphoid tissue involvement of the nasopharynx and Waldeyer's ring. Lymph nodes affected may include those of the axilla, mediastinum, tracheobronchial tree, and abdomen. Fever and enlargement of the spleen and liver may follow infection of these patients. The disease may last from several months to 5 years and usually leads to death, although remission has been described in occasional patients. The blood serum contains proteins with an electrophoretic peak that correspond in mobility to the homogeneous (Bence-Jones negative) protein present in the urine. There is an elevated sedimentation rate, mild anemia, thrombocytopenia, leukopenia, and sometimes eosinophilia. Abnormal plasma cells and lymphocytes may appear in the blood and occasionally be manifested as plasma cell leukemia. To confirm the diagnosis of γ heavy-chain disease, one must demonstrate a spike with the electrophoretic mobility of a fast γ or β globulin reactive with antiserum against γ heavy chains, but not with κ or λ light chains. There are numerous deletions in γ chains that vary in location, but often include most of the variable region and the total C_H1 segment, with continuation of the normal sequence where the hinge region begins. Whereas patients with Franklin's disease demonstrate heavy γ chains in both the blood serum and urine, the other immunoglobulin levels are diminished. There is total failure in light-chain synthesis in all individuals. Thus, γ heavy-chain disease may mimic lymphomas of one type or another (as well as multiple myeloma, toxoplasmosis, and histoplasmosis) and may be associated occasionally with tuberculosis, RA, and various other conditions.

Essential mixed cryoglobulinemia is a condition that is identified by purpura (skin hemorrhages), joint pains, impaired circulation in the extremities on exposure to cold (Raynaud's phenomenon), and glomerulonephritis. Renal failure may result. Polyclonal IgG, IgM, and complement are detectable as granular deposits in the glomerular basement membranes. The cryoprecipitates containing IgG and IgM may also contain hepatitis B antigen, as this condition is frequently a sequela of hepatitis B.

Purpura hyperglobulinemia: Hemorrhagic areas around the ankles or legs in patients whose serum immunoglobulin levels are strikingly increased. Examples include Waldenström's macroglobulinemia, multiple myeloma, and Sjögren's syndrome.

Monoclonal immunoglobulin is a protein formed by an expanding clone of antibody-synthesizing cells in the body of a patient with either a tumor or a benign condition. The classic example is multiple myeloma or Waldenström's macroglobulinemia, in which an expanded clone of cells producing a homogeneous and uniform immunoglobulin product can be demonstrated. A number of tumors may stimulate monoclonal immunoglobulin synthesis. These include adenocarcinoma and carcinoma of the cervix, liver, and bladder, as well as Kaposi's sarcoma and angiosarcoma. Selected infections may also evoke a monoclonal immunoglobulin response. Among these are viral hepatitis, tuberculosis, and schistomiasis. Other conditions that stimulate this response include thalassemia, autoimmune hemolytic anemia, autoimmune diseases such as scleroderma and pemphigus vulgaris, and various other conditions.

A **monoclonal protein** is protein synthesized by a clone of identical cells derived from a single cell.

Bence-Jones (B-J) proteins (Figure 17.126 and Figure 17.127) represent the light chains of either the κ or λ variant excreted in the urine of patients with a paraproteinemia as a result of excess synthesis of such chains

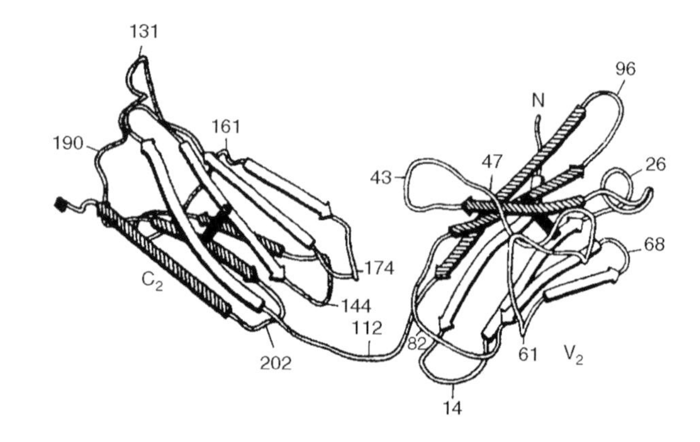

FIGURE 17.126 Bence-Jones protein.

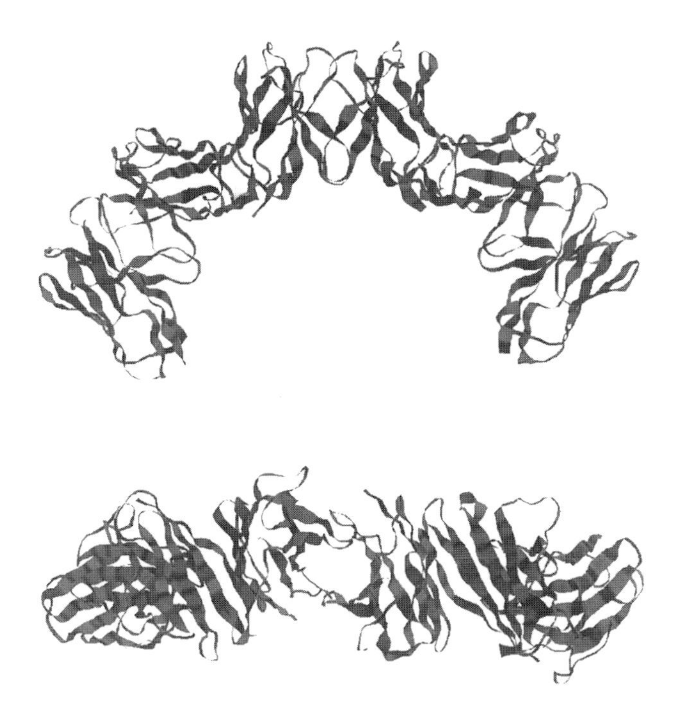

FIGURE 17.127 Bence-Jones protein.

or from mutant cells which make only such chains. Both mechanisms appear operative, and over 50% of patients with multiple myeloma, a plasma cell neoplasm, have B-J proteinuria. The highest frequency of B-J excretion is seen in IgD myeloma; the lowest is seen in IgG myeloma. The daily amount excreted parallels the severity of the disease. The B-J proteins are secreted mostly as dimers and show unusual heat solubility properties. They precipitate at temperatures between 40° to 60°C and redissolve again near 100°C. With proper pH control and salt concentration, precipitation may detect as low as 30 mg/100 ml of urine. Better identification is by protein electrophoresis.

POEMS syndrome is a condition that manifests polyneuropathy, organomegaly, endocrinopathy, monoclonal gammopathy, and skin alterations. It occurs on a background of sclerosing myeloma, Castleman's disease, and occasionally other lymphoproliferative disorders. The neuropathy is initially distal, symmetrical mixed demyelinating and axonal, and involves both motor and sensory fibers. It is progressive and ultimately fatal unless treated. The myeloma is always of the IgG or IgA type and almost always γ. Elevated cytokine levels could account for many of the symptoms including the organomegaly and the endocrine and pigmentary changes. TNF activates the proopiomelanocortin gene. Both IL-1β and TNF-α are osteoclast activating factors. There is a 100-fold increase in vascular endothelial growth factor (VEGF) levels. Radiotherapy of isolated solitary plasmacytomas has proven helpful.

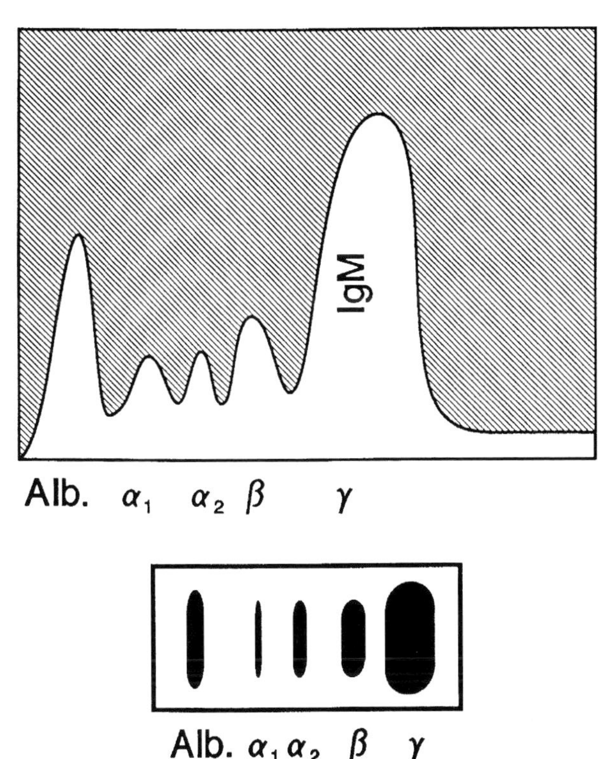

FIGURE 17.128 Waldenström's macroglobulinemia with IgM spike.

POEMS is an abbreviation for polyneuropathy, organomegaly, endocrinopathy, monoclonal gammopathy, and skin alterations. See POEMS syndrome.

Crow-Fucase syndrome: See POEMS syndrome.

Waldenström's macroglobulinemia (Figure 17.128) is a paraproteinemia that is second in frequency only to multiple myeloma, occurring mostly in people over 50 years of age. It may be manifested in various clinical forms. Most of the features of the disease are related to the oversynthesis of monoclonal IgM. Relatively mild cases may be characterized by anemia and weakness or pain in the abdomen resulting from enlargement of the spleen and liver. A major difference from multiple myeloma is a lack of osteolytic lesions of the skeleton, although patients may have peripheral lymphadenopathy. On bone marrow examination, many kinds of cells are found with characteristics of plasma cells and lymphocytes constituting so-called lymphocytoid plasma cells. Many are transitional or intermediate between one type or another. Patients may develop bleeding disorders of some type due to the paraproteins in their circulation. The more severe forms of the disease are characterized by features that resemble chronic lymphocytic leukemia or even lymphosarcoma with a rapidly fatal course. Many individuals may develop anemia. The large molecules of IgM with a molecular weight approaching one million lead to increased viscosity of the blood. Central nervous system and visual difficulties may also be manifested.

Macroglobulinemia of Waldenström is a condition usually of older men in which monoclonal IgM is detected in the serum and elevated numbers of lymphoid cells and plasmacytoid lymphocytes expressing cytoplasmic IgM are found in the bone marrow. However, these subjects do not have the osteolytic lesions observed in multiple myeloma. Due to the high mol wt of the IgM and increased levels of this immunoglobulin, blood viscosity increases, leading to circulatory embarrassment. Patients often develop skin hemorrhages and anemia, as well as neurological problems. This condition is considered less severe than multiple myeloma.

Macroglobulinemia refers to the presence of greater than normal levels of macroglobulins in the blood.

Macroglobulin is a relatively high-mol-wt serum protein. Macroglobulins have a sedimentation coefficient of 18 to 20 S and high carbohydrate content. Each type of macroglobulin belongs to a particular Ig class and is more homogeneous than the Igs produced in immune responses. Elevated levels appear on electrophoresis as a sharp peak in the migration area of the corresponding Ig class. Macroglobulins are monoclonal in origin and restricted to one κ or λ light-chain type. The level of macroglobulins increases significantly in lymphocytic and plasmolytic disorders such as multiple myeloma or leukemia. It also increases in some collagen diseases, reticulosis, chronic infectious states, and carcinoma. The 820- and 900-kDa IgM molecules are both α_2 macroglobulins.

M macroglobulin is an IgM paraprotein that occurs in Waldenström's macroglobulinemia.

Paraproteinemias are malignant diseases in which there is proliferation of a single clone of plasma cells producing monoclonal immunoglobulin. These are commonly grouped as paraproteinemias that may be manifested in several forms. Diseases associated with paraproteinemias include multiple myeloma, Waldenström's macroglobulinemia, cryoglobulinemia, plasmacytoma of soft tissues, amyloidosis, heavy-chain disease, lymphomas, leukemia, sarcomas, gastrointestinal disorders associated with tumors, chronic infection, and some endocrine disorders.

Paraimmunoglobulins: (1) The physical characteristics of some immunoglobulins present in a variety of pathologic conditions or in others of unknown etiology, and (2) the secretory products of neoplastic lymphocytes. One form, the M protein of macroglobulin, is present in the normal serum, but increased levels may result in increased serum viscosity with sluggish blood flow and development of thrombi or in central nervous system lesions. Increased levels of M protein are considered as paraimmunoglobulinopathies.

Pseudoparaproteinemia refers to an elevation of transferrin to levels at least twice normal (200 to 400 mg/dl). Profound iron deficiency anemia leads to this increase. There is only one molecular form of transferrin in the serum that is concentrated into a single band that travels in the β region in serum electrophoresis, giving the appearance of a paraproteinemia. However, a real paraproteinemia is characterized by a spike on serum electrophoresis representing a monoclonal gammopathy.

Beta–gamma bridge: Patients with chronic liver disease such as that caused by alcohol, chronic infection, or connective tissue disease may synthesize sufficient polyclonal proteins whose electrophoretic mobilities are in the beta–gamma range to cause obliteration of the beta and gamma peaks, forming a "bridge" from one to the other.

Paraendocrine syndromes include clinical signs and symptoms induced by hormones synthesized by neoplasms.

M macroglobulin is an IgM paraprotein that occurs in Waldenström's macroglobulinemia.

β-pleated sheet describes a protein configuration in which the b sheet polypeptide chains are extended and have a 35-nm axial distance. Hydrogen bonding between NH and CO groups of separate polypeptide chains stabilize the molecules. Adjacent molecules may be either parallel or antiparallel. The β-pleated sheet configuration is characteristic of amyloidosis and is revealed by Congo red staining followed by polarizing light microscopy that yields an apple-green birefringence and ultrastructurally consists of nonbranching fibrils.

In **amyloid β fibrillosis** all amyloids have a b-pleated sheet structure, which accounts for the ability of Congo red to stain them and the ability of proteolytic enzymes to digest them. See also amyloid and amyloidosis.

Burkitt's lymphoma (Figure 17.129) is an Epstein-Barr virus-induced neoplasm of B lymphocytes. It affects the

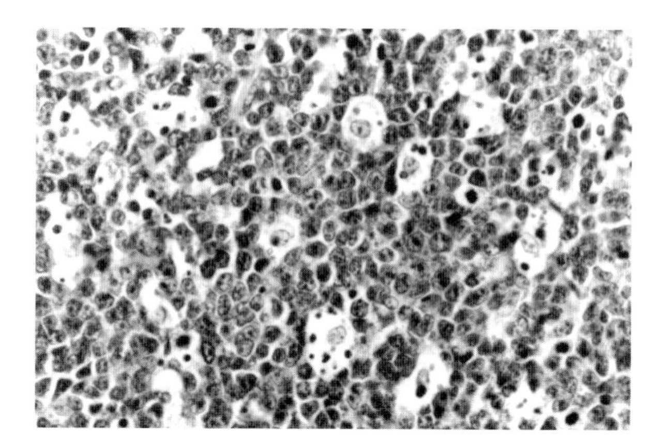

FIGURE 17.129 Burkitt's lymphoma.

jaws and abdominal viscera. It is seen especially in African children. The Epstein-Barr virus is present in tumor cells which may reveal rearrangement between the c-*myc*-bearing chromosome and the immunoglobulin heavy-chain gene-bearing chromosome. Burkitt's lymphoma patients have antibodies to the Epstein-Barr virus in their blood sera. The disease occurs in geographic regions that are hot and humid and where malaria is endemic. It occurs in subjects with acquired immunodeficiency and in other immunosuppressed individuals. There is an effective immune response against the lymphoma that may lead to remission.

BLR-1/MDR-15 (Burkitt's lymphoma receptor-1/ Monocyte-derived receptor-15) is a member of the G-protein-coupled receptor family, the chemokine receptor branch of the rhodopsin family. It is expressed on B cells and memory T cells; chronic B lymphoid leukemia and non-Hodgkin's lymphoma cells; peripheral blood monocytes and lymphocytes. It is expressed in tonsils and secondary lymphoid organs, and in selected other tissues. Most Burkitt's lymphoma cell lines express the receptor. IL-4 and IL-6 downregulate BLR-1. The receptor is also downregulated by monoclonal antibodies against CD40 and CD3. Jurkat cells expressing BLR-1 do not bind known chemokines. Also called BLR-1 and MDR-15.

Myc is an oncogene designated v-*myc* when isolated from the avian myelocytomatosis retrovirus and designated c-*myc* when referring to the cellular homologue. Two others designated N-L-*myc* have been cloned. The *myc* genes are activated by overexpression either by upregulation, caused by transcriptional regulatory signal mutations in the first intron, or by gene amplification. Normal tissues contain c-*myc*. When c-*myc* is in its normal position on chromosome 8, it remains transcriptionally silent, but when it is translocated, as in Burkitt's lymphoma, it may become activated. The protooncogene c-*myc* is amplified in early carcinoma of the uterine cervix, lung cancer, and promyelocytic leukemia.

Epstein-Barr virus (EBV) is a DNA herpes virus linked to aplastic anemia, chronic fatigue syndrome, Burkitt's lymphoma, histiocytic sarcoma, hairy cell leukemia, and immunocompromised patients. EBV may promote the appearance of such lymphoid proliferative disorders as Hodgkin's and non-Hodgkin's lymphoma, infectious mononucleosis, nasopharyngeal carcinoma, and thymic carcinoma. It readily transforms B lymphocytes and is used in the laboratory for this purpose to develop long-term B lymphocyte cultures. Antibodies produced in patients with EBV infections include those that appear early and are referred to as EA, antibodies against viral capsid antigen (VCA), and antibodies against nuclear antigens (EBNA). EBV selectively infects human B cells by binding to complement receptor 2 (CR2 or CD21). It causes infectious mononucleosis and establishes a latent infection of B cells that persists for life and is controlled by T cells.

Epstein-Barr nuclear antigen is a molecule that occurs in B cells before virus-directed protein can be found in nuclei of infected cells. Thus, it is the earliest evidence of Epstein-Barr virus infection and can be found in patients with conditions such as infectious mononucleosis and Burkitt's lymphoma.

c-*myb* is a protooncogene that codes for the 75- to 89-kDa phosphoprotein designated c-Myb in the nucleus, which immature hematopoietic cells express during differentiation. When casein kinase II phosphorylates Myb at an N-terminal site, this blocks the union of Myb to DNA and blocks continued activation. This phosphorylation site is deleted during oncogenic transformation, which permits Myb to combine with DNA.

v-*myb* oncogene: A genetic component of an acute transforming retrovirus that leads to avian myeloblastosis. It represents a truncated genetic form of c-*myb*.

Lymphoma belt is an area across Central Africa on either side of the equator where an increased incidence of Epstein-Barr virus-induced Burkitt's lymphoma occurs. Burkitt's lymphoma is a relatively common childhood cancer in Uganda.

Sezary syndrome (Figure 17.130) is a disease that occurs in middle age, affecting males more commonly than females. It is a neoplasm, i.e., a malignant lymphoma of $CD4^+$ T helper lymphocytes with prominent skin involvement. There is a generalized erythroderma, hyperpigmentation, and exfoliation. Fissuring and scaling of the skin on the palms of the hands and soles of the feet may occur. The peripheral blood and lymph nodes contain the typical cerebriform cells that have a nucleus that resembles the

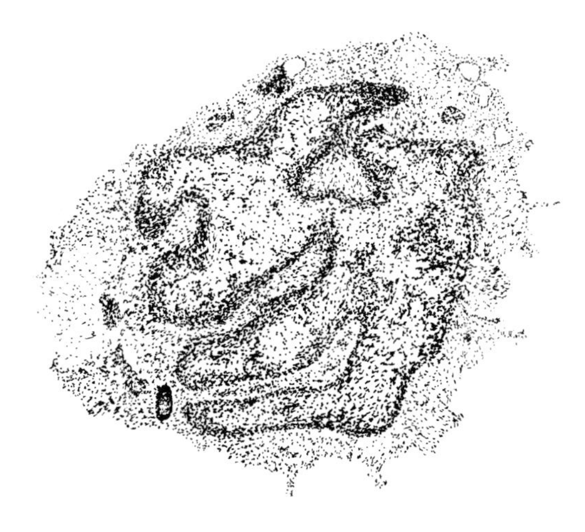

FIGURE 17.130 Lymphocyte from a patient with Sezary syndrome.

brain. There is extensive infiltration of the skin by leukocytes, with prominent clustering in the epidermis forming Pautrier's evidence. Late in the disease, T immunoblasts may appear. The so-called Sezary cells are T lymphocytes.

Sezary cells are T lymphocytes that form E rosettes with sheep red cells and react with anti-T antibodies. Sezary cells from most individuals show a diminished response to plant mitogens, although those from a few demonstrate normal reactivity. Sezary cells are also poor mediators of T cell cytotoxicity, but they can produce a migration inhibitory factor (MIF)-like lymphokine. Sezary cells neither produce immunoglobulin nor do they act as suppressors; however, they have a helper effect for immunoglobulin synthesis by B cells.

Behcet's disease is characterized by oral and genital ulcers, vasculitis, and arthritis that recur as a chronic disease in young men. It is postulated to have an immunologic basis and possibly to be immune complex mediated. Perivascular infiltrates of lymphocytes may occur. The serum contains immune complexes, and immunofluorescence may reveal autoantibodies against the oral mucous membrane. It is associated with HLA-1 in ethnic group, but not Caucasians. The principal immunologic dysfunction in Behcet's disease is neutrophil hyperfunction, shown by augmented chemotactic responsiveness and superoxide production. Behcet's disease is an inflammatory condition of uncertain cause. No autoimmune mechanism has been proven.

Transmissible spongiform encephalopathy (TSE) immunity: Paradoxically, a functional immune system is a requisite for the efficient experimental transmission of scrapie and for TDE agents to pass from one species to another. Immunologically immature mice are less susceptible to scrapie than are older mice, and SCID mice are refractory to infection. PHA and BCG increase susceptibility to infection. Interferon, antilymphocyte serum, neonatal thymectomy, or whole body radiation fail to alter the incubation period.

Mouse hepatitis virus (MHV) is a DNA virus that causes murine hepatitis and encephalitis. MHV infects oligodendrocytes and leads to demyelination without the presence of immune system cells.

Chronic fatigue syndrome (CFS) is a disabling fatigue that persists for at least 6 months. Although the etiology is idiopathic, laboratory studies on patients reveal a consistent observation of immune system dysfunction primarily affecting the cellular immune response. $CD4^+$ T helper cells and $CD8^+$ suppressor/cytotoxic cells may be normal, increased, or decreased, but the CD4/DC8 ratio is usually elevated. This has been attributed to a diminished number of suppressor cells with a concomitant increase in cytotoxic T cell ($CD8^+$, $CD28^+$, $CD11b^-$) numbers. The increased cytotoxic T cells express HLA-DR and/or CD38 activation markers. Manifestations of altered T cell functions also include decreased delayed-type hypersensitivity, diminished responsiveness in mitogen-stimulation assays *in vitro*, increased suppression of immunoglobulin synthesis by T cells, and elevated spontaneous suppressor activity. NK cells may be normal, increased, or decreased, but there may be qualitative alterations in NK cell function. Elevated IgG antibody titers to Epstein-Barr virus (EBV) early antigen and capsid antigen are demonstrable in many CFS patients. Occasionally, increased antibodies against cytomegalovirus (CMV), herpes simplex, HHV-6, coxsackie B, or measles may be observed. Some CFS patients have abnormal levels of IgG, IgM, IgA, or IgD. Approximately one third of CFS patients have antibodies against smooth muscle or thyroid. Laboratory test results from CFS patients should be interpreted as one battery and not as individual tests.

Eosinophilic myalgia syndrome (EMA) is an intoxication syndrome observed in persons in the U.S. that appeared linked to the consumption of L-tryptophan, proposed by some health advocates as an effective treatment for various disorders such as insomnia, premenstrual syndrome, etc. It was associated with a strain of *Bacillus amyloliquefaciens* employed to produce tryptophan commercially. The inducing agent was apparently an altered amino acid, DTAA (ditryptophan aminal acetaldehyde), a contaminant introduced during manufacture. Clinical manifestations of the syndrome include arthralgia, myopathy, angioedema, alopecia, mobiliform rash, oral ulcers, sclerodermoid lesions, restricted lung disease, fever, lymphadenopathy, and dyspnea, among other features. There was a significant eosinophilia. IL-5 was believed to have a role in injury to tissues. Histopathologic examination revealed arteriolitis and sclerosing skin lesions.

Hemolytic anemia is a disease characterized by diminished circulating erythrocytes as a consequence of their destruction based either on an intrinsic abnormality, as in sickle cell anemia and thalassemia, or as a consequence of membrane-specific antibody and complement. Certain infections such as malaria are also associated with hemolytic anemia. Free hemoglobin in serum may lead to renal problems.

PCH is an abbreviation for paroxysmal cold hemaglobinuria.

HAM test: See paroxysmal nocturnal hemoglobulinuria (PNH).

PNH cells are red blood cells of paroxysmal nocturnal hemoglobinuria patients. At a weak acid pH, PNH cells disrupt spontaneously. The ability of complement to lyse these cells is much more pronounced than its action on normal erythrocytes subjected to conditions that promote their lysis by complement.

18 Immunodeficiencies: Congenital and Acquired

Immunodeficiencies are classified as either primary diseases with a genetic origin or diseases that are secondary to an underlying disorder. **X-linked (congenital) agammaglobulinemia** results from a failure of pre-B cells to differentiate into mature B cells. The defect in **Bruton's disease** is in the rearrangement of immunoglobulin heavy-chain genes. This disorder occurs almost entirely in males and is apparent after 6 months of age, following the disappearance of the passively transferred maternal immunoglobulins. Patients have recurrent sinopulmonary infections caused by *Haemophilus influenzae, Streptococcus pyogenes, Staphlococcus aureus*, and *Streptococcus pneumoniae*, with absent or decreased B cells and decreased serum levels of all immunoglobulin classes. The T cell system and cell-mediated immunity appear normal.

Thymic hypoplasia (DiGeorge syndrome) occurs when the immune system in infants is deprived of thymic influence. T cells are absent or deficient in the blood and thymus-dependent areas of lymph nodes and spleen. Infants with this condition are highly susceptible to infection by viruses, fungi, protozoa, or intracellular bacteria due to defective intracellular microbial killing by phagocytic cells with interferon. By contrast, B cells and immunoglobulins are not affected.

Severe combined immunodeficiency (Swiss-type agammaglobulinemia) comprises a group of conditions manifesting variable defects in both B and T cell immunity. In general, there is a lymphopenia with deficiency of T and B cell numbers and function. The thymus is hypoplastic or absent. Lymph nodes and other peripheral lymphoid tissues reveal depleted B and T cell regions. Infants with severe combined immunodeficiency show increased susceptibility to infections by viruses, fungi, and bacteria, and often succumb during the first year.

These three categories of immunodeficiency involving either B cells or T cells or both are presented in greater detail in the sections that follow.

Immunoincompetence is the inability to produce a physiologic immune response. For example, patients with acquired immune deficiency (AIDS) become immunoincompetent as a consequence of the destruction of their helper/inducer ($CD4^+$) T lymphocyte population. Infants born without a thymus or experimental animals thymectomized at birth are immunologically incompetent. Children born with severe combined immunodeficiency due to one or several causes are unable to mount an appropriate immune response. Immunoincompetence may involve either the B cell limb, as in Bruton's hypogammaglobulinemia, or the T cell limb, as in patients with DiGeorge's syndrome.

Secondary immunodeficiencies are more common than the primary forms. The best known is acquired immunodeficiency disease (e.g., AIDS), which results from destruction of the helper/inducer ($CD4^+$) lymphocyte. Most of these individuals develop opportunistic infections caused by viruses, fungi, protozoa, and bacteria that are not commonly pathogenic (Figure 18.1).

Immunological deficiency state: Immunodeficiency.

Immunodeficiency is a failure in humoral antibody or cell-mediated limbs of the immune response. If attributable to intrinsic defects in T and/or B lymphocytes, the condition is termed a primary immunodeficiency. If the defect results from loss of antibody and/or lymphocytes, the condition is termed a secondary immunodeficiency.

Primary immunodeficiency refers to diminished immune reactivity attributable to an intrinsic abnormality of T or B lymphocytes.

Immunodeficiency disorders are conditions characterized by decreased immune function. They may be grouped into four principal categories based on recommendations from a committee of the World Health Organization. They include antibody (B cell) deficiency, cellular (T cell) deficiency, combined T cell and B cell deficiencies, and phagocyte dysfunction. The deficiency can be congenital or acquired. It can be secondary to an embryologic abnormality or an enzymatic defect, or may be attributable to an unknown cause. Types of infections produced in the physical findings are characteristic of the type of immunodeficiency disease. Screening tests identify a number of these conditions whereas others have an unknown etiology. Antimicrobial agents for the treatment of recurrent infections, immunotherapy, bone marrow transplantation, enzyme replacement, and gene therapy are all modes of treatment.

Congenital immunodeficiency refers to a varied group of unusual disorders with associated autoimmune manifestations,

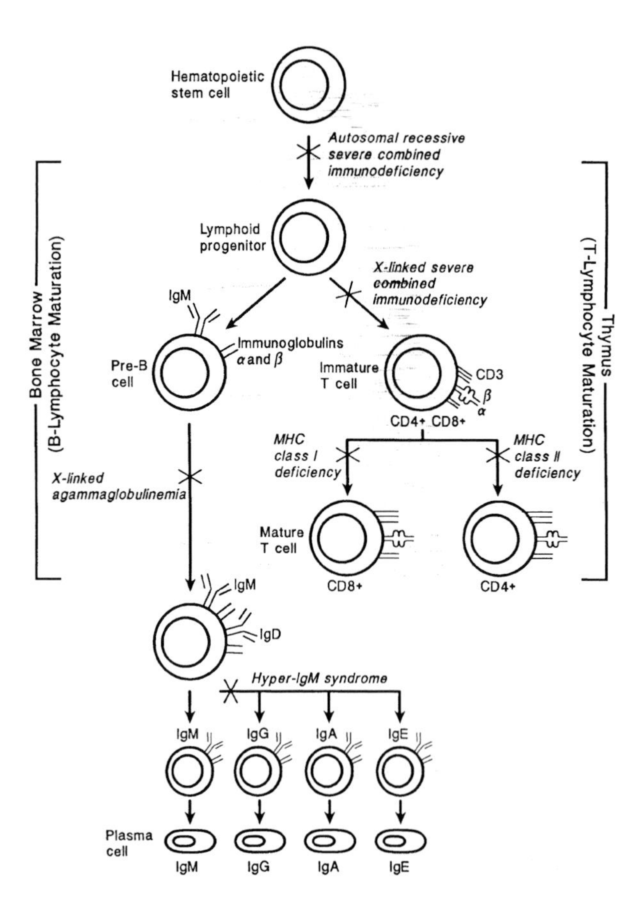

FIGURE 18.1 Maturation of T and B lymphocytes.

increased incidence of malignancy, allergies, and gastrointestinal abnormalities. These include defects in stem cells, B cells, T cells, phagocytic defects, and complement defects. An example is severe combined immunodeficiency due to various causes. The congenital immunodeficiencies are described under the separate disease categories.

Bloom syndrome (BS) is a very rare clinical entity characterized by a strikingly small but well-proportioned body size. It is the consequence of either homozygosity or compound heterozygosity of a mutation of *BLM*, a gene that encodes a phylogenetically highly conserved nuclear protein. It has been suggested by some that this condition be grouped with the genetically determined immunodeficiencies. Respiratory tract infections occur much more commonly in BS than in the general population. If not treated promptly they progress to life-threatening infections. Yet 37% of BS patients manifest no increased tendency to infection of any kind. It is a disease characterized by diminished numbers of T lymphocytes, decreased antibody levels, and increased susceptibility to respiratory infections, tumors, and radiation injury. Mutations in a DNA helicase cause the disorder.

Infantile sex-linked hypogammaglobulinemia refers to an antibody deficiency syndrome that is sex linked and occurs in males following the disappearance of passively transferred antibodies from the mother following birth. Serum immunoglobulin concentrations are relatively low, and there is defective antibody synthesis giving rise to recurring bacterial infections. Cell-mediated immunity remains intact.

Transient hypogammaglobulinemia of infancy is a temporary delay in the onset of antibody synthesis during the first 12 months or even 24 months of life. This leads only to a transient, physiologic immunodeficiency following catabolism of maternal antibodies passed to the infant across the placenta to the fetal circulation. Helper T cell function is impaired, yet B cell numbers are at physiologic levels.

Hypogammaglobulinemia refers to deficient levels of IgG, IgM, and IgA serum immunoglobulins. This may be attributable to either decreased synthesis or increased loss. Hypogammaglobulinemia can be physiologic in neonates. It may be a manifestation of either congenital or acquired antibody deficiency syndromes. Several types are described and include Bruton's disease and a congenital type, as well as an acquired type such as in chronic lymphocytic leukemia. Human γ globulin is used for treatment.

Agammaglobulinemia (hypogammaglobulinemia) (Figure 18.2) was used in earlier years before the development of methods sufficiently sensitive to detect relatively small quantities of gammaglobulin in the blood. Primary agammaglobulinemia was attributed to defective immunoglobulin formation, whereas secondary agammaglobulinemia referred to immunoglobulin depletion, as in loss through inflammatory bowel disease or through the skin in burn cases.

Primary agammaglobulinemia: See antibody deficiency syndrome.

X-linked agammaglobulinemia (Bruton's X-linked agammaglobulinemia) affects males who develop recurrent sino-pulmonary or other pyogenic infections at 5 to 6 months of age after disappearance of maternal IgG. There is defective B-lymphocyte gene (chromosome Xq21.3-22).

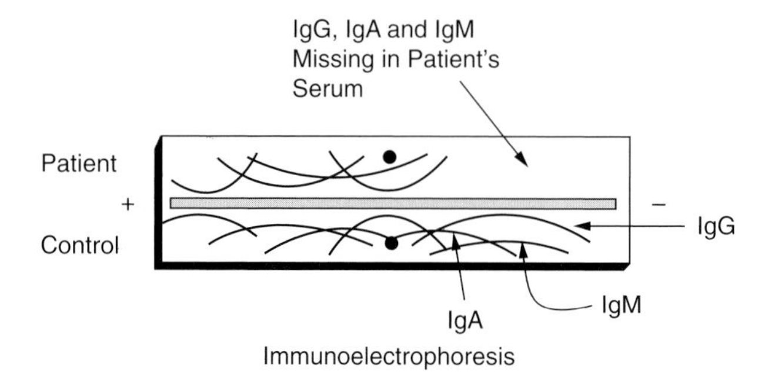

FIGURE 18.2 Agammaglobulinemia.

Whereas B cells and immunoglobulins are diminished, there is normal T cell function. Supportive therapy includes gammaglobulin injections and antibiotics. Repeated infections may lead to death in childhood. The bone marrow of affected children contains pre-B cells with constant regions of immunoglobulin m chains in the cytoplasm. There may be defective VH-D-JH gene rearrangement.

Btk is a protein tyrosine kinase encoded for by the defective gene in X-linked agammaglobulinemia (XLA). B lymphocytes and polymorphonuclear neutrophils express the btk protein. In XLA (Bruton's disease) patients, only the B lymphocytes manifest the defect, and the maturation of B lymphocytes stops at the pre-B cell stage. There is rearrangement of heavy chain genes but not of the light chain genes. The btk protein might have a role in linking the pre-B cell receptor to nuclear changes that result in growth and differentiation of pre-B cells.

Infantile agammaglobulinemia is a synonym for X-linked agammaglobulinemia.

Bruton's agammaglobulinemia is a synonym for X-linked agammaglobulinemia.

Congenital agammaglobulinemia: See X-linked agammaglobulinemia.

Duncan's syndrome: Marked lymphoproliferation and agammaglobulinemia that are associated with Epstein-Barr virus infection. The rapidly proliferating B lymphocytes produce neoplasms which may rupture the spleen. This defect in the immune response with susceptibility to infection is inherited as an X-linked recessive disorder. The individual is not able to successfully resist infection by the Epstein-Barr virus.

Antibody deficiency syndrome: A few patients have been observed with this syndrome, in which normal immunoglobulin levels are present but the ability to mount an immune response to immunogenic challenge is impaired. This condition is associated with several separate disease states and might more properly be considered a syndrome. Some may present clinically as severe combined immunodeficiency with diminished cell-mediated immunity, lymphopenia, and infection by microorganisms of low pathogenicity. There are normal or even elevated numbers of plasma cells, and there may be no demonstrable T cell deficiency, both of which are in contrast to the usual clinical picture of severe combined immunodeficiency. These individuals may develop autoimmune reactions and show reduced numbers of lymphoid cells with surface immunoglobulin in the circulating blood. One possible explanation for normal immunoglobulin levels and an inadequate humoral immune response to antigenic challenge could be accounted for by a defect in clonal diversity, resulting in an antibody response to only a limited number of antigens. Some investigators have associated the defect with T cell clonal diversity. This combined B and T cell system disorder resembles but is less pronounced than severe combined immunodeficiency (SCID). Some patients with this defect develop paraproteins with subsequent agammaglobulinemia and clinical manifestations closely resembling severe combined immunodeficiency.

Epstein-Barr immunodeficiency syndrome: Duncan's X-linked immunodeficiency. This is an X-linked or autosomal recessive condition associated with congenital cardiovascular and central nervous system defects. Patients may develop infectious mononucleosis that is fatal. There is aplasia of the bone marrow, agammaglobulinemia, and agranulocytosis, and the response to mitogens and antigens by B cells is greatly diminished. Natural killer cell activity is decreased and T cells are abnormal. Patients may develop hepatitis, B cell lymphomas, and immune suppression.

ecto-5′-nucleotidase deficiency is a purine metabolism alteration that produces immunodeficiencies of B lymphocytes.

Swiss type agammaglobulinemia is a severe combined immunodeficiency disease. See Swiss type severe combined immunodeficiency.

Swiss type immunodeficiency: See Swiss agammaglobulinemia.

Immunodeficiency with thymoma is an abnormality in B cell development with a striking decrease in B cell numbers and in immunoglobulins in selected patients with thymoma. There is a progressive decrease in their cell-mediated immunity as the disease continues. Most patients develop chronic sinopulmonary infections. Approximately 20% develop thrombocytopenia and about 25% have splenomegaly. Immunoglobulin class alterations are variable. Skin test reactivity and responsiveness to skin allografts are decreased. Patients have few or no lymphocytes expressing surface immunoglobulin. The disease is due to a stem-cell defect. The preferred treatment is gammaglobulin administration.

B cell leukemias can be classified as pre-B cell, B cell, or as plasma cell neoplasms. They include Burkitt's lymphoma, Hodgkin's disease, and chronic lymphocytic leukemia. Neoplasma of plasma cells are associated with multiple myeloma and Waldenström's macroglobulinemia. Many of these conditions are associated with hypogammaglobulinemia and a diminished capacity to form antibodies in response to the administration of an immunogen. In chronic lymphocytic leukemia (CLL), more than 95% of individuals have malignant leukemic cells that are identifiable as B lymphocytes expressing surface immunoglobulin. These patients frequently develop infections and have autoimmune manifestations such as autoimmune hemolytic anemia. **Chronic lymphocytic leukemia/small lymphocytic lymphoma** patients may have secondary immunodeficiency, which affects both B and T limbs of the immune response. Diminished immunoglobulin levels are due primarily to diminished synthesis (Figure 18.3).

Mantle zone lymphoma is a type of follicular lymphoma of intermediate grade. There is small lymphocyte proliferation in the mantle zone encircling benign germinal centers, splenomegaly, and generalized lymphadenopathy. Histopathologically, there are B cells that vary in size from small to relatively large blasts containing clumped chromatin. IgM is usually present.

Selective immunoglobulin deficiency is an insufficient quantity of one of the three major immunoglobulins or a subclass of IgG or IgA. The most common is selective IgA deficiency followed by IgG3 and IgG2 deficiencies. Patients suffering from selective immunoglobulin deficiency may be either normal or manifest increased risk for bacterial infections.

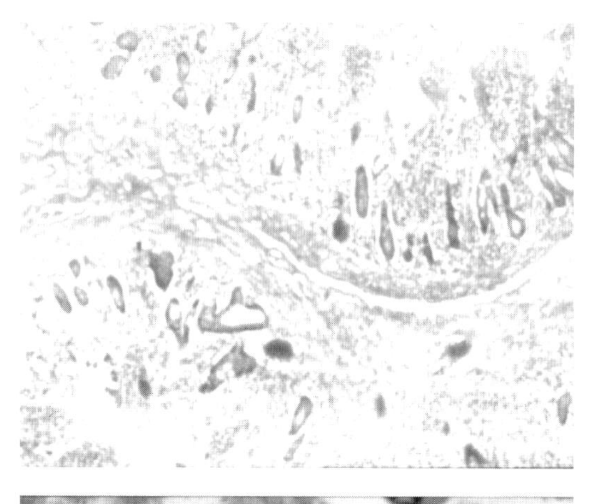

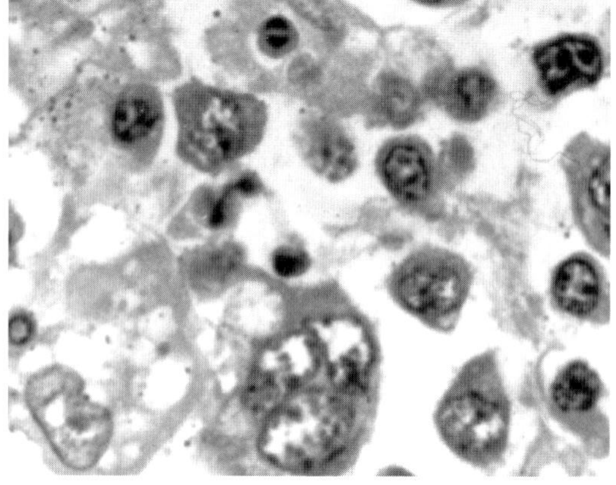

FIGURE 18.3 B cell lymphoma of the gut.

Immunoglobulin A deficiency is the most frequent human immunodeficiency that affects 1 in 600 persons in the U.S. Even though the B cells of these individuals have IgA on the surface, they do not differentiate into plasma cells that secrete the IgA. IgA levels are decreased from a normal value of 76 to 390 mg/dl to a value below 5 mg/dl. Almost half develop anti-IgA antibodies that are subclass specific. Those IgA-deficient individuals with heightened susceptibility to infection by pyogenic microorganisms also are deficient in IgG2 and often IgG4. The administration of a blood transfusion to IgA-deficient patients possessing anti-IgA antibodies can lead to anaphylactic shock due to IgE antibodies specific for IgA or to a fatal hemolytic transfusion reaction. Patients may also have intestinal lymphangiectasia, arthritis, gluten-sensitive enteropathy, allergies, and myotonic dystrophy. They may also develop low-mol-wt IgM antibodies against food substances such as milk. Other clinical features may include sinopulmonary infections, cirrhosis, and autoimmune disease. IgA deficiency is detectable in approximately three fourths of ataxia telangiectasia patients. Intravenous immune globulin with only minute quantities of IgA may be beneficial to these patients.

Selective IgA deficiency is the most frequent immunodeficiency disorder. It occurs in approximately 1 in 600 individuals in the population. It is characterized by nearly absent serum and secretory IgA. The IgA level is less than 5 mg/dl, whereas the remaining immunoglobulin class levels are normal or elevated. The disorder is either familial or it may be acquired in association with measles, other types of virus infection, or toxoplasmosis. The patients may appear normal and asymptomatic, or they may have some form of an associated disease. IgA is the principal immunoglobulin in secretions and is an important part of the defense of mucosal surfaces. Thus, IgA-deficient individuals have an increased incidence of respiratory, gastrointestinal, and urogenital infections. They may manifest sinopulmonary infections and diarrhea. Selective IgA deficiency is diagnosed by the demonstration of less than 5 mg/dl of IgA in serum. The etiology is unknown but is believed to be arrested B cell development. The B lymphocytes are normal with surface IgA and IgM or surface IgA and IgD. Some patients also have an IgG2 and IgG4 subclass deficiency. They are especially likely to develop infections. IgA-deficient patients have an increased incidence of respiratory allergy and autoimmune disease such as systemic lupus erythematosus (SLE) and rheumatoid arthritis (RA). The principal defect is in IgA B lymphocyte differentiation. The 12-week-old fetus contains the first IgA B lymphocytes that bear IgM and IgD as well as IgA on their surface. At birth, the formation of mature IgA B lymphocytes begins. Most IgA B cells express IgA exclusively on their surface, with only 10% expressing surface IgM and IgD in the adult. Patients with selective IgA deficiency usually express the immature phenotype, only a few of which can transform into IgA-synthesizing plasma cells. Patients have an increased incidence of HLA-A1, -B8, and -Dw3. Their IgA cells form, but do not secrete IgA. There is an increased incidence of the disorder in certain atopic individuals. Some selective IgA deficiency patients form significant titers of antibody against IgA. They may develop anaphylactic reactions upon receiving IgA-containing blood transfusions. The patients have an increased incidence of celiac disease and several autoimmune diseases as indicated above. They synthesize normal levels of IgG and IgM antibodies. Autosomal recessive and autosomal dominant patterns of inheritance have been described. Selective IgA deficiency has been associated with several cancers, including thymoma, reticulum cell sarcoma, and squamous cell carcinoma of the esophagus and lungs. Certain cases may be linked to drugs such as phenytoin or other anticonvulsants. Some individuals develop antibodies against IgG, IgM, and IgA. Gammaglobulin should not be administered to selective IgA-deficient patients.

IgA deficiency: See immunoglobulin A deficiency.

Selective IgM deficiency occurs when IgM is absent from the serum. Although IgM may be demonstrable on plasma cell surfaces, it is not secreted. This could be related to an alteration in secretory peptide or due to the action of suppressor T lymphocytes specifically on IgM-synthesizing and secreting cells. Gram-negative microorganisms may induce septicemia in affected individuals, since IgM's major role in protection against infection is in the intravascular compartment rather than in extravascular spaces where other immunoglobulin classes may be active.

IgM deficiency syndrome is an infrequent condition characterized by diminished activation of complement and decreased B lymphocyte membrane 5-ecto nucleotidase. Patients manifest heightened susceptibility to pulmonary infections, septicemia, and tumors.

IgG subclass deficiency refers to decreased or absent IgG2, IgG3, or IgG4 subclasses. Total serum IgG is unaffected because it is 65 to 70% IgG1. Deletion of constant heavy chain genes or defects in isotype switching may lead to IgG subclass deficiency. IgG1 and IgG3 subclasses mature quicker than do IgG2 or IgG4. Patients have recurrent respiratory infections and recurring pyogenic sinopulmonary infections with *Hemophilus influenzae*, *Staphylococcus aureus*, and *Streptococcus pneumoniae*. Some patients may manifest features of autoimmune disease such as SLE. IgG2–IgG4 deficiency is often associated with recurrent infections or autoimmune disease. IgG2 deficiency is reflected as recurrent sinopulmonary infections and nonresponsiveness to polysaccharide antigens such as those of the pneumococcus. In the IgG4 deficient patients, recurrent respiratory infections as well as autoimmune manifestations also occur. The diagnosis is established by the demonstration of significantly lower levels of at least one IgG subclass in the patient compared with IgG subclass levels in normal age-matched controls. Gammaglobulin is the treatment choice. Very infrequently, C-gene segment deletions may lead to IgG subclass deficiency.

Selective IgA and IgG deficiency affect both males and females, and are either X-linked, autosomal recessive, or can be acquired later in life. There may be a genetic defect in the switch mechanism for immunoglobulin-producing cells to change from IgM to IgG or IgA synthesis. Respiratory infections with pyogenic microorganisms or autoimmune states that include hemolytic anemia, thrombocytopenia, and neutropenia may occur. Numerous IgM-synthesizing plasma cells are demonstrable in both lymph nodes and spleen of affected individuals.

Light-chain deficiencies: In addition to deficiencies in heavy chains, one may observe light-chain deficiencies. The ratio of κ to λ light chains may be altered in individuals with immunodeficiency. κ Chain deficiency has been associated with respiratory infections, megaloblastic anemia, and diarrhea. It has also been associated with achlorhydria and pernicious anemia and has even been seen in cases of malabsorption, diabetes, and cystic fibrosis. T cell function in all of these individuals was within normal limits, with only defective B cell immunity observed. Abnormal κ and λ light chain ratios are secondary findings in certain diseases, whereas in others they may be primary etiologic agents.

κ light-chain deficiency is a rare condition in which point mutations in the Cκ gene at chromosome 2p11 cause an absence of κ light chains in the serum and generate B lymphocytes whose surfaces are bereft of κ light chains.

Immunoglobulin deficiency with elevated IgM is antibody deficiency characterized by heightened susceptibility to infection by pyogenic microorganisms. Whereas IgM and IgD levels are elevated in the serum, the IgA and IgG concentrations are greatly diminished or not detectable. Patients often manifest IgM autoantibodies against neutrophils and platelets. The only immunoglobulins secreted by B cells in this X-linked recessive immunodeficiency syndrome are IgM and IgD.

Selective IgA and IgM deficiency is a concomitant reduction in both IgA and IgM concentrations with normal IgG levels. The IgG produced in many individuals may not be protective, and recurrent infections may result. There is an inadequate response to many immunogens in this disease, which occurs in four males to every female affected.

Common variable immunodeficiency (CVID) is a relatively common congenital or acquired immunodeficiency that may be either familial or sporadic. The familial form may have a variable mode of inheritance. Hypogammaglobulinemia is common to all of these patients and usually affects all classes of immunoglobulin, but in some cases only IgG is affected. The World Health Organization (WHO) classifies three forms of the disorder: (1) an intrinsic B lymphocyte defect, (2) a disorder of T-lymphocyte regulation that includes deficient T helper lymphocytes or activated T suppressor lymphocytes, and (3) autoantibodies against T and B lymphocytes. The majority of patients have an intrinsic B cell defect with normal numbers of B cells in the circulation that can identify antigens and proliferate but cannot differentiate into plasma cells. The ability of B cells to proliferate when stimulated by antigen is evidenced by hyperplasia of B cell regions of lymph nodes, spleen, and other lymphoid tissues. Yet, differentiation of B cells into plasma cells is blocked. The deficiency of antibody that results leads to recurrent bacterial infections, as well as intestinal infestation by *Giardia lamblia*, which produces a syndrome that resembles sprue. Noncaseating granulomas occur in many organs. There is an increased incidence of autoimmune diseases such as pernicious anemia, rheumatoid arthritis, and hemolytic anemia. Lymphomas also occur in these immunologically deficient individuals.

Acquired agammaglobulinemia: See common variable immunodeficiency.

Common variable antibody deficiency: See common variable immunodeficiency (CVID).

Transcobalamin II deficiency: Infants deficient in transcobalamin II, the main transport protein for vitamin B_{12}, develop megaloblastic anemia and agammaglobulinemia. B lymphocytes require vitamin B_{12} for terminal differentiation. B_{12} therapy corrects the deficiency, which has an autosomal recessive mode of inheritance.

Lesch-Nyhan syndrome is a deficiency of hypoxanthine-guanine phosphoribosyl transferase that leads to neurological dysfunction and B cell immunodeficiency.

Transcobalamin II deficiency with hypogammaglobulinemia see an association of transcobalamin II deficiency with hypogammaglobulinemia, gastrointestinal, and hematologic disorders. This is inherited as an autosomal recessive trait. Patients with this condition may manifest macrocytic anemia, thrombocytopenia, leukopenia, and malabsorption resulting from small intestinal mucosal atrophy. Transcobalamin II is a protein needed for vitamin B_{12} transport in the blood. Circulating B lymphocytes are normal, but plasma cells are absent from the bone marrow. Affected subjects often fail to produce antibodies following immunogenic challenge. T cell responsiveness to PHA and skin tests are within normal limits. Replacement therapy with vitamin B_{12} given intramuscularly has improved immunoglobulin levels in the blood and rendered immunization against common antigens successful. Thus, the defect in the ability of B cells to undergo clonal expansion and to mature into antibody-producing B cells is related to a deficiency in transcobalamin II needed for the passage of vitamin B_{12} to the cell's internal environment.

DiGeorge syndrome (Figure 18.4) is a T cell immunodeficiency in which there is failure of T cell development but normal maturation of stem cells and B lymphocytes. This is attributable to failure in the development of the thymus, depriving the individual of the mechanism for T lymphocyte development. DiGeorge syndrome is a recessive genetic immunodeficiency characterized by failure of the thymic epithelium to develop. Maldevelopment of the thymus gland is associated with thymic hypoplasia. Anatomical structures derived from the third and fourth pharyngeal pouches during embryogenesis fail to develop. This leads to a defect in the function of both the thymus and parathyroid glands. DiGeorge syndrome is believed to be a consequence of intrauterine malfunction. It is not familial. Tetany and hypocalcemia, both characteristics of hypoparathyroidism, are observed in DiGeorge syndrome in addition to the defects in T cell immunity. Peripheral lymphoid tissues exhibit a deficiency of lymphocytes in

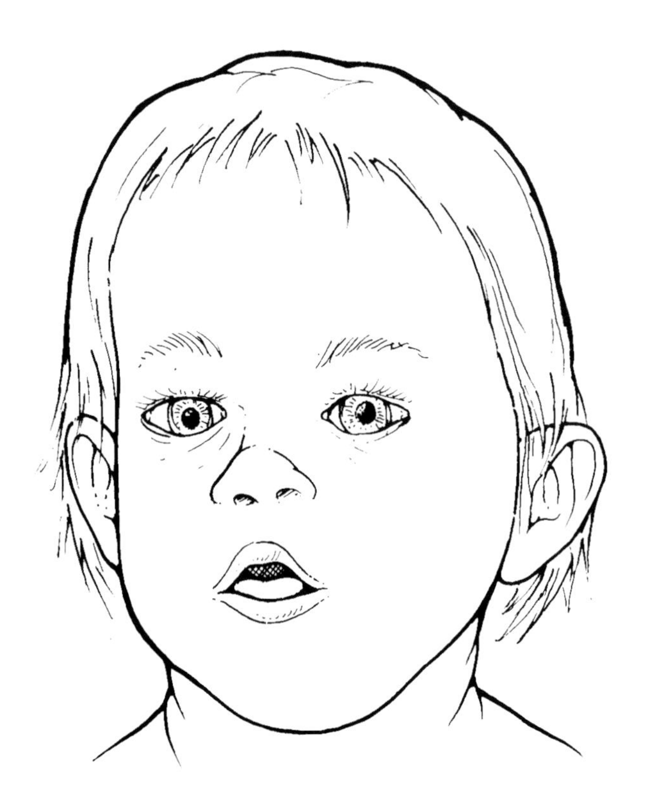

FIGURE 18.4 DiGeorge syndrome.

thymic-dependent areas. By contrast, the B or bursa equivalent–dependent areas, such as lymphoid follicles, show normal numbers of B lymphocytes and plasma cells. Serum immunoglobulin levels are within normal limits and there is a normal immune response following immunization with commonly employed immunogens. A defect in delayed-type hypersensitivity is demonstrated by the failure of affected patients to develop positive skin tests to commonly employed antigens such as candidin or streptokinase and the inability to develop an allograft response. Defective cell-mediated immunity may increase susceptibility to opportunistic infections and render the individual vulnerable to a graft-vs.-host reaction in blood transfusion recipients. There is also minimal or absent *in vitro* responsiveness to T cell antigens or mitogens. The most significant advance has been the identification of microdeletions on human chromosome 22q in most DiGeorge syndrome patients. Considerable success in treatment has been achieved with fetal thymic transplants and by the passive administration of thymic humoral factors.

Thymic hypoplasia is an immunodeficiency that selectively affects the T cell limb of the immune response. Early symptoms soon after birth may stem from associated parathyroid abnormalities leading to hypocalcemia and heart defects, which may lead to congestive heart failure. There is lymphopenia with diminished T cell numbers. T lymphocyte function cannot be detected in peripheral blood T cells. There is variation in antibody levels and function. The condition has been successfully treated by thymic transplantation. Some DiGeorge patients may have normal B cell immunity. All the others may have diminished immunoglobulin levels and may not form specific antibody following immunization. Clinically, DiGeorge patients may have a fish-shaped mouth, and abnormal faces with low-set ears, hypertelorism, and antimongoloid eyes, in addition to the other features mentioned above.

Nezelof's syndrome is a hypoplasia of the thymus leading to a failure of the T lymphocyte compartment with no T cells and no T cell function. By contrast, B lymphocyte function remains intact. Thus, this is classified as a T lymphocyte immunodeficiency.

Cell-mediated immunodeficiency syndrome is a condition in which cell-mediated immunity is defective. This may be manifested as negative skin tests following the application of tuberculin, histoplasmin, or other common skin test antigens; failure to develop contact hypersensitivity following application of sensitizing substances such as DNCB to the skin; or failure to reject an allograft, such as a skin graft. Severe combined immunodeficiency is characterized by defective T lymphocyte as well as B lymphocyte limbs of the immune response. DiGeorge syndrome is characterized by failure of development of the T cell-mediated limb of the immune response.

T cell immunodeficiency syndromes (TCIS) cause decreased immune function as a consequence of complete or partial defects in the function of T lymphocytes. HUETER patients develop recurrent opportunistic infections; may manifest cutaneous anergy, wasting, diminished life expectancy, retardation in growth, and increased likelihood of developing graft-vs.-host disease; and have very serious or even fatal reactions following immunization with BCG or live virus vaccines. They also have an increased likelihood of malignancy. T cell immunodeficiencies are usually more profound than B cell immunodeficiencies. There is no effective treatment. This group of disorders includes thymic hypoplasia known as DiGeorge syndrome; cellular immunodeficiency with immunoglobulins, termed Header syndrome; and defects of T lymphocytes caused by deficiency of purine nucleoside phosphorylase and lack of inosine phosphorylase.

Interleukin-1 receptor deficiency: $CD4^+$ T cells deficient in IL-1 receptors in affected individuals fail to undergo mitosis when stimulated and fail to generate IL-2. This leads to a lack of immune responsiveness and constitutes a type of combined immunodeficiency. Opportunistic infections are increased in affected children who have inherited the condition as an autosomal recessive trait.

Immunodeficiency with T cell neoplasms occurs in almost one third of patients with acute lymphocytic leukemia (ALL), in individuals with Sezary syndrome, and in a very few chronic lymphocytic leukemia (CLL) patients who develop a malignant type proliferation of lymphoid cells. Sezary cells are poor mediators of T cell cytotoxicity, but they can produce migration inhibitory factor (MIF)-like lymphokine. They produce neither immunoglobulin nor suppressor substances, but they do have a helper effect for immunoglobulin synthesis by B cell. In mycosis fungoides, skin lesions contain T lymphocytes, and there is an increased number of null cells in the blood with a simultaneous decrease in the numbers of B and T cells. T cell immunity is decreased in this condition, but IgA and IgE may be elevated. Whereas acute lymphocytic leukemia patients show major defects in cell-mediated or in humoral (antibody) immunity, a few of them manifest profound reduction in their serum immunoglobulin concentration. This has been suggested to be due to malignant expansion of their T-suppressor lymphocytes.

Reticular dysgenesis is the most severe form of all combined immunodeficiency disorders. It is believed to be caused by a cellular defect at the level of hematopoietic stem cells. This leads to a failure in the development of B cells, T cells, and granulocytes. This condition is incompatible with life and leads to early death of affected infants. The only possibility for treatment is bone marrow transplantation. This condition has an autosomal recessive mode of inheritance.

Chronic mucocutaneous candidiasis is an infection of the skin, mucous membranes, and nails by *Candida albicans* associated with defective T cell-mediated immunity that is specific to *Candida.* Skin tests for delayed hypersensitivity to the *Candida* antigen are negative. There may also be an associated endocrinopathy. The selective deficiency in T lymphocyte immunity leads to increased susceptibility to chronic *Candida* infection. T cell immunity to non-*Candida* antigens is intact. B cell immunity is normal, which leads to an intact antibody response to *Candida* antigens. T lymphocytes form migration inhibitor factor (MIF) to most of the antigens, except for those of *Candida* microorganisms. The most common endocrinopathy that develops in these patients is hypoparathyroidism. Clinical forms of the disease may be either granulomatous or nongranulomatous. *Candida* infection of the skin may be associated with the production of granulomatous lesions. The second most frequent endocrinopathy associated with this condition is Addison's disease. The disease is difficult to treat. The antifungal drug ketoconazole has proven effective. Intravenous amphotericin B has led to improvement. Transfer factor has been administered with variable success in selected cases.

Combined immunodeficiency is a genetically determined or primary immunodeficiency that may affect T cell-mediated immunity and B cell (humoral antibody)-mediated immunity. The term is usually reserved for immunodeficiency that is less profound than severe combined immunodeficiency. Combined immunodeficiency may occur in both children and adults.

Severe combined immunodeficiency syndrome (SCID) (Figure 18.5) is a profound immunodeficiency characterized by functional impairment of both B and T lymphocyte limbs of the immune response. It is inherited as an X-linked or autosomal recessive disease. The thymus has only sparse lymphocytes and Hassal's corpuscles, or is bereft of them. Several congenital immunodeficiencies are characterized as SCID. There is T and B cell lymphopenia and decreased production of IL-2. There is an absence of delayed-type hypersensitivity, cellular immunity, and of normal antibody synthesis following immunogenic challenge. SCID is a disease of infancy with failure to thrive. Affected individuals frequently die during the first 2 years of life. Clinically, they may develop a measles-like rash, show hyperpigmentation, and develop severe recurrent (especially pulmonary) infections. These subjects have heightened susceptibility to infectious disease agents such as *Pneumocystis carinii* , *Candida albicans* , and others. Even attenuated microorganisms, such as those used for immunization, (e.g., attenuated poliomyelitis viruses) may induce infection in SCID patients. Graft-vs.-host disease

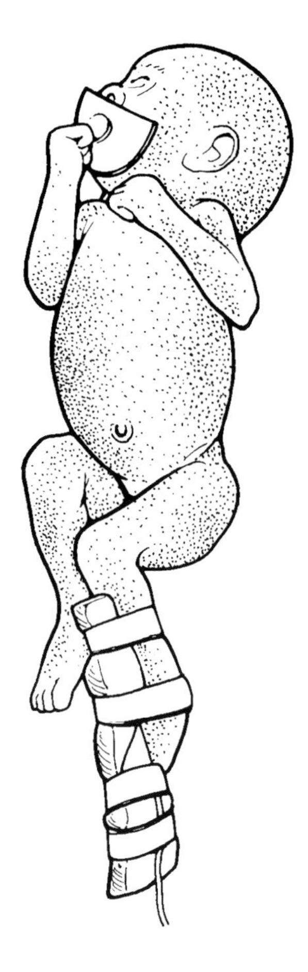

FIGURE 18.5 Severe combined immunodeficiency syndrome (SCID).

is a problem in SCID patients receiving unirradiated blood transfusions. Maternal–fetal transfusions during gestation or at parturition, or blood transfusions at a later date, provide sufficient immunologically competent cells entering the SCID patient's circulation to induce graft-vs.-host disease. SCID may be manifested in one of several forms. SCID is classified as a defect in adenosine deaminase (ADA) and purine nucleoside phosphorylase (PNP) enzymes and in a DNA-binding protein needed for HLA gene expression. Treatment is by bone marrow transplantation or by gene therapy, and enzyme reconstitution in those cases caused by a missing gene, such as adenosine deaminase deficiency.

Human SCID (hu-SCID) mouse: Since the SCID mouse does not have a functioning immune system, xenogeneic transplantation can be accomplished with little or no graft rejection. This has led to the establishment of a chimeric construct in which a functional human immune system can be established within the SCID mouse (hu-SCID). This model permits the evaluation of many important aspects of human immune-mediated pathology. Other applications of the hu-SCID model include the study of human infectious diseases and the validation of novel immune-targeted therapies. Currently, immune-mediated human diseases are being recreated in the hu-SCID mouse.

The **SCID (severe combined immundeficiency) human mouse** is a murine immunodeficiency model in which human immune system elements such as bone marrow and thymic fragments have been introduced. Pluripotent human hematopoietic stem cells differentiate into mature immunocytes in these mice, rendering them useful for the investigation of lymphocyte development.

Immunodeficiency animal models: Common murine models of immunodeficiency include the mouse strains designated SCID, Moth-eaten, Viable moth-eaten, Nude, Lipopolysaccharide, X-linked immunodeficiency, Complement deficient, Beige, Dominant spotting, Steel, and Wasted. Animal models of immunodeficiency are critical for understanding how genetic abnormalities affect immune competence.

A **SCID (severe combined immunodeficiency) mouse** with an autosomal recessive mutation expressed as severe combined immune deficiency has been described in the CB-17Icr mouse strain. These mice do not have serum immunoglobulins, yet their adenosine deaminase (ADA) levels are normal. They lack T and B lymphocytes. Thus, they fail to respond to either T cell-dependent or to T cell-independent antigens when challenged. Likewise, their lymph node or spleen cells fail to proliferate following challenge by T or B lymphocyte mitogens. The lymphoid stroma in their lymph nodes and spleen is normal. Even though there is no evidence of T cell-mediated immunity, they do have natural killer cells and mononuclear phagocytes that are normal in number and function. The mutation likewise does not affect myeloid and erythroid lineage cells. B cell development is arrested at the pro-B cell stage before cytoplasmic or surface immunoglobulins are present. There are also normal numbers of macrophages in the spleen, peritoneum, and liver. The SCID mutation is associated with an intrinsic defect in lymphoid stem cells. The main characteristic of SCID mice is the failure of their lymphocytes to express antigen-specific receptors. This is due to disordered rearrangements of T cell receptors or of immunoglobulin genes. The defect in recombination of antigen-specific receptor genes may be associated with the absence of a DNA recombinase specific for lymphocytes in these mice. This mouse model may be used to investigate the effects of anti-HIV drugs as well as of immunostimulants as a substitute for human experimentation. This model is also useful for investigations of neoplasms in hosts lacking an effective immune response.

A **severe combined immune-deficient (SCID) mouse** developed following a mutation that occurred spontaneously in BALB/c mice and is characterized by impairment of double-stranded DNA break repair. V(D)J rearrangement is mainly affected, leading to a defective coding joint formation that prevents maturation of B and T lymphocytes. At the molecular level, the SCID mutation is a genetically determined deficiency of the DNA-dependent protein kinase (DNA-PK) which induces DNA break repair by forming an activated complex with the DNA end-binding Ku proteins (p80 and p70) upon association with damaged DNA.

SCID (severe combined immunodeficiency): See severe combined immunodeficiency syndrome.

ZAP-70 deficiency is a rare autosomal recessive type of severe combined immunodeficiency syndrome (SCID) in which there is a selective absence of CD8$^+$ T cells and by abundant CD4$^+$ T lymphocytes in the peripheral blood which do not respond to T cell receptor (TCR)-mediated stimuli *in vitro*. Peripheral T cells from patients with the disorder manifest defective T cell signaling attributable to inherited mutations within the kinase domain of the TCR-associated protein tyrosine kinase (PTK) ZAP-70. ZAP-70 deficiency showed that PTKs, and especially ZAP-70, are necessary for the physiologic development and function of T cells in humans. The condition is marked by CD8 lymphocytopenia but presents during infancy with severe, recurrent, frequently fatal infections resembling those in SCID patients. They have normal or elevated numbers of circulating CD3$^+$/CD4$^+$ T lymphocytes, but essentially no CD8$^+$ T cells. These T cells fail to respond to mitogens or to allogeneic cells *in vitro* or to form cytotoxic T lymphocytes. By contrast, NK activity is normal, and they have normal or elevated numbers of B cells, and low to elevated serum immunoglobulin concentrations. The thymus may have normal architecture with normal numbers of double-positive (CD4$^+$/CD8$^+$) thymocytes but no CD8 single-positive thymocytes. The condition results from mutations in the gene encoding ZAP-70, a non-src family protein tyrosine kinase important in T cell signaling.

X-linked severe combined immunodeficiency (XSCID) is a condition characterized by T cell development failure during an early intrathymic stage, resulting in a lack of production of mature T cells or of T cell-dependent antibody. It is a consequence of defect in a gene that encodes portions of receptors for different cytokines, such as inherited mutations in the common γ chain of the receptor for IL-2, IL-4, IL-7, IL-9, and IL-15, resulting in impaired ability to transmit signals from the receptor to intracellular proteins.

JAK3-SCID is a severe combined immunodeficiency attributable to JAK3 deficiency. Immunophenotypic and functional analysis of circulating lymphoid cells permits classification of severe combined immune deficiency

(SCID) into distinct subgroups. Neither immunological nor clinical features can distinguish between X-linked and (AR) T-B$^+$ SCID. Those cytokine receptors that utilize the common γ chain always associate with the intracellular tyrosine kinase designated as JAK3. The major cytokine receptor transducing subunit binds to another JAK kinase, JAK1. Markedly reduced expression of JAK3 protein due to mutations in the JAK3 gene in unrelated infants with T-B$^+$ SCID revealed the critical role of JAK for lymphoid development and function. In the U.S. less than ten SCID patients are characterized by defective expression of JAK3 but the incidence has been greater in Europeans.

MHC class I deficiency is a type of severe combined immunodeficiency in which class I MHC molecules are not expressed on the patient's lymphocyte membranes. The trait has an autosomal recessive mode of inheritance.

Major histocompatibility complex Class II deficiency (MHC II deficiency) is a condition characterized by the lack of MHC II expression, which leads to a severe defect in both cellular and humoral immune responses to foreign antigens and is consequently characterized by an extreme susceptibility to viral, bacterial, fungal, and protozoal infections, primarily of the respiratory and gastrointestinal tracts. Severe malabsorption with failure to thrive ensues, often leading to death in early childhood. Also called bare lymphocyte syndrome type II, **MHC class II deficiency** is a type of combined immunodeficiency in which the patient's lymphocytes and monocytes not only fail to express class II MHC molecules on their surfaces but also have diminished expression of class I MHC antigens. The condition, which appears principally in North African children, has an autosomal recessive mode of inheritance. Patients are able to synthesize the class I invariant chain β_2 microglobulin. Whereas the numbers of B cells and T lymphocytes in the circulating blood are normal, patients have agammaglobulinemia and diminished cell-mediated immunity. Malabsorption in the gastrointestinal tract and diarrhea are commonly associated with this deficiency.

Omenn's syndrome is a severe immunodeficiency characterized by the development soon after birth of a generalized erythroderma desquamation, failure to thrive, protracted diarrhea, hepatosplenomegaly, hypereosinophilia, and markedly elevated serum IgE levels. It has been observed in association with neonatal minimal change nephritic syndrome. Circulating and tissue-infiltrating activated T lymphocytes that do not respond normally to mitogens or antigen *in vitro* are increased. These T cells are both oligoclonal and polyclonal. Circulating B cells are not found and the only hypogammaglobulinemia is that for IgE. Superficial lymph node architecture is grossly abnormal due to a proliferation of interdigitating S-100 protein-positive nonphagocytic reticulum cells and a depletion of B lymphocytes. This is believed to be a Th2-mediated condition even though the fundamental defect has not been discovered. The condition is fatal in the first 5 months of life unless corrected by bone marrow transplantation following chemoablation.

X-linked lymphoproliferative disease (XLP) is a combined immunodeficiency involving T and B lymphocytes, which exacerbates following exposure to Epstein-Barr virus (EBV). In older children and adults, EBV, one of the eight known human herpesviruses, is the causative agent of infectious mononucleosis. This is frequently a self-limiting polyclonal lymphoproliferative disease with an excellent prognosis. In immunosuppressed individuals, EBV infection may lead to life-threatning lymphoproliferative disorders and lymphoma. EBV has been primarily associated with certain malignancies such as endemic Burkitt's lymphoma (BL), nasopharyngeal carcinoma (NPC), certain B and T cell lymphomas, and approximately 50% of Hodgkin's disease. The XLP (*LYP*) gene locus has been mapped to Xq25. In general, XLP has an unfavorable prognosis. However, transplantation of hematopoietic stem cells from a suitable donor may cure this immunodeficiency. XLP is characterized by alteration of the (*SAP/SH2D1A*) gene. The two sets of target molecules for this small SH2 domain-containing protein include (1) a family of hematopoietic cell surface receptors, i.e., the SLAM family, and (2) a second molecule that is a phosphorylated adapter. EAT-2, a SAP-like protein, also interacts with these surface receptors. SAP/SH2D1A is a natural inhibitor of SH2-domain-dependent interactions with SLAM family members. Dysgammaglobulinemia and B cell lymphoma can occur without prior EBV infection. *SAP/SH2D1A* gene controls signaling via the SLAM surface receptors, thereby playing a critical role in T cell and APC interactions during viral infections. **X-linked lymphoproliferative syndrome** is a type of infectious mononucleosis in X-linked immunodeficiency patients who manifest EBNA-positive lymphoid cells and polyclonal B lymphocyte proliferation, as well as plasma blasts in the blood. The disease may be caused by an inability to combat Epstein-Barr virus, leading to a fatal outcome.

Somatic gene therapy is a potential therapy to treat or cure inherited immunodeficiency diseases. Stem cells derived from the patient are transfected with a normal copy of the gene that is defective in the patient. The stem cells that are now equipped with a good copy of the defective gene are reinfused into the patient's circulation.

Swiss agammaglobulinemia is a type of severe combined immunodeficiency (SCID) that has an autosomal recessive mode of inheritance. Patients usually die during infancy as a consequence of severe diarrhea, villous atrophy, and malabsorption with disaccharidase deficiency. Because of severely impaired humoral and cellular immune defense mechanisms, patients have an increased susceptibility to various opportunistic infections such as those induced by *Pneumocystis carinii*, *Candida albicans*, measles, varicella, and cytomegalovirus. Patients are also at increased risk of developing graft-vs.-host disease following blood transfusion. A stem cell defect leads to diminished numbers of T and B lymphocytes. Patients have elevated liver enzymes, lymphopenia, anemia, and diarrhea, causing an electrolyte imbalance. They are usually treated with antibiotics, γ globulin, and an HLA 6-antigen match bone marrow transplant.

Swiss type of severe combined immunodeficiency is a condition that results from a defect at the lymphocytic stem cell level. It results in cellular abnormalities that affect both T and B cell limbs of the immune response. This culminates in impaired cell-mediated immunity and humoral antibody responsiveness following challenge by appropriate immunogens. The mode of inheritance is autosomal recessive.

Adenosine deaminase (ADA) deficiency (Figure 18.6) is a form of severe combined immunodeficiency (SCID) in which affected individuals lack an enzyme, adenosine deaminase (ADA), which catalyzes the deamination of adenosine as well as deoxyadenosine to produce inosine and deoxyinosine, respectively. Cells of the thymus, spleen, and lymph node, as well as red blood cells, contain free ADA enzyme. In contrast to the other forms of SCID, children with ADA deficiency possess Hassall's corpuscles in the thymus. The accumulation of deoxyribonucleotides in various tissues, especially thymic cells, is toxic and is believed to be a cause of immunodeficiency. As deoxyadenosine and deoxy-ATP accumulate, the latter substance inhibits ribonucleotide reductase activity, which inhibits formation of the substrate needed for synthesis of DNA. These toxic substances are especially injurious to T lymphocytes. The autosomal recessive ADA deficiency leads to death. Two-fifths of severe combined immunodeficiency cases are of this type. The patient's signs and symptoms reflect defective cellular immunity with oral candidiasis, persistent diarrhea, failure to thrive, and other disorders, with death occurring prior to 2 years of age. T lymphocytes are significantly diminished. There is eosinophilia and elevated serum, and urine adenosine and deoxyadenosine levels. As bone marrow transplantation is relatively ineffective, gene therapy is the treatment of choice (Figure 18.7).

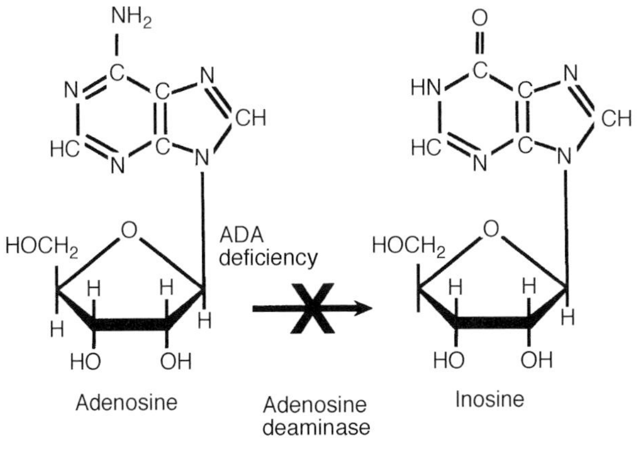

FIGURE 18.6 ADA deficiency.

Purine nucleotide phosphorylase (PNP) is an enzyme involved in purine metabolism. A deficiency of this enzyme leads to the accumulation of purine nucleosides that are toxic for developing T lymphocytes. This enzyme deficiency leads to severe combined immunodeficiency.

Purine nucleoside phosphorylase (PNP) deficiency is a type of severe combined immunodeficiency caused by mutant types of PNP. This results in the retention of metabolites that have a toxic effect on T cells. B lymphocytes appear unaffected and their numbers are normal. All cells of mammals contain PNP, which acts as a catalyst in the phosphorolysis of guanosine, deoxyguanosine, and inosine. Insufficient PNP leads to an elevation in the concentration within cells of deoxyguanosine, guanosine, deoxyguanosine triphosphate (dGTP), and guanosine triphosphate (GTP). dGTP blocks the enzyme ribonucleoside-diphosphate reductase which participates in DNA synthesis. T cell precursors are especially sensitive to death induced by these compounds. PNP is comprised of three 32-kDa subunits. Its gene, located on chromosome 14q13.1, codes for a 289-amino-acid-residue polypeptide chain. Immunologic defects associated with this disorder are characterized by anergy, lymphocytopenia, and diminished T lymphocytes in the blood. By contrast, serum immunoglobulin levels and the response following deliberate immunization with various types of immunogens are within normal limits. There is an autosomal recessive mode of inheritance. Treatment is by bone marrow transplantation.

Deoxyguanosine: See purine nucleoside phosphorylase (PNP) deficiency.

Nucleoside phosphorylase is an enzyme that is only seldom decreased in immunodeficiency patients. It catalyzes inosine conversion into hypoxanthine.

Gnotobiotic is an adjective that describes an animal or an environment where all the microorganisms are known, i.e., either a germ-free animal or an animal contaminated with one microorganism can be considered as gnotobiotic. Besides animal models, the so-called "bubble boy" who suffered severe combined immunodeficiency survived for 8 years in a plastic bubble that enclosed his gnotobiotic, germ-free environment.

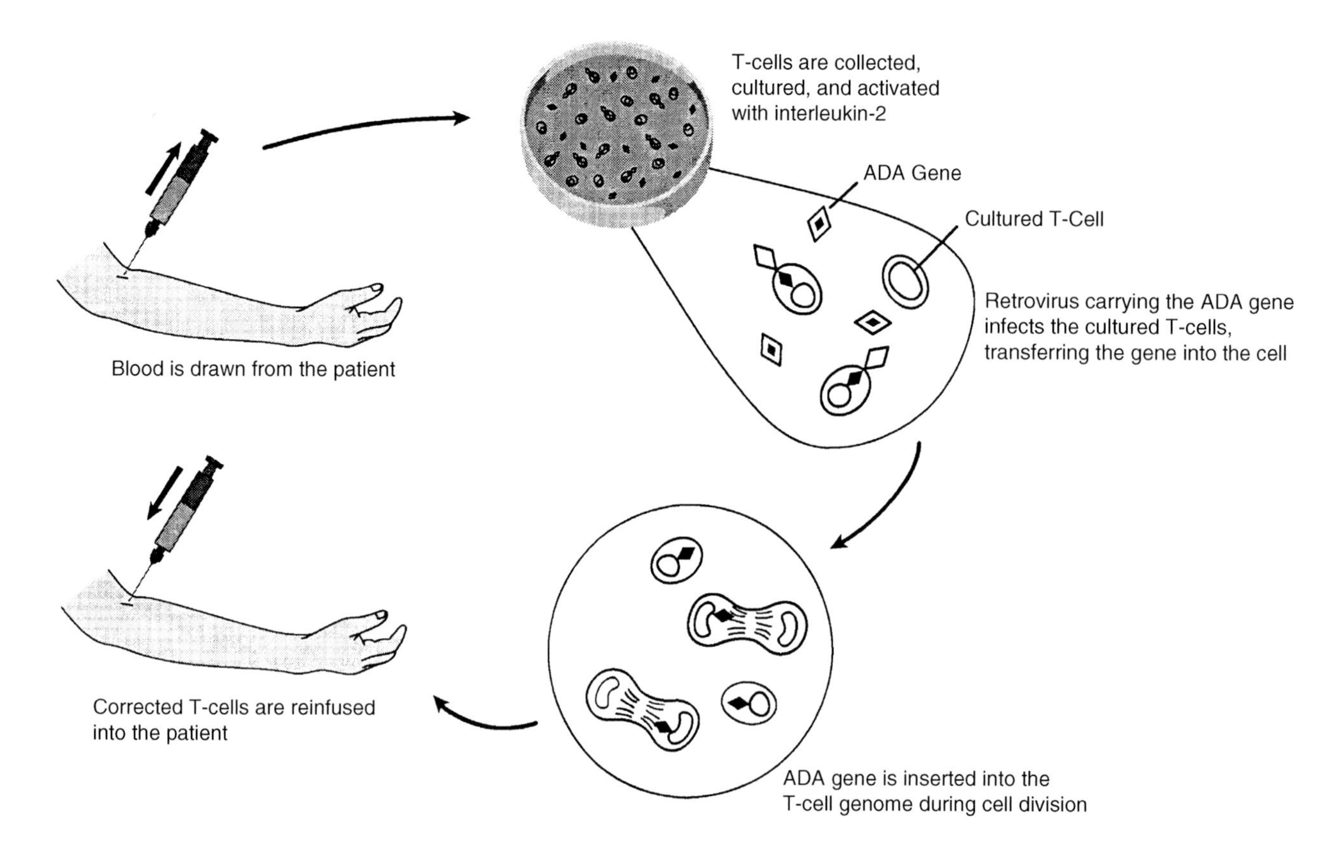

FIGURE 18.7 Gene therapy for ADA deficiency.

The term **"bubble boy"** refers to the 12-year-old male child maintained in a germ-free (gnotobiotic) environment in a plastic bubble from birth because of his severe combined immunodeficiency. A bone marrow transplant from a histocompatible sister was treated with monoclonal antibodies and complement to diminish alloreactive T lymphocytes. He died of a B cell lymphoma as a consequence of Epstein-Barr virus-induced polyclonal gammopathy that transformed into monoclonal proliferation, leading to lymphoma.

Bare lymphocyte syndrome (BLS) causes failure to express class I HLA-A, -B, or -C major histocompatibility antigens due to defective β_2 microglobulin expression on the cell surface. This immune deficiency is inherited as an autosomal recessive trait. In some individuals, the class II HLAODR molecules are likewise not expressed. Patients may be asymptomatic or manifest respiratory tract infections, mucocutaneous candidiasis, opportunistic infections, chronic diarrhea and malabsorption, inadequate responsiveness to antigen, aplastic anemia, leukopenia, decreased T lymphocytes, and normal or elevated B lymphocytes. The mechanism appears to be related to either defective gene activation or inaccessibility of promoter protein. DNA techniques are required for tissue typing.

Wiskott-Aldrich syndrome (Figure 18.8) is an X-linked recessive immunodeficiency disease of infants characterized

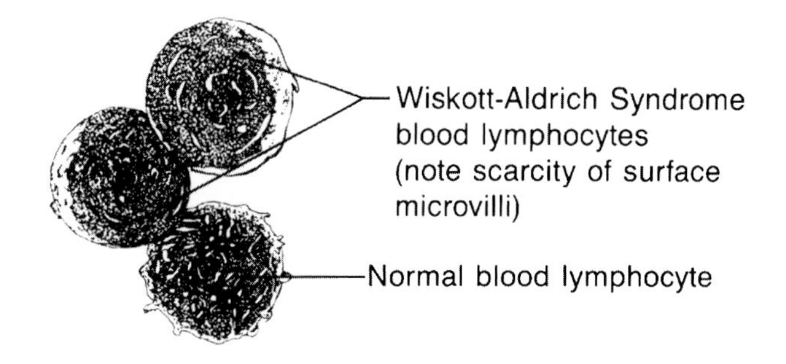

FIGURE 18.8 Wiskott-Aldrich syndrome.

by thrombocytopenia, eczema, and increased IgA and IgE levels. There is decreased cell-mediated immunity (and delayed hypersensitivity), and the antibody response to polysaccharide antigens is defective, with only minute quantities of IgM appearing in the serum. There may be an inability to recognize processed antigen. Male patients may have small platelets with absent surface glycoprotein Ib. Whereas IgA and IgE are increased, IgM is diminished, although IgG serum concentrations are usually normal. By electron microscopy, T lymphocytes appear to be bereft of the markedly fimbriated surface of normal T cells. T lymphocytes have abnormal sialophorin. Patients may have an increased incidence of malignant lymphomas. Bone marrow transplantation corrects the deficiency.

Elevated IgE, defective chemotaxis, recurrent infection, and eczema: Both men and women with the third disorder may manifest several bacterial infections, eczema, abscesses, and pneumonia associated with group A β hemolytic streptococci and *Staphylococcus aureus*. They demonstrate normal function of both B and T cell limbs of the immune response, although their IgE levels are strikingly elevated, reaching values that are tenfold those of normal individuals. Only chemotactic function is defective, with other parameters of phagocytic cell function within normal limits. No treatment other than antibiotic therapy is available.

Immunodeficiency with thrombocytopenia and eczema (Wiskott-Aldrich syndrome) is an X-linked recessive disease characterized by thrombocytopenia, eczema, and susceptibility to recurrent infections, sometimes leading to early death. The life span of young boys is diminished as a consequence of extensive infection, hemorrhage, and sometimes malignant disease of the lymphoreticular system. Infectious agents affecting these individuals include the Gram-negative and Gram-positive bacteria, fungi, and viruses. The thymus is normal morphologically, but there is a variable decline in cellular immunity, which is thymus dependent. The lymph node architecture may be altered as paracortical areas become depleted of cells with the progression of the disease. IgM levels in serum are low, but IgA and IgG may be increased. Isohemagglutinins are usually undetectable in the serum. Patients may respond normally to protein antigens, but show defective responsiveness to polysaccharide antigens. There is decreased immune responsiveness to lipopolysaccharides from enteric bacteria and B blood group substances.

Immunodeficiency with thrombocytopenia is a synonym for Wiskott-Aldrich syndrome.

Thymic alymphoplasia is a severe combined immune deficiency transmitted as an X-linked recessive trait.

Immunodeficiency with partial albinism is a type of combined immunodeficiency characterized by decreased cell-mediated immunity and deficient natural killer cells. Patients develop cerebral atrophy and aggregation of pigment in melanocytes. This disease, which leads to death, has an autosomal recessive mode of inheritance.

Ataxia telangiectasia is a disorder characterized by cerebellar ataxia, oculocutaneous telangiectasis, variable immunodeficiency which affects both T and B cell limbs of the immune response, the development of lymphoid malignancies, and recurrent sinopulmonary infections. Clinical features may appear by 2 years of age. A total of 40% of patients have selective IgA deficiency. The disease has an autosomal recessive mode of inheritance. There may be lymphopenia, normal or decreased T-lymphocyte numbers, and a normal or diminished lymphocyte response to PHA and allogeneic cells. The delayed-type hypersensitivity skin test may not stimulate any response. Some individuals may have an IgG2, IgG4, or IgA2 subclass deficiency. Other patients may reveal no IgE antibody level. There is diminished antibody responsiveness to selected antigens. B cell numbers are usually normal and NK cell function is within physiologic limits. The level of T cell deficiency varies. Defects in DNA repair mechanisms lead to multiple breaks, inversions, and translocations within chromosomes, rendering them highly susceptible to the injurious action of ionizing radiation and radiomimetic chemicals. The chromosomal breaks are especially apparent on chromosomes 7 and 14 in the regions that encode immunoglobulin genes and T cell receptor genes. The multiple chromosomal breaks are believed to be linked to the high incidence of lymphomas in these patients. α-fetoprotein is also elevated. Endocrine abnormalities associated with the disease include glucose intolerance associated with anti-insulin receptor antibodies and hypogonadism in males. Patients may experience retarded growth and hepatic dysfunction. Death may occur in many of the patients related to recurrent respiratory tract infections or lymphoid malignancies.

Hereditary ataxia telangiectasia: See ataxic telangiectasia.

LFA-1 deficiency is an immunodeficiency that is caused by a defect in lymphocyte function-associated antigen, a 95-kDa β-chain linked to CD11a which aids NK-binding, T helper cell reactivity, and cytotoxic T cell-mediated killing. This deficiency is associated with pyogenic mucocutaneous infections, pneumonia, diminished respiratory burst, and abnormal cell adherence in chemotaxis, causing poor wound healing among other features.

Phagocyte disorders are conditions characterized by recurrent bacterial infections that can involve the skin, respiratory tract, and lymph nodes. Evaluation of phagocytosis should include tests of motility, chemotaxis, adhesion, intracellular killing (respiratory burst), enzyme testing, and examination of the peripheral blood smear. Phagocyte disorders include the following conditions: chronic granulomatous disease; myeloperoxidase deficiency; Job syndrome (Hyperimmunoglobulin E syndrome); Chediak-Higashi syndrome; and leukocyte adhesion deficiency, together with less common disorders.

Chronic granulomatous disease (CGD) is a disorder that is inherited as an X-linked trait in two thirds of the cases and as an autosomal recessive trait in the remaining one third. Clinical features are usually apparent before the end of the second year of life. There is an enzyme defect associated with NADPH oxidase. This enzyme deficiency causes neutrophils and monocytes to have decreased consumption of oxygen and diminished glucose utilization by

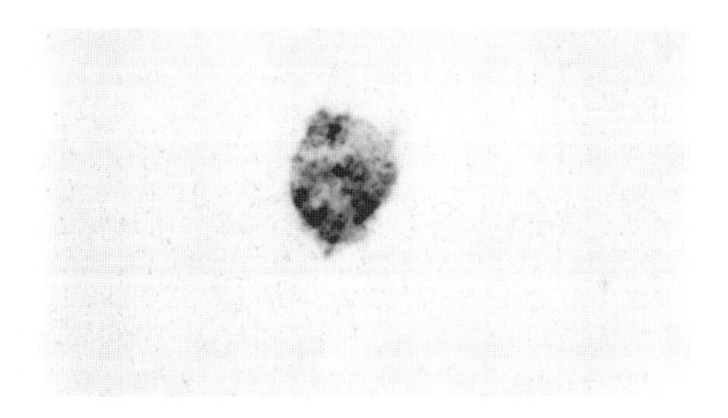

FIGURE 18.9 Nitroblue tetrazolium test (NBT).

the hexose monophosphate shunt. Although neutrophils phagocytize microorganisms, they do not form superoxide and other oxygen intermediates that usually constitute the respiratory burst. Neutrophils and monocytes also form a smaller amount of hydrogen peroxide, have decreased iodination of bacteria, and have diminished production of superoxide anions. All of this leads to decreased intracellular killing of bacteria and fungi. Thus, these individuals have an increased susceptibility to infection with microorganisms that normally are of relatively low virulence. These include *Aspergillus*, *Serratia marcescens*, and *Staphylococcus epidermidis*. Patients may have hepatosplenomegaly, pneumonia, osteomyelitis, abscesses, and draining lymph nodes. The quantitative nitroblue tetrazolium (NBT) test and the quantitative killing curve are both employed to confirm the diagnosis (Figure 18.9). Most microorganisms that cause difficulty in CGD individuals are catalase positive. Therapy includes interferon γ, antibiotics, and surgical drainage of abscesses (Figure 18.10).

Phagocytic cell function deficiencies: Patients with this group of disorders frequently show an increased susceptibility to bacterial infections but are generally able to successfully combat infections by viruses and protozoa. Phagocytic dysfunction can be considered as either extrinsic or intrinsic defects. Extrinsic factors include diminished opsonins that result from deficiencies in antibodies and complement, immunosuppressive drugs or agents that reduce phagocytic cell numbers, corticosteroids that alter phagocytic cell function, and autoantibodies against neutrophil antigens that diminish the number of PMNs in the blood circulation. Complement deficiencies or inadequate complement components may interfere with neutrophil chemotaxis to account for other extrinsic defects. By contrast, intrinsic defects affect the ability of phagocytic cells to kill bacteria. This is related to deficiencies of certain metabolic enzymes associated with the intracellular digestion of bacterial cells. Among the disorders are chronic granulomatous disease, myeloperoxidase deficiency, and defective glucose-6-phosphate-dehydrogenase.

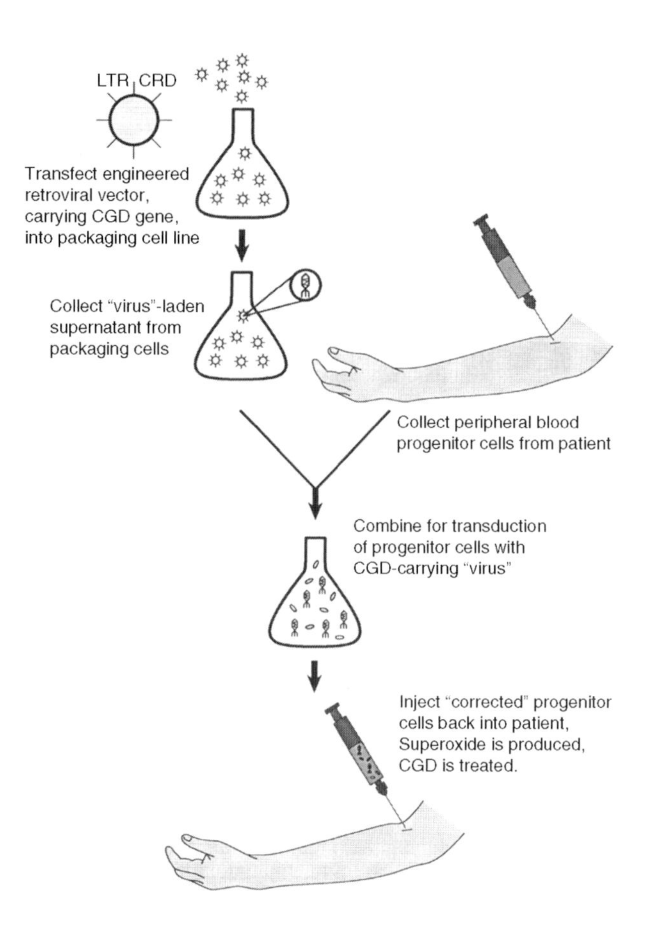

FIGURE 18.10 Strategy for gene therapy for CGD with peripheral blood progenitor cells.

Cytochrome b deficiency: See chronic granulomatous disease.

CGD is an abbreviation for chronic granulomatous disease.

The **phagocytic index (PI)** is an *in vivo* measurement of the ability of the mononuclear phagocyte system to remove foreign particles. It may be represented by the rate at which injected carbon particles are cleared from the blood. The PI is increased in graft-vs.-host disease.

Phagocytic dysfunction refers to the altered ability of macrophages, neutrophils, or other phagocytic cells to ingest microorganisms or to digest them following ingestion. This represents a type of immunodeficiency involving phagocytic function.

Deficiency of secondary granules is a rare disorder in which neutrophils are bereft of secondary granules. This has an autosomal recessive mode of inheritance. Affected individuals show an increased incidence of infection by pyogenic microorganisms.

Chemotactic disorders are conditions attributable to abnormalities of the complex molecular and cellular interactions involved in mobilizing an appropriate phagocytic cell response to injuries or inflammation. This can involve defects in either the humoral or cellular components of chemotaxis that usually lead to recurrent infections. The process begins with the generation of chemoattractants. Among these chemoattractants that act *in vivo* are the anaphylatoxins (C3a, C4a, and C5a), leukotriene B_4 (LTB_4), IL-8, GM-CSF, and platelet activating factors. Once exposed to chemoattractant, circulating neutrophils embark upon a four-stage mechanism of emigration through the endothelial layer to a site of tissue injury where phagocytosis takes place. The four stages include (1) rolling or initial margination by the selectins (L-, P-, E-); (2) stopping on the endothelium by CD18 integrins and ICAM-1; (3) neutrophil–neutrophil adhesion by CD11b/CD18; and (4) transendothelial migration by CD11b/CD18, CD11a/CD18, ICAM-1. Chemotactic defects can be either acquired or inherited. Specific disorders are listed separately.

Chemotactic assays: The chemotactic properties of various substances can be determined by various methods. The most popular is the Boyden technique. This consists of a chamber separated into two compartments by a Millipore® filter of appropriate porosity, through which cells can migrate actively but not drop passively. The cell preparation is placed in the upper compartment of the chamber and the assay solution is placed in the lower compartment. The chamber is incubated in air at 37°C for 3 h, after which the filter is removed and the number of cells migrating to the opposite surface of the filter are counted.

Chronic and cyclic neutropenia is a syndrome characterized by recurrent fever, mouth ulcers, headache, sore throat, and furunculosis occurring every 3 weeks in affected individuals. This chronic agranulocytosis leads to premature death from infection by pyogenic microorganisms in affected children who may have associated pancreatic insufficiency, dysostosis, and dwarfism. Antibodies can be transmitted from the maternal to the fetal circulation to induce an isoimmune neutropenia. This may consist of either a transitory type in which the antibodies are against neutrophil antigens determined by the father or a type produced by autoantibodies against granulocytes.

Neutrophil nicotinamide adenine dinucleotide phosphate oxidase: See chronic granulomotous disease.

Copper and immunity: The trace metal copper is required for a healthy immune system. Copper insufficiency in humans leads to pathologic effects that may include cerebral degeneration in Menkes' syndrome and increased susceptibility to infection among infants who are copper deficient. In neonatal copper deficiency there is a marked neutropenia associated with infections. Insufficient copper intake among domestic animals leads to diminished bactericidal activity, impairment of neutrophil function, and an increased susceptibility to bacterial and fungus infections.

Copper deficiency: Trace amounts of copper are required for the ontogeny and proper functioning of the immune system. Neonates or malnourished children with copper deficiency may have associated neutropenia and an increased incidence of infection. Antioxidant enzyme levels are diminished in copper deficiency. This may render immunologically competent cells unprotected against elevated oxygen metabolism associated with immune activation.

Glucose-6-phosphate dehydrogenase deficiency: Occasionally, individuals of both sexes have been shown to have no glucose-6-phosphate dehydrogenase in their leukocytes. This could be associated with deficient NADPH, with diminished hexose monophosphate shunt activity and reduced formation of hydrogen peroxide in leukocytes, which are unable to kill microorganisms intracellularly as was described in chronic granulomatous disease (CGD). Clinical aspects resemble those in CGD, except that glucose-6-phosphate dehydrogenase deficiency occurs later and affects both sexes, in addition to being associated with hemolytic anemia. Although the NBT test is within normal limits, glucose-6-phosphate dehydrogenase activity is deficient, the killing curve is altered, there is abnormal formation of H_2O_2, and oxygen consumption is inadequate. No effective treatment is available.

Tuftsin deficiency: Tuftsin is a tetrapeptide that stimulates phagocytes. Tuftsin is split from an immunoglobulin by the action of one proteolytic enzyme in the spleen that cleaves the carboxy terminus between residues 292 and 293, and another enzyme that is confined to neutrophil membranes (leukokinase) that splits the molecule between positions 288 and 289. Thus, tuftsin deficiency, which is transmitted as an autosomal recessive trait, results from a lack of this splenic enzyme. Although γ globulin has been used, there is no known treatment.

Lazy leukocyte syndrome is a disease of unknown cause in which patients experience an increased incidence of pyogenic infections such as abscess formation, pneumonia, and gingivitis, and which is linked to defective neutrophil chemotaxis in combination with neutropenia. Random locomotion of neutrophils is also diminished and abnormal. This is demonstrated by the vertical migration of leukocytes in capillary tubes. There is also impaired exodus of neutrophils from the bone marrow.

Leukocyte adhesion deficiency (LAD) is a recurrent bacteremia with staphylococci or *Pseudomonas* linked to defects in the leukocyte adhesion molecules known as integrins. These include the CD11/CD18 family of molecules. CD18 β chain gene mutations lead to a lack of complement receptors CR3 and CR4 to produce a congenital disease marked by recurring pyogenic infections. Deficiency of p150,95, LFA-1, and CR3 membrane proteins leads to diminished adhesion properties and mobility of phagocytes and lymphocytes. There is a flaw in synthesis of the 95-kDa β-chain subunit that all three of these molecules share. The defect in mobility is manifested as altered chemotaxis, defective random migration, and faulty spreading. Particles coated with C3 are not phagocytized and therefore fail to activate a respiratory burst. The CR3 and p150,95 deficiency account for the defective phagocytic activity. LAD patients' T cells fail to respond normally to antigen or mitogen stimulation and are also unable to provide helper function for B cells producing immunoglobulin. They are ineffective in fatally injuring target cells and they do not produce the lymphokine γ interferon. LFA-1 deficiency accounts for the defective response of these T lymphocytes as well as all natural killer cells, which also have impaired ability to fatally injure target cells. Clinically, the principal manifestations are a consequence of defective phagocyte function rather than of defective T lymphocyte function. Patients may have recurrent severe infections, a defective inflammatory response, abscesses, gingivitis, and periodontitis. There are two forms of leukocyte adhesion deficiency. Those with the severe deficiency do not express the three α- and four β-chain complexes, whereas those with moderate deficiency express 2.5 to 6% of these complexes. There is an autosomal recessive mode of inheritance for leukocyte adhesion deficiency.

Job's syndrome (Figure 18.11) refers to cold staphylococcal abscesses or infections by other agents that recur. There is associated eczema, elevated levels of IgE in the serum, and phagocytic dysfunction associated with glutathione reductase and glucose-6-phosphatase deficiencies. The syndrome has an autosomal recessive mode of inheritance.

Hyperimmunoglobulin E syndrome (HIE) is a condition characterized by markedly elevated IgE levels, i.e., greater than 5000 IU/ml. The patients have early eczema and repeated abscesses of the skin, sinuses, lungs, eyes, and ears. *Staphylococcus aureus*, *Candida albicans*, *Hemophilus influenzae*, *Streptococcus pneuomoniae*, and group A hemolytic streptococci are among the more common infectious agents. The principal infection produced by *Staphylococcus aureus* and *C. albicans* is a "cold abscess" in the skin. The failure of IgE to fix complement and therefore cause inflammation at the infection site is characteristic. IgG antibodies against IgE form complexes that bind to mononuclear phagocytes, resulting in

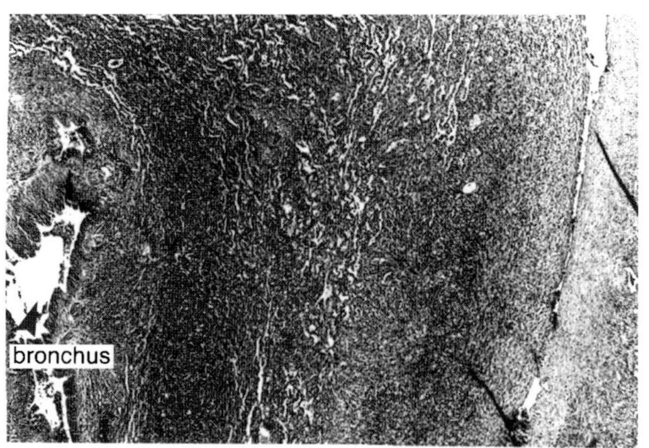

FIGURE 18.11 Job's syndrome with bronchogenic abscess.

monokine release that induces calcium resorption from bone. As calcium is lost from the bone, osteoporosis results, leading to bone fractures. Patients with HIE have diminished antibody responses to vaccines and to major histocompatibility antigens. They may be anergic, and *in vitro* challenge of their lymphocytes with mitogens or antigens leads to diminished responsiveness. There is also a decrease in the $CD8^+$ T lymphocyte population in the peripheral blood. The disease becomes manifest in young infants, shows no predilection for males vs. females, and is not hereditary.

Hyperimmunoglobulin M syndrome is an immunodeficiency disorder in which the serum IgM level is normal or elevated. By contrast, the serum IgG and IgA levels are strikingly diminished or absent. These patients have repeated infections and may develop neoplasms in childhood. This syndrome may be transmitted in an X-linked or autosomal dominant fashion. It may also be related to congenital rubella. The condition is produced by failure of the T lymphocytes to signal IgM-synthesizing B cells to switch to IgG- and IgA-producing cells. In this X-linked disease in boys who are unable to synthesize immunoglobulin isotypes other than IgM, there is a defect in the gene encoding the CD40 ligand. The T_H cells fail to express CD40L. These patients fail to develop germinal centers or displaced somatic hypermutation. They do not form memory B cells and are subject to pyogenic bacterial and protozoal infections.

X-linked hyper-IgM syndrome: See hyperimmunoglobulin M syndrome.

Hyper-IgM syndrome: Monoclonal antibodies against CD40 surface protein of B cells induce isotype switching in the presence of appropriate costimulatory cytokines.

CD40L is a surface molecule transiently expressed on activated T cells mostly of the $CD4^+$ subpopulation. Mutations of CD40L account for the X-linked form of the disease.

Myeloperoxidase is an enzyme present in the azurophil granules of neutrophilic leukocytes that catalyzes peroxidation of many microorganisms. Myeloperoxidase in conjunction with hydrogen peroxidase and halide has a bactericidal effect.

Myeloperoxidase (MPO) deficiency is a lack of 116-kDa myeloperoxidase in both neutrophils and monocytes. This enzyme is located in the primary granules of neutrophils. It possesses a heme ring which imparts a dark-green tint to the molecule. MPO deficiency has an autosomal recessive mode of inheritance. Clinically, affected patients have a mild version of chronic granulomatous disease. *Candida albicans* infections are frequent in this condition.

Chediak-Higashi syndrome (Figure 18.12) is a childhood disorder with an autosomal recessive mode of inheritance that is identified by the presence of large lysosomal granules in leukocytes that are very stable and undergo slow degranulation. Multiple systems may be involved. Repeated bacterial infections with various microorganisms, partial albinism, central nervous system disorders, hepatosplenomegaly, and an inordinate incidence of malignancies of the lymphoreticular tissues may occur. The large cytoplasmic granular inclusions that appear in white blood cells may also be observed in blood platelets and can be seen by regular light microscopy in peripheral blood smears. There is defective neutrophil chemotaxis and an altered ability of the cells to kill ingested microorganisms. There is a delay in the killing time, even though hydrogen peroxide formation, oxygen consumption, and hexose monophosphate shunt are all within normal limits. There is also defective microtubule function, leading to defective phagolysosome formation. Cyclic AMP levels may increase. This causes decreased neutrophil degranulation and mobility. High doses of ascorbic acid have been shown to restore normal chemotaxis, bactericidal activity, and degranulation. Natural killer cell numbers and function are decreased. There is an increased incidence of lymphomas in Chediak-Higashi patients. There is no effective therapy other than the administration of antibiotics for the infecting microorganisms. The disease carries a poor prognosis because of the infections and the neurological complications. The majority of affected individuals die during childhood, although occasional subjects may live longer.

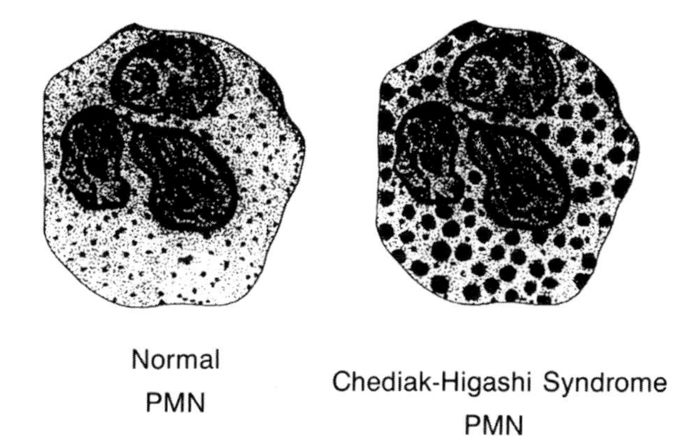

FIGURE 18.12 Chediak-Higashi syndrome.

Beige mice are a mutant strain of mice that develops abnormalities in pigment, defects in natural killer cell function, and heightened tumor incidence. This serves as a model for the Chediak-Higashi disease in man.

Complement deficiency conditions are rare. In healthy Japanese blood donors, only 1 in 100,000 persons had no C5, C6, C7, and C8. No C9 was contained in 3 of 1000 individuals. Most individuals with missing complement components do not manifest clinical symptoms. Additional pathways provide complement-dependent functions that are necessary to preserve life. If C3, factor I, or any segment of the alternative pathway is missing, the condition may be life-threatening with markedly decreased opsonization in phagocytosis. C3 is depleted when factor I is absent. C5, C6, C7, or C8 deficiencies are linked with infections, mainly meningococcal or gonococcal, which usually succumb to complement's bactericidal action. Deficiencies in classical complement pathway activation are often associated with connective tissue or immune complex diseases. SLE may be associated with C1qrs, C2, or C4 deficiencies. Hereditary angioedema (HAE) patients have a deficiency of C1 inactivator. A number of experimental animals with specific complement deficiencies have been described, such as C6 deficiency in rabbits and C5 deficiency in mice. Acquired complement deficiencies may be caused by either accelerated complement consumption in immune complex diseases with a type III mechanism or by diminished formation of complement proteins as in acute necrosis of the liver.

C1 deficiencies: Only a few cases of C1q, C1r, or C1r and C1s deficiencies have been reported. These have an autosomal recessive mode of inheritance. Patients with these defects may manifest SLE, glomerulonephritis, or pyogenic infections. They have an increased incidence of type III (immune complex) hypersensitivity diseases. Half of C1q-deficient persons may contain physiologic levels of mutant C1q that are not functional.

C1 inhibitor (C1 INH) deficiency: The absence of C1 INH is the most frequently found deficiency of the classic complement pathway and may be seen in patients with hereditary angioneurotic edema. This syndrome may be expressed as either a lack of the inhibitor substance or a functionally inactive C1 INH. The patient develops edema of the face, respiratory tract, including the glottis and bronchi, and the extremities. Severe abdominal pain may occur with intestinal involvement. Since C1 INH can block Hagemann factor (factor XII) in the blood clotting mechanism, its absence can lead to the liberation of kinin and fibrinolysis, which results from the activation of plasmin. The disease is inherited as an autosomal dominant trait. When edema of the larynx occurs, the patient may die of asphyxiation. When abdominal attacks occur, there may be watery diarrhea and vomiting. These bouts usually span 48 h and are followed by a rapid recovery. During an attack of angioedema, C1r is activated to produce C1s, which depletes its substrates C4 and C2. The action of activated C1s on C4 and C2 leads to the production of a substance that increases vascular permeability, especially that of post-capillary venules. C1 and C4 cooperate with plasmin to split this active peptide from C2. Of the families of patients with hereditary angioneurotic edema, 85% do not contain C1 INH. Treatment is by preventive maintenance. Patients are given inhibitors of plasmin such as ε-aminocaproic acid and tranexamic acid. Methyl testosterone, which causes synthesis of normal C1 INH in angioneurotic edema patients, is effective by an unknown mechanism.

Hereditary angioedema (HAE) is a disorder in which recurrent attacks of edema, persisting for 48 to 72 h occur in the skin, gastrointestinal, and respiratory tracts. It is nonpitting and can threaten death if laryngeal edema becomes severe enough to obstruct the airway. Edema in the jejunum may be associated with abdominal cramps and bilious vomiting. Edema of the colon may lead to watery diarrhea. There is no redness or itching associated with edema of the skin. Tissue trauma or no apparent initiating cause may induce an attack. It is due to decreased or absent C1 inhibitor (C1 INH). It is inherited in an autosomal dominant fashion. Heterozygotes for the defect develop the disorder. Greatly diminished C1 INH levels (5 to 30% of normal) are found in affected individuals. Activation of C1 leads to increased cleavage of C4 and C2, decreasing their serum levels during an attack. C1 INH is also a kinin system inactivator. C1 INH deficiency in HAE permits a kinin-like peptide produced from C2b to increase vascular permeability leading to manifestations of HAE. Some have proposed that bradykinin may represent the vasopermeability factor. Hereditary angioedema has been treated with aminocaproic acid and transexamic acid, but they do not elevate C1 INH or C4 levels. Anabolic steroids such as danazol and stanozolol, which activate C1 INH synthesis in affected individuals, represent the treatment of choice.

Acquired C1 inhibitor deficiency is a condition in which the C1 inhibitor is inactivated, resulting in elevated C4 and C2 cleavage as C1 is activated. Patients experience repeated laryngeal, intestinal, and subcutaneous tissue swelling. A kinin-like peptide derived from C2b promotes vascular permeability and produces symptoms. Subjects with B lymphocyte or plasma-cell monoclonal proliferation may develop acquired C1 inhibitor deficiency. These syndromes include multiple myeloma, Waldenström's macroglobulinemia, and B cell lymphoma. Subjects may develop antiidiotypic antibodies against membrane immunoglobulins or myeloma proteins. Idiotype–antiidiotype interactions at the cell membrane may result in C1 fixation and may increase utilization of C4 and C2 as well as of C1 inhibitor.

C1q deficiency may be found in association with lupus-like syndromes. C1r deficiency, which is inherited as an autosomal recessive trait, may be associated with respiratory tract infections, glomerulonephritis, and skin manifestations, which resemble an SLE-like disease. C1s deficiency is transmitted as an autosomal dominant trait and patients may again show SLE-like signs and symptoms. Their antigen–antibody complexes can persist without resolution.

C2 deficiency: Rare individuals may demonstrate a failure to express C2. Although no symptoms are normally associated with this trait, which has an autosomal recessive mode of inheritance, autoimmune-like manifestations that resemble features of certain collagen–vascular diseases, such as SLE, may appear. Thus, many genetically determined complement deficiencies are not associated with signs and symptoms of disease. When they do occur, it is usually manifested as an increased incidence of infectious diseases that affect the kidneys, respiratory tract, skin, and joints.

C3 deficiency is an extremely uncommon genetic disorder that may be associated with repeated serious pyogenic bacterial infections and may lead to death. The C3-deficient individuals are deprived of appropriate opsonization, prompt phagocytosis, and the ability to kill infecting microorganisms. There is defective classical and alternative pathway activation. Besides infections, these individuals may also develop an immune complex disease such as glomerulonephritis. C3 levels that are one half normal in heterozygotes are apparently sufficient to avoid the clinical consequences induced by a lack of C3 in the serum.

CR3 deficiency syndrome: See leukocyte adhesion deficiency.

C4 deficiency is an uncommon genetic defect with an autosomal recessive mode of inheritance. Affected individuals have defective classical complement pathway activation. Those who manifest clinical consequences of the defect may develop SLE or glomeruloneophritis. Half of the patients with C4 and C2 deficiencies develop SLE, but deficiencies in these two complement components are not usually linked to increased infections.

C5 deficiency is a very uncommon genetic disorder that has an autosomal recessive mode of inheritance. Affected individuals have only trace amounts of C5 in their plasma. They have a defective ability to form the membrane attack complex (MAC), which is necessary for the efficient lysis of invading microorganisms. C5-deficient individuals have an increased susceptibility to disseminated infections by *Neisseria* microorganisms such as *N. meningitidis* and *N. gonorrhoeae*. Heterozygotes may manifest 13 to 65% of C5 activity in their plasma and usually show no clinical effects of their partial deficiency. C5-deficient mice also have been described.

C6 deficiency is a highly uncommon genetic defect with an autosomal recessive mode of inheritance in which affected individuals have only trace amounts of C6 in their plasma. They are defective in the ability to form an MAC and have increased susceptibility to disseminated infections by *Neisseria* microorganisms, which include gonococci and meningococci. C6-deficient rabbits have been described.

C7 deficiency is a highly uncommon genetic disorder with an autosomal recessive mode of inheritance in which the serum of affected persons contains only trace amounts of C7 in the plasma. They have a defective ability to form an MAC and show an increased incidence of disseminated infections caused by *Neisseria* microorganisms. Some may manifest an increased propensity to develop immune complex (type III hypersensitivity) diseases such as glomerulonephritis or SLE.

C8 deficiency is a highly uncommon genetic disorder with an autosomal recessive mode of inheritance in which affected individuals are missing C8 α, γ, or β chains. This is associated with a defective ability to form an MAC. Individuals may have an increased propensity to develop disseminated infections caused by *Neisseria* microorganisms such as meningococci.

C9 deficiency is a highly uncommon genetic disorder with an autosomal recessive mode of inheritance in which only trace amounts of C9 are present in the plasma of affected persons. There is a defective ability to form the MAC. The serum of C9-deficient subjects retains its lytic and bactericidal activity even though the rate of lysis is decreased compared to that induced in the presence of C9. There are usually no clinical consequences associated with this condition. The disorder is more common in the Japanese than in most other populations.

Secondary immunodeficiency is an immunodeficiency that is not due to a failure or intrinsic defect in the T and B lymphocytes of the immune system. It is a consequence of some other disease process and may be either transient or permanent. The transient variety may disappear following adequate treatment whereas the more permanent type persists. Secondary immunodeficiencies are commonly produced by many effects. For example, those that appear in patients with neoplasms may result from effects of the tumor. Secondary immunodeficiencies may cause an individual to become susceptible to microorganisms that would otherwise cause no problem. They may occur following immunoglobulin or T lymphocyte loss, the administration of drugs, infections, cancer, effects of ionizing radiation on immune system cells, and other causes.

Intestinal lymphangiectasia: Escape of immunoglobulins, as well as other proteins and lymphocytes, into the intestinal tract as a consequence of lymphatic dilation in the intestinal villi. The loss of immunoglobulin leads to secondary immunodeficiency. In addition to primary intestinal telangiectasia, obstruction of lymphatic drainage of the intestine produced by a lymphoma represents a secondary type.

Late-onset immune deficiency is a disease of unknown cause associated with gastric carcinoma, pernicious anemia, atrophic gastric autoimmunity, and several other conditions.

Immunodeficiency from hypercatabolism of immunoglobulin occurs when serum levels of immunoglobulins fluctuate according to their rates of synthesis and catabolism. Although many immunological deficiencies result from defective synthesis of immunoglobulins and lymphocytes, immunoglobulin levels in serum can decline as a consequence of either increased catabolism or loss into the gastrointestinal tract or other areas. Defective catabolism may affect one to several immunoglobulin classes. For example, in myotonic dystrophy, only IgG is hypercatabolized. In contrast to the normal levels of IgM, IgA, IgD, IgE, and albumin in the serum, the IgG concentration is markedly diminished. Synthesis of IgG in these individuals is normal, but the half-life of IgG molecules is reduced as a consequence of increased catabolism. Patients with ataxia telangiectasia and those with selective IgA deficiency have antibodies directed against IgA that remove this class of immunoglobulin. Patients with the rare condition known as familial hypercatabolic hypoproteinemia demonstrate reduced IgG and albumin levels in the serum and slightly lower IgM levels, but the IgA and IgE concentrations are either normal or barely increased. Although synthesis of IgG and albumin in such patients is within normal limits, the catabolism of these two proteins is greatly accelerated.

Immunodeficiency from severe loss of immunoglobulins and lymphocytes occurs when the gastrointestinal and urinary tracts are two sources of serious protein loss in disease processes. The loss of integrity of the renal glomerular basement membrane, renal tubular disease, or both may result in loss of immunoglobulin molecules into the urine. Since the small IgG molecules would pass through in many situations leaving larger IgA molecules in the intravascular space, all immunoglobulins are not lost from the serum at the same rate. More than 90 diseases that affect the gastrointestinal tract have been associated with protein-losing gastroenteropathy. This may be secondary to inflammatory or allergic disorders, or disease processes involving the lymphatics. In intestinal lymphangiectasia associated with lymphatic blockage, lymphocytes as well as protein are lost. Lymphatics in the small intestine are dilated. Intestinal lymphangiectasia patients show defects in both humoral and cellular immune mechanisms. The major immunoglobulins are diminished to less than half of normal. IgG is affected more than IgA and IgM, which are more affected than IgE.

An **immunodeficiency associated with hereditary defective response to Epstein-Barr virus** is an immunodeficiency that develops in previously healthy subjects with a normal immune system who have developed a primary Epstein-Barr virus infection. They develop elevated numbers of natural killer cells in the presence of a lymphopenia. The condition is serious, and its acute stage may lead to B cell lymphoma or failure of the bone marrow or agammaglobulinemia. The disease may be fatal. The condition was first considered to have an X-linked recessive mode of inheritance recurring only in males, but it has now been found in occasional females. The term Duncan's syndrome is often used to describe the X-linked variety of this condition.

An **acquired immunodeficiency** is a decrease in the immune response to immunogenic (antigenic) challenge as a consequence of numerous diseases or conditions that include acquired immune deficiency syndrome (AIDS), chemotherapy, immunosuppressive drugs such as corticosteroids, psychological depression, burns, nonsteroidal antiinflammatory drugs, radiation, Alzheimer's disease, celiac disease, Waldenstrom's macroglobulinemia, multiple myeloma, aplastic anemia, sickle cell disease, malnutrition, aging, neoplasia and diabetes mellitus, and numerous other conditions.

Gene therapy is the introduction of a normal functional gene into cells of the bone marrow to correct a genetic defect. Also termed somatic gene therapy since it does not affect the germline genes. A mechanism to achieve a therapeutic effect by transferring new genetic information either into affected cells or tissues, or into accessory cells. Adenosine deaminase deficiency has been successfully treated by this method. The technique also applies to patients with neoplasms or degenerative syndrome.

Human immune globulin (HIG) is a pooled globulin preparation from the plasma of donors who are negative for HIV, which is used in the treatment of primary immunodeficiency, such as severe combined immunodeficiency, Bruton's disease, and combined variable immunodeficiency, and in cases of idiopathic thrombocytopenic purpura. The method of production is extraction by cold ethanol fractionation at acid pH. Viruses are inactivated, which permits the safe administration of the HIG to patients without risk of HIV, HAV, HBV, or non-A, non-B hepatitis.

19 Acquired Immune Deficiency Syndrome (AIDS)

Acquired immunodeficiency describes a decrease in the immune response to immunogenic (antigenic) challenge as a consequence of numerous diseases or conditions that include acquired immunodeficiency syndrome (AIDS), chemotherapy, immunosuppressive drugs such as corticosteroids, psychological depression, burns, nonsteroidal anti-inflammatory drugs, radiation, Alzheimer's disease, celiac disease, sarcoidosis, lymphoproliferative disease, Waldenström's macroglobulinemia, multiple myeloma, aplastic anemia, sickle cell disease, malnutrition, aging, neoplasia, diabetes mellitus, and numerous other conditions.

AIDS (acquired immune deficiency syndrome) is a disease induced by the human immunodeficiency retrovirus designated HIV-1. Although first observed in homosexual men, the disease affects both males and females equally in central Africa and is beginning to affect an increasing number of heterosexuals with cases in both males and females in the Western countries in North America and Europe. Following exposure to the AIDS virus, the incubation period is variable and may extend to 11 years before clinical AIDS occurs in HIV-positive males in high-risk groups. It is transmitted by blood and body fluids, but is not transmitted through casual contact or through air, food, or other means. Besides homosexual and bisexual males, others at high risk include intravenous drug abusers, hemophiliacs, the offspring of HIV-infected mothers, and sexual partners of any HIV-infected individuals in the above groups.

Acquired immune deficiency syndrome (AIDS) is a retroviral disease marked by profound immunosuppression that leads to opportunistic infections, secondary neoplasms, and neurologic manifestations. It is caused by the human immunodeficiency virus HIV-1, the causative agent for most cases worldwide with a few in western Africa attributable to HIV-2. Principal transmission routes include sexual contact, parenteral inoculation, and passage of the virus from infected mothers to their newborns. Although originally recognized in homosexual or bisexual men in the U.S., it is increasingly a heterosexual disease. It appears to have originated in Africa, where it is a heterosexual disease, and has been reported from more than 193 countries. The CD4 molecule on T lymphocytes serves as a high-affinity receptor for HIV. HIVgp120 must also bind to other cell surface molecules termed coreceptors for cell entry. They include CCR5 and CXCR4 receptors for β chemokines and α chemokines. Some HIV strains are macrophage-tropic whereas others are T cell-tropic. Early in the disease HIV colonizes the lymphoid organs. The striking decrease in $CD4^+$ T cells is a hallmark of AIDS that accounts for the immunodeficiency late in the course of HIV infection, but qualitative defects in T lymphocytes can be discovered in HIV-infected persons who are asymptomatic. Infection of macrophages and monocytes is very important, and the dendritic cells in lymphoid tissues are the principal sites of HIV infection and persistence. In addition to the lymphoid system, the nervous system is the major target of HIV infection. It is widely accepted that HIV is carried to the brain by infected monocytes. The microglia in the brain are the principal cell type infected in that tissue. The natural history of HIV infection is divided in three phases that include (1) an early acute phase, (2) a middle chronic phase, and (3) a final crisis phase. Viremia, measured as HIV-1 RNA, is the best marker of HIV disease progression and it is valuable clinically in the management of HIV-infected patients. Clinically, HIV infection can range from a mild acute illness to a severe disease. The adult AIDS patient may present with fever, weight loss, diarrhea, generalized lymphadenopathy, multiple infections, neurologic disease, an in some cases secondary neoplasms. Opportunistic infections account for 80% of deaths in AIDS patients. Prominent among these is pneumonia caused by *Pneumocystis carnii* as well as other common pathogens. AIDS patients also have a high incidence of certain tumors, especially Kaposi's sarcoma, non-Hodgkin's lymphoma, and cervical cancer in women. No effective vaccine has yet been developed.

Acute AIDS syndrome: Within the first to the sixth week following HIV-1 infection, some subjects develop the flu-like symptoms of sore throat, anorexia, nausea and vomiting, lymphadenopathy, maculopapular rash, wasting, and pain in the abdomen, among other symptoms. The total leukocyte count is slightly depressed with possible CD4 to CD8 ratio inversion. Detectable antibodies with specificity for HIV constituents gp120, gp160, p24, and p41 are not detectable until at least 6 months following infection. Approximately 33% of the infected subjects manifest the acute AIDS syndrome.

AIDS serology: Three to six weeks after infection with HIV-1 there are high levels of HIV p24 antigen in the plasma. One week to three months following infection there is an HIV-specific immune response resulting in the formation of antibodies against HIV envelope protein gp-120 and HIV core protein p24. HIV-specific cytotoxic T lymphocytes are also formed. The result of this adaptive immune response is a dramatic decline in viremia and a clinically asymptomatic phase lasting from 2 to 12 years. As $CD4^+$ T cell numbers decrease, the patient becomes clinically symptomatic. HIV-specific antibodies and cytotoxic T lymphocytes decline, and p24 antigen increases.

AIDS belt (Figure 19.1) refers to the geographic area across central Africa that describes a region where multiple cases of heterosexual AIDS, related to sexual promiscuity, was reported. Nations in this belt include Burundi, Central African Republic, Kenya, the Congo, Malawi, Rwanda, Tanzania, Uganda, and Zambia.

Retrovirus (Figure 19.2 and Figure 19.3) is a reverse transcriptase-containing virus such as the human immunodeficiency virus and human T cell leukemia virus. An RNA virus that can insert and efficiently express its own genetic information in the host cell genome through transcription of its RNA into DNA, which is then integrated into the genome of host cells. Retroviruses are employed in research to deliberately insert foreign DNA into a cell. Thus, they have the potential for use in gene therapy when a host cell gene is either missing or defective. Retroviruses have been used to tag tumor-infiltrating lymphocytes in experimental cancer treatment.

Human immunodeficiency virus (HIV) (Figure 19.4) is the retrovirus that induces acquired immune deficiency syndrome (AIDS) and associated disorders. It was previously designated as HTLV-III, LAV, or ARV. It infects $CD4^+$ T lymphocytes, mononuclear phagocytes carrying CD4 molecules on their surface, follicular dendritic cells, and Langerhans cells. It produces profound immunodeficiency affecting both humoral and cell-mediated immunity. There is a progressive decrease in $CD4^+$ helper/inducer T lymphocytes until they are finally depleted in many patients. There may be polyclonal activation of B lymphocytes with elevated synthesis of immunoglobulins. The immune response to the virus is not protective and does not

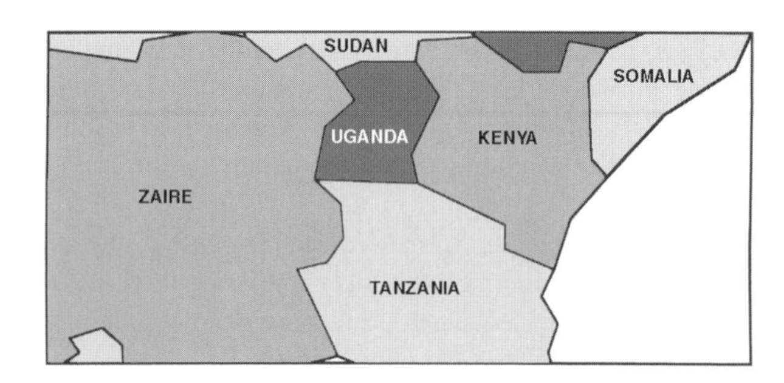

FIGURE 19.1 AIDS belt.

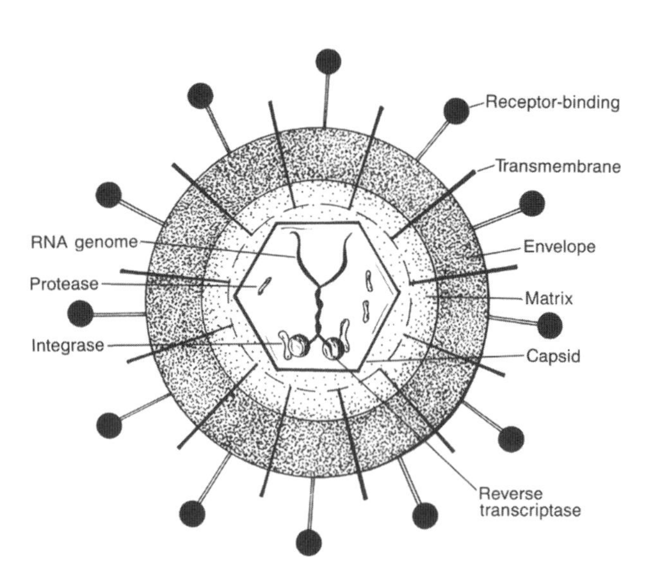

FIGURE 19.2 Retrovirus.

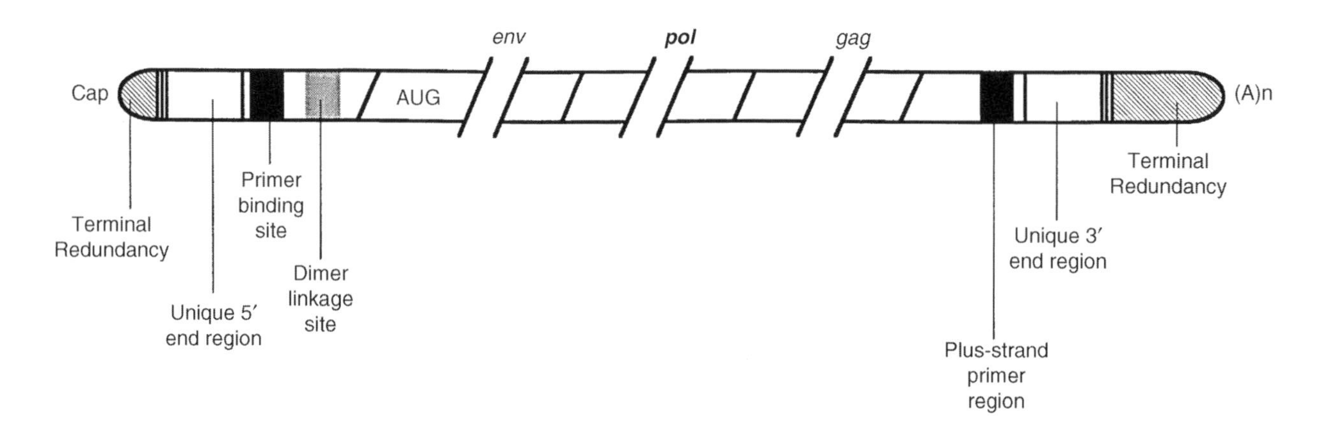

FIGURE 19.3 Retroviral genome.

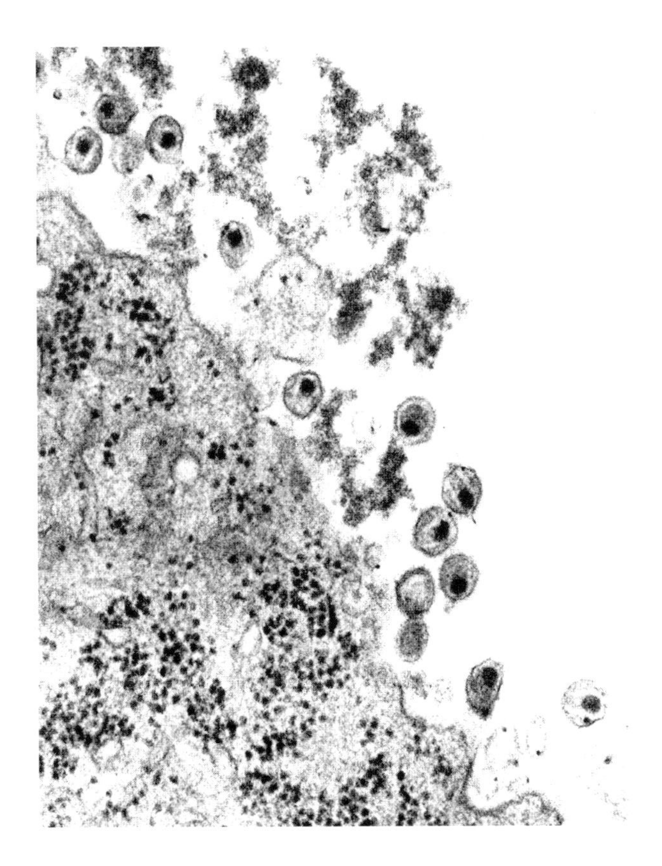

FIGURE 19.4 HIV. Electron micrograph. (Courtesy of Dr. Tom Folks, CDC, Atlanta, GA).

improve the patient's condition. The virus is comprised of an envelope glycoprotein (gp160) which is its principal antigen. It has a gp120 external segment and a gp41 transmembrane segment. CD4 molecules on $CD4^+$ lymphocytes and macrophages serve as receptors for gp120 of HIV. It has an inner core that contains RNA and is encircled by a lipid envelope. HIV contains structural genes designated *env*, *gag*, and *pol* that encode the envelope protein, core protein, and reverse transcriptase, respectively. HIV also possesses at least six additional genes: Gene *tat*, that regulate HIV replication. It can increase production of viral protein several thousand fold. Gene *rev* encodes proteins that block transcription of regulatory genes. Gene *vif (sor)* is the virus infectivity gene whose product increases viral infectivity and may promote cell-to-cell transmission. Gene *nef* is a negative regulatory factor that encodes a product that blocks replication of the virus. Genes *vpr* (viral protein R) and *vpu* (viral protein U) have also been described. No successful vaccine has yet been developed, although several types are under investigation.

AIDS virus: See human immunodeficiency virus (HIV).

Cytokine upregulation of HIV coreceptors: HIV-1 virus strains use the chemokine receptors CCR-5, CXCR-4, or both, to enter cells. Expression of these chemokine receptors may predetermine susceptibility of hematopoietic subsets to HIV-1 infection. Certain cytokines can influence the dynamics of HIV-1 infection by altering chemokine receptor expression levels on hematopoietic cells. During chronic HIV-1 infection, proinflammatory cytokines such as TNF-α and IFN-γ are secreted in excess. IFN-γ increases cell surface expression of CCR-5 by human mononuclear phagocytes and of CXCR-4 by primary hematopoietic cells. In addition, GN-CSF can decrease and IL-10 can increase expression of CCR-5. Further research into cytokine-mediated regulation of chemokine receptors may lead to increased understanding of how these receptors affect the pathogenesis of AIDS.

Fusin is a receptor present on $CD4^+$ T lymphocytes and selected other human cells that is linked to a G protein and is believed to be requisite for HIV fusion with target cells.

HTLV-IV is a human retrovirus isolated from western Africa that is related to HIV-1 and HIV-2, but appears nonpathogenic.

HALV (human AIDS-lymphotropic virus) (historical): Human AIDS-lymphotrophic virus. A designation proposed to replace Montagnier's LAV designation and Gallo's HTLV-III designation for the AIDS virus. However, the designation HIV for human immunodeficiency virus was subsequently chosen instead.

HIV infection: (Figure 19.5) The recognition of infection by the human immunodeficiency virus (HIV) is through seroconversion. Following conversion to positive reactivity in an antibody screening test, a Western blot analysis is performed to confirm the result of positive testing for HIV. HIV mainly affects the immune system and the brain. It affects primarily the $CD4^+$ lymphocytes which are necessary to initiate an immune response by interaction with

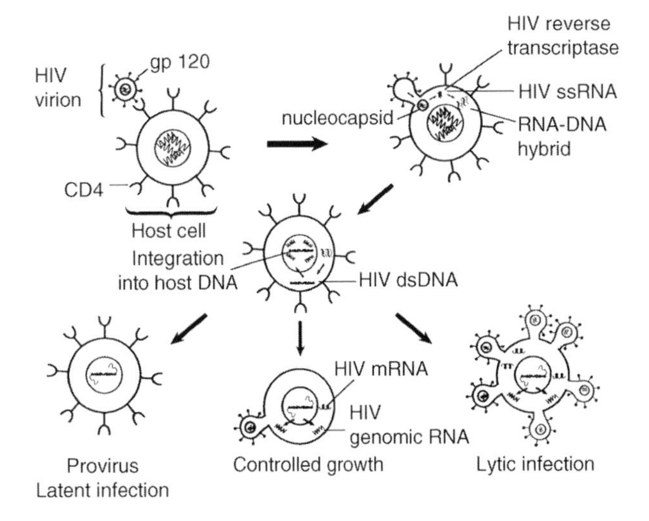

FIGURE 19.5 HIV infection.

antigen-presenting cells. This also deprives other cells of the immune system from receiving a supply of interleukin-2 through CD4+ lymphocyte stimulation, leading to a progressive decline in immune system function. HIV transmission is by either sexual contact, through blood products, or horizontally from mother to young. Although first observed in male homosexuals, it later became a major problem of intravenous drug abusers and ultimately has become more serious in the heterosexual population, affecting an increasing number of women as well as men. Clinically, individuals may develop acute HIV mononucleosis that usually occurs 2 to 6 weeks following infection, although it may occur later. The main symptoms include headache, fever, malaise, sore throat, and rash. Patients may develop pharyngitis; generalized lymphadenopathy; a macular or urticarial rash on the face, trunk, and limbs; and hepatosplenomegaly.

The severity of the symptoms may vary from one individual to another. Acute HIV infection may also induce neurologic diseases, including meningitis, encephalitis, and other neurologic manifestations. Some individuals may not develop symptoms or illness for years. Other individuals develop AIDS-related complex (ARC), which represents progressive immune dysfunction. Symptoms include fever, night sweats, weight loss, chronic diarrhea, generalized lymphadenopathy, herpes zoster, and oral lesions. Individuals with ARC may progress to AIDS or death may occur in the ARC stage. ARC patients do not revert to an asymptomatic condition. Other individuals may develop persistent generalized lymphadenopathy (PGL) characterized by enlarged lymph nodes in the neck, axilla, and groin. The Centers for Disease Control (CDC) has set up criteria for the diagnosis of AIDS. These include the individuals who develop certain opportunistic infections and neoplasms, HIV-related encephalopathy, and HIV-induced wasting syndrome. The most frequent opportunistic infections in AIDS patients include *Pneumocystis carnii* , which produces pneumonia, and *Mycobacterium avium-intracellulare* , among other microorganisms. The most frequent tumor in AIDS patients is Kaposi's sarcoma. The definition of AIDS by the CDC now includes HIV-related encephalopathy and HIV wasting syndrome. At the present time, AIDS is 100% fatal.

Window refers to: (1) The period between exposure to a microorganism and the appearance of serologically detectable antibody. It is observed in hepatitis B as well as in HIV-1 infections. In hepatitis B, there is a "core window" that occurs in active but unidentified hepatitis B infection. The hepatitis B surface antigen (HBsAg) can no longer be detected, and the antibody against hepatitis B surface antigen (anti-HBs) has not reached sufficiently high levels to be detected. (2) The period between the first infection with HIV-1 and synthesis of anti-p24 and anti-p41 antibodies in amounts measurable by the ELISA assay. Use of the polymerase chain reaction to demonstrate the p24 antigen can be useful to indicate infection during the window. The window period in HIV-1 infection may be between 3 and 9 months, or it may reach 36 months. Blood donated for transfusion in the U.S. is assayed for anti-HIV-1 p24 antibody. Thus, these units of blood could be in the HIV-1 infection window.

Anti-p24 antibodies against the viral core protein p24 appear within weeks of acute HIV infection and may have a role in the decrease in plasma viremia associated with primary infection. The decline in anti-p24 antibodies is linked to HIV disease progression.

Gay bowel syndrome describes a constellation of gastrointestinal symptoms in homosexual males related to both infectious and noninfectious etiologies before the AIDS epidemic. Clinical features include alterations in bowel habits, condyloma accuminata, bloating, flatulence, diarrhea, nausea, vomiting, adenomatous polyps, fistulas, fissures, hemorrhoids, and perirectal abscesses, among many other features. Associated sexually transmitted infections include syphilis, herpes simplex, gonorrhea, and *Chlamydia trachomatis.* Numerous other microbial species identified include human papilloma virus, *Campylobacter* organisms, hepatitis A and B, cytomegalovirus, and parasites such as *Entamoeba histolytica.* Treatment varies with the etiology of various gay bowel syndrome manifestations.

Kaposi's sarcoma is a malignant neoplasm that may consist of a discreet intradermal nodule with vascular channels lined by atypical endothelial cells and extravasated erthrocytes with deposits of hemosiderin. This vascular tumor, seen previously largely in elderly patients of Mediterranean and Jewish stock, is now recognized in a more aggressive form as one of the presentations of AIDS. Whereas classical Kaposi's sarcoma is on the lower limbs and is only very slowly progressive, the more aggressive form seen in AIDS of discreet vascular tumors is scattered widely over the body. This tumor is associated with infection by Kaposi's sarcoma–associated herpes virus (human herpesvirus 8).

Progressive multifocal leukoencephalopathy is a central nervous system disease characterized by demyelination, very little inflammation associated with patches of cortical degeneration, oligodendrogliocytes, intranuclear viral inclusions, aberrant large astrocytes, and reactive fibrillary astrocytes. This condition occurs in immunosuppressed individuals such as those with AIDS or latent virus infections such as measles. Papova (DNA type), often JC virus, usually causes it.

HUT 78 is the original designation for a cell line derived from a patient with *mycosis fungoides*, now termed H-9, that was susceptible to infection with HIV-1 virus and has greatly aided in HIV-1 culture *in vitro.*

Intracellular immunization is a recent term used to describe interference with wild-type virus replication by a dominant negative mutant viral gene. This has been suggested to be of possible use in protecting cells against HIV-1 infection because of the easy accessibility of $CD4^+$ cells. By using *tat*, *gag*, and *rev* mutant genes and a mutant CD4 cell which bears the KDEL sequence, HIV envelope protein transport to the cell surfaces is inhibited.

VLIA (virus-like infectious agent) is a mycoplasma, possibly synergistic with HIV-1, leading to profound immunodeficiency. It has been named *Mycoplasma incognitos*.

Retrovirus immunity: See human immunodeficiency virus; AIDS.

Persistent generalized lymphadenopathy (PGL) is a clinical stage of HIV infection.

PCP is an abbreviation for *Pneumocystis carnii* pneumonia.

Neopterin is a guanosine triphosphate metabolite which macrophages synthesize following their stimulation by γ interferon from activated T lymphocytes. Neopterin levels in both serum and urine of HIV-1 infected patients rise as the infection progresses. This, together with diminishing $CD4^+$ lymphocyte levels, reflects progression of HIV-1 infection to clinical AIDS.

Quaternary syphilis is the stage of syphilis that follows tertiary syphilis. It is characterized by a necrotizing encephalitis with tissues rich in spirochetes. End-stage HIV patients who have completely lost cell-mediated immunity against treponemal antigens may show this form of syphilis. Although it has been rare in the past, quaternary syphilis is being seen more and more in AIDS patients also infected with *Treponema pallidum* who show increased susceptibility to neurosyphilis.

AIDS embryopathy is a condition in children born to HIV-infected mothers who are intravenous drug abusers. Affected children have craniofacial region defects that include microcephaly, hypertelorism, a cube-shaped head, a saddle nose, widened palpebral fissures with bluish sclera, triangular philtrum, and widely spreading lips.

HIV-1 virus structure (Figure 19.6) is comprised of two identical RNA strands which constitute the viral genome. These are associated with reverse transcriptase and p17 and p24, which are core polypeptides. These components are all enclosed in a phospholipid membrane envelope that is derived from the host cell. Proteins gp120 and gp41 encoded by the virus are anchored to the envelope.

Lentiviruses are a group of slow retroviruses that have a long incubation period and may take years to become manifest. Human immunodeficiency virus (HIV) is included in this group.

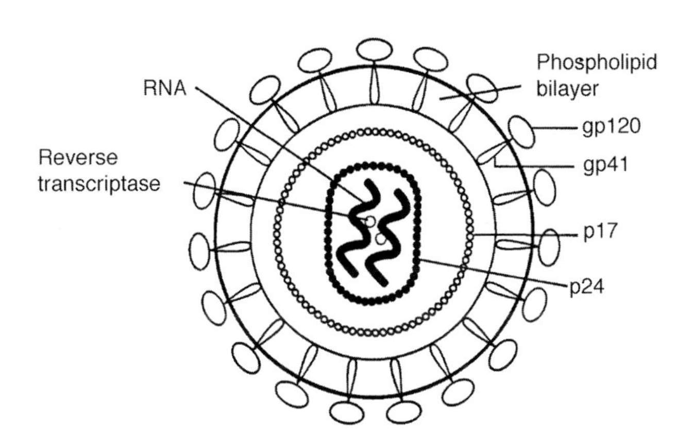

FIGURE 19.6 HIV-1 virus structure.

RNA-directed DNA polymerase (reverse transcriptase) is a DNA polymerase present in retroviruses such as human immunodeficiency virus (HIV) and Rous sarcoma virus that can use an RNA template to produce DNA. The primer needed must contain a free 3′-hydroxyl group that is base paired with the template. This produces a DNA–RNA hybrid. Reverse transcriptase is critical in recombinant DNA techniques since it is employed for first-strand cDNA synthesis.

Reverse transcriptase is an enzyme that is a critical component of retroviruses. It translates the RNA genome into DNA before integration into host cell DNA. It also permits RNA sequences to be converted into complementary DNA (cDNA) and to be cloned. It is encoded by HIV, and the purified form is used to clone complementary DNAs encoding a gene of interest from messenger RNA. Inhibitors of the enzyme have been used as therapy for HIV-1 infection.

p24 antigen is a human immunodeficiency virus type 1 (HIV-1), 24-kDa core antigen that is the earliest indicator of infection with HIV-1. It is demonstrable days to weeks prior to seroconversion to antibody synthesis against HIV-1. Testing for the p24 antigen does not reveal anti-HIV-1 seronegative persons or those with inapparent infections who wish to donate blood.

rev protein is a product encoded by the *rev* gene of the human immunodeficiency virus (HIV). Rev protein facilitates the transport of viral RNA from the nucleus to the cytoplasm during replication of HIV.

LAV: Lymphadenopathy-associated virus (See HIV-1).

gp120 is a surface 120-kDa glycoprotein of human immunodeficiency virus type 1 (HIV-1) that combines with the CD4 receptor on T lymphocytes and macrophages. Synthetic soluble CD4 molecules have been used to block gp120 antigens and spare $CD4^+$ lymphocytes from becoming infected. The *env* gene, which mutates frequently,

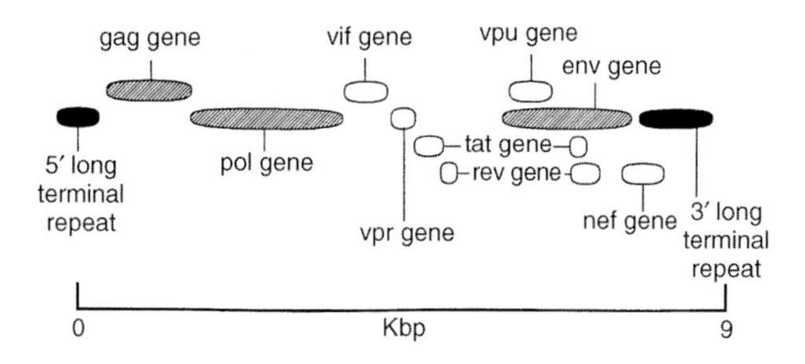

FIGURE 19.7 HIV-1 genes.

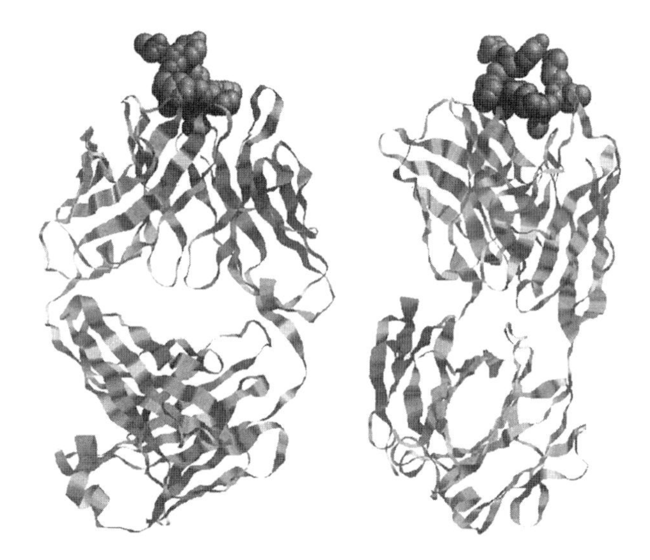

FIGURE 19.8 IgG/gp120 complex.

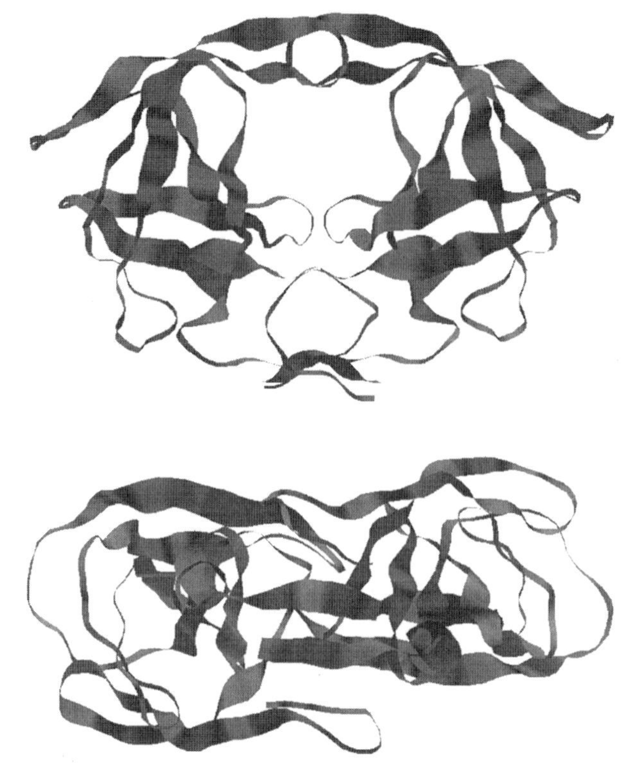

FIGURE 19.9 HIV-1 protease. NMR.

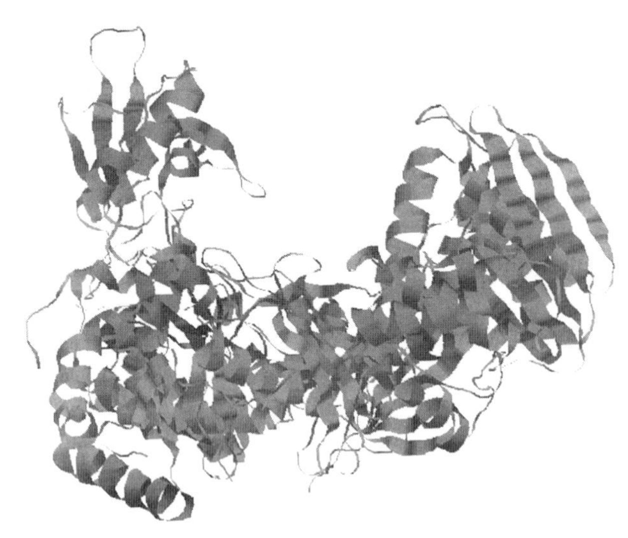

FIGURE 19.10 HIV-1 reverse transcriptase.

encodes gp120, thereby interfering with host efforts to manufacture effective or protective antibodies.

Peptide T is a small HIV-1 envelope polypeptide that was first believed to have potential in treating AIDS but was later withdrawn.

Tat is the protein product of the *tat* gene of HIV. It increases the rate of transcription of viral RNA. The activation of latently infected cells leads to synthesis. Tat protein binds to a transcriptional enhancer in the long terminal repeat of the provirus increasing proviral genome transcription.

HIV-1 genes (Figure 19.7 and Figure 19.8) include the *gag* gene which encodes the structural core proteins p17, p24, p15, and p55 precursor. *pol* encodes a protease (Figure 19.9) that cleaves *gag* precursors. It also encodes reverse transcriptase (Figure 19.10) that produces proviral DNA from RNA and encodes an integrase that is necessary for proviral insertion. *env* encodes gp160 precursor, gp120, and gp41 in mature proteins. gp120 binds CD4 molecules, and gp41 is needed for fusion of the virus with the cell. *vpr's* function is unknown. *vif* encodes a 23-kDa product that is necessary for infection of cells by free virus and is not needed for infection from cell to cell. *tat* encodes a p14 product that binds to viral long-terminal repeat (LTR) sequence and activates viral gene transcription. *rev* encodes a 20-kDa protein product that is needed for post-transcriptional expression of *gag* and *env* genes. *nef* encodes a 27-kDa protein that inhibits HIV transcription and slows viral replication. *vpu* encodes a 16-kDa protein product that may be required for assembly and packaging of new virus particles.

Envelope glycoprotein (*env*) is a gene of retroviruses that codes for *env* envelope glycoprotein. (See HIV-1 genes.) It is present on the plasma membrane of infected cells and on the host cell-derived membrane coat of viral particles. The NV proteins may be required for viral infectivity. HIV *env* proteins include gp41 and gp120 that bind to CD4 and chemokine receptors, respectively, on human T lymphocytes and facilitate fusion of viral and T cell membrane.

Pol is a retrovirus structural gene that codes for reverse transcriptase. The structural genes of HIV-1 also include *gag* and *env.*

gag is the retroviral HIV-1 gene that encodes the heterogeneous p24 protein of the virus core.

HIV-2 is an abbreviation for human immunodeficiency virus-2. Previously referred to as HTLV-IV, LAV-2, and SIV/AGM. This virus was first discovered in West African individuals who showed aberrant reactions to HIV-1 and simian immunodeficiency virus (SIV). It shows greater sequence homology (70%) with SIV/MAC than with HIV-1 (40% sequence homology). HIV-2 has only 50% conservation for *gag* and *pol.* The remaining HIV genes are even less conserved than this. It has p24, gp36, and gp140 structural antigens. Its clinical course resembles that of AIDS produced by HIV-1, but it is confined primarily to western Africa and is transmitted principally through heterosexual promiscuity.

The subject recently infected with HIV may develop either no symptoms or an acute infectious mononucleosis-like condition. The principal symptoms include headache, sore throat, fever, rash, and malaise. This illness is apparent 2 to 6 weeks following infection but may occur between 5 d and 3 months. On examination, there is a macular or urticarial rash on the limbs, face and trunk, hepatosplenomegaly, generalized lymphadenopathy, and pharyngitis. The illness ranges from mild to severe, possibly requiring hospitalization. Acute HIV infection may also involve the nervous system and be associated with encephalitis, meningitis, cranial nerve palsies, peripheral neuropathy, and myopathy. In 3 to 6 weeks after infection with HIV-I there are high levels of HIV p24 antigen in the plasma (Figure 19.11). 1 week to 3 months following infection, there is an HIV-specific immune response resulting in the formation of antibodies against HIV envelope protein gp120 and HIV core protein p24. HIV-specific cytotoxic T lymphocytes are also formed. The result of this adoptive immune response is a dramatic decline in viremia and a clinically asymptomatic phase lasting from 2 to 12 years. As $CD4^+$ T cell numbers decrease, the patient becomes clinically symptomatic. HIV-specific antibodies and cytotoxic T lymphocytes decline, and p24 antigen increases.

HIV-2V possesses *X-ORF* and *VPX*, which are unique to it.

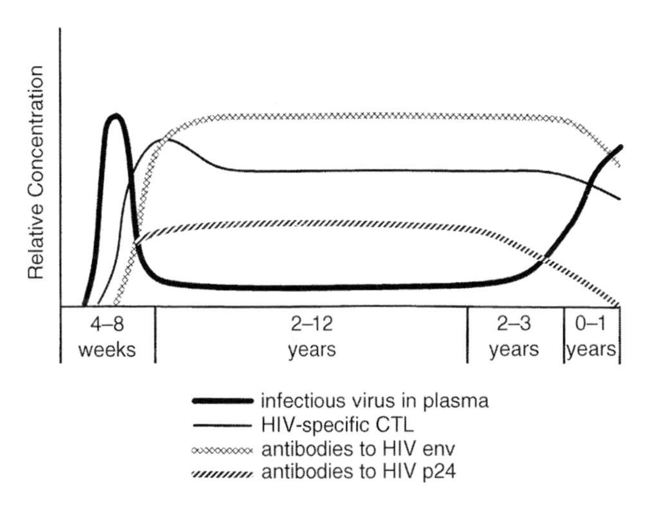

FIGURE 19.11 AIDS serology.

AIDS-related complex (ARC) is the former term for the preamble to AIDS that consists of a constellation of symptoms and signs which include a temperature of greater than 38°C, a greater than 10% loss of body weight, lymphadenopathy, diarrhea, night sweats of greater than a 3-month duration, and fatigue. Laboratory findings include $CD4^+$ T lymphocyte levels of less than 0.4×10^9, a CD4:CD8 T lymphocyte ratio of less than 1.0, leukopenia, anemia, and thrombocytopenia. There may be a decreased response to PHA, principally a T cell mitogen, and anergy, manifested as failure to respond to skin tests. In contrast, there may be a polyclonal gammopathy. A diagnosis of ARC requires at least two of the clinical manifestations and two of the laboratory findings listed above.

Changes in the nomenclature for disease associated with HIV caused AIDS-related complex (ARC) to be termed symptomatic HIV infection without an AIDS-defining condition. Progressive immune dysfunction is defined by manifestations of HIV-related symptoms. Patients experience persistent fever, night sweats, some weight loss, psoriasis, eczema, seborrheic dermatitis, diarrhea, herpes zoster, oral candidiasis, and oral hairy leukoplakia. The latter two indicate the development of AIDS. A few developed blood platelet counts less than 50,000µl.

The CDC criteria for diagnosing AIDS include the development of selected opportunistic infections and tumors, HIV-induced wasting syndrome, HIV-related encephalopathy, and various other diseases indicative of AIDS based on laboratory findings. Subjects with a CD4 lymphocyte count below 200 cells/ml blood or a CD4 T lymphocyte level below 14% irrespective of clinical symptoms are considered to have AIDS. AIDS-defining illnesses also include recurrent bacterial pneumonia, pulmonary tuberculosis, and invasive cervical cancer. Opportunistic infections common in these individuals include cervical cancer, *Pneumocystitis carnii* pneumonia, disseminated toxoplasmosis, cryptococcus, and Mycobacterial disease (including mycobacterium avium complex) and tuberculosis, the current herpes simplex infection, disseminated cytomegalovirus infection, and histoplasmosis. AIDS patients are also prone to developing staphylococcal and pneumococcal infections as well as Salmonella bacteremia. Lymphocytic interstitial pneumonitis and recurrent bacterial infections may have a higher incidence in children with AIDS than in adults. Kaposi's sarcoma is the most frequently associated tumor with AIDS. It involves the endothelium and mesenchymal stroma but is a less frequent presenting illness than in the past. HHV-8 herpes virus may be the causative agent of Kaposi's sarcoma. B cell lymphomas may also occur.

Up to two thirds of AIDS patients may develop CNS signs and symptoms such as sustained cognitive behavior and motor impairment referred to as the AIDS dementia complex. This is believed to be associated with infection of microglial cells with the HIV-1 virus and could be due to the structural similarity of gp120 of HIV-1 to neuroleukin. Patients have memory loss, are unable to concentrate, have poor coordination of gait, and altered psychomotor function, among other symptoms. The subcortical white matter and deep gray matter degenerate; lateral and posterior spinal cord columns show white matter vacuolization; the gp120 of HIV serves as a calcium channel inhibitor causing toxic levels of calcium within neurons.

AIDS enteropathy is a condition that may be seen in AIDS-related complex patients marked by diarrhea, especially nocturnal; wasting; possibly fever; and defective D-xylose absorption, leading to malnutrition. The small intestine may demonstrate atrophy of villi and hyperplasia of crypts. Both small and large intestines may reveal diminished plasma cells, elevated intraepithelial lymphocytes, and viral inclusions.

AIDS encephalopathy refers to AIDS dementia. See AIDS dementia complex.

AIDS dementia complex: Up to two-thirds of AIDS patients may develop CNS signs and symptoms such as sustained cognitive behavior and motor impairment believed to be associated with infection of microglial cells with the HIV-1 virus. This could be due to the structural similarity of gp120 of HIV-1 to neuroleukin. Patients have memory loss, are unable to concentrate, have poor coordination of gait, and have altered psychomotor function, among other symptoms. The subcortical white matter and deep gray matter degenerate; lateral and posterior spinal cord columns show white matter vacuolization; and the gp120 of HIV serves as a calcium channel inhibitor, causing toxic levels of calcium within neurons.

***Pneumocystis carnii* (PCP)** (Figure 19.12) is a protozoan parasite that infects immunocompromised subjects such as AIDS patients, transplant recipients, lymphoma and leukemia patients, and others immunosuppressed for one reason or the other. It is diagnosed in tissue sections stained with the Gomori-methenamine silver stain. A mannose receptor facilitates the organism's uptake by macrophages. Approximately one half of those hospitalized with a first infection by PCP die. The organism has two major forms: a trophozoite and a cyst. A trophic form is the smaller form (1 to 4 μm), is pleomorpic, and is present in clusters. The cyst stage is larger (5 to 8 μm) and contains as many as eight intracystic bodies. There are two groups of *P. carnii* antigens, i.e., a large surface complex, designated major surface glycoprotein (MSG); and gpA or gp120 discovered in organisms derived from human subjects, with a mol wt of 95 to 140 kDa. MSG facilitates the microorganism's interaction with the host. The other major antigen complex is a glycoprotein of 35 to 45 kDa in human *P. carnii*.

Cytomegalovirus (CMV) (Figure 19.13) is a herpes (DNA) virus group that is distributed worldwide and is not often a problem except in individuals who are immunocompromised, such as the recipients of organ or bone marrow transplants, or individuals with acquired immunodeficiency syndrome (AIDS). Histopathologically, typical inclusion bodies that resemble an owl's eye are found in multiple tissues. CMV is transmitted in the blood.

AIDS treatment: Two classes of antiviral drugs are used to treat HIV infection and AIDS. Nucleotide analogs inhibit reverse transcriptase activity. They include azidothymidine-AZT, dideoxyinosine, and dideoxycytidine. They may diminish plasma HIV RNA levels for considerable periods, but often fail to stop disease progression because of development of mutated forms of reverse transcriptase that resist these drugs. Viral protease inhibitors are now used to block the processing of precursor proteins into mature viral capsid and core proteins. Currently, a triple-drug therapy consisting of protease inhibitors (Figure 19.14 to Figure 19.16), together with two separate reverse transcriptase inhibitors, are used to reduce plasma viral RNA to very low levels in patients treated for more than

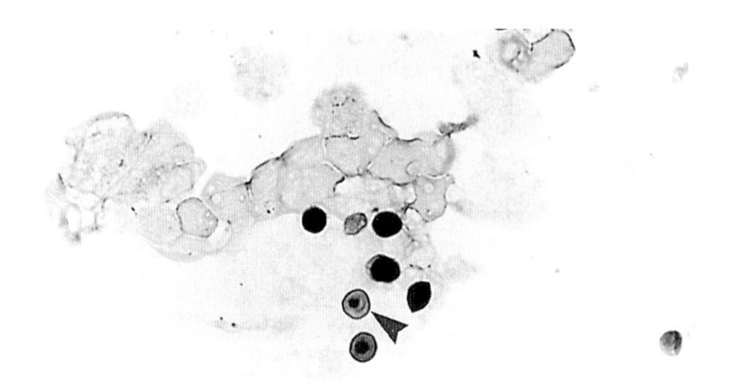

FIGURE 19.12 *Pneumocystis carnii.*

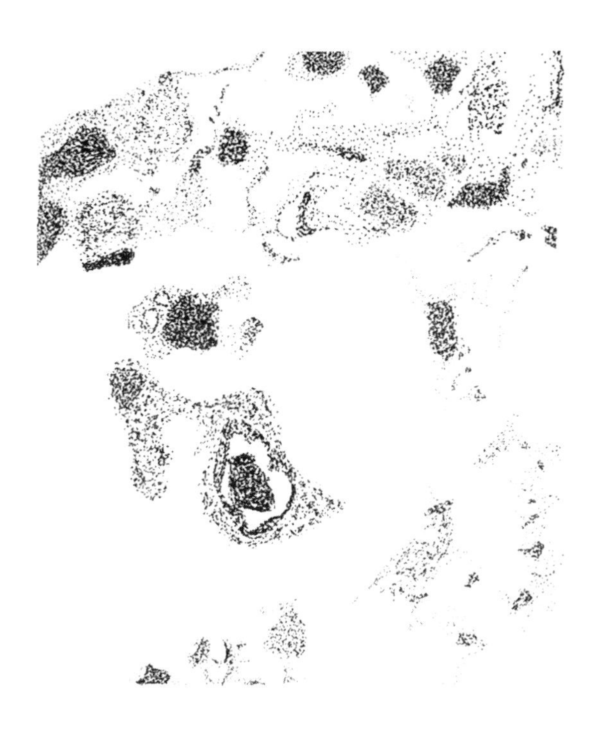

FIGURE 19.13 CMV nuclear and cytoplasmic inclusions in the lung.

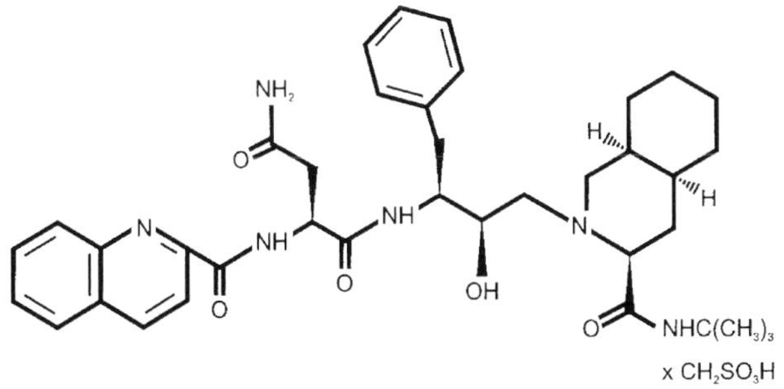

FIGURE 19.14 Saquinavir mesylate.

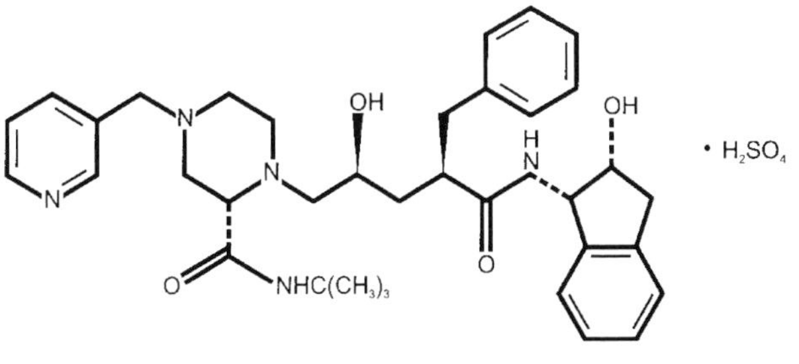

FIGURE 19.15 Indinavir sulfate.

1 year. It remains to be determined whether or not resistance to this therapy will develop. Disadvantages include their great expense and the complexity of their administration. Antibiotics are used to treat the many infections to which AIDS patients are susceptible. Whereas viral resistance to protease inhibitors may develop after only a few days,

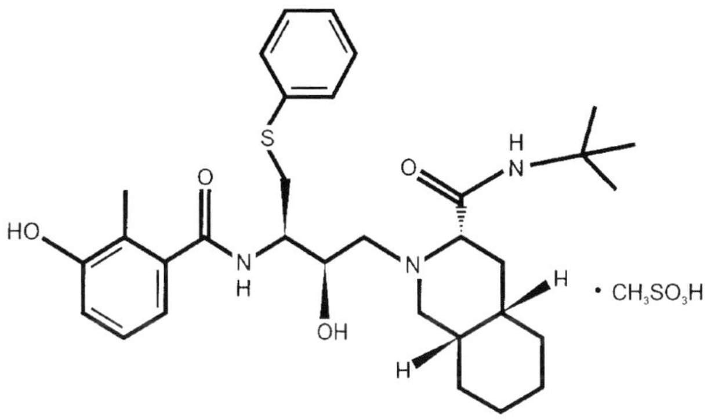

FIGURE 19.16 Nelfinavir mesylate.

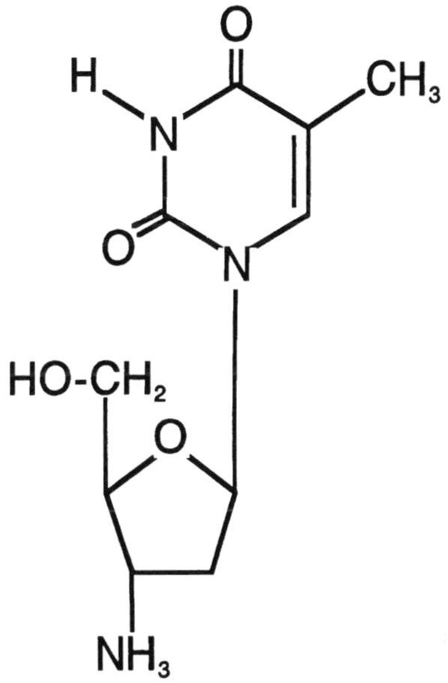

FIGURE 19.17 Zidovudine.

resistance to the reverse transcriptase inhibitor zidovudine may occur only after months of administration. Three of four mutations in the viral reverse transcriptase are necessary for resistance to zidovudine, yet only one mutation can lead to resistance to protease inhibitors.

Many advances have been made in **AIDS treatment**. Although no drug is curative, zidovudine (azidothymidine-AZT), ddC (dideoxycytidine), and ddI (dideoxyinosine) are effective in delaying progression of the disease. Many experimental preparations are under investigation, such as DAB/486 IL-2, which is cytotoxic for high-affinity IL-2 receptors expressed on HIV-infected T lymphocytes.

Zidovudine (3′-azido-3′-deoxythymidine) or AZT (Figure 19.17) is a reverse-transcriptase inhibitor that is a thymidine analog. It is FDA approved for the treatment of AIDS. The mechanism of action includes phosphorylation of the drug *in vivo* to 3′-azido-3′-deoxythymidine triphosphate. This combines with human immunodeficiency virus

(HIV reverse transcriptase), which leads to cessation of DNA elongation. Lamivudine is the (-) enantiomer of a dideoxy analog of cytidine. It has a molecular formula of $C_8H_{11}N_3O_3S$ and a molecular weight of 229.3. Zidovudine and lamivudine are combined as the active ingredients in Combivir®.

Ziagen® is a carbocyclic synthetic nucleoside analog. It is FDA approved for the treatment of HIV. The mechanism of action includes activation of abacavir to carbovir triphosphate, which inhibits the activity of HIV-1 reverse transcriptase both by competing with the natural substrate dGTP and by its incorporation into viral DNA. Abacavir is the active component of the drug.

Azidothymidine is a synonym for zidovudine.

AZT (3′-azido-3′-deoxythymidine). See zidovudine.

BI-RG-587 is a powerful inhibitor of reverse transcriptase in humans. This dipyridodiazepinone can prevent the replication of HIV-1 *in vitro*. It can be used in conjunction with such nucleoside analogs as zidovudine, ddI, and ddC, as well as in subjects whose HIV-1 infection no longer responds to these drugs.

ddC (dideoxycytidine) is an inhibitor of reverse transcriptase used in AIDS treatment. It resembles ddI.

ddI (2′,3′-dideoxyinosine) is a purine analog that blocks HIV-1 *in vivo*. It is transformed into a triphosphorylated substance, ddATP, which blocks HIV reverse transcriptase and suppresses the replication of HIV by inhibiting viral DNA synthesis. Administration of ddI may be followed by an elevation in the $CD4^+$ T helper cells and a significant decrease in p24 antigen, an indicator of HIV activity in the blood. AIDS patients tolerate ddI better than they do zidovudine.

Epivir® is a synthetic nucleoside analog. It is FDA approved for the treatment of HIV. The mechanism of action includes phosphorylation of the drug to its active 5′-triphosphate metabolite, which inhibits reverse transcriptase via DNA chain termination after incorporation of the nucleoside analog. Lamivudine is the active component of the drug.

Highly active anteretroviral therapy (HAART) describes the combined use of reverse transcriptase inhibitors and a viral protease inhibitor for HIV infection. This type of therapy can diminish virus titers to undetectable levels for more than 1 year and slow the progression of HIV disease.

Norvir® is a peptidomimetic inhibitor of both the HIV-1 and HIV-2 proteases. It is FDA approved for the treatment of HIV. The mechanism of action includes inhibition of HIV protease, rendering the enzyme incapable of processing the *gag-pol* polyprotein precursor, which leads to production of noninfectious immature HIV particles. Ritonavir is the active component of the drug.

Retrovir® is a synthetic nucleoside analog of thymidine, in which the 3′-hydroxy group is replaced by an azido group. It is FDA approved for the treatment of HIV. The mechanism of action includes the conversion of zidovudine to its active metabolite zidovudine 5′-triphosphate, which inhibits the activity of the HIV reverse transcriptase by both competing for utilization with the natural substrate and its incorporation into viral DNA. Zidovudine is the active component of the drug.

Sustiva® is an HIV-1 specific, nonnucleoside, reverse-transcriptase inhibitor. It is FDA approved for the treatment of HIV. The mechanism of action includes noncompetitive inhibition of HIV-1 reverse transcriptase by efavirenz. Efavirenz, the active component of the drug, has no inhibitory effect on HIV-2 reverse transcriptase.

Videx® is a synthetic purine nucleoside analog. It is FDA approved for the treatment of HIV. The mechanism of action includes the activation of didanosine to dideoxyadenosine 5′-triphosphate, which inhibits the activity of HIV-1 reverse transcriptase both by competing with the natural substrate and by its incorporation into viral DNA. Didanosine is the active component of the drug.

Viracept® is an inhibitor of the HIV-1 protease that prevents cleavage of the gag-pol polyprotein, resulting in the production of noninfectious virus. It is FDA approved for the treatment of HIV. Nelfinavir mesylate is the active component of the drug.

Serpins are a large family of protease inhibitors.

Viramune® is a nonnucleoside reverse-transcriptase inhibitor. It is FDA approved for the treatment of HIV. Nevirapine, the active component of the drug, binds directly to reverse transcriptase and blocks the RNA-dependent and DNA-dependent polymerase activities by causing disruption of the enzyme's catalytic site. Nevirapine has no inhibitory effect on HIV-2 reverse transcriptase.

Pentamidine isoethionate is a substance useful in the treatment of *Pneumocystis carnii* pneumonia in AIDS patients who have failed to respond to trimethoprimsulfamethoxazole therapy. It is administered by aerosol and has diminished *P. carnii* pneumonia by 65%. Adverse effects include azothemia, arrythmia, hypotension, diabetes mellitus, pancreatitis, and severe hypoglycemia.

Ribavarin (1-8-5-D ribofuranosyl-1,2,4-triazole-3-carboxamide) (Figure 19.18) is a substance that interferes with mRNA capping of certain viruses, thereby restricting the synthesis of viral proteins. It is used as an aerosol to treat severe respiratory syncytial virus infection in children.

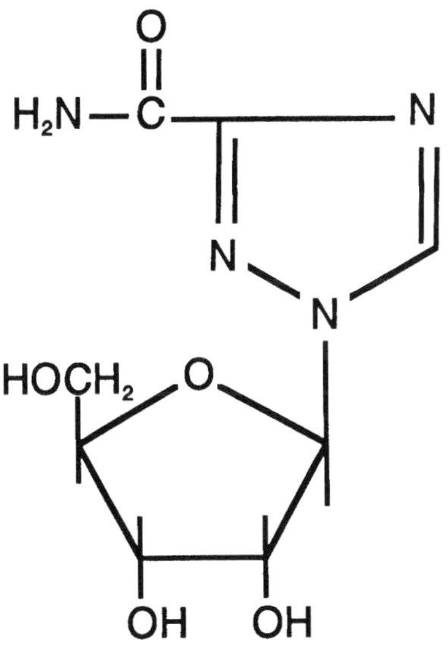

FIGURE 19.18 Ribavirin.

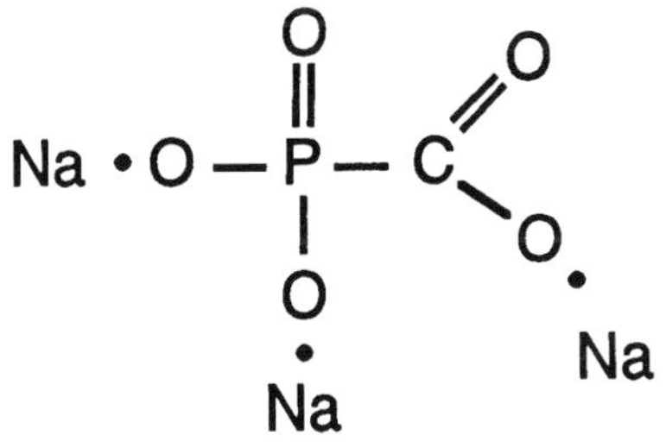

FIGURE 19.19 Foscarnet.

Foscarnet (Figure 19.19) is an investigational drug used to combat cytomegalovirus-induced pneumonia, hepatitis, colitis, and retinitis in AIDS patients rendered nonresponsive to gancyclovir, which is a frequently used treatment for cytomegalovirus infection.

Gancyclovir (9-[2-hydroxy-1(hydroxymethyl) ethoxymethyl] guanine) is an antiviral drug used for the therapy of immunocompromised patients infected with cytomegalic inclusion virus. Five days of gancyclovir therapy has proven effective for clearing CMV from the blood, urine, and respiratory secretions. It has been used successfully to treat CMV retinitis, gastroenteritis, and hepatitis. Drug resistance may develop. The drug may induce neutropenia and thrombocytopenia as side effects. It has not proven very effective in AIDS or bone marrow transplant patients.

Development of an AIDS vaccine always has been made difficult by the genetic potential of HIV for extensive antigenic variations. Whereas many of the viral gene products capable of inducing humoral immunity are known, humoral immunity is insufficient to prevent HIV disease. By contrast, viral products that can induce effective cell-mediated immunity require investigations to determine whether or not they can effectively prevent disease. Several experimental AIDS vaccines are under investigation. HIV-2 inoculation into cynomologus monkeys apparently prevented them from developing simian AIDS following injection of the SIV virus.

ALVAC is an experimental AIDS vaccine developed for the first test of a human AIDS vaccine in Africa. The vaccine has undergone safety testing in the U.S. and France with no serious side effects reported and is being used in Uganda, where AIDS has killed nearly half a million people and left one million children orphaned.

gp160 vaccine is a vaccine that contains a cloned segment of the envelope protein of HIV-1. It activates both humoral and cellular immunity against HIV products during early infection with HIV-1. It diminishes the rate at which $CD4^+$ T lymphocytes are lost.

Pediatric AIDS describes AIDS in infants who are infected vertically, i.e., from mother to young through intrauterine or intrapartum infection. They show symptoms usually between 3 weeks and 2 years of age. They develop lymphadenopathy; fever; increased numbers of B lymphocytes in the peripheral blood; thrombocytopenia; and increased levels of IgG, IgM, and IgD in the serum. They may develop lymphoid interstitial pneumonia, chronic otitis media encephalopathy, recurrent bacterial infections, and *Candida* esophagitis. Epstein-Barr virus infection may produce interstitial pneumonia and salivary gland inflammation. They may also have bloodborne infections with such microorganisms as *Hemophilus influenzae* or pneumococci. The adult age pattern of either opportunistic infections or Kaposi's sarcoma is rarely seen in HIV-1-infected infants. These children have a 50% 5-year mortality.

SIV (simian immunodeficiency virus) is a lentivirus of primates that resembles HIV-1 and HIV-2 in morphology and attraction to cells that bear CD4 molecules such as lymphocytes and macrophages. SIV also shares with these human viruses the additional genes lacking in other retroviruses, which include *vip*, *rev*, *upr*, *tat*, and *nef*. The SIV virus induces the classic cytopathologic alterations of the type produced by HIV, and it can also induce chronic disease following a lengthy latency. SIVmac239 is an SIV clone that induces a disease resembling AIDS in monkeys.

Simian immunodeficiency virus (SIV) causes a disease resembling human AIDS in rhesus monkeys. The SIV sequence reveals significant homology with HIV-2, a cause of AIDS in western Africa.

SAIDS (simian acquired immunodeficiency syndrome) describes an immunodeficiency of rhesus monkeys induced by retrovirus group D. The animals develop opportunistic infections and tumors. Their $CD4^+$ lymphocytes decrease.

They suffer wasting and develop granulomatous encephalitis. The sequence homology between SIV and HIV-1 is minimal, but the sequence homology between SIV and HIV-2 is significant.

SRV-1 is a simian AIDS virus type D that shows little similarity with HIV-1. However, they both contain genes that resemble one another. This strain was responsible for an infection among a colony of macaques in California.

***tat* gene** is a retrovirus gene found in HIV-1. The tat transactivating protein, which this gene encodes, gains access to the nucleus and activates viral proliferation. Additional retroviral genes become activated. Mesenchymal tumors may be induced by the *tat* genes in experimental animals.

MAIDS is an abbreviation for (1) murine acquired immunodeficiency syndrome and (2) monoclonal antiidiotypic antibodies.

Mycoplasma–AIDS link is a mechanism postulated by L. Montagnier for AIDS development. HIV-1 virus binds to cells first activated by *Mycoplasma* infection.

20 Immunosuppression

Clinical immunosuppression has been used to treat immunological diseases, including autoimmune reactions, as well as to condition recipients of solid organ allografts or of bone marrow transplants.

Suppression refers to an immunologic unresponsiveness attributable to either substances produced by selected lymphocytes or by the administration of immunosuppressive drugs such as for organ transplantation.

Nonspecific T cell suppressor factor refers to a $CD8^+$ suppressor T lymphocyte soluble substance that nonspecifically suppresses the immune response.

Immune suppression refers to decreased immune responsiveness as a consequence of therapeutic intervention with drugs or irradiation or as a consequence of a disease process that adversely affects the immune response, such as acquired immune deficiency syndrome (AIDS).

Immunosuppression describes either the deliberate administration of drugs such as cyclosporine, azathioprine, corticosteroids, FK506, or rapamycin; the administration of specific antibody; the use of irradiation to depress immune reactivity in recipients of organ or bone marrow allotransplants; and the profound depression of the immune response that occurs in patients with certain diseases such as acquired immune deficiency syndrome (AIDS) in which the helper-inducer ($CD4^+$) T lymphocytes are destroyed by the HIV-1 virus. In addition to these examples of nonspecific immunosuppression, antigen-induced specific immunosuppression is associated with immunologic tolerance.

Nonspecific suppression is a state induced by soluble molecules such as cytokines, nitric oxide, and prostaglandins released from cells, in a nonantigen-specific manner, that downmodulates immune function. Nonspecific suppressive cell populations such as natural suppressor (NS) cells may induce suppression through soluble mediators. Nonspecific suppressive mechanisms would be of interest for tolerance induction if they could be induced temporarily during a critical period when alloreactive was present.

Immunosuppressive agents include drugs such as cyclosporine, FK506, rapamycin, azathioprine, or corticosteroids; antibodies such as antilymphocyte globulin; and irradiation. These produce mild to profound depression of a host's ability to respond to an immunogen (antigen), as in the conditioning of an organ allotransplant recipient. Substances that inhibit adaptive immune responses may be used also to treat autoimmune diseases.

NONSPECIFIC IMMUNOSUPPRESSION

Corticosteroids (Figure 20.1 to Figure 20.5) are lympholytic steroid hormones, such as cortisone, derived from the adrenal cortex. Glucocorticoids such as prednisone or dexamethasone can diminish the size and lymphocyte content of lymph nodes and spleen while sparing proliferating myeloid or erythroid stem cells of the bone marrow. Glucocorticoids may interfere with the activated lymphocyte's cell cycle. Glucocorticoids are cytotoxic for selected T lymphocyte subpopulations and are also able to suppress cell-mediated immunity and antibody synthesis, as well as the formation of prostaglandin and leukotrienes. Corticosteroids may lyse either suppressor or helper T lymphocytes, but plasma cells may be more resistant to their effects. However, precursor lymphoid cells are sensitive to the drug, which may lead in this way to decreased antibody responsiveness. The repeated administration of prednisone diminishes the concentration of specific antibodies in the IgG class, whose fractional catabolic rate is increased by prednisone. Corticosteroids interfere with the phagocytosis of antibody-coated cells by macrophages (Figure 20.6). Glucocorticoids have been widely administered for their immunosuppressive properties in autoimmune diseases such as autoimmune hemolytic anemia, systemic lupus erythematosus (SLE), Hashimoto's thyroiditis, idiopathic thrombocytopenic purpura, and inflammatory bowel disease. They are also used in the treatment of various allergic reactions and for bronchial asthma. Corticosteroids have been widely used in organ transplantation, especially prior to the introduction of cyclosporine and related drugs. Management of rejection crises without producing bone marrow toxicity has also been achieved with these drugs. Long-term usage has adverse effects that include adrenal suppression.

Glucocorticoids are powerful immunosuppressive and anti-inflammatory drugs. They reduce circulating lymphocytes and monocytes, in addition to suppressing interleukin-1 (IL-1) and IL-2 production. However, their chronic use produces adverse effects, including increased susceptibility to infection, bone fractures, diabetes and cataracts.

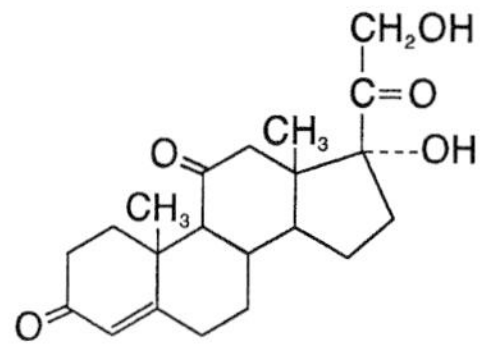

FIGURE 20.1 Structure of cortisone.

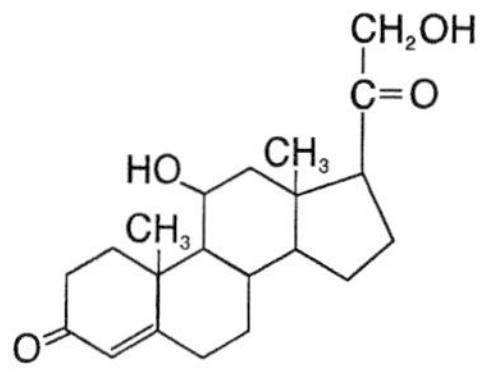

FIGURE 20.2 Structure of corticosterone.

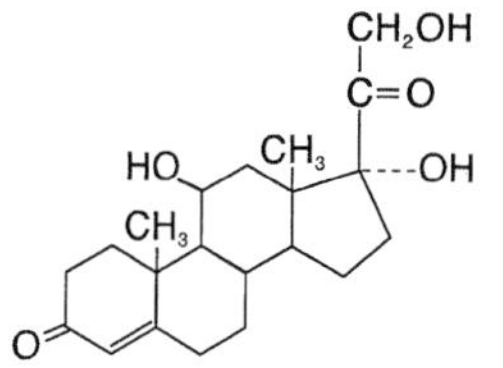

FIGURE 20.3 Structure of cortisol.

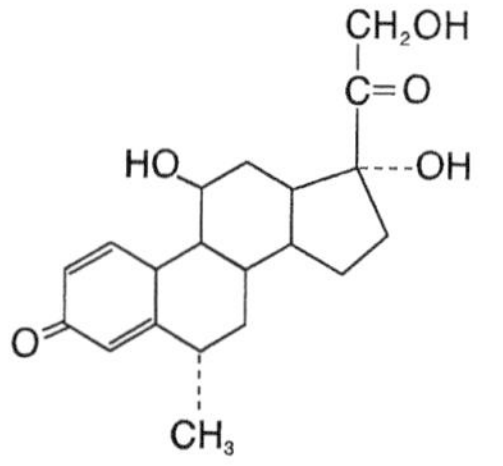

FIGURE 20.4 Structure of 6α-methylprednisolone.

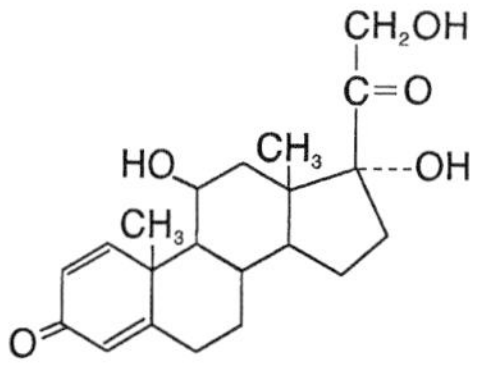

FIGURE 20.5 Structure of prednisolone.

GCs act by binding to the cytoplasmic GC receptor which regulates transcription of cytokine genes associated with inflammation. GCs regulate expression of the cytokines TNF-α, GC-CSF, IL1, IL-2, IL-3, IL-4, IL-5, IL-6, and IL-8. IL-1, IL-6, and TNF-α induce GC release in a prostaglandin–corticotropin-releasing hormone-dependent manner from the hypothalamic–pituitary–adrenal axis. GCs are powerful inhibitors of nitric oxide synthase induction

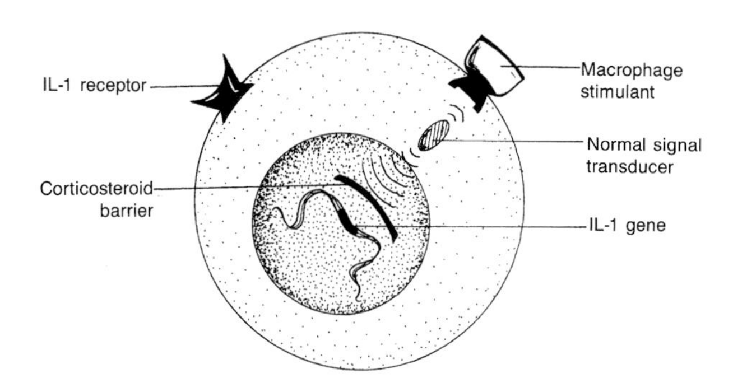

FIGURE 20.6 Mechanism of action of corticosteroids.

and induce the expression of the 37-kDa immunomodulator, lipocortin-1, in PMNL and macrophages. Lipocortin is a powerful inhibitor of phospholipase A_2, which releases arachidonic acid for proinflammatory eicosanoid synthesis. GCs downregulate expession of the eosinophil chemokine RANTES. Endogenous GCs (corticosterone) inhibit neutrophil chemotaxis in inflammatory cholestatis. GC is also a diminished neutrophil extravasation. GCs also induce a prolonged increase in plasma cortisol.

Prednisolone (1,4-Pregnadiene-11β,17α, 21-Triol-3,20-Dione) is a semisynthetic steroid with glucocorticoid action.

Prednisone (1,4-Pregnadiene-17α,21-DIOL-3,11,20-Trione) is a synthetic steroid with glucocorticoid action.

17-hydroxycorticosteroids (17-OHCS): Adrenal steroid hormones synthesized by the action of 17-hydroxylase, including cortisone, cortisol, 11-deoxycortisol, and tetrahydro derivities of 17-hydroxylase. 17-OHCSs in the urine give an indication of the adrenal gland's functional status and catabolic rates. 17-OHCSs are elevated in Cushing's disease, obesity, pregnancy, and pancreatitis, but decreased in hypopituitarism and Addison's disease.

Chemical "splenectomy": Deliberate suppression of the immune system by the administration of high-dose corticosteroids (1 mg/kg/d) or intravenous immunoglobulin (0.4 g/kg/d). This prevents endocytosis of cells or microorganisms opsonized by a coating of immunoglobulin or complement, which blocks Fc receptors. Although the opsonized particles are bound, they are not endocytosed. This procedure has been used in the management of hypersplenism associated with certain immune disorders such as autoimmune hemolytic anemia, Felty's syndrome, or autoimmune neutropenia.

6-mercaptopurine (6-MP) is a powerful immunosuppressive drug used prior to the introduction of cyclosporine in organ transplantation (Figure 20.7). It is also an effective chemotherapeutic agent for the treatment of acute leukemia of childhood as well as other neoplastic

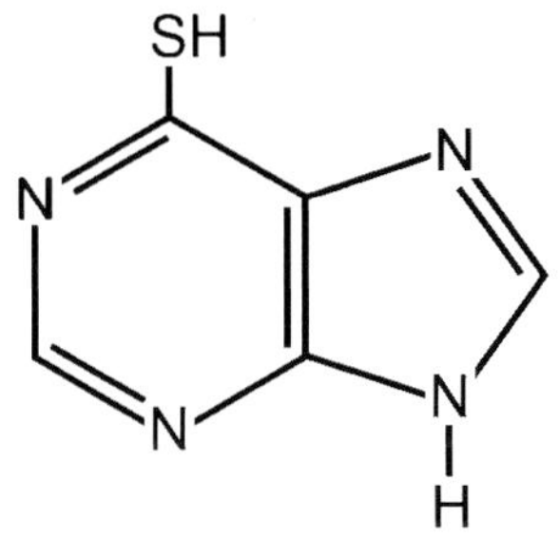

FIGURE 20.7 Structure of 6-mercaptopurine (6MP).

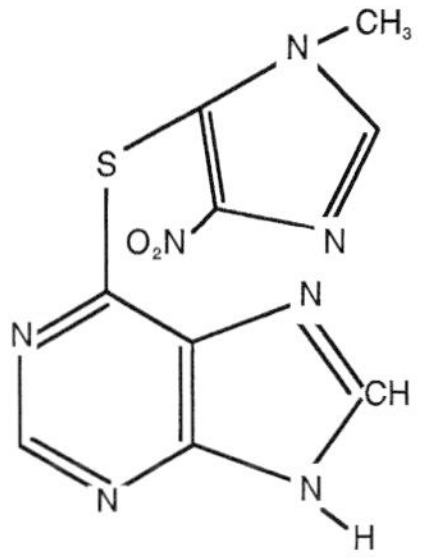

(6-[(1-methyl-4-nitro-1H-imidazol-5-yl)thio]-1H-purine)

FIGURE 20.8 Structure of azathioprine.

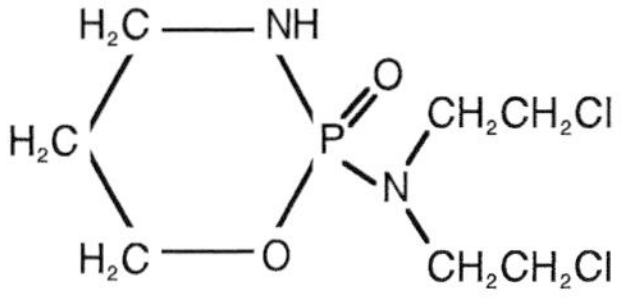

FIGURE 20.9 Structure of cyclophosphamide.

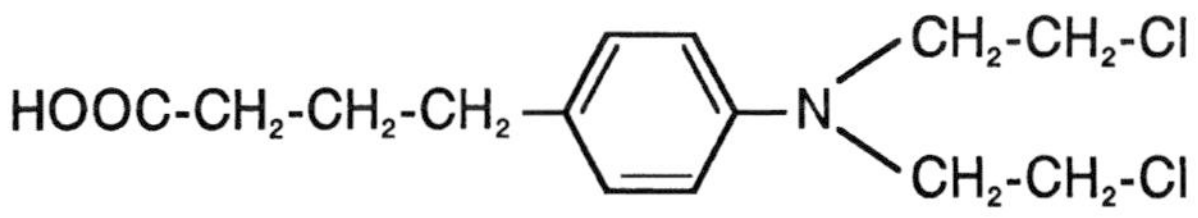

FIGURE 20.10 Structure of chlorambucil (4-[bis(2-chloroethyl)amino-phenylbutyric acid).

conditions. 6-MP is a purine analog in which a thiol group replaces the 6-hydroxyl group. Hypoxanthine-guanine phosphoribosyl transferase (HGPRT) transforms 6-MP to 6-thioinosine-5′-phosphate. This reaction product blocks various critical purine metabolic reactions. 6-MP is also incorporated into DNA as thioguanine.

Azathioprine (Figure 20.8) is a nitroimidazole derivative of 6-mercaptopurine, a purine antagonist. Following administration, it is converted to 6-mercaptopurine *in vivo*. Its principal action is to interfere with DNA synthesis. Of less significance is its ability to impair RNA synthesis. Azathioprine has a greater inhibitory effect on T cell than on B cell responses, even though it suppresses both cell-mediated and humoral immunity. It diminishes circulating NK and killer cell numbers. Azathioprine has been used to treat various autoimmune disorders including rheumatoid arthritis, other connective tissue diseases, autoimmune blood diseases, and immunologically mediated neurological disorders. It is active chiefly against reproducing cells. The drug has little effect on immunoglobulin levels or antibody titers, but it does diminish neutrophil and monocyte numbers in the circulation.

Cyclophosphamide (*N,N*-bis-[2-choroethyl]-tetrahydro-2H-1,3,2-oxazaphosphorine-2-amine-2-oxide) is a powerful immunosuppressive drug that is more toxic for B than for T lymphocytes (Figure 20.9). Therefore, it is a more effective suppressor of humoral antibody synthesis than of cell-mediated immune reactions. It is administered orally or intravenously and mediates its cytotoxic activity by crosslinking DNA strands. This alkylating action mediates target cell death. It produces dose-related lymphopenia and inhibits lymphocyte proliferation *in vitro*. The greater effect on B than on T cells is apparently related to B cells' lower rate of recovery. Cyclosphosphamide is beneficial for therapy of various immune disorders including rheumatoid arthritis (RA); systemic lupus erythematosus (SLE)-associated renal disease; Wegener's granulomatosis and other vasculitides; autoimmune hematologic disorders including idiopathic thrombocytopenic purpura, pure red cell aplasia, and autoimmune hemolytic anemia. It is used also to treat Goodpasture's syndrome and various glomerulonephritises. Its beneficial use as an immunosuppressive agent is tempered by the finding of its significant toxicity, such as its association with hemorrhagic cystitis, suppression of hematopoiesis, gastrointestinal symptoms, etc. It also may increase the chance of opportunistic infections and be associated with an increased incidence of malignancies such as non-Hodgkin's lymphoma, bladder carcinoma, and acute myelogenous leukemia.

Chlorambucil (4-[bis(2-chloroethyl)amino-phenylbutyric acid) (Figure 20.10) is an alkylating and cytotoxic drug. It is not as toxic as is cyclosphosphamide and has served as an effective therapy for selected immunological diseases such as RA, SLE, Wegener's granulomatosis, essential cryoglobulinemia, and cold agglutinin hemolytic anemia. Although chlorambucil produces bone marrow suppression, it has not produced hemorrhagic cystitis and is less irritating to the GI tract than is cyclosphosphamide. Chlorambucil increases the likelihood of opportunistic infections and the incidence of some tumors.

Busulfan (1,4-butanediol dimethanesulfonate) is an alkylating drug that is toxic to bone marrow cells and is used to condition bone marrow transplant recipients (Figure 20.11).

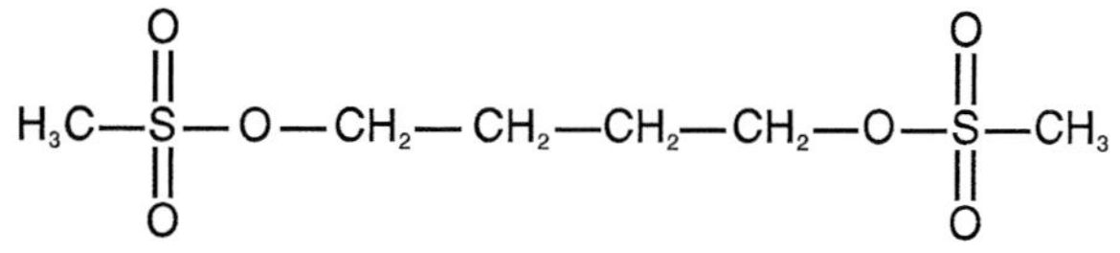

1,4-butanediol dimethanesulfonate

FIGURE 20.11 Structure of busulfan.

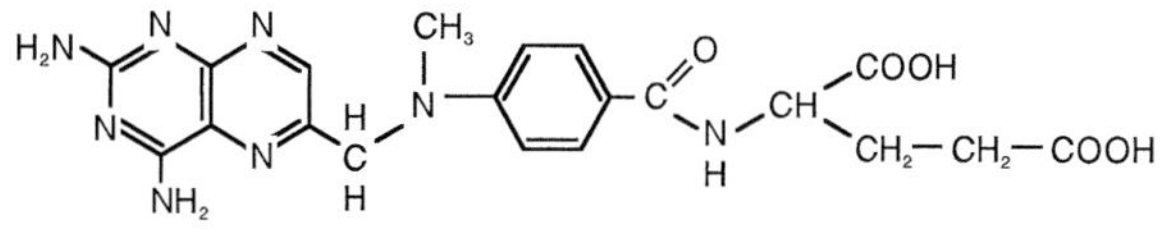

FIGURE 20.12 Structure of methotrexate.

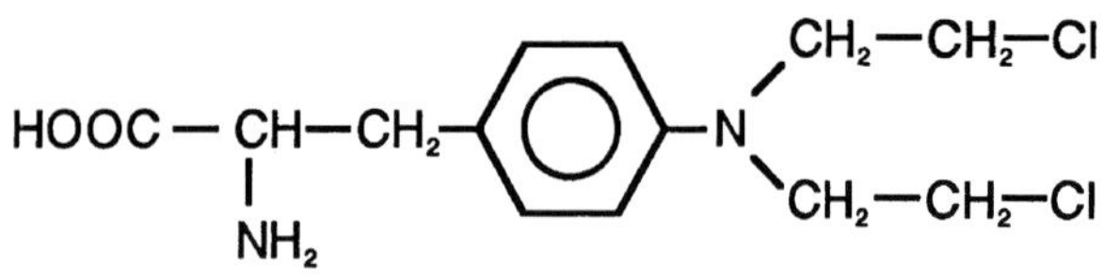

1-phenylalanine mustard

FIGURE 20.13 Structure of melphalan.

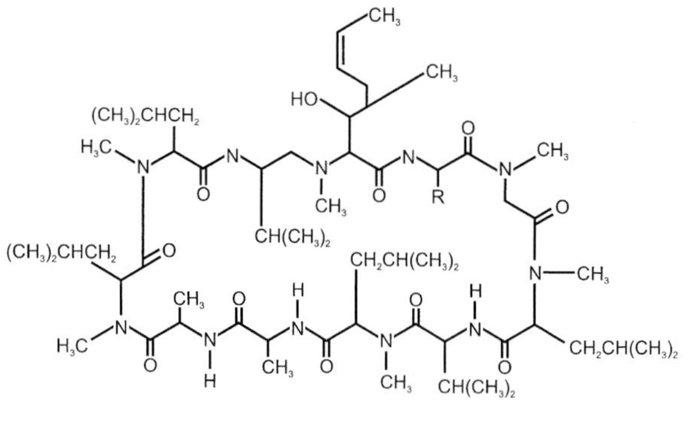

FIGURE 20.14 Structure of cyclosporine.

Methotrexate (*N*-[p-[[2,4-diamino-6-pteridinyl-methyl] methylamino] benzoyl] glutamic acid) is a drug that blocks synthesis of DNA and thymidine in addition to its well-known use as a chemotherapeutic agent against neoplasia (Figure 20.12). It blocks dihydrofolate reductase, the enzyme requisite for folic acid conversion to tetrahydrofolate. Methotrexate has been used to treat cancer, psoriasis, rheumatoid arthritis, polymyositis, Reiter's syndrome, graft-vs.-host disease (GVH), and steroid-dependent bronchial asthma. It inhibits both humoral and cell-mediated immune responses. The major toxicity of methotrexate is hepatic fibrosis, which is dose related. It may also produce hypersensitivity pneumonitis and megaloblastic anemia. It interferes with the binding of IL-1β to its receptor and may function as an anticytokine.

Melphalan (l-phenylalanine mustard) is a nitrogen mustard that is employed for therapy of multiple myeloma patients (Figure 20.13).

Cyclosporine (cyclosporin A) (ciclosporin) is a cyclic endecapeptide of 11 amino acid residues isolated from soil fungi, which has revolutionized organ transplantation (Figure 20.14). Rather than acting as a cytotoxic agent, which defines the activity of a number of currently available immmunosuppressive drugs, cyclosporine (CSA) produces an immunomodulatory effect principally on the helper/inducer (CD4) lymphocytes that orchestrate the generation of an immune response. A cyclic polypeptide, CSA blocks T cell help for both humoral and cellular immunity. A primary mechanism of action is its ability to suppress IL-2 synthesis. It fails to block activation of antigen-specific suppressor T cells and thereby assists development of antigen-specific tolerance. Side effects include nephrotoxicity and hepatotoxicity with a possible increase in B cell lymphomas. Some individuals may also develop hypertension. CSA's mechanism of action appears to include inhibition of the synthesis and release of lymphokines and alteration of expression of MHC gene products on the cell surface. CSA inhibits IL-2 mRNA formation. This does not affect IL-2 receptor expression on the cell surface. Although CSA may diminish the number of low-affinity binding sites, it does not appear to alter high-affinity binding sites on the cell surface. CSA inhibits the early increase in cytosolic-free calcium, which occurs in beginning activation of normal T lymphocytes. It appears to produce its effect in the cytoplasm rather than on the cell surface of a lymphocyte. This could be due to its ability to dissolve in the plasma membrane lipid bilayer. CSA's cytosolic site of action may involve calmodulin and/or cyclophilin, a protein kinase.

Although immunosuppressive action cannot be explained based upon CSA–calmodulin interaction, this association closely parallels the immunosuppressive effect. CSA produces a greater suppressive effect upon class II than upon class I antigen expression in at least some experiments. While decreasing T helper lymphocytes, the T suppressor cells appear to be spared following CSA therapy. Not only sparing, but amplification of T lymphocyte suppression has been reported during CSA therapy. This is a powerful immunosuppressant that selectively affects $CD4^+$ helper T cells without altering the activity of suppressor T cells, B cells, granulocytes, and macrophages. It alters lymphocyte function, but it does not destroy the cells. CSA's principal immunosuppressive action is to inhibit IL-2 production and secretion. Thus, the suppression of IL-2 impairs the development of suppressor and cytotoxic T lymphocytes that are antigen specific. It has a synergistic immunosuppressive action with corticosteroids. Corticosteroids interfere with IL-2 synthesis by inhibiting IL-1 release from monocytes and macrophages. Cyclosporine,

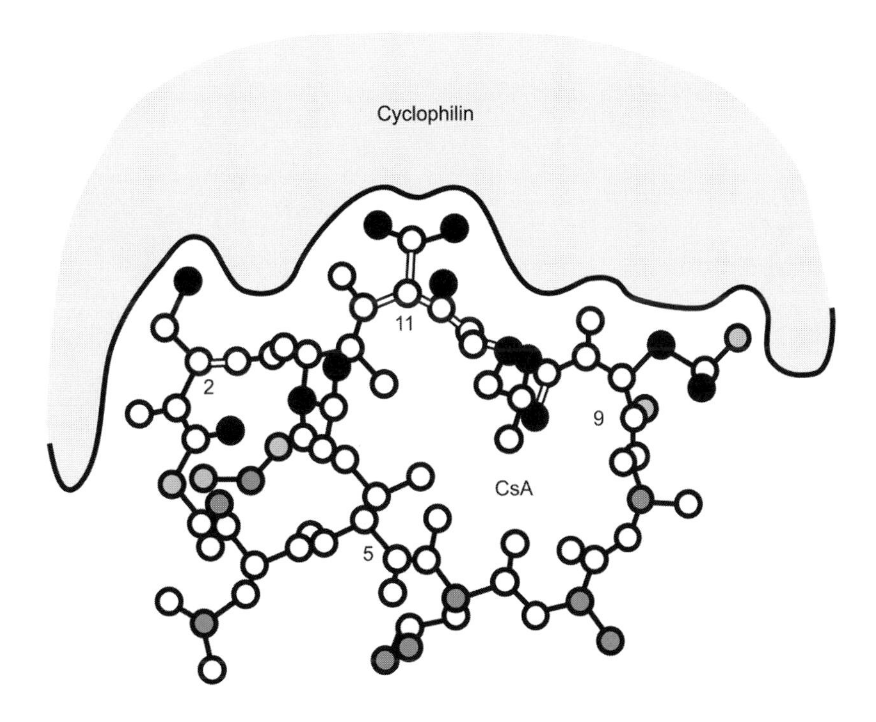

FIGURE 20.15 Cyclosporine (CSA) bound to cyclophilin.

although water insoluble, has been successfully employed as a clinical immunosuppressive agent principally in preventing rejection of organ and tissue allotransplants including kidney, heart, lung, pancreas, and bone marrow. It has also been successful in preventing GVH reactions. The drug has some nephrotoxic properties, which may be kept to a minimum by dose reduction. As with other long-term immunosuppressive agents, there may be an increased risk of lymphoma development such as Epstein-Barr (EBV) associated B cell lymphomas.

Immunophilins are high-affinity receptor proteins in the cytoplasm that combine with such immunosuppressants as cyclosporin A, FK-506, and rapamycin. They prevent the activity of rotamase by blocking conversion between *cis-* and *trans-*rotamers of the peptide and protein substrate peptidyl-prolylamide bond. Immunophilins are important in transducing signals from the cell surface to the nucleus. Immunosuppressants have been postulated to prevent signal transduction mediated by T lymphocyte receptors, which blocks nuclear factor activation in activated T lymphocytes. Cyclophilin- and FK506-binding proteins represent immunophilins. Drug–immunophilin complexes are implicated in the mechanism of action of the immunosuppressant drugs cyclosporine, FK506, and rapamycin.

Cyclophilin an 18-kDa protein in the cytoplasm that has peptidyl–prolyl isomerase functions. It has a unique and conserved amino acid sequence that has a broad phylogenetic distribution. It represents a protein kinase with a postulated critical role in cellular activation. It serves as a catalyst in *cis-trans-*rotamer interconversion. Cyclophilin catalyzes phosphorylation of a substrate, which then serves as a cytoplasmic messenger associated with gene activation. Genes coding for the synthesis of lymphokines would be activated in helper T lymphocyte responsiveness. Cyclophilin has a high affinity for cyclosporine (CSA), which accounts for the drug's immunosuppressive action (Figure 20.15). Inhibition of cyclophilin-mediated activities as a consequence of CSA–cyclophilin interaction (Figure 20.16) could lead to inhibition of the synthesis and release of lymphokines. CSA not only inhibits primary immunization but it may halt an ongoing immune response. This has been postulated to occur through inhibition of continued lymphokine release and by suppression of continued effector cell activation and recruitment.

Tacrolimus: See FK506.

FK506 (Figure 20.17) is a powerful immunosuppressive agent synthesized by *Streptomyces tsukubaensis*. Its principal use is for immunosuppression to prevent transplant rejection. FK506 has been used experimentally in liver transplant recipients. It interferes with the synthesis and binding of IL-2 and resembles cyclosporin, with which it may be used synergistically. Its immunosuppressive properties are 50 times greater than those of cyclosporin. It has been used in renal allotransplantation, but like cyclosporin also produces nephrotoxicity. It also has some neurotoxic effects and may have diabetogenic potential. The drug continues to be under investigation and is awaiting FDA approval. Also called tacrolimus.

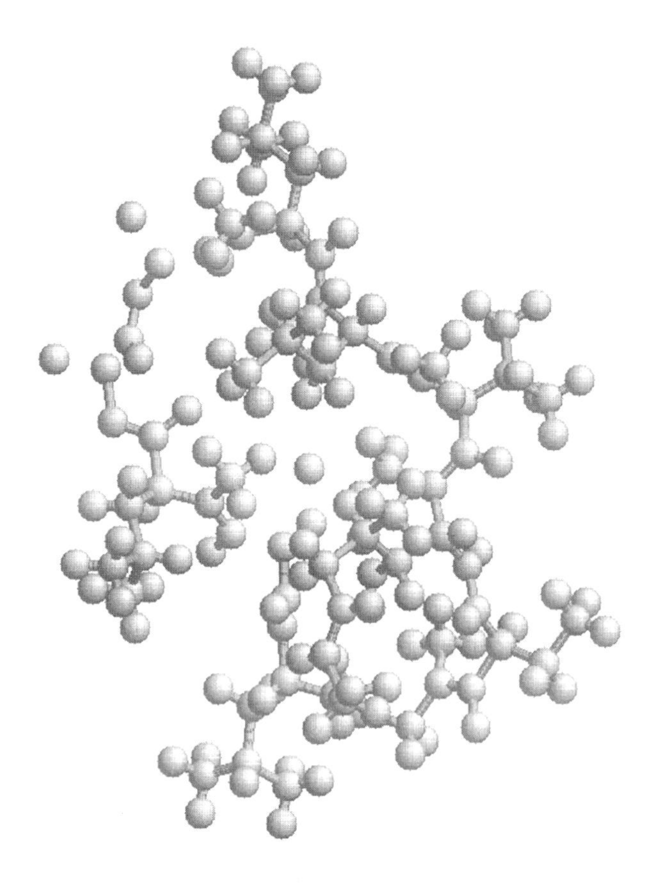

FIGURE 20.16 NMR structure of cyclosporine A as bound to cyclophilin A.

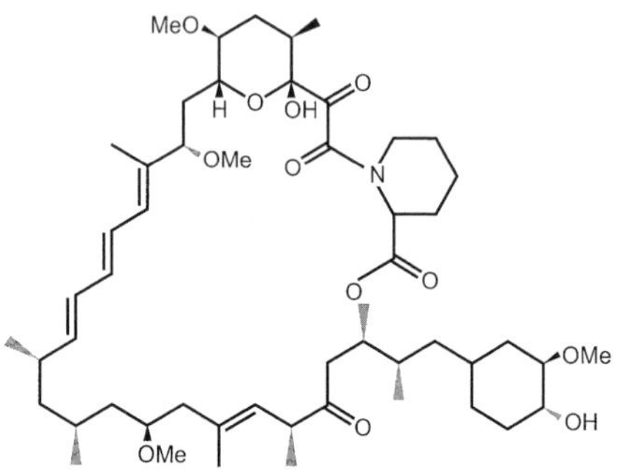

FIGURE 20.17 Structure of FK506.

FKBP (FK-binding proteins) (Figure 20.18) is a protein that binds FK506. It is a rotamase enzyme with an amino-acid sequence that closely resembles that of protein kinase C. It serves as a receptor for both FK506 and rapamycin (Figure 20.19). Cyclophilins that bind tacrolimus (FK506) are FK-binding proteins.

Rapamycin (Figure 20.20) is a powerful immunosuppressive drug derived from a soil fungus on Rapa Nui on Easter Island. It resembles FK506 (Figure 20.21) in structure but has a different mechanism of action. Rapamycin

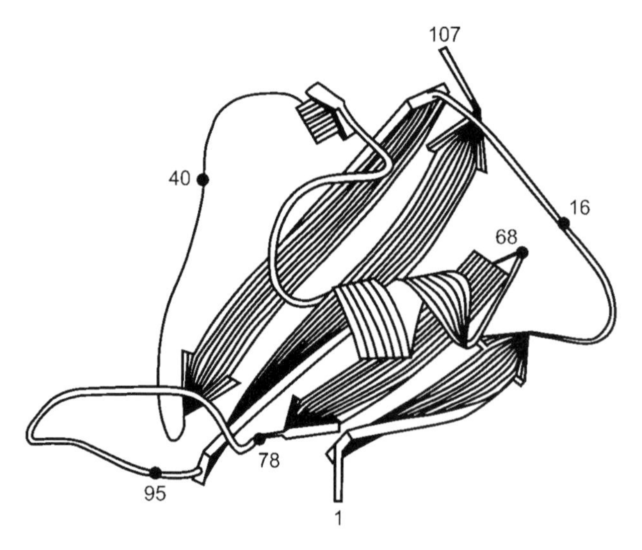

FIGURE 20.18 Ribbon model of human FKBP.

suppresses B and T lymphocyte proliferation, lymphokine synthesis, and T cell responsiveness to IL-2. To achieve clinical immunosuppression, rapamycin is effective at concentrations one eighth those required for FK506 and at 1% of the levels required for cyclosporin (Figure 20.22).

Sirolimus is the drug name for rapamycin.

Sandoglobulin®: See human immune globulin.

Brequinar sodium (BQR) is an antineoplastic and immunosuppressive agent (Figure 20.23). Its major activity is inhibition of the *de novo* biosynthesis of pyrimidine nucleosidases, resulting in inhibition of both DNA and RNA synthesis. BQR has also been shown to interfere with IgM production by IL-6 stimulated SKW6.4 cells, although in a manner independent of DNA synthesis. In transplantation studies, BQR has been shown to inhibit both the humoral and the cellular immune responses of the host, thereby significantly suppressing acute and antibody-mediated graft rejection.

Mycophenolate mofetil (Figure 20.24) is a new immunosuppressive drug that induces reversible antiproliferative effects specifically on lymphocytes, but does not induce renal, hepatic, and neurologic toxicity. Its action is based on adequate amounts of guanosine and deoxyguanosine nucleotides being required for lymphocytes to proliferate following antigenic stimulation. Thus, an agent that reversibly inhibits the final steps in purine synthesis, leading to a depletion of guanosine and deoxyguanosine nucleotides, could induce effective immunosuppression. Mycophenolate mofetil was found to produce these effects. It is the morpholinoethyl ester of mycophenolic acid. *In vivo*, it is hydrolyzed to the active form, mycophenolic acid, which the liver converts to mycophenolic acid glucuronide, which is biologically inactive and is excreted in

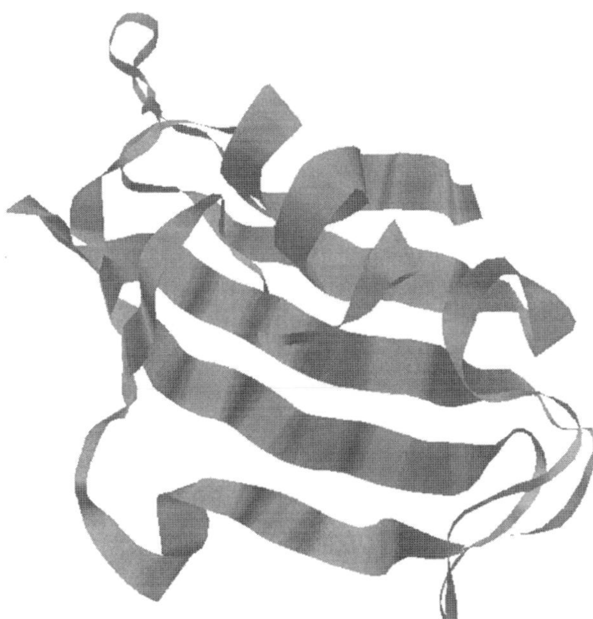

FIGURE 20.19 Atomic structure of FKBP12-rapamycin, an immunophilin–immunosuppressant complex.

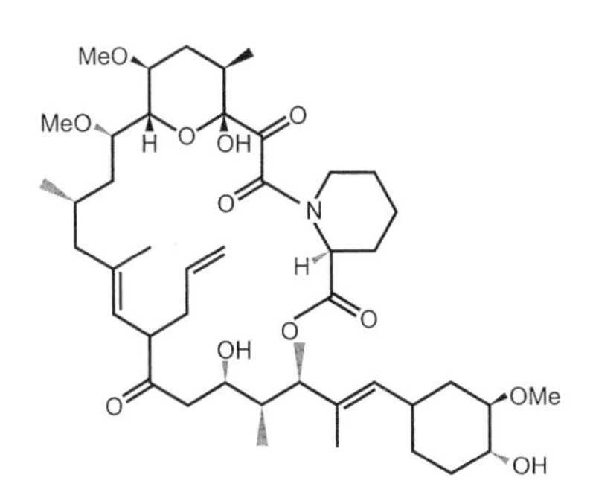

FIGURE 20.20 Structure of rapamycin.

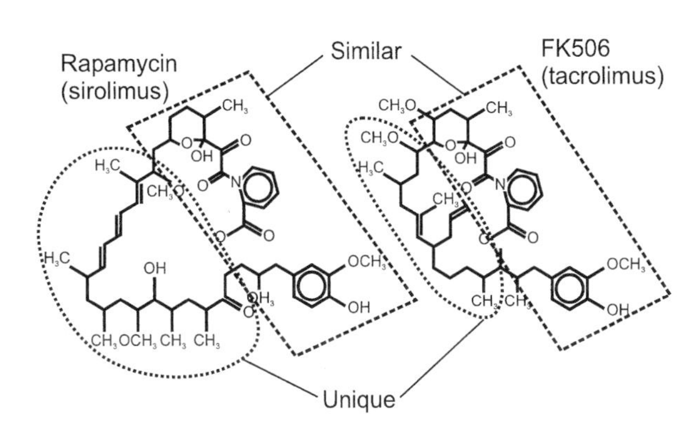

FIGURE 20.21 Uniqueness and similarities of FK506 and rapamycin.

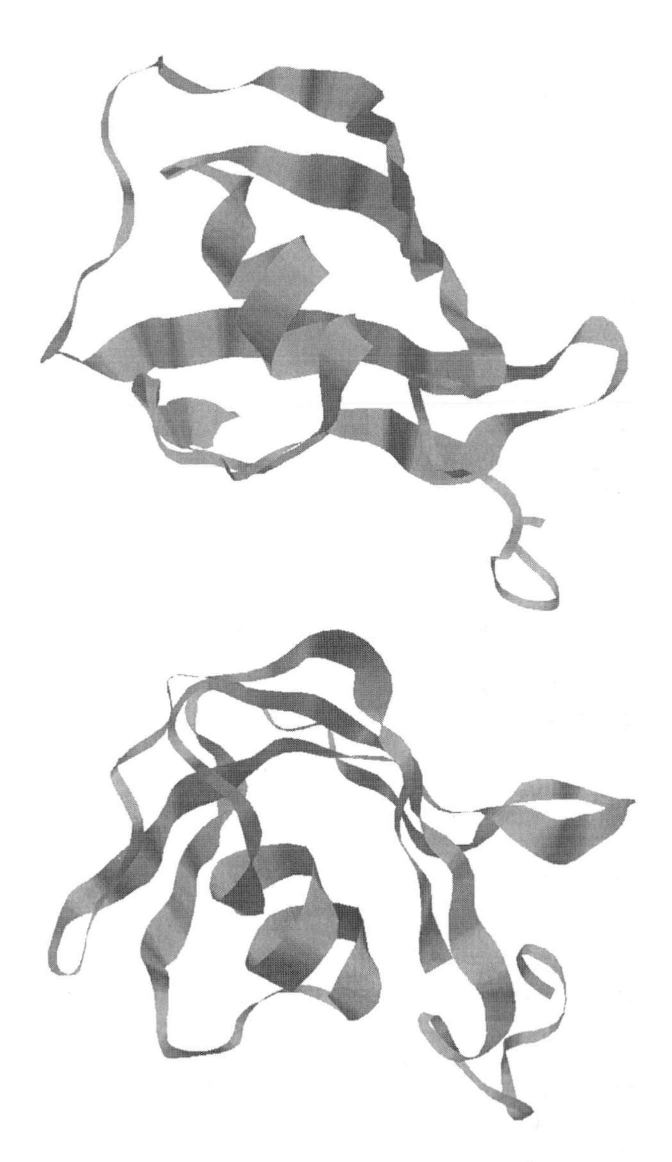

FIGURE 20.22 Human recombinant form of FK506 and rampamycin-binding protein expressed in *E. coli*.

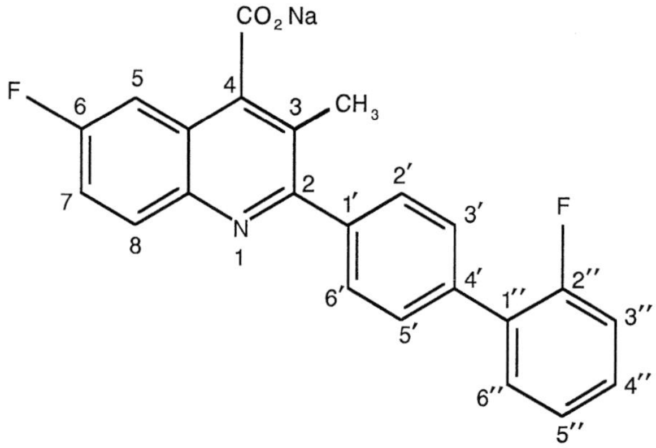

FIGURE 20.23 Structure of brequinar sodium (BQR).

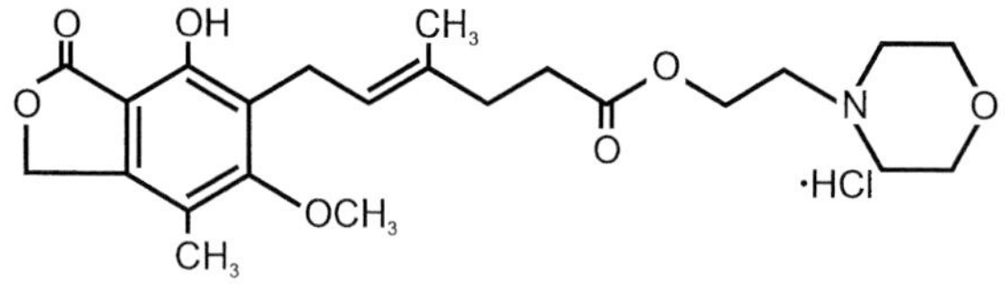

FIGURE 20.24 Structure of mycophenolate mofetil.

the urine. Mycophenolate blocks proliferation of peripheral blood mononuclear cells of humans to both B and T cell mitogens. It is also blocks antibody formation as evidenced by inhibition of a recall response by human cells challenged with tetanus toxoid. Its ability to block glycosylation of adhesion molecule that facilitate leukocyte attachment to endothelial cells and target cells probably diminishes recruitment of lymphocytes and monocytes to sites of rejection or chronic inflammation. Mycophenolate does not affect neutrophil chemotaxis, microbicidal activity, or supraoxide production. *In vivo*, mycophenolate prevents cytotoxic T cell generation and rejection of allergenic cells. It inhibits antibody formation in a dose-dependent manner and effectively prevents allograft rejection in animal models, especially when used in conjunction with cyclosporine. Current data suggest that the use of mycophenolate, together with cyclosporine and prednisone, leads to less allograft rejection than is achieved with combinations of these drugs without mycophenolate.

RS61443 (Mycophenolate mofetil) is an experimental immunosuppressive drug for the treatment of refractory, acute cellular allograft rejection, and possibly chronic rejection, in renal and other organ allotransplant recipients. Derived from mycophenolic acid, RS61443 interferes with guanosine synthesis, thereby blocking both T cell and B cell proliferation. It acts synergistically with cyclosporin in counteracting chronic rejection and has been reported to prevent FK506- and cyclosporin A-induced obliterative vasculopathy.

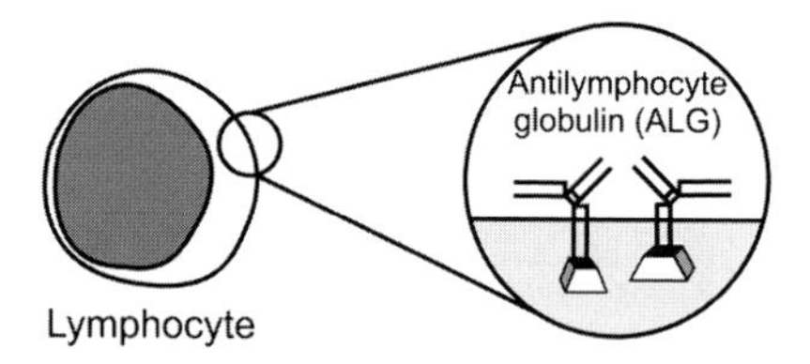

FIGURE 20.25 Antilymphocyte globulin.

Antimetabolite is a drug that interrupts the normal intracellular processes of metabolism, such as those essential to mitosis. Antimetabolite drugs such as azathioprine, mercaptopurine, and methotrexate induce immunosuppression in organ transplant recipients and diminish autoimmune reactivity in patients with selected autoimmune diseases.

Antilymphocyte serum (ALS) or antilymphocyte globulin (ALG) is an antiserum prepared by immunizing one species, such as a rabbit or horse, with lymphocytes or thymocytes from a different species, such as a human (Figure 20.25). Antibodies present in this antiserum combine with T cells and other lymphocytes in the circulation to induce immunosuppression. ALS is used in organ transplant recipients to suppress graft rejection. The globulin fraction known as antilymphocyte globulin (ALG), rather than whole antiserum, produces the same immunosuppressive effect.

Antithymocyte serum (ATS) contains antibody raised by immunizing one species, such as a rabbit or horse, with thymocytes derived from another, such as a human. The resulting antiserum has been used to induce immunosuppression in organ transplant recipients. It acts by combining with the surface antigens of T lymphocytes and suppressing their action.

Antithymocyte globulin (ATG) is the globulin fraction of serum containing antibodies generated through immunization of animals such as rabbits or horses with human thymocytes. It has been used clinically to treat rejection episodes in organ transplant recipients.

OKT®3 (Orthoclone OKT®3) is a commercial mouse monoclonal antibody against the T cell surface marker CD3 (Figure 20.26 and Figure 20.27). It may be used therapeutically to diminish T cell reactivity in organ allotransplant recipients experiencing a rejection episode; OKT3 may act in concert with the complement system to induce T cell lysis or may act as an opsonin, rendering T cells susceptible to phagocytosis. Rarely, recirculating T lymphocytes are removed in patients experiencing rejection crisis by thoracic duct drainage or extracorporeal irradiation of the blood. Plasma exchange is useful for temporary reduction in circulating antibody levels in selective diseases such as hemolytic disease of the newborn,

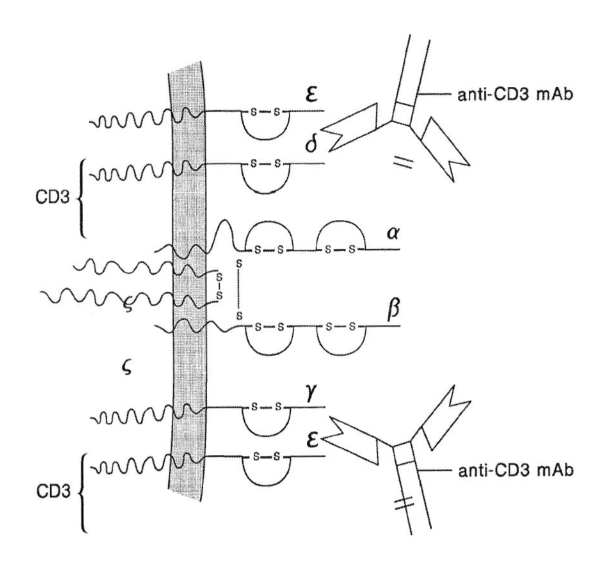

FIGURE 20.26 Anti-CD3 mAb interation with CD3 at T cell surface.

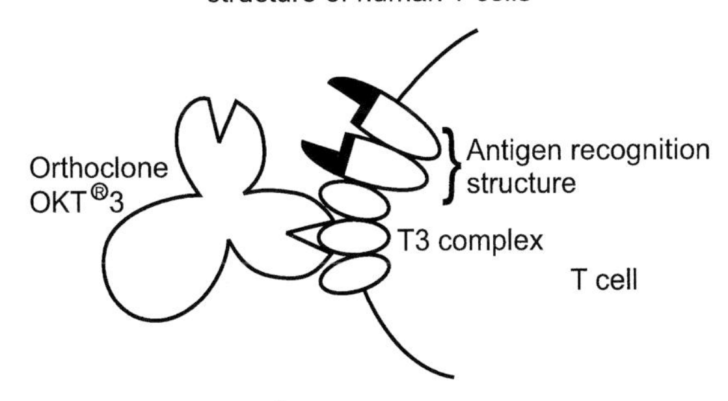

FIGURE 20.27 OKT®3 bound to the T3 complex of a T cell.

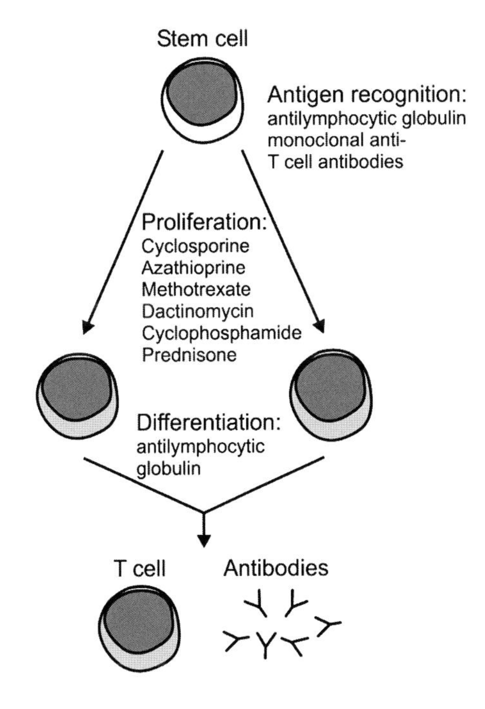

FIGURE 20.28 Summary of the actions of immunosuppressive drugs.

myasthenia gravis, or Goodpasture's syndrome. Immunosuppressive drugs act on all of the T and B cell maturation processes (Figure 20.28).

Radiation and immunity: Ionizing radiation injures DNA and other cellular constituents. The most radiosensitive cells in the body are the lymphoid tissues and recirculating lymphocytes. Small doses of radiation can lead to programmed cell death (apoptosis) of lymphocytes and of myeloid progenitor cells. Immune system sensitivity to radiation is characterized by (1) the magnitude of the suppression of the primary or secondary immune response, (2) reduction in the size of lymphoid organs, and (3) the diminished number of surviving lymphoid cells. Whereas lymphoid cells are very radiosensitive, selected radioresistant immune functions (and cells) are maintained after significant doses of radiation. The effects of radiation on immunity do not depend on their actions on lymphocytes alone but also on the leukocytes that mediate nonspecific host responses. Macrophages and other nondividing cells in the body are relatively radioresistant. They may even be activated following irradiation, revealing enhanced RNA synthesis and enhanced production of cytokines, and increased synthesis and release of lysosomal enzymes. Polymorphonuclear neutrophilic leukocyte phagocytic ability is not decreased by radiation which may even increase intracellular killing of phagocytized microorganisms. Radiation of 80 to 500 rad can rapidly kill most mature resting lymphocytes within hours. Lymphocytes stimulated by mitogen or antigen become more radioresistant and maintain their function and viability even following 1000 rad of radiation. The thymus is the most radiosensitive lymphoid organ. Radiation is usually more harmful to the antibody response than to delayed-type hypersensitivity responses. The irradiation of mice (440 rad) just before antigen exposure leads to immune suppression, whereas the same dose of irradiation 4 d after antigenic challenge leads to increased antibody titers. The *in vivo* secondary immune response is more radioresistant than is the primary response. Delayed-type hypersensitivity (DTH) is relatively radioresistant by comparison with antibody production. Most cytotoxic T lymphocytes are radioresistant to doses of several thousand rad. Many immunodeficient conditions have been associated with increased sensitivity to radiation.

A **radiomimetic drug** is an immunosuppressive drug such as an alkylating agent that is employed in the treatment of cancer. Its effect on DNA mimics that of ionizing radiation.

Total lymphoid irradiation (TLI) is a technique to induce immunosuppression in which lymphoid organs are irradiated, whereas other organs are protected from irradiation. This method has been used in the therapy of lymphomas.

H65-RTA is an immunosuppressant used clinically in the treatment of acute GVH disease and rheumatoid arthritis. H65-RTA is an immunoconjugate made up of anti-CD5 monoclonal antibody coupled to a cytotoxic enzyme. It leads to inhibition of protein synthesis, cell depletion, and lymphocyte activation.

Immunosenescence describes the age-associated decrease of the immune system and host defenses. Cell-mediated immunity declines in the elderly who develop a secondary decrease in humoral immunity. Elderly persons may have defective host defenses that place them at a greater risk for developing an infectious disease and may manifest an increased risk of morbidity and mortality from infectious diseases.

SPECIFIC IMMUNOSUPPRESSION

Anti B and T cell receptor idiotype antibodies interact with antigenic determinants (idiotopes) at the variable N-terminus of the heavy and light chains comprising the paratope region of an antibody molecule where the antigen-binding site is located. The idiotope antigenic determinants may be situated either within the cleft of the antigen-binding region or located on the periphery or outer edge of the variable region of heavy- and light-chain components. Antiidiotypic antibodies also block T cell receptors for antigen for which they are specific.

Anti-target antigen antibodies may be used to block MHC class II molecules to prolong allograft survival or to remove Rh(D) positive cells to prevent sensitization of Rh(D) negative mothers.

Clonal deletion (negative selection) refers to the elimination of self-reactive T lymphocyte in the thymus during the development of natural self-tolerance. T cells recognize self antigens only in the context of MHC molecules. Autoreactive thymocytes are eliminated following contact with self antigens expressed in the thymus before maturation is completed. The majority of $CD4^+$ T lymphocytes in the blood circulation that survived clonal deletion in the thymus failed to respond to any stimulus. This reveals that clonal anergy participates in suppression of autoimmunity. Clonal deletion represents a critical mechanism to rid the body of autoreactive T lymphocytes. It is brought about by minor lymphocyte stimulation (mls) antigens that interact with the V β region of the T lymphocyte receptor, thereby mimicking the action of bacterial superantigen. Intrathymic and peripheral tolerance in T lymphocyte can be accounted for by clonal deletion and functional inactivation of T cells reactive against self.

B cell lymphoproliferative syndrome (BLS): A rare complication of immunosuppression in bone marrow or organ transplant recipients. Epstein-Barr virus appears to be the etiologic agent. It occurs in less than 1% of HLA-identical bone marrow recipients and is more likely in those cases where anti-CD3 monoclonal antibodies were used to treat GVH disease. Clinically, it may be either a relatively mild infectious mononucleosis or a proliferating and relentless lymphoma that produces high mortality. Monoclonal antibodies to the B cell antigens, CD21 and CD24, have proven effective in controlling the B cell proliferation, but further studies are needed.

Polyomavirus immunity: Polyomavirus infections are ususally associated with reactivation of the virus in immunocompromised hosts. Two polyomaviruses, JC virus (JCV) ad BK virus (BKV), infect humans. BKV and JCV infect children. In the U.S., one half of all children acquire antibodies to BKV during the first 3 years of life and to JCV by the age of 10 or 12. Antibody responsiveness is assayed by neutralization, hemagglutination inhibition, and ELISA. Antibody titers often increase during virus activation. IgM antibodies that are virus specific have been found in renal allotransplant patients. Cell-mediated immunity develops against BKV. JCV and BKV remain in the body after primary infections. Cell-mediated immunity is believed to prevent or limit virus reactivation in healthy individuals, but they are reactivated in immunosuppressed subjects.

Wasting disease: Neonatal thymectomy in mice can lead to a chronic and eventually fatal disease characterized by lymphoid atrophy and weight loss. Animals may develop ruffled fur, diarrhea, and a hunched appearance. Gnotobiotic (germ-free) animals fail to develop wasting disease following neonatal thymectomy. Thus, thymectomy of animals that are not germ free may lead to fatal infection as a consequence of greatly decreased cell-mediated immunity. Wasting disease is also called runt disease. Wasting may appear in immunodeficiency states such as AIDS as well as in GVH reactions.

Cytotoxic drugs are agents that kill self-replicating cells such as immunocompetent lymphocytes. Cytotoxic drugs have been used for anticancer therapy as well as for immunosuppression in the treatment of transplant rejection and aberrant immune responses. The four cytotoxic drugs commonly used for immunosuppression include cyclosphosphamide, chlorambucil, azathioprine, and methotrexate.

Cytosine arabinoside is an antitumor substance that is inactive by itself, but following intracellular conversion to the nucleoside triphosphate, acts as a competitive inhibitor with regard to dCPP of DNA polymerase. It has an immunosuppressive effect on antibody formation in both the primary and secondary immune responses and also depresses the generation of cell-mediated immunity.

Reticuloendothelial blockade describes the temporary paralysis of phagocytic cells of the mononuclear phagocytic system with respect to phagocytic ability by the injection of excess amounts of inert particles such as colloidal carbon, gold, or iron. Once the mononuclear phagocytes have expended their entire phagocytic ability taking up these inert particles, they can no longer phagocytize administered microorganisms or other substances that would normally be phagocytized.

The **trophoblast** is a layer of cells in the placenta that synthesizes immunosuppressive agents. These cells are in contact with the lining of the uterus.

Uromodulin Tamm-Horsfall protein is an 85-kDa α_1 acid glycoprotein produced in Henle's ascending loop and distal convoluted tubules by epithelial cells. It is a powerful immunosuppressive protein based upon N-linked carbohydrate residues. It inhibits proliferation of T cells induced by antigen and monocyte cytotoxicity. Uromodulin is a ligand for interleukin-1 α, interleukin-1 β, and tumor necrosis factor (TNF).

Cycle-specific drugs are immunosuppressive and cytotoxic drugs that lead to death of mitotic and resting cells.

21 Transplantation Immunology

Transplantation is the replacement of an organ or other tissue, such as bone marrow, with organs or tissues derived ordinarily from a nonself source such as an allogeneic donor. Organs include kidney, liver, heart, lung, pancreas (including pancreatic islets), intestine, or skin. In addition, bone matrix and cardiac valves have been transplanted. Bone marrow transplants are given for nonmalignant conditions such as aplastic anemia as well as to treat certain leukemias and other malignant diseases.

Transplantation immunology is the study of immunologic reactivity of a recipient to transplanted organs or tissues from a histoincompatible recipient. Effector mechanisms of transplantation rejection or transplantation immunity consist of cell-mediated immunity and/or humoral antibody immunity, depending upon the category of rejection. For example, hyperacute rejection of an organ such as a renal allograft is mediated by preformed antibodies and takes place soon after the vascular anastomosis is completed in transplantation. By contrast, acute allograft rejection is mediated principally by T lymphocytes and occurs during the first week after transplantation. There are instances of humoral vascular rejection mediated by antibodies as a part of the acute rejection in response. Chronic rejection is mediated by a cellular response.

Histocompatibility is tissue compatibility as in the transplantation of tissues or organs from one member to another of the same species, an allograft, or from one species to another, a xenograft. The genes that encode antigens which should match if a tissue or organ graft is to survive in the recipient are located in the major histocompatibility complex (MHC) region. This is located on the short arm of chromosome 6 in humans (Figure 21.1 and Figure 21.2) and of chromosome 17 in the mouse. Class I and class II MHC antigens are important in tissue transplantation. The greater the match between donor and recipient, the more likely the transplant is to survive. For example, a six-antigen match implies sharing of two HLA-A antigens, two HLA-B antigens, and two HLA-DR antigens between donor and recipient. Even though antigenically dissimilar grafts may survive when a powerful immunosuppressive drug such as cyclosporine is used, the longevity of the graft is still improved by having as many antigens match as possible.

A **histocompatibility locus** is a specific site on a chromosome where the histocompatibility genes that encode histocompatibility antigens are located. There are major histocompatibility loci such as HLA in humans and H-2 in the mouse across which incompatible grafts are rejected within 1 to 2 weeks. There are also several minor histocompatibility loci with more subtle antigenic differences, across which only slow, low-level graft rejection reactions occur.

Histocompatibility antigen is one of a group of genetically encoded antigens present on tissue cells of an animal that provoke a rejection response if the tissue containing them is transplanted to a genetically dissimilar recipient. These antigens are detected by typing lymphocytes on which they are expressed. These antigens are encoded in humans by genes at the HLA locus on the short arm of chromosome 6 (Figure 21.2). In the mouse, they are encoded by genes at the H-2 locus on chromosome 17 (Figure 21.3).

Transplantation antigens are histocompatibility antigens that stimulate an immune response in the recipient that may lead to rejection.

The **minor histocompatibility locus** is a chromosomal site of genes that encode minor histocompatibility antigens which stimulate immune responses against grafts containing these antigens.

Minor histocompatibility antigens are molecules expressed on cell surfaces that are encoded by the minor histocompatibility loci, not the major histocompatibility locus. They represent weak transplantation antigens by comparison with the major histocompatibilty antigens. However, they are multiple, and their cumulative effect may contribute considerably to organ or tissue graft rejection. Graft rejection based on a minor histocompatibility difference between donor and recipient requires several weeks compared to the 7 to 10 d required for a major histocompatibility difference. Minor histocompatibility antigens may be difficult to identify by serological methods.

Minor transplantation antigens: See minor histocompatibility antigens.

H-Y is a Y chromosome-encoded minor histocompatibility antigen that may induce male skin graft rejection by females or destruction of lymphoid cells from males by effector cytotoxic T lymphocytes from females.

HY is the male-specific transplantation antigen. Females of some but not all inbred mouse strains can reject skin grafts from males of the same syngeneic strain. By contrast, male-to-male, female-to-female, and female-to-male grafts succeed. This indicated the presence of a minor histocompatibility antigen gene on the Y chromosome that was designated H-Y. H-Y is a weak transplantation antigen compared to the mouse MHC-designed H-2. Several H-Y epitopes have been identified in mice and one in humans. H-Y peptide epitopes are derived from several linked genes.

Minor histocompatibility peptides: H antigens. Among minor antigens thus far identified are H-3 antigens, male-specific H-Y antigen, β_2 microglobulin, and numerous others that have not yet been firmly established.

Minor lymphocyte stimulatory (MIs) loci: See Mls antigens.

Minor lymphocyte-stimulating genes: See Mls genes.

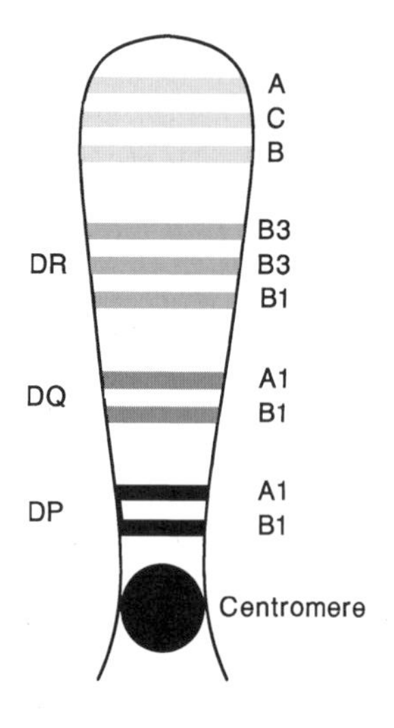

FIGURE 21.1 Human chromosome 6.

Minor lymphocyte-stimulating (Mls) determinants are characterized by their activation of a marked primary mixed-lymphocyte reaction (MLR) between lymphocytes of mice sharing an identical MHC haplotype. MHC class II molecules on various cell surfaces present Mls epitopes to naive T lymphocytes which mount a significant response. V β-specific monoclonal antibodies have facilitated the definition of Mls epitopes. Mls determinants activate T lymphocytes expressing selected β specificities. See also Mls antigens.

Histocompatibility testing is a determination of the MHC class I and class II tissue type of both donor and recipient prior to organ or tissue transplantation. In man HLA-A, HLA-B, and HLA-DR types are determined, followed by cross-matching donor lymphocytes with recipient serum prior to transplantation. A mixed lymphocyte culture (MLC) was formerly used in bone marrow transplantation, but has now been replaced by molecular DNA typing. The MLC may also be requested in living related organ transplants. As in renal allotransplantation, organ recipients have their serum samples tested for percent reactive antibodies, which reveals whether or not they have been presensitized against HLA antigens of an organ for which they may be the recipient.

Human leukocyte antigen (HLA) is the product of the MHC in humans that contains the genes that encode the polymorphic MHC Class I and Class II molecules as well as other important genes.

HLA is an abbreviation for human leukocyte antigen. The HLA histocompatibility system in humans represents a complex of MHC class I molecules distributed on essentially all nucleated cells of the body and MHC class II molecules that are distributed on B lymphocytes, macrophages, and a few other cell types. These are encoded by genes at the MHC. In humans the HLA locus is found on the short arm of chromosome 6. This has now been well defined, and in addition to encoding surface isoantigens, genes at the HLA locus also encode immune

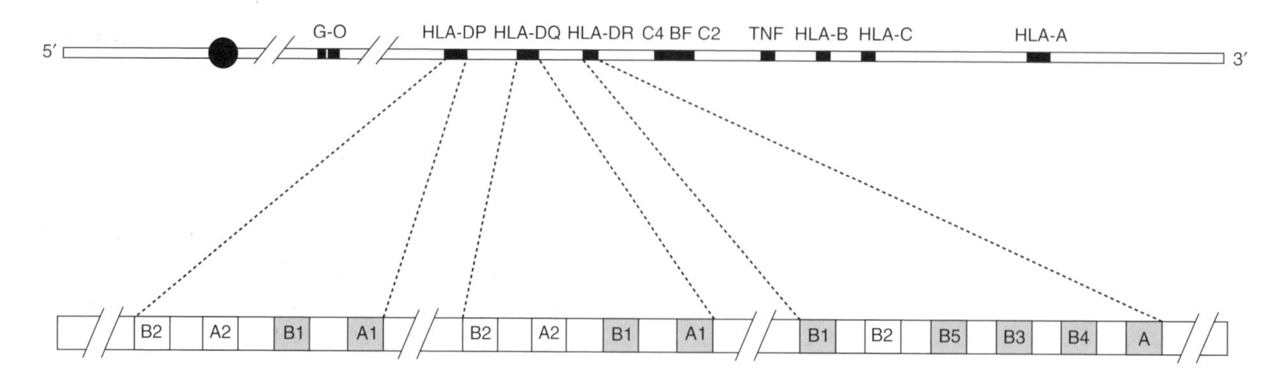

FIGURE 21.2 Short arm of human chromosome 6.

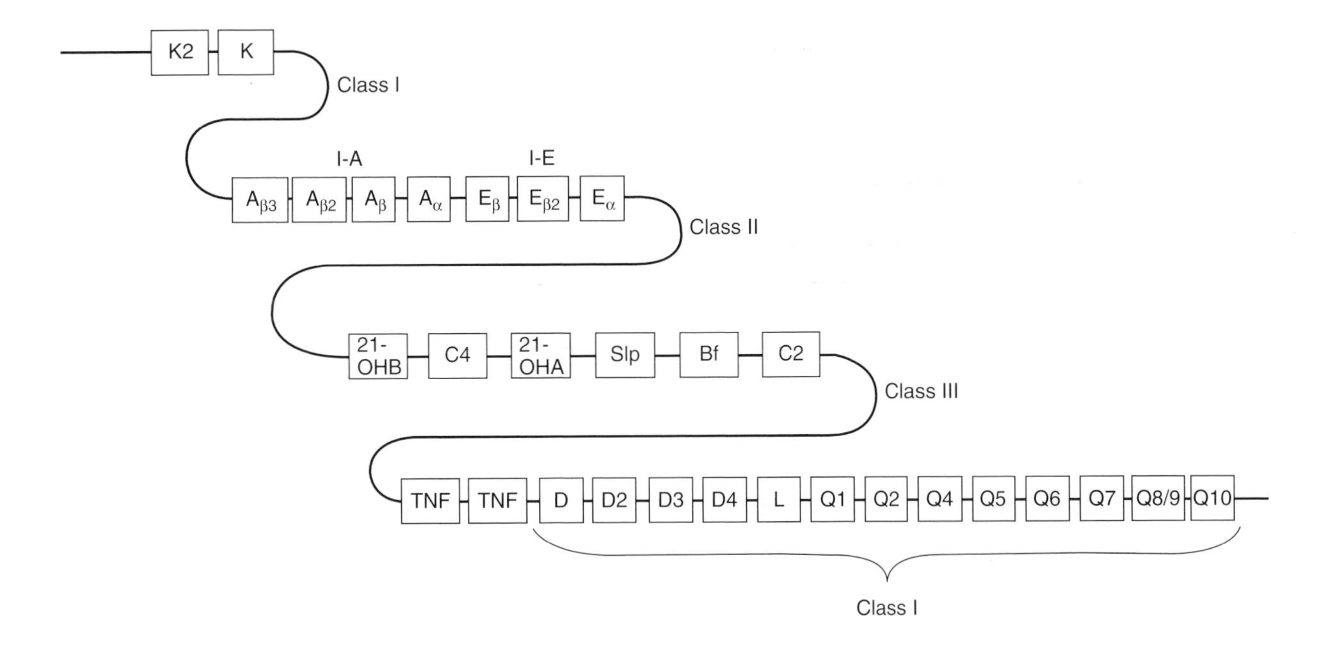

FIGURE 21.3 H-2 complex on chromosome 17 of a mouse.

response (*Ir*) genes. The class I region consists of HLA-A, HLA-B, and HLA-C loci, and the class II region consists of the D region which is subdivided into HLA-DP, HLA-DQ, and HLA-DR subregions. Class II molecules play an important role in the induction of an immune response, since antigen-presenting cells must complex an antigen with class II molecules to present it in the presence of interleukin-1 to $CD4^+$ T lymphocytes. Class I molecules are important in presentation of intracellular antigen to $CD8^+$ T lymphocytes as well as for effector functions of target cells. Class III molecules encoded by genes located between those that encode class I and class II molecules include C2, BF, C4a, and C4b. Class I and class II molecules play an important role in the transplantation of organs and tissues. The microlymphocytotoxicity assay is used for HLA-A, -B, -C, -DR, and -DQ typing. The primed lymphocyte test is used for DP typing. Uppercase letters designate individual HLA loci such as HLA-B and alleles are designated by numbers such as in HLA-B*0701.

HLA Class III: See MHC genes and Class III MHC molecules.

HLA locus refers to the major histocompatibility locus in man.

Immunotyping: See immunophenotyping.

w is the symbol for "workshop" that is used for HLA antigen and cluster of differentiation (CD) designations when new antigenic specificities have not been conclusively decided. Once the specificities have been agreed upon among authorities, "w" is removed from the designation.

Polymorphism indicates the occurrence of two or more forms, such as ABO and Rh blood groups, in individuals of the same species. This is due to two or more variants at a certain genetic locus occurring with considerable frequency in a population. Polymorphisms are also expressed in the HLA system of human leukocyte antigens as well as in the allotypes of immunoglobulin γ and κ chains.

Supratypic antigen: See public antigen.

HLA-A is a class I histocompatibility antigen in humans (Figure 21.4). It is expressed on nucleated cells of the body. Tissue typing to identify an individual's HLA-A antigens employs lymphocytes.

HLA-B is a class I histocompatibility antigen (Figure 21.5) in humans which is expressed on nucleated cells of the body. Tissue typing to define an individual's HLA-B antigens employs lymphocytes.

HLA-C is a class I histocompatibility antigen in humans which is expressed on nucleated cells of the body. Lymphocytes are employed for tissue typing to determine HLA-C antigens. HLA-C antigens play little or no role in graft rejection.

The human MHC class II region is the **HLA-D region,** which is comprised of three subregions designated DR, DQ, and DP. Multiple genetic loci are present in each of these. DN (previously DZ) and DO subregions are each comprised of one genetic locus. Each class II HLA molecule is comprised of one α and one β chain that constitute a heterodimer. Genes within each subregion encode a particular class II molecule's α and β chains. Class II genes

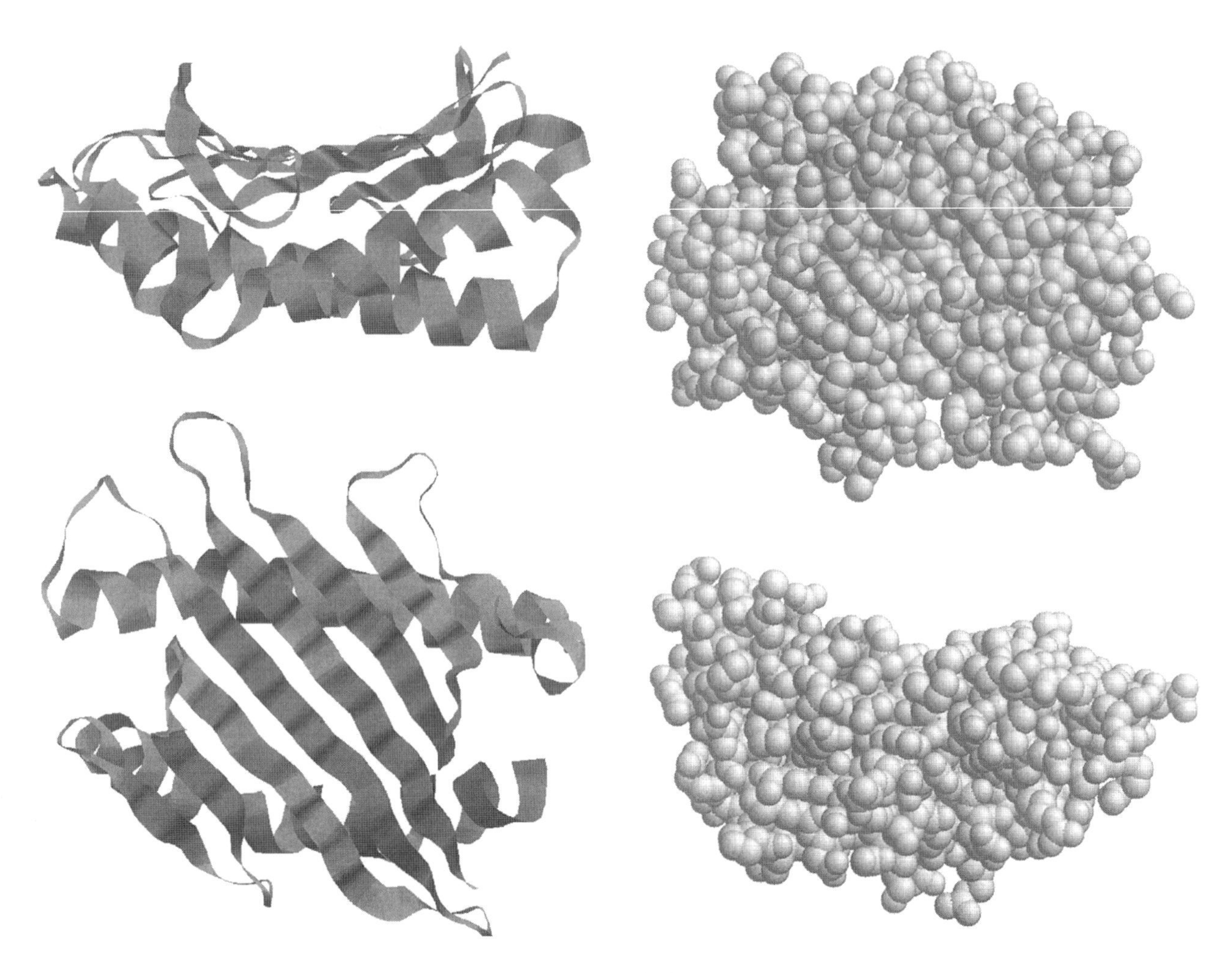

FIGURE 21.4 Human class I histocompatibility antigen (HLA-A0201) complexed with a decameric peptide from calreticulin (HLA-A0201). Human recombinant extracellular fragment expressed in *E. coli*; peptide synthetic based on sequence of human calreticulin.

FIGURE 21.5 Class I histocompatibility antigen HLA-B*2705 complexed with nonapeptide arg-arg-ile-lys = ala-ile-thr-leu-lys (theoretical model).

that encode α chains are designated A, whereas class II genes that encode β chain are designated B. A number is used following A or B if a particular subregion contains two or more A or B genes.

Primed lymphocyte test (PLT): Lymphocytes previously exposed or primed to a certain antigen in a primary mixed lymphocyte culture will divide rapidly when reexposed to the same antigen. Using a primed cell, one can determine whether or not an unknown cell possesses the original stimulating antigen. Cells previously exposed to MHC class II HLA antigens can be used in HLA typing for HLA-D region antigens. It is an assay for the detection of lymphocyte-associated determinants (LAD). For this procedure, lymphocytes donated by a normal person can serve as responder cells against the antigens of a known cell type. The test is based on the secondary stimulation of the primed or sensitized lymphocytes. The original stimulator serves as a positive control. The response of the sensitized cell to other cells measured by the incorporation of tritiated thymidine, by comparison with the control, may suggest sharing of HLA-D-associated antigens with the original stimulator cell if high stimulation values result. The HTC typing procedure, on the other hand, implies an antigenic determinant shared between the two cell types when there is little or no response.

PLT is a positive typing procedure and has the advantage that homozygous donor cells are not required. Primed or sensitized cells can be prepared whenever they are needed and frozen for future use. These cells can be used to type unknowns within a period of 24 h. This eliminates the 5 to 6 d needed for a homozygous cell typing procedure.

Primed lymphocyte typing (PLT) is a method to type for HLA-D antigenic determinants. It is a type of mixed-lymphocyte reaction in which cells previously exposed to allogeneic lymphocytes of known specificity can be reexposed to unknown lymphocytes to determine their HLA-DP type, for example.

The **HLA-DP subregion** is the site of two sets of genes designated HLA-DPA1 and HLA-DPB1 and the pseudogenes HLA-DPA2 and HLA-DPB2. DP α and DP β chains encoded by the corresponding genes DPA1 and DPB1 unite to produce the DPαβ molecule. DP antigen or type is determined principally by the very polymorphic DPβ chain, in contrast to the much less polymorphic DPα chain. DP molecules carry DPw1–DPw6 antigens.

The **HLA-DQ subregion** consists of two sets of genes designated DQA1 and DQB1, and DQA2 and DQB2. DQA2 and DQ B2 are pseudogenes. DQα and DQβ chains, encoded by DQA1 and DQB1 genes, unite to produce the DQαβ molecule. Although both DQα and DQβ chains are polymorphic, the DQβ chain is the principal factor in determining the DQ antigen or type. DQαβ molecules carry DQw1–DQw9 specificities.

The **HLA-DR subregion** is the site of one HLA-DRA gene (Figure 21.6). Although DRB gene number varies with DR type, there are usually three DRB genes, termed DRB1, DRB2, and DRB3 (or DRB4). The DRB2 pseudogene is not expressed. The DR α chain, encoded by the DRA gene, can unite with products of DRB1 and DRB3 (or DRB4) genes which are the DR β-1 and DR β-3 (or DR β-4) chains. This yields two separate DR molecules, DR αβ-1 and DR αβ-3 (or DR αβ-4). The DR β chain determines the DR antigen (DR type) since it is very polymorphic, whereas the DR α chain is not. DR αβ-1 molecules carry DR specificities DR1–DRw18. Yet, DR αβ-3 molecules carry the DRw52, and the DR αβ-4 molecules carry the DRw53 specificity.

W,X,Y boxes (class II MHC promoter) are three conserved sequences found in the promoter region of the HLA-DRα chain gene. The X box contains tandem regulatory sequences designated X1 and X2. Any cell that expresses MHC class II molecules will have all three boxes interacting with binding proteins, and decreased or defective production of some of these binding proteins can result in the "bare lymphocyte syndrome."

HLA-DR antigenic specificities are epitopes on DR gene products. Selected specificities have been mapped to defined loci. HLA serologic typing requires the identification of a prescribed antigenic determinant on a particular HLA molecular product. One typing specificity can be present on many different molecules. Different alleles at the same locus may encode these various HLA molecules. Monoclonal antibodies are now used to recognize certain antigenic determinants shared by various molecules bearing the same HLA typing specificity. Monoclonal antibodies have been employed to recognize specific class II alleles with disease associations.

HLA-DM facilitates the loading of antigenic peptides onto MHC class II moleules. As a result of proteolysis of the invariant chain, a small fragment called the class II-associated invariant chain peptide, or CLIP, remains bound to the MHC class II molecule. CLIP peptide is replaced by antigenic peptides, but in the absence of HLA-DM, this does not occur. The HLA-DM molecule must therefore play some part in removal of the CLIP peptide and in the loading of antigenic peptides.

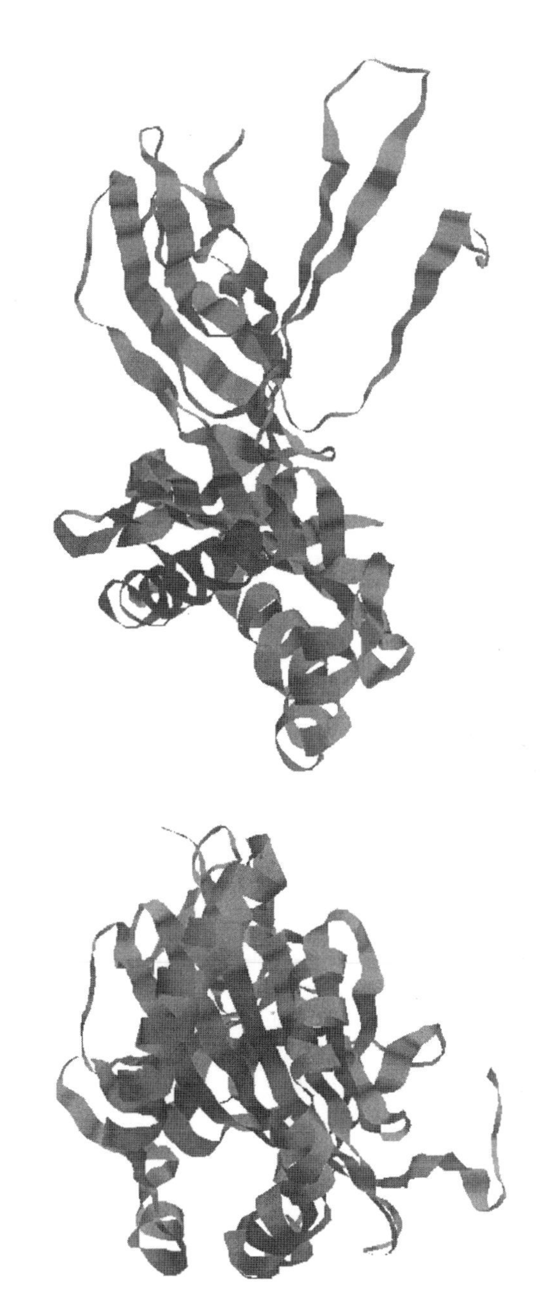

FIGURE 21.6 HLA-DR1 histocompatibility antigen.

Minor lymphocyte stimulatory (MIs) loci: See MIs antigens.

HLA-E is an HLA class I nonclassical molecule.

HLA-F is an HLA class I nonclassical molecule.

HLA-G is a polymorphic class I HLA antigen with extensive variability in the α-2 domain. It is found on trophoblasts, i.e., placenta cells and trophoblastic neoplasms. HLA-G is expressed only on cells such as placental extravillous cytotrophoblasts and choriocarcinoma that fail to express HLA-A, -B, and -C antigens. HLA-G expression is most pronounced during the first trimester of pregnancy. Trophoblast cells expressing HLA-G at the maternal–fetal junction may protect the semiallogeneic fetus from "rejection." Prominent HLA-G expression suggests maternal immune tolerance.

HLA-H is a pseudogene found in the MHC class I region that is structurally similar to HLA-A but is nonfunctional due to the absence of a cysteine residue at position 164 in its protein product and the deletion of the codon 227 nucleotide.

HLA nonclassical class I genes are located within the MHC class I region and encode products that can associate with β2 microglobulin. However, their function and tissue distribution are different from those of HLA-A, -B, and -C molecules. Examples include HLA-E, -F, and -G. Of these, only HLA-G is expressed on the cell surface. It is uncertain whether or not these HLA molecules are involved in peptide binding and presentation like classical class I molecules.

An **extended haplotype** consists of linked alleles in positive linkage disequilibrium, situated between and including HLA-DR and HLA-B of the MHC of man. Examples of extended haplotypes include the association of B8/DR3/SCO1/GLO2 with membranoproliferative glomerulonephritis, and of A25/B18/DR2 with complement C2 deficiency. Extended haplotypes may be a consequence of crossover suppression through environmental influences, together with selected HLA types, leading to autoimmune conditions. The B27 relationship to *Klebsiella* is an example. PCR amplification and direct sequencing help identify a large number of allelic differences and specific associations of extended haplotypes with disease. Extended haplotypes are more informative than single polymorphisms. Some diseases associated with extended haplotypes include Graves' disease, pemphigus vulgaris, type I (juvenile onset) insulin-dependent diabetes mellitus, celiac disease, psoriasis, and autoimmune hepatitis.

Linkage disequilibrium refers to the appearance of HLA genes on the same chromosome with greater frequency than would be expected by chance. This has been demonstrated by detailed studies in both populations and families, employing outbred groups in which numerous different haplotypes are present. With respect to the HLA-A, -B, and -C loci, a possible explanation for linkage disequilibrium is that there has not been sufficient time for the genes to reach equilibrium. However, this possibility is remote for HLA-A, -B, and -D linkage disequilibrium. Natural selection has been suggested to maintain linkage disequilibrium that is advantageous. If products of two histocompatibility loci play a role in the immune response and appear on the same chromosome, they might reinforce one another and represent an advantageous association. An example of linkage disequilibrium in the HLA system of man is the occurrence on the same chromosome of HLA-A3 and HLA-B7 in the Caucasian American population.

Lymphocyte defined (LD) antigens are histocompatibility antigens on mammalian cells that induce reactivity in a mixed-lymphocyte culture (MLC) or mixed-lymphocyte reaction.

HLA disease association: Certain HLA alleles occur in a higher frequency in individuals with particular diseases than in the general population. This type of data permits estimation of the "relative risk" of developing a disease with every known HLA allele. For example, there is a strong association between ankylosing spondylitis, which is an autoimmune disorder involving the vertebral joints, and the class I MHC allele, HLA-B27. There is a strong association between products of the polymorphic class II alleles HLA-DR and -DQ and certain autoimmune diseases, since class II MHC molecules are of great importance in the selection and activation of $CD4^+$ T lymphocytes which regulate the immune responses against protein antigens. For example, 95% of Caucasians with insulin-dependent (type I) diabetes mellitus have HLA-DR3 or HLA-DR4 or both. There is also a strong association of HLA-DR4 with rheumatoid arthritis. Numerous other examples exist and are the targets of current investigations, especially in extended studies employing DNA probes. Calculation of the relative risk (RR) and absolute risk (AR) can be found elsewhere in this book.

Immunoinhibitory genes are selected HLA genes that appear to protect against immunological diseases. Their mechanisms of action are in dispute.

HLA allelic variation is a genomic analysis that has identified specific individual allelic variants to explain HLA associations with rheumatoid arthritis, type I diabetes mellitus, multiple sclerosis, and celiac disease. There is a minimum of six α and eight β genes in distinct clusters, termed HLA-DR, -DQ, and -DP within the HLA class II genes. DO and DN class II genes are related, but map outside DR, DQ, and DP regions. There are two types of dimers along the HLA cell-surface HLA-DR class II molecules. The dimers are made up of either DRα-polypeptide associated with $DR\beta_1$-polypeptide or DR with $DR\beta_2$-polypeptide. Structural variation in class II gene products is linked to functional features of immune recognition leading to individual variations in histocompatibility, immune recognition, and susceptibility to disease.

There are two types of structural variations which include variation among DP, DQ, and DR products in primary amino acid sequence by as much as 35% and individual variation attributable to different allelic forms of class II genes. The class II polypeptide chain possesses domains which are specific structural subunits containing variable sequences that distinguish among class II α genes or class II β genes. These allelic variation sites have been suggested to form epitopes, which represent individual structural differences in immune recognition.

Interallelic conversion refers to genetic recombination between two alleles of a locus in which a segment of one allele is replaced with a homologous segment from another. HLA Class I and HLA Class II alleles are formed in this way.

HLA oligotyping is a recently developed method using oligonucleotide probes to supplement other histocompatibility testing techniques. Whereas serological and cellular methods identify phenotypic characteristics of HLA proteins, oligotyping defines the genotype of the DNA that encodes HLA protein structure and specificity. Thus, oligotyping can identify the DNA type even when there is a failure of expression of HLA genes that render serological techniques ineffective.

HLA tissue typing (Figure 21.7) is the identification of MHC class I and class II antigens on lymphocytes by serological and cellular techniques. The principal serological assay is microlymphocytotoxicity using microtiter plate containing predispensed antibodies against HLA specificities to which lymphocytes of unknown specificity plus rabbit complement and vital dye are added. Following incubation, the wells are scored according to the relative proportion of cells killed. This method is employed for organ transplants such as renal allotransplants. For bone marrow transplants, mixed lymphocyte reaction procedures are performed to determine the relative degree of histocompatibility or histoincompatibility between donor and recipient. Serological tests are largely being replaced by DNA typing procedures employing PCR methodology and DNA or oligonucleotide probes, especially for MHC class II typing.

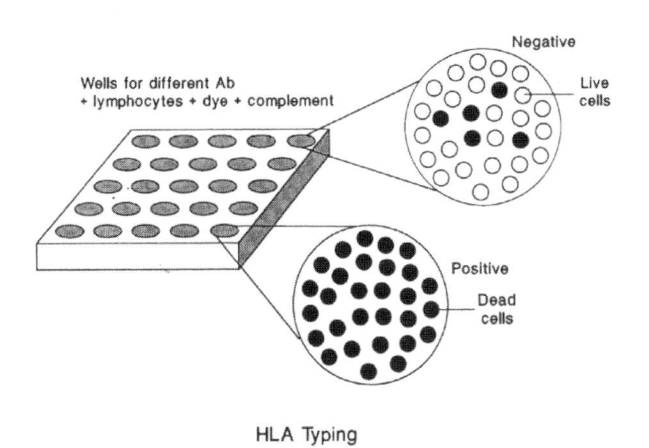

FIGURE 21.7 HLA tissue typing.

Class I typing involves reactions between lymphocytes to be typed with HLA antisera of known specificity in the presence of complement. Cell lysis is detected by phase or fluorescence microscopy. This is important in parentage testing, disease association, transfusion practices, and transplantation. HLA-A, -B, and -C antigens should be defined by at least one of the following: (1) at least two different sera if both are monospecfic, (2) one monospecific and two multispecific antisera, (3) at least three multispecific antisera if all multispecific are used.

Class II typing detects HLA-DR antigens using purified B cell preparations. It is based on antibody-specific, complement-dependent disruption of the cell membrane of lymphocytes. Cell death is demonstrated by the penetration of dye into the membrane. Class II typing is more difficult than class I typing because of the variability of both B cell isolation and complement toxicity. At least three antisera must be used if all are monospecific; at least three antisera, must be used if all are monospecific; at least five antisera must be used for multispecifc sera.

Antibody screening: Candidates for organ transplants, especially renal allografts, are monitored with relative frequency for changes in their percent reactive antibody (PRA) levels. Obviously, those with relatively high PRA values are considered to be less favorable candidates for renal allotransplants than are those in whom the PRA values are low. PRA determinations may vary according to the composition of the cell panel. If the size of the panel is inadequate, it may affect the relative frequency of common histocompatibility antigens found in the population.

Tissue typing is the identification of MHC class I and class II antigens on lymphocytes by serological and cellular techniques. The principal serological assay is microlymphocytotoxicity using microtiter plates containing predispensed antibodies against HLA specificities to which lymphocytes of unknown specificity plus rabbit complement and vital dye are added. Following incubation, the wells are scored according to the relative proportion of cells killed. This method is employed for organ transplants such as renal allotransplants. For bone marrow transplants, mixed lymphocyte culture (MLC), also called mixed lymphocyte reaction, procedures are performed to determine the relative degree of histocompatibility or histoincompatibility between donor and recipient. Serological tests are largely being replaced by DNA typing procedures employing PCR methodology and DNA or oligonucleotide probes, especially for MHC class II typing. Sequence specific primer (SSP) technology is currently the method of choice in molecular typing.

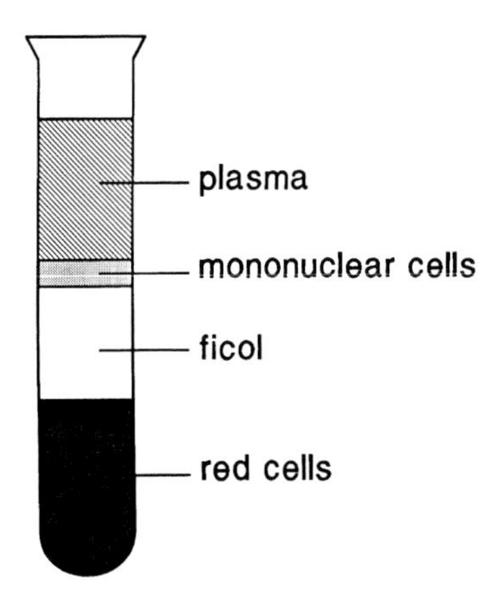

FIGURE 21.8 Schematic representation of Ficoll-hypaque technique of cell separation.

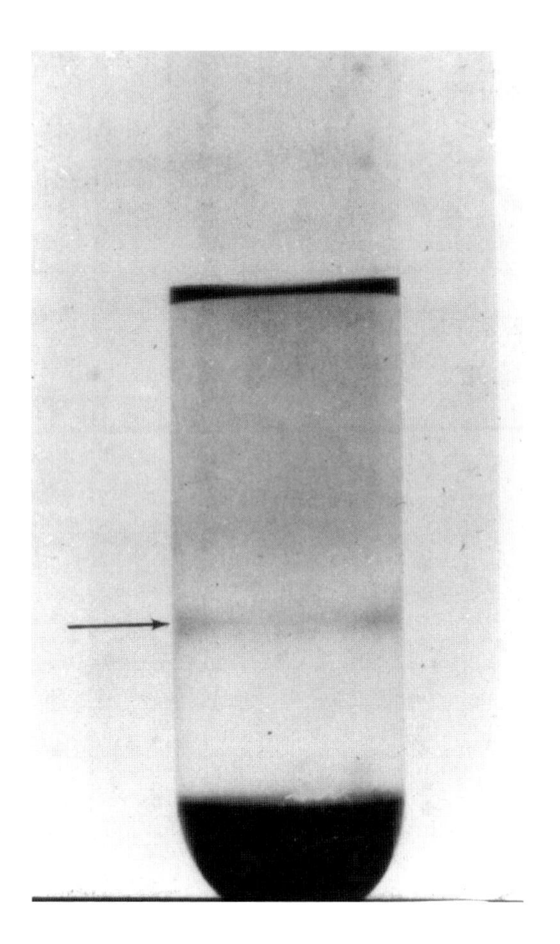

FIGURE 21.9 The separation of lymphocytes from peripheral blood by centrifugation using Ficoll-hypaque.

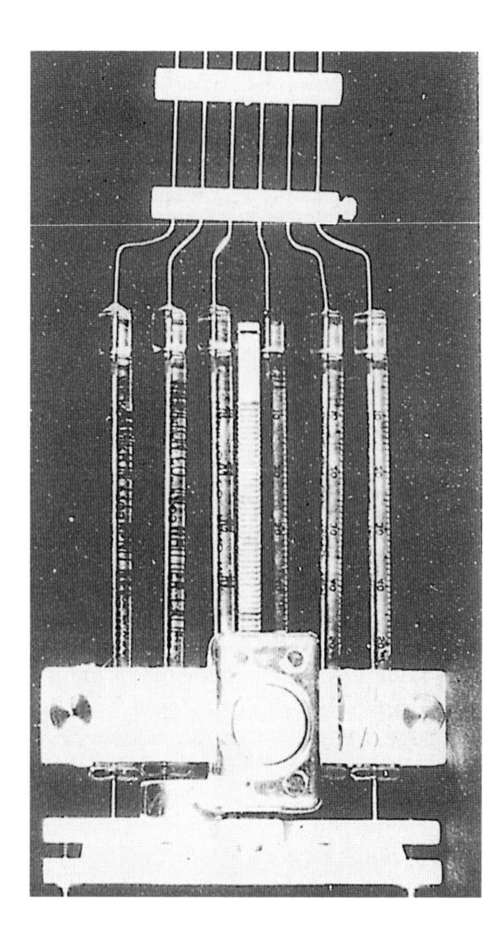

FIGURE 21.10 A Hamilton syringe that is used to dispense lymphocytes into Terasaki plates for tissue typing.

Microlymphocytotoxicity is a widely used technique for HLA tissue typing. Lymphocytes are separated from heparinized blood samples by either layering over Ficoll-hypaque (Figure 21.8 and Figure 21.9), centrifuging and removing lymphocytes from the interface or by using beads. After appropriate washing, these purified lymphocyte preparations are counted, and aliquots are dispensed using a Hamilton syringe (Figure 21.10) into microtiter plate wells (Figure 21.11) containing predispensed quantities of antibody. When used for HLA testing, antisera in the wells are specific for known HLA antigenic specificities. After incubation of the cells and antisera, rabbit complement is added and the plates are again incubated. The extent of cytotoxicity induced is determined by incubating the cells with trypan blue, which enters dead cells and stains them blue, while leaving live cells unstained. The plates are read by using an inverted phase contrast microscope (Figure 21.12). A scoring system from 0 to 8 (where 8 implies >80% of target cells killed) is employed to indicate cytotoxicity. Most of the sera used to date are multispecific, as they are obtained from multiparous females who have been sensitized during pregnancy by HLA antigens determined by their spouse. Monoclonal antibodies are being used with increasing frequency in tissue typing. This technique is useful to identify HLA-A, HLA-B, and HLA-C antigens. When purified B cell preparations and specific antibodies against B cell antigens are employed, HLA-DR and HLA-DQ antigens can be identified.

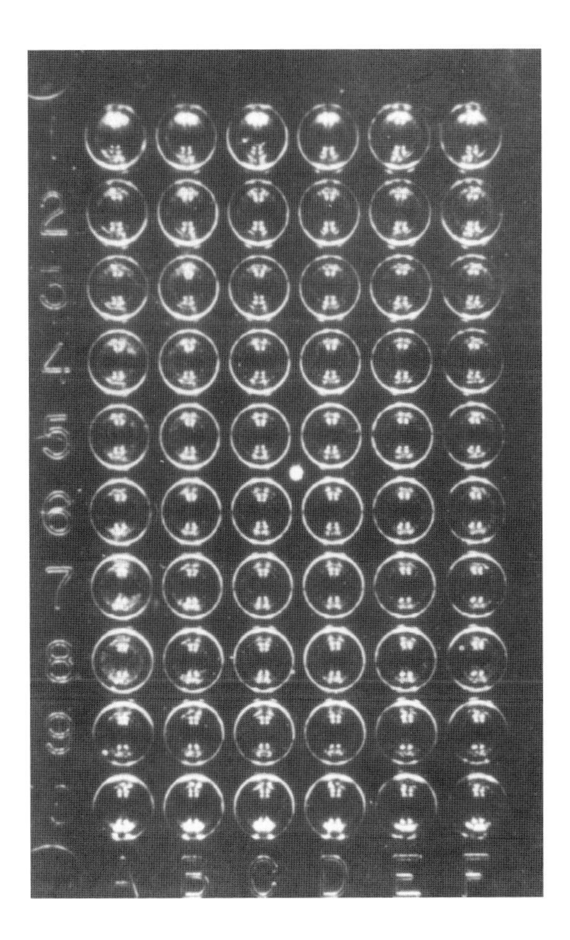

FIGURE 21.11 A Terasaki plate consisting of depressions in a plastic plate that contains predispensed antibodies to HLA antigens of various specificities and into which are placed patient lymphocytes and rabbit complement for tissue typing.

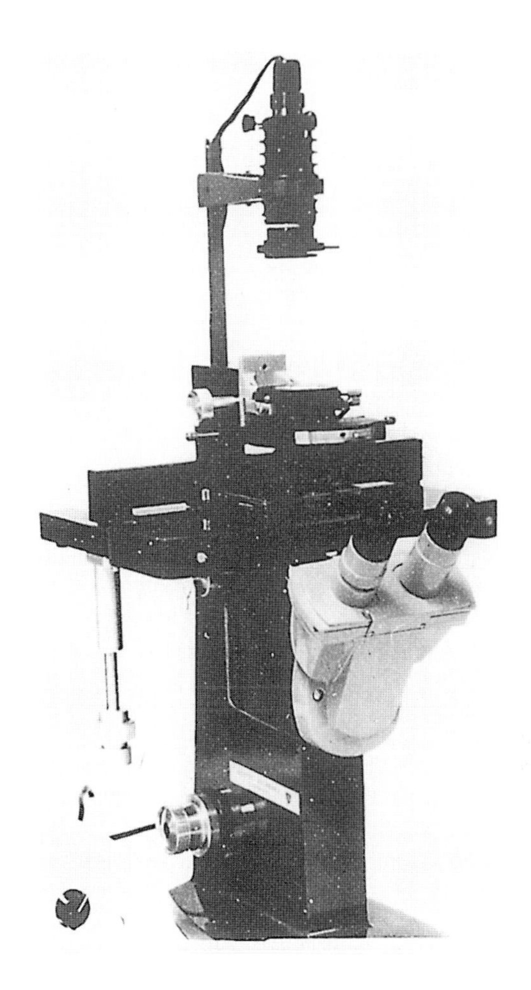

FIGURE 21.12 An inverted light microscope used to read Terasaki plates to determine tissue type.

A **cell tray panel** (Figure 21.13) is used to detect and identify HLA antibodies. Patient serum is tested against a panel of known cells. The panel (or percent) reactive antibody (PRA) is the percent of panel cells reacting with a patient's serum. It is expressed as a percentage of the total reactivity, i.e., %PRA = (No. of positive reactions/No. of cells in panel) × 100. This percentage is a useful indicator of the proportion of HLA antibodies of the patient.

Patients may have preformed antibodies against class I or II HLA antigens. If these patients receive organs that possess the corresponding antigens, they will likely experience hyperacute or delayed rejection for class I or class II incompatibilities, respectively. In order to detect such incompatibilities before transplantation, a **cross-matching procedure** is performed. The conventional cross-matching procedure (Figure 21.14) for organ transplants involves the combination of donor lymphocytes with recipient serum. There are three major variables in the standard cross-match procedure that predominantly affect the reactivity of the cell/sera sensitization. These include (1) incubation time and temperature; (2) wash steps after cell/sera sensitization; and (3) the use of additional

TRAY			CELL		PANEL TYPING							
POS	#CENTRL	TEST	ID	RACE	A		B		C		BW	
1A		8	10571T	H	1	2	8	35	7			6
1B		8	9891T	C	1	2	44	51	1	5	4	
1C		8	9884T	B	1	2	57	82	3	6	4	6
1D		8	9898T	B	1	23	45	49	6	7	4	6
1E		8	10356T	B	1	23	58	72	6		4	6
1F		8	10990T	O	1	24	27	37	2	6	4	
2F		8	10367T	C	1	32	8	51	7		4	6
2E		8	7109T	H	1		13	64	6	8	4	6
2D		1	6606T	C	2	11	18	38	7		4	6
2C		1	10567T	C	2	11	37	60	3	6	4	6
2B		8	10988T	C	2	24	51	55	3		4	6
2A		1	10359T	C	2	25	57	62	5	6	4	6
3A		1	10549T	O	2	26	39	61	1	7		6
3B		1	10361T	O	2	26	54	62	1	3		6
3C		1	10570T	O	2	26	60	65	4	8		6
3D		1	9899T	B	2	30	8	58	7		4	6
3E		1	10352T	O	2	30	13	46	1	6	4	6
3F		1	10547T	C	2	31	35	47	4		4	6
4F		1	6688T	C	2	31	50	60	3	6		6
4E		1	10568T	H	2	32	41	61	2	7		6

FIGURE 21.13 Cell tray panel showing positive reactions (8s) for HLA-A1 at tray positions 1A, 1B, 1C, 1D, 1E, 1F, 2F, and 2E, and a positive reaction for HLA-A24 at position 2B.

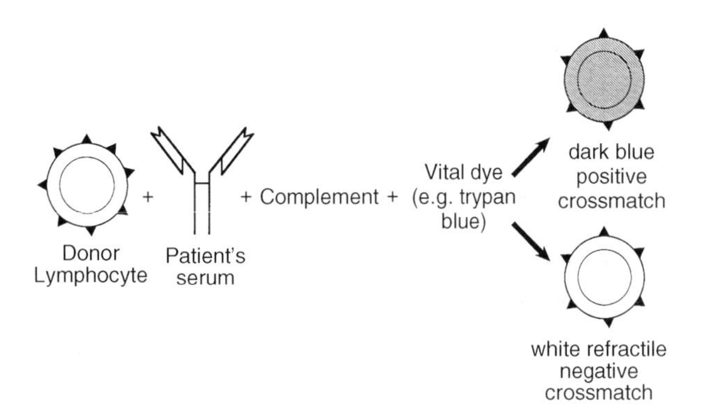

FIGURE 21.14 Crossmatching procedure.

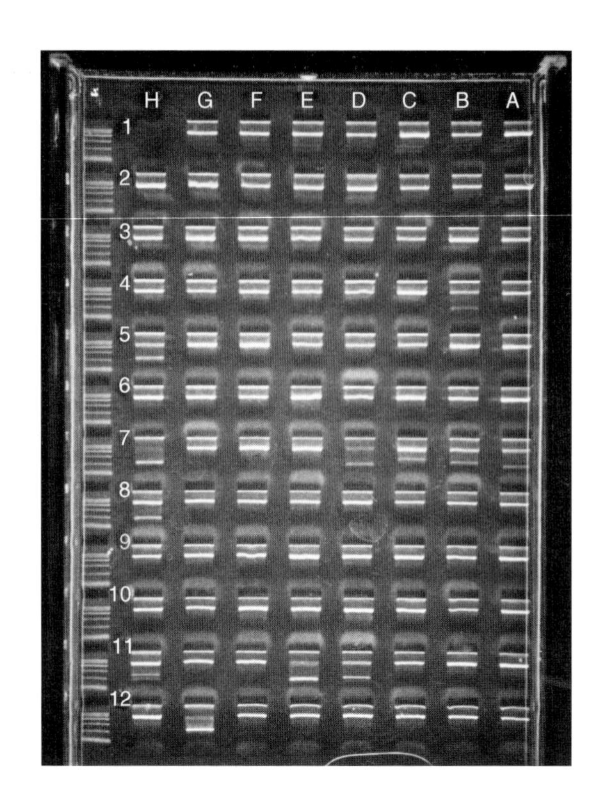

FIGURE 21.14a Example of high-resolution DRBI typing using sequence-specific primer methodology. Molecular weight ladder of known base pairs is in the far left column for base pair sizing.

reagents, such as antiglobulin in the test. Variations in these steps can cause wide variations in results. Lymphocytes can be separated into T and B cell categories for crossmatch procedures that are conducted at cold (4°C), room (25°C), and warm (37°C) temperatures. These permit the identification of warm anti-T cell antibodies that are almost always associated with graft rejection.

Molecular (DNA) typing: sequence-specific priming (SSP) is a method that employs a primer with a single mismatch in the 3′-end that cannot be employed efficiently to extend a DNA strand because the enzyme Taq polymerase, during the PCR reaction, and especially in the first PCR cycles which are very critical, does not manifest 3′-5′ proofreading endonuclease activity to remove the mismatched nucleotide. If primer pairs are designed to have perfectly matched 3′-ends with only a single allele, or a single group of alleles, and the PCR reaction is initiated under stringent conditions, a perfectly matched primer pair results in an amplification product, whereas a mismatch at the 3′-end primer pair will not provide any amplification product. A positive result, i.e., amplification, defines the specificity of the DNA sample. In this method, the PCR amplification step provides the basis for identifying polymorphism. The post-amplification processing of the sample consists only of a simple agarose gel electrophoresis to detect the presence or absence of amplified product. DNA amplified fragments are visualized by ethidium bromide staining and exposure to UV light. A separate technique detects amplified product by color fluorescence. The primer pairs are selected in such a manner that each allele should have a unique reactivity pattern with the panel of primer pairs employed. Appropriate controls must be maintained (Figure 21.14a).

CREGs are crossreactive groups. Public epitope-specific antibodies identify CREGs. Public refers to both similar (cross-reactive) and identical (public) epitopes shared by more than one HLA gene product.

CYNAP antibodies are cytotoxicity negative but absorption-positive antibodies that are concerned with HLA tissue typing. Most alloantibodies to public epitopes display CYNAP when tested in complement-dependent cytotoxicity assays. Most alloantisera contain public or CREG antibodies, but they act operationally as "private" antibodies because of their CYNAP phenomenon. For this reason, the relative insensitivity of standard CDC, due to CYNAP, has been useful for detecting discrete gene products. Standard CDC is not the recommended procedure to define HLA molecule binding specificities. The antiglobulin-augmented CDC (AHG-CDC) more accurately defines the true binding capabilities of alloantisera than do complement-independent assays by overriding the CYNAP phenomenon. CDC is the procedure of choice for HLA antigen detection and HLA antiserum analysis.

Haplotype designates those phenotypic characteristics encoded by closely linked genes on one chromosome inherited from one parent. It frequently describes several MHC alleles on a single chromosome. Selected haplotypes are in strong linkage disequilibrium between alleles of different loci. According to Mendelian genetics, 25% of siblings will share both haplotypes.

CYNAP phenomenon: See CYNAP antibodies.

Phenotype designates observable features of a cell or organism that are a consequence of interaction between the genotype and the environment. The phenotype represents those genetically encoded characteristics that are expressed. Phenotype may also refer to a group of organisms with the same physical appearance and the same detectable characteristics.

MHC haplotype refers to the set of genes in a haploid genome inherited from one parent. Children of parents designated ab and cd will probably be ac, ad, bc, or bd.

Trypan blue is a vital dye used to stain lymphoid cells, especially in the microlymphocytotoxicity test used for HLA tissue typing. Cell membranes whose integrity has been interrupted by antibody and complement permit the dye to enter and stain the cells dark blue. By contrast, the viable cells with an intact membrane exclude the dye and remain as bright circles of light in the microscope. Dead cells stain blue.

Trypan blue dye exclusion test is a test for viability of cells in culture. Living cells exclude trypan blue by active transport. When membranes have been interrupted, the dye enters the cells, staining them blue and indicating that the cell is dead. The method can be used to calculate the percent of cell lysis induced.

The **2-mercaptoethanol agglutination test** is a simple test to determine whether or not an agglutinating antibody is of the IgM class. If treatment of an antibody preparation, such as a serum sample, with 2-mercaptoethanol can abolish the serum's ability to produce agglutination of cells, then agglutination was due to IgM antibody. Agglutination induced by IgG antibody is unaffected by 2-mercaptoethanol treatment and just as effective after the treatment as it was before. Dithiothreitol (DTT) produces the same effect as 2-mercaptoethanol in this test.

Small "blues" are blue aggregates of acellular debris observed in clinical histocompatibility testing using the microlymphotoxicity test. It occurs in the wells of tissue typing trays and is due to an excess amount of trypan blue mixed with protein. This is a technical artifact.

Serologically defined (SD) antigens are mammalian cellular membrane epitopes that are encoded by MHC genes. Antibodies detect these epitopes.

Serological determinants are epitopes on cells that react with specific antibody and complement, leading to fatal injury of the cells. Serological determinants are to be distinguished from lymphocyte determinants, which are epitopes on the cell surface to which sensitized lymphocytes are directed, leading to cellular destruction. Although the end result is the same, antibodies and lymphocytes are directed to different epitopes on the cell surface.

In a **mixed-lymphocyte culture (MLC),** lymphocytes from two members of a species are combined in culture where they are maintained and incubated for 3 to 5 d. Lymphoblasts are formed as a consequence of histoincompatibility between the two individuals donating the lymphocytes. The lymphocyte antigens of these genetically dissimilar subjects each stimulate DNA synthesis by the other, which is measured by tritiated thymidine uptake that is assayed in a scintillation counter. See mixed-lymphocyte reaction (MLR).

MLC: Abbreviation for mixed-lymphocyte culture.

In the **mixed-lymphocyte reaction (MLR)**, lymphocytes from potential donor and recipient are combined in tissue culture. Each of these lymphoid cells has the ability to respond by proliferating following stimulation by antigens of the other cell. In the one-way reaction, the donor cells are treated with mitomycin or irradiation to render them incapable of proliferation. Thus, the donor antigens stimulate the untreated responder cells. Antigenic specificities of the stimulator cells that are not present in the responder cells lead to blastogenesis of the responder lymphocytes. This leads to an increase in the synthesis of DNA and cell division. This process is followed by introduction of a measured amount of tritiated thymidine, which is incorporated into the newly synthesized DNA. The mixed-lymphocyte reaction usually measures a proliferative response and not an effector cell killing response. The test is important in bone marrow and organ transplantation to evaluate the degree of histoincompatibility between donor and recipient. Both $CD4^+$ and $CD8^+$ T lymphocytes proliferate and secrete cytokines in the MLR. Also called mixed-lymphocyte culture.

Mixed leukocyte reaction (MLR): See mixed-lymphocyte reaction (MLR).

Homozygous describes containing two copies of the same allele.

The **homozygous typing cell (HTC) technique** is an assay that employs a stimulator cell that is homozygous at the HLA-D locus. An HTC incorporates only a minute amount of tritiated thymidine when combined with a homozygous cell in the MLR. This implies that the HTC shares HLA-D determinants with the other cell type. By contrast, when an HTC is combined with a nonhomozygous cell, much larger amounts of tritiated thymidine are incorporated. Many variations between these two extremes are noted in actual practice. Homozygous typing cells are frequently obtained from the progeny of marriages between cousins.

Homozygous typing cells (HTCs) are cells obtained from a subject who is homozygous at the HLA-D locus. HTCs facilitate MLR typing of the human D locus.

Lymphocyte determinants are target cell epitopes identified by lymphocytes rather than antibodies from a specifically immunized host.

Cross-match testing is an assay used in blood typing and histocompatibility testing to ascertain whether or not

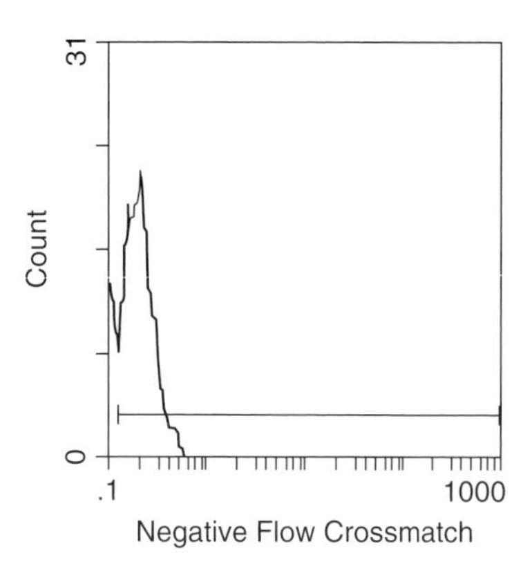

FIGURE 21.15 Negative flow crossmatch.

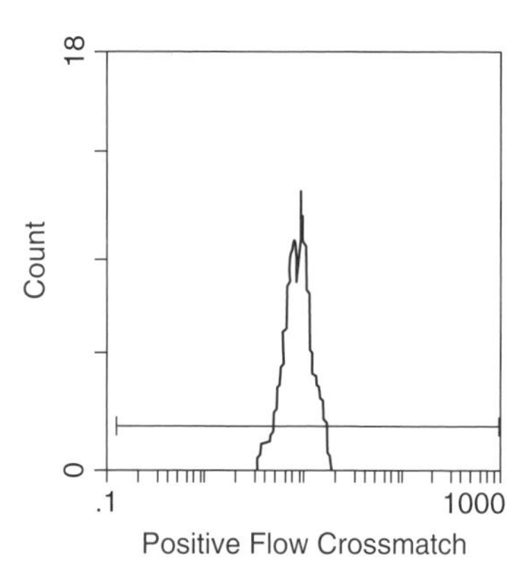

FIGURE 21.16 Positive flow crossmatch.

donor and recipient have antibodies against each other's cells that might lead to transfusion reaction or transplant rejection. Cross-matching reduces the changes of graft rejection by performed antibodies against donor cell surface antigens which are usually MHC antigens. Donor lymphocytes are mixed with recipient serum, complement is added, and the preparation observed for cell lysis.

Flow cytometry can also be used to perform the cross-matching procedure. This method is highly sensitive (considerably more sensitive than the direct cytotoxicity method). Flow cross-matching is also faster and can distinguish antibodies according to class (IgG vs. IgM) and target cell specificity (T cells from B cells). It is a valuable procedure in organ and bone marrow transplantation and is particularly suitable to measuring antibodies against HLA class I antigens on donor T cells. False positives are rare and most errors are due to low sensitivity (lower antibody concentration). Flow cross-matching has the potential to be standardized and automated. The flow cytometry cross-matching method commonly utilizes $F(ab')_2$ antihuman IgG conjugated to fluorescein, and anti-CD3 for T cells conjugated to phycoerythrin. A two parameter display of anti-CD3 vs. IgG is generated. A positive flow crossmatch is defined as median channel shift values >40 (Figure 21.15 and Figure 21.16).

Original Broad Specificities	Splits and Associated Antigens #
A2	A203#, A210#
A9	A23, A24, A2403#
A10	A25, A26, A34, A66
A19	A29, A30, A31, A32, A33, A74
A28	A68, A69
B5	B51, B52
B7	B703#
B12	B44, B45
B14	B64, B65
B15	B62,B63, B75, B76, B77
B16	B38, B39, B3901#, B3902#
B17	B57, B58
B21	B49, B50, B4005#
B22	B54, B55, B56
B40	B60, B61
B70	B71, B72
Cw3	Cw9, Cw10
DR1	DR103#
DR2	DR15, DR16
DR3	DR17, DR18
DR5	DR11, DR12
DR6	DR13, DR14, DR1403#, DR1404#
DQ1	DQ5, DQ6
DQ3	DQ7, DQ8, DQ9
Dw6	Dw18, Dw19
Dw7	Dw11, Dw17

FIGURE 21.17 Splits.

Splits are human leukocyte antigen (HLA) subtypes (Figure 21.17). For example, the base antigen HLA-B12 can be subdivided into the splits HLA-B44 and HLA-B45. The term "split" is used to designate an HLA antigen that was first believed to be a private antigen but later was shown to be a public antigen. The former designation can be placed in parenthesis following its new designation, i.e., HLA-B44(12).

A **private antigen** (Figure 21.18) is (1) an antigen confined to one MHC molecule; (2) an antigenic specificity restricted to a few individuals; (3) a tumor antigen restricted to a specific chemically induced tumor; (4) a low-frequency epitope present on red blood cells of fewer than 0.1% of the population, i.e., Pt^a, By, Bp^a, etc.; and (5) HLA antigen encoded by one allele such as HLA-B27.

A **public antigen (supratypic antigen)** is an epitope which several distinct or private antigens have in common (Figure 21.18). A public antigen such as a blood group antigen is one that is present in greater than 99.9% of a population. It is detected by the indirect antiglobulin (Coombs' test). Examples include Ve, Ge, Jr, Gy^a, and Ok^a. Antigens that occur frequently but are not public antigens include Mns, Lewis, Duffy, P, etc. In blood banking, there is a problem finding a suitable unit of blood for a transfusion to

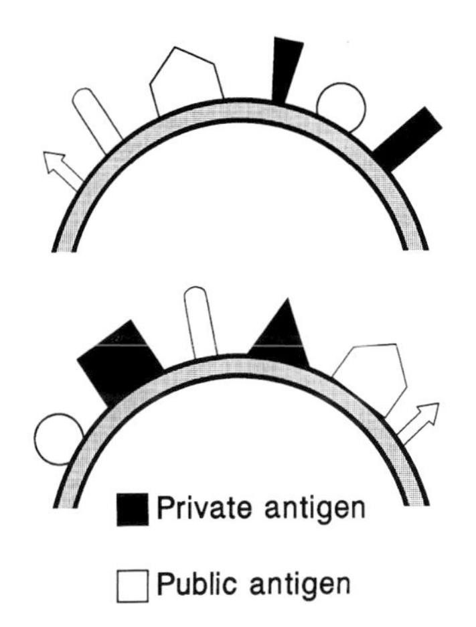

FIGURE 21.18 Public and private antigens.

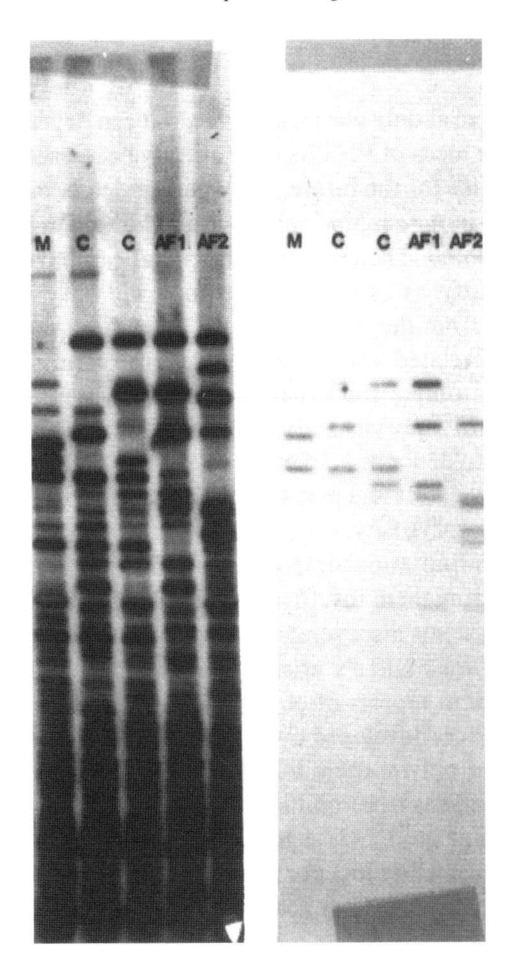

FIGURE 21.19 Multilocus probes.

recipients who have developed antibodies against public antigens.

Multilocus probes (Figure 21.19) are used to identify multiple related sequences distributed throughout each person's genome. Multilocus probes may reveal as many as 20 separate alleles. Because of this multiplicity of alleles, there is only a remote possibility that two unrelated persons would share the same pattern, i.e., about 1 in 30 billion. There is, however, a problem in deciphering the multibanded arrrangement of minisatellite RFLPs, as it is difficult to ascertain which bands are allelic. Mutation rates of minisatellite HVRs remain to be demonstrated, but are recognized occasionally. This method can be used in resolving cases of disputed parentage.

Single locus probes (SLPs) are probes which hybridize at only one locus. These probes identify a single locus of variable number of tandem repeats (VNTRs) and permit detection of a region of DNA repeats found in the genome only once and located at a unique site on a certain chromosome. Therefore, an individual can have only two alleles that SLPs will identify, as each cell of the body will have two copies of each chromosome, one from the mother and the other from the father. When the lengths of related alleles on homologous chromosomes are the same, there will be only a single band in the DNA typing pattern. Therefore, the use of an SLP may yield either a single- or double-band result from each individual. Single-locus markers such as the pYNH24 probe developed by White may detect loci that are highly polymorphic, exceeding 30 alleles and 95% heterozygosity. SLPs are used in resolving cases of disputed parentage.

Immediate spin crossmatch is a test for incompatibility between donor erythrocytes and the recipient patient's serum. This assay reveals ABO incompatibility in practically all cases, but is unable to identify IgG alloantibodies against erythrocyte antigens.

Orthotopic is an adjective that describes an organ or tissue transplant that has been in the site usually occupied by that organ or tissue.

An **orthotopic graft** (Figure 21.20) is an organ or tissue transplant that is placed in the location that is usually occupied by that particular organ or tissue.

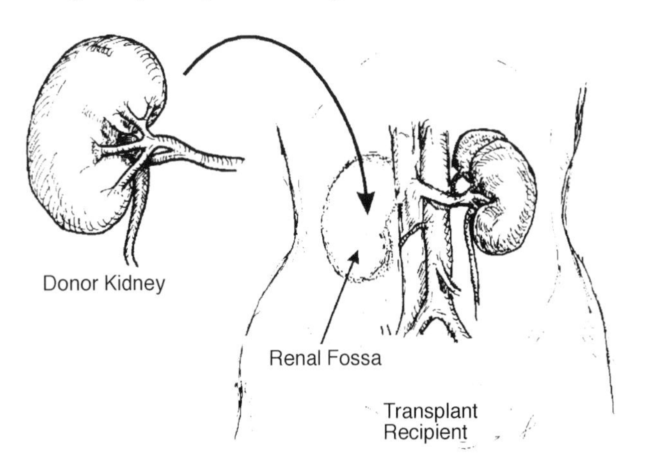

FIGURE 21.20 Orthotopic graft.

Heterotopic is an adjective that describes the placement of an organ or tissue graft in an anatomic site other than the one where it is normally located.

A **heterotopic graft** is a tissue or organ transplanted to an anatomic site other than the one where it is usually found under natural conditions. For example, the anastomosis of the renal vasculature at an anatomical site that would situate the kidney in a place other than the renal fossa where it is customarily found.

A **graft** is the transplantation of a tissue or organ from one site to another within the same individual or between individuals of the same or a different species.

Heart–lung transplantation is a procedure that has proven effective for the treatment of primary respiratory disease with dysfunction of gas exchange and alveolar mechanics, together with a secondary elevation in pulmonary vascular resistance, and in primary high-resistance circulatory disorder associated with pulmonary vascular disease.

A **rescue graft** is a replacement graft for an original graft that failed.

Privileged sites are anatomical locations in the body that are protected from immune effector mechanisms because of the absence of normal lymphatic drainage. Antigenic substances such as tissue allografts may be placed in these sites without evoking an immune response. Privileged sites include the anterior chamber of the eye, the cheek-pouch of the Syrian hamster, and the central nervous system. Tissue allografts in these locations enjoy a period of protection from immunologic rejection, as the diffusion of antigen from graft sites to lymphoid tissues is delayed. Immune privilege alters the induction of immunity to antigens first encountered via privileged sites and also inhibits the expression of certain forms of alloimmunity in these same sites.

Immunologically privileged sites are certain anatomical sites within the animal body that provide an immunologically privileged environment which favors the prolonged survival of alien grafts. The potential for development of a blood and lymphatic vascular supply connecting graft and host may be a determining factor in the qualification of an anatomical site as an area which provides an environment favorable to the prolonged survival of a foreign graft. Immunologically privileged areas include (1) the anterior chamber of the eye, (2) the substantia propria of the cornea, (3) the meninges of the brain, (4) the testis, and (5) the cheek pouch of the Syrian hamster. Foreign grafts implanted in these sites show a diminished ability to induce transplantation immunity in the host. These immunologically privileged sites usually fail to protect alien grafts from the immune rejection mechanism in hosts previously or simultaneously sensitized with donor tissues. The capacity of cells expressing Fas ligand to cause deletion of activated lymphocytes provides a possible explanation for the phenomenon of immune privilege. Animals with a deficiency in either Fas ligand or the Fas receptor fail to manifest significant immune privilege. Both epithelial cells of the eye and Sertoli cells of the testes express Fas ligand. Immune privilege is a consequence not only of the lack of an inflammatory response but also from immune consequences of the accumulation of apoptotic immune cells within a tissue. Immune cell apoptosis may be a signal to terminate inflammation. Apoptotic cell accumulation during an immune response could activate the development of cells that function to downregulate or suppress further immune activation.

Immune privilege: See immunologically privileged sites.

An **immunologic barrier** is an anatomical site that diminishes or protects against an immune response. This refers principally to immunologically privileged sites where grafts of tissue may survive for prolonged periods without undergoing immunologic rejection. This is based mainly on the lack of adequate lymphatic drainage in these areas. Examples include prolonged survival of foreign grafts in the brain.

A **semisyngeneic graft** is a graft that is ordinarily accepted from an individual of one strain into an F_1 hybrid of an individual of that strain mated with an individual of a different strain (Figure 21.21).

Graft facilitation is a prolonged graft survival attributable to conditioning of the recipient with IgG antibody, which is believed to act as a blocking factor. It also decreases cell-mediated immunity. This phenomenon is related to immunologic enhancement of tumors by antibody and has been referred to as immunological facilitation (*facilitation immunologique*).

Immunologic facilitation (*facilitation immunologique*) is the slightly prolonged survival of certain normal tissue allografts, e.g., skin, in mice conditioned with isoantiserum specific for the graft.

Immunologic enhancement is the prolonged survival, conversely the delayed rejection, of a tumor allograft in a host as a consequence of contact with specific antibody. Both the peripheral and central mechanisms have been postulated. In the past, coating of tumor cells with antibody was presumed to interfere with the ability of specifically reactive lymphocytes to destroy them, but today a central effect in suppressing cell-mediated immunity, perhaps through suppressor T lymphocytes, is also possible.

Enhancement is the prolonged survival, conversely the delayed rejection, of tumor or skin allografts in individuals

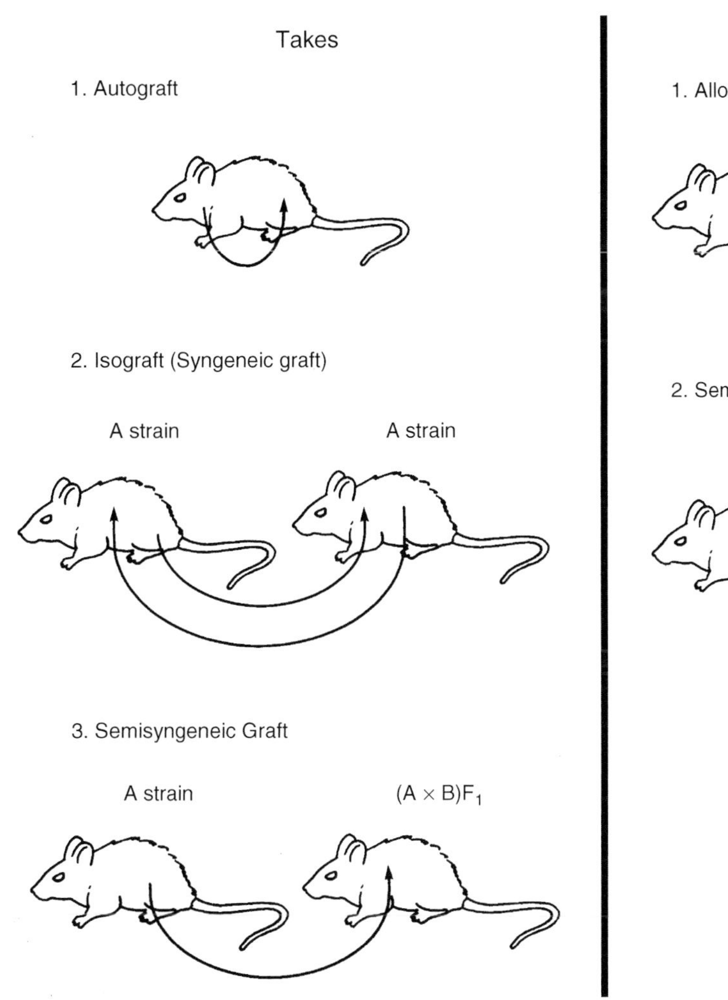

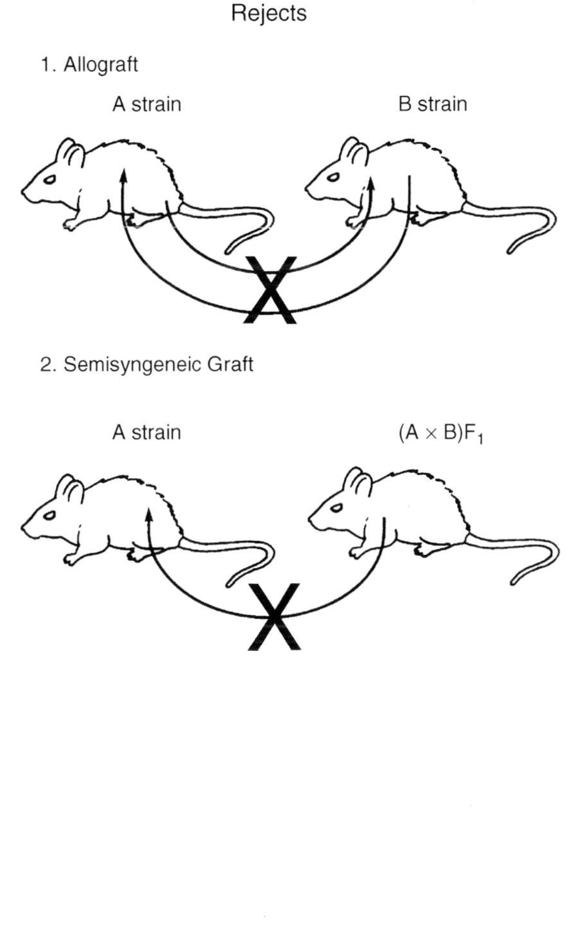

FIGURE 21.21 Semisyngeneic graft.

previously immunized or conditioned by passive injection of antibody specific for graft antigens. This is termed immunological enhancement and is believed to be due to a blocking effect by the antibody.

Enhancing antibodies are blocking antibodies that favor survival of tumor or normal tissue allografts.

An **allograft** is an organ, tissue, or cell transplant from one individual or strain to a genetically different individual or strain within the same species. Allografts are also called homografts (Figure 21.22).

An **allotransplant** refers to the transplantation of an organ or tissue from one individual to another member of the same species.

Fetus allograft: Success of the haplo-nonidentical fetus as an allograft was suggested in the 1950s by Medawar, Brent, and Billingham to rely on four possibilities. This proposal suggested that (1) the conceptus might not be immunogenic, (2) that pregnancy might alter the immune response, (3) that the uterus might be an immunologically privileged site, and (4) that the placenta might represent an effective immunological barrier between mother and fetus. Further studies have shown that transplantation privilege afforded the fetalplacental unit in pregnancy depends on intrauterine mechanisms. The pregnant uterus has been shown not to be an immunologically privileged site. Pregnancies usually are successful in maternal hosts with high levels of preexisting alloimmunity. The temporary status has focused on specialized features of fetal trophoblastic cells that facilitate transplantation protection. Fetal trophoblast protects itself from maternal cytotoxic attack by failing to express on placental villous cytotrophoblast and syncytiotrophoblast any classical polymorphic class I or II MHC antigens. Constitutive HLA expression is also not induced by known upregulators such as interferon y. Thus classical MHC antigens are not expressed throughout gestation. Extravillous cytotrophoblast cells selectively express HLA-G, a nonclassical class I MHC antigen which has limited genetic polymorphism. HLA-G might protect the cytotrophoblast population from MHC-nonrestricted

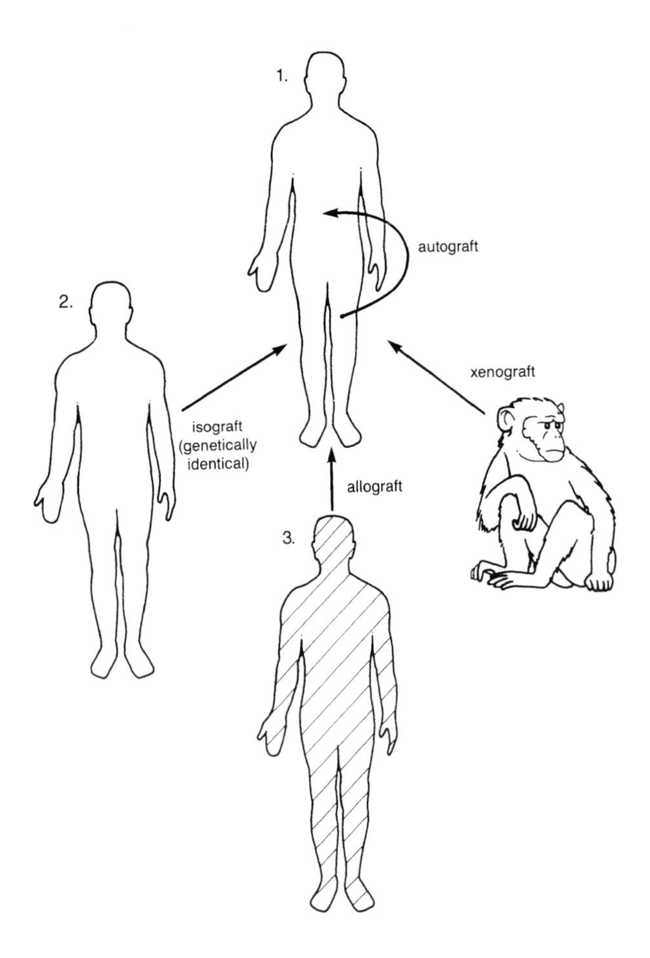

FIGURE 21.22 Types of grafts.

natural killer (NK) cell attack. The trophoblast also protects itself from maternal cytotoxicity during gestation by expressing a high level of complement regulatory proteins on its surface, such as membrane cofactor protein (MCP;CD46), decay accelerating factor (DAF;CD55), and membrane attack complex inhibitory factor (CD59). The maternal immune system recognizes pregnancy, i.e., the fetal trophoblast, in a manner that results in cellular, antibody, and cytokine responses that protect the fetal allograft. CD56 positive large granular lymphocytes may be regulated by hormones in the endometrium that control their function. They have been suggested to be a form of NK cell in arrested maturation possibly due to persistent expression of HLA-G on target invasive cytotrophoblast. Contemporary studies have addressed cytokine interactions at the fetal–maternal tissue interface in pregnancy. HLA-G or other fetal trophoblast antigens have been postulated to possibly stimulate maternal lymphocytes in endometrial tissue to synthesize cytokines and growth factors that act in a paracrine manner beneficial to trophoblast growth and differentiation. This has been called the immunotrophism hypothesis. Other cytokines released into decidual tissue include colony stimulating factors (CSFs), tumor necrosis factor α(TNF-α), IL-6, and transforming growth factor 3 (TGF-β). Fetal syncytiotrophoblast has numerous growth factor receptors. Thus an extensive cytokine network is preset within the uteroplacental tissue that offers both immunosuppressive and growth promoting signals. In humans, IgG is selectively transported across the placenta into the fetal circulation following combination with transporting Fcγ receptors on the placenta. This transfer takes place during the 20th to the 22nd week of gestation. Maternal HLA-specific alloantibody that is specific for the fetal HLA type is bound by nontrophoblastic cells expressing fetal HLA antigens. These include macrophages, fibroblasts, and endothelium within the villous mesenchyme of placental tissue, thereby preventing these antibodies from reaching the fetal circulation. Maternal antibodies against any other antigen of the fetus will likewise be bound within the placental tissues to a cell expressing that antigen. The placenta acts as a sponge to absorb potentially harmful antibodies. Exceptions to placental trapping of deleterious maternal IgG antibodies include maternal IgG antibodies against RhD antigen and certain maternal organ-specific autoantibodies.

An **allogeneic graft** is an allograft consisting of an organ, tissue, or cell transplant from a donor individual or strain to a genetically different individual or strain within the same species.

Homologous is an adjective that describes something from the same source. For example, an organ allotransplant from one member to a recipient member of the same species, i.e., renal allotransplantation in humans.

Allogeneic bone marrow transplantation: Hematopoietic cell transplants are performed in patients with hematologic malignancies, certain nonhematologic neoplasms, aplastic anemias, and certain immunodeficiency states. In allogeneic bone marrow transplantation the recipient is irradiated with lethal doses either to destroy malignant cells or to create a graft bed. The problems that arise include graft-vs.-host (GVH) disease and transplant rejection. GVH disease occurs when immunologically competent cells or their precursors are transplanted into immunologically crippled recipients. Acute GVH disease occurs within days to weeks after allogeneic bone marrow transplantation and primarily affects the immune system and epithelia of the skin, liver, and intestines. Rejection of allogeneic bone marrow transplants appears to be mediated by NK cells and T cells that survive in the irradiated host. NK cells react against allogeneic stem cells that are lacking self MHC Class I molecules and therefore fail to deliver the inhibitory signal to NK cells. Host T cells react against donor MHC antigens in a manner resembling their reaction against solid tissue grafts.

Hemopoietic resistance (HR): Transplantation of allogeneic, parental, or xenogeneic bone marrow or leukemia cells into animals exposed to total body irradiation often results in the destruction of the transplanted cells. The mechanism causing the failure of the transplant appears similar with all three types of cells. This phenomenon, designated hemopoietic resistance (HR), has a genetic basis and mechanism different from conventional transplantation reactions against solid tumor allografts. It does not require prior sensitization and apparently involves the cooperation between NK cells and macrophages, both resistant to irradiation. The NK cells have the characteristics of null cells; macrophages play an accessory cell role. The cooperative activity seems to represent *in vivo* surveillance against leukemogenesis.

Homologous chromosomes are a pair of chromosomes containing the same linear gene sequences, each derived from one parent.

Immunoisolation describes the enclosure of allogeneic tissues such as pancreatic islet cell allografts within a membrane that is semipermeable, but does not itself induce an immune response. Substances of relatively low mol wt can reach the graft through the membrane, while it remains protected from immunologic rejection by the host.

Allogeneic inhibition is the better growth of homozygous tumors when they are transplanted to homozygous syngeneic hosts of the strain of origin than when they are transplanted to F1 hybrids between the syngeneic (tumor) strain and an allogeneic strain. This is manifested as a higher frequency of tumor and shorter latency period in syngeneic hosts. The better growth of tumor in syngeneic than in heterozygous F1 hybrid hosts was initially termed syngeneic preference. When it became apparent that selective pressure against the cells in a mismatching environment produced the growth difference, the phenomenon was termed allogeneic inhibition.

Syngeneic preference is the better growth of neoplasms when they are transplanted to histocompatible recipients than when they are transplanted in histoincompatible recipients. See also allogeneic inhibition.

Incompatibility refers to dissimilarity between the antigens of a donor and recipient as in tissue allotransplantation or blood transfusions. The transplantation of a histoincompatible organ or the transfusion of incompatible blood into a recipient may induce an immune response against the antigens not shared by the recipient in injurious consequences.

Homograft is the earlier term for allograft, i.e., an organ or tissue graft from a donor to a recipient of the same species.

Homograft reaction is an immune reaction generated by a homograft (allograft) recipient against the graft alloantigens. Also called an allograft reaction.

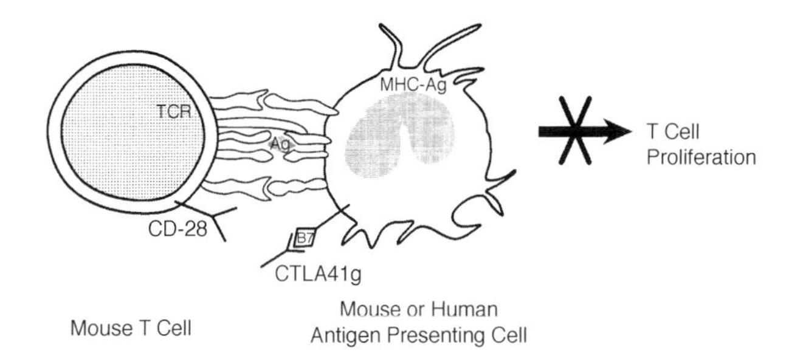

FIGURE 21.23 Induction of tolerance to a xenogenic tissue graft.

Homograft rejection is an earlier term for allograft rejection, i.e., an immune response induced by histocompatibility antigens in the donor graft that are not present in the recipient. This is principally a cell-mediated type of immune response.

Homotransplantation: Homograft, i.e., allograft transplantation.

Heterograft: See xenograft.

Heterogeneic: See xenogeneic.

In transplantation biology, **heterologous** refers to an organ or tissue transplant from one species to a recipient belonging to another species, i.e., a xenogeneic graft. It also refers to something from a foreign source.

A **xenograft** (Figure 21.23) is a tissue or organ graft from a member of one species, i.e., the donor, to a member of a different species, i.e., the recipient. It is also called a heterograft. Antibodies and cytotoxic T cells reject xenografts several days following transplantation.

Xenogeneic is an adjective that refers to tissues or organs transplanted from one species to a genetically different species, e.g., a baboon liver transplanted to a human.

Xenoantigen is an antigen of a xenograft. Also called heteroantigen.

Xenoantibody is an antibody specific for xenoantigen.

Xenoantibodies are antibodies formed in one species that are specific for antigens of a separate species.

Xenoreactive refers to a T cell or antibody response to an antigen of a graft derived from another species. The T lymphocyte may recognize an intact xenogenic MHC molecule or a peptide from a xenogeneic protein bound a self MHC molecule.

Xenotransplantation is organ or tissue transplantation between members of different species. An example of transplantation of tissues or organs from one species to another is a chimpanzee heart transplanted into a human recipient. It represents a possible substitute for the shortage of human organs for clinical transplantation. Xenogeneic transplantation can involve concordant or discordant donors, according to the phylogenetic distance between the species involved. Natural preformed antibodies in a recipient specific for donor endothelial antigens that lead to hyperacute rejection of most vascularized organ transplants now occur in discordant species combinations. The immune response to a xenotransplant resembles the response to an allotransplant. However, there are greater antigenic differences between donor and host in xenotransplantation than in allotransplantation. Previously termed heterotransplantation.

Xenozoonosis is a term that describes transmission of infection that might be the consequence of xenotransplantation. Infections resulting from xenotransplantation might involve infection of recipient cells with endogenous retroviral sequences from donor cells, giving rise themselves or after recombination with human endogenous retroviral sequences to previously unknown pathogenic viruses. Such new viruses might be pathogenic for other human beings in addition to the xenograft recipient.

Zoonosis is a term that describes the general process of cross-species infection.

Xenotype refers to molecular variations based on differences in structure and antigenic specificity. Examples would include membrane antigens of cells or immunoglobulins from separate species.

A **syngraft** is a transplant from one individual to another within the same strain. Syngrafts are also called isografts.

Syngeneic is an adjective that implies genetic identity between identical twins in humans or among members of an inbred strain of mice or other species. It is used principally to see transplants between genetically identical members of a species.

An **isograft** is a tissue transplant from a donor to an isogenic recipient. Grafts exchanged between members of an inbred strain of laboratory animals such as mice are syngeneic rather than isogenic.

Isogeneic (isogenic) is an adjective implying genetic identity such as identical twins. Although used as a synonym for syngeneic when referring to the genetic relationship between members of an inbred strain (of mice), the inbred animals never show the absolute identity, i.e., identical genotypes, observed in identical twins.

Isologous means derived from the same species. Also called isogeneic or syngeneic.

An antigen found in a member of a species that induces an immune response if injected into a genetically dissimilar member of the same species is termed an **isoantigen**. These are antigens carrying identical determinants in a given individual. Isoantigens of two individuals may or may not have identical determinants. In the latter case they are allogeneic with respect to each other and are called alloantigens.

Since the individual red blood cell antigens have the same molecular structure and are identical in different individuals, they have been referred to in the past as isoantigens. This is only a descriptive term and should not be used, because two individuals may be allogeneic by virtue of the assortment of the antigens present on their red blood cells. An isoantigen is an antigen of an isograft.

Isoantibody is an antibody that is specific for an antigen present in other members of the species in which it occurs. Thus, it is an antibody against an isoantigen. Also called alloantibody.

Isoleukoagglutinins are antibodies in the blood sera of multiparous females and of patients receiving multiple blood transfusions that recognizes surface isoantigens of leukocytes and leads to their agglutination.

Leukoagglutinin is an antibody or other substance that induces the aggregation or agglutination of white blood cells into clumps.

A **donor** is one who offers whole blood, blood products, bone marrow, or an organ to be given to another individual. Individuals who are drug addicts or test positively for certain diseases such as HIV-1 infection or hepatitis B, for example, are not suitable as donors. There are various other reasons for donor rejection not listed here. To be a blood donor, an individual must meet certain criteria which include blood pressure, temperature, hematocrit, pulse, and history. There are many reasons for donor rejection, including low hematocrit, skin lesions, surgery, drugs, or positive donor blood tests.

An **organ bank** is a site where selected tissues for transplantation, such as acellular bone fragments, corneas, and bone marrow, may be stored for relatively long periods until needed for transplantation. Several hospitals often share such a facility. Organs such as kidneys, liver, heart, lung, and pancreatic islets must be transplanted within 48 to 72 h and are not suitable for storage in an organ bank.

Organ brokerage, or the selling of an organ such as a kidney from a living related donor to the transplant, recipient, is practiced in certain parts of the world but is considered unethical and is illegal in the U.S., as it is in violation of the National Organ Transplant Act (Public Law 98-507,3 USC).

Adoptive immunity (Figure 21.24) is the term assigned by Billingham, Brent, and Medawar (1955) to transplantation immunity induced by the passive transfer of specifically immune lymph node cells from an actively immunized animal to a normal (previously nonimmune) syngeneic recipient host.

Adoptive immunization is the passive transfer of immunity by the injection of lymphoid cells from a specifically immune individual to a previously nonimmune recipient host. The resulting recipient is said to have adoptive immunity.

Adoptive transfer is a synonym for adoptive immunization. The passive transfer of lymphocytes from an immunized individual to a nonimmune subject with immune system cells such as $CD4^{+}$ T lymphocytes. Tumor-reactive T cells have been adoptively transferred for experimental cancer therapy.

Leukocyte transfer: See adoptive transfer.

Lymphocyte transfer reaction: See normal lymphocyte transfer reaction.

Normal lymphocyte transfer reaction: The intracutaneous injection of an individual with peripheral blood lymphocytes from a genetically dissimilar, allogeneic member of the same species leads to the development of a local,

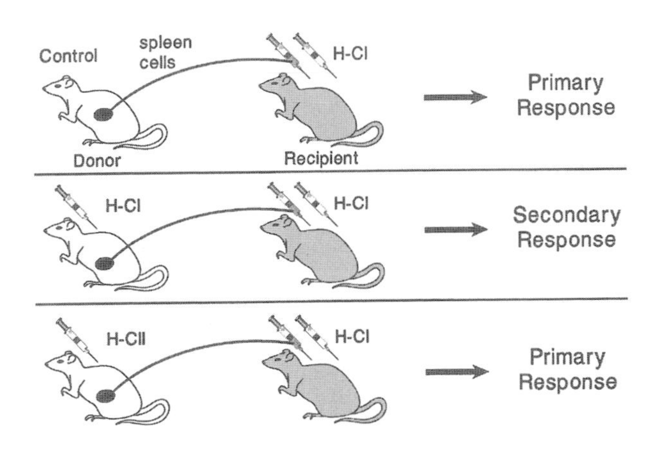

FIGURE 21.24 Adoptive immunity.

erythematous reaction that becomes most pronounced after 48 h. The size of the reaction has been claimed to give some qualitative indication of histocompatibility or histoincompatibility between a donor and recipient. This test is not used in clinical practice.

A **direct reaction** is a skin reaction caused by the intracutaneous injection of viable or nonviable lymphocytes into a host that has been sensitized against donor tissue antigens. This represents a type IV hypersensitivity reaction, which is classified as a delayed-type reaction mediated by T cells. Reactivity is against lymphocyte surface epitopes.

A **skin graft** uses skin from the same individual (autologous graft) or donor skin that is applied to areas of the body surface that have undergone third degree burns. A patient's keratinocytes may be cultured into confluent sheets that can be applied to the affected areas, although these may not "take" because of the absence of type IV collagen 7 S basement membrane sites for binding and fibrils to anchor the graft.

A **skin-specific histocompatibility antigen** is a murine skin minor histocompatibility antigen termed Sk that can elicit rejection of skin but not other tissues following transplantation from one parent into the other parent that has been irradiated and rendered a chimera by the previous injection of F_1 spleen cells. The two parents are from different inbred strains of mice. The rate of rejection is relatively slow. Immunologic tolerance of F_1 murine spleen cells to the skin epitope of the parent in which they are not in residence is abrogated following residence in the opposite parent.

A **split thickness graft** is a skin graft that is only 0.25 to 0.35 mm thick and consists of epidermis and a small layer of dermis. These grafts vascularize rapidly and last longer than do regular grafts. They are especially useful for skin burns, contaminated skin areas, and sites that are poorly vascularized. Thick split thickness grafts are further resistant to trauma, produce minimal contraction, and permit some amount of sensation, but graft survival is poor.

Pancreatic transplantation (Figure 21.25) is a treatment for diabetes. Either a whole pancreas or a large segment

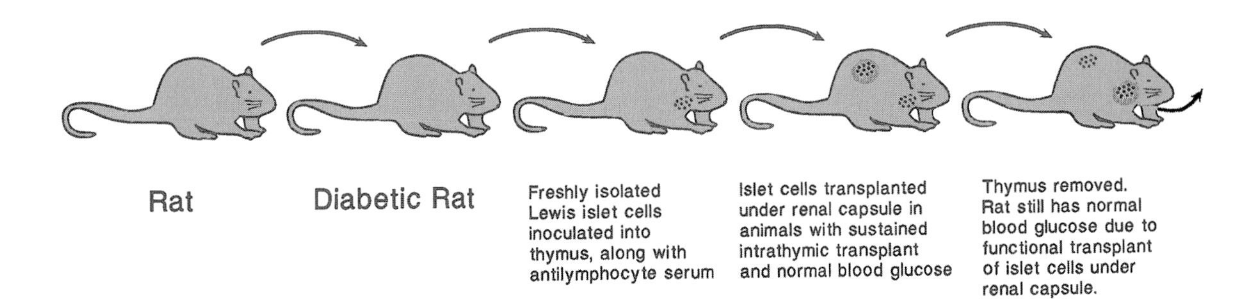

FIGURE 21.25 Protocol for pancreas transplant.

of it, obtained from cadavers, may be transplanted together with kidneys into the same diabetic patient. It is important for the patient to be clinically stable and for there to be as close a tissue (HLA antigen) match as possible. Graft survival is 50 to 80% at 1 year.

Islets of Langerhans are groups of endocrine cells within the exocrine pancreas that consist of α cells that secrete glucagon, β cells that secrete insulin, and δ cells that secrete somatostatin.

Islet cell transplantation is an experimental method aimed at treatment of type I diabetes mellitus. The technique has been successful in rats but less so in man. It requires sufficient functioning islets from a minimum of two cadaveric donors that have been purified, cultured, and shown to produce insulin. The islet cells are administered into the portal vein. The liver serves as the host organ in the recipient who is treated with FK506 or other immunosuppressant drugs.

Bone marrow is a soft tissue within bone cavities that contains hematopoietic precursor cells and hematopoietic cells that are maturing into erythrocytes, the five types of leukocytes, and thrombocytes. Whereas red marrow is hemopoietic and is present in developing bone, ribs, vertebrae, and long bones, some of the red marrow may be replaced by fat and become yellow marrow. Bone marrow cells are stem cells from which the formed elements of the blood, including erythrocytes, leukocytes, and platelets are derived. B lymphocyte and T lymphocyte precursors are abundant. The B lymphocytes and pluripotent stem cells in bone marrow are important for reconstitution of an irradiated host. Bone marrow transplants are useful in the treatment of aplastic anemia, leukemias, and immunodeficiencies. Patients may donate their own marrow for subsequent bone marrow autotransplantation if they are to receive intense doses of irradiation.

Bone marrow cells are stem cells from which the formed elements of the blood, including erythrocytes, leukocytes, and platelets are derived. B lymphocyte and T lymphocyte precursors are abundant. The B lymphocytes and pluripotent stem cells in bone marrow are important for reconstitution of an irradiated host. Bone marrow transplants are useful in the treatment of aplastic anemia, leukemias, and immunodeficiencies. Patients may donate their own marrow for subsequent bone marrow autotransplantation if they are to receive intense doses of irradiation.

Bone marrow transplantation is a procedure used to treat both nonneoplastic and neoplastic conditions not amenable to other forms of therapy. It has been especially used in cases of aplastic anemia, acute lymphocytic leukemia, and acute nonlymphocytic leukemia. A total of 750 ml of bone marrow are removed from the iliac crest of an HLA-matched donor. Following appropriate treatment of the marrow to remove bone spicules, the cell suspension is infused intravenously into an appropriately immunosuppressed recipient who has received whole-body irradiation and immunosuppressive drug therapy. GVH episodes, acute graft-vs.-host disease (GVHD), or chronic GVHD may follow bone marrow transplantation in selected subjects. The immunosuppressed patients are highly susceptible to opportunistic infections.

Autologous bone marrow transplantation (ABMT): Leukemia patients in relapse may donate marrow which can be stored and readministered to them following a relapse. Leukemic cells are removed from the bone marrow which is cryopreserved until needed. Prior to reinfusion of the bone marrow, the patient receives supralethal chemoradiotherapy. This mode of therapy has improved considerably the survival rate of some leukemia patients.

Immunotoxin: The linkage of an antibody specific for target cell antigens with a cytotoxic substance such as the toxin ricin yields an immunotoxin. Upon parenteral injection, its antibody portion directs the immunotoxin to the target and its toxic portion destroys target cells on contact. Among its uses is the purging of T cells from hematopoietic cell preparations used for bone marrow transplantation. Immunotoxin is a substance produced by the union of a monoclonal antibody or one of its fractions to a toxic molecule such as a radioisotope, a bacterial or plant toxin, or a chemotherapeutic agent. The antibody portion is intended to direct the molecule to antigens on a target cell, such as those of a malignant tumor, and the toxic portion of the molecule is for the purpose of destroying the target cell. Contemporary methods of recombinant DNA technology have permitted the preparation of specific hybrid molecules for use in immunotoxin therapy. Immunotoxins may have difficulty reaching the intended target tumor, may be quickly metabolized, and may stimulate the development of antiimmunotoxin antibodies. Crosslinking proteins may likewise be unstable. Immunotoxins have potential for antitumor therapy and as immunosuppressive agents.

Platelet-associated immunoglobulin (PAIgG) is present in 10% of normal individuals, 50% of those with tumors, and 76% of septic patients, and may be induced by GVHD. PAIgG is present in 71% of autologous marrow graft recipients and in 50% of allogeneic marrow graft recipients.

Autologous is an adjective that refers to derivation from self. The term describes grafts or antigens taken from an individual and returned to the same subject from which they were derived.

An **autograft** is a graft of tissue taken from one area of the body and placed in a different site on the body of the same individual, e.g., grafts of skin from unaffected areas to burned areas in the same individual.

Autologous graft refers to the donation of tissue such as skin or bone marrow by the same individual who will subsequently receive it either at a different anatomical site, as in skin autografts for burns, or at a later date, or as in autologous bone marrow transplants.

Bone marrow chimera: The inoculation of an irradiated recipient mouse with bone marrow from an unirradiated donor mouse which ensures that lymphocytes and other cellular elements of the blood will be of donor genetic origin. They have been useful in demonstrating lymphocyte and other blood cell development.

Stem cells have two unique biological features that include self-renewal and multilineage differentiation potential. In the past, stem cells were divided into two types that include the pluripotential stem cell and the committed stem cell. Pluripotential stem cells were the progenitors of many different hematopoietic cells, whereas the progeny of committed stem cells were of one cell type. "Committed stem cell" is now termed "progenitor cell." Stem cells arise from yolk sac blood islands and usually are noncycling. They are not morphologically recognizable. Cell culture studies have yielded much information about hematopoietic precursor cells. Hematopoietic stem cells express the progenitor cell antigen CD34, which can be detected using monoclonal antibodies and by flow cytometry.

A **hematopoietic stem cell** is a bone marrow cell that is undifferentiated and serves as a precursor for multiple cell lineages. These cells are also demonstrable in the yolk sac and later in the liver in the fetus.

Hematopoietic stem cell (HSC) transplants are used to reconstitute hematopoietic cell lineages and to treat neoplastic diseases. A total of 25% of allogeneic marrow transplants in 1995 were performed using hematopoietic stem cells obtained from unrelated donors. Since only 30% of patients requiring an allogeneic marrow transplant have a sibling that is HLA-genotypically identical, it became necessary to identify related or unrelated potential marrow donors. It became apparent that complete HLA compatibility between donor and recipient is not absolutely necessary to reconstitute patients immunologically. Transplantation of unrelated marrow is accompanied by an increased incidence of GVHD. Removal of mature T lymphocytes from marrow grafts decreases the severity of GVHD but often increases the incidence of graft failure and disease relapse. HLA-phenotypically identical marrow transplants among relatives are often successful. HSC transplantation provides a method to reconstitute hematopoietic cell lineages with normal cells capable of continuous self-renewal. The principal complications of HSC transplantation are GVHD, graft rejection, graft failure, prolonged immunodeficiency, toxicity from radiochemotherapy given pre- and posttransplantation, and GVHD prophylaxis. Methrotrexate and cyclosporin A are given to help prevent acute GVHD. Chronic GVHD may also be a serious complication involving the skin, gut, and liver and an associated sicca syndrome. Allogenic HSC transplantation often involves older individuals and unrelated donors. Thus, blood stem cell transplantation represents an effective method for the treatment of patients with hematologic and nonhematologic malignancies and various types of immunodeficiencies. The *in vitro* expansion of a small number of $CD34^+$ cells stimulated by various combinations of cytokines appears to give hematopoietic reconstitution when reinfused after a high-dose therapy. Recombinant human hematopoietic growth factors (HGF) (cytokines) may be given to counteract chemotherapy treatment–related myelotoxicity. HGF increase the number of circulating progenitor and stem cells, which is important for the support of high-dose therapy in autologous as well as allogeneic HSC transplantation.

A **chimera** (Figure 21.26) is the presence in an individual of cells of more than one genotype. This can occur rarely under natural circumstances in dizygotic twins, as in cattle, who share a placenta in which the blood circulation has become fused, causing the blood cells of each twin to circulate in the other. More commonly, it refers to humans or other animals who have received a bone marrow transplant that provides a cell population consisting of donor and self cells. Tetraparental chimeras can be produced by experimental manipulation. The name chimera derives from a monster of Greek mythology that had the body of a goat, the head of a lion, and the tail of a serpent.

Chimerism is the presence of two genetically different cell populations within an animal at the same time.

Hematopoietic chimerism: A successful bone marrow transplant leads to a state of hematological and/immunological

FIGURE 21.26 Chimera.

chimerism in which donor type blood cells coexist permanently with host type tissues, without manifesting alloreactivity to each other. Usually incomplete or mixed hematopoietic chimerism are generated following bone marrow transplantation in which both host type and donor type blood cells can be detected in the recipient. In bone marrow transplantation, not only is immune reactivity against donor type cells an obstacle to bone marrow engraftment, there is also the problem of GVHD-mediated by donor T cells reactive against host antigens. See chimera.

Radiation chimera: See irradiation chimera.

An **irradiation chimera** is an animal or human whose lymphoid and myeloid tissues have been destroyed by lethal irradiation and successfully repopulated with donor bone marrow cells that are genetically different.

Radiation bone marrow chimeras: Mice that have been subjected to heavy radiation and then reconstituted with allogeneic bone marrow cells, i.e., from a different mouse strain. Thus, the lymphocytes are genetically different from the surroundings in which they develop. These chimeric mice have yielded significant data in the investigation of lymphocyte development.

Backcross refers to a crossing of a heterozygous organism and a homozygote. The term commonly refers to the transfer of a particular gene from one background strain/stock to an inbred strain via multigenerational matings to the desired strain. Breeding an F_1 hybrid with either one of the strains that produced it.

Corneal transplants (Figure 21.27) are different from most other transplants in that the cornea is a "privileged site." These sites do not have a lymphatic drainage. The rejection rate in corneal transplants depends on vascularization; if vascularization occurs, the cornea becomes accessible to the immune system. HLA incompatibility increases the risk of rejection if the cornea becomes vascularized. The patient can be treated with topical steroids to cause local immunosuppression.

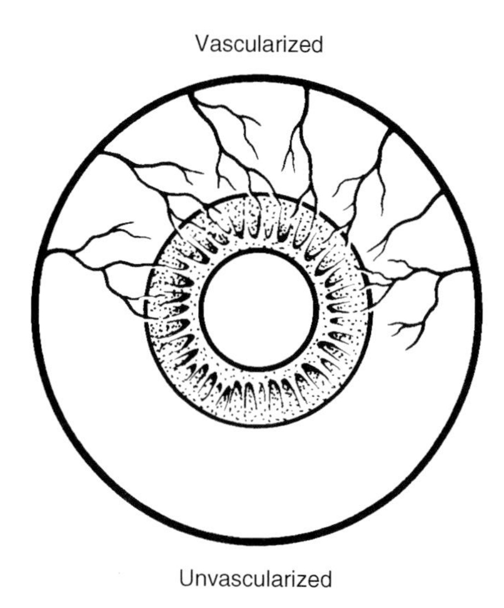

FIGURE 21.27 Corneal transplant.

Certain anatomical sites within the animal body provide an immunologically privileged environment which favors the prolonged survival of alien grafts. The potential for development of a blood and lymphatic vascular supply connecting graft and host may be a determining factor in the qualification of an anatomical site as an area which provides an environment favorable to the prolonged survival of a foreign graft. Immunologically privileged sites include (1) the anterior chamber of the eye, (2) the substantia propoa of the cornea, (3) the menges of the brain, (4) the testis, and (5) the cheek pouch of the Syrian hamster. Foreign grafts implanted in these sites show a diminished ability to induce transplantation immunity in the host. These immunologically privileged sites usually fail to protect alien grafts from the immune refection mechanism in hosts previously or simultaneously sensitized with donor tissues.

Leptin, the antiobesity hormone, is an endothelial cell mitogen and chemoattractant, and it induces angiogenesis in a cornea implant model. Endothelial cells express OB-Rβ, the leptin receptor.

The **allogeneic effect** (Figure 21.28) is the synthesis of antibody by B cells against a hapten in the absence of carrier-specific T cells, provided allogeneic T lymphocytes are present. Interaction of allogeneic T cells with the MHC class II molecules of B cells causes the activated T lymphocytes to produce factors that facilitate B-cell differentiation into plasma cells without the requirement for helper T lymphocytes. There is allogeneic activation of T cells in the GVH reaction.

Alloreactive is the recognition by antibodies or T lymphocytes from one member of a species cell or tissue antigens of a genetically nonidentical member.

Alloreactivity is the stimulation of immune system T cells by non self MHC molecules attributable to antigenic differences between members of the same species. It represents

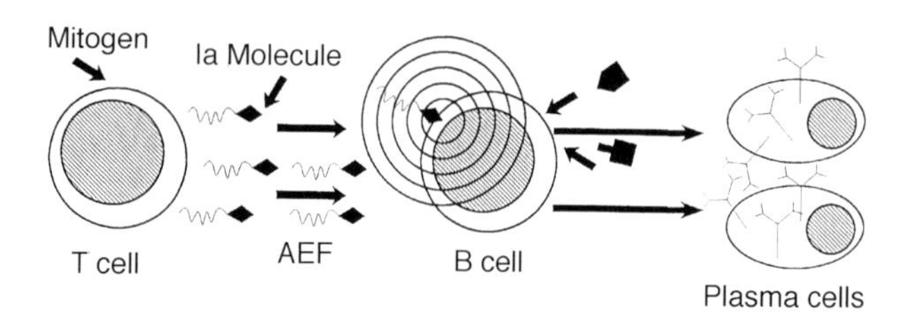

FIGURE 21.28 Allogeneic effect factor.

the immune response to an alloantigen based on recognition of allogeneic MHC.

Allogeneic disease includes the pathologic consequences of immune reactivity of bone marrow allotransplants in immunosuppressed recipient patients as a result of GVH reactivity in genetically dissimilar members of the same species.

Homologous disease: See allogeneic disease and graft-vs.-host disease (GVHD).

Alloimmunization is defined as an immune response provoked in one member or strain of a species with an alloantigen derived from a different member or strain of the same species. Examples include the immune response in man following transplantation of a solid organ graft such as a kidney or heart from one individual to another. Alloimmunization with red blood cell antigens in humans may lead to pathologic sequelae, such as hemolytic disease of the newborn (erythroblastosis fetalis) and in a third $Rh(D)^+$ baby born to an $Rh(D)^-$ mother.

Allogeneic (or **allogenic**) is an adjective that describes genetic variations or differences among members or strains of the same species. The term refers to organ or tissue grafts between genetically dissimilar humans or unrelated members of other species.

Alloantiserum is an antiserum generated in one member or strain of a species not possessing the alloantigen (e.g., histocompatibility antigen), with which they have been challenged, that is derived from another member or strain of the same species.

A **take** is the successful grafting of skin that adheres to the recipient graft site 3 to 5 d following application. This is accompanied by neovascularization as indicated by a pink appearance. Thin grafts are more likely to "take" than thicker grafts, but the thin graft must contain some dermis to be successful. The term "take" also refers to an organ allotransplant that has survived hyperacute and chronic rejection.

Engraftment is the phase during which transplanted bone marrow manufactures new blood cells.

Graft rejection (Figure 21.29) is an immunologic destruction of transplanted tissues or organs between two members or strains of a species differing at the MHC for that species (i.e., HLA in man and H-2 in the mouse). The rejection is based upon both cell-mediated and antibody-mediated immunity against cells of the graft by the histoincompatible recipient. First-set rejection usually occurs within 2 weeks after transplantation. The placement of a second graft with the same antigenic specificity as the first in the same host leads to rejection within 1 week and is termed second-set rejection. This demonstrates the presence of immunological memory learned from the first experience with the histocompatibility antigens of the graft. When the donor and recipient differ only at minor histocompatibility loci, rejection of the transplanted tissue may be delayed, depending upon the relative strength of the minor loci in which they differ. Grafts placed in a hyperimmune individual, such as those with preformed antibodies, may undergo hyperacute or accelerated rejection. Hyperacute rejection of a kidney allograft by preformed antibodies in the recipient is characterized by formation of fibrin plugs in the vasculature as a consequence of the antibodies reacting against endothelial cells lining vessels, complement fixation, polymorphonuclear neutrophil attraction, and denuding of the vessel wall, followed by platelet accumulation and fibrin plugging. As the blood supply to the organ is interrupted, the tissue undergoes infarction and must be removed.

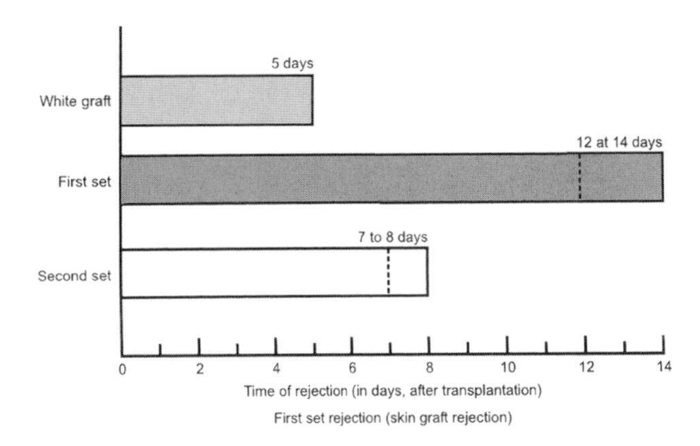

FIGURE 21.29 Types of skin graft rejection.

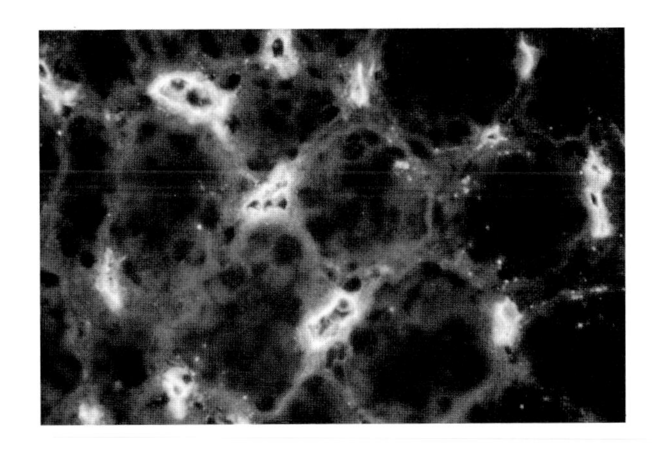

FIGURE 21.29a Immunofluorescent "staining" of C4d in peritubular capillaries.

Immunofluorescent "staining" of C4d in peritubular capillaries of renal allograft biopsies reveals a humoral component of rejection (Figure 21.29a).

First-set rejection is an acute form of allograft rejection in a nonsensitized recipient. It is usually completed in 12 to 14 d and is mediated by type IV (delayed-type) hypersensitivity to graft antigens.

Immunological rejection is the destruction of an allograft or even a xenograft in a recipient host whose immune system has been activated to respond to the foreign tissue antigens.

Rejection is an immune response to an organ allograft such as a kidney transplant. *Hyperacute rejection* is due to preformed antibodies and is apparent within minutes following transplantation. Antibodies reacting with endothelial cells cause complement to be fixed, which attracts polymorphonuclear neutrophils, resulting in denuding of the endothelial lining of the vascular walls. This causes platelets and fibrin plugs to block the blood flow to the transplanted organ, which becomes cyanotic and must be removed. Only a few drops of bloody urine are usually produced. Segmental thrombosis, necrosis, and fibrin thrombi form in the glomerular tufts. There is hemorrhage in the interstitium and mesangial cell swelling; IgG, IgM, and C3 may be deposited in arteriole walls. *Acute rejection* occurs within days to weeks following transplantation and is characterized by extensive cellular infiltration of the interstitium. These cells are largely mononuclear cells and include plasma cells, lymphocytes, immunoblasts, and macrophages, as well as some neutrophils. Tubules become separated, and the tubular epithelium undergoes necrosis. Endothelial cells are swollen and vacuolated. There is vascular edema, bleeding with inflammation, renal tubular necrosis, and sclerosed glomeruli. *Chronic rejection* occurs after more than 60 d following transplantation and may be characterized by structural changes such as interstitial fibrosis, sclerosed glomeruli, mesangial proliferative glomerulonephritis, crescent formation, and various other changes.

Second-set rejection is rejection of an organ or tissue graft by a host who is already immune to the histocompatibility antigens of the graft as a consequence of rejection of a previous transplant of the same antigenic specificity as the second, or as a consequence of immunization against antigens of the donor graft. The accelerated second-set rejection compared to rejection of a first graft is reminiscent of a classic secondary or booster immune response.

Second-set response is a term that describes the accelerated rejection of a second skin graft from a donor that is the same as or identical with the first donor. The accelerated rejection is seen when regrafting is performed within 12 to 80 d after rejection of the first graft. It is completed in 7 to 8 d and is due to sensitization of the recipient by the first graft.

Indirect antigen presentation: In organ or tissue transplantation, the mechanism whereby donor allogeneic MHC molecules present microbial proteins. The recipient professional antigen-presenting cells process allogeneic MHC proteins. The resulting allogeneic MHC peptides are presented, in association with recipient (self) MHC molecules, to host T lymphocytes. By contrast, recipient T cells recognize unprocessed allogeneic MHC molecules on the surface of the graft cells in direct antigen presentation.

White graft rejection is an accelerated rejection of a second skin graft performed within 7 to 12 d after rejection of the first graft. It is characterized by lack of vascularization of the graft and its conversion to a white eschar. The characteristic changes are seen by day 5 after the second grafting procedure. The transplanted tissue is rendered white because of hyperacute rejection, such as a skin or kidney allograft. Preformed antibodies occlude arteries following surgical anastomosis, producing infarction of the tissue graft.

ALG is an **a**bbreviation for antilymphocyte globulin.

ALS (antilymphocyte serum) or **ALG (antilymphocyte globulin):** See antilymphocyte serum.

Antilymphocyte serum (ALS) or antilymphocyte globulin (ALG) is an antiserum prepared by immunizing one species, such as a rabbit or horse, with lymphocytes or thymocytes from a different species, such as a human. Antibodies present in this antiserum combine with T cells and other lymphocytes in the circulation to induce immunosuppression. ALS is used in organ transplant recipients to suppress graft rejection (Figure 21.30). The globulin fraction known as ALG rather than whole antiserum produces the same immunosuppressive effect.

Antithymocyte globulin (ATG): IgG isolated from the blood serum of rabbits or horses hyperimmunized with human thymocytes is used in the treatment of aplastic anemia patients and to combat rejection in organ transplant recipients. The equine ATG contains 50mg/ml of

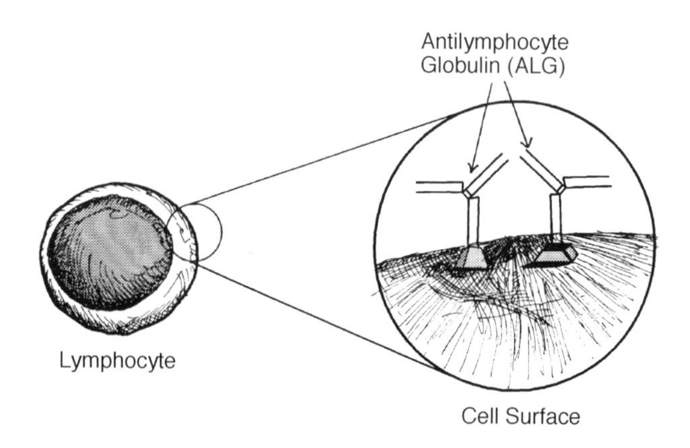

FIGURE 21.30 Antilymphocyte globulin (ALG).

immunoglobulin and has yielded 50% recovery of bone marrow and treated aplastic anemia patients.

ATG is an abbreviation for antithymocyte globulin.

Orthoclone OKT3 is a commercial antibody against the T cell surface marker CD3. It may be used therapeutically to diminish T cell reactivity in organ allotransplant recipients experiencing a rejection episode; OKT3 may act in concert with the complement system to induce T cells lysis, or may act as an opsonin, rendering T cells susceptible to phagocytosis. Rarely, recirculating T lymphocytes are removed in patients experiencing rejection crisis by thoracic duct drainage or extracorporeal irradiation of the blood. Plasma exchange is useful for temporary reduction in circulating antibody levels in selective diseases, such as hemolytic disease of the newborn, myasthenia gravis or Goodpasture's syndrome. Immunosuppressive drugs act on all of the T and B cell maturation processes.

Mouse immunoglobulin antibodies: A total of 40% of human subjects may harbor heteroantibodies that include human antimouse antibodies (HAMA). HAMA in serum may induce falsely elevated results in immunoassays that involve mouse antibodies. This may represent a problem in organ transplant patients who receive mouse monoclonal antibodies such as anti-CD3, anti-CD4, and anti-IL-2R for treatment.

Transplantation rejection (Figure 21.31) is the consequence of cellular and humoral immune responses to a transplanted organ or tissue that may lead to loss of function and necessitate removal of the transplanted organ or tissue. Transplantation rejection episodes occur in many transplant recipients, but are controlled by such immunosuppressive drugs as cyclosporine, rapamycin, or FK506, or by monoclonal antibodies against T lymphocytes.

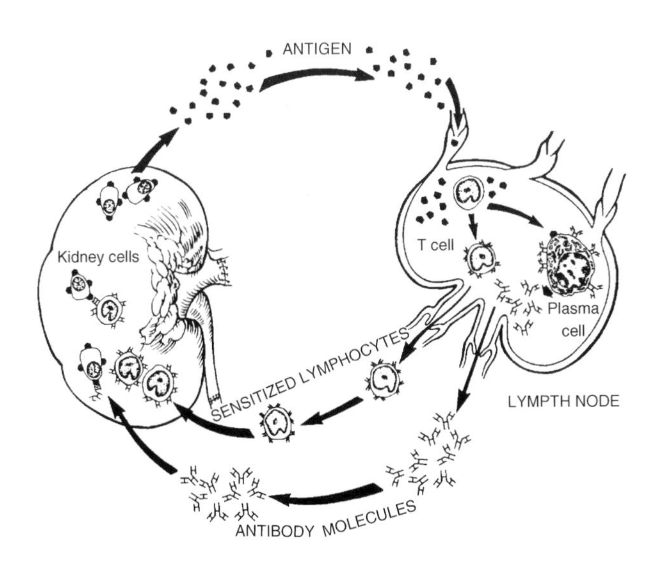

FIGURE 21.31 Rejection.

Host-vs.-graft disease (HVGD) is a consequences of humoral and cell-mediated immune response of a recipient host to donor graft antigens.

Hyperacute rejection (Figure 21.32 to Figure 21.37) is due to preformed antibodies and is apparent within minutes following transplantation. Antibodies reacting with endothelial cells cause complement to be fixed, which attracts polymorphonuclear neutrophils, resulting in denuding of the endothelial lining of the vascular walls. This causes platelets and fibrin plugs to clock the blood flow to the transplanted organ that becomes cyanotic and must be removed. Only a few drops of bloody urine are usually produced. Segmental thrombosis, necrosis, and fibrin thrombi form in the glomerular tufts. There is hemorrhage in the interstitium, mesangial cell swelling, IgG, and IgM, and C3 may be deposited in arteriole walls.

Hyperacute rejection is accelerated allograft rejection attributable to preformed antibodies in the circulation of the recipient that are specific for antigens of the donor.

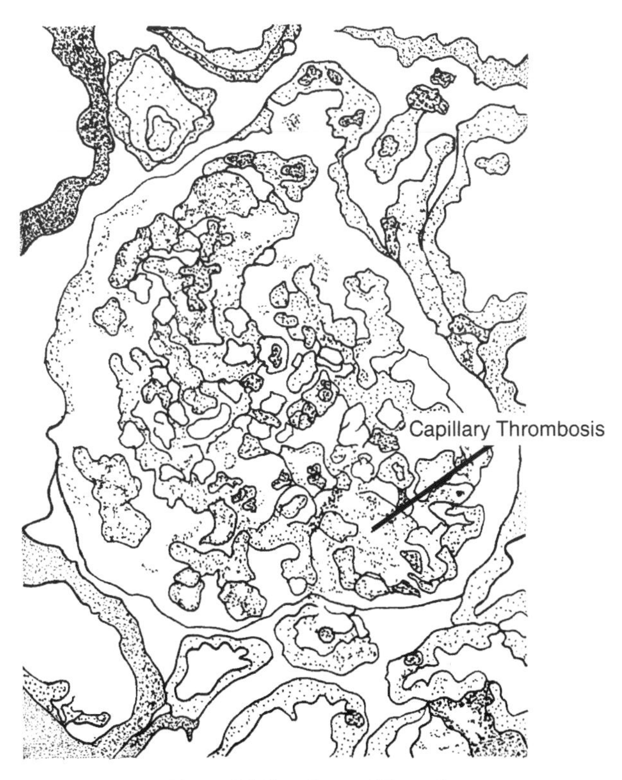

FIGURE 21.32 Schematic representation of hyperacute graft rejection.

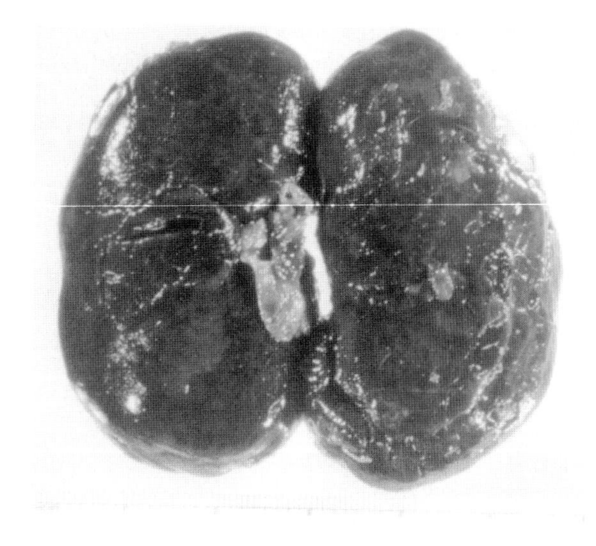

FIGURE 21.33 Hyperacute rejection of renal allotransplant showing swelling and purplish discoloration. This is a bivalved transplanted kidney. The allograft was removed within a few hours following transplantation.

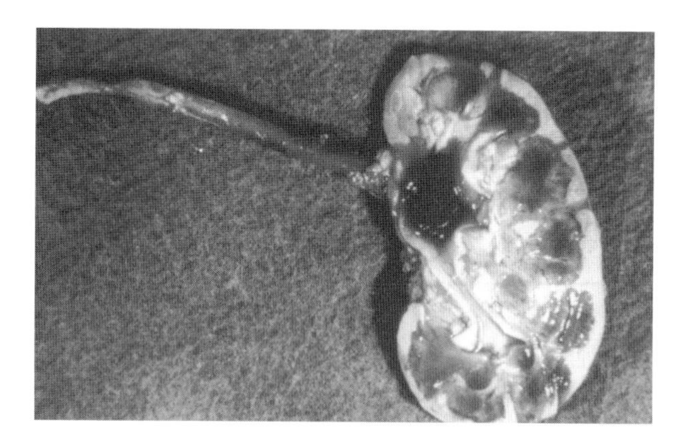

FIGURE 21.34 A bivalved transplanted kidney showing hyperacute rejection. There is extensive pale cortical necrosis. This kidney was removed 5 d after transplantation.

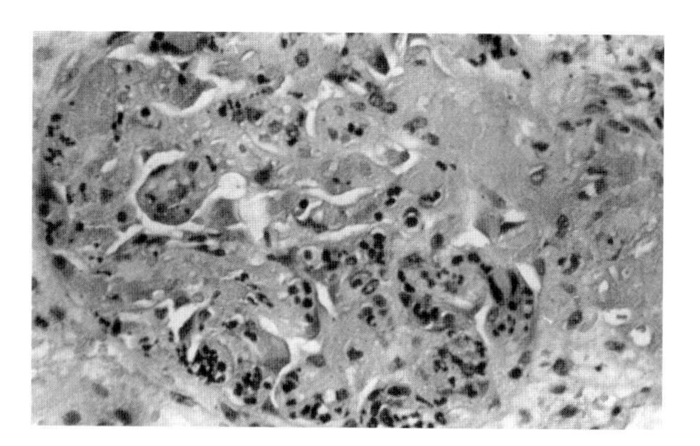

FIGURE 21.35 Microscopic view of hyperacute rejection showing a necrotic glomerulus infiltrated with numerous polymorphonuclear leukocytes. H&E stained section 25X.

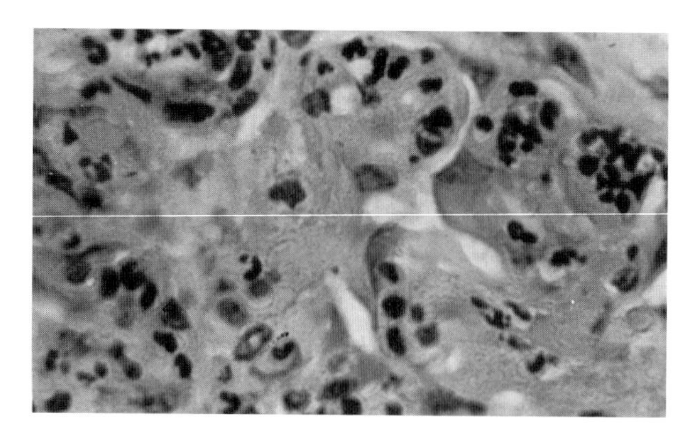

FIGURE 21.36 A high-power view of the same necrotic glomerulus shown in Figure 21.35. There are large numbers of polymorphonuclear leukocytes present. Extensive endothelial cell destruction is apparent. H&E stained section 50X.

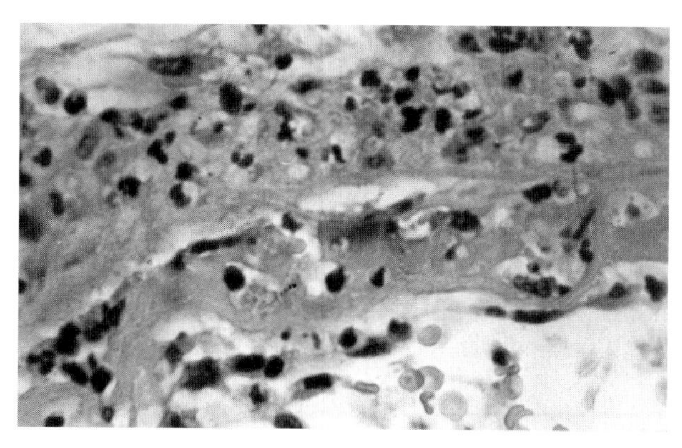

FIGURE 21.37 Microscopic view of hyperacute rejection showing necrosis of the wall of a small arteriole.

These antibodies react with antigens of endothelial cells lining capillaries of the donor organ. It sets in motion a process that culminates in fibrin plugging of the donor organ vessels, resulting in ischemia and loss of function and necessitating removal of the transplanted organ.

Acute rejection (Figure 21.38 to Figure 21.44) occurs within days to weeks following transplantation and is characterized by extensive cellular infiltration of the interstitium. These cells are largely mononuclear cells and include plasma cells, lymphocytes, immunoblasts, and macrophages as well as some neutrophils. Tubules become separated and the tubular epithelium undergoes necrosis. Endothelial cells are swollen and vaculoated. There is vascular edema, bleeding with inflammation, renal tubular necrosis, and sclerosed glomeruli.

Acute rejection is a type of graft rejection in which T lymphocytes, macrophages, and antibodies mediate vascular and tissue injury that may commence a week following transplantation. The response to the graft includes

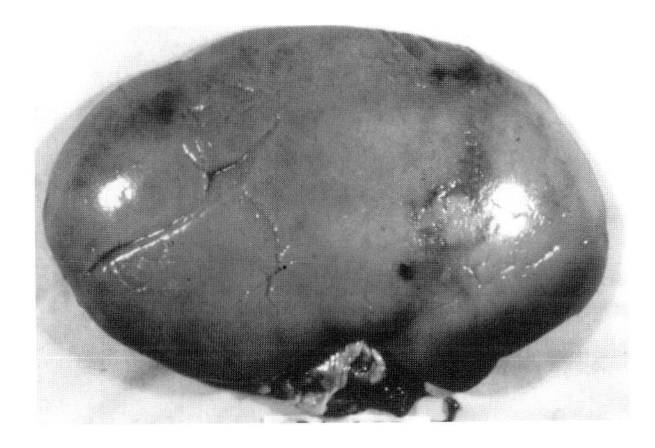

FIGURE 21.38 Acute rejection of a renal allograft in which the capsular surface shows several hemorrhagic areas. The kidney is tremendously swollen.

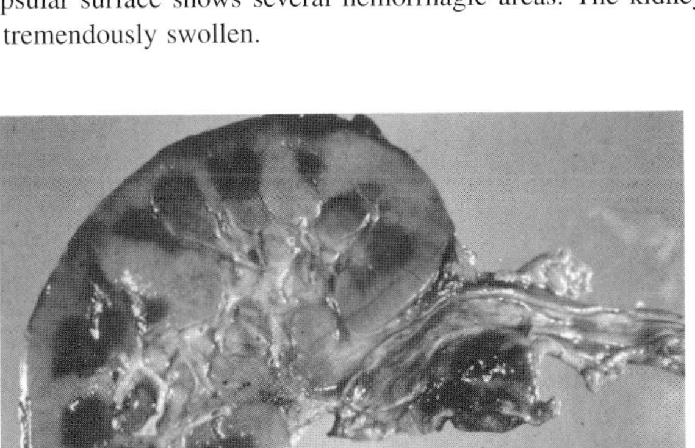

FIGURE 21.39 Acute rejection of a bivalved kidney. The cut surface bulges and is variably hemorrhagic and shows fatty degeneration of the cortex.

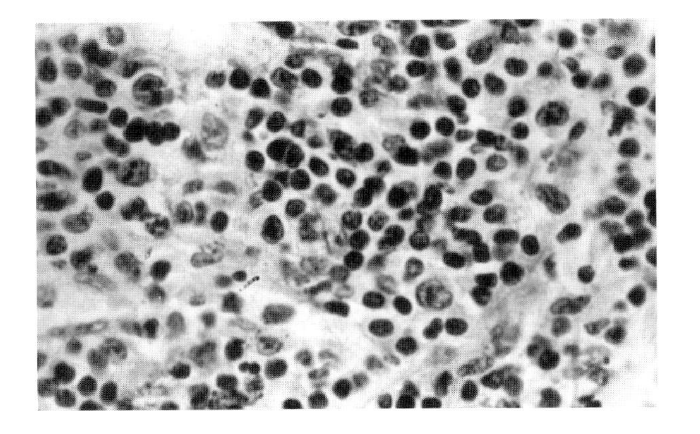

FIGURE 21.40 Microscopic view of the interstitium revealing predominantly cellular acute rejection. There is an infiltrate of variabily sized lymphocytes. There is also an infiltrate of eosinophils.

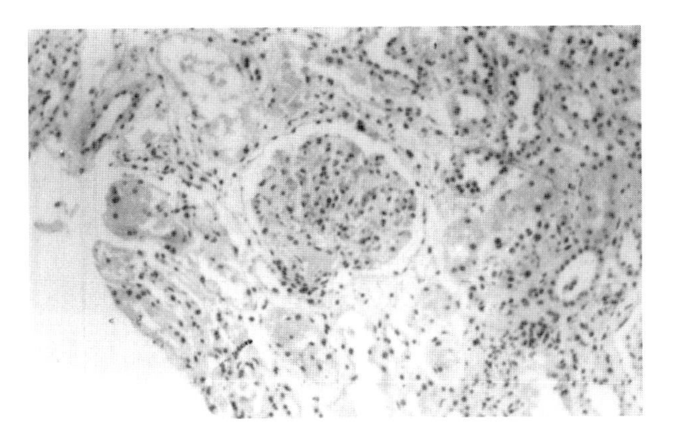

FIGURE 21.41 Microscopic view of acute rejection showing interstitial edema. Mild lymphocytic infiltrate. In the glomerulus, there is also evidence of rejection with a thrombus at the vascular pole.

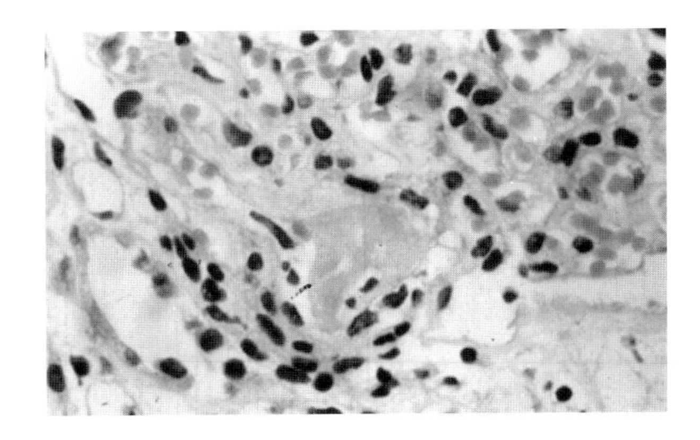

FIGURE 21.42 A higher magnification of the thrombus at the hilus of the glomerulus.

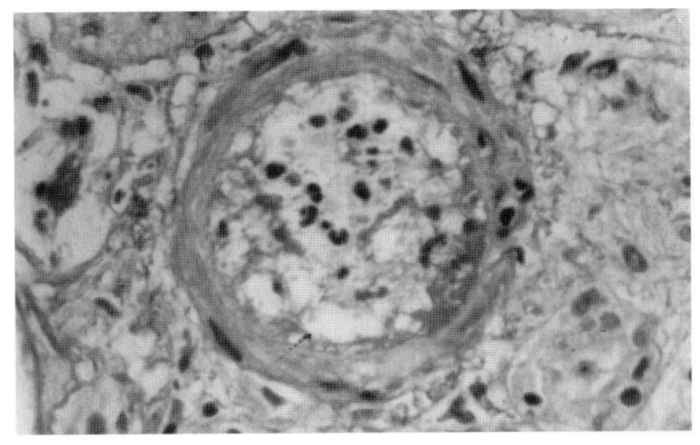

FIGURE 21.43 A trichrome stain of a small interlobular artery showing predominantly humoral rejection There is tremendous swelling of the intima and endothelium with some fibrin deposition and a few polymorphonuclear leukocytes.

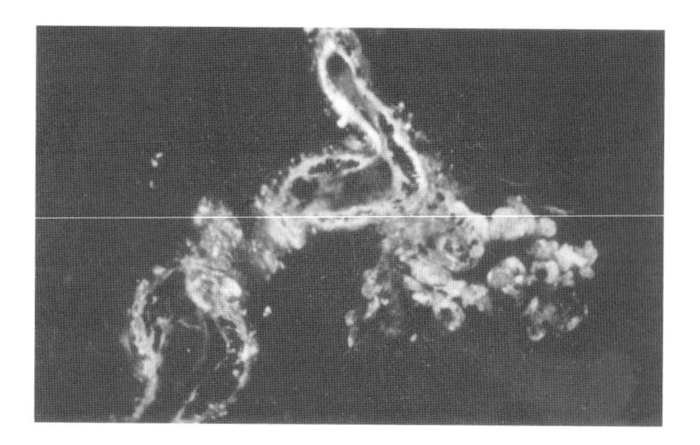

FIGURE 21.44 Immunofluorescence preparation showing humoral rejection with high-intensity fluorescence of arteriolar walls and of some glomerular capillary walls. This pattern is demonstrable in antiimmunoglobulin and anticomplement stained sections.

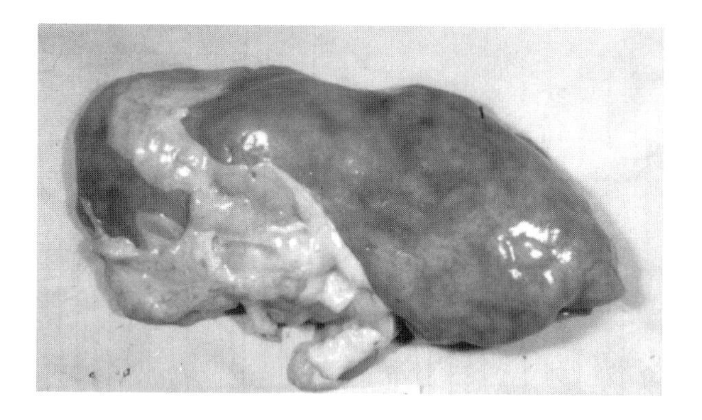

FIGURE 21.45 Renal allotransplant showing chronic rejection. The kidney is shrunken and malformed.

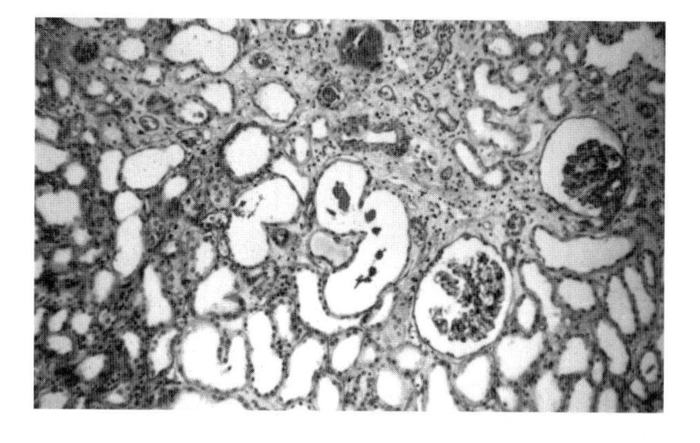

FIGURE 21.46 Microscopic view of chronic rejection showing tubular epithelial atrophy with interstitial fibrosis and shrinkage of glomerular capillary tufts.

formation of antibodies and activation of effector T lymphocytes that mediate the process.

Chronic rejection (Figure 21.45 to Figure 21.47) occurs after more than 60 d following transplantation and may

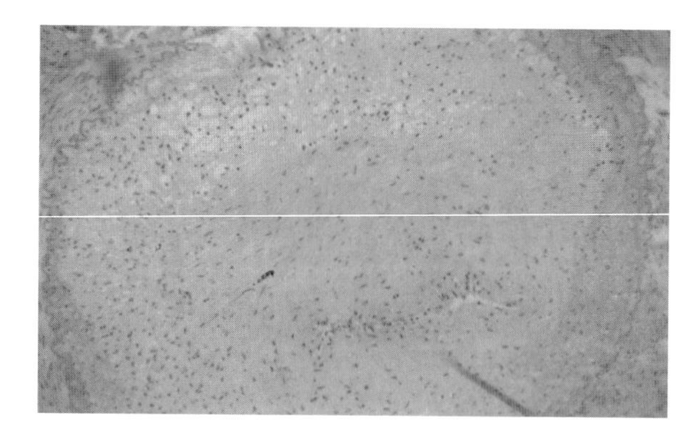

FIGURE 21.47 The wall of an artery in chronic rejection. There is obliteration of the vascular lumen with fibrous tissue. Only a slit-like lumen remains.

be characterized by structural changes such as interstitial fibrosis, sclerosed glomeruli, mesangial proliferative glomerulonephritis, crescent formation, and various other changes.

Chronic rejection is a type of allograft rejection that occurs during a prolonged period following transplantation and is characterized by structural changes such as fibrosis with loss of normal organ architecture. The principal pathologic change is occlusion of arteries linking the graft to the host. This results from intimal smooth muscle cell proliferation and has been referred to as graft arteriosclerosis.

Graft arteriosclerosis is characterized by intimal smooth muscle cell proliferation that occludes graft arteries. It may occur 6 to 12 months following transplantation and leads to chronic rejection of vascularized organ grafts. It is probably attributable to a chronic immune response to alloantigens of the vessel wall. It is also termed accelerated arteriosclerosis.

The **graft-vs.-host reaction (GVHR)** is the reaction of a graft containing immunocompetent cells against the genetically dissimilar tissues of an immunosuppressed recipient. Criteria requisite for a GVHR include (1) histoincompatibility between the donor and recipient, (2) passively transferred immunologically reactive cells, and (3) a recipient host who has been either naturally immunosuppressed because of immaturity or genetic defect, or deliberately immunosuppresed by irradiation or drugs. The immunocompetent grafted cells are especially reactive against rapidly dividing cells. Target organs include the skin, gastrointestinal tract (including the gastric mucosa), and liver, as well as the lymphoid tissues. Patients often develop skin rashes and hepatosplenomegaly and may have aplasia of the bone marrow. GVHR usually develops within 7 to 30 d following the transplant or infusion of the lymphocytes. Prevention of the GVHR is an important procedural step in several forms of transplantation and may be accomplished by irradiating the transplant. The clinical course of GVHR may take a hyperacute, acute, or chronic form as seen in graft rejection.

GVH: See graft-vs.-host reaction and graft-vs.-host disease.

Secondary disease is a condition that occurs in irradiated animals whose cell population has been reconstituted with histoincompatible, immunologically competent cells derived from allogeneic donor animals. Ionizing radiation induces immunosuppression in the recipients, rendering them incapable of rejecting the foreign cells. Thus, the recipient has two cell populations, its own and the one that has been introduced, making these animals radiation chimeras. After an initial period of recovery, the animals develop a secondary runt disease, which is usually fatal within 1 month.

Posttransfusion graft-vs.-host disease is a condition that resembles postoperative erythroderma that occurs in immunocompetent recipients of blood. There is dermatitis, fever, marked diarrhea, pancytopenia, and liver dysfunction.

Graft-vs.-host disease (GVHD) is a disease produced by the reaction of immunocompetent T lymphocytes of the donor graft that are histoincompatible with the tissues of the recipient into which they have been transplanted. For the disease to occur, the recipient must be either immunologically immature, immunosuppressed by irradiation or drug therapy, or tolerant to the administered cells, and the grafted cells must also be immunocompetent. Patients develop skin rash, fever, diarrhea, weight loss, hepatosplenomegaly, and aplasia of the bone marrow. The donor lymphocytes infiltrate the skin, gastrointestinal tract, and liver. The disease may be either acute or chronic. Murine GVH disease is called "runt disease," "secondary disease," or "wasting disease." Both allo- and autoimmunity associated with GVHD may follow bone marrow transplantation. A total of 20 to 50% of patients receiving HLA-identical bone marrow transplants still manifest GVHD with associated weight loss, skin rash, fever, diarrhea, liver disease, and immunodeficiency. GVHD may be either acute, which is an alloimmune disease, or chronic, which consists of both allo- and autoimmune components. The conditions requisite for the GVH reaction include genetic differences between immunocompetent cells in the marrow graft and host tissues, immunoincompetence of the host, and alloimmune differences that promote proliferation of donor cells that react with host tissues. In addition to allogenic marrow grafts, the transfusion of unirradiated blood products to an immunosuppressed patient or intrauterine transfusion from mother to fetus may lead to GVHD.

GVH disease: See graft-vs.-host disease.

Toxic epidermal necrolysis is a hypersensitivity reaction to certain drugs such as allopurinol, nonsteroidal antiinflammatory drugs, barbiturates, sulfonamides such as sulfmethoxazole-trimethoprim, carbamazepine, and other agents. It may closely resemble erythema multiforme. Patients develop erythema, subepidermal bullae, and open epidermal lesions. They become dehydrated, show imbalance of electrolytes, and often develop abscesses with sepsis and shock. Toxic epidermal necrolysis may also be observed in a hyperacute type of graft-vs.-host reaction, especially in some babies receiving bone marrow transplants.

Graft-vs.-leukemia (GVL): Bone marrow transplantation as therapy for leukemia. Partial genetic incompatibility between donor and recipient is believed to facilitate elimination of residual leukemia cells by T lymphocytes from the transplant.

Parabiotic intoxication is the result of a surgical union of allogeneic adult animals. The course of immune reactivity can be modified to take a single direction by uniting parental and F_1 animals. A hybrid recognizes parental cells as its own and does not mount an immune response against them, but alloantigens of F_1 hybrid cells stimulate the parental cells leading to graft-vs.-host disease.

The **Simonsen phenomenon** is a graft-vs.-host reaction in chick embryos that have developed splenomegaly following inoculation of immunologically competent lymphoid cells from adult chickens. Splenic lymphocytes are increased and represent a mixture of both donor and host lymphocytes.

Acute graft-vs.-host reaction, the immunopathogenesis of acute GVHD, consists of recognition, recruitment, and effector phases. Epithelia of the skin (Figure 21.48 to Figure 21.52), gastrointestinal tract (Figure 21.53 to Figure 21.57), small intrahepatic biliary ducts, and liver (Figure 21.58 to Figure 21.60), and the lymphoid system constitute primary targets of acute GVHD. GVHD development may differ in severity based on relative antigenic differences between donor and host and the reactivity of donor lymphocytes against non-HLA antigens of recipient tissues. The incidence and severity of GVHD has been

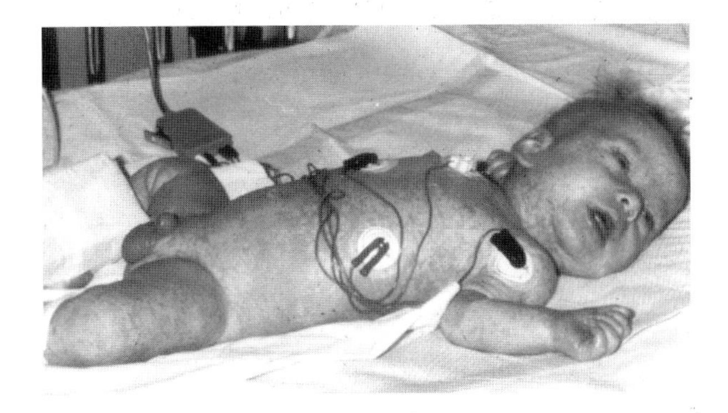

FIGURE 21.48 A diffuse erythematous to morbilliform rash in a child with acute graft-vs.-host disease (GVHD).

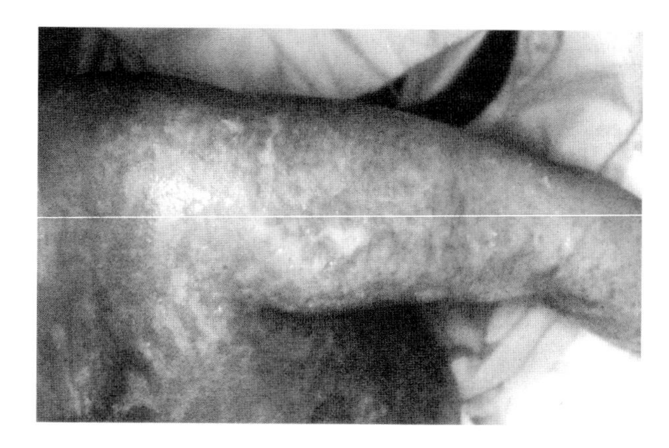

FIGURE 21.49 Diffuse erythematous skin rash in a patient with acute graft-vs.-host reaction (GVHR).

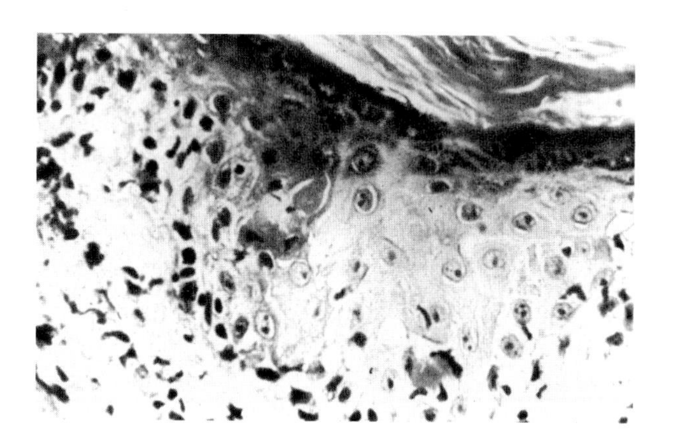

FIGURE 21.50 Histologically, there is an intense interface dermatitis with destruction of basal cells, particularly at the tips of the rete ridges, incontinence of melanin pigment, and necrosis of individual epithelial cells, referred to as apoptosis.

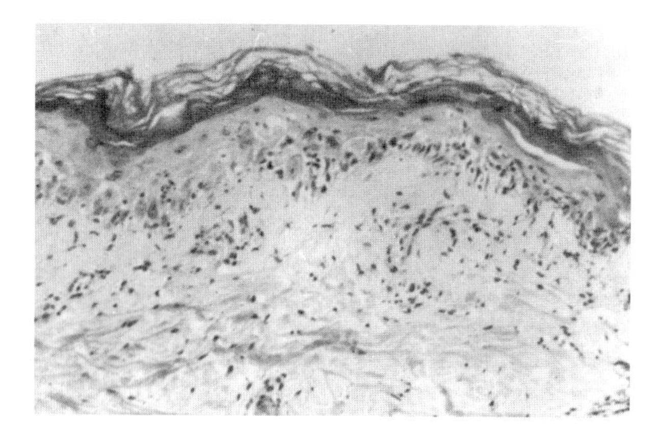

FIGURE 21.51 Histological appearance of the skin in graft-vs.-host disease with disruption of the basal cell layer, hyperkeratosis, and beginning sclerotic change.

FIGURE 21.52 Papulosquamous rash in graft-vs.-host disease.

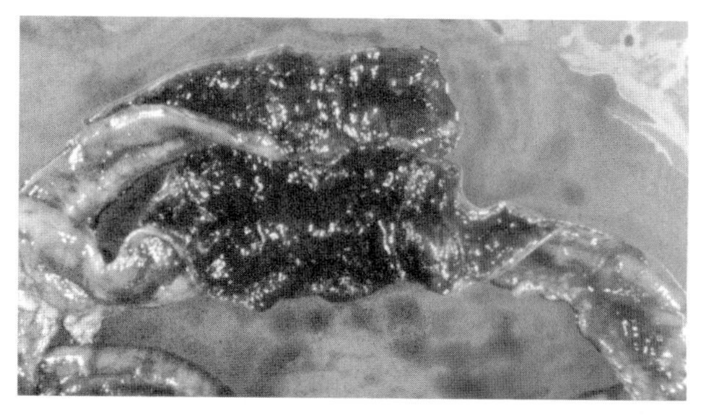

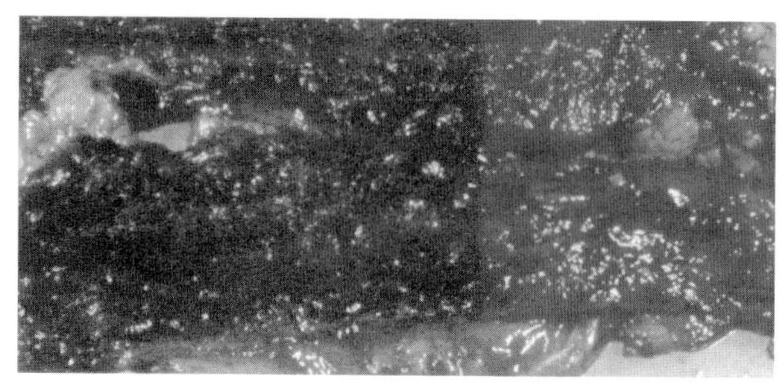

FIGURE 21.53 AND FIGURE 21.54 Gastrointestinal graft-vs.-host disease in which there is a diffuse process that usually involves the ileum and cecum, resulting in secretory diarrhea. Grossly there is diffuse erythema, granularity, and loss of folds, and when severe, there is undermining and sloughing of the entire mucosa, leading to fibrinopurulent clots of necrotic material. Sometimes there is frank obstruction in patients with intractable-graft-vs.-host disease.

ascribed also to HLA-B alleles, i.e., an increased GVHD incidence associated with HLA-B8 and HLA-B35. Epithelial tissues serving as targets of GVHD include keratinocytes, erythrocytes, and bile ducts, which may express Ia antigens following exposure to endogeneous interferon produced by T lymphocytes. When Ia antigens are expressed on nonlymphoid cells, they may become antigen-presenting cells for autologous antigens and aid perpetuation of autoimmunity.

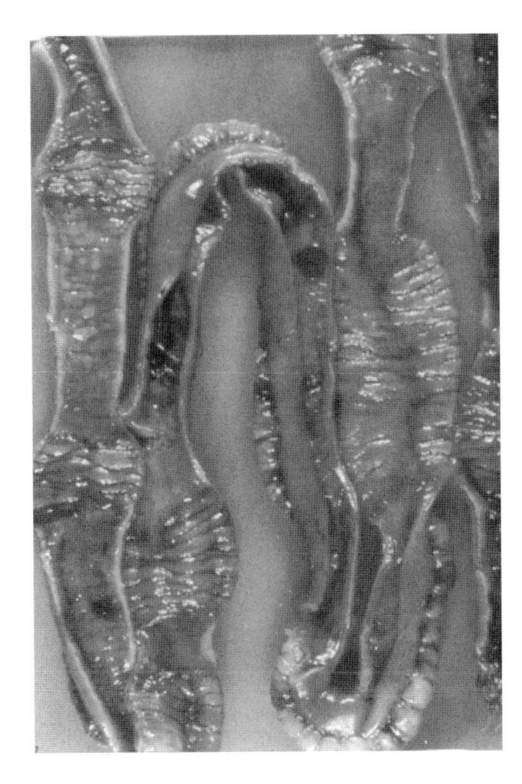

FIGURE 21.55 Stenotic and fibrotic segments alternating with more normal-appearing dilated segments of gut in graft-vs.-host disease.

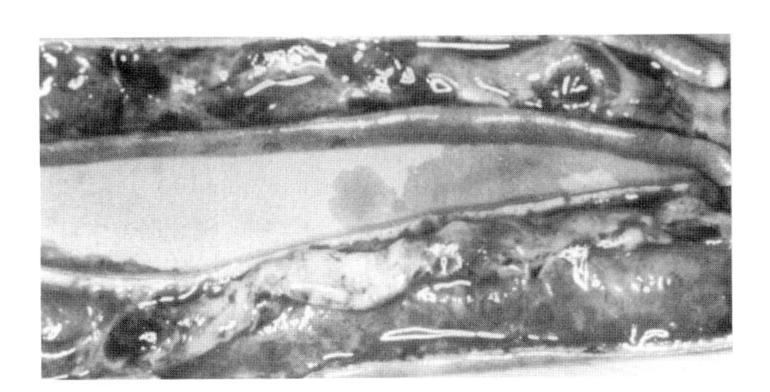

FIGURE 21.56 Sloughing of the mucosal lining of the gut in graft-vs.-host disease.

Cytotoxic T lymphocytes mediate acute GVHD. While most immunohistological investigations have implicated CD8$^+$ (cytotoxic/suppressor) lymphocytes, others have identified CD4$^+$ (T helper lymphocytes) in human GVHD, whereas natural killer (NK) cells have been revealed as effectors of murine but not human GVHD. Following interaction between effector and target cells, cytotoxic granules from cytotoxic T or NK cells are distributed over the target cell membrane, leading to perforin-induced large pores across the membrane and nuclear lysis by deoxyribonuclease. Infection, rather than failure of the primary target organ (other than gastrointestinal bleeding), is the major cause of mortality in acute GVHD. Within the

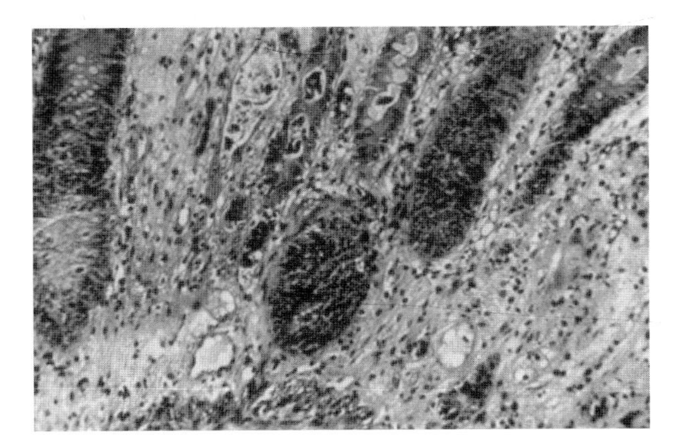

FIGURE 21.57 Histologically, graft-vs.-host disease in the gut begins as a patchy destructive enteritis localized to the lower third of the crypts of Lieberkuhn.

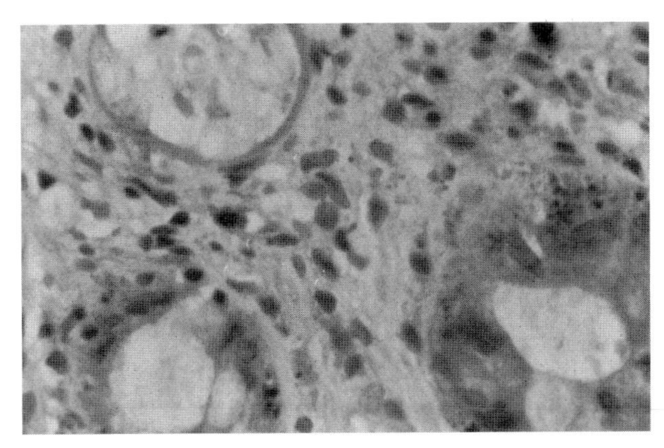

FIGURE 21.58 The earliest lesions are characterized by individual enterocyte necrosis with karyorrhectic nuclear debris, the so-called exploding crypt, which progresses to a completely destroyed crypt as shown in the upper left-hand corner.

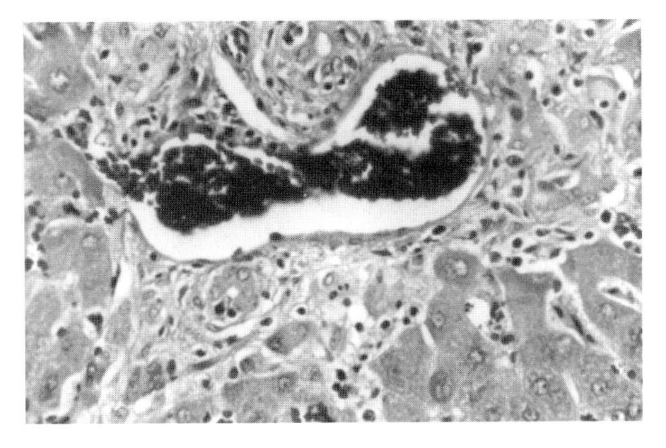

FIGURE 21.59 Hepatic graft-vs.-host disease is characterized by a cholestatic hepatitis with characteristic injury and destruction of small bile ducts that resemble changes seen in rejection. In this section of early acute GVHD, there are mild portal infiltrates with striking exocytosis into bile ducts associated with individual cell necrosis and focal destruction of the bile ducts.

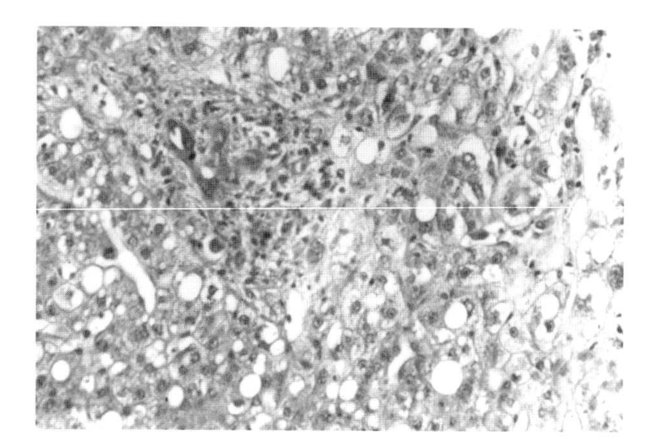

FIGURE 21.60 This liver section from a patient with GVHD demonstrates the cholestatic changes that evolve from hepatocellular ballooning to cholangiolar cholestasis with bile microliths, which signifies prolonged GVHD.

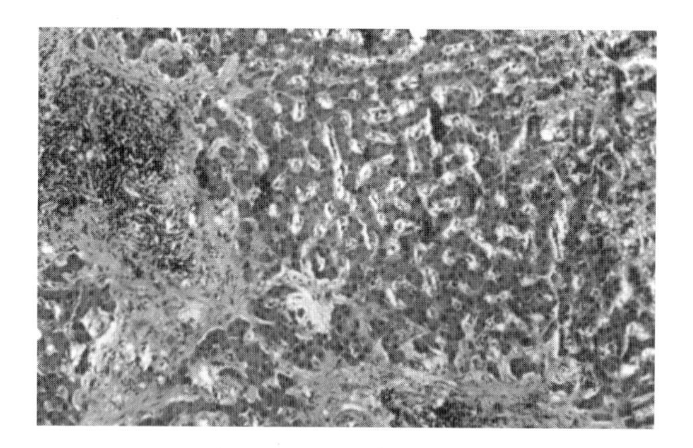

FIGURE 21.61 Chronic GVHD of the liver with pronounced inflammation and portal fibrosis with disappearance of bile ducts.

first few months posttransplant, all recipients demonstrate diminished immunoglobulin synthesis, decreased T helper lymphocytes, and increased T suppressor cells. Acute GVHD patients manifest an impaired ability to combat viral infections. They demonstrate an increased risk of cytomegalovirus (CMV) infection, especially CMV interstitial pneumonia. GVHD may also reactivate other viral diseases such as herpes simplex.

Immunodeficiency in the form of acquired B cell lymphoproliferative disorder (BCLD) represents another serious complication of post-bone marrow transplantation. Bone marrow transplants treated with pan-T cell monoclonal antibody or those in which T lymphocytes have been depleted account for most cases of BCLD, which is associated with severe GVHD. All transformed B cells in cases of BCLD have manifested the Epstein-Barr viral genome.

Chronic graft-vs.-host disease (GVHD) may occur in as many as 45% of long-term bone marrow transplant recipients. Chronic GVHD (Figure 21.61) differs both

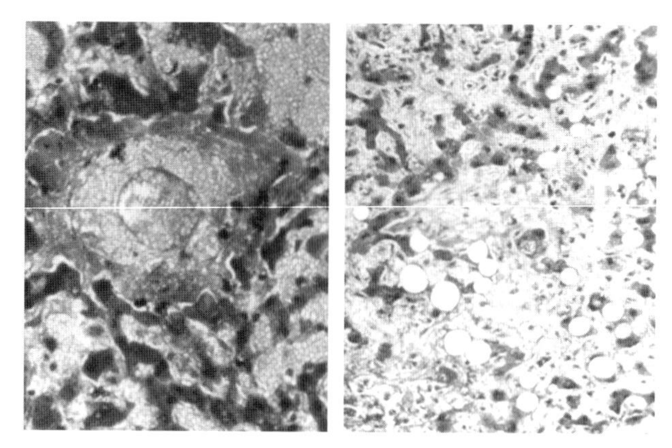

FIGURE 21.62 Venoocclusive disease (VOD) accompanying graft-vs.-host disease of the liver. On the left is early VOD with concentric subendothelial widening and sublobular central venules with degeneration of surrounding pericentral hepatocytes. There is deposition of fibrin and Factor VIII. On the right is a late lesion of VOD showing fibrous obliteration of the central venule and the sinusoids by combination of types 3, 1, and even type 4 collagen.

clinically and histologically from acute GVHD and resembles autoimmune connective tissue diseases. For example, chronic GVHD patients may manifest skin lesions resembling scleroderma; sicca syndrome in the eyes and mouth; inflammation of the oral, esophageal, and vaginal mucosa; bronchiolitis obliterans; occasionally myasthenia gravis; polymyositis; and autoantibody synthesis. Histopathologic alterations in chronic GVHD, such as chronic inflammation and fibrotic changes in involved organs, resemble changes associated with naturally occurring autoimmune disease. The skin may reveal early inflammation with subsequent fibrotic changes. Infiltration of lacrimal, salivary, and submucosal glands by lymphoplasmacytic cells leads ultimately to fibrosis. The resulting sicca syndrome, which resembles Sjögren's syndrome, occurs in 80% of chronic GVHD patients. Drying of mucous membranes in the sicca syndrome affects the mouth, esophagus, conjunctiva, urethra, and vagina. The pathogenesis of chronic GVHD involves the interaction of alloimmunity, immune dysregulation, and resulting immunodeficiency and autoimmunity. The increased incidence of infection among chronic GVHD patients suggests immunodeficiency. The dermal fibrosis is associated with increased numbers of activated fibroblasts in the papillary dermis. T lymphocyte or mast cell cytokines may activate this fibroplasia, which leads to dermal fibrosis in chronic GVHD.

OKT®3 (Orthoclone OKT®3) is a commercial mouse monoclonal antibody against the T cell surface marker CD3. It may be used, therapeutically, to diminish T cell reactivity in organ allotransplant recipients experiencing a rejection episode. OKT3 may act in concert with the complement system to induce T cell lysis, or it may act as an opsonin, rendering T cells susceptible to phagocytosis.

Venoocclusive disease (VOD) is a serious liver complication after marrow transplantation (Figure 21.62). Histopathology of early VOD reveals concentric subendothelial widening and sublobular central venules with degeneration of surrounding pericentral hepatocytes. At this early stage, there is deposition of fibrin and Factor VIII. Late lesions of VOD show fibrous obliteration of the central venule and sinusoids by combination of type 3, 1, and even type 4 collagen. The clinical diagnosis of VOD is reasonably accurate based on the combination of jaundice, ascites, hepatomegaly, and encephalopathy in the first 2 weeks post-transplant. The incidence may be higher among older patients with diagnosis of AML or CML and with hepatitis. The mortality rate of VOD is relatively high at 32%.

22 Tumor Immunology

Biologists have long been fascinated with possible differences between neoplastic cells (Figure 22.1 to Figure 22.3) and their so-called normal counterparts or tissues of origin. This led to the search for antigens on tumors that are absent from normal tissues. The aim of finding such immunologic differences would be both for cancer testing and for cancer treatment purposes. This search has met with varying degrees of success.

A **neoplasm** is any new and abnormal growth that may be either a benign or malignant tumor.

Cancer is an invasive, metastatic, and highly anaplastic cellular tumor that leads to death. Neoplasms are often divided into two broad categories of carcinoma and sarcoma.

Modification of proteins by phosphorylation or specific proteolysis may change their covalent architecture to yield new antigenic determinants or epitopes termed **neoantigens.** The epitope is newly expressed on cells during development or in neoplasia. Neoantigens include tumor-associated antigens. New antigenic determinants may also emerge when a protein changes conformation or when a molecule is split, exposing previously unexpressed epitopes.

A **carcinogen** is any chemical or physical cancer-producing agent. Carcinogens comprise the epigenetic type that does not damage DNA but causes other physiological alterations that predispose to cancer, and the genotoxic type that reacts directly with DNA or with micromolecules that then react with DNA.

A **carcinoma** is a malignant tumor composed of epithelial cells that infiltrate surrounding tissues and lead to metastases.

A **choriocarcinoma** is an unusual malignant neoplasm of the placenta trophoblast cells in which the fetal neoplastic cells are allogeneic in the host. On rare occasions, these neoplasms have been "rejected" spontaneously by the host. Antimetabolites have been used in the treatment of choriocarcinoma.

Sarcomas are tumors arising from connective tissue.

"Sneaking through" is the successful growth of a sparse number of transplantable tumor cells that have been inoculated into a host in contrast to the induction of tumor immunity and lack of tumor growth in the same host if larger doses of the same cells are administered.

Autochthonous is an adjective that indicates "pertaining to self," occuring in the same subject. Also called autologous.

Spontaneous remission is the reversal of progressive growth of the neoplasm with inadequate or no treatment. Spontaneous remission occurs only rarely.

A **benign tumor** is an abnormal proliferation of cells that leads to a growth that is localized and contained within epithelial barriers. It does not usually lead to death, in contrast to a malignant tumor.

Malignant is an adjective that means leading to death, as by a malignant neoplasm.

Metastasis is the transfer of disease from one organ or part to another not directly connected with it. For example, malignant tumors may need anatomical sites distant from the primary tumor's site of origin, leading to the establishment of secondary tumors.

Malignolipin (historical) is a substance claimed in the past to be specific for cancer and to be detectable in the patient's blood early in the course of the disease. This is no longer considered valid. Malignolipin is comprised of fatty acids, phosphoric acid choline, and spermine. When injected into experimental animals, it can produce profound anemia, leukopoiesis, and cachexia.

Oncogenes are genes with the capacity to induce neoplastic transformation of cells. They are derived from either normal genes termed protooncogenes or from oncogenic RNA (oncorna) viruses. Their protein products are critical for regulation of gene expression or growth signal transduction. Translocation, gene amplification, and point mutation may lead to neoplastic transformation of protooncogene. Oncogenes may be revealed through use of viruses that induce tumors in animals or by derivation of tumor-causing genes from cancer cells. There are more than 20 protooncogenes and cellular oncogenes in the human genome. An oncogene alone cannot produce cancer. It must be accompanied by malignant transformation which involves multiple genetic steps. Oncogenes encode four types of proteins that include growth factors, receptors, intracellular transducers, and nuclear transcription factors.

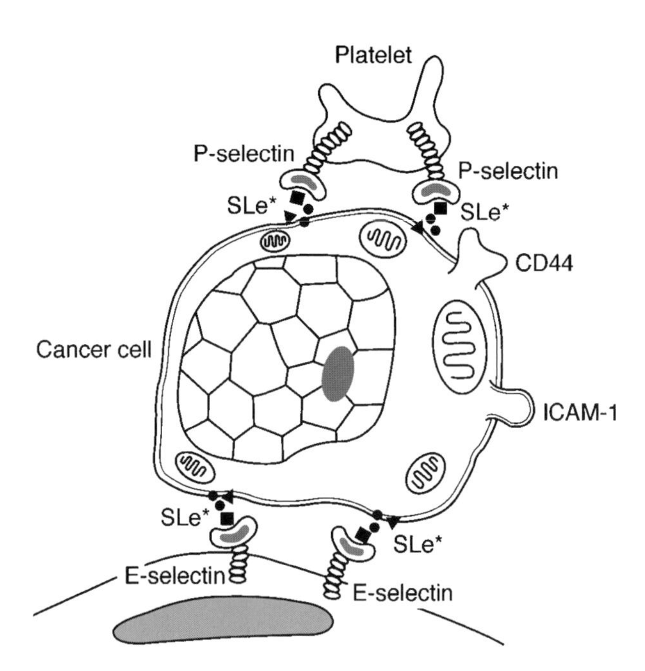

FIGURE 22.1 Schematic representation of a tumor cell attached through E-selectin molecules to an endothelial cell surface.

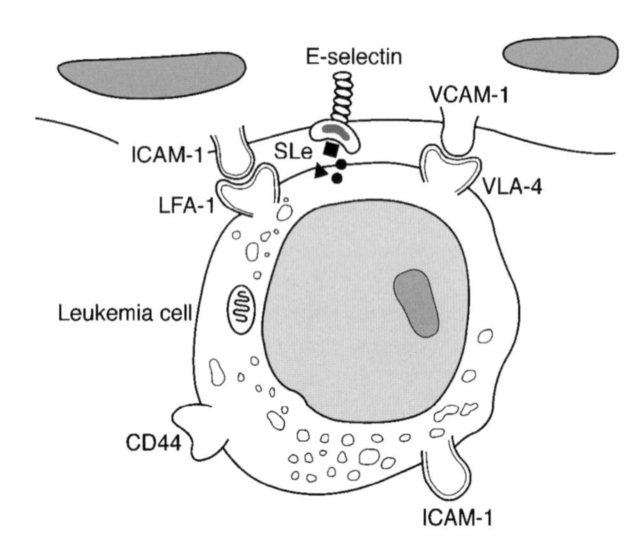

FIGURE 22.2 Schematic representation of a leukemia cell attached to an endothelial cell surface via adhesion molecules.

The **oncogene theory** is a concept of carcinogenesis that assigns tumor development to latent retroviral gene activation through irradiation or carcinogens. These retroviral genes are considered to be normal constituents of the cell. Following activation, these oncogenes are presumed to govern the neoplasm through hormones that are synthesized and even the possible construction of a complete oncogenic virus. This concept states that all cells may potentially become malignant.

Oncogenesis is the process whereby tumors develop.

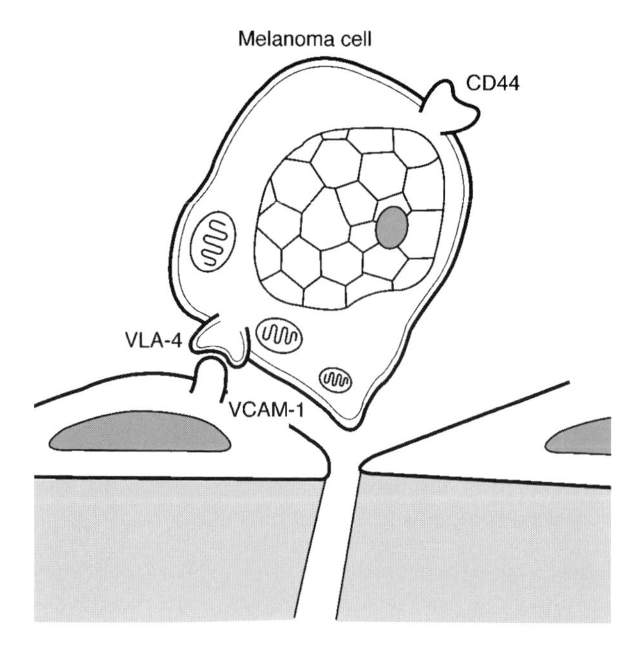

FIGURE 22.3 Schematic representation of a melanoma cell attached to an endothelial cell surface through a VLA-4-VCAM-1 interaction.

Oncomouse is a commercially developed transgenic animal into which human genes have been introduced to make the mouse more susceptible to neoplasia. This transgenic mouse is used for both medical and pharmaceutical research.

A **protooncogene** is a cellular gene that shows homology with a retroviral oncogene. It is found in normal mammalian DNA and governs normal proliferation and probably also differentiation of cells. Mutation or recombination with a viral genome may convert a protooncogene into an oncogene, signifying that it has become activated. Oncogenes may act in the induction and/or maintenance of a neoplasm. Protooncogenes united with control elements may induce transformation of normal fibroblasts into tumor cells. Examples of protooncogenes are c-*fos*, c-*myc*, c-*myb*, c-*ras*, etc. Alteration of the protooncogenes, leading to synthesis of an aberrant gene product, is believed to facilitate its becoming tumorigenic. An elevation in the quantity of gene product produced is also believed to be associated with protooncogenes becoming tumorigenic.

Cellular oncogene: See protooncogene.

Ras is one of a group of 21-kDa guanine nucleotide-binding proteins with intrinsic GTPase activity that participates in numerous different signal transduction pathways in a variety of cells. *Ras* gene mutations may be associated with tumor transformation. Ras is attracted to the plasma membrane by tyrosine phosphorylated adapter proteins during T lymphocyte activation, where GDP–GTP exchange factors are activated. GTP-Ras then activates the

MAP kinase cascade that results in *fos* gene expression and assembly of AP-1 trascription factor.

Ras: See small G proteins.

Rous sarcoma virus (RSV) is an RNA type C oncovirus that is single stranded and produces sarcomas in chickens. It is the typical acute transforming retrovirus. Within its genome are *gag*, *pol*, *env*, and v-*src* genes; *gag* encodes a core protein, *pol* encodes reverse transcriptase, and *env* encodes envelope glycoprotein. V-*scr* is an oncogene associated with the oncogenic capacity of the virus.

A **promoter** is (1) the DNA molecular site where RNA polymerase attaches and the point at which transcription is initiated. The promoter is frequently situated adjacent to the operator, and upstream from it is an operon. A TATA box and a promoter are both required for immunoglobulin gene transcription. (2) In tumor biology, a promoter mediates the second stage or promotion stage in the process of carcinogenesis. It may be a substance that can induce a tumor in an experimental animal that has been previously exposed to a tumor initiator. Yet the promoter alone is not carcinogenic.

A **v-*myb* oncogene** is a genetic component of an acute transforming retrovirus that leads to avian myeloblastosis. It represents a truncated genetic form of c-*myb*.

Tumor promoter: See phorbol ester(s).

Amphiregulin is a glycoprotein member of the epidermal growth factor (EGF) family of proteins. The carboxyl-terminal amino acid residues of amphiregulin positions 46–84 share much sequence homology with the EGF family of proteins. The actions of amphiregulin are wide ranging, including the stimulation of proliferation of certain tumor cell lines, fibroblasts, and various other normal cells. These actions are mediated by binding to EGF receptors possessing intrinsic tyrosine kinase activity.

Tumor imaging is an experimental and clinical medical technique employed to localize neoplastic lesions in a body using a labeled antibody or its fragment. Tumor imaging is based on the presence of an antigen expressed only on a tumor cell or at least has a significant difference in amount and/or distribution between the tumor and normal tissues.

Tumor-associated antigens (Figure 22.4) are antigens designated as CA-125, CA-19-9, and CA195, among others, that may be linked to certain tumors such as lymphomas, carcinomas, sarcomas, and melanomas, but the immune response to these tumor-associated antigens is not sufficient to mount a successful cellular or humoral immune response against the neoplasm. Three classes of tumor-associated antigens have been described. Class I

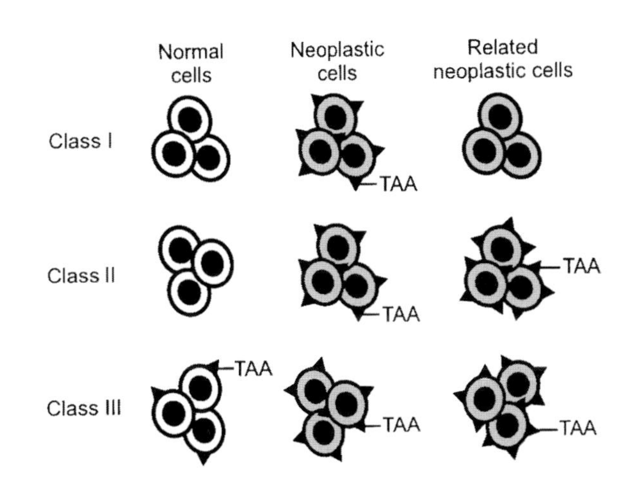

FIGURE 22.4 Schematic representation of tumor-associated antigens (TAA) among normal and neoplastic cells.

antigens are very specific for a certain neoplasm and are absent from normal cells. Class II antigens are found on related neoplasms from separate individuals. Class III antigens are found on malignant as well as normal cells, but show increased expression in the neoplastic cells. Assays of clinical value will probably be developed for Class II antigens, since they are associated with multiple neoplasms and very infrequently found in normal individuals.

Tumor antigens are cell surface proteins on tumor cells that can induce a cell-mediated and/or humoral immune response. See also tumor-associated antigens and tumor-specific antigens.

Thymicleukemia antigen (TL) is an epitope on thymocyte membrane of TL^+ mice. As the T lymphocytes mature, antigen disappears but resurfaces if leukemia develops. TL antigens are specific and are normally present on the cell surface of thymocytes of certain mouse strains. They are encoded by a group of structural genes located at Tla locus, in the linkage group IX, very close to the D pole of the H-2 locus on chromosome 17. There are three structural Tl genes, one of which has two alleles. The TL antigens are numbered from 1 to 4 specifying four antigens: TL.1, TL.2, TL.3, and TL.4. Antigens TL.3 and TL.4 are mutually exclusive. Their expression is under the control of regulatory genes, apparently located at the same Tla locus. Normal mouse thymocytes belong to three phenotypic groups: Tl^-, Tl.2, and TL.1, 2, 3. Development of leukemia in the mouse induces a restructuring of the TL surface antigens of thymocytes with expression of TL.1 and TL.2 in TL^- cells, expression of TL.1 in TL.2 cells, and expression of TL.4 in both TL^- and TL.2 cells. When normal thymic cells leave the thymus, the expression of TL antigen ceases. Thus, thymocytes are TL^+ (except the TL^-) and the peripheral T cells are TL^-. In transplantation experiments TL^+ tumor cells undergo antigenic modulation.

Tumor cells exposed to homologous antibody stop expressing the antigen and thus escape lysis when subsequently exposed to the same antibody plus complement.

CD10 (CALLA) is an antigen, also referred to as common acute lymphoblastic leukemia antigen (CALLA), that has a mol wt of 100 kDa. CD10 is now known to be a neutral endopeptidase (enkephalinase). It is present on many cell types, including stem cells, lymphoid progenitors of B and T cells, renal epithelium, fibroblasts, and bile canaliculi.

Prostate-specific antigen (PSA) is a marker in serum or tissue sections for adenocarcinoma of the prostate. PSA is a 34-kDa glycoprotein found exclusively in benign and malignant epithelium of the prostate. Men with PSA levels of 0 to 4.0 ng/ml and a nonsuspicious digital rectal examination are generally not biopsied for prostate cancer. Men with PSA levels of 10.0 ng/ml and above typically undergo a prostate biopsy. About one half of these men will be found to have prostate cancer. Certain kinds of PSA, known as bound PSA, link themselves to other proteins in the blood. Other kinds of PSA, known as free PSA, float by themselves. Prostate cancer is more likely to be present in men who have a low percentage of free PSA relative to the total amount of PSA. This finding is especially valuable in helping to differentiate between cancer and other benign conditions, thus eliminating unnecessary biopsies among men in that diagnostic gray zone who have total PSA levels between 4.0 and 10.0 ng/ml. The PSA molecule is smaller than prostatic acid phosphatase (PAP). In patients with prostate cancer, preoperative PSA serum levels are positively correlated with the disease. PSA is more stable and shows less diurnal variation than does PAP. PSA is increased in 95% of new cases for prostatic carcinoma compared with a 60% increase for PAP; PSA is increased in 97% of recurrent cases compared with a 66% increase of PAP. PAP may also be increased in selected cases of benign prostatic hypertrophy and prostatitis, but these elevations are less than those associated with adenocarcinoma of the prostate. It is inappropriate to use either PSA or PAP alone as a screen for asymptomatic males. TUR, urethral instrumentation, prostatic needle biopsy, prostatic infarct, or urinary retention may also result in increased PSA values. PSA is critical for the prediction of recurrent adenocarcinoma in postsurgical patients. PSA is also a useful immunocytochemical marker for primary and metastatic adenocarcinoma of the prostate.

Oncofetal antigens (Figure 22.5) are markers or epitopes present in fetal tissues during development but not present, or found in minute quantities, in adult tissues. These cell-coded antigens may reappear in certain neoplasms of adults due to derepression of the gene responsible for their formation. Examples include carcinoembryonic antigen (CEA), which is found in the liver, intestine, and pancreas

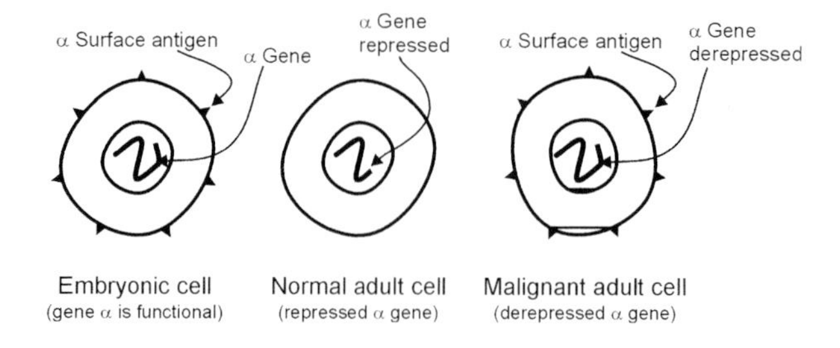

FIGURE 22.5 Oncofetal antigen.

of the fetus, and also in both malignant and benign gastrointestinal conditions. Yet it is useful to detect recurrence of adenocarcinoma of the colon based upon demonstration of CEA in the patient's serum; α-fetoprotein (AFP) is demonstrable in approximately 70% of hepatocellular carcinomas.

A **fetal or oncofetal antigen** that is expressed as a normal constituent of embryos and not in adult tissues. It is reexpressed in neoplasms of adult tissues, apparently as a result of derepression of the gene responsible for its formation.

Carcinoma-associated antigens are self antigens whose epitopes have been changed due to effects produced by certain tumors. Self antigens are transformed into a molecular structure for which the host is immunologically intolerant. Examples include the T antigen, which is an MN blood group precursor molecule exposed by the action of bacterial enzymes, and Tn antigen, which is a consequence of somatic mutation in hematopoietic stem cells caused by inhibition of galactose transfer to *N*-acetyl-D-galactosamine.

Embryonic antigens are protein or carbohydrate antigens synthesized during embryonic and fetal life that are either absent or formed in only minute quantities in normal adult subjects. Fetoproteins (AFP) and carcinoembryonic antigen (CEA) are fetal antigens that may be synthesized once again in large amounts in individuals with certain tumors. Their detection and level during the course of the disease and following surgery to remove a tumor reducing the substance may serve as a diagnostic and prognostic indicator of the disease process. Blood group antigens, such as the iI, which are reversed in their levels of expression in the fetus and in the adult, may show a reemergence of i antigen in adult patients with thalassemia and hypoplastic anemia. Cold autoagglutinins specific for it may be found in infectious mononucleosis patients. Common acute lymphoblastic leukemia antigen (CD10) is rarely found on peripheral blood cells of normal subjects, whereas CALLA cells coexpressing IgM and CD19 molecules may be found in fetal bone marrow and peripheral blood samples. CD10 may be expressed in children with common acute lymphoblastic leukemia.

FIGURE 22.6 Carcinoembryonic antigen (CEA).

Carcinoembryonic antigen (CEA) (Figure 22.6) is a 200-kDa membrane glycoprotein epitope that is present in the fetal gastrointestinal tract in normal conditions. However, tumor cells, such as those in colon carcinoma, may reexpress it. CEA was first described as a screen for identifying carcinoma by detecting nanogram quantities of the antigen in serum. It was later shown to be present in certain other conditions as well. CEA levels are elevated in almost one third of patients with colorectal, liver, pancreatic, lung, breast, head and neck, cervical, bladder, medullarythyroid, and prostatic carcinoma. However, the level may be elevated also in malignant melanoma, lymphoproliferative disease, and smokers. Regrettably, CEA levels also increase in a variety of nonneoplastic disorders, including inflammatory bowel disease, pancreatitis, and cirrhosis of the liver. Nevertheless, determination of CEA levels in the serum is valuable for monitoring the recurrence of tumors in patients whose primary neoplasm has been removed. If the patient's CEA level reveals a 35% elevation compared with the level immediately following surgery, this may signify metastases. This oncofetal antigen is comprised of one polypeptide chain with one variable region at the amino terminus and six constant region domains. CEA belongs to the immunoglobulin superfamily. It lacks specificity for cancer, thereby limiting its diagnostic usefulness. It is detected with a mouse monoclonal antibody directed against a complex glycoprotein antigen present on many human epithelial-derived tumors. This reagent may be used to aid in the identification of cells of epithelial lineage. The antibody is intended for qualitative staining in sections of formalin-fixed, paraffin-embedded tissue. Anti-CEA antibodies specifically bind to antigens located in the plasma membrane and cytoplasmic regions of normal epithelial cells. Unexpected antigen expression or loss of expression may occur, especially in neoplasms. Occasionally, stromal elements surrounding heavily stained tissue and/or cells show immunoreactivity. Clinical interpretation of any staining or its absence must be complemented by morphological studies and evaluation of proper controls.

CEA is the abbreviation for carcinoembryonic antigen.

SV40 (simian virus 40) (Figure 22.7 and Figure 22.8) is an oncogenic polyoma virus. It multiplies in cultures of rhesus monkey kidney and produces cytopathic alterations in African green monkey cell cultures. Inoculation into newborn hamsters leads to the development of sarcomas. SV40 has 5243 base pairs in its genome. It may follow either of two patterns of life cycle according to the host cell. In permissive cells, such as those from African green monkeys, the virus-infected cells are lysed, causing the escape of multiple viral particles. Lysis does not occur in nonpermissive cells infected with the virus. By contrast, they may undergo oncogenic transformation in which SV40 DNA sequences become integrated into the genome of the host cell. Cells that have become transformed have characteristic morphological features and growth properties. SV40 may serve as a cloning vector. It is a diminutive icosahedral papovavirus that contains double-stranded DNA. It may induce progressive multifocal leukoencephalopathy. SV40 is useful for the *in vitro* transformation of cells as a type of "permissive" infection ultimately resulting in lysis of infected host cells.

Oncogenic virus (Figure 22.9) is any virus, whether DNA or RNA, that can induce malignant transformation of cells. An example of a DNA virus would be human papillomavirus, and an RNA virus would be retrovirus.

α-fetoprotein (Figure 22.10) is a principal plasma protein in the α globulin fraction present in the fetus. It bears considerable homology with human serum albumin. It is produced by the embryonic yolk sac and fetal liver and consists of a 590-amino acid residue polypeptide chain structure. It may be elevated in pregnant women bearing fetuses with open neural tube defects, central nervous system defects, gastrointestinal abnormalities, immunodeficiency syndromes, and various other abnormalities. After parturition, the high levels in fetal serum diminish to levels that cannot be detected. α-fetoprotein induces immunosuppression, which may facilitate neonatal tolerance. Based on *in vitro* studies, it is believed to facilitate suppressor T lymphocyte function and diminish helper T lymphocyte action. Liver cancer patients reveal significantly elevated serum levels of α-fetoprotein. In immunology, however, it is used as a marker of selected tumors such as hepatocellular carcinoma. It is detected by the avidin–biotin–peroxidase complex (ABC) immunoperoxidase technique using monoclonal antibodies.

The **melanoma antigen-1 gene (MAGE-1)** in humans was derived from a malignant melanoma cell line. It encodes for an epitope that a cytotoxic T lymphocyte clone

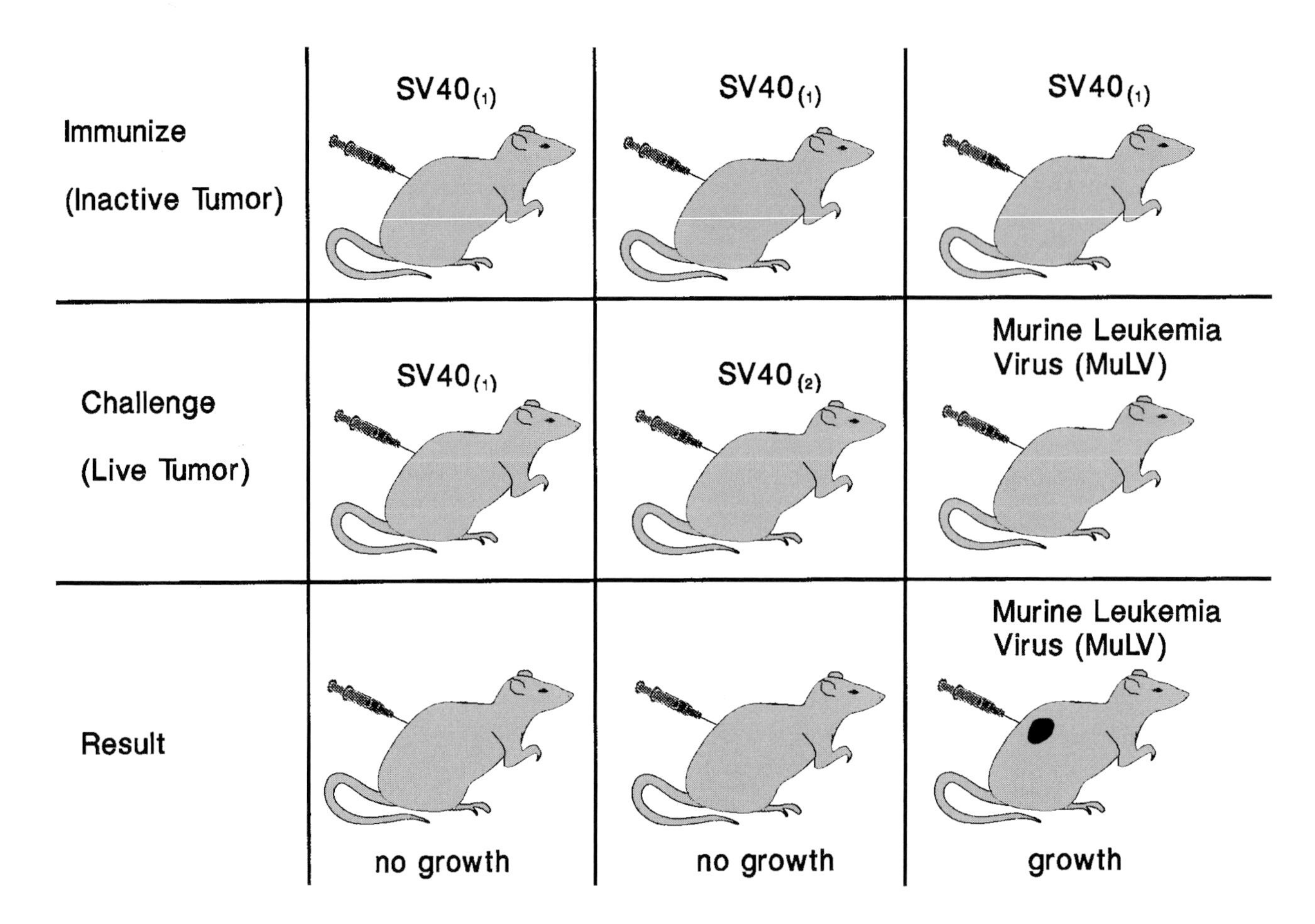

FIGURE 22.7 SV-40.

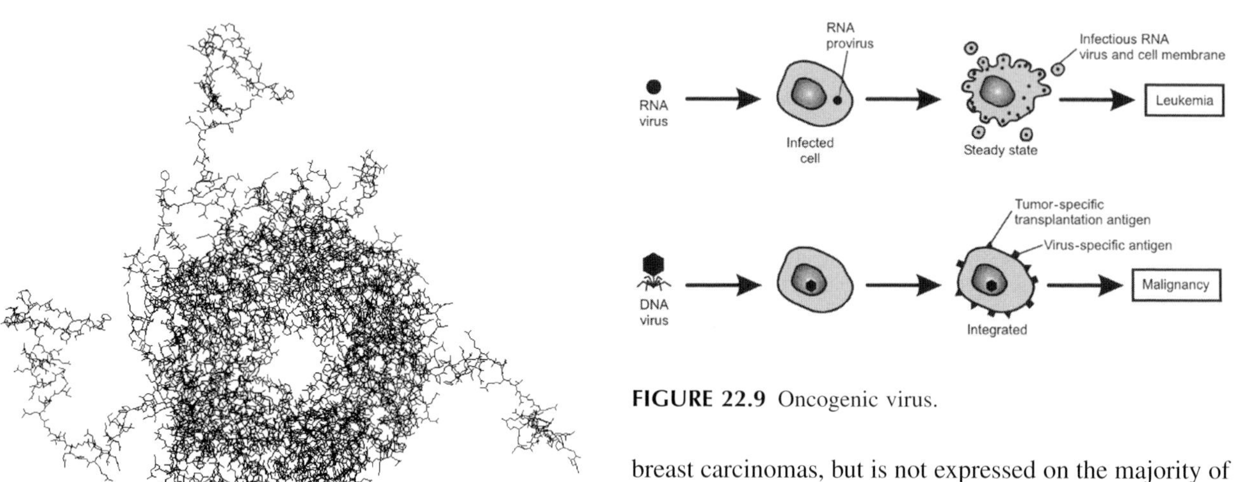

FIGURE 22.8 SV-40. Resolution 3.1 Å.

FIGURE 22.9 Oncogenic virus.

specific for melanoma recognizes. This clone was isolated from a patient bearing melanoma. **MAGE-1 protein** is found on one half of all melanomas and one fourth of all breast carcinomas, but is not expressed on the majority of normal tissues. Even though MAGE-1 has not been shown to induce tumor rejection, cytotoxic T lymphocytes in melanoma patients manifest specific memory for MAGE-1 protein.

Melanoma-associated antigens (MAA) are antigens associated with the aggressive, malignant, and metastatic tumors arising from melanocytes or melanocyte-associated nevus cells. Monoclonal antibodies have identified 40+ separate MAAs. They are classified as MHC molecules,

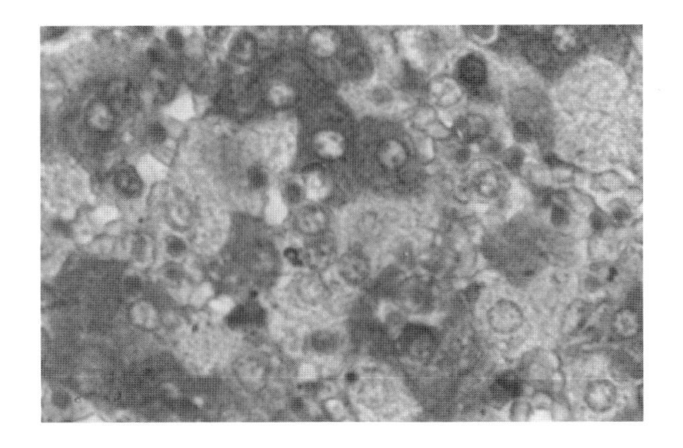

FIGURE 22.10 α-fetoprotein.

cation-binding proteins, growth-factor receptors, gangliosides, high molecular weight extracellular matrix-binding molecules, and nevomelanocyte differentiation antigens. Some of the antigens are expressed on normal cells, whereas others are expressed on tumor cells. Melanoma patient blood sera often contain anti-MAA antibodies which are, regrettably, not protective. Monoclonal antibodies against MAAs aid studies on the biology of tumor progression, immunodiagnosis, and immunotherapy trials.

Modulation: See antigenic modulation.

Tumor cells may be subject to alterations in antigenic structure. **Antigenic transformation** refers to changes in a cell's antigenic profile as a consequence of antigenic gain, deletion, reversion, or other process. **Antigenic gain** refers to nondistinctive normal tissue components that are added or increased without simultaneous deletion of other normal tissue constituents. **Antigenic deletion** describes antigenic determinants that have been lost or masked in the progeny of cells that usually contain them. Antigenic deletion may take place as a consequence of neoplastic transformation or mutation of parent cells resulting in the disappearance or repression of the parent cell genes. **Antigenic modulation** is the loss of epitopes or antigenic determinants from a cell surface following combination with an antibody. The antibodies either cause the epitope to disappear or become camouflaged by covering it. **Antigenic diversion** refers to the replacement of a cell's antigenic profile by the antigens of a different normal tissue cell. Used in tumor immunology, **antigenic reversion** is the change in antigenic profile characteristic of an adult cell to an antigenic mosaic that previously existed in the immature or fetal cell stage of the species. Antigenic reversion may accompany neoplastic transformation.

Tumor cells express **tumor-specific determinants** or **epitopes** present on tumor cells but identifiable also in varying quantities and forms on normal cells. **Tumor-specific antigen (TSA)** (Figure 22.11) are present on tumor cells,

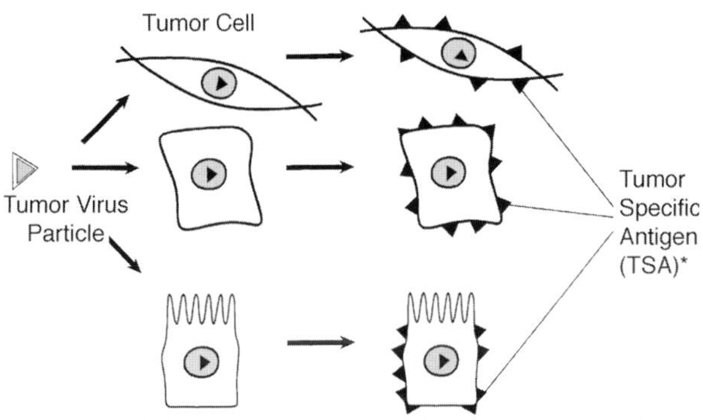

FIGURE 22.11 Tumor-specific antigens (TSA).

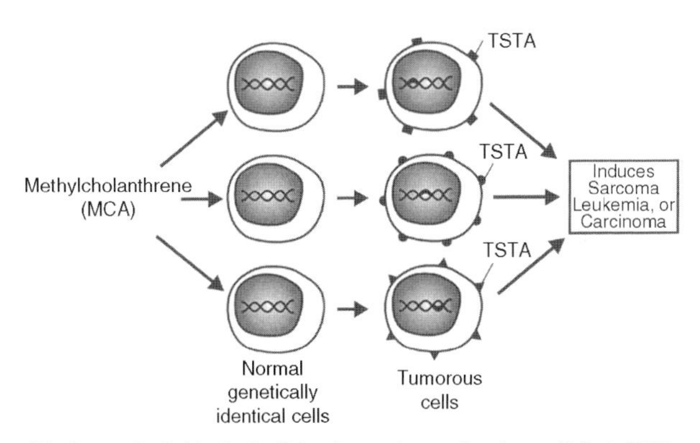

FIGURE 22.12 Tumor-specific transplantation antigens (TSTA).

but not found on normal cells. Murine tumor-specific antigens can induce transplantation rejection in mice. **Tumor-specific transplantation antigen (TSTA)** (Figure 22.12) are epitopes that induce rejection of tumors transplanted among syngeneic (histocompatible) animals.

Tumor rejection antigen is an antigen that is detectable when transplanted tumor cells are rejected. Also called tumor transplant antigen.

TATA is the abbreviation for tumor-associated transplantation antigen.

Macrophages (Figure 22.13 and Figure 22.14) are mononuclear phagocytic cells derived from monocytes in the blood that were produced from stem cells in the bone marrow. These cells have a powerful although nonspecific role in immune defense. These intensely phagocytic cells contain lysosomes and exert microbicidal action against microbes which they ingest. They also have effective tumoricidal activity. They may take up and degrade both protein and polysaccharide antigens and present them to T lymphocytes in the context of MHC class II molecules.

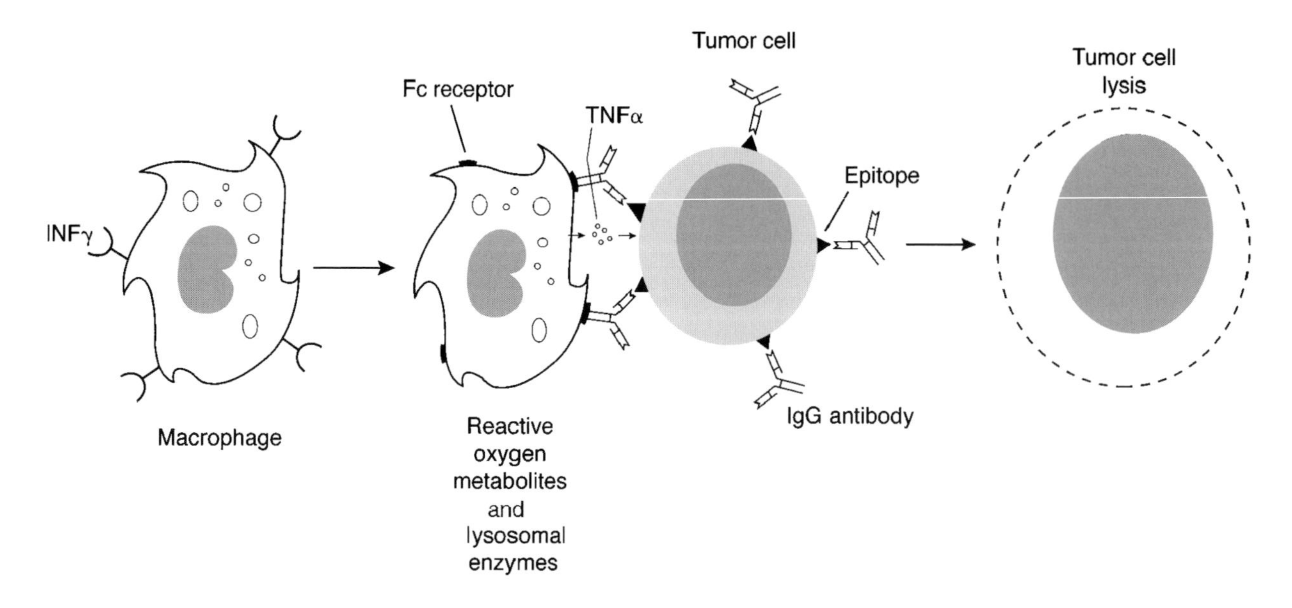

FIGURE 22.13 Macrophage-mediated tumor cell lysis is mediated by several mechanisms. Activated macrophages express Fcγ receptors that anchor IgG molecules attached to tumor cells but not normal cells, resulting in the release of lysosomal enzymes and reactive oxygen metabolites that lead to tumor cell lysis. Another mechanism of macrophage-mediated lysis includes the release of the cytokine tumor necrosis factor α that may unite with high affinity TNFα receptors on a tumor cell surface resulting in its lysis, or the effect of TNFα on the small blood vessels and capillaries of vascularized tumors leading to hemorrhagic necrosis producing a localized Shwartzman-like reaction.

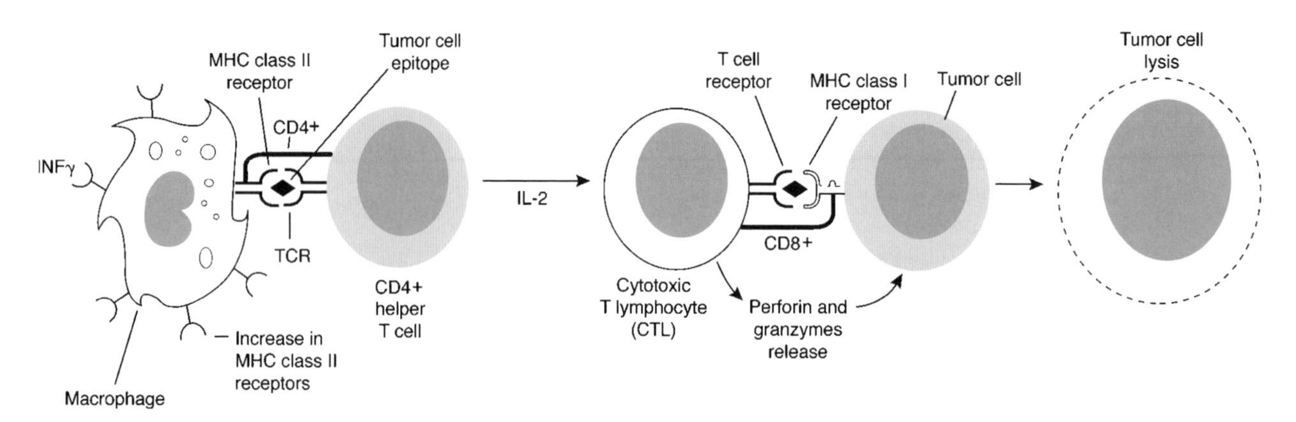

FIGURE 22.14 Macrophage-mediated tumor immunity.

They interact with both T and B lymphocytes in immune reactions. They are frequently found in areas of epithelium, mesothelium, and blood vessels. Macrophages have been referred to as adherent cells since they readily adhere to glass and plastic and may spread on these surfaces and manifest chemotaxis. They have receptors for Fc and C3b on their surfaces, stain positively for nonspecific esterase and peroxidase, and are Ia antigen positive when acting as accessory cells that present antigen to $CD4^+$ lymphocytes in the generation of an immune response. Monocytes, which may differentiate into macrophages when they migrate into the tissues, make up 3 to 5% of leukocytes in the peripheral blood. Macrophages that are tissue-bound may be found in the lung alveoli, as microglial cells in the central nervous system, as Kupffer cells in the liver, as Langerhans cells in the skin, as histiocytes in connective tissues, as well as macrophages in lymph nodes and peritoneum. Multiple substances are secreted by macrophages including complement components C1 through C5, factors B and D, properdin, C3b inactivators, and β-1H. They also produce monokines such as interleukin-1, acid hydrolase, proteases, lipases, and numerous other substances.

Natural killer (NK) cells (Figure 22.15 and Figure 22.16) attack and destroy tumor cells and certain virus-infected cells. They constitute an important part of the natural immune system, do not require prior contract with antigen, and are not MHC restricted by the MHC antigens. NK cells are lymphoid cells of the natural immune system that express cytotoxicity against various nucleated cells,

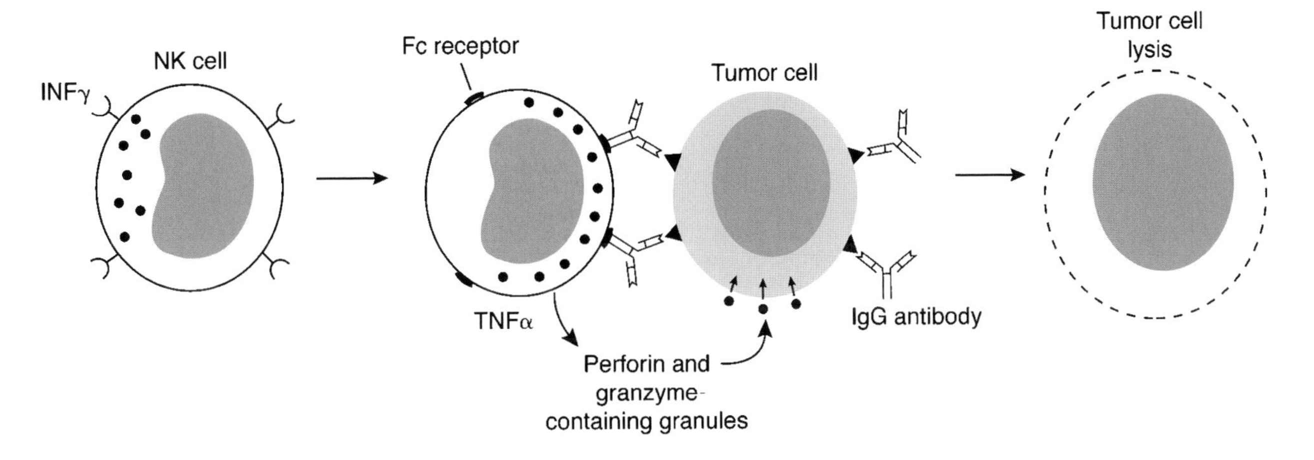

FIGURE 22.15 NK-cell-mediated killing of tumor cells by antibody-dependent cell-mediated cytotoxicity.

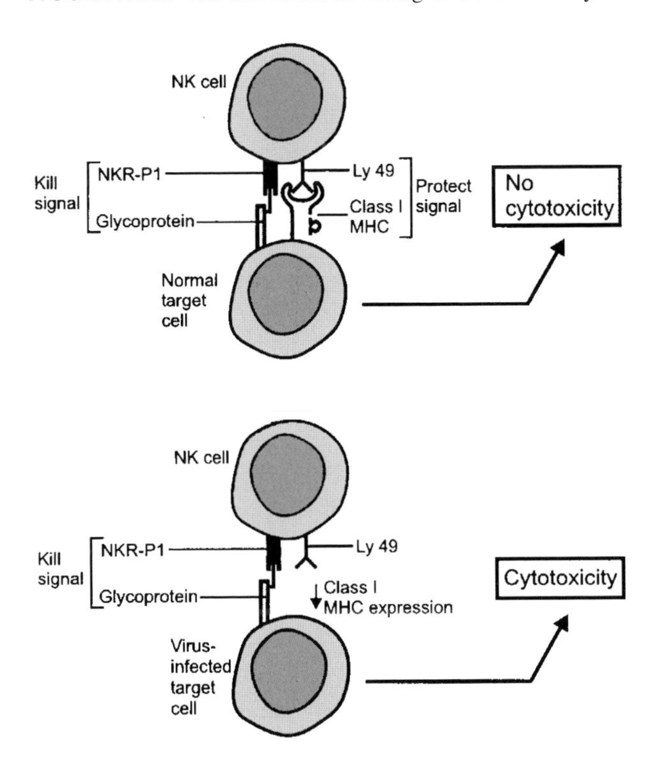

FIGURE 22.16 Proposed mechanism of NK cell cytotoxicity restricted to altered self cells. Kill signal is generated when the NK cell's NKR-PI receptor interacts with membrane glycoprotein or normal and altered self cells. The kill signal can be countermanded by interaction of the NK cells Ly49 receptor with class I MHC molecules. Thus, MHC Class I expression prevents NK cell killing of normal cells. Diminished Class I expression on altered self cells leads to their destruction.

including tumor cells and virus-infected cells. NK cells, killer (K) cells, or antibody-dependent, cell-mediated cytotoxicity (ADCC) cells induce lysis through the action of antibody. Immunologic memory is not involved, as previous contact with the antigen is not necessary for NK cell activity. The NK cell is approximately 15 μm in diameter and has a kidney-shaped nucleus with several, often three, large cytoplasmic granules. The cells are also called large granular lymphocytes (LGLs). In addition to their ability to kill selected tumor cells and some virus-infected cells, they also participate in ADCC by anchoring antibody to the cell surface through an Fc γ receptor. Thus, they are able to destroy antibody-coated nucleated cells. NK cells are believed to represent a significant part of the natural immune defense against spontaneously developing neoplastic cells and against infection by viruses. NK cell activity is measured by a ^{51}Cr release assay employing the K562 erythroleukemia cell line as a target. NK cells secrete IFN-γ and fail to express antigen receptors such as immunoglobulin receptors or T-cell receptors. Cell-surface stimulatory receptors and inhibitory receptors, which recognize self-MHC molecules, regulate their activation.

Designer lymphocytes are lymphocytes into which genes have been introduced to increase the cell's ability to lyse tumor cells. Tumor-infiltrating lymphocytes transfected with these types of genes have been used in experimental adoptive immunotherapy.

Cytotoxic T lymphocytes (CTLs) (Figure 22.17 and Figure 22.18) are specifically sensitized T lymphocytes that are usually $CD8^+$ and recognize antigens through the T cell receptor on cells of the host infected by viruses or that have become neoplastic. $CD8^+$ cell recognition of the target is in the context of MHC class I histocompatibility molecules. Following recognition and binding, death of the target cell occurs a few hours later. CTLs secrete lymphokines that attract other lymphocytes to the area and release serine proteases and perforins that produce ion channels in the membrane of the target, leading to cell lysis. Interleukin-2, produced by $CD4^+$ T cells, activates cytotoxic T cell precursors. Interferon-γ generated from CTLs activates macrophages. CTLs have a significant role in the rejection of allografts and in tumor immunity. A minor population of $CD4^+$ lymphocytes may also be cytotoxic, but they recognize target-cell antigens in the context of MHC class II molecules.

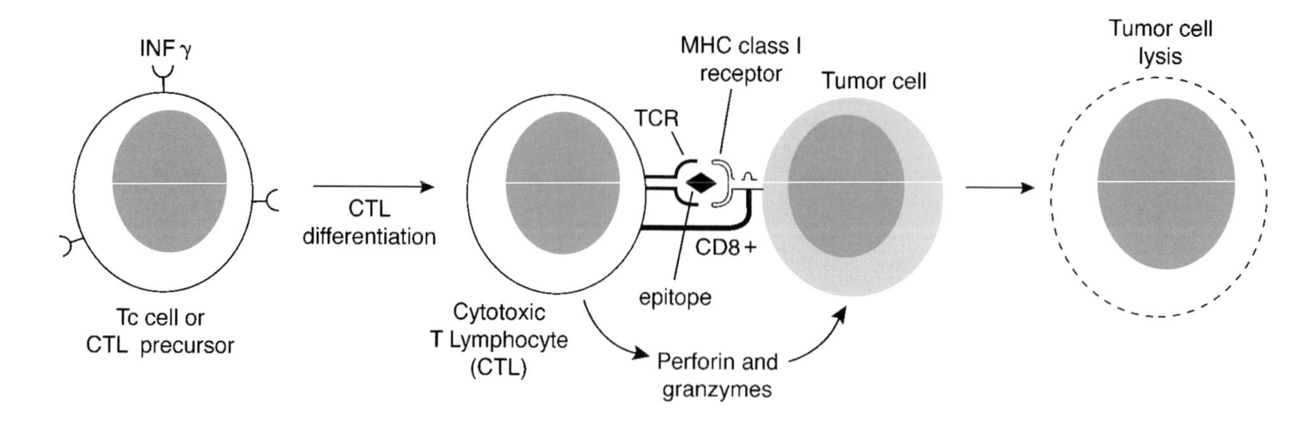

FIGURE 22.17 Cytotoxic T lymphocyte (CTL)-mediated tumor lysis.

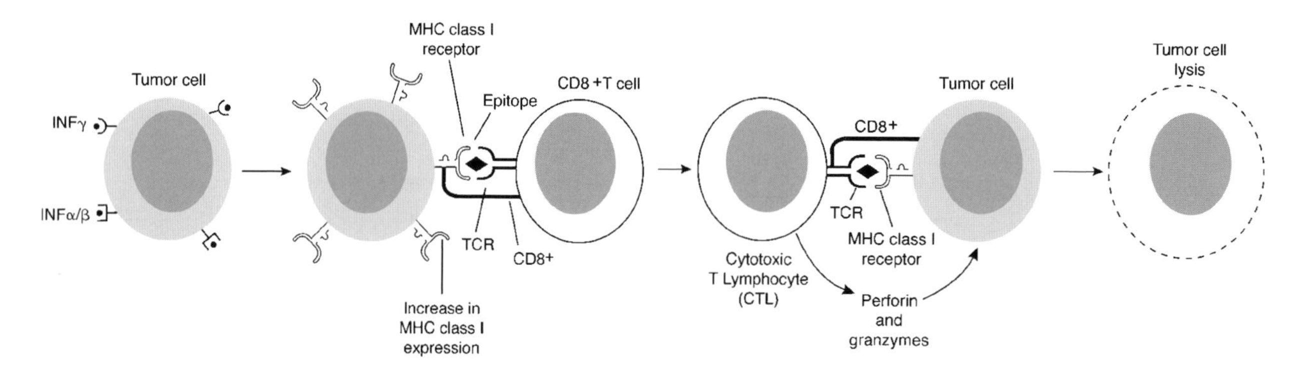

FIGURE 22.18 CTL-mediated killing of tumor cells.

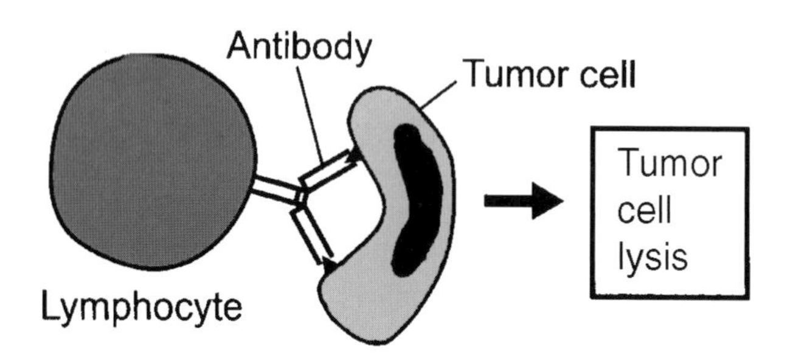

FIGURE 22.19 Antibody-dependent cell-mediated cytotoxicity (ADCC).

Tumor-specific IgG antibodies may act in concert with immune system cells to produce antitumor effects. **Antibody-dependent cell-mediated cytotoxicity (ADCC)** (Figure 22.19) is a reaction in which T lymphocytes and NK cells, including large granular lymphocytes, neutrophils, and macrophages, may lyse tumor cells, infectious agents, and allogeneic cells by combining through their Fc receptors with the Fc region of IgG antibodies bound through their Fab regions to target cell surface antigens. Following linkage of Fc receptors with Fc regions, destruction of the target is accomplished through released cytokines. It represents an example of participation between antibody molecules and immune system cells to produce an effector function.

Antibody-directed enzyme prodrug therapy (ADEPT) is a type of treatment in which an antibody is used to target an enzyme to a tumor and unbound reagent is allowed to clear. A nontoxic prodrug is then given, and this is activated by the enzyme to form a cytotoxic drug at the tumor site. An important part of adept is bystander killing. Since the drugs are activated extracellularly by the antibody–enzyme complex, neighboring cells may also be killed by a mechanism that does not require translocation across intracellular membranes. By contrast, immunotoxins kill only the cell to which they bind.

A **heteroconjugate** is a hybrid of two different antibody molecules.

Heteroconjugate antibodies (Figure 22.20) are antibodies against a tumor antigen coupled covalently to an antibody specific for a natural killer cell or cytotoxic T lymphocyte surface antigen. These antibodies facilitate binding of cytotoxic effector cells to tumor target cells. Antibodies against effector cell surface markers may also be coupled covalently with hormones that bind to receptors on tumor cells.

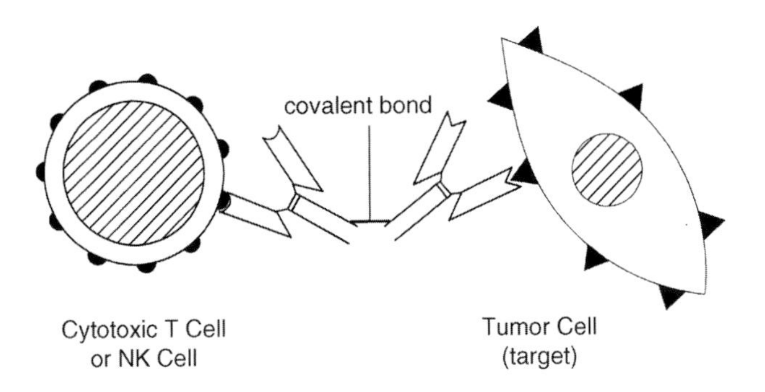

FIGURE 22.20 Heteroconjugate antibodies.

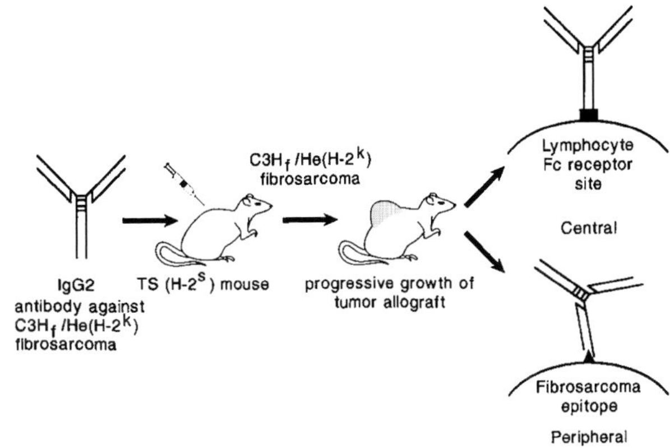

FIGURE 22.21 Immunologic enhancement (tumor enhancement).

Immunosurveillance refers to the policing or monitoring function of immune system cells to recognize and destroy clones of transformed cells prior to their development into neoplasms and to destroy tumors after they develop. Immunosurveillance is believed to be mediated by the cellular limb of the immune response. Indirect evidence in support of the concept includes (1) an increased incidence of tumors in aged individuals who have decreased immune competence, (2) increased tumor incidence in children with T cell immunodeficiencies, and (3) the development of neoplasms (lymphomas) in a significant number of organ or bone marrow transplant recipients who have been deliberately immunosuppressed.

Immunoselection is the selective survival of cells due to their diminished cell surface antigenicity. This permits these cells to escape the injurious effects of either antibodies or immune lymphoid cells.

Immunological escape is a mechanism of escape in which tumors that are immunogenic continue to grow in immunocompetent syngeneic hosts in the presence of a modest *in vivo* antitumor immune response. Escape mechanisms may facilitate tumors in evading a fatal tumoricidal response and render them incapable of inducing such a response. Failure of tumor antigen presentation by MHC class I molecules, and lack of costimulation and downregulation of tumor-destructive immune responses by tumor antigens, immune complexes, and molecules, and such as TGF-β and P15E are all believed to contribute to the inefficiency of tumor immunity.

Immunologic enhancement (tumor enhancement) (Figure 22.21) describes the prolonged survival, conversely the delayed rejection, of a tumor allograft in a host as a consequence of contact with a specific antibody. Antitumor antibodies may have a paradoxical effect. Instead of eradicating a neoplasm, they may facilitate its survival and progressive growth in the host. Both the peripheral and central mechanisms have been postulated. Coating of tumor cells with antibody was presumed, in the past, to interfere with the ability of specifically reactive lymphocytes to destroy them, but today a central effect in suppressing cell-mediated immunity, perhaps through suppressor T lymphocytes, is also possible. Enhancing antibodies are blocking antibodies that favor survival of tumor or normal tissue allografts.

Immunologic facilitation (facilitation immunologique) is the slightly prolonged survival of certain normal tissue allografts, e.g., skin, in mice conditioned with isoantiserum specific for the graft.

Immunotherapy employs immunologic mechanisms to combat disease. These include nonspecific stimulation of the immune response with BCG immunotherapy in treating certain types of cancer, and the IL-2/LAK cell adoptive immunotherapy technique for treating selected tumors.

Biological response modifiers (BRM) are a wide spectrum of molecules, such as cytokines, that alter the immune response. They include substances such as interleukins, interferons, hematopoietic colony-stimulating factors, tumor necrosis factor, B lymphocyte growth and differentiating factors, lymphotoxins, and macrophage-activating and chemotactic factors, as well as macrophage-inhibitory, eosinophils chemotactic, and osteoclast-activating factors, etc. BRM may modulate the immune system of the host to augment antirecombinant DNA technology and are available commercially. An example is α interferon used in the therapy of hairy cell leukemia.

Interferon α (IFN-α) is an immunomodulatory 189-amino acid residue glycoproteins synthesized by macrophages and B cells that are able to prevent the replication of viruses, are antiproliferative, and are pyrogenic, inducing fever. IFN-α stimulates natural killer cells and induces expression of class I MHC antigens. It also has an immunoregulatory effect through alteration of antibody responsiveness. The 14 genes that encode IFN-α are positioned on the short arm of chromosome 9 in man. Polyribonucleotides, as well as RNA or DNA viruses, may induce

IFN-α secretion. Recombinant IFN-α has been prepared and used in the treatment of hairy cell leukemia, Kaposi's sarcoma, chronic myeloid leukemia, human papilloma virus-related lesions, renal cell carcinoma, chronic hepatitis, and other selected conditions. Patients may experience severe flu-like symptoms as long as the drug is administered. They also have malaise, headache, depression, supraventricular tachycardia, and may possibly develop congestive heart failure. Bone marrow suppression has been reported in some patients.

Immunoscintigraphy (Figure 22.22 and Figure 22.23) is the formation of two-dimensional images of the distribution of radioactivity in tissues following the administration of antibodies labeled with a radionuclide that are specific for tissue antigens. A scintillation camera is used to record the images. **Immunolymphoscintigraphy** is a method used to determine the presence of tumor metastasis to lymph nodes. Antibody fragments or monoclonal antibodies against specific tumor antigens are radiolabeled and then detected by scintigraphy.

Radioimmunoscintigraphy is the use of radiolabeled antibodies to localize tumors or other lesions through use of radioactivity scanning following injection *in vivo*.

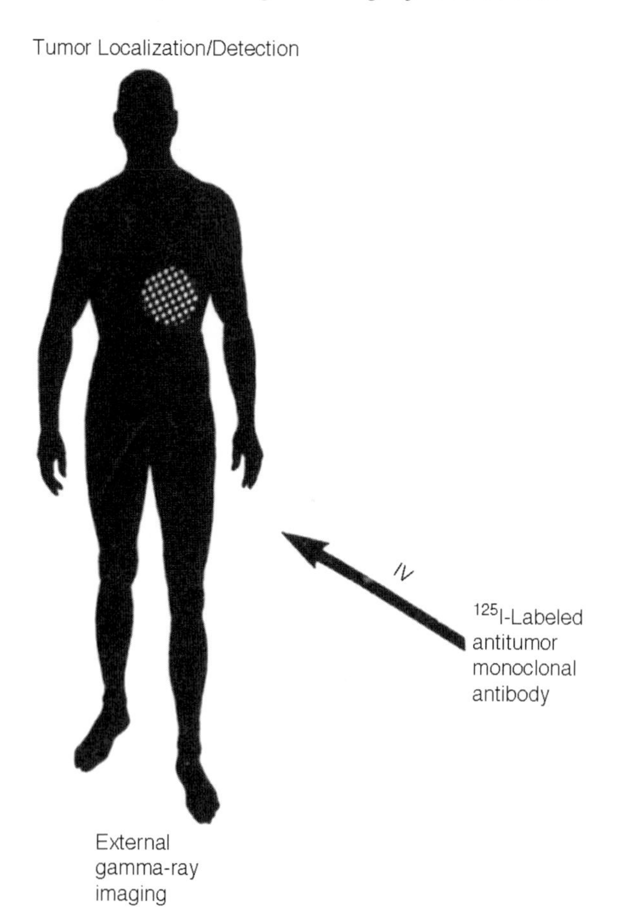

FIGURE 22.22 Immunoscintigraphy.

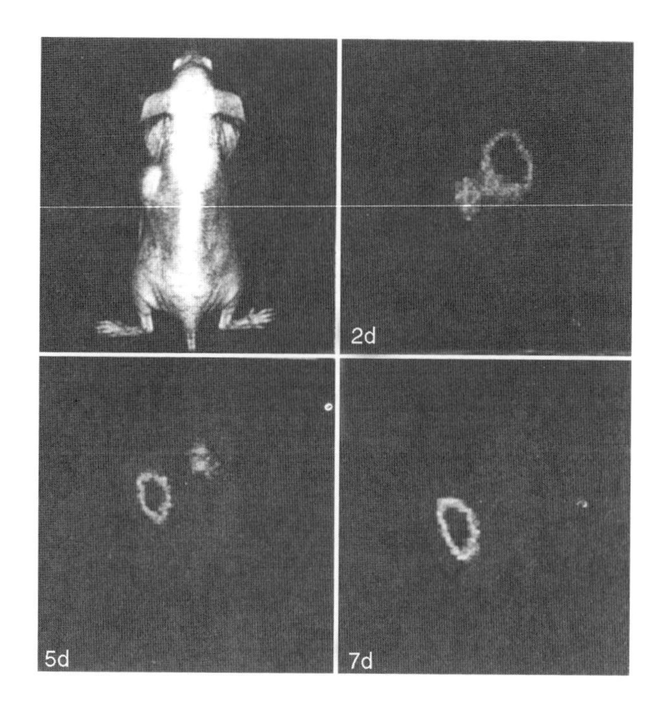

FIGURE 22.23 Immunoscintigraphy (nude mouse) with a ^{131}I-labeled monoclonal antibody. The mouse shown bears a human colon carcinoma in its left flank. The scintigrams were recorded 2, 5, and 7 days postinjection. While the second picture shows mainly the blood pool and little of the tumor, the tumor is the major imaged spot in the body after 5 days; after 7 days, only the tumor is recognizable.

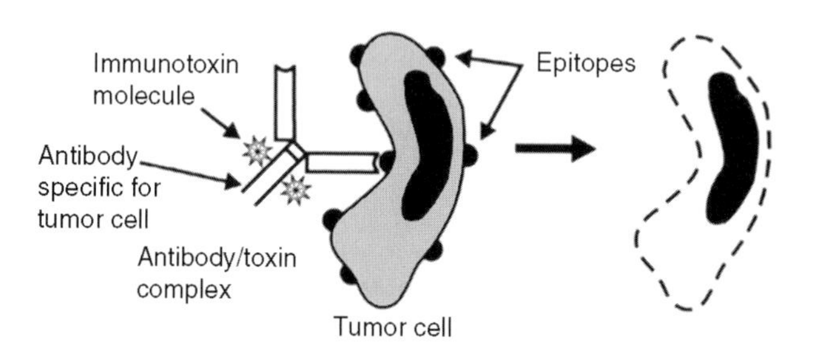

FIGURE 22.24 Immunotoxin.

Immunolymphoscintigraphy is a method to determine the presence of tumor metastasis to lymph nodes. Antibody fragments or monoclonal antibodies against specific tumor antigens are radiolabeled and then detected by scintigraphy.

An immunotoxin (Figure 22.24) is produced by linking an antibody specific for target cell antigens with a cytotoxic substance such as the toxin ricin. Upon parenteral injection, its antibody portion directs the immunotoxin to the target and its toxic portion destroys target cells on contact. An immunotoxin may also be a monoclonal antibody or one of its fractions linked to a toxic molecule such as a radioisotope, a bacterial or plant toxin, or a chemotherapeutic agent. The antibody portion is intended to direct the molecule to antigens on a target cell such as those of a malignant tumor and the toxic portion of the molecule is for the purpose of destroying the target cell. Contemporary methods of recombinant DNA technology have permitted the preparation of specific hybrid molecules for use in immunotoxin therapy. Immunotoxins may have difficulty reaching the intended target tumor, may be quickly metabolized, and may stimulate the development of antiimmunotoxin antibodies. Crosslinking proteins may likewise be unstable.

Among the uses of immunotoxin is the purging of T cells from hematopoietic cell preparations used for bone marrow transplantation. Immunotoxins have potential for antitumor therapy and as immunosuppressive agents.

Ricin is a toxic protein found in seeds of *Ricinus communis* (castor bean) plants. It is a heterodimer comprised of a 30-kDa α chain, which mediates cytotoxicity, and a 30-kDa β chain, which interacts with cell surface galactose residues that facilitate passage of molecules into cells in endocytic vesicles. Ricin inhibits protein synthesis by linkage of a dissociated α chain in the cytosol to ribosomes. The ricin heterodimer or its α chain conjugated to a specific antibody serves as an immunotoxin.

Ricinus communis: See ricin.

Abrin (Figure 22.25) and ricin are examples of immunotoxins. Abrin is a powerful toxin and lectin used in immunological research by Paul Ehrlich (*circa* 1900). It is extracted from the seeds of the jequirity plant and causes agglutination of erythrocytes.

Magic bullet is a term coined by Paul Ehrlich in 1900 to describe what he considered to be the affinity of a drug for a particular target. He developed "606" (salvarsan), an arsenical preparation, to treat syphilis. In immunology, it describes a substance that could be directed to a target by a specific antibody and injure the target once it arrives. Monoclonal antibodies have been linked to toxins such as diphtheria toxin, or ricin, as well as to cytokines for use as magic bullets.

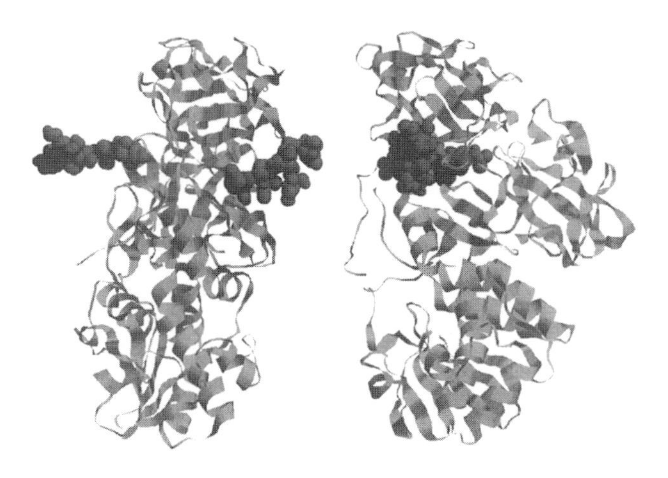

FIGURE 22.25 Abrin – A.

Adoptive immunotherapy (Figure 22.26 and Figure 22.27) is the experimental treatment of terminal cancer patients with metastatic tumors unresponsive to other modes of therapy by the inoculation of lymphokine-activated killer (LAK) cells or tumor-infiltrating lymphocytes (TIL) together with IL-2. This mode of therapy has shown some success in approximately one tenth of treated individuals with melanoma or renal cell carcinoma.

Lymphokine-activated killer (LAK) cells are lymphoid cells derived from normal or tumor patients cultured in medium with recombinant IL-2, which become capable of lysing NK-resistant tumor cells as revealed by ^{51}Cr-release cytotoxicity assays. These cells are also referred to as lymphokine-activated killer cells. Most LAK activity is derived from NK cells. The large granular lymphocytes (LGL) contain all LAK precursor activity and all active NK cells. In accord with the phenotype of precursor cells, LAK effector cells are also granular lymphocytes expressing markers associated with human NK cells. The asialo Gm_1^+ population, known to be expressed by murine NK cells, contains most LAK precursor activity. Essentially all LAK activity resides in the LGL population in the rat. LAK cell and IL-2 immunotherapy has been employed in human cancer patients with a variety of histological tumor types when conventional therapy has been unsuccessful. Approximately one fourth of LAK- and IL-2-treated patients manifested significant responses, and some individuals experienced complete remission. Serious side effects include fluid retention and pulmonary edema attributable to the administered IL-2.

LAK cells are lymphokine-activated killer cells.

IL-2/LAK cells are interleukin-2/lymphokine-activated killer cells. NK cells, which express only the p70 and not the p55 receptor for IL-2, are incubated with IL-2, converting them into an activated form referred to as LAK cells. The IL-2/LAK combination has been used to treat cancer patients through adoptive immunotherapy, which has been successful in inducing transient regression of tumors in selected cases of melanoma, colorectal carcinoma, non-Hodgkin's lymphoma, and renal cell carcinoma, as well as regression of metastases in the liver and lung of some patients. There may be transient defective chemotaxis of neutrophils, and patients often develop "capillary leak syndrome," producing pulmonary edema. Patients may also develop congestive heart failure.

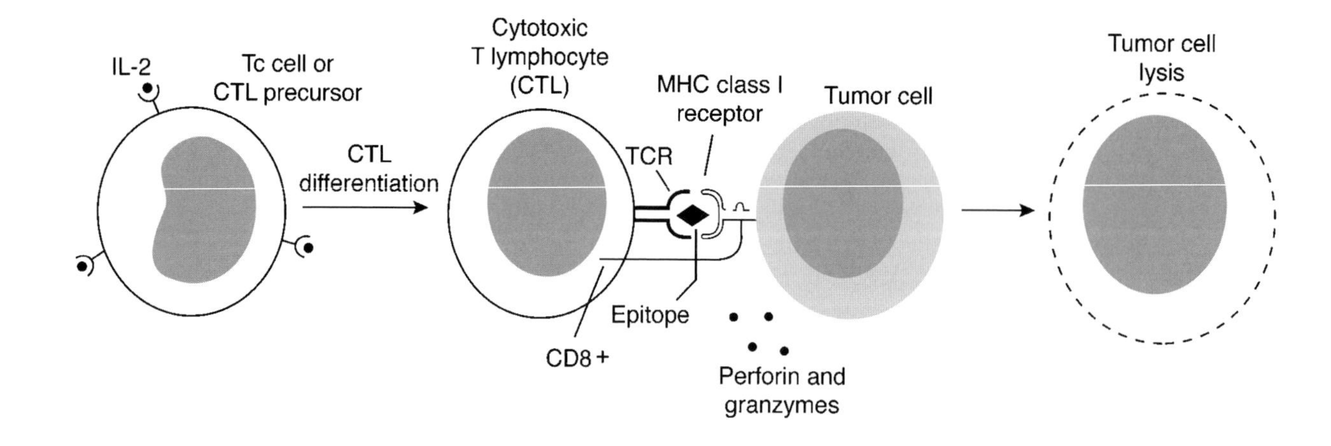

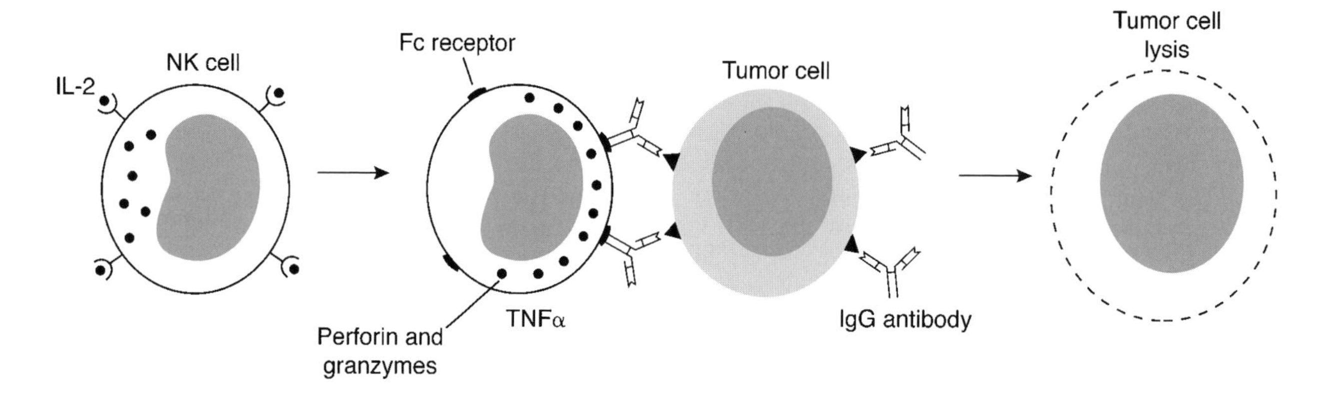

FIGURE 22.26 Interleukin-2 (IL-2) immunotherapy.

Tumor immunity: Numerous experimentally induced tumors in mice express numerous specific transplantation antigens which can induce an immune response that leads to destruction of neoplastic cells *in vivo.* Lymphocytes play a critical role in the immunological destruction of many antigenic tumors. Both cell-mediated and antibody-mediated immune responses to human neoplasms have been identified and their targets characterized in an effort to develop clinically useful immunotherapy.

Tumor-infiltrating lymphocytes (TIL) are lymphocytes isolated from the tumor they are infiltrating. They are cultured with high concentrations of IL-2, leading to expansion of these activated T lymphocytes *in vitro.* TILs are very effective in destroying tumor cells and have proven much more effective than LAK cells in experimental models. TILs have 50 to 100 times the antitumor activity produced by LAK cells. TILs have been isolated and grown from multiple resected human tumors, including those from kidney, breast, colon, and melanoma. In contrast to the non-B-non-T LAK cells, TILs nevertheless are generated from T lymphocytes and phenotypically resemble cytotoxic T lymphocytes. TILs from malignant melanoma exhibit specific cytolytic activity against cells of the tumor from which they were extracted, whereas LAK cells have a broad range of specificity. TILs appear unable to lyse cells of melanomas from patients other than those in whom the tumor originated. TILs may be tagged in order that they may be identified later.

TIL is the abbreviation for tumor-infiltrating lymphocytes.

Reverse immunology is a process that involves computerized algorithms to predict the likelihood of a particular mutation resulting in a strong antigen. Several of the mutant proteins and peptides have been used to examine the possibility of inducing tumor-specific immunity.

Hybrid resistance is the resistance of members of an F_1 generation of animals to growth of a transplantable neoplasm from either one of the parent strains.

Concomitant immunity is resistance to a tumor that has been transplanted into a host already bearing that tumor. Immunity to the reinoculated neoplasm does not inhibit growth of the primary tumor.

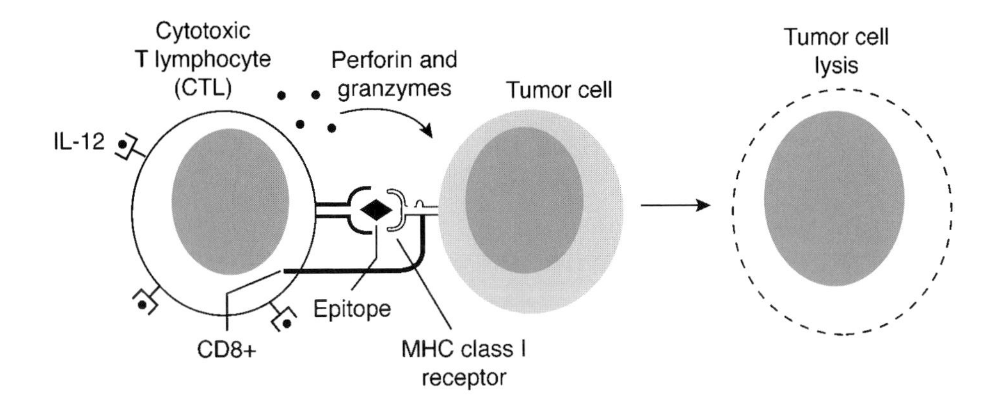

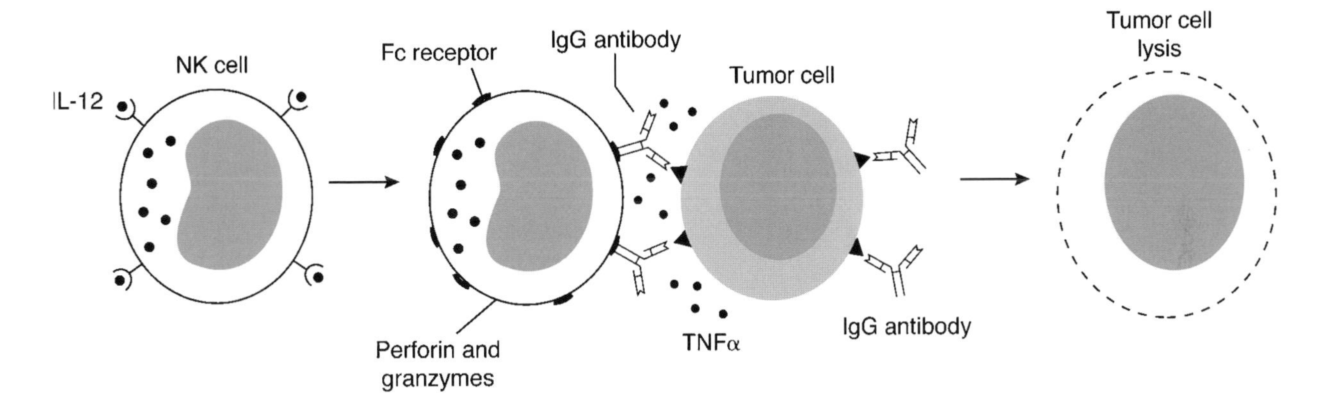

FIGURE 22.27 Interleukin-12 (IL-12) immunotherapy.

CA-15-3 is an antibody specific for an antigen frequently present in the blood serum of metastatic breast carcinoma patients.

CA-19-9 is a tumor-associated antigen found on the Lewis A blood group antigen that is sialylated or in mucin-containing tissues. In individuals whose serum levels exceed 37 U/ml, 72% have carcinoma of the pancreas. In individuals whose levels exceed 1000 U/ml, 95% have pancreatic cancer. Anti-CA-19-9 monoclonal antibody is useful to detect the recurrence of pancreatic cancer following surgery and to distinguish between neoplastic and benign conditions of the pancreas. However, it is not useful for pancreatic cancer screening.

CA-125 is a mucinous ovarian carcinoma cell surface glycoprotein detectable in the patient's blood serum. Increasing serum concentrations portend a grave prognosis. It may also be found in the blood sera of patients with other adenocarcinomas such as breast, gastrointestinal tract, uterine cervix, and endometrium.

CALLA is common acute lymphoblastic leukemia antigen. Also known as CD10.

Calcitonin is a hormone that influences calcium ion transport. Immunoperoxidase staining demonstrates calcitonin in thyroid parafollicular or C cells. It serves as a marker characteristic of medullary thyroid carcinoma and APUD neoplasms. Lung and gastrointestinal tumors may also form calcitonin.

Blocking factors are agents such as immune complexes in the serum of tumor-bearing hosts that interfere with the capacity of immune lymphoid cells to mediate cytotoxicity of tumor target cells.

Antimalignin antibodies are specific for the 10-kDa protein malignin comprised of 89 amino acids. These antibodies are claimed to be increased in cancer patients without respect to tumor cell type. It has been further claimed that antibody levels are related to survival. These claims will require additional confirmation and proof to be accepted as fact.

The **Winn assay** is a method to determine the ability of lymphoid cells to inhibit the growth of transplantable tumors *in vivo*. Following incubation of the lymphoid cells and tumor cells *in vitro*, the mixture is injected into the

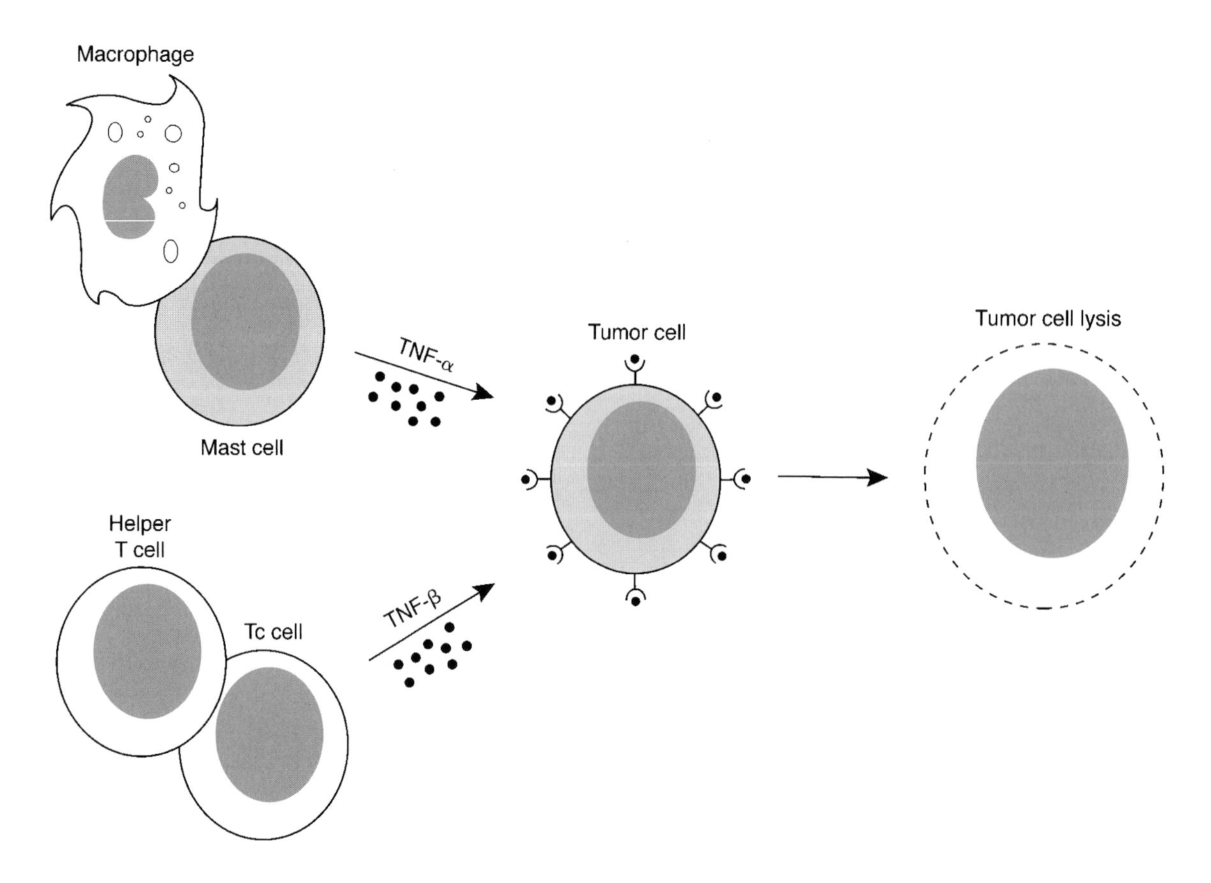

FIGURE 22.28 Tumor necrosis factor (TNF)-mediated immune reaction.

skin of X-irradiated mice. Growth of the transplanted cells is followed. T lymphocytes that are specifically immune to the tumor cells will inhibit tumor growth and provide information related to tumor immunity.

Tumor necrosis factor α (TNF-α) (Figure 22.28 and Figure 22.29) is a cytotoxic monokine produced by macrophages stimulated with bacterial endotoxin. TNF-α participates in inflammation, wound healing, and remodeling of tissue. TNF-α, which is also called cachectin, can induce septic shock and cachexia. It is a cytokine comprised of 157 amino acid residues. It is produced by numerous types of cells including monocytes, macrophages, T lymphocytes, B lymphocytes, NK cells, and other types of cells stimulated by endotoxin or other microbial products. The genes encoding TNF-α and TNF-β (lymphotoxin) are located on the short arm of chromosome 6 in man in the MHC region. High levels of TNF-α are detectable in the blood circulation very soon following administration of endotoxins or microorganisms. The administration of recombinant TNF-α induces shock, organ failure, and hemorrhagic necrosis of tissues in experimental animals including rodents, dogs, sheep, and rabbits, closely resembling the effects of lethal endotoxemia. TNF-α is produced during the first 3 d of wound healing. It facilitates leukocyte recruitment, induces angiogenesis, and promotes fibroblast proliferation. It can combine with receptors on selected tumor cells and induce their lysis. TNF mediates the antitumor action of murine natural cytotoxic (NC) cells, which distinguishes their function from that of NK and cytotoxic T cells. TNF-α was termed "cachectin" because of its ability to induce wasting and anemia when administered on a chronic basis to experimental animals. Thus it mimics the action in cancer patients and in those with chronic infection with HIV or other pathogenic microorganisms. It can induce anorexia which may lead to death from malnutrition.

Tumor necrosis factor-β (TNF-β) is a 25-kDa protein synthesized by activated lymphocytes. It can kill tumor cells in culture, induce expression of genes, stimulate proliferation of fibroblasts, and can mimic most of the actions of TNF-α (cachectin). It participates in inflammation and graft rejection and was previously termed "lymphotoxin." TNF-β and TNF-α have approximately equivalent affinity for TNF receptors. Both 55-kDa and 80-kDa TNF receptors bind TNF-β. TNF-β has diverse effects that include killing of some cells and causing proliferation of others. It is the mediator whereby cytolytic T cells, natural killer

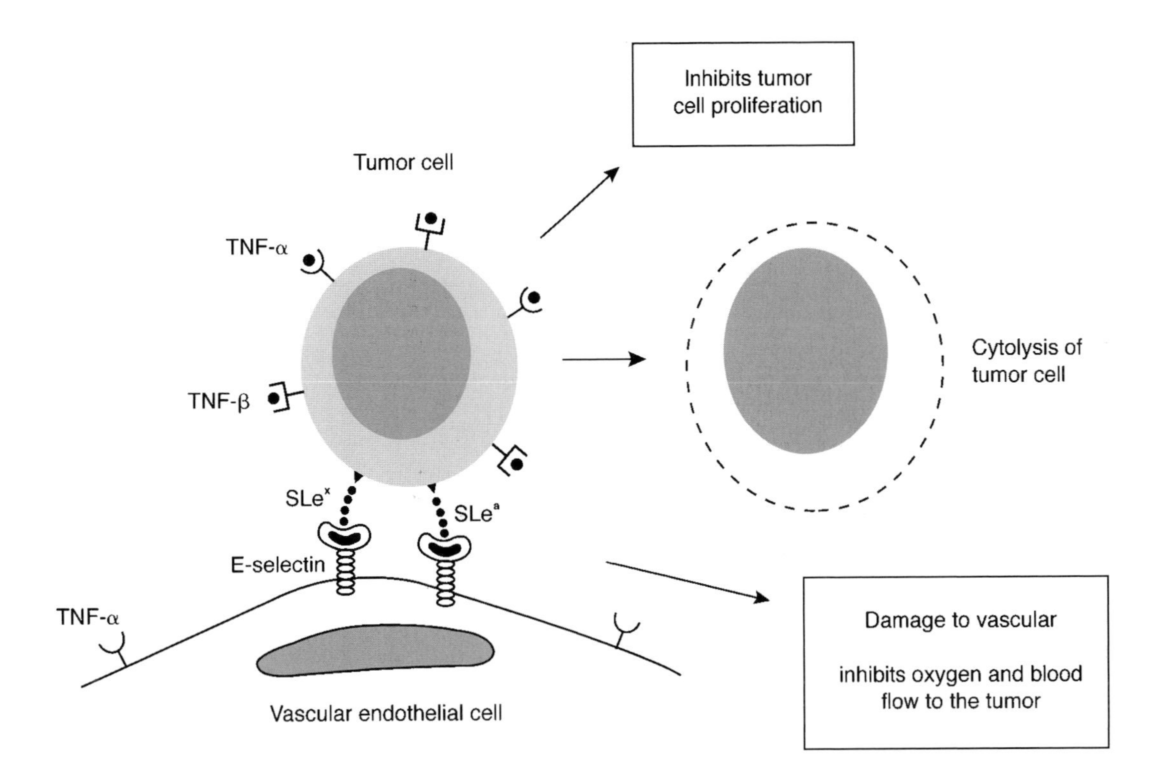

FIGURE 22.29 Tumor necrosis immunotherapy.

cells, lymphokine-activated killer cells, and "helper-killer" T cells induce fatal injury to their targets. TNF-β and TNF-α have been suggested to play a role in AIDS, possibly contributing to its pathogenesis.

Tumor necrosis factor receptor is a receptor for tumor necrosis factor that is comprised of 461 amino acid residues and which possesses an extracellular domain that is rich in cysteine.

23 Immunity against Microorganisms

Natural immunity: Entry of a pathogenic organism into a susceptible host is followed by invasion and colonization of tissues, circumvention of the host immune response, and injury or decreased function of host tissues. Microbial immunity consists of several factors. Natural and acquired immune mechanisms facilitate the body's resistance against microorganisms. Microbes vary in the lymphocyte responsiveness and effector mechanisms they elicit. The skill with which pathogenic microorganisms resist the host's immune defense mechanisms governs their survival and pathogenicity. Paradoxically, the host response to a pathogenic microorganism, rather than the microbe itself, may induce injury to host tissues. Factors that determine the outcome of man's encounter with pathogenic microorganisms include the microbe's virulence and the size of the infecting dose on the one hand and specific defense mechanisms of the host on the other.

Pathogenicity refers to the capacity of a microorganism to induce disease. Factors that contribute to pathogenicity include toxin production, activation of host inflammatory responses, and perturbation of host cell metabolism. If host defenses are decreased significantly, as in the immunocompromised host, opportunistic infections, produced by microorganisms that are not normally pathogenic for the individual, may result. There are multiple causes for diminished host resistance that include accidentally or surgically induced trauma to the mucous membranes or skin, localized lesions, leukocyte defects, complement defects, or defective B or T cell responses. Various drugs such as antibiotics may also alter the normal flora of the body. Microbes that produce opportunistic infections generally are of low virulence, i.e., their level of pathogenicity is low.

Both nonspecific constitutional factors and specific immune mechanisms provide host resistance. Nonspecific resistance mechanisms protect against body surface colonization by microorganisms with pathogenic potential, thereby blocking their penetration of underlying tissues.

Protective immunity consists of both natural, nonspecific immune mechanisms as well as actively acquired specific immunity that results in the defense of a host against a particular pathogenic microorganism. Protective immunity may be induced either by active immunization with a vaccine prepared from antigens of a pathogenic microorganism or by experiencing either a subclinical or clinical infection with the pathogenic microorganism.

The skin, as well as mucous membranes of various anatomical regions such as the conjunctiva, nose, mouth, intestinal tract, and lower genital tract, have a normal commensal flora. Microbial properties, host factors, and exogenous factors determine the nature of colonization. The ability of a microorganism to adhere to mucosa or epithelial cells is a significant factor. Microbes in the normal flora may compete with pathogenic microorganisms for receptors on cell surfaces. Fibronectin on epithelial cells may bind *Staphylococcus aureus* and group A hemolytic streptococci. Microbes in the commensal flora may also synthesize bacteriocins that inhibit other bacteria. They may also compete with them for nutrient substances. Thus, the normal flora serves as an effective mechanism for inducing colonization resistance. This can be interrupted by the use of broad-spectrum antibiotics resulting in colonization of the surface by pathogenic microorganisms. Gram-negative bacteremia may even result in an immunocompromised host. Another consequence of antibiotic therapy may be overgrowth of yeast or of *Clostridium difficile*, a toxin-producing Gram-positive bacillus that is an anaerobic and antibiotic resistant that can lead to diarrhea and colitis. Host age, hormones, nutrition, and diseases such as diabetes mellitus or malignancy may influence the normal flora. For example, the vaginal flora is sparse in both prepubertal and post menopausal females, but is rich in acidophilic lactobacilli during the child-bearing years of life. Lactobacilli convert glycogen to lactic acid, yielding a pH of 4 to 5, which inhibits many potential pathogens that might otherwise colonize the vaginal mucosa. Microorganisms in the normal flora of various anatomical regions may induce natural antibodies that would be active against potential pathogenic microorganisms bearing cross-reacting antigens.

Secretory immunoglobulin A (SigA) can interfere with the attachment of bacteria to host cells by coating the microbes. It may also neutralize their exotoxins, inhibit their motility, and agglutinate them. It is not involved in opsonization or lysis of bacteria through complement. SIgA's ability to prevent adherence of such microorganisms as *Vibrio cholerae, Giardia lamblia*, and selected respiratory viruses to mucosal surfaces represents a significant defense mechanism. Whereas gastric acidity can destroy most microorganisms, *Mycobacterium tuberculosis* and enteroviruses are not destroyed by it. Gram-negative bacteria may colonize the stomach and small intestine

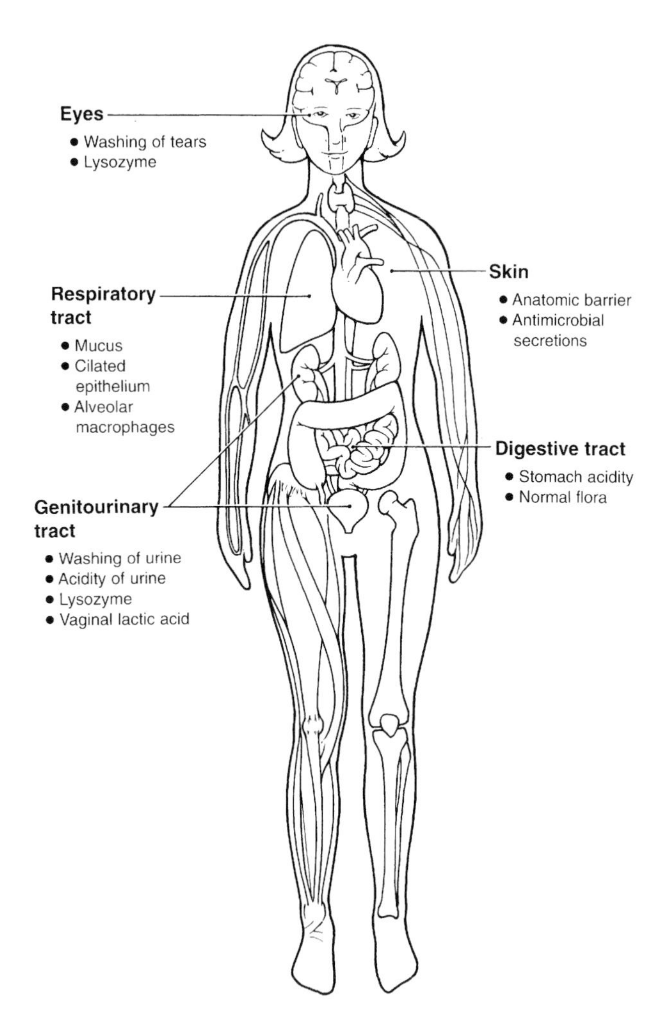

FIGURE 23.1 External defense barriers of the human body.

Intact skin
Mucus Motion of cilia Coughing/sneezing
Cell shedding
Flushing of microbes by tears, saliva, urine, perspiration, other body fluids
Emesis and diarrhea aid microbial elimination

FIGURE 23.2 Mechanical barriers against infection.

in subjects with achlorhydria. Unconjugated bile may prevent bacterial growth in the small gut. Intestinal peristalsis also guards against overgrowth of microorganisms in blind loops.

The skin and mucous membrane serve as **mechanical barriers** to the entrance of microorganisms (Figure 23.1 and Figure 23.2). The papilloma virus and a few other infection agents may penetrate the skin, but most microorganisms are excluded by it. Free fatty acids from sebaceous glands and lactic acid present in perspiration together with an acid pH of 5 to 6 and the dryness of the skin are unfavorable to microorganisms. *Staphylococcus aureus* may colonize hair follicles and sweat glands to produce furuncles, carbuncles, and abscesses. *Pseudomonas aeruginosa* may infect skin injured by burns. Injury to the gastric mucosa by irradiation or cytotoxic drugs may culminate in infection by the normal flora of the intestine.

Lysozyme and lactoferrin are both antimicrobial substances found in **surface secretions** of the mucosa. Lysozyme induces lysis of bacterial cells through breaking the linkage connecting *N*-acetyl muraminic acid and *N*-acetylglucosamine in the walls of Gram-positive bacterial cells. Lactoferrin interrupts metabolism of bacterial iron.

Lysozyme can induce lysis of some Gram-positive bacterial cell walls but not Gram-negative bacteria unless antibody and complement are also present. It accentuates complement activity. Lactoferrin, a protein that binds iron, competes with microorganisms for this substance. By chelating iron, lactoferrin deprives microbes of the free iron they require for growth. Neutrophil secondary granules also contain lysozyme and lactoferrin. Beta lysin is a thrombocyte-derived antibacterial protein that is effective mainly against Gram-positive bacteria. It is released when blood platelets are disrupted, as occurs during clotting. β lysin acts as a nonantibody humoral substance that contributes to nonspecific immunity (Figure 23.3).

Inhaled microorganisms in dust or droplets greater than t μm adhere to the mucosa lining the upper respiratory tract and are swept upward by cilia to the posterior pharynx, followed by expectoration or swallowing; this is called **directional flow**. Particles less than 5 μm reach the alveoli and are phagocytized by alveolar macrophages. Cigarette smoke or other pollutants, as well as bacterial or viral infection such as pertussis and influenza, may diminish the sweeping action of cilia, thereby rendering the subject susceptible to secondary bacterial pneumonia. Intubation and tracheostomy may also decrease normal resistance mechanisms, leading to infection. Tears and blinking actions protect the eyes causing microorganisms to be diluted and flushed out through the nasolacrimal duct into the nasopharynx. The antibacterial action of tears is attributable to lysozyme.

Even though urine can support bacterial growth in the bladder, the acid pH of urine and voiding serve as defensive mechanisms against infection. Ascending infection that is discouraged by the longer male urethra is more common in females with a shorter urethra. Urinary stasis in subjects with posterior urethral valves, prostatic hypertrophy, or calculi facilitates infections.

An **opsonin** is a substance that adheres to the surface of a microorganism and makes it more attractive or delectable

Factor	Function	Source
Lysozyme	Catalyzes hydrolysis of cell wall muco-peptide	Tears, saliva, nasal secretions, body fluids, lysosomal granules
Lactoferrin, transferrin	Binds iron and competes with microorganisms for it	Specific granules of PMNs
Lactoperoxidase	May be inhibitory to many micro-organisms	Milk and saliva
Beta-lysin	Effective mainly against Gram-positive bacteria	Thrombocytes, normal serum
Chemotactic factors	Induce reorientation and directed migration of PMNs, monocytes, and other cells	Bacterial substances and products of cell injury and denatured proteins
Properdin	Activates complement in the absence of antibody-antigen complex	Normal plasma
Interferons	Act as immuno-modulators to increase the activities of macrophages	Leukocytes, fibroblasts, natural killer cells, T cells
Defensins	Block cell transport activities	Polymorphonuclear granules

FIGURE 23.3 Nonspecific humoral defense mechanisms.

to a phagocyte. Opsonins facilitate or enhance phagocytosis of microbes, which constitutes a cornerstone of constituitive defense against infection. Both nonimmune and immune substances may serve as opsonins. C3b, produced during complement activation, forms a covalent bond with the bacterial cell surface thereby rendering it susceptible to phagocytosis by C3b receptor-bearing neutrophils, monocytes, and macrophages. Adherence of opsonized bacteria to the phagocyte cell surface facilitates phagocytosis. Leukocyte receptors for C3b are termed CR1, CR2, CR3, and CR4. Among the pediatric population, CR3 is associated with increased susceptibility to bacterial infections, a condition termed leukocyte adhesion deficiency.

Opsonins bind to bacteria, erythrocytes, or other particles to increase their susceptibility to phagocytosis. Opsonins include antibodies such as IgG3, IgG1, and IgG2 that are specific for epitopes on the particle surface. Following interaction, the Fc region of the antibody becomes anchored to Fc receptors on phagocyte surfaces, thereby facilitating phagocytosis of the particles. In contrast to these so-called heat-stable antibody opsonins are the heat-labile products of complement activation such as C3b or C3bi, which are linked to particles by transacylation with the C3 thiolester. C3b combines with complement receptor 1 and C3bi combines with complement receptor 3 on phagocytic cells. Other substances that act as opsonins include the basement membrane constituent, fibronectin (Figure 23.4).

Fibronectin is an adhesion-promoting dimeric glycoprotein of relatively high mol wt found abundantly in the connective tissue and basement membrane. The tetrapeptide Arg-Gly-Asp-Ser facilitates cell adhesion to fibrin; Clq; collagens; heparin; and type I-, II-, III-, V-, and VI-sulfated proteoglycans. Fibronectin is also present in plasma and on normal cell surfaces. Approximately 20 separate fibronectin chains are known. They are produced from the fibronectin gene by alternative splicing of the RNA transcript. Fibronectin is comprised of two 250-kDa subunits joined near their carboxy terminal ends by disulfide bonds. The amino acid residues in the subunits vary in number from 2145 to 2445. Fibronectin is important in contact inhibition, cell movement in embryos, cell-substrate adhesion, inflammation, and wound healing. It may also serve as an opsonin and function as an adhesion molecule in cellular interactions. Fibronectin may also react with complement components. Intensive care patients often lose fibronectin from their pharynx causing alteration of the normal flora with colonization by coliforms.

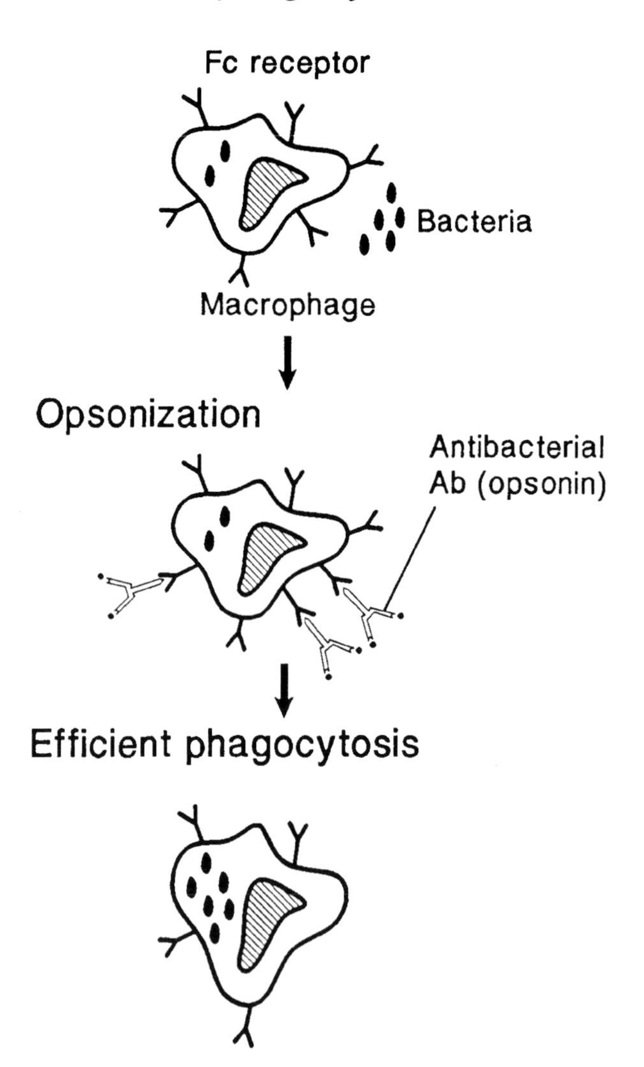

FIGURE 23.4 Opsonization.

Microorganism	Cell Type
Extracellular bacteria	Polymorphonuclear neutrophils (PMNs)
Parasites	Eosinophils
Intracellular microorganisms (i.e., mycobacteria or fungi)	Macrophages
Viruses	Lymphocytes, NK cells

FIGURE 23.5 Nonspecific cellular defense.

Azurophil granules (Primary granules)	Specific granules (Secondary granules)
Bacterial permeability-inducing protein (BPI)	
Cathespin G	
Cationic antimicrobial protein (CAP) 57	
Cationic antimicrobial protein (CAP) 37	
Defensins: HP1 HP2 HP3	
Elastase	
Lysozyme	
Myeloperoxidase	Bacterial chemotaxin receptors
Collagenase	
C5a receptors	
Gelatinase	
Lactoferrin	
Lysozyme	
NADPH	
Vitamin B_{12}-binding protein	

FIGURE 23.6 Substances associated with neutrophils.

PMNs (polymorphonuclear neutrophils), sometimes called the soldiers of the body, are first to arrive at areas of invading and rapidly multiplying bacteria. They contain both primary or azurophilic granules and secondary or specific granules that serve as reservoirs for the digestive and hydrolytic enzymes such as lysozyme (Figure 23.6) before they are delivered to the phagosome. Frequently, the PMNs die after ingesting and destroying the invading microorganisms. Macrophages that serve as scavengers ingesting debris left by neutrophils killed by the microorganisms they phagocytized, are resilient and survive.

A **secondary granule** is a structure in the cytoplasm of polymorphonuclear leukocytes that contains vitamin B_{12}-binding protein, lysozyme, and lactoferrin in neutrophils. Cationic peptides are present in eosinophil secondary granules. Histamine, platelet-activating factor, and heparin are present in the secondary granules of basophils.

Phagocytic cells are polymorphonuclear neutrophils and eosinophils as well as macrophages (the mononuclear phagocytes) which have a critical role in defending the host against microbial infection (Figure 23.5). Polymorphonuclear neutrophils and occasionally eosinophils appear first in areas of acute inflammation followed later on by macrophages. Chemotactic factors, including formylmethionylleucylphenylalanine (f-Met-Leu-Phe) are released by actively multiplying bacteria. This is a powerful attractant for PMNs whose membrane have a specific receptor for it. Different types of infectious agents may stimulate different types of cellular response. When particles greater than 1 mm become attached and engulfed by a cell, the process is known as phagocytosis. Various factors present in the serum and known as oposins coat

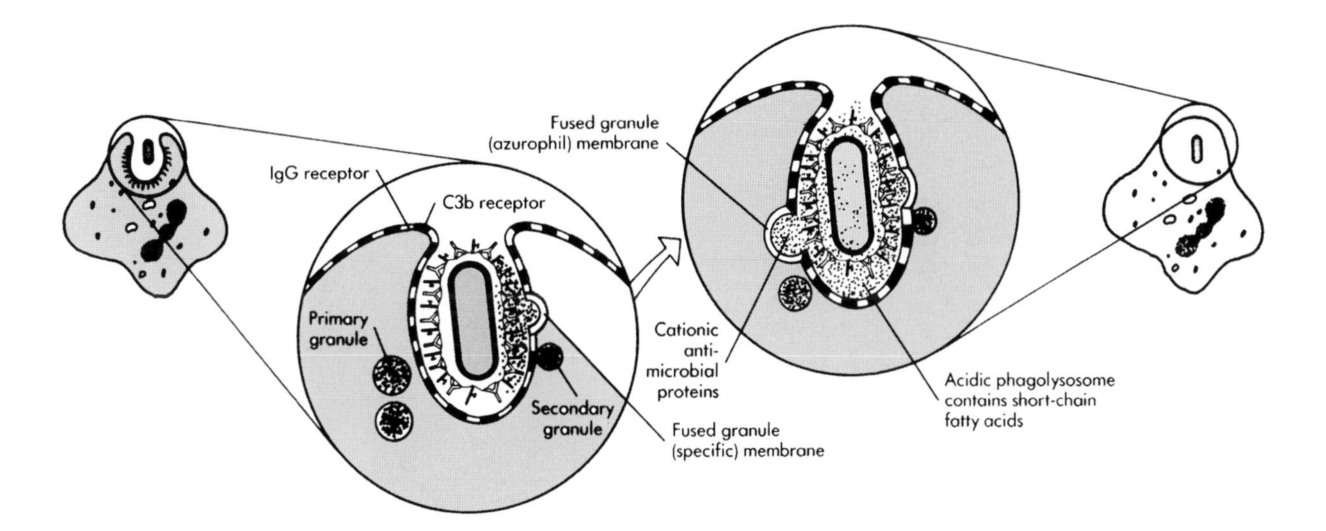

FIGURE 23.7 Steps of phagocytic endocytosis.

microorganisms or other particles and make them more delectable to phagocyte cells. These include nonspecific substances such as complement component C3b, as well as specific antibodies located in the IgG or IgG3 fractions. Capsules enable microorganisms such as pneumococci and *Hemophilus* to resist phagocytosis.

Mononuclear phagocytes are mononuclear cells with pronounced phagocytic ability that are distributed extensively in lymphoid and other organs. "Mononuclear phagocyte system" should be used in place of the previously popular "reticulo-endothelial system" to describe this group of cells. Mononuclear phagocytes originate from stem cells in the bone marrow that first differentiate into monocytes that appear in the blood for approximately 24 h or more with final differentiation into macrophages in the tissues. Macrophages usually occupy perivascular areas. Liver macrophages are termed Kupffer cells, whereas those in the lung are alveolar macrophages. The microglia represent macrophages of the central nervous system, whereas histiocytes represent macrophages of connective tissue. Tissue stem cells are monocytes that have wandered from the blood into the tissues and may differentiate into macrophages. Mononuclear phagocytes have a variety of surface receptors that enable them to bind carbohydrates or such protein molecules as C3 via complement receptor 1 and complement receptor 3, and IgG and IgE through Fcγ and Fcε receptors. The surface expression of MHC class II molecules enables both monocytes and macrophages to serve as antigen-presenting cells to $CD4^+$ T lymphocytes. Mononuclear phagocytes secrete a rich array of molecular substances with various functions. A few of these include interleukin-1; tumor necrosis factor α; interleukin-6; C2, C3, C4, and factor B complement proteins; prostaglandins; leukotrienes; and other substances.

Mononuclear phagocytes include monocytes in the blood and macrophages in the tissues which have cell surface receptors for Fcγ and C3b. They are also able to phagocytize microorganisms coated with opsonins and kill many but not all microorganisms during the process. Some microorganisms, such as mycobacteria, survive and multiply with macrophages which may serve as a reservoir or transport mechanism to help them reach other areas of the body. Other intracellular microorganisms that are not killed by nonimmune macrophages include *Listeria monocytogenes, Brucella* species, *Legionella pneumophilia, Cryptococcus neoformans, Toxoplasma gondii*, and *Pneumocystis carinii*. However, the development of cell-mediated immunity with the production of gamma interferon by lymphocytes is able to activate these macrophages to enable them to kill the intracellular pathogens. Activated macrophages produce interleukin-1 and tumor necrosis factor alpha which promote inflammation.

Phagocytosis (Figure 23.7) is an important clearance mechanism for the removal and disposition of foreign agents and particles or damaged cells. Macrophages, monocytes, and polymorphonuclear cells are phagocytic cells. In special circumstances, other cells such as fibroblasts may show phagocytic properties; these are called facultative phagocytes.

Phagocytosis may involve nonimmunologic or immunologic mechanisms. Nonimmunologic phagocytosis refers to the ingestion of inert particles such as latex beads, or of other particles that have been modified by chemical treatment or coated with protein. Details of the recognition process of such particles are not known. Damaged cells are also phagocytized by nonimmunologic mechanisms. It is believed that in the latter case, damaged cells are also coated with immunoglobulin or other proteins which facilitate their recognition.

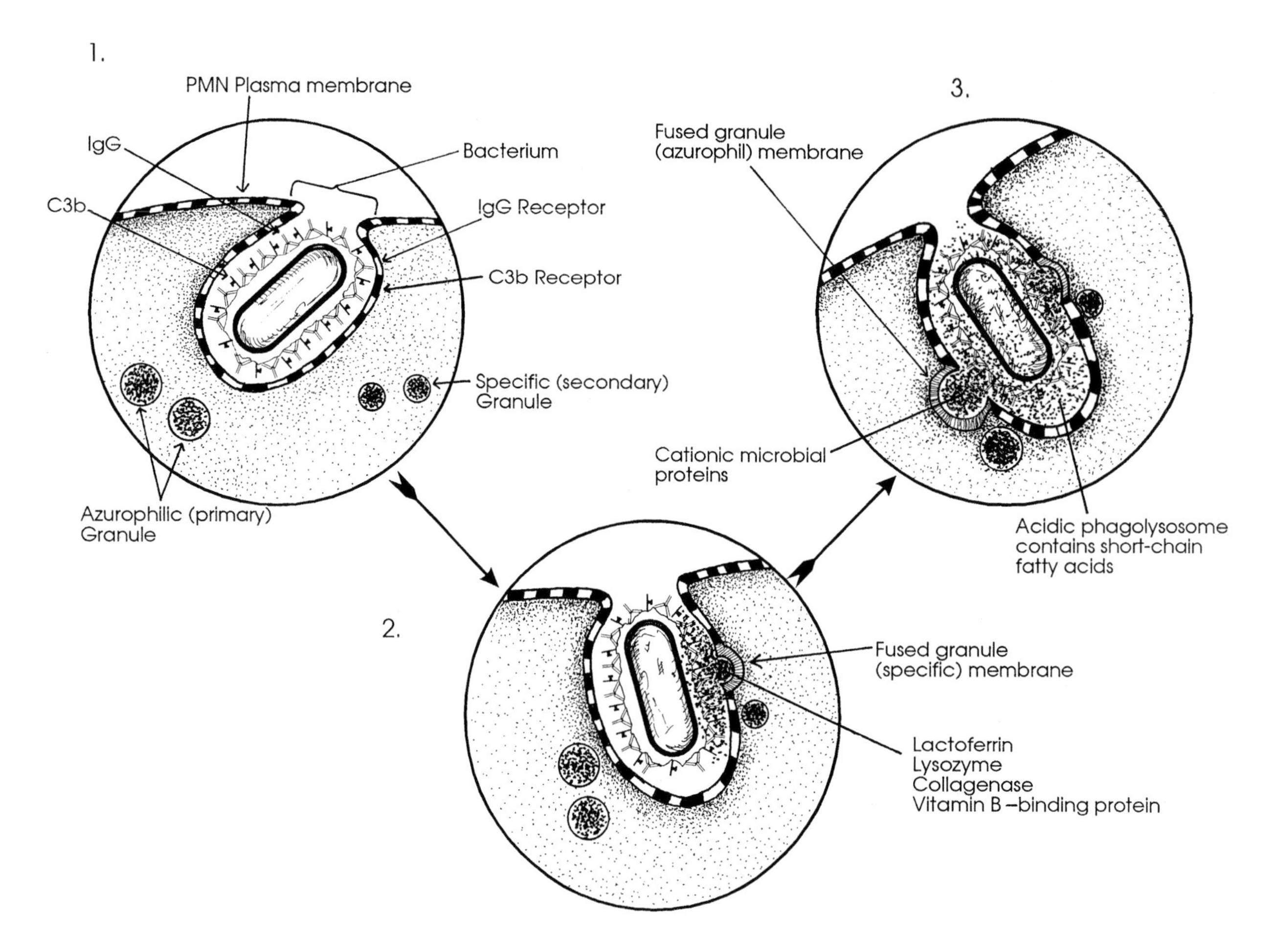

FIGURE 23.8 Phagocytosis.

Phagocytosis involves several steps: attachment, internalization, and digestion (Figure 23.8). The initiation of ingestion is known as the "zipper mechanism." After attachment, the particle is engulfed within a fragment or plasma membrane and forms a phagocytic vacuole. This vacuole fuses with the primary lysosomes to form the phagolysosome in which the lysosomal enzymes are discharged and the enclosed material is digested. Remnants of indigestible material can be recognized subsequently as residual bodies. The process is associated with stimulation of phagocyte metabolism.

Phagocytic dysfunction may be due to either extrinsic or intrinsic defects. The extrinsic variety encompasses opsonin deficiencies secondary to antibody or complement factor deficiencies, suppression of phagocytic cell numbers by immunosuppressive agents, corticosteroid-induced interference with phagocytic function, decreased neutrophils through antineutrophil autoantibody; and abnormal neutrophil chemotaxis as a consequence of complement deficiency or abnormal complement components. Intrinsic phagocytic dysfunction is related to deficiencies in enzymatic deficiencies that participate in the metabolic pathway leading to bacterial cell killing. These intrinsic disorders include chronic granulomatous disease, characterized by defects in the respiratory burst pathway, myeloperoxidase deficiency, and glucose-6-phosphate dehydrogenase deficiency (G6PD). Consequences of phagocytic dysfunction include increased susceptibility to bacterial infections but not to viral or protozoal infections. Selected phagocytic function disorders may be associated with severe fungal infections. Severe bacterial infections associated with phagocytic dysfunction range from mild skin infections to fatal systemic infections.

Phagocytosis may involve nonimmunologic or immunologic mechanisms. Nonimmunologic phagocytosis refers to the ingestion of inert particles such as latex particles or of other particles that have been modified by chemical treatment or coated with protein. Damaged cells are also phagocytized by nonimmunologic mechanisms. Damaged cells may become coated with immunoglobulin or other proteins which facilitate their recognition.

Phagocytosis of microorganisms involves several steps: attachment, internalization, and digestion. After attachment, the particle is engulfed within a membrane fragment and a phagocytic vacuole is formed. The vacuole fuses with the primary lysosome to form the phagolysosome, in which the lysosomal enzymes are discharged and the enclosed material is digested. Remnants of indigestible material can be recognized subsequently as residual bodies. Polymorphonuclear neutrophils (PMNs), eosinophils, and macrophages play an important role in defending the host against microbial infection. PMNs and occasional eosinophils appear first in response to acute inflammation, followed later by macrophages. Chemotactic factors are released by actively multiplying microbes. These chemotactic factors are powerful attractants for phagocytic cells which have specific membrane receptors for the factors. Certain pyogenic bacteria may be destroyed soon after phagocytosis as a result of oxidative reactions. However, certain intracellular microorganisms such as *Mycobacteria* or *Listeria* are not killed merely by ingestion and may remain viable unless there is adequate cell-mediated immunity induced by γ interferon activation of macrophages.

Phagocytic dysfunction may be due to either extrinsic or intrinsic defects. The extrinsic variety encompasses opsonin deficiencies secondary to antibody or complement factor deficiencies, suppression of phagocytic cell numbers by immunosuppressive agents, corticosteroid-induced interference with phagocytic function, neutropenia, or abnormal neutrophil chemotaxis. Intrinsic phagocytic dysfunction is related to deficiencies in enzymatic killing of engulfed microorganisms. Examples of the intrinsic disorders include chronic granulomatous disease, myeloperoxidase deficiency, and glucose-6-phosphate dehydrogenase deficiency. Consequences of phagocytic dysfunction include increased susceptibility to bacterial infections but not to viral or protozoal infections. Selected phagocytic function disorders may be associated with severe fungal infections. Severe bacterial infections associated with phagocytic dysfunction range from mild skin infections to fatal systemic infections.

Chemotaxis is the process whereby chemical substances direct cell movement and orientation. The orientation and movement of cells in the direction of a chemical's concentration gradient is positive chemotaxis, whereas movement away from the concentration gradient is termed negative chemotaxis. Substances that induce chemotaxis are referred to as chemotaxins and are often small molecules, such as C5a, formyl peptides, lymphokines, bacterial products, leukotriene B_4, etc., that induce positive chemotaxis of polymorphonuclear neutrophils, eosinophils, and monocytes. These cells move into inflammatory agents by chemotaxis. A dual chamber device called a Boyden chamber is used to measure chemotaxis, in which phagocytic cells in culture are separated from a chemotactic substance by a membrane. The number of cells on the filter separating the cell chamber from the chemotaxis chamber reflect the chemotactic influence of the chemical substance for the cells.

Chemotaxis is locomotion of cells that may be stimulated by the presence of certain substances in their environment. This locomotion may be random in direction, i.e., it is not oriented with respect to the stimulus although there is a direct cause/effect relationship between stimulus and response. In contrast, the directed locomotion implies an orientation of cell movement with respect to the inducing stimulus. The latter form of cell movement is called chemotaxis and may be positive, in which the stimulus acts as an attractant, or negative, in which the stimulus acts as a repellent.

Substances that may stimulate random cell locomotion are called cytotoxigens; those that stimulate directed migration are called cytotoxins or chemotactic factors. The main element in the effect of chemotactic factors is the presence of a concentration gradient that determines the direction of cell migration. Under these circumstances a chemotactic signal is provided to the cells under consideration. In the absence of such a gradient, chemotactic factors enhance the random migration.

CD230 is a 35-kDa molecule with broad antigenic expression. It is a large membrane prion protein that occurs normally in neurons of the human brain and is thought to be involved in synaptic transmission. In prion diseases, such as Creutzfeld-Jakob disease (CJD) and bovine spongiform encephalopathy (BSE), altered forms of the normal cellular protein (PrPc) can occur upon contact with an infectious prion protein (PrPsc) from another host. The altered PrPsc form differs from the host-encoded PrPc in its conformational structure, but the sequence of both forms is reported to be the same. The altered PrPsc form is resistant to proteolytic degradation and accumulates in cytoplasmic vesicles of diseased individuals, forming lesions, vacuoles, and amyloid deposits. Also referred to as a prion protein.

Phagocytes may kill microorganisms they have ingested by either of two separate mechanisms. One of these, **oxygen-dependent killing**, is activated by a powerful oxidative burst that culminates in the formation of hydrogen peroxide and other antimicrobial substances (Figure 23.9). In addition to this oxygen-dependent killing mechanism, phagocytized intracellular microbes may be the targets of toxic substances released from granules into the phagosome, leading to microbial cell death by an oxygen-independent mechanism.

For oxygen-dependent killing of microbes, membranes of specific granules and phagosomes fuse. This permits interaction of NADPH oxidase with cytochrome b. With the aid of quinone, this combination reduces oxygen to

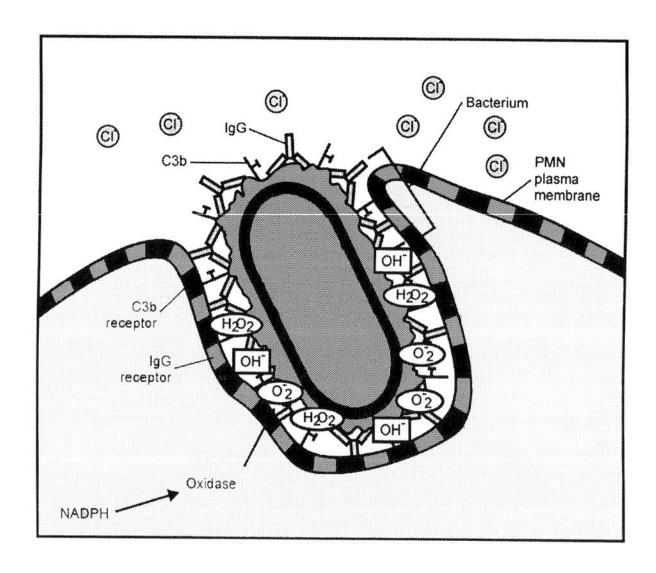

FIGURE 23.9 Formation of bactericide and hydrogen peroxide catalyzed by NADPH oxidase.

superoxide anion, O2. In the presence of a catalyst superoxide dismutase, superoxidase ion is converted to hydrogen peroxide.

The clinical relevance of this process is illustrated by chronic granulomatous disease (CGD) in children who fail to form superoxide anions. They have diminished cytochrome b. Even though phagocytosis is normal, they have impaired ability to oxidize NADPH and destroy bacteria through the oxidative pathway. The oxidative mechanism kills microbes through a complex process. Hydrogen peroxide together with myeloperoxidase transforms chloride ions into hypochlorous ions that kill microorganisms. Azurophil granule fusion releases myeloperoxidase to the phagolysosome. Some microorganisms such as pneumococci may themselves form hydrogen peroxide.

Natural killer (NK) cells (Figure 23.10 and Figure 23.11) attack and destroy tumor cells and certain virus-infected cells. They constitute an important part of the natural immune system, do not require prior contact with antigen, and are not MHC restricted by the major histocompatibility complex (MHC) antigens. NK cells are lymphoid cells of the natural immune system that express cytotoxicity against various nucleated cells including tumor cells and virus-infected cells. NK cells, killer (K) cells, or antibody-dependent cell-mediated cytotoxicity (ADCC) cells induce lysis through the action of antibody. Immunologic memory is not involved, as previous contact with the antigen is not necessary for NK cell activity. The NK cell is approximately 15 μm in diameter and has a kidney-shaped nucleus with several, often three, large cytoplasmic granules. The cells are also called large granular lymphocytes (LGL). In addition to their ability to kill selected tumor cells and some virus-infected cells, they also participate in ADCC by anchoring antibody to the cell surface through an Fcγ receptor. Thus, they are able to destroy antibody-coated nucleated cells. NK cells are believed to represent a significant part of the natural immune defense against spontaneously developing neoplastic cells and against infection by viruses. NK cell activity is measured by a ^{51}Cr release assay employing the K562 erythroleukemia cell-line as a target. NK cells secrete IFN-γ and fail to express antigen receptors such as immunoglobulin receptors or T cell receptors. Cell surface stimulatory receptors and inhibitory receptors that recognize self MHC molecules regulate their activation.

Although not phagocytic, natural killer (NK) cells attack and destroy certain virus-infected cells. They constitute a part of the natural immune system, do not require prior contact with antigen, and are not MHC restricted. On contacting a virus-infected cell, NK cells produce perforin that leads to the formation of pores in the infected cell membrane

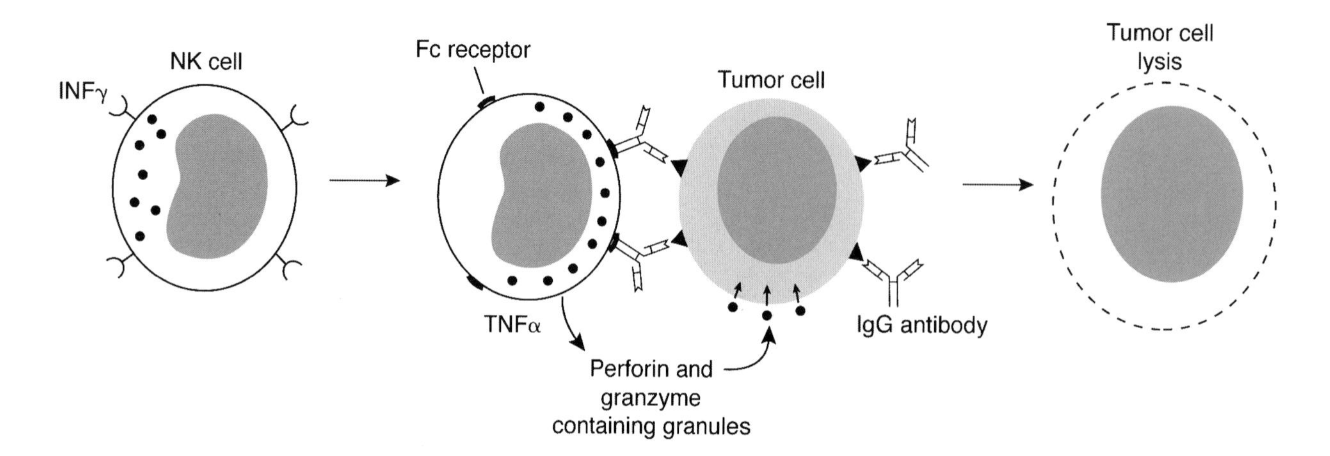

FIGURE 23.10 NK cell-mediated killing of tumor cells by antibody-dependent cell-mediated cytotoxicity.

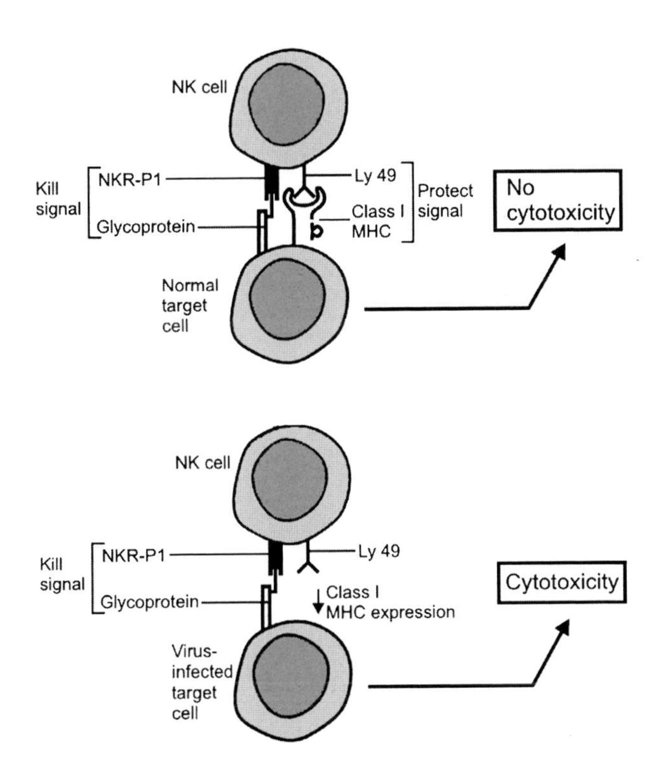

FIGURE 23.11 Proposed mechanism of NK cell cytotoxicity restricted to altered self cells. Kill signal is generated when the NK cell's NKR-PI receptor interacts with membrane glycoprotein of normal and altered self cells. The kill signal can be countermanded by interaction of the NK cells Ly49 receptor with class I MHC molecules. Thus, MHC Class I expression prevents NK cell killing of normal cells. Diminished Class I expression on altered self cells leads to their destruction.

Natural killer (NK) cells
Antibody-dependent cytotoxic cells K cells NK cells
Lymphokine-activated killer (LAK) cells
Tumor-infiltrating lymphocytes (TILS)

FIGURE 23.12 Lymphoid cells participating in nonspecific immunity.

leading to osmotic lysis. Interferon enhances NK cell activity. These cells appear to be large granular lymphocytes and are significant in antiviral defense and in surveillance against the development of neoplasia. NK cells lyse certain virus-infected cells without MHC restriction. Questions remain concerning the phenotype of NK cells, even though several monoclonal antibodies reactive with them are available. The natural immune system, in which the NK cells are key participants, does not involve memory. It does not require sensitization and cannot be enhanced by specific antigens. Other nonmemory cells include polymorphonuclear leukocytes and macrophages (Figure 23.12), which are important in early defense against infectious agents and possibly tumors. NK cells are able to lyse selected tumor target cells without prior sensitization and in the absence of antibody or complement. NK and cytotoxic T cells have been shown to share similar lytic mechanisms. Both cell types have granules that contain perforin or C9-related protein which lyse target cells without antibody or complement.

NK activity is measured by a chromium release assay, employing the K562 erythroleukemia cell line as a target. Whereas NK cells mediate their effect in the absence of antibody or complement, killer **(K) cells** or **ADCC (antibody-dependent cell-mediated cytotoxicity)** cells induce lysis through the action of antibody. With the demonstration of Fc receptors on their surface, NK cells may actually be the killer (K) cells responsible for ADCC activity through attached IgG antibody. They mediate their classic effects via cell-surface receptors for antigen.

Other than NK or ADCC cells, circulating monocytes or macrophages also mediate cell lysis through antibody molecules. Cytotoxic T cells (CTL) apparently recognize specific target cells through interaction with MHC antigens on the cell surface. Whereas either helper or killer T cells are directed to MHC proteins, NK cells apparently do not recognize MHC determinants. NK cell activity is located in the low-density population of lymphocytes which have large granules in their cytoplasm, i.e., large granular lymphocytes (LGL). Even though NK cells are lethal to tumor cells *in vitro*, very little data exists about their *in vivo* activity. Studies in mice suggest NK cells to be important in protection against selected virus infections. NK cells are also believed to play a regulatory role in the immune system, encompassing downregulation of antibody responses.

Leukocyte activation: The first step in activation is adhesion through surface receptors on the cell. Stimulus recognition is also mediated through membrane-bound receptors. An inducible endothelial–leukocyte adhesion molecule that provides a mechanism for leukocyte–vessel wall adhesion has been described.

Studies on H_2O_2 secretion by PMNs exposed to macrophage and lymphocyte products suggest that surface adherent leukocytes undergo a large prolonged respiratory burst. Recombinant TNF-alpha delays H_2O_2 release, demonstrating that soluble factors from macrophages and lymphocytes can affect adherent PMNs with respect to cytotoxic potential. Studies on regulation of neutrophil activation by platelets reveal that platelet-derived growth factor (PDGF) does not alter the resting level of superoxide generation but inhibits the rate and extent of f-Met-Leu-Phe-induced oxidative burst. Intracellular Ca^{++} increases upregulation of ligand-independent cell surface expression of f-Met-Leu-Phe receptors in neutrophils, whereas phorbol myristate acetate (PMA) activates downregulation of these receptors. A pertussis toxin-sensitive GTP-binding protein regulates monocyte phagocytic function.

Complement receptor 3 (CR3) facilitates the ability of phagocytes to bind and ingest opsonized particles. There is a relatively large family of homologous adhesion-promoting receptor proteins, including leukocyte proteins, that identify the sequence Arg-Gly-Asp. Molecules found to be powerful stimulators of PMN activity include recombinant IFN-γ, granulocyte–macrophage colony-stimulating factor, TNF and lymphotoxin. Investigations of storage sites for the several protein receptors have revealed a mobile intracellular storage compartment in human neutrophils. Chemotactic stimuli, such as f-Met-Leu-Phe, may cause translocation of granules acting as storage sites to the cell surface, which could be a requisite for neutrophil adhesion and chemotaxis.

Dephosphorylation pathways for inositol triphosphate isomers culminate in the elevation of intracellular Ca^{++} and protein kinase C activation. NADPH oxidase, which utilizes hexose monophosphate shunt-generated NADPH, catalyzes the respiratory burst. Both Ca^{++} and protein kinase C play a key role in the activated pathway. Activated human neutrophils manifest an elevated expression of complement decay-accelerating factor, which protects erythrocytes from injury by autologous complement. Transduction of decay-accelerating factors to the cell surface following stimulation by chemoattractants may be significant in protecting PMNs from complement-mediated injury. This type of process would permit PMNs to manifest unreserved function in sites of inflammation.

Surface adherent leukocytes undergo a large prolonged respiratory burst. NADPH oxidase, which utilizes hexose monophosphate shunt-generated NADPH, catalyzes the respiratory burst. Both Ca^{2+} and protein kinase C play a key role in the activation pathway. Complement receptor 3 (CR3) facilitates the ability of phagocytes to bind and ingest opsonized particles. Molecules found to be powerful stimulators of PMN activity include recombinant IFN-γ, granulocyte–macrophage colony-stimulating factor, TNF, and lymphotoxin.

The **K cells (killer cells)**, also called null cells, have lymphocyte-like morphology but functional characteristics different from those of B and T cells. They are involved in a particular form of immune response, the antibody-dependent cellular cytotoxicity (ADCC), killing target cells coated with IgG antibodies. A K cell is an Fc-bearing killer cell that has an effector function in mediating antibody-dependent cell-mediated cytotoxicity. An IgG antibody molecule binds through its Fc region to the K cell's Fc receptor. Following contact with a target cell bearing antigenic determinants on its surface for which

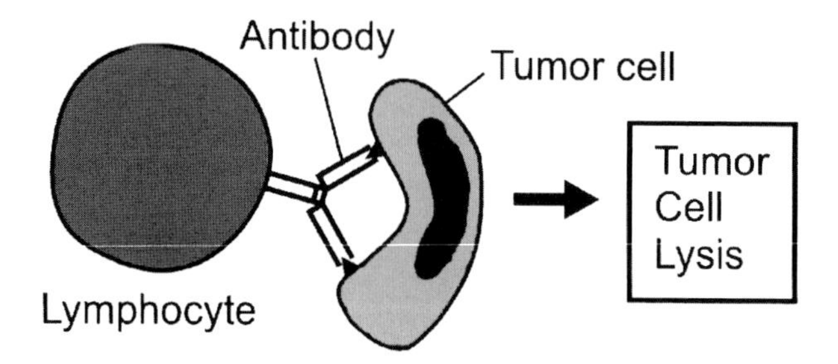

FIGURE 23.13 ADCC.

the Fab regions of the antibody molecule attached to the K cell are specific, the lymphocyte-like K cell releases lymphokines that destroy the target. This represents a type of immune effector function in which cells and antibody participate. Besides K cells, other cells that mediate antibody-dependent cell-mediated cytotoxicity include NK cells, cytotoxic T cells, neutrophils, and macrophages.

Tumor specific IgG Antibodies may act in concert with immune system cells to produce antitumor effects. **Antibody-dependent cell-mediated cytotoxicity (ADCC)** (Figure 23.13) is a reaction in which T lymphocytes, NK cells, including large granular lymphocytes, neutrophils, and macrophages may lyse tumor cells, infectious agents, and allogeneic cells by combining through their Fc receptors with the Fc region of IgG antibodies bound through their Fab regions to target cell surface antigens. Following linkage of Fc receptors with Fc regions, destruction of the target is accomplished through released cytokines. It represents an example of participation between antibody molecules and immune system cells to produce an effector function.

Humans have **innate immunity against extracellular** bacteria. Neutrophil (PMN), monocyte, and tissue macrophage phagocytosis leads to rapid microbicidal action against ingested microbes from the extracellular environment. The capacity of a microorganism to resist phagocytosis and digestion in phagocytic cells is a principal feature of its virulence. Complement activation represents a significant mechanism for ridding the body of invading microorganisms. A **peptidoglycan layer** (Figure 23.14) in the cell walls of Gram-positive bacteria as well as lipopolysaccharide, or LPS (Figure 23.15), in the cell walls of Gram-negative bacteria are able to activate the alternative pathway of complement without antibody. Also associated with LPS is flagellar antigen and somatic antigen (Figure 23.16).

Flagellar antigens, or **H antigens**, are epitopes on flagella of enteric bacteria that are motile and Gram-negative. H is from the German word "*Hauch*," which means breath, and refers to the production of a film on agar plates that resembles breathing on glass. In the Kaufmann-White

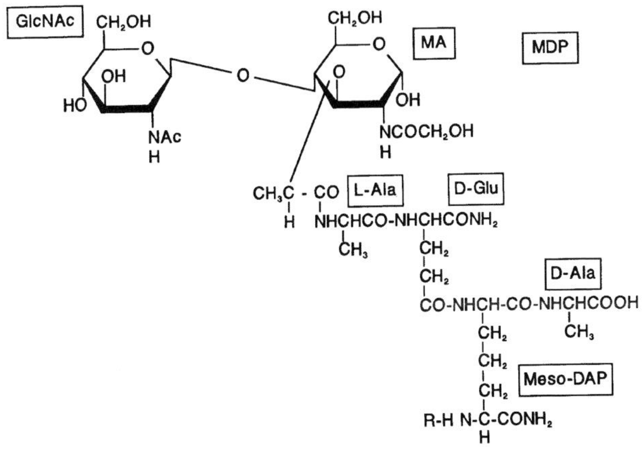

FIGURE 23.14 Peptidoglycan (murein).

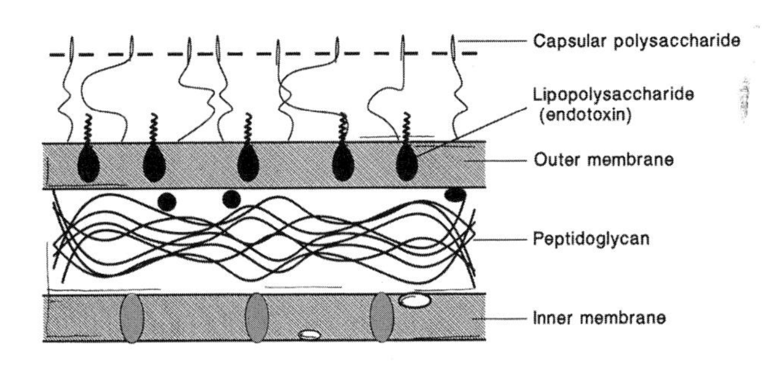

FIGURE 23.15 Cross-section of Gram-negative bacterial cell wall.

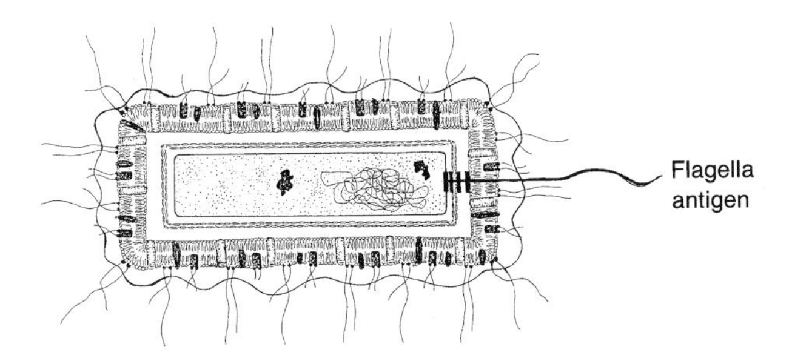

FIGURE 23.16 Bacterial cell.

classification scheme for *Salmonella*, H antigens serve as the basis for the division of microorganisms into Phase I and Phase II, depending on the flagellin. Phase variation may result in a switch to production of the other type that is genetically controlled.

A **somatic antigen,** or **O antigen**, is a lipopolysaccharide-protein antigen of enteric microorganisms which is used for their serological classification. O antigens of the *Proteus* species serve as the basis for the Weil-Felix reaction which is employed to classify *Rickettsia*. O antigens of *Shigella* permit them to be subdivided into 40 serotypes. The exterior oligosaccharide repeating unit side chain is responsible for specificity and is joined to lipid A to form lipopolysaccharide and to lipid B. The O antigen is the most variable part of the lipopolysaccharide molecule.

C3b may deposit on LPS where it is safe from the inactivating effects of factors H and I. LPS may also activate the classical complement pathway in the absence of antibody by combining with Clq. The C3b that results from activation of complement serves as an opsonin when linked to the bacterial surface, making the bacterial cell more attractive to phagocytes. The membrane attack complex (MAC) induces lysis of bacterial cells. Complement reaction products play an active role in inflammation through the attraction and stimulation of leukocytes.

Lipopolysaccharides or **endotoxins** (Figure 23.17) induce macrophages and selected other cells such as endothelial cells of vessels to synthesize cytokines, such as interleukin-1 (IL-1), tumor necrosis factor (TNF), interleukin-6 (IL-6), and interleukin-8 (IL-8) molecules that participate in inflammation (Figure 23.18). Monokines and cytokines from macrophages activate nonspecific inflammation and facilitate lymphocyte activation by bacterial epitopes. PMNs and monocytes adhere to the endothelium of vessels in areas of infection through the action of cytokines. These inflammatory cells migrate, accumulate in local areas, and become activated, enabling them to destroy the microorganisms. Local tissue injury may be an unintended consequence of these resisting processes. Fever and the formation of acute-phase reactants may also be consequences of cytokine action. Some cytokines may facilitate specific immune mechanisms by stimulating both T and B cells. Excessive cytokine synthesis may lead to pathologic sequelae during infection by extracellular microorganisms. Gram-negative bacterial infection can lead to disseminated intravascular coagulation (DIC) and vascular collapse known also as endotoxin shock or septic shock in which it is mediated mainly by TNF.

Flagellar antigens are epitopes of flagella in an organelle that renders some bacteria motile. Also called H antigens.

Flagellin is a protein that is a principal constituent of bacterial flagella. It consists of 25- to 60-kDa monomers that are arranged into helical chains which wind around a central hollow core. Polymeric flagellin is an excellent thymus-independent antigen. Mutations may occur in the central part of a flagellin monomer that has a variable sequence.

A **somatic antigen** is an antigen such as the O antigen which is part of a bacterial cell's structure.

Endotoxin is a Gram-negative bacterial cell wall lipopolysaccharide (LPS) that is heat stable and causes

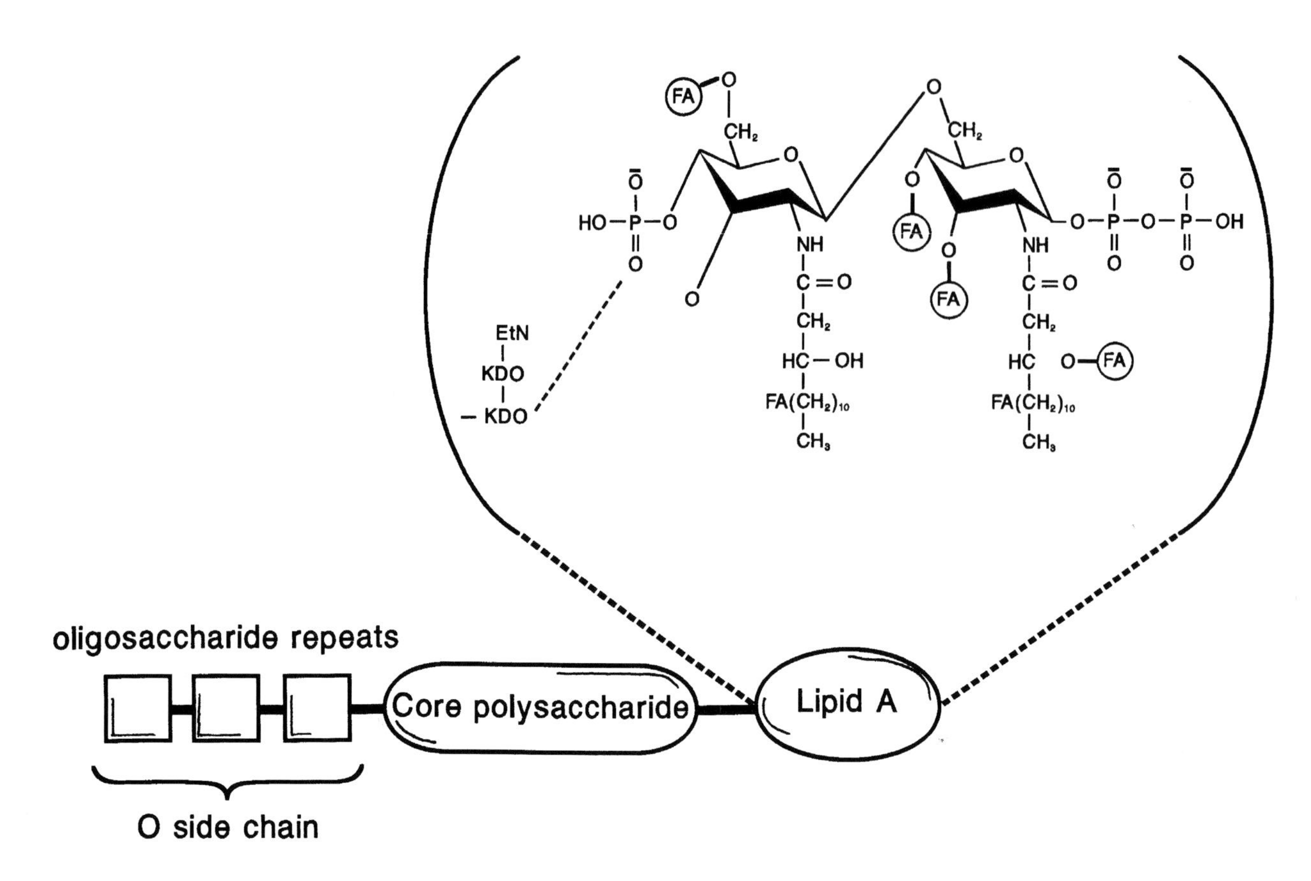

FIGURE 23.17 Lipopolysaccharide or endotoxin.

Cell differentiation factors	Alpha interferon
CSF	Plasma proteins
Cytotoxic factors	Coagulation factors
TNFα	Oxygen metabolites H_2O_2 Superoxide anion
Cachectin	Arachidonic acid metabolites Prostaglandins Thromboxanes Leukotrienes
Hydrolytic enzymes Collagenase Lipase Phosphatase	Complement components C1 to C5 Properdin Factors B, D, I, H
Endogenous pyrogen IL-1	

FIGURE 23.18 Secreted products of macrophages that have a protective effect on the body.

neutrophils to release pyrogens. It may produce endotoxin or hemorrhagic shock and modify resistance against infection. Endotoxins comprise an integral constituent of the outer membrane of Gram-negative microorganisms. They are significant virulence factors and induce injury in a number of ways. Toxicity is associated with the molecule's lipid A fraction, which is comprised of a β-1,6-glucosaminyl-glucosamine disaccharide substituted with phosphate groups and fatty acids. Lipopolysaccharide (LPS) has multiple biological properties that include ability to induce fever, lethal action, initiation of both complement and blood coagulation cascades, mitogenic effect on B lymphocytes; the ability to stimulate production of cytokines such as tumor necrosis factor and interleukin-1; and the ability to clot *Limulus* amebocyte lysate. Relatively large amounts of lipopolysaccharide released from Gram-negative bacteria during Gram-negative septicemia may produce endotoxin shock.

Endotoxin shock occurs following exposure to relatively large amounts of endotoxin produced during bacterial sepsis with *Escherichia coli*, *Pseudomonas aeruginosa*, or meningococci. It is characterized by falling blood pressure and disseminated intravascular coagulation (DIC). DIC leads to the formation of thrombi in small blood vessel, leading to such devastating consequences as bilateral cortical necrosis of the kidneys and blockage of the blood supply to the brain, lungs, and adrenals. When DIC affects the adrenal glands, as in certain meningococcal infections, infarction leads to adrenal insufficiency and death. This is the Waterhouse–Friderichsen syndrome.

An **enterotoxin** is a bacterial toxin that is heat stable and causes intestinal injury.

A **mitogen** is a substance, often derived from plants, that causes DNA synthesis and induces blast transformation and division by mitosis. Lectins, representing plant-derived mitogens or phytomitogens, have been widely used in both experimental and clinical immunology to evaluate T and B lymphocyte function *in vitro*. Phytohemagglutinin (PHA) is principally a human and mouse T cell mitogen, as is concanavalin A (Con A). By contrast, lipopolysaccharide (LPS) induces B lymphocyte transformation in mice, but not in humans. Staphylococcal protein A is the mitogen used to induce human B lymphocyte transformation. Pokeweed mitogen (PWM) transforms B cells of both humans and mice, as well as their T cells.

Natural immunity against viruses occurs when virus-infected host cells sensitize type I interferon. This blocks virus replication. NK cells, which are not MHC restricted, provide early antiviral effects following infection. Type I interferon accentuates their action. Both complement and phagocytosis play significant roles in removal of extracellular viruses.

Phagocytosis is the chief mechanism of **innate immunity against intracellular bacteria** whereby intracellular pathogenic microorganisms should be eliminated. However, this is frustrated by the resistance of many intracellular microbes to intracellular dissolution. Thus, the natural immune mechanism of phagocytosis is of little use in controlling infection by intracellular microorganisms. Bacteria of this category may persist in the tissues leading to chronic infection.

Innate immunity against intracellular bacteria is frustrated by the resistance of many intracellular microbes to intracellular dissolution. Thus, the natural immune mechanism of phagocytosis is of little use in controlling infection by intracellular microorganisms. Bacteria of this category may persist in the tissues, leading to chronic infection.

Parasitic protozoa and helminths are adept at survival within the host through successful resistance of host **innate immune mechanisms against parasites**. Whereas parasitic stages isolated from invertebrates may be lysed through activation of the alternate complement pathway, parasites isolated from humans or other veterbrate hosts are often susceptible to complement lysis. This could be attributable to either disappearance of surface molecules that activates complement or adherence to the surface of decay-accelerating factor (DAF) or other regulatory proteins.

Mechanisms whereby mature adult schistosomes are able to evade the immune response of the host include low surface antigenicity, disguise, host molecule mimicry, surface antigen sequestration and shedding, and reduced surface antigenicity, as well as other evasions mechanisms. Reducing surface antigenicity by host molecule masking, shedding, or sequestration of antigen represents a successful mechanism for parasites to escape the immune system. Whereas macrophages may ingest protozoa, numerous pathogenic parasites may resist intracellular killing or even replicate within the phagocyte. The outer coat of helminths helps to protect against intracellular killing by neutrophils or macrophages.

Specific immune responses can be mounted against parasites. Parasites such as protozoa and helminths elicit a variety of immune responses. Helminthic infections including *Nippostrongylus* schistosomes and filaria evoke titers of IgE that exceed those induced by other infectious agents. Helminths specifically stimulate Cd4^{+} helper t lymphocytes that form IL-4 and IL-5. ADCC involving eosinophils and IgE antibody is believed to be effective in immunity against helminths since the major basic protein in eosinophil granules is toxic to helminths. Coating helminths with IgE-specific antibody followed by eosinophil attachment through the Fc regions leads to ADCC by eosinophils.

Innate immune mechanisms against parasites: Whereas parasitic stages isolated from invertebrates may be lysed through activation of the alternate complement pathway, parasites isolated from humans or other veterbrate hosts are often susceptible to complement lysis. This could be attributable to either disappearance of surface molecules that activates complement or adherence to the surface of decay-accelerating factor (DAF) or other regulatory proteins.

Acquired immunity can develop after previous contact with the organism through infection (overt or subclinical) or by deliberate immunization with a vaccine prepared from the etiologic agent.

Acute-phase serum is a serum sample drawn from a patient with an infectious disease during the acute phase.

An **adaptive immune response** is the response of B and T lymphocytes to a specific antigen and the development of immunological memory. The response involves clonal selection of lymphocytes that respond to a specific antigen. Also called acquired immune response.

Adaptive immunity is protection from an infectious disease agent as a consequence of clinical or subclinical infection with that agent or by deliberate immunization against that agent with products from it. This type of immunity is mediated by B and T lymphocytes following exposure to specific antigen. It is characterized by specificity, immunological memory, and self/nonself recognition. This type of immunity is in contrast to natural or innate immunity.

Cardiolipin is diphosphatidyl glycerol, a phospholipid, extracted from beef heart as the principal antigen in the Wasserman complement fixation test for syphilis used earlier in the century.

Cecropin is an antibacterial protein derived from immunized cecropia moth pupae. It is also found in butterflies. Cecropin is a basic protein that induces prompt lysis of selected Gram-negative and Gram-positive bacteria.

Coagglutination refers to the interaction of IgG antibodies with the surface of protein A-containing *Staphylococcus aureus* microorganisms through their Fc regions, followed by interaction of the Fab regions of these same antibody molecules with surface antigens of bacteria for which they are specific. Thus, when the appropriate reagents are all present, coagglutination will take place in which the Y-shaped antibody molecule will serve as the bridge between staphylococci and the coagglutinated microorganism for which it is specific.

Concomitant immunity is the continued survival of adult worms from a primary infection while the host is demonstrably resistant to reinfection by a secondary challenge of fresh cercariae.

1. In tumor immunology, resistance to a tumor that has been transplanted into a host already bearing that tumor. Immunity to the reinoculated neoplasm does not inhibit growth of the primary tumor.
2. Resistance to reinfection of a host currently infected with that parasite.

Convalescent serum is a patient's blood serum sample obtained 2 to 3 weeks following the beginning of a disease. The finding of an antibody titer to a pathogenic microorganism that is higher than the titer of a serum sample taken earlier in the disease is considered to signify infection produced by that particular microorganism. For example, a fourfold or greater elevation in antibody titer would represent presumptive evidence that a particular virus, for example, had induced the infection in question.

Effector mechanisms refer to the means whereby post innate and adaptive immune responses destroy and eliminate pathogens from the body.

The **effector phase** is that part of an immune response following recognition and activation phases during which a foreign antigen, such as a microbe, is inactivated or destroyed.

The **effector response** is an event that follows antigen recognition and binding by antibody, such as complement-mediated lysis.

Immunological memory: The effectiveness of protective immunity against an infectious agent depends on the ability of the immune system to retain a memory of the original infective agent in order to provide an enhanced immune response on reexposure to the same agent. This usually prevents overt infection and prevents a fatal outcome. However, not all consequences of immune memory are beneficial. A second infection with the dengue virus may be more severe than the first. Another detrimental effect of immunity is sensitization to an allergen, leading to a hypersensitivity reaction. **Immunological memory** is governed by many factors with both B cells and T cells contributing to it. It depends upon interactions between memory T cells and memory B cells.

Systemic inflammatory response syndrome (SIRS): The systemic effects of disseminated bacterial infection. Mild and severe forms have been described. The mild form of SIRS is characterized by fever, neutrophilia, and an increase in acute-phase reactants. LPS or other bacterial products may stimulate these changes that are mediated by innate immune system cytokines. Severe SIRS is characterized by disseminated intravascular coagulation, adult respiratory distress syndrome, and septic shock.

Innate or constitutive defense system: Humans are confronted with a host of microorganisms with the potential to induce serious or fatal infections. Yet, nature has provided appropriate molecules, cells, and receptors that can protect against these microbes. Many of these defenses are general or nonspecific and do not require previous exposure to the offending pathogen (or closely related organism). These important mechanisms constitute the **innate** or **constitutive defense system**. Another important defense system is **acquired immunity**, which can develop after previous contact with the organism through infection (overt or subclinical).

Naturally acquired immunity describes the protection provided by previous exposure to a pathogenic microorganism or antigenically related organism.

In contrast, **artificially acquired immunity** develops as a result of immunization with vaccines — either with attenuated organisms or with killed organisms or subunit components. Toxoids provide excellent immunity against the effects of microorganisms such as *Corynebacterium diphtheriae* and *Clostridium tetani* that produce powerful exotoxins. Active immunization with appropriate booster injections leads to the development of IgG which provides immunity of long duration. Acquired immunity depends upon antibodies and T cells.

Passive immunity involves the transfer of resistance against an infectious disease agent from an immune individual to a previously susceptible recipient. **Natural passive immunity** describes the transfer of IgG antibodies across the placenta

Opsonic–promote ingestion and killing by phagocytic cells (IgG)
Block attachment (IgA)
Neutralize toxins
Agglutinate bacteria–may aid in clearing
Render motile organisms nonmotile
Abs only rarely affect metabolism or growth of bacteria (Mycoplasma)
Abs, combining with antigens of the bacterial surface, activate the complement cascade, thus inducing an inflammatory response and bringing fresh phagocytes and serum Abs into the site
Abs, combining with antigens of the bacterial surface, activate the complement cascade, and through the final sequences the membrane attack complex (MAC) is formed involving C5b-C9

FIGURE 23.19 Antimicrobial actions of antibodies.

from mother to child. IgA secretory antibodies may also be passively transferred from mother to child in breast milk.

Artificially acquired passive immunity describes the transfer of immunoglobulins from an immune individual to a nonimmune, susceptible recipient. Passive immunity of this type is more often used for prophylaxis than for therapy. It provides immediate protection of the recipient for relatively short periods (few weeks). Human sera are preferred for passive immunization to avoid serum sickness induced by foreign serum proteins.

Specific immune response to extracellular bacteria: Antibodies are the primary agents that protect the body against extracellular bacteria. Microbial cell wall polysaccharides serve as thymus-independent antigens that stimulate specific IgM antibody responses. Cytokine production may even permit switching from IgM to IgG production. Protein antigens of extracellular bacteria primarily stimulate activate $CD4^+$ T cells. Toxins of extracellular bacteria may activate multiple $CD4^+$ T lymphocytes. When a bacterial toxin stimulates an entire family of T lymphocytes that express products of a certain family of v_βT lymphocyte receptor genes, it is referred to as a superantigen. Immune stimulation of this type may lead to the production of abundant quantities of cytokines that lead to pathologic sequelae. The resistance mechanism against extracellular bacteria regrettably may include two reactions that produce tissue injury: acute inflammation and endotoxin shock. In addition, late in the course of a bacterial infection, pathogenic antibodies may appear. The multiple lymphocytes clones stimulated by either bacterial endotoxins or superantigens may lead to the production of autoimmunity through overriding specific T cell bypass mechanisms. Autoreactive lymphocytes may also be activated during this process.

Bacterial immunity: Bacteria produce disease by toxicity, invasiveness, or immunopathology, or combinations of these three mechanisms. Immune mechanisms may require the development of a neutralizing antitoxin or mechanisms to destroy the microorganism. The animal body provides both nonspecific and specific defenses. Nonspecific defenses include natural barriers such as the skin, acidity in the gut and vagina, mucosal coverings, etc. Those microorganisms not excluded may be recognized by acute-phase proteins, formol peptide receptors, receptors for bacterial cell-wall components, complement, and receptors that promote cytokine release. Other factors influence the T_H1/T_H2 balance of the T cell response. Cytokines play a protective role during nonspecific recognition and early defense. Bacteria may interact with complement leading to three types of protective function. Antibodies are important in neutralizing bacterial toxins. Secretory IgA can inhibit the binding of bacteria to epithelial cells. This antibody may also sensitize bacterial cells and render them susceptible to injury by complement. Phagocytic cells are important antibacterial defenses. Monocytes and polymorphonuclear cells have both oxygen-dependent and oxygen-independent antimicrobial mechanisms. Oxygen-independent killing may be accomplished by exposure to lysozyme and neutroproteases. Cell-mediated immunity mediated by T lymphocytes is another important antibacterial mechanism. They function through the release of lymphokines that have various types of consequences. Cytotoxic T cells and NK cells also have an important role in antimicrobial immunity. Mechanisms of immunopathology include septisemic shock in the adult respiratory distress syndrome; the Shwartzman reaction; the Koch phenomenon and necrotizing T cell dependent granulomas; and heat shock proteins and the possible development of autoimmunity. Thus, the immune response to bacteria is varied and complex but effective in the animal organism with an intact immune response.

Bacteriolysin is an agent such as an antibody or other substance that lyses bacteria.

Bacteriolysis refers to the disruption of bacterial cells by such agents as antibody and complement or lysozyme, causing the cells to release their contents.

Capsule swelling reaction: Pneumococcus swelling reaction. See Quellung reaction.

Cat scratch disease, or regional lymphadenitis, is common usually in children following a cat scratch. The condition is induced by a small Gram-negative bacillus. Erythematous papules may appear on the hands or forearms at the site of the injury. Patients may develop fever, malaise, swelling of the parotid gland, lymphadenapathy that is regional or generalized, maculopapular rash, anorexia, splenic enlargement, and encephalopathy. There may be hyperplasia of lymphoid tissues, formation of granulomas, and abscesses. A positive skin test together with the history establishes the diagnosis. Gentamycin and ciprofloxacin have been used in treatment.

Chancre immunity describes the resistance to reinfection with *Treponema pallidum* that develops 3 months following a syphilis infection that is untreated.

Cholera toxin is a *Vibro cholerae* enterotoxin comprised of five B subunits that are cell-binding 11. 6-kDa structures that encircle a 27-kDa catalase that conveys ADP-ribose to G protein, leading to continual adenyl cyclase activation. Other toxins that resemble cholera toxin in function include diphtheria toxin, exotoxin A, and pertussis toxin.

Circulatory system infections: The principal infection involving the circulatory system with immunologic sequelae is infective endocarditis. In acute endocarditis attributable to *Staphylococcus aureus*, there is high fever and, if untreated, a fatal rapid destruction of the heart valves. In subacute endocarditis there is a more indolent course and immunologic complications follow. Emboli break off from the infected heart valve. The principal causative microorganisms are streptococci, which include viridans streptococci and the more antibiotic resistant enterococci. Antibody levels, measured by immunoblotting, are greatly increased in enterococcal or streptococcal endocarditis and these are species specific.

***Corynebacterium diphtheriae* immunity:** Toxins produced by all strains of this organism are identical immunologically, which means that antitoxins may neutralize them equally. A single toxoid is used for effective immunization. There is no type-specific immunity. Immunization does not protect against the infection but against the systemic and local effects of the toxin. A high level of immunity is conferred but it is not complete.

Dapsone (diaminodiphenyl sulfone) is a sulfa drug that has been used in the treatment of leprosy. It has also shown efficacy for prophylaxis of malaria and for therapy of *dermatitis herpetiformis.*

DDS syndrome refers to a hypersensitivity reaction that occurs in 1 in 5000 leprosy patients who have been treated with dapsone (DDS, 4,4′-diaminodiphenyl sulfone), a drug that prevents folate synthesis by inhibiting the *p*-aminobenzoic acid condensation reaction. Patients develop hemolysis, agranulocytosis, and hypoalbuminemia, as well as exfoliative dermatitis and life-threatening hepatitis.

Encapsulated bacteria are surrounded by a thick carbohydrate coating or capsule that protects microorganisms such as pneumococci from phagocytosis. Infection-producing encapsulated bacteria cannot be effectively phagocytized and destroyed unless they are first coated with an opsonizing antibody, formed in an adaptive immune response, and complement.

End-binders are selected anticarbohydrate-specific antibodies that bind the ends of oligosaccharide antigens, in contrast to those that bind the sides of these molecules.

The **Fernandez reaction** is an early (24 to 48-h) tuberculin-like delayed-type hypersensitivity reaction to lepromin observed in tuberculoid leprosy; a skin test for leprosy.

FTA-ABS (Fluorescence treponema antibody absorption) is a serological test for syphilis that is very sensitive, i.e., approaching 100% in the diagnosis of secondary, tertiary, congenital, and neurosyphilis. It is an assay for specific antibodies to *Treponema pallidum* in the serum of patients suspected to have syphilis. Before combining the patient's serum with killed *T. pallidum* microorganisms fixed on a slide, the serum is first absorbed with an extract of Reiter's treponemes to remove group-specific antibodies. After washing, the specimen is covered with fluorescein-labeled antihuman globulin. This is followed by examination by fluorescence microscopy equipped with ultraviolet light. Demonstration of positive fluorescence of the target microorganisms reveals specific antibody present in the patient's serum. The greater specificity and sensitivity of this test makes it preferable to the previously used FTA-200 assay.

H antigens are epitopes on flagella of enteric bacteria that are motile and Gram-negative. H is from the German word "*Hauch*," which means breath, and refers to the production of a film on agar plates that resembles breathing on glass. In the Kaufmann–White classification scheme for *Salmonella*, H antigens serve as the basis for the division of microorganisms into phase I and phase II, depending on the flagellins they contain. A single cell synthesizes only one type of flagellin. Phase variation may result in a switch to production of the other type that is genetically controlled.

Halogenation refers to halogen binding to the cell wall of a microorganism with resulting injury to the microbe.

Hib (*Hemophilus influenzae* type b) is a microorganism that induces infection mostly in infants less than 5 years of age. Approximately 1000 deaths out of 20,000 annual cases are recorded. A polysaccharide vaccine (Hib Vac) was of only marginal efficacy and poorly immunogenic. By contrast, anti-Hib vaccine that contains capsular polysaccharide of Hib bound covalently to a carrier protein such as polyribosylribitol-diphtheria toxoid (PRP-D) induces a very high level of protection that reached 94% in one cohort of Finnish infants. PRP-tetanus toxoid has induced 75% protection. PRP-diphtheria toxoid vaccine has been claimed to be 88% effective.

Homozygote describes an organism whose genotype is characterized by two identical alleles of a gene.

Intimin is a bacterial membrane protein expressed on enteropathogenic *Escherichia coli* and can bind to both $\alpha_4\beta_1$ and $\alpha_4\beta_7$ integrins.

Intracellular pathogens are microorganisms, including viruses and bacteria, that grow within cells.

Jarisch-Herxheimer reaction is a systemic reaction associated with fever, lymphadenopathy, skin rash, and headaches that follows the injection of penicillin into patients with syphilis. It is apparently produced by the release of significant quantities of toxic or antigenic substances from multiple *Treponema pallidum* microorganisms.

K antigens are surface epitopes of Gram-negative microorganisms. They are either proteins (fimbriae) or acid polysaccharides found on the surface of *Klebsiella* and *Escherichia coli* microorganisms. K antigens are exterior to somatic O antigens. They are labile to heat and cross-react with the capsules of other microorganisms such as *Hemophilus influenzae, Streptococcus pneumoniae*, and *Neisseria meningitidis*. K antigens may be linked to virulent strains of microorganisms that induce urinary tract infections. Anti-K antibodies are only weakly protective.

Lyme disease is a condition, first described in Lyme, Connecticut, where an epidemic of juvenile rheumatoid arthritis (Still's disease) was found to be due to *Borrelia burgdorferi*. It is the most frequent zoonosis in the U.S. with concentration along the eastern coast. Insect vectors include the deer tick (*Ixodes dammini*), white-footed mouse tick (*I. pacificus*), wood tick (*I. ricinus*), and lone star tick (*Amblyomma americanum*). Deer and field mice are the hosts. In stage I, a rash termed *erythema chronicum migrans* occurs. The rash begins as a single reddish papule and plaque that expands centrifugally to as much as 20 cm. This is accompanied by induration at the periphery with central clearing that may persist months to years. The vessels contain IgM and C3 deposits. Stage II is the cardiovascular stage, which may be accompanied by pericarditis, myocarditis, transient atrial ventricular block, and ventricular dysfunction. Neurological symptoms also ensue and include Bell's palsy, meningoencephalitis, optic atrophy, and polyneuritis. Stage III is characterized by migratory polyarthritis. The diagnosis requires the demonstration of IgG antibodies against the causative agent by Western immunoblotting. Lyme disease is treated with the antibiotics tetracycline, penicillin, and erythromycin. *Borrelia burgdorferi*, the causative agent of this chronic infection, is a spirochete that may evade the immune response.

Mycoplasma immunity: High-titer cold agglutinin autoantibodies against sialo-oligosaccharide of the Ii antigen type are sometimes found during *Mycoplasma pneumoniae* infections. The first line of defense against these microorganisms is phagocytosis. Yet mycoplasmas can survive neutrophil phagocytosis if specific antibodies are not present. Secretory IgA is significant in preventing localized colonization, yet systemic antibodies protect from primary infection and secondary spread from localized colonization. Mycoplasmas can evade the humoral immune response by undergoing antigenic variation of surface antigens. T cells also appear to play a role in immunity to mycoplasma, which needs to be further investigated.

Organism-specific antibody index (OSAI) is the ratio of organism-specific IgG to total IgG in cerebrospinal fluid compared to the ratio of organism-specific IgG in serum to the total serum IgG. This is illustrated in the following formula:

If the index is greater than 1, signifying a greater quantity of organism-specific immunoglobulin in CSF than in the blood serum, this implies that organism-specific IgG is being synthesized in the intra-blood–brain barrier (IBBB) and suggests that the specific organism of interest is producing an infection of the central nervous system. Similar indices can be calculated for IgM and IgA antibody classes.

A **pathogen** is an agent such as a microorganism that can produce disease through infection of the host.

Pathogen-associated molecular pattern (PAMP) are repetitive motifs of molecules such as lipopolysaccharide, peptidoglycan, lipoteichoic acids, and mannans that are broadly expressed by microbial pathogens not found on host tissues. The immune system's pattern recognition receptors (PRRs) makes use of them in differentiating between pathogen antigens and self antigens.

Pattern recognition receptors (PRRs) bind to pathogen-associated molecular patterns (PAMPs). Natural or innate immune system receptors which recognize molecular patterns that comprise frequently encountered structures produced by microorganisms. These receptors enhance natural immune responses against microbes. CD14 receptors on macrophages that bind bacterial endotoxin to activate macrophages, and the mannose receptor on phagocytes that bind microbial glycoproteins or glycolipids, are examples of pattern recognition receptors.

Protein M (M antigen) is Group A streptococcal protein found on fimbriae. It is antiphagocytic and facilitates virulence of streptococci.

A **protoplast** is a bacterial cell from which the cell wall has been removed. It includes the cell protoplasm and the cytoplasmic membrane. Lysozyme digestion of

Gram-positive bacteria that contain a peptidoglycan cell wall yields protoplasts that require hypertonic media for survival, and they do not usually multiply. The hypertonic solution protects them from lysis. Protoplasts can be produced from Gram-positive bacteria also by treatment with penicillin or other antibiotics which inhibit synthesis of the cell wall. Gram-negative bacteria have a cell wall comprised of a thin peptidoglycan layer enclosed by an exterior membrane of lipopolysaccharide. Protoplasts prepared from Gram-negative bacteria are frequently termed spheroplasts.

PRP antigen is polyribosyl-ribitol capsular polysaccharide. An antiphagocytic cell wall constituent of *Hemophilus influenzae* that provides this microorganism with an effective mechanism to induce disease. Type-specific antibodies that facilitate immunization are a requisite for protective immunity against *H. influenzae*. Children less than 2 years old are poor producers of anti-PRP antibodies, making them more susceptible to the infection.

Pyogenic bacteria are microorganisms such as Gram-positive staphylococci and streptococci that induce predominantly polymorphonuclear leukocytes inflammatory responses leading to the formation of pus at sites of infection.

A **pyogenic infection** is infection associated with the generation of pus. Microorganisms that are well known for their pus-inducing or pyogenic potential include *Streptococcal pyogenes, Staphylococcus aureus, Streptococcus pneumoniae,* and *Hemophilus influenzae.* Antibody-deficient patients and those having defective phagocytic cell capacity show increased susceptibility to pyogenic infections. Patients with complement deficiency such as C3 deficiency, factor I deficiency, etc., are also prone to developing pyogenic infections.

Pyogenic microorganisms stimulate a large polymorphonuclear leukocyte response to their presence in tissues.

Pyrogen is a substance that induces fever. It may be either endogenously produced, such as interleukin-1 released from macrophages and monocytes, or it may be an endotoxin associated with Gram-negative bacteria produced exogenously that induces fever.

Phosphocholine antibodies are synthesized during selected bacterial infections especially by streptococcus (S), but also by Mycloplasma, Proteus, Trichinella, and Neisseria, in addition to helminithic parasites. $CD5^+$ B cells form IgM antibodies that have a limited idiotype spectrum V_H/V_L gene usage and provide protective immunity from infection. These antibodies are a cross-react with phophatidycholine, pneumoccoci, DsDNA, and sphingomellin. Phosphocholine antibody affinity diminishes with age.

Pili are structures that facilitate adhesion of bacteria to host cells and are therefore direct determinants of virulence.

The **Quellung phenomenon** refers to the swelling of the pneumococcus capsule following exposure to antibodies against pneumococci. See Quellung reaction.

A **Quellung reaction** is the swelling of bacterial capsules when the microorganisms are incubated with species-specific antiserum. Examples of bacteria in which this can be observed include *Streptococcus pneumoniae, Hemophilus influenzae, Neisseria,* and *Klebsiella* species. The combination of a drop of antiserum with a drop of material from a patient containing an encapsulated microorganism and the addition of a small loopful of 0.3% methylene blue produces the Quellung reaction. The microorganisms are stained blue and are encircled by a clear halo that resembles swelling but is in fact antigen–antibody complex produced at the surface of the organism. The reaction is due to an alteration of the refractory index.

Scarlet fever is a condition associated with production of erythrogenic toxin by group A hemolytic streptococci associated with pharyngitis. Patients develop a strawberry-red tongue and generalized erythematous blanching areas that do not occur on the palms, the soles of the feet, or in the mouth. Patients may also develop Pastia's lines, which are petechiae in a linear pattern.

Septic shock: Hypotension, with a systolic blood pressure of less than 90 mmHg or a decrease in the systolic pressure baseline of more than 40 mmHg, in individuals with sepsis. It may be induced by the systemic release of TNF-α following bacterial infection of the blood, frequently with Gram-negative bacteria. Vascular collapse, disseminated intravascular coagulation, and metabolic disorders occur. Septic shock results from the effects of bacterial LPS and cytokines, including TNF, IL-12, and IL-1. Also termed endotoxin shock.

Seroconversion is the first appearance of specific antibodies against a causative agent in the blood either during the course of an infection or following immunization.

Serotype refers to the use of specific antibodies to classify bacterial subtypes based on variations in the surface epitopes of the microorganism. Serotyping has long been used to classify *Salmonella,* streptococci, *Shigella,* and many other bacteria. May also describe human alloantigens such as HLA and blood group antigens.

***Shigella* immunity:** The host protective immune response to *Shigella* infection is poorly understood. Since M cells of the gut take up the microorganisms, secretory IgA immune responsiveness has been postulated to be protective in Shigellosis, but it has been hard to establish. The immunity induced is type specific with reinfection occurring only within different *Shigella* species or serotypes. This is believed to be associated with an immune response to lipopolysaccharide determinants. *Shigella* may destroy antigen-presenting cells in the host following systemic exposure to *Shigella* antigens and toxins before an immune response can be established. Serum IgG has no protective effect. Oral vaccines with attenuated *Shigella* induce type-specific protection. Previous *Shigella* infection leads to specific IgA secretion in breast milk. Antibodies develop early against somatic *Shigella* antigens. Shiga toxin is a multimeric protein comprised of a single enzymatically active A subunit and 5B subunits needed for toxin binding. It is synthesized only by *Shigella dysenteriae* type I strains. It is an important virulence factor in the pathogenesis of hemolytic–uremic syndrome, which may be a complication of infection. Shiga toxin induces IgM antibody responses but the IgG response is lacking. However, IgG can be raised against Shiga toxin in animal models. Protection against *Shigella* has been associated with the humoral response to LPS or plasmid-encoded protein antigens. T cells also become activated *in vivo* during *Shigella* infection. There may be some correlation between T cell activation and the severity of the disease. Local cytokine synthesis is also significant in Shigellosis. Increased levels of IL-1, IL-6, TNF, and IFNγ have been found in stool specimens and plasma of infected patients. Both T_H1 and T_H2 cytokines are present in Shigellosis. There may be both a humoral and a cytotoxic defense mechanism during infection. High-serum antitoxin titers failed to protect monkeys against intestinal disease following challenge with live *Shigella dysentariae*. Postinfection reactive arthritis may also occur. Heat-killed whole cell *Shigella* vaccines used in the past failed to give protection. Although mucosal secretory IgA is to prevent bacterial attachment to the mucosa and to neutralize toxins, it has not been proved that either mechanism is significant in Shigellosis.

A **spheroplast** is a Gram-negative bacterial protoplast that contains outer membrane remains.

Staphylococcal enterotoxins (Ses) are bacterial cell constituents that cause food poisoning and activate numerous T lymphocytes by binding to MHC class II molecules and the Vβ domain of selected T cell receptors, which qualifies staphylococcal enterotoxins to be classified as superantigens.

***Staphylococcus* immunity:** Most individuals synthesize antibodies reactive with staphylococcal antigens that include the cell wall-associated teichoic acid and the extracellular protein α-hemolysin. These antibodies are not protective against staphylococcal infections. They may be of some use in diagnosis, but it is necessary to show a significant increase in antibody titer. Yet high titers of antibodies against TSST, enterotoxins, and exfoliatins are protective. These toxins are immunomodulatory, mitogenic, and induce cytokine synthesis. No effective staphylococcal vaccine has been developed.

***Streptobacillus* immunity:** Even though an immune response is induced in subjects infected with *S. moniliformis*, both the exact mechanisms and relative contribution of antibody remain to be determined. In mouse experiments, antibody confers only partial immunity to challenge with the microorganism. Inactivated *S. moniliformis* vaccines induce only partial protection against challenge.

***Streptococcus* immunity:** The most effective host resistance against pneumococcal (*Streptococcus penumoniae*) infection is by synthesizing IgG or IgM that interacts with capsular polysaccharide which opsonizes the bacteria for ingestion and killing by professional phagocytes. Anticapsular antibody develops following colonization or infection. Most individuals have low resistance as reflected by their lack of antibody to most of the commonly infecting pneumococcal serotypes even though normal adults may have sufficient antibody against phosphocholine-containing epitopes of pneumococcal cell wall. This antibody does not appear to be protective in human subjects since reaction between this antibody and the bacterial cell wall occurs beneath the capsule. The complement fragments and bound Fc are inaccessible to phagocytic cells. Vaccines to prevent streptococcal infections are hindered by the systemic local and systemic reactions that follow administration of large does of M protein given in an effort to induce type-specific antibody responses. This may be attributable to M proteins serving as superantigens. Heart reactive epitopes have been removed from M proteins, which has made possible immunization with purified M protein preparations to induce type-specific opsonic antibodies that do not cross-react with heart antibodies. Immunization protocols have also included attempts to stimulate antibodies against lipoteichoic acid, the adherence constituent of streptococci. But these efforts have been limited by the poor immunogenicity of this component. Vaccination with capsular polysaccharides is effective in preventing pneumococcal infection.

Streptococcal M protein is a cell-wall protein of virulent *Streptococcus pyogenes* microorganisms which interferes with phagocytosis and also serves as a nephritogenic factor.

Superinfection "immunity" refers to the inability of two related organisms, for example, plasmids, to invade a host cell at the same time.

Toxic shock syndrome is an acute systemic toxic reaction in which the patient manifests shock, skin exfoliation, conjunctivitis, and diarrhea that results from the excessive production of cytokines by $CD4^+$ T cells activated by the bacterial superantigen toxic shock syndrome toxin-1 (TSST-1), secreted by *Staphylococcus aureus*.

Transferrin is a protein that combines with and competes for iron with bacteria.

Tetanus is a disease in which the exotoxin of *Clostridium tetani* produces tonic muscle spasm and hyperreflexia, leading to tris (lock jaw), generalized muscle spasms, spasm of the glottis, arching of the back, seizures, respiratory spasms, and paralysis. Tetanus toxin is a neurotoxin. The disease occurs 1 to 2 weeks after tetanus spores are introduced into deep wounds that provide anaerobic growth conditions.

Tetanus toxin is the exotoxin synthesized by *Clostridium tetani.* It acts on the nervous system, interrupting neuromuscular transmission and preventing synaptic inhibition in the spinal cord. It binds to a nerve cell membrane glycolipid, i.e., disialosyl ganglioside. The effects of tetanus toxin are countered by specific antitoxin.

Treponema immunity: Infection with *Treponema pallidum* induces both cellular and humoral immune responses. The humoral response is characterized by the synthesis of both phospholipid and treponemal antibodies that are detected in the serological diagnosis of syphilis. Flocculation tests have long been used to detect phospholipid or cardiolipin antibodies. They make use of a cardiolipin–lecithin–cholesterol antigen in the VDRL test. Cardiolipin F antibody specific for host cell mitochondrial cardiolipin (autoantibody) is also associated with syphilis but is not used in diagnosis. Treponemal antibodies that appear after infection together with cardiolipin antibodies are detectable by immunofluorescence. Treponemal antibodies are still detectable in the host for many years. Autoantibodies against tissue phospholipids and other tissue components also occur in *T. Pallidum*-infected hosts. The cell-mediated immune response consists of macrophage activation and induction of $CD8^+$ T cells and $CD4^+$ T cells. The cell-mediated immunity appears to be more important than the humoral response in the development of immunity.

The **VDRL (Venereal Disease Research Laboratory) test** is a reaginic screening assay for syphilis. VDRL antigen is combined with heat-inactivated serum from the patient, and the combination is observed for flocculation by light microscopy after 4 min. Reaginic assays are helpful as screening tests for early syphilis and are usually positive in secondary syphilis, although the results are more variable in tertiary syphilis. VDRL is negative in approximately half of the neurosyphilis cases. Reaginic tests for syphilis may be biologically false-positive in such conditions as malaria, lupus erythematosus, and acute infections. Biologic false-positive VDRL tests may also be seen in some cases of hepatitis, infectious mononucleosis, rheumatoid arthritis, or even pregnancy.

A **biological false-positive reaction** results when a positive serological test for syphilis, such as in the VDRL (Venereal Disease Research Laboratory), is produced by the serum of an individual who is not infected with *Treponema pallidium.* It is attributable to antibodies reactive with antigens of tissues such as the heart from which cardiolipin antigen used in the test is derived. The blood sera of patients with selected autoimmune diseases, including SLE, may contain antibodies that give a biological false-positive test for syphilis.

BFPR is an abbreviation for biological false-positive reaction.

Vibrio cholerae **immunity:** *Vibrio cholerae* induces disease by production of virulence factors and toxins. Protective antigens have been identified and progress has been made in vaccine development. Mucosal immunity is mediated by secretory IgA antibodies as necessary for protection against cholera. The V. *cholerae* is noninvasive. Thus, antibodies block adherence and inhibit colonization. They either neutralize the toxin or inhibit their binding to specific receptors. Intestinal immune response begins in the lymphoid follicle where the mononuclear cells are responsible for phagocytic uptake of antigen from the intestine. Live microorganisms are more effective immunogens than are killed ones. An initial infection with cholera yields long-lasting immunity. The most important protective antigen is the O antigen of the LPS. Antibodies to TCP are protective but are biotype-restricted as a consequence of biotype-specific antigenic variation in the major structural subunit. The other protective antigen is the B subunit of CT. The major epidemic strain now appears to be 0139 which increases the significance of anti-LPS immunity since this new strain has the virulence of the original 01 El Tor strains. Immunity against non-LPS antigens is far less important than to the specific LPS type. Non-LPS antigens are less immunogenic than is the LPS antigen.

Virulence genes govern the expression of multiple other genes with changing environmental conditions such as temperature, pH, osmolarity, etc. An example is the *toxR* gene of *Vibrio cholerae* that coordinates 14 other genes of this microorganism.

Yersinia **immunity:** Both humoral and cell-mediated immune responses develop in Yersiniosis. The antibody is directed mainly to lipopolysaccharide. Antibodies in patients with *Yersinia* infection are specific for numerous epitopes. The antibodies persist for months and in some cases even years. Reactive arthritis, a frequent postinfection complication of Yersiniosis, is strongly associated with HLA-B27 antigen. The immune response against the infectious agents is believed to have a significant role in the pathogenesis of *Yersinia*-triggered reactive arthritis. No effective vaccine is available for Yersinia infection.

Diphtheria toxin is a 62-kDa protein exotoxin synthesized and secreted by *Corynebacterium diphtheriae.* The exotoxin, which is distributed in the blood, induces neuropathy and myocarditis in humans. Tryptic enzymes nick the single-chain diphtheria toxin. Thiols reduce the toxin to produce two fragments. The 40-kDa B fragment gains access to cells through their membranes, permitting the 21-kDa A fragment to enter. Whereas the B fragment is not toxic, the A fragment is toxic and it inactivates elongation factor-2, thereby blocking eukaryocytic protein synthesis. Guinea pigs are especially sensitive to diphtheria toxin, which causes necrosis at injection sites, hemorrhage of the adrenals, and other pathologic consequences. Animal tests developed earlier in the century consisted of intradermal inoculation of *C. diphtheriae* suspensions into the skin of guinea pigs that were unprotected, compared to a control guinea pig that had been pretreated with passive administration of diphtheria antitoxin for protection. In later years, toxin generation was demonstrated *in vitro* by placing filter paper impregnated with antitoxin at right angles to streaks of *C. diphtheriae* microorganisms growing on media in Petri plates. Formalin treatment or storage converts the labile diphtheria toxin into toxoid.

In a **Moloney test**, diphtheria toxoid is injected intradermally and the skin response is observed to determine whether or not the subject is hypersensitive to diphtheria prophylactic substances.

***Bacteroides* immunity:** *Bacteroides fragilis* has a capsular polysaccharide that serves as a virulence factor which led to use of this polysaccharide as a vaccine in rats. The vaccine led to excellent antibody levels but did not prevent abscess formation. In the rat model, the findings strongly point to the role of T cells in the immune response to *B. fragilis* in abscess prevention. Murine studies confirmed the rat findings and demonstrated that immune T cells are antigen specific. Abscess formation and prevention in response to *B. fragilis* is mediated by T cell mechanisms. Abscess formation requires a precursor-type T cell. Prevention of abscess requires T cells belonging to the suppressor phenotype which communicate via a small polypeptide factor. These cells are antigen-specific but not MHC-restricted.

Bacterial immunoglobulin-binding proteins are molecules expressed by selected microorganisms that interact with immunoglobulin at sites other than their antigen-combining regions. Thus, the antibody's ability to bind its homologous antigen is not impaired. Gram-positive bacteria express six types of IgG-binding protein, including protein A (type I) which is expressed on *Staphylococcus aureus.* Most human group A steptococci express type II receptors and exhibit great variability in immunoglobulin-binding capacity. Protein G, the type III IgG binding protein, is expressed by most human group C and G streptococcal isolates. Type IV, V, and VI immunoglobulin-binding proteins require further investigation.

Bactericidin is an agent such as an antibody or nonantibody substance in blood plasma which destroys bacteria.

***Bacillus anthracis* immunity:** Protective immunity against anthrax is induced only by the antigen designated PA. The *B. anthracis* polypeptide capsule is only weakly immunogenic and is not believed to contribute to naturally acquired immunity or to induce protection against anthrax. The Ca^-/tx^+ strains of *B. anthracis* are protective. One of these strains is used to prepare a very successful live-spore animal vaccine. It is considered unsuitable in the West to use in humans because it retains some residual virulence. Human vaccine developed half a century ago consists of aluminum hydroxide-adsorbed cell-free filtrates of cultures of a noncapsulating, nonproteolytic derivative of strain V770. Several doses are required. An immune response to PA but not to LF or EF are critical for protection. Immunization with strains synthesizing either PA or EF or PA and LF resulted in higher antibody responses to EF or LF, respectively. There is a synergistic relationship between PA and LF or EF in immunoprotection. Some cellular immune mechanisms are believed to be significant in the induction of protective immunity. PA has been shown with monoclonal antibodies to have at least three nonoverlapping antigenic regions. It is anticipated that the elucidation of protective motifs on PA may lead to the development of a subunit-based vaccine. Passive immunity in the treatment of anthrax has been unsuccessful.

***Klebsiella* immunity:** Most healthy adults have high levels of natural resistance against *Klebsiella pneumoniae.* The organism classically induces lung infection leading to a massive confluent lobar consolidation with polymorphonuclear leukocytes, edema, and abscess formation with extensive cavity formation. Type-specific antibodies to the carbohydrate capsule are critical to recovery. These antibodies that appear within the first two weeks following infection serve as opsonins to facilitate the killing of this microorganism by phagocytes. The immunity to *K. pneumoniae* is long lasting. This microorganism is believed to have a part in the development of ankylosing spondylitis (AS) since these individuals develop high levels of serum IgA against the microbe. AS is associated with HLA-B27 antigen. The mechanism has been claimed to be either molecular mimicry in which antibody against this antigen will bind to self-antigenic determinants leading to destructive immunity or that *Klebsiella* plasmids might encode the formation of a bacterial modifying factor that reacts with HLA-B27, rendering it susceptible to immune attack. The microbe's main virulence factor is the capsule that helps it to resist phagocytosis. Although there have been reports of several enterotoxins, the main pathologic effect is associated with the production of an endotoxin.

The extracellular toxic complex ETC contains endotoxin, capsule, and protein that is lethal when injected into mice, leading to pathologic changes that resemble those produced in *Klebsiella pneumoniae* lobar pneumonia in humans. Vaccines are aimed at developing antibodies against capsule polysaccharides which are able to prevent experimental *Klebsiella pneumoniae* sepsis. An effective vaccine must contain capsular polysaccharide from 25 different serotypes.

Bordetella pertussis is the etiologic agent of whooping cough in children. Killed *B. pertussis* microorganisms are administered in a vaccine together with diphtheria toxoid and tetanus toxoid as DPT. The endotoxin of *B. pertussis* has an adjuvant effect that can facilitate antibody synthesis.

***Bordetella* immunity:** *Bordetella pertussis* produces the respiratory disease whooping cough or pertussis. Both humoral and cell-mediated immune responses follow infection. Neutralizing antibodies are believed to be the principal protective mechanism against infection with *Bordetella pertussis.* Antibodies are formed against *Bordetella pertussis* antigens that include PT, FHA, PRN, and fimbriae, which are associated with protection against pertussis. Pertussis infection can lead to long-lasting immunity against subsequent pertussis. Patients who have recovered from the infection develop anti-*Bordetella pertussis* immunoglobulin A in serum and saliva, pointing to the role of mucosal antibodies. Cell-mediated immunity is also believed to be significant since T_H1 cells specific for the microorganism occur in persons following either infection or vaccination. High-titer IgG antibodies may clear bacteria in respiratory infections revealed by animal studies, but cell-mediated immunity is necessary to completely eliminate the microorganism from mouse lungs. In humans, immunity that follows infection may prevent respiratory colonization, whereas immunity that follows vaccination may protect against toxin-mediated disease. *B. pertussis* antigens suppress the host response to pertussis both *in vitro* and *in vivo*. The whole-cell vaccine comprised of killed whole virulent *B. pertussis* has been replaced with acellular preparation combined with diphtheria and tetanus toxoid in the currently used DTaP vaccine. It provides fewer and milder side effects than the whole-cell vaccine preparation and is more effective in inducing serum antibody responses and protection from pertussis.

***Borrelia* immunity:** Relapsing fever and Lyme disease are produced by members of this genus. *Borrelia burgdorferi* sensu lato spirochetes are the causative agents of Lyme disease. These microorganisms are covered by a slime layer comprised of self-molecules that block immune recognition. The slime layer acting as a capsule prevents phagocytosis, but when incubated with specific antibody in complement, the microbe is killed. The principal surface proteins in the exterior cell membrane include A, B, C, D, E, and F that are designated as Outer surface proteins (Osp). The proteins are heterogeneous but their function remains to be determined. Spirochetes that cause Lyme disease upregulate or downregulate Osp A and Osp C during the course of human infection. Other antigens in the outer membrane of *Borrelia burgdorferi* sensu lato include 16-, 27-, 55-, 60-, 66-, and 83-kDa proteins. Osp A is the principal candidate for a Lyme disease vaccine. Soon after infection with *Borrelia burgdorferi* sensu lato, the microorganisms become refractory to the action of bactericidal antibodies.

***Brucella* immunity:** The immune response to *Brucella* infection is marked by early IgM synthesis followed by a switch to IgG and IgA. IgE is also detected. IgM persists for an unusually long time, possibly responding to the T cell independent antigen LPS. Antibodies are important for the serodiagnosis of *Brucella* infection. Antibody's only protective role is probably as a preexisting mucosal antibody that decreases initial infection. Cell-mediated immune responses confer protective immunity. *Brucellae* are facultative intracellular bacteria that are controlled by macrophage activation and granuloma formation to isolate the infectious agents. Both $CD4^+$ and $CD8^+$ T lymphocytes participate in the experimental infection in mice. The $CD4^+$ T cells synthesize interferon γ, and the $CD8^+$ T cells lyse ineffective macrophages. IL-12 is important in controlling the differentiation of T cells and natural killer cells to produce IFN-γ, which facilitates cell-mediated immunity. IL-1, TNF-α, IL-6, M-CSF, and G-CSF are all produced during experimental infection. Minute granulomata comprised of epithelioid cells, neutrophils, mononuclear leukocytes, and giant cells are produced in the tissues of humans infected with *Brucella* species. Hepatosplenomegaly is a common clinical feature. Delayed-type hypersensitivity, which is a correlate of cell-mediated immunity, induces immunopathological changes. In addition to the granulomata, there is development of a severe generalized delayed-type hypersensitivity response that mimics many features of the infection itself. Diagnosis depends on testing for antibodies in the serum. Most of the antibodies are directed against LPS. The agglutination test is the one most widely used but is being replaced by ELISA. Whereas *B. abortus* strain 19 is employed to immunize cattle and *B. melitensis* stain Rev 1 is used to immunize sheep and goats, human vaccination has been used essentially only in Russia. Other preventive measures include pasteurization of milk products.

***Campylobacter* immunity:** Circulating antibodies develop rapidly in patients with *Campylobacter enteritis.* These antibodies fix complement, are bactericidal, and agglutinate. Following an initial but short-lived IgM response, there is a rapid IgA response that peaks 14 d after onset of symptoms but declines by the fifth week. IgG antibodies appear by the tenth day after infection and are present for several months. Antibodies are believed to limit the infection. Serologic tests for diagnosis depend on an acid-extractable surface antigen that consists mainly of flagellin which is the immunodominant surface antigen. Antiflagellin antibodies appear early during an infection and are believed to be protective. MOMP is also immunogenic. All of these antibodies have specificity for homologous and heterologous strains. LPSs induce variable antibody responses. Half of the cases of *Campylobacter enteritis* develop during Guillian–Barré syndrome. These patients develop increased circulating IgG and IgM that bind to GM1 and GD1 ganglioside epitope and which cross-react with LPS of certain serotypes of *C. Jejuni*. Definitive studies of the T cell response to *Campylobacter* remain to be performed.

***Clostridium* immunity:** Clostridia produce disease by releasing exotoxins. They may produce more than one toxin and each one is immunologically unique. For example, each of the five types of *Clostridium perfringens* produces a different toxin. Clostridia enter the host by many routes to produce disease. *Clostridium perfringens* gains access through traumatic or surgical wounds to produce gas gangrene and wound infections. The microorganism is aided by a poor blood supply in the area of the wound. The clostridia divide and produce toxins that cause disease. When antibiotics upset the normal bowel florae, *C. difficile* may multiply and induce colitis. The toxins released from this organism act on intestinal epithelial cells and produce diarrhea and chronic inflammation. *C. botulinum* does not grow in the host but forms toxins in contaminated food that when ingested lead to the disease. Most clostridial diseases do not induce protective immunity because the amount of toxin required to produce disease is less than that need to induce an immunologic response. Even though systemic immunity does not follow an episode of the disease, tetanus toxoid can induce immunity that may last for 5 years. Tetanus immunoglobulin is also valuable for passive immunization in suspected cases. Antibotulinum toxin antibody is available for laboratory workers. Intravenous administration of gammaglobulin containing high titers of antibody to *C. difficile* toxin has been useful in the therapy of patients with relapsing *C. difficile* diarrhea.

***Escherichia coli* immunity:** IgM and IgG antibodies are formed against O, H, and K antigens of *E. coli* of the diarrhea-producing strains in infants. Secretory IgA specific for *E. coli* diminishes the adherence of diarrhea-producing *E. coli* in the intestines of infants. Secretory IgA in breast milk also offers significant protection of infants through passive immunity. Secretory IgA may be specific for LT enterotoxins and for colonization factor antigens.

***Francisella* immunity:** The causative agent of tularemia, *Francisella tularensis*, may induce two forms of the disease, ulceroglandular tularemia which is borne by vectors or induced by contact with infected animals, and respiratory tularemia which is caused by inhalation of contaminated dust. A powerful antibody response occurs during infection with this microorganism and it is detectable by agglutination, ELISA, or other techniques. The antibodies appear at the end of the second or during the third week of infection. They persist for several years following recovery. IgM antibodies do not appear before IgG antibodies and may even be present for years following recovery. Cell-mediated immunity in tularemia is demonstrated by a delayed-type hypersensitivity test or by *in vitro* activation of T cells. Cell-mediated immunity against *F. tularensis* is a requisite for host protection. The microorganism's capsule protects it against lysis by complement and affords resistance to intracellular killing by polymorphonuclear leukocytes. Attenuated strains of the microorganism, termed *F. tularensis* LVS are easily killed by polymorphonuclear leukocytes and are susceptible to hydrochloric acid and hydrogen peroxide produced as a result of the oxidative burst. No significant toxins are produced by *F. tularensis*. T-cell mediated immunity can prevent fulminating disease. Infection or vaccination with live attenuated bacteria can induce host protection against tularemia.

***Fusobacterium* immunity:** *F. nucleatum* adheres to lymphocytes through lectin-like ligands to facilitate other activity or exert an immunosuppressive effect on them. The latter enhances the pathogenicity of the microorganism. Patients with periodontal disease, peritonsillar cellulitis and abscesses, infectious mononucleosis, and acute streptococcal, nonstreptococcal, and recurrent tonsillitis manifest increased levels of antibodies against protein antigens of F. *nucleatum*. Antibodies against these outer membrane proteins may point to a pathogenic role for this microorganism in these infections. Delayed-type hypersensitivity to *F. necrophorum* has been induced in mice, and leukotoxin-specific antibodies have been found in cattle. Natural infection or vaccination with a toxoid does not always protect against reinfection with this microorganism. It is believed to be a weak immunogenic pathogen.

***Haemophilus* immunity:** Serum bactericidal activity to *Haemophilus influenzae* serotype b(Hib) is associated with protection from infection. This effect is mediated by antibody to the capsule. Antibody to the PRP polysaccharide capsule protects against invasive infection. Two-year-olds have only low titers of anticapsular antibodies and are susceptible to reinfection and episodes of Hib meningitis. NTHI strains of the organism demonstrate antigenic heterogeneity of surface antigens among strains. Infection does not induce protective immunity from infection by

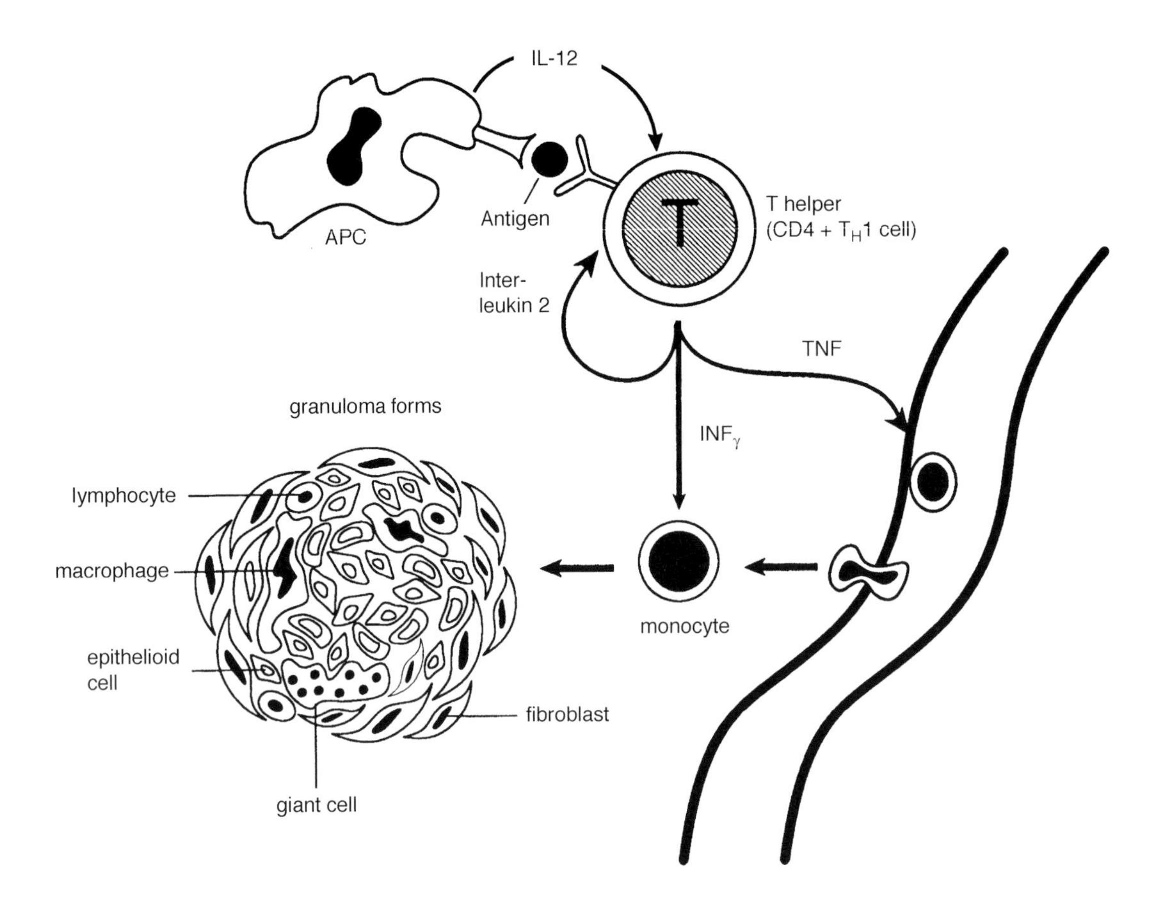

FIGURE 23.20 Granuloma.

NTHI strains following otitis media attributable to an NTHI strain. Serum bactericidal antibody develops and this is associated with protection. The immune response is strain specific, which renders the child susceptible to infection with other strains of NTHI. The immune response is strain specific, with reactivity directed to the immunodominant surface epitopes on the P2 OMP. *H. influenzae* LOS is a principal toxin against which humoral immune responses are directed. The mucosal immune response to NTHI remains to be defined. Conjugate vaccines developed to prevent invasive infections by Hib have been very successful in preventing the infectious disease. High titers of serum IgG are specific for Hib capsular polysaccharide are associated with increased bactericidal and protective activity. Linking the PRP capsular polysaccharide to protein carriers has yielded a conjugate vaccine which is effective in protecting infants against invasive Hib infections.

***Legionella* immunity:** Immunity against *Legionella* pneumophila, a facultative intracellular pathogen that induces Legionnaires' disease in humans, depends on cellular immune mechanisms, including the release of IFNγ. T_H1 CD^+ T cells play a significant role in the development of acquired immunity in mice. Acquired immunity to *Legionella* pneumophila is believed to be a consequence of both humoral and cellular immune responses that facilitate enhanced uptake of the microorganisms by activated mononuclear phagocytic cells.

Frei test is a tuberculin type of delayed hypersensitivity skin test employed to reveal delayed-type hypersensitivity in lymphogranuloma venereum patients. Following intradermal injection of lymphogranuloma venereum virus, an erythematous and indurated papule develops after 4 d.

Lepra cells are foamy macrophages that contain clusters of *Mycobacterium leprae* microorganisms that are not degraded because cell-mediated immunity has been lost. These are found in lepromatous leprosy, but are not observed in tuberculoid leprosy.

Lepromatous leprosy is a chronic granulomatous disease induced by *Mycobacterium leprae*. The condition is contagious and is also known as Hansen's disease. A second form of leprosy is termed tuberculoid, which is a more benign and stable form of leprosy. Both lepromatous and tuberculoid leprosy infect the peripheral nervous system.

***Leptospira* immunity:** Newly isolated leptospires evade the host immune system by not reacting with specific antibody which permits their multiplication. On entering the host, they also evade the host immune system by their sequestration in renal tubules, uterine lumen, eye, or brain. The humoral immune is marked by production of IgM which, together with complement and phagocytic cells, begins to clear leptospires from the host. This is followed by the production of opsonizing and neutralizing IgG. There is only low-grade cell-mediated immunity to *Leptospira* infection. Vaccines for use in animals consist of chemically inactivated whole *Leptospira* cultures which have proven to be somewhat effective. The immunity induced is serovar specific. These antibodies do not afford protection. Virulent leptospires are better immunogens than are avirulent organisms. It is significant that antigens located on the outer envelope maintain their natural configuration. Both IgM and IgG are induced following parenteral administration of a vaccine, as the IgM is a more efficient agglutinating and neutralizing antibody for leptospires than is IgG.

Moro test is a variant of the tuberculin test in which tuberculin is incorporated into an ointment that is applied to the skin to permit the tuberculin to enter the body by inunction.

Mycobacterium is a genus of aerobic bacteria, including *Mycobacterium tuberculosis*, which can survive within phagocytic cells and produce disease. Cell-mediated immunity is the principal host defense mechanism against mycobacteria.

MOTT (mycobacteria other than *Mycobacterium tuberculosis*) is an acronym for mycobacteria other than those that induce tuberculosis. Their recognition is increasing.

TB is the abbreviation for tuberculosis.

***Mycobacteria* immunity:** Immunity to tuberculosis is highly complex and involves cell-mediated mechanisms. The host immune response to tuberculosis is inappropriate, leading to tissue injury through immune mechanisms rather than elimination of the invading microorganism. In the mouse, immunity depends upon TNF-α, a T_H1 cytokine pattern and MHC class II. Mouse murine macrophages activated by IFN-γ inhibit proliferation of *M. tuberculosis*. β_2 Microglobulin is required for immunity to mycobacteria, which may point to the participation of $CD8^+$ MHC Class I restrictor cytotoxic T cells. $\gamma\delta$ T cell receptors identify mycobacterial antigens such as heat shock proteins. These cell types secrete IFN-γ, and are cytotoxic and classified as a type I response. A type II response renders mice more susceptible to tuberculosis. Immunity in humans is also associated with a T_H1 type of response associated with macrophage activation and cytotoxic removal of infected cells. Human tuberculosis patients form specific IgE and IgG4 antibodies, both of which are IL-4 dependent. IL-10 levels may also be increased. A T_H2 response is associated with progressive disease in humans. *M. tuberculosis* can cause release of TNF-α from primed macrophages. This cytokine is required for protection but also has a role in immunopathology. Tuberculosis patients develop necrotic lesions which are believed to help wall off established infections. TNF-α's toxicity in a mycobacterial lesion depends on whether or not the T-cell response is T_H1 or T_H2. Necrosis is not produced when TNF-α is injected into a T_H1 inflammatory site but marked tissue injury results when it is injected into a T_H1-/T_H2-mediated site. Infection by *Mycobacterial leprae* is weakly associated with MHC haplotypes. It appears to determine the type of disease that will develop instead of susceptibility. In tuberculoid leprosy, both *in vivo* and *in vitro* equivalents of T cell mediated responsiveness such as skin test reactivity in lymphoproliferation in response to antigens of *M. leprae* are intact. T_H1 cytokines are produced in these lesions. Unless treated, these will develop into lepromatous leprosy in which the T_H1 response is impaired. In leprosy, immunity probably requires macrophage activation and cytotoxic T cells activated by the T_H1 response. In lepromatous leprosy, there is depressed T cell mediated immunity *in vivo* and *in vitro*. Numerous macrophages packed with bacilli are present in the lesions, but there is no response. There is a continuous heavy antigen load and lack of a T cell mediated response to the microorganism. The T cells express a T_H2 cytokine pattern.

Caseous necrosis (Figure 23.21) is tissue destruction, as occurs in tuberculosis, that has the appearance of cottage cheese.

Caseation necrosis is a type of necrosis present at the center of large granulomas such as those formed in tuberculosis. The white cheesy appearance of the central necrotic area is the basis for the term.

Invasin is a membrane protein derived from *Yersina pseudotuberculosis* that binds to $\alpha_4\beta_1$ integrins and has the capacity to induce T cell costimulatory signals.

MAC, *Mycobacterium avium* complex, is a systemic infection that regularly affects subjects with AIDS, up to 66% of whom still have peripheral blood $CD4^+$ T lymphocytes. Infection with this complex is a clear indication of immunosuppression. MAC is successfully treated with clarithromycin.

MAIS complex: *Mycobacterium avium-intracellulare-scrofulaceum*. Three species of mycobacteria that express the same antigens and lipids on their surfaces and also have the same biochemical reactions, antibiotic susceptibility, and pigment formation. They frequently occur together clinically. MAIS complex is relatively rare, but it occurs in 5 to 8% of AIDS patients when their $CD4^+$ T lymphocyte levels diminish to less than 100 cells per cubic millimeter of blood. Affected patients have

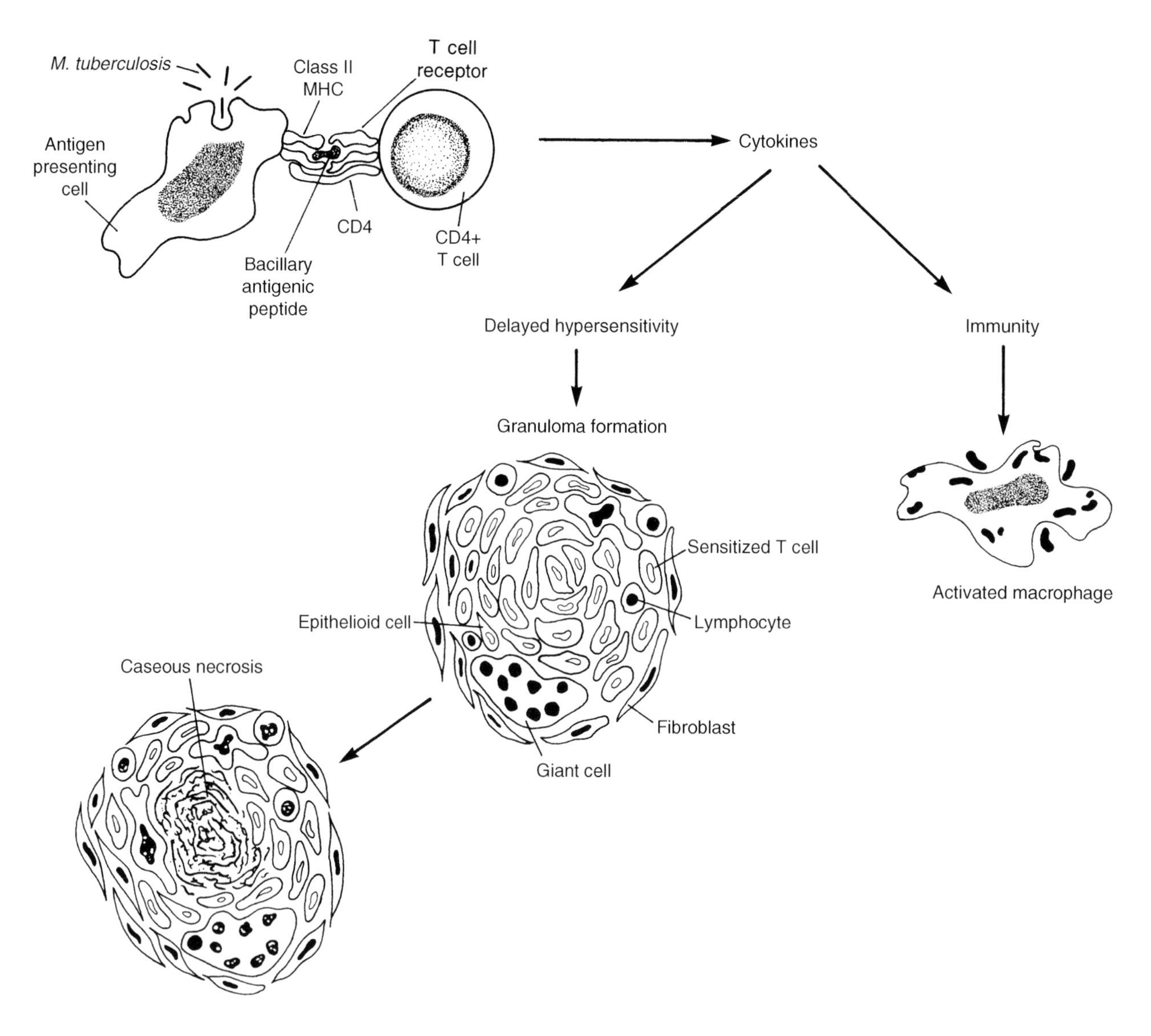

FIGURE 23.21 Caseous necrosis (tuberculosis).

persistent diarrhea, night sweats and fever, abdominal pain, anemia, and extrahepatic obstruction. Ciprofloxacin, clofazimine, ethambutol, and rifampicin, as well as rifabutin, clarithromycin, and azithromycin have been used in treatment.

Mantoux test is a type of tuberculin reaction in which an intradermal injection of tuberculin tests for cell-mediated immunity. A positive test signifies delayed (type IV) hypersensitivity to *Mycobacterium tuberculosis,* which indicates previous or current infection with this microorganism.

The **Mitsuda reaction** is a graduated response to an intracutaneous inoculation of lepromin, a substance used in the lepromin test. A nodule representing a subcutaneous granulomatous reaction to lepromin occurs 2 to 4 weeks after inoculation and is maximal at 4 weeks. It indicates granulomatous sensitization in a leprosy patient. Although not a diagnostic test, it can distinguish tuberculoid from lepromatous leprosy in that this test is positive in tuberculoid leprosy, as well as in normal adult controls, but is negative in lepromatous leprosy patients.

Ghon complex is the combination of a pleural surface-healed granuloma or scar on the middle lobe of the lung, together with hilar lymph node granulomas. The Ghon complex signifies healed primary tuberculosis.

***Neisseria* immunity:** Even though patients with gonococcal infection develop increased levels of serum and mucosal antibody of IgG and IgA classes reactive with gonococcal surface antigens, they develop repeated episodes of urogenital infections. Thus, natural infection does not confer protective immunity. The surface antigens may be antigenically heterogeneous or this may be attributable to the brevity of the mucosal antibody response and the lack of activity of serum antibody in the mucosal infection. Attachment and colonization of the mucosa with gonococci in their encounter with polymorphonuclear neutrophils play an important role. Protective immunity against meningococcal infection is associated with complement-dependent, bactericidal serum antibodies specific for capsular polysaccharide. Cell-mediated immunity does not appear to have a significant role. There is no successful gonococcal vaccine. Polysaccharide vaccines to protect against *N. meningitidis* of serogroup A and C induce protective bactericidal antibodies. The current vaccine contains polysaccharide from serogroups A, C, Y, and W-135. Polysaccharide A and C vaccine has been conjugated to tetanus toxoid or another protein carrier to enhance immunogenicity in young children. Polysaccharide vaccines may not induce protection in serogroup B meningococcal disease. Protection against these strains may be linked to antibody specific for outer membrane proteins which can be used to induce immunity.

***Pasteurella* immunity:** Immunity against pasteurella multocida is mainly humoral antibody-mediated, yet cell-mediated immune responses also occur. Naturally acquired immunity to this organism can develop in unvaccinated cattle and water buffalo. Even though the organism is surrounded by a capsule that contributes to its virulence, antibody acting as opsonin can render these microorganisms readily available for phagocytosis by monocytes, macrophages, and polymorphonuclear neutrophils. Protein toxins from certain stains can induce the formation of neutralizing antitoxins, but purified proteins have not been shown to induce protective immunity. Cattle, buffalo, and poultry vaccinated with bacterins, formalin-killed organisms in a water-in-oil emulsion, form antibodies against lipopolysaccharide (LPS). Anti-LPS antibodies are also associated with naturally acquired immunity. Not only bacterins but also live-attenuated vaccines have been employed to control *P. multicida* infections.

***Proteus* immunity:** *Proteus mirabilis* administered to mice by the transurethral route can induce antibodies that react with purified lipopolysaccharide, flagella, outer membrane protein (OMP), and mannose-resistant *proteus*-like MR/P fimbriae. Murine vaccination leads to partial immunity. Rheumatoid arthritis patients' blood sera also react with *P. mirabilis*. A hexamer peptide within hemolysin is the target of the immunogenic response, probably due to its close similarity to susceptibility sequences of HLA/DR1 and DR4. Several proteins encoded by chromosomal genes serve as *P. mirabilis* virulence factors. These help the host evade immune defenses and lead to cell and tissue injury. These include urease and HPM homolysin. The latter leads to lysis of erythrocytes or epithelial cell membrane injury in culture. The organism also produces four types of fimbriae, surface proteins that have a role in adherence. It also secretes a protease that degrades IgA.

***Pseudomonas aeruginosa* immunity:** *Pseudomonas aeruginosa* infection is followed by the development of antibodies that facilitate opsonophagocytosis in protection against subsequent infections. Thus, an adequate antibody response is required for protection. Antibody alone may be insufficient for protection since cystic fibrosis patients' lungs continue to be chronically colonized even in the presence of potent serum antibody responses to several antigens of this microorganism. Active vaccination is made less desirable by the fact that *P. aeruginosa* infections cannot be reliably predicted. Thus, passively administered hyperimmune intravenous immunoglobulin from immunized volunteers has been used to confer protection. Nevertheless, contemporary investigations show that antibodies against LPS serotypes and to exotoxin A did not significantly protect recipients.

***Salmonella* immunity:** Natural and adaptive immunity are necessary for survival following primary infection with virulent microorganisms of this genus. *Salmonellae* grow exponentially in the reticuloendothelial system and may reach 10^8 microorganisms, which is a lethal number, apparently due to endotoxin as a contributing lethal factor. The balance between strain virulence and host resistance controls the rate at which bacterial cells increase in the reticuloendothelial system. Bone marrow-derived cells TNF-α, IFN-γ, and IL-12 but not T lymphocytes control early growth. This is termed the plateau phase after which an immune response clears the microorganisms from the tissue and provides effective immunity to reinfection. $CD4^+$ and $CD8^+$ T lymphocytes, together with TNF-α, IFN-γ, and IL-12 help to clear bacteria from the tissues, but antibody is also required in addition to cell-mediated immunity. The antibody may be directed against lipopolysaccharides (LPS) in animal infections with these microorganisms. In humans, antibodies to Vi antigen are believed to be significant. The diagnosis of typhoid fever is based on the detection of antibodies to O and H antigen and to Vi antigens in carriers. Immunization with killed samonellae does not induce cell-mediated immunity and confers less protection than immunization with live organisms that induce both cell-mediated and humoral immunity. Antibody alone is also not protective. The protective immunogen is probably LPS O antigen or Vi antigen in *S. typhi*, other protein antigens, or some combination of these. IgA provides partial immunity. Since killed microorganisms used in vaccines in the past do not induce appropriate cell-mediated immunity and are highly reactogenic, live vaccines of superior efficacy in experimental models are being developed.

Vi antigen is a virulence antigen linked to *Salmonella* microorganisms. It is found in the capsule and interferes with serological typing of the O antigen, a heat-stable lipopolysaccharide of enterobacteriaceae.

Cholera toxin is a *Vibro cholerae* enterotoxin comprised of five B subunits which are cell-binding 11.6-kDa structures that encircle a 27-kDa catalase which conveys ADP-ribose to G protein, leading to continual adenyl cyclase activation. Other toxins that resemble cholera toxin in function include diphtheria toxin, exotoxin A, and pertussis toxin.

Exotoxin is an extracellular product of pathogenic microorganisms. Exotoxins are 3- to 500-kDa polypeptides produced by such microorganisms as *Corynebacterium diphtheriae, Clostridium tetani,* and *C. botulinum. Vibrio cholerae* produces exotoxins that elevate cAMP levels in intestinal mucosa cells and increase the flow of water and ions into the intestinal lumen, producing diarrhea. Exotoxins are polypeptides released from bacterial cells and are diffusible, thermolabile, and able to be converted to toxoids that are immunogenic but not toxic. Bacterial exotoxins are either cytolytic, acting on cell membranes, or bipartite (A-B toxins), linking to a cell surface through the B segment of the toxin and releasing the A segment only after the molecule reaches the cytoplasm where it produces injury. Some may serve as superantigens.

Coccidiodin is a *Coccidioides immitis* culture extract that is used in a skin test for cell-mediated immunity against the microorganism in a manner analogous to the tuberculin skin test.

***Listeria*:** A genus of small Gram-positive motile bacilli that have a palisading pattern of growth. The best known is *L. monocytogenes*, which has a special affinity for monocytes and macrophages in which it takes up residence. It may be transmitted in contaminated milk and cheese. Approximately one third of the cases are in pregnant females, resulting in transplacental infection that may induce abortion or stillbirth. Infected infants may develop septicemia, vomiting, diarrhea, cardiorespiratory distress, and meningoencephalitis. Individuals with defective immune reactivity may develop endocarditis, meningoencephalitis, peritonitis, or other infectious processes. Treatment is with ampicillin, erythromycin, gentamycin, or chloramphenicol.

***Listeria* immunity:** Immune responsiveness to *Listeria* has been investigated mostly in a mouse model. The microorganism results are only in macrophages but are also in hepatocytes of infected host. Natural immunity in mice is controlled by Lr1 locus on chromosome II. Mice resistant to *Listeria* respond to inoculated listeriae with large numbers of inflammatory phagocytes. Neutrophils and mononuclear phagocytes kill *L. monocytogenes in vitro* although resident macrophages are less effective. Oxygen-dependent and independent bactericidal mechanisms facilitate destruction of listeriae. Multiple cell types and mediators are involved in resistance against *Listeria*. T cell-mediated immunity is significant but not the only factor in resistance to listeriosis. Neutrophils are also significant to resistance against *Listeria*. Other cells implicated in resistance to *Listeria* infections include granulocytes, NK cells, and cytotoxic T cells. Infection with the microorganism is followed by the expression of several cytokines and include IFN-γ, IL-1β, TNF-α, and GM-CSF among others. The two cytokines most critical for resistance to Listeriosis are IFN-γ and TNF-α. IFN-γ (T_H1) is necessary for resistance, but IL-4 and IL-10 (T_H2) hinder resistance. Recombinant cytokines that can increase resistance to *L. monocytogenes* infection include IFN-γ, TNF-α, IL-1β, and IL-12 but potentiate in inflammation. Experimental animals injected with a sublethal dose of viable listeriae develop resistance to rechallenge for a few months followed by decreased resistance. Killed microorganisms failed to provide effective immunity.

***Listeria monocytogenes*:** Specific immune responses can be mounted against intracellular bacteria and fungi. Some bacteria reproduce inside cells of the host. For example, mycobacteria and *Listeria monocytogenes* (Figure 23.22) are organisms of high pathogenicity that survive in phatocytic cells such as macrophages where they resist dissolution. Within the macrophage, they are not exposed to specific antibody. In addition to mycobacteria and *Listeria* species, a number of fungi are also intracellular pathogens.

Group agglutination: In the serologic classification of microorganisms, the identification of group-specific antigens rather than species-specific antigens by the antibody used for serotyping.

A **virus** is an infectious agent that ranges from 10^6 D for the smallest viruses to 200×10^6 D for larger viruses such

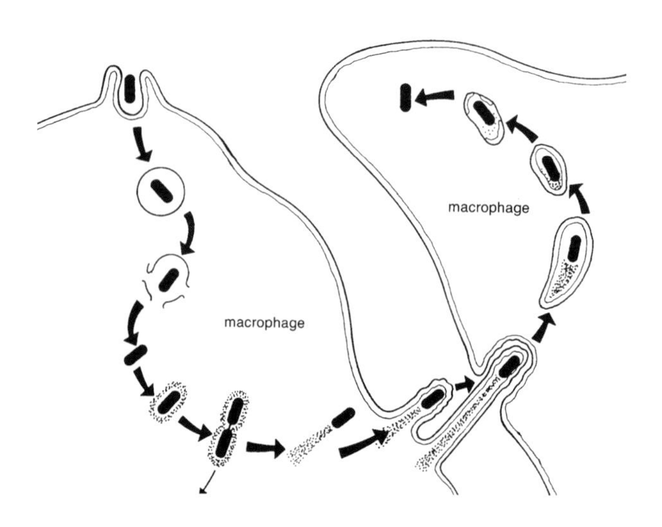

FIGURE 23.22 *Listeria.*

as the poxviruses. Viruses contain single or double-stranded DNA or RNA that is either circular or open and linear. The nucleic acid is enclosed by a protein coat, termed a capsid, comprised of a few characteristic proteins. Most viruses are helical or icosahedral. There may be a lipid envelope which may contain viral proteins. Viruses may be incubated with cells in culture, where they produce characteristic cytopathic effects. Inclusion bodies may be produced in cells infected by viruses. Viruses infect host cells through specific receptors. Examples of this specificity include cytomegalovirus linking to β_2 microglobulin, Epstein-Barr virus linking to C3d receptor (CR2), and HIV-1 binding to CD4.

Virion is complete virus particles.

Viroid is a 100-kDa 300-bp subviral infectious RNA particle comprised of a circular single-stranded RNA segment. It induces disease in certain plants. Viroids may be escaped introns.

Viropathic refers to host tissue injury resulting from infection by a pathogenic virus.

A **capsid** is a virus protein envelope comprised of subunits that are called capsomers.

Viral interference is resistance of cells infected with one virus to infection by a second virus.

Latency is a condition in which the viruses enter a cell but do not replicate. Once reactivated, the virus may replicate and cause disease.

Lysogeny is the condition in which a viral genome (provirus) is associated with the genome of the host in such a manner that the genes of the virus remain unexpressed.

Neuraminidase is an enzyme that cleaves the glycosidic bond between neuraminic acid and other sugars. Neuraminic acid is a critical constituent of multiple cell surface glycoproteins and confers a negative charge on the cells. Cells treated with neuraminidase agglutinate more readily than do normal cells because of the diminished coulombic forces between them. Cells treated with neuraminidase activate the alternate complement pathway. Neuraminidase is produced by myxoviruses, paramyxoviruses, and such bacteria as *Clostridium perfrigens* and *Vibrio cholerae.* Neuraminidase together with hemagglutinin is found on the spikes of the influenza virus.

Viral immunity: Congenital immunodeficiency patients have given insight into the relative significance of various constituents of the immune system. Subjects with isolated defects of cell-mediated immunity contract severe and often fatal viral infections that include measles and chicken pox. By contrast, those individuals with isolated immunoglobulin deficiency usually recover normally from most viral infections except enteroviruses which may lead to chronic infection of the central nervous system. Certain generalizations may be reached concerning viral immunity. These include the following.

Antibodies act mainly by neutralizing virions, rendering them noninfectious. By contrast, cell-mediated immunity is against virus antigens present in infected cells. Antibody prevents primary infection and reinfection through neutralization of viruses on mucosal surfaces and limiting their spread in body fluids, whereas cell-mediated immunity eliminates intracellular infection and limits reactivation of persistent viruses. The immune system may be considered in three phases, i.e., immediate (less than 4 h), early (4 to 96 h), and late (greater than 96 h). It may also be divided into humoral and cell-mediated components that include both specific and nonspecific mechanisms.

Neutralizing antibody: See neutralization and neutralization test.

Viral hemagglutination: Selected viruses may combine with specific receptors on surfaces of red cells from various species to produce hemagglutination. The ability of antiviral antibodies to inhibit this reaction constitutes hemagglutination inhibition, which serves as an assay to quantify the antibodies. One must be certain that inhibition is due to the antibody and not to a nonspecific agent such as mucoproteins with myxoviruses or lipoproteins with arbor viruses. Blood sera to be assayed by this technique must have the inhibitor activity removed by treatment with neuraminidase or acetone extraction, depending upon the chemical nature of the inhibitor. The attachment of virus particles to cells is termed hemadsorption. Among viruses causing hemagglutination are those which induce influenza and parainfluenza, mumps, Newcastle disease, smallpox and vaccinia, measles, St. Louis encephalitis, Western equine encephalitis, Japanese B encephalitis, Venezuelan equine encephalitis, West Nile fever, Dengue viruses, respiratory syncytial virus, and some enteroviruses. Herpes simplex virus can be absorbed to tanned sheep red cells and hemagglutinated in the presence of specific antiserum against the virus. This method is termed indirect virus hemagglutination.

Virus-associated hemophagocytic syndrome is an aggressive hemophagocytic state that occurs in both immunocompromised and nonimmunocompromised individuals, usually those with herpetic infections including CMV, EBV, and herpes virus, and may occur in infections by adenovirus and rubella, as well as in brucellosis, candidiasis, leishmaniasis, tuberculosis, and salmonellosis. There is lymphadenopathy, hepatosplenomegaly, pulmonary infiltration, skin rash, and pancytopenia. The disease is sometimes confused with malignant histiocytosis, lymphoma, sinus histiocytosis, and lymphomatoid granulomatosis.

Virus-neutralizing capacity refers to serum's ability to prevent virus infectivity. Neutralizing antibody is usually of the IgM, IgG, or IgA class.

Neutralization by antibodies is a specific immune responses to viruses. Viruses are obligate intracellular parasites that multiply within host cells whose nucleic acids and protein synthesis capability are subverted and appropriated for virus propagation. Viruses may injure cells they have infected by interfering with the cell's protein synthesis and normal functioning, leading to host cell death. This constitutes a cytopathic effect. Viruses that are not cytopathic may induce a latent infection in which they remain inside host cells and induce synthesis of proteins that provoke a specific immune response. This consists of cytolytic T lymphocytes that are specific for the virus and destroy the virus-infected cell. Viral proteins may induce delayed-type hypersensitivity that leads to cellular injury. Both antibodies from B cells and specifically sensitized T cells confer immunity against viruses. Before host cell invasion, specific antibodies may neutralize virions through a process known as **neutralization** (Figure 23.23). However, following penetration of host cells, T-cell mediated immunity is required for destruction of the virus-infected host cells.t the br

Papovaviruses are minute tumor viruses that are icosahedral and contain double-stranded DNA. Included in the group are SV40 and polyomavirus that may cause malignant and benign tumors. Permissive or nonpermissive infections occur with papovavirus. Following permissive infection of monkey cells, papovavirus replicates, leading to lysis. T antigens, which are early papovavirus proteins that occur in nonpermissive rodent cells, can lead to transformation of the cells that is not reversible if the viral genome is integrated into the host genome. It is reversible if the cell can eliminate the viral genome.

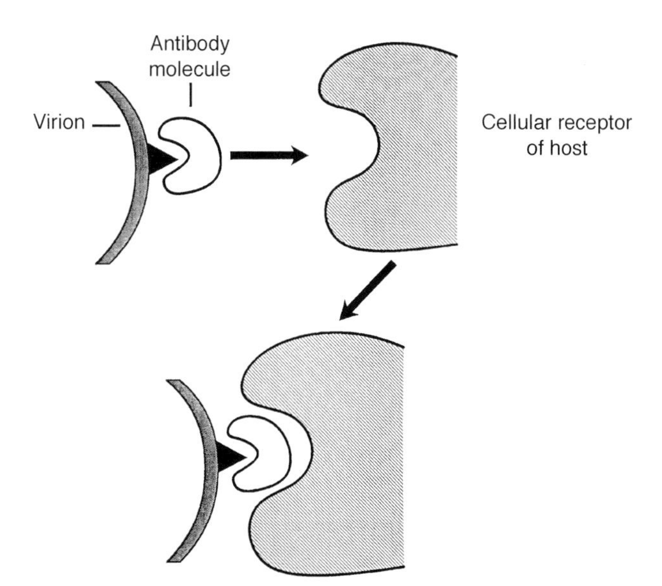

FIGURE 23.23 Neutralizing antibody molecules.

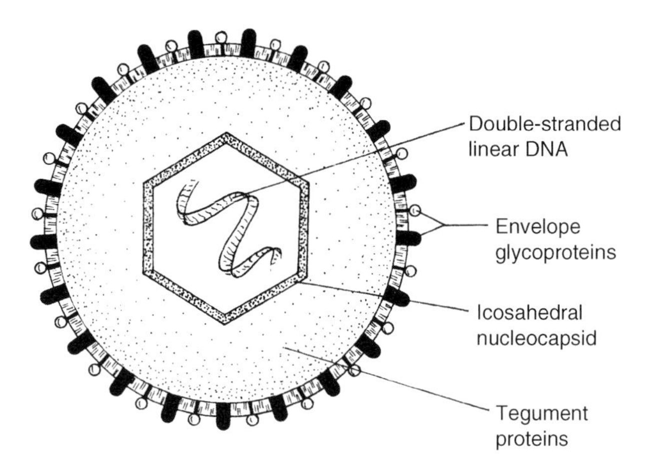

FIGURE 23.24 Herpesvirus.

HPV, human papilloma virus, is a human virus that has the potential to be oncogenic and occurs most frequently in individuals with multiple sexual partners. There are 46 HPV genotypes. HPV can be demonstrated by *in situ* hybridization in proliferations of epithelial cells that are benign, such as condyloma accuminatum, or malignant, such as squamous cell carcinoma of the uterine cervix. Whereas HPV types 6 and 11 are not usually premalignant, HPV types 16, 18, 31, 33, and 35 are linked to cervical intraepithelial neoplasia (CIN), cervical dysplasia, and anogenital cancer. HPV is predicted to induce derepression as a neoplastic mechanism. HPV encodes E6, a viral protein that combines with the tumor suppressor protein p53.

HHV is the abbreviation for human herpesvirus.

Herpesvirus (Figure 23.24) is a DNA virus family that contains a central icosahedral core of double-stranded DNA. There is a lipoprotein envelope that is trilaminar and 100 nm in diameter and a nucleus that is 30 to 43 nm in diameter. Herpes viruses may persist for years in a dormant state. Six types have been described. HSV-1 (herpes simplex virus-1) can account for oral lesions such as fever blisters. HHV-2 (human herpes virus-2) produces lesions below the waistline and is sexually transmitted. It may produce venereal disease of the vagina and vulva, as well as herpetic ulcers of the penis. Both simplex-1 and simplex-2 may infect the brain (Figure 23.25). HHV-3 (herpes varicella-zoster) occurs clinically as either an acute form known as chicken pox or a chronic form termed shingles. HHV-4 (Epstein-Barr virus), HHV-5 (cytomegalovirus), and HHV-6 (human B cell lymphotrophic virus) are the other types of herpes virus.

Herpes simplex virus 1 and 2 (HSV 1 and 2) polyclonal antibody is an antibody used to identify specific and qualitative localization of herpes simplex virus (HSV)

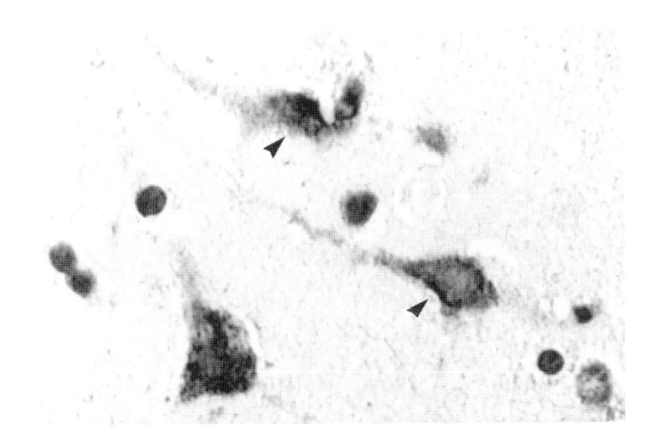

FIGURE 23.25 Herpes simplex in the brain.

types 1 and 2 in formalin-fixed, paraffin-embedded, or frozen tissue sections. HSV-1 is most often acquired during early childhood by nonvenereal means. It causes gingivostomatis (fever blisters). HSV-2 causes *Herpes genitalis.* HSV-2 is usually acquired by venereal contact but can also be acquired by a neonate at the time of delivery from an infected mother's genital tract.

HSV is the abbreviation for herpes simplex virus.

Herpes simplex virus immunity: The two related subtypes of herpes simplex virus are designated type I (HSV-1) and type II (HSV-2). They are presently designated as human herpes viruses 1 and 2, respectively. Immune response to HSV has been investigated mainly in mouse, rabbit, and guinea pig animal models. Genetic resistance of a host animal is by several mechanisms such as the effective killing action of macrophages and natural killer cells and interferon synthesis. NK cell and interferon activity can restrict an infection but cannot clear the virus, which depends on the host-specific immune response. Both antibody and T cell immunity are induced. Antibodies interact with the virus, infected cell glycoproteins, capsid proteins, and selected infected cell polypeptides. In mice, passively transferred antibodies against HSV can protect them from lethal doses of the virus. Antibody alone is insufficient to clear virus from the nervous system or the periphery and may retard virus spread through the nervous system. Antibody activities of HSV-specific antibody *in vitro* include virus neutralization, complement-mediated cytotoxicity, and ADCC. Reactivation and recurrence or reinfection with HSV in humans often occurs in the presence of high titers of neutralizing antibodies, which reveals that the antibody fails to protect. T cells are vitally important in HSV infections. The cytotoxic T lymphocyte response to HSV is mediated by $CD8^+$ T lymphocytes in both humans and mice. $CD4^+$ (MHC) Class II-restricted lymphocytes are involved in the delayed-type hypersensitivity response to HSV, LCTL activity, and T cell help in the antibody production. $CD4^+$ T cells are critical for clearance of the virus. The T_H1 subset is protective. HSV survives for many years in the host through establishment of a latent infection without discernible protein expression. The efficacy of avirulent or inactivated virus and subunit vaccines has not been definitely established in human trials. Animal trials have revealed that protection can be induced with glycoprotein D using recombinants *Salmonella* introduced orally.

Herpesvirus-6 immunity: Humoral immune responses during primary infection include an IgM response. Secondary infection is associated with an increase in IgG titer as well as a recrudescence of IgM reactivity. The humoral immune response is strongly cross-reactive among HHV-6 variants. The T cell response to infection remains to be elucidated.

HBLV (human B lymphotropic virus): Herpesvirus 6.

Herpesvirus-8 immunity: Antibodies specific for a latent HHV-8 antigen and to a recombinant structural antigen present in Kaposi's sarcoma patients or at those at risk for developing this disease. There is strong evidence for a link between HHV-8 in the pathogenesis of Kaposi's sarcoma. Antibodies to HHV-8 antigens occur in HIV-uninfected homosexual men than in the general population including blood donors. Antibodies to undefined structural HHV-8 antigens present in a quarter of all blood donors in North America. Yet antibodies to the recombinant capsid-related and latent HHV-8 proteins are present in 0 to 2% of blood donors in North America and Northern Europe.

Herpes zoster is a viral infection that occurs in a band-like pattern according to distribution in the skin involved nerves. It is usually a reactivation of the virus that causes chickenpox.

Cytomegalovirus (CMV) (Figure 23.26) is a herpes (DNA) virus group that is distributed worldwide and is not often a problem, except in individuals who are immunocompromised, such as the recipients of organ or bone marrow transplants or individuals with acquired immunodeficiency syndrome (AIDS). Histopathologically, typical inclusion bodies that resemble an owl's eye are found in multiple tissues. CMV is transmitted in the blood. Two classes of antiviral drugs are used to treat HIV infection and AIDS. Nucleotide analogues inhibit reverse transcriptase activity. They include azidothymidine (AZT), dideoxyinosine, and dideoxycytidine. They may diminish plasma HIV RNA leveld for considerable periods but often fail to stop disease progression because of development of mutated forms of reverse transcriptase that resist these drugs. Viral protease inhibitors are now used to block the processing of precursor proteins into mature viral capsid and core proteins. Currently, a triple-drug therapy consisting of protease inhibitors are used to reduce plasma viral RNA to very low levels in patients treated for more than 1 year. It remains to be determined whether or not resistance to this therapy will develop. Disadvantages include their

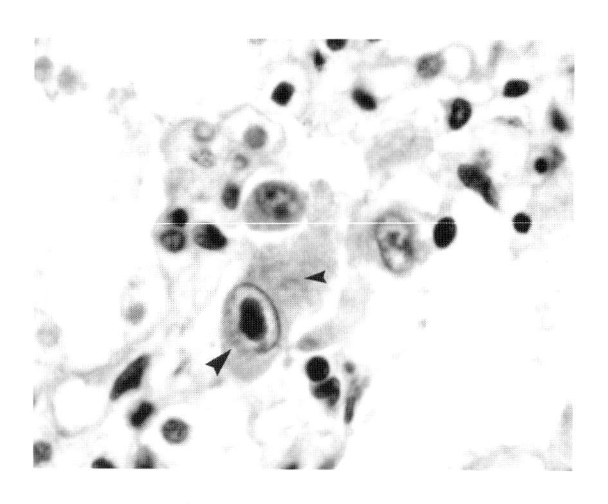

FIGURE 23.26 Cytomegalovirus.

great expense and the complexity of their administration. Antibiotics are used to treat the many infections to which AIDS patients are susceptible. Whereas viral resistance to protease inhibitors may develop after a few days, resistance to the reverse transcriptase inhibitor zidovudine may occur only after moths of administration. Three of four mutations develop in the viral resistance to zidovudine, yet only one mutation can lead to resistance to protease inhibitors.

Cytomegalovirus (CMV) immunity: Cytomegalovirus that induces injury only if the host immune response is impaired, which makes it a significant pathogen for the fetus, allograft recipients, and individuals with acquired immune deficiency syndrome (AIDS). Host immune response to CMV is both cell-mediated and humoral. Cytotoxic T lymphocytes are specific for viral structural phosphoproteins. Cell-mediated immunity appears to be the major mechanism that controls CMV replication in murine CMV, and NK cells are also important. Viral proteins induce a limited humoral response to two surface glycoproteins, gB and gH, which are neutralizing domains. Both cell-mediated and humoral immunity are insufficient to block reactivation of latent virus or to protect against reinfection from an exogenous source. CMV upregulates adhesion molecules. CMV pneumonitis is an immunopathologic. CMV is linked to infection of solid organ grafts and with graft-vs.-host disease after bone marrow transplants. It causes immunosuppression by unknown mechanisms. A live attenuated vaccine strain has no effect on the incidence of CMV infection, and a recombinant gB vaccine is in phase I clinical trials.

Owl eye appearance (Figure 23.27) describes inclusions found by light microscopy in cytomegalovirus (CMV) infection. CMV-infected epithelial cells are enlarged and

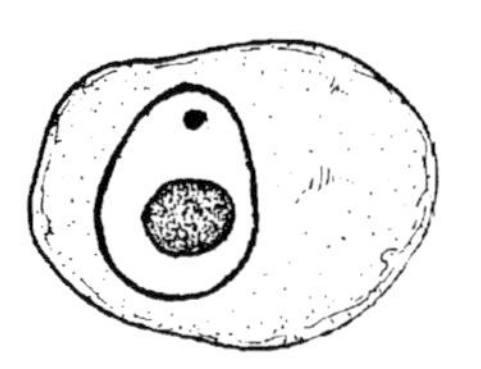

FIGURE 23.27 Owl eye appearance.

exhibit prominent eosinophilic intranuclear inclusions that are half the size of the nucleus and are encircled by a clear halo.

Chicken pox (varicella) is a human herpes virus type 3 (HHV-3) induced in acute infection that occurs usually in children less than 10 years of age. There is anorexia, malaise, low fever, and a prodromal rash following a 2-week incubation period. Erythematous papules appear in crops and intensify for 3 to 4 d. They are very pruritic. Complications include viral pneumonia, secondary bacterial infection, thrombocytopenia, glomerulonephritis, myocarditis, and other conditions. HHV-3 may become latent when chicken pox resolves. Its DNA may become integrated into the dorsal route ganglion cells. This may be associated with the development of herpes zoster or shingles later in life.

A **picornavirus** is a small RNA virus with a naked capsid structure. More than 230 viruses categorized as enteroviruses, rhinoviruses, cardioviruses, and aphthoviruses comprise this family.

ECHO virus (enteric cytopathogenic human orphan virus) is comprised of 30 types within the picornavirus family. It is cytopathic in cell culture and produces clinical manifestations in patients that include upper respiratory tract infections, diarrhea, exanthema, viremia, and sometimes poliomyelitis and viral meningitis.

Abelson murine leukemia virus (A-MuLV) is a B cell murine leukemia-inducing retrovirus that bears the v-*abl* oncogene (Figure 23.28). The virus has been used to immortalize immature B-lymphocytes to produce pre-B cell or less differentiated B cell lines in culture. These have been useful in unraveling the nature of immunoglobulin differentiation, such as H and L chain immunoglobulin gene assembly, as well as class switching of immunoglobulin.

HAV is an abbreviation for hepatitis A virus.

Hepatitis A (Figure 23.29) is a picornavirus which is also called enterovirus 72. It is spread either person to person by the fecal–oral route or by consumption of contaminated water or food.

Hepatitis serology refers to hepatitis B antigens and antibodies against them (Figure 23.30). Core antigen is designated HBc. The HBc particle is comprised of double-stranded DNA and DNA polymerase. It has an association with Hbe antigens. Core antigen signifies persistence of

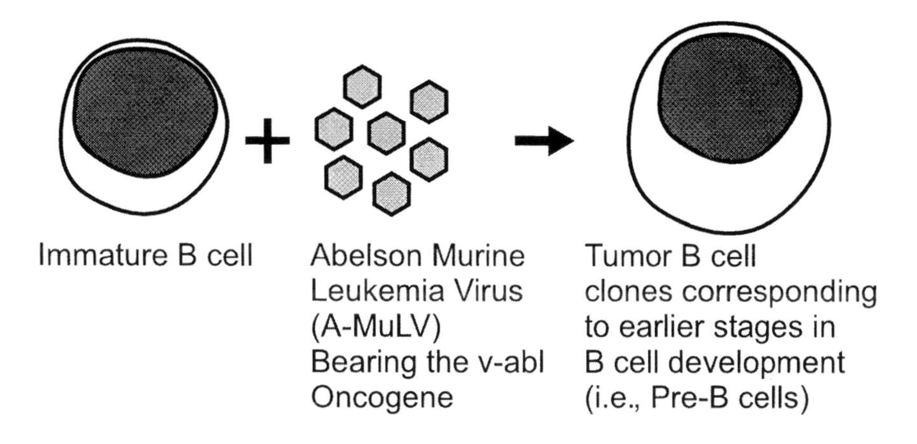

FIGURE 23.28 Abelson murine leukemia virus.

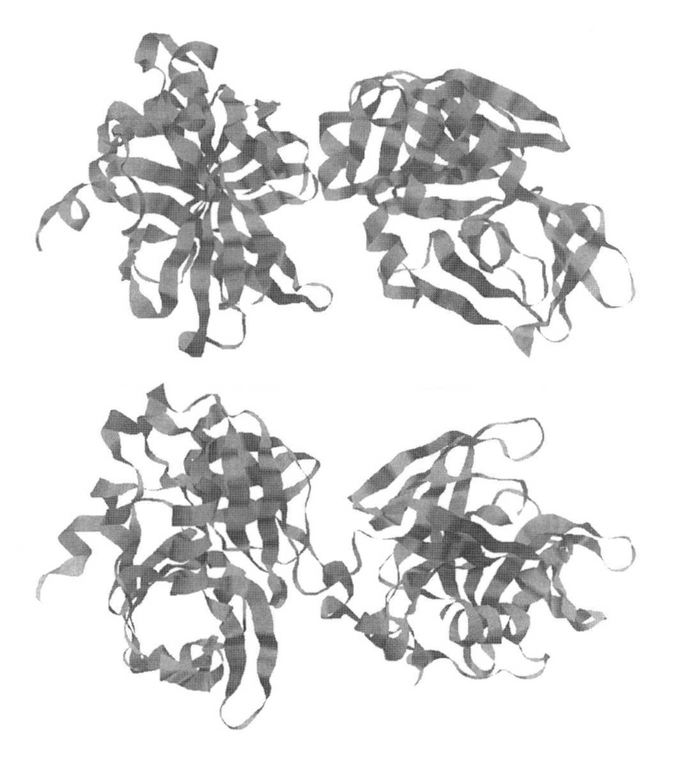

FIGURE 23.29 Hepatitis A virus 3C proteinase.

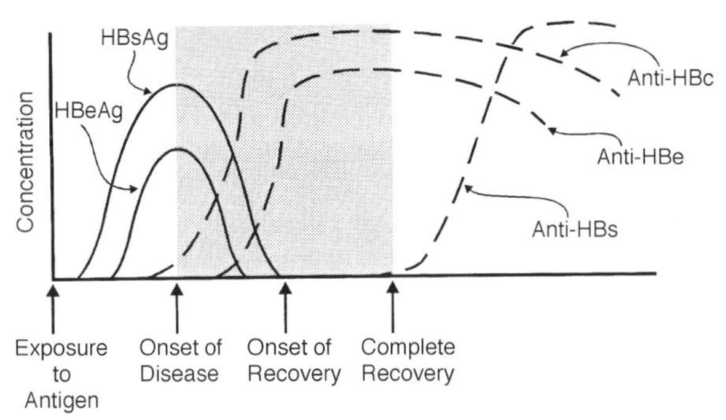

FIGURE 23.30 Hepatitis serology.

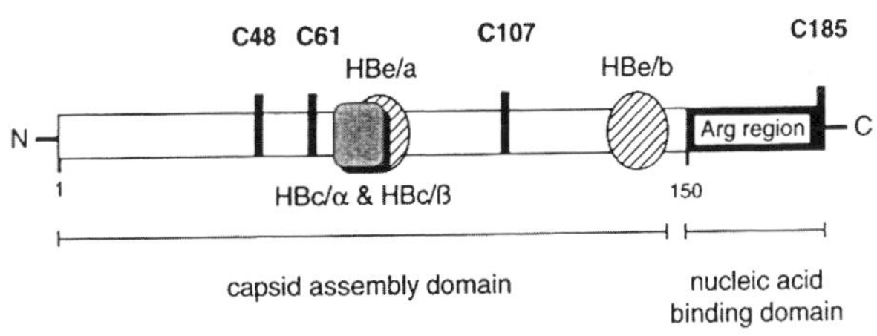

FIGURE 23.31 Schematic structure of the HBV core polypeptide. The 185 residue p21.5 polypeptide (genotype A) is shown with the amino-terminus (N) at the left and the carboxyl-terminus (C) at the right. The open region depicts the hydrophobic assembly domain (a.a.s 1–149; open): the Arg-rich nucleic acid binding region, also known as the protamine domain (a.a.s 150–185), is shown shaded. Hatched ovals indicate the approximate locations of the Hbe/a and Hbe/b antigenic determinants. The shaded rectangle portrays the capsid-specific Hbc/α and Hbc/β epitopes which supposedly overlap Hbe/a. Also indicated are the four Cys residues 48, 61, 107, and 185 (vertical bars).

replicating hepatitis B virus. Anti-HBc antibody is a serologic indicator of hepatitis B (Figure 23.31 and Figure 23.32). It is an IgM antibody that increases early and is still detectable 20 years postinfection. The IgM anti-HBc antibody assay is the one best antibody assay for acute hepatitis B. The Hbe antigen (Hbe) follows the same pattern as HbsAg antigen. When found, it signifies a carrier state. The anti-Hbe increases as Hbe decreases. It appears in patients who are recovering and may last for years after the hepatitis has been resolved. The first antigen that is detectable following hepatitis B infection is surface antigen (HBs). It is detectable a few weeks before clinical disease and is highest with the first appearance of symptoms. This antigen disappears 6 months from infection. Antibody to HBs increases as the HbsAg levels diminish. Anti-HBs often is detectable for the lifetime of the individual.

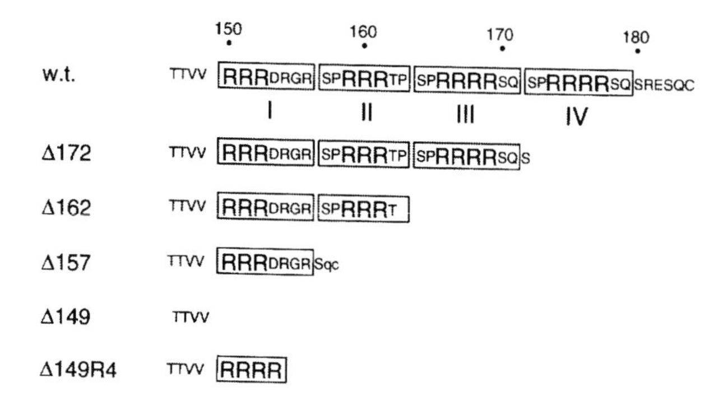

FIGURE 23.32 The protamine region of p21.5 contains four Arg-rich repeats that mediate interactions between the core protein and nucleic acid. Shown are the C-terminal amino acid sequences (residues 150–185) of wild-type (w.t.) p21.5 (top), as well as a series of truncated core proteins with defined endpoints.

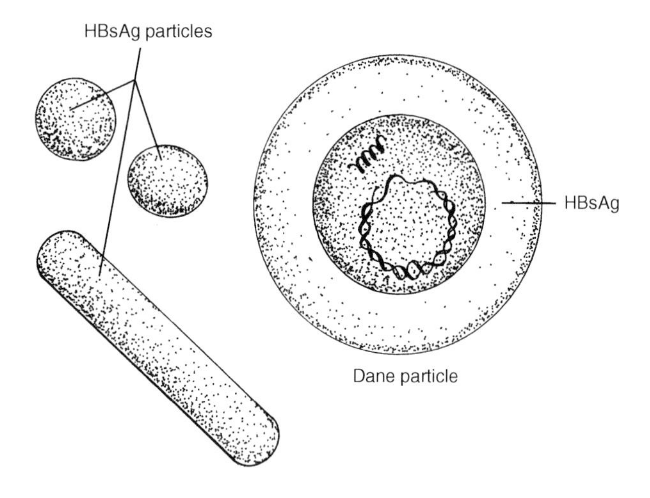

FIGURE 23.33 DANE particle.

Hepatitis B is a DNA virus that is relatively small and has four open reading frames. The S gene codes for HBsAg. The P gene codes for a DNA polymerase. There is an X gene and a core gene that code for HBcAg and the precore area that codes for HBeAg.

HBx is a regulatory gene of hepatitis B that codes for the production of an HBx protein which is a transcriptional transactivator of viral genes. This modifies expression of the host gene. In transgenic mouse-induced hepatomas, it may promote development of hepatocellular carcinoma.

DANE particle (Figure 23.33) is a 42-nm structure identified by electron microscopy in hepatitis B patients in the acute infective stage. The DANE particle has a 27-nm diameter icosahedral core that contains DNA polymerase.

HBV is the abbreviation for hepatitis B virus.

Australia antigen (AA) is hepatitis B viral antigen. The name is derived from detection in an Australian aborigine.

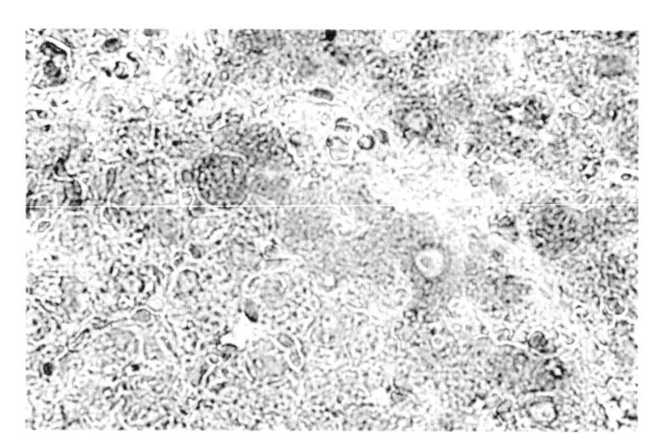

FIGURE 23.34 Hbs antigen in liver cells.

Australia antigen is demonstrable in the cytoplasm of an infected hepatocyte. In early hepatitis B, there is sublobular cell involvement, but later in the disease, only some hepatocytes are antigen positive. There is a positive correlation between the presence of hepatitis B antigen in the liver of a group of people and that group's incidence of hepatocellular carcinoma.

Hepatitis B surface antigen (HbsAg) antibody us a murine monoclonal antibody specific for the HbsAg phenotype.

HbcAg is hepatitis B core antigen. This 27-nm core can be detected in hepatocyte nuclei.

HbsAg (Figure 23.34) is hepatitis B virus envelope or surface antigen.

Hepatitis B virus protein X: See HBx.

HbeAg is hepatitis B nucleocapsid constituent of relatively low molecular weight which signifies an infectious state when it appears in the serum.

Hepatitis B virus immunity: Hepatitis B virus (HBV) infection leads to chronic liver disease and hepatocellular carcinoma (HCC). The virus is not believed to cause direct cytopathic injury of liver cells. Liver injury is likely due to the host immune response. During acute hepatitis B infection, IgM anti-HBc, a thymus-independent response appears in the early phases of the infection together with HBsAg and HBeAg. Anti-pre-S1 also appears early in the infection together with a potent MHC class I restricted cytolytic T lymphocyte (CTL) response specific for the numerous epitopes of the structural and nonstructural proteins of the virus. The CDLs are critical for viral clearance, as are cytokines, interferon, IFN-γ, and TNFα. In the early phase of acute hepatitis B, nucleocapsid antigens, HBcAg, and HBeAg induce a powerful MHC class II restricted T helper cell proliferative response. The MHC class II locus DRB1*-1302 is associated with recovery. The CD4 response has a significant role in recovery. When the virus is eliminated from the liver, HBeAg is lost from the serum and anti-HBe is detected. A short time thereafter, HBsAg is lost and anti-HBs antibodies appear. Immune response fails to eliminate HBV in selected patients, who become chronically infected. HBsAg either isolated from infected serum or created by recombinant DNA technology represents a successful immunogen in the induction of protective immunity against HBV infection.

Hepatitis, non-A, non-B (C) (NANBH) is the principal cause of hepatitis that is transfusion related. Risk factors include intravenous drug abuse (42%), unknown risk factors (40%), sexual contact (6%), blood transfusion (6%), household contact (3%), and health professionals (2%). There are 150,000 cases per year in the U.S. Of those cases, 30 to 50% become chronic carriers, and one fifth develop cirrhosis. Parenteral NANBH is usually hepatitis C, and enteric NANBH is usually hepatitis E.

Hepatitis C virus immunity: HCV-infected subjects develop specific antibodies whose clinical and biological significance remains to be determined. They are not protective. Most anti-HCV positive sera are also RT-PCR positive. Antibodies to the envelope glycoproteins may have neutralizing activity. Cellular immune response against recombinant viral antigens have been investigated in proliferation and cytotoxicity assays. Chronic progressive liver disease may be linked to T_H2 cytokine profiles from liver-derived T cells. HCV-specific cytotoxic $CD8^+$ T cells have been found in HCV-infected patients. HCV escapes surveillance by the immune system by altering its antigenic determinants. Problems for vaccine development include diversity of HCV genotypes and subtypes in the hypervariability of HCV quasi-species within the host.

Hepatitis D virus: See δ agent.

Delta agent (hepatitis D virus [HDV]) is a viral etiologic agent of hepatitis that is a circular single-stranded incomplete RNA virus without an envelope. It is a 1.7-kb virus and consists of a small, highly conserved domain and a larger domain manifesting epitopes. HDV is a subviral satellite of the hepatitis B virus (HBV), on which it depends to fit its genome into virions. Thus, the patient must first be infected with HBV to have HDV. Individuals with the delta agent in their blood are positive for HBsAg, anti-HBC, and often HBe. This agent is frequently present in IV drug abusers and may appear in AIDS patients and hemophiliacs.

Hepatitis E virus (HEV) is enteric non-A, non-B hepatitis. A single-stranded RNA virus that has an oral–fecal route of transmission and can introduce epidemics under poor sanitary conditions where drinking water is contaminated and the population is poorly nourished.

E antigen refers to a hepatitis B virus antigen present in the blood sera of chronic active hepatitis patients.

Hepatitis E virus immunity: Immunity against hepatitis E virus (HEV), a self-limiting disease that resembles hepatitis A, consists of an immunoglobulin M (IgM) antibody response against HEV in the acute phase of the disease. Once a peak titer is reached, it declines to undetectable levels 5 months after the immune response begins. IgG titers reach their height during the early convalescent phase and decline during the following months. Up to 50% of postinfection patients reveal undetectable levels of HEV-specific IgG. Yet, some persons still have IgG antibodies 2 to 14 years following infection. HEV-specific IgG antibodies prevent reinfection and clinical hepatitis E in young adults.

Hepatitis immunopathology panel is a profile of assays that are very useful to establish the clinical and immune status of a patient believed to have hepatitis. The panel for acute hepatitis may include HBsAg, anti-HBc, anti-ABs, antihepatitis A (IgM), anti-HBe, and antihepatitis C. The panel for chronic hepatitis (carrier) includes all of these except for the antihepatitis A test.

Chronic active hepatitis (autoimmune) (Figure 23.35) is a disease that occurs in young females who may develop fever, arthralgias, and skin rashes. They may be of the HLA-B8 and -DR3 haplotype and suffer other autoimmune disorders. Most develop antibodies to smooth muscle, principally against actin, and autoantibodies to liver membranes. They also have other organ- and nonorgan-specific autoantibodies. A polyclonal hypergammaglobulinemia may be present. Lymphocytes infiltrating portal areas destroy hepatocytes. Injury to liver cells produced by these infiltrating lymphocytes produces piecemeal necrosis. The inflammation and necrosis are followed by fibrosis and cirrhosis. The T cells infiltrating the liver are $CD4^+$. Plasma cells are also present, and immunoglobulins may be deposited on hepatocytes. The autoantibodies

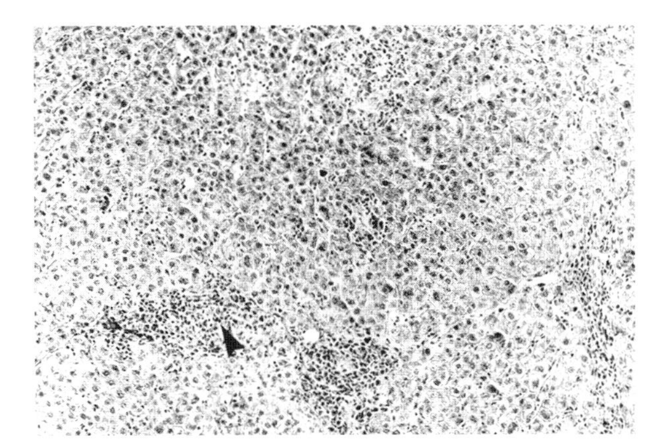

FIGURE 23.35 Chronic active hepatitis.

against liver cells do not play a pathogenetic role in liver injury. There are no serologic findings that are diagnostic. Corticosteroids are useful in treatment. The immunopathogenesis of autoimmune chronic active hepatitis involves antibody, K cell cytotoxicity, and T cell reactivity against liver membrane antigens. Antibodies and specific T suppressor cells reactive with LSP are found in chronic active hepatitis patients, all of whom develop T cell sensitization against asialoglycoprotein (AGR) antigen. Chronic active hepatitis has a familial predisposition.

Shingles (Herpes zoster) is a virus infection that occurs in a band-like pattern according to distribution in the skin of involved nerves. It is usually a reactivation of the virus that causes chickenpox.

Parvovirus is a minute icosahedral virus comprised of single-stranded DNA that may replicate in previously uninfected host cells or in those already infected with adenovirus.

Parvovirus immunity: Specific IgM followed by IgG antibodies that occur in the second week of exposure can effectively clear the infection. Parvovirus B19 targets the erythroid bone marrow cells. Antibodies that develop following first exposure to the virus are specific mainly for VP2 epitopes. This is followed by antibodies that are specific for VP1 epitopes. VP1 unique region linear epitopes are critical to induce effective neutralizing antibodies. IgG antibodies persist for life. Immunodeficient individuals who cannot develop an adequate antibody response may develop persistent parvovirus infection. VP1 specific antibodies are crucial for the control of parvovirus B19 infection. Commercial immunoglobulin preparation contains parvovirus B19-neutralizing antibody and are useful for the treatment of persistent parvovirus infection. Antibody synthesis in immunocompetent subjects presents clinical manifestations as a consequence of direct viral cytotoxicity. Immunodeficient patients who have an impaired capacity to synthesize neutralizing antibodies develop severe anemia as a result of suppressed erythropoiesis by the virus. Empty parvovirus capsids enriched for VP1 epitopes induce protective antibody responses. A recombinant form has shown promise in clinical trials.

Poliovirus is a picornavirus of the genus enteroviridae. There are three polio serotypes. Polio and other enterviruses are spread mainly by the fecal–oral route. Poliomylelitis occurs around the world; however, in the Western Hemisphere the wild-type virus has been eliminated by successful vaccines.

Rotavirus is a double-stranded RNA virus that is encapsulated and belongs to the reovirus family. It is 70 nm in diameter and causes epidemics of gastroenteritis, which are usually relatively mild but may be severe in children less than 2 years of age.

Rabies is an infection produced by an RNA virus following a bite from an infected animal. The virus passes across the neuromuscular junction and infects the nerve from which it reaches the central nervous system. It also reaches salivary glands of lower animals. The virus infection leads to cerebral edema, congestion, round cell infiltration of the spinal cord and grey matter in the brain stem, and profound loss of Purkinje cells. Negri bodies are found prominently in the medulla oblongata, hippocampus, and cerebellum. Clinically, the fury associated with the disease is due to irritability of the central nervous system. There is fever, hyperethesia, and anoxia aggression; it may be paralytic. Human rabies is rare in the U.S. It is more common in other animals, with most of the cases appearing in skunks and raccoons. Fewer cases occur in bats and only 2% each in dogs and cats. The virus is transmitted from one person to another by inhalation or by corneal transplantation, but not by human bites.

Prion (Figure 23.36) is an infectious particle comprised of a protein with appended carbohydrate. It is the most diminutive infectious agent known. The three human diseases in which prions have been implicated include kuru, Creutzfeldt-Jakob disease, and Gerstmann-Straussler syndrome. They have also been implicated in the following animal diseases: sheep and goat scrapie, bovine spongiform encephalopathy, chronic wasting of elk and mule deer, and transmissible mink encephalopathy. Prions do not induce inflammation nor do they stimulate antibody synthesis. They resist formalin, heat, ultraviolet radiation, and other agents that normally inactivate viruses. They possess a 28-kDa, hydrophobic, glycoprotein particle that polymerizes, forming an amyloid-like fibrillar structure.

Slow viruses are agents that induce infectious encephalitis following a lengthy latency. Slow viruses consist of conventional viruses and prions that are comprised of subverted cell proteins. Among the conventional group is measles, which induces subacute sclerosing panencephalitis,

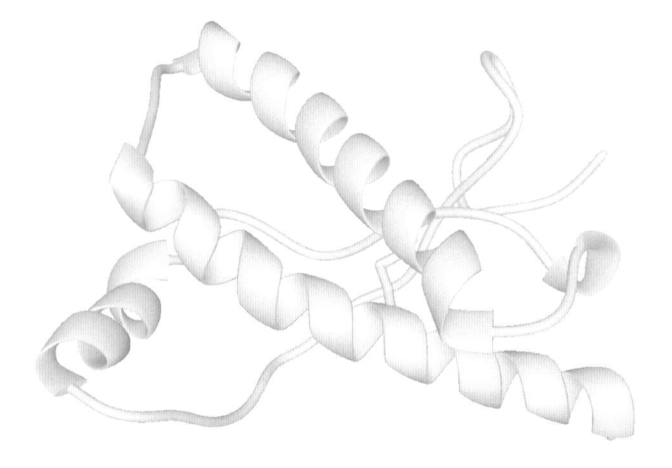

FIGURE 23.36 Prion.

papovavirus, which induces progressive multifocal leukoencephalopathy, and rubella, which induces rare progressive rubella panencephalitis. The agents that cause kuru and Creutzfeldt-Jakob disease are among the nonconventional group of slow viruses.

Viral capsids (Figure 23.37) are envelope antigens, thereby inhibiting adherence and invasion of host cells. Antibodies may also act as opsonins that increase the attractiveness of viral particles to phagocytic cells. Secretory IgA antibody is important in neutralizing viruses on mucosal surfaces. Complement also facilitates phagocytosis and may be significant in viral lysis.

Capsular polysaccharide is a constituent of the protective coating around a number of bacteria such as the pneumococcus (*Streptococcus pneumoniae)* which is a polysaccharide, chemically, and stimulates the production of antibodies specific for its epitopes. In addition to the pneumococcus, other microorganisms such as *Streptococci* and certain *Bacillus* species have polysaccharide capsules.

Antigenic variation represents a mechanism whereby selected viruses, bacteria, and animal parasites may evade the host antibody or T cell immune response, thereby permitting antigenically altered etiologic agents of disease to produce a renewed infection. The variability among infectious disease agents is of critical significance in the development of effective vaccines. Antigenic variation affects the surface antigens of the viruses, bacteria, or animal parasite in which it occurs. By the time the host has developed a protective immune response against the antigens originally present, the latter have been replaced in a few surviving microorganisms by new antigens to which the host is not immune, thereby permitting survival of the microorganism or animal parasite and its evasion of the host immune response. Thus, from these few surviving viruses, bacteria, or animal parasites, a new population of infectious agents is produced. This cycle may be repeated, thereby obfuscating the protective effects of the immune response.

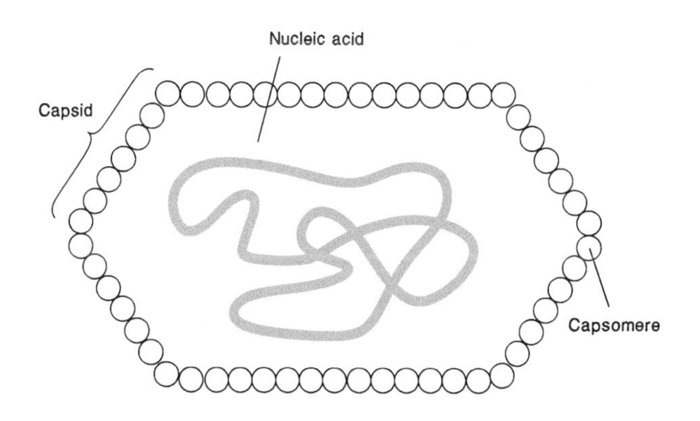

FIGURE 23.37 Capsid.

Viruses are able to evade immune mechanisms. Besides the sanctuary viruses enjoy once they have entered host cells, these disease agents have additional means to escape host immune mechanisms. Viruses are especially adept at **antigenic variation** (Figure 23.38) whereby they may alter their surface antigenic structure once antibodies are formed against their original epitopes. This process may be repeated many times leading to the production of numerous strains of a particular virus that are antigenically and, therefore, serologically distinct. The **influenza** (Figure 23.39 to Figure 23.41) and **AIDS** viruses are especially versatile in this regard.

Viruses may also induce immunosuppression in the host they infect. The AIDS virus (HIV-1) is well known to target $CD4^+$ helper/inducer lymphocytes that are central to mounting any type of immune response. Selected viruses such as the **Epstein-Barr virus** (Figure 23.42) may induce immunosuppression by mechanisms that are yet to be determined but possibly attributable to genes that encode substances which dampen antiviral immune responsiveness.

Epstein-Barr virus (EBV) is a DNA herpes virus linked to aplastic anemia, chronic fatigue syndrome, Burkitt's lymphoma, histiocytic sarcoma, hairy cell leukemia, and immunocompromised patients. EBV may promote the appearance of such lymphoid proliferative disorders as Hodgkin's and non-Hodgkin's lymphoma, infectious

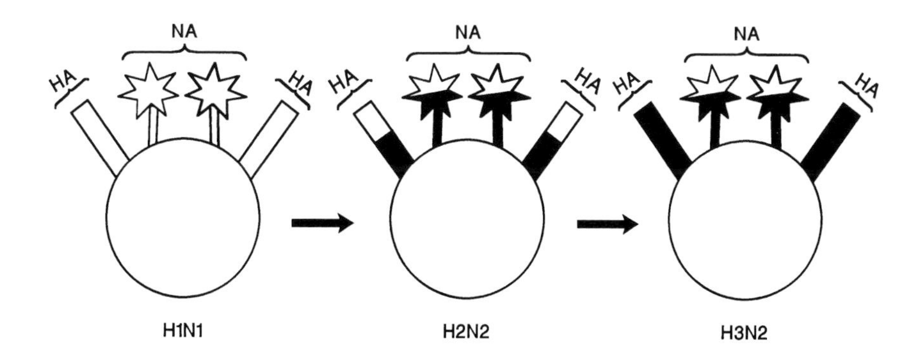

FIGURE 23.38 Antigenic variation.

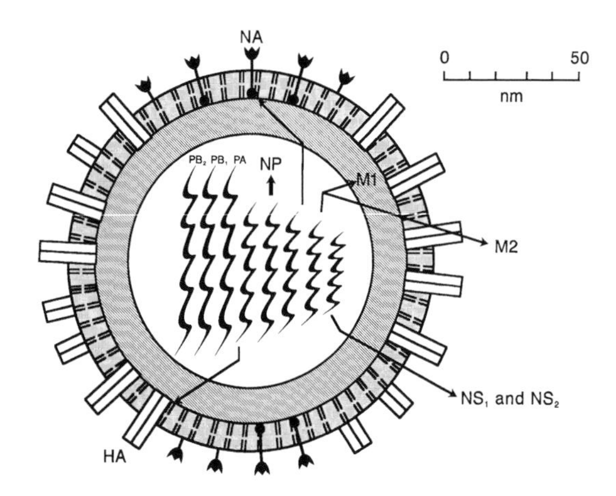

FIGURE 23.39 Influenza virus.

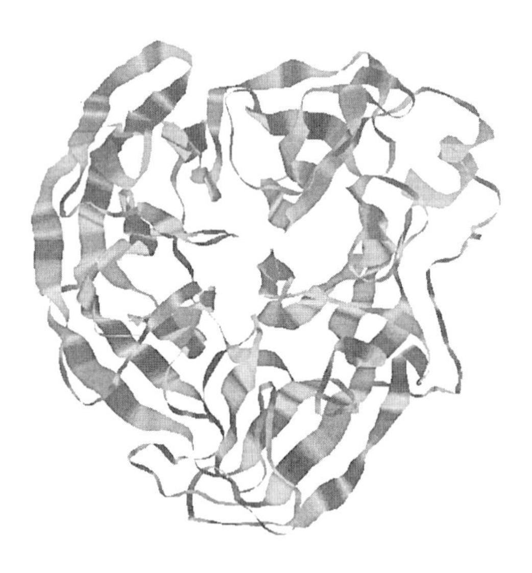

FIGURE 23.40 Influenza A subtype N2 neuraminidase (sialidase). A/Tokyo/3/67.

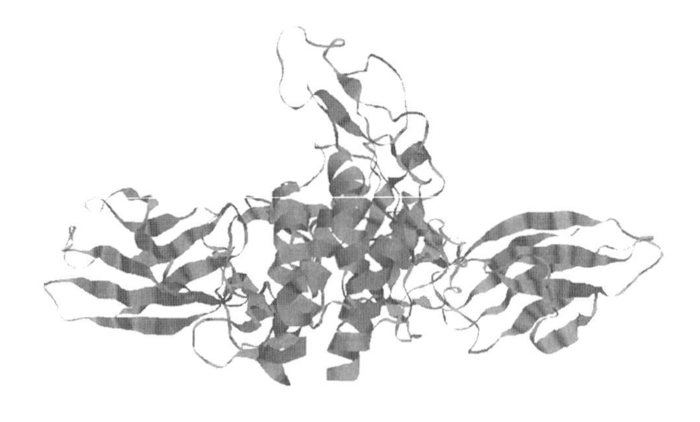

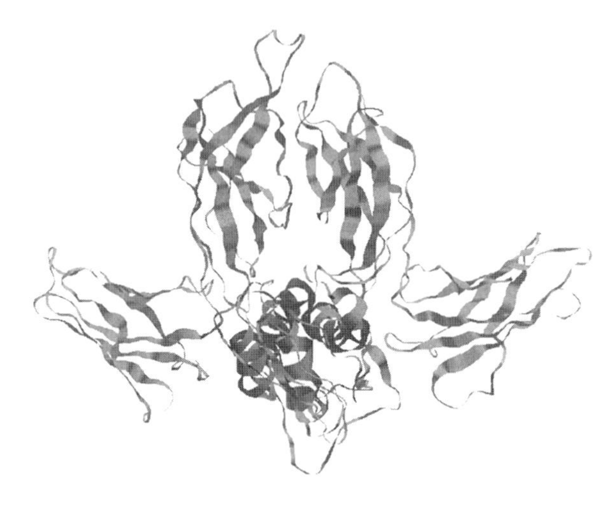

FIGURE 23.41 Influenza B/LEE/40 neuraminidase (sialidase).

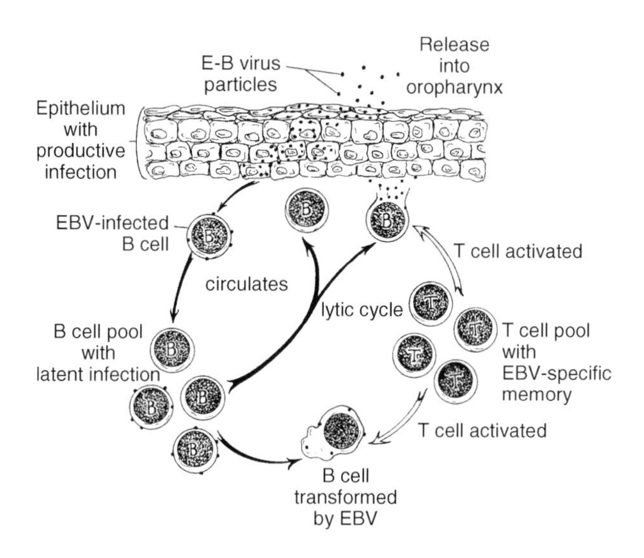

FIGURE 23.42 Epstein-Barr virus–host interaction.

mononucleosis, nasopharyngeal carcinoma, and thymic carcinoma. It readily transforms B lymphocytes and is used in the laboratory for this purpose to develop long-term B lymphocyte cultures. Antibodies produced in patients with EBV infections include those that appear early and are referred to as EA, antibodies against viral capsid antigen (VCA), and antibodies against nuclear antigens (EBNA).

Influenza viruses are infectious agents that induce an acute, febrile, and respiratory illness often associated with myalgia and headache. The classification of influenza A viruses into subtypes is based on their hemagglutinin (H) and neuraminidase (N) antigens. Three hemagglutinin (H1, H2, and H3) and two neuraminidase (N1 and N2) subtypes are the principal influenza A virus antigenic subtypes that produce disease in man. Due to antigenic change (antigenic drift), infection or vaccination by one strain provides little or no protection against subsequent infection by a distantly related strain of the same subtype. Influenza B viruses undergo less frequent antigenic variation.

Influenza virus immunity: Influenza A viruses undergo changes in their surface hemagglutinin (HA) and neuraminidase (NA) glycoproteins, leading to repeated epidemics and pandemics. Mucosal IgA and serum IgG antibodies are specific for the HA molecule-neutralized virus infectivity and mainly confer resistance to reinfection. This is the reason for vaccination against epidemic strains with killed virus. The antibody response to HA is subtype specific but antigenic drift facilitates escape of infectious virus from antibody-mediated destruction. Five protective antigenic sites are present on the globular head of the HA molecule. Antibodies against neuraminidase fail to prevent infection but diminish spread of the virus. Enzyme-active centers on each of the four subunits of the NA are in a central cavity encircled by two antigenic sites. Influenza A viruses undergo variability which includes antigenic drift that yields variance of contemporary epidemic strains that are sufficiently different to avoid neutralization by antibody. The more extensive antigenic shift may be a consequence of dual infection with a human and an animal influenza A virus. This can lead to the development of a novel pandemic strain for which humans have no preexisting antibody. $CD8^+$ virus-immune cytotoxic T lymphocytes can clear influenza A viruses from infected lungs. It is necessary for the effector lymphocyte to come into direct contact with the virus-infected target cell. Immune $CD4^+$ T lymphocytes are involved in lung consolidation in influenza penumonia and have a significant role in clearance of the virus. $CD4^+$ T cells facilitate antibody production. They also facilitate the generation of "helper cytokine" such as IL-4 and IL-5 that promote clonal expansion and differentiation of B cells into antibody-secreting plasma cells. Influenza A virus infection primes the host for a secondary $CD8^+$ T lymphocyte response to other influenza A viruses. Influenza-immune $CD4^+$ helper T cells are specific for viral peptides presented in the context of self MHC class II glycoproteins.

Influenza hemagglutinin is an influenza virus coat glycoprotein that binds selected carbohydrates on human cells, the initial event in virus and viral infection. Antigenic shift results from changes in the hemagglutinin.

A **v-*myb* oncogene** is a genetic component of an acute transforming retrovirus that leads to avian myeloblastosis. It represents a truncated genetic form of c-*myb*.

Creutzfeldt-Jakob syndrome is a slow virus infection of brain cells that reveal membrane accumulations.

Street virus is a natural or genetically unmodified virus such as rabies that can be isolated from animals.

Subacute sclerosing panencephalitis is a slow virus disease that occurs infrequently as a complication of measles and produces progressively destructive injury to the brain through slow replication of defective viruses.

Acyclovir 9 (2-hydroxyethoxy-methylguanine) is an antiviral nucleoside analog that blocks herpes simplex virus-2 (HSV-2), the causative agent of genital herpes. HSV thymidine kinase activates acyclovir, through monophosphorylation, followed by triple phosphorylation with host enzymes to yield a powerful blocking action of the DNA polymerase of HSV-2. Acyclovir is prescribed for the treatment of HSV-2 genital infection.

Q fever is an acute disease caused by the rickettsia *Coxiella burnetii*. Cattle, sheep, goats, and several small marsupials serve as reservoirs. Ticks are the main vector. The microorganism is highly infectious and multiplies readily to produce clinical infection. The onset is abrupt and is accompanied by headaches, high fever, myalgia, malaise, hepatic dysfunction, interstitial pneumonitis, and fibrinous exudate. Q fever may also induce atypical pneumonia, rapidly progressive pneumonia, or be a coincidental finding to a systemic illness. The disease has a relatively low mortality. It is treated successfully with tetracycline and chloramphenicol.

Rickettsia immunity: The immune response in rickettsial infection involves both humoral and cell-mediated immune responses that are both powerful and persistent. Antibodies that act as opsonins render microorganisms susceptible to destruction by macrophages even though complement and antibody are not bactericidal. Antibody-dependent cellular cytotoxicity has also been demonstrated. Interferon γ-activated macrophages effectively destroy rickettsia. Interferon γ and to a lesser degree tumor necrosis factor can control reproduction of rickettsiae in nonprofessional phagocytes. Protection depends upon cell-mediated immunity rather than antibody alone. Rickettsial infections activate NK cells. Both LAK and cytotoxic T cells can kill rickettsia-infected cells. Numerous attenuated rickettsial and killed subunit vaccines are available.

***Chlamydia* immunity:** Chlamydiae infect many animal species and various anatomical sites. No single pattern of host response can be described, but there may be similarities in effector mechanisms. The strain or serovar of the infecting microorganism is also significant. *In vitro* studies and genital respiratory and ocular animal models have provided most of the information about both protective and pathologic host responses. Acute inflammation is the initial response with participation by polymorphonuclear leukocytes (PMNs) that may counteract the microorganisms but also cause pathologic changes. When chlamydiae infect epithelial cells, IL-1 is released as well as IL-8, a potent PMN chemoattractant. When these microorganisms infect macrophages, LPS triggers the synthesis of TNFα, IL-1, and IL-6. They may also activate the alternative pathway of complement. IL-8, TNFα, and complement may induce chemotaxis of PMNs to the local site. ICAM-1, VCAM-1, and MAdCAM-1 are all detectable early in the infection and could be addressins responsible for PMN extravasation at sites of infection. NK cells may appear early after infection of the genital tract. Chlamydial infection produces both humoral and cell-mediated immune responses. Cell-mediated immunity has been found important in both mouse and guinea pig models. The $CD4^+$ T cells are the main subset responsible for protective cell-mediated immunity. In genital infections, the T_H1 subset of $CD4^+$ T cells is the principal cell type leading to the formation of high levels of IFN-γ which is believed to have a protective role. In some models antibody appears to be important in the resolution of infection. Antibody may neutralize chlamydial elementary bodies *in vitro*. Immunity to chlamydial infections is short lived. There is currently no effective vaccine for chlamydia infections in humans, but a veterinary vaccine is available.

Lymphogranuloma venereum (LGV) is a sexually transmitted disease induced by *Chlamydia trachomatis* that is divided into L_1, L_2, and L_3 immunotypes. It is rare in the U.S. but endemic in Africa, Asia, and South America. Clinically, patients develop papulo ulcers that heal spontaneously at the inoculation site. This is followed by development of inguinal and perirectal lymphadenopathy. There is skin sloughing, hemorrhagic proctocolitis, purulent draining, fever, headache, myalgia, aseptic meningitis, arthralgia, conjunctivitis, hepatitis, and erythema nodosum. Various antibody assays used in the diagnosis of LGV include complement fixation, with a titer of greater than 1:32, and immunofluorescence. Also, the Frei test, which consists of the intracutaneous inoculation of a crude antigen into the forearm, is used and can be read after 72 h. It is considered positive if the area of induration is greater than 6 mm.

Adenoviruses, infection, and immunity: Species-specific icosahedral DNA viruses that belong to numerous serotypes. They produce multiple clinical syndromes in humans, including respiratory, genitourinary, gastrointestinal, and conjunctival infections. They are resistant to most antiviral chemotherapy. An oral vaccine has been very successful in preventing acute respiratory disease in military personnel. Humans develop serotype-specific neutralizing antibodies to the structural proteins, thereby preventing reinfection with the same serotype. Early (E) nonstructural proteins produce significant immunologic effects. The virus has a double-stranded linear DNA and more than 12 structural proteins. There is no virus envelope. The viral polypeptides serve as antigens for host immune responses generated as a consequence of infection as a result of immunization with a vaccine. The internal structural proteins are not believed to be involved in humoral or cell-mediated immunity.

Anticytomegalovirus (CMV) antibody is a mouse monoclonal antibody that reacts with CMV-infected cells, giving a nuclear staining pattern with early antigen and a nuclear and cytoplasmic reaction with the late viral antigen. The antibody shows no cross-reactivity with other herpesviruses or adenoviruses.

Arenavirus immunity: Even though arenaviruses may not cause ill effects in carrier rodents, when transmitted to primates they may induce severe encephalitis, hepatitis, or hemorrhagic syndrome. LCMV is a classic member of this group and has been used to investigate virus-induced cell-mediated immunity. Infection of mice may be either acute, which induces strong cell-mediated immunity, or a persistent infection that is associated with little cell-mediated immunity. The immune response to the acute infection either kills the mouse by causing meningitis, or subsides and renders the mouse LCMV-immune. $CD8^+$ T cells not only clear the virus infection but may mediate pathological changes of choriomeningitis when injected intracerebrally. LCMV infection induces interferon γ and tumor necrosis factor α which affect the host immune response. IFN-γ induces NK cell proliferation that eliminates pathogen-infected cells and selected tumors. Arenaviruses persist in long-term carriers through antigenic variation that evades the host immune response. Mice may be protected from lethal LCMV challenge by peptide vaccination or by vaccinia vectors that express viral epitopes.

Bunyaviridae immunity: Bunyaviridae are very immunogenic, stimulating the formation of neutralizing antibodies that are type specific and reveal limited cross-reactivity within the genus. Solid immunity that follows infection protects against reinfection. The immune response in some individuals may be too late to prevent central nervous system or liver invasion. For example, Rift Valley fever virus may lead to liver necrosis and high mortality. Some American hantaviruses produce the greatest effect on the lungs rather than the liver, kidneys, or central nervous system, leading to hantavirus pulmonary syndrome. G1 glycoprotein antibodies neutralize viral infectivity and prevent hemagglutination. The dominant antigen is nucleocapsid protein in complement fixation assays. Little is known concerning the role of cell-mediated immunity in resistance. Formalin-inactivated veterinary vaccines against Rift Valley fever and Nairobi sheep disease are available.

B-type virus (*Aspergillus macaques*) is an Old World monkey virus that resembles herpes simplex. Clinical features include intermittent shedding and reactivation in the presence of stress and immunosuppression. Humans who tend these monkeys may become infected with fatal consequences. B-type viruses possess an eccentric nuclear core.

Calcivirus immunity: Human calciviruses have been shown to cause gastric distress. Antibodies against members of this group such as small round-structured viruses (SRSV) that include Norwalk virus (NV) may cross-react with other viruses of this same form; thus, a rise in antibody titer is insufficient for diagnosis. IgA antibody responses appear more specific. Reinfection by NV is common as preformed antibody correlates with susceptibility to the illness. Therefore, people who recover often become susceptible again on rechallenge. High titers attained after several infections are protective in some studies although this has not been confirmed. Antibody to HuCV is protective and mainly type specific. Vaccines are available for FCV and RHDV.

Canine distemper virus induces disease in dogs and is associated with demyelination, probably induced by myelin-sensitive lymphocytes.

Coronavirus immunity: Neutralizing and fusion-inhibiting antibodies are directed mainly against the protein S antigen. Antibodies against HE, N, M, and sM are also significant in neutralization together with complement factors. Among coronavirus antigens, there is high antigenic variability of the S1 subunit of IBV. This subunit induces neutralizing antibodies that bind to discontinuous epitopes. The S2 subunit contains linear epitopes and an aminodominant region. The S1 subunit of the molecule contains the most immunogenic sites of BCV and MHV. The S2 subunit of MHV also induces neutralizing and fusion-inhibiting antibodies. BCV-specific antibodies against the HE protein participate in neutralization. The M protein of MHV and TGEV stimulate complement-dependent neutralizing antibodies. The TGEV sM protein is involved in neutralization. T lymphocyte responses against SN protein facilitate virus elimination and may confer protection against encephalomyelitis in mice and rats. An immunodominant T cell antigenic site in the N protein of MHV induces neutralizing S-protein-specific antibodies.

Coxsackie is a picornavirus family from Enteroviridiae. Coxsackie A viruses have 23 virotypes and Coxsackie B viruses have six types. Clinical conditions produced by Coxsackie viruses include herpangina, epidemic pleurodynia, aseptic meningitis, summer grippe, and acute nonspecific pericarditis and myocarditis.

Cytopathic effect (of viruses) refers to injurious effects of viruses on host cells produced by various biochemical and molecular mechanisms that are independent of host immunity against the virus. Selected viruses produce disease even though they have little cytopathic effect because the immune system recognizes and destroys the virus-infected cells.

Defective endogenous retroviruses: Partial retroviral genomes that are integrated into host-cell DNA and carried as host genes.

Dengue is an infection produced by the group B arbovirus, flavivirus, which the mosquito (*Stegomyia Aedes aegypti*) transmits. Dengue fever that occurs in the tropical regions of Africa and America may either be benign or produce malignant Dengue hemorrhagic shock syndrome, in which patients experience severe bone pain (break-bone fever). They have myalgia, biphasic fever, headache, lymphadenopathy, and a morbilliform maculopapular rash on the trunk. They also manifest thrombocytopenia and lymphocytopenia.

EBNA (Epstein-Barr virus nuclear antigen): See Epstein-Barr nuclear antigen.

Filovirus immunity: No effective immune responses are associated with fatal filovirus infections that cause fulminating hemorrhagic fever with severe shock syndrome and high mortality in humans and nonhuman primates. Those antibodies that do develop in monkeys against Ebola Reston virus are nonprotective. No significant role for neutralizing antibodies has been found for viral clearance. Extensive alterations of the parafollicular regions in the spleen and lymph nodes lead to destruction of antigen-presenting dendritic cells, pointing to disruption of cell-mediated immunity during filoviral hemorrhagic fever. Besides these cytolytic effects, the high carbohydrate content of these viruses may suppress immune reactivity. A nonstructural glycoprotein secreted from cells infected with the Ebola virus may interfere with an immune response against the virus. A fragment of spike GP released by infected cells may have a similar action. Filovirus GP possesses a sequence motif homologous to an immunosuppressive domain of retroviral glycoproteins. Filoviruses are now known to induce immunosuppression in the infected host, which contributes to the rapid course and severity of the infection. Even though little neutralizing antibody can be detected in human convalescent sera, it is believed that passive immunization with antibodies against Ebola virus afford some benefit in treatment. No vaccines for humans are presently available.

Flavivirus immunity: Yellow fever, dengue, Japanese encephalitis, and tickborne encephalitis are the most important viruses of the flavivirus group. The E protein plays a critical role in infection and immunity since it possesses cellular receptor-binding determinants, a membrane fusion activity, and epitopes for neutralizing antibodies. Macrophages clear the viremia, yet antiviral function may be affected by their state of activation and levels of virus-specific antibodies. West Nile virus infection is associated with impaired NK cell function, which may be related to MHC class I antigen expression on virus-infected cells, which could represent an immune escape mechanism. Virus-specific antibodies provide protection against flavivirus disease. Anti-E protein antibodies are protective in various species and are believed to play a major role in immunity and natural infections. Previous infections are thought to ameliorate or protect from subsequent infections with heterologous viruses by inducing formation of group-specific antibodies. Flavivirus NS 1 protein stimulates protective antibodies. Antibody-dependent cellular cytotoxicity (ADCC) occurs in dengue fever. Cellular immunity is believed to be required for control of infection since T cells adoptively transferred into unimmunized mice can protect them against lethal encephalitis. In dengue fever, there is antibody-enhanced infection of mononuclear phagocytes. Primary dengue infection sensitizes serotype cross-reactive memory T lymphocytes for activation during the secondary infection, leading to inflammatory cytokine release that facilitates the development of capillary leak syndrome. Both CD4 and CD8 cytotoxic lysis of virus-infected cells has been absorbed in dengue fever. Dengue antigen stimulates $CD4^+$ T cells to synthesize interferon γ. Memory responses are primed for major activation during secondary infections. The YF17D strain of yellow fever virus was the first live-attenuated vaccine for this virus family. It has proven safe and highly effective in inducing long-lasting immunity. Other members of this virus group are also candidates for vaccine development. Tick-borne encephalitis virus vaccine is a formalin-inactivated preparation which is highly effective in producing few side effects. Vaccination against Japanese encephalitis has included both inactivated and live attenuated viruses.

Infectious mononucleosis is a disease of teenagers and young adults who have a sore throat, fever, and enlarged lymph nodes. Atypical large lymphocytes with increased cytoplasm, which is also vacuolated, are found in the peripheral blood and have been shown by immunophenotyping to be T cells. They are apparently responding to Epstein-Barr virus-infected B lymphocytes. There is also lymphocytosis, neutropenia, and thrombocytopenia. Patients also develop heterophile antibodies which agglutinate horse, ox, and sheep red blood cells as revealed by the Paul-Bunnell test. Infectious mononucleosis is the most common condition that EBV causes. There may be splenomegaly and chemical hepatitis.

Infectious mononucleosis syndrome(s) include conditions induced by viruses that produce an acute and striking peripheral blood monocytosis and lead to symptoms resembling those of infectious mononucleosis induced by Epstein-Barr virus. Examples include herpes virus, cytomegalovirus, HIV-1, HHV-6, and *Toxoplasma gondii* infections.

Kuru is a slow virus disease of some native tribes of Guinea that practice cannibalism. Transmission is through skin lesions of individuals preparing infected brains for consumption. The virus accumulates in brain cell membranes.

Newcastle disease is follicular conjunctivitis induced by an avian paramyxovirus that blocks the oxidative burst in phagocytes. Cytokines that produce fever are formed. There is recovery in approximately 1 to 2 weeks. In birds the agent induces pneumoencephalitis, which is fatal.

P1 kinase is a serine/threonine kinase activated by interferons α and β. It prevents translation by phosphorylating eIF2, the eukaryotic protein synthesis initiation factor. This facilitates inhibition of viral replicaton.

Papillomavirus immunity: Papillomaviruses induce skin and mucosa neoplasia. Humans who have been infected with this virus develop antibodies that react with papillomavirus capsid proteins. Cervical cancer patients frequently form antibodies against the E7 protein of HPV 16 and HPV 18. Patients' sera have also demonstrated antibodies against E2, E6, and E7 proteins. Chronic infection in immunologically competent hosts point to the possibility that the viral antigens may not be recognized by the immune system. HPV disease occurs in immunosuppressed transplant patients and in acquired immune deficiency syndrome (AIDS). Thus, cell-mediated immunity is significant for the control HPV infection. NK cells are significant in the cellular response to HPV infection. NK cell activity is decreased in patients with HPV-induced neoplasia. Decreased nomers of the potent antigen-presenting cells known as Langerhans cells occur in HPV precancerous lesions. Viral antigens are presented to T lymphocytes via class I MHC molecules which are downregulated in HPV-induced cervical and laryngeal lesions. The papillomavirus evades the immune system through downregulation of MHC class I molecules and Langerhans cells and diminished susceptibility to NK cells. Vaccines have been developed in animal modes, especially cattle. Therapeutic vaccines for subjects already infected with these viruses hold promise.

Parainfluenza virus (PIV) immunity: Immunity against parainfluenza virus infection (presented clinically as croup, upper respiratory infections, and pharyngitis) is manifested as an increase in IgG antibodies to PIV in 93%, 81%, and 80% of PIV type 1, 2, and 3 infections, respectively. IgM antibodies occur in 40 to 90%. They are common cross-reactions of IgG antibodies to PIV 1 and 3 but not of antibodies to PIV 2 and PIV 1 and 3. Cross-reactions are less frequent with IgM antibodies. EIA is more sensitive than complement fixation but is of lower specificity because of cross-reactions of PIV with mumps virus. Type-specific PIV antigens are found in 94 to 100% of culture-positive nasopharyngeal aspirates.

Paramyxovirus immunity: Both serum antibody and cell-mediated immunity are induced by infection with human paramyxoviruses that cause such common childhood diseases as measles, mumps, and respiratory tract infections. Both limbs of the immune response are important for recovery from disease although the relative significance of these varies with the particular virus of this group. Secretory antibody is important in some of them such as respiratory infections, but it is only partially protected. Most all of the virus-encoded proteins induce serum antibody detectable after infection. Antibodies specific for M protein and F protein are usually at low titers. Even though antinucleocapsid antibody is often present in high titer, the only neutralizing antibodies are those specific for the attachment protein and the fusion protein and are thus protected. Antibodies against either F or HN proteins are protective but the greatest protection is induced when both antigens are used for immunization. The cell-mediated immune response to paramyxoviruses remains to be defined. These viruses may evade the host immune response and nonspecifically suppress cell mediated immunity through infection of monocytes and macrophages as observed in measles infection. They also evade the host immunity by establishing a persistent infection. Measles, mumps, Newcastle disease, canine distemper, and rhinderpest virus vaccines are presently available. These are all live attenuated virus vaccines.

Poxvirus immunity: Multiple antibodies are produced in response to poxvirus infection. These antibodies can be detected by numerous assays including complement fixation, virus neutralization, and ELISA among others. Not all of these antibodies are protective. Significant are the antibodies that neutralize enveloped or nonenveloped virus infectivity, those that combine with circulating antigens to facilitate immune clearance by phagocytes and those that together with effector cells and complement lyse infected cells. Neutralizing antibody may diminish thyremia by acting on extracellular virus. The passive transfer of antibodies has been shown to protect mice and sheep from infection by viruses of this group, but the passive protection is brief and of limited effectiveness compared with active immunization with live virus that induces long-lasting immunity. Cross-reactive neutralizing antibodies may be detected within a week following infection and last for a generation. But the level of neutralizing antibody and the host immune status are not directly correlated. The cell-mediated response is the principal mechanism of protection and recovery from poxvirus diseases. Immune T lymphocytes and monocytes and macrophages are necessary for regression of ectromelia infection of mice. They develop delayed-type hypersensitivity that is a T cell mediated response. Vaccinia virus infection of Rhesus monkeys and hampsters is mediated by NK cells, but by cytotoxic T cells in sheep. Vaccinia virus shows interferon sensitivity. Poxvirus synthesis remains intracellular and is transmitted from one cell to another without exposure to neutralizing antibody, which is effective only against virus budding from infected cells. The virus may also survive in the skin, where epidermal Langerhans' cells process and present antigen to lymphocytes to generate a protective immune response. Poxviruses encode functional homologues of host cell factors that regulate the immune system. Edward Jenner in 1798 showed that cowpox cross-protects against smallpox virus infection. Cowpox vaccinia was invaluable in ridding the world of smallpox. Thereafter, it was judged that the effects of vaccination were a liability as the disease smallpox had been eliminated. Vaccinia has also been used in the control of pox in various animal species. Vaccinia is currently used as a vector for genes of other viral pathogens in the development of vaccines.

Provirus is the DNA version of a retrovirus that has been integrated into the host cell genome where it may remain inactive transcriptionally for prolonged periods.

Picornavirus is a small RNA virus with a naked capsid structure. More than 230 viruses categorized as enteroviruses, rhinoviruses, cardioviruses, and aphthoviruses comprise this family.

Picornavirus immunity: Neutralizing antibodies have an important role in protection against picornaviruses as shown by the ability of passively transferred antibodies to block virus replication and disease progression. The early IgM response is less specific than the subsequent IgG and IgA responses. There is considerable cross-reactivity among the different serotypes. Virus neutralization by antibody involves Fc receptor-mediated endocytosis (opsonization) and interactions that prevent virus penetration and uncoating or induce lethal RNA unpackaging. $CD4^+$ T cells have also been shown to be significant in picornavirus infections that induce cell-mediated responses with the production of cytokines. $CD4^+$ and $CD8^+$ T lymphocytes recognize specific epitopes. Picornaviruses evade the immune system through antigenic variation of neutralizing antibody epitopes and may also involve variation at T cell sites/MHC-binding structures, which interferes with help for humoral immune responses and cell-mediated killing of infected cells. Vaccines against picornaviruses depend on their ability to induce neutralizing antibodies in the host following administration of either a live attenuated virus or a chemically inactivated intact virus.

Reovirus immunity: Most humans acquire a reovirus-specific antibody response in infancy or early childhood. Reovirus induces strong humoral and cell-mediated immune responses. Reovirus specific antibodies are directed mainly against the outer capsid protein σ1. Yet serum antibodies to other viral proteins are also induced. Reovirus-specific antibodies to the σ1 hemagglutinin protein are mainly serotype specific. By contrast, reovirus-specific monoclonal antibodies reactive against other external capsid proteins reveal serotype nonspecific neutralizing or hemagglutinin inhibiting properties. Reovirus-specific monoclonal antibodies can block attachment of the virus to host cells as well as inhibit internalization and intracellular proteolytic uncoating. Virus-specific cytotoxic T lymphocytes are also elicited in addition to T cell mediated delayed-type hypersensitivity reactions. The DTH response is serotype specific, whereas the CDLs are mainly serotype nonspecific. These viruses may induce antigen-specific as well as antigen-nonspecific immunosuppression. Reoviruses can induce specific T cell responses in systemic tissues. A virus-specific B cell response occurs in Peyer's patches, and a cytotoxic T lymphocyte response in Peyer's patches. The ability of reovirus to enter Peyer's patches via M cells reveals that these viruses might be used in a mucosal vaccine.

Rhabdovirus immunity: Resistance to rabies is in part genetically controlled, as has been shown in mice where it is controlled by one or two genes. Resistance is a dominant trait. There is a difference among species in susceptibility to rabies virus infection. Immunity may be either nonspecific or specific. Interferon plays a critical role since rhabdoviruses are quite susceptible to interferon action. In rabies there is no serological evidence of infection prior to onset of the disease which is usually fatal. Vaccination studies have yielded the most information concerning specific immunity. Vaccination during the incubation period, if not repeated, can cause the "early death phenomenon." Passively transferred specific antibodies can protect against rabies. The relative significance of cell-mediated immunity as a protective mechanism remains to be demonstrated. Yet antibody titers and protection are closely correlated. Protective mechanisms following postexposure treatment of humans with rabies vaccine involve T lymphocytes. Rabies virus infection progresses silently in the nervous system without inducing any detectable humoral immune response. Anti-rabies vaccination must distinguish between preventive vaccination and postexposure treatment. Several vaccine injections together with specific immunoglobulin inoculation are warranted in postexposure treatment for humans when an individual has been badly exposed. Preventive vaccination is usually carried out in veterinary medicine. Contemporary vaccines confer partial or no protection against selected rabies-related virus infections. Only inactivated vaccines are licensed for use in humans. Those previously used that contain nervous tissue are dangerous because of their myelin content which may induce hypersensitivity reactions that lead to paralysis. Most current vaccines are prepared from virus grown in cell culture. While attenuated virus vaccines have been used in domestic animals in the past, they have been replaced by newer potent inactivated vaccines. Recombinant vaccines make use of a vaccinia recombinant virus containing the rabies virus glycoprotein gene which is able to induce production of virus glycoprotein in infected cells, to induce rabies virus neutralizing antibodies, and to protect susceptible hosts. It is active by oral administration.

Rhinovirus immunity: Although most rhinovirus infections resolve spontaneously, respiratory tract infections may occur in immunocompromised hosts. There are nearly 100 serotypes of rhinovirus. Studies with human rhinovirus 2 (HRV-2) reveal that HRV-2 specific immunoglobulins in blood sera and nasal secretions increase 1 to 2 weeks following inoculation. HRV-2 antibodies peak at 35 d following inoculation. Serum-neutralizing antibodies remain elevated for many years following infection. Local specific antibodies cannot be detected after 2 years. Preinoculation IgA levels in nasal washings are diminished in those who become infected compared with those who do not. Aspirin and acetaminophen suppress serum antibody responses.

Smallpox: See variola.

Theiler's virus myelitis: Murine spinal cord demyelination that is considered to be an immune-based consequence of a viral infection.

***Togavirus* immunity:** Lifelong immunity is induced by infection with a number of the *togaviruses*. Attenuated vaccines have been used to successfully control Venezuelan equine encephalitis virus in horses. Induction of vaccinal immunity in livestock can protect humans through vaccination of the intermediate host. Antibodies against E1 protein and the E2 protein can neutralize and passively protect against alpha virus infection in mice and monkeys. Nonstructural protein antibodies can recognize surface components of infected cells. Anti-NS-1 antibodies are highly efficient in activating complement on cell surfaces, leading to lysis of infected cells. Maturation of these viruses from infected cells is by budding through the cytoplasmic membrane. The recognition of nonstructural proteins on infected cell surfaces by T lymphocytes is a significant immunity mechanism. Vaccinia virus live vaccine expressing nonstructural proteins have been used as experimental vaccines. E1 proteins of alphaviruses participate in cell surface adsorption and fusion. E2 proteins contain significant virulence determinants.

TORCH panel is a general serologic screen to identify antenatal infection. Elevated levels of IgM in a neonate reflect *in utero* infection. The panel may be further refined by determining IgM antibody specific for certain microorganisms. TORCH is an acronym for toxoplasma, other, rubella, cytomegalic inclusion virus, herpes (and syphilis). There are both false-positive and false-negative reactions in the quantitative TORCH screen. If the TORCH panel is positive, it is indicative of *in utero* infection which may have major consequences. Toxoplasmosis may result in microglial nodules, thrombosis, necrosis, and blocking of the foramina, leading to hydrocephalus. Rubella may cause hepatosplenomegaly, congenital heart disease, petechiae and purpura, decreased weight at birth, microcephaly, cataracts, and central nervous system manifestations including seizures and bulging fontanelles. Cytomegalovirus is characterized by hepatosplenomegaly, hyperbilirubinemia, microcephaly, and thrombocytopenia at birth, followed later by deafness, mental retardation, learning disabilities, and other manifestations. Herpes simplex can lead to premature birth. The central nervous system manifestations include seizures, chorioretinitis, paralysis that is either flaccid or spastic, and coma. Syphilis is an addendum to the TORCH designation, but congenital syphilis has increased in recent years and is not associated with specific clinical findings.

Varicella is a human herpes virus type 3 (HHV-3) induced in acute infection that occurs usually in those less than 10 years of age. There is anorexia, malaise, low fever, and a prodromal rash following a 2-week incubation period. Erythematous papules appear in crops and intensify for 3 to 4 d. They are pruritic. Complications include viral pneumonia, secondary bacterial infection, thrombocytopenia, glomerulonephritis, mycocarditis, and other conditions. HHV-3 may become latent when chicken pox resolves. Its DNA may become intergrated into the dorsal route ganglion cells. This may be associated with the development of herpes zoster or shingles later in life.

Varicella-zoster virus immunity: Varicella-zoster virus (VZV) causes two separate illnesses, i.e., chicken pox or varicella and shingles or herpes zoster. Chicken pox is the primary infection, and reactivation of the virus in adulthood causes shingles, a dermatomal exanthem. The immune response to chicken pox includes IgM response at the end of the incubation period when a vesicular rash appears. Immunofluorescence can be used to detect VZV-specific antibodies reacting with outer membrane of live VCV-infected cells. The initial antibodies are specific for VCV gB, followed soon thereafter by antibodies to gH and gE. Chicken pox patients also develop a cellular immune response which reacts with the same viral glycoproteins recognized by the antibody as well as the regulatory protein termed IE62. Chickenpox patients develop lymphocyte proliferative responses to VZV gE, gI, gB, gH, and IE62 antigens. $CD8^+$ class I-restricted T cells and $CD4^+$ class II-restricted T cells mediate VZV-specific cytotoxicity. The VZV cellular immune responses control the severity of the chicken pox exanthem in normal individuals. A latent VZV infection develops in most children in whom the virus remains dormant in the dorsal root ganglia for many years. In senior adult years, the virus may become reactivated and induce herpes zoster (shingles). This is associated with decreased immunity that accompanies increasing age. In the period prior to the development of zoster, anti-VZV glycoprotein antibody is greatly diminished in the serum, and cellular immunity is likewise decreased. Cells synthesizing IFN-γ (T_H1 cells) diminishes more than those producing interleukin-4 (IL-4) (T_H2 cells). Several weeks after the appearance of herpes zoster, high titers of VZV-specific antibody appear and there is an increased lymhoproliferative response to VZV antigens. Varicella-zoster immune globulin (VZIG) of high titer can prevent chicken pox if injected intramuscularly. It is important to give the globulin within 3 to 4 d after exposure to chicken pox. A live attenuated varicella vaccine is available for use in the U.S., Europe, and Japan. Vaccination is followed by the development of both humoral and cellular immune responses.

Variola (smallpox): *Variola major* is a *Poxvirus variolae*-induced disease that has now been eliminated from the worldwide human population. This virus-induced disease caused vesicular and pustular skin lesions, leading to disfigurement. It produced viremia and toxemia. Approximately one third of the people who were unvaccinated succumbed to the disease. *V. minor* (alastrim) is a mild form of smallpox. It was produced by a different strain that was so weak it was unable to induce the formation of pocks on the chick chorioallantoic membrane.

Fungi are single-celled and multicellular eukaryotic microorganisms such as yeast and molds that may induce a host of diseases. Immunity to fungi involves both cell-mediated and humoral immune responses.

Fungal immunity: Nonspecific immune mechanisms of the host that form a first line of defense against fungal infections include the mechanical barrier provided by the skin and mucous membranes, competition by the normal bacterial flora for nutrients, and the respiratory tract's mucociliary clearance mechanism. Fungi are not lysed by the terminal components of the complement system and specific antibody. Yet complement components serving as opsonins facilitate phagocytosis of fungi by phagocytic cells. Fungi are powerful activators of the alternative complement system. Neutrophils are very significant in protection against various mycoses including disseminated candidiasis and invasive aspergillosis. Monocytes and resident macrophages vary in their ability to kill fungi. Few specific anti-fungal activities of activated human macrophages have been demonstrated. Bronchoalveolar macrophages play an important role in the immune response to inhaled fungi. Natural killer cells inhibit the growth of *C. neoformans* and *P. brasiliensis in vitro.* NK cells have also been shown to clear cryptococcus from mice. Specific antibodies are of little use in host defense against most mycoses. But specific cell-mediated immunity is paramount for a protective immune response to *C. neoformans* disease and the dimorphic fungi. It is also important for protection against dermatophyte infections. Cell-mediated immunity plays an important role in protection against mucocutaneous candidiasis. AIDS patients have a high incidence of fungal infections. Cytokines formed in a specific cell-mediated immune response facilitate the antifungal action of natural killer cells, nonspecific T lymphocytes, and neutrophils. Cytokines such as tumor necrosis factor, granulocyte macrophage colony stimulating factor (GM-CSF), and interleukin-12 (IL-12) released during a cell-mediated immune response may activate effector cells to kill fungi. Immunosuppressive cytokines such as IL-10 and transforming growth factor β (TGF-β) are also formed in response to fungal infection. Severe fungal infections usually occur in profoundly immunosuppressed patients who have diminished responses to immunization. Defects in host immunity associated with fungal infections are being identified in order that cytokines such as interferon γ (IFN-γ) may be administered to chronic granulomatous disease patients, and granulocyte colony-stimulating factor and GM-CSF may be administered to neutropenic patients. Monoclonal antibodies against capsular glucuronoxylomannan have been given to patients with cryptococcosis.

***Candida* immunity:** Resistance against *Candida* begins with the nonspecific barriers such as intact skin and mucosal epithelium in addition to the indigenous bacterial flora that competes for binding sites. Once these protective barriers have been breached, neutrophils are the major cellular defense by phagocytosing the *Candida* microorganisms with intracellular killing through oxidative mechanisms. Monocytes and eosinophils also participate in this process. Microabscesses may form in infected tissues. Mononuclear cells constitute the main inflammatory response in more chronic infections. IgG, IgM, and IgA immunoglobulin classes of *Candida*-specific antibodies have been found in infected patients. Local mucosal immunity such as in the vagina is associated with the development of IgA antibodies in secretions. Even though antibody titers were elevated in infected patients, the humoral immune response does not have a principal role in host defense against *Candida.* Patients with defects in cell-mediated immunity, such as AIDS patients, and those with chronic mucocutaneous candidiasis have increased susceptibility to *Candida* infections. Vaccination has been determined to be ineffective in preventing *Candida* infections.

***Coccidioides* immunity:** Immunity against C*occidioides immitis* depends upon T lymphocytes. IFN-γ plays an important role in protection which can be conferred by the recombinant form of this cytokine in experimental mice. Monocytes have a precise role in limiting infection before a specific immune response develops. Spherules and arthroconidia induce the synthesis of TNF-α by human monocytes *in vitro.* TNF-α alone or combined with IFN-γ promotes killing of spherules by human monocytes *in vitro.* TNF-α and IL-6 levels have been shown to be elevated in patients with overwhelming infection by this organism. Antigen overload and specific suppressor T lymphocyte activity from a circulating humoral suppressor substance may sometimes suppress the T lymphocyte. A *Coccidioides*-specific response occurs in some patients. A positive skin test indicates that the patient has been previously infected. This usually remains positive for the patient's lifetime. Up to 90% of all infected individuals develop an antibody response to *C. immitis.* Mycelial phase antigens are most often used in serodiagnosis. IgM forms early but disappears after 6 months, whereas IgG elevated titers may indicate dissemination. Immunity induced by infection is species- and, in some instances, strain-specific, yet immunization with purified antigens may induce heterologous protection. Immunity is mediated by T cells and is far more significant to resistance than is the humoral immune response which is also stimulated. Antibodies act mainly against extracellular parasites to reduce invasion. $CD4^+$ T cells control primary infections whereas $CD8^+$ lymphocytes are more significant in later stages of infection. There is a vaccine for chickens, but it is expensive.

***Cryptosporidium* immunity:** Mucosal immunization might prove useful to prevent cryptosporidiosis in AIDS patients. Little is known of the host immune response to C. *parvum.* The infection is increased in severity and duration in immunosuppressed individuals, indicating that a specific mucosal immune response must be induced in the host. IFN-γ limits the infection whereas $CD4^+$ T cells limit the duration of the infection. Thus both IFN-γ and $CD4^+$ T cells are critical to inducing resistance to and resolution of the infection. IL-12 activates both natural killer cells and cytotoxic T lymphocytes and induces IFN-γ synthesis. The administration of IL-12 prevents C. *parvum* murine infection. This proves that exogenous IL-12 therapy can prevent the infection through an IFN-γ-dependent specific immune mechanism and that endogenous IL-12 synthesis helps to limit C. *parvum* infection.

Cytochalasins are metabolites of various species of fungi that affect microfilaments. They bind to one end of actin filaments and block their polymerization. Thus, they paralyze locomotion, phagocytosis, capping, cytokinesis, etc.

***Cryptococcus neoformans* immunity:** The polysaccharide capsule of *C. neoformans* serves as an anti-phagocytic mechanism. It blocks binding sites recognized by phagocytic receptors for β-glucan and mannan that could mediate phagocytosis and secretion of TNF-α. The capsule also covers IgG bound to the cell wall but it is the site of complement activation in the alternative pathway in which IC3b fragments might facilitate opsonization. Neutrophils, monocytes, and NK cells all show anticryptococcal activity *in vitro*. Nonencapsulated *C. neoformans* generate elevated levels of IL-2 and IFN-γ *in vivo*. The polysaccharide capsule may also induce suppressor T cells that synthesize a factor which inhibits binding of the organism by macrophages. Critical to immunity to this fungus is the recognition of encapsulated *C. neoformans* by antigen-specific mechanisms. A specific immune response is essential to control encapsulated *C. neoformans*. NK and T cells exert their antifungal action against *C. neoformans* independent of oxygen or nitrogen radicals. T cell-mediated immunity is critical for acquired immunity against *C. neoformans*. NK cells have also been shown to play an important role.

***Histoplasma* immunity:** Cell-mediated immunity is the main host defense against infection with *Histoplasma capsulatum*. The specific cell-mediated response in humans occurs in lymphoid organs and other tissues 7 to 18 d following exposure to conidia. This leads to the initiation of healing of lesions and organs with the formation of granulomas that have central necrosis. Lymph nodes that drain sites of infection are enlarged, encapsulated, and may calcify. Delayed-type hypersensitivity responses to histoplasmin are detectable a month after infection. Macrophages and neutrophils may harbor *H. capsulatum*. Yeasts and conidia that are not coated with opsonins bind to epitopes of the lymphocyte function-associated antigen 1 (CD11a/CD18), complement receptor type III (CD11b/CD18), and the P150-95 complex (CD11c/CD18) of adhesion-promoting receptors on human macrophages. Yeast bonding requires diovalent cations. The unopsonized use induces formation of hydrogen peroxide by human macrophages. Yet *Histoplasma* yeasts fail to induce a respiratory burst when phagocytosed. Antigen-specific $CD4^+$ T lymphocytes mediate both protective immunity and delayed-type hypersensitivity. Cytokines that influence the outcome of infection in naïve animals include IL-12 and TNF-α. IL-3, GMCSF, and macrophage CSF activate human microphages to block yeast cell growth. There is no licensed vaccine for *H. capsulatum* but its heat shock protein system may be a promising candidate for a future vaccine.

Ketokonazole is an antifungal drug used to treat chronic mucocutaneous candidias.

Mucocutaneous candidiasis: Cellular immunodeficiency is associated with this chronic *Candida* infection of the skin, mucous membranes, nails, and hair, with about 50% of patients manifesting endocrine abnormalities. Cell-mediated immunity to *Candida* antigens alone is absent or suppressed. The individual manifests anergy following the injection of *Candida* antigen into the skin. Immunity to other infectious agents, including other fungi, bacteria, and viruses, is not impaired. The B cell limb of the immune response, even to *Candida* antigens, does not appear to be affected. The antibody response to *Candida* and other antigens is within normal limits. The relative numbers of both T and B lymphocytes are normal, and immunoglobulins are at normal or elevated levels. Four clinical patterns have been described. The most severe is known as early chronic mucocutaneous candidiasis with granuloma and hyperkeratotic scales on the nails or face. These have an associated endocrinopathy in about 50% of the cases. The second type is late-onset chronic mucocutaneous candidiasis, which involves the oral cavity or occasionally the nails. The third form is transmitted as an autosomal recessive trait and is usually not associated with endocrine abnormalities. It is a mild to moderately severe disorder. The fourth form is known as juvenile familial polyendocrinopathy with candidiasis, which may be associated with hypoparathyroidism with or without Addison's disease. Those individuals in whom endocrinopathy is associated with mucocutaneous candidiasis may demonstrate autoantibodies against the endocrine tissue involved. In addition to the immunologic abnormalities described above, there is diminished formation of lymphokines, e.g., macrophage migration inhibitory factor (MIF), directed against *Candida* antigens. Recommended treatment includes antimycotic agents and immunologic intervention designed to improve resistance of the host.

***Nocardia* immunity:** Infection by microorganisms of this genus usually begin as a pulmonary infection that may be either localized or disseminated. *Nocardia* species have a complex antigenic structure. Immunocompromised patients have an increased likelihood of developing infections by Nocardia. Defects in the mononuclear phagocyte system increase host susceptibility. The microorganisms grow in monocytes and macrophages with the addition of recombinant IFN-γ and TNF-α enhance nocardial growth. T cell of mice immunized against whole cells of *Nocardia* species become immunologically reactive against the microorganism and kill it directly. Whereas T cell-deficient athymic mice show increased susceptibility to nocardial infection, B cell-deficient mice do not. Thus, T cells rather than B cells are critical for host immunity against nocardial infection. No effective vaccine against *Nocardia* is presently available.

Spherulin is an antigen derived from spherules of *Coccidioides immitis* that has been used for the delayed-type hypersensitivity skin test for coccidioidomycosis.

Parasites are organisms that derive sustenance from a living host such as worms and protozoa.

Parasite immunity: The cytokine network is critical in parasitic infection. Contemporary research is attempting to untangle this complex network in order to develop appropriate mechanisms to combat infections. Partial success has been achieved in attempting to control the direction of an immune response by incorporating cytokines into a vaccine against leishmaniasis. IL-12 has been used to prevent granuloma formation in schistosomiasis. A fine balance must be maintained between ensuring protection while reducing the possibilities of counter protection. The primary concern in parasitic infections is not to determine whether an immune response occurs, but whether the interaction between parasite and host will lead to protection or pathological changes or a combination of the two. It is necessary to reveal which antigens induce protection, how this may be induced artificially, and to ascertain what causes the pathological changes and how they may be countered. No commercially available vaccines against any human parasitic diseases exist. There are only a few against parasites of veterinary importance.

***Ascaris* immunity:** The roundworm *Ascaris lumbricoides* infects 1.3 billion people. Infected subjects mount strong IgG antibody responses specific for the parasite, but most individuals respond to only a subset of parasite constituents. Only 20% of individuals respond to the ABA-1 antigen/allergen. Laboratory studies have shown that immune responses to *Ascaris* antigens in laboratory rodents is restricted by the class II MHC region. People may vary in the specificity of their hypersensitivity reactions to *Ascaris lumbricoides*. The IgE response appears to be protective in ascariasis and is believed to be a protective mechanism in other helminth infections. Mouse experiments show that the immune response to *Ascaris* infection is dominated by T_H2 cells, which helps to explain the elevated IgE levels, eosinophilia, and mastocytosis observed in these infections. The T_H2 response is critical for immune elimination of the parasites. Although the parasite is able to alter its surface and secreted antigens, it remains to be proven that this serves as an effective mechanism to evade the host immune system. No vaccine is available.

Babesiosis immunity: The host immune response to babesiosis, a malaria-like disease transmitted by parasitized *Ixodes* ticks, depends in part on the spleen, which has a central role in immune defense. Patients in whom the spleen has been removed are more susceptible to infection by *Babesia* and manifest elevated parasitemia. Complement activation by *Babesia* might lead to formation of TNF-α and IL-1, promoting local defense. Complement levels decrease in babesiosis. Patients develop increased circulating C1q binding activity and decreased C4, C3, and CH50 levels. The formation of TNG-α and IL-1 could account for many of the clinical features of the disease. Besides macrophages, other cellular immune functions are a critical part of the response to *Babesia*. T cell-deficient mice manifest significantly increased parasitemia. Cellular immunity is diminished by the disease itself, which is also associated with increased $CD8^+$ T lymphocytes, diminished monocyte mitogen responsiveness, and polyclonal hypergammaglobulinemia.

Chagas' disease: The immune response effectively controls the high number of parasites in the acute phase, leading to essentially undetectable parasitemia in the chronic phase, yet sterile immunity and complete parasite clearance and cure have not been achieved in humans or in experimental animals infected with *Trypanosoma cruzi*. The immune response does not achieve a cure but maintains a host–parasite balance that lasts for the lifetime of the infected person. Various antigens have been used in vaccine trials but most only reduce the parasitemia during the acute phase of the disease and transform lethal to nonlethal infections. No vaccination has produced complete protection, and the vaccinated animals still become infected. Decreased parasitemia may diminish the incidence and severity of the chronic phase.

***Echinococcus* immunity:** The genus *Echinococcus* includes four species of tapeworm parasites, among which is *E. granulosus*. Its cycle of transmission involves interaction between a definitive host, such as carnivores, and an intermediate host (herbivores/omnivores). Each host may reveal two morphologically distinct parasite stages. Definitive hosts are infected with the tapeworm stage of *Echinococcus*. Immunological methods have been used to diagnose infection in definitive hosts and to develop a recombinant vaccine that is highly effective in protecting sheep from hydatid infection. Ovine hydatid cyst fluid is rich in antigen. The principal parasite antigens are designated antigen 5 and antigen B. Antibodies against these are useful in immunodiagnosis of hydatid infection in humans. Protein antigens of 27 kDa and 94 kDa from protoscoleces are recognized by sera of dogs infected with *E. granulosus*. Oncosphera antigens of 22, 30, and 37 kDa are specific for *E. granulosus* and are stage specific for oncospheres. Little is known regarding cellular responses against *E. granulosus* infection in dogs, but T cells and activated macrophages are believed to play a significant role in cellular immunity against *Echinococcus* in intermediate hosts. With respect to the humoral immune response to infection, dogs form IgA, IgG, and IgM antibodies against *E. granulosus*. IgA antibody against *E. granulosus is* produced in the intestinal mucosa, but some dogs manifest elevated levels of IgE. IgA, IgE, IgG, and IgM antibodies are synthesized in intermediate hosts infected with this microorganism. Immunodiagnosis can be carried out by the ELISA in the definitive host by detecting circulating antibodies specific for the microorganism. Coproantigens of *Echinococcus* may be detected in the feces of infected dogs. The mucosal IgA produced in dogs has little effect on the worms, which have the capacity to suppress cytotoxic and effector activity in the region of the scolex. *E. granulosus* is believed to modulate the immune system of the intermediate host through production of cytotoxic substances, immunosuppressive and immunostimulatory cytokines. Concomitant immunity is critical in ensuring the survival of *E. granulosus* in intermediate hosts. This refers to the capacity of established hydatid cysts to avoid the immune system of the intermediate host while inducing an effective immune response against subsequent infection by this microorganism. Concomitant immunity is mediated by antibody and is directed against oncospheres. Killing is induced through an antibody-dependent, complement-mediated lysis of the parasite.

Elephantiasis is enlargement of extremities by lymphedema caused by lymphatic obstruction during granulomatous reactivity in filariasis.

***Entamoeba histolytica* antibody** is a specific serum antibody that develops in essentially all individuals infected with *E. histolytica*. The antibody responses comprise mainly of IgG and to a lesser degree IgA. IgM declines quickly whereas specific IgG remains increased for months or years. Coproantibodies are found in the feces of amebic dysentery patients, whereas amebic liver abscess patients have secretory IgA in their saliva and colostrum. There is little evidence of cellular immunity and granuloma formation in amebic dysentery ulcers and in amebic liver abscesses. *E. histolytica* trophozoites not only resist normal human leukocytes but in fact kill them. IFN-γ induced by antigenic stimulation activates macrophages to kill trophozoites. The organism itself is potently chemotactic for neutrophils, which are killed on contact with the parasite. Lytic enzymes released from dead cells induce tissue injury. IFN-γ and TNF-α-activated neutrophils are able to kill trophozoites. There have been conflicting reports of the relative complement sensitivity of *E. histolytica*.

Filarial immunity: Increased levels of parasite-specific antibodies are synthesized following filarial infection. Subjects with asymptomatic microfilaremic infections develop high titers of filarial-specific IgG4 antibodies, yet patients with chronic lymphatic obstruction develop mainly IgGl, IgG2, and IgG3. Most infected subjects develop antifilarial IgE antibodies. IgE and IgG4 antibodies are usually directed against the same epitopes and are regulated by IL-4 and IL-13. Little or no proliferative response to parasite antigens occurs in lymphocytes from asymptomatic microfilaremia individuals. This lack of T cell reactivity is parasite antigen-specific since responsiveness to nonparasite antigens and mitogens is unaffected. Asymptomatic microfilaremic subjects are unable to produce IFN or IFN-γ but retain the ability to synthesize IL-4 and IL-5. Subjects with chronic lymphatic pathology synthesize IFN-γ, IL-4, and IL-5 following exposure to the parasite antigen. IL-10 modulates the synthesis of IFN-γ in microfilaremia. Prenatal exposure to microfilarial stage antigens can lead to long-term anergy to filarial antigens once naturally infected. Protective immunity can be induced by attenuated larvae or by repeated infections. Individuals who develop resistance to new infection while maintaining adult parasites acquire concomitant immunity. The few individuals who remain free of infection in spite of long-term residence in high endemic areas are said to have putative immunity.

A **helminth** is a parasitic worm. Infections by helminths induce a T_H2-regulated immune response that is associated with inflammatory infiltrates rich in eosinophils and IgE synthesis.

Hookworm immunity: Hookworms in humans induce various antibody responses that may be assayed by either the ELISA or the radioimmunoassay technique. There is a prominent immune response to excretory–secretory (ES) products as well as to surface cuticular antigens. There is also a sharp rise in specific immunoglobulin isotypes during infection. There is a marked elevation of total serum IgE in human hookworm disease. The remaining additional immunoglobulin in the circulation is not specific for parasite antigens. Serum IgA may be diminished during hookworm disease because hookworm proteases are able to digest host IgA. The greater the burden of worms, the more intense the antibody response to adult antigens. Those subjects who have fewer worms develop higher titers of antilarval antibodies, which increases resistance to larval challenge.

Hookworms induce T_H2 type responses together with specific IgE antibodies against ES products and eosinophilia. Evidence is lacking to support the concept that T_H2-dependent immunity plays a major part in host protective immunity against hookworms. Little is known concerning cellular responses to hookworms. Eosinophilia is a common finding in hookworm infection. There is increased production of superoxide and enhanced chemotaxins of eosinophils from infected donors, yet the eosinophilia has not been linked to host-protective immunity to hookworms. No vaccines are available for immunization against hookworm disease in humans.

Leishmania is an obligate intracellular protozoan parasite with an affinity for infecting macrophages. It produces chronic inflammatory disease of numerous tissues. In mice, T_H1 responses to *Leishmania major* that include IFN-γ-synthesis, control infection. By contrast, T_H2 responses to IL-4 synthesis result in disseminated lethal disease.

Leishmaniasis is a parasitic human infection that can lead to the development of different disease conditions ranging from cutaneous lesions to fatal visceral infection. Mouse models of infection have yielded much information on mechanisms of susceptibility and resistance to this infection, rendering experimental leishmaniasis a fine model system to evaluate T cell subset polarization and its relationship to pathogenesis. These intracellular protozoan parasites affect 12 million people worldwide. *Leishmania major* or *L. tropica* may cause cutaneous leishmaniasis; *L. braziliensas* cause mucocutaneous leishmaniasis; and *L. donovini* or *L. infantum* that induce viseral leishmaniasis produce the clinical disease known as Kala-azar (black disease), dum dum fever, or ponos. This clinical disease follows spread of the parasite from the skin lesion to tissue macrophages in the liver, spleen, and bone marrow. Patients develop fever, malaise, weight loss, coughing, and diarrhea, as well as anemia, darkening skin, and hepatosplenomegaly. Immunity depends on polarization of $CD4^+$ T_H cell subsets. In a T_H2 response to infection, IL-4 and IL-10 correlate with disease susceptibility, whereas a murine T_H1 response is associated with production of interferon γ (IFN-γ) and IL-2, which lead to resolution of lesions in animals that remain refractory to further challenge.

Malaria is a disease induced by protozoan parasites (*Plasmodium* species) with a complex life cycle in a mosquito and a vertebrate host. Four species of the parasite are responsible for human malaria. Numerous immunogenic proteins are formed at each morphologically distinct stage in the life cycle. The asexual stage in the blood stream causes the disease. The parasite employs various mechanisms to evade a protective immune response. However, immunity against the parasite and the disease eventually develops from repeated exposure. Malaria vaccine development is in progress.

Metronidazole is the drug of choice to treat intestinal and extraintestinal amebiasis and neurogenital trichomoniasis, and an alternative drug in *Giardia lamblia*, *Balantidium coli*, and *Blastocystis hominis* and infections.

A **MOTT cell** is a type of plasma cell in which refractile eosinophilic inclusion bodies that resemble Russell bodies are found. It is associated with African sleeping sickness. It is demonstrable in periarteriolar cuffs in the brain of patients in late stages of African trypanosomiasis.

Opisthorchiasis–clonorchiasis immunity: Antibodies synthesized in patients infected with opisthorchiasis or chlonorchiasis react with the various developmental stages of the parasites. Stage-specific and crossreactive common antigens have been detected and identified. IgG is the predominant class of antibodies in the serum but IgE have also been detected to a lesser degree. However, the bile contains secretory IgA as the principal immunoglobulin. The parasite antigens manifest an IgE-potentiating activity. Even though high-antibody titers are achieved in infected patients, the protective ability of these antibodies is doubtful. Some investigations have suggested that complement-fixing antibodies might have a role but *Opisthorchis* may be able to activate complement by way of the alternative pathway leading to lymphocyte killing. Cell-mediated immunity also occurs following natural infection or immunization with the parasite antigen. The role of T cells remains to be determined. Primary infection does not appear to protect against reinfection by the same parasite. Thus there appears to be a lack of protective acquired immunity. These liver flukes survive within the biliary system that may serve as a type of immunologically privileged site. The parasites can shed their surface tegument following injury by immune mechanism, representing yet another mechanism to evade host defenses.

***Onchocerciasis volvulus* immunity:** The filarial parasite that can induce dermal and ocular complications contains antigens that induce antigen-specific IgG, IgM, IgE, and IgA antibodies, in addition to polyclonal bead stimulation. Humoral immunity develops early in chimpanzees, but the exact role of antibodies in protection remains to be determined. Antibodies against microfilariae facilitate adherence of granulocytes *in vitro*, and eosinophils and neutrophils mediate antibody-dependent killing. Massive eosinophil degranulation may lead to tissue injury. Infected subjects develop elevated IgE which may worsen ocular lesions by contributing to acute inflammation. Much of the IgE antibody is *O. volvulus* antigen specific. Immune complexes may also contribute to acute inflammation. Cell-mediated immunity is downregulated to antigens that are specific and nonspecific for the infectious agent. Onchocerciasis is marked by a predominant T_H2 cytokine response in subjects with ocular pathology, whereas a T_H1 response is associated with immunity. Thus the host immune response is significant in both pathology and protection in onchocerciasis. The HLA-D allele influences the pathogenesis of *O. volvulus* infection. Ocular pathology is the most serious effect of this disease affecting both anterior and posterior segments of the eye. Protection against onchocerciasis has been directed to vectal control but development of a protective vaccine would be a better solution. Larval antigens are the best targets for a prophylactic vaccine.

Sabin-Feldman dye test is an *in vitro* diagnostic test for toxoplasmosis. Serial dilutions of patient's serum are combined with *Toxoplasma gondii* microorganisms, and complement is added. If specific antibodies against *Toxoplasma* organisms are present in the serum, complement interrupts the integrity of the toxoplasma membrane admitting methylene blue which has been added to the system and stains the interior of the organism. That dilution of patient's serum in which one half of the *Toxoplasma* organisms have been fatally injured is the titer.

***Schistosoma* immunity:** The immune response to the blood flukes classified as schistosomes is complex. Repeated exposure to schistosome larval antigens may lead to hypersensitivity and cercarial dermatitis (swimmers itch). Exposure to large numbers of *S. mansoni* or *S. japonicum* can lead to a serum sickness or immune complex-like disease, whereas immune reactions to later stages of the infection may be associated with resistance against infection. Many of the pathological changes in schistosome infections are linked to deposition of eggs which induce granulomatous reactions in the tissues resulting in fibrosis. The granuloma is a delayed-type hypersensitivity reaction that is T cell dependent. In addition to the cells expected in a granuloma, eosinophils, lymphocytes, and macrophages are also present. The fibrosis is also egg antigen induced. Multiple immune parameters are activated by these eggs and their antigens leading to a modulation in chronically infected individuals. Egg antigens may induce a protective as well as an immunopathological response. Whereas the eggs induce mostly immunopathologic effects, reactions to the schistosomulum are mostly protective of the host. Adult worms from a primary infection can continue to survive in individuals resistant to reinfection with fresh cercariae through "concomitant immunity." Irradiated cercariae can be used to induce immunization of mice and other experimental animals against cercarial challenge. The main target of destructive immunological attack is the migrating schistosomulum. Humans develop concomitant immunity slowly. Resistance is correlated with peripheral blood eosinophilia in *S. haematobium* infections. Infection-protective immunity may be associated in elderly persons with IgE antibodies against adult worm antigens. More than 90% of the surface antigens of the young schistosomulum are carbohydrates. Anti-egg antibodies may cross-react with these antigens. Adult worms activate T_H1 responses, whereas eggs induce T_H2 responses. Protection in mice is mediated mainly by T_H1 cells and can be potentiated with IL-12. TNF-α is associated with granuloma formation. Adult worms are usually not susceptible to immune attack either by coating themselves with host-derived macromolecules that mask parasite surface antigens or the worms may shed antigenic macromolecules from the outer tegument, rendering their outer surfaces immunologically inert. Even though egg identification in human excreta has long been the method of diagnosis, indirect diagnosis using antibody detected by the ELISA technique is used with increasing frequency. Schistosome-derived carbohydrate antigens in the blood are also helpful in immunodiagnosis. There is no vaccine for protection against schistosomiasis in humans, yet irradiated larvae have been used to immunize cattle.

Antigen masking is the ability of some parasites (e.g., *S. mansoni*) to become coated with host proteins, theoretically rendering them "invisible" to the host's immune system.

***Strongyloides* immunity:** Immunoglobulin G (IgG), IgA, and IgE classes of immunoglobulin form in response to antigens of *Strongyloides stercoralis* filariform larvae. The principal humoral responses by the IgG_4 subclass may be directed to more than 50 different 15- to 100-kDa antigens. The remaining IgG subclasses recognize fewer than 20 antigens. Zinc endopeptidase together with 31- and 28-kDa proteins are antigens that induce specific immune responses; however, none of these antibodies is protective against dissemination in the host. Patients without detectable humoral responses remain asymptomatic. Immunocompromised patients with disseminated infection may manifest high titers of parasite-specific antibodies. Impaired cell-mediated immunity has been claimed by some to facilitate parasite dissemination, but this has not been proven, especially since AIDS patients have not developed this as an opportunistic infection.

***Strongyloides* hyperinfection:** *S. stercoralis* larvae may invade the tissues of immunosuppressed patients with enteric strongyloides infection to produce this condition.

Suramin (Antrypol, 8,8′-(carbonyl-bis -(imino-3,1-phenylenecarbonylimino))-bis-1,3,5-naphthalene trisulfonic acid) is a therapeutic agent for African sleeping sickness produced by trypanosomes. Of immunologic interest is its ability to combine with C3b, thereby blocking factor H and factor I binding. The drug also blocks lysis mediated by complement by preventing attachment of the membrane attack complex of complement to the membranes of cells.

***Taenia solium* immunity:** Preencystment immunity (early immunity) refers to the immune response at the oncosphere penetration site. Late postoncospheral or postencystment immunity refers to the immune response at the final establishment site. Secretory IgA in gut secretions likely attacks invading oncospheres since it is resistant to intestinal enzymes. Mast cells surround invading oncospheres and developing larvae, which suggests that IgE might react with antigen leading to degranulation of these cells whose products cause increased vascular permeability. This would permit IgG antibodies to reach the invading site. Eosinophils also surround invading oncospheres but there is no evidence that they induce injury. When oncospheres that are newly hatched reach their establishment site, they transform from a stage in which they are highly vulnerable to attack to one in which they are completely resistant. Both humoral and cellular responses to *T. solium* is heterogeneous in both pigs and humans. 90% of the serum antibody is IgG. The *T. solium* cysticerci induce a chronic granulomatous reaction in pig muscle with extensive eosinophil infiltration and degranulation. T cell immunity to larval cestodes remains to be demonstrated. Larval cestodes may contain blocking antibodies on the surface. Molecular mimicry is also believed to represent an evasive strategy whereby this parasite avoids the host immune response. Some larvae produce inhibitors of proteolytic enzymes such as trypsin and chymotrypsin. Some cestode larvae may induce immunosuppression of the host. Vaccination with living eggs can induce complete protection against challenge infection, but there is only a limited supply of the eggs. This has been remedied by the development of a recombinant taeniid vaccine antigens which have been able to induce protection in animals.

***Theileria* immunity:** Immunity against these tick-transmitted intracellular protozoan parasites of domestic animals depend upon cell-mediated immune responses for protection. Humoral immune responses directed against schizonts and piroplasms are insignificant in the development of natural protective immunity. Animals that recover may be resistant to homologous challenge and developing immunity that persists for 3 to 5 years. Infection with sporozoites and the development of schizonts are critical for the development of natural immunity. Live attenuated, subunit and recombinant vaccines have been used. *T. annulata* live vaccine is prepared from attenuated cell lines that produce infection in cattle without causing disease. These induce protective immunity. *T. parva* subunit vaccine contains p67 antigen, the recombinant form of which protects 70% of immunized cattle against experimental challenge.

***Toxocara canis* immunity:** The immune response in humans to the domestic dog roundworm or *Toxocara canis* includes the development of antibodies that are useful for immunodiagnosis to detect infection. The seroprevalence rate in the U.S. general population is 2.8% for adults but 23.1% for children. The immune response is also characterized by development of eosinophilic granulomata which may appear throughout the body except for the brain. Larvae within liver granulomata may be killed. *T. canis* infection induces a powerful T_H2 response in experimental animals. There is no available vaccine.

***Toxoplasma gondii* immunity:** Both humoral and cell-mediated responses follow infection with *T. gondii*. The cellular immune response is the principal mediator of resistance to infection, although both cellular and humoral confer resistance. Antibodies activate complement by the classical pathway to lyse extracellular parasites. Tachyzoites coated with immunoglobulin are killed within macrophages. Both monocytes and neutrophils also mediate effective killing. Whereas antibodies mediate only a protective effect that is partial, cell-mediated immunity is critical for survival of the host during acute infection. Activation of macrophages by IFN-γ is a principal effector mechanism in toxoplasma infection. Macrophages kill by both the oxidative and nonoxidative pathways. Both NK and T cells are essential components of the efferent limb of the protective cell-mediated immune response. $CD8^+$ T cells produce gamma interferon to activate macrophages whereas $CD4^+$ T cells synthesize interleukin 2 which facilitates IFN-γ synthesis by $CD8^+$ T cells. NK cells also produce IFN-γ which helps to induce a T_H1-type response (IL-2 and IFN-γ synthesis) of $CD4^+$ T cells. TGF-β and IL-10 downregulate IFN-γ production by NK cells during infection. AIDS patients may develop toxoplasmic encephalitis, pointing to the significance of cell-mediated immunity which they have lost. Tachyzoites in the acute stage of infection are the principal targets for the protective immune response. Tachyzoites induce both antigen-specific and nonspecific suppressor cells to inhibit induction of the immune response to the parasite. No vaccine for *T. gondii* is available.

***Trichuris trichiura* immunity:** Immune responses against these worm infections of humans include specific IgG, IgA, and IgE antibody responses. The IgG level increases with the level of the infection, but elevated IgA levels may be associated with diminished worm burdens. The cellular response remains to be defined. Macrophages release TNF-α and there is IgE-mediated release of mast cell mediators that contribute to inflammation and enteropathy associated with infection. In mice the immune response to *T. muris* consists of a powerful T cell response that leads to premature expulsion of worms from the intestine. Primary infections are followed by the development of resistance to reinfection. Immunity can be passively transferred with either $CD4^+$ T cells or immune sera. Inflammatory responses are not believed to contribute to immunity. Murine stains that effectively eliminate the infection develop responses mediated by T_H2 cells. There is no vaccine for *Trichuris*.

Tropical eosinophilia is a hypersensitivity to filarial worms manifested in the lungs. It has been reported in the Near East and in the Far East. Patients develop wheezing, productive cough, and a cellular infiltrate comprised of eosinophils, lymphocytes, and fibroblasts. Fibrosis may result.

Trypanosome immunity: Vaccination against African sleeping sickness known as trypanosomiasis has thus far been unattainable. Even though the antigen that induces a protective humoral response is well known as a surface glycoprotein that covers the entire trypanosome, the organism repeatedly changes the antigenic structure of this glycoprotein (the variant surface glycoprotein VSG), thereby evading destruction by the host immune response. Only a single VSG is expressed by a trypanosome at a time. Trypanosomes that express the same VSG are classified as belonging to the same variable antigenic type (VAT). Only short-term immunity can be induced by

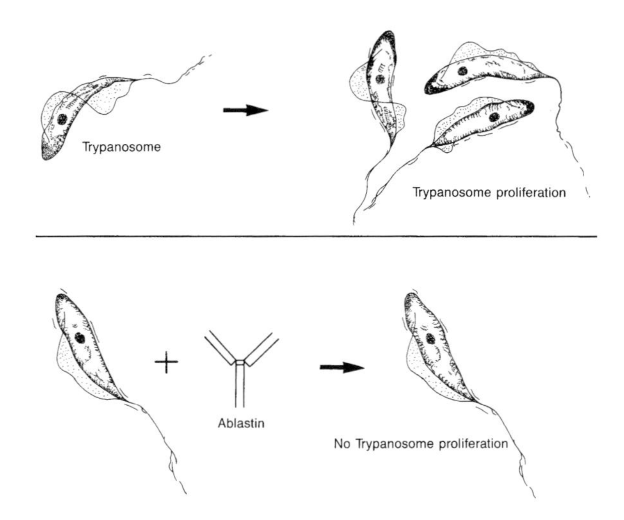

FIGURE 23.43 Ablastin.

allowing cattle to become infected by fly bites and then treating the infection as soon as the parasitemia becomes patent. The powerful humoral response in trypanosomal infections is characterized by the appearance of IgM and IgG antibodies that are associated with the elimination of each VAT. These antibodies kill parasites and clear them from the blood by complement activation by way of the classic pathway and opsonization that results in uptake by the liver's Kupffer cells. The immune response in experimental mice consists of both T-dependent and T-independent components. Lymphocyte responsiveness is profoundly depressed in trypanosome infections.

Released antigen is antigen derived from trypanosomes during an infection that may appear in the patient's serum. It corresponds to the antigenic type of the trypanosome infecting the individual.

Ablastin is an antibody with the exclusive property of preventing reproduction of such agents at the rat parasite *Trypanosoma lewisi* (Figure 23.43). It does not demonstrate other antibody functions.

Parasites such as *Schistosoma mansoni* produce eggs that induce granuloma formation and isolation of the eggs. The development of fibrosis interrupts the venous blood supply to the liver, leading to hypertension and cirrhosis.

Intracellular protozoa often activate specific cytotoxic T cells. They represent a critical mechanism to prevent dissemination of intracellular malarial parasites. Parasite antigen–antibody (immune) complexes may be trapped in the renal microvasculature, leading to immune complex glomerulonephritis. Parasites are capable of evading immune mechanisms. Animal parasites have developed remarkable mechanisms to establish chronic infections in the vertebrate host. Natural immunity to them is weak, and the parasites have devised novel and ingenious mechanisms to circumvent specific immunity. Parasites either camouflage their own antigens or interfere with host immunity. Some parasites mask their antigens by coating themselves with host proteins, which prevents their detection by host immune mechanisms. Some develop resistance by biochemical alterations of their surface coat. They are also adept at changing their surface antigens by antigenic variation, which may frustrate attempts to prophylactic immunization. Other parasites, such as *Entamoeba histolytica*, may shed their antigenic coats.

Schistosomiasis is a schistosome infection that is characteristically followed by a granulomatous tissue reaction.

Fasciola immunity: An inflammatory response in the liver is associated with primary infection of a host with the liver fluke *Fasciola hepatica*. The cellular infiltrate includes neutrophils, eosinophils, lymphocytes, and macrophages. A significant eosinophilia occurs in the peritoneal cavity and the blood. There is no cellular response encircling the juvenile flukes themselves. Numerous macrophages and fibroblasts occur in injured areas in the chronic phase of a primary infection. This is followed by liver fibrosis which is accompanied by numerous $CD8^+$ and y8- TCR^+T cells. Numerous lymphocytes and eosinophils infiltrate larger bile ducts, and in cattle a granulomatous reaction is followed by bile duct calcification. Resistance to reinfection follows primary infection of cattle, rats, and possibly humans, yet this fails to occur in sheep and mice. The failure of sheep to develop resistance against reinfection may be associated with cellular immune deficiencies. Products released from *F. hepatica* may adversely affect the host immune response such as cathepsin proteases which flukes secrete and which may cleave immunoglobulins, thereby inhibiting attachment to host effector cells. Other immunosuppressive molecules may be toxic to host cells.

24 Vaccines and Immunization

A **vaccine** may contain live attenuated or killed microorganisms or parts or products from them capable of stimulating a specific immune response comprised of protective antibodies and T cell immunity. A vaccine should stimulate a sufficient number of memory T and B lymphocytes to yield effector T cells and antibody-producing B cells from memory cells. Viral vaccine should also be able to stimulate high titers of neutralizing antibodies. Injection of a vaccine into a nonimmune subject induces active immunity against the modified pathogens.

Other than macromolecular components, a vaccine may consist of a plasmid that contains a cDNA encoding an antigen of a microorangism. Other vaccines include antiinsect vector vaccines, fertility-control vaccines, peptide-based preparations, antiidiotype preparations and DNA vaccines, among others. There is no antiparasite vaccine manufactured by conventional technology in use at present. Vaccines can be prepared from weakened or killed microorganisms, inactivated toxins, toxoids derived from microorganisms, or immunologically active surface markers extracted from microorganisms. They can be administered intramuscularly, subcutaneously, intradermally, orally, or intranasally; as single agents or in combination. An ideal vaccine should be effective, well tolerated, easy and inexpensive to produce, easy to administer, and convenient to store. Vaccine side effects include fever, muscle aches and injection site pain, but these are usually mild. Reportable adverse reactions to vaccines include anaphylaxis, shock, seizures, active infection, and death.

Vaccination is immunization against infectious disease through the administration of vaccines for the production of active (protective) immunity in humans or other animals.

To **vaccinate** is to inoculate with a vaccine to induce immunity against a disease.

Vaccinable is the capability of being vaccinated successfully.

Multivalent vaccine: See polyvalent vaccine.

Combined prophylactic: See mixed vaccine.

A **mixed vaccine** is a preparation intended for protective immunization that contains antigens of more than one pathogenic microorganism. Thus, it induces immunity against those disease agents whose antigens are represented in the vaccine. It may also be called a polyvalent vaccine.

Dead vaccine: See inactivated vaccine.

Attenuated is an adjective that denotes diminished virulence of a microorganism.

Attenuation is the decrease of a particular effect, such as exposing a pathogenic microorganism to conditions that destroy its virulence, but leave its antigenicity or immunogenicity intact.

An **attenuated pathogen** is one that has been altered to the point that it will grow in the host and induce immunity without causing clinical illness.

To **attenuate** is the process of diminishing the virulence of a pathogenic microorganism, rendering it incapable of causing disease. Attenuated bacteria or viruses may be used in vaccines to induce better protective immunity than would have been induced with a killed vaccine.

Hyperimmune is a descriptor for an animal with a high level of immunity that is induced by repeated immunization of the animal to generate large amounts of functionally effective antibodies, in comparison to animals subjected to routine immunization protocols, perhaps with fewer boosters.

Hyperimmunization is the successive administration of an immunogen to an animal to induce the synthesis of antibody in relatively large amounts. This procedure is followed in the preparation of therapeutic antisera by repeatedly immunizing animals to render them "hyperimmune."

A **bacterial vaccine** is a suspension of killed or attenuated bacteria prepared for injection to generate active immunity to the same microorganism.

Diathelic immunization describes protective immunity induced by injecting antigen into the nipple or teat of a mammary gland.

A **bacterin** is a vaccine comprised of killed bacterial cells in suspension. Inactivation is by either chemical or physical treatment.

In vaccine standardization, the **immunizing dose (ImD_{50})** is that amount of the immunogen required to immunize 50% of the test animal population, as determined by appropriate immunoassay.

Formol toxoid is a toxoid generated by the treatment of an exotoxin such as diphtheria toxin with formalin. Although first used nearly a century ago, it was subsequently modified to contain an adjuvant such as an aluminum compound to boost immune responsiveness to the toxoid. It was later replaced with the so-called triple vaccine of diphtheria, pertussis, and toxoid vaccine.

Anavenom is a toxoid consisting of formalin-treated snake venom which destroys the toxicity but preserves immunogenicity of the preparation.

A **live vaccine** is an immunogen for protective immunization that contains an attenuated strain of the causative agent, an attenuated strain of a related microorganism that cross-protects against the pathogen of interest, or the introduction of a disease agent through an avenue other than its normal portal of entry or in combination with an antiserum.

Passive immunization describes the transfer of a specific antibody or of sensitized lymphoid cells from an immune to a previously nonimmune recipient host. Unlike active immunity, which may be of a relatively long duration, passive immunity is relatively brief, lasting only until the injected immunoglobulin or lymphoid cells have disappeared. Examples of passive immunization include (1) the administration of γ globulin to immunodeficient individuals and (2) the transfer of immunity from mother to young, i.e., antibodies across the placenta or the ingestion of colostrum-containing antibodies.

A **killed vaccine** is an immunizing preparation comprised of microorganisms, either bacterial or viral, that are dead but retain their antigenicity, making them capable of inducing a protective immune response with the formation of antibodies and/or stimulation of cell-mediated immunity. Killed vaccines do not induce even a mild case of the disease which is sometimes observed with attenuated (greatly weakened but still living) vaccines. Although the first killed vaccines contained intact dead microorganisms, some modern preparations contain subunits or parts of microorganisms to be used for immunization. Killed microorganisms may be combined with toxoids, as in the case of the DPT (diphtheria–pertussis–tetanus) preparations administered to children.

International Unit of Immunological Activity refers to the use of an international reference standard of a biological preparation of antiserum or antigen of a precise weight and strength. The potency or strength of biological preparations such as antitoxins, vaccines, and test antigens derived from microbial products and antibody preparations may be compared against such standards to reflect their strength or potency.

A **lapinized vaccine** is a preparation used for immunization that has been attenuated by passage through rabbits until its original virulence has been lost.

The **LD_{50}** is the dose of a substance such as a bacterial toxin or microbial suspension that leads to the death of 50% of a group of test animals within a certain period following administration. This has been employed to evaluate toxicity or virulence and to evaluate the protective qualities of vaccines administered to experimental animals.

LEP (low egg passage) is a type of vaccine for rabies that has been employed for the immunization of dogs and cats.

Live attenuated vaccine is an immunizing preparation consisting of microorganisms whose disease-producing capacity has been weakened deliberately in order that they may be used as immunizing agents. Response to a live attenuated vaccine more closely resembles a natural infection than does the immune response stimulated by killed vaccines. The microorganisms in the live vaccine are actually dividing to increase the dose of immunogen, whereas the microorganisms in the killed vaccines are not reproducing, and the amount of injected immunogen remains unchanged. Thus, in general, the protective immunity conferred by the response to live attenuated vaccines is superior to that conferred by the response to killed vaccines. Examples of live attenuated vaccines include those used to protect against measles, mumps, polio, and rubella. Live attenuated virus vaccines contain live viruses in which accumulated mutations impede their growth in human cells and their disease-causing capacity.

HEP is the abbreviation for "high egg passage," which signifies multiple passages of rabies virus through eggs to achieve attenuation for preparation of a vaccine appropriate for use in immunizing cattle.

Mass vaccination refers to immunization with vaccines during an outbreak of a communicable disease in an effort to prevent an epidemic. For example, mass vaccinations may be carried out in schools and hospitals during meningitis or hepatitis epidemics.

Live attenuated measles (rubeola) virus vaccine is an immunizing preparation that contains live measles virus strains. It is the preferred form except in patients with lymphoma, leukemia, or other generalized malignancies; radiation therapy; pregnancy; active tuberculosis; egg sensitivity; prolonged drug treatment that suppresses the immune response, such as corticosteroids or antimetabolites; or administration of gammaglobulin, blood, or plasma.

Persons in these groups should be administered immune globulin immediately following exposure.

Live measles virus vaccine is a standardized attenuated virus immunizing preparation used to protect against measles.

Live measles and mumps virus vaccine is a standardized immunizing preparation that contains attenuated measles and mumps viruses.

Live measles and rubella virus vaccine is a standardized immunizing preparation that contains attenuated measles and rubella viruses.

Live oral poliovirus vaccine is an immunizing preparation prepared from three types of live attenuated polioviruses. An advisory panel to the Centers for Disease Control and Prevention recommended in 1999 that its routine use be discontinued. It contains a live yet weakened virus that has led to eight to ten cases of polio each year. Now that the polio epidemic has been eliminated in the U.S., this risk is no longer acceptable. Also called Sabin vaccine.

Live rubella virus vaccine is an attenuated virus immunizing preparation employed to protect against rubella (German measles). All nonpregnant susceptible women of childbearing age should be provided with this vaccine to prevent fetal infection and the congenital rubella syndrome, i.e., possible fetal death, prematurity, impaired hearing, cataract, mental retardation, and other serious consequences.

A **challenge** is the deliberate administration of an antigen to induce an immune reaction in an individual previously exposed to that antigen to determine the state of immunity.

MMR vaccine is measles–mumps–rubella vaccine. A live attenuated virus vaccine given at 15 months of age or earlier. A booster injection is given later. The vaccine is effective in stimulating protective immunity in most cases. It might prove ineffective in children younger than 15 months of age if they still have massively transferred antibodies from the mother. This vaccine should not be given to pregnant women, immunodeficient individuals undergoing immunosuppressive therapy, or individuals with acute febrile disease.

A **heterologous vaccine** induces protective immunity against pathogenic microorganisms which the vaccine does not contain. Thus, the microorganisms that are present in the heterologous vaccine possess antigens that cross-react with those of the pathogenic agent absent from the vaccine. Measles vaccine can stimulate protection against canine distemper. Vaccinia virus was used in the past to induce immunity against smallpox because the agents of vaccinia and variola share antigens in common.

Heterotypic vaccine: See heterologous vaccine.

Historically, the intracutaneous inoculation of pus from lesions of smallpox victims into healthy, nonimmune subjects to render them immune to smallpox was known as **variolation**. In China, lesional crusts were ground into a powder and inserted into the recipient's nostrils. These procedures protected some individuals, but often led to life-threatening smallpox infection in others. Edward Jenner's introduction of vaccination with cowpox to protect against smallpox rendered variolation obsolete.

Smallpox vaccine: an immunizing preparation prepared from the lymph of cowpox vesicles obtained from healthy vaccinated bovine animals. This vaccine is no longer used as smallpox has been eradicated throughout the world.

Smallpox vaccination involves the induction of active immunity against smallpox (variola) by immunization with a related agent, vaccinia virus, obtained from vaccinia vesicles on calf skin. Shared, and therefore cross-reactive, epitopes in the vaccinia virus provide protective immunity against smallpox. This ancient disease has now been eliminated worldwide with the only laboratory stocks maintained in Atlanta (U.S.) and Moscow (Russia). These stocks are supposed to have been destroyed by both agencies once the virus was sequenced. Smallpox vaccination was first developed by the English physician Edward Jenner, whose method diminished the mortality from 20% to less than 1%. Following application of vaccinia virus by a multiple pressure method, vesicles occur at the site of application within 6 to 9 d. Maximum reactivity is observed by day 12. Initial vaccinations were given to 1- to 2-year-old infants with revaccination after 3 years. Children with cell-mediated immunodeficiency syndromes sometimes developed complications such as generalized vaccinia spreading from the site of inoculation. Postvaccination encephalomyelitis also occurred occasionally in adults and in babies less than 1 year old. The procedure was contraindicated in subjects experiencing immunosuppression due to any cause.

Vaccination is immunization against infectious disease through the administration of vaccines for the production of active (protective) immunity in man or other animals.

Cowpox (Figure 24.1) is a bovine virus disease that induces vesicular lesions on the teats. It is of great historical significance in immunology because Edward Jenner observed that milkmaids who had cowpox lesions on their hands failed to develop smallpox. He used this principle in vaccinating humans with the cowpox preparation to produce harmless vesicular lesions at the site of inoculation (vaccination). This stimulated protective immunity against smallpox (variola) because of shared antigens between the vaccinia virus and the variola virus.

FIGURE 24.1 James Gillray's cartoon of the cowpox.

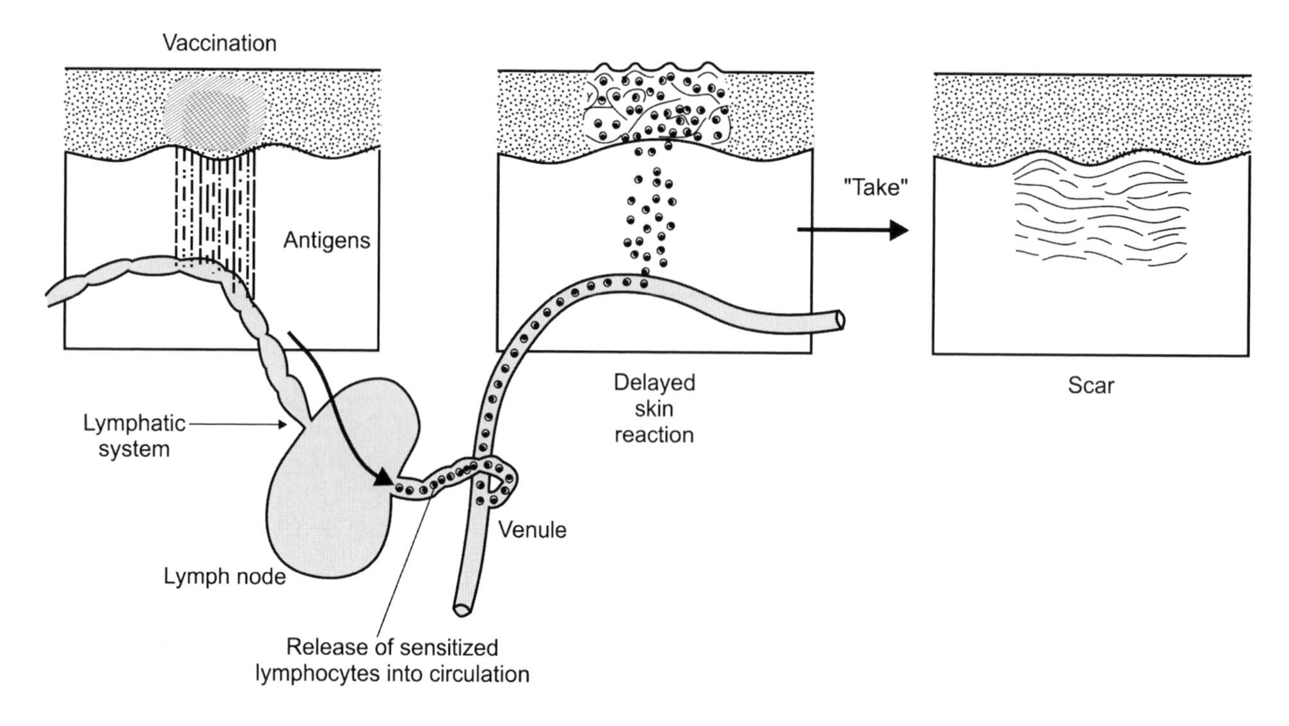

FIGURE 24.2 Vaccination against smallpox.

Vaccinia refers to a virus termed *Poxvirus officinale* derived from cowpox and used to induce active immunity against smallpox through vaccination (Figure 24.2). It differs from both cowpox and smallpox viruses in minor antigens.

Vaccinia gangrenosa: Chronic progressive vaccinia.

Eczema vaccinatum: On occasion, in subjects receiving smallpox vaccination in the past, the virus in the vaccine superinfected areas of skin affected by atopic dermatitis. This led to generalized vaccinia, which was severe and frequently fatal. It was also referred to as Kaposi's varicelliform eruption.

Progressive vaccinia describes an adverse reaction to smallpox vaccination in children with primary cell-mediated immunodeficiency, such as severe combined immunodeficiency. The vaccination lesion would begin to spread from the site of inoculation and cover extensive areas of the body surface, leading to death.

Postvaccinal encephalomyelitis is a demyelinating encephalomyelitis that occurs approximately 2 weeks following vaccination of infants less than 1 year of age, as well as of adults, with vaccinia virus to protect against smallpox. This rare complication of the smallpox vaccination frequently leads to death.

Vaccinia immune globulin is hyperimmune gammaglobulin used to treat dermal complications of vaccination for smallpox, such as eczema vaccinatum and progressive vaccinia. This is no longer used as smallpox is believed to have been eradicated throughout the world.

Varicella (chickenpox) vaccine is an immunizing preparation comprised of attenuated varicella virus.

Chronic progressive vaccinia (vaccinia gangrenosa) (historical): An unusual sequela of smallpox vaccination in which the lesions produced by vaccinia on the skin became gangrenous and spread from the vaccination site to other areas of the skin. This occurred in children with cell-mediated immunodeficiency.

Generalized vaccinia is a condition observed in some children being vaccinated against smallpox with vaccinia virus. There were numerous vaccinia skin lesions that occurred in these children who had a primary immunodeficiency in antibody synthesis. Although usually self-limited, children who also had atopic dermatitis in addition to the generalized vaccinia often died.

Tetanus antitoxin is an antibody raised by immunizing horses against *Clostridium tetani* exotoxin. It is a therapeutic agent to treat or prevent tetanus in individuals with contaminated lesions. Anaphylaxis or serum sickness (type III hypersensitivity) may occur in individuals receiving second injections because of sensitization to horse serum proteins following initial exposure to horse antitoxin. One solution to this has been the use of human antitetanus toxin of high titer. Treatment of the IgG fraction yields (F(ab′)2 fragments which retain all of the toxin-neutralizing capacity but with diminished antigenicity of the antitoxin preparation.

Tetanus vaccine is an immunizing preparation to protect against *Clostridium tetani*. See DTaP vaccine.

A **toxoid** is formed by treating a microbial toxin with formaldehyde to inactivate toxicity but leave the immunogenicity (antigenicity) of the preparation intact. Toxoids are prepared from exotoxins produced in diphtheria and tetanus. These are used to induce protective immunization against adverse effects of the exotoxins in question.

Tetanus toxoid is prepared from formaldehyde-treated toxins of *Clostridium tetani*. It is an immunizing preparation to protect agains tetanus. Individuals with increased likelihood of developing tetanus as a result of a deep penetrating wound with a rusty nail or other contaminated instrument are immunized by subcutaneous inoculation. The preparation is available in both fluid and adsorbed forms. It is included in a mixture with diphtheria toxoid and pertussis vaccine and is known as DTP or triple vaccine. It is employed to routinely to immunize children less than 6 years old.

An immunizing preparation containing **toxoid–antitoxin floccules** has been used to induce active immunity against diphtheria in subjects who show adverse reactions to alum-precipitated toxoid. The preparation consists of diphtheria toxoid combined with diphtheria antitoxin in the presence of minimal excess antigen. It has been used in individuals who are hypersensitive to alum-precipitated toxoid alone. Horse serum in the preparation may induce hypersensitivity to horse protein in some subjects.

Killed virus vaccines are immunogen preparations containing virons deliberately killed by heat, chemicals, or radiation.

An **inactivated vaccine** is an immunizing preparation that contains microorganisms such as bacteria or viruses that have been killed to stop their replication while preserving their protection-inducing antigens. Formaldehyde, phenol, and β-propiolactone have been used to inactivate viruses, whereas formaldehyde, acetone, phenol, or heating have been methods used to kill bacteria to be used in vaccines.

A **polyvalent vaccine** is comprised of multiple antigens from more than one strain of a pathogenic microorganism or from a mixture of immunogens such as the diphtheria, pertussis, and tetanus toxoid preparation.

Immunoprophylaxis describes disease prevention through the use of vaccines to induce active immunization or antisera to induce passive immunization.

An **autogenous vaccine** (Figure 24.3) is prepared by isolating and culturing of microorganisms from an infected subject. The microorganisms in culture are killed and used as an immunogen, i.e., a vaccine, to induce protective immunity in the same subject from which they were derived.

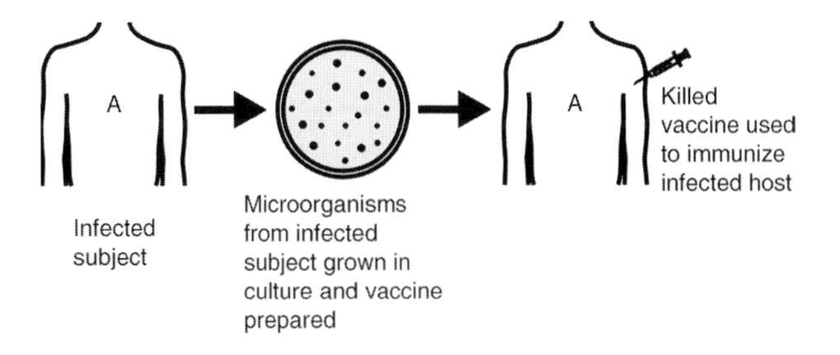

FIGURE 24.3 Autogenous vaccine.

In earlier years, this was a popular method to treat *Staphylococcus aureus*-induced skin infections.

A **caprinized vaccine** is a preparation used for therapeutic immunization, which contains microorganisms attenuated by passage through goats.

An **edible vaccine** is a genetically altered food containing microorganisms or related antigens that may induce active immunity against infection.

A **homologous vaccine** is an autogenous vaccine.

DTaP vaccine (Figure 24.4) is an acellular preparation used for protective immunization that is comprised of diptheria and tetanus toxoids and acellular pertussis proteins that is used to induce protective immunity in children against diptheria, tetanus, and pertussis. Children should receive DTaP vaccine at the ages of 2, 4, 6, and 15 months, with a booster given at 4 to 6 years of age. The tetanus and diptheria toxoids should be repeated at 14 to 16 years of age. The vaccine is contraindicated in individuals who have shown prior allergic reactions to DTaP or in subjects with acute or developing neurologic disease. DTaP vaccine is effective in preventing most cases of the disease it addresses.

DPT vaccine is a discontinued immunizing preparation that consisted of a combination of diphtheria and tetanus toxoids and killed pertussis microorganisms but is no longer used in pediatric immunizations because of the

	Diphtheria, tetanus toxoids, and whole cell pertussis vaccine	Diphtheria, tetanus toxoids and acellular pertussis vaccine	Diphtheria and tetanus toxoid (pediatric)	Tetanus and diphtheria toxoids (adult)	Diphtheria and tetanus toxoids and Hib conjugate and whole cell pertussis vaccines
Synonyms	DTP, DTwP	DTP, DTaP	DT	Td	DTwP-Hib
Manufacturers	several	Connaught, Lederle	several	several	Lederle
Concentration (per 0.5 ml) Diphtheria Tetanus Pertussis Hib	 6.5-12.5 Lf u 5-5.5 Lf u 4 u none	 6.7-7.5 Lf u 5 Lf u either 46.8 mcg or 300 HA u none	 6.6-12.5 Lf u 5-7.6 Lf u none none	 2 Lf u 2-5 Lf u none none	 6.7 or 12.5 Lf u 5 Lf u 4 u 10 mcg
Packaging	5 or 7.5 ml vials	5 or 7.5 vials	5 ml vial, 0.5 ml syringe	5 or 30 ml vials, 0.5 ml syringe	5 ml vial
Appropriate age range	2 months to <7 years	18 months to <7 years	2 months to <7 years	7 years to adult	typically 2 to 15 months
Standard schedule	Five 0.5 ml doses: at 2, 4, 6, and 18 months and 4-6 years of age	For doses 4 and 5: at 18 months and at 4-6 years of age	Three 0.5 ml doses: at 2, 4 and 10-16 months of age	Three 0.5 ml doses: the second 4-8 weeks after the first and the third 6-12 months after the second	Four 0.5 ml doses: at 2, 4, 6, and 15 months of age
Routine additional doses	none	none	none	every 10 years	none
Route	IM	IM	IM	IM, jet	IM

FIGURE 24.4 Diptheria, tetanus, and pertussis — combination summary.

superiority of DTaP vaccine that contains only acellular pertussis microorganisms.

Combination vaccines are immunizing preparations that contain immunogens (antigens) from more than one pathogenic microorganism. They induce protection against more than one disease.

Diphtheria toxin is a 62-kDa protein exotoxin synthesized and secreted by *Corynebacterium diphtheriae.* The exotoxin, which is distributed in the blood, induces neuropathy and myocarditis in humans. Tryptic enzymes nick the single-chain diphtheria toxin. Thiols reduce the toxin to produce two fragments. The 40-kDa B fragment gains access to cells through their membranes, permitting the 21-kDa A fragment to enter. Whereas the B fragment is not toxic, the A fragment is toxic and it inactivates elongation factor-2, thereby blocking eukaryocytic protein synthesis. Guinea pigs are especially sensitive to diphtheria toxin, which causes necrosis at injection sites, hemorrhage of the adrenals, and other pathologic consequences. Animal tests developed earlier in the century consisted of intradermal inoculation of *C. diphtheriae* suspensions into the skin of guinea pigs that were unprotected, compared to a control guinea pig that had been pretreated with passive administration of diphtheria antitoxin for protection. In later years, toxin generation was demonstrated *in vitro* by placing filter paper impregnated with antitoxin at right angles to streaks of *C. diphtheriae* microorganisms growing on media in Petri plates. Formalin treatment or storage converts the labile diphtheria toxin into toxoid.

Diphtheria immunization results from the repeated administration of diphtheria toxoids [as alum precipitated toxoids (APT)]. Toxoid–antitoxin floccules (TAF) are an alternate form for adults who show adverse reaction to APT. Besides this active immunization procedure, diphtheria antitoxin can also be given for passive immunization in the treatment of diphtheria.

Diphtheria toxoid is an immunizing preparation generated by formalin inactivation of *Corynebacterium diphtheriae* exotoxins. This toxoid, which is used in the active immunization of children against diphtheria, is usually administered as a triple vaccine, together with pertussis microorganisms and tetanus toxoid; as purified toxoid which has been absorbed to hydrated aluminum phosphate (PTAP); or as alum-precipitated toxoid (APT). Infants immunized with one of these preparations develop active immunity against diphtheria. Toxoid–antitoxin floccules (TAF) may be administered to adults who demonstrate adverse hypersensitivity reactions to toxoids.

Diphtheria vaccine is an immunizing preparation to protect against *Corynebacterium diphtheriae.* See DTaP vaccine.

	Risk of Occurrence After	
Problem	Vaccination	Disease
Seizures	1:1750	1:25–1:50
Encephalitis	1:100,000	1:1000–1:4000
Severe brain damage	1:310,000	1:2000–1:8000
Death	1:1,000,000	1:200–1:1000

FIGURE 24.5 Pertussis vaccine.

PTAP is a purified diphtheria toxoid which has been adsorbed on hydrated aluminum phosphate. It has been employed to induce active immunity against diphtheria.

CRM 197 is a carrier protein used in vaccines. It is a nontoxic mutant protein related to diphtheria toxin.

Catch-up vaccine (or vaccination) refers to the immunization of unvaccinated children at a convenient time, such as the first day of school, rather than at the optimal time for antibody synthesis. This procedure protects many children who have missed vaccines at regularly scheduled intervals a second opportunity for disease prevention and control.

Pertussis vaccine (Figure 24.5) is a preparation used for prophylactic immunization against whooping cough in children. It consists of virulent *Bordetella pertussis* microorganisms that have been killed by treatment with formalin. It is administered in conjunction with diphtheria toxoid and tetanus toxoid as a so-called triple vaccine. In addition to stimulating protective immunity against pertussis, the killed *Bordetella pertussis* microorganisms act as an adjuvant and facilitate antibody production against the diphtheria and tetanus toxoid components in vaccine. Rarely does a hypersensitivity reaction occur. To reduce the toxic effects of the vaccine, an acellular product is now in use.

Whooping cough vaccine: See pertussis vaccine.

Vaccine extraimmunization refers to the administration of excessive or repeated doses of a vaccine to children or adults, usually as a consequence of poor recordkeeping.

Triple vaccine is an immunizing preparation comprised of three components and used to protect infants against diphtheria, pertussis (whooping cough), and tetanus. It is made up of diphtheria toxoid, pertussis vaccine, and tetanus toxoid. The first of four doses is administered between 3 and 6 months of age. The second dose is administered 1 month later, and the third dose is given 6 months after the second. The child receives a booster injection when beginning school (Figure 24.6).

Pneumococcal polysaccharide vaccine contains a polysaccharide found in the *Streptococcus pneumoniae*

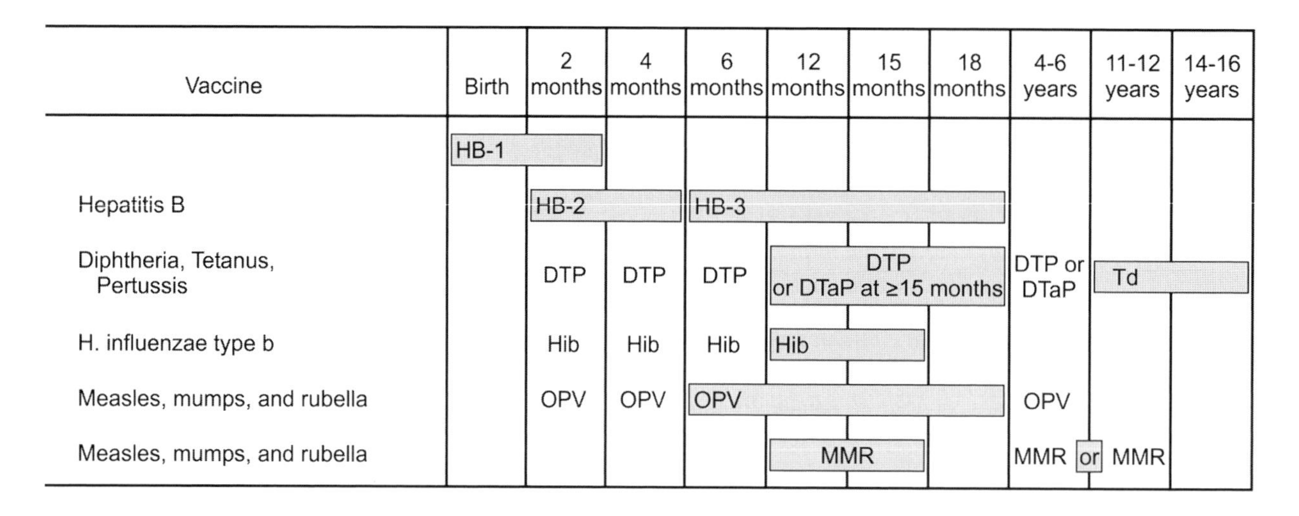

FIGURE 24.6 Childhood immunization schedule recommended by U.S. Public Health Service, January 1996.

capsule that is type specific antigen. It is a virulence factor. Serotypes of this microorganism are based upon different specificities in the capsular polysaccharide which is comprised of oligosaccharide repeating units. Glucose and glucuronic acid are the repeating units in type III polysaccharide.

Pneumococcal 7 valent conjugate vaccine is a pneumococcal vaccine employed to actively immunize infants and small children. The immunizing preparation is comprised of antigens derived from seven capsular serotypes of *Streptococcus pneumoniae*.

Polyvalent pneumococcal vaccine is an immunizing preparation that contains 23 of the known 83 pneumococcal capsular polysaccharides. It induces immunity for a duration of 3 to 5 years. The vaccine is believed to protect against 90% of the pneumococcal types that induce serious illness in patients over 2 years of age. Children at high risk can be vaccinated at 6 months of age and reinoculated at 2 years of age. It is especially indicated in high-risk patients such as those with sickle cell disease, chronic debilitating disease, immunological defects, and the elderly. Vaccination against pneumococcal disease is of increasing significance as *Streptococcus pneumoniae* becomes increasingly resistant to antibiotics.

Rabies vaccine: In humans, significant levels of neutralizing antibody can be generated by immunization with a virus grown in tissue culture in diploid human embryo lung cells. A rabies vaccine adapted to chick embryos, especially egg passage material, is used for prophylaxis in animals prior to exposure. The "historical" vaccine originally prepared by Pasteur made use of rabbit spinal cord preparations to which the virus had become adapted. However, they were discontinued because of the risk of inducing postrabies vaccination encephalomyelitis.

Human diploid cell rabies vaccine (HDCV) is an inactivated virus vaccine prepared from fixed rabies virus grown in human diploid cell tissue culture.

Rabies vaccination: See postrabies vaccination encephalomyelitis.

Postrabies vaccination encephalomyelitis is a demyelinating disease produced in humans actively immunized with rabies vaccine containing nervous system tissue to protect against the development of rabies. Serial injections of rabbit brain tissue containing rabies virus killed by phenol could induce demyelinating encephalomyelitis in the recipient. This method of vaccination was later replaced with a vaccine developed in tissue culture that did not contain any nervous tissue.

TAB vaccine is an immunizing preparation used to protect against enteric fever. It is comprised of *Salmonella typhi* and *S. paratyphi* A and B microorganisms that have been killed by heat and preserved with phenol. The bacteria used in the vaccine are in the smooth specific phase. They also contain both O and Vi antigens. The vaccine is administered subcutaneously. Lipopolysaccharide from the Gram-negative bacteria may induce fever in vaccine recipients. If *Salmonella paratyphi* C is added, the vaccine is referred to as TABC. If tetanus toxoid is added, it is referred to as TABT.

Typhoid vaccination: See TAB vaccine.

Typhoid vaccine: Two forms of immunizing preparation are currently in use. One is a live, attenuated *S. typhi* strain Ty2la that is used as an oral vaccine administered in four doses by adults and children over 6 years of age. It affords protection for 5 years. This vaccine is contraindicated in patients taking antimicrobial drugs and in AIDS patients. A second type of the vaccine, which is for parenteral use, is prepared from the capsular polysaccharide of *Salmonella typhi*. It is administered to 6-month-old children, divided into two doses spaced 4 weeks apart. It is effective 55 to 75% of the time, and lasts for 3 years.

Yellow fever vaccine is a lyophilized attenuated vaccine prepared from the 17D strain of liver-attenuated yellow-fever virus grown in chick embryos. A single injection may confer immunity that persists for a decade.

Salk vaccine is an injectable poliomyelitis virus vaccine, killed by formalin that was used for prophylactic immunization against poliomyelitis prior to development of the Sabin oral polio vaccine.

Inactivated poliovirus vaccine is an immunizing preparation prepared from three types of inactivated polioviruses. Also called Salk vaccine.

Sabin vaccine is an attenuated live poliomyelitis virus vaccine that is administered orally to induce local immunity in the gut, which is the virus's natural route of entry, thereby stimulating local as well as systemic immunity against the causative agent of the disease.

Provocation poliomyelitis is an uncommon consequence of attempted immunization against poliomyelitis in which paralysis follows soon after injection of vaccines such as those that contained *Bordetella pertussis* or alum.

Poliomyelitis vaccines: The three strains of poliomyelitis virus combined into a live attenuated oral poliomyelitis vaccine was first introduced by Sabin. Replication in the gastrointestinal tract stimulates effective local immunity associated with IgA antibody synthesis. Individuals to be immunized receive three oral doses of the vaccine. This largely replaces the Salk vaccine which was introduced in the early 1950s as a vaccine comprised of the three strains of poliovirus that had been killed with formalin. This preparation must be administered subcutaneously.

Influenza virus vaccine is a purified and inactivated immunizing preparation made from viruses grown in eggs. It cannot lead to infection. It contains (H1N1) and (H3N2) type A strains and one type B strain. These are the strains considered most likely to cause influenza in the U.S. Whole virus and split virus preparations are available. Children tolerate the split virus preparation better than the whole virus vaccine. The influenza virus vaccine is a polyvalent immunizing preparation that contains inactivated antigenic variants of the influenza virus (types A and B) either individually or combined for annual use. It protects against epidemic disease and the morbidity and mortality induced by influenza virus, especially in the aged and chronically ill. The vaccine is reconstituted each year to protect against the strains of influenza virus present in the population.

BCG (Bacille Calmette-Guerin) is a *Mycobacterium bovis* strain maintained for more than 75 years on potato, bile glycerine agar, which preserves the immunogenicity but dissipates the virulence of the microorganism. It has long been used in Europe as a vaccine against tuberculosis, although it never gained popularity in the U.S. It has also been used in tumor immunotherapy to nonspecifically activate the immune response in selected tumor-bearing patients, such as those with melanoma or bladder cancer. BCG has been suggested as a possible vector for genes that determine HIV proteins such as *gag*, *pol*, *env*, gp20, gp40, reverse transcriptase, gp20, gp40, and tetanus toxin. In certain countries with a high incidence of tuberculosis, BCG provides effective immunity against the disease, diminishing the risk of infection by approximately 75%. The vaccine has a disadvantage of rendering skin testing for tuberculosis inaccurate, especially in the first 5 years after inoculation, because it induces hypersensitivity to tuberculin.

Challenge stock is an antigen dose that has been precisely measured and administered to an individual following earlier exposure to an infectious microorganism.

Tuberculosis immunization requires the induction of protective immunity through injection of an attenuated vaccine containing BCG. This vaccine was more widely used in Europe than in the U.S. in an attempt to provide protection against development of tuberculosis. A local papule develops several weeks after injection in individuals who were previously tuberculin negative, as it is not administered to positive individuals. It is claimed to protect against development of tuberculosis, although not all authorities agree on its efficacy for this purpose. In recent years, oncologists have used BCG vaccine to reactivate the cellular immune system of patients bearing neoplasms in the hope of facilitating antitumor immunity.

Measles vaccine is an attenuated virus vaccine administered as a single injection to children at 2 years of age or between 1 and 10 years old. Contraindications include a history of allergy or convulsions. Puppies may be protected against canine distemper in the neonatal period by the administration of attenuated measles virus which represents a heterologous vaccine. Passive immunity from the mother precludes early immunization of puppies with live canine distemper vaccine.

Mumps vaccine is an attenuated virus vaccine prepared from virus generated in chick embryo cell cultures. It is a live attenuated immunizing preparation employed to prevent mumps. It should be administered under the same guidelines and restrictions that apply to live attenuated measles virus vaccine.

Rubella vaccine is an attenuated virus vaccine used to immunize girls 10 to 14 years of age. It is used in the MMR

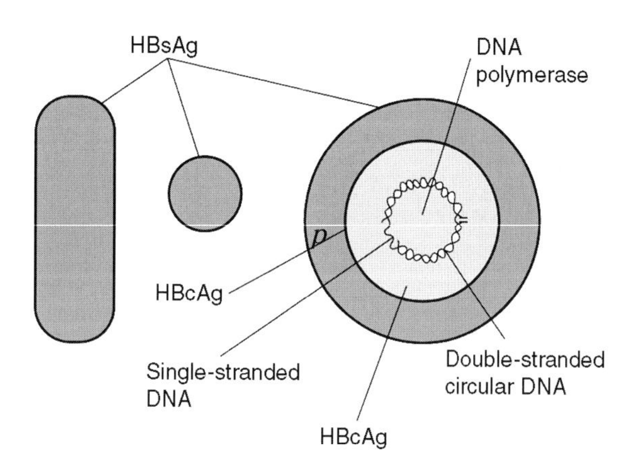

FIGURE 24.7 Hepatitis B virus and its antigens.

combination, or used alone to immunize seronegative women of childbearing age, but it is not to be used during pregnancy.

Although several types of **rinderpest vaccines** have been used in the past, the most satisfactory contemporary vaccine contains a virus adapted to tissue culture.

β propiolactone is a substance employed to inactivate the nucleic acid core of pathogenic viruses without injuring the capsids. This permits the development of an inactivated vaccine, as the immunizing antigens that induce protective immunity are left intact.

Hepatitis vaccine (Figure 24.7) is a vaccine used to actively immunize subjects against hepatitis B virus and contains purified hepatitis B surface antigen. Current practice uses an immunogen prepared by recombinant DNA technology referred to as Recombivax®. The antigen preparation is administered in three sequential intramuscular injections to individuals such as physicians, nurses, and other medical personnel who are at risk. Temporary protection against hepatitis A is induced by the passive administration of pooled normal human serum immunoglobulin which protects against hepatitis A virus for a brief time. Antibody for passive protection against hepatitis must be derived from the blood sera of specifically immune individuals.

Hepatitis B vaccine: Human plasma-derived hepatitis B vaccine (Heptavax-B), which was developed in the 1980s, was unpopular because of the fear of AIDS related to any product for injection derived from human plasma. It was replaced by a recombinant DNA vaccine (Recombivax®) prepared in yeast (*Saccharomyces cervesiae)*. It is very effective in inducing protective antibodies in most recipients. Nonresponders are often successfully immunized by intracutaneous vaccination. An immunizing preparation containing hepatitis B protein antigen produced by genetically engineered yeast.

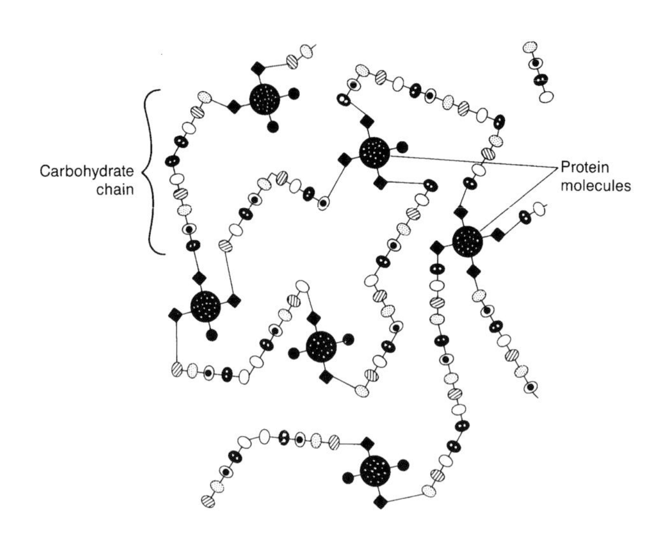

FIGURE 24.8 Conjugate vaccine.

Conjugate vaccine (Figure 24.8) is an immunogen comprised of polysaccharide bound covalently to proteins. The conjugation of weakly immunogenic, bacterial polysaccharide antigens with protein carrier molecules considerably enhances their immunogenicity. Conjugate vaccines have reduced morbidity and mortality for a number of bacterial diseases in vulnerable populations such as the very young or adults with immunodeficiencies. An example of a conjugate vaccine is *Haemophilus influenzae* 6 polysaccharide polyribosyl-ribitol-phosphate vaccine.

Hib (Hemophilus influenzae type b) is a microorganism that induces infection mostly in infants less than 5 years of age. Approximately 1000 deaths out of 20,000 annual cases are recorded. A polysaccharide vaccine (Hib Vac) was of only marginal efficacy and poorly immunogenic. By contrast, anti-Hib vaccine, which contains capsular polysaccharide of Hib bound covalently to a carrier protein such as polyribosylribitol-diphtheria toxoid or PRP-D, induces a very high level of protection that reached 94% in one cohort of Finnish infants. PRP-tetanus toxoid has induced 75% protection. PRP-diphtheria toxoid vaccine has been claimed to be 88% effective.

Haemophilus influenzae type b vaccine (HB) is an immunizing preparation that contains purified polysaccharide antigen from *H. Influenzae* microorganisms and a carrier protein. It diminishes the risks of epiglottitis, meningitis, and other *H. influenzae*-induced diseases during childhood.

Malaria vaccine: Although there is no effective vaccine against **malaria** (Figure 24.9), several vaccine candidates are under investigation, including an immunogenic but nonpathogenic *Plasmodium* sporozoite that has been attenuated by radiation. Circumsporoite proteins combined with sporozoite surface protein 2(SSP-2) are immunogenic.

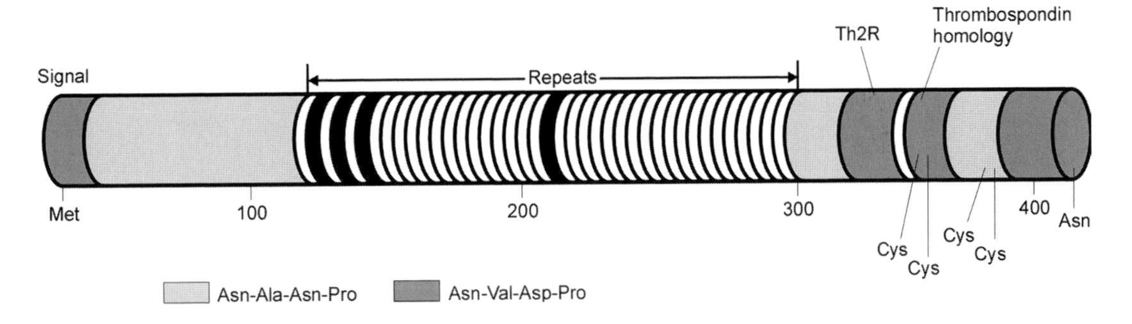

FIGURE 24.9 Circumsporate protein of malaria.

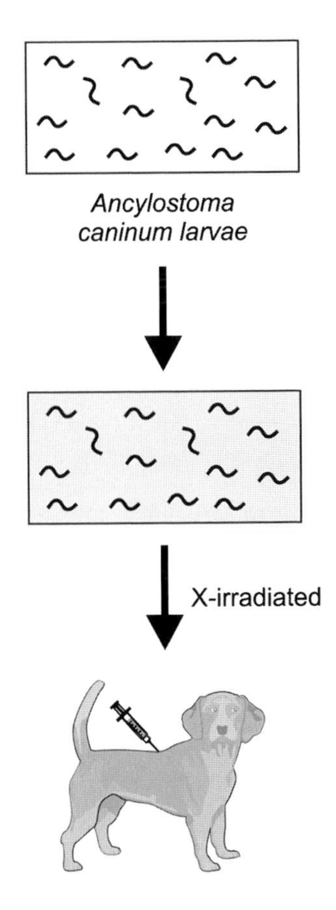

FIGURE 24.10 Hookworm vaccine.

Murine studies have shown the development of transmission-blocking antibodies following immunization with vaccinia into which has been inserted the *Plasmodium falciparum* surface 25-kDa protein designated Pfs25. Attempts have been made to increase natural antibodies against circumsporozoite (CS) protein to prevent the prehepatoinvasive stage. The high mutability of *Plasmodium falciparum* makes prospects for an effective vaccine dim.

Hookworm vaccine (Figure 24.10) is a live vaccine to protect dogs against the hookworm, *Ancylostoma caninum*. The vaccine is comprised of X-irradiated larvae to halt their development to adult forms.

Several experimental **AIDS vaccines** (Figure 24.11) are under investigation. HIV-2 inoculation into cyomologus monkeys apparently prevented them from developing simian AIDS following injection of the SIV virus. Various problems relate to the successful development of an HIV vaccine to protect against AIDS. There is no precise animal model that can be employed to evaluate a vaccine against HIV. Studies of vaccine efficacy in humans are also problematic. The simian immunodeficiency virus (SIV), a close relative of HIV that infects macaques, produces a disease that closely resembles AIDS in Asian macaques but not in the African variety. This variability of SIV points to the problem in attempting to extrapolate results of vaccine trials in macaques to trials in humans. The requirements to establish immunity against HIV infection are unknown. Unlike infections that induce long-lasting protective immunity, HIV coexists with an intensive immune response. HIV infection leads to a progressive immunodeficiency even though there is variability among individuals in resistance to HIV which could be attributable either to their infection by a mutant HIV or to resistance to HIV infection mediated by their $CD8^+$ T cells. The type of immunity induced is also significant. A $CD8^+$ T cell or TH1 cell immune response would probably be most desirable. AIDS patients usually produce TH2 cell cytokines and their $CD4^+$ T cell response to HIV components is affected by TH1 cytokines. Subunit vaccines have also been attempted but these induce immunity to only some proteins of the virus and, when tested in chimpanzees, have induced immunity specific only for the exact virus strain used to make the vaccine. Such subunit vaccines fail to protect against natural infection. In addition, there are ethical issues involved in AIDS vaccine development.

HGP-30 is an experimental vaccine for AIDS that employs a synthetic HIV core protein, p17.

Plague vaccine: *Yersinia pestis* microorganisms killed by heat or formalin are injected intramuscularly to induce immunity against plague. It is administered in three doses 4 weeks or more apart. The duration of the immunity is approximately 6 months. A live attenuated vaccine, used mainly in Java, has also been found to induce protective immunity.

Plague vaccine is an immunizing preparation prepared either from a crude fraction of killed plague microorganisms, *Yersinia pestis*, or synthetically from recombinant proteins. It is rarely used except in a laboratory or for field workers in regions where plague is endemic.

Typhus vaccine is an immunizing preparation that contains killed rickettsiae microorganisms of a strain or strains of epidemic typhus rickettsiae.

Typhus vaccination: Protective immunization against typhus transmitted by lice or fleas and against Rocky Mountain spotted fever is achieved by the administration of inactivated vaccines. Rickettsiae prepared in chick embryo yolk sacs or tissues are treated with formaldehyde to render them inactive. Rather than provide protective immunity that prevents the disease, these vaccines condition the host to experience a milder or less severe form of the disease than that which occurs in a nonvaccinated host.

Meningococcal vaccine is an immunizing preparation that contains bacterial polysaccharides from certain types of meningococci. Meningococcal polysaccharide vaccines A, C, Y, and W135 are available for preventing diseases induced by those serogroups. There is no vaccine for meningococcal serogroup B.

A **DNA vaccine** is an immunizing preparation comprised of a bacterial plasmid containing a cDNA encoding a protein antigen. The mechanism apparently consists of professional antigen-presenting cells transfected *in vivo* by the plasmid, which then express immunogenic peptides that induce specific immune responses. The CpG nucleotides present in the plasmid DNA serve as powerful adjuvants. DNA vaccines may induce powerful cytotoxic T lymphocyte responses. This is an immunizing preparation made by genetic engineering in which the gene that encodes an antigen is inserted into a bacterial plasmid which is injected into the host. Once inside, it employs the nuclear machinery of the host cell to manufacture and express the antigen. In contrast to other vaccines, DNA vaccines may induce cellular as well as humoral immune responses.

CpG nucleotides: Bacterial DNA unmethylated cytidine-guanine sequences that facilitate immune responses acting as an adjuvant. They are believed to enhance DNA vaccine efficacy.

DNA vaccination consists of imunization with plasmid DNA to induce an adaptive immune response against the encoded protein. Bacterial DNA which is rich in unmethylated CpG dinuleotides serves as an adjuvant for this kind of vaccination.

Synthetic vaccines are substances used for prophylactic immunization against infectious disease prepared by artificial techniques such as from cloned DNA or through peptide synthesis.

Genetic immunization consists of the inoculation of plasmid DNA encoding a protein into muscle for the purpose of inducing an adaptive immune response. For reasons yet to be explained, the plasmid DNA is expressed and induces T cell responses and antibody formation to the protein which the DNA encoded.

Distemper vaccine is an attenuated canine distemper virus vaccine prepared from virus grown in tissue culture or chick embryos.

Brucella vaccine is a preparation used for the prophylactic immunization of cattle. It contains live, attenuated *Brucella abortus* microorganisms. A second vaccine, comprised of McEwen strain 45/20 killed microorganisms in a water-in-oil emulsion (adjuvant), has also been used.

Cholera toxin is a *Vibro cholerae* enterotoxin comprised of five B subunits which are cell-binding 11.6-kDa structures encircling a 27-kDa catalase that conveys ADP-ribose to G protein, leading to continual adenyl cyclase activation. Other toxins that resemble cholera toxin in function include diptheria toxin, exotoxin A, and pertussis toxin.

Cholera vaccine is an immunizing preparation comprised of *Vibrio cholerae* smooth strains Inaba and Ogawa in addition to El Tor vibrio, which have been killed by heat or formalin treatment. It is designed to induce protective active immunity against cholera in regions where it is endemic, as well as in travelers to those locations. The immunity induced is effective for only about 12 weeks.

Canine parvovirus vaccine: Initially, a feline enteritis vaccine that was live and attenuated was used based on its cross-reactivity with canine parvovirus. Canine parvovirus may have originated from the feline enteritis organism by mutation. This vaccine was later replaced with attenuated canine parvovirus vaccine.

Newcastle disease vaccines include: (1) an inactivated virus raised in chick embryos that is incorporated into aluminum hydroxide gel adjuvant, and (2) live virus grown in chick embryos and attenuated in a graded manner. Strains with medium virulence are administered parenterally, and those that are less virulent are given to birds either in drinking water or as an aerosol.

Schistosomiasis vaccines: Schistosomiasis (billharziasis), also termed snail fever, is the worst human disease induced by a metazoan parasite. Five different species of *Schistosoma* produce the disease. Two isoforms of the schistosome enzyme glutathione-*S*-transferase have been investigated as candidates for a vaccine. They manifest highly effective immunogens in a rat model. Two surface antigens from the migratory larval stage of *S. mansoni* parasites or other molecules appear promising in animal models. In addition, two different parasite muscle proteins have also been tested. None of these antigens has reached clinical trials, and there is no immediate prospect for development of an effective vaccine even though the search continues.

T cell vaccination (TCV) is a technique to modulate an immune response in which T lymphocytes are administered as immunogens. The vaccine is comprised of T cells specific for the target autoantigen in an autoimmune response to be modulated. For example, antimyelin basic protein (MBP) $CD4^{+}8^{-}$ T cells, serving as a vaccine, were irradiated (1500 R) and injected into Lewis rats. Vaccination with the attenuated anti-MBP T cells induced resistance against subsequent efforts to induce experimental allergic encephalomyelitis (EAE) by active immunization with myelin basic protein in adjuvant. This technique could even induce resistance against EAE adoptively transferred by active anti-MBP T cells. This method was later successfully applied to other disease models of adjuvant arthritis, collagen-induced arthritis, experimental autoimmune neuritis, experimental autoimmune thyroiditis, and insulin-dependent diabetes mellitus (IDDM). TCV was demonstrated to successfully abort established autoimmune disease and spontaneous autoimmune disease in the case of IDDM.

Sensitized vaccine is an immunizing preparation that contains bacteria treated with their homologous immune serum.

Serum virus vaccination is a method no longer used that consisted of administering an immunizing preparation of an infectious agent such as a live vaccine virus together with an antiserum specific for the virus. It was intended to ameliorate the effects of the live infectious agent. This method was abandoned because it was considered a dangerous practice.

A **reassortant vaccine** is an immunizing preparation in which antigens from several viruses or from several strains of the same virus are combined.

A **recombinant vaccine** is an immunogen preparation for prophylactic immunization, comprised of products of recombinant DNA methodology, prepared by synthesizing proteins employing cloned complementary DNA.

Saponin is a glucoside used in the past for its adjuvant properties to enhance immune reactivity to certain vaccine constituents. It was considered to slow the release of immunogen from the site of injection and to induce B cells capable of forming antibody at the site of antigen deposition.

Prophylactic immunization is a procedure to prevent disease through either active immunization or passive immunization. Active immunization usually induces longer lasting protection than does passive immunization.

An **idiotype vaccine** is an antibody preparation that mimics antigens at the molecular level. Such vaccines induce immunity specific for the antigens they mimic. They are not infectious to the recipient, are physiologic, and can be used in place of many antigens, e.g., idiotype vaccine related to *Plasmodium falciparum* circumsporozote (CS) protein.

An **antiidiotypic vaccine** is an immunizing preparation of antiidiotypic antibodies that are internal images of certain exogenous antigens. To develop an effective antiidiotypic vaccine, epitopes of an infectious agent that induce protective immunity must be identified. Antibodies must be identified which confer passive immunity to this agent. An antiidiotypic antibody prepared using these protective antibodies as the immunogen, in some instances, can be used as an effective vaccine.

Antiidiotypic vaccines have effectively induced protective immunity against such viruses as rabies, coronavirus, cytomegalovirus, and hepatitis B; such bacteria as *Listeria monocytogenes, Escherichia coli,* and *Streptococcus pneumoniae*; and such parasites as *Schistosoma mansoni* infections. Antiidiotypic vaccination is especially desirable when a recombinant vaccine is not feasible. Monoclonal antiidiotypic vaccines represent a uniform and reproducible source for an immunizing preparation.

Immunological contraception is a method to prevent an undesired pregnancy. Vaccines that induce antibodies and cell-mediated immune responses against either a hormone or gamete antigen significant to reproduction have been developed. Such vaccines control fertility in experimental animals. They have undergone exhaustive safety and toxicological investigations which have shown the safety and reversibility of some of the vaccines, and with approval of regulatory agencies and ethics commissions have undergone clinical trials in humans. Six vaccines, three in women and three in men, have completed phase I clinical trials showing their safety and reversibility. One vaccine has successfully completed phase II trials in females, proving efficacy. The trials have determined the titers of antibodies and other immunological features. The fertilized egg makes the hormone hCG. Antibodies that inactivate one or more hormones involved in the production of gametes and sex steroids could be expected to impair fertility. Blocking the action of LHRH would also inhibit the synthesis of sex steroids. This might prove useful in controlling fertility of domestic animals but would not be acceptable for contraception in humans. Two vaccines against LHRH are in clinical trials in prostate carcinoma patients. An FSH vaccine would act at the level of male fertility since FSH is required for spermatogenesis in primates.

25 Therapeutic Immunology

ANTIOXIDANTS AND IMMUNITY

Immune function depends on a balance between free radical and the antioxidant status of the body. Exposure of healthy adults to high levels of oxidants leads to diminished immune responses. Exposure to low levels of dietary antioxidants also decreases immune responses such as delayed type hypersensitivity. There is increased oxidative stress and immune dysfunction in rheumatoid arthritis, aging, and cigarette smoking. This leads to damage to lipids and other cellular components by free radicals. Antioxidant status is reduced in arthritic patients and smokers compared with controls. Thus, supplemental antioxidants are useful in diminishing oxidative stress and improving immune function. Increased levels of antioxidants may be needed for elderly individuals to maintain delayed-type hypersensitivity responses.

ARGININE AND IMMUNITY

The dibasic nitrogen-rich amino acid arginine has a marked immunomodulatory function. It is critical for the maintenance of nitrogen balance and physiologic functions in humans. Supplemental administration of arginine in experimental animals has led to increased thymic size, lymphocyte count, and a lymphocyte mitogenic response to mitogens and antigens. There is enhanced IL-2 synthesis. It protects against posttraumatic thymic involution and the impairment of T cell function. It enhances delayed hypersensitivity reactions in animal studies and also promotes host antitumor responses. Arginine is essential for the function of various immunoregulatory proteins that include thymosin, thymopentin, and tuftsin. Arginine has a powerful effect on numerous cells and molecules of the immune system and may have future potential as a pharmacologic agent in the treatment of immunocompromised patients.

Serotherapy is a form of treatment for an infectious disease developed almost a century ago in which antiserum raised by immunizing horses or other animals against exotoxin, such as that produced by *Corynebacterium diphtheriae*, was administered to children with diphtheria. This antitoxin neutralized the injurious effects of the toxin. Thus, serotherapy was intended for prevention and treatment.

Immunotherapy is a treatment mechanism in which therapy is the aim at targets of the immune system that include antigen-presenting cells, activated T cells, macrophages, and B cells. The term also refers to the use of immunological reagents such as antibodies, T cells, or modifications of them in therapy. Both types have been applied to animal models of autoimmune diseases and attempts are in progress to apply immunotherapy to human autoimmune diseases. A principal goal of immunotherapy is to selectively diminish the unwanted immune response but retain protective immune mechanisms. Selective immunosuppression involves the induction of immunologic tolerance which is antigen-induced specific immunosuppression. Oral tolerance also shows promise. Trials are being conducted with peptide-induced tolerance, peptide–MHC complexes, and TCR peptides. Bystander suppression, nonspecific immune suppression, and antibody-dependent immunotherapy are other treatment modalities.

Therapeutic antisera are serum antibody preparations employed to either protect against disease or for disease therapy. They are distinct from antibody or antisera preparations used for the serological identification of microorganisms. Therapeutic antisera, such as horse antitoxin against diphtheria, were widely used earlier in the 20th century. A few specific antisera such as tetanus antitoxin are still used.

Tetanus antitoxin: Antibody raised by immunizing horses against *Clostridium tetani* exotoxin. It is a therapeutic agent to treat or prevent tetanus in individuals with contaminated lesions. Anaphylaxis or serum sickness (type III hypersensitivity) may occur in individuals receiving second injections because of sensitization to horse serum proteins following initial exposure to horse antitoxin. One solution to this has been the use of human antitetanus toxin of high titer. Treatment of the IgG fraction yields $F(ab_2)$ fragments which retain all of the toxin-neutralizing capacity but with diminished antigenicity of the antitoxin preparation.

Diphtheria antitoxin is an antibody generated by the hyperimmunization of horses against *Corynebacterium diphtheriae* exotoxin with injections of diphtheria toxoid and diphtheria toxins. When used earlier in the 20th century to treat children with diphtheria, many of the recipients developed serum sickness. It may be employed for passive immunization to treat diphtheria or for short-term protection during epidemics. Presently, pepsin digestion of the serum globulin fraction of the antitoxin yields

$F(ab')_2$ fragments of antibodies that retain their antigen-binding property but lose the highly antigenic Fc region. This process diminishes the development of serum sickness-type reactions and is called despecification.

Gas gangrene antitoxin contains antibodies found in antisera against exotoxins produced by *Clostridium perfringens*, *C. septicum*, and *C. oedematiens*, bacteria, which may cause gas gangrene. In the past, this antiserum was used together with antibiotics and surgical intervention in the treatment of wounds where gas gangrene was possible.

Intravenous immune globulin (IVIG) is an immunoglobulin preparation comprised principally of IgG derived from the blood plasma of about 1000 donors. This preparation may be effective against hepatitis A and B, cytomegalovirus, rubella, varicella-zoster, tetanus, and various other agents. IVIG is administered to children with common variable immunodeficiency, X-linked agammaglobulinemia, or other defects of the antibody limb of the immune system. Even though AIDS is a $CD4^+$ T cell defect, IVIG may help to protect against microbial infection in HIV-I-infected children. It may also be effective in autoimmune or idiopathic thrombocytopenia, autoimmune neutropenia, or Kawasaki's disease. IVIG may induce anaphylactic reactions and fever, as well as headache, muscle aches, and cardiovascular effects on blood pressure and heart rate.

IVIG is the abbreviation for intravenous immunoglobulin.

Immune serum is an antiserum containing antibodies specific for a particular antigen or immunogen. Such antibodies may confer protective immunity.

Immune serum globulin is an injectable immunoglobulin that consists mainly of IgG extracted by cold ethanol fractionation from pooled plasma from up to 1000 human donors. It is administered as a sterile $16.5 \pm 1.5\%$ solution to patients with immunodeficiencies and as a preventive against certain viral infections including measles and hepatitis A.

Human immune globulin (HIG) is a pooled globulin preparation from the plasma of donors who are negative for HIV that is used in the treatment of primary immunodeficiency, such as severe combined immunodeficiency, Bruton's disease, and combined variable immunodeficiency, and in cases of idiopathic thrombocytopenic purpura. The method of production is extraction by cold ethanol fractionation at acid pH. Viruses are inactivated, which permits the safe administration of the HIG to patients without risk of HIV, HAV, HBV, or non-A, non-B hepatitis.

HIG is the abbreviation for human immune globulin.

HGG is the abbreviation for human γ globulin.

HBIG (hepatitis B immunoglobulin) is a preparation of donor pool-derived antibodies against hepatitis B virus (HBV). It is heat treated and shown not to contain human immunodeficiency virus. HBIG is given at the time when individuals are exposed to HBV and 1 month thereafter. Low-titer (1:128) and high-titer (1:100,000) preparations are available.

Monoclonal antibody (Mab) therapy refers to treatment with monoclonal antibodies to suppress immune function, kill target cells, or treat specific inflammatory diseases. Mab demonstrate highly specific binding to precise cellular or molecular targets. Since monoclonal antibodies are derived from mouse cells, they have the potential to induce allergic reactions in recepients. Monoclonal antibodies have multiple uses in health care. For example, Edrecolomab® is used to treat solid tumors; Enlimomab® is used to treat organ transplant rejection; Infliximab® is used to treat Crohn's disease and rheumatoid arthritis; OKT3® is used to treat organ transplant rejection; Palivizumab® is used for respiratory syncytial virus; Rituxamab® is used to treat leukemias and lymphomas; Rhumabvegf® is used to treat solid tumors; and Transtuzumab® is used to treat metastatic breast cancer. (Table 25.1)

HAMA is human antimouse antibody. A group of murine monoclonal antibodies administered to selected cancer patients in research treatment protocols stimulate synthesis of antimouse antibody in the human recipient. The murine monoclonal antibodies are specific for the human tumor cell epitopes. The antimouse response affects the continued administration of the monoclonal antibody preparation. The hypersensitivity induced can be expressed as anaphylaxis, subacute allergic reactions, delayed-type hypersensitivity, rash, urticaria, flu-like symptoms, gastrointestinal disorders, dyspnea, hypotension, and renal failure.

Campath-1 (CD52) CAMPATH-1M is a rat IgM monoclonal antibody against the CD52 antigen. It is able to lyse cells using human complement, making no other manipulation necessary to deplete T cells other than to add donor serum. This has been used to deplete T cells to prevent GVHD. CD52 is a lipid-anchored glycoprotein with a very small peptide constituent. Anti-CD52 antibodies are potent lytic agents, killing cells *in vitro* with human complement as well as *in vivo*. These antibodies have been used for therapy of leukemia, bone marrow transplantation, organ transplantation, rheumatoid arthritis, vasculitis, and multiple sclerosis. CAMPATH-1H (human IgG1) was the first monoclonal antibody to be humanized. It was formed by transplanting complementarity-determining regions of campath-1G into human heavy and light chain genes. Even though initial binding affinity was decreased, this was corrected by modifying framework residues. Patients receiving this humanized antibody developed a

TABLE 25.1

Antibody Name	Target Antigen	Conditions Treated/Prevented
Abciximab (ReoPro®)	Glycoprotein II_bIII_a receptor	Complications of coronary angioplasty
ABX-CBL	CD147	GVHD
ABX-EGF	EGFr	EGF-dependent human tumor
ABX-IL8	IL-8	Rheumatoid arthritis, psoriasis
AcuTect™		Diagnoses of acute venous thrombosis
AFP-Scan™	AFP	Detection of liver and germ cell cancers
Alemtuzumab (Campath®)	CD52	B cell chronic lymphocytic leukemia, multiple sclerosis, kidney transplant rejection
Apolizumab (Remitogen™)	1D10 antigen	B cell Non-Hodgkin's lymphoma, solid tumors
Arcitumomab (CEA-Scan®) Technetium-99m-labeled	Carcinoembryonic antigen	Presence, location, and detection of recurrent and metastatic colorectal cancer
Anti-CD11a hu1 124	CD11a	Psoriasis
Basiliximab (Simulect®)	CD25 (IL-2 receptor)	Allograft rejection
Bectumomab		Non-Hodgkin's lymphoma
Bevacizumab (Avastin™)	VEGF	Metastatic renal cell carcinoma
Capromab Pendetide (Prostascint®) Indium-111 labeled	Prostate membrane specific antigen (PMSA)	Radioimmunoscintigraphy for prostate cancer
Cetuximab	EGFr	Head and neck, breast, pancreatic, colorectal cancers
CEACide™	Carcinoembryonic antigen	Colorectal cancer
Daclizumab (Zenapax®)	CD25 (IL-2 receptor)	Allograft rejection
Edrecolomab (Panorex®)	17-1A cell surface antigen	Colorectal cancer
Efalizumab (Xanelim™)	CD11a	Rheumatoid arthritis
Enlimomab	CD54 (ICAM-1)	Organ transplant rejection
Epratuzumab (LymphoCide™)	CD22	Non-Hodgkin's lymphoma
Gemtuzumab ozogamicin Mylotarg®	CD33 calicheamicin	Acute myeloid leukemia
Hu23F2G (LeukArrest™)	CD11/18 (leukointegrin)	Ischemic stroke
Hu1124	CD11a	Psoriasis
Ibritumomab tiuxetan (Zevalin™)	CD20	B cell non-Hodgkin's lymphoma
Igovomab (Indimacis 125®)	Tumor-associated antigen CA125	Detection of ovarian adenocarcinoma
Imciromab pentetate (Myoscint®)	Human cardiac myosin	Myocardial infarction imaging
IMC-C225 (ERBITUX™)	EGFR	EGF-dependent human tumor
Infliximab (Remicade®)	TNF-α	Crohn's disease, rheumatoid arthritis
Inolimomab	IL-2 receptor	Organ transplant rejection
LDP-01	β2 integrin	Stroke, kidney transplant rejection
LDP-02	α4β7 integrin receptor	Crohn's disease, ulcerative colitis
LeuTech® 99cTc-Anti-CD15 Anti-Granulocyte Antibody	CD15	Imaging infection sites
Lerdelimumab	TGFb2	Glaucoma, cataract
Lym-1 Yttrium-90 labeled	HLA-DR	Non-Hodgkin's lymphoma
LymphoScan®	CD22	Detection of B cell non-Hodgkin's lymphoma
MAK-195F	TNF-α	Hyperinflammatory response in sepsis syndrome
MDX-33	CD64	Idiopathic thrombocytopenia purpura
MDX-H210	Bispecific HER2 × CD64	Breast, colorectal, kidney, ovarian, prostate cancers
MDX-447	Bispecific EGFR × CD64	Head, neck, renal cancers
Mitumomab (BEC2)	GD3-idiotypic	Small cell lung cancer, melanoma
Muromonab (Orthoclone OKT®3)	CD3	Allograft rejection
Natalizumab (Antegren®)	α-4 integrin (VLA-4)	Multiple sclerosis, Crohn's disease
Nebacumab (Centoxin®)	Bacterial endotoxins	Gram-negative bacteria sepsis
Nofetumomab (Verluma®)	Carcinoma-associated antigen	Detection of small-cell lung cancer
OctreoScan® Indium-111 labeled	Somatostatin receptor	Immunoscintigraphic localization of primary and metastatic neuroendocrine tumors that contain somatostatin receptors

(Continued)

TABLE 25.1 (*Continued*)

Antibody Name	Target Antigen	Conditions Treated/Prevented
Olizumab rhuMAb-E25	Ig-E	Allergic asthma, allergic rhinitis
Oncolym® (131Lym-1) iodine131 labeled	HLA-DA	B cell non-Hodgkin's lymphoma
Omalizumab (Xolair™)	IgE	Allergic asthma, allergic rhinitis
Oregovomab (OvaRex ®)	Tumor-associated antigen CA125	Ovarian cancer
ORTHOCLONE OKT4A	CD4	CD4-mediated autoimmune diseases, allograft rejection
Palivizumab (Synagis®)	Antigenic site of the F protein of respiratory syncytial virus (F gp)	Respiratory syncytial virus infection
Pexelizumab (5G1.1-SC)	Complement C5	AMI, UA, CPB, PTCA
Priliximab	CD4	Crohn's disease, multiple sclerosis
Regavirumab	Cytomegalovirus (CMV)	Acute CMV disease
Rituximab (Rituxan®)	CD20	Non-Hodgkin's lymphoma
Satumomab pendetide (OncoScint® CR/OV)	Tumor-associated glycoprotein-72	Detection of colorectal and ovarian cancers
Sevirumab (Protovir®)	Cytomegalovirus (CMV)	Prevention of CMV infection in bone marrow transplant patients
Siplizumab (MEDI-507)	CD2	Acute GVHD, psoriatic arthritis
Smart™ M195	CD33	Acute myeloid leukemia, myelodysplastic syndrome
Sulesomab (LeukoScan®) Technicium-99m labeled	Surface granulocyte nonspecific cross-reacting antigen	Detection of osteomyelitis, acute atypical appendicitis
Tecnemab K1	High molecular weight melanoma-associate antigen	Diagnosis of cutaneous melanoma lesions
Tositumomab (Bexxar®) Iodine 131 attached	B lymphocyte surface protein	Non-Hodgkin's lymphoma
Trastuzumab (Herceptin®)	Her2/neu	Her2 positive metastatic breast cancer
Visilizumab (Nuvion™, Smart™ anti-CD3)	CD3	GVHD, ulcerative colitis
Vitaxin™	$\alpha v\beta 3$ integrin	Solid tumors
Votumumab (Humaspect®)	Cytokeratin tumor-associated antigen	Detection of carcinoma of colon and rectum
YM-337	GPIIb/IIIa	Prevention of platelet aggregation
Zolimomab	CD5, ricin A-chain toxin	GVHD

Notes: GVHD: graft-vs.-host disease; EGFr: epidermal growth factor receptor; IL-2: interleukin-2; IL-8: interleukin-8; AFP: α fetoprotein; VEGF: vascular endothelial growth factor; ICAM-1: intercellular adhesion molecule-1; TNF-α: tumor necrosis factor-α; TGFb2-transforming growth factor b2; HLA: human leukocyte antigen; VLA-4: very late antigen-4; AMI: acute myocardial infarction; UA: unstable angina; CPB: cardiopulmonary bypass; PTCA: percutaneous transluminal coronary angioplasty; GPIIb/IIIa: glycoprotein II_bIII_a.

very low antiglobulin response compared with the original rat antibody. The CD52 antigen is a glycoprotein with only 12 amino acids. It is a complex carbohydrate consisting of sialylated polylactosamine units with fucosylated mannose core. It is attached to Asn3 at the C-terminus as a glycosylphosphatidylinositol (GPI) anchor. The CAMPATH-1 epitope is comprised of the C-terminal amino acids and part of the GPI anchor, which means that the antibodies bind near the cell membrane which facilitates cell lysis. The antigen is expressed abundantly on all lymphocytes except for plasma cells, and also on monocytes, macrophages, and eosinophils. It is not found on any other tissues except the male reproductive tract where it is expressed strongly on epithelial cells lining the epididymis, vas deferens, and seminal vesicle. The CD52 antigen is a principal membrane protein of sperm.

Trastuzumab is a recombinant DNA-derived humanized monoclonal antibody that selectively binds with high affinity in a cell-based assay to the extracellular domain of the human epidermal growth factor receptor 2 protein, HER2. The antibody is an IgG_1 kappa that contains human framework regions with the complementarity-determining regions of a murine antibody (4D5) that binds to HER2. Commercially distributed as Herceptin®.

Herceptin®: See Trastuzumab.

REMICADE® (infliximab) is a chimeric $IgGl_k$ monoclonal antibody with an approximate mol wt of 149,100 Da. It is composed of human constant and murine variable regions. I nfliximab binds specifically to human tumor necrosis factor alpha (TNFα) with an association constant of 10^{10} M^{-1}. Infliximab is produced by a recombinant cell line cultured by continuous perfusion and is purified by a series of steps that includes measures to inactivate and remove viruses.

ReoPro® (abciximab) is the Fab fragment of the chimeric human–murine monoclonal antibody 7E3. Abciximab binds to the glycoprotein (GP) IIb/IIIa ($\alpha_{IIb}\beta_3$) receptor of human platelets and inhibits platelet aggregation. Abciximab also binds to the vitronectin ($\alpha_v\beta_3$) receptor found on platelets and vessel wall endothelial and smooth muscle cells. The chimeric 7E3 antibody is produced by continuous perfusion in mammalian cell culture. The 47,615-Da Fab fragment is purified from cell culture supernatant by a series of steps involving specific viral inactivation and removal procedures, digestion with papain, and column chromatography.

E5 is a murine monoclonal IgM antibody to endotoxin that has proven safe and capable of diminishing mortality and helping to reverse organ failure in patients with Gram-negative sepsis who are not in shock during therapy. E5 may recognize and combine with lipid A epitope.

HA-1A is a human monoclonal IgM antibody specific for the lipid A domain of endotoxin. It can prevent death of laboratory animals with Gram-negative bacteremia and endotoxemia. In clinical trials, HA-1A has proven safe and effective for the treatment of patients with sepsis and Gram-negative bacteremia, whether or not they are in shock, but not in those with focal Gram-negative infection.

Rituxan (rituximab) is an anticancer monoclonal antibody that serves as a type of biotherapy by binding to tumor cells and triggering the immune system to kill target tumor cells rather than using toxic chemicals to accomplish this result. The antibody has been approved by the FDA for use in patients with non-Hodgkin's lymphoma.

Simulect© (basiliximab) is a chimeric (murine/human) monoclonal antibody (IgG_{1k}), produced by recombinant DNA technology, that functions as an immunosuppressive agent, specifically binding to and blocking the interleukin-2 receptor α-chain (IL-2Rα, also known as CD25 antigen) on the surface of activated T-lymphocytes. Based on the amino acid sequence, the calculated mol wt of the protein is 144 kDa. It is a glycoprotein obtained from fermentation of an established mouse myeloma cell line genetically engineered to express plasmids containing the human heavy and light chain constant region genes and mouse heavy and light chain variable region genes encoding the RFT5 antibody that binds selectively to the 11,2Rα. The active ingredient, basiliximab, is water soluble. The drug product, Simulect, is a sterile lyophilisate which is available in 6-ml colorless glass vials. Each vial contains 20 mg basiliximab, 7.21 mg monobasic potassium phosphate, 0.99 mg disodium hydrogen phosphate (anhydrous), 1.61 mg sodium chloride, 20 mg sucrose, 80 mg mannitol, and 40 mg glycine, to be reconstituted in 5 ml of Sterile Water for Injection, USP. No preservatives are added.

Tumor immunotherapy: Immunological surveillance, first proposed by Paul Ehrlich 100 years ago, is believed to be a protective mechanism against the development of neoplasia in intact healthy subjects. Significant suppression of the immune system from any cause may favor tumor development. Contemporary attempts at therapeutic modification of the immune response include the use of immune response modifiers, monoclonal antibodies, cancer vaccines, and gene therapy. Genetically engineered cytokines such as IL-2 and interferon-α have been used to activate the immune system. Early hopes for success have now abated. Interferons have some proven efficacy in the management of melanoma, renal cell carcinoma, and hairy cell leukemia, and as adjuvant therapy in certain hematological malignancies such as low-grade lymphoma and myeloma. Although IL-2 has been used in the past to treat renal cell carcinoma and melanoma with the generation of tumor-infiltrating lymphocytes and lymphokine-activated killer cells, it has toxic physiologic effects. Monoclonal antibodies as anticancer agents are promising but have problems of delivery to targets, reactions to murine antibodies, and difficulties in linking therapeutic warheads to these "smart bombs." Bispecific antibodies have been used to crosslink targets to immune effector cells to activate a cell-mediated antitumor response. Current cancer vaccines include autologous cell lines, allogeneic cell lines, genetically modified tumor cells, glycoproteins, stripped glycoproteins, peptides, antitumor idiotypes, and polynucleotides encoding tumor antigens. Gene therapy may be used to modify tumor cells to express costimulatory molecules such as B7-1 to help initiate a cell-mediated response against the neoplasm.

Altered peptide ligands (APL) are peptides that may closely resemble an agonist peptide in amino acid sequence. They induce only a partial response from T lymphocytes specific for the agonist peptide. Altered peptide ligands are analogs of immunogenic peptides in which the T cell receptor contact sites have been altered, usually by substitution with another amino acid. Even though these peptides fail to stimulate T cell proliferation, they active some T cell receptor-mediated functions. Antagonist peptides specifically downmodulate the agonist-induced response. APL can act therapeutically by modulating the cytokine pattern of T cells or they may induce a form of anergy in T cells. They stimulate a partial response from T lymphocytes specific for the agonist peptide.

Autolymphocyte therapy (ALT) is an unconfirmed immunotherapeutic treatment for metastatic renal carcinoma. Leukocytes from the patient are isolated and activated with monoclonal antibodies to induce the leukocytes to synthesize and secrete cytokines. Cytokines produced in the supernatant are combined with a sample of the patient's own lymphocytes and reinjected. Preliminary reports claim success, but these are not confirmed.

Dendritic cell immunotherapy: The lack of detectable tumor-specific immune responses in humans led to the use of autologous dendritic cells in active immunotherapy. Dendritic cells are expanded from progenitors *ex vivo*, charged with tumor antigens, and reinfused. Methods are being sought to genetically modify dendritic cells with antigens encoded by viral and nonviral vectors. Dendritic cells might also be used to vaccinate humans at the time a primary tumor is resected. This would permit an immune response to be available to act against metastases not detectable at the time the primary tumor is identified.

Despecification is a method to reduce the antigenicity of therapeutic antisera prepared in one species and used in another. To render immunoglobulin molecules less immunogenic in the heterologous recipient, they may be treated with pepsin to remove the molecules' most immunogenic portion, i.e., the Fc fragment, but leave intact the antigen-bonding regions, i.e., $F(ab')_2$ fragments which retain their antitoxic properties. Had such a treatment been available earlier in the 20th century, serum sickness induced by horse antitoxin administered to human patients as a treatment for diphtheria would have been greatly reduced.

Biologicals are substances used for therapy that include antitoxins, vaccines, products prepared from pooled blood plasma, and biological response modifiers (BRM). BRMs are prepared by recombinant DNA technology and include lymphokines such as interferons, interleukins, tumor necrosis factor, etc. Monoclonal antibodies for therapeutic purposes also belong in this category. Biologicals have always presented problems related to chemical and physical standardization to which drugs are subjected. They are regulated by the Food and Drug Administration (FDA) within the U.S.

Immunomodulation describes therapeutic alteration of the immune system by the administration of biological response modifiers such as lymphokines or antibodies against cell surface markers bound to a toxin such as ricin.

Biological response modifiers (BRM) include a broad spectrum of molecules, such as cytokines, that alter the immune response. They include substances such as interleukins, interferons, hematopoietic colony-stimulating factors, tumor necrosis factor, B lymphocyte growth and differentiating factors, lymphotoxins, and macrophage-activating and chemotactic factors, as well as macrophage inhibitory factor, eosinophils chemotactic factor, osteoclast activating factor, etc. BRM may modulate the immune system of the host to augment antirecombinant DNA technology and are available commercially. An example is α interferon used in the therapy of hairy cell leukemia.

Inosiplex (Isoprinosine) is an immunomodulating agent that increases natural kill (NK) cell cytotoxicity as well as T cell and monocyte functional activites. It has produced slight benefit in AIDS patient and has been used in Europe to treat diverse immunodeficiency diseases but is not yet approved for use in the U.S.

Roquinimex is an immunomodulating drug capable of augmenting NK cell numbers and reactivity and stimulating various T and B cell functions. It has been used to stimulate immune function in post-bone marrow transplant patients. By contrast, it prevents relapses of chronic relapsing allergic encephalomyelitis in animal models of that disease. It also reduces relapses, disease activity, and disease progression in multiple sclerosis patients. If confirmed in further studies, this immunomodulator will have a significant future role in clinical medicine.

B lymphocyte Stimulator (BlyS) is a naturally occurring protein that stimulates the immune system to produce antibodies. It is being tested as a potential drug for the treatment of immunodeficiency that occurs in higher than normal levels in rheumatoid arthritis (RA) and lupus (SLE) patients and could account for their immune systems becoming overeactive and attacking joints in RA or connective tissues in SLE.

Immunoaugmentive therapy (IAT) describes an approach to tumor therapy practiced by selected individuals in Mexico, Germany, and the Bahama Islands that is not based on scientific fact and is of unproven efficacy or safety. Concoctions of tumor cell lysate and blood serum from tumor-bearing patients as well as from normal individuals is injected into cancer patients ostensively for the two to provide tumor antibody and "blocking" and "deblocking" proteins. The method is not based upon any scientifically approved protocol and is of very doubtful value. The treatment preparations are often contaminated with microorganisms.

An **immunoprotein** is an immunologically active protein such as one that serves as a target for immunological probes or therapy.

Immunoablation is the deliberate destruction of a patient's immune competence to condition the patient for organ transplantation or to treat refractory autoimmune diseases.

Bombesin is a neuropeptide of 14 residues that is analogous to a gastrin-releasing peptide that is synthesized in the gastrointestinal tract and induces GI smooth muscle contraction and the release of stomach acid and the majority of GI hormones with the exception of secretin. Bombesin injection into the brain may induce hyperglucagonemia, hyperglycemia, analgesia, and hypothermia. Bombesin facilitates bronchial epithelial cell proliferation, and pancreatic and small-cell carcinoma. Antibombesin antibodies might prove useful in the future for the treatment of small-cell lung carcinoma whose cells bear bombesin receptors.

Botulinum toxin is formed by *Clostridium botulinum*. The 150-kDa type A toxin is available in purified form and is employed to treat a neuromuscular junction disease such as dystonias. It acts by combining with the presynaptic cholinergic nerve terminals where it is internalized and prevents exocytosis of acetylcholine. Subsequently, sprouting takes place, and new terminals are formed which reinnervate the muscle.

Avionics®: Interferon-β-1a preparation approved for the treatment of multiple sclerosis.

Cimetidine is an H_2-receptor antagonist that is capable of over 90% reduction in food-stimulated and nocturnal secretion of gastric acid after a single dose. This histamine H_2-receptor antagonist is of interest to immunologists because of its efficacy in treating common variable immunodeficiency, possibly through suppressor T lymphocyte inhibition. It also has an immunomodulating effect by reducing the activity of suppressor lymphocytes. Other drug immunomodulators include low-dose cyclophosphamide given prior to immunization with a tumor vaccine, which can augment the immune response in both animals and humans. Indomethacin can nullify the effects of suppressor macrophages.

Levamisole is an antihelminthic drug used extensively in domestic animals and birds that was found to also produce immunostimulant effects. The drug may potentiate or restore the function of T lymphocytes and other leukocytes. This agent increases the magnitude to the delayed-type hypersensitivity or T cell-medicated immunity in humans. Although it has some efficacy in the treatment of RA, Levamisole induces severe agranular cytosis, which requires discontinuation of use. It also potentiates the action of fluorouracil in adjuvant therapy of colorectal cancer.

Hydroxychloroquine (2-[[4-[7-chloro-4-quinolyl]amino] ethyl-amino]ethanol sulfate). An antimalarial drug that has also been used in therapy of the connective tissue diseases, such as SLE and RA.

Chlorambucil (4-[bis(2-chloroethyl)amino-phenylbutyric acid) is an alkylating and cytotoxic drug. Chlorambucil is not as toxic as is cyclophosphamide and has served as an effective therapy for selected immunological diseases such as rheumatoid arthritis, SLE, Wegener's granulomatosis, essential cryoglobulinemia, and cold agglutinin hemolytic anemia. Although it produces bone marrow suppression, it has not produced hemorrhagic cystitis and is less irritating to the GI tract than is cyclophosphamide. Chlorambucil increases the likelihood of opportunistic infections and the incidence of some tumors.

Aminophylline: See theophylline.

An **antihistamine** is a substance that links to histamine receptors, thereby inhibiting histamine action. Antihistamine drugs derived from ethylamine block H1 histamine receptors, whereas those derived from thiourea block the H2 variety.

Naproxen (2-naphthaleneacetic acid, 6-methoxy-α-methyl) is an antiinflammatory drug used in the treatment of arthritis, especially RA of children and adults, as well as ankylosing spondylitis.

Naprosyn is a nonsteroidal antiinflammatory drug (NSAID).

Nonsteroidal antiinflammatory drugs (NSAIDs) are a category of drugs used in the treatment of rheumatoid arthritis (RA), gouty arthritis, ankylosing spondylitis, and osteoarthritis. The drugs are weak organic acids. They block prostaglandin synthesis by inhibiting cyclooxygenase and lipooxygenase. They also interrupt membrane-bound reactions such as NADPH oxidase in neutrophils, monocyte phospholipase C, and processes regulated by G protein. They also exert a number of other possible activities such as diminished generation of free radicals and superoxides, which may alter intracellular cAMP levels, diminishing vasoactive mediator release from granulocytes, basophils, and mast cells. Salicylates and other similar drugs are used to treat rheumatic disease through their capacity to suppress the signs and symptoms of inflammation. These drugs also exert antipyretic and analgesic effects. Their antiinflammatory properties make then most useful in the management of disorders in which pain is related to the intensity of the inflammatory process. NSAIDs used for special indications include indomethacin and ketorolac. Gastric irritation caused by some of the original NSAIDs led to the introduction of newer NSAIDs that include phenylbutazone and a host of other compounds for use and treatment of such conditions as RA or osteoarthritis.

NSAIDs, including aspirin, are employed mainly for their antiinflammatory, analgesic, and antipyretic effects. NSAIDs are widely used but are not always completely safe considering their gastrointestinal tract and kidney toxicity. All of these drugs are able to inhibit the enzyme cyclooxygenase that is the principal means by which they diminish pain, inflammation, and fever. Cyclooxygenase catalyzes arachidonic acid conversion to prostaglandins (GH).

Propylthiouracil is an antithyroid drug that blocks TPO activity.

Phenylbutazone (4-butyl-1,2-diphenyl-3,5-pyrazolidene-dione) is a drug that prevents synthesis of prostaglandin and serves as a powerful antiinflammatory drug. It is used in therapy of RA and ankylosing spondylitis.

Quinidine (β-quinine; 6′-methoxycinchonan-9-ol) is a stereoisomer of quinine recognized for its cardiac anti-arrhythmic effects.

Tolmetin (1-methyl-5-*p*-toluoylpyrrole-2-acetic acid) is an antiinflammatory agent effective in the therapy of arthritis, including juvenile RA, RA, and ankylosing spondylitis.

Vinblastine is a chemotherapeutic alkaloid that leads to lysis of rapidly dividing cells by disruption of the mitotic spindle microtubules. It is used for therapy of leukemia, Hodgkin's disease, lymphoproliferative disorders, and malignancies.

Vincristine is a chemotherapeutic alkaloid that lyses proliferating cells through disruption of the mitotic spindle. It is used in the therapy of leukemia and other malignancies.

Intolerance refers to adverse reactivity following the administration of normal doses of a drug.

L-phenlyalanine mustard is a nitrogen mustard that is employed for therapy of multiple myeloma patients.

26 Comparative Immunology

Plant immunity: Immunity in higher plants consists of an immune state of **systemic acquired resistance (SAR)** that follows a local infection by pathogenic microorganisms that leads to lesions with death of host cells. SAR encompasses a broad spectrum of bacterial, viral, and fungal agents, as well as the infective pathogen. Various SAR genes encode numerous microbicidal proteins induced by endogenous chemicals that include salicylic acid, which combines with a catalase to increase H_2O_2, which may facilitate defense. (Figure 26.1)

Phytoimmunity encompasses both active and passive immune-like phenomena in plants. Plant substances active in phytoimmunity include phytonicides and phytoalexins. Plant resistance to many diseases is associated with the presence of antibiotic substances in plant tissues. Antibiotic substances are inherent in both susceptible and resistant varieties of plant species. They may be constitutional inhibitors present in a plant before contact with a parasite or induced antibiotic substances that arise in plants after contact with a parasite. Defense reactions in plants that are associated with the formation and conversion of antibiotic substances include reactions to wounding and the necrotic reaction. Plant resistance to a specific disease is determined by the various antibiotic substances they contain and the synergistic action of these agents with differing roles in phytoimmunity. Plant varieties differ in the quantity of antibiotic substances present in intact tissues and the intensity of their generation in response to infection. They also differ in the nature of subsequent conversions that may produce a marked increase in antibiotic activity.

Phytonicides are substances produced in both traumatized and nontraumatized plants that represent one of the factors active in plant immunity. Phytonicides have bactericidal, fungicidal, and protisticidal properties.

Phytoalexins are plant substances that are active in phytoimmunity.

Phytomitogens are plant glycoproteins that activate lymphocytes through stimulation of DNA synthesis and induction of blast transformation.

Phytohemagglutinin (PHA) is a lectin or carbohydrate-binding protein synthesized by plants. It crosslinks human T lymphocyte surface molecules such as the T cell receptor, thereby causing polyclonal activation and T cell aggregation or agglutination. PHA is used often in T lymphocyte functional studies to investigate T cell activation. Clinically, it is employed in the laboratory to determine whether or not a patient's T lymphocytes are functional or it may be used to induce T lymphocyte mitosis to gather karyotypic data. An extract of the red kidney bean (*Phaseolus vulgaris*) that contains PHA, a powerful cell-agglutinating and mitogenic principle. PHA was the first polyclonal mitogen to be described. It permits lymphocytes to be activated in *in vitro* cultures and to investigate lymphocyte growth in lymphokine synthesis. PHA consists of five tetrameric glycoproteins, each of which contains two different subunits designated ENL and have molecular weights of 33 kDa and 31.6 kDa, respectively. PHA stimulates the proliferation of peripheral blood mononuclear cells, notably T cells, in the presence of monocytes acting as accessory cells. These cells do not respond to PHA. PHA binds to glycoproteins on the T cell surface.

Phagocytosis is significant in all animal species. It may be facilitated by agglutinins and bactericidins that bind to pathogen-associated molecular patterns (PAMPs) on the microorganism's surface to render a basis for nonself recognition.

Protozoa are microscopic animals that must recognize and eat food. Intricate genetic mechanisms control their surface proteins.

Bacteria: Bacterial cells may become infected by viruses known as bacteriophages. Restriction endonucleases are believed to be responsible for the indentification and breakdown of viral DNA without injury to the host bacterium's DNA. Bacteriophages that develop resistance to this enzymatic action infect bacterial cells successfully. (Figure 26.2)

Immunity in prokaryotes: Numerous antimicrobial agents synthesized and released extracellularly by bacteria have a specific effect on the bacteria. They include (1) enzymatically synthesized antibiotics, (2) posttranslationally modified peptide antibiotics, (3) protein antibiotics as bacteriocins, protein exotoxins, and bacteriolytic enzymes, and (4) intemperate or temperate bacterial viruses. Once these agents are inside the bacterial cell they must remain unchanged to exert their actions. Bacteriocins released by bacteria may inhibit the growth of other sensitive, closely

FIGURE 26.1 Plant immunity.

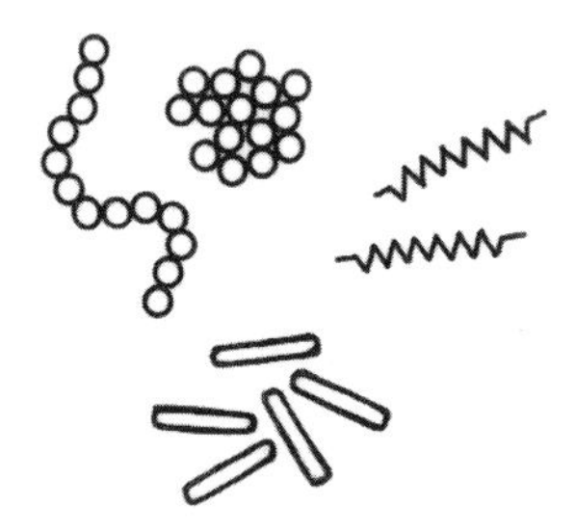

FIGURE 26.2 Bacteria.

related bacteria. Yet the strain producing the bacteriocin is usually able to resist its effect through specific immunity peptides or proteins.

Antibiotic synthesizing strains protect themselves from their own products by forming immunity proteins. Immunity to bacteriophage is the failure of an infected bacterial cell to be reinfected by a phage of the same type. Colicins and bacteriophages together with many other substances facilitate bacterial strain competitiveness. Immunity mechanisms are very specific and depend on protein–protein, protein–DNA, or RNA–DNA interactions. Numerous mechanisms confer immunity to prokaryotic organisms against specific bactericidal agents. "Immunity" also refers to replicons carrying a transposon: a duplicate copy of a transposon fails to insert into a replicon already carrying a copy of the same transposon.

Invertebrate immunity: Invertebrates have various mechanisms to recognize and respond to nonself substances even though they lack a lymphoid immune system. They possess both cellular and humoral components that mediate an immune-like response. Invertebrate internal defense responses include phagocytosis, encapsulation, and nodule formation. The molecular recognition and effector mechanisms involved from one species to another are diverse. Some of the factors involved include α_2 microglobulins, c-reactive proteins, antibacterial peptides, serine proteinases and proteinase inhibitors, C-type lectins and complement-related factors, glucan binding proteins, and some antibacterial peptides. Alloaggressive responses have been observed in noncolonial invertebrates. Invertebrates do form certain members of the immunoglobulin superfamily such as adhesion molecules and receptors for tyrosine kinases. Within the immunoglobulin superfamily, only humulin, a protein isolated from lepidopterans, is induced following bacterial challenge. Humulin has four immunoglobulin-like domains whose primary structure will closely resemble cell adhesion molecules than immunoglobulins. With respect to invertebrate effector molecules, antibacterial peptide are classified into several families that include lysozymes, cecropins, capecins or insect defensins, attacin-like, and proline-rich antibacterial peptides. Invertebrates have cascades of endogenous serine proteinases that are important in defense. Cytokine-like activity possibly mediated by factors equivalent to IL-1 and IL-6 have been recognized. Receptors have also been identified on phagocytic cells. Thus, invertebrate cytokine-like factors may play a role in the regulation of nonspecific responses to tissue injury and infection. Opioid peptides, opiate alkaloids, and other neuropeptides may modulate chemotaxis and cell adhesion. (Figure 26.3)

Sponges: Even the most primitive invertebrate animals, i.e., marine sponges, can discriminate self from nonself, leading to rejection of parabiosed fingers of different colonies within 7 to 9 d. Species-specific glycoproteins of sponge cells are used for identification of self and inhibition of hybrid colony formation. Placed in apposition to one another, nonidentical sponge colonies become necrotic at the interface. Second grafts undergo accelerated rejection.

Corals: Syngrafts, i.e., genetically identical grafts, are successful in corals. By contrast, allografts of genetically nonidentical ones are rejected slowly with injury to both donor and recipient. A vestige of adaptive immunity is revealed by limited evidence of immunological memory of a prior rejection episode.

Worms: Four types of cells are present in the earthworm coelom, all of which are phagocytic. Some cells participate in allograft rejection, whereas others synthesize antibacterial substances. (Figure 26.4)

A **coelomocyte** is a circulating or fixed-ameboid phagocytic leukocyte that participates in the defense of invertebrate animals that have a coelom by phagocytosis and encapsulation.

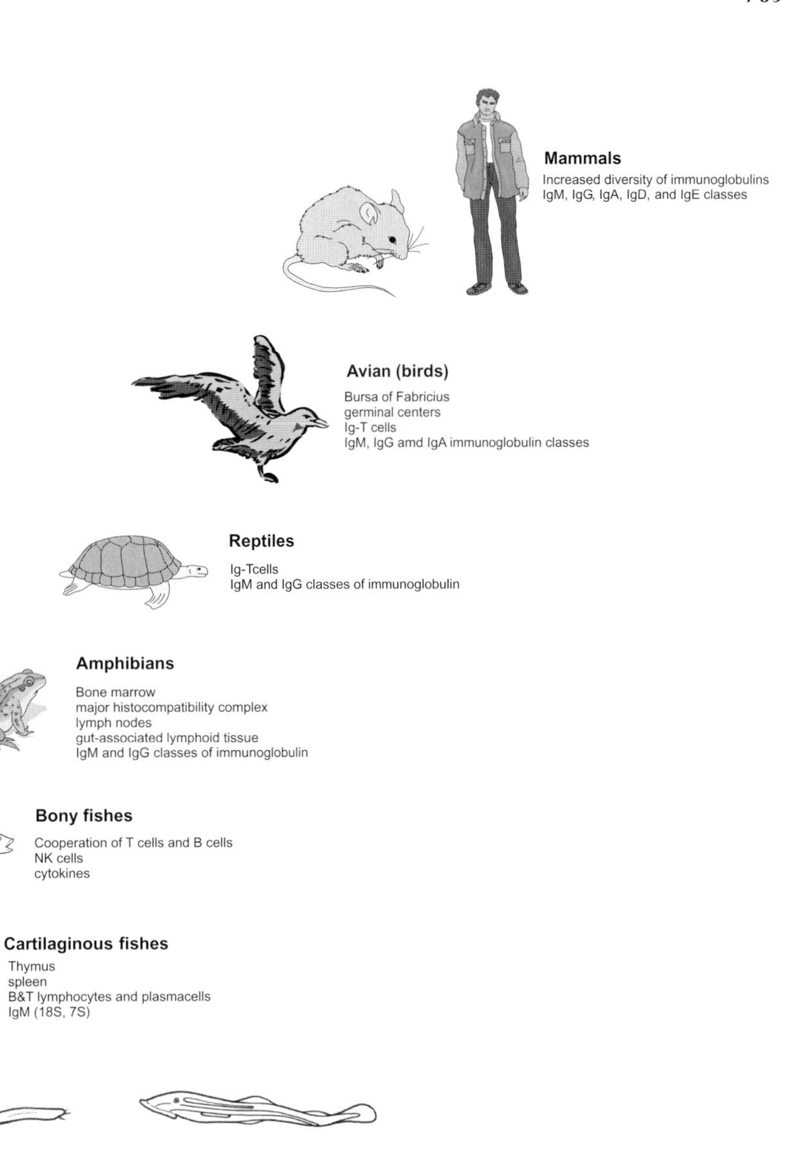

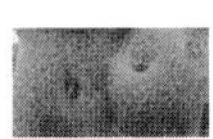
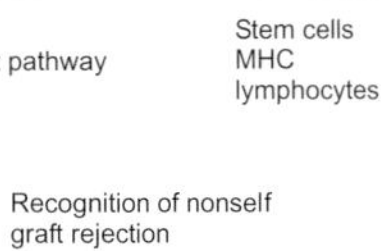

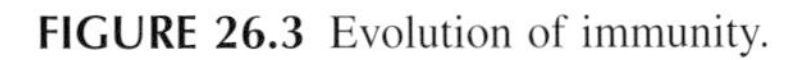

FIGURE 26.3 Evolution of immunity.

FIGURE 26.4 Earthworm.

FIGURE 26.5 Molluscs and arthropods.

Molluscs and Arthropods: Interestingly, molluscs and arthropods fail to manifest graft rejection even though they do possess significant humoral factors. These include what may be the most primitive alternative complement pathway components that might provide resistance to selected parasites. (Figure 26.5)

Insects: Toll receptors in insects induce the formation of antimicrobial proteins in response to molecular motifs present on the insect pathogen surfaces and fungal polysaccharides. Infection of higher insects leads to the rapid production of antimicrobial peptides after activation of transcription factors that link to promoter sequence configurations homologous to regulatory features of the acute phase response in animals. *Drosophila* have toll molecules that serve as a receptor for PAMPs that activate NFκB. A loss of function toll mutation renders these flies

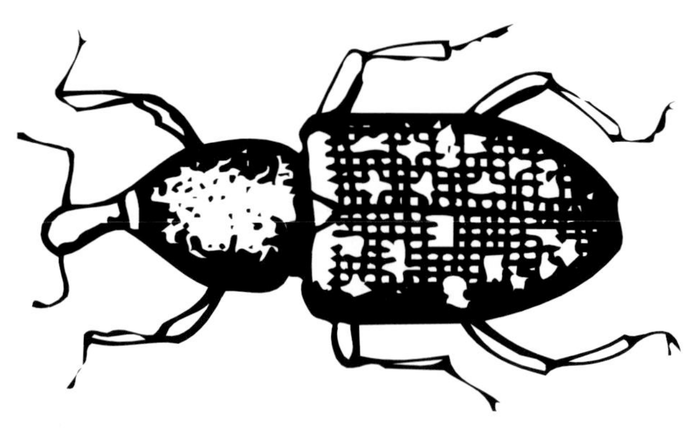

FIGURE 26.6 Insects.

susceptible to infection by fungi. Antimicrobial peptides synthesized by insects include cyclic peptides, including anti-Gram-positive defensins (4 kDa) and drosomycin, a 5-kDa antifungal agent. Infection-induced linear peptides include cecropins, that induce lethal injury of microbial membranes by producing ion channels. Other linear peptides include anti-Gram-negative glycine or proline-rich polypeptides. Integrins in insects may be related to complement receptor CR3. (Figure 26.6)

Iridovirus immunity: Insect immunity against this virus consists mainly of a cellular response including phagocytosis, encapsulation, nodule formation, or coagulation to recognize foreignness. Attacin, cecropins, lysozyme, and phenol oxidase occur in various dipteran and lepidopteran species. These inducible agents may play a toxic role in the defense mechanism. Only nonspecific immunity not associated with antigen antibody reactions or complement or interferon occurs.

Encapsulation is the reaction of leukocytes to foreign material that cannot be phagocytized because of its large size. Multiple layers of flattened leukocytes form a wall surrounding the foreign body and isolate it within the tissues. This type of reaction occurs in invertebrates including annelids, mollusks, and arthropods. In vertebrates, macrophages surround the foreign body, a granuloma is formed, and fibroblasts subsequently appear. A fibrous capsule is formed.

Fertilizin is a glycoprotein present as a jelly-like substance surrounding sea urchin eggs. It behaves as an antigen-like substance from the standpoint of valence. Sperm, which contain proteinaceous antifertilizin, are agglutinated into clumps when combined with soluble fertilizin, which is dissolved from eggs by acidified sea water.

Echinoderms: In the starfish, made famous by Metchinkoff's studies of phagocytosis, allograft rejection occurs and is marked by cellular infiltration. There is a significant specific memory response. Cytokine-like molecules similar to interleukin-1 (IL-1) and tumor necrosis

FIGURE 26.7 Starfish.

factor (TNF) have been recognized in echinoderms as well as other intervertebrates. (Figure 26.7)

Tunicates: Tunicates, including the seq-squirt *Amphioxus*, manifest hemopoietic cells that are self-renewing, lymphoid-type cells and a single MHC that governs rejection of foreign grafts.

A primitive complement system is present in invertebrates, including a protease inhibitor, an α_2-macroglobulin with a structure homologous to C3 with internal thioester, in the horseshoe crab. This could be a primordial form of C3, activated by infection-released proteases, that attaches to the microbial surface where it serves as a ligand for phagocytes. Limulin, produced also by the horseshoe crab, is homologous to acute phase C-reactive protein and is believed to be an evolutionary precursor of Clq, mannose-binding protein and lung surfactant.

Invertebrates are also able to wall off invading microorganisms by proteolytic processes that surround the microbes with a gelled hemolymph coagulum.

Invertebrate cytokine-like molecules may regulate host defense mechanisms in a manner resembling cytokine networks in vertebrates. Cytokine-related molecules in protozoans include the pheromone Er-1 which is structurally and functionally similar to interleukin-2 (IL-2). IL-1α, IL-β-, and tumor necrosis factor (TNF)-like molecules are present in annelids, echinoderms, and tunicates. Invertebrate IL-1 induces proliferation, aggregation, and phagocytic activity by these lower animals' blood cells. Additional invertebrate cytokine-like molecules in insects include a plasmatocyte

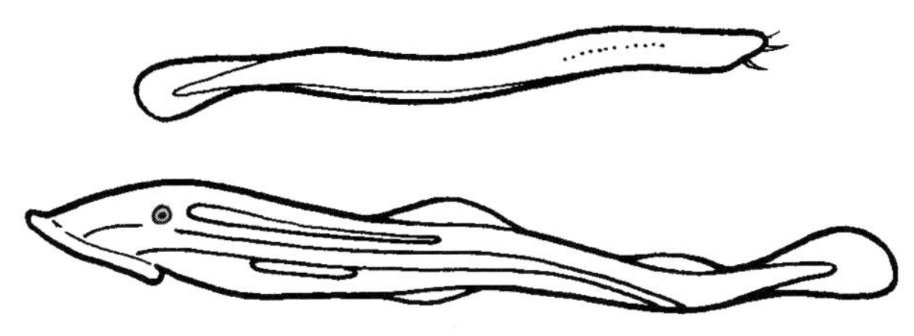

FIGURE 26.8 California hagfish and lamprey.

(leukocyte-type) depletion factor, a leukocyte activator (i.e., hemokinin) and several encapsulation and phagocytosis stimulants.

Echinoderm leukocytes produce a factor that induces mitogenesis of mammalian lymphocytes and starfish white blood cell accumulation. A tunicate inflammatory cytokine affects vertebrate antibody synthesis, phagocytosis, and cell-mediated cytotoxicity and phagocytosis in prawns.

Jawless fishes (cyclostomes, e.g., hagfish and lampreys): The lowest vertebrate investigated, the California hagfish, does not manifest true lymphocytes and adapative T and B cell responses. This species has no thymus, an erythropoietic spleen, or lymphocyte-like cells in the circulation, and gamma M macroglobulin. The hagfish responds to hemocyanin if its body temperature is maintained at 20°C, but true immunoglobulins are not synthesized. (Figure 26.8)

The lamprey has a primitive thymus, lymphopoietic spleen, family of circulating lymphocytes, and gamma globulins.

In the cyclostomes, the most primitive surviving vertebrates, aggregates of lymphoid cells are present in the pharynx and other locations. They develop the first true antibody immunoglobulin with a four-chain unit structure synthesized following challenge by different antigens, a pivotal event in the evolution of immunity. Interestingly, other immunoglobulin superfamily molecules, such as adhesion molecules, had already appeared in invertebrates (e.g., arthropods).

The **T cell system**, which mediates self–nonself recognition, appeared earlier in evolution than did antibody, which is essentially confined to vertebrates. Vertebrate T and B lymphocyte responses are clearly defined and then separate developmental sites have been identified. The appearance of a true thymus in the bony fishes (teleosts), amphibians, reptiles, birds, and mammals was accompanied by MHC molecules. Cytotoxic T cells, cell-mediated immunity and allograft rejection, T cell-dependent, rapid anamnestic antibody responses of high affinity and heterogeneity are confined to warm blooded vertebrates, i.e., mammals and birds.

Cartilaginous fishes: Another major milestone in the evolution of immunity is the development by cartilaginous fishes, including sharks, of the thymus, the anamnestic (secondary) memory antibody response, of plasma cells for antibody synthesis, and the spleen. These vertebrates have both 7S and 18S IgM immunoglobulins that are disulfide linked. The low and high molecular weights are probably attributable to polymerization rather than differences in heavy chain class. The classic complement pathway components also make their debut. (Figure 26.9) The lower elasmobranchs have a more advanced thymus and spleen than do earlier forms. Chondrostean paddlefish manifest involution of the thymus with age and the presence of plasma cells. The higher elasmobranchs also show thymus involution with age and plasma cells.

Conversion of 19S to 7S immunoglobulin occurs at the evolutionary advance from holosteans to teleosteans (bony fish).

Bony fish (teleosts): T and B lymphocyte functions become separate and distinctive, NK cells appear, and important cytokines, such as IL-2 and IFN, appear in the bony fish. A polymorphic major histocompatibility complex (MHC) system, resembling the mammalian MHC, is demonstrable in Zebra fish. (Figure 26.10)

Fish immunity: Primary and secondary lymphoid tissues of the fish are found in the thymus, kidney, and spleen. Immune cells are also found in the skin and mucous membranes. The gills play an important role in antigen uptake, since they contain lymphoid and antibody secreting cells. There are no lymph nodes or gut-associated lymphoid tissues in fish. The thymus is found in the gill chamber as

FIGURE 26.9 Cartilaginous fish.

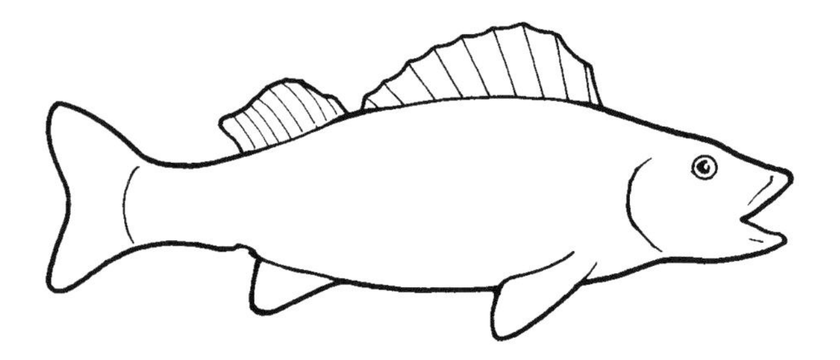

FIGURE 26.10 Bony fish (teleosts).

a paired organ. The fish thymus resembles the mammalian thymus in structure and function. It contains lymphocytes and some epithelial type cells and macrophages. It represents the primary lymphoid organ. Fish thymocytes resemble T lineage cells in mammals. The anterior tip of the kidney consists largely of lymphoreticular and hematopoietic tissues. It is the principal antibody-synthesizing tissue. The fish spleen is comprised mostly of red pulp sinuses, whereas the white pulp is poorly differentiated. The fish kidney and spleen have melanomacrophage centers. These may be analogs of germinal centers in mammals. Fish have T- and B-like lymphocytes. The B-like population possesses surface immunoglobulin similar to mammalian B lymphocytes. Although the fish T-like lymphocyte antigen receptor has not yet been characterized, it is postulated to be homologous to the mammalian T cell receptor. Monocytes and macrophages as well as granulocytes that participate in inflammatory responses have been identified in fish. Mast cells have not been demonstrated but a granulocytic cell with eosinophilic granules that otherwise resembles mast cells has been described together with natural cytotoxic cells. Fish demonstrate both humoral and cell-mediated immunity which is temperature dependent. Some species respond immunologically quite slowly following antigenic challenge at 4°C yet exhibit a maximal response at 12°C, but the optimal temperature varies from one species to another. An IgM-like antibody is the only immunoglobulin class present. It is a tetrameric molecule, and in some species it is monomeric. Secretory IgM has also been described. Two types of light chains comparable to kappa and lambda light chains of vertebrates and heavy chains of different molecular masses have been described. The kinetics of the antibody response in fish resembles that in higher vertebrates. Fish also demonstrate cell-mediated immunity which includes production of lymphokines, yet no fish cytokine has yet been characterized. There is evidence for interferon y, IL-1, IL-2, colony stimulating factor, transforming growth factor P, and tumor necrosis factor in fish. Immunological memory involving both humoral and cell-mediated responses has been described. Memory antibody responses in fish are usually lower than in mammals, and memory is temperature-dependent. Immunological tolerance has also been show in some fish species. Vaccines are available to protect fish against the main bacterial and viral diseases. Three methods of fish vaccination include injection, immersion, and oral. Injection vaccinations stimulate protective immunity. Immersion vaccination involves placing the fish in water containing formalin-killed broth culture vaccines and immunizing them through immersion for one minute in the vaccine which enters through the gills. Vaccine absorption through the hind intestine leads to protective immunity in certain species of fish. (Figure 26.11)

IgW is an immunoglobulin isotype in sharks.

FIGURE 26.11 Fish.

Immunoglobulin evolution: Like mammalian immunoglobulins, dogfish shark immunoglobulins contain two light and two heavy polypeptide chains that express charge diversity. Shark immunoglobulin most closely resembles mammalian IgM. All vertebrates synthesize antibodies that have classic immunoglobulin structure. They consist of two kinds of polypeptide chains that are linked covalently by disulfide bonds and are polydisperse in charge. Immunoglobulins of all classes of vertebrates derived from the primitive jawed fish, the placoderms are now known, such as those from elasmobranchs, which include sharks and rays; teleost fish, which include goldfish; amphibians, which include bullfrogs and *Xenopus*; and reptiles and birds. All of these synthesize antibodies comparable to the IgM isotype of mammals. The IgM molecules of bony fish teleosts occur as tetramers rather than pentemers. Lungfish, amphibians, reptiles, and birds develop other classes of immunoglobulin characterized by distinct heavy chains that resemble γ chains. Distinct immunoglobulin monomers of amphibians, reptiles, and birds have larger heavy chains than the mammalian γ chain and may represent gene duplications. These non-IgM immunoglobulins are designated IgY (amphibians, reptiles, chickens). Other distinct immunoglobulin classes are present in amphibians (IgX), lungfish (IgN), reptiles (IgN), and birds (IgN). Light chains are designated as either κ or λ and their relative frequencies vary from one species to another. Sharks possess a second major immunoglobulin class termed IgW or IgNARC (new antigen receptor from cartilaginous fish) which has a heavy chain that possesses one variable and six constant domains and may be the primitive immunoglobulin in evolution. IgW variable domains are similar to those of heavy chains but are different, showing variation in sequence consistent with diversity through many recognized antigens. Two genes are required early in T and B cell development for recombination of variable, diversity, and joining segments of immunoglobulin light and heavy chains and T cell receptor chains to take place. The most primitive living vertebrates, the cyclostomes, hagfish, and lampreys synthesize antibodies equivalent to those of higher vertebrates in their possession of heavy and light polypeptide chains, disulfide covalent bonding, and charge disparity.

The immunoglobulin domain structure's ability for noncovalent mutual binding led to formation of an extensive **immunoglobulin superfamily** of recognition molecules, including immunoglobulins (ig), T cell receptors (TCRs), MCH class I and II molecules, β2 microglobulin, CD4, CD8, the poly-Ig receptor, and Thy-1. A separate **integrin superfamily** includes LFA-1 and VLA molecules and addressed the binding of leukocytes to endothelial cells and extracellular matrix proteins.

Phylogenetic-associated residues are amino acid residues in immunoglobulin variable regions at a specific position in immunoglobulin (or other protein) molecules in one or more species.

Somatic gene conversion is the nonreciprocal exchange of sequences between genes. A portion of the donor gene or genes is copied into an acceptor gene, leading to alteration only in the acceptor gene. It serves as a mechanism to generate diverse immunoglobulins in nonhuman species.

Amphibian immune system: With the movement from exclusive existence in water by fish onto land by amphibians, a new and second class of immunoglobulin appears, viz., IgG with its class specific γ heavy chain. Lymph nodes, gut-associated lymphoid tissue (GALT), and bone marrow hemopoiesis also make their initial appearance.

Amphibians serve as crucial models for developmental and comparative immunological analysis. The two species used for investigation include the axolotl and *Xenopus*. Hematopoietic precursors rapidly colonize thymus buds in *Xenopus*. The spleen consists of red and white pulp. The axolotl also has a spleen. IgM-synthesizing cells are present in the digestive tract in axolotl. IgM-, IgX-, but no IgY-synthesizing cells are found in the *Xenopus* digestive tract. Amphibians have both B and T like cells in the blood. Monoclonal antibodies specific for different heavy and light chain immunoglobulin chain isotypes can be identified in both *Xenopus* and axolotl. A CD8 membrane glycoprotein has also been found in *Xenopus*. *Xenopus* B and T cells respond to plant mitogens in the same manner observed with mammalian T and B cells. The axolotl has B lymphocytes that proliferate in response to lipopolysaccharide (LPS).

FIGURE 26.12 Amphibian (frog).

FIGURE 26.13 *Xenopus.*

Genes that encode the α and β chains of the axolotl TCR have been cloned. In *Xenopus*, IgM and IgX with high-molecular-weight, and IgY with low-molecular-weight immunoglobulin classes have been identified. The *Xenopus* MHC encodes both Class Ia and Class II molecules. This MHC region also mediates rapid rejection of grafts. Thymectomy in *Xenopus* may diminish or abolish allograft rejection. (Figure 26.12)

Jugular bodies are nodules ventral to external jugular veins that contain lobules of lymphoid cells separated by sinusoids paved with phagocytic cells. Jugular bodies filter the blood but are not a part of the lymphoid system. They may be found in selected amphibian species.

Xenopus is an anuran amphibian that serves as an excellent model for investigations on the ontogeny of immunity. Comparative immunologic research employs isogeneic and inbred families of *Xenopus. Xenopus* is especially useful in studying the role of the thymus in immune ontogeny because early thymectomy of free-living larvae does not lead to runting. (Figure 26.13)

IgX is an immunoglobulin isotype of *Xenopus.*

IgY is an immunoglobulin isotype of *Xenopus.*

Reptile immunity: Environmental conditions affect the structure and functional activity of various reptile organs including those of the immune system. The reptilian thymus is the first lymphoid organ to develop. It is a bilateral organ lying on either side of the neck in lizards and anterior to the heart in snakes. It is frequently multilobulated in turtles. The thymus is well developed in reptiles collected in nature under the correct seasonal temperature and other environmental conditions. Otherwise, it may reveal degrees of degeneration. Reptiles have thymus cell surface molecules that resemble their serum immunoglobulin, but are likely precursors of the T lymphocyte receptor. They also have both IgM and IgG imog classes. The spleen is the most significant peripheral lymphoid organ in reptiles. It becomes lymphopoietic much later than the thymus in lizards, snakes, and tortoises. Of reptile splenic lymphocytes, 50 to 70% are thymus derived and located in thymus-dependent areas. Reptiles do not develop germinal centers and nodules found in mammalian secondary lymphoid organs. T- and B-like cells differ in their sensitivity to environmental conditions. Gut-associated lymphoid tissue (GALT) develops in lizards, snakes, and chelonians. Reptiles do not have tonsils, Peyers's patches, or an appendix, but do have numerous lymphoid aggregates in the mucosa and submucosa along the gut. In lizards, the esophageal lymphoid aggregates containing T-like lymphocytes are greatly affected by seasonal and environmental conditions, whereas the B-like cells and the GALT in reptiles are resistant to changes in season. The spleen and GALT carry out humoral immune mechanisms. Even though reptiles do not have true lymph nodes, their lymphoid system is homologous to that of mammals except for its high sensitivity to ambient conditions. Reptile lymphocytes are strongly responsive to plant mitogens. Lizards and snakes are able to reject skin grafts in a manner similar to mammals.

Reptiles also manifest major functional markers of the MHC. Reptiles produce at least two classes of immunoglobulin that resemble IgM and a second immunoglobulin of lower mol wt that is antigenically related to the 7S Ig of birds and amphibians. All reptiles can mount a powerful immune response to various antigens that include bacteria, protozoa, erythrocytes, and proteins. Body temperature is a critical regulator of reptilian immune function. Reptiles, in contrast to fish and amphibians, are restricted in the temperature range over which immune responses can occur. (Figure 26.14)

FIGURE 26.14 Reptile.

IgR is a second class of immunoglobulin protein in *Raja kenojei*.

Avian (bird) immunity: Birds are unique in forming their B lymphocytes exclusively in a special organ, the bursa of Fabricus in the gut near the cloaca. Their thymus is multilobular and is the site of T cell maturation. No classic lymph nodes are present. They reveal a new class of immunoglobulin, i.e., IgA, in addition to IgM and IgG. The avian complement system is remarkably different from that of mammals as exemplified by Factor B functioning like mammalian C4 and C2.

Although turkeys, ducks, pigeons, and Japanese quail have been investigated, the domestic chicken is the principal representative of this biological group. There is much similarity between the avian and the mammalian immune systems, especially with respect to lymphoid organ structure, the generation of antibody variability, and the arrangements of immunoglobulin and MHC genes. More than a dozen highly inbred chicken strains have been used in research. Chickens are usually excellent antibody producers. Their immune responses develop early. They form IgM prior to IgG and develop poor affinity maturation reflecting the lack of extensive generation of antibody variability in peripheral tissues. Monoclonal antibodies are difficult to prepare from chicken cells. B lymphoid cell development takes place in a distinct primary lymphoid organ, the bursa of Fabricius. Another difference from the mammalian immune system is the multilobed thymus comprised of six to seven distinct lobes on one side of the neck. Chicken T cells develop from precursors, which enter the thymus during development. Avian nonlymphoid hematopoietic cells are well defined as are the chicken immunoglobulin classes, IgM, IgA, and an IgG-like class termed IgY. Chicken β_2-microglobulin and MHC class I α (BF) and class II (BL) molecules are homologous to their mammalian counterparts both structurally and functionally. Chicken T cell surface constituents have been identified that are similar to mammalian T cell TCR, in addition to CD3, CD4, CD5, CD6, CD8, CD28, and CD45 antigens and the interleukin-2 (IL-2) receptor α chain (CD25). Whereas λ and γ chain genes are linked closely,

FIGURE 26.15 Avian immunity (chicken).

light and heavy chain gene complexes are not linked. Chicken TCR $\alpha\beta\gamma$ and δ genes have been cloned and sequenced. The chicken MHC (B complex) is situated on microchromosome 16 which bears the nucleolar organizer region (NOR). The avian immune system is especially susceptible to the avian leukoses, Marek's disease, and infectious bursal disease (IBD). Vaccines against IBD and other diseases of chicks are available. (Figure 26.15)

Mammals: Mammals develop additional immunoglobulin classes, viz., IgD and IgG, as well as subclasses. They also manifest increased diversity of MHC antigens. Three distinct recognition systems are seen in mammals. These include antibody, on B cells only, the T cell receptor, only on T lymphocytes and a spectrum of cells (the MHC), all of whose genes evolved from a common ancestor. There is close similarity between the mouse and human immune systems, which has greatly facilitated elucidation of human immunity through inbreeding and transgenic mouse research. Whereas rats have powerful natural immunity, other mammals, such as whales and Syrian hamsters, manifest little MHC polymorphism. (Figure 26.16)

The **Harderian gland** is a tear-secreting gland in the orbit of the eye in mammals and birds. IgG, IgM, and secretory IgA may originate from this location in birds.

Gene conversion refers to recombination between two homologous genes in which a local segment of one gene is replaced by the homologous segment of a second gene. In avian species and lagomorphs, gene conversion facilitates immunoglobulin receptor diversity principally through

FIGURE 26.16 Mammals (mouse and man).

homologous inactive V gene segments exchanging short sequences with an active, rearranged variable-region gene.

Hm-1 is the designation for the Syrian hamster's MHC. Class II MHC genes have been recognized.

GPLA is the guinea pig MHC. There are two loci that encode class I MHC molecules (*GPLA-B* and *GPLA-S*). The *GPLA-Ia* locus encodes class II MHC molecules. Complement protein factor B, C2, and C4 are encoded by other loci. The genes are *BF* (factor B), C2, C4, *Ia*, *B*, and *S*.

L2C leukemia is a B cell neoplasm of guinea pigs that is transplantable.

Rabbit immunity: The rabbit immune system is quite similar to that of the human with only minor variations. The GALT in the rabbit consists of an appendix, Peyer's patches and diffuse lymphatic nodules. GALT and other peripheral lymphoid tissues of the rabbit contain a permanent lymphatic system with lymph nodes. The rabbit has a prominent spleen and a thymus that undergoes involution in adulthood. Lymphpoiesis originates in the bone marrow and the maturing cells occupy appropriate tissues and organs. Rabbit leukocytes resemble those of other mammalian mononuclear phagocyte systems. Lymphoid cell populations and circulation routes resemble those of other mammals, making the rabbit an excellent model for immunological investigations. Rabbit cytokines include migration inhibition factor (MIF), chemotactic factor, migration stimulation factor (MSF), IL-1, IL-2, and TNF-α. IgM, IgG, IgE, and IgA immunoglobulins and several groups of allotypes have been discovered in rabbits. The rabbit MHC has both Class I and Class II regions. Both B and T cells as well as microphages and polymorphonuclear leukocytes have been described. (Figure 26.17)

FIGURE 26.17 Rabbit.

The **sacculus rotundus** is the lymphoid tissue-rich terminal segments of the ileum in the rabbit. It is a part of the gut-associated lymphoid tissue (GALT).

The **bas** mutation in an mRNA splicing acceptor site in rabbits leads to diminished expression of the principal type of immunoglobulin κ light chain (MO-κ-1). Rabbits with the bas mutation have λ light chains, although a few have the κ-2 isotype of light chains.

TFA antigens in rabbits and rats result from changes in liver cell components as a consequence of exposure to the anesthetic halothane. Rats administered halothane intraperitoneally expressed maximum amounts of the 100-, 76-, 59-, and 57-kDa antigens after 12 h. They were still detectable after 7 days. TFA antigen expression varies in humans as a consequence of variability of hepatic cytochrome P-450 isoenzyme profiles.

Rabbit immunoglobulin allotypes: A rabbit immunoglobulin λ light chain allotype 100 designated as c7 and c21.

An **e allotype** is a rabbit IgG heavy chain allotype. The e allotypes are determined by an amino acid substitution at position 309 in C_H2. The e14 allotype heavy chains possess threonine at position 309, whereas e15 allotype heavy chains possess alanine at that location. They are determined by alleles at the *de* locus that encodes rabbit gamma chain constant regions.

An **f allotype** is a rabbit IgA subclass α heavy chain allotype. Allelic genes at the *f* locus encode five *f* allotypes. These are designated f69 through f73. More than one allotypic determinant is associated with each allotype.

Canine immunity: The canine immune system, using the dog as the classic example, is structurally and functionally very similar to the mouse and human counterparts. As in the human, the dog has numerous natural resistance mechanisms that prevent disease-causing agents by nonimmunologic means. For example, these include the skin and mucous membranes. The dog has the same immunoglobulin classes as described in man. Cell-mediated immunity is essentially no different from the murine and human equivalent. The MHC in the dog is known as DLA, which encodes the DLA class I and class II histocompatibility antigens. Natural killer (NK) cells, killer (K) cells, and T suppressor cells have also been described in the dog. There are also multiple inherited canine immunodeficiencies. Acquired immune deficiencies may also be associated with vitamin and mineral deficiencies, and various autoimmune conditions have been described in the dog. Among these is SLE that is associated with MHC DLA-A7. (Figure 26.18)

FIGURE 26.18 Dog.

Feline immunity: Although peripheral lymphoid tissues and the thymus of cats are comparable to those of other mammals, they do possess a population of pulmonary intravascular macrophages that may render them to manifest increased susceptibility to septic shock mediated by macrophage-derived tumor necrosis factor. A total of 40–45% of feline peripheral blood lymphocytes are B cells, whereas 32–41% are T cells. Of feline peripheral blood cells, 20% are null cells that are considered to be NK cells. Both T helper and T suppressor activities have been described. Feline IL-1 and IL-2 as well as IL-6 have been identified, although the physical chemical properties of the latter are different from those of the human and the mouse. Interferons α, β, and γ and have been characterized and resemble those of other species. Although all immunoglobulin isotype of other species are recognized in the cat, neither IgE nor IgD has been formally identified. Cats also have secretory IgA with a J-chain that resembles equivalent molecules of other species. Although cats do not respond to tuberculin, they do develop a good delayed hypersensitivity response in the skin to dinitrochlorobenzene, called viral antigens, and to BCG vaccine, but the delayed-type response is less intense than in other species. The granulomatous reaction to tuberculosis is essentially the same as in other mammals. The MHC is designated FLA, which is not polymorphic. The FLA system contains 10 to 20 class I gene loci, only two of which are expressed. They manifest five class II gene loci, only three of which are expressed. The lack of MHC polymorphism renders bone marrow transplantation very successful in this species. Cats possess all major complement components at levels equivalent to those in other mammals. The principal target organ for anaphylaxis in the cat is the lung where serotonin, released from mast cells, is the principal mediator.

FIGURE 26.19 Feline immunity.

Flea-bite dermatitis is the most frequent allergic skin disease of cats. Feline blood group antigens are designated A and B. In the U.S., 99% of cats belong to group A and 1% to group B. (Figure 26.19)

Spontaneous autoimmune diseases in cats include hemolytic anemia, hyperthryoidism, thrombocytopenia purport, pemphigus vulgaris, pemphigus foliaceous, myasthenia gravis, systemic lupus erythematosus, and arthritis. Cats only rarely have immunodeficiencies. Chediak-Higashi syndrome is the most common congenital defect in immune function. The leukocytes manifest large lysosomal granules, but there is no increased susceptibility to infection. There is a failure of colostral immunoglobulins to be passively transferred in the cat. Secondary immunodeficiency of cats is attributable usually to FeLV infection which often induces severe immunodeficiency in infected animals. Feline immunodeficiency virus (FIV) is a lentivirus that infects both domestic and wild cats. It leads to depletion of $CD4^+$ T lymphocytes and resembles human AIDS. The virus receptor is mediated through fCD9 rather than fCD4. There is polyclonal stimulation of B cells and of $fCD8^+$ T cells. The disease terminates in an AIDS-like condition. Cats also develop plasmacytomas, although they are not common.

Ungulate immunity: Ungulates develop an immune system similar to that of most other mammalian species. The thymus and bone marrow serve as the sources for generation of T and B cells, respectively. A feature unique to ruminant ungulates is the major role played in primary B cell development by the ileal Peyer's patch which is a single structure in the terminal ileum. Ruminants have a higher proportion of $\gamma\delta$ T cells than do other species. Ungulate immunoglobulins consist of IgG1, IgG2, IgM, IgA (serum), IgA (secretory), and IgE isotypes. IgA antibodies predominate on mucosal surfaces in ruminants, whose mammary glands produce milk that is different in immunoglobulin content compared to other species. There is no immunoglobulin transport across the placenta in ruminants. The colostrum contains very high levels of IgG1 and very little IgA. The IgA in milk is derived from the plasma. A biliary pump removes IgA from serum for delivery by way of the bowel to the gut lumen. Whereas the immunologic tissues are well developed before birth, differentiation of peripheral organs does not occur until there is antigenic stimulation in extrauterine life. This antigenic stimulation is followed by the development of follicles and germinal centers. Since ungulates do not receive immunoglobulin prenatally from the mother, they are born agammaglobulinemic. However, large amounts of maternal IgG are concentrated in the mammary gland just prior to parturition and are ingested and absorbed intact into the circulation by the suckling neonate. In addition to antibodies in the colostrum that are transferred to the ruminant neonate, cells and other soluble factors in milk are also important in passive protection. (Figure 26.20)

FIGURE 26.20 Ungulate.

Porcine immunity: The pig's immune system differs from that of mice and humans in having inverted lymph nodes; lymphocytes leaving lymph nodes in the blood rather than in efferent lymph; pigs having four types of Peyer's patches; pigs having small tonsilar papillae in the tongue together with palatine and nasopharyngeal tonsils and a failure to transfer immunoglobulins across the placenta. The pig MHC encodes the production of both MHC class I and class II products. They have peripheral blood leukocytes of the same type as found in humans. Unique features of porcine blood leukocytes are the relatively high proportion of $\gamma\delta$ T cells, the high $CD4^+/CD8^+$ ratio in adults, and the increased frequencies of $CD4^+/CD8^+$ double-positive T cells. Four CD markers have been described and four classes of immunoglobulin IgM, IgA, IgG, and IgE have been identified. The pig has both $\alpha\beta$ T cell and $\gamma\delta$ T cells as well as NK cells, adhesion molecules of the E selectin type, and cytokines, that include interferons, inflammatory cytokines, and chemokines. Other pig cytokines include IL-2, IL-4, IL-5, IL-10, IL-12, GM-CSF, and G-CSF, among others. (Figure 26.21)

Ovine immune system: Sheep are valuable for investigation of the physiology of the immune system. A technique termed lymphatic cannulation of single lymph nodes in sheep offers a powerful tool to analyze immune cell populations within various immunological microenvironments. The cannulated lymph node maintains intact vascular and neurological connections. Cannulation of the efferent lymphatic permits monitoring of immunological events in a single lymph node. Numerous studies have been conducted on ovine cellular and molecular markers of lymphocyte populations, their subsets and receptors. MHC class I and class II genes and proteins, immunoglobulins, and cytokine have been described. The ILEAL Peyer's patch in the intestine serves as a primary lymphoid organ for B cell lymphopoiesis. Serum IgG is higher in sheep than in humans. The three principal T cell subsets in sheep include the majority of either $CD4^+$ or $CD8^+$ T cells that express the $\alpha\beta$ T cell receptor together with the T3 molecule and other adhesion molecules. The double-negative CD4-/CD8-T cell subset contains the $\gamma\delta$ T cell subset which is a measured subpopulation in rumanants. Both MHC class I and class II molecules are products of the MHC region in sheep. The afferent lymph of sheep is a physiological source of antigen-presenting dendritic cells. (Figure 26.22)

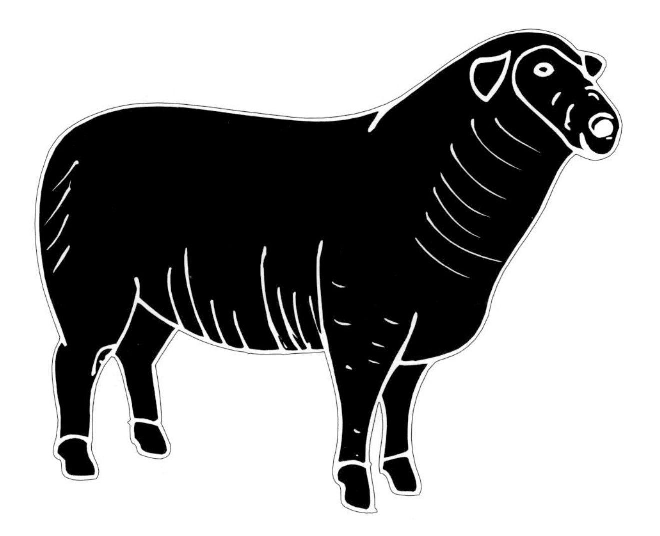

FIGURE 26.22 Sheep.

FIGURE 26.21 Pig.

T globulin is a serum protein found after hyperimmunization of horses and is a $\gamma 1$ 7 S globulin that appears as a prominent band of immunoglobulin when the serum is electrophoresed. It may be a subtype of IgG.

Marsupial immunity: Opossums have been studied as representatives of the marsupialia. At birth, the pouch neonates are immunologically incompetent and inexperienced antigenically. It is necessary for them to develop immune competence without delay. This is facilitated by a rapidly developing thymus in a cervical location. Whereas some species may pass maternal antibodies across the placenta, most of the transfer of maternal antibodies begins with suckling and continues until the young leave the pouch at which time immune competence is fully developed. Adult marsupials possess lymphoid organs that participate in both humoral and cellular immune responses. Some data suggest that the humoral immune response may be initially sluggish and that selective immune responses may be refractory to certain cytokines. (Figure 26.23)

Random breeding refers to the mating of members of a population at random. The genetic diversity produced by random breeding depends upon the size of the population. If it is large, genetic diversity will be maintained. If it is small, genetic uniformity will result in spite of random breeding.

Inbred strain refers to laboratory animals developed by sequential brother–sister matings. After 20 generations, the animals, e.g., mice, are said to be inbred. They are homozygous at approximately 98% of genetic loci. This homogeneity at the histocompatibility loci permits successful grafting, without rejection, among members of the

FIGURE 26.23 Opossum.

inbred strain. Recessive deleterious genes may become homozygous during inbreeding, leading them to manifest their negative effects with respect to such factors as growth rate, susceptibility to disease, or fertility, and thus limiting the number of possible inbred generations. An additional problem is the development of sublines of an inbred strain caused by mutations and evolutionary factors.

Inbreeding describes the mating of animals of a species that are genetically more similar to one another than to members of that same species selected by chance or at random in the population. Deliberate inbreeding of experimental animals is carried out to induce genetic uniformity or homozygosity. Raising inbred strains of mice for laboratory investigation involves brother–sister matings for 20 or more generations. Thereafter, the progeny are said to be "inbred."

Inbred mouse strain: See inbred strain.

C3H/HeJ mice are a mutant substrain of C3H mice that manifests a suppressed response by macrophages and B cells to challenge with lipopolysaccharide. Their macrophages do not produce interleukin-1 and tumor necrosis factor following the lipopolysaccharide challenge. This mutation has an autosomal dominant mode of inheritance and is encoded by genes on chromosome 4. Such immunosuppression leads to an increased incidence of microbial infections in the C3H/HeJ mice.

The **CBA mouse** is a strain of inbred mice from which many substrains have been developed, such as CBA/H-T6.

Balb/c mice are an inbred white mouse strain that responds to an intraperitoneal inoculation of mineral oil and Freund's complete adjuvant with a myoproliferation reaction.

CBA/N mouse is a CBA murine mutant that is incapable of responding immunologically to linear polysaccharides and selected other thymus-independent immunogens. The mutant strain's B cells are either defective or immature. The Lyb3, Lyb5, and Lyb7 B lymphocyte subset is not present in these mice. This mutation is designated *xid*, which has an X-linked recessive mode of inheritance. Their serum IgG concentrations are diminished. CBA/N mice mount only weak immune responses to thymus-dependent immunogens.

Congenic signifies a mouse line that is identical or almost identical to other inbred strains with the exception of the substitution of an alien allele at a single histocompatibility locus that crosses with a second inbred strain permitting introduction of the foreign allele.

Coisogenic: See congenic strains.

Congenic mice: See congenic strains.

Coisogenic strains are inbred mouse strains that have an identical genotype except for a difference at one genetic locus. A point mutation in an inbred strain provides the opportunity to develop a coisogenic strain by inbreeding the mouse in which the mutation occurred. The line carrying the mutation is coisogenic with the line not expressing the mutation. Considering the problems associated with developing coisogenic lines, congenic mouse strains were developed as an alternative. See congenic strains.

Congenic strains: Inbred mouse strains that are believed to be genetically identical except for a single genetic locus difference. Congenic strains are produced by crossing a donor strain and a background strain. Repeated back-crossing is made to the background strain, selecting in each generation for heterozygosity at a certain locus. Following 12 to 14 backcrosses, the progeny are inbred through brother–sister matings to yield a homozygous inbred strain. Mutation and genetic linkage may lead to random differences at a few other loci in the congenic strain. Designations for congenic strains consist of the symbol for the background strain followed by a period and then the symbol for the donor strain.

A **conventional mouse** is one maintained under ordinary living conditions and provided water and food on a regular basis.

Commensal mice are those that associate closely with humans.

Recombinant inbred strains are inbred strains of F_2 generation mice developed by crossing two inbred strains to yield F_1 and then F_2 generations. Progeny of the F_2

generation are inbred until they become homozygous at most loci. The progeny approach complete genetic identity and homozygosity. Recombination occurs during meiosis and consists of crossing over and recombination of parts of two chromosomes. Recombinant inbred stains help to establish genetic linkages. These genetically uniform and homozygous mice offer a means to study the consequences of reassorting various parental genes such as heavy chain genes.

The ***gld*** **gene** is a murine mutant gene on chromosome 1. When homozygous, the *gld* gene produces progressive lymphadenopathy and lupus-like immunopathology.

The designation **"conventional (holoxenic) animals"** refers to experimental animals exclusive of those that have been raised in a gnotobiotic or germ-free environment.

F protein is a 42-kDa protein in the cytoplasm of murine hepatocytes. It occurs in F.1 and F.2 allelic forms in separate inbred murine strains.

Low responder mice are inbred mouse strains that produce a poor immune response to selected antigens in comparison to the response by other inbred mouse strains. This is associated with the low responder's lack of appropriate *Ir* genes. Low responsiveness is governed by class II MHC genes.

Lymphocytic choriomeningitis (LCM) is a murine viral disease that produces inflammatory brain lesions in the affected mouse as a result of delayed-type hypersensitivity to viral antigens on brain cells infected with the LCM virus. This infectious agent, which is classified as an arenavirus, is endemic in the mouse population and occasionally occurs in humans. Only adult mice that become infected develop the lesions, as those infected *in utero* are rendered immunologically tolerant to the viral antigens and fail to develop disease. An adult with an intact immune system exposed to LCM virus either becomes immune or succumbs to the acute infection, which is associated with lymphadenopathy, splenomegaly, and T lymphocyte perivascular infiltration of the viscera, especially the brain. A chronic carrier state can be induced in either neonatal mice or those with an impaired immune system through infection with the virus. Although carriers generate significant quantities of antiviral antibodies, the infection persists, and virus–antibody immune complexes become deposited in the renal glomeruli, walls of arteries, liver, lungs, and heart. The passive transfer of cytotoxic T lymphocytes from an immune animal to a carrier results in specific reactivity against LCM viral epitopes on cell membranes in the brain and meninges, which leads to profound inflammation with death of the animal. Transmission of this disease is through the excrement of rodents and is seen especially in winter when rodents enter dwellings. There is fever, headache, flu-like symptoms, and lymphocytosis in the cerebrospinal fluid. There may be associated leukopenia and thrombocytopenia. This disease must be distinguished from infectious mononucleosis, herpes zoster, and enterovirus infection.

LCM: See lymphocytic choriomeningitis.

Primate (nonhuman) immune system: Nonhuman primates represent the best animal models of many human

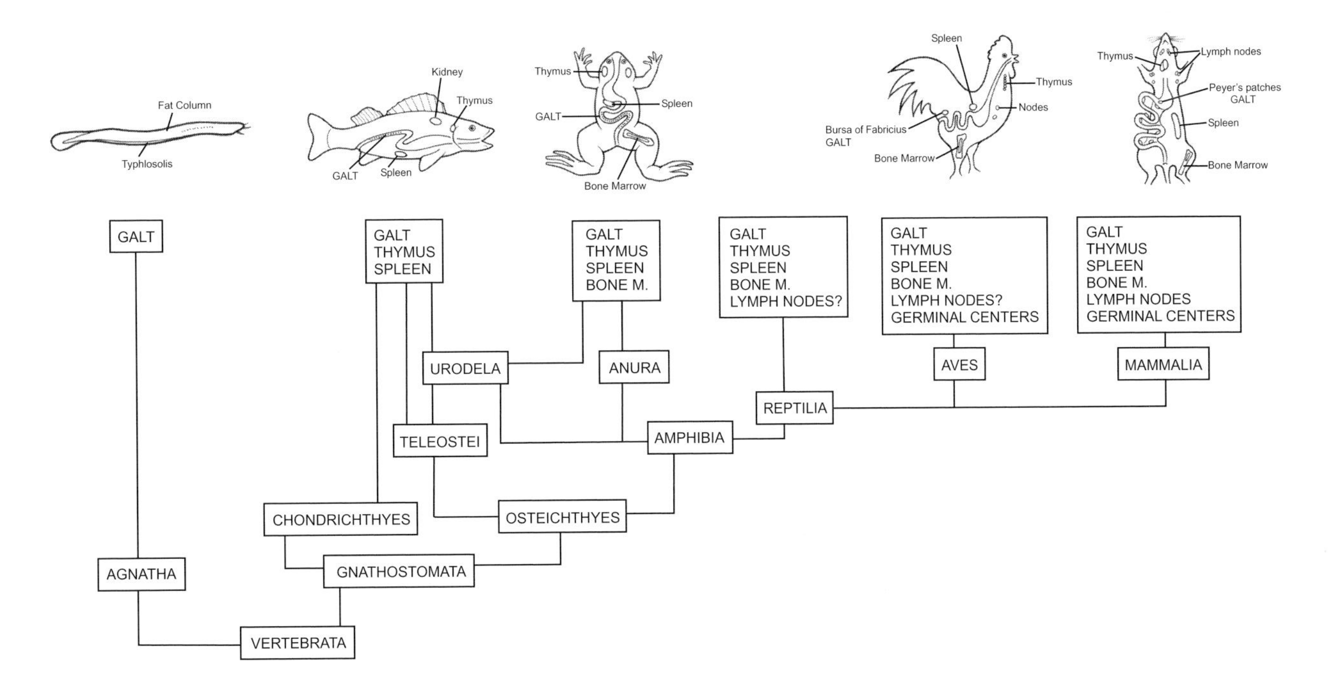

FIGURE 26.24 Evolution of the lymphoid systems.

diseases based on the resemblance of their immune systems to those of humans. Lymphocyte subsets of human and nonhuman primates have shown to be quite similar to the use of leukocyte-specific monoclonal antibodies. There are also similarities between human and nonhuman primate MHC and TCR genes.

An **antiglutinin** is a mammalian seminal plasma substance that prevents washed spermatozoa from spontaneously agglutinating, i.e., autoagglutinating.

Lower vertebrates have cytokines that are similar functionally to those in mammals. Cytokines present in various classes of vertebrates include interleukins, interferons, colony stimulating factors, tumor necrosis factor, and chemokines. T lymphocytes from bony fish, urodele and anuran amphibians, and snakes and amphibians produce T cell growth factors (TCGF) when stimulated. *Xenopus* TCFG- resembles mammalian IL-2 both biochemically and functionally. Bony fish, amphibian, and bird macrophages manifest IL-1-like activity. Ectotherms have interferon-like factors that are antiviral and activate macrophages.

Phylogeny is the evolutionary development of a species.

In immunology, **species specificity** refers to cellular or tissue antigens present in one species only and not found in other species.

27 Immunological Methods and Molecular Techniques

MEASUREMENT OF ANTIGENS AND ANTIBODIES

Just as the eclectic science of immunology intersects essentially all of the basic biological sciences, it makes use of many biochemical techniques such as chromatography and protein fractionation. It also employs the newer methods of molecular genetics such as gene sequencing and related techniques. Advances in technology have armed the immunologist with the powerful tools of PCR technology, immunophenotyping by flow cytometry, hybridomas and monoclonal antibodies, DNA typing, enzyme-linked immunosorbent assay (ELISA), and radiolabeling of immune system molecules. In addition, the time-honored methods of precipitation, agglutination, complement fixation, and related techniques have long been used by the immunologist. Since its inception, immunologic science has not only maintained a unique nomenclature but also special techniques that elucidated some of Nature's most jealously guarded secrets through scientific investigation. Inbred mice, of known genetic constitution, and more recently, transgenic animals including knockout mice, offer new avenues for elucidating some of immunology's most perplexing conundrums. Great tomes are currently available that describe the myriad immunological techniques now available. A representative number of the more commonly used ones are described here (Figure 27.1 to Figure 27.4).

Absorption is the elimination of antibodies from a mixture by adding soluble antigens or the elimination of soluble antigens from a mixture by adding antibodies.

An **antibody absorption test** is a serological assay based upon the ability of a cross-reactive antigen to diminish a serum sample's titer of antibodies against its homologous antigen, i.e., the antigen that stimulated its production. Crossreactive antibodies, as well as crossreactive antigens, may be detected in this way.

An **immunoabsorbent** is a gel or other inert substance employed to absorb antibodies from a solution or to purify them.

Adsorption is the elimination of antibodies from a mixture by adding particulate antigen or the elimination of particulate antigen from a mixture by adding antibodies. The incubation of serum-containing antibodies such as agglutinins with red blood cells or other particles may remove them through sticking to the particle surface.

Gel filtration chromatography (Figure 27.5) is a method that permits the separation of molecules on the basis of size. Porous beads are allowed to swell in buffer, water, or other solutions and are packed into a column. The molecular pores of the beads will permit the entry of some molecules into them but exclude others on the basis of size. Molecules larger than the pores will pass through the column and emerge with the void volume. Since the solute molecules within the beads maintain a concentration equilibrium with solutes in the liquid phase outside the beads, molecular species of a given weight, shape, and degree of hydration move as a band. Gel chromatography using spherical agarose gel particles is useful in the exclusion of IgM, which is present in the first peak. Of course, other molecules of similar size, such as macroglobulins, are also present in this peak. IgG is present in the second peak, but the fractions of the leading side are contaminated by IgA and IgD.

Immunoabsorption is the removal of a selected group of antibodies by antigen or the removal of antigen by interaction with specific antibody.

Immunoadsorbents are specific antibodies that are chemically bound to solid supports. When a mixture containing antigen is poured over the solid support, antigen is bound noncovalently to the immunoadsorbent from which it may be isolated after treating the sorbent–ligand complex with a denaturing agent.

Adsorption chromatography is a method to separate molecules based on their adsorptive characteristics. Fluid is passed over a fixed-solid stationary phase.

Agar gel is a semisolid substance prepared from seaweed agar that has been widely used in the past in bacteriology, but is used in immunology for diffusion of antigen and antibody in Ouchterlony-type techniques, electrophoresis, immunoelectrophoresis, and related methods.

Gel diffusion (Figure 27.7) is a method to evaluate antibodies and antigens based upon their diffusion in gels toward one another and their reaction at the point of contact in the gel.

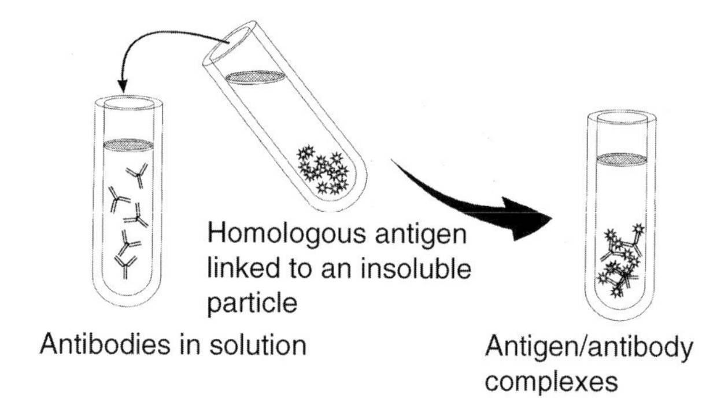

FIGURE 27.1 Absorption is the elimination of antibodies from the mixture by adding soluble antigens or the elimination of soluble antigen from a mixture by adding antibodies.

Affinity chromatography is a method to isolate antigen or antibody based upon antigen–antibody binding. Antibody molecules fixed to a solid support such as plastic or agarose beads in a column, constituting the solid phase, may capture antigen molecules in solution passed over the column. Subsequent elution of the antigen is then accomplished with acetate buffer at pH 3.0 or diethylamine at pH 11.5.

Double immunodiffusion is a precipitation reaction in gel media in which both antibody and antigen diffuse radially from wells toward each other, thereby forming a concentration gradient. A visible line of precipitation forms as equivalence is reached.

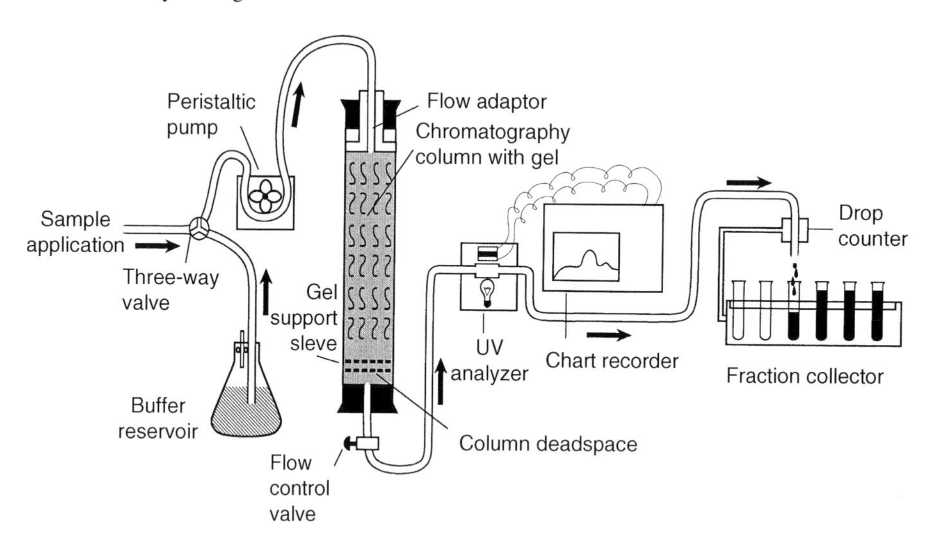

FIGURE 27.2 Column chromatography. Arrows indicate direction of flow. Chromatography refers to a group of methods employed for the separation of proteins.

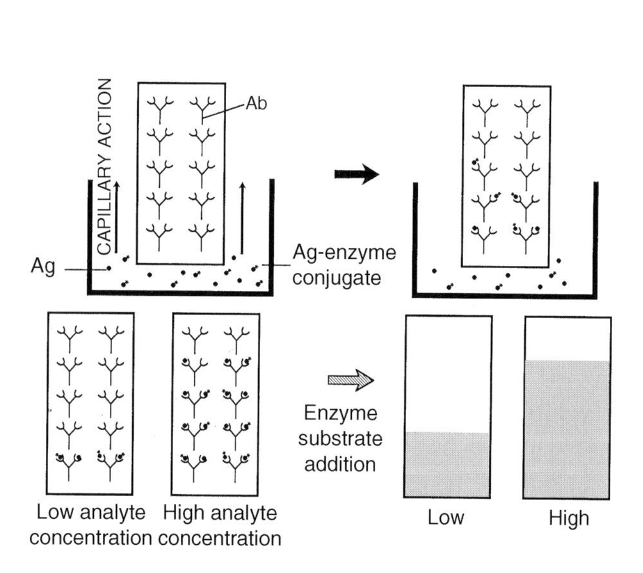

FIGURE 27.3 Immunochromatography.

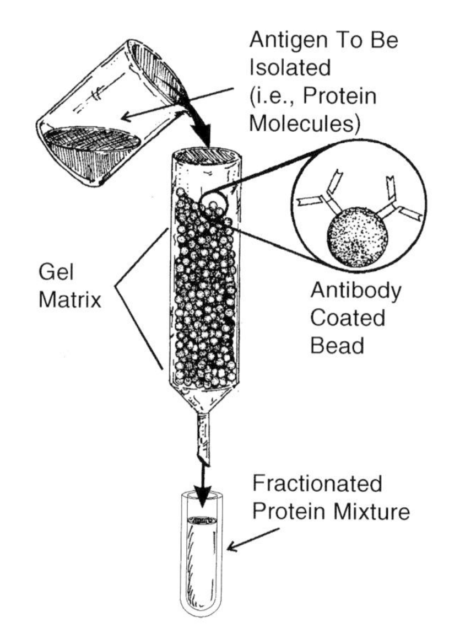

FIGURE 27.4 Absorption chromatography is a method to separate molecules based on their absorptive characteristics. Fluid is passed over a fixed solid stationary phase.

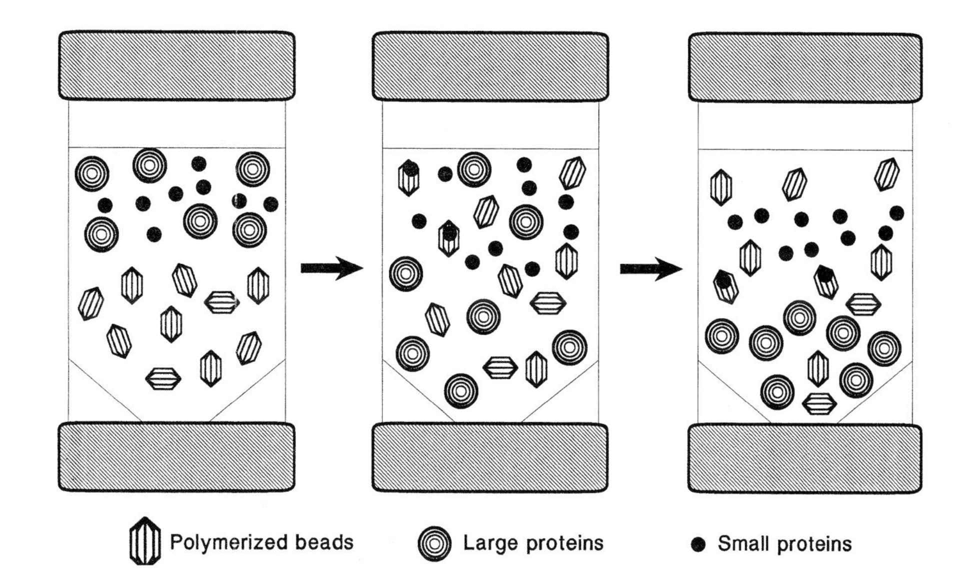

FIGURE 27.5 Gel filtration chromatography.

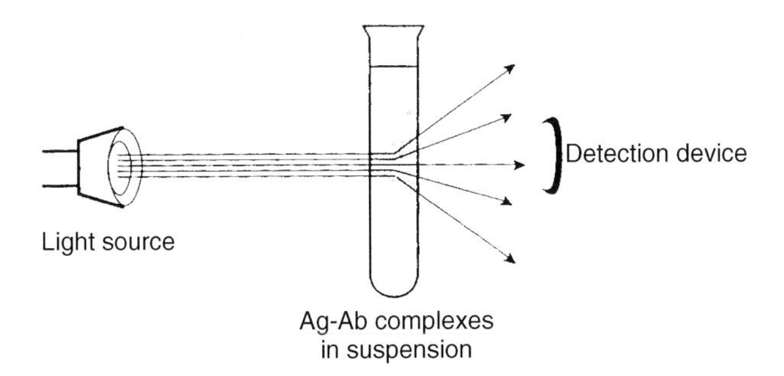

FIGURE 27.6 Turbidimetry is the quantification of a substance in suspension based on the suspension's ability to reduce forward light transmission.

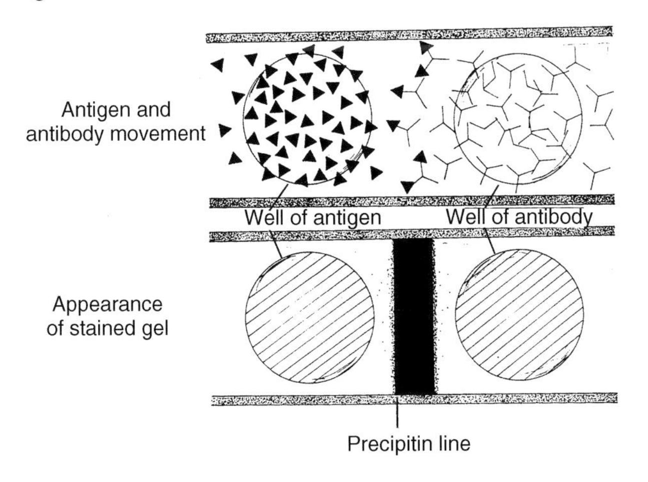

FIGURE 27.7 Gel diffusion is a method to evaluate antibodies and antigens based upon their diffusion in gels toward one another and their reaction at the point of contact in the gel.

Chromatography is a group of methods employed for the separation of proteins.

Immunodiffusion is a method in which antigen and antibody are placed in wells at different sites in agar gel and are permitted to diffuse toward each other in the gel. The formation of precipitin lines at their point of contact in the agar gel signifies a positive reaction, showing that the antibody is specific for the antigen in question. Multiple variations of this technique have been described.

Sephadex® is a trade name for a series of crosslinked dextrans used in chromatography.

FICA (fluoroimmunocytoadherence) refers to the use of column chromatography to capture antigen-binding cells.

Ion exchange chromatography is a method that permits the separation of proteins in a solution, taking advantage of the net charge differences among them. It involves the electrostatic binding of proteins onto a charged resin suspended in buffer and packed into a column. Since serum proteins vary in their charge, binding to or elution from the column is possible by gradually increasing or decreasing the salt concentration (with or without changes in pH). This affects the type of proteins binding to the resin. With buffers of low molarity and pH greater than 6.5, the IgG in solution is not adsorbed on the column and passes through with the first buffer volume.

Dialysis is a method to separate a solution of molecules that differ in molecular weight by employing a semipermeable membrane. Antibody affinity is measured by equilibrium dialysis.

Doubling dilution (Figure 27.6 and 27.8) is a technique used in serology to prepare serial dilutions of serum. A

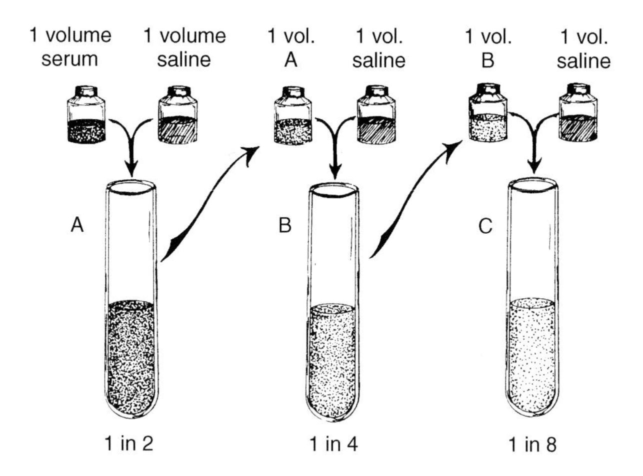

FIGURE 27.8 Doubling dilution.

fixed quantity (one volume) of physiologic saline is added to each of a row of serological tubes, except for the first tube in the row which receives two volumes of serum. One volume of serum from the first tube is added to one volume of saline in the second tube. After thoroughly mixing the contents with the transfer pipette, one volume of the second tube is transferred to the third, and the procedure is repeated down the row. This same volume is then discarded from the final tube after the contents have been thoroughly mixed. Thus, the serum dilution in each tube is double that in the preceding tube. The first tube is undiluted; the second contains a 1:2 dilution; the third a 1:4; the fourth 1:8, etc.

Serial dilution is the successive dilution of antiserum in a row of serological tubes containing physiologic saline solution as diluent to yield the greatest concentration of antibody in the first tube and the least amount in the last tube which contains the highest dilution. For example, a double quantity of antiserum is placed in the first tube, half of which is transferred to the second tube containing an equal volume of diluent. After thorough mixing with a serological pipette, an equivalent amount is transferred to the successive tube, etc. The same volume removed from the last tube in the row is discarded. This represents a doubling dilution.

The **quantitative precipitin reaction** is an immunochemical assay based on the formation of an antigen–antibody precipitate in serial dilutions of the reactants, permitting combination of antigen and antibody in various proportions. The ratio of antibody to antigen is graded sequentially from one tube to the next. The optimal proportion of antigen and antibody is present in the tube that shows the most rapid flocculation and yields the greatest amount of precipitate. After washing, the precipitate can be analyzed for protein content through procedures such as the micro-Kjeldahl analysis to ascertain nitrogen content, spectrophotometric assay, or other techniques. Heidelberger and Kendall used the

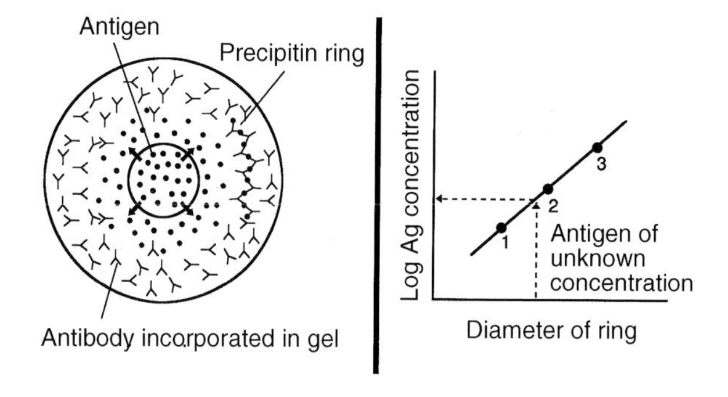

FIGURE 27.9 Single immunodiffusion (Mancini technique).

technique extensively, employing pneumococcus polysaccharide antigen and precipitating antibody in which nitrogen determinations reflected a quantitative measure of antibody content.

Single immunodiffusion (Mancini technique): (1) A technique in which antibody is incorporated into agar gel and antigen is placed in a well that has been cut into the surface of the antibody-containing agar (Figure 27.9). Following diffusion of the antigen into the agar, a ring of precipitation forms at the point where antigen and antibody have reached equivalence. The diameter of the ring is used to quantify the antigen concentration by comparison with antigen standards. (2) The addition of antigen to a tube containing gel into which specific antibody has been incorporated. Lines of precipitation form at the site of interaction between equivalent quantities of antigen and antibody.

Radiolabeling: Radioisotope incorporation into macromolecules has increased the sensitivity of immunoassays 30,000-fold. They permit the detection and quantification of picogram quantities of an antigen or receptor. Radiolabeled antibodies can be employed to detect primary binding of antibody to antigen, providing sensitivity and measurements of affinity. They may be used to identify a tumor site or the presence and topography of molecules in a cell membrane. Radiolabeled antigens demonstrated that B cells bind antigen via cell surface immunoglobulin receptors.

Isotopic labeling (radionuclide labeling) refers to the introduction of a radioactive isotope into a molecule by either external labeling through tagging molecules with ^{125}I or other appropriate isotope or by internal labeling in which ^{14}C or ^{3}H-labeled amino acids are added to tissue culture, which allows the cells to incorporate the isotope. Once labeled, molecules can be easily traced and their fate monitored by measuring radioactivity.

Trace labeling: See isotopic labeling.

Single diffusion test is a type of gel diffusion test in which antigen and antibody are placed in proximity with one

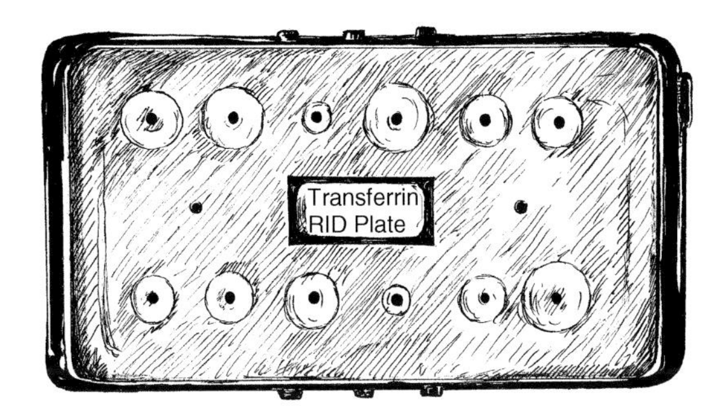

FIGURE 27.10 Single radial immunodiffusion.

another and one of them diffuses into the other, leading to the formation of a precipitate in the gel. This is in contrast to double diffusion in which both antigen and antibody diffuse toward each other.

Single radial immunodiffusion (Figure 27.10) is a technique to quantify antigens. Plates are poured in which antibody is incorporated into agar, wells are cut, and precise quantities of antigen are placed in the wells. The antigen is permitted to diffuse into the agar-containing antibody and produce a ring of precipitation upon interaction with the antibody. As diffusion proceeds, an excess of antigen develops in the area of the precipitate causing it to dissolve only to form once again at a greater distance from the site of origin. At the point where antigen and antibody have reached equivalence in the agar, a precipitation ring is produced. The precipitation ring encloses an area proportional to the concentration of antigen measured 48 to 72 h following diffusion. Standard curves are employed using known antigen standards. The antigen concentration is determined from the diameter of the precipitation ring. This method can detect as little as 1 to 3 μg/ml of antigen. Known also as the Mancini technique.

The **Mancini test** is a single radial diffusion test.

Radioimmunodiffusion is a variation of the immunodiffusion technique which uses a radioactively labeled antibody. This enhances the sensitivity of the results when read by autoradiography. See single radioimmunodiffusion.

Radioimmunodiffusion test: See single radial immunodiffusion.

Reverse radioimmunodiffusion is a technique to quantify antibody levels that varies from the single radial diffusion test in only one detail. Samples are placed in gel containing antigen. As diffusion takes place, precipitation rings that are produced are directly proportional to the antibody concentration.

Reverse Mancini technique: See reverse radioimmunodiffusion.

Ammonium sulfate precipitation: The ammonium sulfate method is a means of measuring the primary antigen-binding capacity of antisera and detects both precipitating and nonprecipitating antibodies. If offers an advantage over equilibrium dialysis in that large, nondializable protein antigens may be used. This assay is based on the principle that certain proteins are soluble in 50% saturated ammonium sulfate, whereas antigen–antibody complexes are not. Thus, complexes may be separated from unbound antigen. Spontaneous precipitation will occur if a precipitating-type antibody is used, until a point of antigen excess is reached where complex aggregation no longer occurs and soluble complexes are formed. Upon the addition of an equal volume of saturated ammonium sulfate solution (SAS), these complexes become insoluble, leaving radiolabeled antigen in solution. SAS fractionation does not significantly alter the stoichiometry of the antibody–antigen reaction and inhibits the release or exchange of bound antigen. Thus, radioactivity of this "induced" precipitate is a measure of the antigen-binding capacity of the antisera as opposed to a measure of the amount of antigen or antibody spontaneously precipitated.

Antigen-binding cell (ABC) assay: The principle of this assay is the binding of cells bearing receptors for antigen to a gelatin dish in which antigen is incorporated. After incubation for a specified time and temperature, the unbound cells are washed out, and the bound cells are collected following melting of the gelatin layer at 37°C. The harvested cells are washed, counted, and used for various other assays.

Antigen capture assay is a method to identify minute quantities of antigen in blood sera or supernatants. Antibodies of high titer are linked to an insoluble solid support and the specimen containing the antigen to be evaluated is passed over the solid phase. This will bind or capture the antigen, making it available for reaction with a separate enzyme-labeled antibody, which reacts with and reveals the captured antigen.

ASLT is the abbreviation for the antistreptolysin O test.

ASO (antistreptolysin O) is a laboratory technique that serves as an indicator of infection by group A b hemolytic streptococci. IgM antibody titers, expressed in Todd units (TU), increase fourfold within 3 weeks after infection in untreated subjects. Penicillin treatment decreases the ASO titer. Less than 166 TU is normal, whereas greater than 333 TU in children and greater than 250 TU in adults suggests recent infection. The ASO assay depends upon hemolysis inhibition. The greatest dilution of a patient's blood combined with 1 U of streptolysin O that prevents the lysis of erythrocytes determines the Todd units, the reciprocal of endpoint dilution.

Autoantibody assays are tests for autoantibodies that can bind to self antigens and include a broad spectrum of techniques. These include agglutination, such as latex agglutination, indirect immunofluorescence, indirect immunoperoxidase, radioimmuno assay, ELISA, hemagglutination, bioassay, binding assay, and immobilization.

A **blocking test** is an assay in which the interaction between an antigen and its homologous antibody is inhibited by the previous exposure of the antigen to a different antibody which has the same specificity as the first one, but does not have the same biological function. In a different situation, a hapten may be used to prevent the reaction of an antibody with its intended antigen. This is referred to as the hapten inhibition test. An example would be blood group substance soluble molecules equivalent to erythrocyte surface isoantigen epitopes found in the body fluids. See ABO blood group substances.

Capture assays are methods to measure antigens or antibodies in which antibodies bound to plastic capture antigens or antigens bound to plastic capture antibodies. Labeled antigens or antiimmunoglobulins may be used to measure antibody binding to a plate-bound antigen. Antigen binding to an antibody bound to a plate can be assayed with an antibody that binds to a different antigenic determinant or epitope on the antigen.

Cold ethanol fractionation is a technique used to fractionate serum proteins by precipitation with cold ethanol. One of the fractions obtained is Cohn fraction II, which contains the immunoglobulins. This method has been largely replaced by more modern and sophisticated techniques.

Cold target inhibition refers to the introduction of unlabeled target cells to inhibit radioisotope release from labeled target cells through the action of antibody or cell-mediated immune mechanisms.

Competitive binding assays are serological tests in which unknowns are detected and quantified by their ability to inhibit binding of a labeled known ligand to a specific antibody. Also termed competitive inhibition assay.

Competitive inhibition assay is a test in which antigens or antibodies are assayed by binding of a known anitbody or antigen to a known amount of lableled antibody or antigen. Known or unknown sources of antibody or antigen are then used as competitive inhibitors. Also termed competitive inhibition assay.

The **conglutinating complement absorption test** is an assay based on the removal of complement from the reaction medium if an antigen–antibody complex develops. This is a test for antibody. As in the complement fixation test, a visible or indicator combination must be added to determine whether any unbound complement is present. This is accomplished by adding sensitized erythrocytes and conglutinin, which is prepared by combining sheep erythrocytes with bovine serum that contains natural antibody against sheep erythrocytes as well as conglutinin. Horse serum may be used as a source of nonhemolytic complement for the reaction. Aggregation of the erythrocytes constitutes a negative test.

Conglutination is the strong agglutination of antigen–antibody–complement complexes by conglutinin, a factor present in normal sera of cows and other ruminants. The complexes are similar to EAC1423 and are aggregated by conglutinin in the presence of Ca^{2+}, which is a required cation. Conglutination is a sensitive technique for detecting complement-fixing antibodies.

The **conglutinin solid phase assay** is a test that quantifies C3bi-containing complexes that may activate complement by either the classical or the alternate pathways.

A **consumption test** is an assay in which antigen or antibody disappears from the reaction mixture as a result of its interaction with the homologous antibody or antigen. By quantifying the amount of unreacted antigen or antibody remaining in the reaction system and comparing it with the quantity of that reagent that was originally present, the result can be ascertained. The antiglobulin consumption test is an example.

A **control** is a specimen of known content used together with an unknown specimen during an analysis in order that the two may be compared. A positive control known to contain the substance under analysis and a negative control known not to contain the substance under analysis are required.

Coprecipitation refers to the addition of an antibody specific for either the antigen portion or the antibody portion of immune complexes to effect their precipitation. Protein A may be added instead to precipitate soluble immune complexes. The procedure may be employed to quantify low concentrations of radiolabeled antigen that are combined with excess antibody. After soluble complexes have formed, antiimmunoglobulin or protein A is added to induce coprecipitation.

***Crithidia* assay** is the use of a hemoflagellate termed *Crithidia luciliae* to measure anti-dsDNA antibodies in the serum of systemic lupus erythematosus (SLE) patients by immunofluorescence methods. The kinetoplast of this organism is an altered mitochondrion that is rich in double-stranded DNA.

Crithidia luciliae is a hemoflagellate possessing a large mitochondrion that contains concentrated mitochondrial DNA in a single large network called the kinetoplast. It is used in immunofluorescence assays to detect the presence of anti-dsDNA antibodies in the blood sera of SLE patients.

Cytotoxicity assays are techniques to quantify the action of immunological effector cells in inducing cytolysis of target cells. The cell death induced is either programmed cell death (apoptosis) in which the dying cell's nuclear DNA disintegrates and the cell membrane increases in permeability, or it leads to necrosis that does not involve active metabolic processes and leads to increased membrane permeability without immediate nuclear disintegration. Cell-mediated cell lysis usually induces apoptosis, whereas antibody and complement usually induce necrosis. Cell death is determined by measurement of increased membrane permeability and by detecting DNA disintegration. The two methods used to determine membrane permeability of cells include dye exclusion in which trypan blue is used to stain dead cells but not viable ones and the other is the chromium-release assay in which target cells are labeled with radioactive ^{51}Cr, which is released from cells that develop increased membrane permeability as a consequence of immune attack. To quantitatively measure either DNA disintegration, ^{125}IUdR or ^{3}H-thymidine can be used to label nuclear DNA.

Cytotoxicity tests: (1) Assays for the ability of specific antibody and complement to interrupt the integrity of a cell membrane, which permits a dye to enter and stain the cell. The relative proportion of cells stained, representing dead cells, is the basis for dye exclusion tests. See microlymphocytotoxicity. (2) The ability of specifically sensitized T lymphocytes to kill target cells whose surface epitopes are the targets of their receptors. Loss of the structural integrity of the cell membrane is signified by the release of a radioisotope such as ^{51}Cr, which was taken up by the target cells prior to the test. The amount of isotope released into the supernatant reflects the extent of cellular injury mediated by the effector T lymphocytes.

EA is an abbreviation for erythrocyte (E) coated with specific antibody (A). This is a technique to measure the activity of Fcγ receptors. Sheep red blood cells with subagglutinating quantities of IgG antibodies are placed in contact with cells at room temperature. IgG Fc receptor-bearing cells will combine with the EA, resulting in rosette formation.

ED_{50} is the 50% effective dose. For example, 50% hemolysis can be determined more accurately than can a 100% endpoint.

Hapten inhibition test is an assay for serological characterization or elucidation of the molecular structure of an epitope by blocking the antigen-binding site of an antibody specific for the epitope with a defined hapten.

Hemadsorption inhibition test: A red blood cell suspension is added to a tissue culture infected with a hemagglutinating virus. Viral hemagglutinin, expressed at the tissue culture cell surfaces, facilitates the hemadsorption of erythrocyte aggregates to the tissue culture surfaces. Antiviral antibody added to the culture prevents this hemadsorption, which serves as the basis for testing for antiviral antibody.

Immunoassay is a test that measures antigen or antibody. When choosing an immunoassay technique, one should keep in mind the differing levels of sensitivity of various methods. Whereas immunoelectrophoresis is relatively insensitive, requiring 5 to 10,000 ng/ml for detection, the enzyme-linked immunoabsorbent assay (ELISA), radioimmunoassay (RIA), and immunofluorescence may detect less than 0.001 ng/ml. Between these two extremes are agglutination that detects 1 to 10,000 ng/ml and complement fixation which detects 5 ng/ml.

Immunoelectroadsorption is a quantitative assay for antibody using a metal-treated glass slide to adsorb antigen followed by antibody from serum. Adsorption is facilitated by an electric current. Measurement of the antibody layer's thickness reflects the serum concentration.

The **inhibition test** is: (1) Blocking an established serological test such as agglutination or precipitation through the addition of an antigen for which the antibody in the test system is specific. It shows the specificity of the reactants. (2) Inhibition of an antigen–antibody interaction through the addition of a hapten for which the antibody is specific. See hapten inhibition test. (3) Preventing the action of a virus through addition of antibody specific for the virus.

Interfacial test: See ring test.

The **leading front technique** is a method to assay chemotaxis or cell migration which evaluates differences in the migration of stimulated and nonstimulated cells.

Mixed agglutination is aggregation (agglutination) produced when morphologically dissimilar cells that share a common antigen are reacted with antibody-specific cells for this epitope. The technique is useful in demonstrating antigens on cells which by virtue of their size or irregular shape are not suitable for study by conventional agglutination tests. It is convenient to use an indicator such as a red cell which possesses the antigen being sought. Thus, the demonstration of mixed agglutination in which the indicator cells are linked to the other cell type suspected of possessing the common antigen constitutes a positive test.

The **mixed-antiglobulin reaction** is a test to demonstrate antibodies adsorbed to cell surfaces. The addition of antiglobulin-coated red cells to a suspension of cells suspected of containing cell surface antibodies results in formation of erythrocyte-test cell aggregates if the test is positive. This is caused by linkage of the antiglobulin to the immunoglobulin on the surface of test cells.

Mixed hemadsorption is the demonstration of antiviral antibody by the mixed-antiglobulin reaction.

Nephelometry is a technique to assay proteins and other biological materials through the formation of a precipitate of antigen and homologous antibody. The assay depends on the turbidity or cloudiness of a suspension. It is based on determination of the degree to which light is scattered when a helium–neon laser beam is directed through the suspension. The antigen concentration is ascertained using a standard curve devised from the light scatter produced by solutions of known antigen concentration. This method is used by many clinical immunology laboratories for the quantification of complement components and immunoglobulins in patients' sera or other body fluids.

P-80 is an assay of an antiserum's ability to precipitate antigen. This test yields data equivalent to that obtained by the quantitative precipitation reaction. A constant quantity of radioisotope-labeled antigen is added to doubling dilutions of antiserum in a row of tubes. The tube in the zone of antigen excess in which precipitation of 80% of the antigen occurs is the end point.

Polyethylene glycol assay for CIC is a method to detect, characterize, and quantitate antigens and antibodies in complexes that sediment in polyethylene glycol. This technique has only borderline clinical utility. In serum sickness conditions, it is able to detect circulating immune complexes (CIC) associated with decreased C4, C3, and CH50.

Protein B is a group B streptococcal protein capable of binding the Fc regions of IgA molecules. It is used in immunoassays and purification techniques for human serum and secretory IgA. It has the unique capacity to bind specifically to human IgA1 and IgA2 subclasses and shows no cross-reactivity with other immunoglobulin classes or serum proteins.

Immunoprecipitation is a method to recover or isolate an antigen molecule from solution by uniting it with an antibody and rendering the antigen–antibody complex insoluble through interaction with an anti-antibody or through coupling the first antibody to an insoluble support such as a particle or bead.

The **radioimmunoprecipitation assay (RIPA)** is a method that demonstrates the presence of antibodies against viral constituents. Virus grown in culture in the presence of radioactive amino acid is disrupted and incubated with a test sample that might contain antibodies specific for viral antigens. This is followed by polyacrylamide gel electrophoresis of the immunoglobulins in the test sample.

The **sia test (historical)** is a former qualitative test for macroglobulinemia in which the patient's serum was placed in water in one tube and in saline in another tube. Precipitation of the serum in water, attributable to IgM's low water solubility, but not in saline, constituted a positive test.

Streptolysin O test: See ASO.

STS: Abbreviation for serological test for syphilis.

TPI: See *Treponema pallidum* immobilization test.

The ***Treponema pallidum* immobilization test** is a diagnostic test for syphilis in which living, motile *T. pallidum* microorganisms are combined with a sample of serum presumed to contain specific antibody. Complement is also present. If the serum sample contains anti-*Treponema* antibody, the motile microorganisms become immobilized. This test is much more specific for the diagnosis of syphilis than is the complement fixation test such as the old Wassermann reaction. It has been rarely used because living *T. pallidum* microorganisms are not readily available and are hazardous to work with.

Trypanosome adhesion test: Selected mammalian red blood cells or bacteria may stick to the surface of trypanosomes when specific antibody and complement are present. This has been used in the past as a test for antibodies to trypanosomes.

The **Rieckenberg reaction** is a trypanosome immune adherence test. Anticoagulated blood of an animal that recovered from trypanosomiasis is combined with live trypanosomes. Provided the same antigenic type of trypanosome that produced the infection is used for the test, blood platelets adhere to the trypanosomes.

Turbidimetry is the quantification of a substance in suspension based on the suspension's ability to reduce forward light transmission.

Radioimmunoassay (RIA) refers to a technique to assay either antigen or antibody which is based on radiolabeled antigen competitively inhibiting the binding of antigen, which is not labeled, to specific antibodies (Figure 27.11 to Figure 27.13). Minute quantities of enzymes, hormones, or other immunogenic substances can be assayed by RIA. Enzyme immunoassays are largely replacing RIAs because of the problems associated with radioisotope regulation and disposal.

RIA: Radioimmunoassay.

Double diffusion test (Figure 27.14) is a test in which solutions of antigen and antibody are placed in separate wells of a gel containing electrolyte. The antigen and antibody diffuse toward one another until their molecules meet at the point of equivalence and precipitate, forming a line of precipitation in the gel. In addition to the two-dimensional technique, double immunodiffusion may be accomplished in a tube as a one-dimensional technique. The Oakley-Fulthorpe test is an example of this type of reaction. This technique may be employed to detect whether antigens are similar or different or share epitopes. It may also be used to investigate antigen and antibody purity. See also reaction of identity, reaction of nonidentity, and reaction of partial identity.

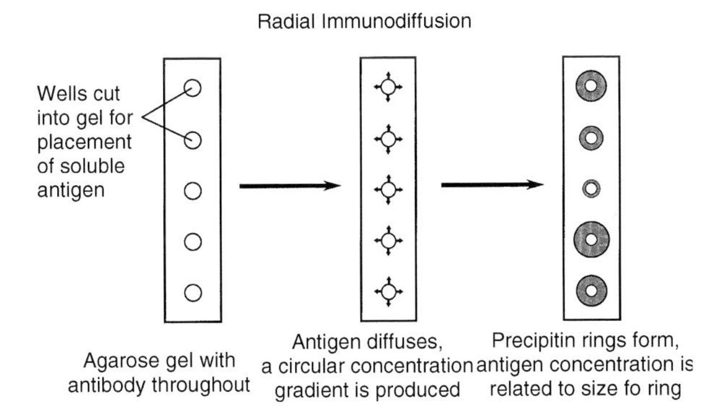

FIGURE 27.11 Radial immunodiffusion.

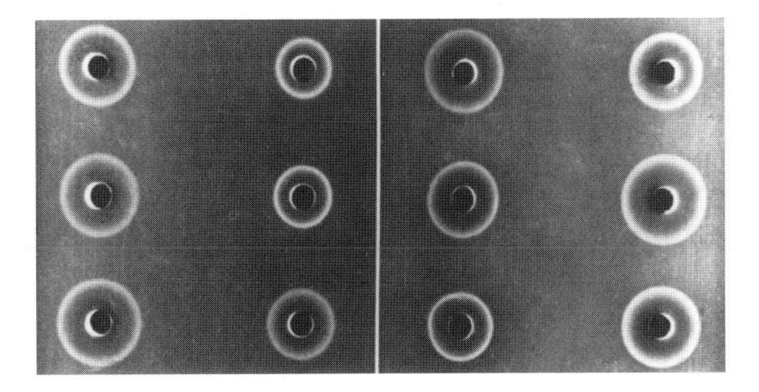

FIGURE 27.12 Single radial immunodiffusion.

The **cloned enzyme donor immunoassay** is a homogeneous enzyme immunoassay (EIA) based on the modulation of enzyme activity by bound fragments of beta-galactosidase.

Hormone immunoassays: Multiple hormones, including thyroid-stimulating hormone, human growth hormone, insulin, glucagons, and many others may be measured by an immunoassay using either radioactively labeled reagents or through enzyme color reactions using the ELISA technique. Labeled and unlabeled hormone are allowed to compete for binding sites with antihormone antibody. This is followed by the separation of bound from unbound hormone by one of several techniques.

The **solid-phase radioimmunoassay** requires the attachment of antigen (or antibody) to an insoluble support, which can be used to capture antibodies (or antigens) in a specimen to be assayed. Antibodies in a serum sample are exposed to excess antigen on an insoluble support and sufficient time is allowed for antigen–antibody interaction. This is followed by washing and the application of radiolabelled anti-Fc antibodies specific for the Fc regions of the captured antibodies. After washing, quantification of the bound antibody is determined from the amount of radioactivity adhering to the insoluble support. Various materials

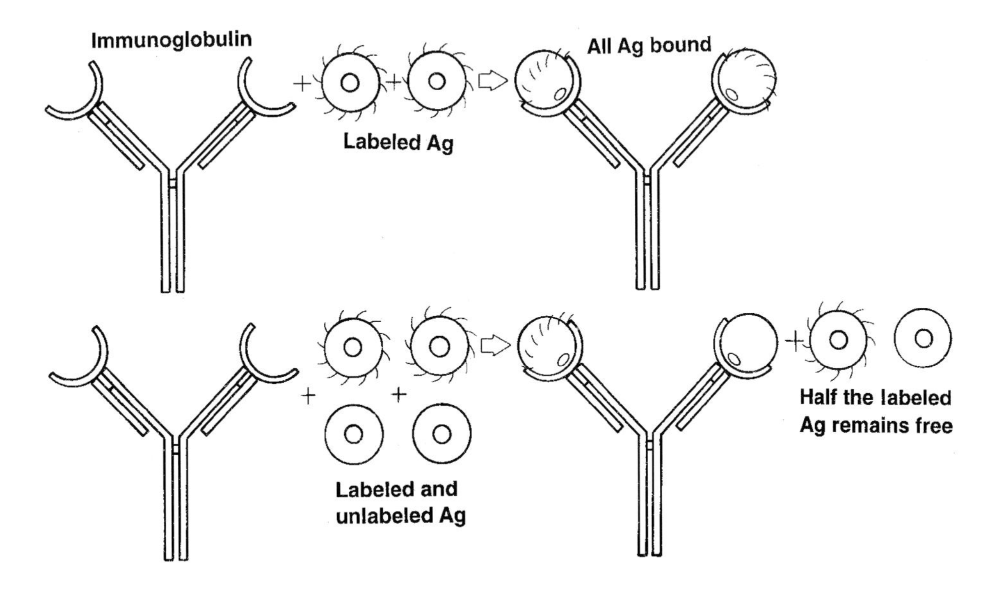

FIGURE 27.13 Radioimmunoassay (RIA).

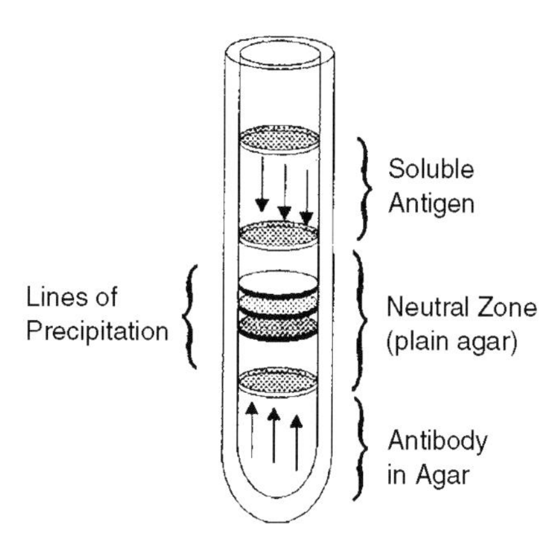

FIGURE 27.14 Double diffusion test.

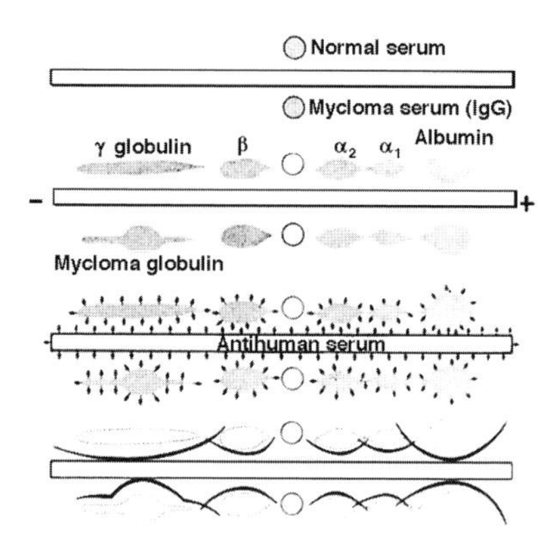

FIGURE 27.15 Immunoelectrophoresis (IEP).

may be used as an appropriate insoluble support. These include Sepharose® beads or tissue culture plate wells. An unrelated protein must be used to coat the insoluble support prior to application of the specific antibody to saturate areas of the insoluble support where antigen is not located.

The **end-point immunoassay** is a test in which the signal is measured as the antigen–antibody complex reaches equilibrium.

Ascoli's test is a ring precipitin assay used in the past to identify anthrax antigen in tissues, skins, and hides of animals infected with *Bacillus anthracis*. The simple test was considered useful in that it could identify anthrax antigen in decaying material from which anthrax bacilli could no longer be cultured.

Two-dimensional gel electrophoresis is a technique to separate proteins by isoelectric focusing in one dimension, followed by SDS-PAGE on a slab gel at right angles to the first dimension. The protein to be analyzed is subjected to isoelectric focusing by placing the soluble protein in a pH gradient and applying an electrical charge. The protein moves to the pH where it has a neutral charge. This is followed by electrophoresis in gel at a 90° angle to separate the proteins according to size. This procedure yields a pattern known as a fingerprint that is very specific. Large numbers of distinct proteins can be separated and identified by this technique.

Immunoelectrophoresis (IEP) (Figure 27.15) is a method to identify antigens on the basis of their electrophoretic mobility, diffusion in gel, and formation of precipitation arcs with specific antibody. Electrophoresis in gel is combined with diffusion of a specific antibody in a gel medium containing electrolyte to identify separated antigenic substances. The presence or absence of immunoglobulin molecules of various classes in a serum sample may be identified in this way. One percent agar containing electrolyte is layered onto microscope slides, allowed to gel, and patterns of appropriate troughs and wells are cut in the solidified medium. Antigen to be identified is placed in the circular wells cut into the agar medium. This is followed by electrophoresis, which permits separation of the antigenic components according to their electrophoretic mobility, and antiserum is placed in a long trough in the center of the slide. After antibody has diffused through the agar toward each separated antigen, precipitin arcs form where the antigen and antibody interact. Abnormal amounts of immunoglobulins result in changes in the shape and position of precipitin arcs when compared with the arcs formed by antibody against normal human serum components. With monoclonal gammapathies, the arcs become broad, bulged, and displaced. The absence of immunoglobulin classes such as those found in certain immunodeficiencies can also be detected with IEP.

IE: Abbreviation for immunoelectrophoresis.

IEP: Abbreviation for immunoelectrophoresis.

Immunoelectroosmophoresis: See counterimmunoelectrophoresis.

CIE: Abbreviation for counterimmunoelectrophoresis or crossed immunoelectrophoresis.

Immunoosmoelectropheresis: See counterimmunoelectrophoresis.

Radioimmunoelectrophoresis is a type of immunoelectrophoresis that employs radiolabeled antibody or antigen to identify individual precipitin arcs by subsequent autoradiography of the arcs.

IFE: Abbreviation for immunofixation electrophoresis.

Countercurrent electrophoresis: See counterimmunoelectrophoresis.

Counter electrophoresis: See counterimmunoelectrophoresis.

Counterimmunoelectrophoresis (CIE) is an immunoassay in which antigen and antibody are placed into wells in agar gel and followed by electrophoresis in which the antigen that carries a negative charge migrates toward the antibody which moves toward the antigen by electroendosomis. Interaction of antigen and antibody molecules in the gel leads to the formation of a precipitin line. The method has been used to identify serotypes of *Streptococcus pneumoniae, Neisseria meningitidis* groups, and *Haemophilus influenzae* type b.

Counter migration electrophoresis: See counterimmunoelectrophoresis.

Crossed immunoelectrophoresis is a gel diffusion method employing two-dimensional immunoelectrophoresis. Protein antigens are separated by gel electrophoresis. This is followed by the insertion of a segment of the gel into a separate gel into which specific antibodies have been incorporated. The gel is then electrophoresed at right angles to the first electrophoresis, forcing the antigen into the gel containing antibody. This results in the formation of precipitin arcs in the shape of a rocket that resembles bands formed in the Laurell rocket technique.

Tandem immunoelectrophoresis is a method that is a variation of crossed immunoelectrophoresis in which the material to be analyzed is placed in one well cut in the gel and the reference antigen is placed in a second well. Following electrophoresis in one direction, it is repeated at a right angle which drives the antigens that have been separated into another gel containing specific antibodies. Planes of precipitation form and are observed to determine whether or not they share identity with the reference antigen.

The **Lancefield precipitation test** is a ring precipitation test developed by Rebecca Lancefield to classify streptococci according to their group-specific polysaccharides. The polysaccharide antigen is derived by treatment of cultures of the microorganisms with HCl, formimide, or a *Streptomyces albus* enzyme. Antiserum is first placed into a serological tube, followed by layering the polysaccharide antigen over it. A positive reaction is indicated by precipitation at the interface.

An **Oudin test** (Figure 27.16) is a type of precipitation in gel that involves single diffusion. Antiserum, incorporated into agar, is placed in a narrow test tube. This is overlaid with an antigen solution which diffuses into the agar to yield precipitation rings. Also called single radial diffusion test. A band of precipitation forms at the equivalence point.

The **Ouchterlony test** (Figure 27.17) is a double diffusion in a gel type of precipitation test. Antigen and antibody solutions are placed in separate wells that have been cut into an agar plate prepared with electrolyte. As the antigen and antibody diffuse through the gel medium, a line of precipitation forms at the point of contact between antigen and antibody. Results are expressed as reaction of identity, reaction of partial identity, or reaction of nonidentity. See those entries for further details.

The **Oakley-Fulthorpe test** (Figure 27.18) is a double-diffusion type of precipitation test performed by incorporating antibody into agar which is placed in the tube followed by a layer of plain agar. A solution of antigen is

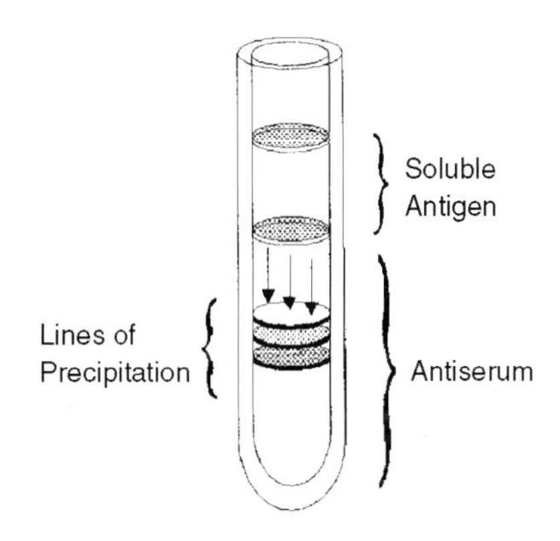

FIGURE 27.16 Oudin test.

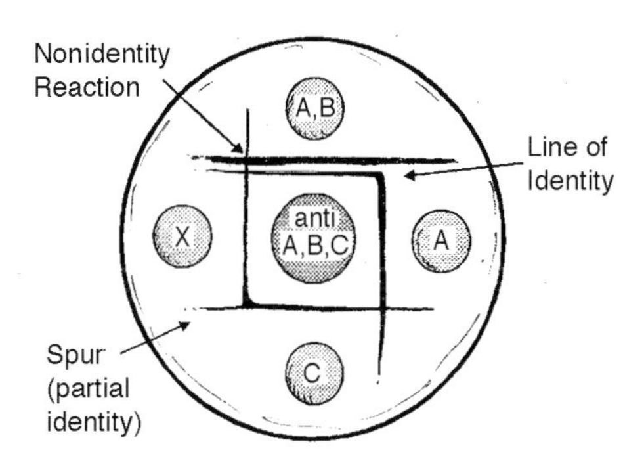

FIGURE 27.17 Ouchterlony test.

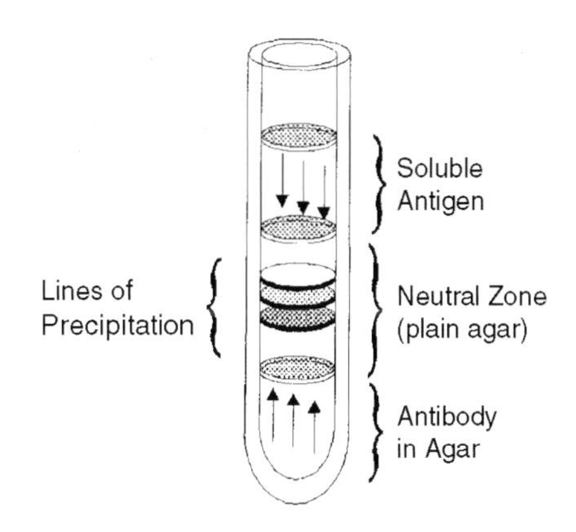

FIGURE 27.18 Oakley-Fulthorpe test.

placed on top of the plain agar in the tube and precipitation occurs where antigen and antibody meet in the plain agar layer.

Elek plate (Figure 27.19) is a method to show toxin production by *Corynebacterium diphtheriae* colonies growing on an agar plate. Diphtheria antitoxin impregnated into a strip of filter paper is placed at a right angle to a streak of the microorganisms on the agar plate. Toxin formation by the growing microbes interacts with antitoxin in the filter paper to form a line of precipitation.

Rocket electrophoresis (Figure 27.20) describes the electrophoresis of antigen into an agar-containing specific antibody. In this electroimmunodiffusion method, lines of precipitation formed in the agar by the antigen–antibody interaction assume the shape of a rocket. The antigen concentration can be quantified since the rocket-like area is proportional to the antigen concentration. This can be deduced by comparing with antigen standards. This technique has the advantage of speed since it can be completed within hours instead of longer periods required for single radial immunodiffusion. Also called Laurell rocket electrophoresis.

FIGURE 27.19 Elek plate.

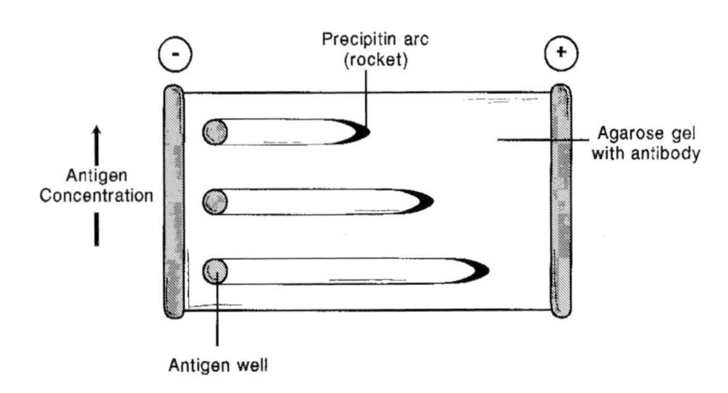

FIGURE 27.20 Rocket electrophoresis.

Agarose is a neutral polygalactoside consisting of alternating *d*-galactose and 3,6-anhydrogalactose linear polymer, the principal constituent of agar. Gels made from agarose are used for the hemolytic plaque assay and for leukocyte chemotaxis assays, as well as for immunodiffusion and nucleic acid/protein electrophoresis.

Electrophoresis is a method for separating a mixture of proteins based on their different rates of migration in an electrical field. Zone electrophoresis represents a technical improvement in which a stabilizing medium such as cellulose acetate serves as a matrix for buffer and as a structure to which proteins can remain attached following fixation. By this technique, plasma proteins are resolved into five or six major peaks. Zone electrophoresis permits a gross evaluation of the levels of immunoglobulins and other proteins in the serum. In cases of increased levels, electrophoresis indicates whether this involves a general proliferation and hypersecretion by lymphocytes derived from multiple individual cells (polyclonal origin; the proteins are heterogeneous) or involves proliferation and hypersecretion by lymphoid cells derived from a limited number of individual cells (monoclonal origin; the proteins are homogeneous).

Electrophoretic mobility is the electrophoretic velocity, v, of a charged particle expressed per unit field strength; hence, $u = v/E$, where E is the field strength. The value of u is positive if the particle moves towards the pole of lower potential and negative in the opposite case. The electrophoretic mobility depends only on molecular parameters.

A **lane** is the path of migration of a molecule of interest from a well or point of application in gel electrophoresis. A substance is propelled within the confines of this path or corridor by an electric current that induces migration and separation of the molecules into bands according to size.

SDS-PAGE refers to polyacrylamide gel electrophoresis in sodium dodecyl sulfate. See also polyacrylamide gel electrophoresis.

Sepharose® is the trade name for agarose gels used in electrophoresis.

The **Laurell rocket test** is a method to quantify protein antigens by rapid immunoelectrophoresis. Antiserum is incorporated in agarose into which wells are cut and protein antigen samples are distributed. The application of an electric current at 90° angles to the antigen row drives antigen into the agar. Dual lines of immune precipitate emanate from each well and merge to form a point where no more antigen is present, producing a structure which resembles a rocket. The amount of antigen can be determined by measuring the rocket length from the well to the point of precipitate. This length is proportional to the total amount of antigen in the preparation.

Pulsed-field gel electrophoresis is a method for separating DNA molecules that vary from a few kilobase pairs (kbp) to 4000 kbp. The direction of the electric field is repeatedly altered, causing the molecules to change direction of migration and to enter new pores in the gel. Thus, both small and large DNA molecules migrate through the gel based on their size. Migration of the smaller molecules is more rapid than that of the larger molecules.

Laurell crossed immunoelectrophoresis: See crossed immunoelectrophoresis.

Fluorography is a method to identify radiolabeled proteins following their separation by gel electrophoresis. A fluor such as diphenyl oxazole is incorporated into the gel where it emits photons on exposure to a radioisotope. After drying, the gel is placed on x-ray film in the dark.

Zone electrophoresis is the separation of proteins on cellulose acetate (or on paper) based upon the charge when an electric current is passed through the gel.

FIGE: Field inversion gel electrophoresis. See pulsed-field gradient gel electrophoresis.

Density gradient centrifugation is the centrifugation of relatively large molecules, such as in a solution of DNA, with a density gradient substance such as cesium chloride. This method also permits the separation of different types of cells as they are centrifuged through a density gradient produced by a substance to which they are impermeable. A commonly used material is Ficoll-Hypaque. Separation of cells is according to size as they progress through the gradient. When they reach the level where their specific gravity is the same as that of the medium, cell bands of different density are produced. This technique is widely employed to separate hematopoietic cells.

Electroimmunodiffusion (Figure 27.21) is a double-diffusion in-gel method in which antigen and antibody are forced toward one another in an electrical field. Precipitation occurs at the site of their interaction. See Laurell rocket test and rocket immunoelectrophoresis. Also called counter immunoelectrophoresis.

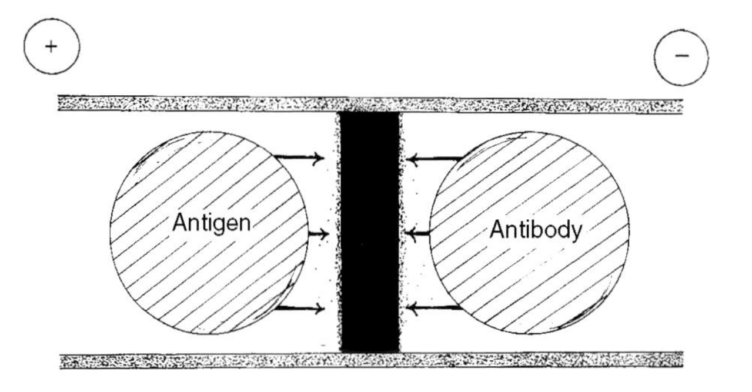

FIGURE 27.21 Electroimmunodiffusion.

Ultrafiltration is the passage of solutions or suspensions through membranes with minute pores of graded sizes.

Ultracentrifugation is the separation of cell components, including organelles and molecules, through high-speed centrifugation reaching 6000 rpm with a gravitational force up to 500,000 g. In differential velocity centrifugation, there is a stepwise increase in gravitational force to remove selected components. Following centrifugation of a cellular homogenate at 600 g for 10 min to isolate the nuclei, further spinning at 15,000 g for 5 min permits isolation of mitochondria, lysozomes, and peroxisomes. Respinning at 100,000 g for 1 h permits isolation through sedimentation of the plasma membrane, microsomal fraction, endoplasmic reticulum, and large polyribosomes. Respinning at 300,000 g for 2 h permits sedimentation of ribosomal subunits and small polyribosomes. This leaves the cytosol, which is the soluble portion of the cytoplasm. Separation can also be achieved by sucrose density gradient ultracentrifugation. Cesium chloride combined with molecules to be analyzed permits the molecules to migrate to a particular density equivalent. Ultracentrifugation may be either an analytical method, which is used to identify proteins that differ in sedimentation coefficient, or as a preparative method to separate proteins based on their densities and shapes.

S is the abbreviation for sedimentation coefficient.

Svedberg unit refers to a sedimentation coefficient unit that is equal to 10^{-13} sec. Whereas most immunoglobulin molecules, such as IgG, sediment at 7 S, the pentameric IgM molecule sediments at 19 S.

The **sedimentation coefficient** is the rate at which a macromolecule or particle sediment that is equivalent to the velocity per unit centrifugal field. $s = (dx/dt)/w^2x$. The sedimentation coefficient is s, the velocity is dx/dt, and the angular velocity is w. The distance from the axis of the centrifuge rotor is x. The size, shape, and weight of the macromolecule in question, as well as the concentration and temperature of the solutions, but not the centrifuge speed, determine the sedimentation coefficient. Measurement is in Svedberg units.

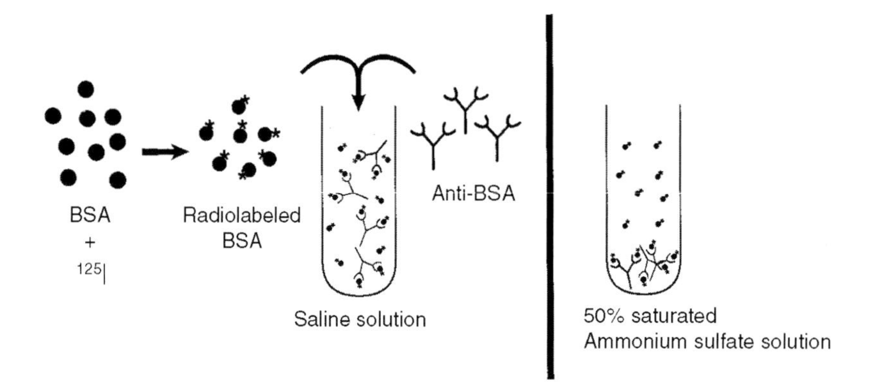

FIGURE 27.22 Farr technique.

Sedimentation pattern refers to the configuration of red blood cells on a test tube or plastic plate bottom at the conclusion of a hemagglutination test. The formation of a covering mat on the curved bottom of the tube or well signifies that agglutination has taken place. The formation of a round button where the red blood cells have settled to the midpoint of the bottom of the tube or well and were not retained on the curvature constitutes a negative reaction with no agglutination.

Zonal centrifugation is the separation of molecules according to size based on molecular mass and centrifugation time.

Percoll® is a density gradient centrifugation medium used to isolate certain cell populations such as natural killer (NK) cells. It is a colloidal suspension.

Salt precipitation is an earlier method to separate serum proteins based on the principle that globulins precipitate when the concentration of sodium sulfate or ammonium sulfate is less than the concentration at which albumin precipitates. Euglobulins precipitate at concentrations that are less than those at which pseudoglobulins precipitate. This method was largely replaced by chromatographic methods using Sephadex® beads and related techniques.

Salting out refers to salt precipitation of serum proteins such as globulins.

S value (Sverdberg unit) refers to the sedimentation coefficient of a protein that is ascertained by analytical ultracentrifugation.

The **Farr technique** (Figure 27.22) is an assay to measure primary binding of antibody with antigen as opposed to secondary manifestations of antibody–antigen interactions such as precipitation, agglutination, etc. It is a quantitation of an antiserum's antigen-binding properties and is appropriate for antibodies of all immunoglobulin classes and subclasses. The technique is limited to the assay of antibody against antigens soluble in 40% saturated ammonium sulfate solution in which antibodies precipitate. Following interaction of antibody with radiolabeled antigen, precipitation in ammonium sulfate separates the bound antigen from the free antigen. The quantity of radiolabeled antigen that reacted with antibody can be measured in the precipitate. The antibody dilution that precipitates part of the ligand reflects the antigen-binding ability.

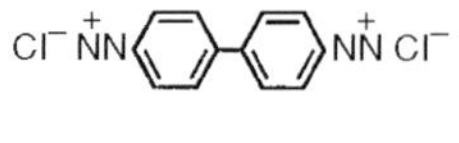

FIGURE 27.23 Bis-diazotized benzidine.

Bis-diazotized benzidine refers to a chemical substance that serves as a bivalent coupling agent that can link to protein molecules (Figure 27.23). This method was used in the past to conjugate erythrocytes with antigens for use in the passive agglutination test.

BDB: See bis-diazotized benzidine.

The **AET rosette test (historical)** is a technique used previously to enumerate human T cells based upon the formation of sheep red cell rosettes surrounding them. The use of sheep red cells treated with aminoethylthiouridium bromide renders the rosettes more stable than when using sheep red cells untreated by this technique. This technique was later replaced by the use of anti-CD2 monoclonal antibodies and flow cytometry.

Erythrocyte agglutination test: See hemagglutination test.

Hemagglutination (Figure 27.24) is the aggregation of red blood cells by antibodies, viruses, lectin, or other substances. The hemagglutination inhibition test is an assay for antibody or antigen based on the ability to interfere with red blood cell aggregation. Certain viruses are able to agglutinate red blood cells. In the presence of antiviral antibody, the ability to agglutinate erythrocytes is inhibited. Thus, this serves as a basis to assay the antibody.

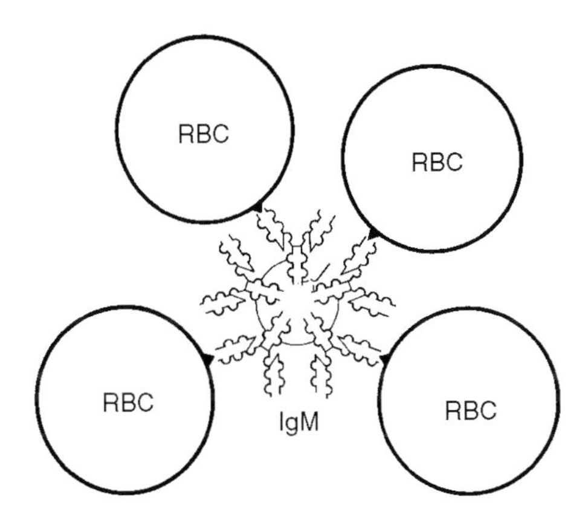

FIGURE 27.24 Hemagglutination.

Minimal hemagglutinating dose (MHD): In the hemagglutination inhibition test for antiviral antibodies, the MHD is the least amount of hemagglutinating virus that will completely agglutinate the red cells in a single volume of a standard suspension.

The **Paul-Bunnell test** is an assay for heterophile antibodies in infectious mononucleosis patients. It is a hemagglutination test in which infectious mononucleosis patient serum induces sheep red blood cell agglutination. Absorption of the serum with guinea pig kidney tissue removes antibody to the Forssman antigen but does not remove the sheep red blood cell agglutinin which can be absorbed with ox cells. This hemagglutinin is distinct from antibodies against the causative agent of infectious mononucleosis, i.e., the Epstein-Barr virus.

The **red cell-linked antigen antiglobulin test** is a passive hemagglutination test in which the red cells serve only as carriers for antigen coated on their surfaces. It can identify either agglutinating antibodies or nonagglutinating (incomplete) antibodies by the aggregation or clumping of antigen-bearing red cells. To perform the assay, the test serum is incubated with red cells treated with antigen, which are then washed and antibody against human globulin is added.

The **sheep red blood cell agglutination test** is an assay in which sheep erythrocytes are either agglutinated by antibody or are used as carrier particles for an antigen adsorbed to their surface, in which case they are passively agglutinated by antibodies specific for the adsorbed antigen.

TPHA: See *Treponema pallidum* hemagglutination assay.

The ***Treponema pallidum*** **hemagglutination assay** is a test for antibodies specific for *T. pallidum* used formerly to diagnose syphilis. *T. pallidum* antigens were coated onto sheep red blood cells treated with tannic acid and formalin. Aggregation of the antigen-coated red cells signified that antibody was present.

Rose-Waaler test: Sheep red blood cells are treated with a subagglutinating quantity of rabbit anti-sheep erythrocyte antibody. These particles may be used to identify rheumatoid factor in the serum of rheumatoid arthritis (RA) patients. Agglutination of the IgG-coated red cells constitutes a positive test and is based upon immunological crossreactivity between human and rabbit IgG molecules. It may be positive in collagen vascular diseases other than RA, but it has still proven beneficial in diagnosis.

The **antiglobulin consumption test** is an assay to test for the presence of an antibody in serum which is incubated with antigen-containing cells or antigen-containing particles. After washing, the cells or particles are treated with antiglobulin reagents and incubated further. If any antibody has complexed with the cells or particles, antiglobulin will be taken up. Antiglobulin depletion from the mixture is evaluated by assaying the free antiglobulin in the supernatant through combination with incomplete antibody-coated erythrocytes. No hemagglutination reveals that the antiglobulin reagent was consumed in the first step of the reaction and shows that the original patient's serum contained the antibody in question.

Tanned red cells are prepared by treating a suspension of erythrocytes with a 1:20,000 to 1:40,000 dilution of tannic acid that renders their surfaces capable of adsorbing soluble antigen. Thus, they have been widely used as passive carriers of soluble antigens in passive hemagglutination reactions. By adding toluene diisocyanate, the protein can become covalently bound to the red cell surface. However, this is not necessary for routine hemagglutination reactions.

The **tanned red cell test** is a passive hemagglutination assay in which red blood cells are used only as carrier particles for soluble antigens. Agglutination of the cells by specific antibody signifies a positive reaction. To render erythrocytes capable of adsorbing soluble protein antigens to their surface, the cells are treated with a weak tannic acid solution. This promotes cell surface attachment of the soluble protein antigen.

The **latex fixation test** (Figure 27.25) is a technique in which latex particles are used as passive carriers of soluble antigens adsorbed to their surfaces. Antibodies specific for the adsorbed antigen then cause agglutination of the coated latex particles. This has been widely used and is the basis of an RA test in which pooled human IgG molecules are coated on the surface of latex particles which are then agglutinated by antiimmunoglobulin antibodies in the sera of RA patients.

Latex particles are inert particles of defined size that are used as carriers of either antigens or antibodies in latex agglutination immunoassays. An example is the RA test

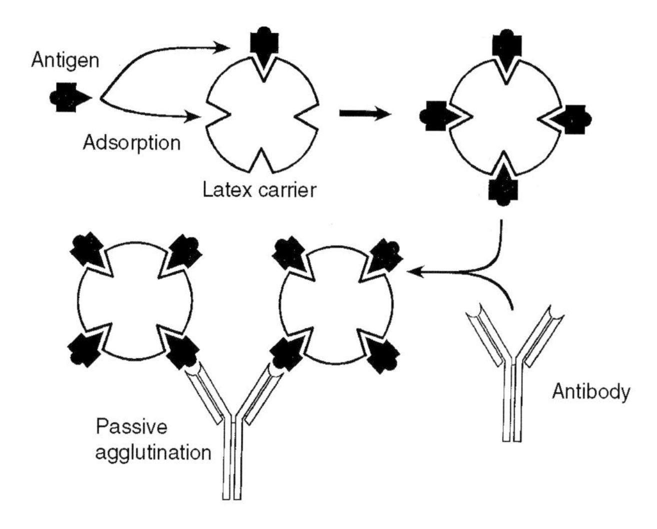

FIGURE 27.25 Latex fixation test.

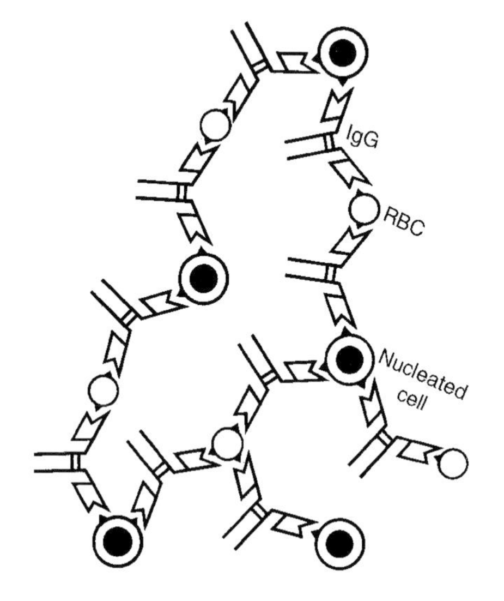

FIGURE 27.26 Mixed agglutination.

in which latex particles are coated with pooled human IgG that serves as antigen. These IgG-coated particles are agglutinated by rheumatoid factor (antiimmunoglobulin antibody) that may be detected in an RA patient's serum.

Mixed agglutination (Figure 27.26) describes the aggregation (agglutination) produced when morphologically dissimilar cells that share a common antigen are reacted with antibody specific for this epitope. The technique is useful in demonstrating antigens on cells that by virtue of their size or irregular shape are not suitable for study by conventional agglutination tests. It is convenient to use an indicator, such as a red cell that possesses the antigen being sought. Thus, the demonstration of mixed agglutination in which the indicator cells are linked to the other cell type suspected of possessing the common antigen constitutes a positive test.

Bentonite ($Al_2O_3 \cdot 4SiO_2 \cdot H_2O$) is aluminum silicate that is hydrated and colloidal. This insoluble particulate substance has been used to adsorb proteins, including antigens. It was used in the past in the bentonite flocculation test.

Biolistics refers to the coating of small particles, such as colloidal gold, with an agent such as a drug, for nucleic acid or other substance that is to be conveyed into a cell. A helium-powered gun is employed to fire particles into the recipient's dermis.

Inert particle agglutination tests are assays that employ particles of latex, bentonite, or other inert materials to adsorb soluble antigen on their surface to test for the presence of specific antibody in the passive agglutination test. The particles coated with adsorbed antigen agglutinate if antibody is present. An example is the RA test in which pooled human IgG is adsorbed to latex particles that agglutinate if combined with a serum sample containing rheumatoid factor, i.e., anti-IgG autoantibody.

Passive agglutination refers to the aggregation of particles with soluble antigens adsorbed to their surfaces by the homologous antibody. The soluble antigen may be linked to the particle surface through covalent bonds rather than by mere adsorption. Red blood cells, latex, bentonite, or collodion particles may be used as carriers for antigen molecules adsorbed to their surfaces. When the red blood cell is used as a carrier particle, its surface has to be altered in order to facilitate maximal adsorption of the antigen to its surface. Several techniques are employed to accomplish this. One is the tanned red blood cell technique that involves treating the red blood cells with a tannic acid solution that alters their surface in a manner that favors the adsorption of added soluble antigen. A second method is the treatment of red cell preparations with other chemicals such as bis-diazotized benzidine.

Indirect agglutination (passive agglutination) is the aggregation or agglutination of a specific antibody with carrier particles such as latex particles or tanned red blood cells to which antigens have been adsorbed or with bis-diazotized red blood cells to which antigens have been linked chemically. See passive agglutination.

With this passive agglutination technique, even relatively minute quantities of soluble antigens may be detected by the homologous antibody agglutinating carrier cells on which they are adsorbed. Since red blood cells are the most commonly employed particle, the technique is referred to as passive hemagglutination. Latex particles are used in the RA test in which pooled IgG molecules are adsorbed to latex particles and reacted with sera of RA patients containing rheumatoid factor (IgM anti-IgG antibody) to produce agglutination. Polysaccharide antigens will stick to red blood cells without treatment. When proteins are used, however, covalent linkages are required.

The **RPR (rapid plasma reagin) test** is an agglutination test used in screening for syphilis. Antilipoidal (nontreponemal) antibodies (reagins) develop in the host usually within 4 to 6 weeks after infection with *Treponema pallidum* . Of patients with primary syphilis, 93% develop a positive RPR.

Slide agglutination test refers to the aggregation of particulate antigen such as red blood cells, microorganisms, or latex particles coated with antigen within 30 sec following contact with specific antibody. The reactants are usually mixed by rocking the slide back and forth, and agglutination is observed both macroscopically and microscopically. The test has been widely used in the past for screening, but is unable to distinguish reactions produced by crossreacting antibodies which can be ruled out in a tube test that allows dilution of the antiserum.

Slide flocculation test: See slide agglutination test.

The **tube agglutination test** is an agglutination assay that consists of serial dilutions of antiserum in serological tubes to which particulate antigen, such as microorganisms, is added.

The **Weil-Felix reaction** is a diagnostic agglutination test in which *Proteus* bacteria are agglutinated by the sera of patients with typhus. The reaction is based upon the cross-reactivity of the carbohydrate antigen shared between *Rickettsiae* and selected *Proteus* strains. Various rickettsial diseases can be diagnosed based upon the reaction pattern of antibodies in the blood sera of rickettsial disease patients with O-agglutinable strains of *Proteus* OX19, OX2, and OX12.

The **Widal reaction** is a bacterial agglutination test employed to diagnose enteric infections caused by *Salmonella*. Doubling dilutions of patient serum are combined with a suspension of microorganisms known to cause enteric fever such as *S. typhi*, *S. paratyphi* B, and *S. paratyphi* A and C. The microorganisms used in the test should be motile, smooth, and in the specific phase. To assay H agglutinins, formalin-treated suspensions are used, and to assay O agglutinin, alcohol-treated suspensions are employed. The Widal test is positive after the tenth day of the disease and may be false-positive if an individual previously received a TAB vaccine. Thus, it is important to repeat the test and observe a rising titer rather than to merely observe a single positive test. Widal originally described the test to diagnose *S. paratyphi* B infection.

Indirect hemagglutination test: See passive agglutination test.

The **passive agglutination test** (Figure 27.27) is an assay to recognize antibodies against soluble antigens that are attached to erythrocytes, latex, or other particles by either adsorption or chemical linkage. In the presence of antibodies specific for the antigen, aggregation of the passenger particles occurs. Examples of this technique include the RA latex agglutination test, the tanned red cell technique, the bentonite flocculation test, and the bis-diazotized benzidine test.

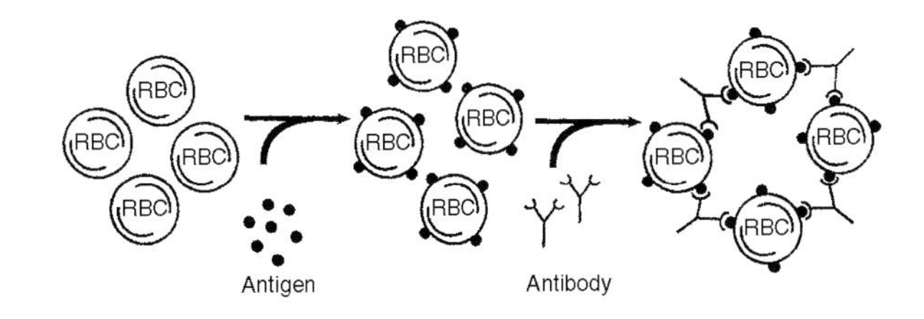

FIGURE 27.27 Passive agglutination test.

$$\begin{array}{ccccc} & H & & SH & \\ & | & & | & \\ H- & C & - & C & -OH \\ & | & & | & \\ & H & & H & \end{array}$$

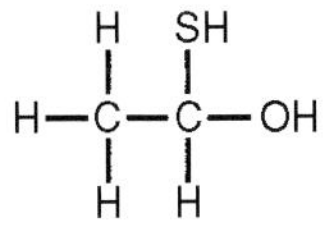

FIGURE 27.28 2-Mercaptoethanol agglutination test.

The **Takatsy method** is a technique that employs tiny spiral loops on the end of a handle that resembles those used for wire loops by bacteriologists. The loops are carefully engineered to retain a precise volume when immersed in a liquid. They are used to prepare doubling dilutions of a test liquid in microtiter wells of test plates. As the loops are passed from one well to the next, a spiral motion helps to discharge the contents into the well diluent and mix it. Several loops can be manipulated by one operator at the same time using a single plastic plate with multiple wells. This method has been applied to hemagglutination assays.

Microtiter technique: See Takatsy method.

The **vaginal mucous agglutination test** is an assay for antibodies in bovine vaginal mucous from animals infected with *Campylobacter fetus, Trichomonas fetus*, and *Brucella abortus*. The mucous can be used in the same manner as serum for a slide or tube agglutination test employing the etiologic microorganisms as antigen.

The **2-mercaptoethanol agglutination test** (Figure 27.28) is a simple test to determine whether or not an agglutinating antibody is of the IgM class. If treatment of an antibody preparation, such as a serum sample, with 2-mercaptoethanol can abolish the serum's ability to produce agglutination of cells, the agglutination was due to IgM antibody. Agglutination induced by IgG antibody is unaffected by 2-mercaptoethanol treatment and just as effective after the treatment as it was before. Dithiothreitol (DTT) produces the same effect as 2-mercaptoethanol in this test.

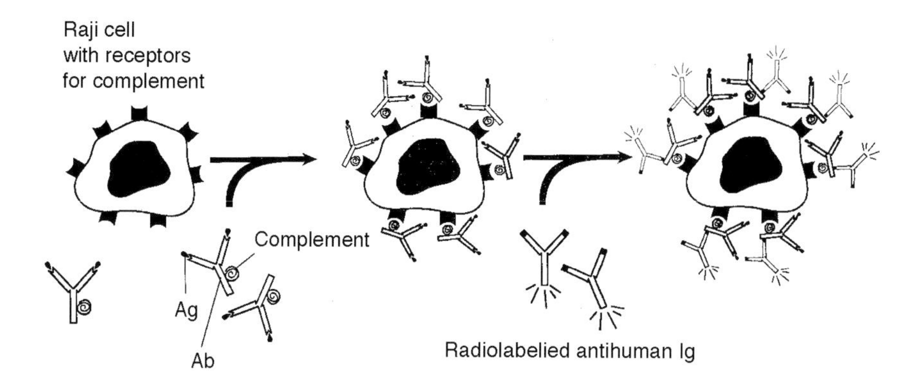

FIGURE 27.29 Raji cell assay.

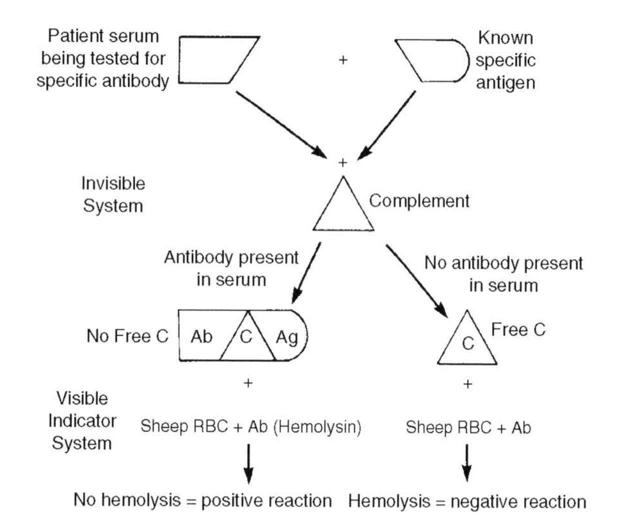

FIGURE 27.30 Complement fixation reaction.

An **immobilization test** is a method for the identification of antibodies specific for motile microorganisms by determining the ability of antibody to inhibit motility. This may be attributable to adhesion or agglutination of the microorganisms' flagella or cell wall injury when complement is present.

The **Raji cell assay** (Figure 27.29) is an *in vitro* assay for immune complexes in serum. The technique employs Raji cells, a lymphoblastoid B lymphocyte tumor cell line that expresses receptors for complement receptor 1, complement receptor 2, FCγ and C1q receptors. The cell line does not express surface immunoglobulins. Following combination of Raji cells with the serum sample, the immune complex is bound and quantified using radiolabeled $F(ab')_2$ fragments of antibodies against IgG.

In the **complement fixation reaction** (Figure 27.30) the primary union of antigen with antibody takes place almost instantaneously and is invisible. A measured amount of complement present in the reaction mixture is taken up by complexes of antigen and antibody. The consumption or binding of complement by antigen–antibody complexes, this serves as the basis for a serologic assay in which antigen is combined with a serum specimen suspected of containing the homologous antibody. Following the addition of a measured amount of complement, which is fixed or consumed only if antibody was present in the serum and has formed a complex with the antigen, sheep red blood cells, sensitized (coated) with specific antibody, are added to determine whether or not the complement has been fixed in the first phase of the reaction, implying that homologous antibody was not present in the serum, and complement remains free to lyse the sheep red blood cells sensitized with antibody. Hemolysis constitutes a negative reaction. The sensitivity of the complement fixation text falls between that of agglutination and precipitation. Complement fixation tests may be carried out in microtiter plates, which are designed for the use of relatively small volumes of reagents. The lysis of sheep red blood cells sensitized with rabbit antibody is measured either in a spectrophotometer at 413 nm or by the release of ^{51}Cr from red cells that have been previously labeled with the isotope. Complement fixation can detect either soluble or insoluble antigen. Its ability to detect virus antigens in impure tissue preparations makes the test still useful in diagnosis of viral infections.

C1q binding assay for circulating immune complexes (CIC): There are two categories of methods to assay circulating immune complexes: (1) The specific binding of CIC to complement components, such as C1q or the binding of complement activation fragments within the CIC to complement receptors, as in the Raji cell assay. (2) Precipitation of large and small CIC by polyethylene glycol. The C1q binding assay measures those CIC capable of binding C1q, a subcomponent of the C1 component of complement and capable of activating the classical complement pathway.

The **gonococcal complement fixation test** is an assay that uses as antigen an extract of *Neisseria gonorrhoea.* It is of little value in diagnosing early cases of gonorrhea that appear before the generation of an antibody response, but may be used to identify late manifestations in untreated individuals.

A **complement fixation assay** is a serologic test based on the fixation of complement by antigen–antibody complexes. It has been applied to many antigen–antibody systems and was widely used earlier in the century as a serologic test for syphilis.

Reiter complement fixation test (historical) is a diagnostic test for syphilis that used an antigen derived from a protein extract of *Treponema pallidum* (the Reiter strain). This test identified antibodies formed against *Treponema* group antigens.

Cardiolipin is diphosphatidyl glycerol, a phospholipid, extracted from beef heart, as the principal antigen in the Wasserman complement fixation test for syphilis used earlier in the century.

The **Wassermann reaction** is a complement fixation assay used extensively in the past to diagnose syphilis. Cardiolipin extracted from ox heart served as antigen which reacts with antibodies that develop in patients with syphilis. Biologic false-positive reactions using this test require the use of such confirmatory tests as the FTA-ABS test, the Reiter's complement fixation test, or the *Treponema pallidum* immobilization test. Both FTA and TPI tests use *T. pallidum* as antigen.

CFT is the abbreviation for complement fixation test.

CH_{50} unit refers to the amount of complement (serum dilution) that induces lysis of 50% of erythrocytes sensitized (coated) with specific antibody. More specifically, the 50% lysis should be of 5×10^8 sheep erythrocytes sensitized with specific antibody during 60 min of incubation at 37°C. To obtain the complement titer, i.e., the number CH_{50} present in 1 ml of serum that has not been diluted, the log $y/1 - y$ (y = % lysis) is plotted against the log of the quantity of serum. At 50% lysis, the plot approaches linearity near $y/1 - y$.

Fluorochrome is a label such as fluorescein isothiocyanate or rhodamine isothiocyanate, which is used to label antibody molecules or other substances. A fluorochrome emits visible light of a defined wavelength upon irradiation with light of a shorter wavelength such as ultraviolet light.

Fluorescein (Figure 27.31) is a yellow dye that stains with brilliant apple-green fluorescence when excited. Its isothiocyanate derivative is widely employed to label proteins such as immunoglobulins that are useful in diagnostic medicine as well as in basic science research.

Fluorescein isothiocyanate (FITC) (Figure 27.32) is a widely used fluorochrome for labeling antibody molecules.

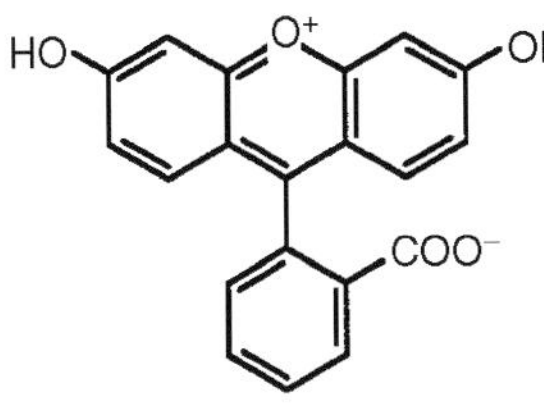

FIGURE 27.31 Fluorescein.

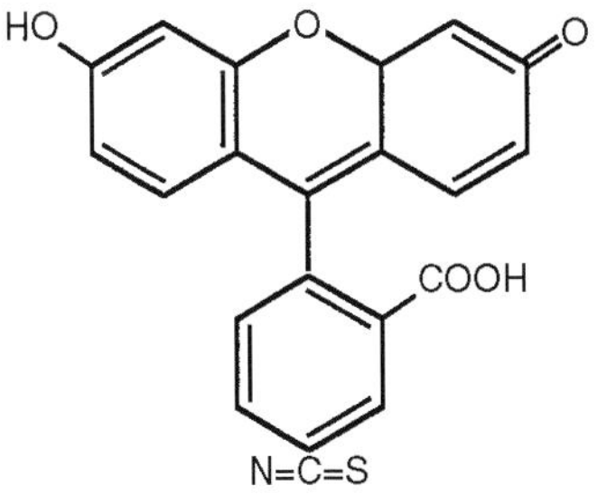

FIGURE 27.32 Fluorescein isothiocyanate (FITC).

It may also be used to label other proteins. Fluorescein-labeled antibodies are popular because they appear apple-green under ultraviolet irradiation, permitting easy detection of antigens of interest in tissues and cells. FITC fluoresces at 490 and 520 nm. FITC-labeled antibodies are useful for the demonstration of immune deposits in both skin and kidney biopsies.

Lissamine rhodamine (RB200) is a fluorochrome that produces orange fluorescence. Interaction with phosphorus pentachloride yields a reactive sulphonyl chloride that is useful for labeling protein molecules to be used in immunofluorescence staining methods.

Fluorescein-labeled antibody is an antibody tagged with a fluorescein derivative such as fluorescein isothiocyanate. These antibodies are useful to localize antigens in tissues and cells by their brilliant apple-green fluorescence under ultraviolet light.

The **F:P ratio** is the fluorescence to protein ratio that expresses the ratio of fluorochrome to protein in an antibody preparation labeled with the fluorochrome.

A **fluorescent antibody** is an antibody molecule to which a fluorochrome has been conjugated, such as fluorescein isothiocyanate.

The **fluorescent antibody technique** (Figure 27.33 to Figure 27.36) is an immunofluorescence method in which antibody labeled with a fluorochrome such as FITC is used to identify antigen in tissues or cells when examined by ultraviolet light used in fluorescence microscopy. Besides the direct technique, antigens in tissue sections treated with unlabeled antibody can be "counterstained" with fluorescein-labeled

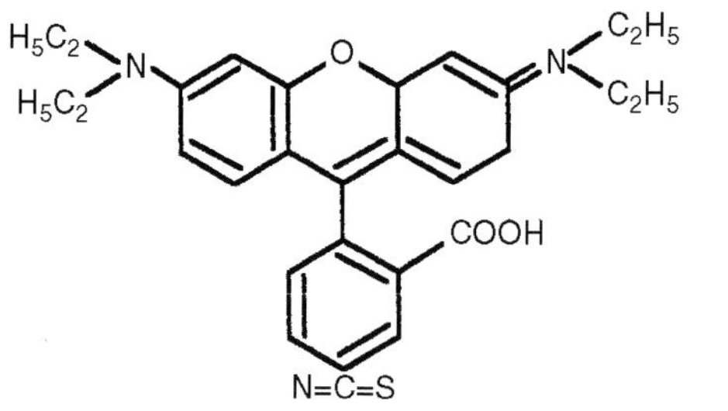

H_5C_2 N C_2H_5 O N H_5C_2 C_2H_5 COOH N=C=S

FIGURE 27.33 Rhodamine B isothiocyanate is a reddish-orange fluorochrome used to label immunoglobulins or other proteins for use in immunofluorescence studies.

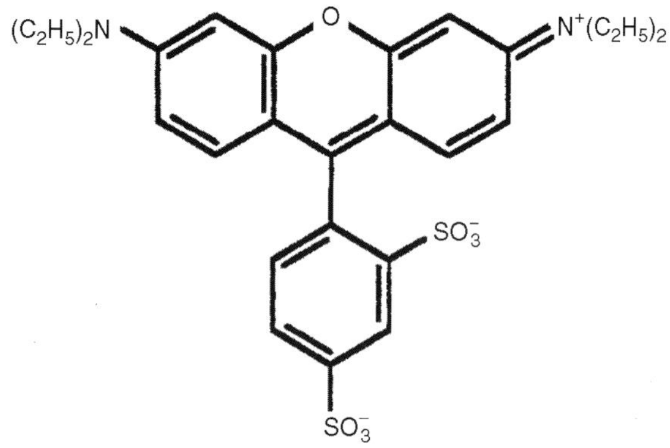

$(C_2H_5)_2N$ O $N^+(C_2H_5)_2$ SO_3^- SO_3^-

FIGURE 27.34 Rhodamine disulfonic acid is a red fluorochrome used in immunofluorescence.

$H_5C_2N^+$ O NC_2H_5 C SO_3^- SO_3Na

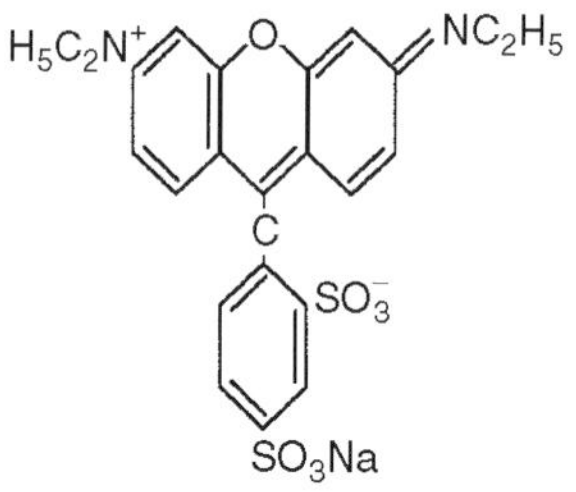

FIGURE 27.35 Lissamine rhodamine (RB200) is a fluorochrome that produces orange fluorescence. Interaction with phosphorus pentachloride yields a reactive sulphonyl chloride that is useful for labeling protein molecules to be used in immunofluorescence staining methods.

antiimmunoglobulin to localize antigen in tissues by the indirect immunofluorescence method. The **indirect fluorescence antibody technique** is a method to identify antibody or antigen using a fluorochrome-labeled antibody which combines with an intermediate antibody or antigen rather than directly with the antibody or antigen being sought. The indirect test has a greater sensitivity than those of the direct fluorescence antibody technique. It is often referred to as the sandwich or double layer method.

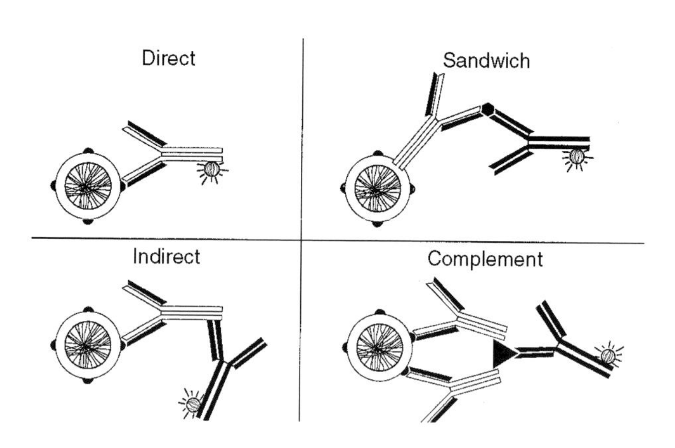

FIGURE 27.36 Fluorescent antibody technique.

Anti-*Toxoplasma gondii* antibody reacts with an epitope of *T. gondii* which is resistant to formalin fixation and paraffin embedding. When tested by indirect immunofluorescene on infected glioma cells, anti-*T. gondii* stains the outer surface of tachyzoites of two different strains of *T. gondii* (RH and T626). Using an immunoperoxidase technique, both tachyzoites and encysted bradyzoites have been labeled in infected lung.

Sandwich methodology: See sandwich technique.

The **sandwich technique** is the identification of antibody or of antibody-synthesizing cells in tissue preparations in which antigen is placed in contact with the tissue section or smear, followed by the application of antibody labeled with a fluorochrome such as fluorescein isothiocyanate (FITC) that is specific for the antigen. This yields a product consisting of antibody layered on either side of an antigen, which accounts for the name "sandwich."

An **excitation filter** is a filter in fluorescent microscopes that permits only light of a specific excitation wavelength, such as ultraviolet light, to pass through.

Fluorescence is the emission of light of one wavelength by a substance irradiated with light of a different wavelength.

Fluorescence microscopy employs a special microscope which uses ultraviolet light to illuminate a tissue or cell stained with a fluorochrome-labeled substance such as an antibody against an antigen of interest in the tissue. When returning from an excited state to a ground state, the tissue emits fluorescent light, which permits the observer to localize an antigen of interest in the tissue or cell.

Natural fluorescence is **autofluorescence**. A **sandwich immunoassay** is a technique in which the analyte is bound to a solid phase and a labeled reagent subsequently bound immunochemically to the analyte.

Fluorescence quenching is a method to ascertain association constants of antibody molecules interacting with ligands. Fluorescence quenching results from excitation energy transfer, where certain electronically excited residues in protein molecules, such as tryptophan and tyrosine, transfer energy to a second molecule which is bound to the protein. Maximum emission is a wavelength of approximately 345 nm. The attachment of the acceptor molecule need not be covalent. This transfer of energy occurs when the absorbence spectrum of the acceptor molecule overlaps with that of the emission spectrum of the donor and takes place via resonance interaction. There is no need for direct contact between the two molecules for energy transfer. If the acceptor molecule is nonfluorescent, diminution of energy occurs through nonradiation processes. On the other hand, if the acceptor molecule is fluorescent, the transfer of radiation to it results in its own fluorescence (sensitized fluorescence). Fluorescence quenching techniques can provide very sensitive quantitative data on antibody–hapten interactions.

Confocal fluorescent microscopy is a technique in which optics are used to form images at very high resolution by having two origins of fluorescent light that join only at one plane of a thicker section.

The **sandwich ELISA** is a method in which surface-bound antibody traps a protein by binding to one of its epitopes. An enzyme-linked antibody specific for a different epitope on the protein surface is employed to detect the trapped protein. This is a highly specific assay.

Fluorescent protein tracing employs fluorescent dyes are used in place of nonfluorescent dyes because they are detectable in a much lower concentration. Radioactive labeling is employed usually if the substance to be detected is present in minute amounts. Fluorescent labeling, however, provides simplicity of technique and precise microscopic observation of fluorescence. Fluorescent microscopic preparations require several hours and permit localization at the cellular level, whereas autoradiograms require a longer period and are localized at the tissue level. Either fluorescein (apple-green fluorescence) or rhodamine (reddish-orange fluorescence) compounds may be used for tracing.

Fluorescence enhancement refers to the increased fluorescence of certain substances after their combination with antibody. This is attributable to changing the substance from an aqueous milieu to the antibody combining site's hydrophobic surroundings.

A **Cryostat®** is a microtome in a refrigerated cabinet used by pathologists to prepare frozen tissue sections for surgical pathologic diagnosis. Immunologists use this method of quick frozen thin sections for immunofluorescence staining by fluorochrome-labeled antibody to identify antigens, antibodies, or immune complexes in tissue sections such as renal biopsies.

Immunofluorescence is a method for the detection of antigen or antibody in cells or tissue sections through the use of fluorescent labels, termed fluorochromes, by fluorescent light microscopic examination. The most commonly used fluorochromes are fluorescein isothiocyanate, which imparts an apple-green fluorescence, and rhodamine B isothiocyanate, which imparts a reddish-orange tint. This method, developed by Albert Coons in the 1940s, has a wide application in diagnostic medicine and research. In addition to antigens and antibodies, complement and other immune mediators may also be detected by this method. It is based on the principle that following adsorption of light by molecules, they dispose of their increased energy by various means, such as emission of light of longer wavelength. Fluorescence is the process whereby emission is of relatively short duration (10^{-6} to 10^{-9} sec) for return of the excited molecules to the ground state. The active groups in protein that allow them to attach fluorochromes include free amino and carboxyl groups at the ends of each polypeptide chain, many free amino groups and lysine side chains, many free carboxyl groups in asparatic and glutamic acid residues, the guanidino group of arginine, the phenolic group of tyrosine, and the amino groups of histidine and tryptophan. Labeling antibody molecules with fluorochromes does not alter their antigen-binding specificity. Several immunofluorescence techniques are available. In the direct test, smears of the substance to be examined are fixed with heat or methanol and followed by flooding with a fluorochrome-antibody conjugate. This is followed by incubating in a moist chamber for 30–60 min at 37°C, after which the smear is washed first in buffered saline for 5–10 min and second in tap water for another 5–10 min. These washing procedures remove uncombined conjugated globulin. After adding a small drop of buffered glycerol and the cover slip, the smear may be examined with the fluorescence light microscope. In the indirect test, which is more sensitive than the direct, a smear or tissue section is first flooded with unlabeled antibody specific for the antigen being sought. After washing, fluorescein-labeled antiimmunoglobulin of the species of the primary antibody is layered over the section. After appropriate incubation and washing, the section is cover slipped and examined as in the direct method. Other variations, such as complement staining, are also available. The indirect method is more sensitive than the direct method and considerably less expensive than one fluorochrome-labeled antiimmunoglobulin may be used with multiple primary antibodies specific for a battery of antigens. The technique is widely used to diagnose and classify renal diseases, bullous skin diseases, and for the study of cells and tissues in connective tissue disorders such as SLE.

Membrane immunofluorescence refers to the reaction of a fluorochrome-labeled antibody with cell surface receptors of viable cells. This reaction of fluorescent antibody with surface antigens rather than internal antigens is the basis for many immunologic assays such as labeling of lymphocytes with reagents for immunophenotyping by flow cytometry, patching, and capping, and to detect changes in surface antigens through antigenic variation.

Nonspecific fluorescence is fluorescence emission that does not reflect antigen–antibody interaction and may confuse interpretation of immunofluorescence tests. Either free fluorochrome or fluorochrome tagging of proteins other than antibody such as serum albumin, α globulin, or β globulin may contribute to nonspecific fluorescence. Nonspecific staining is accounted for in appropriate controls.

Quenching: When immunofluorescent cell or tissue preparations treated with fluorochrome-labeled antibodies are exposed to ultraviolet light under the microscope, the emission of fluorescent radiation from the fluorochrome label diminishes as a result of quenching. The term may also refer to diminished efficiency of assaying radioactivity in a scintillation counter by such agents as ethanol or hydrogen peroxide (H_2O_2).

Rhodamine isothiocyanate is a reddish-orange fluorochrome used to label immunoglobulins or other proteins for use in immunofluorescence studies.

The **double-layer fluorescent antibody technique** is an immunofluorescence method to identify antigen in a tissue section or cell preparation on a slide by first covering and incubating it with antibody or serum containing antibody that is not labeled with a fluorochrome. After appropriate time for interaction, the preparation is washed and a second application of fluorochrome-labeled antibody such as goat or rabbit antihuman immunoglobulin is applied to the tissue or cell preparation and again incubated. This technique has greater sensitivity than does the single-layer immunofluorescent method. Examples include the application of serum from a patient with Goodpasture's syndrome to a normal kidney section acting as substrate followed by incubation and washing, and then covering with fluorochrome-labeled goat antihuman IgG to detect antiglomerular basement membrane antibodies in the patient's serum. A similar procedure is used in detecting antibodies against intercellular substance antigens in the serum of patients with pemphigus vulgaris.

RB200: See lissamine rhodamine.

The **direct fluorescence antibody method** (Figure 27.37) employs antibodies, either polyclonal or monoclonal, labeled with a fluorochrome such as fluorescein isothiocyanate, which yields an apple green color by immunofluorescence microscopy, or rhodamine isothiocyanate, which yields a reddish-orange color, to identify a specific antigen. This technique is routinely used in immunofluorescence evaluation of renal biopsy specimens as well as skin biopsy preparations to detect immune complexes comprised of the various immunoglobulin classes or complement components.

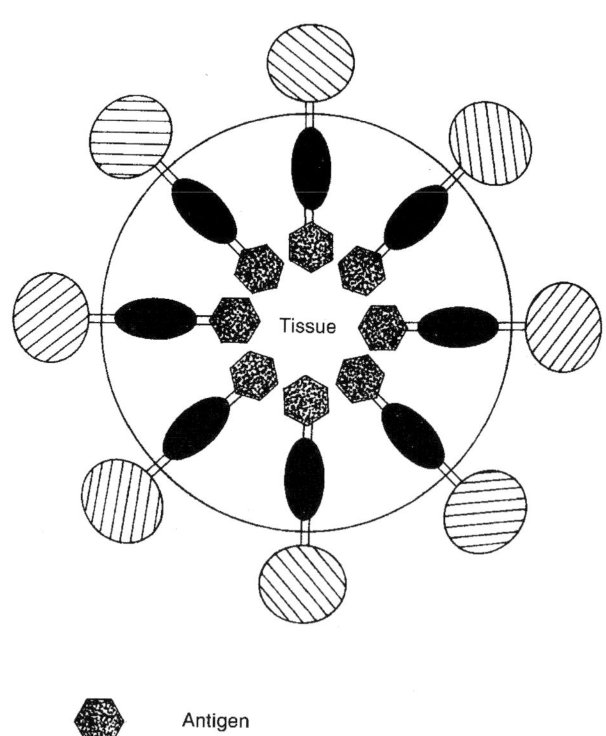

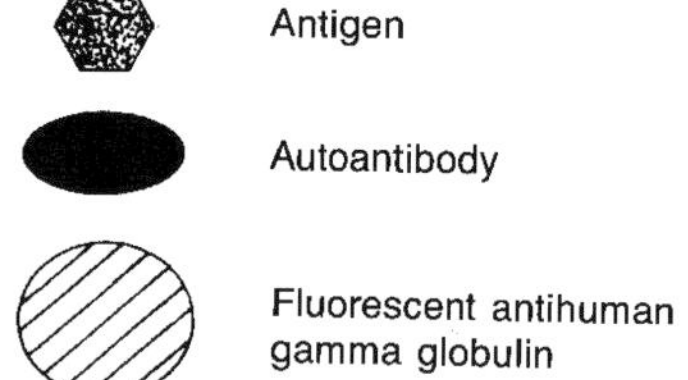

FIGURE 27.37 Direct fluorescence antibody technique.

Direct immunofluorescence refers to the use of fluorochrome-labeled antibody to identify antigens, especially those of tissues and cells. An example is the immunofluorescence evaluation of renal biopsy specimens.

Direct staining is a version of the fluorescent antibody staining technique in which a primary antibody has been conjugated with a fluorochrome and applied directly to a tissue sample containing the antigen in question.

Indirect immunofluorescence refers to the interaction of unlabeled antibody with cells or tissues expressing antigen for which the antibody is specific, followed by treatment of this antigen–antibody complex with fluorochrome-labeled antiimmunoglobulin that interacts with the first antibody, forming a so-called sandwich.

The **indirect fluorescence antibody technique** is a method to identify antibody or antigen using a fluorochrome-labeled antibody which combines with an intermediate antibody or antigen rather than directly with the antibody or antigen being sought. The indirect test has a greater sensitivity than those of the direct fluorescence antibody technique. It is often referred to as the sandwich or double-layer method.

Quin-2 (Figure 27.38) is a derivative of quinoline which combines with free Ca^{++} to accentuate fluorescence intensity. It can be introduced into cells as an ester followed by deesterification. When T or B lymphocytes containing Quin-2 are activated, their fluorescence intensity rises, implying an elevation of free Ca^{++} in the cytosol.

Ferritin is an iron-containing protein that is electron dense and serves as a source of stored iron until it is needed for the synthesis of hemoglobulin. It is an excellent antigen and is found in abundant quantities in horse spleen. Ferritin's electron-dense quality makes it useful to label antibodies or antigens to be identified or localized in electron microscopic preparations.

Ferritin labeling (Figure 27.39) is achieved by conjugating ferritin to antibody molecules to render them visible in histologic or cytologic specimens observed by electron microscopy. Antibodies may be labeled with ferritin by use of a crosslinking reagent such as toluene-2,4-diisocyanate. The ferritin-labeled antibody may be reacted directly with the specimen, or ferritin-labeled antiimmunoglobulin may be used to react with unlabeled specific antibody attached to the target tissue antigen.

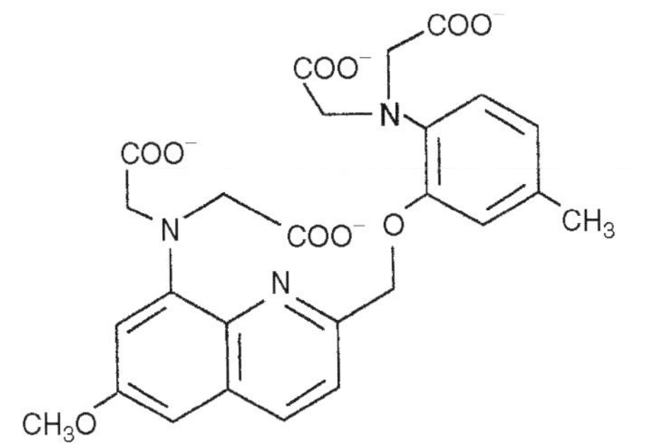

FIGURE 27.38 Quin-2.

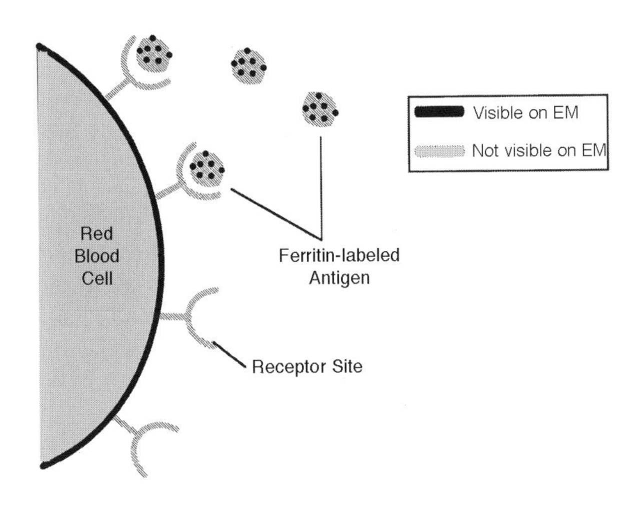

FIGURE 27.39 Ferritin labeling.

Immunoelectron microscopy refers, traditionally, to the use of antibodies labeled with ferritin to study the ultrastructure of subcellular organelles and, more recently, the use of immunogold labeling and related procedures for the identification and localization of antigens by electron microscopy.

The **immunoferritin method** (Figure 27.40) is a technique to aid detection by electron microscopy of sites where antibody interacts with antigen of cells and tissues. Immunoglobulin may be conjugated with ferritin, an electron-dense marker, without altering its immunological reactivity. These ferritin-labeled antibodies localize molecules of antigen in the subcellular areas. Electron-dense ferritin permits visualization of antibody binding to homologous antigen in cells and tissues by electron microscopy. In addition to ferritin, horseradish-peroxidase-labeled antibodies may also be adapted for use in immunoelectron microscopy.

Immunogold labeling (Figure 27.41) is a technique to identify antigens in tissue preparations by electron microscopy. Sections are incubated with primary antibody and followed by treatment with colloidal gold-labeled anti-IgG antibody. Electron-dense particles are localized at sites of antigen–antibody interactions.

Immunohistochemistry is a method to detect antigens in tissues that employs an enzyme-linked antibody specific for antigen. The enzyme degrades a colorless substrate to

A ferritin-conjugated anti-A globulin treated group A red cell. The binding of ferritin particles to the surface of the erythrocyte denotes the site of the antigen.

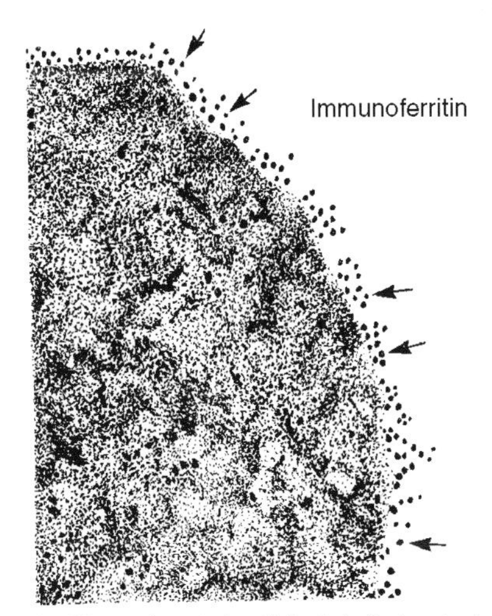

FIGURE 27.40 Immunoferritin method.

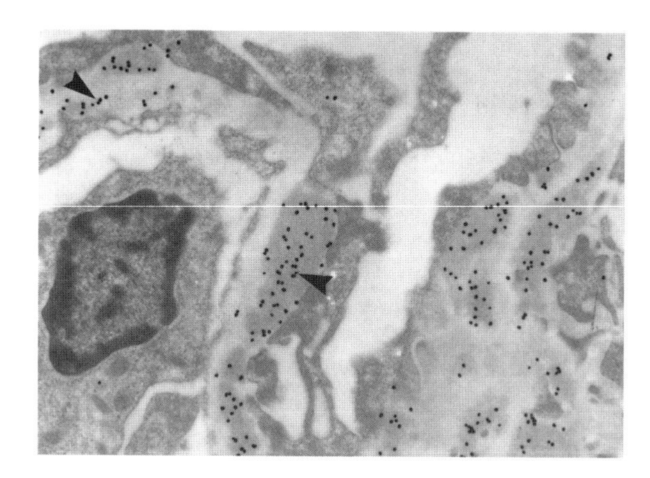

FIGURE 27.41 Immunogold labeling.

a colored insoluble substance that precipitates where the antibody and, therefore, the antigen are located. Identification of the site of the colored precipitate and the antigen in the tissue section is accomplished by light microscopy. Diagnostic pathology services routinely offer approximately 100 antigens identified by immunoperoxidase technology that are used in diagnosis.

Immunogold silver staining (IGSS) is an immunohistochemical technique to detect antigens in tissues and cells by light microscopy. IGSS offers higher labeling intensity than that of most other methods when examined in a bright field or in conjunction with polarized light. The technique successfully stains tissue sections from paraffin wax, resin, or cryostat preparations. It is also effective for cell suspensions or smears, cytospin preparations, cell cultures, or tissue sections. Both 1- and 5-nm gold conjugates are used for light microscopy. The 1-nm particles are advantageous in studies of cell penetration. In immunogold silver staining, primary antibody is incubated with tissues or cells to localize antigens that are identified with gold-labeled secondary antibodies and silver enhanced.

The **immunoperoxidase method** (Figure 27.42) was introduced by Nakene and Pierce in 1966 who proposed that enzymes be used in the place of fluorochromes as labels for antibodies. Horseradish peroxidase (HRP) is the enzyme label most widely employed. The immunoperoxidase technique permits the demonstration of antigens in various types of cells and fixed tissues. This method has certain advantages that include (1) the use of conventional light microscopy, (2) the stained preparations may be kept permanently, (3) the method may be adapted for use with electron microscopy of tissues, and (4) counterstains may be employed. The disadvantages include the following: (1) the demonstration of relatively minute positively staining areas is limited by the light microscope's resolution, (2) endogenous peroxidase may not have been completely eliminated from the tissue under investigation, and (3)

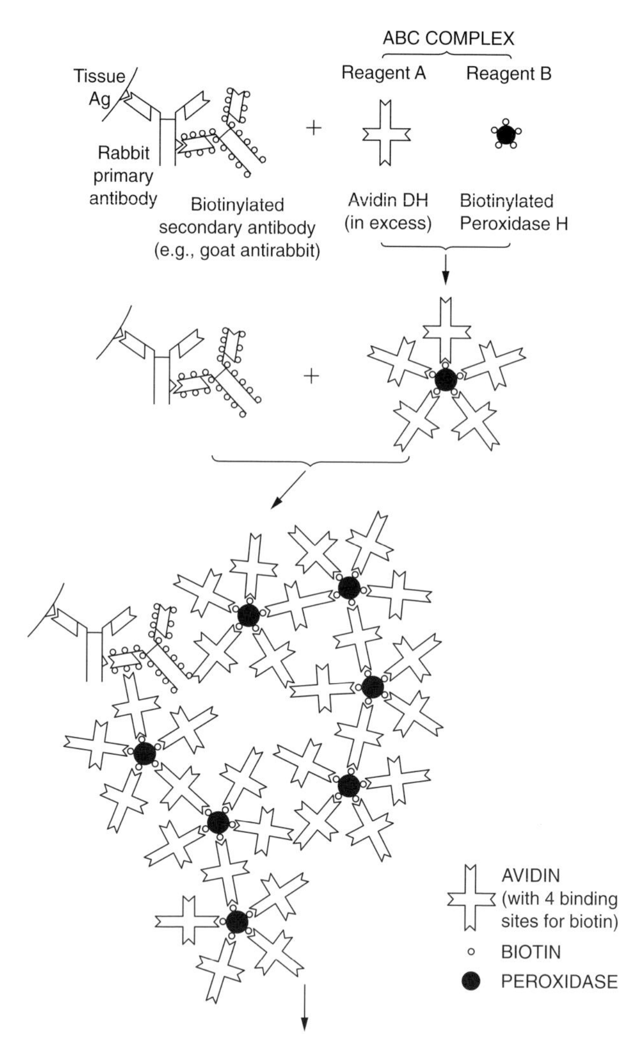

FIGURE 27.42 Immunoperoxidase method.

diffusion of products resulting from the enzyme reaction away from the area where antigen is localized.

The **ABC method** is a unique immunoperoxidase procedure for localizing a variety of histologically significant antigens and other markers. The procedure employs biotinylated antibody and a preformed avidin: biotinylated enzyme complex and has been termed the "ABC" technique. Because avidin has such an extraordinarily high affinity for biotin, the binding of avidin to biotin is essentially irreversible. In addition, avidin has four binding sites for biotin and most proteins including enzymes can be conjugated with several molecules of biotin. These properties allow macromolecular complexes (ABCs) to be formed between avidin and biotinylated enzymes.

Antigen retrieval is a novel method for the rescue of antigens from formalin-fixed paraffin-embedded tissue. It consists of heating sections in a microwave oven or in a pressure cooker in the presence of an antigen retrieval solution. It is designed for use in immunohistochemical staining with certain antibodies. This technique increases staining intensity and reduces background staining of many important markers in formalin-fixed tissue. Its use helps overcome false-negative staining of overfixed tissue, expand the range of antibodies useful for routinely processed tissue, and increases the usefulness of archival materials for retrospective studies. In addition to microwave heating, the pH of the antigen retrieval solution is an important cofactor for some antigens. Three antigen retrieval solutions covering a wide pH range are used. These include a citrate-based neutral pH solution, Tris-based high pH solution, and a glycine-based low pH solution.

The **peroxidase–antiperoxidase (PAP) technique** employs unlabeled antibodies and a PAP reagent. This has proven highly successful for the demonstration of antigens in paraffin-embedded tissues as an aid in surgical pathologic diagnosis. Tissue sections preserved in paraffin are first treated with xylene, and after deparaffinization they are exposed to a hydrogen peroxide solution which destroys the endogenous peroxidase activity in tissue. The sections are next incubated with normal swine serum which suppresses nonspecific binding of immunoglobulin molecules to tissues containing collagen. Thereafter, the primary rabbit antibody against the antigen to be identified is reacted with the tissue section. Primary antibody that is unbound is removed by rinsing the sections which are then covered with swine antibody against rabbit immunoglobulin. This so-called linking antibody will combine with any primary rabbit antibody in the tissue. It is added in excess, which will result in one of its antigen-binding sites remaining free. After washing, the PAP reagent is placed on the section, and the antibody portion of this complex, which is raised in rabbits, will be bound to the free antigen-binding site of the linking antibody on the sections. The unbound PAP complex is then washed away by rinsing. To read the sections microscopically, it is necessary to add a substrate of hydrogen peroxide and aminoethylcarbazole (AEC) which permits the formation of a visible product that may be detected with the light microscope. The AEC is oxidized to produce a reddish-brown pigment that is not water soluble. Peroxidase catalyzes the reaction. Because peroxidase occurs only at sites where the PAP is bound via linking antibody and primary antibody to antigen molecules, the antigen is identified by the reddish-brown pigment. The tissue sections can then be counterstained with hematoxylin or other suitable dye, covered with mounting medium and cover slips, and read by conventional light microscopy. The PAP technique has been replaced, in part, by the avidin–biotin complex (ABC) technique.

The **PAP (peroxidase–antiperoxidase) technique** (Figure 27.43 to Figure 27.47) is a method for immunoperoxidase staining of tissue to identify antigens with antibodies. This method employs unlabeled antibodies and a PAP reagent. The same PAP complex may be used for dozens of different unlabeled antibody specificities. If the primary antibody against the antigen being sought is made in the rabbit, then tissue sections treated with this reagent are exposed to sheep antirabbit immunoglobulin followed by

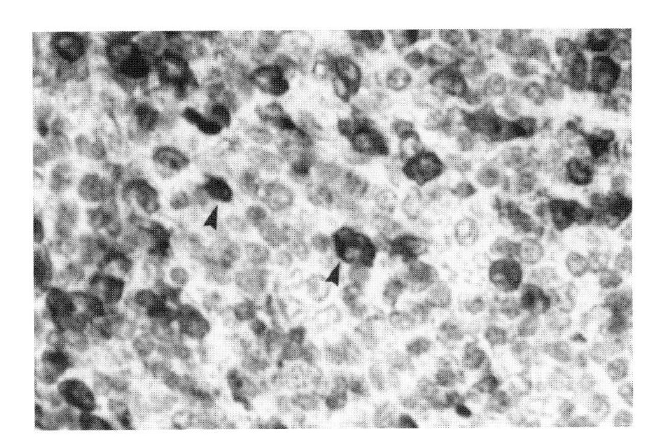

FIGURE 27.43 Plasma cells decorated with antibody.

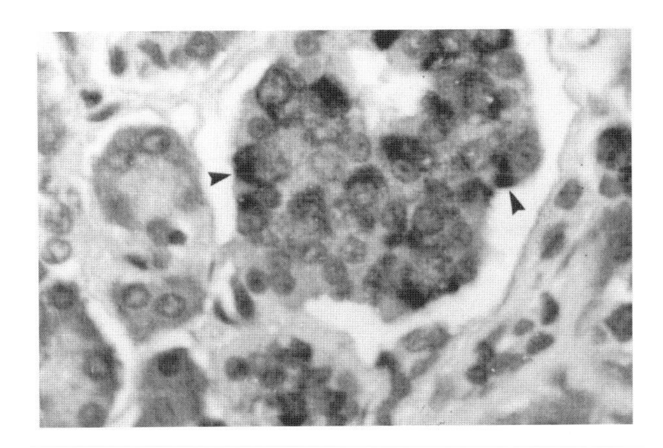

FIGURE 27.44 Insulin in β cells.

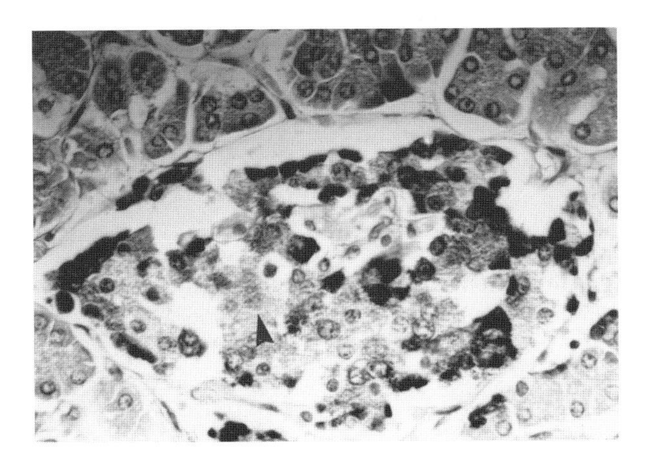

FIGURE 27.45 Chromogranin.

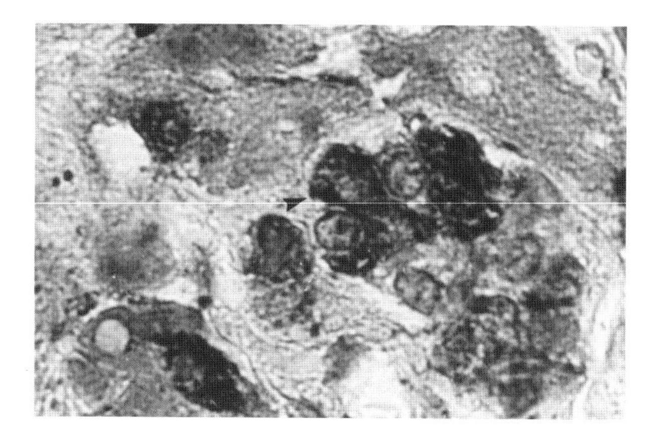

FIGURE 27.46 Prolactin staining in pituitary.

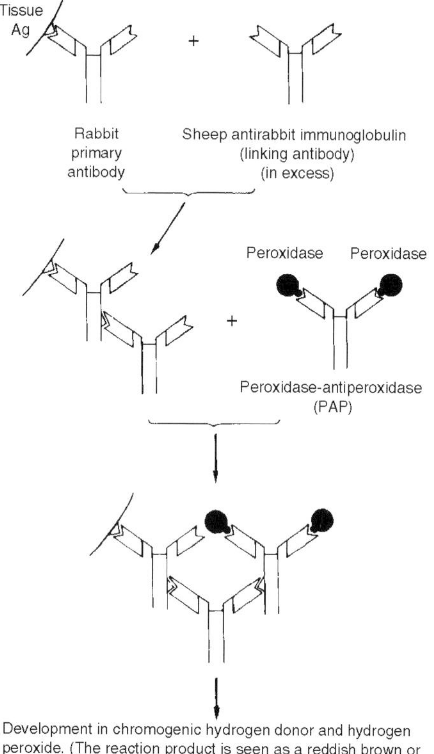

FIGURE 27.47 Peroxidase–antiperoxidase (PAP) technique.

the PAP complex. For human primary antibody, an additional step must link the human antibody into the rabbit sandwich technique. Paraffin-embedded tissue sections are first treated with xylene, and after deparaffinization, they are exposed to hydrogen peroxide to destroy the endogenous peroxidase. Sections are next incubated with normal sheep serum to suppress nonspecific binding of immunoglobulin to tissue collagen. Primary rabbit antibody against the antigen to be identified in combined with the tissue section. Unbound primary antibody is removed by rinsing the sections which are then covered with sheep antibody against rabbit immunoglobulin. This linking antibody will combine with any primary rabbit antibody in the tissue. It is added in excess, which results in one of its antigen-binding sites remaining free. After washing, the PAP is placed in the section, and the rabbit antibody part of this complex will be bound to the free antigen binding site of the linking antibody. The unbound PAP complex is then washed away by rinsing. A substrate of hydrogen peroxide and AEC is placed on the tissue section leading to formation of a visible color reaction product that can be seen by light microscopy. Peroxidase is localized only at sites where the PAP is bound via linking antibody and primary antibody to antigen molecules, permitting the antigen to be identified as an area of reddish-brown pigment. Tissues may be counterstained with hematoxylin.

A **chromogenic substrate** is a colorless substance that is transformed into a colored product by an enzymatic reaction.

In immunoperoxidase staining of paraffin-embedded tissues, tissue drying may produce nonspecific coloring at the periphery, which is an artifact often termed **edge artifact**.

***In-situ* hybridization** is a technique to identify specific DNA or RNA segments in cells or tissues, viral plaques, or colonies of microorganisms. DNA in cells or tissue fixed on glass slides must be denatured with formamide before hybridization with a radiolabeled or biotinylated DNA or RNA probe that is complementary to the tissue mRNA being sought. Proof that the probe has hybridized to its complementary strand in the tissue or cell under study must be by autoradiography or enzyme-labeled probes, depending on the technique being used.

Aminoethylcarbazole (AEC), 3-Amino-9-ethyl carbazole, is used in the ABC immunoperoxidase technique to produce a visible reaction product detectable by light microscopy when combined with hydrogen peroxide. AEC is oxydized to produce a reddish-brown pigment that is not water soluble. Peroxidase catalyzes the reaction. Because peroxidase is localized only at sites where the PAP is bound via linking antibody and primary antibody to antigen molecules, the antigen is identified by the reddish-brown pigment.

Pan keratin antibodies are comprised of a "cocktail" of antibodies reactive with high-mol-wt cytokeratin and low-mol-wt keratin (AE1/AE3). By immunoperoxidase staining, these antibodies identify most epithelial cells and their derived neoplasms, irrespective of the site of origin or the level of differentiation.

Biotin–avidin system: Avidin is an egg-white derived glycoprotein with an extraordinarily high affinity (affinity constant > 10^{15} M^{-1}) for biotin. Streptavidin is similar in properties to avidin but has a lower affinity for biotin. Many biotin molecules can be coupled to a protein, enabling the biotinylated protein to bind more than one molecule of avidin. If biotinylation is performed under gentle conditions, the biological activity of the protein can be preserved. By covalently linking avidin to different ligands, such as fluorochromes, enzymes, or EM markers, the biotin–avidin system can be employed to study a wide variety of biological structures and processes. This system has proven particularly useful in the detection and localization of antigens, glycoconjugates, and nucleic acids by employing biotinylated antibodies, lectins, or nucleic acid probes.

Pancreatic islet cell hormones: Immunoperoxidase staining of islet cell adenomas with antibodies to insulin, glucagon, somatostatin, and gastrin facilitates definition of their clinical phenotype.

Autoradiography is a method employed to localize radioisotopes in tissues or cells from experimental animals injected with radiolabeled substances. The radioisotopes serve as probes bound to specific DNA or RNA segments. Radioactivity is detected by placing the x-ray or photographic emulsion into contact with the tissue sections or nylon/nitrocellulose membranes in which they are localized to record sites of radioactivity. The technique permits the detection of radioactive substances by analytical methods involving electrophoresis, Southern blotting, and Northern blot hybridization.

Intracellular cytokine staining refers to the use of fluorescent labeled anticytokine antibodies to "stain" permeabilized cells that synthesize the cytokine in question.

PAS: (1) Abbreviation for periodic acid Schiff stain for polysaccharides. This technique identifies mucopolysaccharide, glycogen, and sialic acid among other chemicals containing 1,2-diol groups. (2) Abbreviation for *para*-aminosalicylic acid, which is used in the treatment of tuberculosis.

Neuron-specific enolase (NSE) is an enzyme of neurons and neuroendocrine cells, as well as their derived tumors (e.g., oat cell carcinoma of lung), demonstrable by immunoperoxidase staining. NSE also occurs in some neoplasms not derived from neurons or endocrine cells. (Figure 27.48)

Enzyme immunoassay (EIA) is a technique employed to measure immunochemical reactions based on enzyme catalytic properties. The three widely used techniques include a heterogeneous EIA technique, ELISA, and two homogeneous techniques, enzyme-multiplied immunoassay technique (EMIT) and cloned enzyme donor immunoassay (CEDIA).

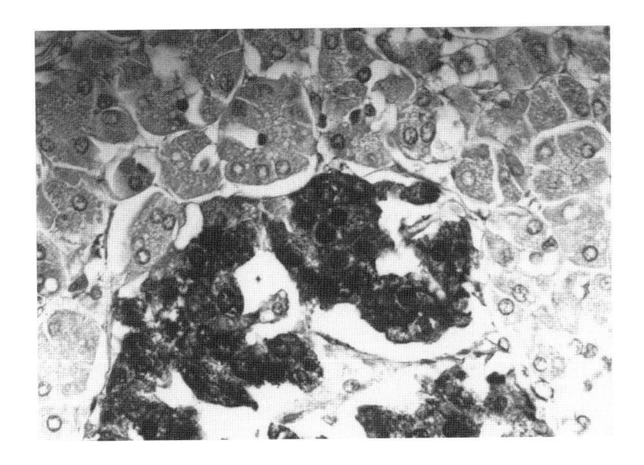

FIGURE 27.48 Neuron-specific enolase.

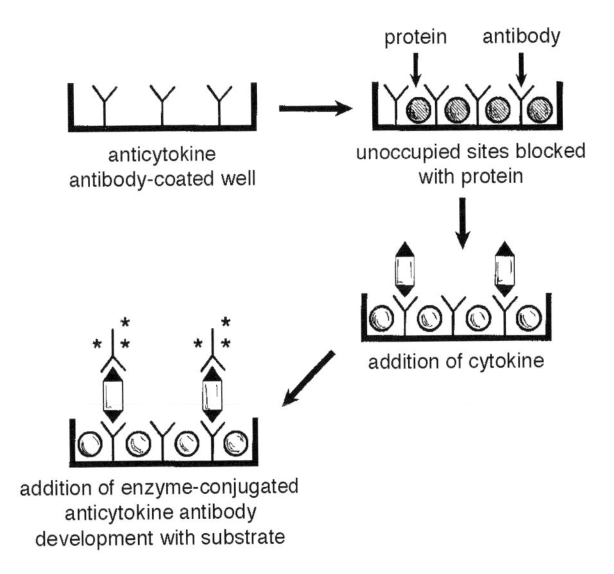

FIGURE 27.49 ELISA.

The **enzyme-linked immunosorbent assay (ELISA)** (Figure 27.49) is an immunoassay that employs an enzyme linked to either antiimmunoglobulin or antibody specific for antigen and detects either antibody or antigen. This method is based on the sandwich or double-layer technique, in which an enzyme rather than a fluorochrome is used as the label. In this method, antibody is attached to the surface of plastic tubes, wells, or beads to which the antigen-containing test sample is added. If antibody is being sought in the test sample, then antigen should be attached to the plastic surface. Following antigen–antibody interaction, the enzyme–antiimmunoglobulin conjugate is added. The ELISA test is read by incubating the reactants with an appropriate substrate to yield a colored product that is measured in a spectrophotometer. Alkaline phosphatase and horseradish peroxidase are enzymes that are often employed. ELISA methods have replaced many radioimmunoassays because of their lower cost, safety, speed, and simplicity in performing.

EIA is an abbreviation for enzyme immunoassay.

Enzyme labeling is a method such as the immunoperoxidase technique that permits detection of antigens or antibodies in tissue sections by chemically conjugating them to an enzyme. Then, by staining the preparation for the enzyme, antigen or antibody molecules can be located. See immunoperoxidase method.

Spot ELISA is an assay that is a variation on standard ELISA. It is used primarily for the detection of immunoglobulin-secreting cells (ISC) or cytokine-secreting cells (CSC), although future applications may include detection of specific hormone secreting cells. As in standard ELISA, the starting point is a plastic or nitrocellulose vessel coated with antigen or capture antibody. The ISC or CSC of interest is added and then removed, following sufficient incubation time for the cell to secrete its immunoglobulin or cytokine. The secreted product binds locally to the capture protein and is subsequently detected by enzyme-linked antibody. Finally, a substrate that yields an insoluble product is added and the resulting colored precipitate is quantified.

EMIT is an abbreviation for enzyme-multiplied immunoassay technique.

The **enzyme-multiplied immunoassay technique** is an immunoassay used to monitor therapeutic drugs such as antitumor, antiepileptic, antiasthmatic, and metabolites of cocaine and of other agents subject to abuse. It is a one-phase, competitive enzyme-labeled immunoassay.

ELISA (enzyme-linked immunosorbent assay): See enzyme-linked immunosorbent assay.

Antihistone antibodies are associated with several autoimmune diseases that include SLE, drug-induced lupus, juvenile RA, and RA. H-1 antibodies are the most common in SLE followed by anti-H2B, anti-H2A, anti-H3, and anti-H4, respectively. Antihistone antibodies are usually assayed by the ELISA technique.

Collagen disease/lupus erythematosus diagnostic panel refers to a battery of serum tests for the diagnosis of collagen vascular disease that yields the most information for the least cost.

The **ELISPOT assay** is a modification of the enzyme-linked immunosorbent assay (ELISA) which involves the capture of products secreted from cells placed in contact with antigen or antibody fixed to a plastic surface. An enzyme-linked antibody is then used to identify the captured products by cleaving a colorless substrate to yield a colored spot.

Western blot (immunoblot) (Figure 27.50) is a method to identify antibodies against proteins of precise molecular weights. It is widely used as a confirmatory test for HIV-1 antibody following the HIV-1 antibody screen test performed by the ELISA assay. Following separation of proteins by one- or two-dimensional electrophoresis, they are blotted or transferred to a nitrocellulose or nylon membrane followed by exposure to biotinylated or radioisotope-labeled antibody. The antigen under investigation is revealed by either a color reaction or autoradiography, respectively.

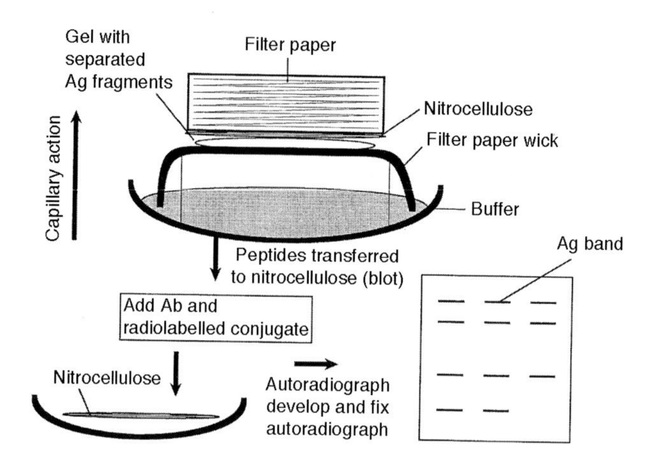

FIGURE 27.50 Western blot (immunoblot).

Immunoblot (Western blot) refers to the interaction between labeled antibodies and proteins that have been absorbed on nitrocellulose paper. See Western blot.

Immunoblotting is a method to identify antigen(s) by the polyacrylamide gel electrophoresis (PAGE) of a protein mixture containing the antigen. PAGE separates the components according to their electrophoretic mobility. After transfer to a nitrocellulose filter by electroblotting, antibodies labeled with enzyme or radioisotope and which are specific for the antigen in question are incubated with the cellulose membrane. After washing to remove excess antibody that does not bind, substrate can be added if an enzyme was used, or autoradiography can be used if a radioisotope was used to determine where the labeled antibodies were bound to homologous antigen. Also called Western blotting.

Protein blotting: See immunoblotting.

Protein separation techniques: Proteins may be purified using both electrophoresis and chromatography. Individual techniques are discussed separately. See affinity chromatography; isoelectric focussing; SDS-polyacrylamide gel electrophoresis (SDS-PAGE).

Southwestern blot is a method that combines Southern blotting that identifies DNA segments, with Western immunoblotting that characterizes proteins. A protein may be hybridized to a molecule of single-stranded DNA bound to the membrane. Southwestern blotting is helpful in delineating nuclear transcription-related proteins.

The **Cleveland procedure** is a form of peptide map in which protease-digested protein products, with sodium dodecyl sulfate (SDS) present, are subjected to SDS-PAGE. This produces a characteristic peptide fragment pattern that is typical of the protein substrate and enzyme used.

Blot refers to the transfer of DNA, RNA, or protein molecules from an electrophoretic gel to a nitrocellulose or nylon membrane by osmosis or vacuum, followed by immersing the membrane in a solution containing a complementary, i.e., mirror-image molecule corresponding to the one on the membrane. This is known as a hybridization blot.

Southern blotting (Figure 27.51) is a procedure to identify DNA sequences. Following extraction of DNA from cells, it is digested with restriction endonucleases to cut DNA at precise sites into fragments. This is followed by separation of the DNA segments according to size by electrophoresis in agarose gel, denaturation with sodium hydroxide, and transfer of the single-stranded DNA to a nitrocellulose membrane by blotting. This is followed by hybridization with a ^{35}S- or ^{32}P-radiolabeled probe of complementary DNA. Alternatively, a biotinylated probe may be used.

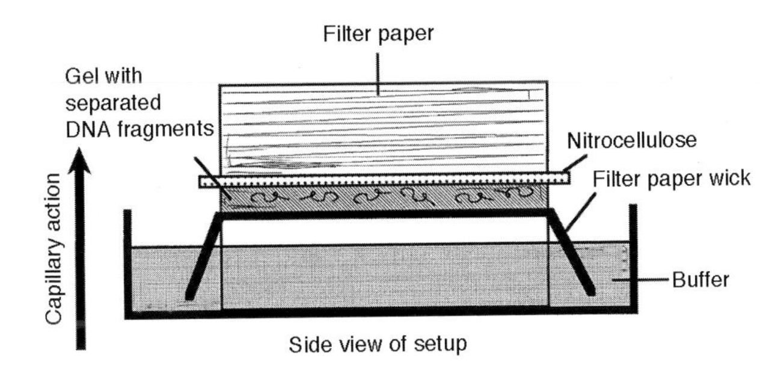

FIGURE 27.51 Southern blotting.

Autoradiography or substrate digestion identifies the location of the DNA fragments that have hybridized with the complementary DNA probe. Specific sequences in cloned and in genomic DNA can be identified by Southern blotting. Whereas DNA analysis is referred to as Southern blot, RNA analysis is referred to as a Northern blot, and protein analysis is referred to as a Western blot. A Northwestern blot is one in which RNA-protein hybridizations are formed.

Northern blotting (Figure 27.52) is a method to identify specific mRNA molecules. Following denaturation of RNA in a particular preparation with formaldehyde to cause the molecule to unfold and become linear, the material is separated by size through gel electrophoresis and blotted onto a natural cellulose or nylon membrane. This is then exposed to a solution of labeled DNA "probe" for hybridization. This step is followed by autoradiography. Northern blotting corresponds to a similar method used for DNA fragments which is known as Southern blotting.

***In-situ* hybridization** (Figure 27.53) is a technique to identify specific DNA or RNA segments in cells or tissues or in viral plaques or colonies of microorganisms. DNA in cells or tissue fixed on glass slides must be denatured with formamide before hybridization with a radiolabeled or biotinylated DNA or RNA probe that is complementary to the tissue mRNA being sought. Proof that the probe has hybridized to its complementary strand in the tissue or cell under study must be by autoradiography or enzyme-labeled, probes depending on the technique being used.

Molecular hybridization probe is a molecule of nucleic acid, which is labeled with a radionuclide or fluorochrome that can reveal the presence of complementary nucleic acid through molecular hybridization such as *in situ*.

FISH (fluorescence *in situ* hybridization) is a method to determine ploidy by examining interphase (nondividing) nuclei in cytogenetic and cytologic samples.

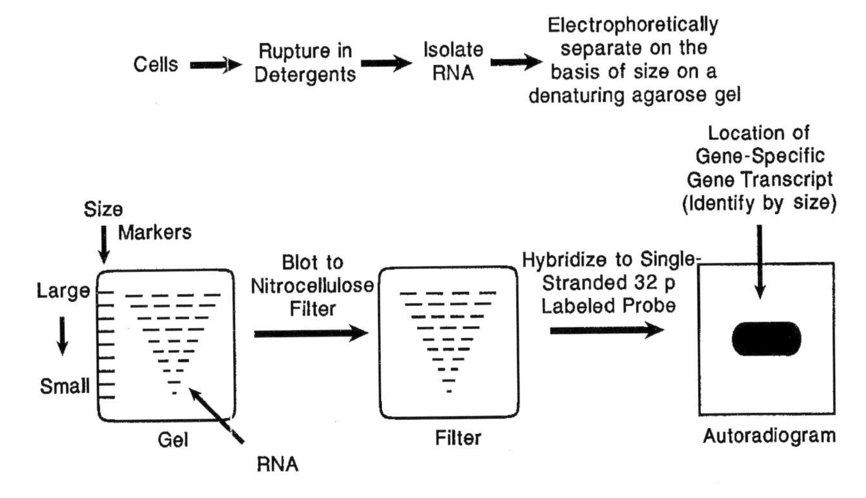

FIGURE 27.52 Northern blotting.

The **polymerase chain reaction (PCR)** (Figure 27.54) is a technique to amplify a small DNA segment beginning with as little as 1 μg. The segment of double-stranded DNA is placed between two oligonucleotide primers through many cycles of amplification. Amplification takes place in a thermal cycler, with one step occurring at a high temperature in the presence of DNA polymerase that is able to withstand the high temperature. Within a few hours, the original DNA segment is transformed into millions of copies. PCR methodology has been used for multiple purposes, including detection of human immunodeficiency virus 1(HIV-1), the prenatal diagnosis of sickle cell anemia, and gene rearrangements in lymphoproliferative disorders, among numerous other applications. The technique is used principally to prepare enough DNA for analysis by available DNA methods and is used widely in DNA diagnostic work. PCR has a 99.99% sensitivity.

PCR is an abbreviation for polymerase chain reaction.

FIGURE 27.53 *In-situ* hybridization.

Reverse transcriptase polymerase chain reaction (RT-PCR) is a technique employed to amplify RNA sequences. Reverse transcriptase is used to convert an RNA sequence into a cDNA sequence that is amplified by PCR using gene-specific primers. This technique is a variation of the polymerase chain reaction (PCR) employed to amplify a complementary cDNA of a gene of interest.

Taq polymerase or *Thermus aquaticus* polymerase. A heat-resistant DNA polymerase that greatly facilitates use of the polymerase chain reaction to amplify minute quantities of DNA from various sources into a sufficiently large quantity that can be analyzed.

DNA fingerprinting (Figure 27.55) is a method to demonstrate short, tandem-repeated highly specific genomic sequences known as minisatellites. There is only a 1 in 30 billion probability that two persons would have the identical DNA fingerprint. It has greater specificity than restriction fragment length polymorphism (RFLP) analysis. Each individual has a different number of repeats. The insert-free wild-type M13 bacteriophage identifies the hypervariable minisatellites. The sequence of DNA that identifies the differences is confined to two clusters of 15-bp repeats in the protein III gene of the bacteriophage. The specificity of this probe, known as the Jeffries probe, renders it applicable to parentage testing, human genome mapping, and forensic science. RNA may also be split into fragments by an enzymatic digestion followed by electrophoresis. A characteristic pattern for that molecule is produced and aids in identifying it.

DNA microarray is a technique in which a different DNA is placed on a small section of a microchip. The microarray

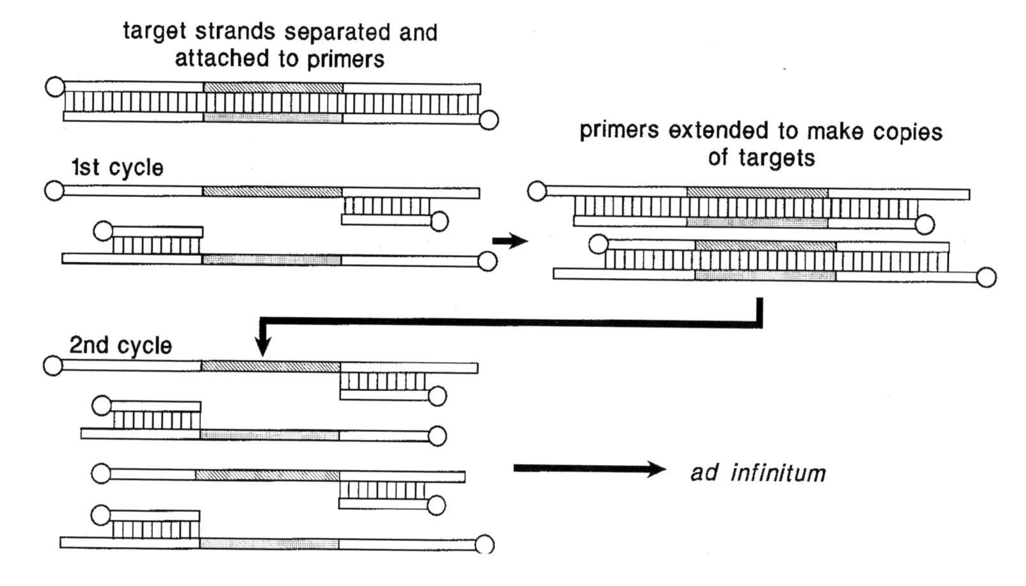

FIGURE 27.54 Polymerase chain reaction (PCR).

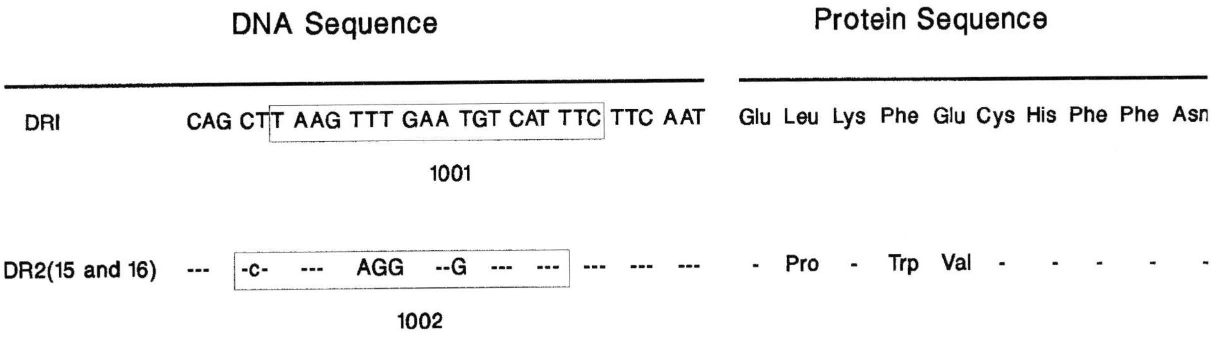

	DNA Sequence	Protein Sequence
DRI	CAG CT T AAG TTT GAA TGT CAT TTC TTC AAT	Glu Leu Lys Phe Glu Cys His Phe Phe Asn
	1001	
DR2(15 and 16)	--- -c- --- AGG --G --- --- --- --- ---	- Pro - Trp Val - - - - -
	1002	

FIGURE 27.55 DNA fingerprinting.

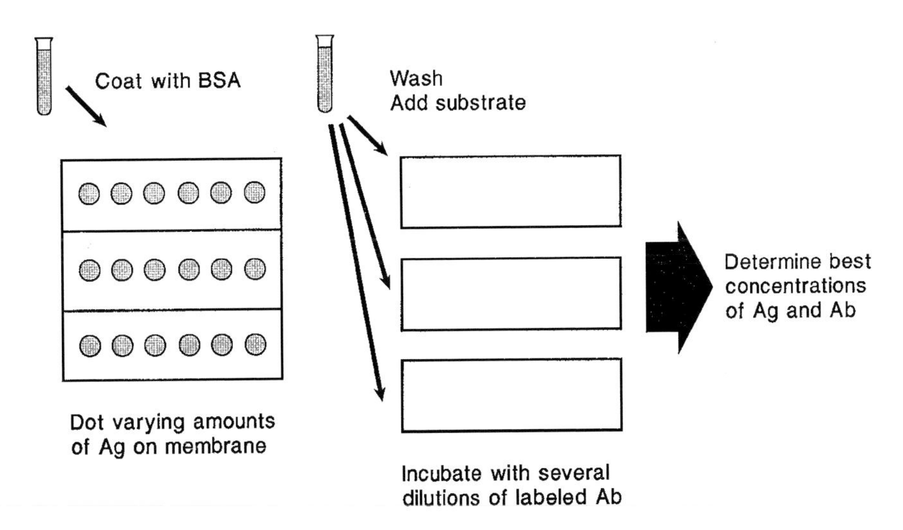

FIGURE 27.56 Dot blot.

is then used to evaluate expression of RNA in normal or neoplastic cells.

Dot blot (Figure 27.56) is a rapid hybridization method to partially quantify a specific RNA or DNA fragment found in a specimen without the need for a Northern or Southern blot. After serially diluting DNA, it is "spotted" on a nylon or nitrocellulose membrane and then denatured with NaOH. It is then exposed to a heat-denatured DNA fragment probe that is believed to be complementary to the nucleic acid fragment whose identity is being sought. The probe is labeled with ^{32}P or ^{35}S. When the two strands are complementary, hybridization takes place. This is detected by autoradiography of the radiolabeled probe. Enzymatic, nonradioactive labels may also be employed.

Multilocus probes (MLPs) (Figure 27.57) are probes used to identify multiple related sequences distributed throughout each person's genome. Multilocus probes may reveal as many as 20 separate alleles. Because of this multiplicity of alleles, there is only a remote possibility that two unrelated persons would share the same pattern, i.e., about 1 in 30 billion. There is, however, a problem in deciphering the multibanded arrangement of minisatellite

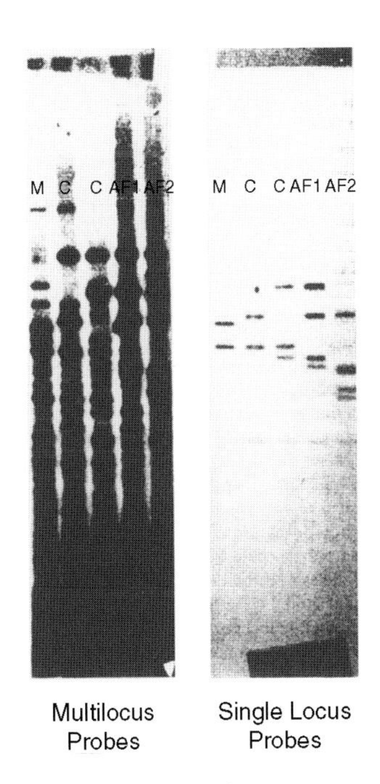

FIGURE 27.57 Multilocus probes (MLPs).

RFLPs, as it is difficult to ascertain which bands are allelic. Mutation rates of minisatellite HVRs remain to be demonstrated but are recognized occasionally. Used in resolving cases of disputed parentage.

Restriction fragment length polymorphism (RFLP) refers to genome diversity in DNA from different subjects revealed by restriction map comparisons. It is based on differences in restriction fragment lengths which are determined by sites of restriction endonuclease cleavage of the DNA molecules. This is revealed by preparing Southern blots using appropriate molecular hybridization probes. Polymorphisms may be demonstrated in exons, introns, flanking sequences, or any DNA sequence. Variations in DNA sequence show Mendelian inheritance. Results are useful in linkage studies and can help to identify defective genes associated with inherited disease.

RFLP (restriction fragment length polymorphism) is a method to identify local DNA sequence variations of humans or other animals that may be revealed by the use of restriction endonucleases. These enzymes cut double-stranded DNA at points where they recognize a very specific oligonucleotide sequence, resulting in DNA fragments of different lengths that are unique to each individual animal or person. The fragments of different sizes are separated by electrophoresis. The technique is useful for a variety of purposes, such as identifying genes associated with neurologic diseases (e.g., myotonic dystrophy) which are inherited as autosomal dominant genes or in documenting chimerism. The fragments may also be used as genetic markers to help identify the inheritance patterns of particular genes.

Single locus probes (SLPs) are probes that hybridize at only one locus. These probes identify a single locus of variable number of tandem repeats (VNTRs) and permit detection of a region of DNA repeats found in the genome only once and located at a unique site on a certain chromosome. Therefore, an individual can have only two alleles that SLPs will identify, as each cell of the body will have two copies of each chromosome, one from the mother and the other from the father. When the lengths of related alleles on homologous chromosomes are the same, there will be only a single band in the DNA typing pattern. Therefore, the use of an SLP may yield either a single- or double-band result from each individual. Single locus markers such as the pYNH24 probe developed by White may detect loci that are highly polymorphic, exceeding 30 alleles and 95% heterozygosity. SLPs are used in resolving cases of disputed parentage.

A **λ cloning vector** is a genetically engineered λ phage that can accept foreign DNA and be used as a vector in recombinant DNA studies. Phage DNA is cleaved with restriction endonucleases, and foreign DNA is inserted. Insertion vectors are those with a single site where phage DNA is cleaved and foreign DNA inserted. Substitution or replacement vectors are those with two sites which span a DNA segment that can be excised and replaced with foreign DNA.

Sequence-specific priming (SSP) is a method that employs a primer with a single mismatch in the 3′-end that cannot be employed efficiently to extend a DNA strand because the enzyme Taq polymerase, during the PCR reaction, and especially in the first PCR cycles which are very critical, does not manifest 3′ -5′ proofreading endonuclease activity to remove the mismatched nucleotide. If primer pairs are designed to have perfectly matched 3′-ends with only a single allele, or a single group of alleles and the PCR reaction is initiated under stringent conditions, a perfectly matched primer pair results in an amplification product, whereas a mismatch at the 3′-end primer pair will not provide any amplification product. A positive result, i.e., amplification, defines the specificity of the DNA sample. In this method, the PCR amplification step provides the basis for identifying polymorphism. The postamplification processing of the sample consists only of a simple agarose gel electrophoresis to detect the presence or absence of amplified product. DNA amplified fragments are visualized by ethidium bromide staining and exposure to UV light. A separate technique detects amplified product by color fluorescence. The primer pairs are selected in such a manner that each allele should have a unique reactivity pattern with the panel of primer pairs employed. Appropriate controls must be maintained.

The **plaque-forming cell (PFC) assay** (Figure 27.58) is a technique for demonstrating and enumerating cells forming antibodies against a specific antigen. Mice are immunized with sheep red blood cells (SRBC). After a specified period of time, a suspension of splenic cells from the immunized mouse is mixed with antigen (SRBC) and spread on a suitable semisolid gel medium. After or during incubation at 37°C, complement is added. The erythrocytes that have anti-SRBC antibody on their surface will be lysed. Circular areas of hemolysis appear in the gel medium. If viewed under a microscope, a single antibody-forming cell can be identified in the center of the lytic area. There are several

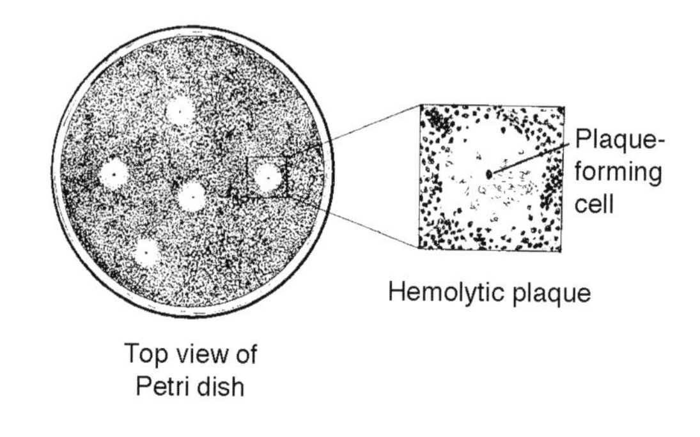

FIGURE 27.58 Plaque-forming cell (PFC) assay.

modifications of this assay, as some antibodies other than IgM may fix complement less efficiently. In order to enhance the effects, an antiglobulin antibody called developing antiserum is added to the mixture. The latter technique is called indirect PFC assay.

The **PFC (plaque-forming cell)** is an *in vitro* technique in which antibody-synthesizing cells derived from the spleen of an animal immunized with a specific antigen produce antibodies that lyse red blood cells coated with the corresponding antigen in the presence of complement in a gel medium. The reaction bears some resemblance to β hemolysis produced by streptococci on a blood agar plate. When examined microscopically, a single antibody-producing cell can be detected in the center of the plaque-forming unit.

Jerne plaque assay (Figure 27.59) is a technique to identify and enumerate cells synthesizing antibodies. Typically, spleen cells from a mouse immunized against sheep red blood cells are combined with melted agar or agarose in which sheep erythrocytes are suspended. After gentle mixing, the suspension is distributed into Petri plates where it gels. This is followed by incubation at 37°C, after which complement is added to the dish from a pipette. Thus, the sheep erythrocytes (SRBC) surrounding cells

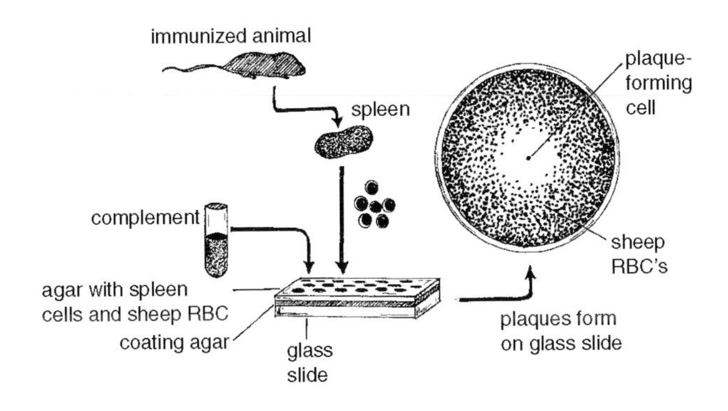

FIGURE 27.59 Jerne plaque assay.

secreting IgM antibody against SRBC are lysed by the added complement, producing a clear zone of hemolysis resembling the effect produced by β hemolytic streptococci on blood agar. IgG antibody against sheep erythrocytes can be identified by adding anti-IgG antibody to aid lysis by complement. Whereas modifications of this method have been used to identify cells producing antibodies against a variety of antigens or haptens conjugated to the sheep red cells, it can also be used to ascertain the immunoglobulin class being secreted. This method is also known as the **hemolytic plaque assay.**

PFU is an abbreviation for plaque-forming unit. An assay of plaques that develop in the hemolytic plaque assay and related techniques.

The **reverse plaque method** (Figure 27.60) is a method to identify antibody-secreting cells regardless of their antibody specificity. The antibody-forming cells are suspended in agarose and incubated at 37°C in Petri plates with sheep red cells coated with protein A. Anti-Ig and complement are also present. Cells synthesizing and secreting immunoglobulin become encircled by Ig–anti-Ig complexes and then link to protein A on the erythrocyte surfaces. This leads to hemolytic plaques (zones of lysis). Thus, any class of immunoglobulin can be identified by this technique through the choice of the appropriate antibody.

The **Cunningham plaque technique** is a modification of the hemolytic plaque assay in which an erythrocyte monolayer between a glass slide and cover slip is used without agar for the procedure.

Nylon wool (Figure 27.61) is a material that has been used to fractionate T and B cells from a mixture of the two based upon the tendency for B cells to adhere to the nylon wool, whereas the T cells pass through. B cells are then eluted from the column. Previously, tissue typing laboratories used this technique to isolate B lymphocytes for

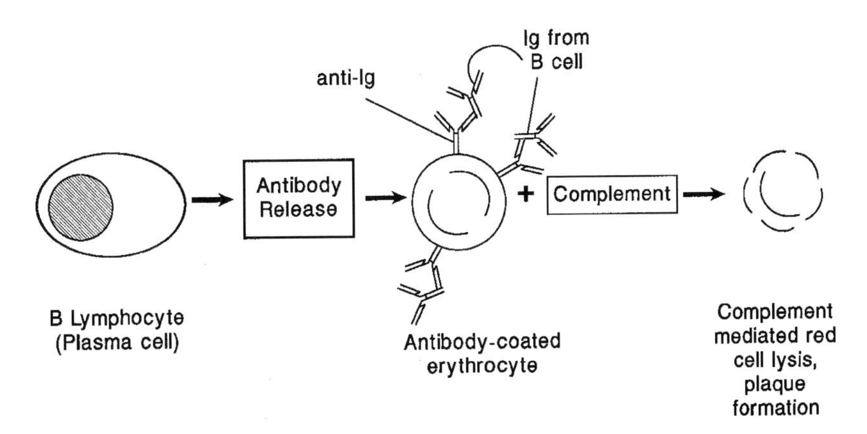

FIGURE 27.60 Reverse plaque method.

MHC class II (B cell) typing. Magnetic beads have replaced nylon wool for lymphocyte T and B cell separation.

Microlymphocytotoxicity (Figure 27.62) is a widely used technique for HLA tissue typing. Lymphocytes are separated from heparinized blood samples by either layering over Ficoll-Hypaque, centrifuging and removing lymphocytes from the interface, or with beads. After appropriate washing, these purified lymphocyte preparations are counted and aliquots dispensed into microtiter plate wells containing predispensed quantities of antibody. When used for human histocompatibility (HLA) testing, antisera in the wells are specific for known HLA antigenic specificities. After incubation of the cells and antisera, rabbit complement is added and the plates are again incubated. The extent of cytotoxicity induced is determined by incubating the cells with trypan blue, which enters dead cells staining them blue but leaves live cells unstained. The plates are read by using an inverted phase contrast microscope. A scoring system from 0 to 8 (where 8 implies >80% of target cells killed) is employed to indicate cytotoxicity. Most of the sera used to date are multispecific, as they are obtained from multiparous females who have been sensitized during pregnancy by HLA antigens determined by their spouse. Monoclonal antibodies are being used with increasing frequency in tissue typing. This technique is useful to identify HLA-A, HLA-B, and HLA-C antigens. When purified, B cell preparations and specific antibodies against B cell antigens are employed, HLA-DR and HLA-DQ antigens can be identified.

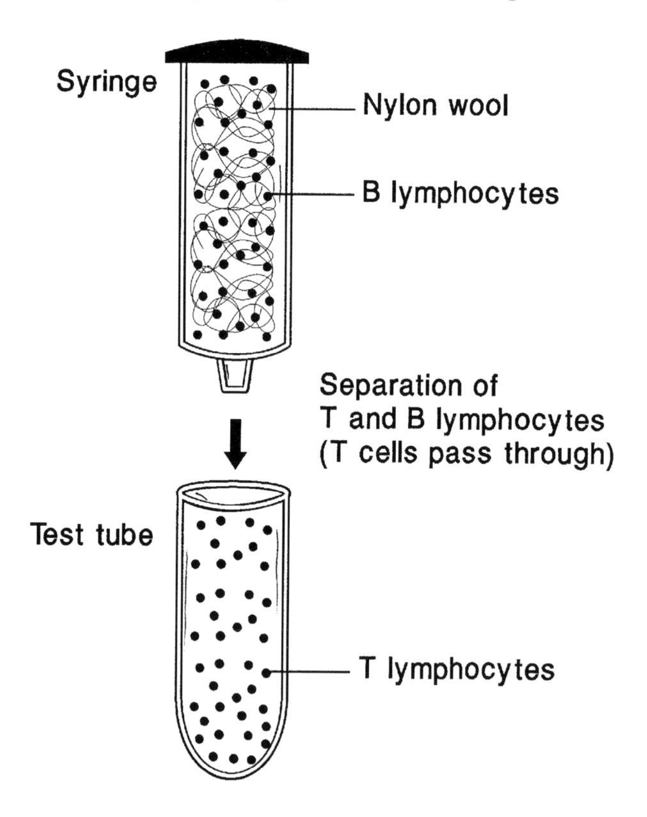

FIGURE 27.61 Nylon wool.

In the **cell-mediated lympholysis (CML) test**, responder (effector) lymphocytes are cytotoxic for donor (target) lymphocytes after the two are combined in culture (Figure 27.63). Target cells are labeled by incubation with ^{51}Cr at 37°C for 60 min. Following combination of effector and target cells in tissue culture, the release of ^{51}Cr from target cells injured by cytotoxicity represents a measure of cell-mediated lympholysis (CML). The CML assay gives uniform results, is relatively simple to perform, and is rather easily controlled. The effector cells can result from either *in-vivo* sensitization following organ grafting or can be induced *in vitro*. Variations in effector to target cell ratios can be employed for quantification.

In the **mixed lymphocyte reaction (MLR),** lymphocytes from potential donor and recipient are combined in tissue

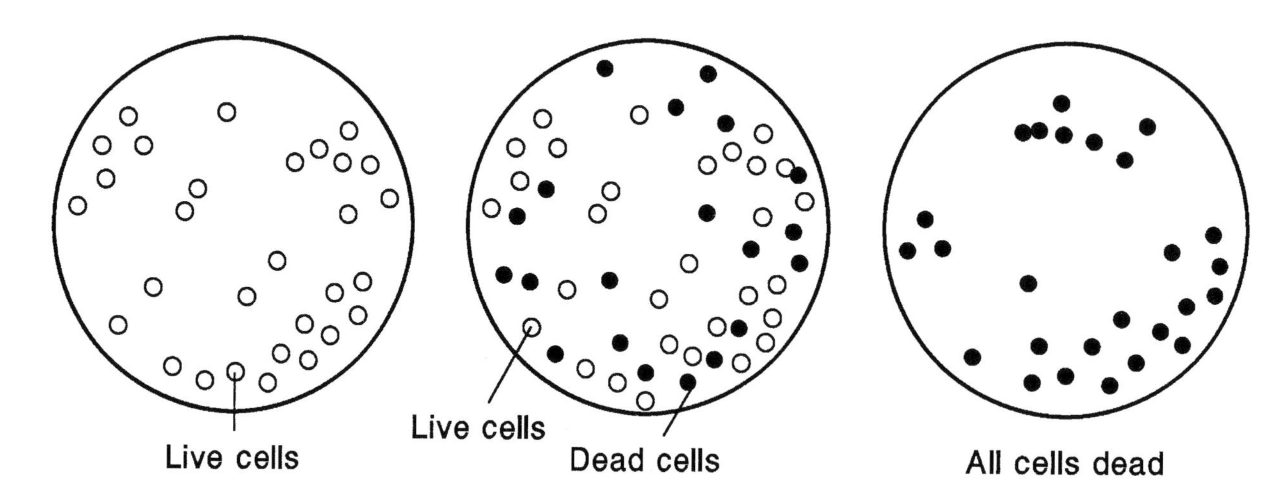

FIGURE 27.62 Microlymphocytotoxicity.

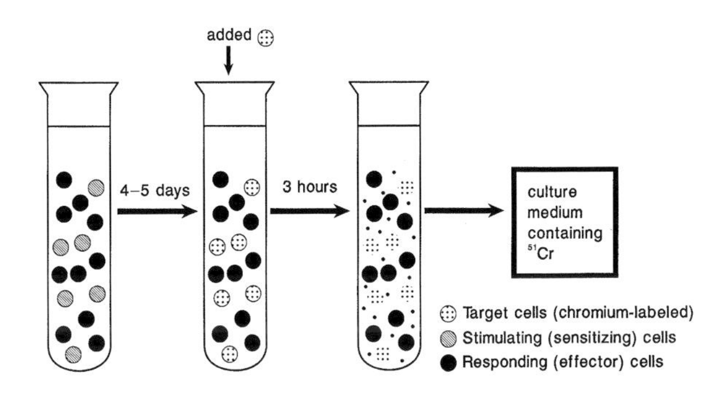

FIGURE 27.63 Cell-mediated lympholysis (CML) test.

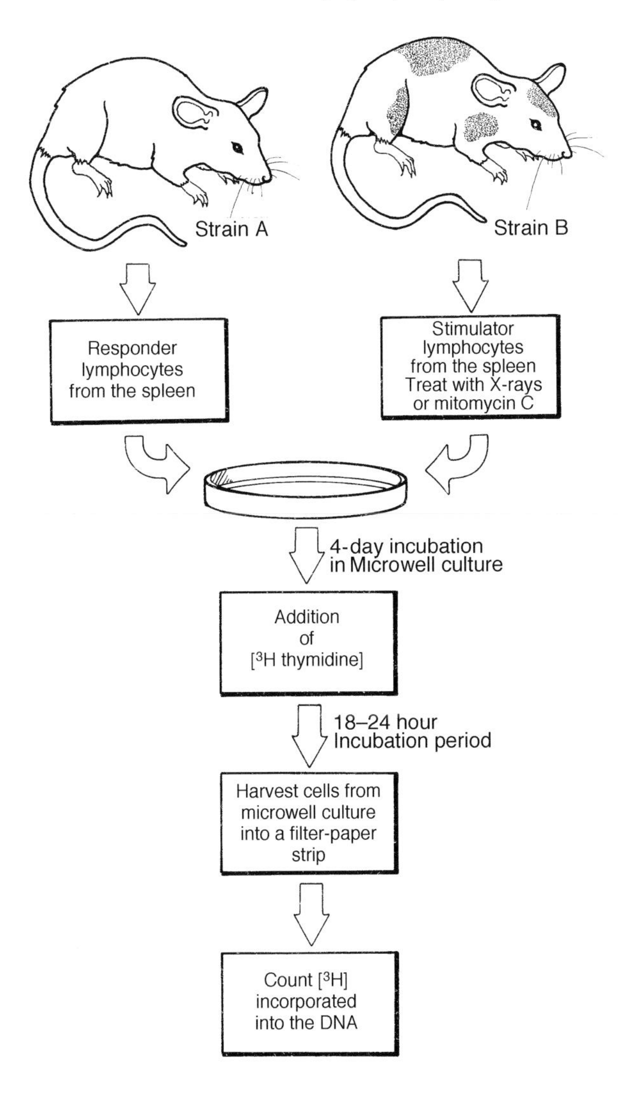

FIGURE 27.64 Mixed lymphocyte reaction (MLR).

culture (Figure 27.64). Each of these lymphoid cells has the ability to respond by proliferating following stimulation by antigens of the other cell. In the one-way reaction, the donor cells are treated with mitomycin or irradiation to render them incapable of proliferation. Thus, the donor antigens stimulate the untreated responder cells. Antigenic specificities of the stimulator cells that are not present in the responder cells lead to blastogenesis of the responder lymphocytes. This leads to an increase in the synthesis of DNA and cell division. This process is followed by introduction of a measured amount of tritiated thymidine, which is incorporated into the newly synthesized DNA. The mixed-lymphocyte reaction usually measures a proliferative response and not an effector-cell-killing response. The test is important in bone marrow and organ transplantation to evaluate the degree of histoincompatibility between donor and recipient. Both $CD4^+$ and $Cd8^+$ T lymphocytes proliferate and secrete cytokines in the MLR. Also called mixed-lymphocyte culture.

Lymphocyte transformation (Figure 27.65) is an alteration in the morphology of a lymphocyte induced by an antigen, mitogen, or virus interacting with a small, resting lymphocyte. The transformed cell increases in size and amount of cytoplasm. Nucleoli develop in the nucleus, which becomes lighter staining as the cell becomes a blast. Epstein-Barr virus transforms B cells, and the human T cell leukemia virus transforms T cells. The lymphocyte transformation test involves activation of lymphocytes with mitigens, antigen, superantigens, and antibodies to components of cell membranes. This leads to their synthesis of proteins that include immunoglobulins, cytokines, and growth factors. The activated lymphocyte enters the cell cycle, synthesizes DNA, and replicates and undergoes metabolic and morphologic changes. The mitogens, phytohemagglutinin and Concanvalin-A, superantigens, anti-CD3 and antigens that are presented by antigen-presenting cells activate T lymphocytes. Antiimmunoglobulin, bacterial lypopolysaccharides, and staphycoccal protein A activate B lymphocytes. The lymphocyte transformation assay is a broadly used *in vitro* test to evaluate lymphocyte function in patients.

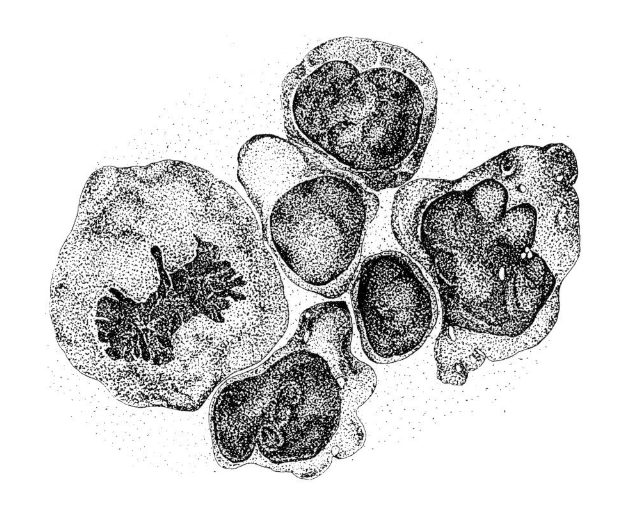

FIGURE 27.65 Lymphocyte transformation.

The **lymphocyte antigen stimulation test** is an assay for the *in vitro* assessment of impaired cell-mediated immunity. This test is useful to evaluate patients with genetic or acquired immunodeficiencies, bacterial and viral infections, cancer, autoimmune disorders, transplantation-related disorders, antisperm antbodies, and individuals with previous exposure to a variety of antigens, allergens, pathogens, and metals/chemicals. Lymphocyte antigen stimulation is assayed by (^{3}H)-thymidine uptake or a flow cytometric assay (based on expression of the activation antigen CD69) with (^{3}H)-thymidine incorporation. Antigen-stimulated culture supernatants can be assessed for cytokine production by EIA.

The **lymphocyte mitogen stimulation test** is an assay used for the *in vitro* assessment of cell-mediated immunity in patients with immunodeficiency, autoimmunity, infectious diseases, cancer, and chemical-induced hypersensitivity reactions. Healthy human lymphocytes have receptors for mitogens such as the plant lectin concanavalin-A (Con A), pokeweed mitogen (PWM), *Staphylococcus* protein A, and chemicals. Lymphocytes respond to these mitogens that stimulate large numbers of lymphocytes, without prior sensitization. In contrast to antigens, mitogens do not require a sensitized host. Mitogens may stimulate both B and T cells, and the inability of lymphocytes to respond to mitogens suggests impaired cell-mediated or homoral immunity.

The **lymphocyte toxicity assay** is a test to evaluate adverse reactions to drugs, especially anticonvulsants. Incubation with liver microsomes is believed to metabolize the drug to the *in vivo* metabolite that kills lymphocytes from sensitized patients but not from controls. Lymphocytes derived from nonreactive individuals do not show significant lymphocyte toxicity.

Macrophage–monocyte inhibitory factor (MIF) (Figure 27.66) is a substance synthesized by T lymphocytes in response to immunogenic challenge that inhibits the migration of macrophages. MIF is a 25-kDa lymphokine. Its mechanism of action is by elevating intracellular cAMP, polymerizing microtubules, and stopping macrophage migration. MIF may increase the adhesive properties of macrophages, thereby inhibiting their migration. The two types of the protein MIF include one that is 65-kDa with a pI of 3 to 4 and another that is 25-kDa with a pI of approximately 5.

The **macrophage migration test** is an *in vitro* assay of cell-mediated immunity. Macrophages and lymphocytes from the individual to be tested are placed into segments of capillary tubes about the size of microhematocrit tubes and incubated in tissue culture medium containing the soluble antigen of interest, with maintenance of appropriate controls incubated in the same medium not containing

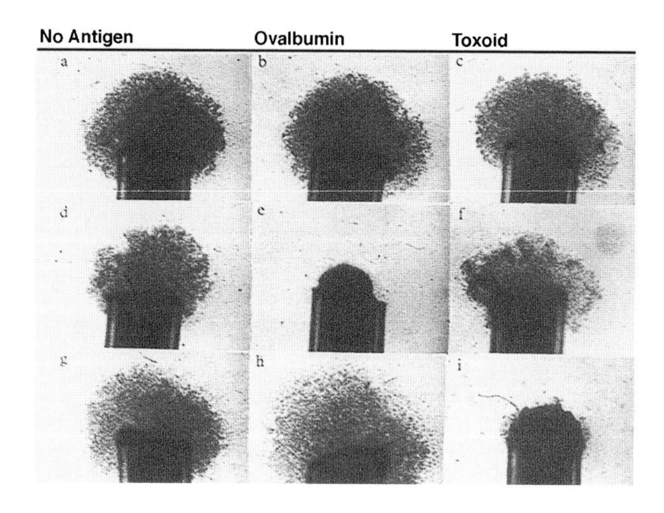

FIGURE 27.66 Macrophage–monocyte inhibitory factor (MIF).

the antigen. Lymphocytes from an animal or human sensitized to the antigen release a lymphokine called migration inhibitory factor that will block migration of macrophages from the end of the tube where the cells form an aggregated mass. The macrophages in the control preparation (that does not contain antigen) will migrate out of the tube into a fan-like pattern.

Macrophage functional assays are tests of macrophage function. (1) Chemotaxis: macrophages are placed in one end of a Boyden chamber and a chemoattractant is added to the other end. Macrophage migration toward the chemoattractant is assayed. (2) Lysis: macrophages acting against radiolabeled tumor cells or bacterial cells in suspension can be measured after suitable incubation by measuring the radioactivity of the supernatant. (3) Phagocytosis: radioactivity of macrophages that have ingested a radiolabeled target can be assayed.

Panning (Figure 27.67) is a technique to isolate lymphocyte subsets through the use of petri plates coated with monoclonal antibodies specific for lymphocyte surface markers. Thus, only lymphocytes bearing the marker being sought bind to the petri plate surface.

Immunobeads (Figure 27.68) are minute plastic spheres with a coating of antigen (or antibody) that may be aggregated or agglutinated in the presence of the homologous antibody. Immunobeads are used also for the isolation of specific cell subpopulations such as the separation of B cells from T cells that is useful in class II MHC typing for tissue transplantation.

The **nitroblue tetrazolium (NBT) test** (Figure 27.69) is an assay that evaluates the hexose monophosphate shunt in phagocytic cells. The soluble yellow dye, nitroblue tetrazolium, is taken up by neutrophils and monocytes

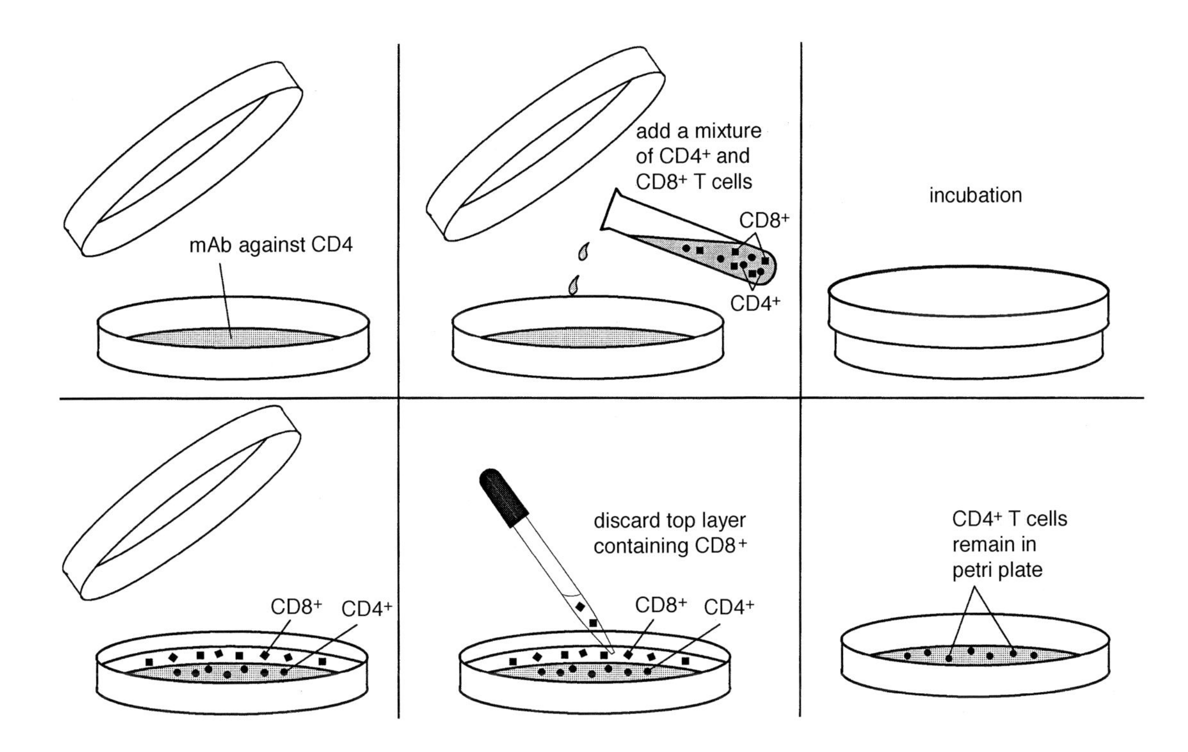

FIGURE 27.67 Panning.

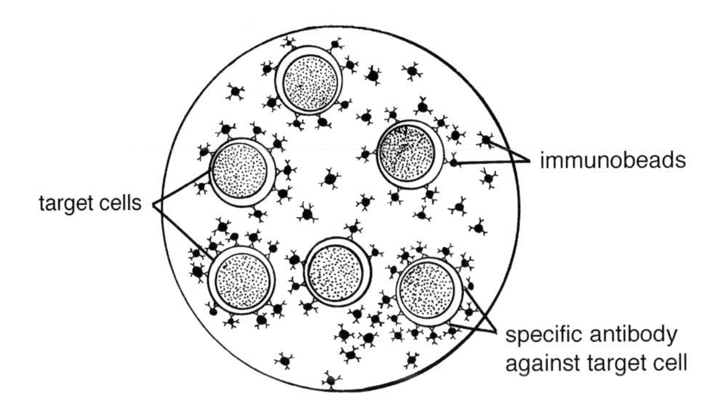

FIGURE 27.68 Immunobeads.

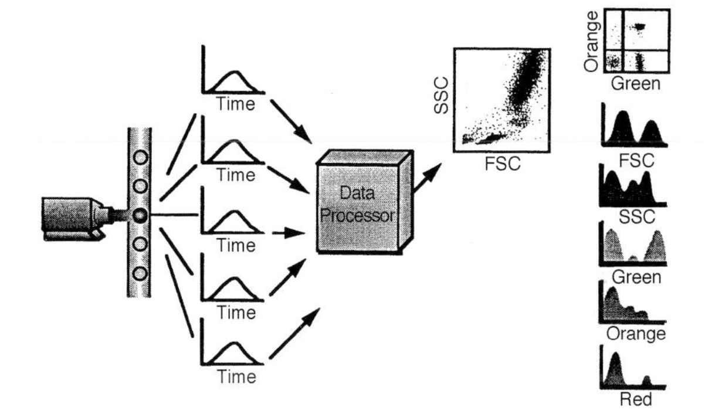

FIGURE 27.70 Flow cytometry is a fast, accurate way to measure multiple characteristics of a single cell simultaneously. These objective measurements are made one cell at a time, at routine rates of 500 to 4000 particles per second in a moving fluid stream. A flow cytometer measures relative size (FSC), relative granularity or internal complexity (SSC), and relative florescence. Use of three-color flow cytometry to analyze blood cells by size, cytoplasmic granularity, and surface markers labeled with different fluorochromes.

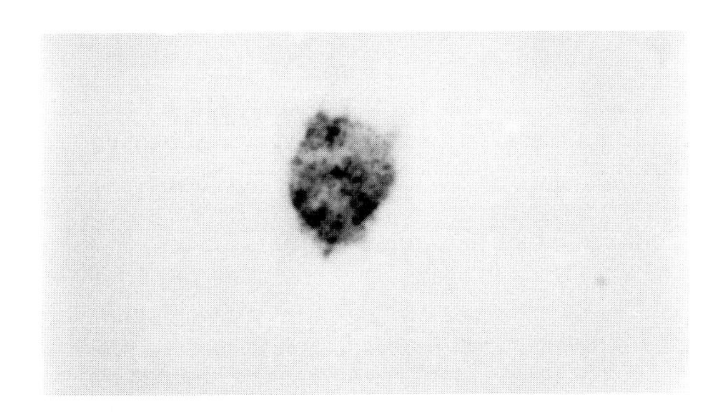

FIGURE 27.69 Nitroblue tetrazolium test (NBT).

during phagocytosis. In normal neutrophils, the NBT is reduced by enzymes to insoluble, dark blue formazan crystals within the cell. Neutrophils from patients with chronic granulomatous disease are unable to reduce the nitroblue tetrazolium. The ability to reduce NBT to the insoluble deep blue formazan crystals depends on the generation of superoxide in the neutrophil being tested.

Flow cytometry (Figure 27.70) is an analytical technique to phenotype cell populations that requires a special apparatus, termed a flow cytometer, that can detect fluorescence on individual cell in suspension and thereby ascertain the number of cells that express the molecule binding a fluorescent probe. Cell suspensions are incubated with fluorescent-labeled monoclonal antibodies or other probes and the quantity of probe bound by each cell in the population is assayed by passing the cells one at a time through a spectrofluorometer with a laser-generated incident beam. Sample cells flow single file past a narrowly focused excitation light beam that is used to probe the cell properties of interest. As the cells pass the focused excitation lightbeam, each cell scatters light and may emit fluorescent light, depending on whether or not it is labeled with a fluorochrome or is autofluorescent. Scattered light is measured in both the forward and perpendicular directions relative to the incident beam. The fluorescent emissions of the cell are measured in the perpendicular directions by a photosensitive detector. Measurements of light scatter and fluorescent emission intensities are used to characterize each cell as it is processed. Flow cytometry is a fast, accurate way to measure multiple characteristics of a single cell simultaneously. These objective measurements are made one cell at a time, at routine rates of 500 to 4000 particles per second in a moving fluid stream. A flow cytometry measures relative size (FSC), relative granularity or internal complexity (SSC), and relative fluorescence. Three-color flow cytometry is used to analyze blood cells by size, cytoplasmic granularity, and surface markers labeled with different fluorochromes. Flow cytometry serves as the basis for numerous very different, highly specialized assays. It is a multifactorial analysis technique and provides the capability for performing many of these assays simultaneously.

The **neutrophil microbicidal assay** is a test that assesses the capacity of polymorphonuclear neutrophil leukocytes to kill intracellular bacteria.

Bright is an adjective used in flow cytometry to indicate the relative fluorescence intensity of cells being analyzed, with bright designating the greatest intensity and dim representing the lowest intensity of fluorescence.

FACS® is an abbrebiation for fluorescence-activated cell sorter.

A **fluorescence-activated cell sorter (FACS®)** is an instrument that measures the size, granularity, and fluorescence of cells attributable to bound fluorescent antibodies, as individual single cells flow in a stream past photodetectors. Single-cell analysis by this method is termed flow cytometry, and the machine used to make these measurements and/or sort cells is termed a flow cytometer or cell sorter.

Dim is an adjective used in flow cytometry to indicate the relative fluorescence intensity of cells being analyzed, with dim representing the lowest intensity and bright designating the greatest intensity of fluorescence.

Immunophenotyping is the use of monoclonal antibodies and flow cytometry to reveal cell surface or cytoplasmic antigens that yield information that may reflect clonality and cell lineage classification. This type of data is valuable clinically in aiding the diagnosis of leukemias and lymphomas through the use of a battery of B cell, T cell, and myeloid markers. However, immunophenotyping results must be used only in conjunction with morphologic criteria when reaching a diagnosis of leukemia or lymphoma.

Texas red is a fluorochrome derived from sulforhodamine 101. It is often used as a second label in fluorescence antibody techniques where fluorescein, an apple-green label, is also used. This provides two-color fluorescence.

Chemiluminescence is the conversion of chemical energy into light by an oxidation reaction. A high-energy peroxide intermediate, such as luminol, is produced by the reaction of a precursor substance exposed to peroxide and alkali. The emission of light energy by a chemical reaction may occur during reduction of an unstable intermediate to a stable form. Chemiluminescence measures the oxidative formation of free radicals such as superoxide anion by polymorphonuclear neutrophils and mononuclear phagocytes. Light is released from these cells after they have taken up luminol (5-amino-2,3-dihydro-1,4-phthalazinedione). This is a mechanism to measure the respiratory burst in phagocytes. The oxidation of luminol increases intracellular luminescence. Chronic granulomatous disease may be diagnosed by this technique.

Chemotactic assays: The chemotactic properties of various substances can be determined by various methods. The most popular is the Boyden technique. This consists of a chamber separated into two compartments by a Millipore® filter of appropriate porosity, through which cells can migrate actively but not drop passively. The cell preparation is placed in the upper compartment of the chamber, and the assay solution is placed in the lower compartment. The chamber is incubated in air at 37°C for 3 h, after which the filter is removed and the number of cells migrating to the opposite surface of the filter are counted.

Phycoerythrin is an extensively used label for immunofluorescence. This light-gathering plant protein absorbs light efficiently and emits a brilliant red fluorescence.

Light scatter refers to light dispersion in any direction which can be useful in the study of cells by flow cytometry (Figure 27.71 to Figure 27.85). A cell passing through a laser beam both absorbs and scatters light. Fluorochrome staining of cells permits absorbed light to be emitted as fluorescence. Forward angle light scatter permits identification of a cell in flow and determination of its size.

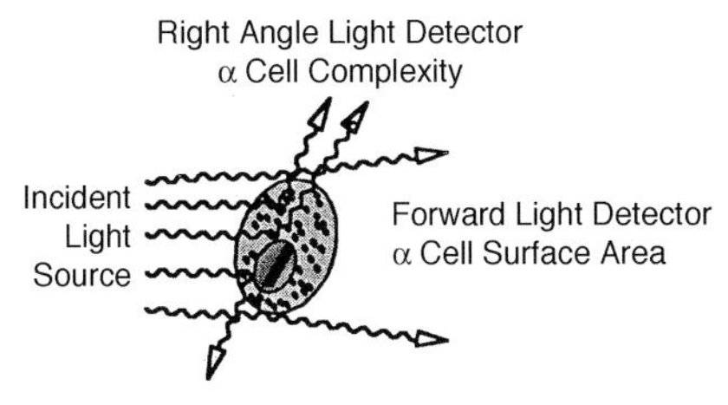

FIGURE 27.71 Properties of Forward Scatter Light (FSC) and Side Scatter Light (SCC) are measured by observing how light disperses when a laser hits the cell.

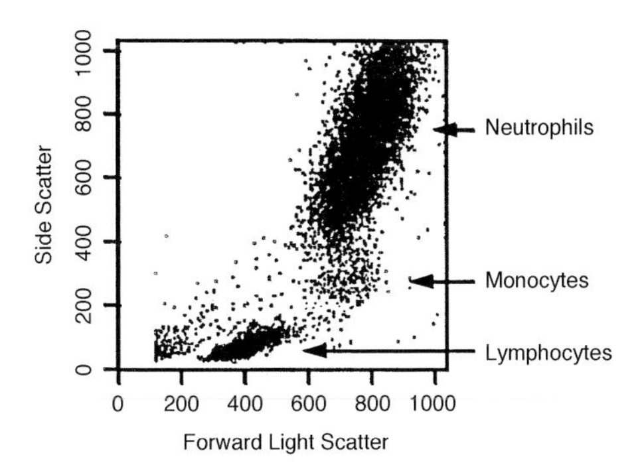

FIGURE 27.72 Each dot represents the data from one cell. The bigger the cell, the larger the FSC signal and the farther to the right the dot will appear on the x-axis. The more complex or granular the cell, the larger the SCC signal and the higher it will appear on the y-axis. It is possible to discern lymphocytes, monocytes, and neutrophils in this plot.

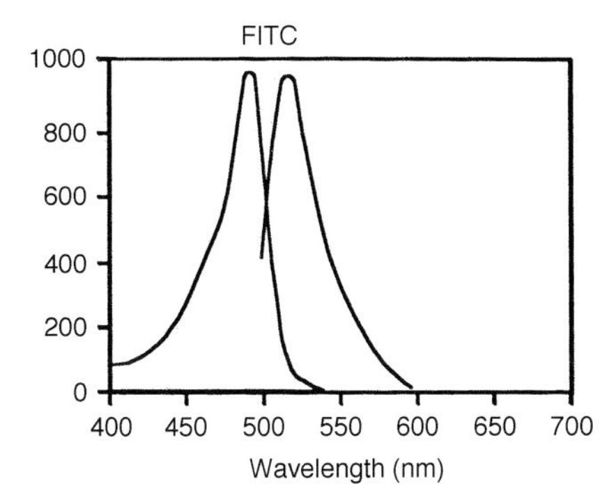

FIGURE 27.73 The absorption and emission spectra for the FITC fluorochrome are shown here. The peak absorption is around 488 nm and the peak emission is around 530 nm.

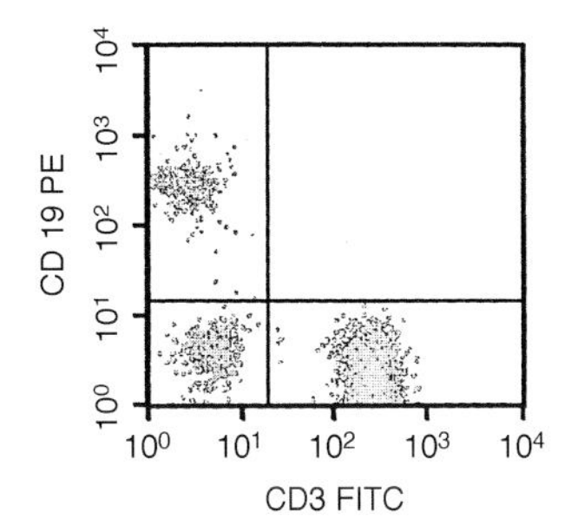

FIGURE 27.74 The FITC-positive cells fall in the lower right quadrant and PE-positive cells fall in the upper left quadrant. Cells that are positive for both FITC and PE are in the upper right quadrant.

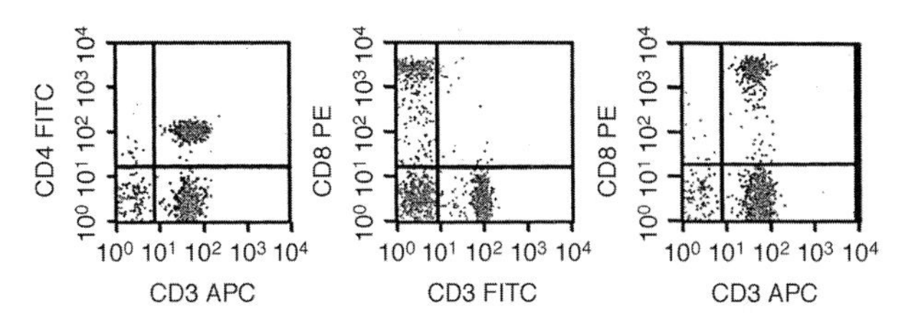

FIGURE 27.75 These are three bivariate plots displaying FITC-, PE-, and APC-stained lymphocytes. All $CD3^+$ cells are stained with APC, the $CD4^+$ cells are stained with FITC, and the $CD8^+$ cells are stained with PE.

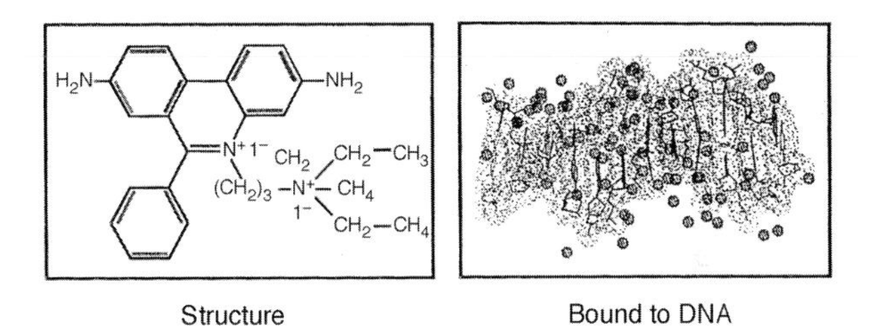

FIGURE 27.76 DNA content can be quantified by use of the fluorescent dye propidium iodide (PI). The dye intercalates between the base pairs to stain double-stranded nucleic acids. The amount of DNA that PI binds is proportional to the DNA content.

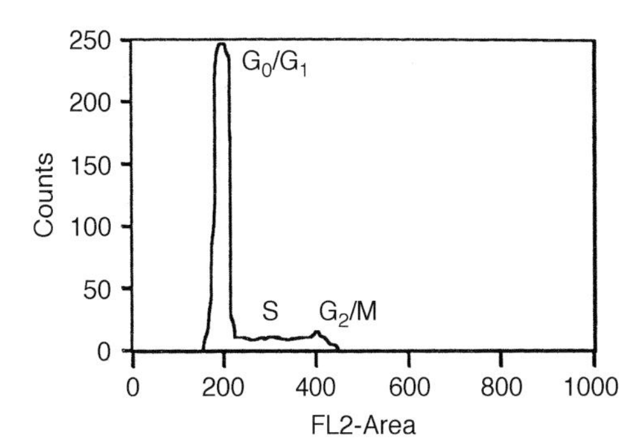

FIGURE 27.77 Staining DNA with PI and analyzing the sample by flow cytometry permits the percentage of cells in each phase of the cell cycle to be determined.

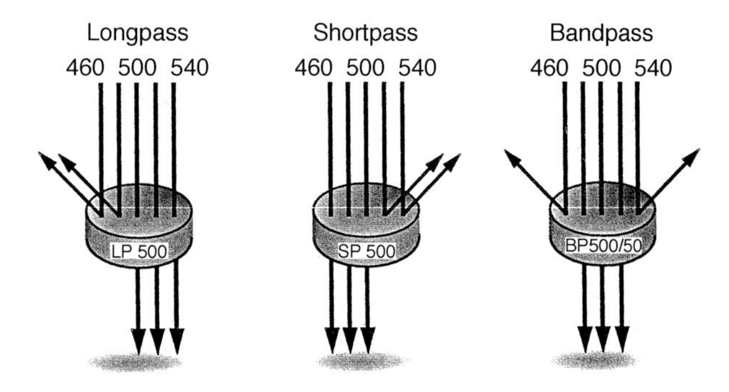

FIGURE 27.78 The specificity of an optical detector for a particular fluorescent dye is optimized by placing a filter in front of the detector which allows a narrow range of wavelengths to pass through the filter.

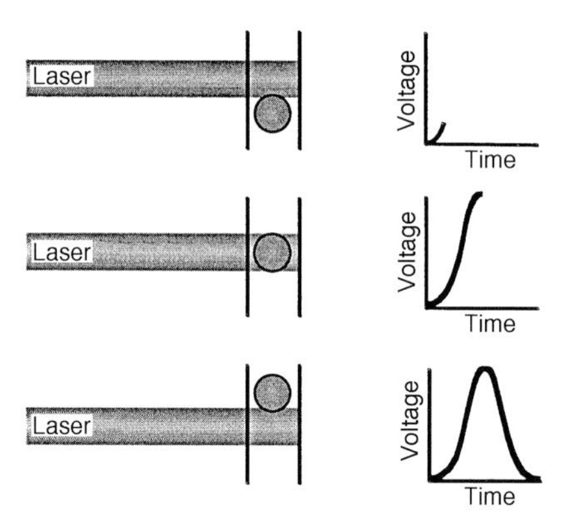

FIGURE 27.79 A pulse is created when the particle enters the laser beam and starts to scatter light. The highest point of the pulse occurs when the particle is in the center of the beam, and the maximum amount of scatter is achieved. As the particle leaves the laser, the pulse comes back down to the baseline.

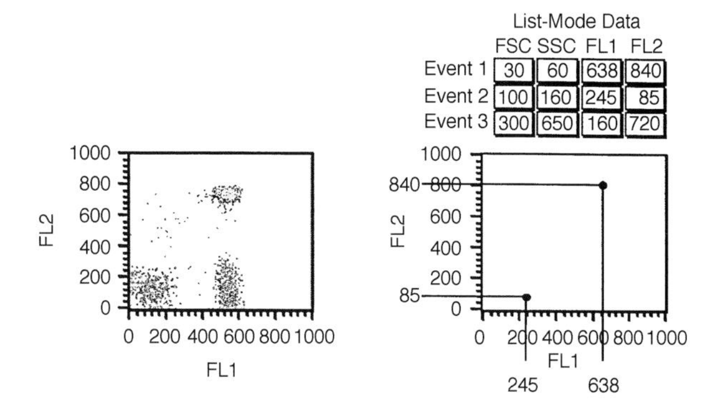

FIGURE 27.80 Useful information can be obtained with the dot plot by determining the percentages for each population.

If higher-angle light scatter is added, some specific cell populations may also be identified. Light scatter measured at 90° to the laser beam and flow stream yields data on cell granularity or fine structure. Light scatter depends on such factors as cell size and shape, cell orientation in flow,

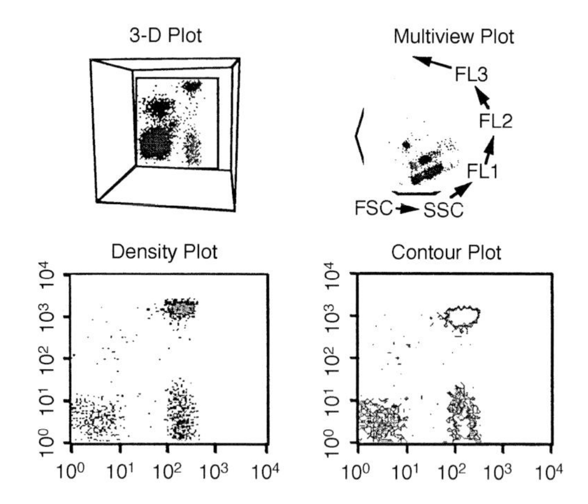

FIGURE 27.81 The different plots can be used to clearly display and analyze populations of interest.

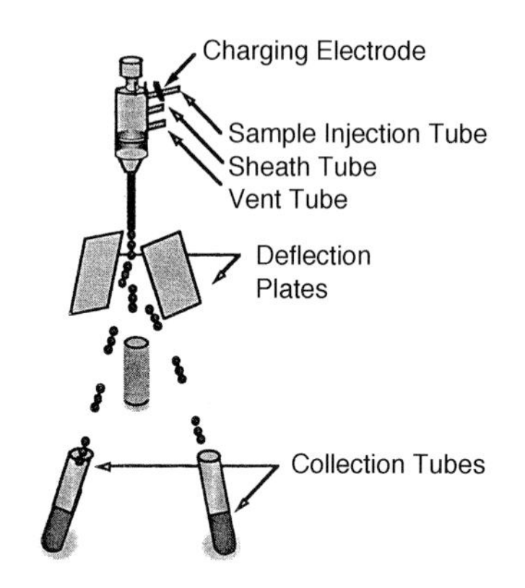

FIGURE 27.82 Particles can be isolated after they are passed through the laser by charging the particle. Depending on the charge, the particle will either travel to the left or right sort tube, repelled or attracted toward the charged plate. All noncharged particles travel to the waste.

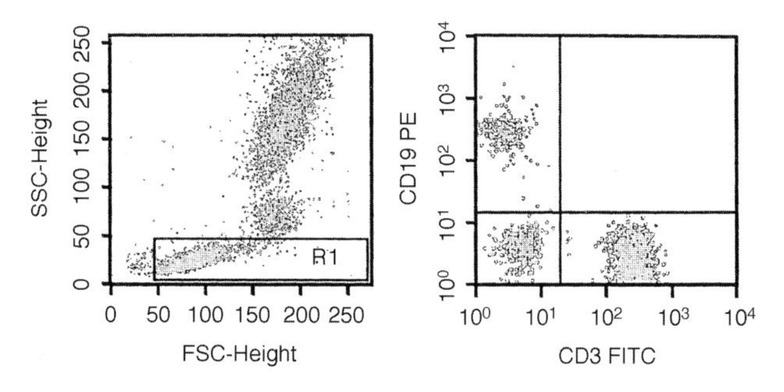

FIGURE 27.83 The conventional method for identifying lymphocyte subsets is light-scatter gating. The lymphocytes are gated, markers are set using a two-color isotype control, then subsequent immunofluorescence analyses of the remaining files are completed.

cellular internal structure, laser beam shape and wavelength, and the angle of light collection.

Limiting dilution (Figure 27.86) is a method of preparing aliquots that contain single cells through dilution to a point

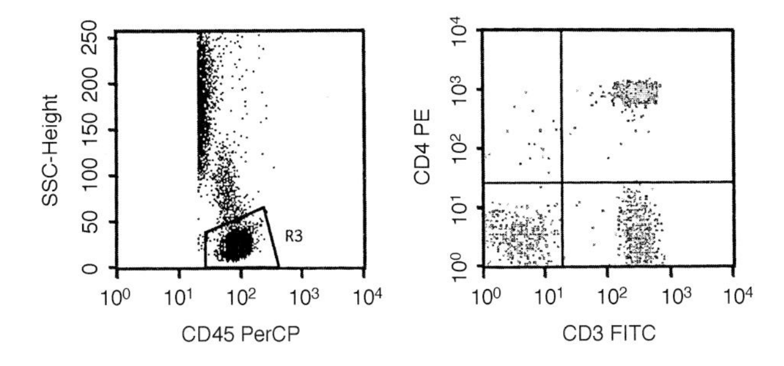

FIGURE 27.84 Unlike traditional methods of light-scatter gating where lymphocyte gate purity and recovery are concerns, TriTEST allows the CD45-positive lymphocyte population to be gated, providing unambiguous identification.

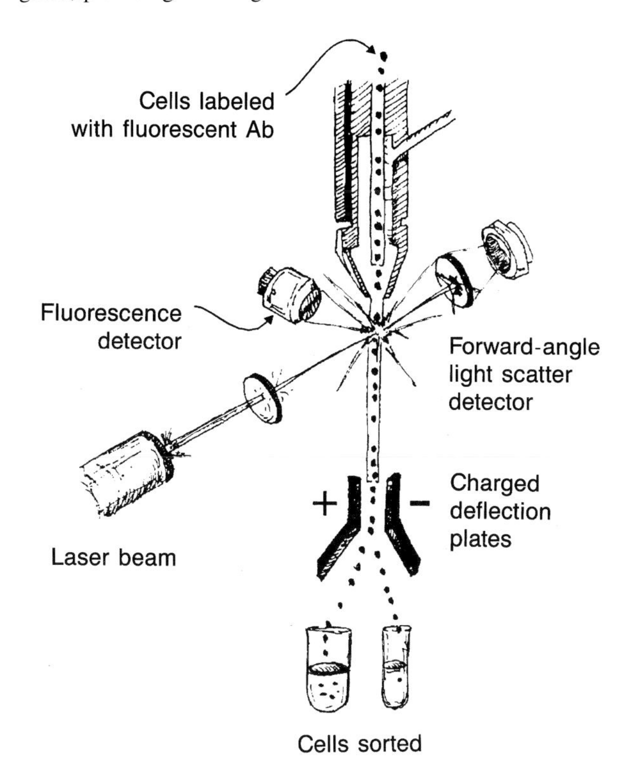

FIGURE 27.85 Light scatter.

where each aliquot contains only one cell. The apportionment of cells by this method follows the Poissonian distribution, which yields 37% of aliquots without any cells and 63% with one or more cells. This technique can be used to estimate a certain cell's frequency in a population. For example, it may be employed to approximate the frequency of helper T lymphocytes, cytotoxic T lymphocytes, or B lymphocytes in a lymphoid cell suspension or to isolate cells for cloning in the production of monoclonal antibodies.

The **prick test** (Figure 27.87) is an assay for immediate (IgE-mediated) hypersensitivity in humans. The epidermal surface of the skin on which drops of diluted antigen (allergen) are placed is pricked by a sterile needle passed through the allergen. The reaction produced is compared with one induced by histamine or another mast-cell secretogogue. The test is convenient, simple, and rapid and produces little discomfort for the patient in comparison with the intradermal test. It may even be used for infants.

The **patch test** (Figure 27.88) is an assay to determine the cause of skin allergy, especially contact allergic (type IV) hypersensitivity. A small square of cotton, linen, or paper impregnated with the suspected allergen is applied to the skin for 24 to 48 h. The test is read by examining the site 1 to 2 d after applying the patch. The development of redness (erythema), edema, and formation of vesicles constitutes a positive test. The impregnation of tuberculin into a patch was used by Vollmer for a modified tuberculin test. There are multiple chemicals, toxins, and other allergens that may induce allergic contact dermatitis in exposed members of the population.

A **skin test** is any one of several assays in which a test substance is either injected into the skin or applied to it to determine the host response. Skin tests have long been used to determine host hypersensitivity or immunity to a particular antigen or product of a microorganism. Examples include the tuberculin test, the Schick test, the Dick test, the patch test, the scratch test, etc.

The **Dick test** (Figure 27.89) is a skin test to signify susceptibility to scarlet fever in subjects lacking protective antibody against the erythrogenic toxin of *Streptococcus pyogenes*. A minute quantity of diluted erythrogenic toxin is inoculated intradermally in the individual to be tested. An area of redness (erythema) occurs at the injection site 6 to 12 h following inoculation of the diluted toxin in individuals who do not have neutralizing antibodies specific for the erythrogenic toxin and who are therefore susceptible to scarlet fever. A heat-inactivated preparation of the same diluted toxin is also injected intradermally in the same individual as a control against nonspecific hypersensitivity to other products of the preparation.

Waaler-Rose test: See Rose-Waaler test.

The **tuberculin test** refers to the 24- to 48-h response to intradermal injection of tuberculin. If positive, it signifies delayed-type hypersensitivity (type IV) to tuberculin and implies cell-mediated immunity to *Mycobacterium tuberculosis*. The intradermal inoculation of tuberculin or of purified protein derivative (PPD) leads to an area of erythema and induration within 24 to 48 h in positive individuals. A positive reaction signifies the presence of cell-mediated immunity to *M. tuberculosis* as a consequence of past or current exposure to this microorganism. However, it is not a test for the diagnosis of active tuberculosis.

Stormont test: A double intradermal tuberculin test.

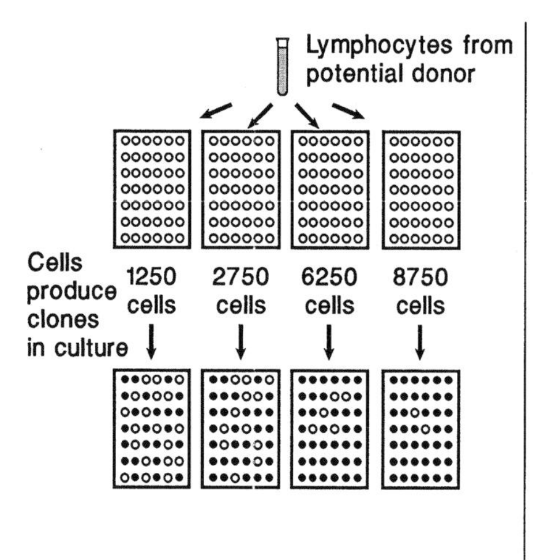

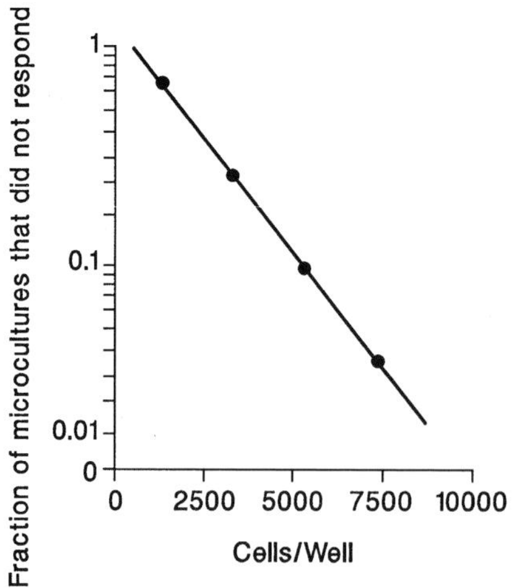

FIGURE 27.86 Limiting dilution.

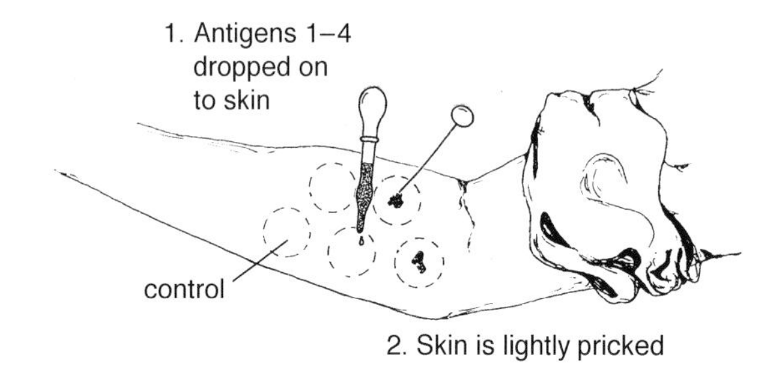

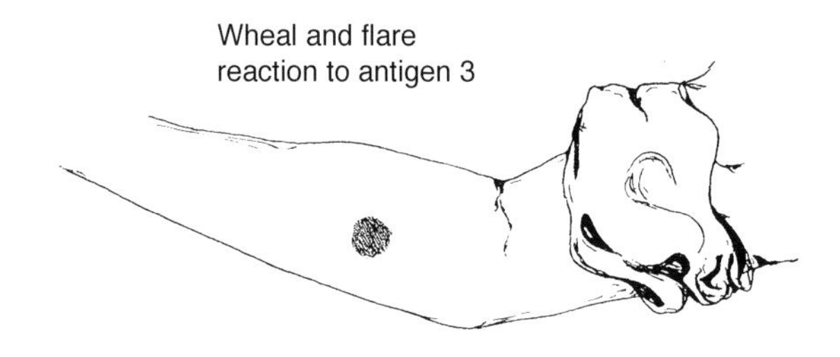

FIGURE 27.87 Prick test.

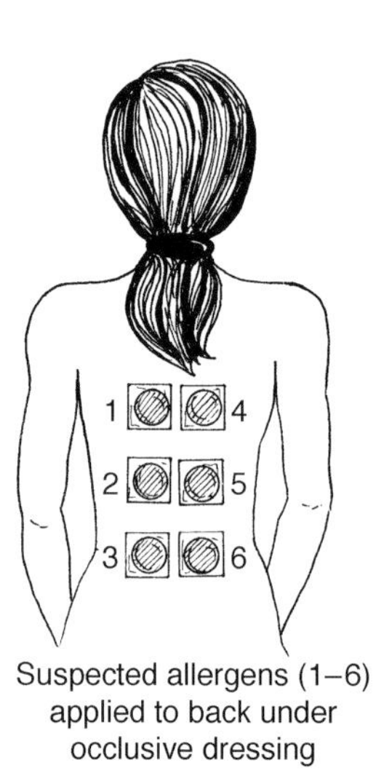

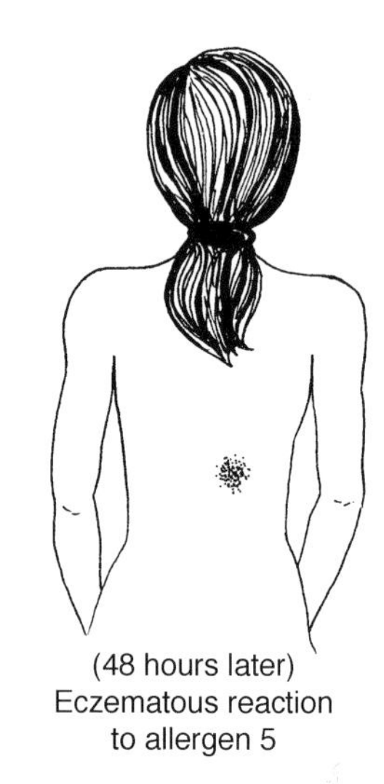

FIGURE 27.88 Patch test.

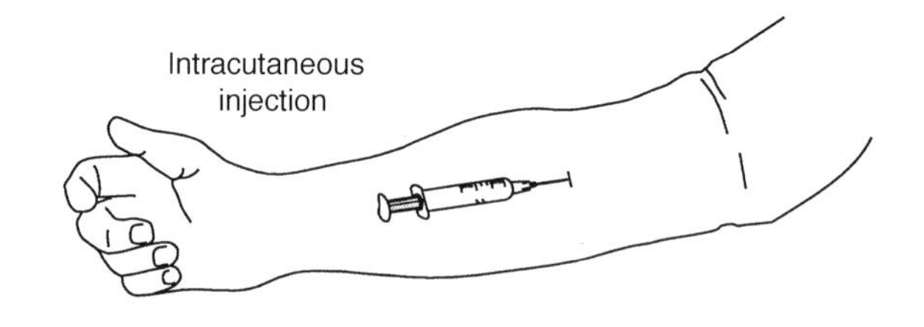

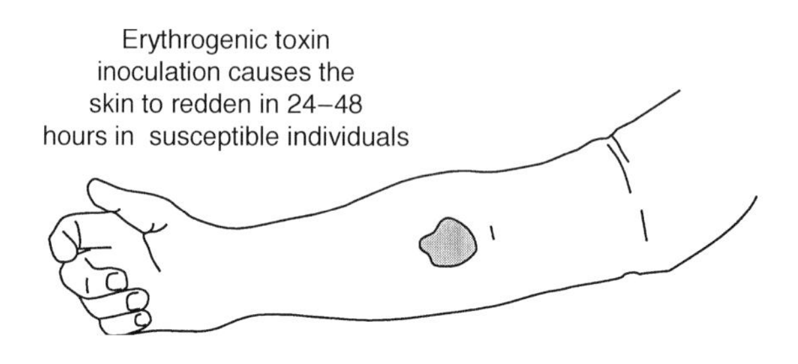

FIGURE 27.89 Dick test.

The **tine test** is a human tuberculin test that involves the intradermal inoculation of dried, old tuberculin using a four-pointed applicator that introduces the test substance 2 mm below the surface.

The **Vollmer test (historical)** is a tuberculin patch test employing gauze treated with tuberculin.

Coccidioidin is a *Coccidioides immitis* culture extract that is used in a skin test for cell-mediated immunity against the microorganism in a manner analogous to the tuberculin skin test.

Histoplasmin is an extract from cultures of *Histoplasma capsulatum* that is injected intradermally, in the same manner as the tuberculin test, to evaluate whether an individual has cell-mediated immunity against this microorganism.

The **histoplasmin test** is a skin test analogous to the tuberculin skin test, which determines whether or not an individual manifests delayed-type hypersensitivity (cell-mediated) immunity to *Histoplasma capsulatum,* the causative agent of histoplasmosis in man. A positive skin test implies an earlier or a current infection with *H. capsulatum.*

Johnin is an extract from culture medium in which *Mycobacterium johnei* is growing. It can be used in a skin test of cattle for the diagnosis of Johne's disease. Its preparation parallels the extraction of old tuberculin or purified protein derivative (PPD) used in the tuberculin test.

Brucellin is a substance similar to tuberculin, but derived from a culture filtrate of *Brucella abortus* that is used to test for the presence of delayed-type hypersensitivity to brucella antigens. The test is of questionable value in diagnosis.

The **Schick test** is a test for susceptibility to diphtheria. Standardized diphtheria toxin is adjusted to contain 1/50 MLD in 0.1 ml, which is injected intracutaneously into the subject's forearm. Development of redness and induration within 24 to 36 h after administration constitutes a positive test if it persists for 4 d or longer. The presence of 1/500 to 1/250 or more of a unit of antitoxin per milliliter of the patient's blood will result in a negative reaction because of neutralization of the injected toxin. Neither redness nor induration appears if the test is negative. An individual with a negative test possesses sufficient antitoxin to protect against infection with *Corynebacterium diphtheriae,* whereas a positive test denotes susceptibility. A control is always carried out in the opposite forearm. For this test, toxin that has been diluted and heated to 70°C for 15 min is injected intracutaneously. Heating destroys the toxin's ability to induce local tissue injury; however, it does not affect the components of the diphtheria bacilli or of the medium that might evoke an allergic response in the individual. If the size and duration of the reaction at the injection site in the control approximates the reaction in the test arm, the result is negative. If the reaction is at least 50% larger and of longer duration on the test arm compared to the control, the individual is both allergic to the materials in the bacilli or in the medium and susceptible to the toxin. A positive Schick reaction suggests that diphtheria immunization is needed.

The **Casoni test** is a diagnostic skin test for hydatid disease in humans induced by *Echinococcus granulosus* infection. In sensitive individuals, a wheal and flare response develops within 30 min following intradermal inoculation of a tapeworm or hydatid cyst fluid extract. This is followed within 24 h by an area of induration produced by this cell-mediated delayed-hypersensitivity reaction.

The **heaf test** is a type of tuberculin test in which an automatic multiple puncture device with six needles is used to administer the test material by intradermal inoculation. The multiple needles advance 2 to 3 mm into the skin. Also called tine test.

The **Montenegro test** is a diagnostic assay for South American leishmaniasis induced by *Leishmania brasiliensis.* The intracutaneous injection of a polysaccharide antigen derived from the causative agent induces a delayed hypersensitivity response in the patient. They are not usually found in myositis, scleroderma, or Sjögren's syndrome.

Old tuberculin (OT) is a broth culture, heat-concentrated filtrate of medium in which *Mycobacterium tuberculosis* microorganisms were grown. It was developed by Robert Koch for use in tuberculin skin tests nearly a century ago.

OT (historical) is old tuberculin.

Romer reaction (historical): Romer in 1909 described erythematous swelling following intracutaneous injection of diphtheria toxin in small quantities. The reaction was found to be neutralized by homologous antitoxin. The smallest amount of diphtheria toxin that produced a definite reaction was defined as the MRD or minimal reaction dose. In general, the MRD of a given toxin is equivalent to about 1/250 to 1/500 of the MLD (minimal lethal dose). The Lr is the smallest amount of toxin which, after mixing with one unit of antitoxin, will produce a minimal skin lesion when injected intracutaneously into a guinea pig.

Passive cutaneous anaphylaxis (PCA) is a skin test that involves the *in vivo* passive transfer of homocytotropic antibodies that mediates type I immediate hypersensitivity (e.g., IgE in man) from a sensitized to a previously nonsensitized individual by intradermally injecting the antibodies, which become anchored to mast cells through their Fc receptors. This is followed hours or even days later by intravenous injection of antigen mixed with a dye such as Evans Blue. Crosslinking of the cell-fixed (e.g., IgE) antibody receptors by the injected antigen induces a type I immediate hypersensitivity reaction in which histamine and other pharmacological mediators of immediate hypersensitivity are released. Vascular permeability factors act on the vessels to permit plasma and dye to leak into the extravascular space, forming a blue area that can be measured with calipers. In humans, this is called the Prausnitz-Küstner (PK) reaction.

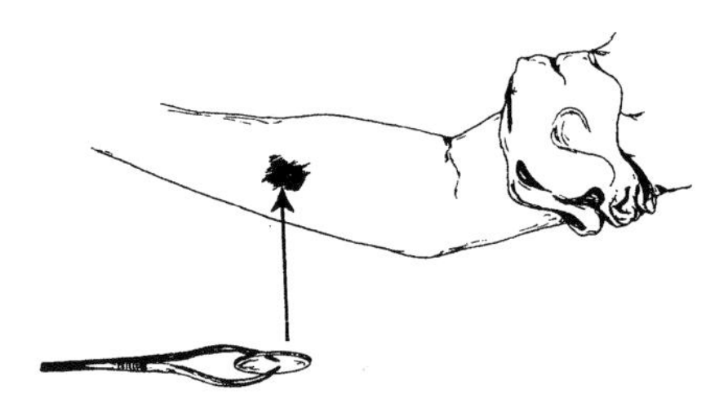

FIGURE 27.90 Rebuck skin window.

Test dosing is a method to determine whether or not an individual has type I anaphylactic hypersensitivity to various drugs, e.g., penicillin, or antisera prior to administration. However, the procedure is not without danger, as even a scratch test with highly diluted penicillin preparations in highly sensitized subjects has been known to produce fatal anaphylactic shock.

Histamine release assay: In 4 to 13% of allergic patients, the IgE-mediated release of histamine from basophils depends on cytokines. These anti-IgE-nonreleasers do respond to f-Met-Leu-Pro (FMLP), which bypasses the IgE receptor pathway. Histamine-released assays are valuable when skin testing in RAST function suboptimally, especially in urticaria and atopic dermatitis patients in whom only a weak correlation between IgE in disease is apparent. Histamine-release assays may be informative in urticaria, asthma, and atopic dermatitis patients.

Rebuck skin window (Figure 27.90) is a clinical method for assessing chemotaxis used *in vivo* by making a superficial abrasion of the skin which is then covered with a glass slide. This is removed several hours later, air dried, and stained for leukocyte content.

A **skin window** is a method to observe the sequential changes in types of cells during the development of acute inflammation. Following superficial abrasion of an area of skin, sterile cover slips are applied, removed at specified intervals thereafter, stained, and observed microscopically for the types of cells present. The first cells to appear are polymorphonuclear neutrophils which comprise most of the cell population within 3 to 4 h of the induced injury. By contrast, the cover slip removed after 12 h reveals the presence of mononuclear cells such as lymphocytes, plasma cells, and monocytes. The cover slip removed after 24 h reveals predominantly monocytes and macrophages. Also termed a Rebuck window, named for the individual who perfected the method.

The **radioallergosorbent test (RAST)** (Figure 27.91) is a technique to detect specific IgE antibodies in a patient's serum. This solid-phase method involves binding of the allergen–antigen complex to an insoluble support such as dextran particles or Sepharose®. The patient's serum is then passed over the allergen-support complex, which permits specific IgE antibodies in the serum to bind with the allergen. After washing to remove nonreactive protein, radiolabeled antihuman IgE antibody is then placed in contact with the insoluble support where it reacts with the bound IgE antibody. Both the allergen and the anti-IgE antibody must be present in excess for the test to be accurate. The amount of radioactivity on the beads is proportional to the quantity of serum antibody that is allergen specific.

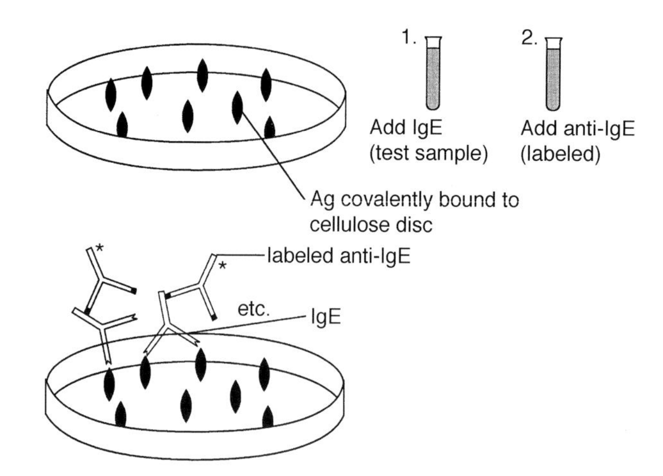

FIGURE 27.91 Radioallergosorbent test (RAST).

Radioallergosorbent test: See RAST.

The **radioimmunosorbent test (RIST)** (Figure 27.92) is a solid-phase radioimmunoassay to determine the serum IgE concentration. A standard quantity of radiolabeled IgE is added to the serum sample to be assayed. The mixture is then combined with Sephadex® or Dextran beads coated with antibody to human IgE. Following incubation and washing, the quantity of radiolabeled IgE bound to the beads is measured. The patient's IgE competes with the radiolabeled IgE or antibody attached to the beads. Therefore, the decrease in labeled IgE attached to the beads compared to a control in which labeled IgE combines with the beads without competition represents the patient's serum concentration of IgE. The radioallergosorbent test by comparison assays IgE levels reactive with a specific allergen.

RIST: See radioimmunosorbent test.

The **paper radioimmunosorbent test (PRIST)** (Figure 27.93) is a technique to assay serum IgE levels. It resembles the radioimmunoabsorbent test except that filter paper discs impregnated with antihuman IgE is used in place of Sephadex® discs.

Immunoradiometry is a radioimmunoassay method in which the antibody rather than the antigen is radiolabeled.

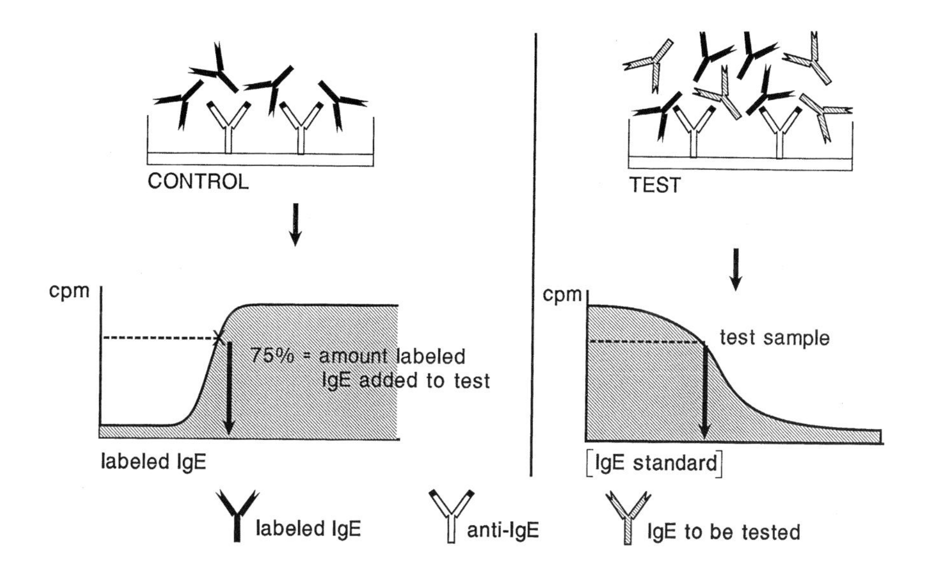

FIGURE 27.92 Radioimmunosorbent test (RIST).

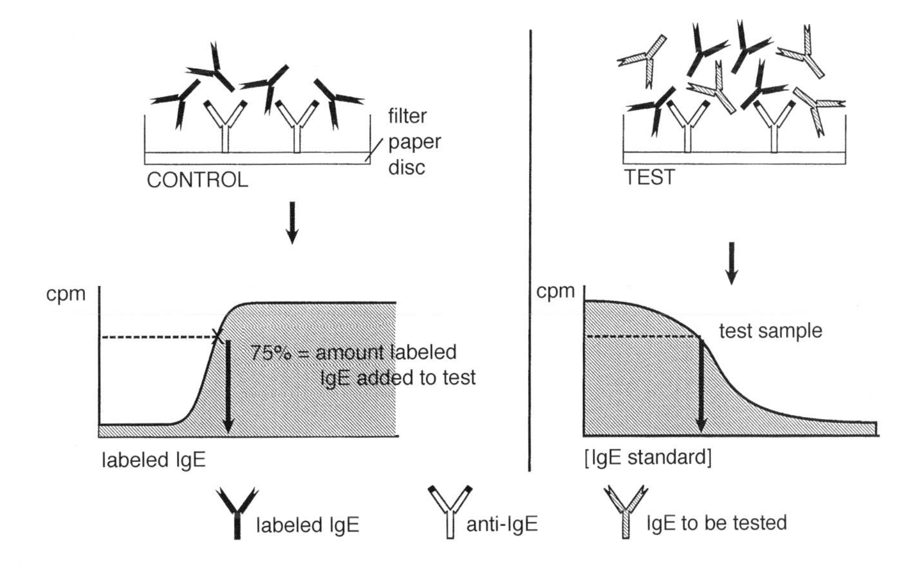

FIGURE 27.93 Paper radioimmunosorbent test (PRIST).

The **immunoradiometric assay (IRMA)** (Figure 27.94) is a quantitative method to assay certain plasma proteins based on a "sandwich" technique using radiolabeled antibody, rather than radiolabeled hormone competing with hormone from a patient in the radioimmunoassay (RIA).

The **hook effect** is an artifact that may be seen in IRMA that occurs when a hormone being assayed is in very high concentration. The excess amount cannot be measured by the detector system since it will have obtained a theoretical limit. The diminished counts with the labeled antibody at the elevated hormone concentration yield spuriously low results. Thus, IRMA is not an appropriate method to assay hormones present in relatively high concentrations, such as gastrin, prolactin, or hCG. The hook effect requires measurement of two separate concentrations to establish linearity.

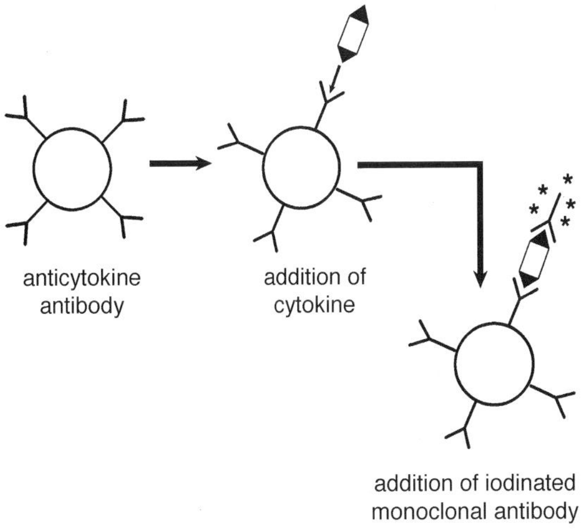

anticytokine antibody
addition of cytokine
addition of iodinated monoclonal antibody

FIGURE 27.94 Immunoradiometric assay (IRMA).

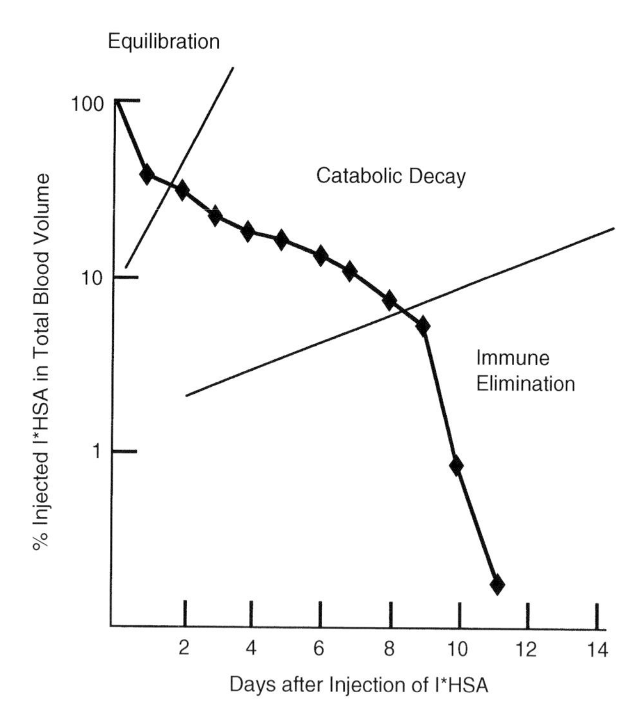

FIGURE 27.95 Immune elimination.

Immune elimination (Figure 27.95) refers to accelerated removal of an antigen from the blood circulation following its interactions with specific antibody and elimination of the antigen–antibody complex through the mononuclear phagocyte system. A few days following antigen administration, antibodies appear in the circulation and eliminate the antigen at a much more rapid rate than occurs in nonimmune individuals. Splenic and liver macrophages express Fc receptors that bind antigen–antibody complexes as well as complement receptors which bind those immune complexes that have already fixed complement. This is followed by removal of immune complexes through the phagocytic action of mononuclear phagocytes. Immune elimination also describes an assay to evaluate the antibody response by monitoring the rate at which a radiolabeled antigen is eliminated in an animal with specific (homologous) antibodies in the circulation.

Immune clearance: See immune elimination.

The **Winn assay** (Figure 27.96) is a method to determine the ability of lymphoid cells to inhibit the growth of transplantable tumors *in vivo.* Following incubation of the lymphoid cells and tumor cells *in vivo*, the mixture is injected into the skin of X-irradiated mice. Growth of the transplanted cells is followed. T lymphocytes that are specifically immune to the tumor cells will inhibit tumor growth and provide information related to tumor immunity.

Strain refers to genetically identical animals such as mice or rats used in medical research.

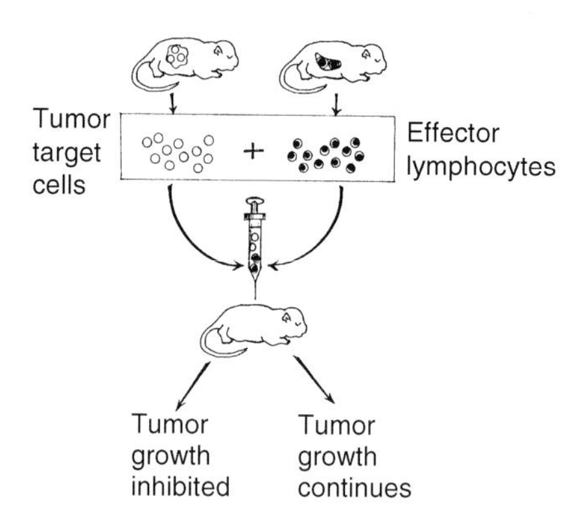

FIGURE 27.96 Winn assay.

A **transgenic animal** is an animal into whose genome a foreign gene has been introduced. Introduction of the exogenous gene into a mouse can be by either microinjection into a pronucleus of an egg that has been fertilized recently or through retroviruses. The egg that has received the foreign gene is transferred to the oviduct of a pseudopregnant female. If the gene becomes integrated into a chromosome, it is passed on to the progeny through the germ line and will be expressed in all cells.

Transgenes are genes that are artificially and deliberately introduced in the germ line and are foreign. The **Foreign gene**. See transgenic mice.

Transgenic is a term that describes an organism that has had DNA from another organism put into its genome through recombinant DNA techniques. These animals are usually made by microinjection of DNA into the pronucleus of fertilized eggs, with the DNA integrating at random.

Transgenic mice (Figure 27.97) carry a foreign gene that has been artificially and deliberately introduced into their germ line. The added genes are termed transgenes. Fertilized egg pronuclei receive microinjections of linearized DNA. These are placed in pseudopregnant female oviducts and development proceeds. About one fourth of the mice that develop following injection of several hundred gene copies into pronuclei are transgenic mice. Transgenic mice have been used to study genes not usually expressed *in vivo* and alterations in genes that are developmentally regulated to express normal genes and cells where they are not usually expressed. Transgenic mice are also used to delete certain populations of cells with transgenes that encode toxic proteins. They are highly significant in immunologic research.

A **transgenic mouse** is a mouse developed from an embryo into which foreign genes were transferred. Transgenic mice have provided much valuable information related to immunological tolerance, autoimmune phenomena, oncogenesis, developmental biology, and related topics. The transgene has

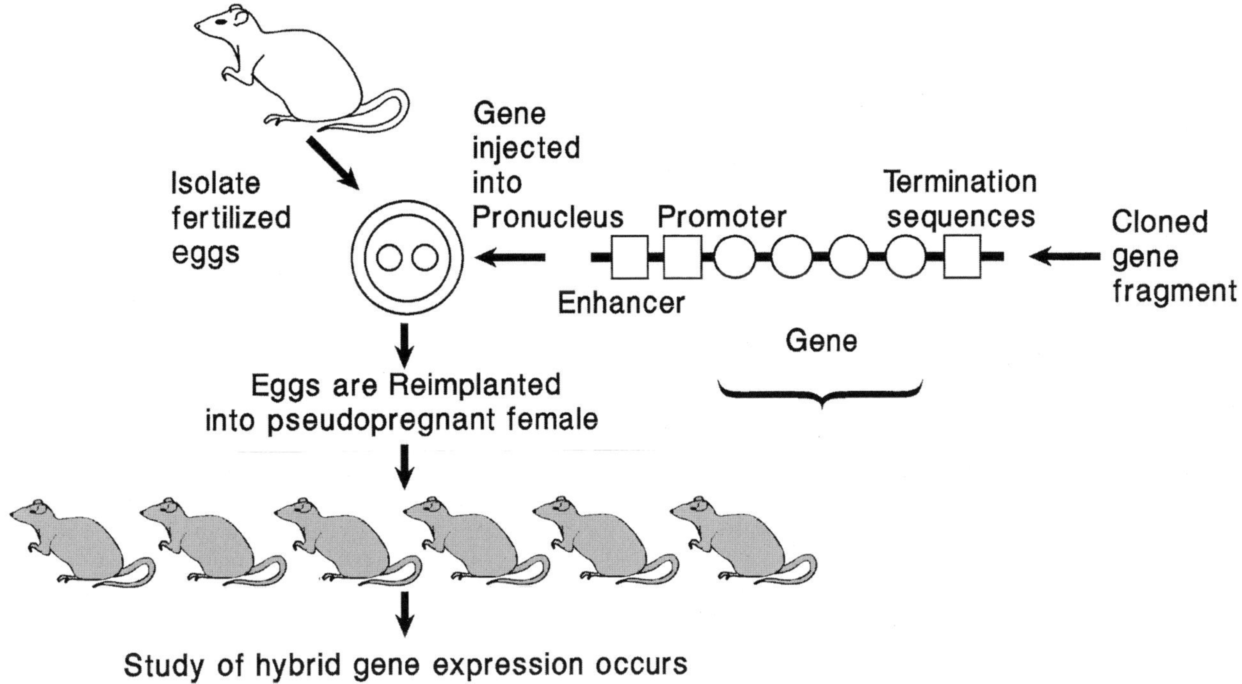

FIGURE 27.97 Transgenic mice.

been introduced and stably incorporated into germ-line cells ensuring that it can be passed on to the progeny. A specific DNA sequence is injected in to the pronuclei of fertilized mouse eggs. Trangenes insert randomly at chromosomal breakpoints and are inherited as simple Mendelian traits. Studies with transgenic mice have yielded much data about cytokines, cell surface molecules, and intracellular signaling molecules.

Transgenic organisms are animals or plants into which foreign genes that encode specific proteins have been inserted. However, controlling the site of gene insertion has not been accomplished yet. Insertion into some positions may even lead to activation of the host's own structural genes.

Transgenic line refers to a transgenic mouse strain in which the transgene is stably integrated into the germ line and therefore inherited in Mendelian fashion by succeeding generations.

Transgenics refers to the transfer of needed genes into an organism for the purpose of providing a missing protein which these genes encode.

A **germ-free animal** is one such as a laboratory mouse, raised under sterile conditions, where it is free from exposure to microorganisms and is not exposed to larger organisms. Germ-free animals have decreased serum immunoglobulin and lymphoid tissues that are not fully developed. The diet may also be controlled to avoid exposure to food antigens. Most difficult is the ability to maintain a virus-free environment for these animals.

Gene knockout is laboratory jargon for gene disruption by homologous recombination. It may refer to a cell or animal in which a specific gene's function has been purposely eliminated by replacing the normal gene with an inactive mutant gene.

A **knockout mouse** is a transgenic mouse in which a mutant allele or disrupted form of the gene replaces a normal gene, leading to the mouse's failure to produce a functional gene product. Much has been learned about the role of cytokines, cell surface receptors, signaling molecules, and transcription factors in the immune system using knockout mice.

Knockout gene is a descriptor for the generation of a mutant organism in which the function of a particular gene has been completely eliminated (a null allele). To successfully knockout a gene, cloned and sequenced genomic DNA and a suitable embryonic (ES) cell line are necessary. A sequence insertion targeting approach may be used. The advantage of an insertion vector is that the frequency of integration is nine-fold higher than with an equal-length replacement-type vector. Homologous recombination techniques can be used to achieve targeted disruption of one or more genes in mice. Knockout mice deprived of functional genes that encode cytokines, cell surface receptors, signaling molecules, and transcription factors are critical for contemporary immunologic research.

Genetic knockout is a technique to introduce precise genetic lesions into the mouse genome to cause "gene disruption" and generate an animal model with a specific genetic defect. Specific defects may be introduced into any murine gene by permitting investigation of this alteration *in vivo*. Technological advances that have made this possible include using homologous recombination to introduce defined changes into the murine genome, and the reintroduction of genetically altered embryonic stem cells into the murine germ line to produce mutant mouse strains.

Outbreeding refers to mating of subjects who showed greater genetic differences between themselves than randomly chosen individuals of a population. This process encourages genetic diversity. It is in contrast to inbreeding and random breeding.

Isoelectric focusing (IEF) is an electrophoretic method to fractionate amphoteric molecules, especially proteins, according to their distribution in a pH gradient in an electric field created across the gradient. Molecular distribution is according to isoelectric pH values. The anode repels proteins that are positively charged and the cathode repels proteins that are negatively charged. Thus, each protein migrates in the pH gradient and bands at a position where the gradient pH is equivalent to the isoelectric pH of the protein. A chromatographic column is used to prepare a pH gradient by the electrolysis of amphoteric substances. A density gradient or a gel is used to stabilize the pH gradient. Proteins or peptides focus into distinct bands at that part of the gradient that is equivalent to their isoelectric point. Isoelectric focusing is a technique that permits the separation of protein substances on the basis of their isoelectric characteristics. Thus, this technique can be employed to define heterogeneous antibodies. It may also be employed to purify homogeneous immunoglobulins from heterogeneous pools of antibody.

IEF is an abbreviation for isoelectric focusing.

Spectrotype: In isoelectric focusing analysis, the arrangement of bands in a gel that is characteristic for either one protein or a category of proteins. An antibody spectrotype on an isoelectric focusing gel may signify that it is the product of a particular antibody-synthesizing clone.

Isoelectric point (pI) is the pH at which a molecule has no charge, as the number of positive and negative charges are equal. At the isoelectric pH, the molecule does not migrate in an electric field. The solubility of most substances is minimal at their isoelectric point.

Viability techniques are methods employed to determine the viability of cells maintained *in vitro*. Of the many dyes that have been employed for viability assays, the anionic trypan blue is the one most frequently used. For example, the viability of leukocytes suspended in balanced salt solution are evaluated for their ability to exclude trypan blue form the cell interior. Leukocytes are counted on a hemacytometer and the percentage of viability is calculated by dividing the number of unstained viable cells by the number of all cells counted (stained and unstained).

Dye exclusion test is an assay for the viability of cells *in vitro*. Vital dyes such as eosin and trypan blue are excluded by living cells; however, the loss of cell membrane integrity by dead cells admits the dye that stains the cell. The dye exclusion principle is used in the microlymphocytotoxicity test employed for HLA typing in organ transplantation.

A **dye test** is an assay to determine whether an individual has become infected with *Toxoplasma*. Antibody in an infected patient's serum prevents living toxoplasma organisms, obtained from an infected mouse's peritoneum, from taking up methylene blue. Therefore, the organisms do not stain blue if antitoxoplasma antibody is present in the serum.

Immunomagnetic technique: The use of magnetic microspheres to sort, isolate, or identify cells with specific antigenic determinants.

Immunonephelometry is a test that measures light that is scattered at a 90° angle to a laser or light source as it is passed through a suspension of minute complexes of antigen and antibody. Measurement is made at 340 to 360 nm using a spectrophotometer.

Plaque-forming cells are the antibody-producing cells in the center of areas of hemolysis observed microscopically when reading a hemolytic plaque assay. The antibodies they form are specific for red blood cells suspended in the gel medium surrounding them. Once complement is added, the antibody-coated erythrocytes lyse, producing clear areas of hemolysis surrounding the antibody-forming cell. The antibody produced may be specific not only for red blood cell surface antigens but also for soluble antigens deliberately coated on their surfaces for assay purposes.

Plaque-forming assay: See hemolytic plaque assay.

The **plaque-forming cell (PFC) assay** is a technique for demonstrating and enumerating cells forming antibodies against a specific antigen. Mice are immunized with sheep red blood cells (SRBC). After a specified period of time, a suspension of splenic cells from the immunized mouse is mixed with antigen (SRBC) and spread on a suitable semisolid gel medium. After or during incubation at 37°C, complement is added. The erythrocytes that have anti-SRBC antibody on their surface will be lysed. Circular areas of hemolysis appear in the gel medium. If viewed under a microscope, a single antibody-forming cell can be identified in the center of the lytic area. There are several modifications of this assay, since some antibodies other than IgM may fix complement less efficiently. In order to enhance the effects, an antiglobulin antibody called developing antiserum is added to the mixture. The latter technique is called indirect PFC assay.

Plaque technique: See hemolytic plaque assay or phage neutralization assay.

Paternity testing refers to tests performed to ascertain the biological (genetic) parentage of a child. In the past, these have included erythrocyte enzymes, red blood cell antigens, HLA antigens, immunoglobulin allotypes, nonimmunoglobulin serum proteins, and more recently, DNA "fingerprinting" (typing). The demonstration of a genetic marker in a child that is not present in either the father or mother or in cases where none of the paternal antigens are present in the child is enough evidence for direct exclusion of paternity. Another case of direct exclusion of paternity is the failure of a child to express a gene found in both the mother and putative father. When a child expresses a gene that only the man can transmit and which the putative father does not express, this is evidence for indirect exclusion of paternity. When a child is homozygous for a marker not present in the mother or putative father, or if the parent is homozygous for a marker not found in the child, then paternity can be excluded as an indirect exclusion. Also called identity testing.

Identity testing: See paternity testing.

A **virus neutralization test** is an assay based upon the ability of a specific antibody to neutralize virus infectivity. This assay can be employed to measure the titer of antiviral antibody. This test may be performed *in vivo* using susceptible animals or chick embryos or it may be done *in vitro* in tissue culture.

Bacteriophage neutralization test: See phage neutralization assay.

The **phage neutralization assay** is a laboratory test in which bacteriophage is combined with antibodies specific for it to diminish its capacity to infect a host bacterium. This neutralization of infectivity may be quantified by showing the decreased numbers of plaques produced when the phage, which has been incubated with specific antibody, is plated on appropriate bacteria. The technique is sensitive and can demonstrate even weak antibody activity.

The **neutralization test** is an assay based on the ability of antibody to inactivate the biological effects of an antigen or of a microorganism expressing it. Neutralization applies especially to inactivation of virus infectivity or of the biological activity of a microbial toxin.

Buffy coat refers to the white cell layer that forms between the red cells and plasma when anticoagulated blood is centrifuged.

Cell separation methods: Cell separation techniques were first based, in the early 1960s, on differences such as cell size and density. Subsequently, membrane receptors or surface antigens were found to be differently expressed by lymphocyte subsets. Currently, the most popular lymphocyte separation techniques include immunoselection procedures that employ monoclonal antibodies. The two methods used for lymphocyte separation based on physical differences include sedimentation separation and density gradient separation. Other methods are based on functional properties of the cell such as adhesive or phagocytic properties. Selected mononuclear cell types can adhere to plastic surfaces or to nylon wool. Other techniques employ selective depletion of cells such as lymphocytes undergoing proliferation. Mitogens in culture can be employed to select given lymphocyte populations based on their ability to respond to these stimulants. Rosetting techniques permit the detection or purification of cells expressing a certain surface receptor for antigen. Immunoselection techniques employ either monoclonal or polyclonal antibodies specific for surface antigens on lymphocyte subsets. Immunotoxicity procedures can be used to induce selective cytolysis of cells expressing a certain antigen at the cell surface by reacting the cell with antibodies. Immunoadhesion procedures make use of antibodies against cell surface antigens bound to a solid support that permits the capture of cells by adherence to the support. Immunomagnetic beads to which antibodies have been attached may also be used. Magnetic cell sorting is based on the use of monoclonal antibodies or lectins that bind specifically to surface antigen/receptor expressed by a certain cell subset. Flow cytometry is a precise and objective method to quantify the number of cells expressing a given surface marker and the extent to which the marker is expressed.

Cell surface molecule immunoprecipitation is a technique to analyze cell surface molecules with monoclonal antibodies and antisera. Immunoprecipitation is based on solubilization of membrane proteins by the use of nonionic detergents; subsequent interaction of specific antibody with solubilized membrane antigen; and recovery of antibody–antigen complexes by binding to an insoluble support which permits washing procedures to remove unbound molecules. Analysis of immunoprecipitates can be accomplished by SDS-page or isoelectric focusing (IEF).

Chromium release assay: The release of chromium (^{51}Cr) from labeled target cells following their interaction with cytotoxic T lymphocytes or antibody and K cells (ADCC) or NK cells. The test measures cell death, which is reflected by the amount of radiolabel released according to the number of cells killed.

Cell line refers to cultured neoplastic cells or normal cells that have been transformed by chemicals or viruses. Transformed cell lines may be immortal, enabling them to be propagated indefinitely in culture.

Cryopreservation is the technique of freezing tissue or cells or other biological materials to remain genetically stable and metabolically inert. Cryopreservation may involve freezers (–80°C), or preservation with dry ice (–79°C) or liquid nitrogen (–196°C).

Ficoll is a 400-kDa water-soluble polymer comprised of sucrose and epichlorohydrin. It is employed in the manufacture of Ficoll-Hypaque, a density gradient substance used to separate and purify mononuclear cells by centrifugation following removal of the buffy coat.

Ficoll-Hypaque is a density gradient medium used to separate and purify mononuclear cells by centrifugation.

Footprints refer to macrophages filled with *Mycobacterium leprae* without caseation necrosis. A similar situation may be observed in anergic Hodgkin's disease patients and in AIDS patients infected with *M. avium-intercellulare.*

Heterokaryon refers to the formation of a hybrid cell by fusion of two separate cells in suspension leading to a cell form with two nuclei. Cell fusion may be accomplished through the use of polyethylene glycol or ultraviolet light-inactivated Sendai virus.

A **hybrid cell** is a cell produced when two cells fuse and their two nuclei fuse to form a heterokaryon. Although hybrid cell lines can be established from clones of hybrid cells, they lack stability and delete chromosomes, which is nevertheless useful for gene mapping. Hybrid cell lines can be isolated by using HAT as a selective medium.

T cell hybridomas: The immortalization of normal T lymphocytes by fusion with continuously replicating tumor cells. Fusion randomly immortalizes T lymphocytes regardless of their antigen specificity and genetic restrictions to form a T cell hybridoma. This represents one of two methods to isolate and propagate T cell lines in clones of defined specificity. The other technique is to span clones of normal immune T lymphocytes stimulated with appropriate antigens and antigen-presenting cells. The hybridoma technique holds the advantage over T cell cloning in the relative ease in securing large numbers of T cells of interest and their biologically active products. Lymphokines and other regulatory molecules together with their nRNA and DNA represent T cell hybridoma products. This technology has also facilitated evaluation of T cell receptors and their antigen recognition mechanisms. The adoptive (passive) transfer of autoantigen-specific T cell hypridomas in mice can induce autoimmune diseases.

Immune cell cryopreservation is accomplished by using glycerol and dimethylsufoxide (DMSO) to cryopreserve bone marrow for transplantation. It first involved freezing at a constant rate of 1°C min^{-1} to –79°C, with storage for a 6-month period. Dye exclusion was used to test cells for viability post thaw. Lymphocytes have been cryopreserved for *in vitro* studies using 15% DMSO and storing in liquid nitrogen (–196°C) for 3 months followed by an assay for viability.

Immunocytoadherence is a method to detect cells with surface immunoglobulin, either synthesized or attached through Fc receptors. Red blood cells coated with the homologous antigen are mixed with the immunoglobulin-bearing cells and result in rosette formation. A laboratory assay employed to identify antibody-bearing cells by the formation of rosettes comprised of red blood cells and antibody-bearing cells.

Leukocyte culture: Whereas mononuclear blood cells have been cultured *in vitro* in a medium containing serum to support growth in the past, culture of cells to be used for patient reinfusion must be grown in media that is free of serum, endotoxin, and antibiotics. Tissue culture vessels for the large-scale expansion of leukocytes include polystyrene flasks that can be stacked on top of one another in an incubator. Gas-permeable cell-culture bags, 30 of which containing 1500 ml of cell culture each, may be placed in an incubator; tissue culture bioreactors include the hollow fiber cell culture bioreactors and the rotary cell culture system, both of which provide a 3-dimensional growth environment. Clinical applications of leukocyte culture include (1) generation of LAK cells and TILs for adoptive immunotherapy; (2) $CD34^+$ cell culture and *in vitro* generation of hematopoietic precursors for bone marrow reconstitution; and (3) culture of dendritic cells for use in active immunization.

Antisperm antibody is an antibody specific for any one of several sperm constituents. Antisperm agglutinating antibodies are detected in blood serum by the Kibrick sperm agglutination test, which uses donor sperm. Sperm-immobilizing antibodies are detected by the Isojima test. The subject's serum is incubated with donor sperm, and motility is examined. Testing for antibodies is of interest to couples with infertility problems. Treatment with relatively small doses of prednisone is sometimes useful in improving the situation by diminishing antisperm antibody titers. One half of infertile females manifest IgG or IgA sperm-immobilizing antibodies which affect the tail of the spermatozoa. By contrast, IgM antisperm head agglutinating antibodies may occur in homosexual males.

Plasmapheresis is a technique in which blood is withdrawn from an individual, the desired constituent is separated by centrifugation, and the cells are reinjected into the patient. Thus, plasma components may be removed from the circulation of an individual by this method. The technique is also useful to obtain large amounts of antibodies from the plasma of an experimental animal.

Plasmapheresis is used therapeutically to rid the body of toxins or autoantibodies in the blood circulation. Blood taken from the patient is centrifuged, the cells are saved, and the plasma is removed. Cells are resuspended in albumin, fresh normal plasma, or albumin in saline, and returned to the patient. The ill effects of a toxin or of an autoantibody may be reduced by 65% by removing approximately 2500 ml of plasma. Removal of twice this amount of plasma may diminish the level of a toxin or of an autoantibody by an additional 20%. This procedure has been used to treat patients with myasthenia gravis, Eaton-Lambert syndrome, Goodpasture's syndrome, hyperviscosity syndrome, post-transfusion purpura, and acute Guillain-Barré syndrome.

Acridine orange is a fluorescent substance that binds nonspecifically to RNA with red fluorescence and to DNA with green fluorescence. It also interacts with polysaccharides, proteins, and glycosaminoglycans. It is a nonspecific tissue stain that identifies increased mitoses and shows greater sensitivity but less specificity than the Gram stain. It is carcinogenic and of limited use in routine histology.

Cloned DNA is a DNA fragment or gene introduced into a vector and replicated in eukaryotic cells or bacteria.

Concatamer integration occurs when the entire genome of vector including the bacterial plasmid is integrated into the host genome.

Electroporation is a technique to insert molecules into cells through use of brief high-voltage electric pulses. It can be used to insert DNA into animal cells. The electrical discharge produces tiny pores that are nanometers in diameter in the plasma membrane. These pores admit supercoiled or linear DNA.

The **Feulgen reaction** is a standard method that detects DNA in tissues.

Methyl green pyronin stain is a stain used in histology or histopathology that renders DNA green and RNA red. It has been widely used to demonstrate plasma cells and lymphoblasts that contain multiple ribosomes containing RNA in their cytoplasm.

Footprinting is a method to ascertain the DNA segment (or segments) that binds to a protein. Radiolabeled double-stranded DNA is combined with the binding protein to yield a complex that is exposed to an endonuclease that cuts the molecules once and at random. The digested DNA is electrophoresed in polyacrylamide gel together with a control DNA sample (which has been treated similarly, but without added protein) to permit separation of fragments differing in length by one nucleotide. Autoradiography of the material reveals a series of bands representing the DNA fragments. In the area of protein binding, the DNA is spared from digestion, and no corresponding bands appear compared to the control. The protected area's specific location can be ascertained by running a DNA sequencing gel in parallel.

***In situ* transcription** is a method in which mRNA acts as a template for complementary DNA for reverse transcription in tissues that have been fixed.

Minisatellite refers to DNA regions comprised of tandem repeats of DNA short sequences.

Nick translation is a technique used to make a radioactive probe of a DNA segment. Nick translation signifies the movement of a nick, i.e., single-stranded break in the double-stranded helix, along a duplex DNA molecule.

A **plasmid** is an extrachromosomal genetic structure that consists of a circular, double-stranded DNA molecule which permits the host bacterial cell to resist antibiotics and produce other effects that favor its survival. Plasmid replication is independent of the bacterial chromosome. Plasmids have been used widely in recombinant DNA technology.

Protoplast fusion is a technique for DNA transfer from one group of bacteria to others, to myeloma cells, or other animal cells in culture. The exposure of plasmid-bearing *Escherichia coli* microorganisms to lysozyme and EDTA yields protoplasts that may be fused with myeloma cells by polyethylene glycol treatment.

Recombinant DNA technology is the technique of isolating genes from one organism and purifying and reproducing them in another organism. This is often accomplished through ligation of genomic or cDNA into a plasmid or viral vector where replication of DNA takes place.

RNAse protection assay is a technique to detect and quantify messenger RNA (mRNA) copies of specific genes based on nRNA hybridization to radiolabeled RNA probes followed by digestion of the unhybridized RNA with

RNAse. Double-stranded RNA duplexes formed as a result of the hybridization are resistant to RNAse degradation. Their size depends on the probe length. Gel electrophoresis is used with their separation. Radioautography is employed for their detection and quantification.

Slot blot analysis is a quick technique to detect gene amplification by determining a solution's DNA content by electrophoresis. The technique is closely similar to dot blot analysis, except that a slot instead of a punched-out hole is cut in the agar.

The **TUNEL assay (TdT-dependent dUTP-biotin nick end labeling)** is a technique that identifies apoptotic cells *in situ* by fragmentation of their DNA. Immunohistochemical staining with enzyme-linked streptavidin identifies biotin-tagged dUTP added to the free 3′ ends of the DNA fragments by the enzyme TdT.

In **TUNEL-based assays**, DNA fragments can be stained *in situ* by using terminal deoxynucleotidyl transferase (TdT) to polymerize labeled nucleotides onto the ends of nicked DNA (TUNEL, TdT-mediated d UTP nick end labeling). For example, following TdT labeling, biotinylated nucleotides may be detected with a chromogenic or fluorometric-conjugated streptavidin, or brominated nucleotides may be detected with a highly sensitive, biotinylated anti-BrdU antibody and chromogenic-conjugated streptavidin.

Transcription refers to RNA synthesis using a DNA template.

Transduction is the use of a virus to transfer genes, such as the use of a bacteriophage to convey genes from one bacterial cell to another one. Other viruses such as retroviruses may also transfer genes from one cell type to another.

Transfection is the transfer of double-stranded DNA extracted from neoplastic cells for the purpose of producing phenotypic alterations of malignancy in the recipient cells.

Spectratyping refers to selected types of DNA gene segments that give a repetitive spacing of three nucelotides, or one codon.

SRY is the protein encoded for by the sex-determining region of the Y chromosome termed the *sry* gene in man. It is equivalent to the Y chromosome's testis-determining gene. The corresponding protein in mice is termed Sry. The murine *sry* gene can cause transgenic female mice to become phenotypic males when the gene is inserted into them.

The **shift assay** is a useful method to identify protein-DNA interactions that may mediate gene expression, DNA repair, or DNA packaging. The assay can also be used to determine the affinity abundance, binding constants, and binding specificity of DNA-binding proteins. The gel shift assay is performed by annealing two labeled oligonucleotides that contain the test binding sequence, then incubating the duplex with the binding protein. The mixture is then separated on a nondenaturing polyacrylamide gel. Duplexes that are bound by protein migrate more slowly than unbound duplexes and appear as bands that are shifted relative to the bands from the unbound duplexes. Also called gel mobility shift assay or gel shift assay or gel retardation or band shift assay.

Zygosity refers to characterization of an individual's heredity traits in terms of gene pairing in the zygote from which it develops.

RNA splicing is the method whereby RNA sequences that are nontranslatable (known as introns) are excised from the primary transcript of a split gene. The translatable sequences (known as exons) are united to produce a functional gene product.

Apheresis is the technique whereby blood is removed from the body, its components are separated, and some are retained for therapeutic or other use, and the remaining elements are recombined and returned to the donor. Also called hemapharesis.

Leukapheresis is a method that removes circulating leukocytes from the blood of healthy individuals for transfusion to recipients with decreased immunity or who are leukopenic. Leukapheresis is also used in leukemia patients who have too many white cells. The procedure leads to temporary relief of symptoms attributable to hyperleukocytosis.

Site-directed mutagenesis is a laboratory procedure that involves the substitution of amino acids in a protein whose function is defined for the purpose of localizing a certain activity.

Capsule swelling reaction: Pneumococcus swelling reaction. See Quellung reaction.

Lethal dose is the amount of a toxin, virus, or any other material that produces death in all members of the species receiving it within a specified period of time following administration.

Minimum lethal dose (MLD) is that dose of a substance or agent that will kill 100% of the population being tested.

The **Ramon test (historical)** was an imprecise method for assaying the activity of any given preparation of diphtheria (or tetanus) toxin. Varying quantities of antitoxin are combined with a constant quantity of toxin *in vitro*. The tubes are placed in a 44 to 46°C water bath and are observed often. The test is read by noting the tube where flocculation occurs first. This is the point of equivalence where antitoxin has neutralized the homologous toxin. However, this assay is based on antigenicity of the toxin with which the antitoxin combines, in contrast to toxicity. Therefore, it is a measure of combining power and provides an indirect idea of toxicity only insofar as toxicity and antigenicity are positively correlated. Since the two are not always closely correlated, this method is less reliable than the *in vivo* technique of Ehrlich that measures the actual toxic effect of the toxin and the ability of antitoxin to combat it. The Ramon test measures toxin in L_f (flocculating) units. The L_f unit is defined as the amount of toxin which flocculates most rapidly with one unit of antitoxin. The L_f value, in contrast to other L values described, must be calculated. To determine the L_f value for a given toxin, the following formula is used:

$$L_f/\text{ml toxin} = \frac{\text{antitoxin units/ml} \times \text{ml of antitoxin}}{\text{ml of toxin}}$$

Thus, the L_f content of a toxin may be determined if the following values are known: (1) antitoxin units per milliliter of antitoxin; (2) milliliter of antitoxin required for most rapid flocculation with toxin; and (3) milliliter of toxin employed. Although the Ramon flocculation test was classically used to determine the L_f value of toxin, it may be carried out in reverse to assay the antitoxin units in each milliliter of antitoxin which has not been previously standardized. The same formula is applicable.

$$L_f/\text{ml toxin} = \frac{\text{antitoxin units/ml} \times \text{ml of antitoxin}}{\text{ml of toxin}}$$

$$\text{Antitoxin units/ml} = \frac{L_f/\text{ml toxin} \times \text{ml of toxin}}{\text{ml of antitoxin}}$$

Varying quantities of toxin of known L_f value are combined with a constant amount of antiserum. The tube where flocculation first occurs is the point of equivalence. Therefore, the amount of toxin in a milliliter is substituted into the formula together with the known values, which include the L_f per milliliter of toxin and the number of milliliters of antitoxin held constant. By simple arithmetic, the antitoxin units per milliliter may then be calculated. In this quantitative precipitin test, antibody dilutions are varied, but antigen dilutions are kept the same. The first tube where precipitation occurs is considered the end point.

Serial passage is a method to attenuate a pathogenic microorganism but retain its immunogenicity by transfer through several animal hosts, growth media, or tissue culture cells.

28 Diagnostic Immunohistochemistry

Immunohistochemistry is a method to detect antigens in tissues that employs an enzyme-linked antibody specific for antigen. The enzyme degrades a colorless substrate to a colored insoluble substance that precipitates where the antibody and, therefore, the antigen are located. Identification of the site of the colored precipitate and the antigen in the tissue section is accomplished by light microscopy. Diagnostic pathology services routinely offer approximately 100 antigens, identified by immunoperoxidase technology, which are used in diagnosis.

Immunoperoxidase method: Nakene and Pierce, in 1966, first proposed that enzymes be used in the place of fluorochromes as labels for antibodies. Horseradish peroxidase (HRP) is the enzyme label most widely employed. The immunoperoxidase technique permits the demonstration of antigens in various types of cells and fixed tissues. This method has certain advantages that include the following: (1) the use of conventional light microscopy; (2) the stained preparations may be kept permanently, (3) the method may be adapted for use with electron microscopy of tissues, and (4) counterstains may be employed. The disadvantages include the following: (1) the demonstration of relatively minute positively staining areas is limited by the resolution of the light microscope; (2) endogenous peroxidase may not have been completely eliminated from the tissue under investigation; and (3) diffusion of products results from the enzyme reaction away from the area where antigen is localized.

The peroxidase–antiperoxidase (PAP) technique (Figure 28.1) employs unlabeled antibodies and a PAP reagent. This has proven highly successful for the demonstration of antigens in paraffin-embedded tissues as an aid in surgical pathologic diagnosis. Tissue sections preserved in paraffin are first treated with xylene, and after deparaffinization they are exposed to a hydrogen peroxide solution that destroys the endogenous peroxidase activity in tissue. The sections are next incubated with normal swine serum, which suppresses nonspecific binding of immunoglobulin molecules to tissues containing collagen. Thereafter, the primary rabbit antibody against the antigen to be identified is reacted with the tissue section. Primary antibody that is unbound is removed by rinsing the sections, which are then covered with swine antibody against rabbit immunoglobulin. This so-called linking antibody will combine with any primary rabbit antibody in the tissue. It is added in excess, which will result in one of its antigen-binding sites remaining free. After washing, the PAP reagent is placed on the section, and the antibody portion of this complex, which is raised in rabbits, will be bound to the free antigen-binding site of the linking antibody on the sections. The unbound PAP complex is then washed away by rinsing. To read the sections microscopically, it is necessary to add a substrate of hydrogen peroxide and aminoethylcarbazole (AEC), which permits the formation of a visible product that may be detected with the light microscope. The AEC is oxidized to produce a reddish-brown pigment that is not water-soluble. Peroxidase catalyzes the reaction. Because peroxidase occurs only at sites where the PAP is bound via linking antibody and primary antibody to antigen molecules, the antigen is identified by the reddish-brown pigment. The tissue sections can then be counterstained with hematoxylin or other suitable dye, covered with mounting medium and cover slips, and read by conventional light microscopy. The PAP technique has been replaced, in part, by the avidin–biotin complex (ABC) technique (Figure 28.2).

Streptavidin is a protein isolated from streptomyces that binds biotin. This property makes streptavidin useful in the immunoperoxidase reaction that is employed extensively in antigen identification in histopathologic specimens, especially in surgical pathologic diagnosis.

Immunodiagnosis involves the use of antibody assays, immunocytochemistry, the identification of lymphocyte markers, and other techniques to diagnose infectious diseases and malignant neoplasms.

Decorate is a term used by immunologists to describe the reaction of tissue antigens with monoclonal antibodies, described as "staining," in the immunoperoxidase reaction. Thus, a tissue antigen stained with a particular antibody is said to be decorated with that monoclonal antibody. Immunoperoxidase techniques give a reddish-brown color to the reaction product that is read by light microscopic observation.

Antibroad-spectrum cytokeratin is a mouse monoclonal antibody that may be used to identify cells of normal and abnormal epithelial lineage and as an aid in the diagnosis of anaplastic tumors. The cytokeratins are a group of intermediate filament proteins that occur in normal and neoplastic cells of epithelial origin. The 19 known human cytokeratins are divided into acidic and basic subfamilies. They occur in pairs in epithelial tissues, the composition

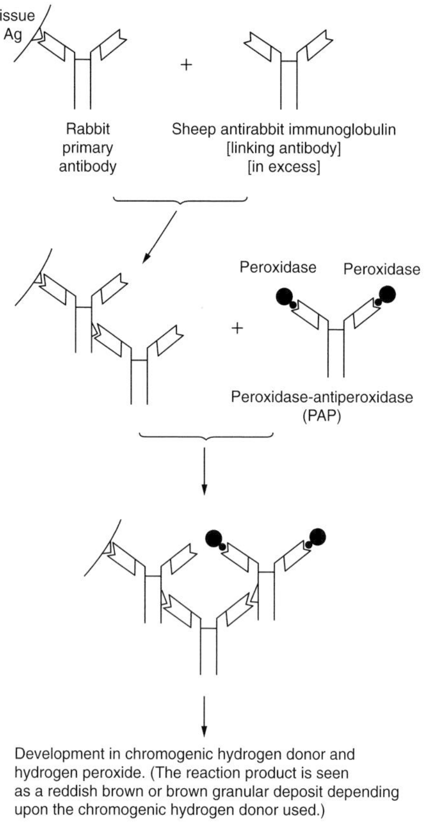

Tissue Ag

Rabbit primary antibody

Sheep antirabbit immunoglobulin [linking antibody] [in excess]

Peroxidase Peroxidase

Peroxidase-antiperoxidase (PAP)

Development in chromogenic hydrogen donor and hydrogen peroxide. (The reaction product is seen as a reddish brown or brown granular deposit depending upon the chromogenic hydrogen donor used.)

FIGURE 28.1 The peroxidase–antiperoxidase (PAP) technique.

of pairs varying with the epithelial cell type, stage differentiation, cellular growth, environment, and disease state. The pankeratin cocktail recognizes most of the acidic and all of the basic cytokeratins, making it a useful general stain for nearly all epithelial tissues and their tumors. This antibody binds specifically to antigens located in the cytoplasmic region of normal simple and complex epithelial cells. The antibody is used to qualitatively stain cytokeratins in sections of formalin-fixed paraffin-embedded tissue. Anti-pankeratin primary antibody contains a mouse monoclonal antibody raised against an epitope found on human epidermal keratins. It reacts with 56.5-kDa, 50-kDa, 48-kDa and 40-kDa cytokeratins of the acidic subfamily, and 65- to 67-kDa, 64-kDa, 59-kDa, 58-kDa, 56-kDa, and 52-kDa cytokeratins of the basic subfamily. In anaplastic tumors, the percentage of tumor cells showing cytokeratin reactivity may be small (less than 5%). Unexpected antigen expression or loss of expression may occur, especially in neoplasms. Occasionally, stromal elements surrounding heavily stained tissue and or cells will show immunoreactivity. The clinical interpretation of any staining or its absence must be complemented by morphological studies and evaluation of proper controls.

AE1/AE3 pan-cytokeratin monoclonal antibody (Figure 28.3) provides the broadest spectrum of keratin reactivity among

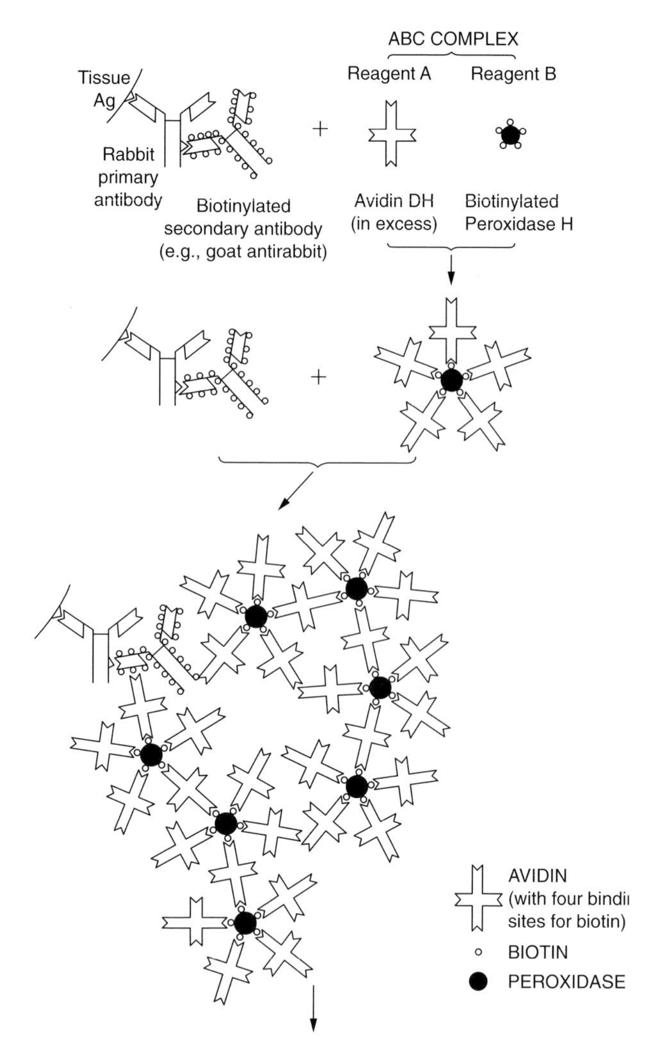

FIGURE 28.2 The avidin–biotin complex (ABC) technique.

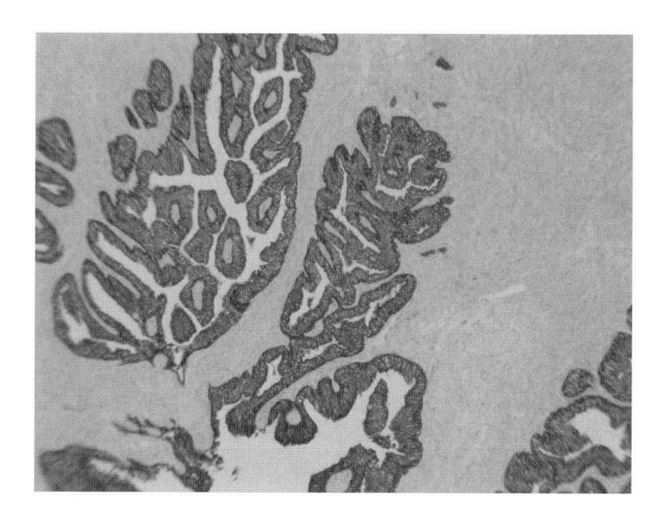

FIGURE 28.3 Cytokeratin cocktail — prostate.

the 19 catalogued human epidermal keratins and produces positive staining in virtually all epithelia.

Anti-high molecular weight human cytokeratin antibodies are mouse monoclonal antibodies that identify keratins of approximately 66 kDa and 57 kDa in extracts of the stratum corneum. The antibody labels squamous, ductal, and other complex epithelia. It is reactive with both squamous and ductal neoplasms and variably with those derived from simple epithelium. Consistently positive are squamous cell carcinomas and ductal carcinomas, most notably those of the breast, pancreas, bile duct, and salivary gland; transitional cell carcinomas of the bladder and nasopharynx; and thymomas and epithelioid mesotheliomas. Adenocarcinomas are variably positive. The antibodies are largely unreactive with adenomas of endocrine organs, carcinomas of the liver (hepatocellular carcinoma), endometrium, and kidney. Mesenchymal tumors, lymphomas, melanomas, neural tumors, and neuroendocrine tumors are unreactive.

Antihuman cytokeratin (CAM 5.2) (cytokeratin 8,18) (Figure 28.4) is a monoclonal antibody against cytokeratins which are polypeptide chains that form structural proteins within the epithelial cell cytoskeleton. Nineteen different molecular forms of cytokeratin have been identified in both normal and malignant epithelial cell lines. Because specific combinations of cytokeratin peptides are associated with different epithelial cells, these peptides are clinically important markers for classifying carcinomas (tumors of epithelial origin) and for distinguishing carcinomas from malignant tumors of nonepithelial origin such as lymphomas, melanomas, and sarcomas. The identification of cytokeratin has gained increasing importance in immunopathology.

Cytokeratin 7 (K72), mouse: Anticytokeratin 7 (K72) mouse monoclonal antibody reacts with proteins that are found in most ductal, glandular, and transitional epithelium of the urinary tract and bile duct epithelial cells. Cytokeratin 7 distinguishes between lung and breast epithelium that stain positive, and colon and prostate epithelial cells that are negative. This antibody also reacts with many benign and malignant epithelial lesions, e.g., adenocarcinomas of the ovary, breast, and lung. Transitional cell carcinomas are positive and prostate cancer is negative. This antibody does not recognize intermediate filament proteins.

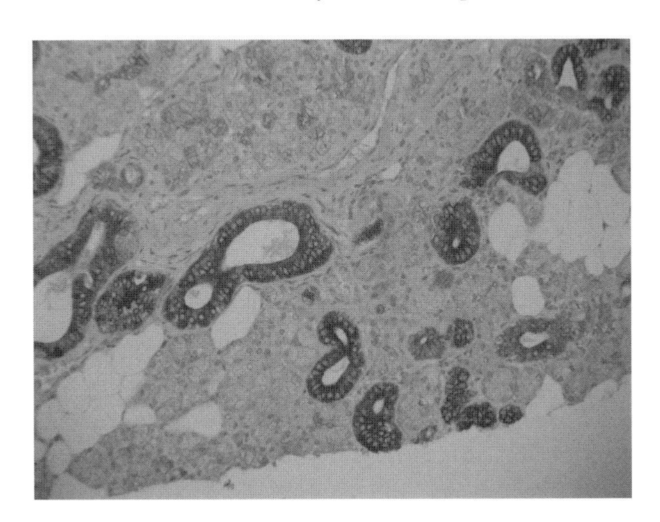

FIGURE 28.4 Cytokeratin 18 — salivary gland.

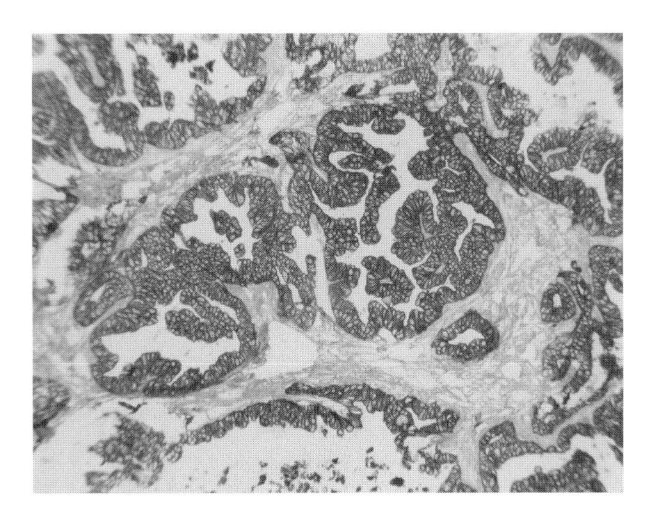

FIGURE 28.5 Cytokeratin 7 — adenocarcinoma of the lung.

Antihuman cytokeratin 7 antibody (Figure 28.5) is a mouse monoclonal antibody directed against the 54-kDa cytokeratin intermediate filament protein identified as cytokeratin 7, a basic cytokeratin found in most glandular epithelia and in transitional epithelia. The antibody reacts with a large number of epithelial cell types including many ductal and glandular epithelia. In general, the antibody does not react with stratified squamous epithelia but is reactive with transitional epithelium of the urinary tract. The antibody reacts with many benign and malignant epithelial lesions. Keratin 7 is expressed in specific subtypes of adenocarcinomas from ovary, breast and lung, whereas carcinomas from the gastrointestinal tract remain negative. Transitional cell carcinomas express keratin 7 whereas prostate cancer is generally negative. The antibody does not react with squamous cell carcinomas, rendering it a rather specific marker for adenocarcinoma and transitional cell carcinoma. In cytological specimens, the antibody permits ovarian carcinoma to be distinguished from colon carcinoma.

Antihuman cytokeratin-20 monoclonal antibody (Figure 28.6) reacts with the 46-kDa cytokeratin intermediate filament protein. It reacts with intestinal epithelium, gastric foveolar epithelium, a number of endocrine cells of the upper portions of the pyloric glands, as well as with the urothelium and Merkel's cells in the epidermis. The antibody has been tested

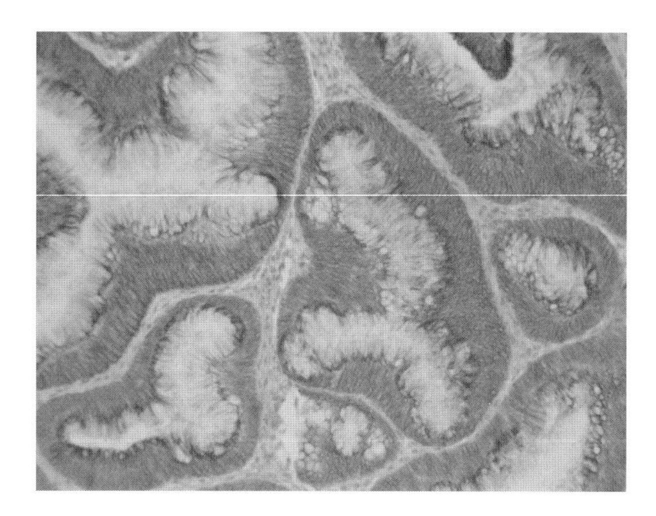

FIGURE 28.6 Cytokeratin 20 — adenoma of the colon.

on a series of carcinomas including primary and metastatic lesions. There is a marked difference in expression of cytokeratin 20 among various carcinoma types. Neoplasia expressing cytokeratin 20 are derived from normal epithelia expressing cytokeratin 20. Colorectal carcinomas consistently express cytokeratin 20, whereas adenocarcinomas of the stomach express cytokeratin 20 to a lesser degree. Adenocarcinomas of the gall bladder and bile ducts, ductal cell adenocarcinomas of the pancreas, mucinous ovarian tumors, and transitional-cell carcinomas have been found to stain positively with the antibody. Most of the carcinomas from other sites were not positive using the antibody to cytokeratin 20, e.g., adenocarcinomas of the breast, lung and endometrium, and nonmucinous tumors of the ovary. Merkel cell carcinomas of the skin stain normally with the anticytokeratin 20 antibody. There was a lack of positivity in small-cell lung carcinomas and in intestinal and pancreatic neuroendocrine tumor cells.

Cytokeratin (34betaE12), mouse: Anticytokeratin (34betaE12) mouse monoclonal antibody detects cytokeratin 34betaE12, a high-molecular-weight cytokeratin that reacts with all squamous and ductal epithelium and stains carcinomas. This antibody recognizes cytokeratins 1,5,10, and 14 that are found in complex epithelia. Cytokeratin 34betaE12 shows no reactivity with hepatocytes, pancreatic acinar cells, proximal renal tubes, or endometrial glands; there has been no reactivity with cells derived from simple epithelia. Mesenchymal tumors, lymphomas, melanomas, neural tumors, and neuroendocrine tumors are unreactive with this antibody. Cytokeratin 34betaE12 has been shown to be useful in distinguishing prostatic adenocarcinoma from hyperplasia of the prostate.

Anti-low molecular weight cytokeratin is a mouse monoclonal antibody directed against an epitope found on human cytokeratins. It may be used to aid in the identification of cells of epithelial lineage. The antibodies are intended for qualitative staining in sections of formalin-fixed paraffin-embedded tissue. Antikeratin primary antibody specifically binds to antigens located in the cytoplasmic regions of normal epithelial cells. Unexpected antigen expression or loss of expression may occur, especially in neoplasms. Occasionally, stromal elements surrounding heavily stained tissue and or cells will show apparent immunoreactivity. The clinical interpretation of any staining, or its absence, must be complemented by morphological studies and evaluation of proper controls.

Nonsquamous keratin (NSK) is a marker, demonstrable by immunoperoxidase staining, that is found in glandular epithelium and adenocarcinomas but not in stratified squamous epithelium.

Epithelial membrane antigen (EMA) (Figure 28.7) is a marker that identifies, by immunoperoxidase staining, most epithelial cells and tumors derived from them, such as breast carcinomas. However, various nonepithelial neoplasms, such as selected lymphomas and sarcomas, may express EMA also. Thus, it must be used in conjunction with other markers in tumor identification and/or classification.

Antiepithelial membrane antigen (EMA) antibody is a mouse monoclonal antibody directed against a mucin epitope present on most human epithelial cells. This antibody reacts with epithelial mucin, a heavily glycosylated molecule with a molecular weight of circa 400 kDa. Epithelial membrane antigen is widely distributed in epithelial tissues and tumors arising from them. Normal glandular epithelium and tissue from nonneoplastic diseases stain in lumen membranes and cytoplasm. Malignant neoplasms of glandular epithelium frequently show a change

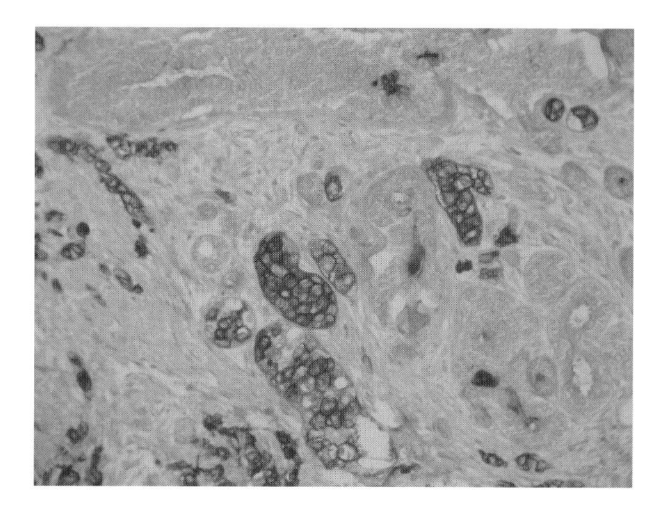

FIGURE 28.7 Epithelial membrane antigen — squamous carcinoma.

in pattern with the appearance of adjacent cell membrane staining. EMA is of value in distinguishing both large-cell anaplastic carcinoma from diffuse histiocytic lymphoma and small-cell anaplastic carcinoma from well and poorly differentiated lymphocytic lymphomas.

An **epithelial cell adhesion molecule (EpCAM)** is considered a pan-carcinoma antigen. It is highly expressed on a variety of adenocarcinomas of different origin such as breast, ovary, colon, and lung, whereas its expression in normal tissue is very limited.

Intermediate filaments are 7- to 11-nm diameter intracellular filaments observed by electron microscopy that are lineage specific. They are intermediate in size between actin microfilaments, which are 6 nm in diameter, and microtubules, which are 25 nm in diameter. They are detected in cell and tissue preparations by monoclonal antibodies specific for the filaments and are identified by the immunoperoxidase method. The detection of various types of intermediate filaments in tumors is of great assistance in determining the histogenetic origin of many types of neoplasms.

Vimentin is a 55-kDa intermediate filament protein synthesized by mesenchymal cells such as vascular endothelial cells, smooth muscle cells, histiocytes, lymphocytes, fibroblasts, melanocytes, osteocytes, chondrocytes, astrocytes, and occasional ependymal and glomerular cells. Malignant cells may express more than one intermediate filament. For example, immunoperoxidase staining may reveal vimentin and cytokeratin in breast, lung, kidney, or endometrial adenocarcinomas.

Antivimentin antibody is a mouse monoclonal antibody raised against purified bovine eye lens vimentin. This antibody reacts with the 57-kDa intermediate filament protein, vimentin. This reagent may be used to aid in the identification of cells of mesenchymal origin. The antibody is intended for qualitative staining in sections of formalin-fixed, paraffin-embedded tissue. It binds specifically to antigens located in the cytoplasm of mesenchymal cells. The clinical interpretation of any staining, or its absence, must be complemented by morphological studies and evaluation of proper controls.

Desmin is a 55-kDa intermediate filament molecule found in mesenchymal cells that include both smooth and skeletal muscle, endothelial cells of the vessels, and probably myofibroblasts. In surgical pathologic diagnosis, monoclonal antibodies against desmin are useful in identifying muscle tumors.

Desmin (D33), mouse: Desmin antibody detects a protein that is expressed by cells of normal smooth, skeletal, and cardiac muscles. The light microscope has suggested that desmin is primarily located at or near the periphery of Z lines in striated muscle fibrils. In smooth muscle, desmin interconnects cytoplasmic dense bodies with membrane-bound dense plaques. Desmin antibody reacts with leiomyomas, rhabdomyomas, and perivascular cells of glomus tumors of the skin (if they are of myogenic nature). This antibody is basically used to demonstrate the myogenic components of carcinosarcomas and malignant mixed mesodermal tumors.

Antidesmin antibody is a mouse monoclonal antibody (clone DE-R-11) raised against purified porcine desmin that reacts with the 53-kDa intermediate filament protein desmin. This reagent may be used to aid in the identification of cells of myocyte lineage. The antibody is intended for qualitative staining in sections of formalin-fixed, paraffin-embedded tissue. Antidesmin primary antibody specifically binds to antigens located in the cytoplasm of myocytic cells. The clinical interpretation of any staining, or its absence, must be complemented by morphological studies and evaluation of proper controls.

Myoglobin is an oxygen-storing muscle protein that serves as a marker of muscle neoplasms, demonstrable by immunoperoxidase staining for surgical pathologic diagnosis.

Myoglobin antibody is a reagent that stains normal striated muscle and striated muscle containing tumor. Using immunohistochemical procedures on formalin-fixed paraffin-embedded tissues, this antibody stains human skeletal and cardiac muscle.

Actin, the principal muscle protein, which together with myosin causes muscle contraction, is used in surgical pathology as a marker for the identification of tumors of muscle origin. Actin is identified through immunoperoxidase staining of surgical pathology tissue specimens.

Antimuscle actin primary antibody is a mouse monoclonal antibody (clone HUC1-1) directed against an actin epitope found on muscle actin isoforms. This reagent may be used to aid in the identification of cells of myocytic lineage. The antibody is intended for qualitative staining in sections of formalin-fixed, paraffin-embedded tissue. Antimuscle actin antibody specifically binds to antigens located in the cytoplasmic regions of normal muscle cells. Unexpected antigen expression or loss of expression may occur, especially in neoplasms. Occasionally, stromal elements surrounding heavily stained tissue and or cells will show immunoreactivity. Clinical interpretation must be complemented by morphological studies and the evaluation of appropriate controls.

Antihuman α-smooth muscle actin is a mouse monoclonal antibody that reacts with the α-smooth muscle isoform of actin. The antibody reacts with smooth muscle cells of vessels and different parenchyma without exception, but with

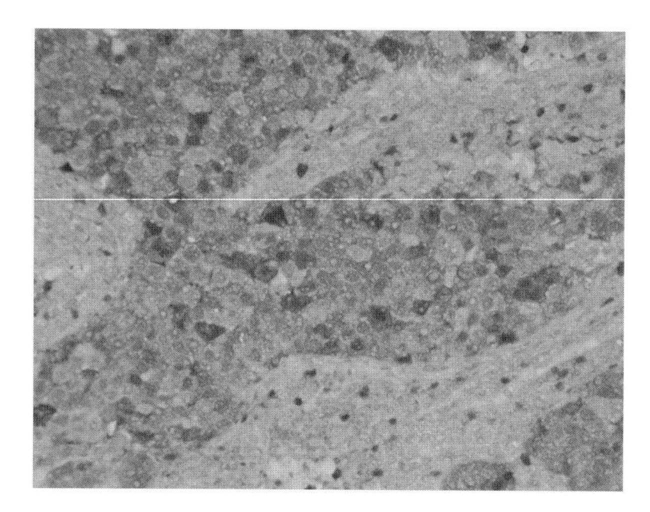

FIGURE 28.8 S-100 — metastatic melanoma — lymph node.

different intensity, according to the amount of α-smooth muscle actin present in smooth muscle cells, myoepithelial cells, pericytes, and some stromal cells in the intestine, testes, breast, and ovary. The antibody also reacts with myofibroblasts in benign and reactive fibroblastic lesions and perisinusoidal cells of normal and diseased human livers.

S-100 (Figure 28.8) is a heterodimeric protein comprised of α and β chains. It is present in a variety of tissues and is especially prominent in nervous system tissue including brain, neural crest, and Schwann cells. It also is positive in breast ducts, sweat and salivary glands, bronchial glands and Schwann cells, serous acini, malignant melanomas, myoepithelium, and neurofibrosarcomas.

S-100 protein is a marker, demonstrable by immunoperoxidase staining, that is extensively distributed in both central and peripheral nervous systems and tumors arising from them, including astrocytomas, melanomas, Schwannomas, etc. Most melanomas express S-100 protein. Such nonneuronal cells as chondrocytes and histiocytes are also S-100 positive.

S-100 protein antibody is a mouse monoclonal antibody specific for S-100 protein that is found in normal melanocytes, Langerhans cells, histiocytes, chrondrocytes, lipocytes, skeletal and cardiac muscle, Schwann cells, epthelial and myoepithelial cells of the breast, salivary and sweat glands, and glial cells. Neoplasms derived from these cells also express S-100 protein, albeit nonuniformly. A large number of well-differentiated tumors of the salivary gland, adipose and cartilaginous tissue, and Schwann cell derived tumors express S-100 protein. Almost all malignant melanomas and cases of histiocytosis X are positive for S-100 protein. Despite the fact that S-100 protein is a ubiquitous substance, its demonstration is of great value in the identification of several neoplasms, particularly melanomas.

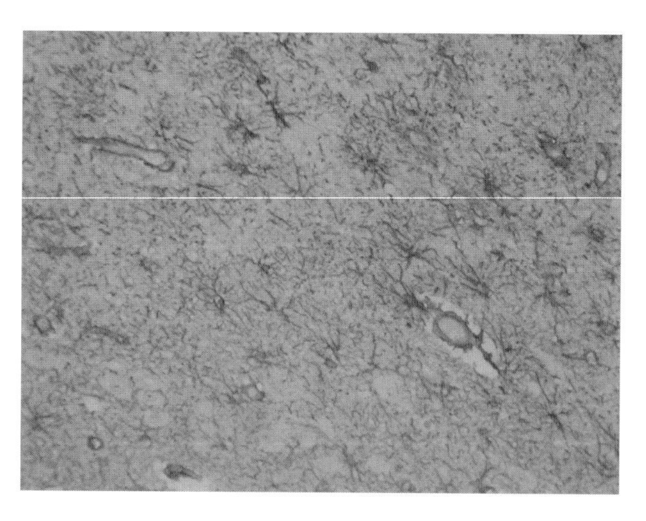

FIGURE 28.9 Glial fibrillary acidic protein — brain.

Glialfibrillary acidic protein (GFAP) (Figure 28.9) is an intermediate filament protein constituent of astrocytes, which is also abundant in glial cell tumors. The immunoperoxidase technique employing monoclonal antibodies against the GFAP is used in surgical pathologic diagnosis to identify tumors based on their histogenetic origin.

Antiglial fibrillary acidic protein (GFAP) antibody is a rabbit polyclonal antibody directed against glial fibrillary acidic protein present in the cytoplasm of most human astrocytes and ependymal cells. This reagent may be used to aid in the identification of cells of glial lineage. The antibody is intended for qualitative staining in sections of formalin-fixed, paraffin-embedded tissue. Anti-GFAP antibody specifically binds to the glial fibrillary acidic protein located in the cytoplasm of normal and neoplastic glial ells. Unexpected antigen expression or loss of expression may occur, especially in neoplasms. Occasionally, stromal elements surrounding heavily stained tissue and or cells will show immunoreactivity. The clinical interpretation of any staining, or its absence, must be complemented by morphological studies and evaluation of proper controls.

Synaptophysin (Figure 28.10) is a neuroendocrine differentiation marker that is detectable by the immunoperoxidase technique used in surgical pathologic diagnosis. Tumors in which it is produced include ganglioneuroblastoma, neuroblastoma, ganglioneuroma, paraganglioma, pheochromocytoma, medullary carcinoma of the thyroid, carcinoid, and tumors of the endocrine pancreas.

Antihuman synaptophysin antibody is a rabbit antibody that reacts with a wide spectrum of neuroendocrine neoplasms of neural type including neuroblastomas, ganglioneuroblastomas, ganglioneuromas, pheochromocytomas, and chromaffin and nonchromaffin paragangliomas. The antibody also labels neuroendocrine neoplasms of epithelial

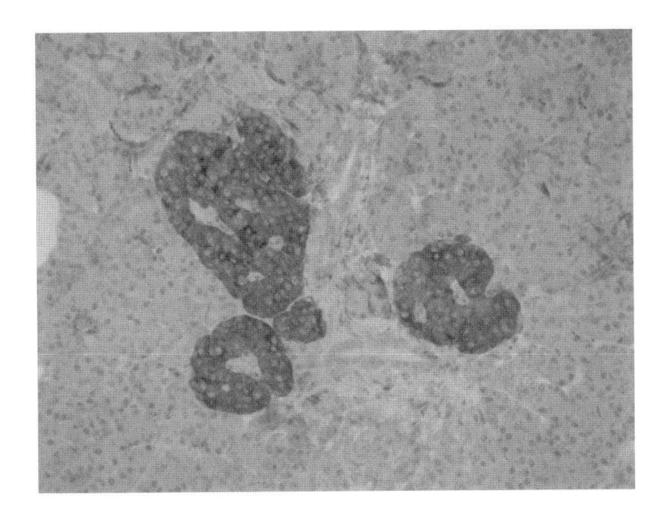

FIGURE 28.10 Synaptophysin — pancreas.

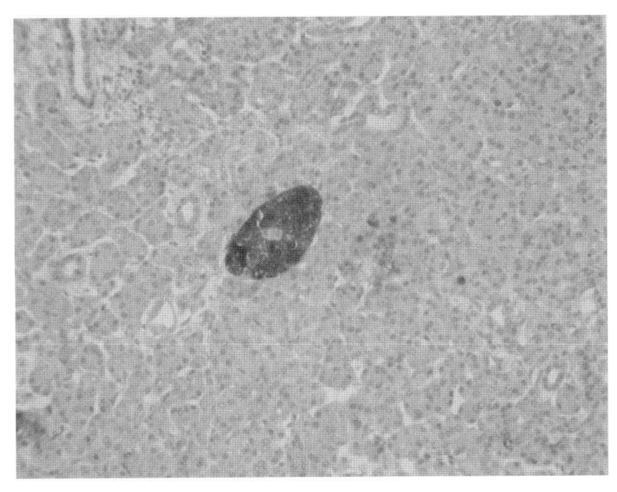

FIGURE 28.11 Neuron-specific enolase (NSE) — pancreas.

type including pituitary adenomas, islet cell neoplasms, medullary thyroid carcinomas, parathyroid adenomas, carcinoids of the bronchopulmonary and gastrointestinal tracts, neuroendocrine carcinomas of the bronchopulmonary and gastrointestinal tracts, and neuroendocrine carcinomas of the skin.

Anti-Purkinje cell antibody has been detected in the circulation of subacute cerebellar degeneration patients and in those with ovarian neoplasms and other gynecologic malignancies.

Anti-GM$_1$ antibodies are antibodies found in 2 to 40% of Guillain-Barré syndrome (GBS) patients. They are mainly IgG$_1$, or IgA, rather than IgM, even though IgM anti-GM$_1$ antibodies have been found in a few GBS cases. Anti-GM$_1$ are more frequent in GBS patients who experienced *C. jejuni* infection (up to 50% of cases). Titers are highest initially and fall as the disease progresses. These antibodies are present in spinal fluid, apparently due to disruption of the blood–nerve barrier. Anti-GM$_1$ antibodies recognize surface epitopes on *Campylobacter* bacteria, stains, and possibly a saccharide identical to the terminal tetrasaccharide of GM$_1$ that has been found in *Campylobacter* lipopolysaccharide. IgG anti-GM$_1$ has been postulated to selectively injure motor nerves.

Anti-Ewing's sarcoma marker (CD99) is a mouse monoclonal antihuman MIC2 gene product. (Ewing's sarcoma marker) antibody reacts only with glioblastoma and ependymoma of the central nervous system and certain islet cell tumors of the pancreas. Because the MIC2 gene products are most strongly expressed on the cell membrane of Ewing's sarcoma (ES) and primitive peripheral neuroectodermal tumors (pPNET), demonstration of the gene products allows for the differentiation of these tumors from other round-cell tumors of childhood and adolescence.

Anti-Ri antibody is an antibody found in serum and spinal fluid of patients with opsoclonus without myoclonus that occurs in conjunction with gait ataxia in women with breast cancer. The anti-Ri antibody reacts with 55-kDa and 80-kDa proteins present in the nuclei of CNS neurons and breast tumor cells. The condition may remit, manifest exacerbations and remissions, and occasionally respond to steroids or other immune interventions.

Neuron-specific enolase (NSE) (Figure 28.11) is an enzyme of neurons and neuroendocrine cells, as well as their derived tumors, e.g., oat cell carcinoma of lung, demonstrable by immunoperoxidase staining. NSE occurs also in some neoplasms not derived from neurons or endocrine cells.

Neuron-specific enolase (NSE) antibody is a murine monoclonal antibody directed against γ-γ enolase present on most human neurons, normal and neoplastic neuroendocrine cells, and some megakarocytes. This reagent may be used to aid in the identification of cells of neural or neuroendocrine lineage. The antibody is intended for qualitative staining in sections of formulin-fixed paraffin-embedded tissue. Anti-NSE antibody specifically binds to the γ-γ enolase located in the cytoplasm of normal and neoplastic neuroendocrine cells. Unexpected antigen expression or loss expression may occur, especially neoplasms. Occasionally, stromal elements surrounding heavily stained tissue and/or cells will show immunoreactivity. The clinical interpretation of any staining, or its absence, must be complemented by morphological studies and evaluation of proper controls.

A **neurofilament** is a marker, demonstrable by immunoperoxidase staining, for neural-derived tumors as well as selected endocrine neoplasms with neural differentiation.

Hanganitziu-Deicher antigen is an altered ganglioside present in certain human neoplasms (CD3, GM1, and terminal 4NAcNeu).

Immunoperoxidase staining of pituitary adenomas with antibodies to the **pituitary hormones** ACTH, GH, prolactin, FSH, and LH facilitates definition of their clinical phenotype.

Adrenocorticotrophic hormone (ACTH) antibody is a polyclonal antibody preparation useful in immunoperoxidase procedures to stain corticotroph cells of the pituitary gland and benign and malignant tumors arising from these cells, in formalin-fixed, paraffin-embedded tissue biopsies.

ALZ-50 is a monoclonal antibody that serves as an early indicator of Alzheimer's disease by reacting with Alzheimer's brain tissue, specifically protein A-68.

Antihuman follicle-stimulating hormone (FSH) antibody (Figure 28.12) is a rabbit antibody that labels gonadotropic cells in the pituitary. Positive staining for adenohypophyseal hormones assists in the classification of pituitary tumors. FSH is an adenohypophyseal glycoprotein hormone found in gonadotropic cells of the anterior pituitary gland of most mammals. Gonadotropic cells average about 10% of anterior pituitary cells. This antibody can be used for immunohistochemical staining.

Chromogranin monoclonal antibody (Figure 28.13) is used to recognize chromogranin A (68 kDa) and other related chromogranin polypeptides from human, monkey, and pig. It is designed for the specific and quantitative localization of human chromogranin in paraffin-embedded and frozen tissue sections. It aids the localization of secretory storage granules in endocrine cells. Chromogranin A is a large, acidic protein present in catecholamine-containing granules of bovine adrenal medulla. It may be widely distributed in endocrine cells and tissues, which share some common characteristics and are known as APUD cells. Dispersed throughout the body, they are also referred to as the diffuse neuroendocrine system (DNES). Chromogranin has been demonstrated in several elements of the DNES, including anterior pituitary, thyroid parafollicular C cells, parathyroid chief cells, pancreatic islet cells, intestinal enteroendocrine cells, and tumors derived from these cells. Chromogranin immunoreactivity has also been observed in the thymus, spleen, lymph nodes, fetal liver, neurons, the inner segment of rods and cones, the submandibular gland, and the central nervous system. Chromogranin is a widespread histological marker for polypeptide producing cells (APUD) and the tumors derived from them.

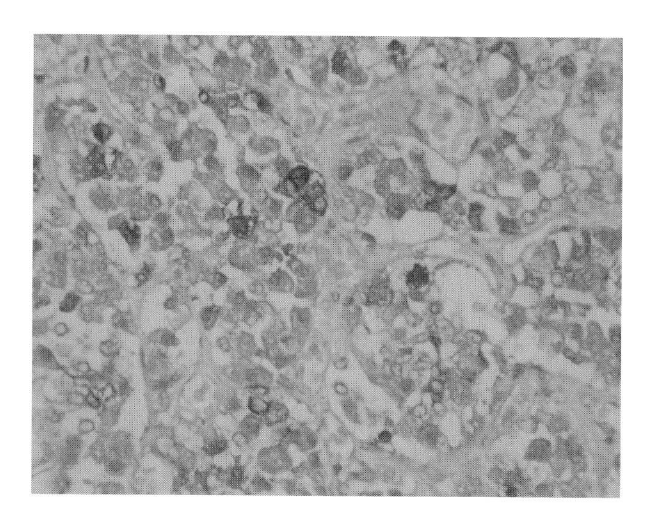

FIGURE 28.12 Follicle-stimulating hormone (FSH) — pituitary.

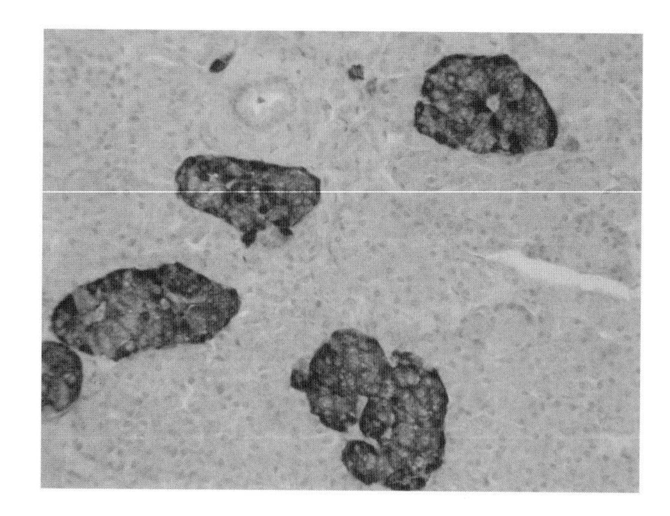

FIGURE 28.13 Chromogranin — pancreas.

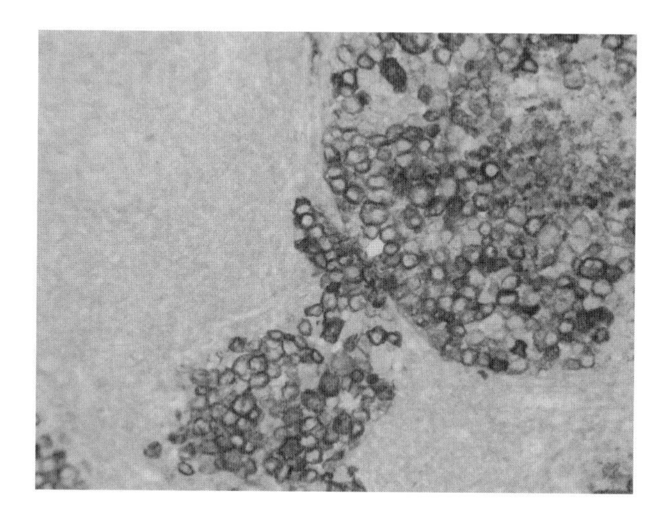

FIGURE 28.14 HMB-45. Melanoma in lymph node.

Antigrowth hormone (GH) antibody is a rabbit polyclonal antibody against human growth hormone that positively stains the growth hormone-producing cells and somatotrophs of the pituitary gland and malignant and benign neoplasms arising from these cells.

Antiprolactin antibody is a rabbit antibody that gives positive staining of the prolactin cells of the anterior pituitary and benign and malignant neoplasms derived from these cells.

Antimelanoma primary antibody is a mouse monoclonal antibody (clone HMB-45) raised against an extract of pigmented melanoma metastases from lymph nodes directed against a glycoconjugate present in immature melanosomes. This antibody may be used to aid in the identification of cells of melanocytic lineage. The antibody is for qualitative staining in sections of formalin-fixed paraffin-embedded tissue. This antibody binds specifically to antigens located on immature melanosomes. Unexpected antigen expression or loss of expression may occur, especially in neoplasms. Clinical interpretation must be complemented by morphological studies and evaluation of proper controls.

Anti-BRST-2 (GCDFP-15) monoclonal antibody is specific for BRST-2 antigen expressed by apocrine sweat glands, eccrine glands (variable), minor salivary glands, bronchial glands, metaplastic epithelium of the breast, benign sweat gland tumors of the skin, and the serous cells of the submandibular gland. Breast carcinomas (primary and metastatic lesions) with apocrine features express the BRST-2 antigen. BRST-2 is positive in extramammary Paget's disease. Other tumors are negative.

Anti-BRST-3 (B72.3) monoclonal antibody recognizes TAG-72, (Figure 28.15) a tumor-associated oncofetal antigen expressed by a wide variety of human adenocarcinomas. This antigen is expressed by 84% of invasive ductal breast carcinoma and 85 to 90% of colon, pancreatic, gastric, esophageal, lung (non-small cell), ovarian, and endometrial adenocarcinomas. It is not expressed by leukemias, lymphomas, sarcomas, mesotheliomas, melanomas, or benign tumors. TAG-72 is also expressed on normal secretory endometrium but not on other normal tissues.

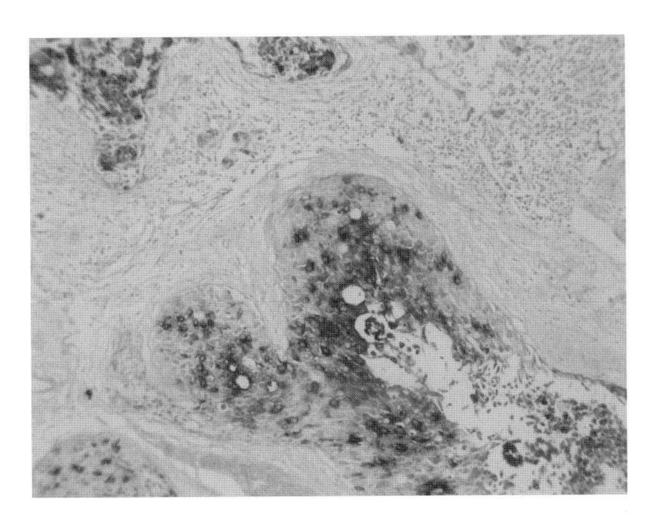

FIGURE 28.15 Tag 72 — carcinoma of the breast.

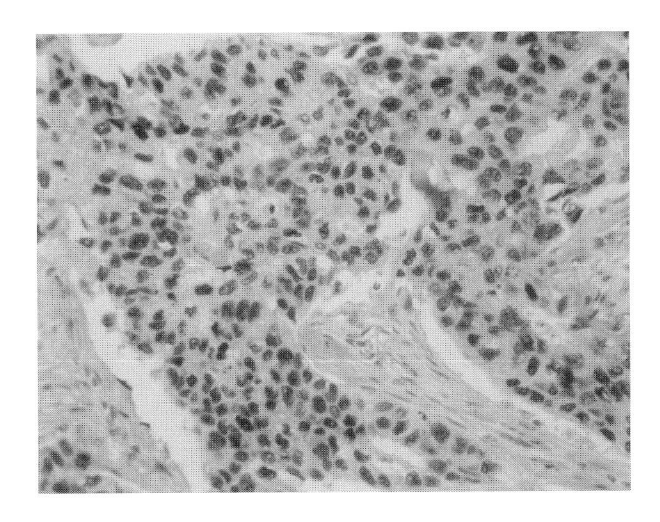

FIGURE 28.16 Estrogen receptor — carcinoma of the breast.

Estrogen/progesterone receptor protein: Monoclonal antibodies against estrogen receptor protein and against progesterone receptor protein permit identification of tumor cells by their preferential immunoperoxidase staining for these markers, whereas stromal cells remain unstained. This method is claimed by some to be superior to cytosol assays in evaluating the clinical response to hormones. (Figure 28.16)

Antiestrogen receptor antibodies are mouse monoclonal specific for estrogen receptors. The estrogen receptor (ER) content of breast cancer tissue is an important parameter in the prediction of prognosis and response to endocrine therapy. Monoclonal antibodies to ER permit the determination of receptor status of breast tumors to be carried out in routine histopathology laboratories. Although monoclonal antibodies that recognize ER were only effective on frozen sections initially, currently available monoclonal antibodies are effective on formalin-fixed, paraffin-embedded tissues to allow the determination of ER in routinely processed and archival material.

CA-15-3 is an antibody specific for an antigen frequently present in the blood serum of metastatic breast carcinoma patients.

CA-125 (Figure 28.17) is a mucinous ovarian carcinoma cell surface glycoprotein detectable in blood serum. Increasing serum concentrations portend a grave prognosis. It may also be found in the blood sera of patients with other adenocarcinomas, such as breast, gastrointestinal tract, uterine cervix, and endometrium.

CA-125 antibody is a mouse monoclonal antibody that reacts with malignant ovarian epithelial cells. The antigen is formalin resistant, permitting the detection of ovarian cancer by immunohistochemistry, although serum assays for this protein are widely used to monitor ovarian cancer.

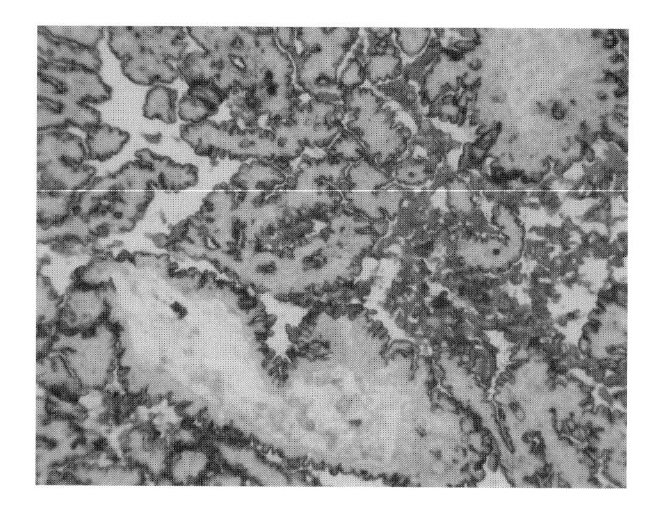

FIGURE 28.17 CA125 — papillary carcinoma of the ovary.

FIGURE 28.18 Placental alkaline phosphatase (PLAP) — placenta.

CA-125 also reacts with antigens in seminal vesicle carcinoma and anaplastic lymphoma.

c-*erb*-B2 murine monoclonal antibody is specific for *c-erb*-B2 oncoprotein which is expressed by tumor cell membranes at a level detectable by immunohistochemistry in up to 20% of adenocarcinomas from various sites including ovary, gastrointestinal tract, and breast. Immunohistochemical staining correlates with gene amplification. In the case of breast cancer, *c-erb*-B2 expression has been shown to be associated with poor prognosis. Between 15 and 30% of invasive ductal cancers are positive for c-*erb*-B2. Almost all cases of Paget's disease and approximately 70% of cases of *in situ* ductal carcinoma are positive.

Cu-18 is a glycoprotein of breast epithelium. Immunoperoxidase staining identifies this marker in most breast tumors and a few tumors of the ovary and lung. Stomach, pancreas, and colon tumors do not express this antigen.

Lactalbumin is a breast epithelial cell protein demonstrable by immunoperoxidase staining that is found in approximately one half to two thirds of breast carcinomas for which it is relatively specific. More than 50% of metastatic breast tumors and some salivary gland and skin appendage tumors stain positively for lactalbumin.

Gross cystic disease fluid protein 15 (GCDFP-15) antigen is a 15-kDa glycoprotein that is demonstrable with immunoperoxidase staining and expressed by primary and metastatic breast carcinomas with apocrine features and extramammary Paget's disease. Normal apocrine sweat glands, eccrine glands (variable), minor salivary glands, bronchial glands, metaplastic breast epithelium, benign sweat gland tumors of skin, and submandibular serous cells express GCDFP-15 antigen.

GCDFP-15 (23A3), mouse: Gross cystic disease fluid protein-15 is a 15,000-Da glycoprotein that was localized in the apocrine metaplastic epithelium lining breast cysts and in apocrine glands in the axilla, vulva, eyelid, and ear canal. Approximately 70% of breast carcinomas stain positive with antibody to GCDFP-15. Colorectal carcinomas, as well as mesotheliomas, do not stain with this antibody. Lung adenocarcinoma rarely stains with this antibody.

Human milk-fat globulin (HMFG) is a human milk glycoprotein on secretory breast cell surfaces. Many breast and ovarian carcinomas are positive for HMFG.

In diagnostic immunology, **estradiol** is a marker identifiable in breast carcinoma tissue by monoclonal antibody and the immunoperoxidase technique that correlates, to a limited degree, with estrogen receptor activity in cytosols from the same preparation.

Antiplacental alkaline phosphatase (PLAP) antibody (Figure 28.18) is normally produced by syncytiotrophoblasts after the twelfth week of pregnancy. Human placental alkaline phosphatase is a member of a family of membrane-bound alkaline phosphatase enzymes and isoenzymes. It is expressed by both malignant somatic and germ-cell tumors. PLAP immunoreactivity can be used in conjunction with epithelial membrane antigen (EMA) and keratin to differentiate between germ cell and somatic tumor metastases. Germ cell tumors appear to be universally reactive for PLAP, whereas somatic tumors show only 15 to 20% reactivity.

Antiprogesterone receptor antibody is a mouse monoclonal antibody against human progesterone receptor. A mouse monoclonal antihuman progesterone receptor antibody that specifically recognizes the A and B forms of the receptor in Western blot purified recombinant receptor, normal endometrium, and cell lysates of the progesterone receptor-rich T47D human breast carcinoma cell line.

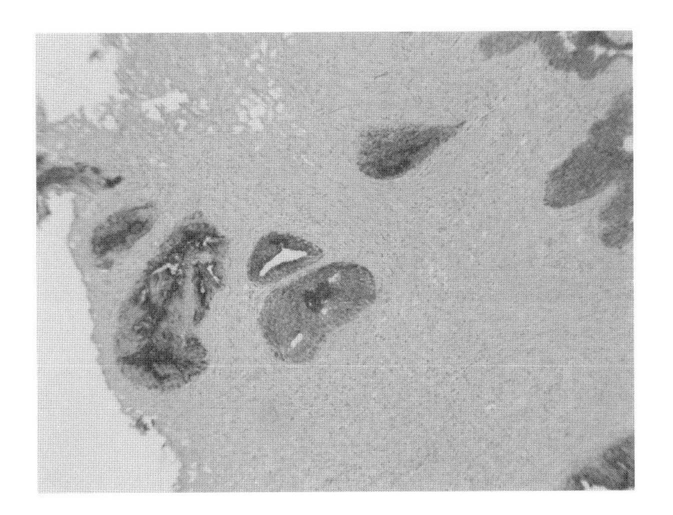

FIGURE 28.19 Prostatic acid phosphatase (PSAP) — prostate.

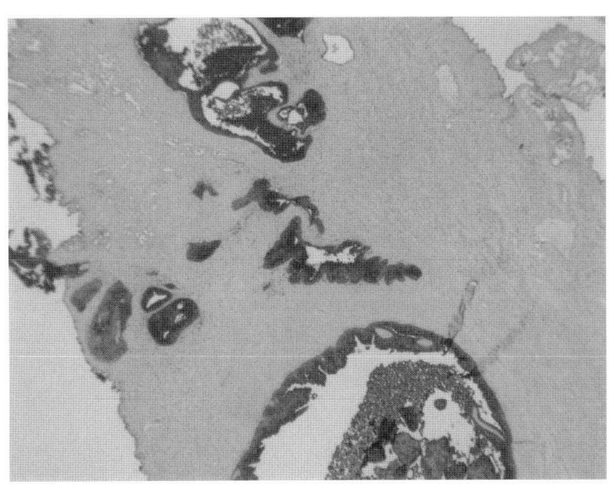

FIGURE 28.20 Prostate-specific antigen (PSA) — prostate.

No reactivity was observed with lysate of the progesterone receptor-negative MDA-MB-231 breast carcinoma cells. No crossreactivity was found with androgen receptor, estrogen receptor, or glucocorticoid receptor. The antibody binds an epitope found between amino acids 165 and 534, in the N-terminal transactivation domain of the progesterone receptor molecule. Various tumors of the female reproductive tract have been shown to express progesterone receptor. Immunoreactivity has been demonstrated in breast carcinoma, uterine papillary serous carcinoma, endometrial carcinoma, ovarian serous borderline tumor, endometrial stromal sarcoma, uterine adenomatoid tumor, and ovarian thecoma. Other tumors that have been shown to stain positively include medullary carcinoma of the thyroid and meningioma.

Prostatic acid phosphatase (PAP)/prostatic epithelial antigen are prostate antigens, identifiable by immunoperoxidase staining, that are prostate-specific and -sensitive. Used together, they detect approximately 99% of prostatic adenocarcinomas.

Antihuman prostatic acid phosphatase (PSAP) (Figure 28.19) is a rabbit antibody that reacts with prostatic ductal epithelial cells — normal, benign hypertrophic, and neoplastic. This antibody labels the cytoplasm of prostatic epithelium, secretions, and concretions.

PSA (prostate-specific antigen) (Figure 28.20) is a substance secreted only by the prostate epithelium and is a 34-kDa glycoprotein serine protease that lyses seminal coagulum. Individuals with benign prostatic hypertrophy have a 30 to 50% elevation in PSA levels, whereas those with prostatic carcinoma have a 25 to 92% elevation. It is a more reliable indicator of prostatic carcinoma than is serum prostatic acid phosphatase (PAP). PSA levels are also valuable in signifying recurrence of prostatic adenocarcinoma. Prostate cancer may occur in 22% of the individuals with PSA levels greater than 4.0 μg/l and 60% of the individuals with PSA levels greater than 10 μg/l.

Prostate-specific antigen (PSA) is a marker in serum or tissue sections for adenocarcinoma of the prostate. PSA is a 34-kDa glycoprotein found exclusively in benign and malignant epithelium of the prostate. Men with PSA levels of 0 to 4.0 ng/ml and a nonsuspicious digital rectal examination are generally not biopsied for prostate cancer. Men with PSA levels of 10.0 ng/ml and above typically undergo prostate biopsy. About one half of these men will be found to have prostate cancer. Certain kinds of PSA, known as bound PSA, link themselves to other proteins in the blood. Other kinds of PSA, known as free PSA, float by themselves. Prostate cancer is more likely to be present in men who have a low percentage of free PSA relative to the total amount of PSA. This finding is especially valuable in helping to differentiate between cancer and other, benign, conditions, thus eliminating unnecessary biopsies among men in that diagnostic gray zone, who have total PSA levels between 4.0 and 10.0 ng/ml. The PSA molecule is smaller than prostatic acid phosphatase (PAP). In patients with prostate cancer, preoperative PSS serum levels are positively correlated with the disease. PSA is more stable and shows less diurnal variation than does PAP. PSA is increased in 95% of new cases of prostatic carcinoma compared with 60% for PAP. It is increased in 97% of recurrent cases compared with 66% of PAP. PAP may also be increased in selected cases of benign prostatic hypertrophy and prostatitis, but these elevations are less than those associated with adenocarcinoma of the prostate. It is inappropriate to use either PSA or PAP alone as a screen for asymptomatic males. Transurethral resection (TUR), urethral instrumentation, prostatic needle biopsy, prostatic infarct, or urinary retention may also result in increased PSA values. PSA is critical for the prediction of recurrent adenocarcinoma in postsurgical patients. PSA

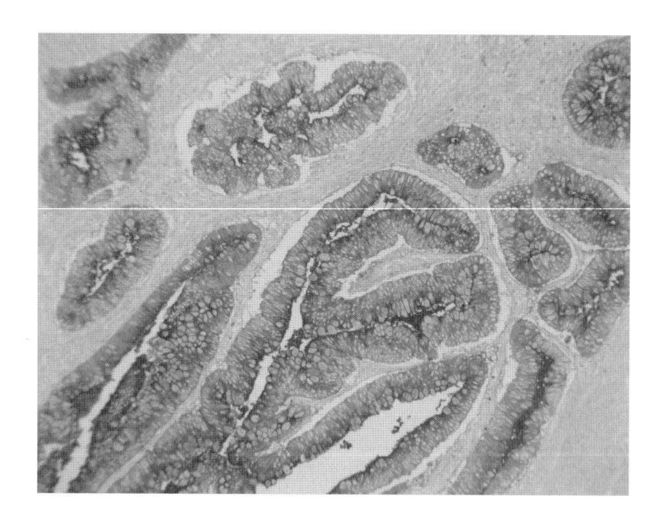

FIGURE 28.21 CEA — carcinoma of the colon.

is also a useful immunocytochemical marker for primary and metastatic adenocarcinoma of the prostate.

Antiprostate specific antigen (PSA) antibody is a rabbit antibody that reacts with prostatic ductal epithelial cells — normal, benign hypertrophic, and neoplastic. This antibody labels the cytoplasm of prostatic epithelium, secretions, and concretions.

Carcinoembryonic antigen (CEA) (Figure 28.21) is a 200-kDa membrane glycoprotein epitope that is present in the fetal gastrointestinal tract in normal conditions. However, tumor cells, such as those in colon carcinoma, may reexpress it. CEA was first described as a screen for identifying carcinoma by detecting nanogram quantities of the antigen in serum. It was later shown to be present in certain other conditions as well. CEA levels are elevated in almost one third of patients with colorectal, liver, pancreatic, lung, breast, head and neck, cervical, bladder, medullarythyroid, and prostatic carcinoma. However, the level may be elevated also in malignant melanoma, lymphoproliferative disease, and smokers. Regrettably, CEA levels also increase in a variety of nonneoplastic disorders, including inflammatory bowel disease, pancreatitis, and cirrhosis of the liver. Nevertheless, determination of CEA levels in the serum is valuable for monitoring the recurrence of tumors in patients whose primary neoplasm has been removed. If the patient's CEA level reveals a 35% elevation compared to the level immediately following surgery, this may signify metastases. This oncofetal antigen is comprised of one polypeptide chain with one variable region at the amino terminus and six constant region domains. CEA belongs to the immunoglobulin superfamily. It lacks specificity for cancer, thereby limiting its diagnostic usefulness. It is detected with a mouse monoclonal antibody directed against a complex glycoprotein antigen present on many human epithelial derived tumors. This reagent may be used to aid in the identification of cells of epithelial lineage. The antibody is intended for qualitative staining in sections of formalin-fixed, paraffin-embedded tissue. Anti CEA antibodies specifically bind to antigens located in the plasma membrane and cytoplasmic regions of normal epithelial cells. Unexpected antigen expression or loss of expression may occur, especially in neoplasms. Occasionally, stromal elements surround heavily stained tissue and/or cells which show immunoreactivity. Clinical interpretation of any staining or its absence must be complemented by morphological studies and evaluation of proper controls.

Serotonin (5-hydroxytryptamine [5-HT]) is a 176-mol-wt catecholamine found in mouse and rat mast cells and in human platelets that participates in anaphylaxis in several species such as the rabbit but not in humans. It induces contraction of smooth muscle, enhances vascular permeability of small blood vessels, and induces large blood vessel vasoconstriction. 5-HT is derived from tryptophan by hydroxylation to 5-hydroxytryptophan and decarboxylation to 5-hydroxytryptamine. In man, gut enterochromaffin cells contain 90% of 5-HT, with the remainder accruing in blood platelets and the brain. 5-HT is a potent biogenic amine with wide species distribution. 5-HT may stimulate phagocytosis by leukocytes and interfere with the clearance of particles by the mononuclear phagocyte system. Immunoperoxidase staining for 5-HT, which is synthesized by various neoplasms, especially carcinoid tumors, is a valuable aid in surgical pathologic diagnosis of tumors producing it.

CA-19-9 is a tumor-associated antigen found on the Lewis A blood group antigen that is sialylated or in mucin-containing tissues. In individuals whose serum levels exceed 37 U/ml, 72% have carcinoma of the pancreas. In individuals whose levels exceed 1000 U/ml, 95% have pancreatic cancer. Anti-CA-19-9 monoclonal antibody is useful to detect the recurrence of pancreatic cancer following surgery and to distinguish between neoplastic and benign conditions of the pancreas. However, it is not useful for pancreatic cancer screening.

Antihuman glucagon antibody is a rabbit antibody that labels A cells of the endocrine mammalian pancreas.

Antihuman chorionic gonadotropin (HCG) antibody is an antibody that reacts with the beta chain of human chorionic gonadotropin (HCG). HCG is a polypeptide hormone synthesized in the syncytiotrophoblastic cells of the placenta and in certain trophoblastic tumors. HCG is a marker for the biochemical differentiation of trophoblastic cells, which often precedes their morphological differentiation. The antibody aids detection of HCG in trophoblastic elements of germ cell tumors of the ovaries, testes, and extragonadal sites. It crossreacts with luteinizing hormone.

O125 (ovarian celomic) is a nonmucinous ovarian tumor antigen demonstrable with homologous antibody by immunoperoxidase staining. Selected mesotheliomas express this antigen as well.

Colon–ovary tumor antigen (COTA) is a type of mucin demonstrable by immunoperoxidase staining in all colon neoplasms and in some ovarian tumors. COTA occurs infrequently in other neoplasms. Normal tissues express limited quantities of COTA.

Antisomatostatin antibody is a rabbit antibody that can be used for the immunohistochemical staining of somatostatin in tumors and hyperplasias of pancreatic islets.

Antipancreatic polypeptide (PP) antibody is a polyclonal antibody that detects pancreatic polypeptide in routinely fixed paraffin embedded or frozen tissue sections. Hyperplasia of pancreatic polypeptide-containing cells (PP cells) is often seen in patients with juvenile diabetes, chronic pancreatitis, and islet cell tumors. Hyperplasia of PP cells (greater than 10% of the islet cell population) in the nontumoral pancreas has been observed in nearly 50% of islet cell tumors. Demonstration of increased numbers of cells secreting pancreatic polypeptide found both within the islets and between the islets is characteristic of type II hyperplasia of pancreatic islets.

Monoclonal antiinsulin antibody is an antibody used for the immunohistochemical localization of the polypeptide hormone insulin that is the most reliable means to accurately characterize the functional repertoire of islet cell tumors. Islet cell neoplasms of the pancreas appear as solitary or multiple circumscribed lesions that contrast sharply with the neighboring pancreatic parenchyma. These tumors are grouped on the basis of their predominant secretory hormone. This monoclonal antibody is used for the specific and qualitative localization of insulin in routinely fixed paraffin-embedded or frozen tissue sections.

Antihuman thyroid-stimulating hormone (TSH) is a rabbit antibody used for the immunochemical detection of thyroid stimulating hormone (TSH) in thyrotrophic cells and in certain pituitary tumors.

Antiparathyroid hormone (PTH) antibody is a polyclonal antibody against parathyroid hormone (PTH). PTH controls the concentration of calcium and phosphate ions in the blood. A decrease in blood calcium stimulates the parathyroid gland to secrete PTH, which acts on cells of bone, increasing the number of osteoclasts and leading to absorption of the calcified bone matrix and the release of calcium into the blood. Hyperparathyroidism may be caused by adenomas, rarely by carcinomas and by ectopic PTH production. PTH is released by renal adenocarcinomas as well as by squamous cell cancers of the bronchus.

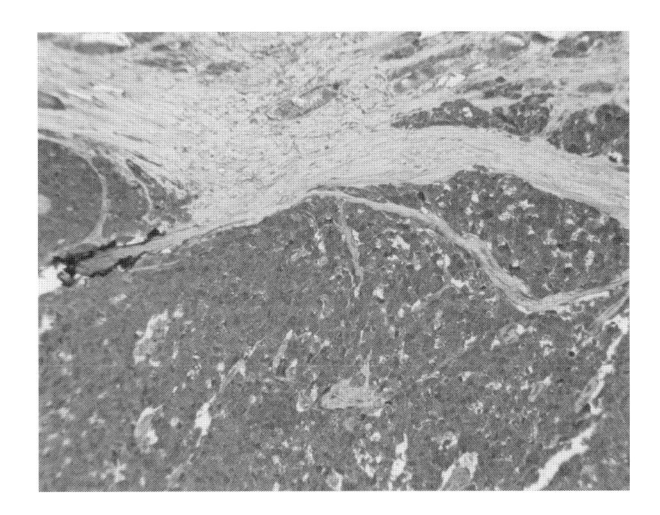

FIGURE 28.22 Calcitonin — medullary carcinoma of the thyroid.

Calcitonin (Figure 28.22) is a hormone that influences calcium ion transport. Immunoperoxidase staining demonstrates calcitonin in thyroid parafollicular or C cells. It serves as a marker characteristic of medullary thyroid carcinoma and APUD neoplasms. Lung and gastrointestinal tumors may also form calcitonin.

Antihuman gastrin is a rabbit antibody that labels G-cells of antropyloric mucosa of the stomach. It permits immunohistochemical detection of gastrin-secreting tumors and G-cell hyperplasia.

Antihuman thyroglobulin is a rabbit antibody that reacts with human thyroglobulin. It labels the cytoplasm of normal and neoplastic thyroid follicle cells. Some staining of colloid may also be observed.

Common leukocyte antigen (LCA) (CD45) is an antigen shared in common by both T and B lymphocytes and expressed, to a lesser degree, by histiocytes and plasma cells. By immunoperoxidase staining, it can be demonstrated in sections of paraffin-embedded tissues containing these cell types. Thus, it is a valuable marker to distinguish lymphoreticular neoplasms from carcinomas and sarcomas (Figure 28.23).

Anti-CD45R (Leukocyte common antigen) is a mouse monoclonal antibody specific for an epitope present on the majority of human leukocytes. This reagent may be used to aid in the identification of cells of lymphocytic lineage. The antibody is intended for qualitative staining in sections of formalin-fixed, paraffin-embedded tissue. It specifically binds to antigens located predominantly in the plasma membrane and to a lesser degree in the cytoplasm of lymphocytes, with variable reactivity to monocytes/histiocytes, and polymorphonuclear leukocytes. Unexpected antigen expression or loss of expression may occur,

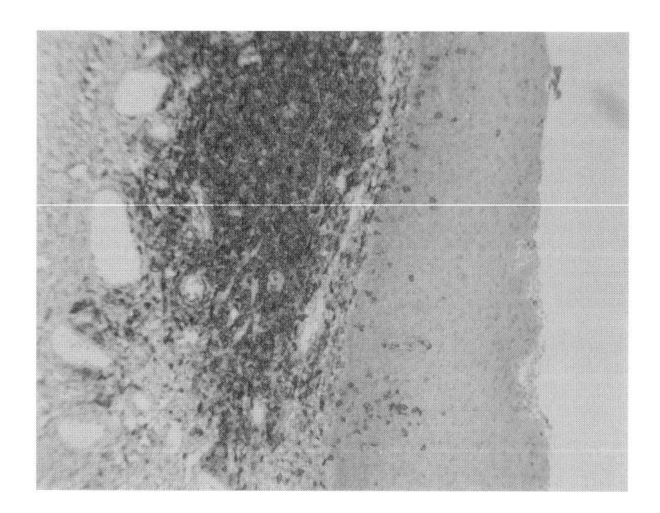

FIGURE 28.23 CD45 — tonsil.

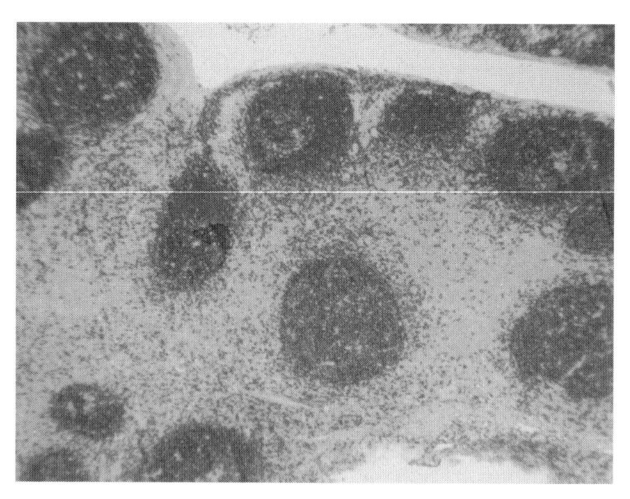

FIGURE 28.24 CD20 — tonsil.

especially in neoplasms. Occasional stromal elements surrounding heavily stained tissues and or cells would show immunoreactivity. The clinical interpretation of any staining or its absence must be complemented by morphological features and evaluation of proper controls.

Anti-T cell (CD45RO) antibody reacts with CD45RO determinant of leukocyte common antigen. It reacts with most T lymphocytes, macrophages, and Langerhans cells of normal tissues. It also reacts with peripheral T cell lymphomas, T cell leukemia, histiocytosis, and monocytic leukemia with mature phenotype. It reacts very rarely with B cell lymphoma and leukemia.

UCHL1 antihuman T cell, CD45RO is a mouse monoclonal antibody that recognizes specifically the 180-kDa isoform of CD45 (leukocyte common antigen). The 180-kDa glycoprotein occurs on most thymocytes and activated T cells, but only a proportion of resting T cells. This antibody and antibodies to the high-molecular-weight form of CD45 (CD45R) seem to define complementary, largely nonoverlapping populations in resting peripheral T cells demonstrating heterogeneity within the CD4 and CD8 subsets. The antibody labels most thymocytes, a subpopulation of resting T cells within both the CD4 and CD8 subsets, and mature activated T cells. Cells of the myelomonocytic series, e.g., granulocytes and monocytes, are also labeled, whereas most normal B cells and NK cells are consistently negative. Weak cytoplasmic staining is however seen in cases of centroblastic and immunoblastic lymphoma.

CD3 antibody has been considered the best all around T-cell marker. This antibody reacts with an antigen present in early thymocytes. The positive staining of this marker may represent a sign of early commitment to the T cell lineage.

Anti-CD1a is a murine monoclonal antibody that reacts with CD1a, a nonpolymorphic MHC class-I related cell surface glycoprotein, expressed in association with β_2 microglobulin. In normal tissues the antibody reacts with cortical thymocytes, Langerhans cells, and interdigitating reticulum cells. It also reacts with thymomas, Langerhans histiocytosis cells (histiocytosis X), and some T cell lymphomas and leukemias. The staining is localized on the membrane.

Anti-CD5 monoclonal antibody detects CD5 antigen, which is expressed in 95% of thymocytes and 72% of peripheral blood lymphocytes. In lymph nodes, the main reactivity is observed in T cells. CD5 antigen is expressed by many T cell leukemias, lymphomas, and activated T cells. CD5 antigen is also expressed on a subset of B cells. CD5 is recommended for the identification of mantle cell lymphomas. Antibodies to CD5 may prove to be of particular use in the detection of T cell acute lymphocytic leukemias (T-ALL), some B cell chronic lymphocytic leukemias (B-CLL), as well as B and T cell lymphomas. CD5 does not react with granulocytes or monocytes.

Ki-67 or -780 are nuclear antigens expressed by both normal and neoplastic-proliferating cells. They are demonstrable by immunoperoxidase staining. A relatively high percentage of positive cells in a neoplasm implies an unfavorable prognosis.

Anti-Ki-67 (MIB) is a mouse monoclonal antibody directed against the Ki-67 nuclear antigen. This reagent may be used to aid in the identification of proliferating cells in normal and neoplastic cell populations. It is intended for qualitative staining in sections of formalin-fixed, paraffin-embedded tissue (some form of antigen enhancement is required for paraffin-embedded samples), frozen tissue, and cytologic preparations. Ki-67 antibody specifically binds to nuclear antigen(s) associated with cell proliferation which is present throughout the active cell cycle (G1, S, G2, and M phases) but absent in resting (G0) cells. Unexpected

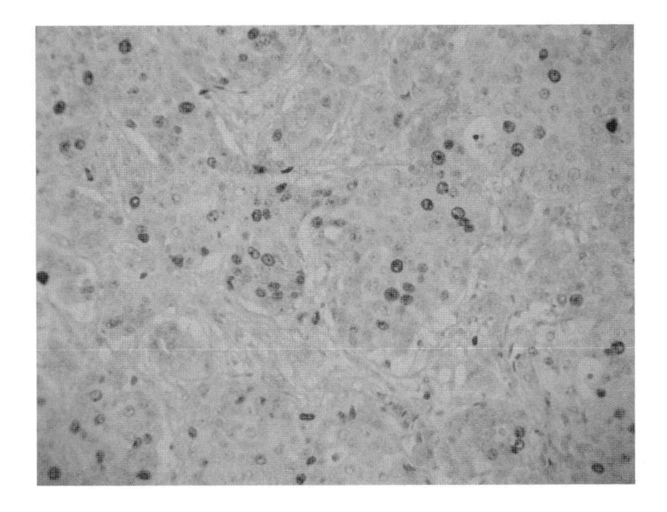

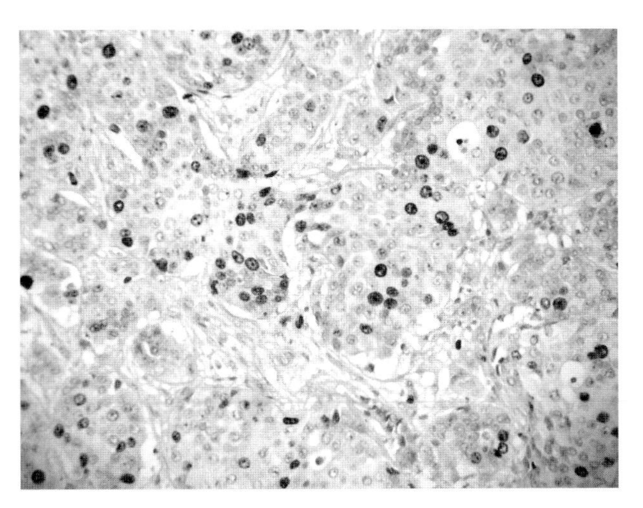

FIGURE 28.25a Ki-67 — Carcinoma of the breast.

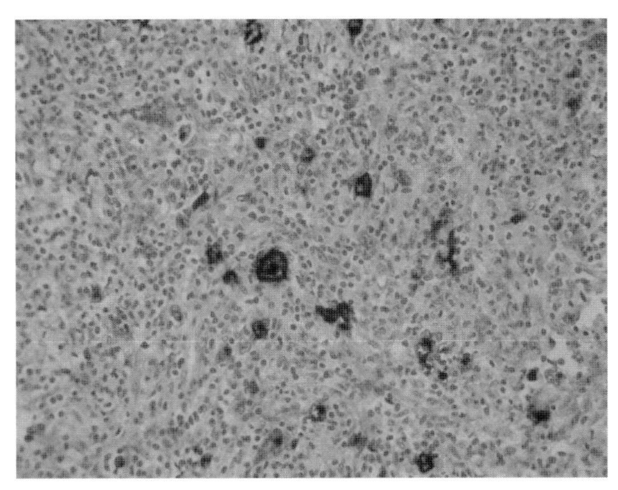

FIGURE 28.26 Hodgkin's disease.

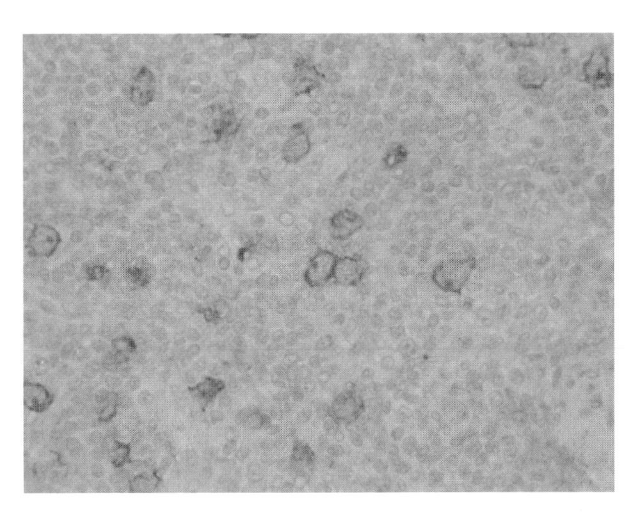

FIGURE 28.27 CD30 — Hodgkin's disease.

antigen expression or loss of expression may occur, especially in neoplasms. Occasionally, stromal elements surrounding heavily stained tissues and or cells will show immunoreactivity. The clinical interpretation of any staining or its absence must be complemented by morphological studies and evaluation of proper controls.

Ki-1 (CD30 antigen) is a marker of Reed-Sternberg cells found in Hodgkin's disease of the mixed-cellularity, nodular-sclerosing, and lymphocyte-depleted types and in selected cases of large-cell non-Hodgkin's lymphomas (Figure 28.27).

Antihuman Ki-1 antigen, CD30 is a mouse monoclonal antibody that reacts with a 595-amino acid transmembrane, 121-kDa glycoprotein. It contains six cysteine-rich motifs in the extracellular domain and is homologous to members of the nerve growth factor receptor superfamily. The CD30 gene was assigned to the short arm of chromosome 1 at position 36. The CD30 antigen was initially designated Ki-1. The antibody detects a formalin-resistant epitope on the 90-kDa precursor molecule. This molecule is processed in the Golgi system into the membrane-bound phosphorylated mature 120-kDa glycoprotein and into the soluble 85-kDa form of CD30, which is released from the supernatant and appears in serum at detectable levels in conditions such as infectious mononucleosis or neoplastically amplified CD30-positive blasts. The CD30 antigen is expressed by Hodgkin's and Reed-Sternberg cells in Hodgkin's disease, by the tumor cells of a majority of anaplastic large-cell lymphomas, and by a varying proportion of activated T and B cells. It is also expressed on embryonal carcinomas.

CD15 (Leu M1) is a monoclonal antibody that recognizes the human myelomonocytic antigen lacto-*N* fucopentose III. It is present on greater than 95% of mature peripheral blood eosinophils and neutrophils and is present at low density on circulating monocytes. In lymphoid tissue, CD15 reacts with Reed-Sternberg cells of Hodgkin's disease and with granulocytes. CD15 reacts with few tissue macrophages and does not react with dendritic cells.

Leu-M1 (CD15) (Figure 28.26) is a granulocyte-associated antigen. Immunoperoxidase staining detects this marker on myeloid cells but not on B or T cells, monocytes, erythrocytes, or platelets. It can be detected in Hodgkin's cells and Reed-Sternberg cells.

CD20 primary antibody (Figure 28.24) is a mouse monoclonal antibody (Clone L26) directed against an intracellular epitope of the CD20 antigen present on human B lymphocytes. This reagent may be used to aid in the identification of cells of B lymphocytic lineage. The antibody is intended for qualitative staining in sections of formalin-fixed paraffin-embedded tissue. Anti-CD20 antibodies specifically bind to antigens located in the plasma membrane and cytoplasmic regions of normal B lymphocytes which may also be expressed in Reed-Sternberg cells. Unexpected antigen expression or loss of expression may occur, especially in neoplasms. Occasionally, stromal elements surrounding heavily stained tissue and/or cells may show immunoreactivity. The clinical interpretation of any staining, or its absence, must be complemented by morphological studies and evaluation of proper controls.

The **CD21 antigen** is a restricted B cell antigen expressed on mature B cells. The antigen is present at high denisty on follicular dendritic cells (FDC), the accessory cells of the B zones. It shows moderate labeling of B cells and a strong labeling of FDC in cryostat sections, whereas the staining of B cells is reduced or abolished in paraffin sections. However, the labeling of FDC in paraffin sections is as strong as on cryostat sections.

The antibody reacts with FDC meshwork in normal and hyperplastic lymph nodes and tonsils. Sharply defined, dense meshwork of FDC in germinal centers is revealed. Follicular mantles of secondary and primary follicles show a loosely textured and ill-defined meshwork of FDC.

Immunohistological analysis of FDC in paraffin sections of non-Hodgkin's lymphomas demonstrates a nodular and usually a dense and sharply defined FDC meshwork in follicular lymphomas (e.g., centroblastic/centrocytic lymphoma) and a loose, ill-defined FDC meshwork of varying size in some diffuse lymphoma types (e.g., centrocytic lymphoma). Precursor B cell lymphomas (lymphoblastic lymphomas), Burkitt's lymphomas, plasmacytomas, and hairy cell leukemias constantly lack FDC. FDC in non-Hodgkin's lymphomas is mainly restricted to peripheral T cell lymphomas of angioimmunoblastic lymphadenopathy (AILD) type and some cases of pleomorphic T cell lymphomas. The FDC meshwork in AILD contains constant hyperplastic venules in contrast to pleomorphic T cell lymphomas. In contrast to B cell lymphomas, the FDC meshworks in T cell lymphomas and AILD contain only a relatively small number of B cells.

CD10 is a mouse monoclonal antibody that reacts with common acute lymphoblastic leukemia antigen (CALLA/CD10) as a useful marker for the characterization of childhood leukemia and B cell lymphomas. This antibody reacts with antigen of lymphoblastic, Burkitt's, and follicular lymphomas; and chronic myelocytic leukemia. Also, CD10 detects the antigen of glomerular epithelial cells and the brush border of the proximal tubules. This characteristic may be helpful in interpreting renal ontogenesis in conjunction with other markers. Other nonlymphoid cells that are reactive with CD10 are breast myoepithelial cells, bile canaliculi, neutrophils, and small population of bone marrow cells, fetal small intestine epithelium, and normal fibroblasts.

Common acute lymphoblastic leukemia antigen (CALLA/CD10) is a useful marker for the characterization of childhood leukemia and B cell lymphomas. This antibody reacts with antigen of lymphoblastic, Burkitt's, and follicular lymphomas; and chronic myelocytic leukemia. Also, CD10 detects the antigen of glomerular epithelial cells and the brush border of the proximal tubules; this characteristic may be helpful in interpreting renal ontogenesis in conjunction with other markers. Other nonlymphoid cells that are reactive with CD10 are breast myoepithelial cells, bile canaliculi, neutrophils, and small population of bone marrow wells, fetal small intestine epithelium, and normal fibroblasts.

Cyclin D1 (polyclonal), rabbit: Anti-Cyclin D1 is a rabbit polyclonal antibody that detects Cyclin D1, one of the key cell-cycle regulators that is a putative protooncogene overexpressed in a wide variety of human neoplasms. Cyclins are proteins that govern transitions through distinct phases of the cell cycle by regulating the activity of the cyclin-dependent kinases. In mid to late G1, Cyclin D1 shows a maximum expression following growth factor stimulation. Cyclin D1 has been successfully employed and is a promising tool for further studies in both cell-cycle biology and cancer-associated abnormalities. This antibody is useful for separating mantle cell lymphomas (Cyclin D1-positive) from SLLs and small cleaved-cell lymphomas (Cyclin D1-negative).

Antihuman kappa light chain (Figure 28.28) is a rabbit antibody that reacts with free kappa light chains as well as kappa chains in intact immunoglobulin molecules. This antibody may be used for typing of free and bound monoclonal light chains by immunoelectrophoresis and immunofixation. It may also be used for immunohistochemistry.

Anti-LN1is a mouse monoclonal antibody against a sialoglycan antigen (CDw75) on cell membranes. In lymphoid tissues, the antibody reacts strongly with the B lymphocytes in the germinal centers, but only faintly with B lymphocytes of the mantle zone. No reaction is observed

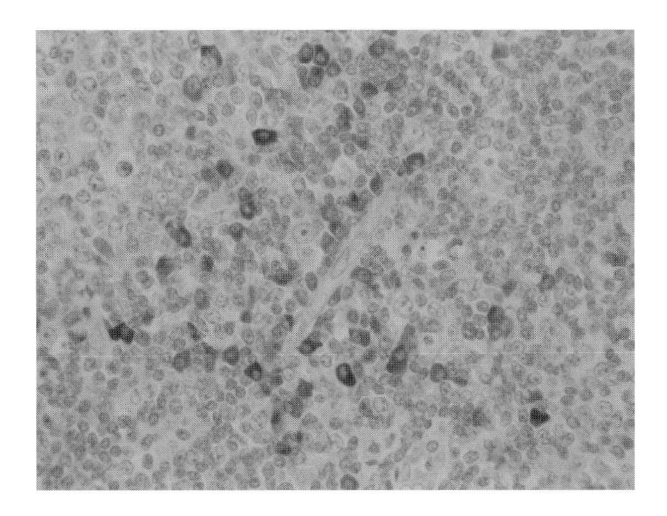

FIGURE 28.28 Kappa light chain — tonsil.

with T lymphocytes. LN1 also reacts with certain epithelial cells, including cells of the distal renal tubules, breast, bronchus, and prostate.

Immunoglobulin is demonstrable by immunoperoxidase staining of plasma cell and B lymphocyte cytoplasm in frozen or paraffin-embedded sections. B-5 fixative is preferable to formalin for demonstration of intracellular IgG or light chains in paraffin sections. Monoclonal cytoplasmic staining for either κ or λ light chains aids the diagnosis of B cell lymphomas.

Antihuman lambda light chain is a rabbit antibody that reacts with free lambda light chains as well as the lambda light chains in intact immunoglobulin molecules.

Anti-bcl-2 primary antibody contains a mouse monoclonal antibody. The bcl-2 oncoprotein expression is inhibited in germinal centers where apoptosis forms a part of the B cell production pathway. In 90% of follicular lymphomas, a translocation occurs which justaposes the bcl-2 gene at 19q21 to an immunoglobulin gene, with subsequent deregulation of protein synthesis and cell proliferation. The bcl-2 product is considered to act as an inhibitor of apoptosis. This observation has turned out to have clinical implications. Distinction of follicular hyperplasia from follicular lymphoma is a common problem in histopathology. Reactive follicles show no staining for bcl-2, whereas the cells in neoplastic follicles exhibit membrane staining.

Anti-BCL-6 (PG-B6p) mouse monoclonal antibody: This is a transcriptional regulator gene which codes for a 706-amino-acid nuclear zinc finger protein. Antibodies to this protein stain the germinal center cells in lymphoid follicles, the follicular cells and interfollicular cells in follicular lymphoma, diffuse large B cell lymphomas, and Burkitt's lymphoma, and the majority of the Reed-Sternberg cells in nodular lymphocyte predominant Hodgkin's disease. In contrast, anti-BCL-6 rarely stains mantle cell lymphoma, and MALT lymphoma bcl-6 expression is seen in approximately 45% of CD30$^+$ anaplastic large cell lymphomas but is consistently absent in other peripheral T cell lymphomas.

CD23(1B12): Anti-CD23 mouse monoclonal antibody is a B cell antibody that is useful in differentiating between B-CLL and B-SLLs that are CD23 positive from mantle cell lymphomas and small cleaved lymphomas that are CD23 negative. This antibody reacts with the antigen that is found on a subpopulation of peripheral blood cells, B lymphocytes, and on EBV transformed B-lymphoblastoid cell lines.

CD31 (JC/70A): Anti-CD31 mouse monoclonal antibody detects CD31 expressed by stem cells of the hematopoietic system and is primarily used to identify and concentrate these cells for experimental studies as well as for bone marrow transplantation. Endothelial cells also express this marker; therefore, antibodies to CD31 have been used as a tool to identify the vascular origin of neoplasms. CD31 has shown to be highly specific and sensitive for vascular endothelial cells. Staining of nonvascular tumors (excluding hematopoietic neoplasms) has not been observed.

Anti-CD34 (Figure 28.29) is a murine monoclonal antibody, raised by immunization with human placental endothelial cells, that has a specificity for the CD34 glycoprotein, which is considered the earliest known CD marker and is expressed on virtually all human hematopoietic progenitor cells.

Anti-CD43 is a murine monoclonal antibody directed against an epitope present on human monocytes, granulocytes, and lymphocytes. This reagent may be used to aid in the identification of cells of lymphoid lineage. It is

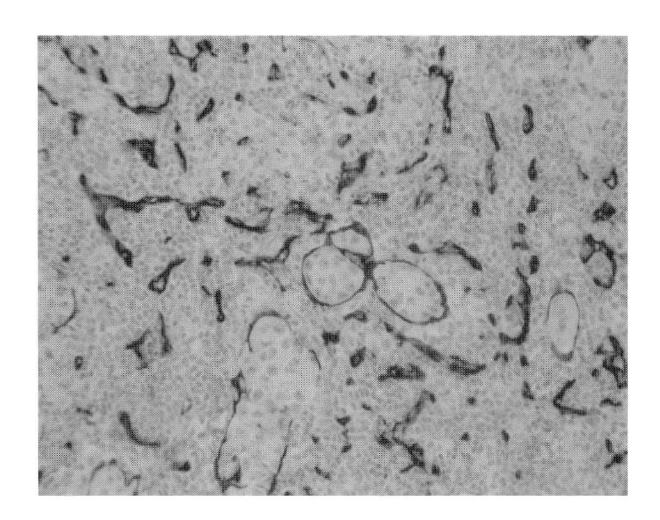

FIGURE 28.29 CD34 — highly vascular tumor.

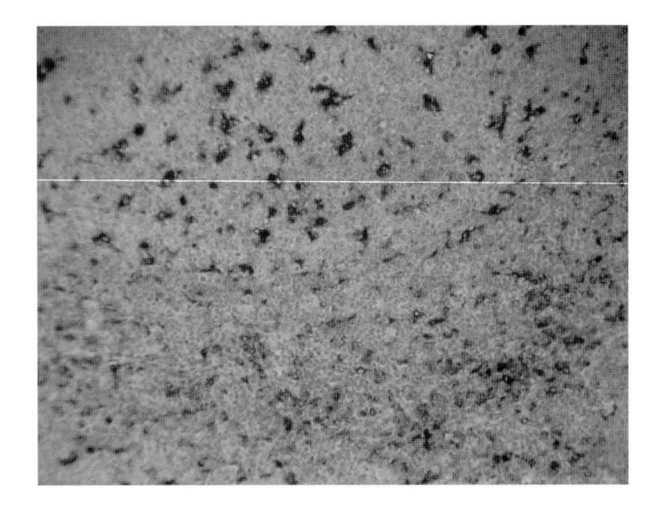

FIGURE 28.30 CD68 — tonsil.

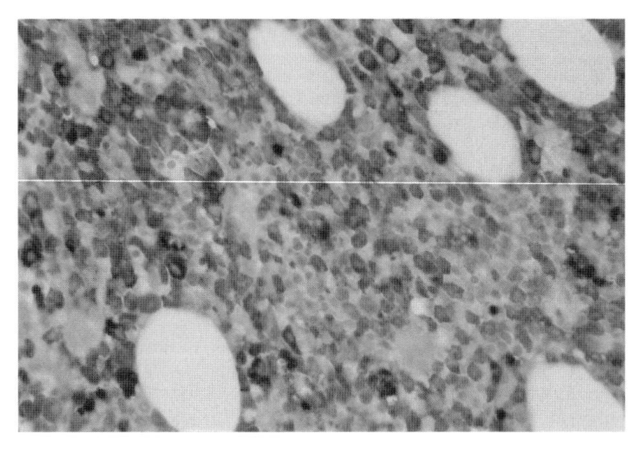

FIGURE 28.31 Myeloperoxidase — bone marrow.

intended for qualitative staining in sections of formalin-fixed, paraffin-embedded tissue. Anti-CD43 antibody specifically binds to antigen located in the plasma membrane and cytoplasmic regions of normal granulocytes or T lymphocytes.

Anti-CD68 (human macrophage marker) (Figure 28.30) is a murine monoclonal antibody that stains macrophages and a wide variety of human tissues, including Kupffer cells and macrophages in the red pulp of the spleen, in the lamina propria of the gut, in lung alveoli, and in bone marrow. Antigen-presenting cells, such as Langerhans cells, are either negative or show weak and/or restricted areas of reactivity, e.g., interdigitating reticulum cells. Resting microglia in the normal white matter of the cerebrum and microglia in areas of infarction react with the antibody. Peripheral blood monocytes are also positive, with a granular staining pattern. The antibody reacts with myeloid precursors and peripheral blood granulocytes. The antibody also reacts with the cell population known as "plasmacytoid T cells" which are present in many reactive lymph nodes and which are believed to be of monocyte/macrophage origin. The antibody stains cases of chronic and acute myeloid leukemia, giving strong granular staining of the cytoplasm of many cells, and also reacts with rare cases of true histiocytic neoplasia. The positive staining of normal and neoplastic mast cells is seen with the antibody as well as staining of a variable number of cells in malignant melanomas. Neoplasms of lymphoid origin are usually negative, although some B cell neoplasms, most frequently small lymphocytic lymphoma and hairy cell leukemia, show weak staining of the cytoplasm, usually in the form of a few scattered granules.

CD99 (HO36-1.1): Anti-CD99 mouse monoclonal antibody reacts with MIC-2 antigen present on the cell membrane of Ewing's sarcoma and primitive peripheral neuroectodermal tumors (PNET). It is also present on some bone marrow, lymph nodes, spleen, cortical thymocytes, granulosa cells of the ovary, most beta cells, CNS ependymal cells, Sertoli's cells of the testis, and a few endothelial cells. MIC-2 has also been identified in lymphoblastic lymphoma, rhabdomyosarcoma, mesenchymal chondrosarcoma, and thymoma.

CD117 (c-kit) (polyclonal), rabbit: Anti-CD117 is a purified immunoglobulin fraction of rabbit antiserum that detects CD117, a tyrosine kinase receptor found on interstitial cells of Cajal, germ cells, bone marrow stem cells, melanocytes, breast epithelium, and mast cells. This receptor is found on a wide variety of tumor cells (follicular and papillary carcinoma of thyroid; adenocarcinomas from endometrium, lung, ovary, pancreas, and breast; and malignant melanoma, endodemal sinus tumor, and small cell carcinoma) but has been particularly useful in differentiating gastrointestinal stromal tumors from Kaposi's sarcoma and tumors of smooth muscle origin.

Antihuman myeloperoxidase antibody (Figure 28.31) is an antibody that is used to discriminate between lymphoid leukemias and myeloid leukemias in formalin-fixed paraffin-embedded tissues.

Antihuman luteinizing hormone (LH) is a rabbit antibody that labels gonadotropic cells of the pituitary. Positive staining for adenohypophyseal hormones assists in classification of pituitary tumors. Lutenizing hormone (LH) is an adenohypophyseal glycoprotein hormone found in gonadotropic cells of the anterior pituitary gland of most mammals. Gonadotropic cells average about 10% of anterior pituitary cells.

α-1 antichymotrypsin is a histiocytic marker. By immunoperoxidase staining, it is demonstrable in tumors derived from histiocytes. It may also be seen in various carcinomas.

Factor VIII (Figure 28.32) is a coagulation protein produced by endothelial cells, which makes it a useful marker

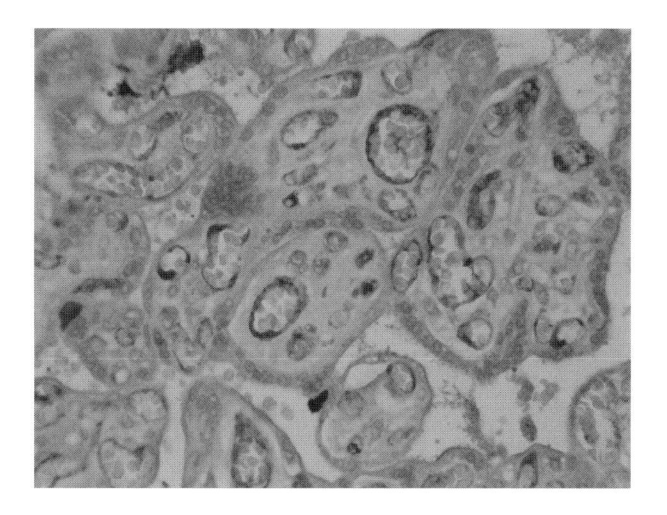

FIGURE 28.32 Factor VIII — placenta.

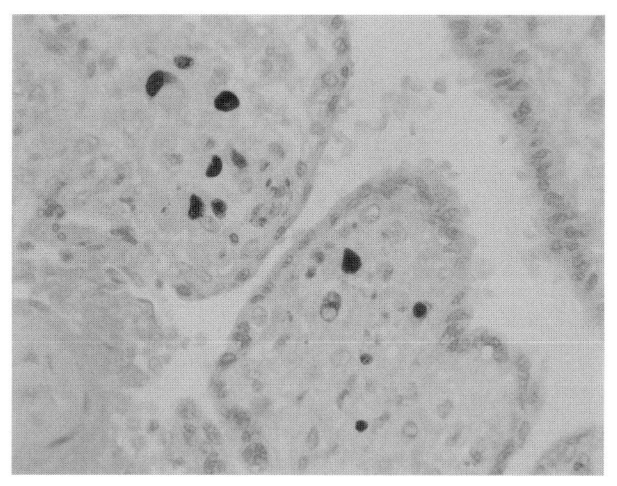

FIGURE 28.33 Cytomegalovirus (CMV) — placenta.

for vascular tumors. It is demonstrable by immunoperoxidase staining. Megakaryocytes and platelets also stain for factor VIII.

Antifactor VIII is a mouse monoclonal antibody that gives positive staining in the cytoplasm of normal vascular endothelial cells of arteries, veins, capillaries, and endocardial cells. Factor VIII related antigen is also present in megakaryocytes and platelets.

Antihuman hemoglobin is a rabbit antibody against hemoglobin A, isolated from erythrocytes of normal adults, that reacts with hemoglobin A and, due to a common alpha chain, also with hemoglobin A_2 and hemoglobin F.

200-4 nuclear matrix protein is a marker expressed preferentially by malignant cells rather than by normal cells. It is demonstrated by immunoperoxidase staining.

Anticytomegalovirus antibody (Figure 28.33) is a mouse monoclonal antibody that reacts with CMV infected cells giving a nuclear staining pattern with early antigen and a nuclear and cytoplasmic reaction with the late viral antigen. The antibody shows no crossreactivity with other herpesviruses or adenoviruses.

Antihepatitis B virus core antigen (HBcAg) antibody labels the nuclei and occasionally cytoplasm of virus infected cells. HBcAg is expressed predominantly in the nuclei of infected liver cells, although variable staining may also be seen in the perinuclear cytoplasm.

Antipapillomavirus is a rabbit antibody against papillomavirus. The structural antigens on this virus can be detected in a variety of proliferative squamous lesions. Only 50 to 60% of lesions caused by papillomavirus will express the structural antigens. This antibody staining is predominantly intranuclear in a focal or diffuse pattern, although perinuclear cytoplasmic staining of koilocytotic cells may also be seen.

Collagen Type IV (CIV22): Anti-Collagen Type IV is a mouse monoclonal antibody that detects Collagen Type IV, the major component of the basal lamina. Antibodies to this molecule confirm its presence and reveal the morphological appearance of the structure. Normal tissue stains with this antibody in a fashion consistent with the sites of mesenchymal elements and epithelial basal laminae. Collagen IV can also be useful in the classification of soft tissue tumors, Schwanomas, and leiomyomas, and their well-differentiated malignant counterparts usually immunoreact to this antibody. The vascular nature of neoplasma, hemangiopericytoma, angiosarcoma, and epitheliod hemangioendothelioma can be revealed by this antibody with greater reliability than nonspecific stains (e.g., silver reticulum).

Polyclonal rabbit anti-calretinin is intended to qualitatively detect normal and malignant mesothelial cells in formalin-fixed, paraffin-embedded tissue sections using light microscopy. Calretinin, a calcium-binding protein with a mol wt of 29 kDa, is a member of the large family of EF-hand proteins that also include S-100 protein. EF-hand proteins are characterized by a helix–loop–helix fold that acts as the calcium-binding site. Calretinin contains six such EF-hand stretches. It is abundantly expressed in central and peripheral neural tissues, especially in the retina and neurons of the sensory pathways. Calretinin is also consistently expressed in normal and reactive mesothelial cell lining of all serosal membranes, eccrine glands of skin, convoluted tubules of kidney, Leydig and Sertoli cells of the testis, endometrium and ovarian stromal cells, and adrenal cortical cells. Calretinin is also a sensitive and specific indicator of normal and reactive mesothelial cells in effusion cytology. This antibody is useful as part of an immunohistochemical marker panel to distinguish mesothelioma from adenocarcinoma. The combination of calretinin and E-Cadherin was shown to have high sensitivity and specificity in differentiating malignant mesothelioma from metastatic adenocarcinoma to the pleura in one study.

P63 (ap53 Homolog at 3q27–29) Ab-4 (cocktail) mouse monoclonal antibody recognizes a 63-kDa protein, identified as p63. Ths p63 gene, a homolog of the tumor-suppressor p53, is highly expressed in the basal or progenitor layers of many epithelial tissues. Protein p63 shows remarkable structural similarity to p53 and to the related p73 gene. Unlike p53, the p63 gene encodes multiple isotypes with remarkable divergent abilities to transactivate p53 reporter genes and induce apoptosis. NeoMarkers' Ab-4 recognizes all known isotypes of p63.

TTF-1 (8G7G3/a), mouse: Thyroid transcription factor-1 is useful in differentiating primary adenocarcinoma of the lung from metastatic carcinomas from the breast and malignant mesothelioma. It can also be used to differentiate small-cell lung carcinoma from lymphoid infiltrates.

Myogenin (F5D), mouse: Anti-Myogenin monoclonal antibody labels the nuclei of myoblasts in developing muscle tissue and is expressed in tumor cell nuclei of rhabdomyosarcoma. Positive nuclear staining may occur in Wilms' tumor, as well as in some myopathies.

Neurofilament (2F11), mouse: Neurofilament antibody stains an antigen localized in a number of neural, neuroendocrine, and endocrine tumors. Neuromas, ganglioneuromas, gangliogliomas, ganglioneuroblastomas, and neuroblastomas stain positively for neurofilament. Neurofilament is also present in paragangliomas and adrenal and extra-adrenal pheochromocytomas. Carcinoids, neuroendocrine carcinomas of the skin, and oat cell carcinomas of the lung also express neurofilament.

Anti-p53 primary antibody (clone Bp53-11) is a mouse monoclonal antibody directed against both the mutant and wild-type of the p53 nuclear phosphoprotein. Very rare normal cells express p53, but alterations in the p53 suppressor gene result in an overproduction of this protein in malignancies. This reagent may be used to aid in the identification of abnormally proliferating cells in neoplastic cell populations. The antibody is intended for qualitative staining in sections of formalin-fixed, paraffin-embedded tissue on a Ventana automated slide staining device. Some form of antigen enhancement is required for paraffin-embedded samples. The p53 antibody specifically binds to nuclear antigen(s) associated with the normal downregulation of cell division. Increased expression of p53 in actively dividing cells is an indication of loss of function due to mutation of the p53 gene.

Inhibin, alpha (R1), mouse: Anti-Inhibin alpha is an antibody against a peptide hormone which has demonstrated utility in differentiation between adrenal cortical tumors and renal cell carcinoma. Sex cord stromal tumors of the ovary, as well as trophoblastic tumors, also demonstrate cytoplasmic positivity with this antibody.

MART-1 (M2-7C10), mouse: MART-1 (also known as Melan A) is a melanocyte differentiation antigen. It is present in melanocytes of normal skin and retina, nevi, and in more than 85% of melanomas. This antibody is very useful in establishing the diagnosis of metastatic melanomas.

BRST-2 (GCDFP-15), monoclonal antibody (murine) detects BRST-2 antigen expressed by apocrine sweat glands, eccrine glands (variable), minor salivary glands, bronchial glands, metaplastic epithelium of the breast, benign sweat gland tumors of the skin, and the serous cells of the submandibular gland. Breast carcinomas (primary and metastatic lesions) with apocrine features express the BRST-2antigen. BRST-2 is positive in extramammary Paget's Disease. Other tumors tested are negative.

E-Cadherin (ECH-6), mouse: Anti-E-Cadherin mouse monoclonal antibody detects E-Cadherin, an adhesion protein expressed in cells of epithelial lineage. It stains positively in glandular epithelium as well as adenocarcinomas of the lung and GI tract and ovary. It has been useful in distinguishing adenocarcinoma from mesothelioma. It has also been shown to be positive in some thyroid carcinomas.

Fascin (55k-2), mouse: Fascin is a very sensitive marker for Reed-Sternberg cells and variants in nodular sclerosis, mixed cellularity, and lymphocyte depletion Hodgkin's disease. It is uniformly negative in lymphoid cells, plasma cells, and myeloid cells. Fascin is positive in dendritic cells. This marker might be helpful in distinguishing between Hodgkin's disease and non-Hodgkin's lymphoma in difficult cases. Also, the lack of expression of fascin in the neoplastic follicles in follicular lymphoma can be helpful in distinguishing these lymphomas from reactive follicular hyperplasia in which the number of follicular dendritic cells is normal or increased.

Index

A

A blood group, 449
α Chain, 192
α-Fetoprotein, 617
α Heavy-chain disease, 529
α helix, 41
α-1 Antichymotrypsin, 794
α-1 Antitrypsin (A1AT), 41
α1-Microglobulin, 503
α2 Macroglobulin (α2M), 503
α2-Plasmin inhibitor-plasmin complexes (α2PIPC), 416, 423
AA Amyloid, 525
AB Blood group, 448
αβ T cell receptor (αβ TCR), 273
αβ T cells, 54, 273
ABC method, 746
Abelson murine leukemia virus (A-MuLV), 171, 662
Ablastin, 683
ABO blood group antigen, 449
ABO blood group substances, 447
ABO blood group system, 448
Abrin, 107, 625
Absorption, 723
Absorption elution test, 449
Abzyme, 215
Acanthosis nigricans, 500
Accessory molecules, 64
Acetaldehyde adduct autoantibodies, 398
Acetylcholine receptor (AChR) antibodies, 407, 485
Acetylcholine receptor (AChR) binding autoantibodies, 407
Acquired agammaglobulinemia, 540
Acquired B antigen, 449
Acquired C1 inhibitor deficiency, 552
Acquired immune deficiency syndrome (AIDS), 555
Acquired immunity, 96, 643, 644
Acquired immunodeficiency, 554, 555
Acquired tolerance, 381
Acridine orange, 773
ACT-2, 41
Actin, 781
Activated leukocyte cell adhesion molecule (ALCAM/ CD166), 28
Activated lymphocyte, 50
Activated lymphocytes, 279
Activated macrophage, 61
Activation, 53
Activation phase, 53
Activation protein-1 (AP-1), 52
Activation unit, 322, 328
Activation-induced cell death (AICD), 54
Active anaphylaxis, 353
Active immunity, 96
Active immunization, 119
Active kinins, 350
Active site, 195
Acute AIDS syndrome, 555
Acute disseminated encephalomyelitis, 509
Acute graft-vs.-host reaction, 607
Acute inflammation, 94
Acute inflammatory response, 94
Acute lymphoblastic leukemia (ALL), 470
Acute myelogenous leukemias (AML), 474
Acute phase proteins, 96
Acute poststreptococcal glomerulonephritis, 503
Acute rejection, 604
Acute rheumatic fever, 493
Acute-phase reactants, 96
Acute-phase response (APR), 95
Acute-phase serum, 643
Acyclic adenosine monophosphate (cAMP), 75
Acyclic guanosine monophosphate (cGMP), 75
Acyclovir 9(2-hydroxyethoxy-methylguanine), 669
Adaptive differentiation, 257
Adaptive immune response, 643
Adaptive immunity, 643
Adaptor proteins, 43, 280
ADCC (antibody-dependent cell-mediated
cytotoxicity), 639
Addison's disease, 433
Addressin, 38
Adenoids, 79
Adenosine, 75
Adenosine deaminase (ADA), 255
Adenosine deaminase (ADA) deficiency, 545
Adenoviruses, infection, and immunity, 670
Adherent cell, 62
Adhesins, 41
Adhesion molecule assays, 28
Adhesion molecules, 28
Adhesion receptors, 29
Adjuvant, 122
Adjuvant disease, 520
Adjuvant granuloma, 122
Adoptive immunity, 596
Adoptive immunization, 597
Adoptive tolerance, 380
Adoptive transfer, 597
Adrenal autoantibody (AA), 404
Adrenergic receptor agonists, 307
Adrenergic receptors, 55
Adrenocorticotrophic hormone (ACTH) antibody, 784
Adsorption, 723
Adsorption chromatography, 723
Adult respiratory distress syndrome (ARDS), 489
Adult T cell leukemia–lymphoma (ATLL), 489
AE1/AE3 pan-cytokeratin monoclonal antibody, 778
AET rosette test (historical), 736
Afferent lymphatic vessels, 82
Affinity, 229
Affinity chromatography, 724
Affinity constant, 230
Affinity maturation, 214

Agammaglobulinemia (hypogammaglobulinemia), 536
Agar gel, 723
Agarose, 734
Agglutination, 245
Agglutination inhibition, 246
Agglutination titer, 246
Agglutinin, 246
Agglutinogen, 246
Aggregate anaphylaxis, 354
Aging and immunity, 416
Agonist, 55
Agonist peptides, 53
Agranulocytosis, 469
Agretope, 153
Agrin, 148
AH50, 341
AIDS, 667
AIDS (acquired immune deficiency syndrome), 555
AIDS belt, 556
AIDS dementia complex, 562
AIDS embryopathy, 559
AIDS encephalopathy, 562
AIDS enteropathy, 562
AIDS serology, 555
AIDS treatment, 562, 563
AIDS vaccines, 695
AIDS virus, 557
AIDS-related complex (ARC), 561
AILA, 476
Alanyl-tRNA synthetase autoantibodies, 409
Albumin agglutinating antibody, 456
Alexine (alexin), 323
ALG, 602
Allele, 102
Allelic dropout, 102
Allelic exclusion, 168, 223
Allelic variants, 140
Allergen, 359
Allergen immunotherapy, 353
Allergic alveolitis, 347
Allergic asthma, 363
Allergic conjunctivitis, 359
Allergic contact dermatitis, 374, 478
Allergic disease immunotherapy, 353
Allergic granulomatosis, 347
Allergic orchitis, 347
Allergic reaction, 360
Allergic response, 360
Allergic rhinitis, 359
Allergoids, 360
Allergy, 360
Alloantibody, 200
Alloantigen, 107
Alloantiserum, 601
Allogeneic (or allogenic), 601
Allogeneic bone marrow transplantation, 594
Allogeneic disease, 601
Allogeneic effect, 600
Allogeneic graft, 594
Allogeneic inhibition, 595
Allograft, 593
Allogroup, 200
Alloimmunization, 466, 601
Allophenic mouse, 103
Alloreactive, 600
Alloreactivity, 600
Allotope, 200, 202
Allotransplant, 593
Allotype, 200
Allotype suppression, 201
Allotypic determinant, 201
Allotypic marker, 202
Allotypic specificities, 202
Allotypy, 202
Alopecia areata, 416, 517
ALS (antilymphocyte serum) or ALG (anti-lymphocyte globulin), 602
Altered peptide ligands (APL), 703
Altered self, 388
Alternaria species, 347
Alternative complement pathway, 333
Alternative pathway C3 convertase, 334
Alum granuloma, 123
Aluminum adjuvant, 123
Aluminum hydroxide gel, 124
Alum-precipitated antigen, 123
Alum-precipitated toxoid, 123
Alums, 122
ALVAC, 565
Alveolar basement membrane (ABM) autoantibodies, 403
Alveolar macrophage, 65
ALZ-50, 784
Am allotypic marker, 202
Amboceptor (historical), 175
Amino acyl tRNA synthetases, 409
Aminoethylcarbazole (AEC), 748
Aminophylline, 705
Ammonium sulfate method, 236
Ammonium sulfate precipitation, 727
Amphibian immune system, 713
Amphiphysin autoantibodies, 411
Amphiregulin, 615
Amyloid, 524
Amyloid β fibrillosis, 532
Amyloid P component, 525
Amyloidosis, 523
Antiidiotypic vaccine, 216
ANA, 418
ANAE (a-naphthyl acetate esterase), 61
AnaINH, 358
Anamnesis, 122
Anamnestic, 122
Anamnestic immune response, 121
Anaphylactic shock, 349
Anaphylactoid reaction, 355
Anaphylatoxin inactivator, 358
Anaphylatoxin inhibitor (Ana INH), 333, 358
Anaphylatoxins, 332, 358
Anaphylaxis, 348
Anavenom, 686
Anchor residues, 146
Anergic B cells, 382
Anergy, 373, 386
Angioedema, 344
Angiogenesis, 91
Angiogenesis factor, 91, 344
Angiogenic factors, 91
Angiogenin, 92
Angioimmunoblastic lymphadenopathy (AILA), 475
Angiopoietins/Tie2, 92
Angry macrophage, 62
Ankylosing spondylitis, 520
ANNA, 483

Annexin V binding, 79
Annexins (lipocortins), 41
Antagonists, 53
Anti B and T cell receptor idiotype antibodies, 576
Anti PM/Scl autoantibodies, 409
Anti U1 RNP autoantibodies, 432
Antiagglutinin, 175
Antianaphylaxis, 353
Antiantibody, 178
Anti-B cell receptor idiotype antibodies, 165
Anti-bcl-2 primary antibody, 793
Anti-BCL-6 (PG-B6p) mouse monoclonal antibody, 793
Antibodies, 173
Antibodies to histidyl t-RNA synthetase (anti-HRS), 430
Antibodies to Mi-1 and Mi-2, 405
Antibody absorption test, 723
Antibody affinity, 213, 230
Antibody deficiency syndrome, 537
Antibody detection, 176
Antibody excess immune complexes (ABICs), 249
Antibody feedback, 216
Antibody fragment, 195
Antibody half-life, 211
Antibody humanization, 177
Antibody repertoire, 175
Antibody screening, 585
Antibody specificity, 175
Antibody synthesis, 176
Antibody titer, 176
Antibody units, 176
Antibody–antigen intermolecular forces, 230
Antibody-binding site, 175
Antibody-dependent cell-mediated cytotoxicity (ADCC), 364, 622, 640
Antibody-directed enzyme pro-drug therapy (ADEPT), 622
Antibody-mediated suppression, 177
Antibody-secreting cells, 163
Antibroad-spectrum cytokeratin, 777
Anti-BRST-2 (GCDFP-15) monoclonal antibody, 785
Anti-BRST-3 (B72.3) monoclonal antibody, 785
Anticardiolipin antibody syndrome, 416
Anti-CD1a, 790
Anti-CD34, 793
Anti-CD43, 793
Anti-CD45R (leukocyte common antigen), 789
Anti-CD5 monoclonal antibody, 790
Anti-CD68 (human macrophage marker), 794
Anticentriole antibodies, 427
Anticentromere autoantibody, 428
Anti-Clq antibody, 325
Anticomplementary, 342
Anticytomegalovirus (CMV) antibody, 670, 795
Anti-D, 454
Anti-desmin antibody, 781
Anti-DEX antibodies, 176
Anti-double-stranded DNA, 416, 517
Anti-endothelial cell autoantibodies, 431
Anti-epithelial membrane antigen (EMA) antibody, 780
Anti-estrogen receptor antibodies, 785
Anti-Ewing's sarcoma marker (CD99), 783
Anti-factor VIII, 795
Antifibrillarin antibodies, 427
Antigen, 107
Antigen capture assay, 727
Antigen clearance, 116
Antigen excess, 248
Antigen masking, 681
Antigen presentation, 145
Antigen processing, 145
Antigen receptors, 55
Antigen recognition activation motif, 271
Antigen retrieval, 746
Antigen unmasking, 116
Antigen–antibody complex, 237
Antigen-binding capacity, 227
Antigen-binding cell (ABC) assay, 727
Antigen-binding site, 116, 176
Antigenic, 111, 619
Antigenic competition, 116
Antigenic deletion, 619
Antigenic determinant, 105, 111
Antigenic diversion, 619
Antigenic drift, 117
Antigenic gain, 619
Antigenic modulation, 619
Antigenic mosaicism, 117
Antigenic peptide, 108, 146
Antigenic profile, 108
Antigenic reversion, 619
Antigenic shift, 117
Antigenic transformation, 619
Antigenic variation, 116, 667
Antigenicity, 111
Antigen-presenting cell (APC), 149
Antigen-specific cells, 56
Antigen-specific suppressor cells, 120
Antigliadin antibodies (AGA), 495
Anti-glial fibrillary acidic protein (GFAP) antibody, 782
Antiglobulin, 455
Antiglobulin consumption test, 737
Antiglobulin inhibition test, 455
Antiglobulin test, 455
Antiglutinin, 722
Anti-GM_1 antibodies, 783
Antigranulocyte antibodies, 464
Anti-growth hormone (GH) antibody, 784
Antiheat shock protein antibodies, 43
Anti-hepatitis B virus core antigen (HBcAg) antibody, 795
Anti-high molecular weight human cytokeratin antibodies, 779
Antihistamine, 355, 705
Antihistone antibodies, 750
Anti-Hu antibodies, 411
Anti-human α-smooth muscle actin, 781
Anti-human chorionic gonadotropin (HCG) antibody, 788
Anti-human cytokeratin (CAM 5.2) (cytokeratin 8,18), 779
Anti-human cytokeratin 7 antibody, 779
Anti-human cytokeratin-20 monoclonal antibody, 779
Anti–human follicle stimulating hormone (FSH) antibody, 784
Anti-human gastrin, 789
Anti-human glucagon antibody, 788
Anti-human hemoglobin, 795
Anti-human kappa light chain, 792
Anti–human Ki-1 antigen, CD30, 791
Anti-human lambda light chain, 793
Anti-human luteinizing hormone (LH), 794
Anti-human myeloperoxidase antibody, 794
Anti-human prostatic acid phosphatase (PSAP), 787
Anti-human synaptophysin antibody, 782
Anti-human thyroglobulin, 789
Anti-human thyroid stimulating hormone (TSH), 789
Anti-I, 462
Antiidiotypic antibody, 216
Antiidiotypic vaccine, 697

Anti-immunoglobulin antibodies, 178, 179
Anti-intrinsic factor autoantibodies, 396
Anti-Ki-67 (MIB), 790
Anti-Ku autoantibodies, 409
Anti-La/SS-B autoantibodies, 429
Anti-LN1, 792
Anti-low molecular weight cytokeratin, 780
Antilymphocyte serum (ALS) or antilymphocyte globulin (ALG), 574, 602
Antimalignin antibodies, 627
Anti-melanoma primary antibody, 785
Antimetabolite, 574
Anti-muscle actin primary antibody, 781
Anti-myelin-associated glycoprotein (MAG) antibodies, 509
Antineutrophil cytoplasmic antibodies (ANCA), 482, 509
Antineutrophil cytoplasmic antibody, 483
Antineutrophil cytoplasmic autoantibodies (pANCA), 483
Antinuclear antibodies (ANA), 417, 516
Antinucleosome antibodies, 419
Anti-p24, 558
Anti-p53 primary antibody (clone Bp53-11), 796
Anti-pancreatic polypeptide (PP) antibody, 789
Anti-papillomavirus, 795
Anti-parathyroid hormone (PTH) antibody, 789
Anti-PCNA, 418
Antiphospholipid antibodies, 421, 434
Antiphospholipid syndrome (APS), 421
Antiplacental alkaline phosphatase (PLAP) antibody, 406, 786
Antiplatelet antibodies, 464
Anti-progesterone receptor antibody, 786
Anti-prolactin antibody, 784
Anti-prostate specific antigen (PSA) antibody, 788
Anti-Purkinje cell antibody, 783
Anti-RA-33, 424
Anti-Ri antibody, 783
Anti-rRNP, 434
Anti-scRNP (Ro/SS-A, La/SS-B), 430
Antiseptic paint, 441
Antiserum, 175
Anti-Sm (Smith) autoantibodies, 419
Anti-Sm autoantibodies, 418
Anti-snRNP (Sm, U1-RNP, U2-RNP), 419
Anti-somatostatin antibody, 789
Antisperm antibody, 512, 772
Anti-SS-A, 429
Anti-SS-B, 430
Anti-T cell (CD45RO), 790
Anti-T cell receptor idiotype antibodies, 272
Anti-target antigen antibodies, 576
Anti-tau antibodies, 510
Antithymocyte globulin (ATG), 574, 602
Antithymocyte serum (ATS), 574
Anti-topoisomerase I (Scl 70), 522
Antitoxin, 175, 178
Antitoxin assay (historical), 176
Antitoxin unit, 176
Anti-*Toxoplasma gondii* antibody, 742
Antivenom, 176
Anti-vimentin antibody, 781
AP-1, 53
APC, 150
APECED (autoimmune polyendocrinopathy-candidiasis-ectodermal dystrophy), 434
Apheresis, 774
Aplastic anemia, 469
APO-1, 82
Apolar or hydrophobic bonding, 229
Apolipoprotein (APO-E), 41
Apolipoprotein E, 62
Apoptosis, 79, 282
Apoptosis and necrosis, 81
Apoptosis, caspase pathway, 80
Apoptosis, suppressors, 82
Appendix, vermiform, 85
APT (alum-precipitated toxoid), 124
Aquaphor®, 124
Aqueous adjuvants, 124
Arachidonic acid (AA) and leukotrienes, 349
Arenavirus immunity, 670
Arlacel® A, 124
Armed effector T cells, 276
Armed macrophages, 63
Arthus reaction, 367
Artificial antigen, 110
Artificial passive immunity, 97
Artificially acquired immunity, 96, 644
Artificially acquired passive immunity, 645
Ascaris immunity, 678
Aschoff bodies, 493
Ascoli's test, 732
Asialoglycoprotein receptor (ASGP R) autoantibodies, 398
ASLT, 727
ASO (antistreptolysin O), 727
Aspergillus species, 491
Aspirin (ASA) acetyl salicylic acid, 360
Aspirin sensitivity reactions, 360
Association constant (K_A), 235
Asthma, 362, 491
Ataxia telangiectasia, 547
ATG, 603
Athymic nude mice, 282
Atopic, 359
Atopic allergy or atopy, 359
Atopic dermatitis, 359, 478
Atopic hypersensitivity, 359
Atopy, 358
Attenuate, 685
Attenuated, 685
Attenuated pathogen, 685
Attenuation, 685
AtxBm, 171
Atypical antineutrophil cytoplasmic antibodies, 483
Auer's colitis, 365
Australia antigen (AA), 664
Autoagglutination, 387
Autoallergy, 387
Autoantibodies against lamin, 400
Autoantibodies against pepsinogen, 398
Autoantibodies to gastric parietal cells, 396
Autoantibody, 387
Autoantibody assays, 728
Autoantigens, 387
Autobody, 204
Autochthonous, 613
Autocrine, 286
Autocrine factor, 286
Autofluorescence, 742
Autogenous vaccine, 689
Autograft, 598
Autoimmune adrenal failure, 404
Autoimmune and lymphoproliferative syndrome, 393
Autoimmune cardiac disease, 409

Autoimmune complement fixation reaction, 388
Autoimmune disease, 390
Autoimmune disease animal models, 390
Autoimmune disease spontaneous animal models, 390
Autoimmune gastritis, 396
Autoimmune hemolytic anemia, 393
Autoimmune hemophilia, 393
Autoimmune hepatitis, 399
Autoimmune lymphoproliferative syndrome (ALPS), 393
Autoimmune neutropenia, 393, 469
Autoimmune polyglandular syndromes, 405
Autoimmune response, 388
Autoimmune skin diseases, 414
Autoimmune thrombocytopenic purpura, 393
Autoimmune thyroiditis, 391, 392
Autoimmune tubulointerstitial nephritis, 402
Autoimmune uveoretinitis, 415
Autoimmunity, 387
Autologous, 598
Autologous bone marrow transplantation (ABMT), 598
Autologous graft, 599
Autolymphocyte therapy (ALT), 703
Autoradiography, 749
Autoreactive T lymphocytes, 388
Autoreactivity, 388
Autosensitization, 388
Avian (bird) immunity, 715
Avidity, 211, 229, 231
Avidity hypothesis, 213, 229
Avionics®, 705
Azathioprine, 569
Azidothymidine, 564
Azoprotein, 116
AZT, 564

B

4-1BB, 30
4-1BB ligand (4-1BBL), 30
B allotype, 200
β barrel, 41
B blood group, 449
B cell activation, 163, 164
B cell antigen receptor, 160
B cell chronic lymphocytic leukemia/small lymphocytic lymphoma (B-CLL/SLL), 471
B cell coreceptor, 161
B cell corona, 89
B cell differentiation and growth factors, 171
B cell leukemias, 537
B cell lymphoproliferative syndrome (BLS), 576
B cell mitogens, 171
B cell specific activator protein (BSAP), 171
B cell tolerance, 160, 382
B cell tyrosine kinase (Btk), 171
B cells, 55, 159
β cells, 41
B cell-stimulating factor 2 (BSF-2), 300
B complex, 129
B genes, 130
B locus, 130
B lymphocyte antigen receptor (BCR) complex, 165
B lymphocyte hybridoma, 160, 210
B lymphocyte receptor, 160
B lymphocyte Stimulator (BlyS), 704
B lymphocyte stimulatory factors, 171
B lymphocyte tolerance, 160
B lymphocytes, 159
β lysin, 78
β propiolactone, 694
B-1 cells, 159
B1a B-cells (CD5), 159
β_1A globulin, 327
β_1C globulin, 327
β_1E globulin, 329
β_1F globulin, 329
β_1H, 336
β_2, 544
B-2 cells, 159
β_2 microglobulin (β_2M), 132
B220, 47
B4,b5,b6 and b9, 225
B7, 54
B7, B7-2, 151
B7.1 costimulatory molecule, 54
B7.2 costimulatory molecule, 54
Babesiosis immunity, 678
Bacillus anthracis immunity, 651
Back typing, 452
Backcross, 600
Bacteria, 707
Bacterial agglutination, 246
Bacterial allergy, 372
Bacterial hypersensitivity, 372
Bacterial immunity, 645
Bacterial immunoglobulin-binding proteins, 651
Bacterial vaccine, 685
Bactericidin, 651
Bacterin, 685
Bacteriolysin, 645
Bacteriolysis, 645
Bacteriophage neutralization test, 771
Bacteroides immunity, 651
β-adrenergic receptor antibodies, 415
Bagassosis, 367, 490
Bak[a], 464
Balb/c mice, 720
BALT, 79
Band test, 515
Bare lymphocyte syndrome (BLS), 546
Bas, 717
Basement membrane antibody, 506
Basophil-derived kallikrein (BK-A), 74
Basophilic, 74
Basophils, 74
BCDF, 171
B-cell growth factor (BCGF), 297
B-cell growth factor I (BCGF-1), 297
B-cell growth factor II (BCGF-2), 298
B-cell stimulating factor 1 (BSF-1), 297
BCG (Bacille Calmette-Guerin), 693
BCGF (B-cell growth factors), 171
Bcl-2, 172
Bcl-2 proteins, 172
Bcl-X_L, 172
BDB, 736
Behcet's disease, 534
Beige mice, 551
Bence-Jones (B-J) proteins, 530
Benign lymphadenopathy, 86
Benign lymphoepithelial lesion, 430

Benign monoclonal gammopathy, 529
Benign tumor, 613
Bentonite ($Al_2O_3 \cdot 4SiO_2 \cdot H_2O$), 738
Bentonite flocculation test, 246
Berger's disease, 505
Berylliosis, 498
Beta-2 glycoprotein-I autoantibodies, 422
Beta-adrenergic receptor autoantibodies, 434
Beta–gamma bridge, 532
BFPR, 650
Biclonality, 477
Bifunctional antibody, 215
Binding constant, 236
Binding protein, 225
Binding site, 195
Biochemical sequestration, 108
Biogenic amines, 350
Biolistics, 738
Biological false-positive reaction, 650
Biological response modifiers (BRM), 319, 623, 704
Biologicals, 704
Biotin–avidin system, 749
Biovin antigens, 109
BiP, 207
Birbeck granules, 67
Bird fancier's lung, 491
BI-RG-587, 564
Chlorambucil, 569, 705
Bis-diazotized benzidine, 736
Bispecific antibody, 214
BLA-36, 477
Blast cell, 76
Blast transformation, 53
Blastogenesis, 53, 280
Blk, 53
Blocking, 217
Blocking antibody, 217
Blocking factors, 627
Blocking test, 728
Blood group antigens, 447
Blood grouping, 447
Blood–thymus barrier, 254
Bloom syndrome (BS), 536
Blot, 751
BLR-1/MDR-15 (Burkitt's lymphoma receptor-1/ Monocyte-derived receptor-15), 533
B-lymphocyte tolerance, 383
BlyS, 310
Bm mutants, 129
Bombay phenotype, 450
Bombesin, 704
Bone marrow, 78, 598
Bone marrow cells, 78, 598
Bone marrow chimera, 599
Bone marrow transplantation, 598
Bony fish (teleosts), 712
Booster, 121
Booster injection, 121
Booster phenomenon, 121
Booster response, 118
Bordetella immunity, 652
Bordetella pertussis, 652
Borrelia immunity, 652
Botulinum toxin, 705
Bovine serum albumin (BSA), 107
Boyden chamber, 40
β-pleated sheet, 41, 532
Bradykinin, 353
Brambell receptor (FcRB), 211
Brequinar sodium (BQR), 572
Brester-Cohn theory, 380
BRMs, 319
Bromelin, 456
Bronchial asthma, 363
Bronchial-associated lymphoid tissue (BALT), 442
Bronchiectasis, 478
BRST-2 (GCDFP-15), monoclonal antibody (murine), 796
Brucella immunity, 652
Brucella vaccine, 696
Brucellin, 765
Brush border autoantibodies, 402
Bruton's agammaglobulinemia, 537
Bruton's disease, 535
BSA, 109
Btk, 537
B-type virus (*Aspergillus macaques*), 671
Bubble boy, 546
Buffy coat, 771
Bullous pemphigoid, 478
Bullous pemphigoid antigen, 479
Bungarotoxin, 408
Bunyaviridae immunity, 670
Burkitt's lymphoma, 532
Bursa equivalent, 157
Bursa of Fabricius, 157
Bursacyte, 157
Bursectomy, 157
Busulfan (1,4-butanediol dimethanesulfonate), 569
Butterfly rash, 515
BXSB mice, 422
Byssinosis, 490
Bystander B cells, 225
Bystander effects, 27
Bystander lysis, 27

C

C gene, 170
C gene segment, 170, 269
C region (constant region), 184
C segment, 170
C1, 324
C1 deficiencies, 343, 551
C1 esterase inhibitor, 324
C1 inhibitor (C1 INH) deficiencies, 343
C1 inhibitor (C1 INH) deficiency, 551
C10, 289
C1q, 325
C1q autoantibodies, 325, 422
C1q binding assay for circulating immune complexes (CIC), 740, 325
C1q deficiency, 344, 552
C1q receptors, 325
C1r, 325
C1s, 326
C2 (complement component 2), 326
C2 and B, 326
C2 and B genes, 139
C2 deficiency, 344, 552
C2A, 326
C2a, 326
C2B, 326

C2b, 326
C2C, 326
C3 (complement component 3), 327
C3 convertase, 322, 323, 327
C3 deficiency, 344, 552
C3 nephritic factor (C3NeF), 335
C3 PA (C3 proactivator), 336
C3 tickover, 334
C3a, 328
C3a receptor (C3a-R), 328
C3a/C4a receptor (C3a/C4a-R), 328, 329
C3b, 328
C3b (inactivated C3b), 334
C3b inactivator, 336
C3b receptors, 339
C3bi (iC3b), 334
C3c, 334
C3d, 335
C3dg, 335
C3e, 335
C3f, 335
C3g, 335
C3H/HeJ mice, 720
C4 (complement component 4), 328
C4 allotypes, 329
C4 deficiency, 344, 552
C4A, 329, 463
C4a, 329
C4B, 329, 463
C4b, 329
C4b inactivator, 336
C4b-binding protein (C4bp), 338
C4bi (iC4b), 338
C4c, 338
C4d, 338
C5 (complement component 5), 329
C5 convertase, 330
C5 deficiency, 345, 553
C5a, 329
C5a receptor (C5a-R), 330
$C5a_{74des\ Arg}$, 330
C5aR (C5 anaphylatoxin receptor), 333
C5b, 330
C6 (complement component 6), 330
C6 deficiency, 345, 553
C7 (complement component 7), 330
C7 deficiency, 345, 553
C8 (complement component 8), 330
C8 deficiency, 345, 553
C9 (complement component 9), 330
C9 deficiency, 345, 553
CA-125, 627, 785
CA-125 antibody, 785
CA-15-3, 627, 785
CA-19-9, 627, 788
Cachectin, 308
Cadherins, 38
Caecal tonsils, 91
Calcineurin, 278
Calcitonin, 627, 789
Calcivirus immunity, 671
CALLA, 471, 627
Calnexin, 127
Calreticulin, 127
CAM (cell adhesion molecules), 25
Campath-1 (CD52) CAMPATH-1M, 700
Campylobacter immunity, 652
Canale-Smith syndrome, 393
Cancer, 613
Candida immunity, 676
Canine distemper virus, 671
Canine immunity, 717
Canine parvovirus vaccine, 696
Capillary leak syndrome, 294
Caplin's syndrome, 498
Capping, 165
Capping phenomenon, 166
Caprinized vaccine, 690
Capsid, 659
Capsular polysaccharide, 667, 108
Capsule swelling reaction, 645, 774
Capture assays, 728
Carbohydrate antigens, 108
Carcinoembryonic antigen (CEA), 617, 788
Carcinogen, 613
Carcinoma, 613
Carcinoma-associated antigens, 616
Carcinomatous neuropathy, 411
Cardiolipin, 644, 741
Cardiolipin autoantibodies, 422
Carrier, 114
Carrier effect, 115
Carrier specificity, 115
Cartilaginous fishes, 711
Cartwheel nucleus, 57
Cascade reaction, 342
Caseation necrosis, 655
Caseous necrosis, 655
Casoni test, 765
Caspase substrates, 80
Caspases, 80
Castleman's disease, 477
Cat scratch disease, 645
Catalase, 72
Catalytic antibodies, 179
Catalytic antibody, 179
Catch-up vaccine, 691
Cathepsins, 150
Cationic proteins, 72
CBA mouse, 720
CBA/N mouse, 720
CC chemokine receptor 1 (CC CKR-1), 289
CC chemokine receptor 2 (CC CKR-2), 289
CC chemokine receptor 3 (CC CKR-3), 289
CC chemokine receptor 4 (CC CKR-4), 289
C-C subgroup, 288
CD (cluster of differentiation), 26
CD antigens, 26
CD molecules, 26
CD1, 146, 262
CD10, 792
CD10 (CALLA), 473, 616
CD11, 36
CD117 (c-kit) (polyclonal), rabbit, 794
CD11a, 36
CD11b, 334
CD13, 64
CD14, 473
CD15, 473
CD15 (Leu M1), 791
CD16, 59
CD19, 161

CD1a, 262
CD1b, 262
CD1c, 262
CD2, 150, 261
CD20, 150, 162
CD20 primary antibody, 792
CD21, 87, 150, 162
CD21 antigen, 792
CD22, 56, 150, 162
CD23, 473
CD23(1B12), 793
CD230, 637
CD25, 295
CD28, 150
CD29, 264
CD2R, 262
CD3, 268, 790
CD3 complex, 269
CD30, 473
CD31 (JC/70A), 793
CD33, 64, 473
CD34, 46, 473
CD35, 339
CD4, 141, 262
CD4 molecule, 141, 263
CD4 T cells, 264
CD40, 53
CD40 ligand, 54, 274
CD40-L, 53
CD41, 474
CD42a, 78
CD42b, 78
CD42c, 78
CD42d, 78
CD44, 37, 474
CD45, 47, 274
CD45R, 48
CD45RA, 48
CD45RB, 47
CD45RO, 48
CD5, 159, 273
CD5 B cells, 159
CD56, 59, 474
CD57, 59, 474
CD59, 331
CD6, 274
CD62E, 37
CD62L, 38
CD62P, 38
CD7, 274
CD8, 141, 150, 265
CD8 molecule, 141, 266
CD8 T cells, 150
CD8 T cells comprise, 266
CD9, 64
CD99 (HO36-1.1), 794
CEA, 617
Cecropin, 644
Celiac disease, 495
Celiac sprue (gluten-sensitive enteropathy), 495
Cell adhesion molecules, 25
Cell line, 772
Cell separation methods, 771
Cell surface molecule immunoprecipitation, 771
Cell surface receptors and ligands, 28
Cell tray panel, 587
Cell-bound antibody (cell-fixed antibody), 191
Cell-mediated hypersensitivity, 369
Cell-mediated immune response, 281
Cell-mediated immunity (CMI), 282
Cell-mediated immunodeficiency syndrome, 541
Cell-mediated lympholysis (CML) test, 756
Cell–surface immunoglobulin, 190
Cellular allergy, 369
Cellular and humoral metal hypersensitivity, 371
Cellular hypersensitivity, 369
Cellular immunity, 96
Cellular immunology, 96
Cellular interstitial pneumonia, 498
Cellular oncogene, 614
CentiMorgan (cM), 129
Central lymphoid organs, 85
Central tolerance, 382
Centriole antibodies, 427
Centroblasts, 87
Centrocytes, 88
Centromere autoantibodies, 428
C-erb-B2 murine monoclonal antibody, 786
Cerebrospinal fluid (CSF) immunoglobulins, 182
CFA, 125
CFT, 741
CFU, 313
CFU-GEMM, 313
CFU-S (colony-forming units, spleen), 46
C_γ, 179
CGD, 548
C_H, 179
C_H1, 179
C_H2, 179
C_H3, 179
C_H4, 179
CH_{50} unit, 341, 741
CHAD, 464
Chagas' disease, 678
Challenge, 119, 687
Challenge stock, 693
Chancre immunity, 646
Chaperones, 41, 207
Charcot-Leyden crystals, 363, 491
Chase-Sulzberger phenomenon, 444
Chediak-Higashi syndrome, 551
Chemical "splenectomy," 568
Chemical adjuvants, 124
Chemiluminescence, 760
Chemoattractant, 41
Chemokine autoantibodies, 433
Chemokine β receptor-like 1, 288
Chemokine receptor, 288
Chemokines, 288
Chemokinesis, 40
Chemotactic assays, 549, 760
Chemotactic deactivation, 40
Chemotactic disorders, 549
Chemotactic factors, 39
Chemotactic peptide, 40
Chemotactic receptors, 40
Chemotaxis, 39, 637
Chickenpox (varicella), 662
Chido (Ch) and Rodgers (Rg) antigens, 462
Chief cell autoantibodies, 414
Chimera, 599
Chimeric antibodies, 179

Chimerism, 599
Chlamydia immunity, 670
Chlorodinitrobenzene, 369
Cholera toxin, 646, 658, 696
Cholera vaccine, 696
Cholinergic urticaria, 362
Chorea, 493
Choriocarcinoma, 613
Chromatography, 725
Chromium release assay, 772
Chromogenic substrate, 748
Chromogranin monoclonal antibody, 784
Chromosomal translocation, 471
Chronic active hepatitis (autoimmune), 497, 665
Chronic and cyclic neutropenia, 549
Chronic fatigue syndrome (CFS), 534
Chronic graft-vs.-host disease (GVHD), 610
Chronic granulomatous disease (CGD), 547
Chronic lymphocytic leukemia (CLL), 472
Chronic lymphocytic leukemia/small lymphocytic lymphoma, 538
Chronic lymphocytic thyroiditis, 486
Chronic mucocutaneous candidiasis, 542
Chronic myelogenous leukemia (CML), 475
Chronic myeloid leukemia, 475
Chronic progressive vaccinia (vaccinia gangrenosa) (historical), 689
Chronic rejection, 606
Chrysotherapy, 433
Churg-Strauss syndrome (allergic granulomatosis), 347, 491
CIC, 366
Cicatrical ocular pemphigoid, 511
CIE, 732
Ciliary neurotrophic factor (CNTF), 299
Cimetidine, 705
Circulating anticoagulant, 422
Circulating dendritic cell, 66, 150
Circulating lupus anticoagulant syndrome (CLAS), 422
Circulating lymphocytes, 51
Circulatory system infections, 646
Cisterna chyli, 83
C_κ, 179
C-kit ligand, 312
C_L, 179
C_λ, 179
Cladosporium species, 359
Class I antigen, 132
Class I MHC molecules, 131
Class I region, 130
Class IB genes, 130
Class II antigens, 134
Class II MHC molecules, 134
Class II region, 134
Class II transactivator (CIITA), 135
Class II vesicle (CIIV), 152
Class III molecules, 135
Class switching (isotype switching), 221
Classes of cytokine receptors, 285
Classic pathway of complement, 321
Classical pathway, 323
Clathrin, 45
Cleveland procedure, 751
CLIP, 153
Clonal anergy, 383
Clonal balance, 383
Clonal deletion (negative selection), 383, 576
Clonal expansion, 384
Clonal ignorance, 382
Clonal restriction, 379
Clonal selection, 208
Clonal selection theory, 208
Clone, 57
Cloned DNA, 773
Cloned enzyme donor immunoassay, 731
Cloned T cell line, 276
Clonotypic, 165, 272
Clostridium immunity, 653
Clotting system, 42
Cluster of differentiation (CD), 26, 261
Clusterin (serum protein SP-40,40), 332
Clustering, 166
C_μ, 179
CMI, 26
c-myb, 533
c-myb gene, 254
Coagglutination, 644
Coagulation system, 42
Coated pit, 45
Coated vesicles, 46
Cobra venom factor (CVF), 328
Cocapping, 166
Coccidioides immunity, 676
Coccidioidin, 658, 764
Coding joint, 169, 270
Codominant, 449
Codominantly expressed, 129
Codon, 102
Coelomocyte, 708
Cogan's syndrome, 511
Cognate interaction, 156
Cognate recognition, 156
Cohn fraction II, 175
Coisogenic, 135, 720
Coisogenic strains, 720
Cold agglutinin, 463
Cold agglutinin syndrome, 464
Cold antibodies, 463
Cold ethanol fractionation, 728
Cold hemagglutinin disease, 463
Cold hypersensitivity, 361
Cold target inhibition, 728
Cold urticaria, 361
Cold-reacting autoantibodies, 394
Collagen, 35
Collagen (types I, II, and III) autoantibodies, 422
Collagen disease and arthritis panel, 422
Collagen disease/lupus erythematosus diagnostic panel, 750
Collagen Type IV (CIV22), 795
Collagen type IV autoantibodies, 422
Collagen vascular diseases, 423
Collectin receptor, 325
Colocalization, 166
Colon antibodies, 494
Colon autoantibodies, 396, 494
Colon–ovary tumor antigen (COTA), 789
Colony stimulating factors (CSF), 311
Colony-forming unit (CFU), 46
Colony-forming units, spleen (CFU-S), 46
Colony-stimulating factors (CSFs), 312
Colostrum, 444
Combination vaccines, 691
Combinatorial diversity, 169, 270
Combinatorial joining, 169
Combined immunodeficiency, 542

Combined prophylactic, 685
Combining site, 176
Commensal mice, 720
Common acute lymphoblastic leukemia antigen (CALLA), 471
Common acute lymphoblastic leukemia antigen (CALLA/CD10), 792
Common leukocyte antigen (LCA) (CD45), 789
Common lymphoid progenitors, 47
Common variable antibody deficiency, 540
Common variable immunodeficiency (CVID), 540
Competitive binding assays, 728
Competitive inhibition assay, 728
Complement (C), 321
Complement activation, 323
Complement deficiency conditions, 343, 551
Complement deviation (Neisser-Wechsberg phenomenon), 342
Complement fixation, 340
Complement fixation assay, 341, 741
Complement fixation inhibition test, 341
Complement fixation reaction, 740
Complement fixing antibody, 341
Complement inhibitors, 338
Complement membrane attack complex, 332
Complement multimer, 331
Complement receptor 1 (CR1), 339
Complement receptor 2 (CR2), 339
Complement receptor 3 (CR3), 339
Complement receptor 4 (CR4), 340
Complement receptor 5 (CR5), 340
Complement receptors, 338
Complement system, 323
Complementarity, 101
Complementarity-determining region (CDR), 179
Complete Freund's adjuvant (CFA), 125
Complex allotype, 201
Complex release activity, 366
Complotype, 135, 327
Concanavalin A (con A), 277
Concatamer integration, 773
Concomitant immunity, 626, 644
Confocal fluorescent microscopy, 743
Conformational determinant, 112
Conformational epitopes, 112
Congenic, 720
Congenic mice, 720
Congenic strains, 720
Congenital agammaglobulinemia, 537
Congenital immunodeficiency, 535
Conglutinating complement absorption test, 345, 728
Conglutination, 345, 728
Conglutinin, 345
Conglutinin solid phase assay, 345, 728
Conjugate, 115
Conjugate vaccine, 694
Conjugated antigen, 115
Connective tissue disease, 423
Connective tissue-activating peptide-III (CTAP-III), 290
Consensus sequence, 102
Constant domain, 179
Constant region, 179
Constitutive defense system, 644
Consumption test, 728
Contact dermatitis, 375
Contact hypersensitivity, 373
Contact sensitivity (CS) (or allergic contact dermatitis), 374
Contact system, 94
Continuous epitopes, 112
Contrasuppression, 268, 379
Contrasuppressor cell, 379
Control, 728
Control tolerance, 382
Convalescent serum, 644
Conventional mouse, 720
Convertase, 323
Coombs' test, 454
Cooperation, 156
Cooperative determinant, 112
Cooperativity, 156
Copolymer, 124
Copper and immunity, 549
Copper deficiency, 549
Coprecipitation, 238, 728
Coproantibody, 444
Corals, 708
Cords of Billroth, 89
Coreceptor, 151
Corneal response, 369
Corneal test, 369
Corneal transplants, 600
Coronavirus immunity, 671
Corticosteroids, 567
Corticotrophin receptor autoantibodies (CRA), 405
Corynebacterium diphtheriae immunity, 646
Costimulator, 151
Costimulatory molecules, 151
Costimulatory signal, 151
Coulombic forces, 228
Counter current electrophoresis, 732
Counter electrophoresis, 732
Counter immunoelectrophoresis (CIE), 733
Counter migration electrophoresis, 733
Cowpox, 687
Coxsackie, 671
CpG nucleotides, 696
CR1, 339
CR2, 339
CR2, Type II complement receptor, 339
CR3 deficiency syndrome, 340, 552
CREGs, 588
CREST complex, 523
CREST syndrome, 428
Creutzfeldt-Jakob syndrome, 669
Crithidia assay, 419, 728
Crithidia luciliae, 419, 728
CRM 197, 691
Crohn's disease, 494
Cromolyn, 354
Cromolyn sodium, 352
Cross-absorption, 246
Crossed immunoelectrophoresis, 733
Cross-match testing, 589
Cross-matching procedure, 587
Cross-priming, 151
Cross-reacting antibody, 177, 246
Crossreacting antigen, 108, 246
Cross-reaction, 246
Cross-reactivity, 246
Cross-sensitivity, 380
Cross-tolerance, 380
Crow-Fucase syndrome, 531
Cryofibrinogenemia, 529
Cryoglobulin, 529
Cryoglobulinemia, 529

Cryopreservation, 772
Cryostat®, 743
Cryptantigens, 463
Cryptococcus neoformans immunity, 677
Cryptodeterminant, 112
Cryptosporidium immunity, 676
Crystallographic antibodies, 520
CSF, 313
CSIF, 303
C-terminus, 184
CTL, 266
CTLA-4, 151
CTLA4-Ig, 151
Cu-18, 786
Cunningham plaque technique, 755
Cutaneous anaphylaxis, 362
Cutaneous basophil hypersensitivity (Jones-Mote hypersensitivity), 376
Cutaneous immune system, 444
Cutaneous lymphocyte antigen, 444
Cutaneous sensitization, 371, 445
Cutaneous T cell lymphoma, 445, 473
C-X-C subgroup, 289
CXCR-4, 289
Cyanogen bromide, 180
Cycle-specific drugs, 577
Cyclin D1 (polyclonal), rabbit, 792
Cyclooxygenase pathway, 349
Cyclophilin, 571
Cyclophosphamide, 569
Cyclosporine (cyclosporin A) (ciclosporin), 570
CYNAP, 588
CYNAP phenomenon, 588
Cytochalasins, 676
Cytochrome b deficiency, 548
Cytochrome c, 80
Cytokeratin (34betaE12), mouse, 780
Cytokeratin 7 (K72), mouse, 779
Cytokine assays, 287
Cytokine autoantibodies, 287, 388
Cytokine inhibitors, 287
Cytokine receptor families, 285
Cytokine receptors, 285
Cytokine synthesis inhibitory factor, 303
Cytokine upregulation of HIV coreceptors, 557
Cytokine-specific subunit, 286
Cytolysin, 267
Cytolytic, 267, 342
Cytolytic reaction, 267, 343
Cytomegalovirus (CMV), 562, 661
Cytomegalovirus (CMV) immunity, 662
Cytopathic effect (of viruses), 671
Cytophilic antibody, 177
Cytoplasmic antigens, 431
Cytosine arabinoside, 577
Cytoskeletal antibodies, 389
Cytoskeletal autoantibodies, 389
Cytoskeleton, 44
Cytosolic aspartate-specific proteases (CASPases), 80
Cytotoxic, 266
Cytotoxic antibody, 177
Cytotoxic CD8 T cells, 266
Cytotoxic drugs, 576
Cytotoxic T cells, 266
Cytotoxic T lymphocyte precursor (CTLp), 266
Cytotoxic T lymphocytes (CTLs), 266, 609, 621
Cytotoxicity, 177, 266, 343
Cytotoxicity assays, 267, 729
Cytotoxicity tests, 177, 267, 729
Cytotoxins, 267
Cytotrophic antibodies, 177
Cytotropic anaphylaxis, 353

D

2,4-Dinitrophenyl (DNP) group, 114
δ Chain, 193
D exon, 169
D exon, 223
D gene, 169
D gene region, 168
D gene segment, 169
D region, 169
D3TX mice, 406
DAF, 340
Dalen-Fuchs nodule, 415
d-amino acid polymers, 110
Dander antigen, 348
DANE particle, 664
Danysz effect, 243
Danysz phenomenon, 243
Dapsone, 646
Dark zone, 87
DAT, 454
ddC (dideoxycytidine), 564
ddI (2′,3′-dideoxyinosine), 564
DDS syndrome, 646
Dead vaccine, 685
Dean and Webb titration, 238
Death domains, 82
Decay-accelerating factor (DAF), 340
Decomplementation, 342
Decorate, 777
Defective endogenous retroviruses, 671
Defensins, 73
Deficiency of secondary granules, 548
Degranulation, 354
Delayed-type hypersensitivity (DTH), 370
Denaturation, 113
Dendritic cell immunotherapy, 704
Dendritic cells (DC), 65
Dendritic epidermal cell, 66
Dengue, 671
Dense-deposit disease, 506
Density gradient centrifugation, 735
Deoxyguanosine, 545
Deoxyribonuclease, 101
Deoxyribonuclease I, 101
Deoxyribonuclease II, 101
Deoxyribonucleoprotein antibodies, 434
Depot-forming adjuvants, 124
Dermatitis herpetiformis (DH), 479
Dermatitis venenata, 375
Dermatographism, 355
Dermatomyositis, 430, 484
Dermatopathic lymphadenitis, 86
Dermatophagoides, 359
Dermatophagoides pteronyssinus, 359
Dermatophytid reaction, 375
Desensitization, 353
Desetope, 153
Designation "conventional" (holoxenic) animals, 721
Designer, 215

Designer lymphocytes, 621
Desmin, 781
Desmin (D33), mouse, 781
Desmoglein, 414
Despecification, 704
Determinant groups (or epitopes), 113
Determinant spreading, 389
Dextrans, 111, 454
Dhobi itch, 372
Diabetes insipidus, 411
Diabetes mellitus, insulin dependent (type I), 500
Diacylglycerol (DAG), 43
Dialysis, 725
Diapedesis, 71
Diathelic immunization, 685
Diazo salt, 116
Diazotization, 115
DIC, 368
Dick test, 763
Differential signaling hypothesis, 259
Differentiation antigen, 109
Differentiation factors, 295
Diffusion coefficient, 245
DiGeorge syndrome, 540
Dilution end point, 247
Dim, 760
Dinitrochlorobenzene (DNCB), 114
Dinitrofluorobenzene (DNFB), 115
Diphtheria antitoxin, 699
Diphtheria immunization, 691
Diphtheria toxin, 651, 691
Diphtheria toxoid, 691
Diphtheria vaccine, 691
Diploid, 101
Direct, 454
Direct agglutination, 451
Direct antigen presentation, 146
Direct antiglobulin test, 455
Direct Coombs' test, 454
Direct fluorescence antibody method, 744
Direct immunofluorescence, 744
Direct reaction, 597
Direct staining, 744
Directional flow, 632
Discoid lupus erythematosus, 518
Discontinuous epitopes, 112
Disodium cromoglycate, 352
Disseminated intravascular coagulation (DIC), 368
Dissociation constant, 236
Distemper vaccine, 696
Distribution ratio, 180
Disulfide bonds, 184
Diversity, 51
Diversity (D) segments, 169, 270
DNA fingerprinting, 103, 752
DNA laddering, 103
DNA library, 100
DNA ligase, 101
DNA nucleotidylexotransferase (terminal deoxynucleotidyl-transferase), 101
DNA nucleotidyltransferase (terminal deoxynucleotidyl transferase [TdT]), 259
DNA microarray, 752
DNA polymerase, 100
DNA polymerase I, 100
DNA polymerase II, 100
DNA polymerase III, 100
DNA vaccination, 696
DNA vaccine, 696
DNA-dependent RNA polymerase, 100
DNBS (2,4-dinitrobenzene sulfonate), 114
DNCB, 114
DNP, 114
DO and DM, 134
Doctrine of original antigenic sin, 210
Domain, 184
Dominant phenotype, 101
Donath-Landsteiner antibody, 458
Donor, 596
Dopamine neuron autoantibodies, 411
Dot blot, 753
Dot DAT, 454
Double diffusion test, 731
Double immunodiffusion, 724
Double-emulsion adjuvant, 125
Double-layer fluorescent antibody technique, 744
Double-negative (DN) cell, 258
Double-negative thymocytes, 258
Double-positive (DP) cell, 258
Double-positive thymocytes, 258
Double-stranded DNA autoantibodies, 420
Doubling dilution, 725
Doughnut structure, 330
DPT vaccine, 690
Draining lymph node, 86
Drakeol 6VR®, 124
Drug allergy, 359
Drug-induced autoimmunity, 389
Drug-induced immune hemolytic anemia, 392
Drug-induced lupus (DIL), 515
Drug-induced lupus erythematosus, 423
DTaP vaccine, 690
DTH, 371
DTH T cell, 371
Duffy antigen/chemokine receptor (DARC), 289
Duffy blood group, 461
Duncan's syndrome, 537
Dye exclusion test, 770
Dye test, 770
Dysgammaglobulinemia, 530

E

E allotype, 717
E antigen, 665
E rosette, 262
E rosette-forming cell, 262
E2A, 169
E32, 172
E5, 703
EA, 729
EAC, 342
EAE, 411
EAMG (experimental autoimmune myasthenia gravis), 408
Early B cell factor (EBF), 171
Early induced responses, 96
EAT, 487
EBF (early B cell factor), 171
EBI1, 289
EBNA (Epstein-Barr virus nuclear antigen), 671
E-Cadherin, 39
E-Cadherin (ECH-6), mouse, 796

ECF-A (eosinophil chemotactic factor of anaphylaxis), 351
Echinococcus immunity, 678
Echinoderms, 710
ECHO virus (enteric cytopathogenic human orphan virus), 662
Eclipsed antigen, 109
ECRF3, 289
Ecto-5′-nucleotidase deficiency, 537
Eczema, 359
Eczema vaccinatum, 688
Eczematoid skin reaction, 359
ED_{50}, 729
Edema, 95
Edge artifact, 748
Edible vaccine, 690
Effector function, 180
Effector lymphocyte, 50
Effector mechanisms, 644
Effector phase, 644
Effector response, 644
Efferent lymphatic vessel, 85
Ehrlich side chain theory (historical), 205
EIA, 350, 750
Eicosanoid, 356
ELAM-1 (endothelial leukocyte adhesion molecule-1), 34
Electroimmunodiffusion, 735
Electrophoresis, 734
Electrophoretic mobility, 734
Electroporation, 773
Elek plate, 734
Elephantiasis, 679
Elevated IgE, defective chemotaxis, recurrent infection, and eczema, 547
ELISA (enzyme-linked immunosorbent assay), 750
ELISPOT assay, 750
Embryonic antigens, 109, 616
Embryonic stem (ES) cells, 46
EMF-1 (embryo fibroblast protein-1), 40
EMIT, 750
Emperipolesis, 51
ENA antibodies, 419
ENA autoantibodies, 420
ENA-78 (epithelial derived neutrophil attractant-78), 40
Encapsulated bacteria, 646
Encapsulation, 710
Encephalitogenic factors, 411
End cell, 57
End piece, 323
End point, 247
End-binders, 108, 646
Endocrine, 42
Endocytic vesicle, 64
Endocytosis, 63
Endogenous, 64
Endogenous pyrogen, 292
Endometrial antibodies, 406
Endometrial autoantibodies, 407
Endomysial autoantibodies, 396
Endophthalmitis phacoanaphylactica, 415
Endoplasmic reticulum, 44, 207
Endoplasmic reticulum autoantibodies, 399
Endosome, 45
Endothelial cell autoantibodies (ECA), 431
Endothelial leukocyte adhesion molecule-1 (ELAM-1), 34
Endothelin, 34
Endotoxin, 641
Endotoxin shock, 642
Endotoxins, 641
End-point immunoassay, 247, 732
End-stage renal disease (ESRD), 506
Engraftment, 601
Enhancement, 592
Enhancer, 168
Enhancing antibodies, 593
Entactin/nidogen, 502
Entactin/nidogen autoantibodies, 402
Entamoeba histolytica antibody, 679
Enteric neuronal autoantibodies, 411
Enterotoxin, 643
Envelope glycoprotein, 561
Enzyme immunoassay (EIA), 749
Enzyme labeling, 750
Enzyme-linked immunosorbent assay (ELISA), 749
Enzyme-multiplied immunoassay technique, 750
Eosinophil and neutrophil chemotactic activities, 74
Eosinophil cationic protein (ECP), 73
Eosinophil chemotactic factors, 73
Eosinophil differentiation factor, 73
Eosinophil granule major basic protein (EGMBP), 74
Eosinophilia, 394
Eosinophilic granuloma, 472
Eosinophilic myalgia syndrome (EMA), 534
Eosinophils, 73
Eotaxin, 288
Eotaxin-1 and eotaxin-2, 289
Eph receptors and ephrins, 53
Ephrin/eph, 53
Epibody, 205
Epithelial cell adhesion molecule (EpCAM), 781
Epithelial membrane antigen (EMA), 780
Epithelial thymic-activating factor (ETAF), 258
Epithelioid cell, 62
Epitope, 112
Epitope spreading, 112
Epitype, 112
Epivir®, 564
Epstein-Barr immunodeficiency syndrome, 537
Epstein-Barr nuclear antigen, 533
Epstein-Barr virus (EBV), 533, 667
Equilibrium dialysis, 235
Ergotype, 411
Erp57, 148
Erythema marginatum, 493
Erythema multiforme, 480
Erythema nodosum, 496
Erythroblastosis fetalis, 457
Erythrocyte agglutination test, 736
Erythrocyte autoantibodies, 451
Erythroid progenitor, 47
Erythropoiesis, 47
Erythropoietin, 47
Escherichia coli immunity, 653
E-selectin (CD62E), 37
Essential mixed cryoglobulinemia, 530
Estrogen/progesterone receptor protein, 785
Euglobulin, 175
Eukaryote, 100
EVI antibodies, 415
Exchange transfusion, 466
Excitation filter, 742
Exercise and immunity, 97
Exercise-induced asthma, 363
Exoantigen, 109
Exocytosis, 75

Exogenous, 75
Exon, 100
Exotoxin, 658
Experimental allergic encephalomyelitis, 508
Experimental allergic neuritis, 509
Experimental allergic orchitis, 509
Experimental allergic thyroiditis, 487
Experimental autoimmune encephalomyelitis, 509
Experimental autoimmune myasthenia gravis (EAMG), 485
Experimental autoimmune neuritis (EAN), 509
Experimental autoimmune oophoritis, 510
Experimental autoimmune sialoadenitis (EAS), 521
Experimental autoimmune thyroiditis (EAT), 487
Experimental autoimmune uveitis (EAU), 511
Extended haplotype, 584
Extravasation, 95
Extrinsic allergic alveolitis, 489
Extrinsic asthma, 363
Exudate, 95
Exudation, 95

F

13-26 Fd′ piece, 198
F protein, 721
F(ab′)$_2$ fragment, 198
Fab fragment, 195
Fab″ fragment, 198
Fabc fragment, 198
Facb fragment, 198
FACS®, 760
F-actin, 42
Factor B, 336
Factor D, 336
Factor D deficiency, 336
Factor H, 336
Factor H deficiency, 336
Factor H receptor (fH-R), 336
Factor I, 336
Factor I deficiency, 337
Factor P (properdin), 336
Factor VIII, 794
Facultative phagocytes, 63
Faenia rectivirgula, 489
FANA, 420
Farmer's lung, 489
Farr technique, 736
Fas (AP0-1/CD95), 82
Fas ligand, 82, 387
Fascin (55k-2), mouse, 796
Fasciola immunity, 683
FasL/Fas toxicity, 82
Fatty acids and immunity, 95
Fb fragment, 199
Fc fragment (fragment crystallizable), 196
Fc piece, 196
Fc receptor, 218
Fc receptors, 218
Fc receptors on human T cells, 276
Fc′ fragment, 198
Fcε receptor (FcεR), 219
Fcγ receptors (FcγR), 218
FcγRI, 219
FcγRII, 219
FcγRIII, 219
Fd fragment, 195
Fd piece, 195
Fd′ fragment, 198
Feline immunity, 717
Felton phenomenon, 386
Fernandez reaction, 646
Ferritin, 745
Ferritin labeling, 745
Fertilizin, 710
Fetal or oncofetal antigen, 616
Fetus allograft, 593
Feulgen reaction, 773
Fibrillarin autoantibodies, 427
Fibrin, 34
Fibrinogen, 34
Fibrinoid necrosis, 367
Fibrinopeptides, 34
Fibronectin, 34, 633
Fibrosis, 95
FICA (fluoroimmunocytoadherence), 725
Ficin, 456
Ficoll, 772
Ficoll-Hypaque, 772
FIGE, 735
Filarial immunity, 679
Filovirus immunity, 671
Final serum dilution, 247
First-set rejection, 602
First-use syndrome, 355
FISH (fluorescence *in situ* hybridization), 751
Fish immunity, 712
Fixed drug eruption, 377
FK506, 571
FKBP (FK-Binding Proteins), 572
Flagellar antigens, 640, 641
Flagellin, 641
Flame cells, 527
Flavivirus immunity, 672
FLIP/FLAM, 80
Flocculation, 241
Flow cytometry, 590, 759
Fluid mosaic model, 44
Fluorescein, 741
Fluorescein isothiocyanate (FITC), 741
Fluorescein-labeled antibody, 741
Fluorescence, 742
Fluorescence enhancement, 743
Fluorescence microscopy, 742
Fluorescence quenching, 236, 742
Fluorescence-activated cell sorter (FACS®), 760
Fluorescent antibody, 741
Fluorescent antibody technique, 741
Fluorescent protein tracing, 743
Fluorochrome, 741
Fluorodinitrobenzene, 369
Fluorography, 735
f-Met peptides, 41
Fog fever, 489
Follicles, 85
Follicular center cells, 85
Follicular dendritic cells, 86
Follicular hyperplasia, 85
Follicular lymphoma, 472
Food allergy, 360
Food and drug additive reactions, 361
Footprinting, 773
Footprints, 772

Forbidden clone theory, 225
Foreign gene, 768
Formol toxoid, 686
Formyl-methionyl-leucyl-phenylalanine (F-Met-Leu-Phe), 41
Forssman antibody, 118
Forssman antigen, 118
Foscarnet, 565
Fractional catabolic rate, 211
Fragmentins, 58
Framework regions (FR), 224
Francisella immunity, 653
Freemartin, 380
Frei test, 654
Freund's adjuvant, 125
Freund's complete adjuvant, 125
Freund's incomplete adjuvant, 125
Front typing, 451
FTA-ABS (fluorescence treponema antibody absorption), 646
Functional affinity, 213
Functional antigen, 110
Functional immunity, 97
Fungal immunity, 675
Fungi, 675
Fusin, 557
Fusobacterium immunity, 653
Fv fragment, 199
Fv region, 195
Fyn, 271

G

γ Globulin, 175
γ Globulin fraction, 175
γ Heavy-chain disease, 530
γ Interferon, 308
γ Macroglobulin, 190
G protein-coupled receptor family, 95
G proteins, 95
GAD-65, 405
Gammopathy, 528
Gancyclovir, 565
Ganglioside autoantibodies, 411
Gas gangrene antitoxin, 700
Gastric cell cAMP stimulating autoantibodies, 397
Gastrin receptor antibodies, 428
Gastrin-producing cell autoantibodies (GPCA), 397
GATA-2 gene, 102
Gatekeeper effect, 34
Gay bowel syndrome, 558
GCDFP-15 (23A3), mouse, 786
γδ T cell receptor, 272
γδ T cell receptor (TCR), 271
γδ T cells, 273
GEF, 95
Gel diffusion, 723
Gel filtration chromatography, 723
Gene bank, 100, 221
Gene cloning, 102, 220
Gene conversion, 101, 715
Gene conversion hypothesis, 131, 220
Gene diversity, 220
Gene knockout, 769
Gene mapping, 101
Gene rearrangement, 168, 221, 270
Gene segments, 168, 270
Gene therapy, 554
Generalized anaphylaxis, 354
Generalized vaccinia, 689
Generative lymphoid organ, 85
Genetic code, 101, 219
Genetic immunization, 696
Genetic knockout, 770
Genetic polymorphism, 102
Genetic switch hypothesis, 221
Genome, 100, 219
Genomic DNA, 100, 219
Genotype, 101, 225
Germ line, 100
Germ-free animal, 769
Germinal centers, 87
Germinal follicle, 85
Ghon complex, 656
Giant cell arteritis, 432
Gld gene, 424, 721
Gliadin autoantibodies, 397
Glial fibrillary acidic protein (GFAP), 782
Globulins, 175
Glomerular basement membrane autoantibodies, 402
Glomerulonephritis (GN), 501
Glucocorticoids, 567
Glucose-6-phosphate dehydrogenase deficiency, 549
Glutamic acid decarboxylase autoantibodies, 412
Gluten-sensitive enteropathy (celiac sprue, nontropical sprue), 494
GlyCAM-1, 36
Glycosylphophatidylinositol (GPI)-linked membrane antigens, 110
Gm allotype, 196, 202
γM globulin, 190
Gm marker, 202
GM_1 autoantibodies, 434
Gnotobiotic, 545
Gold therapy, 424
Golgi apparatus, 45
Golgi autoantibodies, 433
Golgi complex, 45
Gonococcal complement fixation test, 740
Goodpasture's antigen, 404, 507
Goodpasture's syndrome, 404, 506
GOR autoantibodies, 399
gp120, 559
gp160, 565
GPLA, 135, 716
Graft, 592
Graft arteriosclerosis, 606
Graft facilitation, 592
Graft rejection, 601
Graft-vs.-host disease (GVHD), 607
Graft-vs.-host reaction (GVHR), 606
Graft-vs.-leukemia (GVL), 607
Granulocyte, 68
Granulocyte antibodies, 469
Granulocyte autoantibodies, 394
Granulocyte chemotactic protein-2 (GCP-2), 289
Granulocyte colony-stimulating factor (G-CSF), 313
Granulocyte–macrophage colony-stimulating factor (GM-CSF), 313
Granulocyte–monocyte colony-stimulating factor, 313
Granulocyte-specific antinuclear autoantibodies (GS-ANA), 425
Granulocytopenia, 394
Granuloma, 62, 155
Granulomatous hepatitis, 497
Granulopoietin, 312

Granzymes, 58
Graves' disease (hyperthyroidism), 487
Gravity and immunity, 97
Gross cystic disease fluid protein 15 (GCDFP-15) antigen, 786
Group agglutination, 658
Growth factors, 287
Guanine nucleotide exchange factors (GEFs), 95
Gancyclovir, 565
Guillain-Barré syndrome, 412
Gut-associated lymphoid tissue (GALT), 437
GVH, 607
GVH disease, 607

H

17-Hydroxycorticosteroids (17-OHCS), 568
H antigen, 449
H antigens, 640, 646
H chain (heavy chain), 183
H substance, 449
H1 receptors, 351
H1, H2 blocking agents, 351
H-2, 135
H-2 complex, 136
H-2 locus, 136
H2 receptors, 351
H-2 restriction, 136
H-2D and H -2K, 136
H-2I region, 136
H-2L, 136
H65-RTA, 576
H7, 280
HA-1A, 703
Haemophilus immunity, 653
Haemophilus influenzae type b vaccine (HB), 694
Hageman factor (HF), 42
Hairpin loop, 100
Hairy cell leukemia, 472
Half-life ($T_{1/2}$), 211
Halogenation, 646
Halothane antigens, 110
HALV (human AIDS-lymphotropic virus) (historical), 557
HAM, 471
HAM test, 534
HAM-1 and HAM-2 (histocompatibility antigen modifier), 155
HAMA, 700
Hanganitziu-Deicher antigen, 783
Haploid, 101
Haplotype, 142, 225, 588
Hapten, 114
Hapten conjugate response, 115
Hapten conjugates, 115
Hapten inhibition test, 115, 729
Hapten X, 115
Hapten–carrier conjugate, 115
Harderian gland, 715
Hashimoto's disease (chronic thyroiditis), 486
Hassall's corpuscles, 254
HAV, 662
Hay fever, 359
HbcAg, 664
HbeAg, 664
HBIG (hepatitis B immunoglobulin), 700
HBLV (human B lymphotropic virus), 661
HbsAg, 664
HBV, 664
HBx, 664
HCC-1, 289
hCG (human choriogonadotrophic hormone), 44
H-chain disease, 527
HD_{50}, 342
HDN, 457
Delta agent (hepatitis D virus [HDV]), 499, 665
Heaf test, 765
Heart–lung transplantation, 592
Heat inactivation, 342
Heat shock protein antibodies, 43
Heat shock proteins (hsp), 43
Heat-aggregated protein antigen, 110
Heat-labile antibody, 225
Heavy chain, 182
Heavy chain class, 183
Heavy chain class (isotype) switching, 221
Heavy chain subclass, 183
Heavy-chain disease, 529
Heavy-chain diseases, 527
Helicobacter pylori immunity, 441
Helminth, 679
Helper CD4+ T cells, 264
Helper T cells, 264
Helper/suppressor ratio, 277
Hemadsorption inhibition test, 729
Hemagglutination, 450, 736
Hemagglutination inhibition reaction, 451
Hemagglutination inhibition test, 451
Hemagglutination test, 450
Hemagglutinin, 450
Hematogones, 157
Hematopoiesis, 47
Hematopoietic chimerism, 599
Hematopoietic lineage, 47
Hematopoietic stem cell, 599
Hematopoietic stem cell (HSC) transplants, 599
Hematopoietic system, 47
Hematopoietic-inducing microenvironment (HIM), 47
Hematoxylin bodies, 420
Hematuria, 503
Hemocyanin, 110
Hemocytoblast, 47
Hemolysin, 342
Hemolysis, 249
Hemolytic anemia, 534
Hemolytic anemia of the newborn, 457
Hemolytic disease of the newborn (HDN), 457
Hemolytic plaque assay, 755
Hemolytic system, 342
Hemophilia, 42
Hemopoietic resistance (HR), 594
Hemostatic plug, 28
Henoch-Schoenlein purpura, 484
HEP, 686
Heparan sulfate, 30
Heparin, 30
Hepatitis A, 662
Hepatitis B, 498, 664
Hepatitis B surface antigen (HbsAg) antibody, 664
Hepatitis B vaccine, 694
Hepatitis B virus immunity, 664
Hepatitis B virus protein X, 664
Hepatitis C virus immunity, 665
Hepatitis D virus, 665

Hepatitis E virus (HEV), 665
Hepatitis E virus immunity, 665
Hepatitis immunopathology panel, 498, 665
Hepatitis non-A, non-B (C) (NAN BH), 499
Hepatitis serology, 499, 662
Hepatitis vaccine, 694
Hepatitis, non-A, non-B (C) (NANBH), 665
Hepatocyte-stimulating factor, 299
Herbimycin A, 280
Herceptin®, 702
Herd immunity, 96
Hereditary angioedema (HAE), 552
Hereditary angioneurotic edema (HANE), 344
Hereditary ataxia telangiectasia, 547
Hereditary complement deficiencies, 345
Herpes gestationis (HG) autoantibodies, 414
Herpes simplex virus 1 and 2 (HSV 1 and 2) polyclonal antibody, 660
Herpes simplex virus immunity, 661
Herpes zoster, 661
Herpesvirus, 660
Herpesvirus-6 immunity, 661
Herpesvirus-8 immunity, 661
Herxheimer reaction, 366
Heteroantibody, 433
Heteroantigen, 110
Heteroclitic antibody, 173
Heterocliticity, 227
Heteroconjugate, 622
Heteroconjugate antibodies, 622
Heterocytotropic antibody, 173, 225
Heterodimer, 272
Heterogeneic, 595
Heterogeneous nuclear ribonucleoprotein (RA-33) autoantibodies, 425
Heterogenetic antibody, 173
Heterogenetic antigen, 110
Heterograft, 595
Heterokaryon, 772
Heterologous, 595
Heterologous antigen, 110
Heterologous vaccine, 687
Heterophile antibody, 173
Heterophile antigen, 118
Heterotopic, 592
Heterotopic graft, 592
Heterotypic vaccine, 687
Heterozygosity, 102
Heterozygous, 102
Heymann antigen, 503
Heymann glomerulonephritis, 504
Heymann's nephritis, 504
HGG, 700
HGP-30, 695
HHV, 660
Hib (*Hemophilus influenzae* type b), 646, 694
Hidden determinant, 112
HIG, 700
High endothelial postcapillary venules, 88
High endothelial venules (HEV), 88
High-dose tolerance, 381
Highly active anteretroviral therapy (HAART), 564
Highly polymorphic, 102
High-titer, low-avidity antibodies (HTLA), 173
High-zone tolerance, 381
Hinge region, 180
Histaminase, 355
Histamine, 354
Histamine release assay, 766
Histamine-releasing factors (HRF), 355
Histiocyte, 62
Histiocytic lymphoma, 472
Histiocytosis X, 472
Histocompatibility, 127, 579
Histocompatibility antigen, 579, 129
Histocompatibility locus, 127, 579
Histocompatibility testing, 580
Histone (H2A–H2B)–DNA complex autoantibodies (IgG), 420
Histone antibodies, 420
Histone autoantibodies (non-H2A–H2B)–DNA), 420
Histoplasma immunity, 677
Histoplasmin, 765
Histoplasmin test, 765
Histotope, 153
HIV infection, 557
HIV-1, 560
HIV-1 virus structure, 559
HIV-2, 561
HIV-2V, 561
Hives, 351
HLA, 136, 580
HLA allelic variation, 137, 584
HLA Class I, 394
HLA Class II, 394
HLA Class III, 581
HLA class III, 138
HLA disease association, 138, 584
HLA locus, 137, 581
HLA nonclassical class I genes, 132, 139, 584
HLA oligotyping, 585
HLA tissue typing, 585
HLA-A, 137, 394, 581
HLA-B, 394, 581
HLA-B27-related arthropathies, 425
HLA-C, 394, 581
HLA-D region, 138, 581
HLA-DM, 139, 583
HLA-DP subregion, 138, 583
HLA-DQ subregion, 138, 583
HLA-DR antigenic specificities, 139, 583
HLA-DR subregion, 138, 583
HLA-E, 139, 583
HLA-F, 139, 583
HLA-G, 139, 584
HLA-H, 132, 584
Hm-1, 716
Hodgkin's disease, 476
Hof, 57
Homing receptors, 28
Homing-cell adhesion molecule (H-CAM), 29
Homobody, 205
Homocytotrophic antibody, 174
Homodimer, 272
Homograft, 595
Homograft reaction, 595
Homograft rejection, 595
Homologous, 594
Homologous antigen, 110
Homologous chromosomes, 595
Homologous disease, 601
Homologous recombination, 101
Homologous restriction factor (HRF), 340
Homologous vaccine, 690
Homology region, 181

Homology unit, 180
Homopolymer, 107
Homotransplantation, 595
Homozygote, 647
Homozygous, 589
Homozygous typing cell (HTC) technique, 589
Homozygous typing cells (HTCs), 138, 589
Hook effect, 767
Hookworm immunity, 679
Hookworm vaccine, 695
Hormone immunoassays, 731
Horror autotoxicus (historical), 387
Horse serum sensitivity, 360
Host-vs.-graft disease (HVGD), 603
Hot antigen suicide, 110
Hot spot, 180
House dust allergy, 360
HPV, 660
HSA, 108
HSV, 661
HTLV, 471
HTLV-IV, 557
Human diploid cell rabies vaccine (HDCV), 692
Human immune globulin (HIG), 554, 700
Human immunodeficiency virus (HIV), 556
Human leukocyte antigen (HLA), 132, 580
Human milk-fat globulin (HMFG), 786
Human SCID (hu-SCID) mouse, 542
Human T lymphocyte, 36
Humanization, 180
Humanized antibody, 180
Humoral, 55
Humoral antibody, 97
Humoral immune response, 97
Humoral immunity, 175
Humps, 505
HUT 78, 558
H-Y, 579
HY, 580
Hybrid antibody, 214
Hybrid cell, 772
Hybrid hapten, 115
Hybrid resistance, 626
Hydrogen bonds, 229
Hydrophilic, 229
Hydrophobic, 229
Hydrops fetalis, 457
Hydroxychloroquine, 705
Hyperacute rejection, 603
Hypergammaglobulinemia, 528
Hypergammaglobulinemic purpura, 484
Hyper-IgM syndrome, 550
Hyperimmune, 685
Hyperimmunization, 685
Hyperimmunoglobulin E syndrome (HIE), 550
Hyperimmunoglobulin M syndrome, 550
Hyperplasia, 85
Hypersensitivity, 347
Hypersensitivity angiitis, 361, 494
Hypersensitivity diseases, 347
Hypersensitivity pneumonitis, 347, 361, 491
Hypersensitivity vasculitis, 361, 494
Hyperthyroidism, 487
Hypervariable regions, 185
Hypocomplementemia, 323
Hypocomplementemic glomerulonephritis, 503
Hypocomplementemic vasculitis, 481
Hypogammaglobulinemia, 536
Hyposensitization, 353

I

I invariant (Ii), 153
I region, 132, 133
Ia antigens (immune-associated antigen), 132
IBD, 496
Ibuprofen, 425
iC3b-Neo, 334
iC4b, 338
ICA512 (IA-2), 405
ICAM-1 (intercellular adhesion molecule-1), 32
ICAM-2, 32
ICAM-3, 33
Iccosomes (immune complex coated antibodies), 121
ICOS, 280
Id reaction, 375
Identity testing, 771
Idiopathic thrombocytopenic purpura, 396, 469
Idiotope, 203
Idiotype, 202
Idiotype network, 203
Idiotype network theory, 203
Idiotype suppression, 203
Idiotype vaccine, 697
Idiotypic determinant, 203
Idiotypic specificity, 203
IE, 732
IEF, 770
IEP, 732
IFE, 732
IFN, 307
Ig, 181
Ig myeloma subclasses, 526
IgA, 192
Igα and Igβ, 161
IgA deficiency, 539
IgA nephropathy (Berger's disease), 505
IgA paraproteinemia, 527
Igα/Igβ (CD79a/CD79b), 161
IgD, 192
IgE, 193
IgG, 188
IgG index, 188
IgG subclass deficiency, 539
IgG-induced autoimmune hemolysis, 435
IgM, 190
IgM deficiency syndrome, 539
IgM index, 191
IgM paraproteinemia, 527
IgR, 715
IgW, 712
IgX, 714
IgY, 714
Ii antigens, 462
I-J, 132
I-K, 433
Ikaros, 82
IL, 292
IL-1, 292
IL-1 receptor antagonist (IL-1RA), 293
IL-10, 303
IL-11, 304

IL-12, 304
IL-12R, 304
IL-13, 304, 305
IL-13 receptor, 304
IL-13 receptor complex, 304
IL-14, 305
IL-15, 305
IL-15 receptor, 305
IL-16, 306
IL-17, 306
IL-18, 306
IL-2, 294
IL-2 receptor, 305
IL-2 receptor (CD25), 295
IL-2/LAK cells, 625
IL-3, 295
IL-4, 297
IL-5, 298
IL-6, 299
IL-6 receptor, 299
IL-7, 300
IL-9, 302
ImD50, 119
Immature dendritic cells, 66
Immediate hypersensitivity, 347
Immediate spin crossmatch, 466, 591
Immobilization test, 740
Immune, 25
Immune adherence, 339
Immune adherence receptor, 339
Immune and neuroendocrine systems, 384
Immune antibody, 449
Immune cell cryopreservation, 772
Immune cell motility, 25
Immune clearance, 768
Immune complex disease (ICD), 500
Immune complex pneumonitis, 501
Immune complex reactions, 364
Immune costimulatory molecules, 151
Immune cytolysis, 332
Immune deviation, 385
Immune elimination, 237, 768
Immune exclusion, 440
Immune hemolysis, 332
Immune inflammation, 94
Immune interferon, 308
Immune network hypothesis of Jerne, 204
Immune neutropenia, 394
Immune privilege, 592
Immune response, 119
Immune response (Ir) genes, 166
Immune serum, 182, 700
Immune serum globulin, 182, 700
Immune suppression, 567
Immune system, 25
Immune system anatomy, 25
Immune tolerance, 380
Immune–neuroendocrine axis, 384
Immunity, 26
Immunity against extracellular bacteria, 640
Immunity in prokaryotes, 707
Immunization, 119
Immunize, 119
Immunizing dose (ImD_{50}), 686
Immunological suicide, 386
Immunoablation, 704
Immunoabsorbent, 723
Immunoabsorption, 723
Immunoadsorbents, 723
Immunoassay, 729
Immunoaugmentive therapy (IAT), 704
Immunobeads, 758
Immunoblast, 26
Immunoblastic lymphadenopathy, 476
Immunoblastic sarcoma, 476
Immunoblot (Western blot), 750
Immunoblotting, 750
Immunochemistry, 26
Immunocompetent, 26
Immunoconglutination, 346
Immunoconglutinin, 346
Immunocyte, 26, 480
Immunocytoadherence, 772
Immunocytochemistry, 26
Immunodeficiency, 535
Immunodeficiency animal models, 543
Immunodeficiency associated with hereditary defective response to Epstein-Barr virus, 554
Immunodeficiency disorders, 535
Immunodeficiency from hypercatabolism of immunoglobulin, 553
Immunodeficiency from severe loss of immunoglobulins and lymphocytes, 553
Immunodeficiency with partial albinism, 547
Immunodeficiency with T cell neoplasms, 541
Immunodeficiency with thrombocytopenia, 547
Immunodeficiency with thrombocytopenia and eczema (Wiskott-Aldrich syndrome), 547
Immunodeficiency with thymoma, 537
Immunodiagnosis, 777
Immunodiffusion, 725
Immunodominance, 112
Immunodominant epitope, 112
Immunodominant site, 112
Immunoelectroadsorption, 729
Immunoelectron microscopy, 745
Immunoelectroosmophoresis, 732
Immunoelectrophoresis (IEP), 732
Immunoenhancement, 121
Immunoferritin method, 745
Immunofluorescence, 743
Immunofluorescent "staining" of C4d, 601
Immunogen, 111
Immunogenic, 111
Immunogenic carbohydrates, 109
Immunogenicity, 111
Immunoglobulin, 169, 180, 793
Immunoglobulin A (IgA), 191
Immunoglobulin A deficiency, 538
Immunoglobulin alpha (α) chain, 192
Immunoglobulin class, 187
Immunoglobulin class switching, 192
Immunoglobulin D (IgD), 192
Immunoglobulin δ chain, 192
Immunoglobulin deficiency with elevated IgM, 539
Immunoglobulin domain, 184
Immunoglobulin E (IgE), 193
Immunoglobulin epsilon (ε) chain, 193
Immunoglobulin evolution, 713
Immunoglobulin fold, 184
Immunoglobulin fragment, 194
Immunoglobulin function, 182
Immunoglobulin G (IgG), 188

Immunoglobulin gamma (γ) chain, 189
Immunoglobulin gene superfamily, 167
Immunoglobulin genes, 166, 219
Immunoglobulin heavy chain, 182
Immunoglobulin heavy chain, binding protein (BiP), 183, 187
Immunoglobulin κ chain, 187
Immunoglobulin l chain, 186
Immunoglobulin light chain, 186
Immunoglobulin M (IgM), 189
Immunoglobulin mu (μ) chain, 191
Immunoglobulin structure, 181
Immunoglobulin subclass, 188
Immunoglobulin superfamily, 174, 713
Immunoglobulin-like domain, 184
Immunogold labeling, 745
Immunogold silver staining (IGSS), 746
Immunohematology, 447
Immunohistochemistry, 745, 777
Immunoincompetence, 535
Immunoinhibitory genes, 584
Immunoisolation, 595
Immunologic (or immune) paralysis, 386
Immunologic (or immunological), 27
Immunologic adjuvant, 122
Immunologic barrier, 592
Immunologic colitis, 496
Immunologic competence, 379
Immunologic competency, 125
Immunologic enhancement, 386, 592
Immunologic enhancement (tumor enhancement), 623
Immunologic facilitation, 592
Immunologic facilitation (facilitation immunologique), 623
Immunologic tolerance, 380
Immunological contraception, 697
Immunological deficiency state, 535
Immunological escape, 623
Immunological ignorance, 386
Immunological inertia, 379
Immunological infertility, 407
Immunological memory, 644
Immunological reaction, 27
Immunological rejection, 602
Immunological synapse, 147
Immunological unresponsiveness, 379
Immunologically activated cell, 27
Immunologically competent cell, 27
Immunologically privileged sites, 592
Immunologist, 27
Immunology, 27
Immunolymphoscintigraphy, 624
Immunomagnetic technique, 770
Immunomodulation, 704
Immunomodulator, 381
Immunonephelometry, 770
Immunoosmoelectropheresis, 732
Immunoparasitology, 469
Immunopathic, 469
Immunopathology, 469
Immunoperoxidase method, 777
Immunoperoxidase method, 746
Immunophenotyping, 760
Immunophilins, 571
Immunophysiology, 27
Immunopotency, 109
Immunopotentiation, 121
Immunoprecipitation, 730
Immunoproliferative small intestinal disease (IPSID), 477
Immunoprophylaxis, 689
Immunoprotein, 704
Immunoradiometric assay (IRMA), 767
Immunoradiometry, 766
Immunoreactant, 237
Immunoreaction, 237
Immunoreceptor tyrosine-based activation motif (ITAM), 29
Immunoreceptor tyrosine-based inhibition motif (ITIM), 29
Immunoregulation, 379
Immunoscintigraphy, 624
Immunoselection, 623
Immunosenescence, 576
Immunosuppression, 567
Immunosuppressive agents, 567
Immunosurveillance, 623
Immunotactoid glomerulopathy, 502
Immunotherapy, 623, 699
Immunotoxin, 598, 624
Immunotoxin-induced apoptosis, 81
Immunotyping, 581
Inaccessible antigens, 110
Inactivated poliovirus vaccine, 693
Inactivated vaccine, 689
Inactivation, 343
Inbred mouse strain, 720
Inbred strain, 719
Inbreeding, 720
Incompatibility, 466, 595
Incomplete antibody, 457
Incomplete Freund's adjuvant (IFA), 124, 126
Indirect agglutination (passive agglutination), 738
Indirect antigen presentation, 147, 602
Indirect antiglobulin test, 455
Indirect complement fixation test, 341
Indirect Coombs' test, 454
Indirect fluorescence antibody technique, 742, 744
Indirect hemagglutination test, 739
Indirect immunofluorescence, 744
Indirect template theory (historical), 206
Indomethacin, 425
Inducer, 312
Inducer determinant, 112
Inducer T lymphocyte, 265
Inducible NO synthase (iNOS), 62
Inductive phase, 119
Inductive sites, 440
Inert particle agglutination tests, 738
Infantile agammaglobulinemia, 537
Infantile sex-linked hypogammaglobulinemia, 536
Infection allergy, 372
Infection hypersensitivity, 372
Infection or bacterial allergy, 372
Infectious mononucleosis, 672
Infectious mononucleosis syndrome(s), 672
Infectious tolerance, 386
Inflammation, 92
Inflammatory bowel disease (IBD), 495
Inflammatory CD4 T cells, 264
Inflammatory cells, 94
Inflammatory macrophage, 94
Inflammatory mediator, 95
Inflammatory myopathy, 484
Influenza, 667
Influenza hemagglutinin, 669
Influenza virus immunity, 669

Influenza virus vaccine, 693
Influenza viruses, 669
Inhibin, alpha (R1), mouse, 796
Inhibition test, 729
Inhibition zone, 247
Innate, 644
Innate defense system, 96
Innate immune mechanisms against parasites, 643
Innate immunity, 96
Innate immunity against extracellular bacteria, 640, 643
Innate or constitutive defense system, 644
Innocent bystander, 27
Innocent bystander hemolysis, 394
Inoculation, 119
Inosiplex (Isoprinosine), 704
Inositol 1, 4, 5-triphosphate (IP_3), 171, 260
Insects, 710
In-situ hybridization, 748, 751
In-situ transcription, 773
Instructional model, 206
Instructive theory (of antibody formation), 206
Insulin receptor autoantibodies, 405
Insulin resistance, 405
Insulin-dependent (type I) diabetes mellitus (IDDM), 405, 499
Insulin-like growth factor-II (IGF-II), 44
Insulin-like growth factors (IGFs), 43, 315
Insulitis, 405
Intal®, 352
Integrin family of leukocyte adhesive proteins, 30
Integrins, 28
Integrins, HGF/SF activation of, 30
Interallelic conversion, 585
Intercalated cell autoantibodies, 402
Intercellular adhesion molecule-1 (ICAM-1), 31
Intercellular adhesion molecule-2 (ICAM-2), 32
Intercellular adhesion molecule-3 (ICAM-3), 33
Intercrine cytokines, 291
Interdigitating reticular cells, 66
Interfacial test, 729
Interferon α (IFN-α), 623
Interferon β (IFN-β), 307
Interferon γ (IFN-γ), 307
Interferon γ (IFN-γ) inducible protein-10 (IP-10), 307
Interferon γ receptor, 308
Interferon regulatory factors (IRF), 307
Interferon(s) α (IFN-α), 307
Interferons (IFNs), 306
Interleukin 18 (IL-18), 306
Interleukin 6 receptor, 300
Interleukin 8 receptor, type A (IL-8RA), 301
Interleukin 8 receptor, type B (IL-8RB), 302
Interleukin(s) (IL), 291
Interleukin-1 (IL-1), 292
Interleukin-1 (IL-1) receptor, 292
Interleukin-1 receptor antagonist protein (IRAP), 293
Interleukin-1 receptor deficiency, 293, 541
Interleukin-10 (IL-10), 303
Interleukin-10 (IL-10) (cytokine synthesis inhibitory factor), 302
Interleukin-11 (IL-11), 303
Interleukin-11 receptor, 303
Interleukin-12 (IL-12), 304
Interleukin-12 receptor, 304
Interleukin-13 (IL-13), 304
Interleukin-14 (IL-14), 305
Interleukin-15 (IL-15), 305
Interleukin-16 (IL-16), 305
Interleukin-17 (IL-17), 306
Interleukin-19 (IL-19), 306
Interleukin-2 (IL-2), 293
Interleukin-2 receptor (IL-2R), 294
Interleukin-2 receptor α subunit (IL-2Rα), 295
Interleukin-2 receptor αβγ subunit (IL-α2Rβγ), 295
Interleukin-2 receptor β subunit (IL-2Rβ), 295
Interleukin-2 receptor βγ subunit (IL-2Rβγ), 295
Interleukin-2 receptor γ (IL-2Rγ) subunit, 295
Interleukin-20 (IL-20), 306
Interleukin-21 (IL-21), 306
Interleukin-22 (IL-22), 306
Interleukin-23 (IL-23), 306
Interleukin-3 (IL-3), 295
Interleukin-3 receptor (IL-3R), 295
Interleukin-4 (IL-4) (B-cell growth factor), 296
Interleukin-4 receptor (IL-4R), 297
Interleukin-5 (IL-5) (eosinophil differentiation factor), 298
Interleukin-5 receptor complex, 298
Interleukin-6 (IL-6), 298
Interleukin-7 (IL-7), 300
Interleukin-8 (IL-8), 300
Interleukin-8 (IL-8) (neutrophil-activating protein 1), 300
Interleukin-9 (IL-9), 302
Interleukin-9 (murine growth factor P40, T cell growth factor III), 302
Intermediate filaments, 781
Internal image, 204
International Unit of Immunological Activity, 686
Interstitial dendritic cells, 66
Interstitial fluid, 92
Interstitial nephritis, 503
Intervening sequence, 100
Intestinal lymphangiectasia, 553
Intimin, 647
Intolerance, 706
Intrabody, 177
Intracellular cytokine staining, 288, 749
Intracellular immunization, 559
Intracellular pathogens, 647
Intraepidermal lymphocytes, 444
Intraepithelial lymphocytes, 439
Intraepithelial T lymphocytes, 272
Intravenous immune globulin (IVIG), 700
Intrinsic affinity, 236
Intrinsic association constant, 236
Intrinsic asthma, 491
Intrinsic factor, 397
Intrinsic factor antibodies, 397
Intron, 100
Inulin, 333
Inv, 184
Inv allotypes, 184
Inv allotypic determinant, 184
Inv marker, 184
Invariant (Ii) chain, 152
Invasin, 655
Invertebrate immunity, 708
Inverted repeat, 100
Ion exchange chromatography, 725
Ionic, 228
Ir genes, 131
Iridovirus immunity, 710
Iron and immunity, 100
Irradiation chimera, 600
Ischemia, 42
ISCOMs, 124

Islet cell autoantibodies (ICA), 405
Islet cell transplantation, 598
Islets of Langerhans, 598
Isoagglutinin, 200
Isoallergens, 360
Isoallotypic determinant, 200
Isoantibody, 596
Isoelectric focusing (IEF), 770
Isoelectric point (pI), 770
Isoforms, 42
Isogeneic (isogenic), 596
Isograft, 596
Isohemagglutinin, 450
Isoimmunization, 466
Isoleukoagglutinins, 596
Isologous, 596
Isophile antibody, 466
Isophile antigen, 466
Isoproterenol, 363
Isoschizomer, 102
Isotope, 200
Isotopic labeling (radionuclide labeling), 726
Isotype, 199
Isotype switching, 221
Isotypic determinant, 200
Isotypic exclusion, 223
Isotypic specificities, 200
Isotypic variation, 200
ITAMs, 29
ITP, 469
IVIG, 700

J

J chain, 187
J exon, 167
J gene segment, 167, 270
J region, 167, 270
JAK3-SCID, 543
JAK-STAT signaling pathway, 287
Janus kinases (Jaks), 287
Jarisch-Herxheimer, 647
Jawless fishes (cyclostomes, e.g., hagfish and lampreys), 711
Jerne network theory, 204
Jerne plaque assay, 755
Jo-1 autoantibodies, 430
Jo1 syndrome, 431
Job's syndrome, 550
Johnin, 765
Jones criteria, 493
Jones-Mote reaction, 376
Jugular bodies, 714
Junctional diversity, 167, 223
Juvenile onset diabetes, 406
Juvenile rheumatoid arthritis (JRA), 426

K

(K) cells, 639
K (killer) cells, 60
K antigens, 647
K cells (killer cells), 640
κ chain, 183
k light-chain deficiency, 539
K region, 134
K562 cells, 60
Kabat-Wu plot, 196
Kallikrein, 352
Kallikrein inhibitors, 352
Kallikrein–kinin system, 351
Kaposi's sarcoma, 558
Kappa (k), 183
Karyotype, 101
Kawasaki's disease, 493
Kell blood group system, 460
Keratinocyte growth factor (KGF), 319
Keratoconjunctivitis sicca, 415
Kernicterus, 457
Ketokonazole, 677
Keyhole limpet hemocyanin (KLH), 110
Ki autoantibodies, 420
Ki-1 (CD30 antigen), 791
Ki-67 or -780, 790
Kidd blood group system, 461
Killed vaccine, 686
Killed virus vaccines, 689
Killer activatory receptors (KARs), 59
Killer cell (K cell), 60
Killer cell immunoglobulin-like receptors (KIRs), 135
Killer inhibitory receptors (KIRs), 135
Killer T cell, 266
Kilobase (kb), 100
Kinetochore autoantibodies, 429
Kininases, 352
Kininogens, 352
Kinins, 352
Klebsiella immunity, 651
KLH, 110
KM (formerly Inv), 183
Km allotypes, 183
Knockout gene, 769
Knockout mouse, 769
Koch phenomenon, 372
Ku, 421
Ku antibodies, 421
Ku autoantibodies, 409
Kupffer cell, 64
Kuru, 672
Kveim reaction (historical), 490

L

L cell conditioned medium, 315
L chain, 183
λ cloning vector, 754
L+ dose (historical), 249
L2C leukemia, 716
L3T4, 265
L3T4+ T lymphocytes, 265
λ5, 158
l5 B cell development, 158
Lactalbumin, 786
Lactoferrin, 96
Lactoperoxidase, 96
LAF (lymphocyte-activating factor), 292
LAG-3, 59
LAK cells, 625
LAM-1 (leukocyte adhesion molecule-1), 36
Lambda (l) chain, 183
Lamina propria, 437

Laminin, 35
LAMP 1, 45
LAMP 1 and LAMP 2, 50
LAMP 2, 45
Lancefield precipitation test, 733
Landsteiner, 227
Landsteiner's rule (historical), 448
Lane, 734
Langerhans cells, 66, 444
Lapinized vaccine, 686
Large granular lymphocytes (LGL), 58
Large lymphocyte, 50
Large pre-B cells, 157
Large pyroninophilic blast cells, 85
LAT, 280
Latency, 659
Latent allotype, 200
Late-onset immune deficiency, 553
Late-phase reaction (LPR), 92
Latex allergy, 352
Latex fixation test, 737
Latex particles, 737
LATS protector, 392
Lattice theory, 239
Laurell crossed immunoelectrophoresis, 735
Laurell rocket test, 735
LAV, 559
Lazy leukocyte syndrome, 549
LCA (leukocyte common antigen), 47
LCAM, 28
Lck, fyn, ZAP (phosphotyrosine kinases in T cells), 269
LCM, 721
LD_{50}, 686
LDCF (lymphocyte-derived chemotactic factor), 301
LE cell, 421, 516
LE cell "prep,", 421
LE cell test, 421
LE factor, 421
Leader sequence, 208
Leading front technique, 729
Lectin pathway of complement activation, 323
Lectin-like receptors, 42
Lectins, 42
Legionella immunity, 654
Leishmania, 679
Leishmaniasis, 679
Lens-induced uveitis, 415
Lentiviruses, 559
LEP (low egg passage), 686
Lepra cells, 654
Lepromatous leprosy, 654
Leptin, 600
Leptospira immunity, 654
Lesch-Nyhan syndrome, 540
LESTR, 291
Lethal dose, 774
Lethal hit, 267
Letterer-Siwe disease, 472
Leu-CAM, 48
Leukapheresis, 774
Leukemia, 469
Leukemia inhibitory factor (LIF), 316
Leukemia viruses, 470
Leukoagglutinin, 596
Leukocidin, 74
Leukocyte activation, 47, 639
Leukocyte adhesion deficiency (LAD), 549
Leukocyte adhesion molecule-1, 48
Leukocyte adhesion molecules, 47
Leukocyte adhesion proteins, 48
Leukocyte chemotaxis inhibitors, 48
Leukocyte common antigen (LCA, CD45), 275
Leukocyte culture, 772
Leukocyte functional antigens (LFAs), 48
Leukocyte groups, 132
Leukocyte inhibitory factor (LIF), 316
Leukocyte integrins, 48
Leukocyte migration inhibitory factor, 316
Leukocyte transfer, 597
Leukocytes, 47
Leukocytoclastic vasculitis, 481
Leukocytosis, 47
Leukopenia, 47
Leukophysin, 266
Leukotaxis, 40
Leukotriene, 356
Leu-M1 (CD15), 792
Levamisole, 705
Lewis blood group system, 459
Lewis[y]/Sialyl-Lewis[x]CD15/CD15S, 459
L_f dose (historical), 249
L_f flocculating unit (historical), 249
LFA-1 deficiency, 547
LFA-1, LFA-2, LFA-3, 30
LFA-2, 31
LFA-3, 31
LGL (large granular lymphocyte or null cell), 58
LGSP (leukocyte sialoglycoprotein), 254
Liacopoulos phenomenon (nonspecific tolerance), 386
Liberated CR1, 339
Ligand, 28
Light chain, 183
Light chain subtype, 183
Light chain type, 183
Light scatter, 760
Light zone, 88
Light-chain deficiencies, 539
Light-chain disease, 528
Limiting dilution, 762
Lineage infidelity, 474
Linear determinants, 112
Linear epitope, 112
Linear staining, 507
Linkage disequilibrium, 584
Linked recognition, 260
Linker of activation in T cells (LAT), 280
Lipid raft, 148
Lipopolysaccharide (LPS), 109
Lipopolysaccharides, 641
Liposome, 123
Lipoxygenase pathway, 350
Liquefactive degeneration, 515
Lissamine rhodamine (RB200), 741
Listeria, 658
Listeria immunity, 658
Listeria monocytogenes, 658
Live attenuated measles (rubeola) virus vaccine, 686
Live attenuated vaccine, 686
Live measles and mumps virus vaccine, 687
Live measles and rubella virus vaccine, 687
Live measles virus vaccine, 687
Live oral poliovirus vaccine, 687

Live rubella virus vaccine, 687
Live vaccine, 686
Liver cytosol autoantibodies, 400
Liver membrane antibodies, 400
Liver membrane autoantibodies, 400
Liver–kidney microsomal antibodies, 401
Liver–kidney microsomal autoantibodies, 401
Liver–kidney microsome (LKM-1) autoantibodies, 402
Liver–kidney microsome 2 (LKM-2) autoantibodies, 402
Liver–kidney microsome 3 (LKM-3) autoantibodies, 402
LM autoantibodies, 400
LMP genes, 148
LMP-2 and LMP-7, 148
L_o dose (historical), 249
Local anaphylaxis, 354
Local immunity, 437
Loci, 129
Locus, 129
Long homologous repeat, 338
Long-acting thyroid stimulator (LATS), 364, 392, 488
Long-lived lymphocyte, 50
Low responder mice, 721
Low-dose (or low-zone) tolerance, 381
LPAM-1, 38
L-phenlyalanine mustard, 706
L-plastin (LPL), 48
LPR, 94
LPS, 171
LPS-binding protein (LBP), 171
L_r dose (historical), 249
L-Selectin, 439
L-selectin (CD62L), 38
LTα, 266
Lung autoantibodies, 404
Lupoid hepatitis, 400
Lupus anticoagulant, 423, 517
Lupus erythematosus, 423
Lupus erythematosus and pregnancy, 423
Lupus nephritis, 423
Lutheran blood group, 462
Lw antibody, 453
Ly antigen, 171, 276
Ly1 B cell, 171
Ly6, 171, 277
Lyb, 171
Lyb-3 antigen, 171
Lyme disease, 647
Lymph, 84
Lymph gland, 85
Lymph node, 83
Lymphadenitis, 86
Lymphadenoid goiter, 391
Lymphadenopathy, 86
Lymphatic system, 84
Lymphatic vessels, 85
Lymphatics, 85
Lymphoblast, 48
Lymphocyte, 48
Lymphocyte activation, 52, 119
Lymphocyte activation factor (LAF), 292
Lymphocyte anergy, 383
Lymphocyte antigen receptor complex, 165
Lymphocyte antigen stimulation test, 758
Lymphocyte chemokine (BLC), 289
Lymphocyte chemotaxis, 52
Lymphocyte defined (LD) antigens, 584
Lymphocyte determinants, 589
Lymphocyte function-associated antigen-1 (LFA-1), 30
Lymphocyte function-associated antigen-2 (LFA-2), 31
Lymphocyte function-associated antigen-3 (LFA-3), 31
Lymphocyte homing, 52
Lymphocyte maturation, 158, 257
Lymphocyte mitogen stimulation test, 758
Lymphocyte receptor repertoire, 51
Lymphocyte toxicity assay, 758
Lymphocyte trafficking, 51
Lymphocyte transfer reaction, 597
Lymphocyte transformation, 757
Lymphocytic choriomeningitis (LCM), 721
Lymphocytic interstitial pneumonia (LIP), 489
Lymphocytopenic center, 88
Lymphocytosis, 51
Lymphocytotoxic autoantibodies, 394
Lymphocytotoxin, 311
Lymphocytotrophic, 51
Lymphogranuloma venereum (LGV), 670
Lymphoid, 85
Lymphoid cell, 49
Lymphoid cell series, 49
Lymphoid enhancer factor-1(LEF-1), 273
Lymphoid follicle, 85
Lymphoid lineage, 50
Lymphoid nodules (or follicles), 85
Lymphoid organs, 85
Lymphoid progenitor cell, 48
Lymphoid system, 78
Lymphoid tissues, 78
Lymphokine, 288
Lymphoma, 477
Lymphoma belt, 533
Lymphomatoid granulomatosis, 494
Lymphomatosis, 477
Lymphopenia, 49
Lymphopoiesis, 49
Lymphoreticular, 49
Lymphorrhages, 50
Lymphotactin (Ltn), 290
Lymphotoxin (LT), 311
Lysins, 179
Lysis, 249
Lysogeny, 659
Lysosome, 45
Lysozyme, 632
Lyt 1,2,3, 268
Lyt antigens, 268
Lytic granules, 266

M

2-Mercaptoethanol agglutination test, 589, 739
6-Mercaptopurine (6-MP), 706
6′-Methoxycinchonan-9-ol), 568
88 Monocytes, 75
M cell, 439
μ chain, 191
M component, 528
M macroglobulin, 532
M protein, 527
Mab, 210
MAC, 332, 655
MAC-1, 36, 175
Macroglobulin, 532

Macroglobulinemia, 532
Macrophage chemotactic and activating factor (MCAF), 64
Macrophage chemotactic and activating factor (MCAF) (MCP-1), 290
Macrophage chemotactic factor (MCF), 64, 308
Macrophage colony-stimulating factor (M-CSF), 313
Macrophage cytophilic antibody, 65
Macrophage functional assays, 65, 758
Macrophage immunity, 62
Macrophage inflammatory peptide-2 (MIP-2), 65
Macrophage inflammatory protein 1α (MIP-1α), 318
Macrophage inflammatory protein 1β (MIP-1β), 318
Macrophage inflammatory protein-1-α (MIP-1), 64
Macrophage inflammatory protein-1α (MIP-1α), 317
Macrophage inflammatory protein-2 (MIP-2), 318
Macrophage migration inhibitory factor, 288
Macrophage migration test, 758
Macrophage/monocyte chemotaxis, 62
Macrophage-activating factor (MAF), 64, 308
Macrophage–monocyte inhibitory factor (MIF), 758
Macrophages, 60, 619
MadCAM-1, 38, 438
MAF, 308
MAGE-1 protein, 618
MAIDS, 566
MAIS complex, 655
Major basic protein (MBP), 363
Major histocompatibility complex (MHC) molecule, 132
Major histocompatibility complex Class II deficiency (MHC II deficiency), 544
Major histocompatibility complex restriction, 141
Major histocompatibility system, 127
Malaria, 680, 694
Malaria vaccine, 694
Malignant, 613
Malignolipin (historical), 613
Mammals, 715
Mancini test, 727
Mannose receptor, 62
Mannose-binding lectin (MBL), 62
Mantle, 88
Mantle zone, 86
Mantle zone lymphoma, 473, 538
Mantoux test, 656
Marek's disease, 412
Marginal zone, 90
Margination, 71
Marsupial immunity, 719
MART-1 (M2-7C10), mouse, 796
Mass vaccination, 686
Mast cell activation, 350
Mast cell growth factor-1, 295
Mast cell growth factor-2, 297
Mast cell tryptase, 350
Mast cell–eosinophil axis, 351
Mast cells, 74
Masugi nephritis, 503
Maternal immunity, 443
Maternal immunoglobulins, 211
Mature B cell, 159
Mature T cells, 277
MBP, 412
McCleod phenotype, 460
Mcg isotypic determinant, 184
MCP-1 in atherosclerosis, 291
MCTD, 515
MDP, 126
Measles vaccine, 693
Mechanical barriers, 632
Medulla, 85
Medullary cord, 85
Medullary sinuses, 86
Megakaryocytes, 77
MEL-14, 27
MEL-14 antibody, 27
Melanoma antigen-1 gene (MAGE-1), 617
Melanoma growth stimulatory activity (MGSA), 290
Melanoma-associated antigens (MAA), 618
Melphalan (l-phenylalanine mustard), 570
Membrane Attack Complex (MAC), 331
Membrane attack unit, 323
Membrane cofactor of proteolysis (MCP or CD46), 336
Membrane cofactor protein (MCP), 336
Membrane complement receptors, 338
Membrane immunofluorescence, 743
Membrane immunoglobulin, 170
Membranoproliferative glomerulonephritis (MPGN), 502
Membranous glomerulonephritis, 502
Memory, 122
Memory cells, 122
Memory lymphocytes, 122
Memory T cells, 122
Meningococcal vaccine, 696
Mercury and immunity, 403
Metaproterenol, 363
Metastasis, 613
Metatype autoantibodies, 394
Methotrexate, 570
Methyl green pyronin stain, 773
Metronidazole, 680
MGUS, 529
MHC, 127, 129, 151, 153
MHC class I deficiency, 544
MHC class I molecules, 132
MHC class IB molecules, 134
MHC class II compartment (MIIC), 134
MHC class II deficiency, 544
MHC class II region, 134
MHC class II transactivator (CIITA), 134
MHC congenic mice, 135
MHC disease associations, 135
MHC functions, 131
MHC genes, 129
MHC haplotype, 589
MHC molecules, 131
MHC mutant mice, 135
MHC peptide tetramers, 131
MHC peptide-binding specificity, 131
MHC recombinant mice, 131
MHC restriction, 140
MHC-I antigen presentation, 155
MIC molecules, 440
Microenvironment, 27
Microfilaments, 27
Microfold cells, 439
Microglial cell, 63
Microlymphocytotoxicity, 586, 756
Microtiter technique, 739
Microtubules, 45
Mid-piece, 323
MIF (macrophage/monocyte migration inhibitory factor), 288
MIG, 291

Mikulicz's syndrome, 521
Miller-Fisher syndrome, 507
Minimal hemagglutinating dose (MHD), 737
Minimal hemolytic dose (MHD), 342
Minimum lethal dose (MLD), 774
Minisatellite, 773
Minor histocompatibility antigens, 579
Minor histocompatibility locus, 579
Minor histocompatibility peptides, 580
Minor lymphocyte stimulatory (MIs) loci, 580, 583
Minor lymphocyte-stimulating (Mls) determinants, 580
Minor lymphocyte-stimulating genes, 580
Minor transplantation antigens, 579
MIP-1 (macrophage inflammatory protein-1-α), 318
MIP-1α receptor, 318
MIP-1β, 318
MIRL (membrane inhibitor of reactive lysis), 331
Mitochondria, 45
Mitochondrial antibodies, 400
Mitochondrial autoantibodies (MA), 401
Mitogen, 643
Mitotic spindle apparatus autoantibodies, 425
Mitsuda reaction, 656
Mixed agglutination, 729, 738
Mixed hemadsorption, 730
Mixed leukocyte reaction (MLR), 589
Mixed lymphocyte reaction (MLR), 756
Mixed vaccine, 685
Mixed-antiglobulin reaction, 729
Mixed-connective tissue disease (MCTD), 515
Mixed-lymphocyte culture (MLC), 589
Mixed-lymphocyte reaction (MLR), 589
MK-571, 363
MLC, 589
MLNS (mucocutaneous lymph node syndrome), 494
MMR vaccine, 687
MNSs blood group system, 458
Mo1, 334
Modulation, 619
Molecular (DNA) typing, 588
Sequence specific priming (SSP), 588
Molecular hybridization probe, 751
Molecular mimicry, 433
Molluscs and arthropods, 710
Moloney test, 651
Monoclonal, 210
Monoclonal antibody (MAb), 208
Monoclonal antibody (MAb) therapy, 700
Monoclonal anti-insulin antibody, 789
Monoclonal gammopathy, 528
Monoclonal gammopathy of undetermined significance (MGUS), 528
Monoclonal immunoglobulin, 530
Monoclonal immunoglobulin deposition disease (MIDD), 524
Monoclonal protein, 530
Monocyte chemoattractant of a protein-2 (MCP-2), 290
Monocyte chemoattractant protein-1 (MCP-1), 290
Monocyte chemoattractant protein-3 (MCP-3), 290
Monocyte colony-stimulating factor (MCSF), 314
Monocyte-derived neutrophil chemotactic factor, 301
Monocyte–phagocyte system, 76
Monogamous bivalency, 217
Monogamous multivalency, 218
Monokine, 288
Mononuclear cells, 48
Mononuclear phagocyte, 68
Mononuclear phagocyte system, 67
Mononuclear phagocytes, 635
Montenegro test, 765
Mooren's ulcer, 511
Moro test, 655
MOTT (mycobacteria other than *Mycobacterium tuberculosis*), 655
MOTT cell, 680
Mouse hepatitis virus (MHV), 534
Mouse immunoglobulin antibodies, 603
MRL-lpr/lpr mouse, 424
Mucins, 26
Mucocutaneous candidiasis, 677
Mucocutaneous lymph node syndrome, 493
Mucosa homing, 438
Mucosa-associated lymphoid tissue (MALT), 437
Mucosal immune system, 437
Mucosal lymphoid follicles, 439
Mucosal tolerance, 443
Multicatalytic proteinase autoantibodies, 410
Multilocus probes (MLPs), 591, 753
Multiple autoimmune disorders (MAD), 434
Multiple myeloma, 525
Multiple sclerosis (MS), 507
Multiple-emulsion adjuvant, 125
Multivalent, 227
Multivalent antiserum, 227
Multivalent vaccine, 685
Mumps vaccine, 693
Muramyl dipeptide (MDP), 126
Mutant, 169
Mutation, 169
Myasthenia gravis (MG), 485
Myc, 533
Mycobacteria immunity, 655
Mycobacterial adjuvants, 124
Mycobacterial peptidoglycolipid, 125
Mycobacterium, 655
Mycophenolate mofetil, 572
Mycoplasma immunity, 647
Mycoplasma–AIDS link, 566
Mycosis fungoides, 473
Myelin autoantibodies, 412
Myelin basic protein (MBP), 412, 508
Myelin basic protein (MBP) antibodies, 412
Myelin-associated glycoprotein (MAG) autoantibodies, 412
Myeloid antigen, 473
Myeloid cell series, 68
Myeloma, 527
Myeloma protein, 527
Myeloma, IgD, 527
Myelomatosis, 527
Myeloperoxidase, 70, 551
Myeloperoxidase (MPO) deficiency, 551
Myocardial autoantibodies (MyA), 410
Myogenin (F5D), mouse, 796
Myoglobin, 781
Myoglobin antibody, 781
Myoid cells, 254
Myositis-associated autoantibodies, 431
Myositis-specific autoantibodies, 431

N

200-4 Nuclear matrix protein, 795
N region, 169
N-addition, 169
Naïve, 51

Naïve lymphocyte, 51
NAP, 70
NAP-1, 301
NAP-2 (neutrophil activating protein-2), 70
Naprosyn, 705
Naproxen, 705
Nasopharyngeal-associated lymphoreticular tissue (NALT), 84
Native immunity, 96
Natural antibodies, 449
Natural autoantibodies, 177, 389
Natural fluorescence, 742
Natural immunity, 96, 631
Natural immunity against viruses, 643
Natural killer (NK) cells, 58, 620, 638, 639
Natural passive immunity, 644
Naturally acquired immunity, 644
Negative induction apoptosis, 81
Negative phase, 119
Negative selection, 259
Neisseria immunity, 656
Neisser-Wechsberg phenomenon, 342
Neoantigens, 613
Neonatal Fc receptor (FcRn), 177
Neonatal immunity, 177
Neonatal thymectomy syndrome, 253
Neoplasm, 613
Neopterin, 559
Nephelometry, 243, 730
Nephritic factor, 403
Nephritic factor autoantibodies, 403
Nephritic syndrome, 503
Nephrotic syndrome, 503
Network hypothesis, 204
Network theory, 204
Neural cell adhesion molecule-L1 (NCAM-L1), 28
Neuraminidase, 659
Neurofilament, 783
Neurofilament (2F11), mouse, 796
Neurological autoimmune diseases, 412
Neuromuscular junction autoimmunity, 408
Neuron specific enolase (NSE), 749, 783
Neuron specific enolase (NSE) antibody, 783
Neuronal autoantibodies, 412
Neuropeptides, 92
Neuropilin, 28, 43
Neutralization, 660
Neutralization test, 771
Neutralizing antibody, 659
Neutropenia, 70
Neutrophil, 69
Neutrophil activating factor 1, 302
Neutrophil activating peptide 2, 302
Neutrophil activating protein 1 (NAP-1), 302
Neutrophil activating protein 2 (NAP-2), 291
Neutrophil attracting peptide (NAP-2), 302
Neutrophil cytoplasmic antibodies, 395
Neutrophil leukocyte, 69, 394
Neutrophil microbicidal assay, 70
Neutrophil microbicidal assay, 760
Neutrophil nicotinamide adenine dinucleotide phosphate oxidase, 549
Neutrophils chemotaxis, 69
New Zealand black (NZB) mice, 395
New Zealand white (NZW) mice, 424
Newcastle disease, 672
Newcastle disease vaccines, 696
Nezelof's syndrome, 541
NF-AT, 280
NFc (nephritic factor of the classical pathway), 336
NF-kB, 55
N-formyl peptide receptor (FPR), 289
N-formylmethionine, 95
NFt (C3bBb-P stabilizing factor), 336
Nick translation, 773
NIP (4-hydroxy,5-iodo,3-nitrophenylacetyl), 115
Nitric oxide (NO), 61, 62
Nitric oxide synthetase, 62
Nitroblue tetrazolium (NBT) test, 758
NK 1.1, 59
NK cell, 58
NK1.1, 59
NK1-T, 59
NK-T, 59
N-linked oligosaccharide, 44
N-nucleotides, 169
Nocardia immunity, 677
NOD (nonobese diabetic) mouse, 406
NON mouse, 406
Non-A non-B hepatitis, 499
Nonadherent cell, 50
Noncovalent forces, 228
Nonimmunologic classic pathway activators, 324
Nonprecipitating antibodies, 177
Nonproductive rearrangement, 101
Nonresponder, 120
Nonsecretor, 452
Nonsequential epitopes, 112
Nonspecific esterase (α naphthyl acetate esterase), 68
Nonspecific fluorescence, 744
Nonspecific immunity, 98
Nonspecific suppression, 567
Nonspecific T cell suppressor factor, 268, 567
Nonspecific T lymphocyte helper factor, 51, 264
Nonsquamous keratin (NSK), 780
Nonsterile immunity, 96
Nonsteroidal antiinflammatory drugs (NSAIDs), 357, 705
Nontissue-specific antigen, 110
Normal lymphocyte transfer reaction, 597
Northern blotting, 751
Norvir®, 564
NP (4-hydroxy,3-nitrophenylacetyl), 115
N-region diversification, 169
NSAID (nonsteroidal antiinflammatory drug), 706
N-terminus, 181
Nuclear dust (leukocytoclasis), 481
Nuclear matrix proteins (NMPs), 101
Nucleolar autoantibodies, 427
Nucleoside phosphorylase, 545
Nude mouse, 282
Null cell, 57
Null cell compartment, 57
Null phenotype, 449
Nutrition and immunity, 98
Nylon wool, 755
NZB/NZW F_1 hybrid mice, 424

O

θ antigen, 268
"O" phage antibody library, 178
O antigen, 449, 641
O blood group, 449

O125 (ovarian celomic), 789
Oakley-Fulthorpe test, 733
Oct-2, 172
OKT monoclonal antibodies, 178
OKT®3 (Orthoclone OKT®3), 574, 610
OKT4, 178
OKT8, 178
Old tuberculin (OT), 765
Oligoclonal bands, 508
Oligoclonal response, 508
Oligosaccharide determinant, 112
Omenn's syndrome, 544
Onchocerciasis volvulus immunity, 680
Oncofetal antigens, 616
Oncogene theory, 614
Oncogenes, 613
Oncogenesis, 614
Oncogenic virus, 617
Oncomouse, 614
One gene, one enzyme theory (historical), 168
One-hit theory, 343
One-turn recombination signal sequences, 168
Open reading frame (ORF), 101
Opisthorchiasis–clonorchiasis immunity, 680
Opsonin, 72, 632
Opsonins, 633
Opsonization, 72
Opsonophagocytosis, 72
Oral immunology, 442
Oral tolerance, 385, 442, 443
Oral unresponsiveness, 442
Organ bank, 596
Organ brokerage, 596
Organism-specific antibody index (OSAI), 647
Organ-specific antigen, 391
Organ-specific autoimmune diseases, 391
Original antigenic sin, 116
Orthoclone OKT3, 603
Orthotopic, 591
Orthotopic graft, 591
Osteoclast-activating factor (OAF), 319
OT (historical), 765
Ouchterlony test, 733
Oudin test, 733
Outbreeding, 770
Ovalbumin (OA), 108
Ovary antibodies (OA), 406
Ovary autoantibodies, 406
Ovine immune system, 719
Owl eye appearance, 662
Oxazolone, 374
Oxidized low-density lipoprotein autoantibodies against, 434
Oxygen-dependent killing, 70, 637
Oxygen-independent killing, 71
Oz isotypic determinant, 200

P

(Phe,G)AL, 115
P, 334
P antigen, 458
P1 kinase, 672
P1[A1] antibodies, 464
P24 antigen, 559
P63 (ap53 Homolog at 3q27–29) Ab-4 (cocktail) mouse monoclonal antibody, 796
P-80, 730
P-addition, 168
PAF, 77
Palindrome, 100
Pan keratin antibodies, 748
Panagglutination, 451
Pancreatic islet cell hormones, 749
Pancreatic transplantation, 597
Panning, 758
Pannus, 518
Pan-T cell markers, 261
PAP (peroxidase-antiperoxidase) technique, 747
Papain, 194
Papain hydrolysis, 194
Paper radioimmunosorbent test (PRIST), 766
Papillomavirus immunity, 672
Papovaviruses, 660
Parabiotic intoxication, 607
Para-Bombay phenotype, 450
Paracortex, 50
Paracrine, 286
Paracrine factor, 287
Paradoxical reaction (historical), 351
Paraendocrine syndromes, 532
Paraimmunoglobulins, 532
Parainfluenza virus (PIV) immunity, 673
Paralysis, 386
Paramyxovirus immunity, 673
Paraneoplastic autoantibodies, 412
Paraneoplastic autoimmune syndromes, 433
Paraneoplastic pemphigus, 413
Paraprotein, 527
Paraproteinemias, 532
Parasite immunity, 678
Parasites, 678
Parathyroid hormone autoantibodies, 391
Paratope, 176
Parenteral, 119
Parietal cell antibodies, 397, 497
Parietal cell autoantibodies, 398
Paroxysmal cold hemoglobinuria (PCH), 395
Paroxysmal nocturnal hemoglobinuria (PNH), 340
Partial identity, 245
Parvovirus, 666
Parvovirus immunity, 666
PAS, 749
Passive agglutination, 738
Passive agglutination test, 739
Passive anaphylaxis, 353
Passive Arthus reaction, 367
Passive cutaneous anaphylaxis (PCA), 362, 765
Passive hemagglutination, 246
Passive hemolysis, 246
Passive immunity, 97, 644
Passive immunization, 215, 686
Passive sensitization, 347
Passive systemic anaphylaxis, 354
Passive transfer, 347
Pasteurella immunity, 657
Patch test, 369, 763
Patching, 165
Paternity testing, 771
Pathogen, 647
Pathogen-associated molecular pattern (PAMP), 647

Pathogenicity, 631
Pathologic autoantibodies, 389
Pattern recognition receptors (PRRs), 647
Paul-Bunnell test, 737
Pax-5 gene, 168
PBC, 498
PCA, 362
PCH, 534
PCP, 559
PCR, 752
PECAM (CD31), 34
Pediatric AIDS, 565
Pemphigoid, 413
Pemphigus erythematosus (Senear-Usher syndrome), 413
Pemphigus foliaceus, 413
Pemphigus vulgaris, 414, 479
Penicillin hypersensitivity, 377
Pentadecacatechol, 375
Pentamidine isoethionate, 564
Pentraxin family, 96
Pentraxins, 96
Pepsin digestion, 193, 194
Peptide T, 560
Peptide-binding cleft, 146
Peptidoglycan layer, 640
Percoll®, 736
Perforin, 267
Periarteriolar lymphoid sheath, 90
Periarteritis nodosa, 484
Perinuclear antibodies, 519
Perinuclear factor (profillagrin) autoantibodies, 425
Peripheral blood mononuclear cells, 51
Peripheral lymphoid organs, 82
Peripheral tolerance, 382
Permeability factors, 287
Permeability-increasing factor, 287
Pernicious anemia (PA), 396, 496
Peroxidase–antiperoxidase (PAP) technique, 747
Persistent generalized lymphadenopathy (PGL), 559
Pertussis adjuvant, 125
Pertussis vaccine, 691
PFC (plaque-forming cell), 755
pFc′ fragment, 194
Pfeiffer phenomenon (historical), 249
PFU, 755
PHA, 277
Phacoanaphylactic endophthalmitis, 415
Phacoanaphylaxis, 362
Phage antibody library, 168
Phage display, 178
Phage display library, 178
Phage neutralization assay, 771
Phagocyte disorders, 547
Phagocytes, 71
Phagocytic cell function deficiencies, 548
Phagocytic cells, 634
Phagocytic dysfunction, 548
Phagocytic index (PI), 548
Phagocytosis, 63, 635, 707
Phagolysosome, 63, 72
Phagosome, 63
Pharyngeal pouch, 251
Pharyngeal pouch syndrome, 251
Pharyngeal tonsils, 86
Phenotype, 588
Phenylbutazone, 520, 706
Philadelphia chromosome, 475
Phoma species, 347
Phorbol ester(s), 54, 110, 278
Phosphatase, 42
Phosphatidylinositol bisphosphate (PIP2), 54
Phosphocholine antibodies, 648
Phospholipid autoantibodies, 422
Photoallergy, 356
Photoimmunology, 445
Phycoerythrin, 760
Phylogenetic-associated residues, 713
Phylogeny, 722
Phytoalexins, 707
Phytohemagglutinin (PHA), 707
Phytoimmunity, 707
Phytomitogens, 707
Phytonicides, 707
Picornavirus, 673
Picornavirus immunity, 673
Picryl chloride, 374
Piecemeal necrosis, 497
Pigeon breeder's lung, 347
Pili, 648
Pinocytosis, 63
Pituitary autoantibodies, 413
Pituitary hormones, 784
P-K reaction, 351
PK test, 351
Plague vaccine, 695, 696
Plant immunity, 707
Plaque technique, 771
Plaque-forming assay, 770
Plaque-forming cell (PFC) assay, 754, 770
Plaque-forming cells, 770
Plasma, 447
Plasma cell antigen, 163
Plasma cell dyscrasias, 528
Plasma cell leukemia, 528
Plasma cells, 56, 163
Plasma half-life ($T_{1/2}$), 225
Plasma histamine, 355
Plasma pool, 225
Plasmablast, 162
Plasmacyte, 162
Plasmacytoma, 476, 528
Plasmapheresis, 773
Plasmid, 773
Plasmin, 195
Plasminogen, 65
Plasminogen activator, 65
Platelet, 77
Platelet antigens, 464
Platelet autoantibodies, 396
Platelet endothelial cell adhesion molecule-1 (PECAM-1) (CD31), 33
Platelet factor 4 (PF4), 318
Platelet transfusion, 464
Platelet-activating factor (PAF), 350
Platelet-associated immunoglobulin (PAIgG), 598
Platelet-derived growth factor (PDGF), 77
Platelet-derived growth factor receptor (PDGF-R), 78
Pluripotent stem cell, 46
PMN, 69
PMNs (polymorphonuclear neutrophils), 634
PM-Scl autoantibodies, 427
Pneumococcal 7 valent conjugate vaccine, 692
Pneumococcal polysaccharide, 108

Pneumococcal polysaccharide vaccine, 691
Pneumocystis carnii (PCP), 562
PNH cells, 534
P-nucleotides, 167
POEMS, 531
POEMS syndrome, 531
Poison ivy, 375
Poison ivy hypersensitivity, 375
Poliomyelitis vaccines, 693
Poliovirus, 666
Pollen hypersensitivity, 360
Polyagglutination, 451
Polyarteritis nodosa, 483
Polyarthritis, 493
Polyclonal, 57
Polyclonal activators, 110
Polyclonal antibodies, 210
Polyclonal hypergammaglobulinemia, 529
Polyclonal rabbit anti-calretinin, 795
Polyclone proteins, 225
Polyendocrine deficiency syndrome (polyglandular autoimmune syndrome), 405
Polyendocrine autoimmunity, 435
Polyethylene glycol assay for CIC, 730
Polygenic, 102
Polygenic inheritance, 120
Polyimmuoglobulin receptor, 437
Polyimmunoglobulin receptor, 219
Polymerase chain reaction (PCR), 752
Polymers, 194
Polymorphism, 453, 581
Polymorphonuclear leukocytes (PMNs), 69
Polymyositis, 484
Polynucleotide, 124
Polyomavirus immunity, 576
Polyspecific antihuman globulin (AHG), 455
Polyspecificity, 227
Polyvalent, 227
Polyvalent antiserum, 178
Polyvalent pneumococcal vaccine, 692
Polyvalent vaccine, 689
Porcine immunity, 719
Positive induction apoptosis, 81
Positive selection, 259
Postcapillary venules, 86
Postcardiotomy syndrome, 410
Postinfectious encephalomyelitis, 413, 510
Postinfectious iridocyclitis, 415
Postrabies vaccination encephalomyelitis, 692
Poststreptococcal glomerulonephritis, 504
Posttransfusion graft-vs.-host disease, 607
Postvaccinal encephalomyelitis, 689
Postzone, 248
Poxvirus immunity, 673
PPD, 372
PPLO (pleuropneumonia-like organisms), 463
P ratio, 741
Prausnitz-Küstner (P-K) reaction (historical), 362
Pre-B cell receptor, 157
Pre-B cells, 157
Precipitating antibody, 176
Precipitation, 237
Precipitation reaction, 237
Precipitin, 176
Precipitin curve, 239
Precipitin reaction, 237
Precipitin test, 237
Prednisolone, 568
Prednisone, 568
Preemptive immunity, 97
Prekallikrein, 42
Preprogenitor cells, 159
Pre-T cell receptor (Pre-TCR), 258
Pre-T cells, 258
Pre-T lymphocyte, 258
Prick test, 352, 763
Primary agammaglobulinemia, 536
Primary allergen, 360
Primary biliary cirrhosis (PBC), 498
Primary follicle, 86
Primary granule, 70
Primary immune response, 120
Primary immunodeficiency, 535
Primary interaction, 227
Primary lymphoid organs, 84
Primary lysosome, 45
Primary nodule, 85
Primary reaction, 227
Primary response, 120
Primary sclerosing cholangitis (PSC), 401
Primary structure, 113
Primate (nonhuman) immune system, 721
Primed, 120
Primed lymphocyte, 120
Primed lymphocyte test (PLT), 582
Primed lymphocyte typing (PLT), 582
Prion, 666
Private antigen, 118, 590
Private idiotypic determinant, 203
Private specificity, 118
Privileged sites, 592
Pro-C3, 327
Pro-C4, 329
Pro-C5, 329
Procomplementary factors, 323
Productive rearrangement, 51
Professional antigen-presenting cells, 150
Progenitor cell, 48
Programmed cell death, 79
Progressive multifocal leukoencephalopathy, 558
Progressive systemic sclerosis (scleroderma), 522
Progressive transformation of germinal centers (PTGC), 477
Progressive vaccinia, 688
Promoter, 615
Properdin (factor P), 333
Properdin deficiency, 334
Properdin pathway, 334
Properdin system, 334
Prophylactic immunization, 697
Propylthiouracil, 706
Prostaglandins (PG), 357
Prostate-specific antigen (PSA), 616, 787
Prostatic acid phosphatase (PAP)/prostatic epithelial antigen, 787
Pro-T cell, 257
Proteasome, 146
Proteasome genes, 148
Protectin (CD59), 332
Protective antigens, 97
Protective immunity, 97, 631
Protein A, 171

Protein AA, 525
Protein B, 730
Protein blotting, 750
Protein kinase C (PKC), 42, 278
Protein M (M antigen), 647
Protein P, 525
Protein S, 42
Protein SAA, 525
Protein separation techniques, 750
Proteinuria, 503
Proteolipid protein autoantibodies, 413
Proteus immunity, 657
Prothrombin antibodies, 396
Prothymocyte, 258
Protooncogene, 614
Protoplast, 647
Protoplast fusion, 773
Protozoa, 707
Provirus, 673
Provocation poliomyelitis, 693
Prozone, 247
Prozone phenomenon, 247
PRP antigen, 648
PSA (prostate-specific antigen), 787
P-selectin (CD62P), 38
Pseudoalleles, 102
Pseudoallergic reaction, 356
Pseudoallergy, 356
Pseudogene, 101
Pseudolymphoma, 477
Pseudolymphomatous lymphadenitis, 527
Pseudomonas aeruginosa immunity, 657
Pseudoparaproteinemia, 532
Pseudopodia, 72
Psoriasis vulgaris, 480
Psychoneuroimmunology, 384
PTα, 258
PTAP, 691
Public antigen (supratypic antigen), 118, 456, 590
Public idiotypic determinant (IdX or CRI), 203
Public specificity, 130
Pulmonary vasculitis, 491
Pulsed-field gel electrophoresis, 735
Purified protein derivative (PPD), 369
Purine nucleoside phosphorylase (PNP) deficiency, 545
Purine nucleotide phosphorylase (PNP), 545
Purpura, 484
Purpura hyperglobulinemia, 530
Pyogenic bacteria, 648
Pyogenic infection, 648
Pyogenic microorganisms, 648
Pyrogen, 292, 648
Pyroglobulins, 175
Pyroninophilic cells, 163

Q

Q fever, 669
Qa, 133
Qa antigens, 134
Qa region, 133
Qa-2 antigen, 134
Quantitative gel diffusion test, 239
Quantitative precipitin reaction, 238, 726
Quaternary structure, 113
Quaternary syphilis, 559
Quellung phenomenon, 648
Quellung reaction, 648
Quenching, 744
Quin-2, 745
Quinidine (β-quinine), 706

R

12/23 Rule, 167
RA, 519
RA cell, 426
RA-33, 519
Rabbit immunity, 716
Rabbit immunoglobulin allotypes, 717
Rabies, 666
Rabies vaccination, 692
Rabies vaccine, 692
Rac, 280
Radial immunodiffusion, 245
Radiation and immunity, 575
Radiation bone marrow chimeras, 600
Radiation chimera, 600
Radioallergosorbent test (RAST), 766
Radioimmunoassay (RIA), 730
Radioimmunodiffusion, 727
Radioimmunodiffusion test, 727
Radioimmunoelectrophoresis, 732
Radioimmunoprecipitation assay (RIPA), 730
Radioimmunoscintigraphy, 624
Radioimmunosorbent test (RIST), 766
Radiolabeling, 726
Radiomimetic drug, 575
RAG-1 and RAG-2, 167
Ragg, 519
Ragocyte, 519
Ragweed, 361
Raji cell assay, 740
Ramon test (historical), 774
RANA (rheumatoid arthritis-associated nuclear antigen), 519
RANA autoantibodies, 427
Random breeding, 719
RANTES, 290
Rapamycin, 572
Ras, 102, 614, 615
Raynaud's phenomenon, 515
RB200, 744
RCA, 323
RCA locus (regulator of complement activation), 323
Reaction of identity, 244
Reaction of nonidentity, 244
Reaction of partial identity, 244
Reactive lysis, 343
Reactive nitrogen intermediates, 72
Reactive oxygen intermediates (ROIs), 30
Reactive oxygen species (ROS), 61
Reagin (historical), 181
Reassortant vaccine, 697
Rebuck skin window, 766
Receptor, 55
Receptor editing, 168
Receptor-associated tyrosine kinases, 52
Receptor-mediated endocytosis, 166
Recirculating pool, 52
Recirculation of lymphocytes, 52

Recognition phase, 86
Recognition unit, 322, 325
Recombinant DNA, 102
Recombinant DNA technology, 102, 773
Recombinant inbred strains, 720
Recombinant vaccine, 697
Recombination activating genes (RAG-1 and RAG-2), 225
Recombination activating genes 1 and 2 (RAG-1 and RAG-2), 167
Recombination recognition sequences, 167
Recombination signal sequences (RSSs)B, 225
Recombinatorial germ-line theory, 207
Red cell-linked antigen antiglobulin test, 737
Red pulp, 89
Reed-Sternberg cells, 477
Regional enteritis, 496
Regulation of complement activation (RCA) cluster, 327
Regulators of complement activity (RCA), 327
Regulatory T cells, 54
Reiter complement fixation test (historical), 741
Rejection, 602
Relapsing polychondritis, 512
Relative risk (RR), 135
Released antigen, 683
REMICADE, 702
REOPRO®, 703
Reovirus immunity, 674
Repeating units, 109
Reptile immunity, 714
Rescue graft, 592
Resident macrophage, 62
Respiratory burst, 70
Responder animals, 120
Resting lymphocytes, 50, 156, 279
Restitope, 153
Restriction endonuclease, 102
Restriction fragment length polymorphism (RFLP), 220, 754
Restriction map, 102
Reticular cells, 86
Reticular dysgenesis, 541
Reticulin autoantibodies, 398
Reticuloendothelial blockade, 577
Reticuloendothelial system (RES), 67
Reticulosis, 477
Reticulum cell, 86
Reticulum cell sarcoma, 473
Retina autoantibodies, 415
Retrovir®, 564
Retrovirus, 556
Retrovirus immunity, 559
Rev protein, 559
Reverse anaphylaxis, 367
Reverse immunology, 626
Reverse Mancini technique, 727
Reverse passive Arthus reaction, 367
Reverse passive cutaneous anaphylaxis (RPCA), 362
Reverse plaque method, 755
Reverse radioimmunodiffusion, 727
Reverse transcriptase, 559
Reverse transcriptase polymerase chain reaction (RT-PCR), 752
RF, 519
RFLP (restriction fragment length polymorphism), 102, 754
Rhabdovirus immunity, 674
Rhemuatoid arthritis cell (RA cell), 519
Rhemuatoid factor (RF), 519
Rhesus antibody, 453
Rhesus antigen, 452
Rhesus blood group system, 452
Rhesus incompatibility, 454
Rheumatic fever (RF), 492
Rheumatoid arthritis (RA), 426, 518
Rheumatoid arthritis cell (RA cell), 426
Rheumatoid factor (RF), 426
Rheumatoid nodule, 426, 519
Rheumatoid pneumonitis, 426
Rhinovirus immunity, 674
RhLA locus, 135
Rh_{null}, 452
Rh_oD immune globulin, 453
Rhodamine isothiocyanate, 744
RhoGAM, 453
RIA, 730
Ribavarin, 564
Ribosomal P protein autoantibodies (RPP), 424
Ribosome, 45
Ricinus communis, 625
Rickettsia immunity, 669
Rieckenberg reaction, 730
Rinderpest vaccines, 694
Ring precipitation test, 238
Ring test, 238
RIST, 766
Rituxan, 703
RNA polymerase, 427
RNA polymerases I, II, and III autoantibodies, 427
RNA splicing, 774
RNA-directed DNA polymerase (reverse transcriptase), 559
RNAse protection assay, 773
Ro/SS-A and La/SS-B, 429
Rocket electrophoresis, 734
Rodgers (Rg) antigens, 462
Romer reaction (historical), 765
Roquinimex, 704
Rosette, 54, 261
Rose-Waaler test, 737
Rotavirus, 666
Round cells, 51
Rous sarcoma virus (RSV), 615
RPR (rapid plasma reagin) test, 739
RS61443 (Mycophenolate mofetil), 574
Rubella vaccine, 693
Runt disease, 282
Runting syndrome, 282
Russell body, 57

S

19 S antibody, 190
S antibody, 184
S protein, 332
S region, 135
S value (Sverdberg unit), 736
S-100, 782
S-100 protein, 782
S-100 protein antibody, 782
Sabin vaccine, 693
Sabin-Feldman dye test, 680
Saccharated iron oxide, 68
Sacculus rotundus, 441, 717
Sago spleen, 524
SAIDS (simian acquired immunodeficiency syndrome), 565

Saline agglutinin, 450
Salk vaccine, 693
Salmonella immunity, 657
Salt precipitation, 736
Salting out, 736
Sanarelli-Shwartzman reaction, 368
Sandoglobulin®, 572
Sandwich ELISA, 743
Sandwich immunoassay, 742
Sandwich methodology, 742
Sandwich technique, 742
Saponin, 697
SAR, 707
Sarcoidosis, 490
Sarcomas, 613
SCAB (single chain antigen-binding proteins), 181
SCID (severe combined immundeficiency) human mouse, 543
Scarlet fever, 648
Scatchard, 230
Scatchard analysis, 230
Scatchard equation, 230
Scavenger receptors, 68
ScFv, 181
Schick test, 765
Schistosoma immunity, 681
Schistosomiasis, 683
Schistosomiasis vaccines, 696
Schlepper, 115
Schultz-Dale test (historical), 351
SCID (severe combined immunodeficiency), 543
SCID (severe combined immunodeficiency) mouse, 543
Scl-70 (topoisomerase I) autoantibody, 428
Scl-70 antibody, 522
Scleroderma, 428
Scr homology-3 (SH-3) domain, 52
SCR1 (soluble complement receptor type 1), 339
Scratch test, 351
SDS-PAGE, 734
Second messengers (IP_3 and DAG), 272
Second signals, 53
Secondary allergen, 360
Secondary antibody response, 120
Secondary disease, 607
Secondary follicle, 88
Secondary granule, 70, 634
Secondary immune response, 120
Secondary immunodeficiency, 553
Secondary lymphoid organs, 86
Secondary lymphoid tissues, 86
Secondary lysosome, 72
Secondary reactions, 227
Secondary response, 120
Secondary structure, 113
Second-set rejection, 602
Second-set response, 602
Secreted immunoglobulin (sIg), 192
Secretor, 452
Secretory component (T piece), 441
Secretory component deficiency, 441
Secretory IgA, 440
Secretory immune system, 440
Secretory immunoglobulin A (SigA), 631
Secretory piece, 192, 441
Sedimentation coefficient, 735
Sedimentation pattern, 736
Sedormid® purpura (historical), 435
Selectins, 26
Selective IgA and IgG deficiency, 539
Selective IgA and IgM deficiency, 539
Selective IgA deficiency, 538
Selective IgM deficiency, 539
Selective immunoglobulin deficiency, 538
Selective theory, 206
Self marker hypothesis (historical), 206
Self-MHC restriction, 272
Self-peptides, 381
Self-restriction, 272
Self-tolerance, 381
Semisyngeneic graft, 592
Senescent cell antigen, 465
Sensitization, 347
Sensitized lymphocyte, 280
Sensitized vaccine, 697
Sephadex®, 725
Sepharose®, 734
Septic shock, 648
Sequence-specific priming (SSP), 754
Sequential determinant, 112
Sequestered antigen, 389
Serial dilution, 726
Serial passage, 775
Seroconversion, 648
Serological determinants, 589
Serologically defined (SD) antigens, 589
Serology, 227
Serotherapy, 699
Serotonin (5-hydroxytryptamine), 356, 788
Serotype, 648
Serpins, 564
Serum albumin, 109, 456
Serum amyloid A component (SAA), 524
Serum amyloid P component (SAP), 524
Serum hepatitis (hepatitis B), 498
Serum sickness, 366
Serum virus vaccination, 697
Severe combined immune-deficient (SCID) mouse, 543
Severe combined immunodeficiency (Swiss-type agammaglobulinemia), 535
Severe combined immunodeficiency syndrome (SCID), 542
Sex hormones and immunity, 390
Sex-limited protein, 327
Sezary cells, 534
Sezary syndrome, 533
SH-2 domain, 52
Sheep red blood cell agglutination test, 737
Shift assay, 774
Shigella immunity, 648
Shingles (Herpes zoster), 666
Shock organ, 351
Shocking dose, 351
Short-lived lymphocytes, 51
Shwartzman (or Shwartzman-Sanarelli) reaction, 368
Sia test (historical), 730
Sialophorin (CD43), 254
Sicca complex, 430
Side chain theory, 205
Side effect, 435
Signal hypothesis, 207
Signal joint, 167, 270
Signal peptide, 163
Signal recognition particle autoantibodies against SRP, 431
Signal sequence, 207

Signal transducers and activators of transcription (STATs), 287
Signal transduction, 52
Silencer sequence, 273
Silencers, 273
Silica adjuvants, 125
Silicate autoantibodies, 432
Silicosis, 497
Simian immunodeficiency virus (SIV), 565
Simonsen phenomenon, 607
Simple allotype, 200
Simulect©, 703
Single cysteine motif-1 (SCM-1), 319
Single diffusion test, 726
Single domain antibodies, 199
Single hit theory, 324
Single immunodiffusion (Mancini technique), 726
Single locus probes (SLPs), 591, 754
Single radial immunodiffusion, 727
Single-chain Fv fragment, 199
Single-positive thymocytes, 256
Sips plot, 236
Sirolimus, 572
Site-directed mutagenesis, 774
SIV (simian immunodeficiency virus), 565
Sjögren's syndrome (SS), 429, 520, 521
Skin autoantibodies, 414
Skin graft, 597
Skin immunity, 444
Skin test, 763
Skin window, 766
Skin-fixing antibody, 179
Skin-reactive factor (SRF), 287
Skin-sensitizing antibody, 348
Skin-specific histocompatibility antigen, 597
Slide agglutination test, 739
Slide flocculation test, 739
Slot blot analysis, 774
Slow viruses, 666
Slow-reacting substance of anaphylaxis (SRS-A), 357
Slp, 327
Small "blues", 589
Small G proteins, 42
Small lymphocyte, 50
Smallpox, 674
Smallpox vaccination, 687
Smallpox vaccine, 687
Smooth muscle antibodies, 401, 498
SNagg, 426
Sneaking through, 613
Snell-Bagg mice, 283
SOD, 71
SODD-Silencer of Death Domains, 309
Solid-phase radioimmunoassay, 731
Solubilized water-in-oil adjuvant, 125
Soluble antigen, 107
Soluble complex, 248
Soluble cytokine receptors, 286
Soluble liver antigen antibodies, 498
Somatic antigen, 641
Somatic gene conversion, 713
Somatic gene therapy, 544
Somatic hypermutation, 208
Somatic mutation, 208
Somatic recombination, 167, 258
Southern blotting, 751
Southwestern blot, 750
SP-40,40, 332
Species specificity, 722
Specific granule, 70
Specific immune response to extracellular bacteria, 645
Specific immunity, 97
Specificity, 161, 270
Speckled pattern, 432
Spectratyping, 774
Spectrotype, 770
Sperm antibodies, 407
Sperm autoantibodies, 407
Spheroplast, 649
Spherulin, 677
Spleen, 88
Spliceosomal snRNP autoantibodies, 424
Split thickness graft, 597
Split tolerance, 385
Splits, 590
Sponges, 708
Spontaneous autoimmune thyroiditis (SAT), 392
Spontaneous remission, 613
Spot ELISA, 750
Sprue, 496
Spur, 245
Squalene, 125
SRBC, 262
Src homology-2 (SH-2) domain, 52
SRS-A, 350
SRV-1, 566
SRY, 774
Ss protein, 327
SS-A, 521
SS-A Ro, 429
SS-A/Ro antibodies, 429
SS-B, 522
SS-B La, 430
SS-B/La antibodies, 430
SSPE, 521
SSS III, 108
St. Vitus dance (chorea), 493
Staphylococcal enterotoxins (Ses), 649
Staphylococcal protein A, 172
Staphylococcus immunity, 649
Status asthmaticus, 363
Status thymolymphaticus (historical), 254
Stem cells, 46, 599
Stem-cell factor (SCF), 46
Steric hindrance, 236
Steric repulsion, 230
Steroid cell antibodies, 405
Stiff man syndrome (SMS), 408
Stimulated macrophage, 61
Stormont test, 763
Strain, 768
Street virus, 669
Streptavidin, 777
Streptobacillus immunity, 649
Streptococcal M protein, 649
Streptococcus immunity, 649
Streptolysin O test, 730
Stress and immunity, 385
Stress proteins, 42
Striational antibodies, 408

Striational autoantibodies (StrAb), 408
Stromal cell, 78
Stromal cell derived factor-1 (SDF-1), 319
Stromal cells, 78, 253
Strongyloides hyperinfection, 681
Strongyloides immunity, 681
STS, 730
Subacute sclerosing panencephalitis, 669
Subset, 277
Substance P, 350
Substrate adhesion molecules (SAM), 29
Sugarcane worker's lung, 490
Sulfite sensitivity, 351
Sulzberger-Chase phenomenon, 385, 444
Superantigen, 107, 154
Superinfection "immunity," 649
Superoxide dismutase, 62
Suppressin, 267
Suppression, 567
Suppressor cell, 267
Suppressor macrophage, 73
Suppressor T cell factor (TSF), 268
Suppressor T cells (Ts cells), 268
Suppressor/inducer T lymphocyte, 267
Supratypic antigen, 107, 581
Suramin, 332, 681
Surface antigens, 110
Surface immunoglobulin, 170
Surface phagocytosis, 71
Surface plasmon resonance (SRP), 236
Surface secretions, 632
Surrogate light, 170
Sustiva®, 564
SV40 (simian virus 40), 617
Svedberg unit, 735
Sweet's syndrome (acute febrile neutrophilic dermatosis), 481
Swiss agammaglobulinemia, 544
Swiss type agammaglobulinemia, 537
Swiss type immunodeficiency, 537
Switch, 208
Switch cells, 208
Switch defect disease, 208
Switch region, 208
Switch site, 208
Syk PTK, 160
Sympathetic nervous system autoantibodies, 414
Sympathetic ophthalmia, 415, 510
Synaptophysin, 782
Synergism, 287
Syngeneic, 596
Syngeneic preference, 595
Syngraft, 596
Synthetic antigen, 105
Synthetic polypeptide antigens, 111
Synthetic vaccines, 696
Systemic acquired resistance (SAR), 707
Systemic anaphylaxis, 354
Systemic autoimmunity, 416
Systemic immunoblastic proliferation, 476
Systemic inflammatory response syndrome (SIRS), 644
Systemic lupus erythematosus (SLE), 416, 512, 521
Systemic lupus erythematosus, animal models, 424
Systemic sclerosis, 428

T

(TG)AL, 115
T activation, 465
T agglutinin, 465
T antigen(s), 464
T cell, 54
T cell antigen receptors, 271
T cell antigen-specific suppressor factor, 268
T cell clonal expansion, 276
T cell development, 256
T cell domains, 261
T cell hybridomas, 772
T cell immunodeficiency syndromes (TCIS), 541
T cell leukemia, 473
T cell leukemia viruses, 473
T cell lymphoma (TCL), 473
T cell maturation, 257
T cell migration, 257
T cell nonantigen-specific helper factor, 264
T cell receptor (TCR), 270
T cell receptor complex, 271
T cell receptor genes, 270
T cell replacing factor (TRF), 277
T cell rosette, 276
T cell specificity, 276
T cell system, 711
T cell vaccination (TCV), 432, 697
T cell-dependent (TD) antigen, 260
T cell-independent (TI) antigen, 261
T cells, 54, 276
T globulin, 719
T lymphocyte (T cell), 275
T lymphocyte clone, 277
T lymphocyte hybridoma, 277
T lymphocyte receptor, 271
T lymphocyte subpopulation, 277
T lymphocyte–B lymphocyte cooperation, 156
T lymphocyte-conditioned medium, 287
T lymphocyte–T lymphocyte cooperation, 225
T piece, 192
T1 antigen, 274
T-200, 274
T3 antigen, 269
T4 antigen, 262
T8 antigen, 266
TAB vaccine, 692
Tac, 280
Tac antigen, 280
TACI (transmembrane activator and CAML-interactor), 309
Tacrolimus, 571
Taenia solium immunity, 681
TAF, 248
Tail peptide, 181
Takatsy method, 739
Takayasu's arteritis, 493
Take, 601
Tandem immunoelectrophoresis, 733
Tanned red cell test, 737
Tanned red cells, 737
TAP 1 and TAP 2 genes, 149
TAPA-1, 161
Tapasin (TAP-associated protein), 148, 149
Tapioca adjuvant (historical), 124
Taq polymerase, 752
Target cell, 343

Tat, 560
Tat gene, 566
TATA, 619
TATA box, 100
TB, 655
Tc lymphocyte, 266
T-cell growth factor (TCGF), 294
T-cell growth factor 1, 294
T-cell growth factor 2, 297
T-cell tolerance, 382
TCGF (T-cell growth factor), 294
TD antigen, 260
T-dependent antigen, 108, 260
TdT, 103
Tec kinase, 160, 279
Telencephalin, 181
Template theory (historical), 206
Tenascin, 35
Terminal complement complex (TCC), 332
Terminal complement complex deficiency, 332
Terminal complement components, 332
Terminal deoxynucleotidyl transferase (TdT), 158, 259
Terminal transferase, 103
Termination of tolerance, 380
Tertiary granule, 70
Tertiary immune response, 122
Tertiary immunization, 122
Tertiary reactions, 227
Tertiary response, 122
Tertiary structure, 113
Test dosing, 766
Testicular autoimmunity, 407
Tetanus, 650
Tetanus antitoxin, 689, 699
Tetanus toxin, 650
Tetanus toxoid, 108, 689
Tetanus vaccine, 689
Tetraparental chimera, 381
Tetraparental mouse, 381
Texas red, 760
TFA antigens, 108, 717
TGF (transforming growth factor, 315
TGF-βs (transforming growth factor-βs), 315
T_H0 cells, 264
T_H1 cells, 264
T_H2 cells, 265
Theiler's virus myelitis, 674
Theileria immunity, 682
Theliolymphocytes, 51
Theophylline, 363
Therapeutic antisera, 699
Thermoactinomyces species, 490
Thoracic duct, 83
Thoracic duct drainage, 83
Thorotrast (thorium dioxide 32THOT), 68
Threonyl-transfer RNA synthetase antibodies, 431
Threonyl-transfer RNA synthetase autoantibodies, 410
Thrombocyte, 77
Thrombocytopenia, 464
Thrombocytosis, 464
Thromboxanes, 357
Thy (θ), 268
Thy 1 antigen, 268
Thy-1, 268
Thy-1+ dendritic cells, 268, 444
Thymectomy, 282
Thymic alymphoplasia, 547
Thymic epithelial cells, 255
Thymic hormones, 280
Thymic hormones and peptides, 280
Thymic humoral factor(s) (THFs), 280
Thymic hypoplasia (DiGeorge syndrome), 535, 541
Thymic leukemia antigen (TL), 256
Thymic medullary hyperplasia, 485
Thymic nurse cells, 255
Thymic stromal-derived lymphopoietin (TSLP), 268, 298
Thymic-leukemia antigen (TL), 615
Thymin, 282
Thymocyte, 255
Thymoma, 253
Thymopentin (TP5), 281
Thymopoietin, 281
Thymosin, 281
Thymosin α-1 (thymopoietin), 281
Thymosine, 280
Thymulin, 268
Thymus, 79, 251
Thymus cell differentiation, 259
Thymus cell education, 260
Thymus-dependent (TD) antigen, 117, 260
Thymus-dependent areas, 86, 260
Thymus-dependent cells, 261
Thymus-independent (TI) antigen, 117, 261
Thymus-replacing factor (TRF), 298
Thyroglobulin, 391
Thyroglobulin autoantibodies, 392
Thyroid antibodies, 391
Thyroid autoantibodies, 487
Thyroid autoimmunity animal models, 392
Thyrotoxicosis, 391, 487
Thyrotropin, 392
Thyrotropin receptor autoantibodies, 392
Tight skin-1 mouse (Tsk1), 522
Tight skin-2 mouse (Tsk2), 522
TIL, 626
T-independent antigen, 108, 261
Tine test, 764
Tingible body, 63
Tingible body macrophages, 63
Tissue transglutaminase autoantibodies, 398
Tissue typing, 585
Tissue-fixed macrophage, 62
Tissue-specific antigen, 391
Titer, 176, 246
TL (thymic-leukemia antigen), 133
Tla antigen, 133
Tla complex, 132
TLR1-10, 72
TNF, 308
TNF receptor-associated factors (TRAFs), 310
TNF-related activation-induced cytokine (TRANCE) (RANK Ligand), 310
TNP, 114
Togavirus immunity, 674
Tolerance, 379
Tolerogen, 381
Tolerogenic, 380
Toll-like receptors, 72
Tolmetin, 706
Tonsils, 91
Topoisomerase I, 428
TORCH panel, 675

Total lymphoid irradiation (TLI), 576
Totipotent, 46
Toxic complexes, 366
Toxic epidermal necrolysis, 607
Toxic shock syndrome, 649
Toxin neutralization (by antitoxin), 248
Toxins, 107
Toxocara canis immunity, 682
Toxoid, 689
Toxoid–antitoxin floccules, 689
Toxoplasma gondii immunity, 682
Tp44 (CD28), 150
Tp44 (CD29), 273
TPA, 65
TPHA, 737
TPI, 730
T_R1, 276
Trace labeling, 726
Traffic area, 261
TRAFs, 310
TRAIL (TNF-related apoptosis-inducing ligand), 310
TRALI (transfusion-related acute lung injury), 467
Trangenic, 768
Transcobalamin II deficiency, 540
Transcobalamin II deficiency with hypogammaglobulinemia, 540
Transcription, 774
Transcytosis, 441
Transduction, 774
Transfection, 774
Transfectoma, 208
Transfer factor (TF), 315
Transferrin, 650
Transferrin receptor (T9), 280
Transformation, 119
Transforming growth factor-α (TGF-α), 315
Transforming growth factor β (TGF-β), 316
Transfusion, 465
Transfusion reaction(s), 466
Transfusion-associated graft-vs.-host disease (TAGVHD), 466
Transgenes, 768
Transgenic animal, 768
Transgenic line, 769
Transgenic mice, 768
Transgenic mouse, 768
Transgenic organisms, 769
Transgenics, 769
Transient hypogammaglobulinemia of infancy, 536
Transmissible spongiform encephalopathy (TSE) immunity, 534
Transplantation, 579
Transplantation antigens, 579
Transplantation immunology, 579
Transplantation rejection, 603
Transport piece, 192
Transporter associated with antigen processing (TAP), 148
Transporter in antigen processing (TAP) 1 and 2 genes, 149
Transudation, 92
Trastuzumab, 702
Treponema immunity, 650
Treponema pallidum hemagglutination, 737
Treponema pallidum immobilization test, 730
TRF, 298
Trichuris trichiura immunity, 682
Trinitrophenyl (picryl) group, 114
Triple response of Lewis, 348
Triple vaccine, 691
Triton X-100, 123
Trophoblast, 76, 577
Tropical eosinophilia, 682
Trypan blue, 589
Trypan blue dye exclusion test, 589
Trypanosome adhesion test, 730
Trypanosome immunity, 682
Tryptic peptides, 225
Ts, 267
Ts1, Ts3 lymphocytes, 267
TsF, 267
TTF-1 (8G7G3/a), mouse, 796
Tube agglutination test, 739
Tuberculid, 372
Tuberculin, 372
Tuberculin hypersensitivity, 372
Tuberculin reaction, 373
Tuberculin test, 373, 763
Tuberculin-type reaction, 373
Tuberculosis immunization, 693, 373
Tubular basement membrane autoantibodies, 403
Tuftsin, 63
Tuftsin deficiency, 549
Tumor antigens, 615
Tumor imaging, 615
Tumor immunity, 626
Tumor immunotherapy, 703
Tumor necrosis factor (TNG) family, 308
Tumor necrosis factor α (TNF-α), 308, 628
Tumor necrosis factor β (TNF-β), 310
Tumor necrosis factor receptors (TNF receptors), 310, 629
Tumor promoter, 615
Tumor rejection antigen, 619
Tumor-associated antigens, 615
Tumor-infiltrating lymphocytes (TIL), 626
Tumor-specific antigen (TSA), 619
Tumor-specific determinants, 619
Tumor-specific transplantation antigen (TSTA), 619
TUNEL assay (TdT-dependent dUTP-biotin nick end labeling), 774
TUNEL-based assays, 774
Tunicates, 711
Turbidimetry, 730
Tween, 124
Tween 80®, 124
Two-dimensional gel electrophoresis, 732
Two-signal hypothesis, 119
Type I anaphylactic hypersensitivity, 348
Type I cytokine receptors, 286
Type I interferons (IFN-α, IFN-β), 307
Type II antibody-mediated hypersensitivity, 363
Type II interferon, 307
Type III immune complex-mediated hypersensitivity, 364
Type IV cell-mediated hypersensitivity, 369
Typhoid vaccination, 692
Typhoid vaccine, 692
Typhus vaccination, 696
Typhus vaccine, 696
Tyrosine kinase, 53

U

U antigen, 459
U1 snRNP autoantibodies, 432
U2 snRNP autoantibodies, 411
Ubiquitin, 43
Ubiquitin autoantibodies, 424
Ubiquitination, 43

UCHL1 anti-human T cell, CD45RO, 790
Ulcerative colitis (immunologic colitis), 495
Ultracentrifugation, 735
Ultrafiltration, 735
Umbrella effect, 529
Undifferentiated connective tissue disease, 433
Ungulate immunity, 718
Unidentified reading frame (URF), 101
Unitarian hypothesis, 207, 249
Univalent, 176
Univalent antibody, 176
Universal donor, 466
Universal recipient, 466
Unprimed, 159
Unproductive rearrangements, 167
Unresponsiveness, 379
Uromodulin Tamm-Horsfall protein, 577
Uropod, 54
Urticaria, 361
Urushiols, 375
US28, 291
Usual interstitial pneumonitis, 488
Uveitis, 511

V

V gene, 167, 170
V gene, 223
V gene segment, 167, 170
V gene segment, 223
V region subgroups, 224, 225
V(D)J recombinase, 168
V(D)J recombination class switching, 167
V(J) recombination, 167
V28, 291
Vaccinable, 685
Vaccinate, 685
Vaccination, 685, 687
Vaccine, 685
Vaccine extraimmunization, 691
Vaccinia, 688
Vaccinia immune globulin, 689
Vaginal mucous agglutination test, 739
Valence, 227
van der Waals forces (London forces), 230
Variability plot, 196
Variable region, 224
Varicella, 675
Varicella (chickenpox) vaccine, 689
Varicella-zoster virus immunity, 675
Variola (smallpox), 675
Variolation, 687
Vascular addressins, 38
Vascular cell adhesion molecule-1 (VCAM-1), 33
Vascular permeability factors, 356
Vasculitis, 365
Vasectomy, 407
Vasoactive amines, 350
Vasoactive intestinal peptide (VIP), 92
Vasoconstriction, 351
Vasodilatation, 351
VDRL (Venereal Disease Research Laboratory) test, 650
Vector, 168
Veiled cell, 65
Venom, 108, 361
Venoocclusive disease (VOD), 611
Vermiform appendix, 91
Very late activation antigens (VLA molecules), 33
Vesiculation, 375
Veto cells, 55
V_H region, 185
Vi antigen, 657
Viability techniques, 770
Vibrio cholerae immunity, 650
Videx®, 564
Vimentin, 781
Vinblastine, 706
Vincristine, 706
Vinyl chloride (VC), 515
Viracept®, 564
Viral capsids, 667
Viral hemagglutination, 659
Viral immunity, 659
Viral interference, 659
Viramune®, 564
Virgin B cells, 164
Virion, 659
Viroid, 659
Viropathic, 659
Virulence genes, 650
Virus, 658
Virus infection associated autoantibodies, 434
Virus neutralization test, 771
Virus-associated hemophagocytic syndrome, 659
Virus-neutralizing capacity, 660
Viscosity, 92
Vitamin A, 98
Vitamin A and immunity, 98
Vitamin B and immunity, 98
Vitamin C and immunity, 98
Vitamin D and immunity, 99
Vitamin E and immunity, 99
Vitiligo, 415
Vitronectin, 35
Vλ, 185
V_L region, 185
VLA receptors, 33
VLIA (virus-like infectious agent), 559
V-MYB oncogene, 533, 615, 669
Vogt-Koyanagi-Harada (VKH) syndrome, 510
Vollmer test (historical), 764
Voltage-gated-calcium channel autoantibodies, 408
Von Krough equation, 343
Vpre-B, 158
V_T region, 271

W

W,X,Y boxes (class II MHC promoter), 583
Waaler-Rose test, 763
Waldenström's macroglobulinemia, 531
Waldeyer's ring, 91
Warm antibody, 462
Wassermann reaction, 741
Wasting disease, 576
Wax D, 125
Wegener's granulomatosis, 481
Weibel-Palade bodies, 38
Weil-Felix reaction, 739

Western blot (immunoblot), 750
Wheal and flare reaction, 94, 361
Whipple's disease, 495
White graft rejection, 602
White pulp, 89
White pulp disease, 472
Whooping cough vaccine, 691
Widal reaction, 739
Window, 558
Winn assay, 627, 768
Wire loop lesion, 515
Wiskott-Aldrich syndrome, 546
Witebsky's criteria, 388
Worms, 708
Wu-Kabat plot, 196

X

X cell, 206
Xenoantibodies, 595
Xenoantibody, 595
Xenoantigen, 595
Xenobiotics, 433
Xenogeneic, 595
Xenograft, 595
Xenopus, 714
Xenoreactive, 595
Xenotransplantation, 595
Xenotype, 596
Xeno-zoonosis, 596
Xg^a, 463
Xid gene, 118
X-linked (congenital) agammaglobulinemia, 535
X-linked agammaglobulinemia (Bruton's X-linked agammaglobulinemia), 536
X-linked hyper-IgM syndrome, 550
X-linked lymphoproliferative disease (XLP), 544
X-linked lymphoproliferative syndrome, 544
X-linked severe combined immunodeficiency (XSCID), 543
XYZ cell theory (historical), 206

Y

Y cell, 206
Yellow fever vaccine, 693
Yersinia immunity, 650

Z

ζ (zeta) chain, 272
Z cell, 206
ZAP-70 deficiency, 543
Zeta potential, 456
Zeta-associated protein of 70 kDa (ZAP-70), 269
Ziagen®, 564
Zidovudine, 563
Zinc, 99
Zinc and immunity, 99
Zippering, 71
Zirconium granuloma, 369
Zonal centrifugation, 736
Zone electrophoresis, 735
Zone of equivalence, 237
Zoonosis, 596
Zygosity, 774
Zymogen, 43
Zymosan, 333